Laboratory Diagnosis of Infectious Diseases
Principles and Practice

Senior Editors
A. Balows W.J. Hausler, Jr. E.H. Lennette

Laboratory Diagnosis of Infectious Diseases
Principles and Practice

VOLUME II
Viral, Rickettsial, and Chlamydial Diseases

Editors
E.H. Lennette   P. Halonen   F.A. Murphy

With 191 Illustrations, 7 in Full Color

Springer-Verlag New York Berlin Heidelberg
London Paris Tokyo

A. Balows, Center for Infectious Diseases, Centers for Disease Control, Atlanta, GA 30333, USA

W. J. Hausler, Jr., Hygienic Laboratory, Oakdale Hall, University of Iowa, Iowa City, IA 52242, USA

E. H. Lennette, Viral and Rickettsial Disease Laboratory (Emeritus Director), California Department of Health Services, and California Public Health Foundation, Berkeley, CA 94704, USA; and Institut Virion, CH-8803 Ruschlikon/Zurich, Switzerland

P. Halonen, Department of Virology, University of Turku, Turku, Finland

F. A. Murphy, Center for Infectious Diseases, Centers for Disease Control, Atlanta, GA 30333, USA

Library of Congress Cataloging-in-Publication Data
Laboratory diagnosis of infectious diseases: principles and
    practices.
        p. cm.
    Includes bibliographies and index.
    Contents: v. 1. Bacterial, mycotic, and parasitic diseases / A.
Balows . . . [et al.], editors—v. 2. Viral, rickettsial, and
chlamydial diseases / E. H. Lennette, P. Halonen, F. A. Murphy,
editors.
    1. Diagnostic microbiology—Laboratory manuals. I. Balows.
Albert.
    [DNLM: 1. Communicable Diseases—diagnosis. 2. Diagnosis,
Laboratory. 3. Virus Diseases—diagnosis. WC 25 L1226]
QR67.L328 1988
616.9'0475—dc19
DNLM/DLC
                                                        88-6581

Typeset by Bi-Comp, York, Pennsylvania.

9 8 7 6 5 4 3 2 1

ISBN-13:978-1-4612-8394-2   e-ISBN-13:978-1-4612-3900-0
DOI:10.1007/978-1-4612-3900-0

Foreword

This two volume work stems from the belief of the Editors that infectious diseases are not only very much with us today but, more importantly, that they will continue to play a significant global role in morbidity and mortality in all people. A continuing need for an informed and knowledgeable community of laboratory scientists is fundamental. Data describing the global impact of infectious diseases are difficult to come by. Fortunately, a recent thoughtful and provocative publication by Bennett et al. (1987) provides us with data derived from several consultants that clearly delineate the impact of infectious diseases on the United States today. In the mid 1980s, almost three-quarters of a billion cases of infectious diseases occurred annually in the United States, resulting in mortality approaching 200,000. These cases represent over two million years lost before age 65 and almost two billion days lost from work, school, or other usual daily activities. It would be very helpful if similar data exist or could be obtained from other developed or developing nations.

Infectious diseases or, more specifically, the agents that cause them usually have little regard for age or sex, but nutritional or immune status, underlying or pre-existing disease, and habitat frequently contribute to the establishment of acute and often chronic infections. Without question, the scientific advances in biology and medicine that have occurred during the past 75 years and the successes that will take place between now and the end of this century will be major forces in keeping the above-cited numbers from increasing. In all likelihood, these numbers should continue a worldwide downward trend.

A master plan to diagnose, treat, control, and prevent infectious diseases has evolved over the past several decades. Ingredients of this master plan include in-depth study and research in all aspects of infectious diseases to completely characterize the etiologic agents and fully describe the natural history of each disease. Inclusive aspects, although the list is by no means complete, are host–parasite interactions, immunology, pathology, microbiology, parasitology, virology, and epidemiology. The most important resource in gaining the necessary knowledge in these areas is well-informed people—particularly those who deal with infectious diseases on a daily basis.

There are several excellent textbooks dealing with medical microbiology, and there are equally well-recognized books devoted to infectious diseases. The Editors of this work, on the other hand, were persuaded that there was a need for a publication that would bring together the most pertinent and relevant information on the principles and practice of the laboratory diagnosis of infectious diseases and include clinical relationships. While this two volume text is directed toward the role of the laboratory in infections—after all, that is where the definitive diagnosis of an infectious disease is most often made—ample consideration is given to the disease, its symptomatology, pathogenesis, epidemiology, treatment, prevention, and control.

This work is designed to keep the clinician and the clinical laboratory scientist informed on nearly all the commonly known infectious diseases and their etiologic agents about which there are useful and reliable data. To achieve this goal, international authorities were enlisted as authors. The international approach using a chapter format designed to give the reader access to both a synopsis and a more detailed discussion on each disease or group of related diseases is unique. Dividing the work into two volumes was a convenient and practical way to make the information readily available in a useful manner, and these volumes therefore constitute the complete work. The Editors welcome comments from the users of this set.

Bennett, J. V., S. D. Holmberg, M. F. Rogers, and S. L. Solomon. 1987. Infections and parasitic diseases in closing the gap: the burden of unnecessary illness, p. 102–104. R. W. Amler and H. B. Dull (ed.). Oxford University Press, New York.

The Senior Editors ALBERT BALOWS
WILLIAM J. HAUSLER, JR.
EDWIN H. LENNETTE

The Associate Editors PEKKA HALONEN
FREDERICK A. MURPHY
MAKOTO OHASHI
ADOLFO TURANO

Preface

The past four decades have witnessed remarkable advances in the field of medical virology, rickettsiology, and chlamydiology, all of which have had great impact on the clinician, epidemiologist, immunologist, pathologist, and especially the clinical microbiologist. To take full advantage of the advances in these fields of medical science, there must be substantial communication and collaboration between all disciplines that are involved in disease prevention and control as well as indirect patient care and management. In this volume, we have included clinical manifestations, pathology, epidemiology, treatment, prevention, and control of viral, rickettsial, and chlamydial diseases to complement the extensive coverage of the principles and practice of laboratory diagnosis. The diagnosis of a viral, rickettsial, or chlamydial disease is not simply the product of laboratory tests, but rather an integrated conclusion with contributions from clinicians and microbiologists, and often from epidemiologists and pathologists. Therefore, it is hoped that the information provided in this volume on laboratory diagnosis will be as helpful to the clinician and others interested in infectious diseases as it should be to the laboratory diagnostician.

The clinical laboratory scientist of today takes for granted the power and diversity of diagnostic methodologies at his or her disposition, and probably gives little thought to how different the situation was only a few decades ago when medical virology, rickettsiology, and chlamydiology were constrained by a sense of mystery and intractability in working with the organisms; the greatest obstacle in working with all of these organisms lays in their strict intracellular life cycles.

Dependence on animal host systems, which continued through the first half of this century, was essentially ended when Enders, Weller, and Robbins, in 1949, demonstrated the feasibility of propagating viruses in cell culture. This epochal work ushered in a whole new era of virology, allowing the discovery of many new pathogenic viruses, and also contrib-

uted greatly to the advance of work on *Rickettsiae* and *Chlamydiae.*

An important basic benefit was the availability of a means for accurately titrating infectivity. In the past two decades, methods for the isolation, cultivation, and identification of viruses, *Rickettsiae,* and *Chlamydiae* have been simplified and made readily accessible to a greater number of laboratories. Polyclonal antisera and, most recently, monoclonal antibodies have helped to make serologic diagnosis even more sensitive and specific.

Isolation and identification still play major roles in diagnostic virology, rickettsiology, and chlamydiology, although in recent years the increasing demands for simpler, quicker, more specific, more sensitive, and less expensive methods for diagnosis have led to further changes in the approaches used in diagnostic laboratories. Methodologies for rapid diagnosis by direct detection of antigens and nucleic acids in clinical specimens are being developed with increasing innovation. The demands on such tests are great because in most cases the amount of antigen and nucleic acid in clinical specimens is very small and the background material in clinical specimens is extremely complex. Nevertheless, the application of rapid antigen and nucleic acid detection tests is progressing rapidly. Currently, the ultimate diagnostic approach applicable to many viral infections may be an initial attempt to detect viral antigens or nucleic acids; if such approaches are unsuccessful, virus isolation methods are used. Parallel trends are seen in the diagnosis of rickettsial and chlamydial infections. In some cases, amplification of the infectious agent to yield more antigen or nucleic acid in a clinical specimen is attempted and followed by one of the rapid diagnostic methods. For example, indirect immunofluorescence using monoclonal antibody reagents has been used to detect the presence of adenoviruses and influenza viruses after clinical specimens have been inoculated into cell cultures for short periods. Although this approach has reduced the time for final diagnosis by one-half, it requires comprehensive sets of reagents and cell culture substrates so that all common organisms associated with a particular disease syndrome may be diagnosed by parallel sets of tests.

Commercial production of everything from cell cultures to complete antigen and nucleic acid detection kits has simplified much of the preparative work that at one time was a great burden in diagnostic laboratories. Commercial supplies of antibodies, both polyclonal and monoclonal, have also made it possible for diagnostic laboratories to more clearly focus on the conduct of tests and their clinical interpretation. The convenience of the current use of prepackaged test components is obvious, but cost containment has become a major factor that each laboratory must address.

Finally, new on the scene is the harbinger of what may be the most rapidly growing segment of microbiological diagnostics, namely, the appearance of "over-the-counter" tests for home use. We believe this will add to the complications of providing confirmatory and reference diagnostic services and in the ultimate diagnosis of infections; only time will tell how rapidly this development will proceed. It is quite likely that this subject will be the focus of much more extensive coverage in the next edition of this and similar publications.

The Editors

EDWIN H. LENNETTE
FREDERICK A. MURPHY
PEKKA HALONEN

Contents: Volume II

Contents: Volume I

Section III   Mycotic Infections

Contributors

D. G. AHEARN, Laboratory for Microbial and Biochemical Sciences, Department of Biology, Georgia State University, Atlanta, GA 30303, USA

L. AJELLO, Division of Mycotic Diseases, Center for Infectious Diseases, Centers for Disease Control, Atlanta, GA 30333, USA

WIDAD AL-NAKIB, MRC Common Cold Unit, Harvard Hospital, Salisbury, Wiltshire SP2 8BW, England

W. L. ALBRITTON, Department of Microbiology, University of Saskatchewan, Saskatoon, Canada S7N 0W0

DANIEL AMSTERDAM, Division of Clinical Microbiology and Immunology, Erie County Laboratory and Medical Center, and Departments of Microbiology and Immunology, State University of New York School of Medicine, Buffalo, NY 14215, USA

LARRY J. ANDERSON, Respiratory and Enterovirus Branch, Division of Viral Diseases and Hospital Infections Program, Centers for Disease Control, Atlanta, GA 30333, USA

M. J. ANDERSON, Department of Medical Microbiology, Faculty of Clinical Sciences, University College and Middlesex School of Medicine, London WC1E 6JJ, England

RAY R. ARTHUR, Department of Immunology and Infectious Diseases, The Johns Hopkins University, School of Hygiene and Public Health, Baltimore, MD 21205, USA

PERTTI P. ARSTILA, Department of Virology, University of Turku, Kiinamyllynkatu 13, SF-20520, Turku, Finland

GEORGE M. BAER, Rabies Laboratory, Viral and Rickettsial Zoonoses Branch, Division of Viral Diseases, Center for Infectious Diseases, Centers for Disease Control, Lawrenceville, GA 30246, USA

MARILYN S. BARTLETT, Department of Pathology, Indiana University School of Medicine, Indianapolis, IN 46223, USA

GAURI BAZAZ-MALIK, Department of Pathology, Vallabhbhai Patch Chest Institute, University of Delhi, Delhi 110 007, India

C. M. BECK-SAGUE, Sexually Transmitted Diseases Laboratory Program, Center for Infectious Diseases, Centers for Disease Control, Atlanta, GA 30333, USA

R. W. BENNETT, Division of Microbiology, Food and Drug Administration, Washington, D.C. 20204, USA

JENNIFER M. BEST, Department of Virology, United Medical and Dental Schools of Guys' and St. Thomas' Hospitals, St. Thomas' Campus, London SE1 7EH, England

JEFFREY W. BIER, Division of Microbiology, Center for Food Science and Applied Nutrition, Food and Drug Administration, Washington, D.C. 20204, USA

WALTER W. BOND, Division of Viral Diseases and Hospital Infections Program, Center for Infectious Diseases, Centers for Disease Control, Atlanta, GA 30333, USA

ANN CALI, Department of Biological Sciences, Boyden Hall, Rutgers University, Newark, NJ 07102, USA

CHARLES H. CALISHER, Division of Vector-Borne Viral Diseases, Center for Infectious Diseases, Centers for Disease Control, Fort Collins, CO 80522, USA

MARIO E. CAMARGO, Laboratorio de Immunologia e Soroepidemiologia, Instituto de Medicina Tropical de São Paulo, São Paulo, Brazil

MARIO CAMPA, Institute of Microbiology, University of Pisa, 56100 Pisa, Italy

JOSEPH C. CHAN, Division of Infectious Diseases, Department of Medicine, University of Miami School of Medicine, Miami, FL 33101, USA

TIMOTHY J. CLEARY, Department of Pathology, University of Miami School of Medicine, Miami, FL 33101, USA

JOHN H. CROSS, Division of Tropical Public Health, Department of Preventive Medicine and Biometrics, Uniformed Services University of the Health

Sciences, F. Edward Hebert School of Medicine, Bethesda, MD 20814-4799, USA

WILLIAM L. CURRENT, Eli Lilly Research Laboratories, Lilly Corporate Center, Indianapolis, IN 46285, USA

RICHARD F. D'AMATO, Division of Microbiology, The Catholic Medical Center of Brooklyn and Queens, Jamaica, NY 11432, USA

A. F. DISALVO, Bureau of Laboratories, South Carolina Department of Health and Environmental Control, Columbia, SC 29202, USA

FRANCES W. DOANE, Department of Microbiology, Faculty of Medicine, University of Toronto, Toronto, Ontario M5S 1A8, Canada

GARY V. DOERN, Department of Clinical Microbiology, University of Massachusetts Medical Center, Worcester, MA 01605, USA

W. LAWRENCE DREW, Clinical Microbiological Laboratory, Mt. Zion Hospital and Medical Center, San Francisco, CA 94611, USA

RICHARD W. EMMONS, Viral and Rickettsial Disease Laboratory, Division of Laboratories, California Department of Health Services, Berkeley, CA 94704, USA

G. FADDA, Institute of Microbiology, University of Sassari, 07100 Sassari, Italy

S. FAINE, Department of Microbiology, Monash University, Clayton, Victoria, Australia 3168

GIUSEPPE FALCONE, Institute of Microbiology, University of Pisa, 56100 Pisa, Italy

W. EDMUND FARRAR, Infectious Diseases and Immunology Division, Department of Medicine, Medical University of South Carolina, Charleston, SC 29425, USA

FRANK FENNER, The John Curtin School of Medical Research, The Australian National University, Canberra 2601, Australia

BERNARD N. FIELDS, Department of Microbiology and Molecular Genetics and the Shipley Institute of Medicine, Harvard Medical School, and Division of Infectious Diseases, Department of Medicine, Brigham & Womens Hospital, Boston, MA 02115, USA

SYDNEY M. FINEGOLD, Research Service and Infectious Disease Section, West Los Angeles Wadsworth V.A. Medical Center, Los Angeles, CA 90073; and Departments of Medicine and Microbiology and Immunology, UCLA School of Medicine, Los Angeles, CA 90024, USA

DANIEL B. FISHBEIN, Viral and Rickettsial Zoonoses Branch, Division of Viral Diseases, Center for Infectious Diseases, Centers for Disease Control, Atlanta, GA 30333, USA

SUSAN FISHER-HOCH, Division of Viral Diseases and Hospital Infections Program, Center for Infectious Diseases, Centers for Disease Control, Atlanta, GA 30333, USA

D. O. FREEDMAN, Laboratory of Parasitic Diseases, National Institute of Allergy and Infectious Disease, National Institutes of Health, Bethesda, MD 20892, USA

MORRIS L. V. FRENCH, Department of Pathology, James Whitcomb Riley Hospital for Children, Indiana University Medical Center, Indianapolis, IN 46223, USA

DAMIEN A. GABIS, Silliker Laboratories, Inc., Chicago Heights, IL 60411, USA

H. R. GAMBLE, Helminthic Diseases Laboratory, Animal Parasitology Institute, Agricultural Research Service, USDA, Beltsville, MD 20705, USA

LYNNE S. GARCIA, Department of Clinical Microbiology, UCLA Medical Center, Los Angeles, CA 90024, USA

ROBERT M. GENTA, Department of Pathology and Laboratory Medicine, V.A. Hospital, University of Cincinnati Medical Center, Cincinnati, OH 45267, USA

W. LANCE GEORGE, Medical and Research Services, Wadsworth Division, West Los Angeles V.A. Medical Center, Los Angeles, CA 90073, and Department of Medicine, UCLA School of Medicine, Los Angeles, CA 90024, USA

MICHAEL A. GERBER, Department of Pediatrics, University of Connecticut School of Medicine, Farmington, CT 06032, USA

MARY ANN GERENCSER, Department of Microbiology, West Virginia University Medical Center, Morgantown, WV 26506, USA

M. J. R. GILCHRIST, Microbiology Laboratory, Veterans Administration Medical Center, Cincinnati, OH 45220, USA

A. GONZÁLEZ-MENDOZA, División de Patológia Experimental de Occidente, Unidad de Investigación Biomédica, Instituto Mexicano del Seguro Social, Centro Medico de Occidente, Guadalajara, Jalisco, Mexico

ROBERT C. GOOD, Division of Bacterial Diseases, Center for Infectious Diseases, Centers for Disease Control, Atlanta, GA 30333, USA

MONICA GRANDIEN, Department of Virology, National Bacteriological Laboratory, S-105 21 Stockhom, Sweden

I. D. GUST, Virus Laboratory, Fairfield Hospital for Communicable Diseases, Fairfield, Victoria 3078, Australia

STEPHEN C. HADLER, Division of Viral Diseases and Hospital Infections Program, Center for Infectious Diseases, Centers for Disease Control, Atlanta, GA 30333, USA

PEKKA HALONEN, Department of Virology, University of Turku, Kiinamyllynkatu 13, SF-20520, Turku, Finland

MAURICE W. HARMON, Division of Viral Diseases,

Center for Infectious Diseases, Centers for Disease Control, Atlanta, GA 30333, USA

S. M. HARMON, Division of Microbiology, Food and Drug Administration, Washington, D.C. 20204, USA

CHARLES L. HATHEWAY, Division of Bacterial Diseases, Center for Infectious Diseases, Centers for Disease Control, Atlanta, GA 30333, USA

KENNETH L. HERRMANN, Division of Viral Diseases, Center for Infectious Diseases, Centers for Disease Control, Atlanta, GA 30333, USA

JOHN C. HIERHOLZER, Respiratory and Enteric Viruses Branch, Division of Viral Diseases, Centers for Disease Control, Atlanta, GA 30333, USA

ANTHONY E. HILGER, Department of Allied Health Professions, University of North Carolina, Chapel Hill, NC 27514, USA

GEORGE V. HILLYER, Department of Pathology, University of Puerto Rico, Medical Science Campus, San Juan, Puerto Rico 00936, USA

YORIO HINUMA, Institute for Virus Research, Kyoto University, Kyoto 606, Japan

LARRY A. HOLCOMB, Hygienic Laboratory, Oakdale Hall, University of Iowa, Iowa City, IA 52242, USA

B. HOLMES, Public Health Laboratory Service, National Collection of Type Cultures, Central Public Health Laboratory, London NW9 5HT, England

IAN H. HOLMES, Department of Microbiology, University of Melbourne, Parkville 3052, Victoria, Australia

DONALD R. HOPKINS, Center for Infectious Diseases, Centers for Disease Control, Atlanta, GA 30333, USA

HENRY D. ISENBERG, Department of Microbiology, Long Island Jewish Medical Center, New Hyde Park, NY 11042, and Professor of Clinical Pathology, State University of New York at Stony Brook, Stony Brook, NY 11794, USA

WILLIAM R. JARVIS, Division of Viral Diseases and Hospital Infections Program, Center for Infectious Diseases, Centers for Disease Control, Atlanta, GA 30333, USA

STIG JEANSSON, Department of Clinical Virology, Institute of Medical Microbiology, University of Göteborg, 41346 Göteborg, Sweden

RUSSELL C. JOHNSON, Department of Microbiology, School of Medicine, University of Minnesota, Minneapolis, MN 55455, USA

JERRY E. JONES, Department of Family Practice, College of Medicine, University of Kentucky, Lexington, KY 40536, USA

S. L. JOSEPHSON, Hygienic Laboratory, Oakdale Hall, University of Iowa, Iowa City, IA 52242, USA

H. KAMO, Department of Medical Zoology, Tottori University School of Medicine, Yonago 683, Japan

JACOBA G. KAPSENBERG, Rijksinstituut voor Volksgezondheid en Milieuhygiene, Laboratorium voor Virologie, 3720 BA Bilthoven, The Netherlands

D. S. KELLOGG, Sexually Transmitted Diseases Laboratory Program, Center for Infectious Diseases, Centers for Disease Control, Atlanta, GA 30333, USA

ALAN KENDAL, Influenza Branch, Center for Infectious Diseases, Centers for Disease Control, Atlanta, GA 30333, USA

MICHAEL P. KILEY, Division of Viral Diseases, Center for Infectious Diseases, Centers for Disease Control, Atlanta, GA 30333, USA

MICHAEL M. KLIKS, Division of Comparative Medicine, John A. Burns School of Medicine, University of Hawaii, Honolulu, HA 96822, USA

T. KOYAMA, Department of Parasitology, National Institute of Health, Shinagawa-ku, Tokyo 141, Japan

DONALD J. KROGSTAD, Departments of Pathology and Internal Medicine Division of Laboratory Medicine, Washington University School of Medicine, St. Louis, MO 63110, USA

YASUO KUDOH, The First Division of Bacteriology, Department of Microbiology, Tokyo Metropolitan Research Laboratory of Public Health, Sinjuku-ku Tokyo 160, Japan

MONIQUE LAFON, Rabies Unit, Institut Pasteur, 75724 Paris Cedex 15, France

S. A. LARSEN, Sexually Transmitted Diseases Laboratory Program, Center for Infectious Diseases, Centers for Disease Control, Atlanta, GA 30333, USA

WILLIAM K. K. LAU, Department of Medicine, John A. Burns School of Medicine, University of Hawaii, Honolulu, HA 96822, USA

DIANE S. LELAND, Department of Pathology, James Whitcomb Riley Hospital for Children, Indiana University Medical Center, Indianapolis, IN 46223, USA

DAVID A. LENNETTE, Virolab, Inc., Berkeley, CA 94710, USA

EVELYNE T. LENNETTE, Virolab, Inc., Berkeley, CA 94710, USA

JAY A. LEVY, Department of Medicine and Cancer Research Institute, Universtiy of California School of Medicine, San Francisco, CA 94143, USA

C. C. LINNEMANN, JR., Division of Infectious Diseases, University of Cincinnati College of Medicine, Cincinnati, OH 45229, USA

S. A. LOCARNINI, Virus Laboratory, Fairfield Hospital for Communicable Diseases, Fairfield, Victoria 3078, Australia

E. G. LONG, Division of Parasitic Diseases, Centers for Disease Control, Atlanta, GA 30333, USA

CARLOS LOPEZ, Division of Viral Diseases and Hospital Infections Program, Center for Infectious Diseases, Centers for Disease Control, Atlanta, GA 30333, USA

ERIK LYCKE, Department of Clinical Virology, Institute of Medical Microbiology, University of Göteborg, 41346 Göteborg, Sweden

S. E. MADDISON, Division of Parasitic Diseases, Center for Infectious Diseases, Centers for Disease Control, Atlanta, GA 30333, USA

CHARLES RICHARD MADELEY, Department of Virology University of Newcastle upon Tyne and Royal Victoria Infirmary, Newcastle upon Tyne NE1 4LP, England

EL SHEIKH MAHGOUB, Department of Medical Microbiology and Parasitology, Faculty of Medicine, University of Khartoum, Khartoum, Republic of Sudan

D. D. MARTIN, Life Sciences Division, U.S. Army Dugway Proving Ground, Dugway, UT 84022, USA

HENRY M. MATHEWS, Division of Parasitic Diseases, Center for Infectious Diseases, Centers for Disease Control, Atlanta, GA 30333, USA

JOSEPH B. MCCORMICK, Special Pathogens Branch, Division of Viral Diseases, Center for Infectious Diseases, Centers for Disease Control, Atlanta, GA 30333, USA

JOSEPH E. MCDADE, Viral and Rickettsial Zoonoses Branch, Division of Viral Diseases, Center for Infectious Diseases, Centers for Disease Control, Atlanta, GA 30333, USA

MICHAEL R. MCGINNIS, Department of Pathology, University of Texas Medical Branch, Galveston, TX 77550, USA

PATRICIA A. MERZ, Department of Virology and Pathology, N.Y. State Office of Mental Retardation and Developmental Disabilities, Institute for Basic Research in Developmental Disabilities, Staten Island, NY 10314, USA

F. MEUNIER, Service de Médecine et Laboratoire d'Investigation Clinique, Clinique H. J. Tagnon, Institut Jules Bordet, 1000 Brussels, Belgium

M. MIYAJI, Department of Fungal Infections, Research Center for Pathogenic Fungi and Microbial Toxicoses, Chiba University, Inohana, Chiba 280, Japan

THOMAS P. MONATH, Division of Vector-Borne Viral Diseases, Center for Infectious Diseases, Centers for Disease Control, Fort Collins, CO 80522, USA

NELSON P. MOYER, Hygienic Laboratory, Oakdale Hall, University of Iowa, Iowa City, IA 52242, USA

FREDERICK A. MURPHY, Center for Infectious Diseases, Centers for Disease Control, Atlanta, GA 30333, USA

PATRICK R. MURRAY, Division of Laboratory Medicine, Department of Pathology, Washington University School of Medicine, St. Louis, MO 63110, USA

K. DARWIN MURRELL, Helminthic Diseases Laboratory, Animal Parasitology Institute, Agricultural Research Service, USDA, Beltsville, MD 20705, USA

KAZUNARI NAKAMURA, The First Department, National Institute for Leprosy Research, Higashimurayama, Tokyo 189, Japan

JAMES H. NAKANO, Center for Infectious Diseases, Centers for Disease Control, Atlanta, GA 30333, USA

THOMAS R. NAVIN, Division of Parasitic Diseases, Centers for Disease Control, Atlanta, GA 30333, USA

RICARDO NEGRONI, Centro de Micrología de la Cátedra de Microbiología, Parasitología, e Immunología, Facultad de Medícine, Universidad de Buenos Aires, Buenos Aires, Argentina

ERLING NORRBY, Department of Virus Research, Karolinska Institutet, S-105 21 Stockholm, Sweden

T. B. NUTMAN, Laboratory of Parasitic Diseases, National Institute of Allergy and Infectious Disease, National Institutes of Health, Bethesda, MD 20892, USA

MAKOTO OHASHI, Tokyo Metropolitan Research Laboratory of Public Health, Shinjuku-ku, Tokyo 160, Japan

CLAES ÖRVELL, Department of Virology, National Bacteriological Laboratory and Karolinska Institute, School of Medicine, S-105 21 Stockholm, Sweden

SIOBHAN O'SHEA, Department of Virology, United Medical and Dental Schools of Guys' and St. Thomas' Hospitals, St. Thomas' Campus, London SE1 7EH, England

ROBERT L. OWEN, Departments of Medicine, Epidemiology, and International Health, University of California–San Francisco, and Cell Biology and Aging Section, V.A. Medical Center, San Francisco, CA 94121, USA

A. A. PADHYE, Division of Mycotic Diseases, Center for Infectious Diseases, Centers for Disease Control, Atlanta, GA 30333, USA

NICHOLAS E. PALUMBO, Department of Medicine, John A. Burns School of Medicine, University of Hawaii, Honolulu, HA 96822

DEMOSTHENES PAPPAGIANIS, Department of Medical Microbiology and Immunology, University of California-Davis School of Medicine, Davis, CA 95616, USA

HERBERT PFISTER, Institut für Klinische und Molekulare Virologie, Universität Erlangen-Nürnberg, D-8520 Erlangen, West Germany

PETER PIOT, Department of Microbiology, Institute of Tropical Medicine (Prince Leopold), B-2000 Antwerp, Belgium

F. PIRALI, Institute of Microbiology, Brescia University, 25100 Brescia, Italy

HELEN M. POLLOCK, Department of Pathology, University of South Alabama Hospitals and Clinics, Mobile, AL 36617, USA

THOMAS J. QUAN, Plague Branch, Division of Vector-Borne Viral Diseases, Center for Infectious Diseases, Centers for Disease Control, Ft. Collins, CO 80522, USA

H. S. RANDHAWA, Department of Medical Mycology, Vallabhbhai Patel Chest Institute, University of Delhi, Delhi 110 007, India

M. RANKI, Orion Genetic Engineering Laboratory, Orion Corporation Ltd., SF-00380 Helsinki, Finland

ANNETTE C. REBOLI, Infectious Diseases and Immunology Division, Department of Medicine, Medical University of South Carolina, Charleston, SC 29425, USA

H. B. REES, JR., 3584 Millstream Lane, Salt Lake City, UT 84109, USA

MARCIA L. RHOADS, Helminthic Diseases Laboratory, Animal Parasitology Institute, Agricultural Research Service, USDA, Beltsville, MD 20705, USA

F. O. RICHARDS, JR., Division of Parasitic Diseases, Center for Infectious Diseases, Centers for Disease Control, Atlanta, GA 30333, USA

LEE W. RILEY, Divisions of Infectious Diseases and Geographic Medicine, Stanford University Medical Center, Stanford, CA 94305, USA

HAN-JONG RIM, Department of Parasitology, Institute for Tropical Endemic Diseases, College of Medicine, Korea University, Chongno-ku, Seoul 110, Korea

MICHAEL G. RINALDI, Department of Pathology, University of Texas Health Science Center at San Antonio and Veterans Administration Mycology Reference Laboratory, Laboratory Service, Audie L. Murphy Memorial Veterans' Hospital, San Antonio, TX 78284, USA

ALLAN R. RONALD, Department of Medicine, Faculty of Medicine, University of Manitoba, Winnipeg, Manitoba R3E 0Z3, Canada

TRENTON K. RUEBUSH, II, Division of Parasitic Diseases, Center for Infectious Diseases, Centers for Disease Control, Atlanta, GA 30333, USA

RAYMOND W. RYAN, Department of Laboratory Medicine, Division of Clinical Microbiology, University of Connecticut School of Medicine, Farmington, CT 06032, USA

G. SATTA, Institute of Microbiology, University of Siena Medical School, Siena, Italy

JULIUS SCHACHTER, Department of Laboratory Medicine, University of California-San Francisco, San Francisco, CA 94143, USA

PETER M. SCHANTZ, Helminthic Diseases Branch, Division of Parasitic Diseases, Center for Infectious Diseases, Centers for Disease Control, Atlanta, GA 30333, USA

H. J. SHADOMY, Department of Microbiology and Immunology, Virginia Commonwealth University, Richmond, VA 23298, USA

KEERTI V. SHAH, Department of Immunology and Infectious Diseases, The Johns Hopkins University, School of Hygiene and Public Health, Baltimore, MD 21205, USA

ROBERT E. SHOPE, Yale Arbovirus Research Unit, Department of Epidemiology and Public Health, School of Medicine, Yale University, New Haven, CT 06510, USA

JOHN H. SILLIKER, Silliker Laboratories of California, Inc., Carson, CA 90746, USA

JAMES W. SMITH, Department of Pathology, Indiana University School of Medicine, Indianapolis, IN 46223, USA

M. A. SMITH, Environmental Technology, Utah Technical College, Orem, UT 84058, USA

H. SÖDERLUND, Orion Genetic Engineering Laboratory, Orion Corporation Ltd., SF-00380 Helsinki, Finland

S. J. STEWART, Laboratory of Microbial Structure and Function, Rocky Mountain Laboratory, NIAID, National Institutes of Health, Hamilton, MT 59841, USA

PIERRE SUREAU, Rabies Unit, Institut Pasteur, 75724 Paris Cedex 15, France

NORMAN S. SWACK, Hygienic Laboratory, Oakdale Hall, University of Iowa, Oakdale Hall, Iowa City, IA 52242, USA

A.-C. SYVÄNEN, Orion Genetic Engineering Laboratory, Orion Corporation Ltd., SF-00380 Helsinki, Finland

MICHIAKI TAKAHASHI, Department of Virology, Research Institute for Microbial Diseases, Osaka University, Suita, Osaka, Japan

GREGORY A. TANNOCK, Respiratory and Enteric Viruses Branch, Divsion of Viral Diseases, Center for Infectious Diseases, Centers for Disease Control, Atlanta, GA 30333, USA

A. A. TERRENI, Bureau of Laboratories, South Carolina Department of Health and Environmental Control, Columbia, SC 29202, USA

RICHARD C. TILTON, Division of Microbiology, University of Connecticut School of Medicine, Farmington, CT 06032, USA

M. TSUBOKURA, Department of Veterinary Microbiology, Faculty of Agriculture, Tottori University, Koyama-cho, Tottori 680, Japan

A. TURANO, Institute of Microbiology, Brescia University, 25100 Brescia, Italy

KENNETH L. TYLER, Department of Microbiology and Molecular Genetics, Harvard Medical School, and Department of Neurology, Massachusetts General Hospital, Boston, MA 02115, USA

DAVID A. J. TYRRELL, MRC Common Cold Unit, Harvard Hospital, Salisbury, Wiltshire SP2 8BW, England

P. VAN DER AUWERA, Service de Médecine et Laboratoire d'Investigation Clinique, Clinique H. J. Tagnon, Institut Jules Bordet, 1–1000 Brussels, Belgium

M. H. V. VAN REGENMORTEL, Institute de Biologie Moléculaire et Cellulaire de CNRS, 67084 Strasbourg Cedex, France

P. E. VARALDO, Institute of Microbiology, University of Ancona Medical School, 60131 Ancona, Italy

NEYLAN A. VEDROS, Department of Biomedical and Environmental Health Sciences, School of Public Health, Department of Biomedical and Environmental Health Sciences, Earl Warren Hall, University of California, Berkeley, CA 94720, USA

G. S. VISVESVARA, Division of Parasitic Diseases, Center for Infectious Diseases, Centers for Disease Control, Atlanta, GA 30333, USA

G. WADELL, Department of Virology, University of Umea, S-901 87, Umea, Sweden

DUARD L. WALKER, Department of Medical Microbiology, University of Wisconsin School of Medicine, Madison, WI 53706, USA

KENNETH W. WALLS, 4006 Northlake Creek Court, Tucker, GA 30084, USA

N. G. WARREN, Division of Consolidated Laboratory Services, Microbiological Sciences Bureau, Richmond, VA 23298, USA

HARUO WATANABE, Department of Bacteriology, National Institute of Health, Kamiosaki, Shinagawa-ku, Tokyo 141, Japan

IRENE WEITZMAN, Bureau of Laboratories, Department of Health, City of New York, New York, NY 10016, USA

HAZEL W. WILKINSON, Division of Bacterial Diseases, Center for Infectious Diseases, Centers for Disease Control, Atlanta, GA 30333, USA

HENRYK M. WISNIEWSKI, New York State Office of Mental Retardation and Developmental Disabilities, Institute for Basic Research in Developmental Disabilities, Staten Island, NY 10314, USA

H. YAMAGUCHI, Institute of Applied Microbiology, University of Tokyo, Bunkyo-ku, Tokyo 113, Japan

NAOKI YAMAMOTO, Department of Virology and Parasitology, Yamaguchi University School of Medicine, Ube, Yamaguchi 755, Japan

YUTAKA ZINNAKA, Department of Microbiology, Toho University School of Medicine, Ohta-ku, Tokyo 143, Japan

ARIE J. ZUCKERMAN, Department of Medical Microbiology and WHO Collaborating Centre for Reference and Research on Viral Hepatitis, London School of Hygiene and Tropical Medicine, London WC1E 7HT, England

Color Plate 1

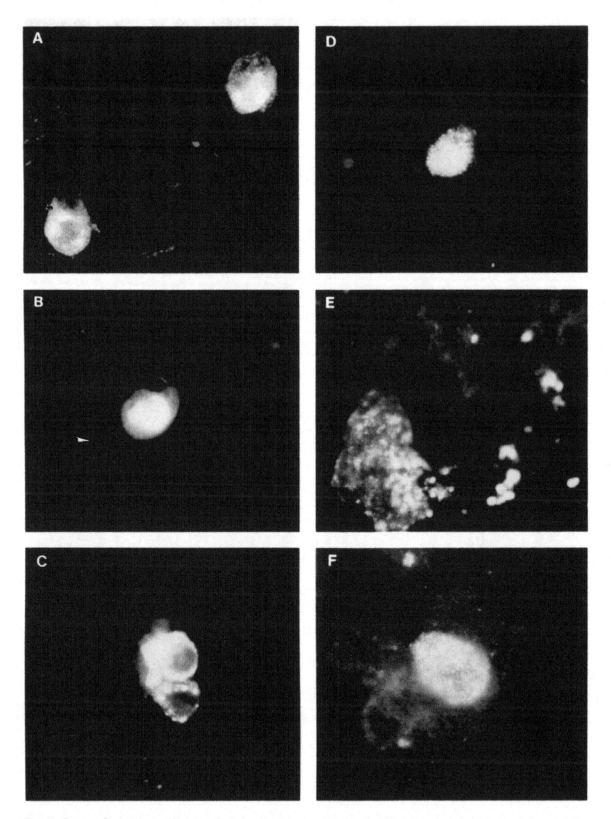

FIG. 1. Immunofluorescence of virus-infected cells from nasopharyngeal aspirates (A to E) and of an impression smear from a brain biopsy specimen (F). A: Respiratory syncytial virus. B: Influenza A virus. C: Parainfluenza 3. D: Adenovirus. E: Nonspecific staining with mucus. F: Herpes simplex virus. See page 64.

Color Plate 2

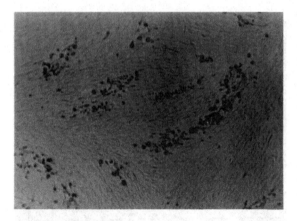

FIG. 1. Cytopathogenic effect due to cytomegalovirus in WI-38 cells. See page 252.

FIG. 2. Cytomegalovirus-infected cells positive by immunofluorescence. See page 253.

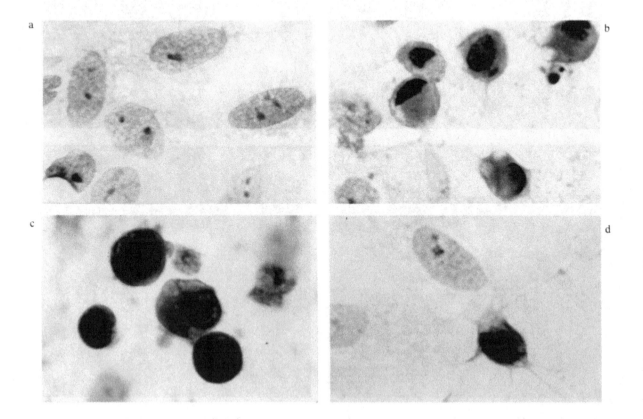

FIG. 3. Enterovirus cytopathic effects in tertiary monkey kidney cells stained after 2 days. a: Cell control; b: echovirus 6; c: echovirus 22; and d: enterovirus 71. See page 699.

SECTION I
General Principles of Viral Diagnostics

CHAPTER 1

Specimen Collection and Transport

CHARLES RICHARD MADELEY, DAVID A. LENNETTE, and PEKKA HALONEN

Introduction

Diagnostic virology is expensive and time consuming. It is often practiced in laboratories not immediately adjacent to the patient, although it is now likely that some tests that are simple to perform will be available for use in the doctor's office or surgery in the next few years. However, no matter where it is done, who it is done by, or what technique(s) are used, it will be wasted time, effort, and resources if the specimens taken are poor, ill timed, or incorrectly handled between the time of collection and the time laboratory procedures are begun.

Many viral infections present as syndromes rather than specific illnesses. Consequently, only a minority of such infections can be readily diagnosed on clinical grounds alone and when the need to establish the diagnosis is important (e.g., prior to chemotherapy, isolating the patient, or contact tracing), it is essential to give the laboratory the best possible chance of identifying the virus concerned by isolation or by detection of virus or viral antigen, viral nucleic acid(s), or induction of specific antibody. In this chapter it is assumed that specimens are being taken to make a diagnosis in an individual. Hence, the probability of a particular virus being found is not relevant, with each specimen being taken and handled to give the best chance of identifying any virus present.

Viruses require living cells to replicate and will not increase in amount in a diagnostic specimen after it has been taken until the virus has been put in contact with suitable cells in the laboratory. In the interval the amount of intact virus can only decrease, at a rate that will depend greatly on the virus concerned and the conditions of storage and transport. Whether any survives at a detectable level will be influenced strongly by the amount of virus in the original specimen and the care with which it is looked after en route to the laboratory.

An understanding of these constraints is essential to any diagnostic virologist and to the clinician. It is more important to the virologist because his work will be impossible if the quality of specimens received is poor. He has therefore a strong interest in influencing the clinician's approach to specimen taking and transport, and the staff of any successful laboratory will spend much time and effort encouraging local clinicians. They will never be able to take a supply of good specimens for granted and ensuring a reasonable proportion will mean telephone calls, preparing circulars and handouts, lectures and pestering colleagues over lunch and in the corridors. It is never time wasted, but the effort can never be relaxed.

Specimen taking and transport is therefore central to all diagnostic virology. Conditions and experience will dictate local variations in the fine details, but it is essential that the principles discussed in this chapter should be understood and applied as appropriate.

Taking the Specimen

From the foregoing it should be apparent that the more virus obtained from the patient, the more likely it is to survive to the laboratory. The best specimens to be taken are discussed in the next section, but some initial comments about specimen quantities are necessary.

With adults, the amount of feces or urine that can be readily obtained is usually far more than is needed to make the diagnosis, and laboratories have objected in the past to large quantities being sent on the grounds that they are not disposal units. This can indeed be true, but the specimen taker should remember that, with viruses that cannot be grown to high titer routinely, the specimen provides the laboratory with all the virus, antigen, or both they will get. This will have to provide enough material both to make the initial diagnosis and, where appropriate, also the raw material for research. If in doubt, it is better to send a larger quantity; the laboratory can discard what it does not need. Arrangements for collection will have to be made, particularly with adults, but acquisition of the specimen is painless to the patient. He has no further need of it, but may be embarrassed over providing feces, and tact may be needed.

Although providing a blood specimen is not contingent on the excretory activities of the patient and can be taken at any time, the quantity available is limited by practical reality. Obtaining more than a small quantity from a neonate is all but impossible and may be dangerous. However, much larger amounts than the normal 2 to 3 ml of venous blood can be obtained as cord blood from the placenta or by a femoral stab into the external femoral artery as it emerges from beneath the inguinal ligament. This is a site of last resort; if viremia is suspected or diagnosis is essential (in, for example, congenital hepatitis B), however, the risk may be justified.

In older children and adults, larger quantities may be taken routinely (although there is an upper limit partly provided by physiology and partly by the patients' willingness to stay in the same room with an operator with a large syringe in his hand). In deciding what size of syringe to use, take into account all the likely needs by considering the following questions:

1. How much is needed for antibody studies? Do not forget that several tests may be necessary on one specimen of blood and that some or all of them may have to be repeated, sometimes several times.
2. How much is needed for viral isolation or detection of viral antigen? These tests may also require repetition, and it may be necessary to send both heparinized and clotted blood specimens.
3. For what other purposes is blood needed at this moment? These purposes may include blood counts, hemoglobin estimation, biochemistry, and the like.

Adding up the necessary quantities can mean taking a larger syringe than originally intended, but it is better to take one large amount (with suitable prior explanation, if possible) than several small ones.

The taking of other specimens will, with very few exceptions, involve some discomfort for the patient. This must be clearly recognized and accepted before starting. Once the decision has been made, the actual taking of the specimen should not be half-hearted; the specimen must be firmly and properly taken even though the patient loathes the experience. Anything less does no service to anyone, least of all to the patient, and this should be explained to him or his relatives beforehand. With infants, the cooperation of the parents in holding the patient while the specimen is taken can be essential for success, but older children frequently cooperate better if the parents are asked to wait in an adjacent room. There are no rules other than getting an adequate specimen, and the operator will have to use his judgment in individual cases.

The specimens appropriate to various clinical syndromes are listed in the next section, but some practical hints on obtaining particular types of specimens follow:

Respiratory Specimens

NASOPHARYNGEAL SECRETIONS

The basic necessary equipment for collection of nasopharyngeal secretions consists of a suction pump connected to a catheter through a mucus trap, with airtight connections between them.

The pump must be portable, and suitable ones are often carried in ambulances for clearing airways in patients on their way to hospital (Fig. 1). They can be electrically or battery-powered and should be neither too large nor too small. A typical specification would be a pump capable of reducing the pressure in a closed container to about 600 mm Hg and create a flow rate of water of about 500 ml in 5 s through a French gauge 8 catheter. It should be connected to the shorter of the connections on the mucus trap, the lid of which should be removable.

The catheter is the critical part of the apparatus. It should be long and thin enough to enter the nose and reach as far as the posterior pharyngeal wall with

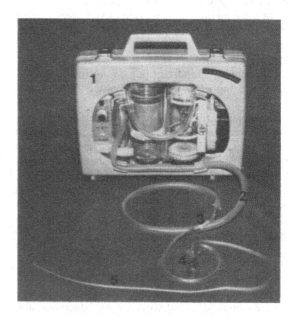

FIG. 1. Apparatus for taking nasopharyngeal secretions. (1) Portable suction pump with battery or mains operation. (2) Wide-bore (9 mm internal diameter) connecting tubing. (3) Y-piece. Note that one leg is not connected; it is variably occluded by one thumb in use to control the amount of suction through the catheter. (4) Sterile plastic mucus trap with detachable airtight lid. (5) Sterile catheter (French gauge 8 for infants and small children, French gauge 12 for adults).

ease. Other than to provide enough for the operator to hold, it needs to be no longer, and cutting off the surplus will shorten the distance for the mucus to travel before reaching the trap. The end should be rounded with a terminal hole. There is no advantage in having side holes as well, because they may allow dilution of the specimen with anterior secretions or anterior squamous cells.

The gauge of the catheter is important and should be selected to suit the size of the patient's nasal passages. Sucking the secretions off the surface of an inflamed mucosa will be unpleasant, but the degree of discomfort appears to be related to the speed of the airflow into the catheter. Hence, the procedure can be made less unpleasant by using as large a gauge as possible. With infants only a small size (e.g., French gauge 8) can be used, but with adults a larger size (e.g., French gauge 12) will prove more acceptable.

The specimen should be taken from the posterior part of the nasal mucosa, which is lined with respiratory epithelium, and not from the anterior part, which is lined with squamous epithelium. *Both* nasal passages should be thoroughly harvested, with the catheter inserted as far into the nostrils as possible. The harvesting should be both prolonged and thorough; success in diagnosis depends very much on the quantity obtained. It is impossible to collect too large a quantity of secretions, but it is very easy to collect too little.

After collecting as much of the nasopharyngeal secretions as possible, they should be sucked into the trap. With thick tenacious mucus this will be difficult, but the temptation to wash the secretions into the trap with a little saline should be resisted. It will dilute the virus-protective properties of the mucus. Beware impatience!

The suction line to the trap should now be detached, the catheter looped around, and the end pushed onto the vacated connection; it may be necessary to snip off the tip with scissors to make it fit. The trap–catheter combination, which is now airtight and therefore leak-proof, should be sent on ice to the laboratory in toto and without delay (see Transporting Specimens to the Laboratory, below, for further details). No transport medium should be added. However, if the laboratory uses only enzyme immunoassays or similar tests to detect viral antigen and does not use immunofluorescence and viral isolation, the specimen can be sent without cooling (Halonen et al., 1985).

NOSE AND THROAT SWABS

Nose and throat swabs must be taken with equal thoroughness. Nose swabs should be taken from the turbinates, *not* from the anterior nares. This will require a hard heart on the part of the operator; swab-

bing an inflamed mucosa is uncomfortable. Both nasal passages should be sampled using plain sterile cotton-wool swabs. Alginate or charcoal swabs should *not* be used; the former can inactivate the infectivity of some viruses and chlamydiae, and the latter are overabsorptive.

The throat swab must be taken from the posterior pharyngeal wall, *not* the tonsils, the tongue, or the palate. The patient should be made comfortable, in a good light, and told to relax. Even with a good light it may be essential to have an assistant with a pen-sized flashlight to ensure that the operator can see clearly.

The posterior wall is then swabbed firmly, briskly, and thoroughly (as indicated by the alternative name for this specimen, a "cough" swab). A good throat swab has been defined as the one removed half a second before the vomit comes up. It is not necessary to make the patient literally sick, but risking it indicates the degree of thoroughness necessary.

Both swabs should be broken off into a suitable viral transport medium (see below); excessively long swab-stick residues may cause the vials to leak by preventing complete closure.

BRONCHOALVEOLAR LAVAGE

Bronchoalveolar lavage specimens are superior for detection of viral infections of the lung and lower respiratory tract. Details of their collection have been reported by Springmeyer et al. (1986). A high proportion of the cells obtained will be leukocytes and squamous cells, but some respiratory cells are normally recovered, and any virus recovered from this level is likely to be significant. At least 60 ml of warmed saline or viral transport medium is introduced through a bronchoscope and removed by suction into a suitable container. There is evidence (Kelly et al., 1987) that a second 60-ml lavage will penetrate more peripherally and may prove a more suitable specimen for virology, although this has yet to be demonstrated. It is also important that the bronchus selected for washout is the one serving an affected portion of lung. The fluid (and cells) should be sent on ice rapidly to the laboratory without further addition.

Neurologic Specimens

CEREBROSPINAL FLUID

For adequate diagnosis, enough cerebrospinal fluid is needed for both viral isolation and serology. This may conflict with usual teaching, that is, it is only safe to take a minimum quantity of cerebrospinal fluid without the risk of "coning" the cerebellum into the foramen magnum. Good judgment is clearly necessary, but the quantities sent to the laboratory are often too small to be useful and at least 2 ml should be obtained if possible. This quantity is for virology alone and is in addition to that needed for bacteriology, cytology, or clinical biochemistry.

Virus or antibody is usually present in low concentration. Accordingly, cerebrospinal fluid should not be diluted in viral transport medium, but sent in a sterile dry container without delay. For serology it is essential to take a serum specimen for testing in parallel, to establish serum–cerebrospinal fluid antibody ratios.

Skin Specimens

VESICLE FLUID

Fluid from vesicular lesions can be taken with a 1-ml syringe and fine (25-gauge) needle. The vesicle fluid can be used for isolation or electron microscopy; for the latter, a minute quantity will often be sufficient, literally only enough to moisten the butt of the needle. This should be left in the needle and syringe and the combination sent to the laboratory, with the plunger still withdrawn and, if necessary, fixed with adhesive tape. No viral transport medium should be added.

Done carefully, this procedure is totally painless, and it is usually worth telling the patient so beforehand. Most people are nervous about "needles" and expect any procedure involving them to hurt. If the needle is held parallel to the skin and gently inserted into the vesicle, the patient may not even be aware that vesicle fluid is being taken. True vesicle fluid from viral lesions is usually water clear, and 1 to 2 μl can be obtained from a lesion without difficulty.

Although the vesicles of orf (paravaccinia) appear large and juicy the amount of fluid, which can be thick and turgid, may be insufficient even for electron microscopy. However, virus is usually plentiful in crusts or the fleshy parts of the lesion.

INDIVIDUAL NONVESICULAR LESIONS

Individual nonvesicular lesions are the only instance in which virus-induced lesions are sampled directly. Virus is usually present where the lesion is developing; specimens should be taken from the active edge of ulcers, by scraping for immunofluorescence, and by swabbing for isolation. The scrapings should be teased out in saline using two mounted needles on areas measuring about 1 cm^2, marked out with a diamond on a microscope slide(s). The spread cells are then allowed to dry. Swabs must be put in viral transport medium without delay.

Specimens of warts or molluscum contagiosum should be snipped from the lesion and sent to the laboratory in a clean sterile vial without viral transport medium. In the case of genital warts, in which the amount of virus will be small, a bigger biopsy specimen should be obtained under local anesthesia.

There is insufficient or no virus in the lesions of maculopapular rashes (measles, rubella, "fifth" disease, and the like) to make sampling worthwhile.

Urogenital Specimens

Specimens may be collected from lesions of the external genitals as for any other skin lesions. For patients with urethritis, an extrafine swab (Micro Diagnostics MD160 or equivalent) should be used to obtain urethral specimens by gentle insertion to a depth of 2 cm, if possible. Recovery of viruses or chlamydiae from urethral exudate is poor compared with results obtained with a properly collected swab. Cervical specimens are best obtained by the use of several swabs (Embil et al., 1982; Mårdh et al., 1981). If the first is used to clean the cervix of mucus and pus, better results will be obtained from subsequent swabs taken as specimens for isolation: the greater the number of cells that are recovered, the higher will be the isolation rate of viruses and chlamydiae. If any lesions are present, they should be swabbed; otherwise, a swab should be inserted about 1 cm into the cervical canal and rotated. Semen is a good specimen for recovery of cytomegalovirus (Howell et al., 1986); a sample of 1 to 2 ml should be diluted in *Chlamydia* transport medium (see Table 4) for best results. Seminal fluid is toxic to cell cultures; the specimen should be diluted and centrifuged, and only the cell pellet inoculated for testing.

Eye Specimens

Swabs taken from the conjunctivae or cornea (from ulcers) are suitable for isolation of viruses, although the yield may be generally low. They are not suitable for other diagnostic techniques, but have no serious side effects. As with other swabs they should be taken thoroughly to pick up as much virus as possible and put into viral transport medium without delay. Each eye should be swabbed separately and the swabs put into separate vials of VTM.

Corneal or conjunctival scrapings provide better specimens for diagnosis and are the only ones suitable for rapid techniques such as immunofluorescence. They are uncomfortable for the patient and carry a risk of symblepharon (in which the eyelids become welded together and may have to be separated surgically). For these reasons some ophthalmologists are unwilling to take them, but if obtained they should be prepared as outlined for skin scrapings.

Gut Specimens

Obtaining stool from infants still in diapers and from bedridden hospital patients presents few problems. Stool can be collected directly from the diaper or from the bedpan whenever they are passed; from ambulatory patients of all ages, however, collection can be more difficult. Containers of various kinds suspended over the toilet are unsuccessful at worst and inhibitory to the patient at best. A possible solution is to put a layer of toilet paper several sheets thick in the toilet beforehand. This will prevent the stool from sinking before a specimen is taken, and cross-contamination can be kept to a minimum by flushing immediately before the procedure is begun.

The specimen can be collected in a clean container (plastic or waxed paper). It need not be sterile, and no viral transport medium should be added. With fluid feces, the lid must be watertight.

Rectal swabs make a very poor alternative specimen to feces. The quantity is inevitably smaller and the failure to isolate a virus from a swab when one is readily isolated from a subsequent stool makes it seem likely that the material recovered on a swab is not, in any case, representative of colonic contents. Unsatisfactory for isolation, a rectal swab is totally useless for electron microscopy.

Postmortem/Autopsy Specimens

Viral titers decline rapidly after the death of the patient, whereas bacteria will multiply and interfere with attempts to isolate virus. These processes will be slowed by refrigeration, but the postmortem should be done as soon as possible. A guide to suitable specimens is given in Table 1.

Where a complete postmortem is done, the collection of satisfactory quantities of relevant tissues is not difficult, although a different set of sterile instruments should be used for each tissue to avoid cross-contamination. To keep bacterial contamination to a minimum, the tissues should be taken in a planned sequence. Wherever there is a choice of site for sampling, the most virus will be found at the active edge of the lesion rather than in a necrotic center or in a region of normal tissue. Exceptions to this rule will be found, but careful selection of the best area to sample is needed with large organs such as the liver, lungs, or brain. Cubes measuring 1 cm^3 or less should be taken and put in viral transport medium on ice for transit to the laboratory. Random sampling is a wasted opportunity, and the laboratory staff should be prepared to advise the pathologist before or during the postmortem over the taking of optimum specimens.

TABLE 1. Postmortem specimens

Syndrome	Specimens required[a]
Respiratory disease	R & L lung,[b] tracheal swab,[b] R & L bronchial swabs,[b] blood[c]
Disease of central nervous system	Brain[b] spinal cord,[b] meningeal swab,[b] cerebrospinal fluid, large bowel contents, blood
Cardiovascular disease	Myocardium,[b] large bowel contents, pericardial fluid (if any), blood[c]
Skin disease	Vesicular/pustular fluid,[d] crusts, ulcer scrapings or tissue, large bowel contents, blood[c] Respiratory specimens may also be appropriate
Hepatitis	Liver,[b] large bowel contents, blood[c]
Abortion (therapeutic or accidental)	Fetus, placenta, cord blood
Undiagnosed fever	Brain,[b] liver,[b] R & L lung,[b] tracheal swab,[b] R & L bronchial swabs,[b] spleen,[b] kidney,[b] pleural fluid, peritoneal fluid, cerebrospinal fluid, large bowel contents, blood[c]
Acquired immunodeficiency syndrome	Blood,[e] other systems for opportunist infections as necessary

[a] Specimens of choice in bold type.
[b] Tissue put in cold (4°C) viral transport medium for transit to laboratory.
[c] Postmortem blood will be lysed to some degree; this may limit its usefulness for viral isolation and some serologic tests.
[d] Obtained with 1-ml syringe and 25-gauge needle; send as syringe and contents.
[e] For antibody estimation.

Specimens of some tissues may be taken by syringe and needle or by Tru-Cut needle (Baxter Health Care Inc., Deerfield, IL) immediately after death. These *may* yield viral isolations but, because they are more or less random small samples of the whole organ, they provide a less than satisfactory isolation–diagnostic rate compared with a full postmortem.

Specimens for Routine Tests

Specimens used in routine testing are listed in Table 2. There are four important rules to observe:

1. Understanding the disease process allows specimens to be taken from accessible sites rather than from the site of maximum damage, which may be less accessible.

TABLE 2. Specimens for routine tests by body system affected

System or illness	Specimens required[a]		
	For direct examination	For viral isolation	For serology
Respiratory	Nasopharyngeal aspirate (IF, EIA) (throat washings, EM)[b] Bronchoalveolar lavage (IF, EIA), lung biopsy (IF, EIA)	Nasopharyngeal aspirate, nose and throat swabs,[c] bronchoalveolar lavage, lung biopsy	Paired sera
Gut	Stool (not rectal swab[d]) (EM, EIA) (vomitus, EM)[b]	Stool (rectal swab)[b]	(Paired sera)[e]
Pyrexial illness	Nasopharyngeal aspirate (IF, EIA) Fresh urine (EM and IF)	Nasopharyngeal aspirate, nose and throat swabs,[c] fresh urine, stool, heparinized blood (for arboviruses and arenaviruses)	Paired sera
Central nervous system	CSF (IF, EIA, and EM), corneal impression smears (IF)[f] (brain biopsy, IF and EM)[g]	Stool, heparinized blood (for arboviruses and HIV), CSF, nose and throat swabs,[c] urine (brain biopsy)[g]	Paired sera, CSF[h]
Skin	Vesicle fluid (EM), ulcer scrapings (as smears, IF) Crusts (EM), biopsy (EM)	Vesicle fluid, ulcer swabs,[c] stool, nose and throat swabs,[c] heparinized blood	Paired sera[i]
Eye	Conjunctival scrapings (as smears, IF)	Conjunctival swabs[c] Corneal swabs[c,j]	Paired sera
Liver	Serum[k]	Heparinized blood (for yellow fever)	Single serum
Cardiovascular system	Cardiac biopsy (EM)[g]	Stool, nose and throat swabs,[c] cardiac biopsy, pericardial fluid[g]	Paired sera
Congenital infections or abnormalities	Ulcer scrapings (as smears, IF) Fresh urine	Nose and throat swabs,[c] (nasopharyngeal aspirate) Fresh urine, stool	Single serum from both mother and infant[l]
Glandular fever	Nasopharyngeal aspirate (IF, EIA)[m]	Nasopharyngeal aspirate,[m] nose and throat swabs,[c,m] fresh urine	Paired sera
Genital	Vesicle fluid (EM), scraping (IF), bubo pus (LM)[n]	Urethral swab, endocervical swab, heparinized blood,[o] bubo pus	Paired sera

[a] IF = Immunofluorescence; EM = electron microscopy; EIA = enzyme immunoassay; CSF = cerebrospinal fluid; HIV = human immunodeficiency virus.
[b] Parentheses indicate specimens of second choice, yielding virus less commonly.
[c] In viral transport medium.
[d] Rectal swabs rarely contain detectable levels of virus.
[e] There are no serologic tests for most viruses associated with diarrhea.
[f] For rabies only.
[g] Rarely taken, but useful if available.
[h] Both serum and CSF should be obtained, for testing in parallel.
[i] First specimen can be tested by EM for parvovirus.
[j] If corneal ulceration is present.
[k] For detection of HB_sAg, HB_eAg, and other nonantibody hepatitis markers.
[l] Both are necessary, but a cord blood specimen from the infant is satisfactory.
[m] For detection of viruses other than EB virus.
[n] For light microscopy.
[o] For lymphocytes.

2. If in doubt whether to send a particular specimen, it is better to err on the side of sending too many rather than too few. "Send everything from every case" is a maxim that must be considered in all cases, even if it is modified later.
3. If in doubt, the laboratory staff will be willing to discuss which specimens should be taken and how they should be sent to the laboratory.
4. Viral infections are usually cleared within 5 days. Therefore, nothing is gained by a delay in taking a specimen.

Viral Transport Medium

The constituents of suitable viral transport media are indicated in Table 3 and of *Chlamydia* transport media in Table 4. They are designed to provide an isotonic solution containing protein, antibiotics to control any bacterial contamination, and a buffer to control the pH.

Viral transport medium is primarily used for transporting swabs, which should be plain cotton ones broken off into the medium, and small pieces of tissue. It may be added to other specimens where inactivation is more likely and where the resultant dilution is acceptable.

A variety of commercially produced transport media are available. They are more expensive than media prepared in the laboratory and there is no good evidence that they are better. The use of semi-solid media, such as Stuart's, has been suggested particularly for multidisciplinary laboratories where the same specimen could be used to isolate bacteria and viruses. Such a compromise solution will reduce the efficiency of viral recovery, however, and it is better to take additional specimens specifically for virology. Transport media have a finite shelf-life, which should not be exceeded.

Transporting Specimens to the Laboratory

General Considerations

This section deals only with transporting specimens from the patient to a local laboratory. If they must be sent further to a more distant laboratory, whether in the same country or abroad, see Transport Between Laboratories.

In transporting the specimens, two important aspects must be considered:

1. Keeping the viral material in the specimen as intact (and as infective) as possible; and

2. preventing others from becoming infected by the specimen.

The amount of virus in the specimen will not increase after it has been taken, but will decline at a rate that will vary depending on several factors.

TABLE 3. Four modifications of viral transport media

Modification	Constituents
A[a]	Hanks balanced salt solution (HBSS), 86 ml
	Bovine albumin solution (Armour fraction V in distilled water), 10%, 10 ml
	Sodium bicarbonate ($NaHCO_3$), 5.6%, 1.5–2.0 ml w/v in distilled water
	Penicillin (10^4 U/ml) and streptomycin (10 mg/ml) in phosphate-buffered saline solution A (PBSA), 1 ml
	Nystatin (2,500 IU/ml) in PBSA, 1 ml
	Phenol red, 0.4% in distilled water, 0.5 ml Total, 100 ml
B[a]	Phosphate-buffered saline A (Dulbecco A, PBSA), 86 ml
	Skimmed milk, pasteurized, 10 ml
	Sodium bicarbonate ($NaHCO_3$), 5.6%, 1.5–2.0 ml w/v in distilled water
	Gentamicin, 5 mg in 1 ml PBSA
	Fungizone, 5 mg in 1 ml PBSA
	Phenol red, 0.4% in distilled water, 0.5 ml Total, 100 ml
C[b]	HBSS with 5% gelatin
D[c]	Tryptose phosphate broth
	Bovine albumin fraction V, 0.5%
	Gentamicin, 50 μg/ml
	Fungizone, 50 μg/ml
	Vancolin, 50 μg/ml
	Cefuroxime sodium, 60 μg/ml

[a] The constituents are prepared separately and sterilized by autoclaving (HBSS or PBSA), filtration, or pasteurization. They are combined using an aseptic technique and distributed in 2-ml quantities in firmly closed screw-cap glass vials. This medium has a shelf-life of about 3 weeks at 4°C. The medium should be a pale orange color; it should be rejected if it is purple, yellow, or cloudy or contains "floaters."

[b] To make 1 liter of medium, dissolve 5.0 g of gelatin in a small volume of very hot water, then add the gelatin solution to HBSS sufficient to make 1 liter, with rapid stirring to ensure that the gelatin is well mixed. Dispense 2.5-ml volumes into screw-cap vials, then autoclave the vials at 121°C for 15 min. The vial caps must be tightened to prevent pH changes unless a nonvolatile buffer, such as 10 mM Hepes, pH 7.4, has been added as a supplement. The sterile viral transport medium can be kept at room temperature for at least 1 year with no loss in efficacy.

[c] Tryptose phosphate broth is sterilized by autoclaving. Stock solution of 4% bovine albumin is made in Eagle's minimal essential medium and sterilized by filtration. The constituents are combined using an aseptic technique and distributed in 2-ml quantities in firmly closed screw-cap vials. This medium has a shelf-life of about 2 months at 4°C.

TABLE 4. Three modifications of *Chlamydia* transport media

Modification	Constituents
A[a]	Sucrose, 68.5 g Potassium phosphate, dibasic, 2.1 g Potassium phosphate, monobasic, 1.1 g Water, 1,000 ml
B	Complete cell culture medium used in isolation of *Chlamydia* Gentamicin, 10 μg/ml Vancomycin, 100 μg/ml Fungizone, 4 μg/ml
C[b]	As A and supplemented with 3% fetal bovine serum Nystatin, 25 U/ml Gentamicin, 50 μg/ml

[a] Dissolve in cell-culture grade water and adjust the pH to 7.2, as necessary, with 1 N sodium hydroxide or hydrochloric acid. Dispense 2.5-ml volumes into screw-cap vials, then autoclave the vials at 121°C for 15 min. Adding 2 to 3% fetal bovine serum (pretested for absence of antibodies to *Chlamydiae*) improves the ability of the chlamydial medium to maintain infectivity, but requires that it be dispensed aseptically after filter sterilization, as the serum cannot be autoclaved. The sterile medium can be kept at room temperature for at least 1 year with no loss in efficacy.
[b] Sucrose–phosphate solution is sterilized by autoclaving and stored at −20°C. The constituents are combined using an aseptic technique and distributed in 2-ml quantities in firmly closed screw-cap vials containing 2 or 3 sterile small glass beads to loosen and break infected cells in the specimen before inoculation. This medium is stored at −20°C.

VIRUS SPECIMENS

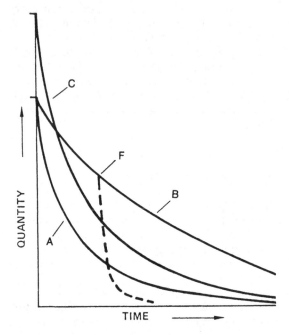

FIG. 2. Decay of virus in diagnostic specimens with time. (A) Nominal decay at room temperature. (B) Effect of cooling to 4°C. Note longer survival. At point (F) the drastic effect of freezing is superimposed, showing a rapid loss of virus until most of the specimen is frozen. (C) Effect of taking a more thorough specimen containing more virus, with storage at room temperature. Note that more virus remains at each point in time. Cooling would further prolong survival.

These factors include temperature and time in transit, the amount of virus originally present, the amount and type of protein, the pH, and even the amount of shaking en route.

Different viral properties will not decay at the same rate and the critical time after which attempts at detection are in vain will depend on whether infectivity, viral antibody, viral nucleic acid (whole or part), or viral morphology is sought. Antigens or nucleic acids generally survive better than infectivity or morphology, but the virologist should always be conscious that a specimen from the patient is a diminishing asset that should reach the laboratory as quickly as possible.

Temperature, time, and quantity are interrelated, as indicated in Figure 2. It is clear that the person taking the specimen should try to take as comprehensive a specimen as possible (i.e., containing as much viral material as can be obtained) and should keep it cold (but *not* frozen) during as brief a transit as can be arranged. Viral titers diminish rapidly as the temperature rises, particularly above 50°C, at which the decline is so rapid that subsequent attempts to re-

cover live virus are a waste of effort. Similarly, severe fluctuations in temperature should be avoided.

Most clinical specimens do not survive slow freezing, but respiratory viruses are particularly affected. Putting the specimen in the freezing compartment of refrigerators and sending the specimen on dry ice (solid carbon dioxide) or on frozen pads ("cold dogs") taken straight from a −20°C freezer have all been done with the best of intentions, and all were disastrous. The best solution to keeping a suitable temperature is to transport the specimens on wet ice in a vacuum flask. When specimens are being sent routinely, it is worth the expense to have a number of vacuum flasks available on the ward(s) or in the doctor's office or surgery. Otherwise, some improvisation may be necessary. A temporary cold box can be improvised from a cardboard box or similar box containing ice cubes in a sealed or knotted polythene bag surrounding and supporting the specimen vial.

Long-term storage can be achieved at −70°C or below, but the specimen should be snap-frozen in a dry-ice/isopentane freezing mixture. This is impractical for routine specimens and, even so, considerable losses occur. Many viruses in specimens frozen at −20°C will rapidly lose infectivity.

Within the flask the specimens should be sent in screw-cap vials (*not* plug closures) with the caps taped to prevent them from unscrewing. If two or more glass vials are sent together, they should be individually wrapped to prevent breakage.

Viruses will survive high centrifugal forces, but are less resistant to frothing caused by shaking. This finding is particularly true of enveloped viruses.

Practical Hints for Transporting Specimens

Some specimens require, or tolerate, a variation in the conditions outlined previously.

Blood and Serum

Antibodies are more temperature stable than viruses, but blood or serum provides a good culture medium for bacteria, and lysed blood is more toxic for cells and will produce a less satisfactory "serum" for some tests. Lysis may be due to inherent red cell fragility (in some blood dyscrasias), frothing, wet specimen containers (or even syringes), some drug treatments, or delay in removing the blood clot. Most of the virus present will be intracellular in the white cells, and its recovery will be more likely if an anticoagulant (e.g., preservative-free heparin, 10 U/ml of blood) is added and the leukocytes are separated and added to the culture. Separation of the serum/plasma from the cells before transport to the laboratory avoids lysis but risks contamination, and plasma is not the specimen of choice for serologic tests.

Stool

The majority of viruses isolated or observed in stool are more resistant to inactivation compared with those from the respiratory tract, and stool material itself provides some protection. Refrigeration is desirable, but not essential.

Urine

Viruses survive poorly in urine, especially cytomegalovirus. Preferably, it should arrive still "smoking" in the laboratory, but if this is not possible it should be cooled to and kept at 4°C.

Smears

Smears should be made on clean microscope slides and thoroughly air-dried. They should not be fixed in any way and, once dry, they can be wrapped individually in clean tissue, metal foil, or the inside sheets of a newspaper. Spacers should not be necessary, but two slides should not be sent face to face because it may be hazardous, and even impossible, to separate them later. Alternatively, they can be sent in cardboard, wooden, or hard plastic slide boxes. They should be carefully and clearly labeled before dispatch.

More Dangerous Specimens

All infectious material is potentially hazardous and poses some risk to those handling it. This hazard should be recognized, and sensible and safe working practices should be routine.

Nonetheless, some specimens can present additional dangers because they come from fatal infections and may cause severe or incurable illness in otherwise healthy subjects, or present unusual or unknown hazards from exotic or uncommon diseases.

Compared with routine specimens, such samples are more hazardous, but quantifying the size of the additional risk can be impossible. When there is some evidence, it will usually suggest that the risk is small; however, the temptation to "play safe" and treat them with exaggerated care will be strong. This temptation will often be reinforced by others who will urge avoidance of all risk, a condition impossible to fulfill. The main purpose of diagnostic virology should not be forgotten and the needs of the patient should not be neglected. The best protection against hazards great or small is a sound technique coupled with an awareness of the dangers from aerosols created indiscriminately. It is rarely necessary to use elaborate precautions and, when they are particularly cumbersome, may fail to create better protection by making the procedures more awkward.

This is not a subject where absolutes exist. Various codes of practice prepared in different countries to provide guidelines should be observed where relevant; the hazard is unknown in some situations, however, because diagnostic procedures are being undertaken for precisely that reason. Here the clinician has to balance his duty to the patient with that to the laboratory staff. This conflict of interests can rarely be completely resolved, but the situation can certainly be helped by a discussion with the director of the laboratory. It is clear that overanxiety can prevent early diagnosis and delay treatment and that there is an obligation on the part of all concerned not to overreact. Clear thinking and a recognition of responsibilities is necessary: the decision to accept a paid job in a diagnostic laboratory implies an acceptance of some risk, and it must be kept within a commonsense perspective. The laboratory staff have a right to share the clinician's assessment of the patient, but this will not be given frankly if they react negatively to sharing some of the risk run by the clinician and his ward staff.

Suitable Containers

Specimens in transit are liable to be dropped or inverted; occasionally they may be crushed. The primary container should therefore be strong and watertight. Thick-walled glass is better than thin, screw caps are better than plugs, and, with blood, the clot will stick to plastic surfaces, preventing proper retraction and release of serum. If the cap alone is plastic, the clot can adhere to it and can be pulled out when the cap is removed, a potential hazard. Plug-type closures can cause aerosols when they are removed.

Although it is not difficult to obtain good specimen containers, several common problems lead to leakage in transit. The use of containers that do not seal well usually results in extensive leakage, and tape or Parafilm will not stop leakage past a poorly fitting cap. Proper containers do not require tape, except as a safeguard against loosening by vibration; improper containers do not benefit from its use. Transportation to distant facilities requires additional care: large containers containing small specimens often allow dessication of the specimen in transit. Cardboard collection containers are unsuitable for transporting specimens and should not be used. Even appropriate containers will leak on rare occasions, usually from inadequate tightening of the closure. Thus, it is prudent to enclose the specimen contained in a sealable ("ziplock") plastic bag. Such bags are available with an exterior pocket in which to place the accompanying requisition form, which will not be soiled in the event of specimen leakage.

Local conditions and availability will influence the final choice of container, but some thought is necessary to ensure that all the carefully obtained specimens arrive safely at the laboratory without leakage and without putting the staff at risk of either physical injury or virologic hazard.

Transport Between Laboratories

Specimens may be sent from one laboratory to another to take advantage of specialized techniques for reference purposes, for research, or for other reasons.

Because the dispatching laboratory is sending a specimen as a deliberate act and will have resources unavailable to the individual doctor or nurse on the ward or in his/her surgery or office, there should be few problems in completing the transfer successfully and safely. Nonetheless, an unacceptably high proportion of these specimens fail to arrive intact, usually because the specimen has leaked in transit. The usual cause is a screw-cap vial with a cap that has not been taped to prevent unscrewing. This is important to the staff of the sending laboratory because most receiving laboratories will automatically discard a leaking specimen, and failure to recognize the obligation to prepare a specimen properly for dispatch wastes time, money, and sometimes, an irreplaceable specimen. The previous suggestion to enclose the specimen in a sealable plastic bag ("ziplock" or heat sealed) also applies here.

Sometimes using a leak-proof and suitable container for the specimen and taping it, if necessary, to ensure it remains leak-proof is only the first part of the process of dispatch. Further wrapping will usually be required.

At one end of the scale, a specimen delivered by hand within one building will not need special packing; at the other, a highly pathogenic virus sent from one country to another will require prior clearance and special packing to conform both to common-sense precautions and to the international regulations of the two countries concerned and of the carrier (airline, post office, and the like). Between these two extremes some variations are acceptable depending on circumstances, but the paramount obligation is not to put at risk the staff handling the specimen en route or those at the receiving end.

Detailed guidance on complete packaging can be found in the text by Madeley (1977) or in current International Air Transport Association regulations. Briefly, the requirements are that the specimen should be contained in a leak-proof container surrounded by enough absorptive material to contain all the fluid should it leak. This should be surrounded by a second leak-proof container (e.g., a heat-sealed polythene bag) surrounded by further shock absorbing material, all of which is placed in a strong, leakproof outer tin, box, or other container capable of withstanding the International Air Transport Association's specified tests of watertightness, heat resistance, and shock resistance. Refrigeration materials (wet or dry ice, cold dogs, and the like) will be put outside this outer container and will require another outer layer of packing to keep everything together. If the specimen has already been stored at $-70°C$, then dry ice (to keep it frozen) is acceptable; however, dry ice generates CO_2 gas, which can make the parcel explode as it sublimates if no route of escape is included in the packaging, and it will make the fluid in the specimen very acidic if it penetrates into it.

The other obligation on the staff of the sending laboratory is to send the necessary information to give the staff of the receiving laboratory sufficient knowledge to process the specimen, handle it appropriately, and place the results of testing into their clinical context. In practice this will mean including

TABLE 5. Checklist of information to be sent with a specimen to a reference laboratory

Details of patient
  Name: Surname/family name
          First name(s)
  Age
  Sex
  Date of birth
  Address
  Unit or personal number
  Occupation
  For females: Whether pregnant
  (if relevant) Date of last period
                 Gravida/parity
  Family history (if relevant)

Details of illness
  Presenting signs/symptoms
  Duration/date of onset
  Other relevant factors
  Immunodeficiency/immunosuppression
  Underlying conditions and/or treat-
    ment
  Other similar or related cases
  Recent travel and/or immunizations
  Relevant previous history

Laboratory details
  Laboratory number
  Date of specimen
  Type of specimen
  Place where specimen was obtained
  Laboratory findings
    Results
    Passage history
    Method of identification
    Method of confirmation
  Other specimens obtained and findings

Investigation requested

considerable details, a checklist of which is provided in Table 5.

It is better to err on the side of too much rather than too little information, but it should be written with an awareness of how it may be interpreted by the recipient. Ambiguities and alarmist statements should be avoided, but adequate warning will be needed with some specimens.

It is also wise, particularly when important specimens are involved, to warn the receiving laboratory that the specimen is being sent, with details of the route, carrier, waybill number (if appropriate), dates of dispatch, and probable arrival. It is also essential to remember that recipients of air freight packages will have to pay handling charges and would prefer some warning.

Conclusions

Specimen collection and delivery to the laboratory without delay and with minimal losses are vital components of a good diagnostic service. Observation of the principles outlined above will not happen without previous effort in planning and constant vigilance on the part of the virologist. Good diagnosis is a partnership between clinician and laboratory staff that must be cultivated carefully, with appropriate husbandry and adding fertilizer from time to time in the form of visits, telephone calls, and even letters. The letter will, however, have very limited effect. The direct personal approach is always better.

Literature Cited

Embil, J. A., H. J. Thiebaux, F. R. Manuel, L. H. Pereira, and S. W. MacDonald. 1982. Sequential cervical specimens and the isolation of chlamydia trachomatis: Factors affecting detection. Sex. Transm. Dis. 10:62–66.

Halonen, P., G. Obert, and J. C. Hierholzer. 1985. Direct detection of viral antigens in respiratory infections by immunoassays: A four year experience and new developments, p. 65–83. In L. M. de la Maza and E. M. Peterson (ed.), Medical virology IV. Lawrence Erlbaum Associates, Publishers, Hillsdale, New Jersey.

International Air Transport Association–Dangerous Goods Regulations. 1987. 28th ed. International Air Transport Association, Montreal.

Kelly, C. A., J. Kotre, C. Ward, D. J. Hendrick, and E. H. Walters. 1987. Anatomical distribution of broncho-alveolar lavage as assessed by subtraction radiography. Thorax 42:624–628.

Madeley, C. R. 1977. Guide to the collection and transport of virological specimens, 40 p. World Health Organization, Geneva.

Mårdh, P. A., L. Weström, S. Colleen, and P. Wolner-Hanssen. 1981. Sampling, specimen handling, and isolation techniques in the diagnosis of chlamydial and other genital infections. Sex. Transm. Dis. 8:280–285.

Springmeyer, S. C., R. C. Hackman, R. Holle, G. M. Greenberg, C. E. Weems, D. Myerson, J. D. Meyers, and E. D. Thomas. 1986. Use of bronchoalveolar lavage to diagnose acute diffuse pneumonia in the immunocompromised host. J. Infect. Dis. 154:604–610.

CHAPTER 2

Nosocomial Viral Infections

LARRY J. ANDERSON, STEPHEN C. HADLER, CARLOS LOPEZ,
WILLIAM R. JARVIS, SUSAN FISHER-HOCH, WALTER W. BOND, and
DANIEL B. FISHBEIN

Disease: Viral infections and their associated illnesses acquired in the hospital or other health care facility.

Etiologic Agents: Gastroenteritis viruses (Norwalk and Norwalk-like viruses, Rotavirus); viral hemorrhagic fever viruses (Lassa virus, Ebola and Marburg viruses, Congo-Crimean hemorrhagic fever virus); hepatitis viruses (hepatitis A virus, hepatitis B virus, non-A and non-B viruses; herpes viruses (cytomegalovirus, Epstein-Barr viruses, herpes simplex viruses, varicella zoster virus); human immunodeficiency virus; respiratory viruses (adenoviruses, coronaviruses, influenza viruses, measles virus, parainfluenza viruses, respiratory syncytial virus, rhinoviruses); other viruses (enteroviruses, human parvovirus B19, mumps virus, papilloma virus, rabies virus, rubella virus); Creutzfeldt-Jakob disease agent; and disinfectants.

Source: Most nosocomial viral infections are introduced from the community by patients, visitors, or health care workers.

Clinical Manifestations, Pathology, and Laboratory Diagnosis: See chapters for the individual viruses.

Epidemiology: Nosocomial viral infections are probably common in all groups of hospitalized patients, but are especially important clinically in infants and young children, probably the elderly, and immunocompromised patients. Health care workers are important in the nosocomial spread of many viruses.

Treatment: See chapters on the individual viruses.

Prevention and Control: Good infection control practices will prevent the nosocomial spread of many viral infections and should be practiced at all times. Persons with unsuspected infections are often the source of spread. The viral diagnostic laboratory should be an active partner in the infection control team.

Introduction

Viruses have recently become appreciated as important causes of nosocomial infections. One group demonstrated that viruses are responsible for 1 to 5% of nosocomial infections in medical, surgical, and gynecologic services, 35% in pediatric services, and 77% in psychiatric services (Valenti et al., 1980). The highest rate of nosocomial viral infections was identified in the pediatric service (67 of 10,000 admissions). Some viruses, such as hepatitis B virus (HBV), are primarily a risk to staff; others, such as respiratory syncytial virus (RSV) and rotavirus, are primarily a risk to patients; still others, such as the cytomegalovirus (CMV), can reactivate from a latent infection and cause disease in immunosuppressed, hospitalized patients. Under the right circumstances any human pathogenic virus can probably be a nosocomial pathogen.

New viruses, changes in the epidemiology of known viruses, new vaccines, antiviral drugs, better diagnostic tests, and increasing numbers of immunosuppressed patients have made nosocomial viral infections an increasingly recognized and ever-changing problem. For example, the increasing rate of HBV infection has led to an increased risk to hospital and laboratory staff, whereas the risk to patients receiving blood transfusions has decreased because of routine screening of blood. Vaccination could eliminate most of the risk of nosocomial HBV infection in hospital and laboratory staff (Immunization Practices Advisory Committee, 1985).

With a better understanding of nosocomial viral infections has come an opportunity to prevent some of them. Good infection control practices can decrease the risk of nosocomial RSV infections (Gala et al., 1986; Gardner et al., 1973; Hall et al., 1978). Infection control practices and vaccination have markedly decreased the risk of transmitting HBV in renal dialysis units (Alter et al., 1986). Unfortunately, these and other control measures are not applied widely enough.

The virology laboratory can be used to help improve the application of infection control measures. Bacteriology laboratories have played an important role in controlling nosocomial infections by providing surveillance data on specific organisms, alerting staff to the emergence of potential nosocomial pathogens, and participating in investigations of nosocomial outbreaks (Weinstein and Mallison, 1978). The viral diagnostic laboratory should follow this example. The techniques are now available for virology laboratories to identify many viral infections (often rapidly), monitor their presence in the hospital, and alert staff of isolates that might represent nosocomial infections. Identifying nosocomial viral infections is an important step toward educating hospital staff about their importance.

Furthermore, molecular techniques provide the opportunity to clarify the epidemiology of nosocomial viral infections. For many viruses, monoclonal antibodies, endonuclease restriction studies, and oligonucleotide fingerprint studies can be used to differentiate among strains and use these differences to infer the source of an infection and determine which infections are related to an outbreak and which are not.

The purpose of this chapter is to highlight the importance of nosocomial viral infections and to point out the responsibility that hospital staff, including laboratory staff, have to be active participants in their control. The sections in this chapter on each virus are intended to provide information on their importance as nosocomial pathogens, their spread, and their control. The last section gives general information on disinfection of the viruses. The chapters on the specific viruses should be consulted regarding information about their biologic, clinical, epidemiologic, and diagnostic features.

Gastroenteritis Viruses

Norwalk and Norwalk-Like Viruses

Although not recognized as common nosocomial pathogens, the Norwalk and Norwalk-like viruses can cause outbreaks of nosocomial gastroenteritis. Outbreaks have been reported from acute- and chronic-care facilities (Appleton and Higgins, 1975; Judson et al., 1975; Pether et al., 1983; Spender et al., 1986) and presumably occur when endemic or epidemic transmission in the community spreads to a hospital or other institution. The community outbreaks occur throughout the year and involve schools, camps, recreational areas, restaurants, nursing homes, hospitals, and cruise ships (Appleton and Higgins, 1975; Baron et al., 1982; Baron et al., 1984; Griffin et al., 1982; Gunn et al., 1982; Judson and Molk, 1975; Kaplan et al., 1982a; Kaplan et al., 1982b; Kappus et al., 1982; Koopman et al., 1982; Morens et al., 1979; Murphy et al., 1979; Oshiro et al., 1981; Taylor et al., 1981; Vogt and Satnik, 1986). Nosocomial outbreaks often involve both patients and staff and their family members. The outbreaks can be explosive, with many persons developing gastroenteritis over a matter of several days to a week.

The illness is characterized by acute onset of a gastroenteritis, usually 24 to 48 h after exposure, that lasts 1 to 3 days. The Norwalk and Norwalk-like viruses are highly infectious and readily spread by either person-to-person transmission (which probably includes direct contact and fomites) or through a common source such as contaminated food or water (Greenberg et al., 1979). In one outbreak in an infant ward, asymptomatic excretion among the infants facilitated transmission (Spender et al., 1986). A common source has usually been implicated in large outbreaks.

Prevention of nosocomial transmission depends on promptly identifying patients with gastroenteritis and instituting enteric precautions and/or eliminating any common source for transmission (Garner and Simmons, 1983).

ROTAVIRUSES

Group A rotaviruses are important nosocomial pathogens in pediatric hospitals and also have been occasionally reported to cause nosocomial outbreaks among adults in hospitals or nursing homes (Bryden et al., 1982; Champsaur et al., 1984a; Champsaur et al., 1984b; Cubitt and Holzel, 1980; Dearlove et al., 1983; Harris et al., 1983; Krause et al., 1981; Marrie et al., 1982; Middleton et al., 1977; Rodriguez et al., 1982; Santosham et al., 1982; Srinivasan et al., 1984). In one children's hospital, rotavirus was found to be the second most common nosocomial pathogen and accounted for 12% of the nosocomial infections (Welliver and McLaughlin, 1984). In other studies, up to 20% of the infants and children admitted without gastroenteritis developed nosocomial rotavirus infection during their hospitalization (Middleton et al., 1977; Ryder et al., 1977). Nosocomial rotavirus infections usually occur during community outbreaks that occur in the winter months in temperate climates each year.

Rotavirus gastroenteritis is often severe and may account for up to 50% of each years cases of childhood gastroenteritis requiring hospitalization. It occasionally results in death from severe dehydration or electrolyte imbalances (Carlson et al., 1978; Gurwith et al., 1981; Koopman et al., 1981). Increased morbidity and mortality and prolonged excretion occur in immunodeficient patients (Jarvis et al., 1983; Saulsbury et al., 1980). Normal persons usually excrete rotavirus from day 3 to day 8 of illness; neonates can excrete virus for several months.

Rotaviruses are usually spread through close person-to-person contact and by fomites, and staff are likely to be involved in transmission by inadvertently carrying contaminated material from one patient to another. The virus can persist for prolonged periods in stools and the environment and is relatively resistant to commonly used disinfectants. Using effective disinfectants to clean environmental surfaces should reduce the likelihood of transmission (Sattar et al., 1983; Tan and Schnagi, 1981).

Control of nosocomial rotavirus infections requires identifying infected patients, cohorting patients excreting the virus, instituting enteric precautions, and emphasizing handwashing (Garner and Simmons, 1983). In neonatal nurseries control has been complicated by asymptomatic excretion in as many as 11 to 50% of infected neonates (Krause et al., 1981; Rodriguez et al., 1982). Asymptomatic excretion of rotavirus is more common in neonates and children <6 months old than in older children (Champsaur et al., 1984a,b).

Rotavirus vaccines are being developed and may become an important factor in preventing rotavirus gastroenteritis (Kapikian et al., 1986).

Hemorrhagic Fever Viruses

Nosocomial outbreaks of viral hemorrhagic fever can occur in both endemic (primarily the tropics) and nonendemic regions. Hospital-based outbreaks have been reported with high mortality rates involving Lassa, Ebola, Marburg, and Crimean-Congo hemorrhagic fever (CCHF) viruses. With the increasing frequency of international air travel, cases can easily be imported to nonendemic areas and patients be admitted to hospitals distant from the source of exposure,

particularly to hospitals near major airports. It is important to consider the diagnosis in febrile patients in or coming from areas endemic for these viruses. Nosocomial transmission has usually occurred when a diagnosis of hemorrhagic fever was not considered and appropriate infection control measures were not instituted. For example, nosocomial transmission has occurred after surgical and obstetric procedures and attending a critically ill patient in an emergency room (Burney et al., 1980; Fraser et al., 1974; Monath et al., 1973). Newly acquired data on mode and risk of transmission and prospects of effective treatment have simplified the approach to safe management of patients potentially infected with these viruses (Centers for Disease Control, 1988; Helmick et al., 1986).

Lassa Virus

Lassa virus is the only arenavirus that has been implicated in nosocomial infections (Fraser et al., 1974; Monath et al., 1973). It is endemic in West Africa, where most infections derive from close contact with the natural host, *Mastomys natalensis,* a rat that commonly infests village homes. It is an important infectious disease in West Africa; in one hospital in Sierra Leone, 15% of all adult medical admissions and 30% of adult medical deaths are due to Lassa fever (McCormick et al., 1987b). The initial symptoms of Lassa fever are similar to those of many other viral infections; they include fever, headache, sore throat, dry cough, and malaise. In some patients the illness progresses to include severe edema of the head and neck, respiratory distress, hypovolemic shock, encephalopathy, and death (McCormick et al., 1987a).

Person-to-person spread occurs but requires intimate contact with blood or body secretions. In a hospital-based study in a Lassa virus-endemic area of Sierra Leone, 496 hospital staff were followed between 1972 and 1983 (Hemlick et al., 1987). This study showed that strict but simple-barrier nursing techniques, such as wearing gloves, gowns, and masks and using a disinfectant—0.5% sodium hypochlorite, a 10% solution of household bleach—liberally prevented transmission to hospital staff.

It is now recommended that patients be in a single room with an anteroom and staff should use strict barrier nursing techniques (Centers for Disease Control, 1987c). The number of medical personnel attending the patient should be limited, and they should be experienced in infection control practices. Special care must be taken in handling needles and instruments and in performing invasive procedures. When an accidental exposure involving inoculation of blood or secretions or contamination of mucous membranes occurs, oral ribavirin should be offered as postexposure prophylaxis. Ribavirin is effective in treating infection, especially if given within 7 days of onset of illness (McCormick et al., 1986).

Ebola and Marburg Viruses

Marburg disease and Ebola hemorrhagic fever are uncommon diseases, but several outbreaks have centered around hospitals or laboratories (Baron et al., 1983; Martini and Siegert, 1971; Report of an International Commission, 1978). Both viruses derive from central and southern Africa; sporadic cases of Marburg disease have occurred in travelers in rural areas, and epidemics of Ebola have occurred in Sudan and Zaire (Baron et al., 1983). The natural host and the ecology of these viruses remains obscure.

Infection is manifested by acute onset of fever, headache, and malaise followed by profuse vomiting, often a morbilliform or maculopapular rash, and a bleeding diathesis (McCormick and Johnson, 1984). In some outbreaks of Ebola hemorrhagic fever, mortality has been very high; for example, in the outbreak in Zaire in 1976 it was 88% (Report of an International Commission, 1978).

As for Lassa virus, spread of Marburg and Ebola viruses requires close contact with infected persons, their blood, or secretions (Baron et al., 1983; Martini and Siegert, 1971). The epidemic in Zaire was promptly stopped when needles were no longer shared among patients and simple barrier-nursing was instituted (Report of an International Commission, 1978).

No specific postexposure prophylaxis or treatment is available. However, it is reasonable to give exposed or infected persons immunoglobulin, although it has not been proven efficacious. The plasma should be checked for the presence of HBV and human immunodeficiency virus (HIV).

Crimean-Congo Hemorrhagic Fever Virus (CCHF)

There have been several hospital-based outbreaks of CCHF with high mortality rates associated with failure to observe barrier-nursing techniques (Burney et al., 1980; Swanapoel et al., 1987). Crimean-Congo hemorrhagic fever is a tick-borne bunyavirus with wide geographic distribution that extends from the tip of Africa through parts of Eastern Europe and southern USSR and as far as northwest China (McCormick and Johnson, 1984). Apparently human cases only occur sporadically, but the true incidence and the epidemiology of this disease is unknown. The natural host is probably a small rodent or hare, but many mammalian species may become infected,

including domestic animals. The illness is characterized by abrupt onset of fever and chills and joint, back, and epigastric pain followed by diarrhea, a petechial rash, and in some cases, a bleeding diathesis, hypovolemic shock, and death. Overt bleeding is common in severely ill patients and may be an important factor in the spread of CCHF.

Nosocomial infection occurs by close contact in a fashion similar to that for Lassa fever virus infections. Ribavirin should be considered for postexposure prophylaxis and for treatment (Centers for Disease Control, 1988). Ribavirin is effective in vitro against CCHF but has not been studied in humans.

Hepatitis Viruses

Hepatitis A Virus (HAV)

Transmission of HAV in hospitals was uncommon in the past but is now recognized with increasing frequency (Centers for Disease Control, 1977c; Goodman et al., 1982; Orenstein et al., 1981). The disease is rarely spread from patients hospitalized with clinical hepatitis A, presumably due to decreasing virus excretion at onset of illness and use of isolation precautions (Dienstag et al., 1975; Favero et al., 1979; Papavangelou et al., 1981; Purcell et al., 1984). Spread in virtually all outbreaks has been from a patient who is either incubating infection or has asymptomatic infection, is not suspected of having HAV infection, and is incontinent of feces (Azimi et al., 1986; Goodman et al., 1982; Klein et al., 1984; Krober et al., 1984; Noble et al., 1984; Reed et al., 1984; Seeberg et al., 1981). Index patients acquire infection in the community as sporadic cases or as part of outbreaks that most often occur in settings with crowding or poor hygiene and sometimes are associated with contaminated food or water (Francis and Maynard, 1979). Although unusual, transfusion-associated HAV infection in a hospitalized infant is usually asymptomatic and, therefore, can readily spread to susceptible adults (Hollinger et al., 1983; Noble et al., 1984; Seeberg et al., 1981).

Nosocomial outbreaks usually involve transmission from patient to staff via the fecal-oral route; nursing staff are at highest risk, but other staff, including physicians and housekeepers, and family members may also become infected. Laboratory workers and hospital staff not exposed to patients are not at risk of infection. Other patients occasionally become infected, possibly by inadvertent transmission of HAV from one patient to another by hospital staff (Klein et al., 1984). Food-borne outbreaks also have been reported among hospital staff, but secondary transmission within the hospital has not been a problem (Eisenstein et al., 1963; Meyers et al., 1974). Hepatitis A virus infection is rarely fulminant, does not progress to chronic hepatitis, and is usually asymptomatic in children <5 years old. Excretion of virus is greatest before onset of symptoms.

The key to preventing nosocomial spread of HAV is to use good infection-control practices when handling feces and equipment and objects contaminated with feces (Favero et al., 1979; Garner et al., 1983). Enteric precautions for human HAV cases are recommended during the first 7 days of illness; isolation is recommended for patients incontinent of feces. Because most episodes of nosocomial hepatitis A involve unsuspected index patients, the best prevention is to follow good handwashing practices with all patients and particularly those incontinent of feces.

When an outbreak occurs, diagnostic tests can be used to ascertain which patients and staff are infected. Infected patients should be placed on enteric precautions and can be cohorted. Exposed susceptible persons should be considered for immunoglobulin prophylaxis. If given before or within 2 weeks of exposure, immunoglobulin is effective in preventing or moderating hepatitis A infection (Immunization Practices Advisory Committee, 1985).

Hepatitis B Virus (HBV)

Hepatitis B virus is the primary cause of nosocomial hepatitis in this country (Maynard, 1981). Nosocomial transmission of HBV from patients to persons in healthcare occupations is well documented (Dienstag et al., 1982; Hadler et al., 1985; Hirschowitz et al., 1980; Levy et al., 1977; Lewis et al, 1973; Pattison et al., 1975). In the United States, most groups of patients have a low rate of HBV infection (0.1 to 0.5% are virus carriers), but some, such as intravenous drug abusers, homosexuals or promiscuous heterosexuals, institutionalized developmentally disabled persons, and hemophiliacs, have a much higher rate of infection. Hepatitis B virus is highly endemic in most developing countries, and 5 to 20% of immigrants from Africa and Asia may be chronically infected with the virus. Transmission of HBV from infected healthcare workers to patients is uncommon but can occur (Alter et al., 1975; Collaborative Study, 1980; Kane and Lettau, 1985; Lettau et al., 1986; Meyers et al., 1978; Snydman et al., 1976b) and nosocomial HBV transmission from patient to patient is rare (primarily of concern in hemodialysis units) (Centers for Disease Control, 1977d; Garibaldi et al., 1973; Syndman et al., 1976a; Szmuness et al., 1974).

The risk from nosocomial HBV infection is important not only due to the acute illness but also since 5 to 10% of adults with HBV infection will develop a persistent, viremic infection, and 10 to 33% of persistently infected persons develop chronic liver disease that can progress to cirrhosis or hepatocellular carcinoma.

Hepatitis B virus is transmitted by percutaneous or mucosal exposure to blood or serum-derived body fluids or by sexual contact with persons with acute or chronic HBV infection. Transmission in the hospital setting most commonly occurs by exposure to blood from asymptomatic carriers (Francis and Maynard, 1979; Immunization Practices Advisory Committee, 1985). Blood and serous fluids contain the highest titers of virus and are the most likely vehicles for virus transmission; other body fluids (saliva, semen) are potentially infectious but contain 0.001 as much virus. Exposure to extremely small quantities of HBV surface antigen (HBsAg) positive blood may transmit infection (Francis and Maynard, 1979). The virus can remain viable for at least 1 week on environmental surfaces at ambient conditions but is not unusually resistant to chemical germicides (Bond et al., 1983).

The risk of acquiring HBV by healthcare workers is most closely related to degree of contact with blood- or serum-derived body fluids, frequency of accidents with blood-contaminated needles or sharp objects, and the rate of HBV infection in the patient population (Dienstag et al., 1982; Hadler et al., 1985; Harris et al., 1984; Osterholm et al., 1979; Pattison et al., 1975). The usual types of hospital exposures are accidents with contaminated needles or other sharp instruments and inapparent parenteral exposures. Inapparent exposures can occur when infectious material contaminates cuts or abrasions directly or through a fomite and when infectious material contaminates mucous membranes during pipetting or splash accidents (Favero et al., 1973; Garibaldi et al., 1972; Maynard, 1981; Pattison et al., 1974). Hospital staff at high risk of acquiring nosocomial HBV infection include medical technologists, operating room staff, phlebotomists and intravenous therapy nurses, surgeons, pathologists, and oncology and dialysis staff. Dental professionals, laboratory and blood bank technicians, emergency medical staff and technicians, and morticians are also at risk.

Education is an important part of control measures for nosocomial transmission of HBV. Healthcare personnel should understand how HBV is transmitted and the importance of asymptomatic carriage in its transmission. They should treat blood and body fluids from all patients as if they were contaminated with HBV and other bloodborne viruses such as HIV (Favero et al., 1979; Snydman et al., 1975). In the clinical laboratory, the appropriate practices and procedures include (1) use of mechanical pipettors (no mouth pipetting); (2) prohibition of eating, drinking, or smoking in the laboratory; (3) use of laboratory coats and gloves and protective eye wear; (4) daily (or more frequent) cleanup of work surfaces; and (5) use of proper techniques when handling phlebotomy needles. Visible spills should be promptly cleaned with absorbent material followed by disinfecting the area with hypochlorite solution having at least 500 mgm/l free available chlorine (1% solution of household bleach) or another suitable disinfectant (Bond et al., 1983).

Hepatitis B vaccination, coupled with the safety measures outlined, can essentially eliminate the risk to healthcare personnel of acquiring HBV in the workplace. The vaccine is safe, immunogenic, and highly effective in preventing hepatitis B for 5 or more years in those who respond to the three-dose series (Hadler et al., 1986; Szmuness et al., 1980). Hepatitis B virus vaccine should be given to all healthcare personnel who have direct contact with human blood or body secretions or risk of needle accidents. Ideally, vaccination should be offered to such persons while they are in training (Immunization Practices Advisory Committee, 1985).

Persons hospitalized with acute hepatitis B infection or known to be HBV carriers should be placed on blood and body fluid precautions (Favero et al., 1979; Garner and Simmons, 1983). A private room is not required, although it is advisable for patients who have a bleeding diathesis.

Control of HBV infection in dialysis units has been achieved by regular monthly serologic screening for HBsAg, room isolation of carriers, elimination of shared equipment between patients, and vaccination programs for both staff and patients (Alter et al., 1983; Alter et al., 1986; Centers for Disease Control, 1985c; Stevens et al., 1984).

Hepatitis B virus surface antigen positive health workers need not be restricted from patient contact, but should be educated regarding risks of transmitting infection and take precautions to minimize these risks (Lettau et al., 1986). If HBV transmission to patients is documented, decisions about further patient contact should be made on a case-by-case basis.

When susceptible staff are exposed to HBsAg-positive blood, hepatitis B immunoglobulin (HBIG) is effective and should be given for postexposure prophylaxis (Immunization Practices Advisory Committee, 1985; Grady et al., 1978; Seeff et al., 1978). The combination of HBIG with HBV vaccine increases efficacy after exposure to more than 90% and does not decrease the immunogenicity of vaccine (Beasley et al., 1983; Wong et al., 1984).

Non-A, Non-B Hepatitis Viruses

Non-A, non-B hepatitis is a diagnosis of exclusion that includes several distinct virus diseases of grow-

ing importance (Dienstag, 1983; Hollinger et al., 1980). One of the diseases is posttransfusion non-A, non-B hepatitis. An estimated 3 to 8% of blood donors appear to chronically carry the virus that causes this disease, and 90% of all posttransfusion hepatitis is now due to this virus. The virus apparently can also be transmitted by needlestick exposure (Ahtone et al., 1980), and persons working in healthcare occupations with patient or blood exposure are at elevated risk of infection (Alter et al., 1982). Parenteral drug abusers are another group at high risk of being infected and can be a source for nosocomial transmission. Non-A, non-B transmission from patients to other patients and to staff may occur in hemodialysis units (Centers for Disease Control, 1985e; Galbraith et al., 1979).

Non-A, non-B agents cause between 15 to 35% of acute hepatitis cases in the United States, primarily in adults, and up to 40% of cases may be followed by persistent viremia and chronic hepatitis (Alter et al., 1982; Francis et al., 1984).

The principal mode of transmission in this country is blood-borne, but about half of the sporadic cases cannot be attributed to parenteral transmission, and other pathways for transmission probably exist. The mechanisms for transmission of posttransfusion non-A, non-B hepatitis in the hospital are probably similar to those for HBV, but the low titer of viremia (estimated to be between 10^2 and 10^6 infectious doses/ml) probably makes the risk lower.

The recommended precautions to prevent transmission of non-A, non-B hepatitis are identical to those for hepatitis B (Favero et al., 1979; Garner et al., 1983). If an exposure does occur it is reasonable to give immunoglobulin (0.06 ml/kg), though efficacy studies have given equivocal results (Immunization Practices Advisory Committee, 1985; Knodell et al., 1976; Seeff et al., 1977).

Herpes Viruses

Cytomegalovirus (CMV)

Cytomegalovirus infection in the hospital is primarily a concern for the patient with an immature or compromised immune system. Hospital-acquired CMV infection usually occurs as either reactivation of latent infection or transmission of virus through blood or blood products (Adler, 1986; Meyers et al., 1986; Tolpin et al., 1985). Most adults have been infected with CMV and may excrete virus at anytime without symptoms (Adler, 1986; Alford et al., 1981; Meyers et al., 1986; Peterson et al., 1980). In immunosuppressed, hospitalized patients, both primary and reactivated CMV infection can produce life-threatening disease. The fetus and neonate are also at risk of serious congenital anomalies or life-threatening disease with CMV infection (Hanshaw, 1971; Stagno and Whitley, 1985).

Transmission of CMV occurs in 10 to 25% of seronegative recipients of seropositive blood donors (Adler, 1986; Meyers et al., 1986). Cytomegalovirus is a common infection in patients with organ transplants and this infection usually results from immunosuppressive therapy leading to reactivation of latent infection (Meyers et al., 1986). Transplant recipients, including those who are seropositive, also can acquire exogenous infection (Grundy et al., 1986). Cytomegalovirus is readily transmitted within households and day-care centers, suggesting that close contact can result in transmission. The presence of virus in genital secretions at the time of delivery and in the breast milk after delivery is associated with transmission to the neonate (Stagno and Whitley, 1985).

Prevention of hospital-acquired CMV infection has focused on patients with immature or compromised immune systems. The use of blood or blood products from seronegative donors in seronegative infants has dramatically reduced the incidence of CMV infections (Adler, 1986; Yeager et al., 1981). Frozen, deglyceralized red blood cells appear to lack viable leukocytes and do not transmit CMV infection. As most CMV disease in organ transplant recipients is reactivated disease, prevention has been directed toward prophylactic therapy with hyperimmune serum, antiviral drugs, and interferon (Grundy et al., 1986; Meyers et al., 1986). Each of these treatments has shown some promise.

Because most adults are latently infected and can excrete virus at any time, it is impractical to identify and isolate persons excreting the virus. One group, however, has recommended that congenitally infected infants (who can excrete high titers of virus for up to 3 months) should be placed in isolation and staff, especially pregnant women, should strictly adhere to good handwashing practices (Plotkin, 1986).

A live, attenuated CMV vaccine is being evaluated in seronegative renal transplant recipients as a means of minimizing primary CMV infection in that group of patients (Plotkin et al., 1984).

Epstein-Barr Virus (EBV)

Epstein-Barr virus does not appear to be a major nosocomial pathogen. Because most adults have been infected with EBV and most primary infections become persistent, the source of EBV infection in the hospitalized patient is usually endogenous. Reactivation and asymptomatic excretion in saliva can occur in 20% or more of patients (Yao et al., 1985). Immunosuppression, however, may allow EBV in-

fection to progress to a lymphoproliferative disease or malignancy (Klein and Klein, 1984). Although rare, transmission has occurred after blood transfusion (Turner et al., 1972).

Because many normal persons periodically shed virus, there are no practical hospital infection control measures other than good personal hygiene.

Herpes Simplex Virus

Nosocomial herpes simplex virus type 1 and 2 (HSV-1 and -2) infections can occur by either reactivation of latent infection or by exogenous infection. Because reactivation of latent HSV infection occurs spontaneously in normal persons and exogenous infection can occur in previously infected persons, it is often difficult to distinguish between the two (Whitley, 1985a). Despite this difficulty, there are reports of nosocomial transmission of HSV infections in neonates, immunosuppressed patients, and staff (Foley et al., 1970; Greaves et al., 1980; Linneman et al., 1978; Naragi et al., 1976). The immunosuppressed patient and the neonate are at greatest risk of serious disease with either type of infection (Saral et al., 1981; Siegal et al., 1981). Herpes simplex virus-1 and -2 infections of organ transplant recipients and patients with acquired immunodeficiency syndrome (AIDS) often disseminate and can be life threatening. Newborns, with their immature immune system, are also susceptible to life-threatening, disseminated disease (Nahmias and Roizman, 1973; Whitley, 1985a).

Spread of HSV-1 and -2 requires close personal contact (Corey, 1985; Whitley, 1985a). Persons with lesions are infectious until the vesicles have completely healed. Asymptomatic shedders can also transmit virus and probably represent an important source of infection in the general population. Neonatal herpes can result from vertical transmission, transmission during delivery, or transmission after delivery (Douglas et al., 1983; Katz et al., 1980; Sullivan-Bolyai et al., 1983; Van Dyke and Spector, 1984; Yeager et al., 1983). Most newborns with neonatal herpes are born of mothers who are asymptomatic excreters of the virus (Whitley et al., 1980).

There is currently no way to prevent reactivation of latent HSV infection, but good infection control practices may prevent spread from a known excreter of HSV (Garner and Simmons, 1983). The risk of infection in infants born of women excreting HSV in the genital tract can be diminished, but not eliminated, by delivering the child by Cesarean section before or within 6 h of the rupture of membranes (Stagno and Whitley, 1985). When infection does occur, vidarabine or acyclovir can be used to treat the disease (Whitley, 1985b).

Varicella Zoster Virus (VZV)

Nosocomial VZV, like the other herpes viruses, can occur as a primary infection, chicken pox, reactivation of latent infection and herpes zoster (shingles) (Brunell 1981; Hope-Simpson, 1965; Weller, 1983). Varicella zoster virus is very contagious, and its introduction into the hospital can be a serious problem. Varicella infection is most common in children (more than 80% of cases in the United States occur in children <10 years old [Preblud and D'Angelo, 1979; Weller, 1983]), and consequently children with (or incubating) varicella and persons with herpes zoster are the primary source of VZV infections in the hospital (Weitekamp et al., 1985).

The risk of serious illness with nosocomial varicella infection is in patients with compromised immune systems (Weller, 1983). The mortality rate in immunodeficient patients varies with the extent of compromise of the cell-mediated immune system and ranges from 7 to 50%. Severe recrudescent varicella zoster also occurs more often in immunodeficient patients. Varicella infection can also be a serious illness in otherwise healthy adults, in pregnant women, and in neonates (Enders, 1984; Paryani and Arvin, 1986; Preblud, 1981).

Varicella zoster virus is transmitted very efficiently to susceptible persons both by small particle aerosol and by contact with infected skin lesions. In households, the secondary attack rate for varicella among susceptible persons is almost 90% (Weller, 1983) and for herpes zoster it is 25% (Berlin and Campbell, 1970). Patients may shed virus as early as 2 days before the onset of rash and be contagious until all skin lesions are dried and crusted (Gordon, 1962). Patients with a primary or secondary immunodeficiency appear to shed more virus than their otherwise normal counterparts do and generally for a longer time.

When VZV is introduced into the hospital it must be addressed rapidly and efficiently to reduce spread of infection, especially to immunosuppressed patients. Weitekamp et al. (1985) have developed an algorithm for the prevention of nosocomial VZV infections. All cases including the index case must be identified and sent home or isolated. Atypical cases with few or unusual lesions may require laboratory tests, such as the Tzanck test, for confirmation. Patients too severely ill to be discharged must be kept in strict isolation until all lesions have crusted. Air should flow into the patient's room under negative pressure, and masks, gowns, and gloves should be worn in the patient's room. Only seropositive staff should care for VZV-infected patients. Because an outbreak of VZV in areas with immunosuppressed patients requires rapid decisions about staffing patterns based on susceptibility to VZV, all employees

working in these areas should be screened for VZV antibodies.

All persons exposed to VZV infected patients (those who have been in the same room as a patient with varicella or in contact with a patient with varicella zoster) and who have no history of having had varicella should be tested for VZV antibodies using a sensitive test. Exposed, seronegative persons are candidates for varicella zoster immunoglobulin (VZIG) (Zaia, et al., 1983). If given within 3 days of exposure, VZIG will usually ameliorate varicella. Immunodeficient patients with severe varicella or with disseminated, recrudescent zoster have been treated effectively with both vidarabine and acyclovir (Prober et al., 1982; Whitley et al., 1982). A live, attenuated VZV vaccine has been developed for use in high-risk patient populations and may decrease the risk of serious disease with nosocomial VZV (Gershon, 1980; Takahashi et al., 1974).

Human Immunodeficiency Virus (HIV)

Nosocomial transmission of HIV from infected patients to healthcare workers has been rare and limited to situations in which direct parenteral exposure to contaminated blood occurred (Anonymous, 1984; Centers for Disease Control, 1985d; Hirsch et al., 1985; Neisson-Vernant et al., 1986; Oksenhendler et al., 1986; Stricof and Morse, 1986; Weiss et al., 1985a,b). Most such infections have followed needlestick exposures but there has been one report of HIV seroconversion following blood contamination of skin or mucous membranes (Centers for Disease Control, 1987c). Prospective studies have shown the risk of disease transmission by a needlestick injury from a patient with AIDS is low (less than 1%), in contrast to the much higher risk (14 to 26%) of disease transmission of HBV by a needlestick accident (McCray et al., 1986). There is no evidence of HIV transmission from infected healthcare workers to patients, or of patient-to-patient transmission in the hospital setting. Studies of family contacts of AIDS patients indicate that casual contact does not result in disease transmission and support the conclusion that nosocomial transmission of HIV infection is a low risk for exposures other than parenteral inoculation of HIV-contaminated material. The risk of transmission through blood transfusions or administration of blood products has been nearly eliminated by screening blood for HIV antibody (Centers for Disease Control, 1985b).

Clinical symptoms are not a good indicator of the presence of HIV infection because most infections, at least initially, are asymptomatic. Within 5 years of infection an estimated 15 to 30% of persons will develop AIDS (Centers for Disease Control, 1986b; Curran et al., 1985). The best predictor of HIV infection is having a risk factor. More than 90% of infected patients are homosexual men, parenteral drug abusers, hemophiliacs, recipients of contaminated blood transfusions, sexual partners of someone in the above-noted risk groups, or infants born to infected mothers. Human immunodeficiency virus can be acquired from blood, semen, vaginal secretions, and possibly breast milk; the virus has been isolated from saliva, tears, and urine, but these secretions have not been implicated in transmission.

Preventive measures for nosocomial HIV transmission are similar to those recommended for hepatitis B virus, and focus on blood and body fluid precautions (Centers for Disease Control, 1982, 1983e, 1985a; Immunization Practices Advisory Committee, 1985; Conte et al., 1983; American Hospital Association, 1984). Healthcare workers should be aware that many persons with HIV infection will be asymptomatic and that all blood or body fluid specimens should be handled as if they contained HIV and HBV. Precautions in the clinical laboratory setting are identical to those recommended for hepatitis B (Centers for Disease Control, 1982). Procedures in which blood or body fluids may become aerosolized or spattered should be avoided or done only in a protective hood. Persons performing autopsies should wear protective clothing and waterproof aprons, and use hand instruments which minimize risk of aerosols. Recommendations for other settings have been also published (Centers for Disease Control, 1985a, 1985d, 1985f, 1986a).

For episodes of known needlestick or other percutaneous or mucosal exposure to HIV-contaminated blood, no specific prophylaxis can be offered. The HIV-contaminated blood may also contain HBV, however, for which prophylaxis is available and should be given. It is recommended that the exposed person be evaluated clinically and serologically for evidence of HIV infection as soon as possible after exposure, and, if seronegative, retested periodically thereafter (for example, 6 weeks and 3, 6, and 12 months after exposure) to determine if transmission has occurred (Centers for Disease Control, 1985a). Exposed persons should be advised that the risk of acquiring HIV from the exposure is low and also be counseled to avoid the risk of transmitting HIV to others.

The standard blood and body fluid precautions and disinfection and sterilization strategies routinely practiced in dialysis centers are adequate to prevent transmission of HIV in these centers (Centers for Disease Control, 1986c). To date, there is no evidence of HIV transmission from patient to patient,

from patient to staff, or from staff to patient in this setting.

Healthcare workers infected with HIV and who have only casual contact with patients pose no risk of disease transmission. All healthcare workers, regardless of HIV infection status, should wear gloves when touching mucosal membranes or open tissues and should avoid patient contact if they have exudative dermatitis (Centers for Disease Control, 1985a). Professionals who routinely perform invasive procedures, such as surgeons and oral surgeons, and who are infected with HIV should be handled just like they would be handled if they had any other illness that might effect their physical or mental competence. They can continue to work if physically and mentally able and if they can follow basic infection control procedures (Centers for Disease Control, 1986a). In general, decisions regarding continued practice should be made by informed hospital and local and state public health officials in conjunction with the personal physician of the infected practitioner.

Respiratory Viruses

Adenoviruses

Nosocomial adenovirus infections have been identified among all age groups. In children, nosocomial infections have been most often associated with outbreaks of acute respiratory illness, sometimes with significant morbidity and mortality (Centers for Disease Control, 1978b, 1983a; Harris et al., 1972; Herbert et al., 1977; Pingleton et al., 1978; Straube et al., 1983). There also have been several reports of nosocomial enteric adenovirus infections, and in immunosuppressed patients these infections have been associated with significant mortality (Chiba et al., 1983; Flewett et al., 1975; Yolken et al., 1982). Among adults, the most common nosocomial adenovirus infection is probably epidemic keratoconjunctivitis (EKC) transmitted in an ophthalmology office, but outbreaks of nosocomial pharyngitis and conjunctivitis also have been reported among adults (Buehler et al., 1984; Keenlyside et al., 1983; Kemp et al., 1983; Larsen et al., 1986; Levandowski and Rubenis, 1981; Richmond et al., 1984; Warren et al., 1986). Hospital staff have often become infected from patients during these nosocomial outbreaks. Infants and young children and immunosuppressed patients are at greatest risk from nosocomial infection (Shields, 1985; Yolken et al., 1982).

Unlike the other respiratory viruses the adenoviruses usually do not cause seasonal, community-wide outbreaks, but, are present throughout the year and sometimes cause focal outbreaks. Nosocomial infections may be associated with either endemic disease or focal outbreaks. Examples of adenovirus outbreaks include acute respiratory disease in military recruits (Dudding et al., 1972; Takafuji et al., 1979) and acute pharyngoconjunctival fever (Caldwell et al., 1974; D'Angelo et al., 1979). Infection is most common in children, among whom up to 50% of infections are asymptomatic (Brandt et al., 1969; Fox et al., 1969).

Adenovirus probably can be spread by aerosol (Couch et al., 1966) and by close contact and fomites. The virus can be isolated for up to 1 week from respiratory secretions, up to 2 weeks from eye swabs of persons with conjunctivitis, and up to 12 weeks and more from stool specimens (Fox et al., 1969).

The importance of good infection control measures in stopping nosocomial transmission has been clearly demonstrated in two outbreaks. One outbreak of conjunctivitis in a hospital occurred after the staff stopped using gloves, gowns, and masks while caring for a patient with adenovirus pneumonia (Levandowski and Rubenis, 1981). The other, an outbreak of EKC in a large ophthalmology clinic, continued until the recommended infection control measures (triaging patients and cohorting patients and staff, using good handwashing practices, postponing procedures on patients with EKC, and disinfecting equipment) were strictly enforced (Warren et al., 1986). Contact isolation is recommended for patients with adenovirus infections in the hospital (Gerner and Simmons, 1983). Procedures for controlling outbreaks of EKC in an ophthalmology office have been outlined by Buehler et al. (1984).

Coronaviruses

Coronaviruses do not appear to be important nosocomial pathogens, although there are reports of coronavirus nosocomial gastroeteritis (Caul and Clarke, 1975; Chany et al., 1982). They are common respiratory pathogens (Hamre and Beem, 1972; Larson et al., 1980) and probably cause nosocomial upper respiratory infections as well; but our ability to study nosocomial coronavirus infections is limited by difficulties in identifying infections. Coronaviruses are presumably transmitted by close contact in a fashion similar to that for rhinovirus and RSV.

Influenza Viruses

Nosocomial outbreaks of influenza A and B virus infections occur during community outbreaks, can spread through hospitals, nursing homes, and chronic-care facilities in an explosive fashion, and involve infants, children, and adults (Bean et al., 1983; Blumenfeld et al., 1959; Centers for Disease Control, 1981b, 1983b; Gardner et al., 1973; Hall and Douglas, 1975; Kapila et al., 1977; Mathur et al., 1980; Meibalane et al., 1977; Sims, 1981; Van Voris et al., 1982; Wenzel et al., 1977). Influenza C virus causes sporadic cases of mild illness and has been associated with one outbreak in a children's home (Katagiri et al., 1983; Mogabgab, 1963). Community-wide epidemics of influenza A or B occur each year in the winter or spring in temperate climates and are the only infectious diseases that are regularly associated with an increase in national mortality during outbreaks (Noble, 1982). The severity of illness associated with nosocomial influenza depends on the age and health of the patients and their levels of immunity to the outbreak strain. Serious illness with nosocomial influenza infection occurs most often in the elderly and members of high-risk groups such as persons with chronic cardiovascular or pulmonary disease (requiring regular medical follow-up), diabetes mellitus, renal dysfunction or immunosuppression, and residents of nursing homes and chronic-care facilities (Barker and Mullooly, 1980; Immunization Practices Advisory Committee, 1986).

Transmission occurs by respiratory aerosol and possibly by close contact and fomite spread (Alford et al., 1966; Moser et al., 1979). In nosocomial outbreaks, the hospital staff may spread influenza by becoming infected themselves, either from the community or the hospital, and directly transmitting virus to patients (Bauer et al., 1973; Van Voris et al., 1982). Influenza virus is shed in respiratory secretions beginning at or shortly before onset of illness and often persists for 5 days and sometimes longer (Frank et al., 1981; Togo et al., 1970). Shedding persists for longer in children (Brocklebank et al., 1972; Hall and Douglas, 1975).

Nosocomial influenza virus infections can be controlled by vaccination and amantadine or rimantadine prophylaxis. Several studies have demonstrated that vaccination can reduce infection rates among staff and, with sufficient levels, even prevent an outbreak from occurring (Arden et al., 1986; Blumenfeld et al., 1959; Budnick et al., 1984; Centers for Disease Control, 1983b, 1987b; Patriarca et al., 1985, 1986). Amantadine and rimantadine are effective only against influenza A viruses and when used prophylactically, have reduced the rate and severity of nosocomial infections (Atkinson et al., 1986; Dolan et al., 1982; O'Donoghue et al., 1973). During community outbreaks of influenza, adults with influenza infection can be cohorted and children should be placed on contact isolation for the duration of their illness or cohorted with other infected children (Garner and Simmons, 1983; Hoffman and Dixon, 1977). Amantadine prophylaxis should be considered for patients at high risk who have not been immunized and for nonimmunized staff. Preferably, high-risk patients and hospital employees (or employees of other institutions) who have extensive contact with high-risk patients should be immunized before the influenza season begins (Immunization Practices Advisory Committee, 1986).

Measles Virus

With the widespread vaccination of children, nosocomial measles is uncommon in developed countries (Bloch et al., 1985; Davis et al., 1986). Between 1980 and 1984, approximately 241 persons acquired measles in medical facilities (Davis et al., 1986). Most cases (76%) occurred in patients or visitors, and the remainder occurred in medical personnel (24%). Patient-to-patient (50%) and patient-to-staff (37%) transmission were most common. Medical personnel were rarely the source of disease for others. Severe illness is most common in adults, children <2 years old, and immunosuppressed patients.

Measles is highly communicable; the virus can be spread by droplets, direct or indirect contact with infectious excretions or secretions, and aerosol; patients are communicable from the prodromal stage to 4 days after the onset of the rash.

Prevention of nosocomial measles requires a high index of suspicion. Suspected infected persons should be separated from susceptible persons regardless of the medical setting (whether office, emergency room, clinic, or hospital ward) and especially from infants and the immunosuppressed who are at high risk from serious illness with infection. Hospitalized measles patients should be placed in respiratory isolation until 4 days after onset of rash (Garner and Simmons, 1983). Immunosuppressed patients should remain on respiratory isolation for the duration of their illness.

Personnel at high risk of exposure to measles should be immune; measles immunity can be documented by history of live measles vaccination at ≥12 months of age, history of physician-diagnosed measles, or laboratory confirmed seropositivity to measles. Vaccination of personnel who are serosusceptible is strongly encouraged. Approximately 5 to 15% of medical personnel are serosusceptible (Dales et al., 1985; Krause et al., 1979; Preblud et al., 1982). If susceptible personnel are exposed to measles, they should be removed from direct patient contact from either the 5th to the 21st day after exposure or for 7 days after rash onset (Williams, 1983). If a measles outbreak occurs, consideration should be given to the vaccination of all susceptible patients up to 72 h after exposure or, if vaccine is contraindicated, post-exposure measles immunoglobulin should be given within 6 days of exposure.

Parainfluenza Viruses

The parainfluenza viruses are important nosocomial respiratory pathogens in pediatric hospitals (Gardner et al., 1973; Meissner et al., 1984; Mufson et al., 1973; Sims, 1981; Wenzel et al., 1977). They also have been reported to cause outbreaks of serious respiratory illness among the elderly in nursing homes and among patients receiving immunosuppressive therapy (Centers for Disease Control, 1978a; DeFabritus et al., 1979). Nosocomial outbreaks usually occur in association with community outbreaks. Each serotype has somewhat different clinical and epidemiologic characteristics in the community and the hospital (Chanock et al., 1963; Denny and Clyde, 1986; Foy et al., 1973; Glezen et al., 1984; Monto et al., 1973). Parainfluenza virus 1 and 2 are common causes of croup and pneumonia in children and in recent years in the United States, have caused outbreaks in the autumn of alternating years. Parainfluenza virus 3 is a common cause of pneumonia and bronchiolitis in infants and young children, is often present throughout the year, and periodically causes outbreaks. Parainfluenza viruses 4a and 4b usually do not cause serious illness and are not considered to be important nosocomial pathogens (Gardner, 1969; Killgore and Dowdle, 1970).

Parainfluenza viruses are probably spread in a fashion similar to that for RSV, that is, through close contact and fomites. Staff are likely to play a major role in transmission by carrying virus from one patient to another or by becoming infected and directly infecting patients. Virus can be isolated for 1 to 2 days before onset of illness, often for 5 to 7 days after onset, and sometimes for several weeks after onset of illness (Chanock et al., 1961; Frank et al., 1981; Hall et al., 1977).

Control of parainfluenza virus infections depends on identifying infection, placing children on contact isolation, and using good handwashing practices, cohorting patients and staff, and wearing gowns and gloves when soiling is likely (Garner and Simmons, 1983; Williams, 1983). Although unproven, it is likely that, as for RSV, wearing goggles by staff would help prevent nosocomial transmission (Gala et al., 1986).

Respiratory Syncytial Virus (RSV)

Respiratory syncytial virus is an important nosocomial pathogen in infants and young children. Nosocomial RSV usually occurs during the yearly community outbreaks that begin in the late fall or winter in temperate climates and last for 2 to 5 months (Bruckova et al., 1979; Ditchburn et al., 1971; Gardner et al., 1973; Goldson et al., 1979; Gouyon et al., 1985; Hall et al., 1975b; Meissner et al., 1984; Mintz et al., 1979; Neligan et al., 1970; Sims, 1981; Valenti et al., 1982; Wenzel et al., 1977). In one study 45% of children in the hospital for >6 days became infected with RSV, all those hospitalized for >4 weeks became infected, and infected children were hospitalized for twice as long as those not infected (Hall et al., 1975b). The mortality associated with nosocomial RSV has been reported to be as high as 5% (MacDonald et al., 1982) but probably is closer to 0.5% in otherwise healthy children (Hall et al., 1986). In infants and young children with compromised cardiac, pulmonary, or immune systems, RSV mortality rates of 10 to 37% have been reported (Hall et al., 1986; MacDonald et al., 1982). Nosocomial outbreaks also have been reported among the elderly in nursing homes (Centers for Disease Control, 1977b; Garvie and Gray, 1980; Hart, 1984; Mathur et al., 1980; Morales et al., 1983; Sorvillo et al., 1984) and among older children and young adults in an institution for the mentally retarded (Finger et al., 1987).

Studies by Hall and others have shown that RSV is spread by direct contact, fomites, and perhaps large particle aerosol (Hall and Douglas, 1981; Hall et al., 1980, 1981). Nosocomial transmission appears to often involve the staff who inadvertently transfer virus between patients or become infected themselves and directly infect patients (Ditchburn et al., 1971; Hall et al., 1975b; Meissner et al., 1984; Valenti et al., 1982). Data from investigations of two outbreaks suggest that staff may have infected patients while manipulating orotracheal respiratory tubes or gastric feeding tubes (Meissner et al., 1984; Valenti et al., 1982). Patients can shed virus and potentially infect others for several days before onset of illness and usually for 5 to 7 days (but sometimes for 3 to 4 weeks) after onset of illness (Frank et al., 1981; Hall et al., 1976; Hall et al., 1975a). Prolonged shedding is more common in patients with altered immune systems and in newborns (Hall et al., 1986; Hall et al., 1979; Jarvis et al., 1983).

Diagnosing RSV infection in a hospitalized patient has become an important part of patient care and management. There is some evidence that infection control measures can decrease the nosocomial spread of RSV (Gardner et al., 1973; Hall et al., 1978); therefore, it is important to know which patients are infected and to institute contact isolation, use good handwashing practices, cohort patients and staff (during outbreaks), and wear gowns if soiling is likely (Garner and Simmons, 1983; Williams, 1983). Two recent studies have shown that wearing goggles by staff further decreases the rate of nosocomial infections in patients as well as staff (Agah et al., 1985; Gala et al., 1986).

The antiviral drug ribavirin has been effective both in decreasing shedding of virus and speeding recovery in patients with RSV infection (Hall and McBride, 1986); its use may also decrease the risk of nosocomial spread.

Rhinoviruses

Although common respiratory pathogens, rhinoviruses have infrequently been identified as nosocomial infections, probably because infection is usually associated only with upper respiratory illness (Gwaltney et al., 1966; Horn et al., 1975; Person et al., 1970; Sims et al., 1981; Valenti et al., 1980, 1982). Rhinoviruses, however, have been reported to cause lower respiratory tract illness, especially in patients who are immunosuppressed (George and Mogabgab, 1969; Halperin et al., 1983; Krilov et al., 1986; Valenti et al., 1982; Wood and Corbitt, 1985). Rhinovirus is usually present in respiratory secretions for 4 to 6 days after onset of illness, but can be present for up to 4 weeks, and is found in the highest titer in nasal secretions (Douglas et al., 1966; Fox et al., 1975; Gwaltney et al., 1966).

Transmission and methods of interrupting transmission have been well studied for the rhinoviruses; these results probably apply to most other respiratory viruses. Rhinoviruses are most easily transmitted by autoinoculation of the mucosa of the eyes or nose by hands contaminated through direct person-to-person contact of fomites (D'Alessio et al., 1976; Gwaltney and Hendley, 1982; Gwaltney et al., 1978; Hendley et al., 1973; Pancic et al., 1980; Reed, 1975). Aerosols do not appear to be important in their transmission. Spraying environmental surfaces with an antiviral substance (Gwaltney and Hendley, 1982), washing hands with an antiviral subtance (Gwaltney et al., 1980), and using virucidal paper handkerchiefs (Dick et al., 1986; Hayden et al., 1985) have decreased rhinovirus transmission in controlled experiments; but the effectiveness of these measures under natural conditions has not been demonstrated.

The recommended infection control measures include contact isolation for infants and young children, but no isolation for adults (Garner and Simmons, 1983).

Other Viruses

Enteroviruses

Nosocomial enterovirus infections with significant morbidity and some mortality have been reported in infant nurseries (Brightman et al., 1966; Cramblett et al., 1973; Davies et al., 1979; Faulkner and van Rooyen, 1973; Javette et al., 1956; Kinney et al., 1986; Lapinleimu and Hakulinen, 1972; McDonald et al., 1971; Mertens et al., 1982; Nagington et al., 1978, 1983: Swender et al., 1974), in staff in a surgical unit, and in older children and staff in a chronic-care hospital (Johnston and Burke, 1986; Morgante et al., 1971). Where vaccination has controlled polio-myelitis, it is not an important nosocomial pathogen.

Nosocomial enterovirus infections usually occur during the yearly enterovirus outbreaks of late summer and early fall in temperate climates (Moore, 1982; Nelson et al., 1979; Phillips et al., 1980; Strikas et al., 1986). Enterovirus infections occur in all age groups and lead to a wide range of clinical illnesses, from asymptomatic infection to life-threatening disseminated disease (Elveback et al., 1966; Kogon et al., 1969). Serious infection is most common in the newborn, and it is in the newborn that nosocomial infection is of greatest concern (Eichenwald et al., 1958; Grist et al., 1978; Jenista et al., 1984; Kaplan et al., 1983; Lake et al., 1976; Modlin, 1986; Morens, 1978). Persons with compromised immune systems are also at high risk for serious illness with enterovirus infections (Jarvis et al., 1983; Wilfert et al., 1977; Wood and Corbitt, 1985; Yolken et al., 1982).

Enteroviruses are transmitted by the oral-oral or fecal-oral route. Staff can inadvertently carry virus from one patient to another or become infected themselves and directly infect patients. In one outbreak, echovirus 11 infection was associated with gavage and mouth care (Kinney et al., 1986). Viruses can be shed in respiratory secretions for several days to several weeks after onset of illness and in stools for up to 4 weeks (Karzon et al., 1961; Kogan et al., 1969).

Patients with enterovirus infections should be placed on enteric precautions and patients and staff can be cohorted during outbreaks (Garner and Simmons, 1983). During several outbreaks in newborns, neonatal nurseries have been closed to new admissions during the outbreak (Cramblett et al., 1973; Davies et al., 1979; Kinney et al., 1986; Nagington et al., 1978, 1983). In one outbreak prophylactic immunoglobulin appeared to protect but not in another (Kinney et al., 1986; Nagington et al., 1983).

Human Parvovirus B19 (B19)

Relatively little is known about nosocomial B19 infection, although aplastic crisis (AC) in a patient with a chronic hemolytic anemia has been reported to occur as a nosocomial infection (Evans et al., 1983), and B19 infection has been associated with receiving blood products (Mortimer et al., 1983). The most likely source of B19 infection in the hospital is school-aged children who acquired infection during a community outbreak of fifth disease or AC (Anderson et al., 1984c; Chorba et al., 1986; Pattison et al., 1981; Serjeant et al., 1981). The two most severe complications of B19 infection are AC and fetal death (Anand et al., 1987; Saarinen et al., 1986); therefore, patients with chronic hemolytic anemias and pregnant women are at greatest risk from nosocomial infection.

The mode of transmission of B19 has not been clearly defined; however, B19 DNA has been found in respiratory secretions of patients in the acute phase of their infection, suggesting that the respiratory secretions are involved in transmission (Anderson et al., 1985; Chorba et al., 1986; Plummer et al., 1985). B19 also has been frequently found in acute-phase serum samples of patients with AC and in the urine of several patients with AC while they were viremic (Anderson et al., 1985, 1986; Chorba et al., 1986; Pattison et al., 1981; Serjeant et al., 1981). Studies of the timing of viremia with B19 infection suggest that patients with erythema infectiosum are likely to be infectious before but not after onset of the rash, whereas patients with AC are likely to still be infectious at the onset of their illness and for several days thereafter. A secondary attack rate of about 50% in susceptible family members suggests that close contact leads to efficient transmission.

Respiratory isolation precautions are recommended to prevent the spread of B19 infection (Garner and Simmons, 1983).

Mumps Virus

Nosocomial outbreaks of mumps are rare (Brunell et al., 1968; Sparling, 1969). In these outbreaks, the virus has been introduced by a patient or an employee incubating the illness. Since the introduction of the mumps vaccine, the incidence of mumps has decreased by 97% in the United States (Centers for Disease Control, 1984a) and the risk of nosocomial mumps has similarly decreased.

Mumps virus is believed to be transmitted by droplet spread, direct contact, or fomites. The virus is shed in saliva for from 6 days before to 9 days after salivary gland involvement and can be excreted in the urine for as long as 14 days after onset of the illness.

Control of nosocomial mumps depends on (1) placing ill or exposed persons on respiratory isolation for 9 days after onset of swelling, or until swelling resolves; and (2) vaccination of susceptible persons (Garner and Simmons, 1983). Control of outbreaks is complicated by the fact that excretion of the virus is greatest in the 48 h before the onset of clinical illness and many persons develop asymptomatic infection.

Human Papillomavirus (HPV)

Very little is known about the epidemiology of HPV infections in the community or in hospitals. It is likely that most infections in the hospital represent reactivation of latent virus infection (Ferenczy et al., 1985; Steinberg et al., 1983). For example, patients receiving immunosuppressive therapy often develop multiple warts (Kirchner, 1986; Spencer and Anderson, 1970). If transmission of HPV occurs in the hospital, it probably requires close contact. Sexual transmission is important for the transmission of some serotypes of HPV in the community (Oriel, 1971).

Rabies Virus

Nosocomial transmission of rabies virus is rare, but concern about potential transmission is great. Nosocomial infection has only been documented to occur after transplantation of corneas from humans dying of undiagnosed rabies encephalitis (Centers for Disease Control, 1980, 1981a; Hough et al., 1979).

Many of the recent cases of rabies in the United States were not suspected as being rabies until late in the patient's illness or after death because the classic symptoms associated with human rabies (difficulty swallowing, excitement or agitation, pain or paresthesia at the exposure site, hydrophobia or aerophobia, or hypersalivation) and history of a rabies exposure were not present (Anderson et al., 1984a). The one common clinical feature is the rapid progression of encephalomyelitis. Late diagnosis of rabies has led to many persons being potentially exposed and therefore given rabies postexposure prophylaxis (Helmick et al., 1987).

Almost all cases of human rabies have occurred after bites, but some have developed after the type of nonbite exposures that may occur in the hospital (Anderson et al., 1984b; Helmick et al., 1987). For example, rabies has been reported to develop after infectious material contacted open cuts or mucous membranes. The risk for these types of exposures is low, probably <0.1% (Babes, 1912) but high enough to justify postexposure prophylaxis with a safe vaccine. Rabies also has developed after aerosol exposure (Centers for Disease Control, 1977a; Humphrey et al., 1960; Irons et al., 1957; Winkler et al., 1973).

The virus is most often isolated from saliva and respiratory secretions and less often from cerebrospinal fluids urine sediment, tears, and body tissues from humans with rabies (Anderson et al., 1984a; Duenas et al., 1973; Helmich et al., 1987). The virus has been isolated for up to 5 weeks after onset of illness.

Suspecting rabies early in the illness is critical to reducing potential exposures; it should be considered in any person with a rapidly progressing encephalomyelitis. A patient suspected of having rabies should be placed on contact isolation and persons with close contact with the patient should wear masks, gloves, gowns, and protective eyewear (Remington et al., 1985). Exposed persons, including those previously vaccinated for rabies, should be given the recommended postexposure prophylaxis (Immunization Practices Advisory Committee, 1984).

Rubella Virus

Nosocomial rubella, which infrequently occurs where vaccination coverage is good, is primarily a concern for pregnant women and their unborn fetuses. Serious disease with infection is rare except during the early stages of pregnancy, when it often leads to a series of serious congenital anomalies that make up the congenital rubella syndrome (CRS) (Centers for Disease Control, 1983c). Outbreaks of rubella have been reported from general hospitals (McLaughlin et al., 1979; Polk et al., 1980; Strassburg et al., 1981), maternity units (Carne et al., 1973), and obstetric clinics (Gladstone et al., 1981); they usually occur in the spring when community outbreaks traditionally occur. The most common source of nosocomial rubella is a child admitted while incubating rubella; but medical personnel with symptomatic or asymptomatic disease have also introduced rubella into hospitals and clinics.

Transmission is believed to occur by droplet aerosol, direct contact, or fomites. Adults and children infected with rubella can excrete the virus in respiratory secretions from 7 days before until 4 days after onset of the rash. Infants with CRS can often excrete the virus for months and sometimes up to 4 years after birth (Shewmon et al., 1982).

Vaccination of high-risk susceptible persons is the key to preventing nosocomial rubella. This should be done before an exposure occurs. Because 12 to 14% of hospital employees lack antibody to rubella (McLaughlin et al., 1979; Orenstein et al., 1981; Weiss et al., 1979), those with direct or indirect patient contact should be screened regarding a history of rubella or rubella immunization, and if necessary, tested for rubella antibody before employment. Everyone who is antibody-negative, men and women, and exposed to high-risk areas should be immunized with the rubella vaccine unless contraindicated (for example by pregnancy or immunodeficiency). Female employees should be advised about the potential risk of vaccination during pregnancy, vaccinated, and maintained on an appropriate contraceptive for at least 3 months after the vaccination (Centers for Disease Control, 1983d).

When rubella is introduced into the hospital, susceptible patients and staff should be immunized unless contraindicated (Centers for Disease Control, 1983d, 1984b). Infected patients should be placed on contact isolation (Garner and Simmons, 1983).

Creutzfeldt-Jakob Disease (CJD) Agent

The risk of nosocomial transmission of CJD agent appears to be extremely low and has occurred under only four circumstances: (1) through transplantation of a cornea from a person with CJD (Duffy et al., 1974); (2) through use of contaminated deep stereotactic needles (inserted directly into the brain) after having been sterilized by conventional techniques (Bernoulli et al., 1977); (3) through transplantation of cadaveric dura mater (Centers for Disease Control, 1987a); and (4) through treatment with human growth hormone derived from pools of autopsy-obtained pituitary glands (Brown et al., 1985). The incubation period in the first three groups ranged from 15 to 24 months, but was between 4 to 15 years in the last. There is no evidence of unusual risk of disease in healthcare occupations, and no cases of CJD have been reported among general pathologists, neuropathologists, neurologists, laboratory technicians, autopsy technicians, morticians, or virologists (Masters et al., 1979).

The disease occurs primarily in middle-aged adults and is manifested by a progressive dementia leading to death over several years (Masters et al., 1979).

The only well-documented mode of transmission is parenteral exposure to neurologic tissue from patients with the disease (Bernoulli et al., 1977; Brown et al., 1985; Duffy et al., 1974). Cases infrequently cluster in families, and person-to-person transmission appears to be rare (Masters et al., 1979). Nevertheless, special precautions in dealing with patients and with potentially infective material are prudent. Through studies of CJD and other, related slow virus diseases in animals, tissues derived from the central nervous system, including optic tissues and cerebrospinal fluid, are considered to have the highest infectious titers. Other tissues, such as liver, lung, blood, lymph node, kidney, and urine, may also be infectious (Brown et al., 1982; Committee on Health Care Issues, ANA, 1986). Patients with known or suspected CJD should be handled in a fashion similar to that for patients with hepatitis B (Committee on Health Care Issues, ANA, 1986; Garner and Simmons, 1983; Jarvis, 1982). All body fluids and tissues, particularly brain tissue, cerebrospinal fluid, and blood, and instruments contaminated with these fluids should be handled with care. Patients with CJD do not need to be placed in isolation, but gloves should be worn when handling potentially infectious

material and gowns should be worn if gross exposure to such materials is anticipated. Specimens sent to the laboratory should be labeled as biohazard. Objects contaminated with patient blood or tissues should be handled with added precautions and sterilized before being discarded or cleaned. Creutzfeldt-Jakob disease appears to be unusually resistant to disinfection and requires special procedures to ensure that it is inactivated (See section on Disinfection and Inactivation of Viruses) (Brown et al., 1982; Committee on Health Care Issues, ANA, 1986).

Disinfection and Inactivation of Viruses

The method used for disinfection or sterilization must be considered carefully; all potential pathogens likely to be present and the level of disinfection required for the specific surface, instrument, or piece of equipment must be considered. Microorganisms to be considered include bacteria, fungi, mycobacteria, spores, and viruses; it is not appropriate to consider only one of these pathogens. Guidelines for determining the level of disinfection required for different purposes are outlined by Favero (1985).

Methods of inactivating microorganisms can be separated into sterilization and disinfection procedures. Sterilization procedures are designed to inactivate all microorganisms, including highly resistant bacterial spores, whereas disinfection procedures inactivate most but not all microorganisms. The most commonly used sterilization procedures are steam under pressure (autoclaving: 121°C or above per cycle), dry heat (121°C for 16 h, 160°C for 2 h, or 171°C for 1 h), or ethylene oxide gas (450 to 500 mg/liter at 55 to 60°C) (Favero, 1985).

Disinfection is generally less lethal than sterilization; its effectiveness depends on (1) the numbers and innate resistance of the organisms, (2) the presence of organic material such as blood or feces, (3) the type and concentration of the disinfectant chemical, (4) the time of exposure to the chemical, (5) the temperature during the exposure, and (6) whether or not the material is dry. The number of organisms and presence of contaminating materials are key variables in the disinfection process; therefore, cleaning before final disinfection is critical.

Viruses can be placed into two groups based on the ease with which they are likely to be inactivated by disinfectant chemicals, (1) lipid-containing viruses, and (2) nonlipid viruses (Favero, 1985). The lipid-containing viruses include all the enveloped viruses plus the nonenveloped hepatitis B virus (HBV); the nonlipid viruses include the remaining nonenveloped viruses. The lipid-containing viruses are the easiest to inactivate, whereas the nonlipid

TABLE 1. Susceptibility of viruses to
disinfectant chemicals

| Activity of disinfectant | Virus susceptibility | |
	Lipid-containing viruses	Nonlipid viruses
High	+	+
Intermediate	+	+/−[a]
Low	+	−

[a] Some nonenveloped viruses are resistant to intermediate
level disinfectants such as certain alcohols and phenolic
compounds.

viruses are comparatively more resistant to a spectrum of disinfectants (Table 1). Hepatitis B virus is
inactivated by disinfectants with an intermediate
level of activity but has not yet been tested with the
low-level disinfectants which inactivate other lipid-
containing viruses (Bond et al., 1983). Among the
nonlipid viruses, the smaller ones are more resistant
than the larger ones. The enteroviruses are the most
resistant of the viruses tested (Klein and Deforest,
1983).

Disinfectants can be placed into three groups
based on their ability to inactivate microorganisms
(Table 2). Three commonly available intermediate-
level disinfectants, 70% ethyl alcohol, 95% isopropyl

alcohol, or 0.05 to 0.5% sodium hypochlorite (500 to
5000 mg of free available chlorine per liter or a 1/100
to 1/10 dilution of household bleach) are effective
against a wide range of organisms and can be expected to inactivate lipid-containing and nonlipid viruses in 10 min under the appropriate conditions
(Klein and Deforest, 1983). The commercial disinfectants classified as *tuberculocidal* "hospital disinfectants" [under the nomenclature system of the Environmental Protection Agency (Block, 1983)] are also
intermediate-level disinfectants with broad-spectrum
virucidal activity, as well as activity against a wide
range of other microorganisms (Centers for Disease
Control, 1985a). Some of the most commonly used
disinfectants (certain phenolics and quaternary ammonium compounds) may fail to inactivate enteroviruses. If the conditions are not appropriate (for
example, if a high concentration of organic material
is present) any disinfectant may be ineffective.

The major exception to the previous discussion is
the CJD (or related) agents which are commonly
called "unconventional viruses." Limited studies to
date on the physical and chemical inactivation kinetics of the CJD (or scrapie) agent have shown inconsistent results but have suggested that slow viruses
as a group have rather unusual levels of resistance
(Asher et al., 1986; Brown et al., 1982). These data
suggest that the CJD agent in tissue samples (not on
surfaces or instruments) is resistant to 3 to 10% for-

TABLE 2. Characteristics of disinfectant chemicals

Method	Concentration or level[f]	Activity level
Glutaraldehyde, aqueous[a]	2% or variable	High
Hydrogen peroxide, stabilized	3–6%	High
Formaldehyde, aqueous[b]	1–8%	High to low
Iodophors[c]	30–50 mg of free iodine per liter; 70–150 mg of available iodine per liter	Intermediate
Chlorine compounds[d]	500–5,000 mg of free available chlorine per liter	Intermediate
Alcohol (ethyl; isopropyl)[e]	70%	Intermediate
Phenolic compounds, aqueous	0.5–3%	Intermediate to low
Quaternary ammonium compounds, aqueous	0.1–0.2%	Low

[a] There are several glutaraldehyde-based proprietary formulations on the US market, that is, low-, neutral-, or high-pH
formulations recommended for use at normal or raised temperatures with or without ultrasonic energy, and also a formulation containing 2% glutaraldehyde and 7% phenol. Instructions of the manufacturer regarding use as a sterilant or disinfectant or regarding anticipated dilution during use should be closely followed.
[b] Because of the ongoing controversy of the role of formaldehyde as a potential occupational carcinogen, the use of
formaldehyde is recommended only in limited circumstances under carefully controlled conditions, that is, for the disinfection of certain hemodialysis equipment.
[c] Only those iodophors registered with the EPA as hard-surface disinfectants should be used, and the instructions of the
manufacturer regarding proper use dilution and product stability should be closely followed. Antiseptic iodophors are not
suitable for use as disinfectants.
[d] There currently is a formulation registered with the EPA as a sterilant and disinfectant, depending on contact time, whose
active ingredient is chlorine dioxide. The instructions of the manufacturer regarding use as a sterilant or disinfectant or
regarding anticipated dilution during use should be closely followed.
[e] With volatile products such as alcohols, careful attention should be given to proper contact time during a disinfection
protocol.
[f] For disinfection, exposure times should be 20 to 30 min.

malin, 70% ethanol, ethylene oxide, quaternary ammonium compounds, and ultraviolet and ionizing radiation but is inactivated by immersion in a sodium
hypochlorite solution for 2 h or autoclaving at 121°C
for 1 h. These controversial data have led to rather
unconventional recommendations for disinfection
and sterilization (Committee on Health Care Issues,
ANA, 1986). These guidelines for terminal processing of CJD-contaminated surfaces far exceed the
level of treatment necessary to kill large numbers of
bacterial spores. However, until the inactivation
properties of CJD are studied further, it would be
prudent for healthcare workers to follow the published guidelines (within reason) and handle potentially contaminated objects with extreme caution.

Acknowledgment. The authors thank Novella Goss
for her help in preparing this chapter.

Literature Cited

Adler, S. P. 1986. Nosocomial transmission of cytomegalovirus. Pediatr. Infect. Dis. **5**:239–246.

Agah, R., J. Cherry, A. Garakian, and M. Chapin. 1985. A
study of the respiratory syncytial virus (RSV) infection
rate in personnel caring for children with RSV infection
in which the routine isolation procedure was compared
with the routine procedure supplemented with the use of
masks and goggles, p. 174. Program and Abstracts of the
Twenty-Fifth Interscience Conference on Antimicrobial
Agents and Chemotherapy.

Ahtone, J., D. P. Francis, D. W. Bradley, et al. 1980. Non-
A, non-B hepatitis in a nurse after percutaneous needle
exposure. Lancet **1**:1142.

Alford, C. A., S. Stagno, R. F. Pass, and E. S. Huang.
1981. Epidemiology of cytomegalovirus, p. 159–171. *In*
A. J. Nahmias, W. R. Dowdle, R. F. Schinazzi (ed.),
The human herpesvirus. Elsevier, New York.

Alford, R. H., J. A. Kasel, P. J. Gerone, and V. Knight.
1966. Human influenza resulting from aerosol inhalation. Proc. Soc. Exp. Biol. Med. **122**:800–804.

Alter, H. J., T. C. Chalmers, B. M. Freeman, et al. 1975.
Health care workers positive for hepatitis B surface antigen: are their contacts at risk? N. Engl. J. Med.
292:454–457.

Alter, M. J., M. S. Favero, and J. E. Maynard. 1986. Impact of infection control strategies on the incidence of
dialysis associated hepatitis in the United States. J. Infect. Dis. **153**:1149–1151.

Alter, M. J., M. S. Favero, and D. P. Francis. 1983. Cost
benefit of vaccination for hepatitis B in hemodialysis
centers. J. Infect. Dis. **148**:770–771.

Alter, M. J., R. J. Gerety, L. A. Smallwood, et al. 1982.
Sporadic non-A, non-B hepatitis: frequency and epidemiology in an urban U.S. population. J. Infect. Dis.
145:886–893.

American Hospital Association. 1984. A hospitalwide approach to AIDS. Recommendations of the Advisory
Committee of Infections within hospitals. Infect. Control **5**:242–248.

Anand, A., E. S. Gray, T. Brown, J. P. Clewley, and B. J.
Cohen. 1987. Human parvovirus infection in pregnancy
and hydrops fetalis. N. Engl. J. Med. **316**:183–186.

Anderson, L. J., K. G. Nicholson, R. V. Tauxe, and W. G.
Winkler. 1984a. Human rabies in the United States
1960–79: Epidemiology, diagnosis, and prevention.
Ann. Intern. Med. **100**:728–735.

Anderson, L. J., C. Tsou, R. A. Parker, T. L. Chorba, H.
Wulff, P. Tattersall, and P. P. Mortimer. 1986. Detection of antibodies and antigens of human parvovirus B19
by enzyme-linked immunosorbent assay. J. Clin. Microbiol. **24**:522–526.

Anderson, L. J., L. P. Williams, J. B. Layde, F. R. Dixon,
and W. G. Winkler. 1984b. Nosocomial rabies: Investigation of contacts of human rabies cases associated with
a corneal transplant. Am. J. Pub. Health **74**:370–372.

Anderson, M. J., P. G. Higgins, L. R. Davis, J. S. Williams,
S. E. Jones, I. M. Kidd, J. R. Pattison, and D. A. J.
Tyrrell. 1985. Experimental parvoviral infection in hu-
mans. J. Infect. Dis. **152**:257–265.

Anderson, M. J., E. Lewis, I. M. Kidd, S. M. Hall, and
B. J. Cohen. 1984c. An outbreak of erythema infec-
tiosum associated with human parvovirus infection. J.
Hyg. **93**:85–93.

Anonymous. 1984. Needlestick transmission of HTLV-III
from a patient infected in Africa. Lancet **2**:1376–1377.

Appleton, H., and P. G. Higgins. 1975. Viruses and gastroenteritis in infants. Lancet **1**:1297.

Arden, N. H., P. A. Patriarca, and A. P. Kendal. 1986.
Experiences in the use and efficacy of inactivated influ-
enza vaccine in nursing homes. *In* A. P. Kendal, P. A.
Patriarca, (ed.), Options for the control of influenza.
Alan R. Liss, Inc., New York.

Asher, D. M., C. J. Gibbs, and D. C. Gajdusek. 1986. Slow
viral infections: safe handling of agents of subacute
spongiform encephalopathics, p. 59–71. *In* B. M. Miller,
D. H. M. Groschel, J. H. Richardson, D. Vesley, J. R.
Songer, R. D. Housewright, and W. E. Barkley (ed.),
Laboratory safety: principles and practices. American
Society for Microbiology, Washington, D.C.

Atkinson, W. L., N. H. Arden, P. A. Patriarca, N. Leslie,
K-J Lui, and R. Gohd. 1986. Amantadine prophylaxis
during an institutional outbreak of type A (H1N1) influenza. Arch. Intern. Med. **146**:1751–1756.

Azimi, P. H., R. R. Roberto, J. Guralnik, et al. 1986. Transfusion-acquired hepatitis A in a premature infant with
secondary nosocomial spread in an intensive care nursery. Am. J. Dis. Child **140**:23–27.

Babes, V. 1912. Traite de la rage. Paris: Librarie J.-B. Balil-
liare et fils, p. 81–84.

Barker, W. H., and J. P. Mullooly. 1980. Influenza vaccination of elderly persons. Reduction in pneumonia and
influenza hospitalizations and deaths. J. Am. Med. Assoc. **244**:2547–2549.

Baron, R. C., H. B. Greenberg, G. Cukor, and N. R. Blacklow. 1984. Serological responses among teenagers after
natural exposure to Norwalk virus. J. Infect. Dis.
150:531–534.

Baron, R. C., J. B. McCormick, and O. A. Zubeir. 1983.
Ebola virus disease in southern Sudan: hospital dissemination and intrafamilial spread. Bull. W.H.O. **61**:997–
1003.

Baron, R. C., F. D. Murphy, H. B. Greenberg, C. E. Davis,
D. J. Bregman, G. W. Gary, J. M. Hughes, and L. B.
Schonberger. 1982. Norwalk gastrointestinal illness: an
outbreak associated with swimming in a recreational
lake and secondary person-to-person transmission. Am.
J. Epidemiol. **115**:163–172.

Bauer, C. R., K. Elie, L. Spence, and L. Stern. 1973. Hong

Kong influenza in a neonatal unit. J. Am. Med. Assoc. **223:**1233–1235.

Bean, B., F. S. Rhame, R. S. Hughes, M. D. Weiler, L. R. Peterson, and D. N. Gerding. 1983. Influenza B: hospital activity during a community epidemic. Diagn. Microbiol. Infect. Dis. **1:**177–183.

Beasley, R. P., L. Y. Hwang, G. C. Lee, et al. 1983. Prevention of perinatally transmitted hepatitis B virus infection with hepatitis B immune globulin and hepatitis B vaccine. Lancet **2:**1099–1102.

Berlin, B. S., and T. Campbell. 1970. Hospital-acquired chickenpox following exposures to herpes zoster. J. Am. Med. Assoc. **211:**1831–1833.

Bernoulli, C., J. Siegfried, G. Baumgartner, et al. 1977. Danger of accidental person-to-person transmission of Creutzfeld-Jakob disease by surgery. Lancet **1:**478–479.

Bloch, A. B., W. A. Orenstein, H. C. Stetler, S. G. Wassilak, R. W. Amler, K. J. Bart, C. D. Kirby, and A. R. Hinman. 1985. Health impact of measles vaccination in the United States. Pediatrics **76:**524–532.

Block, S. S. 1983. Federal regulation of disinfectants in the United States, p. 831–844. *In* S. S. Block, (ed.), Disinfection, sterilization and preservation, (3rd ed.) Lea & Febiger, Philadelphia.

Blumenfeld, H. L., E. D. Kilbourne, D. B. Louria, and D. E. Rogers. 1959. Studies on influenza in the pandemic of 1957–1958. I. An epidemiologic, clinical and serologic investigation of an intrahospital epidemic, with a note on vaccination efficacy. J. Clin. Invest. **38:**199–212.

Bond, W. W., M. S. Favero, N. J. Petersen, and J. W. Ebert. 1983. Inactivation of hepatitis B virus by intermediate-to-high-level disinfectant chemicals. J. Clin. Microbiol. **18:**535–538.

Bockman, J. M., D. T. Kingbury, M. P. McKinley, P. E. Bendheim, and S. B. Prusiner. 1985. Creutzfeld-Jakob disease prion proteins in human brains. N. Engl. J. Med. **312:**73–78.

Brandt, C. D., H. W. Kim, A. J. Vargosko, B. C. Jeffries, J. O. Arrobio, B. Rindge, R. H. Parrott, and R. M. Chanock. 1969. Infections in 18,000 infants and children in a controlled study of respiratory tract disease. I. Adenovirus pathogenicity in relation to serologic type and illness syndrome. Am. J. Epidemiol. **90:**484–500.

Brightman, V. J., T. F. M. Scott, M. Westphal, and T. R. Boggs. 1966. An outbreak of coxsackie B-5 virus infection in a newborn nursery. J. Pediatr. **69:**179–192.

Brocklebank, J. T., S. D. M. Court, J. McQuillin, and P. S. Gardner. 1972. Influenza-A infection in children. Lancet **2:**497–500.

Brown, P., C. Gajdusek, C. J. Gibbs, and D. M. Asher. 1985. Potential epidemic of Creutzfeld-Jakob disease from human growth hormone therapy. N. Engl. J. Med. **313:**728–731.

Brown, P., C. J. Gibbs, H. L. Amyx, et al. 1982. Chemical disinfection of Creutzfeld-Jakob disease virus. N. Engl. J. Med. **306:**1279–1281.

Bruckova, M., L. Kunzova, Z. Jezkova, and J. Vocel. 1979. Incidence of RS virus infections in premature children's ward. J. Hyg. Epidemiol. Microbiol. Immunol. **23:**389–396.

Brunell, P. A. 1981. Epidemiology of varicella-zoster virus infections, p. 153–158. *In* A. J. Nahmias, W. R. Dowdle, R. F. Schinazzi, (ed.), The human herpesvirus. Elsevier, New York.

Brunell, P. A., A. Brickman, D. O'Hare, and S. Steinberg. 1968. Ineffectiveness of isolation of patients as a method of preventing the spread of mumps. N. Engl. J. Med. **279:**1357–1361.

Bryden, A. S., M. E. Thouless, C. J. Hall, T. H. Flewett, B. A. Wharton, P. M. Mathew, and I. Crag. 1982. Rotavirus infections in a special-care baby unit. J. Infect. **4:**43–48.

Budnick, L. D., R. L. Stricof, and F. Ellis. 1984. An outbreak of influenza A in a nursing home, 1982. NY State J. Med. **84:**235–238.

Buehler, J. W., R. J. Finton, R. A. Goodman, K. Choi, J. C. Hierholzer, R. K. Sikes, and W. R. Elsea. 1984. Epidemic keratoconjunctivitis: report of an outbreak in an ophthalmology practice and recommendations for prevention. Infect. Control **5:**390–394.

Burney, M. I., A. Ghafoor, M. Saleen, P. A. Webb, and J. Casals. 1980. Nosocomial outbreak of viral hemorrhagic fever caused by Crimean hemorrhagic fever-Congo virus in Pakistan, January 1976. Am. J. Trop. Med. Hyg. **29:**941–947.

Caldwell, G. G., N. J. Lindsey, H. Wulff, D. D. Donnelly, and F. N. Bohl. 1974. Epidemic of adenovirus type 7 acute conjunctivitis in swimmers. Am. J. Epidemiol. **99:**230–234.

Carlson, J. A. K., P. J. Middleton, M. T. Szymanski, J. Huber, and M. Petric. 1978. Fatal rotavirus gastroenteritis. Am. J. Dis. Child **132:**477–479.

Carne, S., C. J. Dewhurst, and R. Hurley. 1973. Rubella epidemic in a maternity unit. Br. Med. J. **1:**444–446.

Carolane, D. J., A. M. Long, P. A. McKeever, S. J. Hobbs, and A. P. Roome. 1985. Prevention of speed of echovirus 6 in a special care baby unit. Arch. Dis. Child **60:**674–676.

Caul, E. O., and S. K. R. Clarke. 1975. Coronavirus propagated from patient with non-bacterial gastroenteritis. Lancet **2:**953–954.

Centers for Disease Control. 1988. Management of the patient with suspected viral hemorrhagic fever. **37:**Suppl 3.

Centers for Disease Control. 1987a. Rapidly progressive demention in a patient who received cadaveric dura mater graft. Morbid. Mortal. Weekly Rep. **36:**49–50, 55.

Centers for Disease Control. 1987b. Influenza A(H1N1) associated with mild illness in a nursing home—Maine. Morbid. Mortal. Weekly Rep. **36:**57–59.

Centers for Disease Control. 1987c. Update: Human immunodeficiency virus infections in health-care workers exposed to blood of infected patients. Morbid. Mortal. Weekly Rep. **36:**285–289.

Centers for Disease Control. 1986a. Recommendations for preventing transmission of infection with human T-lymphotropic virus type III/lymphadenopathy-associated virus during invasive procedures. Morbid. Mortal. Weekly Rep. **35:**221–223.

Centers for Disease Control. 1986b. Update: acquired immunodeficiency syndrome—United States. Morbid. Mortal. Weekly Rep. **35:**17–21.

Centers for Disease Control. 1986c. Recommendations for providing dialysis treatment to patients infected with human T-lymphotropic virus type III/lymphadenopathy-associated virus. Morbid. Mortal. Weekly Rep. **35:**376–378, 383.

Centers for Disease Control. 1985a. Recommendations for preventing transmission of infection with human T-lymphotropic virus type III/lymphadenopathy-associated virus in the workplace. Morbid. Mortal. Weekly Rep. **34:**681–686; 691–695.

Centers for Disease Control. 1985b. Update: public health service workshop on human T-lymphotropic virus type III antibody testing—United States. Morbid. Mortal. Weekly Rep. **34:**477–478.

Centers for Disease Control. 1985c. Routine screening for viral hepatitis in chronic hemodialysis centers. Hepatitis Surv. Rep. 49:5–6.

Centers for Disease Control. 1985d. Evaluation of human T-lymphotrophic virus type III/lymph adenopathy-associated virus infection in health care personnel—United States. Morbid. Mortal. Weekly Rep. 34:575–578.

Centers for Disease Control. 1985e. What control measures should be taken when hemodialysis units are suspected of having non-A, non-B hepatitis? Hepatitis Surv. Rep. 49:3–4.

Centers for Disease Control. 1985f. Recommendations for assisting in the prevention of perinatal transmission of human T-lymphotropic virus type III/lymphadeno-pathy-associated virus and acquired immunodeficiency syndrome. Morbid. Mortal. Weekly Rep. 34:721–726, 731–732.

Centers for Disease Control. 1984a. Mumps—United States, 1983–1984. Morbid. Mortal. Weekly Rep. 33:534–535.

Centers for Disease Control. 1984b. Rubella prevention. Morbid. Mortal. Weekly Rep. 33:301–318.

Centers for Disease Control. 1983a. Adenovirus type 7 outbreak in a pediatric chronic-care facility—Pennsylvania, 1982. Morbid. Mortal. Weekly Rep. 32:258–260.

Centers for Disease Control. 1983b. Impact of influenza on a nursing home population—New York. Morbid. Mortal. Weekly Rep. 32:32–34.

Centers for Disease Control. 1983c. Rubella and congenital rubella—United States, 1980–1983. Morbid. Mortal. Weekly Rep. 32:505–510.

Centers for Disease Control. 1983d. Rubella vaccination during pregnancy—United States, 1971–1982. Morbid. Mortal. Weekly Rep. 32:430–437.

Centers for Disease Control. 1983e. Acquired immunodeficiency syndrome (AIDS): precautions for health care workers and allied professionals. Morbid. Mortal. Weekly Rep. 32:450–451.

Centers for Disease Control. 1983f. Viral hemorrhagic fever: initial management of suspected and confirmed cases. Morbid. Mortal. Weekly Rep. 32(suppl):275–395.

Centers for Disease Control. 1982. Acquired immunodeficiency syndrome (AIDS): precautions for clinical and laboratory staff. Morbid. Mortal. Weekly Rep. 31:577–580.

Centers for Disease Control. 1981a. Human-to-human transmission of rabies virus by a corneal transplant—Thailand. Morbid. Mortal. Weekly Rep. 30:473–474.

Centers for Disease Control. 1981b. Influenza A in a hospital—Illinois. Morbid. Mortal. Weekly Rep. 30:79,80,85.

Centers for Disease Control. 1980. Human-to-human transmission of rabies virus by a corneal transplant—France. Morbid. Mortal. Weekly Rep. 29:25–26.

Centers for Disease Control. 1978a. Parainfluenza outbreaks in extended-care facilities—United States. Morbid. Mortal. Weekly Rep. 27:475–476.

Centers for Disease Control. 1978b. Nosocomial outbreak of pharyngoconjunctival fever due to adenovirus, type 4—New York. Morbid. Mortal. Weekly Rep. 27:49.

Centers for Disease Control. 1977a. Rabies in a laboratory worker, New York. Morbid. Mortal. Weekly Rep. 26:183–184.

Centers for Disease Control. 1977b. Respiratory syncytial virus—Missouri. Morbid. Mortal. Weekly Rep. 26:351.

Centers for Disease Control. 1977c. Outbreak of viral hepatitis in the staff (of a pediatric ward—California). Morbid. Mortal. Weekly Rep. 26:77–78.

Centers for Disease Control. 1977d. Hepatitis control measures for hepatitis B in dialysis centers. Viral hepatitis investigations and control series.

Champsaur, H., M. Henry-Amar, D. Goldszmidt, J. Provot, M. Bourjouane, E. Questiaux, C. Bach. 1984a. Rotavirus carriage, asymptomatic infection, and disease in the first two years of life. II. Serologic response. J. Infect. Dis. 149:675–682.

Champsaur, H., E. Questiaux, J. Provot, M. Henry-Amar, D. Goldszmidt, M. Bourjouane, and C. Bach. 1984b. Rotavirus carriage, asymptomatic infection, and disease in the first two years of life. I. Virus shedding. J. Infect. Dis. 149:667–674.

Chanock, R. M., J. A. Bell, and R. H. Parrott. 1961. Natural history of parainfluenza infection, p. 126–138. In M. Pollard, (ed.), Perspectives in virology. II. Burgess Publishing Co., Minneapolis.

Chanock, R. M., R. H. Parrott, K. M. Johnson, A. Z. Kapikian, and J. A. Bell. 1963. Myxoviruses: parainfluenza. Am. Rev. Respir. Dis. 88:152–166.

Chany, C., O. Moscovici, P. Lebon, and S. Rousset. 1982. Association of coronavirus infection with neonatal necrotizing enterocolitis. Pediatrics 69:209–214.

Chiba, S., I. Nakamura, S. Urasawa, S. Nakata, K. Taniguchi, K. Fujinaga, and T. Nakao. 1983. Outbreak of infantile gastroenteritis due to type 40 adenovirus. Lancet 2:954–957.

Chorba, T., R. Coccia, R. C. Holman, T. Tattersall, L. J. Anderson, J. Sudman, N. S. Young, E. Kurczynski, U. M. Saarinen, R. Koir, D. N. Lawrence, J. M. Jason, and B. Evatt. 1986. The role of parvovirus B19 in aplastic crisis and erythema infectiosum (fifth disease). J. Infect. Dis. 154:383–393.

Collaborative Study. 1980. Acute hepatitis B associated with gynaecological surgery. Lancet 1:1–6.

Committee on Health Care Issues, American Neurological Association. 1986. Precautions in handling tissues, fluids, and other contaminated materials from patients with documented or suspected Creutzfeld-Jakob disease. Ann. Neurol. 19:75–77.

Conte, J. E., W. E. Hadley, M. Sande, et al. 1983. Infection control guidelines for patients with the acquired immunodeficiency syndrome. N. Engl. J. Med. 309:740–744.

Corey, L. 1985. The natural history of genital herpes simplex virus. Perspectives on an increasing problem, p. 1–35. In B. Roizman and C. Lopez, (ed.), The herpesviruses: immunobiology and prophylaxis of human herpesvirus infections. Plenum Press, New York.

Corey, L., and K. K. Holmes. 1983. Genital herpes simplex virus infection: current concepts in diagnosis, therapy, and prevention. Ann. Intern. Med. 98:973–978.

Couch, R. B., T. R. Cate, W. F. Fleet, P. J. Gerone, and V. Knight. 1966. Aerosol-induced adenoviral illness resembling the naturally occurring illness in military recruits. Am. Rev. Respir. Dis. 88:529–535.

Cramblett, H. G., R. E. Haynes, P. H. Azimi, M. D. Hilty, and M. H. Wilder. 1973. Nosocomial infection with echovirus type 11 in handicapped and premature infants. Pediatrics 51:603–607.

Cubitt, W. D., and H. Holzel. 1980. An outbreak of rotavirus infection in a long-stay ward of a geriatric hospital. J. Clin. Pathol. 33:306–308.

Curran, J. W., W. M. Morgan, A. M. Hardy, H. W. Jaffe, W. W. Darrow, and W. R. Dowdle. 1985. The epidemiology of AIDS: current status and future prospects. Science 229:1362–1367.

Dales, L. G., and K. W. Kizer. 1985. Measles transmission in medical facilities. West. J. Med. 142:415–416.

D'Alessio, D. J., J. A. Peterson, C. R. Dick, and E. C. Dick. 1976. Transmission of experimental rhinovirus colds in volunteer married couples. J. Infect. Dis. 133:28–36.

D'Angelo, L. J., J. C. Hierholzer, R. A. Keenlyside, L. J. Anderson, and W. J. Martone. 1979. Pharyngoconjunctival fever caused by adenovirus type 4: report of a swimming pool-related outbreak with recovery of virus from pool water. J. Infect. Dis. 140:42–47.

Davies, D. P., C. A. Hughes, J. Mac Vicar, P. Hawkes, and H. J. Mair. 1979. Echovirus-11 infection in a special-care baby unit. Lancet 1:96.

Davis, R. M., W. A. Orenstein, J. A. Frank, J. J. Sacks, L. G. Dales, S. R. Preblud, K. J. Bart, N. M. Williams, and A. R. Hinman. 1986. Transmission of measles in medical settings. J. Am. Med. Assoc. 255:1295–1298.

Dearlove, J., P. Latham, B. Dearlove, K. Pearl, A. Thomson, I. G. Lewis. 1983. Clinical range of neonatal rotavirus gastroenteritis. Br. Med. J. 286:1473–1475.

DeFabritus, A. M., R. R. Riggio, D. S. David, L. B. Senterfit, J. S. Cheigh, and K. H. Stenzel. 1979. Parainfluenza type 3 in a transplant unit. J. Am. Med. Assoc. 241:384–386.

Denny, F. W., and W. A. Clyde, Jr. 1986. Acute lower respiratory tract infections in nonhospitalized children. J. Pediatr. 108:635–646.

Dick, E. C., S. U. Hossain, K. A. Mink, C. K. Meschievitz, S. B. Schultz, W. J. Raynor, and S. L. Inhorn. 1986. Interruption of transmission of rhinovirus colds among human volunteers using virucidal paper handkerchiefs. J. Infect. Dis. 153:352–356.

Dienstag, J. L. 1983. Non-A, non-B hepatitis. Recognition, epidemiology, and clinical features. Gastroenterology 85:439–462.

Dienstag, J. L., S. M. Feinstone, A. Z. Kapikian, et al. 1975. Fecal shedding of hepatitis A antigen. Lancet 1:765–767.

Dienstag, J. L., and D. M. Ryan. 1982. Occupational exposure to hepatitis B virus in hospital personnel. Infection or immunization? Am. J. Epidemiol. 115:26–39.

Ditchburn, R. K., J. McQuillin, P. S. Gardener, and S. D. M. Court. 1971. Respiratory syncytial virus in hospital cross-infection. Br. Med. J. 3:671–673.

Dolin, R., R. C. Reichman, H. P. Madore, R. Maynard, P. N. Linton, and J. Webber-Jones. 1982. A controlled trial of amantadine and rimantadine in the prophylaxis of influenza A infection. N. Engl. J. Med. 307:580–584.

Douglas, J., O. Schmidt, and L. Corey. 1983. Acquisition of neonatal HSV-1 infection from a parental source contact. J. Pediatr. 103:908–910.

Douglas, R. G., Jr., T. R. Cate, P. J. Gerone, and R. B. Couch. 1966. Quantitative rhinovirus shedding patterns in volunteers. Am. Rev. Respir. Dis. 93:159–167.

Dowdle, W. R., A. J. Nahmias, R. W. Harwell, and F. P. Pauls. 1967. Association of antigenic type of herpesvirus hominis with site of viral recovery. J. Immunol. 99:974–980.

Dudding, B. A., F. H. Top, Jr., P. E. Winter, E. L. Buescher, T. H. Lamson, and A. Leibovitz. 1973. Acute respiratory disease in military trainees. The adenovirus surveillance program, 1966–1971. Am. J. Epidemiol. 97:187–198.

Duenas, A., M. A. Belsey, J. Escobar, P. Medina, and C. Sanmartin. 1973. Isolation of rabies virus outside the human central nervous system. J. Infect. Dis. 127:702–704.

Duffy, P., J. Wolf, G. Collins, et al. 1974. Possible person-to-person transmission of Creutzfeld-Jakob disease. N. Engl. J. Med. 290:692–693.

Eichenwald, H. F., A. Ababio, A. M. Arky, and A. P. Hartman. 1958. Epidemic diarrhea in premature and older infants caused by echo virus type 18. J. Am. Med. Assoc. 166:1563–1566.

Eisenstein, A. B., R. D. Aach, W. Jacobsohn, and A. Goldman. 1963. An epidemic of infectious hepatitis in a general hospital: probable transmission by contaminated orange juice. J. Am. Med. Assoc. 185:171–174.

Elveback, L. R., J. P. Fox, A. Ketler, C. D. Brandt, F. E. Wassermann, and C. E. Hall. 1966. The virus watch program: a continuing surveillance of viral infections in metropolitan New York families. III. Preliminary report on association of infections with disease. Am. J. Epidemiol. 83:436–454.

Enders, C. 1984. Varicella-zoster virus infections in pregnancy. Prog. Med. Virol. 29:166–196.

Endreson, K., K. Gjesdal, I. Orstavik, E. Sivertssen, A. Reikvam, J. C. Ulstrup, and O. O. Aalen. 1985. Primary cytomegalovirus infection following open heart surgery. Acta Med. Scand. 218:423–428.

Evans, A. S. 1981. Epidemiology of Epstein-Barr virus infection and disease, p. 172–183. In A. J. Nahmias, W. R. Dowdle, and R. F. Schinazzi (ed.), The human herpesviruses. Elsevier, New York.

Evans, J. P. M., M. A. Rossiter, T. O. Kumaran, G. W. Marsh, and P. P. Mortimer. 1984. Human parvovirus aplasia: case due to cross infection in a ward. Br. Med. J. 288:681.

Faulkner, R. S., and C. E. van Rooyen. 1973. Echovirus type 17 in the neonate. Can. Med. Assoc. J. 108:878–882.

Favero, M. S. 1985. Sterilization, disinfection and antisepsis in the hospital, p. 129–137. In E. H. Lennette, A. Balows, W. J. Hausler, H. T. Shadomy (ed.), Manual of clinical microbiology (4th ed.). American Society for Microbiology, Washington, D.C.

Favero, M. S., J. E. Maynard, R. T. Leger, D. R. Graham, R. E. Dixon. 1979. Guidelines for the care of patients hospitalized with viral hepatitis. Ann. Intern. Med. 91:872–876.

Favero, M. S., J. E. Maynard, and N. J. Peterson. 1973. Hepatitis B antigen on environmental surfaces. Lancet 2:1455.

Feldman, S., W. T. Hughes, and H. Y. Kim. 1973. Herpes zoster in children with cancer. Am. J. Dis. Child 126:178–184.

Ferenczy, A., M. Mitao, N. Nagai, S. J. Silverstein, and C. P. Crum. 1985. Latent papillomavirus and recurring genital warts. N. Engl. J. Med. 313:784–788.

Finger, F., L. J. Anderson, R. C. Dicker, B. Harrison, R. Doan, A. Downing, and L. Corey. 1987. Epidemic respiratory syncytial virus infection in institutionalized young adults. J. Infect. Dis. (in press).

Fisher-Hoch, S. P., G. S. Platt, G. Lloyd, D. I. H. Simpson, G. H. Neild, and A. J. Barrett. 1983. Haematological and biochemical monitoring of Ebola infection in rhesus monkeys: implications for patient management. Lancet 2:1055–1058.

Flewett, T. H., A. S. Bryden, H. Davies, and C. A. Morris. 1975. Epidemic viral enteritis in a long-stay children's ward. Lancet 1:4–5.

Foley, F. D., K. A. Greenawald, G. Nash, B. A. Pruitt, Jr. 1970. Herpesvirus infection in burned patients. N. Engl. J. Med. 282:652–656.

Fox, J. P., C. D. Brandt, F. E. Wassermann, C. E. Hall, I. Spigland, A. Kogon, and L. R. Elveback. 1969. The virus watch program: a continuing surveillance of viral infections in metropolitan New York families. VI. Observations of adenovirus infections: virus excretion patterns, antibody response, efficiency of surveillance, patterns of infection, and relation to illness. Am. J. Epidemiol. 89:25–50.

Fox, J. P., M. K. Cooney, and C. E. Hall. 1975. The Seattle virus watch. V. Epidemiologic observations of rhinovirus infections, 1965–1969, in families with young children. Am. J. Epidemiol. **101:**122–143.

Foy, H. M., M. K. Cooney, A. J. Maletzky, and J. T. Grayston. 1973. Incidence and etiology of pneumonia, croup and bronchiolitis in preschool children belong to a prepaid medical care group over a four-year period. Am. J. Epidemiol. **97:**80–92.

Francis, D. P., S. C. Hadler, T. J. Prendergast, et al. 1984. Occurrence of hepatitis A, B, and non-A, non-B hepatitis in the United States. Am. J. Med. **76:**69–74.

Francis, D. P., and J. E. Maynard. 1979. The transmission and outcome of hepatitis A, B, and non-A, non-B: a review. Epidemiol. Rev. **1:**17–31.

Frank, A. L., L. H. Taber, C. R. Wells, J. M. Wells, W. P. Glezen, and A. Paredes. 1981. Patterns of shedding of myxoviruses and paramyxoviruses in children. J. Infect. Dis. **144:**433–441.

Fraser, D. W., C. C. Campbell, and T. P. Monath. 1974. Lassa fever in the Eastern Province of Sierra Leone 1970–1972. I. Epidemiologic studies. Am. J. Trop. Med. Hyg. **23:**1131–1139.

Gala, C. L., C. B. Hall, K. C. Schnabel, P. H. Pincus, P. Blossom, S. W. Hildreth, R. F. Betts, R. G. Douglas, Jr. 1986. The use of eye-nose goggles to control nosocomial respiratory syncytial virus infection. J. Am. Med. Assoc. **256:**2706–2708.

Galbraith, R. M., J. L. Dienstag, R. H. Purcell, et al. 1979. Non-A, non-B hepatitis associated with chronic liver disease in a hemodialysis unit. Lancet **1:**951–953.

Gardner, P. S., S. D. M. Court, J. T. Brocklebank, M. A. P. S. Downham, and D. Weightman. 1973. Virus cross-infection in paediatric wards. Br. Med. J. **2:**571–575.

Gardner, S. D. 1969. The isolation of parainfluenza 4 subtypes A and B in England and serological studies of their prevalence. J. Hyg. **67:**545–550.

Garibaldi, R. A., J. A. Bryan, J. N. Forrest, et al. 1973. Hemodialysis associated hepatitis. J. Am. Med. Assoc. **225:**384–389.

Garibaldi, R. A., C. M. Rasmussen, A. W. Holmes, et al. 1972. Hospital-acquired serum hepatitis. Report of an outbreak. J. Am. Med. Assoc. **219:**1577–1580.

Garner, J. S., and B. P. Simmons. 1983. CDC guidelines for isolation precautions in hospitals. Infect. Control. **4(suppl):**245–325.

Garvie, D. G., and J. Gray. 1980. Outbreak of respiratory syncytial virus infection in the elderly. Br. Med. J. **281:**1253–1254.

Gershon, A. 1980. Live attenuated varicella-zoster vaccine. Rev. Infect. Dis. **2:**393–405.

George, R. B., and W. J. Mogabgab. 1969. Atypical pneumonia in young men with rhinovirus infections. Ann. Intern. Med. **71:**1073–1078.

Gladstone, J. S., and S. J. Millian. 1980. Rubella exposure in an obstetric clinic. Obstet. Gynecol. **57:**182–186.

Glezen, W. P., A. L. Frank, L. H. Taber, and J. A. Kasel. 1984. Parainfluenza virus type 3: seasonality and risk of infection and reinfection in young children. J. Infect. Dis. **150:**851–857.

Goldson, E. J., J. T. McCarthy, M. A. Welling, and J. K. Todd. 1979. A respiratory syncytial virus outbreak in a transitional care nursery. Am. J. Dis. Child **133:**1280–1282.

Goodman, R. A., C. C. Carder, J. R. Allen, W. A. Orenstein, and R. J. Finton. 1982. Nosocomial hepatitis A transmission by an adult patient with diarrhea. Am. J. Med. **73:**220–226.

Gordon, J. E. 1962. Chickenpox: an epidemiological review. Am. J. Med. Sci. **244:**362–389.

Gouyon, J. B., P. Pothier, F. Guignier, H. Portier, H. P. Pujol, A. Kazmierczak, P. Chatelain, and M. Alison. 1985. Outbreak of respiratory syncytial virus in France. Eur. J. Clin. Microbiol. **4:**415–416.

Grady, G. F., V. A. Lee, A. M. Prince, et al. 1978. Hepatitis B immune globulin for accidental exposures among medical personnel: final report of a multicenter controlled trial. J. Infect. Dis. **138:**625–638.

Greaves, W. L., A. B. Kaiser, R. H. Alford, and W. Schaffner. 1980. The problem of herpetic whitlow among hospital personnel. Infect. Control. **1:**381–385.

Greenberg, H. B., J. Valdesuso, R. H. Yolken, E. Gangarosa, W. Gary, R. G. Wyatt, J. Konno, H. Suzuki, R. M. Chanock, and A. Z. Kapikian. 1979. Role of Norwalk virus in outbreaks of nonbacterial gastroenteritis. J. Infect. Dis. **139:**564–568.

Griffin, M. R., J. J. Surowiec, D. I. McCloskey, B. Capuano, B. Pierzynski, M. Quinn, R. Wojnarski, W. E. Parkin, H. Greenberg, and G. W. Gary. 1982. Foodborne Norwalk virus. Am. J. Epidemiol. **115:**178–184.

Grist, N. R., E. J. Bell, and F. Assaad. 1978. Enteroviruses in human disease. Prog. Med. Virol. **24:**114–157.

Grundy, J. E., M. Super, and P. O. Griffiths. 1986. Reinfection of a seropositive allograft recipient by cytomagalovirus from donor kidney. Lancet **1:**159–160.

Gunn, R. A., H. T. Janowski, S. Lieb, E. C. Prather, and H. B. Greenberg. 1982. Norwalk virus gastroenteritis following raw oyster consumption. Am. J. Epidemiol. **115:**348–351.

Gurwith, M., W. Wenman, D. Hinde, S. Feltham, and H. Greenberg. 1981. A prospective study of rotavirus infection in infants and young children. J. Infect. Dis. **144:**218–224.

Gwaltney, J. M., Jr., and J. O. Hendley. 1982. Transmission of experimental rhinovirus infection by contaminated surfaces. Am. J. Epidemiol. **116:**828–833.

Gwaltney, J. M., Jr., J. O. Hendley, G. Simon, W. S. Jordan, Jr. 1966. Rhinovirus infections in an industrial population. I. The occurrence of illness. N. Engl. J. Med. **275:**1261–1268.

Gwaltney, J. M., Jr., P. B. Moskalski, and J. O. Hendley. 1980. Interruption of experimental rhinovirus transmission. J. Infect. Dis. **142:**811–815.

Gwaltney, J. M., Jr., P. B. Moskalski, and J. O. Hendley. 1978. Hand-to-hand transmission of rhinovirus colds. Ann. Intern. Med. **88:**463–467.

Hadler, S. C., I. L. Doto, J. E. Maynard, et al. 1985. Occupational risk of hepatitis B infection in hospital workers. Infect. Control **6:**24–31.

Hadler, S. C., D. P. Francis, J. E. Maynard, et al. 1986. Long-term immunogenicity and efficacy of hepatitis B vaccine in homosexual men. N. Engl. J. Med. **315:**209–214.

Hall, C. B., and R. G. Douglas, Jr. 1981. Modes of transmission of respiratory syncytial virus. J. Pediatr. **99:**100–103.

Hall, C. B., and R. G. Douglas, Jr. 1975. Nosocomial influenza infection as a cause of intercurrent fevers in infants. Pediatrics **55:**673–677.

Hall, C. B., R. G. Douglas, Jr., and J. M. Geiman. 1980. Possible transmission by fomites of respiratory syncytial virus. J. Infect. Dis. **141:**98–102.

Hall, C. B., R. G. Douglas, Jr., and J. M. Geiman. 1976. Respiratory syncytial virus infections in infants: quantitation and duration of shedding. J. Pediatr. **89:**11–15.

Hall, C. B., R. G. Douglas, Jr., and J. M. Geiman. 1975a. Quantitative shedding patterns of respiratory of syncytial virus in infants. J. Infect. Dis. **132:**151–156.

Hall, C. B., R. G. Douglas, Jr., J. M. Geiman, and M. K. Messner. 1975b. Nosocomial respiratory syncytial virus infections. N. Engl. J. Med. **293:**1343–1346.

Hall, C. B., R. G. Douglas, Jr., K. C. Schnabel, J. M. Geiman. 1981. Infectivity of respiratory syncytial virus by various routes of inoculation. Infect. Immun. 33:779–783.

Hall, C. B., J. M. Geiman, B. B. Breese, and R. G. Douglas. 1977. Parainfluenza viral infections in children: Correlation of shedding with clinical manifestations. J. Pediatr. 91:194–198.

Hall, C. B., J. M. Geiman, R. G. Douglas, Jr., and M. P. Meagher. 1978. Control of nosocomial respiratory syncytial viral infections. Pediatrics 62:728–732.

Hall, C. B., A. E. Kopelman, R. G. Douglas, Jr., J. M. Geiman, and M. P. Meagher. 1979. Neonatal respiratory syncytial virus infection. N. Engl. J. Med. 330:393–396.

Hall, C. B., and J. T. McBride. 1986. Ribavirin and respiratory syncytial virus. Am. J. Dis. Child 140:331–332.

Hall, C. B., K. R. Powell, N. E. MacDonald, C. L. Gala, M. E. Menegus, S. C. Suffin, and H. J. Cohen. 1986. Respiratory syncytial viral infection in children with compromised immune function. N. Engl. J. Med. 315:77–81.

Halperin, S. A., P. A. Eggleston, J. O. Hendley, P. M. Suratt, D. H. M. Groschel, and J. M. Gwaltney, Jr. 1983. Pathogenesis of lower respiratory tract symptoms in experimental rhinovirus infection. Am. Rev. Respir. Dis. 128:806–810.

Hamre, D., and M. Beem. 1972. Virologic studies of acute respiratory disease in young adults. V. coronavirus 229E infections during six years of surveillance. Am. J. Epidemiol. 96:94–106.

Hanshaw, J. B. 1971. Congenital cytomegalovirus infection: a fifteen year perspective. J. Infect. Dis. 123:555–561.

Harris, D. J., H. Wulff, C. G. Ray, J. D. Poland, T. D. Y. Chin, and H. A. Wenner. 1972. Viruses and disease: III. An outbreak of adenovirus type 7A in a children's home. Am. J. Epidemiol. 93:399–402.

Harris, J. C., R. F. Finger, and J. M. Kobayashi. 1984. The low risk of hepatitis B in rural hospitals. J. Am. Med. Assoc. 252:3270–3272.

Harris, J. S., W. D. Kundin, M. F. Lenahan, and A. Bischone. 1983. Outbreak of rotavirus diarrhea in a fullterm nursery. 23rd Interscience Conference on Antimicrobial Agents and Chemotherapy. Abstract 417.

Hart, R. J. C. 1984. An outbreak of respiratory syncytial virus infection in an old people's home. J. Infect. 8:259–261.

Hayden, G. F., J. O. Hendley, and J. M. Gwaltney, Jr. 1985. The effect of placebo and virucidal paper handkerchiefs on viral contamination of the hand and transmission of experimental rhinoviral infection. J. Infect. Dis. 152:403–407.

Haynes, K., D. M. Danks, H. Gibas, and I. Jack. 1972. Cytomegalovirus in human milk. N. Engl. J. Med. 287:177–178.

Helmick, C. G., R. V. Tauxe, and A. A. Vernon. 1987. Is there a risk to contacts of patients with rabies? Rev. Infect. Dis. 9:511–518.

Helmick, C. G., P. A. Webb, C. L. Scribner, J. W. Krebs, and J. B. McCormick. 1986. No evidence for increased risk of Lassa fever infection in hospital staff. Lancet 2:1202–1205.

Hendley, J. O., R. P. Wenzel, and J. M. Gwaltney, Jr. 1973. Transmission of rhinovirus colds by self-inoculation. N. Engl. J. Med. 288:1361–1364.

Herbert, F. A., D. Wilkinson, E. Burchak, and O. Morgante. 1977. Adenovirus type 3 pneumonia causing lung damage in childhood. Can. Med. Assoc. J. 116:274–276.

Hirsch, M. S., G. P. Wormser, R. Schooley, et al. 1985.

Risk of nosocomial infection with human T-cell lymphotropic virus III (HTLV-III). N. Engl. J. Med. 312:1–4.

Hirschowitz, B. I., C. A. Dasher, F. J. Whitt, et al. 1980. Hepatitis B antigen and antibody and test of liver function—a prospective study of 310 hospital laboratory workers. Am. J. Clin. Pathol. 73:63–68.

Hoffman, P. C., and R. E. Dixon. 1977. Control of influenza in the hospital. Ann. Intern. Med. 87:725–728.

Hollinger, F. B., N. C. Khan, P. E. Oefinger, et al. 1983. Posttransfusion hepatitis type A. J. Am. Med. Assoc. 250:2313–2317.

Hollinger, F. B., J. W. Mosley, W. Szmuness, et al. 1980. Transfusion-transmitted viruses study: Experimental evidence for two non-A, non-B agents. J. Infect. Dis. 142:400–407.

Hope-Simpson, R. E. 1965. The nature of herpes zoster: a long-term study and a new hypothesis. Proc. Roy. Soc. Med. 58:9–20.

Horn, M. E. C., E. Brain, I. Gregg, S. J. Yealland, and J. M. Inglis. 1975. Respiratory viral infection in childhood. A survey in general practice, Roehampton 1967–1972. J. Hyg. 74:157–168.

Houff, S. A., R. C. Burton, and R. W. Wilson. 1979. Human-to-human transmission of rabies virus by corneal transplant. N. Engl. J. Med. 300:603–604.

Humphry, G. L., G. E. Kemp, and E. G. Wood. 1960. A fatal case of rabies in a woman bitten by an insectivorous bat. Publ. Health Rep. 75:317–326.

Immunization Practices Advisory Committee. 1986. Prevention and control of influenza. Morbid. Mortal. Weekly Rep. 35:317–331.

Immunization Practices Advisory Committee. 1985. Recommendations for protection against viral hepatitis. Morbid. Mortal. Weekly Rep. 34:313–324, 329–335.

Immunization Practices Advisory Committee. 1984. Rabies prevention—United States. Morbid. Mortal. Weekly Rep. 33:393–402, 407–408.

Irons, J. V., R. B. Eads, J. E. Grimes, and A. Conklin. 1957. The public health importance of bats. Texas Rep. Biol. Med. 15:292–298.

Jarvis, W. R. 1982. Precautions for Creutzfeld-Jakob disease. Infect. Control 3:238–239.

Jarvis, W. R., P. J. Middleton, and E. W. Gelfand. 1983. Significance of viral infections in severe combined immunodeficiency disease. Pediatr. Infect. Dis. 2:187–191.

Javett, S. N., S. Heymann, B. Mundel, W. J. Pepler, H. I. Lurie, J. Gear, V. Measroch, and Z. Kirsch. 1956. Myocarditis in the newborn infant. A study of an outbreak associated with coxsackie group B virus infection in a maternity home in Johannesburg. J. Pediatr. 48:1–22.

Jenista, J. A., K. R. Powell, and M. A. Menegus. 1984. Epidemiology of neonatal enterovirus infection. J. Pediatr. 104:685–690.

Johnston, J. M., and J. P. Burke. 1986. Nosocomial outbreak of hand-foot-and-mouth disease among operating suite personnel. Infect. Control 7:172–176.

Judson, F. N., and L. Molk. 1975. Epidemic acute infectious nonbacterial gastroenteritis at the Children's Research Institute and Hospital. Am. J. Epidemiol. 102:251–256.

Kane, M. A., and L. A. Lettau. 1985. Transmission of HBV from dental personnel to patients. J. Am. Dent. Assoc. 110:634–636.

Kapikian, A. Z., J. Flores, Y. Hoshino, R. Glass, K. Midthun, M. Gorziglia, and R. M. Channock. 1986. Rotavirus: the major etiologic agent of severe infantile diarrhea may be controlled by a "Jennerian" approach to vaccination. J. Infect. Dis. 153:815–823.

Kapila, R., D. I. Lintz, F. T. Tecson, L. Ziskin, D. B.

Louria. 1977. A nosocomial outbreak of influenza A. Chest **71**:576–579.

Kaplan, J. E., G. W. Gary, R. C. Baron, N. Singh, L. B. Schonberger, R. Feldman, H. B. Greenberg. 1982a. Epidemiology of Norwalk gastroenteritis and the role of Norwalk virus in outbreaks of acute nonbacterial gastroenteritis. Ann. Intern. Med. **96**:756–761.

Kaplan, J. E., R. A. Goodman, L. B. Schonberger, E. C. Lippy, and G. W. Gary. 1982b. Gastroenteritis due to Norwak virus: an outbreak associated with a municipal water system. J. Infect. Dis. **146**:190–197.

Kaplan, M. H., S. W. Klein, J. McPhee, and R. G. Harper. 1983. Group B Coxsackievirus infections in infants younger than three months of age: a serious childhood illness. Rev. Infect. Dis. **5**:1019–1032.

Kappus, K. D., J. S. Marks, R. C. Holman, J. K. Bryant, C. Baker, G. W. Gary, and H. B. Greenberg. 1982. An outbreak of Norwalk gastroenteritis associated with swimming in a pool and secondary person-to-person transmission. Am. J. Epidemiol. **116**:834–839.

Karzon, D. T., G. L. Eckert, A. L. Barron, N. S. Hayner, and W. Winkelstein, Jr. 1961. Aseptic meningitis epidemic due to echo 4 virus. Am. J. Dis. Child **101**:610–622.

Katagiri, S., A. Ohizumi, and M. Homma. 1983. An outbreak of type C influenza in a children's home. J. Infect. Dis. **148**:51–56.

Katz, M., M. A. Greco, L. Antony, and B. K. Young. 1980. Neonatal herpesvirus sepsis following internal monitoring. Int. Gynaecol. Obstet. **17**:631–633.

Keenlyside, R. A., J. C. Hierholzer, and L. J. D'Angelo. 1983. Keratoconjunctivitis associated with adenovirus type 37: an extended outbreak in an ophthalmologist's office. J. Infect. Dis. **147**:191–197.

Kemp, M. C., J. C. Hierholzer, C. P. Cabradilla, and J. F. Obijeski. 1983. The changing etiology of epidemic keratoconjunctivitis: antigenic and restriction enzyme analyses of adenovirus types 19 and 37 isolated over a 10-year period. J. Infect. Dis. **148**:24–33.

Killgore, G. E., and W. R. Dowdle. 1970. Antigenic characterization of parainfluenza 4A and 4B by the hemagglutination-inhibition test and distribution of HI antibody in human sera. Am. J. Epidemiol. **91**:303–316.

Kim, H. W., C. D. Brandt, J. O. Arrobio, B. Murphy, R. M. Chanock, and R. H. Parrott. 1979. Influenza A and B virus infection in infants and young children during the years 1957–1976. Am. J. Epidemiol. **109**:464–479.

Kinney, J. S., E. McCray, J. E. Kaplan, D. E. Low, G. W. Hammond, G. Harding, P. F. Pinsky, M. J. Davi, S. F. Kovnats, P. Riben, W. J. Martone, L. B. Schonberger, and L. J. Anderson. 1986. Risk factors associated with echovirus 11 infection in a hospital nursery. Pediatr. Infect. Dis. **5**:192–197.

Kirchner, H. 1986. Immunobiology of human papillomavirus infection. Prog. Med. Virol. **33**:1–41.

Klein, M., and A. Deforest. 1983. Principles of viral inactivation, p. 422–434. In S. S. Block, (ed.), Disinfection, sterilization and preservation, (3rd ed.). Lea & Febiger, Philadelphia.

Klein, G., and E. Klein. 1984. The changing faces of EBV research. Prog. Med. Virol. **30**:87–106.

Klein, B. S., J. A. Michaels, M. W. Rytel, K. G. Berg, and J. P. Davis. 1984. Nosocomial hepatitis A: a multinursery outbreak in Wisconsin. J. Am. Med. Assoc. **252**:2716–2721.

Knodell, R. G., M. E. Conrad, A. L. Ginsburg, C. J. Bell, and E. P. Flannery. 1976. Efficacy of prophylactic gammaglobulin in preventing non-A, non-B post-transfusion hepatitis. Lancet **1**:557–561.

Kogon, A., I. Spigland, T. E. Frothingham, L. Elveback,

C. Williams, C. E. Hall, and J. P. Fox. 1969. The virus watch program: a continuing surveillance of viral infections in metropolitan New York families. VII. Observations on viral excretion seroimmunity, intra-familial spread and illness association in coxsackie and echovirus infections. Am. J. Epidemiol. **89**:51–61.

Koopman, J. S., E. A. Eckert, H. B. Greenberg, B. C. Strohm, R. E. Isaacson, and A. S. Monto. 1982. Norwalk virus enteric illness acquired by swimming exposure. Am. J. Epidemiol. **115**:173–177.

Koopman, J. S., V. J. Turkish, A. S. Monto, V. Gouvea, S. Srivastava, and R. E. Issacson. 1981. Patterns and etiology of diarrhea in three clinical settings. Am. J. Epidemiol. **119**:114–123.

Krause, P. S., M. Ballows, and B. W. Klemas. 1981. Nosocomial rotavirus infection in a neonatal intensive care unit. Twenty-first Interscience Conference on Antimicrobial Agents and Chemotherapy. Abstract 704.

Krause, P. S., J. D. Cherry, S. Desada-Tous, J. G. Champion, M. Strassburg, C. Sullivan, M. J. Spencer, Y. J. Bryson, R. C. Welliver, and K. M. Boyer. 1979. Epidemic measles in young adults: clinical, epidemiologic and serologic studies. Ann. Intern. Med. **90**:873–876.

Krilov, L., L. Pierik, E. Keller, K. Mahan, D. Watson, M. Hirsch, V. Hamparian, and K. McIntosh. 1986. The association of rhinoviruses with lower respiratory tract disease in hospitalized patients. J. Med. Virol. **19**:345–352.

Krober, M. S., J. W. Bass, J. D. Brown, S. M. Lemon, and K. J. Rupert. 1984. Hospital outbreak of hepatitis A: risk factors for spread. Pediatr. Infect. Dis. **3**:296–299.

Lake, A. M., B. A. Lauer, J. C. Clark, R. L. Wesenberg, and K. McIntosh. 1976. Enterovirus infections in neonates. J. Pediatr. **89**:787–791.

Lapinleimu, K., and A. Hakulinen. 1972. A hospital outbreak caused by echo virus Type 11 among newborn infants. Ann. Clin. Res. **4**:183–187.

Larsen, R. A., J. T. Jacobson, J. A. Jacobson, R. A. Strikas, and J. C. Hierholzer. 1986. Hospital-associated epidemic of pharyngitis and conjunctivitis caused by adenovirus (21/H21+35). J. Infect. Dis. **154**:706–709.

Larson, H. E., S. E. Reed, and D. A. J. Tyrrell. 1980. Isolation of rhinoviruses and coronaviruses from 38 colds in adults. J. Med. Virol. **5**:221–229.

Lettau, L. A., J. D. Smith, D. Williams, et al. 1986. Transmission of hepatitis B with resultant restriction of surgical practice. J. Am. Med. Assoc. **255**:934–937.

Levandowski, R. A., and M. Rubenis. 1981. Nosocomial conjunctivitis caused by adenovirus type 4. J. Infect. Dis. **143**:28–31.

Levy, B. S., J. C. Harris, J. L. Smith, et al. 1977. Hepatitis B in ward and clinical laboratory employees of a general hospital. Am. J. Epidemiol. **106**:330–335.

Lewis, T. L., H. J. Alter, T. C. Chalmers, et al. 1973. A comparison of the frequency of hepatitis B antigen and antibody in hospital and non-hospital personnel. N. Engl. J. Med. **1289**:647–651.

Linnemann, C. C. Jr., T. G. Buchman, I. J. Light, J. L. Ballard, and B. Roizman. 1978. Transmission of herpessimplex virus type 1 in a nursery for the newborn: identification of viral isolates by D.N.A. "fingerprinting." Lancet **1**:964–966.

MacDonald, N. E., C. B. Hall, S. C. Suffin, C. Alexson, P. J. Harris, and J. A. Manning. 1982. Respiratory syncytial viral infection in infants with congenital heart disease. N. Engl. J. Med. **307**:397–400.

Marrie, T. J., S. H. L. Spencer, R. S. Faulkner, J. Ethier, and C. H. Young. 1982. Rotavirus infection in a geriatric population. Arch. Intern. Med. **142**:313–316.

Martini, G. A., and R. Siegert. 1971. Marburg virus disease, p. 230. Springer-Verlag, New York, Heidelberg, Berlin.

Masters, C. L., J. D. Harris, C. Gajdusek, et al. 1979. Creutzfeld-Jakob disease: patterns of worldwide occurrence and the significance of familial and sporadic clustering. Ann. Neurol. **5**:177–188.

Mathur, U., D. W. Bentley, and C. B. Hall. 1980. Concurrent respiratory syncytial virus and influenza A infections in the institutionalized elderly and chronically ill. Ann. Intern. Med. **93**:49–52.

Maynard, J. E. 1981. Nosocomial viral hepatitis. Am. J. Med. **70**:439–444.

McCormick, J. B., and K. M. Johnson. 1984. Viral hemorrhagic fevers, p. 676–697. *In* K. S. Warren, and A. A. F. Mahmoud (ed.). Tropical and geographical medicine. McGraw-Hill, New York.

McCormick, J. B., I. J. King, and P. A. Webb. 1986. Lassa fever: effective therapy with ribavirin. N. Engl. J. Med. **314**:20–26.

McCormick, J. B., I. J. King, P. A. Webb, K. M. Johnson, R. O'Sullivan, E. S. Smith, S. Trippel, and T. C. Tong. 1987a. A case-control study of the clinical diagnosis and course of Lassa fever. J. Infect. Dis. **155**:437–444.

McCormick, J. B., P. A. Webb, J. W. Krebs, K. M. Johnson, and E. S. Smith. 1987b. A prospective study of the epidemiology and ecology of Lassa fever. J. Infect. Dis. **155**:445–455.

McCray, E., Collaborative Needlestick Surveillance Group. 1986. Occupational risk of the acquired immunodeficiency syndrome among health care workers. N. Engl. J. Med. **314**:1127–1132.

McDonald, L. L., J. W. St. Geme, Jr., and B. H. Arnold. 1971. Nosocomial infection with echo virus type 31 in a neonatal intensive care unit. Pediatrics **47**:995–999.

McDougal, B. A., G. R. Hodges, H. D. Lewis, J. W. Davis, and S. A. Caldwell. 1977. Nosocomial influenza A infection. S. Med. J. **70**:1023–1024.

McLaughlin, M. C., and L. H. Gold. 1979. The New York rubella incident: a case for changing hospital policy regarding rubella testing and immunization. Am. J. Public Health **69**:287–289.

Meibalane, R., G. V. Sedmak, P. Sasidharan, P. Garg, and J. P. Grausz. 1977. Outbreak of influenza in a neonatal intensive care unit. J. Pediatr. **91**:974–976.

Meissner, H. C., S. A. Murray, M. A. Kiernan, D. R. Snyderman, and K. McIntosh. 1984. A simultaneous outbreak of respiratory syncytial virus and parainfluenza virus type 3 in a newborn nursery. J. Pediatr. **104**: 680–684.

Mertens, Th., H. Hager, and H. J. Eggers. 1982. Epidemiology of an outbreak in a maternity unit of infections with an antigenic variant of echovirus 11. J. Med. Virol. **9**:81–91.

Meyers, J. D., N. Fluornoy, and E. D. Thomas. 1986. Risk factors for cytomegalovirus infection after human marrow transplantation. J. Infect. Dis. **153**:478–488.

Meyers, J. D., F. J. Romm, W. S. Tihen, and J. A. Bryan. 1974. Food-borne hepatitis A in a general hospital: epidemiologic study of an outbreak attributed to sandwiches. J. Am. Med. Assoc. **231**:1049–1053.

Meyers, J. D., W. E. Stamm, M. Kerr, et al. 1978. Lack of transmission of hepatitis B after surgical exposure. J. Am. Med. Assoc. **240**:1725–1727.

Middleton, P. J., M. T. Szymanski, and M. Petric. 1977. Viruses associated with acute gastroenteritis in young children. Am. J. Dis. Child **131**:733–737.

Mintz, L., R. A. Ballard, S. H. Sniderman, R. S. Roth, and W. L. Drew. 1979. Nosocomial respiratory syncytial virus infections in an intensive care nursery: rapid diagnosis by direct immunofluorescence. Pediatrics **64**:149–153.

Modlin, J. F. 1986. Perinatal echovirus infection: insights from a literature review of 61 cases of serious infection

and 16 outbreaks in nurseries. Rev. Infect. Dis. **8**:918–926.

Mogabgab, W. J. 1963. Viruses associated with upper respiratory illnesses in adults. Ann. Intern. Med. **59**:306–322.

Monath, T. P., P. E. Mertens, R. Patton, C. R. Moser, J. J. Baum, L. Pinneo, G. W. Gary, and R. E. Kissling. 1973. A hospital epidemic of lassa fever in Zorzor, Liberia, March–April 1972. Am. J. Trop. Med. Hyg. **22**:773–779.

Monto, A. S. 1973. The Tecumseh study of respiratory illness: V. Patterns of infection with the parainfluenzaviruses. Am. J. Epidemiol. **97**:338–348.

Monto, A. S., J. S. Koopman, and E. R. Bryan. 1986. The Tecumseh study of illness. XIV. Occurrence of respiratory viruses, 1976–1981. Am. J. Epidemiol. **124**:359–367.

Morales, F., M. A. Calder, J. M. Inglis, P. S. Murdoch, and J. Williamson. 1983. A study of respiratory infections in the elderly to assess the role of respiratory syncytial virus. J. Infect. **7**:236–247.

Moore, M. 1982. Enteroviral disease in the United States, 1970–1979. J. Infect. Dis. **146**:103–108.

Morens, D. M. 1978. Enteroviral disease in early infancy. J. Pediatr. **92**:374–377.

Morens, D. M., R. M. Zweighaft, T. M. Vernon, G. W. Gary, J. J. Eslien, B. T. Wood, R. C. Holman, and R. Dolin. 1979. A waterborne outbreak of gastroenteritis with secondary person-to-person spread. Lancet **1**:964–966.

Morgante, O., D. Wilkinson, E. C. Burchak, M. Bruce, and M. Richter. 1972. Outbreak of hand-foot-and-mouth disease among Indian and Eskimo children in a hospital. J. Infect. Dis. **125**:587–594.

Mortimer, P. P., N. L. C. Luban, J. F. Kelleher, and B. J. Cohen. 1983. Transmission of serum parvovirus-like virus by clotting-factor concentrates. Lancet **2**:482–484.

Moser, M. R., T. R. Bender, H. S. Margolis, G. R. Noble, A. P. Kendal, and D. G. Ritter. 1979. An outbreak of influenza aboard a commercial airliner. Am. J. Epidemiol. **110**:1–6.

Mufson, M. A., H. E. Mocega, and H. E. Krause. 1973. Acquisition of parainfluenza 3 virus infection by hospitalized children. I. Frequencies, rates, and temporal data. J. Infect. Dis. **128**:141–147.

Murphy, A. M., G. S. Grohmann, P. J. Christopher, W. A. Lopez, G. R. Davey, and R. H. Millsom. 1979. An Australia-wide outbreak of gastroenteritis from oysters caused by Norwalk virus. Med. J. Aust. **2**:329–333.

Nagington, J., G. Gandy, J. Walker, and J. J. Gray. 1983. Use of normal immunoglobulin in an echovirus 11 outbreak in a special-care baby unit. Lancet **2**:443–446.

Nagington, J., T. G. Wreghitt, G. Gandy, N. R. C. Roberton, and P. J. Berry. 1978. Fatal echovirus 11 infections in outbreak in special-care baby unit. Lancet **2**: 725–728.

Nahmias, A. J., B. Roizman. 1973. Infection with herpes-simplex viruses 1 and 2. N. Engl. J. Med. **289**:667–674, 719–725, 781–789.

Naraqi, S., G. G. Jackson, and O. M. Jonasson. 1976. Viremia with herpes simplex type 1 in adults: four nonfatal cases, one with features of chicken pox. Ann. Intern. Med. **85**:165–169.

Neisson-Vernant, C., S. Arfi, D. Mathez, J. Leibowitch, and N. Monplaisir. 1986. Needlestick HIV seroconversion in a nurse. Lancet **2**:814.

Neligan, G. A., H. Steiner, P. S. Gardner, and J. McQuillin. 1970. Respiratory syncytial virus infection of the newborn. Br. Med. J. **3**:146–147.

Nelson, D., H. Hiemstra, T. Minor, and D. D'Alessio. 1979. Non-polio enterovirus activity in Wisconsin based on a 20-year experience in a diagnostic virology laboratory. Am. J. Epidemiol. **109**:352–361.

Noble, G. R. 1982. Epidemiological and clinical aspects of influenza, p. 11–51. In A. S. Beare (ed.), Basic and applied influenza research. CRC Press, Boca Raton, FL.

Noble, R. C., M. A. Kane, S. A. Reeves, and I. Roeckel. 1984. Posttransfusion hepatitis A in a neonatal intensive care unit. J. Am. Med. Assoc. 152:2711–2715.

O'Donoghue, J. M., C. G. Ray, D. W. Terry, Jr., and H. N. Beaty. 1973. Prevention of nosocomial influenza infection with amantadine. Am. J. Epidemiol. 97:276–282.

Oksenhendler, E., M. Harzic, J. Le Roux, C. Rabian, and J. P. Clauvel. 1986. HIV infection with seroconversion after a superficial needlestick injury to the finger. N. Engl. J. Med. 315:582.

Orenstein, W. A., P. N. R. Heseltine, S. J. LeGagnoux, and B. Portnoy. 1981. Rubella vaccine and susceptible hospital employees. J. Am. Med. Assoc. 245:711–713.

Orenstein, W. A., E. Wue, J. Wilkins, et al. 1981. Hospital-acquired hepatitis A: report of an outbreak. Pediatrics 67:494–497.

Oriel, J. D. 1971. Natural history of genital warts. Br. J. Vener. Dis. 47:1–13.

Oshiro, L. S., C. E. Haley, R. R. Roberto, J. L. Riggs, M. Croughan, H. Greenberg, and A. Kapikian. 1981. A 27-nm virus isolated during an outbreak of acute infectious nonbacterial gastroenteritis in a convalescent hospital: a possible new serotype. J. Infect. Dis. 143:791–795.

Osterholm, M. S., and J. S. Andrews. 1979. Viral hepatitis in hospital personnel in Minnesota—report of a statewide survey. Minn. Med. 62:683–689.

Pancic, F., D. C. Carpentier, and P. E. Came. 1980. Role of infectious secretions in the transmission of rhinovirus. J. Clin. Microbiol. 12:567–571.

Papavangelou, G. J., A. J. Roumeliotou-Karayannis, and P. C. Contoyannis. 1981. The risk of nosocomial hepatitis A and B virus infections from patients under care without isolation precaution. J. Med. Virol. 7:143–148.

Parrott, R. H. 1957. The clinical importance of group A coxsackie viruses. Ann. NY. Acad. Sci. 67:230–240.

Paryani, S. G., and A. M. Arvin. 1986. Intrauterine infection with varicella-zoster virus after maternal varicella. N. Engl. J. Med. 314:1542–1546.

Patriarca, P. A., J. A. Weber, R. A. Parker, W. N. Hall, A. P. Kendal, D. J. Bregman, and L. B. Schonberger. 1985. Efficacy of influenza vaccine in nursing homes. Reduction in illness and complication during an influenza A (H3N2) epidemic. J. Am. Med. Assoc. 253:1136–1139.

Patriarca, P. A., J. A. Weber, R. A. Parker, W. A. Orenstein, W. N. Hall, A. P. Kendal, and L. B. Schonberger. 1986. Risk factors for outbreaks of influenza in nursing homes: a case control study. Am. J. Epidemiol. 124:114–119.

Pattison, C. P., K. P. Boyer, J. E. Maynard, and P. C. Kelly. 1974. Epidemic hepatitis in a clinical laboratory. J. Am. Med. Assoc. 230:854–856.

Pattison, J. R., S. E. Jones, J. Hodgson, L. R. Davis, J. M. White, C. E. Stroud, and L. Murtaza. 1981. Parvovirus infections and hypoplastic crisis in sickle-cell anemia. Lancet 1:664–665.

Pattison, C. P., J. E. Maynard, K. R. Berquist, et al. 1975. Epidemiology of hepatitis B in hospital personnel. Am. J. Epidemiol. 101:59–64.

Person, D. A., and E. C. Herrmann. 1970. Experiences in laboratory diagnosis of rhinovirus infections in routine medical practice. Mayo Clin. Proc. 45:517–526.

Peterson, P. K., H. H. Balfour, S. C. Marker, D. S. Fryd, R. J. Howard, and R. L. Simmons. 1980. Cytomegalovirus disease in renal allograft recipients: a prospective study of the clinical features, risk factors, and impact on renal transplantation. Medicine 59:283–300.

Pether, J. V. S., and E. O. Caul. 1983. An outbreak of food-borne gastroenteritis in two hospitals associated with a Norwalk-like virus. J. Hyg. 91:343–350.

Phillips, C. A., M. D. Aronson, J. Tomkow, and M. E. Phillips. 1980. Enteroviruses in Vermont, 1969–1978: An important cause of illness throughout the year. J. Infect. Dis. 141:162–164.

Pingleton, S. K., W. W. Pingleton, R. H. Hill, A. Dixon, R. E. Sobonya, and J. Gertzen. 1978. Type 3 adenoviral pneumonia occurring in a respiratory intensive care unit. Chest 73:554–555.

Plotkin, S. A. 1986. Cytomegalovirus in hospitals. Pediatr. Infect. Dis. 5:177–178.

Plotkin, S. A., H. M. Friedman, G. R. Fleisher, D. C. Dafoe, R. A. Grossman, M. L. Smiley, S. E. Starr, C. Woldaver, A. D. Friedman, and C. F. Barker. 1984. Towne-vaccine-induced prevention of cytomegalovirus disease after renal transplants. Lancet 1:528–530.

Plummer, F. A., G. W. Hammond, K. Forward, L. Sekla, L. M. Thompson, S. E. Jones, I. M. Kidd, and M. J. Anderson. 1985. An erythema infectiosum-like illness caused by human parvovirus infection. N. Engl. J. Med. 313:74–79.

Polk, B. F., J. A. White, P. C. DeGirolami and J. F. Modlin. 1980. Outbreak of rubella among hospital personnel. N. Engl. J. Med. 303:541–545.

Preblud, S. R. 1981. Age-specific risks of varicella complications. Pediatrics 68:14–17.

Preblud, S. R., and L. J. D'Angelo. 1979. Chickenpox in the United States, 1972–1977. J. Infect. Dis. 140:257–260.

Preblud, S. R., F. Gross, N. A. Halsey, A. R. Hinman, K. L. Herrmann, and J. P. Koplan. 1982. Assessment of susceptibility to measles and rubella. J. Am. Med. Assoc. 247:1134–1137.

Prober, C. C., L. E. Kirk, and R. E. Keeney. 1982. Acyclovir therapy of chickenpox in immunosuppressed children—a collaborative study. J. Pediatr. 101:622–625.

Purcell, R. H., S. M. Feinstone, J. R. Ticehurst, R. J. Daemer, and B. M. Baroudy. 1984. Hepatitis A virus, p. 9–22. In G. N. Vyas, J. L. Dienstag, and J. H. Hoofnagle (ed.), Viral hepatitis and liver disease. Grune and Stratton, Orlando, FL.

Reed, C. M., T. L. Gustafson, J. Siegel, and P. Duer. 1984. Nosocomial transmission of hepatitis A from a hospital-acquired case. Pediatr. Infect. Dis. 3:300–303.

Reed, S. E. 1975. An investigation of the possible transmission of rhinovirus colds through indirect contact. J. Hyg 75:249–258.

Remington, P. L., T. Shope, and J. Andrews. 1985. A recommended approach to the evaluation of human rabies exposure in an acute care facility. J. Am. Med. Assoc. 254:67–69.

Report of an International Commission. 1978. Ebola haemorrhagic fever in Zaire, 1976. Bull. W.H.O. 56:271–293.

Richmond, S., R. Burman, E. Crosdale, L. Cropper, D. Longson, B. E. Enoch, and C. L. Dodd. 1984. A large outbreak of keratoconjunctivitis due to adenovirus type 8. J. Hyg. 93:285–291.

Rodriguez, W. J., H. W. Kim, C. D. Brandt, A. B. Fletcher, and R. H. Parrott. 1982. Rotavirus: a cause of nosocomial infection in the nursery. J. Pediatr. 101:274–277.

Roizman, B., P. G. Spear, and E. D. Kieff. 1973. Herpes simplex viruses I and II: A biochemical definition, p. 129–166. In Persistent virus infections. Academic Press, New York.

Ryder, R. W., J. E. McGowen, M. H. Hatch, and E. L. Palmer. 1977. Reovirus-like agent as a cause of nosocomial diarrhea in infants. J. Pediatr. 90:698–702.

Saarinen, U. A., T. L. Chorba, P. Tattersall, N. S. Young,

L. J. Anderson, E. Palmer, and P. F. Coccia. 1986. Human parvovirus B19 induced epidemic red-cell aplasia in patients with hereditary hemolytic anemia. Blood **67**:1411–1417.

Santosham, M., A. Pathak, S. Kottapalli, J. Vergara, S. Wong, J. Frochlick, and R. B. Sack. 1982. Neonatal rotavirus infection. Lancet **1**:1070–1071.

Saral, R., W. H. Burns, O. L. Laskin, G. W. Santos, and P. S. Lietman. 1981. Acyclovir prophylaxis of herpes-simplex-virus infections: a randomized, double-blind, controlled trial in bone-marrow-transplant recipients. N. Engl. J. Med. **305**:63–67.

Sattar, S. A., R. A. Raphael, H. Locknan, and V. S. Springthorpe. 1983. Rotavirus inactivation by chemical disinfectants and antiseptics used in hospitals. Can. J. Microbiol. **29**:1464–1469.

Saulsbury, F. T., J. A Winkelstein, and R. H. Yolken. 1980. Chronic rotavirus infection in immunodeficiency. J. Pediatr. **97**:61–65.

Seeberg, S., A. Brandberg, H. Svante, P. Larsson, and S. Lundgren. 1981. Hospital outbreak of hepatitis A secondary to blood exchange in a baby. Lancet **1**:1155–1156.

Seeff, L. B., E. C. Wright, H. J. Zimmerman, et al. 1978. Type B hepatitis after needle-stick exposure: prevention with hepatitis B immune globulin. Final report of the Veterans Administration Cooperative Study. Ann. Intern. Med. **88**:285–293.

Seeff, L. B., J. H. Zimmerman, E. L. Wright, et al. 1977. A randomized double-blind controlled trial of the efficacy of immune serum globulin for the prevention of post-transfusion hepatitis. A Veterans Administration cooperative study. Gastroenterology **72**:111–121.

Serjeant, G. R., J. M. Topley, K. Mason, B. E. Serjeant, J. R. Pattison, S. E. Jones, and R. Mohamed. 1981. Outbreak of aplastic crises in sickle cell anemia associated with parvovirus-like agent. Lancet **2**:595–597.

Shewmon, D. A., J. D. Cherry, and S. E. Kirby. 1982. Shedding of rubella virus in a 4-1/2 year old boy with congenital rubella. Pediatr. Infect. Dis. **1**:342–343.

Shields, A. F., R. C. Hackman, K. H. Fife, L. Corey, and J. D. Meyers. 1985. Adenovirus infections in patients undergoing bone-marrow transplantation. N. Engl. J. Med. **312**:529–533.

Siegal, F. P., C. Lopez, G. S. Hammer, A. E. Brown, S. J. Kornfeld, J. Gold, J. Hassett, S. Z. Hirschman, S. Cunningham-Rundles, and D. Armstrong. 1981. Severe acquired immunodeficiency in male homosexuals, manifested by chronic perianal ulcerative herpes simplex lesions. N. Engl. J. Med. **305**:1439–1444.

Sims, D. G. 1981. A two year prospective study of hospital-acquired respiratory virus infection on pediatric wards. J. Hyg. **86**:335–342.

Snydman, D. R., J. A. Bryan, and R. E. Dixon. 1975. Prevention and control of nosocomial viral hepatitis, type B (hepatitis B). Ann. Intern. Med. **83**:838–845.

Snydman, D. R., J. A. Bryan, E. J. Macon, et al. 1976a. Hemodialysis associated hepatitis: report of an epidemic with further evidence on mechanisms of transmission. Am. J. Epidemiol. **104**:563–570.

Snydman, D. R., S. H. Hindman, M. D. Wineland, J. A. Bryan, and J. E. Maynard. 1976b. Nosocomial viral hepatitis B. A cluster among staff with subsequent transmission to patients. Ann. Intern. Med. **85**:573–577.

Sorvillo, F. J., S. F. Huie, M. A. Strassburg, A. Butsumyo, W. X. Shandera, and S. L. Fannin. 1984. An outbreak of respiratory syncytial virus pneumonia in a nursing home for the elderly. J. Infect. **9**:252–256.

Sparling, D. 1969. Transmission of mumps. N. Engl. J. Med. **270**:276.

Spencer, E. S., and H. K. Anderson. 1970. Clinically evi-

dent, non-terminal infections with herpes viruses and the wart virus in immunosuppressed renal allograft recipients. Br. Med. J. **1**:251–254.

Spender, Q. W., D. Lewis, and E. H. Price. 1986. Norwalk-like viruses: study of an outbreak. Arch. Dis. Child. **61**:142–147.

Spigland, I., J. P. Fox, L. R. Elveback, F. E. Wassermann, A. Ketler, C. D. Brandt, and A. Kogon. 1966. The virus watch program: a continuing surveillance of viral infections in metropolitan New York families. II. Laboratory methods and preliminary report on infections revealed by virus isolation. Am. J. Epidemiol. **83**:413–435.

Srinivasan, G., E. Azarcon, M. R. L. Muldoon, G. Jenkins, S. Polavarapu, C. A. Kallick, and R. S. Pildes. 1984. Rotavirus infection in normal nursery: epidemic and surveillance. Infect. Control **10**:478–481.

Stagno, S., R. J. Whitley. 1985. Herpesvirus infections of pregnancy. Part I: cytomegalovirus and Epstein-Barr virus infections. N. Engl. J. Med. **313**:1270–1274.

Steinberg, B. M., W. C. Topp, P. S. Schneider, and A. L. Abramson. 1983. Laryngeal papillomavirus infection during clinical remission. N. Engl. J. Med. **308**:1261–1264.

Stevens, C. E., H. J. Alter, P. E. Taylor, E. A. Zang, E. J. Harley, and W. Szmuness. 1984. Hepatitis B vaccine in patients receiving hemodialysis. Immunogenicity and efficacy. N. Engl. J. Med. **311**:496–501.

Strassburg, M. A., D. T. Imagawa, S. L. Fannin, J. A. Turner, A. W. Chow, R. A. Murray, and J. D. Cherry. 1981. Rubella outbreaks among hospital employees. Obstet. Gynecol. **57**:283–288.

Straube, R. C., M. A. Thompson, R. B. Van Dyke, G. Wadell, J. D. Connor, D. Wingard, and S. A. Spector. 1983. Adenovirus type 7b in a children's hospital. J. Infect. Dis. **147**:814–819.

Straus, S. E. 1986. Oral acyclovir for recurrent herpesvirus infections, p. 165–175. *In* C. Lopez and B. Roizman (ed.), Human herpesvirus infections. Raven Press, New York.

Straus, S. E., W. Reinhold, H. A. Smith, W. T. Ruzechan, D. K. Henderson, R. M. Blaese, and J. Haz. 1984. Endonuclease analysis of viral DNA from varicella and subsequent zoster infections in the same patient. N. Engl. J. Med. **311**:1362–1364.

Stricof, R. C., and D. L. Morse. 1986. HTLV-III/LAV seroconversion following a deep intramuscular needlestick injury. N. Engl. J. Med. **314**:1115.

Strikas, R. A., L. J. Anderson, and R. A. Parker. 1986. Temporal and geographic patterns of isolates of nonpolio enterovirus in the United States, 1970–1983. J. Infect. Dis. **153**:346–351.

Sullivan-Bolyai, J. Z., K. H. Fife, R. F. Jacobs, Z. Miller, and L. Corey. 1983. Disseminated neonatal herpes simplex virus type 1 from a maternal breast lesion. Pediatrics **71**:455–457.

Swanepoel, R., A. J. Shepherd, P. A. Leman, S. P. Shepherd, G. M. McGillivray, M. J. Erasmus, L. A. Searle, and D. E. Gill. 1987. Epidemiologic and clinical features of Crimean-Congo hemorrhagic fever in southern Africa. Am. J. Trop. Med. Hyg. **36**:120–132.

Swender, P. T., R. J. Shott, and M. L. Williams. 1974. A community and intensive care nursery outbreak of coxsackievirus B5 meningitis. Am. J. Dis. Child **127**:42–45.

Szmuness, W., A. M. Prince, G. F. Grady, et al. 1974. Hepatitis B infection: a point prevalence study in 15 US hemodialysis centers. J. Am. Med. Assoc. **227**:901–906.

Szmuness, W., C. E. Stevens, E. J. Harley, et al. 1980. Hepatitis B vaccine: demonstration of efficacy in a controlled clinical trial in a high risk population in the United States. N. Engl. J. Med. **303**:833–841.

Takafuji, E. T., J. C. Gaydos, R. G. Allen, and F. H. Top,

Jr. 1979. Simultaneous administration of live enteric-coated adenovirus types 4, 7, and 21 vaccines: safety and immunogenicity. J. Infect. Dis. **141:**48–53.

Takahashi, M., T. Otsuka, Y. Okuno, Y. Asano, T. Yazaki, and S. Isomura. 1974. Live varicella vaccine used to prevent the spread of varicella in children in hospital. Lancet **2:**1288–1290.

Tan, J. A., and R. D. Schnagl. 1981. Inactivation of a rotavirus by disinfectants. Med. J. Aust. **1:**19–23.

Taylor, J. W., G. W. Gary, Jr., and H. B. Greenberg. 1981. Norwalk-related viral gastroenteritis due to contaminated drinking water. Am. J. Epidemiol. **114:**584–592.

Togo, Y., R. B. Hornick, V. J. Felitti, M. L. Kaufman, A. T. Dawkins, Jr., V. E. Kilpe, and J. L. Claghorn. 1970. Evaluation of therapeutic efficacy of amantadine in patients with naturally occurring A2 influenza. J. Am. Med. Assoc. **211:**1149–1156.

Tolpin, M. D., J. A. Stewart, D. Warren, B. A. Majica, M. A. Collins, S. A. Doveikio, C. Cabradilla, V. Schauf, T. N. K. Rajii, and K. Nelson. 1985. Transfusion transmission of cytomegalovirus confirmed by restriction endonuclease analysis. J. Pediatr. **107:**953–956.

Turner, A. R., R. N. MacDonald, and B. A. Cooper. 1972. Transmission of infectious mononucleosis by transfusion of pre-illness serum. Ann. Intern. Med. **77:**751–753.

Valenti, W. M., T. A. Clarke, C. B. Hall, M. A. Menegus, and D. L. Shapiro. 1982. Concurrent outbreaks of rhinovirus and respiratory syncytial virus in an intensive care nursery: epidemiology and associated risk factors. J. Pediatr. **100:**722–726.

Valenti, W. M., C. B. Hall, R. G. Douglas, Jr., M. A. Menegus, and P. H. Pincus. 1980. Nosocomial viral infections: I. Epidemiology and significance. Infect. Control **1:**33–37.

Van Dyke, R. B., S. A. Spector. 1984. Transmission of herpes simplex virus type 1 to a newborn infant during endotracheal suctioning for meconium aspiration. Pediatr. Infect. Dis. **3:**153–156.

Van Voris, L. P., R. B. Belshe, and J. L. Shaffer. 1982. Nosocomial influenza B virus infection in the elderly. Ann. Intern. Med. **96:**153–158.

Vogt, R. L., and F. Satink. 1986. Clam-associated gastroenteritis. N. Engl. J. Med. **315:**583.

Warren, D., J. Farrar, E. Hurwitz, K. Nelson, J. Hierholzer, E. Ford, and L. Anderson. 1986. A large outbreak of epidemic keratoconjunctivitis (EKC): problems in controlling nosocomial spread, p. 233. Program and Abstracts of the Twenty-sixth Interscience Conference on Antimicrobial Agents and Chemotherapy. American Society for Microbiology, Washington, D.C.

Weinstein, R. A., and G. F. Mallison. 1978. The role of the microbiology laboratory in surveillance and control of nonsocomial infections. Am. J. Clin. Pathol. **69:**130–136.

Weiss, K. E., C. E. Falvo, and E. Buimovici-Klein. 1979. Evaluation of an employee health service as a setting for a rubella screening and immunization program. Am. J. Public Health **69:**281–283.

Weiss, S. H., J. J. Goedart, M. G. Sarngadharan, et al. 1985a. Screening test for HTLV-III (AIDS agent) antibodies. J. Am. Med. Assoc. **253:**221–225.

Weiss, S. H., W. C. Saxinger, D. Rechtman, et al. 1985b. HTLV-III infection among health care workers. J. Am. Med. Assoc. **254:**2089–2093.

Weitekamp, M. R., P. Schan, and R. C. Aber. 1985. An algorithm for the control of nosocomial varicella-zoster virus infection. Am. J. Infect. Control **13:**193–198.

Weller, T. H. 1983. Varicella and herpes zoster. Changing concepts of the natural history, control and importance of a not-so-benign virus. N. Engl. J. Med. **309:**1362–1368, 1434–1440.

Weller, T. H. 1971. The cytomegloviruses: ubiquitous agents with protean manifestations. N. Engl. J. Med. **285:**203–214, 267–274.

Welliver, R. C., and S. McLaughlin. 1984. Unique epidemiology of nosocomial infection in a children's hospital. Am. J. Dis. Child **138:**131–135.

Wenzel, R. P., E. C. Deal, and J. O. Hendley. 1977. Hospital-acquired viral respiratory illness on a pediatric ward. Pediatrics **60:**367–371.

Whitley, R. J. 1985a. Epidemiology of herpes simplex viruses, p. 1–44. In B. Roizman, (ed.), The herpesviruses. (Vol. 3). Plenum Press, New York.

Whitley, R. J. 1985b. A perspective on the therapy of human herpesvirus infections, p. 339–369. In B. Roizman, C. Lopez, (ed.), The herpesviruses. Immunobiology and prophylaxis of human herpesvirus infections. Plenum Press, New York.

Whitley, R., M. Hilty, R. Haynes, Y. Bryson, J. D. Connor, S. J. Soong, and C. A. Alford. 1982. Vidarabine therapy of varicella in immunosuppressed patients. J. Pediatr. **101:**125–131.

Whitley, R. J., A. J. Nahmias, A. M. Visintine, C. L. Fleming, and C. A. Alford. 1980. The natural history of herpes simplex virus infection of mother and newborn. Pediatrics **66:**489–494.

Wilfert, C. M., R. H. Buckley, T. Mohanakumar, J. F. Griffith, S. L. Katz, J. K. Whisnant, P. A. Eggleston, M. Moore, E. Treadwell, M. N. Oxman, and F. S. Rosen. 1977. Persistent and fatal central-nervous-system echovirus infections in patients with agammaglobulinemia. N. Engl. J. Med. **296:**1485–1489.

Williams, W. W. 1983. CDC guideline for infection control in hospital personnel. Infect. Control **4**(suppl):326–349.

Winkler, W. G., T. R. Fashinelli, L. Leffingwell, P. Howard, and J. P. Conomy. 1973. Airborne transmission of rabies in a laboratory worker. J. Am. Med. Assoc. **226:**1219–1221.

Wong, V. C. W., H. M. H. Ip, H. W. Reesink, et al. 1984. Prevention of the HBsAg carrier state in newborn infants of mothers who are chronic carriers of HBsAg and HBeAg by administration of hepatitis B vaccine and hepatitis B immunoglobulin: double-blind randomized placebo-controlled study. Lancet **1:**921–926.

Wood, D. J., and G. Corbitt. 1985. Viral infections in childhood leukemia. J. Infect. Dis. **152:**266–273.

Yao, Q. Y., A. B. Rickinson, and M. A. Epstein. 1985. A re-examination of the Epstein-Barr virus carrier state in healthy seropositive individuals. Intern. J. Cancer **35:**35–42.

Yeager, A. S., R. C. Ashley, and L. Corey. 1983. Transmission of herpes simplex virus from father to neonate. J. Pediatr. **103:**905–907.

Yeager, A. S., F. C. Grumet, E. B. Hafleigh, A. M. Arvin, J. S. Bradley, and C. G. Prober. 1981. Prevention of transfusion-acquired cytomegalovirus infections in newborn infants. J. Pediatr. **98:**281–287.

Yolken, R. H., C. A. Bishop, T. R. Townsend, E. A. Bolyard, J. Barlett, G. W. Santos, and R. Saral. 1982. Infectious gastroenteritis in bone-marrow-transplant recipients. N. Engl. J. Med. **306:**1009–1012.

Zaia, J. A., M. J. Levin, and S. R. Preblud. 1983. Evaluation of varicella-zoster immune globulin: protection of immunosuppressed children after household exposure to varicella. J. Infect. Dis. **147:**737–743.

SECTION II
General Technological Approaches to Diagnostics

CHAPTER 3

Virus Isolation and Identification

DIANE S. LELAND and MORRIS L. V. FRENCH

Cell Cultures: Monolayer cultures of primary, diploid, and continuous cell lines are the hosts of choice for virus isolation. Quality cell cultures are available commercially and are conveniently maintained in the laboratory. After proper decontamination and purification, each clinical sample is inoculated into several types of cell cultures; the preferred lines vary from virus to virus.

Alternative Cell Culture Techniques: New cell culture techniques have been developed to enhance virus isolation. Shell vials containing a monolayer of cells growing on a cover slip provide a virus isolation system that can be placed in the centrifuge and spun to enhance virus infection. Following incubation, the infected cover slip is withdrawn and stained to confirm virus identification. Shell vials are helpful in cytomegalovirus isolation. Herpes simplex isolation and identification systems are commercially available. A culture of herpes-susceptible cells and an immunostaining system are included. Inclusion of microspheres in cell culture has been shown to provide increased surface area for attachment of cells. Increased cell numbers provide for increased virus production. Several viruses can be isolated in suspensions of human lymphocytes rather than in traditional monolayer cultures.

Virus Quantitation: Viruses are quantitated by titration and plaquing methods. Formulas by Reed-Muench (1938) and Karber (1931) are used to determine the end point from titration data.

Virus Detection and Identification: Both immunologic and nonimmunologic methods are used in identification of isolated viruses and for detection of viral antigen within cell cultures or within clinical materials.

Nonimmunologic methods include histologic staining, electron microscopy, hemadsorption (HAD), hemagglutination (HA), challenge interference (CI), and DNA probes. HAD, HA, and CI are used primarily to detect non-cytopathogenic effect-producing viruses within cell cultures. DNA probes provide the most specific viral identification and will probably replace many traditional techniques.

Immunologic methods include immunofluorescence (IF), immunoperoxidase (IP), enzyme-linked immunosorbent assay (ELISA), neutralization (NEUT), hemagglutination inhibition (HAI), passive agglutination (PA), radioimmunoassay, and immune adherence hemagglutination. IF, IP, ELISA, and PA provide rapid and accurate virus identification, whereas NEUT and HAI, although accurate, are more costly and time-consuming.

Virus Isolation and Identification

In the past 5 to 10 years, progress in clinical diagnostic virology has been remarkable. Traditional techniques for virus isolation have been refined, and new alternative methods have been developed. All virus isolation and identification services have been simplified and enhanced by the commercial availability of quality cell cultures and antisera. Techniques previously reserved for research facilities are now performed on a routine basis in clinical virus-isolation facilities. New diagnostic tests for direct detection and identification of viral antigens in clinical samples are now available. These methods provide rapid, accurate, and cost-effective identification of many common viral pathogens, as well as confirmation of infections due to several viruses such as rotavirus and hepatitis, which, owing to their lack of proliferation in standard cell lines, have previously eluded the clinical virologist.

In this chapter, standard viral diagnostic methods are described. Guidelines are included for virus isolation as well as for direct methods in which virus isolation is not required. Traditional methods and contemporary technology are described; techniques useful in routine viral identification are emphasized, and those methods which are best reserved for full-service virology laboratories are outlined.

Virus Isolation

Cell Cultures

Because viruses lack ribosomes and systems for synthesis of energy and proteins, they are not capable of reproducing on artificial media. Viruses are obligate intracellular parasites and can replicate only within living cells. Several living cell systems have been used for viral proliferation. These include cell cultures, embryonated eggs, and animals. No single system will support the growth of all viruses; hence, the laboratory is faced with the problem of selecting culture environments suitable for isolation of the largest number of viral agents.

Cell cultures are the host system most frequently used for virus cultivation because of their broad spectrum of susceptibility and the relative ease with which cell cultures are maintained in the laboratory. Cell cultures are prepared with suspensions of cells which, if necessary, have been dissociated from the parent tissue by action of proteolytic enzymes and chelating agents. Dissociated cells, when placed in a culture vessel, adhere to the surface of the vessel and replicate to form a layer of cells (monolayer); adherence to the surface is an integral part of survival and

subsequent cell proliferation. This is the mode of culture common to most normal cells other than hemopoietic cells (Freshney, 1983). Other types of cells such as mature hemopoietic cells, transformed cell lines, or cells from malignancies may replicate in suspension. These cells can survive and proliferate without attachment (Freshney, 1983). Both monolayer and suspended cell cultures are useful in the isolation of viruses in the laboratory.

Monolayer cultures are observed microscopically without removing the monolayer from the wall of the vessel and without staining. Many types of virus-induced changes, collectively referred to as the cytopathogenic effect (CPE) of a virus and including any alteration in cellular characteristics that is virus induced, are visible when unstained cell monolayers are examined with the $10\times$ objective of a standard light microscope. Other manifestations of CPE, such as inclusion bodies or chromosomal alterations, can be evaluated only in stained preparations. Suspended cell preparations are more difficult to evaluate microscopically, and the presence of a viral infection must be confirmed by alternative techniques.

Three types of monolayer cell cultures—primary, diploid, and established or continuous—are used in virus isolation. Primary cells are the first generation of cells that grow from the tissue of origin. Primary cell preparations contain predominantly epithelial cells, and the cells have normal diploid chromosomes of the same number as the parent tissue. Primary cell preparations are susceptible to a variety of human viral pathogens. One problem encountered occasionally with primary cultures is that of indigenous viral pathogens that may have infected the animal host prior to cell culture preparation. These indigenous viruses may produce CPE during cell culturing or may remain largely undetectable while reducing the susceptibility of the cultured cells to viral infection. Primary cell types include primary rhesus monkey kidney (pRMK) and primary human embryonic kidney (pHEK).

Diploid cell lines, as their name suggests, are those which continuously maintain their diploid chromosome number throughout serial passages; these cells usually die out after the 50th passage. Some well-known diploid cell strains are human lung fibroblasts (MRC-5 or WI-38) and human foreskin (MRHF).

Continuous (heteroploid, established) cell lines demonstrate heteroploid or aneuploid chromosome numbers during repeated subculturing. Continuous cell lines originate from malignancies; they replicate vigorously and are usually not difficult to culture, being capable of indefinite passaging. Continuous cell lines include human cervical carcinoma (HeLa) and human laryngeal carcinoma (HEp-2).

Primary cell cultures may be initiated in the virus laboratory from a desired tissue by allowing cells to migrate from fragments of tissue or by disaggregating the tissue mechanically or enzymatically to produce a suspension of cells. The enzymes most frequently used are trypsin, collagenase, elastase, hyaluronidase, DNase, pronase, or combinations thereof. Each tissue requires a different set of conditions, although certain requirements, summarized by Freshney (1983), are shared by most cells: 1) fat and necrotic tissue is best eliminated prior to culturing; 2) tissue should be chopped fine with minimum damage; 3) enzymes used for disaggregation should be removed by gentle centrifugation afterward; 4) the concentration of cells initially should be increased over that required for normal subculture, since the proportion of cells that survive primary culture may be quite low; 5) a rich medium should be used, and fetal bovine sera, rather than calf or horse sera, may give better survival; and 6) embryonic tissue disaggregates more readily and proliferates more rapidly than adult tissue in primary culture.

In-house preparation of cell cultures allows for utilization of cultured cells while cells are young and actively metabolizing. This state, rather than the steady-state metabolism of mature, confluent, monolayer cultures, has been shown to be superior for viral isolation (Aurelian, 1969). In-house preparation need not begin with procurement of a primary culture. Cell lines may be purchased commercially; then cells can be maintained within the laboratory for preparation of culture tubes as required. Most diploid and heteroploid lines can be transferred with excellent quality of cells maintained (see cell subculture procedure, below). In-house preparation may be especially desirable with continuous cells such as Vero (African green monkey kidney) or HEp-2, which tend to proliferate vigorously and overgrow their monolayers.

Many types of cell cultures are available commercially, and supply houses will prepare primary, diploid, or continuous cells in tubes or flasks and in the medium specified by the purchaser. Cultures are delivered to the purchaser when monolayers are confluent. Busy laboratories with limited staff may favor commercial purchase of cell cultures. Some of the burden of cell culture maintenance, subculturing, aliquoting, and feeding is relieved in this way and allows staff to concentrate on virus isolation.

Cell Culture Media

Cell culture media consist of balanced salt solutions enriched with amino acids, vitamins, and animal sera. Basic media enriched with 10% serum serve as outgrowth media for freshly subcultured or newly established cell cultures when rapid cell proliferation is desired. The amino acid L-glutamine is also added in a concentration of 3% to outgrowth media to stimulate cell growth. The serum content is reduced to 2% in maintenance media used for cultures that have reached confluency and are to be maintained in a steady state. Incorporation of a buffering system in the medium is mandatory to aid in maintaining proper pH. Media with Hanks balanced salt solution contain a low concentration of sodium bicarbonate and function effectively in a closed system where vessels are incubated with caps tightly closed in an ambient air incubator. Media with Earles balanced salt solution contain a higher concentration of sodium bicarbonate and function well in an open system where vessels are incubated with caps loosely in place in an incubator with a 5% CO_2 environment.

All media, both outgrowth and maintenance, contain a pH indicator, usually phenol red. A salmon-pink color indicates a 6.9 to 7.2 pH, optimal for cell culture maintenance. The color changes to bright pink, indicating increased pH, when the buffering system is disturbed (as when caps are not tightly closed) or when cells are not metabolizing well. A color change to bright yellow indicates excess acid production and usually is seen in cultures infected by bacteria or fungi or in cultures that proliferated excessively and are in need of refeeding or subculturing. The color indicator allows the virologist to evaluate the condition of the cell culture with a macroscopic glance. Although a color change may be observed in cultures extensively infected by virus, virus proliferation may not produce a change in color of the medium.

Antibiotics are routinely added to cell culture media to protect the culture from bacterial or fungal contamination. Solutions containing combinations of antibiotics suitable for cell culturing are available commercially; these contain antibiotics in concentrations appropriate for convenient addition to cell culture media. Antibiotics commonly added to cell cultures include penicillin/streptomycin and gentamicin sulfate. The addition of an antifungal agent such as amphotericin B is also common practice and is helpful in the culturing of clinical samples from the throat or female genital tract, which are commonly contaminated with fungi.

Subculturing of Adherent Cells

Subculturing or transferring of cultured cells is required when a culture has reached maturity and has overgrown its present vessel or when additional vessels or vessels of different volume are desired. The transfer of cells from one vessel to another is accomplished by detaching the cells from their existing vessel and transferring them to new, sterile vessels. A subculturing procedure is given below.

CELL SUBCULTURE PROCEDURE

1. Remove and discard medium from the cell culture vessel by aspirating with a pipette under vacuum. Rinse monolayer with 3 to 4 ml of phosphate-buffered saline (PBS). Discard PBS.
2. Add a chelating agent such as EDTA or a trypsin–EDTA mixture (pre-warmed to 35°C) as follows:
    a. 0.5 ml for 16 × 125-mm screw-cap tubes.
    b. 2.0 ml for 25-cm² flask.
    c. 4.0 ml for 75-cm² flask.
    d. 6.0 ml for 150-cm² flask.
3. Incubate the culture vessel for 5 to 10 min at 35°C, with the chelating agent covering the cell sheet. Observe after 5 min to see whether the cell sheet is breaking loose from the vessel surface. Tap vessel sharply against palm of hand to aid in loosening tissue.
4. When tissue has loosened completely, add a cell culture outgrowth medium as follows:
    a. 1.0 ml for 16 × 125-mm screw-cap tubes.
    b. 4.0 ml for 25-cm² flask.
    c. 8.0 ml for 75-cm² flask.
    d. 12.0 to 16 ml for 150-cm² flask.
    This medium will stop the action of the chelating agent.
5. Disperse cells by drawing cells and fluid up and down in a pipette.
6. Make a cell count if desired (refer to *Procedure for Enumeration of Cells*, below). A "split ratio" is known for most cell types. This is the ratio at which the cells have been found to replicate adequately.
    Example: If the split ratio is 3 for 1, three new flasks of the same size (divide cell mixture evenly among the three flasks) or one of triple size could be made from the starting flask.
7. Label appropriate new culture vessel(s) with the name of the cell line, date subcultured, and with the generation level, that is, generation level from old flask +1.
8. Aliquot the resuspended cell mixture into the new culture vessel(s). Add cell culture outgrowth medium to make the *final volumes* indicated below (0.2 to 0.5 ml of medium per cm² is recommended) (Freshney, 1983):
    a. 1.0 ml for 16 × 125-mm screw-cap tubes.
    b. 6.0 ml for 25-cm² flasks.
    c. 15.0 ml for 75-cm² flasks.
    d. 30.0 ml for 150-cm² flasks.
9. Gently swirl the culture vessel to disperse cells evenly throughout medium.
10. Incubate the culture vessel at 35°C. Growth of some cell types is enhanced by an initial 1 h of incubation in a 5% CO_2 incubator. Loosen caps slightly during this incubation. Close caps snugly following this incubation.
11. Observe daily for outgrowth of the cells and for changes in the pH of the medium. If the medium becomes acid or basic, replace it with fresh medium.
12. When a confluent cell sheet is obtained, replace outgrowth medium with maintenance medium.

Procedure for Enumeration of Cells

Use a hemocytometer (Neubauer) with cover slip (Fig. 1). This hemocytometer has two platforms which support the cover slip. The cover slip covers two identical areas which are ruled into 9 squares of equal size (Fig. 2). Each square is 1 mm × 1 mm. The depth between the cover slip and the ruled surface is 0.1 mm. The volume of one ruled square is 0.1 mm³ (1 mm length × 1 mm width × 0.1 mm depth).

1. If desired, make a dilution of cells in stain (trypan blue stain for viability counts) or in a diluent such as sterile saline. Record the dilution factor; it will be used in later calculations.
2. Draw up a suspension of well-mixed cells into a pipette or capillary tube. Position the cover slip in place on the support platforms of the counting chamber and fill one side of the cleaned and dried counting chamber by touching the tip of the pipette or capillary tube to the side of the cover slip (Fig. 1). Some counting chambers will have a V notch for this purpose. The mixture will seep under the cover slip to fill the counting area; do not flood the chamber or allow fluid to flow into the channels on each side of the counting area. Fill the other side of the chamber in the same manner.

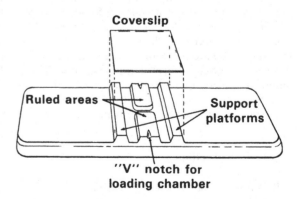

FIG. 1. Neubauer hemocytometer with cover slip, support platforms, ruled areas, and "V" notch for loading the chamber.

3. Allow the cells to settle for 2 to 3 min. Place counting chamber on the stage of a microscope.

4. Focus on the ruled area of the counting chamber with the low-power objective. The light must be reduced (lower the condenser or partially close the iris diaphragm) for the ruled lines to be observed.

5. Count the cells in the five large squares indicated in Fig. 2. Move to the other side of the chamber and count the cells in the five large squares. The total number of cells in these 10 squares is divided by 10 to obtain the mean count per square. This is the number of cells per 0.1 mm^3 of suspension. Multiply this number by 10,000 to determine the number of cells per cc^3 or per ml of suspension. If a dilution was made in step no. 1, multiply the final number by this dilution factor.

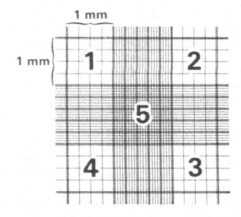

FIG. 2. Ruled area for the Neubauer hemocytometer. There are 9 large squares, each measuring 1 mm × 1 mm. The distance from the cover slip to the ruled surface is 0.1 mm, giving each square a volume of 0.1 mm^3. Cells within the 5 numbered squares are counted on each of the two ruled areas of the hemocytometer.

Specimen Collection, Processing, and Inoculation

Virus isolation is enhanced by:

1. Collection of samples during the acute stage of illness when virus shedding is greatest; this insures acquisition of the largest quantity of virus.

2. Procurement of intact cells in the sample; some viruses lose infectivity unless they are maintained within intact cells.

3. Prompt delivery of the sample to the virology laboratory; viruses are inactivated by heat, adverse pH, and other environmental conditions to which samples may be subjected during transportation to the laboratory. Samples may be overgrown by usual flora type bacteria and fungi during prolonged periods of transportation.

Samples collected from contaminated sites such as mouth, vagina, or skin are best placed in a liquid transport medium such as Hanks balanced salt solution containing antibiotics and 2% calf serum. The site of sampling is dictated by patient symptoms. Culture lesions in cutaneous and mucous membrane disease, and throat and rectum in enteric illnesses; sample the throat or nose in upper respiratory infections; and collect sputum in lower respiratory infections. In central nervous system infections, submit cerebral spinal fluid along with anticoagulated peripheral blood, throat, and rectal cultures. Guidelines for selection of appropriate specimen types and for proper sample collection and transport are described elsewhere in this volume.

In the virology laboratory, samples collected from sites contaminated with bacterial or fungal flora are decontaminated by centrifugation, filtration, and addition of antibiotics. Routine processing of swab samples received in the laboratory in 2.5-ml aliquots of transport medium includes: 1) vortexing of swab in transport medium; 2) removal of swab from transport medium; 3) centrifugation of transport medium at 1,500 × g for 10 min; and 4) use of clear supernatant fluid for inoculation of viral cultures. Samples collected from sites devoid of normal flora, such as cerebral spinal fluid, are inoculated into cell cultures directly without decontamination processing. Anticoagulated peripheral blood samples are processed by density centrifugation in ficoll-hypaque to separate mononuclear cells from the remainder of the sample; the mononuclear cells are used in viral culture inoculation. Separation of mononuclear cells can also be accomplished with commercially available separator tubes that contain a layer of polysaccharide and sodium diazoate and a layer of polyester gel (Becton-Dickinson, 1986).

If samples will be stored less than 24 h prior to delivery to the virus laboratory, they may be refrigerated at 4°C. Storage for periods exceeding 24 h should be at −70°C or in liquid nitrogen. If the sample will be transported via mail or messenger to a reference laboratory, dry ice is used during transport, and the sample must be sealed in a container impervious to CO_2.

Clinical materials that have been processed for viral culturing may be introduced into the cell culture in one of two ways. Addition of 0.2 to 0.3 ml of processed sample into a cell culture tube with medium in place is one accepted method of inoculation. A second technique, adsorption inoculation, involves removal of liquid medium from the cell culture tube. The processed sample, 0.2 to 0.3 ml per tube, is placed directly on the monolayer, where it remains for a 1-h incubation period at 37°C. After this time, the remaining sample is aspirated and discarded, and fresh culture medium is added to the culture tube. The adsorption method is recommended for recovery of viruses such as cytomegalovirus, which spread via cell-to-cell transfer; however, certain specimens such as urine or feces may be toxic to the cells so that adsorption inoculation is destructive to the monolayer.

If excess processed sample remains following inoculation of cell cultures, it is wise to store it at −70°C. Occasionally, because of laboratory accidents or contaminated cell cultures, restarting of the viral isolation process is required; the stored, processed sample is used for this purpose. Also, additional assays, requested retrospectively by the physician, can be done on stored material. Excess sample can be stored in 2.0-ml cryovials; addition of 0.2 ml of a 5% glycerol solution in phosphate-buffered saline or of 70% sorbitol is recommended. A storage period of 3 months or more is usually satisfactory. Approximately 1,000 tubes can be stored per cubic foot of freezer space by using an 81-cell divider in a cardboard box measuring 5⅛" (width) × 5½" (depth) × 2⅝" (height) (Fig. 3). Boxes are numbered consecutively, and the position of each specimen is identified by row and position. For example, Box 207, Row H, position 8 (207H8). This storage system is compatible with a computer system, but is also easy to use manually.

Effective Utilization of Cell Cultures in Diagnostic Virology

Two tubes each of primary and diploid cells are routinely inoculated for each clinical sample. Primary

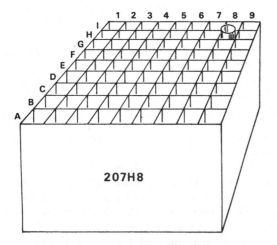

FIG. 3. Cardboard storage box with 81 compartments. Boxes are numbered consecutively, with each sample identified by letter of row and number of position in row. Example shown is Box 207, Row H, Position 8.

monkey kidney cells such as pRMK or cynomolgus monkey kidney cells are strongly recommended owing to their wide range of susceptibility. A continuous line may be included, if desired. This schema provides for isolation of most human viral pathogens that replicate in standard cell cultures. Inoculated cultures are incubated at 36°C, preferably in rotating drums, for a period of 14 days with daily examination for viral CPE. Routine use of several cell lines allows the virologist to predict viral identity early on by comparing the time required for CPE production, the cell line(s) affected, and the appearance of CPE. A table such as Table 1 may aid in determining presumptive virus identification.

Enhanced virus recovery may result when the standard schema is modified, as described below, when certain specific viral suspects have been indicated. Each of these agents, along with specific directions for their isolation, is described, in detail, in the chapters which follow.

CYTOMEGALOVIRUS

Decrease primary and continuous cell culture tubes. Add extra tubes of diploid fibroblasts, preferably of more than one type, and hold fibroblasts for an extended period of 30 days. Cytomegalovirus isolation is enhanced by low-speed centrifugation during inoculation of specimen into shell vials containing cover slip monolayers of diploid fibroblasts (Gleaves et al., 1984, 1985a). (See Alternative Cell Culture Techniques, below.)

RUBELLA

Inoculate 2 tubes each of primary African green monkey kidney cells (pAGMK) and/or rabbit cornea cells (SIRC) in addition to routine culture types. Incubate one tube of AGMK and one of SIRC at 33°C. Incubate all other cultures under standard conditions.

HERPES SIMPLEX

Inoculate pHEK, primary rabbit kidney cells (RK), or mink lung (ML) cells and decrease the number of other cell culture tubes. If "herpes simplex only" is specified, inoculate one tube of at least two of the following cell types: pHEK, RK, or ML. Cell culture systems, specifically designed for isolation and identification of herpes simplex virus, are available commercially. (See Alternative Cell Culture Techniques, page 45).

HUMAN IMMUNODEFICIENCY VIRUS (HIV) AND EPSTEIN-BARR VIRUS (EBV)

These viruses are not isolated in standard cell cultures. Both require culturing in suspensions of human lymphocytes. (See Alternative Cell Culture Techniques.)

Alternative Cell Culture Techniques

SHELL VIALS

Monolayers of adherent cells grown on 12-mm round cover slips contained in 1-dram shell vials have provided a viral isolation system that uses centrifugation to enhance virus isolation. The vials are prepared by adding dispersed cells to sterile shell vials containing cover slips; the vials are incubated in an upright position until the cells form a monolayer on the cover slip. For virus isolation, culture medium is decanted from the vial, suspect clinical material is placed directly on the cell monolayer, and the vial is placed in the centrifuge. Following centrifugation, culture medium is added to each vial, and vials are incubated for the desired time period. The cover slip is stained and examined to confirm identification of the specific agent (Fig. 4).

This technique has been used extensively for culturing biovars of *Chlamydia trachomatis* that do not grow well in standard cell culture systems. For chlamydia isolation, inoculated monolayers of iododeoxyuridine or cycloheximide-treated cells are centrifuged at 3,000 × g for 1 h. Following incubation, the infected cover slip monolayers are stained and examined for intracellular inclusions (Schacter, 1980).

The infectivity of cytomegalovirus (CMV), which often requires 7 to 21 days to produce characteristic CPE in cell cultures, is enhanced by centrifugation (Hudson et al., 1976). Using shell-vial cultures of

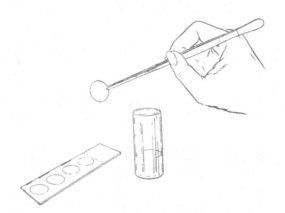

FIG. 4. Shell vial. Cells are grown in a monolayer on a cover slip in the bottom of the vial. Sample material is inoculated directly onto the monolayer, and the vial is spun in the centrifuge. Following a proper incubation period, the cover slip is stained with specific antiserum, mounted on a microscope slide, and examined to confirm virus identification.

MRC-5 diploid fibroblasts and centrifuging the inoculated monolayers for 1 h at 700 × g, Gleaves et al. (1984) were able to demonstrate early nuclear protein of CMV by indirect immunofluorescent staining in 100% of CMV-infected cultures at 36 h after infection. In further studies Gleaves et al. (1985a) compared this technique with standard cell culture isolation of CMV and found the shell-vial culture assay to be as specific as and more sensitive than conventional tube culture isolation.

Gleaves et al. (1985b), using commercially available, fluorescein-labeled, monoclonal antisera, identified 100% of herpes simplex (HSV)-positive samples by serotype 16 h after inoculation in spin-inoculated, cover slip monolayers of MRC-5 cells. Salmon et al. (1986) used spin-inoculated, cover slip monolayers of ML cells and an immunoperoxidase staining method to identify 96% of HSV-positive cultures after overnight incubation; only 58% of these cultures were positive in standard cell cultures examined at the same time after inoculation.

The use of shell-vial cover slip cultures and centrifugation has also been reported for rapid detection of adenovirus. In spin-inoculated cover slip monolayers of HEp-2 cells, even low titer adenovirus samples were identified at 24 h after inoculation by immunofluorescent staining (Espy et al., 1986).

HSV ISOLATION AND IDENTIFICATION SYSTEMS

Because HSV type 1 and type 2 account for 80% of clinical virology isolates (Moseley et al., 1981), considerable attention has been devoted to isolation of these agents. Several commercially produced isolation systems are designed specifically for rapid detection and identification of HSV. These systems include a conveniently packaged cell culture of HSV-susceptible cells such as Vero, ML, RK, or Medical Research Human Foreskin (MRHF). The cells are usually growing on the flat side of a plastic tube. After the monolayer has been inoculated with a clinical sample and incubated for an appropriate period, a confirmatory test, usually an immunoperoxidase staining technique provided with the culture system, is performed directly on the cell monolayer.

Results obtained with these systems have varied considerably. Hayden et al. (1983) reported 73% sensitivity for HSV detection at 24 h after inoculation, using a commercial system. Fayram et al. (1983), Sewell et al. (1984), Rubin and Rogers (1984), and Phillips et al. (1985) reported 65%, 79%, 78%, and 92% sensitivity respectively for HSV isolation at 48 h after inoculation with the same system.

HSV isolation systems routinely provide confirmation of HSV infection more rapidly than do standard cell cultures. This is not unexpected in that

TABLE 1. Laboratory identification of some common viruses[a]

Virus	Cytopathogenic effect			Additional information	HAD[b]	HA[b]
	Primary monkey kidney	Human diploid fibroblasts	Continuous lines			
Adenovirus		Less sensitive	Enlarged, rounded, grape-like clusters in HEp-2 or HeLa in 7–10 days	Proliferate well in pHEK in 5–7 days	0	+ Hu + Ra + Rh
Arboviruses (includes EEE, WEE, SLE, etc.)		No accepted monolayer cell culture system		Intracerebral mouse inoculation for isolation	0	+ Goose erythrocytes
Cytomegalovirus		Elongated foci of small, rounded cells in 10–30 days in WI-38, MRHF, MRC-5		Use shell vials with centrifugation inoculation for rapid (2–3 day) isolation	0	0
Enteroviruses (includes polio, echo, coxsackie)	Cells shrink, detach, lyse, remain attached by filopodia, 2–5 days	Less sensitive	Less sensitive	Some Coxsackie A strains best isolated in suckling mice	0	+ Hu
Epstein-Barr virus		No accepted monolayer cell culture system		Coculture with human fetal cord blood lymphocytes—look for cell transformation	0	0
Hepatitis		No accepted monolayer cell culture system		Best diagnosed serologically or by immunologic method for antigen detection	0	0
Herpes simplex	Less sensitive	Enlarged-ballooned cells, multinucleate giant cells in 3–5 days	Less sensitive	Best isolated in pHEK, ML, or pRK in 1–3 days	0	0
Human immunodeficiency virus		No accepted monolayer cell culture system		Coculture with stimulated normal human lymphocytes	0	0
Influenza Type A	CPE rare			Some strains require embryonated eggs	+ GP + Hu	+ GP + Hu + Ch
Type B	Vacuolation, lysis, 3–5 days				+GP + Hu	+ GP + Hu + Ch
Measles (rubeola)	Multinucleate giant cells, 7–10 days			Proliferate well in pHEK	+ Rh	+ African green monkey

Virus	CPE	Comments		
Mumps	Large syncytia 6–7 days. No CPE in some strains		+ GP	+ Hu + GP
Papilloma virus (human)	No accepted monolayer cell culture system	Diagnosis by histologic exam or test for viral antigen	0	0
Parainfluenza	Minimal CPE: dark granular cells and syncytia in 10 days	For all parainfluenzas, the majority of strains produce no CPE in cell culture	+ GP + Hu	+ GP + Ch
Reovirus	Granular, nonsloughing cells in 7–12 days (resembles cellular toxicity)		0	+ Hu
Respiratory syncytial virus	Some show syncytia in 5–10 days. Indistinct rounding and syncytia in 4–7 days in HEp-2 and HeLa		0	0
Rhinovirus	Some show CPE in 5–10 days. Rounding and refractility in 2–7 days	Some proliferate in pHEK	0	0
Rotavirus (human)	No accepted monolayer cell culture system	Some proliferate following trypsin pretreatment of samples and incubation in trypsin-supplemented medium in several cell lines	0	0
Rubella	Replicate within pAGMK. Seldom produce CPE.	May replicate in pRK or rabbit cornea cells. Detect using challenge interference	0	+ Hu + Ch
Vaccinia-variola	Cell fusion leading to plaques in 2–4 days. Some show CPE in 3–6 days		0	+ Ch
Varicella zoster	Small foci of rounded, swollen cells in 5–7 days		0	0

[a] Reactions may vary depending on virus strain and quantity and cell culture type and condition.

[b] Abbreviations: Ch = chicken erythrocytes, CPE = cytopathogenic effect, EEE = eastern equine encephalitis, GP = guinea pig erythrocytes, HA = hemagglutination, HAD = hemadsorption, HeLa = human cervical carcinoma cells, HEp-2 = human laryngeal carcinoma cells, Hu = human group 0 erythrocytes, ML = mink lung cells, MRC-5 = Medical Research human lung fibroblasts, MRHF = Medical Research Human Foreskin cells, pAGMK = primary African green monkey cells, pHEK = primary human embryonic kidney cells, pRK = primary rabbit kidney cells, Ra = rat erythrocytes, Rh = rhesus monkey erythrocytes, SLE = St. Louis encephalitis, WEE = Western equine encephalitis, WI-38 = Wistar Institute human lung fibroblasts, 0 = negative, + = positive.

HSV isolation systems rely on immunologic identification of HSV antigen in the cell monolayer, whereas traditional HSV detection depends upon CPE production by the virus. HSV isolation systems provide a convenient and useful alternative when full-service viral isolation facilities are not available. The systems require some expertise, but, in general, can be managed without extensive virology training. Such systems generally represent an increased cost compared with that of standard cell culture methods (Rubin and Rogers, 1984).

MICROCARRIER TECHNOLOGY

In 1967 van Wezel described a new technique for expanding tissue cultures of anchorage-dependent cells in which microscopic particles, rather than the wall of a culture vessel, were used to serve as a surface for cell attachment. The particles called microcarriers were kept in suspension to increase the area available for cell growth. Various types of microcarriers, including swollen, collagen-coated dextran beads (Widell et al., 1984), DEAE Sephadex (Van Wezel et al., 1978), and glass beads (Thornton et al., 1985), have been used.

Microcarriers were used in large-scale production of polio virus (Van Wezel et al., 1978), foot-and-mouth-disease virus (Meignier et al., 1980), HSV (Thornton et al., 1985), and hepatitis A (Widell et al., 1984) and have proven suitable for vaccine production. Meignier et al. (1980) estimated that a single technician could alone manage microcarrier cultures equivalent in production to that achieved with 600 to 900 rolling flasks. Media changing is simplified with microcarriers, and opportunities for contamination of both laboratory personnel and culture contents are minimized. The microcarrier system also facilitates frequent sampling of aliquots for monitoring of cell and virus levels and for sequential harvesting.

HUMAN LYMPHOCYTE SUSPENSION CULTURES

Viruses such as HIV or EBV, which replicate within lymphocytes, are best isolated via inoculation of infected clinical materials into laboratory cultures of human lymphocytes. For HIV isolation, lymphocytes from infected individuals are co-cultivated with stimulated, normal lymphocytes (Levy et al., 1984). Stimulated lymphocyte suspensions are prepared from blood collected from normal human donors. The lymphocytes are separated via Ficoll-Hypaque density separation and activated for 24 h by incubation with phytohemagglutinin. Fluorescent antibody testing and enzyme immunoassays, as well as DNA probes and viral reverse transcriptase assays, have been used to demonstrate HIV antigen or to confirm HIV proliferation in infected cells. For EBV, culturing of throat washings or saliva in suspensions of human fetal cord blood lymphocytes is recommended (Reedman and Klein, 1973). The EBV-transformed lymphocytes will proliferate, whereas non-transformed cells disintegrate. Immunofluorescence testing for EBV nuclear antigen confirms the infection.

Contamination of Cell Cultures

All cell cultures, whether commercially purchased or originated and maintained in-house, may be subject to contamination by bacteria, fungi, or mycoplasma. The sources for contamination are many and include the usual throat flora of laboratory personnel, clinical specimens, or contaminated cell culture reagents and additives such as bovine sera.

Bacterial and fungal contaminants usually produce gross changes in the appearance of the culture. Culture media become turbid, the pH of the media becomes acid, cells appear granular and may detach from the culture vessel, and, microscopically, organisms may be observed floating in the culture media.

Bacterial and fungal contamination are not always obvious. Inoculate samples from suspicious cultures onto blood and/or Sabourauds agar and into thioglycollate or similar broth medium to test for growth of contaminating bacteria or fungi. Cultures contaminated with bacteria or fungi are usually discarded because contamination is difficult to eliminate.

The least obvious cell culture contaminant is mycoplasma. Mycoplasma contamination may go undetected through many generations of cells. It is our recommendation that all cell cultures carried in-house be checked every 4 to 6 weeks for mycoplasma. Although commercial suppliers of cell cultures routinely check each generation of cells for mycoplasma, it is wise to monitor an occasional sample of each commercial cell line. Mycoplasma isolation requires the use of broth or plate media that are enriched with yeast extract and horse serum. Formulas for mycoplasma isolation media have been published previously (Hay, 1985). An incubation period of 48 h to 7 days may be required for growth of the organisms on agar, and subculturing of broth media to agar plates at 10 to 14 days is recommended for detection of small quantities of the organisms.

Culturing for mycoplasma may not represent the ideal method for monitoring of cell culture contamination in that more than 60% of *Mycoplasma hyorhinis* cell culture contaminants fail to grow on mycoplasma isolation media (Arata et al., 1972). A cost-effective and apparently accurate alternative for detection of mycoplasma contamination involves the principle of nucleic acid hybridization. A labeled DNA probe that hybridizes to *Mycoplasma* and *Acholeplasma* ribosomal RNA can be used to detect the six mycoplasma species that have been shown to cause 98% of laboratory infections (McGarrity, 1982); the probe will also detect other *Mycoplasma* and *Spiroplasma* species. The DNA probe technique requires a 1-h time period. One DNA probe for detection of mycoplasma contamination in cell cultures is commercially available at present (Gen-Probe, 1986). This technique will probably prove superior to culturing for detection of mycoplasma contamination in cell cultures.

A cell line contaminated with mycoplasma is usually lost to further use in the virology laboratory. Attempts to "cure" mycoplasma contamination have not been encouraging in the past. New combination antibiotics (Boehringer Mannheim, 1986) have been developed for the purpose of decontaminating mycoplasma-infected cell lines. Although these agents are new and are marketed at this time for research purposes only, they may be the only viable alternative for salvaging a valuable and otherwise irreplaceable cell line.

Embryonated Eggs and Laboratory Animals

Early virologists used embryonated eggs and laboratory animals, especially mice, extensively prior to the 1950s, when cell culture technology was not well established. Because of the expense of animals and eggs, the cumbersome nature of proper animal and egg maintenance, and the broad range of virus susceptibility provided by cell cultures, the use of animals and eggs in clinical diagnostic virology has declined such that these systems are seldom, if ever, used. Certain viruses, including the arboviruses and Coxsackie A viruses, do not proliferate well in standard cell cultures, so isolation in an alternate system such as suckling mice is recommended. Work with agents of spongiform encephalopathy relies totally on animals, and hamsters are most useful in studies of virus-induced tumors and cerebral malformations (Johnson, 1982). Inoculation of mice is usually by the intraperitoneal or intracranial route. Inoculation, observation, and harvesting techniques for use with suckling mice have been published previously (Landry and Hsiung, 1985).

Embryonated eggs remain the host of choice for certain strains of influenza and for some arboviruses. Chlamydial agents of lymphogranuloma venereum, trachoma, and oculogenital infections as well as psittacosis have been isolated in eggs (Schachter and Dawson, 1979). Although culturing in eggs produces higher yields of chlamydial agents, tissue culture isolation is preferred owing to its ease of performance. Chlamydia isolation in eggs is particularly well established in laboratories that do not use cell cultures regularly or have problems in obtaining cell culture reagents. Chlamydia isolation is described in detail elsewhere in this volume.

Egg inoculation, descriptions of which have been published previously (Hawkes, 1979), may facilitate viral proliferation in the amniotic cavity, allantoic cavity, the yolk sac, or the chorioallantoic membrane. Because CPE cannot be easily observed in egg and animals, other signs of viral proliferation such as animal death or paralysis, plaques of virus visible on egg membranes, or viral hemagglutination by egg fluid must serve to indicate viral presence. Definitive virus identification must then be confirmed with a specific method.

Virus Quantitation

Many viral identification procedures cannot be performed until the quantity or "dose" of the virus has been determined. Quantitation is accomplished by preparing serial, tenfold dilutions of the virus-infected material; an aliquot from each dilution serves as inoculum for susceptible hosts. Ideally, a large number of hosts is inoculated with each virus dilution. This approach, though best scientifically, is impractical in clinical virology; the cost of large numbers of animals or cultures is high, and variations may still be encountered (Lennette, 1969). Most laboratories utilize a modified titration method to predict the viral dose within acceptable accuracy limits.

Most titration procedures are quantal rather than quantitative in that results do not indicate the number of infectious virus particles in the inoculum. These assays provide a result defined in terms of the highest viral dilution in which the virus affects 50% of the hosts. With animal or egg death, the unit is expressed as 50% lethal dose (LD_{50}); for animal paralysis, the unit is the paralyzing dose (PD_{50}); and with cell culture infectivity, the unit is expressed as tissue culture infective doses ($TCID_{50}$). Estimation of viral dose by quantal techniques is satisfactory for most clinical laboratory procedures. Virus quantities required in common identification techniques are defined by manufacturers of viral antisera in terms of 50% infectivity end points.

TITRATION PROCEDURE FOR USE WITH CELL CULTURES

1. Label 7 sterile tubes (12 × 75 mm) with dilutions 10^{-1}, 10^{-2} . . . through 10^{-7} and pipette 0.9 ml of cell culture medium into each tube.
2. Pipette 0.1 ml of virus suspension into the first dilution tube. Mix well.
3. With a fresh pipette, transfer 0.1 ml of dilution from tube No. 2 and mix well. Continue this serial dilution through tube No. 7 with a fresh pipette for each transfer. Note: In preparation of viral

TABLE 2. Data for 50% infectivity end point determinations

| A | Observed at individual dilution | | | Cumulative values | | |
Virus dilution	B No. affected	C No. unaffected	D Affected ratio (%)	E Cumulative no. affected	F Cumulative no. unaffected	G Affected ratio (%)
10^{-1}	4	0	4/4 (100)	8	0	8/8 (100)
10^{-2}	3	1	3/4 (75)	4	1	4/5 (80)
10^{-3}	1	3	1/4 (25)	1	4	1/5 (20)
10^{-4}	0	4	0/4 (0)	0	8	0/8 (0)

dilutions, it is important to discard pipettes between dilutions in order to avoid carrying virus particles on the pipette to the next dilution; failure to do so results in misleadingly high infectivity end points.

4. Label several tubes of a susceptible type of cell culture to correspond with each of the serial dilutions:

    No. 1 = 10^{-1} No. 4 = 10^{-4} No. 7 = 10^{-7}
    No. 2 = 10^{-2} No. 5 = 10^{-5}
    No. 3 = 10^{-3} No. 6 = 10^{-6}

5. Using a clean pipette for each dilution, transfer 0.1 ml of each dilution into the corresponding cell culture tubes.

6. Incubate cell cultures at 35°C in a slanted cell culture tube rack. Observe microscopically at intervals over a 7-day period (or longer for more slowly growing viruses).

7. The end point of the titration is the highest dilution showing evidence of viral infection in 50% of the cell cultures. This end point can be calculated with the Karber or Reed-Muench equations.

REED-MUENCH CALCULATIONS OF 50% INFECTIVITY END POINTS

To facilitate easy calculation of 50% infectivity end points with the Reed-Muench method (Reed and Muench, 1938), prepare a table as shown in Table 2, which includes:

Column	Data
A	= Virus dilution tested.
B	= Number of hosts affected (killed, paralyzed, showing CPE) by virus at each dilution.
C	= Number of hosts not affected by virus at each dilution.
D	= Ratio of affected hosts (%): Column B divided by the total of Columns B + C. Calculate percentage.
E	= Cumulative number of hosts affected. Calculated by adding numbers starting at the bottom of Column B and adding up.

Column	Data
F	= Cumulative number of hosts unaffected. Calculated by adding numbers starting at the top of Column C and adding down.
G	= Cumulative ratio of affected hosts (%): number from Column E divided by the total of Columns E + F. Calculate percentage.

Use of cumulative values aids in equalizing chance variations. Using these data, three steps are required for calculation of the 50% end point:

Step No. 1. The proportionate distance between the two dilutions nearest 50% is calculated by use of the equation below (use values from column G):

$$\frac{(\% \text{ Affected next above 50\%}) - 50\%}{(\% \text{ Affected next above 50\%}) - (\% \text{ Affected next below 50\%})}$$

Step No. 2. The proportionate distance (from step No. 1) is corrected for the dilution factor by multiplying the proportionate distance by the logarithm$_{10}$ of the dilution increment used. For twofold dilutions this factor is 0.3, for fivefold dilutions this factor is 0.7, and for tenfold dilutions this factor is 1.0. In procedures such as this, this factor is understood to be negative.

Step No. 3. The 50% end point is calculated by adding the corrected proportionate distance to the negative logarithm of the dilution next above 50% affected.

An example is provided below, from data of Table 2.

Step No. 1. Calculate the proportionate distance:

$$\frac{80\% - 50\%}{80\% - 20\%} = \frac{30}{60} = 0.5$$

Step No. 2. Correct the proportionate distance for dilution factor.

The dilutions are tenfold, so use -1:

$$-1 \times 0.5 = -0.5$$

Step No. 3. Calculate the 50% end point by adding the corrected dilution factor to the negative $\log_{10}$ of the dilution next above 50%:

$$-2 + (-0.5) = -2.5$$

Therefore, the 50% end point dilution is $10^{-2.5}$

Karber Method for Calculation of 50% End Points

The 50% end point titer can also be calculated by the method of Karber that follows (Karber, 1931). Cumulative values are not required for this method, although they can be used.

Karber equation:

Negative logarithm of the 50% end point =

Negative log of the highest virus concentration used −

$$\left[\left(\frac{\text{Sum of \% affected at each dilution}}{100} - 0.5 \right) \right.$$

$$\left. \times (\text{Log of dilution}) \right]$$

Using data from Table 2, the 50% end point is calculated as follows using the Karber equation:

$$-1 - \left[\left(\frac{100 + 75 + 25 + 0}{100} - 0.50 \right) \times 1 \right]$$

$$-1 - \left[\left(\frac{200}{100} - 0.5 \right) \times 1 \right]$$

$$-1 - [(2.0 - 0.5) \times 1]$$

$$-1 - 1.5 = -2.5$$

Therefore, the 50% end point is $10^{-2.5}$

Plaque Assays

A more precise method of virus quantitation is provided by plaque assays. Plaque assays, first described by Dulbecco (1952), involve adsorption of tenfold serial dilutions of infected material onto monolayer cell cultures. The monolayers are overlaid with agar-containing nutrient medium, which prevents free virus spread and limits infection to cell-to-cell spread. A circular plaque of infected cells is formed around each infectious unit. The cultures are stained with vital dye, and plaques are counted to determine the number of plaque-forming units. Plaque assays for herpesvirus (Wentworth and French, 1969) and for cytomegalovirus (Wentworth and French, 1970) have been described.

Virus Detection and Identification

Success in viral isolation is only half the battle in providing clinical diagnostic virology services; the isolation must be followed by confirmation of viral identity. In some cases, an identification method also serves as a virus detection method. Several common human viral pathogens produce little, if any, visual evidence of their presence; therefore, an identification/detection method must be applied to a possibly infected cell culture in order to confirm that a virus is proliferating within. Several methods used for detection of viruses in cell cultures may also be used to analyze clinical materials directly to provide identification of viral antigens within the sample; virus isolation is not required. Nonimmunologic techniques rely upon structural or physiologic characteristics of the viruses, while immunologic methods involve the reaction of known virus-specific antibodies with unknown viral antigen within the infected clinical material or cell culture. An overview of methods for use in detection and identification of cell culture isolates as well as for direct detection of viral antigen in clinical samples is presented below.

Collection of Specimens for Direct Viral Antigen Identification

Viability of the virus is not required for direct detection methods, which somewhat simplifies specimen collection and transport. For direct methods involving microscopic examination, clinical samples can be collected from the site of interest and placed directly on glass microscope slides. Frosted-end glass slides are preferred and should be labeled with the patient's name, the site from which the sample originated, and the date of collection. Cotton swabs can be used to collect material from nose, throat, lesions, etc., taking care to obtain cellular material. The swab is then rubbed, with a circular motion, on the surface of the slide; the slide is air dried and forwarded to the laboratory.

For direct methods in which the sample is assayed in suspension, the sample may be collected with a swab. The swab is placed in appropriate transport medium. Sputum, urine, or stool samples are sent in suitable containers to the laboratory, where appropriate suspensions are prepared. On all samples submitted for direct examinations, the suspected virus must be identified to insure that appropriate testing is performed.

Nonimmunologic Testing

Histologic Staining

The most popular application of direct staining methods in viral diagnosis is the Tzank test. Cells scraped from the base of vesicular lesions are smeared on a microscope slide, where they are fixed and stained with Wright or Giemsa stain (Rawls, 1980). Syncytial giant cells and ballooning of cytoplasm can be observed in lesions produced by HSV type 1 and type 2 and varicella-zoster (VZ). These viruses cannot be distinguished in the Tzank test. Likewise, Cowdry type A intranuclear inclusions may be found in HSV or VZ-infected cells stained with hematoxylin and eosin. Inclusions are amphophilic to eosinophilic and usually occupy the entire nucleus. Inclusions progress to become eosinophilic and more hyaline and are surrounded by a distinct and wide halo (Strano, 1976). Although direct stains may indicate infection by certain viruses, a definitive identification cannot be made on this basis. Viral antigen identification methods such as immunofluorescence and ELISA (see below) can be performed almost as rapidly as the simple direct stains, and the results from immunologic techniques provide definitive virus identification. Immunologic detection, rather than direct staining, is recommended when appropriate laboratory facilities are available.

Electron Microscopy

Few clinical virus isolation facilities offer electron microscopy (EM) as an option for viral diagnosis. The technology has been applied in identification of viruses such as hepatitis and rotavirus, which do not proliferate in standard cell cultures. Although both hepatitis and rotavirus are now detectable in clinical materials by methods such as ELISA or passive agglutination (see below), EM is the only method that allows observation of the actual viral particles.

Negative contrast staining allows specimens to be prepared and examined within 1 h. Although rapid, this technique does not provide specific identification. With this method for samples collected from skin lesions, poxviruses can be distinguished from members of the Herpetovirus family; however, the Herpetoviruses HSV type 1 and type 2 and varicella-zoster cannot be differentiated.

Immunoelectron microscopy (IEM) confers specificity on EM. Samples can be stained with known specific antibody labeled with electron-dense markers such as ferritin or with an enzyme; substrate is then added (Weakley, 1981). Unlabeled antibodies can also be used for staining. Results are evaluated by examining the sample for antibody-trapped, negatively stained virus particles or viral antigen (Doane, 1986). A relatively new technique, solid-phase immunoelectron microscopy (SPIEM), has been shown to increase sensitivity of EM. In SPIEM, known specific antibodies are bound to the EM grid, and the grid is inverted on a drop of specimen. The antibodies bind viral antigen within the sample (Doane, 1986).

HEMADSORPTION

The term hemadsorption is used to describe the phenomenon of adherence of erythrocytes to the surface of cells infected by certain viruses. Not all viruses are capable of inducing this phenomenon, and not all species of erythrocytes will be hemadsorbed (Table 1). The mechanism of hemadsorption is not defined but is known to be a direct reaction of virus or viral antigen present on the cell surface with the erythrocytes. Hemadsorption can be used as a screening method to identify infected CPE-negative cultures. The procedure is routinely performed in the clinical laboratory on cultures that fail to demonstrate CPE. The medium is decanted from these cultures and is replaced with a dilute (0.08%) suspension of guinea pig erythrocytes. Cultures are refrigerated for 30 min and examined for attachment of erythrocytes to the cell monolayer, which indicates that a virus is present within the cells (Fig. 5). The test does not provide a definitive virus identification. Several species of virus—namely, influenza, parainfluenza, measles, and mumps—have been shown to hemadsorb. Additional testing with a method for definitive virus identification is required after a positive hemadsorption test.

HEMAGGLUTINATION

Hemagglutinating viruses, including the enteroviruses, influenza, parainfluenza, adenovirus, mumps, reovirus, rubella, measles, vaccinia, and variola, have the ability to agglutinate certain species of erythrocytes. The species of erythrocyte hemagglutinated varies from virus to virus and includes human, chicken, monkey, rat, guinea pig, and others (Table 1). In hemagglutination testing, fluid from virus-infected cell cultures is mixed with erythrocytes; the mixtures sit undisturbed until the erythrocytes have settled. Unagglutinated cells settle in a button in the

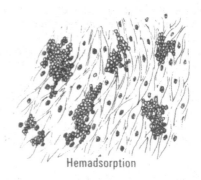

Hemadsorption

FIG. 5. Hemadsorption. Shown is a diagram representing erythrocytes that have been hemadsorbed to a virus-infected monolayer. Erythrocytes adhere to the monolayer in clumps and do not float free when the monolayer is moved.

bottom of the tube, whereas hemagglutinated cells produce a layer or shield of small clumps that covers the bottom of the container. Hemagglutination, like hemadsorption, indicates that a virus is present but does not provide a specific identification; additional specific testing is required.

CHALLENGE INTERFERENCE

Viruses that produce CPE poorly or not at all may be detected by the challenge interference (CI) technique. The basic premise of this method is that virus-infected cells will be unable to support replication of a second infecting virus, that is, the initial virus interferes with the ability of the challenge virus to produce infection. The suspect cell culture is infected with known CPE-producing virus such as echovirus. During an incubation period of 2 to 4 days, the culture is evaluated for the appearance of CPE. If CPE fails to appear, the presence of a preexisting virus is indicated. If CPE develops, it is unlikely that a prior viral infection existed. This technique does not provide a definitive identification of the virus, so additional testing is required. The CI method is not widely utilized at present because new cell lines are available that allow CPE production by viruses previously thought incapable of producing CPE.

DNA PROBES

New technology has provided clinical virology with a powerful diagnostic tool for detection and identification of viruses. DNA probes, small segments of complementary DNA labeled with enzymes or radioactive substances, are now available commercially. These probes are prepared from areas of each viral genome that contain unique sequences. The probe is capable of hybridizing (attaching to) complementary nucleic acid strands of DNA or RNA to form stable double strands. To perform an analysis with a DNA probe, clinical samples or infected cells from cell cultures are treated to cause disruption of cells and separation of double-stranded DNA. The unknown target nucleic acid is immobilized by fixation to a microscopic slide or in a semi-solid medium such as polyacrylamide gel. The DNA probe is applied and hybridizes to complementary viral DNA or RNA segments in the sample. Following the hybridization period, unattached probe is removed by rinsing, or the nucleic acid hybrids are harvested with hydroxy-apatite and centrifugation. The hybridized probe is then measured. Enzyme-labeled probes are detected by the action of their enzymes on a substrate solution, and radiolabeled probes are measured by scintillation counter or by autoradiography. The steps in performance of an *in situ* DNA probe technique are illustrated in Figure 6.

Fung et al. (1985) used a biotin-labeled DNA probe for direct detection of HSV in 162 clinical samples collected from various sites. The probe identified 71.4% of HSV culture-positive samples. Forghani et al. (1985) tested 30 stored human brain specimens, using a biotin-labeled HSV DNA probe. Of 17 samples originally positive for HSV isolation, 16 (94%) were identified as positive by DNA probe. Of 13 HSV culture-negative brain samples, all were HSV negative by DNA probe. DNA probe technology has also been applied for detection of hepatitis B surface antigen (Berninger et al., 1982), adenovirus (Hyypia, 1985), enteroviruses (Hyypia et al., 1984), cytomegalovirus (Enzo Biochem, 1985), and papillomavirus (Gissmann et al., 1986).

False-positive DNA probe results have been reported due to hybridization of the plasmid portions of one probe to plasmid-containing bacteria contaminating the test samples (Diegutis et al., 1986). Nonspecific cytoplasmic staining has also been reported for one biotin-labeled HSV DNA probe (Fung et al., 1985). Improvements in probe and sample preparation are expected to eliminate questionable reactions and to enhance both sensitivity and specificity of DNA probe methods. DNA probe technology offers the following advantages:

1. Probe techniques are usually not affected by contaminating organisms.
2. Hybridization is not reversible.
3. The reaction is not blocked by coating antibodies.
4. Viability of the organism is not required.
5. The end point may be objectively evaluated.

Eventually, DNA probe technology is likely to replace many standard viral identification methods.

Immunologic Methods

IMMUNOFLUORESCENCE

Immunofluorescence techniques, based on the reaction of antigen and fluorescein-labeled antibodies, form the basis for a wide range of viral diagnostic procedures. Direct immunofluorescence (DIF) is used to detect and identify viral antigen in clinical materials and in infected cell cultures. This is the least complicated immunofluorescence method and involves the application of labeled, known, specific antibodies to material that has been fixed to a microscope slide. Following an incubation period of 20 to 30 min, the slide is washed to remove unattached antibodies and is examined with a fluorescence microscope. Bright yellow-green fluorescence is seen in a positive test. The entire DIF procedure from fixation to evaluation of the final result requires approximately 1 h and provides a definitive viral

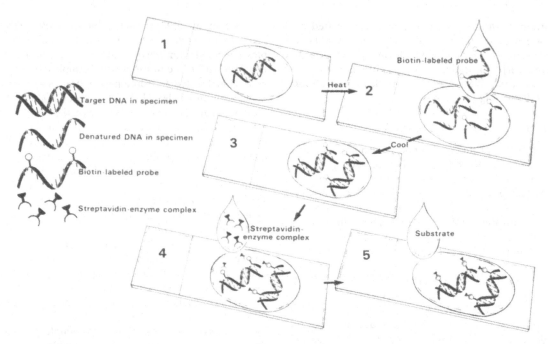

FIG. 6. A sample in situ DNA probe method. (1) The specimen is fixed to a glass microscope slide. Target viral DNA within the specimen is shown. (2) The specimen is heated to denature the target DNA, and biotin-labeled DNA probe is added. (3) Labeled DNA probe hybridizes to denatured target DNA as sample cools. Nonhybridized probe is rinsed away. (4) A detection complex of streptavidin and enzyme is added, which binds to hybridized probe. Unattached detection complex is rinsed away. (5) A substrate is added, and the enzyme acts on the substrate to produce a color change.

identification. Quality fluorescein isothlocyanate (FITC)-labeled antisera are available commercially for identification of many viral antigens, including adenovirus group antigen, herpes simplex, influenza A, mumps, parainfluenza types 1 to 3, reovirus 3, respiratory syncytial virus, and measles. FITC-labeled monoclonal antibodies for herpes simplex types 1 and 2 are also available commercially and provide a rapid and accurate method for differentiating the two virus types, a differentiation not easily made with traditional methods. The accuracy of DIF staining depends on the quality of the antisera; in general, commercially available antisera yield sensitive and specific results. DIF staining is the most rapid but least sensitive of the immunofluorescence staining techniques.

Indirect immunofluorescence (IIF) staining is also used for identification of viral antigen, both in clinical specimens and in virus-infected cell cultures. The technique is a two-step procedure including application of fluorescein-labeled antispecies globulin in the second step. For identification of unknown viral antigen, a smear of infected material is prepared on a microscope slide. Following fixation, this smear is covered with a drop of known, specific viral antiserum and incubated for 20 to 30 min. The smear is rinsed to remove unattached antibodies and is then covered with FITC-labeled antispecies globulin (antisera against globulin of the animal species in which the known, specific viral antiserum was produced), which will bind to immunoglobulin bound to the smear. Following a 20 to 30-min incubation period and rinsing, the smear is examined with a fluorescence microscope. Fluorescence confirms the presence of viral antigen homologous with the known, specific antiserum.

The two-step IIF staining method requires 2.5 to 3.0 h to complete. IIF testing requires longer to perform but offers increased sensitivity compared with DIF staining; more false-positive results are encountered in IIF techniques. Nonspecific adherence of the labeled antispecies globulin may be responsible for false-positive results in some cases. Also, an Fc binding receptor known to be produced in cells infected with herpesvirus types 1 and 2 (Spear et al., 1979) and cytomegalovirus (Keller et al., 1976) may bind antisera to produce aberrant results.

IMMUNOPEROXIDASE STAINING

Immunoperoxidase staining, like immunofluorescent staining, may be performed both as a direct (DIP) and indirect (IIP) technique and has been applied in a modified peroxidase-antiperoxidase (PAP) stain. Immunoperoxidase staining is identical with immuno-fluorescent staining with the exception of the labeling method used. In immunoperoxidase staining the label is peroxidase enzyme rather than fluorescein. Immunoperoxidase techniques depend upon the action of the active peroxidase enzyme upon a suitable substrate to produce a color change. This change is sometimes visible macroscopically or can be evaluated microscopically with a standard light microscope. DIP and IIP methods are applied in the same clinical situations as the DIF and IIF techniques, and time required and expense of reagents are comparable. Although DIF and IIF methods enjoy greater popularity than the immunoperoxidase staining and although more fluorescein-labeled reagents are available commercially, immunoperoxidase staining has some advantages. A standard light microscope can be used to evaluate immunoperoxidase results, and the results are stable and do not fade as does immunofluorescence staining.

The peroxidase-antiperoxidase (PAP) staining technique involves a three-step procedure including,

in order, application of rabbit antiviral antibody, sheep anti-rabbit globulin, and a peroxidase-rabbit-antiperoxidase complex, which binds to the sheep anti-rabbit globulin. This staining is used extensively in commercially available herpesvirus isolation and identification systems (see above).

Enzyme-Linked Immunosorbent Assay

Enzyme-linked immunosorbent assays (ELISA), also called enzyme immunoassays (EIA), are prominent in viral antigen detection and identification. ELISAs are performed in microtiter plates or test tubes. A solid support, such as the wall of the test tube or the surface of metal or plastic beads, is used for attachment of known virus-specific antibody (Fig. 7). The specimen containing the unknown antigen is exposed to the antibody-coated solid support; if homologous antigen is present in the sample, it will be bound by the antibody. Following rinsing to remove unattached components, a preparation of enzyme-labeled, virus-specific antibody is added; this

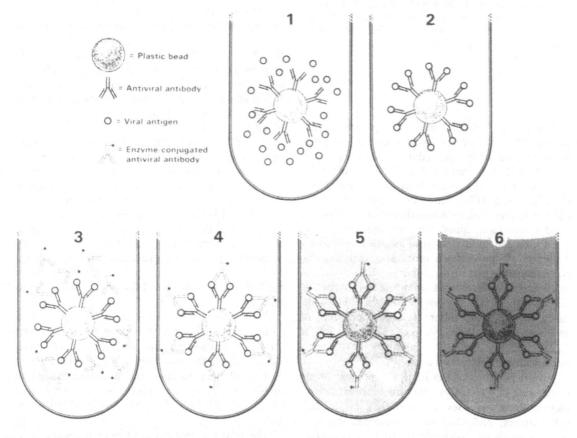

FIG. 7. Solid-phase enzyme-linked immunosorbent assay for detection of viral antigen. (1) Antibody-coated bead is incubated in a suspension of sample, which contains viral antigen. (2) Antigen–antibody complexes form. (3) Bead is placed in a solution containing enzyme-conjugated antiviral antibody. (4) Enzyme-labeled antibodies attach to bound antigen. (5) Bead is placed in a substrate solution. (6) The enzyme of the bound enzyme-labeled antibodies acts on the substrate to produce a color change.

NEUTRALIZATION IN IDENTIFICATION OF UNKNOWN VIRUS

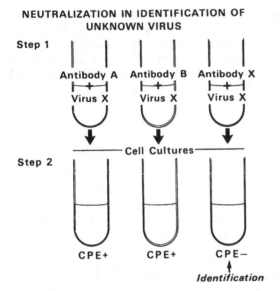

FIG. 8. Neutralization for identification of an unknown virus. In Step 1, the unknown virus (Virus X) is mixed with several known antibodies (Antibody A, B, and X), and the mixtures are incubated. In Step 2, an aliquot from each mixture is inoculated into susceptible cell cultures. The cell cultures *not* showing evidence of viral proliferation (CPE negative) signal neutralization of the virus and confirm the viral identity.

will bind to viral antigen bound in the preceding step. The action of the enzyme on the substrate, added at the conclusion of the procedure, signals a positive test. Most ELISAs require 2 to 3 h to complete.

Commercial ELISA systems are available for identification of HSV, rotavirus, respiratory syncytial virus, and various hepatitis antigens. Most of these systems are suitable for identification of virus in infected cell culture material as well as for direct detection of viral antigen in clinical samples. Reported sensitivity of ELISA for direct HSV detection in clinical samples ranged from 52.5% (Warford et al., 1984) to 78.3% (Morgan and Smith, 1984), compared with 7-day cell culture isolation. Sensitivity of 88% and specificity of 94% have been reported for ELISA direct detection of respiratory syncytial virus in clinical samples (Lauer et al., 1985).

Rotavirus detection via ELISA is popular owing to the lack of alternative rotavirus diagnostic techniques. Rotavirus does not proliferate in standard cell cultures, and electron microscopy, the only method previously available for rotavirus observation, is not available in most clinical virology laboratories. Commercially available rotavirus ELISAs are reportedly as sensitive as, if not more sensitive than, electron microscopy (Chernesky et al., 1985). Al-

though a low incidence of false-positive results has been documented (Sambourg et al., 1985), the technique has shown excellent specificity (Chernesky et al., 1985; Sambourg et al., 1985; Miotti et al., 1985).

Most commercial ELISA systems are modular and include devices for rinsing or bead transfer, a spectrophotometric readout device, and a computer that interprets spectrophotometric readings and prints results. Each system is expensive and none, at this writing, is as yet set up for identification of all of the antigens described in this section. The lack of versatility of individual ELISA systems is a practical consideration in selection of this method for clinical virology laboratories.

NEUTRALIZATION AND HEMAGGLUTINATION INHIBITION TESTING

Several testing methods used in identification of unknown viral agents rely on the ability of known specific viral antibodies to neutralize or inhibit characteristic viral activities. In neutralization testing, incubation of the virus with its specific antibody prior to inoculation of the virus into susceptible cell cultures neutralizes the virus and renders it incapable of infecting susceptible cells. The capacity to hemagglutinate erythrocytes (see above), characteristic of some groups of viruses, can be inhibited by specific antibody. Loss of viral activity through neutralization of viral infectivity (NEUT) or by hemagglutination inhibition (HAI) provides confirmation of the identity of the virus.

Both NEUT (Fig. 8) and HAI tests are performed in two stages. In the first stage, live unknown virus and known viral antibodies are reacted. In stage 2, aliquots of the mixture are inoculated into a system that will allow the condition of the virus to be evaluated. For HAI testing, a suspension of erythrocytes is added, and the capacity of the virus to hemagglutinate is evaluated 2 to 3 h later. For NEUT, the virus is inoculated into susceptible cell cultures or hosts. Following a suitable incubation period of 5 to 7 days, the host tissue is examined for the presence of cytopathogenic effect.

NEUT procedures provide the most accurate results and are the reference standard against which viral identification methods are compared. Groups of virus, such as the enterovirus, which include many closely related serotypes, can be differentiated only by NEUT testing. Protocols for NEUT identification of viruses have been published previously (Schmidt, 1979).

The HAI test method has been a popular one for quantitation of rubella, measles, and mumps antibodies but is not widely used for virus identification. This serologic testing is being replaced in many laboratories by less laborious methods such as ELISA.

Passive Agglutination

Passive agglutination methods, often casually termed latex agglutinations, are the least technically demanding test methods. These use latex particles coated with viral antibodies of known specificity. The suspension of coated particles is mixed with virus-infected cell culture material or with the patient's sample that contains the unknown viral antigen. If this antigen is homologous with the known antibody coating the latex particles, agglutination of the particles occurs. Latex agglutination testing requires only 2 to 20 min to perform, and results of most latex test methods compare favorably with more technically difficult methods.

Latex testing kits, available commercially for identification of herpesvirus antigen in clinical samples, have been reported as 92% sensitive and 98% specific (Wampole Laboratories, 1986) compared with 7-day cell culture isolation of the virus. Latex agglutination products are also marketed for detection of rotavirus in stool samples. Rotavirus latex test sensitivity was reported as 69 to 79% compared with ELISA in one study (Sambourg et al., 1985); in a second study, latex testing showed greater sensitivity than ELISA (Pai et al., 1985). Latex testing, compared with electron microscopy, has shown 82% sensitivity and 99.5% specificity (Pai et al., 1985). Rotavirus latex testing is a convenient alternative to the more sophisticated ELISA.

Other Immunologic Techniques

Radioimmunoassay (RIA)

RIA methods are performed much like ELISA testing (see above); the major difference is in the indicator system. In RIA testing, radiolabeled substances are included to provide indication of test reactivity. RIA methods were the first techniques used extensively in hepatitis antigen detection. Results were excellent, and the RIA technique is still used in the study of hepatitis. The method is losing popularity in clinical laboratories owing to the presence of radiolabeled substances in the testing systems. Most RIA methods are being replaced with ELISA methods, which provide comparable results and do not require radiolabeled reagents.

Immune Adherence Hemagglutination (IAHA)

The IAHA assay is based upon complement fixation by Ag–Ab complexes; the presence of bound complement is demonstrated by addition of primate erythrocytes, which bind to the complement by C'3b receptors present on the erythrocyte surface. IAHA assays require 4 to 5 h for completion compared with 18 h for the traditional complement fixation (CF) test. This technique has been shown to be more sensitive than CF (Lennette and Lennette, 1978) and has been applied to detection of a variety of viral antibodies. Use of IAHA has also been reported in detection of viral proliferation of HFRS (hemorrhagic fever with renal syndrome) virus in cell cultures (Matsuura et al., 1984) and for detection of Australia antigen (Mayami, et al., 1971).

Literature Cited

Aurelian, L. 1969. Factors affecting the growth of canine herpesvirus in dog kidney cells. Appl. Microbiol. 17:179–181.

Becton Dickinson and Company. 1986. Package insert for LeucoPREP™ Tubes.

Berninger, M., M. Hammer, B. Hoyer, and J. L. Gerin. 1982. An assay for the detection of the DNA genome of hepatitis B virus in serum. J. Med. Virol. 9:57–68.

Boehringer Mannheim Biochemica. 1986. Package insert for BM-Cycline.

Chernesky, M., S. Castriciano, J. Mahony, and D. DeLong. 1985. Examination of the rotazyme II enzyme immunoassay for the diagnosis of rotavirus gastroenteritis. J. Clin. Microbiol. 22:462–464.

DelGiudice, R. A., and H. E. Hopps. 1978. Microbiological methods and fluorescent microscopy for direct demonstration of mycoplasma infection of cell cultures, p. 156–169. In G. J. McGarrity, D. G. Murphy, and W. W. Nichols (ed.), Mycoplasma infection of cell cultures. Plenum Press, New York.

Diegutis, P. S., E. Keirnan, L. Burnett, B. N. Nightingale, and Y. E. Cossart. 1986. False-positive results with hepatitis B virus DNA dot-hybridization in hepatitis B surface antigen-negative specimens. J. Clin. Microbiol. 23:797–799.

Doane, F. W. 1986. Electron microscopy and immunoelectron microscopy, p. 71–88. In S. Spector and G. J. Lancz (ed.), Clinical virology manual. Elsevier Science Publishing, Inc., New York.

Dulbecco, R. 1952. Production of plaques in monolayer tissue cultures by single particles of an animal virus, Proc. Natl. Acad. Sci. USA 38:747–752.

Enzo Biochem. 1985. Cytomegalovirus Patho-Gene® kit product information.

Espy, M. J., J. C. Hierholzer, and T. F. Smith. 1986. The effect of centrifugation on the rapid detection of adenovirus in shell vials. Abstracts of the Annual Meeting, American Society for Microbiology, C-65.

Fayram, S. L., S. Aarnaes, and L. M. de la Maza. 1983. Comparison of Cultureset to a conventional tissue culture-fluorescent-antibody technique for isolation and identification of herpes simplex virus. J. Clin. Microbiol. 18:215–216.

Forghani, B., K. W. Dupis, and N. J. Schmidt. 1985. Rapid detection of herpes simplex virus DNA in human brain tissue by in situ hybridization. J. Clin. Microbiol. 22:656–658.

Freshney, I. R. 1983. Culture of animal cells: a manual of basic technique. Alan R. Liss, Inc., New York.

Fung, J. C., J. Shanley, and R. C. Tilton. 1985. Comparison of the detection of herpes simplex virus in direct clinical specimens with herpes simplex virus-specific DNA probes and monoclonal antibodies. J. Clin. Microbiol. 22:748–753.

Gen-Probe, Inc. 1986. Mycoplasma T.C. detection kit, package insert. San Diego, California.

Gissman, L., B. Forbes, M. Pawlita, and A. Schneider. 1986. Filter in situ hybridization; a sensitive method to detect papillomavirus DNA in single cells, p. 157–162. *In* L. S. Lerman, (ed.), Current communication in molecular biology. Cold Spring Harbor Laboratory, Cold Spring Harbor, New York.

Gleaves, C. A., T. F. Smith, E. A. Shuster, and G. R. Pearson. 1984. Rapid detection of cytomegalovirus in MRC-5 cells inoculated with urine specimens by using low-speed centrifugation and monoclonal antibody to an early antigen. J. Clin. Microbiol. **19**:917–919.

Gleaves, C. A., T. F. Smith, E. A. Shuster, and G. R. Pearson. 1985a. Comparison of standard tube and shell vial cell culture techniques for the detection of cytomegalovirus in clinical specimens. J. Clin. Microbiol. **21**:217–221.

Gleaves, C. A., D. J. Wilson, A. D. Wald, and T. F. Smith. 1985b. Detection and serotyping of herpes simplex virus in MRC-5 cells by use of centrifugation and monoclonal antibodies 16 h postinoculation. J. Clin. Microbiol. **21**:29–32.

Hawkes, R. A. 1979. General principles underlying laboratory diagnosis of viral infections, p. 3–48. *In* E. H. Lennette and N. J. Schmidt (ed.), Diagnostic procedures for viral, rickettsial, and chlamydial infections. American Public Health Association, Washington, D.C.

Hay, R. J. 1985. ATCC quality control methods for cell lines, 1st ed. American Type Culture Collection, Rockville, Maryland.

Hayden, F. G., A. S. Sorensen, and J. A. Bateman. 1983. Comparison of the Immulok Cultureset and virus isolation for detection of herpes simplex virus in clinical specimens. J. Clin. Microbiol. **18**:222–224.

Hudson, J. B., V. Misra, and T. R. Mosmann. 1976. Cytomegalovirus infectivity: analysis of the phenomenon of centrifugal enhancement of infectivity. Virology **72**:235–243.

Hyypia, T. 1985. Detection of adenovirus in nasopharyngeal specimens by radioactive and nonradioactive DNA probes. J. Clin. Microbiol. **21**:730–733.

Hyypia, T., P. Stalhandske, R. Vainionpaa, and U. Petterson. 1984. Detection of enteroviruses by spot hybridization. J. Clin. Microbiol. **19**:436–438.

Johnson, R. T. 1982. Viral infections of the nervous system. Raven Press, New York.

Karber, G. 1931. Beitrag zur kollektiven Behandlung pharmakologischer Reihenversuche. Arch. Exp. Pathol. Pharmakol. **162**:480–483.

Keller, R., R. Peitchel, J. N. Goldman, and M. Goldman. 1976. An IgG-Fc receptor induced in cytomegalovirus-infected human fibroblasts. J. Immunol. **116**:772–777.

Landry, M. L., and G. D. Hsiung. 1986. Primary isolation of viruses, p. 31–51. *In* S. Specter and G. J. Lancz (ed.), Clinical virology manual. Elsevier Science Publishing, Inc., New York.

Lauer, B. A., H. A. Masters, C. G. Wren, and M. J. Levin. 1985. Rapid detection of respiratory syncytial virus in nasopharyngeal secretions by enzyme-linked immunosorbent assay. J. Clin. Microbiol. **22**:782–785.

Lennette, E. H. 1969. General principles underlying laboratory diagnosis of viral and rickettsial infections, p. 1–65. *In* E. H. Lennette and N. J. Schmidt (ed.), Diagnostic procedures for viral and rickettsial infections, 4th ed. American Public Health Association, Inc., New York.

Lennette, E. T., and D. A. Lennette. 1978. Immune adherence hemagglutination: alternative to complement-fixation serology. J. Clin. Microbiol. **7**:282–285.

Levy, J. A., A. D. Hoffman, S. M. Kramer, J. A. Landis, and J. M. Shimabukuro. 1984. Isolation of lymphocyto-

pathic retroviruses from San Francisco patients with AIDS. Science **225**:840–842.

Matsuura, Y., K. Sugiyama, C. Morita, S. Morikawa, S. Shiga, T. Komatsu, Y. Akao, and T. Kutamura. 1984. Infectivity titration of hemorrhagic fever with renal syndrome virus: use of immune adherence hemagglutination for detection of virus growth. J. Clin. Microbiol. **20**:483–485.

Mayami, M., K. Okochi, and K. Nishioka. 1971. Detection of Australia antigen by means of immune adherence hemagglutination test. Vox Sang. **20**:178.

McGarrity, G. J. 1982. Detection of mycoplasmal infection of cell cultures, p. 99–131. *In* Advances in cell culture. Academic Press, New York.

Meignier, B., H. Mougeot, and H. Favre. 1980. Foot and mouth disease virus production of microcarrier-grown cells. Dev. Biol. Stand. **46**:249–256.

Miotti, P. G., J. Eiden, and R. H. Yolken. 1985. Comparative efficiency of commercial immunoassays for the diagnosis of rotavirus gastroenteritis during the course of infection. J. Clin. Microbiol. **22**:693–698.

Morgan, M. A., and T. F. Smith. 1984. Evaluation of an enzyme-linked immunosorbent assay for the detection of herpes simplex virus antigen. J. Clin. Microbiol. **19**:730–732.

Moseley, R. C., L. Corey, D. Benjamin, C. Winter, and M. L. Remington. 1981. Comparison of viral isolation, direct immunofluorescence, and indirect immunoperoxidase techniques for detection of genital herpes simplex virus infection. J. Clin. Microbiol. **13**:912–918.

Pai, C. H., M. S. Shahrabad, and B. Ince. 1985. Rapid diagnosis of rotavirus gastroenteritis by a commercial latex agglutination test. J. Clin. Microbiol. **22**:846–850.

Phillips, L. E., R. A. Magliolo, M. L. Stehlik, P. A. Whiteman, S. Faro, and T. E. Rogers. 1985. Retrospective evaluation of the isolation and identification of herpes simplex virus with cultureset and human fibroblasts. J. Clin. Microbiol. **22**:255–258.

Rawls, W. E. 1980. Herpes simplex viruses, p. 783–789. *In* E. H. Lennette (ed.), Manual of clinical microbiology, 3rd ed., American Society for Microbiology, Washington, D.C.

Reed, L. J., and H. Muench. 1938. A simple method of estimating fifty percent endpoints. Am. J. Hyg. **27**:493–497.

Reedman, B. M., and G. Klein. 1973. Cellular localization of an Epstein-Barr virus (EBV)-associated complement-fixing antigen in producer and non-producer lymphoblastoid cell lines. Int. J. Cancer **11**:499–520.

Rubin, S. J., and S. Rogers. 1984. Comparison of cultureset and primary rabbit kidney cell culture for the detection of herpes simplex virus. J. Clin. Microbiol. **19**:920–922.

Salmon, V. C., R. B. Turner, M. J. Speranza, and J. C. Overall, Jr. 1986. Rapid detection of herpes simplex virus in clinical specimens by centrifugation and immunoperoxidase staining. J. Clin. Microbiol. **23**:683–686.

Sambourg, M., A. Goudeau, C. Courant, G. Pinon, and F. Denis. 1985. Direct appraisal of latex agglutination testing, a convenient alternative to enzyme immunoassay for the detection of rotavirus in childhood gastroenteritis, by comparison of two enzyme immunoassays and two latex tests. J. Clin. Microbiol. **21**:622–625.

Schacter, J. 1980. Chlamydiae (psittacosis-lymphogranuloma venerum-trachoma group), p. 357–364. *In* E. H. Lennette (ed.), Manual of clinical microbiology, 3rd ed. American Society for Microbiology, Washington, D.C.

Schacter, J., and C. R. Dawson. 1979. Psittacosis, lymphogranul organuloma venereum agents/TRIC agents, p. 1021–1059. *In* E. H. Lennette and N. J. Schmidt (ed.),

Diagnostic procedures for viral, rickettsial and chlamydial infections. American Public Health Association, Washington, D.C.

Schmidt, N. J. 1979. Cell culture techniques for diagnostic virology, p. 65–140. *In* E. H. Lennette and N. J. Schmidt (ed.), Diagnostic procedures for viral, rickettsial, and chlamydial infections, 5th ed. American Public Health Association, Washington, D.C.

Sewell, D. L., S. A. Horn, and P. W. Dilbeck. 1984. Comparison of cultureset and Bartels Immunodiagnostics with conventional tissue culture for isolation and identification of herpes simplex virus. J. Clin. Microbiol. **19:**705–706.

Spear, P. G., M. F. Para, and R. B. Baucke. 1979. The Fc-binding receptor induced by herpes simplex virus, p. 186–189. *In* A. S. Nahmias, W. R. Dowdle, and R. F. Schinazi (ed.), The human herpesviruses. Elsevier Science Publishing, Inc., New York.

Strano, A. J. 1976. Viral pneumonias, p. 57–64. *In* C. H. Binford and D. H. Connor, (ed.), Pathology of tropical and extraordinary diseases. Armed Forces Institute of Pathology, Washington, D.C.

Thornton, B., I. D. McEnter, and B. Griffiths. 1985. Production of herpes simplex virus from MRC-5 cells grown in a glass bead culture system. Dev. Biol. Stand. **60:**475–481.

Van Wezel, A. L. 1967. Growth of cell strains on microcarriers in homogeneous culture. Nature (London) **216:**64–65.

Van Wezel, A. L., G. Van Steenis, C. A. Hannik, and H. Cohen. 1978. New approach to the production of concentrated and purified inactivated polio and rabies tissue culture vaccines. Dev. Biol. Stand. **41:**159–167.

Wampole Laboratories. 1986. Product information. Latex agglutination test for HSV antigen—Virogen™.

Warford, A. L., R. A. Levy, and K. A. Rekrut. 1984. Evaluation of a commercial enzyme-linked immunosorbent assay for detection of herpes simplex virus antigen. J. Clin. Microbiol. **20:**490–493.

Weakley, B. S. 1981. A beginner's handbook in biological transmission electron microscopy. Churchill Livingstone, New York.

Wentworth, B. B., and L. French. 1969. Plaque assay of herpesvirus hominis on human embryonic fibroblasts, Proc. Soc. Exp. Biol. Med. **131:**588–592.

Wentworth, B. B., and L. French. 1970. Plaque Assay of Cytomegalovirus strains of human origin. Proc. Soc. Exp. Biol. Med. **135:**253–258.

Widell, A., B. G. Hansson, and E. Nordenfelt. 1984. A microcarrier cell culture system for large scale production of hepatitis A virus. J. Virol. Methods **8:**63–71.

CHAPTER 4

Direct Antigen Detection

PERTTI P. ARSTILA and PEKKA E. HALONEN

A direct detection of viral particles, antigens or nucleic acids in clinical samples is the most straightforward strategy for specific viral diagnosis. Almost always, it is also the most rapid diagnostic method. Detection of virus particles requires electron microscopy, and detection of nucleic acids is done by hybridization methods, whereas antigen detection is done by immunologic methods using specific hyperimmune sera or monoclonal antibodies.

In antigen detection, complete virions are not necessarily required in the specimen. In fact, structural and even nonstructural proteins are usually the antigens that are detected since they are produced in large excess compared with full virion production. In immunofluorescence (IF), the specimen must contain infected cells, but in solid-phase immunoassays and in latex agglutination excreted antigens are also assayed. The detectable level of viral antigen varies according to immunoassay type used and the sensitivity of the specific assay. As an example, in IF, a single infected cell can be seen and the diagnosis made. Also, any immunoassay can be only as good as the quality of antibodies used in the test. In addition, because viral antigens are detected by specific antibodies, one has to know what he or she is looking for. If etiologic candidates are numerous (i.e., 100 different rhinovirus types in common cold), a type-specific antigen assay is not practical. A contrasting example is bronchiolitis of infants, where only a few viruses are candidates.

In diagnostic virology, the use of antigen detection has expanded steadily. Reagents have improved and quality control has been organized. The evolvement of monoclonal antibodies has, in particular, had great impact in the development of antigen assays, and numerous commercial kits are now available. The tests are usually simple to perform and often the results are provided in printout forms which are quick and relatively straight-forward to evaluate.

Immunofluorescence

Principles of Immunofluorescence in Antigen Detection

Since the introduction of fluorescein-labeled antibodies in 1941 (Coons et al., 1941), the method was applied to most areas of virology, and only recently has been partly replaced by methods using other labels. There are several recent reviews and books describing immunofluorescence methodology in great detail (Gallo, 1983; Gardner and McQuillin, 1980; Hanson, 1985). In this context only, some general outlines and examples of the most recent applications will be described.

The most frequently used fluorochrome dyes are fluorescein isothiocyanate (FITC) and rhodamine B. The emission maximum for FITC is 517 nm, and for rhodamine 595 nm. The fluorescence of FITC is apple green. The fluorescence of rhodamine is red when proper filter systems are used. The different colors make it possible to use double-labeling of the same specimen in some instances.

In antigen detection, a direct or indirect immunofluorescent method can be used. In a direct assay, antigen in infected cells is detected by virus-specific immunoglobulins conjugated with fluorochrome. In an indirect assay, virus-specific immunoglobulins are followed by anti-species-specific antibodies, which are labeled with fluorochrome. The direct assay is quicker, but on the other hand the sensitivity is better in the indirect assay because of the higher number of binding sites for the conjugate. The indirect method is also more convenient in practice if several viruses are searched for in the same specimen. For direct assays, each antiviral immunoglobulin must be separately conjugated. In indirect methods, only a few conjugated anti-species immunoglobulins are needed.

Sensitivity of IF assay may be further amplified by using biotin-avidin complexes (Nerurkar et al., 1983). Avidin anti-species conjugates are commercially available. Virus-specific biotin-labeled antibodies are more difficult to obtain.

Antibodies Used in Immunofluorescence

As in any antigen detection based on an immunologic reaction, in immunofluorescence the quality of antisera used is the most important single factor affecting the results. Conventional polyclonal antisera are still most commonly used, and are usually produced in rabbits or guinea pigs, but also in mice, hamsters, calves, goats, sheep, monkeys, horses, and even the egg yolk of chickens. In many instances, polyclonal antisera are still the only ones available, but monoclonal antibodies are now rapidly replacing them. In the case of a high quality polyclonal antiserum, a monoclonal antibody offers hardly any benefits in the actual assay, but the mass production and basically unchanged specificity of various batches of the same monoclonal antibody make the use of them practical. Difficulties may arise in a too narrow specificity of monoclonal antibody, and may require the use of pools. In that case, the situation resembles that of polyclonal antisera.

Despite the source of the antibody, the intended reagent must be carefully tested before use. Cross-reactions must be checked against related viruses, tissue culture cells in which immunizing antigen has been grown, other microbes or antigens possibly present in the clinical specimen, and negative clinical specimens. At least in case of polyclonal antisera, removal of undesired reactions by absorption is often necessary.

Conjugates Used in Immunofluorescence

The quality of fluorescein-labeled anti-species immunoglobulins ("conjugates") is just as important as the quality of antiviral antibodies. For indirect assays, conjugates are commercially available and can be used in almost any virus laboratory. In the indirect fluorescence assays, usually only a few conjugates are needed, and they are now commercially available.

As in the case of primary antiviral antisera, conjugates must always be checked by the final user. In principle, the controls are the same as those mentioned previously. Extensive check-ups may not be necessary between different batches. If problems arise, however, they very often are caused by the conjugate.

The preparation of a good conjugate, as noted, usually requires absorption and extensive testing to avoid nonspecific cross-reactions. Therefore, there are only a few commercial conjugates available for direct assays. Theoretically, monoclonal antibodies should make excellent conjugates because of their narrow specificity, which minimizes cross-reactions.

Nonspecific Fluorescence

In an ideal case, specific apple green fluorescence of FITC is seen against a practically dark background. Unfortunately, nonspecific fluorescence is avoided only in IF assays of highest quality. Nonspecific fluorescence is, as mentioned previously, generally caused by antiserum or conjugate that cross-reacts with the specimen or related antigen. Extensive quality controls and absorptions, if necessary, overcome this problem.

In some cases, nonspecific fluorescence might be caused by bacteria in the specimen. This is particularly the case with staphylococci in which protein A binds nonspecifically to the IgG class of immunoglobulins. Another common problem is the Fc receptors of *Herpes viridae*.

Autofluorescence of the clinical specimen itself is also possible. The color and intensity of autofluorescence may vary and cause some difficulties in interpretation. Therefore, when examining clinical material, it is necessary to use counter stains to diminish background fluorescence. Several different dyes are used for this purpose. Naphthalene black, for instance, produces an almost black background, but like Evans blue, it may give a reddish color to the noninfected cells, producing a good contrast to the green color of FITC (Gardner and McQuillin, 1980). Counter stain can be added as the last step of the staining procedure, but it is more convenient to add it directly to the conjugate.

Immunofluorescence Microscope

The development of the IF microscope since the early days of immunofluorescence has been huge. Improvement of filter and lens systems is still taking place, partly as a result of modern computer technology. Major manufacturers have developed several models, which are fully ready for use. The best models have built-in interference and barrier filters for one or more fluorescence systems so that they can be changed by one movement of the revolver.

For most applications, incident light is recommended. A brighter and sharper fluorescent pattern is achieved because the light beam does not go through the specimen. The price of mercury bulbs has become reasonable and their life-time lengthened so that they are superior to halogen bulbs.

Opinions differ concerning the use of objectives. In many cases, dry objectives are more practical, but some experts prefer oil objectives. In either case, one should not be content with standard objectives, but obtain at least plain-corrected ones.

Interpretation of Results

Immunofluorescence is regarded as a demanding technique that requires careful attention to detail throughout the procedure. As always, the collection of the specimen is the first critical point. When possible, the same trained personnel should take the samples. Handling and transportation requirements vary according to the specimen, but in general for tissues or nasopharyngeal aspirates, it is good practice to pack them on ice if the virus laboratory is far away. Antigenicity of the virus may be readily lost if the specimen is stored at room- or higher temperatures. Various smear-type specimens are preferable to be spread directly on the microscope slide and fixed with acetone before shipment. It should also be noted that all microscope slides used for immunofluorescence must be acid-washed before use to remove grease.

The quality of antisera and conjugates has already been emphasized. Any reagent giving nonspecific reactions should be disregarded or further improved (that is, by absorptions).

When interpreting the results, both positive and negative controls must naturally work in a desired way. It is not uncommon to find that the positive control is no longer positive. The reason for this lies most probably in the conjugate, which has lost its activity. Reasons include bacterial growth, aggregation of globulin molecules, or repeated freezing and thawing (among others). It is good practice to sonicate and/or centrifuge the conjugate at least once a week.

The next step is to evaluate the specimen and make sure that it contains a sufficient amount of cells. If this is not the case, no result should be given and a new specimen should be requested. Even when the specimen has been correctly taken, cells can be lost during washing steps as a result of poor drying and fixation or a greasy objective glass.

When the specimen seems to be positive, the microscopist must evaluate the intensity, amount of the fluorescence, and background. In addition, the fluorescent pattern must be specific for the virus being looked for.

Immunofluorescence microscopy must be done by a trained and experienced microscopist. If the strict qualifications of the IF test itself are fulfilled, it does not take an excessively long time to train a microscopist. However, everybody involved in the entire procedure, including the microscopist, should continuously keep up their skills. This can be achieved only by ensuring a sufficient and steady flow of specimens.

Applications of Immunofluorescence in Antigen Detection

APPLICATIONS UTILIZING CONVENTIONAL ANTISERA

Since the first description of immunofluorescence by Coons et al. (1941), it took almost three decades before the technique really took wing. In research laboratories, most viruses have been examined with immunofluorescence, but for diagnostic purposes, its golden period really started in the late 1960s. Recent developments, however, have led to the replacement of immunofluorescence with other methods, such as radioimmunoassay and enzyme immunoassay, which can be automated and are far less labor-consuming. On the other hand, the measuring of light (fluorescence) is still one of the most sensitive detection systems, and methods such as time-resolved fluoroimmunoassay that are measured by special equipment will be seen in the future for viral diagnostic purposes. The major applications of immunofluorescence in the diagnosis of viral diseases can be divided into three categories on the basis of the sample (Table 1).

The first important application utilizing infected tissue was the demonstration of rabies in the salivary glands of rabid animals (Goldwasser et al., 1959). In the diagnosis of rabies, IF has maintained its position to the present day. In the case of a suspected rabid animal, both brain and salivary gland tissue can be collected postmortem. A reliable diagnosis is achieved in a simple way by preparing an impression smear of a small (1 to 2 mm) piece of the tissue and squeezing it between two objective slides (Gardner and McQuillin, 1980), then fixing and staining in a routine way.

The same impression smear technique can be readily used for any tissue specimen taken at autopsy and biopsy. In pneumonia or other lower respiratory tract infections, a lung biopsy may be considered,

TABLE 1. Examples of immunofluorescence in viral antigen detection of clinical specimens

Specimen	Type of viral infection
Tissue (biopsy or sample from autopsy)	Encephalitis, pneumonia, carditis, hepatitis, pancreatitis, exanthemas, etc.
Exfoliated cells	Respiratory infections
Vesicles	Herpes and varicella

although a positive result is not always obtained even in a case of real viral infection.

Encephalitis caused by herpes simplex virus is a life-threatening situation. A rapid diagnosis is needed to determine the specific treatment, but the diagnostics are hampered by several other differential possibilities. A brain biopsy, although possible only in the largest hospitals, seems to be the only way to obtain the specific diagnosis in the early phase of the illness. Again, a negative result does not exclude the possibility of herpes simplex infection.

A special skin punch biopsy technique has been used for diagnosis of varicella zoster and other erythematous viral infections (Olding-Stenkvist and Grandien, 1976). This technique also includes a simple and rapid preparation of the clinical specimen.

One of the most extensive uses of immunofluorescence is in the rapid diagnosis of viral respiratory infections. Gardner and McQuillin (pioneers in this field) have established the techniques (Gardner and McQuillin, 1968; McQuillin et al., 1970), that are widely applied in virus laboratories worldwide (Ørstavik et al., 1984). Viral antigens are looked for in exfoliated epithelial cells of nasopharyngeal aspirate, which is collected by suction through the nostrils. For an ideal result, the cells must be extensively washed to rid them of mucus, which may cause autofluorescence or physically cover the antigenic sites and, thus, diminish the specific fluorescence. The method is especially useful for respiratory syncytial viral infections where a common problem in infected children is the considerable excretion of mucus in the respiratory tract. Other viruses include influenza viruses A and B, parainfluenza viruses 1 to 4, adenoviruses, and measles virus. Some examples of fluoresceing cells from nasopharyngeal aspirates are shown in Figure 1A to D.

For the first time, it was also possible to make a relatively rapid viral diagnosis in the specimens sent from small and distant laboratories. Technical personnel were trained to collect the sample, wash the cells, and fix them on the objective slide, which could then be mailed to a special laboratory for staining and evaluation of the immunofluorescent pattern. The results from our laboratory for the first 5 years of the use of immunofluorescence in the diagnosis of respiratory infections are summarized in Table 2. An

important factor behind this success is the availability of controlled, high quality reagents either commercially or through the European and Pan American Groups for Rapid Viral Laboratory Diagnosis (W.H.O. Scientific Group; 1981).

Although immunofluorescence is still the most widely used method for respiratory viral infections and it has some advantages over other methods, in many laboratories it has been replaced by enzyme immunoassays for practical reasons.

In its early stages of development, it was already possible to examine the vesicles on the skin of herpes simplex or varicella zoster virus-infected patients by immunofluorescence for the presence of antigen (Biegeleisen et al., 1959; Schmidt et al., 1965). An optimal specimen is taken by scraping off cells from the bottom of the vesicle on the drop of saline solution on the objective slide, air-drying it, and processing it normally. Immunofluorescence of the herpes virus family is challenging. It is not always easy to collect enough cells to make a representative sample. In addition to common cross-reactions between different herpes viruses, Fc receptors may also cause problems. Monoclonal antibodies, also commercially available (Balkovic and Hsiung, 1985; Fung et al., 1985), should make the interpretation easier and give the specific type as well.

APPLICATIONS UTILIZING MONOCLONAL ANTIBODIES

Only 10 years after the original report of Köhler and Milstein (1975), monoclonal antibodies had found a permanent place in all fields of biological sciences. Great expectations have also been placed for their use in diagnostic virology. In theory, monoclonal antibodies make an excellent reagent, but many problems in their use remain to be solved. Some benefits and pitfalls are listed in Table 3.

Although monoclonal antibodies have several beneficial features, they may cause problems when used as diagnostic reagents. Obviously, restricted specificity may result in the loss of some prevalent viral subtypes. This might necessitate the use of a pool of two or more monoclonal antibodies. It is not possible to produce a cocktail of reagents with exactly the same characteristics from batch to batch.

TABLE 2. Results of respiratory viral antigen detection by indirect immunofluorescence in nasopharyngeal aspirates from 1978 to 1982, Department of Virology, University of Turku, Finland

No. of specimens	Positive		RSV	Influenza		Parainfluenza			Adenovirus
	No.	%		A	B	1	2	3	
6,190	1,388	22	826	91	13	80	53	101	224

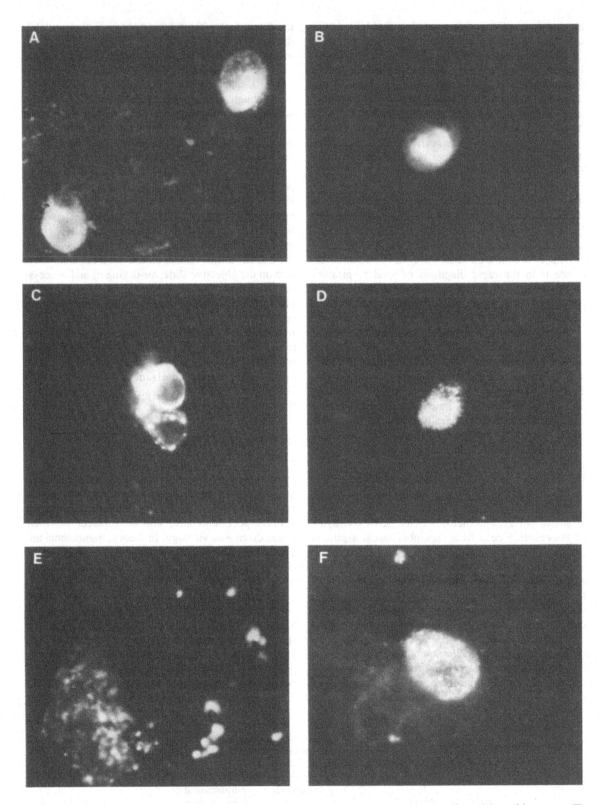

FIG. 1. Immunofluorescence of virus-infected cells from nasopharyngeal aspirates (A to E) and of an impression smear from a brain biopsy specimen (F). (A) Respiratory syncytial virus. (B) Influenza A virus. (C) Parainfluenza 3. (D) Adenovirus. (E) Nonspecific staining with mucus. (F) Herpes simplex virus. A color plate of this figure appears in Color Plate 1 at the beginning of this volume.

TABLE 3. General properties of monoclonal antibodies in immunoassays

Benefits	Pitfalls
Monospecificity due to monoclonality	Too narrow specificity; problems in group detection or due to antigenic variation
Homogenicity	Need for pooled reagents
Good typing reagent	Random success
In vitro production	Problems in characterization, labeling, and stability
Mass production	Unpredictable properties
Low production costs	Low avidity
	Variations between ascites batches

Even the repetitive production of a single monoclonal antibody with an identical character may be impossible because of variation in the animal stock used for raising ascites fluid.

Above all, there is concern about the availability of monoclonal antibodies. Only rather advanced laboratories are able to produce their own reagents, and commercially there is a tendency toward complete kits, not just single reagents. For the same reasons, it may prove difficult to freely exchange information on the quality control of the reagents.

Monoclonal antibodies have been raised for practically all important human viruses, but for pure diagnostic purposes, only a few monoclonal antibodies have been used in immunofluorescence. In the diagnostics of respiratory syncytial viruses, monoclonal antibodies have proved to be equal to polyclonal antibodies (Bell et al., 1983; Routledge et al., 1985). By using monoclonal antibodies, it has been shown that subtypes of respiratory syncytial viruses may occur alternately or concomitantly (Åkerlind and Norrby, 1986). Influenza A and B viruses can also be detected by monoclonal antibodies (McQuillin et al., 1985; Shalit et al., 1985). Antigenic changes of both viruses may cause problems in subsequent epidemic seasons.

Cytomegalovirus (CMV) pneumonia is a growing problem in immunocompromised patients. Monoclonal antibodies have been used successfully in the diagnostics of CMV pneumonitis in bronchoalveolar lavage material and in open-lung biopsies (Emanuel et al., 1986; Hackman et al., 1985).

Reliable typing of herpes simplex viruses is now available as a result of monoclonal antibodies (Pereira et al., 1982). Most laboratories use enzyme immunoassay (Ziegler et al., 1985) for this, but the reagents work equally well in immunofluorescence (Balkovic and Hsiung, 1985). Herpes viral antigen has been detected directly by immunofluorescence in

erythema multiforme patients (Orton et al., 1984). Commercial FITC-conjugated monoclonal antibodies for the detection of herpes virus are also now available (Fung et al., 1985).

Solid-Phase Immunoassays

Principles and Solid-Phase Immunoassays

Solid-phase immunoassays are highly sensitive in the direct detection of viral antigens in clinical specimens, and are widely used both in commercial kits and in homemade tests. They have many advantages including: 1) stability of viral antigens in clinical specimens and accordingly no special requirement for transportation to the laboratory; 2) the possibility of semiautomated bulk testing of specimens; 3) tests can be standardized, resulting in acceptable daily variations; and 4) availability of technologically advanced measuring equipment with printout results and the possibility of analyzing the results by flexible computer programs. The applications in diagnostic virology include hepatitis B (HBsAg) antigen in serum (Lander et al., 1971, Ling and Overby, 1972), gastroenteritis viruses (rotaviruses and adenoviruses) in stool specimens (Halonen et al., 1979; Kalica et al., 1977; Middleton et al., 1977; Sarkkinen et al., 1979; Sarkkinen et al., 1980), respiratory viruses (RSV, influenza A and B, parainfluenza 1, 2, and 3, adenoviruses, and coronaviruses) in nasopharyngeal aspirates (Berg et al., 1980; Chao et al., 1979; Halonen et al., 1985; Sarkkinen et al., 1981a, 1981b, 1981c), herpes simplex and varicella zoster in vesicle fluid (Forghani et al., 1974; Ziegler, 1984; Ziegler and Halonen, 1985), and HIV in blood (Allain et al., 1986; Goudsmit et al., 1986; Paul and Falk, 1986).

The principle of the assays is simple. The catching antibody is a solid phase, which is usually a polystyrene microtitration well, bead, or tube. The specimen is incubated with solid-phase antibody, the nonbound material is washed away, and labeled or nonlabeled (or biotinylated) antibody (followed by labeled anti-immunoglobulin or enzyme-labeled avidin) is added. After another wash, the bound label is measured directly (RIA) or through a color (EIA) or enhancement (TR-FIA) reaction.

Earlier polyclonal immunoreagents were used in solid-phase immunoassays, and one of the most commonly used test configurations is a four-layer or antispecies principle (Fig. 2). The sensitivity of 0.1 ng of viral protein/ml of the specimen can be reached if the first incubation of the specimen with solid-phase antibody is extended to overnight (16 h) at 37°C. This assay principle is simplified by labeling the highly purified secondary antibody. More recently, mono-

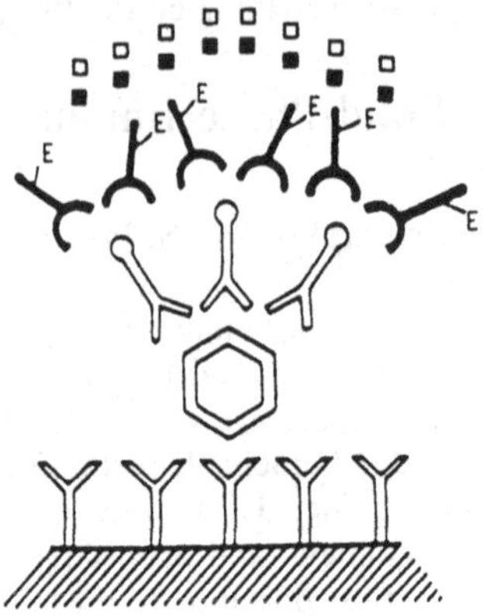

LAYER		REAGENT	INCUBATION
STOPPING SOLUTION		1N HCL	
SUBSTRATE		O-PHENYLDIAMINE	
			30 MIN AT RT
ANTI-SPECIES INDICATOR ANTIBODY		SWINE ANTI-RABBIT IGG ANTIBODY; HRPO-LABELLED	
			1 HOUR AT 37°C
SECONDARY ANTIBODY		IGG FRACTION OF RABBIT ANTI-VIRAL HYPERIMMUNE SERUM	
			1 HOUR AT 37°C
SPECIMEN		VIRUS ANTIGEN (STRUCTURAL PROTEIN)	
			16 HOURS AT 37°C
PRIMARY CATCHING ANTIBODY		IGG FRACTION OF GUINEA PIG ANTI-VIRAL HYPERIMMUNE SERUM	
SOLID PHASE		POLYSTYRENE MICROTITER STRIP	

FIG. 2. Principle of the indirect anti-species enzyme immunoassay for the detection of respiratory viruses in sonicated nasopharyngeal aspirates from patients with acute respiratory disease.

clonal antibodies have been used as immunoreagents and one-incubation (one-wash) assays can be built (Fig. 3). In these 1-h assays, the specimen and the labeled antibody are incubated simultaneously. If the monoclonal antibodies are of high quality and screened for this particular purpose, the sensitivity of the assay can be 0.1 ng of viral protein/ml. However, the very high sensitivity of the assay is not always a critical factor. For instance, gastroenteritis viruses are usually excreted in stool in large quantities, often more than 1 μg/g of stool, and 0.1-ng sen-

sitivities are not required. On the other hand, HSV can occur in lower concentrations, particularly in genital specimens. Respiratory viral antigens are often excreted in hundreds of nanograms in nasopharyngeal secretions, but they can also be found in lower concentrations. Minimal concentrations of HBsAg can be detected in blood. Finally, one of the most critical assays in terms of sensitivity is HIV detection directly in blood specimens.

The specificity of the antigen assay is not usually a problem when immunoreagents have been pre-

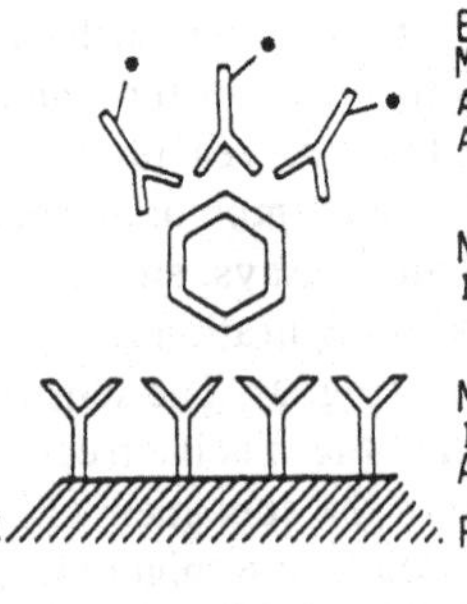

LAYER	REAGENT		INCUBATION
	TYPE	CONCENTRATION	
ENHANCEMENT SOLUTION	15 UM 2-NAPHTOYLTRIFLUOROACETONE 50 UM TRI-N-OCTYLPHOSPHINO OXIDE, 0.1% TRITON X-100 IN 0.1M ACETATE BUFFER, PH 3.2 WITH POTASSIUM HYDROGEN PHTALATE		
			15 MIN AT RT
ANTI-VIRAL INDICATOR ANTIBODY	EU-CHELATE-LABELLED MOUSE MONOCLONAL ANTI-RSV (NUCLEOPROTEIN) ANTIBODY (CLONE A1)	10 NG/ASSAY	
			1 HOUR AT 37°C
SPECIMEN	NUCLEOPROTEIN OF INFLUENZA A	0.1 - 10,000 NG/ML	
CATCHING ANTIBODY	MOUSE MONOCLONAL ANTI-INFLUENZA A (NUCLEOPROTEIN) ANTIBODY (CLONE A3)	500 NG/WELL	
SOLID PHASE	POLYSTYRENE STRIP		

FIG. 3. Principle of the monoclonal one-incubation time-resolved fluoroimmunoassay for the detection of influenza A virus.

pared from monoclonal antibodies. In contrast, extensive evaluations must be made to ensure that the particular epitope on viral antigen being assayed is well conserved in all clinical "isolates" and may require continuous monitoring.

Immunoreagents

Immunoreagents are prepared from antibody-positive human serum, hyperimmune serum, and monoclonal antibodies. Purified IgG or total immunoglobulin fractions (Ig) are better solid-phase catching antibody preparations than are diluted serum or ascites fluid, and IgG or Ig are equally effective. A simple method to prepare Ig from hyperimmune serum or ascites fluid is to precipitate by sodium sulphate, followed by Sepharose G-200 chromatography for desalting. Protein A-sepharose adsorption of IgG is also often used. Mouse immunoglobulins in ascites fluid are efficiently purified by high pressure liquid chromatography (HPLC) (Burchiel et al., 1984).

Total immunoglobulin fraction (Ig) is adsorbed on a polystyrene microtitration well, tube, or bead in carbonate buffer, pH 9.6 (Voller, 1986) by overnight incubation at ambient temperature. However, the binding reaction is not critical for pH or temperature, and binding may be complete in a few hours. The coated plates, strips, or beads can be stored in carbonate buffer for many weeks at 4°C. If longer storage is required, a postcoating with gelatin or bovine albumin, sometimes with carbohydrates, increases the stability.

The optimal concentration of Ig in a solid-phase immunoassay is always pretitrated against increasing concentrations of labeled or secondary antibodies and antigens. The concentrations required depend on the specific activity of the IgG preparation. In optimal conditions, it is about 250 ng of Ig per well, but it may vary from 100 to 1,000 ng per well. Often, several optimal concentrations can be used; if a low concentration of Ig is used in solid phase, an increased concentration of the labeled antibody is required (and vice versa).

Microtitration strips with 8 or 12 wells are convenient in daily routine diagnosis. Depending on the number of specimens tested each day, the correct number of precoated strips are included in the test. An additional advantage is that the immunoreactions are more homogeneous in each well of the strips than in 96-well plates.

The purity of the labeled or secondary antibody is more critical than that of the catching antibody on solid phase. High pressure liquid chromatography-purified monoclonal antibody is an ideal preparation. Polyclonal IgG of high-titered hyperimmune serum

can be used, but often an optimal test requires immunosorbent purified polyclonal antibody.

Labeling of antibody by iodine-125 for radioimmunoassay is usually done by modifications of the Hunter and Greenwood (1962) method. Use of antigen-bound antibody for labeling may have some advantages, particularly with polyclonal antibodies, when immunosorbent purification is combined with labeling (Pelkonen, 1982). Enzyme labeling is usually done according to Wilson and Nakane (1978) (horseradish peroxidase) or Engvall and Perlmann (1972) (alkaline phosphatase). An isothiocyanate reaction is used for labeling with Europium chelate (Hemmilä et al., 1984).

Monoclonal Antibodies

The use of monoclonal antibodies in the detection of viral antigens by solid-phase immunoassays is rapidly increasing and they may soon replace polyclonal antibodies. The problem in their use both on solid phase and as a labeled antibody has been their weak binding capacity. This has been at least partially overcome by using new screening assays in selecting the hybridomas. The screening assay must be as close as possible to the final use of antibodies. As an example, when hybridomas are screened for the production of monoclonal antibodies used in one-step (one-wash) assays, the well is coated with polyclonal antiviral antibody and the screening antigen and the diluted hybridoma cultures are added into the well simultaneously without washing in between (Fig. 4). About half of the monoclonal antibodies are highly reactive in this screening assay and can be used as labeled indicator antibody. The yield of the hybrido-

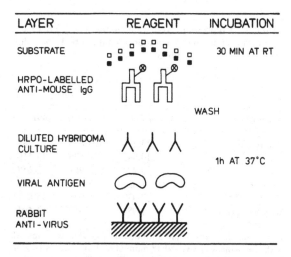

FIG. 4. The screening assay of hybridoma cultures for selecting monoclonal antibodies in one-incubation (one-wash) assays.

mas producing these highly reactive monoclonal antibodies per fusion experiment is considerably lower than that with a standard screening assay using antigen bound directly on solid phase. After cloning, there may finally be only 0 to 5 reactive clones per 1,000 hybridoma wells of the initial fusion experiment.

Primary screening of monoclonal antibodies for solid phase is more difficult, but fortunately monoclonal antibodies, which are highly reactive in the screening assay for the indicator antibody, are often reactive on solid phase as catching antibody. Once the prospective indicator antibody has been labeled, it must be tested against several nonlabeled antibodies on solid phase, which have been reactive in the screening assay (Table 4). In these optimization tests, positive and negative specimens and a purified reference viral preparation must be included.

Comparison of Solid-Phase Immunoassays

When the sensitivity and specificity of radioimmunoassay, enzyme immunoassay, and time-resolved fluoroimmunoassay are compared, the differences are minimal when polyclonal immunoreagents are used in these tests. Actually, the sensitivities and specificities are dictated by the quality of the antibodies used to prepare the immunoreagents. The label material (radioisotope, enzyme, euchelate) has only a marginal effect on the final performance of the tests. Other factors, such as practicality, safety considerations in the laboratory, and availability of commercial kits or reagents, must be taken into account when these tests are selected for the antigen assays in daily diagnosis.

TABLE 4. A representative parainfluenza type 2 TR-FIA test for selecting an optimal monoclonal antibody on solid phase as a catching antibody

Controls		Monoclonal antibodies in solid phase		
		4-2D/2g	1-6D/9H	10-4F/8E
Reference	1,000 ng/ml	73.238[b]	103.676	217.288
virus	100	10.043	16.576	19.510
	10	2.224	2.961	2.657
	0	1.827	1.953	873
Positive NPS		7.479	8.329	15.982
Negative NPS		1.984	1.975	1.591

[a] Eu-labeled monoclonal antibody (1-6D/9H) was tested with the same and two other monoclonal antibodies (4-2D/2g and 10-4F/8E) on solid phase. Three dilutions of reference virus and positive and negative nasopharyngeal specimens (NPS) were included in the optimization test.
[b] Counts per second value.

Radioimmunoassay (RIA)

This technique is still the test with the highest reproducibility. It has less daily variations than do the other solid-phase immunoassays, and the variations between laboratories are minimal and the counting equipment (gamma counters) are reliable. Automation of RIA tests is easier than with EIA tests, and this is one reason why many blood banks with a large number of specimens to be tested for HBsAg still use RIA technology.

The major disadvantages, of course, are the health hazard of handling radioactive material and problems in waste disposal. Actually, in many countries, government regulations rather than real health hazards have limited the use of RIA in diagnostic virology. The short expiration time in commercial RIA kits and the need for frequent labeling of your own reagents are additional factors that have decreased the use of RIA in viral antigen detection.

Enzyme Immunoassay

The enzyme immunoassay (EIA) is the most commonly used solid-phase immunosassay for viral antigen detection. Commercial kits are available for HBsAg and HIV antigen detection in blood, and for rotavirus detection in stool. One of the most advanced tests is the AUSZYME monoclonal test (Abbott Laboratories, North Chicago, Ill.), which is based on monoclonal antibodies both on solid phase and as the HRPO-labeled indicator antibody. The test can be done in one incubation; the serum specimen and the labeled antibody are incubated simultaneously with a bead coated with the catching antibody. One of the benefits of EIA is the possibility for quick visual evaluation of the test.

Many laboratories have reported the development of EIA tests for viral antigen detection, including rotavirus, adenovirus, Norwalk agent, caliciviruses, and astroviruses in stool; herpes simplex in vesicle fluid and genital specimens; varicella zoster in vesicle fluid; cytomegalovirus in urine; arboviruses in blood; and respiratory viruses in nasopharyngeal aspirates and throat swabs. However, the use of these tests in daily diagnostic work has not always been indicated, although this is the most important challenge for the new tests. In our diagnostic unit, rotavirus and adenovirus detection by polyclonal EIA (Halonen et al., 1979; Sarkkinen, 1981) has been performed since 1982, herpes simplex and varicella zoster virus since 1985, and respiratory viruses (influenza A and B, parainfluenza 1, 2, and 3, respiratory syncytial virus, and group-reacting hexon antigen of adenovirus) during 1984 to 1986. Representative results of these EIA tests are shown in Table 5, 6, and 7, and the principle of the indirect anti-species

TABLE 5. A representative indirect enzyme immunoassay for the detection of rotavirus and adenovirus in 1/20 diluted stool specimens from 10 patients with acute gastroenteritis

Controls and specimen no.	Rotavirus	Adenovirus	Conclusion	
Reference antigen[b]	100 ng/ml	1.066[a]	1.271	
	10	0.185	0.205	
	1	0.074	0.087	
	0	0.060	0.054	
Specimen no.				
11340		2.407	0.055	Rotavirus
11341		1.372	0.055	Rotavirus
11345		1.832	0.054	Rotavirus
11429		0.059	0.054	Negative
11430		0.065	0.053	Negative
11450		0.072	0.056	Negative
1048		0.041	0.755	Adenovirus
1431		0.042	0.040	Negative
1470		0.040	1.470	Adenovirus
1471		0.041	0.058	Negative

[a] Optical density.
[b] Purified Nebraska calf diarrhea virus (rotavirus) and crystallized hexon (adenovirus).

TABLE 6. A representative enzyme immunoassay for the detection of herpes simplex virus (HSV) and varicella zoster virus (VZV) in vesicle swabs from patients with vesicular rash

Controls and specimen no.	HSV		VZV		Conclusion
	Normal serum	Immune serum	Normal serum	Immune serum	
Reference virus	0.049	0.977	0.040	0.333	
Control antigen	0.044	0.045	0.044	0.040	
Specimen no.					
3310	0.039	0.047	0.166	1.150	VZV
3311	0.046	0.051	0.059	1.001	VZV
3314	0.042	0.051	0.057	0.057	Negative
1879	0.042	0.046	0.052	0.062	Negative
2065	0.068	0.314	0.059	0.064	HSV

TABLE 7. A representative indirect enzyme immunoassay for the detection of respiratory syncytial virus (RSV), influenza A and B, parainfluenza types 1, 2, and 3, and adenovirus in nasopharyngeal aspirates from seven patients with acute respiratory disease

Reference antigen and specimen no.	RSV	Influenza		Parainfluenza			Adenovirus	Conclusion
		A	B	1	2	3		
Reference								
1,000[a] ng/ml	0.985[d]			0.893	1.790	1.894		
100[b]		1.517	0.870				1.995	
0[c]	0.149	0.036	0.087	0.059	0.039	0.085	0.087	
Specimen no.								
101911	0.129	0.049	0.127	0.090	0.053	0.111	0.083	Negative
101920	0.141	0.080	0.123	0.162	0.098	0.125	0.135	Negative
101942	1.906	0.041	0.090	0.273	0.087	0.104	0.116	RSV
101948	0.087	0.046	0.100	0.079	0.062	1.918	0.105	Parainfl. 3
101954	1.751	0.112	0.131	0.181	0.121	0.129	0.099	RSV
101992	0.116	0.244	0.073	0.093	0.092	0.085	0.084	Influenza A
NIT 39	0.134	0.082	0.121	0.127	0.094	0.098	0.086	Negative

[a] Nonpurified cell lysate antigen.
[b] Purified virus.
[c] Pool of negative specimens.
[d] O.D. value.

TABLE 8. Monthly results of respiratory viral antigen detection in nasopharyngeal aspirates by enzyme immunoassay during the respiratory epidemic period from September 1985 to August 1986, Department of Virology, University of Turku, Finland

Month	No. of specimens tested	Positive No.	Positive (%)	RSV	Influenza A	Influenza B	Parainfluenza 1	Parainfluenza 2	Parainfluenza 3	Adenovirus
September 1985	168	15	(9)	4						11
October	199	33	(17)	22			2		1	8
November	269	75	(28)	63					1	11
December	365	146	(40)	137					1	8
January 1986	361	128	(35)	113	2		3		1	9
February	310	72	(23)	64		1				7
March	280	88	(31)	21	38	3	5		12	9
April	254	55	(22)	8	4	4	2	2	22	13
May	190	31	(16)	1			2		22	6
June	138	19	(14)		1			1	4	13
July	111	15	(14)				1	1	1	12
August	109	5	(5)							5
Total	2,754	682	(25)	433	45	8	15	4	65	112

EIA used in the respiratory virus antigen detection is shown in Figure 3. Monthly results of respiratory virus EIA tests during the respiratory epidemic year from September 1985 to August 1986 are shown in Table 8.

Time-Resolved Fluoroimmunoassay

A new interesting immunoassay, time-resolved fluoroimmunoassay (TR-FIA) (Soini and Kojola, 1983), combines many advantages of RIA and EIA. Some of the limitations of the earlier solid-phase immunoassays are avoided in TR-FIA, but some new requirements, such as being very labor intense, have resulted with this new technology.

The measurement principle of time-resolved fluorescence is shown in Figure 5. The principle is based on new probes (Europium chelate) with a long fluorescent decay time and on a short pulsed excitation of the probes. One-step monoclonal TR-FIAs have been developed for the detection of HBsAg (Siitari et al., 1983) and respiratory viruses (Walls et al., 1986). The principles of this test are presented in Figure 3. In the assay, two monoclonal antibodies are used, each against the same structural protein, but with a different epitope specificity. In some assays, the same monoclonal antibody used as the catching antibody in solid phase can be used as the labeled monoclonal antibody. The catching antibody is coated on the wells of polystyrene microtitration strips, and the

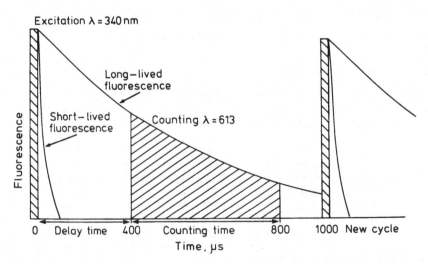

FIG. 5. The detection principle of time-resolved fluorescence. Shown are fluorescence of Eu-chelate and the short decay background fluorescence. Actual decay time is less than 1 μs.

FIG. 6. The 1230 Arcus fluorometer with a disc unit and printer.

indicator antibody is labeled with Europium chelate (Hemmilä et al., 1984). The specimen is simultaneously incubated with the indicator antibody in the well coated with the catching antibody. The incubation time is 1 h at 37°C, followed by a six-cycle washing and the addition of enhancement solution, which dissociates the Europium ion from the antigen-bound antibody and forms a new, highly fluorescent chelate in solution. Fluorescence is measured for 1 s with a single-photon counting fluorometer. Fluorescence activity is expressed as counts per second (cps), and these values are printed out (Fig. 6).

The one-step TF-FIAs of respiratory viruses have been in daily use in our diagnostic unit since January 1, 1987. Typical test results are shown in Table 9, and all results in 1987 are shown in Table 10. The sensitivity of the one-step monoclonal TR-FIA is usually higher than those of the indirect EIAs with

polyclonal antibodies when measured with purified viral preparations. However, the clinical sensitivities have been improved only marginally, except with the influenza A virus TR-FIA in which the number of positive specimens has increased by 20% compared with polyclonal EIA.

The advantage of TR-FIA is clearly in the assays that use monoclonal antibodies. After labeling, monoclonal antibodies are in the monomer form, and the small molecular weight (400) of Eu-chelate ensures the biological activity of antibody molecules (Waris et al., in press). In contrast, horseradish peroxidase labeling of monoclonal antibodies results in aggregates of IgG and HRPO molecules, with a molecular weight of 500,000 to 1,000,000 resulting in decreased biological activity.

Time-resolved fluoroimmunoassays require high working standards. The six-cycle washing is critical because the high input activity in labels (millions of cps) must be reduced to a few hundred of cps in negative specimens. The label must be pipetted precisely into the right place in the well to avoid nonspecific binding of the label. In addition, the relatively expensive fluorometer and difficulties in obtaining the Eu-chelate for labeling have limited the use of TR-FIA.

Enzyme Immunofiltration Staining Assay

In this assay, the infected cells in the clinical specimens are filtered on glass Fiber disk placed in a 48-well plate with a vacuum manifold (Cleveland and

TABLE 9. A representative monoclonal one-step time-resolved fluoroimmunoassay for the detection of respiratory syncytial virus (RSV), influenza A and B, parainfluenza types 1, 2, and 3, and adenovirus in nasopharyngeal aspirates from five patients with acute respiratory disease

Reference antigen, cut-off value, and specimen no.	RSV	Influenza		Parainfluenza			Adenovirus	Conclusion
		A	B	1	2	3		
Reference								
1,000 ng/ml[a]		140.534[c]						
100 ng/ml[b]	8.712		12.251	15.467	10.490	22.670	33.320	
0 ng/ml	534		496	1.454	455	767	721	588
Cut-off	1.157		1.391	4.544	1.632	2.169	1.998	1.360
Specimen no.								
6233	486	441	2.880	387	945	701	585	Negative
6235	513.760	662	1.558	10.72	1.823	789	496	RSV
6237	540	508	1.518	1.195	770	769	672	Negative
6243	8.584	399	1.041	439	663	813	426	RSV
6318	369	431	356.536	1.009	667	873	399	Influenza B

[a] Nonpurified cell lysate antigen.
[b] Purified virus.
[c] Mean counts per second value of duplicates.

TABLE 10. Monthly results of respiratory viral antigen detection by monoclonal one-incubation time-resolved fluoroimmunoassay in nasopharyngeal aspirates

Month	No. of specimens tested	Positive No.	Positive (%)	RSV	Influenza A	Influenza B	Parainfluenza 1	Parainfluenza 2	Parainfluenza 3	Adenovirus
January 1987	334	44	(13)	1	32			1	1	9
February	284	36	(13)	16	3		1		6	10
March	323	76	(24)	41		1	2		22	10
April	230	64	(28)	37			6		15	6
May	246	55	(22)	21			8		13	13
June	191	52	(27)	38			2		4	8
July	189	32	(17)	15			2	1	1	13
August	223	24	(11)	2			4			18
September	226	11	(5)	3			1			7
October	380	29	(8)	17				3		9
November	471	106	(23)	94				1		11
December	672	253	(38)	225				4		24
Total	3,769	782	(21)	510	35	1	26	10	62	138

Richman, 1987). Viral antigens are detected by short incubations with biotinylated monoclonal antibodies, streptavidin-HRPO conjugate, and the substrate aminoethylcarbazole. Infected cells and cell debris are stained red, and the test is read with a microscope. The total time required is 30 min. As few as two virus-infected cells per swab can be detected. Many viruses including herpes simplex, varizella zoster, adenoviruses, and chlamydial antigens have been detected in clinical specimens or cell cultures.

Latex Agglutination Tests

The sensitivities of latex agglutination test used for viral antigen detection are not as high as those of solid-phase immunoassays. However, if the specimen contains large amounts of viral antigen, latex agglutination tests can be used. These simple and very rapid tests are based on small beads coated with viral antibody. The antigen, preferably virus particles in the specimen, forms bridges between the beads, which are agglutinated, and the reaction can be read macroscopically. Such tests have been developed for the detection of rotaviruses (Cevenini et al., 1983; Haikala et al., 1983; Julkunen et al., 1985; Pai et al., 1985; Sanekata et al., 1981) and adenoviruses (Bricout et al., 1987; Grandien et al., 1987) in stool. The sensitivities are 80 to 90% compared with the most sensitive EIAs and 90 to 95% compared with electron microscopy. The specificity is almost 100% when calculated from the specimens that can be tested by this method. However, about 5 to 15% of stool specimens agglutinate control beads coated by normal serum and these specimens must be tested by EIA. New modifications in some commercial kits have reduced the agglutination reactions with control beads.

Recent developments in commercial latex agglutination tests include slides where beads are dried on the spot, and the only step in the test is addition of diluted stool specimens. Previously, the reading of the weak positive specimens required considerable expertise, but in the new tests the evaluation of the results is easier. These tests are quite practical in smaller hospital outpatient laboratories and can even be practiced in the doctor's office.

Future Prospects

A question often asked is whether viral antigen assays or nucleic acid hybridization will be used in the future for the detection of virus directly in clinical specimens. In the next few years, antigen detection, when available and most likely with monoclonal one-incubation (one-wash) assays, will find increasing use. However, the type-specific assay of papillomaviruses is an example where DNA hybridization has no competition from antigen assays. In group- or subgroup-specific enterovirus assays, the prospect that it will be done with nucleic acid hybridization is better than with antigen detection. In addition, we must realize that immunoassays have been used in viral antigen detection for almost 20 years with constant improvements in the tests as compared with only a few years in the use of hybridization technology. We can expect similar improvements in the DNA assays both in the form of increased sensitivity and in more simple and faster tests.

Other questions many investigators ask concern the future developments in rapid diagnosis of respiratory infections. Should we use immunofluorescence or solid phase immunoassays in the detection of respiratory viral antigens? What is the role of antibody assays in the retrospective diagnosis of these infections in relation to antigen detection?

We suggest that choice depends on the previous experience of the laboratory, the number of specimens tested daily, how the transportation of specimens is organized, and the available expertise for IF microscopy. If the laboratory has a large number of specimens and previous experience with EIAs, solid-phase immunoassays may be more practical, whereas in small hospital laboratories, IF may be the better choice. The diagnostic efficiency of these two techniques is almost the same if expertise and good reagents are available. Further developments in respiratory viral antigen detection will include coronavirus and rhinovirus assays, preferably group- or subgroup-specific rhinovirus assays.

Serologic diagnosis of respiratory viral infections by IgG EIA in paired serum specimens is slightly more efficient than antigen detection, with the exception of in adenovirus infections. We suggest that antigen detection is the primary diagnostic method, but that EIA IgG serology should be performed in selected cases if antigen detection is negative and the continued search for an etiologic agent is justified.

Further improvements in antigen detection will be the addition of calicivirus, astrovirus, and coronavirus antigen detections in gastroenteritis virus assays; the addition of a cytomegalovirus assay in antigen detection in genital swabs; increased sensitivity of herpes simplex virus assays (particularly for screening of nonsymptomatic pregnant women); and increased sensitivity of HIV antigen assays.

Latex tests do not compete with the more sensitive assays, but they may find increased use in small hospital laboratories, doctor's offices, and field studies.

Monoclonal antibodies may replace polyclonal immune sera in the near future, but only larger diagnostic centers can afford to produce their own monoclonal antibodies even though they have the expertise to build the immunoassays or immunofluorescence tests if reagents are available. For this reason, the commercial availability of antigen detection kits based on monoclonal antibodies as well as monoclonal reagents is highly desirable.

Laboratories that produce their own monoclonal antibodies should realize that for each specific use, monoclonal antibodies must be produced and screened separately. Reagents that have been produced for immunofluorescence usually are not suitable for solid-phase immunoassays. The screening assays must be designed to be as close as possible to the final use of the monoclonal antibodies. A common mistake is that monoclonal antibodies that will be used in solid-phase antigen assays are screened against viral antigen bound on solid phase. In the final assays, the viral antigen in the specimen is never bound directly on solid phase; there is always a catching antibody first on solid phase. Once the right type of monoclonal antibodies are available, viral antigen detection will be greatly improved independently of the technology used.

Literature Cited

Åkerlind, B., and E. Norrby. 1986. Occurrence of respiratory syncytial virus subtypes A and B strains in Sweden. J. Med. Virol. **19**:241–247.

Allain, J-P., Y. Laurian, D. A. Paul, D. Senn, D. Bastit, C. Bosser, C. Gazengel, C. Guerois, J. J. Larrier, M. F. Torchet, and F. Verroust. 1986. Serological markers in early stages of human immunodeficiency virus infection in haemophilials. Lancet **2**:1233–1236.

Balkovic, E. S., and G. D. Hsiung. 1985. Comparison of immunofluorescence with commercial monoclonal antibodies to biochemical and biological techniques for typing herpes simplex virus isolates. J. Clin. Microbiol. **22**:870–872.

Bell, D. M., E. E. Walsh, J. F. Hruska, K. C. Schnabel, and C. B. Hall. 1983. Rapid detection of respiratory syncytial virus with a monoclonal antibody. J. Clin. Microbiol. **17**:1099–1101.

Berg, R. A., S. I. Rennard, B. R. Murphy, R. H. Yolken, R. Dolin, and S. E. Straus. 1980. New enzyme immunoassays for measurement of influenza A/Victoria/3/75 virus in nasal washes. Lancet **1**:851–853.

Biegeleisen, J. Z. Jr., L. V. Scott, and V. Lewis, Jr. 1959. Rapid diagnosis of herpes simplex virus infection with fluorescent antibody. Science **129**:640–641.

Bricout, F., A. Garbarg-Chenon, and J. C. Nicolas. 1987. Essais de détection des adénovirus humains dan les selles au moyen du test adenolex. Feuillets de biologie **28**:31–33.

Burchiel, S. W., J. R. Billman, and T. R. Alber. 1984. Rapid and efficient purification of mouse monoclonal antibodies from ascites fluids using HPLC. J. Immunol. Methods **69**:33–42.

Cevenini, R., F. Rumpianesi, R. Mazzaracchio, M. Donati, E. Falcieri, and R. Lazzari. 1983. Evaluation of a new latex agglutination test for detecting human rotavirus in faeces. J. Infect. **7**:130–133.

Chao, R. K., M. Fishaut, J. D. Schwartzman, and K. McKintosh. 1979. Detection of respiratory syncytial virus in nasal secretions from infants by enzyme-linked immunosorbent assay. J. Infect. Dis. **139**:483–486.

Cleveland, P. H., and D. D. Richman. 1987. Enzyme filtration staining assay for immediate diagnosis of herpes simplex virus and varicella-zoster virus directly from clinical specimens. J. Clin. Microbiol. **25**:416–420.

Coons, A. H., H. J. Creech, and R. N. Jones. 1941. Immunological properties of an antibody containing a fluorescent group. Proc. Soc. Exp. Biol. Med. **47**:200–202.

Emanuel, D., J. Peppard, P. Stover, J. Gold, D. Amstrong, and V. Hammerling. 1986. Rapid immunodiagnosis of cytomegalovirus pneumonia by bronchoalveolar lavage using human and murine monoclonal antibodies. Ann. Intern. Med. **104**:476–481.

Engvall, E., and P. Perlmann. 1972. Enzyme-linked immunosorbent assay, ELISA: quantitation of specific antibodies by enzyme-labelled anti-immunoglobulin in antigen coated tubes. J. Immunol. **109**:129–135.

Forghani, B., N. J. Schmidt, and E. H. Lennette. 1974. Solid-phase radioimmunoassay for identification of herpesvirus hominis types 1 and 2 from clinical materials. Appl. Microbiol. **28**: 661–667.

Fung, J. C., J. Shanley, and R. C. Tilton. 1985. Comparison of the detection of herpes simplex virus in clinical specimens with herpes simplex virus-specific DNA probes and monoclonal antibodies. J. Clin. Microbiol. **22**:748–753.

Gallo, D. 1983. Uses of immunofluorescence in diagnostic virology. Am. J. Med. Technol. **49**:157.

Gardner, P. S., and J. McQuillin. 1968. Application of the immunofluorescent antibody technique in the rapid diagnosis of respiratory syncytial virus infection. Br. Med. J. **3**:340–343.

Gardner, P. S., and J. McQuillin. 1980. Rapid virus diagnosis: applications of immunofluorescence, 2nd ed. Butterworths, London.

Gardner, P. S., J. McQuillin, M. M. Black, and J. Richardson. 1968. The rapid diagnosis of herpes virus hominis infections in superficial lesions by immunofluorescent antibody techniques. Br. Med. J. **4**:89–92.

Goldwasser, R. S., R. E. Kissling, T. R. Carski, and T. S. Hosty. 1959. Fluorescent antibody staining of rabies virus antigens in the salivary glands of rabid animals. Bull. W.H.O. **20**:579–588.

Goudsmit J., F. de Wolf, D. A. Paul, L. G. Epstein, J. M. Lange, W. J. Krone, H. Speelman, E. Ch. Wolters, J. Van Der Noorda, J. M. Oleske, H. Van Der Helm, and R. A. Coutinho. 1986. Expression of human immunodeficiency virus antigen (HIV-Ag) in serum and cerebrospinal fluid during acute and chronic infection. Lancet **2**:177–180.

Grandien, M., C-A. Pettersson, L. Svensson, and I. Uhnoo. 1987. Latex agglutination test for adenovirus diagnosis in diarrheal disease. J. Med. Virol. **23**:311–316.

Hackman, R. C., D. Myerson, J. D. Meyers, H. M. Shulman, G. E. Sale, L. C. Goldstein, M. Rastetter, N. Flournoy, and E. D. Thomas. 1985. Rapid diagnosis of cytomegaloviral pneumonia by tissue immunofluorescence with a murine monoclonal antibody. J. Infect. Dis. **151**:325–329.

Haikala, O. J., J. O. Kokkonen, M. K. Leinonen, T. Nurmi, R. Mäntyjärvi, and H. K. Sarkkinen. 1983. Rapid detection of rotavirus in stool by latex agglutination: comparison with radioimmunoassay and electron microscopy and clinical evaluation of the test. J. Med. Virol. **11**:91–97.

Halonen, P., H. Bennich, E. Torfason, T. Karlsson, B. Ziola, M-T. Matikainen, E. Hjertsson, and T. Wesslen. 1979. Solid-phase radioimmunoassay of serum immunoglobulin A antibodies to respiratory syncytial virus and adenovirus. J. Clin. Microbiol. **10**:192–197.

Halonen, P., G. Obert, and J. C. Hierholzer. 1985. Direct detection of viral antigens in respiratory infections by immunoassays: a four year experience and new developments in medical virology IV. (L. M. de la Manza and E. M. Peterson eds.), Lawrence Erlbaum Associates, Hillsdale, New Jersey.

Hanson, C. V. 1985. Immunofluorescence and related techniques, p. 119–133. *In* E. H. Lennette (ed.). Laboratory diagnosis of viral infections. Marcel Dekker, New York and Basel.

Hemmilä, I., S. Dakubu, V-M. Mukkula, H. Siitari, and T. Lövgren. 1984. Europium as a label in time-resolved immunofluorometric assays. Analyt. Biochem. **137**:335–343.

Hunter, W. M., and F. C. Greenwood. 1962. Preparation of Iodine-131 labeled human growth hormone of high specific activity. Nature **194**:495–496.

Julkunen, I., J. Savolainen, A. Hautanen, and T. Hovi. 1985. Detection of rotavirus in foecal specimens by enzyme immunoassay, latex agglutination and electron microscopy. Scand. J. Infect. Dis. **17**:245–249.

Kalica, A. R., R. H. Purcell, M. M. Sereno, R. G. Wyatt, H. W. Kim, R. M. Chanock, and A. Z. Kapikian. 1977. A microtiter solid phase radioimmunoassay for detection of the human reovirus-like agent in stools. J. Immun. **118**:1275–1279.

Köhler, G., and C. Milstein. 1975. Continuous cultures of fused cells secreting antibody of predefined specificity. Nature **256**:495–497.

Lander, J. J., H. J. Alter, and R. H. Purcell. 1971. Frequency of antibody to hepatitis-associated antigen as measured by a new radioimmunoassay technique. J. Immun. **106**:1116–1171.

Ling, C. M., and L. R. Overby. 1972. Prevalence of hepatitis B virus antigen revealed by direct radioimmune assay with ^{125}I-antibody. J. Immun. **109**:834–841.

McQuillin, J., P. S. Gardner, and R. McGuckin. 1970. Rapid diagnosis of influenza by immunofluorescent techniques. Lancet **2**:690–694.

McQuillin, J., C. R. Madeley, and A. P. Kendal. 1985. Monoclonal antibodies for the rapid diagnosis of influenza A and B virus infections by immunofluorescence. Lancet **2**:911–914.

Middleton, P. J., M. D. Holdaway, M. Petric, M. T. Szymanski, and J. S. Tam. 1977. Solid-phase radioimmunoassay for the detection of rotavirus. Infect. Immun. **16**:439–444.

Nerurkar, L. S., A. J. Jacob, D. L. Madden, and J. L. Sever. 1983. Detection of genital herpes simplex infections by a tissue culture-fluorescent-antibody technique with biotin-avidin. J. Clin. Microbiol. **17**:149–154.

Olding-Stenkvist, E. and M. Grandien. 1976. Early diagnosis of virus-caused vesicular rashes by immunofluorescence on skin biopsies. I. Varicella zoster and herpes simplex. Scand. J. Infect. Dis. **8**:27–35.

Ørstavik, I., M. Grandien, P. Halonen, P. Arstila, C. H. Mordhost, A. Hornsleth, T. Popow-Kraupp, J. McQuillin, P. S. Gardner, J. Almeida, F. Bircoit, and A. Marques. 1984. Viral diagnosis using the rapid immunofluorescence technique and epidemiological implications of acute respiratory infections among children in different European countries. Bull. W.H.O. **62**(2)307–313.

Orton, P. W., J. C. Huff, M. G. Tonnesen, and W. Weston. 1984. Detection of herpes simplex viral antigen in skin lesions of erythema multiforme. Ann. Intern. Med. **101**:48–50.

Pai, C. H., M. S. Shahrabadi, and B. Ince. 1985. Rapid diagnosis of rotavirus gastroenteritis by a commercial latex agglutination test. J. Clin. Microbiol. **22**:846–850.

Paul, D. A., and L. A. Falk. 1986. Detection of HTLV-III antigens in serum. J. Cell Biochem. Supplement 10A.

Pelkonen, J. 1982. A gene linked to the heavy-chain allotype 1α (1a) controls for the expression of idiotype OX-R1 in the rat. Immunol. Lett. **5**:19–22.

Pereira, L., D. Dondero, D. Gallo, V. Devlin, and J. D. Woodie. 1982. Serological analysis of herpes simplex virus types 1 and 2 with monoclonal antibodies. Infect. Immun. **35**:363–367.

Routledge, E. G., J. McQuillin, A. C. R. Samson, and G. L. Toms. 1985. The development of monoclonal antibodies to respiratory syncytial virus and their use in diagnosis by indirect immunofluorescence. J. Med. Virol. **15**:305–320.

Sanekata, T., Y. Yoshida, and H. Okada. 1981. Detection of rotavirus in faeces by latex agglutination. J. Immun. Methods 41:377–385.

Sarkkinen, H. K. 1981. Human rotavirus antigen detection by enzyme-immunoassay using antisera against Nebraska calf diarrhea virus. J. Clin. Pathol. 34:680–685.

Sarkkinen, H. K., P. E. Halonen, and P. P. Arstila. 1979. Comparison of radioimmunoassay and electron microscopy for the detection of human rotavirus. J. Med. Virol. 4:255–260.

Sarkkinen, H. K., P. E. Halonen, P. P. Arstila, and A. A. Salmi. 1981a. Detection of respiratory syncytial, parainfluenza type 2, and adenovirus antigens by radioimmunoassay and enzyme immunoassay on nasopharyngeal specimens from children with acute respiratory disease. J. Clin. Microbiol. 13:258–265.

Sarkkinen, H. K., P. E. Halonen, and A. A. Salmi. 1981b. Detection of influenza A virus by radioimmunoassay and enzyme-immunoassay from nasopharyngeal specimens. J. Med. Virol. 7:213–220.

Sarkkinen, H. K., P. E. Halonen, and A. A. Salmi. 1981c. Type-specific detection of parainfluenza viruses by enzyme-immunoassay and radioimmunoassay in nasopharyngeal specimens of patients with acute respiratory disease. J. Gen. Virol. 56:59–57.

Sarkkinen, H. K., H. Tuokko, and P. E. Halonen. 1980. Comparison of enzyme immunoassay and radioimmunoassay for detection of human rotaviruses and adenoviruses from stool specimens. J. Virol. Methods 1:331–341.

Schmidt, N. J., E. H. Lennette, J. D. Woodie, and H. H. Ho. 1965. Immunofluorescent staining in the laboratory diagnosis of varicella-zoster virus infections. J. Lab. Clin. Med. 66:403–412.

Shalit, I., P. A. McKee, H. Beauchamp, and J. L. Waner. 1985. Comparison of polyclonal antiserum versus monoclonal antibodies for the rapid diagnosis of influenza A virus infections by immunofluorescence in clinical specimens. J. Clin. Microbiol. 22:877–879.

Siitari, H., I. Hemmilä, E. Soini, T. Lövgren, and V. Koistinen. 1983. Detection of hepatitis B surface antigen using time-resolved fluoroimmunoassay. Nature 301:258–260.

Soini, E., and H. Kojola. 1983. Time-resolved fluorometer for lanthanide chelates—a new generation of nonisotopic immunoassays. Clin. Chem. 29:65–68.

Walls, H. H., K. H. Johansson, M. W. Harmon, P. E. Halonen, and A. P. Kendal. 1986. Time-resolved fluoroimmunoassay with monoclonal antibodies for rapid diagnosis of influenza infections. J. Clin. Microbiol. 24:907–912.

Waris, M., S. Nikkari, P. Halonen, I. Kharitonenkov, and A. Kendal. In press. Eu-chelate and horseradish peroxidase labelled monoclonal antibodies in detection of influenza viruses. Proceedings of Fifth International Symposium on Rapid Methods and Automation in Microbiology and Immunology, Florence, 4-6 November 1987.

W.H.O. Scientific Group. 1981. Rapid laboratory techniques for the diagnosis of viral infections. Report of a scientific group. Technical Report Series No. 661.

Voller, A. 1986. Enzyme immunoassays for parasitic diseases. Trans. Roy. Soc. Trop. Med. Hyg. 70:98–106.

Yolken, R. H., H. W. Kim, T. Clem, R. G. Wyatt, A. R. Kalica, R. M. Chanock, and A. Z. Kapikian. 1977. Enzyme-linked immunosorbent assay (ELISA) for detection of human reovirus-like agent of infantile gastroenteritis. Lancet 2:263–266.

Ziegler, T. 1984. Detection of Varicella-zoster viral antigens in clinical specimens by solid-phase enzyme immunoassay. J. Infect. Dis. 150:149–154.

Ziegler, T., and P. Halonen. 1985. Rapid detection of herpes simplex and varicella-zoster virus antigens from clinical specimens by enzyme immunoassay. Antiviral Res. Suppl. 1:107–110.

Ziegler, T., V. Hukkanen, P. Arstila, P. Auvinen, A. Jalava, and T. Hyypiä. 1985. Typing of herpes simplex virus isolates with monoclonal antibodies and by nucleic acid spot hybridization. J. Virol. Methods 12:169–177.

CHAPTER 5

Antibody Detection

KENNETH L. HERRMANN

Subject: Serodiagnosis (use of known antigens as reagents to detect antibodies in patient sera for diagnosing recent or past viral infection).

Serodiagnostic Principles: Mechanism of antibody production, response patterns to primary and recurrent infections, and significance of antibody to immunity in viral diseases.

Laboratory Methods: Neutralization, complement fixation, hemagglutination inhibition, immune adherence hemagglutination, passive agglutination, hemolysis in gel, radioimmunoassay, enzyme immunoassay, indirect immunofluorescence/fluoroimmunoassay, and Western immunoblotting.

Interpretation of Test Results: Significance of presence of antibody or change in antibody level. Value of immunoglobulin type-specific assays in viral serodiagnosis.

Problems and Pitfalls: Insensitive and nonspecific serodiagnostic tests resulting from heterologous cross-reactions, immunologic interference, substandard reagents, or improper test performance.

Introduction

Antibody detection plays an important role in the investigation of viruses as causes of human illness. Viral serology provides both clinicians and epidemiologists with a powerful tool for diagnosing infection in an individual patient, for determining the susceptibility of a person or population to a specific virus, and for assessing the prevalence of a particular virus in a community. Technologic advances during the past several years have led to more rapid, sensitive, and accurate tests for detecting and measuring antiviral antibodies. The development of automated technology has made these tests easier to perform and has encouraged more clinical laboratories to use them.

The term *serodiagnosis* denotes diagnosis based on procedures that include known antigens as reagents to detect and measure antibodies in patient serum. To use such tests to diagnose an infectious disease, one must suspect a specific causative agent based on clinical or epidemiologic data. One also must have the proper antigen to detect antibody to this agent.

Immunoglobulin Specificity

The most basic feature of the immune response of the host to a foreign antigen is its specificity for that particular antigen. Our understanding of antibody specificity at the structural and the genetic level has been enhanced greatly during the past few years with the application of monoclonal antibodies to the antigenic and structural characterization of viral proteins and with immunochemical or genetic mapping of the antibody-combining sites on the immunoglobulin (Ig) molecule.

The mechanism by which antibodies are formed has been debated for years. It is now generally accepted that each person is endowed with a large pool of B lymphocytes that possess surface receptors capable of responding to or being triggered by an antigen (Mims and White, 1984). The initial step in antibody formation is phagocytosis of the invading pathogen, the viral particle, by macrophages and the presentation of specific viral antigens in some form to B cells. When a B cell encounters the antigen, it binds the antigens with receptors complementary to any of the several antigenic determinants (epitopes) on that antigen. After receiving the appropriate antigen-specific and nonspecific signals from helper T cells, the B cell responds by dividing and differentiating into antibody-secreting plasma cells. Antibodies are either secreted into the blood or lymphatic fluid by plasma cells or remain attached to the surface of lymphocytes or other effector cells. Each clone of plasma cells secretes antibody of only a single Ig class corresponding to the particular receptors it expresses and specific to the epitope that triggered it. The binding strength of a specific antibody receptor with its corresponding epitope is referred to as the affinity of the reaction.

A single virus may contain many different antigens that the host will recognize as foreign, so infection with one virus may cause many different antibodies to appear. In addition, some antigenic determinants of a virus may not be available for recognition by the host until the virus has been disrupted by enzymatic action. Certain internal, or core, antigens of a virus may not be "visible" to the host immune system until the virion has been broken down to reveal these internal components.

When large amounts of antigen are present, as in the early stages of an acute infection, the antigen-reactive B cells may be triggered even if their receptors fit the epitope with relatively poor binding strength. Antibody secreted by the resulting plasma cell clones are thus likely to have low affinity for the antigen. Later, when only small amounts of antigen are present, B cells with receptors of high antigen binding strength are selected; hence the affinity of the antibody secreted increases correspondingly, sometimes 100-fold.

Each antibody-producing cell makes only one specificity of antibody. The spleen and lymph nodes receive foreign antigens via the blood or lymphatic system and synthesize humoral antibodies, mainly of the IgM and IgG classes. On the other hand, the interstitial lymphoid follicles of the respiratory and alimentary tracts, such as the tonsils and Peyer's patches, receive antigens directly from the overlying epithelial cell and secrete local antibodies mainly of the IgA class. Viremic infections lead to the production of both specific IgM and IgG antibodies in most persons. Localized virus infections, however, often stimulate little or no increase of humoral IgM or IgG antibody levels. The larger number of antigen binding sites on the IgM molecules may help to clear the invading pathogen more quickly, even though each individual antigen-binding site may not be the most

efficient for attaching the antigen. Over time, the cells producing IgM switch to producing IgG. The mechanism for this switch involves genes controlling the variable regions of the heavy chain of the IgM molecule; these genes recombine with genes controlling the constant regions of IgG so that the IgG produced is of the same specificity as the IgM. The IgG antibody often has greater avidity for the antigen and is able to bind complement.

Antibody Responses to Viral Infections

Traditional serodiagnosis is based on the detection of specific antibody in the serum of the infected person. Circulating antiviral antibodies, particularly neutralizing antibodies, are widely accepted as prima facie evidence of past infection with the particular virus. Current or very recent infection is usually detected by an initial antibody response or an increase in the level of specific antibody. Most viruses induce detectable levels of specific antibodies after a primary infection and may boost the titer after reinfection or reactivation of latent infection. Serodiagnosis of a viral infection generally requires paired acute- and convalescent-phase serum samples. Confirmation of a suspected virus infection usually requires demonstrating a significant rise of specific antibody, not merely the presence of antibody to the viral agent.

In some situations, however, an antibody rise may not be detected in conjunction with an acute viral infection. Such exceptions include 1) cases in which antibodies had reached peak levels before the acute-phase serum sample was collected, 2) infections in immunocompromised hosts, 3) superficial infections, such as respiratory infections, that fail to induce a humoral antibody response despite significant illness, and 4) acute infections acquired in the presence of passively transferred antibody. Congenital and neonatal infections may be difficult to diagnose by antibody rise because high levels of placentally transferred maternal antibody are often present in the newborn's serum and may mask the development of specific antibodies by the newborn infant.

As variables inherent in serologic procedures can result in substantial differences in the results obtained from one sample tested at different times, acute- and convalescent-phase samples must be tested together in the same run for differences in antibody levels to be meaningfully interpreted. In tests that use doubling dilutions of the serum, a fourfold or greater rise in antibody titer has traditionally been considered significant, whereas a twofold difference is considered within the technical error of the technique (Hall and Felker, 1970). In tests using other methods of quantitating antibody, significant differences in antibody levels must be established independently (Wood and Durham, 1980). Frequently, a critical ratio of convalescent- to acute-phase antibody levels is used to define a significant antibody rise.

As mentioned previously, each virus generally possesses several antigens, some of which are specific for the particular strain of virus and others that are shared by related viruses. Infection with a particular virus commonly stimulates antibody responses to several viral antigens, and these responses may occur at different times. Such temporal differences in antibody responses may be of considerable diagnostic value for certain viral infections such as hepatitis A and B.

Primary Infection

There are considerable differences between the humoral immune response following a primary, or initial, infection with a particular virus and that following a secondary infection or reinfection with the virus. In general, IgM antibodies appear earlier than IgG antibodies in the primary response (Fig. 1). In most primary viral infections, specific antibodies appear in the blood within several days, then rise in titer, and reach a plateau usually within a few weeks. Specific IgM has a half-life of only about 5 days and generally does not persist longer than 2 to 3 months after the acute viral infection. IgG antibodies, on the other hand, usually persist for many years, often for life. Because of the transient nature of the IgM response, the detection of specific IgM antibodies has often been used as an indication of recent or active infection with the virus. Some investigators, however, have reported the persistence of IgM antibody for several months or even years after the onset of primary infection (Al-Nakib et al., 1975; Pass et al., 1983). Such reports are more common for viruses that characteristically produce chronic or persistent infections, such as cytomegalovirus. Others have reported the appearance of specific IgM antibody to agents other than the one responsible for the current infection (Morgan-Capner et al., 1983). Such true heterologous IgM responses are probably rare, but should be considered when the results of IgM tests are used in determining clinical management.

Exogenous Reinfection

The antibody response to viral reinfection is often quite brisk and usually begins within a day or two after onset of the infection (Fig. 2). Antigen-reactive B lymphocytes, on reexposure to the same antigen, produce larger amounts of specific antibody, mainly IgG, after a delay of only a day or two. Little, if any, specific IgM antibody response is detected following most viral reinfections.

Reactivation of Latent Infection

The antibody response following reactivation of a latent viral infection is less predictable than that following either a primary acute infection or an exogenous reinfection. Examples of reactivated infections are herpes zoster (or shingles) and herpes labialis (or fever blisters). In some instances, antibody increases are brisk. However, antibody levels may not change significantly in others; thus, serodiagnostic results may be misleading.

Measurement of Viral Antibodies

Methods used to detect antiviral antibodies can generally be separated into three groups: 1) methods that measure the ability of an antibody to interfere with some special viral function; 2) methods that depend on the antibody, when reacted with antigen, to perform some non-virus-related function; and 3) methods that directly detect the interaction of the antibody with antigen. The first group includes neutralization and hemagglutination inhibition as-

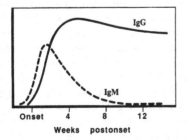

FIG. 1. Immune response in primary virus infection. Note early appearance and rapid decline of IgM compared with the appearance and persistence of IgG antibody.

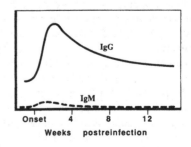

FIG. 2. Immune response in virus reinfection. Note the near absence of an IgM antibody response compared with the rapid booster response of IgG antibody.

says; the second includes complement fixation, passive (or indirect) hemagglutination, immune adherence hemagglutination, and latex agglutination tests; and the third includes enzyme immunoassays, radioimmunoassays, immunofluorescence tests (including fluoroimmunoassays), and Western immunoblotting.

An antibody to a particular antigen may be measurable by a number of different assays. However, different methods for the same agent may measure a different antibody; thus, the results of an antibody assay for a particular virus as measured by one method may not correlate with the results from another method. Each assay may show quantitative as well as temporal differences in antibody response following infection, especially when the assay is run with many different serum samples, representing several variables.

A general overview of the basic principles of the more commonly used methods for detecting and quantifying antiviral antibodies follows in the next sections. Immunodiagnostic tests commonly used for specific viral diseases are listed in Table 1. See the chapters dealing with each virus for additional information regarding the application, performance, and interpretation of the serodiagnostic tests.

Neutralizing Antibody Assay

The neutralization (Nt) test is used to measure antibody to a wide variety of viral agents and represents one of the most sensitive, specific, and clinically important serologic tests available. Nt antibodies persist well beyond the initial illness, and when detected in the absence of symptoms of recent infection with a specific virus, are widely accepted as evidence of immunity to the virus. As Nt antibody can persist in the host for many years, Nt tests are useful in seroepidemiologic studies to determine which viral agents have infected a given population in the past.

Classic viral neutralization occurs when antibody binds to the virion and prevents infection of a susceptible host cell. The precise mechanism of neutralization is not fully understood. Nt antibodies are directed against particular epitopes on accessible surface proteins of the outer capsid or envelope of the virion. Nt antibodies may interfere with viral infectivity by either blocking attachment of the virus to the host cell receptors or by preventing penetration or uncoating of the virion (Mims and White, 1984). The avidity of Nt antibodies may be substantially influenced by the presence of complement and by the degree of glycosylation of the viral glycoprotein antigens (Alexander and Elder, 1984; Sissons and Oldstone, 1980).

The basic procedure for assaying the virus-neutralizing activity of a serum consists of mixing the serum with known virus, incubating the mixture under appropriate conditions, and injecting the mixture into a susceptible host system in which the presence of residual unneutralized virus can be detected. Most virus Nt antibody assays are carried out in cell cultures in vitro, although in vivo hosts, such as mice or embryonated eggs, may be required for some viruses that do not grow well in cell culture.

Serum specimens are generally assayed for Nt antibody content by testing serial dilutions of the serum against a standard dose (usually 100 $TCID_{50}$) of the virus. This is known as the "constant virus-varying serum" technique; it is more useful for demonstrating significant increases in Nt antibody than is the "constant serum-varying virus" procedure, in which a fixed dilution of serum is tested against serial dilutions of virus. In the constant virus-varying serum system, the antibody titer is expressed as the highest serum dilution that neutralizes the test dose of virus. Results of the constant serum-varying virus Nt test are expressed logarithmically as a neutralization index (NI). The NI is the difference in virus neutralizing capacity (in $\log_{10} LD_{50}$) between the unknown and normal control serum samples or between acute- and convalescent-phase serum samples. A log NI of 1.7 is generally accepted as the lower limit for a "positive" result.

The following general procedures can be adapted for detecting Nt antibodies to many human viral pathogens.

NEUTRALIZING ANTIBODY ASSAYS IN TUBE MONOLAYER CELL CULTURES

1. The serum specimen (either known immune serum or unknown serum to be assayed for Nt antibody) is inactivated at 56°C for 30 min to destroy heat-labile, nonspecific viral inhibitory substances. As some Nt reactions are complement dependent, a source of complement (usually fresh guinea pig serum) is added to the maintenance medium for the cultures.
2. Appropriate serial serum dilutions are prepared in either Hanks basic salt solution (BSS) or the maintenance medium to be used on the cell cultures.
3. The virus is diluted to contain 100 $TCID_{50}$ in a volume of 0.1 ml (as determined by previous titration of the virus in the selected cell system). Virus endpoint titers are calculated by the methods of Reed and Muench (1938) or Karber (1931). Virus dilutions are prepared in the same diluent used for the serum dilutions. Filtering the test virus before titration will remove aggregated virus and may substantially enhance virus neutralization.

TABLE 1. Serodiagnostic tests for common viral, rickettsial, and chlamydial diseases of humans[a]

Disease agent	Available tests	Comments
Arenaviruses		
Lassa	IFA, Nt	Testing limited to selected reference centers
Lymphocytic choriomeningitis	CF, IFA, Nt	IFA recommended over CF for diagnosis
Chlamydiae		
Psittacosis–LGV	CF, IFA, EIA	IFA more sensitive than CF or EIA
Enteric viruses		
Poliovirus	CF, Nt	For confirmation of diagnosis
Nonpolio enteroviruses	Nt, EIA	Heterotypic responses often make serologic tests difficult to interpret
Rotavirus	CF, IFA, EIA	Confirmation of diagnosis; antigen detection preferred for diagnosis
Norwalk-like agents	EIA	Sources of test antigen limited
Exanthem viruses		
Rubella	HAI, EIA, IFA, LA, PHA	For diagnosis or immune screening
Rubeola	CF, HAI, IFA, EIA, Nt	For diagnosis or immune screening
Parvovirus	EIA	Sources of diagnostic antigen limited
Poxviruses	CF, HAI, EIA, Nt	Viral isolation preferred for diagnosis
Hepatitis viruses		
Hepatitis A	RIA, EIA	IgM anti-HAV for diagnosis
Hepatitis B	RIA, EIA	IgM anti-HBc diagnostic; antigen detection preferred for diagnosis
Herpesviruses		
Herpes simplex 1 and 2	CF, PHA, EIA, IFA, Nt	Serologic tests of limited diagnostic value
Cytomegalovirus	CF, PHA, EIA, IFA	Serologic tests of limited diagnostic value
Epstein–Barr virus	Heterophil, IFA	IFA tests for anti-VCA, anti-EBNA, and anti-EA of limited diagnostic value
Varicella-zoster	CF, IFA, FAMA	CF for diagnosis only; FAMA generally acknowledged to be most sensitive; antigen detection preferred for diagnosis
Retroviruses		
HIV (HTLV-III, LAV, ARV)	EIA, immunoblot	EIA for immune screening only
HTLV-I	EIA, immunoblot	Heterologous responses common with retroviruses
Rhabdoviruses		
Rabies	RFFIT, IFA	For determining immune status following rabies vaccination and for confirming diagnosis
Respiratory viruses		
Adenovirus	CF, HAI, EIA, Nt	CF and EIA group-specific; good for diagnosis
Coronavirus	CF, HAI, Nt	
Influenza	CF, HAI, EIA	Need antigen of current strain for optimal sensitivity; EIA most specific
Mumps	CF, HAI, IFA, EIA	Heterologous responses common with other paramyxovirus infections
Parainfluenza	CF, HAI, EIA	Heterotypic responses common
Respiratory syncytial	CF, HAI, EIA	EIA more sensitive than CF or HAI for diagnosis
Rhinovirus	Nt	Heterotypic responses occur
Rickettsiae		
RMSF	CF, IFA	
Typhus	CF, IFA	
Q fever	CF, IFA	
Arboviruses		
SLE, EEE, VEE, WEE	CF, HAI, Nt	Heterologous reactions common with CF and HAI tests

[a] CF = Complement fixation; EIA = enzyme immunoassay; FAMA = fluorescent antibody to membrane antigen; HAI = hemagglutination inhibition; IF = indirect fluorescent antibody; LA = latex agglutination; Nt = neutralization; PHA = passive hemagglutination; RFFIT = rabies fluorescent focus inhibition test; RIA = radioimmunoassay.

4. Equal volumes of the serum dilutions and the test virus dilution are mixed. The volume of serum-virus prepared depends on the number of tube cell cultures to be inoculated (0.1 ml each × number of tubes). For a "virus control," the test virus dilution is mixed with an equal volume of diluent, or known "normal" serum of the same species as the test serum, and incubated under the same conditions as the serum–virus mixtures. For Nt antibody assays, concurrent titration of the virus is necessary to confirm that the test dose actually contains approximately 100 $TCID_{50}$.

5. The conditions recommended for incubation of serum-virus mixtures vary widely. For certain viruses, preliminary incubation increases the neutralizing capacity of the serum. However, it is important to avoid incubation conditions under which a labile virus might be destroyed. For most Nt tests, serum–virus mixtures are incubated for 0.5 to 1 h at 37°C.

6. After the incubation period, the serum–virus mixtures and virus controls are inoculated in a volume of 0.2 ml into monolayer tube cultures of appropriate susceptible cells. At least two cultures are employed for each serum–virus mixture.

7. The inoculated cultures are incubated under conditions most suitable for optimal growth of the virus and examined microscopically to see if the serum inhibits the cytopathic effect (CPE) of the virus. Final readings are usually made when the virus control shows that 100 $TCID_{50}$ are present in the test. Alternatively, the cell cultures may be examined for virus growth by direct immunofluorescence, enzyme-linked immunosorbent assay (EIA), or immunoperoxidase staining using labeled anti-viral antibody conjugates. In the case of noncytopathic orthomyxoviruses or paramyxoviruses, the cultures may be tested for hemadsorption after a suitable incubation period; neutralization of the virus is evidenced by failure of the cultures inoculated with serum–virus mixtures to hemadsorb the erythrocytes. The antibody titer is expressed as the reciprocal of the highest serum dilution that neutralizes the virus.

MICRONEUTRALIZATION TESTS IN CELL CULTURES

Neutralization tests performed in flat-bottomed wells of disposable plastic microtitration plates are more economical than tube Nt tests; smaller volumes of cell cultures and reagents are needed and there is no loss of sensitivity or reproducibility. Microtitration plates processed specially for cell culture growth are commercially available.

Cell monolayers may be prepared in the microtitration plate wells before they are inoculated with the serum–virus mixtures, or more conveniently, the cells, serum, and virus may be added at the same time. In the latter method, unneutralized virus can infect and destroy the cells before attachment, and monolayers will not be formed, whereas in cultures containing neutralized virus, the cells may attach and proliferate into a confluent monolayer. Endpoints of microneutralization tests may be based on microscopic observation of viral CPE or hemadsorption or on colorimetrically determined metabolic inhibition, direct immunofluorescence or immunoperoxidase staining, or EIA using specific antiviral conjugates to detect unneutralized virus. As with the tube test, the endpoint titer is expressed as the reciprocal of the highest serum dilution that neutralizes the virus.

PLAQUE REDUCTION NEUTRALIZATION TESTS

Plaque reduction Nt tests are very sensitive for virus Nt antibodies. Enumerating virus plaques in monolayer cell cultures is a more precise means of quantifying viral infectivity than $TCID_{50}$ endpoints in tube or microtitration plate cultures; therefore, demonstrating reduction in virus plaque counts provides a very sensitive means for measuring virus Nt capacity of a serum.

1. The serum specimen is heat inactivated, and appropriate serum dilutions are prepared as described for the tube Nt test.

2. The test virus is diluted to contain sufficient plaque-forming units (pfu) in the inoculum volume to allow precise plaque counting in the monolayer culture.

3. Equal volumes of the diluted test virus and each serum dilution are mixed.

4. After an appropriate incubation period, monolayer cell cultures in bottles or wells of polystyrene cell culture plates are inoculated with serum–virus mixtures, and the cultures are incubated for 1 to 1.5 h at 37°C to permit virus adsorption. The monolayers are then washed with Hanks BSS and covered with an appropriate nutrient overlay containing a solidifying agent such as purified agar or agarose.

5. After a suitable incubation period at 36 to 37°C, plaques are stained with a histochemical stain such as neutral red or crystal violet, or the overlay is removed and plaques are identified by fluorescent antibody, immunoperoxidase, or other immunochemical reagents. An example test plate stained with crystal violet (Fig. 3) demonstrates the reduction or neutralization of plaques by an antibody-positive serum.

6. The neutralizing capacity of a test serum is determined by its ability to reduce the number of virus plaques in the test culture as compared with control cultures inoculated with only the virus. The

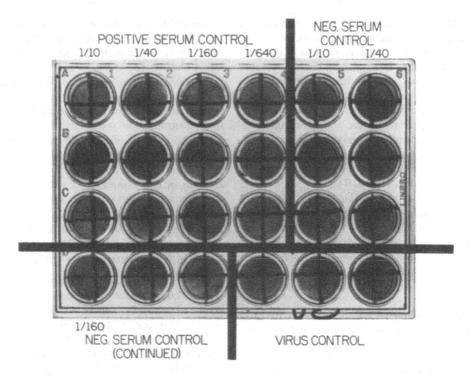

FIG. 3. Results of a plaque reduction neutralization test for poxvirus antibody performed using 24-well polystyrene cell culture trays and LLC-MK₂ cell monolayers. Note the gradual increase in plaque breakthrough for the positive serum, showing none at 1:10, very few at 1:40, few at 1:160, and about 26 plaques per 3 wells at 1:640. The 50% plaque reduction neutralization titer for this positive serum is greater than 640. The negative serum shows no plaque reduction with dilutions from 1:10 through 1:160.

reciprocal of the highest dilution of test serum that reduces the plaque count by 50% or more is the plaque-neutralizing (PNt) antibody titer. The 50% plaque reduction endpoint is most frequently calculated using the probit transformation method (Finney, 1971). The percentage of plaque breakthrough for each serum dilution is calculated and plotted on probit graph paper, showing the percentages of plaque breakthrough on the ordinate and the serum dilutions in negative $\log_{10}$ on the abscissa. A best-fitting straight line is drawn through the points, and a vertical line is dropped from the intersection of the plot line; the 50% probit graph line determines the serum dilution endpoint that gives the 50% plaque reduction.

Neutralizing antibody endpoint titers can be determined for almost any virus that can be grown in cell culture by detecting breakthrough virus with direct immunofluorescent staining of the cell monolayer rather than by CPE. This procedure is referred to as the *fluorescent focus inhibition test*. It is particularly useful for viruses that do not produce recognizable CPEs or plaques.

Neutralization testing for serum antibody has been applied to almost all known human viral patho-gens. Virus Nt tests, although relatively simple to conduct, are expensive in time and materials and may be difficult to interpret because of variability in the titration endpoint and difficulties in standardizing the test method. References to protocols giving details and conditions for specific Nt antibody assays may be found in other chapters of this volume. Most viral antibodies, however, can be more easily and relatively accurately assayed by using other less expensive and less cumbersome methods.

Complement Fixation Test

One of the classic methods for demonstrating the presence of antiviral antibody in a patient's serum is the complement fixation (CF) test. The complex complement system plays an important role in mediating and amplifying immune and inflammatory reactions as part of the natural immunopathologic process. Complement is activated after combining with antigen–antibody complexes. This activation sequence is used in the complement fixation test for antibody.

The conventional CF test consists of two stages. In the first stage, serum (antibody) and known antigen are mixed in the presence of a measured amount of complement (usually a pretitrated dilution of guinea pig serum). If the serum antibody and antigen react specifically and form antigen–antibody complexes, the complement will be fixed and depleted. If the serum antibody and antigen do not form antigen–antibody complexes, the complement will not be fixed and will remain free in the reaction mixture. In the second stage, sheep erythrocytes sensitized (coated) with anti-sheep erythrocyte antibody (hemolysin) are added to the reacted mixture. As complement is a lytic agent, if any active complement remains in the test system, the sheep erythrocytes will be lysed (hemolyzed). Conversely, the absence of hemolysis indicates that the complement had been depleted or "fixed" and therefore an antigen–antibody reaction had occurred in the first stage.

For the successful performance of this assay, all reagents, with the exception of the unknown serum, must be carefully titrated and used in precisely measured amounts. Interpretation of the test depends on complete fixation of complement in the presence of the antibody in the unknown serum sample and complete hemolysis in the absence of the unknown antibody. Protocols for preparing and standardizing all reagents for performing the standard diagnostic complement fixation test have been published (Hawkes, 1979; Palmer and Casey, 1981; U.S. Public Health Service, 1965). The following is a condensed version of the standardized Laboratory Branch Complement Fixation (LBCF) procedure.

LABORATORY BRANCH COMPLEMENT FIXATION TEST PROCEDURE

Serum samples may be tested with one or more antigens. Each test must include positive control serum of known reactivity. The titer of an antiserum is defined as the reciprocal of the highest dilution of serum showing 30% or less hemolysis with the optimal antigen dilution. The serum is considered anticomplementary if less than 75% hemolysis occurs in the serum control of any dilution greater than two dilutions below the titer.

Hemolysin Titration

Sheep red blood cells (RBCs) are sensitized with seven dilutions of hemolysin: 1:1000, 1:1500, 1:2000, 1:2500, 1:3000, 1:4000, and 1:8000. Following addition of complement (C'), the hemolysin-cell mixtures are incubated at 37°C for 1 h, centrifuged, and read with color standards for percent hemolysis. Readings are plotted on arithmetic graph paper, and the graph is examined for a plateau. The second dilution on the plateau is considered the optimal dilution for use in the LBCF.

Complement Titration

Based on the hemolysin titration results, sheep RBCs are sensitized using the optimal dilution of hemolysin. The C' titration is performed in duplicate using the volumes shown in Table 2. Tube titration, rather than microplate titration, is recommended for the C' titration to permit greater precision in estimating the percentage of hemolysis for each dilution in the assay. After the tubes are incubated for 30 min at 37°C, they are centrifuged and read for percentage of hemolysis, and the readings are plotted using log–log graph paper. Working dilutions of C' for the LBCF antigen titration and serum assay are calculated and expressed as the dilution needed to contain 5 $C'H_{50}$ in 0.4 ml. Complement titers are calculated and expressed as the number of $C'H_{50}$ units per ml. One $C'H_{50}$ is the amount of complement that produces 50% lysis in the indicator system.

Antigen Titration (Micro Method)

Antigens are evaluated using at least one reference animal antiserum for specificity and one human reference antiserum (if available) for reactivity. The potency or optimal antigen dilution is normally defined as the dilution that gives the greatest amount of fixation at the highest (or expected) serum antibody titer.

Box titrations are performed in microtitration plates using 25 μl of serially diluted antiserum, 25 μl of serially diluted antigen, and 50 μl C' containing 5 $C'H_{50}$ units. Complement controls for each dilution of test antigen, reference antigen, tissue control antigen, and Veronal-buffered diluent (VBD) are included. The volumes used to prepare the controls are given in Table 3. Acceptable limits of reactivity for the C' controls are given in Table 4. If the VBD controls are not within acceptable limits, the entire test is invalid. Antigen and tissue dilutions that give less than 85% hemolysis in the 2.5-unit control are considered anticomplementary (AC).

In interpreting each antigen titration, it is convenient to draw a line, known as the optimal dilution curve, through the 30% hemolysis endpoints (interpolated if necessary) of each antigen dilution that is not AC. In the sample titration (Table 5), the 1:2 dilution of test antigen is AC, and all other C' con-

TABLE 2. Volumes (ml) of reagents required for complement titration

Reagent	Tube no.			
	1	2	3	4
Veronal buffered diluent	0.60	0.55	0.50	0.40
1:400 dilution of complement	0.20	0.25	0.30	0.40
Sensitized cells	0.20	0.20	0.20	0.20

TABLE 3. Volumes (μl) of reagents required to prepare complement controls

	Assay well		
Reagents	5-unit $C'H_{50}$	2.5-unit $C'H_{50}$	1.25-unit $C'H_{50}$
Antigen[a]			
VBD[b]	25	25	25
Antigen	25	25	25
Complement	50	50	50
Tissue control			
VBD	25	25	25
Tissue	25	25	25
Complement	50	50	50
VBD			
VBD	50	50	50
Complement	50	50	50

[a] Test and reference preparations are incubated overnight (15 to 18 h) at 4°C, then 25 μl of sensitized cells is added and the mixture incubated at 37°C for 30 min. After centrifugation, the tubes are read and the percentage of hemolysis is recorded. A completed antigen titration is illustrated in Table 5.
[b] VBD = Veronal buffered diluent.

trols are within acceptable limits. The optimal dilution curve is drawn from the 1:4 to 1:64 test dilutions. To the right of the curve, the results indicate the greatest fixation of C' at the 1:8 antigen dilution and this dilution would be selected as optimal (the test antigen titer).

Serum Antibody Titration (Micro Method)

A 1:8 dilution of each unknown serum with VBD is prepared in sufficient quantity for tests and controls. The diluted serum is inactivated for 30 min at 56°C.

TABLE 4. Acceptable percentage of hemolysis of complement controls in antigen titration

	Assay well		
Type of control	5-unit $C'H_{50}$	2.5-unit $C'H_{50}$	1.25-unit $C'H_{50}$
Veronal buffered diluent	100	90–100	40–75
Antigen	100	85–100	0–75
Tissue	100	85–100	0–75
Serum	75–100	—	—

Rows of wells on microtitration plates are labeled for the diagnostic test as indicated in Table 6. Measured 50μl volumes of the 1:8 dilutions of unknown, known negative, and known positive serum are then added to their appropriately labeled wells, and serial twofold dilutions of each serum are made in VBD. Test antigen is then added to the wells for the tests and for the complement–antigen controls. Normal tissue antigen is added to the rows of tissue control wells and complement-tissue control wells, and cold VBD to the rows of serum control wells. Diluted complement containing 5C'H_{50} is added to each assay well and to each 5-unit complement control well and mixed thoroughly. Complement diluted to 2.5 units and 1.25 units is added to the appropriate control wells, as indicated in Table 6, and all plates are incubated overnight at 4°C. Following the overnight incubation, the test plates are removed from the refrigerator and allowed to warm to room temperature. Freshly sensitized 2.8% sheep RBCs are added, and the plates are incubated for 30 min at 37°C. Following the incubation period, the plates are centrifuged and read, and the percentage of hemolysis of all test and control wells is recorded.

TABLE 5. Sample antigen titration[a]

	Antigen dilution	Percentage hemolysis: reference antiserum dilutions[b]						Complement controls		
		1:16	1:32	1:64	1:128	1:256	1:512	5-unit $C'H_{50}$	2.5-unit $C'H_{50}$	1.25-unit $C'H_{50}$
Test antigen	1:2	0	0	30	70	100	100	100	75	0
	1:4	0	0	10	50	90	100	100	90	20
	1:8	0	0	0	20	60	90	100	95	60
	1:16	0	0	0	20	70	100	100	95	60
	1:32	0	0	30	80	100	100	100	95	60
	1:64	10	30	50	90	100	100	100	95	70
Reference antigen	Optimal	0	0	0	20	60	95	100	95	65
Test tissue control	1:2	100	100	100				100	85	10
	1:4	100	100	100				100	90	30
Reference tissue control	1:2	100	100	100						
Serum control	None	100	100	100						
VBD-C'[c] controls								100	95	75

[a] Optimal antigen dilution for this example is 1:8.
[b] The titer of the reference antiserum in this example is 128.
[c] VBD-C' = Veronal buffered diluent–complement.

TABLE 6. Diagnostic laboratory branch complement fixation test: micro method

A. Serum titration			Serum dilution (μl)						
			1:8	1:16	1:32	1:64	1:128	1:256	
Unknown serum assay	Test	Unknown serum	25						
		Test antigen	25						→
		5 C'H$_{50}$[a]	50						
	Serum controls	Unknown serum	25						
		VBD[b]	25						→
		5 C'H$_{50}$	50						
	Tissue controls[c]	Unknown serum	25						
		Normal tissue antigen	25						→
		5 C'H$_{50}$	50						
Known positive serum assay	Test	Known positive serum	25						
		Test antigen	25						→
		5 C'H$_{50}$	50						
	Serum controls	Known positive serum	25						
		VBD	25						→
		5 C'H$_{50}$	50						
	Tissue controls	Known positive serum	25						
		Normal tissue antigen	25						→
		5 C'H$_{50}$	50						
Known negative serum assay	Test	Known negative serum	25						
		Test antigen	25						→
		5 C'H$_{50}$	50						
	Serum controls	Known negative serum	25						
		VBD	25						→
		5 C'H$_{50}$	50						
	Tissue controls	Known negative serum	25						
		Normal tissue antigen	25						→
		5 C'H$_{50}$	50						
	Cell control	VBD	100						→

B. Complement controls			5-unit C'H$_{50}$ (μl)	2.5-unit C'H$_{50}$[d] (μl)	1.25-unit C'H$_{50}$[d] (μl)
C'-antigen controls		Test antigen	25	25	25
		VBD	25	25	25
		Complement	50	50	50
C'-tissue controls[c]		Normal tissue antigen	25	25	25
		VBD	25	25	25
		Complement	50	50	50
C'-VBD controls		VBD	50	50	50
		Complement	50	50	50

[a] C'H$_{50}$ = 50% hemolysis unit of complement.
[b] VBD = Veronal buffered diluent.
[c] Tissue controls are used with viral or rickettsial antigens. A normal tissue antigen is used at the same dilution as the test antigen.
[d] For 2.5-unit controls, this is a 1:2 dilution of the complement containing 5-unit C'H$_{50}$ in 50 μl; for 1.25-unit controls, a 1:4 dilution is used.

Interpretation of Results

In serum antibody assays, the complement control readings must meet the same criteria as outlined for antigen titrations in Table 4. If the VBD controls are not within acceptable limits, the test results are not valid. The serum titer is the reciprocal of the highest dilution showing 30% or less hemolysis with the optimal dilution of antigen. A serum is AC if the lowest dilution of the serum without antigen indicates less than 75% hemolysis. When serum-AC activity is encountered, usually another serum sample from the patient is requested. AC reactions are usually limited to the lower serum dilutions (i.e., 1:8 or 1:16); thus high titers (64 or greater) may be demonstrable by CF even in serum with low levels of AC activity.

The CF test has been used widely over the years to detect many antiviral antibodies. In large laboratories where CF is performed routinely, this assay serves as an accurate diagnostic method for many viral infections. The procedure however, is complex, involves a relatively large number of reagents, requires rigid standardization, and takes 18 to 24 h to be completed. For most viral diseases, alternative test systems have been developed that are less technically demanding, more automated, quicker, and more sensitive than the CF test. Laboratories without experience with CF tests should not adopt this method for routine diagnostic testing when other less demanding procedures are available.

Hemagglutination Inhibition Test

Many viruses can hemagglutinate RBCs of one or more animal species under certain conditions (Table 7). This reaction occurs directly between specific virus antigens, called *hemagglutinins*, and receptor

TABLE 7. Conditions in standardized hemagglutination inhibition tests[a]

Virus	Erythrocyte Species	Concentration (%)	Incubation temperature (°C)	Serum treatment
Enteroviruses[b]				
Coxsackie A-20, A-21, A-24; echovirus 3, 11, 13, 19; enterovirus 68	Human O	0.4	4	56°C, 30 min, kaolin adsorption
Coxsackie B-1, B-3, B-5, B-6; echovirus 6, 7, 12, 20, 21, 24, 29, 30, 33	Human O	0.4	37	56°C, 30 min, kaolin adsorption
Coxsackie A-7	Chicken[c]	0.5	24	56°C, 30 min, kaolin adsorption
Reovirus 1, 2, 3	Human O[c]	0.4	24	56°C, 30 min, kaolin
Influenza A and B (fresh isolates)	Guinea pig[d] Human O	0.4	24	RDE-heat, RBC adsorption for guinea pig cells
Influenza A and B (laboratory-adapted)	Chicken	0.5	24	RDE-heat
Influenza C	Chicken	0.5	4	RDE-heat
Parainfluenza 1, 2, 3	Guinea pig Human O	0.4	24	RDE-heat
Parainfluenza 4A and 4B	Rhesus, guinea pig	0.4	24	RDE-heat, RBC adsorption
Mumps	Chicken	0.4	24	RDE-heat
Coronavirus OC 38	Chicken	0.5	24	56°C, 30 min
	Rat[c]	0.4	24	56°C, 30 min, RBC adsorption
Rubeola (measles)	Vervet or rhesus[c]	0.4	37	56°C, 30 min, RBC adsorption
Rubella	Baby chick	0.25	4	Heparin-MnCl₂
	Human O	0.25	24	RBC adsorption
Smallpox, vaccinia	Chicken[c]	0.5	24	56°C, 30 min
Adenoviruses				
Group I	Rhesus	0.4	37	56°C, 30 min, RBC adsorption
Group II, III	Rat[c,d]	0.4	37	56°C, 30 min, RBC adsorption
Arboviruses	Goose	0.4	4	Kaolin, RBC adsorption
Polyoma BK, JC	Guinea pig	0.4	4	RDE-heat

[a] Modified from Hierholzer et al. (1969).
[b] The hemagglutinating enteroviruses may require different pH levels for optimal hemagglutination titers. In addition, many wild strains of these enteroviruses do not hemagglutinate at any pH or temperature.
[c] Animals must be selected for erythrocytes sensitive to viral hemagglutination.
[d] Erythrocytes should be used within 2 days after collection.

sites on the red cell surface and results in visible clumping of the cells. Antibodies to these virus hemagglutinins react with the virus and prevent the hemagglutination. This principle forms the basis of the hemagglutination inhibition (HAI) test.

The HAI test has been widely used for diagnosing current infection and determining immune status for viruses with a hemagglutinin component. The test is relatively easy to perform. However, because most serum contains nonspecific viral inhibitors of hemagglutination, diagnostic serum specimens must be treated before testing for specific antibody. Failure to remove the nonspecific inhibitors may result in false-positive results. Natural agglutinins to the RBCs, if present, must also be removed before assaying for specific antibody.

The sensitivity and specificity of HAI tests depend on many variables, and tests must be carefully controlled if accurate and reproducible results are to be expected. The efforts to standardize viral HAI test procedures are reflected in the following section.

Standardized Hemagglutination and Hemagglutination Inhibition Tests

Standardized microtitration methods for the HA and HAI tests have been described (Hierholzer and Suggs, 1969; Hierholzer et al., 1969). These techniques have been used for detecting and titrating antibody to adenoviruses, myxoviruses, paramyxoviruses, rubeola virus, reoviruses, and vaccinia. The following is a condensed version of this procedure.

Preparation of Red Blood Cells

Red blood cells should be obtained from the proper animal species (see Table 7). They should be washed and standardized on the day of the test.

Titration of Hemagglutinin

The antigens used are titrated by HA in microtitration plates to determine the proper dilution needed to give 4 HA units per 25 μl. Table 7 lists the major variations in the standardized HAI tests. The plates are read for hemagglutination when the cells in the control wells have formed compact buttons or rings of nonagglutinated cells (usually after 1 to 2 h).

To facilitate reading exact endpoints, the plate may be slanted for approximately 30 s so that nonagglutinated cells will flow in a teardrop pattern. Because many viruses elute and give falsely low titers, the agglutination patterns should be read immediately. The last well that shows complete agglutination is considered the endpoint and equals 1 HA unit.

Hemagglutination Inhibition Test

The virus antigen is diluted to contain 4 HA units per 25 μl. To ensure that the proper test dilution is used,

a back titration should be performed before and during the test. If the test dilution is accurate, agglutination will occur in the first three wells but not in the remainder of the back titration series. If too few or too many positive wells are observed, the test antigen concentration may be adjusted by adding more virus or more diluent.

The serum treatments necessary for the HAI tests are given in Table 7. The RBC adsorption is done with 50% RBC in phosphate-buffered saline (PBS) by mixing 0.1 ml of RBC suspension with 1 ml of 1:10 dilution of serum and incubating at 4°C for 1 h. For kaolin adsorption, equal volumes of a 1:5 dilution of serum and a 25% suspension (wt/vol) acid-washed kaolin in PBS are incubated at room temperature for 20 min. Other serum treatment methods have been recommended in the literature for specific HAI tests.

The HAI test is performed in a total volume of 100 μl (25 μl of serum, 25 μl of antigen, and 50 μl of RBC suspension) with suitable reference controls and is read when the control wells show compact buttons or rings (usually after 1 to 2 h). As in the HA test, to facilitate reading, plates may be slanted for 30 s so that nonagglutinated (inhibited) cells will "tear." Cell control wells and serum control wells should show no agglutination. The HAI titer of each serum is defined as the reciprocal of the highest dilution of serum that completely inhibits hemagglutination. If nonspecific agglutination is seen in the 1:20 dilution or above, the serum should be readsorbed with a 50% RBC suspension and the HAI repeated.

Unique HAI test protocols have been developed for some viruses such as rubella and other togaviruses. See other chapters for details and references for these special methods.

Immune Adherence Hemagglutination

Immune adherence hemagglutination (IAHA) is a variation of the CF test in which agglutination rather than lysis of erythrocytes is measured (Lennette and Lennette, 1978; Nelson, 1963). Both methods detect antibodies by demonstrating bound complement to indicate the presence of antigen–antibody complexes. In IAHA, complement that has been bound by antigen–antibody complexes is then bound to the C'3b receptors located on the surface of primate erythrocytes, resulting in hemagglutination. This technique has been used to detect antibodies to influenza, herpes simplex, Epstein–Barr, varicella-zoster, hepatitis B, and rubeola viruses, cytomegalovirus, human rotavirus, and is reportedly up to 20 times more sensitive than CF tests.

Although IAHA offers advantages of increased sensitivity for antibody detection, reduced performance time, and less sensitivity to anticomplementary activity compared with CF tests, few clinical laboratories have adopted this method in place of CF tests for routine viral diagnosis.

Reagents and equipment required for IAHA are similar to those needed for the CF test. The following brief outline for the IAHA procedure has been summarized from that published by Lennette and Lennette (1986).

IMMUNE ADHERENCE HEMAGGLUTINATION MICROTITRATION PROCEDURE

1. Each test serum is pretreated by diluting the serum 1 : 4 with Veronal-buffered saline (VBS) and incubating it at 56°C for 30 min to inactivate natural complement in the serum.
2. VBS (25 μl) containing 0.1% bovine serum albumin is added to each well of the plate.
3. Then 25 μl of each inactivated serum is added to wells 1 and 8 of the corresponding row on the test plate.
4. Using microdiluters, serial twofold dilutions (in sets of seven and five wells) are made for each test serum.
5. Optimally diluted test antigen (25 μl) is added to wells 1 through 7 of each row; 25 μl of identically diluted negative (control) antigen is added to wells 8 through 12. The plates are shaken to mix contents of wells. The optimal antigen dilution is determined by prior block titration with control reference serum.
6. The plates are covered and incubated for 30 min at 37°C.
7. Diluted guinea pig complement (25 μl of 1 : 100 C' in VBS) is added to all test wells. The plates are again shaken and incubated for 40 min at 37°C.
8. Following incubation, the reaction is stopped by adding 25 μl of VBS containing disodium ethylenediaminetetraacetic acid (EDTA) and dithiothreitol (DTT) to each well, and 25 μl of 0.4% human O erythrocytes is added to the mixture.
9. The hemagglutination patterns are read after 1 h. A positive agglutination reaction (i.e., granular or coarse pattern of the settled cells) indicates the presence of antibody; a negative reaction appears as a smooth compact cell button. The reciprocal of the highest dilution of serum resulting in a clearly positive agglutination pattern is considered the endpoint titer.

Passive Agglutination and Passive Hemagglutination

Measuring the agglutination of a particulate antigen by its specific antibody (direct agglutination) is the simplest way to estimate the quantity of that antibody in serum. Very small amounts of antibody can be detected by this method, because only a small number of antibody molecules are necessary to form the antigen–antibody lattice.

The agglutination reaction has been extended to include a wide variety of antigens by attaching soluble antigens to the surface of inert particles, such as latex, bentonite, or RBCs. The role of these particles, once coated or sensitized, is passive; they react as if they themselves possess the antigenic specificity of the coating antigen.

RBCs are extremely convenient carriers of antigen. When specific antibody is added to antigen-coated RBCs, antibody bridges are formed between neighboring cells, and large aggregates of RBCs are produced that are visible to the naked eye. This agglutination is designated passive hemagglutination (PHA) or indirect hemagglutination. The following PHA procedure is adopted and summarized from that recommended by the Centers for Disease Control for toxoplasma, herpes simplex, and cytomegalovirus antibody assay (Palmer et al., 1977).

PASSIVE HEMAGGLUTINATION TEST PROCEDURE

Preparation of 2.5% Sheep Red Blood Cells

Sheep blood is washed three times in PBS (pH 7.2), centrifuging for 5 min at 900 × g between each wash. After the third wash, the cells are packed for 10 min. A 2.5% sheep RBC suspension is then prepared from the packed cells (39 volumes of PBS, pH 7.2, to 1 vol of packed cells).

Tanning of Standardized Cells

Fresh tannic acid solution is prepared by dissolving tannic acid in PBS (pH 7.2) to yield a final dilution of 1 : 1000 (e.g., for each milligram of tannic acid, add 1 ml of PBS, pH 7.2). This 1 : 1000 stock solution may be further diluted 20-fold to obtain the 1 : 20,000 dilution for use in the test. Five milliliters of 2.5% SRBC is mixed with 5 ml of the 1 : 20,000 tannic acid solution in a 15-ml centrifuge tube, and the tube is incubated in a 37°C water bath for 10 min. The mixture is then centrifuged for 5 min, the supernatant fluid is removed, and the cells are resuspended in 10 ml of PBS (pH 7.2). Following a final centrifugation, the supernatant fluid is removed and the cells are reconstituted to a 2.5% suspension by adding PBS (pH 6.4).

Sensitization of Cells

Antigen is diluted to its optimal dilution in PBS, pH 6.4 (the optimal dilution as determined by block titrations is the lowest dilution of antigen yielding the highest specific serum titer). One milliliter of the 2.5% tanned cells is mixed with 1.0 ml of optimal diluted antigen and incubated for 15 min at 37°C.

Control cells are prepared in parallel by mixing 1.0 ml of 2.5% tanned cells with 1.0 ml of PBS (pH 6.4) and incubating for 15 min at 37°C. After centrifugation and removal of the supernatant, the cells are washed several times in rabbit serum diluent (PBS, pH 7.2, containing 1% normal rabbit serum). Following the washings, the cells are adjusted to a 1.0% suspension by adding normal rabbit serum diluent. The 1.0% suspensions of tanned, sensitized cells and tanned, unsensitized cells are now ready for use in the test.

Serum Antibody Assay (Microtitration Adaptation)

Each test serum is inactivated for 30 min at 56°C and diluted 1 : 10 in the rabbit serum diluent. Two rows of eight wells marked for each serum on microtitration U plates, and two wells are marked for the cell controls (one for sensitized and one for unsensitized cells). With a multichannel pipetter or microdropper, 50 μl of normal rabbit serum diluent is added to all wells in which serum dilutions will be made except for the first well in each row, and to the two wells that will serve as the cell controls. One-tenth milliliter of the 1 : 10 dilution of each serum is pipetted into the first wells of the two rows, and twofold dilutions of serum are made with 50-μl microtitration loops. Tanned, sensitized sheep RBCs are added to the wells of the first row of each serum and to the well marked for the sensitized cell control; tanned, unsensitized sheep RBCs are added to wells of the second row of each serum and to the well marked for the unsensitized cell control. The plates are covered and incubated in a 37°C water bath for 2 to 3 h or left at room temperature until the controls show well-formed buttons.

The reciprocal of the highest initial dilution yielding a clearly positive agglutination of sensitized red cells is considered an endpoint. If agglutination of control cells in the second row is equal to that of the sensitized cells, the serum must be adsorbed as follows and the test repeated: 1) Inactivate serum and dilute 1 : 10. 2) Adsorb two or more times with 0.1 ml of 50% sheep RBC per ml of diluted serum for 30 min at 4°C. If this treatment fails to remove nonspecific agglutination, the serum must be adsorbed in the same manner with sheep RBCs sensitized with rabbit serum.

Only in the past few years have passive agglutination and PHA become widely accepted in viral serology. Latex agglutination and PHA tests are currently the most widely used procedures in the United States for rubella immunity screening.

Hemolysis in Gel

Hemolysis in gel (HIG), sometimes referred to as single radial hemolysis, is a serologic method based on lysis of antigen-sensitized RBCs in a gel by antibody in the presence of guinea pig complement. This technique has been widely accepted in some countries to detect or measure antibodies to several hemagglutinating viruses, including rubella, influenza, mumps, and parainfluenza 3 viruses.

The reaction is carried out on RBCs to which viral antigen has been bound and takes place with the cells suspended in an agarose gel containing complement. The serum specimen, previously heated to inactivate native complement, is allowed to diffuse into the gel from a well, and if specific antibody is present, causes a zone of hemolysis around the well. The diameter of the concentric zone of hemolysis is proportional to the concentration of specific antibody in the serum.

Several procedures for performing HIG tests have been reported (Neumann and Weber, 1983; Russell et al., 1975). The method described here has been adapted from the protocol recommended by Public Health Laboratory Service of Great Britain for rubella antibody testing (Pattison, 1982).

HEMOLYSIS IN GEL PROCEDURE

Gel Preparation

One test and one control gel can be prepared in 100-by 100-mm trays or petri dishes as follows. Two 15-ml quantities of 1% agarose in veronal-buffered saline (VBS), pH 7.3, are melted in a 44°C water bath. Into each of two 30-ml screw-cap centrifuge bottles, one marked "test" and the other "control," 0.3 ml of a 15% (vol/vol) suspension of sheep RBCs in VBS is pipetted. The "test" cells are combined with 0.3 ml of the test antigen (hemagglutinin) diluted to contain approximately 128 HA units (see Titration of Hemagglutinin). After 30 min of incubation at room temperature, both bottles are filled with VBS and centrifuged at $900 \times g$ for 10 min. The supernatants are discarded, and the cells are resuspended in 0.5 ml of VBS. One-half milliliter of undiluted guinea pig complement is then added to each bottle, both bottles are placed in a 44°C water bath, and 15 ml of the melted agarose is immediately added to each. After mixing, the contents of each are poured into appropriately labeled 100-mm^2 dishes or trays placed on a level surface. After the gels have thoroughly set at room temperature, they may be stored at 4°C in a humid chamber for up to 5 days before use.

Test Procedure and Controls

Serum to be tested must be inactivated by heating for 20 min at 60°C. Wells with diameters of 2 to 3 mm (capacity, 8 to 11 μl) are cut in the gels using a template, maintaining a minimum of 10 mm between wells. The agar cylinders are sucked out with a Pasteur pipette attached to a vacuum line and trap. Each inactivated serum is added to the appropriate well in the test and control gels. The wells should not be overfilled. Known positive and negative control sera must be tested on each gel tray. If antibody levels are to be quantitated by this method, serial dilutions of a reference "calibrator" serum of known titer must be included. One positive control should be adjusted to have a minimal (breakpoint) potency to define positive (immune) from negative (nonimmune). The gels are incubated in a humid chamber at 37°C overnight. Zones of hemolysis are best read by transilluminating the gels against a black background. The diameter of the zones of hemolysis are measured to the nearest 0.1 mm and compared with the controls.

Interpretation

Test zones larger than that of the minimal potency (breakpoint) control may be reported positive. Test zones smaller than that of the breakpoint control should be reported negative for purposes of immune screening.

Although the HIG test has primarily been used as a qualitative or semiquantitative antibody assay, when accurately calibrated, it can be used in place of other quantitative tests for determining antibody titers for diagnostic purposes.

HIG has several theoretical and practical advantages over other viral antibody assays. It is simple to perform. It is unaffected by most nonspecific inhibitors in human serum that interfere with the virus hemagglutinins, and is therefore more specific than corresponding HI tests, which require special serum treatment to remove nonspecific HA inhibitors before testing. On the other hand, the limited "shelf life" of prepared gels requires that clinical laboratories in countries where the time between kit production and delivery is greater than 2 weeks prepare their own gels.

Radioimmunoassay

Radioimmunoassay (RIA) offers a very sensitive technique for detecting and quantitating viral antibodies and is currently one of the most widely applied of all immunoassays. During the past decade, many RIA systems have been developed (Mushahwar and Brawner, 1986; Mushahwar and Overby, 1983). However, because of the relative instability of the labeled reagents used and the special conditions and equipment required for handling radioisotopes, RIA systems in virology have been limited primarily to the research laboratory, and only a few of these procedures, notably hepatitis antigen and antibody assays, have successfully made the transition from research procedures to commercially available diagnostic tests.

Most radioimmunoassays for detecting anti-viral antibodies are based on either the traditional indirect solid-phase RIA (SPRIA), solid-phase competitive binding RIA, or radioimmunoprecipitation (RIP) systems.

INDIRECT SOLID-PHASE RADIOIMMUNOASSAY PROCEDURES

The indirect SPRIA for antibody detection uses a purified or semi-purified viral antigen adsorbed to plastic beads, tubes, or microtitration wells to create a solid surface capable of capturing and binding specific antibody in the test serum sample. Removable "strips" or rows of microtitration wells are commonly used as the solid phase for RIA and EIA test systems, are available commercially (Titertek FB strips, Flow Laboratories, McLean, Va.), and allow greater economy in use of antigen coated microwells when a full 96-well plate is not needed for an assay. Although a variety of commercially available viral antigens have been used successfully in indirect SPRIAs, the specificity of the assay depends primarily on the quality and purity of the antigen preparation, that is, freedom from contaminating nonviral proteins (Parratt et al., 1982). Control antigen (i.e., antigen that contains all components in the test antigen except the virus) must be available and used whenever possible to detect nonspecific reactivity.

Two variations of the indirect SPRIA will be described in this section. The principles of these two methods are represented schematically in Figures 4A and B. Both use the traditional two-stage "sandwich" technique.

In method A, the serum to be assayed is incubated with the immobilized solid-phase antigen in stage 1, allowing specific antibody, if present, to be captured and bound to the solid surface. Removing the unreacted components of the reaction mixture and washing the surface separates bound and free antibodies. After the surface has been washed, ^{125}I-labeled antigen is added in stage 2 as a probe to react with the remaining exposed binding sites of the captured antibody molecules. Again following incubation and washing, the amount of radioisotope bound through antigen–antibody–antigen linkage to the surface is measured in counts per minute in a gamma scintillation counter. The amount of radioactivity counted on the surface after the final wash is proportional to the concentration of specific anti-viral antibody in the test serum.

The stage 1 reaction of method B is identical to method A. However, ^{125}I-labeled anti-IgG directed against the species of the test serum is added in stage 2 as the probe to detect and bind to antibody captured by the immobilized antigen in stage 1. If recombinant proteins or synthetic peptides are used as the

FIG. 4. Indirect solid-phase radioimmunoassay for detecting antibody. Known antigen bound to the solid surface is reacted with the sample containing antibody. Unbound proteins are washed away and the antibody is detected by a second radiolabeled antigen probe (A) or by a radiolabeled anti-human immunoglobulin probe (B).

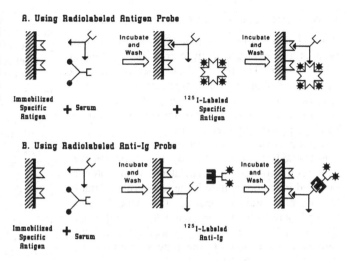

capture antigen and labeled monoclonal anti-human immunoglobulin as the probe, method B may be simplified to one stage by allowing the addition of the probe immediately following the test serum without the need for an intervening wash step. Again as described for method A, the amount of radioactivity binding to the solid phase through the antigen-antibody-anti-IgG complex is proportional to the concentration of specific anti-viral antibody in the test serum.

These indirect SPRIA procedures have been used in the development of commercial test kits for detecting antibody to hepatitis B surface antigen.

COMPETITIVE BINDING SOLID-PHASE RADIOIMMUNOASSAYS

Solid-phase competitive binding RIA for detecting antibody (Fig. 5) uses radiolabeled antibody to compete with the test serum antibody for the finite number of antigen binding sites available on the immobilized solid-phase antigen. As with the indirect procedures described under Indirect Solid-Phase Radioimmunoassay Procedures, the test serum specimen is incubated in stage 1 with the solid-phase antigen, allowing specific antibody, if present, to be captured and bound by the immobilized antigen. Then in stage 2, added ^{125}I-labeled antibody directed against the solid-phase antigen allows radiolabeled antibody to bind to any remaining available binding sites on the immobilized antigen. When specific antibody is present in the test serum, the radiolabeled antibody will have fewer antigen binding sites with which to react. The radioactivity count of the solid phase will therefore be inversely proportional to the amount of specific antibody in the test serum. Competitive binding solid-phase RIA methods have been applied commercially to detect antibodies to hepatitis B core and e antigens.

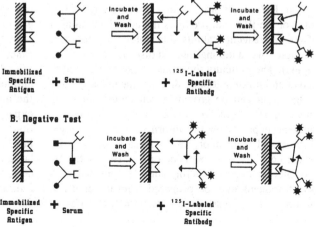

FIG. 5. Competitive binding solid-phase radioimmunoassay. This two-step procedure involves capture of serum antibody by the known antigen bound to the solid-phase surface followed by the addition of radiolabeled probe antibody to combine with all residual available binding sites on the solid-phase antigen. The radioactivity count of the washed solid phase will be inversely proportional to the amount of antibody in the assay specimen.

Radioimmunoprecipitation

The radioimmunoprecipitation (RIP) assay for antibody is a competitive assay carried out in a liquid phase system. The test is usually performed using single, purified proteins as antigen and thus is highly specific. For this test, the viral protein antigen is radiolabeled, added to the test serum, and after incubation, anti-IgG directed against the species of the test serum is added. The anti-Ig cross-links the antigen–antibody complexes already in solution and precipitates the complexes. Radiolabeled viral protein is included in the precipitate if the test serum contains homologous viral antibodies. The precipitate is washed, and the level of radioactivity in the precipitate counted. The amount of radioactivity in the precipitate is proportional to the concentration of specific antibody in the serum.

The accuracy of the RIP test for viral antibody depends on the antigen used. The extreme sensitivity and potential specificity of RIP can be a disadvantage as well as an advantage. Theoretically, the test can detect very low levels of specific antibody; however, it may also detect unwanted cross-reacting antibodies or antibodies to nonviral components of the antigen. For many other types of antibody assays, satisfactory antigens can be purchased or obtained from colleagues without much difficulty. However, the investigator must frequently prepare his own test antigens for an RIP assay. Purifying the antigen, which may be specific viral protein, is often tedious work. RIP tests, therefore, have had only limited success outside the research laboratory.

Enzyme Immunoassay

Among the most useful and widely applied methods of detecting antiviral antibodies are the indirect enzyme immunoassays (EIA or ELISA). The basic principles of EIA are similar to the indirect solid-phase RIA methods described under Indirect Solid-Phase Radioimmunoassay Procedures.

The EIA incorporates many of the desirable features of the RIA and provides the added advantages of greater reagent stability and potential safety (for example, no handling of radioisotope-labeled reagents). Theoretically these methods can be used to detect and titrate antibody directed against any virus antigen that can be grown in cell culture or in an animal host. The degree of purity required for EIA antigens depends on the information being sought (Kenny and Dunsmoor, 1983). Lysates of virus infected cells are often satisfactory as EIA antigens, especially if a broad-reacting antigen is needed. Cell lysate antigens may be prepared from infected cell cultures by harvesting and washing the infected cells, disrupting them by sonication or freeze-thawing, clarifying the suspension by low-speed centrifugation, and semipurifying the antigen by pelleting the viral proteins at high speed centrifugation followed by resuspending the pellets in PBS or buffer. When preparing cell lysate antigens, it is important that the lysate be as free as possible of serum or other protein supplements from the cell culture medium that may contribute adversely to nonspecific reactivity of the antigen. Such cell lysate antigens should always be used together with control antigens consisting of uninfected cells processed in the same way as the virus-infected cells.

If greater specificity is required, gradient-purified virus or viral subunits may be prepared and used as antigen. Control antigens for these more purified materials may not always be available because uninfected cell cultures processed in the same manner as the antigen often contain no residual protein to bind to the solid phase.

The solid phase used in the EIA is critical to the precision and accuracy of the assay. Although polystyrene microtitration plates or strips and tubes have been the most commonly used, other supports such as beads, disks, sticks, and cuvettes have also been used. Reproducibility of the assay depends on well-to-well consistency in the coating of the solid phase with the antigen.

The sensitivity of any enzyme immunoassay is directly related to the properties of the enzyme label used. A suitable enzyme for EIA is one that can be readily linked to the detector antibody without losing functional activity. To provide sensitivity, the enzyme must have a sufficiently high turnover rate to permit adequate amplification of the serologic reaction within a desirable time. The enzyme should have substrates that are degradable to stable and easily measured products. Finally, the enzyme should be relatively inexpensive, nontoxic, and readily available in purified form. To date, horseradish peroxidase and alkaline phosphatase have been the two most widely used enzymes for EIA systems. Methods for conjugating these enzymes with antihuman gamma globulin have been thoroughly described elsewhere (Avrameas, 1969; Shekarchi and Sever, 1986; Wilson and Nakane, 1978).

Indirect EIAs have been used to measure specific antibodies in a wide variety of viral infections, and the number of applications of the test continues to expand rapidly. In the process, many different variations in both the procedure and in the reagents used have developed. All indirect EIA systems, however, have certain common features. Indirect EIA is similar to the indirect SPRIA (method B) described under Indirect Solid-Phase Radioimmunoassay Procedures. All EIA systems require careful washing between steps to remove unbound reactants, which, if not completely removed, might result in nonspecific or false-positive results. A variety of automated and semiautomated devices and readers are available for EIA. These include pipetting devices, plate washers, and spectrophotometric plate readers.

As with test methodology, many systems have been developed for reporting EIA test results. At one extreme, a single serum dilution is tested, and antibody is measured quantitatively in a spectrophotometer, absorbance is relative to the quantity of antibody present (Voller and Bidwell, 1979). Others have used an "index" or ratio of a spectrophotometric reading for the unknown specimen compared with that of a reference calibrator reading. At the other extreme is the use of serial twofold dilutions with a visual or spectrophotometric reading of the test endpoint (Bullock and Walls, 1977).

EIA technology continues to evolve at a rapid rate. New solid-phase carriers, more sensitive and specific reagents, and improved instruments and readers continue to be introduced. Over the past several years, EIA has replaced other traditional test methods for the serodiagnosis of many common viral infections.

Indirect Immunofluorescence Assays

Indirect fluorescent antibody (IFA) tests and indirect solid-phase fluoroimmunoassays (FIA) have been widely used in many clinical and research laboratories for detecting and quantitating antiviral antibodies. The IFA test is a two-step procedure similar in principle to the indirect EIA and RIA (see sections on radioimmunoassay and enzyme immunoassay and Fig. 4) involving application of patient serum to the antigen preparation (usually virus-infected cells fixed on glass microscope slides) followed by incubation of the slide with fluorochrome-labeled anti-human globulin. Although the IFA procedure is rapid and may be useful in detecting antibodies that are difficult or even impossible to demonstrate by any other means, nonspecific fluorescence frequently occurs and results are open to subjective interpretation. This has resulted in less than optimum sensitivity and specificity in many laboratories.

INDIRECT FLUORESCENT ANTIBODY METHOD

IFA test procedures vary slightly according to the specific antigen and test system used. Specific IFA test protocols and references may be found in some of the other chapters of this book. In general, however, the following procedure can be used to identify antibodies to almost any virus that can be grown in cell culture.

Preparation of Antigen Slides

Slides may be prepared using infected cell monolayers showing 2+ to 3+ CPE. The infected cells are suspended and mixed with an equal number of cells from an uninfected culture. After thorough rinsing in PBS, the cells are sedimented by centrifugation and resuspended in a small amount of PBS containing 1 to 2% fetal bovine serum or 0.05% gelatin. The cell suspension is then spotted on glass slides, allowed to air-dry at room temperature, and fixed in acetone for 5 to 10 min. In addition, uninfected cells are prepared separately to serve as an uninfected cell control. Slides with 10 pre-etched wells are routinely used for antigen spots for IFA testing. After fixation, the antigen slides may be stored frozen in airtight envelopes until needed for testing. Antigen and control slides for viral antibody IFA assays are available commercially.

Test Procedure and Controls

Serial dilutions (beginning at 1:2) of each test serum specimen and of the positive control serum are prepared in PBS. One drop of each serum dilution is placed on a corresponding antigen spot on the slide. One drop of undiluted negative control serum and one drop of PBS are placed on corresponding antigen spots as a negative serum control and a nonspecific staining control for the test. The slides are then incubated in a covered moist chamber for 30 min at 35°C. Following the incubation, each slide is gently rinsed and emersed in a staining dish with PBS for 10 min. The slides are then dipped in distilled water, blotted gently, and air-dried for 10 min at 35°C. A drop of fluorescein-labeled anti-human globulin (conjugate) is placed on each antigen spot, and the slides are again incubated in a moist chamber for 30 min at 35°C. After incubating, the slides are again rinsed thoroughly and air-dried, and a cover slip is added using several drops of buffered glycerol (pH 9.0) or Elvanol mounting medium on the slide.

Interpretation

The stained spots are examined as soon as possible using a fluorescence microscope. Fluoresence of infected cells is recorded as follows: 3 to 4+, bright green; 2+, dull yellow-green; 1+, dim yellow-green; Neg, no yellow-green. The controls are also examined and results recorded. The positive control serum should give the expected endpoint titer and the negative control serum should be negative. The endpoint titer of each patient serum is recorded as the highest dilution showing a 1+ or greater fluorescence.

Commercial reagents and kits are available for IFA testing for some viral antibodies. These kits provide antigen slides (for the desired virus), positive and negative control sera, buffered diluent, and fluorescein-labeled anti-globulin conjugate.

ANTICOMPLEMENT IMMUNOFLUORESCENCE METHOD

The anticomplement immunofluorescence (ACIF) test first described by Goldwasser and Shepard (1958) is a modification of the traditional IFA test that involves incubating patient serum on the antigen slide and adding human complement and then fluorescein-labeled anti-human C'3. The ACIF test avoids the nonspecific staining in the IFA test that results from Fc receptors binding human immunoglobulin nonspecifically (Rao et al., 1977). The ACIF assay also gives brighter staining than the IFA test does because of the amplifying effect resulting from the many C'3 molecules bound by each IgG molecule.

The ACIF test is performed by reacting the diluted immune serum with the virus antigen on a slide for 1 h at 35°C or for 30 min at 35°C followed by overnight refrigeration at 4°C. After washing, the appropriately diluted complement is added and allowed to react for 45 min at 35°C. After further washing, fluorescein-labeled anti-human C'3 is added and allowed to react for 45 min. at 35°C. After washing once more, the preparation is rinsed, air-dried, mounted, and examined as described for the IFA test.

SOLID-PHASE FLUORESCENCE IMMUNOASSAY

The principle of the solid-phase fluorescence immunoassay (FIA) is similar to the IFA, except that the viral antigen is immobilized on an opaque solid-phase surface rather than on a glass slide and fluorescence is measured by a fluorometer rather than by a fluorescence microscope. A commercially available assay system using this principle is the FIAX system marketed by M.A. Bioproducts (Walkersville, Md.). The FIAX test has been used to detect and quantitate antibodies to rubella, measles, mumps, and herpes simplex viruses and cytomegalovirus, as well as nonviral antibodies.

Western Immunoblotting

Western immunoblotting provides an extremely sensitive and specific analytical tool for characterizing antiviral antibodies. Gel electrophoresis is used to separate antigens and identify antibodies to specific viral proteins (Bers and Garfin, 1985). The process of transferring proteins from polyacrylamide gels to an immobilizing matrix, usually nitrocellulose paper, is commonly referred to as blotting. Protein blots can

then be subjected to immunologic or biochemical analysis.

Western blotting was originally described by Towbin and colleagues (1979) to detect specific antibodies against ribosomal proteins of *Escherichia coli*. The procedure has since been modified to allow sensitive and specific confirmation of other viral antibodies including human immunodeficiency virus (HTLV-III/LAV) antibodies in serum samples (Sarngadharan et al., 1984; Schupback et al., 1985). Figure 6 summarizes the principal steps of the assay.

1. The proteins from semipurified (density gradient-banded) virus are separated by electrophoresis on a 12% polyacrylamide gel in the presence of sodium dedecyl sulfate as described by Laemmli (1970).
2. The protein bands are then transferred electrophoretically to a nitrocellulose sheet. Proteins can be readily electroeluted out of the gel onto the nitrocellulose matrix. A wet nitrocellulose filter is placed on the gel, and the filter and gel are then sandwiched between supportive porous pads. The supported "gel and filter sandwich" is submerged in a tank containing transfer buffer and placed between two electrodes. Johnson et al. (1984) recommend a transfer buffer consisting of phosphate buffered saline (PBS) containing 5% non-fat dry milk and 0.001% merthiolate. The choice of transfer buffer depends on the type of gel, the particular immobilizing matrix, and the physical characteristics of the proteins of interest.
3. After completing the electrophoretic transfer, the nitrocellulose filter is removed, rinsed, placed between two layers of parafilm, and cut into 3- to 5-mm wide strips; these can be used immediately or stored at -20°C.
4. For testing, the nitrocellulose strips are freed from their parafilm covers, placed in tubes with 2.4 ml of transfer buffer containing 100 ml of normal goat serum, and incubated for 1 h. Test serum is then added in a dilution of 1:50 (50 μl) or 1:100 (25 μl), and the strips are incubated overnight at 4°C; the strips are then thoroughly washed in PBS containing 0.05% Tween 20 to remove all unreacted serum proteins.
5. To detect the bound serum antibodies, a second labeled anti-human immunoglobulin probe, appropriately diluted in transfer buffer, is incubated with the strips. After final washings, the strips are examined for bound labeled probe. Excellent results have been obtained with affinity-purified and ^{125}I-labeled anti-IgG antibody as well as with biotinylated antibody; with the latter, additional incubations with avidin D–horseradish peroxidase are followed by an appropriate substrate. Both detection systems are very sensitive. Direct

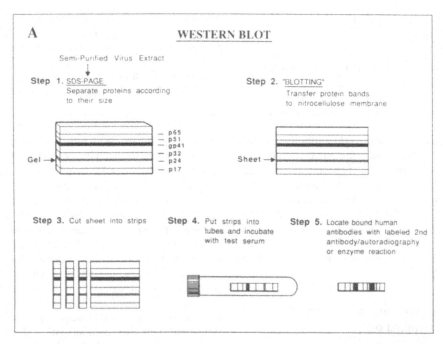

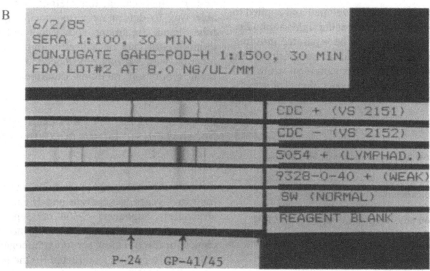

FIG. 6. Western blotting with electrophoretically separated viral proteins for detecting specific antiviral antibodies. (A) Principal steps of the assay. (B) Representative test strips from patients and controls assayed for antibodies to human immunodeficiency virus proteins.

enzyme-labeled anti-immunoglobulins are much less sensitive and should not be used for this procedure.

Strips can be prepared in advance and stored frozen for several months before use. This can substantially shorten the time needed to perform this assay. Interpreting the test results requires comparison with positive and negative sera to establish the precise location of the specific viral proteins on the test strip. Interpretation is easy in cases where antibodies to a variety of viral antigens are present and where the bands are strong. In other cases, however, antibody levels may be weak and only one or two bands may be visible. In addition, bands of cellular proteins may be very close to those of viral proteins. Interpreting Western blot assays in such difficult cases thus depends heavily on the expertise of the reader.

Specific Immunoglobulin M Antibody Detection

IgM antibodies are usually the first immunoglobulins to appear after a primary antigenic stimulus. They are often detectable for only a few weeks and then are replaced by IgG antibodies, which persist for a longer period.

The transient nature of the IgM antibody response appears to be characteristic for most primary viral infections; the determination of specific antiviral IgM antibodies is now a well established method for rapidly diagnosing recent viral infections. For this approach to be successful, the IgM antibody response must be specific, measurable with adequate reliability and sensitivity, and transient (i.e., present only with recent active infection with the specific virus) (Herrmann, 1986).

Methods Used for Immunoglobulin M Antibody Determination

Several methods have been developed and applied to detecting and assaying of specific IgM antibodies (Table 8). These methods can be separated into four general groups: 1) those based on comparing titers before and after chemical inactivation of serum IgM proteins, 2) those based on physicochemical separation of IgM from the other serum immunoglobulin

TABLE 8. Methods for immunoglobulin M (IgM) antibody assay

Methods based on chemical inactivation of IgM
   Alkylation–reduction by mercaptans
Methods based on physicochemical separation of IgM
   Chromatographic gel filtration
   Sucrose density gradient ultracentrifugation
   Ion-exchange chromatography
   Affinity chromatography
   Protein A absorption
Solid-phase indirect immunoassays using labeled anti-human IgM
   Immunofluorescence (IFA)
   Radioimmunoassay (RIA)
   Enzyme immunoassay (EIA)
Reverse "capture" solid-phase IgM assays
   IgM antibody capture EIA
   IgM antibody capture RIA (MACRIA)
   Solid-phase immunosorbent technique (SPIT)
Other assays
   Radioimmunodiffusion
   Radioimmunoprecipitation
   Counterimmunoelectrophoresis
   Anti-IgM hemagglutination
   Latex-IgM agglutination

---

classes, 3) those based on solid-phase indirect immunoassays using labeled anti-human IgM, and 4) reverse "capture" solid-phase IgM assays.

METHODS BASED ON CHEMICAL INACTIVATION OF IMMUNOGLOBULIN M

Mercaptans split the IgM molecule into immunologically inactive parts by breaking disulfide bonds between the polypeptide chains. This simple technique can be used to detect virus-specific IgM antibodies (Banatvala et al., 1967). Briefly, serum antibody titers are measured by standard serologic methods before and after mercaptan treatment. The presence of specific IgM antibody is indicated by a significant decrease in titer of the treated serum aliquot.

The method is simple, but very insensitive. For this test to be positive (i.e., to demonstrate a fourfold or greater decrease in titer between treated and untreated serum), at least 75% of the total virus-specific antibody must be of the IgM class. This would be true only during the early stages of a few viral infections, and therefore, its diagnostic value is very limited. For this reason, this method is not an acceptable way to determine virus-specific IgM. On the other hand, treatment with 2-mercaptoethanol or dithiothreitol may be a useful control step in conjunction with various physicochemical immunoglobulin separation methods described later in this chapter (Caul et al., 1974).

METHODS BASED ON PHYSICOCHEMICAL SEPARATION OF IMMUNOGLOBULIN M

Gel Filtration

Column chromatography methods have been used for many years to separate and isolate serum IgM antibodies. Sephacryl S-300 or Sephadex G-200 (Pharmacia AB, Uppsala, Sweden) are the gels of choice for fractionating serum immunoglobulins. Gel columns are packed as directed by the gel manufacturer. Serum lipoproteins and nonspecific cell agglutinins will elute from these gel columns in the same fractions as IgM; they must be removed before the fractionation if they will interfere with the reliability of the assay of specific antibody activity in the fractions. With both Sephacryl S-300 and Sephadex G-200 columns, IgM is eluted in the first protein peak and IgG in the second. IgA may also be present in the first peak eluted from the Sephadex G-200 column, but not from the Sephacryl S-300 column. Both columns have approximately the same sensitivity in detecting specific viral IgM antibodies.

The specificity of the gel filtration IgM test for diagnosing viral infections is very high, provided several factors that can cause false-positive results are recognized (Pattison et al., 1976). Prolonged storage of serum at -20°C and bacterial contamination of the serum may make the serum pretreatment ineffective. Also, if the serum has been heated to 56°C or higher, IgG will aggregate and may elute in the IgM fractions after gel filtration. To minimize misinterpretation of the gel fractionation results, any presumptive IgM antibody activity in the first peak should be shown to be sensitive to 2-mercaptoethanol.

Sucrose Density Gradient Ultracentrifugation

The most commonly used method of fractionating serum for detecting virus-specific IgM antibodies is high-speed ultracentrifugation of the serum on a sucrose density gradient. Since IgM proteins have a higher sedimentation coefficient (19S) than do other immunoglobulin proteins (7 to 11S), IgM antibodies can be separated from other antibodies by rate-zonal centrifugation. The technique was introduced in 1968 for the rapid diagnosis of recent rubella infection by demonstration of the IgM antibodies (Vesikari and Vaheri, 1968). Since then, modifications of the method have been published (Caul et al., 1976; Forghani et al., 1973), and the test has been applied to other viral infections (Al-Nakib, 1980; Hawkes et al., 1980).

Studies have shown that sucrose density gradient ultracentrifugation may not be as sensitive as some of the more recently developed indirect immunoassays or IgM capture immunoassays. However, because of its high degree of specificity and overall reliability, this method is generally considered the standard for comparison of other new IgM antibody tests. Sucrose gradient ultracentrifugation is a rather laborious procedure, and the high cost of the necessary equipment places it out of financial reach of most clinical diagnostic laboratories. Using vertical rotors with reorienting gradients, however, makes it possible to reduce the centrifugation time from 16 h to only 2 h, making it feasible to test considerably larger numbers of specimens in a given period of time.

Other Physicochemical Separation Methods

Other less frequently used methods for physically separating IgM from the other serum immunoglobulins include ion-exchange chromatography (Johnson and Libby, 1980), affinity chromatography (Barros and Lebon, 1975), and staphylococcal protein A absorption (SPA-Abs) (Ankerst et al., 1974). Ion-exchange chromatography is based on the differential binding of IgM and IgG to anion-exchange resins. Affinity chromatography employs columns of anti-human IgM covalently bound to Sepharose beads to isolate the serum IgM for subsequent assay for specific viral antibodies. Neither of these two

methods has received much attention for IgM antibody assay.

SPA-Abs, on the other hand, has attracted considerable interest as a simple and rapid screening method for IgM antibody. SPA, a cell wall protein present in some *Staphylococcus aureus* strains, binds to the Fc receptor of the IgG molecule and can be used to absorb and remove the IgG component of serum. Antibody detected in serum after SPA-Abs suggests the presence of specific IgM antibody. SPA-Abs does not remove all subgroups of IgG, however, and up to 5% of the original IgG antibody activity may still remain following absorption. This low-level residual antibody activity must not be mistakenly interpreted as representing IgM antibody. SPA-Abs is also used to pretreat serum to remove excess IgG and possible IgG-IgM immune complexes before testing by one of the solid-phase indirect immunoassays using labeled anti-human IgM.

SOLID-PHASE INDIRECT IMMUNOASSAYS

The availability of class-specific antiglobulins has led to the adaptation of several serologic techniques, including IFA (Baublis and Brown, 1968), EIA (Voller and Bidwell, 1976), and RIA (Chau et al., 1983; Knez et al., 1976), for detecting virus-specific IgM antibodies. In these methods, the test serum is incubated with viral antigens bound to a solid-phase surface, and specific IgM antibodies bound to the antigen are subsequently detected with anti-human IgM antibody labeled with a suitable marker (Fig. 7).

Because of the technical simplicity of these methods, commercial indirect immunoassay IgM kits for several viruses, including rubella, herpes simplex, and hepatitis viruses, cytomegalovirus, Epstein–Barr virus, and rotavirus have become available. However, several pitfalls in these methods limit their sensitivity and specificity. These pitfalls can be grouped into three general categories: 1) the quality of available reagents; 2) interference by IgM-class rheumatoid factor; and 3) competition between specific IgG and IgM antibodies in patient serum specimens for available antibody binding sites on the solid-phase antigen. Each of these factors can play

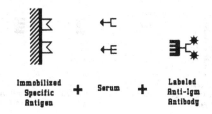

FIG. 7. Indirect solid-phase immunoassay for IgM antibody.

an important role in the reliability of an indirect IgM antibody assay system.

Substantial improvements in the quality of reagents for these assays have occurred during the past few years. More highly purified antigens and highly specific anti-IgM globulin, including monoclonal antibody, conjugates are now available.

False-positive results may occur in these IgM antibody assays if IgM with anti-IgG activity but with no antiviral specificity (i.e., rheumatoid factor) becomes attached to complexes of specific IgG and the bound solid-phase antigen (Fig. 8). On the other hand, failure to detect specific antiviral IgM may occur in serum with high levels of specific antiviral IgG due to the competition for antigen binding sites. Both of these problems can be minimized by preadsorption of the serum with staphylococcal protein A or protein A-Sepharose. Such preadsorption has been shown to effectively eliminate nonspecific IgM activity from serum with known rheumatoid factor and to significantly increase the sensitivity of the specific IgM assay by removing most competing IgG (Kronvall and Williams, 1969).

REVERSE "CAPTURE" SOLID-PHASE IMMUNOGLOBULIN M ASSAYS

Another way to avoid the problems of competitive interference and nonspecific reactivity inherent in indirect immunoassays is by using the reverse IgM antibody capture method (Fig. 9). This method of detecting antiviral IgM antibodies employs a solid phase coated with anti-IgM antibody to "capture" and bind the IgM antibodies in a serum specimen, after which IgG and any immune complexes in the specimen are washed away (Duermeyer and Van der Veen, 1978). Exposure of the bound IgM antibody to specific viral antigen, followed by the addition of a second, labeled anti-viral antibody, completes the test. This approach is attracting considerable support, and it has been used to detect IgM antibodies to hepatitis A, herpes simplex, and rubella viruses, cytomegalovirus, and hepatitis B surface antigen.

The reverse capture solid-phase IgM assays have proved to be very sensitive and specific (Mortimer et al., 1981). Our experience confirms that the IgM cap-

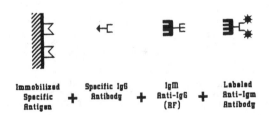

FIG. 8. False-positive results due to rheumatoid factor (RF) interference in the indirect solid-phase immunoassay.

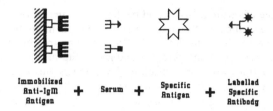

FIG. 9. Reverse capture solid-phase immunoassay for IgM antibody.

ture assays are potentially more sensitive than the standard assay based on sucrose density gradient fractionation. Because the first step in the IgM capture separates IgM antibodies from other serum components, competition between IgG and IgM does not occur. Rheumatoid factor (RF) interference, however, may occur in these IgM capture assays, as it does with the indirect IgM immunoassays described earlier. IgM–RF may be "captured" and bound on the solid phase and then in turn bind labeled antiviral IgG antibodies. Usually, however, interference by RF in these IgM capture assays can be controlled by two simple methods. First, because RF binds to the Fc portion of the IgG molecule, false-positive results can be avoided by using labeled F(ab')₂ fragments as the indicator antibody; and second, since the affinity of RF for aggregated Ig is much greater than for native IgG, the binding sites of the RF can be blocked by adding aggregated Ig to the serum dilution buffer. In general, the reverse capture IgM assays are less susceptible to interference by RF than are traditional indirect solid-phase IgM assays. Because RF interference is relatively simple to eliminate in the reverse capture IgM assay, this assay is potentially more specific than are the indirect immunoassays that use labeled human anti-IgM conjugates.

Interpretation of Immunoglobulin M Antibody Test Results

The presence of specific antiviral IgM antibodies indicates a recent or current infection with the virus in question only if the IgM response is specific (i.e., if these IgM antibodies are not produced by any other infection or condition). The absence of specific IgM antibodies, on the other hand, can rarely be used as evidence to exclude a recent infection with a given virus. Variations in the temporal appearance of IgM antibodies, including the occurrence of prolonged IgM antibody responses, can result in difficulties in interpreting the significance of the test results in relation to the clinical illness in question.

False-positive IgM antibody results may occur because of cross-reactions between closely related viruses. Such cross-reactions have been reported in arbovirus (Wolff et al., 1981) and coxsackie B virus infections (Schmidt et al., 1968). In general, the heterologous IgM antibody responses are low compared with the homologous titers.

Evidence suggesting the occurrence of true polyclonal IgM production in cases of acute infectious mononucleosis has been reported by Morgan-Capner and colleagues (1983). Their report suggests that production of various IgM antibodies may result from Epstein–Barr virus-induced stimulation of B lymphocytes already committed by prior antigenic stimulation. These results emphasize the importance of carefully considering assay results together with the full clinical picture.

In infections with viruses belonging to groups of closely related strains or serotypes (e.g., enteroviruses, adenoviruses, parainfluenza viruses, or togaviruses), serodiagnosis using specific IgM testing may be complicated by the possible absence of a specific IgM response as well as by possible false-positive reactions to related viruses. Specific IgM antibody responses are generally absent in reinfections or reactivations of latent virus infections, and may be very weak or absent in certain immunocompromised patients.

Finally, the expected duration of the specific IgM response must be considered when interpreting the significance of observed specific IgM antibody. Generally, the IgM antibody response following an acute viral infection is of limited duration, usually 1 to 2 months. However, prolonged elevated IgM antibody titers have been observed in some arbovirus infections, in complicated infections, chronic infections, congenital infections, and in some immunosuppressed patients. The persistence of specific IgM in these cases appears usually to be related to the persistence of viral antigen (or even replicating virus) in the patient. Occasionally, prolonged IgM antibody responses have been observed without any apparent reason. Also, as more sensitive methods are developed for detecting specific antiviral IgM antibodies, the time following an acute infection during which specific IgM is detectable will be extended. For the diagnosis of an acute infection, the ideal maximum duration of specific IgM antibodies should be 2 to 3 months. It may therefore be necessary to limit the sensitivity of some assays to retain the optimal diagnostic usefulness of the methods.

The diagnostic value of specific IgM antibody assay is variable and depends on the virus and the infection in question. Generally, transient IgM responses are characteristic of acute viral infections caused by viruses that elicit long-lasting immunity, such as rubella, measles, mumps, and hepatitis A. In these infections, a reliable diagnosis can usually be made or excluded by specific IgM antibody testing of a single serum specimen taken early in the illness.

For other virus infections, such as herpes simplex or cytomegalovirus, the diagnostic usefulness of such tests is much more limited.

Literature Cited

Alexander, S., and J. H. Elder. 1984. Carbohydrate dramatically influences immune reactivity of antisera to viral glycoprotein antigens. Science 226:1328–1330.

Al-Nakib, W. 1980. A modified passive-haemagglutination technique for the detection of cytomegalovirus and herpes simplex virus antibodies: application in virus-specific IgM diagnosis. J. Med. Virol. 5:287–293.

Al-Nakib, W., J. M. Best, and J. E. Banatvala. 1975. Rubella-specific serum and nasopharyngeal immunologic responses following naturally acquired and vaccine-induced infection—prolonged persistence of virus-specific IgM. Lancet 1:182–185.

Ankerst, J., P. Christensen, L. Kjellen, and G. Kronvall. 1974. A routine diagnostic test for IgA and IgM antibodies to rubella virus: absorption of IgG with Staphylococcus aureus. J. Infect. Dis. 130:268–273.

Avrameas, S. 1969. Coupling of enzymes to proteins with glutaraldehyde. Use of the conjugates for detection of antigens and antibodies. Immunochemistry 6:43–52.

Banatvala, J. E., J. M. Best, E. A. Kennedy, E. E. Smith, and M. E. Spence. 1967. A serological method for demonstrating recent infection by rubella virus. Br. Med. J. 3:285–286.

Barros, M. F., and P. Lebon. 1975. Separation des anticorps IgM anti-rebeole par chromatographie d'affinite. Biomed. Express (Paris) 23:184–188.

Bers, G., and D. Garfin. 1985. Protein and nucleic acid blotting and immunobiochemical detection. Bio. Tech. 3:276–288.

Baublis, J. V., and G. C. Brown. 1968. Specific responses of the immunoglobulins to rubella infection. Proc. Soc. Exp. Biol. Med. 128:206–210.

Bullock, S. L., and K. W. Walls. 1977. Evaluation of some of the parameters of the ELISA. J. Infect. Dis. (suppl.) 136:S279–S279.

Caul, E. O., S. J. Hobbs, P. C. Roberts, and S. K. R. Clarke. 1976. Evaluation of simplified sucrose gradient method for the detection of rubella-specific IgM in routine diagnostic practice. J. Med. Virol. 2:153–163.

Caul, E. O., G. W. Smyth, and S. K. R. Clarke. 1974. A simplified method for the detection of rubella-specific IgM employing sucrose density fractionation and 2-mercaptoethanol. J. Hyg. 73:329–340.

Chau, K. H., M. P. Hargie, R. H. Decker, I. K. Mushahwar, and L. R. Overby. 1983. Serodiagnosis of recent hepatitis B infection by IgM class anti-HBc. Hepatology 3:142–149.

Duermeyer, W., and J. Van der Veen. 1978. Specific detection of IgM antibodies by ELISA, applied in hepatitis A. Lancet 2:684–685.

Finney, D. J. 1971. Probit analysis, 3rd ed., p. 20–49. Cambridge University Press, Cambridge.

Forghani, B., N. J. Schmidt, and E. H. Lennette. 1973. Demonstration of rubella IgM antibody by indirect fluorescent antibody staining, sucrose density gradient centrifugation and mercaptoethanol reduction. Intervirology 1:48–59.

Goldwasser, R. A., and C. D. Shepard. 1958. Staining of complement and modifications of fluorescent antibody procedures. J. Immunol. 80:122–131.

Hall, E. C., and M. B. Felker. 1970. Reproducibility in the serological laboratory. Health Lab. Sci. **7:**63–68.

Hawkes, R. A. 1979. General principles underlying laboratory diagnosis of viral infections, p. 3–48. *In* E. H. Lennette and N. J. Schmidt (ed.), Diagnostic procedures for viral, rickettsial and chlamydial infections. American Public Health Association, Washington, D.C.

Hawkes, R. A., C. R. Boughton, V. Ferguson, and N. I. Lehmann. 1980. Use of immunoglobulin M antibody to hepatitis B core antigen in diagnosis of viral hepatitis. J. Clin. Microbiol. **11:**581–583.

Herrmann, K. L. 1986. IgM determinations, p. 219–228. *In* S. Specter and G. J. Lancz (ed.), Clinical virology manual. Elsevier Science Publishing Co., New York.

Hierholzer, J. C., and M. T. Suggs. 1969. Standardized viral hemagglutination and hemagglutination-inhibition tests. I. Standardization of erythrocyte suspensions. Appl. Microbiol. **18:**816–823.

Hierholzer, J. C., and M. T. Suggs, and E. C. Hall. 1969. Standardized viral hemagglutination and hemagglutination-inhibition tests. II. Description and statistical evaluation. Appl. Microbiol. **18:**824–833.

Johnson, D. A., J. W. Gautsch, J. R. Sportsman, and J. H. Elder. 1984. Improved technique utilizing non-fat dry milk for analysis of proteins and nucleic acids transferred to nitrocellulose. Gene Anal. Tech. **1:**3–8.

Johnson, R. B., Jr., and R. Libby. 1980. Separation of immunoglobulin (IgM) essentially free of IgG from serum for use in systems requiring assay of IgM-type antibodies without interference from rheumatoid factor. J. Clin. Microbiol. **12:**451–454.

Kärber, G. 1931. Beitrag zur kollektiven behandlung pharmakologischer reihenversuche. Arch. Exp. Pathol. Pharmakol. **162:**480–483.

Kenny, G. E., and C. L. Dunsmoor. 1983. Principles, problems, and strategies in the use of antigenic mixtures for enzyme-linked immunosorbent assay. J. Clin. Microbiol. **17:**655–665.

Knonvall, G., and R. C. Williams, Jr. 1969. Differences in anti-protein A activity among IgG subgroups. J. Immunol. **103:**828–833.

Laemmli, V. K. 1970. Cleavage of standard proteins during the assembly of the head of bacteriaphage T4. Nature (London) **227:**680–685.

Lennette, E. T., and D. A. Lennette. 1978. Immune adherence hemagglutination: alternative to complement-fixation serology. J. Clin. Microbiol. **7:**282–285.

Lennette, E. T., and D. A. Lennette. 1986. Immune adherence hemagglutination, p. 209–218. *In* S. Specter and G. L. Lancz (ed.), Clinical virology manual. Elsevier Science Publishing Co., New York.

Mims, C. S., and D. O. White. 1984. The immune response to viral infection, p. 87–131. *In* C. S. Mims and D. O. White (ed.), Viral pathogenesis and immunology. Blackwell Scientific Publishers, Palo Alto, Calif.

Morgan-Capner, P., R. S. Tedder, and J. E. Mace. 1983. Rubella-specific IgM reactivity in sera from cases of infectious mononucleosis. J. Hyg. **90:**407–413.

Mortimer, P. P., R. S. Tedder, M. H. Hambling, M. S. Shafi, F. Burkhardt, and U. Schilt. 1981. Antibody capture radioimmunoassay for anti-rubella IgM. J. Hyg. **86:**139–153.

Mushahwar, I. K., and T. A. Brawner. 1986. Radioimmunoassay, p. 111–131. *In* S. Specter and G. J. Lancz (ed.), Clinical virology manual. Elsevier Science Publishing Co., New York.

Mushahwar, I. K., and L. R. Overby. 1983. Radioimmune assays for diagnosis of infectious diseases, p. 167–194. *In* F. S. Ashkar (ed.), Radiobioassays. CRC Press, Boca Raton, Fla.

Nelson, D. S. 1963. Immune adherence. Adv. Immunol. **3:**131–180.

Neumann, P. W., and J. M. Weber. 1983. Single radial hemolysis test for rubella immunity and recent infection. J. Clin. Microbiol. **17:**28–34.

Palmer, D. F., and H. L. Casey. 1981. A guide to the performance of the standardized diagnostic complement fixation method and adaptation to micro test. Centers for Disease Control, Atlanta.

Palmer, D. F., J. J. Cavallaro, K. Herrmann, J. A. Stewart, and K. W. Walls. 1977. Procedural guide for the serodiagnosis of toxoplasmosis, rubella, cytomegalic inclusion disease, and herpes simplex, pp. 24–56, 90–99. Immunology Series No. 5. U.S. Department of Health, Education, and Welfare, Centers for Disease Control, Atlanta.

Parratt, D., H. McKenzie, K. H. Nielsen, and S. J. Cobb. 1982. Radioimmunoassay of antibody and its clinical applications, p. 19–31. John Wiley & Sons Ltd., Chichester, England.

Pass, R. F., P. D. Griffiths, and A. M. August. 1983. Antibody response to cytomegalovirus after renal transplantation: comparison of patients with primary and recurrent infections. J. Infect. Dis. **147:**40–46.

Pattison, J. R. (ed.). 1982. Laboratory investigation of rubella, p. 19–25. Public Health Laboratory Service, Monograph Series No. 16, Her Majesty's Stationery Office, London.

Pattison, J. R., J. E. Mace, and D. S. Dane. 1976. The detection and avoidance of false-positive reactions in tests for rubella-specific IgM. J. Med. Microbiol. **9:**355–357.

Rao, N., D. T. Waruszewski, J. A. Armstrong, R. W. Atchison, and M. Ho. 1977. Evaluation of anticomplement immunofluorescence test in cytomegalovirus infection. J. Clin. Microbiol. **6:**633–638.

Reed, L. J., and H. Muench. 1938. A simple method of estimating fifty per cent endpoints. Am. J. Hyg. **27:**493–497.

Russell, S. M., D. McCahon, and A. S. Beare. 1975. A single radial hemolysis technique for the measurement of influenza antibody. J. Gen. Virol. **27:**1–10.

Sarngadharan, M. G., M. Popovic, L. Bruch, J. Schupbach, and R. C. Gallo. 1984. Antibodies reactive with a human T lymphotropic retrovirus (HTLV-III) in the sera of patients with acquired immune deficiency syndrome. Science **224:**506–508.

Schmidt, N. J., E. H. Lennette, and J. Dennis. 1968. Characterization of antibodies produced in natural and experimental coxsackievirus infections. J. Immunol. **100:**99–106.

Schupbach, J., O. Haller, M. Vogt, R. Luthy, H. Joller, O. Oelz, M. Popovic, M. G. Sarngadharan, and R. C. Gallo. 1985. Antibodies to HTLV-III in Swiss patients with AIDS and pre-AIDS and in groups at risk for AIDS. N. Engl. J. Med. **312:**265–270.

Shekarchi, I. C., and J. L. Sever. 1986. Enzyme immunoassay, p. 133–146. *In* S. Specter and G. L. Lancz (ed.), Clinical virology manual. Elsevier Science Publishing Co., New York.

Sissons, J. G., and M. B. A. Oldstone. 1980. Antibody-mediated destruction of virus infected cells. Adv. Immunol. **29:**209–260.

Towbin, H., T. Staehelin, and J. Gordon. 1979. Electrophoretic transfer of proteins from polyacrylamide gels to nitrocellulose sheets: procedure and some applications. Proc. Natl. Acad. Sci. USA **76:**4350–4354.

U.S. Public Health Service. 1965. Standardized diagnostic complement fixation method and adaptation to micro test. Public Health Service Publication no. 1228 (public health monograph no. 74). U.S. Government Printing Office, Washington, D.C.

Vesikari, T., and A. Vaheri. 1968. Rubella: a method for rapid diagnosis of a recent infection by demonstration of the IgM antibodies. Br. Med. J. **1:**221–223.

Voller, A., and D. E. Bidwell. 1975. A simple method for detecting antibodies to rubella. Br. J. Exp. Pathol. **56:**338–339.

Voller, A., and D. E. Bidwell. 1976. Enzyme immunoassays for antibodies in measles, cytomegalovirus infections and after rubella vaccination. Br. J. Exp. Pathol. **57:**243–247.

Wilson, M. B., and P. K. Nakane. 1978. Recent developments in the periodate method of conjugating horseradish peroxidase, p. 215–224. *In* W. Knass, K. Holubar and G. Wick (ed.), Immunofluorescence and related staining techniques. Elsevier/North Holland, Amsterdam.

Wolff, K. L., D. J. Muth, B. W. Hudson, and D. W. Trent. 1981. Evaluation of the solid-phase radioimmunoassay for diagnosis of St. Louis encephalitis infection in humans. J. Clin. Microbiol. **14:**135–140.

Wood, R. J., and T. M. Durham. 1980. Reproducibility of serological titers. J. Clin. Microbiol. **11:**541–545.

CHAPTER 6

Immunoreagents

M. H. V. VAN REGENMORTEL

Polyclonal Antisera: Immunization of animals; titer and specificity of antisera; cross-reactions between viral serotypes.

Preparation of Antisera: Adeno-, arbo-, arena-, entero-, herpes, influenza, paramyxo-, reo-, and rhinoviruses.

Commercial Sources of Reagents: Antisera and viral antigens.

Monoclonal Antibodies: Production and selection of hybridoma; influence of immunoassay on antibody specificity.

Molecular Structure of Viral Antigens: Types of antigenic determinants in viruses.

Viral Antigens: Purification of virions, preparation of viral proteins by dissociation of virions and by recombinant DNA technology.

Synthetic Peptides: Use of peptides as antigen in viral diagnosis; synthetic peptides as immunogens for raising antibodies that cross-react with viral antigens.

Introduction

In recent years, the laboratory diagnosis of viral infections by isolation and cultivation of the virus is increasingly being replaced by new immunoassay methods that permit detection of viral antigens directly in clinical specimens. These new methods, which avoid the delays that occur when virus is grown in cell culture, are usually referred to as methods of rapid viral diagnosis (Grandien, 1983; Halonen et al., 1984; Schmidt, 1983). They are particularly useful when the virus cannot be cultivated or when it is no longer viable in the specimen (Lennette, 1985). Since the viral antigen is present in the specimen primarily at a very low concentration, it is essential to use highly sensitive immunoassays as well as high-quality reagents. In practice, this means that the level of contamination or interfering substances that can be tolerated in immunoreagents is lower today than in the early days of classical serology.

It is possible to increase the specificity of viral antisera by affinity chromatography or immunoadsorption (Stott and Tyrrell, 1986). Either the antigen of interest is used to adsorb antibodies from the anti-

serum and these are subsequently obtained in pure form by dissociation from the immune complex, or the heterologous antigen is used to remove unwanted antibodies from the antiserum (Dean et al. 1985). Such fractionation can render an antiserum highly specific and increase the relevant effective antibody concentration. However, because of limited supplies of purified antigens, it is not always possible to improve the quality of an antiserum by absorption procedures. Fortunately, the advent of hybridoma technology has made it possible nowadays to completely overcome problems caused by contaminated reagents. When a virus preparation used for preparing monoclonal antibodies (McAb) is contaminated with host antigens or with other viruses, it is still possible during the screening of positive clones to select McAbs that will react only with the virus of interest.

The purpose of this chapter is to review the principles that govern the preparation of reagents suitable for the serological diagnosis of virus infections. No attempt will be made to describe in detail the precise experimental protocols used for purifying viruses, synthesizing peptides, preparing monoclonal antibodies, or immunizing experimental animals.

These techniques have been described at length in recent texts, and the reader will be referred to the specialized literature covering these topics. Comprehensive reviews of viral serodiagnosis are available (Gardner and McQuillin, 1980; Kurstak and Kurstak, 1977; Lennette and Schmidt, 1979; Lennette et al., 1985; Rose et al., 1986; Van Regenmortel and Neurath, 1985).

Polyclonal Antisera

Immunization of Experimental Animals

Few systematic studies have been published concerning the most efficient immunization procedures for obtaining viral antisera from experimental animals. There are several reasons for the lack of reliable information concerning the relative merits of different immunization protocols. Since the immune response measured in individual outbred animals submitted to the same immunization schedule can vary greatly, few workers have attempted to collect the extensive data that would be required to demonstrate the superiority of any particular procedure. Large differences have been recorded in the titer and specificity of antisera obtained from different animals submitted to the same immunization regimen, as well as in antisera of the same animal taken at different stages of immunization (Van Regenmortel and Von Wechmar, 1970). The magnitude of these differences is such that small variations in the effectiveness of different protocols are likely to be overshadowed, especially if the number of animals used for comparison is small.

Another reason for the lack of extensive comparative studies is simply that a wide variety of immunization procedures have been found to give satisfactory results (Dresser, 1986; Horwitz and Scharff, 1969; Stott and Tyrrell, 1986). Viral antigens are mostly good immunogens, and the experience of many workers shows that adequate antisera can be obtained by several different methods. However, because of genetic and physiological animal variability, it is important always to use several animals for antiserum production. In view of the influence of age on immune responsiveness (Makinodan and Kay, 1980), animals should be used neither too young nor too old. Rabbits are the most commonly used animals for antiserum production because of the ease with which they can be injected and substantial volumes of blood collected. Volumes of 30 ml of blood can be obtained repeatedly by puncture of the marginal ear vein of a rabbit, in contrast to a maximum of 1.0 ml obtainable from a mouse by heart puncture.

After repeated injections of a viral antigen, the antibody concentration in the serum will reach a plateau level, indicating that the immune response to an antigen is self-limiting. This phenomenon is brought about by regulatory mechanisms such as antibody feedback (Uhr and Moller, 1968), cell-mediated suppression, and tolerance (Dresser, 1986; Dresser and Mitchison, 1968; Ivanyi and Cerny, 1969).

Using a larger dose of immunogen than is required for eliciting an adequate antibody level is wasteful and can even be counterproductive because of the phenomenon of immunologic tolerance. Furthermore, since contaminants may then be present at a level at which they will also elicit an immune response, it is advisable to inject the minimum quantity of viral antigen that will induce an antiserum of satisfactory titer. Since no preparation of viral antigen is ever likely to be absolutely pure, this precaution is important in all cases. An empirical approach, whereby a series of animals is immunized with a range of doses of antigen, offers the best chance of success.

The amount of viral antigen administered to animals in different published procedures varies widely. The reported dose of antigen used for injection is not proportional to the body weight of the animal, since quantities of 1 to 10 mg of purified antigen per inoculation have been used with rabbits, goats, and horses. It seems that doses of 100 to 200 μg of purified virus are in fact sufficient to immunize rabbits and goats and that larger doses do not lead to proportionally higher antibody levels. In the case of mice, good results have been reported with immunizing doses of 2 to 20 μg of purified virus (Reddecliff and Ludwig, 1966). As little as 10 to 100 ng of virus has been found to trigger a secondary response in the mouse (Marbrook and Matthews, 1966). Mice can be injected intravenously in the tail, intraperitoneally, or intramuscularly and can be bled either from the tail or by heart puncture under anesthesia. When mice are injected with Ehrlich ascites tumor cells about a week after the last immunizing dose of antigen, the animals will develop distended abdomens from which volumes of up to 10 ml of ascitic fluid can be drained. Detailed accounts of procedures used for handling mice and other experimental animals are available (Garvey et al., 1977; Herbert and Kristensen, 1986).

Routes of immunization reported in the literature for different types of animals vary widely (Polley, 1977). Satisfactory results have been obtained by intravenous, intradermal, subcutaneous, intraperitoneal, and intramuscular injections. One of the few generalizations that can be drawn from the vast amount of published information on immunization protocols is that the use of adjuvant economizes on the amount of antigen needed (Warren et al., 1986). The most widely used adjuvant is Freund's adjuvant, which contains mineral oil (known under various commercial names such as Bayol F, Drakeol F) and an emulsifier such as mannide monooleate (Arlacel A) in ratios varying from 2:1 to 9:1. The so-called Freund's complete adjuvant contains, in addition, killed mycobacteria (2 mg dry weight cells/ml); without the bacteria, it is known as Freund's incomplete adjuvant. Both types of adjuvant are available commercially or can be prepared more economically from the various ingredients (Dresser, 1986; McLean, 1982; Van Regenmortel, 1982). A virus–adjuvant emulsion can be prepared by repeated aspirations of a mixture of 3 parts adjuvant and 1 part virus preparation through a syringe (Herbert, 1978).

One method for obtaining large quantities of viral antibodies that obviates the need to bleed animals consists in immunizing laying hens. The hens are readily immunized by a series of intramuscular injections of 50 to 200 μg antigen emulsified in Freund's adjuvant per ml. Viral antibody is conveniently obtained from the egg yolk collected from such immunized animals. The passage of antibodies from the serum of hens to egg yolk is a well-documented phenomenon (Brambell, 1970). One egg yolk contains approximately 100 mg of immunoglobulin. The average yield of purified immunoglobulin obtained by a simple extraction method with polyethylene glycol precipitation (Polson et al., 1980) was found to be 75 mg per egg (Van Regenmortel, 1982). Avian globulins extracted from egg yolk have been shown to be a cheap and convenient source of pure immunoreagent suitable for detecting a variety of viruses (myxo- and paramyxoviruses, rotaviruses, plant viruses) by precipitation (Polson et al., 1980), ELISA (Obert et al., 1981; Van Regenmortel et al., 1983; Yolken et al., 1984; Zrein et al., 1986b), radioimmunoassay (Katz et al., 1985), immunosorbent electron microscopy (Nicolaïeff et al., 1980), immunofluorescence (Gardner and Kaye, 1982), and complement fixation (Van Regenmortel and Burckard, 1985). Since chicken antibodies do not bind to protein A, it is possible to use yolk globulins as trapping antibodies in a solid phase radioimmunoassay. The antigen, bound by the chicken antibodies, can be detected by rabbit antibodies followed by radiolabeled protein A (Katz et al., 1985). Some of the advantages of using yolk antibody in the serodiagnosis of virus infections are: 1) ease of preparation of large quantities of reagent; 2) low cost; 3) anti-host antibodies are avoided when virus grown in embryonated eggs is used for immunizing laying hens; 4) absence of antigenic cross-reactivity between avian and mammalian globulins allows the reagent to be used in multi-layer ELISA procedures (Al Moudallal et al., 1984); and 5) absence of interference by rheumatoid factors in immunoassays, since avian antibody does not bind to mammalian Fc receptors.

Titer and Specificity of Antisera

The suitability of a viral antiserum for any particular immunoassay depends on the concentration of relevant antibodies it contains and on their affinity and specificity. The antibody concentration is rarely known or even measured, and usually it is recorded only in terms of antiserum titer. In general, this refers to the maximum extent to which the antiserum can be diluted while still performing adequately in a given assay. Since the sensitivity of different immunoassays varies considerably, the titer of an antiserum can be defined only with respect to a particular serological test. The same antiserum may, for instance, have a precipitation titer of 10^{-3}, corresponding to the last dilution at which a precipitate is still visible, and a radioimmunoassay titer of 10^{-5}, corresponding to the dilution that will bind 50% of the labeled antigen. Expressing antiserum titers as dilution endpoints is not very satisfactory, since this procedure relies solely on experimental points obtained when the measured signal is very small. A curve-fitting program that incorporates all experimental points can be used to increase the reliability of a calculated titer expressed as dilution endpoint (Kurstak, 1985; Zrein et al., 1986a). Alternatively, in an assay such as ELISA, the antiserum titer can be expressed as the dilution that produces an absorbance of, say, 1.0. Obviously, it is essential to adhere to strictly standardized assay conditions to ensure the validity of comparative measurements.

The degree of antigenic similarity between two viruses is usually estimated from the homologous and heterologous antiserum titers measured in reciprocal tests, with antisera used against each of the two viruses. It has been found expedient to express the degree of antigenic cross-reactivity between two viruses by the number of twofold dilution steps separating homologous from heterologous antiserum titers. This number, which is called the serological differentiation index or SDI (Van Regenmortel and Von Wechmar 1970), can be derived by any technique such as precipitation or ELISA (Jaegle and Van Regenmortel, 1985), but is reliable only if it represents an average value calculated from several different antisera. Average SDI values between two viruses obtained in reciprocal tests have been shown to agree closely (Van Regenmortel, 1982). In the tobamovirus group, the extent of antigenic cross-reactivity measured in this fashion was found to correlate with the extent of sequence similarity in the coat protein of different viruses (Paul et al., 1980; Van Regenmortel, 1985).

The quality of an antiserum depends not only on the antibody concentration, but also on the relative affinity of the different antibodies present in the serum. In the case of multivalent antigens such as viruses, the binding strength of an antibody should be expressed as its avidity rather than by a classical affinity constant derived from the law of mass action (Fazekas de St Groth, 1979; Van Regenmortel, 1982; Yolken, 1985). In polyclonal situations, the mathematical relationships defining the interaction of individual antibody molecules of different avidity with multivalent virions are complex and poorly understood. In practice, it is possible to estimate the relative avidity of different antisera from the comparative slopes in dose-response curves of signal versus antigen or antibody concentration. For instance, antiserum avidity has been estimated from standard curves obtained by nephelometry, by comparing the antigen concentrations that correspond to 100% and 50% of the maximum nephelometric response (Hudson et al., 1981).

The specificity of an antiserum is related to the absence of cross-reactions with heterologous antigens, this is, with antigens other than the one used for preparing the antibody. It is important to distinguish genuine, intrinsic cross-reactions due to a particular antibody reacting with more than one antigen (Berzofsky and Schechter, 1981) from contaminating reactions that arise because the antiserum contains a mixture of antibodies specific for different unrelated antigens. Any antiserum obtained from an experimental animal after a course of immunization will always contain, in addition to the "relevant" antibodies, a wide range of "natural" antibodies against a variety of antigens. For instance, preimmunization rabbit antisera often contain antibodies against reoviruses induced by an earlier infection unbeknown to the investigator. It is important, therefore, always to collect a control preimmunization serum from an animal before embarking on an immunization program. Clearly, the presence in antiserum of antibodies against unrelated antigens does not interfere with the usefulness of the serum, unless the corresponding antigens are also found in the clinical sample to be studied.

Antibodies to host tissues are the most common type of contamination that has to be avoided in order to secure the specificity of serological testing. This can be achieved by raising antiviral antibodies in the same species of host as the one used for propagating the virus intended for immunization. For instance, crude preparations of arboviruses propagated in infant mice are commonly used to immunize adult mice, since no antibodies against the host antigens will then be elicited. Another example is the immunization of hens with influenza viruses grown in eggs (Gardner and Kaye, 1982). When laying hens are used, the viral antibodies can be collected from the yolk of eggs laid by the animal. Another, less commonly used method, which also avoids the need to immunize animals with highly purified viral antigens,

is to render the animals tolerant to contaminating antigens by immunizing them beforehand with high doses of the contaminants (Hansen et al., 1981).

Antisera from hyperimmune animals, that is, animals submitted to multiple doses of the same immunogen, tend to cross-react with a wider range of related viral serotypes than antisera obtained after a single exposure of viral antigen. If a reagent of narrow specificity is required, it is thus preferable to limit the number of injections given to the animal. Alternatively, it is also possible to obtain reagents of high specificity by absorbing the antiserum with one or more related viral serotypes (Baxby, 1982; Johansson et al., 1979). Methods to absorb virus antibodies or host antibodies from antisera have been described by Stott and Tyrrell (1986). A convenient way to remove antibodies reacting with heterologous serotypes is by intragel cross-absorption (Van Regenmortel, 1982).

Antisera used for hemagglutination tests usually have to be treated by a variety of procedures to remove nonspecific inhibitors that interfere with the hemagglutination reaction. Methods suitable for removing these inhibitors have been listed in earlier reviews (Stott and Tyrrell, 1985; Van Regenmortel, 1981).

Procedures Used with Different Groups of Viruses

ADENOVIRUSES

Purified preparations of whole virus or of hexons are desirable to produce high-quality antiserum suitable for most immunoassays. Animals are inoculated by the intramuscular route with Freund's complete adjuvant. Blood is collected 3 to 4 weeks after a booster dose. Antisera raised against different adenovirus serotypes have been extensively used for classifying members of the Adenoviridae (Hierholzer et al., 1975; Stevens et al., 1967).

ARBOVIRUSES

Antisera can be prepared in mice by a series of intraperitoneal injections of freshly prepared suspensions of infected mouse brain (Shope and Sather, 1979). A single subcutaneous injection in hamsters has also been shown to be effective (Karabatsos and Mathews, 1980). The preparation of arbovirus antigens has been described by Monath (1986).

ARENAVIRUSES

Detailed procedures for obtaining antisera to arenaviruses in mice, hamsters, and guinea pigs have been described by Casals (1979).

ENTEROVIRUSES

Antisera have been obtained from mice, guinea pigs, and rabbits (Melnick et al., 1979), as well as from large domestic animals such as horses, sheep, and goats (Hampil and Melnick, 1968; Hampil et al., 1965).

HERPESVIRUSES

Antisera can be raised in rabbits, guinea pigs, horses, and sheep. Detailed immunization protocols have been described by Rawls (1979). Antiserum to the varicella-zoster virus has been raised in rabbits by intramuscular inoculation of purified virus emulsified with Freund's complete adjuvant (Kissling et al., 1968).

INFLUENZA VIRUSES

Rabbits can be immunized by repeated intravenous inoculations of 10^5 to 10^6 infectious doses of virus. Chickens have been immunized by intravenous inoculation of 5 ml of allantoic fluid containing at least 160 virus hemagglutination units (Dowdle et al., 1979). In this case, no antibodies will be produced to egg host components in the immunized chicken. Large quantities of virus antibodies can be obtained from the yolk of eggs laid by such immunized hens (Gardner and Kaye, 1982; Van Regenmortel et al., 1983). Although it is often said that such avian reagents cannot be used for complement fixation tests with myxo- or paramyxoviruses, it is possible to carry out such tests by using a partially purified preparation of the first component of chicken complement (C1) together with a guinea pig serum reagent free of C1 but containing the remaining complement components (Stolfi et al., 1971; Van Regenmortel and Burckard, 1985). Ferrets, which are highly susceptible to infection with influenza viruses, are also good animals for raising antiserum but should be shown to be seronegative before immunization. Ferrets are usually infected by intranasal instillation (Stott and Tyrrell, 1986).

When chemically purified viral polypeptides are available (Schild, 1970, 1972), best results are obtained by immunizing rabbits with a series of intramuscular injections of antigen emulsified in Freund's complete antigen.

PARAMYXOVIRUSES

Antiserum can be obtained from guinea pigs immunized by intranasal infection under light anesthesia (Van der Veen and Sonderkamp, 1965). Each nostril is instilled with 0.1 ml of a virus suspension. Antiserum to Newcastle disease virus can be obtained from infected chickens or from vaccinated birds.

Antisera to various measles antigens have been prepared in rabbits injected intramuscularly with antigen emulsified in Freund's adjuvant (Norrby and Hammerskjold, 1972).

REOVIRUSES

Antiserum can be obtained from guinea pigs infected intranasally. Since the serum of guinea pigs and rabbits often contains reovirus antibodies, preimmunization sera should always be tested before using any animal for the production of antiserum. Rotavirus antibodies have been obtained from immunized rabbits and goats, as well as from the egg yolk of laying hens immunized with purified virus (Obert et al., 1981).

RHINOVIRUSES

Antisera can be obtained from rabbits immunized with purified virus emulsified in Freund's complete or incomplete adjuvant (Cooney and Kenny, 1970). Cross-immunization of rabbits with different serotypes has been shown to lead to the production of antisera that detect a broad range of antigenic cross-reactions among different rhinoviruses (Cooney et al. 1975).

Commercial Sources of Antisera and Viral Antigens

The Linscott Directory of Immunological and Biological reagents (40 Glen Drive, Mill Valley, CA 94941) provides a comprehensive list of worldwide suppliers of viral reagents. Some of the companies and organizations that distribute a wide range of materials for viral diagnosis are:

Accurate Chemical & Scientific Corporation
300 Shames Drive
Westbury, NY 11590, USA

American Type Culture Collection
Sales Department
12301 Parklawn Drive
Rockville, MD 20852, USA

BIOMERIEUX
Marcy-L'Etoile
69620 Charbonnieres-les-Bains, France

Centers for Disease Control
Reagents Division
Atlanta, GA 30333, USA

Connaught Laboratories
1755 Steeles Avenue West
Willowdale, Ontario, Canada M2R3T4

Electro-Nucleonics, Inc., Viral Science Lab
12050 Tech Road
Silver Springs, MD 20904, USA

Flow Laboratories, Inc.
7655 Old Springhouse Road
McLean, VA 22102, USA

Institut Pasteur Production
36 rue du Dr Roux
75725 Paris 15, France

M.A. Bioproducts
Bldg. 100, Biggs Ford Road
Walkersville, MD 21793, USA

Meloy Laboratories, Inc.
Life Sciences Division
6715 Electronic Drive
Springfield, VA 22151, USA

Microbiological Associates
5221 River Road
Bethesda MD 20816, USA

National Institutes of Health
Research Resources Branch, Extramural
    Activities Program
NIAID, Bethesda, MD 20892, USA

Nordic Immunological Laboratories b.v.
Langestraat 57–61
Tilburg, The Netherlands

Polysciences, Inc.
Paul Valley Industrial Park
Warrington, PA 18976, USA

Southwest Foundation for Viral Science
Lab Research & Education
P.O. Box 28147
San Antonio, TX 78284, USA

Wellcome Reagents, Ltd.
303 Hither Green Lane
London SE13 6TL, England

Monoclonal Antibodies

Within 10 years after Kohler and Milstein (1975) developed the hybridoma technology, monoclonal antibodies (McAb) were prepared against all major groups of viruses (Carter and ter Meulen, 1984; Prabhakar et al., 1984; Sander and Dietzgen, 1984; Van Regenmortel, 1984a; Yewdell and Gerhard, 1981; Yolken, 1983). Compared with conventional polyclonal antisera, the main attraction of McAbs is their chemical homogeneity and the fact that they can be produced in practically unlimited quantities. This ensures that uniform serological results can now be obtained in all laboratories, in contrast to the earlier situation in which the inherent variability of individual antisera, available mostly in only small quantities, often led to discrepant serological results. The other major advantages of McAbs are their increased specificity compared with antisera and the fact that they can be produced even when highly purified viral antigen is not available for immunizing animals. Mice and rats needed for fusion experiments can be immunized with quantities of viral antigen of the order of 10 to 100 μg. If the antigen preparation is contaminated with host antigens or with other viruses, it is nevertheless possible to select McAbs that will react only with the particular antigen of interest.

It is important to realize that McAbs are specific for one epitope of a viral antigen rather than for one particular virus, strain, or serotype. If two viral strains differ by only a single epitope and McAbs directed to this epitope are used to discriminate between them, the two strains will be clearly differentiated or may even appear to be unrelated. On the other hand, if two very dissimilar strains are analyzed with a McAb directed against the only common epitope present in the two strains, they may appear to be antigenically identical. It is thus possible to select McAbs that emphasize either what is common between two viruses or what is different between them. When McAbs are used to estimate the degree of antigenic similarity between two viruses, the results will always be highly reproducible, although the apparent degree of similarity revealed by different McAbs may differ considerably. To assess the degree of relatedness between antigens that possess a large number of different epitopes, it is thus necessary to utilize many different McAbs.

A preparation of McAbs consists of chemically homogeneous immunoglobulin molecules which all react in an identical manner with the antigen. Each McAb shows binding activity only in a particular range of experimental conditions—pH, ionic strength, and temperature (Mosman et al., 1980; Underwood and Bean, 1985)—and it may or may not neutralize viral infectivity, precipitate the virions, fix complement, or inhibit hemagglutination by the virus. In contrast, viral antisera are mixtures of many antibodies of varying classes, specificity, and chemical stability, and therefore they usually possess all of the above-mentioned functional activities. Because of its heterogeneous composition, the reactivity of an antiserum is usually preserved over a wider range of experimental conditions than is that of a McAb.

Production of Hybridoma

Technical details concerning the production of hybridoma will not be presented here. Exhaustive descriptions of the hybridoma technology are available (Campbell, 1984; Fazekas de St Groth and Scheidegger, 1980; Galfre and Milstein, 1981; Goding, 1983; Langone and Van Vunakis, 1986), and the fusion protocols used for obtaining McAbs directed against viruses (Prabhakar et al., 1984) do not differ from the classical procedures used with other antigens. The goal of a fusion experiment is to immortalize a cell that secretes a specific antibody, and this is achieved by fusing spleen cells from an immunized mouse or rat with cells of a suitable myeloma line by the use of polyethylene glycol. The fused hybrid cells or hybridomas are allowed to grow in a selective medium that does not support the growth of nonfused cells, and they are subcloned to ensure that the cell lines are truly monoclonal. Since these hybrids are dependent upon products of their own metabolism in order to grow, it is important to use feeder cells when the cultures are started from low numbers of cells.

The state of immunization of the animal providing the spleen cells is an important factor in determining the success of the fusion. It is usually assumed that the animal presenting the highest titer after immunization is likely to give the best results. Selecting such animals for the fusion experiment provides a comforting guideline to the investigator, although it should be remembered that the fused hybridomas do arise from actively dividing parent cells and not from the plasmacytes that produced the antibodies found in the immunized animal. The McAbs will thus reflect the population of immature lymphoblastic B cells present in the spleen and not necessarily the population of antibodies found in the serum of the animal. A wide variety of immunization protocols have been found to give satisfactory results (Prabhakar et al., 1984), and there is little information available regarding the relative merits of different procedures. Usually, viral antigen at a dose of 10 to 100 μg is first administered in complete Freund's adjuvant; subsequent injections are given in incomplete Freund's adjuvant, followed by one or a few intravenous or intraperitoneal injections about 3 days before the fusion.

The immunogen may be live virus (Collins et al., 1982; Gonzales-Scorano et al., 1982), inactivated virus (Pereis et al., 1982), crude lysate of virus-infected cells (Balachandran et al., 1981), or purified fractions such as nucleocapsids or glycoproteins (Heinz et al., 1982; Volk et al., 1982).

Selection Procedures

One of the most critical steps in the production of McAbs is the immunological screening of culture supernatants after fusion. The choice of immunoassay is crucial, since it will affect the type of antibody that is selected. Many investigators use solid-phase radioimmunoassay (RIA) or enzyme-linked immunosorbent assay (ELISA) (Prabhakar et al., 1984), although immunoprecipitation (Cepko et al., 1981),

immunofluorescence, and dot-blot immunoassays (Towbin and Gordon, 1984) are also commonly used. Before embarking on a fusion experiment, one should decide on the screening strategy and on the immunoassay that will be used. It is often useful to test for the presence of antibodies that are able to recognize a variety of serotypes and not only the antigen used to immunize the mouse. It is well established that an antibody may be capable of binding to several related epitopes (Berzofsky and Schechter, 1981) and that it may bind less well to the immunogen than to other antigens (Al Moudallal et al., 1982; Underwood, 1985). This phenomenon, known as heterospecificity, may be put to good use in hybridoma work, since it is possible, from a single fusion experiment, to select clones that react preferentially with different viral serotypes. If only the immunizing viral strain were used for selecting positive hybridoma clones, one might inadvertently discard clones that have a low avidity for the immunogen but react strongly with other serotypes.

It is important to adjust the parameters of the immunoassay to ensure maximum sensitivity of antibody detection. A variety of ELISA procedures have been used for screening hybridoma cell lines (Al Moudallal et al., 1984; Brennand et al., 1986), and multi-layered sandwich tests have been found to be particularly sensitive. The use of an avian antibody in one of the layers is advantageous, since the absence of cross-reaction between avian and mammalian globulins tends to keep background readings low (Al Moudallal et al., 1984; Leslie and Clem, 1969).

The use of antigen-coated plates for selecting the hybridomas that produce appropriate antibodies can introduce an unexpected bias (Al Moudallal et al., 1984; Van Regenmortel, 1984a). This is owing to the fact that the adsorption of proteins or virions to the plastic surface of microtiter plates leads to a certain degree of protein denaturation and disruption of the particles (Altschuh et al., 1985; Friguet et al., 1984; Kennel, 1982; McCullough et al., 1985; Mierendorf and Dimond, 1983; Vaidya et al., 1985). McAbs selected by an assay in which they bind to the "denatured" antigen on the plate may fail to recognize the same antigen in solution, where it possesses its native conformation.

When hybridomas are screened by a double-antibody-sandwich ELISA in which the antigen is attached to the plate through a first layer of antibodies, the selected McAbs tend to be specific for the native configuration of the antigen. This indicates that, when antigens are trapped by antibody-coated plates, their native conformation is preserved. McAbs selected by this procedure are often specific for the quaternary structure of virions and do not react with monomeric subunits of the viral coat protein. Such McAbs specifically recognize neotopes, that is, conformational epitopes that exist only in the assembled virions (Neurath and Rubin, 1971; Van Regenmortel, 1966, 1982). When the virus preparation is directly adsorbed to the solid phase in ELISA at alkaline pH (for instance, pH 9.6), the neotope conformation may be disrupted (Halk et al., 1984; Van Regenmortel, 1984a). Viral subunits, if present, may become preferentially adsorbed to the solid phase and this could prevent intact virions from becoming trapped. This means that McAbs selected by a test with antibody-coated plates may fail to show any activity in a test with virus-coated plates. It is thus important to select McAbs by the same immunoassay that will be used subsequently in diagnostic or other testing.

Relative Merits of Monoclonal Antibodies and Polyclonal Antisera

The main advantages of McAbs over conventional antisera are 1) permanent supply of chemically defined reagents; 2) increased specificity for the antigen at the level of its elementary unit, the epitope; and 3) simplified immunization requirements. Large quantities of McAb preparations can be obtained from tissue culture supernatant fluid or more easily by growing the hybridoma in mice as ascites tumors. As much as 10 to 20 mg of McAb can be collected per milliliter of ascites fluid. If necessary, McAbs can be purified by a variety of methods such as ion-exchange chromatography (Carlsson et al., 1985) and high-pressure liquid chromatography (HPLC) (Burchiel et al., 1984). Purification by affinity chromatography is not always an advantage, since the antibodies with the highest avidity may be lost because of lack of dissociation from the complex or by the denaturation caused by the harsh procedures needed to achieve dissociation.

The many advantages of McAbs should not obscure the fact that these reagents also possess limitations. McAbs are sometimes too specific for many practical purposes. For instance, they may reveal subtle antigenic differences between variants or between native and denatured particles that have little relevance in diagnostic work. It is thus always necessary to select the clones that are best suited for the task at hand.

Another limitation is that McAbs are assay-specific, that is, they may function well in one type of assay but fail altogether in others (Haaijman et al., 1984). For instance, some McAbs lose their activity when conjugated to enzymes or to radioactive markers; some are denatured on the plastic surface when used as capturing antibodies in a solid-phase assay (Suter and Butler, 1986), while others do not precipi-

tate the antigen or do not function well in immunoelectron microscopy.

The lack of stability of many McAbs to treatments that appear not to affect the activity of antisera (freeze-drying, storage, pH and temperature changes) is not due to any particular sensitivity of McAbs, but rather to their intrinsic chemical individuality. The apparent stability of a polyclonal mixture of antibodies simply reflects the fact that, whereas some of the antibody components in the mixture are inactivated, others are likely to resist the treatment. The average stability of the mixture will thus be greater than that of the individual antibody components.

The affinity or avidity of McAbs is usually not so high as that of many antisera obtained from hyperimmunized animals. However, the proportion of high-affinity antibodies in an antiserum is usually very small, and on a statistical basis the number of high-affinity McAbs obtainable from a fusion experiment can be expected to be equally small. A significant number of high-affinity McAbs is likely to be obtained only when the number of clones derived from a fusion is very high. Furthermore, it has been shown that synergistic effects may increase the combined avidity of a mixture of McAbs (Moyle et al., 1983; Thompson and Jackson, 1984).

Antigens

Molecular Structure of Viral Antigens

The antigenic reactivity of viruses resides in restricted parts of the virion surface that are known as antigenic determinants or epitopes. For viruses of a diameter smaller than about 80 nm, the particles consist of nucleocapsids built up of a layer of protein molecules enclosing the nucleic acid. Larger viruses usually have an additional membrane-like structure known as the envelope, which contains projections or spikes made up of glycoproteins. These glycoproteins consist of a hydrophilic moiety that can be removed by digestion with proteolytic enzymes and a hydrophobic segment that remains associated with the lipid bilayer and anchors the protein in the membrane. The lipids of the viral envelope are derived from preexisting cellular membranes and are devoid of antigenic activity. In the myxo- and rhabdoviruses, a layer of nonglycosylated matrix protein is present directly below the lipid bilayer.

Because of their exposed surface location, glycoproteins interact preferentially with antibody, and their antigenic properties are thus of particular interest. The ability of certain envelope glycoproteins such as hemagglutinins to agglutinate erythrocytes is linked to the role they play in the entry of the virus into the host cell. Although the reactivity of many antibodies specific for viral glycoproteins is influenced by the attached carbohydrate chains (Alexander and Elder, 1984), the exact way in which the oligosaccharide side chains contribute to the antigenicity is rarely known (Kaluza et al., 1980; Portetelle et al., 1980). One particular instance in which this role was elucidated concerns the hemagglutinin of a Hong Kong influenza virus mutant, in which the substitution of one amino acid generated a new glycosylation site. The presence of the additional oligosaccharide chain in the mutant was shown to mask an underlying epitope and to inhibit the binding of a monoclonal antibody specific for this region (Skehel et al. 1984).

Most of our knowledge of viral antigenicity concerns the protein epitopes of nucleocapsids. Each epitope consists of a number of amino acid residues that make contact with the binding site (or paratope) of a complementary antibody molecule. Epitopes made up of a series of contiguous residues in the polypeptide chain are called continuous epitopes, whereas discontinuous epitopes consist of residues that are not contiguous in the sequence but are brought together by the folding of the peptide chain. Many linear peptide fragments have been given the label "continuous epitope" because they were found to bind to antibodies raised against the intact molecule. However, it cannot be excluded that such peptides are actually reacting with antibodies directed against a larger discontinuous epitope of which the linear fragment represents only a part. Methods used to fragment viral proteins into peptides of various sizes have recently been reviewed (Doel, 1982).

The current paradigm regarding protein antigenicity states that the vast majority of epitopes are likely to be discontinuous and that the entire accessible surface of proteins consists of a series of overlapping epitopes (Barlow et al., 1986; Benjamin et al., 1984; Van Regenmortel, 1984b). The number of different epitopes present on a protein has been estimated to correspond to the number of different McAbs that can be raised against it (Schroer et al., 1983), a view in sharp contrast to the earlier paradigm which stated that each protein contains only a few discrete epitopes (Atassi 1975, 1984). Regulatory mechanisms leading to immunodominance have been invoked to explain the predominance in each individual antiserum of antibodies specific for only a small number of epitopes.

The majority of epitopes in a native viral protein possess a unique conformation gained from long-range stabilizing interactions that are absent in short fragments excised from the molecule. In particular,

interactions at the level of the quaternary structure of capsids give rise to a new type of epitope called neotope that is absent in the constituent monomeric subunits (Van Regenmortel, 1966). Neotopes owe their existence either to conformational changes of the protein induced by intersubunit bonds or to the juxtaposition of residues from neighboring subunits. On the other hand, depolymerization or denaturation of capsids leads to the appearance of hidden epitopes, or cryptotopes, that are not expressed at the surface of intact virions. The existence of cryptotopes and neotopes was demonstrated initially by the use of cross-absorbed antisera (see Van Regenmortel, 1982), but nowadays they can be studied much more easily by means of specific McAbs.

Methods used in the localization of protein and viral epitopes have recently been reviewed elsewhere (see Chapter 1 in Van Regenmortel and Neurath, 1985; Van Regenmortel, 1984b, 1986a). Most of our knowledge on epitopes has been derived from cross-reactivity studies between proteins and linear peptide fragments, by use of either antiprotein or antipeptide antibodies. Since the exact structure of an epitope is rarely maintained after it has been excised from the protein, the extent of cross-reactivity measured in these experiments is mostly very low. This means that most epitopes that have been "identified" so far correspond only to cross-reactive structures that do not reproduce exactly the antigenic surface patches to which antibodies bind.

Another approach used to localize viral epitopes consists in studying the influence of amino acid substitutions in the viral protein on the binding of McAbs (Carter and ter Meulen, 1984; Yewdell and Gerhard, 1981). If the substitution leads to a considerable reduction in antibody binding, it is assumed that the corresponding residue is directly part of the epitope. However, since the number of available substitutions is limited and since epitopes can be modified by residue exchanges occurring far from the protein surface (Al Moudallal et al., 1982), the information obtained by this approach is often difficult to interpret.

Considerable information on neutralizable viral epitopes has been obtained by attempting to grow virus in the presence of McAbs that neutralize the viral infectivity and selecting nonneutralizable virus mutants. In most cases, the mutants that are no longer neutralized were found to be altered by a single amino acid substitution (Blondel et al., 1986; Caton et al., 1982; Colman et al., 1983; Emini et al., 1982; Minor et al., 1985; Wiley et al., 1981). If the three-dimensional structure of the viral antigen is known, as is the case for influenza hemagglutinin (Wilson et al., 1981), rhinovirus (Rossmann et al., 1985), and poliovirus (Hogle et al., 1985), this type of study can provide detailed information on the location of neutralization epitopes.

Preparation of Virions

The degree of purity required of viral antigens depends on whether they are to be used as diagnostic reagents, immunogens, or for immunochemical studies. Good-quality antigens in lyophilized form suitable for complement fixation and hemagglutination assays can be purchased from various commercial sources (see above section on Commercial Sources of Antisera and Viral Antigens).

In many cases, viral antigen is prepared by infecting cell cultures and harvesting both cells and supernatant after a single cycle of freezing and thawing. Methods suitable for obtaining routine diagnostic antigens for most virus diseases have been reviewed by Polley (1977). A fairly crude preparation obtained from infected cells may be adequate when the antigen is needed for complement fixation, hemagglutination, or neutralization assays. In a solid-phase binding assay, however, the presence of contaminants is more likely to give rise to nonspecific reactions, and highly purified viral antigen may be required, especially if the antigen is to be labeled. This is the case, for instance, when labeled rubella antigen is used in a "capture" type immunoassay in which viral IgM antibodies of a test serum are first trapped on the solid phase by an anti-μ chain reagent and are then revealed by the labeled antigen (Bonfanti et al., 1985). Incidentally, in such assays the specificity of the anti-μ chain reagent is also crucial. Inasmuch as many polyclonal commercial reagents are of inadequate specificity (Zrein et al., 1986a), it may be preferable to use McAbs specific for human μ chains (Forghani et al., 1983).

Viral antigens can be obtained from embryonated eggs (Appleyard and Zwartouw, 1978), from organs of infected laboratory animals (Lennette and Schmidt, 1979; Schmidt and Lennette, 1967) or from biological fluids of infected animals or patients. For instance, hepatitis B virus antigens are derived from serum or plasma, rotaviruses are obtained mostly from fecal samples, and arboviruses from infected mouse brain.

Since the use of infectious antigens in diagnostic serology is hazardous for personnel and can lead to cross-contamination in the laboratory, it is important to inactivate virus antigens with chemical agents such as formaldehyde or β-propiolactone or by γ-irradiation (Polley, 1977).

General methods used for purifying animal viruses have been described extensively elsewhere (Schmidt and Lennette, 1967; Habel and Salzman,

1969; Burrell, 1986). The purification of adeno-, myxo-, and poxviruses has been reviewed by Appleyard and Zwartouw (1978), and detailed accounts of purification procedures used with picorna-, toga-, rhabdo-, myxo-, reo-, and herpesviruses were recently brought together in a handbook of virological procedures (Mahy, 1985). Most purification protocols consist of the following three steps: clarification, concentration, and actual purification. When the initial virus-containing extract is heavily contaminated, it is usual to clarify the material by means of organic solvents such as chloroform, freon, or fluorocarbon and by low-speed centrifugation. The next step is to concentrate the clarified suspension with, for instance, polyethylene glycol precipitation (Juckes, 1971; Polson, 1977), salt precipitation, or high-speed centrifugation to pellet the virus. The precipitate or pellet of virus material is resuspended in a fairly small volume of a suitable buffer solution, and the virions are then separated from the remaining contaminating substances by procedures such as rate-zonal or isopycnic gradient centrifugation on sucrose or cesium chloride gradients (Brakke, 1967), controlled-pore glass-bead chromatography (Heyward et al., 1977), zone electrophoresis (Polson and Russell, 1967), or absorption on erythrocytes (Laver, 1969).

The antigenic properties of nonenveloped viruses such as picornaviruses can be modified during purification procedures by changes in the ionic and chemical environment that influence the conformation of polypeptide chains (see section above on Molecular Structure of Viral Antigens). Such modifications are more easily observed when McAbs instead of antisera are used for antigenic analysis. It is well known, for instance, that the antigenic properties of poliovirus particles are altered when the virions are submitted to various physical or chemical treatments (heating to 56°C, alkaline pH, presence of urea, and so forth). As a result, the antigenic state of the native infectious poliovirus, known as the D state (because of its dense appearance in electron micrographs), is changed to the noninfectious C state (coreless appearance). The D to C transition is brought about by conformational changes in the VP1, VP2, and VP3 polypeptides and by the loss of VP4 from the particles (Koch and Koch, 1985). Procapsids devoid of RNA are in the D or C antigenic state, depending on the pH of the extraction buffer (Rombaut et al., 1982).

The immunogenicity of intact virus particles is usually greater than that of viral subunits (Neurath and Rubin, 1971; Cowan, 1973), although instances have been reported in which soluble antigens or virions disrupted by detergents (for example, rabies and vesicular stomatitis viruses) were equally or more immunogenic than intact virus. As a rule, it is possible to enhance the immunogenicity of unstable virions by chemically stabilizing the capsid structure, for instance by aldehyde treatment (Van Regenmortel, 1982).

Preparation of Viral Proteins

Methods for disrupting virions into their components have been reviewed previously (Neurath and Rubin, 1971; Ralph and Bergquist, 1967). Structurally complex viruses can be disrupted by treatment with Tween ether, saponin, or detergents such as sodium deoxycholate, Triton X-100, or Nonidet. Simple viruses such as picornaviruses can be dissociated at acid or alkaline pH. After dissociation, the viral components are usually separated by SDS polyacrylamide gel electrophoresis, sometimes in the presence of 6 to 8 M urea, and they can be extracted from the gel by electroelution. Carbohydrate-containing proteins can be separated from a mixture of subviral components by adsorption to specific lectins (Hayman et al., 1973; Lund and Salmi, 1981) and elution with appropriate sugars.

As discussed above, viral subunits in the monomeric or oligomeric state possess distinct epitopes (cryptotopes) that are not expressed at the surface of polymerized capsids or whole virions. Even the dissociation of trimeric influenza hemagglutinin into monomers has been shown to alter the specificity of its epitopes (Nestorowicz et al., 1985). Antibodies prepared against individual viral polypeptides usually reveal the existence of more extensive antigenic similarities between related viral serotypes than are apparent from antigenic comparisons made with antibodies directed to whole virions (Blondel et al., 1982; Meloen and Briaire, 1980).

The additional epitopes that become expressed in dissociated capsid proteins are valuable for raising group- or type-specific antibodies which usually have greater discriminatory power than antibodies raised against intact virions. On the other hand, virions with an intact quaternary structure are usually better inducers of neutralizing antibody than dissociated protein subunits (Cartwright et al., 1982).

The use of certain detergents can sometimes lead to unexpected complications in immunoassays. When solid-phase assays are applied to the routine diagnosis of infections caused by enveloped viruses, it is advisable to use membrane antigens of the virions that have been extracted with sodium deoxycholate (Jeansson et al., 1982), a detergent which, contrary to Triton-X, does not interfere with the adsorption of antigen to the solid phase. Another difficulty arises when viral proteins extracted in the presence of SDS are used as immunogen, since antibodies may be produced against the SDS-protein

complex (Lompré et al., 1979). In view of this possibility, controls must be included to ascertain the specificity of any reaction observed with viral antigens treated with SDS.

In recent years, recombinant DNA technology has been used to produce large quantities of certain viral polypeptides, mainly with the purpose of developing new recombinant subunit vaccines. However, this approach will also find many applications in the production of diagnostic reagents. A recent study showed that it is feasible to use expression vectors to direct bacterial synthesis of fusion proteins containing both a viral epitope and a marker enzyme suitable for use in ELISA tests (Offensperger et al., 1985). The pre-S2 region of HBV envelope protein was inserted into a plasmid and expressed in *Escherichia coli,* leading to the production of large amounts of a pre-S2 peptide-β-galactosidase fusion protein suitable as a diagnostic reagent.

It has also been found possible to express the recombinant HBV surface antigen in yeast (Valenzuela et al., 1982; Wampler et al., 1985) and in mammalian cells (Moriarty et al., 1981; Pourcel et al., 1982). Influenza virus hemagglutinin (Gething and Sambrook, 1981; Sveda and Lai, 1981) and glycoprotein D of herpes simplex virus (Berman et al., 1983) have also been expressed in mammalian cells.

As most of the important antigens of envelope viruses are integral membrane proteins, it is necessary to convert by genetic engineering the membrane-bound proteins into secreted proteins. By removing the membrane-binding-domain sequence, it is possible to obtain the synthesis of truncated proteins that are secreted from transfected cells. Truncated forms of glycoproteins of herpes simplex virus (Berman et al., 1985) and human immunodeficiency virus type 1 (Lasky et al., 1986) have been obtained in this manner.

Synthetic Peptides

Most of our knowledge concerning the antigenic structure of proteins has been obtained by studying the immunological cross-reactivity between proteins and certain peptide fragments (see above). The method most commonly used consists in testing natural or synthetic fragments of the protein for their ability to bind to antibodies raised against the whole molecule. Alternatively, the presence of cross-reactivity between peptide fragments and the whole molecule can also be established by using antibodies raised against the peptide (Lerner, 1984; Van Regenmortel, 1986a; Walter, 1986). Both approaches have been successful and have led to the view that it is possible to mimic certain antigenic sites of proteins by means of synthetic peptides ranging in length from 6 to about 20 amino acid residues. Advances in automated solid-phase peptide synthesis have made it possible to obtain synthetic peptides with considerable ease (Kent and Clark-Lewis, 1985; Merrifield, 1986; Sheppard, 1986), and recently, new methods for producing very rapidly large numbers of synthetic peptides have been developed (Geysen et al., 1984; Houghten, 1985).

These technological developments have made it feasible to replace the virion or viral protein used in a diagnostic assay by a synthetic peptide corresponding to one of the epitopes of the viral antigen.

SYNTHETIC PEPTIDES AS ANTIGENS

Viral antigens are sometimes difficult to obtain in purified form and in sufficient quantity. When the position of continuous epitopes in such antigens has been established by immunochemical analysis, it may be possible to replace the viral protein or virion by synthesizing a peptide at a fraction of the cost of producing equivalent quantities of the whole antigen. Another advantage of this approach is linked to the fact that such peptides represent a single epitope of the viral antigen, and they will, therefore, detect only certain classes of viral antibodies in the antiserum. This may result in a more specific diagnosis than is possible with a multispecific antigen. In this respect, it can be said that peptides increase the specificity of antibody detection in the same manner that monoclonal antibodies, compared with polyclonal antisera, increase the specificity of antigen detection.

A typical example of the use of synthetic peptides for detecting specific viral antibodies in serum concerns the diagnosis of hepatitis B virus (HBV) infections. The envelope of HBV consists of three overlapping protein species possessing the same C terminal region of 226 residues (the S protein); the M protein contains, in addition, 55 residues (the pre-S2 region) at the N-terminal end; while the L protein contains a further 119 residues (the pre Sl region) at the N-terminal end (Neurath and Kent, 1985). Recently, the pre S region of HBV has been shown to possess important biological functions, and the sera of patients who recovered from hepatitis B were found to contain antibodies recognizing residues 12 to 32 and 120 to 145 of the L protein (Neurath et al., 1985). Synthetic peptides corresponding to these residues of the pre-S region were used to develop immunoassays for detecting anti-pre-S antibodies in serum early during acute type hepatitis B (Neurath et al., 1986). These peptides were chosen as probes because they had been found to correspond to immunogenic epitopes expressed in the native envelope protein of HBV. In addition to the use of peptides in the diagnosis of HBV, synthetic peptides of Epstein-Barr nuclear antigen (Rhodes et al., 1985) and of

human immunodeficiency virus (Kennedy et al., 1986; Wang et al., 1986) have also been shown to be recognized by antibodies present in patients infected with the respective viruses.

Several recent studies have demonstrated that it is possible to predict the position of continuous epitopes in the primary structure of viral proteins by making use of normalized scales of hydrophilicity, segmental mobility, or accessibility for each of the 20 amino acids (Hopp, 1986; Van Regenmortel, 1986a). When these parameters are averaged over sequence segments of 6 or 7 residues at a time, one obtains a series of plots describing how properties such as hydrophilicity or mobility vary along the polypeptide chain (Hopp and Woods, 1981; Tainer et al., 1985). Local maxima in such plots have been found in many cases to correspond to continuous epitopes in proteins. However, as discussed elsewhere (Van Regenmortel, 1986a, b), all methods to predict the location of epitopes in proteins suffer from the limited antigenicity data base as well as from the fact that all definitions of antigenicity are operationally biased.

SYNTHETIC PEPTIDES AS IMMUNOGENS

Instead of peptides being used as surrogate viral antigens in a diagnostic test, a more common utilization of synthetic peptides exploits their immunogenicity, that is, their ability to induce antibodies that will react with the viral antigen from which the peptide was derived. This powerful application of synthetic peptides has gained considerable popularity only in recent years, although it was already shown more than 20 years ago (Anderer and Schlumberger, 1965) that there was no difficulty in raising antipeptide antibodies that would react with a virus particle and even neutralize its infectivity. Two widely held misconceptions are responsible for the fact that this useful methodology did not come into general use for another 15 years (Lerner, 1984; Walter, 1986). First, it was assumed that proteins possessed only very few epitopes and that these had to be identified before the synthetic approach could be attempted (Atassi, 1984; Benjamin et al., 1984). Second, it was believed that synthetic peptides would very rarely assume the conformation found in the corresponding region of the protein, and thus that it was unlikely that antipeptide antibodies would recognize the protein (Lerner, 1984).

Because of rapid advances in nucleotide sequencing, the primary structures of viral proteins are nowadays deduced mostly from genome sequences. The resulting explosion in protein sequence data has induced many investigators to synthesize short stretches of viral proteins in order to raise antipeptide antibodies. In the majority of cases, these antibodies were found to be capable of reacting with the corresponding viral protein. As a result, it has become widely accepted that antipeptide antibodies are powerful tools for identifying putative gene products (Sutcliffe et al., 1980), purifying viral proteins (Walter et al., 1980, 1982), developing new diagnostic reagents (Kennedy et al., 1986; Klinkert et al., 1986), and assessing the potential of synthetic viral vaccines (Arnon et al., 1983; Shinnick et al., 1983).

The use of synthetic peptides corresponding to the N- and C-termini of proteins for raising antipeptide antibodies that cross-react with the complete molecule has been particularly successful (Walter, 1984; Walter et al., 1980; Harvey et al., 1982; Nigg et al., 1982a). The terminal segments of proteins are predominantly surface oriented (Thornton and Sibanda, 1983) and are less constrained and more mobile than other segments of the polypeptide chain (Westhof et al., 1984). As a result, terminal peptides of 6 to 15 residues resemble fairly closely the configuration present in the corresponding region of the protein, and this probably explains why the antigenic cross-reactivity observed between such peptides and the protein is often relatively high.

It has been claimed (Lerner, 1984) that immunization with peptides leads to a very high frequency of induction of antibodies able to recognize the "native" protein, even when the peptides do not correspond to protein termini. The validity of such a claim would have to rest on the demonstration that the antipeptide antibodies truly recognize the native protein conformation, and not a molecule partly or fully denatured by labeling or by the type of solid-phase immunoassay used (Van Regenmortel, 1986a; Walter, 1986; see also section on Selection Procedures of Monoclonal Antibodies). The requirement that antipeptide antibodies must recognize the native viral antigen is probably of considerable importance in the development of new synthetic vaccines, but it may be less crucial when peptides are used as diagnostic reagents. Viral antigens used, for instance in solid-phase immunoassays or in immunoblotting will often be denatured by the conditions of the test, and in addition some denaturation could be induced on purpose. Since virtually any antipeptide antibody will react with the corresponding denatured protein molecule, the selection of suitable peptide regions for synthesis is considerably simplified. In addition to using the terminal regions, it seems that the synthesis of peptides of 20 to 30 residues comprising stretches predicted to be particularly hydrophilic, accessible, and mobile offers the best chance of success.

The use of 20 to 30 residue-long peptides has the advantage that a protein-like conformation is more likely to be present in them compared with peptides of 5 to 10 residues. Furthermore, the longer peptides can be used for immunization in the free form, that is, without conjugating them to carrier proteins such as bovine serum albumin, ovalbumin, or keyhole limpet hemocyanin. This avoids the numerous potential problems linked to the various conjugation methods (Briand et al., 1985). The different coupling reagents used for conjugation—that is, glutaraldehyde, carbodiimides, bis-diazotized benzidine, and so on (Harvey et al., 1982; Tamura and Bauer, 1982; Tamura et al., 1983)—change the antigenicity of the protein carrier and induce the formation of antibodies specific for chemically modified residues of the carrier. These antibodies will then cross-react with unrelated carrier proteins treated with the same coupling agent (Briand et al., 1985). This means that the specificity of peptide antibodies present in an antiserum raised against a peptide–carrier conjugate must be established either with the unconjugated peptide as antigen, or with the peptide coupled to another carrier by means of another coupling agent. In published reports describing the specificity of antisera prepared against peptide conjugates, this essential control is often omitted.

There is evidence that antipeptide antibodies often show some cross-reactivity with various proteins other than the homologous one from which the peptide was derived (Nigg et al., 1982b). This may be owing to the fact that antipeptide antibodies tend to be more sequence-specific than conformation-specific, especially if the peptide is relatively short (Leach 1984).

The wider range of cross-reactivities observed with peptide antibodies as opposed to protein antibodies is an advantage, since it enhances the likelihood of obtaining a suitable reagent that reacts with the protein of interest. All antibodies possess intrinsic cross-reactivity (Van Regenmortel, 1986a, 1987), and suitable controls are always necessary to establish that the relative specificity of any particular antibody or antiserum is adequate for the task at hand.

Literature Cited

Alexander, S., and J. H. Elder. 1984. Carbohydrate dramatically influences immune reactivity of antisera to viral glycoprotein antigens. Science **226**:1328–1330.

Al Moudallal, Z., D. Altschuh, J. P. Briand, and M. H. V. Van Regenmortel. 1984. Comparative sensitivity of different ELISA procedures for detecting monoclonal antibodies. J. Immunol. Methods. **68**:35–43.

Al Moudallal, Z., J. P. Briand, and M. H. V. Van Regenmortel. 1982. Monoclonal antibodies as probes of the antigenic structure of tobacco mosaic virus. EMBO J. **1**:1005–1010.

Altschuh, D., Z. Al Moudallal, J. P. Briand, and M. H. V. Van Regenmortel. 1985. Immunochemical studies of tobacco mosaic virus. VI. Attempts to localize viral epitopes with monoclonal antibodies. Mol. Immunol. **22**:329–337.

Anderer, F. A., and H. D. Schlumberger. 1965. Properties of different artificial antigens immunologically related to tobacco mosaic virus. Biochim. Biophys. Acta **97**:503–509.

Appleyard, G., and H. T. Zwartouw. 1978. Preparation of antigens from animal viruses, p. 3.1–3.18. *In* D. M. Weir (ed.), Handbook of experimental immunology, 3rd ed., Blackwell, Oxford.

Arnon, R., M. Shapira, and C. O. Jacob. 1983. Synthetic vaccines. J. Immunol. Methods **61**:261–273.

Atassi, M. Z. 1975. Antigenic structure of myoglobin: the complete immunochemical anatomy of a protein and conclusions relating to antigenic structures of proteins. Immunochemistry **12**:423–438.

Atassi, M. Z. 1984. Antigenic structures of proteins. Their determination has revealed important aspects of immune recognition and generated strategies for synthetic mimicking of protein binding sites. Eur. J. Biochem. **145**:1–20.

Balachandran, N., D. Harnish, R. A. Killington, S. Bacchetti, and W. E. Rawls. 1981. Monoclonal antibodies to two glycoproteins of herpes simplex virus type 2. J. Virol. **30**:438–446.

Barlow, D. J., M. S. Edwards, and J. M. Thornton. 1986. Continuous and discontinuous protein antigenic determinants. Nature **322**:747–752.

Baxby, D. 1982. The surface antigens of orthopoxviruses detected by cross-neutralization tests on cross-absorbed antisera. J. Gen. Virol. **58**:251–262.

Benjamin, D. C., J. A. Berzofsky, I. J. East, F. R. N. Gurd, C. Hannum, S. J. Leach, E. Margoliash, J. G. Michael, A. Miller, E. M. Prager, M. Reichlin, E. E. Sercarz, S. J. Smith-Gill, P. A. Todd, and A. C. Wilson. 1984. The antigenic structure of proteins: a reappraisal. Annu. Rev. Immunol. **2**:67–101.

Berman, P. W., D. Dowbenko, L. A. Lasky, and C. C. Simonsen. 1983. Detection of antibodies to herpes simplex virus with a continuous cell line expressing cloned glycoprotein D. Science **222**:524–527.

Berman, P. W., T. Gregory, D. Crase, and L. A. Lasky. 1985. Protection from genital herpes simplex virus type 2 infection by vaccination with cloned type 1 glycoprotein D. Science **227**:1490–1492.

Berzofsky, J. A., and A. N. Schechter. 1981. The concepts of cross-reactivity and specificity in immunology. Mol. Immunol. **18**:751–763.

Blondel, B., R. Crainic, O. Akacem, P. Bruneau, M. Girard, and F. Horodniceanu. 1982. Evidence for common, intertypic antigenic determinants on poliovirus capsid polypeptides. Virology **123**:461–463.

Blondel, B., R. Crainic, O. Fichot, G. Dufraisse, A. Candrea, D. Diamond, M. Girard, and F. Horaud. 1986. Mutations conferring resistance to neutralization with monoclonal antibodies in type 1 poliovirus can be located outside or inside the antibody-binding site. J. Virol. **57**:211–220.

Bonfanti, C., O. Meurman, and P. Halonen. 1985. Detection of specific immunoglobulin M antibody to rubella virus by use of an enzyme-labeled antigen. J. Clin. Microbiol. **21**:963–968.

Brakke, M. K. 1967. Density-gradient centrifugation, p. 93–118. *In* K. Maramorosch and H. Koprowski (ed.), Methods in virology, vol. 2. Academic Press, New York.

Brambell, F. W. R. 1970. The transmission of passive immunity from mother to young. North-Holland Publishing Co., Amsterdam.

Brennand, D. M., M. J. Danson, and D. W. Hough. 1986. A comparison of ELISA screening methods for the production of monoclonal antibodies against soluble protein antigens. J. Immunol. Methods **93**:9–14.

Briand, J. P., S. Muller, and M. H. V. Van Regenmortel.

1985. Synthetic peptides as antigens: pitfalls of conjugation methods. J. Immunol. Methods **78**:59–69.

Burchiel, S. W., J. R. Billman, and T. R. Alber. 1984. Rapid and efficient purification of mouse monoclonal antibodies from ascites fluid using high performance liquid chromatography. J. Immunol. Methods **69**:33–42.

Burrell, C. J. 1986. Preparation of viral antigens, p. 5.1–5.9. *In* D. M. Weir, L. A. Herzenberg, C. Blackwell, and L. A. Herzenberg (ed.), Handbook of experimental immunology, 4th ed., vol. 1. Blackwell, Oxford.

Campbell, A. M. 1984. Monoclonal antibody technology. Elsevier Biomedical Press, Amsterdam.

Carlsson, M., A. Hedin, M. Inganas, B. Harfast, and F. Blomberg. 1985. Purification of in vitro produced mouse monoclonal antibodies. A two-step procedure utilizing cation exchange chromatography and gel filtration. J. Immunol. Methods **79**:89–98.

Carter, M. J., and V. ter Meulen. 1984. The application of monoclonal antibodies in the study of viruses. Adv. Virus Res. **29**:95–130.

Cartwright, B., D. J. Morrell, and F. Brown. 1982. Nature of the antibody response to the foot-and-mouth disease virus particle, its 12S protein subunit and the isolated immunizing polypeptide VP1. J. Gen. Virol. **63**:375–381.

Casals, J. 1979. Arenavirus, p. 815–841. *In* E. H. Lennette and N. J. Schmidt (ed.), Diagnostic procedures for viral, rickettsial, and chlamydial infections, 5th ed. American Public Health Association, Inc., Washington, D.C.

Caton, A. J., G. G. Brownlee, J. W. Yewdell, and W. Gerhard. 1982. The antigenic structure of the influenza virus A/PR/8/34 hemagglutinin (H1 subtype). Cell **31**:417–427.

Cepko, C. L., P. S. Changelian, and P. A. Sharp. 1981. Immunoprecipitation with two-dimensional pools as a hybridoma screening technique: production and characterization of monoclonal antibodies against adenovirus 2 proteins. Virology **110**:385–401.

Collins, A. R., R. L. Knobler, H. Powell, and M. J. Buchmeir. 1982. Monoclonal antibodies to murine hepatitis virus 4 (strain JHM) define the viral glycoprotein responsible for attachment and cell–cell fusion. Virology **119**:358–371.

Colman, P. M., J. N. Varghese, and W. G. Laver. 1983. Structure of the catalytic and antigenic sites in influenza virus neuraminidase. Nature **303**:41–44.

Cooney, M. K., and G. E. Kenny. 1970. Immunogenicity of rhinoviruses. Proc. Soc. Exp. Biol. Med. **133**:645–650.

Cooney, M. K., J. A. Wise, G. E. Kenny, and J. P. Fox. 1975. Broad antigenic relationships among rhinovirus serotypes revealed by cross-immunization of rabbits with different serotypes. J. Immunol. **114**:635–639.

Cowan, K. M. 1973. Antibody response to viral antigens. Adv. Immunol. **17**:195–253.

Dean, P. D. G., W. S. Johnson, and F. A. Middle. 1985. Affinity chromatography. A practical approach. IRL Press, Oxford.

Doel, T. R. 1982. Peptide analysis of viral proteins, p. 143–183. *In* C. R. Howard (ed.), New developments in practical virology. Alan R. Liss, Inc., New York.

Dowdle, W. A., A. P. Kendal, and G. R. Noble. 1979. Influenza viruses, p. 585–609. *In* E. H. Lennette and N. J. Schmidt (ed.), Diagnostic procedures for viral, rickettsial, and chlamydial infections, 5th ed. American Public Health Association, Inc., Washington, D.C.

Dresser, D. W. 1986. Immunization of experimental animals, p. 8.1–8.21. *In* D. M. Weir, L. A. Herzenberg, C. Blackwell, and L. A. Herzenberg (ed.), Handbook of experimental immunology, 4th ed., vol. 1. Blackwell, Oxford.

Dresser, D. W., and N. A. Mitchison. 1968. The mechanism of immune paralysis. Adv. Immunol. 8:129–181.

Emini, E. A., B. A. Jameson, A. J. Lewis, G. R. Larsen, and E. Wimmer. 1982. Poliovirus neutralizing epitopes: analysis of localization with neutralizing monoclonal antibodies. J. Virol. 43:997–1005.

Fazekas de St Groth, S. 1979. The quality of antibodies and cellular receptors, p. 1–42. In I. Lefkovits, and B. Pernis (ed.), Immunol methods, vol. 1. Academic Press, New York.

Fazekas de St Groth, S., and D. Scheidegger. 1980. Production of monoclonal antibodies: strategy and tactics. J. Immunol. Methods 35:1–21.

Forghani, B., C. K. Myoraku, and N. J. Schmidt. 1983. Use of monoclonal antibodies to human immunoglobulin M in "capture" assays for measles and rubella immunoglobulin M. J. Clin. Microbiol. 18:652–657.

Friguet, B., L. Djavadi-Ohaniance, and M. E. Goldberg. 1984. Some monoclonal antibodies raised with a native protein bind preferentially to the denatured antigen. Mol. Immunol. 21:673–677.

Galfre, G., and C. Milstein. 1981. Preparation of monoclonal antibodies: strategies and procedures. Methods Enzymol. 73:3–46.

Gardner, P. S., and S. Kaye. 1982. Egg globulins in rapid virus diagnosis. J. Virol. Methods 4:257–262.

Gardner, P. S., and J. McQuillin. 1980. Rapid virus diagnosis, application of immunofluorescence, 2nd ed. Butterworths, London.

Garvey, J. S., N. E. Cremer, and D. H. Sussdorf. 1977. Methods immunol, 3rd ed. W. A. Benjamin, Reading, Mass.

Gething, M.-J., and J. Sambrook. 1981. Cell-surface expression of influenza haemagglutinin from a cloned DNA copy of the RNA gene. Nature 293:620–625.

Geysen, H. M., R. H. Meloen, and S. J. Barteling. 1984. Use of peptide synthesis to probe viral antigens for epitopes to a resolution of a single amino acid. Proc. Natl. Acad. Sci. USA. 81:3998–4002.

Goding, J. W. 1983. Monoclonal antibodies: principles and practice. Academic Press, New York.

Gonzales-Scorano, F., R. E. Shope, C. Calisher, and N. Nathanson. 1982. Characterization of monoclonal antibodies against the G1 and N proteins of LaCrosse and Tahyna, two California serogroup bunyaviruses. Virology 120:42–53.

Grandien, M. 1983. Activities of the European Group for Rapid Viral Diagnosis: their experience and their achievements, p. 379–384. In L. M. de la Maza and E. M. Peterson (ed.), Medical virology II. Elsevier Biomedical Press, New York.

Haaijman, J. J., C. Deen, C. J. M. Krose, J. J. Zijlstra, J. Coolen, and J. Radl. 1984. Monoclonal antibodies in immunocytology: a jungle full of pitfalls. Immunology Today 5:56–58.

Habel, K., and N. P. Salzman. 1969. Fundamental techniques in virology. Academic Press, New York.

Halk, E. L., H. T. Hsu, T. Aebig, and J. Franke. 1984. Production of monoclonal antibodies against three ilarviruses and alfalfa mosaic virus and their use in serotyping. Phytopathology 74:367–372.

Halonen, P., O. Meurman, U. G. Petterson, M. Rnaki, and T. N.-E. Lovgren. 1984. New developments in diagnosis of virus infections, p. 501–520. In E. Kurstak and R. G. Marusyk (ed.), Control of virus diseases. Marcel Dekker, New York.

Hampil, B., and J. L. Melnick. 1968. WHO collaborative studies on enterovirus reference antisera: second report. Bull. W.H.O. 38:577–593.

Hampil, B., J. L. Melnick, C. Wallis, R. W. Brown, E. T. Braye, and R. R. Adams, Jr. 1965. Preparation of antiserum to enteroviruses in large animals. J. Immunol. 95:895–908.

Hansen, B. L., B. F. Vestergaard, and G. N. Hansen. 1981. Production of monospecific antibodies to varicella zoster virus antigen in rabbits tolerant to human IgG and immunized with antigen immunoprecipitated with human IgG from zoster convalescent sera. J. Immunol. Methods 43:283–290.

Harvey, R., R. Faulkes, P. Gillett, N. Lindsay, E. Paucha, A. Bradbury, and A. E. Smith. 1982. An antibody to a synthetic peptide that recognises SV40 small-t antigen. EMBO J. 4:473–477.

Hayman, M. J., J. J. Skehel, and M. J. Crumpton. 1973. Purification of virus glycoproteins by affinity chromatography using Lens Culinaris phytohaemagglutinin. FEBS Lett. 29:185–188.

Heinz, F. X., R. Berger, O. Majdic, W. Knapp, and C. Kunz. 1982. Monoclonal antibodies to the structural glycoprotein of tick-borne encephalitis virus. Infect. Immun. 37:869–874.

Herbert, W. J. 1978. Mineral-oil adjuvants and the immunization of laboratory animals, p. A3.1–A3.15. In D. M. Weir (ed.), Handbook of experimental immunology, 3rd ed., appendix 3. Blackwell, Oxford.

Herbert, W. J., and F. Kristensen. 1986. Laboratory animal techniques for immunology, p. 133.1–133.36. In D. M. Weir, L. A. Herzenberg, C. Blackwell, and L. A. Herzenberg (ed.), Handbook of experimental immunology, 4th ed., vol. 4. Blackwell, Oxford.

Heyward, J. T., R. A. Klimas, M. D. Stapp, and J. F. Obijeski. 1977. The rapid concentration and purification of influenza virus from allantoic fluid. Arch. Virol. 55:107–119.

Hierholzer, J. C., W. C. Gamble, and W. R. Dowdle. 1975. Reference equine antisera to 33 human adenovirus types: homologous and heterologous titers. J. Clin. Microbiol. 1:65–74.

Hogle, J. M., M. Chow, and D. J. Filman. 1985. The three dimensional structure of poliovirus at 2.9 Å resolution. Science 229:1358–1365.

Hopp, T. P. 1986. Protein surface analysis. Methods for identifying antigenic determinants and other interaction sites. J. Immunol. Methods 88:1–18.

Hopp, T. P., and K. R. Woods. 1981. Prediction of protein antigenic determinants from amino acid sequences. Proc. Natl. Acad. Sci. USA 78:3824–3828.

Horwitz, M. S., and M. D. Scharff. 1969. The production of antiserum against viral antigens, p. 253–263. In K. Habel and N. P. Salzmann (ed.), Fundamental techniques in virology. Academic Press, New York.

Houghten, R. A. 1985. General method for the rapid solid-phase synthesis of large numbers of peptides: Specificity of antigen–antibody interaction at the level of individual amino acids. Proc. Natl. Acad. Sci. USA 82:5131–5135.

Hudson, G. A., R. F. Ritchie, and J. E. Haddow. 1981. Method for testing antiserum titer and avidity in nephelometric systems. Clin. Chem. 27:1838–1844.

Ivanyi, J., and J. Cerny. 1969. The significance of the dose of antigen in immunity and tolerance. Curr. Top. Microbiol. Immunol. 49:114–150.

Jaegle, M., and M. H. V. Van Regenmortel. 1985. Use of ELISA for measuring the extent of serological cross-reactivity between plant viruses. J. Virol. Methods 11:189–198.

James, J. G., S. S. Wang, R. Wisniewolski, and C. Y. Wang. 1986. Detection of antibodies to human T-lymphotropic virus type III by using a synthetic peptide of 21 amino acid residues corresponding to a highly antigenic segment of gp41 envelope protein. Proc. Natl. Acad. Sci. USA 83:6159–6163.

Jeansson, S., H. Elwing, H. Nygren, and S. Olofsson. 1982. Evaluation of solubilized herpes simplex virus membrane antigens in diffusion in gel-enzyme-linked immunosorbent assay (DIG-ELISA). J. Virol. Methods 4:167–176.

Johansson, M. E., M. Grandien, and L. Arro. 1979. Preparation of sera for subtyping of influenza A viruses by immunofluorescence. J. Immunol. Methods 27:263–272.

Juckes, I. R. M. 1971. Fractionation of proteins and viruses with polyethylene glycol. Biochim. Biophys. Acta 229:535–546.

Kaluza, G., R. Rott, and R. T. Schwarz. 1980. Carbohydrate-induced conformational changes of Semliki forest virus glycoproteins determine antigenicity. Virology 102:286–299.

Karabatsos, N., and J. H. Mathews. 1980. Serological reactions of fractionated hamster immunoglobulins with California group viruses. Am. J. Trop. Med. Hyg. 29:1420–1427.

Katz, D., S. Lehrer, and A. Kohn. 1985. Use of chicken and rabbit antibodies in a solid phase protein A radioimmunoassay for virus detection. J. Virol. Methods 12:59–70.

Kennedy, R. C., R. D. Henkel, D. Pauletti, J. S. Allan, T. H. Lee, M. Essex, and G. R. Dreesman. 1986. Antiserum to a synthetic peptide recognizes the HTLV-III envelope glycoprotein. Science 231:1556–1559.

Kennel, S. 1982. Binding of monoclonal antibody to protein antigen in fluid phase and bound to solid supports. J. Immunol. Methods 55:1–12.

Kent, S., and I. Clark-Lewis. 1985. Modern methods for the chemical synthesis of biologically active peptides, p. 29–57. In K. Alitalo, P. Partanen, and A. Vaheri (ed.), Synthetic peptides in biology and medicine. Elsevier Science Publishers, Amsterdam.

Kissling, R. E., H. L. Casey, and E. L. Palmer. 1968. Production of specific varicella antiserum. Appl. Microbiol. 16:160–162.

Klinkert, M.-Q., L. Theilmann, E. Pfaff, and H. Schaller. 1986. Pre-S1 antigens and antibodies early in the course of acute hepatitis B virus infection. J. Virol. 58:522–525.

Koch, F., and G. Koch. 1985. The molecular biology of poliovirus. Springer-Verlag, New York.

Kohler, G., and C. Milstein. 1975. Continuous cultures of fused cells secreting antibody of predefined specificity. Nature 256:495–497.

Kurstak, E. 1985. Progress in enzyme immunoassays: production of reagents, experimental design, and interpretation. Bull. W.H.O. 63:793–811.

Kurstak, E., and C. Kurstak. 1977–1981. Comparative diagnosis of viral diseases, vol. 1–4. Academic Press, New York.

Langone, J. J., and H. Van Vunakis (ed.). 1986. Immunochemical techniques, part I. Hybridoma technology and monoclonal antibodies. Methods Enzymol. 121:1–947.

Lasky, L. A., J. E. Groopman, C. W. Fennie, P. M. Benz, D. J. Capon, D. J. Dowbenko, G. R. Nakamura, W. M. Nunes, M. E. Renz, and P. W. Berman. 1986. Neutralization of the AIDS retrovirus by antibodies to a recombinant envelope glycoprotein. Science 233:209–212.

Laver, W. G. 1969. Purification of influenza virus, p. 82. In K. Haber and N. P. Salzman (ed.), Fundamental techniques in virology. Academic Press, New York.

Leach, S. J. 1984. Antigenicity of proteins and peptides. Ann. Sclavo 2:21–45.

Lennette, D. A. 1985. Collection and preparation of specimens for virological examination, p. 687–693. In E. H. Lennette, A. Balows, W. J. Hausler, Jr., and H. J. Shadomy (ed.), Manual of clinical microbiology, 4th ed. American Society for Microbiology, Washington, D.C.

Lennette, E. H., A. Balows, W. J. Hausler, Jr., and H. J. Shadomy. 1985. Manual of clinical microbiology, 4th ed. American Society for Microbiology, Washington, D.C.

Lennette, E. H., and N. J. Schmidt. 1979. Diagnostic procedures for viral, rickettsial, and chlamydial infections, 5th ed. American Public Health Association, Inc., Washington, D.C.

Lerner, R. A. 1984. Antibodies of predetermined specificity in biology and medicine. Adv. Immunol. 36:1–44.

Leslie, G. A., and L. W. Clem. 1969. Phylogeny of immunoglobulin structure and function. III. Immunoglobulins of the chicken. J. Exp. Med. 130:1337–1352.

Lompré, A. M., P. Bouveret, J. Leger, and K. Schwartz. 1979. Detection of antibodies specific to sodium dodecyl sulfate-treated proteins. J. Immunol. Methods 28:143–148.

Lund, G. A., and A. A. Salmi. 1981. Purification and characterization of measles virus haemagglutinin protein G. J. Gen. Virol. 56:185–193.

Mahy, B. W. J. (ed.) 1985. Virology: a practical approach. IRL Press, Oxford.

Makinodan, T., and M. M. B. Kay. 1980. Age influence on the immune system. Adv. Immunol. 29:287–330.

Marbrook, J., and R. E. F. Matthews. 1966. The differential immunogenicity of plant viral protein and nucleoproteins. Virology 28:219–228.

McCullough, K. C., J. R. Crowther, and R. N. Butcher. 1985. Alteration in antibody reactivity with foot-and-mouth disease virus (FDMV) 146S antigen before and after binding to a solid phase or complexing with specific antibody. J. Immunol. Methods 82:91–100.

McLean, D. M. 1982. Immunological investigations of human virus diseases. Churchill Livingstone, New York.

Melnick, J. L., H. A. Wenner, and C. A. Phillips. 1979. Enteroviruses, p. 471–534. In Lennette, E. H. and N. J. Schmidt (ed.), Diagnostic procedures for viral, rickettsial, and chlamydial infections, 5th ed. American Public Health Association, Inc., Washington, D.C.

Meloen, R. H., and J. Briaire. 1980. A study of the cross-reacting antigens on the intact foot-and-mouth disease virus and its 12S subunits with antisera against the structural proteins. J. Gen. Virol. 51:107–116.

Merrifield, B. 1986. Solid phase synthesis. Science 232:341–347.

Mierendorf, R. C., Jr., and R. L. Dimond. 1983. Functional heterogenicity of monoclonal antibodies obtained using different screening assays. Anal. Biochem. 135:221–229.

Minor, P. D., D. M. A. Evans, M. Ferguson, G. C. Schild, G. Westrop, and J. W. Almond. 1985. Principal and subsidiary antigenic sites for VP1 involved in the neutralization of poliovirus type 3. J. Gen. Virol. 65:1159–1165.

Monath, T. P. 1986. Alphaviruses, flaviviruses, bunyaviruses and Colorado tick fever, p. 541–547. In N. R. Rose, H. Friedman, and J. L. Fahey (ed.), Manual of clinical laboratory immunology, 3rd ed. American Society for Microbiology, Washington, D.C.

Moriarty, A. M., B. H. Hoyer, J. W.-K. Shih, J. L. Gerin, and D. H. Hamer. 1981. Expression of the hepatitis B virus surface antigen gene in cell culture by using a simian virus 40 vector. Proc. Natl. Acad. Sci. USA 78:2606–2610.

Mosman, T. R., M. Gallatin, and B. M. Longenecker. 1980. Alteration of apparent specificity of monoclonal (hybridoma) antibodies recognizing polymorphic histocompatibility and blood group determinants. J. Immunol. **125**:1152–1156.

Moyle, W. R., C. Lin, R. L. Corson, and P. H. Ehrlich. 1983. Quantitative explanation for increased affinity shown by mixtures of monoclonal antibodies: importance of a circular complex. Mol. Immunol. **20**:439–452.

Nestorowicz, A., D. O. White, and D. C. Jackson. 1985. Conformational changes in influenza virus haemagglutinin and its monomer detected by monoclonal antibodies. Vaccine **3**:175–181.

Neurath, A. R., and S. B. H. Kent. 1985. Antigenic structure of human hepatitis viruses, p. 325–366. *In* M. H. V. Van Regenmortel and A. R. Neurath (ed.), Immunochemistry of viruses. The basis for serodiagnosis and vaccines. Elsevier Biomedical Press, Amsterdam.

Neurath, A. R., S. B. H. Kent, and N. Strick. 1986. Hepatitis B vaccine trial study groups. Detection of antiviral antibodies with predetermined specificity using synthetic peptide-β-lactamase conjugates: Application to antibodies specific for the preS region of the hepatitis B virus envelope proteins. J. Gen. Virol. **67**:453–461.

Neurath, A. R., S. B. H. Kent, N. Strick, P. Taylor, and C. E. Stevens. 1985. Hepatitis B virus contains pre-S gene-encoded domains. Nature **315**:154–156.

Neurath, A. R., and B. A. Rubin. 1971. Viral structural components as immunogens of prophylactic value. Karger, Basel.

Nicolaïeff, A., G. Obert, and M. H. V. Van Regenmortel. 1980. Detection of rotavirus by serological trapping on antibody-coated electron microscopic grids. J. Clin. Microbiol. **12**:101–104.

Nigg, E. A., B. M. Sefton, T. Hunter, G. Walter, and S. J. Singer. 1982a. Immunofluorescent localization of the transforming protein of Rous sarcoma virus with antibodies against a synthetic src peptide. Proc. Natl. Acad. Sci. USA **79**:5322–5326.

Nigg, E. A., G. Walter, and S. J. Singer. 1982b. On the nature of crossreactions observed with antibodies directed to defined epitopes. Proc. Natl. Acad. Sci. USA **79**:5939–5943.

Norrby, E., and B. Hammerskjold. 1972. Structural components of measles virus. Microbios **5**:17–29.

Obert, G., R. Gloeckler, J. Burckard, and M. H. V. Van Regenmortel. 1981. Comparison of immunosorbent electron microscopy, enzyme immunoassay and counterimmunoelectrophoresis for detection of human rotavirus in stools. J. Virol. Methods **3**:99–107.

Offensperger, W., S. Wahl, A. R. Neurath, P. Price, N. Strick, S. B. H. Kent, J. K. Christman, and G. Acs. 1985. Expression in *Escherichia coli* of a cloned DNA sequence encoding the pre-S2 region of hepatitis B virus. Proc. Natl. Acad. Sci. USA **82**:7540–7544.

Paul, H. L., A. Gibbs, and B. Wittmann-Liebold. 1980. The relationships of certain tymoviruses assessed from the amino acid composition of their coat proteins. Intervirology **13**:99–109.

Pereis, J. S. M., J. S. Porterfield, and J. T. Roehrig. 1982. Monoclonal antibodies against the flavivirus West Nile. J. Gen. Virol. **58**:283–289.

Polley, J. R. 1977. Viral diagnostic reagents, p. 347–402. *In* E. Kurstak and C. Kurstak (ed.), Comparative diagnosis of viral diseases, vol. II. Academic Press, New York.

Polson, A. 1977. A theory for the displacement of proteins and viruses with polyethylene glycol. Prepar. Biochem. **7**:129–154.

Polson, A., and B. Russell. 1967. Electrophoresis of viruses, p. 391–426. *In* K. Maramorosch, and H. Koprowski (ed.), Methods in virology, vol. II. Academic Press, New York.

Polson, A., M. B. Von Wechmar, and M. H. V. Van Regenmortel. 1980. Isolation of viral IgY antibodies from yolks of immunized hens. Immunol. Commun. **9**:475–493.

Portetelle, D., C. Bruck, M. Mammerickx, and A. Burny. 1980. In animals infected by bovine leukemia virus (BLV) antibodies to envelope glycoprotein gp51 are directed against the carbohydrate moiety. Virology **105**:223–233.

Pourcel, C., E. Sobzack, M. F. Dubois, M. Gervais, J. Drouet, and P. Tiollais. 1982. Antigenicity and immunogenicity of hepatitis B virus particles produced by mouse cells transfected with cloned viral DNA. Virology **121**:175–183.

Prabhakar, B. S., M. V. Haspel, and A. L. Notkins. 1984. Monoclonal antibody techniques applied to viruses, p. 1–18. *In* K. Maramorosch and H. Koprowski (ed.), Methods in virology, vol. VII. Academic Press, New York.

Ralph, R. K., and P. L. Bergquist. 1967. Separation of viruses into components, p. 464–545. *In* K. Maramorosch and H. Koprowski (ed.), Methods in virology, vol. II. Academic Press, New York.

Rawls, W. 1979. Herpes simplex virus types 1 and 2 and herpesvirus simial, p. 309–373. *In* E. H. Lennette, and N. J. Schmidt (ed.), Diagnostic procedures for viral, rickettsial, and chlamydial infections, 5th ed. American Public Health Association, Inc., Washington, D.C.

Reddecliff, J. M., and E. H. Ludwig. 1966. Ascites fluid from immunized mice as a source of plant virus antibodies. Appl. Microbiol. **14**:834–835.

Rhodes, G., D. A. Carson, J. Valbracht, R. Houghten, and J. H Vaughan. Human immune responses to synthetic peptides from the Epstein-Barr nuclear antigen. J. Immunol. **134**:211–216.

Rombaut, B., R. Vrijsen, P. Brioen, and A. Boeye. 1982. A pH-dependent antigenic conversion of empty capsids of poliovirus studied with the aid of monoclonal antibodies to N and H antigen. Virology **122**:215–218.

Rose, N. R., H. Friedman, and J. L. Fahey. 1986. Manual of clinical laboratory immunology, 3rd ed. American Public Health Association, Inc., Washington, D.C.

Rossmann, M. G., E. Arnold, J. W. Erickson, E. A. Frankenberger, J. P. Griffith, H. J. Hecht, J. E. Johnson, G. Kamer, M. Luo, A. G. Mosser, R. R. Rueckert, B. Sherry, and G. Vriend. 1985. Structure of a human common cold virus and functional relationship to other picornaviruses. Nature **317**:145–153.

Sander, E., and R. G. Dietzgen. 1984. Monoclonal antibodies against plant viruses. Adv. Virus Res. **29**:131–169.

Schild, G. C. 1970. Studies with antibody to the purified haemagglutinin of an influenza AO virus. J. Gen. Virol. **9**:191–200.

Schild, G. C. 1972. Evidence for a new type-specific structural antigen of the influenza virus particle. J. Gen. Virol. **15**:99–103.

Schmidt, N. J. 1983. Rapid viral diagnosis. Med. Clin. North Am. **67**:953–972.

Schmidt, N. J., and E. H. Lennette. 1967. The preparation of animal viruses for use as antigens, p. 87–102. *In* C. A. Williams and M. W. Chase (ed.), Methods in immunology and immunochemistry, vol. I. Academic Press, New York.

Schroer, J. A., T. Bender, R. J. Feldmann, and K. J. Kim. 1983. Mapping epitopes on the insulin molecule using monoclonal antibodies. Eur. J. Immunol. **13**:693–700.

Sheppard, R. C. 1986. Modern methods of solid-phase peptide synthesis. Science Tools **33:**9–16.

Shinnick, T. M., J. G. Sutcliffe, N. Green, and R. A. Lerner. 1983. Synthetic peptide immunogens as vaccines. Annu. Rev. Microbiol. **37:**425–446.

Shope, R. E., and G. E. Sather. 1979. Arboviruses, p. 767–814. *In* E. H. Lennette and N. J. Schmidt (ed.), Diagnostic procedures for viral, rickettsial, and chlamydial infections, 5th ed. American Public Health Association, Inc., Washington, D.C.

Skehel, J. J., D. J. Stevens, R. S. Daniel, A. R. Douglas, M. Knossow, I. A. Wilson, and D. C. Wiley. 1984. A carbohydrate side chain on hemagglutinins of Hong Kong influenza viruses inhibits recognition by a monoclonal antibody. Proc. Natl. Acad. Sci. USA **81:**1779–1783.

Stevens, D. A., M. Schaeffer, J. P. Fox, C. D. Brandt, and N. Romano. 1967. Standardization and certification of reference antigens and antisera for 30 human adenovirus serotypes. Am. J. Epidemiol. **86:**617–633.

Stolfi, R. L., R. A. Fugmann, J. J. Jensen, and M. M. Sigel. 1971. A C1-fixation method for measurement of chicken anti-viral antibody. Immunology **20:**299–306.

Stott, E. J., and D. A. J. Tyrrell. 1986. Applications of immunological methods in virology, p. 120.1–120.25. *In* D. M. Weir, L. A. Herzenberg, C. Blackwell, and L. A. Herzenberg (ed.), Handbook of experimental immunology 4th ed., vol. 4. Blackwell, Oxford.

Sutcliffe, J. G., T. M. Shinnick, N. Green, F.-T. Liu, H. L. Niman, and R. A. Lerner. 1980. Chemical synthesis of a polypeptide predicted from nucleotide sequence allows detection of a new retroviral gene product. Nature **287:**801–805.

Suter, M., and J. E. Butler. 1986. The immunochemistry of sandwich ELISAs. II. A novel system prevents the denaturation of capture antibodies. Immunol. Lett. **13:**313–316.

Sveda, M. M., and C.-J. Lai. 1981. Functional expression in primate cells of cloned DNA coding for the hemagglutinin surface glycoprotein of influenza virus. Proc. Natl. Acad. Sci. USA **78:**5488–5492.

Tainer, J. A., E. D. Getzoff, Y. Paterson, A. J. Olson, and R. A. Lerner. 1985. The atomic mobility component of protein antigenicity. Annu. Rev. Immunol. **3:**501–535.

Tamura, T., and H. Bauer. 1982. Monoclonal antibody against the carboxy terminal peptide of pp60^src of Rous sarcoma virus reacts with native pp60^src. EMBO J. **1:**1479–1485.

Tamura, T., H. Bauer, C. Birr, and R. Pipkorn. 1983. Antibodies against synthetic peptides as a tool for functional analysis of the transforming protein pp60^src. Cell **34:**587–596.

Thompson, R. J., and A. P. Jackson. 1984. Cyclic complexes and high avidity antibodies. Trends Biochem. Sci. **9:**1–3.

Thornton, J. M., and B. L. Sibanda. 1983. Amino and carboxy-terminal regions in globular proteins. J. Mol. Biol. **167:**443–460.

Towbin, H., and J. Gordon. 1984. Immunoblotting and dot immunobinding. Current status and outlook. J. Immunol. Methods **72:**313–340.

Uhr, J. W., and E. Moller. 1968. Regulatory effect of antibody on the immune response. Adv. Immunol. **8:**81–127.

Underwood, P. A. 1985. Theoretical considerations of the ability of monoclonal antibodies to detect antigenic differences between closely related variants, with particular reference to heterospecific reactions. J. Immunol. Methods **85:**295–307.

Underwood, P. A., and P. A. Bean. 1985. The influence of methods of production, purification and storage of monoclonal antibodies upon their observed specificities. J. Immunol. Methods **80:**189–197.

Vaidya, H. C., D. N. Dietzler, and J. H. Ladenson. 1985. Inadequacy of traditional ELISA for screening hybridoma supernatants for murine monoclonal antibodies. Hybridoma **4:**271–276.

Valenzuela, P., A. Medina, W. J. Rutter, G. Ammerer, and B. D. Hall. 1982. Synthesis and assembly of hepatitis B virus surface antigen particles in yeast. Nature **298:**347–350.

Van der Veen, J., and H. J. A. Sonderkamp. 1965. Secondary antibody response of guinea pigs to parainfluenza and mumps viruses. Arch. Gesamte Virusforsch. **15:**721–734.

Van Regenmortel, M. H. V. 1966. Plant virus serology. Adv. Virus Res. **12:**207–271.

Van Regenmortel, M. H. V. 1981. Serological methods in the identification and characterization of viruses, p. 183–243. *In* H. Fraenkel-Conrat and R. R. Wagner (ed.), Comprehensive virology, vol. 17. New York: Plenum Publishing Corp.

Van Regenmortel, M. H. V. 1982. Serology and immunochemistry of plant viruses. Academic Press, New York.

Van Regenmortel, M. H. V. 1984a. Monoclonal antibodies in plant virology. Microbiol. Sci. **1:**73–78.

Van Regenmortel, M. H. V. 1984b. Molecular dissection of antigens with monoclonal antibodies, p. 43–82. *In* N. J. Stern and H. R. Gamble (ed.), Hybridoma technology in agricultural and veterinary research. Rowman and Allanheld, Totowa, New Jersey.

Van Regenmortel, M. H. V. 1985. Antigenic structure of plant viruses, p. 467–478. *In* M. H. V. Van Regenmortel and A. R. Neurath (ed.), Immunochemistry of viruses. The basis for serodiagnosis and vaccines. Elsevier Biomedical Press, Amsterdam.

Van Regenmortel, M. H. V. 1986a. Definition of antigenicity in proteins and peptides, p. 81–86. *In* H. Peeters (ed.), Protides of the biological fluids, vol. 34. Pergamon Press, Oxford.

Van Regenmortel, M. H. V. 1986b. Which structural features determine protein antigenicity? Trends Biochem. Sci. **11:**36–39.

Van Regenmortel, M. H. V. 1987. Antigenic cross-reactivity between proteins and peptides: new insights and applications. Trends Biochem. Sci. **12:**237–240.

Van Regenmortel, M. H. V., and J. Burckard. 1985. Quantitative microcomplement fixation tests using chicken anti-viral antibody extracted from egg yolk. J. Virol. Methods **11:**217–223.

Van Regenmortel, M. H. V., and A. R. Neurath. 1985. Immunochemistry of viruses. The basis for serodiagnosis and vaccines. Elsevier Biomedical Press, Amsterdam.

Van Regenmortel, M. H. V., and M. B. Von Wechmar. 1970. A re-examination of the serological relationship between tobacco mosaic virus and cucumber virus 4. Virology **41:**330–338.

Van Regenmortel, M. H. V., R. Yolken, G. Obert, and J. Burckard. 1983. Use of viral antibody extracted from egg yolk for detecting rota- and influenza viruses by ELISA, p. 291–294. *In* S. Avrameas, P. Druet, R. Masseyeff, and G. Feldmann (ed.), Immunoenzymatic techniques. Elsevier Biomedical Press, Amsterdam.

Volk, W. A., R. M. Snyder, D. C. Benjamin, and R. R. Wagner. 1982. Monoclonal antibodies to the glycoprotein of vesicular stomatitis virus: comparative neutralizing activity. J. Virol. **42:**220–227.

Walter, G. 1984. Antibodies against synthetic peptides as a tool in tumor virology. Ann. Sclavo 2:61–70.

Walter, G. 1986. Production and use of antibodies against synthetic peptides. J. Immunol. Methods 88:149–161.

Walter, G., M. A. Hutchinson, T. Hunter, and W. Eckhart. 1982. Purification of polyoma virus medium-size tumor antigen by immunoaffinity chromatography. Proc. Natl. Acad. Sci. USA 79:4025–4029.

Walter, G., K.-H. Scheidtmann, A. Carbone, A. P. Laudano, and R. F. Doolittle. 1980. Proc. Natl. Acad. Sci. USA 77:5197–5200.

Wampler, D. E., E. D. Lehman, J. Boger, W. J. McAleer, and E. M. Scolnick. 1985. Multiple chemical forms of hepatitis B surface antigen produced in yeast. Proc. Natl. Acad. Sci. USA 82:6830–6834.

Warren, H. S., F. R. Vogel, and L. A. Chedid. 1986. Current status of immunological adjuvants. Annu. Rev. Immunol. 4:369–388.

Westhof, E., D. Altschuh, D. Moras, A. C. Bloomer, A. Mondragon, A. Klug, and M. H. V. Van Regenmortel. 1984. Correlation between segmental mobility and the location of antigenic determinants in proteins. Nature 311:123–126.

Wiley, D. C., I. A. Wilson, and J. J. Skehel. 1981. Structural identification of the antibody-binding sites of Hong Kong influenza haemagglutinin and their involvement in antigenic variation. Nature 289:373–378.

Wilson, I. A., J. J. Skehel, and D. C. Wiley. 1981. Structure of the haemagglutinin membrane glycoprotein of influenza virus at 3 Å resolution. Nature 289:366–373.

Yewdell, J. W., and W. Gerhard. 1981. Antigenic characterization of viruses by monoclonal antibodies. Annu. Rev. Microbiol. 35:185–206.

Yolken, H. Y. 1985. Solid phase immunoassays for the detection of viral diseases, p. 121–138. In M. H. V. Van Regenmortel and A. R. Neurath (ed.), Immunochemistry of viruses. The basis for serodiagnosis and vaccines. Elsevier Biomedical Press, Amsterdam.

Yolken, H. Y., S. B. Wee, and M. H. V. Van Regenmortel. 1984. The use of beta-lactamase in enzyme immunoassays for detection of microbial antigens. J. Immunol. Methods 73:109–123.

Yolken, R. H. 1983. Use of monoclonal antibodies for viral diagnosis. Curr. Top. Microbiol. Immunol. 104:177–195.

Zrein, M., G. De Marcillac, and M. H. V. Van Regenmortel. 1986a. Quantitation of rheumatoid factors by enzyme immunoassay using biotinylated human IgG. J. Immunol. Methods 87:229–237.

Zrein, M., G. Obert, and M. H. V. Van Regenmortel. 1986b. Use of egg-yolk antibody for detection of respiratory syncytial virus in nasal secretions by ELISA. Arch. Virol. 90:197–206.

CHAPTER 7

Electron Microscopy

FRANCES W. DOANE

Safety Precautions: Ultraviolet radiation exposure of specimen on EM grid; disposal of grids; virus-inactivation with glutaraldehyde.

Direct Examination of Fluid Specimens: Negative staining methods, including direct application, water drop, agar diffusion, pseudoreplica, ultracentrifugation.

Direct Examination of Tissues and Cells: Preparation of thin sections by standard and rapid embedding methods; collection of cell suspensions by hematocrit centrifugation.

Immunoelectron Microscopy Methods: Negative staining methods, including direct IEM, solid phase IEM, serum-in-agar, immunogold.

Mycoplasma Detection by SEM: Processing of monolayer cell cultures by critical point drying procedure.

Introduction

Electron microscopy (EM), when used in combination with the simple negative staining technique, offers one of the most rapid methods for virus detection, a result often being possible within minutes of receipt of the specimen. Although EM has long been recognized as a useful diagnostic tool (Nagler and Rake, 1948; van Rooyen and Scott, 1948), virologists have been rather slow to use it for detecting viruses in clinical specimens, owing possibly to the following concerns: (a) purchase and maintenance of an electron microscope is expensive; (b) only one specimen can be examined at a time; and (c) the sensitivity of direct EM examination may not be as high as other diagnostic methods.

Some of these problems have now been overcome. Many laboratories have their own electron microscope, or have access to one. Unfortunately, most electron microscopes cannot readily be used with multiple specimen holders. But with respect to the important area of sensitivity of virus detection, there have been impressive advances. With the introduction of immunoelectron microscopy (IEM) to diagnostic virology, extremely sensitive methods for detecting viral antigens and antibodies are now available.

Despite these advances, EM remains a selective tool to be applied in specific situations. It is essential that the EM operator be skilled in the techniques required for preparing specimens. It is equally important that he/she be familiar with viral ultrastructure and with the normal cellular artefacts routinely encountered in clinical specimens and cell culture harvests. Some of the EM techniques that we have found to be useful in diagnostic virology are presented in this chapter. However, for more in-depth coverage of the methods and for detailed descriptions of individual viruses, one would be wise to consult recent review articles and atlases dealing with viral ultrastructure (Chernesky and Mahony, 1984; Cheville, 1975; Davies, 1982; Doane and Anderson, 1977, 1987; England and Reed, 1980; Field, 1982; Hsiung, 1982; Kjeldsberg, 1980; Miller, 1986; Oshiro, 1985; and Palmer and Martin, 1982).

Direct Examination of Specimens

Among the variety of clinical specimens received in the diagnostic laboratory, most are in fluid form and can thus be processed by the very simple and rapid technique of negative staining. When tissue specimens are to be processed, they can be transformed into a suspension in preparation for negative staining, or they can be processed intact for histopathological examination.

Electron microscopy can also be used to detect and identify viral isolates in cell culture harvests. Here again, the fluid nature of the sample enables it to be processed by one of the negative staining procedures.

The sensitivity of EM detection may be increased impressively by the addition of antibody to the clinical specimen, viz., by immunoelectron microscopy (described below). Source of antibody can be specific polyclonal or monoclonal, convalescent serum, or gamma globulin.

Recommended procedures for EM examination of clinical specimens are presented in Table 1.

Safety Precautions

One might imagine that an infectious virus preparation on an EM specimen grid would quickly become inactivated after a few minutes of irradiation under the electron beam; unfortunately, such is not always the case. Thus the EM operator should observe safety precautions not only during the preliminary

stages of specimen preparation, but also in handling and disposing of the prepared specimen grid.

In our laboratory, EM grids containing freshly prepared, negatively stained specimens are exposed to ultraviolet (UV) radiation (900 μW/cm^2 at 15 to 20 cm) for at least 10 min before EM examination. Forceps used to handle grids are rinsed in alcohol (or wiped with an alcohol swab) and flamed *briefly* immediately after use. Once a grid has been examined by EM, it should either be returned to its container (e.g., a small snap-lock plastic Petri dish) or be discarded in a small, wide-base, capped container.

The 2.5% glutaraldehyde solution used to fix tissues for histopathological examination has been shown to be effective in inactivating poliovirus and hepatitis B virus after 30 min (Boudouma, Enjalbert and Didier, 1984; Kobayashi et al., 1984).

Fluid Clinical Specimens

Included in this category are respiratory tract secretions, cerebrospinal fluid, urine, feces, vesicle fluid, and scrapings. Dense specimens such as vesicle scrapings and feces should be mixed with a small volume of diluent, such as distilled water or 1% am-

TABLE 1. Recommended procedures for direct EM examination of clinical specimens

Specimen	Preliminary preparation	EM method[a]	Viruses detected
Blood cells	Fix cell pellet	TS	HIV
Brain tissue	(i) Prepare suspension in water by grinding or by several cycles of freezing and thawing	DA; WD	Herpes simplex/varicella zoster; PML; SSPE
	(ii) Fix small (~1 mm) cubes	TS	
CSF	None required	AD; AF; IEM	Herpes simplex/varicella zoster; mumps; entero (?)
Feces	(i) Centrifuge bacteria from 20% suspension	DA; AD; AF	Adeno; astro; calici; corona; entero; mini-reo; Norwalk; parvo; reo; rota
	(ii) Prepare small suspension in water on slide	IEM	
Liver tissue	Fix small (~1 mm) cubes	TS	Adeno; arena; hepatitis B; herpes; Marburg
Lung biopsy	Prepare suspension in water by grinding or by several cycles of freezing and thawing	DA; WD	Herpes simplex/varicella zoster; CMV; measles; parainfluenza; RSV
Nasopharyngeal secretions	Place drop on slide; if viscous, dilute with equal quantity of water	DA; AD	Corona; herpes simplex; influenza; measles; parainfluenza; RSV
Serum	May require clarification by centrifugation	AD; IEM	Hepatitis B; parvo
Skin tumors	(i) Prepare suspension in water by grinding or by several cycles of freezing and thawing	WD; DA; IEM	Molluscum; orf; papilloma; vaccinia; variola
	(ii) Fix small (~1 mm) cubes	TS	
Tears	None required	AD; PR	Adeno; herpes simplex
Urine	None required	AD; PR; AF	Adeno; CMV; entero; herpes simplex; mumps; papova
Vesicle fluid	Collect in capillary tube or fine needle; dilute with equal quantity of water	DA; AD	Coxsackie; pox; herpes simplex/varicella zoster
Vesicle scab	Grind, underside down, in drop of water on slide		

[a] AD = Agar diffusion; AF = Airfuge; DA = direct application; IEM = immunoelectron microscopy; PR = pseudoreplica; TS = thin section; WD = water drop.

monium acetate, prior to negative staining. A recommended approach is to prepare specimens by two methods, the direct application method and the agar diffusion method. Specimen grids prepared by the direct application method can be examined within a few minutes. Provided there is not an excessive amount of salt in the suspension and there is a sufficiently high concentration of virus in the specimen, virus particles should be visible. While these grids are being examined, those prepared by the slower but more sensitive agar diffusion method can be air-dried. If the first set of grids does not yield results, the second set can then be examined.

If an Airfuge ultracentrifuge is readily available and if large numbers of specimens are being processed, it may be more convenient to use this instrument routinely to prepare specimen grids.

DIRECT APPLICATION METHOD

This is the simplest of the EM methods presented here. It requires approximately 10^8 virus particles per ml in the specimen, a concentration that may be present in a variety of specimens (Doane and Anderson, 1977; Doane et al., 1967; Joncas et al., 1969). It cannot be used for very salty specimens, however, as dried salt crystals tend to obliterate the virus particles.

Materials Required

1. 300-mesh copper EM specimen grids coated with Formvar or parlodion film, preferably stabilized with a light coat of evaporated carbon.
2. Phosphotungstic acid (2%) negative stain, prepared with filtered distilled water, and adjusted to pH 6.5 with 1 N KOH. Best stored in 1-ml syringes at 4°C.
3. Fine-bore Pasteur pipettes; EM forceps; filter paper.

Procedure

1. Place a small drop of specimen on a coated specimen grid held securely with EM forceps. Alternatively, float a grid for 1 to 2 min on a drop of specimen placed on a waxed surface (e.g., a sheet of Parafilm).
2. Add a drop of negative stain.
3. After 30 to 60 sec, touch the fluid with the torn edge of a piece of filter paper, leaving only a moist layer on the grid surface.
4. Air-dry under UV for 10 to 15 min; examine in EM.

WATER DROP METHOD

This is a simple and effective method for removing salt from fluid specimens. It yields a clean, nicely stained preparation (Fig. 1), but it requires a starting

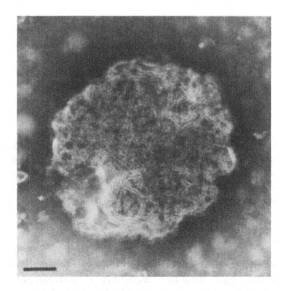

FIG. 1. Paramyxovirus particle in negatively stained nasopharyngeal secretions, prepared for direct electron microscopy by the water drop method. Bar = 100 nm.

concentration of approximately 10^9 virus particles per ml. (Anderson and Doane, 1972).

Materials Required

1. 300-mesh copper EM specimen grids coated with Formvar or parlodion film, preferably stabilized with a light coat of evaporated carbon.
2. Phosphotungstic acid (2%) negative stain, prepared with filtered distilled water, and adjusted to pH 6.5 with 1 N KOH. Best stored in 1-ml syringes at 4°C.
3. Fine-bore Pasteur pipettes; EM forceps; filter paper.
4. A waxed surface (e.g., a sheet of Parafilm).
5. Distilled water.

Procedure

1. Place a drop of filtered distilled water on a waxed surface.
2. Place a small drop of specimen on top of the water drop.
3. Briefly touch the coated surface of a specimen grid to the top of the drop.
4. Withdraw the grid and add to it a drop of negative stain.
5. Air-dry under UV as described above for the direct application method.

AGAR DIFFUSION METHOD

This is a modification of the methods described by Kelenberger and Arber (1957) and Kelen et al. (1971). It is useful for salty specimens, especially

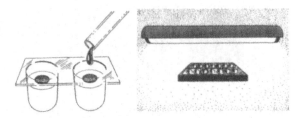

FIG. 2. Preparation of negatively stained specimen by the agar diffusion method. Specimen drops are placed on agar-supported grids (left) and are air-dried under UV light (right). (Reproduced, with permission, from Doane and Anderson, 1987.)

when the virus concentration is low. For small viruses it has a limit of detectability of approximately 10^7 virus particles per ml (Anderson and Doane, 1972) (see Fig. 2).

Materials Required

1. 300-mesh copper EM specimen grids coated with Formvar or parlodion film, preferably stabilized with a light coat of evaporated carbon.
2. Phosphotungstic acid (2%) negative stain, prepared with filtered distilled water, and adjusted to pH 6.5 with 1 N KOH. Best stored in 1-ml syringes at 4°C.
3. Fine-bore Pasteur pipettes; EM forceps; filter paper.
4. Plate of flexible microtiter cups filled approximately 3/4 full with 1% agar or agarose, covered with transparent sealing tape and stored at 4°C until needed.

Procedure

1. For each specimen to be examined, cut a pair of agar-filled cups from the microtiter plate. Remove the sealing tape and dry the agar surface at room temperature for approximately 5 min.
2. Place a coated specimen grid on the surface of each cup.
3. Add a drop of specimen to each grid.
4. Air-dry at room temperature under UV (30 to 60 min).
5. Add a drop of negative stain; remove the grid and draw the bottom lightly across a piece of filter paper. Air-dry on filter paper for 1 to 2 min.

PSEUDOREPLICA METHOD

Although more involved than the agar diffusion method described above, this method utilizes the same principles (Boerner et al., 1981, Lee et al., 1978; Smith, 1967).

Materials Required

1. 300-mesh copper EM specimen grids, *uncoated*.
2. Phosphotungstic acid (2%) negative stain, prepared with filtered distilled water, and adjusted to pH 6.5 with 1 N KOH. Best stored in 1-ml syringes at 4°C.
3. Fine-bore Pasteur pipettes; EM forceps; filter paper.
4. 10 mm × 10 mm × 5 cm squares of 2% agar or agarose.
5. Glass slides; paper toweling.
6. Formvar (0.5%) in ethylene dichloride.

Procedure

1. Place a square of agar on a glass slide. Place a drop of specimen in the center of the square.
2. Air-dry the specimen under UV for 10 to 15 min.
3. When the surface appears dry, flood with 1 to 2 drops of Formvar solution; drain off any excess with absorbent paper.
4. When the Formvar has dried completely, trim the block slightly on all four sides and move it to the extreme end of the slide.
5. Slowly dip the end of the slide into a container of negative stain, at a slight angle, until the Formvar film floats off.
6. Place a bare EM grid in the center of the floating film.
7. Turn up the corner of a piece of filter paper that is 2 to 3 times the size of the floating film; holding this corner, gently place the paper on top of the film. As soon as the paper becomes wet, quickly flip the paper (plus film and grid) 180° onto paper toweling.
8. Air-dry; examine in EM.

AIRFUGE ULTRACENTRIFUGATION

Hammond et al. (1981) reported that they routinely use this method for all clinical specimens submitted to the virus laboratory, increasing the sensitivity of detection by as much as 3 $logs_{10}$.

Materials Required

1. Beckman Airfuge ultracentrifuge with EM-90 rotor.
2. 300-mesh copper EM specimen grids coated with Formvar or parlodion film, preferably stabilized with a light coat of evaporated carbon.
3. Phosphotungstic acid (2%) negative stain, prepared with filtered distilled water, and adjusted to pH 6.5 with 1 N KOH. Best stored in 1-ml syringes at 4°C.
4. Fine-bore Pasteur pipettes; EM forceps; filter paper.

5. Parafilm.
6. Glutaraldehyde (2.5%); filtered distilled water.

Procedure

1. Place a coated specimen grid into the end of each of the six rotor sectors.
2. Add 90 μl of specimen into each sector.
3. Centrifuge at approximately 90,000 rpm for 30 min.
4. To stain each grid, invert it briefly on a drop of negative stain resting on Parafilm.
5. Air-dry under UV as described above.
6. After use, decontaminate the rotor and rotor cover by immersing in glutaraldehyde for 15 to 30 min. Rinse thoroughly in tap water, brushing the sectors with a cotton-tipped swab; soak briefly in 90% alcohol; air-dry.

Tissue Specimens and Intact Cell Suspensions

Tissue samples and cell suspensions are processed intact when a histopathological examination is called for (e.g., when looking for subacute sclerosing panencephalitis virus in brain tissue). When speed is vital (e.g., suspected herpes simplex encephalitis), the specimen can be processed by a rapid embedding method such as the one given below, which will yield sections within 2 to 3 hours. The quickest results, however, are obtained through negative staining. The tissue is ground in 1% ammonium acetate and negatively stained by one of the methods described above for fluid specimens. Although this approach has been successfully employed to detect herpes simplex virus in brain tissue (K. Chia, Toronto General Hospital, personal communication), it does not permit the same degree of methodical examination possible with thin sections (Fig. 3).

The hematocrit method described below has been found to be useful for histopathological examination of small quantities of cells in suspension.

STANDARD FIXATION AND EMBEDDING METHOD

There are many well-established procedures for fixation and embedding of tissue samples and cell pellets (Mackay, 1981; Trump and Jones, 1978; Weakley, 1981). One example is given below.

Materials Required

1. Glutaraldehyde (2.5%) in 0.13 *M* phosphate buffer; final pH 7.3.

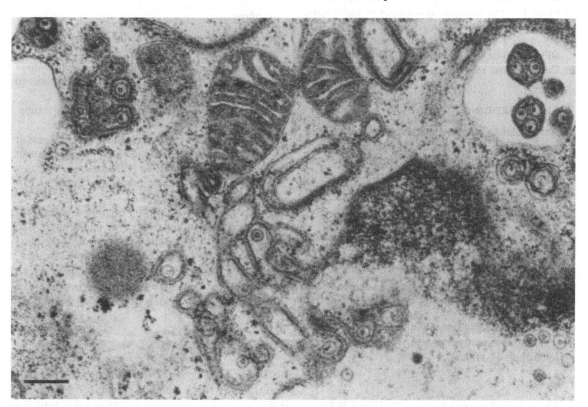

FIG. 3. Thin section of herpes simplex-infected cells. Naked nucleocapsids lie within the nucleoplasm at lower right. Note enveloped particles (upper right), some containing multiple nucleocapsids. Bar = 300 nm.

2. Osmium tetroxide (1%) in 0.13 M phosphate buffer; final pH 7.3.
3. Ethanol, propylene oxide, epoxy embedding medium.
4. 200-mesh copper grids.
5. Miscellaneous equipment and reagents for embedding and sectioning (e.g., oven, ultramicrotome, knives, stains).

Procedure

1. Fix 1-mm cubes of tissue in glutaraldehyde at 4°C for at least 30 min (longer if necessary).
2. Rinse in 3 changes of phosphate buffer, for a total of at least 1 h. Tissue can be stored in buffer at 4°C for prolonged periods if necessary.
3. Fix in osmium tetroxide at room temperature for 30 min.
4. Dehydrate through alcohol as follows: 2 changes of 70% ethanol (5 min each); 70% ethanol (5 min); 95% ethanol (5 min); 3 changes of 100% ethanol (5 min each).
5. Take tissue through 2 changes of propylene oxide (5 min each), and place in a 1 : 1 mixture of propylene oxide and embedding medium at room temperature for 1/2 to 1 h (in a rotating device).
6. Transfer to 100% embedding medium at room temperature for 1 to 24 h (in a rotating device).
7. Place in fresh embedding medium in embedding capsule at 37°C overnight.
8. Polymerize medium in oven (Epon–Araldite at 60°C for 24 to 48 h; Spurr's at 70°C for 8 to 24 h).
9. Cut sections; stain with uranyl acetate and lead citrate.

Rapid Embedding Method

This is similar to standard methods, but has shorter processing times and uses acetone instead of alcohol for dehydrating. Ultrastructural preservation and resolution are only slightly less than that obtained by standard methods (Doane et al., 1974).

Materials Required

1. Glutaraldehyde (2.5%) in 0.13 M phosphate buffer; final pH 7.3.
2. Osmium tetroxide (1%) in 0.13 M phosphate buffer; final pH 7.3.
3. Acetone, epoxy embedding medium.
4. 200-mesh copper grids.
5. Miscellaneous equipment and reagents for embedding and sectioning (e.g., oven, ultramicrotome, knives, stains).

Procedure

1. Fix 1-mm cubes of tissue in glutaraldehyde for 15 min at 4°C.
2. Rinse in 3 changes of phosphate buffer (1 min each).
3. Fix in osmium tetroxide for 15 min at room temperature.
4. Dehydrate through acetone as follows: 2 changes of 70% (total 5 min); 3 changes of 100% (total 5 min).
5. Place in 1 : 1 mixture of 100% acetone and embedding medium for 10 min at room temperature.
6. Transfer through 2 changes of 100% embedding medium (5 min each).
7. Place in embedding capsule in fresh embedding medium and heat at 95°C for 60 min.
8. Cool block; cut and stain sections as usual.

Hematocrit Method

This method can be used for collecting isolated cells, such as blood cells or suspensions of cell cultures, in preparation for embedding and sectioning (Doane et al., 1974).

Materials Required

1. Hematocrit centrifuge and 1.3 mm × 75 mm capillary tubes.
2. Parafilm; plasticine; metal paper clip.
3. Glutaraldehyde (2.5%) in 0.13 M phosphate buffer; final pH 7.3.

Procedure

1. Collect cells into a pellet by light centrifugation in a conical-tip tube, for 3 to 5 min.
2. Withdraw the medium and replace with 2 to 3 drops of glutaraldehyde. Transfer the cell pellet to a sheet of Parafilm.
3. Draw the cells and fixative into a capillary tube; seal one end with a small plug of plasticine.
4. Centrifuge in a hematocrit centrifuge for 3 min at 12,500 rpm. The cells should now form a compact pellet immediately above the plasticine.
5. Score the glass tube and break at a distance 6 to 7 mm above the cell pellet.
6. Invert the tube so that the open end is directed toward a drop of glutaraldehyde on Parafilm. Use a straightened paper clip to push against the plasticine plug, forcing the cell pellet into the fixative.
7. Transfer the cell pellet to a vial containing fresh fixative, and process by a standard or rapid embedding method (as described above).

Cell Culture Harvests

Because it is not possible to examine by EM, on a daily basis, a large number of clinical specimens, many laboratories routinely screen out the positives by inoculating all specimens into cell cultures. The viral isolates can then be identified by electron microscopy (Fig. 4), either by negative staining or by

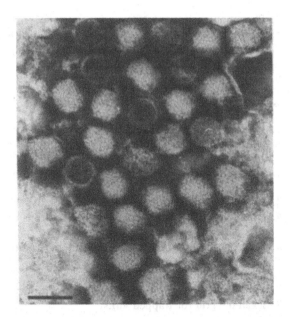

FIG. 4. Negatively stained adenovirus in cell culture lysate. Bar = 100 nm.

thin sectioning (by one of the methods described above).

PREPARATION FOR NEGATIVE STAINING

Procedure

1. Withdraw medium from the culture to be examined; store the medium temporarily.
2. Add 2 to 3 drops of filtered distilled water to the culture; resuspend cells.
3. After 2 to 3 min, process cell lysate by one of the negative staining methods described above, or by the serum-in-agar method of immunoelectron microscopy, with human immune serum globulin in the agar (see below).
4. If no virus is found in the lysate, negatively stain the temporarily stored medium, using the Airfuge ultracentrifugation method or the serum-in-agar method.

Note: Some enveloped RNA viruses such as togaviruses or retroviruses may disrupt in the negative stain. Ultrastructure can be preserved by brief fixation (15 min) in 2.5% buffered glutaraldehyde prior to negative staining (M. Szymanski, Hospital for Sick Children, Toronto, personal communication).

PREPARATION FOR THIN SECTIONING

Procedure

1. Gently resuspend the cultured cells into the overlying medium, using a rubber policeman or a Pasteur pipette.

2. Transfer the cells to a conical-tipped centrifuge tube; centrifuge at 1,500 rpm for 3 to 5 min.
3. Withdraw the medium and replace with 2 to 3 drops of 2.5% buffered glutaraldehyde. Process by a standard or rapid embedding method, with or without the aid of a hematocrit centrifuge (see above).

Note: If difficulty is encountered in keeping the cells in a tightly packed pellet, they can be enrobed in agar as follows. After the cells have been fixed in osmium tetroxide and transferred to buffer, centrifuge them lightly to produce a pellet. Gently withdraw the buffer and add a small drop of molten (43°C) 1% agar. Quickly mix the cells and agar, and pipette the mixture onto a glass slide. Allow the agar to solidify; cut into 1-mm^3 blocks with a clean razor blade. The enrobed cells will remain in place during transfer through subsequent solutions.

Immunoelectron Microscopy Methods Used in Connection with Negative Staining Methods

Immunoelectron microscopy (IEM) has several applications in virus diagnosis (Doane, 1974; Kapikian et al., 1975, 1976, 1979). It was used effectively to identify rubella virus and provided the first morphological characterization of this elusive virus (Best et al., 1967). It can also be used to serotype viruses, especially nonenveloped viruses such as adenoviruses (Luton, 1973; Svensson and von Bonsdorf, 1982; Vassall and Ray, 1974); enteroviruses (Anderson and Doane, 1973; Petrovicova and Juck, 1977); papovaviruses (Gardner et al., 1971; Penny et al., 1972; Penny and Narayan, 1973); and rotaviruses (Gerna et al., 1984).

Perhaps its most important application, however, is in the detection of viral antigen and antibody. The incorporation of antibody into the EM detection system has increased its sensitivity several-hundredfold, especially through the use of methods employing colloidal gold as an antibody label, making IEM one of the most sensitive detection methods available (Hopley and Doane, 1985).

Like other immunoassays, IEM procedures require careful attention to controls. It is essential that all reactants first be titrated to determine optimum test concentrations, and each test must be run in parallel with the appropriate positive and negative controls.

Direct Immunoelectron Microscopy Method

The simple procedure of mixing together equal volumes of the two reactants of viral antigen and antibody was first recommended by Almeida and Waterson (1969). Although these authors included an overnight incubation after step 1, followed by centrifugation at 10,000 to 15,000 rpm for 1/2 h, we tend to omit these steps.

Materials Required

As for agar diffusion method of negative staining (above).

Procedure

1. Prepare 1 : 1 mixtures of specimen and antiserum; incubate at 37°C for 30 min.
2. Process the mixtures by the agar diffusion method (above).
3. Examine grids for the presence of virus–antibody aggregates.

Solid-Phase Immunoelectron Microscopy Method

Long used by plant pathologists (Derrick, 1973), this method is particularly good for increasing the sensitivity of detection of viruses in feces (Pegg-Feige and Doane, 1983, 1984; Svensson et al., 1982, 1983) (see Fig. 5). Antibody-coated grids are most sensitive when used soon after preparation. Although they can be stored at 4°C, their efficiency drops by 50% after 4 to 5 weeks.

Materials Required

1. Rotavirus antiserum (e.g., 1/100 dilution).
2. Protein A, 1 mg/ml.
3. Tris buffer (0.05 M) pH 7.2.

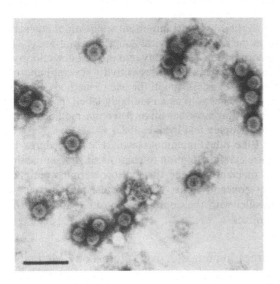

FIG. 5. Rotavirus in stool specimen prepared by solid-phase IEM. Bar = 200 nm. (Micrograph courtesy of Kathryn Pegg Feige.)

4. Phosphotungstic acid (2%) negative stain, pH 6.5.
5. 300-mesh copper EM specimen grids coated with Formvar and carbon, pretreated with UV light (1,700 μW/cm^2 for 30 min prior to use).
6. Parafilm.

Procedure

1. Invert a freshly pretreated grid on a drop of protein A on Parafilm. Leave at room temperature for 10 min.
2. Drain grid briefly on edge; transfer across 3 separate drops of Tris buffer (total 1 to 2 min).
3. Invert on a drop of rotavirus antiserum for 10 min.
4. Drain briefly; transfer across 3 separate drops of Tris buffer (1 to 2 min).
5. Invert on a drop of specimen at room temperature for 30 min.
6. Rinse on 3 drops of Tris buffer (1 to 2 min).
7. Float briefly on a drop of negative stain; air-dry.
8. Examine grid for trapped virus.

Serum in Agar

This is a modification of the agar diffusion method (see above), incorporating antiserum in the agar (Anderson and Doane, 1973). When human immune serum globulin is used in place of specific antiserum, a broad, nonselective detection system is produced (Juneau, 1979; Berthiaume et al., 1981). Viruses detected by this preliminary screening step can be assigned to a virus family on the basis of their morphology. Further serotyping of viruses such as picornaviruses can then be performed by the serum-in-agar method with pools of enterovirus antisera. Antisera should be titrated to determine the IEM endpoint (highest dilution producing an immune complex) and can then be used at 10 to 100 times this value. We routinely use immune serum globulin at a final dilution of 1/25 to 1/50.

Materials Required

1. 300-mesh copper EM specimen grids coated with Formvar or parlodion film, preferably stabilized with a light coat of evaporated carbon.
2. Phosphotungstic acid (2%) negative stain, pH 6.5.
3. Flexible plastic microtiter plates.
4. Agar or agarose.
5. Antiviral antibodies (monoclonal or polyclonal, type-specific or pooled).

Procedure

1. Prepare a 1% molten solution of agar (or agarose) by heating to boiling.
2. Allow the molten agar to cool to approximately 45°C, and add antiserum.

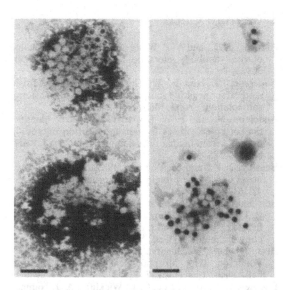

FIG. 6. Enterovirus immune complexes seen in specimens prepared by the serum-in-agar method. *Left:* In the presence of homologous antibody; virus–antibody aggregates are much easier to detect on the grid than single isolated virus particles. *Right:* In the presence of homologous antibody and protein A-gold. An aggregate is clearly identified by the presence of the electron-dense gold label. Bar = 100 nm.

3. Pipette the mixture into microtiter cups (approximately 3/4 full); allow it to solidify at room temperature. Plates can now be covered with plastic sealing tape and stored at 4°C.
4. For use, cut a pair of cups from the plate, remove the tape, and air-dry the agar surface at room temperature for approximately 5 min.
5. Place a coated specimen grid on the surface of each cup.
6. Add a drop of specimen to each grid.
7. Air-dry at room temperature (30 to 60 min).
8. Add a drop of negative stain; remove the grid and draw the bottom lightly across a piece of filter paper. Place on filter paper until dry (1 to 2 min). Examine for the presence of virus–antibody aggregates (Fig. 6).

Immunogold Labeling Method

The immunoelectron microscopy methods described above rely on the presence of antibody-bound virus particles, in the form of either virus–antibody aggregates or virus particles trapped to an antibody-coated grid. By use of antibody labeled with an electron-dense marker, not only virus particles, but also soluble viral antigen can be detected by IEM. Colloidal gold has been shown to be a superior marker for electron microscopy (Faulk and Taylor, 1971), especially when used as an immunocytochemical label

(Polak and Varndell, 1984). When applied to fluid specimens in combination with the negative staining technique, it provides a rapid and extremely sensitive immunoassay (Beasley and Betts, 1985; Hopley and Doane, 1985; Kjeldsberg, 1985; Patterson and Oxford, 1986; Stannard et al., 1982). Although used primarily for detecting viral antigen, colloidal gold should also be useful for detecting small quantities of viral antibody (Hopley and Doane, 1985). The immunogold method described below was developed in our laboratory by Miss Nan Anderson.

Careful attention should be paid to controls, which should include a positive virus control, a negative virus control, and a normal serum control.

Materials Required

1. Reagents and supplies for the serum-in-agar method (see above).
2. 16 nm colloidal gold particles complexed to protein A (protein A-gold).

Procedure

1. Mix equal volumes of protein A-gold and the test specimen; leave at room temperature for 5 to 10 min.
2. Place a drop of mixture on a grid on a serum-in-agar cup. Allow the mixture to air-dry (30 to 60 min).
3. Rinse the grids by immersing sequentially in 5 separate drops on Parafilm; the first 3 drops consist of the buffer used to dilute the protein A-gold (0.5 M tris[hydroxymethyl]-aminomethane hydrochloride, pH 7.0; 0.15% NaCl; 0.5 mg/ml polyethylene glycol 20,000, 0.1% NaN₃); the final 2 drops consist of filtered distilled water.
4. Place grids on a drop of negative stain for 30 sec; remove, air-dry, and examine for the presence of gold label (Fig. 6).

Mycoplasma Screening of Cell Cultures by Scanning Electron Microscopy

The presence of contaminating mycoplasmas in cell cultures (Fig. 7) can severely reduce the virus susceptibility of the culture isolation system. Unfortunately, these insidious microorganisms are often difficult to detect. A screening system that has been used with some success employs the scanning electron microscope (Doane and Anderson, 1987; Ho and Quinn, 1977). The method used in our laboratory is presented here.

Materials Required

1. Fixatives and dehydrating alcohol as required for processing tissues for thin sections (see above).

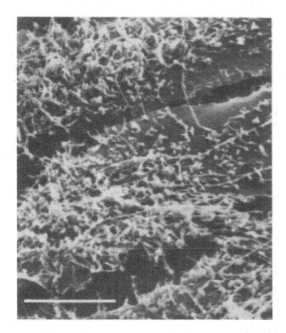

Fig. 7. Scanning electron micrograph showing three cells heavily contaminated with mycoplasma. Bar = 10 μm.

2. Source of CO_2 or freon.
3. Critical-point dryer.

Procedure

1. The culture to be examined should be grown on a cover slip, as long as possible in antibiotic-free medium. Ideally, the cell monolayer should not be confluent, so that open spaces on the cover slip (where mycoplasmas tend to attach) can be observed.
2. Remove the nutrient medium and replace it with 2.5% glutaraldehyde. Fix at 4°C for at least 1 h, preferably overnight.
3. Wash the cover slip 3 times with buffer; fix in 1% osmium tetroxide for 30 min.
4. Rinse in buffer and dehydrate through an alcohol series of 1 change in each of 50%, 70%, and 95%, and 3 changes of absolute alcohol (10 min per change).
5. Critical-point dry the culture in CO_2 or freon; attach the cover slip to a specimen stub, coat the specimen with gold/palladium, and examine in a scanning electron microscope. (If preparations cannot be examined immediately, store in a desiccator).

Acknowledgments. I am indebted to my colleagues Miss Nan Anderson, who developed many of the methods presented in this chapter, and Dr. Peter Middleton, who advised on Table 1.

Literature Cited

Almeida, J. D., and A. P. Waterson. 1969. The morphology of virus–antibody interaction. Adv. Virus Res. **15**:307–338.

Anderson, N., and F. W. Doane. 1972. Agar diffusion method for negative staining of microbial suspensions in salt solutions. Appl. Microbiol. **24**:495–496.

Anderson, N., and F. W. Doane. 1973. Specific identification of enteroviruses by immuno-electron microscopy using a serum-in-agar diffusion method. Can. J. Microbiol. **19**:585–589.

Beasley, J. E., and M. P. Betts. 1985. Virus diagnosis: a novel use for the protein A-gold probe. Med. Lab. Sci. **42**:161–165.

Berthiaume, L., R. Alain, B. McLaughlin, P. Payment, and P. Trepanier. 1981. Rapid detection of human viruses in faeces by a simple and routine immune electron microscopy technique. J. Gen. Virol. **55**:223–227.

Best, J. M., J. E. Banatvala, J. D. Almeida, and A. P. Waterson. 1967. Morphological characteristics of rubella virus. Lancet **2**:237–239.

Boerner, C. F., F. K. Lee, C. L. Wickliffe, A. J. Nahmias, H. D. Cavanagh, and S. E. Strauss. 1981. Electron microscopy for the diagnosis of ocular viral infections. Ophthalmology **88**:1377–1380.

Boudouma, M., L. Enjalbert, and J. Didier. 1984. A simple method for the evaluation of antiseptic and disinfectant virucidal activity. J. Virol. Methods **9**:271–276.

Chernesky, M. A., and J. B. Mahony. 1984. Detection of viral antigens, particles, and early antibodies. Yale J. Bio. Med. **57**:757–776.

Cheville, N. F. 1975. Cytopathology in viral diseases. *In* J. L. Melnick (ed.), Monographs in virology, vol. 10. Karger, Basel.

Davies, H. A. 1982. Electron microscopy and immune electron microscopy for detection of gastroenteritis viruses, p. 37–49. *In* D. A. Tyrrell and A. Z. Kapikian (ed.), Virus infections of the gastrointestinal tract. Marcel Dekker, New York.

Derrick, K. S. 1973. Quantitative assay for plant viruses using serologically specific electron microscopy. Virology **56**:652–653.

Doane, F. W. 1974. Identification of viruses by immunoelectron microscopy, p. 237–255. *In* E. Kurstak and R. Morisset (ed.), Viral immunodiagnosis. Academic Press, New York.

Doane, F. W., and N. Anderson. 1977. Electron and immunoelectron microscopic procedures for diagnosis of viral infections, p. 505–539. *In* E. Kurstak and C. Kurstak (ed.), Comparative diagnosis of viral diseases, vol. II, part B. Academic Press, New York.

Doane, F. W., and N. Anderson. 1987. Electron microscopy in diagnostic virology. A practical guide and atlas. Cambridge University Press, New York.

Doane, F. W., N. Anderson, J. Chao, and A. Noonan. 1974. Two-hour embedding procedure for intracellular detection of viruses by electron microscopy. Appl. Microbiol. **27**:407–410.

Doane, F. W., N. Anderson, K. Chatiyanonda, R. M. Banatyne, D. M. McLean, and A. J. Rhodes. 1967. Rapid laboratory diagnosis of paramyxovirus infections by electron microscopy. Lancet **2**:751–753.

England, J. J., and D. E. Reed. 1980. Negative contrast microscopic techniques for diagnosis of viruses of veterinary importance. Cornell Vet. **70**:125–136.

Faulk, W. P., and G. M. Taylor. 1971. An immunocolloid method for the electron microscope. Immunochemistry **8**:1081–1083.

Field, A. M. 1982. Diagnostic virology using electron microscopic techniques. Adv. Virus Res. **27**:1–69.

Gardner, S. D., A. M. Field, D. V. Coleman, and B. Hulme 1971. A new human papovavirus (BK) isolated from urine after renal transplantation. Lancet **1**:1253–1257.

Gerna, G., N. Passarani, M. Battaglia, and E. Percivalle. 1984. Rapid serotyping of human rotavirus strains by solid-phase immune electron microscopy. J. Clin. Microbiol. **19**:273–278.

Hammond, G. W., P. R. Hazelton, I. Chuang, and B. Klisko. 1981. Improved detection of viruses by electron microscopy after direct ultracentrifuge preparation of specimens. J. Clin. Microbiol. **14**:210–221.

Ho, T. Y., and P. A. Quinn. 1977. Rapid detection of mycoplasma contamination in tissue cultures by SEM, p. 291–299. In O. Johari and R. P. Becker (ed.), Scanning electron microscopy, vol. II. IIT Research Institute, Chicago.

Hopley, J., and F. W. Doane. 1985. Development of a sensitive protein A-gold immunoelectron microscopy method for detecting viral antigens in fluid specimens. J. Virol. Methods **12**:135–147.

Hsiung, G. D. 1982. Diagnostic virology. Yale University, New Haven.

Joncas, J. H., L. Berthiaume, R. Williams, P. Beaudry, and V. Pavilanis. 1969. Diagnosis of viral respiratory infections by electron microscopy. Lancet **1**:956–959.

Juneau, M. L. 1979. Role of the electron microscope in the clinical diagnosis of viral infections from patients' stools. Can. J. Med. Technol. **41**:53–57.

Kapikian, A. Z., J. L. Dienstag, and R. H. Purcell. 1976. Immune electron microscopy as a method for the detection, identification, and characterization of agents not cultivable in an in vitro system, p. 467–480. In N. R. Rose and H. Friedman (ed.), Manual of clinical immunology. American Society for Microbiology, Washington, D.C.

Kapikian, A. Z., S. M. Feinstone, R. H. Purcell, R. G. Wyatt, T. S. Thornhill, A. R. Kalica, and R. M. Chanock. 1975. Detection and identification by immune electron microscopy of fastidious agents associated with respiratory illness, acute nonbacterial gastroenteritis, and hepatitis A. Perspect. Virol. **9**:9–47.

Kapikian, A. Z., R. H. Yolken, H. B. Greenberg, R. G. Wyatt, A. R. Kalica, R. M. Chanock, and H. W. Kim. 1979. Gastroenteritis viruses, p. 927–995. In E. H. Lennette and N. J. Schmidt (ed.), Diagnostic procedures for viral, rickettsial and chlamydial infections. American Public Health Association, Washington, D.C.

Kelen, A. E., A. E. Hathaway, and D. A. McLeod. 1971. Rapid detection of Australia/SH antigen and antibody by a simple and sensitive technique of immunoelectron microscopy. Can. J. Microbiol. **17**:993–1000.

Kelenberger, E., and W. Arber. 1957. Electron microscopical studies of phage multiplication. 1. A method for quantitative analysis of particle suspensions. Virology **3**:245–255.

Kjeldsberg, E. 1980. Application of electron microscopy in viral diagnosis. Pathol. Res. Pract. **167**:3–21.

Kjeldsberg, E. 1985. Specific labelling of human rotaviruses and adenoviruses with gold-IgG complexes. J. Virol. Methods **12**:47–57.

Kobayashi, H., M. Tsuzuki, K. Koshimizu, H. Toyama, N. Yoshihara, T. Shikata, K. Abe, K. Mizuno, N. Otomo, and T. Oda. 1984. Susceptibility of hepatitis B virus to disinfectants or heat. J. Clin. Microbiol. **20**:214–216.

Lee, F. K., A. J. Nahmias, and S. Stagno. 1978. Rapid diagnosis of cytomegalovirus infection in infants by electron microscopy. N. Engl. J. Med. **299**:1266–1270.

Luton, P. 1973. Rapid adenovirus typing by immunoelectron microscopy. J. Clin. Pathol. **26**:914–917.

Mackay, B. 1981. Introduction to diagnostic electron microscopy. Appleton-Century-Crofts, New York.

Miller, S. E. 1986. Detection and identification of viruses by electron microscopy. J. Electron Microsc. Techn. **4**:265–301.

Nagler, F. P. O., and G. Rake. 1948. The use of the electron microscope in diagnosis of variola, vaccinia and varicella. J. Bacteriol. **55**:45–51.

Oshiro, L. S. 1985. Application of electron microscopy to the diagnosis of viral infections, p. 55–72. In E. H. Lennette (ed.), Laboratory diagnosis of viral infections. Marcel Dekker, New York.

Palmer, E. L., and M. L. Martin. 1982. An atlas of mammalian viruses. CRC Press, Boca Raton.

Patterson, S., and J. S. Oxford. 1986. Analysis of antigenic determinants on internal and external proteins of influenza virus and identification of antigenic subpopulations of virions in recent field isolates using monoclonal antibodies and immunogold labelling. Arch. Virol. **88**:189–202.

Pegg-Feige, K., and F. W. Doane. 1983. Effects of specimen support film in solid phase immunoelectron microscopy. J. Virol. Methods **7**:315–319.

Pegg-Feige, K., and F. W. Doane. 1984. Solid phase immunoelectron microscopy for rapid diagnosis of enteroviruses, p. 226–227. In G. W. Bailey (ed.), 42nd Annual Proceedings of the Electron Microscopy Society of America. San Francisco Press, Inc., San Francisco, CA.

Penny, J. B., and O. Narayan. 1973. Studies of the antigenic relationships of the new human papovaviruses by electron microscopy agglutination. Infect. Immun. **8**:299–300.

Penny, J. B., L. P. Weiner, R. M. Herndon, O. Narayan, and R. T. Johnson. 1972. Virions from progressive multifocal leukoencephalopathy; rapid serological identification by electron microscopy. Science **178**:60–62.

Petrovicova, A., and A. S. Juck. 1977. Serotyping of coxsackieviruses by immune electron microscopy. Acta Virol. **21**:165–167.

Polak, J. M., and I. M. Varndell (ed.) 1984. Immunolabelling for electron microscopy. Elsevier, Amsterdam.

Smith, K. O. 1967. Identification of viruses by electron microscopy, p. 545–572. In H. Busch (ed.), Methods in cancer research, vol. 1. Academic Press, New York.

Stannard, L. M., M. Lennon, M. Hodgkiss, and H. Smuts. 1982. An electron microscopic demonstration of immune complexes of hepatitis B e-antigen using colloidal gold as a marker. J. Med. Virol. **9**:165–175.

Svensson, L., M. Grandien, and C. A. Pettersson. 1983. Comparison of solid-phase immune electron microscopy by use of protein A with direct electron microscopy and enzyme-linked immunosorbent assay for detection of rotavirus in stool. J. Clin. Microbiol. **18**:1244–1249.

Svensson, L., and C. D. von Bonsdorff. 1982. Solid-phase immune electron microscopy (SPIEM) by use of protein A and its application for characterization of selected adenovirus serotypes. J. Med. Virol. **10**:243–253.

Trump, B. F., and R. T. Jones (ed.) 1978. Diagnostic electron microscopy, vol. 1. Wiley, New York.

van Rooyen, C. E., and G. D. Scott. 1948. Smallpox diagnosis with special reference to electron microscopy. Can. J. Public Health **39**:467–477.

Vassall, J. H., and C. G. Ray. 1974. Serotyping of adenoviruses using immune electron microscopy. Appl. Microbiol. **28**:623–627.

Weakley, B. S. 1981. A beginner's handbook in biological electron microscopy. Churchill Livingstone, Edinburgh.

Nucleic Acids in Viral Diagnosis

M. RANKI, A.-C. SYVÄNEN, AND H. SÖDERLUND

Hybridization Principle: Single-stranded nucleic acids reform a double-stranded structure only if complementary.

Probe: A defined fragment of the nucleic acid to be identified carrying a label.

Characteristics of Hybridization: Exquisite specificity, sensitivity of 10^5 for target nucleic acid molecules.

Prerequisites of a Practical Test: Easy specimen pretreatment and easy separation of hybrid from unreacted probe.

Practical Methods: Sandwich hybridization is a three-component reaction with a capture probe attached on a solid matrix, and a labeled probe in solution together with the specimen nucleic acid. Labeled hybrids are formed on the solid carrier. All components can be in solution, the capture probe carrying an affinity label. Hybrids are collected onto the solid carrier based on affinity of the capture to the matrix.

Detection of Hybrids: Radioactive labels allow the highest sensitivity. Enzymatic methods for detection of biotin-labeled probes are frequently used. Other nonisotopic detection systems are emerging.

Applications: Infections with irregular genome expression such as human papillomavirus and human immunodeficiency virus infections. Infections with hepatitis B virus where accurate measure of virus quantity cannot be done by other methods, and cytomegalovirus that poses difficulties by standard techniques.

Introduction

The biological information of any organism is carried in its genome, either as DNA or RNA, as a sequence of the four nucleotides. The information is expressed in the synthesis of proteins, the properties of which are determined by their amino acid sequence. The proteins act either as structural components of the organism or as enzymes catalyzing or controlling physiological reactions. The proteins thus create the phenotype typical for the organism. In diagnosis a number of methods are available to recognize this phenotype (Fig. 1).

The methods to diagnose viruses include propagation in cell cultures in which case identification can be based on susceptible cell types, morphological changes induced, growth rate, and so on. More directly, the virus particles may be identified by physical methods (e.g., electron microscopy), or their structural components can be recognized using immunochemical methods.

Theoretically, microbial diagnosis could be based entirely on direct identification of the genotype of the organism, independently of whether it is a viroid, a virus, a bacterium, or a eukaryotic cell. By targeting the identification toward the actual source of biological diversity and/or similarity (i.e., the genetic information), the diagnosis can be made at any level of specificity.

Viral genomes are relatively small, and their sizes are distinct. In a few cases, the virus can be identified by direct isolation of its genome by electrophoresis. Alternatively, the genome may be cleaved enzymatically, and certain fragments can be identified. Such direct methods require relatively high concentrations of virus, and their use is therefore limited.

Universal methods are obtained when focusing on recognizing given nucleotide sequences. These tests

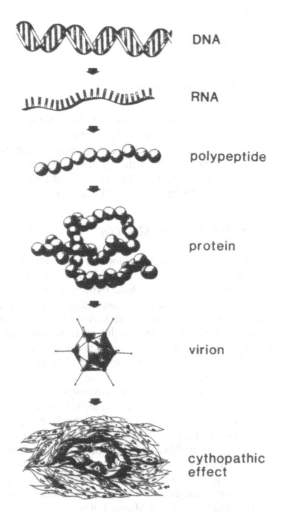

DNA

RNA

polypeptide

protein

virion

cythopathic
effect

FIG. 1. Diagnostic possibilities at different levels of genome expression: DNA and RNA by hybridization; polypeptide by immunochemistry; protein by enzymatic activity or immunochemistry; virion by physical methods, immunochemistry, or hybridization; cytopathic effect by microscopy or any of the above methods.

are based on the hybridization reaction, in which the intrinsic property of nucleic acids to form double-stranded helices by base-pairing is utilized. Since molecular cloning allows the isolation and amplification of any polynucleotide fragment, sequences complementary to any viral DNA or RNA can be obtained. Sequence identities or homologies are indicative of evolutionary relationship, and hybridization is therefore an extraordinarily specific tool for diagnosis.

The Hybridization Reaction

In the hybridization reaction, two single-stranded polynucleotides (DNA or RNA) react with each other, forming a double-stranded structure when their nucleotide sequence homology is high enough (Fig. 2). The reaction kinetics are thus of second order, and the rate is dependent on the concentration of both reactants. Because of the large size of nucleic acids, the reaction can be seen as a two-stage event—a primary collision to initiate the pairing, followed by a rapid zippering reaction completing the helix formation (Wetmur and Davidson, 1968). The specificity of the reaction is a result of the hydrogen bonding between the base pairs adenine to thymine and cytosine to guanosine. A major stabilizing force is the hydrophobic "stacking energy" obtained in the helix formation when the bases pile up to lie flat on top of each other (Marmur et al., 1963).

What is needed to transform the principle of polynucleotide duplex formation into a test detecting specific genes? First, a probe is required. The probe is a DNA or RNA molecule that has a sequence complementary to the genome to be detected. When added

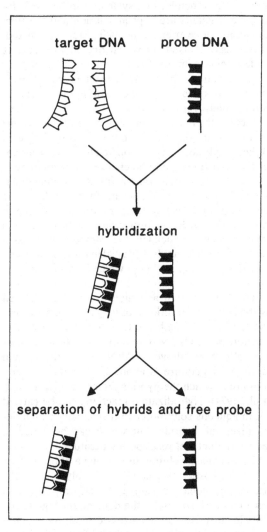

target DNA   probe DNA

hybridization

separation of hybrids and free probe

FIG. 2. Principle of a hybridization assay.

to a sample, the probe will base-pair with its target polynucleotide sequence. There are different possibilities to obtain probes, the most common being the molecular cloning of a fragment of the viral genome. Alternatively, oligonucleotides 20 to 50 bases long can be synthetized by chemical means. Cloning, nucleotide sequence determination, and oligonucleotide synthesis are outside the scope of this chapter. We want, however, to stress that probes with desired specificity can be selected so that they detect either a specific strain or a larger family of viruses.

The hybrid formation is detected with the aid of a label in the probe. Thus, a suitable label should be selected. The choices are essentially the same as in immunodiagnostics—radioactivity, enzymes, or fluorescent markers. Their advantages/disadvantages will be described below.

The probe is usually added in a large excess in hybridization reactions. Free probe must consequently be removed from hybrids prior to detection (Fig. 2). In principle, two systems can be used. Either the reactants are kept in solution and the fractionation is done after the annealing step, or a mixed-phase assay is used, in which one of the reacting nucleic acids is immobilized on a solid support prior to hybridization. The separation step is crucial. How this is performed distinguishes different hybridization methods from each other.

Finally, the reaction conditions must be set. Some treatments are required to release the nucleic acid of the sample and render it single-stranded, whereafter the conditions are to be changed to favor annealing. Sample treatment varies with the starting material and with the test format. In general, strong detergents and elevated temperatures are enough to release DNA or RNA from virions or from infected cells. Boiling dissociates the strands of double-stranded nucleic acids. They denature at a given "melting temperature" (T_m) which is in the range of 85 to 95°C.

The hybridization annealing reaction is affected by the reaction conditions, of which the temperature and the ionic strength are the most important. The reaction proceeds most rapidly at high concentrations of cations (above 0.4 M Na^+), and the optimal reaction temperature is about 25°C below the temperature at which the hybrids dissociate $(T_m$; Britten et al., 1974). Hybridization reactions can be carried out in aqueous solution or in formamide, which destabilizes the hybrids, but which can be added to reduce the optimal reaction temperature.

The technical solution to creating a hybridization based diagnostic test should preferably fulfill three criteria: it should be simple to perform, fast, and sensitive. Unfortunately, the demand for high sensitivity in most cases antagonizes demands for simplicity and rapidity (Syvänen, 1986). In the following

section, various compromises that are in use or under development are described.

Hybridization Methods

To be identifiable by hybridization, the viral nucleic acid is released from the virus, and the host cell and is rendered single-stranded. Depending on the nature of the specimens and on the hybridization method, additional purification of the nucleic acid from biological material may be required (Table 1).

Filter Hybridization Methods

Filter hybridization methods have gained widespread use both in basic research and in routine medical diagnosis. In this mixed-phase hybridization, one of the reacting nucleic acids is immobilized on a filter, which provides a convenient way to separate the formed hybrids from the reaction mixture. Nitrocellulose, to which the nucleic acids are noncovalently bound, is the most frequently used filter material. However, nucleic acid fragments smaller than 200 to 300 base pairs bind poorly to nitrocellulose, and loss of nucleic acids from the filters during hybridization has been demonstrated. These drawbacks have led to an increasing use of nylon-based membranes as solid support in filter hybridization. Nucleic acids can be covalently bound to nylon membranes. Therefore repeated hybridizations are possible after removal of hybridized probe molecules (Anderson and Young, 1985; Reed and Mann, 1985).

SPOT HYBRIDIZATION

Spot or dot blot hybridization is used for semiquantitative demonstration of the presence of specific DNA or RNA sequences (Denhardt, 1966; Gillespie and Spiegelman, 1965; Kafatos et al., 1979). The nucleic acids to be analyzed are immobilized in single-stranded form as spots on a filter. The immobilization prevents self-annealing of double-stranded molecules and allows the target sequences to be identified by hybridization to a labeled nucleic acid probe. After the reaction, unhybridized probe molecules are removed from the filter by washing. The hybrids formed are then detected on the filter by autoradiography if an isotopically labeled probe is used (Fig. 3) or by colored precipitates formed in an enzymatic detection procedure.

A specificity problem is often encountered in spot hybridization when using probes cloned in plasmids or bacteriophages. They will also hybridize to sequences homologuous to the vector part frequently present in bacteria in the clinical samples to be analyzed (Ambinder et al., 1986). This problem can be

TABLE 1. Procedure for the analysis of biological samples by spot hybridization

Crude samples[a]	Serum or cells

Release of the DNA
  Treatment with enzyme and detergent
  ↓
Purification of the DNA
  Phenol extraction
  Ethanol precipitation
  ↓
Denaturation of the DNA
  Boiling
  ↓
Immobilization of the DNA
  Spotting under pressure
  Baking at 80°C (nitrocellulose) or UV irradiation (nylon)
	Immobilization of samples
	  Spotting under pressure
	  ↓
	Release of the DNA
	  Treatment with enzyme and detergent
	  ↓
	Denaturation of the DNA
	  Alkaline treatment
	  Baking or UV irradiation

Prehybridization ↓ Hybridization to the labeled probe ↓ Washing ↓ Detection	

[a] Feces, mucus, urine, biopsies, for example.

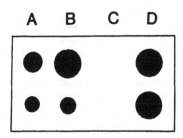

FIG. 3. Detection of human papillomavirus type 11 (HPV 11) by spot hybridization followed by autoradiography. Lane A: Standard samples containing 10^7 (top) and 5×10^6 (bottom) molecules of cloned HPV 11 DNA. Lanes B, D: HPV 11–positive cervical scrapings. Lane C: Human cells as negative controls. Hybridization was carried out in 50% formamide, $6 \times$ SSC (SSC = 0.15 M sodium chloride, 15 mM sodium citrate), $1 \times$ Denhardt's solution (0.2% BSA, 0.2% polyvinyl pyrrolidone, and 0.2% Ficoll), 1% SDS, 0.1 mg/ml RNA at 42°C overnight. The probe DNA was cloned HPV 11 DNA labeled by nick translation to a specific activity of 10^8 cpm/μg and used at 10^6 cpm/ml (Parkkinen et al., 1986).

avoided by using hybridization probes that do not contain vector sequences. Such probes are obtained by isolating the specific insert from the vector, by in vitro transcription of specific inserts with the SP6 RNA polymerase (Melton et al., 1984), or by using synthetic oligonucleotide probes.

Unspecific binding of the probe to biological material in the specimens is a second frequently occurring problem in spot hybridization. To avoid this, it is often necessary to purify the nucleic acid before immobilization. A typical procedure for the purification of a nucleic acid from biological material involves extractions with organic solvents (phenol, chloroform, diethyl ether) followed by precipitation with ethanol. The attachment of the nucleic acid to the filter requires additional steps. These measures, summarized in Table 1, add up to a tedious procedure for the preparation of the samples for spot hybridization. To some extent unspecific binding of probe DNA to the filters can be avoided by adding to the hybridization reaction components mimicking the chemical properties of nucleic acids (Denhardt 1966).

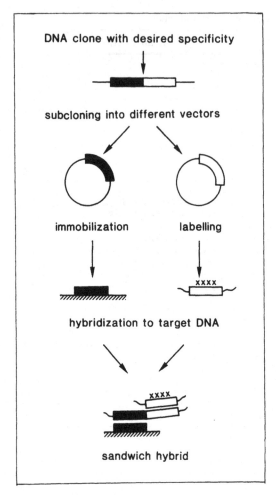

DNA clone with desired specificity

subcloning into different vectors

immobilization   labelling

xxxx

hybridization to target DNA

xxxx

sandwich hybrid

FIG. 4. Principle of the sandwich hybridization technique. Two identifying probes are cloned into nonhomologous vectors. One of the probes is immobilized, and the second is labeled. Target DNA, present in a sample, hybridizes to both probes, resulting in the formation of sandwich hybrids detectable on the solid support with the aid of the labeled probe.

In viral diagnosis, the most convenient application of spot hybridization is the analysis of samples that can be immobilized without extensive pretreatment—for example, serum (Scotto et al., 1983). In some cases, the target nucleic acid is released and denatured in situ on the filter. This approach has been used for the detection of parvovirus (Anderson et al., 1985) and hepatitis B virus (Yoffe et al., 1986) in human serum. The same approach can be utilized when infected cells are collected on filters. This "filter in situ hybridization" (FISH) has been used for detection of Epstein-Barr virus (Brandsma and Miller, 1980) and papilloma virus (Schneider et al., 1985).

SANDWICH HYBRIDIZATION

The sandwich hybridization technique (Ranki et al., 1983) allows more convenient identification of specific nucleic acids from crude biological samples than spot hybridization. In this method the sample is added to the reaction in solution, which allows unpurified biological material to be analyzed without background problems (Palva et al., 1984; Parkkinen et al., 1986; Virtanen et al. 1983). In sandwich hybridization, the target nucleic acid is identified by two separate probe fragments (Fig. 4). The fragments have been derived from adjacent sequences in the target nucleic acid and are nonhomologous to each other and cloned into nonhomologous vectors—for example, a derivative of the plasmid pBR 322 and a M13 phage vector. One of the probes has been immobilized on a small filter to act as capturing reagent, and the other probe is labeled. During the reaction the target nucleic acid hybridizes to the probe in solution and to the immobilized capturing fragment, thereby mediating labeling of the filter.

Because complementarity between the target and the two probes is required, background from sequences homologous to vector sequences is avoided in sandwich hybridization. After hybridization, the filters are washed, and the bound label is quantified by counting the radioactivity or by nonisotopic measurement. In sandwich hybridization, the results are most conveniently obtained as numeric values, which facilitates quantification of the results with the aid of a standard curve as well as objective definition of the limit of positivity (Fig. 5).

SOUTHERN BLOT HYBRIDIZATION

The Southern blot hybridization technique yields information on the size of specific nucleic acid fragments (Southern, 1975) and can be used to detect minor differences due to point mutations, deletions, or translocations within a DNA of interest. For Southern hybridization, purified target DNA is digested with one or several restriction endonucleases recognizing known sequences. Altered restriction enzyme cleavage patterns are revealed by separation of the fragments formed according to their size by gel electrophoresis. The fragments are then transferred from the gel to a filter for identification by hybridization to a labeled probe. The hybridization and detection procedures used in Southern blot hybridization are in principle identical to those used in spot hybridization.

In virology, Southern hybridization is useful in epidemiological studies where the detection of minor differences in the nucleotide sequences of the genomes is required to distinguish between different virus strains. This approach has been used in epidemiological studies of cytomegalovirus, for instance

(Spector et al., 1985). Southern hybridization can also be used to determine whether a viral genome is free in the cell or integrated into the genome of the host cell (see below).

Kinetics of Filter Hybridization

The rate of a hybridization reaction is decreased when one of the reactants is immobilized on a filter (Flavell et al., 1974). For filter hybridization, different protocols are used depending on the method and the purpose of the analysis (Anderson and Young, 1985). In most filter hybridization assays, overnight reactions are required to obtain maximal sensitivity. In direct filter hybridization techniques, the reaction rate can be accelerated by using probes at high molar excess over the target nucleic acid (Leary et al., 1983). This is possible without background problems using nonradioactively labeled probes or with small oligonucleotide probes. The addition of inert polymers (dextran sulfate, polyethylene glycol) acting as volume excluders to the reaction mixture can be used to increase the rate of filter hybridization reactions (Amasino 1986; Palva, 1985; Wahl et al., 1979).

Sandwich hybridization reaches 75% of the maximal reaction in 18 h (Ranki et al., 1983). The annealing of the target nucleic acid to the immobilized capturing fragment is the rate-limiting reaction. Thus, the reaction rate cannot be significantly increased by raising the concentration of the labeled probe. The required overnight reaction is a drawback in diagnostic applications of sandwich hybridization.

Hybridization in Solution

The most favorable reaction kinetics are obtained when all components of a hybridization reaction are present in solution. Solution hybridization methods have conventionally been used in the study of the structure of genes. These methods either require large amounts of material or involve complicated procedures for distinguishing hybrids from unhybridized molecules (Britten et al., 1974).

At defined conditions, hydroxyapatite binds specifically double-stranded nucleic acids (Kohne and Britten, 1971). This property has been utilized for the development of a solution hybridization assay suitable for routine diagnosis. In this method hybrids of target RNA molecules with a labeled single-stranded DNA probe formed in solution are detected after collection on hydroxyapatite. This method has been applied to the detection of ribosomal RNA, which is present in very high concentrations in bacterial samples, allowing short (15 min) reaction times (Edelstein, 1986; Wilkinson et al., 1986). In virology, this approach could be used for

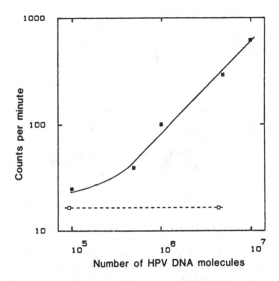

FIG. 5. Standard curve of sandwich hybridization for the detection of human papillomavirus type 11 DNA. Increasing amounts (10^5 to 10^7 molecules) of cloned HPV 11 DNA (■) or no DNA (□) were added to the reaction with the capturing probe (M13 clones, 500 ng/filter), immobilized on nitrocellulose filter disks, and ^{32}P-labeled probe DNA (pBR 322 clones, 4×10^5 cpm/assay) in a 0.4-ml solution consisting of $4 \times$ SSC, $1 \times$ Denhardt's solution, 0.2% SDS, 0.2 mg/ml denatured salmon sperm DNA. Hybridization was at 65°C for 16 h. After washing, the radioactivity bound to the filters was measured in a beta counter. The detection limit was 10^5 molecules of target DNA (Parkkinen et al., 1986).

the detection of messenger RNA as an amplified target RNA molecule.

In the affinity-based hybrid collection method (Syvänen et al., 1986a), the target nucleic acid is allowed to form hybrids with a pair of probes in solution. One of the probes has been modified with a ligand to enable separation of the formed hybrids from the reaction mixture on an affinity matrix. The other probe is labeled to allow quantification of the collected hybrids (Fig. 6). Biotin in conjunction with streptavidin immobilized on agarose beads or a hapten with an immobilized antibody is suitable as affinity pairs. In the affinity-based hybrid collection method, 70% of the maximal reaction is reached in 2 h, and the reaction can be further enhanced by volume excluders. Unpurified biological samples can be analyzed without background problems by this procedure (Syvänen and Korpela, 1986).

In Situ Hybridization

Hybridization in situ is used for the detection and localization of specific nucleic acid sequences within intact cells or tissues (Gall and Pardue, 1969). Since the analysis is done at the single-cell level, the distribution of specific sequences throughout a tissue can

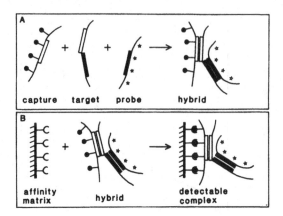

FIG. 6. The affinity-based hybrid collection procedure. (A) The target DNA is allowed to hybridize in solution to two probes. (B) The hybrids formed are collected on an affinity matrix, using the ligand on the capturing probe, and the hybrids are quantified with the aid of the label on the second probe (Syvänen et al., 1986a).

be studied. In the diagnosis of viral infections, it is the method of choice when only a small part of a cell population is infected, and the sensitivity of detection does not allow the sequence of interest to be found from the total nucleic acid pool of a sample (McDougall et al., 1986).

Preparation of the samples for in situ hybridization includes deposition of the cells or tissue on cover slips by cytological methods, fixation, treatment of the preparation to allow diffusion of the hybridization probe into the cells, and denaturation of double-stranded target DNA (Haase et al., 1984; Pardue, 1985) (Table 2). In the sample preparation procedure, a compromise usually has to be made between retainment of morphological structures and hybridization efficiency.

The principle of the in situ hybridization reaction is basically the same as in other mixed-phase hybridization methods, except that the final detection of the formed hybrids is done by microscopic examination of autoradiographic grains (Fig. 7) or of colored or fluorescent end products. The tedious sample preparation procedure and long hybridization and autoradiographic exposure times required in in situ hybridization are inconvenient for routine purposes with the technology of today.

Labeling of Hybridization Probes

Isotopic Labeling

Isotopes are still the most commonly used labels for hybridization probes, because they are sensitive and can be detected conveniently by autoradiography or by counting of radioactive disintegrations. A summary of the isotopes used to label nucleic acids is presented in Table 3. ^{32}P used as label in filter hybridization can be detected with high sensitivity by autoradiography. ^{35}S also gives good resolution in autoradiography and has the advantage of a longer half-life

TABLE 2. Procedure for in situ hybridization (modified from Haase et al., 1984)

Deposition of cells or tissue on glass slides
↓
Fixation
↓
Pretreatment: acid/base, heat, proteolytic enzyme —————————┐
↓   ↓
Hybridization (for RNA)   Digestion with RNAase
                                                                    ↓
                                                  Postfixation with paraformaldehyde
                                                                    ↓
Hybridization (for single-stranded DNA)   Denaturation of DNA
                                                                    ↓
                                                  Hybridization (for double-stranded DNA)
↓
Washing
↓
Coating with emulsion
↓
Autoradiographic exposure
↓
Development and staining
↓
Microscopy

than ^{32}P. ^{125}I and ^{32}P are about equally sensitive by counting. ^{3}H is widely used in in situ hybridization, because it has low energy and travels less in autoradiographic emulsions resulting in high resolution, but very long (months) exposure times are required with ^{3}H. ^{125}I and ^{35}S can be used with shorter (in the range of 1 week) exposure times in in situ hybridization and are thus more suitable for diagnostic applications.

The isotopes are usually introduced into the nucleic acids enzymatically as labeled nucleotide analogues (Table 3). DNA polymerases are used for nick translation of double-stranded DNA (Rigby et al., 1977) or for synthesis of complementary strands starting from a primer (Feinberg and Vogelstein, 1983; Hu and Messing, 1982; Studencki and Wallace, 1984). Cloned DNA can be transcribed into RNA using the SP6 RNA polymerase system (Melton et al., 1984). Nucleic acids can be labeled in the 5' end using T4 polynucleotide kinase (Maxam and Gilbert, 1980) and in the 3' end with terminal deoxyribonucleotide transferase (Jackson et al., 1972). A chemical method for labeling single-stranded nucleic acids directly with ^{125}I is also available (Commerford, 1971). A thorough review of the principles and methods of radioactive labeling of nucleic acids is given in (Arrand, 1985). Probes with specific activities between 10^8 and 10^9 cpm/μg nucleic acid can be obtained with all these procedures.

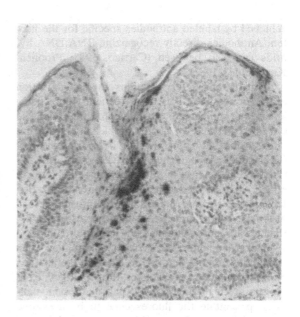

FIG. 7. Detection of human papillomavirus type 11 in a penile papillary condyloma by in situ hybridization with ^{35}S-labeled cloned HPV 11 DNA as probe. Hybridization was for 40 h at 42°C under stringent conditions. The condensation of black autoradiography grains indicates the presence of HPV 11 DNA confined to nuclei of epithelial cells. (Picture by Stina Syrjänen, University of Kuopio, Kuopio, Finland.)

Nonisotopic Labeling and Detection

During the last few years much effort has been put into the development of stable nonradioactive labels, which would facilitate routine applications of nucleic acid hybridization. Most of the nonradioactive detection procedures available today are indirect. The probe molecule is modified with a marker group, which is identified after hybridization with the aid of a labeled detector molecule. A prerequisite when modifying nucleic acids with chemical groups is that the hybridization properties of the molecules remain unaltered.

The most frequently used nonradioactive marker for hybridization probes is biotin. Biotin is introduced into nucleic acids as biotinylated nucleotide analogues using nucleic acid polymerases (Langer et al., 1981; Murasugi and Wallace, 1984) or by chemical reactions (Chollet and Kawashima, 1985; Forster et al., 1985; Viscidi et al., 1986). After hybridization, the biotinylated probes are detected with labeled streptavidin or avidin, two proteins with an extremely strong affinity for biotin ($K_d = 10^{-15}$ M).

In addition to biotin, haptens detectable by antibodies are suitable as markers for hybridization probes. The haptens (e.g., fluorene, sulfone, or p-nitrophenyl groups) are introduced by chemical modification of the DNA (Landegent et al., 1984; Tchen et al., 1984; Vincent et al., 1982). Mercurated nucleic acid probes to which haptens can be bound after the hybridization via sulfhydryl groups represent a modification of this approach (Hopman et al., 1986). Detection of the hybridized hapten-modified probes is

TABLE 3. Isotopes most commonly used for the labeling of nucleic acid hybridization probes

Isotope	Half-life	Type of decay	Energy	Specific activity of nucleotide precursor (Ci/mmol)
^{32}P	14 days	β	High	3,000–6,000
^{125}I	60 days	γ	Medium	1,500
^{35}S	87 days	β	Medium	1,000
^{3}H	12 years	β	Low	45–60

achieved by labeled antibodies specific for the hapten. Antibodies directly recognizing RNA:DNA hybrids can also be used (Carrico, 1985; Prooijen-Knegt et al., 1982).

In mixed-phase hybridizations, the final detection of the formed hybrids is most frequently done using an enzyme conjugated to the detector molecule. The enzyme catalyzes the formation of colored precipitates visible as spots on the hybridization filter (Leary et al., 1983; Tchen et al., 1984) or in the microscope in the case of in situ hybridization (Brigati et al., 1983). In sandwich and solution hybridization techniques, enzyme substrates forming soluble end products yielding numeric results from a spectrophotometer are more suitable (Syvänen et al., 1986a).

Background fluorescence is often a problem when using conventional fluorescent labels in the analysis of biological material. Chelates of the lanthanide europium are characterized by long fluorescent half-lives, permitting the fluorescence to be measured after a delay time during which background fluorescence disappears (Soini and Kojola, 1983). Europium-labeled antibodies have been used for sensitive detection of hapten-modified DNA in spot and sandwich hybridization (Syvänen et al., 1986b). Enhanced chemiluminescence is another sensitive method that has been applied as the final step in indirect detection of filter hybridization. The light emission from the chemiluminescent reaction catalyzed by the enzyme-labeled detector molecule is measured in a photomultiplier tube or recorded from the filter on instant photographic film (Matthews et al., 1985). A summary of indirect, nonradioactive detection methods is presented in Table 4.

So far, only a few direct approaches for nonisotopic detection of nucleic acid hybrids have been applied. Enzymes (peroxidase, phosphatase) modified with positively charged groups have been crosslinked to polynucleotides (Renz and Kurz, 1984), or alkaline phosphatase has been linked to a chemically modified base in an oligonucleotide probe (Jablonski et al., 1986). These enzymatically labeled probes are used in filter hybridization, followed by direct detection in one step by the addition of a substrate.

The main advantages of nonradioactive hybridization probes are their stability and the fact that they can be used at high concentrations to increase the rate of hybridization reactions. Drawbacks of nonradioactive probes are the multistep detection procedures and lower sensitivity of detection than with isotopes. When used in in situ hybridization, an additional advantage is the faster detection procedure than autoradiography of radiolabeled probes (Unger et al., 1986).

Sensitivity of Detection

The detection sensitivity of a hybridization assay is determined by the specific activity (detectable signal/mass unit) of the labeled probe used. In practice, however, the actual sensitivity is determined by the level of unspecific background signals (signal/noise ratios).

In filter hybridization, the high sensitivity achieved with ^{32}P as label depends entirely on long autoradiographic exposure times. The lowest limit of detection with overnight hybridization and 24-h autoradiography is in the range of 10^4 molecules (10^{-20} moles) of target nucleic acid. With direct measurement in beta or gamma counters, the limit of detection is 10^5 to 5×10^5 molecules of target using probes with specific activities in the range of 10^8 to 10^9 cpm/μg.

The enzymatic detection procedures commonly used today are about 10 times less sensitive than counting of radioactivity. Increased signals can be obtained in these methods using polymerized enzymes as labels on the detector molecule (The "Blue-Gene" system from Bethesda Research Laboratories, MD) or by enzymatic amplification of the primary product of the detection reaction (Stanley et

TABLE 4. Comparison of nonradioactive procedures for detection of immobilized target DNA (modified from Syvänen, 1986)

Marker	Detector label	End point	Amount of sequences detected[a]	Reference
Biotin	Avidin, alkaline phosphatase	Precipitating color	4 pg	Leary et al. (1983)
Biotin	Avidin, β-galactosidase	Fluorescence	8 pg	Nagata et al. (1985)
Biotin	Streptavidin, peroxidase	Chemiluminescence	1 pg	Matthews et al. (1985)
Fluorene groups	Antibodies, alkaline phosphatase	Precipitating color	5 pg	Tchen et al. (1984)
Fluorene or sulfone groups	Antibodies, europium	Time-resolved fluorescence	0.3 pg	Syvänen et al. (1986)
Enzyme	—	Precipitating color	4 pg	Renz and Kurz (1984)

[a] In these experiments the probe and the immobilized target DNA were identical, so the pg quantities reflect the sensitivity differences. In Syvänen et al. (1986), M13 clones also containing unhybridizing vector sequences were used as probes.

al., 1985). The sensitivity of detection also depends on the size of the nucleic acid probe used. Small oligonucleotide probes generally yield lower detection sensitivities than large probes.

Despite the fact that the actual sensitivity of detection in hybridization assays is high and amounts of target DNA in the attomole or picogram range can be detected, a limitation of nucleic acid hybridization methods today is their insufficient sensitivity for many applications. This problem originates from the fact that often only one or two copies of the target nucleic acid are present per organism.

Amplification of the Target DNA

A promising approach for increasing the sensitivity of detection in nucleic acid hybridization takes advantage of the unique property of nucleic acids as the only self-duplicable molecules known. The idea is to amplify the number of target molecules in the sample before detection. This is done in vitro with oligonucleotide primers and DNA polymerase in the so-called polymerase chain reaction (Saiki et al., 1985, 1986). The target DNA is denatured, primers are hybridized to both strands of the target, and new strands are synthesized by primer extension with DNA polymerase. This cycle is repeated, resulting in an exponential increase in copies of the target DNA (Fig. 8).

Hartley et al., 1986 have used another approach to amplify the target. In their "probe-vector" system, the amplification is performed in vivo. A plasmid carrying a probe region and a phenotypic marker (e.g., an antibiotic resistance gene) is linearized within the probe region, and the ends are made single-standed. Hybridization to the target will lead to recircularization of the plasmid (Fig. 9). The hybridization mixture is then used to transform *Escherichia coli* cells. Only circular plasmids will transform the

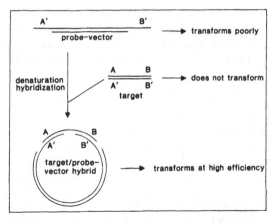

FIG. 9. Principle of hybrid detection by the probe-vector system.

bacterium to an antibiotic-resistant phenotype, and therefore transformants will arise only as a response to hybrid formation. The number of bacterial colonies obtained should reflect the concentration of target in the sample. The "probe-vector" system is in theory extremely sensitive, but it remains to be seen how it will perform in clinical practice.

Clinical Applications

Criteria for Applying Nucleic Acid Hybridization for Viral Diagnosis

Even if nucleic acid hybridization is a universal identification method, it will, in the first place, become useful in solving diagnostic problems that cannot be answered by other methods. Such problems include (1) microbial infections where the genome is not regularly expressed (nevertheless, the presence of the genome, even in a latent form, is significant); (2) estimation of the quantity of infectious virus when no fast biological infectivity assay is available; (3) specific identification of problematic microorganisms the diagnosis of which by conventional methods is time-consuming or inadequate; and (4) determining the identity or dissimilarity of two virus isolates that have minor differences in their DNA genomes.

In addition to defining the diagnostic question, some parameters concerning the specimen should be taken into account when choosing the appropriate hybridization technique. One is the crudeness of the specimen, determining the pretreatment need (Table 1), and the other is the sensitivity of the method as opposed to the quantity of target nucleic acids in the specimen. For instance, more virus is usually produced during primary childhood infections than dur-

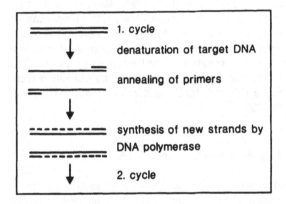

FIG. 8. Amplification of the target DNA by the polymerase chain reaction.

ing secondary infections with partial preexisting immunity. This was observed, for example, when quantitating adenovirus in the respiratory secretions from children and adults. The virus quantitites differed by 1 or 2 orders of magnitude (Lehtomäki et al., 1986; Virtanen et al., 1983). The problem is analogous when detecting cytomegalovirus from congenitally infected children as opposed to detection of CMV from immunocompromised patients.

In the following, we will concentrate on some illustrative examples, where nucleic acid hybridization has been of decisive importance in the development of an adequate diagnosis. We will not review the applications where the preexisting diagnostic methods are fully satisfactory and where hybridization is merely a complementary technique.

Detection of Human Papillomavirus in Genital Specimens by Nucleic Acid Hybridization

Human papillomaviruses (HPV) replicate in differentiating human epithelial cells and therefore cannot be grown in cell culture. More than 40 different virus types have been isolated. In addition to causing benign warts, they are associated with malignancies (Gissmann, 1984). Several HPV types have been isolated from genital warts or flat condylomas (HPV 6, 11, 16, 18, and 31), of which HPV 16 and 18 are considered the risk types, because they are found in a high proportion of the highergrade cervical intraepithelial neoplasias (CIN) and cervical cancer (Boshart et al., 1984; Crum et al., 1984; Durst et al., 1983; Gissmann, 1984; Syrjänen et al., 1985b). Even if the association of certain HPV types with malignancy seems to be strong, the proportion of the infections ultimately leading to progression is not clear (Syrjänen et al., 1985b).

The diagnostic problems connected with HPV infections are the following: (1); virus does not grow in cell culture; (2) virus can cause latent infections without constantly expressing its genome, so antigen detection is successful only in about 50% of infections (Syrjänen et al., 1985a); (3) detection of viral antigen does not allow virus typing (Syrjänen, et al. 1985a); (4) histology and cytology indicate the presence of HPV but are not type-specific; and (5) serological methods are noninformative. Nucleic acid hybridization, allowing type-specific detection of HPV genomes, has become the method of choice in HPV diagnosis in combination with cytology and histology.

The most common diagnostic questions are listed in Table 5. If only the possible presence or absence of the virus in a given specimen is studied, many different hybridization methods will provide the answer as summarized. Spot test requires purification of specimen DNA, whereas in the filter in situ technique, the specimen cells are collected onto the filter and treated there (Table 1). For the sandwich hybridization, specimens are boiled in the presence of detergents before subjecting to the test. Papillomavirus genomes are present in hundreds of copies per cell in a productive infection, implying that the sensitivity of each of the techniques will be satisfactory in most cases.

Detection of a latent infection with a few genome copies is more difficult. The efficacies of the methods are between 40 and 90% for specimens taken from patients with a suspected HPV infection. More important than the technique itself is the quality of the specimen. The specimen must represent the whole ectocervical epithelium and therefore contain a large number of cells in order to be suitable for screening (Burk et al., 1986; Parkkinen et al., 1986). When taken properly, such as by cervicovaginal lavage, the exfoliated cells provide a more adequate sample than a punch biopsy taken under colposcopic control. Burk et al. (1986) found HPV DNA in the exfoliated cells of 20 of 29 biopsied patients, whereas only 14 of the biopsies were positive.

Nucleic acid hybridization has also revealed the HPV DNA in cervical scrapes taken from women with a normal Pap smear. The percentage of positives reported varies from 2 to 29% (Burk et al., 1986; Schneider et al., 1985); suggesting that screening by hybridization might be more sensitive than by Pap smears. HPV DNA was also found in biopsies taken from histologically normal tissue when studying the frequency of recurrence after surgical treatment (Ferenczy et al., 1985) or women attending the gynecological clinic for reasons other than HPV (Cox et al., 1986).

The state of the HPV genome in the cell is of importance with regard to the prognosis of the infection in case of CIN. An HPV genome integrated into the host genome is more often associated with malignancy than when present in the episomal state (Gissmann 1984; Millan et al., 1986). Because the restriction pattern of the viral DNA will differ if episomal or integrated, Southern blot analysis is required to answer this question (Fig. 10). The extent of infection and the correlation of cell morphology to infection have been studied by the in situ hybridization technique (Fig. 7) (McDougall et al., 1986; Syrjänen et al., 1986).

Hepatitis B Virus and DNA Detection

Because no cell culture method is available, the presence of HBV is diagnosed with the aid of a variety of immunochemical methods detecting the presence of HBV surface antigen (sAg), core antigen (cAg), and

TABLE 5. Application of nucleic acid hybridization techniques to diagnosis of genital human papillomavirus infections

Question	Test	Clinical problem/implications	Usefulness	Reference
Presence and type of HPV	Spot	Screening of cervical cells for HPV infections and thus for potential risk of developing cervical malignancy	Allows type-specific diagnosis; reveals HPV infections with normal cytology also	Wickenden et al. (1985)
	Filter in situ	Screening of cervical cells for HPV infections and thus for potential risk of developing cervical malignancy		Wagner et al. (1984), Schneider et al. (1985)
	Sandwich	Screening of cervical cells for HPV infections and thus for potential risk of developing cervical malignancy		Parkkinen et al. (1986)
HPV type in cervical cells or biopsy	Spot	Assessment of risk for cervical malignancy in patients with HPV suggestive cytology	Possible only by DNA detection	Gissmann (1984)
For sandwich the clinical problem/ implication mentioned for spot should be repeated	Sandwich	Assessment of risk for cervical malignancy in patients with HPV suggestive cytology		Parkkinen et al. (1988)
Physical state of the HPV genome in biopsy or cervical cells	Southern blot	Assessment of risk for cervical malignancy in patients with HPV suggestive cytology		Burk et al. (1986), Millan et al. (1986)
Correlation of presence of HPV genome to viral antigen expression and histology in biopsy or exfoliated cells	In situ hybridization	Location of HPV infection; activity of infection	Research tool	Syrjänen et al. (1986)

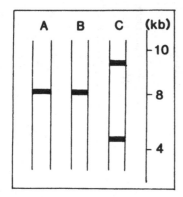

FIG. 10. Principle of determining the physical state of papillomavirus DNA in infected cells by Southern blot. The DNA preparations are digested with a restriction enzyme cleaving the DNA only once. (A) Marker DNA. (B) Episomal linearized DNA in infected cells. (C) Integrated DNA.

the e antigen (eAg), which is considered a marker for the aggressiveness of the disease. Immunoresponses against each of the antigens defined above are also measured. The availability of HBV-specific probes has allowed hybridization experiments and has considerably increased our knowledge on the biology of the HBV infection (for review, see Tiollais et al., 1985). Consequently, new needs for diagnosis are emerging. The diagnostic questions that could be answered by applying DNA hybridization are listed in Table 6.

By far the most important diagnostic task is the reliable prevention of HBV transmission in blood transfusions. Today all donated blood is screened for the presence of HBsAg. Nevertheless, 10 to 15% of hepatitis associated with transfusion is caused by HBV (Aach and Kahn, 1980). Nucleic acid hybridization, quantifying DNA in serum, gives a direct measure of the number of virions present and is thus a measure of the activity of the viral infection. DNA was detected in 66 to 74% of sAg-positive and in 83 to 100% of eAg-positive chronic carriers (Lieberman et al., 1983; Scotto et al., 1983).

Recent studies, using hybridization have revealed that infectious HBV is, however, present in unexpected situations, which would have been considered uninfectious by the serological criteria. First, 18 to 30% of anti-HBe-positive carriers were found DNA-positive (Scotto et al., 1983; Wu et al., 1986). Second, HBV DNA was frequently present in patients suffering from HBsAg-negative chronic liver disease (Brechot et al., 1985). DNA was present in 10% of the sera studied and frequently in the liver. In six of 10 DNA-positive sera, DNA was the only marker for HBV infection.

The range of target cells susceptible to HBV infection has been shown to be wider than initially predicted. In addition to hepatocytes, the viral genome is found in peripheral blood leukocytes of asymptomatic and sick chronic carriers either in free or integrated form (Pontisso et al., 1984; Shen et al., 1986; Yoffe et al., 1986). The frequency correlates to the presence of HBeAg in serum—that is, the replicative activity of the virus. The finding of HBV DNA in leukocytes is of importance, because the peripheral blood mononuclear cells might transmit the infection even in a situation where the virus production by hepatocytes has ceased.

DNA is mostly detected by the spot test from 10 to 100 μl serum. The sensitivity of the technique allows, at its best, detection of 10^4 target DNA molecules. Is this sufficient? Krogsgaard et al. (1986) had the opportunity to analyze the serum of 13 donor and recipient pairs after transfusion of HBVsAg-positive blood. Ten of the donors were DNA negative. Nevertheless, nine of the recipients became infected. This study shows that the absence of demonstrable DNA does not preclude infectivity.

Detection of HBV DNA is a useful technique today when estimating the activity of a sAg-positive infections, such as for determining need for antiviral therapy or risk of transmission of infection from mother to child during pregnancy. The quantity of virus present is usually more than 10^7 virus particles per milliliter, which is sufficient for positive identification (Wu et al., 1986).

The second major diagnostic question relates to the prognosis of the disease evaluated on the basis of the HBV genome in the infected cell. Two different forms of chronic carriership have been described (Brechot et al., 1981). One is characterized by free viral DNA in the liver, and the other by viral DNA integrated into the hepatocyte genome. Free DNA is transcribed into two distinct mRNAs, encoding the viral surface and the core proteins. HBV DNA is found in the serum, indicating an active infection. The integrated DNA only expresses the mRNA coding for the viral surface protein. The infection is quiescent with positive HBsAg, but rarely with positive HBeAg (Brechot et al., 1981; Yokosuka et al., 1986). In hepatocellular carcinoma, HBV DNA is often found integrated into the host cell genome (Blumberg and London, 1981; Shafritz et al., 1981).

Latent Human Immune Deficiency Virus Genome in Infected Cells

HIV is a retrovirus that has an RNA genome in virions but a DNA genome in infected cells. In the cells it either exists as free proviral DNA or is integrated into the host cell chromosome. HIV infects lymphocytes, at least T cells and macrophages, in which the infection can lead to cell death or to persistent infec-

TABLE 6. DNA hybridization in HBV diagnosis

Question	Test	Implication	Usefulness in diagnosis	Reference
Presence and quantity in serum	Spot	Activity of the infection in chronic carriers	Sensitive enough in active chronic infections	Scotto et al. (1983), Lieberman et al. (1983), Pontisso et al. (1985)
Screening for HBV in serum	Spot	Prevention of blood transfusion hepatitis	Reveals HBV marker negative cases; not sensitive enough	Brechot et al. (1985), Krogsgaard et al. (1986)
	Probe-vector method		Not known	Hartley et al. (1986)
State of viral genomes in cells	Southern	Prognosis of the disease	Informative with representative biopsy	Yokosuka et al. (1986), Brechot et al. (1981)
	Southern	Infection of PBMC/transmission of infection by blood	Not known	Pontisso et al. (1984), Yoffe et al. (1986)
	Southern	Diagnosis of hepatocellular carcinoma	Useful	Shafritz et al. (1981)

tion. The mechanism of persistence is not known. The virus can be recovered by culture in most of the seropositive infected individuals regardless of symptoms (for review, see Wong-Staal and Gallo, 1985).

Because of the long symptom-free latency period typical to HIV, it is important to screen blood donors to prevent HIV transmission by transfusions. Testing is usually done by immunoassays measuring antibodies to the viral structural proteins. Even if these assays perform well, it has become apparent that antibodies are not always present. Four of 96 patients positive by virus isolation were seronegative as studied by Salahuddin et al. (1984). Several possible explanations for this finding can be presented: (1) the individuals had not had time to seroconvert; (2) these individuals will never produce antibodies, because HIV replicates slowly, and the antigens do not get exposed to the humoral immune system; (3) antibodies are present, but they are complexed with viral antigens and therefore not detected. Whatever the reason, additional tests detecting either viral antigen or the genome are thus indicated.

It was recently reported that HIV Ag was detectable in 25% of infected persons. The antigen was found as soon as 2 weeks after exposure and lasted for 3 to 5 months (Allain et al., 1986). As has been described above for HPV and HBV, the latent viral infections can be most reliably detected by nucleic acid hybridization. The difficulty with HIV is the small number of infected lymphocytes both in peripheral blood and in lymphnodes. Using in situ hybridization, Harper et al. (1986) showed that cells expressing virus RNA could be shown in 86% of

lymphnode and 50% of blood samples studied. The RNA was present in about 0.01% of the lymphocytes in peripheral blood in a fairly low copy number. The above-cited studies indicate that more sensitivity is required before hybridization or antigen detection methods can be used in blood bank screening.

The in situ hybridization technique has been extremely valuable in studies correlating virus replication to symptoms associated with AIDS. Virus is detected in brain explaining the encephalopathy symptoms (Koenig et al., 1986) and in lung, indicating its direct involvement in the lymphocytic interstitial pneumonitis (Chayt et al., 1986).

Detection of Cytomegalovirus by Nucleic Acid Hybridization

Cytomegalovirus (CMV) infections are most often asymptomatic and harmless. Problems arise in two situations: congenital CMV infection, and CMV infection in an immunocompromised transplant recipient. In the latter case, an accurate and rapid diagnosis, distinguishing the reactivated CMV infection from symptoms of rejection, is of utmost value for appropriate treatment and consequently the outcome of the patient. Diagnosis is usually based on detection of virus from urine.

Several studies demonstrating the efficacy of the hybridization method as compared to culture are summarized in Table 7. In general, the hybridization methods detect 70 to 90% of the infections from which CMV has been isolated. The true efficacy of

TABLE 7. Diagnosis of cytomegalovirus by nucleic acid hybridization

Specimens		Culture result	Clinical diagnosis	Positive by hybridization[a]		Reference
No.	Source			No.	(%)	
24	Urine	+	12 congenital or primary infection; 12 transplant patients or others	22	(92)	Spector et al. (1984)
26	Urine	−	23 transplant patients; 3 others	3	(11)	Spector et al. (1984)
14	Buffy coat	+	Bone marrow transplantation	13	(93)	Spector et al. (1984)
53	Buffy coat	−	Bone marrow transplantation	21	(40)	Spector et al. (1984)
41	Total leukocytes	ND	Seropositive blood donors	3	(7)	Spector and Spector (1985)
30	Urine	+	Congenital infection and transplant patients	25	(83)	Schuster et al. (1986)
65	Urine	−	Congenital infection and transplant patients	4	(6)	
48	Urine	+	Congenital infection and transplant patients	34	(70)	Chou and Merigan (1983)
57	Urine	−	Congenital infection and transplant patients	0	(0)	
51	Urine	+	Congenital infection and transplant patients	47	(92)	Buffone et al. (1986)
4	Urine	+	Congenital or other primary infection	4	(100)	Virtanen et al. (1984)

[a] Spot hybridization expect in Virtanen et al. (1984) where sandwich hybridization was used.

the methods is difficult to evaluate, because the specimens studied all represent selected materials, and the numbers are restricted. There is fairly good agreement that a spot hybridization technique detects CMV from 10 ml of urine provided the virus titer is over 100 $TCID_{50}$ U/ml, which seems also to be reached in patients with reactivated CMV after organ transplant. When starting with urine, the specimen DNA needs to be concentrated by DNA purification and subsequent precipitation. This is commonly achieved by first pelleting the virus by ultracentrifugation followed by phenol extraction and ethanol precipitation.

The transplant patients have a particularly high frequency of microbial infections in their urinary tract, which means that culture cannot be performed. In the conventional spot test, a high number of false positives can be obtained because of the plasmid sequences in the labeled probe which hybridize to specific homologous sequences present in bacteria encountered in the urine. The frequency of false positivity was reported to be as high as 10 to 25% (Kahan and Landers, 1985). This problem is circumvented in the nucleic acid sandwich hybridization technique (Virtanen et al., 1984).

Hybridization seems to be a powerful tool to detect CMV from peripheral blood leukocytes (Spector and Spector, 1985; Spector et al., 1984). In recipients of bone marrow transplants, hybridization revealed the presence of viral DNA in 40% of the specimens, even in the absence of cultivable virus. The result was of predictive value, since CMV pneumonitis developed in some patients that were positive only by the DNA test (Spector et al., 1984). However, the presence of latent virus does not necessarily mean that virus production is activated, and therefore caution is needed when interpreting the results. This has become apparent when looking at the high frequency of virus positivity in bronchial alveolar lavation specimens in individuals without any sign of cytomegalovirus disease (Volin et al., 1985).

Hybridization permits distinction between productive and latent infection. Schrier et al. (1985) applied in situ hybridization to detect early viral transcripts with a specific probe from the peripheral blood mononuclear cells in naturally infected seropositive individuals without signs of cytomegalovirus disease. The number of infected cells was 0.3 to 2% of the total number of cells studied. Virus could not be recovered by cocultivating these cells with permissive cells. Also, a probe detecting late viral transcripts or an antiserum detecting viral structural proteins would allow distinction between productive and latent infection. In situ hybridization has also been applied to detection of CMV in biopsies of tumor specimens—for example, Kaposi's sarcoma (Spector and Spector, 1985).

Other Herpes Viruses

All herpesviruses remain latent in the host after primary infection. Recently the significance of Epstein-Barr virus (EBV) has been stressed. During infectious mononucleosis, it is usually isolated from throat washings and circulating B lymphocytes, from which it can also be rescued in the latent phase of the infection. EBV has now been shown to replicate in vivo in the ectocervical epithelium of the uterine cervix (Sixbey et al., 1986). If EBV is a sexually transmitted disease, and if it can remain latent in the cervix, its significance in the interplay of infections and development of malignancies will have to be taken into account. In terms of diagnosis, it means that demonstration of EBV DNA from the cervical epithelium will be included in the pattern of hybridization assays performed.

Identification of Nucleic Acids Without Hybridization

Genome Identification

In a few cases, clinical specimens contain so much virus that their genomes can be identified directly by physical methods. In these cases, immunochemical methods also work well and are easier to perform. However, the genome isolation may give information preferable to that obtained by immunological methods. This is exemplified by diarrheal feces samples from which rotavirus is detected (Sanders, 1985). The genome of rotavirus is composed of 11 double-stranded RNA molecules (Estes et al., 1983). The distribution of these fragments according to size gives rise to typical patterns after electrophoresis through polyacrylamide or agarose gels (Fig. 11). These patterns vary in a manner diagnostic for different viral strains (Lourenco et al., 1981). Electrophoretic analysis of viral RNA is probably the most convenient way for epidemiological studies.

Detection of Minor Differences

Diagnostic questions related to similarity or dissimilarity of two virus isolates are sometimes relevant. The answer is easiest obtained by comparing the actual nucleotide sequence of the isolates. This can be done by direct sequencing over a relevant region or by "fingerprint" analysis. RNA fingerprints are obtained after nuclease digestion, which gives rise to a specific set of oligonucleotides. Technically, the fingerprint analysis is complicated and involves the propagation and purification of ^{32}P-labeled RNA from the isolate. It has been used, for instance, to

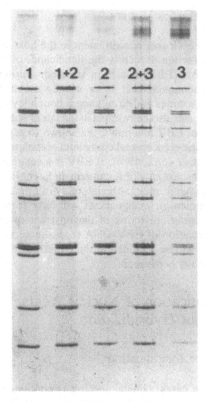

FIG. 11. Three patient stool samples (1, 2, and 3) analyzed in polyacrylamide gel electrophoresis. The 11 RNA segments of samples 1 and 2 run differently as seen in lane (1 + 2), where these samples are coelectrophoresed. Samples 2 and 3 are identical by the above criteria. The genotype differences do not necessarily correlate to antigenic differences.

study the rapid evolution of the poliovirus genome (Nottay et al., 1981). Rapid development of the nucleotide sequencing methodology has outdated the use of fingerprints in many cases. If the approximate location of the expected changes in the RNA genome is known, it is possible to determine the sequence at that site with the aid of specific primers and reserve transcriptase. In this way the molecular changes in the genome of the poliovirus type 3 that led to an outbreak of poliomyelitis in Finland in 1984 and 1985 were characterized (Magrath et al., 1986).

Conclusions

Nucleic acid hybridization fulfills the criteria of a diagnostic method that is universally applicable to the detection of any living organism. Properties making hybridization to an outstanding tool are (1) the genome is always present whether expressed or not, and can therefore always be identified; (2) identification by hybridization has exquisite and predetermined specificity due to the strict base pairing rule;

and (3) nucleic acids can be specifically amplified before detection.

The decisive question in the acceptance of the principle of nucleic acid detection for viral diagnosis will be the technical performance. An enormous effort is being put into the development of these techniques to achieve simplicity, rapidity, and sensitivity. The goal is clear—a development from a research tool to a practical method. Some of the diagnostic problems that can be solved by genome identification are so important that pressure to get the tests into use is heavy. The technical development needed is linked to sample treatment, test format, nonradiometric detection methods, and instrumentation.

It is not likely that methods for nucleic acid hybridization will replace the well-performing conventional diagnostic methods in the near future. Rather, hybridization will be applied in fields where other diagnostic methods either are nonexisting or do not perform satisfactorily.

Literature Cited

Aach, R. D., and R. A. Kahn. 1980. Post-transfusion hepatitis: current perspectives. Ann. Intern. Med. **92**:539–546.

Allain, J.-P., D. A. Paul, Y. Laurian, and D. Senn. 1986. Serological markers in early stages of human immunodeficiency virus infection in haemophiliacs. Lancet **2**:1233–1236.

Amasino, R. M. 1986. Acceleration of nucleic acid hybridization rate by polyethylene glycol. Anal. Biochem. **152**:304–307.

Ambinder, R. F., P. Charache, S. Staal, P. Wright, M. Forman, S. D. Hayward, and G. S. Hayward. 1986. The vector homology problem in diagnostic nucleic acid hybridization of clinical specimens. J. Clin. Microbiol. **24**:16–20.

Anderson, M. J., S. E. Jones, and A. C. Minson. 1985. Diagnosis of human parvovirus infection by dot-blot hybridization using cloned viral DNA. J. Med. Virol. **15**:163–172.

Anderson, M. L. M., and B. D. Young. 1985. Quantitative filter hybridization, p. 73–111. In B. D. Hames and S. J. Higgins (ed.), Nucleic acid hybridization, a practical approach. IRL Press, Oxford, U.K.

Arrand, J. E. 1985. Preparation of nucleic acid probes, p. 17–45. In B. D. Hames and S. J. Higgins (ed.), Nucleic acid hybridization, a practical approach. IRL Press, Oxford, U.K.

Blumberg, B. S., and W. T. London. 1981. Hepatitis B virus and the prevention of primary hepatocellular carcinoma. N. Engl. J. Med. **304**:782–784.

Boshart, M., L. Gissmann, H. Ikenberg, A. Kleinheinz, W. Scheurlen, and H. Zur Hausen. 1984. A new type of papillomavirus DNA, its presence in genital cancer biopsies and in cell lines from cervical cancer. EMBO J. **3**:1151–1157.

Brandsma, J., and G. Miller. 1980. Nucleic acid spot hybridization: rapid quantitative screening of lymphoid cell lines for Epstein-Barr viral DNA. Proc. Natl. Acad. Sci. USA **77**:6851–6855.

Brechot, C., J. Scotto, P. Charnay, M. Hadchouel, F. Degos, C. Trepo, and P. Tiollais. 1981. Detection of hepatitis B virus DNA in liver and serum: a direct appraisal of the chronic carrier state. Lancet 2:765–768.

Brechot, C., F. Degos, C. Lugassy, V. Thiers, S. Zafrani, D. Franco, H. Bismuth, C. Trepo, J.-P. Benhamou, J. Wands, K. Isselbacher, P. Tiollais, and P. Berthelot. 1985. Hepatitis B virus DNA in patients with chronic liver disease and negative tests for hepatitis B surface antigen. N. Engl. J. Med. 312:270–276.

Brigati, D. J., D. Myerson, J. J. Leary, B. Spalholz, S. Z. Travis, C. K. Y. Fong, and G. D. Hsuing. 1983. Detection of viral genomes in cultured cells and paraffin-embedded tissue sections using biotin-labeled hybridization probes. Virology 126:32–50.

Britten, R. J., D. E. Graham, and B. R. Neufeld. 1974. Analysis of repeating DNA sequences by reassociation. Methods Enzymol. 29E:363–418.

Buffone, G. J., C. M. Schimbor, G. J. Demmler, D. R. Wilson, and G. J. Darlington. 1986. Detection of cytomegalovirus in urine by nonisotopic DNA hybridization. J. Infect. Dis. 154:163–166.

Burk, R. D., A. S. Kadish, S. Calderin, and S. L. Romney. 1986. Human papillomavirus infection of the cervix detected by cervicovaginal lavage and molecular hybridization: correlation with biopsy results and Papanicolaou smear. Am. J. Obstet. Gynecol. 154:982–989.

Carrico, R. J. 1985. Nucleic acid hybridization assay employing detectable anti-hybrid antibodies. Eur. Pat. Appl. No. 85106127.5.

Chayt, K. J., M. E. Harper, L. M. Marselle, E. B. Lewin, R. M. Rose, J. M. Oleske, L. G. Epstein, F. Wong-Staal, and R. C. Gallo. 1986. Detection of HTLV-III RNA in lungs of patients with AIDS and pulmonary involvement. J. Am. Med. Assoc. 256:2356–2359.

Chollet, A., and E. H. Kawashima. 1985. Biotin-labeled synthetic oligodeoxyribonucleotides: chemical synthesis and uses as hybridization probes. Nucleic Acids Res. 13:1529–1541.

Chou, S., and T. C. Merigan. 1983. Rapid detection and quantitation of human cytomegalovirus in urine through DNA hybridization. N. Engl. J. Med. 308:921–925.

Commerford, S. L. 1971. Iodination of nucleic acids in vitro. Biochemistry 10:1993–2000.

Cox, M. F., C. A. Meanwell, N. J. Maitland, G. Blackledge, C. Scully, and J. A. Jordan. 1986. Human papillomavirus type 16 homologous DNA in normal human ectocervix. Lancet 2:157–158.

Crum, C. P., H. Ikenberg, R. M. Richard, and L. Gissmann. 1984. Human papillomavirus type 16 and early cervical neoplasia. N. Engl. J. Med. 310:880–883.

Denhardt, D. T. 1966. A membrane-filter technique for the detection of complimentary DNA. Biochem. Biophys. Res. Commun. 23:641–646.

Durst, M., L. Gissmann, H. Ikenberg, and H. zur Hausen. 1983. A papillomavirus DNA from a cervical carcinoma and its prevalence in cancer biopsy samples from different geographic regions. Proc. Natl. Acad. Sci. USA 80:3812–3815.

Edelstein, P. H. 1986. Evaluation of the Gen-probe DNA probe for the detection of Legionellae in culture. J. Clin. Microbiol. 23:481–484.

Estes, M. K., E. L. Palmer, and J. F. Obijeski. 1983. Rotaviruses: a review. Curr. Top. Microbiol. Immunol. 105:123–184.

Feinberg, A. P., and B. A. Vogelstein. 1983. A technique for radiolabelling DNA restriction endonuclease fragments to high specific activity. Anal. Biochem. 132:6–13.

Ferenczy, A., M. Mitao, N. Nagai, S. J. Silverstein, and C. P. Crum. 1985. Latent papillomavirus and recurring genital warts. N. Engl. J. Med. 313:784–788.

Flavell, R. A., E. J. Birfelder, P. M. Sanders, and P. Borst. 1974. DNA-DNA hybridization on nitrocellulose filters. 1. General considerations and non-ideal kinetics. Eur. J. Biochem. 47:535–543.

Forster, A. C., J. L. McInnes, D. C. Skingle, and R. H. Symons. 1985. Nonradioactive hybridization probes prepared by the chemical labelling of DNA and RNA with a novel reagent, photobiotin. Nucleic Acids Res. 13:745–761.

Gall, J., and M. L. Pardue. 1969. Formation and detection of RNA-DNA hybrid molecules in cytological preparations. Proc. Natl. Acad. Sci. USA 63:378–383.

Gillespie, D., and S. Spiegelman. 1965. A quantitative assay for DNA-RNA hybrids with DNA immobilized on a membrane. J. Mol. Biol. 12:829–842.

Gissmann, L. 1984. Papillomaviruses and their association with cancer in animals and in man. Cancer Surv. 3:161–181.

Haase, A., M. Brahic, L. Stowring, and H. Blum. 1984. Detection of viral nucleic acids by in situ hybridization. Methods Virol. 7:189–225.

Harper, M. E., L. M. Marselle, R. C. Gallo, and F. Wong-Staal. 1986. Detection of lymphocytes expressing human T-lymphotropic virus type III in lymphnodes and peripheral blood from infected individuals by in situ hybridization. Proc. Natl. Acad. Sci. USA 83:772–776.

Hartley, J. L., M. Berninger, J. A. Jessee, F. R. Bloom, and G. F. Temple. 1986. Bioassay for specific DNA sequences using a non-radioactive probe. Gene 49:295–302.

Hopman, A. H. N., J. Wiegant, G. I. Tesser, and P. Van Dujin. 1986. A nonradioactive in situ hybridization method based on mercurated nucleic acid probes and sulphydryl-hapten ligands. Nucleic Acids Res. 14:6471–6488.

Hu, N., and J. Messing. 1982. The making of strand-specific M13 probes. Gene 17:271–277.

Jablonski, E., E. W. Moomaw, R. H. Tullis, and J. L. Ruth. 1986. Preparation of oligodeoxynucleotide-alkaline phosphatase conjugates and their use as hybridization probes. Nucleic Acids Res. 14:6115–6128.

Jackson, D., R. Symons, and P. Berg. 1972. Biochemical methods for inserting new genetic information into DNA of simian virus 40: circular SV40 DNA molecules containing lambda phage genes and the galactose operon of Escherichia coli. Proc. Natl. Acad. Sci. USA 69:2904–2909.

Kafatos, F. C., and C. W. Jones, and A. Efstratiadis. 1979. Determination of nucleic acid sequence homologies and relative concentrations by a dot hybridization procedure. Nucleic Acids Res. 7:1541–1552.

Kahan, B. D., and T. A. Landers. 1985. Rapid detection of cytomegalovirus infection using a DNA probe. Transplant. Proc. 17:989–992.

Koenig, S., H. E. Gendelman, J. M. Orenstein, M. C. Dal Canto, G. H. Pezeshkpour, M. Yungbluth, F. Janotta, A. Aksamit, M. A. Martin, and A. S. Fauci. 1986. Detection of AIDS virus in macrophages in brain tissue from AIDS patients with encephalopathy. Science 233:1089–1093.

Kohne, D. E., and R. J. Britten. 1971. Hydroxyapatite techniques for nucleic acid reassociation. Proc. Nucleic Acids Res. 2:500–512.

Krogsgaard, K., P. Wantzin, J. Andershvile, P. Kryger, P. Andersson, and J. O. Nielsen. 1986. Hepatitis B virus DNA in hepatitis B surface antigen-positive blood donors: relation to the hepatitis B e system and outcome in recipients. J. Infect. Dis. 153:298–303.

Landegent, J. E., N. Jansen in de Wal, R. A. Baan, J. H. J. Hoeijmakers, and M. van der Ploeg. 1984. 2-Acetylaminofluorene-modified probes for the indirect hybrido-cytochemical detection of specific nucleic acid sequences. Exp. Cell Res. **153**:61–72.

Langer, P. R., A. A. Waldrop, and D. C. Ward. 1981. Enzymatic synthesis of biotin-labeled polynucleotides: novel nucleic acid affinity probes. Proc. Natl. Acad. Sci. USA **78**:6633–6637.

Leary, J. J., D. J. Brigati, and D. C. Ward. 1983. Rapid and sensitive colorimetric method for visualizing biotin-labeled DNA probes hybridized to DNA or RNA immobilized on nitrocellulose: bioblots. Proc. Natl. Acad. Sci. USA **80**:4045–4049.

Lehtomäki, K., I. Julkunen, K. Sandelin, J. Salonen, M. Virtanen, M. Ranki, and T. Hovi. 1986. Rapid diagnosis of respiratory adenovirus infections in young adult men. J. Clin. Microbiol. **24**:108–111.

Lieberman, H. M., D. R. LaBrecque, M. C. Kew, S. J. Hadziyannis, and D. A. Shafritz. 1983. Detection of hepatitis B virus DNA directly in human serum by a simplified molecular hybridization test: comparison to HBeAg/anti-HBe status in HBsAg carriers. Hepatology **3**:285–291.

Lourenco, M. H., J. C. Nicolas, J. Cohen, R. Scherrer, and F. Bricout. 1981. Study of human rotavirus genome by electrophoresis: attempt of classification among strains isolated in France. Ann. Virol. (Paris) **132E**:161–173.

Magrath, D. I., P. D. Minor, M. Ferguson, D. M. A. Evans, G. C. Schild, F. Horaud, R. Crainic, M. Stenvik, and T. Hovi. 1986. Antigenic and molecular properties of type 3 poliovirus responsible for an outbreak of poliomyelitis in a vaccinated population. J. Gen. Virol. **67**:899–905.

Marmur, J., R. Rownd, and C. L. Schildkraut. 1963. Denaturation and renaturation of DNA. Prog. Nucleic Acids Res. **1**:232–300.

Matthews, J. A., A. Batki, C. Hynds, and L. J. Kricka. 1985. Enhanced chemiluminescent method for the detection of DNA dot-hybridization assays. Anal. Biochem. **151**:205–209.

Maxam, A. M., and W. Gilbert. 1980. Sequencing end-labeled DNA with base-specific chemical cleavages. Methods Enzymol. **65**:499–560.

McDougall, J. K., D. Myerson, and A. M. Beckman. 1986. Detection of viral DNA by in situ hybridization. J. Histochem. Cytochem. **34**:33–38.

Melton, D. A., P. A. Krieg, M. R. Rebagliati, T. Maniatis, K. Zinn, and M. R. Green. 1984. Efficient in vitro synthesis of biologically active RNA and RNA hybridization probes from plasmids containing a bacteriophage SP6 promotor. Nucleic Acids Res. **12**:7035–7056.

Millan, D. W. M., J. A. Davis, T. E. Torbet, and M. S. Campo. 1986. DNA sequences of human papillomavirus types 11, 16, and 18 in lesions of the uterine cervix in the west of Scotland. Br. Med. J. **293**:93–96.

Murasugi, A., and R. B. Wallace. 1984. Biotin-labelled oligonucleotides: enzymatic synthesis and use as hybridization probes. DNA **3**:269–277.

Nagata, Y., H. Yokota, O. Kosuda, K. Yokoo, K. Takemura, and T. Kikuchi. 1985. Quantification of picogram levels of specific DNA immobilized in microtiter wells. FEBS Lett. **183**:379–382.

Nottay, B. K., O. M. Kew, M. H. Hatch, J. P. Heyward, and J. F. Obijeski. 1981. Molecular variation of type 1 vaccine related and wild polioviruses during replication in humans. Virology **108**:405–423.

Palva, A. 1985. Nucleic acid spot hybridization for detection of Chlamydia trachomatis. FEMS Microbiol. Lett. **28**:85–91.

Palva, A., H. Jousimies-Somer, P. Saikku, P. Väänänen, H. Söderlund, and M. Ranki. 1984. Detection of Chla-mydia trachomatis by nucleic acid sandwich hybridization. FEMS Microbiol. Lett. **23**:83–89.

Pardue, M. L. 1985. In situ hydridization, p. 179–202. In B. D. Hames and S. J. Higgins (eds.), Nucleic acid hybridization, a practical approach. IRL press, Oxford, U.K.

Parkkinen, S., R. Mäntyjärvi, K. Syrjänen, and M. Ranki. 1986. Detection of human papillomavirus DNA by the nucleic acid sandwich hybridization method from cervical scraping. J. Med. Virol. **20**:279–288.

Parkkinen, S., S. Syrjänen, K. Syrjänen, M. Yliskoski, J. Tenhunen, R. Mäntyjärvi, and M. Ranki. 1988. Screening of premalignant cervical lesions for HPV 16 DNA by sandwich and in situ hybridization techniques. Gynecol. Oncol. **30**:251–264.

Pontisso, P., M. C. Poon, P. Tiollais, and C. Brechot. 1984. Detection of hepatitis B virus DNA in mononuclear blood cells. Br. Med. J. **288**:1563–1566.

Pontisso, P., L. Chemello, G. Fattovich, A. Alberti, G. Realdi, and C. Brechot. 1985. Relationship between HBcAg in serum and liver and HBV replication in patients with HBsAg-positive chronic liver disease. J. Med. Virol. **17**:145–152.

Prooijen-Knegt, A. C. van, J. F. M. Hoek, J. G. J. Bauman, P. van Dujin, I. G. Wool, and M. van der Ploeg. 1982. In situ hybridization of DNA sequences in human methaphase chromosomes visualized by an indirect fluorescent immunocytochemical procedure. Exp. Cell Res. **141**:397–407.

Ranki, M., A. Palva, M. Virtanen, M. Laaksonen, and H. Söderlund. 1983. Sandwich hybridization as a convenient method for the detection of nucleic acids in crude samples. Gene **21**:77–85.

Reed, K. C., and D. A. Mann. 1985. Rapid transfer of DNA from agarose gels to nylon membranes. Nucleic Acids Res. **13**:7207–7221.

Renz, M., and C. Kurz. 1984. A colorimetric method for DNA hybridization. Nucleic Acids Res. **12**:3435–3444.

Rigby, P. W. J., M. Dieckmann, C. Rhodes, and P. Berg. 1977. Labeling deoxyribonucleic acid to high specific activity in vitro by nick translation with DNA polymerase I. J. Mol. Biol. **113**:237–251.

Saiki, R. K., S. Scharf, F. Faloona, K. B. Mullis, G. T. Horn, H. A. Erlich, and N. Arnheim. 1985. Enzymatic amplification of β-globin genomic sequences and restriction site analysis for diagnosis of sickle cell anemia. Science **230**:1350–1354.

Saiki, R. K., T. L. Bugawan, G. T. Horn, K. B. Mullis, and H. A. Erlich. 1986. Analysis of enzymatically amplified β-globin and HLA-DQ DNA with allele-specific oligonucleotide probes. Nature **324**:163–166.

Salahuddin, S. Z., P. D. Markham, R. R. Redfield, J. E. Groopman, M. G. Sarngadharan, M. F. McLane, A. Sliski, and R. C. Gallo. 1984. HTLV-III in symptom-free seronegative persons. Lancet **2**:1418–1420.

Sanders, R. C. 1985. Molecular epidemiology of human rotavirus infections. Eur. J. Epidemiol. **1**:19–32.

Schneider, A., H. Kraus, R. Schuhmann, and L. Gissmann. 1985. Papilloma infection of the lower genital tract: detection of viral DNA in gynecological swabs. Int. J. Cancer **35**:443–448.

Schrier, R. D., J. A. Nelson, and M. B. Oldstone. 1985. Detection of human cytomegalovirus in peripheral blood lymphocytes in a natural infection. Science **230**:1048–1051.

Schuster, V., B. Matz, H. Wiegand, B. Traub, D. Kampa, and D. Neumann-Haefelin. 1986. Detection of human cytomegalovirus in urine by DNA-DNA and RNA-DNA hybridization. J. Infect. Dis. **154**:309–314.

Scotto, J., M. Hadchouel, C. Hery, J. Yvart, P. Tiollais, and C. Brechot. 1983. Detection of hepatitis B virus DNA in serum by a simple spot hybridization technique: comparison with results for other viral markers. Hepatology 3:279–284.

Shafritz, D. A., D. Shouval, H. I. Sherman, S. J. Hadziyannis, and M. C. Kew. 1981. Integration of hepatitis B virus DNA into the genome of liver cells in chronic liver disease and hepatocellular carcinoma. N. Engl. J. Med. 305:1067–1073.

Shen, H.-D., K.-B. Choo, S.-D. Lee, Y.-T. Tsai, and S.-H. Han. 1986. Hepatitis B virus DNA in leukocytes of patients with hepatitis B virus-associated liver diseases. J. Med. Virol. 18:201–211.

Sixbey, J. W., S. M. Lemon, and J. S. Pagano. 1986. A second site for Epstein-Barr virus shedding: the uterine cervix. Lancet 2:1122–1124.

Soini, E., and H. Kojola. 1983. Time-resolved fluorometer for lanthanide chelates—a new generation of nonisotopic immunoassays. Clin. Chem. 29:65–68.

Southern, E. M. 1975. Detection of specific sequences among DNA fragments separated by gel electrophoresis. J. Mol. Biol. 98:503–517.

Spector, S. A., and D. H. Spector. 1985. The use of DNA probes in studies of human cytomegalovirus. Clin. Chem. 31:1514–1520.

Spector, S. A., J. A. Rua, D. H. Spector, and R. McMillan. 1984. Detection of human cytomegalovirus in clinical specimens by DNA-DNA hybridization. J. Infect. Dis. 150:121–126.

Spector, S. A., T. R. Neuman, and K. K. Hirata. 1985. Rapid determination of molecular relatedness of isolates of human cytomegalovirus. J. Infect. Dis. 152:755–759.

Stanley, C. J., F. Paris, A. Plumb, A. Webb, and A. Johansson. 1985. Enzyme amplification. A new technique for enhancing the speed and sensitivity of enzyme immunoassays. Int. Biotech. Lab. 3:46–51.

Studencki, A. B., and R. B. Wallace. 1984. Allele-specific hybridization using oligonucleotide probes of very high specific activity: discrimination of the human β^a- and β^s-globin genes. DNA 3:7–15.

Syrjänen, K., R. Mäntyjärvi, M. Väyrynen, H. Holopainen, S. Saarikoski, S. Syrjänen, S. Parkkinen, and O. Castren. 1985a. Structural protein expression in biological behavior of human papillomavirus infections in uterine cervix. Cervix 3:249–266.

Syrjänen, K., S. Parkkinen, R. Mäntyjärvi, M. Väyrynen, S. Syrjänen, H. Holopainen, S. Saarikoski, and O. Castren. 1985b. Human papillomavirus (HPV) type as an important determinant of the natural history of HPV infections in uterine cervix. Eur. J. Epidemiol. 1:180–187.

Syrjänen, S., K. Syrjänen, R. Mäntyjärvi, S. Parkkinen, M. Väyrynen, S. Saarikoski, and O. Castren. 1986. Human papillomavirus (HPV) DNA sequences demonstrated by in situ DNA hybridization in serial paraffin-embedded cervical biopsies. Arch. Gynecol. 239:39–48.

Syvänen, A.-C. 1986. Nucleic acid hybridization: from research tool to routine diagnostic method. Med. Biol. 64:313–324.

Syvänen, A.-C., and K. Korpela. 1986. Detection of ampicillin and tetracycline resistance genes in uropathogenic Escherichia coli strains by affinity-based nucleic acid hybrid collection. FEMS Microbiol. Lett. 36:225–229.

Syvänen, A.-C., M. Laaksonen, and H. Söderlund. 1986a. Fast quantification of nucleic acid hybrids by affinity-based hybrid collection. Nucleic Acids Res. 14:5037–5048.

Syvänen, A.-C., P. Tchen, M. Ranki, and H. Söderlund.
1986b. Time-resolved fluorometry: a sensitive method to quantify DNA-hybrids. Nucleic Acids Res. 14:1017–1028.

Tchen, P., R. P. P. Fuchs, E. Sage, and M. Leng. 1984. Chemically modified nucleic acids as immunodetectable probes in hybridization experiments. Proc. Natl. Acad. Sci. USA 81:3466–3470.

Tiollais, P., C. Pourcel, and A. Dejean. 1985. The hepatitis B virus. Nature 317:489–495.

Unger, E. R., L. R. Budgeon, D. Myerson, and D. J. Brigati. 1986. Viral diagnosis by in situ hybridization. Description of a rapid-simplified colorimetric method. Am. J. Surg. Pathol. 10:1–8.

Vincent, C., P. Tchen, M. Cohen-Solal, and P. Kourilsky. 1982. Synthesis of 8-(2-4 dinitrophenyl 2-6 aminohexyl)-amino-adenosine 5' triphosphate: biological properties and potential uses. Nucleic Acids Res. 10:6787–6796.

Virtanen, M., A. Palva, M. Laaksonen, P. Halonen, H. Söderlund, and M. Ranki. 1983. Novel test for rapid viral diagnosis: detection of adenovirus in nasopharyngeal mucus aspirates by means of nucleic-acid sandwich hybridisation. Lancet 1:381–383.

Virtanen, M., A.-C. Syvänen, J. Oram, H. Söderlund, and M. Ranki. 1984. Cytomegalovirus in urine: detection of viral DNA by sandwich hybridization. J. Clin. Microbiol. 20:1083–1088.

Viscidi, R. P., C. J. Connelly, and R. H. Yolken. 1986. Novel chemical method for the preparation of nucleic acids for nonisotopic hybridization. J. Clin. Microbiol. 23:311–317.

Volin, L., R. Leskinen, E. Taskinen, P. Tukiainen, P. Ruutu, P. Häyry, and T. Ruutu. 1986. Bronchoalveolar lavage in the diagnosis of pulmonary complications in bone marrow transplant recipients. Transplant. Proc. 18:130–131.

Wagner, D., H. Ikenberg, N. Boehm, and L. Gissmann. 1984. Identification of human papillomavirus in cervical swabs by deoxyribonucleic acid in situ hybridization. Obstet. Gynecol. 64:767–772.

Wahl, G. M., M. Stern, and G. R. Stark. 1979. Efficient transfer of large DNA fragments from agarose gels to diazobenzyloxymethyl-paper and rapid hybridization by using dextran sulfate. Proc. Natl. Acad. Sci. USA 76:3683–3687.

Wetmur, J. G., and N. Davidson. 1968. Kinetics of renaturation of DNA. J. Mol. Biol. 31:349–370.

Wickenden, C., A. Steele, A. D. B. Malcolm, and D. V. Coleman. 1985. Screening for wart virus infection in normal and abnormal cervices by DNA hybridization of cervical scrapes. Lancet 1:65–67.

Wilkinson, H. W., J. S. Sampson, and B. B. Plikaytis. 1986. Evaluation of a commercial gene probe for identification of Legionella cultures. J. Clin. Microbiol. 23:217–220.

Wong-Staal, F., and R. C. Gallo. 1985. Human T-lymphotropic retroviruses. Nature 317:395–403.

Wu, J.-C., S.-D. Lee, J.-Y. Wang, L.-P. Ting, Y.-T. Tsai, K.-J. Lo, B. N. Chiang, and M. J. Tong. 1986. Analysis of the DNA of hepatitis B virus in the sera of Chinese patients infected with hepatitis B. J. Infect. Dis. 153:974–977.

Yoffe, B., C. A. Noonan, J. L. Melnick, and F. B. Hollinger. 1986. Hepatitis B virus DNA in mononuclear cells and analysis of cell subsets for the presence of replicative intermediates of viral DNA. J. Infect. Dis. 153:471–477.

Yokosuka, O., M. Omata, F. Imazeki, Y. Ito, and K. Okuda. 1986. Hepatitis B virus RNA transcripts and DNA in chronic liver disease. N. Engl. J. Med. 315:1187–1192.

SECTION III
The Viruses, Viral Infections, and Nomenclature

CHAPTER 9

Virus Taxonomy and Nomenclature

FREDERICK A. MURPHY

The ICTV Universal System of Virus Taxonomy: Virus families, genera, and species.
Virus Nomenclature: Formal nomenclature usage for unambiguous identification, vernacular usage.
Structural, Physicochemical, and Replicative Characteristics Used in Taxonomy: A listing.
Classification of Viruses on the Basis of Epidemiologic Criteria: Groupings useful in the diagnostic setting.
The Place of Taxonomy in Diagnostic Virology: "Differential diagnosis" and "rule-out diagnosis," causal relationship between virus and disease, the adequate description of new viruses.
A Taxonomic Description of the Families of Viruses Containing Human and Animal Pathogens: Family characteristics and listings of member viruses.

"Viruses should be considered as viruses because viruses are viruses."

—Andre Lwoff, 1957

Introduction

The diagnostician might ask, "What is the need for virus taxonomy in my work?" The answer, of course, is that a system for classification and nomenclature, that is, a system of taxonomy, is a practical necessity whenever large numbers of distinct organisms are being dealt with. This practical necessity becomes clear whenever one compares an "unknown" with known organisms in an attempt to identify it. In this case, the organization of the list of known organisms usually takes the form of a taxonomic scheme. The practical necessity for taxonomy also becomes clear whenever candidate organisms must be considered in a differential diagnosis. Again, the list of candidate organisms usually takes the form of a taxonomic scheme. As in each of these cases, another value of taxonomic organization and standardized nomenclature lies in the avoidance of ambiguity in communicating information about the identification of organisms; taxonomic nomenclature is

chosen for its precision, its avoidance of confusing synonyms, and its universality in all languages (Murphy, 1985).

The earliest taxonomic experiments involving viruses were designed to separate them from microbes that could be seen in the light microscope and that usually could be cultivated on rather simple media. In the experiments that led to the first discoveries of viruses by Beijerinck and Ivanovski (tobacco mosaic virus), Loeffler and Frosch (foot-and-mouth disease virus), and Reed and Carroll (yellow fever virus) at the turn of the century, one single physicochemical characteristic was measured—filterability (Waterson and Wilkinson, 1978). No other physicochemical measurements were possible at that time, and most studies of viruses centered on their ability to cause infections and diseases. The earliest efforts to classify viruses, therefore, were based upon common pathogenic properties, common organ tropisms, and common ecological and transmission characteristics. For example, viruses that share the pathogenic property of causing hepatitis (e.g., hepatitis A virus, hepatitis B virus, yellow fever virus, Rift Valley fever virus) were brought together as "the hepatitis viruses."

The first substantial studies of the nature of viruses were begun in the 1930s, but it was not until about 1950 that biochemical and morphological information on virion composition and structure was comprehensive enough to influence classification concepts. In the 1950s, the first groupings of viruses on the basis of common virion properties emerged—the myxovirus group (Andrewes et al., 1955), the poxvirus group (Fenner and Burnet, 1957), and the herpesvirus group (Andrewes, 1954). At the same time, there was an explosion of discovery of new viruses affecting humans and animals. Prompted by this rapidly growing mass of data, several individuals and committees advanced their own classification schemes. The result was confusion over competing, conflicting schemes, and for the first but not the last time it became clear that virus classification and nomenclature are topics that give rise to very strongly held opinions.

Against this background, in 1966 the International Committee on Nomenclature of Viruses (ICNV)[1] was established at the International Congress of Microbiology in Moscow. At that time, virologists already sensed a need for a single, universal taxonomic scheme. There was little controversy that the hundreds of viruses being isolated from humans and animals, as well as from plants, insects, and bacteria, should be classified in a single system separately from all other biological entities, but there was much controversy over the taxonomic system to be used. Lwoff and his colleagues argued for the adoption of an all-embracing scheme for the classification of viruses into subphyla, classes, orders, suborders, and families (Matthews, 1983). These descending hierarchical divisions would have been based upon nucleic acid type, capsid symmetry, presence or absence of an envelope, etc. Opposition to this scheme was based on the arbitrariness of decisions as to the relative importance of different virion characteristics and upon the feeling that not enough was known about the characteristics of most viruses to warrant an elaborate hierarchy. The hierarchical issue was defeated in the ICNV, but otherwise the Lwoff–Horne–Tournier (1962) scheme became the basis of the universal taxonomy system that has been built upon ever since (Wildy, 1971). In this system, as many virion characteristics as possible are considered and weighted as criteria for making divisions into *families,* in some cases *subfamilies, genera,* and *species.* The relative order and weight assigned to each characteristic are in fact set arbitrarily and are influenced by prejudgments of relationships that "we would like to believe [from an evolutionary standpoint], but are unable to prove" (Fenner, 1974). The system does not use any hierarchical levels higher

than *families,* and the system does not imply any phylogenetic relationships beyond those "proven" experimentally (e.g., by nucleic acid hybridization, nucleotide sequencing, gene reassortment and recombination). The system is quite different from that used for the taxonomy of bacterial and other microbial organisms. The system is pragmatic—it is useful and is being used widely. It has replaced all competing classification schemes for all viruses. It eliminates ambiguity for all users, diagnosticians and clinicians, teachers and students, and laboratory researchers and epidemiologists. It brings order and precision to diagnostic reports, to textbooks, and to the research literature (Brown, 1987; Fenner, 1976; Matthews, 1979, 1982; Wildy, 1971).

The ICTV Universal System of Virus Taxonomy

In the most recent ICTV report (Brown, 1987), about 1,400 viruses have been assigned to approved taxa, and about 500 more have been designated as "probable" or "possible" members of these taxa. There is a sense that, at least for the viruses that cause clinical disease in humans and domestic animals, a significant fraction of all existing viruses have already been isolated and entered into the taxonomic system. This sense is based upon the infrequency in recent years of discoveries of viruses that do not fit into present taxa. The present universal system of virus taxonomy is set arbitrarily at hierarchical levels of *family,* in some cases *subfamily, genus,* and *species.* Lower hierarchical levels, such as subspecies, strain, variant, etc., are established and used by international specialty groups and by culture collections.

Virus Families

Virus families are designated by terms with the suffix -*viridae.* Families represent clusters of viruses, organized into genera, that share common characteristics and that are distinct from other major clusters of viruses. Despite the arbitrariness of early criteria for creating families, this highest level in the taxonomic hierarchy now seems soundly based and rather stable. All the families of viruses containing members that infect humans or other vertebrates have quite distinct virion morphologic characteristics and distinct genomes and strategies of replication; this indicates phylogenetic independence or at least great phylogenetic separation. In fact, although the taxonomic hierarchy does not necessarily have any evolutionary implications, it now seems, on the one hand, that the member viruses of a family do share a common evolutionary origin, and, on the other hand,

---

[1] The International Committee on Nomenclature of Viruses (ICNV) became the International Committee on Taxonomy of Viruses (ICTV) in 1973.

that the member viruses of different families usually do not share a common evolutionary origin. For example, it seems unlikely that the many similar structural and replicative characteristics of all the many diverse poxviruses could stem from other than a common ancestor. On the other hand, the recently discovered genomic similarities between the member viruses of the families *Rhabdoviridae* and *Paramyxoviridae* are being used by some virologists to argue that the two families should be brought together in some way. These similarities will be weighed against the many differences between the member viruses of the two families, differences that were used to construct the two families in the first place. Where new data contradict the sense of common ancestry of viruses now placed in the same family, or contradict the sense of distinct ancestry of viruses now placed in separate families, there are likely to be changes in the present taxonomic structure, with some further division or merger of present families.

So far, in three families—the *Retroviridae*, the *Poxviridae*, and the *Herpesviridae*—subfamilies have been introduced to allow a more complex hierarchy of taxa, in keeping with the apparent intrinsic complexity of the relationships among member viruses. Subfamilies are designated by terms with the suffix -*virinae*.

Virus Genera

Virus genera are designated by terms with the suffix -*virus*. Virus genera represent groups of species that share certain common characteristics and that are distinct from other groups of species. This level in the hierarchy of taxa also seems pragmatically based and useful. The criteria used for creating genera differ from family to family, and as more viruses are discovered and studied, there will be continuing pressures to use smaller and smaller structural, physicochemical, or serological differences to create new genera in many families. As with the family taxon, no commonality in phylogenetic origin is necessarily implied in the genus taxon; however, more and more evidence is accumulating that the members of genera do have a common evolutionary origin.

The subgenus taxon is used in one virus family, the *Retroviridae*, to divide the unnamed genus comprising the type C oncoviruses into three taxa, for the mammalian, avian, and reptilian type C oncoviruses.

Virus Species

Although the species taxon is widely regarded as the most important level in any taxonomic hierarchy, it is proving to be the most difficult to apply to viruses. In fact, virus species have not yet been formally de-

fined, nor have associated nomenclature problems been resolved. By extension from the definitions of virus families and genera, it might be presumed that virus species would simply be defined as clusters of strains of viruses with certain properties that separate one cluster from others. However, some virologists have argued that virus species must be defined as species are defined in other taxonomic systems. For example, in the taxonomy of animals, species are defined as "groups of interbreeding natural populations that are reproductively isolated from other such groups" (Matthews, 1983). This latter definition cannot be applied to all viruses, nor can criteria used to define species in other taxonomic systems. The issue continues to be debated.

In any case, it is clear that the species term will eventually be defined as an equivalent (or near equivalent) of the present vernacular term *virus*. Just as the term *virus* is defined differently in different virus families, so the term *species* will likely be defined in some cases by structural or physicochemical criteria and in other cases by serological or biological criteria. For example, Sindbis virus, mumps virus, polio 1 virus, and vaccinia virus each meet the definition of the vernacular term *virus*, and will likely be designated *species*. These examples, however, do not reflect the difficulty that will be encountered in differentiating species and strains, nor do they reflect the difficulty in species nomenclature caused by attempts to adapt multi-word vernacular terms (e.g., eastern equine encephalitis virus, Colorado tick fever virus, etc.). These problems probably will be overcome in the next few years, but in the meantime, the ICTV lists virus species only under the heading "English vernacular name" (Brown, 1987).

Virus Nomenclature

Formal Nomenclature Usage for Unambiguous Identification

Unambiguous virus identification is one value of the universal system of taxonomy (Murphy, 1983). Despite the incompleteness of the system, journal editors will soon require unambiguous virus identification somewhere in each publication. For example, under *Materials and Methods*, each virus could be identified by family, genus, and species terms—perhaps in combination with precise strain designation terms as developed by specialty groups. It is unlikely that such formal virus identification would ever be required in a diagnostic report, but it is likely that more precise virus identification will be required for labeling diagnostic reagents, seed virus stocks, etc.

Even in the absence of formal species designations, there is value in using the vernacular species names listed in the most current report of the ICTV—*The Classification and Nomenclature of Viruses* (Brown, 1986, 1987; Matthews, 1982)—rather than any synonyms or transliterated common virus names. Likewise, there is value in using strain and variant designations developed by recognized international specialty groups and culture collections, rather than any local laboratory-coded terms.

The matter of deciding how type species and strains are chosen and designated remains the responsibility of international specialty groups, some of which operate under the auspices of the World Health Organization and other agencies. In the past, some confusion was caused by the ICTV's identification of type species as part of descriptions of taxa; however, the ICTV has never been responsible for the identification of type species with the kind of precision that must be used by specialty groups and culture collections for nontaxonomic purposes. There has also been confusion caused by conflicting claims by individuals having personal interests in the choice of prototypes. This has especially been the case where prototype strains have become valuable as substrates for diagnostic reagents, vaccines, etc. In this regard, the designation of prototype species and strains for noncommercial purposes must be seen as a primary responsibility of international specialty groups.

The best model for the kind of virus description necessary to avoid all ambiguity in identification is that of the *American Type Culture Collection* in its frequently updated *Catalogue of Animal and Plant Viruses, Fifth Edition* (1986). For example, St. Louis encephalitis virus is listed as:

> St. Louis encephalitis virus   Class III
> *ATCC VR-80 Strain:* Hubbard. *Original source:* Brain of patient, Missouri, 1937. *Reference:* Mc-Cordock, H. A., et al., *Proc. Soc. Exp. Biol. Med. 37:*288, 1937. *Preparation:* 20% SMB in 50% NIRS infusion broth; supernatant of low speed centrifugation. *Host of Choice:* sM (i.c.); M (i.c.). *Incubation:* 3–4 days. *Effect:* Death. *Host Range:* M, Ha, CE, HaK, CE cells. *Special Characteristics:* Infected brain tissue will have a titer of about 10^7. Agglutinates goose and chicken RBC. Cross reacts with many or all members of Group B arboviruses. Requests for this agent must carry a signed statement assuming all risks and responsibility for lab handling.

In formal usage, the first letter of virus family, subfamily, and genus terms are capitalized, and the terms are underlined or printed in italics. Species terms, which, as stated above, are used in English vernacular form, are not capitalized, underlined, or italicized. In formal usage, the name of the taxon should precede the term for the taxonomic unit; for example . . . "the family *Paramyxoviridae*" . . . "the genus *Morbillivirus*." Further, it was decided years ago that virus nomenclature would not be Latinized. For example, terms such as *Flavivirus fabricius, Orthopoxvirus variolae,* and *Herpesvirus varicellae,* which were used at one time, should not be perpetuated.

The following represent examples of full, formal, taxonomic terminology:

1. Family *Poxviridae,* subfamily *Chordopoxvirinae,* genus *Orthopoxvirus,* vaccinia.
2. Family *Herpesviridae,* subfamily *Alphaherpesvirinae,* genus *Simplexvirus,* herpes simplex 2.
3. Family *Picornaviridae,* genus *Enterovirus,* polio 1.
4. Family *Rhabdoviridae,* genus *Lyssavirus,* rabies.

Vernacular Usage of Virus Nomenclature

In informal vernacular usage, virus family, subfamily, genus, and species terms are written in lower case Roman script—they are not capitalized, underlined, or printed in italics. In informal usage, the name of the taxon should not include the formal suffix, and the name of the taxon should follow the term for the taxonomic unit; for example, "the paramyxovirus family" or "the morbillivirus genus."

Use of vernacular terms for virus taxonomic units and virus names should not lead to unnecessary ambiguity or loss of precision in virus identification. The formal family, subfamily, and genus terms and the English vernacular species terms listed in the most current report of the ICTV—*The Classification and Nomenclature of Viruses* (Brown, 1987)—should be used as the basis for choosing vernacular terms, rather than any synonyms or transliterations.

Structural, Physicochemical, and Replicative Characteristics Used in Taxonomy

Laboratory techniques that have been used for taxonomic purposes over the past 20 years have included measurements of virion morphology (by electron microscopy), virion stability (by varying pH and temperature, adding lipid solvents and detergents, etc.), virion size (by filtration through fibrous and porous microfilters), and virion antigenicity (by many different serologic methods). Taxonomic placement of viruses by these conventional means has represented more a repetition of historical steps in virus characterization than the use of any rational sequential identification protocol. These conventional means have worked, because after large numbers of viruses had been studied and their characteristics worked into the universal taxonomic scheme, it was necessary in most cases to measure only a few characteristics to place a new virus, especially a new variant from a well studied source, in its proper taxonomic niche. For example, a new adenovirus, isolated from the human respiratory tract and identified by serologic means, should be easy to place in its niche in the family *Adenoviridae*. The exception to this easily applied system is seen when a new virus is found that does not have a usual pattern of properties. Such a virus becomes a candidate prototype for a new taxon—a new family or genus. In such cases, comprehensive characterization of all virion properties is called for, but if techniques are used that have only an empirical basis, then taxonomic constructions that follow may be faulted. Over the years, more and more physicochemical data and serological studies have been called for before new viruses have been placed in taxa.

We now understand many of the fundamental molecular bases for those empirical, physical-property measurements that were originally used to construct most of the virus families and genera. Moreover, through the use of monoclonal antibodies, synthesized peptides (by prediction of amino acid sequence from virus nucleotide sequence), and epitopic mapping, we have begun to understand the molecular basis for those serological methods that were used taxonomically in the original constructions of families and genera. In addition, we have added "strategy of replication" and genetic considerations to taxonomic decisions. Many of the characteristics that are used in deciding taxonomic constructions are listed in Table 1.

The single most important technological development underpinning the development of modern virus taxonomy was the invention by Brenner and Horne (1959) of the negative–staining technique for electron microscopic examination of virions. The impact of this technique was immediate: 1) virions could be characterized with respect to size, shape, surface structure, and, often, symmetry; 2) the method could be applied to viruses infecting all kinds of hosts, including humans, experimental animals, cell cultures, arthropods, etc.; and 3) virions could be characterized in unpurified material, including diagnostic specimens. Negative staining facilitated the rapid accumulation of data about the physical properties of many viruses. Thin-section electron microscopy of virus-infected cell cultures and tissues of infected humans and experimental animals provided complementary data on virion morphology, mode and site of virion morphogenesis, presence or absence of an envelope, etc. Thus, today, in most cases, viruses can

TABLE 1. Properties of viruses used in taxonomic constructions

A. *Properties of virions*
1. Virion size
2. Virion shape
3. Presence or absence of an envelope and peplomers
4. Capsomeric symmetry and structure

B. *Properties of genome*
1. Type of nucleic acid—DNA or RNA—and size
2. Strandedness—single-stranded or double-stranded
3. Linear or circular
4. Sense—positive, negative, or ambisense
5. Number of segments
6. Size of genome or genome segments
7. Presence or absence and type of 5'-terminal cap
8. Presence or absence of 5'-terminal, covalently linked polypeptide
9. Presence or absence of 3'-terminal poly (A) tract
10. Nucleotide sequence

C. *Properties of proteins*
1. Number of proteins
2. Size of proteins
3. Functional activities of proteins (especially virion transcriptase, virion reverse transcriptase, virion hemagglutinin, virion neuraminidase, virion fusion protein)
4. Amino acid sequence

D. *Replication*
1. Strategy of replication of nucleic acid
2. Characteristics of transcription
3. Characteristics of translation and posttranslational processing
4. Site of accumulation of virion proteins, site of assembly, site of maturation and release
5. Cytopathology, inclusion body formation

E. *Physical properties*
1. pH stability
2. Thermal stability
3. Cation (Mg^{++}, Mn^{++}) stability
4. Solvent stability
5. Detergent stability
6. Radiation stability

F. *Biological properties*
1. Serologic relationships
2. Host range, natural and experimental
3. Pathogenicity, association with disease
4. Tissue tropisms, pathology, histopathology
5. Transmission
6. Vector relationships
7. Geographic distribution

be placed in their appropriate family, and often in their appropriate genus, after visualization and measurement by negative-stain or thin-section electron microscopy. Because virion morphology is such a valuable summary characteristic in virus taxonomy, representative viruses of each of the 21 families (and some morphologically distinct genera) containing human and animal pathogens are illustrated in this chapter.

Classification of Viruses on the Basis of Epidemiologic Criteria

Separate from the formal universal taxonomic system and the formal and vernacular nomenclature that stems from it, there are other classifications of viruses that are useful in clinical, epidemiological, and diagnostic settings. These are based upon virus tropisms and modes of transmission. Most viruses of humans and animals are transmitted by inhalation, ingestion, injection (including via arthropod bites), close contact (including sexual contact), or transplacental passage (White and Fenner, 1986).

Enteric viruses are usually acquired by ingestion (fecal-oral transmission) and replicate primarily in the intestinal tract. The term is usually restricted to viruses that remain localized in the intestinal tract, rather than causing generalized infections. Enteric viruses are included in the families *Reoviridae* (genera *Rotavirus* and *Reovirus*), *Coronaviridae*, *Picornaviridae* (genus *Enterovirus*), *Adenoviridae,* and *Caliciviridae*.

Respiratory viruses are usually acquired by inhalation (respiratory transmission) or by fomites (hand-to-nose/mouth/eye transmission) and replicate primarily in the respiratory tract. The term is usually restricted to viruses that remain localized in the respiratory tract, rather than causing generalized infections. Respiratory viruses are included in the families *Orthomyxoviridae, Paramyxoviridae* (genera *Paramyxovirus* and *Pneumovirus*), *Coronaviridae, Adenoviridae,* and *Picornaviridae* (genus *Rhinovirus*).

Arboviruses (from "*arthropod-bo*rne viruses") multiply in their hematophagous (blood-feeding) arthropod hosts and are then transmitted by bite to a vertebrate host, in which virus multiplication produces a viremia of sufficient magnitude to infect another blood-feeding arthropod. Thus, the cycle is perpetuated. Part of the cycle can be bypassed by the virus via vertical transmission, in which transovarial infection passes virus directly from one arthropod generation to the next. Arboviruses are included in the families *Togaviridae, Flaviviridae, Bunyaviridae, Reoviridae* (genus *Orbivirus*), *Rhabdoviridae,* and the unnamed family containing African swine fever virus.

Oncogenic viruses are acquired by close contact (including sexual contact), injection, and by unknown means. The viruses usually infect only specific target tissues, where they usually become persistent and may evoke transformation of host cells, which may in turn progress to malignancy. Viruses that have demonstrated the capacity to be oncogenic, in experimental animals or in nature (but not in all cases in humans or domestic animals), are included in the families *Herpesviridae, Adenoviridae, Papovaviridae, Hepadnaviridae,* and *Retroviridae* (subfamilies *Oncovirinae* and possibly *Lentivirinae*).

The Place of Taxonomy in Diagnostic Virology

"Differential Diagnosis" and "Rule-Out Diagnosis"

Because the etiologic diagnosis of a viral disease usually requires laboratory tests, two-way communication is necessary between the clinician and the laboratory diagnostician. The clinician usually makes a preliminary diagnosis on the basis of four kinds of evidence: 1) *Clinical features* allow recognition with varying certainty in typical cases of many viral diseases (e.g., varicella exanthem and systemic disease, measles exanthem and systemic disease, paralytic poliomyelitis). 2) *Epidemic behavior* of a disease, in a typical population, may allow recognition (e.g., influenza, arbovirus diseases such as dengue and yellow fever, enterovirus exanthems). 3) *Epidemiologic circumstances* may indicate probable etiology (e.g., respiratory syncytial virus as the primary cause of croup and bronchiolitis in infants; hepatitis B or non-A, non-B viruses as the likely cause of hepatitis following blood transfusion). 4) *Organ involvement* may suggest a probable etiology (e.g., mumps virus as the cause of parotitis, viruses in general as the cause of 80–90 percent of acute respiratory infections) (Evans, 1982).

Shortcomings in the predictive value of these kinds of evidence suggest that the laboratory diagnostician as well as the clinician must appreciate the range of possible etiologic agents in particular disease syndromes (McIntosh, 1985). There is value in initially assembling an inclusive "long list" of possible etiologic agents, so that no candidate agent is overlooked. In most cases this is done informally, and the process is adjusted to the complexity of the case. In general, two forms of review of the "long list" are used—the approach of "differential diagnosis" and that of "rule-out diagnosis." The difference between the two is small, involving only how the "long list" is handled. In "differential diagnosis," possible etiologies are listed in descending order of likelihood. In "rule-out diagnosis," only etiologies that are consistent with the data at hand are listed. In each case, the process of arriving at a diagnosis is dynamic, with increasing evidence leading to a shortening of the list. There is no attempt to shorten the list arbitrarily nor to decide on the etiologic agent without confirmatory evidence.

The universal system of virus taxonomy may be used as the source of the "long list" of candidate etiologic agents. The system serves to organize the "long list" logically, and because the system is comprehensive, it is unlikely that known viruses will be overlooked. The kinds of evidence that serve to place an etiologic agent in its proper family and genus should also play a major role in shortening the "differential diagnosis" or "rule-out diagnosis" list, and should in most cases provide the etiologic information needed to focus immunologic (serologic) identification techniques. For example, the "long list" of possible etiologies for a slowly progressive central nervous system disease would include many viruses that are difficult or impossible to cultivate. However, the identification by electron microscopy of spherical, 45-nm, nonenveloped virions in the nuclei of cells in a brain biopsy from a patient with such a disease would go far toward shortening the list to the family *Papovaviridae,* genus *Polyomavirus,* thereby suggesting a diagnosis of JC or SV 40, virus-induced, progressive multifocal leukoencephalopathy. In this case, many viruses known to invade the brain and cause slowly progressive neurologic disease would be eliminated from the differential or rule-out diagnostic list; e.g., member viruses of the families *Herpesviridae, Adenoviridae, Togaviridae, Flaviviridae, Paramyxoviridae, Rhabdoviridae, Bunyaviridae, Arenaviridae, Retroviridae,* and *Picornaviridae.*

Causal Relationship Between Virus and Disease

One of the landmarks in the study of infectious diseases was the development of the *Henle-Koch* postulates of causation. They were originally drawn up for bacteria and protozoa, but were revised in 1937 by Rivers and again in 1976 by Evans in attempts to accommodate the special problem of proving disease causation by viruses (Evans, 1976) (Table 2).

The problem is still difficult, especially when viruses are considered as causative of chronic diseases, neoplastic diseases, and slowly progressive neurological diseases. Because most such diseases cannot be reproduced in experimental animals, virologists have had to evaluate causation indirectly, via "guilt by association," an approach that relies to a large degree on epidemiologic data and patterns of serologic reactions in populations. The framework of virus taxonomy, again, plays a role, this time in helping to evaluate some of the critieria for causation. This is especially the case in evaluating the likelihood that particular kinds of viruses might be etiologically, rather than coincidentally or opportunistically, associated with a given disease. For example, very early in the investigation of the acquired immunodeficiency syndrome (AIDS), when many kinds of viruses were being isolated from patients and many candidate etiologic agents were being advanced publicly, several virologists working on the disease predicted that the etiologic agent would turn out to be a member of the family *Retroviridae.* This

TABLE 2. Criteria for causation: a unified concept appropriate for viruses as causative agents, based on the Henle-Koch postulates (from Evans, 1976)

1. *Prevalence* of the disease should be significantly higher in those exposed to the putative virus than in controls not so exposed.
2. *Incidence* of the disease should be significantly higher in those exposed to the putative virus than in controls not so exposed (prospective studies).
3. *Exposure* to the putative virus should be present more commonly in those with the disease than in controls without the disease.
4. *Temporally,* the disease should follow exposure to the putative virus, with a distribution of incubation periods following a normal pattern.
5. *A spectrum* of effects on hosts should follow exposure to the putative virus, presenting a logical biological gradient of responses, from mild to severe.
6. *A measurable host response,* such as an antibody response, should follow exposure to the putative virus. In those individuals lacking prior experience, the response should appear regularly, and in those individuals with prior experience, the response should be anamnestic.
7. *Experimental reproduction* of the disease should follow deliberate exposure of animals or humans to the putative virus, but nonexposed controls should remain disease-free. Deliberate exposure may be in the laboratory or in the field, as with sentinel animals.
8. *Elimination* of the putative virus or its vector should decrease the incidence of the disease.
9. *Prevention or modification* of infection, via immunization or drugs, should decrease the incidence of the disease.
10. *"The whole thing should make biologic and epidemiologic sense."*

prediction was based upon review of the biological and pathogenetic properties of member viruses of each family, the ruling out of most, and the recognition of similarities between characteristics of the diseases caused by known retroviruses and the disease AIDS. This prediction guided some of the early experimental approaches toward finding the etiologic agent. Later, after the discovery of the etiologic agent, human immunodeficiency virus (HIV), taxonomic considerations predicted its placement in the genus *Lentivirus* of the family *Retroviridae*. This taxonomic prediction, in turn, has been guiding experimental design in many areas, including cell biology, diagnostics, and epidemiology.

The Adequate Description of New Viruses

Over the years, as thousands of viruses have been isolated from human and animal specimens, there have been many errors of duplication—that is, viruses isolated in different laboratories have been given different names and the chance for coexistence in various virus lists (Murphy, 1983). Viruses have also been placed in the wrong lists—the most notable instance involving the emergence of the family *Reoviridae* (genus *Reovirus*) from the initial placement of its prototype members in the list of human enteroviruses. More recently, several named but serologically "ungrouped" viruses listed in the *International Catalogue of Arboviruses* (Karabatsos, 1985) were found actually to be unrecognized isolates of Lassa virus and Rift Valley fever virus, viruses that must be handled under maximum biocontainment conditions. The reasons for these and similar problems have been 1) the inadequate characterization and description of viruses by those who isolate them, and 2) the inadequate review of data by international specialty groups. The number of instances of such problems is declining, but because there are serious consequences, further attention is warranted.

When an "unknown" is first worked on in the diagnostic laboratory, its initial characterization may involve only standardized protocols. That is, only a few characteristics may be determined as the lead-in for specific (usually serologic) identification procedures. Only when an "unknown" fails to yield to routine procedures is there a call for more extensive study. One key to simplifying and rationalizing such study is to set useful techniques into a proper sequence based upon taxonomic characteristics. This sequence of procedures should include logical "short-cuts," so as to avoid the extra effort and expense of "filling in all the boxes." The only situation in which there should be no "short-cuts" is in the characterization of a virus that may stand as the prototype of a new family or genus; otherwise, tech-

niques and the sequence for their use should be streamlined. For example, negative-stain electron microscopy represents a logical "short-cut" for the initial placement of unknowns that emerge from diagnostic protocols. If an "unknown" is shown by electron microscopy to be a rhabdovirus, what is the value of checking whether its genome is DNA or RNA, or whether it is sensitive to low pH or to solvents? What is the value of doing serology against other than the known rhabdoviruses (except perhaps in testing for the presence of contaminant viruses)? Comprehensive testing is the key to discovery of novel viruses (such as a hypothetical bullet-shaped virus with a double-stranded DNA genome!), but

TABLE 3. Families containing human and animal viruses

Dividing characteristics	Virus family names
DNA Viruses	
dsDNA, enveloped	*Poxviridae*
	Iridoviridae
	Herpesviridae
dsDNA, nonenveloped	*Adenoviridae*
	Papovaviridae
	Hepadnaviridae
ssDNA, nonenveloped	*Parvoviridae*
RNA Viruses	
dsRNA, nonenveloped	*Reoviridae*
	Birnaviridae
ssRNA, enveloped, no DNA step in replication, positive-sense genome	*Togaviridae*
	Flaviviridae
	Coronaviridae
negative-sense genome, nonsegmented genome	*Paramyxoviridae*
	Rhabdoviridae
	[*Filoviridae*][a]
negative-sense genome, segmented genome ambisense genome, segmented genome	*Orthomyxoviridae*
	Bunyaviridae
	Arenaviridae
DNA step in replication, positive-sense genome, nonsegmented genome	*Retroviridae*
ssRNA, nonenveloped	*Picornaviridae*
	Caliciviridae

[a] Brackets are used in this chapter to identify taxa and names that have not yet been approved by the ICTV. Abbreviations and terms used: ds = double-stranded; ss = single-stranded; enveloped = possessing an outer lipid-containing bilayer partly derived from host cell membrane; positive-sense genome = for RNA viruses, genome with sequence the same as mRNA, = for DNA viruses, genomes with sequence analogous to mRNA; negative-sense genome = genome with sequence complementary to mRNA; ambisense genome = genome with some genes having positive and other genes having negative sense.

"filling in the boxes" is so demanding of laboratory resources that it detracts from needed, focused studies of viruses that do exhibit charcteristics of an unusual nature.

Assuring the adequacy of characterization and description of new viruses is a particular responsibility of reference laboratories, international reference centers, international specialty groups, and culture collections.

A Taxonomic Description of the Families of Viruses Containing Human and Animal Pathogens

Of the more than 61 families of viruses recognized by the ICTV, 21 contain viruses that infect humans and animals (Table 3). Although the rest of this volume is concerned only with human viruses, families containing animal viruses that do not cause disease in humans are included here because there are frequent questions about such viruses infecting humans. An unresolved issue concerns the order of listing the virus families; this issue again points out the arbitrariness of weighting virion characteristics for taxonomic purposes (Matthews, 1983). The listing of the families containing human and animal viruses in Table 3 is in the order currently used by the ICTV (Matthews, 1982). The order is set by nucleic acid type; the presence or absence of an envelope; genome replication strategy; genome sense (positive, negative, or ambisense); and genome segmentation. An alphabetical listing of the common pathogenic viruses of humans is also provided for cross-referencing purposes (Table 4).

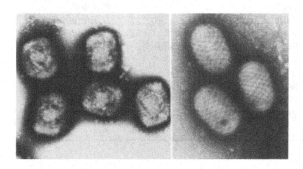

FIG. 1. *Poxviridae, Chordopoxvirinae, Orthopoxvirus,* vaccinia (left), and *Parapoxvirus,* orf (right). Magnification ×55,000. (Micrographs provided by the courtesy of E. L. Palmer, M. L. Martin, A. K. Harrison, S. G. Whitfield, L. D. Pearson, T. Baker, E. H. Cook, C.-H. von Bonsdorff, J. Esposito, C. Smale, M. V. Nermut, C. Mebus, V. I. Heine, and F. X. Heinz.)

Poxviridae

Family: *Poxviridae* (the poxviruses) (Matthews, 1982)
Subfamily: *Chordopoxvirinae* (the poxviruses of vertebrates)
Genus: *Orthopoxvirus* (vaccinia, smallpox)
Genus: *Parapoxvirus* (orf)
Genus: *Avipoxvirus* (fowlpox)
Genus: *Capripoxvirus* (sheeppox)
Genus: *Leporipoxvirus* (myxoma)
Genus: *Suipoxvirus* (swinepox)
Genus: [*Yatapoxvirus*] (yabapox and tanapox)
Genus: [*Molluscipoxvirus*] (molluscum contagiosum)
Subfamily: *Entomopoxvirinae* (the poxviruses of insects)

CHARACTERISTICS

Poxvirus virions are large, brick-shaped (or ovoid in the case of the genus *Parapoxvirus*), 300 to 450 × 170 to 260 nm in size, with an external envelope, a complex coat of tubular structures, and an internal structure made up of a DNA-containing core and one or two lateral bodies. The genome consists of a single molecule of dsDNA with covalently closed ends ("hairpins"), size 130 to 280 kbp. The viruses have more than 100 proteins and several enzymes (including a DNA-dependent DNA transcriptase and a DNA-dependent RNA polymerase). Little is known about poxvirus DNA replication. Initial transcription occurs within the virus core, leading to the synthesis of early proteins, some of which lead to uncoating of viral DNA, allowing further transcription and DNA replication. Replication and assembly occur in cytoplasm in viroplasm ("viral factories"), and virions are released by budding (enveloped virions) or by cell destruction (nonenveloped virions). There is a common poxvirus group antigen (nucleoprotein antigen), and member viruses of the genus *Orthopoxvirus* produce a nonvirion hemagglutinin.

HUMAN PATHOGENS

Orthopoxvirus: variola (smallpox, alastrim) virus, vaccinia virus, monkeypox virus, cowpox virus. *Parapoxvirus:* orf (contagious pustular dermatitis) virus, pseudocowpox (milker's nodule) virus. [*Yatapoxvirus*]: yabapox virus, tanapox virus, [*Molluscipoxvirus*]: molluscum contagiosum virus.

ANIMAL PATHOGENS

Orthopoxvirus: cowpox virus, ectromelia (mousepox) virus, rabbitpox virus, monkeypox virus. *Parapoxvirus:* orf (contagious pustular dermatitis) virus,

TABLE 4. Important pathogenic viruses of humans

Virus	Family	Virus	Family
Adenovirus, human 1 to 43	*Adenoviridae*	Marburg	[*Filoviridae*]
Astrovirus	Unclassified	Mayaro	*Togaviridae*
B-virus (Cercopithecus herpesvirus)	*Herpesviridae*	Measles	*Paramyxoviridae*
BK	*Papovaviridae*	Mokola	*Rhabdoviridae*
Bunyamwera	*Bunyaviridae*	Molluscum contagiosum	*Poxviridae*
Bwamba	*Bunyaviridae*	Monkeypox	*Poxviridae*
California encephalitis	*Bunyaviridae*	Mumps	*Paramyxoviridae*
Central European encephalitis	*Flaviviridae*	Murray Valley encephalitis	*Flaviviridae*
Chandipura	*Rhabdoviridae*	Norwalk (and related viruses)	Unclassified
Chikungunya	*Togaviridae*	O'nyong-nyong	*Togaviridae*
Colorado tick fever	*Reoviridae*	Omsk hemorrhagic fever	*Flaviviridae*
Coronavirus, human OC 43 and 229-E	*Coronaviridae*	Orf (contagious pustular dermatitis)	*Poxviridae*
Coronavirus, human enteric	*Coronaviridae*	Oriboca	*Bunyaviridae*
Cowpox	*Poxviridae*	Oropouche	*Bunyaviridae*
Coxsackie A 1 to 22, A 24	*Picornaviridae*	Orungo	*Reoviridae*
Coxsackie B 1 to 6	*Picornaviridae*	Papillomavirus, human 1 to 40	*Papovaviridae*
Creutzfeldt-Jakob	Unclassified	Parainfluenza 1 to 4	*Paramyxoviridae*
Crimean-Congo hemorrhagic fever	*Bunyaviridae*	Parvovirus, human B 19	*Parvoviridae*
Cytomegalovirus, human	*Herpesviridae*	Piry	*Rhabdoviridae*
Delta	Unclassified	Polio 1 to 3	*Picornaviridae*
Dengue 1 to 4	*Flaviviridae*	Pseudocowpox (milker's nodule)	*Poxviridae*
Duvenhage	*Rhabdoviridae*	RA-1	*Parvoviridae*
Eastern equine encephalitis	*Togaviridae*	Respiratory syncytial	*Paramyxoviridae*
Epstein-Barr	*Herpesviridae*	Rabies	*Rhabdoviridae*
Ebola	[*Filoviridae*]	Rhinovirus, human 1 to 113	*Picornaviridae*
Echovirus, human 1 to 9, 11 to 27, 29 to 34	*Picornaviridae*	Rift Valley fever	*Bunyaviridae*
		Rocio	*Flaviviridae*
Enterovirus, human 68 to 71	*Picornaviridae*	Ross River	*Togaviridae*
Hantaan (hemorrhagic fever with renal syndrome)	*Bunyaviridae*	Rotavirus, human	*Reoviridae*
		Rubella	*Togaviridae*
Hepatitis A	*Picornaviridae*	Russian spring-summer encephalitis	*Flaviviridae*
Hepatitis B	*Hepadnaviridae*	SV 40	*Papovaviridae*
Hepatitis, non-A, non-B	Unclassified	Sandfly fever-Naples	*Bunyaviridae*
Herpes simplex 1 to 2	*Herpesviridae*	Sandfly fever-Sicilian	*Bunyaviridae*
Immunodeficiency virus, human (HIV = LAV, HTLV-III, ARV)	*Retroviridae*	St. Louis encephalitis	*Flaviviridae*
		T-lymphotropic 1, human (HTLV-I)	*Retroviridae*
Influenza A, B, and C	*Orthomyxoviridae*	T-lymphotropic 2, human (HTLV-II)	*Retroviridae*
Isfahan	*Rhabdoviridae*	Tahyna	*Bunyaviridae*
JC	*Papovaviridae*	Tanapox	*Poxviridae*
Japanese encephalitis	*Flaviviridae*	Vaccinia	*Poxviridae*
Junin (Argentine hemorrhagic fever)	*Arenaviridae*	Varicella-zoster	*Herpesviridae*
Kemerovo	*Reoviridae*	Variola	*Poxviridae*
Kuru	Unclassified	Venezuelan equine encephalitis	*Togaviridae*
Kyasanur Forest	*Flaviviridae*	Vesicular stomatitis	*Rhabdoviridae*
La Crosse	*Bunyaviridae*	West Nile	*Flaviviridae*
Lassa	*Arenaviridae*	Western equine encephalitis	*Togaviridae*
Louping-ill	*Flaviviridae*	Yabapox	*Poxviridae*
Lymphocytic choriomeningitis	*Arenaviridae*	Yellow fever	*Flaviviridae*
Machupo (Bolivian hemorrhagic fever)	*Arenaviridae*		

pseudocowpox (milker's nodule) virus, bovine papular stomatitis virus. *Avipoxvirus:* many specific bird poxviruses. *Capripoxvirus:* sheeppox virus, goatpox virus, lumpy skin disease virus (of cattle). *Lepori-* *poxvirus:* myxoma virus (of rabbits), rabbit fibroma virus, hare fibroma virus, squirrel fibroma virus. *Suipoxvirus:* swinepox virus. Ungrouped: tanapox virus and yabapox virus (of monkeys).

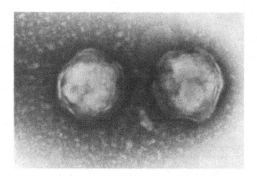

FIG. 2. African swine fever virus. Magnification ×150,000.

Iridoviridae and the Unnamed Family for African Swine Fever Virus

Family: *Iridoviridae* (the iridoviruses) (Matthews, 1982)
Genus: *Iridovirus* (small iridescent insect viruses)
Genus: *Chloriridovirus* (large iridescent insect viruses)
Genus: *Ranavirus* (frog viruses)
Genus: *Lymphocystivirus* (lymphocystis viruses—fish)
[Family]:   [unnamed] African swine fever virus (Brown, 1986)

CHARACTERISTICS

Iridovirus and African swine fever virions consist of a lipid-containing envelope (missing on some insect viruses), surrounding a distinctly hexagonal icosahedral nucleocapsid. Overall virion diameter is 125 to 300 nm (African swine fever virus is 175 to 215 nm). The genome consists of one molecule of dsDNA, size 150 to 350 kbp (African swine fever has a genome size of 150 kbp). The viruses have more than 20 structural proteins and several virion-associated enzymes. Little is known about viral DNA replication and RNA transcription. Replication occurs in the cytoplasm (although the nucleus is needed for viral DNA synthesis), and virions are released by budding or cell destruction. Antibodies raised against African swine fever virus virions do not neutralize, but do react in binding assays.

HUMAN PATHOGENS

None known.

ANIMAL PATHOGENS

African swine fever virus and several frog viruses.

Herpesviridae

Family: *Herpesviridae* (the herpesviruses) (Roizman et al., 1982)
Subfamily: *Alphaherpesvirinae* (the herpes simplex-like viruses)
Genus: *Simplexvirus* (herpes simplex-like viruses)
Genus: [*Varicellovirus*] (varicella and pseudorabies-like viruses)
Subfamily: *Betaherpesvirinae* (the cytomegaloviruses)
Genus: *Cytomegalovirus* (human cytomegaloviruses)
Genus: [*Muromegalovirus*] (murine cytomegaloviruses)
Subfamily: *Gammaherpesvirinae* (the lymphocyte-associated viruses)
Genus: *Lymphocryptovirus* (EB-like viruses)
Genus: [*Thetalymphocryptovirus*] (Marek's disease-like viruses)
Genus: *Rhadinovirus* (saimiri-ateles-like viruses)

CHARACTERISTICS

Herpesvirus virions consist of 1) an envelope with surface projections, 2) a tegument consisting of amorphous material, 3) an icosahedral nucleocapsid, 100 nm in diameter with 162 prismatic capsomeres, and 4) a core consisting of a fibrillar spool on which the DNA is wrapped. Virus particles have an overall diameter of 150 to 200 nm. The genome consists of one molecule of dsDNA, size 120 to 220 kbp. Genomic DNA contains terminal and internal reiterated sequences, usually forming two covalently linked components (L and S), arranged in five different patterns. Virions have more than 30 structural proteins, including, in some cases, an Fc receptor in their envelope. DNA replication is complex, involving several viral enzymes in a rolling-circle mechanism. Transcription and translation are coordinately regu-

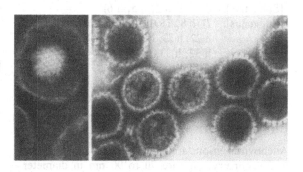

FIG. 3. *Herpesviridae, Alphaherpesvirinae, Simplexvirus,* herpes simplex 1, enveloped (left), and naked capsids (right). Magnification ×210,000.

lated and sequentially ordered in a cascade with three major stages. Replication takes place in the nucleus, and capsids acquire their envelopes via budding through the inner lamella of the nuclear envelope. Virions are released via transport across the cytoplasm in membranous vesicles, which then fuse with plasma membrane. Some herpesviruses induce neoplasia, and many or all persist for the lifetime of their natural host.

HUMAN PATHOGENS

Simplexvirus: herpes simplex virus 1 and 2, cercopithecine herpesvirus 1 (B-virus). [*Varicellovirus*]: varicella-zoster virus. *Cytomegalovirus:* human cytomegalovirus. *Lymphocryptovirus:* EB virus.

ANIMAL PATHOGENS

Simplexvirus: infectious bovine rhinotracheitis virus, bovine mammillitis virus, cercopithecine herpesviruses 1 (B-virus) and 2. [*Varicellovirus*]: pseudorabies virus (of swine), equine rhinopneumonitis and coital exanthema viruses. [*Muromegalovirus*]: murine cytomegalovirus. *Lymphocryptovirus:* baboon herpesvirus, pongine (chimpanzee) herpesvirus. [*Thetalymphocryptovirus*]: Marek's disease herpesvirus (of fowl), turkey herpesvirus. *Rhadinovirus:* herpesvirus ateles, herpesvirus saimiri. Hundreds of other herpesviruses are associated with diseases of animals; most of these have not yet been assigned to subfamilies or genera.

Adenoviridae

Family: *Adenoviridae* (the adenoviruses) (Wigand et al., 1982)
Genus: *Mastadenovirus* (mammalian adenoviruses)
[Subgenus]: *A* (h 12, h 18, h 31)
[Subgenus]: *B* (h 3, h 7, h 11, h 14, h 16, h 21, h 34, h 35)
[Subgenus]: *C* (h 1, h 2, h 5, h 6)
[Subgenus]: *D* (h 8, h 9, h 10, h 13, h 15, h 17, h 19, h 22, h 23, h 24, h 26, h 27, h 29, h 30, h 32, h 33, h 36, h 37)
[Subgenus]: *E* (h 4)
[Subgenera]: [unnamed, for animal adenoviruses]
Genus: *Aviadenovirus* (avian adenoviruses)

CHARACTERISTICS

Adenovirus virions are nonenveloped, have icosahedral symmetry, and are 70 to 90 nm in diameter. Virions have 252 capsomeres; of these, the 12 vertex capsomeres (penton bases) have fiber projections that carry major species-specific and minor subge-

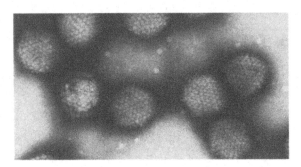

FIG. 4. *Adenoviridae, Mastadenovirus,* h5 (human adenovirus type 5). Magnification ×260,000. (Human adenoviruses are designated by the letter *h*, followed by serial numbers. Mammalian and avian adenoviruses are designated by a three-letter code for their host genus, followed by serial numbers. Subgenera have only been established for human adenoviruses.)

nus-specific epitopes, and the 240 nonvertex capsomes (hexons) carry major genus and subgenus-specific and minor species-specific epitopes. The genome consists of one molecule of dsDNA, size 36 to 38 kbp. The viruses have at least 10 structural proteins. DNA replication is complex and distinct from that of other viruses; it involves covalently bound proteins at the ends of the DNA and circularization. Transcription is also complex, involving early and late genes distributed randomly along both strands of the DNA. Replication and assembly occur in the nucleus, and virions are released via cell destruction. Most adenoviruses have a very narrow host range. Several of the viruses cause tumors in newborns of heterologous species.

HUMAN PATHOGENS

Mastadenovirus: h 1 to h 43.

ANIMAL PATHOGENS

Mastadenovirus: equ 1 (equine adenovirus), can 1 (infectious canine hepatitis), can 2 (canine adenovirus 2). Many other adenoviruses of mammals and birds are pathogenic.

Papovaviridae

Family: *Papovaviridae* (the papovaviruses) (Melnick et al., 1974)
Genus: *Papillomavirus* (papillomaviruses)
Genus: *Polyomavirus* (polyomaviruses)

CHARACTERISTICS

Papovavirus virions are nonenveloped, have icosahedral symmetry, and are 45 nm (polyomaviruses)

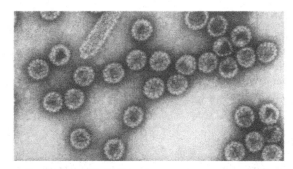

FIG. 5. *Papovaviridae, Papillomavirus,* human papillomavirus (untyped). Magnification ×150,000.

or 55 nm (papillomaviruses) in diameter. Virions have 72 capsomeres in a skewed (T = 7) arrangement. The genome consists of one molecule of circular dsDNA, size 5 kbp (polyomaviruses) or 8 kbp (papillomaviruses). The genome may persist in infected cells in an integrated form (polyomaviruses) or in an episomal form (papillomaviruses). The viruses have 5 to 7 structural proteins. DNA replication starts at a fixed point on the genome, which remains in circular configuration throughout the process. Transcription is complex, with early and late phases, involving both DNA strands. There is mRNA splicing and overlapping translation. Replication and assembly occur in the nucleus, and virions are released via cell destruction. Most of the viruses have a narrow host range. Host cell transformation and oncogenesis in experimentally inoculated animals are characteristic of particular viruses.

HUMAN PATHOGENS

Papillomavirus: human papillomaviruses (HPV) 1 to 46 (particular viruses are the cause of papillomas at different sites; HPV 6, 11, 16, and 18 are associated with genital condylomas and cervical carcinoma). *Polyomavirus:* JC, SV 40 (progressive multifocal leukoencephalopathy), BK (kidney infection and mild respiratory disease).

ANIMAL PATHOGENS

Papillomavirus: Shope papillomavirus (of rabbit), bovine papillomaviruses, and papillomaviruses of many animal species. *Polyomavirus:* SV 40 (from rhesus monkey) and several other polyomaviruses are oncogenic in many experimentally inoculated animals.

Hepadnaviridae

Family: *Hepadnaviridae* (the hepatitis B-like viruses) (Gust et al., 1986)
Genus: *Hepadnavirus* (hepatitis B-like viruses)

CHARACTERISTICS

Hepadnavirus virions consist of a lipid-containing envelope (containing hepatitis B surface antigen [HBsAg]) surrounding a nucleocapsid (containing hepatitis B core antigen [HBcAg]). Virions are 42 nm (the Dane particle) and nucleocapsids are 18 nm in diameter. The genome consists of one molecule of DNA, which is circular, nicked, and mainly double-stranded, with a large, single-stranded gap, size 3.2 kbp when fully double-stranded. Virions have two major polypeptides making up the HBsAg, a single polypeptide making up the HBcAg, and there are other important structural polypeptides (HBeAg). DNA replication involves repair of the single-standed gap, conversion to a supercoiled helix, and transcription of two classes of RNA, mRNA for protein synthesis and genomic RNA, which is transcribed in turn by a virus-specific reverse transcriptase to make genomic DNA. Replication in hepatocytes takes place in the nucleus (with HBcAg accumulation); HBsAg production occurs in the cytoplasm in massive amounts and is formed into 22-nm HBsAg particles, which are shed into the bloodstream, producing antigenemia. Persistence is common and is associated with chronic disease and neoplasia.

HUMAN PATHOGENS

Hepadnavirus: hepatitis B virus.

ANIMAL PATHOGENS

Hepadnavirus: hepatitis B-like viruses of Eastern woodchuck (*Marmota monax*), ground squirrel (*Spermophilus beecheyi*), red-bellied squirrel (*Callosciurus erythràcus*), and Pekin duck (*Anas domesticus*).

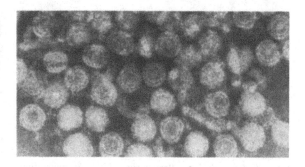

FIG. 6. *Hepadnaviridae, Hepadnavirus,* hepatitis B. Magnification ×250,000.

Parvoviridae

Family: *Parvoviridae* (the parvoviruses) (Siegl et al., 1985)
Genus: *Parvovirus* (parvoviruses of mammals and birds)
Genus: *Dependovirus* (adeno-associated viruses—AAVs)
Genus: *Densovirus* (insect parvoviruses)

CHARACTERISTICS

Parvovirus virions are nonenveloped, have icosahedral symmetry, and are 18 to 26 nm in diameter. Virions have 32 capsomeres in a T = 3 arrangement. The genome consists of one molecule of ssDNA, size 5 kb. Viruses of the genus *Parvovirus* preferentially encapsidate negative-sense DNA, whereas members of the other two genera encapsidate positive- and negative-sense DNA equally. The viruses have three major polypeptides. DNA replication is complex, involving the formation of a hairpin structure; extension to form a complete double-stranded, intermediate, endonuclease cleavage; and repetition of these cycles. Replication and assembly take place in the nucleus; replication requires host cell functions of the late S phase of the cell division cycle. Virions are very stable (pH, heat). The viruses have narrow host ranges.

HUMAN PATHOGENS

Parvovirus: human parvovirus B19 (aplastic crisis in hemolytic anemias; erythema infectiosum—fifth disease). A parvovirus, RA-1, has recently been associated with rheumatoid arthritis. Norwalk virus and related viruses that cause gastroenteritis may be parvoviruses.

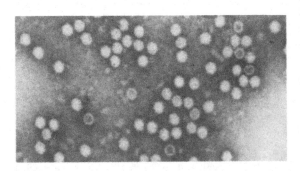

FIG. 7. *Parvoviridae, Parvovirus,* H-1. Magnification ×190,000.

ANIMAL PATHOGENS

Parvovirus: feline panleukopenia virus, canine parvovirus, mink enteritis virus, Aleutian mink disease virus, bovine parvovirus, goose parvovirus, porcine parvovirus, minute virus of mice, and other murine parvoviruses.

Reoviridae

Family: *Reoviridae* (the reoviruses) (Matthews, 1982)
Genus: *Reovirus* (reoviruses of humans and animals)
Genus: *Orbivirus* (orbiviruses)
Genus: *Rotavirus* (rotaviruses)
Genus: *Phytoreovirus* (plant reovirus subgroup 1)
Genus: *Fijivirus* (plant reovirus subgroup 2)
Genus: *Cypovirus* (cytoplasmic polyhedrosis viruses)
[Genus]: [proposed, for Colorado tick fever virus]

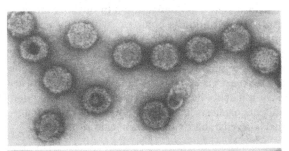
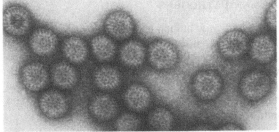
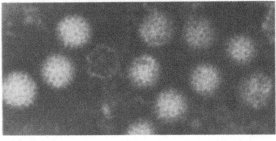

FIG. 8. *Reoviridae, Reovirus,* reovirus 2 (top); *Rotavirus,* human rotavirus (middle); *Orbivirus,* bluetongue 11 (bottom). Magnification ×150,000 (top and middle); Magnification ×270,000 (bottom).

CHARACTERISTICS

Viruses of each genus differ in morphologic and physicochemical details. In general, virions are non-enveloped, have two nucleocapsid shells, each with icosahedral symmetry, and have a diameter of 60 to 80 nm. The genome is linear dsRNA and divided into 10 (genus *Reovirus* and *Orbivirus*), 11 (genus *Rotavirus*), or 12 (Colorado tick fever virus) segments, size 22 (total genome; genus *Reovirus*), 18 (genus *Orbivirus*), 16 to 21 (genus *Rotavirus*), or 27 (Colorado tick fever virus) kbp. The viruses have 10 to 12 structural polypeptides; viruses carry a transcriptase and other enzymes. RNA replication is coordinated with translation: each genome segment is transcribed into mRNA; some molecules are used in translation, others are assembled into virion precursors, where they serve as templates for double-stranded RNA synthesis. Replication and assembly take place in cytoplasm, often in association with granular or fibrillar inclusion bodies. Rotaviruses and reoviruses are spread by direct contact (fecal-oral transmission) and fomites; orbiviruses and Colorado tick fever virus are transmitted via arthropods.

HUMAN PATHOGENS

Reovirus: reoviruses 1, 2, and 3 have been isolated from humans, but a causative relationship to illness has not been proven. *Orbivirus:* Orungo virus (febrile illness in Nigeria and Uganda), Kemerovo virus (febrile illness in USSR and Egypt). *Rotavirus:* human rotaviruses. Proposed genus: Colorado tick fever virus.

ANIMAL PATHOGENS

Reovirus: reoviruses 1, 2, and 3 (mice and other species). *Orbivirus:* bluetongue viruses, African horsesickness viruses, epizootic hemorrhagic disease of deer viruses, and other viruses affecting various animal species. *Rotavirus:* rotaviruses are important causes of diarrhea in most animal species.

Birnaviridae

Family: *Birnaviridae* (the 2-segment dsRNA viruses) (Matthews, 1982)
Genus: *Birnavirus* (2-segment dsRNA viruses)

CHARACTERISTICS

Birnavirus virions are noneveloped, have icosahedral symmetry, and are 60 nm in diameter. Virions have 92 capsomeres in a T = 9 arrangement. The genome consists of two segments of linear dsRNA,

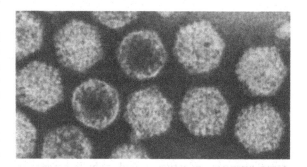

FIG. 9. *Birnaviridae, Birnavirus,* infectious bursal disease. Magnification ×270,000.

size 7 kbp (total genome). The viruses have four major structural polypeptides including a transcriptase. RNA replication has not been studied in detail. Replication and assembly take place in cytoplasm, and virus release occurs via cell destruction.

HUMAN PATHOGENS

None known.

ANIMAL PATHOGENS

Birnavirus: infectious bursal disease virus of chickens, infectious pancreatic necrosis virus of fish.

Togaviridae

Family: *Togaviridae* (the togaviruses) (Porterfield et al., 1978)
Genus: *Alphavirus* ("group A" arboviruses)
Genus: *Rubivirus* (rubella virus)
Genus: *Pestivirus* (mucosal disease viruses)
Genus: *Arterivirus* (equine arteritis virus)

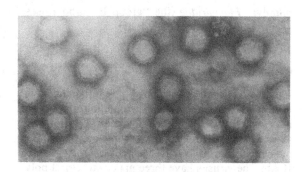

FIG. 10. *Togaviridae, Rubivirus,* rubella. Magnification ×150,000.

CHARACTERISTICS

Togavirus virions consist of a lipid-containing envelope, with fine peplomers, surrounding a nucleocapsid with icosahedral symmetry. Virions are 60 to 70 nm in diameter. The genome consists of one molecule of positive-sense, infectious, ssRNA, size 12 kb. The viruses have 3 to 4 major structural polypeptides, 1 or 2 of which are glycosylated. Some of the viruses exhibit pH-dependent hemagglutinating activity. RNA replication involves the synthesis of a complementary negative strand, which serves as template for genomic RNA synthesis. In alphaviruses, a subgenomic mRNA is synthesized in excess, from which a viral polyprotein is translated and cleaved into the structural proteins. Replication takes place in cytoplasm, and assembly involves budding through host cell membranes. All alphaviruses, but not rubella or the pestiviruses, replicate in arthropods as well as in vertebrates.

HUMAN PATHOGENS

Alphavirus: eastern equine encephalitis virus, western equine encephalitis virus, Venezuelan equine encephalitis virus, chikungunya virus, o'nyong-nyong virus, Ross River virus, Mayaro virus. *Rubivirus:* rubella virus.

ANIMAL PATHOGENS

Alphavirus: eastern equine encephalitis virus, western equine encephalitis virus, Venezuelan equine encephalitis virus. *Pestivirus:* mucosal disease virus (bovine virus diarrhea virus), hog cholera virus, border disease virus of sheep. *Arterivirus:* equine arteritis virus. Ungrouped: lactic dehydrogenase virus of mice.

Flaviviridae

Family: *Flaviviridae* (the "group B" arboviruses) (Westaway et al., 1985)
Genus: *Flavivirus* ("group B" arboviruses)

CHARACTERISTICS

Flavivirus virions consist of a lipid-containing envelope, with fine peplomers, surrounding a spherical nucleocapsid with unknown symmetry. Virions are 40 to 50 nm in diameter. The genome consists of one molecule of positive-sense, infectious, ssRNA, size 10 kb. The viruses have three major structural polypeptides, one of which is glycosylated. The viruses exhibit pH-dependent hemagglutinating activity. RNA replication involves the synthesis of a comple-

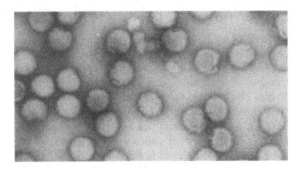

FIG. 11. *Flaviviridae, Flavivirus,* tick-borne encephalitis. Magnification ×140,000.

mentary negative-stranded intermediate that serves as template for the synthesis of genomic RNA. The genomic RNA also serves as mRNA, having one long open reading frame for the translation of a polyprotein which is cleaved into all viral proteins. Replication takes places in cytoplasm, and assembly involves passage through host cell membranes, at which time virions become enveloped by an unknown process. Most flaviviruses replicate in arthropods as well as in vertebrates.

HUMAN PATHOGENS

Flavivirus: yellow fever virus, dengue viruses, West Nile virus, St. Louis encephalitis virus, Japanese encephalitis virus, Murray Valley encephalitis virus, Rocio virus, tick-borne encephalitis viruses, and others.

ANIMAL PATHOGENS

Flavivirus: yellow fever virus, Japanese encephalitis virus, tick-borne encephalitis viruses, Wesselsbron virus. Ungrouped: simian hemorrhagic fever virus.

Coronaviridae

Family: *Coronaviridae* (the coronaviruses) (Siddell et al., 1983)
Genus: *Coronavirus* (coronaviruses of mammals and birds)

CHARACTERISTICS

Coronavirus virions consist of a pleomorphic lipid-containing envelope, with large club-shaped peplomers, within which is coiled a helical nucleocapsid

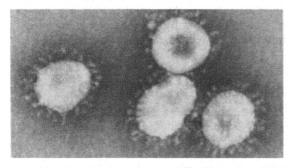

FIG. 12. *Coronaviridae, Coronavirus,* human coronavirus OC 43. Magnification ×200,000.

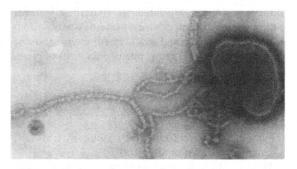

FIG. 13. *Paramyxoviridae, Paramyxovirus,* human parainfluenza 1. Magnification ×80,000.

with a diameter of 10 to 20 nm. Virions have a diameter of 80 to 160 nm. The genome consists of one molecule of positive-sense, infectious, ssRNA, size 16 to 21 kb. The viruses have three major structural polypeptides, two of which are glycosylated. RNA replication involves the synthesis of a negative-stranded template for the synthesis of genomic RNA, but coordinated with this is the synthesis of six mRNA species that represent a nested set of sequences—with the sequence of each RNA contained in the 3' end of each larger RNA species. Replication takes place in cytoplasm, and assembly involves budding, usually through intracytoplasmic membranes; release occurs via membrane fusion and exocytosis and by cell destruction.

HUMAN PATHOGENS

Coronavirus: human coronaviruses 229-E and OC43 (common cold, upper respiratory tract infection, probably pneumonia, and possibly gastroenteritis). Human enteric coronavirus (probably diarrhea).

ANIMAL PATHOGENS

Coronavirus: infectious bronchitis virus of fowl, transmissible gastroenteritis virus of swine, hemagglutinating encephalomyelitis virus of swine, turkey bluecomb virus, calf coronavirus (neonatal diarrhea), feline infectious peritonitis virus.

Paramyxoviridae

Family: *Paramyxoviridae* (the paramyxoviruses) (Kingsbury et al., 1978)
Genus: *Paramyxovirus* (paramyxoviruses)
Genus: *Morbillivirus* (measles-like viruses)
Genus: *Pneumovirus* (respiratory syncytial viruses)

CHARACTERISTICS

Paramyxovirus virions consist of a pleomorphic lipid-containing envelope, with large peplomers, within which is coiled a helical nucleocapsid with a diameter of 18 nm (genera *Paramyxovirus* and *Morbillivirus*) or 12 to 15 nm (genus *Pneumovirus*). Virions are 150 to 300 nm in diameter. The genome consists of one molecule of negative-sense ssRNA, size 16 to 20 kb (some virus particles contain a positive-sense RNA strand). The viruses have 6 to 10 major polypeptides including a transcriptase. The surface glycoproteins of some viruses have neuraminidase activity (genus *Paramyxovirus*), hemagglutinating activity (genera *Paramyxovirus* and *Morbillivirus*), and fusion activity (all genera). Hemagglutinating and fusion activities are based in separate surface proteins (HN or HA, and F proteins). RNA replication first involves mRNA transcription from the genomic RNA via a virion transcriptase; later, with the protein products of this transcription, there is production of full-length, positive-stranded template, which in turn is used for the synthesis of genomic RNA. Replication takes place in cytoplasm, and assembly occurs via budding upon plasma membrane. Viruses have a narrow host range. Morbilliviruses may cause persistent infections.

HUMAN PATHOGENS

Paramyxovirus: parainfluenza 1, 2, 3, 4 viruses (upper respiratory disease, bronchitis/bronchiolitis, pneumonia), mumps virus. *Morbillivirus:* measles virus. *Pneumovirus:* respiratory syncytial virus.

ANIMAL PATHOGENS

Paramyxovirus: Newcastle disease virus, parainfluenza 1 virus (Sendai virus of mice), parainfluenza 3 virus (cattle). These and the other parainfluenza viruses cause respiratory disease in other animals and birds. *Morbillivirus:* rinderpest virus of cattle, canine distemper virus, peste-des-petits-ruminants virus of sheep and goats. *Pneumovirus:* bovine respiratory syncytial virus, pneumonia virus of mice.

Rhabdoviridae

Family: *Rhabdoviridae* (the rhabdoviruses) (Brown et al., 1979)
Genus: *Vesiculovirus* (vesicular stomatitis-like viruses)
Genus: *Lyssavirus* (rabies-like viruses)
[Genera]: [proposed, for the many ungrouped rhabdoviruses of mammals, birds, fish, arthropods, and plants]

CHARACTERISTICS

Rhabdovirus virions are bullet-shaped (plant rhabdoviruses are often bacilliform), 70 to 85 nm in diameter and 130 to 380 nm long (most common length is 180 nm). Virions consist of a lipid-containing envelope, with large peplomers, within which a helical nucleocapsid is coiled into a precise cylindrical structure. The genome consists of one molecule of negative-sense ssRNA, size 13 to 16 kb. The viruses have 4 to 5 major polypeptides including a transcriptase. Some viruses exhibit hemagglutinating activity via the peplomer glycoprotein (G). Serologic cross-reactions involve primarily the nucleoprotein (N). RNA replication first involves mRNA transcription from the genomic RNA via a virion transcriptase; later, with the protein products of this transcription, there is production of full-length, positive-stranded template, which in turn is used for the synthesis of genomic RNA. Replication takes place in cytoplasm, and assembly occurs via budding upon plasma (vesicular stomatitis viruses) or intracytoplasmic (rabies

viruses) membranes. Many of the viruses replicate in and are transmitted by arthropods; rabies virus is transmitted between mammals by bite.

HUMAN PATHOGENS

Vesiculovirus: vesicular stomatitis-Indiana, -New Jersey, -Cocal viruses, Chandipura virus, Piry virus, Isfahan virus. *Lyssavirus:* rabies virus, Mokola virus, Duvenhage virus.

ANIMAL PATHOGENS

Vesiculovirus: vesicular stomatitis-Indiana, -New Jersey, -Cocal, -Alagoas viruses, Chandipura virus, Piry virus, Isfahan virus. *Lyssavirus:* rabies virus, Mokola virus, Duvenhage virus, Lagos bat virus, Kotonkan virus, Obodhiang virus. Ungrouped: bovine ephemeral fever virus, Egtved and infectious hematopoietic necrosis virus of fish.

Filoviridae

Family: [*Filoviridae*] (Marburg and Ebola viruses) (Kiley et al., 1982)
Genus: [*Filovirus*] (Marburg and Ebola viruses)

CHARACTERISTICS

Marburg and Ebola virions appear as long filamentous forms, sometimes with branching, and sometimes as "U"-shaped, "6"-shaped or circular forms. Virions have a uniform diameter of 80 nm and vary greatly in length (up to 14,000 nm). The infectious virion unit length is 790 nm (Marburg virus) or 970 nm (Ebola virus). Virions consist of a lipid-containing envelope, with large peplomers, surrounding a rather rigid helical nucleocapsid. The genome consists of one molecule of negative-sense ssRNA, size

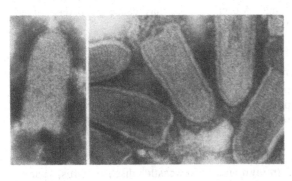

FIG. 14. *Rhabdoviridae, Lyssavirus,* rabies (left); *Vesiculovirus,* vesicular stomatitis—Indiana (right). Magnification ×240,000.

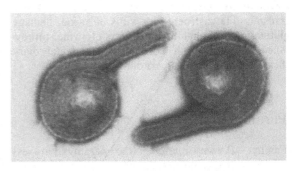

FIG. 15. [*Filoviridae*], [*Filovirus*], Ebola. Magnification ×66,000.

12.7 kb. The viruses have 5 major polypeptides. RNA replication has not been studied in detail. Replication takes place in the cytoplasm, and assembly involves envelopment via budding of preformed nucleocapsids. Marburg and Ebola viruses are "Biosafety Level 4" pathogens; they must be handled in the laboratory under maximum containment conditions.

HUMAN PATHOGENS

[*Filovirus*]: Marburg and Ebola viruses.

ANIMAL PATHOGENS

[*Filovirus*]: Marburg and Ebola viruses (experimental disease in monkeys, guinea pigs, hamsters, and mice).

Orthomyxoviridae

Family: *Orthomyxoviridae* (the influenza viruses) (Dowdle et al., 1975)
Genus: *Influenzavirus* (influenza A and B viruses)
[Genus]: [unnamed] (influenza C virus)

CHARACTERISTICS

Orthomyxovirus virions consist of a pleomorphic lipid-containing envelope, with large peplomers, within which is a helical nucleocapsid (arrangement uncertain) with a diameter of 9 to 15 nm. Virions are 90 to 120 nm in diameter, but may be filamentous with lengths up to several micrometers. The genome consists of 7 (influenzavirus C) or 8 (influenzaviruses A and B) segments of linear, negative-sense ssRNA, size 13.6 kb (total genome). The viruses have 7 to 9 major polypeptides including a transcriptase and a neuraminidase (NA). The viruses exhibit hemagglutinating activity via one peplomer glycoprotein complex (HA1-HA2). RNA replication involves a primary mRNA transcription from each segment of the genomic RNA via a virion transcriptase; later, with the protein products of this transcription, there is production of full-length, complementary RNA for each segment, each of which in turn is used as template for the synthesis of genomic RNA segments. Replication takes place in the nucleus (N protein accumulation) and cytoplasm, and assembly occurs via budding upon plasma membrane. The viruses can reassort genes during mixed infections.

HUMAN PATHOGENS

Influenzavirus: influenza A, B, and C viruses.

ANIMAL PATHOGENS

Influenzavirus: influenza A viruses infect swine, horses, seals, fowl, and many other species of birds. Unnamed genus: influenza C virus infects swine.

Bunyaviridae

Family: *Bunyaviridae* (the bunyaviruses) (Bishop et al., 1980)
Genus: *Bunyavirus* (Bunyamwera supergroup)
Genus: *Phlebovirus* (sandfly fever viruses)
Genus: *Nairovirus* (Nairobi sheep disease-like viruses)
Genus: *Uukuvirus* (Uukuniemi-like viruses)
[Genus]: [*Hantavirus*] (hemorrhagic fever with renal syndrome viruses)

CHARACTERISTICS

Bunyavirus virions consist of a lipid-containing envelope, with fine peplomers, within which are three loosely helical, circular, nucleocapsid structures

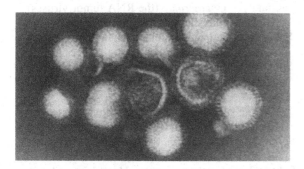

FIG. 16. *Orthomyxoviridae, Influenzavirus,* influenza A. Magnification ×150,000.

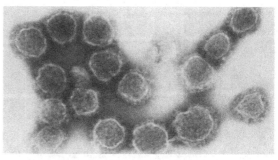

FIG. 17. *Bunyaviridae, Bunyavirus,* LaCrosse. Magnification ×130,000.

with a diameter of 2 to 2.5 nm. Virions are 90 to 120 nm in diameter. The genome consists of 3 molecules of "circular" (ends hydrogen-bonded), negative-sense and in the genus *Phlebovirus* ambisense ssRNA, size 13.5 to 21 kb (total genome). The viruses have 4 major polypeptides including a transcriptase. The viruses exhibit hemagglutinating activity via one of the two peplomer glycoproteins. RNA replication involves a primary transcription of mRNA from each segment of the genomic RNA via a virion transcriptase; later, with the protein products of this transcription, there is production of full-length, complementary RNA for each segment, each of which in turn is used as template for the synthesis of genomic RNA segments. Replication takes place in the cytoplasm, and assembly occurs via budding upon smooth membranes of the Golgi system. Closely related viruses can reassort genes during mixed infections.

HUMAN PATHOGENS

Bunyavirus: Bunyamwera virus, Bwamba virus, Oriboca virus, Oropouche virus, Guama virus, California encephalitis virus, LaCrosse virus, Tahyna virus. *Phlebovirus:* sandfly fever-Naples virus, sandfly fever-Sicilian virus, Rift Valley fever virus. *Nairovirus:* Crimean-Congo hemorrhagic fever virus. *Hantavirus:* Hantaan virus (Korean hemorrhagic fever, hemorrhagic fever with renal syndrome, nephropathia epidemica)

ANIMAL PATHOGENS

Phlebovirus: Rift Valley fever virus. *Nairovirus:* Nairobi sheep disease virus.

Arenaviridae

Family: *Arenaviridae* (the arenaviruses) (Pfau et al., 1974)
Genus: *Arenavirus* (arenaviruses)

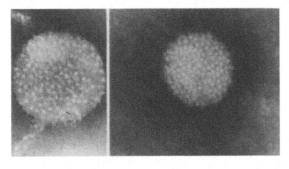

FIG. 18. *Arenaviridae, Arenavirus,* lymphocytic chorio-meningitis (left); Tacaribe (right). Magnification ×250,000.

CHARACTERISTICS

Arenavirus virions consist of a pleomorphic lipid-containing envelope, with large peplomers, within which are two loosely helical, circular, nucleocapsid structures and a variable number of ribosome particles. Virions are 50 to 300 nm (mean 110 to 130 nm) in diameter. The genome consists of two molecules of "circular" (ends hydrogen-bonded), negative and ambisense ssRNA, size 10 to 14 kb (total genome). The viruses have 3 major polypeptides including a transcriptase. Peplomer glycoproteins are involved in type-specific serologic reactivity, and the nucleo-protein is responsible for cross-reactivities. RNA replication involves a primary transcription of mRNA from each segment of the genomic RNA via a virion transcriptase; later, with the protein products of this transcription, there is production of full-length, complementary RNA for each segment, each of which in turn is used as template for the synthesis of genomic RNA segments. Replication takes place in cyptoplasm, and assembly occurs via budding from plasma membrane. The human pathogens, Lassa, Machupo, and Junin viruses, are *"biosafety level 4"* pathogens and can be worked with in the laboratory only under maximum containment conditions.

HUMAN PATHOGENS

Arenavirus: lymphocytic choriomeningitis (LCM) virus, Lassa virus, Machupo virus (Bolivian hemorrhagic fever), Junin virus (Argentine hemorrhagic fever).

ANIMAL PATHOGENS

Arenavirus: LCM virus of mice and hamsters.

Retroviridae

Family: *Retroviridae* (the retroviruses) (Vogt, 1976; Matthews, 1982)
Subfamily: *Oncovirinae* (the RNA tumor viruses)
Genus: [unnamed] (type C oncoviruses)
Subgenus: [unnamed] (mammalian type C oncoviruses)
Subgenus: [unnamed] (avian type C oncoviruses)
Subgenus: [unnamed] (reptilian type C oncoviruses)
Genus: [unnamed] (type B oncoviruses)
[Genus]:   [unnamed] (type D oncoviruses)
Subfamily: *Spumavirinae* (the foamy viruses)
Genus: *Spumavirus* (foamy viruses)
Subfamily: *Lentivirinae* (the Maedi/visna-like viruses)
Genus: *Lentivirus* (Maedi/visna-like viruses)

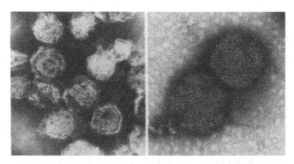

FIG. 19. *Retroviridae, Oncovirinae,* [unnamed genus, type B oncovirus] mouse mammary tumor (B-particle) (left); [unnamed genus, type C oncovirus] Rauscher murine leukemia (C-particle) (right). Magnification ×150,000 (left); Magnification ×120,000 (right).

CHARACTERISTICS

Retrovirus virions consist of a lipid-containing envelope, with peplomers, surrounding an icosahedral capsid which in turn contains a probably helical nucleocapsid. Virions are 80 to 130 nm in diameter. The genome consists of two identical molecules of positive-sense ssRNA, bound noncovalently; each RNA subunit is 3.5 to 9 kb in size, depending on the virus. The viruses have 7 major polypeptides, including a reverse transcriptase. Peplomer glycoproteins contain type-specific epitopes, and internal proteins express cross-reactive epitopes. RNA replication is unique: genomic RNA serves as template for the synthesis of viral DNA via the action of virion reverse transcriptase; this is followed by digestion of the RNA strand from the DNA/RNA hybrid, and synthesis of complementary DNA. This DNA is circularized, integrated into the host DNA, and then used for transcription, including transcription of full-length genomic RNA species. Virion assembly occurs via budding upon plasma membranes. Most of the viruses are oncogenic, causing leukemias, lymphomas, carcinomas, and sarcomas.

HUMAN PATHOGENS

Oncovirinae [type C oncovirus]: human T-lymphotropic virus 1 (HTLV-I) (adult T-cell leukemia), human T-lymphotropic virus 2 (HTLV-II) (associated with hairy-cell leukemia). *Lentivirinae, Lentivirus:* human immunodeficiency viruses 1 and 2 (HIV-1, HIV-2) (formerly called lymphadenopathy-associated virus—LAV, human T-lymphotropic virus 3—HTLV-III, or AIDS-related virus—ARV) (acquired immunodeficiency syndrome—AIDS).

ANIMAL PATHOGENS

Oncovirinae [type C oncovirus]: bovine leukemia (leukosis) virus, feline sarcoma and leukemia vi-

ruses, baboon, gibbon ape, and wooly monkey leukemia/sarcoma viruses, avian reticuloendotheliosis, sarcoma, and leukemia viruses, reptilian oncoviruses, mouse mammary tumor virus. *Lentivirinae, Lentivirus:* Maedi, progressive pneumonia and visna viruses of sheep, equine infectious anemia virus, caprine arthritis-encephalitis virus.

Picornaviridae

Family: *Picornaviridae* (the picornaviruses) (Cooper et al., 1978)
Genus: *Enterovirus* (enteroviruses)
Genus: *Cardiovirus* (EMC-like viruses)
Genus: *Rhinovirus* (rhinoviruses)
Genus: *Aphthovirus* (foot-and-mouth-disease viruses)

CHARACTERISTICS

Picornavirus virions are nonenveloped, have icosahedral symmetry, and are 30 nm in diameter. The genome consists of one molecule of positive-sense, infectious, ssRNA, size 7.2 to 8.4 kb. Virions have 4 major polypeptides. RNA replication involves the synthesis of a complementary RNA which serves as template for genome RNA synthesis. Genome RNA also serves as mRNA, being translated into a polyprotein that is cleaved into all the viral proteins, including those proteins that serve as enzymes for specific cleavages. Replication involves translation of a large precursor polyprotein and posttranslational cleavage into functional polypeptides. Replication and assembly take place in cytoplasm, and virus is released via cell destruction. The viruses have a narrow host range.

HUMAN PATHOGENS

Enterovirus: polioviruses 1, 2, 3, coxsackieviruses Al-22,[4] A24, coxsackieviruses Bl-6, human echoviruses 1–9, 11–27, 29–34, human enteroviruses 68–

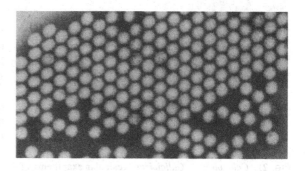

FIG. 20. *Picornaviridae, Enterovirus,* polio 1. Magnification ×200,000.

71, hepatitis A virus (Gust et al., 1983). *Rhinovirus:* human rhinoviruses 1–113. (*Note:* coxsackievirus A23 = echovirus 9; echovirus 10 = reovirus 1; echovirus 28 = human rhinovirus 1A; human enterovirus 72 = hepatitis A virus.)

ANIMAL PATHOGENS

Enterovirus: swine vesicular disease virus, murine poliovirus (Theiler's virus), porcine enteroviruses 1–11, bovine enteroviruses 1–7, simian enteroviruses 1–18. *Cardiovirus:* EMC virus. *Rhinovirus:* bovine rhinoviruses 1–2, equine rhinoviruses 1–2. *Aphthovirus:* foot-and-mouth-disease viruses.

Caliciviridae

Family: *Caliciviridae* (the caliciviruses) (Schaffer et al., 1980)
Genus: *Calicivirus* (caliciviruses)

CHARACTERISTICS

Calicivirus virions are nonenveloped, have icosahedral symmetry, and are 35 to 40 nm in diameter. Virions have 32 cup-shaped surface depressions in a T = 3 arrangement. The genome consists of one molecule of positive-sense ssRNA, size 8 kb. The viruses have one major polypeptide (and two minor polypeptides). RNA replication has not been studied in detail. Replication and assembly take place in cytoplasm, and virus is released via cell destruction. The viruses have narrow host ranges.

HUMAN PATHOGENS

Calicivirus: possibly Norwalk virus and similar viruses, which are proven causes of gastroenteritis, are caliciviruses. Other calicivirus-like agents have been identified as possible causes of gastroenteritis.

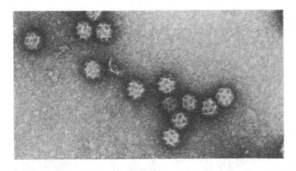

FIG. 21. *Caliciviridae, Calicivirus,* vesicular exanthema of swine. Magnification ×250,000.

ANIMAL PATHOGENS

Calicivirus: vesicular exanthema viruses 1 to 12 of swine, San Miguel sea lion viruses 1 to 8, feline caliciviruses, possibly caliciviruses of calves, swine.

Unclassified Viruses

The *spongiform encephalopathies*, which include Creutzfeldt-Jakob disease (CJD) and kuru of humans, scrapie of sheep and goats, and chronic wasting disease of mule deer and elk, are caused by transmissible infectious agents, the nature of which remains enigmatic. The diseases all have very long incubation (or preclinical) phases, followed by progressive disease marked by destruction of neurons, neurological dysfunction, leading inexorably to death. The etiological agents are highly resistant to inactivation by many physical and chemical agents, and they seem to be non-immunogenic. As their nature is unraveled, they may stretch the definition of the term *virus* as we use it today.

The agent of delta hepatitis, *delta virus*, is a defective satellite of hepatitis B virus, requiring the presence of hepatitis B virus for its replication and assembly. The virion is 35 to 37 nm in diameter and consists of an envelope made of HBsAg surrounding a structure containing delta antigen and a ssRNA genome, size 1.75kb. The virus has not yet been classified.

The agents of *non-A, non-B hepatitis* fall into at least two categories, one transmitted in a fecal-oral cycle and having properties somewhat like hepatitis A, and the second transmitted parenterally by blood transfusion, sexual contact, etc., and having some properties like hepatitis B. Despite repeated claims for the isolation and characterization of these viruses, their natures have not yet been resolved, and they have not been classified.

Norwalk and related viruses are the proven cause of diarrhea in humans, but have not been classified. Evidence has been presented, over several years, that the viruses are caliciviruses or parvoviruses, but because the viruses cannot be propagated in cell culture it has been difficult to obtain enough material from feces to characterize the nucleic acid or the structural proteins.

The name *Toroviridae* has been proposed for hitherto undescribed viruses associated with diarrhea in horses (Berne virus) and calves (Breda virus). The virions are enveloped and disk-shaped (35 × 170 nm) and contain a nucleocapsid of presumed helical symmetry. The genome is a single molecule of positive-sense ssRNA, size 20 kb.

The name *Astrovirus* has been accorded unofficially to viruses visualized in the electron microscope in fecal specimens from humans, calves, and lambs. The virions are spherical, 40 to 60 nm in diameter, and have a characteristic star-shaped outline. The genome is a single molecule of ssRNA about the same size as that of picornaviruses, but there are only two virion proteins.

Borna disease virus causes an encephalomyelitis of horses and sheep; the virus contains RNA, but is otherwise unclassified.

Literature Cited

American Type Culture Collection. 1986. Catalogue of animal and plant viruses. American Type Culture Collection, Washington, D.C.

Andrewes, C. H. 1954. Nomenclature of viruses. Nature (London) 173:260–261.

Andrewes, C. H., F. B. Bang, and F. M. Burnet. 1955. A short description of the Myxovirus group (influenza and related viruses). Virology 1:176–180.

Bishop, D. H. L., C. H. Calisher, J. Casals, M. P. Chumakov, S. Y. Gaidamovich, C. Hanoun, D. K. Lvov, I. D. Marshall, N. Oker Blom, R. F. Pettersson, J. S. Porterfield, P. K. Russell, R. E. Shope, and E. G. Westaway. 1980. *Bunyaviridae*. Intervirology 14:125–143.

Brenner, S., and R. W. Horne. 1959. A negative staining method for high resolution electron microscopy of viruses. Biochim. Biophys. Acta 34:103–110.

Brown, F. 1986. The classification and nomenclature of viruses: summary of results of meetings of the ICTV in Sendai, September 1984. Intervirology 25:141–143.

Brown, F. 1987. Classification and nomenclature of viruses (Fifth report of the ICTV). Intervirology, in press.

Brown, F., D. H. L. Bishop, J. Crick, R. I. B. Francki, J. J. Holland, R. Hull, K. M. Johnson, G. Martelli, F. A. Murphy, J. F. Obijeski, D. Peters, C. R. Pringle, M. E. Reichmann, L. G. Schneider, R. E. Shope, D. I. H. Simpson, D. F. Summers, and R. R. Wagner. 1979. *Rhabdoviridae*. Intervirology 12:1–17.

Cooper, P. D., V. I. Agol, H. L. Bachrach, F. Brown, Y. Ghendon, A. J. Gibbs, J. H. Gillespie, K. Lonberg-Holm, B. Mandel, J. L. Melnick, S. B. Mohanty, R. C. Povey, R. R. Rueckert, F. L. Schaffer, and D. A. J. Tyrrell. 1978. *Picornaviridae*: second report. Intervirology 10:165–180.

Dowdle, W. R., F. M. Davenport, H. Fukumi, G. C. Schild, B. Tumova, R. G. Webster, and X. Zakstelskaja. 1975. *Orthomyxoviridae*. Intervirology 5:245–251.

Evans, A. S. 1976. Criteria for causation: a unified concept appropriate for viruses as causative agents. Yale J. Biol. Med. 49:175–186.

Evans, A. S. 1982. Viral infections of humans. Plenum Publishing Corp., New York.

Fenner, F. 1974. The classification of viruses; why, when and how. Aust. J. Exp. Biol. Med. Sci. 52:223–231.

Fenner, F. 1976. Classification and nomenclature of viruses (second report of the ICTV). Intervirology 7:1–115.

Fenner, F., and F. M. Burnet. 1957. A short description of the Poxvirus group (vaccinia and related viruses). Virology 4:305–310.

Gust, I. D., C. J. Burrell, A. G. Coulepis, W. S. Robinson, and A. J. Zuckerman. 1986. Taxonomic classification of human hepatitis B virus. Intervirology 25:14–29.

Gust, I. D., A. G. Coulepis, S. M. Feinstone, S. A. Locarnini, Y. Moritsugu, R. Najera, and G. Siegl. 1983. Taxonomic classification of hepatitis A virus. Intervirology 20:1–7.

Karabatsos, N. (ed.) 1985. International catalogue of arboviruses, 3rd ed. American Society of Tropical Medicine and Hygiene, San Antonio, Texas.

Kiley, M. P., E. T. W. Bowen, G. A. Eddy, M. Isaacson, K. M. Johnson, F. A. Murphy, S. R. Pattyn, D. Peters, O. W. Prozesky, R. L. Regnery, D. I. H. Simpson, W. Slenzca, P. Sureau, G. vanderGroen, and P. Webb. 1982. *Filoviridae*: a taxonomic home for Marburg and Ebola viruses? Intervirology 18:24–32.

Kingsbury, D. W., M. A. Bratt, P. W. Choppin, R. P. Hanson, Y. Hosaka, V. terMeulen, E. Norrby, W. Plowright, R. Rott, and W. H. Wunner. 1978. *Paramyxoviridae*. Intervirology 10:137–152.

Lwoff, A., R. Horne, and P. Tournier. 1962. A system of viruses. Cold Spring Harbor Symp. Quant. Biol. 27:51–55.

Matthews, R. E. F. 1979. Classification and nomenclature of viruses (third report of the ICTV). Intervirology 12:132–296.

Matthews, R. E. F. 1982. Classification and nomenclature of viruses (fourth report of the ICTV). Intervirology 17:1–199.

Matthews, R. E. F., (ed.) 1983. A critical appraisal of viral taxonomy, p. 256. CRC Press Inc., Boca Raton, Florida.

McIntosh, K. 1985. Diagnostic virology, p. 309–322. In B. N. Fields (ed.), Virology. Raven Press, New York.

Melnick, J. L., A. C. Allison, J. S. Butel, W. Eckhart, B. E. Eddy, S. Kit, A. J. Levine, J. A. R. Miles, J. S. Pagano, L. Sachs, and V. Vonka. 1974. *Papovaviridae*. Intervirology 3:106–120.

Murphy, F. A. 1983. Current problems in vertebrate virus taxonomy, p. 37–61. In R. E. F. Matthews (ed.), A critical appraisal of viral taxonomy. CRC Press Inc., Boca Raton, Florida.

Murphy, F. A. 1985. Virus taxonomy, p. 7–25. In B. N. Fields (ed.), Virology. Raven Press, New York.

Pfau, C. J., G. H. Bergold, J. Casals, K. M. Johnson, F. A. Murphy, I. R. Pedersen, W. E. Rawls, W. P. Rowe, P. A. Webb, and M. C. Weissenbacher. 1974. *Arenaviridae*. Intervirology 4:207–218.

Porterfield, J. S., J. Casals, M. P. Chumakov, S. Y. Gaidamovich, C. Hanoun, I. H. Holmes, M. C. Horzinek, M. Mussgay, N. Oker Blom, P. K. Russell, and D. W. Trent. 1978. *Togaviridae*. Intervirology 9:129–148.

Roizman, B., L. E. Carmichael, G. Deinhardt, G. deThe, A. J. Nahmias, W. Plowright, F. Rapp, P. Sheldrick, M. Takahashi, and K. Wolf. 1982. *Herpesviridae*. Definition, provisional nomenclature and taxonomy. Intervirology 16:201–217.

Schaffer, F. L., H. L. Bachrach, F. Brown, J. H. Gillespie, J. N. Burroughs, S. H. Madin, C. R. Madeley, R. C. Povey, F. Scott, A. W. Smith, and M. J. Studdert. 1980. *Caliciviridae*. Intervirology 14:1–6.

Siddell, S., R. Anderson, D. Cavanaugh, K. Fukiwara, H. D. Klenk, M. R. Macnaughton, M. Pensaert, S. A. Stohlman, L. Sturman, B. A. M. van der Zeijst. 1983. *Coronaviridae*. Intervirology 20:181–190.

Siegl, G., R. C. Bates, K. I. Berns, B. J. Carter, D. C. Kelly, E. Kurstak, and P. Tattersall. 1985. Characteristics and taxonomy of *Parvoviridae*. Intervirology 23:61–73.

Vogt, P. K. 1976. The *Oncovirinae*—a definition of the group, p. 327–339. *In* Report No. 1 of WHO Collaborating Centre for Collection and Evaluation of Data on Comparative Virology, UNI-Druck, Munich.

Waterson, A. P., and L. Wilkinson. 1978. An introduction to the history of virology. Cambridge University Press, London.

Westaway, E. G., M. A. Brinton, S. Y. Gaidamovich, M. C. Horzinek, A. Igarashi, L. Kaariainen, D. K. Lvov, J. S. Porterfield, P. K. Russell, and D. W. Trent. 1985. *Flaviviridae*. Intervirology **24:**183–192.

White, D. O., and F. Fenner. 1986. Medical virology. Academic Press, New York.

Wigand, R., A. Bartha, R. S. Dreizin, H. Esche, H. S. Ginsberg, M. Green, J. C. Hierholzer, S. S. Kalter, J. B. McFerran, U. Pettersson, W. C. Russell, and G. Waddell. 1982. *Adenoviridae:* second report. Intervirology **18:**169–176.

Wildy, P. 1971. Classification and nomenclature of viruses (first report of the ICNV). Monogr. Virol. **5:**181.

CHAPTER 10

Poxviridae: The Poxviruses

FRANK FENNER AND JAMES H. NAKANO

Diseases: Human monkeypox, vaccinia, cowpox, parapoxvirus infections (orf, milker's nodules), tanapox, molluscum contagiosum.

Etiologic Agents: Monkeypox, vaccinia, cowpox, parapox, tanapox, and molluscum contagiosum viruses.

Sources: Monkeypox—African squirrels and primates; vaccinia—nosocomial; cowpox—cows, cats, ?rodents; orf—sheep and goats; milker's nodules—cattle; tanapox—African wildlife; molluscum contagiosum—man.

Clinical Manifestations: Monkeypox—generalized pustular rash, lymphadenopathy, fever; 11% case fatality rate. Vaccinia and cowpox—localized pustular skin lesions, slight fever. Orf, milker's nodules, and tanapox—localized nodular skin lesions. Molluscum contagiosum—multiple chronic skin nodules.

Pathology: Monkeypox, vaccinia, and cowpox—vesiculopustular skin lesions; parapoxvirus infections—proliferative skin lesions; tanapox—thickening of epidermis, with characteristic inclusion bodies; molluscum contagiosum—hyperplasia of epidermis with large cells containing "molluscum bodies."

Laboratory Diagnosis: All poxvirus diseases: electron microscopic demonstration of characteristic virions. Monkeypox, vaccinia, and cowpox viruses can be isolated on chorioallantoic membrane and produce characteristic lesions. Parapoxviruses have a distinctive morphology and can be isolated in tissue culture. Tanapox—electron microscopy, clinical signs, and epidemiology. Molluscum contagiosum—electron microscopy and biopsy of lesion.

Epidemiology: Human monkeypox—rare zoonosis, confined to central and western Africa; 70% of cases are infected from wildlife sources, 30% by person-to-person infection. Vaccinia—worldwide, infection by inoculation or, rarely, contact with human cases. Cowpox—rare zoonosis, confined to Europe, from lesions on cow or cat or ?rodent source. Parapoxvirus infections—worldwide, from contact with lesions on sheep or goats (orf) or cattle (milker's nodules). Tanapox—rare zoonosis, confined to central and eastern Africa, from wildlife sources, ?insect-borne. Molluscum contagiosum—worldwide, infection by contact infection with other humans, including sexual contact.

Treatment: Symptomatic.

Prevention and Control: Human monkeypox—vaccination with vaccinia virus is protective but is not used. Vaccinia, cowpox, parapoxvirus infections—avoidance of contact with infectious animals. Tanapox—avoidance of enzootic areas. Molluscum contagiosum—avoidance of physical contact with cases.

Introduction

Long before the nature of viruses was understood, several diseases—smallpox, cowpox, horsepox, sheeppox, and, mistakenly, chickenpox and the "great pox" (syphilis)—were grouped together as "pox diseases" because of the similar lesions they produced in the skin. With advances in light microscopy, virus particles of a similar size and appearance were seen in stained extracts of lesions of several of these diseases, and the viruses were grouped together as pox viruses (Aragao, 1927; Goodpasture, 1933). Electron microscopy confirmed this grouping, and when virus classification was embarked upon by a committee of the International Association of Microbiological Societies in the 1950s, Fenner and Bur-

net (1957) were commissioned to write a description of the group. Subsequently, in the First Report of the International Committee on Nomenclature of Viruses (Wildy, 1971), the "poxvirus group" was defined on the basis of the morphology and nucleic acid of the virions. Five "subgroups" were recognized, as were several unclassified poxviruses. Five years later (Fenner, 1976), the subgroup was given family status (*Poxviridae*), and the subgroups were categorized as genera. The Third Report (Matthews, 1979) saw the addition of one genus and the erection of two subfamilies: *Chordopoxvirinae* (poxviruses of vertebrates) and *Entomopoxvirinae* (poxviruses of insects). Currently, the larger of these subfamilies, *Chordopoxvirinae*, contains eight named families and some unclassified poxviruses (Table 1). Only two of the many viruses in these eight genera, variola virus and molluscum contagiosum virus, are associated with a specifically human disease (Table 2), and one of these diseases smallpox, is now extinct. Eight other poxviruses, belonging to three genera, can cause disease in man, usually as a zoonosis. The distribution of some of these diseases is worldwide; others are localized to particular parts of the world.

Four orthopoxviruses can cause human infection. Variola virus caused smallpox, and monkeypox virus, known to occur naturally only in western and central Africa, is an infection of wildlife that on rare occasions causes in man a generalized exanthematous disease clinically very similar to smallpox; it is occasionally transmitted from person to person. Vaccinia virus is a laboratory product that used to be inoculated in the skin to protect humans against

TABLE 1. Classification of poxviruses of vertebrates (family *Poxviridae;* subfamily *Chordopoxvirinae*)

Genus	Prototype virus
Orthopoxvirus[a]	Vaccinia
Parapoxvirus[a]	Pseudocowpox
Capripoxvirus	Sheeppox
Suipoxvirus	Swinepox
Leporipoxvirus	Myxoma
Avipoxvirus	Fowlpox
Yatapoxvirus[a]	Tanapox
Molluscipoxvirus[a]	Molluscum contagiosum

[a] Genus includes viruses that infect humans.

TABLE 2. Poxviruses that can cause disease in man

Genus and species	Disease	Geographic distribution	Hosts naturally infected
Orthopoxvirus			
Variola virus	Smallpox; now extinct	Formerly worldwide	Man
Monkeypox virus	Generalized disease: a rare zoonosis	Central and western Africa	Monkeys, squirrels
Vaccinia virus	Used for smallpox vaccination	Formerly worldwide	Laboratory product
Cowpox virus	Localized lesion: a rare zoonosis	Europe	Rodents, cats, cows
Parapoxvirus			
Orf virus	Localized lesion: a rare zoonosis	Worldwide	Sheep, goats
Pseudocowpox virus	Milker's nodules: a rare zoonosis	Worldwide	Dairy cows
Bovine papular stomatitis virus	Localized lesion: a rare zoonosis	Worldwide	Calves, beef cattle
Yatapoxvirus[a]			
Tanapox virus	Localized lesion: a rare zoonosis	Kenya, Zaire, USA[b]	Monkeys, ?rodents
Yabapox virus	Localized lesion: very rare accidental infections	Western Africa, USA[b]	Monkeys, baboons
Molluscipoxvirus[a]			
Molluscum contagiosum	Many skin lesions; specifically human disease	Worldwide	Man

[a] Proposed generic names.
[b] In the United States, in laboratory colonies of primates during 1960s.

smallpox, producing a localized lesion and, very rarely, a generalized rash or encephalitis. Cowpox virus is probably a virus of rodents which man may acquire from that source or via cows or cats, which are also probably incidental hosts for the virus. Among the parapoxviruses, orf virus causes skin lesions in sheep, goats, and wild ungulates that can be transmitted to man; milker's nodules in man are caused by contact with lesions on the teats of infected cows, and humans can occasionally be infected with bovine papular stomatitis virus from calves. Two serologically related African poxviruses that comprise the proposed genus *Yatapoxvirus,* tanapox virus and yaba poxvirus, can infect man. Tanapox is probably mechanically transmitted by insects and produces single or sometimes multiple skin lesions; yaba poxvirus infections of man are known only as a result of deliberate or accidental inoculation. Molluscum contagiosum is a specifically human infection, which is relatively common in children in some tropical countries and also occurs as a sexually transmitted disease. Molluscum contagiosum virus is the sole member of the proposed genus *Molluscipoxvirus.*

The Viruses

The properties of viruses of the 10 viruses listed in Table 2 differ according to the genus to which they belong, but many properties are shared by all poxviruses (Table 3).

Morphology and Physicochemical Properties

All poxviruses except members of the genus *Parapoxvirus* have a similar morphology, although there are differences in detail. Vaccinia virus, the prototype *Orthopoxvirus,* can be taken as the model.

MORPHOLOGY OF VIRION OF VACCINIA VIRUS

The virion is brick-shaped with rounded corners, measuring 235 to 280 × 165 to 225 nm. Unlike viruses of most families, the nucleocapsid has neither icosahedral nor helical symmetry. Figure 1, which is based on electron microscopic studies of vaccinia virus using thin sections, negative staining, and freeze-etching, represents the virion as consisting of four major elements: core, lateral bodies, outer membrane, and, as an inconstant component, an envelope. The well-defined central core contains the viral DNA, and on each side of the core there is an oval mass called the lateral body. The core and lateral bodies are enclosed within a well-defined outer membrane, which has a charateristic ribbed surface structure (see Fig. 2) and is composed of a large

TABLE 3. Properties of poxviruses that infect man

Most genera: brick-shaped virion, 235–280 × 165–225 nm, irregular arrangement of short surface tubules on outer membrane

Parapoxviruses: ovoid, 250–295 × 160–190 nm, with regular spiral arrangement of long "tubule" on outer membrane

Complex structure with core, lateral bodies, outer membrane, and sometimes envelope

Linear dsDNA, 140 kbp (parapoxvirus); 165–210 kbp (orthopoxvirus); 180 kbp (molluscivirus); 160 kbp (yatapoxvirus)

Transcriptase and several other enzymes in virion

Cytoplasmic replication

Common "subfamily" antigen; extensive serologic cross-reactions and cross-protection within but not between genera; some species-specific antigens

number of "surface tubules." The viral DNA within the core, which is associated with at least four different proteins, is maintained in a superhelical configuration and occurs in globular structures interconnected by DNA-protein fibers, resembling the nucleosome structures of eukaryotic chromatin (Soloski and Holowczak, 1981). Virions released spontaneously from cells are often enclosed within a lipoprotein envelope which contains the vaccinia hemagglutinin and several other virus-specified polypeptides (Payne, 1978; Payne and Norrby, 1976). Virions released by cellular disruption are infectious but lack an envelope (Appleyard et al., 1971); their outer surface is then composed of the outer membrane.

The envelope probably plays a role in the spread of virions within the animal body and thus in pathogenesis (Payne, 1980), and the low protective power of inactivated poxvirus vaccines is due in part to the fact that they consist of inactivated nonenveloped virions (Boulter and Appleyard, 1973), whereas live virus vaccines produce envelope proteins in the process of replication (Payne and Kristenson, 1979).

MORPHOLOGY OF POXVIRUSES OF OTHER GENERA (EXCEPT PARAPOXVIRUS)

The virions of other orthopoxviruses are indistinguishable from those of vaccinia virus, although the presence of an envelope has not been demonstrated in all other species—probably because extracellular released particles have not been examined. Virions of yatapoxviruses are comparable to those of vaccinia, but an envelope is almost always present in specimens obtained from human lesions, and the surface tubules appear to be somewhat more prominent than in virions of vaccinia virus (see Fig. 2E). The virions of molluscum contagiosum virus are similar

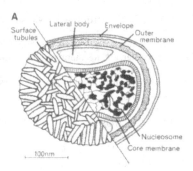
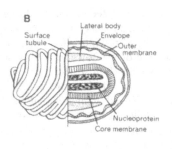

FIG. 1. Structure of the virions of vaccinia and orf viruses. Right-hand side, section of enveloped virion; left-hand side, surface structure of nonenveloped particle. (A) Vaccinia virus, an orthopoxvirus. The viral DNA and several proteins within the core are organized as a "nucleosome." The core has a 9-nm-thick membrane, with a regular subunit structure. Within the virion, the core assumes the shape of a dumbbell because of the large lateral bodies, which are, in turn, enclosed in a protein shell about 12 nm thick—the outer membrane, the surface of which consists of irregularly arranged surface tubules, which in turn consist of small globular subunits. Virions released through the cytoplasmic membrane are enclosed within an envelope which contains host cell lipids and several virus-specific polypeptides, including the hemagglutinin; they are infectious. Most virions remain cell-associated and are released by cellular disruption. These particles lack an envelope, so the outer membrane constitutes their surface; like the enveloped particles, they are also infectious. (B) Orf virus, a parapoxvirus. The outer membrane consists primarily of a single long tubule that appears to be wound around the particle. In negatively stained electron microscopes (see Fig. 2C), both sides are visible, giving a characteristic crisscross appearance. In section, the lateral bodies are less apparent than in vaccinia virions, and the detailed structure of the core membrane and contents has not been elucidated. The envelope is usually closely applied to the surface of the outer membrane.

to those of orthopoxviruses, but they also have somewhat more prominent surface tubules (see Fig. 2G).

MORPHOLOGY OF THE PARAPOXVIRUS VIRION

The virions of parapoxviruses are distinctly different from those of other chordopoxviruses. They are ovoid, measuring 250 to 295 × 160 to 190 nm, with an axial ratio of 1.6, compared with 1.3 for vaccinia virus. The surface structure is distinctive, consisting of what appears to be a single, long, spirally wound tubule (Fig. 1B). Since both sides of this tubule are seen in negatively stained electron micrographs (Fig. 2C), the virions then have a criss-cross appearance (Nagington et al., 1964). Many particles, especially in clinical specimens, appear to have a loose envelope (Nagington and Horne, 1962; Rosenbusch and Reed, 1983). The lateral bodies are less well developed than those seen in vaccinia virus (Peters et al., 1964).

GENOME OF ORTHOPOXVIRUSES

The basic structure of the genome of all poxviruses is the same, but there are differences among genera. The genome of orthopoxviruses has been studied in greatest detail. It consists of a single linear molecule of double-stranded DNA, M_r for different species, 110 million to 140 million, comprising between 165 and 210 kbp. Orthopoxvirus DNAs contain no unusual bases, and the guanine + cytosine ratio is very low—about 36%.

The DNA of vaccinia virus behaves in an anomalous way when it is denatured. Instead of separating, the two sister strands form a large, single-stranded circular molecule, being attached at or near each end of the genome by covalent links (Geshelin and Berns, 1974). For the most part, the DNA sequences in the vaccinia genome are unique, but the two terminal fragments cross-hybridize with each other (Wittek et al., 1977) and with the termini of other species of orthopoxvirus (Mackett and Archard, 1979). This inverted terminal repetition is about 10 kbp long in the strains of vaccinia virus used by Wittek, but its length varies considerably in other orthopoxviruses. For example, in a series of mutants of cowpox virus studied by Archard et al. (1984), the length of the terminal repetition varied from 4.5 to 41 kbp, almost 20% of the entire genome. Within each terminal repeat of vaccinia virus, there are 30 reiterations of a 70-bp sequence arranged in tandem and grouped into two discrete groups of 17 and 13 units (Wittek and Moss, 1980). Garon et al. (1978) visualized the terminal repetition by electron microscopy, which showed that the opposite ends of each strand are complementary to each other. The continuity of the DNA chain around the single-strand hairpin loop at the end of the molecule was shown in a variety of ways, culminating in the determination of the base sequence by Baroudy et al. (1982).

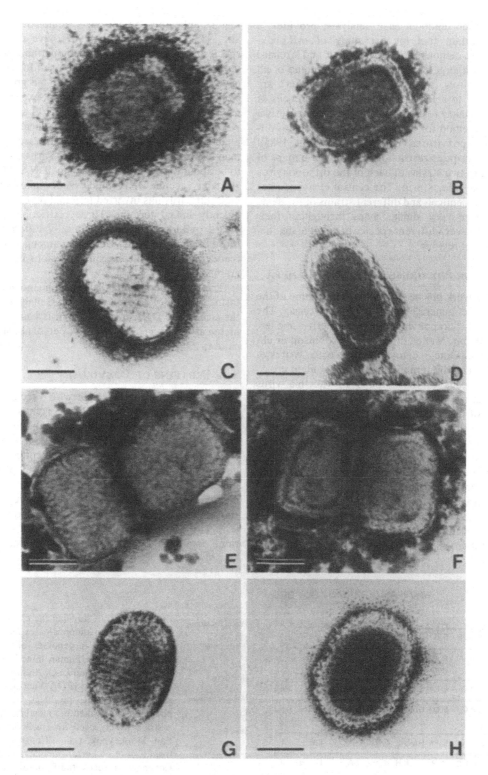

FIG. 2. Virions of poxviruses, as seen in negatively stained preparations of material obtained from lesions in humans. Bars = 100 nm. Left panel: M forms, illustrating the structure of the outer membrane. Right panel: C forms, pene-trated by phosphotungstate. (A,B) Variola virus; all other orthopoxviruses are similar. (C,D) Orf virus; all other para-poxviruses are similar. (E,F) Tanapox virus; yabapox virus is similar. (G,H) Molluscum contagiosum virus.

Restriction endonuclease analysis provides a most important tool for the study of orthopoxviruses. The demonstration by Mackett and Archard (1979) that a large central part of the genome of all orthopoxviruses is very similar but that there are species and to a lesser extent strain differences at each end, makes restriction endonuclease analysis a powerful method for taxonomic comparisons of different orthopoxviruses (Esposito and Knight, 1985). Figure 3 illustrates restriction endonuclease maps of the DNA of two strains of each of the orthopoxvirus species that infects man. The central cleavage sites are almost identical, and different strains of the same species have very similar maps throughout their length; however, different species have quite distinctive terminal regions.

GENOME OF POXVIRUSES OF OTHER GENERA

Much less work has been done on the genome of the poxviruses of genera other than *Orthopoxvirus*. The genome of parapoxviruses is distinctly smaller, about 140 kbp (Menna et al., 1979; Robinson et al., 1982), the guanine + cytosine ratio is 63% (Wittek et al., 1979), and the terminal fragments contain crosslinks analogous to those found in vaccinia virus (Menna et al., 1979). The restriction endonuclease patterns of different parapoxvirus isolates, particularly when using *Eco*RI and *Hin*dIII, appear to be more heterogeneous than those of orthopoxviruses (Raffii and Burger, 1985), even within species—for example, bovine papular stomatitis virus (Wittek et al., 1980) and orf virus (Robinson et al., 1982). However, DNA fragments from internal parts of the genome show cross-hybridization between species (orf, pseudocowpox, bovine papular stomatitis viruses), but cross-hybridization between the terminal

regions appears to be species-specific (Gassmann et al., 1985). Robinson et al. (1987) have shown that species-specific patterns are demonstrable with orf virus when *Bam*HI and *Kpn*I are used for digestion of the DNA (Fig. 4). The New Zealand group have cloned the entire genome of orf virus, except for the termini, into various vectors (Mercer et al., 1987).

Molluscum contagiosum virus replicates incompletely in cell culture, but DNA extracted from human lesion material was shown by Parr et al. (1977) to be about the same size as vaccinia DNA (M_r 118 × 10^6); like vaccinia DNA, the DNA of molluscum contagiosum virus has inverted terminal repeats and covalently linked termini. Three restriction endonuclease patterns, which were totally different from those of vaccinia virus, were seen among 11 isolates recovered from clinically similar lesions in one geographic area.

The DNA of yabapoxvirus has a guanosine + cytosine content of 32.5% (Yohn and Gallagher, 1969); the size of the DNA of tanapox virus has been estimated as 160 kbp (J. J. Esposito, personal communication, 1986).

THE PROTEINS OF POXVIRUSES

Most studies have been made with vaccinia virus. Two-dimensional gel electrophoresis of disrupted virions revealed over 110 distinct polypeptides (Essani and Dales, 1979); virus-infected cells contain another 50 nonstructural, virus-specific polypeptides. Many virus-specific enzymes are found in orthopoxvirus-infected cells (Dales and Pogo, 1981), including thymidine kinase, the DNA sequence of which is very similar in vaccinia, variola, and monkeypox viruses (Esposito and Knight, 1984), and several in the virion, notably a DNA-dependent RNA polymerase

Hind III A'AGCTT *Sma* I CCC'GGG

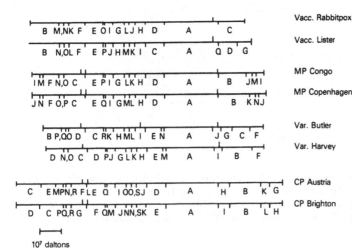

FIG. 3. Diagram showing the cleavage sites of the restriction endonucleases *Hin*dIII and *Sma*I in the DNA genomes of orthopoxviruses that cause human infections. (Data from M. Mackett and L. C. Archard, J. Gen. Virol. 45:683–701, 1979.) Vaccinia strains: rabbitpox Utrecht, produces hemorrhagic pocks on chorioallantoic membrane; Lister, a strain widely used for smallpox vaccination. Monkeypox strains: Congo, recovered from a case of human monkeypox in 1970; Copenhagen, recovered from infected cynomolgus monkey in Copenhagen in 1958. Variola strains: Butler, reference strain of variola minor virus; Harvey, reference strain of variola major virus. Cowpox strains: Austria; Brighton, reference strain of cowpox virus.

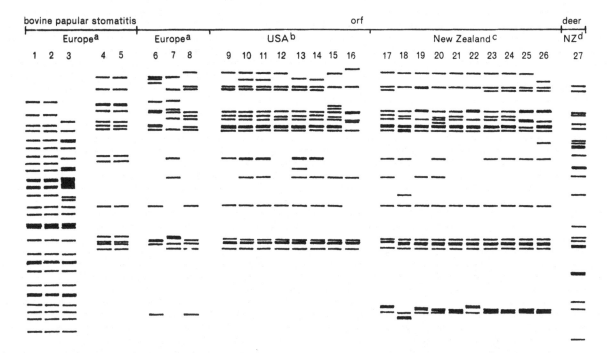

FIG. 4. Diagram illustrating electropherograms of DNA of a number of strains of parapoxvirus after digestion with *Kpn*I. Although there are greater differences between strains than found with orthopoxviruses, orf viruses from New Zealand (17–26) show a general resemblance to orf viruses from Europe (6–8) and the United States (9–16), and they are clearly different from strains of bovine papular stomatitis virus (1–5) and a newly discovered parapoxvirus of deer (27). Sources of data: (a) Wittek et al., 1980; (b) Rafii and Burger, 1985; (c) Robinson et al., 1987; (d) Horner et al., 1987; A. J. Robinson, unpublished data, 1986. (Courtesy, Dr. A. J. Robinson.)

(transcriptase). Proteins that are important in serological and immunological reactions are discussed below.

Replication

Replication of poxviruses occurs in the cytoplasm and can be demonstrated in enucleate cells. However, a function associated with the nucleus appears to be required for efficient replication (Silver et al., 1979). This may be a component of cellular RNA polymerase II, which is associated with the viral RNA polymerase during replication (Morrison and Moyer, 1986).

After fusion of the virion with the plasma membrane or entry via endocytosis, the viral core is released into the cytoplasm (Dale and Siminovitch, 1961). There are at least three cycles of transcription (Fig. 5), the transcripts being translated directly into proteins (i.e., without cleavage or splicing), some of which undergo posttranslational cleavage to yield the functional molecules. Immediate early transcription is initiated by the viral transcriptase within the core. Functional capped and polyadenylated mRNAs, produced within minutes after infection, are translated into enzymes that release the viral DNA from the core. Transcription of about 100 genes, distributed throughout the genome, occurs before viral DNA synthesis begins.

Early proteins include thymidine kinase, DNA polymerase, and several other enzymes. With the onset of DNA replication 1.5 to 6 h after infection, there is a dramatic shift in gene expression, and almost the entire genome is transcribed, but transcripts from the early genes (i.e., those transcribed before DNA replication begins) are not translated.

Virion formation occurs in circumscribed areas of the cytoplasm ("viral factories"). Spherical immature particles can be visualized by electron microscopy (Fig. 6); their outer bilayer becomes the outer membrane of the virion, and the core and lateral bodies differentiate within it. Some of these mature particles move to the vicinity of the Golgi complex, acquire an envelope, and are released from the cell. However, most particles are not enveloped and are released by cell disruption.

Antigenic Composition

The large and complex virions of poxviruses contain many antigens. Some antigens show cross-reactivity across the whole subfamily *Chordopoxvirinae*; many

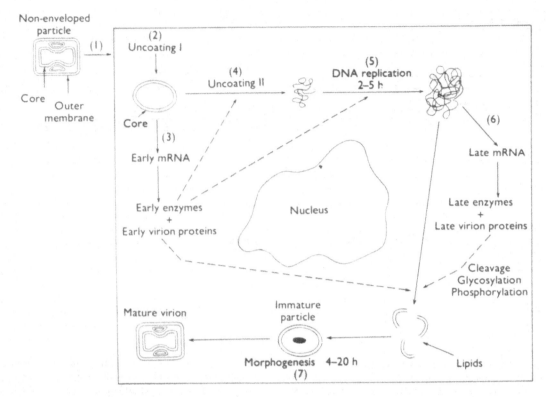

Fig. 5. Replication cycle of vaccinia virus. Both enveloped and nonenveloped particles are infectious, but they differ in their attachment to cells and mode of entry (not shown). The sequence of events is (1) attachment and entry; (2) uncoating I, whereby the outer membrane is removed by cellular enzymes, leaving the core; (3) immediate early transcription from DNA within the core by the viral transcriptase, leading to the production of early enzymes which include the enzyme that produces (4) uncoating II and release of the viral DNA into the cytoplasm. Early transcription continues, and (5) DNA replication simultaneously occurs, after which (6) late transcription occurs from the newly synthesized DNA, followed by translation and cleavage, glycosylation, and phosphorylation of some of the late proteins. Morphogenesis (7) goes through stages of cupules, immature particles, and mature virion (see Fig. 6). Envelopment (see text) not shown. (From B. Moss, in Virology; B. N. Fields (ed.), p. 690; New York: Raven Press, 1985; with permission.)

antigens, including those important in generating a protective immune response, show cross-reactivity within each of the genera *Orthopoxvirus*, *Parapoxvirus*, and *Yatapoxvirus* but not between genera. Some antigens of orthopoxviruses are species-specific; those of other genera have not yet been adequately studied in this respect.

ANTIGENS COMMON TO THE SUBFAMILY *Chordopoxvirinae*

Investigations carried out some years ago and using crude chemical and serological methods showed that one or several antigens were shared by all members of the subfamily *Chordopoxvirinae* (Takahashi et al., 1959; Woodroofe and Fenner, 1962). More recently, Ikuta et al. (1979) demonstrated by radioimmunoprecipitation that among the 30 antigenic polypeptides found in one-dimensional gels prepared from cells infected with vaccinia, cowpox, and Shope fibroma viruses (a *Leporipoxvirus*), there were four that showed cross-reactivity between the orthopoxviruses and Shope fibroma virus. There were, as expected, additional cross-reactive polypeptides shared by vaccinia and cowpox viruses but not found in fibroma virus.

SEROLOGIC CROSS-REACTIVITY BETWEEN VIRUSES OF THE SAME GENUS

Cross-protection in experimental animals and cross-neutralization in tests with sera from recovered animals are the classical ways of allocating poxviruses to their genera. Thus vaccination with vaccinia virus protects experimental animals against infection with any other orthopoxvirus, and vaccinia-immune sera neutralize the infectivity of all other species of orthopoxvirus almost as well as they neutralize vaccinia virus infectivity.

The antigens that elicit the production of neutralizing antibodies to orthopoxviruses fall into two classes: those located on the surface tubular ele-

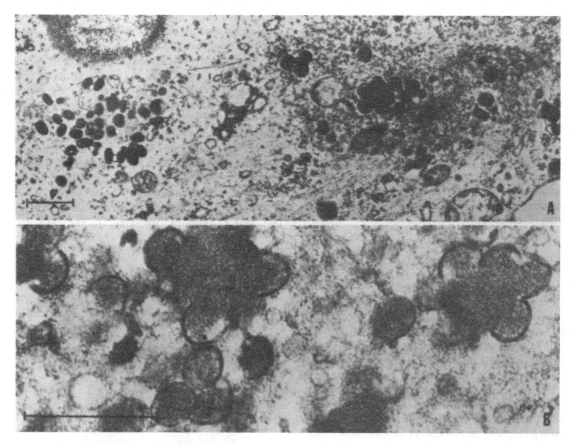

FIG. 6. Morphogenesis of vaccinia virus. (A) Infected cell showing viral "factory" area with immature particles on right; mature naked intracellular virions on left. Bar = 1,000 nm. (B) "Factory" area showing the "caps" (cupules) of developing immature viral particles. Bar = 100 nm. ([A] from L. G. Payne and K. Kristensson, J. Virol. 32:614–622, 1979; [B] from S. Dales and L. Siminovich, J. Biophys. Biochem. Cytol. 10:475–503, 1961; by courtesy of Dr. S. Dales and Dr. L. G. Payne.)

ments of the outer membrane of the virion, and some of the virus-specific antigens in the viral envelope. A 58K protein polymerizes to form the surface tubules that are a prominent feature of the outer membrane of vaccinia virions. Purified surface tubular elements elicit neutralizing antibody to nonenveloped but not to enveloped virions and block the neutralizing capacity of antibody to nonenveloped virions (Stern and Dales, 1976; Payne, 1980). Other protective antigens are located in the viral envelope, which is found only in virions that are released naturally from cells and contains 10 virus-specified polypeptides, nine of which are glycosylated (Payne, 1978). Antibody to the isolated envelopes neutralizes the infectivity of enveloped forms of vaccinia virus, as demonstrated by the "anticomet" test of Appleyard et al. (1971), whereas antibody to inactivated nonenveloped virions does not (Payne, 1980).

Members of the same genus show a high level of cross-reactivity in all other serologic tests, but some differences can be found by agar gel precipitation and other tests that involve the detection of several different antigens. Among the poxviruses, only orthopoxviruses agglutinate red cells; hemagglutination-inhibition is a convenient serologic test for detecting past infection with orthopoxviruses but not for differentiating between species. Some serologic tests, such as the complement-fixation test, remain positive for only a few months after infection; this test has been used to detect recent subclinical smallpox in vaccinated subjects (Heiner et al. 1971).

SPECIES-SPECIFIC SEROLOGIC TESTS

There are so many ways of distinguishing between species of *Orthopoxvirus* by biological tests and genome analysis (see below) that serologic tests are not used for this purpose. However, for some investigations, such as studies of the ecology of monkeypox virus (for review see Fenner et al., 1988), it was important to be able to determine what species of *Orthopoxvirus* was involved in the infection of animals whose serum contained orthopoxvirus-specific antibodies. Species-specific diagnosis of recent infection with monkeypox, vaccinia, and variola viruses can be made with hyperimmune or other highly potent sera by absorption with appropriate viral suspensions and tests for residual antibody by gel precipitation (Gispen and Brand-Saathof, 1974), immunofluorescence (Gispen et al., 1974), radioimmunoassay (Hutchinson et al., 1977), or ELISA (Marennikova et al., 1981). The radioimmunoassay adsorption test, using anti-gamma-globulin or staphylococcus A protein, has proved useful in serological tests on sera of a variety of wild animals, including camels, primates, squirrels, and some rodents (see below).

Genetics

Genetic studies have been carried out with members of the genus *Orthopoxvirus* but not with other poxviruses. Within this genus, there is a considerable amount of data on white pock mutants of rabbitpox virus (a strain of vaccinia virus), cowpox virus, and monkeypox virus. Most such mutants are caused by deletions of variable and rather large segments of DNA, sometimes associated with transpositions (Archard et al., 1984; Dumbell and Archard, 1980; Moyer and Rothe 1980). White pock mutants that involve deletions from different termini show recombination in doubly infected cells (Fenner and Sambrook, 1966; Lake and Cooper, 1980). Recombination also occurs between different species of *Orthopoxvirus*, producing "hybrid" viruses (Woodroofe and Fenner, 1960; Bedson and Dumbell, 1964a,b).

Reactions to Physical and Chemical Reagents

The most important feature of poxviruses in this respect is that they are relatively resistant to environmental conditions. Thus, infectivity may persist in dry scabs from cases of smallpox for months or even years (Wolff and Croon, 1968), and infectivity with parapoxviruses may persist for months in contaminated fields or on contaminated skins.

Pathogenesis and Pathology

Pathogenesis of Infections with a Generalized Rash

Before its eradication, smallpox was by far the most important and severe generalized poxvirus disease of man, but now only monkeypox virus and, very rarely, vaccinia and orf viruses produce human dis-

ease with a generalized skin rash, although molluscum contagiosum and occasionally tanapox may be associated with multiple skin lesions.

The mode of infection in monkeypox is unknown, but it may be via minute abrasions on the skin or, perhaps, in person-to-person infections (Jezek et al., 1986a) by the respiratory route. During the incubation period of 12 to 14 days, the virus probably replicates within cells of the reticuloendothelial system, notably in lymph nodes, which may be enlarged and tender several days before the rash appears. There is then a viremia, probably cell-associated, and the epidermal cells in various parts of the body are infected, to produce a rash, which is most extensive on the face and extremities (see Fig. 7). The rash goes through stages of macule, papule, vesicle, pustule, and scab over a period of about 2 weeks. The extent

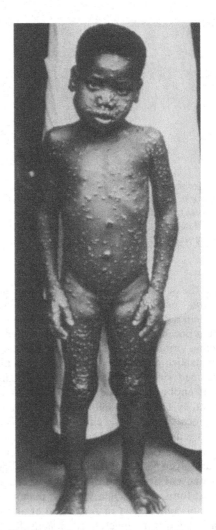

FIG. 7. Human monkeypox. Lesions in a 7-year-old girl from Zaire, on day 8 after the appearance of the rash, which is indistinguishable from that of smallpox. Note enlargement of lymph nodes, not seen in smallpox. (Courtesy of Dr. Z. Jezek.)

of general signs (fever, prostration, toxemia) is roughly proportional to the extent of the rash.

Based on studies with animal models, notably mousepox, the cell-mediated immune response plays a major role in recovery from generalized poxvirus infections (Blanden, 1971; Cole and Blanden, 1982).

Pathogenesis of Localized Infections

Cowpox and vaccinia viruses ordinarily produce infections that remain localized to the site of entry of virus through the skin; either deliberately as an inoculation or accidentally via an abrasion. All poxviruses of genera other than *Orthopoxvirus* that infect man (see Table 2) produce localized skin lesions. The parapoxvirus infections are usually occupational diseases associated with handling infected sheep, cows, calves, or wild ungulates; tanapox virus is probably mechanically transmitted by arthropods, and molluscum contagiosum is spread by contact, sometimes sexual and sometimes via swimming or playing. In infections with vaccinia and cowpox viruses, the local lymph node usually becomes tender and swollen a few days after infection occurs, and there may be a transient viremia. Lesions elsewhere in the body usually occur only in persons with eczema or cell-mediated immunedeficiency, although generalized vaccinia, a self-limited disease, occurs as a very rare complication in persons without immunological deficiencies.

The duration of immunity to reinfection by poxviruses that remain localized appears to be shorter than in generalized infections, but there is usually an accelerated response for years after infection. Reinfection with parapoxviruses is relatively common among occupationally exposed subjects.

Pathology of Skin Lesions Produced by Orthopoxviruses

Only the orthopoxviruses produce pustular lesions; their pathology is much the same in local infections and in generalized rashes. The evolution of the lesions in cases of smallpox has been described by Bras (1952a). In generalized infections, the earliest change is dilation of the capillaries in the papillae of the corium with swelling of the lining endothelium and infiltration of lymphocytes and histiocytes. Changes then appear in the overlying epidermis. In localized infections, the virus is delivered directly into cells of the Malpighian layer of the epidermis. A few cells in the middle layer of the epidermis become enlarged and vacuolated, and the nuclei become condensed. Cells in the Malpighian layer proliferate and swell, leading to thickening of the epithelium. Intercellular edema develops, the cell membranes rupture, and a loculated vesicle develops, with septa

formed by the remains of incompletely destroyed cells. In the fully developed vesicle, the roof is formed of compressed cells of the stratum spinosum, the keratohyaline layer, and the horny layer, and the base is formed by degenerate cells of the Malpighian layer. Soon after formation of the vesicle, polymorphonuclear leukocytes enter from the dermis to produce a pustule. Eventually, fluid is absorbed from the pustule, the contents dry up, and epithelial cells from the sides of the cavity grow under the residual mass of the exudate, leaving a crust of degenerated epithelial cells, leukocytes, and debris. After healing, the sites of lesions are hypo- or hyperpigmented, and there may be permanent scars (pockmarks). Pockmarks are most severe on the face, because of destruction of the numerous sebaceous glands found there followed by organization and subsequent shrinking of granulation tissue (Bras, 1952b).

In early lesions, cytoplasmic inclusions are readily found in sections stained by hematoxylin and eosin as round or oval, faintly basophilic or acidophilic bodies lying in the cytoplasm, usually close to the nucleus (B-type inclusions; Guarnieri bodies). Cowpox virus–infected cells contain two types of cytoplasmic inclusion bodies—the rather faint, irregular, B-type inclusions found in all orthopoxvirus infections, and numerous large, homogeneous, acidophlic, A-type inclusion bodies (Kato et al., 1959). Depending on genetic factors, mature virions may or may not be occluded within the A-type inclusion bodies (Shida et al., 1977), which are composed of a single species of high-molecular-weight protein (Patel et al., 1986).

Pathology of Skin Lesions Produced by Parapoxviruses

The skin lesions of orf and milker's nodules are markedly proliferative (Johannessen et al., 1975; Kluge et al., 1972). Changes that occur early in the proliferating keratinocytes include nucleolar enlargement and focal lysis of keratin fibrils. Extreme swelling of the cells results in ballooning degeneration which, when accompanied by pale eosinophilic B-type cytoplasmic inclusions and nuclear shrinkage, is pathognomonic of orf. Dermal infiltration with monocytes and lymphoid cells is prominent around hyperemic capillaries and venules, and infection of the endothelial cells may produce endothelial proliferation.

Pathology of Skin Lesions Caused by Yatapoxviruses

In human tanapox lesions, there is a marked thickening of the epidermis with extensive ballooning degeneration of the prickle cell layer. The swollen epidermal cells contain large, pleomorphic, granular eosinophilic, cytoplasmic B-type inclusion bodies. The nuclei are swollen, and there is peripheral concentration of the chromatin and large nucleoli.

After subcutaneous or intradermal inoculation of monkeys or man, yaba monkey tumor poxvirus produces tumors that are composed of masses of histiocytes, which are later infiltrated with lymphocytes and polymorphonuclear cells. No true neoplastic proliferation occurs, and the lesions regress as the immune response develops (Grace and Mirand, 1963).

Pathology of Molluscum Contagiosum Lesions

The typical molluscum contagiosum lesion consists of a localized region of hypertrophied and hyperplastic epidermis extending down into the underlying dermis without breaking the basement membrane and projecting above the adjacent skin as a visible tumor (see Fig. 8). There is an increase in mitotic figures in the germinal layer, above which pathological changes in the nuclei and cytoplasm are seen. Near the surface, the lesions become more and more pronounced. Eventually, each cell is many times larger than normal, and the cytoplasm is filled with a large hyaline acidophilic granular mass known as the molluscum body, which pushes the nucleus to the edge of the cell. The core of the lesion consists of degenerating epidermal cells, which contain B-type cytoplasmic inclusion bodies, and keratin, which uninfected cells continue to produce. The fully developed lesion is loculated, and there is very little inflammatory reaction in the corium unless secondary bacterial infection has occurred.

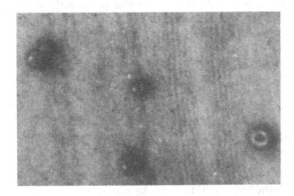

FIG. 8. Lesions of molluscum contagiosum in human skin. (Courtesy of Dr. D. Lowy, Institute for Dermatologic Communication and Education and the American Academy of Dermatology.)

Clinical Features

Each poxvirus infection of man has distinctive features, except for infections with the parapoxviruses, which are alike except that their location depends on the particular occupational exposure.

Clinical Features of Human Monkeypox

The clinical features of human monkeypox are indistinguishable from those of what was called ordinary-type smallpox (Fenner et al., 1988), except that enlargement of the lymph nodes is prominent and occurs early (Arita et al., 1985). There is a prodromal illness lasting 1 to 3 days with fever, prostration, and usually lymph node enlargement, which may occur only in the neck or inguinal lesion but may be generalized (Fig. 7). Skin lesions then develop more or less simultaneously all over the body and evolve together at the same rate through papules, vesicles, and pustules before umbilicating, drying, and desquamating. This process usually takes about 2 to 3 weeks, depending on the severity of the disease. The distribution of the rash is mainly peripheral. Severe eruptions can cover the entire body, including the palms and soles. Lesions have been noted on the mucous membranes, tongue, and genitalia. The case fatality rate is about 11%. Following recovery, pockmarks may develop, usually on the face, and occasionally corneal lesions produce unilateral blindness.

Clinical Features of Vaccinia

Although smallpox vaccination of civilians has been discontinued, service personnel in several countries are still vaccinated, and there are proposals to reintroduce vaccination for the delivery of genes for the antigens of a variety of other infectious agents (Quinnan, 1985). It is therefore appropriate to describe the clinical features and complications of vaccination as seen with strains used prior to the eradication of smallpox, with the proviso that it is likely that strains used as vectors for foreign genes will be genetically engineered to reduce their virulence and propensity to produce complications.

PRIMARY VACCINATION

Ordinarily, vaccinia virus is deliberately introduced into the skin; occasionally cases occur in contacts, especially in those with eczematous lesions. In primary vaccination, a papule appears at the vaccination site 4 or 5 days after vaccination; 2 or 3 days later, this becomes vesicular and constitutes the umbilicated and multilocular "Jennerian vesicle." The contents rapidly become turbid because of the infiltration of inflammatory cells, and the central lesion is surrounded by erythema and induration, which reach their maximum diameter on the ·9th or 10th day. At this time, the draining lymph nodes in the axilla are enlarged and tender, and many patients have a mild fever. The pustule dries from the center outward, and the brown scab falls off after about 3 weeks, leaving a scar by which vaccination can be recognized for many years.

REVACCINATION

Since primary vaccination induces cell-mediated as well as humoral immunity, revaccination is associated with an accelerated skin response to viral proteins. Such an accelerated response can be produced by noninfectious vaccine; the development of vesicles (indicating that viral replication has occurred) is necessary if revaccination is to enhance immunity.

COMPLICATIONS OF VACCINATION

Very rarely, primary vaccination may be followed by complications. The most serious is progressive vaccinia (vaccinia necrosum), which occurs only in persons with immunological deficiencies, usually of the cell-mediated immune system (Fulginiti et al., 1968). Eczema vaccinatum may occur in eczematous children who are vaccinated or are infected by contact with vaccinees. Neither progressive vaccinia nor eczema vaccinatum should occur in developed countries, since vaccination is absolutely contraindicated in cases of immunological deficiencies and should be carried out in cases of eczema only if absolutely essential, and then under cover of vaccinia-immune globulin.

When smallpox vaccination was conducted on a large scale, postvaccinial encephalitis was a serious and unpredictable complication. During the 1930s to 1960s, it occurred with widely differing frequencies in different countries, ranging from 8 in 650,000 primary vaccinations in the United States in 1968 (Lane et al., 1970) to 1 in 8,000 (80 in 640,000) in some European countries in the 1930s (Wilson, 1967). Its pathogenesis and pathology resemble those of other types of postinfection encephalitis (Johnson, 1982), perivenular inflammation and demyelination being the principal lesions. The case fatality rate was 25 to 40%.

Clinical Features of Cowpox

In cases acquired by contact with infected cows, one or more lesions appear on the hands (Fig. 9D). The thumbs, the first interdigital cleft, and the forefinger are especially liable to attack. Scratches or abrasions of the skin may determine localization of the lesions elsewhere on the hands, forearms, or face. The lesions resemble those of primary vaccination, passing through stages of vesicle and pustule before a scab forms. Local edema may be more pronounced than in vaccination, and there are lymphangitis, lymphadenitis, and often fever for a few days. Baxby (1977) noted that in children, cowpox was sometimes rather more severe than vaccinia. There are sometimes multiple primary lesions, but a generalized rash has not been observed, probably because cases of human infection with cowpox virus are rare.

Clinical Features of Parapoxvirus Infections

Parapoxviruses cause occupational diseases of man acquired by contact with infected sheep or other ungulates (orf), cows (milker's nodules), or calves and beef cattle with bovine papular stomatitis. Infection occurs through abrasions of the skin, and localized lesions are found on the hands or sometimes the face. The lesions of orf are rather large nodules that may be multiple, the surrounding skin being inflamed (Johannessen et al., 1975) (Fig. 9F). There may be a low fever and swelling of the draining lymph nodes; the local lesion is rather painful. Milker's nodules are hemispheric, cherry-red papules, which appear 5 to 7 days after exposure and gradually enlarge into firm, elastic, purple, smooth, hemispheric nodules varying in size up to 2 cm in diameter, and when fully developed may be umbilicated (Fig. 9E). They are relatively painless, but they may itch. They are highly vascular, which explains the purple color, but they do not ulcerate. The granulation tissue that makes up the mass of the nodule gradually becomes absorbed, and the lesions disappear after 4 to 6 weeks. The only evidence of generalization is occasional slight swelling of the draining lymph nodes.

Clinical Features of Yatapoxvirus Infections

Tanapox is a relatively common human disease in certain parts of Africa and has been observed among persons who handled laboratory primates in the United States. Human infection with yabapoxvirus has only been observed after deliberate or accidental inoculation.

CLINICAL FEATURES OF HUMAN TANAPOX

The appearance of the characteristic skin lesion has been described by Jezek et al. (1985). Initially there is a small nodule, without any central abrasion such as is often seen with an insect bite, which soon becomes papular and gradually enlarges to reach a maximum diameter of ~15 mm by the end of the second week but never becomes pustular (Fig. 10).

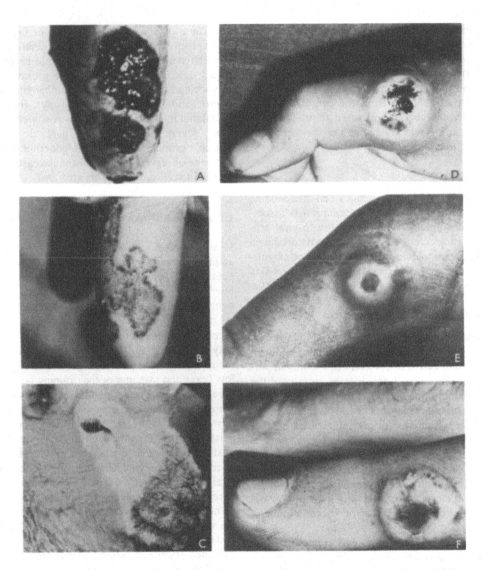

FIG. 9. Cowpox, pseudocowpox, and orf in animals and in humans. (A,B,C) In animals. (A) Cowpox ulcer on teat of cow, 7 days after onset of symptoms. (B) Pseudocowpox (milker's nodule virus) on teat of cow. (C) Scabby mouth caused by orf virus, in a lamb. (D,E,F) Lesions caused by these viruses on the hands. (D) Cowpox; (E) milker's nodule (pseudocowpox); (F) orf. (A and B, courtesy of Dr. E. P. J. Gibbs; C, courtesy of Dr. A. Robinson; D, courtesy of Dr. A. D. McNae; E and F, courtesy of Dr. J. Nagington.)

The nodule is surrounded by an edematous zone and a large erythematous areola. The draining lymph nodes are enlarged and tender from about day 5 after the appearance of the skin lesion, which may remain nodular but usually ulcerates during week 3 and then gradually heals within 5 to 6 weeks, leaving a scar. In Kenya, Downie et al. (1971) noted that the lesions were almost always solitary and on the upper arm, face, neck, and trunk. In Zaire, however, Jezek et al. (1985) observed that 22% of patients had multiple lesions—usually two but sometimes three or more, the maximum number seen on one patient being 10.

Multiple lesions were often close to each other and usually evolved together, although they differed in size. In Zaire, the distribution of lesions was different from that seen in Kenya, 72% being on the lower limbs, 17% on the upper limbs, 7% on the trunk, and 5% on the head.

CLINICAL FEATURES OF INFECTION WITH YATAPOXVIRUS

Lesions in *Cercopithecus aethiops* are flat nodules with a maximum diameter of about 3 cm, but large

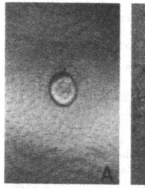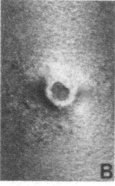

Fig. 10. Tanapox. Lesions on the thigh of a Zairean woman aged 27 years (A) 14 days and (B) 45 days after its appearance. (Courtesy of Mr. M. Szczeniowski.)

protuberant tumors, which appear to be histiocytomas, are produced in cynomolgus and rhesus monkeys. In inoculated cancer patients and in one case of accidental laboratory infection in man, the virus produced local skin nodules after an incubation period of about a week, reaching a maximum diameter of 2 cm and regressing after 3 to 4 weeks (Grace and Mirand, 1963).

Clinical Features of Molluscum Contagiosum

The lesions of molluscum contagiosum are pearly, flesh-colored, raised, firm, umbilicated nodules in the epidermal layer of the skin, usually 2 to 5 mm in diameter (Fig. 8). The incubation period, as determined by inoculation of human volunteers, varies from 14 to 50 days. The lesions may appear anywhere on the skin of the body except the palms and soles, often as crops in localized areas, probably due to multiple simultaneous infections or localized mechanical spread. It seems unlikely that generalization of the virus by the bloodstream occurs. The nodules are painless, and at the top of each there is often an opening through which a small white core can be seen. The disease is chronic, and lesions often persist for months or even a few years but ultimately resolve spontaneously, or following trauma or bacterial infection. However, recurrences may occur, probably owing to reinfection. Lesions in genital regions may be mistaken for genital herpes.

Diagnosis

Smallpox, now extinct, was the poxvirus disease for which laboratory diagnosis was first and most extensively developed. Where it was endemic, diagnosis was made on the basis of clinical and epidemiologic criteria, but laboratory confirmation was essential for suspected cases of importations into industrialized countries and in the recently endemic countries during the final stages of the eradication campaigns. However, many techniques once used for smallpox diagnosis, such as stained smears and agar gel precipitation, have long been superseded for the laboratory diagnosis of poxvirus infections, and they will not be described here.

For all poxvirus infections of man, the method of choice for laboratory diagnosis is electron microscopy of lesion material, supplemented by cultivation of the virus on the chorioallantoic membrane or in tissue culture. Restriction endonuclease digestion of poxvirus DNA is useful for positively identifying species of *Orthopoxvirus* and *Parapoxvirus*. Serologic methods have no place in routine diagnosis but have been essential for epidemiologic and ecologic investigations of monkeypox.

Biosafety Precautions

Before describing methods for the laboratory diagnosis of poxvirus infections, it is useful to recall that some of the viruses are hazardous agents for which special precautions are required during the collection and handling of specimens, as well as in the diagnostic laboratory. Laboratory requirements range from biosafety level 2 to biosafety level 4 (Table 4).

TABLE 4. Biosafety levels for handling poxviruses in the laboratory

Virus	Biosafety level[a]	Other safety measures
Variola virus (including suspected smallpox)	4 (3½)	Vaccination with vaccinia virus every 3 years
Monkeypox virus	2 (3 for infected animals)	Vaccination with vaccinia virus every 3 years
Cowpox and vaccinia viruses	2	Vaccination with vaccinia virus every 3 years
Parapoxviruses	2	Nil
Yatapoxviruses	2	Nil
Molluscum contagiosum virus	2	Nil

[a] As described in Centers for Disease Control–National Institutes of Health Biosafety Manual (1984).

VARIOLA VIRUS

At present, only the two WHO Collaborating Centers for Smallpox and Other Poxvirus Infections, located at the Centers for Disease Control, Atlanta, and the Research Institute for Viral Preparations, Moscow, are authorized by the World Health Organization to handle variola virus and maintain variola virus stocks. There is, however, a remote possibility that in some other laboratory there may be material containing variola virus stored in a freezer, forgotten because the person who had stored it had died or left the establishment. Unlabeled specimens, or containers that might contain variola virus discovered in such a situation, should immediately be destroyed.

Endemic smallpox has been eradicated from the world since 1977, but suspected cases are still being reported (Khodakevich and Arita, 1985). In the United States and in other countries of the western hemisphere, any suspected case of smallpox is reportable by telephone to the respective state, territorial, or national health department. If after review the case is still regarded as suspected smallpox, it should be reported in the United States to the International Health Program Office, Centers for Disease Control, Atlanta, GA 30333, telephone (404) 329–3145. Specimens for laboratory diagnosis should be sent to Poxvirus Laboratory, Viral Exanthems and Herpesvirus Branch, Division of Viral Diseases, Center for Infectious Diseases, CDC, Atlanta, GA 30333. In other parts of the world, any suspected case of smallpox should be reported to Smallpox Eradication Unit, World Health Organization, 1211 Geneva 27, Switzerland, and specimens should be sent to that address.

Laboratory facilities for handling and storing variola virus should be at biosafety level 4 (actually 3½) as described by the Centers for Disease Control–National Institutes of Health (CDC-NIH) Biosafety Manual (1984), but the special pressurized suit is not necessary, and class II rather than class III safety cabinets are adequate. Otherwise, the standard microbiological practices and special practices are those prescribed by that manual for biosafety level 4. In addition, anyone who enters the smallpox laboratory must have been vaccinated with smallpox vaccine within the previous 3 years.

MONKEYPOX VIRUS

Laboratory facilities for handling and storing monkeypox virus should be at biosafety level 2, as described in the CDC-NIH manual. Standard microbiological practices, special practices, and containment equipment are as described for biosafety level 2, except that all who enter the laboratory must have been vaccinated with smallpox vaccine within the previous 3 years. Animals infected with monkeypox virus should be kept and handled in a laboratory facility designed for biosafety level 3.

VACCINIA AND COWPOX VIRUSES

The laboratory facility for handling and storing vaccinia and cowpox viruses should be at biosafety level 2. Large volumes of high-titered virus materials should be handled in a class II biosafety cabinet.

All those who work with or might come in direct contact with these viruses should be vaccinated every 3 years, including those who do not work with poxviruses but work in the same laboratory or with equipment such as a centrifuge in which large volumes of high-titered virus are handled. Visitors and service personnel who do not come in direct contact with vaccinia or cowpox virus do not require vaccination, but access to the laboratory for nonvaccinated individuals should be restricted during periods of intensive laboratory activity.

Parapoxvirus, Yatapoxvirus, and Molluscivirus

Viruses belonging to these genera (orf, milker's nodule, bovine papular stomatitis, tanapox, yabapox, and molluscum contagiosum viruses) should be handled under biosafety level 2 in practices, containment equipment, and laboratory facilities.

Specimen Collection

Poxvirus diseases are unusual in that all material for laboratory diagnosis is taken from skin lesions. As described earlier in this chapter, the kinds of lesions differ according to the disease (Figs. 7 to 10). Generalized vesiculopustular rashes occur only in smallpox, monkeypox, and the very rare cases of generalized vaccinia or eczema vaccinatum; local pustular lesions occur in human infections with vaccinia and cowpox viruses. In all other poxvirus infections of humans, the lesions are single (rarely multiple) nodules.

Orthopoxvirus SPECIMENS (SMALLPOX, HUMAN MONKEYPOX, VACCINIA, AND COWPOX)

Orthopoxvirus lesions go through papular, vesicular, pustular, and scabbing stages. Appropriate material for laboratory diagnosis differs according to the stage of the lesion. It must be emphasized that adequate amounts of material must be provided to permit effective testing.

Maculopapular Stage

Cleanse the skin with an alcohol sponge, scrape the lesion with a scalpel blade, reaching deeply to the lower epithelial layers, and make a thick smear (not spread out) on a glass slide; take at least four glass slide samples. Do not attempt to fix the smears in any way, since they will be reconstituted for electron microscopy and culture. Pack slides in a plastic slide holder. Before the eradication of smallpox, it was important to collect specimens at this stage, so as to be able to diagnose smallpox as early as possible. For other diseases, it is better to defer collection until a later stage, when more material can be collected.

Vesiculopustular Stage

Remove the top of the vesicle or pustule with a scalpel blade or forceps, and collect as much vesicular or pustular fluid as possible in four capillary tubes, filling each to a height of at least 10 mm. Seal the ends of the capillary tubes with plastic glue or clay, and place them in a screw-capped tube or vial. When the fluid is too viscous to collect by a capillary tube, scrape the base of the lesions, and make thick smears on glass slides. The amount of vesicular fluid that can be collected from the single lesions of vaccinia and cowpox may be very small.

Crusting Stage

Collect no fewer than three crusts, preferably 10, and place them in a screw-capped vial. Do not add any liquid to the crusts. Only fragments of crusts may be available for collection from cases of vaccinia and cowpox.

Parapoxvirus SPECIMENS (ORF, MILKER'S NODULES, AND BOVINE PAPULAR STOMATITIS)

Collect vesicular fluid, if available, and crusts and fragments of crusts as described for orthopoxvirus specimens. Also collect biopsy material if possible; this is more likely to give positive results when examined with the electron microscope. Do not add any liquid to the crusts or biopsy material.

Yatapoxvirus SPECIMENS (TANAPOX AND YABAPOX)

For tanapox, collect crusts, if present, and biopsies of the lesions. For yabapox, collect biopsies of the tumors, or excise a whole tumor, and place in a screw-capped vial. Do not add liquid to the specimens.

Molluscipoxvirus SPECIMENS (MOLLUSCUM CONTAGIOSUM)

Collect expelled material from the lesions on swabs, and place them in a screw-capped tube, or transfer the lesion material directly into a screw-capped vial with an appropriate stick or a scalpel blade. Do not add liquid to the specimens. Biopsy material is also useful, some of which should be fixed, since the histologic picture is pathognomonic.

Shipment of Specimens

Detailed instructions for shipment are found in Chapter 1, "Specimen Collection and Transport." Specimens should be packed in a primary container (a screw-capped vial or test tube), tightly closed, and sealed with waterproof tape. The primary container should then be packed in a durable, watertight secondary container. The space between the primary and secondary containers should have sufficient nonparticulate absorbent material, such as paper towels, to absorb the contents of the primary container should it leak. The secondary container is placed in a tertiary corrugated fiberboard or cardboard container for shipping. A copy of the specimen information should be enclosed between the secondary and tertiary containers.

Specimens from Possible Cases of Smallpox in Archaeologic Material

Although no cases of smallpox have occurred since 1978, concern is sometimes expressed by public health officials and archaeologists about the risk of contracting smallpox from handling archaeologic materials, including mummified human remains, in situations in which deaths from smallpox may have occurred. This concern is based on the stability of viability of variola virus under favorable conditions of storage (Wolff and Croon, 1968). However, there is a vast difference between the potential viability of virus in crusts stored in a laboratory in Holland for a few years and corpses subject to autolysis and bacterial attack for hundreds of years. The only situation that might conceivably present a risk is the disinterment of bodies buried deep in ice or permanently frozen soil (Ewart, 1983). Personnel handling such material should be adequately protected by vaccination, and specimens should be collected, shipped, and examined as described for variola virus.

Given the remarkable state of preservation of some mummified tissues (Lewin, 1967, 1977), it is possible that poxvirus particles could be morphologically preserved for hundreds, even thousands, of years. Poxvirus particles were recently visualized by electron microscopy in skin taken from the mummified corpse of a child who died from smallpox in Naples during the 16th century (Fornaciari and Marchetti, 1986). It is important that such material be examined by a virologist skilled in the electron microscopic examination of poxviruses, since many artifacts occur that bear some resemblance to poxvirus particles (Nakano, 1985b).

Diagnosis on the Basis of Epidemiologic and Clinical Information

The diagnosis of most of the poxvirus diseases listed in Table 2 can be made by a process of elimination, using some well-established epidemiologic facts, including the knowledge that except for smallpox, vaccinia, and molluscum contagiosum, poxvirus diseases are zoonoses. Therefore, before undertaking any laboratory diagnosis, one should consider the epidemiologic data outlined below for each disease.

SMALLPOX

1) The last case of endemic smallpox in the world occurred in October 1977. 2) The infection was perpetuated by person-to-person transmission. 3) There is no animal reservoir of smallpox virus, and no known source of the virus except for the two smallpox virus repositories at the Centers for Disease Control, Atlanta, and the Research Institute for Viral Preparations, Moscow. 4) There should be a convincing epidemiologic association of the suspected case with smallpox virus. 5) Smallpox was often mistaken for chickenpox, and chickenpox was often mistaken for smallpox. Virtually every case of suspected smallpox investigated by the laboratory at the Centers for Disease Control after eradication was eventually diagnosed as chickenpox. If after considering these facts one is still convinced that the case in question may be smallpox, one should contact the appropriate national, state, or territorial public health authority, as outlined above.

HUMAN MONKEYPOX

1) This is strictly an African disease and is transmitted to man from infected monkeys and squirrels. 2) Although it could be imported into western countries as an infection of an African monkey, no cases have been recorded in monkeys in Europe or the United States since 1968, and no human cases have ever been recognized other than in the enzootic regions in western and central Africa. 3) The rate of person-to-person transmission is about 15%, compared with about 60% for smallpox. 4) As with smallpox, human monkeypox is sometimes diagnosed as chickenpox, and chickenpox is sometimes diagnosed as human monkeypox.

VACCINIA

1) This is a "man-made" infection produced by deliberate inoculation of vaccinia virus as smallpox vaccine. 2) A vaccinee may by direct contact transmit the infection to another person or to an animal (buffalo, rabbit, cow), which may in turn transmit the infection to man.

COWPOX

1) This disease has been reported in humans, cows, cats, and zoo animals in Great Britain, western Europe, and the Soviet Union. 2) Cowpox has never existed in the United States. Reports of "cowpox" in man from the United States (Nakano, 1986) and many of those in European countries involved patients who were infected by cows suffering with either vaccinia virus infection (cows previously infected by vaccinated persons) or pseudocowpox, or they were not due to poxvirus infection at all. 3) Cowpox virus may by transmitted to man from cows or cats, or there may be no recognized animal source (Baxby, 1977).

ORF

1) The infection is found worldwide and is usually transmitted to man from domestic sheep and goats and, in some cases from wild ungulates. 2) In the United States and Australia, farmers and shearers working closely with sheep and goats are the most likely persons to contract orf, but in New Zealand the majority of cases occur in abattoir workers handling sheep (Robinson and Petersen, 1983). In North America and Europe, infection may occasionally occur among hunters or taxidermists who handle infected material from wild ungulates.

MILKER'S NODULE

The infection is found worldwide and is usually transmitted to man from dairy cows.

BOVINE PAPULAR STOMATITIS

The infection is found worldwide and is usually transmitted to man from calves and beef cattle.

TANAPOX

1) As with monkeypox, this disease is confined to Africa, and so far it has been found only in Kenya and Zaire, especially near rivers. 2) It is thought to be transmitted to man from monkeys and other animals by blood-sucking insects. 3) The lesion (Fig. 10) is fairly characteristic.

YABAPOX

1) This is an African disease. 2) No natural infections of man have been reported, but humans are susceptible, as shown by accidental and deliberate inoculation.

MOLLUSCUM CONTAGIOSUM

1) This infection is found worldwide. 2) In children, the infection is transmitted by nonsexual direct con-

tact or fomites, but in young adults it is commonly sexually transmitted. 3) The lesions (Fig. 8) are very characteristic.

Direct Virus Detection

Skin lesions produced by most poxvirus infections, unlike manifestations of infection by many other viruses, contain a relatively high concentration of virions and of viral antigens. At the Centers for Disease Control, it was found that a 10% suspension of a smallpox crust usually contains 10^6 to 10^7 virions per milliliter. Other workers (Mitra et al., 1974) reported finding $10^{4.7}$ to $10^{7.5}$ pock-forming units per gram of scabs. Thus, skin lesions are well suited for direct virus detection by electron microscopy.

Electron Microscopy

Electron microscopy for direct virus detection is the fastest and most dependable method for the laboratory diagnosis of all poxvirus infections. Procedures for electron microscopy are described in Chapter 7; detailed descriptions for preparation of specimens and grids and the electron microscopic examination of poxvirus specimens are given in Nakano (1979). It is important to be able to distinguish between poxvirus and herpesvirus virions and artifacts (nonvirus particles that resemble them). Some examples of such artifacts are illustrated in Nakano (1985b).

Orthopoxvirus Infections

It is impossible to differentiate between the virions of variola, monkeypox, vaccinia, and cowpox viruses. The disease most commonly confused with smallpox in the past, and with human monkeypox in central Africa at the present time, is chickenpox. Electron microscopy has proved to be a rapid and dependable method for differentiating between these poxvirus and herpesvirus infections.

The efficiency of electron microscopy for finding poxvirus particles in specimens from cases of smallpox and human monkeypox that had been shipped to the Centers for Disease Control from endemic countries in Africa and Asia, via WHO in Geneva, was 98.6% (Nakano, 1982), whereas herpesvirus particles were found in only 60 to 70% of specimens from cases that were eventually diagnosed as chickenpox. Thus, when no poxvirus or herpesvirus particles were found by electron microscopy in specimens from a patient suspected to have smallpox or monkeypox, the patient was probably not suffering from either of these diseases but may have had chickenpox or some other disease.

As visualized in negatively stained preparations, two forms of poxvirus particle can be distinguished,

depending on penetration of the virion by phosphotungstate. For vaccinia virus, these were designated as M and C particles (Westwood et al., 1964), M for mulberry, because of the similarity of the surface structure of the M form, which is produced by the delineation of the surface tubules of the outer membrane (see Fig. 1A), to a mulberry, and C for capsule, seen when stain penetrates the virion, revealing the capsule and other internal structures. These two forms of poxvirus virions can be found in specimens from patients with smallpox, monkeypox, vaccinia, and cowpox. M forms are usually found in specimens of vesicular fluid, whereas the C forms are in the majority of specimens from dried scabs. Figure 2A illustrates M forms of variola virus found in the last case of endemic smallpox in the world, which was diagnosed in Somalia on October 26, 1977; Fig. 2B illustrates a C form of variola virus. All other orthopoxviruses have a similar morphology.

Parapoxvirus Infections

The virions of orf, pseudocowpox, and bovine papular stomatitis viruses are indistinguishable from one another in size and morphology but are easily differentiated from those of all other poxviruses, being rather narrow and having rounded ends (Figs. 1B, 2). As shown in Fig. 2C, an M form of orf virus (the designation is not appropriate, for the virion resembles a skein of wool) has tubules that are arranged in parallel (see Fig. 1B) but present a crisscross pattern when both sides of the particle are visualized. In specimens from lesions, parapoxviruses are often surrounded by an envelope (Fig. 2C). Figure 2D illustrates the C form of the virion of orf virus.

Yatapoxvirus Infections

Lesions caused by tanapox and yabapox viruses are characteristic enough that, given epidemiologic and clinical histories, these infections can be diagnosed quite easily; a positive result by electron microscopy will confirm the diagnosis. Although most specimens collected from patients with smallpox and monkeypox contain a large number of poxvirus particles, those collected from tanapox patients, when examined with the electron microscope, appear to contain even more single particles and more aggregates of particles per field. Tanapoxvirus M forms are enveloped and show a prominent pattern of tubules on the surface (Fig. 2E). Virions (both M and C forms) of tanapoxvirus are surrounded by an envelope (Fig. 2F), whereas those of orthopoxviruses are not, and the M forms of tanapoxvirus usually have more prominent surface tubules than are seen in orthopoxviruses. These differences are found only in clinical specimens, not in material obtained from cultured cells (Nakano, 1985a).

Experience at the Centers for Disease Control with yabapox virus virions viewed by electron microscopy is limited to specimens obtained from a small number of nonhuman primates; most were enveloped.

Molluscipoxvirus Infections

Molluscum contagiosum virus produces characteristic lesions (Fig. 8), so the infection is usually diagnosed clinically. However, because the infection has been noted increasingly as being sexually transmitted in young adults, the lesions may be in the moist genital areas, and if they become inflamed and ulcerated, they may be confused with those produced by herpesviruses (Dennis et al., 1985). In such cases, electron microscopy can provide a rapid and accurate diagnosis. As in tanapox, lesions of molluscum contagiosum contain an abundance of virions, which usually occur as single particles rather than as aggregates. They are similar to orthopoxvirus virions in shape and size but can usually be differentiated from them by the fact that the M forms of molluscum contagiosum virions have more prominently visible surface tubules (Fig. 2G). Enveloped forms are not seen in clinical material.

Virus Isolation and Identification

The chick embryo chorioallantoic membrane and cell cultures are used for isolation of poxviruses. All species of orthopoxvirus can be isolated on the chorioallantoic membrane and in selected cell cultures, but parapoxviruses and yatapoxviruses can only be isolated in cell cultures, and it has not yet proved possible to culture the virus of molluscum contagiosum satisfactorily.

ISOLATION OF ORTHOPOXVIRUSES ON THE CHICK EMBRYO CHORIOALLANTOIC MEMBRANE

The standard procedures for the preparation of the chorioallantoic membrane for the inoculation of viruses and the subsequent examination and harvesting of these membranes are described in Chapter 3, but aspects peculiar to growing poxviruses are described here.

Before inoculation, fertile chicken eggs must be incubated at 38.0°C to 39.0°C and used at 11 to 13 days. Eggs incubated for less than 11 days may be refractory to infection with poxviruses, and by harvest time the embryos in those incubated for more than 13 days are too large to handle. Temperatures lower than 38.0 to 39.0°C for incubation prior to inoculation may render the membranes less susceptible, and during the summer months, in spite of correct temperature and correct length of time for incubation, chorioallantoic membranes may vary in sensitivity. Sometimes growth is normal. At other times vaccinia virus, used as a control, may not grow on such membranes, or it may grow, but with atypically small pocks, or normal size vaccinia virus pocks may be produced, but the pock titer may be lower than expected by 0.5 to 1.0 $\log_{10}$ units. After inoculation, the eggs are incubated at 35.5°C for 72 h before examining the chorioallantoic membrane. The easiest and quickest way to differentiate variola, monkeypox, vaccinia, and cowpox viruses from one another and to identify each of these viruses is by the characteristic morphology of the pocks they produce on the chorioallantoic membrane (Table 5; Fig. 11).

Variola Virus

Pocks of variola virus are about 1 mm in diameter, grayish white to white, opaque, and round. They

TABLE 5. Biological characteristics of species of *Orthopoxvirus* that produce infections in man

Characteristic	Variola virus	Monkeypox virus	Vaccinia virus	Cowpox virus
Host range	Narrow	Broad	Broad	Broad
Pocks on the chorioallantoic membrane (CAM)	Small, opaque, white	Small, opaque, hemorrhagic	Strains vary, large opaque white or hemorrhagic	Large hemorrhagic
Ceiling temperature CAM	37.5–38.5°C	39°C	41°C	40°C
Rabbit skin lesion	Small, transient, nontransmissible	Indurated, hemorrhagic	Strains vary, indurated nodule, sometime hemorrhagic	Large, indurated, hemorrhagic
Lethality for				
Suckling mice	Low	High	High to very high	Variable
Chick embryo	Low	Medium	Very high	High
Type A inclusion bodies	−	−	−	+
Type B inclusion bodies	+	+	+	+
Size of DNA (kbp)	182	180	186	218

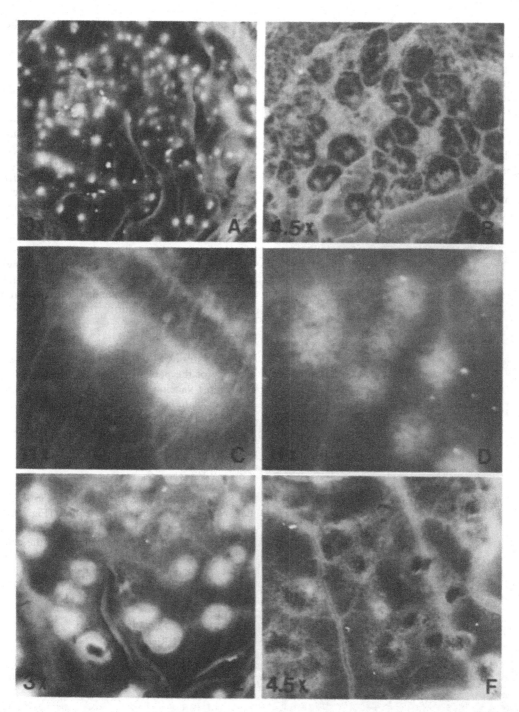

FIG. 11. Pocks produced on the chorioallantoic membrane after 72-h incubation. Magnifications noted were in original prints. (A) Variola virus; pocks dense white, never hemorrhagic, whatever the temperature of incubation postinoculation. (B) Cowpox virus; bright red hemorrhagic pocks. (C,D) Monkeypox virus; pocks ulcerated after incubation at 35.5 to 37°C, but often hemorrhagic after incubation at 33.5 to 34°C. (E,F) Vaccinia virus. Vaccine strain Wyeth (E) produces large white pocks, sometimes with a central ulcer. Neurovaccinia strains, e.g., rabbitpox virus (F), produce hemorrhagic pocks.

have a regular periphery and a smooth convex outer surface, the central region of the pocks being raised above the surface of the chorioallantoic membrane (Fig. 11A). The pocks are never hemorrhagic, and they are all nearly the same size. The central area is opaque and surrounded by a halo, resembling a sunny-side-up fried egg.

According to data collected at the Centers for Disease Control between 1966 and 1978, the efficiency of isolation of variola virus on the chorioallantoic membrane from material submitted from patients with smallpox was 89%. If the specimens could have been more rapidly delivered to the laboratory or ade-

quately refrigerated during transportation, the efficiency rate would have been close to 100%.

Monkeypox Virus

Pocks of monkeypox virus are about 1 mm in diameter, but unlike those of variola virus they are flat, ridged at the periphery, and not raised above the surface of the membrane. There is some variation in appearance, but most pocks produced at 35.5 to 37.0°C have a centrally located crater (Fig. 11C), and those produced at 33.5 to 34.0°C are hemorrhagic (Fig. 11D). The hemorrhagic appearance is due to the deposit of erythrocytes at the surface of the chorioallantoic membrane, not in the pock itself (Fig. 12A).

The efficiency of isolation of monkeypox virus on the chorioallantoic membrane, according to data collected between 1970 and 1986, was 90%. However, all monkeypox material was some weeks at ambient temperature during transit, and virus may sometimes have been inactivated during transit.

Vaccinia Virus

Different strains of vaccinia virus differ in the morphology of the pocks they produce (Fenner, 1958). With most vaccine strains, pocks are 3 to 4 mm in diameter, flattened, with central necrosis and ulceration (Fig. 11E). Pocks of some strains, e.g., rabbitpox virus, have a hemorrhagic appearance (Fig. 11F).

Cowpox Virus

Pocks of cowpox virus are 2 to 4 mm in diameter, flat, round, with a bright red center (Fig. 11B). A cross section of a cowpox virus pock (Fig. 12B) shows that the erythrocytes lie beneath the ectodermal layer, which contains the infected cells, and not in the surface tissue layer of the chorioallantoic membrane, as in pocks produced by monkeypox virus.

ISOLATION OF POXVIRUSES IN TISSUE CULTURE

Except for molluscum contagiosum virus, all poxviruses that infect man can be isolated in cultured cells.

Orthopoxvirus Infections

Variola, monkeypox, vaccinia, and cowpox viruses grow well in primary or stable lines of human and nonhuman primate cells. Those commonly used are primary rhesus and African green monkey kidney cells, stable lines such as Vero, LLC-MK2, F-L, and diploid cells such as human embryonic lung fibroblasts. With a few exceptions, orthopoxviruses can

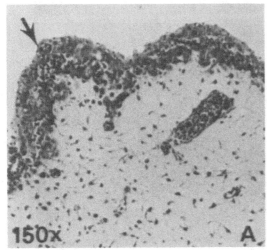

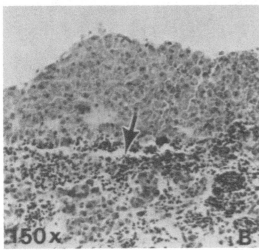

FIG. 12. Sections of the hemorrhagic pocks produced by monkeypox and cowpox viruses, showing the location of dark-staining nucleated erythrocytes. Magnification ×150. (A) Monkeypox virus; the erythrocytes are on the surface of the central area of the pock. (B) Cowpox virus; the erythrocytes are beneath the ectodermal cells on the surface of the pock.

grow also in cells of other mammals and birds, such as rabbit, mouse, hamster, and embryonic chick fibroblasts. However, variola virus does not grow well in rabbit cells but can be adapted by passage (Dumbell and Bedson, 1966).

Variola Virus. Variola virus from clinical specimens produces a cytopathic effect in 1 to 3 days in LLC-MK2 or Vero cells, which begins with rounding of the cells and the formation of hyperplastic foci. This is followed by the formation of small plaques, 1 to 3 mm in diameter, when the infected cells slough off. When high-titered virus inoculum is used, the cytopathic effect may spread so rapidly that the formation of plaques can be missed if the plates are not observed carefully.

Monkeypox, Vaccinia, and Cowpox Viruses. These three orthopoxviruses produce similar cytopathic effects in LLC-MK2 or Vero cells and cannot be differentiated from one another by plaque morphology. A specimen containing any one of these viruses produces a cytopathic effect in cell culture in 1 to 3 days by fusing cells and forming foci, which become plaques measuring 2 to 6 mm in diameter in 2 to 3 days; that is, they are much larger than those of variola virus. They usually have cytoplasmic bridging.

Although monkeypox virus cannot be differentiated from vaccinia and cowpox by the cytopathic effect produced in LLC-MK2 or Vero cells, it can be differentiated from these viruses and variola virus by the fact that it does not grow in a continuous line of pig embryonic kidney (PEK) cells (Marennikova et al., 1972).

Parapoxvirus Infections

Orf Virus. Although ovine embryonic kidney and testis cells are the first choice for initial isolation of orf virus from human infections, bovine embryonic and testis cells are a very close second, and some strains of orf can be isolated in primary or secondary monkey kidney cells (rhesus and African green monkeys) and primary human amnion cells. For initial isolation of orf virus from infections of sheep, however, ovine cells are preferable (Nagington, 1968). Once the virus is isolated from either humans or sheep, it can grow in any of the cells mentioned and also in human embryonic lung fibroblasts and LLC-MK2 cells. The cytopathic effect produced by the virus appears in 3 to 6 days (Fig. 13), but on initial isolation the cytopathic effect may not appear for 6 days or more.

Milker's Nodule and Bovine Papular Stomatitis Viruses. These viruses are isolated from human lesions in bovine embryonic kidney and testis cells, or in ovine embryonic kidney and testis cells, or pri-

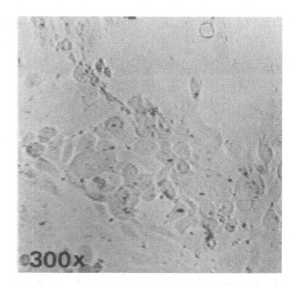

Fig. 13. Cytopathic effect of orf virus in ovine kidney cells 3 days after inoculation. Unstained; magnification ×300.

mary human amnion cells (Nagington, 1968). Once isolated, they grow well in human embryonic fibroblasts and LLC-MK2 cells.

Yatapoxvirus Infections

Tanapox Virus. Tanapox virus was first isolated from human lesions in human thyroid cells. Once isolated, the virus grows well in primary vervet monkey and patas monkey kidney cells, WI-38, HEP-2, and Vero cells (Downie et al., 1971). At the Centers for Disease Control, we were able to isolate the virus from human tanapox in Zaire in LLC-MK2 cells, a stable line of rhesus monkey kidney cells, but only after the incubation temperature was lowered from 35.5 to 33.5°C (Nakano, 1982). Once isolated, the virus grows well in human embryonic lung fibroblast cells, primary African green monkey kidney cells, and CV-1 cells.

The cytopathic effect in LLC-MK2 cells initially appears in 6 to 8 days after inoculation and usually takes 1 to 2 additional passages, because of the "toxic" reaction in the cell cultures caused by the heavily contaminated condition of many specimens. The titer attained is generally $10^{4.5}$ PFU/ml. Figure 14A shows an overall view of a tanapox virus plaque on primary African green monkey kidney cells 13 days after inoculation, and Fig. 14B shows the cytopathic effect at a higher magnification, with several giant cells.

Yabapox Virus Infections. The virus has been grown in BSC-1, CV-1, and LLC-MK2 cells, but the titer is usually very low.

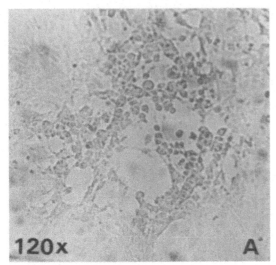

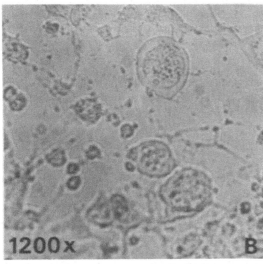

FIG. 14. Cytopathic effect of tanapox virus in primary African green monkey kidney cells 13 days after inoculation. Unstained. (A) General view, magnification ×120. (B) Appearance of infected cells at higher magnification, magnification ×1,200. The magnifications noted were in the original prints.

Molluscum Contagiosum

Molluscum contagiosum virus has been reported to show a cytopathic effect in primary human amnion, primary monkey kidney, mouse embryo, HeLa, BSC-1, WI-38, MRC, F-L, and human foreskin cells (Chang and Weinstein, 1961; Neva, 1962; Raskin, 1963; La Placa, 1966; Francis and Bradford, 1976; McFadden et al., 1979). The virus initially produces rounding and clumping of the cells, but then they resume their original appearance. The cytopathic effect can be passaged but is eventually lost. It has been suggested that "replication is blocked in vitro because the agent is unable to pass through the un-

coating phase" (McFadden et al., 1979). This means that for practical purposes the virus cannot be grown in the laboratory, although virions can be visualized by electron microscopy of first-passage human fetal diploid kidney fibroblast cells (Dennis et al., 1985).

ANIMAL INOCULATION

Although animals have never been used routinely to isolate the poxviruses considered in this chapter, inoculation of rabbits, suckling mice, and chick embryos has been used to differentiate one orthopoxvirus from another, especially when human monkeypox was first discovered, and its differentiation from variola virus was a matter of great importance. The differences in virulence of the four orthopoxviruses that infect humans for rabbit, suckling mice, and chick embryos are summarized in Table 5. However, electropherograms of DNA fragments produced by restriction endonuclease digestion of orthopoxvirus DNA is a much more satisfactory way than biological tests of determining the species to which an orthopoxvirus belongs, and even for comparing strains (Dumbell and Kapsenberg, 1982). For this reason, animal tests other than the appearance of pocks produced by orthopoxviruses on the chorioallantoic membrane, already described, will not be further discussed.

ANALYSIS OF VIRAL DNA

The general procedures for the analysis of viral DNA by electrophoresis of fragments obtained after digestion with restriction endonucleases are described in Chapter 8. The method has been extensively applied to the study of DNA from various species of *Orthopoxvirus* and to a more limited extent with parapoxviruses, but not yet to other genera. Tests have usually been carried out with DNA extracted from purified virions. However, utilizing the fact that poxviruses replicate in the cytoplasm, Esposito et al. (1981) have developed a method whereby the viral DNA can be extracted directly from infected monolayer cultures.

Orthopoxviruses

The most useful enzymes are *Hind*III, *Sma*I, *Xho*I, *Sac*I, and *Sal*I. The differentiation of species of *Orthopoxvirus* by restriction endonuclease analysis of the DNA has already been described. The method has been of great value in research (Esposito and Knight, 1985) and is an essential component in the construction of vectored vaccines, but little use has yet been made of it for diagnosis. However, two studies reveal its value in situations where the accurate diagnosis of the strains of virus involved has been of critical importance (Fig. 15). The first in-

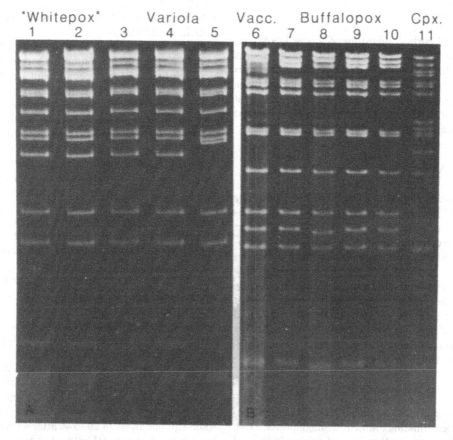

FIG. 15. Use of electropherograms of DNA fragments of orthopoxviruses for species and strain diagnosis. (A) Comparison of digestion products of DNA of "whitepox" viruses and selected variola strains, after digestion with *Sal*I. 1, 2 = Utrecht "whitepox" virus strains; 3, 4 = strains of variola virus from Vellore, India; 5 = variola virus, Harvey strain. The "whitepox" viruses were contaminants of normal cynomolgus kidney cells with one of the Vellore strains of variola virus (4). (From Dumbell and Kapsenberg, 1982; with permission.) (B) Comparison of strains of "buffalopox" virus recovered from outbreaks in India (7, 8, 9, 10) with vaccinia virus (6) and cowpox virus (11) after digestion with *Hin*dIII. The outbreaks of "buffalopox" were due to vaccinia virus. (Courtesy of Dr. K. R. Dumbell.)

volved explanation of the origin of "whitepox" virus in "uninfected" cynomolgus kidney cells in Utrecht, the Netherlands, in 1964 (Gispen and Brand-Saathof, 1972). By comparing the electrophopherograms after *Sal*I digestion of the DNAs of the two "whitepox" viruses, two strains of variola virus from Vellore, India, that had been isolated in the Utrecht laboratory in 1964 and a standard strain of variola virus, Dumbell and Kapsenberg (1982) (Fig. 15A) demonstrated that the "whitepox" viruses were due to inadvertent contamination of uninoculated cultures of cynomolgus kidney cells with one of the Vellore strains of variola virus.

The other problem involved the nature of buffalopox virus. After study of the biological properties of four strains obtained from India, Baxby and Hill (1971) concluded that three of these were strains of vaccinia virus but that the fourth was a "new" orthopoxvirus which should be designated "buffalopox virus." However, K. R. Dumbell (personal communication, 1983) concluded from restriction endonuclease analysis that all four were strains of vaccinia virus. The question became of greater importance when outbreaks of buffalopox were reported in India in 1984 through 1986, long after cessation of routine vaccination of the general public. Again, the strains were very similar to vaccinia virus in their biological characteristics, but they differed in some properties (D. Baxby, personal communication, 1986). However, electropherograms after digestion of the DNA with *Hin*dIII (Fig. 15B) showed unequivocally that they were strains of vaccinia virus (K. R. Dumbell, personal communication, 1986).

This technique is the method of choice for making a definitive species diagnosis of viruses of the genus *Orthopoxvirus*, and if suitable material is available, it may be used to identify strains of a particular species of *Orthopoxvirus*.

Parapoxviruses

As described earlier, digestion of virion DNA with *Bam*HI and *Kpn*I can be used to distinguish orf virus from pseudocowpox and bovine papular stomatitis viruses. This technique was used by A. J. Robinson (personal communication, 1986) to demonstrate that a parapoxvirus recovered from deer in New Zealand (Horner et al., 1987) was not a strain of orf virus.

Antibody Assays

Serologic diagnosis is now used for the laboratory diagnosis of poxvirus infections only in research projects, when no virologic specimen is available and an accurate specific diagnosis is needed, as in investigations of monkeypox in Zaire (Jezek et al., 1986b). In addition, antibody assays have proved useful in ecologic and epidemiologic investigations of poxvirus infections. Several serologic tests have been used at various times (Table 6); only those currently used (ELISA, radioimmunoassay, and radioimmunoassay absorption) or of historical importance (hemagglutinin inhibition) will be further discussed.

HEMAGGLUTININ INHIBITION TEST (HI)

Among the poxviruses, only orthopoxviruses produce hemagglutinating antigen [hemagglutinin; HA), which agglutinates erythrocytes of certain individual chickens and those of every individual turkey. The HI test is therefore applicable only to orthopoxviruses, and it does not differentiate between different species.

The method of performing the standard HI microtiter test is described in Chapter 5. The reagents specific for orthopoxvirus HI tests are 1) chicken erythrocytes that can be agglutinated by vaccinia HA antigen (erythrocytes from every new chicken must be pretested to select only those birds whose erythrocytes are agglutinated by the vaccinia HA antigen)

and 2) HA antigen, produced by growing vaccinia virus in tissue culture.

In addition to specific antibody, two relevant nonspecific factors are often found in serum: nonspecific hemagglutinin and nonspecific hemagglutination inhibitors. The first can be eliminated by adsorbing the serum with a 50% chicken erythrocyte suspension, and the latter can usually be eliminated by treating the serum with periodate. Details of production and usage of these reagents are found in Nakano (1979).

In the HI test, two hemagglutination units are used, and since this makes the test very sensitive, an HI titer of 5 was formerly considered negative. Experience with the test in serologic surveys of monkeypox cases in Africa, including comparison with the RIA test (see below), led to the conclusion that an HI titer of 5 should be considered positive. Therefore, although some false-positive cases are encountered, the HI test, as now used for screening, is apt to detect more cases of past infection with monkeypox virus.

It was formerly believed that HI antibody persisted for only a few months after the onset of smallpox. However, since 1970, sera from a number of monkeypox patients have been tested at increasing intervals. In a few cases, the HI antibody disappeared in 2 to 3 months after onset, but many subjects had titers of 20 to 40 beyond 8 months, and some still had titers of 20 to 40 after 4 years. Thus, at least in monkeypox, HI antibody sometimes persists much longer than a few months.

HI antibody is detectable in smallpox and human monkeypox patients within 4 to 7 days of the onset of rash; the titers are generally 80 or greater and may reach 1,000 or more. A patient with a dependable clinical history may be diagnosed to have variola virus or monkeypox virus infection with an HI titer of > 40 in a blood specimen taken 4 weeks after onset.

For a person who is vaccinated every 3 years, however, HI is not a good method to monitor the antibody response to revaccination, since the titer

TABLE 6. Serologic tests that have been used for assay of antibodies in poxvirus infections

Serologic test	Orthopoxvirus	Parapoxvirus	Yatapoxvirus	Molluscipoxvirus
Hemagglutination inhibition	+[a]	−	−	−
Complement fixation	+	+	+	+
Serum neutralization	+	+	+	−
Immunofluorescence	+	+	+	−
Agar gel precipitation	+	+	+	−
ELISA	+[a]	+[a]	+[a]	−
Radioimmunoassay	+[a]	−	−	−
Radioimmunoassay adsorption	+[b]	−	−	−

[a] Used for screening tests.
[b] Used for species-specific diagnosis of previous orthopoxvirus infection.

may increase very little, for example, from < 5 to 5, or from 5 to 10, and only rarely to 20. In many repeatedly vaccinated persons, whose postvaccination HI titer reaches only 10, the titer after 6 months to 1 year may be 5 or even < 5, so the test cannot be used to evaluate the immune status of that individual. If a person has not been vaccinated for over 5 years, however, the titer may increase from 5 to 40 or greater. If vector vaccinia virus becomes widely used and people are repeatedly vaccinated, some test other than HI should be used for monitoring the vaccinee's antibody responses.

The HI antibody response after cowpox virus infection may be quite good even if the person has been vaccinated previously, similar to that of vaccinees who had not been vaccinated for over 5 years.

ENZYME-LINKED IMMUNOSORBENT ASSAY (ELISA)

A standardized procedure for the indirect ELISA is described in Chapter 5. For poxvirus antibody assay, the method described by Voller et al. (1976) was modified by changing the coating buffer for poxvirus antigen and normal cell antigen from carbonate bicarbonate buffer (pH 9.6) to phosphate-buffered saline (pH 7.2). A serum is determined positive when, for example:

Mean optical density (OD) of test serum + monkeypox virus antigen	= 0.20
Mean OD of test serum + normal cell antigen	= 0.03
Mean OD of (diluent + monkeypox virus antigen) and (diluent + normal cell antigen)	= 0.012
Mean OD from calculation above (0.012) + 2 SD	= 0.025
Mean OD of (test serum + monkeypox virus antigen) − mean OD of (test serum + normal cell antigen)	= 0.20 − 0.03 = 0.17

Since 0.17 is greater than 0.025, the serum is positive. *Note:* Mean OD plus 2 SD gives 95% confidence. Each dilution of a test serum is tested and calculated as above to determine its end-dilution titer.

The ELISA test is used to assay antibodies to orthopoxviruses, parapoxviruses, and yatapoxviruses, but it has not been used for molluscum contagiosum.

Orthopoxvirus Infections

The indirect ELISA for orthopoxviruses does not identify specific antibodies produced against variola, monkeypox, vaccinia, or cowpox viruses. Monkeypox virus is used at the Centers for Disease Control

as the antigen for the assay of antibodies for all orthopoxviruses. It has been most extensively used for diagnosis and serologic surveys of cases of human monkeypox in Africa. Occasionally, ELISA can detect monkeypox antibody in a serum collected 1 day after the onset of rash, but most patients do not show a significant titer ($\geq$ 320) until day 3. The antibody reaches its maximum titer ($\geq$ 1,280) during the second week after onset. Most patients' sera retain titers of 320 or greater even after 8 or 9 years.

ELISA is useful for assaying the vaccinia antibody response in persons who are repeatedly vaccinated, because not only can it detect an antibody response in significant titers in vaccinees with a "good" skin reaction, but it is positive in vaccinees with "poor" skin reactions, whereas all other serologic tests are negative or the titers are too low to be evaluated in such cases (Table 7).

Parapoxvirus Infections

ELISA appears to be sensitive for parapoxviruses, but its use has been limited, and hence the method cannot be properly evaluated.

Tanapox

Although ELISA detects tanapox antibody during the first week after onset of the lesion(s), results are apt to be negative in sera collected during this time. The antibody titer apparently reaches its maximum (5,120) during the second week and is maintained during the third week. Titers begin to decline about a month after the onset. Tanapox antibody is still detected by ELISA with titers of 160 to 320 at 1 to 1½ years after the onset.

RADIOIMMUNOASSAY (RIA)

The indirect RIA to detect orthopoxvirus antibodies by an antiglobulin labeled with radioactive ^{125}I (Ziegler et al., 1975) has been used at the Centers for Disease Control for diagnosis and serologic surveys of monkeypox virus infection in man, nonhuman primates, squirrels, rabbits, and rodents. It has also been used to assay tanapox antibody, but not parapoxviruses or molluscum contagiosum virus. The procedure is described in Chapter 5.

Orthopoxvirus Infections

At 1 to 3 days after the onset of rash in patients with smallpox and monkeypox, RIA does not detect the antibody response as well as the HI test, but toward the end of the first week and during the second week after onset, the RIA titers rise and range from 3,000 to a maximum of 20,000 or greater. Vaccinated patients may show an anamnestic response, producing titers of 70,000 or greater. The titers remain high for

TABLE 7. Serologic responses to recent vaccination in persons who have had repeated vaccinations, by different serologic tests and in relation to the nature of the response to the most recent vaccination[a]

Serologic test	Skin reaction		
	Good take	Moderate take	Poor take
Hemagglutinin inhibition	10–20	5–10	5–5
Complement fixation	4–16	4–8	4
Serum neutralization	300–5,000	200–1,000	100–600
Immunofluorescence	8–64	4–16	4–4
ELISA	640–5,120	320–1,280	160–320
Radioimmunoassay	1,000–3,000	800–1,500	300–600

[a] Responses indicate range (as reciprocal titers) found in different individuals after recent revaccination, not increases in titers.

1 or 2 months. The test appears to be the most sensitive available, but because of this high sensitivity, titers less than 150 are of doubtful significance for African sera, and for sera from the United States, titers less than 80 are of questionable significance. Antibody detectable by RIA apparently persists for a long time, because titers of 700 to 2,000 are found in monkeypox patients 8 to 9 years after recovery.

RIA, like ELISA, is useful for assaying antibody response in persons who are vaccinated every year or every 3 years, because sera from such multiple revaccinees, especially those with "moderate" or "poor" takes when tested by other tests, are often negative, or, at best, the titers are so low that the antibody response cannot be evaluated. These vaccinees often do not get a "good take" even if they are revaccinated with a potent vaccine and with special care taken in the vaccination technique. Table 7 compares the serologic responses measured by each test according to the vaccinees' degree of skin reaction; the titers obtained by RIA and ELISA, but not those obtained with other tests, are high enough for evaluation. Thus, for all multiple revaccinees, the "adequate" level of antibody response is determined by comparing their titers (by ELISA and RIA) with the lowest titers produced by a number of multiple vaccinees who had a "good" take. Consequently, regardless of how low their titers are by other tests, if their titers are equal to or greater than 640 by ELISA and 900 by RIA, we consider the antibody level adequate. When the titers are less, we recommend revaccination. If on revaccination with a potent vaccine and good vaccination technique, the vaccinee still does not have a "good take," we must assume that the vaccinee has a high level of cell-mediated immunity.

Tanapox

Indirect RIA has been used at the Centers for Disease Control to assay tanapox antibody and has been found to be as useful as ELISA.

RADIOIMMUNOASSAY ADSORPTION (RIAA)

The RIAA (Hutchinson et al., 1977; Walls et al., 1980, 1981) has been used at the Centers for Disease Control to identify monkeypox-specific antibody. Serum specimens are first screened for those with high titers by RIA, usually >300, but >500 is preferred for sera collected from African natives. A serum specimen is first split, then one of the split portions is absorbed with normal chorioallantoic membrane and the other with vaccinia virus antigen. The residual antibody in each portion is reacted against chorioallantoic membrane and vaccinia, monkeypox, and variola virus antigens. The numerical values from the difference of the two portions produce a pattern from which it can be determined whether the serum contains specific monkeypox antibody, vaccinia antibody, or variola antibody. In a blood specimen collected during the first month after onset of monkeypox, it may be difficult to identify monkeypox-specific antibody, probably because of a higher proportion of less avid antibody globulin in the serum. In a serum of a monkeypox patient who had previously been vaccinated, the test sometimes identifies vaccinia antibody instead of monkeypox antibody, probably because of original antigenic sin. However, such cases can be correctly diagnosed as monkeypox, because their sera invariably have extraordinarily high titers by RIA and other serologic tests. Some African sera, and to a lesser extent some sera from the United States, have a residual titer which cannot be removed by adsorption with the vaccinia virus antigen. This unadsorbable "RIA titer" is probably nonspecific.

Using the parameters established to determine the titers by RIA, it was noted that a number of sera from unvaccinated children from Zaire aged 1 year or less had a mean titer of 200, and a number of sera from unvaccinated children from the United States aged 5 years or less had a mean titer of 96. Thus, we can conclude that an African serum with an RIA titer of less than 200 and a United States serum of less than 100 may be negative. However, since some African sera and many sera from U.S. subjects had no titer, we cannot conclude that all African sera with a titer of less than 200 and all U.S. sera with titer of less than 100 are essentially negative. The interpretation of each serum with "low" RIA titer depends on the results of other serologic tests, or the RIA titer must be verified by the adsorption test with vaccinia virus antigen.

Radioimmunoassay Adsorption with Staphlococcus Protein A

The capacity of staphyloccocus protein A (SpA) labeled with radioactive ^{125}I to bind with immunoglobulin of a number of wild animals in Africa has been reported (Richman et al., 1982; Lindmark et al., 1983). At the Centers for Disease Control, RIA using staphylococcus protein A labeled with radioactive ^{125}I has served well to assay orthopoxvirus antibody in several wild animals found in Zaire and other African countries. The use of labeled staphylococcus protein A was extended to RIAA for identification of specific monkeypox antibody in wild animals. The RIAA test with ^{125}I-labeled staphylococcus protein A identified monkeypox-specific antibody in man, monkeys, squirrels, and rabbits, and it identified vaccinia-specific antibody in man, monkeys, and rabbits.

Epidemiology and Natural History

Except for vaccinia and molluscum contagiosum, poxvirus infections of man are zoonoses—indeed, even vaccinia is sometimes contracted from an animal source. The epidemiology and natural history of such infections are therefore quite different from those of specifically human viral diseases, and they are different from the commonest kinds of viral zoonoses, since none of the chordopoxviruses are arboviruses.

Monkeypox

Human monkeypox is a zoonosis that is restricted to the inhabitants of villages in the tropical rain forests of certain countries in central and western Africa. It is a rare disease, and most cases are acquired from a wildlife source; a few are due to person-to-person spread. It is primarily a disease of children, with a case fatality rate of about 10%; there is no difference in incidence associated with season. Wildlife sources include squirrels (*Funisciurus anerythrus, Funisciurus lemniscatus,* and *Heliosciurus rufobrachium*), chimpanzees, and several species of monkeys. Squirrels are believed to constitute the reservoir host in some places. Infection of humans probably occurs through contact with infected animals or their organs. Person-to-person spread occurs with difficulty, the secondary attack rate in unvaccinated household contacts being 15.7%, compared with 58.4% in smallpox (Fenner et al., 1988). Serial transmission for four generations has been reported (Jezek et al., 1986a).

Vaccinia

Man is infected with vaccinia virus from one of three sources: deliberate inoculation, contact infection from an inoculated person, or infection from an animal source. Deliberate vaccination is still practiced among personnel of the armed forces in some countries, and it may be reintroduced more widely if genetically engineered vaccinia virus comes into use as a vector for genes of other infectious agents. Currently, the major risk of contact infection occurs when nonvaccinated persons come into close physical contact with vaccinated armed forces personnel; the consequences are most serious in eczematous children.

Infection of man from an animal source used to occur when vaccination was more widespread and cows and buffalo became infected by milkers; infection could then spread from lesions on affected cows or buffalo to man. It has been reported that buffalo still constitute a source of infection in India; it is not clear whether this is due to periodic reintroduction of vaccinia virus from human sources or whether "buffalopox" has become an enzootic disease of buffalo. In such outbreaks, infections of humans are commoner than formerly because of the cessation of routine vaccination against smallpox.

Cowpox

The traditional mode of infection of human beings with cowpox virus was (and is) by "inoculation" of the hands of milkers by contact with ulcers on the teats of cattle caused by cowpox virus (see Fig. 9A,B). This used to be the most commonly reported mode of infection, and over the years many such episodes have been reported. Cowpox due to cowpox virus was never a common disease, even in Jenner's time, and the occurrence of lesions of "spurious cowpox" on cow's teats gave rise to much confusion (Gibbs et al., 1970) (see below).

In recent years, Baxby (1977 and personal communication, 1983) has pointed out that cows were directly implicated as a source of cowpox virus in only three of 16 virologically confirmed cases in human beings in England between 1969 and 1981. Other human cases have been traced to infected zoo animals (Marennikova et al., 1977) and infected domestic cats (Bennett et al., 1986).

Neither cows (Gibbs et al., 1973), cats, nor zoo animals maintain cowpox virus enzootically; they are all, like man, incidental hosts. The reservoir hosts probably differ in different parts of Europe. The only virologically confirmed evidence that rodents are the reservoir hosts came from eastern Europe and the Soviet Union, where white rats (Marennikova et al., 1978a) and wild gerbils and susliks (Marennikova et al., 1978b) are enzootically infected. Cowpox virus does not occur outside of Europe and the Soviet Union; reports of "cowpox" in the Americas and Australia are due to confusion with vaccinia (Lum et al., 1967) or pseudocowpox, which is enzootic in cattle in the Americas.

Parapoxvirus Infections

In contrast to human orthopoxvirus infections, in which the reservoir host is unknown but infection of man may occur from any of a variety of incidental hosts (as well, presumably, as from the unknown reservoir host), the source of each of the three parapoxvirus infections of man is well known. Orf is contracted by persons (shearers, abattoir workers) who handle sheep or goats suffering from scabby mouth (see Fig. 9C); milker's nodules, by persons who milk cows with active lesions of pseudocowpox on their teats (see Fig. 9B); and bovine papular stomatitis, by veterinarians and others by handling infected calves.

In contrast to cowpox, pseudocowpox is an enzootic infection of cows, and infection may persist indefinitely in relatively small herds (Gibbs and Osborne, 1974). Infection of cows occurs mechanically via milkers' hands or the teat cups of milking machines, through small abrasions on the teats. Chronic lesions occur, seen as relatively mild scabbing, and immunity appears to be short-lived. Cyclical waves of reinfection occur in herds, especially in autumn and spring.

Bovine papular stomatitis is primarily a disease of beef cattle and their calves, caused by another species of parapoxvirus. Lesions on the muzzles of calves may produce lesions resembling those of pseudocowpox on their mothers' teats.

Orf (contagious pustular dermatitis, contagious ecthyma) is a common and cosmopolitan disease of sheep and goats, in which it causes papules and vesicles on the skin of the lips (Fig. 9C) and sometimes around the nostrils and eyes. Interference with suckling due to the facial lesions may lead to rapid emaciation. Ewes suckling affected lambs may develop lesions on their udders. Healing occurs, without scarring, in about a month. Persistence of the virus in flocks is due in large part to the persistent infectivity of virions in scabs that fall on pasture plants or the soil. Recovered animals are solidly immune, and vaccination of lambs or kids with a live, nonattenuated vaccine is practiced in some countries.

In North America and Europe, contagious pustular dermatitis also occurs in a variety of wild ungulates: reindeer, musk ox, Alaskan mountain goat, bighorn sheep, chamois, and Dall sheep (Robinson and Balassu, 1981). Such animals constitute possible sources of infection for hunters and taxidermists.

Tanapox

Human tanapox has been recognized in Kenya and Zaire but probably occurs much more widely in eastern, western and central Africa. In studies of 264 virologically confirmed cases in Zaire, Jezek et al. (1985) noted that cases occurred throughout the year, but mainly in the period November–March. There was considerable variation in the frequency between years and seasons, and cases occurred most frequently during the period when bloodsucking insects were most active. Cases were clustered geographically, and the vast majority occurred in children and other persons who spent a considerable amount of time on or near the river. Only a few cases occurred in hunters, traditional farmers, and plantation workers.

Tanapox virus produced epizootics in macacus monkeys in three primate centers in the United States in 1966 (McNulty et al., 1968). In each of these outbreaks, some animal handlers were infected, apparently through abrasions of the skin (references in Nakano, 1977).

After intradermal inoculation, tanapox virus produces skin lesions in both African and Asian monkeys (Downie et al., 1971), and because of its specificity for primates when laboratory animals were tested, it was suggested that monkeys might be the reservoir host. Positive sera have been found in a variety of species of African primates (Downie and Espana, 1973; Tsuchiya and Tagaya, 1970), but no investigations have been made of other wild animals, and it is not clear whether monkeys, like man, are only incidental hosts. Man may be infected by bloodsucking insects acting as mechanical vectors.

Molluscum Contagiosum

Molluscum contagiosum has a worldwide distribution. In some places—Papua New Guinea (Sturt et al., 1971), for example, and Zaire (Jezek, personal communication, 1985)—it is very common in children. It may occur sporadically or in small epidemics. Direct or indirect contagion appears to be the mode of spread, and in Western countries public baths and swimming pools may be implicated.

Among young adults it may be a sexually transmitted disease (Brown et al.,1981).

Prevention and Control

Orthopoxvirus infections, including human monkeypox, could be prevented by vaccination of exposed populations with vaccinia virus. However, human monkeypox is such a rare disease, occurring in areas and in a country in which resources are limited, that it is considered that the difficulties and expense of maintaining vaccination, and the risks associated with it, outweigh the benefits.

No method of prevention or control is known for molluscum contagiosum. For the other zoonotic poxvirus infections, most of which produce only trivial lesions, the only method of prevention is by avoidance of contact.

Literature Cited

Appleyard, G., A. J. Hapel, and E. A. Boulter. 1971. An antigenic difference between intracellular and extracellular rabbitpox virus. J. Gen. Virol. **13**:9–17.

Aragao, H. de B. 1927. Myxoma of rabbits. Mem. Inst. Oswaldo Cruz **20**:237–247.

Archard, L. C., and M. Mackett. 1979. Restriction endonuclease analysis of red coxpox virus and its white pock variant. J. Gen. Virol. **45**:51–63.

Archard, L. C., M. Mackett, D. E. Barnes, and K. R. Dumbell. 1984. The genome structure of cowpox virus white pock variants. J. Gen. Virol. **65**:875–886.

Arita, I., Z. Jezek, L. Khodakevich, and Kalisa Ruti. 1985. Human monkeypox: a newly emerged orthopoxvirus zoonosis in the tropical rain forests of Africa. Am. J. Trop. Med. Hyg. **34**:781–789.

Baroudy, B. M., S. Venkatesan, and B. Moss. 1982. Incompletely base-paired flip-flop terminal loops link the two DNA strands of the vaccinia virus genome into one uninterrupted polynucleotide chain. Cell **28**:315–324.

Baxby, D. 1977. Is cowpox misnamed? A review of 10 human cases. Br. Med. J. **1**:1379–1381.

Baxby, D., and B. J. Hill. 1971. Characteristics of a new poxvirus isolated from Indian buffaloes. Arch. Ges. Virusforsch. **35**:70–79.

Bedson, H. S., and K. R. Dumbell. 1964a. Hybrids derived from the viruses of alastrim and rabbit pox. J. Hyg. (Lond.) **62**:141–146.

Bedson, H. S., and K. R. Dumbell. 1964b. Hybrids derived from the viruses of variola major and cowpox. J. Hyg. (Lond.) **62**:147–158.

Bennett, M., C. J. Gaskell, R. M. Gaskell, D. Baxby, and T. J. Gruffydd-Jones. 1986. Poxvirus infection in the domestic cat: some clinical and epidemiological observations. Vet. Rec. **118**:387–390.

Blanden, R. V. 1971. Mechanisms of recovery from a generalized viral infection: mousepox III. Regression of infectious foci. J. Exp. Med. **133**:1090–1104.

Boulter, E. A., and G. Appleyard. 1973. Differences between extracellular and intracellular forms of poxvirus and their implications. Prog. Med. Virol. **16**:86–108.

Bras, G. 1952a. The morbid anatomy of smallpox. Doc. Med. Geog. Trop. **4**:303–351.

Bras, G. 1952b. Observations on the formation of smallpox scars. Arch. Pathol. **54**:149–156.

Brown, S. T., J. F. Nalley, and S. J. Kraus. 1981. Molluscum contagiosum. Sex. Trans. Dis. **8**:227–234.

Centers for Disease Control–National Institutes of Health. 1984. Biosafety in microbiology and biomedical laboratories, HHS publication No. (CDC) 84-8395. U.S. Government Printing Office, Washington, D.C.

Chang, T., and L. Weinstein. 1961. Cytopathic agents isolated from lesions of molluscum contagiosum. J. Invest. Dermatol. **37**:433–439.

Cole, G. A., and R. V. Blanden. 1982. Immunology of poxviruses, p. 1–19. *In* A. J. Nahmias, and R. J. O'Reilly (ed.), Comprehensive immunology, vol. 9. Immunology of human infection, part II: viruses and parasites; immunodiagnosis and prevention of infectious diseases. Plenum Press, New York.

Dales, S., and B. G. T. Pogo. 1981. Biology of poxviruses. Virol. Monogr. **18**:1–109.

Dales, S., and L. Siminovitch. 1961. The development of vaccinia virus in Earle's strain L cells as examined by electron microscopy. J. Biophys. Biochem. Cytol. **10**:475–503.

Dennis, J., L. S. Oshiro, and J. W. Bunter. 1985. Molluscum contagiosum, another sexually transmitted disease: its impact on the clinical virology laboratory. J. Infect. Dis. **151**:376.

Downie, A. W., and C. Espana. 1973. A comparative study of tanapox and yaba viruses. J. Gen. Virol. **19**:37–49.

Downie, A. W., C. H. Taylor-Robinson, A. E. Caunt, G. S. Nelson, P. E. C. Manson-Bahr, and T. H. C. Matthews. 1971. Tanapox: a new disease caused by a poxvirus. Br. Med. J. **1**:363–368.

Dumbell, K. R., and L. C. Archard. 1980. Comparison of white pock mutants of monkeypox virus with parental monkeypox and with variola-like viruses isolated from animals. Nature **286**:29–32.

Dumbell, K. R., and H. S. Bedson. 1966. Adaptation of variola virus to growth in the rabbit. J. Pathol. Bacteriol. **91**:459–465.

Dumbell, K. R., and J. G. Kapsenberg. 1982. Laboratory investigation of two "whitepox" viruses and comparison with two variola strains from southern India. Bull. WHO **60**:381–387.

Esposito, J. J., and J. C. Knight. 1984. Nucleotide sequence of the thymidine kinase gene region of monkeypox and variola viruses. Virology **135**:561–567.

Esposito, J. J., J. C. Knight. 1985. Orthopoxvirus DNA: a comparison of restriction profiles and maps. Virology **143**:230–251.

Esposito, J. J., R. Condit, and J. Obijeski. 1981. The preparation of orthopoxvirus DNA. J. Virol. Methods **2**:175–179.

Essani, K., and S. Dales. 1979. Biogenesis of vaccinia: evidence for more than 100 polypeptides in the virion. Virology **95**:385–394.

Ewart, W. B. 1983. Causes of mortality in a subarctic settlement (York Factory, Man.) 1713–1946. Can. Med. Assoc. J. **129**:571–574.

Fenner, F. 1958. The biological characters of several strains of vaccinia, cowpox and rabbitpox viruses. Virology **5**:502–529.

Fenner, F. 1976. Classification and nomenclature of viruses. Second report of the International Committee on Taxonomy of Viruses. Intervirology **7**:1–115.

Fenner, F., and F. M. Burnet. 1957. A short description of the poxvirus group (vaccinia and related viruses). Virology **4**:305–314.

Fenner, F., and J. F. Sambrook. 1966. Conditional lethal

mutants of rabbitpox virus. II. Mutants (*p*) that fail to multiply in PK-2a cells. Virology **28**:600–609.

Fenner, F., D. A. Henderson, I. Arita, Z. Jezek, and I. D. Ladnyi. 1988. Smallpox and its eradication. World Health Organization, Geneva.

Fornaciari, G., and A. Marchetti. 1986. Intact smallpox virus particles in an Italian mummy of sixteenth century. Lancet **2**:625.

Francis, R. D., and H. B. Bradford, Jr. 1976. Some biological and physiological properties of molluscum contagiosum virus propagated in cell culture. J. Virol. **19**:382–388.

Fulginiti, V. A., C. H. Kempe, W. E. Hathaway, D. S. Pearlman, O. F. Serber, Jr., J. J. Eller, J. J. Joyner, Sr., and A. Robinson. 1968. Progressive vaccinia in immunologically deficient individuals. Birth Defects **4**:129–145.

Garon, C. F., E. Barbosa, and B. Moss. 1978. Visualization of an inverted terminal repetition in vaccinia virus DNA. Proc. Natl. Acad. Sci. USA **75**:4863–4867.

Gassman, U., R. Wyler, and R. Wittek. 1985. Analysis of parapoxvirus genomes. Arch. Virol. **83**:17–31.

Geshelin, P., and K. I. Berns. 1974. Characterization and localization of the naturally occurring cross-links in vaccinia virus DNA. J. Mol. Biol. **88**:785–796.

Gibbs, E. P. J., and A. D. Osborne. 1974. Observations on the epidemiology of pseudocowpox in south-west England and south Wales. Br. Vet. J. **130**:150–159.

Gibbs, E. P. J., R. H. Johnson, and A. D. Osborne. 1970. The differential diagnosis of viral skin infections of the bovine teat. Vet. Rec. **87**:602–609.

Gibbs, E. P. J., R. H. Johnson, and D. F. Collings. 1973. Cowpox in a dairy herd in the United Kingdom. Vet. Rec. **92**:56–64.

Gispen, R., and B. Brand-Saathof. 1972. "White" poxvirus from monkeys. Bull. WHO **46**:585–592.

Gispen, R., B. Brand-Saathof. 1974. Three specific antigens produced in vaccinia, variola and monkeypox infections. J. Infect. Dis. **129**:289–295.

Gispen, R., J. Huisman, B. Brand-Saathof, and A. C. Hekker. 1974. Immunofluorescence test for persistent poxvirus antibodies. Arch. Ges. Virusforsch. **44**:391–395.

Goodpasture, E. W. 1933. Borreliotoses: fowl-pox, molluscum contagiosum, variola-vaccinia. Science **77**:119–121.

Grace, J. T., Jr., and E. A. Mirand. 1963. Human susceptibility to a simian tumor virus. Ann. N.Y. Acad. Sci. **108**:1123–1128.

Heiner, G. G., N. Fatima, R. W. Daniel, J. L. Cole, R. L. Anthony, and F. R. McCrumb, Jr. 1971. A study of inapparent infection in smallpox. Am. J. Epidemiol. **94**:252–268.

Horner, G. W., A. J. Robinson, R. Hunter, B. T. Cox, and R. Smith. 1987. Parapoxvirus infections in New Zealand red deer (*Cervus elaphus*). N.Z. Vet. J. **35**:41–45.

Hutchinson, H. D., D. W. Ziegler, D. E. Wells, and J. H. Nakano. 1977. Differentiation of variola, monkeypox, and vaccinia antisera by radioimmunoassay. Bull. WHO **55**:613–623.

Ikuta, K., H. Miyamoto, and S. Kato. 1979. Serologically cross-reactive polypeptides in vaccinia, cowpox and Shope fibroma viruses. J. Gen. Virol. **44**:557–563.

Jezek, Z., I. Arita, M. Szczeniowski, M. Paluku, R. Kalisa, and J. H. Nakano. 1985. Human tanapox in Zaire: clinical and epidemiological observations on cases confirmed by laboratory examination. Bull. WHO **63**:1027–1035.

Jezek, Z., I. Arita, M. Mutombo, C. Dunn, J. H. Nakano, and M. Szczeniowski. 1986a. Four generations of proba-

ble person-to-person transmission of human monkeypox. Am. J. Epidemiol. **123**:1004–1012.

Jezek, Z., S. S. Marennikova, M. Mutumbo, J. H. Nakano, K. M. Paluka, and M. Szczeniowski. 1986b. Human monkeypox: a study of 2510 contacts of 214 patients. J. Infect. Dis. **154**:551–555.

Johannessen, J. V., H.-K. Krogh, I. Solberg, A. Dalen, H. van Wijngaarden, and B. Johansen. 1975. Human orf. J. Cutan. Pathol. **2**:265–283.

Johnson, R. T. 1982. Viral infections of the nervous system. Raven Press, New York.

Kato, S., M. Takahashi, S. Kameyama, and J. Kamahora. 1959. A study on the morphological and cyto-immunological relationship between the inclusions of variola, cowpox, rabbitpox, vaccinia (variola origin) and vaccinia IHD and a consideration of the term "Guarnieri body." Biken J. **2**:353–363.

Khodakevich, L., and I. Arita. 1985. Investigating rumours of smallpox. World Health Forum **6**:171–173.

Kluge, J. P., N. F. Cheville, and T. M. Peery. 1972. The pathogenesis of cantagious ecthyma. Am. J. Vet. Res. **33**:1191–1200.

Lake, J. R., and P. D. Cooper. 1980. Deletions of the terminal sequences in the genomes of the white pock (*u*) and host-restricted (*p*) mutants of rabbitpox virus. J. Gen. Virol. **48**:135–147.

Lane, J. M., F. L. Ruben, J. M. Neff, and J. D. Millar. 1970. Complications of smallpox vaccination, 1968: result of ten statewide surveys. J. Infect. Dis. **122**:303–309.

La Placa, M. 1966. On the mechanism of the cytopathic changes produced in human amnion cell culture by the molluscum contagiosum virus. Arch. Ges. Virusforsch. **18**:374–378.

Lewin, P. K. 1967. Palaeo-electron microscopy of mummified tissue. Nature **213**:416–417.

Lewin, P. K. 1977. Mummies that I have known. Am. J. Dis. Child. **131**:349–350.

Lindmark, R., K. Thoren-Tolling, and J. Sjoquist. 1983. Binding of immunoglobulins to protein A and immunoglobulin levels in mammalian sera. J. Immunol. Methods **62**:1–13.

Lum, G. S., F. Soriano, A. Trejos, and J. Llerana. 1967. Vaccinia epidemic and epizootic in San Salvador. Am. J. Trop. Med. Hyg. **16**:332–338.

Mackett, M., and L. C. Archard. 1979. Conservation and variation in *Orthopoxvirus* genome structure. J. Gen. Virol. **45**:683–701.

Marennikova, S. S., E. M. Seluhina, N. N. Mal'ceva, K. L. Cimiskjan, and G. R. Macevic. 1972. Isolation and properties of the causal agent of a new variola-like disease (monkeypox) in man. Bull. WHO **46**:599–611.

Marennikova, S. S., N. N. Maltseva, V. I. Korneeva, and N. M. Garanina. 1977. Outbreak of pox disease among carnivora (Felidae) and Edentata. J. Infect. Dis. **135**:358–366.

Marennikova, S. S., E. M. Shelukhina, and V. A. Fimina. 1978a. Pox infection in white rats. Lab. Anim. **12**:33–36.

Marennikova, S. S., I. D. Ladnyj, Z. I. Ogorodnikova, E. M. Shelukhina, and N. N. Maltseva. 1978b. Identification and study of a poxvirus isolated from wild rodents in Turkmenia. Arch. Virol. **56**:7–14.

Marennikova, S. S., N. N. Malceva, and N. A. Habahpaseva. 1981. ELISA—a simple test for detecting and differentiating antibodies to closely related orthopoxviurses. Bull. WHO **59**:365–369.

Matthews, R. E. F. 1979. Classification and nomenclature of viruses. Third report of the International Committee on the Taxonomy of Viruses. Intervirology **12**:132–296.

McFadden, G., W. E. Pace, J. Purres, and S. Dales. 1979.

Biogenesis of poxviruses: transitory expression of molluscum contagiosum early functions. Virology **94**:297–313.

McNulty, W. P., Jr., W. C. Lobitz, F. Hu, C. A. Maruppo, and A. S. Hall. 1968. A pox disease in monkeys transmitted to man. Arch. Dermatol. **97**:286–293.

Menna, A., R. Wittek, P. A. Bachmann, A. Mayr, and R. Wyler. 1979. Physical characterization of a stomatitis papulosa genome: a cleavage map for the restriction endonucleases *Hind*III and *Eco* RI. Arch. Virol. **59**:145–156.

Mercer, A. A., K. Fraser, G. Barns, and A. J. Robinson. 1987. The structure and cloning of orf virus DNA. Virology **157**:1–12.

Mitra, A. C., J. K. Sarkar, and M. K. Mukherjee. 1974. Virus content of smallpox scabs. Bull. WHO **51**:106–107.

Morrison, D. K., and R. W. Moyer. 1986. Detection of a subunit of cellular pol II within highly purified preparations of RNA polymerase isolated from rabbit pox virus. Cell **44**:587–596.

Moyer, R. W., and C. T. Rothe. 1980. The white pock mutants of rabbit poxvirus. I. Spontaneous host range mutants contain deletions. Virology **102**:119–132.

Nagington, J. 1968. The growth of paravaccinia viruses in tissue culture. Vet. Rec. **82**:477–482.

Nagington, J., and R. W. Horne. 1962. Morphological studies of orf and vaccinia viruses. Virology **16**:248–260.

Nagington, J., A. A. Newton, and R. W. Horne. 1964. The structure of orf virus. Virology **23**:461–472.

Nakano, J. H. 1977. Comparative diagnosis of poxvirus diseases, p. 287–339. *In* E. Kurstak and C. Kurstak (ed.), Comparative diagnosis of viral diseases, vol. 1. Academic Press, New York.

Nakano, J. H. 1979. Poxviruses, p. 257–308. *In* E. H. Lennette and N. J. Schmidt (ed.), Diagnostic procedures for viral, rickettsial and chlamydial infections, 5th ed. American Public Health Association, Washington, D.C.

Nakano, J. H. 1982. Human poxvirus diseases and laboratory diagnosis, p. 125–147. *In* L. M. de la Maza and E. M. Peterson (ed.), Medical virology. Elsevier, New York.

Nakano, J. H. 1985a. Human poxvirus diseases, p. 401–424. *In* E. H. Lennette (ed.), Laboratory diagnosis of viral infections. Marcel Dekker, New York.

Nakano, J. H. 1985b. Poxviruses, p. 733–741. *In* E. H. Lennette, A. Balows, W. J. Housler, Jr., and H. J. Shadomy (eds.), Manual of clinical microbiology, 4th ed. American Society for Microbiology, Washington, D.C.

Nakano, J. H. 1986. Poxviruses, p. 417–435. *In* S. Specter, and G. C. Lancz (ed.), Clinical virology manual. Elsevier, New York.

Neva, F. A. 1962. Studies on molluscum contagiosum. Arch. Intern. Med. **110**:720–725.

Parr, R. P., J. W. Burnett, and C. F. Garon. 1977. Structural characterization of the molluscum contagiosum virus genome. Virology **81**:247–256.

Patel, D. D., D. J. Pickup, and W. K. Joklik. 1986. Isolation of cowpox A-type inclusions and characterization of their major protein component. Virology **149**:174–189.

Payne, L. G. 1978. Polypeptide composition of extracellular enveloped vaccinia virus. J. Gen. Virol. **27**:28–37.

Payne, L. G. 1980. Significance of extracellular enveloped virus in the in vitro and in vivo dissemination of vaccinia. J. Gen. Virol. **50**:89–100.

Payne, L. G., and K. Kristensson. 1979. Mechanism of vaccinia virus release and its specific inhibition by N_1-isonicotinoyl-N_2-3-methyl-4-chlorobenzoylhydrazine. J. Virol. **32**:614–622.

Payne, L. G., and E. Norrby. 1976. Presence of haemagglutinin in the envelope of extracellular vaccinia virus particles. J. Gen. Virol. **32**:63–72.

Peters, D., G. Muller, and D. Buttner. 1964. The fine structure of paravaccinia viruses. Virology **23**:609–611.

Quinnan, G. V., Jr. (ed.). 1985. Vaccinia viruses as vectors for vaccine antigens. Elsevier, New York.

Rafii, F., and D. Burger. 1985. Comparison of ecthyma virus genomes by restriction endonucleases. Arch. Virol. **84**:283–289.

Raskin, J. 1963. Molluscum contagiosum. Arch. Dermatol. **87**:552–559.

Richman, D. D., P. H. Cleveland, M. N. Oxman, and K. M. Johnson. 1982. The binding of staphylococcal protein A by the sera of different animal species. J. Immunol. **128**:2300–2305.

Robinson, A. J., and T. C. Balassu. 1981. Contagious pustular dermatitis (orf). Vet. Bull. **51**:771–782.

Robinson, A. J., and G. V. Peterson. 1983. Orf virus infection of workers in the meat industry. N.Z. Med. J. **96**:81–85.

Robinson, A. J., G. Ellis, and T. Balassu. 1982. The genome of orf virus: restriction endonuclease analysis of viral DNA isolated from lesions of orf in sheep. Arch. Virol. **71**:43–55.

Robinson, A. J., G. Barns, K. Fraser, E. Carpenter, and A. A. Mercer. 1987. Conservation and variation in orf virus genomes. Virology **157**:13–23.

Rosenbusch, R. F., and D. F. Reed. 1983. Reaction of convalescent bovine antisera with strain-specific antigens of parapoxviruses. Am. J. Vet. Res. **44**:875–878.

Shida, H., K. Tanabe, and S. Matsumoto. 1977. Mechanism of virus occlusion into A-type inclusion during poxvirus infection. Virology **76**:217–233.

Silver, M., G. McFadden, S. Wilton, and S. Dales. 1979. Biogenesis of poxviruses: role for the DNA dependent RNA polymerase II of the host cell during expression of late functions. Proc. Natl. Acad. Sci. USA **76**:4122–4125.

Soloski, M. J., and J. A. Holowczak. 1981. Characterization of supercoiled nucleoprotein complexes released from detergent-treated vaccinia virions. J. Virol. **37**:770–783.

Stern, W., and S. Dales. 1976. Biogenesis of vaccinia: isolation and characterization of a surface component that elicits antibody suppressing infectivity and cell-cell fusion. Virology **75**:232–241.

Sturt, R. J., H. K. Muller, and G. D. Francis. 1971. Molluscum contagiosum in villages of the West Sepik district of New Guinea. Med. J. Aust. **2**:751–754.

Takahashi, M., S. Kameyama, S. Kato, and J. Kamahora. 1959. The immunological relationship of the poxvirus group. Biken J. **2**:27–29.

Tsuchiya, Y., and I. Tagaya. 1970. Plaque assay of variola virus in a cynomolgus monkey kidney cell line. Arch. Ges. Virusforsch. **32**:73–81.

Voller, A., D. Bidwell, and A. Bartlett. 1976. Microplate enzyme immunoassays for the immunodiagnosis of virus infections, p. 506–512. *In* N. E. Rose and H. Friedman (ed.), Manual of clinical immunology. American Society for Microbiology, Washington, D.C.

Walls, H. H., D. W. Zeigler, and J. H. Nakano. 1980. A study of the specificities of sequential antisera to variola and monkeypox viruses by radioimmunoassay. Bull. WHO **58**:131–138.

Walls, H. H., D. W. Ziegler, and J. H. Nakano. 1981. Characterization of antibodies to orthopoxviruses in human sera by radioimmunoassay. Bull. WHO **59**:253–262.

Westwood, J. C. N., W. J. Harris, H. T. Zwartouw, O. H. J. Titmuss, and G. Appleyard. 1964. Studies on the structure of vaccinia virus. J. Gen. Microbiol. **34**:67–78.

Wildy, P. 1971. Classification and nomenclature of viruses. First report of the International Committee on Nomenclature of Viruses. Monogr. Virol. **5:**1–81.

Wilson, G. S. 1967. The hazards of immunization. Athlone, London.

Wittek, R., and B. Moss. 1980. Tandem repeats within the inverted terminal repetition of vaccinia virus DNA. Cell **21:**277–284.

Wittek, R., A. Menna, D. Schumperli, S. Stoffel, H. K. Muller, and R. Wyler. 1977. HindIII and SstI restriction sites mapped on rabbit poxvirus and vaccinia virus DNA. J. Virol. **23:**669–678.

Wittek, R., C. C. Kuenzle, and R. Wyler. 1979. High C + G content in parapoxvirus DNA. J. Gen. Virol. **43:**231–234.

Wittek, R., M. Herlyn, D. Schumperli, P. A. Bachmann, A. Mayr, and R. Wyler. 1980. Genetic and antigenic heterogeneity of different parapoxvirus strains. Intervirology **13:**33–41.

Wolff, H. L., and J. J. A. B. Croon. 1968. The survival of smallpox virus (variola minor) in natural circumstances. Bull. WHO **38:**492–493.

Woodroofe, G. M. 1960. The heat inactivation of vaccinia virus. Virology **10:**379–382.

Woodroofe, G. M., and F. Fenner. 1960. Genetic studies with mammalian poxviruses. IV. Hybridization between several different poxviruses. Virology **12:**272–282.

Woodroofe, G. M., and F. Fenner. 1962. Serological relationships within the poxvirus group: an antigen common to all members of the group. Virology **16:**334–341.

Yohn, D. S., and J. F. Gallagher. 1969. Some physicochemical properties of yaba poxvirus deoxyribonucleic acid. J. Virol. **3:**114–118.

Ziegler, D. W., H. D. Hutchinson, J. P. Koplan, and J. H. Nakano. 1975. Detection by radioimmunoassay of antibodies in human smallpox patients and vaccinees. J. Clin. Microbiol. **1:**311–317.

CHAPTER 11

Herpesviridae: Herpes Simplex Virus

ERIK LYCKE and STIG JEANSSON

Disease: Herpes simplex virus infections.

Etiologic Agents: Herpes simplex virus types 1 and 2.

Source: Human to human spread of infection.

Clinical Manifestations: Oropharyngeal infections, eczema herpeticum, keratoconjunctivitis, genital infections, neonatal infections, encephalitis, meningoencephalitis, nosocomial infections, and infections in the immunocompromised patient.

Pathology: Primary and recurrent infections, the latter originating from activation of infection in latently infected sensory neurons. Commonly, a restricted mucocutaneous vesiculation and ulceration; occasionally, a necrotizing systemic disease. Type 1 virus is commonly associated with oropharyngeal infections, type 2 is commonly associated with genital infections.

Laboratory Diagnosis: Virus isolation or demonstration of viral antigens in secretions or infected tissue specimens. Serology by observations of specific IgG, IgM, and/or IgA antibody response. Typing of virus by type-specific, usually monoclonal antibodies; demonstration of virus-type-related antibodies by means of type-specific viral antigen preparations.

Epidemiology: Worldwide, most commonly subclinical infection, transmission essentially by virus-containing secretions from oropharyngeal or genitourinary tracts. Circulation of virus is maintained by spread of virus from activated latent infections.

Treatment: Antiviral drugs such as acyclovir (Zovirax).

Prevention and Control: Active immunization with experimental vaccines and passive immunization with hyperimmune immunoglobulin preparations are under evaluation.

Introduction

The approximately 80 different herpesviruses recognized today are widespread throughout the animal kingdom and have a ubiquitous and worldwide distribution. Almost every class of animal harbors one or more members of the herpesvirus family. During the course of evolution, the herpesviruses have become markedly aligned to their different hosts. Because of this association of the virus with its particular host, the herpesviruses in their natural environment mainly cause subclinical infections with symptom-free shedding of virus, whereas severe disease is relatively uncommon.

Although disease is a less common consequence of infection, the number of severe and life-threatening conditions induced by herpes simplex virus (HSV) infection is significant, primarily because of its high frequency. Thus, HSV is the most important cause of sporadic encephalitis and blindness in the western world and is responsible for one of the most widespread sexually transmitted diseases. Indications of a gradually increasing spread of genital infections associated with an apparently growing incidence of neonatal infections have attracted both interest and concern (Jarrat, 1983; Mindel and Sutherland, 1983; Nilsen, 1985).

In the Hippocratic medicine diseases are described that clearly bear the hallmarks of HSV, suggesting that these infections have probably followed humanity since the earliest period of our history (Wildy, 1973). The name *herpes* has its origin in the

Greek word *herpein* and alludes to a slowly progressing (creeping) vesicular rash of the skin. The infectious character of herpetic skin eruptions was probably first shown in 1873 by Vidal, but the experimental transfer of infection from cases of human herpetic keratitis and labial eruptions to the rabbit cornea is generally attributed to Löwenstein (1919). In 1920, Gruter reported on the successful transmission of the infection from human being to rabbit and from an experimentally infected rabbit back to the normal cornea of a blind man. Through experimental studies in laboratory animals in the 1930s and 1940s, the viral nature of the infectious agent gradually became apparent, and when production of large amounts and an accurate quantification of the infective virus became feasible with the chick embryo and cell culture techniques, the biologic characteristics of the viruses and the medical importance of herpetic infections could be elucidated.

The demonstration of neutralizing antibodies against HSV (Andrewes and Carmichael, 1930; Zinsser and Tang, 1929) offered the first means of distinguishing between primary and reactivated HSV infections. However, it was not until it became possible to differentiate between antibodies induced by HSV type 1 and type 2 infections that primary and recurrent infections could be distinguished by serologic methods. In fact, this differentiation is a fairly recent achievement, although the basis for type-distinguishing serology and typing of isolates was formed at the end of the 1960s (Nahmias and Dowdle, 1968).

In the 1970s, molecular biology provided the final documentation for the existence of genetic and biochemical differences between the two types of HSV (Nahmias and Roizman, 1973a,b,c). It became evident that both types of HSV could cause primary as well as recurrent genital herpes infections (Reeves et al., 1981) and that genital herpes was not exclusively a manifestation of HSV type 2. Epidemiologic studies revealed that genital HSV infections constituted one of the most important sexually transmitted diseases. For this and other reasons, genital HSV infection was selected as the main target for developing antiviral therapy.

The Virus

The gross anatomy of the deoxyribonucleic acid (DNA) genome is the same for all herpesviruses (Roizman, 1982). More than 80% of the genome is formed by two large unique components, designated U_l (unique long) and U_s (unique short). These components are linked to each other by smaller repeated nucleotide sequences, giving the impression that during evolution two primarily free genes might have been linked to each other. The small repeated sequences appear in an inverted order at one end relative to their order at the other end of the unique components and, because they can function as molecular links between the larger genome components, this arrangement facilitates genome circularization (Hayward et al., 1975). The DNA is replicated from the circularized genome.

Nucleotide sequencing of the HSV genome (molecular weight approximately 100 million) has revealed more than 50% homology between the two types, and that certain sequences are homologous for several herpesviruses. For instance, the genes of glycoprotein B (gB) of HSV and varicella-zoster virus show about 70% homology, and the gB gene of HSV-1 and HSV-2 share about 90% nucleotide and amino acid homology (Bzik et al., 1984; Snowden et al., 1985). This homology is one probable reason for some of the cross-reactions in serologic tests that may be encountered with certain types of antigens from different herpesviruses.

The HSV genes and their products are divided into three different classes: alpha, beta, and gamma genes corresponding to immediate early, delayed early, and late proteins. Approximately 50 polypeptides coded for by the HSV genome have been detected. The alpha genes are responsible for regulating gene functions and for enhancing and counteracting other gene activities. The beta proteins include the viral enzymes, such as viral thymidine kinase and DNA polymerases. Like these enzymes, two of the glycoproteins (gB and gD) are formed essentially before synthesis of virus DNA is initiated, and they are therefore also classified as beta proteins. Other structural proteins of the virion (i.e., capsid proteins and the other envelope glycoproteins) are represented among the gamma proteins. The interdependence in expression and control of the alpha, beta, and gamma genes via their corresponding proteins is often termed *cascade phenomenon,* as synthesis of beta and gamma proteins is dependent on preceding expression of alpha and beta genes, respectively, and shutoff of alpha and beta gene activities in turn is controlled via beta and gamma proteins.

The diameter of the virus nucleocapsid is 100 nm, and the enveloped virion measures from 150 to 250 nm. The double-stranded DNA is coiled around a proteinaceous toroid, which is surrounded by 12 pentameric and 150 hexameric capsomeres arranged in icosahedral symmetry. The space between the nucleocapsid and the envelope is called the *tegument* and seems to contain fiberlike structures that connect the nucleocapsid with the envelope. This latter structure is composed of a lipid membrane, derived from the inner lamina of the infected cell's nuclear membrane and cytoplasmic membranes, and of glycoproteins coded for by the virus.

At least six different glycoproteins, designated B, C, D, E, G, and H, form projections on the surface of the envelope. These glycoproteins are important for attachment, penetration, and cell fusion reactions as well as for the egress of newly formed virions (Spear, 1985). In addition, they constitute important surface antigens of the virus. Because glycoproteins C and E can function as receptors for complement factor (C3b) and F_c fragments of immunoglobulin (IgG), respectively, it is possible that these glycoproteins can modify the immune response of the host. The Fc receptor can sometimes create problems in diagnostic virology.

Viral Replication

Glycoproteins C and D have been implicated in the attachment of HSV to cellular receptors specific for either type 1 or type 2 (Vahlne et al., 1979). In the electron microscope, adsorbed virus is found attached to the plasma membrane. However, unlike most other enveloped viruses, HSV seems to be internalized mainly by fusion reactions between the envelope and plasma membrane rather than by a receptor-mediated endocytosis. The nucleocapsid is transported to the cell nucleus and the DNA is released at the nuclear pore. Translation and DNA replication take place in the cell nucleus. Within 4 h after infection, the protein synthesis of the infected cell is shut off. This shutoff is manifested primarily by degradation of host ribosomal ribonucleic acid (RNA) and seems to be initiated by a structural protein of the infecting virus, perhaps also controlled via one of the immediate, early formed viral proteins (Read and Frenkel, 1983).

Transcription of viral DNA depends on cellular RNA polymerase II. The HSV messenger (m)RNA transcripts are produced and translated in a way not unlike that of cellular mRNA. Splicing does not seem to be as common with HSV as with other herpesviruses. Viral proteins formed are transported back to the nucleus, and when alpha and beta protein synthesis is completed, the viral DNA is replicated. Two kinds of specific viral enzymes, thymidine kinase and DNA polymerase, have become the main targets for research and industrial efforts to create antiviral drugs.

The DNA molecule is packed into preformed capsids in the cell nucleus, and the nucleocapsids bind to the inner lamella of the nuclear membrane. It is uncertain whether the virus is permanently enveloped when it passes through the nuclear membrane or if it

is first enveloped, then de-enveloped. The final envelopment occurs in the endoplasmic reticulum and the Golgi apparatus, where the envelope proteins become glycosylated.

Both N-linked and O-linked oligosaccharides have been demonstrated in HSV-induced glycoproteins. The sugars seem important as stabilizers, but may also have other functions (e.g., during intracellular transport of the virus). For instance, viral release seems to be achieved by a process that mimics exocytosis and seems dependent on the function of glycoprotein H (Spear, 1985). In the course of infection, the viral glycoproteins appearing on the surface of the plasma membrane become the main targets for the immune reactions induced by the infection.

Pathogenesis and Pathology

Latency

Latency is a well-recognized, but still poorly defined phenomenon. It is demonstrable by transforming the nonproductive state of the latent infection into viral production (i.e., reactivation). Latency plays a considerable role in the epidemiology of HSV infections and circulation of virus, as the reactivation of latent infections is responsible for the appearance of recurrent infections. Latent HSV infections are established in the ganglionic sensory neurons. Probably most HSV infections of mucocutaneous tissues lead to neuritic uptake and retrograde axonal transport of the virus to ganglionic nerve cells (Kristensson et al., 1972, Lycke et al., 1984). How the HSV infection is restricted and maintained during latency remains incompletely understood. Whether the virus genome is integrated into cellular DNA or the genome exists as a plasmidlike structure, as in latent Epstein–Barr virus infection, is uncertain. Recent findings would favor the latter hypothesis, as virus DNA extracted from latently infected cells has been reported to exist in an endless (i.e., circular) form (Rock and Frazer, 1983). Moreover, antiviral drugs with the ability to block specific viral enzymes have no effect on established latent infections, suggesting that viral genomes in latently infected dividing cells are probably replicated by cellular polymerases (Klein and Friedmann-Kien, 1977; Svennerholm et al., 1981). In fact, several observations indicate that the restriction of the infection during latency might be controlled by the infected cell. The restricted turnover of DNA in mature neurons might be of importance; however, it is also possible that the nerve cells, upon exposure to virus, actively develop the antiviral state by forming factors blocking virus replication.

Reactivation of the latent viral genome with resulting viral replication depends on several different stimuli, including mechanical, chemical, hormonal, and radiative. Some of these treatments can obviously interfere with a possible cell-mediated restriction of infection, whereas others might primarily affect the immunologic surveillance of reactivated infections. Reactivation appears to be relatively common. Most often, however, the virus produced is eliminated rapidly by immune reactions before the infection spreads and clinical symptoms appear. Therefore, the only manifestation of reactivation of a latent infection might sometimes be asymptomatic shedding of virus (Centifanto et al., 1972). As the sensory ganglion can be infected with more than one virus, isolates from reactivated infections in an individual are not necessarily of the same virus strain (Strand et al., 1987), and presumably the epidemiologic picture is further complicated by recombination of strains.

Cellular Changes

Mucous membranes and skin erosions are the most common portals of entry of HSV infections, and epithelial cells seem resistant to infection only as long as their cornified surface is intact. Intraepidermal vesicles are formed when the underlying epithelial cells, particularly the prickle cells, become infected. Foci with ballooning cell degeneration and a gradually developing inflammatory reaction constitute the vesicular and subsequent pustular changes.

Productively infected cells die as a result of metabolic and dysfunctional changes induced by viral replication. Insertion of viral glycoproteins into cellular membranes is probably important for some of the alterations of cellular membrane functions, revealed, for example, by changes in ion transport and increased permeability (Frietz and Nahmias, 1972). In the infected organism, cell death is caused by immune reactions directed against the viral antigens exposed on the cell surface, as well as changes induced by viral replication.

The delayed type of hypersensitivity mediated by certain subsets of T cells and class II restricted cells is important not only for protection against HSV infection (Nash et al., 1981), but also because of the deleterious immunopathologic reactions provoked. For instance, the stromal type of keratitis associated with ocular HSV infection seems to be elicited by cellular immune mechanisms.

Since the discovery that HSV carries nucleotide sequences with apparent homology to pieces of cell DNA (Puga et al., 1982) and potential oncogenes (Camacho and Spear, 1978; Reyes et al., 1979), the discussion about the involvement of HSV in cell transformation and induction of tumors has been renewed. Transformation of permissive cells by HSV can occur only seldom, as normally almost all infected cells will die of the infection. With partially inactivated virus or defective virus with a reduced ability to replicate, transformation is encountered with greater frequency.

Absence of specific protein markers in transformed cells, the considerable variability in the amounts of viral DNA retained in transformed cells, and a gradual loss of viral DNA when HSV-transformed cells are cultured have established the basis of the so-called "hit and run" theory. According to this hypothesis, the HSV infection is regarded as an effect triggering transformation but is less important in maintaining the state of transformation. Thus, so far it cannot be completely excluded that under certain conditions HSV, and presumably defective virus, might be able to activate cellular oncogenes. Whether this activity is a result of gene translocation, transfer of cellular oncogenes, effects on cellular control mechanisms, or mutations is unknown.

Clinical Features

Generally, it is rational to describe the clinical signs of HSV infections by distinguishing between infections caused by type 1 and those caused by type 2, taking into account the primary or recurrent nature of the infection, and by differentiating diseases of mucocutaneous tissues from infections of other organ systems; sometimes, however, this basis for subdivision will seem too great a simplification for describing a complex clinical condition. Usually, HSV infections are transmitted with virus-containing secretions and the symptoms localized to the primary site of infection, be it the oropharynx, the conjunctiva, the skin, or the genital mucosa. As a rule, symptoms of primary infections are more severe than those of recurrences and show a greater tendency to generalize, particularly in small children. Most infections in the oropharyngeal area are of HSV type 1 origin, and recurrent genital infections are usually associated with shedding of type 2 virus.

However, as mentioned previously, the clinical picture and the pathogenesis sometimes appear more complex. For example, half of all patients with encephalitis demonstrate serologic signs of a primary infection, with no mucocutaneous symptoms associated with onset of or preceding the central nervous system disease (Whitley, 1985). In other HSV-induced infections an obvious immunopathologic element is involved, with increased severity of symptoms in recurrences. Such is the case, for example, with HSV-induced stromal keratoconjunctivitis. Some of the primary genital infections are caused by

HSV type 1 (Ashley et al., 1985), and severe genital type 1 infections have been reported, particularly in women.

Primary Infections of the Oropharynx

Primary infection of the oral mucous membranes with HSV-type 1 is usually a disease of children less than 5 years old. After an incubation of 2–7 days, gingivostomatitis associated with fever, malaise, and swollen, tender cervical lymph nodes is present. In children this condition may be accompanied by symptoms of gastrointestinal involvement with vomiting and diarrhea. The gums are swollen, red, and painful, and the oral mucosa is covered with numerous small vesicles that erode and become ulcerated. Healing is spontaneous and is completed in about 2 weeks.

In other cases, particularly in older children and young adults, the most prominent symptom may be ulcerating rhinitis or tonsillitis. Leukemic children and immunocompromised patients frequently develop generalized disease, with involvement of the respiratory tract and esophagus.

Recurrent Herpes Labialis

The most characteristic symptom of recurrent HSV type 1 infections is herpes labialis, with distinctly restricted eruptions appearing on the lips, the nostrils, or the skin around the mouth. The development of vesicles is usually preceded by a prodromal stage of itching or other sensations a few hours to a day before the eruption. Upon healing the vesicles become crusts, and the infection is over within 8–14 days.

Other Recurrent Skin Eruptions

Erosions or other traumata of the skin can be portals for primary infections as well as triggers for recurrence of a latent infection (e.g., herpes gladiatorum in athletes, infections in atopic eczemas, and postsurgical eruptions). In a few cases the clinical picture suggests a recurrent form of zosteriform herpes, with a zosterlike, unilateral arrangement of the vesicles over all or part of a dermatome.

Laboratory and other infections contracted on fingers and hands frequently become recurrent (e.g., herpetic whitlows). Infections on the fingers are often complicated by lymphangitis and lymphadenopathy, and occasionally peripheral infections are followed by episodes of a self-limiting meningitis.

The influence of hypersensitivity reactions against viral proteins has been regarded as important in the erythema multiforme-like eruption occasionally associated with herpes infections. The pathogenetic mechanisms involved are not obvious, although infectious virus is frequently isolated from the efflorescences.

Eye Infections

Of great medical importance are the herpetic eye infections, mainly caused by HSV type 1, which are among the most common causes of blindness in Europe and North America. In primary infections conjunctivitis is the most prominent clinical feature, often complicated by spread to the cornea and leading to classic dendritic infection with pain, photophobia, lacrimation, and reduced vision. Occasionally the cornea becomes perforated. The disease is self-limiting, but the 3 wk generally necessary for healing can be reduced with antiviral drugs. When the primary infection is healed, however, recrudescence may follow (Table 1). In some syndromes of recurrent eye diseases the immunopathologic influence on the pathogenesis is, as mentioned, apparent. Infectious virus cannot be isolated from the eye, and T-cell-dependent immune reactions monitor the clinical picture.

Genital Herpes

Statistics from several countries indicate an increasing prevalence of genital HSV infections during the 1970s, with spread of the infection to middle-class and more privileged socioeconomic groups that previously demonstrated a relatively low incidence. It is uncertain, however, whether these observations directly reflect an increased incidence of genital HSV infections, a consequence of more liberal views on sexual attitudes, or are attributable to improved diagnostic methods and a greater medical awareness of the importance of sexually transmitted HSV infections.

Both type 1 and type 2 viruses cause primary infections. The reported prevalence of type 1 infections can vary considerably in selected populations and from one country to another, but is probably not less than 10%. In patients with partial immunity in-

TABLE 1. Herpes simplex virus-induced recurrent ocular disease

Infectious epithelial herpes simplex keratitis
Noninfectious trophic epithelial herpes simplex keratitis
Infectious stromal herpes simplex keratitis
Noninfectious disciform stromal herpes simplex keratitis
Uveitis

duced by a previous oral type 1 infection, the genital infection is clinically less severe. Itching, paresthesia, pain, dysuria, and vaginal discharge, accompanied sometimes by fever and usually by inguinal lymphadenopathy, are the most common symptoms in women. Multiple vesicles appear on the vulva, in the vagina, and on the portio. The mean duration (12 days) and severity of clinical symptoms are usually greater in women than in men. In men, the eruptions may be localized to any part of the penis, but often the vesicles are found on the sulcus of the glans and on the preputium.

The recurrent clinical infection seems almost invariably a manifestation of type 2 infection; however, studies on patients shedding virus between two episodes of clinical recurrence (Yeager, 1984) have revealed the presence of HSV type 1 as well as type 2 (Strand et al., 1987). In recurrent infections approximately half of all patients have prodromal sensory symptoms 1–2 days before the eruption of vesicles, and the infection heals within a few days. On the other hand, the number and duration of recurrences can vary from less than one episode per year to frequent, almost continuous clinical symptoms (Harger et al., 1983). As in the primary infection, the symptoms of disease seem less severe and of shorter duration in men than in women.

Neonatal Infection

The most severe complication of genital HSV infection is the neonatal disease, which has been reported to increase in association with higher prevalences of genital infections (Sullivan-Boyai et al., 1983). Women with a virus-producing genital infection transmit the virus to their children at delivery. As inapparent genital infections are relatively common, neonatal infections can also occur in the children of asymptomatic mothers, but the risk is low. The risk of infection for the child of a mother with clinical symptoms of infection is nearly 100%. The children of mothers with an ongoing primary infection are most severely affected, whereas the risk of a fatal outcome is considerably less when the mother has a recurrent infection. An evaluation made by Nahmias and Visintine (1976) showed that in one-third of such children severe central nervous system and abdominal infections were already present at the clinical debut of disease and that two of three neonatally infected children with skin injuries later developed generalized systemic infection. Other common symptoms are those of the eyes and mucocutaneous bleeding.

Approximately 75% of neonatal infections are caused by HSV type 2, and in 90% of cases a maternal source of infection can be found. Case reports and studies on nursery outbreaks have also emphasized the risk of nosocomial neonatal infection by transmission of virus from personnel with acute cutaneous HSV infections. This risk seems particularly high in wards with children suffering from complicating immunosuppressive diseases.

Neurologic Disease

Herpes simplex virus type 1 is the most common etiology for sporadic encephalitis in the western world. The incidence has been estimated at one to two cases per million inhabitants. As mentioned previously, this severe neurologic complication can occur in association with both the primary and the reactivated latent form of infection. In experimental animals HSV infection of the central nervous system can be shown to follow distinct neuronal pathways, passing one or more synapses (Kristensson et al., 1982). The characteristic localization of the encephalitis in human beings, the temporal lobes, probably reflects the transfer of virus via the trigeminal ganglion, trigeminal nerve root, and trigeminal pathways of the brain stem. Virus is usually absent from the cerebrospinal fluid, and therefore specimens from brain biopsy must be used for viral isolation (Table 2).

Initially the patients may exhibit, in addition to

TABLE 2. Laboratory diagnosis of herpes simplex virus encephalitis

Type of assay	Method	Comment
Examination of biopsy material	Viral isolation Antigen detection Histology and/or electron microscopy	With properly collected specimens, a combination of these methods will yield very high detection rates
Examination of cerebrospinal fluid	Serology Antigen detection Viral isolation	Virus or viral antigens are demonstrable only in cases of meningitis; indication of intrathecal antibody production has significant diagnostic importance
Mapping of infectious foci	Testing with labeled antiviral drug	Experimental, but no clinical experience available; demonstration of selective and discrete uptake of specific drug might be an important noninvasive diagnostic tool

symptoms of acute infection, personality changes such as excitation and aggression. Later, symptoms of a rapidly progressing intracranial process appear, including electroencephalographic changes unilaterally located to the frontotemporal lobe and increased intracranial pressure. Finally, a focal necrotizing encephalitis develops. The mortality is about 70% in untreated patients. Invalidating sequelae, common in untreated survivors, are also demonstrable in some patients who survive as a result of treatment with antiviral drugs (Sköldenberg et al., 1984; Whitley et al., 1981, 1986).

Mortality is considerably lower in patients with HSV-induced meningoencephalitis associated with type 2 infection. In adults this form of leptomeningitis occasionally becomes recurrent and, in contrast to patients with type 1-induced encephalitis, they demonstrate significantly increased cell counts in their cerebrospinal fluid.

Herpes simplex virus infections have been implicated in parainfectious neurologic syndromes such as Guillain-Barré syndrome, in paretic conditions of unknown etiology such as Bell paresis, and in multiple sclerosis. The general neurobiologic and neurologic aspects of HSV infections have been reviewed recently (Johnson, 1982; Lycke, 1985).

The Immunocompromised Host

Patients with congenital or acquired immune defects are at risk of contracting severe primary HSV infections and, depending on the degree of immunosuppression, might fail to cope with reactivated latent infections. Various clinical symptoms have been reported, including progressive mucocutaneous infections, hepatitis, and proctitis.

Diagnosis

The clinical demand for more rapid and informative laboratory diagnosis of HSV infections has drastically increased during the last 10 years. This growing interest is mainly due to widespread clinical use and testing of antiviral drugs, which not only creates a need for accurate and rapid diagnostic methods but generates questions about reasons for treatment failures and the possible development of drug-resistant strains. Moreover, in association with the apparently increasing incidence of genital and neonatal HSV infections during the 1960s and 1970s, the lack of reliable information about their epidemiology became obvious and prompted the production of HSV type-specific antigens for serologic screening. In addition, specific clinical problems, such as selection of pregnant women at risk of transferring the virus to their babies and encephalitic patients for treatment with

antiviral drugs, has prompted the use of more rapid and accurate diagnostic methods. The following sections describe some techniques that have been used routinely in most diagnostic laboratories and others that, although promising, currently remain in limited use.

Specimen Collection

FOR ISOLATION OF VIRUS

It is important to collect specimens as early as possible in the course of the illness. The highest isolation rate from cutaneous lesions is obtained when the efflorescences have reached the vesicle stage only; the rate of viral isolation decreases as the herpetic lesions evolve from vesicles or ulcerations to crusts. If the skin requires cleaning, sterile saline should be used. Alcohol and other antiseptics should be avoided.

Vesicle fluid can be collected by lancing the vesicles and drawing up the fluid that oozes out with an open-ended capillary tube or a syringe. Fluid may also be collected on a moistened swab, which is subsequently broken off into a tube of transport medium. Specimens from ulcerations are collected by firmly rubbing the efflorescences with swabs moistened with transport medium. The swab is then immersed in transport medium; the fluid and cells are released into the medium by repeatedly pressing the swab against the walls of the container.

Specimens from the urethra are collected by inserting a swab 2 cm into the urethra and rotating it. Specimens from the female cervical area are collected in a similar fashion; however, mucopurulent secretions should be removed before sampling.

Cerebrospinal fluid should be collected in all cases with suspected neurologic involvement and, because demonstration of virus in brain biopsy material is the most conclusive means of diagnosing HSV encephalitis (Baumann et al., 1985; Nahmias et al., 1982), brain tissue should be studied in such cases. Specimens should be taken from the frontotemporal lobe, the most suitable site for specimen collection indicated by computed tomography scanning. Tissue specimens obtained at surgery or postmortem should be handled with care to avoid bacterial contamination.

Herpes simplex virus is relatively labile and can only be stored for 2–3 days at 4°C with maintained infectivity. Specimens should therefore be transported the day of sampling or the following day. Specimens that require a longer period of storage should be kept at −70°C. Moreover, when collecting samples from patients with recurrent infections, it is essential to avoid blood contamination as much as possible to prevent neutralization of the virus.

Several types of transport media exist (Bishal and

Labzoffsky, 1974; Nahmias et al., 1971), and some commercially available media have been reported to be efficient stabilizers, reducing loss of viral infectivity (Johnson et al., 1984; Warford et al., 1984).

A useful way to maintain infective virus before and during transportation is to inoculate cell cultures directly with the specimen and use the cell cultures as transport vehicles (Strand et al., 1987). The virus can then be stored at room temperature for several days without inactivation, even when the cell culture shows signs of degeneration. Upon arrival at the laboratory, the inoculated cells are transferred to fresh virus-sensitive cell cultures for viral isolation.

FOR IMMUNOFLUORESCENCE

Infected cells in vesicles or other efflorescences are collected by lifting away the top of the vesicle or lesion and collecting the cells from the bottom. Cells are then smeared evenly onto glass slides and fixed in methanol for light microscopy and in acetone for immunofluorescence techniques. An alternative is to suspend the cells in medium, sediment the cells by centrifugation, and apply the sediment on glass slides. The specimen is then air-dried and fixed.

FOR ELECTRON MICROSCOPY

A sufficient number of virus particles might be present in certain types of specimens, such as vesicle fluid and infected tissues, to make use of the electron microscope feasible for diagnostic purposes. At least one million particles per 0.1 ml of suspension are required, however, to allow detection of virions by electron microscopy. Vesicle fluid is collected by aspiration with a thin needle attached to a small syringe containing 0.1 ml of distilled water.

FOR SEROLOGY

Generally, acute and convalescent sera are required for serologic diagnosis. The first blood sample, collected as soon as possible after the onset of illness, should be followed by a second after an interval of 2–3 weeks. Blood samples, collected aseptically in dry sterile tubes without additives, are kept under refrigeration to produce clotting, and serum separated from the clot is stored at $-20°C$. Cerebrospinal fluid, used diagnostically to demonstrate intrathecally produced antibodies (Alestig et al., 1984; Skóldenberg et al., 1984; Vandvik et al., 1982), is stored at the same temperature.

Viral Isolation and Identification

Viral isolation is the most sensitive method of demonstrating HSV. Practically all primary cell cultures and established cell lines of mammalian origin sup-

port multiplication of HSV. Minor differences in susceptibility exist between different cell lines, but play no major role in the diagnostic result. For example, human diploid fibroblasts, Vero cells, rabbit kidney cells, and certain cell culture–fluorescent antibody sets, all commercially available, can be used with essentially similar results (Fayram et al., 1983; Hayden et al., 1983; Rubin and Rogers, 1984).

Cell cultures with confluent monolayers of cells are inoculated with 0.1 ml of the specimen to be studied. Sensitivity is somewhat increased if the fluid medium is drained from the culture before inoculation. The virus in the specimen is allowed to adsorb to the cells at 37°C for 1 h. Maintenance medium is then added, and the cultures are incubated at 37°C and observed daily in the microscope for viral cytopathic effects. Herpes simplex virus causes a characteristic swelling and rounding of infected cells, and certain virus strains form syncytia of various sizes (Smith et al., 1973). Usually the cytopathic changes are visible within 3 days, and cultures seldom require following for more than 1 week. In fact, about 50% of virus-positive samples are detectable within 24 h.

Several methods for identification and typing of isolates have been described. Most of these methods are based on the use of HSV-specific and/or type-specific antibodies directed against discrete epitopes on HSV antigens, i.e., the viral glycoproteins. The specificity, type selection, and antigen-discriminating capacities of the antibody preparation used will determine the results of each particular method. It is sometimes sufficient to type the isolated virus as HSV, but frequently determination of the type specificity of the isolate gives more valuable diagnostic and epidemiologic information.

Nilheden et al. (1983) infected monolayers of green monkey kidney cells in microtiter plates with HSV isolates. The plates were incubated at 37°C overnight and fixed with glutaraldehyde, and viral antigens were identified and typed by type-specific monoclonal antibodies in a peroxidase enzyme-linked antibody assay. Other techniques are based on antigens extracted from lysed infected cells and adsorbed to wells of microtiter plates coated with catching anti-HSV antibodies. Monoclonal antibodies against HSV are subsequently used for characterization and identification of adsorbed specific antigens (Clayton et al., 1985; Frame et al., 1984; Grillner and Landquist, 1983; Yolken, 1982). The enzyme-linked antibody assays demonstrate several advantages for identification and typing of HSV, as these techniques are neither technically demanding nor expensive and should therefore minimize any reader bias.

Monoclonal antibodies are also used routinely in the typing of HSV with immunofluorescence microscopy. Infected cells from a virus-positive culture are transferred to slides, and specific sera are used in direct or indirect immunofluorescence techniques. Immunofluorescence permits the infecting virus to be thoroughly characterized with a panel of monoclonal antibodies, revealing the intratypic variation of the isolated virus (Sutherland et al., 1986).

Viral isolation in cell culture combined with biotin–avidin fluorescent staining seems to be a rapid, sensitive, and accurate method of detecting and identifying HSV in clinical samples (Nerurkar et al., 1984a,b). After incubation of inoculated cultures for 24 h, the cells are stained with a biotin-linked polyclonal antibody directed against type 2 virus and a fluorescein–avidin conjugate. The method seems as sensitive as conventional isolation and typing techniques, and results are available 1 day after inoculation of cultures. Gleaves and co-workers (1985) have reported success with enhancing adsorption by centrifuging virus onto cells, incubating the cultures for 16 h, and using type-specific antibodies with indirect immunofluorescence for demonstrating and typing HSV.

Neutralization is the classic, although time-consuming and expensive, serologic technique for typing isolated virus. As a rule, rabbit antisera produced by immunization with prototype laboratory strains of HSV are used (Rawls, 1979). The neutralization is performed according to standard procedures, although several dilutions of the typing sera or varying incubation times are employed frequently to allow the kinetics of viral neutralization to be studied. Type relatedness between virus and typing serum is revealed by the rate or degree of viral neutralization in the presence of antiserum against type 1 or type 2. Antisera prepared against HSV type 2 usually contain relatively more antibodies cross-reactive with type 1 than there are anti-type 1 sera reacting against type 2 virus. As very few available antisera do not show intertypic cross-reactions, the methods are laborious to perform and sometimes yield inconsistent results. Therefore, neutralization is no longer the method of choice for diagnostic typing of isolates.

The highest resolution for typing and characterization of strains is achieved by endonuclease restriction enzyme digestion and electrophoretic analysis of HSV DNA. The endonuclease restriction enzymes most commonly used are *Bam*HI, *Eco*RI, and *Hin*dIII. The technique not only produces accurate typing, but gives the "fingerprint" of each particular strain. The analysis has been used to study nosocomial outbreaks of HSV infection in a pediatric intensive care unit (Buchman et al., 1978) and to demonstrate genital reinfection with exogenous HSV type 2 (Smith et al., 1985). Strains that produce inconclusive results with serologic typing should be typed with endonuclease restriction enzymes.

Direct Detection of HSV or HSV Antigens

IMMUNOFLUORESCENCE MICROSCOPY

Light microscopy has been used for detection of HSV antigens on Papanicolaou-stained cell smears (Thin et al., 1975). However, many virus-positive specimens are not detected histologically, and other Herpesviridae produce cell changes that are morphologically identical to those of HSV. Immunofluorescence efficiently increases both the sensitivity and specificity of light microscopy; nonetheless, only about 80% of virus-positive specimens will be demonstrable. Thus the number of virus-infected cells in the smears is critical.

The cells to be studied should be evenly smeared onto a glass slide, air-dried, and fixed with acetone. When the acetone has evaporated, the slide is dipped in buffered saline (pH 7.2) and covered with appropriately diluted conjugated antibody. The slides are incubated in a humid atmosphere for 30 min, washed in phosphate-buffered saline, mounted with buffered glycerol or elvanol, and examined in a fluorescence microscope. Each batch of slides tested should include positive and negative controls (i.e., HSV type 1 and 2 infected and uninfected cells). Fixed cells can be stored at −70°C. Simple methods for large-scale production of monoclonal antibodies have been published (Sjögren-Jansson and Jeansson, 1985); fluorescein isothiocyanate-conjugated monoclonal antibodies are commercially available, however, and can be used for both detection and typing of HSV antigens (see below and Pereira [1982] for a review).

Monoclonal antibodies are also used for indirect immunofluorescence (Sutherland et al., 1986). However, with this technique the Fc receptor induced by HSV might create certain problems. On cells infected with HSV, a membrane receptor with affinity for the Fc portion of IgG is induced (Watkins, 1965; Westmoreland and Watkins, 1974). This receptor has been identified as HSV glycoprotein E (gE) (Baucke and Spear, 1979; Para et al., 1982). In indirect immunofluorescence sera known to lack HSV, antibodies can give a positive reaction because of low-avidity binding between the Fc portion of IgG and HSV-induced gE. This reaction is especially prominent for HSV type 1 infected cells, whereas type 2 infected cells give a weaker reaction (Feorino et al., 1977). Of human immunoglobulins of various classes, only IgG reacts with gE. When IgG subclass specificity was studied, it was found that gE did not bind to IgG3 (Johansson et al., 1984) or to mouse IgG (Johansson et al., 1985). Recent findings indicate that the Fc receptor induced by HSV type 1 is probably a complex of two different glycoproteins; however, only one of these is detected in type 2 infected cells. This finding might explain the differences in Fc-receptor activity observed between type 1 and type 2 infected cells.

Enzyme-linked immunoassay (ELISA) and enzyme-linked immunosorbent assay (EIA) are techniques often employed for direct antigen detection. Wells of a microtiter plate are coated with purified HSV-specific antibodies. After the plate is washed the specimens are added, the plate is incubated and washed again, and enzyme-linked HSV-specific antibody is added. After final incubation, washings, and addition of relevant enzyme substrate, the enzyme activity remaining in the wells is recorded.

Several studies comparing EIA-based direct antigen detection and viral isolation procedures have been published (Alexander et al., 1985; Clayton et al., 1985; Grillner and Landquist, 1983; Nerurkar et al., 1984a,b; Ziegler et al., 1983). In general, the immunoassays are somewhat less sensitive than viral isolation for diagnosing HSV infections, but demonstrate advantages such as rapidity of assay, relative inexpensiveness, and insensitivity to loss of viral infectivity.

Other tests are based on radioimmunoassay (RIA). Except for the specific problems involved in the use of radioactive tracers, the advantages and limitations of RIA are the same as those for EIA-based methods.

Production of monoclonal antibodies has been of paramount importance in the development of antigen detection tests. A high degree of test specificity is granted; however, because monoclonal antibodies are usually of low avidity, test sensitivity is not always optimal (Brock and van Heyningen, 1983; Campbell, 1984; Thompson, 1984). In addition to their low affinity, monoclonal antibodies are often difficult to link to the solid phase or to enzymes without destroying their antigen-binding properties. Moreover, there are problems associated with the production of monoclonal antibodies against certain HSV antigens. Thus, HSV type 2 antigens seem able to suppress antibody formation (Kampe et al., 1985; Nick et al., 1986), and sex-related factors of the immunized animal influence the result (Knoblich et al., 1983). Because of their high specificity, some monoclonal antibodies may react only with a particular subset of type 1 or type 2 virus. The nonreactivity of some isolates underscores the risk of using a single monoclonal antibody to demonstrate HSV infection (Pereira et al., 1982; Swierkosz et al., 1985; Zheng et al., 1983).

Nick-translated ^{32}P-labeled probes have been used to demonstrate HSV DNA in clinical specimens. By means of nucleic acid hybridization about 10,000 homologous sequences, corresponding to two to eight infected cells, can be detected in assays, some of which can be completed in only 24 h (Redfield et al., 1983). These assays include Southern blotting, spot or slot techniques, and in situ procedures (Fung et al., 1985). HSV-specific probes prepared from the BamHI fragment and the U_l–U_s junction regions of the type 1 and type 2 genomes demonstrated a sensitivity corresponding to 90% and 80% of type 1 and type 2 isolations, respectively (Redfield et al., 1983).

The short half-life of ^{32}P is an obvious disadvantage, and the substitution of the isotope with a nonradioactive tracer was an important improvement (Kulski and Norval, 1985; Langer et al., 1981). Using the biotin–avidin–alkaline phosphatase complex for detection of biotinylated probes, it was possible to detect as little as 1 pg of DNA (Leary et al., 1983). Another advantage of the biotin-labeled probes is that they can be used in high concentrations, thus reducing the hybridization time to 1 h. As the enzyme reaction only requires a few minutes, the method produces rapid results. When hybridization is performed on nitrocellulose filters, DNA extracted from swabs of the eye and of vesicle fluid can be directly filtered. However, specimens containing copious mucous material require phenol extraction to eliminate proteins that compete with nucleic acids for binding sites on the filters (Richman et al., 1984).

Available methods for direct identification of viral antigens in clinical specimens are rapid and relatively inexpensive, but the user must have a thorough knowledge of how to produce and evaluate the quality of the pertinent reagents. These methods are generally less sensitive than HSV isolation methods; moreover, subclinical virus shedding cannot be studied, and biologic characterization of viral strains requires isolation of the virus. Direct identification techniques and viral isolation methods therefore do not exclude but complement each other (Corey, 1986).

ELECTRON MICROSCOPY

The herpes simplex virions are readily recognized in the electron microscope by their characteristic structure, and in some specimens (e.g., vesicle fluid) the concentration of virus may be sufficient to allow detection. Information regarding preparation of specimens, negative staining, immune electron microscopy, and the like are presented by Kjeldsberg (1980), by Almeida (1983), and in Chapter 7 (Electron Microscopy) in this volume.

Tracing HSV-Infected Tissue In Vivo by Virus-Specific and Labeled Markers

Mapping of virus-infected regions with noninvasive diagnostic tests has been attempted with both labeled anti-HSV monoclonal antibodies and radiolabeled antiviral drugs that selectively interact with virus-specific enzymes. The rationale for the mapping is that the labeled antiviral drug will accumulate in demonstrable concentrations only in regions where the corresponding viral enzyme is present. The need for this kind of test is obvious, particularly for diagnosis and management of HSV encephalitis, and an early diagnosis that obviates brain biopsy would undoubtedly save many patients. At present no clinically evaluated tests are available, but in experimental animals with focal encephalitides radioactively labeled anti-HSV drugs have yielded some promising results (Saito et al., 1982).

Antibody Assays

The standard procedure for obtaining a serologic diagnosis of viral diseases comprises the demonstration of specific IgM antibodies and a significant rise in antibody titers from the acute to the convalescent phase of the infection. In HSV infections this serologic pattern is complicated by two conditions, primarily 1) the recurrence of infection and 2) the antigenic relatedness between different Herpesviridae, particularly HSV type 1 and HSV type 2. Primary infection with type 1 or type 2 virus induces conventional antibody responses, including both IgM and IgG, but in patients with recurrent infections the recrudescence is often not reflected in changes of antibody titers. In recurrent infections IgM antibodies may or may not appear; the presence of IgM therefore does not exclude the recurrence of infection. In tests employing antigen preparations that do not adequately differentiate between antibodies directed against type 1 and type 2 virus, pre-existing antibodies against type 1 in individuals with genital primary type 2 infections can mask the antibody response against type 2 virus. In addition, recurrent mucocutaneous HSV infections often have only limited tissue engagement, and therefore the humoral antibody responses are not pronounced. Demonstration of IgA antibodies in secretions can sometimes provide the necessary information.

In some tests cross-reactivity of antibodies between HSV and varicella-zoster virus has been observed (Schmidt, 1982; Schmidt and Gallo, 1984); however, as mentioned previously, the most marked cross-reactions have been noted between HSV type 1 and HSV type 2. The degree of cross-reactivity in the test mainly reflects the specificity of the antigen employed, and the old saying that "no serologic test is better than the antigen used" remains highly relevant.

Many methods previously in common use are no longer those of first choice (e.g., neutralization, complement fixation, and immunofluorescence tests; see Rawls [1979]). Only occasionally are these tests used routinely for serologic diagnosis of HSV infections in viral diagnostic laboratories. Hemadsorption and hemagglutination have never been widely utilized, as Herpesviridae induce no hemagglutinins and techniques with HSV antigen-sensitized erythrocytes are often unpredictable and produce nonspecific results.

The assays most commonly used are based on EIA or RIA solid-phase techniques. The HSV antigen is adsorbed to the wells of plastic microtiter plates or to polystyrene balls (Cremer et al., 1982; Vestergaard, 1980). After washing, addition of the samples to be tested, incubation, and a second washing procedure, the detecting anti-human antibodies labeled with an enzyme or a radioactive tracer are added. Finally, the plates are washed a third time and, in the case of EIA, the enzyme substrate is added. When crude antigens prepared from infected and uninfected cell cultures are used, a relatively high background will be obtained and the sensitivity of the test becomes correspondingly low. With purified antigens the background absorbance will be lowered and the sensitivity increased. In addition, pure or purified antigen preparations can be highly diluted before use to lower the cost of testing (Coleman et al., 1983; Jeansson et al., 1983).

Conjugated antibodies against the different immunoglobulin classes make demonstration of the immunoglobulin class specificity of the antibody response possible. However, efficient separation of the different immunoglobulins or an immunoglobulin class capture technique is required. For IgM antibodies the development of μ-capturing antibodies has solved some problems, including, to some extent, those associated with antibodies to aggregated IgG or so-called rheumatoid factor (Duermeyer and van der Veen, 1978). The solid phase is coated with antibodies to μ chains of human IgM, enabling it to capture IgM antibodies, and labeled viral antigen is added for visualization of the captured antibody (Schmitz et al., 1980; van Loon et al., 1981, 1985). A reagent labeled with IgG can thus be avoided, and the risk of false-positive results due to interference by rheumatoid factor is markedly reduced. On the other hand, both the sensitivity and the specificity of the anti-μ antibody preparation can create problems. These problems may be overcome with the use of recently developed monoclonal antibodies against heavy chains of human immunoglobulin. However, rheumatoid factor that may have bound antiviral IgG in the patient's serum might attach to the solid phase, and the IgG antibodies thus present might in turn bind the labeled virus antigen. Testing for the presence of rheumatoid factor, therefore, must be included. The sensitivity of the μ-capturing technique depends to a large extent on the proportion of virus-specific antibodies in the total IgM.

The production of monoclonal antibodies against IgG subclasses has made evaluation of the subclass specificity of anti-HSV antibodies possible. For these tests, monoclonal antibodies specific for human IgG1, IgG2, IgG3, and IgG4 have been included in EIA-based tests (Coleman et al., 1985; Gilljam et al., 1985). It appears that IgG1 is formed in response to almost all HSV infections, whereas IgG2, IgG3, and IgG4 are detected more often in sera from patients with recurrent infections than in sera from those with primary HSV infections.

In serum, anti-HSV antibodies of the IgA class are present in almost all patients with seroimmunity (Friedman and Kimmel, 1982; Morris et al., 1985). More useful diagnostic information can be obtained, however, if local IgA production is studied. The IgA antibodies are found in lacrimal fluid from patients with primary or acute recurrent eye infections (Pedersen et al. 1982; Shani et al., 1985) and in cervicovaginal secretions from patients with genital type 2 virus infections (Merriman et al., 1984).

As mentioned previously, the extensive cross-reactivity between HSV type 1 and type 2 has created specific problems in determining the viral type specificity of serum antibodies. With recently developed assays such antibodies are demonstrable, however, by removing antibodies via adsorption of sera on cultures infected with one of the two viral types before testing in EIA (Hampar et al., 1985) or by using a type 2-specific antigen, gG-2 (Roizman et al., 1984). Glycoprotein G of HSV type 2, evaluated in ELISA for determination of antibodies against type 2, apparently binds only gG-2 antibodies without any cross-reactions against HSV type 1 induced antibodies (Lee et al., 1985; Svennerholm et al., 1984). With this antigen the prevalence of seropositivity against type 1 and 2 virus can be re-evaluated, presumably with new and interesting information about the epidemiology of HSV infections.

Testing for Development of Drug-Resistant HSV Strains

The activity of most antiviral drugs that inhibit HSV replication is directed against virus-coded DNA polymerase. Some drugs inhibit the DNA polymerase directly, whereas others, such as nucleoside analogues, block the DNA polymerase after phosphorylation by nucleoside kinases (Furman et al., 1984). Acycloguanosine (acyclovir) is an example of this latter group of drugs (Fyfe et al., 1978). The major mechanism that produces resistance to acyclovir, and the only one observed to yield resistant isolates in patients, is the development of thymidine kinase negative (TK$^-$) mutants (Schnipper and Crumpacker, 1980). Normally, TK$^-$ mutants are produced at a frequency of 1/10,000 during HSV multiplication (Parris and Harrington, 1982). Such TK$^-$

mutants usually demonstrate reduced infectivity and pathogenicity, as well as the ability to establish latency (Burns et al., 1982), but some mutants may display unchanged infectivity and pathogenicity (Field and Darby, 1980).

Of the methods used for demonstrating the development of resistance, the dye uptake method or colorimetric assay (Finter, 1969; McLaren et al., 1983) estimates the inhibitory effect of a drug on viral cell destruction by assaying remaining cells vitally stained. In plaque reduction tests (McLaren et al., 1982), the numbers of plaques produced by a standardized viral suspension in the absence and presence of a drug are assayed. More precise, but also more laborious, is the inhibition of virus production assay (Svennerholm et al., 1985). With this technique the amounts of virus produced in the presence and absence of a drug are compared, and the difference indicates the degree of viral sensitivity to the drug. In another type of test the amounts of viral antigens produced in the presence and absence of the drug are studied by measuring the viral antigens with ELISA (Wahren et al., 1983). The drug resistance mediated by TK$^-$ has also been assayed by determining the uptake of ^{125}I-labeled iododeoxycytidine in HSV-infected cell cultures. In this test, TK$^+$ mutants increase cellular uptake of the labeled compound 10-fold compared with cells infected with TK$^-$ mutants or uninfected cells (van Dyke and Connor, 1985).

Strains of HSV less sensitive to antiviral drugs have been observed in immunocompromised patients treated for long periods. However, there are naturally occurring HSV strains that show relative resistance to a particular drug without exposure to the drug, and in clinical trials strains lacking drug sensitivity have been isolated from patients treated with placebo (McLaren et al., 1982; Svennerholm et al., 1979). Moreover, clinical correlates supporting the importance of differences in drug sensitivity between viral strains are still essentially lacking (Lehrman et al., 1986; Svennerholm et al., 1985). In evaluating drug sensitivity, therefore, any conclusions regarding an induced drug resistance or even an altered sensitivity level of HSV strains isolated from drug-treated patients must be drawn very cautiously.

Interpretation of Laboratory Data

For a correct interpretation of laboratory findings, available results must be evaluated in the light of clinical observations. It goes without saying that viral isolation or detection of viral antigens or DNA in a patient must not be interpreted as if HSV were the disease-causing agent. In the adult population most individuals have been exposed to HSV; many are latently infected, and shedding of virus asymptomatically or in association with insignificant symptoms is relatively common. In contrast, demonstration of HSV in specimens from the central nervous system or in young infants has a highly significant diagnostic value. Likewise, HSV infections in patients with eye diseases, eczematous lesions, or immunosuppressive processes probably represent infections influencing the course of the disease.

Several reasons for technical failures to demonstrate HSV have been discussed. It should also be re-emphasized that, in the subacute phase of HSV infection, immunopathologic reactions may be important in the pathogenesis of disease and may cause even greater deterioration than the tissue damage caused by viral cytopathogenicity. These pathologic immune reactions are often associated with little or no production of infectious virus.

In patients with primary HSV infections seroconversion occurs regularly, and serologic findings are usually interpreted without difficulty. Serum antibodies of the IgG class appearing in high titers during the early phase of acute HSV infection, therefore, probably indicates a recurrent type of infection. In many patients with recurrent infections it is often difficult to detect significant changes in serum antibodies during the course of the disease. The IgG levels might reveal only insignificant changes from one episode of recurrence to another. The IgM antibodies may be present or absent. Thus, the persistence of IgM antibodies shows great individual variation, and IgM serum positivity can be prolonged (e.g., in immunosuppressed patients). In cases of suspected recurrence, it might be helpful to test with antigens of different origin (e.g., compare responses to glycoprotein and nucleocapsid antigens, respectively). It seems that the greater the proportion of antibodies to nucleocapsid antigens, the longer the history of infection. As mentioned previously, detection of type 2 virus infections in patients previously immunized against HSV type 1 requires the use of a type-specific antigen to distinguish between type-specific and cross-reacting antibodies. With these antigens, and also with the many commercially available antibody preparations, it is highly advisable that the type specificity of the reagents be thoroughly evaluated before routine testing is initiated. Significant cross-reactivity has been demonstrated with some reagents claimed to be type specific.

The demonstration of antibodies in cerebrospinal fluid samples is essential for serologic diagnosis of neurologic HSV infections, particularly if there is access to more than one sample; changes in titers can be traced and may indicate intrathecal production of antibodies. Comparison of antibody titers in blood samples collected in parallel with those of cerebrospinal fluid may in turn support the assumption of intrathecal antibody production. A fourfold or

TABLE 3. Presence of herpes simplex virus or viral antigens in biopsy/necropsy material and herpes simplex virus antibody in samples of cerebrospinal fluid indicative of intrathecal antibody production[a]

Brain biopsy/necropsy (*herpes simplex virus/antigen*)	Intrathecal antibody production	
	Yes	No
Positive/positive	12	2
Negative/positive	5	3
Negative/negative	6	19
Not examined	25	55
Total	48	79

[a] Adapted from Sköldenberg et al. (1984).

greater increase of cerebrospinal fluid/serum antibody ratio against HSV antigen, not accompanied by a corresponding change in the ratio for an unrelated antigen (e.g., measles antigen), is usually regarded as indicative of intrathecal production of HSV antibodies (Table 3). However, misinterpretation may result from admixture of serum with cerebrospinal fluid and polyclonal stimulation of antibody production, demonstrable in patients with diseases such as multiple sclerosis.

Laboratory Safety

Laboratory infections with HSV are not uncommon and, although most people are immunized, certain precautions are advisable. All laboratory workers should be tested for naturally acquired serum immunity to HSV. Individuals with skin lesions, eczema, and the like should not handle materials containing HSV. Gloves should always be used during preparation of infectious reagents, inoculation of cell cultures with specimens, and similar procedures. The eyes should be protected when there is any risk of aerosol or other droplet contamination. Bench surfaces and other possibly contaminated areas should be regularly disinfected with alcohol or other cleansers effective against HSV.

Epidemiology and Natural History

HSV infections are transmitted through virus-containing secretions and essentially spread by virus in saliva or secretions from the genitourinary tract. There are other means for transmission, but these play no significant role in the general outline of HSV epidemiology. Consequently, social conditions and individual behavior that increase the chances for contact with virus-contaminated secretions greatly influence its spread and epidemiologic patterns. The circulation of HSV is maintained by reactivation of latent infections and shedding of virus from clinically overt or inapparent infections. Primary oropharyngeal infections in children are regularly due to type 1 while the type 2 infections are acquired when an individual becomes sexually active during adolescence. The epidemiologic importance of genetic or molecular differences between the two viral types is obscure. A few cases of pharyngeal type 2 virus infections have been observed in adults, and it is now well established that some primary genital infections are caused by HSV type 1.

Like the oropharyngeal infections, diseases of the eye are usually caused by type 1, and because recurrent infections seem to induce severe clinical conditions more frequently than primary infections, the clinically treated cases of herpetic keratitis among adults outnumber the cases of primary herpetic keratitis in children 5- to 10-fold (Ribaric, 1976).

In the neonate, the infection is transmitted on passage through an infected birth canal, and in most cases the source is a maternal genital type 2 infection. In older children the primary viral infection acquired as frequently from the parents or other adults as from siblings or playmates (Juretic, 1966). In more than half of these cases no symptoms of infection are noticed, and by 15 years of age the prevalence of HSV type 1 antibody has reached 20% in industrialized countries and 50 to 70% in developing countries. When the primary infection is associated with clinical disease, the symptoms are often severe and the duration of disease may exceed 2 weeks. In contrast, the recurrent HSV type 1 infection is short-lived, and healing is often noted within 2 days (Spruance et al., 1977). Almost one-third of all infected individuals experience a recurrence of the type 1 infection.

Genital infections are spread more often by contacts who are unaware that they have transmissible HSV infection than by those who report a history consistent with previous recurrent herpes infection (Mertz et al., 1985). Thus, asymptomatic shedding of the virus is probably the most important factor in the spread of genital infections. Immunity induced by a type 1 infection in childhood, which seems to mitigate substantially the symptoms associated with genital type 2 infection acquired later in life, help to induce asymptomatic type 2 shedding. A genital HSV infection seems to induce protection locally, and reinfection is uncommon in patients with symptomatic recurrent infections (Schmidt et al., 1984; Lakeman et al., 1986). The overall dependence on immune reactions for control of HSV infections (recently reviewed by Rawls [1985]) is dramatically illustrated by the progressive and life-threatening diseases caused by HSV in immunocompromised patients.

In the United States, a steadily increasing incidence of genital HSV infections was registered for 1965 through 1979, and in 1979 an estimated 270,000 new infections occurred (Chuang et al., 1983). The spread of genital infection has raised concern about its effect on the incidence of neonatal infections (Nahmias et al., 1977). Asymptomatic viral excretion seems three to five times more common in pregnant women who experience their first symptomatic genital infection during pregnancy than in women with known recurrent infections (Brown et al., 1985), suggesting that anamnestic information alone will not adequately trace the group at risk. Follow-up of pregnant women with consecutive sampling of HSV from the cervix and vulvar lesions to determine whether vaginal delivery is advisable seems promising, however, and may reduce the need for cesarean section (Harger et al., 1983).

Prevention and Control

Immunization of human volunteers against HSV infection using active infectious HSV strains has been attempted experimentally; however, insufficient information regarding its protective effect against subsequent natural HSV infections, the safety of the vaccine strain, and the possible risk of oncogenic effects have been major hindrances to general acceptance. Subunit vaccines without viral DNA but containing relevant viral protein to induce immunity are currently under development in several laboratories. The cloned truncated forms of HSV glycoproteins have been used for some of these experimental vaccines, and the immunogens tested have provided significant protection against infection in some animal models (Berman et al., 1985).

Passive immunization with immunoglobulin preparations containing HSV neutralizing antibodies has also been studied in animal models and found to prevent neonatal and central nervous system infections efficiently (Davis et al., 1979; Georgiades et al., 1982). Preparations of anti-HSV hyperimmune globulins tested in immunocompromised children have suggested a protective effect (G. Lidin-Jansson, personal communication); evaluation is difficult, however, as the study for ethical reasons lacks controls. As mentioned previously, cesarean section is the otherwise generally accepted means of preventing neonatal infection when the pregnant woman has an ongoing virus-producing genital infection at the time of delivery.

Finally, prophylaxis with the antiviral drug acyclovir has been attempted in patients receiving toxic chemotherapy or immunosuppressive drugs who are thus susceptible to severe infections if HSV is contracted or activated (Prentice, 1983; Saral et al., 1983; Strand et al., 1984). Long-term prophylactic use of acyclovir in patients with disseminated muco-cutaneous (Mindel, 1984) or genital HSV infections (Bryson et al., 1985) has arrested the progress of disease effectively, but has been found insufficient in preventing recurrences.

Literature Cited

Alestig, K., L. Burman, A. Forkman, K. Lövgren, M. Forsgren, T. Bergström, E. Dahlqvist, and A. Fryden. 1984. Acyclovir versus vidarabine in herpes simplex encephalitis. Randomized multicentre study in consecutive Swedish patients. Lancet 2:707–711.

Alexander, I., C. R. Ashley, K. J. Smith, J. Harbour, A. P. Roome, and J. M. Darville. 1985. Comparison of ELISA with virus isolation for the diagnosis of genital herpes. J. Clin. Pathol. 38:554–557.

Almeida, J. D. 1983. Uses and abuses of diagnostic electron microscopy, p. 147–158. In P. Bachman (ed.), Current topics in microbiology and immunology. New developments in diagnostic virology, vol. 104. Springer-Verlag KG, Berlin.

Andrewes, C. H, and E. A. Carmichael. 1930. A note on the presence of antibodies to herpes virus in post-encephalitis and other human sera. Lancet 1:857–858.

Ashley, R., J. Benedetti, and L. Corey. 1985. Humoral immune response to HSV-1 and HSV-2 proteins in patients with primary genital herpes. J. Med. Virol. 17:153–166.

Baucke, R. B., and P. G. Spear. 1979. Membrane proteins specified by herpes simplex viruses V. Identification of an Fc-binding glycoprotein. J. Virol. 32:779–789.

Baumann, R. J., J. W. Walsh, R. L. Gilmore, C. Lee, P. Wong, H. D. Wilson, and W. R. Markesbery. 1985. Brain biopsy in cases of neonatal herpes simplex encephalitis. Neurosurgery 16:619–624.

Berman, P. W., T. Gregory, D. Crase, and L. A. Lasky. 1985. Protection from genital herpes simplex virus type 2 infection by vaccination with cloned type 1 glycoprotein D. Science 227:1490–1492.

Bishal, F. R., and N. A. Labzoffsky. 1974. Stability of different viruses in a newly developed transport medium. Can. J. Microbiol. 20:75–80.

Brock, D. J., and S. van Heyningen. 1983. A simple method for ranking the affinities of monoclonal antibodies. J. Immunol. Methods 62:147–153.

Brown, Z. A., L. A. Vontver, J. Benedetti, C. W. Critchlow, D. E. Hackok, C. J. Sells, S. Berry, and L. Corey. 1985. Genital herpes in pregnancy: risk factors associated with recurrences and asymptomatic viral shedding. Am. J. Obstet. Gynecol. 153:24–30.

Bryson, Y., M. Dillon, M. Lovett, D. Bernstein, E. Garratty, and J. Sayre. 1985. Treatment of first episode genital HSV with oral acyclovir: long term follow-up of recurrences. Scand. J. Infect. Dis. Suppl. 47:70–79.

Buchman, T. G., B. Roizman, G. Adams, and B. Hewitt-Stover. 1978. Restriction endonuclease fingerprinting of herpes simplex virus DNA: a novel epidemiological tool applied to a nosocomial outbreak. J. Infect. Dis. 138:488–498.

Burns, W. H., R. Saral, G. W. Santos, O. L. Laskin, P. S. Lietman, C. McLaren, and D. W. Barry. 1982. Isolation and characterization of resistant herpes simplex virus after acyclovir therapy. Lancet 1:421–423.

Bzik, D. J., B. A. Fox, N. A. DeLuka, and S. Person. 1984. Nucleotide sequence specifying the glycoprotein gene, gB of herpes simplex virus type 1. Virology 133:301–314.

Camacho, A., and P. G. Spear. 1978. Transformation of

hamster embryofibroblasts by a specific fragment of the herpes simplex virus genome. Cell 15:993–1002.

Campbell, A. M. 1984. The production and characterization of human and rodent hybridomas, p. 99–101. In R. H. Burdon and P. H. van Knippenberg (ed). Monoclonal antibody technology. Elsevier Science Publishing Co., New York.

Centifanto, Y. M., D. M. Drylic, S. L. Deardourff, and H. E. Kaufman. 1972. Herpesvirus type 2 in male genito-urinary tract. Science 178:318–319.

Chuang, T. Y., W. P. D. Su, H. O. Perry, D. M. Ilstrup, and L. T. Kurland. 1983. Incidence and trend of herpes progenitalis. Clin. Proc. 58:436–441.

Clayton, A.-L., U. Beckford, C. Roberts, S. Sutherland, A. Druse, J. Best, and S. Chantler. 1985. Factors influencing the sensitivity of herpes simplex virus detection in a simultaneous enzyme-linked immunosorbent assay using monoclonal antibodies. J. Med. Virol. 17:275–282.

Coleman, R. M., A. J. Nahmias, S. C. William, D. J. Phillips, C. M. Black, and C. B. Reimer. 1985. IgG subclass antibodies to herpes simplex virus. J. Infect. Dis. 151:929–936.

Coleman, R. M., L. Pereira, P. D. Bailey, D. Dondero, C. Wickliffe, and A. J. Nahmias. 1983. Determination of herpes simplex virus type-specific antibodies by enzyme-linked immunosorbent assay. J. Clin. Microbiol. 18:287–291.

Corey, L. 1986. Laboratory diagnosis of herpes simplex virus infections. Principles guiding the development of rapid diagnostic tests. Diagn. Microbiol. Infect. Dis. 4(Suppl. 3):1118–1198.

Cremer, N. E., C. K. Cossen, C. V. Hansen, and G. R. Shell. 1982. Evaluation and reporting of enzyme immunoassay determinations of antibody to herpes simplex virus in sera and cerebrospinal fluid. J. Clin. Microbiol. 15:815–823.

Davis, W. B., J. A. Taylor, and J. E. Oakes. 1979. Ocular infection with herpes simplex virus type 1: prevention of acute herpetic encephalitis by systemic administration of virus-specific antibody. J. Infect. Dis. 140:534–539.

Duermeyer, W. Z., and J. van der Veen. 1978. Specific detection of IgM antibodies by ELISA, applied in hepatitis A. Lancet 2:684–685.

Fayram, S. L., S. Aarnaes, and L. M. de la Maza. 1983. Comparison of Culturset to a conventional tissue culture–fluorescent-antibody technique for isolation and identification of herpes simplex virus. J. Clin. Microbiol. 18:215–216.

Feorino, P. M., S. L. Shore, and C. B. Reimer. 1977. Detection by indirect immunofluorescence of Fc receptors in cells acutely infected with herpes simplex virus. Int. Arch. Allergy Appl. Immunol. 53:222–233.

Field, H. J., and G. Darby. 1980. Pathogenicity in mice of strains of herpes simplex virus which are resistant to acyclovir in vitro and in vivo. Antimicrob. Agents Chemother. 17:209–216.

Finter, N. B. 1969. Dye uptake methods for assessing viral cytopathogenicity and their application to interferon assays. J. Gen. Virol. 5:419–427.

Frame, B., J. B. Mahony, N. Balachandran, W. E. Rawls, and M. A. Chernensky. 1984. Identification and typing of herpes simplex virus by enzyme immunoassay with monoclonal antibodies. J. Clin. Microbiol. 20:162–166.

Friedman, M. G., and N. Kimmel. 1982. Herpes simplex virus specific serum immunoglobulin A: detection in patients with primary or recurrent herpes infections and in healthy adults. Infect. Immun. 37:374–377.

Frietz, M. E., and A. J. Nahmias. 1972. Reversal polarity in transmembrane potentials of cells infected with herpesviruses. Proc. Soc. Exp. Biol. Med. 139:1159–1161.

Fung, J. C., J. Shanley, and R. C. Tilton. 1985. Comparison

of the detection of herpes simplex virus in direct clinical specimens with herpes simplex virus-specific DNA probes and monoclonal antibodies. J. Clin. Microbiol. **22:**748–753.

Furman, P. A., M. H. St. Clair, and T. Spector. 1984. Acyclovir triphosphate is a suicide inactivator of the herpes simplex virus DNA polymerase. J. Biol. Chem. **259:**9575–9579.

Fyfe, J. A., P. M. Keller, P. A. Furman, R. L. Miller, and G. B. Elion. 1978. Thymidine kinase from herpes simplex virus phosphorylates the new antiviral compound 9-(2-hydroethoxymethyl) guanine. J. Biol. Chem. **253:**8721–8727.

Georgiades, J. A., J. Montgomery, T. K. Hughes, D. Jensen, and S. Baron. 1982. Determinants of protection by human immune globulin against herpes neonatorum. Proc. Soc. Exp. Biol. **170:**291–297.

Gilljam, G., V. A. Sundqvist, A. Linde, P. Pihlstedt, A. E. Eklund, and B. Wahren. 1985. Sensitive analytic ELISAs for subclass herpes virus IgG. J. Virol. Methods **10:**203–214.

Gleaves, C. A., D. J. Wilson, A. D. Wold, and T. F. Smith. 1985. Detection and serotyping of herpes simplex virus in MRC-5 cells by use of centrifugation and monoclonal antibodies 16 h postinoculation. J. Clin. Microbiol. **21:**29–32.

Grillner, L., and M. Landquist. 1983. Enzyme-linked immunosorbent assay for detection and typing of herpes simplex virus. Eur. J. Clin. Microbiol. **2:**39–42.

Gruter, W. 1920. Experimentelle und Klinische Untersuchungen uber den sogenannten Herpes corneae. Ber. Versam. Deutsch. Ophtalm. Ges. **42:**162–166.

Hampar, B., M. Zweig, D. Showalter, V. Bladen, and C. W. Riggs. 1985. Enzyme-linked immunosorbent assay for determination of antibodies against herpes simplex virus types 1 and 2 in human sera. J. Clin. Microbiol. **21:**496–500.

Harger, J. H., G. J. Pazin, J. A. Armstrong, M. C. Breinig, and M. Ho. 1983. Characteristics and management of pregnancy in woman with genital herpes simplex virus infection. Am. J. Obstet. Gynecol. **145:**784–791.

Hayden, F. G., A. S. Sörensen, and J. A. Bateman. 1983. Comparison of the Immulok Culturset kit and virus isolation for detection of herpes simplex virus in clinical specimens. J. Clin. Microbiol. **28:**222–224.

Hayward, G. S., R. J. Jacob, S. C. Wadsworth, and B. Roizman. 1975. Anatomy of herpes simplex virus DNA; evidence for four populations of molecules that differ in the relative orientations of their long and short segments. Proc. Natl. Acad. Sci. USA **72:**4243–4247.

Jarrat, M. 1983. Herpes simplex infection. Arch. Dermatol. **119:**99–103.

Jeansson, S., M. Forsgren, and B. Svennerholm. 1983. Evaluation of solubilized herpes simplex virus membrane antigen by enzyme-linked immunosorbent assay. J. Clin. Microbiol. **18:**1160–1166.

Johansson, P. J. H., T. Hallberg, V.-A. Oxelius, A. Grubb, and J. Blomberg. 1984. Human immunoglobulin class and subclass specificity of Fc receptors induced by herpes simplex virus type 1. J. Virol. **50:**796–804.

Johansson, P. J. H., E. B. Myhre, and J. Blomberg. 1985. Specificity of Fc receptors induced by herpes simplex virus type 1: comparison of immunoglobulin G from different animal species. J. Virol. **56:**489–494.

Johnson, I. B., R. W. Leavitt, and D. F. Richards. 1984. Evaluation of the Virocult transport tube for isolation of herpes simplex virus from clinical specimens. J. Clin. Microbiol. **20:**120–122.

Johnson, R. T. 1982. Viral infections of the nervous system. Raven Press, New York.

Juretic, M. 1966. Natural history of herpetic infection. Helv. Paediatr. Acta **21:**356–368.

Kampe, P., A. Knoblich, M. Dietrich, and D. Falke. 1985. Differences in humoral immunogenicity between herpes simplex virus types 1 and 2. J. Gen. Virol. **66:**2215–2223.

Kjeldsberg, E. 1980. Application of electron microscopy in viral diagnosis. Pathol. Res. Pract. **167:**3–21.

Klein, R. J., and A. E. Friedmann-Kien. 1977. Latent herpes simplex virus infections in sensory ganglia of mice after topical treatment with adenine arabinoside and adenine arabinoside monophosphate. Antimicrob. Agents Chemother. **12:**577–581.

Knoblich, A., P. Kampe, V. Harle-Grupp, and D. Falke. 1983. Kinetics and genetics of herpes simplex virus-induced antibody formation in mice. Infect. Immun. **39:**15–23.

Kristensson, K., E. Lycke, and J. Sjöstrand. 1972. Spread of herpes simplex virus in peripheral nerves. Acta Neuropathol. (Berlin) **17:**44–53.

Kristensson, K., I. Nennesmo, L. Persson, and E. Lycke. 1982. Neuron to neuron transmission of herpes simplex virus. Transport of virus from skin to brain stem nuclei. J. Neurol. Sci. **54:**149–156.

Kulski, J. K., and M. Norval. 1985. Nucleic acid probes in diagnosis of viral diseases of man. Brief review. Arch. Virol. **83:**3–15.

Lakeman, A. D., A. J. Nahmias, and R. J. Whitley. 1986. Analysis of DNA from recurrent genital herpes simplex virus isolates by restrictive endonuclease digestion. Sex. Transm. Dis. **13:**61–66.

Langer, P. R., A. A. Waldrop, and D. C. Ward. 1981. Enzymatic synthesis of biotin-labelled polynucleotides: novel nucleic acid affinity probes. Proc. Natl. Acad. Sci. USA **78:**6633–6637.

Leary, J. J., D. J. Brigati, and D. C. Ward. 1983. Rapid and sensitive colorimetric method for visualizing biotin-labelled DNA probes hybridized to DNA or RNA immobilized on nitrocellulose: bioblots. Proc. Natl. Acad. Sci. USA **80:**4045–4049.

Lee, F. K., R. M. Coleman, L. Pererira, P. D. Bailey, M. Tatsuno, and A. J. Nahmias. 1985. Detection of herpes simplex virus type 2-specific antibody with glycoprotein G. J. Clin. Microbiol. **22:**641–644.

Lehrman, S. N., J. N. Douglas, L. Corey, and D. W. Barry. 1986. Recurrent genital herpes and suppressive oral acyclovir therapy. Relation between clinical outcome and in-vitro drug sensitivity. Ann. Intern. Med. **104:**786–790.

Lycke, E. 1985. Virus-induced changes in neural cells, p. 509–529. In A. Lajhta (ed.), Handbook of neurochemistry. vol. 10. Plenum Publishing Co., New York.

Lycke, E., K. Kristensson, B. Svennerholm, A. Vahlne, and R. Ziegler. 1984. Uptake and transport of herpes simplex virus in neurites of rat dorsal root ganglia cells in culture. J. Gen. Virol. **65:**55–64.

Löwenstein, A. 1919. Aetiologische Untersuchungen uber den fieberhaften Herpes. Munch. Med. Wochenschr. **66:**769–770.

McLaren, C., M. N. Ellis, and G. A. Hunter. 1983. A colorimetric assay for the measurement of the sensitivity of herpes simplex virus to antiviral agents. Antiviral Res. **3:**223–234.

McLaren, C., C. D. Sibrack, and D. W. Barry. 1982. Spectrum of sensitivity to acyclovir of herpes simplex virus clinical isolates. Am. J. Med. **73:**376–379.

Merriman, H., S. Woods, C. Winter, A. Fahnlander, and L. Corey. 1984. Secretory IgA antibody in cervicovaginal secretions from women with genital infection due to herpes simplex infection. J. Infect. Dis. **149:**505–510.

Mertz, G. J., O. Schmidt, J. L. Jourden, M. E. Guinan, M. L. Remington, A. Fahnlander, and L. Corey. 1985. Frequency of acquisition of first-episode genital infection with herpes simplex virus from symptomatic and asymptomatic source contacts. Sex. Transm. Dis. **12:**33–39.

Mindel, A. 1984. Long term oral acyclovir in disseminated mucocutaneous herpes simplex: a case report. Br. J. Vener. Dis. **60:**125–126.

Mindel, A., and S. Sutherland. 1983. Genital herpes—the disease and its treatment including intravenous acyclovir. J. Antimicrob. Chemother. **12**(Suppl. B):51–59.

Morris, G. E., R. M. Coleman, B. B. Benetato, and A. J. Nahmias. 1985. Persistence of serum IgA antibodies to herpes simplex, varicella-zoster, cytomegalovirus, and rubella virus detected by enzyme-linked immunosorbent assays. J. Med. Virol. **6:**343–349.

Nahmias, A. J., and W. R. Dowdle. 1968. Antigenic and biologic differences in herpesvirus hominis. Prog. Med. Virol. **10:**110–159.

Nahmias, A. J., and B. Roizman. 1973a. Infection with herpes simplex viruses 1 and 2. N. Engl. J. Med. **289:**667–674.

Nahmias, A. J., and B. Roizman. 1973b. Infection with herpes simplex viruses 1 and 2. N. Engl. J. Med. **289:**719–725.

Nahmias, A. J., and B. Roizman. 1973c. Infection with herpes simplex viruses 1 and 2. N. Engl. J. Med. **289:**781–789.

Nahmias, A. J., and A. M. Visintine. 1976. Herpes simplex, p. 156–190. *In* J. Remington and J. Klein (ed.), Infectious diseases of the fetus and newborn. The W. B. Saunders Co., Philadelphia.

Nahmias, A. J., A. M. Visintine, C. B. Reimer, I. Del Buano, S. L. Shore, and S. E. Starr. 1977. Herpes simplex virus infection of the fetus and newborn. p. 63–77. *In* S. Krugman and A. Gershon (ed.), Symposium on infection of the fetus and newborn infant. Progress in clinical and biological research. Alan R. Liss, New York.

Nahmias, A., C. Wickliffe, J. Pipkin, A. Leibowitz, and R. Hutten. 1971. Transport media for herpes simplex viruses types 1 and 2. Appl. Microbiol. **22:**451–454.

Nahmias, A. J., R. J. Whitley, A. N. Visintine, Y. Takei, C. A. Alford, Jr., and the Collaborative Antiviral Study Group. 1982. Herpes simplex encephalitis. Laboratory evaluations and their diagnostic significance. J. Infect. Dis. **145:**829–836.

Nash, A. A., J. Phelan, and P. Wildy. 1981. Cell mediated immunity in herpes simplex infected mice: H2 mapping of the delayed type hypersensitivity response and the antiviral T cell response. J. Immunol. **126:**1260–1268.

Nerurkar, L. S., M. Namba, G. Brashears, A. J. Jacob, Y. J. Lee, and J. L. Sever. 1984a. Rapid detection of herpes simplex virus in clinical specimens by use of a capture biotin–streptavidin enzyme-linked immunosorbent assay. J. Clin. Microbiol. **20:**109–114.

Nerurkar, L. S., M. Namba, and J. L. Sever. 1984b. Comparison of standard tissue culture, tissue culture plus staining, and direct staining for detection of genital herpes simplex virus infection. J. Clin. Microbiol. **19:**631–633.

Nick, S., P. Kampe, A. Knoblick, B. Metzger, and D. Falke. 1986. Suppression and enhancement of humoral antibody formation by herpes simplex virus types 1 and 2. J. Gen. Virol. **67:**1015–1024.

Nilheden, E., S. Jeansson, and A. Vahlne. 1983. Typing of herpes simplex virus by an enzyme-linked immunosorbent assay with monoclonal antibodies. J. Clin. Microbiol. **17:**677–680.

Nilsen, A. 1985. Genital herpes. Scand. J. Infect. Dis. Suppl. **47:**51–57.

Para, M. F., L. Goldstein, and P. G. Spear. 1982. Similarities and differences in the Fc-binding glycoprotein (gE) of herpes simplex viruses types 1 and 2 and tentative mapping of the viral gene for this glycoprotein. J. Virol. **41:**137–144.

Parris, D. H., and J. E. Harrington. 1982. Herpes simplex virus variants resistant to high concentrations of acyclovir exists in clinical isolates. Antimicrob. Agents Chemother. **22:**71–77.

Pedersen, B., S. Moller-Andersen, A. Klauber, E. Ottoway, J. U. Pranse, C. Zhong, and B. Norrild. 1982. Secretory IgA specific herpes simplex virus in lacrimal fluid from patients with herpes keratitis—a possible diagnostic parameter. Br. J. Ophthalmol. **66:**648–653.

Pereira, L. 1982. Monoclonal antibodies to herpes simplex viruses 1 and 2, p. 119–138. *In* J. G. R. Hurrell (ed.), Monoclonal hybridoma antibodies: techniques and applications. CRC Press, Boca Raton, Fla.

Pereira, L., D. V. Dondero, D. Gallo, V. Devlin, and J. D. Woodie. 1982. Serological analysis of herpes simplex virus types 1 and 2 with monoclonal antibodies. Infect. Immun. **35:**363–367.

Prentice, H. G. 1983. Use of acyclovir for prophylaxis of herpes infections in severely immunocompromised patients. J. Antimicrob. Chemother. **12**(Suppl. B):153–159.

Puga, A., E. M. Cantin, and A. L. Notkins. 1982. Homology between murine and human cellular DNA sequences and the terminal repetition of S component of herpes simplex virus type 1 in DNA. Cell **31:**81–89.

Rawls, W. E. 1979. Herpes simplex virus types 1 and 2 and herpesvirus simiae, p. 336–346. *In* E. H. Lennette and N. J. Schmidt (ed.), Diagnostic procedures for viral, rickettsial and chlamydial infections. American Public Health Association, Washington, D.C.

Rawls, W. E. 1985. Herpes simplex virus, p. 537–540. *In* B. N. Fields, D. M. Knipe, J. L. Melnick, R. M. Chanock, B. Roizman, and R. E. Shope (ed.), Virology. Raven Press, New York.

Read, G. S., and N. Frenkel. 1983. Herpes simplex virus mutants defective in the virion associated shut-off of host polypeptide synthesis and exhibiting abnormal synthesis of alpha (immediate early) viral polypeptides. J. Virol. **46:**498–512.

Redfield, D. C., D. D. Richman, S. Albanil, M. N. Oxmanil, and G. M. Wahl. 1983. Detection of herpes simplex virus in clinical specimens by DNA hybridization. Diagn. Microbiol. Infect. Dis. **1:**117–128.

Reeves, W. C., L. Corey, H. G. Adams, L. A. Vontver, and K. K. Holmes. 1981. Risk of recurrence after first episode of genital herpes. N. Engl. J. Med. **305:**315–319.

Reyes, G. R., R. La Femina, S. D. Hayward, and G. S. Hayward. 1979. Morphological transformation by DNA fragments of human herpesviruses: evidence for two distinct transforming regions in herpes simplex virus types 1 and 2 and lack of correlation with biochemical transfer of thymidine kinase gene. Cold Spring Harbor Symp. Quant. Biol. **44:**629–641.

Ribaric, V. 1976. The incidence of herpetic keratitis among population. Ophthalmologica **173:**19–22.

Richman, D., N. Schmidt, S. Plotkin, R. Yolken, M. Cherensky, K. McIntosh, and M. Mattheis. 1984. Summary of a workshop on new and useful methods in rapid viral diagnosis. J. Infect. Dis. **150:**941–951.

Rock, D. L., and N. W. Frazero. 1983. Detection of HSV-1 genome in central nervous system of latently infected mice. Nature (London) **302:**523–525.

Roizman, B. 1982. The family Herpesviridae: general description, taxonomy and classification, p. 1–23. *In* B. Roizman (ed.), The herpesviruses. vol. 1. Plenum Publishing Co., New York.

Roizman, B., B. Norrild, C. Chah, and L. Pereira. 1984. Identification and preliminary mapping with monoclonal antibodies of a herpes simplex virus type 2 glycoprotein lacking a known type 1 counter part. Virology 133:242–247.

Rubin, S. J., and S. Rogers. 1984. Comparison of Culturset and primary rabbit kidney cell culture for the detection of herpes simplex virus. J. Clin. Microbiol. 19:920–922.

Saito, Y., R. W. Price, D. A. Rottenberg, J. J. Fox, T.-L. Su, and K. A. Watanabe. 1982. Quantitative autoradiographic mapping of herpes simplex virus encephalitis with a radiolabeled antiviral drug. Science 217:1151–1153.

Saral, R., R. F. Ambinder, W. H. Burns, C. A. Angelopulos, D. E. Griffin, P. J. Burke, and P. S. Lietman. 1983. Acyclovir prophylaxis against herpes simplex virus infection in patients with leukemia. Ann. Intern. Med. 99:773–779.

Schmidt, N. J. 1982. Further evidence for common antigens in herpes simplex and varicella-zoster viruses. J. Med. Virol. 9:27–36.

Schmidt, N. J., and D. Gallo. 1984. Class-specific antibody responses to early and late antigens of varicella and herpes simplex viruses. J. Med. Virol. 13:1–12.

Schmidt, O. W., K. H. Fife, and L. Corey. 1984. Reinfection is an uncommon occurrence in patients with symptomatic recurrent genital herpes. J. Infect. Dis. 149:645–646.

Schmitz, H., U. von Deimling, and B. Flehmig. 1980. Detection of IgM antibodies to cytomegalovirus (CMV) using an enzyme-labelled antigen (ELA). J. Gen. Virol. 50:59–68.

Schnipper, L. E., and C. S. Crumpacker. 1980. Resistance of herpes simplex virus to acycloguanosine: role of virus thymidine kinase and DNA polymerase loci. Proc. Natl. Acad. Sci. USA 77:2270–2273.

Shani, L., R. Szanton, R. David, Y. Yassur, and I. Sarov. 1985. Studies on HSV specific IgA antibodies in lacrimal fluid from patients with herpes keratitis by solid phase radioimmunoassay. Exp. Res. 4:103–111.

Sjögren-Jansson, E. and S. Jeansson. 1985. Large-scale production of monoclonal antibodies in dialysis tubing. J. Immunol. Methods 84:359–364.

Sköldenberg, B., K. Alestig, L. Burman, A. Forkman, K. Lövgren, R. Norrby, G. Stjernstedt, M. Forsgren, T. Bergström, E. Dahlqvist, A. Frydén, K. Norlin, E. Olding-Stenkvist, and I. Uhnoo. 1984. Acyclovir versus vidarabine in herpes simplex encephalitis. Randomised multicenter study in consecutive Swedish patients. Lancet 2:707–711.

Smith, I. W., B. Barr, K. Slatford, and D. H. H. Robertson. 1985. Restriction enzyme analysis of herpes simplex virus isolates from known contacts of patients with genital herpes. Lancet 1:979.

Smith, I. W., J. F. Pedutherer, and D. H. H. Robertson. 1973. Characterization of genital strains of herpesvirus hominis. Br. J. Vener. Dis. 49:385–390.

Snowden, B. W., P. R. Kinchington, K. L. Powell, and I. W. Halliburton. 1985. Antigen and biochemical analysis of gB of herpes simplex virus type 1 and type 2 and of cross-reacting glycoproteins induced by bovine mammilitis virus and equine herpesvirus type 1. J. Gen. Virol. 66:231–247.

Spear, P. G. 1985. Glycoproteins specified by herpes simplex viruses, p. 315–356. *In* B. Roizman (ed.), The herpesviruses, vol. 3. Plenum Press, New York.

Spruance, S. L., J. C. Overall, E. R. Kern, G. G. Kreuger, V. Pliam, and W. Miller. 1977. The natural history of recurrent herpes labialis. Implications of antiviral therapy. N. Engl. J. Med. 297:69–75.

Strand, A., A. Vahlne, B. Svennerholm, J. Wallin, and E. Lycke. 1986. Asymptomatic viruses shedding in men with genital herpes infection. Scand. J. Infect. Dis. 18:195–197.

Strand, S. E., H. E. Takiff, M. Seidlin, S. Bachrach, L. Lininger, J. J. DiGiovanni, K. A. Westerha, H. A. Smith, S. N. Lehrman, T. Creagh-Kirk, and D. W. Alling. 1984. Suppressing of frequently recurring genital herpes. A placebo controlled double-blind trial of oral acyclovir. N. Engl. J. Med. 310:1545–1550.

Sullivan-Boyai, J., H. Hull, C. Wilson, and L. Corey. 1983. Neonatal herpes simplex virus infection in King County, Washington. J.A.M.A. 250:3059–3062.

Sutherland, S., B. Morgan, A. Mindel, and W. L. Chan. 1986. Typing and subtyping of herpes simplex virus isolates by monoclonal fluorescence. J. Med. Virol. 18:235–245.

Svennerholm, B., S. Olofsson, S. Jeansson, A. Vahlne, and E. Lycke. 1984. Herpes simplex virus type-selective enzyme-linked immunosorbent assay with *Helix pomatia* lectin purified antigens. J. Clin. Microbiol. 19:235–239.

Svennerholm, B., A. Vahlne, G. B. Löwhagen, A. Widell, and E. Lycke. 1985. Sensitivity of HSV strains isolated before and after treatment with acyclovir. Scand. J. Infect. Dis. Suppl. 47:149–154.

Svennerholm, B., A. Vahlne, and E. Lycke. 1979. Inhibition of herpes simplex virus infection in tissue culture by trisodium phosphonoformate. Proc. Soc. Exp. Med. Biol. 161:115–118.

Svennerholm, B., A. Vahlne, and E. Lycke. 1981. Persistent reactivable latent herpes simplex virus infection in trigeminal ganglia of mice treated with antiviral drugs. Arch. Virol. 69:43–48.

Swierkosz, E. M., M. Q. Arens, R. R. Schmidt, and T. Armstrong. 1985. Evaluation of two immunofluorescence assays with monoclonal antibodies for typing of herpes simplex virus. J. Clin. Microbiol. 21:643–644.

Thin, R. N., W. Aha, J. D. Parker, C. S. Nicol, and G. Canti. 1975. Value of Papanicolaou-stained smears in the diagnosis of trichomoniasis, candidiasis and cervical herpes simplex virus infection in women. Br. J. Vener. Dis. 51:116–118.

Thompson, R. J. 1984. Are monoclonal antibodies the end of radioimmunoassay? Trends Biochem. Sci. 7:419–420.

Vahlne, A., B. Svennerholm, and E. Lycke. 1979. Evidence for herpes simplex virus type-selective receptors on cellular plasma membranes. J. Gen. Virol. 44:217–225.

Vandvik, B., F. Vartdal, and E. Norrby. 1982. Herpes simplex encephalitis: intrathecal synthesis of oligoclonal virus-specified IgG, IgA and IgM antibodies. J. Neurol. 228:25–38.

van Dyke, R. B., and J. D. Connor. 1985. Uptake of I-125-iododeoxycytidine by cells infected with herpes simplex virus: a rapid screening test for resistance to acyclovir. J. Infect. Dis. 152:1206–1211.

van Loon, A. M., F. M. V. Heessen, M. van der Logt, and J. van der Veen. 1981. Direct enzyme-linked immunosorbent assay that uses peroxidase labelled antigen for determination of immunoglobulin M antibody to cytomegalovirus. J. Clin. Microbiol. 13:416–422.

van Loon, A. M., M. van der Logt, F. M. V. Heessen, and J. van der Veen. 1985. Use of labelled antigen for the detection of immunoglobulin M and A antibody to herpes simplex virus in serum and cerebrospinal fluid. J. Med. Virol. **15**:183–195.

Vestergaard, B. F. 1980. Herpes simplex virus antigens and antibodies: a survey of studies based on quantitative immunoelectrophóresis. Rev. Infect. Dis. **2**:899–913.

Wahren, B., J. Harmenberg, V.-A. Sundqvist, B. Levén, and B. Sköldenberg. 1983. A novel method for determining the sensitivity of herpes simplex virus to antiviral compounds. J. Virol. Methods. **6**:141–149.

Warford, A. L., W. G. Eveland, C. A. Strong, R. A. Levy, and K. A. Rekrut. 1984. Enhanced virus isolation rate by use of the transporter for a regional laboratory. J. Clin. Microbiol. **19**:561–562.

Watkins, J. F. 1965. Adsorption of sensitized sheep erythrocytes to HeLa cells infected with herpes simplex virus. Nature **202**:1364–1365.

Westmoreland, D., and J. F. Watkins. 1974. The IgG receptor induced by herpes simplex virus: studies using radioiodinated IgG. J. Gen. Virol. **24**:167–178.

Whitley, R. J. 1985. Epidemiology of herpes simplex viruses, p. 1–44. *In* B. Roizman (ed.), The herpesviruses, vol. 3. Plenum Press, New York.

Whitley, R. J., C. A. Alford, M. S. Hirsch, R. T. Schooley, J. P. Luby, F. Y. Aok, D. Hanley, A. J. Nahmias, and S. J. Soong. 1986. Vidarabine versus acyclovir therapy in herpes simplex encephalitis. N. Engl. J. Med. **314**:144–149.

Whitley, R. J., S. J. Soong, M. S. Hirsch, A. W. Karchmer, R. Dolin, G. Galasso, J. K. Dunnick, C. A. Alford, and the NIAD Colloborative Antiviral Study Group. 1981. Herpes simplex encephalitis: vidarabine therapy and diagnostic problems. N. Engl. J. Med. **304**:313–318.

Wildy, P. 1973. Herpes history and classification, p. 1–25. *In* A. S. Kaplan (ed.), The herpesvirus. Academic Press, New York.

Yeager, A. S. 1984. Genital herpes simplex infections: effect of asymptomatic shedding and latency on management of infections in pregnant women and neonates. J. Invest. Dermatol. **83**:53–56.

Yolken, R. H. 1982. Enzyme immunoassays for the detection of infectious antigens in body fluids: current limitations and future prospects. Rev. Infect. Dis. **4**:35–68.

Zheng, Z. M., D. R. Mayo, and G. D. Hsuing. 1983. Comparison of biological, biochemical, immunological and immunochemical techniques for typing of herpes simplex virus isolates. J. Clin. Microbiol. **17**:396–399.

Ziegler, T., O. H. Meurman, P. P. Arstila, and P. E. Halonen. 1983. Solid-phase enzyme-immunoassay for the detection of herpes simplex virus antigens in clinical specimens. J. Virol. Methods **7**:1–9.

Zinsser, H., and F. F. Tang. 1929. Further experiments on agent of herpes. J. Immunol. **17**:343–355.

CHAPTER 12

Herpesviridae: Epstein-Barr Virus

EVELYNE T. LENNETTE

Disease: Infectious mononucleosis (IM), Burkitt's lymphoma (BL), undifferentiated nasopharyngeal carcinoma (NPC).

Etiologic Agent: Epstein-Barr virus—etiologic in IM; cofactor in BL and NPC.

Source: Human saliva and blood.

Clinical Manifestations: Acute or insidious onset with fever, pharyngitis, lymphadenopathy. May be associated with varying degrees of splenomegaly, hepatomegaly, or frank hepatitis. Rare central nervous system involvement.

Pathology: Marked lymphocytosis involving most of the lymphoreticular tissues; hematologic abnormalities include >10,000 lymphocytes/ml with >10 to 15% atypical lymphocytes; elevated liver enzymes.

Laboratory Diagnosis: Positive Paul-Bunnell heterophile antibodies within the first 4 weeks of illness in adolescent and older patients. EBV serology may be needed for heterophile-negative patients; EBV isolation not useful.

Epidemiology: Worldwide; infections mostly asymptomatic in children; morbidity increases with age of primary infection; infections occur in children younger than 5 years of age in lower socioeconomic communities; acute disease is mostly sporadic with no seasonal variation among older patients in more affluent communities.

Treatment: Anecdotal report of efficacy with acyclovir and its derivatives.

Prevention and Control: Avoidance of salivary exchange.

Introduction

In 1958, Dennis Burkitt, a British missionary surgeon in Uganda, presented for the first time his pioneer studies on a little-known African childhood lymphoma which was later to bear his name (Burkitt, 1958). Three years later, tumor biopsies from several patients were successfully propagated by his colleague Epstein, who then discovered with electron microscopy a herpes-like virus in the continuous tumor line designated as Epstein-Barr 1 (Epstein and Barr, 1964; Epstein et al., 1964). Later, the Henles established that the virus was a hitherto unknown member of the *Herpesviridae,* on the basis of its unique biologic and immunologic properties (Henle and Henle, 1966). This new and distinct herpes virus became known as Epstein-Barr virus (EBV), after the cell line in which it was first observed.

Using the immunofluorescence technique, Henle and Henle quickly demonstrated that antibodies to EBV were present in all of the Burkitt's lymphoma (BL) patients, thus establishing an association between the virus and the patients (Henle et al., 1969). However, to their surprise, anti-EBV antibodies were found not only in sera of BL patients, but also in a large number of sera collected from many parts of the world. In addition, similar if not identical viruses were found in quick succession in lymphoblastoid lines initiated from peripheral blood of normal donors, patients with leukemia, other cancers, and infectious mononucleosis (Diehl et al., 1968; Ikawata and Grace, 1964; Moore et al., 1966, 1967).

The worldwide distribution and high incidence of antibodies to EBV led the Henles to suspect the virus to be the etiologic agent of a common disease. They intensified their search of the putative disease with serologic testing on paired sera from patients with infections of unknown etiology in the Philadelphia area. Their efforts were rewarded when seroconversion to EBV was detected in the sera of their young laboratory technician during her mononucleosis illness (Henle et al., 1968). More significantly, lymphoblastoid lines established from her peripheral blood were also positive for the virus. These observations were confirmed several times in their own laboratory among staff members undergoing similar illnesses (Diehl et al., 1968). Thus, the first evidence that EBV may be the causal agent of infectious mononucleosis (IM) was obtained. In the intervening years, extensive serosurveys have led to the identification of EBV as the etiologic agent of IM, as well as its role as a cofactor of malignancies such as Burkitt's lymphoma (BL), nasopharyngeal carcinoma (NPC), and a number of B-cell lymphomas (see below).

Viral Properties

Virus Morphology

As with all herpesviruses, the maturation of EBV involves the budding of immature particles from the nucleus of infected cells, through both the nuclear and cell membranes. Intact immature particles in the nuclei of infected cells are icosahedrons with 162 capsomeres surrounding a dense core (Hummeler et al., 1966). Upon completion of budding, the mature EBV virions appear as 120 nm particles in diameter with a dense DNA (100×10^6 daltons) core of 45 nm, surrounded by a double-membrane envelope (Epstein et al., 1964; Schulte-Holthausen and zur Hausen, 1970).

Replication

Although the full range of permissible target cells has not been determined, EBV is generally considered B-lymphotropic as well as epitheliotropic. Specific EBV receptors, required for infection, were shown to be identical with the CR2 receptors on B lymphocytes, which bind activated complement C3d fragment (Fingeroth et al., 1984; Klein et al., 1978). Recent studies with monoclonal antibodies have extended the previous finding to include C3d receptors on epithelial cells, such as those from the oropharynx and the ectocervix (Young et al., 1986). It is now known that CR2 regions on the target cells can bind both EBV and C3d fragment by virtue of their shared amino acid sequences (Nemerow et al., 1986).

Most of the information concerning EBV replication was derived from observations made in vitro on continuous lymphoblastoid lines. EBV infection of B lymphocytes invariably leads to transformation (Henle et al., 1967; Nilsson et al., 1971). The transformed cells undergo blast formation and cell surface alterations; they acquire the ability to grow into EBV-genome-positive, permanent lines, with concomitant expression of viral antigens (Henle et al., 1967; Menezes et al., 1976; Robinson et al., 1977).

EBV-Induced Antigens

Infection of the target cells leads to two forms of viral cycles: 1) latent, nonproductive and 2) productive, replicative infections. In both cycles, one of the earliest antigens expressed is LYDMA (lymphocyte-detected membrane antigen), a cell-surface antigen that is recognized by T cells (thymus-dependent lymphocytes) of IM patients specifically and is the target antigen of T-cell-mediated killing reaction (Svedmyr and Jondal, 1975). Expression of EBNA (EBV-associated nuclear antigen) either follows or parallels LYDMA at 12 to 24 h post infection (Aya and Osato, 1974; Crawford et al., 1978). EBNA is found as a nonstructural, intranuclear antigen(s), present in all EBV-transformed cell lines and in tumors from BL and NPC patients (Huang et al., 1974; Reedman and Klein, 1973; Reedman et al., 1974). The average concentration of EBNA per cell is proportional to the number of viral genomes present (Ernberg et al., 1977). The latter varies widely from cell line to cell line, although it is stably maintained and is characteristic for the line (Thorley-Lawson and Strominger, 1976). EBNA binds to double-stranded and, to a lesser extent, single-stranded DNA with no apparent specificity (Luka et al., 1977). It has recently become apparent that EBNA is probably not a single antigenic moiety, but a multicomponent antigen complex, on the basis of reactivities of sera from different classes of patients (Grogan et al., 1983). The major component EBNA 1, detected by the majority of the human sera, has been purified and sequenced in its entirely. As detected by immunoblotting, EBNA 1 is a polypeptide of 65 to 72 kilodaltons (kDa) (Strnad et al., 1981). A second component, 82 kDa in molecular weight, designated as EBNA 2, may be important for transformation (Skare et al., 1985). The roles of other minor EBNA components are still unclear (Sculley et al., 1984).

Viral expressions among the transformed B cell lines vary widely. They range from those cell lines that are found only in the latent cycle, with LYDMA and EBNA often the only two detectable viral expressions, to those that contain 2 to 20% of the cells with spontaneous, fully productive cycles. In addition, the degree of inducibility from latent to productive infection, either spontaneous or drug induced, is also a stable trait of the individual cell line (Klein and Dombos, 1973). Some cell lines are totally resistant to induction by halogenated pyrimidines, whereas others can readily enter into the productive cycle with exposure of the same drugs (Gerber and Lucas, 1972). Once the productive cycles have begun, replication of the viruses is often aborted. The exact blockage of the viral cycle is also determined by the particular cell line (Pritchett et al., 1976).

In the fully productive, replicative cycle, the synthesis of EA (early antigen complex) follows EBNA, but precedes viral DNA synthesis (Gergely et al., 1971). Induction of EA always results in a fatal outcome of the infected cell. Antigenically, it is composed of at least two subcomponents, EA-D (EA-diffuse) and EA-R (EA-restricted), which can be differentiated by the morphology of immunofluorescence staining and by their resistance to alcohols (Henle et al., 1970). The former is resistant to methanol, while EA-R is sensitive to both ethanol and methanol.

The viral capsid antigen complex (VCA) appears late in the replicative cycle and is a structural component of the viral nucleocapsids (Henle and Henle, 1966; Silvestre et al., 1971). Two other structural components of the virion are the early and late membrane antigen complexes (EMA and LMA). EMA is synthesized prior to EA and DNA synthesis and can be localized in the viral envelope. LMA is synthesized after viral DNA synthesis; it is also found on the virus envelope and is the target antigen for neutralizing antibodies (Pearson et al., 1970; Silvestre et al., 1974).

Of the above-mentioned antigens, the clinically important ones include VCA, EA-D, EA-R, and EBNA. Humoral responses against these antigens follow a known, distinct, but overlapping course such that antibody patterns obtained from a single serum from various patient populations can determine the patients' immune and disease status in relation to EBV (see below).

EBV Biotypes

Elucidation of EBV virotypes has been hampered by the lack of a fully permissive, susceptible cell system. Currently, virus isolation can be accomplished reliably only with the lymphocyte transformation assay with freshly fractionated, cord-blood lymphocytes. Depending on the target cells, every naturally occurring isolate has the capacity to transform B lymphocytes and to undergo a productive viral cycle if the appropriate indicator cells were used. Isolates from a variety of sources, including saliva and EBV-infected peripheral blood B lymphocytes of patients with various EBV-associated diseases, do not show

major strain variations based on biochemical and antigenic characteristics (Menezes et al., 1975a; Miller et al., 1974). However, the biologic properties of the viral isolates in vitro are, in large part, affected by host factors (Miller and Lipman, 1973). Since much of the earlier work has relied exclusively on the lymphocytic indicator cell, it was not possible to determine how much of the homogeneity observed among the isolates was host determined. More recent work with cultured epithelial cells (Sixbey et al., 1983) to study fresh salivary EBV isolates showed that some isolates can replicate in epithelial cells, whereas laboratory strains with transforming activity could not. These intriguing findings suggest the possibility of biologic variations among EBV isolates, the variations depending on the replication sites.

Although most of the natural isolates and laboratory strains show no major variations, one laboratory strain does not conform to the norm. The virus, the P3HR-1 strain, lost the transforming ability after prolonged passages in the laboratory (Hinuma et al., 1967) and contains 15% more DNA sequences than its transforming counterparts (Pritchett et al., 1975). P3HR-1 virus is considered a laboratory-derived mutant because of its unique biologic traits.

Pathology and Pathogenesis

The hallmark feature of IM is the marked lymphocytosis involving almost all of the lymphoreticular tissues, leading to lymphadenopathy, splenomegaly, hepatomegaly, and lymphoid hyperplasia of the oronasopharynx. Enlarged lymph nodes show striking follicular hyperplasia, with numerous macrophages, atypical lymphocytes, and nodular arrangement of basophilic lymphoblasts. Increased numbers of plasmacytoid cells and, occasionally, Reed-Sternberg cells are seen. Infiltration of mononuclear, atypical, pleiomorphic cells occurs in many organs. In the peripheral blood, atypical lymphocytes are most frequently of the Downey II type, and less frequently of the III and I types. They are strikingly pleiomorphic and can vary in both size and shape. The cytoplasm is usually vacuolated and foamy in appearance, with the nucleus either round or lobulated.

Although the source of infections is uncertain, it is suspected to be either free virus or infected monocytes found in the saliva of virus shedders. The oropharynx is probably the initial site of viral replication, since infectious virus can be recovered from the saliva of nearly all IM patients (Miller et al., 1973). Specific sites that have been proposed include the parotid salivary glands (Morgan et al., 1979) and pharyngeal squamous epithelial cells (Lemon et al., 1977). More recently, the tongue epithelium was shown to sustain highly active EBV replication, adding another potential site to the list (Greenspan et al., 1985). From the oropharynx, the infection spreads via the circulatory route. However, neither free infectious virus nor virus-producing cells have been detected in the peripheral blood of IM patients. During the first week of illness, approximately 1 in 1000 B lymphocytes is infected, decreasing to 1 in a million or more lymphocytes after convalescence (Rocchi et al., 1977). As a result, EBV-positive lymphoblastoid lines can be readily established from the acutely ill patient, only infrequently from the seropositive individuals with past infection, and never from the susceptible (uninfected) individuals.

During the acute disease, there is a marked lymphoproliferation of B lymphocytes, with accompanying hyperimmunoglobulinemia. There is also an activation of cytotoxic T cells and macrophages. Monoclonal staining of the cells has shown the T cells to belong in the suppressor or T8 subset (Tosato et al., 1979), EBV-infected B lymphocytes, whether from allogenic or autochthonous origin, are targets of specific killing action by the effector T lymphocytes (Svedmyr and Jondal, 1975). Presumably, these activated T cytotoxic cells play a role in controlling the lymphoproliferative process in the primary disease.

As is the case with other herpesviruses, EBV infections always result in a persistent, latent infection, which may be intermittently reactivated. Recent reports suggest that infectious, transforming virus can be recovered from nearly all individuals, and not just from 20%, as previously believed. With more sensitive techniques, Yao et al. (1985) could detect a low, persisting, and unchanging level of virus in the majority of subjects, while 20% shed larger and more readily detectable titer. The low level of virus shed in the oropharynx is the most likely source of transmitted viral infections. Under immunosuppressive conditions, the incidence of high-titer virus shedding increases markedly, from 20% observed in healthy controls, to 50% in patients with immunosuppressive therapy, to more than 80% in patients with Hodgkin's disease (Lange et al., 1978; Strauch et al., 1974).

Clinical Features

Transmission of Clinical Course

The most frequent mode of transmission of EBV is via salivary contact, either through kissing or by exposure to contaminated eating implements (Hoagland, 1955). Because of the insidious onset, the incubation period has not been firmly established but is estimated to be between 4 and 7 weeks. Primary EBV infections are mostly asymptomatic. The risk of accompanying disease varies to a large extent with the age of the patient, ranging from less than 1% in the age group below 2 years of age, to 60% at 15 years or older (Niederman et al., 1970). The classical presentation of infectious mononucleosis includes cervical lymphadenopathy, fever, and pharyngitis. Splenomegaly occurs in about half of the cases, and hepatitis in about 20%. Typically, IM is a self-limiting illness of 2 to 4 weeks' duration, but may have a protracted course lasting several months. There is usually a transient leukopenia followed by a marked increase in both normal and atypical lymphocytes and monocytes. There is an absolute as well as a relative leukocytosis. By the second week from onset, the number of total leukocytes typically ranges from 4,500 to 20,000 cells per cubic millimeter. In both children and adult patients, 50 to 80% of the circulating leukocytes are monocytes. For diagnostic significance, the atypical lymphocytes should number at least 10 to 15% of the total white cell count. Although the atypical lymphocytes are found in greatest number in IM, they are also seen in other diseases such as rubella, hepatitis, cytomegalovirus infections, as well as some allergic conditions (Pejme, 1964).

Some degree of liver involvement probably occurs in every case as detectable increases in both the serum level of SGGT (gamma glutamyl transpeptidase) or SGOT (glutamic-oxalacetic transamidase). Peak levels of SGOT do not exceed 500 mU/ml in IM, although bilirubin may reach levels above 2.0 ng in 15% of the cases. Liver inflammation is transient and usually resolves by the 5th week.

Complications of Primary Infection

Although IM is generally a self-limiting, lymphoproliferative illness, it can cause severe, protracted illness, sometimes with fatal outcome. Lymphocytic infiltration of the spleen may vary from scattered foci to extensive invasion of the sinuses, leading to hemorrhage and even rupture. Frank hepatitis occurs in about 8 to 20% of the patients between 2 and 4 weeks after onset. A number of other complications have been associated with primary IM. Thrombocytopenia of some degree is usually noted. Severe thrombocytopenia purpura, sometimes associated with hemolytic anemia, can also occur. Hyperplastic infiltrations of the corresponding tissues and organs can lead to pericarditis, myocarditis, pneumonitis, Reye's syndrome, encephalitis, and other neurologic syndromes, such as Guillain-Barré syndrome and Bell's palsy.

Although IM occurs frequently in adults of childbearing age, there have been very few congenital anomalies attributable to EBV infections. In a rare case of adequately documented EBV infection in the first trimester of pregnancy, micrognathia, cataracts, hypotonia, and petechiae were noted in the infant (Goldberg et al., 1981).

Patients with an unusual genetic immunodefect known as the X-linked lymphoproliferative syndrome (XLP) frequently have overwhelming EBV primary infections, which can be fatal in the majority of the cases (Bar et al., 1974; Purtilo et al., 1978). Their humoral and cellular immune dysfunctions predispose them to erythrocyte aplasia, aplastic anemia, hypogammaglobulinemia, and lymphoproliferative diseases. The 15% of the XLP patients who survive IM often succumb to B cell lymphomas (Purtilo et al., 1982). Fatal cases have also been described in patients with no apparent predisposing conditions, whose EBV primary infections were complicated by a histiocytic, hemophagocytic syndrome involving the bone marrow (Wilson et al., 1981).

In addition to complications associated with the acute phase of IM, it is now evident, in rare patients, the EBV primary infection can trigger the onset of a series of serious, protracted, chronic conditions involving multiple organs (Schooley et al., 1986). In these patients, underlying immune deficiencies are suspected of exacerbating the disease, while the EBV infection appears to be the precipitating factor.

EBV Reactivations

While intermittent reactivations in immunocompetent, seropositive individuals are usually asymptomatic, EBV reactivations were suggested as a possible etiologic agent in a group of patients, mostly young adults and middle-aged patients, whose illnesses were characterized by low-grade fever, chronic fatigue, and recurrent lymphadenopathy (Jones et al., 1985; Straus et al., 1985). Although the serologic findings on many of these patients were indeed compatible with EBV reactivations, the causative link between the patients' illnesses and EBV reactivity is still tenuous at best.

With immunosuppressed patients, it is even more difficult to assess the clinical effects of EBV reactivations independently from the patients' underlying illnesses and associated therapies. Several studies on EBV reactivations in renal transplant and Hodgkin's disease patients did not report increased morbidity associated with the increased rate of viral shedding in these patients (Cheeseman et al., 1980; Lange et al., 1978).

The most striking complication of EBV reactivations is observed in patients at risk for the acquired immunodeficiency syndrome (AIDS) who develop oral hairy leukoplakia (Greenspan et al., 1984). The latter is a flat, poorly demarcated, usually painless,

recurrent lesion of the tongue epithelium. Cryostat sections of lesions showed epithelial cells expressing high levels of EBV-induced antigens (Greenspan et al., 1985), with intact virions visible by electron microscopy. This lesion has not been described in non-AIDS patients.

EBV-Associated Malignancies

While the exact role of EBV in malignancies is not well understood, it is associated with tumors of both lymphocytic origin (Burkitt's lymphoma, BL) and of epithelial origin (undifferentiated nasopharyngeal carcinoma, NPC) (Henle and Henle, 1974; Klein, 1975). In both of these malignancies, EBV appears to be a necessary, but not a sufficient factor. In biopsies of both tumors, EBV-induced viral antigens and DNA can be readily demonstrated (Huang et al., 1974; Reedman et al., 1974; zur Hausen et al., 1970). Extensive chromosomal analyses of BL monoclonal tumors show chromosomal reciprocal translocations between chromosomes 8 and 14 (Manalov and Manolova, 1972; Zech et al., 1976), 8 and 2 (Miyoshi et al., 1979), and 8 and 22 (Magrath et al., 1983).

EBV has been increasingly implicated in the etiology of B-cell lymphomas in immunodeficient populations, both in transplant patients as well as in patients with acquired immunodeficiency syndrome (AIDS) (Hanto et al., 1981; Ziegler et al., 1982). As in BL and NPC, EBV-induced antigens and genomes were detectable in these undifferentiated B-cell lymphomas (Hanto et al., 1985; Ziegler et al., 1982). The characteristic t(8;14) translocation found in BL has not been reported for lymphomas of the immunodeficient patients. While the exact role of EBV in these malignancies is not well understood, it appears that the T-cell dysfunctions of these patients predispose them to uncontrolled proliferation of EBV-transformed lymphocytes and lead to the development of lymphomas.

Diagnosis

Collection and Storage of Specimens

The majority of the primary infections can be serodiagnosed on a single serum, preferably collected during the acute phase of the illness. For the Paul-Bunnell heterophile antibodies assay, the serum must be collected within the first 3 weeks of illness. For EBV-specific serology, only an acute serum is required. Convalescent serum collected 2 to 3 months after onset may be needed only infrequently for confirmation. Aseptically collected sera can be stored at 5°C for several months and at −20°C indefinitely.

For isolation of excreted virus, 2 to 5 ml of saliva can be aspirated into a clean vial. Alternatively, a throat gargle can be collected from adult patients with 5 to 10 ml of sterile saline. For both types of specimens, fetal bovine serum (2 to 5%) should be added as stabilizer. Broad-spectrum antibiotics such as gentamicin may be added to inhibit microbial contaminants. The specimens should be refrigerated promptly. As soon as possible, the samples are filtered through 0.2-nm filters to remove bacterial and fungal contaminants.

Tissues to be examined for viral antigens are collected aseptically and refrigerated in saline or in tissue-culture medium. Smear preparations can be made from the tissues on microscope slides for direct detection of viral antigens by immunostaining. Alternatively, thin cryostat sections can be prepared for immunostaining if the histopathology of the infection is to be determined. All the above preparations for immunostaining should be air dried and fixed prior to storage or testing. Depending on the antigens to be tested, the specimens should be fixed with either acetone (for VCA and EA) or with an equal mixture of acetone and methanol if EBNA is the antigen under test. For the detection of EBV nucleic acid, fresh or frozen tissues can be used in the Southern blot or dot blot procedures, whereas the cryosections or paraffin sections can be used for in situ hybridizations.

For cultivation of EBV-infected peripheral blood lymphocytes, 10 to 30 ml of heparinized (5 to 10 U/ml) blood is needed. The sample should be processed as soon as possible, although refrigeration is adequate for up to 24 h.

Virus Detection

Electron microscopy was instrumental in the initial discovery of EBV in cultured Burkitt's lymphoblastoid lines. Although this technique is historically important, it is not suitable for general detection of EBV. The majority of infected lymphoid cells and tumors are latently infected. It is unusual to find intact EBV particles in biopsy specimens, thus limiting the usefulness of electron microscopy. It is useful, however, in the diagnosis of EBV-associated oral hairy leukoplakia (Greenspan et al., 1985). Cryosections of tongue lesion biopsies of these patients contain large numbers of fully mature EB virions.

Direct Antigen Detection

Direct antigen detection is not useful for the diagnosis of primary infection. It is very useful, however, in the diagnosis of EBV-associated malignancies. As mentioned previously, EBV-associated malignancies are mostly monoclonal in origin, with every cell expressing EBNA. EBNA, as well as other EBV-induced antigens, can be detected readily by immunostaining techniques. For detection of the various EBV antigens in tissues, smears or touch preparations of infected tissues or cells are suitable specimens.

For tissue specimens, anti-complementary indirect immunofluorescence (ACIF) has been the most useful staining procedure, with human reference sera of known specificities as the intermediate reagents (Reedman et al., 1974). Human sera, used with appropriate controls, are still the preferred reagents for the detection of various EBV antigens. Indirect immunofluorescence is usually not suitable on tissues because of the high background of endogenous immunoglobulins in the specimens. Recently, mouse monoclonal anti-EBV antibodies have become commercially available; however, the performance characteristics of these antisera for clinical diagnosis and tissue staining have not been adequately evaluated.

Direct Viral Nucleic Acid Detection

Nucleic acid hybridization methods were used with excellent results in the earlier phase of EBV research and were instrumental in linking EBV with the associated malignancies (Nonoyama and Pagano, 1973; zur Hausen et al., 1970). Although this particular approach remains a research procedure, the availability of nonradioactive probes will clearly put the hybridization techniques within the feasibility of the reference viral diagnostic laboratories. Depending on the type and quantity of specimens available, the sensitivity required, the number of specimens to be tested, and the level of information needed, Southern blot transfer, DNA dot blot, in situ hybridization on filter and in cytospin preparations can all be used for the detection of EBV nucleic acids.

The time required for the completion of the above procedures depends on the nature of the label (isotopes vs. nonradiolabel) on the probes, the genomic content (number of copies of the particular nucleic sequence to be studied) of the probes, and the type of clinical specimens to be tested. It can range from a few days, as in the in situ cytohybridization with biotin-labeled recombinant probe containing the large internal repeated sequence within the EBV genome, to several weeks if autoradiography is needed.

Virus Isolation and Identification

The presence of cell-free virus in saliva and throat wash can be detected by the ability of the virus to transform cord blood lymphocytes. While this virus assay has been instrumental in EBV research, isolation is not useful for clinical diagnosis of EBV infections for the following reasons: 1) the lymphocyte transformation assay requires freshly fractionated cord blood lymphocytes, a reagent not routinely available; 2) the 3 to 4 weeks required for the completion of the assay is too long to be clinically useful, and 3) the uncertain clinical significance of a positive isolation, when the majority of the seropositive individuals are shedding virus in the oropharynx. For these reasons, serological diagnosis is the preferred approach.

Paul-Bunnell Heterophile Antibodies

Even before the discovery of EBV, reliable diagnosis of IM can be made since 1932 by the detection of the Paul-Bunnell (PB) heterophile antibodies (PB-ab) (Paul and Bunnell, 1932). PB-ab are among a large number of transient antibodies produced during primary IM. These antibodies include a number with undetermined target antigens as well as those with autoimmune specificities such as anti-nuclear, anti-lymphocyte antibodies and rheumatoid factors (Carter, 1966). The IgM Paul-Bunnell antibodies specific to IM are present in 85 to 95% of the patients during the first month of their illness, with peak levels at 2 weeks (Davidsohn and Lee, 1969). 2 to 3% of healthy individuals may have low level of Pb-Ab for extended period of time (Horwitz et al., 1976). Anamnestic PB-ab responses are not known to occur, either with EBV reactivations or with other viral infections (Kano and Milgrom, 1977).

A number of assays have been developed to measure PB-Ab, all based on the antibodies' ability 1) to agglutinate sheep or horse erythrocytes; 2) to bind to bovine erythrocytes and their associated antigenic components, and 3) not to react with guinea pig kidney tissue. In contrast, other agglutinins of sheep and horse erythrocytes, such as Forssman heterophile antibodies, can be absorbed with guinea pig kidney tissue only, whereas Hanganutziu-Diecher heterophile antibodies can be absorbed with both guinea pig and bovine erythrocytes (Kano and Milgrom, 1977). Of all the heterophile assays, the Paul-Bunnell-Davidsohn differential has been the best documented reference test (Davidsohn and Lee, 1969), with only rare false positives seen in patients with rheumatoid arthritis and various malignant conditions (Davidsohn and Lee, 1962; Horwitz et al., 1976). However, this test has fallen out of use because of its requirement for involved absorption steps.

Currently, there are two principal procedures for the detection of Pb-ab. The first, the ox-cell hemoly-

sin, is quantitative and can be performed in microtiter plates (Leyton, 1952). It requires no absorption steps and is based on complement-induced hemolysis of PB-Ab bound to bovine erythrocytes. A titer of 1:40 or above is considered positive. Its specificity and sensitivity are comparable to the Paul-Bunnell-Davidsohn (PBD) assay (Davidsohn and Lee, 1964). Although the ox cell hemolysis procedure is much simplified compared with PBD, it is not performed in most laboratories.

The most widely available tests are the commercial, qualitative, slide agglutination assays. They require 1 to 2 min to complete and are simple to perform. Typically, one drop of serum on a glass slide is first mixed separately with a drop of bovine or guinea pig kidney absorbent; this is followed by the addition of a drop of finely suspended horse or sheep erythrocytes to each of the above mixture. Positive IM sera would not agglutinate the indicator erythrocytes after absorption with bovine stromas, whereas the guinea pig treated mixture would show rapid clumping of the indicator erythrocytes. The rapid slide tests have virtually replaced all other forms of heterophile testing because of their combined advantages of convenience, rapidity, and economy. The overall performance of the various slide kits varies with the source of indicator cells and preservative used. In general, those with horse erythrocytes have the fewest false positives or negatives, whereas those with formalinized sheep erythrocytes have the largest number, up to 3% of sera tested (Galloway, 1969).

There are a number of limitations with PB-ab testing. The first disadvantage is the short interval during which PB-ab are present in the serum. Many patients come to the attention of their physicians after the PB-ab are no longer detectable. The second disadvantage is the low level of PB-ab encountered in pediatric patients. Only half of the children under five have PB-ab detectable by the usual, rapid slide tests, although more than 90% of them have actual measurable Pb-ab by another, more sensitive assay, the immune adherence agglutination procedure (Fleisher et al., 1979).

The current recommendation for the serologic diagnosis of IM is summarized in Table 1. 1) Positive PB-ab slide tests are considered diagnostic in sera of patients with clinical and hematologic features consistent with IM. No further testing is needed. 2) If the clinical presentation is characteristic of IM and the heterophile test is negative, EBV specific serology should be considered in order to overcome the possibility of late collection of the serum sample. If the patient has a prior history of mononucleosis-like illness, other infectious agents should be included in the differential diagnosis, as cytomegalovirus, human immunodeficiency virus, adenoviruses, and

TABLE 1. Approaches to EBV-specific serology

Clinical picture	Paul-Bunnell heterophile antibodies	Recommendations
Mononucleosis	Positive	Diagnostic of IM; no further testing needed.
Mononucleosis	Negative	1. EBV-specific serology.
		2. Testing for other agents.
Atypical for IM	Negative or positive	EBV-specific serology to rule out false heterophile results.

Toxoplasma gondii are all capable of inducing mononucleosis-like illnesses. 3) If the illness is not characteristic of IM and yet the heterophile test is positive, EBV specific serology is needed to rule out a false-positive heterophile result.

EBV Serology for Diagnosis of Primary Infection

Paired sera are usually not necessary with EBV serology. Interpretations can usually be made on the basis of the distinctive profiles obtainable against a panel of EBV antigens rather than quantitative differences between paired sera. Table 2 summarizes the most commonly observed patterns.

Humoral response to primary EBV infection is quite rapid. Most symptomatic patients have a peak titer of IgG-VCA by the time they consult their physicians. Fourfold titer increases with subsequent sera cannot be demonstrated in 80% of the IM patients (Henle et al., 1974). The presence of IgM-VCA can be demonstrated in parallel with IgG-VCA for only about 4 weeks after onset. While IgM-VCA antibodies can be useful for confirmation of primary infection, they are not as reliable a marker of current infection as one would wish. Approximately 15% of the patients no longer have demonstrable IgM on their first physician visit. In addition to the false-

negative problem, false-positive IgM-VCA titers can arise from serum containing rheumatoid factor (Henle et al., 1979).

During convalescence, IgG-VCA starts to decline after reaching some peak level; it is maintained at a lower, stable level for life. The peak levels of IgG-VCA attained during primary IM vary widely with the individual's age and immune functions. Single, high cut-off IgG-VCA titer cannot be used to indicate primary infection.

Neutralizing antibodies increase in level during early convalescence and also persist for life. In practice, complex neutralization assays are rarely used to test for immunity. Anti-VCA titers are often used as indicators of immunity, as nearly everyone with measurable anti-VCA titers would also have neutralizing antibodies.

IgG antibodies to EA-D show a transient rise in 85% of IM patients during the acute phase, particularly in those with marked lymphadenopathy. Anti-EA/D usually become undetectable within 3 to 6 months from onset (Henle et al., 1974). In young children under 2 years of age and in adults with asymptomatic seroconversion, anti-EA/R, instead of anti-EA/D, are present during the acute phase (Biggar et al., 1978a). In most cases, however, anti-EA/R are not present in the acute phase but appear transiently during late convalescence (Horwitz et al., 1985). Anti-EBNA antibodies are rarely detectable in

TABLE 2. Serologic profiles for EBV infections

	Antibodies to					
Clinical status	VCA-IgM	VCA-IgG	EA-D	EA-R	EBNA	Hallmark antibodies
Susceptible	−[a]	−	−	−	−	
Current primary	+	+	+ or −	−	−	IgM-VCA
Recent primary	+ or −	+	+ or −	+	± or −	
Past infection	−	+	−	−	+	
Reactivated	−	+	+ or −	+ or −	+	IgG EA-D or EA-R
Burkitt's lymphoma	−	++	−	++	+	IgG-EA-R
NPC	−	++[b]	++[b]	−	+	IgG-EA-D and IgA-EA-D IgA-VCA

[a] Negative (−): <1:5. Positive (+): ≥1:5.
[b] IgA as well as IgG.

the acute phase sera. They usually appear during the 2nd to 3rd month after onset, reach peak titers at 4 to 6 months, and are maintained for life (Henle et al., 1974).

The extent of the humoral responses to the various EBV antigens varies with the patients' ages and immune functions. The exact titers attained and the time needed to develop a spectrum of antibodies are variable. In addition, the titers obtained among laboratories are not comparable, partly owing to the variable quality of the increasingly available commercial reagents and also of the fluorescence microscopes in use. It is, therefore, not acceptable to use a single titer of antibodies to any single antigen as a basis for diagnosis of primary EBV infection. The profile of the various antibodies is predictable and is subject to few variations. If the complete testing panel is not feasible, the following alternative minimum combination should be done: IgG-VCA with anti-EBNA, in combination with the heterophile test. With the combination of the three, the diagnosis of the majority of primary EBV infections need not depend on the measurement of IgM-VCA, which suffers both from false-negative and false-positive reactions. Reliance on anti-EBNA, which does not appear until late convalescence, effectively extends the period during which the diagnosis of primary infection can be made. A definitive diagnosis can be made in 90 to 95% of the acute sera if measurements of antibodies to at least VCA and EBNA are obtained.

EBV Serology in Reactivated Infections

Under normal conditions, the antibody responses to EBV are tightly regulated by the host. Serologic profiles of healthy individuals are surprisingly stable. Conversely, patients with various immunodeficiencies, whether iatrogenic or genetic, develop gradual changes in the overall EBV profile during viral reactivations (Henle and Henle, 1980). These changes vary according to the exact nature of the patient's condition. They may be transient in the case of immunosuppressive viral infection such as CMV, or can be longer lasting, such as in patients with various malignancies or in patients undergoing transplantations, whose EBV serologic profile is characterized by elevated levels of anti-EA/D or -EA/R with or without concomitant change in level of antibodies to other EBV antigens, e.g., VCA and EBNA (Cheeseman et al., 1980). In the majority of the patients with reactivated EBV serology, anti-EBNA antibodies are present at a fairly normal level. Infrequently, the absence of the latter has been noted in some immunedeficiencies, particularly in patients with T-cell dysfunctions (Henle and Henle, 1980, 1981).

EBV Serology in Associated Malignancies

In general, IgG-VCA levels in BL patients are eight- to tenfold above the geometric mean of a healthy, age-matched population (Henle et al., 1969). Unusually high EA/R antibodies are found in these patients and can be correlated quantitatively with the tumor burden (Henle and Henle, 1977). Similarly, anti-VCA titers are also eight- to tenfold higher in NPC patients than in healthy controls (Henle et al., 1970). However, anti-EA/D, not EA/R, are characteristically high in NPC. Increased levels of IgA-VCA and IgA-EA/D are also useful markers, as they increase and decrease with the tumor burden and can be used as prognostic indicators (Henle et al., 1974).

Interpretation of EBV Serology

Antibody profiles against the four EBV antigens can be used (Table 2) to classify the patients as 1) susceptible; 2) current primary infection; 3) recent infection (within last 6 to 8 weeks); 4) past infection (at least 6 months prior); 5) past infection, with possible reactivation.

Antibody profiles associated with categories 1, 2, and 4 are rarely ambiguous. Interpretations for categories 3 and 5 are less straight-forward and may require testing of a second serum collected 3 months later. For example, in the typical primary infection, testing of serum collected during late convalescence would show the presence of IgG but not IgM to VCA, and absence or low level of anti-EBNA. One should be aware that this pattern can also arise in patients undergoing T-cell suppression who have declining anti-EBNA titers. Testing of a later serum should be able to differentiate between the two types of conditions. In the recently infected patients, anti-EBNA should show significant increase, whereas in the immunosuppressed patients, anti-EBNA would remain unchanged. In rare instances, late convalescent sera may still contain anti-EA/D antibodies when anti-EBNA are at low to moderate titers, hence can be confused with that from a patient with reactivated serology. Again, retesting 3 months later should distinguish the former condition by the subsequent disappearance of anti-EA/D, whereas the titer in the latter condition would most often remain unchanged.

Epidemiology

EBV has a worldwide distribution, affecting more than 95% of the adult population. In areas with poor sanitary conditions and crowding, EBV infections occur as soon as maternal antibodies disappear (Biggar et al., 1978b). The mean age for EBV infections is less than 2 years, when nearly all of the primary infections are silent. In socioeconomically privileged areas, the mean age of primary infection is delayed to 14 to 16 years, when an estimated one-third or more of the individuals will have symptomatic infections (Niederman et al., 1970). Transmission of the virus is very efficient, based on the high prevalence of seropositive individuals in all parts of the world. However, IM rarely occur in outbreaks, presumably owing to the requirement of salivary exchange among young adults, when the disease is more apparent.

BL is primarily a disease of African children and New Guineans. The time and space clustering seen in the African cases suggested an infectious agent, such as malaria, as a necessary co-factor (Burkitt, 1969). Cases of EB-associated BL outside Africa are rare and occur mainly in adults. In parts of Africa where EBV primary infections occur at infancy, Burkitt's early epidemiologic investigations had shown that BL is associated with well-defined regions that were also known to be hyperendemic for malaria. Burkitt also pointed out that the only other area of the world with identical temperature and rainfall conditions and where malaria is hyperendemic coincides precisely with the only other BL endemic region, New Guinea. In both regions, EBV affects 100% of the children, malaria more than 50%. These observations led to the hypothesis that BL is the result of multifactorial causations, with genetic predisposition playing a significant part. Supporting evidence for this hypothesis includes the low prevalence of BL in areas not hyperendemic for malaria, the decrease in incidence of BL in areas with aggressive antimalarial eradication programs, and the lowered incidence of BL among sickle-cell children, a trait that confers partial protection against malaria.

As the first known human virus to be associated with neoplasms, EBV was the subject of an intensive, multinational collaboration in Africa, including a prospective serosurvey of 42,000 children for the development of BL (Kafuko et al., 1972). Valuable information was gained on the distribution of EBV-associated BL and of EBV antibodies pre and post malignancy. For example, the serosurveys showed that the interval between primary infection and malignancy averaged between 3 and 5 years. Patients who developed BL had substantially higher anti-VCA months prior to the development of BL, compared with an age-matched control group. Calculation of risk factor attached to high anti-VCA titer for subsequent development of BL in the children had shown it to be 30 times higher than in the controls.

Two forms of BL were discovered, one EBV associated and the second not, based on the presence of EBNA and EBV genome in the tissues. The former is endemic in Africa and only sporadic else-where; the non-EBV-associated BL is sporadic everywhere, including Africa. EBV-BL occurs primarily in children in Africa. In the United States, it has been reported primarily in adults (Judson et al., 1977; Ziegler et al., 1982).

With EBV-associated NPC, a combination of genetic and environmental factors may be important, as this disease is found predominantly in Southern Chinese (Clifford, 1970; Ho, 1975). The clear climatic and environmental factors seen with BL are absent in the epidemiology of NPC. In Southern China, East and North Africa, and Alaska, where NPC is prevalent, EBV primary infections occur very early in life. The interval between primary EBV infection and NPC is much longer than for BL, ranging from 30 to 50 years. The incidence of NPC in first-generation Chinese immigrants is similar to native Cantonese, while children of interracial as well as interethnic Chinese have an intermediate risk of acquiring the malignancy (Ho, 1975).

Laboratory Procedures

Cell Lines

For EBV serologic assays, several lymphoblastoid lines are needed to provide suitable antigens for serology. Table 3 lists alternative cell lines that may be used. They can be divided into three groups:

1) PRODUCER LINES

There are a limited number of lines that continuously produce infectious viruses in culture. The proportion of VCA-positive cells for each of these lines appears to be determined largely by virus and host factors, and to a lesser extent by culture conditions. The fraction of cells expressing late antigen, such as VCA, at any time is generally low, ranging from 2% in the EB3 line to 10 to 20% in the P3HR-1 and B95-8 lines. Under optimal conditions, a stable ratio of positive to negative cells can be maintained for long periods (12 to 18 months or longer) if cells are grown at 32 to 34°C. However, VCA expression in these lines can gradually decrease to less than 1% over time, rendering them unsuitable for immunoassays. The host regulatory factors that switch on the expression of EBV genome are not understood. Hence, there are no known methods for continually boosting the rate at which VCA-negative cells convert to VCA-positive status. Chemical inducers can be used to increase the level of positive cells for the purpose of making antigen preparations. However, as discussed previously, induction of the genome invariably leads to cell death and hence to the loss of positive cells from the culture.

TABLE 3. Commonly used lymphoblastoid lines and EBV antigen expression

	Origin	EBNA	EA	VCA	Virus production	Reference
Producer lines						
P3HR1	BL	+	+	+	Lytic	Hinuma et al. (1967)
EB3	BL	+	+	+	Lytic	Epstein et al. (1965)
B95-8	BL	+	+	+	Transforming	Miller and Lipman (1973)
Nonproducer lines						
Raji	BL	+	±[a]	−	Negative	Pulvertaft (1965)
EBV-negative lines						
BJAB	B lymphoma	−	−	−	Negative	Menezes et al. (1975)
Ramos	B lymphoma	−	−	−	Negative	Klein et al. (1975)
Molt-4	T leukemia	−	−	−	Negative	Minowada and Ohnuma (1972)

[a] Less than 0.1% of total cells.

2) NONPRODUCER LINES

These are EBV-transformed lines expressing primarily EBNA. While any EBV-infected cell line can be used for the preparation of EBNA for immunoassays, the line most commonly used is a Burkitt's lymphoma line designated as Raji. It is one of the first tumor monoclonal BL lines to be established and characterized.

3) EBV-NEGATIVE LYMPHOBLASTOID LINES

These are lymphoblastoid lines that do not contain detectable EBV genomes. They are of both B- and T-cell origins and are useful as negative antigen controls, which are necessary to exclude the presence of anti-nuclear and anti-lymphocytic antibodies unrelated to EBV in patients with autoimmune conditions.

The general procedures and requirements for the maintenance of these lines are fairly uncomplicated. Although various cell culture media can be used, the most widely used medium for lymphoid cultures is RPMI 1640 supplemented with 10% bovine serum. Supplemental glutamine and bovine serum can be added for rapid growth, but they are not essential for the long-term maintenance of the lines. The lines should be subcultured at weekly intervals, with seeding density of about 2×10^5 cells/ml. Subculturing can be accomplished simply by diluting the cultures in fresh medium to the appropriate cell density. The majority of the dividing cells are in unattached clusters during the logarithmic phase. Once in the stationary phase, dying cells detach from the clusters, eventually yielding single-cell suspensions at 6 to 7 days. The peak density of the cultures is usually attained on the 4th or 5th day of growth at 34°C at approximately 2 to 4×10^6 cells/ml for most of the EBV-infected lines.

For laboratories planning to prepare their own reagents, it is important to have cell preservation facilities at liquid nitrogen temperature. In addition to the accidental loss due to microbial contaminations, replacement cultures are needed because of the slow decline in the number of VCA-positive cells frequently observed in the producer lines. Ample numbers of ampules or vials of the lines showing the desired number of positive cells should be frozen as early as permissible. To freeze the cultures, cells in their growth phase should be resuspended in fresh growth medium containing 10% DMSO. A satisfactory, inexpensive freezing apparatus can be made from a heavy-walled (2.5 cm or thicker) styrofoam container. Two pin holes are made with an ice pick or nail on the outer wall of the box. A second styrofoam block with the desired number of hollows can serve as the interior vial holder. Sealed vials in this apparatus can then be cooled very gradually overnight to −70°C in an ultralow temperature freezer and transferred to liquid nitrogen temperature the next day. The vials can remain at −70°C for several weeks prior to transfer to liquid nitrogen storage. For revival of the cells, frozen vials are quickly thawed in water at 37°C. The cell suspension is promptly diluted in prewarmed growth medium. After low-speed centrifugation, the fluid containing DMSO is discarded and replaced with fresh growth medium. Within a few hours of incubation, viable cells should be visible as clusters.

Virus Preparation

TRANSFORMING VIRUS

High-titered transforming EBV stock can be prepared from B95-8, a marmoset lymphoblastoid line transformed with human EBV. Titers of 10^6 transforming units/ml can be obtained from 7- to 10-day-old culture fluid by first centrifuging the spent medium at $1000 \times g$ to remove cells, followed by filtration through 450-nm pore size filter to remove cell debris. Virus preparations can be frozen in small vials at −70°C or at liquid nitrogen temperature for long-term storage.

For quantitation of B95-8 virus, 0.5 ml of serial, tenfold dilution of the virus stock is added to 10^6 lymphocytes, which have been separated from heparinized cord blood by density gradient centrifugation through suitable polymers. After 30 min of incubation at 37°C with occasional mixing, 1.5 ml of fresh medium is added. Clusters of transformed cord lymphocytes should appear after 1 to 4 weeks, depending on the inoculating concentration. Uninfected cell controls should be included to exclude rare, spontaneous transformation events. Confirmation of the presence of EBV can be accomplished with detection of EBNA in the infected cells. (See below staining procedure.)

LYTIC VIRUS

The P3HR-1 cell line is the most common source of lytic virus. Because of the low yield of virus, a 100- to 1000-fold concentration of the culture medium is needed to reduce the storage space requirement. Culture medium (grown at 32 to 34°C for 10 to 14 days) is first clarified with low-speed centrifugation. The virus is then sedimented from the medium by high-speed centrifugation at $50,000 \times g$ for 30 min, resuspended in the desired volume of growth medium, and frozen at −70°C in small vials.

For quantitation of the P3HR-1 virus, 0.5 ml of serial, tenfold dilutions of the virus preparation is added to 10^7 Raji cells and incubated at 37°C for 1 h with occasional mixing. Twenty ml of medium is then added to yield a cell density of 5×10^5 cells/ml. Forty eight hours post infection, when maximal number of cells are expressing EA antigens, the infected cells are collected by centrifugation and smeared on glass slides for immunofluorescence staining. (See below for EA staining procedures.)

Smear Preparations

Table 4 is a summary of culture conditions and appropriate fixatives for preparation of cell smears suitable for indirect immunofluorescence assays. For preparation of cell smears of the various specific EBV antigens, the following methods can be used to maximize the number of cells expressing the desired antigen:

1. VCA: Culture of P3HR1 cells is grown at 32 to 34°C for 7 to 10 days, air dried smears are fixed for 1 to 3 min in acetone. If EB3 line is used, starvation in arginine-free medium for 48 h would increase the VCA-positive cells to 10% (Henle and Henle, 1968). Preparations of either lines are suitable for VCA IgG and IgA determinations, whereas P3HR1 cells are required for the determination of IgM, as EB3 cells contain endogenous IgM.

2. EA: Two general methods are available. The first involves superinfection of Raji cells with P3HR1 lytic virus, at a multiplicity calculated to yield 20 to 30% EA-positive cells. In the second method, a nonproducer line such as Raji is treated with chemical inducers for 3 to 5 days, thereby increasing EA-positive cells from less than 0.1% up to 10 to 20% of the treated culture. Of the inducers, the most effective one appears to be 12-O-tetradecanoylphorbol-13 acetate (zur Hausen et al., 1978), while others such as sodium butyrate and iododeoxyuridine have also been successfully used (Luka et al., 1979; Sugawara et al., 1973). Each lot of chemicals needs to be titrated for optimal concentration prior to use. With the first method, consistent lots of smears can be prepared from any single preparation of virus. Reproducibility with the second method is more difficult to achieve. The degree of induction can vary with temperature, the growth phase of the cells, and other general growth conditions, resulting in greater lot-to-lot variation. In addition, chemically induced cells tend to aggregate, thus affecting the overall quality of the smear preparation. For both methods, the infected cells are harvested at 72 to 96 h posttreatment, washed in buffer to remove the drug, and smeared. Air-dried smears fixed with acetone (1 to 3 min) will contain both EA-D- and EA-R-positive cells, whereas fixation with methanol (1 min) will yield cells positive only

TABLE 4. Culture conditions and antigen preparations for EBV serology

Cell line	Temperature (°C)	Length (days)	Antigen	Fixative, fixing time
P3HR-1	32–34	7–10	VCA	Acetone, 1–3 min
EA (+) Raji	34–37	2–3 post infection or induction	EA-D and EA-R	Acetone, 1 min
		2–3 post infection or induction	EA-D only	Methanol, 1 min
Raji	34–37	≤4–5	EBNA	Acetone : methanol (1 : 1 mixture), 1 min

for EA-D. There are currently no known procedures to prepare cell smears positive only for EA-R.

3. EBNA: For preparation of smears suitable for immunofluorescence staining, cell suspensions are centrifuged at $700 \times g$, the cells are resuspended in phosphate-buffered saline containing 0.25% bovine serum albumin to a density of 10^7 cells/ml. The cell suspension is then smeared as a very thin film on glass slides, air dried, and fixed with an equal mixture of acetone and methanol. It is critical that the smears be dried as rapidly as possible; hence, the thinner the cells are spread and the faster the air flow during drying, the better the results.

4. EBNA-negative control: Smears of EBV-negative lymphoblastoid lines are prepared and fixed exactly as described for EBNA-positive lines.

Fixed smears may be kept at refrigerated temperature for short-term storage (up to 2 to 3 months). For longer storage, they may be frozen at $-20°C$ or lower temperature in sealed containers or wrapping. Moisture or condensation is generally deleterious to most antigens and should be avoided.

Immunofluorescence Procedures

Indirect immunofluorescence staining (IFA) is used to determine the various antibody subclasses to VCA and EA antigens, while ACIF is required to detect antibodies to EBNA. For both procedures, all incubations are at 37°C for 30 min, followed by two rinses in PBS for antisera removal.

Sera to be tested are diluted 1:5 in PBS and inactivated at 56°C for 30 min. Serial, fourfold dilutions of the inactivated sera are then prepared. In the IFA procedure, cell smears containing the desired antigens are incubated with each dilution of the serum, followed by a second incubation of the desired fluorescein isothiocyanate (FITC) anti-human immunoglobulin subclasses. After a final rinse in distilled water to remove residual salts, the slides should be air dried and mounted for microscopic examinations. Known positive and negative controls should be included in each run. For IgM and IgA determinations, it is important to include VCA-IgM or IgA-negative but VCA-IgG-positive serum to check for the specificities of the FITC conjugates.

For the ACIF procedure, there are three incubation steps. In the first, the cell smears are incubated with the dilutions of the test serum. In the second, guinea pig complement at optimal dilution (generally at 1:50 of whole serum) is used, followed by a final incubation with FITC anti-guinea pig C3. The complement-fixation step increases both the specificity and the sensitivity of the assay.

False-positive IgM-VCA and anti-EBNA reactions cause the most problems for both technical and interpretational reasons. Every IgM-VCA-positive serum should be checked undiluted for the presence of rheumatoid factor (RF). If the latter is present, various methods may be used to remove the interfering factor. The most practical ones include column chromatography, absorption with *Staphylococcus aureus* protein A, with latex beads linked to human IgG, and with heat-aggregated IgG (Ankerst et al., 1974; Henle et al., 1979; Johnson and Libby, 1980; Shirodaria et al., 1973). The treated serum can then be retested for EBV-specific IgM. False-positive anti-EBNA reactions are usually due to anti-nuclear antibodies (ANA) present in patients with various viral infections or with autoimmune diseases (Faber and Elling, 1966).

Table 5 gives a check list of the controls needed for the detection of each component of the EBV serologic panel. It is important that they be included in the test to serve as reference staining patterns. As all the lines are lymphoblastoid in nature, they have characteristic surface immunoglobulins as well as immunoglobulin-producing cells which would react with the FITC conjugates. With each test run, the

TABLE 5. Controls for immunofluorescence procedures for EBV serology

Ab to	Assay	FITC antibodies against	Range of dilutions	Serum controls	
				Positive	Negative
VCA	IFA	IgG (γ chain)	20–1,280	VCA-IgG(+)	VCA-IgG(−)
		IgM (μ chain)	5–160	VCA-IgM(+)/IgG(+)	VCA-IgM(−)/IgG(+)
		IgA (α chain)	5–320	VCA-IgA(+)/IgG(+)	VCA-IgA(−)/IgG(+)
EA-D	IFA	IgG (γ chain)	5–320	EA-D IgG(+)	EA-D IgG(−)
		IgA (α chain)	5–80	EA-D-IgA(+)/IgG(+)	EA-D IgA(−)/IgG(+)
EA-R		IgG (γ chain)	5–80	EA-R IgG(+)	EA-R IgG(−)
EBNA	ACIF	Guinea pig C3	5–640	EBNA(+)	EBNA(−)
ANA or lymphocyte	ACIF	Guinea pig C3	5–80	ANA(+) EBNA(−)	ANA(−) EBNA(−)

following features of the stained slides must be kept in mind: the ratio of positive to negative cells, the size and the staining morphology of the positive cells, and the number of globulin-producing cells, usually discernible when negative serum controls are included. With training and practice, the microscopist can learn to differentiate acceptable variations from unacceptable deviations of these patterns.

Literature Cited

Ankerst, J., P. Christensen, L. Kjellen, and G. Kronvall. 1974. A routine diagnostic test for IgA and IgM antibodies to rubella virus: absorption of IgG with *Staphylococcus aureus*. J. Infect. Dis. **130**:268–273.

Aya, T., and T. Osato. 1974. Early events in transformation of human cord leukocytes by Epstein-Barr virus: induction of DNA-synthesis, mitosis and virus associated nuclear antigen synthesis. Int. J. Cancer **14**:341–347.

Bar, R. S., C. J. Delor, K. P. Clausen, P. Hurtubise, W. Henle, and J. Hewetson. 1974. Fatal infectious mononucleosis in a family. N. Engl. J. Med. **290**:363–367.

Biggar, R. J., G. Henle, J. Böcker, E. T. Lennette, G. Fleisher, and W. Henle. 1978a. Primary Epstein-Barr virus infections in African infants. II. Clinical and serological observations during seroconversion. Int. J. Cancer **22**:244–250.

Biggar, R. J., W. Henle, G. Fleisher, J. Böcker, E. T. Lennette, and G. Henle. 1978b. Primary Epstein-Barr virus infections in African infants. I. Decline of maternal antibodies and time of infection. Int. J. Cancer **22**:239–243.

Burkitt, D. 1958. A sarcoma involving the jaws in African children. Br. J. Surg. **46**:218–223.

Burkitt, D. P. 1969. Etiology of Burkitt's lymphoma—an alternative hypothesis to a vectored virus. J. Natl. Cancer Inst. **42**:19–28.

Carter, R. L. 1966. Antibody formation in infectious mononucleosis. II. Other 19 S antibodies and false-positive serology. Br. J. Haematol. **12**:268–275.

Cheeseman, S. H., W. Henle, R. H. Rubin, N. E. Tolkoff-Rubin, B. Cosimi, K. Cantell, S. Winkle, J. T. Herrin, P. Black, P. Russell, and M. S. Hirsch. 1980. Epstein-Barr virus infection in renal transplants. Ann. Intern. Med. **93**:39–42.

Clifford, P. 1970. On the epidemiology of nasopharyngeal carcinoma. Int. J. Cancer **5**:287–309.

Crawford, D. H., A. B. Rickinson, S. Finerty, and M. A. Epstein. 1978. Epstein-Barr (EB) virus genome-containing, EB nuclear antigen-negative B-lymphocyte populations in blood in acute infectious mononucleosis. J. Gen. Virol. **38**:449–460.

Davidsohn, I., and C. L. Lee. 1962. The laboratory in the diagnosis of infectious mononucleosis. Med. Clin. North Am. **46**:225–244.

Davidsohn, I., and C. L. Lee. 1964. Serologic diagnosis of infectious mononucleosis. A comparative study of five tests. Am. J. Clin. Pathol. **41**:115–125.

Davidsohn, I., and C. L. Lee. 1969. The clinical serology of infectious mononucleosis. p. 177–200. *In* R. L. Carter and H. G. Penman (ed.), Infectious mononucleosis. Blackwell Scientific Publications, Oxford.

Diehl, V., G. Henle, W. Henle, and G. Kohn. 1968. Demonstration of a herpes group virus in cultures of peripheral leukocytes from patients with infectious mononucleosis. J. Virol. **2**:663–669.

Epstein, M. A., B. G. Achong, and Y. M. Barr. 1964. Virus particles in cultured lymphoblasts from Burkitt's lymphoma. Lancet **1**:702–703.

Epstein, M. A., and Y. M., Barr. 1964. Cultivation in vitro of human lymphoblasts from Burkitt's malignant lymphoma. Lancet **1**:252–253.

Epstein, M. A., Y. M. Barr, and B. G. Achong. 1965. Studies with Burkitt's lymphoma. Wistar Inst. Symp. Monogr. **4**:69–82.

Ernberg, I., D. Killander, M. Andersson-Anvret, G. Klein, and L. Lundin. 1977. Relationship between the amount of Epstein-Barr virus (EBV) determined nuclear antigen (EBNA)/cell and the number of EBV DNA copies. Nature (London) **266**:269–271.

Faber, V., and P. Elling. 1966. Leukocyte-specific antinuclear factors in patients with Felty's syndrome, rheumatoid arthritis, systemic lupus erythematosus and other diseases. Acta Med. Scand. **179**:257–267.

Fingeroth, J. D., J. J. Weiss, T. F. Tedder, J. L. Strominger, P. A. Biro, and D. T. Fearon. 1984. Epstein-Barr virus receptor of human B-lymphocytes is the C3d receptor CR2. Proc. Natl. Acad. Sci. USA **81**:4510–4514.

Fleisher, G., E. T. Lennette, G. Henle, and W. Henle. 1979. Incidence of heterophile antibody responses in children with infectious mononucleosis. J. Pediatr. **94**:723–728.

Galloway, E. 1969. Comparison of the three slide tests for infectious mononucleosis with Davidsohn's presumptive and differential heterophil test. Can. J. Med. Technol. **31**:197–206.

Gerber, P., and S. Lucas. 1972. Epstein-Barr virus-associated antigens activated in human cells by 5-bromodeoxyuridine. Proc. Soc. Exp. Biol. Med. **141**:431–435.

Gergely, L., G. Klein, and I. Ernberg. 1971. The action of DNA antagonists on Epstein-Barr virus (EBV) associated early antigens (EA) in Burkitt's lymphoma lines. Int. J. Cancer **7**:293–302.

Goldberg, G., V. A. Fulginit, G. Ray, P. Ferry, J. Jones, H. Cross, and L. Minnich. 1981. In utero Epstein-Barr virus (infectious mononucleosis) infection. J. Am. Med. Assoc. **246**:1579–1581.

Greenspan, D., J. S. Greenspan, M. Conant, V. Petersen, S. Silverman, Jr., and Y. de Souza. 1984. Oral "hairy" leucoplakia in male homosexuals: evidence of association with both papillomavirus and a herpes-group virus. Lancet **2**:831–834.

Greenspan, J. S., D. Greenspan, E. T. Lennette, D. I. Abrams, M. A. Conant, V. Petersen, and U. K. Freese. 1985. Replication of Epstein-Barr virus within the epithelial cells of oral "hairy" leukoplakia and AIDS-associated lesion. N. Engl. J. Med. **313**:1564–1571.

Grogan, E. A., W. P. Summers, S. Dowling, D. Shedd, L. Gradoville, and G. Miller. 1983. Two Epstein-Barr viral nuclear neoantigens distinguished by gene transfer, serology and chromosome binding. Proc. Natl. Acad. Sci. USA **80**:7650–7653.

Hanto, D. W., G. Frizzera, K. J. Gajl-Peczalska, and R. L. Simmons. 1985. EB virus, immunodeficiency and B cell lymphoproliferation. Transplantation **39**:461–472.

Hanto, D. W., K. Sakamoto, D. T. Purtilo, R. L. Simmons, and J. S. Najarian. 1981. The Epstein-Barr virus in the pathogenesis of posttransplant lymphoproliferative disorders: clinical, pathologic and virologic correlation. Surgery **90**:204–213.

Henle, G., and W. Henle. 1966. Studies on cell lines derived from Burkitt's lymphoma. Trans. N.Y. Acad. Sci. **29**:71–79.

Henle, W., and G. Henle. 1968. Effect of arginine-deficient media on the herpes-type virus associated with cultured Burkitt tumor cells. J. Virol. **2**:182–191.

Henle, G., W. Henle, P. Clifford, V. Diehl, G. W. Kafuko, B. Kirya, G. Klein, R. H. Morrow, G. M. R. Munube, M. C. Pike, P. M. Tukei, and J. L. Ziegler. 1969. Antibodies to EB virus in Burkitt's lymphoma and control groups. J. Natl. Cancer Inst. 43:1147–1157.

Henle, G., W. Henle, and V. Diehl. 1968. Relation of Burkitt tumor associated herpes-type virus to infectious mononucleosis. Proc. Natl. Acad. Sci. USA 59: 94–101.

Henle, G., E. T. Lennette, M. A. Alspaugh, and W. Henle. 1979. Rheumatoid factor as a cause of positive reactions in tests for Epstein-Barr virus specific IgM antibodies. Clin. Exp. Immunol. 36:415–422.

Henle, W., V. Diehl, G. Kohn, H. zur Hausen, and G. Henle. 1967. Herpes-type virus and chromosome marker in normal leukocytes after growth with irradiated Burkitt cells. Science 157:1064–1065.

Henle, W., and G. Henle. 1974. Epstein-Barr virus and human malignancies. Cancer 34:1368–1374.

Henle, W., and G. Henle. 1977. Antibodies to the R component of Epstein-Barr virus-induced early antigens in Burkitt's lymphoma exceeding in titer antibodies to Epstein-Barr viral capsid antigen. J. Natl. Cancer Inst. 58:785–786.

Henle, W., and G. Henle. 1980. Consequences of persistent Epstein-Barr virus infections, p. 3–9. In M. Essex, G. Todaro, and H. zur Hausen (ed). Viruses in naturally occurring cancers, Cold Spring Harbor Conferences on Cell Proliferation, Vol 7. Cold Spring Harbor Laboratory, Cold Spring Harbor, NY.

Henle, W., and G. Henle. 1981. Epstein-Barr virus-specific serology in immunologically compromised individuals. Cancer Res. 411:4222–4225.

Henle, W., G. Henle, and C. A. Horwitz. 1974. Epstein-Barr virus specific diagnostic tests in infectious mononucleosis. Human Pathol. 5:551–565.

Henle, W., G. Henle, B. A. Zajac, G. Pearson, R. Waubke, and M. Scriba. 1970. Differential reactivity of human serums with early antigens induced by Epstein-Barr virus. Science 169:188–190.

Hinuma, Y., M. Kohn, J. Yamaguchi, D. J. Wudarski, J. R. Blakesee, Jr., and J. T. Grace, Jr. 1967. Immunofluorescence and herpes-type virus particles in the P3HR-1 Burkitt lymphoma cell line. J. Virol. 1:1045–1051.

Ho, H. C. 1975. Epidemiology of nasopharyngeal carcinoma. J. R. Coll. Surg. Edinb. 20:223–235.

Hoagland, R. J. 1955. The transmission of infectious mononucleosis. Am. J. Med. Sci. 229:262–272.

Horwitz, C. A., W. Henle, G. Henle, H. Polesky, H. Wexler, and P. Ward. 1976. The specificity of heterophil antibodies in patients and healthy donors with no or minimal signs of infectious mononucleosis. Blood 47:91–98.

Horwitz, C. A., W. Henle, G. Henle, H. Rudnick, and E. Latts. 1985. Long term serological follow-up of patients for Epstein-Barr virus after recovery from infectious mononucleosis. J. Infect. Dis. 151:1150–1153.

Huang, D. P., J. H. C. Ho, W. Henle, and G. Henle. 1974. Demonstration of Epstein-Barr virus-associated nuclear antigen in nasopharyngeal carcinoma cells from fresh biopsies. Int. J. Cancer 14:580–588.

Hummeler, K., G. Henle, and W. Henle. 1966. Fine structure of a virus in cultured lymphoblasts from Burkitt lymphoma. J. Bacteriol. 91:1366–1368.

Ikawata, S., J. T. Grace, Jr. 1964. Cultivation in vitro of myeloblasts from human leukemia. N.Y. State J. Med. 64:2279–2282.

Johnson, R. B., Jr., and R. Libby. 1980. Separation of immunoglobulin M (IgM) essentially free of IgG from serum for use in systems requiring assay of IgM type antibodies with interference from rheumatoid factor. J. Clin. Microbiol. 12:451–454.

Jones, J. F., C. G. Ray, L. L. Minnich, M. J. Hicks, R. Kibler, and D. O. Lucas. 1985. Evidence of active Epstein-Barr virus infection in patients with persistent, unexplained illnesses: elevated anti-early antigen antibodies. Ann. Intern. Med. 102:1–7.

Judson, S. C., W. Henle, and G. Henle. 1977. A cluster of Epstein-Barr virus associated American Burkitt's lymphoma. N. Engl. J. Med. 297:464–468.

Kafuko, G. W., B. E. Henderson, B. G. Kirya, G. M. R. Munube, P. M. Tukei, N. E. Day, G. Henle, W. Henle, R. H. Morrow, M. C. Pike, P. G. Smith, and E. H. Williams. 1972. Epstein-Barr virus antibody levels in children from the West Nile district of Uganda. Report of a field study. Lancet 1:706–709.

Kano, K., and F. Milgrom. 1977. Heterophile antigens and antibodies in Medicine, p. 43–69. In W. Arber, W. Henle, P. H. Hofsneider, et al. (ed.), Current topics in microbiology and immunology. Springer-Verlag, New York.

Klein, G. 1975. The Epstein-Barr virus and neoplasia. N. Engl. J. Med. 293:1353–1356.

Klein, G., and L. Dombos. 1973. Relationship between the sensitivity of EBV-carrying lymphoblastoid line to superinfection and the inducibility of the resident viral genome. Int. J. Cancer 11:327–337.

Klein, G., B. Giovanella, A. Westman, J. Stehlin, and D. Mumford. 1975. An EBV-genome-negative cell line established from an American Burkitt lymphoma. Receptor characteristics, EBV infectability and permanent conversion into EBV-positive sublines by in vitro infection. Intervirology 5:319–334.

Klein, G., E. Yefenof, K. Falk, and A. Westman. 1978. Relationship between Epstein-Barr virus (EBV)-production and the loss of the EBV receptor/complement receptor complex in a series of sublines derived from the same original Burkitt's lymphoma. Int. J. Cancer 21:552–560.

Lange, B., A. Arbeter, J. Hewetson, and W. Henle. 1978. Longitudinal study of Epstein-Barr virus antibody titers and excretion in pediatric patients with Hodgkin's disease. Int. J. Cancer 22:521–527.

Lemon, S. M., L. M. Hutt, J. E. Shaw, J. L. Li, and J. S. Pagano. 1977. Replication of EBV in epithelial cells during infectious mononucleosis. Nature (London) 268:268–270.

Leyton, G. B. 1952. Ox cell hemolysins in human serum. J. Clin. Pathol. 5:324–328.

Luka, J., G. Kallin, and G. Klein. 1979. Induction of the Epstein-Barr virus (EBV) cycle in latently infected cells by N-butyrate. Virology 94:228–231.

Luka, J., W. Siegert, and G. Klein. 1977. Solubilization of the Epstein-Barr virus determined nuclear antigen and its characterization as a DNA-binding protein. J. Virol. 22:1–8.

Magrath, I., J. Erikson, J. Whang-Peng, H. Sieverts, G. Armstrong, D. Benjamin, T. Triche, O. Alabaster, and C. M. Croce. 1983. Synthesis of kappa light chains by cell lines containing an 8;22 chromosomal translocation derived from a male homosexual with Burkitt's lymphoma. Science 222:1094–1098.

Manalov, G., and Y. Manolova. 1972. Marker band in one chromosome 14 from Burkitt's lymphomas. Nature (London) 237:33–34.

Menezes, J., M. Jondal, W. Leibold, and G. Dorval. 1976. Epstein-Barr virus (EBV) interactions with human lymphocyte subpopulations, virus adsorptions, kinetics of expression of Epstein-Barr virus-associated nuclear antigen, and lymphocyte transformation. Infect. Immun. **13**:303–310.

Menezes, J., W. Leibold, and G. Klein. 1975a. Biological differences between Epstein-Barr virus (EBV) strains with regard to lymphocyte transforming ability, superinfection and antigen induction. Exp. Cell Res. **92**:478–484.

Menezes, J., W. Leibold, G. Klein, and G. Clements. 1975b. Establishment and characterization of an Epstein-Barr (EBV)-negative lymphoblastoid B cell lines (BJA-B) from an exceptional, EBV-genome-negative African Burkitt's lymphoma. Biomedicine **22**:276–284.

Miller, G., and M. Lipman. 1973. Release of infectious Epstein-Barr virus by transformed marmoset leukocyte. Proc. Natl. Acad. Sci. USA **70**:190–194.

Miller, G., J. C. Niederman, and L. Andrews. 1973. Prolonged oropharyngeal excretion of EB virus following infectious mononucleosis. N. Engl. J. Med. **137**:140–147.

Miller, G., J. Robinson, L. Heston, and M. Lipman. 1974. Differences between laboratory strains of Epstein-Barr virus based on immortalization, abortive infection and interference. Proc. Natl. Acad. Sci. USA **71**:4006–4010.

Minowada, J., and T. Ohnuma. 1972. Rosette-forming human lymphoid cell lines. I. Establishment and evidence for origin of thymus-derived lymphocytes. J. Natl. Cancer Inst. **49**:891–895.

Miyoshi, I., S. Hiraki, I. Kimura, K. Miyamoto, and J. Sato. 1979. 2/8 translocation in a Japanese Burkitt's lymphoma. Experientia **35**:742–743.

Moore, G., J. T. Grace, Jr., P. Citron, R. Gerner, and A. Barns. 1966. Leukocyte cultures of patients with leukemia and lymphomas. N.Y. State J. Med. **66**:2757–2764.

Moore, G. E., R. E. Gerner, and H. A. Franklin. 1967. Culture of normal human leukocytes. J. Am. Med. Assoc. **199**:519–524.

Morgan, D. G., J. C. Niederman, G. Miller, H. W. Smith, J. M. Dowaliby. 1979. Site of Epstein-Barr virus replication in the oropharynx. Lancet **2**:1154–1157.

Nemerow, G. R., D. Mold, V. Keivens-Schwend, and N. R. Cooper. 1986. The Epstein-Barr (EBV) gp 350/300 envelope protein mediates EBV attachment to the EBV/C3d receptor of B cells. Abstract, 11th Int Herpesvirus Workshop, Leeds, England, p. 281.

Niederman, J. C., A. S. Evans, M. S. Subramanyan, and R. W. McCollum. 1970. Prevalence, incidence and persistence of EB virus antibody in young adults. N. Engl. J. Med. **282**:361–365.

Nilsson, K., G. Klein, W. Henle, and G. Henle. 1971. The establishment of lymphoblastoid lines from adult and fetal lymphoid tissue and its dependence on EBV. Int. J. Cancer **8**:443–450.

Nonoyama, M., and J. S. Pagano. 1973. Homology between Epstein-Barr virus DNA and viral DNA from Burkitt's lymphoma and nasopharyngeal carcinoma determined by DNA–DNA reassociation kinetics. Nature (London) **242**:44–47.

Paul, J. R., and W. W. Bunnell. 1932. The presence of heterophile antibodies in infectious mononucleosis. Am. J. Med. Sci. **183**:80–104.

Pearson, G., F. Dewey, G. Klein, G. Henle, and W. Henle. 1970. Relationship between neutralization of Epstein-Barr virus and antibodies to cell membrane antigens induced by the virus. J. Natl. Cancer Inst. **45**:989–997.

Pejme, J. 1964. Infectious mononucleosis. A clinical and haematological study of patients and contacts and a comparison with healthy subjects. Acta Med. Scand. (Suppl) **413**:34.

Pritchett, R. F., S. D. Hayward, and E. D. Kieff. 1975. DNA of Epstein-Barr virus. I. Comparative studies of the DNA of EBV from HR-1 and B95-8 cells. Size, structure and relatedness. J. Virol. **15**:556–584.

Pritchett, R., M. Pedersen, E. Kieff. 1976. Complexity of EBV homologous DNA in continuous lymphoblastoid cell lines. Virology **74**:227–231.

Pulvertaft, R. J. V. 1964. Cytology of Burkitt's tumour (African lymphoma). Lancet **1**:238–240.

Purtilo, D. T., K. Sakamoto, V. Barnabei, J. Seeley, T. Bechtold, G. Rogers, J. Yetz, and S. Harada. 1982. Epstein-Barr virus induced diseases in males with the X-linked lymphoproliferative syndrome (XLP). Am. J. Med. **73**:49.

Purtilo, D. T., I. Szymankski, J. Bhawan, J. P. S. Yang, L. M. Hutt, W. Boto, L. DeNicola, R. Maier, and D. Thorley-Lawson. 1978. Epstein-Barr virus infections in the X-linked recessive lymphoproliferative syndrome. Lancet **1**:798–801.

Reedman, B. M., and G. Klein. 1973. Cellular localization of an Epstein-Barr virus (EBV)-associated complement-fixing antigen in producer and non-producer lymphoblastoid cell lines. Int. J. Cancer **11**:499–520.

Reedman, R. M., G. Klein, J. H. Pope, M. K. Walters, J. Hilgers, and B. Johansson. 1974. Epstein-Barr virus-associated complement-fixing and nuclear antigens in Burkitt's lymphoma biopsies. Int. J. Cancer **13**:755–763.

Robinson, J. E., W. A. Andiman, E. Henderson, and G. Miller. 1977. Host determined differences in expression of surface marker characteristics on human and simian lymphoblastoid cell lines transformed by Epstein-Barr virus. Proc. Natl. Acad. Sci. USA **74**:749–753.

Rocchi, G., A. deFelic, G. Ragona, and A. Heiviz. 1977. Quantitative evaluation of Epstein-Barr virus infected mononuclear peripheral blood leukocytes in infectious mononucleosis. N. Engl. J. Med. **296**:132–134.

Schooley, R. T., R. W. Carey, G. Miller, W. Henle, R. Eastman, E. J. Mark, K. Kenyon, E. O. Wheeler, and R. Rubin. 1986. Chronic Epstein-Barr virus infection associated with fever and interstitial pneumonitis. Clinical and serologic features and response to antiviral chemotherapy. Ann. Intern. Med. **104**:636–643.

Schulte-Holthausen, H., and H. zur Hausen. 1970. Partial purification of the Epstein-Barr virus and some properties of its DNA. Virology **40**:776–779.

Sculley, T. B., P. J. Walker, D. J. Moss, and J. H. Pope. 1984. Identification of multiple Epstein-Barr virus-induced nuclear antigens with sera from patients with rheumatoid arthritis. J. Virol. **52**:88–93.

Shirodaria, P. D., K. B. Fraser, and F. Stanford. 1973. Secondary fluorescent staining of viral antigens by rheumatoid factor and fluorescein-conjugated anti-IgM. Ann. Rheum. Dis. **32**:53–57.

Silvestre, D., I. Ernberg, C. Neauport-Sautes, F. M. Kourilsky, and G. Klein. 1974. Differentiation between early and late membrane antigen on human lymphoblastoid cell lines infected with Epstein-Barr virus. II. Immunoelectron microscopy. J. Natl. Cancer Inst. **53**:67–74.

Silvestre, D., F. M. Kourilsky, G. Klein, Y. Yata, C. Neauport-Sautes, and J. P. Levy. 1971. Relationship between the EBV-associated membrane antigen on Burkitt lymphoma cells and the viral envelope demonstrated by immunoferritin labelling. Int. J. Cancer **8**:222–233.

Sixbey, J. W., E. H. Vesterinen, J. G. Nedrud, N. Raab-Traub, L. A. Walton, and J. S. Pagano. 1983. Replication of Epstein-Barr virus in human epithelial cells infected in vitro. Nature (London) **306**:480–483.

Skare, J., J. Farley, J. L. Strominger, K. O. Fresen, M. S. Cho, and H. zur Hausen. 1985. Transformation by Epstein-Barr virus requires DNA sequences in the region of BAM HI fragments Y and H. J. Virol. **55**:286–297.

Strauch, B., L. Andrew, N. Siegel, and G. Miller. 1974. Oropharyngeal excretion of Epstein-Barr virus by renal transplant recipients and other patients treated with immunosuppressive drugs. Lancet **1**:234–237.

Straus, S. E., G. Tosato, G. Armstrong, T. Lawley, O. T. Preble, W. Henle, R. Davey, G. Pearson, J. Epstein, I. Brus, and R. M. Blaese. 1985. Persisting illness and fatigue in adults with evidence of Epstein-Barr virus infection. Ann. Intern. Med. **102**:7–16.

Strnad, B. C., T. C. Schuster, R. F. Hopkins, R. H. Neubaeur, and H. Rabin. 1981. Identification of an Epstein-Barr virus nuclear antigen by fluoro-immunoelectrophoresis and radio-immunoelectrophoresis. J. Virol. **38**:996–1004.

Sugawara, K., F. Mizuno, and T. Osato. 1973. Induction of Epstein-Barr virus-related membrane antigens by 5-iododeoxyuridine in non-producer human lymphoblastoid cells. Nature (London) **246**:70–72.

Svedmyr, E., and M. Jondal. 1975. Cytotoxic cell specific for B cell lines transformed by Epstein-Barr virus are present in patients with infectious mononucleosis. Proc. Natl. Acad. Sci. USA **72**:1622–1626.

Thorley-Lawson, D., and J. L. Strominger. 1976. Transformation of human lymphocytes by Epstein-Barr virus is inhibited by phosphonoacetic acid. Nature (London) **263**:332–334.

Tosato, G., I. Magrath, I. Koski, N. Dooley, and M.

Blaese. 1979. Activation of suppressor T cells during Epstein-Barr virus-induced infectious mononucleosis. N. Engl. J. Med. **301**:1133–1137.

Wilson, E. R., A. Malluh, S. Stagno, and W. M. Crist. 1981. Fatal Epstein-Barr virus-associated hemophagocytic syndrome. J. Pediatr. **98**:260–262.

Yao, Q. Y., A. B. Rickinson, and M. A. Epstein. 1985. A re-examination of the Epstein-Barr virus carrier state in healthy seropositive individuals. Int. J. Cancer **35**:35–42.

Young, L. S., J. W. Sixbey, D. Clark, and A. B. Rickinson. 1986. Epstein-Barr virus receptors on human pharyngeal epithelia. Lancet **1**:240–242.

Zech, L., U. Haglund, K. Nilsson, and G. Klein. 1976. Characteristic chromosomal abnormalities in biopsies and lymphoid cell lines from patients with Burkitt and non-Burkitt lymphomas. Int. J. Cancer **17**:47–56.

Ziegler, J. L., W. L. Drew, R. C. Miner, L. Mintz, E. Rosenbaum, J. Gershow, E. T. Lennette, J. Greenspan, E. Shillitoe, J. Beckstead, C. Casavant, and K. Yamamoto. 1982. Outbreak of Burkitt's like lymphoma in homosexual men. Lancet **2**:631–633.

zur Hausen, H., F. J. O'Neill, and U. K. Freese. 1978. Persisting oncogenic herpesvirus induced by the tumour promoter TPA. Nature (London) **272**:373–375.

zur Hausen, H., H. Schulte-Holthausen, G. Klein, W. Henle, G. Henle, P. Clifford, and L. Sanntesson. 1970. EBV DNA in biopsies in Burkitt's tumours and anaplastic carcinomas of the nasopharynx. Nature (London) **228**:1056–1958.

CHAPTER 13

Herpesviridae: Cytomegalovirus

W. LAWRENCE DREW

Disease: Cytomegalic inclusion disease, congenital; cytomegalovirus mononucleosis; cytomegalovirus infection in the immunosuppressed.

Etiologic Agent: Human cytomegalovirus (HCMV), a member of the herpesvirus family.

Source: Human, by 1) direct contact, including sexual and parturition; 2) vertical; 3) breast milk; 4) blood or organ transplantation.

Clinical Manifestations: Clinical manifestations depend on the syndrome. Congenital: hepatosplenomegaly, petechial rash, microcephaly, intracerebral calcifications, hearing loss. Mononucleosis: fever, malaise. Reactivation or primary infection in immunosuppressed: pneumonia, hepatitis (chemical), retinitis, encephalitis, colitis, esophagitis, adrenalitis.

Pathology: Cytomegalic cells containing a dense central intranuclear inclusion with or without associated mononuclear cell inflammatory response.

Laboratory Diagnosis: Viral isolation from urine, blood, lung, semen, or tissue. Seroconversion is diagnostic, but increase in immunoglobulin G antibody may not be significant. Immunoglobulin M antibody indicates active infection. Detection of antigen or genome may provide rapid diagnosis of infection.

Epidemiology: Worldwide, especially underdeveloped countries where infection is universal by childhood. In developed areas, approximately one-half of adults are infected. Rates of antibody prevalence increase with age and with lowering of socioeconomic status. Spread by close personal contact, including sexual. Also spread vertically, by natal exposure to infected cervix, by breast milk, and by blood transfusion or organ transplantation. Virus remains latent in host and can be reactivated, especially by immunosuppression.

Treatment: DHPG [9-(1,3-dihydroxy-2-propoxymethy)guanine] is an effective antiviral agent that suppresses CMV during treatment, but does not eradicate it.

Prevention and Control: Infection can be prevented by avoiding close personal contact (including sexual) with an actively infected individual. Transmission by blood transfusion or organ transplantation can be avoided by using blood or organs from seronegative donors.

History

Characteristic enlarged "cytomegalic" cells with intranuclear inclusions were detected in tissues as early as 1881. Disseminated disease with cytomegalic inclusions was first described in 1921 by Goodpasture and Talbot, who reported a fatal case of this entity in a 6-week-old infant with lung, kidney, and liver involvement. Some investigators thought that these cells were protozoa, but Goodpasture and Talbot (1921) suspected that they were viral, as did Lipschuetz (1921). Diagnosis of cytomegalic inclusion disease (CID) by a cytologic examination of urine was first described in 1952 by Fetterman. Finally, in 1956 and 1957, three different laboratories reported the isolation of human cytomegalovirus (CMV) from urine and tissues (Rowe et al., 1956; Smith, 1956; Weller et al., 1957). Ironically, one of these isolates resulted from efforts to grow *Toxoplasma* in culture and another was found during efforts to grow adenovirus from adenoids.

Description of Agent

Cytomegalovirus is a member of the Herpesviridae. The virion contains a core of double-strained linear DNA with a molecular weight of 100 to 150×10^6. The surrounding capsid is an icosahedron of 162 capsomeres with 5 capsomeres along each triangular edge of the icosahedron. Cytomegalovirus is quite labile and is destroyed rapidly by heat (56°C for 30 min), low pH, ether, and cycles of freezing and then thawing. If freezing a specimen is necessary, 25 to 35% sorbitol provides stabilization (Weller and Hanshaw, 1962). Strain differences are identified by restriction endonuclease techniques which indicate that most human strains differ from each other (Huang et al., 1980).

Pathology

The histologic hallmark of CMV infection in vivo is the cytomegalic cell which is enlarged and contains a dense central intranuclear inclusion. These cells may be found in any area of the body and are thought to be epithelial in origin. Cytomegalic cells are not merely a host cell reaction to virus challenge; they actually contain virion and/or nucleoside and virus-specific antigen. In addition to the large type A intranuclear inclusion, there may also be granular inclusions in the cytoplasm. There may or may not be an inflammatory response in the vicinity of cytomegalic cells—if present, it is usually mononuclear cell in character.

In any one organ, cytomegalic cells are rarely numerous, raising a question about the contribution of CMV to the overall pathology. This dilemma most frequently arises in the lung, where lung biopsy may show only rare cytomegalic cells, yet there is considerable lung pathology. Moreover, other pathogens such as *Pneumocystis carinii* may be detected, adding further doubt as to the pathogenic role of CMV. Indeed, these two microorganisms are found together so frequently that a synergistic relationship may exist between them (Follansbee et al., 1982). Studies of tissue with monoclonal antibodies and in situ hybridization techniques suggest that significant CMV infection can exist in cells and organs in the absence of these cytomegalic cells (Goldstein et al., 1982; Myerson et al., 1984). It thus appears that typical inclusion-bearing cytomegalic cells are a crude and insensitive measure of CMV pathology and that many cells may be infected but not demonstrate these histologic characteristics. In the absence of another explanation, the finding of even a few such cells may implicate CMV as the cause of the entire pathology present in the tissue specimen.

Pathogenesis

In general, the pathogenesis of CMV is similar to that of other herpesviruses. It shares with them the capacity for 1) cell-to-cell spread even in the presence of circulating antibody, 2) establishing a latent state in the host, 3) reactivation under conditions of immunosuppression, and 4) inducing at least transient immunosuppression in the recipient.

Cell-to-Cell Spread in the Presence of Antibody

This property is especially notable during reactivation of CMV infection and in perinatal infection. In the former, latent virus is able to reactivate from the cell of latency (see below) and spread widely within tissues despite the presence of circulating neutralizing antibody in patient serum. In the neonate, infection is acquired during birth through exposure to the virus in the cervix, vagina, or both. It may also be transmitted via breast milk. In either case these infections come only from CMV seropositive mothers who have passively transferred antibody to the fetus, yet exogenous virus is able to infect. Presumably these infections can spread because the virus is highly cell associated and spreads via coalescing cells.

Latency

Following primary CMV infection, the virus becomes latent in the host. The exact site(s) of latency and the mechanisms of persistence are not completely understood. Leukocytes, presumably mononuclear, appear to be infected latently and account for transmission of the virus via blood and leukocyte transfusions. Also, organs such as kidneys and heart harbor the virus, but the exact cell is not known (Bowden et al., 1986; Peterson et al., 1983). The mechanism of latency may be partially explained by the finding that lymphocytes can be abortively infected by the virus, with the expression of immediate early but not late gene products and no progeny virus (Schrier et al., 1985).

Reactivation

Latent CMV infection appears to be reactivated by immunosuppression (e.g., corticosteroids, human immunodeficiency viral infection) and possibly by allogeneic stimulation (i.e., the host response to transfused or transplanted cells).

Induction of Immunosuppression

Primary CMV infection, even when asymptomatic, induces a marked reversal of the ratio of helper to suppressor T-lymphocyte subsets (Carney et al., 1981; Drew et al., 1985). This reversal is primarily the result of an increase in the suppressor cell population, but there is also a reduction in the number of helper T lymphocytes. Over a period of months these ratios return to or near preinfection levels. Lymphocyte function is also diminished during the acute infection (e.g., proliferative responses to CMV antigens and mitogens), but returns to normal during convalescence (Rinaldo et al., 1980). Given the propensity of CMV to induce immunologic abnormality and the high prevalence of CMV infection in homosexual men (Drew et al., 1981), there has been speculation that CMV might be a cofactor in the acquired immune deficiency syndrome (AIDS).

Epidemiology, Natural History, and Clinical Manifestations

Congenital Infection

Cytomegalovirus is a much more common cause of congenital viral infection than is rubella. Nearly 1% of all infants born in the United States are infected with this virus at birth, although only 1 in 10 shows any clinically evident stigmata such as microcephaly, intracerebral calcification, hepatosplenomegaly, and rash (Ho, 1982). Altogether 6,000 to 7,000 newborns per year may suffer unilateral or bilateral hearing loss and mental retardation as a result of congenital CMV infection.

It may be difficult to clinically distinguish an infant symptomatic at birth from one infected by other members of the TORCH complex (toxoplasmosis, rubella, and herpes simplex virus), and laboratory studies play a definitive role in proving the exact etiology. Congenital CMV infection is best documented by isolating the virus from the infant's urine during the first week of life.

Fetuses are infected by virus ascending from the cervix or by virus in the maternal blood. The cervix is probably the route in immune mothers, the bloodstream in mothers with primary infection. Primary infection in the mother causes more severe disease in the infant. In populations with a high prevalence of CMV antibody, few mothers have primary infection; thus, although there are many congenitally infected babies most, if not all, appear clinically normal. High antibody prevalence (e.g., 90%) is found in underdeveloped countries and in lower socioeconomic groups in the United States. In other populations in the United States, however, only about 50% of mothers are seronegative and are therefore susceptible to primary CMV infection.

Infection During Birth

In the United States, up to 20% of pregnant women at term harbor CMV in the cervix (Stagno et al., 1975). Approximately one-half of the neonates born through an infected cervix acquire CMV infection and become excretors of the virus at 3 to 4 weeks of age. Neonates may also acquire CMV infection from maternal milk or colostrum, because 14% of seroimmune mothers shed CMV in these secretions (Stagno et al., 1980).

Another means of CMV acquisition by neonates is through blood transfusions. Yeager et al. (1981) have shown that if babies born of seronegative mothers are given only blood from seronegative donors, they do not acquire CMV infection. However, if seronegative babies are exposed to blood from seropositive donors, 13.5% acquire CMV infection in the immediate postneonatal period.

Full-term and otherwise healthy babies who acquire CMV at birth usually show no ill effects, although the eventual impact on intellectual and neurologic function is not known. For instants requiring hospitalization for premature birth or other conditions, the impact is more immediate and certain. After three or more weeks of hospitalization, up to 37% may become culture positive (Ballard et al., 1979). Of 16 infected infants in our nursery, 14 demonstrated a recognizable complex of clinical findings. The most frequent sign was hepatosplenomegaly (13 [93%]), then sepsis (12 [86%]), respiratory deterioration (10 [71%]), and fever above 38°C (5 [36%]). Ten (71%) had a peculiar gray pallor when they began excreting virus. Three infants died.

Laboratory findings that helped identify the syndrome included atypical lymphocytosis of greater than 8% (93% of infants). Twenty-nine percent also had platelet counts of $75,000/mm^3$ or less.

Cytomegalovirus in Adulthood

In low socioeconomic populations and in underdeveloped countries, postneonatal CMV infection is common, apparently as a result of crowded living conditions. If such conditions are not present, only 10 to 15% of adolescents are seroimmune. During young adulthood, however, the rate of seropositivity rapidly increases so that by age 35 approximately 50% show serologic evidence of past infection. Although most CMV infections acquired in young adulthood are asymptomatic, patients may develop clinical illness resembling infectious mononucleosis with atypical lymphocytosis and a negative heterophile antibody test but with less pharyngitis and lymphadenopathy. Indeed, when this syndrome is encountered in patients over age 25, it is more apt to be due to CMV than to Epstein–Barr virus (EBV). There is increasing evidence that many of these infections are sexually transmitted or at least require very close personal contact. The evidence for sexual transmission of CMV infection is as follows: 1) The virus does not spread readily among adults by ordinary personal contact, even during prolonged exposure to people who are excreting the virus (Wenzel et al., 1973). 2) Cytomegalovirus has been isolated from the cervix of 13 to 23% of women at venereal disease clinics (Jordan et al., 1973). 3) The virus has also been isolated from semen and is present there in the highest titer of any bodily secretion. 4) Chretien and co-workers (1977) reported CMV mononucleosis in two men after sexual contact with a woman whose cervix and urine were CMV positive. Evidence of recent infection was also found in another sexual contact of one of the men. 5) Antibody prevalence is high among homosexual men. Ninety-three percent of homosexual men attending a venereal disease clinic had serologic evidence of past infection, whereas male blood donors and heterosexual men in the same age group had antibody prevalence rates of 36% and 47%, respectively (Drew et al., 1981). Also, urinary excretion of CMV was found in 14 of 90 homosexual men (15.5%), but in none of 101 heterosexual men ($P < 0.005$).

Post-Transfusion Infection

Post-transfusion mononucleosis infection (PTM) was described frequently in the past as a complication of surgery requiring cardiopulmonary bypass, but it may follow transfusions in neonates and in patients undergoing other types of surgery, especially splenectomy (Drew and Miner, 1982). Transmission of CMV by blood is most often asymptomatic; if symptoms are present, they typically resemble infectious mononucleosis. Fever, splenomegaly, and atypical lymphocytosis usually begin 3 to 6 weeks postoperatively, but the heterophile antibody test is negative. In recent years PTM has decreased in frequency for at least two reasons: 1) Fewer units of blood are used in cardiac surgery and 2) many nurseries use only CMV-seronegative blood, especially if the infant is seronegative.

Post-transfusion infection can be prevented by giving only CMV-antibody-negative blood to seronegative patients. Some evidence suggests that the use of frozen, deglycerolized blood (which eliminates leukocytes) can prevent transmission. Conversely, the use of leukocyte transfusions may increase the risk of CMV infection. As mentioned previously, Yeager (1975) has shown that if infants born of seronegative mothers are given blood only from seronegative donors, they do not acquire CMV infection. However, if seronegative babies are given blood from seropositive donors, 13.5% immediately acquire CMV infection.

Infection Following Transplantation

Cytomegalovirus may also be transmitted by kidneys or bone marrow used for transplantation (Bowden et al., 1986; Peterson et al., 1983). This primary type of infection can apparently be prevented by giving only organs from seronegative donors to recipients who are seronegative (Bowden et al., 1986; Peterson et al., 1983). On the other hand, transplant recipients who are seropositive may reactivate CMV during periods of intense immunosuppression, but these episodes are less apt to be associated with severe clinical disease (see below).

Infection in the Immunocompromised Host

If a patient becomes immunocompromised by either a disease or its therapy, the latent CMV may reactivate and cause clinical illness. Primary infection with CMV may also occur in these patients, and this form is more likely to give symptomatic illness. The most common manifestations of infection are fever and pneumonia, but esophagitis, colitis, mild hepatitis, chorioretinitis, and even encephalitis may occur. In some immunocompromised patients, CMV is a primary etiologic agent; in others, it coexists with other organisms such as *Pneumocystis carinii*.

Chorioretinitis

Ocular disease due to CMV occurs only in patients with severe immunodeficiency and is common in patients with AIDS. Blurring of vision or visual loss are common complaints and ophthalmologic examination reveals large yellowish-white granular areas with perivascular exudates and flame-shaped hemorrhages ("cottage cheese and catsup"). Histologically, coagulation necrosis and microvascular abnormalities are present.

Pneumonia

Some patients with CMV infection and pulmonary disease have no other pathogens present on diagnostic bronchoscopy, and CMV is the presumptive cause of their pneumonia. In others, organisms such as *P. carinii* are seen and the presence of CMV may be only incidental. Diagnosis of CMV pneumonia is most certain when there is a combination of factors, including positive CMV culture from lung tissue or pulmonary secretions, the presence of pathognomic cells with intranuclear or intracytoplasmic inclusion bodies, and the absence of other pathogenic organisms.

When CMV causes pulmonary disease in immunocompromised patients, the presenting syndrome is that of an interstitial pneumonia. Patients complain of gradually worsening shortness of breath, dyspnea on exertion, and a dry, nonproductive cough. The heart and respiratory rate are elevated, and auscultation of the lungs often reveals minimal findings with no evidence of consolidation. Chest x-ray reveals diffuse interstitial infiltrates similar to those present in *P. carinii* pneumonia (PCP). Hypoxemia is invariably present.

Central Nervous System Infection

Although it is not considered a neurotropic virus, cytomegalovirus can, on occasion, produce central nervous system infection. Personality changes, difficulty in concentrating, headaches, and sommolence are frequent findings. Diagnosis can be confirmed only by brain biopsy, with evidence of perventricular necrosis, giant cells, intranuclear and intracytoplamic inclusions, and isolation or identification of the virus.

Gastrointestinal Infection

Colonic ulceration, lower gastrointestinal bleeding, and mucosal vasculitis have been attributed to CMV in the presence of positive CMV cultures and histopathology suggestive of CMV enterocolitis.

Patients with CMV colitis present with diarrhea, weight loss, anorexia, and fever. Sigmoidoscopy reveals diffuse submucosal hemorrhages and diffuse mucosal ulcerations. Biopsy specimens may reveal vasculitis with neutrophilic infiltration, CMV-infected endothelial cells, and nonspecific inflammation. Cytomegalovirus esophagitis may also occur and may mimic candidal esophagitis. Cytomegalovirus may be found in any gastrointestinal ulceration, but its contribution to the pathogenesis of these ulcerations is uncertain.

Specimens for Culture

URINE

Clean voided first morning urine is preferred. During holding and transportation specimens should be kept at 4°C; there is no appreciable drop in virus titer at 4°C for up to 7 days (Feldman, 1962). Alternatively, viral titer in specimens may be stabilized by freezing at −70 to −80°C in suspensions of 30% sorbitol (Weller and Hanshaw, 1962).

The recovery of CMV from urine is increased two to threefold by processing several specimens. Urine can be directly inoculated into tissue culture in 0.2-ml aliquots and then centrifuged and the sediment resuspended in 0.5 to 1.0 ml of supernatant urine also to be inoculated in 0.2-ml aliquots. Urine may be treated with antibiotics before inoculation, but if the specimen is fresh this procedure is usually unnecessary.

BLOOD

Viremia with CMV has been associated with the acute phase of infection in nonimmunocompromised patients and has been correlated with lethal pneumonia in bone marrow transplant recipients.

To culture blood, 5 ml of heparinized blood is used. Cytomegalovirus is associated with the leukocyte fraction or "buffy coat" that can be obtained by simple sedimentation or dextran sedimentation with Ficoll–Hypaque separation, or more effectively by using a Ficoll–Paque/Macrodex density-gradient system (Howell et al., 1979). In this two-step technique, lymphocytes are collected in a layer on top of the Ficoll–Paque. Neutrophils, which sediment below the Ficoll–Paque, can be separated from erythrocytes by the use of Macrodex, a dextran polymer. The neutrophil and lymphocyte fractions are then combined for inoculation into human diploid cell cultures. Up to 28 days are usually required before the cytopathic effects of CMV are detected and a positive report can be made. Multiple blood cultures, such as those recommended for bacterial cultures, are not required for CMV cultures.

TISSUE

Tissue explants and cells grown in cell cultures after dispersal of the cells from tissue fragments have provided higher rates of viral isolation than homogenized specimens. Presumably, viral inhibitors may be released into the homogenate, resulting in lower recovery rates in the disrupted cells. Lung, liver, and brain are the tissues that most frequently yield CMV. As with urine sediment, the cytologic detection of CMV inclusions is at least three to six times less sensitive than viral isolation from the tissue. If homogenates are used, they should be centrifuged and the supernatant fluid used as the inoculum.

Culture Procedure

Cytomegalovirus grows only in diploid fibroblast cell cultures, so specimens for the isolation of this virus must be inoculated into cultures such as human embryonic lung or foreskin fibroblasts or cell lines such as WI-38, MA-184, or Flow 2000. Cultures must be maintained for at least 4 to 6 weeks because the characteristic cytopathogenic effect (CPE) (Fig. 1) develops very slowly in specimens with very low titers of the virus. In infants, however, the urine culture requires only an average of 6 days for detection of CMV.

To maintain viability of cell cultures for this period of time, it is necessary to change the maintenance medium weekly.

Identification of Isolate

A urine culture with characteristic inclusions appearing only in diploid fibroblast cells and only after days to weeks is highly suggestive of CMV. However, certain adenoviruses that have been encountered in urine, especially from immunocompromised patients, may slowly develop a CPE indistinguishable from that of CMV. These adenoviruses are apt to grow in human embryonic kidney cultures, thereby providing a clue to their real identity. Because of these adenoviruses, it may be necessary to use immunologic procedures to confirm a urine isolate as CMV. When specimens other than urine are cultured, viruses such as varicella-zoster and even herpes simplex may produce a CMV-like CPE. Therefore, for complete accuracy, isolates from these sites

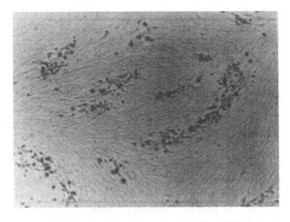

FIG. 1. Cytopathogenic effect due to cytomegalovirus in WI-38 cells. The color illustration for this figure can be found on Color Plate 2 at the beginning of this volume.

suspected to be CMV should be routinely confirmed. Indirect fluorescent-antibody (IFA) testing is the most commonly used confirmation procedure and suitable antibodies, both monoclonal and polyclonal, are becoming commercially available for this purpose.

If passage is required, it is necessary to pass cells rather than supernatant fluid because CMV is a very cell-associated virus.

Rapid culture results may be achieved by "culture amplification" of a specimen. In this procedure, specimens are inoculated by centrifuging them in a shell vial seeded with MRC-5 cells. After overnight incubation, the coverslips are stained for antigen by indirect immunofluorescence using commercially available monoclonal antibody. Of 770 specimens processed in this manner by Gleaves et al. (1984), 124 (16%) were positive for CMV versus 88 (11%) by standard viral culture. Moreover, the 88 specimens positive by standard detection of cytopathogenic effect required an average of 8 days in culture. Further evaluations are necessary to determine if this procedure is equally effective for cultures from other sites.

Interpretation of Cultures for CMV

The diagnosis of CMV *infection* is substantiated by a positive culture from any site or by seroconversion. The diagnosis of CMV *disease* is much more difficult. Patients may excrete virus in urine, in semen, or from the cervix for years following acquisition of the virus. Thus, the recovery of CMV in the urine does not *prove* that CMV is causing a patient's present illness. Recovery of the virus from blood is suggestive of active disease due to CMV, but patients may be asymptomatic even when viremic. Recovery of the virus from a lung culture supports a diagnosis of CMV pneumonitis, especially if characteristic viral inclusions are seen in the tissue. In the absence of inclusions, a positive lung culture may only be the equivalent of a positive blood culture and not explain a pneumonia. On the other hand, if CMV is recovered from the lung, and no other pathogen or pathologic process is identified, CMV may be the cause of the pneumonia even when inclusions are absent. Sorting out a contributing role for CMV in pneumonia is frequently complicated by the identification of coexisting pathogens such as *P. carinii*. Indeed, these two particular agents are found together so frequently that a synergistic relationship between them is suspected.

Similar interpretive problems pertain to recovery of CMV from intestinal and esophageal biopsy specimens. Again, if no other pathogen is identified, CMV may explain a colitis/esophagitis, especially when viral inclusions are present. When CMV is recovered from other organs such as brain, liver, and adrenal gland, it is likely that the virus is responsible for disease in these sites, especially when inclusions are also seen. Viral inclusions are not a sine qua non for proving disease because they can be quite rare and can be missed due to sampling error.

CMV Antigen Detection

Antibodies, especially monoclonal, have been used to attempt direct detection of CMV antigens in tissue by immunofluorescence (IF) (Fig. 2). Goldstein et al. (1982) compared IF with culture on lung samples obtained by open lung biopsy or autopsy. The IF was performed on frozen sections of lung using a single monoclonal antibody. Of 15 with positive lung culture for CMV, 12 (80%) were also positive by IF. In another report using a mixture of 9 monoclonal antibodies that recognize at least 16 CMV antigens, Volpi et al. (1983) detected CMV antigen in 7 of 7 culture-positive biopsy specimens.

Most recently, similar attempts have been made with bronchoalveolar lavage specimens. The sensitivity of this procedure compared with culture has been as low as 34% (Martin and Smith, 1986) and as high as 100% in identifying patients who had CMV pneumonia by histologic and radiologic criteria (Emmanuel et al., 1986).

Several attempts have been made to develop antigen detection systems for use in identifying CMV in urine (Yolken, 1982). These systems have generally utilized enzyme-linked immunosorbent assay (ELISA) techniques, but the results have been poor due to inadequate sensitivity as well as specificity.

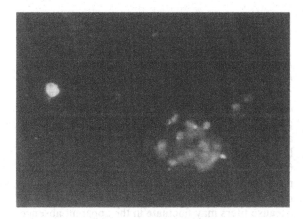

FIG. 2. Cytomegalovirus-infected cells positive by immunofluorescence. The color illustration for this figure can be found on Color Plate 2 at the beginning of this volume.

Detection of CMV by DNA Hybridization

Cytomegalovirus has been successfully identified in urine by dot blot hybridization, but the method is less sensitive than standard tissue culture (Chou and Merigan, 1983). No specimens with less than 500 PFU/ml were positive in the dot hybridization assay (a false-negative rate of 30%). Spector et al. (1984) reported a false-negative rate of 8% on urine and blood; in contrast to results with urine, however, dot blot hybridization on blood samples was positive in 40% of samples negative by culture. The explanation for this major discrepancy is not apparent.

Recent efforts have been directed toward detecting CMV by in situ hybridization. In this procedure formalin-fixed, paraffin-embedded tissue sections were reacted with a mixture of cloned CMV probes that were partially substituted with biotinylated deoxyuridine. Of 14 CMV culture-positive lung specimens, in situ hybridization was positive in 13 (93%) whereas histologic examination was positive in only 9 (64%) (Myerson et al., 1984). Recent modifications of this procedure enable laboratories to perform it within 8 h of cutting the tissue section (Unger et al., 1986).

Until the sensitivity is improved and the significance of a positive assay/negative culture is clarified, these procedures will remain investigative rather than "state of the art" for clinical laboratories.

Cytology and Histology

Cytologic examination of urine has been utilized to diagnose congenital CMV infection, but is negative in up to one-third of symptomatic congenitally infected viruric infants (Weller and Hanshaw, 1962).

Cytologic examination may be applied to other specimens, such as lung lavage, cervical secretions, or tracheal secretions. In each instance the cells are stained with hematoxylin and eosin or the Papanicolaou technique. Characteristic enlarged cells (cytomegaly) with dense central inclusions (owl-eye cells) are seen. Similar but smaller cells may be seen in other herpesvirus infections or as a result of adenovirus infection. Bronchoalveolar lavage specimens contain inclusion-bearing cells in only one-half of the specimens from which CMV is recovered in culture.

Histologic examination has been used to speed the diagnosis of CMV pneumonia. Hackman et al. (1985) reported that 21 of 27 marrow transplant recipients whose lung biopsy was CMV positive by either culture or immunofluorescence had cytologic or histologic findings characteristic of CMV.

Serologic Tests

Complement Fixation Test

The complement fixation (CF) test has been widely used as a method for measuring antibody to CMV. The AD 169 laboratory adapted strain is most frequently used to produce antigen, and the antibody titer of a serum sample is recorded as the highest serum dilution that gives 30% or higher fixation. The CF test measures total antibody and does not discriminate between immunoglobulin (Ig)G and IgM. As is true with several herpesviruses, the CF method for measuring antibody is less sensitive than procedures such as indirect fluorescent antibody (IFA), anticomplement immunofluroescence (ACIF), ELISA, and the like. Mintz et al. (1980) found that the CF test was 90% sensitive in detecting seropositive individuals compared to 97.5% by an ACIF procedure. This lack of CF antibody sensitivity was true despite using glycine-extracted antigen, which is more sensitive than freeze-thawed antigen (Kettering et al., 1977). In another comparative evaluation, Phipps et al. (1983) reported that the CF test was only 75% sensitive in comparison to several other methods (indirect hemagglutination, enzyme immunoassay, IFA). Moreover, the CF test was considered to make the highest technical demands and have the slowest turnaround time (Phipps et al., 1983). Complement fixation is suitable for detecting seroconversion, as patients usually develop antibody of sufficient titer to overcome the insensitivity. The CF antibody is specific and does not cross-react with herpes simplex or varicella-zoster antigens (Starr et al., 1967).

It has been reported that in a number of patients CF antibody titers may disappear after primary infection even when tested using glycine-extracted antigen (Waner et al., 1973). Such patients would subsequently be serologically negative, leading to an incorrect interpretation of susceptibility to primary infection. The number of such individuals is probably not high.

Indirect Fluorescent Antibody and Anticomplement Immunofluorescence Tests

Indirect fluorescent antibody tests using CMV-infected fibroblast cultures are prone to false positivity because CMV induces cytoplasmic Fc receptors on the surface of infected cells that may nonspecifically bind IgG molecules (Keller et al., 1976). To overcome this problem, an ACIF procedure has been developed (Kettering et al., 1977). In this assay, as in the standard IFA test, CMV-infected fibroblasts grown in suspension or monolayer are used. These cells are then incubated with test serum, followed by a guinea pig complement and then fluorescein-tagged anti-guinea pig C_3 antibody. In theory, complement should be bound only if specific CMV immune globulin is present in the test serum. In a further attempt to eliminate false-positive ACIF results, nuclei, but not cytoplasm, from infected cells can be used in the assay (Mintz et al., 1980). In this method, nuclei are isolated by three cycles of centrifuging CMV-infected cells in sterile distilled water at $500 \times g$ for 10 min.

Other Serologic Tests

Enzyme immunoassays (EIA, ELISA), immune adherence hemagglutination (IAHA), and indirect hemagglutination (IHA) are other serologic tests that are used to measure CMV antibody. The details of these procedures are described in Chapter 5.

Cytomegalovirus Antibody to Early Antigen

A test for measuring antibody to CMV-induced early antigens has been proposed as a measure of primary or active CMV infection (The et al., 1974). Early antigen is expressed in fibroblast cells infected by CMV in the presence of cytosine arabinoside. Griffiths et al. (1980), however, were unable to confirm a correlation between antibody to this antigen and transmission of virus to infants. At present, the significance of CMV antibody to early antigen is unclear.

Interpretation of Tests for Cytomegalovirus Immunoglobulin G Antibody

Although viral isolation supports a diagnosis of CMV infection, serologic tests help to prove that the patient is actively or recently infected. In acute acquired infection of the older child or adult, it is often possible to demonstrate the development of CMV antibody by studying paired acute and late or convalescent sera. Although the demonstration of a fourfold increase in antibody titer is the classic requirement for confirming an infection, this criterion may not be suitable for substantiating a CMV infection, because titers may fluctuate in the apparent absence of active infections. Thus, it is of much greater diagnostic significance to show that antibody was absent in the acute specimen but appeared in the second specimen.

A rise in antibody titer (fourfold or greater) between paired or multiple serum specimens may identify those persons undergoing 1) a recent CMV infection due to reactivation or reinfection, or 2) primary infection where the acute-phase specimen was obtained too late to observe seroconversion. However, such an increase in antibody titer basically reflects recent antigenic stimulation, and in itself is not evidence of either a primary infection or reactivation of latent virus. The greater the increase (e.g., 16-fold, 32-fold, or greater), the more likely is active CMV infection. As mentioned previously, fourfold or greater fluctuations in titer may occur in the absence of active infection.

The lack of a significant increase in antibody titer does not necessarily exclude the possibility of CMV infection, as the patient may have developed a maximal titer of antibody by the time clinical illness develops and medical attention is sought. Stationary antibody titers may also result from high IgM antibody levels in the first of paired or multiple serum specimens, thus preventing detection of increases in total antibody. Test results should always be interpreted in the light of the patient's clinical condition (i.e., the results should be used in conjunction with information available from clinical evaluation and from other diagnostic tests). Whenever possible, serologic diagnoses of CMV infection, especially titer rises rather than seroconversions, should be confirmed by viral isolation.

A negative test on newborn cord blood may be useful in excluding possible infection, whereas a positive test on cord blood should be interpreted with caution. The presence of antibody may only represent transplacental passage; its significance should be determined by viral isolation. A positive test (1:8 or greater) by CF, IF, ELISA, and the like for antibody on a single serum specimen in patients older than 6 to 9 months indicates previous infection with CMV, but may also be found in an individual undergoing current infection. The relationship of any single determination to a current infection cannot be assessed unless additional serum specimens are examined for evidence of seroconversion or an antibody titer rising dramatically. High antibody titers may persist for a year or longer after infection, and so cannot unequivocally be regarded as evidence of a current infection. A negative test (less than 1:8) in children and adults without overt symptoms of infection suggests that the individual is susceptible to primary infection. In the presence of suspected CMV illness, the possibility that the specimen was taken too early to contain antibodies should be ruled out by examining a second specimen taken 10 to 14 days later. Both sera should be examined simultaneously in the same test for seroconversion.

Excellent detailed discussions on interpreting the results of CMV serologic tests are available (Arvin, 1985; Reynolds et al., 1979).

Cytomegalovirus Immunoglobulin M Antibody

Numerous attempts have been made to develop a reliable assay for CMV IgM antibody. In general, the procedure involves reacting patient serum with CMV antigen and following with an anti-IgM antibody labeled with fluorescein or an enzyme. If IgM antibody to CMV is present, the anti-IgM antibody will bind and be detectable by fluorescence or an enzyme reaction.

Cytomegalovirus-specific IgM antibody may be helpful in indicating recent or active infection, especially if seroconversion has already occurred by the time the first blood specimen is obtained. This antibody is rarely absent in an adult who has been clinically ill with primary CMV infection for 2 weeks or more.

In many instances these assays have lacked sensitivity, specificity, or both. With solid-phase assays in particular, false-positive IgM assays may result from the presence of rheumatoid factor, an IgM antibody directed against any IgG antibody. If rheumatoid factor, produced in the acute phase of many viral infections, is present in serum along with CMV IgG antibody, it will bind to the IgG antibodies, which are, in turn, complexed with CMV antigen. When fluorescein- or enzyme-tagged antibody to human IgM antibody is then added, it will bind and give a false-positive result. Rheumatoid factor is a problem in IFA, ACIF, and ELISA tests, but may be removed by 1) column fractionation, 2) absorption with staphylococcal protein A, or 3) latex beads coated with human IgG (Kangro et al., 1984). False-positive CMV IgM antibodies have been reported in 24% of infants who were not congenitally infected by the virus (Stagno et al.; 1980).

False-negative CMV IgM antibody assays may occur when there is a great excess of CMV IgG antibody, because IgG molecules are more avid for antigen. This result is especially likely to occur in congenitally infected newborns, who may have large amounts of passively acquired maternal CMV IgG antibody and relatively low levels of viral IgM produced by the fetus. False-negative CMV IgM antibodies have been reported in 21% of congenitally infected infants (Stagno et al., 1980).

Other pitfalls in utilizing this test include 1) even though the incubation period of CMV infection is several weeks, IgM antibody does not usually develop until the patient has been clinically ill for at least a week and often longer; 2) IgM antibody appears not only during primary CMV infection, but also may reappear during reactivations; 3) patients may continue to produce this antibody for 200 days or more following primary infection; and 4) heterotypic IgM responses may occur (e.g., EBV IgM).

Prevention and Control

As discussed in the section on epidemiology, there is strong evidence suggesting that CMV infections are sexually transmitted and that semen is the major vector in such transmission. Among homosexual men only passive anal intercourse correlated either with the initial presence of antibody to CMV or with seroconversion to CMV during the course of the study (Mintz et al., 1983). Cytomegalovirus antibody was present in 57 (96.6%) of 59 men who engaged in passive anal intercourse, but in only 73.7% of men who did not. Antibody prevalence in the latter was not significantly different from that in heterosexual men. These data suggest that exposure of the anorectal mucosa to CMV-infected semen may constitute a major mode of infection by CMV in homosexual men. Increased rates of CMV infection in this population would be reduced by refraining from passive/receptive anal intercourse. In addition, condoms have been shown to prevent transmission of CMV, even after trauma simulating sexual intercourse (Katznelson et al., 1984). Similarly, transmission of CMV via semen to female sex partners may be an important mode of spread from actively infected men (Handsfield et al., 1985). Condoms may therefore be useful in minimizing the spread of the virus to women from heterosexual partners, which is especially important if the women are both seronegative for CMV and pregnant.

Vaccine

The prime targets for vaccination would be seronegative women of childbearing age. Live attenuated CMV vaccines have been developed that induce antibody formation as well as cell-mediated immune responses (Elek and Stern, 1974; Plotkin et al., 1984), but it is not certain that they protect against natural viral infection. Furthermore, there are concerns that vaccine strains of CMV might be oncogenic, although they are probably no more so than wild CMV.

Another question is whether vaccine virus becomes latent in the host and could reactivate during pregnancy, thereby leading to congenital infection. Current evidence suggests that at least the Towne strain of vaccine virus does not induce latency (Plotkin et al., 1984). Some workers have proposed a subunit vaccine using only CMV proteins. Such a product could not become latent and, lacking viral DNA, is also unlikely to be oncogenic.

Finally, it is not clear that vaccines—or, for that matter, natural infections—entirely prevent new exogenous infections. This question would have to be resolved before any attenuated-virus vaccine strategy is employed.

For now, the best protection against severe congenital disease may be acquisition of CMV infection by women during their childhood. Such is the case in underdeveloped countries, where nearly 100% of women are seropositive at an early age. These women may give birth to infants with high rates of congenital infection due to reactivation of virus; however, because they do not experience primary infections during pregnancy, severe congenital infection is largely avoided.

Treatment

Because of its in vitro activity, idoxuridine was studied in patients with CMV infection as early as 1968 (Conchie, 1968). Idoxuridine is an analog of thymidine that is incorporated into viral DNA, causing inhibition of DNA synthesis. This drug has potential activity against all DNA viruses, including herpes simplex, varicella-zoster, Epstein–Barr, and cytomegalovirus. Conchie (1968) and later Barton and Tobin (1970) used idoxuridine to treat neonatal CMV infection. All of the treated infants experienced a reduction in CMV titer in their urine, but there was no clinical improvement and severe toxicity was noted. Idoxuridine is presently considered too toxic and insufficiently effective for systemic administration as an antiviral drug.

The anticancer drug cytosine arabinoside (Ara-C) inhibits the synthesis of the DNA precursor deoxycytidine (Cher, 1965). Treatment of congenital CMV infection with Ara-C was studied in three trials. A moderate decrease in the titer of urinary CMV found in one study of four infants (Rasmussen et al., 1984) was accompanied by both hematologic and hepatic toxicity, and no clinical improvement was found in any of the patients. Plotkin and Stetler (1969) used Ara-C to treat three infants; all showed clinical improvement, but severe thrombocytopenia limited the use of the drug. A third study of two infants showed variable antiviral effects, but again there was substantial toxicity in the forms of vomiting, neutropenia, and thrombocytopenia (Kraybill et al., 1972).

Adenine arabinoside (Ara-A or vidarabine) was synthesized in the early 1960s as an anticancer drug. It is converted to the triphosphate intracellularly and acts by inhibition of DNA polymerase (Whitley et al., 1980). After demonstrating efficacy for treatment of herpetic keratitis and encephalitis as well as herpes zoster infection, vidarabine was studied in patients with CMV infection. Marker and associates

(1980) treated 14 renal transplant patients with CMV infection, 4 in an open protocol and the subsequent 10 in a randomized, placebo-controlled fashion. There was no evidence of clinical or virologic efficacy in the 9 patients who received vidarabine. Pollard et al. (1980) treated 7 organ transplant recipients with CMV retinitis. Of the 7 patients treated with vidarabine, 5 had a favorable clinical response; this result, however, was not significantly different from the improvement noted in 3 of 6 placebo recipients. In an open study of treatment of various cytomegalovirus syndromes in both normal and immunosuppressed patients, Ch'ien and colleagues (1974) found that vidarabine was well tolerated, but had only a transient antiviral effect in the normal hosts; immunosuppressed patients showed only a slight reduction in urine viral titers, and viremia persisted.

Human leukocyte (α) interferon is part of the natural response to a variety of viral infections. Administration of exogenous α interferon has been shown to be effective for prevention and treatment of a variety of viral infections. A double-blind, placebo-controlled trial of interferon for prevention of CMV infections was performed in patients receiving standard immunosuppressive therapy with or without antithymocyte globulin (Cheesemen et al., 1979). There was significantly less CMV viremia in the group receiving interferon, but this finding was limited to those patients who did not receive antithymocyte globulin. There was also a delay in the onset of CMV viruria in the treated patients. α-Interferon therapy was studied in four patients with CMV retinitis (Chou et al., 1984). Three of these patients suffered from AIDS, and one had Hodgkin's disease. The urine viral titers increased in two patients and decreased in the other two. All three patients with AIDS experienced progression of the CMV retinitis; the retinal lesions stabilized in the subject with Hodgkin's disease.

Bone marrow transplant patients with CMV pneumonia showed no clinical improvement when treated with interferon (Meyers et al., 1980) or interferon combined with vidarabine (Meyers et al., 182).

DHPG

DHPG is an acyclic nucleoside analog of thymidine, which is structurally similar to acyclovir. Because DHPG was developed simultaneously by several pharmaceutical manufacturers, it has also been termed BW B759U, 2'-NDG (2'-nordeoxyguanosine], and BIOLF-62. In vitro, DHPG shows activity toward all human herpesviruses at concentrations equivalent to those at which acyclovir is active toward herpes simplex virus (Field et al., 1983; Mar et al., 1983). Tocci et al. (1984) found DHPG to cause

50% inhibition of DNA synthesis by CMV strains at concentrations of 0.1 to 2.0 M. In studies conducted in three different laboratories, similar ED_{50} values were found for numerous strains of CMV (Plotkin et al., 1985). DHPG itself is inactive, and it must be converted to the triphosphate to inhibit viral DNA synthesis. The drug is selectively phosphorylated to the monophosphate in virus-infected cells, then further converted to the triphopshate (the active form) by cellular enzymes, presumably deoxyguanosine kinases (Field et al., 1983). Herpes simplex and varicella-zoster viruses induce their own thymidine kinase in infected cells, and this enzyme efficiently phosphorylates DHPG. However, the mechanism by which monophosphorylation occurs in cells infected with CMV and Epstein–Barr virus is obscure, as these viruses do not specify a thymidine kinase. Cytomegalovirus infection stimulates synthesis of cellular enzymes, including thymidine kinase, but this enzyme does not activate DHPH sufficiently rapidly to explain the levels observed in infected cells (Freitas et al., 1985). Additionally, infection of cells with mutants of CMV resistant to DHPG stimulate cellular levels of thymidine kinase comparable to those seen in cells infected with susceptible strains. Although DHPG functions like acyclovir as a selective inhibitor of viral DNA polymerase, unlike acyclovir it has a 5' hydroxy group and thus can be incorporated into DNA and act as a terminator of DNA chain formation. It might therefore be expected to be more toxic than acyclovir, a supposition that has been borne out in animal and clinical studies (Freitas et al., 1985).

Successful treatment with DHPG of life- or sight-threatening CMV infections in immunocompromised patients has been reported (Bach et al., 1985; Felsenstein et al., 1985; Masur et al., 1986). A total of 13 patients with AIDS and CMV retinitis treated with DHPG were reported to have favorable clinical responses. Viral cultures of urine, blood, or throat for CMV usually reverted to negative concomitantly with decreased retinal inflammation, hemorrhage, and often improvement in visual acuity. A collaborative group recently reported the results of treating 26 immunocompromised patients having life- or sight-threatening CMV infection with intravenous DHPG (Collaborative DHPG Treatment Study Group, 1986). Of the 22 patients with virologically confirmed CMV infection, 17 showed a favorable clinical response to DHPG at dosages of 10 to 15 mg/kg per day administered for 10 to 21 days. Of the 13 patients with CMV retinitis studied, 11 had a favorable clinical response. Clearing or reduction of CMV shedding was observed in all 17 patients for whom serial CMV cultures were available. Taken as a whole, these clinical experiences suggest that intravenous DHPG may have significant antiviral activity against CMV in infected patients. Unfortunately, bone marrow transplant patients have not fared so well with DHPG therapy. Shepp et al. (1985) reported treating 10 bone marrow transplant patients having culture-documented CMV pneumonia with DHPG (7.5 or 15 mg/kg per day). Although CMV cultures of urine and blood became negative and the amount of infectious CMV in lung tissue decreased from pretreatment titers, all 10 patients died (Shepp et al., 1985).

After the initial, favorable clinical and virologic response of patients with AIDS to a full therapeutic course of DHPG (7.5 mg/kg per day or more), most patients have experienced a clinical and often a viral relapse. In a report on 26 patients, 11 (79%) of 14 patients with AIDS who had adequate follow-up experienced clinical relapse (Collaborative DHPG Treatment Study Group, 1986). A majority (57%) of patients who were then placed on a dosage of 30 mg/kg per week as a maintenance regimen appear to have achieved control of retinitis and viral shedding. The reason for continued clinical progression, and in some cases continued viral "breakthrough" in the remaining patients on maintenance DHPG therapy, is not known. A limiting factor in the use of this drug is bone marrow toxicity. Neutropenia (less than 1,000 granulocytes/μl) occurred in 5 of the 14 patients during maintenance DHPG therapy (Collaborative DHPG Treatment Study Group, 1986).

Laboratory Safety

As detailed previously, CMV is transmitted by blood transfusion and by very close personal contact. On the other hand, health care workers caring for patients known to be excreting CMV are not at increased risk for acquiring the virus. Several early papers suggested such a risk (Haneberg et al., 1980; Yeager, 1975), but a recent, more carefully controlled study refuted this conclusion (Dworsky, 1983). Laboratory workers per se were not studied. Intravenous drug abusers do not have an unusually high prevalence of CMV antibody (Brodie et al., 1984), further suggesting that CMV is unlikely to be acquired by accidental needle sticks or exposure to blood in the laboratory. Bodily fluids such as urine, throat secretions, and especially semen have high titers of CMV, but transmission of the virus due to handling these specimens has not been documented. Laboratory workers handling any bodily fluid should observe standard handwashing, glove usage, or both. Masking or gowning is not necessary.

Literature Cited

Arvin, A. M. 1985. Cytomegaloviruses, p. 227. *In* E. H. Lennette (ed.), Laboratory diagnosis of viral infections. Marcel Dekker, Inc., New York.

Bach, M. C., S. P. Bagwell, N. P. Knapp, K. M. Davis, and P. S. Hedstrom. 1985. 9-(1,3-Dihydroxy-2-proppoxy-methyl)guanine for cytomegalovirus infections in patients with the acquired immunodeficiency syndrome. Ann. Intern. Med. **103**:381–382.

Ballard, R. B., W. L. Drew, K. G. Hufnagle, and P. A. Riedel. 1979. Acquired cytomegalovirus infection in pre-term infants. Am. J. Dis. Child. **133**:482–485.

Barton, B. W., and J. Tobin. 1970. The effects of idoxuridine on the excretion of cytomegalovirus in congenital infection. Ann. N.Y. Acad. Sci. **173**:90–95.

Bowden, R. A., et al. 1986. Cytomegalovirus immune globulin and seronegative blood products to prevent primary cytomegalovirus infection after bone marrow transplantation. N. Engl. J. Med. **314**:1006–1010.

Brodie, H., W. L. Drew, and S. Maayan. 1984. Prevalence of Kaposi's sarcoma in AIDS patients reflects differences in rates of cytomegalovirus infection in high risk groups. AIDS Memorandum (NIAID) **1**:12.

Carney, W. P., et al. 1981. Analysis of T lymphocyte subsets in cytomegalovirus mononucleosis. J. Immunol. **126**:2114–2116.

Cheeseman, S. H., et al. 1979. Controlled clinical trial of prophylactic human-leukocyte interferon in renal transplantation. N. Engl. J. Med. **300**:1345–1349.

Ch'ien, L. T., et al. 1974. Effect of adenine arabinoside on cytomegalovirus infections. J. Infect. Dis. **130**:32–39.

Chou, S., J. Dylewski, M. Gaynon, P. Egbert, and T. C. Merigan. 1984. Alpha-interferon administration in cytomegalovirus retinitis. Antimicrob. Agents Chemother. **25**:25–28.

Chou, S., and T. C. Merigan. 1983. Rapid detection and quantitation of human cytomegalovirus in urine through DNA hybridization. N. Engl. J. Med. **308**:921–925.

Chretien, J. H., C. G. McGinnis, and A. Muller. 1977. Venereal causes of cytomegalovirus mononucleosis. J. Am. Med. Assoc. **238**:1644–1645.

Collaborative DHPG Treatment Study Group. 1986. Treatment of serious cytomegalovirus infections using 9-(1,3-dihydroxy-2-propoxymethyl)guanine in patients with AIDS and other immunodeficiencies. N. Engl. J. Med. **314**:801–805.

Conchie, A. F., B. W. Barton, and J. O. Tobin. 1968. Congenital cytomegalovirus infection treated with idoxuridine. Br. Med. J. **4**:162–163.

Drew, W. L., and R. C. Miner. 1982. Transfusion-related cytomegalovirus infection following non-cardiac surgery. J. Am. Med. Assoc. **247**:2389–2391.

Drew, W. L., L. Mintz, R. C. Miner, M. Sands, and B. Ketterer. 1981. Prevalence of cytomegalovirus infection in homosexual men. J. Infect. Dis. **143**:188–192.

Drew, W. L., E. S. Mocarski, E. Sweet, and R. C. Miner. 1984. Multiple infections with CMV in AIDS patients: Documentation by southern blot hybridization. J. Infect. Dis. **150**:952–953.

Drew, W. L., et al. 1985. Cytomegalovirus infection and abnormal T-lymphocyte subset ratios in homosexual men. Ann. Intern. Med. **103**:61–63.

Dworsky, M. E., K. Welch, G. Cassady, and S. Stagno. 1983. Occupational risk for primary cytomegalovirus infection among pediatric health-care workers. N. Engl. J. Med. **309**:950–953.

Elek, S. D., and H. Stern. 1974. Development of a vaccine against mental retardation caused by cytomegalovirus infection in utero. Lancet **1**:1–5.

Emanuel, D., et al. 1986. Rapid immunodiagnosis of cytomegalovirus pneumonia by bronchoalveolar lavage using human and murine monoclonal antibodies. Ann. Intern. Med. **104**:476–481.

Feldman, R. A. 1968. Cytomegalovirus in stored urine specimens. J. Pediatr. **73:**611–614.

Felsenstein, D., et al. 1985. Treatment of cytomegalovirus retinitis with 9-[2-hydroxy-1-(hydroxymethyl) ethoxymethyl]guanine. Ann. Intern. Med. **103:**377–380.

Fetterman, G. H., F. E. Sherman, N. S. Fabazio, and F. M. Studnicki. 1968. Generalized cytomegalic inclusion disease of the newborn: Localization of inclusions in the kidney. Arch. Pathol. **86:**86–94.

Field, A. K., et al. 1983. 9-([2-Hydroxy-1-(hydroxymethyl) ethoxymethyl]guanine: A selective inhibitor of herpes group virus replication. Proc. Natl. Acad. Sci. USA **80:**4139–4143.

Follansbee, S. E., D. F. Busch, C. B. Wofsy, D. L. Coleman, J. Gullett, G. P. Aurigemma, T. Ross, W. K. Hadley, and W. L. Drew. An outbreak of *Pneumocystis carinii* pneumonia in homosexual men. Ann. Intern. Med. 1982;**96(part 1):**705–713.

Freitas, V. R., et al. 1985. Activity of 9-(1,3-dihydroxy-2-propoxymethyl)guanine compared with that of acyclovir against human, monkey and rodent cytomegalovirus. Antimicrob. Agents Chemother. **28:**240–245.

Gleaves, C. A., T. F. Smith, E. A. Shuster, and G. R. Pearson. 1984. Rapid detection of cytomegalovirus in MRC-5 cells inoculated with urine specimens by using low-speed centrifugation and monoclonal antibody to an early antigen. J. Clin. Microbiol. **19:**917–919.

Goldstein, L. C., et al. 1982. Monoclonal antibodies to cytomegalovirus: Rapid identification of clinical isolates and preliminary use in diagnosis of cytomegalovirus pneumonia. Infect. Immun. **38:**273–281.

Goodpasture, E. W., and F. B. Talbot. 1921. Concerning the nature of "protozoan-like" cells in certain lesions of infancy. Am. J. Dis. Child. **21:**415–425.

Griffiths, P. D., K. J. Buie, and R. B. Heath. 1980. Persistence of high titre antibodies to the early antigens of cytomegalovirus in pregnant women. Arch. Virol. **604:**303–309.

Hackman, R. C., et al. 1985. Rapid diagnosis of cytomegaloviral pneumonia by tissue immunofluorescence with a murine monoclonal antibody. J. Infect. Dis. **151:**325–329.

Handsfield, H. H., et al. 1985. Cytomegalovirus infection in sex partners: Evidence for sexual transmission. J. Infect. Dis. **151:**344–348.

Haneberg, B., E. Bertnes, and G. Haukenes. 1980. Antibodies to cytomegalovirus among personnel at a children's hospital. Acta Paediatr. Scand. **69:**407–409.

Howell, C. L., M. J. Miller, W. J. Martin. 1979. Comparison of rates of virus isolation from leukocyte populations separated from blood by conventional and Ficoll-Paque/Macrodex methods. J. Clin. Microbiol. **10:**533–537.

Huang, E. S., S. M. Huong, G. E. Tegtmeier, and C. A. Alford, Jr. 1980. Cytomegalovirus: Genetic variation of viral genomes. Ann. N.Y. Acad. Sci. **354:**332–346.

Jordan, M. C., W. E. Rousseau, G. R. Noble, J. A. Stewart, and T. D. Y. Chinn. 1973. Association of cervical cytomegaloviruses with venereal disease. N. Engl. J. Med. **288:**932–934.

Kangro, H. O., J. C. Booth, T. M. F. Bakir, Y. Tryhorn, and S. Sutherland. 1984. Detection of IgM antibodies against cytomegalovirus: Comparison of two radioimmunoassays, enzyme-linked immunosorbent assay and immunofluorescent antibody test. J. Med. Virol. **14:**73–80.

Katznelson, S., W. L. Drew, and L. Mintz. 1984. Efficacy of the condom as a barrier to the transmission of cytomegalovirus. J. Infect. Dis. **150:**155–157.

Keller, R., R. Peitchel, J. N. Goldman, and M. Goldman.

1976. An IgG-Fc receptor induced in cytomegalovirus-infected human fibroblasts. J. Immunol. **116:**772

Kettering, J. D., N. J. Schmidt, D. Gallo, and E. H. Lennette. 1977. Anti-complement immunofluorescence test for antibodies to human cytomegalovirus. J. Clin. Microbiol. **6:**627–632.

Kraybill, E. N., J. L. Sever, G. B. Avery, and N. Movassaghi. 1972. Experimental use of cytosine arabinoside in congenital cytomegalovirus infection. J. Pediatr. **80:**485–487.

Lipscheutz, B. 1921. Unterschungen uber die aetiologie der krankheiten der Herpes genitalis. Arch. Dermatol. Syph. **136:**428–482.

Mar, E. C., Y. C. Cheng, and E.-S. Huang. 1983. Effect of 9-(1,3-dihydroxy-2-propoxymethyl)guanine on human cytomegalovirus replication *in vitro*. Antimicrob. Agents Chemother. **24:**518–521.

Marker, S. G., R. J. Howard, K. E. Groth, A. R. Mastry, R. L. Simmons, and H. H. Balfour. 1980. A trial of vidarabine for cytomegalovirus infection in renal transplant patients. Arch. Intern. Med. **140:**1441–1444.

Martin, W. J., and T. F. Smith. 1986. Rapid detection of cytomegalovirus in bronchoalveolar lavage specimens by a moncolonal antibody method. J. Clin. Microbiol. **23:**1006–1008.

Masur, H., et al. 1986. Effect of 9-(1,3-dihydroxy-2-propoxymethyl)guanine on serious cytomegalovirus disease in eight immunosuppressed homosexual men. Ann. Intern. Med. **104:**41–46.

McCracken, G. J., Jr., and J. P. Luby. 1972. Cytosine arabinoside in the treatment of congenital cytomegalic inclusion disease. J. Pediatr. **80:**488–495.

McKeating, J. A., S. Stagno, P. R. Stirk, and P. D. Griffiths. 1985. Detection of cytomegalovirus in urine samples by enzyme-linked immunosorbent assay. J. Med. Virol. **16:**367–373.

Meyers, J. D., R. McGuffin, Y. Bryson, K. Cantell, and E. Thomas. 1982. Treatment of cytomegalovirus pneumonia after marrow transplant with combined vidarabine and human leukocyte interferon. J. Infect. Dis. **146:**80–84.

Meyers, J. D., R. McGuffin, P. Neiman, J. Singer, and E. Thomas. 1980. Toxicity and efficacy of human leukocyte interferon for the treatment of cytomegalovirus pneumonia after marrow transplantation. J. Infect. Dis. **141:**555–562.

Mintz, L., W. L. Drew, R. C. Miner, and E. H. Braff. 1983. Cytomegalovirus infections in homosexual men: An epidemiological study. Ann. Intern. Med. **99:**326–329.

Mintz, L., R. C. Miner, and A. S. Yeager. 1980. Anticomplement immunofluorescence test that uses isolated fibroblast nuclei for detection of antibodies to human cytomegalovirus. J. Clin. Microbiol. **12:**562–566.

Myerson, D., R. C. Hackman, and J. D. Meyers. 1984. Diagnosis of cytomegaloviral pneumonia by in situ hybridization. J. Infect. Dis. **150:**272–277.

Peterson, P. K., et al. 1983. Risk factors in the development of cytomegalovirus-related pneumonia in renal transplant recipients. J. Infect. Dis. **148:**1121.

Phipps, P. H., L. Gregoire, E. Rossier, and E. Perry. 1983. Comparison of five methods of cytomegalovirus screening of blood donors. J. Clin. Microbiol. **18:**1296–1300.

Plotkin, S. A., W. L. Drew, D. Felsenstein, and M. S. Hirsch. 1985. Sensitivity of clinical isolates of human cytomegalovirus to 9-(1,3-dihydroxy-2-propoxymethyl)-guanine. J. Infect. Dis. **152:**833–834.

Plotkin, S. A., et al. 1984. Prevention of cytomegalovirus disease by Towne strain live attenuated vaccine, p. 271–287. *In* S. A. Plotkin, et al (ed.), CMV: Pathogenesis and prevention of human infection. Alan R. Liss, Inc., New York.

Plotkin, S. A., and H. Stetler. 1969. Treatment of congenital cytomegalic inclusion disease with antiviral agents. Antimicrob. Agents Chemother. 9:372–379.

Pollard, R. B., P. Egvert, J. Gallagher, and T. C. Merigan. 1980. Cytomegalovirus retinitis in immunosuppressed hosts. I. Natural history and effects of treatment with adenine arabinoside. Ann. Intern. Med. 93:655–664.

Rasmussen, L., P. T. Chen, J. G. Mullenax, and T. C. Merigan. 1984. Inhibition of human cytomegalovirus replication by 9-(1,3-dihydroxy-2-propoxymethyl)-guanine alone and in combination with human interferons. Antimicrob. Agents Chemother. 26:441–445.

Reynolds, D. W., S. Stagno, and C. A. Alford. 1979. Laboratory diagnosis of cytomegalovirus infections, In E. H. Lennette and N. J. Schmidt (ed.), Diagnostic procedures for viral, rickettsial, and chlamydial infections, ed. 5. American Public Health Association, Washington, D.C.

Rice, G. P. A., R. D. Schrier, and M. B. A. Oldstone. 1984. Cytomegalovirus infects human lymphocytes and monocytes: Virus expression is restricted to immediate early gene products. Proc. Natl. Acad. Sci. USA 81:6134–6138.

Rinaldo, C. R., Jr., W. P. Carney, B. S. Richter, P. H. Black, and M. S. Hirsch. 1980. Mechanisms of immunosuppression in cytomegalovirus mononucleosis. J. Infect. Dis. 141:488–495.

Rowe, W. P., J. W. Hartley, S. Waterman, H. C. Turner, and R. J. Huebner. 1956. Cytopathogenic agent resembling human salivary gland virus recovered from tissue cultures of human adenoids. Proc. Soc. Exp. Biol. Med. 92:418–424.

Schrier, R. D., J. A. Nelson, and M. B. A. Oldstone. 1985. Detection of human cytomegalovirus in peripheral blood lymphocytes in a natural infection. Science 230:1048–1051.

Shepp, D. H., et al. 1985. Activity of 9-[2-hydroxy-1-(hydroxymethyl)ethoxymethyl]guanine in the treatment of cytomegalovirus pneumonia. Ann. Intern. Med. 103:368–373.

Smith, M. G. 1956. Propagation in tissue cultures of a cytopathogenic virus from human salivary gland virus (SGV) disease. Proc. Soc. Exp. Biol. Med. 92:424–430.

Spector, S. A., J. A. Rua, D. H. Spector, and R. McMillan. 1984. Detection of human cytomegalovirus in clinical specimens by DNA-DNA hybridization. J. Infect. Dis. 150:121–126.

Stagno, S., et al. 1975. Cervical cytomegalovirus excretion in pregnant and nonpregnant women: Suppression in early gestation. J. Infect. Dis. 131:522–527.

Stagno, S., et al. 1980. Comparative study of diagnostic procedures for congenital cytomegalovirus infection. Pediatrics 65:251–257.

Stagno, S., D. W. Reynolds, R. F. Pass, and C. A. Alford, Jr. 1980. Breast milk and the risk of cytomegalovirus infection. N. Engl. J. Med. 302:1073–1074.

Starr, J. G., D. Calafiore, and H. L. Casey. 1967. Experience with a human cytomegalovirus complement fixing antigen. Am. J. Epidemiol. 86:507–512.

Starr, S. E., J. P. Glazer, H. M. Friedman, J. D. Farquhar, and S. A. Plotkin. 1981. Specific cellular and humoral immunity after immunization with live Towne strain cytomegalovirus vaccine. J. Infect. Dis. 143:585–589.

The, T. H., G. Klein, and M. M. A. C. Langenhuysen. 1974. Antibody reactions to virus-specific early antigens (EA) in patients with cytomegalovirus (CMV) infection. Clin. Exp. Immunol. 16:1–12.

Tocci, M. J., et al. 1984. Effects of the nucleoside analogue 2'NDG on cytomegalovirus replication. Antimicrob. Agents Chemother. 25:247–252.

Unger, E. R., L. R. Budgeon, D. Myerson, and D. J. Brigati. 1986. Viral diagnosis by in situ hybridization. Description of a rapid simplified colorimetric method. Am. J. Surg. Pathol. 10:1–8.

Volpi, A., R. J. Whitley, R. Ceballos, S. Stagno, and L. Pereira. 1983. Rapid diagnosis of pneumonia due to cytomegalovirus with specific monoclonal antibodies. J. Infect. Dis. 147:1119–1120.

Waner, J. L., T. H. Weller, and S. V. Kevy. 1973. Patterns of cytomegalovirus complement-fixing antibody activity: A longitudinal study of blood donors. J. Infect. Dis. 127:538–543.

Weller, T. H., et al. 1971. The cytomegaloviruses: Ubiquitous agents with protean clinical manifestations (first of two parts). N. Engl. J. Med. 285:203–214.

Weller, T. H., and J. B. Hanshaw. 1962. Virologic and clinical observations on cytomegalic inclusion disease. N. Engl. J. Med. 266:1233–1244.

Weller, T. H., J. C. Macauley, J. M. Craig, and P. Wirth. 1957. Isolation of intranuclear inclusion producing agents from infants with illnesses resembling cytomegalic inclusion disease. Proc. Soc. Exp. Biol. Med. 94:4–12.

Whitley, R., et al. 1980. Vidarabine: A preliminary review of its pharmacological properties and therapeutic use. Drugs 20:267–282.

Yeager, A. S. 1975. Longitudinal, serological study of cytomegalovirus infections in nurses and in personnel without patient contact. J. Clin. Microbiol. 2:448–452.

Yeager, A. S., et al. 1981. Prevention of transfusion-acquired cytomegalovirus infections in newborn infants. J. Pediatr. 98:281–287.

Yolken, R. H. 1982. Enzyme immunoassays for the detection of infectious antigens in body fluids: Current limitations and future prospects. Rev. Infect. Dis. 4:35–68.

CHAPTER 14

Herpesviridae: Varicella-Zoster Virus

MICHIAKI TAKAHASHI

Diseases: Varicella (chickenpox), herpes zoster (shingles).

Etiologic Agent: Varicella-zoster virus.

Source: Individuals with varicella or herpes zoster, by direct or indirect contact or airborne transmission.

Clinical Manifestations: Acute onset and progression of rash, leading to vesicles is most common feature of varicella. A wide variety of sequelae, including purpura, arthritis, pneumonitis, and involvement of the central nervous system, may occur in rare instances. In herpes zoster, eruption is characteristically unilateral and limited to the cutaneous innervation of a single sensory ganglion. Pain and paresthesia are usually associated with eruption.

Pathology: The cutaneous lesion begins with infection of the capillary endothelial cells with subsequent spread to the epithelial cells of the epidermis. Focal lesions are found in the mucosa of respiratory, gastrointestinal, and other organs. The lung is usually the organ most severely involved.

Laboratory Diagnosis: Viral antigen detection by immunofluorescence, enzyme immunoassay, or isolation from vesicles is most convenient and reliable. A fourfold rise in antibody by enzyme immunoassay, complement fixation, immune-adherence hemagglutination, or fluorescent antibody to membrane antigen is considered diagnostic.

Epidemiology: Varicella is highly communicable in children worldwide. Herpes zoster is also infectious, although the risk is lower than with varicella.

Treatment: Acyclovir.

Prevention and Control: A live varicella vaccine (Oka strain) has been developed and used particularly for high-risk children.

Description of Disease

Zoster has been described in very early medical literature, whereas chickenpox was confused with smallpox. Clinical differentiation was apparently made by Heberden in the late 18th century (Weller, 1965). Von Bokay (1909), reporting the occurrence of chickenpox following cases of zoster in two families, was the first to draw attention to this association. There is now no doubt that zoster is another clinical manifestation of the varicella viral infections that cause chickenpox. Subsequent clinical observations, experimental infections, and laboratory studies have confirmed this observation. Lipschutz (1929) noted the histologic similarity between the skin lesions of both diseases. Kundratitz (1925) and Bruusgaard

(1932) inoculated children with zoster vesicle fluid and produced chickenpoxlike disease. Secondary transmission of typical varicella was often observed among the susceptible household contacts of these individuals following a typical incubation period. The viral agent that causes chickenpox and zoster was first isolated in cell culture by Weller (1953) and Weller and Witton (1958). The viruses recovered were identical morphologically and serologically.

Subsequently, the isolation of varicella virus from vesicle fluid and its propagation in cultured cells of human and simian origin was described. Under these conditions, however, the virus remained strictly cell associated, and cell-free infectious virus could not be obtained and was not suitable for detailed studies of its physical or biochemical properties. The finding that cell-free virus could be obtained under suitable conditions from primary culture of infected human cells made more extensive studies of the virus possible, including serologic tests and the development of attenuated strain suitable for immunization. Detailed studies of varicella-zoster virus (VZV), and its replication, antigen, and genome are now possible owing to advances in various laboratory techniques, including gene cloning.

The Viruses

Varicella-zoster virus has been regarded as a member of the herpesvirus group. According to the third report of the International Committee on Taxonomy of Viruses, varicella-zoster virus (human herpesvirus 3) and herpes simplex virus (HSV) 1,2 have been assigned to the subfamily α-herpes viridae based on their biologic properties (Roizman et al., 1981). This virus is characterized by a relatively short replication cycle and by frequent latent infection exclusive to ganglia.

Physicochemical Properties and Morphology

Varicella-zoster virions are round or polygonal with central dots. Virions from both zoster and varicella vesicles appeared identical. The nucleocapsid is approximately 100 nm in diameter and consists of 162 hexagonal capsomeres with central axial hollows organized in an icosahedron with $5:3:2$ axial symmetry (Almeida et al., 1962). The intact VZV particle measures 180 to 200 nm in diameter. At least 30 polypeptides with different molecular weights have been detected in the VZV virion, 5 to 6 of which are glycosylated (Shemer et al., 1980; Shiraki et al., 1982). Varicella-zoster viral DNA is $80 + 3 \times 10^6$ daltons, 124884 base pairs (Davison and Scott, 1986), in length with inverted terminal sequences present internally that result in two isomeric forms of DNA molecules (Dumas et al., 1981; Ecker and Hyman, 1982; Straus et al., 1982).

Replication

The multiplication cycle for VZV has not well been defined because of obstacles relating to the low titer of input virus or to the use of infected cells as inoculum. In a study of VZV replication in human embryonic fibroblasts infected with cell-free VZV (Yamanishi et al., 1980), viral antigen was detected in the cytoplasm as early as 2 h postinfection; diffuse fluorescence became brighter and cytoplasmic fluorescence was observed at 14 h postinfection. The spread of virus to the neighboring cells was recog-

nized at 18 h postinfection. These results suggest that 8 to 14 h is required for viral maturation and that 4 additional hours is needed for the virus to spread to neighboring cells. Herpes simplex virus replicates and progeny virus is released by 8 hours postinfection, whereas human cytomegalovirus requires 4 days (Smith and de Harven, 1973). Thus, the maturation of VZV resembles that of HSV.

Varicella-zoster virus is known to induce virus-specific deoxythymidine kinase (Dobersen et al., 1976; Ogino et al., 1977) and DNA polymerase (Mar et al., 1978; Miller and Rapp, 1977), which have been studied indirectly through the action of antiviral compounds such as acyclovir (Elion et al., 1977).

Antigenic Composition and Genetics

Recently, new nomenclature was formulated for VZV glycoproteins (gpI,II,III,IV) that play a major role in the immune reaction (Davison et al., 1986). A physical map of genes for VZV glycoproteins (Keller et al., 1984) and thymidine kinase (Sawyer et al., 1986) is presented in Figure 1. Similarity is evident between VZV and HSV-1 in the layout of the genes of those proteins in the genetic map.

The gpI group of glycoproteins (45 to 100 kilodaltons [kDa]) are the most abundant and immunogenic of the VZV envelope glycoproteins. They elicit the formation of complement-dependent neutralizing antibodies and mediate antibody-dependent cellular cytotoxicity. The gpI gene is mapped to the 72-kDa open reading frame at the right-hand end of the VZV Us region and displays a small degree of sequence homology to HSV-1 gE.

The gpII group glycoproteins have two size ranges, 115 to 140 and 57 to 66 kDa. These glycoproteins, which elicit the formation of complement-dependent and independent neutralizing antibodies, are the second most abundant and immunogenic of the VZV envelope glycoproteins. The gpII gene has been mapped 0.47 in the center of the VZV UL region and displays a significant amount of serologic cross-reactivity and sequence homology to HSV-1 gB.

The gpIII group consists of 105 to 118-kDa polypeptides, which elicit the formation of complement-independent neutralizing antibodies, and is the third most abundant and immunogenic of the VZV envelope glycoproteins.

The gpIV glycoproteins are 45- and 55-kDa polypeptides. The genes for these minor glycoproteins are the 39-kDa open reading frame at the center of the VZV Us region, which displays a small degree of sequence homology to the HSV-1 Us7 gene.

Very recently, VZV deoxythymidine kinase gene was mapped by transfection of cloned viral DNA

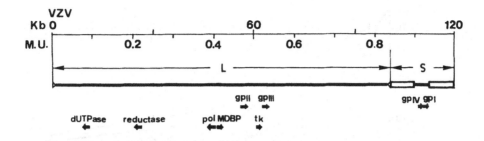

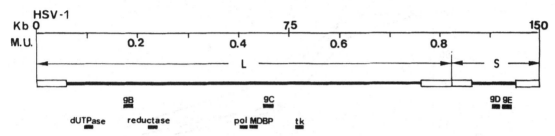

Fig. 1. Physical map of varicella-zoster viral (VZV) genes in comparison with herpes simplex viral (HSV-1) genes.

fragment into thymidine kinase-deficient mouse L(LTK⁻) cells and subsequent biochemical transformation of these cells to the LTK⁻ phenotype. Such transforming activity was between map units 0.50 and 0.52.

The immunologic relationship of VZV to other herpesviruses, particularly HSV, has been investigated not only for its taxonomic interest, but for its clinical significance. Regarding the immunologic relationship of VZV and HSV, a rise in the titer of preexisting antibody to the heterologous virus has been observed after viral infection (Kaspenberg, 1964; Schaap and Huisman, 1968; Schmidt et al., 1969; Svedmyr, 1965; Weller and Witton, 1958).

Harbour and Caunt (1979) and Shiraki et al. (1982) have investigated in detail the immunologic relationship between VZV and HSV, particularly cross-reactions of polypeptides between the viruses. For this study, specific VZV antibody raised in green monkeys or guinea pigs by repeated inoculation of VZV-infected monkey kidney cells or VZV-infected guinea pig embryo cells was employed. Rabbit antisera specific to HSV prepared by inoculation of HSV-infected rabbit kidney cells were also used. There was no cross-neutralization between the viruses, whereas cross-staining was noted on indirect immunofluorescent staining using human serum complement. When cross-reactions were investigated by immunoprecipitation, a few cross-reactive polypeptides were detected (Shiraki et al., 1982). Recently, VZV gpII was found to display a significant

amount of serologic cross-reactivity and sequence homology to HSV-1 gB (Hosler et al., 1985; Kitamura et al., 1986).

These results indicate that VZV and HSV have a few common polypeptides with common antigenic determinants that are not responsible for neutralization. There is also a possibility that preexisting antibody is reinforced by infection with heterologous virus.

Reactions to Physical and Chemical Agents

VZV is one of the most labile viruses; therefore, particular caution is required in selecting suitable suspending medium to preserve viral infectivity. SPGA (0.218 M sucrose, 0.0038 M KH₂PO₄, 0.0072 M K₂HPO₄, 0.0049 M sodium glutamate, and 1% bovine serum albumin) has been reported to be suitable for the preservation of VZV after thermal inactivation (Hondo et al., 1976). PSGC medium (phosphate-buffered saline [PBS](−) containing 5% sucrose, 0.1% sodium glutamate, and 10% fetal calf serum), a simplified form of SPGA, is almost comparable to SPGC in preserving the infectivity of VZV (Asano and Takahashi, 1978) (Fig. 2). With these media, decreases in viral infectivity are minimal with storage at −70°C, and no decrease in infectivity has been noticed after storage for 1 year.

For cryopreservation of VZV, sugar in the suspending medium seems essential to minimize loss of

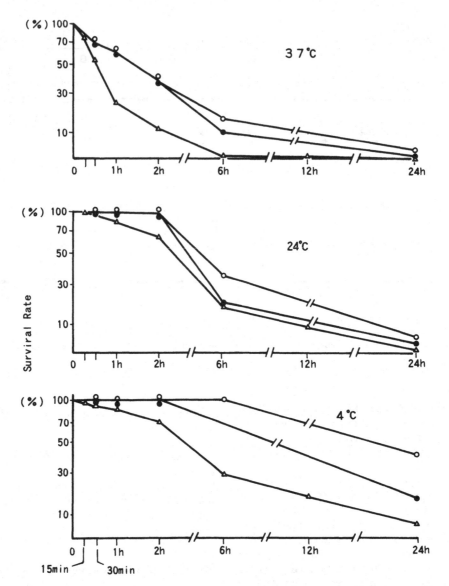

FIG. 2. Comparison of thermal inactivation of virus in various media. ○ = SPGA medium (0.218 M sucrose, 0.0038 M KH$_2$PO$_4$, 0.0072 M K$_2$HPO$_4$, 0.0049 M sodium glutamate, 1% bovine albumin); ● = PSGC medium (phosphate-buffered saline [−] containing 5% sucrose, 0.1% sodium glutamate, and 10% fetal calf serum); △ = medium 199 containing 4% fetal calf serum. (Adapted from Asano and Takahashi [1978], with permission.)

infectivity. Grose et al. (1981) reported that VZV frozen in solutions lacking sugar had little or not residual infectivity after vacuum sublimation was completed. Visualization by electron microscopy demonstrated large numbers of enveloped virions in viral preparations lyophilized in media containing sucrose. In marked contrast, VZV subjected to freeze-drying in buffered solutions without sugar consisted mainly of naked nucleocapsids, suggesting that residual moisture retained by the sugar seemed to prevent viral disenvelopment.

As expected for an enveloped virus, infectivity is rapidly lost when virion preparations are treated with organic solvents or trypsin.

Pathogenesis and Pathology

Pathogenesis

Recently, the pathogenesis of chickenpox was thoroughly reviewed by Grose (1982), and a probable schema is illustrated in Figure 3. Based on epidemio-

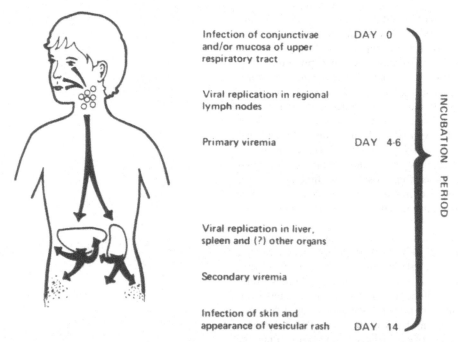

Infection of conjunctivae
and/or mucosa of upper
respiratory tract   DAY 0

Viral replication in regional
lymph nodes

Primary viremia   DAY 4-6

Viral replication in liver,
spleen and (?) other organs

Secondary viremia

Infection of skin and
appearance of vesicular rash   DAY 14

INCUBATION PERIOD

FIG. 3. Pathogenesis of chickenpox. (From Grose [1981], with permission.)

logic evidence, VZV is assumed to spread by droplet nuclei or airborne droplets. Airborne transmission of chickenpox in hospitals has been demonstrated (Asano et al., 1980; Leclair et al., 1980). The initial site of infection may be the conjunctiva, the upper respiratory tract, or both. Conjunctival and respiratory infection was demonstrated in animal model experiments using guinea pigs (Matsunaga et al., 1982; Myers et al., 1980). The virus then replicates at a local site, probably the regional cervical lymph tissue. Large multinuclear giant cells have been found in adenoidal and tonsillar tissue 3 days before the onset of chickenpox (Tomlinson, 1939). The duration of viral replication at the local site may be estimated from observations on direct injection into human subjects of vesicle fluid from patients with varicella or zoster (Bruusgaard, 1932; Kundratitz, 1925). In these studies, the incubation period in the 33 cases that developed symptoms of varicella was usually 8 to 10 days, with a range of 7 to 12 days. As the incubation period of natural VZV infection is usually 14 to 15 days, the duration of viral replication at the local site may be estimated to be 4 to 6 days.

Primary viremia, which may occur after release of virus that has replicated at the local site of infection, may be explained by the observation of neonatal chickenpox on day 1 or 2 after birth following maternal varicella in the last 2 weeks of gestation (Grose, 1982). The major sites of viral replication are not yet clear, but may be deduced from the results of autopsies of neonates who died of chickenpox (Cheatham

et al., 1956; Ehrlich et al., 1958; Miliauskas and Webber, 1984). Although data on fluorescent-antibody staining were not available, pathologic changes including formation of intranuclear inclusions were found in the liver, spleen, lungs, and other organs, but not the thyroid gland, heart, bladder, and ureter.

Secondary viremia may occur after viral replication in some or all of the above sites. The virus will quickly invade cutaneous tissue, where viral replication results in vesicle formation.

Recently, secondary viremia was detected by a sensitive culture technique for viral isolation in immunocompetent children after close exposure to wild-type VZV. Naturally infected children who developed clinical varicella had viremia between 5 days before and 1 day after development of rash (Asano et al., 1985a,b; Ozaki et al., 1984).

Zoster

It has been believed that VZV becomes latent in dorsal ganglion cells. However, information regarding the presence of the virus in nervous tissue is extremely meager, despite the regular demonstration of virus in skin lesions. It may be that latent virus differs from infectious virus and that its presence in tissue culture is extremely difficult to demonstrate. Herpeslike viral particles have been visualized by electron microscopy in ganglia removed from patients who died of VZV infection (Esiri and Tomlinson, 1972; Ghatak and Zimmerman, 1973; Naga-shima et al., 1975). The isolation of VZV from a patient with zoster was reported not only from the skin lesion but also from the corresponding spinal ganglion, which had typical histologic lesions. Viral antigen was also demonstrated in the same spinal ganglion by immunofluorescence (Shibuta et al., 1974). It has been shown recently by DNA–DNA and DNA–RNA in situ hybridization techniques using cloned VZV DNA that VZV DNA resides in the trigeminal ganglia of most adults examined who died of diseases other than zoster; this finding suggests that VZV persists in latent form in most adults after natural infection (Gilden et al., 1983; Hyman et al., 1983).

It has also been reported that no variation in the mobility of digested DNA fragments was detected among strains isolated from the same patient or from household contacts or roommates, whereas some variation was detected among strains isolated from independent epidemics (Hayakawa et al., 1986). No variation in the mobility of digested DNA fragments was detected between strains isolated from a patient who developed varicella and then zoster, indicating that no alteration in the VZV genome occurred during latency (Hayakawa et al., 1986; Straus et al., 1984).

There is no conclusive evidence on the route by which VZV travels to the ganglia. Hope-Simpson (1965) noted that the pattern of incidence for zoster on the individual sensory ganglia was similar to the distribution of the rash in chickenpox, and may bear a direct relationship to it. An area with dense chickenpox rash may establish more latent virus in the related ganglia, and in turn cause more attacks of zoster in later life. Based on this observation, Hope-Simpson proposed that VZV in vesicles on the skin or mucosal surface travels to the ganglia via the sensory nerve fibers. In explaining why sensory ganglia, not motor ganglia, are selected for viral lodgement, this theory is persuasive (Fig. 4).

It has been reported that considerably high titers of neutralizing antibodies are detectable at the onset or 1 or 2 weeks before onset of rash (Asano and Takahashi, 1978; Brunell et al., 1975), and that there is no fall in geometric mean titers of fluorescent antibody to membrane antigen (FAMA) with aging (Gershon and Steinberg, 1981). In contrast, cell-mediated immunity assessed by lymphocyte transformation or VZV skin testing is significantly depressed at the onset of rash and in the elderly (Burke et al., 1982; Hata, 1980; Russel et al., 1973). Therefore, depression of cell-mediated immunity against VZV appears more closely correlated with the development of zoster than with a decline in humoral immunity.

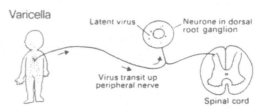

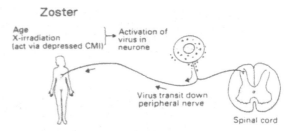

FIG. 4. Pathogenesis of zoster. (Adapted from Mims and White [1984], with permission.)

Pathology

The cutaneous lesions of varicella begin with infection of the capillary endothelial cells, with subsequent spread to epithelial cells of the epidermis. Intracellular (balloon degeneration) and extracellular edema develop, eventually producing separation of the epidermis and a delicate clear vesicle. The roof is formed by the uninvolved stratum corneum and the base by the infected basal layer of the epidermis. Coincident with vesicle formation, perivascular infiltration of round cells occurs in the dermis. Initially, the vesicle contains degenerated epithelial cells and abundant cell-free VZV. Polymorphonuclear leukocytes and macrophages later invade from the surrounding dermis and the fluid turns cloudy. Eventually the fluid is resorbed, leaving a flat adherent crust that is detached by the regrowth of adjacent epithelial cells. Lesions can proceed from papule to crust in 8 to 12 h. Uncomplicated varicella generally heals without scarring. Mucosal lesions develop similarly, but the vesicle roof quickly breaks down, leaving a rapidly healing, shallow ulcer. Fatal cases of varicella have widespread vascular damage with typical intranuclear inclusion in cells lining the small blood and lymphatic vessels (Cheatham et al., 1956; Ehrlich et al., 1958; Weller, 1965).

Clinical Features

Mode of Transmission

Airborne transmission of chickenpox in hospitals has been demonstrated (Asano et al., 1980; Leclair et al., 1980). Vesicles in the mucosa of the respiratory tract may rupture readily, thereby spreading infectious virus in the vesicular fluid from patients to susceptible contacts by airborne transmission.

Incubation Period and Period of Infectivity

The incubation period of chickenpox is usually 14 to 15 days (Gordon and Meader, 1929; Hope-Simpson, 1952). The range is generally considered to be 10 to 20 days. The infectivity of children with preeruptive chickenpox has been determined by careful analysis of hospital outbreaks in which one or more children were by chance transferred out of an open ward before the occurrence of disease there. In these circumstances, children in the late incubation period of chickenpox were infectious 1, 2, and possibly 4 days before the appearance of vesicular rash (Evans, 1940; Gordon and Meader, 1929). Children with chickenpox remained infectious for the first 4 to 5 days after onset of the exanthems.

Clinical Symptoms

The occurrence of recognizable preeruptive symptoms is not a constant feature, and in young children the appearance of rash is often the first clinical event. In older children and adults, however, headache, pains in the limbs, nausea, or vomiting may occur as prodromal symptoms.

The rash of varicella, the most typical feature of the disease, is characterized by a rapid progression from macules to papules to vesicles to crusts. Vesicles develop on the mucous membranes of the mouth as well as on the skin. They occur most commonly over the palate and usually rupture so rapidly that the vesicular stage may be missed. The palpebral conjunctiva, pharynx, larynx, trachea, and rectal and vaginal mucosa may also be involved. The cutaneous eruption appears early and most profusely on the back, chest, and abdomen. Chickenpox usually has little or no apparent extracutaneous manifestations. A wide variety of sequelae have occurred in rare instances, however, including purpura, arthritis, appendicitis, pneumonitis, hepatitis, pericarditis, glomerulonephritis, and orchitis, as well as involvement of the nervous system.

Chickenpox in children with primary immunodefi-ciency may lead to a more severe illness called *progressive chickenpox*. Children with deficient cell-mediated immunity are especially susceptible, as well as children immunosuppressed due to malignancy and anticancer therapy. The persistent viral replication and viremia has been documented in severe disseminated varicella (Feldman and Epp, 1979; Myers, 1979). Chickenpox in adults is often a more severe disease than in healthy children. In general, fever is higher and more prolonged, constitutional symptoms are more severe, rash is more profuse, and complications are more frequent. Primary varicella pneumonia has been observed in 14% of adult patients (Krugman and Katz, 1981).

Congenital varicella syndrome has been reported, but very rarely. Infants born following maternal varicella in early pregnancy (8 to 15 weeks of gestation) showed a syndrome consisting of low birth weight for gestational age, cicatrical skin lesions, a hypotrophic limb, eye abnormalities (chorioretinitis, microophthalmia, optic atrophy), and evidence of brain damage and mental retardation (Gershon, 1975).

When maternal VZV infection occurs shortly before or after birth, a severe or fatal disseminated infection may occur in newborn infants. At this time it is unlikely that passively acquired varicella-zoster antibody will be present to modify the infection, and the immature immunocompetence of such infants may cause severe syndromes. The disseminated varicella infection is characterized by hemorrhagiclike lesions and involvement of the lungs and liver (Gershon, 1975).

Diagnosis

Specimen Collection

In collecting specimens for viral isolation, the labile infectivity of VZV should be taken into consideration. To minimize loss of infectivity, it is important to choose suitable media for collecting vesicular fluid or pharyngeal secretions. SPGA and PSGC media (see Reactions to Physical and Chemical Agents) may be suitable for preserving the infectivity of VZV. When these media are used, 15 to 50% of VZV survive for 24 h at 4°C. Furthermore, comparison of viral recovery from sucrose phosphate (0.2 M SP) and from 70% sorbitol at −70, 4, and 20°C showed 0.2 M SP to be superior in preserving the stability of VZV (Howell and Miller, 1983). From these results, media containing 0.2 M sucrose and fetal calf serum or albumin are recommended. Virus from vesicle fluid is reasonably stable even without added diluent at −65 to −70°C, but rapidly loses infectivity at

−20°C. A detailed study of factors influencing quantitative isolation of VZV has been reported (Levin et al., 1984).

Some preservation of infectivity was obtained by simply freezing the infected cells as monolayers with this medium. Rapid freezing of infected cells with 5 to 10% glycerol preserved infectivity for long periods. No particular caution is needed for biosafety in collecting specimen from suspected cases of varicella-zoster.

Direct Viral Detection

Vesicular lesions resulting from VZV infection can usually be differentiated from all but herpes simplex lesions by Tzanck preparation. Cells are scraped from the base of an early vesicle and stained with Giemsa, Papanicolou, or hematoxylin-eosin. Smears made from the lesions will contain multinucleated giant cells and intranuclear inclusion bodies. These features are absent in smears from smallpox and other nonherpetic vesicles, although the presence of degenerated cells can make interpretation difficult (Barr et al., 1977; Blank et al., 1951).

Punch biopsies to detect for intranuclear inclusions and multinuclear giant cells provide earlier and more reliable material for examination, but suffer from the same limitations and lack of specificity as Tzanck smears. Electron microscopic examination of vesicle fluid or biopsy material can differentiate herpetic infection from smallpox, but cannot separate VZV from HSV infection and is usually not readily available.

Direct Antigen Detection

Gel Precipitation

Gel diffusion tests were developed in the 1960s to differentiate VZV infection from smallpox. These tests were applied clinically and permitted definitive diagnosis of infection within 12 to 24 h (Brunell et al., 1971; Uduman et al., 1972).

Gel is prepared in 0.5% Ionagar (Oxoid Ltd., London, England) in Sorensen phosphate buffer at pH 8.2 containing 1 μg/ml of free protamine and 0.1% of sodium azide. The final pH of the agar is 8.5. Wells 2 mm in diameter are prepared so that their centers are 5 mm apart. Incubation is routinely conducted at 25°C, although lines can frequently be detected earlier if the incubation temperature is 37°C.

Counterimmunoelectrophoresis

Gel diffusion has now been largely supplanted by counterimmunoelectrophoresis (CIE), which can provide results within 2 h (Frey et al., 1981). CIE employs vesicle fluid as the antigen and appears to be both highly sensitive (100% positive in documented VZV infection versus 85% by culture) and specific, with no false-positive results or cross-reactions with HSV.

In brief, 4 ml of 1% agarose in 0.05 M barbital buffer, pH 8.6, is poured onto microscope slides. Wells 2 mm in diameter are made 8 mm apart. Cathodal wells are filled with VZV antigen or vesicular fluid and anodal wells are filled with concentrated antiserum to VZV. Antiserum is obtained from convalescing patients with zoster, varicella immunoglobulin, or VZV monoclonal antibody, which were screened for precipitating antibody to VZV by CIE. Plates are read for visible precipitation lines immediately after electrophoresis and, if negative, read after an additional 45 min at 4°C.

DIRECT IMMUNOFLUORESCENCE STAINING FOR DETECTION OF VARICELLA-ZOSTER VIRAL ANTIGEN

Direct immunofluorescence staining of a vesicular cell smear or tissue specimen has also proved useful in the rapid diagnosis of VZV infections (Schmidt et al., 1965). Smears of cells from vesicular lesions are prepared by collecting cellular scrapings from the base of lesions. Smears are air-dried for transport to the laboratory, fixed in acetone for 10 min at room temperature, and dried. Hyperimmune serum produced in the rhesus monkey by immunization with rhesus monkey kidney cell-grown VZV or preferentially monoclonal antibody is conjugated with a fluorescent marker and used to test clinical specimens after appropriate dilution to the point where no cross-staining is detected in HSV-infected cells. Immunofluorescence of VZV is apparently quite specific and considerably more sensitive than viral isolation, particularly late in infection. The ability to detect VZV antigen in tissue even after vesicle-fluid cultures are no longer positive makes immunofluorescence particularly valuable.

In laboratory testing of 47 patients by viral culture and direct immunofluorescences, the immunofluorescence test established the diagnosis in 24 of 28 patients (86%) strongly suspected on clinical grounds of having varicella-zoster infection. Viral culture was positive in only 10 to 28 patients (86%). Skin lesions in 19 patients who had other diagnoses were negative by both tests. Thus, direct immunofluorescence appears to be highly sensitive and specific for the rapid diagnosis of cutaneous varicella-zoster infection.

Direct immunofluorescence staining was further compared with viral isolation for detection of VZV in clinical materials. Immunofluorescence staining was markedly more sensitive than viral isolation for demonstration of VZV in lesion and tissue specimens, detecting all of the specimens positive for VZV (45 of 45), whereas isolation detected only 23% (13 of 42) (Schimidt et al., 1980).

SOLID-PHASE ENZYME IMMUNOASSAY

Detection of VZV antigen in clinical specimens by solid-phase enzyme immunoassay was developed for this purpose (Ziegler, 1984; Ziegler and Halonen, 1985; Ziegler et al., 1983). Guinea pig and rabbit immunoglobulin to purified VZV nucleocapsids was used as immunoreagent. In preliminary experiments, the optimal dilutions of guinea pig and rabbit immunoglobulin are determined by checkerboard titrations against the immunizing antigen. Micotiter plates are coated with guinea pig immunoglobulin in carbonate buffer (pH 9.6) at a concentration of 2.5 μg/ml (0.25 μg per well) and the plates are incubated at room temperature overnight. The VZV antigen is inoculated, and after overnight incubation at 37°C the plates are washed with PBS containing 0.1% Tween 20 and further incubated for 1 h at 37°C with rabbit immunoglobulin to VZV. After washing, peroxidase-conjugated guinea pig antibodies to rabbit immunoglobulin (Ig)G are added. After incubation for 1 h at 37°C, the plates are washed and 1,2-phenylenediamine (3 μg/ml) in citrate buffer (pH 5.5) containing 0.03% H_2O_2 is added. The plates are then kept at room temperature in the dark for 60 min before the color reaction is stopped by adding 150 μl of 1 N HCl per well. The absorption values are measured with a photometer.

Very few (1 of 60) specimens from which HSV had been isolated gave a weak positive reaction in this VZV assay. Of 24 specimens from which VZV isolation was attempted, the virus was isolated from 8, but 21 were shown to contain VZV antigen. Thus, solid-phase enzyme immunoassay enables rapid and specific diagnosis of VZV infections.

RAPID DETECTION OF VZV ANTIGEN WITH VZV ANTIBODY AND *Staphylococcus aureus*

Rapid detection of VZV antigen in clinical specimens by the use of VZV antibody and *Staphylococcus aureus* CowanI strain, rich in protein A (SRA), was developed (Dishon and Morgensen, 1983). This method is based on the biologic character of protein A, which combines with the Fc portion of IgG 1, 2, and 4. Cytospin preparations from suspected cases are incubated with VZV antisera. After being rinsed, the slide is covered with a 1% suspension of SRA and further incubated. After washing the slide is examined for cells with adhering SRA under a microscope equipped with an interference contrast device. By this method, all of four cases of VZV infection from which VZV was isolated were diagnosed correctly. This method is easy to perform and results can be

obtained within 3 h from the time specimens are delivered to the laboratory.

Direct Viral Nucleic Acid Detection

Dot-blot hybridization using a molecular cloned viral DNA probe has been developed (Seidlin et al., 1984). *Eco*RI fragments of VZV are cloned in a phage λ vector and DNA is purified from phage that was amplified in *Eco*li LE392. The clone mixture is nick-translated with ^{32}PLdCTP to a specific activity of at least 10^8 dpm/μg. The probe is then heated to 95°C for 5 min, chilled immediately, and used the same day for hybridization. Extracted DNA from the sample is used for each blot. For hybridization and autoradiography, nitrocellulose filters are rinsed with distilled water then applied to a Hybri-Dot Manifold (BRL, Bethesda, Md.). Each well is washed with 1 M ammonium acetate. Unknown specimens in volumes of 10 to 50 μl are then applied. The nitrocellulose is baked at 80°C for 3 h in a vacuum oven. Heat-denatured nick-translated VZV cloned probe and denatured calf thymus DNA are added to the filter, and hybridization is conducted at 42°C in thermally sealed plastic bags for 16 to 24 h with gentle shaking. At the end of the hybridization period the filters are washed, and after air-drying autoradiography is performed for 2 to 48 h at room temperature. Films are then developed and read.

The assay can be completed in 36 to 48 h and carried out successfully with amounts of viral DNA ranging from 10 pg to 10 ng. It was reported that in analyses of 38 specimens from patients with a clinical diagnosis of VZV infection, viral isolation yielded positive results in 58%, whereas this assay yielded positive results in 76%. The specificity of viral isolation was 100%, whereas that of this assay was 94%. Thus, this technique appears to be sensitive, specific, and useful for analysis of tissue and body fluids.

Viral Isolation and Identification

ISOLATION OF VZV FROM VESICLE FLUID

Varicella-zoster virus can usually be recovered from vesicle fluid during the first 3 days of rash by inoculation onto actively growing human cell culture (Meurisse, 1969). The lesions of herpes zoster yield virus for a longer interval than those of varicella. As the virus is extremely labile, specimens should be inoculated onto tissue culture immediately or stored at −70°C. Freezing at −20°C for 24 h resulted in less than 1% of the original viral titer, as compared with 70 to 90% when samples were frozen at −70°C for 24 h (Levin et al., 1984).

ISOLATION OF VZV FROM PHARYNGEAL SECRETION

Isolation of VZV from pharyngeal secretions is usually difficult. During the first 3 days of exanthem, VZV was isolated from 12 of 12 vesicle-fluid samples, from 3 of 23 nasal swabs, and from 2 of 22 pharyngeal swabs. No virus was isolated during the incubation period (Trlifajova et al., 1984).

ISOLATION OF VZV FROM BLOOD

It is generally difficult to establish VZV viremia in naturally infected hosts, except in those who are immunocompromised (Feldman and Epp, 1979; Myers, 1979). Recently a sensitive culture technique for isolation of VZV from blood has been developed (Ozaki et al., 1984).

Approximately 2 ml of blood is drawn into a syringe containing 0.1 ml of heparin (1,000 units/ml), which is diluted 1 : 2 with balanced salt solution, layered onto a Ficoll–Hypaque gradient, and centrifuged ($400 \times g$) for 30 min at room temperature. The monolayer cell layer is aspirated and washed twice in culture medium. The aspirated cells are inoculated into cell culture within 5 h of blood sampling. The ratio of mononucleocytes and HEL cells is approximately 1 : 1. The cultures are observed for 7 days and, even if no cytopathic effect is observed, the cells are treated with trypsin and suspended in 2 ml of maintenance medium. The cell suspension then is mixed with 5 ml of newly prepared HEL suspension (2×10^5/ml) in growth medium and plated. Observation is continued for an additional 3 weeks. By this method, VZV could be isolated from the blood of immunocompetent children between 5 days before and 1 day after the appearance of rash (Asano et al., 1985a,b; Ozaki et al., 1984).

ISOLATION FROM SPINAL GANGLION

Varicella-zoster virus has been isolated at autopsy from the ganglia of a case with developed zoster (Shibuta et al., 1974).

VIRAL IDENTIFICATION

Viral isolates are primarily identified by characteristic cytopathic effects (CPE) to VZV and then by VZV-specific immunofluorescence staining. The infected cells are treated with trypsin and transferred onto HEL cells that were grown on a slide. About 2 days later, the medium is decanted and the slides are rinsed in PBS, air-dried, and fixed at −20°C with acetone. Human varicella convalescent serum or preferentially monoclonal antibody to VZV is used for indirect immunofluorescence staining.

Antibody Assay

Several methods have been developed for antibody assay for VZV. Among these, complement fixation (CF), immune-adherence hemagglutination (IAHA), fluorescent antibody to membrane antigen (FAMA), and enzyme immunoassay (EIA) appear suitable for routine use in the diagnostic laboratory for serodiagnosis, and FAMA and EIA should produce better results when assessing the immunity of individuals to clinical VZV infection.

COMPLEMENT FIXATION TEST

Extracts prepared from VZV-infected cells disrupted by freezing and thawing (Brunnell and Casey, 1964) or by sonication (Gold and Godek, 1965; Schmidt et al., 1964) are satisfactory antigens.

When VZV-infected cell cultures in 500-ml bottles (10×12 cm^2) exhibit extensive CPE, cells are collected, washed three times with PBS by light centrifugation ($1,000 \times g$ for 5 min), and suspended in 5 ml of PBS(−). Cells are then sonicated for 3 min at 20 kc/s and centrifuged at 3,000 rpm for 15 min. The supernatant usually has an antigen titer of 1 : 32 to 1 : 64, and four units are used for the reaction.

When sera from individuals with chickenpox are assayed for CF antibody with crude infected-cell extract, almost all sera drawn on day 8 or later after onset of the rash have detectable antibody. After about 3 months the titer falls, becoming undetectable in some at approximately 1 year, and no antibody is detected in adults with a past history of chickenpox more than 20 years before (Asano and Takahashi, 1978; Gold and Godek, 1965; Weller and Witton, 1958). In patients with zoster, the titer of CF antibody rises more rapidly and to higher levels, but falls to low levels several months after onset. Thus, the CF test may be applicable for diagnosis of acute or recent VZV infection, but is of little value in assessing the immune status of individuals to VZV infection.

NEUTRALIZATION TEST

As with other viruses, the neutralization (NT) test has been presumed a reliable method for serologic diagnosis and for assessing the immune status of individuals to VZV. Cell-free virus produced by sonic disruption of VZV-infected cells is used in a plaque reduction assay (Caunt and Shaw, 1969; Schmidt et al., 1965). Cell-free VZV can be obtained by ultrasonic disruption of infected human cells (Brunnell, 1967; Schmidt and Lennette, 1976). The following steps are recommended to obtain high-titer cell-free virus from infected cells: 1) Cultured cells in growth phase are used for inoculation of virus. 2) High-input multiplicity is used as inoculum. For this purpose, cells are inoculated at a ratio of 1 infected cell to 5 to 10 uninfected cells in monolayer. 3) Infected cell monolayers are harvested by treatment with ethylenediaminetetraacetate before advanced cytopathic change is apparent.

Conditions for using the NT test with VZV were investigated in detail by Asano and Takahashi (1978). A 50% plaque reduction NT test is usually performed. Samples of the viral preparation diluted to approximately 100 to 200 plaque-forming units (PFU) per 0.1 ml are mixed with equal amounts of serial two- or fourfold dilutions of serum inactivated by heating at 56°C for 30 min; the mixture is then incubated at 37°C for 30 min with occasional shaking. Subsequently, 0.2-ml volumes of serum–virus mixture are inoculated onto cell cultures in 36-mm plastic dishes, which are incubated at 37°C for 2 h to permit adsorption of unneutralized virus. Then medium is added and incubation is continued for 5 to 6 days. Cell cultures are fixed with Formalin and stained with methylene blue. Foci are counted with a binocular microscope.

In serologic follow-up of patients with varicella and zoster by NT, an increase in NT antibody is detected 7 days after the onset of rash, and adults who had had chickenpox previously had 1:16 to 1:32 NT antibody. In patients with zoster, it was noted that all 4 cases examined had NT antibody (1:16 to 1:64) 1 to 2 weeks before onset of the disease.

Schmidt and Lennette (1975) reported that neutralization of VZV by human sera and immune rhesus monkey sera was enhanced by fresh guinea pig complement. The enhancement was not restricted to either the IgM or the IgG class of immunoglobulin. Grose et al. (1979) have also reported on the complement-enhanced neutralizing antibody response to VZV (C/NT). Generally the C/NT antibody titer was two- to fourfold higher than the FAMA titer. The absence of C/NT activity at a serum dilution of 1:4 indicated susceptiblity of the children to VZV infection and correlated with an absence of FAMA (less than 1:2). A survey of susceptible leukemic children exposed to chickenpox revealed that several recipients of zoster immune globulin had a subclinical infection, as manifested by seroconversion and persistence of C/NT antibody to VZV.

Potentiation of VZV neutralization by anti-human immunoglobulin has been reported. Heterologous anti-human immunoglobulin added to the virus–serum mixture was utilized to potentiate the neutralization reaction. The anti-human immunoglobulin enhanced the sensitivity of the NT test 7- to 107-fold as compared with the conventional neutralization test (Asano et al., 1982a). The specificity of the test was confirmed by seroconversion of recipients of live varicella vaccine.

Thus the NT test, including C/NT and the method potentiated by anti-human immunoglobulin, although technically cumbersome, may be used reliably in the diagnosis of subclinical VZV infection and for evaluation of individual humoral immunity.

Direct and Indirect Immunofluorescence Using Fixed Infected Cells

Infected human cells showing 90% CPE are scraped in one-tenth original volume of media, and smears are prepared by placing small drops of suspension on microscope slides. The smears are allowed to dry at room temperature and are then fixed in cold acetone at −20°C for 18 to 24 h.

For direct fluorescent-antibody staining, the rinsed smears are flooded with the working dilutions of the anti-VZV conjugate.

For indirect fluorescent-antibody staining, acute and convalescent serum specimens are first inactivated at 50°C for 30 min and twofold dilutions are prepared. A drop of each serum dilution is applied to a smear of infected cells, and the slides are incubated at 37°C for 20 min. The slides are washed twice in PBS and a working dilution of fluorescein-conjugated anti-human globulin is added to each smear. After incubation at 37°C for 20 min, the slides are washed twice, drained, and mounted in buffered glycerol–saline solution.

The titer is determined by the degree of cytoplasmic (rather than membrane) fluorescence. When the indirect immunofluorescence assay was compared with CF test, the two assays were roughly comparable in their ability to demonstrate fourfold or greater rises in antibody. It is uncertain whether the indirect immunofluorescence assay may be used to detect humoral immunity to VZV in individuals who have had chickenpox during childhood (Schmidt et al., 1969).

Anticomplement Immunofluorescence Test

The anti-complement immunofluorescence (ACIF) test has been developed for assay of VZV antibodies (Kettering et al., 1977). VZV-infected cells on glass slides are fixed with acetone at room temperature and treated with serial dilutions of test serum. After incubation for 20 min at 37°C, the slides are rinsed. An optimal dilution of complement (usually 1:40) is added and incubation conducted for 20 min at 37°C. After rinsing with PBS, an optimal dilution of fluorescein-conjugated anti-guinea pig complement is added and incubation conducted for 20 min at 37°C. After they are rinsed with PBS, the slides are examined under an ultraviolet illuminator. The ACIF test was found to be comparable to the FAMA test in sensitivity and can be used for examining sera at low dilutions of 1:2 and 1:4 (Gallo and Schmidt, 1981).

FLUORESCENT ANTIBODY TO VZV-INDUCED MEMBRANE ANTIGEN

General

When a cell is infected with VZV, viral antigens are induced in the plasma membrane (Ito and Barron, 1973). Antibodies to these virus-specific antigens can be detected by an indirect fluorescence technique that utilizes unfixed infected cells (Brunnell et al., 1975; Gershon and Krugman, 1975; Williams et al., 1974). Human serum to be tested is incubated with a suspension of infected cells, after which a fluorescent anti-human globulin is added to the mixture. If serum specimens contain antibody to the virus, a fluorescent halo is noted around the infected cells owing to the reaction of specific antibody with varicella-zoster membrane antigen.

The procedures are as follows: Human fibroblasts are inoculated with VZV-infected cells, usually at a ratio of 1 infected cell to 5 to 10 uninfected cells. When CPE has developed in about 90% of cells, the cells are scraped into PBS(−) and centrifuged at $1,000 \times g$ for 5 min. The supernatant is discarded and the pelleted cells are resuspended in PBS to yield a concentration of approximately 10^5 cells/ml.

Aliquots of 0.025 ml of cell suspension are inoculated with an equal volume of diluted test sera in microtiter plates. After 30 min at 25°C in a humidified atmosphere, the cells are washed three times with PBS by centrifugation of microtiter plates at $1,000 \times g$ for 10 min. Supernatant fluid is then decanted. A dilution of fluorescein-conjugated goat antibody to human globulin, 0.025 ml per well, is incubated at 25°C for 30 min with the washed cells. The cells are washed an additional three times with PBS. Approximately 10 μl of the resuspended cells is added to one drop of a 1:9 solution of PBS and glycerol on a glass slide.

Compared with the CF test, FAMA antibody was detected earlier during the course of the disease and rose to higher titers. In addition, adults who had chickenpox during childhood also had VZV antibody at a serum dilution of 1:2 or higher by FAMA assay. The FAMA assay was proved sensitive enough to assess the VZV humoral immune status of children with malignancies who were enrolled in the clinical trials of zoster immune globulin (ZIG) and to diagnose subclinical infection in subjects who received ZIG within 72 h of exposure to chickenpox (Gershon et al., 1974). The subclinical infection was not detected by CF assay. Therefore, FAMA can be utilized to diagnose not only acute VZV infection but also to evaluate the immune status of high-risk children.

FAMA Test Using Monodisperse Glutaraldehyde-Fixed Target Cells

Controlled trypsinization and glutaraldehyde fixation were employed to prepare a monodisperse suspension of noninfectious VZV membrane antigen positive target cells that can be stored indefinitely at −70°C. A microtiter immunofluorescence assay utilizing these target cells was shown to provide a sensitive and specific means for detecting and quantitating antibody to VZV (Zaia and Oxman, 1977).

Simplified FAMA Procedure

A simplified FAMA procedure has been developed (Baba et al., 1984b). The test procedure was simplified by using Terasaki tissue culture plates (60 flat-bottom wells) for the reaction and for direct observation by fluorescence microscopy. For this procedure, preparation of VZV-infected Vero or human embryo cells stored in liquid nitrogen can be used as antigen.

Aliquots of 0.005 ml of test sera diluted in microtiter plates are transferred to Terasaki plates and incubated with an equal volume of VZV-infected cell suspension in a humidified atmosphere for 60 min at 37°C. The plates are dipped into 180 ml of Hanks balanced salt solution to remove the supernatant from the cells adhering to the bottom of the plate. The washed cells are incubated at 37°C for 30 min with 0.005 ml per well of a dilution of fluorescein-conjugated goat antisera to heavy-chain human IgG, IgA, and IgM. The plates are dipped into Hanks balanced salt solution to remove the remaining conjugates from the cells, and the cells are examined directly in a inverted fluorescence microscope.

The serologic responses of patients after the onset of varicella examined by the simplified FAMA, NT, and IAHA tests are presented in Figure 5. The antibody response detected by the FAMA test and the NT test are markedly similar. In contrast, IAHA antibody showed a rapid rise within 1 week and a rapid decline within 2 months after the onset of varicella infection. These results indicate that FAMA antibody remains high during the convalescent phase and may persist at a level comparable to that of NT antibody for a long time. The kinetics of the appearance of serum antibody IgG, IgA, and IgM and nasopharyngeal (secretory) IgA as measured by the FAMA test and skin-reactive immunity in patients with clinical varicella are presented in Figure 6 (Baba et al., 1984a).

IMMUNE-ADHERENCE HEMAGGLUTINATION TEST

The IAHA technique has been adapted to detect VZV antibody (Gershon et al., 1976; Kalter et al.,

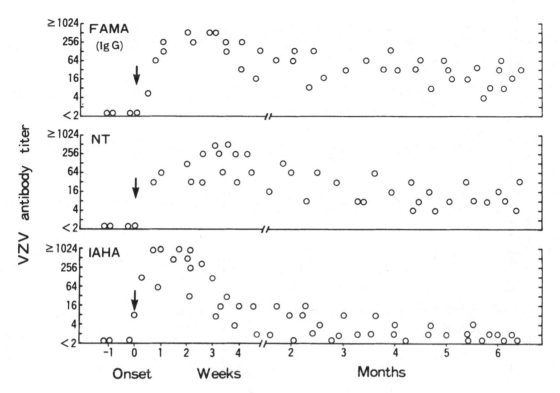

FIG. 5. Kinetics of the appearance of varicella-zoster viral (VZV) antibody in patients who developed varicella as measured by fluorescent antibody to VZV-induced membrane antigen (FAMA), neutralizing test (NT), and immune adherence hemagglutination (IAHA) test. Arrows at day 0 indicate onset of rash. (From Baba et al. [1984b], with permission.)

1977). This reaction is dependent on the activation of the C3 component (C3a, C3b) of complement by the antigen–antibody complex. On the molecule of C3b, stable binding sites appear, which become the receptor sites for human erythrocytes leading to hemagglutination.

Varicella-zoster antigen is prepared in the same way as CF antigen. Guinea pig complement stored at −70°C is diluted for use in gelatine Veronal buffer (GVB). Usually a dilution of 1:80 to 1:100 is used.

Human type O positive whole blood is collected in twice the volume of Alsver solution and kept at 4°C. Red blood cells (RBCs) can be kept for up to 3 weeks. Before use, RBCs are washed three times with GVB and adjusted to a 1.5% suspension in GVB. Not all RBCs are suitable for the test. A donor whose cells give 4+ agglutination with varicella-zoster antigen and antibody (but not with the control preparation) is used.

The procedure for IAHA testing of VZV is as follows: Serial serum dilutions of 1:2 to 1:256 are made in GVB in microtiter plates. Varicella-zoster antigen, 0.025 ml (or control antigen), is added to each well containing serum dilution. Plates are incubated at 37°C for 1 h. Complement at the approxi-

mate dilution in GVB (0.025 ml) is added to all wells. The plates are mixed for 10 s and incubated at 37°C for 40 minutes, and dithiothreitol-GVB-ethylenediaminetetraacetate (0.025 ml) is added to all wells. Then a 1.5% suspension of RBC in GVB (0.025 ml) is added to each well. The plates are mixed for 5 min and left at room temperature for 3 h for hemagglutination to develop.

When serum specimens for chickenpox or zoster were assayed by the IAHA, FAMA, and CF tests, titers obtained with IAHA were comparable to or greater than those with FAMA and consistently higher than those with CF (Kalter et al., 1977). The IAHA test has been successfully applied to detect antibody responses in children who received a live varicella vaccine (Yamada et al., 1979). Antibody responses could readily be detected by IAHA in the vaccinees even when the CF test failed. The IAHA test is rapid and convenient for detecting antibody responses either by natural infection or after vaccination. However, some adults who had had chickenpox in the distant past showed negative results by IAHA, whereas positive results were obtained by FAMA (Kalter et al., 1977). Therefore, IAHA appears less sensitive than FAMA for determination of

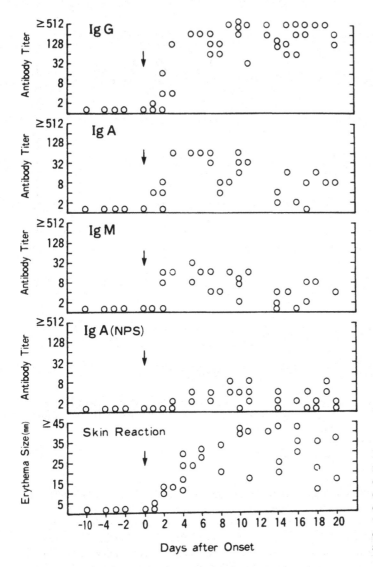

FIG. 6. Kinetics of the appearance of humoral (blood) and nasopharyngeal (NPS [secretory]) immunity as measured by fluorescent antibody to membrane antigen, and cellular immunity as assessed by skin test with varicella-zoster viral antigen. Arrows at day 0 indicate onset of rash. (From Baba et al. [1984a], with permission.)

humoral immune status. When humoral immunity declines (i.e., the number of IgG antibody molecules decreases), humoral antibody tends to become undetectable by IAHA or CF, probably because two molecules of IgG in close proximity at the particulate antigen are required to fix one molecule of the first component of complement (Borsos and Rapp, 1965).

ENZYME IMMUNE ASSAY

General

An enzyme-linked immunosorbent assay (ELISA) has been adapted for detection of antibody to VZV. Trays of cuvettes are coated with 0.25 ml of the viral antigens in 0.005 M PBS (pH 7.2) at dilutions previously determined to be optimal by titration with standard reference sera. The cuvettes are dried by a fan

at 37°C overnight. They are then packaged in airtight polyethylene bags containing silica gel desiccant and stored at 5°C until use.

Duplicate serum samples at a 1 : 100 dilution in PBS–Tween 20 containing 0.5% bovine serum albumin are incubated in the coated cuvettes (0.25 ml per cuvette) for 90 min at room temperature. After the cuvettes are washed, the F(ab)² fragment of goat anti-human IgG labeled with peroxidase, alkaline phosphatase, or β-galactosidase is added. Immunoglobulin labeled with β-galactosidase is most specific but expensive. The F(ab)² fragment is recommended because the reaction of some serum samples with control antigen is less with the labeled F(ab)² fragment than with the labeled IgG molecule. After incubation for 90 min at room temperature, the cuvettes are washed and the substrate (O-phenylenediamine, P-nitrophenyl phosphate, or O-nitrophenyl-β-D-galactopyranoside) is added. In samples, high-positive, low-positive and negative serum should be included. The optical density of the reaction with the control antigen should be subtracted from that with the viral antigen for VZV.

The sensitivity and specificity of the ELISA were compared with those of NT, IAHA, and CF (Forghani et al., 1978). Test sera showed little non-specific reactivity in the ELISA system, and valid results could usually be obtained at serum dilutions as low as 1:8. Demonstration of the presence or absence of VZV antibody by ELISA showed 94% correlation with results obtained in NT tests, but ELISA titers were 2- to 16-fold higher than NT antibody titers. Results by IAHA showed 87% correlation with those obtained by NT. No false-positive IAHA results were seen, but a number of false-negative IAHA results were seen at 1:8 serum dilution, particularly in older individuals. It was also shown that with increasing age (more than 40 years), and presumably increased time from varicella infection, NT antibody levels generally decline to 1:8 or 1:16, ELISA levels fall to 1:128 or 1:256, and IAHA and CF antibody titers are usually 1:8 or less. All tests demonstrated rises of heterotypic varicella-zoster antibody titer in selected patients with initial HSV infection, but fewer heterotypic responses were seen by ELISA than by the other methods.

Gershon et al. (1981) compared ELISA for antibody to VZV with FAMA and found that the two tests produced comparable results, although FAMA was slightly more sensitive. Shanley et al. (1982) reported that the two tests are similar in both sensitivity and specificity and that the absence of antibody detected by ELISA correlated with susceptibility to VZV infection.

Cremer et al. (1985) compared results obtained with an EIA performed at one serum dilution with those of indirect immunofluorescence and NT. When results of NT were taken as a true indicator of immunity, predictive values of 93.3% for a positive result and 100% for a negative result by EIA were presented. By anticomplement immunofluorescence testing, the predictive value of a positive result was 97.7% and that of a negative result was 88.5%.

Shehab and Brunnell (1983) reported that the results of EIA correlated well with those of FAMA. No positive reaction was found in EIA for three sera from children susceptible to the virus, whereas a nonspecific reaction was found by FAMA in one of the three sera. The results of EIA clearly distinguished immune from susceptible individuals. Heterologous reactions to herpes simplex virus were found in one of seven sera and to cytomegalovirus in two of five sera. However, EIA is prompt, sensitive, and reproducible and can be used for routine screening for susceptibility to varicella.

Competitive Enzyme Immunoassay

The competitive EIA test was compared with competitive radioimmunoassay (RIA), CF, and indirect fluorescent-antibody testing for the detection of antibody to VZV (Wreghitt et al., 1984). In the competitive EIA, anti-VZV IgG horseradish peroxidase conjugate was used instead of caprine antiserum to human IgG conjugated with alkaline phosphatase. Competitive EIA and RIA were the most sensitive tests, being 10-fold more sensitive than indirect immunofluorescence and 20 times as sensitive as CF.

Enzyme Immunoassay for Antibodies to Membrane-Associated Antigen of VZV

An in situ EIA to viral membrane antigen has been developed (Cox et al., 1984). In this method, VZV membrane antigen is fixed with glutaraldehyde in the wells, in which EIA is performed subsequently. The results of this assay correlate well with those of FAMA, and both were far more specific than CF.

Assay of Virus-Specific Immunoglobulin Subclasses

Sundqvist et al. (1984) reported that IgG3-specific antiviral antibodies were predominant in varicella, whereas IgG1 was the dominant subclass in zoster. In assaying subclass immunoglobulins, microplates coated with antigen are incubated with patient serum. After being washed, the subclass-specific monoclonal antisera are added and incubated. The plates are washed, and 100 μl of peroxidase-conjugated rabbit anti-mouse IgG is added. After incubation, the plates are washed and the substrate orthophenyldiaminine is added. After 60 min at room temperature, the reaction is terminated with 2 M H_2SO_4 and the absorbance at 450 nm is measured.

Antibody Class Capture Assay

Antibody class capture assay was developed by Forghani et al. (1984). Cups in microtiter plates are sensitized by the addition of goat antibodies specific for the heavy chains of human IgM, IgA, or IgG. After overnight incubation at room temperature, the contents of the wells are aspirated and the wells are washed. Serum dilutions are prepared in the plates. After incubation for 1 h at 37°C, the plates are washed and an optimal dilution of viral antigen is added. After overnight incubation at room temperature, the plates are washed and the optimal dilution of the VZV monoclonal antibody pool is added. Plates are incubated for 2 h at 37°C, then washed, and alkaline phosphatase-labeled goat anti-mouse serum is added and incubated for 1 h. The plates are washed and the fluorogenic substrate is added.

VZV-specific IgA and IgG antibody titers demonstrable by antibody class capture assay are comparable with those measured by solid-phase indirect enzyme immunofluorescence assay. However, antibody class capture assay is totally insensitive to VZV IgG antibody in individuals with past infection, and thus would not be suitable for determination of immune status to VZV.

Significance of VZV-Specific IgA in the Early Diagnosis

Wittek et al. (1983) found that primary infection with VZV was associated with early production of IgA antibodies. Of 36 subjects with varicella tested 1 to 5 days after onset, 22 had detectable IgA, and all of the negative sera were obtained before day 3 of the varicella exanthem. In addition, VZV IgA was found in only 3 of 23 sera from adults who had varicella more than 20 years before.

Subsequently, Tovi et al. (1985) reported that VZV-specific IgG and IgA antibodies were determined by immunoperoxidase assay in 90 serial serum samples of 26 patients with cephalic zoster. VZV-specific IgG antibodies were found in the first serum sample from all patients who showed fourfold rises or decreases in VZV-specific IgG antibody during the illness. In 17 (80.9%) of 21 patients from whom a serum sample was available as early as days 1 to 5 after onset of illness, VZV-specific IgA antibodies were detected. The second serum sample, taken during the convalescent period, was positive for VZV-specific IgA in all patients and showed a marked rise in titer. Although all of the healthy controls matched by sex and age had VZV-specific IgG antibodies, VZV-specific IgA antibodies were detected in low titer (1:2) in only 10.4% (3 of 26) of subjects.

By using a VZV-specific IgA antibody titer of more than 2 and more than 4 immunoperoxidase assay as a "cutoff" for young and adult patients with zoster, respectively, one can obtain an early diagnosis of VZV infection in 83% of the young and 50% of the adult patients with a single serum determination of VZV-specific IgA antibody titer. Accordingly, determination of VZV-specific antibodies by immunoperoxidase assay could be a helpful adjunct in early diagnosis of patients with zoster.

RADIOIMMUNOASSAY

General

A sensitive solid-phase radioimmunoassay for detection of antibodies to VZV has been described (Arvin and Koropchak, 1980; Friedman et al., 1979). VZV antigen (0.025 ml) or uninfected control antigen (0.025 ml) is added to each well; the plate is allowed to dry overnight at room temperature and the sera are added. The plates are incubated at 37°C for 1 h and washed with PBS containing 0.05% Tween 20.

^{125}I-labeled anti-human IgG or IgM is added to each well and the plates are incubated for 1 h at 37°C. The plates are then rinsed and allowed to dry. The wells are separated by passing a "hot wire" below the surface of the plate, transferred to counting vials, and counted in a gamma scintillation counter.

Rapid Radioimmunoassay Using ^{125}I-Labeled Staphylococcal Protein A for Antibody to VZV

A rapid radioimmunoassay using ^{125}I-labeled staphylococcal protein A for antibody VZV has also been developed (Richman et al., 1981).

The principle of the method is to determine the amount of ^{125}I-labeled staphylococcal protein A that bound to VZV antibody combined with VZV antigen; thus, the antibody titer can be assayed. The serum to be tested is serially diluted and each dilution in the well is incubated with target VZV-infected cells. After remaining unbound antibody is removed, ^{125}I-labeled staphylococcal protein A is added to each well. After incubation, unbound protein A is washed off and the amount of bound protein A is determined. As the nonspecific binding of radiolabeled protein appears somewhat large, particular caution may be required technically in radioimmunoassay.

DIAGNOSIS OF VZV INFECTION BY DETECTION OF VIRAL DEOXYTHYMIDINE KINASE IN SERUM AND VESICLE FLUID

A sensitive enzyme assay utilizing [^{125}I]iododeoxyuridine as the substrate in combination with isozyme-specific antisera was used for direct detection and typing of herpesvirus deoxythymidine kinase in clinical specimens (Kallander et al., 1983). The VZV deoxythymidine kinase specific antisera used were human convalescent sera.

An investigation of 16 coded vesicle-fluid specimens, taken in connection with VZV and HSV infections, revealed viral deoxythymidine kinase activity in 14 samples. All positive samples except one were taken within 5 days after the onset of illness. Serologic typing of the deoxythymidine kinase activities easily established whether the vesicles were caused by VZV, HSV-1, or HSV-2. The results were obtainable within 5 h and were in agreement with those of immunofluorescence tests or viral isolation when positive.

Diagnosis of Susceptibility of Individuals to VZV Infection by VZV Skin Test

A varicella skin test has been developed to assess susceptibility to VZV (Kamiya et al., 1977). A varicella-zoster skin antigen was prepared by harvesting infected human diploid cells, washing with PBS, sonication, and centrifugation at 3,000 rpm for 20 min. The supernatant was used as skin antigen after heating at 56°C for 30 min, which destroyed infectivity but not antigenicity. A volume of 0.1 ml of viral or control antigen was injected intracutaneously in the forearm. The reaction was read, usually after 48 h, by measuring the diameters of erythematous changes. An improved skin antigen prepared from fluid from VZV-infected culture, containing a minimal amount of cellular component, was later developed (Asano et al., 1981).

The skin test was successfully applied prospectively to identify the susceptible children exposed to a varicella patient in an institution for children (Baba et al., 1978; Kamiya et al., 1977). The results of the varicella skin test correlate closely with susceptibility to clinical varicella. In animal experiments using guinea pigs, it was demonstrated that there was no cross-reaction between VZV and HSV-1 in the skin test (Shiraki et al., 1984).

Recently, the skin test was further evaluated by LaRussa et al. (1985) in 16 adults known to be immune or susceptible to varicella and in 109 adults with no history of varicella. The skin test was well tolerated, compared favorably with established methods of determining immunity to varicella (i.e., FAMA, ELISA, and lymphocyte proliferation assay), and accurately predicted which subjects would develop clinical varicella after close exposure. Similar results were reported by Florman et al. (1985). Thus, the skin test with VZV antigen is a convenient method for assessing the immune status of individuals to VZV.

Interpretation of Laboratory Data

The definitive diagnostic procedure for identification of VZV infection is isolation of the etiologic agent. Punch biopsies, looking for intranuclear inclusions and multinuclear giant cells, provide earlier and more reliable material for examination, but this method cannot differentiate VZV infection from HSV infection.

Detection of VZV-specific antigen in biopsy material by immunfluorescence is reliable. For this purpose, antiserum specific for VZV and not cross-reactive with HSV should be used. Usually human anti-VZV antiserum, anti-VZV sera from monkeys and guinea pigs are cross-reactive to HSV in various serologic tests except NT. However, their antibody titers are much higher to homologous VZV than to heterologous HSV. Therefore, antiserum should be diluted until it is no longer cross-reactive to HSV before it is used to detect VZV antigen.

As described in the section Antigenic Composition and Genetics, an antigenic relationship between VZV and HSV has been demonstrated. Heterotypic booster immune responses were occasionally demonstrated in CF, NT, IAHA, FAMA, and EIA, but such a heterotypic response is less pronounced in EIA than in the other tests (Forghani et al., 1978) and is limited to the IgG response (Schmidt and Gallo, 1984). Heterotypic IgG antibody responses in HSV and VZV infections have generally been demonstrated only in individuals with serologic evidence of a previous infection with the viral heterotype (Gallo and Schmidt, 1981; Schmidt, 1982; Schmidt et al., 1969). Therefore, demonstration of VZV antibody in single serum tested by any serologic method indicates the presence of immunity to VZV in the subject from whom the serum was taken. When paired sera show a rise in antibody titer to VZV, clinical observation should be included in the diagnosis. If a greater than fourfold rise in antibody titer is observed to VZV despite no apparent clinical symptoms of chickenpox, or zoster, or herpes simplex, IgM antibody should be measured by using sensitive and specific radioimmunoassay (Arvin and Koropchak, 1980) or capture enzyme immunoassay (Forghani et al., 1984). The specificity of IgM antibody responses to VZV was evidenced by the fact that IgM antibody to VZV was not demonstrated in primary or secondary HSV infections where heterotypic rises in VZV IgG antibody occurred (Schmidt and Gallo, 1984). Therefore, if a rise in IgM antibody to VZV is detected for paired sera, VZV infection, not HSV infection, is strongly suspected even without any clinical symptom.

Epidemiology and Natural History

Varicella is a highly communicable disease, primarily of childhood, with secondary infection rates in susceptible household contacts approaching 90 to 100%. Since virtually everyone acquires varicella by adulthood at steady-state conditions, the actual number of cases that occurs annually should approximate the birth number.

In industrial societies in temperate regions, varicella is endemic in the total population, but epidemic in susceptible clustered subgroups, and exhibits a characteristic seasonal fluctuation in incidence. Annual incidence is lowest in September and peaks in

March and April. Considering that this disease is mainly endemic in schools for younger children, summer vacation may account for the low incidence during the summer.

A seroepidemiologic survey of varicella made in the Nagoya in Japan revealed that, of 1,473 recorded cases of varicella, 81.4% were in children less than 6 years old and 9.6% in those less than 1 year old. Of the 168 recorded cases in children less than 1 year old, about 30% were less than 5 months old. Serologically transferred maternal antibody decreases rapidly, becoming undetectable in infants 4 to 6 months old. Then, with increased age, the percentage of seropositive children gradually increases, reaching 53% at 4 to 5 years old and almost 100% in those more than 9 years old (Ozaki et al., 1980). In agreement with this observation, 90% of the reported cases in Massachusetts during 1952 to 1961 were in children less than 10 years old, and the maximum observed incidence was at the age of 6 years (Gordon, 1962). In tropical climates, chickenpox has been observed to occur more frequently in older age groups and a high rate has been documented in immigrant nurses in England and the United States (Hastie, 1980).

Herpes zoster is also infectious, although analysis of household contacts suggests that the risk is lower than that of varicella. It is noteworthy that in a semiclosed small community in downtown Osaka, 4 of 13 outbreaks of varicella were initiated by patients with zoster (Baba et al., 1984a). In that community, the occurrence of varicella outbreaks among children younger than 6 months was not uncommon. The observation that maternal varicella antibody generally disappears before the child is 6 months old and that the preexisting immunity does not give complete protection against clinical illness (Baba et al., 1982) may explain why infants younger than 6 months old develop varicella postnatally more frequently than measles. Furthermore, it has been observed that such infants are at increased risk of severe varicella. Of 53 infants younger than 12 months of age exposed during domiciliary institutional outbreaks, 20 (38%) contracted severe varicella, defined as more than 400 vesicles or a confluent rash (Baba et al., 1978).

Prevention and Control

Chemotherapy

Many antiviral drugs used to treat herpes group viruses have analogues of natural purine and pyrimidine nucleosides. These compounds are incorporated into viral DNA or phosphorylated products inhibiting virus-specified DNA synthesizing enzyme, thereby blocking new viral DNA synthesis. Cytosine arabinoside and adenine arabinoside have been tested in patients with zoster, and adenine arabinoside was shown effective in clearance of virus from vesicle, cessation of new vesicle formation, and shortening time to pustulation (Whitley et al., 1976). However, some side effects have also been reported.

Acyclovir (acycloguanosine) has a unique specificity against viruses of the herpes group. It is phosphorylated by virus-induced thymidine kinase, and the final product, acycloguanosine triphosphate, is a competitive inhibitor of the virus-induced DNA polymerase. The compound thus formed is 10 to 30 times more active against viral than cellular DNA polymerase; therefore, this drug is remarkably nontoxic to cells. It may also act as a DNA chain terminator (Biron and Elion, 1980).

In clinical use, acyclovir is reported to have significantly improved the rate of healing of skin lesions and shortened the period of pain in the acute phase of zoster (Peterslund et al., 1981). No adverse effects due to acyclovir have been observed in the kidney, liver, or hematopoietic function. Thus, acyclovir has been used widely throughout the world.

Recently bromovinyldeoxyuridine, fluoroiodoaracytosine, and fluoromethylarauracil have been developed and reported to be effective against varicella or zoster. Bromovinyldeoxyuridine is particularly promising because it is far more (1,000 times) effective against VZV than acyclovir in vitro (Machida et al., 1982). Results of double-blind tests are awaited.

Passive Immunization

Passive immunization has been attempted with immune serum globulin, usually administered to persons who were closely exposed to VZV. Results have been inconsistent, probably owing to variation in varicella-zoster antibody titer in different batches of immune serum globulin (Gershon, 1980). However, varicella-zoster immune globulin, prepared from plasma with a high antibody titer to VZV, has been found effective in modifying varicella in high-risk varicella-susceptible persons (Gershon et al., 1974; Zaia et al., 1983).

The clinical attack rate of varicella among recipients of varicella-zoster immune globulin has varied from about 20 to 65%, but illness has usually been extremely mild when the immune globulin has been administered within 72 h after contact. Moreover, subclinical varicella infection has been found in some of the recipients of varicella-zoster immune globulin. Thus, varicella-zoster immune globulin appears to be effective for prophylaxis against varicella. A serious problem, however, has been associated with this treatment. Because of the unpredictable nature of nosocomial infection with varicella, there are some children whose exposure to varicella is not recognized and therefore not protected. In practice, in cases of outbreaks in hospitals or institutions, the spread of varicella is usually prolonged and difficulties are encountered in deciding the proper time to give passive immunization to those susceptible.

Active Immunization

A live varicella vaccine has been developed recently (Takahashi et al., 1974, 1975, 1985a,b). The Oka strain of VZV was serially passaged 11 times in human embryonic lung cells at 34°C and 12 times in guinea pig embryo cells at 37°C, then passaged in human diploid cells (WI-38 and MRC-5 cells) several times to prepare the vaccine. The vaccine virus was found to be slightly temperature sensitive at 39°C and showed higher susceptibility to cultured guinea pig cells than wild-type viruses. In the cleavage profile of viral DNA, a distinctive difference was noticed between Oka strains and other wild-type strains (Hayakawa et al., 1984, 1986; Martin et al., 1982).

The results of extensive clinical use of the Oka strain live varicella vaccine for 10 years (1974 to 1984) in Japan (Takahashi et al., 1985b) indicate that children with malignancies can be immunized safely and effectively as long as their disease is in remission, show a normal cell-mediated immunity as assessed by phytohemagglutination or other reagents, and whose anticancer therapy, except for 6-mercaptopurine, is suspended for 2 weeks. An immune response was observed in a high percentage of the vaccinees. Some vaccinees acquired exogenous varicella infection, but symptoms were usually mild. In high-risk children other than those with malignancies, such as children with nephrosis, autoimmune diseases, and bronchial asthma receiving immunosuppressive drugs, vaccination resulted in few clinical reactions with a good immune response when vaccination was conducted with the same or fewer precautions than those outlined for leukemic patients. In this group, the decrease in immunity is less and the protective effect is greater than in those with malignancies.

For non-high-risk children with disease, the precautions necessary for vaccination of high-risk children are not needed, although children with high fever or extreme malnutrition must be excluded. After vaccination, clinical reactions are minimal and the immune response is good. The rate of exogenous varicella infection is low with mild clinical symptoms.

On vaccination of normal children, no or few clinical reactions, if any, are observed. In accordance

with a high immune response (greater than 95%), exogenous varicella infection was low (less than 5%). It has been shown that this vaccine, if administered within 72 h of exposure, apparently is able to prevent clinical varicella in normal susceptible children (Asano et al., 1977, 1982b). Presumably, VZV-specific cell-mediated or humoral responses develop sufficiently quickly to abort naturally acquired infection (Takahashi and Baba, 1984). Therefore, this vaccine is effective for postexposure as well as for preexposure prophylaxis, and long-term (7 to 10 years) protective immunity was proved in the follow-up study of vaccine recipients (Asano et al., 1985).

In the United States and Europe, results similar to those seen in Japan have been obtained with Oka strain varicella vaccine (André, 1985; Arbeter et al., 1982; Brunell et al., 1982; Gershon et al., 1984; Weibel et al., 1984). At the end of 1984, this vaccine was approved in several European countries, and in 1986 it was approved in Japan for administration to high-risk children and others.

Literature Cited

Almeida, J., A. Howatson, and M. G. Williams. 1962. Morphology of varicella virus. Virology **16:**353–355.

André, F. E. 1985. Worldwide experience with the Oka-strain live varicella vaccine. Postgrad. Med. J. **61**(suppl. 4):113–120.

Arbeter, A. M., S. E. Starr, R. E. Weibel, and S. A. Plotkin. 1982. Live attenuated varicella vaccine: immunization of healthy children with the OKA strain. J. Pediatr. **100:**886–893.

Arvin, A. M., and C. M. Koropchak. 1980. Immunoglobulin M and G to varicella-zoster virus measured by solid-phase radioimmunoassay: antibody response to varicella and herpes zoster infections. J. Clin. Microbiol. **12:**367–374.

Asano, Y., P. Albrecht, S. Stagno, and M. Takahashi. 1982a. Potentiation of varicella-zoster virus neutralization by anti-immunoglobulin. J. Infect. Dis. **146:**524–529.

Asano, Y., S. Hirose, S. Iwayama, T. Miyata, T. Yazaki, and M. Takahashi. 1982b. Protective effect of immediate inoculation of a live varicella vaccine in household contacts in relation to the viral dose and interval between exposure and vaccination. Biken J. **25:**43–45.

Asano, Y., N. Itakura, Y. Hiroishi, S. Hirose, T. Nagai, T. Ozaki, T. Yazaki, K. Yamanishi, and M. Takahashi. 1985a. Viremia is present in incubation period in nonimmunocompromised children with varicella. J. Pediatr. **106:**69–71.

Asano, Y., N. Itakura, Y. Hiroishi, S. Hirose, T. Ozaki, K. Kuno, T. Nagai, T. Yazaki, K. Yamanishi, and M. Takahashi. 1985b. Viral replication and immunologic response in children naturally infected with varicella-zoster virus and in varicella vaccine recipients. J. Infect. Dis. **152:**862–868.

Asano, Y., S. Iwayama, T. Miyata, T. Ozaki, K. Tsuzuki, S. Ito, and M. Takahashi. 1980. Spread of varicella in hospitalized children having no direct contact with an indication zoster case and its prevention by a live vaccine. Biken J. **23**:157–161.

Asano, Y., T. Nagai, T. Miyata, T. Yazaki, S. Ito, K. Yamanishi, and M. Takahashi. 1985. Long-term protective immunity of recipients of the Oka strain of live varicella vaccine. Pediatrics **75**:667–671.

Asano, Y., H. Nakayama, T. Yazaki, R. Kato, S. Hirose, K. Tsuzuki, S. Ito, S. Isomura, and M. Takahashi. 1977. Protection against varicella in family contacts by immediate inoculation with live vaccine. Pediatrics **59**:3–7.

Asano, Y., K. Shiraki, M. Takahashi, T. Nagai, T. Ozaki, and T. Yazaki. 1981. Soluble skin test antigen of varicella-zoster virus prepared from the fluid of infected cultures. J. Infect. Dis. **143**:684–692.

Asano, Y., and M. Takahashi. 1978. Studies on neutralization of varicella-zoster virus and serological follow-up of cases of varicella and zoster. Biken J. **21**:15–23.

Baba, K., H. Yabuuchi, H. Okuni, and M. Takahashi. 1978. Studies with live varicella vaccine and inactivated skin test antigen: Protective effect of the vaccine and clinical application of the skin test. Pediatrics **61**:550–555.

Baba, K., H. Yabuuchi, M. Takahashi, A. A. Gershon, and P. L. Ogra. 1984a. Seroepidemiologic behavior of varicella zoster virus infection in a semiclosed community after introduction of VZV vaccine. J. Pediatr. **105**:712–716.

Baba, K., H. Yabuuchi, M. Takahashi, and P. Ogra. 1982. Immunologic and epidemiologic aspects of varicella infection acquired during infancy and early childhood. J. Pediatr. **100**:881–885.

Baba, K., M. Yoshida, A. Tawa, H. Yabuuchi, K. Maeda, and M. Takahashi. 1984b. A simplified immunofluorescence technique for antibody to varicella-zoster membrane antigen (FAMA). Biken J. **27**:23–29.

Barr, R. J., R. J. Herten, and J. H. Graham. 1977. Rapid method for Tzanck preparation. J. Am. Med. Assoc. **237**:1119–1120.

Biron, K. K., and G. B. Elion. 1980. In vitro susceptibility of varicella-zoster virus to acyclovir. Antimicrob. Agents Chemother. **18**:443–447.

Blank, H., C. Burgeon, C. D. Baldridge, P. L. McCarthy, and F. Urbach. 1951. Cytologic smears in diagnosis of herpes simplex, herpes zoster and varicella. J. Am. Med. Assoc. **146**:1410–1912.

Borsos, T., and H. J. Rapp. 1965. Complement fixation on cell surface by 19S and 7S antibodies. Science **150**:505–506.

Brunell, P. A. 1967. Separation of infectious varicella-zoster virus from human embryonic lung fibroblasts. Virology **31**:732–734.

Brunnell, P. A., and H. L. Casey. 1964. Crude tissue culture antigen for determination of varicella-zoster complement fixing antibody. Public Health Rep. **79**:839.

Brunell, P. A., B. H. Cohen, and M. Granat. 1971. A gel-precipitation test for the diagnosis of varicella. Bull. W.H.O. **44**:811–814.

Brunell, P. A., A. A. Gershon, S. A. Uduman, and S. Steinberg. 1975. Varicella-zoster immunoglobulin during varicella, latency and zoster. J. Infect. Dis. **132**:49–54.

Brunell, P. A., Z. Shehab, C. Geiser, and J. E. Waugh. 1982. Administration of live varicella vaccine to children with leukemia. Lancet **2**:1069–1072.

Bruusgaard, E. 1932. The mutual relation between zoster and varicella. Br. J. Dermatol. Syph. **44**:1–24.

Burke, B. L., R. W. Steel, O. W. Beard, J. S. Wood, T. D. Chain, and D. J. Marmer. 1982. Immune responses to varicella-zoster in the aged. Arch. Intern. Med. **142**:291–293.

Caunt, A. E., and D. G. Shaw. 1969. Neutralization tests with varicella-zoster virus. J. Hyg. **67**:343–352.

Cheatham, W. J., T. H. Weller, T. F. Dolan, and J. C. Dower. 1956. Varicella: report of two fatal cases with necropsy, virus isolation and serologic studies. Am. J. Pathol. **32**:1015–1035.

Cox, J. C., M. B. Moloney, R. W. Herrington, A. W. Hampson, and J. G. Hurrell. 1984. Enzyme immunoassay for antibodies to membrane associated antigen of varicella zoster virus. J. Virol. Methods **8**:137–145.

Cremer, N. E., C. K. Cossen, G. Shell, J. Diggs, D. Gallo, and N. J. Schmidt. 1984. Enzyme immunoassay versus plaque neutralization and other methods for determination of immune status to measles and varicella-zoster viruses and versus complement fixation for serodiagnosis of infection with those viruses. J. Clin. Microbiol. **21**:869–874.

Davison, A. J., C. M. Edson, R. W. Ellis, B. Forghani, D. Gilden, C. Grose, P. M. Keller, A. Vafai, Z. Wroblewska, and K. Yamanishi. 1986. New common nomenclature for glycoprotein genes of varicella-zoster virus and their glycosylated products. J. Virol. **57**:1195–1197.

Davison, A. J., and J. E. Scott. 1986. The complete DNA sequence of varicella-zoster virus. J. Gen. Virol. **67**:1759–1816.

Dishon, T., and S. C. Morgensen. 1983. Rapid and direct detection of herpes simplex virus and varicella-zoster virus antigen in clinical specimens by staphylococcal reagent and membrane filtration. Eur. J. Clin. Microbiol. **2**:581–587.

Dobersen, M. J., M. Jerkofsky, and S. Greer. 1976. Enzymatic basis for the selective inhibition of varicella-zoster virus by 5-halogenated analogues of deoxycytidine. J. Virol. **20**:478–486.

Dumas, A. H., J. L. M. C. Geelen, N. W. Weststrate, P. Werthein, and J. van der Noordaa. 1981. XbaI, Pst I and BglII restriction enzyme maps of the two orientations of the varicella-zoster virus genome. J. Virol. **39**:390–400.

Ecker, J. R., and R. W. Hyman. 1982. Varicella zoster virus DNA exists as two isomers. Proc. Natl. Acad. Sci. USA **79**:156–160.

Ehrlich, R. M., J. A. P. Turner, and M. Clarke. 1958. Neonatal varicella case report with isolation of the virus. J. Pediatr. **53**:139–147.

Elion, G. B., P. A. Furman, J. A. Fyfe, P. de Miranda, L. Beauchamp, and H. J. Schaeffer. 1977. Selectivity of action of an antiherpetic agent, 9-(2-hydroxy-ethoxymethyl) guanine. Proc. Natl. Acad. Sci. USA **74**:5716–5720.

Esiri, M. M., and A. H. Tomlinson. 1972. Demonstration of virus in trigeminal nerve and ganglion by immunofluorescence and electronmicroscopy. J. Neurol. Sci. **15**:35–48.

Evans, P. 1940. An epidemic of chickenpox. Lancet **2**:239–340.

Feldman, S., and E. Epp. 1979. Detection of viremia during incubation of varicella. J. Pediatr. **94**:746–748.

Florman, A. L., E. T. Umland, D. Ballow, A. H. Cushing, L. C. McLaren, T. J. Gribble, and M. H. Duncan. 1985. Evaluation of a skin test for chickenpox. Infect. Control (Thorofare) **6**:314–316.

Forghani, B., C. K. Myoraku, K. W. Dupuis, and N. J. Schmidt. 1984. Antibody class captures assays for varicella-zoster virus. J. Clin. Microbiol. **19**:606–609.

Forghani, B., J. Schmidt, and J. Dennis. 1978. Antibody assays for varicella-zoster virus: Comparison of enzyme immunoassay with neutralization, immune adherence hemagglutination and complement fixation. J. Clin. Microbiol. **8**:545–552.

Frey, H. M., S. P. Steinberg, and A. A. Gershon. 1981. Rapid diagnosis of varicella-zoster virus infections by countercurrent immunoelectrophoresis. J. Infect. Dis. **143:**274–280.

Friedman, M. G., S. Leventon-Kriss, and I. Sarow. 1979. Sensitive solid-phase radioimmunoassay for detection of human immunoglobulin G antibodies to varicella-zoster virus. J. Clin. Microbiol. **9:**1–10.

Gallo, D., and N. J. Schmidt. 1981. Comparison of anticomplement immunofluorescence and fluorescent antibody-to-membrane antigen tests for determination of immunity status to varicella-zoster virus and for serodifferentiation of varicella-zoster and herpes simplex virus infections. J. Clin. Microbiol. **14:**539–543.

Gershon, A. A. 1975. Varicella in mother and infant: a problem old and new, p. 79–95. *In* S. Krugman and A. A. Gershon (ed.), Infections of the fetus and the newborn infant. Alan R. Liss, New York.

Gershon, A. A. 1980. Live attenuated varicella-zoster vaccine. Rev. Infect. Dis. **2:**393–407.

Gershon, A. A., H. M. Frey, S. P. Steinberg, M. D. Seeman, D. Bidwell, and A. Voller. 1981. Determination of immunity to varicella using an enzyme-linked-immunosorbent assay. Arch. Virol. **70:**169–172.

Gershon, A. A., Z. G. Kalter, and S. Steinberg. 1976. Detection of antibody to varicella-zoster virus by immune adherence hemagglutination. Proc. Soc. Exp. Biol. Med. **151:**762–765.

Gershon, A. A., and S. Krugman. 1975. Seroepidemiologic survey of varicella: value of specific fluorescent antibody test. Pediatrics **56:**1005–1008.

Gershon, A. A., A. Steinberg, and P. A. Brunell. 1974. Zoster immune globulin A: further assessment. N. Engl. J. Med. **290:**243–245.

Gershon, A. A., and S. P. Steinberg. 1981. Antibody responses to varicella-zoster virus and the role of antibody in host defense. Am. J. Med. Sci. **282:**12–17.

Gershon, A. A., S. P. Steinberg, L. Gelb, G. Galasso, W. Borkowsky, P. LaRussa, A. Ferrara, and the N.I.H. Varicella Vaccine Collaborative Study Group. 1984. Live attenuated varicella vaccine. Efficacy for children with leukemia in remission. J. Am. Med. Assoc. **252:**355–362.

Ghatak, N. R., and H. M. Zimmerman. 1973. Spinal ganglion in herpes zoster. Arch. Pathol. **95:**411–455.

Gilden, D. H., A. Vafai, V. Shtram, Y. Becker, M. Devlin, and M. Wellish. 1983. Varicella-zoster virus DNA in human sensory ganglia. Nature (London) **306:**478–480.

Gold, E., and G. Godek. 1965. Complement fixation studies with varicella-zoster antigen. J. Immunol. **95:**692–695.

Gordon, J. E. 1962. Chickenpox: An epidemiological review. Am. J. Med. Sci. **244:**362–389.

Gordon, J. E., and F. M. Meader. 1929. The period of infectivity and serum prevention of chickenpox. J. Am. Med. Assoc. **93:**2013–2015.

Grose, C. 1981. Variation on a theme by Fenner: The pathogenesis of chickenpox. Pediatrics **68:**735–737.

Grose, C. 1982. Varicella-zoster virus infections chickenpox (varicella) and shingles (zoster), p. 85–150. *In* R. Glaser and T. Gotlieb-Stematsky (ed.), Human herpesvirus infections. Clinical aspects. Marcel Dekker, Inc., New York.

Grose, C., B. J. Edmond, and P. A. Brunell. 1979. Complement-enhanced neutralizing antibody response to varicella-zoster virus. J. Infect. Dis. **139:**432–437.

Grose, C., W. E. Friedrichs, and K. O. Smith. 1981. Cryoprecipitation of varicella-zoster virus without loss of structural integrity or infectivity. Intervirology **15:**154–160.

Harbour, D. A., and A. E. Caunt. 1979. The serological relationship of varicella-zoster virus to other primate herpesviruses. J. Gen. Virol. **45:**469–477.

Hastie, I. R. 1980. Varicella-zoster affecting immigrant nurses. Lancet **2:**154–155.

Hata, S. 1980. Skin test with varicella-zoster virus antigen on herpes zoster patients. Arch. Dermatol. Res. **268:**65–70.

Hayakawa, Y., S. Torigoe, K. Shiraki, K. Yamanishi, and M. Takahashi. 1984. Biologic and biophysical markers of a live varicella vaccine strain (Oka): identification of clinical isolates from vaccine recipients. J. Infect. Dis. **149:**956–963.

Hayakawa, Y., T. Yamamoto, K. Yamanishi, and M. Takahashi. 1986. Analysis of varicella-zoster virus DNAs of clinical isolates by endonuclease HpaI. J. Gen. Virol. **67:**1817–1829.

Hondo, R., H. Shibuta, and M. Matsumoto. 1976. An improved plaque assay for varicella virus. Arch. Virol. **51:**355–359.

Hope-Simpson, R. E. 1952. Infectiousness of communicable diseases in the household (measles, chickenpox and mumps). Lancet **2:**549–554.

Hope-Simpson, R. E. 1965. The nature of herpes zoster: A long-term study and a new hypothesis. Proc. R. Soc. Med. **58:**9–20.

Hosler, A., R. A. Respess, D. J. Waters, and D. A. Thorley-Lawson. 1985. Cross-reactivity between herpes simplex virus glycoprotein B and a 63,000 dalton varicella-zoster virus envelope glycoprotein. J. Virol. **56:**333–336.

Howell, C. L., and M. J. Miller. 1983. Effect of sucrose phosphate and sorbitol on infectivity of enveloped viruses during storage. J. Clin. Microbiol. **18:**658–662.

Hyman, R. W., J. R. Ecker, and R. B. Tensor. 1983. Varicella-zoster virus RNA in human trigeminal ganglia. Lancet **2:**814–816.

Ito, M., and A. L. Barron. 1973. Surface antigens produced by herpesviruses: Varicella-zoster virus. Infect. Immun. **8:**48–52.

Kallander, C. F. R., J. S. Gronowitz, and E. Olding-Stenkvist. 1983. Rapid diagnosis of varicella-zoster virus infection by detection of viral deoxythymidine kinase in serum and vesicle fluid. J. Clin. Microbiol. **17:**280–287.

Kalter, Z. G., S. Steinberg, and A. A. Gershon. 1977. Immune adherence hemagglutination: further observation on demonstration of antibody to varicella-zoster virus. J. Infect. Dis. **135:**1010–1013.

Kamiya, H., T. Ihara, A. Hattori, T. Iwasa, M. Sakurai, T. Izawa, A. Yamada, and M. Takahashi. 1977. Diagnostic skin test reactions with varicella virus antigen and clinical application of the test. J. Infect. Dis. **136:**784–788.

Kaspenberg, J. G. 1964. Possible antigenic relationship between varicella zoster virus and herpes simplex virus. Arch. Gesamte Virusforsch. **15:**67–73.

Keller, P. M., B. J. Neff, and R. W. Ellis. 1984. Three major glycoprotein genes of varicella-zoster virus whose products have neutralization epitopes. J. Virol. **52:**293–297.

Kettering, J. D., N. J. Schmidt, D. Gallo, and E. H. Lennette. 1977. Anti-complement immunofluorescence test for antibodies to human cytomegalovirus. J. Clin. Microbiol. **6:**627–632.

Kitamura, K., J. Namazue, H. Campo-Vera, T. Ogino, and K. Yamanishi. 1986. Induction of neutralizing antibody against varicella-zoster virus (VZV) by VZV gp3 and cross reactivity between VZV gp3 and herpes simplex gB. Virology **149:**74–82.

Krugman, S., and S. L. Katz. 1981. Varicella-zoster infections, p. 486–506. *In* Infectious diseases of children. The C. V. Mosby Co., St. Louis.

Kundratitz, K. 1925. Experimentelle Übertragen von Herpes Zoster auf Menschen und die Beziehungen von Herpes Zoster zu Varicellen. Z. Kinderheikd. **39:**379–387.

LaRussa, P., S. P. Steinberg, M. D. Seeman, and A. A. Gershon. 1985. Determination of immunity of varicella-zoster virus by means of an intradermal skin test. J. Infect. Dis. **152:**869–875.

Leclair, J. M., J. A. Zaia, M. J. Levine, R. G. Congdon, and D. A. Goldman. 1980. Airborne transmission of chickenpox in a hospital. N. Engl. J. Med. **302:**450–453.

Levin, M. J., S. Leventhal, and H. A. Masters. 1984. Factors influencing quantitative isolation of varicella-zoster virus. J. Clin. Microbiol. **19:**880–883.

Lipschutz, B. 1929. Untersuchungen uber die Atiologie der Krankheiten der Herpes-gruppe (Herpes zoster, Herpes genitalis, Herpes febrilis). Arch. Dermatol. Syphilol. **136:**428–482.

Machida, H., A. Kuninaka, and H. Yoshino. 1982. Inhibitory effects of antiherpesviral thymidine analogs against varicella-zoster virus. Antimicrob. Agents Chemother. **21:**358–361.

Mar, E. C., Y. S. Huang, and E. S. Huang. 1978. Purification and characterization of varicella-zoster virus-induced DNA polymerase. J. Virol. **26:**249–256.

Martin, J. H., D. E. Dohner, W. J. Wellinghof, and L. D. Gelb. 1982. Restriction endonuclease analysis of varicella-zoster vaccine virus and wild type DNAs. J. Med. Virol. **9:**69–76.

Matsunaga, Y., K. Yamanishi, and M. Takahashi. 1982. Experimental infection and immune response of guinea pigs with varicella-zoster virus. Infect. Immun. **37:**407–412.

Meurisse, E. V. 1969. Laboratory studies on the varicella-zoster virus. J. Med. Microbiol. **2:**317–325.

Miliauskas, J. R., and B. L. Webber. 1984. Disseminated varicella at autopsy in children with cancer. Cancer **53:**1518–1525.

Miller, R. L., and F. Rapp. 1977. Varicella-zoster virus-induced DNA polymerase. J. Gen. Virol. **36:**515–524.

Mims, C. A., and D. O. White. 1984. Persistent infections, p. 200–253. *In* Viral pathogenesis and immunology. Blackwell Scientific Publications, Ltd., Oxford.

Myers, M. G. 1979. Viremia caused by varicella-zoster virus: association with malignant progressive varicella. J. Infect. Dis. **140:**229–233.

Myers, M. G., H. L. Duer, and C. K. Hausler. 1980. Experimental infection of guinea pigs with varicella-zoster virus. J. Infect. Dis. **142:**414–420.

Nagashima, K., M. Nakagawa, and H. Endo. 1975. Pathology of the human spinal ganglion in varicella-zoster virus infection. Acta Neuropathol. **33:**105–117.

Ogino, T., T. Otsuka, and M. Takahashi. 1977. Induction of deoxythymidine kinase activity in human embryonic lung cells infected with varicella-zoster virus. J. Virol. **21:**1232–1235.

Ozaki, T., T. Ichikawa, Y. Matsui, T. Nagai, Y. Asano, K. Yamanishi, and M. Takahashi. 1984. Viremic phase nonimmunocompromised children with varicella. J. Pediatr. **104:**85–87.

Ozaki, T., T. Nagai, T. Kimura, T. Ichikawa, S. Suzuki, H. Kito, and Y. Asano. 1980. The age distribution of neutralizing antibodies against varicella-zoster virus in healthy individuals. Biken J. **23:**9–14.

Peterslund, N. A., K. Seyer-Hansen, J. Ipsen, V. Esmann, H. Schonheyder, and H. Juhl. 1981. Acyclovir in herpes zoster. Lancet **2:**827–830.

Richman, D. D., P. H. Cleveland, M. N. Oxman, and J. A. Zaia. 1981. A rapid radioimunoassay using ^{125}I-labeled staphylococcal protein A for antibody to varicella-zoster virus. J. Infect. Dis. **143:**693–699.

Roizman, B., L. E. Carmichael, F. Deinhardt, G. de-The, A. J. Nahmias, W. Plowright, F. Rapp, P. Schledrick, M. Takahashi, and K. Wolf. 1981. Herpes viridae: definition, provisional nomenclature, and taxonomy. Intervirology **16:**201–217.

Russel, A. S., R. A. Maini, M. Baily, and D. C. Dumonde. 1973. Cell-mediated immunity to varicella-zoster antigen in acute herpes zoster (shingles). Clin. Exp. Immunol. **14:**181–185.

Sawyer, M. H., J. M. Ostrove, J. M. Felser, and S. E. Straus. 1986. Mapping of the varicella zoster virus deoxypyrimidine kinase gene and preliminary identification of its transcript. Virology **149:**1–9.

Schaap, C. J. P., and J. Huisman. 1968. Simultaneous rise in complement fixing antibodies against herpesvirus hominis and varicella-zoster virus in patient with chickenpox and shingles. Arch. Gesamte Virusforsch. **25:**52–57.

Schmidt, N. J. 1982. Further evidence for common antigens in herpes simplex and varicella-zoster viruses. J. Med. Virol. **9:**27–36.

Schmidt, N. J., and D. Gallo. 1984. Class-specific antibody responses to early and late antigens of varicella and herpes simplex viruses. J. Med. Virol. **13:**1–12.

Schmidt, N. J., D. Gallo, V. Delvin, J. D. Woodie, and R. W. Emmons. 1980. Direct immunofluorescence staining for detection of herpes simplex and varicella-zoster virus antigen in vesicular lesions and certain tissue specimens. J. Clin. Microbiol. **12:**651–655.

Schmidt, N. J., and E. H. Lennette. 1975. Neutralizing antibody response to varicella-zoster virus. Infect. Immun. **12:**606–613.

Schmidt, N. J., and E. H. Lennette. 1976. Improved yields of cell-free varicella-zoster virus. Infect. Immun. **14:**709–715.

Schmidt, N. J., E. H. Lennette, and R. L. Magoffin. 1969. Immunological relationship between herpes simplex and varicella-zoster viruses demonstrated by complement fixation, neutralization and fluorescent antibody tests. J. Gen. Virol. **4:**321–328.

Schmidt, N. J., E. H. Lennette, C. W. Shon, and T. K. Shinomono. 1964. A complement fixing antigen for varicella-zoster derived from infected cultures of human fetal diploid cells. Proc. Soc. Exp. Biol. Med. **116:**144–149.

Schmidt, N. J., E. H. Lennette, J. D. Woodie, and H. H. Ho. 1965. Immunofluorescent staining in the laboratory diagnosis of varicella virus infections. J. Lab. Clin. Med. **66:**403–412.

Seidlin, M., H. E. Takiff, H. A. Smith, J. Hay, and S. E. Straus. 1984. Detection of varicella-zoster virus by dot-blot hybridization using a molecularly cloned viral DNA probe. J. Med. Virol. **13:**53–61.

Shanley, J., M. Myers, B. Edmond, and R. Steele. 1982. Enzyme-linked immunosorbent assay for detection of antibody to varicella-zoster virus. J. Clin. Microbiol. **15:**208–211.

Shehab, Z., and P. A. Brunell. 1983. Enzyme-linked immunosorbent assay for susceptibility to varicella. J. Infect. Dis. **148:**472–476.

Shemer, Y., S. Leventon-Kriss, and I. Sarov. 1980. Isolation and polypeptides characterization of varicella-zoster virus. Virology **106:**133–140.

Shibuta, H., T. Ishikawa, R. Hondo, T. Aoyama, K. Kurata, and M. Matsumoto. 1974. Varicella virus isolation from spinal ganglion. Arch. Gesamte Virusforsch. **45:**382–385.

Shiraki, K., T. Okuno, K. Yamanishi, and M. Takahashi. 1982. Polypeptides of varicella zoster virus (VZV) and immunological relationship of VZV and herpes simplex virus (HSV). J. Gen. Virol. **61:**255–269.

Shiraki, K., K. Yamanishi, and M. Takahashi. 1984. Biologic and immunologic characterization of the soluble skin test antigen of varicella-zoster virus. J. Infect. Dis. **149**:501–504.

Smith, J. D., and E. de Harven. 1973. Herpes simplex virus and cytomegalo virus replication in WI-38 cells. I. Sequence of viral replication. J. Virol. **12**:919–930.

Straus, S. E., J. Owens, W. T. Ruyechan, H. E. Takiff, T. A. Casey, G. F. Vande Woude, and J. Hay. 1982. Molecular cloning and physical mapping of varicella-zoster virus DNA. Proc. Natl. Acad. Sci. USA **70**:993–997.

Straus, S. E., W. Reinhold, H. A. Smith, W. T. Ruyechan, D. K. Henderson, R. Michael Blaser, and J. Hay. 1984. Endonuclease analysis of viral DNA from varicella and subsequent zoster infections in the same patient. N. Engl. J. Med. **22**:1362–1364.

Sundqvist, V. A., A. Linde, and B. Wahren. 1984. Virus specific immunoglobulin G subclass in herpes simplex and varicella-zoster virus infections. J. Clin. Microbiol. **20**:94–98.

Svedmyr, A. 1965. Varicella virus in Hela cells. Arch. Gesamte Virusforsch. **17**:495–503.

Takahashi, M., and K. Baba. 1984. A live varicella vaccine: Its protective effect and immunologic aspect of varicella-zoster virus infection, p. 255–278. *In* L. de la Maza and E. M. Peterson (ed.), Medical virology III. Elsevier Science Publishing Co., Inc., New York.

Takahashi, M., Y. Hayakawa, K. Shiraki, K. Yamanishi, Y. Asano, and T. Ozaki. 1985a. Attenuation and laboratory markers of the Oka-strain varicella-zoster virus. Postgrad. Med. J. **61**(suppl. **4**):37–46.

Takahashi, M., H. Kamiya, K. Baba, Y. Asano, T. Ozaki, and K. Horiuchi. 1985b. Clinical experience with Oka live varicella vaccine in Japan. Postgrad. Med. J. **61**(suppl. **4**):61–67.

Takahashi, M., Y. Okuno, T. Otsuka, J. Osame, A. Takamizawa, T. Sasada, and T. Kubo. 1975. Development of a live attenuated varicella vaccine. Biken J. **18**:25–33.

Takahashi, M., T. Otsuka, Y. Okuno, Y. Asano, T. Yazaki, and S. Isomura. 1974. Live vaccine used to prevent the spread of varicella in children in hospital. Lancet **2**:1288–1290.

Tomlinson, T. H. 1939. Giant cell formation in the tonsils in the prodromal stage of chickenpox. Am. J. Pathol. **15**:523–526.

Tovi, F., T. Hadar, J. Sidi, B. Sarov, and I. Sarov. 1985. The significance of specific IgA antibodies in the serum in the early diagnosis of zoster. J. Infect. Dis. **152**:230.

Trlifajova, J., D. Bryndova, and M. Ryc. 1984. Isolation of varicella-zoster virus from pharyngeal and nasal swabs in varicella patients. J. Hyg. Epidemiol. Microbiol. Immunol. (Prague) **28**:201–206.

Uduman, S. A., A. A. Gershon, and P. A. Brunell. 1972. Rapid diagnosis of varicella zoster infections by agar-gel diffusion. J. Infect. Dis. **126**:193–195.

Von Bokay, J. 1909. Uber den atiologischen Zuzammenhang der varicellen mit gewissen Fallen von Herpes Zoster. Wien. Klin. Wochenschr. **22**:1323–1326.

Weibel, R. E., B. J. Neff, B. Kuter, H. A. Guess, C. A. Rothenberger, A. J. Fitzgerald, K. A. Conner, A. A. McLean, M. R. Hilleman, E. B. Buynak, and E. M. Scolnick. 1984. Live attenuated varicella virus vaccine

efficacy trial in healthy children. N. Engl. J. Med. **310**:1409–1415.

Weller, T. H. 1953. Serial propagation in vitro of agents producing inclusion bodies derived from varicella and herpes zoster. Proc. Soc. Exp. Biol. Med. **83**:340–346.

Weller, T. H. 1965. Varicella-herpes zoster virus, p. 915–925. *In* F. L. Horsfall and I. Tamm (ed.), Viral and rickettsial infections of man, 4th ed. J. B. Lippincott, Philadelphia.

Weller, T. H., and H. M. Witton. 1958. The etiologic agents of varicella and herpes zoster. Serologic studies with the viruses as propagated in vitro. J. Exp. Med. **108**:869–890.

Whitley, R. J., L. T. Chien, R. Dolin, G. J. Galasso, C. A. Alford, Jr., and Collaborative Study Group. 1976. Adenine arabinoside therapy of herpes zoster in the immunosuppressed. NIAID Collaborative Antiviral Study. N. Engl. J. Med. **294**:1193–1199.

Williams, V., A. A. Gershon, and P. A. Brunell. 1974. Serologic response to varicella-zoster membrane antigens measured by indirect immunofluorescence. J. Infect. Dis. **136**:669–672.

Wittek, A. E., A. M. Arvin, and C. M. Koropchak. 1983. Serum immunoglobulin A antibody to varicella-zoster virus in subjects with primary varicella and herpes zoster infections and immune subjects. J. Clin. Microbiol. **18**:1146–1149.

Wreghitt, T. G., R. S. Tedder, J. Nagington, and R. B. Ferns. 1984. Antibody assay for varicella-zoster virus: Comparison of competitive enzyme-linked immunosorbent assay (ELISA), competitive radioimmunoassay (RIA), complement fixation, and indirect immunofluorescence assays. J. Med. Virol. **13**:361–370.

Yamada, A., S. Ogino, Y. Asano, T. Otsuka, M. Takahashi, K. Baba, and H. Yabuuchim. 1979. Comparison of 4 serological test-complement fixation, neutralization, fluorescent antibody to membrane antigen and immune adherence hemagglutination for assay of antibody to varicella-zoster (V-Z) virus. Biken J. **22**:55–60.

Yamanishi, K., Y. Matsunaga, T. Ogino, M. Takahashi, and A. Takamizawa. 1980. Virus replication and localization of varicella-zoster virus antigens in human embryonic fibroblast cells infected with cell-free virus. Infect. Immun. **28**:536–541.

Zaia, J. A., M. J. Levin, S. R. Preblud, J. Leszczynski, G. C. Wright, R. J. Ellis, A. C. Curtis, M. A. Valerino, and J. LeCore. 1983. Evaluation of varicella-zoster immune globulin: protection of immunosuppressed children after household exposure to varicella. J. Infect. Dis. **147**:737–743.

Zaia, J. A., and M. N. Oxman. 1977. Antibody to varicella-zoster virus-induced membrane antigen: immunofluorescence assay using monodisperse glutaraldehyde-fixed target cells. J. Infect. Dis. **136**:519–530.

Ziegler, T. 1984. Detection of varicella-zoster viral antigens in clinical specimens by solid-phase enzyme immunoassay. J. Infect. Dis. **150**:149–154.

Ziegler, T., and P. E. Halonen. 1985. Rapid detection of herpes simplex and varicella-zoster virus antigens from clinical specimens by enzyme immunoassay. Antiviral Res. Suppl. **1**:107–110.

Zeigler, T., O. H. Meurman, P. P. Arstila, and P. E. Halonen. 1983. Solid-phase enzyme-immunoassay for the detection of herpes simplex virus antigens in clinical specimens. J. Virol. Methods **7**:1–9.

CHAPTER 15

Adenoviridae: The Adenoviruses

G. WADELL

Disease and Clinical Manifestations: Acute respiratory diseases, pharyngoconjunctival fever, infections of the eye, acute cystitis, gastroenteritis, meningitis.

Etiologic Agents: Adenovirus, 42 human serotypes; several serotypes in animals.

Pathology: Epithelial cells and lymphatic tissue in respiratory tract, gastrointestinal tract, eye, and urinary bladder are most commonly affected.

Laboratory Diagnosis: Virus isolation in cell cultures, virion detection by electron microscopy; antigen detection by immunofluorescence, enzyme immunoassays, and latex agglutination; nucleic acid detection by DNA hybridization; serology in paired serum specimens by complement fixation tests and enzyme immunoassays. Virus isolation, antigen detection, and serology are in daily use in diagnostic laboratories.

Epidemiology: Worldwide; different serotypes cause disease in different age groups and show distinctly different organ tropism; severity of disease varies in different geographic areas; certain serotypes prevalent in army training centers.

Treatment: No antiviral chemotherapeutic agent in clinical use.

Prevention and Control: Oral vaccine is used in the U.S. Army; no vaccine available for the civilian population; the serotypes causing childhood diseases have not been attenuated.

In an attempt to evaluate adenoids as a tissue for growth of polio viruses, Rowe et al. (1953) detected adenoviruses. Two-thirds of their cultures showed a slowly progressive cytopathic effect. The agents isolated from these cultures were consequently designated the adenoid-degenerating (AD) agents. Adenoviruses were also soon established as etiologic agents of acute respiratory disease by Hilleman and Werner (1954). Adenoviruses have been isolated from every species of placental mammals, marsupials, birds, and amphibians studied.

In industrialized countries, adenoviruses are characterized as causing acute febrile pharyngitis in young children, adenoconjunctival fever in school children, and pneumonia in infants and military recruits who sleep in dormatories. Adenoviruses have been described as the cause of epidemic follicular conjunctivitis and epidemic keratoconjunctivitis (Jawetz, 1959), acute hemorrhagic cystitis (Numazaki et al., 1968), and diarrhea in infants (Uhnoo et al., 1984). Adenoviruses are also an increasing problem in immunocompromised hosts, including bone marrow transplant recipients (Shields et al., 1985).

The impact of adenoviruses on morbidity in Asia, Latin America, and Africa is not completely known. Adenovirus-associated respiratory infections are of major concern in China (Jiang et al., 1984). Respiratory adenovirus infections have been recognized as a significant cause of permanent lung damage after viral pneumonia, which can be particularly dangerous during measles epidemics (Becroft, 1979; Chiu et al., 1977; Kaschula et al., 1983; Spiegelblatt, 1983; Teng, 1960). In Japan and Taiwan, seasonal outbreaks of adenovirus-associated keratoconjunctivitis occur. Enteric adenoviruses have been recognized as etiologic agents in infantile diarrhea in South Africa and Japan (Kidd et al., 1986, Shinozaki et al., 1987b).

Classification

The adenovirus family consists of two genera—*Avi* adenovirus and *Mast* adenovirus. The number of members of each host species identified mirrors the diagnostic effort. Several adenovirus serotypes have been detected in pigs, sheep, cattle, and nonhuman primates. In human beings, 42 serotypes (Table 1) have been recognized (Wigand et al., 1987). The scheme for the classification of adenoviruses has been outlined for human adenoviruses by Rosen (1958). He described the capacity of human adenoviruses to agglutinate rat and monkey erythrocytes, and suggested that human adenoviruses could be divided into four different subgroups (subgenera) based on differential hemagglutinating properties (Rosen, 1960). The receptor binding structures are localized on the tip of the fibers, and their conformation could be of decisive importance in determining the tropism of different subgenera of human adenoviruses. The genes for the fiber protein differ substantially between members of different sub-

TABLE 1. Human adenovirus serotypes

Type	Prototype	Source	Diagnosis
1	Ad71	Adenoid	Hypertrophied tonsils and adenoids
2	Ad6	Adenoid	Hypertrophied tonsils and adenoids
3	GB	Nasal washing	Common cold (volunteer)
4	RI-67	Throat washing	Primary atypical pneumonia
5	Ad75	Adenoid	Hypertrophied tonsils and adenoids
6	Ton99	Tonsils	Hypertrophied tonsils and adenoids
7	Gomen	Throat washing	Pharyngitis
7a	S-1058	Throat swab	Undifferentiated respiratory infection
8	Trim	Eye swab	Epidemic keratoconjunctivitis
9	Hicks	Stool	Rheumatoid arthritis?
10	JJ	Eye swab	Conjunctivitis
11	Slobitski	Stool	Paralytic polio (type 1 poliovirus also recovered)
12	Huie	Stool	Nonparalytic polio?
13	AA	Stool	Healthy child
14	DeWitt	Throat swab	Acute respiratory disease
15	Ch38	Eye swab	Conjunctivitis (early trachoma?)
16	Ch79	Eye swab	Conjunctivitis (early trachoma?)
17	Ch22	Eye swab	Conjunctivitis (early trachoma?)
18	DC	Anal swab	Niemann-Pick disease?
19	587	Conjunctiva	Trachoma
20	931	Conjunctiva	Early trachoma?
21	1645	Conjunctiva	Trachoma
22	2711	Conjunctiva	Trachoma
23	2732	Conjunctiva	Trachoma
24	3153	Conjunctiva	Trachoma
25	BP-1	Anal swab	No specific illness
26	BP-2	Anal swab	No specific illness
27	BP-4	Anal swab	No specific illness
28	BP-5	Anal swab	No specific illness
29	BP-6	Anal swab	No specific illness
30	BP-7	Anal swab	No specific illness
31	1315/63	Stool	Healthy child
32	HH	Anal swab	Healthy child
33	DJ	Anal swab	Healthy child
34	Coimpton	Urine	Renal transplant recipient
35	Holden	Lung and kidney	Renal transplant recipient
36	275	Stool	Enteritis
37	GW	Eye	Keratoconjunctivitis
38	LJ	Anal swab	Bronchopneumonia
39	D335	Stool	Bronchitis
40	Dugan/Hovi	Stool	Gastroenteritis
41	Tak	Stool	Gastroenteritis
42	54/82	Stool	Healthy child

genera coding for fiber proteins with a characteristic length difference (Green et al., 1983; Norrby, 1969; Signäs et al., 1985).

Huebner (1967) suggested that human adenoviruses should be divided into subgenera based on their oncogenicity in newborn hamsters. Subgenus A contains adenovirus serotypes that cause tumors in every animal within 2 months. Subgenus B contains adenoviruses that induce tumors in a few animals after an observation period of 1 year. The nononcogenic adenoviruses can transform rodent cells in vitro. They are divided into subgenera C or D based on the antigenic reactivity of their T antigen (McAllister et al., 1969).

Analysis of the adenovirus proteins by SDS-polyacrylamide gel electrophoresis (PAGE) (Maizel et al., 1968a,b) demonstrated that the virion contained at least nine unique polypeptides. Among these, the internal polypeptides are expected to be conserved and offer a means of classifying human adenoviruses. SDS-PAGE was consequently used to divide 41 human adenovirus serotypes into 6 subgenera. The procedure was compatible with the previous methods and extended these as regards Ad4 (subgenus E) and Ad40 and Ad41 (subgenus F) (Wadell, 1979, 1984; Wadell et al., 1980). The DNA homology of adenoviruses has been studied by filter hybridization, liquid hybridization (Green, 1970; Green et al., 1979), and by use of DNA restriction endonucleases (Wadell et al., 1980). It is concluded that the degree of DNA homology between members of the same subgenus is greater than 50%, whereas the homology between members of different subgenera is less than 20% based on liquid hybridization (Table 2).

More precise information on the genetic distance between moderately to closely related adenoviruses (that is, members of the same subgenus) can be obtained by analysis of the pairwise comigrating restriction fragments (PCRF) after digestion with 12 to 15 different restriction endonucleases (Li and Wadell, 1986, 1988; Wadell et al., 1980). A catalogue of the restriction profiles obtained after digestion of DNA from the 42 human adenovirus prototypes by use of restriction endonucleases Bam HI, Bgl II, Bst EII, Hind III, and Sam I is now available (Adrian et al., 1986). This procedure is of value for identifying new adenovirus isolates and, furthermore, reveals a pronounced genetic variability within each analyzed serotype. Eighteen genome types of Ad3 and 15 genome types of Ad7 have been detected by use of 12 different restriction enzymes (Li and Wadell, 1986, 1988).

Taxonomic Terminology

The members of each *genus* of adenoviruses are characterized by sharing common epitopes on the hexon.

A *subgenus* is defined by a DNA homology of more than 50% between members within the subgenus and less than 20% between members of different subgenera.

A *genomic cluster* is a group of genome types that are significantly more closely related to each other than to any other genome type.

The *genome type* denotes a distinct viral entity within a genomic cluster that is identified by use of DNA restriction enzymes.

The designation *recombinant* should be used only when the two parental genomes have been identified.

An *evolutionary variant* describes a genetic alternation generated via insertion or intragenomic recombination in progeny of the same strain.

The *serotype* is defined by neutralization of the infectivity of the virus by hyperimmune sera. The ratio of homologous to heterologous neutralization titers must be greater than 16. The epitopes specific for one serotype are usually shared by genome types from a few different genomic clusters.

Strain corresponds to the progeny of each wild type isolate.

Physicochemical Properties and Morphology of the Adenovirus Particle

The adenovirus virion is composed of at least 10 different structural polypeptides. The virus particles are nonenveloped icosahedrons, with a corner to corner diameter of 82 nm. The capsid is formed by 252 capsomers, of which 240 are designated hexons arranged in a hexagonal symmetry. Each hexon is composed of three copies of polypeptide II. The vertex capsomers (the base of the penton) are arranged in a pentagonal symmetry. The vertex capsomer has been suggested to be composed of four copies of polypeptide III.

An antenna-like projection (the fiber) extends from each vertex capsomer. The fiber is a glycoprotein that varies in length. The fiber protein contains a repeating sequence of 15 amino acid residues initially containing two short α-strands and two short β-bends. The fiber is stabilized by dimer formation of the peptides. Fibers of members of different subgenera have been reported to display characteristic size differences (Norrby, 1969). The fibers of different lengths are formed by a variation in the number of fifteen amino acid residues repeating motives (Green et al., 1983; Signäs et al., 1985). The knob of the fiber contains a number of epitopes collectively described as the type-specific γ-determinant and also the receptor seeking conformation of the adenovirus particle.

Polypeptide IIIa is associated with the vertex region. Five molecules of polypeptide IIIa are avail-

TABLE 2. Properties of human adenovirus serotypes of subgenera A to F

Subgenus	Serotype	DNA Homology (%)		G + C (%)	Number of SmaI[a] fragments	Apparent molecular mass of the major internal polypeptides (kD)			Hemagglutination pattern[b]	Length of fibers (nm)	Oncogenicity in newborn hamsters	Tropisms/symptoms
		Intrageneric	Intergeneric			V	VI	VII				
A	12, 18, 31	48–69	8–20	48	4–5	51.0–51.5, 46.5–48.5[c]	25.5–26.0	18	IV	28–31	High (tumors in most animals in 4 months)	Cryptic enteric infection
B:1	3, 7, 16, 21	89–94	9–20	51	8–10	53.5–54.5	24	18	I	9–11	Weak (tumors in few animals in 4–18 months)	Respiratory disease
B:2	14,[d] 11, 34, 35											Persistent infections of the kidney
C	1, 2, 5, 6	99–100	10–16	58	10–12	48.5	24	18.5	III	23–31	Nil	Respiratory disease persists in lymphoid tissue
D	8–10, 13, 15, 17, 19, 20, 22–30, 32, 33, 36, 37, 38, 39, 42[e]	94–99	4–17	58	14–18	50.0–50.5[f]	23.2	18.2	II	12–13	Nil	Keratoconjunctivitis
E	4		4–23	58	16–19	48	24.5	18	III	17	Nil	Conjunctivitis
F	40, 41	62–69	15–22	52	9–12	46.0–48.5	25.5	17.5	IV	28–33	Nil	Respiratory disease Infantile diarrhea

Modified from Wadell (1984).

[a] The restricted DNA fragments were analyzed on 0.8 to 1.2% agarose slab gels. DNA fragments smaller than 400 bp were not resolved.

[b] I = Complete agglutination of monkey erythrocytes; II = complete agglutination of rat erythrocytes; III = partial agglutination of rat erythrocytes; IV = agglutination of rat erythrocytes discernible only after addition of heterotypic antisera.

[c] Polypeptide V of Ad31 was a single band of 48 K.

[d] Members of subgenus B are divided into two clusters of DNA homology based on pronounced differences in DNA restriction sites.

[e] Only DNA restriction and polypeptide analysis have been performed with Ad32–39 and Ad42.

[f] Polypeptides V and VI of Ad8 showed apparent molecular mass of 45 and 22 K, respectively.

able for each vertex capsomer. Polypeptide IIIa is exposed on the surface of the virion and extends to the interior of the capsid. Hexons and vertex capsomers are connected with polypeptide V, VIII, and IX to form a tight capsid (Everitt et al., 1973). Polypeptide VII is abundant, amounting to 1,070 copies per particle, and the molecular mass ratio of these polypeptides to DNA is approximately 1:1. Polypeptide VII is arginine-rich and resembles histons. The complex between DNA and polypeptide VII is extremely stable, forming a compact core structure. There are 180 molecules of polypeptide V. This is a moderately basic peptide that is less tightly bound to the DNA and protein VII. The linear Ad2 DNA molecule is a 35.937 ± 9 base-pair complex, the ends of which are covalently bound to a 55K protein. The internal adenovirus polypeptides including the terminal protein are processed during viral assembly by an adenovirus-specific endopeptidase. The adenovirus capsid is stable to exposure to low pH, bile, and proteolytic enzymes. Adenoviruses can consequently replicate to high titers in the gut.

Hemagglutination

The hemagglutination of adenoviruses was first demonstrated by Rosen (1958). Most adenovirus prototypes can be demonstrated to agglutinate erythrocytes (deJong et al., 1983; Hierholzer, 1973; Norrby, 1969; Rosen, 1960). The adenovirus receptors are expressed to a various extent on erythrocytes of different host species. The adenovirus 19a genome type has been responsible for epidemic outbreaks of conjunctivitis (Wadell and deJong, 1980). This genome type agglutinates guinea pig and dog erythrocytes (Kemp et al., 1983), whereas the adenovirus 19 prototype is devoid of this type of hemagglutinating capacity.

The adenoviruses of different subgenera agglutinate erythrocytes in the following way:

Subgenus A: incomplete agglutination of rat erythrocytes

Subgenus B: complete agglutination of monkey erythrocytes

Subgenus C: incomplete agglutination of rat erythrocytes

Subgenus D: complete agglutination of rat and human erythrocytes

Subgenus E: incomplete agglutination of rat erythrocytes

Subgenus F: hardly discernable agglutination of rat erythrocytes.

The number of erythrocyte receptors for different adenovirus serotypes varies for members of different subgenera. Rat erythrocytes carry higher numbers of receptors for members of subgenus D than for members from subgenus C (Wadell, 1969), whereas monkey erythrocytes can carry 5 times more receptors for Ad3 subgenus B than rat erythrocytes for Ad2 (Inouye and Norrby, 1973).

Hemagglutination mediated by virions is characterized by formation of a bridge between neighboring erythrocytes, allowing complete agglutination to occur. Since pentons and fibers are monovalent hemagglutinins, they may interfere with the virion-associated hemagglutination; if they occur in excess, a characteristic incomplete hemagglutination pattern will be noted (Wadell, 1969). The presence of subgenus-specific epitopes of the shaft of the fiber can be used to create divalent hemagglutinins and has also been a versatile technique for initial epitope mapping of human adenoviruses. The method that reveals adenovirus hemagglutination by use of heterotypic antisera has been named hemagglutination enhancement (Norrby, 1969; Wadell and Norrby, 1969). This can be demonstrated for all adenoviruses with the exception of the members of subgenus B. The short fibers of this subgenus cannot be enhanced by heterotypic antisera since they only express the type-specific epitopes on the tip of the fiber. Antibodies directed against the γ-epitopes of the fiber can be assayed by hemagglutination inhibition tests, thus allowing typing of the adenovirus serotype by a simple procedure.

Elaborations of the classification of adenoviruses based on hemagglutination originally proposed by Rosen (1958) have been suggested by Hierholzer (1973) and Norrby (1968).

Replication

Replication of DNA and transcription and maturation of adenoviruses take place in the cell nucleus. Several promoter regions support transcription from the early regions, whereas one major late promoter directs transcription of the structural proteins. The spliced mRNA is translated in the cytoplasm.

Infection is initiated by the attachment of a fiber, the virus attachment protein (VAP), to a receptor on the surface of permissive cells. This initial step of the infection cycle has been suggested to account for the distinctly different tropisms of adenoviruses of different subgenera. Penetration of virions occurs by an incompletely understood mechanism. Cell-associated adenovirus particles are taken into endosomes by adsorptive endocytosis. An ATP-driven proton pump lowers the pH within the endosome. This results in a hydrophobic alteration of the surface of the adenovirus capsid. Penetration of the lipid bilayer by the virion appears to cause disruption of the endosome, allowing the capsid to be transported to the cell nucleus. The remaining capsid is lost during the deposition of the virus core into the cell nucleus.

Six different early transcription regions have been identified. Transcription is initiated by early region E1A, which contains enhancer-like structures. Both regions E1A and E1B are needed for the expression of other viral genes and for adenovirus transformation. Regions E2A and E2B encode several proteins which are required for viral DNA replication. The 19 KD, virus-specific glycoproteins encoded by the E3 region, and several other early virus-specific proteins are formed before viral DNA synthesis can be initiated. This occurs 7 h after infection. Five hours later, the synthesis of viral structural proteins can be initiated. The synthesis of the cellular proteins then gradually ceases.

Within a few minutes, newly formed viral structural polypeptides are transported by means of interaction with virus-specific transportation peptides into the nucleus, where the assembly of virus particles starts 7 h after the initiation of DNA synthesis. Up to 10^5 virus particles are assembled in each cell within 30 h after infection with the most rapidly growing types of human adenoviruses. The proportion of infectious particles among these varies with different serotypes from 1 in 10 to 1 in 10,000. The translation of viral mRNA coding for structural proteins continues uninterrupted for about 40 h. This results in an accumulation of a 10-fold excess of viral polypeptides. No virus-specific function has been shown to be responsible for the release of infectious viral particles from the cell nucleus. Adenoviruses are, therefore, considered to be released in connection with the disruption of the cells.

Pathogenesis

Adenoviruses are characterized by the ability to shut off expression of the host mRNA and by the unregulated excess synthesis of adenovirus structural proteins. This results in the accumulation of viral proteins as intranuclear inclusion bodies, which are incompatible with normal cell function. In severe adenovirus pneumonia, the intranuclear inclusions in the alveolar cells are characteristic. In the upper respiratory tract cells, ciliary abnormalities and microtubular aberrations lead to defective mucociliary clearance (Carson et al., 1985).

Differences in target cell specificity between the adenovirus fibers have been suggested to be responsible for the distinctly different tropisms of members of the six different adenovirus subgenera. Furthermore, a toxin-like activity has been associated with vertex capsomers (the base of the penton). Isolated penton bases can cause cell detachment of monolayer cell cultures after 2 h of incubation. This activity is particularly strong in preparations from adenoviruses of subgenera B and C (Wadell and Norrby, 1969). In addition, the adenovirus capsid exerts a direct effect on the lipid bilayer of the endosome, allowing the release of endosome content into the cytosol. However, the gene products of the adenovirus genome responsible for the adenovirus-induced symptoms in the host are, on the whole, not defined.

Several proteins of the E3 region have a structure compatible with membrane proteins. The 19K glycoprotein binds to and hampers the glycosylation of the MHC class I heavy chain. This results in impaired transport of MHC class I to the cell membrane.

Ad1, Ad2, and Ad5 (members of subgenus C) persist in tonsils for several years through a low grade replication. Whether only a subset of lymphocytes is infected is not known. The shedding of infectious virus in stools for at least 2 years has also been documented (Fox and Hall, 1980). The source may be Peyer's patches of lymphoid tissue in the gut, but this has not been demonstrated.

Members of subgenus B:2 (Ad11, Ad34 and Ad35) cause persistent infections of the kidney. Silent shedding of Ad11 for 6 months into the urine of a healthy pregnant woman has been documented by Gardner (personal communication). Ad11 has also been demonstrated to be transmitted via the kidney, causing hemorrhagic cystitis in a transplant recipient. These serotypes are also activated in patients with AIDS and bone marrow transplant recipients. A certain cluster of Ad11 genome types causes outbreaks of respiratory disease.

Clinical Syndromes

Respiratory Infections

Adenoviruses are responsible for 5% of the acute respiratory infections in children under the age of 4 years, whereas they account for about 10% of the hospitalized respiratory infections in this age group. Characteristic symptoms are pharyngitis with an exudative tonsillitis (which may be difficult to distinguish from streptococcal tonsillitis) and frequently conjunctivitis together with nasal congestion and cough. Children usually have an elevated temperature, myalgia, and headache.

Adenoviruses can also cause laryngotracheobronchitis, but the pneumonias that occur in young children are the most serious clinical manifestations. These may occur as a consequence of infection with the endemic Ad2 and Ad5, among which certain strains (genome types) can be more aggressive than others. Ad3 and Ad7, which appear in epidemic outbreaks due to the low herd immunity, may also create severe problems. The clinical symptoms are characterized by fever, cough, dyspnea, and respiratory wheezing. The severity of the infections is related to overcrowding. In the winter of 1959, 3,398 cases of adenovirus pneumonia with a fatality rate of 15.5% were seen at a Peking children's hospital. Fatal cases among children younger than 2 years of age have also been reported from France, Germany, The Netherlands, Finland, Canada, and Australia (review by Wadell et al., 1980b).

Among the survivors of severe respiratory infection, in particular among Inuits in Canada, Indians in the United States, and Polynesians in New Zealand, residual lung damage caused by secondary obliterative bronchiolitis has been reported. Furthermore, a so-called hyperlucent lung disease has been traced back to adenovirus infections in reports from Canada. The most common serotypes involved in this condition are Ad3, Ad4, Ad7, and Ad21 (representatives of subgenera B and E).

In a prospective 10 year follow-up study from Finland, Similä et al. (1981) reported on sequelae after Ad7 pneumonia: 22 children were examined 10.7 years after the pneumonia; 12 had an abnormal chest X-ray (6 of these had bronchiectasis) and in 16 the results of pulmonary function tests were abnormal. This is in agreement with the sequelae reported among North American Indians or Inuits. In all these studies, children younger than 2 years of age were more prone to develop both severe primary infections, occasionally with a fatal outcome, and long-term sequelae.

Adenoviruses can also be a severe threat to children in developing countries who are exposed to measles infections. In a series of autopsies on children dying from respiratory disease after measles in South Africa, adenovirus pneumonias were considered to be the cause of death in 25% (Kaschula et al., 1983).

Ad7 sometimes appears in clear-cut outbreaks that involve school-age children. In this case, a clinical entity with high fever and a mean duration of 7 days, bronchopneumonia, gastroenteritis, and meningitis may occur. The outcome is generally favorable.

Pharyngoconjunctival Fever

This disease, characterized by conjunctivitis, fever, pharyngitis, and adenoidal enlargements, was first reported in association with swimming pools. Ad3 and Ad7 are most frequently associated with this syndrome, whereas Ad4 is common in Asia and Latin America but unusual among children in western countries. Adequate levels of chlorine are usu-

ally sufficient to inhibit outbreaks connected with the use of swimming pools.

Acute Respiratory Disease in Military Recruits

Acute respiratory disease (ARD) is usually caused by Ad4, Ad7, and Ad21. Ad14 has been reported from Holland. Ad3 is rarely reported, probably because of the more extensive herd immunity against this virus. In general, outbreaks do not involve seasoned troops, but cause a high morbidity among newly enlisted recruits. Adenovirus infections among healthy civilian adults are much less common. Crowding of people from widely different parts of the country, allowing repeated exposure to highly infectious doses, and the strenuous physical exercise may account for the unusually high degree of severe infections, some of which may have a fatal outcome. It usually appears during the third week of training. Characteristic symptoms include fever for at least 4 days, malaise, headache, nasal congestion, sore throat, hoarseness, and cough. About 10% develop patchy pulmonary infiltrates.

Infections of the Eye

Pharyngoconjunctival fever is a characteristic adenovirus syndrome. The acute follicular conjunctivitis that is part of this syndrome can also appear as a separate entity. After an incubation period of about 8 days, swelling of both the bulbar and the palpebral conjunctiva is seen, with involvement of the cornea in 5 to 10% of patients. In addition, a characteristic preauricular lymphadenopathy can usually be found.

In western countries, the disease is frequently reported as "swimming pool conjunctivitis" and recently also as a nosocomial infection. The most frequently encountered types are Ad3 (subgenus B) and Ad4 (subgenus E), but Ad11, Ad2 and Ad6 (subgenus C), Ad9, Ad10, Ad15, Ad17, Ad20, Ad22, and Ad29 (subgenus D), and Ad16 (subgenus B) have also been reported to cause conjunctivitis. In Asia, there is a significantly higher prevalence of neutralizing antibodies to Ad3 and Ad4 than in western countries, and infection usually presents as conjunctivitis.

Epidemic adenovirus keratoconjunctivitis is a distinctly different syndrome that affects males in western countries and children in Asia and may last for 4 to 6 weeks. A typical slow clinical progression starts after an incubation period of 8 days, with foreign body sensation, prominent edema of both conjunctivae and eye lids, and subconjunctival hemorrhage.

Preauricular adenopathy is frequently found; pharyngitis may be seen. The keratitis first develops as a diffuse epithelial engorgement, which proceeds to punctuate epithelial lesions. Two weeks after the initiation of the disease, the conjunctivitis may heal, and small round subepithelial opacities can be seen in the cornea that can last for several years. Ad8, Ad19a, and Ad37 are now the predominant causes of adenovirus-associated keratoconjunctivitis. In Japan, Ad3, Ad4, Ad8, Ad19, and Ad37 have been demonstrated to cause 39% of the viral infections treated at an eye hospital (Aoki et al., 1981).

Acute Hemorrhagic Cystitis

Acute dysuria with hematuria, which occurs predominantly among 6 to 15 year-old boys, has been described from Japan (Numazaki et al., 1968). Ad11 was the most frequently isolated adenovirus serotype. Similar observations have been reported from the United States, but adenoviruses were incriminated less frequently as causative agents. It is important not to confuse this condition with glomerulonephritis. Hemorrhagic cystitis caused by Ad11 has also been diagnosed in kidney transplant recipients.

Diarrhea

Adenoviruses have been associated with 3.6 to 15% of all hospitalized children with viral gastroenteritis. Gastroenteritis may be a sign of a systemic infection such as that caused by Ad3 or Ad7. They may cause both respiratory symptoms and diarrhea in a child with high fever. Ad31 is frequently isolated from stools of children with diarrhea, and may be of etiologic importance. However, the enteric adenoviruses Ad40 and Ad41, which are fastidious and grow only in selected heteroploid cell lines but best in primary or tertiary monkey kidney cells or 293 cells, have been demonstrated to account for two-thirds of diarrhea cases caused by adenoviruses. The incubation period of infections with enteric adenoviruses was 8 to 10 days, viral shedding 3 to 13 days, and the mean age of the patients 15 months (Ad40) and 28 months (Ad41). Ad40 and Ad41 were rarely associated with respiratory symptoms. Enteric adenoviruses cause diarrhea that is less acute than typical rotavirus diarrhea, but of longer duration. This is usually the reason for parents seeking medical attention and hospital care. The mean duration of diarrhea caused by established adenoviruses, Ad40, and Ad41 was 6.2, 8.6, and 12.2 days, respectively (Uhnoo et al., 1984). It is apparent that the enteric adenoviruses are important causes of infantile diarrhea in western countries, second only to rotaviruses. In view of the protracted diarrhea, it is important to analyze the frequency of lactose intolerance and celiac disease as a consequence of infections with enteric adenoviruses.

Intussusception

Invagination or intussusception is a condition that appears as a consequence of a telescopic introduction of a proximal into a distal intestinal segment. It has been suggested that adenitis in the intestinal wall or nearby mesenteric nodes is responsible for this condition. The adenitis should frequently be associated with adenovirus infection. In addition, reports have appeared indicating the presence of adenovirus inclusions in appendices that have been removed under the diagnosis of acute appendicitis.

Meningitis

Adenoviruses may be an infrequent cause of meningitis. Ad3 and Ad7 account for two-thirds of all adenovirus-associated cases of meningitis. Subacute focal encephalitis and even meningoencephalitis have been reported. Frequently, the patients are very small children with an underlying disease or severe combined immunodeficiency (SCID) and immunocompromised hosts.

IMMUNOCOMPROMISED HOSTS

Children with SCID are prone to develop severe infections with the most frequently occurring persistent adenoviruses (that is, Ad1, Ad2, and Ad5 of subgenus C, and Ad12, Ad18, and Ad31 of subgenus A). Furthermore, epidemic outbreaks of Ad3, Ad4, or Ad7 may also cause life-threatening infections. A sequential analysis of 15 immunocompromised patients with adenovirus infection has been reported by Zahradnik et al. (1980). Twelve of these patients had pneumonia, and nine died from respiratory or hepatic failure; Ad1, Ad2, and Ad6 of subgenus C occurred among the children, whereas Ad4 and Ad5 caused severe infection (four of six were fatal) in the adults.

Acquired Immunodeficiency Syndrome (AIDS)

DNA homology cluster 2 of subgenus B (that is, adenoviruses 11, 34 and 35) are apparently prone to establish latent infections of the kidney. Ad34 and 35 are highly overrepresented among urine specimens from AIDS patients and recipients of bone marrow or kidney transplants.

Bone Marrow Transplant Recipients

These patients are vulnerable to activation of all latent DNA viruses. Cytomegaloviruses are the best known and the most important threat to these patients. In Seattle, adenovirus infections have been demonstrated in 8% of bone marrow transplant recipients. It is also of interest that infection with Ad5 and the closely related Ad11, Ad34 and Ad35 is relatively frequent among these patients (Shields et al., 1985).

A prospective study of the etiology of severe enteric infections in 78 bone marrow recipients (mean age 21 years) was performed in Baltimore (Yolken et al., 1982). Adenoviruses were found in 12 of the 31 cases in which human enteric pathogens could be identified. The mortality among infected and uninfected patients was 55 and 13%, respectively (p < 0.001). Six of the patients died within 11 days of their first positive stool culture for enteric virus. Five of these six patients were infected with adenoviruses. These adenovirus isolates were not typed.

Extensive prospective studies are required to ascertain the role of different adenoviruses in primary and reactivated infections in the immunocompromised host in order to evaluate the need for appropriate means of immunoprophylaxis.

Laboratory Diagnosis

Specimen Collection

The most reliable source of adenovirus is stool specimens. Adenovirus may, however, be isolated from the conjunctiva, nasopharynx, pharynx, leucocytes, urine, and cerebrospinal fluid. Samples of stools are preferable to rectal swabs. Adenoviruses are thermostable and do not need refrigeration during transport. The specimens should be transported in tissue culture medium with fetal calf serum, bovine serum albumin, or gelatine. The medium should be supplemented with penicillin and streptomycin or gentamycin to prevent the growth of bacteria. Specimens from urine can be obtained by pelleting desquamated cells and resuspending the cells in a small volume of tissue culture medium.

Bacterial contamination can be controlled by shaking specimens with chloroform and performing a careful centrifugation to create a sharp interphase so that the supernatant can be removed without including any of the chloroform.

Virus Isolation

CELL LINES

Adenoviruses are strictly host cell-specific. Human cells should preferably be used. However, primary to tertiary cynomologous kidney cells display permissiveness for human adenoviruses in general.

Primary human embryo kidney (HEK) cells and the established cell line A-549 cells derived from oat cell carcinoma are permissive for adenovirus serotypes 1 through 39, but enteric adenoviruses Ad40 and Ad41 cannot be isolated in these cells. For the production of large amounts of adenoviruses, the continuous epithelial cell line Hep-2 derived from an epidermoid carcinoma, the HeLa cells derived from human cervical carcinoma, or KB cells derived from carcinoma of an oral cavity are useful. In general, a blind passage may be required for isolation of the adenoviruses of subgenus A and Ad8 of subgenus D.

The 293 cells and primary monkey kidney cells are particularly useful for the isolation of the enteric adenoviruses, Ad40, and Ad41. The 293 cells are HEK cells immortalized by the E1A-E1B region of Ad5 DNA. These cells can complement the host-dependent step in the replication of the enteric adenoviruses. Care should be taken to use a high multiplicity of infection and preferential incubation in a suspension of freshly trypsinized 293 cells (Shinozaki et al., 1987a).

ISOLATION PROCEDURE

Adsorption of the virus inoculum to cells by rocking or rolling increases the efficiency of virus isolation. It is essential that the volume is kept at a minimum. The inoculum can be removed after 1 h (if it is expected to be toxic) and maintenance medium added to the tube.

On inspection of the cell cultures, the adenovirus-specific cytopathic effect (CPE) should first be detected at the margin of the cell culture. The characteristic rounding of the cells with refractile intranuclear inclusions will be noted. Adenoviruses can also induce glycolysis in continuous cell lines. The lowered pH in the maintenance medium can be used to monitor the presence of adenoviruses. Adenoviruses (in particular, members of subgenera B and C) produce large surplus amounts of pentons. The vertex capsomer part of the penton has a tendency to interact with the host cell membrane and induce rounding of the cells and subsequent cell detachment from the support. This may occur two to four hours after infection, but this effect should not be mistaken for productive virus replication. The complete synthesis of adenovirus particles requires at least 30 to 36 h. The release of infectious virions is slow, and may be the function of degradation of the infected cells. A primary isolation of an unknown serotype may thus require at least 10 days' incubation, and a blind passage is often required.

IDENTIFICATION OF ISOLATES

After replication in tissue culture, the presence of adenovirus is suggested by the characteristic cytopathic effect (that is, either isolated rounded cells with intranuclear inclusions or the formation of grape-like clusters of cells that have been detached from the support). A tentative identification of adenovirus isolation can be confirmed by immunofluorescence analysis, complement fixation, or hemagglutination.

TYPING OF ISOLATES

Adenovirus strains are identified as serotypes. This requires that the isolate can grow in tissue culture to an infectious titer that is sufficiently high to allow clearcut neutralization when assayed after incubation for 1 h at 20°C and preferably overnight at +4°C with specific reference antisera. A new serotype can be designated only if the ratio of homologous to heterologous titers is above or equal to 16. Several serotypes of subgenera B and C induce surplus production of pentons that are responsible for an early CPE. This toxin-like effect can be neutralized in a subgenus-specific fashion (Wadell and Norrby, 1969) that should not be confused with the serotype-specific neutralization of infectivity. Monoclonal antibodies have been applied to type enteric adenoviruses Ad40 and Ad41 (Herrman et al., 1987). In theory, but to date not in practice, monoclonal antibodies recognizing type-specific epitopes inducing neutralizing antibodies of all the human adenovirus types could be used for typing of isolates.

Sources of Reference Adenovirus Antisera

Reference antisera can be obtained from American Type Culture Collection: rabbit antisera specific for serotypes 1 to 33; Centers for Disease Control, Atlanta, Georgia: horse antisera specific for serotypes 1 to 39 (Hierholzer et al., 1975); and Abteilung für Virologie, 6650 Homburg, F.R.G.: rabbit antisera specific for serotypes 1 to 42.

Hemagglutination Inhibition

Most human adenoviruses agglutinate rat, human, or monkey erythrocytes. This hemagglutination can be inhibited in a serotype-specific fashion. The sera to be assayed should be absorbed with the erythrocytes to be used. This procedure is simple and efficient. However, certain strains share epitopes that induce

neutralizing and hemagglutination-inhibiting antibodies from two different serotypes. Serum neutralization assays should, therefore, not be omitted when designing procedures to serotype human adenoviruses.

Typing of Adenoviruses of DNA Restriction Analysis

This procedure allows identification of the fastidious adenovirus serotypes 40 and 41. It is a swift and efficient procedure. It should be emphasized that this procedure provides information on genome type rather than serotype (Wadell et al., 1980a).

The DNA restriction patterns of the prototypes of Ad1 to Ad41 are available (Adrian et al., 1986). However, most prototypes no longer circulate, most likely because of continuous evolution. Serotype assignment can frequently be performed since several RE patterns are unaltered. Furthermore, DNA restriction patterns of the most frequently circulating genome types have been published. Members of subgenera A and C may create problems as a result of a large expressed genetic variability. RE patterns of the following genome types are available: Ad1, Ad2, and Ad5 (Adrian et al., 1985; Aird et al., 1983; Bruckova et al., 1980; Fife et al., 1985); Ad3 and Ad7 (Bailey and Richmond, 1986; Li and Wadell, 1986, 1988; O'Donnell et al., 1986; Wadell and Varsanyi, 1978); Ad14 and Ad16 (Adrian et al., 1985); Ad11, Ad34, and Ad35 (Li and Wadell, 1988; Valerama-Leon et al., 1985); Ad4 (Li and Wadell, 1988b; Wadell et al., 1980); Ad8, Ad19, and Ad37 (Fujii et al., 1984; Wadell and deJong, 1980; Wadell et al., 1981); and Ad40 and Ad41 (Kidd, 1984; Kidd et al., 1984; Uhnoo et al., 1983; Wadell et al., 1980, 1987).

Direct Virus Detection

ELECTRONMICROSCOPY

This technique is ideal for detecting the characteristic 80 nm icosahedral adenovirus particle with fibers protruding from each of the 12 verticles. An infection caused by enteric adenoviruses can yield 10^{11} virions per gram of stool. This fact stimulated Brandt et al. (1984) to distinguish enteric adenoviruses from all other adenoviruses by an estimation of the relative amount of adenoviruses particles in stool specimens as "one or more adenovirus particles identified per minute of viewing."

IMMUNE ELECTRON MICROSCOPY

Aggregation of adenovirus particles mediated by specific hyperimmune sera can be efficiently visualized by electron microscopy (Norrby et al., 1969; Wadell,

1972). This procedure is particularly useful for detection of enteric adenoviruses Ad40 and Ad41 (Wood and Bailey, 1987).

SOLID PHASE IMMUNE ELECTRONMICROSCOPY (SPIEM)

This technique is a modification of the electronmicroscopy (EM) technique described above and allows detection of adenoviruses in stool specimens, and even provides information on the adenovirus serotype. This method is designed so that hyperimmune adenovirus-specific IgG is adsorbed to the EM grid coated with *Staphylococcus aureus* protein. The conditions for the detection of adenoviruses have been elaborated by Svensson and von Bonsdorff (1982).

ANTIGEN DETECTION

The adenovirus genome encodes more than 50 different polypeptides. Several of the viral structural proteins are synthesized in a 10-fold excess. The detection of adenoviruses would rely predominantly on the identification of the major structural proteins, the hexon, the penton, or the fiber. The traditional method for detection of the adenovirus antigens is the complement fixation assay. However, the amount of antigen is so high that a whole array of methods have been used including the gel diffusion, single radial immunodiffusion (Pereira et al., 1972), crossed immunoelectrophoresis, (Boulanger and Puvion, 1974), immunoelectroosmophoresis (Jacobsson et al., 1979), radioimmunoassay (Halonen et al., 1980), and ELISA (Johansson et al., 1980; Roggendorff et al., 1982). Latex agglutination can also be used (Grandien et al., 1987).

SDS-POLYACRYLAMIDE GEL ELECTROPHORESIS

Virus particles can be directly isolated from stools and after centrifugation through discontinuous and continuous cesium chloride gradients submitted to SDS polyacrylamide gel electrophoresis. This analysis yields information on the apparent molecular weight of the structural viral polypeptides, which can be used to identify and classify adenovirus strains (Wadell, 1979; Wadell et al., 1980). However, restriction analysis of viral DNA will provide more detailed information concerning the genetic variability.

NUCLEIC ACID DETECTION

Virus particles have to be purified from stool specimens either by centrifugation on cesium chloride gradients (Wadell et al., 1973) or by precipitation using polyethylene glycol (Beards, 1982). Several methods have been elaborated to extract viral DNA after one round of replication in tissue culture. These methods represent modifications of the Hirt supernatant procedure, allowing extraction of nonencapsidated viral DNA and, thus, large amounts of DNA from infected cells.

Preparation of viral DNA can be performed in the following way: Lyse of the infected cell by addition of 0.6% sodium dodecyl sulphate (SDS) to the infected monolayer. Pronase B is then predigested for 90 min at 37°C or proteinase K to a final concentration of 2 mg/ml or 100 μg/ml, respectively, and incubated for 60 min at 37°C. NaCl is added to a final concentration of 1 molar. After gentle mixing, the lysate is stored over night at 4°C. After sedimentation at 17,000 × g for 30 min, ribonuclease A and T1 are added to a final concentration of 30 mg/ml and 80 units, respectively. These preparations should have been preincubated at 80°C for 30 min. After digestion, the second addition of pronase is followed by incubation for 4 h. Ultimately, the DNA is extracted three times with phenol and ether, and precipitated with two volumes of ethanol at −20°C.

In the procedure by Shinagawa et al. (1983), the phenol extraction is performed prior to pronase digestion. This procedure results in denaturation of the terminal viral protein, covalently bound to the linear viral, which is then trapped in the phenol aqueous interphase. This procedure can be performed in microscale in Eppendorf tubes, and results in DNA of high quality. Several methods (Darville, 1985; Fife et al., 1985) miniaturized the procedure by labeling the adenovirus infected cells with organic phosphate, allowing sufficient amount of p^{32}-labeled viral DNA to perform DNA restriction analysis from cells in tissue culture tubes. Alternatively, silver staining (Brown et al., 1984) can be applied to viral DNA extracted from microplate cultures, according to Shinozaki et al., 1987).

The presence of viral DNA can also be detected by dot blot or slot blot analysis (Allard et al., 1985; Hyypiä, 1985; Kidd et al., 1985; Takiff et al., 1985) or sandwich (Ranki et al., 1983) southern blot (Allard et al., 1985) hybridization. DNA restriction endonucleases of the obtained DNA yields the most detailed information, providing an opportunity to ascertain whether an isolate represents one of the 42 different adenovirus prototypes or one of the numerous recognized adenovirus genome types. Agarose electrophoresis analysis of the DNA restriction fragments is preferentially performed in submerged, horizontal agarose gels, allowing resolution of restriction fragments to 400 kilobase pairs. Higher resolution can be obtained by increasing the agarose concentration or, alternatively, polyacrylamid gel electrophoresis can be applied. DNA restriction patterns of adenovirus prototypes can be found in the report by Wadell et al. (1980), and a complete catalogue of the Bam HI, Bgl II, Bst EII, Hind III, and Sma I profiles in the study of Adrian et al. (1986).

Antibody Assays

Seroresponse is a sign of an adenovirus infection that might have been symptomatic or asymptomatic. Between 47 and 55% of all adenovirus infections are subclinical. On the other hand, based on seroconversion, Fox and Hall concluded (1980) that the real contribution of adenoviruses to disease was double of what is estimated from data from virus isolation only.

Screening of adenovirus-specific antibodies will consequently yield information on the prevalence of adenovirus infections in different geographic regions and populations. Hexons of mammalian adenoviruses share group-specific epitopes. An adenovirus group-specific immune response can, therefore, be assayed by complement fixation (CF), ELISA, or RIA techniques. An efficient antigen preparation should contain a pool of antigen from serotypes representing each subgenus. Alternatively, serotypes representing subgenera B, C, and F (for example, Ad2, Ad7, and Ad41) should be selected. An infected cell harvest that is sonicated and clarified is a feasible source of antigen because of the 10-fold excess synthesis of structural viral proteins that are not used in the assembly of virions.

Complement fixation has been most frequently used to ascertain seroconversions against adenoviruses. A four-fold or greater increase in CF antibody is a significant sign of a current adenovirus infection. The ELISA method is superior to the CF assay with regards to detection of immune responses in children.

The immune response to each serotype must be monitored by use of serotype-specific assays. Hemagglutination inhibition is an efficient method for the assay of serotype-specific responses. If unfractionated hemagglutinin preparations are to be used, a heterotypic serum should preferably be added to allow conversion of the monovalent hemagglutinins (that is, pentons fibers to divalent complete hemagglutinins). We frequently use anti-Ad6 serum for this purpose.

Serum neutralization assays are more laborious, but represent the most reliable methods to detect type-specific antibodies against each adenovirus serotype.

Interpretation of Laboratory Results

Recovery of adenoviruses from throat swabs, tears, blood, or cerebrospinal fluid is, as a rule, associated with a recent infection that more frequently is symptomatic than asymptomatic. On the other hand, stool isolates belonging to subgenera A, C, and D are frequently the result of prolonged shedding that may be extended for over 2 years (Fox and Hall, 1980). Members of DNA homology cluster 2 of subgenus B may also be shed in the urine for prolonged periods. However, the large amounts of virions observed by electron microscopy or quantitated by antigen detection assays in stools from children with diarrhea that are characteristic for the enteric adenoviruses have not, to our knowledge, been seen in healthy children. Differential growth characteristics as a means to distinguished between enteric and nonenteric adenoviruses in A549 and 293 cells must be interpreted with caution (Brown, 1984).

TYPING

Both hemagglutination inhibition (HI) and serum neutralization (SN) may yield ambiguous results. The cross-reactivity of HI is extensive in subgenera A and F. Furthermore, so-called "intermediate strains" sharing SN and HI epitopes from two different serotypes may cause spurious typing results if only one of the two methods is applied. They can be demonstrated in several subgenera (that is, Ad3-7, Ad3-16, and Ad14-16 of subgenus B) (Adrian et al., 1985), but are most frequent among members of subgenus D.

The SN assay is the method that defines an adenovirus serotype. The interpretation of SN results can be complicated by the occurrence of cross-reactivity (for example, subgenus A, Ad12/Ad31; subgenus B Ad7/Ad11; Ad11/Ad34/Ad35; subgenus D Ad19/Ad37, and subgenus F Ad40/Ad41) (Wigand, 1987). This implies that a wild type isolate has to be quantitatively analyzed against reference sera specific for several prototypes known to display SN cross-reactivity.

RESTRICTION ENDONUCLEASE (RE) ANALYSIS OF VIRAL DNA

Wild type isolates that display extensive SN cross-reactivity or that are hard to type because of their fastidious properties are now efficiently assigned to the correct subgenus by DNA restriction. If several REs are used, correct identification may be achieved by using the catalogue of RE patterns published by Adrian et al. (1986). It should be emphasized that several or, more likely, most prototypes no longer circulate, but have been replaced by more recently appearing genome types. Several papers previously mentioned must be consulted to identify the genome types that circulate during the 1980s.

The use of REs cannot altogether substitute for SN. In particular, members of subgenus C may best be identified by an SN assay that can follow the RE analysis.

ANTIBODY ASSAYS

Attempts to elaborate IgM assays for the detection of adenovirus infections have, on the whole, been unsuccessful. Although the high CF titer of single serum may be indicative, the seroconversion is the sine qua non for the verification of a recent adenovirus infection.

Epidemiology

General

Reports of adenovirus isolation to the WHO Register have been evaluated over a 10 year period (1967 to 1976) (Schmitz et al., 1983). During this period, adenoviruses accounted for 13% of a total of 135,702 reported isolations. Adenovirus infections were second only to influenza A, which represented 28% of the reported isolates. The reported adenovirus isolations are distributed according to age in the following way: 22% (<1 year); 42% (1 to 4 years); 18% (5 to 14 years); 10% (15 to 24 years); 7% (25 to 59 years); and 1% (>60 years).

On the whole, members of different adenovirus subgenera show distinctly different organ tropisms. The epidemic distribution of different serotypes is, therefore, presented according to subgenus.

Subgenus A

Ad12, Ad18, and Ad31 represent only 0.5% of the reported typed virus isolates. Members of subgenus A share several properties with the enteric adenoviruses and are otherwise distinguished from all other adenovirus serotypes by three characteristics: (1) the majority of isolates have been obtained from infants (0 to 11 months); (2) 91% of the reported isolates were recovered from stools; and (3) 60% of the children had gastrointestinal disease. A study of the prevalence of neutralizing antibody in children in Rome demonstrated that Ad31 and Ad18 were second only to members of subgenus C and Ad3.

Subgenus B

DNA HOMOLOGY CLUSTER B:1

Ad3 and Ad7 account for 13 and 19.7%, respectively, of all adenovirus isolates typed and reported to WHO. Both Ad3 and Ad7 cause respiratory infections and show an epidemic appearance. Both serotypes are most frequently isolated from children younger than the age of 4 years.

Three different epidemic patterns of Ad7 infection have been identified. The first pattern consists of outbreaks among infants, mainly in the winter season. These infections may be severe with a fatal outcome, particularly if they occur among children living under crowded conditions. The second pattern consists of epidemic outbreaks of respiratory disease among school children. Severe infections with fever for 7 days are usually encountered. A fatal outcome is rare. The third pattern is seen as outbreaks among newly enlisted military recruits.

In Japan, the frequency of reported Ad7 isolates and the antibody prevalence were highly discordant. Ad7-specific antibodies were detected in 30% of the children and in 50% of adults. However, in Japan, only 2% of the adenovirus strains were typed as Ad7, whereas 52% of the isolates were typed as Ad3.

The situation in Europe has been different. In West Germany, Ad3 and Ad7 accounted for 11 and 25%, respectively, of the typed adenoviruses. Three Japanese strains were genome-typed as the Ad7 P1 genome types, which may be of low virulence. This information indicates that isolation frequency is influenced by the severity of virus-associated diseases and by the tendency of virus strains to cause persistent infection with shedding of infectious virus over extended periods.

DNA restriction analysis of adenovirus strains can reveal the genetic variability within each serotype (Wadell, 1984; Adrian and Wigand; 1986). This is best exemplified by the analysis of Ad3 and Ad7, in which 18 and 15 genome types, respectively, were identified using 12 restriction endonucleases (Li and Wadell, 1986, 1988).

The distribution of the different Ad7 genome types is worldwide. In Africa, type Ad7c predominates. In Australia, the distribution of the Ad7 genome types is similar to that in Europe, the difference being that a shift from Ad7c to Ad7b took place in 1975. In Brazil, Ad7e has been the only genome type detected. This genome type was unique to Brazil. In China, severe outbreaks of Ad7-associated respiratory disease with mortality among infants have been reported. Ad7d and Ad7g are unique to China. In Europe, more than 90% of all isolates from patients are genome-typed as Ad7b and Ad7c. These two appear to be mutually exclusive, a shift from

Ad7c to Ad7b having occurred in 1969. In the United States, Ad7b strains have predominated among the Ad7 isolates collected on the west coast since 1962. Ad7p$_1$ and Ad7a$_2$ strains have been detected among Ad7 isolates recovered through the Virus Watch Program in Seattle, Washington. Seventeen genome types of Ad3 have also been identified. The predominant Ad3 genome types in China, Europe, and North America were Ad3a$_2$, Ad3p$_1$, and Ad3a, respectively. Ad16 can be associated with conjunctivitis, whereas Ad21 can cause outbreaks of severe respiratory disease. Ad21 has been isolated from military recruits at all major basic training centers in the U.S. Army, which has prompted the introduction of a live Ad21 vaccine. A shift in the appearance of Ad21 genome types was noted in Europe during the last decades (Van der Avoort et al., 1986).

DNA Homology Cluster B:2

Ad11, Ad14, Ad34, and Ad35 account for less than 1% of reported adenovirus isolates. The prevalence of antibodies against Ad11 among Italian children is around 2% but Ad11 is more common in China. Ad14 differs from the other members and has been reported to cause outbreaks of respiratory disease among military recruits. Ad11, Ad34, and Ad35 are closely related and can cause persistent infections of the kidneys. Ad11 has been demonstrated to cause hemorrhagic cystitis. The persistent nature of these adenoviruses is also well illustrated by the fact that they are the most common adenoviruses to be isolated from AIDS patients or bone marrow transplant recipients. One of the four genomic clusters of Ad11 contains genome types that are exclusively associated with respiratory outbreaks.

Subgenus C

The subgenus C members account for 59% of all adenovirus strains reported to WHO. The relative frequencies are 25, 4, 11, and 2.4% for Ad2, Ad1, Ad5, and Ad6, respectively. They are most frequently isolated in children younger than the age of 4 years. Distinct epidemics of genome types of these adenoviruses occur.

The subgenus C adenoviruses have been designated endemic adenoviruses because of their propensity to cause persistent infections and shedding for years in stools. The infection can be perpetuated within families through spread to the newly born child of the family. Because these adenoviruses are harbored in the tonsils and shed from the gut, double infections occur frequently. Ad5 is of particular interest since this virus can cause severe infections in immunocompromised hosts.

Subgenus D

Subgenus D contains 24 serotypes (that is, more than half of all recognized human adenoviruses). However, only 4.1% of all typed adenovirus strains reported to WHO belong to subgenus D. Infections caused by members of subgenus D are more common in Asia and Africa. Ad8 is the classic cause of epidemic keratoconjunctivitis (Jawetz, 1959).

Ad19 was isolated in Saudi Arabia in 1955 from a child with trachoma. This type was not reported again until 1973 when several epidemic outbreaks of keratoconjunctivitis caused by Ad19 appeared in both Europe and the United States. These outbreaks were caused by a new genome type, Ad19a, which was different from the original Ad19 prototype. In 1976, a new adenovirus type Ad37 related to Ad19a appeared. Ad37 is now an important cause of keratoconjunctivitis. This type may also appear as sporadic infections. Furthermore, Ad37 can be sexually transmitted since Ad37 isolates from penile lesions and cervicitis have been reported (Wadell et al., 1981). These three serotypes are the predominant pathogens of adenovirus-associated keratoconjunctivitis in western countries (Kemp et al., 1983).

Subgenus E

Ad4 is the only member of this subgenus. It is rarely isolated from children in Europe or the United States, being detected in only 4 of 1,800 children with proven adenovirus disease. Ad4 accounts for 2.4% of the adenovirus isolates reported to WHO. Most isolates were obtained from adults, and the frequency of Ad4 from adults was superseded only by the serotypes of subgenus D. The prevalence of Ad4-specific antibodies among adults is 30, 50, and 60% in the United States, Japan, and Taiwan, respectively. Three separate genomic clusters of Ad4 have been identified (Li and Wadell, 1988b).

The Ad4 prototype has been isolated from military recruits and identified as a cause of the characteristic epidemic outbreaks of respiratory disease. The Ad4a genome type is commonly isolated in Asia. In 1977, a 5 year old boy died with disseminated Ad4 infection after treatment at an intensive care unit in Buffalo, New York. A nosocomial outbreak of pharyngoconjunctival fever affected several of his attendants at the hospital (Faden et al., 1978). All the Ad4 strains were genome-typed as Ad4a. During recent years, several reports on Ad4-associated conjunctivitis have appeared.

Subgenus F

Enteric adenoviruses Ad40 and Ad41 are the only members of subgenus F (deJong et al., 1983). They are shed in large amounts (10^{11} virus particles/g feces) from children with diarrhea. In contrast to rotaviruses, enteric adenoviruses cause diarrhea in children throughout the year. Outbreaks have been reported from hospitals and boarding schools. Enteric adenoviruses cannot be found in stools from healthy controls. In a prospective study in Uppsala, New York, in 1981, Ad40 and Ad41 accounted for 7.9% of the children seeking hospital care for diarrhea. Seroconversion could be demonstrated in 70% of the patients (Uhnoo et al., 1984; Wadell et al., 1987).

Ad40 and/or Ad41 have been isolated in stool specimens from Africa, Asia, Europe, Latin America, and North America. About 50% of 6 to 8 year old children in Africa, Asia, and Europe have neutralizing antibodies against Ad41. On the basis of seroconversion, Shinozaki et al. (1987b) showed that 19% of cases of diarrhea in Japanese children were caused by enteric adenoviruses.

Prevention and Control

An oral enteric-coated adenovirus vaccine of Ad4 and Ad7 has been used in the U.S. Army for more than 15 years (Top, 1975; Top et al., 1971), and millions of army trainees have been vaccinated. The attenuated vaccine strains replicate in the intestine and cause an asymptomatic enteric infection with good neutralizing antibody response and a high rate of protection. These vaccines have not been used in civilian populations and the adenovirus types that are pathogenic to children have not been attenuated.

Literature Cited

Adrian, T., G. Wadell, J. C. Hierholzer, and R. Wigand. 1986. DNA restriction analysis of adenovirus prototypes 1 to 41. Arch. Virol. 91:277–290.

Adrian, T., R. Wigand, and J. C. Hierholzer. 1985. Immunological and biochemical characterization of human adenoviruses from subgenus B. II. DNA restriction analysis. Arch. Virol. 84:79–89.

Aird, F., J. J. King, and H. B. Younghusband. 1983. Identification of a new strain of adenovirus type 2 by restriction endonuclease analysis. Gene 22:133–134.

Allard, A. K., G. Wadell, K. M. Evander, and G. K. K. Lindman. 1985. Specific properties of two enteric adenovirus 41 clones, mapped within the E1A-region. J. Virol. 54:151–160.

Aoki, K., M. Kato, H. Ohtsuka, H. Tokita, T. Obara, K. Ishii, N. Nakazono, and H. Sawada. 1981. Clinical and etiological study of viral conjunctivitis during six

years 1974–1979, in Sapporo, Japan. Ber. Deutsch. Ophtalmol. Ges. 78:383–391.

Bailey, A. S., and S. J. Richmond. 1986. Genetic heterogeneity of recent isolates of adenovirus types 3, 4, and 7. J. Clin. Microbiol. 24:30–35.

Beards, G. 1982. A method for purification of rotaviruses and adenoviruses. J. Virol. Methods 4:343–352.

Becroft, D. M. O. 1979. Pulmonary sequeale of epidemic type 21 adenovirus infection: 1–13 year follow-up. Arch. Dis. Child. 54:155–156.

Boulanger, P. A., and G. Puvion. 1974. Adenovirus assembly: cross-linking of adenovirus type 2 hexons in vitro. Eur. J. Biochem. 43:465.

Brandt, D. D., W. J. Rodriquez, H. W. Kim, J. D. Arobio, B. C. Jeffries, and R. H. Parrott. 1984. Rapid presumptive recognition of diarrhea associated adenoviruses. J. Clin. Microbiol. 20:1008–1009.

Brown, M. 1984. Selection of nonfastidious adenovirus species in 293 cells inoculated with stool specimens containing adenovirus 40. J. Clin. Microbiol. 22:205–209.

Brown, M., M. Petric, and P. J. Middleton. 1984. Silver-staining of DNA restriction fragments for the rapid identification of adenovirus isolates: application during nosocomial outbreaks. J. Virol. Methods 9:87–98.

Bruckova, M., G. Wadell, G. Sundell, L. Syrucek, and L. Kunzova. 1980. An outbreak of respiratory disease due to a type 5 adenovirus identified as genome type 5a. Acta Virol. 24:161–165.

Carson, J. L., A. M. Collier, and S.-C. S. Hu. 1985. Acquired ciliary defects in nasal epithelium of children with acute viral upper respiratory infections. N. Engl. J. Med. 312:463–468.

Chiu, F. H., K. Y. Li, H-Y Wang, H. Y. Chang, and L. Shao. 1977. Etiology of virus pneumonia among children in Peking 1973–1975. China Med. J. 3:125–130.

Darville, J. M. 1985. Simplified restriction endonuclease method for typing and subtyping adenoviruses. J. Clin. Pathol. 38:331–335.

deJong, J. C., R. Wigand, A. H. Kidd, G. Wadell, J. G. Kapsenberg, C. J. Muzerie, A. G. Wermenbol, and R. G. Firtzlaff. 1983. Candidate adenovirus 40 and 41: fastidious adenovirus from human infantile stool. J. Med. Virol. 11:215–231.

Everitt, E., L. Lutter, and L. Philipson. 1973. Structural proteins of adenovirus. XII. Location and neighbor relationship among proteins of adenovirion type 2 as revealed by enzymatic iodination, immunoprecipitation and chemical cross-linking. Virology 67:197.

Faden, H., M. Gallagher, P. Ogra, and S. McLaughlin. 1978. Nosocomial outbreak of pharyngoconjunctival fever due to adenovirus type 4, New York. Morbid. Mortal. Weekly Rep. 27:49.

Fife, K. H., R. Ashley, A. F. Shields, D. Salter, J. D. Meyers, and L. Corey. 1985. Comparison of neutralization and DNA restriction enzyme methods for typing clinical isolates of human adenovirus. J. Clin. Microbiol. 23:95–100.

Fujii, S.-I., N. Nakazono, K. Ishii, C. C. Lin, M. M. Sheu, C. W. Chen, and K. Fujinaga. 1984. Molecular epidemiology of adenovirus type 8 (Ad8) in Taiwan: four subtypes recovered during the period of 1980–1981 from patients with epidemic keratoconjunctivitis. Jpn. J. Med. Sci. Biol. 37:161–169.

Fox, J. P., and C. E. Hall. 1980. Viruses in families, PSG, Littleton, MD.

Grandien, M., C.-A. Pettersson, L. Svensson, and I. Uhnoo. 1987. Latex agglutination test for adenovirus diagnosis in diarrheal disease. J. Med. Virol. 23:311–316.

Green, M. 1970. Oncogenic viruses. Annu. Rev. Biochem. 39:701–756.

Green, M., J. K. Mackey, W. S. M. Wold, and P. Rigden. 1979. Thirty-one human adenovirus serotypes (Ad1–Ad31) from five groups (A–E) based upon DNA genome homologies. Virology **93**:481–492.

Green, N. M., N. G. Wrigley, W. C. Russell, S. R. Martin, and A. D. McLachlan. 1983. Evidence for a repeating cross-sheet structure in the adenovirus fiber. EMBO J. **2**:1357.

Halonen, P., H. Sarkkinen, P. Arstila, E. Hjertsson, and E. Torfason. 1980. Four-Layer Radioimmunoassay for detection of adenovirus in stool. J. Clin. Microbiol. **11**:614–617.

Herrman, J. E., D. M. Perron-Henry, and N. R. Blacklow. 1987. Antigen detection with monoclonal antibodies for the diagnosis of adenovirus gastroenteritis. J. Infect. Dis. **155**:1167–1171.

Hierholzer, J. 1973. Further subgrouping of human adenoviruses by differential hemagglutination. J. Infect. Dis. **128**:541–550.

Hierholzer, J. C., W. C. Gamble, and W. R. Dowdle. 1975. Reference equine antisera to 33 human adenovirus types: homologous and heterologous titers. J. Clin. Microbiol. **1**:65–74.

Hilleman, M. R., and J. H. Werner. 1954. Recovery of new agents from patients with acute respiratory illness. Proc. Soc. Exp. Biol. Med. **85**:183–188.

Huebner, R. J. 1967. p. 147–166. *In* M. Pollard (ed.), Perspectives in virology, Vol. 5. Academic Press, New York.

Hyypiä, T. 1985. Detection of adenovirus in nasopharyngeal specimens by radioactive and non-radioactive DNA probes. J. Clin. Microbiol. **21**:730–733.

Inouye, S., and E. Norrby. 1973. Kinetics of the attachment of radioisotope-labelled adenovirus hemagglutinins to red blood cells and their detachment by antibody and by nonlabelled hemagglutinins of homologous and heterologous types. Arch. Gesamte Virusforsch. **42**:388–398.

Jacobsson, P. Å., M. E. Johansson, and G. Wadell. 1979. Identification of an enteric adenovirus by immuno-electroosmophoresis (IEOP) technique. J. Med. Virol. **3**:307–312.

Jawetz, E. 1959. The story of shipyard eye. Br. Med. J. **1**:873–878.

Jiang, Z.-F., F.-T. Chu, and Z.-Y. Xu. 1984. Adenoviral pneumonia in infants and children in the Beijing area, p. 153. *In* R. M. Douglas and E. Kerby- Eaton (ed.). Acute respiratory infections in childhood. Proceedings of an International Workshop, Sydney.

Johansson, M. E., I. Uhnoo, A. H. Kidd, C. R. Madeley, and G. Wadell. 1980. Direct identification of enteric adenovirus, a candidate new serotype associated with infantile gastroenteritis. J. Clin. Microbiol. **12**:95–100.

Kaschula, R. O. C., J. Druker, and A. Kipps. 1983. Late morphological consequences of measles: A lethal and debilitating lung disease among the poor. Rev. Infect. Dis. **5**:395–404.

Kemp, M. C., J. C. Hierholzer, C. P. Cabbradilla, and J. F. Obijeski. 1983. The changing etiology of epidemic keratoconjunctivitis: antigenic and restriction enzyme analyses of adenovirus types 19 and 37 isolated over a ten year period. J. Inf. Dis. **148**:24–33.

Kidd, A. H. 1984. Genome variants of adenovirus 41 (subgroup G) from children with diarrhoea in South Africa. J. Med. Virol. **14**:49–59.

Kidd, A. H., F. E. Berkowitz, P. J. Blaskovic, and B. D. Schoub. 1984. Genome variants of human adenovirus 40 (subgroup F). J. Med. Virol. **14**:235–246.

Kidd, A. H., E. H. Harley, and M. J. Erasmus. 1985. Specific detection and typing of adenovirus types 40 and 41 in stool specimens by dot blot hybridization. J. Clin. Microbiol. **22**:934–939.

Kidd, A. H., A. Rosenblatt, T. G. Besselaar, et al. 1986. Characterization of rotaviruses and subgroup F adenoviruses from acute summer gastroenteritis in South Africa. J. Med. Virol. **18**:159–168.

Li, Q-g, and G. Wadell. 1986. Analysis of 15 different genome types of adenovirus 7 isolated on five continents. J. Virol. **60**:331–335.

Li, Q-g, and G. Wadell. 1988a. Comparison of 18 genome types of adenovirus type 3 identified among strains recovered from six continents. J. Clin. Microbiol. **26**:1009–1015.

Li Q-g, and G. Wadell. 1988. Genetic variability of adenovirus 4. Arch. Virol. (in press).

Maizel, J. V. Jr., D. O. White, and M. D. Scharff. 1968a. The polypeptides of adenovirus. I. Evidence for multiple protein components in the virion and comparison of types 2, 7A, and 12. Virology **36**:115–125.

Maizel, J. V. Jr., D. O. White, and M. D. Scharff. 1968b. The polypeptides of adenovirus. II. Soluble proteins, cores, top components and the structure of the virion. Virology **36**:126–136.

McAllister, R. M., M. O. Nicolson, G. Reed, J. I. Kern, R. V. Gilden, and R. J. Huebner. 1969. J. Natl. Cancer Inst. **43**:917–923.

Norrby, E. 1968. Biological significance of structural adenovirus components. Curr. Top. Microbiol. Immunol. **43**:1.

Norrby, E. 1969. The structural and functional diversity of adenovirus capsid components. J. Gen. Virol. **5**:221–236.

Norrby, E., H. L. Marnsyk, and M.-L. Hammarskjöld. 1969. The relationship between the soluble antigens and the virion of adenovirus type 3. V. Identification of antigen specificities available at the surface of virions. Virology **38**:477–482.

Numazaki, Y., S. Shigeta, T. Kumasaka, T. Miyazawa, M. Yamanaka, N. Vano, S. Takai, and N. Ishida. 1968. Acute hemorrhagic cystitis in children: isolation of adenovirus type 11. N. Engl. J. Med. **278**:700–704.

O'Donnell, B., E. Bell, S. B. Payne, V. Mautner, and U. Desselberger. 1986. Genome analysis of species 3 adenoviruses isolated during summer outbreaks of conjunctivitis and pharyngoconjunctival fever in the Glasgow and London areas in 1981. J. Med. Virol. **18**:23–227.

Pereira, H. G., R. D. Machado, and G. C. Schild. 1972. Study of adenovirus hexon antigen-antibody reactions by single radial diffusion techniques. J. Immunol. Methods **2**:121–128.

Ranki, M., M. Virtanen, A. Palva, M. Laaksonen, R. Pettersson, L. Kääriäinen, P. Halonen, and H. Söderlund. 1983. Nucleic acid sandwich hybridization in adenovirus diagnosis, Curr. Top. Microbiol. Immunol. **104**:307–318.

Roggendorf, M., R. Wigand, F. Deinhardt, and G. G. Frösner. 1982. Enzyme-linked immunosorbent assay for acute adenovirus infection. J. Virol. Methods **4**:27–35.

Rosen, L. 1958. Hemagglutination by adenoviruses. Virology **5**:574.

Rosen, L. 1960. Hemagglutination-inhibition technique for typing adenovirus. Am. J. Hyg. **71**:120–128.

Rowe, W. P., R. J. Huebner, L. K. Gillmore, R. H. Parrot, and T. G. Ward. 1953. Isolation of a cytopathogenic agent from human adenoids undergoing spontaneous degeneration in tissue culture. Proc. Soc. Exp. Biol. Med. **84**:570.

Schmitz, H., R. Wigand, and W. Heinrich. 1985. Worldwide epidemiology of human adenovirus infections. Am. J. Epidemiol. **117**:455–466.

Shields, A. F., R. C. Hackman, K. H. Fife, L. Corey, and J. D. Meyers. 1985. Adenovirus infections in patients undergoing bone marrow transplantation. N. Engl. J. Med. **312**:529–533.

Shinagawa, M., A. Matsuda, T. Ishigama, H. Goto, and G. Sato. 1983. A rapid and simple method for preparation of adenovirus DNA from infected cells. Microbiol. Immunol. **27**:817–822.

Shinozaki, T., K. Araki, H. Ushijima, R. Fuji, and Y. Eshita. 1987a. Use of Graham 293 cells in suspension for isolating enteric adenoviruses from the stools of patients with acute gastroenteritis. J. Infect. Dis. **156**:146.

Shinozaki, T., K. Araki, H. Ushijima, and R. Fuji. 1987b. Antibody response to enteric adenovirus types 40 and 41 in sera from people in various age groups. J. Clin. Microbiol. **25**:1679–1682.

Signäs, C., G. Akusjärvi, and U. Pettersson. 1985. Adenovirus 3 fiber polypeptide gene; implications from the structure of the fiber protein. J. Virol. **53**:672–678.

Similä, S., O. Linna, and P. Lanning. 1981. Chronic lung damage caused by adenovirus type 7: a ten-year follow-up study. Chest **80**:127–131.

Spiegelblatt, L., and R. Rosenfeld. 1983. Hyperlucent lung: Long-term complication of adenovirus type 7 pneumonia. Can. Med. Assoc. J. **128**:47–49.

Svenssson, L., and C. H. von Bonsdorff. 1982. Solid-phase immune electronmicroscopy (SPIEM) by use of protein A and its application for characterization of selected adenovirus serotypes. J. Med. Virol. **10**:243–253.

Takiff, H. E., M. Seidlin, P. Krause, et al. 1985. Detection of enteric adenoviruses by dot-blot hybridization using a molecularly cloned viral DNA probe. J. Med. Virol. **16**:107–118.

Teng, C-H. 1960. Adenovirus pneumonia epidemic among Peking infants and pre-school children in 1958. China Med. J. **80**:331–339.

Top, F. H. Jr. 1975. Control of adenovirus acute respiratory disease in U.S. Army trainees. Yale J. Biol. Med. **48**:185–195.

Top, F. H. Jr., E. L. Buescher, W. H. Bancroft, and P. K. Russell. 1971. Immunization with live types 7 and 4 adenovirus vaccines. II. Antibody response and protective effect against acute respiratory disease due to adenovirus type 7. J. Infect. Dis. **124**:155–160.

Uhnoo, I., G. Wadell, L. Svensson, and M. Johansson. 1983. Two new serotypes of adenoviruses causing infantile diarrhea. Dev. Biol. Stand. **55**:311–318.

Uhnoo, J., G. Wadell, L. Svensson, and M. E. Johansson. 1984. Importance of enteric adenoviruses 40 and 41 in acute gastroenteritis in infants and young children, Journal of Clin. Microbiol. **20**:365–372.

Valderama-Leon, G., P. Flomenberg, and M. S. Horwitz. 1985. Restriction endonuclease mapping of adenovirus 35, a type isolated from immunocompromised hosts. J. Virol. **56**:647–650.

Van der Avoort H., T. Adrian, R. Wigand, A. Wermenbol, T. Zomerdijk, and J. C. deJong. 1986. Molecular epidemiology of adenovirus type 21 in the Netherlands and the Federal Republic of Germany from 1960 to 1985. J. Clin. Microbiol. **24**:1084–1088.

Wadell, G. 1969. Hemagglutination with adenovirus serotypes belonging to Rosen's subgroup II and III. Proc. Soc. Exp. Biol. Med. **132**:413–421.

Wadell, G. 1972. Sensitization and neutralization of adenovirus by specific sera against capsid subunits. J. Immunol. **108**:622–632.

Wadell, G. 1979. Classification of human adenoviruses by SDS polyacrylamide gel electrophoresis of structural polypeptides. Intervirology **11**:47–57.

Wadell, G. 1984. Molecular Epidemiology of human adenoviruses. Curr. Top. Microbiol. Immun. **110**:191–120.

Wadell, G., A. Allard, M. Johansson, L. Svensson, and I. Uhnoo. 1987. Enteric adenoviruses, p. 63–91. E. R. Bishop (ed.), Novel diarrhoea viruses. Ciba Symposium No. 128.

Wadell, G., and J. G. deJong. 1980. Use of restriction endonucleases for identification of a genome type of adenovirus 19 associated with kerato-conjunctivitis. Infect. Immun. **27**:292–296.

Wadell, G., M.-L. Hammarskjöld, and T. Varsanyi. 1973. Incomplete particles of adenovirus type 16. J. Gen. Virol. **20**:287–302.

Wadell, G., M.-L. Hammarskjöld, G. Winberg, T. Varsanyi, and G. Sundell. 1980a. Genetic variability of adenoviruses. Ann. N.Y. Acad. Sci. **354**:16–42.

Wadell, G., and E. Norrby. 1969. Immunological and other biological characteristics of pentons of human adenoviruses. J. Virol. **4**:671–680.

Wadell, G., G. Sundell, and J. C. deJong. 1981. Characterization of candidate adenovirus 37 by SDS polyacrylamide gel electrophoresis of virion polypeptides and DNA restriction site mapping. J. Med. Virol. **7**:119–125.

Wadell, G., and T. Varsanyi. 1978. Demonstration of three different subtypes of adenovirus type 7 by DNA restriction site mapping. Inf. Immun. **21**:238–246.

Wadell, G., T. Varsanyi, A. Lord, and R. N. P. Sutton. 1980b. Epidemic outbreaks of adenovirus 7 with special reference to the pathogenicity of adenovirus genome type 7b. Am. J. Epidemiol. **112**:619–628.

Wigand, R. 1987. Pitfalls in the identification of adenoviruses. J. Virol. Methods **16**:161–169.

Wigand, R., T. Adrian, and F. Bricout. 1987. A new human adenovirus of subgenus D: candidate adenovirus type 42. Arch. Virol. **94**:283–286.

Wood, D. J., and A. S. Bailey. 1987. Detection of adenovirus types 40 and 41 in stool specimens by immune electron microscopy. J. Med. Virol. **21**:191–199.

Yolken, R. H., C. A. Bishop, T. R. Townsend, E. A. Bolyard, J. Bartlett, G. W. Santos, and R. Saral. 1982. Infectious gastroenteritis in bone-marrow-transplant recipients. N. Engl. J. Med. **306**:1009–1012.

Zahradnik, J. M., M. J. Spencer, and D. D. Porter. 1980. Adenovirus infection in the immunocompromised patient. Am. J. Med. **68**:725–732.

CHAPTER 16

Papovaviridae: The Papillomaviruses

HERBERT PFISTER

Disease: Skin warts, anogenital warts (condyloma acuminatum, cervical intraepithelial neoplasia, bowenoid papulosis), oral and laryngeal papillomas, conjunctival papillomas, focal epithelial hyperplasia Heck; anogenital cancer, laryngeal carcinomas, tongue carcinomas, skin carcinomas of patients with epidermodysplasia verruciformis.

Etiologic Agents: Human papillomavirus genotypes 1 to 46.

Source: Man by direct contact.

Clinical Manifestations: Circumscribed benign tumors of skin and mucosa, which frequently regress spontaneously. Specific lesions may progress to cancer. Subclinical infections frequent.

Pathology: Epithelial tumors, showing acanthosis and variable degrees of papillomatosis and hyperkeratosis.

Laboratory Diagnosis: Demonstration of virus particles in the electron microscope and, more convenient, of virus capsid antigen by means of a genus-specific antiserum in a PAP test. Identification and typing of viral DNA by nucleic acid hybridization.

Epidemiology: Worldwide incidence at least 20% of the population. Some human papillomavirus types restricted to ethnic groups.

Treatment: Surgical therapy (excision, curettage, laser), application of caustic agents, cryotherapy, interferon.

Prevention and Control: Disinfection, prevention of sexual transmission. No vaccine available yet.

Introduction

Papillomaviruses form one genus of the *Papovaviridae* family. The acronym "papova" means *pa*pillomavirus, *po*lyomavirus, and *va*cuolating agent (SV40). The latter two viruses are members of the second genus (Matthews, 1982). The genera were grouped together on the basis of shared ultrastructural features and oncogenic potential. There is otherwise no evidence for genetic relatedness.

Papillomaviruses cause benign tumors in skin and mucosa of man. The etiologic role was demonstrated at the beginning of this century by transmission with cell-free filtrates of warts (Ciuffo, 1907). Some primary lesions may undergo malignant conversion into invasively growing squamous cell carcinomas (Zur Hausen, 1977). Viral DNA persists in cancers and is partially expressed, which may suggest that the virus is involved in maintenance of the malignant state (Pfister, 1987). The association of papillomaviruses with cervical cancer is especially remarkable, as this tumor accounts worldwide for 15% of all malignancies in females.

During the past 10 years, papillomaviruses turned out to be very heterogeneous. At present, 46 types are differentiated on the basis of limited or lacking cross-hybridization of viral DNAs. The type largely determines the clinical picture and probably the natural history of the lesions. Using DNAs of newly characterized types for screening, it was possible to associate further clinical syndromes with papillomavirus infection such as bowenoid papulosis or cervical dysplasia, oral leukoplakia or conjunctival papilloma.

The Viruses

Papillomaviruses are nonenveloped icosahedral virions about 55 nm in diameter with double-stranded, covalently closed circular DNA of 7.9 kb ± 10% (Rowson and Mahy, 1967). The capsid consists only of proteins forming 72 capsomers, which are arranged on a $T = 7$ surface lattice. Because of the lack of lipids, they are resistant to ether and other lipid solvents. The sedimentation coefficient is 300S, and the bouyant density in cesium chloride 1.34 g/ml. Empty particles have a sedimentation coefficient of 168S and a density of 1.29 g/ml. Tubular capsids may be detected at this density, which are assumed to result from aberrant maturation.

The capsid consists of at least two structural proteins. The species with a relative molecular mass in the range of 53,000 to 59,000 represents 80% of the total viral protein (Favre et al., 1975; Gissmann et al., 1977; Orth et al., 1977) and carries type-specific and genus-specific antigenic determinants (Nakai et al., 1986). Immunization of animals with native capsids usually triggers a type-specific response (Gissmann et al., 1977; Orth et al., 1977, 1978), whereas genus-specific sera are obtained by injection of detergent-disrupted particles (Jenson et al., 1980). A minor protein component of the capsid has an average molecular mass of 70,000. The DNA of human papillomavirus (HPV) 1 and bovine papillomavirus (BPV) 1 was shown to be associated with cellular histones H2a, H2b, H3, and H4 to form a chromatinlike complex (Pfister, 1984). Cellular histones H3 and H4 appear modified.

The DNA in virions exists in three forms: as supercoiled molecule, as open circle with at least one single-strand nick, and as linear molecule. The average guanine-plus-cytosine content of sequenced papillomavirus genomes is 42.6%, the extreme values being 36.5% for HPV 16 and 47.5% for deer papillomavirus. Nucleotide sequence data revealed a rather uniform genome organization. All significant open reading frames, which code for viral proteins, are located on one DNA strand (sense strand), appear in comparable positions, and are of similar size (Pettersson et al., 1987). Based on gene expression in different wart layers and in transformed cells, one differentiates an early region (open reading frames E1 to 8), comprising 4 kb, and a 2.5- to 3-kb late part (open reading frames L1 and 2), coding for structural proteins. The so-called early part (expressed prior to DNA replication) encodes genes that play a role in replication, transcription, and oncogenic transformation. Upstream to the 5' end of the early region there are sequences (0.5 to 1 kb) without major open reading frames that display control elements for transcription and the origin of DNA replication. All papillomaviruses show considerable DNA sequence homology throughout their genomes and up to 60% within open reading frames E1, E2 and L1. This is reflected by significant cross-hybridization under less stringent conditions, which allow reannealing of DNA single strands with some mismatch (e.g., 40°C below the melting temperature (T_m) of the DNA). Cross-hybridization of HPV DNAs under relaxed conditions provides an important tool to screen for HPV infection without type-specific reagents.

The replication of papillomaviruses is tightly linked to the differentiation of keratinocytes (Pfister, 1984). Cells from the basal layer of the epidermis are not permissive for HPV. It is generally assumed, however, that they can be persistently infected. Infection of cultured foreskin keratinocytes with virions from plantar warts resulted in replication and episomal persistence of viral DNA (La Porta and Taichman, 1982) without evidence of cytopathic effects or particle maturation. Productive DNA replication occurs in prickle cells, actually already in the first suprabasal layer, and viral DNA can be disclosed in these cells by in situ hybridization. As keratinocyte differentiation proceeds, capsid proteins are synthesized, and virions appear in the nuclei of cells from the granular layer and in the keratin layer. Virus particles and viral antigens are found exclusively in the nucleus as long as the nuclear membrane is intact.

Papillomaviruses cannot be propagated in tissue culture. Virions or viral DNA must be isolated from biopsy material, and the yield is usually too low to obtain enough antigen for a serological characterization. Classification of human papillomaviruses is therefore exclusively based on a comparison of the nucleic acids, which are molecularly cloned in bacterial plasmids. Different types show per definition less than 50% cross-hybridization when tested by reassociation in liquid phase (Coggin and Zur Hausen, 1979). Isolates, which show more than 50% cross-hybridization but differ in their restriction endonuclease cleavage patterns, are regarded as subtypes. Based on this criterion 46 HPV types have been differentiated up to now (Table 1). The DNAs of some types, such as HPV 1, 4, 30, or 34, do not cross-hybridize with DNAs of other types when tested under stringent conditions ($T_m - 20°C$). The remaining types form groups, members of which cross-react between 1 and 40%. The value of cross-hybridization should not be confused with actual homology. As a matter of fact, one determines that part of the genome which shows sufficient homology to allow hybridization under stringent conditions. The overall homology is usually significantly higher than the percentage of cross-hybridization. Closely related viruses like HPV 5 and 8 or HPV 16 and 31 share 70 to 80% nucleotide sequence homology but show only 23 or 38% cross-hybridization, respectively (Fuchs et

TABLE 1. Human papillomavirus types grouped according to DNA sequence homology[a]

A	B[b]	C	D[b]	D[b]	D[b]	E[b]	F	G	H	I	J	K	L	M	N	O	P
1	2	4	5	9	24	6	7	16	18	26	30	33	34	35	39	41	43
	27		8	15		11	40	31	32								
			12	17													
	29		14	37		13			42								
			19	38		44			45								
	3		20														
	10		21														
	28		22														
			23														
			25														
			36														
			46														

[a] Based on Beaudenon et al., 1986; Kahn et al., 1986; Kawashima et al., 1986; Lorincz et al., 1986; A. Lorincz, G. Orth, and H. zur Hausen, personal communication; H. Pfister, unpublished; Pfister et al., 1986a; Scheurlen et al., 1986.

[b] Subgroups can be differentiated within B, D, and E, which comprise rather closely related viruses.

al., 1986; Lorincz et al., 1986a, personal communication; Seedorf et al., 1985; K. Zachow et al., personal communication).

Viruses, which are related on the DNA level, generally reveal common pathogenic properties. All viruses from group B, for example, induce common or flat skin warts, and members of group D lead to flat, macular lesions in patients with epidermodysplasia verruciformis. Pathogenic differences were noted for isolates with 30% cross-hybridization or less. HPV 5, for example, is distinguished from closely related viruses like HPV 25 or HPV 12 by its regular association with skin carcinomas of patients with epidermodysplasia verruciformis. In contrast, the relatives have only been detected in benign tumors so far (Orth, 1986).

Some isolates disturb the correlation between genetic overall relationship and pathogenesis. HPV 13 and 32 both induce focal epithelial hyperplasias of the oral mucosa but belong to different groups according to their preferential cross-hybridization with HPV 6 and HPV 18, respectively. The discrepancies may be explained by the uneven distribution of homologies throughout the HPV genomes (Pfister et al., 1986a). In some cases, classification is obviously based on a preferential homology in subgenomic regions, which are not very relevant to pathogenic properties. This can be revised as soon as we know more about biological functions of individual genes.

Pathogenesis and Pathology

Papillomaviruses affect cutaneous and mucosal squamous epithelium and induce benign circumscribed tumors, which tend to regress spontaneously after a period of time (Pfister, 1984). Some lesions develop into carcinomas, however, mainly on the basis of long persistence and/or exposure to chemical or physical carcinogens. The malignant tumors harbor HPV DNA in which a few specific HPV types were shown to prevail. This may provide a basis for a prognostic evaluation of diagnostic HPV typing in precursor lesions.

All human papillomaviruses induce pure epithelial proliferations, which are characterized by acanthosis and by an intact basal membrane. There is a spectrum from exuberant exophytic growths to extremely flat, inconspicuous lesions, which can be hardly detected with the naked eye. The epithelial proliferation may extend downward, leading to inverted papillomas with accentuated rete pegs. Warts vary widely in papillomatosis and hyperkeratosis. Extensive papillomatosis is noted with plantar warts, common warts, exophytic genital warts (condylomata acuminata), and laryngeal papillomas. Capillaries are often drawn into papillary lesions, which is visible as abnormal vascular pattern when looking at mucosal proliferations. Hyperkeratosis is a common feature of most skin warts.

The mechanism of tumor induction by papillomaviruses is largely unknown. The persisting viral genome may either increase the proliferation rate or prolong the normal life-span of keratinocytes. An abnormal differentiation was demonstrated for laryngeal papilloma cells and was discussed as possible cause for the hypertrophy of the epithelium (Steinberg, 1986). Common warts were reported to be monoclonal in origin, but genital warts seem to be of multicellular origin.

Virus-specific cytopathogenic effects are most prominent in the stratum granulosum. It contains foci of vacuolated cells with wrinkled, pyknotic nuclei and prominent keratohyalin granules. Especially in mucosal lesions, these cells are referred to as koilocytes. The histopathology is rather type-specific (Gross et al., 1982; Jablonska et al., 1980). Viral capsid antigen and particles appear in some of the koilocytes, but others are negative, suggesting that the cytopathogenic effects are mainly due to virus replication before particle maturation.

There is mounting evidence for clinically inapparent infections with papillomaviruses. HPV DNA was detected in clinically and histologically normal mucosa of the larynx of former papilloma patients who were in remission for 2 years (Steinberg et al., 1983) and in healthy mucosa surrounding condylomas or HPV associated intraepithelial neoplasias (Ferenczy et al., 1985). HPV DNA was also disclosed in 36% of nonglycogenated metaplastic epithelia from the cervix uteri of routinely screened women (Fuchs et al., 1987).

Mode of Transmission

Skin warts are likely to be transmitted by direct skin contact or by contact with HPV-contaminated surfaces—for example, bathroom floors. Autoinoculation by scratching was described. Genital HPV infections occur by sexual intercourse and represent a venereal disease. Many papillomas of the oral cavity and the respiratory tract are caused by HPV types that infect the genital mucosa (De Villiers, et al., 1986b; Gissmann et al., 1983; Naghashfar et al., 1985). An infection during passage through an HPV-infected birth canal or by oral-genital sex was discussed for juvenile-onset laryngeal papillomas and for warts of the oral cavity, respectively. Some lesions of the oral mucosa harbor HPV 2 (Lutzner et al., 1982), which is typically found in common warts. It may be acquired by habits like chewing warts, sometimes reported from children. Infection presumably depends on microlesions or local abrasion of the skin or the mucosa. At the cervix uteri, proliferating cells are constantly exposed at the squamocolumnar border, and the majority of cervical HPV infections indeed occur at this site.

The experimental transmission of papillomaviruses is rather inefficient. Nevertheless, it turned out that 50 to 60% of sexual partners of patients with genital HPV infections develop condylomas (Pfister, 1987). Transmission of some HPV types like HPV 2, HPV 3, or HPV 4 seems to be facilitated in the case of impaired immunity of the patient (Jablonska et al., 1980). Other HPVs were detected only in very special conditions. Types from group D are mainly restricted to patients with epidermodysplasia verruciformis, and even family members who

have intimate contact do not develop warts (Jablonska et al., 1980). HPV 7 occurs preferentially in butchers and meat handlers, and HPV 13 shows ethnic restrictions. The reason for these specificities is unknown.

Based on experimental transmission of human warts, the incubation period varies between 3 and 18 months (Rowson and Mahy, 1967). A follow-up of natural transmission by sexual contact revealed incubation periods in the range of 4 to 6 weeks (Barrett et al., 1954). Warts must be regarded as infectious throughout their clinical course, although long-standing tumors tend to produce fewer virus particles. Regarding different HPV types, there is a wide variation in virus yields. Plantar warts and common warts are particularly rich in mature particles, whereas laryngeal papillomas appear extremely poor.

Clinical Symptoms

The association of HPV types with specific syndromes has been detailed in a number of recent reviews, which also provide comprehensive reference material (Jablonska et al., 1985; Mounts and Shah, 1984; Pfister, 1984, 1987). References in the following will be mainly confined to the latest publications.

Skin Warts

Eleven different HPV types were isolated from skin warts. The proliferations are differentiated into common, plantar, and flat warts based on localization, morphology, and histology. HPV 1 induces mainly solitary deep plantar warts, which may be very painful. Histology reveals large, confluent, eosinophilic cytoplasmic inclusions and basophilic nuclear inclusions corresponding to aggregates of virus particles. HPV 2 infection leads to frequently multiple exophytic, papillomatous common warts, usually on the hands, and to endophytic mosaic lesions on the sole of the foot. HPV 4 can be isolated from small, endophytic, hardly elevated plantar or common warts. HPV 3, 10, and 28 induce flat proliferations, which are almost always multiple (juvenile warts). HPV 26 and 27 were detected only in patients with severely impaired immunity, and HPV 7 was isolated almost exclusively from common warts of butchers or meat handlers. The skin warts virtually never develop into malignant tumors.

Epidermodysplasia Verruciformis (EV)

Epidermodysplasia verruciformus is based on an inherited susceptibility to infection with selected HPV types. Patients often have a history of parental consanguinity. The analysis of familial cases suggests that inheritance is either autosomal-recessive or X-linked (Androphy et al., 1985; Orth, 1986), indicating that the disease results from defects in either of at least two different genes. EV patients frequently show T-cell defects (Glinski et al., 1981) but are nevertheless able to fight virus infections in general. Even regarding papillomaviruses, the increased susceptibility appears confined to specific types. HPV 2 occasionally induces common warts. HPV 3 and 10 were readily isolated from flat warts similar to those found in the general population. The most commonly isolated HPV types, however, were HPV 5, 8, 20, and 17, which induce very flat, red, reddish-brown or achromatic macules (Orth, 1986). A large number of more or less closely related types was detected in morphologically and histologically similar lesions (Table 2), showing clusters of large, swollen cells with pale-staining cytoplasm in the prickle and granular layer. The warts arise first during childhood, gradually spread all over the body, and persist lifelong.

Malignant tumors develop in about one-third of the patients between 2 and 60 years after the onset of verrucosis, mainly at light-exposed sites, which suggests a role of ultraviolet irradiation. A high incidence of skin carcinomas (80%) was observed in the southwest of Japan (Tanigaki et al., personal communication), whereas Africans have a much better prognosis. The carcinomas are often of the in situ type but may grow invasively. Metastases are rare unless triggered by irradiation, which is absolutely contraindicated for that reason. The DNA of HPV 5 or HPV 8 persists extrachromosomally in high copy number in more than 90% of the primary carcinomas and in metastasis and is partially transcribed into 1.4 and 1.9 mRNAs covering open reading frames E6 and E7, which are supposed to play a role in oncogenic transformation (Orth, 1986). Virus particles and capsid antigen were not detected in carcinoma cells. The prevalence of HPV 5 and HPV 8 in malignant tumors seems to indicate that these types share an increased carcinogenic potential.

HPV types from group D are seldom recovered from normal patients. They sometimes appear in lesions of immunosuppressed patients, and HPV 5 DNA was detected in two skin carcinomas, which may indicate that HPV 5 contributes to the relatively high incidence of skin cancer in transplant recipients (Gassenmaier et al., 1986; Lutzner et al., 1983; Tanigaki et al., 1986). Group D HPVs were recently observed in keratoacanthomas, in solar keratosis, in a basalioma, and in a melanoma (Orth, 1986; Pfister et al., 1986b; Scheurlen et al., 1986). It is not possible at the moment to decide whether the viruses are etiologically related to these lesions or are just passengers.

TABLE 2. Occurrence of HPV types in benign and malignant tumors of man

Tumor	HPV types	
	Frequent	Rare
Skin warts		
Plantar wart	1	2, 4
Common wart	2, 4	1, 7, 26, 29
Flat wart	3, 10	27, 28, 41
Epidermodysplasia verruciformis specific macules	5, 8, 17, 20, 36	9, 12, 14, 15, 19, 21–25, 38, 46
Anogenital warts		
Condyloma acuminatum	6, 11	1, 2, 10, 16, 30, 44, 45
Buschke-Löwenstein tumor	6, 11	
Cervical intraepithelial neoplasia	6, 16, 31	11, 18, 33, 35, 42, 43, 44
Bowenoid papulosis	16	6, 34, 39, 40, 42
Benign head and neck tumors		
Laryngeal papilloma	6, 11	
Oral papilloma	6, 11	2, 16
Focal epithelial hyperplasia Heck	13, 32	1
Conjunctival papilloma	11	
Malignant tumors		
Squamous cell skin carcinomas of patients with epidermodysplasia	5, 8	3, 14, 17, 20
Bowen's disease of the skin		2, 16, 34
Cervical cancer	16, 18	6, 10, 11, 31, 33, 35, 39, 45
Penile and vulval cancer	6, 16, 18	
Laryngeal carcinoma		16, 18, 30
Tongue carcinoma		2, 16

Anogenital HPV Infections

The exophytic condylomata acuminata are a well-known manifestation of HPV infection in the anogenital area. They occur on penis and vulva, in the urethra and vagina, at the uterine cervix, and in the perianal region and are mainly induced by HPV 6 or 11 (Gissmann et al., 1983). Condylomata acuminata may become large and invasive (Buschke-Löwenstein tumors), still showing typical histologic features of condylomas. They do not usually metastasize, however, and true malignant conversion of condylomata acuminata is very rare (Zur Hausen, 1977). The genus-specific structural antigen was detected in cases of pruritic vulvar squamous papillomatosis (Growdon et al., 1985).

About 70% of HPV-induced lesions in the cervix are extremely flat and frequently not visible unless magnified by colposcopy and/or treated with acetic acid, which makes them turn white. Histology shows acanthotic epithelium and koilocytotic atypia in the superficial cells (Meisels et al., 1982). The association with HPV is proved by demonstration of viral particles, DNA, and capsid antigen. A number of synonyms are currently used for these lesions such as flat condyloma, noncondylomatous cervical wart virus infection, subclinical papillomavirus infection, mild dysplasia, or cervical intraepithelial neoplasia (CIN) of grade I, depending on the intention to emphasize the character as a primarily benign viral lesion or the potential to develop into cancer. About 40% of these proliferations are induced by HPV 6 or 11. HPV 16–associated lesions are frequently distinguished by nuclear atypia and abnormal mitoses (Crum et al., 1984; Gross et al., 1985), thus showing features of more severe intraepithelial neoplasia (atypical condyloma, CIN II). Koilocytes are less frequent or even absent, which indicates that koilocytosis cannot be regarded as a necessary parameter of genital HPV infection. The characteristics of flat condylomas associated with other HPV types are not yet evaluated in detail.

Cervical dysplasias progress from moderate to severe forms (CIN III or carcinoma in situ) and are regarded as precursors of cervical cancer. They may regress or persist unchanged at any stage, but progression is the more likely the more advanced the process. The percentage of capsid antigen-positive lesions decreases with increasing severity, but HPV DNA can be demonstrated in 80 to 90 % of CIN of all stages and in invasive cancers and metastases (Boshart et al., 1984; De Villiers et al., 1986a; Dürst et al., 1983; Fuchs et al., 1987; Gissmann et al., 1983; Green et al., 1982; Lancaster et al., 1986; Lorincz et al., 1986b; McCance et al., 1985; Yoshikawa et al., 1985). DNA of HPV 16 prevails worldwide in carcinomas (35 to 65%). This was interpreted as evidence for a special oncogenic potential of this virus type. HPV 18 DNA appears rather frequently in Africa and South America (25%). The incidence of HPV 10, 6/11, 31, 33, 39, and 45 in carcinomas is very similar—close to 5%. The DNA of HPV 16 and 18 is partially or fully integrated into host DNA of cancer cells, whereas other viruses seem to persist extrachromosomally. HPV 16–specific transcripts were demonstrated in cervical carcinoma biopsies covering mainly open reading frames E6 and E7 (Schwarz et al., 1985; Smotkin and Wettstein, 1986).

Flat condylomas also appear at external genital sites of both sexes (Gross et al., 1985). HPV 6 and 11 again induce proliferations of low atypia. HPV 16 was described in cases of severe dysplasia (Ikenberg et al., 1983), which are referred to as bowenoid papulosis (multicentric), Bowen's disease (solitary), or VIN or PIN (vulval or penile intraepithelial neoplasia). The lesions are very inconspicuous and may be only detected as acido white epithelium. Bowenoid papulosis occurs in young, sexually active adults and usually ends by spontaneous regression in spite of the severe histologic picture. HPV infections at external genital sites in general have a considerably better prognosis than infections of the cervix. HPV types associated with cervical cancer (HPV 10, 16, 18) were also found in penile, vulval, and anal carcinomas (Beckmann et al., 1985; Dürst et al., 1983; Green et al., 1982; McCance et al., 1986).

All genital HPV infections in females worsen during pregnancy and improve after delivery, which indicates a hormonal activation.

Head and Neck Tumors

HPV types 13 and 32 induce focal epithelial hyperplasia of the oral mucosa (Heck's disease), which is very prevalent in natives of Greenland and in American Indians (Orth, personal communication; Pfister et al., 1983). The lesions are benign and disappear spontaneously.

Other HPV-induced lesions of the oral cavity resemble common warts or condylomas and were partially shown to harbor HPV 2, 6, 11, or 16 DNA. Not further specified HPV DNA was detected in four out of five leukoplakias (Löning et al., 1985). Oral carcinomas were associated with DNA of HPV 2, 11, and 16, and leukoplakias were observed in some cases at the same time, representing potential precursor lesions (De Villiers et al., 1985).

Laryngeal papillomatosis is a rare disease. Induced in most cases by HPV 6 or 11 (Gissmann et al., 1983; Mounts and Shah, 1984), the papillomas appear most frequently at the vocal cords and cause hoarseness and voice change. In children, they are usually multiple, show exuberant growth, and may lead to obstruction of the airways. The papillomas may extend down the trachea and into the bronchi. Preliminary data suggest an association between this more aggressive behavior and HPV type, but this has to be substantiated by larger numbers. Because of frequent recurrences after surgical removal, laryngeal papillomas represent a severe clinical problem. Malignant conversion of juvenile-onset lesions is extremely rare—they rather disappear after puberty—but carcinoma may develop after radiation therapy (Zur Hausen, 1977). In contrast, laryngeal papillomas of adults should be regarded as carcinoma precursors, 20% of which may become malignant. HPV DNA—namely, HPV 16, 18 and 30—has been detected in a few laryngeal cancers (Kahn et al., 1986; Scheurlen et al., 1987; Zur Hausen, personal communication). In analogy with infections of the genital mucosa, the benign precursor lesions induced by HPV 16 and 18 may be very inconspicuous. Verrucous carcinomas of the larynx harbor HPV 16–related sequences (Brandsma et al., 1986).

DNA of HPV 11 was detected in a conjunctival papilloma (Lass et al., 1983). Another lesion was induced by an HPV type whose DNA did not hybridize to other HPV DNAs under stringent conditions (Pfister et al., 1985). HPV 16–related sequences were demonstrated in a nasal papilloma. The genus-specific structural antigen was detected in koilocytotic lesions of the esophagus (Winkler et al., 1985).

Diagnosis

A viral wart is seldom misdiagnosed by the clinician, and there is usually no need for confirmatory tests. However, in view of the enlarged spectrum of papillomavirus-associated disorders, it became important to demonstrate viral footprints to establish the correct diagnosis. Over and above that, the supposed correlation between infection by certain virus types and an increased risk of malignant conversion may provide a basis for prognosis, which will require diagnostic virus classification. This field is just beginning to evolve. The following is therefore rather an attempt to outline prospects of diagnosis than a catalog of established methods and criteria for evaluation.

Specimen Collection

Diagnosis of papillomavirus infections is mainly aimed at the detection of viral antigen and nucleic acid in tumors and exfoliated epithelial cells. If tumors are surgically removed by excision or curettage, part of the material can be spared for these tests. Punch biopsies can be taken just for diagnostic

purposes. Especially from mucosa lesions, sufficient material for diagnosis may be obtained by scraping with a wooden spatula or by swabbing. The resulting cells are then transferred to buffered saline. Burk et al. (1986) collected exfoliated cervicovaginal cells by lavage and detected HPV DNA in 94% of women with a class III or IV Pap smear, in 45% of the patients with a class II smear, and in 29% of the women with a normal Pap result. The cervix was lavaged with approximately 7 to 8 ml of saline solution, and the washings were aspirated from the posterior vaginal fornix.

For extraction of virus particles and isolation of viral DNA the biopsy material must be kept frozen at $-20°C$ until use. It can be dispatched in dry ice. Native material must be regarded as infectious. Safe handling should be guaranteed by wearing gloves and a laboratory coat and by employing aseptic techniques. Formalin-fixed, paraffin-embedded samples may be processed for antigen detection and in situ hybridization.

Direct Virus Detection

Virus particles can be detected by the electron microscope in supernatants of homogenized tumors or tumor cells and in ultrathin sections (Almeida et al., 1962; Della Torre et al., 1978; Lutzner et al., 1982). The tumors are minced and thoroughly ground in a mortar with a little sand and a few drops of buffer. The preparation is clarified by low-speed centrifugation and used for microscopy after negative staining with phosphotungstic acid or uranyl acetate. Alternatively, the biopsy is fixed in glutaraldehyde and osmium tetroxide, dehydrated in ethanol, embedded in Epon, and cut with an Ultramicrotome. The sections can be examined after staining with uranyl acetate and lead citrate. It is possible to reprocess conventional paraffin sections for thin-section electron microscopy by the method of Coleman et al. (1977), which allows for a correlation of light microscopy findings with electron microscopy data. Mature virus particles appear in the nuclei of cells from the upper epidermal layers. They are first seen in association with the nucleoli of cells in the stratum spinosum (Almeida et al., 1962). In lesions with low particle production, they are confined to the uppermost cell layers (Della Torre et al., 1978).

The amount of virus particles in various tumor types differs dramatically, ranging from 10^{13} particles in some plantar and common warts to fewer than 10^5 particles in laryngeal papillomas. The latter is at the limits of detectability in the electron microscope, and it may take hours of screening to detect a virion or a positive nucleus. In contrast, there are plenty of particles throughout the layers of plantar warts, usually forming huge paracrystalline arrays, filling the entire nucleus and pushing the chromatin to the margin.

The direct detection of virus is of limited significance as a practical screening technique. It depends on the availability of an electron microscope, is time-consuming in the case of lesions with low particle production, and provides no information on the virus type. This method was successfully replaced by immunological and DNA hybridization techniques.

Direct Antigen Detection

Capsid antigen can be disclosed by indirect immunofluorescence tests in frozen wart sections using type-specific antisera (Jablonska et al., 1980; Orth et al., 1977, 1978; Pfister et al., 1979). Such sera are only available, however, for HPV 1, 2, 3, 4, and 5. They were obtained by immunization of rabbits or guinea pigs with purified virions and are not commercially distributed. Monoclonal antibodies to the major capsid protein of HPV 1 were described recently (Roseto et al., 1984).

A hyperimmune rabbit serum and the peroxidase-antiperoxidase (PAP) technique were used to identify HPV 1 antigens in formalin-fixed tissue (Jenson et al., 1982). The access to formalin-fixed material offers a great advantage for routine diagnostic purposes, and the method is comparable to immunofluorescence in both sensitivity and specificity. HPV 1 antigens were identified in half of the plantar warts and in about 10% of common warts. Immunofluorescence complement fixation tests were shown to be more sensitive (Jablonska et al., 1982), which is important in cases of limited antigen production.

Further type-specific antisera could theoretically be obtained in the meantime for all virus types. The open reading frame L1, which codes for the major capsid antigens, can be expressed in bacteria, and protein preparations may be used for immunization of animals. An application in practice must be preceded, however, by systematic studies on cross-reactivity.

A genus-specific antiserum could be obtained by immunization of animals with detergent-disrupted papillomavirus particles (Jenson et al., 1980). Such a serum is commercially available and is widely used to demonstrate productive HPV infection in formalin-fixed, paraffin-embedded sections (Fig. 1). As discussed above, viral antigens appear mostly in the superficial layers of the epithelium and are absent from cells of the germinal layer. When screening a variety of papillomavirus-induced lesions from skin and mucosa, only 40 to 70% will be positive for capsid antigen. A negative result obviously does not exclude an HPV infection. In the case of a low number of antigen-producing cells—such as in most laryn-

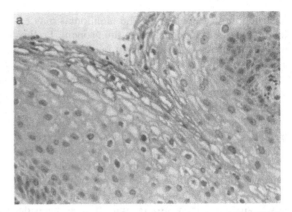

FIG. 1. Formalin-fixed, paraffin-embedded sections of a vulvar condyloma. (a) Stained with hematoxylin and eosin. (b) Adjacent section, uncounterstained immunocytochemistry for papillomavirus genus-specific antigen. Note only rare positive nuclei in the superficial cells (arrowheads). Photographs kindly provided by M. Stoler.

geal papillomas—several serial sections should be screened for final evaluation. When consecutive recurrences of the same lesion were examined, negative results were observed to change to positive and back to negative (Lack et al., 1980). This points to some modulation of capsid protein synthesis in the course of papilloma development, which may be crucial if antigen levels are close to the limit of test sensitivity.

Concerning cervical intraepithelial neoplasias (CIN), the percentage of antigen-positive lesions decreases with increasing severity (Guillet et al., 1983; Kurman et al., 1983; Syrjänen, 1983; Walker et al., 1983). Only 25 to 3% of moderate or severe dysplasias show the group-specific antigen, whereas 50 to 80% harbor viral DNA (see below). The presence of structural antigens is therefore suggestive of but not proof a lower-grade dysplasia.

The group-specific antiserum was recently used to detect and quantitate bovine papillomavirus in wart suspensions by means of an enzyme-linked immunosorbent assay (ELISA) (El Shazly et al., 1985). Virus particles were partially purified by low-speed and high-speed centrifugation. The ELISA detected 1 ng/ml of viral protein. Apart from the possibility to quantify, the test was qualitatively as sensitive as immunohistochemistry and will therefore not replace PAP staining in routine diagnosis.

For diagnosis of tumors that do not allow productive HPV infection, reagents would be desirable that detect early viral proteins expressed in nonpermissive cells. Antisera were raised in the meantime against genetically engineered early proteins. Using a hyperimmune serum specific for the E7 protein of HPV 16, a 20,000-dalton protein could be precipitated from extracts of the cervical cancer–derived cell lines CaSki and SiHa (Smotkin and Wettstein, 1986). No protein was precipitated from the HPV 18 DNA-positive HeLa cell line, which may reflect either a different gene expression of HPV 18 or a lack of serological cross-reactivity between E7 proteins of HPV 16 and HPV 18. These observations are very stimulating, but it is so far impossible to reveal early papillomavirus antigens in tumor material, probably because of very small amounts of specific proteins.

Direct Viral Nucleic Acid Detection

Nucleic acid hybridization up to now represents the method of choice for detection and typing of papillomaviruses. The DNAs of all known HPV types are cloned in plasmids and serve as reference reagents. Southern blot hybridization is certainly the most sensitive and reliable technique, which detects 0.1 copy or even less of HPV DNA per cell. Since the binding of the radioactive probe is confined to distinct bands, it is possible to identify extremely faint signals as specific. Native biopsy material is needed, however, for phenol extraction of DNA, which limits the application in practice. For routine laboratories, in situ hybridization techniques that skip the DNA extraction step are much more convenient. One may use touch smears or frozen sections and also Formalin-fixed, paraffin-embedded material. In addition, cytology or histology can be evaluated at the same time.

Two test systems were described, based on Southern blotting. The standard procedure starts with a restriction enzyme digest of total DNA from tumors or exfoliated cells, followed by gel electrophoresis, transfer to nitrocellulose filter, and hybridization with radiolabeled HPV DNA, which discloses HPV-specific DNA fragments (Lancaster et al., 1983, 1986; Lorincz et al., 1986; McCance et al., 1985; Yoshikawa et al., 1985).

To screen for the presence of HPV DNA in general, hybridization is carried out under relaxed conditions (i.e., for example at $T_m - 40°C$) arbitrarily

using DNA of one HPV type as labeled probe or preferably a DNA cocktail, comprising several distantly related types, to increase the pool of cross-hybridizing sequences. These conditions allow the detection of HPV DNA sequences regardless of the HPV type. It should be noted, however, that relaxed hybridization is at least 10 times less sensitive than stringent hybridization, which may be crucial in cases of low DNA copy number. Relaxed conditions should therefore be replaced by stringent ones whenever homologous probes are available.

HPV typing primarily depends on specific hybridization to individual reference DNAs under stringent conditions at $T_m - 20°C$ (Fig. 2). The cleavage pattern of the viral DNA in question, which is characteristic for the enzyme and for the virus type, provides additional and sufficient information for final classification of those HPV types, which show extensive cross-reactivity with several HPV DNAs (see Table 1).

Most previous studies confined themselves to a rather small number of HPV types. A preselection was made based on the established association between HPV types and clinical symptoms (see Table 2) and on special interest in viruses, which are frequently detected in malignant tumors. Such restrictions seem to be justified and necessary for practical purposes. A comprehensive diagnosis dealing with all 46 types known at present must be reserved for

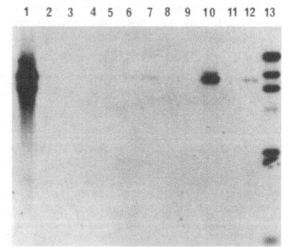

FIG. 2. Autoradiogram of a Southern blot hybridization of ^{32}P-labeled HPV 16 DNA to DNAs isolated from a cervical carcinoma (1) from CIN III (4, 10); CIN II (5, 6, 11, 12); CIN I (7, 8, 9); and normal epithelium (2, 3). The biopsy DNAs were cleaved with BamHI, which linearizes HPV 16 prototype DNA. Off-size bands in lane 1 are due to integration into cellular DNA. Note different strength of signals and very weak, specific bands in lanes 2, 6, and 7. Lane 13 shows radiolabeled size standard DNA.

special questions. In this case, the HPV type is best narrowed down by a series of hybridizations using several cocktails followed by hybridizations with type-specific probes from the positive cocktail. The procedure is obviously rather time-consuming and extremely difficult if small amounts of tumor DNA are only sufficient for one or two blots. Filters must then be used repeatedly, starting with relaxed hybridization, followed by a stringent wash and a series of subsequent stringent hybridizations. In this case, the filters should be kept moist all the time. The use of nylon membranes, which are more stable and bind DNA more tightly, is recommended.

The problem can be better solved by a reverse assay (De Villiers et al., 1986a; Pfister et al., 1986). One microgram of tumor DNA is radiolabeled in this case and hybridized to a Southern blot containing reference DNAs of all known HPV types (Fig. 3). The blots are hybridized and washed under stringent conditions. Applying 100 ng each of the reference DNAs, 10 HPV genome copies per cell can be detected (De Villiers et al., 1986a). Three or four frozen sections of tumor material will provide enough DNA for this analysis.

Two in situ hybridization techniques were applied to papillomavirus diagnosis. In one case, a cell suspension is spotted onto nitrocellulose filters. Alkali treatment lyses the cells and denatures the DNA. After neutralization and baking, the filters are ready for hybridization, which is obviously very fast (Schneider et al., 1985; Wagner et al., 1984). In cases of low amounts of viral DNA, however, it is difficult to prove specificity, and closely related HPV types

such as HPV 6 and 11 or HPV 16 and 31 cannot be differentiated. Hybridization under relaxed conditions is not possible owing to background problems.

The second approach uses paraffin sections of tumor material, which offers access to samples, routinely processed for pathology. The probes can be made by nick translation of cloned HPV genomes using either biotinylated (Beckmann et al., 1985) or ^{35}S-labeled (Gupta et al., 1985) deoxyribonucleotide triphosphates. DNA hybrids in sections are consequently visualized by immunocytochemical reactions or by autoradiography. Stoler and Broker (1986) used single-stranded tritium-labeled RNA probes for in situ hybridization. These were transcribed from HPV genomes, which were cloned in Gemini vectors, where they are flanked by the SP6 and T7 promoters. Starting from either of the two promoters, one obtains a sense strand probe or an antisense probe, which hybridizes to viral mRNA. The sense orientation probe, in contrast, reacts only with HPV DNA denatured prior to hybridization. When testing genital tumors, much higher signals were observed with the antisense probes for mRNAs than with probes for DNA (Fig. 4). Some signals were even detected in basal cells of condylomas, and the invasive areas of a cervical carcinoma displayed a low-level labeling with HPV 16 antisense RNA. The sensitivity of detection was estimated to be 25 genomic mRNA equivalents per cell. Because of the stability of RNA-RNA duplexes, the slides can be washed at high stringency after hybridization, which results in low background. Closely related HPV types could be distinguished by relative signal

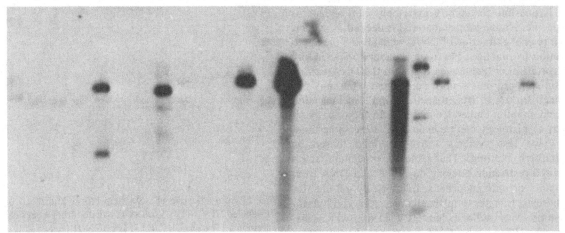

FIG. 3. Southern blot hybridization of reference DNAs for HPV 1 to 26 with ^{32}P-labeled DNA from flat, macular lesions of a patient with Hodgkin's disease. Reference DNAs were cleaved with appropriate restriction enzymes to separate viral from vector DNA. The biopsy DNA hybridized to a number of related HPV types belonging to group D. After cloning of the viral DNA from the biopsy, it was characterized as representing a new HPV type, tentatively designated HPV 46 (Ellinger, Gross, Fuchs, and Pfister, unpublished).

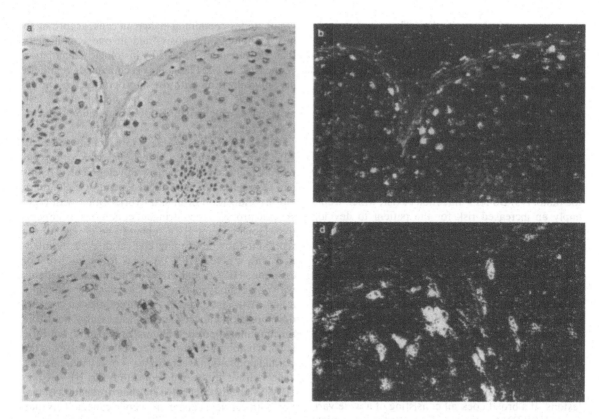

FIG. 4. Same vulvar condyloma as in Figure 1. (a) In situ hybridization with an HPV 6 RNA sense probe for viral DNA. Note the signal is limited to the cell nuclei. Hematoxylin counterstain. (b) Same section as (a); darkfield op- tics accentuate the signal. (c) Adjacent section to (a). Anti- sense RNA probe for HPV 6 mRNA. The signal pattern now has a cytoplasmic distribution. (d) Same field as (c); darkfield optics. Photographs kindly provided by M. Stoler.

strength obtained with the different probes in adjacent sections, and this might be improved by using type-specific, subgenomic RNA probes from less conserved areas of the HPV genome. Taken together, the method combines satisfactory sensitivity and specificity and surpasses immunocytochemistry by far (for comparison see Figs. 1 and 4).

Virus Isolation and Antibody Assays

These techniques play no role in diagnosis of papillomavirus infections. Virus isolation fails because there exists no animal or cell culture system susceptible to productive infection by HPV. On the other hand, the antibody response of the patients is slow and weak and seems to have little impact on the course of the disease (Jablonska et al., 1982; Pfister, 1984). It may take several months to develop low-titer antibodies after the onset of warts. IgM antibodies appear first and prevail for a long time without a detectable influence on tumor growth. The appearance of IgG antibodies was discussed as a frequent attendant symptom of wart regression. The studies were carried out with HPV antigen from pooled warts and are probably mainly representative for HPV 1. A type-specific humoral immune response was also detected with HPV 2, 3, and 8 and a pool of HPV 5, 8, 9 virions as antigens, but IgM and IgG antibodies were not differentiated and have not been correlated with wart regression so far. HPV 3–induced flat warts of a renal allograft recipient did not regress in spite of anti–HPV 3 antibodies of the IgG class with a titer of 1 : 80 in an indirect immunofluorescence test. A protective role of antibodies is not well established.

The percentage of anti–HPV 1 antibody-positive sera in the unselected population varies with age. Antibodies were detected by radioimmunoassay in 50% of young adults and in 35% of people between 30 and 60 years old (Pfister and Zur Hausen, 1978). Specific antibodies to HPV 1 were found by immunodiffusion in 10% of healthy adults and in about 20% of patients with common or flat warts (Jablonska et al., 1982). HPV 3 antibodies were detectable in 3 out of 21 patients with flat warts, and HPV 2 antibodies in 3 out of 12 patients with common warts. No patients with flat warts ($n = 21$) and no healthy adults ($n = 30$)

revealed anti–HPV 2 antibodies. These data point to differences in intensity and persistence of the antibody response to various HPV types. Humoral immunity to most HPV types was not tested, because type-specific antigens were not available.

Interpretation of Laboratory Data

Papillomavirus diagnosis has to address at least three questions: (1) Is a given lesion induced by HPV? (2) Does HPV persist in normal epithelium surrounding the tumor? (3) Do HPV infections exist, which would imply an increased risk for the patient to develop cancer? Question 1 serves differential diagnosis of tumors of skin and mucosa. The easiest approach is the demonstration of viral structural antigens in thin sections of fixed tissues or in exfoliated cells by means of the group-specific antiserum using the peroxidase-antiperoxidase technique. A positive result confirms HPV infection, but the absence of detectable antigen does not rule out an HPV etiology, because structural antigen is produced in only half of the lesions. Demonstration of viral DNA by Southern blot hybridization under relaxed conditions seems more promising, especially with precancerous lesions. If a broad cocktail consisting of a wide variety of HPV DNAs was used for screening, one might even apply stringent conditions and switch to in situ hybridizations with thin sections of Formalin-fixed, paraffin-embedded material, which would bring the test within reach of routine laboratories. One can also ask if a given patient is infected by HPV—for example, by screening routine swabs for the presence of HPV DNA. This may help to follow up and to prevent transmission (see Prevention and Control).

Question 2, regarding persistence of HPV in healthy epithelium, may be relevant to the treatment of recalcitrant warts. It is well known that warts recur frequently after surgical removal or laser treatment. Ferenczy and colleagues (1985) have recently identified HPV DNA in normal skin adjacent to lesions at a distance of 15 mm. Condylomas or intraepithelial neoplasm recurred after laser treatment in six of nine patients with inapparently infected skin margins but in only 1 of 11 patients without detectable HPV DNA in the surrounding tissue. Thus, screening of the periphery of a lesion for HPV DNA may provide valuable information regarding the extent of laser therapy. Cells could be removed from the area adjacent to the lesion by a swab and hybridized in situ with a DNA representative for the HPV type within the lesion prior to treatment.

The detection of HPV DNA may be a special issue in screening lymph nodes for metastasis of cervical cancer. HPV DNA was demonstrated in six of six lymph nodes containing malignant cells, according to histopathologic diagnosis, and in one lymph node that showed no metastasis by histologic examination (Lancaster et al., 1986). This recommends Southern blot hybridization with the appropriate DNA probe as sensitive test system for the detection of early metastasis, which might be used in addition to histologic screening.

The most intriguing question concerns the early detection of those HPV infections, which may impose a higher risk of cancer development. To this end a precise typing, which depends on Southern blot hybridization or in situ hybridization with type-specific probes under stringent conditions, is imperative. However, for HPV typing to be of benefit, the risk factor for developing clinical disease by infection with a given HPV type has to be known. For example, HPV 5 and HPV 8 infections of patients with epidermodysplasia verruciformis seem to imply a high risk, since one out of three infected patients develops skin cancer. Similar values may hold true for immunosuppressed allograft recipients, but little is known about the prevalence of these viruses in the normal population where HPV 5– or HPV 8–positive cancers have not been detected thus far. HPV 16 and HPV 18 are distinguished by their frequent association with cervical cancer and penile cancer. This may indicate that infections with these viruses increase the cancer risk five- to ten-fold relative to other HPV infections. However, exact data have to await prospective follow-up studies, which are just beginning and will be evaluated within a few years. Regarding the high prevalence of HPV 16 in the normal population (up to 10%; Fuchs et al., 1987; H. zur Hausen and W. D. Lancaster, personal communication), the absolute risk to develop a cervical cancer is probably lower than 1 carcinoma per 100 infections. It should be also noted that in general, HPV 16 infections of external genital sites of both sexes have a better prognosis than infections of the cervix uteri. Taken together, the studies to date suggest good reason for more frequent monitoring of HPV 16–infected patients. Radical therapy, on the other hand, would be inappropriate unless supported by a severe clinical picture.

The situation may change in the future if prospective studies reveal types or subtypes that imply a significantly higher cancer risk in comparison with other types. In parallel, however, one should look for parameters of HPV infection that are closer to malignant conversion of the precursor lesions. Integration of HPV 16 or HPV 18 DNA into the host genome may be one candidate. A predominant expression of open reading frames E6 and E7 also seems to be a consistent feature of malignant tumors, which could be tested, for example, by in situ hybridization with E6/E7-specific, antisense RNA probes.

The time point when these events happen during progression and the relevance for prognosis remain to be established.

Epidemiology

Plantar, common, and flat warts are most prevalent in children and young adults (Pfister and Zur Hausen, 1978; Bunney, 1982). The low incidence later in life is probably due to immune mechanisms. Anogenital HPV infections are frequently seen in sexually promiscuous adults and show the age distribution typical of venereal disease. Flat condylomas of the cervix were observed in 2% of a large number of unselected women (Meisels et al., 1982). Demonstration of viral DNA in healthy epithelium suggests that genital HPV infections occur more frequently than predicted by cytologic screening, and the incidence is estimated to be at least 20% (Fuchs et al., 1987; Lorincz et al., 1986b). Inconspicuous lesions like bowenoid papulosis will significantly contribute to virus spreading. As estimated from the number of consultations for genital warts and severe dysplastic cervical lesions (Scholl et al., 1985), the incidence of HPV infections has steadily increased in the past 10 years.

Laryngeal papillomas have a bimodal age distribution, with a first peak between 2 and 5 years and a second one at 40 years, but occur basically at any age. The high risk of children may be explained by primary infection during delivery from a mother with genital HPV infection. Regarding the etiologic agents, there is no obvious difference between juvenile-onset and adult-onset lesions. The latter could therefore represent late reactivations of perinatal infections. In contrast to genital tract warts, laryngeal papillomas are a rare disease with an annual incidence of about 1 per 1 million population (Mounts and Shah, 1984). This implies a rather inefficient transmission during delivery or at least a low risk to develop overt disease. Demonstration of HPV DNA should therefore be no indication for cesarean section unless the birth canal is heavily obstructed by florid condylomas.

The restriction of some HPV infections to certain ethnic groups or to people with an impaired cell-mediated immunity was mentioned above. Lesions that are induced by HPV types from group D are mainly confined to patients with epidermodysplasia verruciformis, which is by itself a very rare disorder. This raises questions about how these patients become infected with the large spectrum of group D viruses. Clinically inapparent, widespread infections in the general population seem to be the most likely reservoir for these viruses. A more recent screening of skin lesions indeed revealed HPV 5–related viruses in keratoacanthomas, solar keratosis, basaliomas and one melanoma (Orth, 1986; Pfister et al., 1986; Scheurlen et al., 1986), which demonstrates the occurrence in the general population irrespective of the question about the etiology of these tumors. Epidermodysplasia seems to be more frequent in Japan, showing an increasing northeast-southwest gradient (Tanigaki et al., personal communication). The relative importance of consanguinity and virus prevalence is not yet examined.

Control and Prevention

Papillomaviruses must be assumed to be rather stable, although this was not analyzed in detail owing to the lack of a quantitative test system. Transmission of plantar warts may be reduced by thorough disinfection of public bathroom floors. In view of the high incidence of genital HPV infections, care should be taken of disinfection of medical instruments to avoid iatrogenic transmission by routine gynecologic examinations. If an HPV infection is diagnosed, the patient's sexual partner(s) should be examined and precautions should be taken to avoid (further) sexual transmission until treatment is successful. The detection of HPV infections in males may well serve as a hint for the detection of cervical intraepithelial neoplasias in the female partners.

Transmission of papillomaviruses during the process of birth was discussed above. It is likely that delivery by cesarean section prior to rupture of the fetal membranes would decrease the risk of contracting laryngeal papilloma, and cesarean delivery should be recommended in the case of florid condylomas.

A vaccine for human papillomaviruses is not yet available. A recombinant DNA vaccine against BPV 1 was tested in preliminary field trials (Pilacinski et al., 1985). Significantly fewer animals developed warts after virus challenge in the vaccinated groups than in the control or placebo groups. One should also try to identify early viral proteins, if any, that are exposed at the surface of tumor cells. They could trigger an immune response of both preventive and therapeutic value.

Recalcitrant warts may by removed by application of caustic agents, by cryotherapy, or by surgery (Bunney, 1982). Leukocyte (alpha) interferon proved effective in the treatment of laryngeal papillomas and condylomas. In the case of laryngeal papillomas, treatment generally fails to eradicate the virus infection, as demonstrated by recurrences after halt of therapy or dose reduction (Pfister, 1984). Interferon therapy may be more efficient with condylomas. Complete responses were obtained in 70 to 80% of the patients after IM application of 5 MU (daily) to

0.6 MU (twice weekly) interferon for 6 to 8 weeks (Friedman-Kien et al., 1985; Gall et al., 1985).

Literature Cited

Almeida, J. D., A. F. Howatson, and M. G. Williams. 1962. Electron microscope study of human warts: sites of virus production and nature of the inclusion bodies. J. Invest. Dermatol. **38**:337–345.

Androphy, E. J., I. Dvoretzky, and D. R. Lowy. 1985. X-linked inheritance of epidermodysplasia verruciformis. Arch. Dermatol. **121**:864–868.

Barrett, T. J., J. D. Silbar, and J. P. McGinley. 1954. Genital warts—a venereal disease. J. Am. Med. Assoc. **154**:333–334.

Beaudenon, S., D. Kremsdorf, O. Croissant, S. Jablonska, S. Wain-Hobson, and G. Orth. 1986. A novel type of human papillomavirus associated with genital neoplasias. Nature **321**:246–249.

Beckmann, A. M., D. Myerson, J. R. Daling, N. B. Kiviat, C. M. Fenoglio, and J. K. McDougall. 1985. Detection and localization of human papillomavirus DNA in human genital condylomas by in situ hybridization with biotinylated probes. J. Med. Virol. **16**:265–273.

Boshart, M., L. Gissmann, H. Ikenberg, A. Kleinheinz, W. Scheurlen, and H. zur Hausen. 1984. A new type of papillomavirus DNA, its presence in genital cancer biopsies and in cell lines derived from cervical cancer. EMBO J. **3**:1151–1157.

Brandsma, J. L., B. M. Steinberg, A. L. Abramson, and B. Winkler. 1986. Presence of human papillomavirus type 16 related sequences in verrucous carcinoma of the larynx. Cancer Res. **46**:2185–2188.

Bunney, M. H. 1982. Viral warts: their biology and treatment. Oxford University Press, Oxford.

Burk, R. D., A. S. Kadish, S. Calderin, and S. L. Romney. 1986. Human papillomavirus infection of the cervix detected by cervicovaginal lavage and molecular hybridization: correlation with biopsy results and papanicolaou smear. Am. J. Obstet. Gynecol. **154**:982–989.

Ciuffo, G. 1907. Innesto positivo con filtrato di verrucae volgare. Giorn. Ital. Mal. Venereol. **48**:12–17.

Coggin, J. R. Jr., and H. zur Hausen. 1979. Workshop on papillomaviruses and cancer. Cancer Res. **39**:545–546.

Coleman, D. V., W. J. I. Russell, J. P. T. Hodgson, and J. F. Mowbray. 1977. Human papova virus in papanicolaou smear of urinary sediment detected by transmission electron microscopy. J. Clin. Pathol. **30**:1015–1020.

Crum, C. P., H. Ikenberg, R. M. Richart, and L. Gissman. 1984. Human papillomavirus type 16 and early cervical neoplasia. N. Engl. J. Med. **310**:880–883.

Della Torre, G., S. Pilotti, G. De Palo, and F. Rilke. 1978. Viral particles in cervical condylomatous lesions. Tumori **64**:549–553.

De Villiers, E. M., H. Weidauer, H. Otto, and H. zur Hausen. 1985. Papillomavirus DNA in human tongue carcinomas. Int. J. Cancer **36**:575–578.

De Villiers, E. M., A. Schneider, G. Gross, and H. zur Hausen. 1986a. Analysis of benign and malignant urogenital tumors for human papillomavirus infection by labelling cellular DNA. Med. Microbiol. Immunol. **174**:281–286.

De Villiers, E. M., C. Neumann, J.-Y. Le, H. Weidauer, and H. zur Hausen. 1986b. Infection of the oral mucosa with defined types of human papillomaviruses. Med. Microbiol. Immunol. **174**:287–294.

Dürst, M., L. Gissmann, H. Ikenberg, and H. zur Hausen. 1983. A new type of papillomavirus DNA from a cervical carcinoma and its prevalence in cancer biopsies from different geographic regions. Proc. Natl. Acad. Sci. USA **80**:3812–3815.

El Shazly, M. O., J. P. Sundberg, T. McPherron, and R. D. Smith. 1985. Enzyme-linked immunosorbent assay for the detection of bovine papillomavirus. Am. J. Vet. Res. **46**:1737–1739.

Favre, M., F. Breitburd, O. Croissant, and G. Orth. 1975. Structural polypeptides of rabbit, bovine, and human papilloma viruses. J. Virol. **15**:1239–1247.

Ferenczy, A., M. Mitao, N. Nagai, S. J. Silverstein, and C. P. Crum. 1985. Latent papillomavirus and recurring genital warts. N. Engl. J. Med. **313**:784–788.

Friedman-Kien, A. E., T. F. Plasse, P. Cremin, B. Castro, H. Badiak, J. R. Geffen, D. Fedorczyk, and J. R. Trout. 1985. Natural leukocyte interferon for treatment of condylomata acuminata. A randomized, double-blind, placebo controlled study, p. 217–233. In P. M. Howley and T. R. Broker (ed.), Papillomaviruses: molecular and clinical aspects. Alan R. Liss, New York.

Fuchs, P. G., T. Iftner, J. Weninger, and H. Pfister. 1986. Epidermodysplasia verruciformis–associated human papillomavirus 8: genomic sequence and comparative analysis. J. Virol. **58**:626–634.

Fuchs, P. G., F. Girardi, and H. Pfister. 1987. Papillomavirus infection in cervical tumors of Austrian patients, p. 297–300. In Cancer cells 5. Cold Spring Harbor Laboratory, Cold Spring Harbor, NY.

Gall, S. A., C. E. Hughes, K. Trofatter, Jr., J. Whisnant, and P. K. Weck. 1985. Lymphoblastoid interferon for the therapy of condylomata acuminata: an inter-study comparison of two dose regimens, p. 201–215. In P. M. Howley and T. R. Broker (ed.), Papillomaviruses: molecular and clinical aspects. Alan R. Liss, New York.

Gassenmaier, A., P. Fuchs, H. Schell, and H. Pfister. 1986. Papillomavirus DNA in warts of immunosuppressed renal allograft recipients. Arch. Dermatol. Res. **278**:219–223.

Gissmann, L., H. Pfister, and H. zur Hausen. 1977. Human papilloma viruses (HPV): characterization of 4 different isolates. Virology **76**:569–580.

Gissmann, L., L. Wolnik, H. Ikenberg, U. Koldovsky, H. G. Schnürch, and H. zur Hausen. 1983. Human papillomavirus types 6 and 11 DNA sequences in genital and laryngeal papillomas and in some cervical cancers. Proc. Natl. Acad. Sci. USA **80**:560–563.

Glinski, W., S. Obalek, S. Jablonska, and G. Orth. 1981. T cell defect in patients with epidermodysplasia verruciformis due to human papillomavirus type 3 and 5. Dermatologica **162**:141–147.

Green, M., K. H. Brackmann, P. R. Sanders, P. M. Loewenstein, J. H. Freel, M. Eisinger, and S. A. Switlyk. 1982. Isolation of a human papillomavirus from a patient with epidermodysplasia verruciformis: presence of related viral DNA genomes in human urogenital tumors. Proc. Natl. Acad. Sci. USA **79**:4437–4441.

Gross, G., H. Pfister, L. Gissmann, and M. Hagedorn. 1982. Correlation between human papillomavirus type (HPV) and histology of warts. J. Invest. Dermatol. **78**:160–164.

Gross, G., H. Ikenberg, L. Gissmann, and M. Hagedorn. 1985. Papillomavirus infection of the anogenital region: correlation between histology, clinical picture, and virus type. Proposal of a new nomenclature. J. Invest. Dermatol. **85**:147–152.

Growdon, W. A., T. B. Lebherz, A. Rapkin, G. D. Mason, and G. Parks. 1985. Pruritic vulvar squamous papillomatosis: evidence for human papillomavirus etiology. Obstet. Gynecol. **66**:564–568.

Guillet, G., L. Braun, K. Shah, and A. Ferenczy. 1983. Papillomavirus in cervical condylomas with and without associated cervical intraepithelial neoplasia. J. Invest. Dermatol. **81:**513–516.

Gupta, J. W., H. E. Gendelman, Z. Naghashfar, P. Gupta, N. Rosenshein, E. Sawada, J. E. Woodruff, and K. V. Shah. 1985. Specific identification of human papillomavirus type of cervical smears and paraffin sections by in situ hybridization with radioactive probes. Int. J. Gynecol. Pathol. **4:**211–218.

Ikenberg, H., L. Gissmann, G. Gross, E.-I. Grussendorf-Conen, and H. zur Hausen. 1983. Human papillomavirus type 16–related DNA in genital Bowen's disease and in bowenoid papulosis. Int. J. Cancer **32:**563–565.

Jablonska, S., G. Orth, G. Glinski, S. Obalek, M. Jarzabek-Chorzelska, O. Croissant, M. Favre, and G. Rzesa. 1980. Morphology and immunology of human warts and familial warts, p. 107–131. In P. A. Bachmann (ed.), Leukaemias, lymphomas and papillomas: comparative aspects. Taylor and Francis, London.

Jablonska, S., G. Orth, and M. A. Lutzner. 1982. Immunopathology of papillomavirus-induced tumors in different tissues. Springer Semin. Immunopathol. **5:**33–62.

Jablonska, S., G. Orth, S. Obalek, and O. Croissant. 1985. Cutaneous warts. Clinical, histologic, and virologic correlations. Clin. Dermatol. **3**(4):71–82.

Jenson, A. B., J. R. Rosenthal, C. Olson, F. Pass, W. D. Lancaster, and K. Shah. 1980. Immunological relatedness of papillomaviruses from different species. JNCI **64:**495–500.

Jenson, A. B., S. Sommer, C. Payling-Wright, F. Pass, C. C. Link, Jr., and W. D. Lancaster. 1982. Human papillomavirus: frequency and distribution in plantar and common warts. Lab. Invest. **47:**491–497.

Kahn, T., E. Schwarz, and H. zur Hausen. 1986. Molecular cloning and characterization of the DNA of a new human papillomavirus (HPV30) from a laryngeal carcinoma. Int. J. Cancer **37:**61–65.

Kawashima, M., S. Jablonska, M. Favre, S. Obalek, O. Croissant, and G. Orth. 1986. Characterization of a new type of human papillomavirus found in a lesion of Bowen's disease of the skin. J. Virol. **57:**688–692.

Kurman, R. J., A. B. Jenson, and W. D. Lancaster. 1983. Papillomavirus infection of the cervix. II. Relationship to intraepithelial neoplasia based on the presence of specific viral structural proteins. Am. J. Surg. Pathol. **7:**39–52.

Lack, E. E., A. B. Jenson, H. G. Smith, G. B. Healy, F. Pass, and G. F. Vawter. 1980. Immunoperoxidase localization of human papillomavirus in laryngeal papillomas. Intervirology **14:**148–154.

Lancaster, W. D., R. J. Kurman, L. E. Sanz, S. Perry, and A. B. Jenson. 1983. Human papillomavirus: detection of viral DNA sequences and evidence for molecular heterogeneity in metaplasias and dysplasias of the uterine cervix. Intervirology **20:**202–212.

Lancaster, W. D., C. Castellano, C. Santos, G. Delgado, R. J. Kurman, and A. B. Jenson. 1986. Human papillomavirus deoxyribonucleic acid in cervical carcinoma from primary and metastatic sites. Am. J. Obstet. Gynecol. **154:**115–119.

La Porta, R. F., and L. B. Taichman. 1982. Human papilloma viral DNA replicates as a stable episome in cultured epidermal keratinocytes. Proc. Natl. Acad. Sci. USA **79:**3393–3397.

Lass, J. H., A. S. Grove, J. J. Papale, D. M. Albert, A. B. Jenson, and W. D. Lancaster. 1983. Detection of human papillomavirus DNA sequences in conjunctival papilloma. Am. J. Ophthalmol. **96:**670–674.

Löning, T., H. Ikenberg, J. Becker, L. Gissmann, I. Hoepfer, and H. zur Hausen. 1985. Analysis of oral papillomas, leukoplakias, and invasive carcinomas for human papillomavirus type related DNA. J. Invest. Dermatol. **84:**417–420.

Lörincz, A. T., W. D. Lancaster, and G. F. Temple. 1986a. Cloning and characterization of the DNA of a new human papillomavirus from a woman with dysplasia of the uterine cervix. J. Virol. **58:**225–229.

Lörincz, A. T., G. F. Temple, J. A. Patterson, A. B. Jenson, R. J. Kurman, and W. D. Lancaster. 1986b. Correlation of cellular atypia and human papillomavirus deoxyribonucleic acid sequences in exfoliated cells of the uterine cervix. Obstet. Gynecol. **68:**508–512.

Lutzner, M., R. Kuffer, C. Blanchet-Bardon, and O. Croissant. 1982. Different papillomaviruses as the causes of oral warts. Arch. Dermatol. **118:**393–399.

Lutzner, M., G. Orth, V. Dutronquay, M.-F. Ducasse, H. Kreis, and J. Crosnier. 1983. Detection of human papillomavirus type 5 DNA in skin cancers of an immunosuppressed renal allograft recipient. Lancet **2:**422–424.

Matthews, R. E. F. 1982. Classification and nomenclature of viruses. Intervirology **17:**1–199.

McCance, D. J., M. J. Campion, P. K. Clarkson, P. M. Chesters, D. Jenkins, and A. Singer. 1985. Prevalence of human papillomavirus type 16 DNA sequences in cervical intraepithelial neoplasia and invasive carcinoma of the cervix. Br. J. Obstet. Gynaecol. **92:**1101–1105.

McCance, D. J., A. Kalache, K. Ashdown, L. Andrade, F. Menezes, P. Smith, and R. Doll. 1986. Human papillomavirus types 16 and 18 in carcinomas of the penis from Brazil. Int. J. Cancer **37:**55–59.

Meisels, A., C. Morin, and M. Casas-Cordero. 1982. Human papillomavirus infection of the uterine cervix. Int. J. Gynecol. Pathol. **1:**75–94.

Mounts, P., and K. Shah. 1984. Respiratory papillomatosis: etiological relation to genital tract papillomaviruses. Prog. Med. Virol. **29:**90–114.

Naghashfar, Z., E. Sawada, M. J. Kutcher, J. Swancar, J. Gupta, R. Daniel, H. Kashima, J. D. Woodruff, and K. Shah. Identification of genital tract papillomaviruses HPV-6 and HPV-16 in warts of the oral cavity. J. Med. Virol. **17:**313–324.

Nakai, Y., W. D. Lancaster, L. Y. Lim, and A. B. Jenson. 1986. Monoclonal antibodies to genus- and type-specific papillomavirus structural antigens. Intervirology **25:**30–37.

Orth, G. 1986. Epidermodysplasia verruciformis: a model for understanding the oncogenicity of human papillomaviruses, p. 157–174. In D. Evered and S. Clark (ed.), Papillomaviruses. Wiley, Chichester, U.K.

Orth, G., M. Favre, and O. Croissant. 1977. Characterization of a new type of human papilloma virus that causes skin warts. J. Virol. **24:**108–120.

Orth, G., S. Jablonska, M. Favre, O. Croissant, M. Jarzabek-Chorzelska, and G. Rzesa. 1978. Characterization of two new types of human papilloma viruses in lesions of epidermodysplasia verruciformis. Proc. Natl. Acad. Sci. USA **75:**1537–1541.

Pettersson, U., H. Ahola, A. Stenlund, and J. Moreno-Lopez. 1987. Organization and expression of papillomavirus genomes. In N. P. Salzmann and P. M. Howley (eds.), The *Papovaviridae:* the papillomaviruses. Plenum Press, New York.

Pfister, H. 1984. Biology and biochemistry of papillomaviruses. Rev. Physiol. Biochem. Pharmacol. **99:**111–181.

Pfister, H. 1987. Human papillomaviruses and genital cancer. Adv. Cancer Res. **48:**113–147.

Pfister, H., and H. zur Hausen. 1978. Seroepidemiological studies of human papilloma virus (HPV 1) infections. Int. J. Cancer **21:**161–165.

Pfister, H., G. Gross, and M. Hagedorn. 1979. Characterization of human papillomavirus 3 in warts of a renal allograft patient. J. Invest. Dermatol. **73**:349–353.

Pfister, H., I. Hettich, U. Runne, L. Gissmann, and G. N. Chilf. 1983. Characterization of human papillomavirus type 13 from focal epithelial hyperplasia Heck lesions. J. Virol. **47**:363–366.

Pfister, H., P. G. Fuchs, and H. E. Völcker. 1985. Human papillomavirus DNA in conjunctival papilloma. Graefe's Arch. Clin. Exp. Ophthalmol. **223**:164–167.

Pfister, H., J. Krubke, W. Dietrich, T. Iftner, and P. G. Fuchs. 1986a. Classification of the papillomaviruses—mapping the genome, p. 3–22. *In* D. Evered and S. Clark (ed.), Papillomaviruses. Wiley, Chichester, U.K.

Pfister, H., A. Gassenmaier, and P. G. Fuchs. 1986b. Demonstration of human papillomavirus DNA in two keratocanthomas. Arch. Dermatol. Res. **278**:243–246.

Pilacinski, W. P., D. L. Glassman, K. F. Glassman, D. E. Reed, M. A. Lum, R. F. Marshall, and C. C. Muscoplat. 1985. Development of a recombinant DNA vaccine against bovine papillomavirus infection in cattle, p. 257–271. *In* P. M. Howley and T. R. Broker (ed.), Papillomaviruses: molecular and clinical aspects. Alan R. Liss, New York.

Roseto, A., P. Pothier, M.-C. Guillemin, J. Peries, F. Breitburd, N. Bonneaud, and G. Orth. 1984. Monoclonal antibodies to the major capsid protein of human papillomavirus type 1. J. Gen. Virol. **65**:1319–1324.

Rowson, K. E. K., and B. W. J. Mahy. 1967. Human papova (wart) virus. Bacteriol. Rev. **31**:110–131.

Scheurlen, W., L. Gissmann, G. Gross, and H. zur Hausen. 1986. Molecular cloning of two new HPV types (HPV 37 and HPV 38) from a keratoacanthoma and a malignant melanoma. Int. J. Cancer **37**:505–510.

Scheurlen, W., A. Stremlau, L. Gissmann, D. Höhn, H.-P. Zenner, and H. zur Hausen. 1987. Rearranged HPV 16 molecules in an anal carcinoma and in a laryngeal carcinoma. Int. J. Cancer (in press).

Schneider, A., H. Kraus, R. Schuhmann, and L. Gissmann. 1985. Papillomavirus infection of the lower genital tract: detection of viral DNA in gynecological swabs. Int. J. Cancer **35**:443–448.

Scholl, S. M., E. M. Kingsley Pillers, R. E. Robinson, and P. J. Farrell. 1985. Prevalence of human papillomavirus type 16 DNA in cervical carcinoma samples in east anglia. Int. J. Cancer **35**:215–218.

Schwarz, E., U. K. Freese, L. Gissmann, W. Mayer, B. Roggenbuck, A. Stremlau, and H. zur Hausen. 1985. Structure and transcription of human papillomavirus sequences in cervical carcinoma cells. Nature **314**:111–114.

Seedorf, K., G. Krämmer, M. Dürst, S. Suhai, and W. G. Röwekamp. 1985. Human papillomavirus type 16 DNA sequence. Virology **145**:181–185.

Smotkin, D., and F. O. Wettstein. 1986. Transcription of human papillomavirus type 16 early genes in a cervical cancer and a cancer-derived cell line and identification of the E7 protein. Proc. Natl. Acad. Sci. USA **83**:4680–4684.

Steinberg, B. M. 1986. Laryngeal papillomatosis is associated with a defect in cellular differentiation, p. 208–220. *In* D. Evered and S. Clark (ed.), Papillomaviruses. Wiley, Chichester, U.K.

Steinberg, B. M., W. C. Topp, P. S. Schneider, and A. L. Abramson. 1983. Laryngeal papillomavirus infection during clinical remission. N. Engl. J. Med. **308**:1261–1264.

Stoler, M. H., and T. R. Broker. 1986. In situ hybridization detection of human papillomavirus DNAs and messenger RNAs in genital condylomas and a cervical carcinoma. Hum. Pathol. **17**:1250–1258.

Syrjänen, K. J. 1983. Human papillomavirus lesions in association with cervical dysplasias and neoplasias. Obstet. Gynecol. **62**:617–624.

Tanigaki, T., R. Kanda, and K. Sato. 1986. Epidermodysplasia verruciformis (L-L, 1922) in a patient with systemic lupus erythematosus. Arch. Dermatol. Res. **278**:247–248.

Wagner, D., H. Ikenberg, N. Boehm, and L. Gissmann. 1984. Identification of human papillomavirus in cervical swabs by deoxyribonucleic acid in situ hybridization. Obstet. Gynecol. **64**:767–772.

Walker, P. G., A. Singer, J. L. Dyson, K. V. Shah, A. To, and D. V. Coleman. 1983. The prevalence of human papillomavirus antigen in patients with cervical intraepithelial neoplasia. Br. J. Cancer **48**:99–101.

Winkler, B., V. Capo, W. Reumann, A. Ma, R. La Porta, S. Reilly, P. M. R. Green, R. M. Richart, and C. P. Crum. 1985. Human papillomavirus infection of the esophagus—a clinicopathologic study with demonstration of papillomavirus antigen by the immunoperoxidase technique. Cancer **55**:149–155.

Yoshikawa, H., T. Matsukura, E. Yamamoto, T. Kawana, M. Mizuno, and K. Yoshiike. 1985. Occurrence of human papillomavirus types 16 and 18 DNA in cervical carcinomas from Japan: age of patients and histological type of carcinomas. Jpn. J. Cancer Res. **76**:667–671.

Zur Hausen, H. 1977. Human papilloma viruses and their possible role in squamous cell carcinomas. Curr. Top. Microbiol. Immunol. **78**:1–30.

CHAPTER 17

Papovaviridae: The Polyomaviruses

RAY R. ARTHUR and KEERTI V. SHAH

Disease: Progressive multifocal leukoencephalopathy (PML). Ureteral stenosis. Hemorrhagic cystitis.

Etiologic Agents: JC virus (JCV) of PML. BK virus (BKV) of urinary tract disease. A human B-cell lymphotropic polyomavirus is suspected to exist on the basis of serologic evidence. Simian virus 40 was a contaminant of early poliovirus vaccines.

Sources: JCV and BKV transmitted person to person probably by the respiratory route. Viruses excreted in urine. Donor kidneys may introduce infection in recipients of renal transplants.

Clinical Manifestations: Occur mainly in immunosuppressed populations as a result of virus reactivation. PML is a rare, fatal disease with an insidious onset and symptoms reflecting multifocal involvement of the brain by JCV. Ureteral stenosis is a rare, late complication of renal transplantation; some cases are associated with BKV. BKV-associated hemorrhagic cystitis occurs frequently in recipients of bone marrow transplants and rarely in immunologically normal individuals.

Pathology: Viruses produce basophilic intranuclear inclusions in infected cells. Urinary tract and central nervous system are affected.

Laboratory Diagnosis: Depends largely on demonstration of viral particles, viral antigen, or viral genome either directly in clinical specimens or after cultivation of the viruses in cell lines of human origin. Serologic tests not often helpful.

Epidemiology: Worldwide prevalence. Primary infections occur in childhood and are largely asymptomatic. Viruses are reactivated in times of immunologic impairment.

Treatment: Specific antiviral therapy is not available. Reduction in immunosuppression, when possible, may ameliorate disease.

Prevention and Control: No measures are in use or under investigation.

Introduction

Progressive multifocal encephalopathy (PML), a rare demyelinating disease of the central nervous system which occurs on a background of immunodeficiency, has been described in the literature since 1930 under a variety of names. In 1961, Richardson correctly proposed that the pathognomonic nuclear changes in oligodendrocytes and astrocytes in PML were "consistent with the cytopathologic effects of a virus, which, in the presence of impaired immunologic responses, could result in atypical lesions" (Richardson, 1961). Particles of polyomavirus morphology were identified in the oligodendrocyte nuclei in PML lesions in 1965 (Silverman and Rubinstein, 1965; ZuRhein and Chou, 1965) and JC virus (JCV), the causative agent of PML, was isolated in 1971 after inoculation of a PML brain extract into primary human fetal glial (PHFG) cells (Padgett et al., 1971). In the same year, a second human polyomavirus, BK virus (BKV), was isolated from a renal transplant recipient who was excreting in his urine large numbers of cytologically abnormal epithelial cells with enlarged nuclei (Gardner et al., 1971). The virus was recovered after inoculation of the urine of the patient into Vero cells, a cell line of African green monkey kidney origin. BKV infection is associated with urinary tract disease in immunosuppressed patients.

The existence of a third human polyomavirus has been inferred on the basis of cross-reactivity of human sera with a B-cell lymphotropic virus of African green monkeys (zur Hausen et al., 1980), but this virus has not been isolated.

Primary infections with BKV and JCV occur in childhood and are largely asymptomatic. The viruses remain latent after primary infection and can be reactivated in subsequent years, especially when there is an impairment of T-cell functions. Almost all the pathogenic effects of JCV and BKV occur in immunologically compromised groups.

Millions of individuals in the United States were inadvertently exposed to small amounts of live simian virus 40 (SV40) between 1955 and 1961, when this virus of Asian macaques was a contaminant of the inactivated poliovirus (Salk) vaccines (Shah and Nathanson, 1976). Simian virus 40 is oncogenic for laboratory animals and is capable of transforming mammalian cells. Therefore, there was concern that some individuals inoculated with SV40-contaminated vaccine may develop tumors. However, there is no persuasive evidence that SV40 in the vaccine produced any ill effect in the exposed population.

Initially, polyomaviruses and papillomaviruses were grouped together as papovaviruses on the basis of virion morphology, circular double-stranded DNA genome, and nuclear site of multiplication. The two groups are currently classified as different genera of the family *Papovaviridae* (Melnick et al., 1974). Nucleic acid hybridization and immunologic studies have revealed no evidence of an evolutionary relationship between the two genera. Furthermore, the genetic organization of the two groups is quite dissimilar. The polyomavirus genome codes for about one half of the genetic information on each DNA strand, whereas all of the information in the papillomavirus genome is located on one strand. It is probable that the two genera will be separated into unrelated virus groups and the term "papovavirus" will cease to have any taxonomic significance.

The Viruses

Polyomaviruses are widely distributed in vertebrates (Table 1). Recognized hosts of polyomaviruses include humans, monkeys, cattle, rabbit, mouse, hamster, rat, and budgerigar.

Morphology and Physicochemical Properties

The virion is nonenveloped, has an icosahedral symmetry, and 72 capsomers (Fig. 1). It is small with a diameter of about 45 nm and a virion molecular weight of about 27×10^6 daltons (Brady and Salzman, 1986). The viral genome constitutes about 12% of the virion mass and is a double-stranded, covalently closed, circular, supercoiled molecule with about 5×10^3 base pairs and a molecular weight of about 3×10^6 daltons. The genomes of JCV and BKV are completely sequenced as are the genomes of SV40, African green monkey lymphotropic virus, and polyoma viruses. The genomes of JCV, BKV, and SV40 have a nearly identical genetic organization and a high degree of nucleotide sequence homology; overall, the JCV genome shares 75% of the sequences with the BKV genome and 69% of the sequences with the SV40 genome (Walker and Frisque, 1986). The viral genome is functionally divided into an early region (2.3 kb) which codes for large and small T proteins, a late region (2.3 kb) which codes for viral capsid proteins VP1, VP2, and VP3 and a noncoding regulatory region (0.4 kb) which contains the large T antigen-binding sites, origin of DNA replication, and transcription control sequences. The early and late regions are transcribed from different strands of the DNA molecule.

Six proteins are coded by BKV, JCV, and SV40 genomes. Large and small T are nonstructural early proteins that mediate a variety of functions including

TABLE 1. Polyomaviruses: natural hosts and affected body sites

Host	Virus	Site of multiplication
Human	JC virus (JCV)	Urinary tract, brain
	BK virus (BKV)	Urinary tract
Monkey	Simian virus 40 (SV40) (Asian macaques)	Kidney, brain, lung
	Simian agent 12 (SA12) (baboon)	Kidney
	Lymphotropic papovavirus (LPV) (African green monkey)	B lymphoblast
Cattle	Bovine polyomavirus (BPoV)	Kidney
Rabbit	Rabbit kidney vacuolating virus (RKV)	Kidney
Mouse	Polyoma virus	Kidney, gut, lung
	K virus	Lung endothelial cells
Hamster	Hamster papovavirus (HaPV)	Skin
Athymic rat	Rat polyoma virus	Parotid glands
Budgerigar	Budgerigar fledgling disease virus (BFDV)	Many organs

initiation of viral DNA replication, stimulation of host DNA synthesis, modulation of early and late transcription, and the establishment and maintenance of cellular transformation (Brady and Salzman, 1986). VP1, VP2, and VP3 are late proteins that form the virion capsid. The sixth protein is a small late protein termed agnoprotein; its function is not well understood. In addition to these six viral-coded proteins, the viral DNA is complexed with histones of cellular origin.

Replication

Both JCV and BKV have a restricted host range. In in vitro studies, they efficiently infect only cells of human origin. JCV infects PHFG cells derived from fetal brain and urothelial cells derived from infant urine. BKV infects primary human embryonic kidney cells and diploid human fibroblasts.

The virion enters the cell by pinocytosis and travels to the nucleus where viral uncoating and multiplication takes place. The molecular mechanisms of viral transcription, DNA replication, and viral assembly are known in great detail (Livingston and Bikel, 1985) and are briefly summarized here. After uncoating, the early strand of the input viral DNA (which codes for large T and small t proteins) is transcribed. The mRNAs for large T and small t proteins are generated from the early strand by differential splicing, and the proteins are synthesized. The large T antigen is a phosphorylated multifunctional protein. It binds to one or more of the antigen binding sites on the noncoding region of the viral DNA to stimulate viral transcription and to initiate viral DNA replication. It also interacts with specific cellular proteins to stimulate the formation of host enzymes required for the replication and metabolism of DNA. Viral DNA replicates bidirectionally from the origin of replication, with one strand synthesized in a continuous manner and the other as fragments that are later joined. The replicated DNA is complexed with cellular histones. After new viral DNA is synthesized, the large T antigen binds to the origin region to down regulate early transcription. This is accompanied by the transcription of the late message, the generation of mRNAs for the late proteins, the synthesis of viral capsid proteins and the agnoprotein, assembly of virions, and cell lysis.

In nonpermissive cells, for example, rodent cells, JCV and BKV infection results in the production of early transcripts and T antigens, but late transcripts and late proteins are not synthesized. Some of the infected cells may become transformed. Transformed cells maintain and express the large T antigen and contain copies of the viral genome that are most often integrated into the host cell genome.

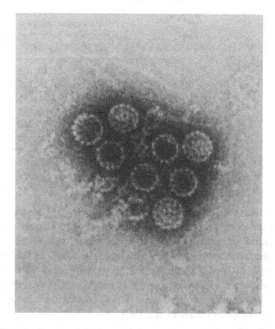

Fig. 1. Immune electron microscopic preparation showing BKV virions in urine from a patient. Particles reacted with BKV antiserum are aggregated. (Courtesy of Magdelena Reissig.)

Antigenic Composition

Capsids of polyomaviruses are resolved into six polypeptides (VP1 to VP6) by SDS-PAGE. VP1 is the major capsid protein and accounts for 70% of the virion mass. VP2 and VP3 are minor capsid proteins. VP4, VP5, and VP6 are cellular histones that are complexed with the viral genome. In addition to the three capsid proteins, the viral genome also codes for large and small T antigens and for an agnoprotein. The molecular weights of these six virus-coded proteins of JCV and BKV as deduced from nucleotide sequence data (Walker and Frisque, 1986) are shown in Table 2. The homologous proteins of the two viruses share large numbers of amino acids ranging from a high of 83% for the large T antigen, to 75 to 79% for the capsid proteins and small T antigen, to a low of 59% for the agnoprotein.

JCV and BKV have both type-specific and type-common antigenic determinants in the virion. Antibodies produced after natural infection are nearly all type specific (Gardner, 1977; Padgett and Walker, 1976). Little or no cross-reactivity is seen between JCV and BKV when human sera with high antibody titers to one virus are examined for antibodies to the other virus by immunofluorescence, hemagglutination-inhibition, neutralization, or ELISA. A low degree of cross-reactivity is found in hyperimmune rabbit sera prepared by inoculation of multiple doses of

TABLE 2. A comparison of viral proteins of JCV and BKV[a]

| Protein | Molecular weight | | No. of amino acids in JCV | Percent of BKV amino acids shared with JCV |
	JCV	BKV		
VP1	39,606	40,106	354	78
VP2	37,366	38,345	344	79
VP3	25,743	26,718	225	75
Large T	79,305	80,499	688	83
Small T	20,236	20,469	172	78
Agnoprotein	8,081	7,396	71	59

[a] Modified from Table IV of Walker and Frisque (1986).

intact virus (Penny and Narayan, 1973). Shared, type-common antigenic determinants are largely located internal to the virion surface. Antibodies to these determinants, made by immunization of rabbits with disrupted viral capsids or with purified VP1 of any one polyomavirus are strongly cross-reactive with capsid antigens of all polyomaviruses in infected cells (Shah et al., 1977).

Reactions to Physical and Chemical Reagents

Because they are nonenveloped, JCV and BKV are relatively stable viruses (Melnick et al., 1974). They are resistant to treatment by ether and to sodium desoxycholate. Infectivity is destroyed by autoclaving, by treatment with detergents, and at alkaline pH.

Pathogenesis and Pathology

The viruses have a high degree of species, tissue, and cell specificity. They multiply efficiently only in their natural hosts. Both JCV and BKV multiply in the epithelial cells of the urinary tract. In addition, JCV exhibits neurotropism as evidenced by its ability to infect the brain in PML cases, its requirement of fetal glial cells for in vitro multiplication, and its ability to induce brain tumors in experimentally inoculated nonhuman primates (London et al., 1978).

Primary infections with JCV and BKV occur in childhood and are largely asymptomatic. The viruses have been recovered rarely from immunologically normal individuals. The study of animal polyomaviruses suggests that the following sequence of events occurs in the course of infection. The viruses probably enter the body by the respiratory tract and multiply at the site of primary infection. They are then transported to kidneys, the target organ, by viremia. They multiply in the urinary tract and may produce transient viruria. In the immunocompetent host, the viruses remain latent for an indefinite time without any ill effect. Genomic sequences of JCV

and BKV are found in cadaver kidneys (Chesters et al., 1983). The viruses are reactivated in a variety of conditions in which T cell functions are impaired, for example, organ transplants, AIDS, immunosuppressive therapy for malignant or chronic diseases, primary immunodeficiency diseases, pregnancy, diabetes, and old age. Reactivation is manifested by virus multiplication in the urinary tract and viruria. The pathologic consequences of the infections appear to be the result of cytolytic action of the viruses on infected cells. For example, the lesions and symptoms of PML are fully explained on the basis of destruction of oligodendrocytes as a result of infection by JCV. Immunopathologic factors are not significantly involved in diseases caused by JCV and BKV.

JCV and BKV, like other polyomaviruses, multiply only in the nucleus. Productive infection of the cell is accompanied by increased nuclear size, nuclear inclusions, and cell death. In the infected host, the affected organs show characteristic, often pathognomonic changes, for example, the altered oligodendrocytes in PML lesions and the inclusion-bearing urothelial cells during JCV and BKV viruria.

Both BKV and JCV are oncogenic when inoculated subcutaneously or intracerebrally into newborn hamsters (Padgett and Walker, 1976; ZuRhein, 1983; Yoshiike and Takemoto, 1986). JCV also induces cerebral neoplasms in owl and squirrel monkeys after intracerebral inoculation of the virus (London et al., 1978; London et al., 1983).

Clinical Features

The precise manner of transmission, the incubation period, and the period of infectivity are not known for either JCV or BKV. Transmission of the viruses is efficient as evidenced by the observation that nearly one half of the children have antibodies to BKV by the age of 3 to 4 years (Shah et al., 1973). The virus is excreted in urine.

The etiology of PML is clearly established as being due to opportunistic JCV infection in immunocompromised individuals. In addition, BKV infection may be responsible for some cases of hemorrhagic or nonhemorrhagic cystitis and for some cases of ureteral stenosis. Viruria without associated disease occurs frequently as a result of virus reactivation (Table 3).

Primary Infection

In immunocompetent hosts primary infections are essentially asymptomatic. The only evidence of illness associated with primary infection is provided by a prospective study of children in which antibody conversion to BKV was associated with mild respiratory disease (Goudsmit et al., 1982). Although transmission by the respiratory route is consistent with the rapid acquisition of antibodies in childhood, BKV has not been recovered from respiratory secretions of either ill or healthy individuals. No illness has been associated with primary JCV infection in immunocompetent hosts.

Progressive Multifocal Leukoencephalopathy

Progressive multifocal leukoencephalopathy (PML) is a rare, subacute, demyelinating disease of the central nervous system which occurs on a background of conditions that are known to impair T-cell functions (Johnson, 1982; Walker and Padgett, 1983). Most PML cases occur at older ages, probably as a result of JCV reactivation. However, in recent years, PML has been recorded in young children with immunodeficiency diseases and in young adults who have renal allografts or are suffering from AIDS (Miller et al., 1982; Padgett and Walker, 1983). The disease in immunodeficient children may be a result of unchecked primary infection with JCV. PML is a rare complication of a wide variety of conditions that include lymphoproliferative diseases such as Hodgkin's disease, chronic lymphocytic leukemia, and lymphosarcoma; chronic diseases such as sarcoidosis and tuberculosis; and diseases such as rheumatoid arthritis and

polymyositis which require prolonged immunosuppressive therapy. In a small number of cases of PML, no underlying disease is identified.

The pathologic lesion of PML is unique (Fig. 2). The brain contains multiple foci of demyelination. These contain at their edges enlarged oligodendrocytes (PML cells) that have enlarged, deeply basophilic nuclei with inclusion bodies. The centers of the foci of demyelination show a loss of myelin and of oligodendrocytes. A majority of the lesions also contain greatly enlarged, bizarre, giant astrocytes, which have pleomorphic hyperchromatic nuclei (Fig. 2). These giant astrocytes resemble the malignant astrocytes of pleomorphic glioblastomas. The affected oligodendrocytes represent lytic productive infection with JCV and contain abundant amounts of viral particles (Figs. 3 and 4), viral antigen, and viral DNA. The changes in astrocytes are thought to be the result of transformation of the cells by JCV; affected astrocytes seldom display viral particles or viral antigen but contain JCV DNA. Virus and viral antigen are absent in "burnt-out" cases of PML, which are devoid of altered oligodendrocytes. Demyelinating foci occur most commonly in the subcortical white matter. The cerebrum is almost always affected. Inflammatory response is absent in most cases.

The disease has an insidious onset. The symptoms reflect the multifocal involvement of the central nervous system. Impaired speech and vision are common, and mental deterioration occurs early in the disease. The disease progresses rapidly and paralysis of limbs, cortical blindness, and sensory abnormalities are common in the late stages of the disease. Death occurs within 3 to 6 months after onset. Throughout the disease, the patient remains afebrile and the cerebrospinal fluid is normal. Electroencephalograms show nonspecific changes. The lymphocytes of the patient have a pronounced inability to respond to JCV antigen in lymphoproliferative assays (Willoughby et al., 1980).

The diagnosis of PML can be conclusively established only by pathologic examination of tissue taken by brain biopsy or postmortem. JCV particles, antigen, and DNA are readily demonstrable in affected

TABLE 3. Illnesses[a] associated with JCV and BKV

Virus	Infection	Immunodeficient host	Immunocompetent host
JCV	Primary	Asymptomatic; rarely PML in children	Asymptomatic
	Reactivation	Asymptomatic; PML	Symptomless viruria
BKV	Primary	Asymptomatic; rarely tubulonephritis	Asymptomatic; mild respiratory disease; rarely hemorrhagic or nonhemorrhagic cystitis
	Reactivation	Asymptomatic; ureteral stenosis in renal transplant recipients; hemorrhagic cystitis in bone marrow transplant recipients	Symptomless viruria

[a] Transient hepatic and renal dysfunction, pancreatitis, and pericardial effusion have been recorded during JCV or BKV infections.

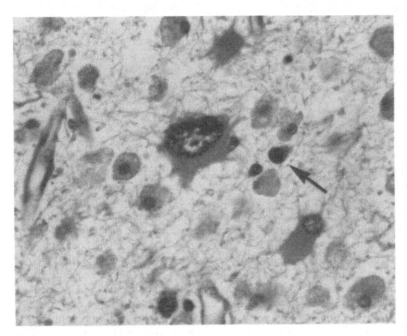

FIG. 2. Characteristic cytopathology of PML: bizarre giant astrocytes and oligodendrocyte with enlarged deeply basophilic nucleus (*arrow*). Also seen are round macrophages and nonspecific reactive astrocytes with fine fibrillary processes (hematoxylin and eosin stain). (Courtesy of Gabriele M. ZuRhein.)

oligodendrocytes. In contrast, JCV has never been demonstrated in normal brains or in nondiseased areas of PML brains. Viral DNA recovered from any individual PML brain is homogeneous in size (Grinnell et al., 1983a). Viral genomes cloned from different PML brains show variations in restriction enzyme digest patterns. Small amounts of unintegrated JCV DNA may be found at some extraneural sites such as kidney, liver, lung, lymph node, and spleen of PML patients (Grinnell et al., 1983b). The widest dissemination of JCV DNA was found in two children with PML, suggesting that the disease in these cases was the result of unchecked primary infection.

Infections in Renal Transplant Recipients

Both BKV and JCV infections occur commonly in renal transplant recipients. Viruria has been demonstrated in 10 to 40% and antibody rise in 20 to 50% of the recipients (Hogan et al., 1984). Viruria may be transient or prolonged and may occur at any time after transplant (Gardner et al., 1984). Infections may be the result of reactivation of latent viruses in the recipients or may be brought in with the kidneys of seropositive donors (Andrews et al., 1982). Some cases of ureteral stenosis, which is a late and rare complication of renal transplantation, are associated with BKV infections (Hogan et al., 1984). There is no clear evidence that infection triggers rejection of the transplant or that it leads to renal failure (Gardner et al., 1984).

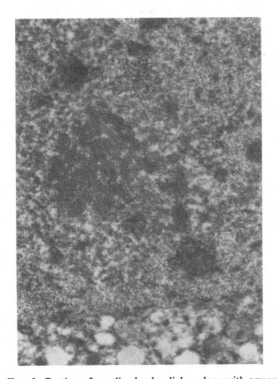

FIG. 3. Portion of an oligodendroglial nucleus with arrays of virus particles between the chromatin. The cytoplasm (*bottom*) contains no virus particles. (Courtesy of Gabriele M. ZuRhein.)

Infections in Bone Marrow Transplant Recipients

Bone marrow transplant (BMT) recipients are severely immunocompromised because the immune system of the recipient is completely destroyed by

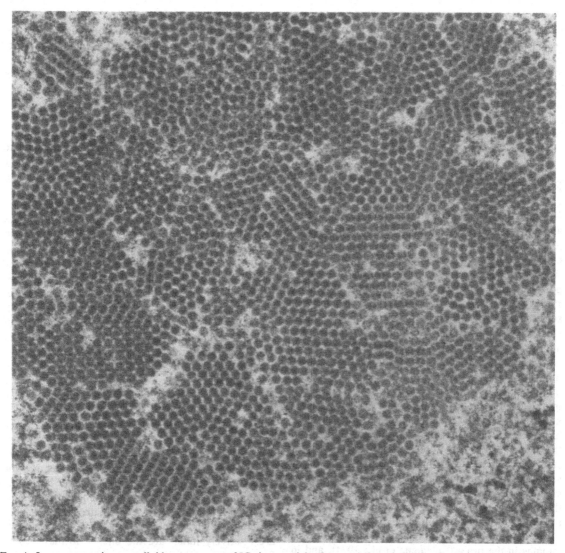

FIG. 4. Large composite crystalloid arrangement of JC virus particles from an infected oligodendroglial nucleus (from brain of JC, source of original virus isolation). (Courtesy of Gabriele M. ZuRhein.)

drugs and radiation in preparation for the transplantation. These patients are subject to reactivation of many latent viral and bacterial pathogens. BK viruria occurs in more than half of the BK-seropositive patients, and is far more common than JC viruria, which occurs in less than 10% of JCV-seropositive patients (Arthur et al., 1986). Viruria is not seen in patients who are seronegative at the time of transplantation. Most of the reactivations have onset between 3 and 7 weeks after transplant. BK viruria may be transient or may last for up to several weeks. BKV infection probably contributes to the occurrence of late-onset hemorrhagic cystitis, a frequent and potentially life-threatening complication in this group of patients. There is a strong temporal association between BK viruria and hemorrhagic cystitis (Arthur et al., 1986; Apperley et al., 1987). The

disease is probably precipitated by the combined effects of BK viruria, immunosuppression, and the toxicity of some of the drugs for the bladder mucosa.

Infections in Pregnancy

Pregnancy reactivates JCV and BKV infections. Viruria has been demonstrated in about 3% and antibody rise in about 15% of pregnant women (Coleman et al., 1980; Daniel et al., 1981). There is no evidence that reactivated viruses are transmitted transplacentally to infect the fetus. Early reports of detection of BKV-specific IgM antibodies in umbilical cord sera (Rziha et al., 1978; Taguchi et al., 1975) have not been confirmed (Borgatti et al., 1979).

Cystitis and Kidney Disease in Children

Primary BKV infection and viruria have been linked to occasional cases of cystitis in otherwise healthy children (Hashida et al., 1976; Padgett et al., 1983). Irreversible tubulointerstitial nephritis has been documented in an immunodeficient child who suffered prolonged BKV infection (Rosen et al., 1983). These reports indicate that BKV infection may occasionally cause severe urinary tract disease.

Role in Human Malignancies

JCV and BKV are oncogenic in laboratory animals and are capable of transforming human cells. Therefore, evidence for the roles of these viruses in the etiology of human tumors has been sought. The viruses, viral antigens, and viral DNAs have been reported from a number of human tumors (Howley, 1983), but the viruses have not been consistently associated with any specific human tumor. In the hamster, BKV produces brain tumors (ependymomas and choroid plexus papillomas), tumors of pancreatic islets, and osteosarcomas. In a study of human tumors of similar histologic types, BKV DNA and infectious BKV have been recovered from brain tumors and from insulinomas (Caputo et al., 1983; Takemoto et al., 1974). A BK variant recovered from a human insulinoma was found to have a deletion in the early region that abolished the expression of small T antigen and an insertion in the regulatory region that produced rearrangements in the enhancer elements (Pagnani et al., 1986). It has been proposed that viruses with variant genomes may produce human malignancies (Pagnani et al., 1986).

In one patient, PML coexisted with multifocal gliomas in the brain (Castaigne et al., 1974), and in another patient multiple malignant astrocytomas were found with PML (Sima et al., 1983). The topographic correlation between the areas of demyelination and the tumors in these two patients suggested that the tumors may have arisen in the PML lesions.

Epidemiology and Natural History of JCV and BKV

Most of the infections with both JCV and BKV occur in childhood. Fifty percent of the children in a suburban population in the United States, acquired antibodies to BKV by the age of 3 to 4 years and to JCV by the age of 10 to 14 years (Padgett and Walker, 1973; Shah et al., 1973). The antibody prevalence to BKV reaches nearly 100% by the age of 10 to 11 years and then declines to about 70 to 80% at older ages. Antibody prevalence to JCV reaches a peak of about 75% by adult age.

No animal reservoirs are known for JCV or BKV. The exact manner of transmission of these viruses from an infected to a susceptible individual is not known. JCV and BKV are excreted in the urine of infected individuals, but it is not clear if the urine-disseminated virus serves as the source of infection of susceptible individuals. The acquisition of BKV antibodies in the early years of life is suggestive of a spread by the respiratory route. A mild respiratory illness has been associated with seroconversion to BKV, but the virus has not been recovered from the respiratory secretions of either ill or healthy individuals (Goudsmit et al., 1982).

After primary infection the viruses remain latent for an indefinite time. Reactivation of the viruses may be brought about not only by overt immunosuppression as in organ transplantation but also by more subtle changes such as pregnancy. The duration and intensity of viruria and the proportion of seropositive individuals who display reactivation probably varies according to the degree of immunosuppression. BKV and JCV do not respond similarly with respect to reactivation. Although both viruses are commonly reactivated in recipients of renal transplants, BKV is reactivated far more frequently than JCV in recipients of bone marrow transplants.

Control and Prevention

Attempts to treat PML have been generally unsuccessful. Nucleic acid base analogs such as adenine arabinoside and cytosine arabinoside have been tried in an attempt to inhibit viral multiplication. Patients in whom it is possible to reduce iatrogenic immunosuppression and patients whose immune defenses are relatively intact are the ones who may benefit from therapy and have a less rapidly progressive disease. However, remissions are rare, and death is the expected outcome. Hemorrhagic cystitis in BMT recipients appears to be a result of the combined effects of viruria, immunosuppression, and toxicity of the drugs for the urothelium. Improved treatments that reduce immunosuppression and drug toxicity may be expected to reduce the incidence of this disease.

There are no measures recommended, or considered to be necessary, for the prevention of BKV and JCV infections in the general population.

Diagnosis

Until recently, BKV and JCV were identified most often by immune electron microscopy and virus isolation in cell cultures. Both of these methods have limitations. They are also not suitable for testing large numbers of samples. Assays for direct detection of viral antigen and viral DNA, which have been described recently, are more suitable for laboratory diagnosis of BKV and JCV. Serologic methods are of limited value in routine diagnosis.

Specimen Collection and Transport: Biosafety

Diagnosis of active polyomavirus infection is usually based on examination of urine. Examination of brain tissue is required for the diagnosis of PML. Detailed instructions for specimen shipment are found in Chapter 1, "An Introduction to Viral Diagnostics." Specimen handling and diagnostic laboratory procedures do not require any additional measures above those used in prudent laboratory practice. Laboratory requirements for routine diagnostic procedures are biosafety level 2, as described by the Centers for Disease Control Biosafety Manual (Richardson and Barkley, 1984).

Direct Virus Detection

CYTOLOGY

Cytologic examination of urinary epithelial cells is a widely used technique to screen for polyomavirus infections. The discovery of BKV was a result of cytologic observation made on a Papanicolaou smear of urinary sediment from a renal allograft recipient (Gardner et al., 1971). Urine specimens are prepared and stained for examination by standard cytologic procedures (Traystman et al., 1980). A typical polyomavirus-infected urothelial cell is enlarged, with a prominent deeply staining basophilic nucleus containing a single inclusion (Fig. 5). The intranuclear inclusions in some cells contract, separate from the nuclear membrane, and have an indistinct halo. These cytologic abnormalities need to be distinguished from neoplastic changes (Kahan et al., 1980). It is not possible to differentiate BKV from JCV by cytologic examination. Viruria is often present without any recognized cytologic abnormality and cytologic changes due to other virus infections, for example, cytomegalovirus infection, are not always distinguishable from those due to polyomavirus infections. Therefore, virologic confirmation of a cytologic diagnosis is recommended.

ELECTRON MICROSCOPY

Electron microscopy is a rapid and frequently used technique for the detection of BKV and JCV. Papovavirus particles were visualized in brain tissue from patients with PML (Silverman and Rubinstein, 1965; ZuRhein and Chou, 1965) years before the isolation of JCV in cell culture was reported (Padgett et al., 1971). Procedures for electron microscopy are described in Chapter 7. In negatively stained prepa-

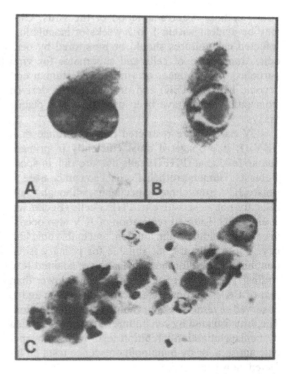

FIG. 5. Urinary epithelial cells dysplaying cytologic changes due to polyomavirus infection. (A and B) Cells with enlarged nuclei occupying two thirds of cell, and a comet-like appearance of the cytoplasm. The nucleus is filled with a large gelatinous mass that is frequently attached to the nuclear membrane and is sometimes surrounded by an irregular halo (B). (C) A cast formation consisting of urinary epithelial cells showing nuclear changes due to BKV infection. (Courtesy of Monica D. Traystman.)

rations from urines, virions are seen as spherical (icosahedral), nonenveloped particles and are about 44 nm in diameter (Fig. 1). Particles of similar morphology (Figs. 3 and 4) are seen by examination of ultrathin sections of diseased brain tissue from patients with PML.

BKV and JCV can be specifically identified by immune electron microscopy (Chapter 7). Virions reacted with a specific antiserum may either show aggregation or may display antibody on the surface (Traystman et al., 1980).

Virus Isolation and Identification

Although BKV was first isolated in monkey cells (Vero), BKV replication is most efficient in cells of human origin. Human diploid lung fibroblasts (WI-38), embryonic kidney (HEK), embryonic lung, fetal brain, and urothelial cells have been used for primary isolation of BKV from clinical specimens (Hogan et al., 1984). Cytopathic effect (CPE) is variable and dependent on the cell line. Observation periods of several weeks to months are required for primary viral isolation. After adaptation of the virus, CPE may be evident within 1 to 2 weeks of inoculation. Infected cell cultures should be monitored by periodic examination of cells and supernates for viral particles, viral antigen, or viral DNA. Human embryonic kidney (HEK) and urothelial cells derived from infant urine have been used for BKV plaque assays.

JCV has a more restricted host cell range than BKV. It is propagated most efficiently in primary human fetal glial (PHFG) cells that are rich in spongioblasts. Human urothelial cells also can be used to isolate JCV from clinical specimens, although these cells are not as sensitive as PHFG cells (Beckmann et al., 1982). Limited replication of JCV also occurs in cells from adult brain, amnion, and embryonic kidney. These cells are not suitable for primary isolation, but are used for the propagation of adapted JCV strains. The growth of JCV is somewhat slower than that of BKV. Cytopathic effect is nonspecific and is observed several weeks or months after inoculation. The virus isolated by cell culture is usually identified by hemagglutination-inhibition tests with virus-specific antisera, or by immunofluorescence tests on infected cells (Fig. 6).

Antigen Detection

IMMUNOLOGIC STAINING OF VIRUS INFECTED CELLS

BKV and JCV can be identified by immunofluorescent staining of infected cells with virus-specific, flu-orescein-conjugated antiserum. This method has been used to identify BKV and JCV in exfoliated urothelial cells in urine (Hogan et al., 1980). Urinary cells are sedimented by centrifugation, placed on slides, fixed, stained, and examined for nuclear fluorescence. This method is suitable for specific virus identification in urine samples that have previously been shown to contain inclusion-bearing cells. It may be a sensitive method for detecting mixed BKV and JCV infections (Hogan et al., 1980). Immunofluorescent staining also has been applied for detection of JCV antigens in PML brains (Bauer et al., 1973; Budka and Shah, 1983). Immunohistochemical staining of paraffin sections by an immunoperoxidase technique may be used for detecting polyomavirus antigens in infected tissues (Gerber et al., 1980; Fig. 7).

ELISA FOR ANTIGEN DETECTION

ELISA is a sensitive and specific method for detection of either BKV or JCV in urine (Arthur et al., 1985). It is particularly suited for testing large numbers of samples. A standard procedure for the sandwich technique is described in Chapter 5. The following method is a double-antibody indirect assay that is used to test supernatants from centrifuged urines (1500 × g, 15 min) for either BKV or JCV antigens. Briefly, microtiter wells are coated with nonimmune and virus-specific rabbit immune serum, followed by a second coating step with 1% bovine serum albumin (wt/vol). The urine sample is added to both nonimmune and immune-coated wells and allowed to react overnight at 4°C. The detector antibody is a human serum with high-titered antibody to either BKV or JCV. Goat antihuman alkaline-phosphatase conjugated antibody and paranitrophenyl phosphate are

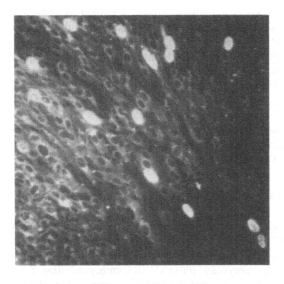

FIG. 6. Immunofluorescent staining of polyomavirus-infected cells showing antigen localized in the nucleus. BKV-infected WI-38 cells stained with virus-specific rabbit antiserum.

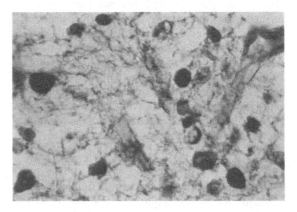

FIG. 7. Immunoperoxidase staining of swollen nuclei of oligodendrocytes in brain tissue from a patient with PML. The brain section was reacted with a genus-specific rabbit antiserum produced by immunization with disrupted SV40 virions.

used as the enzyme-labeled antiglobulin and substrate, respectively. Plates are incubated at room temperature for up to 24 h before optical densities (OD) are determined. The OD difference between values of immune wells and nonimmune wells is calculated for both negative controls and samples. A positive sample is defined as one having a mean OD difference between immune wells and nonimmune wells that is greater than 3 SD above the mean OD difference between immune wells and nonimmune wells of the negative controls. There is little or no cross-reactivity between BKV and JCV in ELISA for antigen.

Detection of Viral DNA

Two nucleic acid hybridization techniques have been used for the detection and identification of BKV and JCV in clinical samples. In the first method, in situ hybridizations are performed on filters on which either urine sediments have been directly spotted and denatured (Arthur et al., 1985) or on which boiled urines have been applied (Gibson et al., 1985). In the second method, Southern hybridization, DNA is extracted from cells, treated with restriction enzymes, electrophoresed, denatured, transferred to filters, and hybridized. Details of hybridization techniques are described in Chapter 8, "Nucleic Acids in Viral Diagnostics."

The Southern hybridization is the "gold standard" for specific diagnosis because it permits determination of the sizes of the hybridizing fragments. This information is valuable in establishing the specificity of the test and in the identification of the infecting viral type. The filter in situ (dot blot) method is suitable for screening a large number of samples because it obviates the need for DNA extraction and gel electrophoresis.

BKV and JCV genomes have a nucleotide sequence homology of 75%. Therefore, the genomic probe will cross-hybridize with the heterologous virus extensively if tests are performed under conditions of low stringency (e.g., at Tm-26°C; 0.66 M Na+, 68°C) and to a lesser extent when tests are performed at conditions of high stringency (e.g., Tm-16°C; 0.16 M Na+, 68°C) or Tm-10°C (0.08 M Na+, 68°C). Parallel hybridization tests with both BKV and JCV probes may be required to establish the diagnosis with certainty.

FILTER IN SITU (DOT BLOT) HYBRIDIZATION

This procedure detects BKV and JCV in urine sediments spotted directly on nitrocellulose (Arthur et al., 1985). After centrifugation (1500 × g, 15 min), 10 μl of urine sediments are spotted directly on a nitrocellulose filter using a 96-well manifold. Cells are lysed and DNA is denatured and fixed in situ by sequential washes in 0.5 N NaOH for 5 min, followed by washings of 2 min each in 1.0 M Tris-HCl (pH 7.4), 1.5 M NaCl/0.5 M Tris-HCl (pH 7.4), and 1 × SSCP (0.15 M NaCl, 0.015 M sodium citrate, 0.02 M sodium phosphate buffer (pH 6.8). Filters are washed in 95% ethanol for 3 min, allowed to air dry, and then baked at 80°C in a vacuum oven. The molecularly cloned viral genome is separated from the vector sequences of the plasmid and purified before radiolabeling. Hybridizations are performed under conditions of low stringency (Tm-37°C) using a mixture of 32P-labeled, molecularly cloned BKV and JCV DNAs. After hybridization, filters are washed at conditions of high stringency (Tm-16°C) and autoradiography is performed. Specimens that are positive with the mixed BKV plus JCV probe are then examined separately with BKV and JCV probes. If both probes hybridize with the sample DNA, the diagnosis is made by a comparison of the intensities of hybridization. Figure 8 illustrates the results obtained by this method using a mixed BKV plus JCV probe.

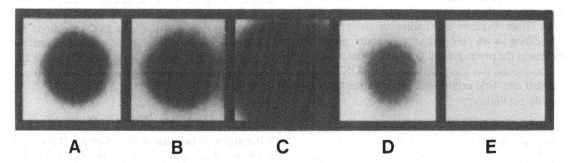

A   B   C   D   E

FIG. 8. Dot blot filter in situ hybridization (using a mixed BKV and JCV 32P-labeled probe) of sequential urine specimens from a bone marrow transplant recipient who developed hemorrhagic cystitis 16 days after transplantation. Symptoms resolved on the 35th day after transplantation. Specimens collected relative to onset of cystitis were (A) 6 days before; (B) 1 day before; (C) 1 day after; (D) 8 days after; and (E) negative control. An equal volume (10 μl) of urine sediment from each specimen was tested. BKV was identified by Southern hybridization as the infecting virus.

Alternatively, specimens may be tested separately with each probe.

The in situ filter hybridization test is capable of detecting 5 pg of viral DNA. However, signals of low intensity are difficult to distinguish from nonspecific background. In mixed infections, if one of the viruses is in low amounts, it may not be diagnosed because its low intensity signal with the homologous probe would be difficult to distinguish from cross-hybridization with the heterologous probe.

SOUTHERN TRANSFER HYBRIDIZATION

Southern transfer hybridization is the method of choice for examination of tissue specimens for BKV and JCV genomic sequences. It permits the differentiation between BKV and JCV on the basis of their distinctive restriction enzyme digest patterns, as well as by a comparison of the intensity of signals upon sequential hybridizations with BKV and JCV viral probes under conditions of high stringency.

DNA is extracted (Hirt, 1967) from tissue specimens or from sediments of centrifuged urine (5 to 50 ml). A maximum of 10 μg of sample DNA is digested first with BamHI (both BKV and JCV have a single BamHI site) to linearize any polyomavirus DNA present in the sample, and then with HhaI (BKV has a single HhaI site, whereas JCV is not cleaved by this enzyme) and loaded into a single lane of an agarose gel. After electrophoresis and transfer to a nylon membrane, the hybridizations are performed sequentially with ^{32}P-labeled BKV and JCV probes, as follows: (1) hybridization with BKV at Tm-26°C; (2) wash at Tm-10°C; (3) boil the filter to remove the probe; (4) hybridization with JCV at Tm-26°C; and (5) wash at Tm-10°C. Autoradiography is performed after each of the above steps. Figure 9 illustrates the identification of BKV and JCV in urine specimens tested in this manner.

Hybridizing bands of sizes other than those described may be due to partial digestion of the DNA samples or due to variability in the number of restriction sites on the genomes. Occasionally, bacterial DNA (from bacteria contaminating the specimen) hybridizing with the plasmid vector sequence present in the probe gives a false positive signal. Hybridization of the filter with a vector DNA probe (without any viral inserts) may be required to evaluate this possibility.

Antibody Assays

Serologic techniques used to measure the BKV and JCV antibodies include hemagglutination inhibition (HI) (Gardner, 1973; Padgett and Walker, 1973), indirect immunofluorescence assay (Shah et al., 1973), neutralization (Shah et al., 1973; Padgett et al.,

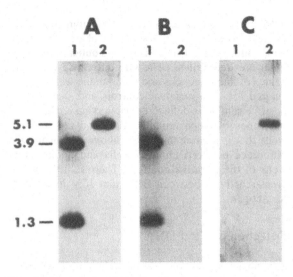

FIG. 9. Southern hybridization of urine specimens from a BKV-positive bone marrow transplant recipient (*lane 1*) and a JCV-positive renal transplant recipient (*lane 2*). DNA was extracted from urine sediments and digested with BamHI and HhaI. (A and B) Filter probed with ^{32}P-labeled BKV DNA and washed under conditions of low stringency (Tm-26°C) (A) and under conditions of high stringency (Tm-10°C) (B). (C) Specimens probed with JCV DNA and washed under conditions of high stringency (Tm-10°C). Digestion of BKV with BamHI and HhaI produces fragments of 3.9 and 1.3 kb in size, whereas JC is not cleaved by HhaI and appears as a single 5.1 kb band.

1977), immune electron microscopy (Gardner et al., 1971), complement fixation (Gardner et al., 1971), immunoelectro-osmophoresis (Dei and Urbano, 1982), ELISA (Burguiere et al., 1980; Iltis et al., 1983; Flaegsted and Traavik, 1985a; Flaegsted and Traavik, 1985b) and radioimmunoassay (RIA) (Brown et al., 1984; Zapata et al., 1984). Both virus-specific IgG and IgM can be detected by most of these methods. Preparative procedures for the separation of IgM from other immunoglobulins present in serum are required for some of these assays. The seroepidemiology of BKV and JVC infections has been studied most extensively by HI. This method will be presented, as will a description of ELISA assays for the detection of virus-specific IgG and IgM.

HEMAGGLUTINATION INHIBITION ASSAY

Viral antigen for hemagglutination assays is prepared from infected cell cultures by treating lysed cells (repetitive freezing and thawing) with neuraminidase (receptor destroying enzyme) (Takemoto et al., 1979). The number of hemagglutinating units of virus present in the stock antigen preparation is estimated by mixing 25 μl volumes of 0.5% washed human O cells with equal volumes of doubling dilutions of the antigen in 96-well plates. The test is incubated at 4°C (or on wet ice) and read when cells in the negative control have settled. The highest dilution of the antigen suspension exhibiting complete hemagglutination is defined as containing one hemagglutinating unit of the antigen. For the hemagglutination inhibition (HI) test, 4 to 8 hemagglutinating units of antigen in 25-μl volume is mixed with equal volumes of doubling dilutions of serum and incubated at 37°C for 1 h. The mixtures are then placed on wet ice and 50 μl of 0.5% human O cells in cold phosphate buffered saline are added. The test is read when the cells in the negative control serum have settled (approximately 1 h). The reciprocal of the highest dilution of serum that completely inhibits hemagglutination is the HI titer of the serum.

Human sera contain low levels of nonspecific inhibitors of hemagglutination. Therefore, HI titers of 1 : 40 or less in untreated sera cannot be interpreted with confidence. These inhibitors are removed by pretreating the serum with receptor-destroying enzyme (Gardner et al., 1973), acetone (Shope and Sather, 1979), or NaIO$_4$ (Dei et al., 1982). The acetone extraction procedure consistently removes inhibitors but may also reduce titers of specific antibodies. Antibody titers of 1 : 20 or greater in treated sera are indicative of past polyomavirus infection. IgM-containing fractions of serum can be tested for the presence of virus-specific IgM antibody by HI. Some human sera also contain low levels of nonspecific agglutinins. They can be removed by absorption with red blood cells before testing.

ELISA

ELISA techniques are sensitive assays for the detection of BKV-specific and JCV-specific antibodies. Comparison of ELISA to other serologic assays for BKV-specific antibodies indicate that ELISA has equal or greater sensitivity than HI and greater specificity than indirect immunofluorescence assays (Flaegstead and Traavik, 1985a; Iltis et al., 1983). Antigens used in ELISA include infected cells, freeze-thawed extracts of infected cells, or purified virions. Antibody titers are determined by either endpoint dilution or by extrapolation from a curve generated with dilutions of a high-titered serum standard. We have chosen the latter approach for the development of BKV- and JCV-specific antibody assays and have used purified virions as antigen. Briefly, duplicate wells of a 96-well microtiter plate (Immulon II, Dynatech, Alexandria, VA) are coated with purified virions in 0.06 M carbonate buffer (pH 9.6). Wells containing only carbonate buffer serve as controls. The plates are incubated overnight at 4°C and then washed with phosphate-buffered saline con-

taining 0.05% Tween 20 (PBST). Dilutions of serum specimens (1 : 100 and 1 : 1000) in PBST containing 2% fetal bovine serum are added to separate sets of control and antigen wells. The plates are then incubated for 1 h at 37°C and then washed. Goat antihuman IgG conjugated to alkaline phosphatase is diluted in the same buffer that was used for the test serum, added to wells and incubated for 1 h at 37°C. The plates are again washed and paranitrophenyl phosphate is added to all wells. Optical densities for all wells are determined spectrophotometrically (405 nm) after 1 h incubation at room temperature. The value for each serum dilution is the sum of the OD in antigen wells minus the sum of OD in control wells. Any serum sample whose value is greater than the mean plus 3 SD of the value of three negative sera at the same dilution as the test sera is considered positive. Titers of test sera are determined by comparing their OD values to those of a high-titered serum standard. The antibody titer of this standard serum is known as a result of earlier conventional endpoint titration tests. The standard serum is titrated in each test and the resulting ODs are plotted. Titers of test sera are estimated by dividing the serum dilution that has an OD value in the linear part of the curve (either the 1 : 100 or 1 : 1000 dilution), by the dilution of the standard serum that has the same OD value, and multiplying this quotient by the previously determined endpoint titer of the standard serum. When both serum dilutions have an OD value in the linear portion of the curve, the higher dilution is used to estimate the titer. Titers derived by this method, using only two dilutions of sera, are comparable to titers obtained by endpoint titration.

BKV- and JCV-specific IgM are present in the serum after primary infections, as well as in some individuals after reactivation of the latent viruses. Measurement of virus-specific IgM in serum usually requires the removal of IgG that may compete for the same epitopes and therefore compromise sensitivity, or may interact with rheumatoid factor and yield false-positive results. The separation of IgM and IgG is usually accomplished by centrifugation in sucrose gradients or by column chromatography. A BKV-specific IgM antibody ELISA assay has been described that produces reliable results without serum pretreatment (Flaegstad and Traavik, 1985b). Although these authors reported no interference of BKV-specific IgG in measuring virus-specific IgM by a solid phase antigen assay, they preferred a capture IgM assay because it was not influenced by rheumatoid factor and produced higher signal-to-noise ratios.

Acknowledgments. This work was supported in part by Public Health Service grants AI 16959 and CA 44962.

Addendum

Polymerase Chain Reaction

A recently described technique, gene amplification by polymerase chain reaction (PCR) (Saiki et al., 1985), may provide the most sensitive as well as the most specific method for diagnosis of polyomaviruses. In this method, the target DNA is amplified more than a 100,000-fold by repeated cycles of heat denaturation, annealing with a pair of oligomer primers, and primer extension with a heat stable DNA polymerase (Saiki et al., 1988). The primers flank the target DNA, are complementary to opposite DNA strands, and extend in opposite directions to synthesize the target DNA. Each cycle, which is completed in a few minutes, results in a two-fold increase in the target DNA. The procedure is simplified by the availability of a DNA Thermal Cycler (Perkin-Elmer/Cetus Corp., Norwalk, Connecticut) which automates the temperature changes required for denaturation, annealing, and primer extension. The DNA amplified in 30 to 40 cycles is then identified.

We have applied the PCR technique for the detection of BKV and JCV genomes in urine specimens (Arthur and Shah, unpublished data). A single set of 20 base oligonucleotide primers capable of amplifying both BKV and JCV DNA was employed to synthesize a 175 base segment in the early region of the genome. BKV and JCV were differentiated by hybridization of the PCR products with ^{32}P end-labeled oligonucleotide probes (40-mers) specific for the sequences flanked by the respective primers. The amplified region of JCV, but not of BKV, contains a BamHI restriction site. Therefore, the PCR products could be also identified by gel electrophoresis following treatment with BamHI.

Preliminary studies show that less than 200 copies of the viral genome may be detected by this method. PCR was more sensitive than filter in situ hybridization and Southern hybridization for the detection of BKV and JCV in DNAs extracted from urines of renal allograft recipients. DNA amplification could be performed directly on boiled urine or on paraffin sections of routinely collected pathological specimens (Shibata et al., 1988).

Literature Cited

Andrews, C., K. V. Shah, R. Rubin, and M. Hirsch. 1982. BK papovavirus infections in renal transplant recipients: contribution of donor kidneys. J. Infect. Dis. 145:276.

Apperley, J. F., S. J. Rice, J. A. Bishop, Y. C. Chia, T. Krausz, S. D. Gardner, and J. M. Goldman. 1987. Late-onset hemorrhagic cystitis associated with urinary excretion of polyomaviruses after bone marrow transplantation. Transplantation 43:108–112.

Arthur, R. R., A. M. Beckmann, C. C. Li, R. Saral, and K. V. Shah. 1985. Direct detection of the human papovavirus BK in urine of bone marrow transplant recipients: comparison of DNA hybridization with ELISA. J. Med. Virol. 16:29–36.

Arthur, R. R., K. V. Shah, S. J. Baust, G. W. Santos, and R. Saral. 1986. Association of BK viruria with hemorrhagic cystitis in recipients of bone marrow transplants. N. Engl. J. Med. 315:230–234.

Bauer, W. R., A. P. Turel, Jr., and K. P. Johnson. 1973. Progressive multifocal leukoencephalopathy and cytarabine. J. Am. Med. Assoc. 226:174–176.

Beckmann, A. M., K. V. Shah, and B. L. Padgett. 1982. Propagation and primary isolation of papovavirus JC in epithelial cells derived from human urine. Infect. Immunol. 38:774–777.

Borgatti, M., F. Costanzo, M. Portolani, C. Vollo, L. Osti, M. Masi, and G. Barbanti-Brodano. 1979. Evidence for reactivation of persistent infection during pregnancy and lack of congenital transmission of BK virus, a human papovavirus. Microbiologia 2:173–178.

Brady, J. N., and N. P. Salzman. 1986. The papovaviruses: general properties of polyoma and SV40, p. 1–26. In N. P. Salzman, (ed.), The papovaviridae: the polyomavirus, Vol. I, Plenum Press, New York.

Brown, D. W. G., S. D. Gardner, P. E. Gibson, and A. M. Field. 1984. BK virus specific IgM responses in cord sera, young children and healthy adults detected by RIA. Arch. Virol. 82:149–160.

Budka, H., and K. V. Shah. 1983. Papovavirus antigens in paraffin sections of PML brains, p. 299–309. In J. L. Sever and D. L. Madden, (eds.), Papovaviruses and human neurological disease, Alan R. Liss, New York.

Burguiere, A. M., B. Fortier, F. Bricout, and J. M. Huraux. 1980. Control of BK virus antibodies in contacts of patients under chronic hemodialysis or after renal transplantation (by an enzyme-linked immunosorbent assay). Pathol. Biol. 28:541–544.

Caputo, A., A. Corallini, M. P. Grossi, L. Carra, P. G. Balboni, M. Negrini, G. Milanesi, G. Federspil, and G. Barbanti-Brodano. 1983. Episomal DNA of a BK virus variant in a human insulinoma. J. Med. Virol. 12:37–49.

Castaigne, P., P. Rondot, R. Escourolle, J. Ribadeau, F. Dumas, F. Cathala, and J.-J. Hauw. 1974. Leucoencephalopathie multifocale progressive et 'gliomes' multiples. Rev. Neurol. (Paris) 130:379–392.

Chesters, P. M., J. Heritage, and D. J. McCance. 1983. Persistence of DNA sequences of BK virus and JC virus in normal human tissues and in diseased tissues. J. Infect. Dis. 147:676–684.

Coleman, D. V., M. R. Wolfendale, R. A. Daniel, N. K. Dhanjal, S. D. Gardner, P. E. Gibson, and A. M. Field. 1980. A prospective study of human polyomavirus infection in pregnancy. J. Infect. Dis. 142:1–8.

Daniel, R., K. Shah, D. Madden, and S. Stagno. 1981. Serological investigation of the possibility of congenital transmission of papovavirus JC. Infect. Immunol. 33:319–321.

Dei, R., F. Marmo, D. Corte, M. G. Sampietro, E. Franceschini, and P. Urbano. 1982. Age related changes in the prevalence of precipitating antibodies to BK virus in infants and children. J. Med. Microbiol. 15:285–291.

Dei, R., and P. Urbano. 1982. Immunoelectro-osmophoresis for the human papovavirus BK. J. Virol. Methods 3:311–318.

Flaegstad, T., and T. Traavik. 1985a. Detection of BK virus antibodies measured by enzyme-linked immunosorbent assay (ELISA) and two haemagglutination inhibition methods: a comparative study. J. Med. Virol. **16:**351–356.

Flaegstad, T., and T. Traavik. 1985b. Detection of BK virus IgM antibodies by two enzyme-linked immunosorbent assays (ELISA) and a hemagglutination inhibition method. J. Med. Virol. **17:**195–204.

Gardner, S. D. 1973. Prevalence in England of antibody to human polyomavirus (BK). Br. Med. J. **i:**77–78.

Gardner, S. D. 1977. The new human papovaviruses: their nature and significance, p. 93–115. *In* A. P. Waterson, (ed.), Recent advances in clinical virology, Churchill Livingstone, Edinburgh.

Gardner, S. D., A. M. Field, D. V. Coleman, and B. Hulme. 1971. New human papovavirus (BK) isolated from urine after renal transplantation. Lancet **1:**1253–1257.

Gardner, S. D., E. F. D. MacKenzie, C. Smith, and A. A. Porter. 1984. Prospective study of the human polyomaviruses BK and JC and cytomegalovirus in renal transplant recipients. J. Clin. Pathol. **37:**578–586.

Gerber, M. A., K. V. Shah, S. N. Thung, and G. ZuRhein. 1980. Immunohistochemical demonstration of common antigen of polyomaviruses in routine histological tissue sections of animal and man. Am. J. Clin. Pathol. **73:**794–797.

Gibson, P. E., S. D. Gardner, and A. A. Porter. 1985. Detection of human polyomavirus DNA in urine specimens by hybridot assay. Arch. Virol. **84:**233–240.

Goudsmit, J., P. Wertheim-van Dillen, A. van Strein, and J. van der Noordaa. 1982. The role of BK virus in acute respiratory tract disease and the presence of BKV DNA in tonsils. J. Med. Virol. **10:**91–99.

Grinnell, B. W., B. L. Padgett, and D. L. Walker. 1983a. Comparison of infectious JC virus DNAs cloned from human brain. J. Virol. **45:**299–308.

Grinnell, B. W., B. L. Padgett, and D. L. Walker. 1983b. Distribution of nonintegrated DNA from JC papovavirus in organs of patients with progressive multifocal leukoencephalopathy. J. Infect. Dis. **147:**669–675.

Hashida, Y., P. C. Gaffney, and E. J. Yunis. 1976. Acute hemorrhagic cystitis of childhood and papovavirus-like particles. J. Pediatr. **89:**85–87.

Hirt, B. 1967. Selective extraction of polyoma DNA from infected mouse cell cultures. J. Mol. Biol. **26:**365–369.

Hogan, T. F., B. L. Padgett, and D. L. Walker. 1984. Human polyomaviruses, p. 969–995. *In* R. B. Belshe, (ed.), Textbook of human virology, PSG Publishing Co., Inc., Littleton, MA.

Hogan, T. F., B. L. Padgett, D. L. Walker, E. C. Borden, and J. A. McBain. 1980. Rapid detection and identification of JC virus and BK virus in human urine using immunofluorescence microscopy. J. Clin. Microbiol. **11:**178–183.

Howley, P. 1983. Papovaviruses: search for evidence of possible association with human cancer, p. 253–305. *In* L. A. Phillips, (ed.), Viruses associated with human cancer, Marcel Dekker, New York.

Iltis, J. P., C. S. Cleghorn, D. L. Madden, and J. L. Sever. 1983. Detection of antibody to BK virus by enzyme-linked immunosorbent assay compared to hemagglutination inhibition and immunofluorescent antibody staining, p. 157–168. *In* J. L. Sever and D. L. Madden, (eds.), Polyomaviruses and human neurological disease, Alan R. Liss, New York.

Johnson, R. 1982. Progressive multifocal leukoencephalopathy, p. 255–263. *In* R. T. Johnson, (ed.), Viral infections of the nervous system, Raven Press, New York.

Kahan, A., D. Coleman, and L. Koss. 1980. Activation of human polyomavirus infection—detection by cytologic technics. Am. J. Clin. Pathol. **74:**326–332.

Livingston, D. M., and I. Bikel. 1985. Replication of papovaviruses, p. 393–410. *In* B. N. Fields, (ed.), Virology, Raven Press, New York.

London, W., S. Houff, D. Madden, D. Fuccillo, M. Gravell, W. Wallen, A. Palmer, J. Sever, B. Padgett, D. Walker, G. ZuRhein, and T. Ohashi. 1978. Brain tumors in owl monkeys inoculated with a human polyomavirus (JC virus). Science **201:**1246–1249.

London, W. T., S. A. Houff, P. E. McKeever, W. G. Wallen, J. L. Sever, B. L. Padgett, and D. L. Walker. 1983. Viral-induced astrocytomas in squirrel monkeys, p. 227–237. *In* J. L. Sever and D. L. Madden, (eds.), Polyomaviruses and human neurological diseases, Alan R. Liss, New York.

Melnick, J. L., A. C. Allison, J. S. Butel, W. Eckhart, B. E. Eddy, S. Kit, A. L. Levine, J. A. Miles, J. S. Pagano, L. Sach, and V. Vonka. 1974. Papovaviridae. Intervirology **3:**106–120.

Miller, J., R. Bartlett, C. Briton, M. Tapper, G. Gahr, P. Bruno, M. Marquardt, M. Hays, J. McMurtry, J. Weissman, and M. Bruno. 1982. Progressive multifocal leukoencephalopathy in a male homosexual with T-cell deficiency. N. Engl. J. Med. **307:**1436–1438.

Padgett, B. L., C. M. Rogers, and D. L. Walker. 1977. JC virus, a human polyomavirus associated with progressive multifocal leukoencephalopathy: additional biological characteristics and antigenic relationships. Infect. Immunol. **15:**656–662.

Padgett, B. L., and D. L. Walker. 1973. Prevalence of antibodies in human sera against JC virus, an isolate from a case of progressive multifocal leukoencephalopathy. J. Infect. Dis. **127:**467–470.

Padgett, B. L., and D. L. Walker. 1976. New human papovaviruses. Prog. Med. Virol. **21:**1–35.

Padgett, B. L., and D. L. Walker. 1983. Virologic and serologic studies of progressive multifocal leukoencephalopathy, p. 107–117. *In* J. L. Sever and D. L. Madden, (eds.), Polyomaviruses and human neurological disease, Alan R. Liss, New York.

Padgett, B. L., D. L. Walker, G. M. ZuRhein, R. J. Eckroade, and B. H. Dessel. 1971. Cultivation of papovalike virus from human brain with progressive multifocal leukoencephalopathy. Lancet **1:**1257–1260.

Padgett, B. L., D. L. Walker, M. M. Desquitado, and D. U. Kim. 1983. BK virus and non-haemorrhagic cystitis in a child. Lancet **1:**770.

Pagnani, M., M. Negrini, P. Reschiglian, A. Corallini, P. G. Balboni, S. Scherneck, G. Macino, G. Milanesi, and G. Barbanti-Brodano. 1986. Molecular and biological properties of BK virus-IR, a BK virus variant isolated from a human tumor. J. Virol. **59:**500–505.

Penny, J. B., Jr., and O. Narayan. 1973. Studies of the antigenic relationships of the new human papovaviruses by electron microscopy agglutination. Infect. Immunol. **8:**299–300.

Richardson, E. P., Jr. 1961. Progressive multifocal leukoencephalopathy. N. Engl. J. Med. **265:**815–823.

Richardson, J. H., W. E. Barkley. (eds.). 1984. Biosafety in microbiological and biomedical laboratories 1984. HHS Publication no. (CDC) 84-8395. U.S. Department of Health and Human Services, Public Health Service, Washington, D.C.

Rosen, S., W. Harmon, A. Krensky, P. J. Edelson, B. L. Padgett, B. W. Grinnell, M. J. Rubino, and D. L. Walker. 1983. Tubulo-interstitial nephritis associated with polyomavirus (BK type) infection. N. Engl. J. Med. **308:**1192–1196.

Rziha, H. J., G. W. Bornkamm, and H. zur Hausen. 1978. BK virus: I. Seroepidemiologic studies and serologic response to viral infection. Med. Microbiol. Immunol. (Berlin) **165**:73–81.

Saiki, R. K., S. Scharf, F. Faloona, K. B. Mullis, G. T. Horn, H. A. Erlich, and N. Arnheim. 1985. Enzymatic amplification of β-globin genomic sequences and restriction site analysis for diagnosis of sickle cell anemia. Science **230**:1350–1354.

Saiki, R. K., D. H. Gelfand, S. Stoffel, S. J. Scharf, R. Higuchi, G. T. Horn, K. B. Mullis, and H. A. Erlich. 1988. Primer-directed enzymatic amplification of DNA with a thermostable DNA polymerase. Science **239**:487–491.

Shah, K. V., R. W. Daniel, and R. Warszawski. 1973. High prevalence of antibodies to BK virus, an SV40 related papovavirus, in residents of Maryland. J. Infect. Dis. **128**:784–787.

Shah, K., and N. Nathanson. 1976. Human exposure to SV40: review and comment. Am. J. Epidemiol. **103**:1–12.

Shah, K. V., H. L. Ozer, H. N. Ghazey, and T. J. Kelly, Jr. 1977. Common structural antigen of papovaviruses of the simian virus 40-polyoma subgroup. J. Virol. **21**:179–186.

Shibata, D. K., N. Arnheim, and W. J. Martin. 1988. Detection of human papilloma virus in paraffin-embedded tissue using the polymerase chain reaction. J. Exp. Med. **167**:225–230.

Shope, R. E., and G. E. Sather. 1979. Arboviruses, p. 767–814. *In* E. H. Lennett and N. J. Schmidt, (eds.), Diagnostic procedures for viral, rickettsial and chlamydial infections (5th ed.), American Public Health Assoc., Washington, D.C.

Silverman, L., and L. J. Rubinstein. 1965. Electron microscopic observations on a case of progressive multifocal leukoencephalopathy. Acta Neuropathol. **5**:215–224.

Sima, A. F., D. S. Finkelstein, and D. R. McLachlin. 1983. Multiple malignant astrocytomas in a patient with spontaneous progressive multifocal leukoencephalopathy. Ann. Neurol. **14**:183–188.

Taguchi, F., D. Nagaki, M. Saito, C. Haruyama, K. Iwasaki, and T. Suzuki. 1975. Transplacental transmission of BK virus in humans. Jpn. J. Microbiol. **19**:395–398.

Takemoto, K. K., P. M. Howley, and T. Miyamura. 1979. JC human papovavirus replication in human amnion cells. J. Virol. **30**:384–389.

Takemoto, K. K., A. S. Rabson, M. R. Mullarkey, M. F. Blaese, C. F. Garon, and D. Nelson. 1974. Isolation of papovavirus from brain tumor and urine of a patient with Wiskott-Aldrich syndrome. J. Natl. Cancer Inst. **53**:1205–1207.

Traystman, M. D., P. K. Gupta, K. V. Shah, M. Reissig, L. T. Cowles, W. D. Hillis, and J. K. Frost. 1980. Identification of viruses in the urine of renal transplant recipients by cytomorphology. Acta Cytol. **24**:501–510.

Walker, D., and B. Padgett. 1983. Progressive multifocal leukoencephalopathy, p. 161–193. *In* H. Fraenkel-Conrat and R. R. Wagner, (eds.), Comprehensive virology, Vol. 18, Plenum Press, New York.

Walker, D. L., and R. J. Frisque. 1986. The biology and molecular biology of JC virus, p. 327–377. *In* N. P. Salzman, (ed.), The papovaviridae: the polyomavirus, Vol. I, Plenum Press, New York.

Willoughby, E., R. W. Price, B. L. Padgett, D. L. Walker, and B. Dupont. 1980. Progressive multifocal leukoencephalopathy (PML): in vitro cell-mediated immune responses to mitogens and JC virus. Neurology **30**:256–262.

Yoshiike, K., and K. T. Takemoto. 1986. Studies with BK virus and monkey lymphotropic papovavirus, p. 295–326. *In* N. P. Salzman, (ed.), The papovaviridae: the polyomavirus, Vol. I, Plenum Press, New York.

Zapata, M., J. B. Mahoney, and M. A. Chernesky. 1984. Measurement of BK papovavirus IgG and IgM by radioimmunoassay (RIA). J. Med. Virol. **14**:101–114.

zur Hausen, H., L. Gissmann, A. Mincheva, and J. F. Bocker. 1980. Characterization of a lymphotropic papovavirus, p. 365–373. *In* M. Essex, G. Todaro, and H. zur Hausen, (eds.), Viruses in naturally occurring cancer, Vol. VIIa, Cold Spring Harbor Laboratory Press, New York.

ZuRhein, G. M. 1983. Studies of JC virus-induced nervous system tumors in the Syrian hamster: a rev. p. 205–221. *In* J. L. Sever and D. L. Madden, (eds.), Polyomaviruses and human neurological disease, Liss, New York.

ZuRhein, G. M., and S.-M. Chou. 1965. Particles resembling papova viruses in human cerebral demyelination disease. Science **148**:1477–1479.

CHAPTER 18

Parvoviridae: The Parvoviruses

M. J. ANDERSON

Disease: Erythema infectiosum (fifth disease, slapped-cheek disease); aplastic crisis.

Etiologic Agent: Autonomous human parvovirus B19.

Source: Respiratory tract secretions of infected persons (often asymptomatic), by inhalation; blood/blood products.

Clinical Manifestations: Erythema infectiosum; mild, rubellalike erythematous rash illness, accompanied in adults (especially women) by arthropathy. Arthropathy may occur without rash. Aplastic crisis; in chronic hemolytic anemias only. Rapid fall in hemoglobin, with reticulocytopenia.

Pathology: Generally a mild condition, thus histologic studies are rare. In aplastic crisis, examination of bone marrow shows absence of erythroid precursors or recovery from erythropoietic block. Leucopenia.

Laboratory Diagnosis: Detection of virus (antigen by immunoassay or DNA by hybridization) in acute-phase sera, especially in aplastic crisis. Detection of specific IgM by immunoassay up to 12 weeks after the onset of symptoms.

Epidemiology: Worldwide. Peaks of infection occur in spring and early summer in temperate zones. Infection most common in children, although up to 40% of adults may be susceptible. Infection occurs both in outbreaks (especially in primary school students) and sporadically.

Treatment: No specific antiviral agent. Symptomatic relief of arthropathy in erythema infectiosum by nonsteroid antiinflammatory agents (e.g., ibuprofen). In aplastic crisis, transfusion to restore hemoglobin concentrations.

Prevention and Control: No vaccine available. The majority of infectious persons are asymptomatic, although in some cases virus excretion may persist a short time after onset of rash in erythema infectiosum. Chronic hemolytic anemics suffering aplastic crisis are infectious for some days after onset of symptoms and should be isolated from susceptible hemolytic anemics and pregnant women.

Introduction

There are three genera within the *Parvoviridae* and, among them parvoviruses are found infecting a wide range of host species in nature. Densonucleosis viruses infect only hosts of the order *Insecta* and are found in neither man nor vertebrate animals. Defective dependoviruses have been described in a number of mammalian species, including man. These viruses are dependent for their replication on coinfection with a helper virus; first described in association with adenovirus infections, in the past

these viruses were called "adeno-associated" viruses (AAVs). However, it is now clear that members of the genus *Adenoviridae* are not unique in providing the necessary help for dependovirus replication; herpesviruses may also act as helper viruses. Moreover, recent work suggests that host cells treated with certain carcinogens may support the replication of dependoviruses (Schlehofer et al., 1986). No causal link has been shown between dependovirus infection and disease in either man or animals. The third group within the *Parvoviridae* is formed of autonomous parvoviruses capable of independent replication. It is with the human viruses in this group that this chapter is concerned.

History

The first parvoviruses infecting man were discovered by electron microscopy in stool specimens (Paver et al., 1973). These fecal parvoviruses, often called "small round viruses" (SRVs), have been linked with gastrointestinal symptoms, most often in outbreaks associated with the consumption of shellfish (Dunnet et al., 1984). However, since these viruses may also be found in the feces of asymptomatic individuals (Paver et al., 1973), the precise pathogenic role of SRVs remains unclear. It should be noted that viruses such as Norwalk, and Ditchling agent, which are often referred to as parvoviruslike, are not parvoviruses; their size, 27 mm, is larger than that of parvoviruses, and their surface structure appears more to resemble caliciviruses than parvoviruses. Definitive classification of these agents awaits the characterization of their genomes.

In 1975, a second type of parvovirus was discovered in the serum of asymptomatic blood donors. The virus was discovered by chance as an agent responsible for false-positive results in the counterimmunoelectrophoresis (CIE) tests then in use for the detection of hepatitis B virus surface antigen and which, at the time, used human immune serum as a source of antibody. Electron microscopy of sera giving these false-positive results revealed uniform naked icosahedral particles with a diameter of 20 nm. Preliminary studies of the physicochemical nature of this agent suggested its probable identity as a member of the *Parvoviridae* (Cossart et al., 1975). In the following years reports concerning this virus named it variously as human parvoviruslike agent, (Anderson, 1982), serum parvoviruslike virus (Cohen et al., 1983), and B19, after one of the original isolates (Shneerson et al., 1980). In 1983, the chemical properties of this virus were described, permitting its classification as a parvovirus (Clewley, 1984; Summers et al., 1983), and subsequent reports referred to the "human parvovirus" or "human serum parvovirus" to denote its separate antigenic identity from the fecal parvoviruses. However, following the decision of the International Committee for the Taxonomy of Viruses the virus is now named B19 (Brown, 1988).

For a number of years after its discovery, B19 infection appeared to be asymptomatic, or at most associated with mild nonspecific febrile illness which was sometimes accompanied by self-limiting leukopenia (Shneerson et al., 1980). However, in 1981, the central role of B19 in the etiology of aplastic crisis in chronic hemolytic anemias was identified (Duncan et al., 1983; Kelleher et al., 1983; Rao et al., 1983; Serjeant et al., 1981). Two years later, erythema infectiosum (fifth disease) was recognized as the common clinical manifestation of B19 infection (Anderson et al., 1984).

Most recently, a third type of serologically distinct parvovirus has been described. RA-1, an autonomous parvovirus, was isolated in suckling mice inoculated with synovial tissue from cases of rheumatoid arthritis (Simpson et al., 1984). At present little is known of the biology, epidemiology, or pathogenesis of this agent in man.

The Virus

As discussed above, three serologically distinct types of autonomous parvovirus are found infecting man. However, only one of these, B19, has been causally associated with specific disease syndromes, and it is with this human serum parvovirus B19 that the remainder of this chapter is concerned.

Morphology and Physicochemical Properties

B19 virus is a naked icosahedral virus with a diameter of 22 nm and buoyant density in CsCl of 1.3 to 1.40 g/dl. SDS–polyacrylamide gel electrophoresis reveals two proteins of molecular weight 83 kDa (VP-1) and 58 kDa (VP2), of which the smaller is the predominant species (Tattersall and Cotmore, 1987). These two capsid proteins make up between 63 and 81% of the virion mass, with the DNA making up the bulk if not all of the remainder. Parvovirus particles appear not to contain lipids, carbohydrates, histone-type proteins, or either cellular or virally coded enzymes.

Parvovirus particles are classically regarded as highly stable; they are resistant to lipid solvents, heating to 56°C for 60 min, pH ranges between 3 and 9, and high salt concentrations (e.g., those used in isopycnic cesium chloride gradients) (Ward and Tattersall, 1982). Tattersall and Cotmore (1987) state that prolonged storage and repeated freeze-thawing may compromise virus integrity so that some parvoviruses are no longer stable to extremes of heat or high salt concentration, although in the absence of these insults they retain their infectivity.

The above properties are common to members of the genus *Parvovirus*. At present, no sensitive culture system has been described for B19, so the physicochemical properties of the virus may only be inferred. However, the ability of B19 to inhibit the in vitro development of erythropoietic colonies is believed to depend on infection of erythroid precursor cells by the virus (see below), and this method of assaying B19 infectivity has been used to investigate the physicochemical properties of B19 (Young et al., 1984a). These authors reported resistance of infectivity to ether, chloroform, DNase, and RNase but little resistance to acid (0.5 N HCl) and alkali (0.05 N NaOH). Infectivity was completely destroyed by heating at 56°C for 5 min but was unaffected by heating at 45°C for 30 min. This report does not detail the duration of storage of the material used or the number of freeze-thaw cycles to which it had been exposed; the relatively poor resistance to pH and heat may have been due to suboptimal storage conditions of the sample of B19 tested.

Antigenic Composition

Parvovirus B19 is currently regarded as existing as a single serotype, antigenically unrelated to either the human fecal parvoviruses or RA-1. Unlike the majority of animal-borne, autonomous parvoviruses, B19 has not been found to agglutinate erythrocytes from any animal species, and as was noted above, no convenient sensitive virus isolation system exists, so neutralization studies are not practicable.

By analogy with studies of feline and canine parvoviruses (Parrish and Carmichael, 1983), the existence of several different, overlapping, neutralizing sites on the capsid, each made up of a number of overlapping neutralizing epitopes, may be predicted. Minor antigenic drift has been noted in these viruses, but these existence or nonexistence of subtypes of B19 has yet to be investigated. Epidemiological data would suggest, however, that infection with B19 confers lifelong immunity (see below).

Genetics

The genome of B19 is a single piece of single-stranded DNA, some 5.5 kb in length. Unlike other autonomous parvoviruses, B19 packages plus and minus DNA strands into separate virions in approximately equal proportions (Clewley, 1984; Summers et al., 1983). Parvovirus DNA is organized as a linear coding region bounded at each end by terminal palindromic sequences which fold into hairpin duplexes (Bourguignon et al., 1976). In B19, these hairpins are relatively long (~330 nucleotides vs. 100 to 250 nucleotides in animal parvoviruses) and are probably inverted terminal repeats (Shade et al., 1986). This latter property, like the packaging of both plus and minus strands, is shared by B19 and the dependoviruses. Although all the evidence available suggests that B19 is an autonomous parvovirus, the similarity in terminal genome structure with the dependoviruses suggests that like these helper-de-

pendent viruses, B19 may also be capable of integration into the host chromosome (Cheung et al., 1980).

The coding regions of B19 are confined to the plus strand and comprise two open reading frames, coding at the left-hand, 3' end, for nonstructural proteins, and at the right-hand, 5' end, for structural proteins. Restriction endonuclease analysis of 17 isolates of B19 obtained between 1972 and 1984 indicates that the genome of B19 is relatively stable with only two changes (additional KpnI and XbaI sites) occurring in the right-hand portion of the genome in two of the isolates. One of these additional sites was present in virus isolated both in France in 1972 and in England in 1975; the other was present only in the 1972 French isolate (Morinet et al., 1986).

Replication

The end result of productive autonomous parvovirus replication is the lysis of mitotically active cells. Mitosis is required for DNA replication and gene expression, and lysis for the release of progeny virus.

Because of the inability of common cell culture systems to support the replication of B19, little is known of the process of replication of this virus. However, by analogy with other autonomous parvoviruses, of which the replication of minute virus of mice (MVM) is the best studied, an outline can be inferred.

The first stage involves adsorption of the virus to specific receptors on the host cell. These may be very numerous (e.g., 105 sites per cell of MVM) and either comprise or involve protein. The expression of these receptors would appear to be under developmental control. Differentiating cultures of murine teratoma cells yield only a single cell type supporting replication of MVM (Tattersall, 1978), and a similar window of susceptibility in erythroid precursor cells has been shown for B19. Following adsorption, the virions are internalized into the cell through coated pits (Richards et al., 1977), and the genome is transported to the nucleus, most probably within the virion. These steps may proceed regardless of the stage of the host cell in the mitotic cycle.

Within the nucleus the DNA is uncoated, and in procedures dependent upon some host cell functions found only in late S phase, DNA synthesis occurs so that a duplex is formed of one parent, virion strand, and one new strand. DNA replication occurs, and at the same time transcription leading to the formation of both nonstructural and structural proteins begins. The precise mechanism of DNA replication and transcription of B19 remains to be elucidated; the structural similarity of the genome of B19 and that of defective dependoviruses, coupled with the observation that B19 replication does not require coinfection with a helper virus, suggests that caution should be applied in the use of analogy with either system.

Capsid proteins begin to accumulate before the peak of virion DNA synthesis, but it would seem that these are not cytotoxic (long-term cultures synthesizing MVM capsid proteins may be sustained, but this has not proved possible in cells expressing nonstructural proteins (Pintel et al., 1984). The newly formed DNA is excised from the double-stranded replicative form and packaged within capsids, and eventually the nuclear and then cytoplasmic membranes degenerate, releasing progeny virus.

To date, the only cell system capable of supporting the replication of B19 is that of primary cultures of human erythroid precursor cells obtained either from the bone marrow or peripheral blood. These cultures, originally described on a methylcellulose semisolid medium (Duncan et al., 1983; Young et al., 1984a) but more recently adapted to suspension culture (Ozawa et al., 1986) require the hormone erythropoietin for the growth and differentiation of erythroid precursor cells (erythroblasts and normoblasts). Forty-eight hours after the addition of B19 virus to these cultures, viral antigen is demonstrable by immunofluorescence, and crystalline arrays of parvoviruslike particles can be demonstrated within the nucleus by electron microscopy. Infected cells show ultrastructural signs of damage, including abnormal mitochondrial structure, vacuolization, nucleolar degeneration, and chromatin margination (Young et al., 1984b) similar to those described in autonomous animal parvovirus-infected cells (Siegl, 1984). At this time, between 5,000 and 8,000 B19 genome copies are present in each infected cell, the majority in the nucleus (Ozawa et al., 1986).

Pathogenesis

The pathogenesis of parvovirus B19 disease involves two separate components. The first is due to the lytic infection of susceptible dividing cells, and the second to interaction with the products of the immune response. As was stated earlier, autonomous parvoviruses may only replicate in dividing cells. Thus infection of an organ or tissue where a significant proportion of the cells are dividing may give rise to organ-specific disease. This is seen in canine and feline parvovirus infections where virus replication in the crypt cells of the intestine gives rise to a severe enteritis (Hammon and Enders, 1939; Macartney et al., 1984).

B19 virus has not been successfully transmitted to any experimental animal. However, inoculation of human volunteers has permitted the study of B19 infection, the host responses, and the clinical consequences of these under controlled conditions (Anderson et al., 1985a; Potter et al., 1987).

Course of B19 Infection

Following intranasal instillation of B19 virus, a systemic infection occurs; virus is first detectable in peripheral blood 6 days after inoculation and rises to attain peak titers of some 10^{11} virus particles per milliliter blood 2 to 3 days later. Thereafter the viremia declines, and virus becomes undetectable within 2 weeks of inoculation. The source of this circulating virus remains to be determined. Coincident with the peak of the viremia, virus is excreted from the throat and may be detected in throat swab and gargle specimens for 4 to 5 days (Fig. 1).

Immune Response

B19-specific IgM antibody is first detectable as the viremia wanes, 9 to 10 days after inoculation, and rises rapidly to reach peak concentrations within 1 week. During the early part of the IgM response, much of the antibody circulates as immune complexes, bound to virions. These complexes do not fix complement. B19-specific IgG does not become detectable until the viremia has cleared, 2 weeks after the initiation of infection. The concentration of IgG rises more slowly than that of IgM, and peak titers may not be attained until 3 to 4 weeks after the initiation of infection.

Hematologic Effects

B19 virus is able to replicate in mitotically active erythroid precursor cells. Bone marrow aspirates obtained 10 days after inoculation of the virus show a complete absence of erythroid cells of all developmental stages. This loss is reflected in the peripheral blood picture where retriculocyte (immature red cell) numbers fall to undetectable levels for a period of 7 to 10 days. Associated with this arrest of red cell production by the bone marrow, is a fall in peripheral hemoglobin concentrations of the order of 0.2 g/dl/day. In response to the cessation of erythrocyte production, the concentration of the hormone erythropoietin rises sharply to stimulate development of erythroid precursors, returning to base-line levels 7 to 10 days later.

In vitro B19 has no effect on myelopoietic cells. However, in vivo transient but significant reductions in first lymphocyte, and 2 days later neutrophil and platelet counts, are a consistent feature of infection. The mechanisms operating to bring about these reductions are uncertain. However, coincident with the fall in neutrophil and platelet numbers, there is a rise in β-2-microglobulin concentrations, suggesting damage to these cells. In contrast, the reduction in lymphocyte numbers seen equally in helper and suppressor subsets occurs prior to the increase in β-2-

FIG. 1. Schematic representation of the virologic, hematologic, and clinical events in B19 virus infection. (Reproduced with permission from Principles and Practice of Clinical Virology, © 1987 by John Wiley and Sons Ltd., Chichester, England.)

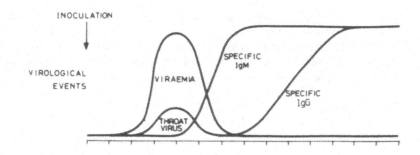

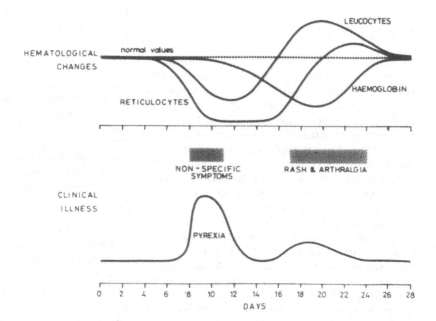

microglobulin and is coincident with peak viremia; this may be due to sequestration of lymphocytes to the site of viral replication.

Biochemical Effects

Although no significant changes were noted in biochemical parameters in infected volunteers, abnormalities in liver enzyme concentrations are occasionally noted (author's unpublished observations). Transient increases in alanine aminotransferase indicate a mild degree of liver cell damage. Whether this is due to the lytic viral replication or to an interaction with components of the immune response is unknown.

Clinical Effects

The clinical features of B19 infection are described in detail in the next section. However, the duality of B19 disease should be considered in the context of pathogenesis.

As was stated earlier, B19 disease involves two separate components. Chronologically the first, the lytic action of B19 replication in erythroid precursors, gives rise to transient arrest of erythrocyte production. In the hematologically normal individual, the resultant slight drop in hemoglobin levels is rarely clinically apparent. However, in chronic hemolytic anemias where the life-span of the erythrocyte is reduced from 120 days to less than 20, hemoglobin levels fall rapidly to very low, often life-threatening concentrations. This precipitate self-limiting anemia is termed the aplastic crisis and is discussed in detail in the next section.

Coincident with the viremia, infected individuals suffer a mild flulike illness with pyrexia and chills, lasting 1 to 3 days. The pathogenesis of this pyrexia has not been investigated, although it is of interest to note that these symptoms occur predominantly in the latter part of the viremia, when immune complexes are detectable in the peripheral blood.

The second phase of B19 disease occurs about 1 week after the peak of the viremia, at a time when virus is only rarely detectable (Anderson, Grilli,

Hoskins, and Davies, unpublished observations; Chorba et al., 1986), but IgM is present in high titer. This second phase is characterized by an erythematous maculopapular rash, accompanied in the early stages by a mild pyrexia and accompanied by arthropathy, especially in adult women. This rash illness, erythema infectiosum, is the commonly observed manifestation of B19 infection.

Clinical Features

Transmission

Parvovirus B19 is transmitted by droplets via the respiratory tract. This is in marked contrast to parvovirus infections in animals, which are spread by the fecal-oral route. B19 can be detected in throat swab, gargle, or, less readily, nasal washings, but it has not been detected in stool specimens (Anderson et al., 1985a). Virus is excreted from the respiratory tract for some 5 days, beginning 1 week after inoculation, at a time when infected individuals are either asymptomatic or suffering mild, nonspecific symptoms. This pattern of excretion gives rise to case-to-case intervals of 6 to 12 days.

In addition to this "natural route," B19 may occasionally be transmitted in blood and blood products; Mortimer et al. (1983) found a significantly higher rate of seropositivity among hemophiliac children than among untransfused children. However, the frequency of blood-borne transmission is likely to be very low, since chronic viremia has not been described. Indeed, the frequency of viremia among voluntary blood donors in England has been estimated at 1 in 40,000 (Anderson, 1982).

On two occasions, viral DNA has been detected in urine samples (Anderson et al., 1985a; Chorba et al., 1986). In the first of these reports, the samples concerned were contaminated with menstrual blood, and the virus was assumed to have been present in this blood rather than in the urine. In the second report, the age and sex of the two individuals concerned were not stated.

Prodromal Symptoms

Although a search of the literature reveals a conspicuous absence of prodromal symptoms in erythema infectiosum, volunteer studies (Anderson et al., 1985a) supported early findings (Shneerson, 1980) of a febrile episode with nonspecific symptoms of headache, chills, myalgia, and malaise accompanying the viremic phase of B19 infection (Fig. 1). During this illness, virus may be detected in respiratory tract secretions. A recent study (Anderson, Grilli,

Hoskins, and Davies, unpublished observations) has shown that virus replication in the respiratory tract may be associated with local symptoms of sore throat, so that the clinical picture at this stage may be indistinguishable from influenza infection.

Erythema Infectiosum

Following the prodromal illness, infected individuals are symptom-free for about 1 week before the onset of the exanthematous phase of illness, erythema infectiosum (EI).

The exanthem in classical EI occurs in three stages (Wadlington, 1957). The first begins some 18 days after the acquisition of infection and is characterized by a bright red rash on the cheeks (slapped-cheek appearance), the edges of which may be slightly raised. The second phase of the exanthem occurs 1 to 4 days after the onset of facial involvement, with the appearance of an erythematous, maculopapular rash on the trunk and limbs. The rash is initially discrete but spreads to involve large areas. Toward the end of this stage, there is central clearing of the rash to give the characteristic lacy or reticular pattern.

The third stage of the exanthem is highly variable in duration, lasting from 1 to 3 or more weeks, and is characterized by changes in the intensity of the rash with periodic complete evanescence and recrudescence. These fluctuations are related to environmental factors such as exposure to sunlight and temperature.

Although EI has long been recognized as a distinct clinical entity, of probably infectious etiology, it was only in 1983 that B19 was recognized as the etiological agent (Anderson et al., 1984). Following this first report, outbreaks of EI occurring throughout the world have been investigated and found to share this etiology (Chorba et al., 1986; Okabe et al., 1984; Plummer et al., 1985). These cases were all diagnosed by the detection of B19-specific IgM (see next section). The use of such tests has revealed a spectrum of disease in which classical cases of EI occupy a central position.

Subclinical, nonexanthematous infection occurs (Anderson et al., 1985a) and is especially common among children (Anderson, Grilli, Hoskins, Davies, unpublished observations; Plummer et al., 1985).

The most common complication of EI is joint involvement. This is relatively rare in children (< 10% of cases), but in adults it is the norm, occurring in 80% or more cases of exanthematous infection. There is a range of severity from mild arthralgia to frank arthritis throughout the age groups, although age- and sex-related differences are beginning to emerge.

In children, the sexes are affected equally. The symptoms are often asymmetric and may be more severe and of longer duration than those in adults (Reid et al., 1985). Among adults, the vast majority of the affected are women. The most common presentation is a sudden onset of symmetric arthritis affecting the small joints of the hand, wrists, ankles, knees, or any combination of these. More than two-thirds of cases resolve within 2 weeks, and the majority have recovered in 4. In many respects B19 arthropathy is similar to that seen in association with acute rubella infection, and like rubella, the arthropathy may occur in the absence of a rash.

Aplastic Crisis

The most serious illness so far associated with B19 virus infection is the aplastic crisis that occurs in patients with underlying chronic hemolytic anemia. In these patients, who have in common a shortened red cell survival time, the profound reticulocytopenia of B19 infection results in the depression of hemoglobin concentrations to critical levels. With the resolution of infection, reticulocytes reappear in the peripheral blood, and hemoglobin concentrations return to the normal steady-state values for these patients. This transient profound anemia may occur in any individual whose erythrocytes have a shortened life-span. Examples of such conditions include sickle cell anemia, hereditary spherocytosis, beta-thalassemia intermedia, pyruvate kinase deficiency (Anderson and Pattison, 1984), and HEMPAS (a diserythropoeitic anemia) (West et al., 1986).

B19 infection does not invariably result in aplastic crisis in patients with chronic hemolytic anemia. Some individuals escape this effect if they have been recently transfused (Anderson et al., 1982a); this may be due either to a protective effect of transfused antibody (more than 60% of donors are immune) or to the substitution of longer-lived, donated erythrocytes for the patient's own, fragile ones or to a combination of these two. Certainly among individuals suffering the more severe form of beta-thalassemia, aplastic crisis is a rare complication. In such cases the anemia is so severe that the patient is maintained by regular, frequent transfusions.

Among populations where aplastic crisis does occur, the severity of the episode varies between individuals. This may reflect the variation in erythrocyte life span between patients.

Recently, an aplastic crisis–like episode has been reported in association with B19 infection in a patient with iron deficiency anemia (Lefrere and Bougeois, 1986). In this case, B19-related cessation of erythropoiesis brought about a clinically significant drop in hemoglobin concentration in a patient already anemic owing to iron deficiency.

Complcations of B19 Infection

Although both arthropathy and aplastic crisis may be regarded as complications of B19 infection, these conditions develop as the norm in groups at risk (adult women and chronic hemolytic anemics, respectively) and so have been included in the foregoing description of the normal clinical consequences of B19 infection.

Reported complications of EI include cases of transient hemolytic anemia (Wadlington and Riley, 1968), encephalitis with full recovery (Balfour et al., 1970), and encephalopathy in a 9-month-old boy resulting .in permanent sequelae (Hall and Horner, 1977). However, these cases occurred prior to the appreciation of B19 as the causative agent of EI; in view of the difficulty of diagnosing sporadic cases of EI on clinical grounds, it is not certain that these cases were due to B19 infection.

INFECTION IN PREGNANCY

The properties of B19–production of a high-titer viremia and a requirement for dividing host cells— suggest a poor outlook if infection occurs during pregnancy. In the largest documented clinical study of EI, teratogenicity was not observed (Ager et al., 1966).

However, B19 virus has been shown to cross the placenta. In a pregnancy terminated at 17 weeks' gestation because of fetal hydrops, there was evidence of B19 infection in the mother about 1 month prior to termination, and B19 virus DNA was detected by hybridization in the placenta, kidney, adrenals, heart, thymus, and liver of the fetus (Brown et al., 1984). In a second hydropic fetus delivered stillborn 12 weeks after maternal B19 infection, viral DNA was detected in fetal liver, lung, brain, and adrenal (Bond et al., 1986). In another pregnancy, a stillbirth of a macerated fetus with ascites occurred at 39 weeks' gestation after a normal pregnancy. Maternal and fetal blood were B19-specific IgM-positive, indicating recent maternal infection and transplacental infection of the fetus (Knott et al., 1984).

The overall incidence of the effects of B19 virus in pregnancy cannot be stated with certainty. Undoubtedly, many pregnancies continue to full-term delivery of normal infants. However, there are a number of reports indicating a higher than expected rate of fetal loss (Anand et al., 1986; Mortimer et al., 1985a; author's unpublished observations).

Moreover, EI appears to be significantly less common in pregnant women than in nonpregnant women of the same age (Anderson et al., 1985b). Whether this is due to a modifying effect of pregnancy on the clinical consequences of B19 infection or B19 infection leads to very early fetal loss before a pregnancy is diagnosed is not known.

In contrast to the possible fetal loss that may be associated with B19 infection, there is no evidence that the virus causes congenital abnormalities. Neither B19 virus nor specific IgM could be detected in sera from 283 abnormal infants born in Manchester, England (95), or Bern, Switzerland (153) (Mortimer et al., 1985a). The definitive assessment of the effects of B19 virus in pregnancy can only be made by large-scale prospective studies; these are under way.

CYTOPENIAS

Studies of B19 infection in volunteers indicate that in addition to reticulocytopenia, transient lymphopenia, neutropenia, and thrombocytopenia are common results of B19 infection. Although these appear to date not to be of sufficient severity or duration to cause the patient distress, they may be detected during detailed investigation of patients in whom B19 infection is not suspected and thence lead to clinical concern and overinvestigation (Neild et al., 1986; Saunders et al., 1986).

NEPHROPATHY

In the latter phases of B19 infection, B19 virus–IgM antibody immune complexes may be detected in peripheral blood, affording the possibility of renal damage. Moreover, virus may be excreted in urine (Chorba et al., 1986). It is therefore to be expected that renal involvement may occur in B19 infection.

Only two documented instances, however, are known to the author. The first occurred in 1985, in a young boy who experienced microscopic hematuria for some weeks following B19 infection which presented with rash and arthropathy (unpublished observations). The second occurred in 1986, in a 44-year-old woman who presented with a typical B19 illness of EI, with rash and arthropathy, and 3 weeks later suffered acute nephritis. Her urinary output dropped to 600 ml per 24 h, and the urine contained blood and protein. At the same time, her blood pressure was raised and she suffered edema, especially of the face and lower legs. The patient required Frusemide for some 4 weeks to maintain a satisfactory urinary output and suffered microscopic hematuria for some months more (Anderson and Cresswell, unpublished observations).

It is likely that this patient suffered an unusually severe B19 infection, as her liver was affected too. Since urinary protein determinations are not routine investigations in patients with rash illness, the incidence of renal involvement is unknown.

HEPATITIS

In 1976, the discovery of B19 virus in the serum of a patient suffering acute hepatitis raised the possibility that B19 might be one of the elusive non-A, non-B hepatitis viruses. However, examination of sera from cases of non-A non-B hepatitis for both B19 virus (P. P. Mortimer and B. J. Cohen, unpublished observations) and IgM antibody (author's unpublished observations) suggests that this virus is not a common cause of hepatitis.

Biochemical analysis of the blood of infected volunteers did not reveal any abnormality during B19 infection (Anderson et al., 1985a). In some patients, however, slight liver damage does occur, although this is not severe enough to manifest as frank jaundice and is only found by accident during detailed investigation of patients with unusually severe B19 infection (author's unpublished observations).

PURPURA

Vascular purpura has been noted in a number of cases of B19 infection. In the majority of such cases, the nonnecrotic, petechial purpura develops during viremia and is not associated with abnormally low platelet counts. Occasionally such cases are diagnosed as Henoch-Schonlein purpura, and the petechae are accompanied by abdominal pain and large joint arthralgia (Lefrere et al., 1985a, 1986). More rarely, the purpura may be characterized by bruising rather than petechiae, and in such cases it is likely to be associated with a low platelet count (Mortimer et al., 1985b).

Diagnosis

It is clear from Fig. 1 that the diagnosis of aplastic crisis may be made early in the course of the illness by the detection of virus. In such patients and in cases presenting with upper respiratory tract involvement, this may be accomplished by testing a throat swab or, preferably, serum specimen. No special conditions are required for either of these specimens, although if detection of virus is to be attempted by DNA:DNA hybridization, storage of specimens at $-70°C$ is to be preferred. Serum specimens giving negative results when examined for virus should be tested for B19-specific IgM.

In marked contrast to the situation of aplastic crisis, EI occurs late in the course of the infection, when virus is seldom detectable in serum. Although this explains in part the historic difficulty of elucidating the agent of EI, it also means that diagnosis must be made by B19-specific IgM detection.

Virus Detection

B19 virus may be detected in blood by electron microscopy, counterimmunoelectrophoresis, radio- or enzyme-immunoassay for antigen, or DNA:DNA hybridization. Because of the relatively low titer of virus present in throat swab specimens, only immunoassay and hybridization are sufficiently sensitive to detect virus in these specimens.

ELECTRON MICROSCOPY

Immune electron microscopy of serum specimens confers the dual advantages of specificity and sensitivity. The small size (21 nm) of B19 necessitates examination of grids at high power ($\times 100,000$), but the addition of antibody-positive serum permits scanning at lower power for clumps of virus which may then be examined in more detail at $\times 100,000$.

The serum specimen should be mixed with an equal volume of serum containing a high concentration of B19-specific IgG serum and incubated at 37°C for 30 min. A parallel specimen is incubated with B19 antibody-negative serum as a control preparation. Following this incubation, the serum/antibody mixtures are diluted 1 in 10 in PBS and centrifuged at $40,000 \times g$ for 1 h. The supernatant is discarded, and the pellet is resuspended in distilled water to a volume one-tenth that of the starting serum specimen volume.

Phosphotungstic acid, pH 6.3, is added to give a final concentration of 3%, and the suspension is touched to a Formvar-coated grid.

COUNTERCURRENT IMMUNOELECTROPHORESIS (CIE)

Although this is probably the least sensitive method among those described here for the detection of B19, it is simple to perform and gives results in 1 h. It uses the opposite charges on virus antigen and antibody molecules to speed migration of the two toward each other in agarose gel. Where antigen and antibody meet and complex, a precipitin line is formed. Depending on whether the antibody or antigen is of known specificity, the test will detect either virus or its specific antibody.

Preparation of Agarose Gel

One percent agarose gel is prepared by dissolving agarose (of a grade suitable for electrophoresis) in barbitone acetate buffer, pH 8.6, in a boiling water bath or microwave oven. The gel can be used immediately or stored at 4°C and heated until liquid immediately prior to use.

The agarose is poured to a depth of at least 1 mm onto a clean glass slide on a level surface and allowed to solidify at room temperature (glass lantern slides, 85 × 85 mm, are ideal and require 12 ml agarose). Slides prepared in this way may be stored at 4°C in a moist box for 1 week.

Immediately prior to use, a row of holes 3 mm in diameter with centers 6 mm apart is punched in the gel to receive antigen. A second, parallel row is punched below the first row to receive antibody; the centers of the upper and lower rows should be 5 mm apart. A corner is then cut off the gel for orientation (Fig. 2).

Loading and Running the Gel

The wells in the upper row are filled with 6-μl volumes of sera to be tested, and the lower row with 6-μl volumes of B19 antibody-positive serum, care being taken not to overfill the wells or introduce air bubbles.

The gel is now ready for electrophoresis, which should be carried out with minimum delay. Contact between the gel and the electrode troughs is made with blotting paper (3MM or similar) wetted in barbitone acetate buffer. Electrophoresis is carried out at 120 V (17 mA) for 1 h with the gel orientated such that antigen migrates toward the anode.

Reading Results

At the end of the electrophoretic run, the slide is removed from the tank and examined for precipitin lines. This is most conveniently accomplished by carefully wiping moisture from the underside of the slide before placing it over a light box. (An egg-candling box fitted with a low-wattage bulb can be used. Alternatively, an X-ray viewing light box largely masked with black paper is convenient.) Precipitin lines of antigen-antibody complex form between the wells which were filled with antigen and antibody. The sensitivity of this test may be increased by holding gels at 4°C for 1 h after electrophoresis, prior to examination.

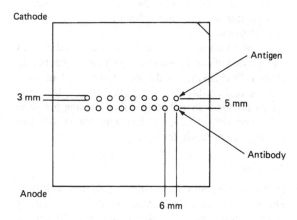

FIG. 2. Gel for countercurrent immunoelectrophoresis (CIE).

Interpretation and Validation of Results

CIE testing for antigen is insensitive, detecting virus at a concentration of 10^9 particles per milliliter. The likelihood of false-positive results is therefore low. Positive results should be confirmed be retesting against a bank of B19 antibody-positive and B19 antibody-negative sera. None of these sera should contain antibody to HBsAg, since this antigen too may be present in sufficiently high concentration in serum to give precipitin lines with antisera containing anti-HBsAg.

Positive results should ideally be confirmed by electron microscopy using either a fresh aliquot of the specimen (see above) or the precipitin line emulsified in a drop of water (Kelleher et al., 1983).

RADIOIMMUNOASSAY

The most widely used assays for B19 antibody are based on the class-specific antibody capture technique (see below). These tests have been adapted for antigen detection by substituting the material under test for B19 antigen and known high concentration B19-specific IgM or IgG positive serum for the "test" antibody. The procedure described here is for an antigen detection test based on IgM antibody captive. Clearly the test may be based on IgG antibody capture also.

Preparation of Solid Phase

In the author's laboratory, etched polystyrene beads 6.4 mm in diameter are used exclusively for solid-phase radioimmunoassays. The reaction volume in this system is 200 μl per bead, and all tests are carried out in duplicate. The methods described could be adapted for microtiter wells. Beads are coated with anti-μ chain antibody by gentle agitation at room temperature in anti-μ antibody diluted in 0.02 M Tris pH 7.5 for 1 h. Beads are stored at 4°C in the same mixture for at least 24 h prior to use. (Beads may be stored under anti-μ for 1 month without loss of sensitivity.)

Immediately prior to use, beads are rinsed once in phosphate-buffered saline (PBS) containing 0.05% Tween 20 (PBST). Beads are then coated in B19-specific IgM antibody by incubation at 37°C for 3 h in serum diluted 1 in 100 in PBST containing 5% fetal calf serum (PST-FCS). The beads are then washed in PBST and are ready for use.

The Test

The material under test should be diluted in PBST. The author finds a dilution of 1 in 10 to be ideal for serum. Throat swab transport medium, PBS nasal washes, and gargles may be tested diluted 1 in 2 in PBS with 0.1% Tween 20 and 10% FCS. The beads are added to the diluted test material and incubated at room temperature overnight. The beads are then washed in PBST, and bound antigen is detected by sequential incubation in monoclonal anti-B19 antibody diluted in PBST-FCS for 3 h at 37°C followed by incubation in ^{125}I-labeled antimouse immunoglobulin, diluted in PBST-FCS with 10% antibody-negative human serum, to contain 50,000 cpm/200 ml. Beads are then washed, and the bound radioactivity is detected in a gamma counter, for 5 min.

Interpretation

Controls to be used in this test should approximate as nearly as possible the material under test: thus, virus-negative serum for test serum, virus transport medium for swab media, and PBS for PBS gargles and nasal washes. A test sample binding at least twice as much radiolabel as its control may be considered positive.

Positive samples should be confirmed by repeating the test, using both B19-specific, IgM-positive and -negative sera. This is particularly important in testing materials for which negative control samples are not readily available—joint aspirates, CSF, and the like.

ENZYME IMMUNOASSAY

In the author's laboratory and throughout the United Kingdom, solid-phase B19 immunoassays utilize ^{125}I-labeled antibody exclusively. However, such assays may be adapted to utilize enzyme labels in place of radiolabel. Indeed, a recent publication (Anderson et al., 1986) describes the use of enzyme-linked immunosorbent assays for the detection of B19-specific IgM, IgG, and antigen. These assays are identical in principal to the antibody capture radioimmunoassays, differing only in the solid phase used (flat-bottomed microtiter plates vs. polystyrene beads), the reaction volume (75 vs. 200 μl), the diluents, and the end-stage label. The following method for B19 antigen detection in serum specimens is that described by Anderson et al. (1986).

Preparation of Solid Phase

Flat-bottomed, 96-well Immulon-2 microtiter plates are coated with antibody to human IgM diluted in 0.01 M carbonate buffer (pH 9.6) overnight at 4°C. The plate is then washed, and the remaining free protein binding sites are blocked by incubation with PBS with 0.5% gelatin and 2% normal goat serum (PBS-G-NGS) for 30 min at 35°C. The wells are then coated with anti-B19 human IgM antibody by incubation with a high-titer serum diluted 1 in 100 in PBS-G with 0.15% Tween 20 (PBS-GT) for 1.5 h at 37°C. Control wells incubated with diluent alone are also prepared. After washing, the wells are ready for use.

The Test

The serum to be tested is diluted 1 in 10 in PBS-GT and incubated overnight at room temperature in four replicate antibody-coated wells and two control wells. After washing, bound antigen is detected in two of the antibody-coated wells and the control wells by incubation with monoclonal anti-B19 antibody diluted in PBS-GT for 1.5 h at 35°C. The remaining two wells are reacted with monoclonal antibody to respiratory syncytial virus, as further controls.

Bound monoclonal antibody is detected by sequential incubation with peroxidase-conjugated antimouse IgM antibody diluted in PBS-GT-NGS for 1 h at 35°C and substrate (O-phenylenediamine with 0.015% hydrogen peroxide in 0.15 M citrate buffer pH 5.5). The reaction is stopped after 30 to 45 min by the addition of 3.5 M HCl and the absorbance at 490 is measured.

Interpretation

Positive : negative (P : N) ratios are calculated using (1) the wells reacted with anti-RSV antibody and (2) the control wells, which are not exposed to human anti-B19 IgM. The authors state that P : N values are usually higher for the former type of control. A specimen giving P : N values of greater than 2.8 for both controls is considered positive.

It is likely that a serum specimen containing both B19 antigen and B19-specific IgM will give high absorbance values in the second type of control: these wells have been coated with antihuman IgM but not deliberately treated with anti-B19 IgM. Thus, the IgM binding sites are available to capture the IgM antibody in the test serum specimen. If some of this IgM is B19-specific and B19 antigen is also present in the sample, the monoclonal anti-B19 detector antibody will be bound to the solid phase. The use of this type of control is therefore likely to give rise to false-negative results in antigen tests where the criterion for positivity is a T : N greater than 2.8 for both types of control.

DNA : DNA Hybridization

Although the genome of B19 is a single piece of single-stranded DNA, the fact that positive and negative strands are packaged with approximately equal efficiency has facilitated the production of cloned segments of viral genome. A number of such clones are in existence and may be obtained from the au-

thors' laboratories (Anderson et al., 1985c; Clewley, 1985; Cotmore and Tattersall, 1984). The method described below is that used in the author's laboratory, where the 500-base B19 genome insert cloned in plasmid PVTM I is used as a DNA probe (Anderson et al., 1985c). For a full discussion of the variations possible, the reader is referred to Chapter 8.

Preparation of the Probe

Double-stranded cloned B19 genome is labeled with ^{32}P by nick translation (Rigby et al., 1977) using reagents supplied by Amersham International. Briefly, each microgram of DNA is incubated with approximately 200 μg [d-^{32}P]dCTP in the presence of 20 mM dATP, dGTP, and dTTP with 2.5 units DNA polymerase I and 50 pg DNase, all in TE buffer (10 mM Tris, 1 mM EDTA, pH 8.0), at 15°C for 5 h. At the end of the reaction, 20 μg denatured calf thymus carrier DNA is added, and DNA is separated from unincorporated nucleotides by fractionation on a column of Sephadex G-50 (Pharmacia, Sweden) using TE buffer as eluant. The relative ^{32}P content of the fractions is estimated with a monitor. Specific activities of 1 to 8 $\times$ 10^7 dpm/μg DNA are obtained.

Dot-Blot Test

Five- or ten-microliter volumes of the specimens are spotted onto nitrocellulose (Schleicher and Schull, Germany) or Genescreen (New England Nuclear) sheets and allowed to dry. Specimens are denatured by floating on alkali (1 M NaCl with 0.1 M NaOH) and then neutralized and rinsed in Tris-buffered saline at pH 7.4. The sheets are then blotted dry between sheets of 3MM filter paper and fixed by baking at 80°C under vacuum for 2 h (Mason et al., 1982).

To reduce nonspecific binding of probe DNA, the nitrocellulose sheets are incubated with 25 ml/100 cm^2 of 0.9 M NaCl with 90 mM sodium citrate containing 0.5% SDS, 100 μg/ml carrier DNA, and 2 $\times$ Denhardt's solution (0.04% each of bovine serum albumin, Ficoll 400, and polyvinyl pyrrolidone [Denhardt, 1966]) at 65°C for 6 h in a shaking water bath. This incubation and the following one are carried out in polythene bags, constructed to fit the filters. The bags are made from a "home freezer" kit, incorporating a tube of polythene and a heat sealer.

The prehybridization mixture is removed from the filter by cutting off a corner of the bag and pouring off the fluid. The mixture is then replaced with 6 ml/100 cm^2 60 mM sodium citrate, 0.6 M NaCl, 3 $\times$ Denhardt's, 0.5% SDS, 100 μg/ml carrier DNA, 10% dextran sulfate, and 20 to 40 μCi ^{32}P-labeled probe DNA that has been boiled for 3 min immediately prior to use to denature the two strands. The bag is resealed and placed within a second sealed polythene bag to prevent contamination of the water bath should the inner bag rupture. The sheets are then incubated at 65°C overnight in a shaking water bath. Following this hybridization incubation, the hybridization mixture is decanted from the filter. If stored at 4°C, this mixture can be reused provided it is thoroughly boiled for 5 min immediately prior to use to denature the probe DNA. Because of the short half-life of ^{32}P, the mixture should only be stored for a few days.

The top is then cut off the bag, and the filter is floated onto washing buffer (1) containing 90 mM sodium citrate, 0.9 M NaCl, 1 × Denhardt's, 0.5% SDS and 10 μg/ml carrier DNA, preheated to 65°C. Filters are then washed sequentially in buffer (1) (3 washes of 20 min); (2) 15 mM sodium citrate, 0.9 M NaCl, 0.5% SDS (3 washes of 20 min); (3) 9 mM sodium citrate, 0.09 M NaCl, 0.5% SDS (1 wash of 10 min).

The sheets are then blotted between sheets of 3MM paper and dried. The sheets are wrapped in Saran wrap, and bound ^{32}P is detected by exposure to Kodak X-Omat S film with an image-intensifying screen at −70°C overnight or for periods up to 5 days.

Interpretation

Dot-blot hybridization may be used to detect virus in any fluid sample. Detection of virus in tissue samples may be accomplished by extracting DNA from the sample and dot-blotting the extract.

Positive results are shown by blackening of the X-ray film above the sample spot on the filter. The position of a putative positive spot should therefore correspond exactly to the position of the test sample. Occasionally, black spots in irrelevant positions may be seen. These may usually be readily distinguished from the positive reactions by their small size and relatively high intensity when they are due to "dirty" hybridization mixture binding to the filter.

Confirmation of positive results may be obtained by analysis of the size of the DNA binding the probe. Briefly, DNA is extracted from the sample and subjected to electrophoresis in 0.8% agarose gel. Electrophoretically separated DNA molecules migrate different distances in the gel depending on their size. The DNA is transferred from the agarose to nitrocellulose or Genescreen filters, fixed by baking at 80°C, and detected by hybridization exactly as described for a dot blot. The genome of B19 is 5.5 kb in length, so a positive signal band corresponding to 5.5 kb confirms the dot-blot result.

Dot-blot hybridization is not directly quantitative, although some idea of the amount of virus genome in the sample may be gained by the size and intensity of the spot (Fig. 3). Titration of the sample along with a

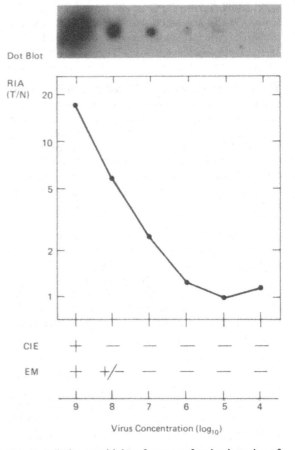

FIG. 3. Relative sensitivity of systems for the detection of B19.

preparation of known virus content permits more accurate quantitation.

The test described here is capable of detecting virus DNA at a concentration of 10^5 to 10^6 particles per milliliter of specimen. The sensitivity may be increased by extracting DNA from the sample by incubation in proteinase K to disrupt the virion, extracting the DNA with phenol and chloroform and precipitation of the DNA in ethanol. By resuspending the resultant pellet in a small volume of TE buffer, the DNA concentration of the sample is effectively increased.

Aqueous samples may more conveniently be concentrated onto the filter by use of a negative-pressure manifold, such as the hybridot (BioRad) or slot blot (Scheiller Schull) apparatus (see Chapter 8).

COMPARISON OF TESTS FOR B19 VIRUS

Figure 3 illustrates the relative sensitivity of EM, CIE, radioimmunoassay for antigen, and DNA : DNA hybridization for viral genome.

The rapidity with which the relatively insensitive

tests of EM and CIE detect virus means that, in spite of their low sensitivity, they are of value in diagnosis where a rapid result is required. Radioimmunoassay and DNA:DNA hybridization are 10^3 to 10^5 times more sensitive, but they require one or two overnight incubations. Where a specimen gives a positive result of EM or CIE, the result can be acted upon, but negative results should be confirmed by testing in either radioimmunoassay or hybridization. Radioimmunoassay and DNA hybridization give results that correlate well in most situations, although hybridization is somewhat more sensitive.

Radioimmunoassay will theoretically detect viral antigen in the absence of whole virus. The author knows of no situation in which this has occurred to date, although with increasing interest in B19, this capability may prove important in, for example, the investigation of arthropathy. In contrast, hybridization has given positive results in specimens containing virus complexed with antibody, where RIA gave negative results owing to the antigenic sites being masked by antibody (Anderson et al., 1985c).

In the context of screening serum or plasma from donated units of blood for B19 to be used as a reagent in antibody tests, CIE, RIA, and hybridization may all be used. The low sensitivity of CIE is of no importance, since the detection of a unit of blood containing high-titer virus is the object. The choice of test will thus depend on the equipment and reagents available; both CIE and RIA require a source of high-titer antibody-positive serum, and RIA requires monoclonal anti-B19 antibody in addition. Hybridization, on the other hand, requires only cloned viral DNA, and this may provide a positive control, obviating the need for using precious virus in the search.

RIA can readily be automated to facilitate the processing of large numbers of sera and could potentially be carried out in the transfusion center alongside "routine" screening. Hybridization is eminently suitable for large-scale screening, especially if the loading of filters with samples can be automated. (The author has made use of the 96-headed microtiter dispensers used in transfusion centers. 3MM filters are stapled to nitrocellulose filters cut to the size of a microliter plate; the 3MM paper draws the plasma through the nitrocellulose filter so that the 25-μl sample forms a dot some 3 to 4 mm in diameter.) The filters are processed in large, partially segmented bags and exposed to large (chest X-ray size) films, significantly reducing the handling time per specimen.

Antibody Detection

Theoretically, with the exception of DNA:DNA hybridization, any of the methods described above may be adapted for the detection of specific antibody by substituting B19 antigen for the unknown "test" material and the serum under test for B19 antibody-positive serum. Historically, immune electron microscopy (IEM) and CIE were the first tests to be used for the detection of B19 antibody. Though rapid, these tests are insensitive and require relatively large quantities of virus and hence have more recently been largely superseded by solid-phase radio- and enzyme-labeled immunoassays based on the antibody capture principle.

Serological diagnosis of recent B19 infection may be made in principle by the detection of seroconversion, or increasing antibody concentration in paired acute and convalescent serum samples, and the first serological diagnoses of B19 infection employed these principles. The desirability of making a rapid diagnosis relevant to the management of the patient's illness has been widely perceived by clinical virologists. The inherent delay in diagnoses requiring a convalescent serum specimen has led to the development of diagnostic techniques requiring only a single serum sample, taken during the patient's illness. Thus, the detection of virus-specific IgM antibody indicating current or recent infection has become a cornerstone of diagnostic virology.

The short-lived nature of the IgM antibody response means, of course, that these tests are wholly unsuitable for the determination of immunity or susceptibility to infection. There is thus a requirement for a different type of test, a sensitive assay for virus-specific IgG antibody, to determine these parameters.

B19 serology rapidly took on the challenge of diagnosis of infection with the advent of B19-specific IgM tests (Anderson et al., 1982; Cohen et al., 1983). However, in the absence of a vaccine, there has been little pressure for tests for IgG antibody, and this is reflected in the poor sensitivity of the tests currently available.

ELECTRON MICROSCOPY

Electron microscopy may be used to detect anti-B19 antibody by the method described above, and by testing acute and convalescent sera, seroconversion may be demonstrated. Although immune electron microscopy (IEM) is not usually regarded as a technique suitable for the determination of the class of antiviral antibody present in serum, this can be accomplished.

In IEM preparations, the antibody molecules linking virus particles are commonly visible. Occasionally, especially when antibody is in excess, the morphology of the molecules may be discernible, permitting the distinction of pentameric IgM molecules from dimeric antibody.

IgM and IgG class antibodies may be physically separated by either sucrose density gradient centrifugation or gel filtration, and the IgM- and IgG-rich fractions can be examined by IEM. Where gel filtration is used, it is desirable to concentrate the fractions to be tested by dialysis, membrane filtration, or precipitation of the protein with ammonium sulfate.

COUNTERCURRENT IMMUNOELECTROPHORESIS

The use of CIE for the detection of B19 antigen is described above. By substituting B19 antigen for the "test" material, CIE provides a rapid method for the detection of anti-B19 antibody.

Provided the serum containing B19 virus does not also contain HBsag, false-positive results are highly unlikely. The use of CsCl gradient-purified virus removes even this small possibility, providing a highly specific test.

Since the detection of antibody by CIE depends on the formation of immune complexes that form a line of precipitate in the gel, IgM antibody cannot be distinguished from IgG. However, the physical separation of IgM from IgG prior to testing provides IgM or IgG class-specific tests (Kelleher et al., 1983). As with IEM, whereas sucrose density gradient fractions may be tested without further treatment, fractions obtained by gel filtration require concentration.

RADIOIMMUNOASSAY FOR B19-SPECIFIC IgM

This assay is based on the antibody capture principle, and in the author's laboratory it is carried out using a solid phase of etched polystyrene beads. The assay may also be performed in microtiter wells, using either polysytrene Removawells (Dynatech) or flexible polyvinyl chloride microtiter trays (Dynatech) from which the wells can be clipped at the end of the assay for radioactivity determination. The reaction volume is 100 μl per bead or 50 to 200 μl per well, and tests are carried out in duplicate.

The concentration of B19-specific IgM in test sera is determined relative to the concentration of this antibody in a reference serum; a highly reactive serum specimen is designated as containing 100 arbitrary units (AU) of B19-specific IgM. Serial dilutions of this reference serum are made in B19 antibody-negative serum to contain 30, 9.2, 2.8, 0.8, and 0.25 AU. In each assay, these five "standard" sera, together with the antibody-negative serum, are reacted in parallel with the sera under test.

Preparation of the Solid Phase

Etched polystyrene beads, 6.4 mm in diameter (Precision Ball Co., Chicago), are coated with rabbit anti-body to human μ-chain diluted in 0.02 M Tris, pH 7.5, by stirring gently for 1 h at room temperature and incubating overnight at 4°C. Beads coated in this way may be stored at 4°C under anti-μ antibody for 1 month without loss of sensitivity. Immediately prior to use, the beads are removed from the coating mixture and rinsed in PBS containing 0.05% Tween 20 (PBST).

Antigen Preparation

As noted before, the only significant source of B19 antigen at present is the serum or plasma of infected individuals. This may be used, appropriately diluted, in the test without prior treatment. Alternatively, immediately prior to use, virus may be extracted with chloroform; equal volumes of serum and chloroform are mixed by vortex in a microcentrifuge tube, and the mixture is then centrifuged in a microcentrifuge for 1 min. After centrifugation, two phases are apparent, and the supernatant, suitably diluted, is used in the test. With some "isolates" of B19, chloroform treatment results in higher test positive : negative ratios. The reasons for this are unclear, and for any isolate, the efficacy or otherwise of chloroform treatment, as well as the optimum dilution of antigen, must be determined. Serum or plasma should be diluted to contain approximately 10^7 particles per milliliter; for samples that contain B19 readily detectable by CIE, a dilution of between 1 in 50 and 1 in 1,000 will be used.

The Test

Sera to be tested are diluted 1 in 100 in PBST with 10% fetal calf serum (PBST-FCS) and incubated in duplicate on anti-μ-coated beads for 3 h at 37°C. The beads are then washed in PBST, and B19 antigen is added, diluted in PBST-FCS to contain some 10^7 particles per milliliter. The beads are incubated under antigen overnight at room temperature, and the unbound antigen is aspirated from the beads. (Note: To conserve antigen, the unbound virus may be recovered from each IgM assay by ultracentrifugation. The resultant pellet is resuspended in PBS with 10% FCS to the original serum volume and stored at -20°C. After two or three assays, the virus content should be checked by titration and the working dilution modified as necessary.) The beads are washed with PBST and incubated with monoclonal anti-B19 antibody diluted in PBST-FCS for 3 h at 37°C. (In the author's laboratory, monoclonal antibody is produced as ascitic fluid, and the IgG fraction is isolated by ion exchange column chromatography. Tissue culture fluid from monoclone cultures may also be used.) Unbound antibody is washed from the beads with PBST, and ^{125}I-labeled antimouse Ig antibody* is added, diluted in PBST-FCS with 5% B19 antibody-negative human serum to contain 50,000 cpm/200 μl. After 3 h incubation at 37°C, the beads are washed in PBST, and the bound radioactivity is estimated.

Treatment of Results

The amount of ^{125}I-labeled antimouse antibody (and hence anti-B19 monoclonal antibody and B19 antigen) bound to each bead is estimated by counting gamma emission for 5 min. The background counts are subtracted from the total counts for each bead, and the mean counts for duplicate beads are calculated. For each serum specimen under test, the test:negative ratio is computed by the following:

$$T:N = \frac{\text{Mean cp 5 min beads incubated with test serum}}{\text{Mean cp 5 min beads incubated with neg. serum}}$$

The T:N values for each of the five standard sera are computed (Table 1), and these T:N values are plotted against the B19-specific IgM concentration in arbitrary units (Fig. 4). The T:N for the serum containing 100 AU should be > 20.0, and the T:N for the serum containing 0.25 AU must be > 1.0. T:N ratios are then calculated for each serum specimen under test, and by interpolation from the graph, the concentration of B19-specific IgM is determined.

Interpretation

In common with all laboratory findings, the results of B19-specific IgM tests should be interpreted with reference to the clinical details of the patient's illness, paying particular attention to the interval between the onset of symptoms, and the drawing of the serum specimen.

In general terms, sera containing 10 AU or more of specific IgM may confidently be associated with recent B19 infection. Parvovirus-specific IgM is detectable for 2 to 3 months after infection in the assay described here (Cohen et al., 1983). However, difficulties in interpreting results may occur with sera taken either very early after the onset of symptoms or 12 or more weeks later.

---

* ^{125}I-labeled rabbit antimouse Ig antibody can be purchased from Amersham International. Alternatively, the IgG fraction of antimouse antibody may be labeled with ^{125}I by either the iodogen method of Salacinski et al. (1979) or using chloramine T (Greenwood et al., 1963). Labeled antimouse Ig preparations should be stabilized by the addition of BSA to 5% and sodium azide to 0.1% and may be stored at 4°C for 1 month. The preparation of ^{125}I-labeled antibodies is described in detail in Public Health Laboratory Service Monograph No. 16 (Pattison, 1982).

TABLE 1. Computation of results for standard sera in B19 IgM antibody capture RIA

Antibody concentration (AU)	[125]Iodine bound (counts/5 min)	T:N
0	345	NA
0.25	499	1.45
0.9	830	2.4
2.8	2,108	6.1
9.2	5,374	15.5
30.0	9,497	27.5
100.0	10,978	31.8

Sera taken from a patient infected with B19 within 3 days of the onset of symptoms may give either negative or equivocal results. In the case of a negative result (< 0.25 AU), the specimen should be examined for B19 virus by either RIA or ELISA for antigen or by dot-blot hybridization for DNA. This situation is most likely to arise in the investigation of aplastic crisis (Fig. 1), but it has recently been noted in erythema infectiosum (author's unpublished observations). Indeed, in a number of laboratories providing a diagnostic service for B19 infection, acute sera taken within 10 days of the onset of symptoms are routinely examined for B19 DNA as well as specific IgM (B. J. Cohen, personal communication, and author's current practice). Alternatively, a second, convalescent serum specimen taken 10 to 14 days later may be examined for B19-specific IgM.

Where early sera give equivocal results (< 3.0 AU), a second specimen is required. It has been

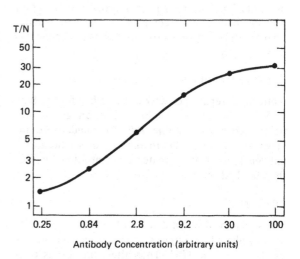

FIG. 4. Standard curve of B19-specific IgM detected by IgM antibody capture radioimmunoassay.

noted that sera containing high concentrations of rubella-specific IgM, taken within 7 days of the onset of illness, may give low-level false-positive results when tested for B19 IgM, and vice versa (Kurtz and Anderson, 1985).

Sera taken 12 weeks or more after the onset of symptoms may contain low concentrations of B-19 specific IgM, which may be difficult to detect with certainty. Where such a specimen gives an equivocal result (< 3.0 AU), a second specimen taken 6 months later should be sought, and the two should be tested in parallel for B19-specific IgG.

Radioimmunoassay for B19-Specific IgG

Like the RIA for specific IgM, this assay is based on the antibody capture principle and may be carried out using a solid phase of polystyrene beads, Removawells, or flexible polyvinyl chloride microtiter trays. The test for specific IgG is conducted using the same methods as those described for the IgM test, with only minor modifications. The reader is therefore referred to the section above for a full description. The modifications converting the test to detect IgG are given in the following sections.

The concentration of B19-specific IgG in test sera is determined relative to the concentration of this antibody in a reference serum. A highly reactive serum specimen is designated as containing 100 AU of B19-specific IgG. Serial dilutions of this reference serum are made in B19 antibody-negative serum to contain 30, 9.2, 2.8, 0.8, and 0.25 AU. In each assay, these five standard sera together with the antibody-negative serum are reacted in parallel with the sera under test.

Etched polystyrene beads are coated with rabbit antibody to human gamma chain, diluted in 0.02 M Tris, pH 7.5, by stirring for 1 h at room temperature and incubating overnight at 4°C. The anti-gamma-coated beads may be stored for 1 month without loss of sensitivity.

Antigen Preparation

Antigen is prepared exactly as described for the IgM RIA (see above). For use in the test, antigen should be diluted such that the 100-AU standard serum gives a T : N of > 20. In the author's laboratory, antigen is used at a greater concentration for IgG than for IgM tests.

The Test

The sera to be tested are diluted 1 in 30 in PBST-FCS and incubated in duplicate on anti-gamma-coated beads for 3 h at 37°C. Thereafter, the test is conducted exactly as for IgM testing.

However, since a higher proportion of test sera

are likely to contain B19-specific IgG than will contain B19-specific IgM, a significant portion of viral antigen will be bound to the beads in each assay. It is therefore not recommended that antigen aspirated from the beads is recycled.

Treatment of Results

The amount of ^{125}I-labeled antimouse antibody (and hence anti-B19 monoclonal antibody and B19 antigen) bound to each bead is estimated by counting gamma emission for 5 min. The background counts are subtracted from the total counts for each bead, and the mean of counts for duplicate beads is calculated. For each serum specimen under test, the test : negative ratio is computed by the following:

$$T : N = \frac{\text{Mean cp 5 min bead incubated with test serum}}{\text{Mean cp 5 min beads incubated with neg. serum}}$$

The T : N values for the five standard sera are computed (Table 2), and these T : N values are plotted against the B19-specific IgG concentration in AU (Fig. 5). The T : N for the serum containing 0.25 AU must be > 1.0. For each serum under test, the T : N ratio is calculated, and the concentration of B19-specific IgG is determined by interpolation from the graph.

Interpretation

The test for specific IgG may be conducted either to diagnose recent infection or to determine previous exposure to B19. The interpretation of results is different in these two circumstances.

Serological diagnosis of recent infection is normally made by the detection of B19-specific IgM (see above). However, detection of specific IgG may be of value if no specimen was obtained within 10 to 12

TABLE 2. Computation of results for standard sera in B19 IgG antibody capture RIA

Antibody concentration (AU)	^{125}Iodine bound (counts/5 min)	T : N
0	150	NA
0.25	180	1.2
0.9	315	2.1
2.8	960	6.4
9.2	1845	12.3
30.0	3903	26.0
100.0	6606	44.0

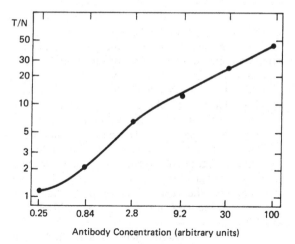

FIG. 5. Standard curve of B19-specific IgG detected by IgG antibody capture radioimmunoassay.

weeks of the onset of symptoms, when specific IgM is detectable. Where a specimen taken before the onset of symptoms is available (e.g., a specimen taken during a routine clinic visit), this may be tested in parallel with a specimen taken more than 10 weeks after the illness. Seroconversion from < 0.25 AU to > 10.0 AU indicates B19 infection in the intervening period. If no preillness specimen is available, examination of two specimens obtained 6 months to 1 year apart may be helpful. Sera taken 2 to 12 months after infection normally contain a high concentration of B19-specific IgG. This concentration falls rapidly in the following year to some 50% of the convalescent concentration. Thereafter, the decline in concentration is slower. Thus, detection of a twofold drop in B19-specific IgG concentration over 6 to 12 months is indicative, though not diagnostic, of recent B19 infection. Since some individuals maintain high concentrations of B19-specific IgG for long periods, a single serum specimen containing high concentration antibody cannot be regarded as indicative of recent infection.

In attempts to determine previous exposure (and thus immunity) to B19 infection, detection of > 1.0 AU in a serum specimen giving a T : N value of > 2 may be considered indicative of immunity. A negative result, however, is not synonymous with susceptibility. This is due to the relative insensitivity of the test. The underlying principle of antibody capture tests, where a representative population of all of the IgG specificities present in a sample is bound to the solid phase, means that only those sera containing a certain minimum *proportion* of IgG of a particular specificity will give a positive signal. Thus, sera containing low concentrations of B19-specific IgG will not bind sufficient B19 antigen to be detected by the

indicator monoclonal antibody/^{125}I label complex, giving false-negative results.

The determination of susceptibility or immunity to B19 is at present of significance in the management of chronic hemolytic anemics exposed to B19 infection, seronegative individuals being at risk of aplastic crisis.

ENZYME IMMUNOASSAYS FOR IgM AND IgG

As noted above, the principle of antibody capture tets may be applied either to radioimmunoassay or enzyme immunoassay (EIA). Although B19 antibody immunoassays in use in England utilize ^{125}I-labeled indicator antibody, Anderson et al. (1986) have described EIAs for B19-specific IgM and IgG. These are described below.

Preparation of Solid Phase

Flat-bottomed 96-well Immulon-2 microtiter plates are coated with antibody to human IgM or IgG diluted in 0.01 M carbonate buffer (pH 9.6) overnight at 4°C. The wells are then washed, and remaining free protein binding sites are blocked by incubation in PBS with 0.5% gelatin and 2% normal goat serum (PBS-G-NGS) for 30 min at 35°C. After washing, the wells are ready for use.

The Test

Test sera are diluted 1 in 100 in PBS with 0.5% gelatin and 0.15% Tween 20 (PBS-G-T) and incubated for 1.5 h at 35°C in each of four replicate wells. The wells are then washed, and B19 antigen in the form of viremic serum, diluted in PBS-G-T added to two of the wells, and control antigen consisting of B19 antigen and antibody-negative serum similarly diluted in PBS-G-T are added to the remaining two wells. The wells are incubated under these antigen overnight at room temperature, and unbound antigen is removed by washing. Monoclonal antibody to B19 in the form of ascitic fluid is diluted in PBS-G-T and reacted with the wells for 1.5 h at 35°C. After washing, peroxidase-conjugated goat antimouse IgM antibody diluted in PBS-G-T for an IgG test, or PBS-G-T-NGS for an IgM test, is added, and the wells are incubated at 35°C for 1 h. The wells are then washed, and the substrate O-phenylenediamine and 0.015% hydrogen peroxide in 0.15 M citrate phosphate buffer (pH 5.5) are added. The reaction is stopped after 30 to 45 min, by the addition of 3.5 M HCl, and the absorbance at 490 is measured.

Treatment of Results

A positive : negative ratio is calculated for each serum by the following:

$$P:N = \frac{\text{Mean absorbance 2 wells reacted with B19 antigen}}{\text{Mean absorbance 2 wells reacted with control antigen}}$$

For the IgM test, the authors state that sera giving P:N values of > 1.5 are considered positive (this figure was derived by testing a number of sera from well persons, computing the mean (1.13) and standard deviation (0.14) of these results, and calculating the mean plus 3 SD at 1.47). For the IgG test, sera giving P:N values > 2.1 together with an absolute absorbance of > 0.030 are considered positive.

Interpretation

Anderson et al. (1986) report that when 25 sera were tested by EIA and RIA similar to the test described above, there was good agreement between the tests, although for both IgM and IgG one specimen gave a positive result in RIA but a negative result in EIA, suggesting that for both tests EIA may be less sensitive than RIA.

The authors refer to "titers" of B19-specific IgM detected in the EIA described (Anderson et al., 1986). It should be noted that titer, defined by Anderson et al. (1986) as the last dilution with a P:N value > 1.5, has no meaning in antibody capture tests. This is because the reactivity of a serum specimen in such a test depends on the *proportion* of the total (IgM) antibody molecules with a particular specificity. If a serum specimen is serially diluted in a human serum-free assay diluent, the proportion of IgM specific for B19 does not alter. It is for this reason that the use of a set of standard sera, as described in the section on RIA for antibody testing, is to be recommended as a means of quantifying specific IgM or IgG antibody.

Epidemiology

Since B19 cannot be grown in tissue culture, antigenic material with which to conduct large-scale serological surveys is not widely available. The only source of virus is the serum of infected individuals who are either asymptomatic or cannot be clinically identified. (The serum of patients undergoing aplastic crisis may also be rich in virus; however, taking large volumes of blood from these severely anemic patients is contrary to their management.) Donations of blood from voluntary asymptomatic, viremic blood donors have provided the bulk of virus to date. By screening donated blood for B19, the frequency of viremia in the donor population of southeast England aged 18 to 65 years has been estimated at 1 in 40,000 during epidemic periods (Anderson, 1982). Much of our knowledge of the epidemiology of B19 has therefore been gained by inference—by analysis of the patterns of B19-associated disease.

B19 is present throughout the year, although in temperate climates outbreaks of infection are more common in the late winter, spring, and early summer months. These outbreaks of infection are often centered on elementary schools when up to 40% of the school may be clinically affected by the rash illness of erythema infectiosum. During these outbreaks, adults (e.g., parents and teachers of cases) may also be affected (Ager et al., 1966). This is in accord with an early serological survey conducted by CIE which indicated that the peak age of antibody acquisition lay between the ages of 4 and 10 years (Edwards et al., 1981).

B19 virus is common: 60% of a north London blood donor population were found to have antibody by a more sensitive radioimmunoassay (Cohen et al., 1983), although even this may be an underestimate. Among the elderly, the rate of seropositivity approaches 80% (E. O. Caul, personal communication).

In addition to seasonality, the virus exhibits longer-term cycles. In Jamaica, for example, peaks of incidence (monitored as cases of aplastic crisis) occur every 3 to 4 years (Serjeant et al., 1981; G. R. Serjeant, unpublished observations). In the United Kingdom, the cycle seems somewhat longer, with peaks occurring every 4 to 5 years.

Case-to-case intervals, determined by the time elapsing between acquisition and excretion of the virus, are independent of the type of disease. Volunteer studies (Anderson et al., 1985a) predict case-to-case intervals of 6 to 12 days, which accords well with case-to-case intervals observed in outbreaks of erythema infectiosum (Ager et al., 1966; Anderson et al., 1984; Plummer et al., 1985) and of aplastic crisis (Lefrere et al., 1985b; Mortimer, 1983).

Control and Prevention

B19 cannot be grown in tissue culture, and no vaccine is available. The virus is spread via the respiratory route, but cases of erythema infectiosum are no longer infectious, so isolation of exanthematous patients serves no useful role. Cases of aplastic crisis may be excreting virus at presentation: in view of this, they would be isolated from patients susceptible to this more severe manifestation of infection—namely, seronegative chronic hemolytic anemics—and from seronegative pregnant women.

It may be possible to protect susceptible chronic hemolytic anemics with normal immunoglobulin and/or whole blood transfusion. This may be considered when B19 infection (indicated by erythema infectiosum) is prevalent in the community.

Blood donations are not routinely screened for B19 antigen. In view of the low incidence (< 1 in 40,000) of virus-positive donations coupled with the fact that, unlike hepatitis B virus, B19 circulates widely in the community, large-scale blood donation screening is of relatively little importance. However, screening of blood for certain individuals (e.g., sero-negative chronic hemolytic anemics, pregnant women, neonates, and immunosuppressed patients) may be considered. This may be achieved by either screening donated units for virus (by immunoassay or DNA : DNA hybridization) or by using blood from known B19-specific IgG-positive donors.

Literature Cited

Ager, E. A., T. D. Y. Chin, and J. P. Poland. 1966. Epidemic erythema infectiosum. N. Engl. J. Med. **275:**1326–1331.

Anderson, M. J. 1982. The emerging story of a human parvovirus-like agent. J. Hyg. **89:**1–8.

Anderson, M. J., and J. R. Pattison. 1984. The human parvovirus. Arch. Virol. **82:**137–148.

Anderson, M. J., L. R. Davis, J. Hodgson, S. E. Jones, L. Murtaza, J. R. Pattison, C. E. Stroud, and J. M. White. 1982a. Occurrence of infection with a parvovirus-like agent in children with sickle cell anaemia during a two-year period. J. Clin. Pathol. **35:**744–749.

Anderson, M. J., L. R. Davis, S. E. Jones, and J. R. Pattison. 1982b. The development and use of an antibody capture radioimmunoassay for specific IgM to a human parvovirus-like agent. J. Hyg. **88:**309–324.

Anderson, M. J., E. Lewis, I. M. Kidd, S. M. Hall, and B. J. Cohen. 1984. An outbreak of erythema infectiosum associated with human parvovirus infection. J. Hyg. **93:**85–93.

Anderson, M. J., P. G. Higgins, L. R. Davis, J. S. Willman, S. E. Jones, I. M. Kidd, J. R. Pattison, and D. A. J. Tyrrell. 1985a. Experimental parvoviral infection in man. J. Infect. Dis. **152:**257–265.

Anderson, M. J., I. M. Kidd, and P. Morgan-Capner. 1985b. Human parvovirus and rubella-like illness. Lancet **2:**663.

Anderson, M. J., S. E. Jones, and A. C. Minson. 1985c. Diagnosis of human parvovirus infection by dot-blot hybridisation using cloned viral DNA. J. Med. Virol. **15:**163–172.

Anderson, L. J., C. Tsou, R. A. Parker, T. L. Chorba, H. Wulff, P. Tattersall, and P. P. Mortimer. 1986. Detection of antibodies and antigens of human parvovirus B19 by enzyme-linked immunosorbent assay. J. Clin. Microbiol. **24:**522–526.

Anand, A., E. S. Gray, T. Brown, J. P. Clewley, and B. S. Cohen. 1986. Human parvovirus infection in pregnancy and hydrops fetalis. N. Engl. J. Med. **316:**183–186.

Balfour, H. H., Jr., G. M. Schiff, and J. E. Bloom. 1970. Encephalitis associated with erythema infectiosum. J. Pediatr. **77:**133–136.

Bond, P. R., E. O. Caul, J. Usher, B. J. Cohen, J. P. Clewley, and A. M. Field. 1986. Intrauterine infection with human parvovirus. Lancet **1:**448–449.

Bourguignon, G. J., P. J. Tattersall, and D. C. Ward. 1976. DNA of minute virus of mice: self-priming, non-permuted, single-stranded genomes with a 5′ terminal hairpin duplex. J. Virol. **20:**290–306.

Brown, F. 1988. The classification and nomenclature of viruses. (Fifth report of the ICTV.) Intervirology (in press).

Brown, T., A. Anand, L. D. Ritchie, J. P. Clewley, Reid, T. M. S. 1984. Intrauterine human parvovirus infection and hydrops fetalis. Lancet **2:**1033–1034.

Cheung, A. K. M., M. D. Hoggan, W. W. Hauswirth, and K. I. Berns. 1980. Integration of the adeno-associated virus genome into cellular DNA in latently infected human Detroit 6 cells. J. Virol. **33:**739–748.

Chorba, T., P. Coccia, R. C. Holman, P. Tattersall, L. J. Anderson, J. Sudman, N. S. Young, E. Kurczynski, U. M. Saarinen, R. Moir, D. N. Lawrence, J. M. Jason, and B. Evatt. 1986. The role of parvovirus B19 in aplastic crisis and erythema infectiosum (fifth disease). J. Infect. Dis. **154:**383–393.

Clewley, J. P. 1984. Biochemical characterisation of a human parvovirus. J. Gen. Virol. **65:**241–244.

Cohen, B. J., P. P. Mortimer, and M. S. Pereira. 1983. Diagnostic assays with monoclonal antibodies for the human serum parvovirus-like virus (SPLV) J. Hyg. **91:**113–130.

Cossart, Y. E., A. M. Field, B. Cant, and D. Widdows. 1975. Parvovirus-like particles in human sera. Lancet **1:**72–73.

Denhardt, D. T. 1966. Membrane filter technique for the detection of complementary DNA. Biochem. Biophys. Res. Commun. **23:**641–646.

Duncan, J. R., M. D. Capellini, M. J. Anderson, C. G. Potter, J. B. Kurtz, and D. J. Weatherall. 1983. Aplastic crisis due to parvovirus infection in pyruvate kinase deficiency. Lancet **2:**14–16.

Dunnet, W. N., B. T. Thorm, R. G. Ayling. 1984. Food poisoning from oysters. C.D.R. **36:**3.

Greenwood, F. C., W. M. Hunter, and J. S. Glover. 1963. The preparation of 131I-labelled human growth hormone of high specific radioactivity. Biochem. J. **89:**114.

Hall, C. B., and F. A. Horner. 1977. Encephalopathy with erythema infectiosum. Am. J. Dis. Child. **131:**65–67.

Hammon, W. D., and J. F. Enders. 1939. A virus disease of cats, principally characterised by aleucocytosis, enteric lesions and the presence of intranuclear inclusion bodies. J. Exp. Med. **69:**327–352.

Kelleher, J. H., N. L. C. Luban, P. P. Mortimer, and T. Kamimura. 1983. The human serum "parvovirus." A specific cause of aplastic crisis in hereditary spherocytosis. J. Paediatr. **102:**720–722.

Knott, P. D., G. A. C. Welply, and M. J. Anderson. 1984. Serologically proven intrauterine infection with parvovirus. Br. Med. J. **289:**1660.

Kurtz, J. B., and M. J. Anderson. 1985. Cross-reactions in rubella and parvovirus specific IgM tests. Lancet **2:**1356.

Lefrere, J. J., and H. Bougeois. 1986. Human parvovirus associated with erythroblastopenia in iron deficiency anaemia. J. Clin. Pathol. **39:**1277–1278.

Lefrere, J. J., A.-M. Courouce, J. Y. Muller, M. Clark, and J. P. Soulier. 1985a. Human parvovirus and purpura. Lancet **2:**730.

Lefrere, J. J., P. Boutard, A.-M. Courouce, T. Lacaze, and R. Girot. 1985b. Familial human parvovirus infections. Lancet **2:**333–334.

Lefrere, J. J., A.-M. Courouce, J. P. Soulier, M. P. Cordier, M. C. Guesne Girault, C. Polonovski, and A. Bensman. 1986. Henoch Schonlein purpura and human parvovirus infection. Pediatrics **78:**183–184.

Macartney, L., I. A. P. McClandish, H. Thompson, and H. J. C. Cornwell. 1984. Canine parvovirus enteritis 2: pathogenesis. Vet. Rec. **115:**453–460.

Mason, W. S., C. Aldrich, J. Summers, and J. M. Taylor.

1982. A symmetric replication of duck hepatitis B virus DNA in liver cells: free minus-strand DNA. Proc. Natl. Acad. Sci. USA **79**:3997–4001.

Morinet, F., J.-D. Tratschin, Y. Perol, and G. Siegl. 1986. Comparison of 17 isolates of the human parvovirus B19 by restriction enzyme analysis. Bret report. Arch. Virol. **90**:165–172.

Mortimer, P. P. 1983. Hypothesis: the aplastic crisis of hereditary spherocytosis is due to a single transmissible agent. J. Clin. Pathol. **36**:445–448.

Mortimer, P. P., N. L. C. Luban, J. F. Kelleher, and B. J. Cohen. 1983. Transmission of serum parvovirus-like virus by clotting factor concentrates. Lancet **2**:482–484.

Mortimer, P. P., B. J. Cohen, M. M. Buckley, J. E. Cradock-Watson, M. K. S. Ridehalgh, F. Burkhardt, and U. Schilt. 1985a. Human parvovirus and the fetus. Lancet **2**:1012.

Mortimer, P. P., B. J. Cohen, M. A. Rossiter, S. M. Fairhead, and A. F. M. S. Rahman. 1985b. Human parvovirus and purpura. Lancet **2**:730–731.

Neild, G., M. J. Anderson, S. Hawes, and B. T. Colvin. 1986. Parvovirus infection after renal transplant. Lancet **2**:1226–1227.

Okabe, N., S. Kobyashi, O. Tatsuzawa, P. P. Mortimer. 1984. Detection of antibodies to human parvovirus in erythema infectiosum (fifth disease). Arch. Dis. Child. **59**:1016–1019.

Ozawa, K., G. Kurtzman, and N. Young. 1986. Replication of the B19 parvovirus in human bone marrow cell cultures. Science **233**:883–886.

Parrish, C. R., and L. E. Carmichael. 1983. Antigenic structure and variation of canine parvovirus type-Z, feline panleukopenia virus and mink enteritis virus. Virology **129**:401–414.

Pattison, J. R. 1982. Laboratory investigation of rubella. No. 16 in Public Health Laboratory Service Monograph Series. Her Majesty's Stationery Office, London.

Paver, W. K., E. O. Caul, C. R. Ashley, and S. K. R. Clarke. 1973. A small virus in human faeces. Lancet **1**:664–665.

Pintel, L. D., M. J. Merchlinsky, and D. C. Ward. 1984. Expression of minute virus of mice structural proteins in murine cell lines transformed by bovine papillomavirus–minute virus of mice plasmid chimera. J. Virol. **52**:320–327.

Plummer, F. A., G. W. Hammond, K. Forward, L. Sekla, L. M. Thomson, S. E. Jones, I. M. Kidd, and M. J. Anderson. 1985. An erythema infectiosum-like illness caused by human parvovirus infection. N. Engl. J. Med. **313**:74–79.

Potter, C. G., A. C. Potter, C. S. R. Hatton, M. J. Anderson, J. R. Pattison, D. A. J. Tyrrell, P. G. Higgins, J. S. Willman, P. M. Cotes, H. F. Parry, and H. M. Chapel. 1987. Variation of erythoid and myeloid precursors in the bone marrow and peripheral blood of volunteer subjects infected with human parvovirus (B19). J. Clin. Infect. **79**:1486–1492.

Rao, K. R. P., A. R. Patel, M. J. Anderson, J. Hodgson, S. E. Jones, and J. R. Pattison. 1983. Infection with a parvovirus-like virus and aplastic crisis in chronic haemolytic anaemia. Ann. Intern. Med. **98**:930–932.

Reid, D. M., T. M. S. Reid, T. Brown, J. A. N. Rennie, and C. J. Eastmond. 1985. Human parvovirus-associated arthritis: a clinical and laboratory description. Lancet **2**:422–425.

Richards, R., P. Linser, and R. W. Armentrout. 1977. Kinetics of assembly of a parvovirus minute virus of mice, in synchronized rat brain cells. J. Virol. **22**:778–793.

Rigby, P. W. S., M. Diekmann, C. Rhodes, and P. Berry. 1977. Labelling deoxyribonucleic acid to high specific activity in vitro by nick translation with DNA polymerase I. J. Mol. Biol. **113**:237–251.

Salacinski, P., J. Hope, C. McClean, V. Clement-Jones, J. Sykes, J. Price, and P. J. Lowry. 1979. A new simple method which allows theoretical incorporation of radioiodine into proteins and peptides without damage. J. Endocrinol. **81**:131–137.

Saunders, P. W. G., M. M. Reid, and B. J. Cohen. 1986. Human parvovirus induced cytopenias: a report of five cases. Br. J. Haematol. **63**:407–410.

Schlehofer, J. R., M. Ehrbar, and H. Zur Hausen. 1986. Vaccinations, herpes simplex virus and carcinogens induce DNA amplification in a human cell line and support replication in a human cell line and support replication of a helpervirus dependent parvovirus. Virology **152**:110–117.

Serjeant, G. R., K. Mason, J. M. Topley, B. Serjeant, J. R. Pattison, S. E. Jones, and R. Mohammed. 1981. Outbreak of aplastic crisis in sickle cell anaemia associated with parvovirus-like agent. Lancet **2**:595–597.

Shade, R. L., M. C. Blundell, S. F. Cotmore, P. Tattersall, and C. R. Astell. 1986. Nucleotide sequence and genome organisation of human parvovirus B19 isolated from an aplastic crisis. J. Virol. **58**:921–927.

Shneerson, J. M., P. P. Mortimer, and E. M. Vandervelde. 1980. Febrile illness due to a parvovirus. Br. Med. J. **2**:1580.

Siegl, G. 1984. The biology and pathogenicity of autonomous parvoviruses, p. 297–348. In K. I. Berns (ed.), The parvoviruses. Plenum, New York.

Simpson, R. W., L. McGinty, L. Simon, C. A. Smith, C. W. Godzeski, and R. J. Boyd. 1984. Association of parvoviruses with rheumatoid arthritis of humans. Science **223**:1425–1428.

Summers, J., S. E. Jones, and M. J. Anderson. 1983. Characterisation of the agent of erythrocyte aplasia as a human parvovirus. J. Gen. Virol. **64**:2527–2532.

Tattersall, P. 1978. Susceptibility to minute virus of mice as a function of host cell differentiation, p. 131–149. In D. C. Ward and P. Tattersall (ed.), Replication of mammalian parvoviruses. Cold Spring Harbor Laboratory, Cold Spring Harbor, NY.

Tattersall, P., and S. Cotmore. 1987. The nature of parvoviruses. In J. R. Pattison (ed.), Parvoviruses and human disease. CRC Press, Boca Raton, FL (in press).

Wadlington, W. B. 1957. Erythema infectiosum: report of an epidemic. J. Tenn. St. Med. Assoc. **50**:1–5.

Wadlington, W. B., and H. D. Riley, Jr. 1968. Arthritis and haemolytic anaemia following erythema infectiosum. JAMA **203**:473–475.

Ward, D. C., and P. Tattersall. 1982. Minute virus of mice, p. 313–332. In H. L. Foster, J. D. Small, and J. G. Fox (ed.), The mouse in biomedical research, Vol. 2. Academic Press, New York.

West, N. C., R. E. Meigh, M. Mackie, and M. J. Anderson. 1986. Parvovirus infection associated with aplastic crisis in a patient with HEMPAS. J. Clin. Pathol. **39**:1019–1020.

Young, N. S., P. P. Mortimer, J. G. Moore, and R. K. Humphries. 1984a. Characterisation of a virus that causes transient aplastic crisis. J. Clin. Invest. **73**:224–230.

Young, N. S., M. Harrison, J. G. Moore, P. P. Mortimer, and R. K. Humphries. 1984b. Direct demonstration of the human parvovirus in erythroid progenitor cells infected in vitro. J. Clin. Invest. **74**:2024–2032.

CHAPTER 19

Reoviridae: The Reoviruses

KENNETH L. TYLER and BERNARD N. FIELDS

Disease: Reovirus. Typically produces asymptomatic or subclinical illnesses resulting in seroconversion. Possibly associated mild gastrointestinal or upper respiratory infections. Association with neonatal biliary atresia and with central nervous system infections has been reported, but remains controversial.

Etiologic Agents: Reoviruses (three serotypes).

Source: Isolation from feces or anal swab specimens, less commonly throat or nasopharyngeal swabs, and rarely cerebrospinal fluid, urine, or tissue specimens.

Clinical Manifestations: Infection typically occurs in childhood, with the majority of individuals becoming seropositive by 20 years of age. The clinical illness associated with seroconversion has not been defined, but may be associated with mild diarrhea and fever or, alternatively, with mild symptoms of upper respiratory infection including rhinorrea, cough, coryza, pharyngitis, headache, and malaise.

Pathology: No definite pathologic lesions occur in man.

Laboratory Diagnosis: Virus isolation in tissue culture; seroconversion or four-fold or greater increase in neutralizing or hemagglutination-inhibiting antibody. Detection of viral antigen in tissues by immunocytochemical methods.

Epidemiology: Worldwide distribution, with no clear pattern of seasonal incidence. Infection occurs primarily in infants and children.

Treatment: No specific treatment is available.

Prevention and Control: Simple hygienic measures designed to minimize fecal-oral and aerosol transmission, including proper disposal of excreta and proper hand-washing measures.

Introduction and History

The first documented isolation of what would later be classified as a reovirus (type 3) was made by Stanley, Dorman, and Ponsford in 1951 from a fecal specimen collected from a 2-½-year-old aboriginal child (Stanley et al., 1953, 1954; Stanley, 1961). This child had a complex medical history of repeated bronchopulmonary infections, bronchiectasis, malnourishment, conjunctivitis, and alopecia. A fecal sample that had been stored at −23°C for 9 months (!) was inoculated intraperitoneally into suckling mice, and the process was repeated with a specimen that had been stored for 15 months. The inoculated mice developed, after a 16-day incubation period, a peculiar syndrome of emaciation, alopecia, jaundice, peritoneal exudation, and encephalitis. In addition, an "oily substance" covered the entire body of the affected mice and matted their hair together into strands, especially over the head, belly, and sacral area ("Oily hair effect" or "OHE"). Jaundice first became evident in the ears of the mice and later became evenly distributed over the skin and often involved the feet, tail, and nose. A blood-stained or "bright yellow" fluid collected in the peritoneal cavity, often preceding other symptoms. Symptoms suggesting central nervous system (CNS) infection occurred in the majority of mice and included incoordination, tremors, paralysis and ataxia. Alopecia occurred only in mice recovering from OHE and encephalitis, it produced bald patches over the head, neck, and near the hind legs. Other symptoms included a "condition resembling conjunctivitis" and in some mice a late-appearing "secondary" jaundice.

After intraperitoneal inoculation, the virus could be demonstrated in skeletal muscle, liver, lung, peritoneal exudate fluid, and brain. The most prominent histopathologic lesions involved the liver and CNS, prompting Stanley and his colleagues to call the new viral isolate "hepato-encephalomyelitis virus" or "HEV." In the CNS the findings included meningitis and perivascular cuffing. The liver often contained focal areas of hepatic necrosis. The bile ducts were surrounded by inflammatory cells and often appeared stenosed.

In the summer of 1953 a virus was isolated from a rectal swab taken from a healthy child ("Lang") living in Cincinnati (Ramos-Alvarez and Sabin, 1954). This virus initially became the prototype for "ECHO 10" and later the prototypic reovirus serotype 1 strain (Sabin, 1959). Although this isolate was classified as an Echovirus on the basis of its source (stool), method of isolation (tissue culture), and an erroneous belief that it was nonpathogenic for newborn mice, it was recognized almost immediately that the pattern of cytopathic effect (CPE) produced by ECHO 10 in monkey kidney tissue culture differed from that produced by other known enteroviruses. Two other viruses were subsequently isolated by the same investigators from children admitted to hospitals in Cincinnati with "summer diarrheal illnesses" (Ramos-Alvarez, 1957; Ramos-Alvarez and Sabin 1958). One of these, referred to initially as "D5" because it was the fifth unclassifiable isolate from a child ("Jones") with diarrheal illness, would subsequently become the prototype for reovirus serotype 2 (Sabin 1959). Another isolate from the same series from a child "Dearing" would subsequently become the prototype for reovirus serotype 3 (Sabin, 1959). Another virus, recovered in 1957 from an anal swab specimen taken from a 17-month-old baby ("Abney") living in an institution in Washington, D.C., became the second prototypic reovirus type 3 strain (Rosen et al., 1960). At the time the virus was isolated the child had a febrile illness diagnosed as an "upper respiratory infection."

A number of other early reovirus isolates were included in the "SV" series of Hull (Hull et al., 1956, 1958) and included SV4, SV12, SV28, and SV59. Interestingly, Hull had grouped all of these isolates together, because he recognized that they produced a similar and completely distinctive pattern of CPE in monkey kidney tissue culture. The monkey kidney cells became granular and vacuolized in appearance and did not slough off the glass tissue culture vials as readily as when they were infected by other enteroviruses.

In 1959, Albert Sabin proposed that all of the viruses discussed above be included in a new group of respiratory and enteric viruses for which he suggested the name REOVIRUSES (R = respiratory, E = enteric, O = orphan) (Sabin, 1959). The name was intended to stress the frequent isolation of these viruses from the respiratory and enteric tracts, as well as to emphasize the lack of a clear-cut association of these viruses with any known disease (hence "orphan"). Sabin noted that these viruses were much larger than the other known enteroviruses, that they produced a distinctive pattern of CPE which differed from that of other enteroviruses, that they were in fact pathogenic for newborn mice (unlike other Echoviruses), and finally that their hemagglutination properties differed from those of the other Echoviruses.

The Viruses

Classification

The family *Reoviridae* is composed of six genera. Viruses representing three of these genera (the orthoreoviruses, rotaviruses, and orbiviruses) can infect vertebrate animals, including man. Orbiviruses and rotaviruses are discussed in detail in Chapters 20 and 21, respectively, and therefore this chapter will be devoted exclusively to the orthoreoviruses ("reovirus").

Reoviruses are classified in the family *Reoviridae* on the basis of certain physicochemical properties, the nature of the viral genome, and the general replicative strategy employed by the virus in infected cells (see Joklik, 1983, for review). The virions are nonenveloped and have a double-capsid shell. The capsid has icosahedral symmetry and an approximate outer diameter of 75 nm. The viral genome is composed of ten discrete segments of double-stranded RNA (total molecular weight, circa 15×10^6 daltons), each of which is transcribed into one mRNA. Replication occurs exclusively in the cytoplasm and involves removal of the outer capsid shell of the virion, although complete uncoating does not occur. The virion contains all of the enzymes required for replication and transcription, including an RNA-dependent RNA polymerase and enzymes responsible for 5'-capping of the viral mRNAs.

Reoviruses isolated from mammals, including man, can be grouped into three serotypes (Rosen, 1960; Sabin, 1959). These three serotypes all appear to share a common group antigen detectable either by complement fixation (Sabin, 1959) or immunodiffusion (Leers et al., 1968). The original prototype for the ECHO 10 virus strain, "Lang," became the prototype for reovirus type 1. An isolate obtained from the stool of a child with a diarrheal illness, "Jones," became the prototype for reovirus type 2. Reovirus serotype 2 shows a greater degree of antigenic heterogeneity than the other serotypes, and it has been suggested that four specific subtypes can be recognized by hemagglutination–inhibition tests (Hartley et al., 1962). A virus isolated from a child with a diarrheal illness, "Dearing," and one from a child with an upper respiratory illness, "Abney," became the prototypes for reovirus type 3 (Rosen, 1960; Sabin, 1959). A brief report has appeared from Japan suggesting that two "new" serotypes of reovirus have been isolated from cattle (Kurogi et al., 1974), but confirmation of this report has been lacking, and there has been no enthusiasm for expanding the number of mammalian serotypes beyond the original three.

A large number of reoviruses have been isolated from birds. These do not appear to fall into the serotypic groups described above for mammalian reoviruses (Deshmukh et al., 1968). The avian reoviruses do appear to share common group antigens (Kawamura and Tsubahara, 1966). These viruses can be grouped into a least five serotypes based on their reactions with specific neutralizing antibodies (Kawamura et al., 1965; Kawamura and Tsubahara, 1966). Most avian reoviruses also differ from their mammalian counterparts in a number of biologic properties. The avian viruses have an abortive replication cycle in mammalian cell lines, most isolates do not hemagglutinate erythrocytes, and many have the capacity to induce cell fusion in certain cell lines (Glass et al., 1973; Kawamura et al., 1965; Spandidos and Graham 1976a,b).

A reovirus referred to as "Nelson Bay Virus" has been isolated from a flying fox (*Pteropus poliocephalus*). The Nelson Bay Virus shares the same group antigen as other mammalian reoviruses, but resembles the avian reoviruses in possessing cell-fusion activity (Gard and Compans, 1970; Gard and Marshall, 1973; Wilcox and Compans, 1982).

Physicochemical Properties and Morphology

Intact viral particles have a buoyant density in cesium chloride of 1.36 to 1.41 g/ml, depending on the procedure used to purify the virus (Joklik, 1983; Smith et al., 1969). The molecular weight of the intact particle, as estimated from its diffusion coefficient, is 129.5×10^6 daltons (Farrell et al., 1974). Approximately 85% of the total molecular weight consists of protein, and the remainder of RNA.

Reoviruses are stable under conditions of ambient temperature. In general, reovirus type 1 appears to be somewhat more heat stable then reovirus type 3. The time to loss of 50% of initial infectivity has been

determined at various temperatures. For reovirus type 1 it was 19 h at 37°C and 3.7 days at 4°C (Rhim et al., 1961). By contrast, it took less than 3 h (157 min) for a strain of reovirus type 3 to lose 50% of its initial infectivity at 37°C (Gomatos et al., 1962). For certain isolates, the loss of reovirus infectivity at high temperature results from removal of the sigma 1 outer capsid protein (Drayna and Fields 1982b,c).

Reovirus type 3 is resistant to antibiotics commonly included in tissue culture media, including streptomycin and penicillin (Stanley et al., 1953, 1954). It is also relatively resistant to formaldehyde (3%), phenol (1%), and hydrogen peroxide (1%) as tested by exposure for 1 h at room temperature. Ethanol (70%) will completely inactivate reovirus type 3 under the identical conditions (Stanley et al., 1953, 1954). Reovirions are not enveloped and hence are resistant to treatment with lipid solvents including ether (Stanley et al., 1954).

Reoviruses are sensitive to inactivation by ultraviolet light (McClain and Spendlove, 1966; Rauth, 1965). Viruses can also be inactivated by the combination of treatment with certain dyes (e.g., neutral red, toluidine blue) and exposure to light in the visible wavelength spectrum (Hiatt, 1960; Wallis and Melnick, 1964).

The proteins that compose the outer capsid of the virion can be disrupted by a variety of physical and chemical manipulations (Drayna and Fields, 1982b,c). Alkaline pH and high temperature (55°C) result in the removal of the viral hemagglutinin, sigma 1. Detergents, such as SDS (1%), can remove the sigma 3 outer capsid protein (Drayna and Fields, 1982b,c). At higher doses or concentrations, disruption of the entire virus particle occurs. Proteases, such as chymotrypsin, can sequentially remove the outer capsid proteins from the virion, although the viral core is relatively resistant to proteolytic digestion (Borsa et al., 1973a,b; Joklik, 1972; Shatkin and Sipe, 1968b; Smith et al., 1969). Proteolytic digestion of the outer capsid can result in loss of infectivity for some strains of reovirus and enhancement of infectivity for others (Rubin and Fields, 1980; Spendlove and Schaffer, 1965; Spendlove et al., 1970).

As will be discussed in more detail later, the physicochemical properties of reovirions have important implications for the storage of specimens suspected to contain virus and for the disinfection of laboratory equipment, glassware, and other items exposed to virus. In general, the viruses are relatively stable at usual environmental temperatures, and storage at normal refrigerator temperatures (4°C) results in only gradual loss of infectivity. Exposure of specimens designed for viral isolation to ultraviolet light, which may be present in settings such as operating rooms or isolation hoods, should be avoided. Of the easily available laboratory disinfectant solutions, ethanol (70% or greater concentration) seems to provide adequate inactivation of the virus, especially if applied for long periods of time (e.g., 1 h).

MORPHOLOGY

Reovirions are composed of an inner protein capsid containing ten double-stranded RNA segments ("the core"), surrounded by an outer protein shell. The outer capsid forms a nearly spherical icosahedron (Joklik, 1983; Jordan and Mayor, 1962; Luftig et al., 1972; Vasquez and Tournier, 1962, 1964) with five-, three-, and twofold axes of rotational symmetry and 12 vertices. It has an approximate diameter of 76 nm (Luftig et al., 1972). Subunit components (capsomers) of the outer capsid can be identified, although their exact number and spatial arrangement has not been definitively established. The capsomers, in turn, appear to be composed of either six (hexagonal capsomers) or five (pentagonal capsomers) smaller subunits (Amano et al., 1971; Luftig et al., 1972; Metcalf, 1982; Palmer and Martin, 1977; Vasquez and Tournier, 1964). Three proteins (σ1, σ3, and μ1c) form the outer capsid shell (see Joklik et al., 1983, for review). The μ1c and σ3 appear to be located in close proximity to each other in the outer capsid of the virus (Lee et al., 1981) in a ratio of 2:1. The σ1 molecules are located at the icosahedral vertices in close proximity to the λ2 core spike (Joklik, 1983; Lee et al., 1981) and appear to be in the form of a multimer (Bassel-Duby et al., 1985). The individual σ1 molecules appear to consist of a long α-helical tail, which may be inserted into a 5-nm-wide central channel in the middle of the λ2 core spike, and a globular head which extends above the virion surface (Bassel-Duby et al., 1985).

The core also has icosahedral symmetry and, like the outer capsid, is composed of subunits. The core diameter has been estimated at 52 nm (Luftig et al., 1972). Complexes composed of five molecules of the λ2 viral protein form "core spikes" (Ralph et al., 1980) that project from the surface of the core at each of its twelve vertices. The core spikes project for a distance of 5 to 6 nm from the outer surface of the core. This corresponds to about half the distance between the surface of the core and the inner surface of the outer capsid (Luftig et al., 1972). The outer capsid surface contains slight indentations that correspond to the location of the core spike (Joklik, 1983; Palmer and Martin, 1977). Three proteins (λ1, λ2, σ2) are the major components of the core and core spike (λ2); σ2 and λ1 appear to be complexed together in a ratio of 2:1 (see Joklik, 1983, for review). σ2 may lie primarily on the inner surface of the core (White and Zweerink, 1976). Three proteins (μ1, μ2, λ3) are also present in small quantities (less than 20 molecules each) in the core, although their precise location and structural arrangement remains unknown (Joklik, 1983).

The observation that reovirus-infected cells stain orthochromatically with the dye acridine orange led directly to the discovery that the reovirus genome was composed of double-stranded RNA (dsRNA) (Gomatos et al., 1962; Gomatos and Tamm, 1963). It had been previously established that reovirus had an RNA genome, and only double-stranded nucleic acid, but not single-stranded RNA, shows an orthochromatic staining pattern with acridine orange. Subsequent studies confirmed the presence of double-stranded RNA (Arnott et al., 1966, 1967; Gomatos and Tamm, 1963; Langridge and Gomatos, 1963) and led to the identification, first, of three distinct size classes (*Large*, *Medium*, *Small*) of dsRNA (Bellamy and Joklik, 1967; Vasquez and Kleinschmidt, 1968; Watanabe and Graham, 1967) and ultimately of ten unique dsRNA segments (Banerjee and Shatkin, 1971; Millward and Graham, 1970; Shatkin et al., 1968). The plus and minus strands of each dsRNA segment are complementary and exactly colinear. They form right-handed double helices with 10 to 11 nucleotides per turn and a pitch of 3 nm (Arnott et al., 1966, 1967; Langridge and Gomatos, 1963; Muthukrishnan and Shatkin, 1975; Smith et al., 1981). The dsRNA strands appear to be packed closely together within the viral core, with adjacent double helices parallel to each other (Harvey et al., 1981). The individual gene segments are not linked together (Joklik. 1983). A number of the viral double-stranded RNA segments have now had their complete nucleotide sequence established (Bassel-Duby et al., 1985; Cashdollar et al., 1985; Richardson and Furuichi, 1983).

Replication

Studies of one-step growth cycles at high MOIs (multiplicity of infection) in reovirus-infected L cells indicate that maximal viral yield occurs at about 15 to 18 h postinfection (Gomatos et al., 1962; Silverstein and Dales, 1968). An eclipse phase, during which viral titer either declines or stays the same, typically lasts for about 6 h (Silverstein and Dales, 1968; Zarbl and Millward, 1983). At low MOI and at temperatures below 37°C, completion of replication may be considerably delayed (Rhim et al., 1961). Typical yield is about 200 to 3,000 PFU/cell, or approximately 2×10^4 to 3×10^5 particles/cell (assuming a particle to PFU ratio of 100:1) (Gomatos et al., 1962; Joklik, 1974, Rhim et al., 1961; Silverstein and Dales, 1968; Zarbl and Millward, 1983). It has been estimated that 70 to 80% of virus may remain cell associated (Gomatos et al., 1962; Rhim et al., 1961), and disruption of cells is necessary for recovery of maximal amounts of virus during assay procedures.

In vitro studies indicate that reovirus adsorbs rapidly to cultured cells. More than 50% of a viral inoculum has adsorbed to cells within 15 min, and 60 to 80% within 1 h (Ramig and Fields, 1977; Silverstein and Dales, 1968). Adsorption rate does not vary significantly between 4°C and 37°C, although at 4°C or below viral penetration does not occur (Dales et al., 1965; Silverstein and Dales, 1968). The reovirus hemagglutinin (sigma 1) is the viral cell attachment protein (Lee et al., 1981; Weiner et al., 1977, 1980a,b). A receptor for reovirus type 3 has been identified on lymphocytes and neural cells and is structurally and physicochemically similar to the β-adrenergic receptor (Co et al., 1985a,b). The best synthesis of available information would suggest that reoviruses enter cells via a process of receptor-mediated endocytosis after binding to specific cell-surface receptors, which for reovirus type 3 may be related to the beta-adrenergic receptor.

Following adsorption, reovirus particles appear within phagocytic vacuoles (endosomes) within cells. These vacuoles subsequently fuse with lysosomes and, within 1 h of penetration, 80% of viral particles are intralysosomal (Silverstein and Dales, 1968). Reovirions with loss of most of the outer capsid proteins (intermediate subviral particles, ISVPs) are still able to penetrate L cells and are fully infectious. It has been suggested that ISVPs may use a "second pathway," distinct from the receptor-mediated endocytosis of complete virions, to enter cells (Borsa et al., 1979).

Reovirus particles inside intracellular vacuoles undergo partial uncoating. Nearly all of the outer capsid protein sigma 3 is removed. Outer capsid proteins μ1c undergoes a series of cleavage reactions, leaving a smaller fragment still attached to the outer capsid ("delta protein") (Chang and Zweerink, 1971; M. Nibert and B. Fields, unpublished; Silverstein et al., 1972). Removal of the outer capsid proteins, and particularly the cleavage of μ1c, coincides with activation of the viral transcriptase (Borsa et al., 1973b; Chang and Zweerink, 1971; Joklik, 1972; Silverstein et al., 1972).

The best available evidence suggests that reovirus particles remain within lysosomes for the duration of the replicative cycle (Dales et al., 1965; Silverstein and Dales, 1968), although other alternatives have also been suggested (see Zarbl and Millward, 1983). As noted above, transcriptional activation seems to be linked to the cleavage of μ1c. Genetic studies have indicated that the viral M2 gene, which encodes the μ1 protein, determines the sensitivity of reovirions to proteolytic digestion in vitro (Rubin and Fields, 1980). In this way this gene may indirectly influence transcriptional activation (Drayna and

Fields, 1982a). Similarly, the L1 gene, which encodes the λ3 protein, determines the pH optimum of the transcriptase (Drayna and Fields, 1982). Modification of the λ2 core-spike protein can also inhibit transcription (Morgan and Kingsbury, 1981; White and Zweerink, 1976). These studies indicate that transcriptional regulation is a complex process that is influenced by the protein products of a number of the reovirus genes.

During transcription both strands of the reovirus dsRNA remain within the uncoated viral particle (Levin et al., 1970; Schonberg et al., 1971; Silverstein et al., 1976). The $(-)$ strand of the dsRNA is transcribed end-to-end, producing full-length copies of each gene segment (Hastings and Millward, 1981; Shatkin and Kozak, 1983; Skehel and Joklik, 1969). Transcription does not start in all dsRNA segments simultaneously, and "early" and "late" mRNAs can be detected (for review, see Joklik, 1974; Watanabe et al., 1967, 1968; Zarbl and Millward, 1983). Once transcription of all dsRNA segments has begun, the transcription frequency appears to be inversely related to the size of the gene segment, with small (S) segments being transcribed with 10 to 20 times the frequency of the large (L) gene segments (Joklik et al., 1981; Zweerink and Joklik, 1970).

Replication of the viral dsRNA segments appear to begin before translation of the gene segments into mRNAs (Lau et al., 1975; Nonoyama et al., 1974; Watanabe et al., 1968). ssRNA transcripts of at least four reovirus genes (L1, M3, S3, S4) are characteristically detected even before the early capped mRNAs (see Zarbl and Millward, 1983, for review). Replication of the remaining six dsRNA segments appears to require the synthesis of at least some viral proteins. It has been suggested that some as yet unidentified viral protein either directly facilitates transcription of these six gene segments or acts indirectly to inhibit a host cell repressor protein (Drayna and Fields, 1982a; Spandidos et al., 1976b; Watanabe et al., 1968). The mechanism that ensures that the 10 viral dsRNA segments are assembled together correctly within each new viral particle remains unknown, although it appears clear that free dsRNA is not found in the cytoplasm of infected cells. It is possible that σ NS, which is known to have RNA-binding activity, plays a role in this process (Gomatos et al., 1981; Huismans and Joklik, 1976). The σ3 protein has strong affinity for dsRNA and may also play a role in viral morphogenesis (Huismans and Joklik, 1976).

Shortly after cells are infected with reovirus, there is a progressive decrease in host cell protein synthesis, and by 10 h postinfection protein synthesis primarily involves viral proteins. Both host cell mRNAs and viral mRNAs are capped. Capping of viral mRNAs appears to occur shortly after transcription (see Shatkin and Kozak, 1983, for review). It has been suggested that it is primarily the "early" viral mRNAs that are capped and that at a certain point in the infectious cycle translation shifts from being cap-dependent to cap-independent, which favors translation of the late uncapped viral mRNAs (see Zarbl and Millward, 1983). However, several investigators have failed to find evidence of a shift from cap-dependent to cap-independent translation (e.g. Detjen et al., 1982). This area still remains controversial, and further studies are required.

One protein product has been identified for each of the reovirus mRNAs with the exception of the mRNA transcribed from the S1 gene segment. The S1 mRNA is dicistronic and encodes two proteins in separate out-of-phase reading frames (Ernst and Shatkin, 1985; Jacobs and Samuel, 1985; Sarkar et al., 1985). As the complete nucleotide sequences for all the reovirus dsRNA segments become available, it is possible that other dicistronic mRNAs may be identified. As noted earlier, there is variation in the transcriptional frequency of the different reovirus dsRNA segments (see above). The resulting variability in the number of mRNAs derived from each dsRNA segment produces some variability in the amount of protein that is synthesized (Lau et al., 1975; Zweerink and Joklik, 1970). In addition, there is variability in the translation efficiency of the different reovirus mRNAs (Bellamy and Joklik, 1967; Graziadei and Graziadei et al., 1973; Joklik, 1981; Levin and Samuel, 1980; McDowell and Joklik, 1971; McDowell et al., 1972; Skup and Millward, 1977; Skup et al., 1981; Zweerink and Joklik, 1970; Zweerink et al., 1971). This may result from differences in mRNA structure and from differences in the ability of mRNAs to utilize the ribosomal translational machinery efficiently (Brendler et al., 1981a,b; Walden et al., 1981; Zarbl and Millward, 1983).

The mechanism by which reovirus particles are assembled and ultimately released from infected host cells is largely unknown. Reovirus produces a lytic infection, and virus release occurs in association with destruction of the host cell. During infection, cytoplasmic inclusions are seen within infected cells (Dales, 1963; Dales et al., 1965; Fields et al., 1971; Mayor and Gomatos et al., 1962; Jordan, 1965; Rhim et al., 1962; Sharpe et al., 1982). These inclusions initially appear as a series of faint granules scattered throughout the cytoplasm; they subsequently enlarge and coalesce to form a "collar" around the nucleus of infected cells. Inclusion body formation is typically maximal at 48 to 72 h postinfection in mouse L cells, although at low temperature (e.g., 31°C) and in other cell types (e.g., monkey kidney cells) this may be delayed (Fields et al., 1971; Rhim et al., 1962). The inclusions are "viral factories" composed of both immature and mature progeny virions in close relationship to parallel arrays of microtubles (Babiss et al., 1979; Dales, 1963). Studies with temperature-sensitive mutants of reovirus provide evidence that the viral factories are the site of viral morphogenesis (Fields et al., 1971).

Viral replication is associated with inhibition of host cell protein synthesis (Sharpe and Fields, 1982; Shatkin and Kozak, 1983; Zarbl and Millward, 1983; Zweerink and Joklik, 1970). Genetic studies have shown that the S4 gene product plays an important role in this inhibition (Sharpe and Fields, 1982); however, the mechanism by which this effect is exerted remains unclear. Reovirus type 3 (and to a lesser extent Reovirus type 1) interferes with host cell DNA synthesis in certain cell lines (Cox and Shaw, 1974; Ensminger and Tamm, 1969a,b, 1970; Gomatos and Tamm, 1963; Hand and Kasupski, 1978; Hand and Tamm, 1974; Roner and Cox, 1985; Sharpe and Fields, 1981; Shaw and Cox, 1973). This inhibition begins as early as 6 to 8 h postinfection, and by 12 h nearly 90% inhibition of host cell DNA synthesis has occurred in Reovirus type 3-infected cells. The effect on DNA replication is apparently owing to inhibition of the initiation step of DNA replication, which in turn results from a reversible effect of reovirus infection on the host cell DNA replication complex (Roner and Cox, 1985). Genetic studies have indicated that the S1 gene product(s) is responsible for the inhibition of DNA synthesis, although the precise mechanism has not been elucidated (Sharpe and Fields, 1981).

Pathogenesis and Pathology

A number of recent reviews are available that discuss the pathogenesis of reovirus infections in experimental animal systems in detail (Sharpe and Fields, 1983, 1985; Tyler & Fields, 1985). Our understanding of reovirus pathogenesis is based largely on experimental studies in mice, although it is assumed that the insights gained from these studies will also be reflective of both natural infections and infections involving other species of animals.

Reoviruses enter the host through the respiratory and gastrointestinal tracts. Information about the sequence of events that follow viral entry in vivo is limited almost exclusively to the gastrointestinal tract. Following oral inoculation, Reoviruses penetrate the intestinal mucosa through microfold (M) cells. These are a specialized type of intestinal epithelial cell that overlie the dome of Peyer's patches (Wolf et al., 1981, 1983). Sequential electron microscopic studies have shown that virus adheres to the luminal (apical) surface of M cells, is endocytosed into intracellular vesicles, and is then transported across the M cell to be released into the extracellular space of the Peyer's patch. Reovirus type 3 can also infect goblet and absorptive cells in the intestine. After inoculation of extremely high doses of Reovirus type 1 into mice, ileal pathology can be produced (Rubin et al., 1985). Virus selectively replicates in the crypt epithelial cells of the ileum, and viral antigen can be detected in these cells by immunocytochemistry. After administration of massive doses of virus, pathologic changes appear in the small intestine. These changes include hyperplasia of the crypts of Lieberkuhn, villus blunting, loss of nuclear polarity in the epithelial cells, and inflammation in the lamina propria. In some cases, transmural perforation of the ileum occurs with the development of a secondary peritonitis (Rubin et al., 1985, 1986). Intestinal pathology has not been reported after oral administration of reovirus type 3 but has been described after massive intravenous inoculation of virus (Rubin et al., 1986). Under these circumstances, hemorrhagic lesions can be seen in the lamina propria of the duodenum and jejenum, associated with an inflammatory infiltrate, and an ulcerative colitis may develop (Rubin et al., 1986). It has been suggested that the reovirus S1 gene plays an important role in determining site-specific patterns of intestinal disease in this experimental system (Rubin et al., 1986).

Reoviruses are able to spread from the gastrointestinal tract to extraintestinal organs (Kauffman et al., 1983a) and the central nervous system after peroral inoculation (Kaye et al., 1986; Rubin and Fields, 1980). Following transport across M cells, reovirus type 1 can be detected sequentially in Peyer's patches, mesenteric lymph nodes, and ultimately the spleen. It has been proposed that this pattern of spread reflects local lymphatic spread followed by bloodstream invasion (Kauffman et al., 1983a). The viral S1 gene appears to be the major determinant of the capacity of reovirus type 1 but not type 3 to utilize the lymphatic–bloodstream pathway to spread to an extraintestinal organ, such as the spleen, after peroral inoculation (Kauffman et al., 1983a).

The pathways utilized by reovirus to spread from the gastrointestinal tract to the CNS is not known. However, spread of reovirus from peripheral inoculation sites, such as footpad and muscle, to the CNS has been extensively investigated. Following intramuscular or footpad inoculation, reovirus type 3 spreads to the spinal cord via nerves, and this spread can be inhibited by nerve section (Tyler et al., 1986). Use of selective inhibitors of fast and slow axonal transport has shown that the neural spread of reovirus type 3 is mediated via the microtubule-associated system of fast axonal transport (Tyler et al., 1986). Following intramuscular or footpad inoculation, reovirus type 1 spreads to the spinal cord via

hematogenous pathways, and its spread is not inhibited by local nerve section or by inhibitors of axonal transport (Tyler et al., 1986). Genetic studies with reassortant viruses have shown that the reovirus S1 gene, which encodes the viral hemagglutinin as well as a small nonstructural protein, determines the capacity of reoviruses to spread to the CNS via these distinct pathways (Tyler et al., 1986).

The factors that determine the capacity of reoviruses to selectively infect certain organs or particular subsets of cells within these organs ("tropism") has been extensively investigated. The reovirus $\sigma 1$ outer capsid protein functions as the viral cell attachment protein (Weiner et al., 1977, 1980a,b; Lee et al., 1981). It is this protein which determines the differential pattern of hemagglutination of reoviruses belonging to different serotypes (Weiner et al., 1978). Studies with monoclonal antibodies have indicated that, in the case of reovirus type 3, a specific functional domain on the $\sigma 1$ protein is responsible for hemagglutination (Burstin et al., 1982).

After intracerebral inoculation into neonatal mice, reovirus type 3 produces a lethal encephalitis. There is extensive cortical necrosis and severe injury to neurons in the limbic system, including Ammon's horn, the septal nuclei, and the mammillary body. The thalamus, basal ganglia, upper brainstem, and cerebellum (Purkinje cells) are involved to variable extents. Microscopic examination of neurons shows that they contain cytoplasmic inclusions. Perivascular cuffing occurs around leptomeningeal and subpial blood vessels. White matter injury does not occur, and an intraparenchymal inflammatory response is usually not prominent (Margolis et al., 1971; Raine and Fields, 1973; Spriggs et al., 1983; Walters et al., 1963; Weiner et al., 1977, 1980b). Reovirus type 3 also infects and severely injures the retinal ganglion cells of the eye after intracerebral inoculation (Tyler et al., 1985).

The pattern of CNS injury produced by reovirus type 1 is distinct from that of reovirus type 3. Reovirus type 1 does not infect neurons or glial cells. It does infect the ependymal cells lining the ventricles and the aqueduct of Sylvius, resulting in the production of hydrocephalus (Jenson et al., 1965; Kilham and Margolis, 1969; Margolis and Kilham, 1969; Masters et al., 1977; Nielsen and Baringer, 1972; Phillips et al., 1970; Walters et al., 1965; Weiner et al., 1977, 1980b). Reovirus type 1 also infects cells in the anterior lobe of the pituitary gland, particularly the growth hormone-producing cells. This results in a particular type of runting syndrome in which there are serum antibodies to growth hormone and extremely low growth hormone levels in infected mice (Onodera et al., 1981).

Reovirus type 2 can also produce hydrocephalus

in neonatal mice. Injury involves the cortex, thalamic nuclei, and cerebellum, although studies are somewhat limited, and direct comparisons with the pattern of injury produced by type 1 and type 3 are difficult to make (Walters et al., 1965).

Genetic studies have clearly shown that the reovirus S1 gene, which encodes the viral hemagglutinin, is responsible for determining the tropism of reovirus within the CNS (Weiner et al., 1977, 1980b; Onodera et al., 1981; Tyler et al., 1985). Further confirmation has come from studies of antigenic variants of reovirus type 3 selected on the basis of their resistance to neutralization by monoclonal antibodies against the hemagglutinin (Bassel-Duby et al., 1986; Kaye et al., 1986, Spriggs and Fields, 1982; Spriggs et al., 1983). These type 3 variants contain single amino acid substitutions in the viral hemagglutinin (Bassel-Duby et al., 1986) and show an altered pattern of CNS tropism (Bassel-Duby et al., 1986; Kaye et al., 1986; Spriggs et al., 1983). A reassortant virus containing the S1 gene of one of these variants replicates its pattern of altered CNS tropism, clearly indicating that this property is due to the altered hemagglutinin (Kaye et al., 1986).

The pattern of injury produced by reoviruses in tissues other than the CNS has been discussed in recent reviews (Tyler & Fields, 1985, 1986). Reoviruses of all three serotypes can produce myocardial injury (Hassan et al., 1965; Walters et al., 1963, 1965). This injury is characterized by gross epicardial lesions. Microscopically there are large areas of coagulative necrosis of the myocardium with associated dystrophic calcification (Hassan et al., 1965). Virus particles can be seen by electron microscopy (EM), and viral antigen can be detected by immunocytochemistry within myocardial cells (Hassan et al., 1965). An inflammatory infiltrate is not prominent, but when it occurs it consists predominantly of lymphocytes and macrophages (Walters et al., 1963).

Reoviruses of all three serotypes can also produce hepatobiliary disease (Bangaru et al., 1980; Papadimitriou, 1965, 1966, 1968; Phillips et al., 1969; Stanley and Joske, 1975a,b; Stanley et al., 1964; Walters et al., 1963). In the liver there are focal regions of necrosis, which can appear grossly as yellow depressed lesions on the surface. Microscopically, there is necrosis of hepatocytes and an inflammatory response consisting of aggregates of mononuclear cells around the central vein and portal tracts. Hyperplasia of the Kupffer cells also occurs. Reovirus type 3 can also infect the mucosal cells of the bile ducts, resulting in chronic obstructive jaundice and producing a histopathologic picture similar to that seen with biliary atresia.

Reovirus-induced injury to the endocrine pancreas has been extensively studied as a potential model for virally induced diabetes mellitus (Onodera

et al., 1978, 1981). Areas of focal necrosis occur in the islets of Langerhans associated with a mononuclear cell infiltrate. Reovirus type 3 primarily infects the insulin-producing beta cells of the islets, whereas reovirus type 1 spares these cells and instead infects the non-beta islet cells. Studies with reassortant reoviruses indicate that these specific patterns of pancreatic tropism are determined by the reovirus Sl gene (Onodera et al., 1981).

Reovirus type 1 and type 3 also produce injury to skeletal muscle. Reovirus type 1 produces a pattern of interstitial inflammation with muscle fiber necrosis confined primarily to the area of intense inflammation. Reovirus type 3 produces a severe necrotizing myositis; inflammation is not a prominent feature of the initial lesion. The genetic bases for these distinct patterns of injury have been studied with reovirus reassortants. The reovirus Sl gene appears to play the predominant role in determining the pattern of injury to skeletal muscle induced by reoviruses (Tyler et al., 1984).

Clinical Features

Human Disease

Seroepidemiologic studies have indicated that the majority of individuals (50 to 80%) have developed antibodies to reovirus by the time they reach their late teens or early twenties, and in some cases substantially earlier (Jackson and Muldoon, 1973; Leers and Royce, 1966; Lerner et al., 1962; Stanley et al., 1964). The pattern of antibody acquisition appears to be very similar in a number of diverse geographic locations and cultural settings (Stanley., 1977). Most of the available serologic evidence is consistent with the fact that infection with reoviruses and subsequent seroconversion typically occur in infancy or early childhood (<5 years). This pattern is evident in a study performed at Boston City Hospital between 1959 and 1962. The incidence of hemagglutination–inhibition (HI) antibodies to reovirus was found to rise from <25% in children under 1 year of age to over 70% in those >$3\frac{1}{2}$ years of age (Lerner et al., 1962).

Despite the apparently ubiquitous nature of reoviruses and the frequency with which humans are infected with them, it has been difficult to produce convincing evidence linking reoviruses to any human disease. In this regard reoviruses remain as much "orphans" as when Sabin originally suggested their name nearly 30 years ago (Sabin, 1959). The majority of reovirus infections are certainly either totally asymptomatic or of such mild severity as to blend imperceptibly into the minor undiagnosed ailments of everyday life (Jackson and Muldoon, 1973; Leers and Royce, 1966; Sprigland et al., 1968).

One of the best documented examples of an "outbreak" of reovirus (type 1) infection occured in a group of nursery children (ages 6 months to 4 years) in a Washington, D.C., welfare institution in 1957. The outbreak was clinically unrecognized and was detected by routinely attempting to isolate viruses from weekly anal swab specimens collected from the children. From 22 children (out of a susceptible population of 73) reovirus type 1 was isolated from stool specimens over a 9-week period. The duration of viral shedding was at least 1 week in 54% and at least 2 weeks in 21% of those tested. One child continued to shed virus in the stool for at least 5 weeks. All of the children with positive stool isolates also showed a rise in anti-reovirus type 1 HI antibody titer.

In an attempt to find clinical correlates for the viral shedding, a retrospective analysis of the children's medical records was made. It appeared that a "mild febrile illness" (temperature, 100.6° to 101.5°F) was temporally associated with the period of virus isolation in many cases, although this was felt to be of only borderline statistical significance. The other clinical features of the affected children concurrent with virus isolation included rhinorrhea (81%), pharyngitis (56%), diarrhea (19%), and otitis media (19%). It was stressed that these symptoms "did not differ significantly in either character or frequency from those of other febrile illnesses which occurred in the same children in the period immediately before and after the interval bracketing their virus isolations" (Rosen, 1960).

Additional evidence concerning the nature of clinically apparent illnesses due to reovirus infections has come from the studies of adult human volunteers inoculated with virus. In two of the available studies (both involving reovirus type 1 only), although most volunteers showed serologic and virologic evidence of viral infection, there were no notable clinical findings (Jackson et al., 1961, 1962; Kasel et al., 1963). The third study (Rosen et al., 1963) involved male prisoners (ages 21 to 38 years) given intranasal inocula of reoviruses belonging to each of the three serotypes. The dose of virus (2-ml inocula) was 10^5 to $10^{5.5}$ TCID$_{50}$/ml (50% tissue culture infective dose in one test and 10^6 to $10^{7.7}$ TCID$_{50}$/ml in a second test. Reovirus type 1 could be isolated from anal swabs of 8/9 volunteers challenged (one patient had pre-existing antibody levels and no isolable virus). A significant increase in homotypic HI antibody titers was documented in 6/8 individuals with positive stool isolates. Three individuals developed clinical illness with features including malaise, rhinorrhea, cough, sneezing, pharyngitis, and headache. An additional volunteer noted "loose stools" but no other symptoms. The illnesses typically began 1 to 2 days after virus ingestion and lasted for 4 to 7 days.

Of the nine volunteers who received reovirus type 2, only two had virus isolated from throat or anal swabs. Both these individuals developed HI antibodies when the homologous virus strain was used in the HI tests. Two other volunteers developed a rise in HI antibody titer without concurrent virus isolation. None of the volunteers had significant clinical symptoms.

Eight of the nine volunteers who received reovirus type 3 had positive anal swab virus isolates. All eight of these individuals also showed a rise in homotypic HI antibody titer. One of the patients with positive virus isolates and a rise in HI antibody titer developed "mild rhinitis"; the other individuals remained well.

The studies cited above, and the frequency with which reoviruses can be isolated from stool or, less commonly, from upper respiratory tract specimens, have provided suggestive evidence for the association between reoviruses and diarrheal illnesses and mild upper respiratory illnesses. More recently, interest has centered around a possible association between reovirus type 3 infection and biliary atresia. In an initial study, paired sera from 12 infants with extrahepatic biliary atresia and from their mothers were examined for neutralizing antibodies to reovirus type 3. Two of the 12 infants had elevated and rising titers (1:30 to 1:120 and 1:40 to 1:120) without detectable maternal antibodies (Bangaru et al., 1980). In a subsequent study it was shown that 68% of 25 infants with biliary atresia (compared with only 8% of 37 controls) had anti-reovirus type 3 antibodies detectable (>1:10) with an indirect-immunofluorescent-antibody technique. Five of the infants showed a significant (>4×) rise in antibody titer (Morecki et al., 1982). An expanded version of this study, with the same antibody detection system, also found that 52% of infants with "idiopathic neonatal hepatitis" (versus 6% of controls) had detectable antibodies to reovirus type 3 (Glaser et al., 1984). The same authors also reported that reovirus type 3 antigens could be detected, by use of avidin-biotin–complex peroxidase immunohistochemistry, in an operative specimen of the porta hepatis from an infant with biliary atresia (Morecki et al., 1984). Almost all of the reports describing an association between reovirus type 3 infection and diseases such as biliary atresia and neonatal hepatitis have come from the same group of investigators, and the status of this association must still be considered sub judice until additional confirmation is forthcoming.

Although a reasonable body of information has been generated concerning the relationship of reoviruses to the diseases discussed above, there also exists a substantial body of literature, consisting largely of anecdotal reports, associating reoviruses with a wide variety of other illnesses. perhaps the most intriguing subset of this literature is the descriptions purporting to document "fatal" cases of reovirus infection. The first of these reports (Krainer and Aronson, 1959) was in a presentation to the American Association of Neuropathologists and never appeared except in abstract form. The case involved a 29-year-old woman who died after an 11-week illness and was found to have disseminated encephalomyelitis. A virus was recovered from two ante mortem cerebrospinal fluid (CSF) specimens and from brain and spinal cord tissue obtained post mortem. No serologic data were presented. It was noted that the virus obtained was pathogenic for adult mice after intracerebral inoculation, a feature that has not typically been described for reovirus isolates (Tardieu et al., 1983). Although the authors felt that their isolate "appeared to be immunologically related to the HEV of Stanley . . .," Dr. Albert Sabin, discussing the case, justifiably concluded that this was an extremely uncertain attribution.

The second fatal case was reported in 1964 (Joske et al., 1964). A 10-month-old infant developed an upper respiratory infection followed by drowsiness, encephalitic symptoms, hemorrhagic skin blebs, and circulatory failure leading to death. At post mortem examination there was evidence of interstitial pneumonia, focal myocarditis and hepatitis, and isolated neuronal "lysis" in restricted regions of the CNS. Reovirus type 1 was isolated from the child's feces and from post mortem brain tissue. No serologic studies were performed. From the available evidence, it is impossible to be sure of the causal role of reovirus in this child's illness, although isolation of virus from tissues such as brain must be considered strong evidence in favor of a possible role for reovirus in this illness.

The third reported case of fatal reovirus infection involved a 5-year-old girl who died of severe pneumonia associated with a generalized maculopapular exanthem (Tillotson and Lerner, 1967). A virus producing CPE typical of reovirus was isolated from post mortem specimens of lung, adrenal glands, liver, spleen, lymph nodes, heart, blood, and brain inoculated onto monkey kidney tissue culture. The isolated virus agglutinated human type O red blood cells but not bovine erythrocytes. Agglutination was inhibited by antisera to reovirus type 3 (and to a lesser degree by antisera to types 1 and 2). On inoculation into mice, the virus isolate produced illness felt to be typical of reovirus type 3 infection. The patient did not have detectable antibodies to reovirus in her serum. Although it does appear that the virus isolated from this patient was a reovirus, the absence

of seroconversion makes the association somewhat tenuous (especially in light of the frequency with which reoviruses can be isolated from monkey kidney cultures).

The association of reoviruses with neurological illnesses has been a recurring theme in several case reports including two of the fatal cases alluded to above (Krainer and Aronson, 1959; Joske et al., 1964). Joske et al. (1964) described two nonfatal cases of meningoencephalitis in children (ages 28 months and 6 years) which they attributed to reovirus. In both cases reovirus (type 2) was isolated from stool specimens of the affected children, but in neither case was neutralizing antibody found in either acute or convalescent sera. A putative role for reoviruses in CNS disease has been revived by a recent report that "reovirus-like particles" can be seen by EM in the brains of both normal individuals (!) and the brains of patients with chronic mental illnesses (e.g., Parkinson's disease, senile and presenile dementia, and schizophrenia). In what is also apparently a unique observation, "reovirus type 3 antigen" was detected by immunofluorescence and immunoperoxidase techniques in "neuronal nuclear inclusion bodies" in the cortex, hippocampus, hypothalamus amygdala and substantia nigra (Averback, 1982). These studies await independent confirmation.

Lerner et al. (1962) tried to determine whether reoviruses were a significant cause of exanthemal illnesses in a pediatric patient population. They studied a large group of children presenting with exanthemal illnesses or nonbacterial pneumonia and found seven patients (ages 4 months to 9 years) with acute and convalescent serum specimens demonstrating a $>4\times$ increase in HI antibody titer to reovirus type 2. Two of the seven patients also had reovirus type 2 isolated from throat or rectal swabs. The clinical findings in these six children included a maculopapular (5) or vesicular (1) rash, mild or moderate fever (6), malaise (4), anorexia (4), pharyngitis (3), cervical adenopathy (2), and diarrhea (1). Unfortunately, when the incidence of HI antibody and virus isolation were compared with an asymptomatic age-matched population, there was no difference, this indicated that "although an etiologic relation between the exanthem and Type 2 reovirus infection in the six patients in whom these antibodies developed is suggested, a definite relation cannot be inferred from the present data. . . ."

In addition to the illnesses discussed above, reports of reovirus infection have also appeared in association with cases of keratoconjunctivitis (Jarudi et al., 1973), asymptomatic viruria (Edmonson et al., 1970; Stanley, 1977), and Burkitt's lymphoma (Bell et al., 1966a, 1967; Levy et al., 1968; Massie et al., 1964).

In summary, mild upper respiratory and gastrointestinal diseases seem the most likely candidates for reovirus-induced illnesses. Evidence is accumulating for an association between reoviruses and neonatal biliary atresia although virus has still not been isolated from patients with this disease. The role of reoviruses in neurologic illnesses, exanthema, conjunctivitis, and the other diseases noted must still be considered speculative until further data are available.

Animal Diseases

No attempt will be made to fully review the numerous studies that have dealt with reovirus infections in animals (see Thein and Scheid, 1981; for a comprehensive review), and no attempt will be made to review avian reoviral infections (see Thein and Scheid, 1981 for review). The majority of reported infections either have been asymptomatic or have involved the respiratory and enteric tracts. Naturally occurring reovirus infections in mice appear to be largely limited to type 3 (Cook, 1963; Hartley et al., 1961; Nelson and Collins, 1961; Nelson and Tarnowski, 1960), which can produce diarrhea, runting, "oily hair effect," jaundice, and neurologic symptoms. In horses, reovirus infections (types 1, 3) take the form of upper and lower respiratory tract disease with symptoms including rhinitis, pharyngitis, laryngitis, conjunctivitis, and cough (Thein and Scheid, 1981). In calves and cattle, infection with reoviruses of all three serotypes has been documented with the production of either respiratory illness or, less commonly, diarrhea (Abinanti, 1963; Lamont et al., 1968; Rosen et al., 1963; Trainor et al., 1966). In sheep and lambs, reovirus types 1 and 3 have been assocatied with upper respiratory infections and enteric illness (Belak and Palfi, 1974; McFerran et al., 1973), and reovirus type 2 has been isolated from asymptomatic animals (Snodgrass et al., 1976). In swine and piglets, reovirus type 1 has been found in animals with diarrhea (enteritis) and respiratory problems (Baskerville et al., 1971; Elazhary et al., 1978; Kasza, 1970; McFerran and Connor, 1970). Reoviruses of all three serotypes have been associated with canine illnesses including cough, rhinitis, conjunctivitis, pneumonia, diarrheal illness, and rare cases of encephalitis (Holzinger and Griesemer, 1966; Lou and Wenner, 1963; Massie and Shaw, 1966; Binn et al., 1970, 1977; Thein and Scheid, 1981). In cats, reovirus has been related to cases of ataxia in kittens (Csiza, 1974) and conjunctivitis (Scott et al., 1970). A particularly interesting category of disease is that reported in subhuman primates. These animals are obviously closely related to man, and the spectrum of illnesses described may

have potential implications for the problem of the relationship of reoviruses to human diseases. Cases of both naturally acquired and experimentally induced respiratory illnesses, including pneumonia, have been reported in *Macaca* sp., *Cercopithecus* sp., chimpanzees, and marmosets (Hull et al., 1956, 1958; Sabin, 1959; Stanley et al., 1954; Thein and Scheid, 1981). Reovirus type 2 has been linked to an outbreak of rhinitis in laboratory chimpanzees (Sabin, 1959, 1960), and a case of interstitial pneumonitis in a wild macaques monkey (Hull et al., 1956). Isolated reports of hepatitis following oral inoculation of reovirus type 3 and meningitis (type 2), and necrosis of ependymal cells and choroid plexus (type 1) have followed intramuscular or intracerebral inoculation of reoviruses of the designated serotypes.

Diagnosis

Specimen Collection

As discussed above under Clinical Features, reoviruses have not been definitively identified as human pathogens. However, diseases in which reoviruses might be suspected as possible etiologic agents include infections of the upper and lower respiratory tract, enteritis and diarrheal illnesses, hepatobiliary diseases including biliary atresia and hepatitis, cardiac and skeletal myositis and myopathies, neurologic diseases (meningitis, encephalitis, myelitis, hydrocephalus), renal diseases, and Burkitt's lymphoma. The nature of these diseases defines the specimens that could conceivably prove useful in viral isolation. Tissue or body-fluid specimens might include nasopharyngeal or tracheal aspirates and washings, throat swabs (respiratory illnesses), stool or fecal swab specimens (intestinal illnesses), cerebrospinal fluid (neurologic illnesses), and urine (renal diseases). However, from a practical point of view it is worth recognizing that the vast majority of reovirus isolates have been from fecal specimens and less commonly from throat swabs. Biopsy material from a wide variety of tissues could potentially be used in an attempt to isolate virus or detect viral antigens by immunocytochemistry (see below).

Throat or nasopharyngeal swabs or sterilely aspirated material from the respiratory tract should be transmitted promptly to the laboratory. To obtain nasal or pharyngeal swab specimens, a dry cotton swab is inserted into the nose or nasopharynx and allowed to sit for several seconds to absorb secretions. Throat swabs are obtained by rubbing the tonsils and posterior pharyngeal area. Nasal or throat washings can be collected in addition to swab specimens. For nasal washings, a few milliliters of sterile saline is placed into each nostril after the patient tilts his head backwards to prevent fluid runoff. The fluid should be allowed to sit for a few seconds, and then the patient can express the fluid by tilting the head forward. The fluid is collected in a clean paper cup and then transferred to an appropriate receptacle. Throat washings can be obtained by having the patient gargle with 10 to 20 ml of sterile saline or Hanks balanced salt solution (BSS). Swabs can be transported in Hanks BSS or an equivalent physiologic solution. Penicillin (500 units/ml) and streptomycin (500 μg/ml) can be added to the transport medium without interfering with subsequent virus isolation.

Fresh stool specimens are generally preferable to rectal swab specimens for the isolation and identification of enteroviruses and rotaviruses, and this would presumably be the case for reovirus isolation as well. Stool specimens that cannot be transported immediately to a diagnostic virology laboratory should ideally be kept frozen at −70°C. If low temperature freezers are not available, storage at refrigerator temperature (4°C) is acceptable. Similar recommendations apply to the storage of other tissue material for reovirus isolation. It is probably advisable to store stool specimens in a glass or plastic screw-cap container rather than the conventional cardboard cup, inasmuch as the latter allows for more desiccation. If fresh stool is not easily available, a rectal swab specimen can be used for viral isolation. A sterile cotton swab is inserted 2 to 5 cm beyond the anal sphincter and rotated. Ideally, some fecal material should be visible on the swab when it is removed. The swab can be transported in Hanks BSS or other equivalent medium.

Preparation of stool specimens or material obtained from rectal swabs should follow standard practices used for enterovirus isolation. Typically a 2 to 10% suspension of feces in sterile water or physiologic saline is made by shaking the stool plus the diluting fluid vigorously with sterile glass beads. It may be desirable to repeat the entire process a second time. The supernatant fluid is then clarified by centrifugation at circa 3000 rpm for 30 min at 4°C. The supernatant from this step is then used for tissue culture inoculation. Tissue specimens (possible examples include lung, liver, bowel, heart, muscle, nervous tissue) should be sterilely macerated, frozen (−70°C), and thawed (37°C) three times and then disrupted by sonication or extensive homogenization before being inoculated onto the tissue culture. Some organ specimens may be toxic to tissue culture if applied directly and may require dilution in sterile physiologic medium. Extremely acidic stool specimens can be titrated to a neutral pH with sodium bicarbonate.

Direct Virus Detection

Reovirus-induced cytopathic effect (CPE) can be detected in a wide variety of cultured cell lines. The most widely used cells include mouse L cell fibroblasts, *Macaca* or *Cercopithecus* kidney cells, primary human kidney cells, and HeLa cells (Tyler and Fields, 1985; Rosen, 1979; Stanley, 1977). Primary *Macaca* kidney cell cultures are generally considered to be the cell line of choice for initial virus isolation (Rosen, 1979). When mouse L cells or kidney cells are used, a confluent monolayer of cells is first prepared. If animal serum is used in the maintenance medium for cultured cells, it must be shown to be free of anti-reovirus antibodies, since these are commonly found in bovine sera and will interfere with virus isolation. If serum not known to be free of antibody to reovirus has been used to grow these cells, the cells should be carefully washed free of serum prior to inoculation. Hanks BSS can be used to wash off serum-containing growth medium from cultures (Rosen, 1979). Serum-free maintenance medium (e.g., 199) or medium with reovirus antibody-free serum can be used to maintain cultures. It is preferable not to change the original maintenance medium; therefore, a number of control vessels should be prepared to monitor the degree of nonspecific cellular degeneration. A 0.1-ml inoculum is generally sufficient for use in 35-mm culture dishes, and 0.2 ml can be used in 60-mm dishes. Culture vials are incubated at 37°C with 5% CO_2 and observed under a tissue culture microscope every 2 or 3 days over a 3-week time span for signs of CPE. If no signs of CPE are seen after 3 weeks, an aliquot of the fluid from the tissue culture vessels should be blindly subcultured into another vial and observed for an additional week (Rosen, 1979; Stanley, 1977; Tyler and Fields, 1985).

The CPE produced by reovirus infection in mouse L cells and monkey kidney cells is usually distinguishable from that caused by other viral agents. Reovirus-infected cells develop pronounced cytoplasmic granularity, and cytoplasmic inclusions can be seen either by phase-contrast microscopy on unstained specimens or after fixed specimens are stained with acridine orange, congo red, or Lendrum's phloxin-tartrazine (Dukes et al., 1969; Gomatos et al., 1962; Lendrum, 1947; Margolis et al., 1971; Rhim et al., 1962). The specific nature of the infection can be confirmed by the use of anti-reovirus antibodies and indirect immunofluorescent or immunocytochemical staining (see below). Plaques may be visible in infected monolayers of mouse L cells or monkey kidney cells, and their visibility is enhanced by staining of these monolayers with neutral red or a similar vital dye.

Newborn mice are particularly susceptible to infection with reoviruses. Under certain circumstances it may be advisable to inoculate 1-day-old mice with tissue specimens (prepared as described above) via the intracerebral, intraperitoneal, and intramuscular routes and to observe them for signs of reovirus-induced illness. The cardinal features of murine illness due to reovirus have been outlined (Stanley et al., 1964; Walters et al., 1963, 1965) and are reviewed under Pathogenesis and Pathology (see also Tyler and Fields, 1985, for review).

Direct Detection of Viral Antigens

Reovirus antigens can be detected in infected cells in vitro or in tissue specimens by either immunofluorescent or immunohistochemical techniques. For indirect immunofluorescence of tissue culture cells infected in vitro, we have generally found the following procedure to be suitable. Cells are washed free of buffer and then fixed with either cold methanol (−20°C) alone for 5 min, cold methanol–acetone (1:1) (−20°C) for 5 min or cold methanol (−20°C) for 5 min followed by cold acetone (−20°C) for 2 min. The cells are allowed to air dry and are then washed with phosphate-buffered saline (PBS) at room temperature. As the primary antiserum, we typically use polyclonal rabbit antiserum to purified reovirus of either type 1 or type 3. After being washed with PBS, a fluorescein-conjugated anti-rabbit IgG (made in sheep, goats, or swine) is used as the secondary antiserum. We have generally found that for most antisera incubation times of 45 to 60 min in a humidified chamber at room temperature suffice. Coverslips with infected cells can be mounted on glass slides, or coverslips can be sealed onto Microtek multichamber slides containing infected cells, with either gelvatol (Monsanto) or glycerol gelatin (Sigma). The dilutions of the primary and secondary antisera required for optimum staining are quite variable and are best determined for each batch of antiserum by checkerboard titrations.

Stained cells are examined under a microscope equipped with epifluorescence and planapochromat objectives. Early in the infectious cycle (circa 12 h) positive immunofluorescence typically takes the form of small inclusions located in the periphery of the cell cytoplasm. Later in the infectious cycle (24 to 48 h) the inclusions become larger and tend to assume a perinuclear location. Nuclear fluorescence does not occur.

We have found that immunocytochemical techniques are also eminently suited for detecting reovirus antigens in vitro. Fixation procedures are identical with those described for immunofluorescence. Tris (0.05 M) (pH 7.6) with 0.05% Triton X-100, to which a small amount (1%) of normal serum from the same species of animal as the secondary antiserum is added, is an excellent buffer. We use polyclonal rabbit antiserum against purified reovirus

type 1 or type 3 as the primary serum. This can generally be diluted into the 1:100 to 1:500 range (in 0.1 M Tris, pH 7.6). The secondary antiserum is an anti-rabbit IgG (made in sheep, goats, or swine) typically diluted 1:20 to 1:30. In the next step, rabbit peroxidase–antiperoxidase (PAP) is then used (typically diluted 1:200). We have found that diaminobenzidine (DAB) is the best substrate for peroxidase reactions. We use a 0.02% solution in 0.05 M Tris (pH 7.6) with 0.03% H_2O_2. This will generally result in good color of the reaction products after 5 to 10 min. The reactions with the primary antiserum, secondary antiserum, and PAP can generally be run for 45 to 60 min at room temperature in a humidity chamber. Commercially available kits (Vector Laboratories, Burlingame, CA) based on avidin–biotin complex (ABC) peroxidase or glucose–oxidase (GO) immunohistochemical staining also work well on in vitro material. We have generally used the buffers described above and have found DAB (for peroxidase reactions) and nitroblue tetrazolium (for GO reactions) to be the most suitable chromogenic substrates.

Immunofluorescent and immunocytochemical procedures can both be used to detect reovirus antigens in infected tissues (Kundin et al., 1966; Hassan et al., 1965; Weiner et al., 1980b; Rubin et al., 1985). We have generally found frozen sections to be superior to paraffin-embedded or epon-impregnated tissues for antigen detection. For nonfrozen sections, brief tissue fixation in Bouin's fixative may be preferable to formalin fixation. Either polyclonal rabbit antiserum to purified reovirus type 1 or type 3 or the IgG fraction of this antiserum can be used as the primary antiserum. We have found that adsorption of most antisera with mouse liver powder (Cappel) decreases the amount of nonspecific staining. Frozen sections can be fixed in cold acetone (4° C) for 3 min and then allowed to air dry. The fixative is washed off with 0.05 M Tris (pH 7.6) for 5 min at room temperature. A subsequent wash in the same buffer containing 2% normal serum from the same species as the secondary antiserum also helps to decrease nonspecific staining. The primary antiserum (polyclonal rabbit anti-reovirus) can generally be diluted into the 1:100 to 1:300 range (in 0.1 M Tris, pH 7.6, with 1% normal serum). The secondary antiserum is an anti-rabbit IgG and is generally diluted 1:20 to 1:30 (e.g., swine immunoglobulins to rabbit immunoglobulins). For peroxidase–antiperoxidase immunohistochemistry, rabbit PAP (Accurate Chemicals, Westbury, NY) is generally diluted 1:200. The reactions with the primary antiserum, secondary antiserum, and PAP can all be run for 30 to 60 min at room temperature in a humidified chamber. Between steps, specimens should be thoroughly washed with buffer (e.g.,. 0.05 M Tris, pH 7.6, with 1% normal serum). 3′-3′-Diaminobenzidine tetrahydrochloride dihydrate (DAB) (Aldrich) made as a 0.02% solution in 0.05 M Tris (pH 7.6) with 0.03% H_2O_2 is a suitable chromogenic substrate. This reaction can generally be allowed to run for 5 to 10 min and is stopped by drenching the slides in water.

Direct Viral Nucleic Acid Detection

Although sequence data are now available for all of the reovirus S (small) genes (Bassel-Duby et al., 1985; Cashdollar et al., 1985; Richardson and Furuichi, 1983), and extensive sequence information is now becoming available for some of the M (medium) and L (large) genes, a reliable procedure involving the use of in situ hybridization for the detection of reovirus messenger RNA or double-stranded genomic RNA in infected cells in vitro or in vivo is not currently available.

Identification of Viral Isolates

Once a virus has been identified as a reovirus by the isolation techniques described previously, it must be further characterized as to serotype. The most reliable method of serotyping involves determination of the hemagglutination (HA) properties of the isolate. The basic procedure outlined is that of Rosen (Rosen, 1960, 1979), as modified in our laboratory. Isolates are grown in mouse L cells or monkey kidney tissue culture. When complete destruction of the cell monolayer has occurred, the resulting material (supernatant fluid and cell debris) is harvested. (For research purposes we employ purified reovirus for HA tests, but this is obviously not practical for typing of clinical isolates.) This material can be stored frozen (−70°C) for prolonged periods without substantial loss of HA activity. The material should be thawed and frozen three times to maximize the release of cell-associated virus. Serial 2× dilutions (1:2 to 1:1,024) of this material ("virus") are made in 0.85% sodium chloride or PBS (pH 7.4); 0.4-ml aliquots of each dilution are placed into 12-mm × 75-mm tubes (with appropriate volume adjustment, V-bottom microtiter plates are also suitable).

Human type O RBCs (erythrocytes) are the preferred substrate for HA reactions. Blood can be collected in heparinized tubes or into an equal volume of Alsever's solution (Gibco, Grand Island, NY). Prior to use, the blood is washed three times in PBS and then resuspended in PBS. For assays in microtiter plates or test tubes, a 0.5 to 0.75% suspension (packed RBC volume: PBS or 0.85% NaCl volume) is usually suitable. The concentration of RBCS should be tested to insure that they produce a suitable button after being allowed to sediment at room temperature for at least one 1 h. Once the RBC suspension has been prepared, 0.2 ml is added to the test

tubes, which are then shaken once and allowed to sit undisturbed at room temperature. The highest dilution of "virus" that shows a faint "button" (i.e., loss of full agglutination) is taken as the end point of the titration and is considered to contain 1 HA unit (in 0.4 ml or whatever volume was employed). Tissue culture stocks of nonpurified virus can contain up to 2,000 HA units/ml. It is not difficult to demonstrate hemagglutination with isolates of reovirus type 1 and type 2, although reovirus type 3 isolates may show low or extremely variable levels of HA (S. Lynn, unpublished; Rosen, 1979). Reovirus type 3 can agglutinate bovine erythrocytes (Eggers et al., 1962) whereas reoviruses of serotype 1 and 2 do not (Weiner et al., 1978).

The hemagglutination–inhibition (HAI) test is used to serotype reovirus isolates. Type serum of known specificity is diluted in 2× steps over the range 1:10 to 1:1,024 in either PBS (pH 7.4) or 0.85% NaCl. For assays performed in 12-mm × 75-mm tubes, 0.2-ml aliquots of diluted antiserum are placed in each tube. An equal amount of "virus" (prepared as described above or, if available, purified virus is preferable) diluted in either PBS or 0.85% NaCl to contain 4 HA units is then added to the antiserum. The mixture is shaken and then allowed to stand at room temperature for 60 min; 0.2 ml of the 0.5 to 0.75% type O RBC suspension is then added, and the mixture is allowed to stand at room temperature for 1 to 4 h (or until complete sedimentation has occurred in a control tube without virus). The HAI titer of an antiserum for the test isolate is defined as the reciprocal of the highest dilution of the antiserum that completely inhibits hemagglutination (i.e., produces a "button"). If one of the typing sera has an HAI titer of >1:40 for the isolate and the other sera have HAI titers of <1:10, the isolate is typed as belonging to the serotype homologous to the high-titer antisera (Rosen, 1979).

Typing of reovirus isolates with HAI is generally preferable to serotyping with plaque-reduction-neutralization assays. It is often difficult to obtain type-specific neutralizing antisera in animals such as rabbits, especially if hyperimmunization protocols requiring multiple booster immunizations are employed. Antisera with more impressive type-specific neutralizing activity can often be produced in birds such as chickens or geese (Behbehani et al., 1966; Bruggeman & Versteeg, 1973; Rosen, 1960, 1979; Stanley, 1977).

Antibody Assays

Serologic studies of reovirus infections in man have typically involved the detection of HAI anibodies, neutralizing antibodies, complement-fixing antibodies, or indirect-immunofluorescent antibodies to reovirus (Stanley, 1977; Rosen, 1979). The methods for detecting each of these types of antibodies and the general applicability and utility of each technique will be discussed separately.

HAI ANTIBODIES

This has generally been the most widely preferred and generally used technique for serologic studies in humans. HAI is often considered to be a simpler procedure and more sensitive than other antibody tests (Rosen, 1985). The general technique is identical with that described earlier for serotyping reovirus isolates except that, instead of use of sera of known specificity to type unknown viral isolates, virus of known serotype is used to detect serotype-specific HAI antibodies in test sera. The test is ideally performed with paired specimens representing sera collected during the acute and convalescent phases of illness. The test sera are diluted through the range from 1 : 10 to 1 : 320 in either PBS (pH 7.4) or 0.85% sodium chloride. If 12-mm × 75-mm test tubes are used, 0.2 ml of each serum dilution is mixed with 0.2 ml of virus solution diluted to contain 4 HA units/0.2 ml. The test sera should be evaluated simultaneously in parallel tests against reovirus of all three serotypes. The virus–serum mixtures are shaken once and then allowed to stand at room temperature for 1 h, 0.2 ml of a 0.5 to 0.75% solution of type O human RBCs, prepared as described earlier, is then added. The combined serum–virus–RBC mixture is then allowed to sit at room temperature until complete sedimentation has occurred (typically 1 to 4 h). The HAI titer of the test serum against each reovirus serotype is the highest serum dilution that completely inhibits hemagglutination (i.e., produces a "button" rather than a "shield"). When serum titers are reported, it should be remembered that the initial serum dilution factor should be corrected for the further dilution of the serum by the addition of both RBCs and virus (e.g., in the assay described above, 0.2 ml of serum is added to 0.2 ml of virus and 0.2 ml of RBCs, resulting in a further 3× dilution of each original serum dilution). A fourfold or greater rise in HAI titers between acute and convalescent sera is considered to be diagnostic of acute reovirus infection.

Nonspecific inhibitors of reovirus-induced HA may be present in human sera and may be mistakenly interpreted as HAI antibody. This problem has been extensively analyzed, and specific techniques for the elimination of these nonspecific inhibitors have been evaluated (Schmidt et al., 1962a, 1965). Heating of the serum to 56° or 60°C, although frequently employed, does not seem to reduce nonspecific inhibitors of reovirus type 1 hemagglutination (Schmidt et al., 1962a, 1965). The most effective technique is to adsorb serum with either kaolin or bentonite. A

1 : 4 (for kaolin) or 1 : 5 (for bentonite) dilution of serum in PBS is mixed with an equal volume of either 25% acid-washed kaolin or 1% bentonite. The mixture is kept at room temperature with occasional shaking for about 20 min. The kaolin or bentonite is removed by centrifugation for 15 to 20 min at 2000 rpm. The resulting supernatant is considered to be a 1 : 8 serum dilution (kaolin) or a 1 : 10 serum dilution (bentonite). It should be noted that in most studies the use of either kaolin- or bentonite-absorbed serum has resulted in a 2 to 4× decrease in HAI antibody titer when compared with nonabsorbed sera (Mann et al., 1967; Schmidt et al., 1962a, 1965; Stanley, 1977).

Serum should also be checked for the presence of isoagglutinins to type O RBC's. An adequate control involves incubating an aliquot of the test serum with an aliquot of the RBC suspension, but without virus. If hemagglutination occurs, isoagglutinins are presumed to be present in the test serum. If isoagglutinins are detected, they can be removed by adsorbing a 1 : 10 dilution of the test serum with a 5 to 10% suspension of the test RBCs at 4°C for 1 h (see Rosen, 1979; Thein and Scheid, 1981).

Individuals infected with reovirus types 1 or 2 typically show both a homologous HAI antibody response toward the infecting serotype and also a heterologous antibody response to the other serotypes. Reovirus type 3 infection typically produces only homologous HAI antibodies (Kasel et al., 1963; Rosen, 1960, 1963, 1979). Heterologous antibody titers are typically lower than homologous titers, although exceptions do occur (Rosen, 1979). A fourfold or greater rise in HAI antibody titer between acute and convalescent sera is considered to be diagnostic of acute reovirus infection. Studies by Rosen et al. have indicated that HAI antibodies are generally present within 3 weeks of an acute infection and that they persist with little diminution in titer for at least 1 year (and presumably longer) (Rosen, 1960, 1979).

NEUTRALIZING ANTIBODIES

Although it has been suggested that neutralization tests are useful for reovirus type differentiation, this is obviously true only if antiserum with high type-specific neutralization (and low heterologous neutralization) titer is available. We have generally found the plaque-reduction-neutralization test to be satisfactory. For assaying serum neutralizing antibody titers, 0.2 ml of serial 2× dilutions of serum in PBS is incubated for 1 h at room temperature with 0.2 ml of reovirus of each serotype diluted (in PBS) to contain 200 plaque-forming units (PFU) per 0.2 ml (= 1,000 PFU/ml). Of the resulting mixture, 0.1 ml is then inoculated onto each of two wells of a Costar six-well plate containing a confluent monolayer of mouse L-cell fibroblasts. (Our laboratory grows L cells in spinner culture with Joklik's modification of Eagle Minimal Essential Medium supplemented with glutamine and fetal calf serum. If 2.5 ml of cells diluted to 4.8 × 10^5 cells/ml are plated into each well of a Costar plate, a confluent monolayer of cells is formed within 12 to 24 h). The inoculum is allowed to adsorb at 37°C in 5% CO_2 incubator for 1 h. The wells are then overlaid with 3 ml of a 1:1 mixture of 2× Medium 199 (with glutamine and fetal calf serum) and agar. After 72 h, a second overlay of the same composition (2 ml) is added. After a further 72 h, the monolayers are stained with neutral red (1:1 mixture of 2× Medium 199 and agar, to which 10 ml of a 2% wt:vol solution of neutral red is added for each 500 ml of 2× 199 and agar). The number of plaques are counted 24 h after neutral red staining. The 80% plaque-reduction antibody titer is the highest dilution of serum that reduces the number of plaques by at least 80% when compared with control (serum-free) levels. A 4× or greater increase in serum neutralizing antibody titer between acute and convalescent sera tested concurrently is indicative of acute infection. As noted earlier, both HAI and neutralizing antibodies persist for at least several years and are in fact presumed to be lifelong, although specific studies directly addressing this issue are not currently available.

COMPLEMENT FIXATION

Although infection with reovirus does produce complement-fixing antibodies that are group-specific (*not* type-specific), assay of complement-fixing antibodies appears to be of limited utility (Stanley, 1977). The complement-fixing antibody response also appears to be less sensitive than either HAI or neutralization tests (Stanley, 1977). The methodology for preparation of reovirus complement-fixing antigen can be found in the original papers on reovirus ("hepatoencephalomyelitis virus") by Stanley (1953).

INDIRECT FLUORESCENT ANTIBODIES

A method for detecting "indirect fluorescent antibodies" to reovirus type 3 in human sera has been described by Morecki et al. (1982). Mouse L-cell fibroblasts are infected with reovirus type 3. Unfortunately the authors do not specifically describe the multiplicity of infection or the duration of infection required for subsequent optimum staining. Infected cells are centrifuged into a pellet, which is sectioned in a cryostat. Sections are fixed in acetone for 3 to 5 min and then incubated for 45 min with serial 2× dilutions (in PBS) of sera. The sections are then washed in PBS and incubated with a fluorescein-conjugated f(ab′)2 fragment of goat anti-human immunoglobulin. The sections are then washed in PBS

and examined under a fluorescent microscope. Uninfected L cells incubated with the test serum are used as controls. The presence of intracytoplasmic fluorescence is scored as positive. This antibody system has not been rigorously tested against more conventional serologic techniques, and its diagnostic utility must be considered uncertain until more information is available.

Interpretation of Laboratory Data

The best evidence of acute reovirus infection is a fourfold or greater increase in either HAI or neutralizing antibody titer against reovirus of a specific serotype. Direct virus isolation from appropriate clinical specimens also provides direct evidence of virus infection. Because of the lack of clear evidence associating reoviruses with human illnesses, care must be taken not to overlook other potential pathogens, since reovirus infection may be coincident with, rather than causal in a particular illness.

Epidemiology and Natural History

The basic features of the epidemiology of reovirus infections have been included as part of the discussion of the Clinical Features of reovirus infection (see above). As noted, reoviruses appear to be of ubiquitous worldwide distribution (Stanley, 1977). Most individuals are initially infected with reovirus in infancy or early childhood, and by the second or third decade of life the majority of individuals (50 to 80%) are seropositive. In the vast majority of cases there are no identifiable clinical correlates to this infection, which appears therefore to be either asymptomatic or at least unrecognized.

Reoviruses have been isolated from a wide variety of animals including chimpanzees, African monkeys (*Cercopithecus*), Asian monkeys (macaque), horses, cattle, sheep, pigs, dogs, cats, mice, a number of marsupials, and birds (see Stanley, 1977). Infection with virus isolation has also been induced in rats, hamsters, and ferrets (Kilham and Margolis, 1969; Nielsen and Baringer, 1972; Raine and Fields, 1973). Antibodies to reovirus have been found in all these animals and in several others, including rabbits, hares, guinea pigs, rats, bats, camels, reptiles, and fish (Stanley, 1977). Infection with reovirus type 2 appears to be more restricted than infection with types 1 and 3, and isolation has apparently been limited to chimpanzees, macaques, horses, cattle, and mice (Stanley, 1977; Thein and Scheid, 1981). The role, if any, of animals as possible reservoirs or carriers of reoviruses that can subsequently infect man is unknown.

Control and Prevention

Since reoviruses have not been clearly implicated in symptomatic human infections, little interest has focused on control or prevention of human reovirus infections. Despite the lack of clear-cut evidence that reoviruses are important animal pathogens, either in economic terms or in terms of the incidence of diseases producing significant morbidity or mortality, reoviruses have been incorporated into a number of polyvalent vaccines for veterinary use (reviewed in Thein and Scheid, 1981).

A trivalent vaccine containing beta-propiolactone-inactivated adenovirus (type 3), parainfluenza (type 3), and reovirus (type 1) is commercially available under the trade name "Pneumovac" for use in the prevention of viral pneumonia in calves (Shefki and Benge, 1969; Thein and Scheid, 1981). Another vaccine, designed for prophylaxis against bovine enzootic bronchopneumonia, which contains formalin-inactivated adenovirus (types 1, 3, and 5), parainfluenza (type 3), and reovirus (types 1 and 3), is also available in Europe (Thein and Scheid, 1981). A polyvalent vaccine for prevention of respiratory tract infections in horses has been marketed under the trade name "Resequin". It contains formalin-inactivated influenza virus-A-equine, reovirus (types 1 and 3), and inactivated rhinopneumonia virus (Thein and Scheid, 1981). Finally, a polyvalent vaccine containing formalin-inactivated reovirus (type 3), human influenza virus (H_3N_2), and parainfluenza virus (type 2) has been developed for the prevention of "kennel cough" in dogs and puppies (Thein and Scheid, 1981). Most of these vaccines have undergone field testing and have been shown to be effective in reducing the incidence and associated morbidity of the specific diseases noted. In several cases (e.g., the horse vaccine), the vaccine has been shown to induce high titer hemagglutination-inhibiting and neutralizing antibodies to the included reovirus serotype, and to prevent clinical illness after challenge with live virus (Thein and Scheid, 1981).

Literature Cited

Abinanti, F. R. 1963. Respiratory disease of cattle and observations on reovirus infections in cattle. Am. Rev. Respir. Dis. **88**:290.

Amano, Y., S. Katagari, N. Ishida, and Y. Watanabe. 1971. Spontaneous degradation of reovirus capsid into subunits. J. Virol. **8**:805.

Arnott, S., F. Hutchinson, M. Spencer, and M. H. F. Wilkins. 1966. X-ray diffraction studies of double helical ribonucleic acid. Nature (London) **211**:227.

Arnott, S., M. H. F. Wilkins, W. Fuller, and R. Langridge. 1967. Molecular and crystal structures of double-helical RNA. J. Mol. Biol. **27**:525.

Averback, P. 1982. Reovirus and pathogenesis of some forms of chronic mental illness. Med. Hypothesis **8**:383.

Babiss, L. E., R. B. Luftig, J. A. Weatherbee, R. R. Weihing, U. R. Ray, and B. N. Fields. 1979. Reovirus serotypes 1 and 3 differ in their in vitro association with microtubules. J. Virol. **30**:863.

Banerjee, A. K., and A. J. Shatkin. 1971. Guanosine-5'-diphosphate at the 5' termini of reovirus RNA: evidence for a segmented genome within the virion. J. Mol. Biol. **61**:643.

Bangaru, B., R. Morecki, J. H. Glaser, L. M. Gartwer, and M. S. Horwitz. 1980. Comparative studies of biliary atresia in human newborn and reovirus-induced cholangitis in weanling mice. Lab. Invest. **43**:456.

Baskerville, A., J. B. McFerran, and T. Connor. 1971. The pathology of experimental infection of pigs with type 1 reovirus of porcine origin. Res. Vet. Sci. **12**:172.

Bassel-Duby, R., A. Jayasuriye, D. Chatterjee, N. Sonenberg, J. V. Maizel, Jr., and B. N. Fields. 1985. Sequence of reovirus hemagglutinin predicts a coiled-coil structure. Nature (London) **315**:421–423.

Bassel-Duby, R., D. R. Spriggs, K. L. Tyler, and B. N. Fields. 1986. Identification of attenuating mutations on the reovirus type 3 sl double-stranded RNA segment with a rapid sequencing technique. J. Virol. **60**:64–67.

Behbehani, A. M., L. C. Foster, and H. A. Wenner. 1966. Preparation of typespecific antisera to reoviruses. Appl. Microbiol. **14**:1051–1053.

Belak, S., V. Palfi. 1974. Isolation of reovirus type 1 from lambs showing respiratory and intestinal symptoms. Arch. Gesamte Virusforsch. **44**(3):177.

Bell, T. M. 1967. Viruses associated with Burkitt's tumor. Prog. Med. Virol. **9**:1–34.

Bell, T. M., A. Massie, and M. G. R. Ross. 1966a. Further isolations of reovirus type 3 from cases of Burkitt's lymphoma. Br. Med. J. **2**:1514.

Bellamy, A. R., and W. K. Joklik. 1967. Studies on reovirus RNA. II. Characterization of reovirus messenger RNA and of the genome RNA segments from which it is transcribed. J. Mol. Biol. **29**:19.

Binn, L. N., E. C. Lazar, J. Helms, and R. E. Cross. 1970. Viral antibody patterns in laboratory dogs with respiratory disease. Am. J. Vet. Res. **31**:697.

Binn, L. N., R. H. Marchwicki, K. P. Keenan, A. J. Strand, and W. F. Engler. 1977. Recovery of reovirus type 2 from an immature dog with respiratory tract disease. Am. J. Vet. Res. **38**(7):927.

Borsa, J., T. P. Copps, M. D. Sargent, D. G. Long, and J. D. Chapman. 1973a. New intermediate subviral particles in the in vitro uncoating of reovirus virions by chymotrypsin. J. Virol. **11**:552.

Borsa, J., B. D. Morash, M. D. Sargent, T. P. Copps, P. A. Lievaart, and J. G. Szekely. 1979a. Two modes of entry of reovirus particles into L cells. J. Gen. Virol. **45**:161.

Borsa, J., M. D. Sargent, T. P. Copps, D. G. Long, and J. D. Chapman. 1979b. Specific monovalent cation effects on modification of reovirus infectivity by chymotrypsin digestion in vitro. J. Virol. **11**:1017.

Borsa, J., M. D. Sargent, D. G. Long, and J. D. Chapman. 1973b. Extraordinary effects of specific monovalent cations on activation of reovirus transcriptase by chymotrypsin in vitro. J. Virol. **11**:207.

Brendler, T., T. Godefroy-Colburn, R. D. Carlill, and R. E. Thach. 1981a. The role of mRNA competition in regulating translation. II. Development of a quantitative in vitro assay. J. Biol. Chem. **256**:11,747.

Brendler, T., T. Godefroy-Colburn, S. Yu, and R. E. Thach. 1981b. The role of mRNA competition in regulating translation. III. Comparison of in vitro and in vivo results. J. Biol. Chem. **256**:11,755.

Bruggeman, C., and J. Versteeg. 1973. Studies on reovirus-antigens. I. Preparation of reovirus-specific immune serum in the rabbit by means of active immunization through the scarified skin. Arch. Gesamte Virusforsch. 42:371–377.

Burstin, S. J., D. R. Spriggs, and B. N. Fields. 1982. Evidence for functional domains on the reovirus type 3 hemagglutinin. Virology 117:146.

Cashdollar, L. W., R. A. Chmelo, J. R. Wiener, and W. K. Joklik. 1985. Sequences of the S1 genes of the three serotypes of reovirus. Proc. Natl. Acad. Sci. USA 82:24–28.

Chang, C.-T., and H. J. Zweerink. 1971. Fate of parental reovirus in infected cell. Virology 46:544.

Co, M. S., G. N. Gaulton, B. N. Fields, and M. I. Greene. 1985a. Isolation and biochemical characterization of the mammalian reovirus type 3 cell-surface receptor. Proc. Natl. Acad. Sci. USA 82:1494–1498.

Co, M. S., G. N. Gaulton, A. Tominaga, C. J. Homcy, B. N. Fields, and M. I. Greene. 1985b. Structural similarities between the mammalian B-adrenergic and reovirus type 3 receptors. Proc. Natl. Acad. Sci. USA 82:5315–5318.

Cook, I. 1963. Reovirus type 3 infection in laboratory mice. Aust. J. Exp. Biol. Med. Sci. 41:651.

Cox, D. C., and J. E. Shaw. 1974. Inhibition of the initiation of cellular DNA synthesis after reovirus infection. J. Virol. 13:760–761.

Csiza, C. K. 1974. Characterization and serotyping of three feline reovirus isolates. Infect. Immun. 9(1):159.

Dales, S. 1963. Association between the spindle apparatus and reovirus. Proc. Natl. Acad. Sci. USA 50:268.

Dales, S., P. Gomatos, and K. C. Hsu. 1965. The uptake and development of reovirus and strain L cells followed with labelled viral ribonucleic acid and ferritin-antibody conjugates. Virology 25:193.

Deshmukh, D. R., H. I. Sayed, and D. S. Pomeroy. 1968. Avian reoviruses. IV. Relationship to human reoviruses. Avian Dis. 13:16.

Detjen, B. M., W. E. Walden, and R. E. Thach. 1982. Transitional specificity in reovirus-infected mouse fibroblasts. J. Biol. Chem. 257:9855–9860.

Drayna, D., and B. N. Fields. 1982a. Activation and characterization of the reovirus transcriptase: genetic analysis. J. Virol. 41:110–118.

Drayna, D., and B. N. Fields. 1982b. Genetic studies on the mechanism of chemical and physical inactivation of reovirus. J. Gen. Virol. 63:149.

Drayna, D., and B. N. Fields. 1982c. Biochemical studies on the mechanism of chemical and physical inactivation of reovirus. J. Gen. Virol. 63:161.

Dukes, C. D., J. L. Parsons, and C. A. L. Stephens, Jr. 1969. Use of acridine orange in lymphocyte transformation test. Proc. Soc. Exp. Biol. Med. 131:1168–1170.

Edmonson, J. H., S. J. Millian, M. Goodenom, and S. L. Lee. 1970. Persistent viruria with reovirus in a patient treated for Hodgkin's disease in a protected environment. J. Infect. Dis. 121:438.

Eggers, H. J., P. J. Gomatos, and I. Tamm. 1962. Agglutination of bovine erythrocytes: a general characteristic of reovirus type 3. Proc. Soc. Exp. Biol. Med. 110:879.

Elazhary, M. A., M. Morin, J. B. Derbyshire, A. Lagac'e, L. Berthiaume, and M. Corbeil. 1978. The experimental infection of piglets with a porcine reovirus. Res. Vet. Sci. 25(1):16.

Ensminger, W. D., and I. Tamm. 1969a. Cellular DNA and protein synthesis in reovirus-infected cells. Virology 39:357.

Ensminger, W. D., and I. Tamm. 1969b. The step in cellular DNA synthesis blocked by reovirus infection. Virology 39:935.

Ensminger, W. D., and I. Tamm. 1970. Inhibition of synchronized cellular deoxyribonucleic acid synthesis during Newcastle disease virus, mengovirus or reovirus infection. J. Virol. 5:672.

Ernst, H., and A. J. Shatkin. 1985. Reovirus hemagglutinin mRNA codes for two polypeptides in overlapping reading frames. Proc. Natl. Acad. Sci. USA 82:48–52.

Farrell, J. A., J. D. Harvey, and A. R. Bellamy. 1974. Biophysical studies of reovirus type 3. I. The molecular weight of reovirus and reovirus cores. Virology 62:145.

Fields, B. N., C. S. Raine, and S. G. Baum. 1971. Temperature-sensitive mutants of reovirus type 3: defects in viral maturation as studied by immunofluorescence and electron microscopy. Virology 43:569.

Gard, G., and R. W. Compans. 1970. Structure and cytopathic effects of Nelson Bay virus. J. Virol. 6:100.

Gard, G., and I. D. Marshall. 1973. Nelson Bay virus: a novel reovirus. Arch. Virol. 43:34.

Glaser, J. M., W. F. Balistreri, and R. Morecki. 1984. Role of reovirus type 3 in persistent infantile cholestasis. J. Pediatr. 105:912–915.

Glass, S. E., S. A. Naqui, C. F. Hall, and K. M. Kerr. 1973. Isolation and characterization of a virus associated with arthritis of chicken. Avian Dis. 17:415.

Gomatos, P. J., O. Prakash, and N. M. Stamatos. 1981. Small reovirus particles composed solely of sigma NS with specificity for binding different nucleic acids. J. Virol. 39:115.

Gomatos, P. J., and W. Stoeckenius. 1964. Electron microscope studies on reovirus RNA Proc. Natl. Acad. Sci. USA 52:1449.

Gomatos, P. J., and I. Tamm. 1963. Base composition of the RNA of a reovirus variant. Science 140:997–1963.

Gomatos, P. J., and I. Tamm. 1963. Macromolecular synthesis in reovirus-infected L Cells. Biochim. Biophys. Acta 72:651.

Gomatos, P. J., I. Tamm, S. Dales, and R. M. Franklin. 1962. Reovirus type 3 physical characteristics and interactions with L cells. Virology 17:441.

Graziadei, W. D., III, and P. Lengyel. 1972. Translation of in vitro synthesized reovirus messenger RNA into proteins of the size of reovirus capsid proteins in a mouse L cell extract. Biochem. Biophys. Res. Commun. 46:1816.

Graziadei, W. D., III, D. Roy, W. Konigsberg, and P. Lengyel. 1973. Translation of reovirus RNA synthesized in vitro into reovirus proteins in a mouse L cell extract. Arch. Biochem. Biophys. 158:266.

Hand, R., and G. J. Kasupski. 1978. DNA and histone synthesis in reovirus-infected cells. J. Gen. Virol. 39:437–448.

Hand, R., and I. Tamm. 1974. Initiation of DNA replication in mammalian cells and its inhibition by reovirus infection. J. Mol. Biol. 82:175–183.

Hartley, J. W., W. R. Rowe, and J. B. Austin. 1962. Subtype differentiation of reovirus type 2 strains by hemagglutination-inhibition with mouse antisera. Virology 96:94.

Hartley, J. W., W. P. Rowe, and R. J. Huebner. 1961. Recovery of REO viruses from wild and laboratory mice. Proc. Soc. Exp. Biol. Med. 108:390.

Harvey, J. D., A. R. Bellamy, W. C. Earnshaw, and C. Schutt. 1981. Biophysical studies of the reovirus type 3. Virology 112:240–249.

Hassan, S. A., E. R. Rabin, and J. L. Melnick. 1965. Reovirus myocarditis in mice: an electron microscopic immunofluorescent and virus assay study. Exp. Mol. Pathol. 4:66.

Hastings, K. E. M., and S. Millward. 1981. Similar sets of

terminal oligonucleotides from reovirus double-stranded RNA and viral messenger RNA synthesized in vitro. Can. J. Biochem. **59**:151.

Hiatt, C. W. 1960. Photodynamic inactivation of viruses. Trans. N.Y. Acad. Sci. **23**:66.

Holzinger, E. A., and R. A. Griesemer. 1966. Effects of reovirus, type I on germ free and disease free dogs. Am. J. Epidemiol. **84**:426.

Huismans, H., and W. K. Joklik. 1976. Reovirus-coded polypeptides in infected cells: isolation of two native monomeric polypeptides with high affinity for single-stranded and double-stranded RNA, respectively. Virology **70**:411.

Hull, R. N., J. R. Minner, and C. C. Mascoli. 1958. New viral agents recovered from tissue cultures of monkey kidney cells. III. Recovery of additional agents both from cultures of monkey tissues and directly from tissues and excreta. Am. J. Hyg. **68**:31.

Hull, R. N., J. R. Minner, and J. W. Smith. 1956. New viral agents recovered from tissue cultures of monkey kidney cells. I. Origin and properties of cytopathogenic agents SV1, SV2, SV4, SV5, SV6, SV11, SV12, and SV15. Am. J. Hyg. **63**:204.

Jackson, G. G., and R. L. Muldoon. 1973. Viruses causing respiratory infection in man. IV. Reoviruses and adenoviruses. J. Infect. Dis. **128**:811.

Jackson, G. G., R. L. Muldoon, and R. S. Cooper. 1961. Reovirus type 1 as an etiologic agent of the common cold. J. Clin. Invest. **40**:1051.

Jackson, G. G., R. L. Muldoon, G. C. Johnson, and M. F. Dowling. 1962. Am. Rev. Resp. Dis. **88**:120–127.

Jacobs, B. L., and C. E. Samuel. 1985. Biosynthesis of reovirus-specified polypeptides: The reovirus S1 mRNA encodes two primary translation products. Virology **143**:63–74.

Jarudi, N. I., D. O. Huggett, and B. Golden. 1973. Reovirus keratoconjunctivitis. Canad. J. Ophthal. **8**:371–373.

Jenson, A. B., E. R. Rabin, D. C. Bentinck, and F. Rapp. 1965. Reovirus viremia in newborn mice. Am. J. Pathol. **49**:1171–1183.

Joklik, W. K. 1972. Studies on the effect of chymotrypsin on reovirions. Virology **49**:700.

Joklik, W. K. 1974. Reproduction of Reoviridae, p. 231–334. *In* H. Fraenkel-Conrat and R. R. Wagner (ed.), Comprehensive virology, vol. 2. Plenum Publishing Corp., New York.

Joklik, W. K. 1981. Structure and function of the reovirus genome. Microbiol. Rev. **45**:483.

Joklik, W. K. 1983. The reovirus particle, p. 9–78. *In* W. K. Joklik (ed.) The Reoviridae. Plenum Publishing Corp., New York.

Jordan, L. E., and H. D. Mayor. 1962. The fine structure of reovirus, a new member of the icosahedral series. Virology **17**:597.

Joske, R. A., D. D. Keall, P. J. Leak, N. F. Stanley, and M. D. Walters. 1964. Hepatitis–encephalitis in humans with reovirus infections. Arch. Intern. Med. **113**:811.

Kasel, J. A., L. Rosen, and H. Evans. 1963. Infection of human volunteers with a reovirus of bovine origin. Proc. Soc. Exp. Biol. Med. **112**:979.

Kasza, L. 1979. Isolation and characterization of a reovirus from pigs. Vet. Rec. **87**(22):681.

Kauffman, R. S., R. Ahmed, and B. N. Fields. 1983b. Selection of a mutant S1 gene during reovirus persistent infection of L cells: role in maintenance of the persistent state. Virology **131**:79.

Kauffman, R. S., J. L. Wolf, R. Finberg, J. S. Trier, and B. N. Fields. 1983a. The sigma 1 protein determines the

extent of spread of reovirus from the gastrointestinal tract of mice. Virology **124**:403.

Kawamura, H., F. Shimizu, M. Maeda, and H. Tsubahara. 1965. Avian reovirus: its properties and serological classification. Natl. Inst. Anim. Health Q. (Tokyo) **5**:115.

Kawamura, H., and H. Tsubahara. 1966. Common antigenicity of avian reoviruses. Natl. Inst. Anim. Health Q. (Tokyo) **6**:187.

Kaye, K. M., D. R. Spriggs, R. Bassel-Duby, B. N. Fields, and K. D. Tyler. 1986. Genetic basis for altered pathogenesis of an immune-selected antigenic variant of reovirus type 3 (Dearing). J. Virol. **59**:90–97.

Kilham, L., and G. Margolis. 1969. Hydrocephalus in hamsters, ferrets, rats and mice following inoculations with reovirus type 1. Lab. Invest. **21**:183.

Krainer, L., and B. E. Aronson. 1959. Disseminated encephalomyelitis in humans with recovery of hepatoencephalitis virus. J. Neuropathol. **18**:339.

Kundin, W. D., C. Liu, and J. Gigstad. 1966. Reovirus infection in suckling mice immunofluorescent and infectivity studies. J. Immunol. **97**:393–401.

Kurogi, H., Y. Inaba, E. Takahashi, K. Sato, Y. Goto, T. Omori, and M. Matumoto. 1974. New serotypes of reoviruses isolated from cattle. Arch. Gesamte Virusforsch. **45**(1/2):157.

Lamont, P. H., J. H. Darbyshire, P. S. Dawson, A. R. Omar, and A. R. Jennings. 1968. Pathogenesis and pathology of infection in calves with strains of reovirus types 1 and 2. J. Comp. Pathol. **78**:23.

Langridge, R., and P. J. Gomatos. 1963. The structure of RNA. Science **141**:694.

Lau, R. Y., D. Van Alstyne, R. Berckmans, and A. F. Graham. 1975. Synthesis of reovirus-specific polypeptides in cells pretreated with cycloheximide. J. Virol. **16**:470.

Lee, P. W. K., E. C. Hayes, and W. K. Joklik. 1981. Protein σ1 is the reovirus cell attachment protein. Virology **108**:156.

Leers, W. D., and K. R. Royce. 1966. A survey of reovirus antibodies in sera of urban children. Can. Med. Assoc. J. **94**:1040.

Leers, W. D., K. R. Rozee, and H. C. Wardlow. 1968. Immunodiffusion and immunoelectrophoretic studies of reovirus antigens. Canad. J. Microbiol. **14**:161–164.

Lendrum, A. C. 1947. The phloxin–tartrazine method as a general histological strain and for the demonstration of inclusion bodies. J. Pathol. Bacteriol. **59**:399.

Lerner, A. M., J. D. Cherry, J. O. Klein, and M. Finland. 1962. Infections with reoviruses. N. Engl. J. Med. **267**:947.

Levin, D. H., N. Mendelsohn, M. Schonberg, H. Klett, S. Silverstein, and A. M. Kapuler. 1970. Properties of RNA transcriptase in reovirus subviral particles. Proc. Natl. Acad. Sci. USA **66**:890.

Levin, K. H., and C. E. Samuel. 1980. Biosynthesis of reovirus-specified polypeptides: purification and characterization of the small-sized class mRNAs of reovirus type 3: coding assignments and translational efficiencies. Virology **106**:1.

Levy, J. A., E. Tanabe, and E. C. Curnen. 1968. Occurrence of reovirus antibodies in healthy African children and in children with Burkitt's lymphoma. Cancer **21**:53.

Lou, T. Y., and H. A. Wenner. 1963. Natural and experimental infection of dogs with reovirus type 1. Pathogenicity of the strain for other animals. Am. J. Hyg. **77**:293.

Luftig, R. B., S. Kilham, A. J. Hay, H. J. Zweerink, and

W. K. Joklik. 1972. An ultrastructure study of virions and cores of reovirus type 3. Virology **48**:170.

Mann, J. J., R. D. Rosen, J. R. Lehrich, and J. A. Kasel. 1967. The effect of kaolin on immunoglobulins: an improved technique to remove the non-specific serum inhibitor of reovirus hemagglutination. J. Immunol. **98**:1136.

Margolis, G., and L. Kilham. 1969. Hydrocephalus in hamsters, ferrets, rats, and mice following inoculations with reovirus type 1. II. Pathologic studies. Lab. Invest. **21**:189.

Margolis, G., L. Kilham, and N. Gonatos. 1971. Reovirus type III encephalitis: observations of virus–cell interactions in neural tissues. I. Light microscopy studies. Lab. Invest. **24**:91.

Massie, A., M. G. Ross, and M. C. Williams. 1964. Isolation of a reovirus from a case of Burkitt's lymphoma. Br. Med. J. **1**:1212.

Massie, E. L., and E. D. Shaw. 1966. Reovirus type 1 in laboratory dogs. Am. J. Vet. Res. **27**:783.

Masters, C., M. Alpers, and B. Kakulas. 1977. Pathogenesis of reovirus type 1 hydrocephalus in mice: significance of aqueductal changes. Arch. Neurol. **34**:18.

Mayor, H. D., and L. E. Jordan. 1965. Studies on reovirus. I. Morphologic observations on the development of reovirus in tissue culture. Exp. Mol. Pathol. **4**:40–50.

McClain, M. E., and R. S. Spendlove. 1966. Multiplicity reactivation of reovirus particles after exposure to ultraviolet light. J. Bacteriol. **92**:1422.

McDowell, M. J., and W. K. Joklik. 1971. An in vitro protein synthesizing system from mouse L fibroblasts infected with reovirus. Virology **45**:724.

McDowell, M. J., W. K. Joklik, L. Villa-Komaroff, and H. F. Lodish. 1972. Translation of reovirus messenger RNAs synthesized in vitro into reovirus polypeptides by several mammalian cell-free extracts. Proc. Natl. Acad. Sci. USA **69**:2649.

McFerran, J. B., and T. Connor. 1970. A reovirus isolated from a pig. Res. Vet. Sci. **11**:388.

McFerran, J. B., T. J. Connor, and R. M. McCracken. 1976. Isolation of adenoviruses and reoviruses from avian species other than domestic fowl. Avian Dis. **20**(3):519.

McFerran, J. B., R. Nelson, and J. K. Clarke. 1973. Isolation and characterization of reoviruses isolated from sheep. Arch. Gesamte Virusforsch. **40**(1/2):72.

Metcalf, P. 1982. The symmetry of the reovirus outer shell. J. Ultrastruct. Res. **78**:292.

Millward, S., and A. F. Graham. 1970. Structural studies on reovirus: discontinuities in the genome. Proc. Natl. Acad. Sci. USA **65**:422.

Morecki, R., J. H. Glaser, S. Cho, W. F. Balistreri, and M. S. Horwitz. 1982. Biliary atresia and reovirus type 3 infection. N. Engl. J. Med. **307**:481.

Morecki, R., J. H. Glaser, A. B. Johnson, and Y. Kress. 1984. Defection of reovirus type 3 in the Porta Hepatis of an infant with extrahepatic biliary atresia: ultrastructural and immunocytochemical study. Hepatology **4**:1137–1142.

Morgan, E. M., and D. W. Kingsbury. 1981. Reovirus enzymes that modify messenger RNA are inhibited by perturbation of the lambda proteins. Virology **113**:565.

Muthukrishnan, S., and A. J. Shatkin. 1975. Reovirus genome RNA segments: resistance to S1 nuclease. Virology **64**:96.

Nelson, J. B., and G. R. Collins. 1961. The establishment and maintenance of a specific pathogen-free colony of Swiss mice. Proc. Anim. Care Panel **11**:65–72.

Nelson, J. B., and G. S. Tarnowski. 1960. An oncolytic virus covered from Swiss mice during passage of an ascites tumour. Nature (London) **188**:866.

Nielsen, S. L., and J. R. Baringer. 1972. Reovirus-induced aqueductal stenosis in hamsters phase contrast and electron microscopic studies. Lab. Invest. **27**:531.

Nonoyama, M., S. Millward, and A. F. Graham. 1974. Control of transcription of the reovirus genome. Nucleic Acids Res. **1**:373.

Onodera, T., A. G. Jenson, J.-W. Yoon, and A. L. Notkins. 1978. Virus-induced diabetes mellitus: reovirus infection pancreatic φ cells in mice. Science **201**:529.

Onodera, T., A. Toniolo, U. R. Ray, A. B. Jensen, R. A. Knazek, and A. L. Notkins. Virus-induced diabetes mellitus. XX. Polyendocrinopathy and autoimmunity. J. Exp. Med. **153**:1457.

Palmer, E. L., and M. L. Martin. 1977. The fine structure of the capsid of reovirus type 3. Virology **76**:109.

Papadimitriou, J. M. 1965. Electron micrographic features of acute murine reovirus hepatitis. Am. J. Pathol. **47**:565.

Papadimitriou, J. M. 1966. Ultrastructural features of chronic murine hepatitis after reovirus type 3 infection. Br. J. Exp. Pathol. **47**:624.

Papadimitriou, J. M. 1968. The biliary tract in acute murine reovirus 3 infection. Am. J. Pathol. **52**:595.

Phillips, P. A., M. P. Alpers, and N. F. Stanley. 1970. Hydrocephalus in mice inoculated neonatally by the oronasal route with reovirus type 1. Science **168**:858.

Phillips, P. A., D. Keast, J. M. Papadimitriou, M. N. Walters, and N. F. Stanley. 1969. Chronic obstructive jaundice induced by reovirus type 3 in weanling mice. Pathology **1**:193.

Raine, C. S., and B. N. Fields. 1973. Ultrastructural features of reovirus type 3 encephalitis. J. Neuropathol. Exp. Neurol. **32**:19.

Ralph, S. J., J. D. Harvey, and A. R. Bellamy. 1980. Subunit structure of the reovirus spike. J. Virol. **36**:894.

Ramig, R. F., and B. N. Fields. 1977. Reoviruses, p. 383–433. In D. P. Nayak (ed.) The molecular biology of animal virus, vol I. Marcel Dekker, New York.

Ramos-Alvarez, M. 1957. Cytopathogenic enteric viruses associated with undifferentiated diarrheal syndromes in early childhood. Ann. N.Y. Acad. Sci. **67**:326–331.

Ramos-Alvarez, M., and A. B. Sabin. 1954. Characteristics of poliomyelitis and other enteric viruses recovered in tissue culture from healthy American children. Proc. Soc. Exp. Biol. Med. **87**:655.

Ramos-Alvarez, M., and A. B. Sabin. 1958. Enteropathogenic viruses and bacteria. Role in summer diarrheal diseases of infancy and early childhood. J. Am. Med. Assoc. **167**:147–156.

Rauth, A. M. 1965. The physical state of viral nucleic acid and the sensitivity of viruses to ultraviolet light. Biophys. J. **5**:257.

Rhim, J. S., L. E. Jordan, and H. D. Mayor. 1962. Cytochemical, fluorescent-antibody and electron microscopic studies on the growth of reovirus (ECHO 10) in tissue culture. Virology **17**:342.

Rhim, J. S., K. O. Smith, and J. L. Melnick. 1961. Complete and coreless forms of reovirus (ECHO 10): ratio of number of virus particles to infective units in the one-step growth cycle. Virology **15**:428.

Richardson, M. A., and Y. Furuichi. 1983. Nucleotide sequence of reovirus genome segment S_3, encoding unstructural protein sigma NS. Nucleic Acids Res. **11**:6399.

Roner, M. R., and D. C. Cox. 1985. Cellular integrity is required for inhibition of initiation of cellular DNA synthesis by reovirus type 3. J. Virol. **53**:350–359.

Rosen, L. 1960. Serologic groupings of reovirus by hemagglutination–inhibition. Am. J. Hyg. **71:**242.

Rosen, L. 1979. Reoviruses, p. 577–584. *In* E. H. Lennette and N. J. Schmidt (ed.), Diagnostic procedures for viral, rickettsial and chlamydial infections, 5th ed. American Public Health Association, Washington, D.C.

Rosen, L., F. R. Abinanti, and J. F. Hovis. 1963. Further observations on the natural infection of cattle with reoviruses. Am. J. Hyg. **77:**38.

Rosen, L., J. F. Hovis, F. M. Mastrota, J. A. Bell, and R. J. Huebner. 1960. An outbreak of infection with a type 1 reovirus among children in an institution. Am. J. Hyg. **71:**266.

Rubin, D. H., M. A. Eaton, and A. O. Anderson. 1986. Reovirus infection in adult mice: the virus hemagglutinin determines the site of intestinal disease. Microb. Pathogen. **1:**79–87.

Rubin, D. H., and B. N. Fields. 1980. Molecular basis of reovirus virulence: role of the M2 gene. J. Exp. Med. **152:**853.

Rubin, D. H., M. J. Kornstein, and A. O. Anderson. 1985. Reovirus serotype 1 intestinal infection: a novel replicative cycle with Ileal Disease. J. Virol. **53:**391–398.

Sabin, A. B. 1959. Reoviruses, a new group of respiratory and enteric viruses formerly classified as ECHO type 10 is described. Science **130:**1387–1389.

Sabin, A. B. 1960. Role of the ECHO viruses in human diseases, p. 78–100. *In:* Rose (ed.), Viral infections of infancy and childhood. Hoeber-Harper, New York.

Sarkar, G., J. Pelletier, R. Bassel-Duby, A. Jayasuriya, B. Fields, and N. Sonenberg. 1985. Identification of a new polypeptide coded by reovirus gene S1. J. Virol. **54:**720.

Schmidt, J., C. Tauchnitz, and O. Kuhn. 1965. Untersuchungen über das Vorkommen Hämagglutinations-hemmender Antikörper gegen die REO-Virustypen 1 und 2 in der Bevolkerung. Z. Hyg. Infekt. Kr. **150:**269.

Schmidt, N. J., J. Dennis, J. Hagens, and E. H. Lennette. 1962a. Studies on hemagglutination inhibition tests for identification of ECHO viruses. Am. J. Hyg. **75:**74–85.

Schmidt, N. J., J. Dennis, J. Hagens, E. M. Lennette. 1962b. Studies on the antibody responses of patients infected with ECHO viruses. Am. J. Hyg. **75:**168–182.

Schonberg, M., S. C. Silverstein, D. H. Levin, and G. Acs. 1971. Asynchronous synthesis of the complementary strands of the reovirus genome. Proc. Natl. Acad. Sci. USA **68:**505.

Scott, F. W., D. E. Kahn, and J. H. Gillespie. 1970. Feline viruses: isolation, characterization and pathogenicity of a feline reovirus. Am. J. Vet. Res. **31:**11.

Sharpe, A. H., L. B. Chen, and B. N. Fields. 1982. The interaction of mammalian reoviruses with the cytoskeleton of monkey kidney CV-1 cells. Virology **120:**399.

Sharpe, A. H., and B. N. Fields. 1981. Reovirus inhibition of cellular DNA synthesis: role of the S1 gene. J. Virol. **38:**389.

Sharpe, A. H., and B. N. Fields. 1982. Reovirus inhibition of cellular RNA and protein synthesis; role of the S4 gene. Virology **122:**381.

Sharpe, A. H., and B. N. Fields. 1900. Pathogenesis of reovirus infection, p. 229–285. *In* W. K. Joklik (ed.), The Reoviridae. Plenum Publishing Corp., New York.

Sharpe, A. H., and B. N. Fields. 1985. Pathogenesis of viral infection: basic concepts derived from the reovirus model. N. Engl. J. Med. **312:**486–497.

Shatkin, A. J., and M. Kozak. 1983. Biochemical aspects of reovirus transcription and translation, p. 79–106. *In* W. K. Joklik (ed.), The Reoviridae. Plenum Publishing Corp., New York.

Shatkin, A. J., and J. D. Sipe. 1968a. Single-stranded adenine-rich RNA from purified reoviruses. Proc. Natl. Acad. Sci. USA **59:**246.

Shatkin, A. J., and J. D. Sipe. 1968b. RNA polymerase activity in purified reoviruses. Proc. Natl. Acad. Sci. USA **61:**1462.

Shatkin, A. J., J. D. Sipe, and P. C. Loh. 1968. Separation of 10 reovirus genome segments by polyacrylamide gel electrophoresis. J. Virol. **2:**986.

Shaw, J. E., and D. C. Cox. 1973. Early inhibition of cellular DNA synthesis by high multiplicities of infections and UV-irradiated reovirus. J. Virol. **12:**704.

Shefki, M. D., and W. P. J. Benge. 1969. A trivalent vaccine against virus pneumonia of calves. Vet. Rec. **85:**582.

Silverstein, S. C., C. Astell, D. H. Levin, M. Schonberg, and G. Acs. 1972. The mechanisms of reovirus uncoating and gene activation in vivo. Virology **47:**797.

Silverstein, S. C., J. D. Christman, and G. Acs. 1976. The reovirus replication cycle. Annu. Rev. Biochem. **45:**375.

Silverstein, S. C., and S. Dales. 1968. The penetration of reovirus RNA and initiation of its genetic function in L-strain fibroblasts. J. Cell Biol. **36:**197.

Skehel, J. J., and W. K. Joklik. 1969. Studies on the in vitro transcription of reovirus RNA catalyzed by reovirus cores. Virology **39:**822.

Skup, D., and S. Millward. 1977. Highly efficient translation of messenger RNA in cell-free extracts prepared from L-cells. Nucleic Acids Res. **4:**3581.

Skup, D., H. Zarbl, and S. Millward. 1981. Regulation of translation in L-cells infected with reovirus. J. Mol. Biol. **151:**35.

Smith, R. E., M. A. Morgan, and Y. Furuichi. 1981. Separation of the plus and minus strands of cytoplasmic polyhedrosis virus and human reovirus double-stranded genome RNAs by gel electrophoresis. Nucleic Acids Res. **9:**5269.

Smith, R. E., H. J. Zweerink, and W. K. Joklik. 1969. Polypeptide components of virions top component and cores of reovirus 3. Virology **39:**791.

Snodgrass, D. R., C. Burrells, and P. W. Wells. 1976. Isolation of reovirus type 2 from respiratory tract of sheep. Arch. Virol. **52(1):**143.

Spandidos, D. A., and A. F. Graham. 1976a. Nonpermissive infection of L cells by an avian reovirus: restricted transcription of the viral genome. J. Virol. **19:**964.

Spandidos, D. A., and A. F. Graham. 1976b. Physical and chemical characterization of an avian reovirus. J. Virol. **19:**968.

Spendlove, R. S., E. H. Lennette, C. O. Knight, and J. H. Chin. 1963. Development of viral antigen and infectious virus on HeLa cells infected with reovirus. J. Immunol. **90:**548.

Spendlove, R. S., M. E. McClain, and E. H. Lennette. 1970. Enhancement of reovirus infectivity by extracellular removal or alterations of the virus capsid by proteolytic enzymes. J. Gen. Virol. **8:**83.

Spendlove, R. S., and F. L. Schaffer. 1965. Enzymatic enhancement of infectivity of reovirus. J. Bacteriol. **89:**597.

Spigland, I., J. P. Fox, L. R. Elveback, F. E. Wasserman, A. Kelter, C. D. Brandt, and A. Kogan. 1968. The virus watch program: a continuing surveillance of viral infections in metropolitan New York families. Am. J. Epidemiol. **83:**413.

Spriggs, D. R., R. T. Bronson, and B. N. Fields. 1983. Hemmaglutinin variants of reovirus type 3 have altered central nervous system tropism. Science **220:**505.

Spriggs, D. R., and B. N. Fields. 1982. Attenuated reovirus

type 3 strains generated by selection of hemagglutinin antigenic variants. Nature (London) **297**:68.

Spriggs, D. R., K. Kaye, and B. N. Fields. 1983. Topological analysis of the reovirus type 3 hemagglutinin. Virology **127**:220.

Stanley, N. F. 1961. Relationship of hepato-encephalomyelitis virus and reoviruses. Nature (London) **189**:687.

Stanley, N. F. 1977. Diagnosis of reovirus infections: comparative aspects, p. 385–421. *In* E. Kurstak and K. Kurstak (ed.), Comparative diagnosis of viral diseases. Academic Press, New York.

Stanley, N. F., D. C. Dorman, and J. PonsFord. 1953. Studies on the pathogenesis of a hitherto undescribed virus (hepato-encephalomyelitis) producing unusual symptoms in suckling mice. Aust. J. Exp. Biol. Med. Sci. **31**:147.

Stanley, N. F., D. C. Dorman, and J. PonsFord. 1954. Studies on the hepato-encephalomyelitis virus. Aust. J. Exp. Biol. **32**:543–562.

Stanley, N. F., and R. A. Joske. 1975a. Animal model: chronic murine hepatitis induced by reovirus type 3. Am. J. Pathol. **80**:181.

Stanley, N. F., and R. A. Joske. 1975b. Animal model: chronic biliary obstruction caused by reovirus type 3. Am. J. Pathol. **80**:185.

Stanley, N. F., P. J. Leak, M. N. Walters, and R. A. Joske. 1964. Murine infection with reovirus. II. The chronic disease following reovirus type 3 infection. Br. J. Exp. Pathol. **45**:142.

Tardieu, M., L. Powers, and H. L. Weiner. 1983. Age dependent susceptibility to reovirus type 3 encephalitis: role of viral and host factors. Ann. Neurol. **13**:602–607.

Thein, P., and R. Scheid. 1981. Reoviral infections, p. 191–216. *In* J. H. Steele (ed.), CRC handbook series in zoonoses, section B: viral zoonoses, volume II. CRC Press, Boca Raton.

Tillotson, J. R., and A. M. Lerner. 1967. Reovirus type 3 associated with fatal pneumonia. N. Engl. J. Med. **276**:1060.

Trainor, P. D., S. B. Mohanty, and F. M. Hetrick. 1966. Experimental infection of calves with reovirus type 1. Am. J. Epidemiol. **83**:217.

Tyler, K. L., R. T. Bronson, K. B. Byers, and B. N. Fields. 1985. Molecular basis of viral neurotropism: Experimental reovirus infection. Neurology **35**:88.

Tyler, K. L., and B. N. Fields. 1985. Reovirus and its replication, p. 823–862. *In* B. N. Fields (ed.), Virology. Raven Press, New York.

Tyler, K. L., and B. N. Fields. 1986. Reovirus infection in laboratory rodents. *In* P. Bhatt, R. Jacoby, H. Morse, and A. New (ed). Viral and mycoplasmal infections of laboratory rodents. Academic Press, Orlando, FL.

Tyler, K. L., D. A. McPhee, and B. N. Fields. 1986. Distinct pathways of viral spread in the host determined by reovirus S1 gene segment. Science **233**:770–774.

Tyler, K. L., W. C. Schoene, and B. N. Fields. 1984. A single viral gene determines distinct patterns of muscle injury: genetics of reovirus myositis. Neurology (NY), **34**(Suppl 1):191.

Vasquez, C., and A. K. Kleinschmidt. 1968. Electron microscopy of RNA strands released from individual reovirus particles. J. Mol. Biol. **34**:137.

Vasquez, C., and P. Tournier. 1964. New interpretation of the reovirus structure. Virology **24**:128.

Walden, W. E., T. Godefroy-Colburn, and R. E. Thach. 1981. The role of mRNA competition in regulating translation. I. Demonstration of competition in vivo. J. Biol. Chem. **256**:11,739.

Wallis, C., and J. L. Melnick. 1964. Irreversible photosensitization of viruses. Virology **23**:520–527.

Walters, M. N., R. A. Joske, P. J. Leak, and N. F. Stanley. 1963. Murine infection with reovirus. I. Pathology of the acute phase. Br. J. Exp. Pathol. **44**:427.

Walters, M. N., P. J. Leake, R. A. Joske, N. F. Stanley, and D. H. Perret. 1965. Murine infection with reovirus. III. Pathology of infection with types I and II. Br. J. Exp. Pathol. **46**:200.

Watanabe, Y., and A. F. Graham. 1967. Structural units of reovirus ribonucleic acid and their possible functional significance. J. Virol. **1**:665.

Watanabe, Y., S. Millward, and A. F. Graham. 1968. Regulation of transcription of the reovirus genome. J. Mol. Biol. **36**:107.

Watanabe, Y., L. Prevec, and A. F. Graham. 1967. Specificity in transcription of the reovirus genome. Proc. Natl. Acad. Sci. USA **58**:1040.

Weiner, H. L., K. A. Ault, and B. N. Fields. 1980a. Interaction of reovirus with cell surface receptors. I. Murine and human lymphocytes have a receptor for the hemagglutinin of reovirus type 3. J. Immunol. **124**:2143.

Weiner, H. L., D. Drayna, D. R. Averill, Jr., and B. N. Fields. 1977. Molecular basis of reovirus virulence: role of the S1 gene. Proc. Natl. Acad. Sci. USA **74**: 5744.

Weiner, H. L., M. L. Powers, and B. N. Fields. 1980b. Absolute linkage of virulence and central nervous system tropism of reoviruses to viral hemagglutinin. J. Infect. Dis. **141**:609.

Weiner, H. L., R. F. Ramig, T. A. Mustoe, and B. N. Fields. 1978. Identification of the gene coding for the hemagglutinin of reovirus. Virology **86**:581.

White, C. K., and H. J. Zweerink. 1976. Studies on the structure of reovirus cores: selective removal of polypeptide λ2. Virology **70**:171.

Wilcox, G. E., and R. W. Compans. 1982. Cell fusion induced by Nelson Bay virus. Virology **123**:312.

Wolf, J. L., R. S. Kauffman, R. Finberg, R. Dambrauskas, and B. N. Fields. 1983. Determinants of reovirus interaction with the intestinal M cells and absorptive cells of murine intestine. Gastroenterology **85**:291.

Wolf, J. L., D. H. Rubin, R. Finberg, R. S. Kauffman, A. H. Sharpe, J. S. Trier, and B. N. Fields. 1981. Intestinal M cells: a pathway for entry of reovirus into the host. Science **212**:471.

Zarbl, H., and S. Millward. 1983. The reovirus multiplication cycle, p. 107–196. *In* W. K. Joklik (ed.) The Reoviridae. Plenum Publishing Corp., New York.

Zweerink, H. J., and W. K. Joklik. 1970. Studies on the intracellular synthesis of reovirus specified proteins. Virology **41**:501.

Zweerink, H. J., M. J. McDowell, and W. K. Joklik. 1971. Essential and non-essential noncapsid reovirus proteins. Virology **45**:716.

CHAPTER 20

Reoviridae: The Orbiviruses (Colorado Tick Fever)

RICHARD W. EMMONS

Disease: Colorado tick fever.

Etiologic Agent: Colorado tick fever virus.

Source: *Dermacentor andersoni* ticks, possibly other ixodid tick species; certain wild rodents (tissues and blood); possibly human blood (by tick bite) or by inoculation or percutaneous contact with infected tick fluids, blood, or infected tissues of rodent hosts or human cases.

Clinical Manifestations: Acute onset of fever, becoming biphasic; chills; headache; photophobia; myalgias; arthralgias; lethargy; nausea and occasionally vomiting; rarely (especially in children) hemorrhagic tendency or signs of meningoencephalitis; fatalities very rare.

Pathology: Leucopenia, mild anemia, persistent viremia; rarely thrombocytopenia and hemorrhagic manifestations (petechial rash, intestinal bleeding); focal necrosis in liver, myocardium, spleen, lymph nodes, intestine, and brain.

Laboratory Diagnosis: Isolation of virus from blood cells (particularly erythrocytes) by suckling mouse or cell culture inoculation; direct fluorescent antibody staining of viral antigen in blood cells in peripheral blood smears; demonstration of significant antibody titer rise and/or of specific IgM antibody by indirect immunofluorescence staining, indirect enzyme immunoassay, neutralizing antibody or complement fixation methods.

Epidemiology: Limited to 11 western and northwestern states of the United States and the southern parts of Alberta and British Columbia provinces in Canada, above 4,000 ft elevation at sites of natural tick-rodent foci (coinciding closely with distribution of the main tick vector); more common in males due to more frequent occupational or recreational exposure; seasonal occurrence (March to October) coinciding with tick activity; closely related viruses have been recognized elsewhere, but have not yet been shown to cause human disease.

Treatment: Supportive only, no specific antiviral treatment known.

Prevention and Control: Avoid tick bites in endemic areas; experimental vaccine for laboratory workers; cases should avoid donating blood for 6 months because of prolonged viremic period.

Introduction

The genus *Orbivirus* is classified under the family *Reoviridae* and includes at least 12 antigenic groups, with at least 61 viruses registered in the *International Catalogue of Arboviruses* (Karabatsos, 1985) and 121 virus types or subtypes in the summary by Gorman and co-authors (1983). The genus and subgroups are likely to undergo further revision as additional information about them is gathered. In addition to the Colorado tick fever (CTF) group, a number of other viruses have been isolated from human beings or otherwise suspected or shown to cause human disease: Kemerovo, Lipovnik, Changuinola, Eyach, and Orungu. Additional members cause important diseases of animals: bluetongue, African horse sickness, epizootic hemorrhagic disease, and Ibaraki. The remaining viruses, in the Corriparta, Eubenangee, Palyam, Wallal, and Warrego groups and in an ungrouped category, are of little importance or need further study to determine their significance.

This chapter is a brief review of the CTF group, which currently comprises CTF and Eyach viruses and possibly additional members described below. A detailed description of the virologic, ecologic, clinical, and diagnostic features will be given only for CTF.

Physicians in the northwestern United States during the latter half of the 19th century described cases of "mountain fever," but could not clearly differentiate them from cases of typhus, typhoid fever, malaria, or other febrile diseases. Several authors described a disease that followed tick bite but was not identical to Rocky Mountain spotted fever, and the term *Colorado tick fever* gradually gained acceptance. There are conflicting claims for the first recognition and accurate description of the disease as a new entity. Topping and co-workers (1940) gave the first detailed clinical description, and Florio and collaborators (Florio and Miller, 1948; Florio et al., 1944, 1950) are credited with the first isolation of the causative virus from human cases and from *Dermacentor andersoni* ticks. The virus was adapted to mice and chick embryos by Koprowski, Cox, and collaborators during the late 1940s (Koprowski and Cox, 1946; Koprowski et al., 1950), allowing development of a vaccine and determination of serologic responses in humans.

A technique for isolating the virus in suckling mice soon replaced the less sensitive and less specific hamster test (Oliphant and Tibbs, 1950). Field studies by Eklund, Kohls, Philip, Burgdorfer, Jellison, and co-workers at the U.S. Public Health Service's Rocky Mountain Laboratory in Montana during the 1950s and 1960s clarified the tick vectors, vertebrate hosts, and ecologic features and added much knowledge about the clinical and epidemiologic features of the human disease (Drevets, 1957; Eklund and Kennedy, 1961; Eklund et al., 1959; Emmons, 1981; Lloyd, 1951; Philip et al., 1975).

Colorado tick fever, although rarely serious or fatal, ranks as one of the most common vector-borne viral diseases in the United States. Case reporting is voluntary and very incomplete. About 100–200 cases are reported annually, but many times that number undoubtedly occur. Its similarity to the more feared, tick-transmitted Rocky Mountain spotted fever results in excessive concern, unnecessary hospitalization and medical costs, and unneeded antibiotic treatment, and it therefore poses a significant problem beyond the actual morbidity it causes (Bowen, 1987).

A virus isolated from *Ixodes ricinus* ticks collected in the Federal Republic of Germany in 1972 was found to be closely related to CTF virus but distinct enough to warrant designation as a separate agent, Eyach virus (Rehse-Küpper et al., 1976, 1979). The virus has also been isolated from ticks in France and may have a role in human disease in Europe, although further studies are needed (Chastel et al., 1984).

Another virus, very similar to both CTF and Eyach, was isolated from a western tree squirrel (*Sciurus griseus*) in 1965 and from a jackrabbit (*Lepus californicus*) in 1976 in California (Lane et al., 1982). Studies indicate that it may be sufficiently different to be classified as a separate virus, but more work is needed (Karabatsos et al., 1987). Further studies are also needed to elucidate the natural cycle and possible clinical significance of this virus.

Other CTF-related viruses may eventually be found elsewhere in the world, and what was originally thought to be a peculiar and unique American virus may be found to have numerous relatives. As information about Eyach virus and other CTF-related viruses is still being developed or verified, and their importance for human disease is unknown, the remaining discussion will deal only with CTF.

Viral Properties

Several studies in the past two decades have clarified the physicochemical and morphologic characteristics of CTF virus (Knudson, 1981; Murphy et al., 1968, 1971; Oshiro and Emmons, 1968). The virion is spherical and apparently has icosahedral architecture with 32 capsomeres on the surface. It has an 80-nm outer diameter and a 50-nm inner core diameter. The viral genome is double-stranded RNA with 12 segments, in contrast to most other orbiviruses, which have 10 or 11 segments. The apparent total molecular weight is 18×10^6, larger than most members of the genus. The virion is nonenveloped, partially resistant to lipid solvents, and acid labile, being readily inactivated at pH 3.0. It is relatively stable at room temperature and readily preserved by freezing or by lyophilization when protected by a protein-containing diluent. However, as for most viruses, repeated freezing–thawing is deleterious. The optimum pH for preservation of infectivity is 7.5 to 7.8.

Many cell culture types can be infected, with or without cytopathic effect, including cells of mammalian, avian, and arthropod origin. Viral replication results in cytoplasmic, granular matrices associated with mature virions and results in accumulation in the cytoplasm (and, to some extent, in the nucleus) of distinctive bundles of filaments or fiberlike strands. These filaments have also been described for Eyach virus (Chastel et al., 1984). The CTF virus-related filaments are made up of virus-specific, antigenic material as shown by ferritin-tagged immune serum treatment (Emmons et al., 1972; Oshiro and Emmons, 1968).

Pathogenesis

In natural vertebrate hosts, laboratory animals, and humans, the virus infects hematopoietic cells, most of which are apparently not lysed or damaged completely but continue to develop and mature, resulting in a prolonged viremia of weeks' or months' duration. The intracellular (largely intraerythrocytic) location of the virus protects it from antibodies or other host defenses (Emmons, 1966, Emmons, 1967; Emmons et al., 1972; Oshiro et al., 1978). This viral persistence and involvement of lymphocytes and platelets may help explain the delayed antibody response noted in natural or experimental infection. Experimental infection of natural hosts or laboratory animals results in viremia titers of up to 10^6 suckling mouse intracerebral median lethal doses per milliliter. In humans the cell-associated viremia can persist for up to 4 months, at gradually diminishing titer, even though the patient has fully recovered from symptoms. Person-to-person transmission via blood transfusion is thus a potential hazard (Emmons et al., 1972; Hughes et al., 1974; Philip et al., 1975) and has in fact been documented (Randall et al., 1975).

As CTF has rarely been known to be fatal, there is little information about pathologic changes in humans. In one fatal case in a 4-year-old boy, purpura and petechiae, gross intestinal hemorrhage, alveolar hyaline membrane, and endothelial damage in lymph node capillaries were seen (Eklund et al., 1959). A 10-year-old girl died with acute renal failure, epistaxis, hematemesis, petechial rash, and disseminated intravascular coagulation syndrome. Focal necrotic changes were seen in the liver, myocardium, spleen, intestine, and brain (Dawson et al., 1972). Hemorrhage secondary to thrombocytopenia and encephalitis are the most serious pathologic abnormalities.

Pathologic changes or clinical illness have not been shown in natural vertebrate hosts, suggesting a long-term adaptation of the virus to them. In laboratory animals the virus can be isolated from many tissues, but this ubiquitousness may merely be due to the presence of blood in the tissue rather than to true tissue involvement. Viral infection and pathologic changes in the brain, spleen, liver, lymphoid tissue, bone marrow, and heart muscle have been described (Black et al., 1947; Eklund and Kennedy, 1961; Hadlow, 1957; Miller et al., 1961). The pathogenesis of infection in human cases may in some ways parallel that seen in animal hosts, such as involvement of the myocardium (Emmons and Schade, 1972) and

liver (Loge, 1985) and the encephalitis and hemorrhagic diathesis mentioned previously.

Clinical Features

The onset of CTF is typically abrupt, beginning 3 to 6 days after tick bite, although the patient is often unaware of the bite. The virus can apparently be transmitted during brief attachment by the tick, whereas in Rocky Mountain spotted fever a day or more of attachment and blood feeding seems necessary to facilitate "reactivation" and transfer of *Rickettsia rickettsii*. Mild or subclinical infections may occur, but most commonly a significant illness follows, with chills, high fever, severe headache, photophobia, retroocular pains, myalgias, arthralgias, and lethargy. The fever is usually biphasic, with a brief remission after 2 to 3 days, then a second 2- to 3-day febrile period. The fever, headache, and other symptoms are often worse during the second phase. Occasionally there is a third febrile period. Anorexia, nausea, and vomiting may be prominent, especially in children. Hemorrhagic tendency and manifestations of cloudy sensorium, disorientation, hallucinations, meningeal signs, and other signs of diffuse encephalitis are unusual and more likely to occur in children than adults, as mentioned previously.

Physical findings are sparse and nonspecific. A transient, petechial, or macular rash is sometimes noted, but it does not have the appearance, the distribution on face, limbs, hands, and feet, or the persistence characteristic of that seen in Rocky Mountain spotted fever. The spleen is sometimes palpable. The most dramatic and helpful laboratory finding is the abrupt drop in the peripheral leukocyte count, often to 2000 to 3000/mm³ or less. There may be a relative lymphocytosis and a left shift in the granulocytic series, with myelocytes and metamyelocytes appearing. Mild anemia may be noted and a transient, sometimes severe thrombocytopenia can occur, leading to hemorrhage. These manifestations are apparently due to a rapid amplification of the infection in bone marrow and lymphoid tissue, with depression of all blood cell elements. Prompt recovery of the hematopoietic system usually occurs, and there is no evidence of persistent bone marrow infection. The unusually persistent viremia described previously is caused predominantly by the presence of virus in the erythrocytes, which survive a normal life span before removal from the circulation.

At least one case of inadvertent transmission of virus by blood transfusion has been reported (Randall et al., 1975), and accidental transmission via needle sticks to hospital or laboratory staff is a possibility, but otherwise the disease is not contagious. Human cases are not important in the natural maintenance cycle, as they are rarely fed on by nymphal ticks (see below), and mosquitoes and other blood-feeding vectors are not involved in the CTF cycle.

Recovery from CTF is generally prompt, although some patients have a prolonged convalescence and slow return of normal physical and mental functioning. Known fatalities are rare, as stated above, but may have been erroneously attributed to Rocky Mountain spotted fever or other diseases. Lasting immunity is the rule, but at least one case with a second documented infection was reported (Goodpasture et al., 1978). At least 12 cases during pregnancy have been followed; one ended in spontaneous abortion, one in multiple fetal abnormalities thought not to be due to CTF, one in infection of the newborn but no sequelae, and the rest in healthy, full-term infants (Eklund et al., 1959; J. D. Poland, personal communication, 1986).

Other diagnostic entities that should be considered and differentiated from CTF in the known endemic area include tick-transmitted diseases such as Rocky Mountain spotted fever, tularemia, tick-borne relapsing fever, Q fever (which can be tick borne, but is most often acquired by the respiratory route), and various non-vector-borne diseases (influenza, enterovirus infection, typhoid fever, and the like). Visitors to the endemic area may have returned home by the time illness begins and may consult a physician who is not familiar with CTF. As specific laboratory tests are also less readily available outside of the endemic area, the diagnosis is likely to be missed. Although the outcome is not likely to be serious, anxiety about the cause, excessive diagnostic tests, inappropriate antibiotic therapy, and high medical costs can be significant problems.

Laboratory Diagnosis

A typical illness following tick bite in an endemic area, along with severe leukopenia, is almost certain to be CTF, but a specific, laboratory-verified diagnosis is preferable. Diagnosis is made by viral isolation, serologic tests, or both. These methods are generally available only in public health or special reference laboratories, as commercial test reagents are not available and the test services are too specialized or not cost-effective for the usual clinical laboratory.

Specimen Collection

The only specimens required are acute-phase and convalescent-phase (2 to 4 weeks after onset) blood samples. However, to detect other agents that could mimic CTF, additional samples should be considered (for example, throat swab or throat washing, stool, and cerebrospinal fluid for attempted isolation of respiratory virus or enterovirus; blood smears stained for *Borrelia* spirochetes; blood cultures for bacteria). Clotted whole blood collected aseptically in sterile tubes without preservatives can be mailed or delivered to the laboratory unrefrigerated, for convenience, as CTF virus is relatively heat stable. The blood clot is used for viral isolation or direct detection of viral antigen. Transporting the blood at 4°C would preserve the virus better, but whole blood should not be frozen because freezing lyses the cells and interferes with viral isolation and with some serologic tests. Separated serum or plasma for serologic tests may be submitted frozen, but the virus can be isolated from these specimens only during the first few days of illness, before neutralizing antibody is formed. To isolate virus from a convalescent blood sample, heparinized or otherwise anticoagulated blood is needed, so that the antibody-containing plasma can be removed and the washed cells used as the inoculum. In cases with suspected involvement of the central nervous system, viral isolation from fresh or frozen cerebrospinal fluid should be attempted.

Precautions appropriate for a Biosafety Level 2 agent should be used in handling specimens, cultures, and inoculated laboratory animals. Numerous laboratory accidents with CTF virus have occurred due to needle-stick injury. Experimental vaccines have been developed (Thomas et al., 1963, 1967) but are not commercially available or generally used.

Direct Antigen Detection

Direct detection of viral antigen by immunofluorescence staining of blood smears can be used for rapid, early diagnosis. Antigen can be demonstrated in infected erythrocytes from the first few days of onset up to 3 to 4 months later, as discussed previously. Expertise and a careful search of stained slides is required, especially during the first few days when infected cells are sparse. Confirmation of CTF alleviates concern about the more serious Rocky Mountain spotted fever and avoids unnecessary antibiotic treatment. The method has been described in detail previously (Emmons and Lennette, 1966). Briefly, hyperimmune serum is prepared in hamsters immunized with stock virus prepared in suckling hamster brains. The globulin fraction is conjugated with fluorescein isothiocyanate, and a working dilution (usually 1:40) is determined and prepared in 20% normal mouse brain suspension in 0.1 M phosphate-buffered saline (PBS) to minimize background staining. Slip smears of a small piece of blood clot or thick smears of washed blood cells (as for a malaria smear) are airdried and fixed in acetone for 30 min at room temperature or at −20°C overnight. Positive and negative control blood smears are prepared similarly from human blood or experimentally infected mouse blood. The conjugate is then added for 20 min at 37°C in a wet chamber, and the slides are washed twice for 10 min with PBS (pH 7.2 to 7.4), then rinsed in distilled water, air-dried, mounted with a cover slip on 25% glycerol–PBS, and examined under an ultraviolet microscope. The CTF viral antigen appears as one or several larger fluorescent foci plus numerous small particles within occasional erythrocytes, often only one in several microscopic fields. The pale white to yellowish, nonspecific fluorescence (autofluorescence) of polymorphonuclear leukocyte granules is readily distinguished from specific fluorescence by an experienced reader.

Immunoperoxidase staining has been evaluated and appears useful for detection of viral antigen in infected cell cultures and mouse tissues, but has not yet found practical application for diagnosis of human cases (Desmond et al., 1979).

Viral Isolation and Identification

Viral isolation is most reliably accomplished by inoculation of suckling mice (Oliphant and Tibbs, 1950). Adult mice and other laboratory animals are not susceptible or do not show signs of illness and are therefore unsuitable for this purpose. Although the virus can be easily adapted to replicate in various cell cultures, none has yet been shown to be sufficiently sensitive for primary viral isolation or been adopted for routine use. The standard method of viral isolation in the Viral and Rickettsial Disease Laboratory, California Department of Health Services, utilizes a 10% suspension of blood clot or washed erythrocytes in 0.75% bovine albumin in PBS containing 500 units of penicillin and 500 units of streptomycin per milliliter of final dilution. Treatment with 0.5% trypsin in Hanks medium has been recommended as a supplement or substitute for multiple saline washings to remove antibody-containing plasma from the blood cells and thus enhance the probability of isolating virus in late convalescence (Hughes et al., 1974). The prepared sample is inoculated intracerebrally (0.02 ml) and intraperitoneally (0.03 ml) into a litter of 1- to 3-day-old mice, which are then observed for 14 days. Signs of illness appear in 4 to 5 days, and brains of mice harvested while ill or after death are used to prepare slip smears for specific identification by immunofluorescence, as described above for blood smears (Emmons and Lennette, 1966). Alternatively, subpassage and identification of the virus by the neutralization index test in mice, using CTF hyperimmune serum, can be done. As mentioned previously, the virus and viral antigen are unusually stable, and

viral isolation or direct immunofluorescence staining of antigen can be successful even after weeks of storage of the blood at room temperature or many months of storage at 4°C.

Antibody Assays

Antibody tests are required if viral isolation has been unsuccessful or if only serum samples are available. As for most viral diseases, paired sera are preferable so that a significant rise (fourfold or greater) in antibody titer can be demonstrated. Antibody persists for many years, presumably a lifetime; thus, the mere presence of antibody does not prove an etiologic relationship to the illness unless rising antibody titer or specific immunoglobulin (Ig)M is demonstrated (see below).

The complement fixation test, utilizing infected mouse brain antigen, was the standard method for many years (Thomas and Eklund, 1960). Adaptation of the virus to hamster kidney cell cultures permitted preparation of a comparable complement fixing antigen with some advantages (Emmons et al., 1969). The microtiter procedure (Hawkes, 1979) is satisfactory and is standard in most laboratories and thus will not be described in detail. As the complement fixing antibody response of patients is often poor or delayed, serologic diagnosis by this test is less reliable.

Neutralizing antibody develops during the first week of illness, earlier than complement fixing antibody, and it therefore may be more difficult to detect a rise in antibody titer by this method. Also, as the techniques are more expensive and cumbersome, they are not used routinely. The standard neutralization test in mice, utilizing either a serum dilution method or the neutralization index method, may be used. A neutralization test in KB cell cultures has been described (Gerloff and Eklund, 1959), and the technique was adapted to a plaque-reduction neutralization test assay (Emmons et al., 1969) that is briefly summarized here. Hamster kidney cells (0-853 line, J. L. Riggs) in culture plates maintained at 35°C in a carbon dioxide incubator are grown to confluency (4 to 5 days). Stock virus prepared from a human strain of plaque-purified CTF virus adapted to the cell line is diluted to contain about 300 plaque-forming units/ml, mixed with equal volumes of twofold dilutions of heated (56°C, 30 min) serum. The serum–virus mixtures are incubated for 1 h at 36°C, then added to the plates, using three plates per dilution. After a 2-h absorption period, the inoculum is aspirated, 2.5% methyl cellulose medium is added, an additional 5-day incubation period under carbon dioxide is used, and a final overlay medium containing neutral red is added. After a final 24-h incubation, plaques are counted in the test and control plates. The antibody titer is the highest serum dilution causing an 80% reduction in plaque number.

A simpler, accurate, and more practical test is the indirect immunofluorescence test (Emmons et al., 1969). Hamster kidney cells or other suitable cell line are infected with stock virus adapted to grow to high titer in the cells. Cells are harvested when about half of them contain antigen, as determined by direct immunofluorescence staining, or heavily infected cell cultures are trypsinized and mixed with normal cells in a 1:3 proportion so that cell spots will show an easily readable pattern of staining cells among nonstaining cells, thus allowing recognition of any nonspecific reactions if present. Numerous test slides and positive and negative control slides can be prepared, with two or more circular smears per slide, or various commercially available slides can be used to prepare the test and control slides. Slides are acetone-fixed for 10 min at room temperature and dried, then stored at $-70°C$ or lower in an electric or liquid nitrogen freezer, where antigenicity will be preserved for a year or more. To assay for immunofluorescence antibody, serum samples are heat inactivated (56°C, 30 min), prepared in serial twofold dilutions, usually from 1:4 to 1:128 in 20% beef brain or mouse brain suspension in PBS (pH 7.3), then added to the virus-infected and control slides. Slides are incubated at 36°C for 20 min, washed vigorously and thoroughly twice in PBS for 10 min each time, rinsed with distilled water, and treated with a predetermined appropriate dilution of anti-human gamma globulin made from rabbit serum and conjugated to fluorescein isothiocyanate (commercially available or prepared in the laboratory). After a further 20-min incubation at 36°C, the slides are thoroughly washed, as described above, and mounted with coverslips on 25% glycerol–saline, then examined under the ultraviolet microscope. The antibody titer is the highest dilution of serum giving at least 1+ specific staining reaction (on a brightness scale of 1 to 4+) and no nonspecific reaction with uninfected cells. Positive and negative control tests are included in parallel with the patient sera being tested. As the immunofluorescence antibody tends to appear earlier than the neutralizing and complement fixing antibodies and usually reaches a higher titer, an early acute-phase serum sample is needed to detect a significant rise in titer during the course of illness. However, the method is readily adapted to detect CTF-specific IgM antibody, allowing a diagnosis to be made serologically on a single early serum sample. In this procedure, fluorescein isothiocyanate conjugated to μ-chain-specific anti-human gamma globulin is used in the last step of the procedure described above.

Use of an enzyme immunoassay procedure for detection of IgG and IgM antibody in CTF (IgM capture enzyme immunoassay) has recently been described (Calisher et al., 1985) and appears to be a simple and rapid technique, although it was found to be less sensitive than a serum-dilution plaque reduction neutralization test (PRNT) in Vero cells. Viral antigen is prepared from infected Vero cells. Microtiter plates are coated with a dilution of pretitrated goat anti-human μ-chain-specific antibody and are incubated overnight at room temperature. A single 1:100 dilution of test serum for screening, or serial twofold dilutions beginning at 1:100 for titrations, are added to the plates, along with suitable controls, then incubated for 2 h at 38°C. The CTF antigen is added, the plates are incubated for 18 h at 4°C, and CTF hyperimmune mouse ascitic fluid is added for a further 1-h incubation at 37°C. Horseradish peroxidase-conjugated goat anti-mouse IgG is then added, followed by further incubation for 1 h at 37°C. After the plates are washed, 2,2'-azino-di-(3-ethylbenzthiazoline sulfonate) (ABTS) substrate with 30% peroxide is added for 5 to 15 min. The plates are read automatically, including appropriate controls, and a ratio of the optical density of the test serum to the mean optical density of a normal serum panel less than 2.0 is considered positive. For detection of IgG antibody, plates are coated with CTF viral antigen, test sera are added for an 18-h, 4°C incubation, and horseradish peroxidase-conjugated goat antibody to human IgG is added. After an additional incubation for 1 h at 37°C, ABTS substrate is added, the plates are incubated further, and the optical density ratio is read and interpreted as for IgM determination. Antibody titers were found to be higher with enzyme immunoassay than with the neutralization test; however, these titers did not appear until 1 to 2 weeks after onset, somewhat later than with the neutralization test, and had nearly disappeared by 2 months. Both the immunofluorescence and enzyme immunoassay methods seem preferable to the neutralization and complement fixation tests, although the lack of a large market for the reagents will inhibit commercial reagent development and thus limit the availability of such tests in clinical laboratories.

Various other serologic techniques have been tried or could be adapted for CTF (e.g., indirect hemagglutination inhibition [Gaidamovich et al., 1974] or a fluorescent focus inhibition test as performed for rabies antibody [Lennette and Emmons, 1972]), but have not yet been found to be practical. As CTF virus has no hemagglutinin, a direct hemagglutination inhibition test cannot be used.

Epidemiology and Natural History

Colorado tick fever virus occurs in at least 11 western and northwestern states (South Dakota, Montana, Wyoming, Colorado, New Mexico, Utah,

Idaho, Nevada, Washington, Oregon, and California) and in the southern parts of Alberta and British Columbia in Canada, at elevations of about 122 m (4000 ft) to more than 305 m (10,000 ft), coinciding very closely to the distribution of the tick vector, *D. andersoni.* Human cases have been identified in nearly all of these areas. Serologic evidence suggests occurrence of the virus in Ontario, but further work is needed for confirmation (Newhouse et al., 1964). As mentioned elsewhere in this chapter, the possibility of CTF-related viruses in new areas must also be studied.

The epidemiology of CTF is typical for a zoonotic, vector-borne disease acquired from specific, focal ecologic niches. Most cases occur in residents of the endemic areas, but many also occur in visitors and vacationers from distant sites where the disease may not be recognized and properly diagnosed. Occasionally, cases may result because infected ticks are transported in clothing, bedding, or some other way, then find a host on which to feed. All ages and races and both sexes are presumed equally susceptible; however, factors influencing tick exposure determine the case distribution. Thus, male cases consistently outnumber female cases two- to threefold, and occupational and recreational pursuits (lumbering, hunting, fishing, camping, and the like) increase the risk of exposure. Seasonal occurrence reflects the activity and host seeking of adult *D. andersoni* ticks. Cases occur from late February or early March to mid-October, with the majority from May through July. The reported incidence of cases varies somewhat from year to year, depending on the climate, the changing population levels of ticks and rodent hosts, and the intensity of case surveillance, laboratory testing, and interest in reporting by physicians. No major fluctuations or cyclic changes, as are seen in many other vector-borne diseases, have been recorded. The epidemiologic features of CTF have been summarized in several reports and large case series, which should be consulted for more details (Bowen, 1987; Earnest et al., 1971; Eklund et al., 1959; Goodpasture et al., 1978; Spruance and Bailey, 1973).

The natural history and ecologic factors determining the maintenance of CTF virus can only be briefly summarized here. Important early contributions to this knowledge have been described in the Introduction, and numerous more recent reviews and careful field studies can be consulted for a more detailed treatment of the subject (Bowen, 1987; Bowen et al., 1981; Carey et al., 1980; Emmons, 1981; McLean et al., 1981; Sonenshine et al., 1976).

The virus cycle involves infection of various rodent and other mammalian species. The prolonged viremia that follows infection facilitates viral transmission to ticks feeding on the hosts, and the virus is transmitted to new hosts after the ticks develop to the next life stage and take another blood meal. The principal or most frequently found hosts thus far have been the golden-mantled ground squirrel (*Spermophilus lateralis*), Columbian ground squirrel (*Citellus columbianus columbianus*), yellow pine chipmunk (*Eutamias amoenus*), and least chipmunk (*E. minimus*). Numerous other species of rodents, rabbits and hares, deer, and other mammals have occasionally been found to have viremia or viral antibody, or are hosts for one or more of the three life stages of the tick vectors and thus also contribute to the maintenance cycle. The natural hosts are not apparently harmed by the infection.

At least eight tick species have been found to be infected, but *D. andersoni* is the predominant one and the only proven vector for human disease. Other species found infected within or close to the geographic range of *D. andersoni* include *D. parumapertus, D. occidentalis, D. albipictus, Otobius lagophilus, Haemaphysalis leporispalustris, Ixodes spinipalpus,* and *Ixodes sculptus.*

There is no transovarial transmission of the virus, but infected ticks remain so through all further life stages. Both male and female ticks can acquire infection by feeding on a viremic host and can transmit the virus via the saliva after they develop to the next stage and feed on a new, susceptible host. The life cycle of *D. andersoni* typically covers 2 years and involves three hosts. Larvae and nymphs that feed in the spring or summer may become infected, then develop to the nymphal or adult stage, respectively, and survive the winter in a dormant state, carrying the virus with them and transmitting it further when they become active the following spring. Overwintering of the virus in viremic, hibernating ground squirrels and chipmunks is also a possibility, although it probably plays a minor role, if any (Emmons, 1966).

Control and Prevention

The only practical means of prevention is avoiding tick-infested areas and tick bites. Protective clothing, frequent inspection of the body, removal of ticks before they become attached and begin to feed, and to some extent the use of tick repellants (such as diethyltoluamide or dimethylphthalate) on clothing are helpful. An experimental formalin-killed vaccine was developed for protection of laboratory workers, as mentioned previously (Thomas et al., 1963, 1967), but immunization is not practical for the general public. There is no specific treatment for the disease. Rest, antipyretics, analgesics, and general supportive care are helpful. The patient should not donate blood for 6 months after recovery to ensure that the prolonged viremic state does not result in transmission of the virus to others. The disease is not otherwise contagious, and hospitalization or isolation of the patient is not usually needed.

The virus cannot be eradicated from its major endemic foci, but it might be suppressed or temporarily eliminated from small geographic areas that pose a special risk, such as campgrounds. Use of acaricides for environmental control would require consultation with state or federal officials regarding currently approved chemicals and methods for their application. Clearing of brush and grass from selected areas might help suppress rodent and tick populations. However, these methods are too difficult or expensive to be of much practical use.

Health education should emphasize the main features of the disease, its clinical and laboratory differentiation from Rocky Mountain spotted fever, the need for better recognition and reporting of cases, and how to prevent tick bites and properly removed attached ticks (Emmons, 1981). In addition to transmitting CTF and Rocky Mountain spotted fever, ticks in the endemic areas of CTF can transmit tularemia, Q fever, and the toxin that causes tick paralysis. Antibiotic prophylaxis simply for a tick bite is not indicated, but prompt recognition and specific antibiotic treatment of Rocky Mountain spotted fever, Q fever, and tularemia are important.

In addition to reporting of human cases, active surveillance by viral isolation from ticks or mammalian hosts and antibody testing of suspected host species would help to delineate more accurately the geographic distribution of CTF virus and its relatives. Much remains to be learned about this relatively recently discovered virus and disease.

Literature Cited

Black, W. C., L. Florio, and M. O. Stewart. 1947. A histologic study of the reaction in the hamster spleen produced by the virus of Colorado tick fever. Am. J. Pathol. 23:217–224.

Bowen, G. S. 1988. Colorado tick fever. 1988. In T. P. Monath (ed.), The epidemiology of arthropod-borne viral diseases. CRC Press, Inc., Boca Raton, Fla. In preparation.

Bowen, G. S., R. G. McLean, R. B. Shriner, D. B. Francy, K. S. Pokorny, J. M. Trimble, R. A. Bolin, A. M. Barnes, C. H. Calisher, and D. J. Muth. 1981. The ecology of Colorado tick fever in Rocky Mountain National Park in 1974. II. Infection in small mammals. Am. J. Trop. Med. Hyg. 30:490–496.

Calisher, C. H., J. D. Poland, S. B. Calisher, and L. A. Warmoth. 1985. Diagnosis of Colorado tick fever virus infection by enzyme immunoassays for immunoglobulin M and G antibodies. J. Clin. Microbiol. 22:84–88.

Carey, A. B., R. G. McLean, and G. O. Maupin. 1980. The structure of a Colorado tick fever ecosystem. Ecol. Monogr. 50:131–151.

Chastel, C., A. J. Main, G. Couatarmanac'h, G. LeLay, D. L. Knudson, M. C. Quillian, and J. C. Beaucournu. 1984. Isolation of Eyach virus (Reoviridae, Colorado

tick fever group) from Ixodes ricinus and I. ventalloi ticks in France. Arch. Virol. 82:161–171.

Dawson, D. L., T. M. Vernon, and an EIS Officer. 1972. Colorado tick fever. Colorado. Morbid. Mortal. Weekly Rep. 21:44.

Desmond, E. P., N. J. Schmidt, and E. H. Lennette. 1979. Immunoperoxidase staining for detection of Colorado tick fever virus, and a study of congenital infection in the mouse. Am. J. Trop. Med. Hyg. 28:729–732.

Drevets, C. C. 1957. Colorado tick fever. Observations on eighteen cases and review of the literature. J. Kansas Med. Soc. 58:448–455,462.

Earnest, M. P., J. C. Breckinridge, R. J. Barr, D. B. Francy, and C. S. Mollohan. 1971. Colorado tick fever. Clinical and epidemiologic features and evaluation of diagnostic methods. Rocky Mountain Med. J. 68:60–62.

Eklund, C. M., and R. C. Kennedy. 1961. Preliminary studies of pathogenesis of Colorado tick fever virus infection of mice, p. 286–293. In H. Libíková (ed.), Biology of viruses of the tick-borne encephalitis complex. Academic Press, Inc., New York.

Eklund, C. M., G. M. Kohls, W. L. Jellison, W. Burgdorfer, R. C. Kennedy, and L. Thomas. 1959. The clinical and ecological aspects of Colorado tick fever, p. 197–203. In Proceedings of the 6th International Congress on Tropical Medicine and Malaria, Lisbon, vol. 5. Instituto de Medicina Tropical. Vol. 16, Suppl. No. 9.

Emmons, R. W. 1966. Colorado tick fever: prolonged viremia in hibernating Citellus lateralis. Am. J. Trop. Med. Hyg. 15:428–433.

Emmons, R. W. 1967. Colorado tick fever along the Pacific slope of North America. Jap. J. Med. Sci. Biol. 20(suppl.):166–170. 1967.

Emmons, R. W. 1981. Colorado tick fever, p. 113–124. In J. H. Steele (ed.), CRC handbook series in zoonosis. Section B: viral zoonoses, vol. 1. CRC Press, Inc., Boca Raton, Fla.

Emmons, R. W., D. V. Dondero, V. Devlin, and E. H. Lennette. 1969. Serologic diagnosis of Colorado tick fever. A comparison of complement-fixation, immunofluorescence, and plaque-reduction methods. Am. J. Trop. Med. Hyg. 18:796–802.

Emmons, R. W., and E. H. Lennette. 1966. Immunofluorescent staining in the laboratory diagnosis of Colorado tick fever. J. Lab. Clin. Med. 68:923–929.

Emmons, R. W., L. S. Oshiro, H. N. Johnson, and E. H. Lennette. 1972. Intraerythrocytic location of Colorado tick fever virus. J. Gen. Virol. 17:185–195.

Emmons, R. W., and H. I. Schade. 1972. Colorado tick fever simulating acute myocardial infarction. J. Am. Med. Assoc. 222:87–88.

Florio, L., and M. S. Miller. 1948. Epidemiology of Colorado tick fever. Am. J. Public Health 38:211–213.

Florio, L., M. S. Miller, and E. R. Mugrage. 1950. Colorado tick fever. Isolation of the virus from Dermacentor andersoni in nature and a laboratory study of the transmission of the virus in the tick. J. Immunol. 64:257–263.

Florio, L., M. O. Stewart, and E. R. Mugrage. 1944. The experimental transmission of Colorado tick fever. J. Exp. Med. 80:165–187.

Gaidamovich, S. Y., G. A. Klisenko, and N. K. Shanoyan. 1974. New aspects of laboratory techniques for studies of Colorado tick fever. Am. J. Trop. Med. Hyg. 23:526–529.

Gerloff, R. K., and C. M. Eklund. 1959. A tissue culture neutralization test for Colorado tick fever antibody and use of the test for serologic surveys. J. Infect. Dis. 104:174–183.

Goodpasture, H. C., J. D. Poland, D. B. Francy, G. S.

Bowen, and K. A. Horn. 1978. Colorado tick fever: clinical, epidemiologic, and laboratory aspects of 228 cases in Colorado in 1973–1974. Ann. Intern. Med. **88**:303–310.

Gorman, B. M., J. Taylor, and P. J. Walker. 1983. Orbiviruses, p. 287–357. W. K. Joklik (ed.), The Reoviridae. Plenum Publishing Corp., New York.

Hadlow, W. J. 1957. Histopathologic changes in suckling mice infected with the virus of Colorado tick fever. J. Infect. Dis. **101**:158–167.

Hawkes, R. A. 1979. General principles underlying laboratory diagnosis of viral infections, p. 3–48. *In* E. H. Lennette and N. J. Schmidt (ed.), Diagnostic procedures for viral, rickettsial and chlamydial infections, 5th ed. American Public Health Association, Washington, D.C.

Hughes, L. E., E. A. Casper, and C. M. Clifford. 1974. Persistence of Colorado tick fever virus in red blood cells. Am. J. Trop. Med. Hyg. **23**:530–532.

Karabatsos, N. (ed.). 1985. International catalogue of arboviruses, 3rd. ed. American Society of Tropical Medicine and Hygiene, San Antonio, Tex.

Karabatsos, N., J. D. Poland, R. W. Emmons, J. H. Mathews, C. H. Calisher, and K. L. Wolff. 1987. Antigenic variants of Colorado tick fever virus. J. Gen. Virol. **68**:1463–1469.

Knudson, D. L. 1981. Genome of Colorado tick fever virus. Virology **112**:361–364.

Koprowski, H., and H. R. Cox. 1946. Adaptation of Colorado tick fever virus to mouse and developing chick embryo. Proc. Soc. Exp. Biol. Med. **62**:320–322.

Koprowski, H., H. R. Cox, M. S. Miller, and L. Florio. 1950. Response of man to egg adapted CTF virus. Proc. Soc. Exp. Biol. Med. **74**:126–131.

Lane, R. S., R. W. Emmons, V. Devlin, D. V. Dondero, and B. C. Nelson. 1982. Survey for evidence of Colorado tick fever virus outside of the known endemic area in California. Am. J. Trop. Med. Hyg. **31**:837–843.

Lennette, E. H., and R. W. Emmons. 1972. The laboratory diagnosis of rabies: review and prospective, p. 77–90. *In* Y. Nagano and F. M. Davenport (ed.), Rabies. Proceedings of Working Conference on Rabies, U.S.–Japan Cooperative Medical Science Program, Tokyo, October 12–14, 1970, 1971. University Park Press, Baltimore.

Lloyd, L. W. 1951. Colorado tick fever. Med. Clin. North Am. **2**:587–592.

Loge, R. V. 1985. Acute hepatitis associated with Colorado tick fever. West. J. Med. **142**:91–92.

McLean, R. G., D. B. Francy, G. S. Bowen, R. E. Bailey, C. H. Calisher, and A. M. Barnes. 1981. The ecology of Colorado tick fever in Rocky Mountain National Park in 1974. I. Objectives, study design, and summary of principal findings. Am. J. Trop. Med. Hyg. **30**:483–489.

Miller, J. K., V. N. Tompkins, and J. C. Sieracki. 1961. Pathology of Colorado tick fever in experimental animals. Arch. Pathol. **72**:149–157.

Murphy, F. A., E. C. Borden, R. E. Shope, and A. Harrison. 1971. Physicochemical and morphological relationships of some arthropod-borne viruses to bluetongue virus—a new taxonomic group. Electron microscopic studies. J. Gen. Virol. **13**:273–288.

Murphy, F. A., P. H. Coleman, A. K. Harrison, and G. W. Gary, Jr. 1968. Colorado tick fever virus: an electron microscopic study. Virology **35**:28–40.

Newhouse, V. F., J. A. McKiel, and W. Burgdorfer. 1964. California encephalitis, Colorado tick fever and Rocky Mountain spotted fever in Eastern Canada. Can. J. Public Health **55**:257–261.

Oliphant, J. W., and R. O. Tibbs. 1950. Colorado tick fever. Isolation of virus strains by inoculation of suckling mice. Public Health Rep. **65**:521–522.

Oshiro, L. S., D. V. Dondero, R. W. Emmons, and E. H. Lennette. 1978. The development of Colorado tick fever virus within cells of the haemopoietic system. J. Gen. Virol. **39**:73–79.

Oshiro, L. S., and R. W. Emmons. 1968. Electron microscopic observations of Colorado tick fever virus in BHK21 and KB cells. J. Gen. Virol. **3**:279–280.

Philip, R. N., E. A. Casper, J. Cory, and J. Whitlock. 1975. The potential for transmission of arboviruses by blood transfusion with particular reference to Colorado tick fever, p. 175–195. *In* T. J. Greenwalt and G. A. Jamieson (ed.), Transmissible disease and blood transfusion. Grune and Stratton, Inc., New York.

Randall, W. H., J. Simmons, E. A. Casper, and R. N. Philip. 1975. Transmission of Colorado tick fever virus by blood transfusion—Montana. Morbid. Mortal. Weekly. Rep. **24**:422.

Rehse-Küpper, B., J. Casals, V. Danielova, and R. Ackermann. 1979. Eyach virus: the first relative of Colorado tick fever virus isolated in Germany, p. 233–238. *In* J. G. Rodriguez (ed.), Recent advances in acarology, vol. II. Academic Press, Inc., New York.

Rehse-Küpper, B., J. Casals, E. Rehse, and R. Ackermann. 1976. Eyach—an arthropod-borne virus related to Colorado tick fever virus in the Federal Republic of Germany. Acta Virol. (Praha) **20**:339–342.

Sonenshine, D. E., C. E. Yunker, C. M. Clifford, G. M. Clark, and J. A. Rudbach. 1976. Contributions to the ecology of Colorado tick fever virus. 2. Population dynamics and host utilization of immature stages of the Rocky Mountain woodtick, *Dermacentor andersoni*. J. Med. Ent. **12**:651–656.

Spruance, S. L., and A. Bailey. 1973. Colorado tick fever. A review of 115 laboratory confirmed cases. Arch. Intern. Med. **131**:288–293.

Thomas, L. A., and C. M. Eklund. 1960. Use of the complement fixation test as a diagnostic aid in Colorado tick fever. J. Infect. Dis. **107**:235–240.

Thomas, L. A., C. M. Eklund, R. N. Philip, and M. Casey. 1963. Development of a vaccine against Colorado tick fever for use in man. Am. J. Trop. Med. Hyg. **12**:678–685.

Thomas, L. A., R. N. Philip, E. Patzer, and E. Casper. 1967. Long duration of neutralizing-antibody response after immunization of man with a formalinized Colorado tick fever vaccine. Am. J. Trop. Med. Hyg. **16**:60–62.

Topping, N. H., J. S. Cullyford, and G. E. Davis. 1940. Colorado tick fever. Public Health Rep. **55**:2224–2237.

CHAPTER 21

Reoviridae: The Rotaviruses

IAN H. HOLMES

Disease: Causes vomiting, diarrhea, and gastroenteritis in animals and man.

Biology of Rotaviruses: Generally a disease of newborn animals, the major symptom is acute diarrhea which may be accompanied by fever and vomiting. A wide variety of mammals and at least some birds are susceptible hosts, with transmission occurring by the fecal-oral route.

Properties of Virions: Virus particles are isometric, 65–75 mm in diameter, with a double-layered protein capsid. The genome consists of 11 segments of double-stranded RNA. The two surface proteins VP7 and VP3 are neutralization antigens, with VP7 determining the virus serotype. The important antigenic sites on VP7 have been located.

Cell Specificity and Cultivation: The virus is adapted to infect the highly differentiated epithelial cells of the small intestine. Growth in cell culture requires the use of susceptible cell lines and the addition of trypsin to activate virions.

Immunity: Protection against rotaviral diarrhea depends on the presence of antibodies in the lumen of the gut. While early subclinical infections can confer protection from disease later, the degree of cross protection against other serotypes is unclear. The role of VP7 and VP3 proteins is currently being investigated.

Diagnosis: The dynamics of virus excretion and antibody responses have been well studied, and the knowledge used to devise numerous diagnostic tests. Many of these are discussed in detail, including methods for detecting virus particles, antigens, or dsRNA, as well as measurement of secretory or serum antibodies. Current problems in subgrouping and cell culture adaptation are also presented and discussed.

Introduction

Although various kinds of bacteria had been linked with diarrheal disease both in humans and in domestic animals, no etiologic agent could be identified in many (often a majority) of cases until the last decade (Holmes, 1979). Viral causation was suspected, but, despite many comprehensive studies and countless unreported attempts, the conventional techniques of cell culture isolation and examination for cytopathology failed to close the gap (Yow et al., 1970). Ironically, pioneering studies on what we now know as rotaviruses affecting human infants, calves, and mice had already been published, but their significance went unrecognized.

Light and Hodes (1949) had produced diarrhea in newborn calves by oral inoculation of fecal filtrates from infants suffering from hospital outbreaks of gastroenteritis. Their results were confirmed, but the use of calves was too inconvenient for most medical researchers, and conventional laboratory animals did not appear to be susceptible. However, it was known that laboratory mouse colonies were sometimes affected by a highly contagious disease called epizootic diarrhea of infant mice (EDIM). This was difficult to study, because uninoculated controls frequently developed the disease when they were kept in rooms used for earlier experiments. Kraft (1958) realized that the agent must spread via dust through the air and established disease-free colonies and filter-top cages which made infectivity titrations possible, though laborious. She demonstrated that the EDIM agent was a virus and visualized it by electron microscopy in the cytoplasm of intestinal epithelial cells (Adams and Kraft, 1967). Although in retrospect her work proved highly relevant to later studies on epidemiology and maternal immunity in human rotaviral disease (Holmes, 1979), it was ahead of its time, and its grant support ran out.

Bovine rotaviruses were shown to be a cause of calf diarrhea (scours) originally in Nebraska by Mebus et al. (1969), who also developed the first rapid and practical diagnostic techniques for rotavirus infections: direct electron microscopy and immunofluorescence applied to fecal extracts. For the first time, rotavirus strains were adapted to growth in cell cultures, although it proved useful to use immunofluorescence to detect virus-infected cultures, since cytopathic effects were sometimes subtle (Fernelius et al., 1972).

Rotaviruses were first detected in humans in Australia in 1973 by electron microscopy of duodenal biopsies from infants with acute "nonbacterial" diarrhea and soon afterward in the United Kingdom, Australia, Canada, and the United States by negative-staining electron microscopy of fecal samples (Bishop et al., 1973, 1974; Flewett et al., 1973; Kapikian et al., 1974; Middleton et al., 1974). As it rapidly became evident that rotaviruses were the major "missing" agents of human infantile diarrhea and they were found to be common in both developing and developed countries, both the tempo and volume of rotavirus research increased to such an extent that it is no longer possible to survey the whole field, and earlier comprehensive reviews should be consulted for access to the older literature (Cukor and Blacklow 1984; Estes et al., 1983; Flewett and Woode, 1978; Holmes, 1979, 1983; McNulty, 1978).

The first electron microscopic studies of rotaviruses all suggested a resemblance in morphology and/or morphogenesis to reoviruses or orbiviruses, but Flewett et al. (1974) soon showed that although human and calf rotaviruses shared common antigens, they were serologically unrelated to reoviruses. Rotavirus particles are icosahedral, are 65 to 75 nm in diameter, and have two concentric capsids with an obvious subunit structure, so they are very easy to recognize by electron microscopy after negative staining. This was fortunate, because the early workers had to rely heavily on electron microscopy for diagnosis, and several years passed before techniques were developed for isolation of human strains in cell culture (Sato et al., 1981; Urasawa et al., 1981; Wyatt et al., 1980).

Thus from the beginning, direct (and relatively rapid) diagnostic techniques were developed for rotavirus detection, because conventional cell culture isolation methods initially failed. Necessity was indeed the mother of invention, and in addition to the established diagnostic methods, which will be described later, a quite remarkable range of ingenious tests have been described. The development of serological tests was facilitated by the facts that most rotaviruses share common (group) antigens and that some more readily cultivable strains were encountered at an early stage. Cultivation of the Nebraska bovine strains has already been mentioned; other bovine strains were isolated in the United Kingdom, and it was realized that the previously unclassified simian virus SA11 was also a rotavirus. The latter had been isolated from an apparently healthy vervet monkey during routine testing related to poliovirus vaccine production (Malherbe and Strickland-Cholmley, 1967) and is now one of the best-characterized rotaviruses.

In recent years both the magnitude of the global problem of acute diarrheal disease and the importance of rotaviruses as major contributors to morbidity and mortality in infants and young children have been established. This has led to considerable current interest in the development of rotavirus vaccines (De Zoysa and Feachem, 1985; Kapikian et al., 1980; WHO, 1980). This research activity has generated a considerable demand for rotavirus diagnosis, and as a result, improved methods and a number of commercially available diagnostic kits have become available. In this review I will attempt to explain the scientific basis and operation of the methods from first principles, but I will also refer to published comparisons with commercial systems where possible. For reasons that will emerge later, rotavirus serotyping is still difficult (and not yet commercial), but promising methods are being developed. Since serotyping has become so important in relation to vaccine development and evaluation, this area represents a major future challenge.

Biology of Rotaviruses

Diseases and Pathogenesis

Rotaviruses cause a remarkably similar type of disease in a wide range of hosts. The symptom is acute diarrhea with or without vomiting and is most common in infants or young animals. Susceptibility to symptomatic disease depends markedly on age, and the age range affected apparently varies with the physiology and development of the intestine and immune systems in different hosts. For example, symptomatic EDIM is restricted to suckling mice from 3 days to 3 weeks of age, while calf or piglet scours due to rotavirus occurs mainly in the first month of life, but human infants remain in danger of dehydration due to rotaviral diarrhea up to 2 years of age. For more details on animal rotavirus infections, see Woode (1982) or Holmes (1983), since we will concentrate here on human infections.

The clinical features of human rotaviral enteritis have been the subject of numerous reports (Davidson et al., 1975; Flewett, 1982; Rodriguez et al., 1977). The incubation period is brief (1 to 2 days), and onset is sudden with vomiting and then diarrhea, which usually lasts for 4 to 5 days. Mild fever (38 to 39°C) is common. In infants under 2 years of age, rotavirus infections tend to be more serious and more likely to cause moderate to severe dehydration than diarrheas caused by other agents (e.g., enterotoxigenic *Escherichia coli*) (Black et al., 1981). Dehydration can generally be corrected by oral rehydration, but it is sometimes so rapid that deaths occur even in areas where parenteral fluid therapy is available (Carlson et al., 1978). Clinically, it is difficult to distinguish rotaviral from bacterial enteritis.

Most studies of pathogenesis of rotaviral disease have been carried out in calves, piglets, or mice, but the limited human studies support a common interpretation. Infection (detected by immunofluorescence) appears to be limited to intestinal epithelial cells and progresses downward from the upper part of the small intestine (McNulty, 1978; Mebus and Newman, 1977). Infected cells are vacuolated and prematurely shed, so the upper parts of villi are denuded and the villi shorten and finish up covered by immature cuboidal epithelial cells. Although there have been a number of reports suggesting an association of respiratory symptoms with rotavirus infections, rotaviral infection of the upper respiratory tract or lung has never been detected in animals despite careful searches. One report of rotavirus antigen in human tracheal washings from cases of pneumonia has appeared (Santosham et al., 1983). Whether rotaviruses occasionally cause infections of the central nervous system (Salmi et al., 1978) remains to be established.

Host Range

It seems very probable that rotaviruses infect all kinds of mammals and certainly at least some species of birds. Most of the more thoroughly investigated species such as humans, calves, pigs, mice, and horses have even been shown to be hosts to more than one serotype, so rotaviruses are remarkably ubiquitous (Estes et al., 1983; Holmes, 1983). Antibodies directed against the common group A rotavirus antigen are extremely prevalent in animal sera, and this must be kept in mind when one is choosing, for example, antiimmunoglobulin sera for quantitation of antirotaviral antibodies in other sera.

Rotaviruses appear to have a quantitative rather than absolute host specificity, so experimentally "human" strains (by definition, rotaviruses isolated from naturally infected humans) have been shown to be capable of infecting piglets, monkeys, calves, lambs, and dogs (Estes et al., 1983; Holmes, 1983). Mice were once thought to be susceptible only to their own murine strains, but diarrhea can be produced in sucklings by large doses of simian or human or even larger doses of bovine rotaviruses (Gouvea et al., 1986; Offit et al., 1986a,b).

It is suspected but not known whether natural cross-species infections occur or in particular whether some human infections are zoonoses. The occurrence of certain common serotypes (see below) in humans, pigs, calves, and dogs certainly supports this possibility, but only a few studies designed to test it have been reported (Garbarg-Chenon et al., 1986)

Epidemiology

Rotavirus infectivity is quite stable to drying and survives particularly well in feces (Flewett, 1982). Since the particles are often excreted in high concentrations, infection doses can readily be acquired by contact with contaminated hands, utensils, or furniture (Holmes, 1983). The primary mechanism of spread of rotaviruses in the community is probably by fecal-oral contamination within families, but nosocomial infections in pediatric wards also appear to be common, and high rates of cross-infection have been reported in preschools and orphanages (Bishop, 1986; Davidson et al., 1975).

During epidemics, rotaviruses have been observed to spread more rapidly than could be accounted for by direct contacts. Respiratory transmission was proposed (Foster et al., 1980), but as was mentioned above, actual upper respiratory tract infection has not been found. It seems more likely that, as in the case of mouse rotavirus infection, dust containing dried feces is suspended in the air and inhaled, then trapped in mucus and swallowed to infect the intestine (Flewett, 1982).

In general, symptomatic rotavirus infections are most common in infants, but is is believed that repeated, often asymptomatic infections occur at intervals throughout life (Holmes, 1983). Although symptoms in normal neonates are often relatively mild, it now appears that rotaviruses are the major etiologic agents of serious diarrheal illness in infants and children under 2 years of age throughout the world (Kapikian and Chanock, 1985). In developing countries, severe infections occur at a slightly younger age (6 to 12 months) than is common in developed countries (12 to 18 months), perhaps indicating more frequent or heavier exposure to rotavirus (Bishop, 1986).

Symptomatic adult rotavirus infections are not uncommon, just underreported. They occur in family members or nursing or medical staff looking after diarrheic infants, in travelers, and even in geriatric patients (Echeverria et al., 1981, 1983; Halvorsrud and Orstavik, 1980; Kim et al., 1977; von Bonsdorff et al., 1976; Wenman et al., 1979). In isolated populations, or as a result of water-borne contamination, adults have also been involved in epidemics (Foster et al., 1980; Hung et al., 1984; Linhares et al., 1981; Lycke et al., 1978).

Properties of Virions

Morphology

Rotaviruses are isometric and 65 to 75 nm in diameter, depending on whether they are incomplete (inner capsid only) or intact, with the outer capsid in place. Both single-shelled (inner capsid) and double-shelled (intact) particles are normally seen by electron microscopy in both fecal and cell culture derived samples (Fig. 1). Although the particles superficially resemble reoviruses, the intact rotaviruses have a sharply defined rim which distinguishes them (Flewett et al., 1974; Holmes et al., 1975). The subunit structure of the inner capsid is shown clearly by negative staining, but it proved hard to define until Roseto et al. (1979) showed by shadowing freeze-dried particles that the structure is icosahedral with a triangulation number T = 13, which gives rise to a skewed structure probably composed of 260 trimeric subunits (Holmes, 1983).

During morphogenesis, rotavirus single-shelled particles bud into the rough endoplasmic reticulum and acquire a temporary pseudoenvelope, but this is lost during maturation, so that infectious virions are not enveloped (Holmes, 1983).

Chemical Composition

Rotavirus particles consist of a genome of 11 segments of double-stranded RNA surrounded by two protein capsids. Complete (double-shelled) rotavirus particles can be separated from single-shelled particles by density gradient centrifugation in cesium chloride (Newman et al., 1975; Rodger et al., 1975). The outer capsid layer is necessary for infectivity (Bridger and Woode, 1976) and is dissociated from the inner capsid when the surrounding calcium concentration is less than about 1 mM. This activates transcription of mRNA and is normally a stage in the process of infection. In vitro uncoating is produced by exposure of the particles to chelating agents such

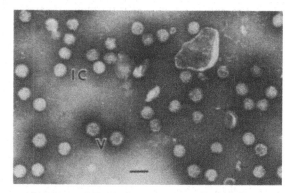

FIG. 1. Single- and double-capsid rotavirus particles and some debris from a human fecal sample negatively stained with ammonium molybdate. The "rough" outline of the inner capsid (IC) and smooth outline of the intact virions (V) are characteristic of rotaviruses and distinguish them morphologically from other double-stranded RNA viruses. Bar represents 100 nm.

as ethylenediaminetetraacetic acid (EDTA) and results in loss of infectivity (Cohen, 1977) but maximum exposure of the group antigen (see below).

Rotavirus infectivity is stable in the pH range 3 to 9 and is resistant to ether, chloroform, and deoxycholate, but 0.1% sodium dodecyl sulfate and alcohols destroy infectivity. For more information on physicochemical properties, see Holmes (1983).

The 11 segments of double-stranded RNA that comprise the genome represent separate genes and range in size from about 3,000 to about 700 base pairs (Holmes, 1983). By polyacrylamide gel electrophoresis (PAGE), most or all of the 11 segments can be resolved: the bands form a characteristic pattern consisting of four large segments, then two, then three running close together, and then the smallest two. Superimposed on this general pattern are minor variations in the mobility of each of the segments, so it is rare to find different strains of rotavirus with identical patterns (Kalica et al., 1978; Rodger and Holmes, 1979). The mobility differences apparently reflect differences in base sequence of genes, so electrophoresis of genome RNA ("electropherotyping") has become a powerful method for distinguishing rotavirus strains and will be discussed in detail in the Diagnosis section.

Technical problems in the early days and genuine variations in apparent molecular weights generated many minor disagreements in the literature on rotavirus proteins, but most of these are now resolved (Estes et al., 1983). Biochemical, immunological, genetic, and molecular approaches have resulted in general agreement on the rotavirus gene coding assignments or genome mapping. These have been reviewed recently by Cukor and Blacklow (1984) and so will not be discussed in detail here, but Fig. 2 will serve to summarize the findings and the nomenclature of RNA segments and proteins. It is based on the work of Kantharidis et al. (1983) and Mason et al. (1983). It will be noted that both RNA segments and proteins are numbered from largest to smallest and that virion structural proteins are prefixed VP and nonstructural proteins are designated NS.

Groups and Subgroups

The rotaviral proteins of most antigenic importance are VP6, VP3, and VP7. VP6 is the major component of the inner capsid and represents 60% of the total protein in intact virions (Novo and Esparza, 1981), and it carries the common (group A) antigen shared by most rotaviruses. This was demonstrated by immunoprecipitation with group-reactive monoclonal antibodies, and in fact the majority of monoclonal antibodies obtained following immunization of mice

with rotaviruses are directed against VP6 (Greenberg et al., 1983a; Sonza et al., 1983).

Some monoclonal antibodies reacting with VP6 show greater specificity and can be used to subdivide group A rotaviruses into subgroups I and II (Greenberg et al., 1983b). These subgroups were originally distinguished using cross-absorbed conventional antisera by complement fixation (CF), immune adherence haemagglutination (IAHA), and enzyme immunoassays (ELISA) and thought to represent serotype differences (Kapikian et al., 1981; Yolken et al., 1978c; Zissis and Lambert, 1978). The situation was clarified by genetic and antigenic analysis of a number of reassortants between a human and a bovine strain that differed in both serotype and subgroup (Kalica et al., 1981). Subgroup specificity segregated with RNA segment 6, which was already known to encode VP6 (Smith et al., 1980). Not all group A rotaviruses can be subgrouped with the available antibodies; for example, murine rotavirus (EDIM) and avian isolates are not classifiable.

Since 1981 reports have gradually accumulated of rotaviruses that are morphologically typical but generally uncultivable in cell culture and that have group antigens different from the predominant group A rotaviruses. Such viruses have been found in pigs, humans, calves, sheep, and chickens. Despite the technical difficulties, some progress has been made, and groups B, C, and D have been characterized by serology (immunofluorescence) and their "atypical" RNA gel patterns (Pedley et al., 1983; Snodgrass et al., 1984). These will be discussed further in the section on electropherotyping, which is at present the only generally available technique capable of detecting such strains in humans.

Serotypes

Rotavirus serotypes are defined by neutralization tests and designated by Arabic numerals (WHO, 1984). The first four numbers were assigned to the first distinct serotypes found in rotaviruses of human origin (Beards and Flewett, 1984; Beards et al., 1980; Wyatt et al., 1983), but rotaviruses of animal origin fit in to the same scheme, some having distinct serotypes (only 2 of which have so far been assigned numbers) and others having the same serotype as human strains (Hoshino et al., 1984, 1985b). A fifth human serotype (Matsuno et al., 1985) also awaits a serotype number. Table 1 lists some prototype or representative cultivable strains of the defined serotypes.

The serotype was originally thought to be dependent only on VP7, the major outer capsid glycoprotein (Bastardo et al., 1981; Killen and Dimmock, 1982), but studies of neutralizing monoclonal anti-

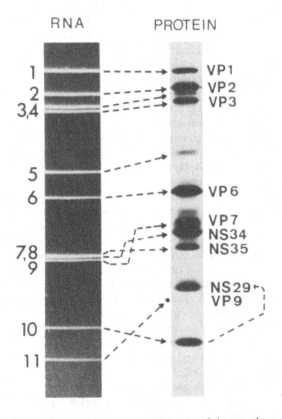

FIG. 2. RNA-protein coding assignments of the rotavirus SA11. Both RNA segments and [^{35}S]methionine-labeled viral proteins are shown as they appear after SDS-PAGE in 10% acrylamide gels using the Laemmli (1970) discontinuous buffer system. Direction of migration is from top to bottom.

TABLE 1. Some cultivable rotavirus strains of defined serotype

Serotype[a]	Strain	Origin	Reference
1	Wa	Human	Wyatt et al. (1983a)
	Ku	Human	Urasawa et al. (1982)
	RV-4	Human	Albert and Bishop (1984)
2	DS-1	Human	Wyatt et al. (1983a)
	S2	Human	Urasawa et al. (1982)
	KUN	Human	Kutsuzawa et al. (1982)
	RV-5	Human	Dyall-Smith and Holmes (1984)
3	YO	Human	Urasawa et al. (1982)
	Ito	Human	Sato et al. (1981)
	MO	Human	Kutsuzawa et al. (1982)
	RV-3	Human (neonate)	Albert and Bishop (1984)
	SA11	Vervet monkey	Malherbe and Strickland-Cholmley (1967)
	RRV (MMU 18006)	Rhesus monkey	Stuker et al. (1980)
4	ST-4	Human (neonate)	Hoshino et al. (1984)
	VA-70	Human	Gerna et al. (1985)
	Gottfried	Porcine	Hoshino et al. (1984)
5	OSU	Porcine	Theil et al. (1977)
6	NCDV	Bovine	Mebus et al. (1971a)
	UKBo	Bovine	Bridger and Woode (1975)
7	Ch2	Avian (chicken)	McNulty et al. (1980)
Unassigned	69M	Human	Matsuno et al. (1985)
	EDIM	Murine	Kraft (1958)

[a] As designated by Hoshino et al. (1984) and WHO (1984).

bodies have shown that although VP7 carries the major neutralization determinants, antibodies specific for VP3 (the minor outer capsid protein) also neutralize rotaviruses (Greenberg et al., 1983c; Sonza et al., 1983; Taniguchi et al., 1985). Since two serotypes are currently defined as distinct if their representative viruses show a 20-fold difference between homologous and heterologous titers in two-way neutralization tests with polyclonal antisera and antibodies against VP3 can contribute 10 to 25% of such titers, it follows that whereas in most rotaviruses that have been successfully serotyped, particular serotypic versions of VP7 must generally be associated with distinct serotypic versions of VP3, an "unusual" combination that could arise by reassortment would result in a virus that appeared to belong to two serotypes.

Hoshino et al. (1985b) recently analyzed the serotype antigens of such a strain (M37) isolated from human neonates by making reassortants with a third (bovine) rotavirus and showed that VP7 of M37 was like that of a prototype type 1 strain (Wa), but its VP3 resembled that of a human type 4 strain (ST3). Thus they pointed out that it would be necessary in future to define serotypes of VP3 and VP7 separately. Although this is not yet possible in practice, serotype-specific (neutralizing) monoclonal antibodies directed against VP7 are available, and some VP7 serotyping by ELISA has already been achieved (Coulson et al., 1985, 1986; Shaw et al., 1985; Thouless et al., 1982), but type-specific anti-VP3 monoclonal antibodies are quite rare, and only a small amount of VP3 antigen is present on virus particles, so VP3 serotyping still looks difficult.

One-way cross-reactions in neutralization tests that have been observed by Hoshino et al. (1984) and other workers cannot be adequately explained by sharing of one or other of VP3 and VP7 between the strains concerned: they seem more likely to be due to genetic drift and antibody selection as with influenza virus (Hamre et al., 1958; Knowlton and Ward, 1985), but proof of this will have to await further sequencing studies.

Genome Reassortment

Other aspects of rotavirus genetics have been mentioned above or are reviewed elsewhere (Holmes, 1983; Kapikian and Chanock, 1985), but reassortment deserves special mention because of its utility in rotavirus research and diagnosis. When cells are coinfected with two different rotaviruses, both parental types and reassortant clones can be isolated by plaque assay of the progeny, provided that both strains grow well in cell culture. Such reassortment of genome segments occurs at high frequency, both in vitro and in vivo, and seems to be a general property of viruses with segmented genomes (Gombold and Ramig, 1986). In the work just cited, mice were simultaneously infected with the simian rotaviruses SA11 and rhesus rotavirus (RRV), and no selection other than that imposed by growth in a "foreign" host was applied. Of the theoretically possible 2,048 (2^{11}) genotypes, 122 were found in the 252 reassortants among a total of 662 clones analyzed. In this case the parental origin of each RNA segment could be distinguished by electrophoretic mobility. In general, segregation of individual genome segments among the reassortants was random, although segments 3 and 5 of SA11 appeared frequently to cosegregate.

Since dual infection with different strains is not uncommon in animals or even in humans (Nicolas et al., 1984; Spencer et al., 1983), it is very likely that reassortment contributes to the variety of electropherotypes observed in epidemiological surveys (Estes et al., 1984a; Rodger et al., 1981).

In vitro, selection can be applied to produce practically useful reassortants. In cell culture, bovine rotaviruses or the simian SA11 and rhesus rotaviruses grow and plaque much better than human rotaviruses. Greenberg et al. (1981, 1982) were able to "rescue" what were at the time uncultivable human strains by coinfecting cells with human rotavirus samples and cultivable UK bovine rotavirus and selecting reassortants from plaques produced by the progeny in the presence of antiserum against bovine rotavirus, Lincoln Nebraska strain (NCDV). In retrospect, there was some luck involved, because cell culture cultivability was found to depend on VP3 of the bovine virus and thus the desired reassortants possessed VP3 from UK bovine rotavirus and VP7 from a human strain (Greenberg et al., 1983a). The selection would not have succeeded if antiserum prepared against UK bovine rotavirus had been used, but, although the VP7 specificity of UK and NCDV bovine strains is identical, they apparently differ in VP3 serotype, and the NCDV antiserum happened to be the only one available (Greenberg et al. 1983c).

More generally, readily cultivable reassortant rotaviruses possessing VP7 proteins of human serotype and VP3 derived from readily cultivable strains such as SA11, RRV, or bovine rotaviruses can now be isolated by selection with neutralizing monoclonal antibodies directed against VP7 of the simian or bovine strains (Greenberg et al., 1983c; Midthun et al., 1985). Such reassortants are very useful for production of serotyping reagents (and potentially as vaccine candidates).

Rotavirus Gene Sequences

Since 1982, most of the RNA segments of a number of rotaviruses have been copied into DNA, cloned, and sequenced. Apart from providing a wealth of new information about the rotavirus proteins, this activity has provided the basis for development of hybridization probes for diagnosis and serotyping. Table 2 is designed to provide a source of basic references for readers interested in this aspect. It is not a complete résumé but does include all available information on the serotypically variable genes 4 and 9. There is a very high degree of base sequence homology (almost 90%) between viruses in the genes encoding nonstructural proteins (segments 7, 8, and 10 in Table 2) and segment 6, so these appear very suitable for the derivation of generic probes for detection of group A rotaviruses, and work along these lines is proceeding (A. R. Bellamy, personal communication).

Even the RNA segments encoding serotypically variable proteins such as VP7 show considerable homology (about 75%), but when the sequences are analyzed in detail, it becomes obvious that within the sequences there are a number of variable regions interspersed with almost constant ones (Dyall-Smith and Holmes, 1984; Richardson et al., 1984). These variable regions offer the possibility of producing serotype-specific probes, based on synthetic oligonucleotides (A. Haralambous, personal communication).

Location of Antigenic Determinants

Although the sequence data for VP7 cited above indicate a high degree of sequence homology, it is remarkable that homology at the level of predicted amino acid sequences is even greater, but still there are regions within which amino acid sequence varies with serotype, and these could represent antigenic determinants. To localize these more definitely, a strategy devised by Caton et al. (1982) was employed. Using a series of serotype 3–specific neutralizing monoclonal antibodies derived by S. Sonza, A. Breschkin, and B. Coulson, antigenic variants of SA11 rotavirus were isolated, their VP7-coding segments were sequenced, and amino acid changes were located (Dyall-Smith et al., 1986; Lazdins et al., 1985). The amino acid changes in the antibody-resistant variants occurred in three regions—near amino acids 87 to 96 (A region), 145 to 150 (B), and 211 to 223 (C). All these regions show considerable variation between serotypes and thus appear likely to be associated with antigenic determinants (Dyall-Smith et al., 1986). It is notable that variants with mutations near the C region are even markedly resistant to neu-

TABLE 2. Rotavirus gene sequences

Segment[a]	Protein	Virus	Base pairs	Amino acids	Reference
4	VP3	Simian SA11	2,244[b]	747[b]	Lopez et al. (1985)
		Human RV-5	2,359	775	Kantharidis et al. (1986)
6	VP6	Simian SA11	1,356	397	Estes et al. (1984b)
		Human Wa	1,356	397	Both et al. (1984)
7	NS34	Bovine UK[c]	1,076	310	Ward et al. (1984)
8	NS35	Simian SA11	1,059	317	Both et al. (1982)
		Bovine UK[d]	1,059	317	Dyall-Smith et al. (1983)
9	VP7	Simian SA11	1,062	326[e]	Both et al. (1983a)
		Human Wa	1,062	326[e]	Richardson et al. (1984)
		Human RV-5	1,062	326[e]	Dyall-Smith and Holmes (1984)
10	NS29	Simian SA11	751	175[f]	Both et al. (1983b)
		Human Wa	750	175[f]	Okada et al. (1984)
11	VP9	Bovine UK	667	198	Ward et al. (1985)
		Human Wa	663	197	Imai et al. (1983)

[a] Numbered as in SA11 (Fig. 2).
[b] Incomplete.
[c] S7 of SA11 corresponds to S9 of UKBo.
[d] S8 of SA11 corresponds to S7 of UKBo.
[e] Glycosylated at one or two sites.
[f] Glycosylated in signal sequence.

tralization by hyperimmune antisera. When more sequences and monoclonal antibodies become available, it should be possible to similarly localize antigenically variable regions of VP3 and thus select nucleotide sequences for hybridization "serotyping" of VP3.

Cell Specificity and Cultivation

Since nearly 20 years of attempts to isolate these very common viruses in cell culture resulted in continual failures, it is instructive to analyze the problems retrospectively. These were twofold: first, restricted occurrence of cellular receptors, and second, a requirement for trypsin activation of infectivity of the virus.

In vivo, rotaviruses appear to infect only the differentiated (mature) columnar epithelial cells covering the villi of the small intestine. This was recognized by electron microscopy and immunofluorescent staining of infected cells in experimentally infected mice and calves and later in humans (Adams and Kraft, 1967; Davidson et al., 1975b; Mebus et al., 1971b; Wilsnack et al., 1969). Undifferentiated, replicating enterocytes in the crypts and lower portions of the villi do not appear to be susceptible. Susceptibility appears to parallel the appearance of brush border receptors, and recent studies suggest that the activity of virus receptors depends on the configuration of carbohydrate moieties on disac-

charidases such as lactase, maltase, and sucrase (Holmes et al., 1976; G. Raghu, personal communication).

Fortuitously, primary kidney cell cultures and some cell lines derived from monkey, bovine, or canine kidney have receptors that function, if not very efficiently, to allow rotavirus adsorption. Even so, it is advisable to rely on immunofluorescent staining rather than cytopathic effects to monitor the degree of rotavirus replication (Mebus et al., 1971a). SA11 and bovine rotavirus strains were adapted to serial passage in such cell cultures, but for improved growth and adaptation of porcine and human rotaviruses, it was finally realized that trypsin was required to activate virus infectivity (Babiuk et al., 1977; Matsuno et al., 1977; Sato et al., 1981; Theil et al., 1977; Urasawa et al., 1981). In the normal habitat of rotaviruses, of course, trypsin is always present. When it is added to cell culture media, serum must be omitted, because sera contain high levels of trypsin inhibitors (McLean and Holmes, 1981).

The essential action of trypsin is specific cleavage of the outer capsid protein VP3 (Espejo et al., 1981; Estes et al., 1981; Lopez et al., 1985). This cleavage activates infectivity, probably by facilitating outer capsid uncoating inside the cell (Clark et al., 1981). Although rotaviruses such as SA11 and most bovine strains will give quite good yields in cells such as the MA104 monkey kidney line without added trypsin (some trypsinlike protease must be associated with the cells), titers are increased in the presence of added trypsin, and the latter is essential in plaque assays (Estes et al., 1983).

Cell culture isolation of human rotaviruses has rarely been achieved without trypsin. Generally samples are activated by treatment with 5 to 10 μg/ml of crystalline trypsin, then infected cultures are maintained in 0.5 μg/ml. Even so, several passages may be required before titers build up, and rolling of cultures may be beneficial (Sato et al., 1981; Urasawa et al., 1981; Wyatt et al., 1983a).

Immunity

The literature on immunity to rotaviral infections and disease, especially in relation to humans, is rather confusing, and many discrepancies remain to be resolved by further research. A number of reasons for this can be suggested. Much earlier work was done before the serotypic variety of rotaviruses was appreciated, and even more where serotyping was not technically feasible, or at any rate was not attempted. Then there seems to be a common preference (reinforced by some overemphasized methodological problems) for studies on sera rather than intestinal secretions or coproantibodies.

The ubiquity of rotaviruses and frequency of asymptomatic infections in young animals (and especially in human infants) is such that it is very difficult to ascertain whether an infection and/or immune response is primary, secondary, or subsequent. Thus neutralizing antibody responses in humans or older animals have often appeared to be "heterospecific," whereas both responses and protection have proved much more predictably serotype-specific in experiments on gnotobiotic or isolated laboratory animals (Gaul et al., 1982; Snodgrass et al., 1984b).

Finally, there is a quantitative rather than an all-or-none element in the outcome of intestinal encounters with rotaviruses: both the dose of virus ingested and the level of intraluminal neutralizing antibodies can vary over a wide range. In the absence of antibodies, replication of even a small infectious dose will soon affect a sufficient area of intestinal epithelium to cause obvious diarrhea. In the presence of moderate levels of neutralizing antibodies, a moderate virus dose may cause a limited infection and excretion but no diarrhea (Little and Shadduck, 1982; Snodgrass et al., 1977). Less commonly, very high doses of virus may overcome immune protection, or conversely unusually high titers of specific antibody may completely block infection (Lecce et al., 1978; McLean and Holmes, 1981).

Animal studies have clearly shown that protection against rotaviral diarrhea depends on the presence of antibodies in the lumen of the gut rather than in the circulation (Offit and Clark, 1985; Snodgrass and Wells, 1976; Woode et al., 1975). Calves or lambs fed

colostrum on day 1 after birth, and thus having moderate levels of humoral antibodies, developed scours when challenged with their respective rotaviruses in the absence of colostrum, 1 or a few days later. However, when colostrum or immune serum was fed shortly before and after the virus challenge, symptoms were suppressed. It remains possible that very high levels of serum antibodies may be protective because of exudation into the gut (Snodgrass and Wells, 1978).

In humans, of course, serum immunoglobulin G (IgG) is transferred transplacentally from mother to infant, and lacteal antibody is predominantly IgA (Cukor et al., 1979; McLean and Holmes, 1980; Yolken et al., 1978a). Varying levels of maternal antibodies and variations in feeding regimes make it difficult to analyze the effect of breast feeding on rotavirus susceptibility, and to complicate interpretations still further, it has been found that trypsin inhibitors in breast milk, whose levels vary independently of antibody levels, can also protect neonates against rotavirus infection (McLean and Holmes, 1981). Even artificially fed neonates appear to have some resistance to rotavirus disease and frequently have subclinical infections (Bishop, 1986; Kapikian and Chanock, 1985), so there have been some disagreements about the importance of breast feeding for reducing rotavirus infections. The topic is reviewed in more detail by Kapikian and Chanock (1985). In general, in humans as in animals, serum antibodies per se do not appear to confer resistance to infection or illness due to rotaviruses.

Both in animals and in humans, there is evidence that early subclinical infections can confer protection against disease later (Bishop et al., 1983; Bridger and Brown, 1981; Snodgrass and Wells, 1976). In the former prospective study, 81 babies were studied from birth up to the age of 3 years. During the first 14 days of life, 44 excreted rotavirus and 37 did not. Subsequently, each group showed a similar frequency of reinfection (54 to 55%), but symptoms were significantly less frequent and less severe in the group who had experienced a neonatal infection than in those who had not. Note that protection was against disease but did not prevent reinfection. In other circumstances, symptomatic reinfections have been recorded, but generally the serotypes of rotavirus involved are not yet known. On the whole, the frequency of reinfections detected is correlated with the frequency of sampling (Gurwith et al., 1981; Mata et al., 1983). Current studies suggest that intensive monitoring can reveal that multiple infections during the first 2 years of life are as typical in Melbourne as in a Guatemalan village (Grimwood, 1986).

Snodgrass et al. (1984) carried out a very important study on the serotype specificity of immune responses to rotavirus immunization. Both in rabbits and in adult cows, they showed that following sequential infections or immunizations, neutralizing antibodies against every serotype to which the animal had previously been exposed were produced, but the response did not broaden to include serotypes that had not been encountered. It seems quite possible that the initial apparent success of a bovine rotavirus vaccine for protection of infants against serotypically distinct human rotavirus infections (Vesikari et al., 1984) may have been due to boosting of antibody responses to serotypes previously experienced in asymptomatic and unsuspected infections.

Nevertheless I should not leave the impression that all is understood in this area. On the one hand, it has been shown that even minor serotypic variation in bovine rotavirus strains can result in failure of cross-protection (Woode et al., 1983), but, on the other hand, well-authenticated examples of heterologous protection have been documented by equally experienced researchers (Wyatt et al., 1979). The explanation probably depends on the fact that the serotypic specificity of rotavirus neutralizing immune responses is not absolute, for reasons that are still obscure. In the study just cited, calves were immunized in utero with bovine rotavirus NCDV strain (serotype 6) and were protected from disease due to postnatal challenge with a human serotype 1 rotavirus. It was subsequently shown that the immunized calves had produced homologous neutralizing antibodies to NCDV but also 5- to 10-fold lower, but still significant, titers of neutralizing antibodies to human serotype 1 virus in their sera (Wyatt et al., 1983b). Similarly, mouse dams hyperimmunized with various human or bovine serotypes of rotavirus produced sufficient levels of heterologous neutralizing antibodies in their sera and (when measured) milk to protect their litters against disease following challenge with SA11 (serotype 3) rotavirus (Offit and Clark, 1985b). Although the immune responses of the dams would be rated as highly type-specific, with a ratio of homologous/heterologous neutralizing antibody titer of about 1,000, heterologous neutralizing titers of the order of 1 to 200 against SA11 virus were apparently protective. It is not yet possible to determine whether such cross-protection depends mainly on antibodies directed against VP7 or VP3 or both.

Diagnosis

Dynamics of Virus Excretion and Antibody Responses

In patients with normal immune systems rotavirus infections are almost always acute, and high titers of particles (or antigen) are excreted for a limited period. In neonates, rotavirus excretion has been detected as early as the first day after birth, indicating a short incubation period (Murphy et al., 1977). In another study of neonates, virus shedding was most often detected around days 4 to 7. Excretion preceded diarrhea by 1 to 3 days and persisted for a few days after (Cameron et al., 1978).

In somewhat older children, maximal virus shedding is detected 3 to 5 days after onset of gastrointestinal symptoms, with lower titers up to about 8 days (Davidson et al., 1975; Konno et al., 1977; Vesikari et al., 1981). More protracted excretion does occur sometimes, but for maximum probability of rotavirus antigen detection, fecal samples need to be obtained within a week after onset of symptoms, and the general rule is the earlier the better. Prolonged rotavirus excretion, accompanying chronic diarrhea, has been reported in children with various primary immune deficiencies (Saulsbury et al., 1980).

For reasons of simplicity and sensitivity, most studies of immune responses to rotavirus infections have in fact measured antibodies directed against the common (group A) antigen, but neutralizing antibodies have more recently been studied also. Since there is no reason to suppose that the two kinds of response do not occur simultaneously, no distinction will be made in this section. Unless otherwise stated, it can be assumed that acute-phase samples have been collected within 1 week of detectable infection and convalescent-phase samples 2 to 6 weeks after infection.

Since it is unusual (if not impossible) to find rotavirus-seronegative adults even in the most isolated populations (Linhares et al., 1981), human infants are born with transplacentally acquired serum antirotavirus IgG levels reflecting (in fact, slightly higher than) those of their mothers (McLean and Holmes, 1980). IgA class antibodies, though present in maternal sera, are not transferred. Possibly because of this background level, it is generally not possible to detect an IgG seroconversion in infants infected during their first 3 months of age, and such infants do not develop persistent rotavirus serum antibodies (Bishop, 1986; Bryden et al., 1982; Grimwood, 1986). An antirotaviral IgM response is detectable in serum and feces within 5 days of the infection, and IgA coproantibody is found from day 4 (or earlier) postinfection but is transient (Grimwood, 1986).

In infants over 3 months of age experiencing a rotavirus infection, IgG seroconversion is reliably detectable whether or not the infection is symptomatic (Bishop et al., 1983; Davidson et al., 1983; Grimwood, 1986; Yolken et al., 1978a), and rotavirus IgG levels in serum persist for at least a year. Rotavirus IgM in serum is detectable by 5 days and peaks 1 to 2 weeks after primary infection but is generally undetectable at 4 months postinfection (Davidson et al., 1983; Grimwood, 1986; Orstavik and Haug, 1976; Yolken et al., 1978b). It is probably not produced after subsequent reinfections (Grim-

wood, 1986; Orstavik et al., 1976). An IgA seroresponse is detectable in most cases and persists for 4
but usually < 12 months (Davidson et al., 1983;
Grauballe et al., 1981; Grimwood, 1986).

Antirotaviral IgM and IgA are detectable in duodenal fluids and feces. In duodenal secretions, IgM is
found in the acute phase, whereas IgA peaks at 3 to 4
weeks postinfection, and remains detectable for
about 4 months (Grimwood, 1986). The coproantibody response is similar but of shorter duration. IgA
is detectable in feces from about day 4 postinfection,
peaks about day 10, and is usually undetectable after
6 weeks (Grimwood, 1986; Simhon et al., 1985;
Sonza and Holmes, 1980). However, persistence of
IgA coproantibody as observed by Grauballe et al.
(1981a) is found in infants who have experienced a
series of either symptomatic or subclinical rotavirus
infections (Grimwood, 1986).

Specimen Collection and Storage

If possible, samples of at least 1 g (or 1 ml) of feces
should be obtained in screw-capped containers. In
the case of diarrheic infants, the liquid phase
squeezed from napkins or absorbent paper is satisfactory. The amount of sample on fecal swabs is generally insufficient for optimal virus detection.

Samples should be chilled to 4°C as soon as possible. Storage or transport conditions depend on the
tests to be carried out. For electron microscopy
(EM), RNA, or antigen detection, freezing is ideal,
but storage at 4°C and transport on wet ice is satisfactory. If virus isolation or serotyping is to be undertaken, samples should be frozen at −70°C (or at
least below −30°C) and transported on dry ice. Domestic freezers (−10 to −20°C) are not satisfactory
for this purpose: rotavirus infectivity is more stable
in feces at 4°C than at −20°C, especially if samples
are stored in well-filled, airtight containers to inhibit
mould growth (G. N. Woode, personal communication). Detailed suggestions for packaging and transport are supplied by Madeley (1977).

For coproantibody detections, stool specimens
should preferably be frozen, but −20°C would be satisfactory in this case. Sera for antibody testing can
be kept at 4°C. For long-term storage, freezing or
storage at 4°C in 50% glycerol is recommended.

Direct Detection of Rotavirus or Viral Antigen

ELECTRON MICROSCOPY

Electron microscopy of stool suspensions was the
first "routine" method for diagnosis of rotavirus infections (Bishop et al., 1974; Flewett et al., 1973;
Kapikian et al., 1974; Mebus et al., 1969). It is still
frequently included in comparisons as a reference

standard against which serological techniques are
checked, although a glance through the literature
shows that there is no "standard" technique for negative staining and electron microscopy of fecal specimens. Nevertheless it is objective and can thus signal
serological false positives (Brandt et al., 1981), and it
picks up rotaviruses of all serogroups and other enteritis viruses that would be missed by the usual rotavirus serology (Holmes, 1979; Hung et al., 1983;
McNulty et al., 1984).

A suggested technique based on Flewett et al.
(1973) and used in our own laboratory is as follows.
A 10% suspension of feces is made in distilled water
and clarified by centrifugation for 15 min at 3,000 × g
in a bench centrifuge. It is then concentrated by ultracentrifugation (say 60 min at 80,000 × g), and the
pellet is resuspended in a few drops of distilled water. Resuspension is easier if the centrifuge tubes can
be left overnight at 4°C at this stage, or sonication
can be used to obtain an even suspension. A drop of
the suspension is placed on a Formvar-carbon-
coated grid for 30 sec, or alternatively the grid can be
floated face down on a drop of specimen on a sheet of
parafilm or wax. Excess liquid is then blotted from
the edge of the grid using a torn edge of a piece of
filter paper, and the grid is placed on a drop of 2%
potassium phosphotungstate (pH 6 to 7) or one-tenth
saturated ammonium molybdate in water. After a
few seconds, excess fluid is again blotted off, and the
grid is dried and examined in the electron microscope. Rotaviruses can be identified on the screen at
a magnification of 20,000 to 30,000×. Usually, both
single-shelled and double-shelled particles are seen
(Fig. 1).

Various variations of the basic procedure have
been recommended. As an alternative to ultracentrifugation, precipitation in 60% ammonium sulfate for
1 h at 4°C, followed by centrifugation for 10 min at
10,000 × g, effectively precipitates rotaviruses (Caul
et al., 1978). In this procedure a distilled-water rinse
of the grid is needed after adsorption of the specimen
and before negative staining. Concentration can also
be effected without centrifugation by adding granules
of a dried gel (Lyphogel) to the sample. Water and
salts are taken up as the granules swell, leaving virus
and other particles in a much reduced volume of liquid (Rogers et al., 1981).

A pseudoreplica procedure for specimen preparation (Smith, 1967) also results in some particle concentration and desalting and appears to be more sensitive for rotavirus detection than direct negative
staining (El-Mekki et al., 1984; Portnoy et al., 1977).

Immune electron microscopy, based on aggregation of virus-antibody complexes in solution followed by centrifugation for 1 h at 15,000 × g and
negative staining of the pellet (Almeida, 1980) increases the sensitivity of rotavirus detection to approximately that of standard enzyme immunoassays

(Brandt et al., 1981; Morinet et al., 1984; Pereira et al., 1983) and in earlier studies was used to demonstrate relationship between different rotaviruses (Flewett et al., 1974) and the location of group antigens (Woode et al., 1976).

More recently solid-phase immune electron microscopy for specific detection of group A rotaviruses has been developed (Nicolaieff et al., 1980; Obert et al., 1981; Svensson et al., 1983). Grids are first coated with protein A, then with antirotavirus antibodies by floating on a diluted hyperimmune antiserum. They are then rinsed and floated on virus specimens before negative staining. Compared with direct negative staining, this method both enhances trapping of rotavirus particles on the grids and decreases adsorption of unrelated debris, so it has been rated as 30 times more sensitive than direct electron microscopy and 10 times more sensitive than ELISA (Svensson et al., 1983). It is of course more labor-intensive for a number of specimens. Gerna et al. (1984, 1985) have further extended this method for subgrouping and serotyping of rotaviruses, as will be discussed in more detail below.

COMPLEMENT FIXATION AND COUNTERIMMUNOELECTROPHORESIS

Continuing approximately chronologically, the first kinds of serological test applied to rotavirus diagnosis were complement fixation (CF) and counterimmunoelectroosmophoresis (CIEOP). Complement fixation is reliable, cheap, and practical for testing large numbers of samples (Kapikian et al., 1974; Thouless et al., 1977; Zissis et al., 1978). It is about as sensitive as direct electron microscopy but less sensitive than solid-phase immunoassays, which are now in more common use.

Once very popular, CIEOP was developed by a number of groups (Grauballe et al., 1977; Middleton et al., 1976; Spence et al., 1977; Tufvesson and Johnsson, 1976). It still has its exponents, because it is quick and is regarded as highly specific, but its sensitivity is low even in comparison with electron microscopy (Clementi et al., 1981; Hammond et al., 1984). Tufvesson and Johnsson (1976) found that the limit of detection was approximately 5×10^7 particles per milliliter, although the precipitates also contain much fine particulate material, probably inner capsid subunits (Mathan et al., 1977). Such precipitates provide an excellent source of antigen for preparation of rotavirus- (group A) specific hyperimmune sera (Grauballe et al., 1977).

RADIOIMMUNOASSAY

Next to be developed were a number of radioimmunoassays for rotavirus detection (Cukor et al., 1978; Kalica et al., 1977; Middleton et al., 1977; Sarkkinen et al., 1979). These were sensitive but limited in application because of the short shelf life of radioactively labeled immunoglobulins and problems with obtaining or disposing of radioisotopes. Thus they were soon supplanted by enzyme immunoassays, which are very similar in principle and in performance (Blacklow and Cukor, 1985; Sarkkinen et al., 1980). Thus I will leave discussion of both principles and practice to the next section but refer anyone desiring a detailed RIA protocol and instructions on preparation of reagents to Blacklow and Cukor (1985).

ENZYME-LINKED IMMUNOSORBENT ASSAY

Many varieties of enzyme immunoassay or ELISA have been developed for rotavirus antigen detection, and this is now the most commonly used technique. The basic design of the assay varies little. A solid phase, usually a polystyrene microtiter plate or ball, is coated with antirotavirus antibodies, then the antigen-containing sample is added, and unbound material is washed away. Bound antigen is then quantitated by adding either enzyme-conjugated antirotaviral antibody (direct test) or a second unlabeled antirotaviral antibody (which must be from a species different from the one in which the coating antibody was made) followed by an enzyme-labeled antiimmunoglobulin specific for the second antibody (indirect test). In either case, a chromogenic substrate for the enzyme is finally added, and bound enzyme (indicating bound antigen underneath) is detected by a color change.

The first rotavirus ELISAs described were of the direct type (Scherrer and Bernard, 1977; Yolken et al., 1977). These have the advantage that only one antiserum has to be prepared: part of it is enzyme-conjugated. The test is also slightly quicker than an indirect assay, since there is one less incubation and one less wash, and a number of commercially available ELISA kits follow this format. In indirect assays the second antibody has an amplifying effect, so these are capable of greater sensitivity, and in laboratories carrying out ELISAs for various antigens it is often convenient to prepare all the second antibodies in the same animal species, then only a single kind of enzyme-labeled antiimmunoglobulin is needed. Generally, some compromise has to be struck between speed and sensitivity: the choice of procedure will depend on whether the aim of the test is a bedside diagnosis or a definitive epidemiological study. Instead of giving another actual protocol, as several are already available (Beards et al., 1984; Blacklow and Cukor, 1985; Grauballe et al., 1981b; Yolken, 1982; Yolken et al., 1986), I will attempt to provide background information to guide those who wish to develop their own test or choose a commercial kit. The World Health Organization assay (Beards et al., 1984) is an example of a very sensitive but lengthy procedure; later I will mention some more rapid commercial approaches.

The type and quality of plastic affects binding of the coating antibody and thus the sensitivity and reproducibility of the whole assay. Gamma-irradiated polystyrene appears to give the best results (Beards et al., 1984; Grauballe et al., 1981b). With such plates it is no longer necessary to avoid using the outside row of wells, which was very uneconomical.

The quality of the coating antibodies is also most important. High-titered hyperimmune antisera prepared by immunizing rabbits or guinea pigs with purified particles or inner capsids of SA11, NCDV bovine or human rotavirus strains, or immunoelectrophoretically purified human rotavirus antigens have all been used successfully (Blacklow and Cukor, 1985; Grauballe et al., 1977). Immunization of hens, followed by extraction of immunoglobulins from the yolks of their eggs, is also efficient (Bartz et al., 1980). When only subgroup I viruses such as SA11 and bovine strains are used as immunizing antigens, there is a theoretical possibility of a bias against subgroup II detection, so Beards et al. (1984) immunized also with subgroup II particles, but the effect may be marginal (Sarkkinen, 1981).

Group A–specific monoclonal antibodies, which for general diagnostic purposes must not be subgroup-specific, offer increased sensitivity and reproducibility (Beards et al., 1984; Cukor et al., 1984; Herrmann et al., 1985; Taniguchi et al., 1984). For coating polystyrene, hyperimmune sera are diluted (usually 1 : 2,000 to 1 : 20,000) in 0.1 M carbonate buffer pH 9.6, but individual monoclones vary, and suitable pH and salt concentrations must be experimentally determined (Beards et al., 1984). Optimal dilutions of coating antibodies (and later, other antibodies and conjugates) are determined by checkerboard titrations using a standard low level of antigen, say a 1 : 100 dilution of a human rotavirus stock. Coating antibody is allowed to adsorb overnight at 4°C or for 2 h at 37°C.

Washing is carried out using phosphate-buffered saline (pH 7.4) containing 0.1% v/v Tween 20 (PBS-Tween) to reduce nonspecific protein absorption. For polystyrene balls, some sort of aspiration system is essential, but plates can be quickly and efficiently washed by immersion in PBS-Tween, making sure all wells are free of bubbles, then flicking the liquid out into a sink. This is repeated six times. Erratic results can frequently be traced to inadequate washing technique.

Samples containing a lot of antigen can exhibit some kind of prozone effect, so 10% fecal extracts should be diluted a further 1 : 4 (final 1 : 40) in PBS-Tween containing 0.01 M EDTA (Beards et al., 1984). This uncoats double-shelled rotavirus particles to expose a maximal amount of group antigen. It is advantageous to pretreat stool samples from neonates with even higher concentrations of EDTA (Coulson and Holmes, 1984), because such samples have frequently caused problems (Krause et al., 1983). Samples can be adsorbed overnight at 4°C for maximum sensitivity, or for 1 to 2 h at 37°C.

Diluent for the second antibody and/or enzyme conjugate often consists of PBS-Tween plus 1% bovine serum albumin, but 2.5% skim milk powder is at least as efficient and much cheaper than the albumin (Johnson et al., 1984). Both bovine serum albumin and supposedly fetal calf serum have on occasion been found to be contaminated with rotaviral antibodies (Offit et al., 1984).

Calf intestinal alkaline phosphatase and horseradish peroxidase are the most popular enzymes for antibody conjugates. Peroxidase is less expensive and easy to conjugate to immunoglobulins (Kurstak, 1985). A wide variety of commercial antiimmunoglobulin conjugates are available and also even antirotavirus conjugates (Dakopatts, Copenhagen, Denmark). A variety of substrates are available; details of the most commonly used ones are given in Table 3. If necessary, results can be read visually, especially if peroxidase and ABTS or TMB substrates are used, the latter before addition of sulfuric acid. Spectrophotometers are much better for judging shades of yellow, and if a commercial ELISA reader is not available, simpler and less expensive versions have been described by Clem and Yolken (1978) and Rook and Cameron (1981).

Each series of tests should include known positive and negative samples as controls. In the commercial kits, these are usually inactivated SA11 or bovine rotavirus preparations and sample diluent, respectively. To determine cutoff values for rotavirus positivity, it is common practice to average a number of negative control readings and to take this mean plus 2 or 3 standard deviations as the cutoff value. This value is then subtracted from test readings. Alternatively, each test reading can be divided by the mean of the negative control readings; the sample is considered positive if this P/N ratio $\geq$ 2.

False-positive reactions can occur in rotavirus ELISAs, owing either to particular reagents or to rheumatoid factor or similar substances in stools (Yolken and Stopa, 1979). One confirmatory system that controls the latter problem has been described by Brandt et al. (1981). At the coating stage, alternate rows of wells in a microtiter plate can be coated with the usual capture antibody and, ideally, a preimmunization serum from the same animal at the same dilution, or if that is not available, a rotavirus antibody-negative serum from the same kind of animal. Each sample is tested in each row, and samples giving a color reaction in both rows are considered to be false positives.

TABLE 3. ELISA substrates

Enzyme	Substrate	Stopping reagent	Visual color	Absorbance read at
Alkaline phosphatase	p-nitrophenyl phosphate 1 mg/ml in 1 M diethanolamine pH 9.8, 0.001 M $MgCl_2$ $6H_2O$	½ vol 3 M NaOH	Yellow	405 nm
Peroxidase	o-phenylene diamine (OPD) 2 mg/ml in 0.1 M citric acid–NaOH pH 5.0 with 0.02% H_2O_2	¼ vol 2 M H_2SO_4	Orange	492 nm
	ABTS[a] 0.4 mg/ml in 0.1 M citrate-phosphate pH 4.0 with 0.003% H_2O_2	½ vol 0.046 M NaF or 0.1% SDS	Green	405 or 650 nm
	TMB[b] dihydrochloride 0.13 mg/ml in 0.1 M acetate-citrate pH 6.0–6.1 plus 0.0036% H_2O_2	½ vol 2 M H_2SO_4	Yellow (blue before stopping)	450 nm

[a] ABTS is 2,2'-azino-bis(3-ethylbenzothiazoline-6-sulfonic acid) diammonium salt.
[b] TMB is 3,3',5,5'-tetramethyl benzidine. The dihydrochloride is water-soluble. If using TMB itself, dilute to 0.1 mg/ml from a stock 10 mg/ml in dimethyl sulfoxide.

Alternatively, a blocking confirmatory test can be performed (Beards et al., 1984). All samples positive in the original screening test are retested after incubation of sample aliquots in parallel with preimmune and postimmune antirotavirus sera at high concentration (say, 1:50 dilution). If the sample contains rotavirus, the aliquot pretreated with immune serum should be blocked; that is, it should produce less than 50% of the color reaction obtained with the other aliquot. This is a more economical procedure if there is not a high frequency of rotavirus positives in the population being tested.

Most commercially available rotavirus detection test kits are direct immunoassays, and some details of a few of them are given in Table 4. This is a rapidly developing field, and potential users will have to make sure that the data are not out of date as well as check the prices and availability in their areas. Precoated solid phases and simultaneous incubation of specimens and conjugates shorten the duration of the newer tests. Numerous comparisons between these commercial tests and previously published ELISAs, electron microscopy, and latex agglutination tests (see below) have been made (Chernesky et al., 1985; Knisley et al., 1986; Miotti et al., 1985; Morinet et al., 1984; Othman et al., 1986; Sambourg et al., 1985; Yolken and Leister, 1981).

LATEX AGGLUTINATION

Latex particles sensitized with antirotaviral IgG or, as control, nonimmune IgG or albumin, offer the possibility of a very rapid (2 to 5 min) diagnosis of rotavirus, provided that the sample contains a moderate amount of antigen. Hughes et al. (1984) have described how to produce the necessary reagents, but diagnostic kits are also available commercially

from at least four sources. Sample preparation (extraction and centrifugation) originally took much longer than the test itself, but small disposable filter units have recently solved this problem.

Drops of stool extract are mixed with equal volumes of test and control latex suspensions on glass slides or test plates (latex may adhere nonspecifically to plastics). Agglutination is read visually after 2 to 5 min. The droppers supplied dispense rather generous drops (Moosai et al., 1985; Othman et al., 1986), and the tests can be done successfully (and economically) by using measured 25-μl volumes of sample and latex.

A number of comparisons with ELISAs and other test methods have been published, and most agree that although the speed and specificity of latex tests are excellent, they are 4 to 10 times less sensitive than reference ELISAs or PAGE when tested against dilute samples (Miotti et al., 1985; Moosai et al., 1985; Morinet et al., 1984; Othman et al., 1986; Sambourg et al., 1985).

REVERSE PASSIVE HEMAGGLUTINATION

Instead of latex particles, erythrocytes can be fixed or tanned and coated with antirotaviral antibodies for antigen detection. Sheep or bovine erythrocytes are usually employed. Some stool extracts have hemagglutinating activity, so it is necessary to preabsorb samples with uncoated erythrocytes and to titrate out samples in parallel using cells coated with antirotaviral and nonimmune immunoglobulins, or coated cells in the presence of excess (blocking) antirotaviral antibody (Nakagomi et al., 1982; Sanekata and Okada, 1983).

At least one commercial version of the test is available in Japan, but the reagents have a short shelf

TABLE 4. Some commercially available ELISA systems

Component	Rotazyme II, Abbott Laboratories (USA)	Enzygnost, Rota Behring (W. Germany)	Rota ELISA, Dakopatts (Denmark)	Pathfinder, Kallestad Laboratories (USA)	Bio Enzabead, Litton Bionetic (USA)
Solid phase	Polystyrene bead	Polystyrene plate	Irradiated polystyrene plate	Polystyrene tube	Plastic-coated metal bead
Coating Ab	Guinea pig anti-rotavirus (pre-coated)	Rabbit anti-SA11 (precoated)	Second generation polyspecific rabbit antirotavirus	Rabbit antirotavirus IgG (precoated)	Rabbit anti-SA11 (precoated)
Conjugate	Peroxidase antirotavirus	Alkaline phosphatase anti-NCDV bovine rotavirus	Peroxidase rabbit antirotavirus	Peroxidase murine monoclonal anti-EDIM rotavirus	Peroxidase IgG of rabbit antirotavirus
Chromogen	OPD	p-nitrophenyl phosphate	OPD	TMB	ABTS
Confirmation	Blocking possible (own sera)	As Rotazyme	Normal rabbit serum available for coating or blocking	As Rotazyme	Negative control beads supplied
Approximate total duration	3 h	4 h	4 h	80 min	4 h

life unless kept below 10°C, and they are thus difficult to export. The commercial kit was faster (2 h) but slightly less sensitive than a reference ELISA (Tsuchie et al., 1983).

It is possible to stabilize coated erythrocytes by glutaraldehyde cross-linking and freeze-drying (Cranage et al., 1985). The stable reagent performed very well: using monoclonal antibodies for coating, Cranage et al. (1985) obtained good correlation with the WHO reference ELISA and comparable sensitivity.

OTHER ANTIGEN DETECTION TESTS

Immune adherence hemagglutination has been applied to the detection and quantitation of rotaviruses in stools and was important for a time as a method of subgrouping (Kapikian et al., 1981; Matsuno and Nagayoshi, 1978). It is a variation on the complement fixation test and is considerably more sensitive, but the fact that suitable erythrocytes are only produced by about 20% of group O blood donors has limited its application.

Foster et al. (1975) introduced the fluorescent virus precipitin test, in which aggregates of rotavirus antigen with fluorescein-tagged antibodies are observed under a fluorescence microscope. Both stool samples and fluorescent antibodies had to be prefiltered (0.45-μm pores) before use, and the test was somewhat laborious. A mixed agglutination, with rotaviral antibodies adsorbed on to protein A–carrying *Staphylococcus aureus* cells, has also been tested successfully (Hebert et al., 1981).

The solid-phase aggregation of coated erythrocytes (SPACE) test of Bradburne et al. (1979) resembled an RIA or ELISA but used erythrocytes coupled with rotaviral antibodies as the detection system. It is simple to read but lacks the amplification characteristics of the ELISA, so it is less sensitive.

Rotaviral RNA Detection

ELECTROPHEROTYPING

As was mentioned earlier (in the section on chemical composition of virions), analysis of rotaviral dsRNA by PAGE results in characteristic migration patterns of the 11 genome segments. These patterns, which are easily recognized and reproducible for individual strains, have for convenience been called electropherotypes or electrophoretypes. Although these cannot be directly correlated with serotype, they provide the finest discrimination available between rotavirus strains and are thus invaluable for epidemiological studies. An extensive review of such studies has been published by Estes et al. (1984). Here I will concentrate more on technical aspects.

Because in the 1970s cultivation and serotyping of rotaviruses were so difficult, most workers who characterized rotaviral RNA soon realized that electropherotyping was very useful for strain identification in the laboratory, but it was Espejo et al. (1977) who first suggested that it "could become a routine diagnostic procedure." At the time, this appeared very optimistic, but as a result of a number of technical improvements, the prophecy has certainly been fulfilled, and RNA analysis is displacing electron microscopy as an objective and sensitive technique for monitoring the performance of serological methods, and it provides additional useful information.

Early methods for RNA extraction were somewhat cumbersome, as they generally included an initial partial purification of the virus from the sample, or of the RNA, or both (Clarke and McCrae, 1981; Espejo et al., 1977; Kalica et al., 1978; Rodger and Holmes, 1979; Theil et al., 1981). Herring et al. (1982) introduced a simpler but very efficient extraction technique which is much more practical for processing large numbers of samples. Fecal samples are diluted 1 : 4 with 0.1 M sodium acetate buffer, pH 5.0, containing 1% w/v sodium dodecyl sulfate (SDS), then extracted once with a phenol-cresol-chloroform mixture. After centrifugation, viral RNA is found in the upper, aqueous phase and is ready for electrophoresis.

The procedure can be readily scaled down slightly and carried out even more rapidly using an Eppendorf or similar centrifuge (Holmes, 1985). Nicolas et al. (1983) have evolved a similar Eppendorf-based procedure with a preliminary fluorocarbon extraction, but this requires a special high-speed vortex mixer. It is possible to omit phenol extraction and to rely on detergents alone, but this sacrifices some sensitivity (Croxson and Bellamy, 1981; Dolan et al., 1985; Moosai et al., 1984).

At the electrophoretic stage, good resolution is obtained by using the sample buffer and discontinuous buffer system of Laemmli (1970) with a 10% polyacrylamide separating gel. Although RNA patterns are generally a stable viral characteristic, slight band variations were sometimes noted when the same samples were reanalyzed, leading to the advice that for reliable sample comparisons, RNA samples must be mixed and electrophoresed together (Rodger et al., 1981). This remains true, but such variations have been recently explained by Espejo and Puerto (1984) and are attributable to changes in buffer system, apparatus, or running current altering the temperature of the gel. Thus a standard methodology is needed, and for 10% Laemmli gels 0.75 mm thick, an overnight run at 10 v/cm or less (about 8 to 10 mA) is recommended.

The second, more important innovation of Herring et al. (1982) was the introduction of silver staining for visualization of the RNA bands. Previously, ethidium bromide staining was the standard method: an ultraviolet transilluminator was required, and Espejo et al. (1980) estimated that the RNA from about 10^{10} rotavirus particles was needed to produce a visible pattern (0.25 μg RNA, generally requiring 1 to 2 g of stool sample). Although ethidium bromide staining was sufficiently sensitive to allow completion of the first large-scale epidemiological studies based on electropherotyping (Buitenwerf et al., 1983; Lourenco et al., 1981; Rodger et al., 1981); not all EM-positive samples produced visible RNA patterns, and more sensitive methods were sought for localizing the bands. Clarke and McCrae (1981) had evolved a ^{32}P-labeling method, but it was very expensive. Silver staining turned out to be the answer.

SDS interferes with silver staining, so while it is still used in the sample buffer for the Laemmli (1970) electrophoretic system, for electropherotyping purposes it is omitted from both the gels and the electrode buffer. In the Herring et al. (1982) method, gels are washed for 30 min in 10% ethanol–0.5% acetic acid, then soaked in 0.011 M silver nitrate for 30 min to 2 h. After a brief rinse with distilled water, the gel is "developed" in 0.75 M sodium hydroxide containing 0.1 M formaldehyde for up to 10 min. The distilled water used throughout must be of good quality. Degassing of the silver and hydroxide solutions and addition of 0.0023 M sodium borohydride to the developer sometimes improves staining. The method is simple, inexpensive, and so sensitive that a result is usually obtainable using the RNA from only 0.01 ml of feces. For diagnosis it is about as sensitive as most ELISA systems (Herring et al., 1982; Pereira et al., 1983; Shinozaki et al., 1985).

The migration patterns of rotaviral RNA are so characteristic that false positives in diagnosis are unlikely, provided that spillover of samples from adjacent wells in the polyacrylamide gel can be ruled out. It is very unusual to find entirely unrelated strains with identical electropherotypes, although two strains with indistinguishable gel patterns yet different serotypes have been reported (Beards et al., 1982). Until recently, the only correlation between electropherotypes and serology of group A rotaviruses was that human serotype 2, subgroup I strains have always been found to have a "short" electropherotype (Fig. 3b), whereas other human serotypes have a "long" pattern (Fig. 3a) (Kalica et al., 1981a). The difference is due to a marked difference in migration of segment 11 of the long electropherotype (Dyall-Smith and Holmes, 1981). Since segment 11 is not directly related to either serotype or subgroup antigens, the correlation appears to be due only to chance. Such correlations are nevertheless useful, for a "supershort" electropherotype (Fig. 3c) drew attention to some human rotavirus samples from Indonesia which appear to belong to a new, fifth

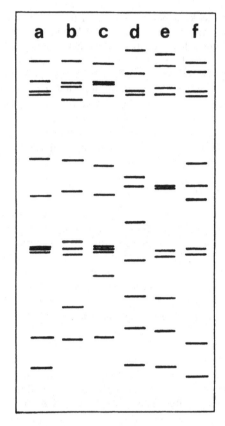

Fig. 3. Diagrammatic representation of rotavirus electropherotypes; direction of electrophoretic migration from top to bottom. (a) Group A "long" pattern, actually SA11, serotype 3. (b) Group A "short" pattern, human RV-5, serotype 2. (c) Group A "supershort" pattern, human rotavirus strain 69M. (d) Group B human, Chinese "adult diarrhea rotavirus." (e) Group B, porcine "rotaviruslike agent." (f) Group C, human or porcine "pararotavirus."

human serotype (Hasegawa et al., 1984; Matsuno et al., 1985).

Electropherotyping can provide much information not available from other diagnostic tests. It provided the first insights into the degree of variety among rotavirus strains (Estes et al., 1984) and the occurrence of different strains within a single epidemic. It has been particularly useful in elucidating patterns of infection in nosocomial outbreaks (Forster and Pastor, 1983; Nicolas et al., 1984; Rodriguez et al., 1983). It has also shown that dual or mixed infections of infants by two rotaviral strains are not uncommon, at least within pediatric hospitals (Nicolas et al., 1984; Spencer et al., 1983). Even single plaques obtained during cell culture adaptation of rotaviruses can yield "mixed" patterns which segregate into different electropherotypes following further single-plaque passage (Garbarg-Chenon et al., 1984; Sabara et al., 1982).

A few reports of atypical RNA patterns produced by group A rotaviruses have now appeared, but these are clearly uncommon. In strains found in some South African neonates, segment 11 had moved up near segment 5 (Besselaar et al., 1986), which was reminiscent of the behavior of segment 5 in bovine strains passaged at high multiplicity of infection in cell culture (Hundley et al., 1985). Concatemers of lower-molecular-weight segments have also been identified in abnormal positions in RNA patterns of samples obtained from immunodeficient children with chronic rotavirus infections (Pedley et al., 1984), but the molecular basis of such occasional rearrangements is not understood.

Electropherotyping has also led to the discovery of a number of rotavirus strains with even more atypical RNA patterns, which on further study have been found to lack the usual group A common antigen and instead have distinct group antigens. At least four such groups are currently being characterized (Chasey et al., 1986; Pedley et al., 1983; Snodgrass et al., 1984). None of these viruses has so far been adapted to grow in cell culture, so serology is very difficult. Fecal samples that appear rotavirus-positive by EM but negative by ELISA may contain such strains, but electropherotyping provides a more definitive diagnosis (Fig. 3) and often a suggestion of the probable group. The group must always be confirmed by serology or hybridization, however, because as more strains are characterized, the distinctions between group patterns are becoming less clear.

Group B rotaviruses were first identified in calves and piglets, but a strain with a slightly different RNA pattern (Fig. 3d) has caused widespread waterborne outbreaks in human adults as well as infants in China (Chen et al., 1985; Hung et al., 1984; Wang et al., 1985). It is possible that the human disease is a zoonosis, because it has not been observed in epidemics in other countries, but rats in the United States were found to be affected by a similar strain, and this apparently caused some human infections (Eiden et al., 1985).

Group C rotaviruses have been found in pigs (Pedley et al., 1983; Snodgrass et al., 1984a) and rather rarely (1% or less of rotavirus infections) in humans (Dimitrov et al., 1983; Nicolas et al., 1983; Pereira et al., 1983b; Rodger et al., 1982). In the human group C RNA pattern (Fig. 3f), one segment from the (group A) 7–9 triplet appears to have moved up closer to segment 6, whereas in group B animal strains (Fig. 3e), one moves down, toward segment 10.

Hybridization

RNA hybridization was first applied at a research level for investigating the sequence relatedness of genome segments among various rotaviruses, using electrophoresis and Northern blotting (Flores et al., 1982; Schroeder et al., 1982; Street et al., 1982). For routine diagnostic purposes, most interest is currently being shown in dot (or spot) hybridization assays ("dot blots"), where denatured viral nucleic acid is immobilized on a solid support and probed with labeled complementary RNA or DNA.

No standard procedure has yet evolved for rotavirus detection by dot hybridization, but several approaches are being tried. Viral RNA extracted from fecal samples is denatured by boiling, quench-cooled on ice, and spotted on to a nitrocellulose or nylon membrane. After air-drying, filters are baked in vacuo for 2 h at 80°C. As probes, Flores et al. (1983) used ^{32}P- or ^{125}I-labeled mRNA transcripts prepared from various cultivable rotaviruses. Other workers have labeled cloned cDNA copies of rotavirus genes with ^{32}P by nick translation (Dimitrov et al., 1985; Lin et al., 1985; Pedley and McCrae, 1984). For precise hybridization conditions, the original articles should be consulted.

For general rotavirus detection, probes based on genome segments encoding nonstructural proteins (see Fig. 2, Table 2) appear most likely to be useful, since these generally show maximal sequence homology between strains—for example, segments 7, 8, 10, or 11. By using probes derived from segments 5, 6, or 9, more strain discrimination is attainable by hybridization under more stringent conditions (Dimitrov et al., 1985; Pedley and McCrae, 1984).

Although none of the methods have yet been tested on a large scale, specificity of rotavirus detection appears to be good, and sensitivity so far approximates that of electropherotyping with silver staining (i.e., 1 to 10 ng of RNA). Flores et al. (1983) achieved even higher sensitivity, but nobody has yet reported making RNA probes from cloned rotaviral genes. When suitable probes become available, dot hybridization may become a very useful technique for identifying rotaviruses of the "atypical" groups (Chen et al., 1985).

Cell Culture

Single-Cycle Immunofluorescence Assay

Even before the infectivity-enhancing effects of trypsin were recognized, limited (single-cycle) growth of human rotaviruses in cell culture was achieved and developed as a diagnostic method. Banatvala et al. (1975) centrifuged stool filtrates on to porcine kidney cell monolayers and after 24 h incubation detected infected cells by immunofluorescent staining. Bryden et al. (1977) developed this into a routine diagnostic procedure using the monkey kidney cell line LLC-MK2 in microtiter trays. Another continuous monkey kidney cell line, MA104, was found preferable by Matsuno et al. (1977), and most later workers agree with their choice.

An updated version of the test has been published by Morinet et al. (1984); the following procedure is based on this and our own experience. Twenty percent fecal extracts in cell culture medium without serum are clarified by low-speed centrifugation and may also be filtered (0.45-μm pore size), but this is not essential if antibiotics are used. Samples are treated with 10 μg/ml trypsin (crystalline, porcine) for 30 min at 37°C, and fetal calf serum is added to 2% to counteract further trypsin action. Washed monolayers of MA104 cells in microtiter trays are inoculated with 100-μl aliquots and incubated or centrifuged for 1 h at 37°C. After removal of inocula, maintenance medium (no serum or trypsin, but with antibiotics) is added, and the plates are incubated overnight at 37°C in 5% CO_2. The original monolayers should not be too old (2-day, just-confluent layers are ideal), and incubation should not be extended beyond 18 to 20 h, or infected cells will be preferentially lost.

After careful removal of the medium (infected cells can easily detach at this stage), the monolayers are fixed with 80% acetone for 10 min at room temperature. After draining and a brief rinse, infected cells are exposed to a diluted hyperimmune rabbit antiserum (e.g., prepared against SA11 rotavirus inner capsids) for 1 h at 37°C, washed, and stained with a fluorescein-conjugated sheep antirabbit globulin. After a final wash, plates are drained and examined through the base using a low-power (16×) objective on an epifluorescence microscope. Infected cells show granular or sometimes almost confluent, cytoplasmic fluorescence.

This test is less sensitive than ELISA for routine diagnosis (Morinet et al., 1984) but is useful if an indication of infectivity is required—for example, in monitoring waste water treatment (Bates et al., 1984) or selecting samples for virus isolation attempts.

ISOLATION AND CULTIVATION

Pretreatment of inocula with high concentrations of trypsin, inclusion of a lower concentration (and no serum) in maintenance medium, and rolling of cell cultures during incubation all seem to assist primary isolation of rotaviruses in cell culture. First, routine isolation of bovine and porcine strains was achieved (Babiuk et al., 1977; Theil et al., 1977), then later, almost routine isolation of human strains (Sato et al., 1981; Urasawa et al., 1981). One human strain, Wa, had previously been isolated after multiple passages through piglets (Wyatt et al., 1980), but Wyatt et al. (1983a) soon switched to using trypsin also.

Although MA104 cells are readily available and capable of growing a range of human strains (Albert and Bishop, 1984), primary vervet or cynomolgus monkey kidney cultures are considered more sensitive for initial isolation of human rotaviruses (Hasegawa et al., 1982; Ward et al., 1984; Wyatt et al., 1983a). Direct isolation of cytopathic strains of human rotaviruses in LLC-MK2 cells without the use of trypsin has been reported (Agliano et al., 1985), but such occurrences always lead us to carry out RNA analyses to rule out the possibility of "reisolation" of a stock cultivable strain, as we have found that cell cultures can be just as susceptible as suckling mice to airborne rotaviruses.

Isolation of human rotaviruses is usually undertaken as a preliminary to serotyping by neutralization. It is very laborious, as four or five passages are needed before success of the attempt can be judged. When dealing with large numbers of samples, it is very useful to electropherotype them first and then select a few representatives of each RNA pattern for further study.

Subgrouping

Rotavirus subgrouping depends on a minor antigenic difference expressed on the major inner capsid protein VP6, which must be detected against a background of shared common (group A) antigens. Originally it was carried out by CF or ELISA using antisera especially prepared for the purpose in guinea pigs; sera usually required absorption with purified viral inner capsids of the alternate subgroup to reduce cross-reactivity (Yolken et al., 1978c; Zissis and Lambert, 1978). Kapikian et al. (1981) used postinfection gnotobiotic calf sera and immune adherence hemagglutination as well as an indirect ELISA and distinguished subgroup from serotype (neutralization) antigens. Gerna et al. (1984) showed that it was possible to subgroup rotaviruses by solid-phase (protein A) immune electron microscopy with cross-absorbed guinea pig antisera, but they found ELISA easier for numbers of samples.

Now subgroup-specific monoclonal antibodies are available, and ELISAs similar to that described above for antigen detection are the method of choice for subgrouping. Some monoclonal antibodies can be used for coating plates, whereas others function better as detection ("topping") antibodies: polyclonal group-specific antibodies are used as the other part of the sandwich (Greenberg et al., 1983b; Nakagomi et al., 1985; Taniguchi et al., 1984; White et al., 1984). Samples are tested in parallel in assays involving subgroup I and subgroup II monoclonal antibodies, results are read by spectrophotometer, and the

ratio of reactivities is determined. The minimum ratio required for subgroup assignment will depend on
the antibodies employed and has to be determined by
testing a number of rotaviruses of known subgroup,
but it is usually about twofold.

Serotyping

FLUORESCENT CELL (FOCUS) NEUTRALIZATION TEST

Rotavirus serotyping was first carried out by serum
neutralization using the single-cycle immunofluorescence assay (Flewett et al., 1978; Thouless et al.,
1977). Pretitrated samples are diluted so that 50 μl
inocula produce about 100 fluorescent cells per microtiter well, then mixed with equal volumes of serial
dilutions of antisera and incubated 1 h at 37°C, before
inoculation of the cell monolayers. The neutralization titer is taken as the reciprocal of the greatest
dilution of antiserum that reduces the fluorescent cell
count by at least 50%.

The test can be applied directly to fecal samples
(i.e., to rotaviruses that have not been adapted to cell
culture), but there are practical limitations. Beards et
al. (1980) found that successful serotyping was possible only if the fecal virus had an infectivity titer of at
least 10^4 focus-forming units per milliliter, which in
their series meant only 1 in 3 of rotavirus-positive
samples. Furthermore, many strains were untypable
because of very marked cross-reactions.

The same test, with the same typing sera, gives
clear-cut results when applied to cell culture adapted
strains (Beards and Flewett, 1984). Twentyfold ho-
mologous/heterologous titer ratios are required for
distinguishing rotavirus serotypes with hyperimmune sera, or eightfold differences if postinfection
sera are used (WHO, 1984; Wyatt et al., 1982). Neutralizing monoclonal antibodies can be detected and
characterized easily and economically using this
technique (Coulson et al., 1985; Sonza et al., 1983).
It is also possible to carry out neutralization tests
using interference with coxsackievirus B1 instead of
immunofluorescence for reading the result (Suzuki et
al., 1984).

PLAQUE REDUCTION ASSAY

With rotavirus strains that are cell culture–adapted
and grow well enough to form plaques, serotyping
can be carried out using conventional plaque reduction assays (Bohl et al., 1984; Hoshino et al., 1984;
Matsuno et al., 1977, 1985; Urasawa et al., 1982;
Wyatt et al., 1982). For human rotavirus isolates that
do not form clear plaques, a radioimmunofocus assay has been suggested (Liu et al., 1984).

ELISA

Because of the limitations mentioned above regarding application of the fluorescent focus neutralization
test to fecal samples and the amount of work required for cell culture adaptation of human rotaviruses as well as that in the neutralization tests
themselves, an ELISA procedure for serotyping has
long been a dream. The high level of immunogenicity
and cross-reactivity of the rotavirus group antigens
frustrated all efforts for some time, but Thouless et
al. (1982) demonstrated that such typing was possible
if antisera were very thoroughly cross-absorbed.

The situation has changed now that serotype-specific monoclonal antibodies are becoming available.
As in the case of subgrouping by ELISA, the monoclones may be used as coating (Shaw et al., 1985) or
detection antibodies (Coulson et al., 1987). Since serotype-specific antigens are located on the outer capsid of rotavirus particles, in serotyping ELISAs,
EDTA must *not* be included in the sample diluent,
since single-capsid particles would be untypable. For
the same reason, if hyperimmune sera are used in the
test, they must contain neutralizing antibodies
against each (or all) of the serotypes to be detected.
Coulson et al. (1987) prepared separate hyperimmune sera for four human serotypes by immunizing
rabbits with purified, double-shelled particles of cultivable, prototype strains (see Table 1). Rows of
wells in microtiter plates are coated with each of
these sera in turn, samples diluted in PBS-Tween
plus 2.5% skim milk powder are added to sets of four
wells, and the corresponding type-specific monoclonal antibodies are used to detect bound particles.
The whole test is modified from Beards et al. (1984),
as described earlier, but the conjugate is peroxidase-
labeled antimouse Ig, and TMB substrate is used.

The monoclonal antibodies used for ELISA serotyping to date are all VP7-specific. In fecal samples
the amount of group-specific VP6 greatly exceeds
that of VP7, so serotyping ELISAs are never likely
to be as sensitive as general (group) antigen-detect-
ing ELISAs. Nevertheless, they are already a great
improvement on infectivity-based tests.

OTHER METHODS

Serotyping by solid-phase immune electron microscopy is also possible, though very labor-intensive.
Gerna et al. (1985) have succeeded in serotyping epidemiologically significant numbers of strains using
cross-absorbed polyclonal antisera and will no doubt
also be able to utilize monoclonal antibodies in the
future, but there must be an easier way.

Now that some serotype-specific sequence differences have been identified in the RNA segment en-

coding VP7 (see the section on rotavirus gene sequences) and further studies on RNA segment 4 promise to provide similar information relevant to VP3 serotypic differences, distinguishing VP3 and VP7 serotypes by hybridization appears to be an attainable goal. Pedley and McCrae (1984), Dimitrov et al. (1985), and Lin et al. (1985) have started to look at this approach. Synthetic oligonucleotide probes are showing much promise for serotyping purposes (A. Haralambous, personal communication).

Antibody Measurement and Serodiagnosis

Although rotavirus infections are usually diagnosed directly by antigen detection, for epidemiologic purposes it is also useful to monitor serum antibody levels, because asymptomatic infections may produce seroconversion even though virus excretion goes undetected.

It is possible to measure type-specific neutralizing antibodies, but this is rarely attempted because of the amount of work involved. Ongoing studies of this type are producing very important information relevant to protective immunity (Chiba et al., 1986). Usually, however, "rotavirus antibodies" means antibodies directed against the group A antigen, since these are predominant in sera and easier to measure. Infections by any group A serotype regularly produce a group-specific seroresponse except in infants less than 3 months old (see section on dynamics of antibody response). Such responses have been measured by CF, immunofluorescence, and radioimmunoassays (Blacklow and Cukor, 1985), but ELISA is the most popular current choice because of its sensitivity and the facility with which different immunoglobulin classes can be distinguished.

Antibody measurement by ELISA has been practiced as long as antigen detection, and many variations of the basic assay have been described. Any cultivable rotavirus can be used as a source of antigen—for example, human Wa strain, bovine NCDV, or monkey SA11. It is convenient to process batches of infected and uninfected cell cultures at the same time; the latter provide control antigen (Ghose et al., 1978). Cultures are frozen and thawed, fluorocarbon extracted in a homogenizer, and the aqueous phase is separated by low-speed centrifugation at 4°C. Supernatant fluids are then centrifuged for 1 h at 90,000 × g, and the pellets are resuspended in 1/100 to 1/20 of the initial volume of phosphate-buffered saline.

Polyvinyl or polystyrene microtiter plates (or beads) are used as the solid phase. Viral antigen and control preparations in alternate rows may be adsorbed directly to the plastic after dilution in carbonate buffer pH 9.6 (Ghose et al., 1978; McLean et al., 1980), or plates may first be coated with a diluted

rabbit or guinea pig antirotaviral serum, then antigens (Grauballe et al., 1981a; Yolken et al., 1978a,b). Optimal dilutions of viral antigen are determined by checkerboard titrations with convalescent sera, and control antigens are used at the same dilution. Sera can be diluted in PBS with 0.2% bovine serum albumin or skim milk powder. After incubation and washing, bound antibodies are detected using appropriate antihuman Ig sera conjugated to peroxidase or alkaline phosphatase.

Antibody levels may be measured by endpoint titration or by measurement at a single dilution and reference to a standard curve (Bishop et al., 1984; Kurstak, 1985). Because antibody-negative human sera are not readily obtainable, levels of binding of positive sera to control antigen are used to establish cutoff levels.

For detecting antirotaviral IgM, direct measurement with an anti-IgM conjugate is possible, but it is probably best to remove excess IgG antibodies by preabsorption of samples with staphylococcal protein A (Cowan strain cells, or Sepharose-bound) before testing (Orstavik and Haug, 1976) or to use an IgM capture system.

Rises in serum IgA antibodies following rotavirus infection are almost as frequent as IgG rises in both infants and adults (Davidson et al., 1983; Grimwood, 1986; Sheridan et al., 1981). Rotaviral IgA coproantibodies can also be measured by ELISA (McLean et al., 1980; Simhon et al., 1985; Sonza and Holmes, 1980). Both intestinal secretions and feces contain significant levels of proteolytic enzymes, and these reduce sensitivity of IgA detection, apparently mainly by removing antigens from the solid phase. Either addition of a protease inhibitor (phenylmethyl-sulfonyl chloride) to the samples (Hohmann et al., 1983) or inclusion of 2.5% skim milk in the sample diluent and sample absorption overnight at 4°C (B. Coulson, personal communication) solves this problem. Inouye et al. (1984) increased sensitivity of coproantibody detection by dissociating their antigen (purified virus particles) into subunits by treatment with chaotropic agents, sodium thiocyanate or guanidine hydrochloride, but with a less highly purified antigen this might not be necessary.

I previously discussed the dynamics of antibody responses in some detail, at least as far as they are known, to provide a background for interpretation of immunodiagnosis. In summary, the most reliable evidence of a recent rotavirus infection is an IgG seroconversion, and next best is detection of IgA and/or IgM coproantibodies 1 to 2 weeks after onset of illness. Serum IgM, optimally detected 7 to 10 days after onset, definitely indicates a recent rotavirus infection but may be detectable only after *primary* infection. In infants less than 3 months old, seroconversion may not be detectable, but IgA or IgM coproantibodies or IgM serum antibodies indicate a recent infection.

Literature Cited

Adams, W. R., and L. M. Kraft. 1967. Electron microscopic study of the intestinal epithelium of mice infected with the agent of epizootic diarrhea of infant mice (EDIM virus). Am. J. Pathol. **51**:39–44.

Agliano, A. M., A. Rossi, and A. Sanna. 1985. Isolation from fecal specimens of new strains of human rotavirus primarily cytopathic for stationary cell cultures without trypsin. Arch. Virol. **84**:119–127.

Albert, M. J., and R. F. Bishop. 1984. Cultivation of human rotaviruses in cell culture. J. Med. Virol. **13**:377–383.

Almeida, J. D. 1980. Practical aspects of diagnostic electron microscopy. Yale J. Biol. Med. **53**:5–18.

Babiuk, L. A., K. Mohammed, L. Spence, M. Fauvel, and R. Petro. 1977. Rotavirus isolation and cultivation in the presence of trypsin. J. Clin. Microbiol. **6**:610–617.

Banatvala, J. E., B. Totterdell, I. L. Chrystie, and G. N. Woode. 1975. In-vitro detection of human rotaviruses. Lancet **2**:821.

Bartz, C. R., R. H. Conklin, C. B. Tunstall, and J. H. Steele. 1980. Prevention of murine rotavirus infection with chicken egg yolk immunoglobulin. J. Infect. Dis. **142**:439–441.

Bastardo, J. W., J. L. McKimm-Breschkin, S. Sonza, L. D. Mercer, and I. H. Holmes. 1981. Preparation and characterization of antisera to electrophoretically purified SA11 virus polypeptides. Infect. Immun. **34**:641–647.

Bates, J., M. R. Goddard, and M. Butler. 1984. The detection of rotaviruses in products of wastewater treatment. J. Hyg. **93**:639–643.

Beards, G. M. 1982. Polymorphism of genomic RNAs within rotavirus serotypes and subgroups. Arch. Virol. **74**:65–70.

Beards, G. M., and T. H. Flewett. 1984. Serological characterization of human rotaviruses propagated in cell cultures. Arch. Virol. **80**:231–237.

Beards, G. M., J. N. Pilfold, M. E. Thouless, and T. H. Flewett. 1980. Rotavirus serotypes by serum neutralization. J. Med. Virol. **5**:231–237.

Beards, G. M., A. D. Campbell, N. R. Cottrell, J. S. M. Peiris, N. Rees, R. C. Sanders, J. A. Shirley, H. C. Wood, and T. H. Flewett. 1984. Enzyme-linked immunosorbent assays based on polyclonal and monoclonal antibodies for rotavirus detection. J. Clin. Microbiol. **19**:248–254.

Besselaar, T. G., A. Rosenblatt, and A. H. Kidd. 1986. Atypical rotavirus from South African neonates. Arch. Virol. **87**:327–330.

Bishop, R. F. 1986. Epidemiology of diarrhoeal disease caused by rotavirus, p. 158–170. *In* J. Holmgren, A. Lindberg, and R. Mollby (ed.), Development of vaccines and drugs against diarrhoea. Nobel Conference 11. Studentlitteratur, Lund, Sweden.

Bishop, R. F., G. P. Davidson, I. H. Holmes, and B. J. Ruck. 1973. Virus particles in epithelial cells of duodenal mucosa from children with acute nonbacterial gastroenteritis. Lancet **2**:1281–1283.

Bishop, R. F., G. P. Davidson, I. H. Holmes, and B. J. Ruck. 1974. Detection of a new virus by electron microscopy of faecal extracts from children with acute gastroenteritis. Lancet **1**:149–151.

Bishop, R. F., G. L. Barnes, E. Cipriani, and J. S. Lund. 1983. Clinical immunity after neonatal rotavirus infec-

tion. A prospective longitudinal study in young children. N. Engl. J. Med. **309**:72–76.

Bishop, R. F., E. Cipriani, J. S. Lund, G. L. Barnes, and C. S. Hosking. 1984. Estimation of rotavirus immunoglobulin G antibodies in human serum samples by enzyme linked immunosorbent assay: expression of results as units derived from a standard curve. J. Clin. Microbiol. **19**:447–452.

Black, R. E., M. H. Merson, I. Huq, A. R. Alim, and M. Yunus. 1981. Incidence and severity of rotavirus and *Escherichia coli* diarrhoea in rural Bangladesh. Implications for vaccine development. Lancet **1**:141–143.

Blacklow, N. R., and G. Cukor. 1985. Viral gastroenteritis agents, p. 805–812. *In* E. H. Lennette (ed.), Manual of clinical microbiology, 4th ed., American Society for Microbiology, Washington, D.C.

Bohl, E. H., K. W. Theil, and L. J. Saif. 1984. Isolation and serotyping of porcine rotaviruses and antigenic comparison with other rotaviruses. J. Clin. Microbiol. **19**:105–111.

Both, G. W., A. R. Bellamy, J. E. Street, and L. J. Siegman. 1982. A general strategy for cloning double-stranded RNA: nucleotide sequence of the simian-11 rotavirus gene 8. Nucleic Acids Res. **10**:7075–7088.

Both, G. W., J. S. Mattick, and A. R. Bellamy. 1983a. Serotype-specific glycoprotein of simian 11 rotavirus: coding assignment and gene sequence. Proc. Natl. Acad. Sci. USA **80**:3091–3095.

Both, G. W., L. J. Siegman, A. R. Bellamy, and P. H. Atkinson. 1983b. Coding assignment and nucleotide sequence of simian rotavirus SA11 gene segment 10: location of glycosylation sites suggests that the signal peptide is not cleaved. J. Virol. **48**:335–339.

Both, G. W., L. J. Siegman, A. R. Bellamy, N. Ikegami, A. J. Shatkin, and Y. Furuichi. 1984. Comparative sequence analysis of rotavirus genomic segment 6: the gene specifying viral subgroups 1 and 2. J. Virol. **51**:97–101.

Bradburne, A. F., J. D. Almeida, P. S. Gardner, R. B. Moosai, A. A. Nash, and R. R. A. Coombs. 1979. A solid phase system for the detection and quantification of rotavirus in faeces. J. Gen. Virol. **44**:615–623.

Brandt, C. D., H. W. Kim, W. J. Rodriguez, L. Thomas, R. H. Yolken, J. O. Arrobio, A. Z. Kapikian, R. H. Parrott, and R. M. Chanock. 1981. Comparison of direct electron microscopy, immune electron microscopy, and rotavirus enzyme-linked immunosorbent assay for detection of gastroenteritis viruses in children. J. Clin. Microbiol. **13**:976–981.

Bridger, J. C., and J. F. Brown. 1981. Development of immunity to porcine rotavirus in piglets protected from disease by bovine colostrum. Infect. Immun. **31**:906–910.

Bridger, J. C., and G. N. Woode. 1975. Neonatal calf diarrhoea: identification of a reovirus-like (rotavirus) agent in faeces by immunofluorescence and immune electron microscopy. Br. Vet. J. **131**:528–535.

Bridger, J. C., and G. N. Woode. 1976. Characterization of two particle types of calf rotavirus. J. Gen. Virol. **31**:245–250.

Bryden, A. S., H. A. Davies, M. E. Thouless, and T. H. Flewett. 1977. Diagnosis of rotavirus infection by cell culture. J. Med. Microbiol. **10**:121–125.

Buitenwerf, J., M. Muilwijk–van Alphen, and G. J. P. Schaap. 1983. Characterization of rotaviral RNA isolated from children with gastroenteritis in two hospitals in Rotterdam. J. Med. Virol. **12**:71–78.

Cameron, D. J. S., R. F. Bishop, A. A. Veenstra, G. L. Barnes, I. H. Holmes, and B. J. Ruck. 1978. Pattern of shedding of two non-cultivable viruses in stools of newborn babies. J. Med. Virol. **2**:7–13.

Carlson, J. A. K., P. J. Middleton, M. T. Szymanski, J. Huber, and M. Petric. 1978. Fatal rotavirus gastroenteritis—analysis of 21 cases. Am. J. Dis. Child. **132**:477–479.

Caton, A. J., G. G. Brownlee, J. W. Yewdell, and W. Gerhard. 1982. The antigenic structure of the influenza virus A/PR/8/34 hemagglutinin (H1 subtype). Cell **31**:417–427.

Caul, E. O., C. R. Ashley, and S. I. Egglestone. 1978. An improved method for the routine identification of faecal viruses using ammonium sulphate precipitation. FEMS Microbiol. Lett. **4**:1–4.

Chasey, D., J. C. Bridger, and M. A. McCrae. 1986. A new type of atypical rotavirus in pigs. Arch. Virol. **89**:235–243.

Chen, G.-M., T. Hung, J. C. Bridger, and M. A. McCrae. 1985. Chinese adult rotavirus is a group B rotavirus. Lancet **2**:1123–1124.

Chernesky, M., S. Castriciano, J. Mahony, and D. DeLong. 1985. Examination of the Rotazyme II enzyme immunoassay for the diagnosis of rotavirus gastroenteritis. J. Clin. Microbiol. **22**:462–464.

Chiba, S., S. Nakata, T. Urasawa, S. Urasawa, T. Yokoyama, Y. Morita, K. Taniguchi, and T. Nakao. 1986. Protective effect of naturally acquired homotypic and heterotypic rotavirus antibodies. Lancet **2**:417–421.

Clark, S. M., J. R. Roth, M. L. Clark, B. B. Barnett, and R. S. Spendlove. 1981. Trypsin enhancement of rotavirus infectivity: mechanism of enhancement. J. Virol. **39**:816–822.

Clem, T. R., and R. H. Yolken. 1978. Practical colorimeter for direct measurement of microplates in enzyme immunoassay systems. J. Clin. Microbiol. **7**:55–58.

Clementi, M., P. Pauri, P. Bagnarelli, G. Carloni, and L. Calegari. 1981. Diagnosis of human rotavirus infections: comparison of an electrophoretic method, a modified complement fixation test and electron microscopy for rotavirus detection. Arch. Virol. **67**:341–344.

Cohen, J. 1977. Ribonucleic acid polymerase activity associated with purified calf rotavirus. J. Gen. Virol. **36**:395–402.

Coulson, B. S., and I. H. Holmes. 1984. An improved enzyme-linked immunosorbent assay for the detection of rotaviruses in faeces of neonates. J. Virol. Methods **8**:165–179.

Coulson, B. S., K. J. Fowler, R. F. Bishop, and R. G. H. Cotton. 1985. Neutralizing monoclonal antibodies to human rotavirus and indications of genetic drift among strains from neonates. J. Virol. **54**:14–20.

Coulson, B. S., J. M. Tursi, W. J. McAdam, and R. F. Bishop. 1986. Derivation of neutralizing monoclonal antibodies to human rotaviruses and evidence that an immunodominant neutralization site is shared between serotypes 1 and 3. Virology **154**:302–312.

Coulson, B. S., L. E. Unicomb, G. A. Pitson, and R. F. Bishop. 1987. Simple and specific enzyme immunoassay using monoclonal antibodies for serotyping human rotaviruses. J. Clin. Microbiol. **25**:509–515.

Cranage, M. P., A. D. Campbell, J. L. Venters, S. Mawson, R. R. A. Coombs, and T. H. Flewett. 1985. Detection and quantitation of rotavirus using monoclonal antibody coupled red blood cells. J. Virol. Methods **11**:273–287.

Croxson, M. C., and A. R. Bellamy. 1981. Extraction of rotaviruses from faeces by treatment with lithium dodecyl sulfate. Appl. Environ. Microbiol. **41**:255–260.

Cukor, G., and N. R. Blacklow. 1984. Human viral gastroenteritis. Microbiol. Rev. **48**:157–179.

Cukor, G., M. K. Berry, and N. R. Blacklow. 1978. Simplified radioimmunoassay for detection of human rotavirus in stools. J. Infect. Dis. **138**:906–910.

Cukor, G., N. R. Blacklow, F. E. Capozza, Z. F. K. Panjvani, and F. Bednarek. 1979. Persistence of antibodies to rotavirus in human milk. J. Clin. Microbiol. **9**:93–96.

Cukor, G., D. M. Perron, R. Hudson, and N. R. Blacklow. 1984. Detection of rotavirus in human stools by using monoclonal antibody. J. Clin. Microbiol. **19**:888–892.

Davidson, G. P., R. F. Bishop, R. R. W. Townley, I. H. Holmes, and B. J. Ruck. 1975a. Importance of a new virus in acute sporadic enteritis in children. Lancet **1**:242–245.

Davidson, G. P., I. Goller, R. F. Bishop, R. R. W. Townley, I. H. Holmes, and B. J. Ruck. 1975b. Immunofluorescence in duodenal mucosa of children with acute enteritis due to a new virus. J. Clin. Pathol. **28**:263–266.

Davidson, G. P., R. J. Hogg, and C. P. Kirubakaran. 1983. Serum and intestinal immune response to rotavirus enteritis in children. Infect. Immun. **40**:447–452.

De Zoysa, I., and R. G. Feachem. 1985. Interventions for the control of diarrhoeal diseases among young children: rotavirus and cholera immunization. Bull. WHO **63**:569–583.

Dimitrov, D. H., M. K. Estes, S. M. Rangelova, L. M. Shindarov, J. L. Melnick, and D. Y. Graham. 1983. Detection of antigenically distinct rotaviruses from infants. Infect. Immun. **41**:523–526.

Dimitrov, D. H., D. Y. Graham, and M. K. Estes. 1985. Detection of rotaviruses by nucleic acid hybridization with cloned DNA of simian rotavirus SA11 genes. J. Infect. Dis. **152**:292–300.

Dolan, K. T., E. M. Twist, P. Horton-Slight, C. Forrer, L. M. Bell, Jr., S. A. Plotkin, and H. F. Clark. 1985. Epidemiology of rotavirus electropherotypes determined by a simplified diagnostic technique with RNA analysis. J. Clin. Microbiol. **21**:753–758.

Dyall-Smith, M. L., and I. H. Holmes. 1981. Gene-coding assignments of rotavirus double-stranded RNA segments 10 and 11. J. Virol. **38**:1099–1103.

Dyall-Smith, M. L., and I. H. Holmes. 1984. Sequence homology between human and animal rotavirus serotype-specific glycoproteins. Nucleic Acids Res. **12**:3973–3982.

Dyall-Smith, M. L., T. C. Elleman, P. A. Hoyne, I. H. Holmes, and A. A. Azad. 1983. Cloning and sequence of UK bovine rotavirus gene segment 7: marked sequence homology with simian rotavirus gene segment 8. Nucleic Acids Res. **11**:3351–3362.

Dyall-Smith, M. L., I. Lazdins, G. W. Tregear, and I. H. Holmes. 1986. Location of the major antigenic sites involved in rotavirus serotype-specific neutralization. Proc. Natl. Acad. Sci. USA **83**:3465–3468.

Echeverria, P., N. R. Blacklow, L. B. Sanford, and G. G. Cukor. 1981. Travellers diarrhoea among American Peace Corps volunteers in rural Thailand. J. Infect. Dis. **143**:767–771.

Echeverria, P., N. R. Blacklow, G. G. Cukor, S. Vibulbandhitkit, S. Changchawalit, and P. Boonthai. 1983. Rotavirus as a cause of severe gastroenteritis in adults. J. Clin. Microbiol. **18**:663–667.

Eiden, J., S. Vonderfecht, and R. H. Yolken. 1985. Evidence that a novel rotavirus-like agent of rats can cause gastroenteritis in man. Lancet **2**:8–11.

El-Mekki, A., W. Al-Nakib, S. K. Sethi, D. A. El-Khalik, and M. Al-Wuhaib. 1984. Pseudoreplica electron microscopy for the detection of rotavirus: comparison with high-speed centrifugation electron microscopy and ELISA. J. Virol. Methods **9**:79–85.

Espejo, R. T., and F. Puerto. 1984. Shifts in the electrophoretic pattern of the RNA genome of rotaviruses under different electrophoretic conditions. J. Virol. Methods **8:**293–299.

Espejo, R. T., E. Calderon, and N. Gonzalez. 1977. Distinct reovirus-like agents associated with acute infantile gastroenteritis. J. Clin. Microbiol. **6:**502–506.

Espejo, R. T., O. Munoz, F. Serafin, and P. Romero. 1980. Shift in the prevalent human rotavirus detected by ribonucleic acid segment differences. Infect. Immun. **27:**351–354.

Espejo, R. T., S. Lopez, and C. Arias. 1981. Structural polypeptides of simian rotavirus SA11 and the effect of trypsin. J. Virol. **37:**156–160.

Estes, M. K., D. Y. Graham, and B. B. Mason. 1981. Proteolytic enhancement of rotavirus infectivity: molecular mechanisms. J. Virol. **39:**879–888.

Estes, M. K., E. L. Palmer, and J. F. Obijeski. 1983. Rotaviruses: a review. Curr. Top. Microbiol. Immunol. **105:**123–184.

Estes, M. K., D. Y. Graham, and D. H. Dimitrov. 1984a. The molecular epidemiology of rotavirus gastroenteritis. Prog. Med. Virol. **29:**1–22.

Estes, M. K., B. B. Mason, S. Crawford, and J. Cohen. 1984b. Cloning and nucleotide sequence of the simian rotavirus gene 6 that codes for the major inner capsid protein. Nucleic Acids Res. **12:**1875–1887.

Fernelius, A. L., A. E. Ritchie, L. G. Classick, J. O. Norman, and C. A. Mebus. 1972. Cell culture adaptation and propagation of a reovirus-like agent of calf diarrhea from a field outbreak in Nebraska. Arch. Ges. Virusforsch. **37:**114–130.

Flewett, T. H. 1982. Clinical features of rotavirus infections, p. 125–145. *In* D. A. J. Tyrrell and A. Z. Kapikian (ed.), Virus infections of the gastrointestinal tract. Marcel Dekker, New York.

Flewett, T. H., and G. N. Woode. 1978. The rotaviruses. Arch. Virol. **57:**1–23.

Flewett, T. H., A. S. Bryden, and H. Davies. 1973. Virus particles in gastroenteritis. Lancet **2:**1497.

Flewett, T. H., A. S. Bryden, H. Davies, G. N. Woode, J. C. Bridger, and J. M. Derrick. 1974. Relation between viruses from acute gastroenteritis of children and newborn calves. Lancet **2:**61–63.

Flewett, T. H., M. E. Thouless, J. N. Pilford, A. S. Bryden, and J. A. N. Candeias. 1978. More serotypes of human rotavirus. Lancet **2:**632.

Flores, J., I. Perez, L. White, M. Perez, A. R. Kalica, R. Marquina, R. G. Wyatt, A. Z. Kapikian, and R. M. Chanock. 1982. Genetic relatedness among human rotaviruses as determined by RNA hybridization. Infect. Immun. **37:**648–655.

Flores, J., E. Boeggeman, R. H. Purcell, M. Sereno, I. Perez, L. White, R. G. Wyatt, R. M. Chanock, and A. Z. Kapikian. 1983. A dot hybridization assay for detection of rotavirus. Lancet **1:**555–558.

Forster, J., and S. Pastor. 1983. Epidemiology of human rotaviruses as determined by electrophoresis of genome RNA. Eur. J. Clin. Microbiol. **2:**141–147.

Foster, L. G., M. W. Peterson, and R. S. Spendlove. 1975. Fluorescent virus precipitin test. Proc. Soc. Exp. Biol. Med. **150:**155–160.

Foster, S. O., E. L. Palmer, G. W. Gary, Jr., M. L. Martin, K. L. Herrmann, P. Beasley, and J. Sampson. 1980. Gastroenteritis due to rotavirus in an isolated Pacific Island group: an epidemic of 3439 cases. J. Infect. Dis. **141:**32–39.

Garbarg-Chenon, A., F. Bricout, and J. C. Nicolas. 1984. Study of genetic reassortment between two human rotaviruses. Virology **139:**358–365.

Garbarg-Chenon, A., J. C. Nicolas, M. Bouvier, G. Desjouis, G. Molinier, D. Repiquet, M. H. Baptista-Lourenco, J. P. Gomant, F. Bricout, and J. M. Huraux. 1986. Epidemiologic and genomic study of rotavirus strains infecting young children and calves in the same rural environment. Eur. J. Epidemiol. **2:**108–111.

Gaul, S. K., T. F. Simpson, G. N. Woode, and R. W. Fulton. 1982. Antigenic relationships among some animal rotaviruses: virus neutralization in vitro and cross protection in piglets. J. Clin. Microbiol. **16:**495–503.

Gerna, G., M. Torsellini, N. Passarani, M. Battaglia, E. Percivalle, A. Sarasini, D. Torre, and P. Ferrante. 1984. Subgrouping of human rotavirus strains by complement fixation, indirect double-antibody sandwich enzyme-linked immunosorbent assay and solid phase immune electron microscopy. Arch. Virol. **81:**193–203.

Gerna, G., N. Passarani, A. Sarasini, and M. Battaglia. 1985. Characterization of serotypes of human rotavirus strains by solid-phase immune electron microscopy. J. Infect. Dis. **152:**1143–1151.

Ghose, L. H., R. D. Schnagl, and I. H. Holmes. 1978. Comparison of an enzyme-linked immunosorbent assay (ELISA) for quantitation of rotavirus antibodies with complement fixation in an epidemiological survey. J. Clin. Microbiol. **8:**268–276.

Gouvea, V. S., A. A. Alencar, O. M. Barth, L. de Castro, A. M. Fialho, H. P. Araujo, S. Majerowicz, and H. G. Pereira. 1986. Diarrhoea in mice infected with a human rotavirus. J. Gen. Virol. **67:**577–581.

Grauballe, P. C., J. Genner, A. Meyling, and A. Hornsleth. 1977. Rapid diagnosis of rotavirus infections: comparison of electron microscopy and immunoelectro-osmophoresis for the detection of rotavirus in human infantile gastroenteritis. J. Gen. Virol. **35:**203–218.

Grauballe, P. C., K. Hjelt, P. A. Krasilnikoff, and P. O. Schitz. 1981a. ELISA for rotavirus-specific secretory IgA in human sera. Lancet **2:**588–589.

Grauballe, P. G., B. F. Vestergaard, A. Meyling, and J. Genner. 1981b. Optimized enzyme-linked immunosorbent assay for determination of human and bovine rotavirus in stools: comparison with electron microscopy, immunoelectro-osmophoresis and fluorescent antibody techniques. J. Med. Virol. **7:**29–40.

Greenberg, H. B., A. R. Kalica, R. G. Wyatt, R. W. Jones, A. Z. Kapikian, and R. M. Chanock. 1981. Rescue of noncultivatable human rotavirus by gene reassortment during mixed infection with ts mutants of a cultivable bovine rotavirus. Proc. Natl. Acad. Sci. USA **78:**420–424.

Greenberg, H. B., R. G. Wyatt, A. Z. Kapikian, A. R. Kalica, J. Flores, and R. Jones. 1982. Rescue and serotypic characterization of noncultivable human rotavirus by gene reassortment. Infect. Immun. **37:**104–109.

Greenberg, H. B., J. Flores, A. R. Kalica, R. G. Wyatt, and R. Jones. 1983a. Gene coding assignments for growth restriction, neutralization and subgroup specificities of the W and DS-1 strains of human rotavirus. J. Gen. Virol. **64:**313–320.

Greenberg, H., V. McAuliffe, J. Valdesuso, R. Wyatt, J. Flores, A. Kalica, Y. Hoshino, and N. Singh. 1983b. Serological analysis of the subgroup protein of rotavirus, using monoclonal antibodies. Infect. Immun. **39:**91–99.

Greenberg, H. B., J. Valdesuso, K. van Wyke, K. Midthun, M. Walsh, V. McAuliffe, R. G. Wyatt, A. R. Kalica, J. Flores, and Y. Hoshino. 1983c. Production and preliminary characterization of monoclonal antibodies directed at two surface proteins of rhesus rotavirus. J. Virol. **47:**267–275.

Grimwood, K. 1986. Serum and mucosal responses to rotavirus infections in infants and young children. M.D. thesis, University of Melbourne, Melbourne, Australia.

Gurwith, M., W. Wenman, D. Hinde, S. Feltham, and H. Greenberg. 1981. A prospective study of rotavirus infection in infants and young children. J. Infect. Dis. 144:218–224.

Halvorsrud, J., and I. Ørstavik. 1980. An epidemic of rotavirus-associated gastroenteritis in a nursing home for the elderly. Scand. J. Infect. Dis. 12:161–164.

Hammond, G. W., G. S. Ahluwalia, B. Klisko, and P. R. Hazelton. 1984. Human rotavirus detection by counterimmunoelectrophoresis versus enzyme immunoassay and electron microscopy after direct centrifugation. J. Clin. Microbiol. 19:439–441.

Hamre, D., C. G. Loosli, and P. Gerber. 1958. Antigenic variants of influenza A virus (PR8 strain). III. Serological relationships of a line of variants derived in sequence in mice given homologous vaccine. J. Exp. Med. 107:829–844.

Hasegawa, A., S. Matsuno, S. Inouye, R. Kono, Y. Tsurukubo, A. Mukoyama, and Y. Saito. 1982. Isolation of human rotaviruses in primary cultures of monkey kidney cells. J. Clin. Microbiol. 16:387–390.

Hasegawa, A., S. Inouye, S. Matsuno, K. Yamaoka, R. Eko, W. Suharyono. 1984. Isolation of human rotaviruses with a distinct RNA electrophoretic pattern from Indonesia. Microbiol. Immunol. 28:719–722.

Hebert, J. P., R. Caillet, B. Hacquard, and B. Fortier. 1981. Use of Staphylococcus aureus protein "A" to detect rotavirus in stools. Pathol. Biol. (Paris) 29:101–104.

Herring, A. J., N. F. Inglis, C. K. Ojeh, D. R. Snodgrass, and J. D. Menzies. 1982. Rapid diagnosis of rotavirus infection by the direct detection of viral nucleic acid in silver stained polyacrylamide gels. J. Clin. Microbiol. 16:473–477.

Herrmann, J. E., N. R. Blacklow, D. M. Perron, G. Cukor, P. J. Krause, J. S. Hyams, H. J. Barrett, and P. L. Ogra. 1985. Enzyme immunoassay with monoclonal antibodies for the detection of rotavirus in stool specimens. J. Infect. Dis. 152:830–832.

Hohmann, A., J. T. La Brooy, G. P. Davidson, and D. J. C. Shearman. 1983. Measurement of specific antibodies in human intestinal aspirate: effect of the protease inhibitor phenyl-methyl sulfonyl fluoride. J. Immunol. Methods 64:199–204.

Holmes, I. H. 1979. Viral gastroenteritis. Prog. Med. Virol. 25:1–36.

Holmes, I. H. 1983. Rotaviruses, p. 359–423. In W. K. Joklik (ed.), The Reoviridae. Plenum Press, New York.

Holmes, I. H. 1985. Epidemiology of rotavirus infections based on analysis of genome RNA, p. 195–200. In S. Tzipori (ed.), Infectious diarrhoea in the young. Elsevier, Amsterdam.

Holmes, I. H., B. J. Ruck, R. F. Bishop, and G. P. Davidson. 1975. Infantile enteritis viruses: morphogenesis and morphology. J. Virol. 16:937–943.

Holmes, I. H., S. M. Rodger, R. D. Schnagl, B. J. Ruck, I. D. Gust, R. F. Bishop, and G. L. Barnes. 1976. Is lactase the receptor and uncoating enzyme for infantile enteritis (rota) viruses? Lancet 1:1387–1389.

Hoshino, Y., R. G. Wyatt, H. B. Greenberg, J. Flores, and A. Z. Kapikian. 1984. Serotypic similarity and diversity of human and animal rotaviruses as studied by plaque reduction neutralization. J. Infect. Dis. 149:694–702.

Hoshino, Y., M. M. Sereno, K. Midthun, J. Flores, A. Z. Kapikian, and R. M. Chanock. 1985. Independent segregation of two antigenic specificities (VP3 and VP7) involved in neutralization of rotavirus infectivity. Proc. Natl. Acad. Sci. USA 82:8701–8704.

Hughes, J. H., A. V. Tuomari, D. R. Mann, and V. V. Hamparian. 1984. Latex immunoassay for rapid detection of rotavirus. J. Clin. Microbiol. 20:441–447.

Hundley, F., B. Biryahwaho, M. Gow, and U. Desselberger. 1985. Genome rearrangements of bovine rotavirus after serial passage at high multiplicity of infection. Virology 143:88–103.

Hung, T., G. Chen, C. Wang, Z. Chou, T. Chao, W. Ye, H. Yao, and K. Meng. 1983. Rotavirus-like agent in adult non-bacterial diarrhoea in China. Lancet 2:1078–1079.

Hung, T., G. Chen, C. Wang, H. Yao, Z. Fang, T. Chao, Z. Chou, W. Ye, X. Chang, S. Den, X. Liong, and W. Chang. 1984. Waterborne outbreak of rotavirus diarrhoea in adults in China caused by a novel rotavirus. Lancet 1:1139–1142.

Imai, M., M. A. Richardson, N. Ikegami, A. J. Shatkin, and Y. Furuichi. 1983. Molecular cloning of double-stranded RNA virus genomes. Proc. Natl. Acad. Sci. USA 80:373–377.

Inouye, S., S. Matsuno, and H. Yamaguchi. 1984. Efficient coating of the solid phase with rotavirus antigens for enzyme-linked immunosorbent assay of immunoglobulin A antibody in feces. J. Clin. Microbiol. 19:259–263.

Johnson, D. A., J. W. Gautsch, J. R. Sportsman, and J. H. Elder. 1984. Improved technique utilizing nonfat dry milk for analysis of proteins and nucleic acids transferred to nitrocellulose. Gene Anal. Tech. 1:3–8.

Kalica, A. R., R. H. Purcell, M. M. Sereno, R. G. Wyatt, H. W. Kim, R. M. Chanock, and A. Z. Kapikian. 1977. A microtiter solid-phase radioimmunoassay for detection of the human reovirus-like agent in stools. J. Immunol. 118:1275–1279.

Kalica, A. R., M. M. Sereno, R. G. Wyatt, C. A. Mebus, R. M. Chanock, and A. Z. Kapikian. 1978. Comparison of human and animal rotavirus strains by gel electrophoresis of viral RNA. Virology 87:247–255.

Kalica, A. R., H. B. Greenberg, R. T. Espejo, J. Flores, R. G. Wyatt, A. Z. Kapikian, and R. M. Chanock. 1981a. Distinctive ribonucleic acid patterns of human rotavirus subgroups 1 and 2. Infect. Immun. 33:958–961.

Kalica, A. R., H. B. Greenberg, R. G. Wyatt, J. Flores, M. M. Sereno, A. Z. Kapikian, and R. M. Chanock. 1981b. Genes of human (strain Wa) and bovine (strain UK) rotaviruses that code for neutralization and subgroup antigens. Virology 112:385–390.

Kantharidis, P., M. L. Dyall-Smith, and I. H. Holmes. 1983. Completion of the gene coding assignments of SA11 rotavirus: gene products of segments 7, 8 and 9. J. Virol. 48:330–334.

Kapikian, A. Z., and R. M. Chanock. 1985. Rotaviruses, p. 863–906. In B. N. Fields, D. N. Knipe, R. M. Chanock, J. L. Melnick, B. Roizman, and R. E. Shope (ed.), Virology. Raven Press, New York.

Kapikian, A. Z., H. W. Kim, R. G. Wyatt, W. J. Rodriguez, S. Ross, W. L. Cline, R. H. Parrott, and R. M. Chanock. 1974. Reoviruslike agent in stools: association with infantile diarrhea and development of serologic tests. Science 185:1049–1053.

Kapikian, A. Z., R. G. Wyatt, H. B. Greenberg, A. R. Kalica, H. W. Kim, C. D. Brandt, W. J. Rodriguez, R. H. Parrott, and R. M. Chanock. 1980. Approaches to immunization of infants and young children against gastroenteritis due to rotaviruses. Rev. Infect. Dis. 2:459–469.

Kapikian, A. Z., W. L. Cline, H. B. Greenberg, R. G. Wyatt, A. R. Kalica, C. E. Banks, H. D. James, Jr., J. Flores, and R. M. Chanock. 1981. Antigenic characterization of human and animal rotaviruses by immune adherence hemagglutination assay (IAHA): evidence for distinctness of IAHA and neutralization antigens. Infect. Immun. **33:**415–425.

Killen, H. M., and N. J. Dimmock. 1982. Identification of a neutralization-specific antigen of a calf rotavirus. J. Gen. Virol. **62:**297–311.

Kim, H. W., C. D. Brandt, A. Z. Kapikian, R. G. Wyatt, J. O. Arrobio, W. J. Rodriguez, R. M. Chanock, and R. H. Parrott. 1977. Human reovirus-like agent infection. Occurrence in adult contacts of pediatric patients with gastroenteritis. J. Am. Med. Assoc. **238:**404–407.

Knisley, C. V., A. J. Bednarz-Prashad, and L. K. Pickering. 1986. Detection of rotavirus in stool specimens with monoclonal and polyclonal antibody-based assay systems. J. Clin. Microbiol. **23:**897–900.

Knowlton, D. R., and R. L. Ward. 1985. Effect of mutation in immunodominant neutralization epitopes on the antigenicity of rotavirus SA11. J. Gen. Virol. **66:**2375–2381.

Konno, T., H. Suzuki, A. Imai, and N. Ishida. 1977. Reovirus-like agent in acute epidemic gastroenteritis in Japanese infants: faecal shedding and serologic response. J. Infect. Dis. **135:**259–266.

Kraft, L. M. 1958. Observations on the control and natural history of epidemic diarrhea of infant mice (EDIM). Yale J. Biol. Med. **31:**121–137.

Krause, P. J., J. S. Hyams, P. J. Middleton, V. C. Herson, and J. Flores. 1983. Unreliability of Rotazyme ELISA test in neonates. J. Pediatr. **103:**259–262.

Kurstak, E. 1985. Progress in enzyme immunoassays: production of reagents, experimental design, and interpretation. Bull. WHO **63:**793–811.

Kutsuzawa, T., T. Konno, H. Suzuki, A. Z. Kapikian, T. Ebina, and N. Ishida. 1982. Isolation of human rotavirus subgroups 1 and 2 in cell culture. J. Clin. Microbiol. **16:**727–730.

Laemmli, U. K. 1970. Cleavage of structural proteins during the assembly of the head of bacteriophage T4. Nature **227:**680–685.

Lazdins, I., S. Sonza, M. L. Dyall-Smith, B. S. Coulson, and I. H. Holmes. 1985. Demonstration of an immunodominant neutralization site by analysis of antigenic variants of SA11 rotavirus. J. Virol. **56:**317–319.

Lecce, J. G., M. W. King, and W. E. Dorsey. 1978. Rearing regimen producing piglet diarrhea (rotavirus) and its relevance to acute infantile diarrhea. Science **199:**776–778.

Light, J. S., and H. L. Hodes. 1949. Isolation from cases of infantile diarrhea of a filtrable agent causing diarrhea in calves. J. Exp. Med. **90:**113–135.

Lin, M., M. Imai, A. R. Bellamy, N. Ikegami, Y. Furuichi, D. Summers, D. L. Nuss, and R. Deibel. 1985. Diagnosis of rotavirus infection with cloned cDNA copies of viral gene segments. J. Virol. **55:**509–512.

Linhares, A. C., F. P. Pinheiro, R. B. Freitas, Y. B. Gabbay, J. A. Shirley, and G. M. Beards. 1981. An outbreak of rotavirus diarrhea among a nonimmune, isolated South American Indian community. Am. J. Epidemiol. **113:**703–710.

Little, L. M., and J. A. Shadduck. 1982. Pathogenesis of rotavirus infection in mice. Infect. Immun. **38:**755–763.

Liu, S., C. Birch, A. Coulepis, and I. Gust. 1984. Radioimmunofocus assay for detection and quantitation of human rotavirus. J. Clin. Microbiol. **20:**347–350.

Lopez, S., C. F. Arias, J. R. Bell, J. H. Strauss, and R. T. Espejo. 1985. Primary structure of the cleavage site associated with trypsin enhancement of rotavirus SA11 infectivity. Virology **144:**11–19.

Lourenco, M. H., J. C. Nicolas, J. Cohen, R. Scherrer, and F. Bricout. 1981. Study of human rotavirus genome by electrophoresis: attempt of classification among strains isolated in France. Ann. Virol. (Paris) **132E:**161–173.

Lycke, E., J. Blomberg, G. Bag, A. Eriksson, and L. Madsen. 1978. Epidemic acute diarrhoea in adults associated with infantile gastroenteritis virus. Lancet **2:**1056–1057.

Madeley, C. R. 1977. Guide to the collection and transport of virological specimens. World Health Organization, Geneva.

Malherbe, H. H., and M. Strickland-Cholmley. 1967. Simian virus SA11 and the related O agent. Arch. Ges. Virusforsch. **22:**235–245.

Mason, B. B., D. Y. Graham, and M. K. Estes. 1983. Biochemical mapping of the simian rotavirus SA11 genome. J. Virol. **46:**413–423.

Mata, L., A. Simhon, J. J. Urrutia, R. A. Kronmal, R. Fernandez, and B. Garcia. 1983. Epidemiology of rotaviruses in a cohort of 45 Guatemalan Mayan Indian children observed from birth to the age of three years. J. Infect. Dis. **148:**452–461.

Mathan, M., J. D. Almeida, and J. Cole. 1977. An antigenic subunit present in rotavirus infected faeces. J. Gen. Virol. **34:**325–329.

Matsuno, S., and S. Nagayoshi. 1978. Quantitative estimation of infantile gastroenteritis virus antigens in stools by immune adherence hemagglutination test. J. Clin. Microbiol. **7:**310–311.

Matsuno, S., S. Inouye, and R. Kono. 1977. Plaque assay of neonatal calf diarrhea virus and the neutralizing antibody in human sera. J. Clin. Microbiol. **5:**1–4.

Matsuno, S., A. Hasegawa, A. Mukoyama, and S. Inouye. 1985. A candidate for a new serotype of human rotavirus. J. Virol. **54:**623–624.

McLean, B., and I. H. Holmes. 1980. Transfer of antirotaviral antibodies from mothers to their infants. J. Clin. Microbiol. **12:**320–325.

McLean, B. S., and I. H. Holmes. 1981. Effects of antibodies, trypsin and trypsin inhibitors on susceptibility of neonates to rotavirus infection. J. Clin. Microbiol. **13:**22–29.

McNulty, M. S. 1978. Rotaviruses. J. Gen. Virol. **40:**1–18.

McNulty, M. S., G. M. Allan, D. Todd, J. B. McFerran, E. R. McKillop, D. S. Collins, and R. M. McCracken. 1980. Isolation of rotaviruses from turkeys and chickens: demonstration of distinct serotypes and RNA electropherotypes. Avian Pathol. **9:**363–375.

McNulty, M. S., D. Todd, G. M. Allan, J. B. McFerran, and J. A. Greene. 1984. Epidemiology of rotavirus infection in broiler chickens: recognition of four serogroups. Arch. Virol. **81:**113–121.

Mebus, C. A., and L. E. Newman. 1977. Scanning electron, light, and immunofluorescent microscopy of intestine of gnotobiotic calf infected with reovirus-like agent. Am. J. Vet. Res. **38:**553–558.

Mebus, C. A., N. R. Underdahl, M. B. Rhodes, and M. J. Twiehaus. 1969. Calf diarrhea (scours): reproduced with a virus from a field outbreak. Univ. Neb. Agric. Exp. Stn. Res. Bull. **233:**1–16.

Mebus, C. A., M. Kono, N. R. Underdahl, and M. J. Twiehaus. 1971a. Cell culture propagation of neonatal calf diarrhea (scours) virus. Can. Vet. J. **12:**69–72.

Mebus, C. A., E. L. Stair, N. R. Underdahl, and M. J. Twiehaus. 1971b. Pathology of neonatal calf diarrhea induced by a reo-like virus. Vet. Pathol. **8:**490–505.

Middleton, P. J., M. T. Szymanski, G. D. Abbott, R. Bortolussi, and J. R. Hamilton. 1974. Orbivirus: acute gastroenteritis of infancy. Lancet **1:**1241–1244.

Middleton, P. J., M. Petric, C. M. Hewitt, M. T. Szymanski, and J. S. Tam. 1976. Counter-immunoelectroosmophoresis for the detection of infantile gastroenteritis virus (orbi-group) antigen and antibody. J. Clin. Pathol. **29**:191–197.

Middleton, P. J., M. D. Holdaway, M. Petric, M. T. Szymanski, and J. S. Tam. 1977. Solid-phase radioimmunoassay for the detection of rotavirus. Infect. Immun. **16**:439–444.

Midthun, K., H. B. Greenberg, Y. Hoshino, A. Z. Kapikian, R. G. Wyatt, and R. M. Chanock. 1985. Reassortant rotaviruses as potential live rotavirus vaccine candidates. J. Virol. **53**:949–954.

Miotti, P. G., J. Eiden, and R. H. Yolken. 1985. Comparative efficiency of commercial immunoassays for the diagnosis of rotavirus gastroenteritis during the course of infection. J. Clin. Microbiol. **22**:693–698.

Moosai, R. B., M. J. Carter, and C. R. Madeley. 1984. Rapid detection of enteric adenovirus and rotavirus: a simple method using polyacrylamide gel electrophoresis. J. Clin. Pathol. **37**:1404–1408.

Moosai, R. B., R. Alcock, T. M. Bell, F. R. Laidler, J. S. M. Pieris, A. P. Wyn-Jones, and C. R. Madeley. 1985. Detection of rotavirus by a latex-agglutination test, Rotalex: comparison with electron microscopy, immunofluorescence, polyacrylamide gel electrophoresis, and enzyme linked immunosorbent assay. J. Clin. Pathol. **38**:694–700.

Morinet, F., F. Ferchal, R. Colimon, and Y. Perol. 1984. Comparison of six methods for detecting human rotavirus in stools. Eur. J. Clin. Microbiol. **3**:136–140.

Murphy, A. M., M. B. Albrey, and E. B. Crewe. 1977. Rotavirus infections of neonates. Lancet **2**:1149–1150.

Nakagomi, O., A. Nakagomi, T. Suto, H. Suzuki, T. Kutsuzawa, F. Tazawa, T. Konno, and N. Ishida. 1982. Detection of human rotavirus by reversed passive hemagglutination (RPHA) using antibody against a cultivable human rotavirus as compared with electron microscopy (EM) and enzyme-linked immunosorbent assay (ELISA). Microbiol. Immunol. **26**:747–751.

Nakagomi, O., T. Nakagomi, H. Oyamada, and T. Suto. 1985. Relative frequency of human rotavirus subgroups I and II in Japanese children with acute gastroenteritis. J. Med. Virol. **17**:29–34.

Newman, J. F. F., F. Brown, J. C. Bridger, and G. N. Woode. 1975. Characterization of a rotavirus. Nature **258**:631–633.

Nicolaieff, A., G. Obert, and M. H. V. van Regenmortel. 1980. Detection of rotavirus by serological trapping on antibody-coated electron microscope grids. J. Clin. Microbiol. **12**:101–104.

Nicolas, J. C., M. H. Lourenco, S. Marchal, J. Cohen, R. Scherrer, and F. Bricout. 1983. Description of a very simplified method for rotavirus ds RNA extraction. Ann. Virol. (Paris) **134**:135–139.

Nicolas, J. C., P. Pothier, J. Cohen, M. H. Lourenco, R. Thompson, P. Guimbaud, A. Chenon, M. Dauvergne, and F. Bricout. 1984. Survey of human rotavirus propagation as studied by electrophoresis of genomic RNA. J. Infect. Dis. **149**:688–693.

Novo, E., and J. Esparza. 1981. Composition and topography of structural polypeptides of bovine rotavirus. J. Gen. Virol. **56**:325–335.

Obert, G., R. Gloeckler, J. Burckard, and M. H. van Regenmortel. 1981. Comparison of immunosorbent electron microscopy, enzyme immunoassay and counterimmunoelectrophoresis for detection of human rotavirus in stools. J. Virol. Methods **3**:99–107.

Offit, P. A., and H. F. Clark. 1985a. Protection against rotavirus-induced gastroenteritis in a murine model by passively acquired gastrointestinal but not circulating antibodies. J. Virol. **54**:58–64.

Offit, P. A., and H. F. Clark. 1985b. Maternal antibody-mediated protection against gastroenteritis due to rotavirus in newborn mice is dependent on both serotype and titer of antibody. J. Infect. Dis. **152**:1152–1158.

Offit, P. A., H. F. Clark, A. H. Taylor, R. G. Hess, P. A. Bachmann, and S. A. Plotkin. 1984. Rotavirus-specific antibodies in fetal bovine serum and commercial preparations of serum albumin. J. Clin. Microbiol. **20**:266–270.

Offit, P. A., G. Blavat, H. B. Greenberg, and H. F. Clark. 1986a. Molecular basis of rotavirus virulence: role of gene segment 4. J. Virol. **57**:46–49.

Offit, P. A., R. D. Shaw, and H. B. Greenberg. 1986b. Passive protection against rotavirus-induced diarrhea by monoclonal antibodies to VP3 and VP7. J. Virol. **58**:700–703.

Okada, Y., M. A. Richardson, N. Ikegami, A. Nomoto, and Y. Furuichi. 1984. Nucleotide sequence of human rotavirus genome segment 10, an RNA encoding a glycosylated virus protein. J. Virol. **51**:856–859.

Ørstavik, I., and K. W. Haug. 1976. Virus-specific IgM antibodies in acute gastroenteritis due to a reovirus-like agent (rotavirus). Scand. J. Infect. Dis. **8**:237–240.

Ørstavik, I., K. W. Haug, and A. Søvde. 1976. Rotavirus-associated gastroenteritis in two adults probably caused by virus reinfection. Scand. J. Infect. Dis. **8**:277–278.

Othman, R. Y., M. Y. Jaliha, N. B. Rasool, and S. K. Lam. 1986. Asian group for rapid viral diagnosis—evaluation of two commercial kits for the detection of rotavirus in faecal samples. Virus Information Exch. Newsl. **3**:5–6.

Pedley, S., and M. A. McCrae. 1984. A rapid screening assay for detecting individual RNA species in field isolates of rotaviruses. J. Virol. Methods **9**:173–181.

Pedley, S., J. C. Bridger, J. F. Brown, and M. A. McCrae. 1983. Molecular characterization of rotaviruses with distinct group antigens. J. Gen. Virol. **64**:2093–2101.

Pedley, S., F. Hundley, I. Chrystie, M. A., McCrae, and U. Desselberger. 1984. The genomes of rotaviruses isolated from chronically infected immunodeficient children. J. Gen. Virol. **65**:1141–1150.

Pereira, H. G., R. S. Azeredo, F. Sutmoller, J. P. G. Leite, V. de Farias, O. M. Barth, and M. N. P. Vidal. 1983a. Comparison of polyacrylamide gel electrophoresis (PAGE), immunoelectron microscopy (IEM) and enzyme immunoassay (EIA) for the rapid diagnosis of rotavirus infection in children. Mem. Inst. Oswaldo Cruz **78**:483–490.

Pereira, H. G., J. P. G. Leite, R. S. Azeredo, V. de Farias, and F. Sutmoller. 1983b. An atypical rotavirus detected in a child with gastroenteritis in Rio de Janeiro, Brazil. Mem. Inst. Oswaldo Cruz **78**:245–250.

Portnoy, B. L., R. Conklin, M. Menn, J. Olarte, and H. L. Du Pont. 1977. Reliable identification of reovirus-like agent in diarrheal stools. J. Lab. Clin. Med. **89**:560–563.

Richardson, A., A. Iwamoto, N. Ikegami, A. Nomoto, and Y. Furuichi. 1984. Nucleotide sequence of the gene encoding the serotype specific antigen of human (Wa) rotavirus: comparison with the homologous genes from simian SA11 and UK bovine rotaviruses. J. Virol. **51**:860–862.

Rodger, S. M., and I. H. Holmes. 1979. Comparison of the genomes of simian, bovine, and human rotaviruses by gel electrophoresis and detection of genomic variation among bovine isolates. J. Virol. **30**:839–846.

Rodger, S. M., R. D. Schnagl, and I. H. Holmes. 1975. Biochemical and biophysical characteristics of diarrhea viruses of human and calf origin. J. Virol. **16**:1229–1235.

Rodger, S. M., R. F. Bishop, C. Birch, B. McLean, and I. H. Holmes. 1981. Molecular epidemiology of human rotaviruses in Melbourne, Australia, from 1973 to 1979, as determined by electrophoresis of genome ribonucleic acid. J. Clin. Microbiol. **13:**272–278.

Rodger, S. M., R. F. Bishop, and I. H. Holmes. 1982. Detection of a rotavirus-like agent associated with diarrhoea in an infant. J. Clin. Microbiol. **16:**724–726.

Rodriguez, W. J., H. W. Kim, J. O. Arrobio, C. D. Brandt, R. M. Chanock, A. Z. Kapikian, R. G. Wyatt, and R. H. Parrott. 1977. Clinical features of acute gastroenteritis associated with human reovirus-like agent in infants and young children. J. Pediatr. **91:**188–193.

Rodriguez, W. J., H. W. Kim, C. D. Brandt, M. K. Gardner, and R. H. Parrott. 1983. Use of electrophoresis of RNA from human rotavirus to establish the identity of strains involved in outbreaks in a tertiary care nursery. J. Infect. Dis. **148:**34–40.

Rogers, F. G., S. Chapman, and H. Whitby. 1981. A comparison of lyphogel, ammonium sulphate and ultra-centrifugation in the concentration of stool-viruses for electron microscopy. J. Clin. Pathol. **34:**227.

Rook, G. A. W., and C. H. Cameron. 1981. An inexpensive, portable, battery-operated photometer for the reading of ELISA tests in microtitration plates. J. Immunol. Methods **40:**109–114.

Roseto, A., J. Escaig, E. Delain, J. Cohen, and R. Scherrer. 1979. Structure of rotaviruses as studied by the freeze-drying technique. Virology **98:**471–475.

Sabara, M., D. Deregt, L. A. Babiuk, and V. Misra. 1982. Genetic heterogeneity within individual bovine rotavirus isolates. J. Virol. **44:**813–822.

Salmi, T. T., P. Arstila, and A. Koivikko. 1978. Central nervous system involvement in patients with rotavirus gastroenteritis. Scand. J. Infect. Dis. **10:**29–31.

Sambourg, M., A. Goudeau, C. Courant, G. Pinon, and F. Denis. 1985. Direct appraisal of latex agglutination testing, a convenient alternative to enzyme immunoassay for the detection of rotavirus in childhood gastroenteritis, by comparison of two enzyme immunoassays and two latex tests. J. Clin. Microbiol. **21:**622–625.

Sanekata, T., and H. Okada. 1983. Human rotavirus detection by agglutination of antibody-coated erythrocytes. J. Clin. Microbiol. **17:**1141–1147.

Santosham, M., R. H. Yolken, E. Quiroz, L. Dillman, G. Oro, W. G. Reeves, and R. B. Sack. 1983. Detection of rotavirus in respiratory secretions of children with pneumonia. J. Pediatr. **103:**583–585.

Sarkkinen, H. K. 1981. Human rotavirus antigen detection by enzyme-immunoassay with antisera against Nebraska calf diarrhoea virus. J. Clin. Pathol. **34:**680–685.

Sarkkinen, H. K., P. E. Halonen, and P. P. Arstila. 1979. Comparison of four-layer radioimmunoassay and electron microscopy for detection of human rotavirus. J. Med. Virol. **4:**255–260.

Sarkkinen, H. K., H. Tuokko, and P. E. Halonen. 1980. Comparison of enzyme immunoassay and radioimmunoassay for detection of human rotaviruses and adenoviruses in stool specimens. J. Virol. Methods **1:**331–341.

Sato, K., Y. Inaba, T. Shinozaki, R. Fujii, and M. Matsumoto. 1981. Isolation of human rotavirus in cell cultures. Arch. Virol. **69:**155–160.

Saulsbury, F. T., J. A. Winkelstein, and R. H. Yolken. 1980. Chronic rotavirus infection in immunodeficiency. J. Pediatr. **97:**61–65.

Scherrer, R., and S. Bernard. 1977. Application of enzyme-linked immunosorbent assay (ELISA) to the detection of calf rotavirus and rotavirus antibodies. Ann. Microbiol. (Paris) **128A:**499–510.

Schroeder, B. A., J. E. Street, J. Kalmakoff, and A. R. Bellamy. 1982. Sequence relationships between the genome segments of human and animal rotavirus strains. J. Virol. **43:**379–385.

Shaw, R., D. L. Stoner-Ma, M. K. Estes, and H. B. Greenberg. 1985. Specific enzyme-linked immunoassay for rotavirus serotypes 1 and 3. J. Clin. Microbiol. **22:**286–291.

Sheridan, J. F., L. Aurelian, G. Barbour, M. Santosham, R. B. Sack, and R. W. Ryder. 1981. Traveller's diarrhea associated with rotavirus infection: analysis of virus-specific immunoglobulin classes. Infect. Immun. **31:**419–429.

Shinozaki, T., H. Ushijima, T. Tajima, B. Kim, K. Araki, and R. Fujii. 1985. Evaluation of four tests for detecting human rotavirus in feces. Eur. J. Pediatr. **143:**238.

Simhon, A., L. Mata, M. Vives, L. Rivera, S. Vargas, G. Ramirez, L. Lizano, G. Catarinella, and J. Azofeifa. 1985. Low endemicity and low pathogenicity of rotaviruses among rural children in Costa Rica. J. Infect. Dis. **152:**1134–1142.

Smith, K. O. 1967. Identification of viruses by electron microscopy, p. 545–567. *In* H. Busch (ed.), Methods in cancer research, Vol. 1. Academic Press, New York.

Smith, M. L., I. Lazdins, and I. H. Holmes. 1980. Coding assignments of dsRNA segments of SA11 rotavirus established by in vitro translation. J. Virol. **33:**976–982.

Snodgrass, D. R., and P. W. Wells. 1976. Rotavirus infection in lambs: studies on passive protection. Arch. Virol. **52:**201–206.

Snodgrass, D. R., and P. W. Wells. 1978. The immunoprophylaxis of rotavirus infections in lambs. Vet. Rec. **102:**146–148.

Snodgrass, D. R., C. R. Madeley, P. W. Wells, and K. W. Angus. 1977. Human rotavirus in lambs: infection and passive protection. Infect. Immun. **16:**268–270.

Snodgrass, D. R., A. J. Herring, I. Campbell, J. M. Inglis, and F. D. Hargreaves. 1984a. Comparison of atypical rotaviruses from calves, piglets, lambs and man. J. Gen. Virol. **65:**909–914.

Snodgrass, D. R., C. K. Ojeh, I. Campbell, and A. J. Herring. 1984b. Bovine rotavirus serotypes and their significance for immunization. J. Clin. Microbiol. **20:**342–346.

Sonza, S., and I. H. Holmes. 1980. Coproantibody response to rotavirus infection. Med. J. Aust. **2:**496–499.

Sonza, S., A. M. Breschkin, and I. H. Holmes. 1983. Derivation of neutralizing monoclonal antibodies against rotavirus. J. Virol. **45:**1143–1146.

Spence, L., M. Fauvel, R. Petro, and S. Bloch. 1977. Comparison of counterimmunoelectrophoresis and electron microscopy for laboratory diagnosis of human reovirus-like agent-associated infantile gastroenteritis. J. Clin. Microbiol. **5:**248–249.

Spencer, E. G., L. F. Avendano, and B. I. Garcia. 1983. Analysis of human rotavirus mixed electropherotypes. Infect. Immun. **39:**569–574.

Street, J. E., M. C. Croxson, W. F. Chadderton, and A. R. Bellamy. 1982. Sequence diversity of human rotavirus strains investigated by Northern blot hybridization analysis. J. Virol. **43:**369–378.

Stuker, G., L. S. Oshiro, and N. J. Schmidt. 1980. Antigenic comparisons of two new rotaviruses from rhesus monkeys. J. Clin. Microbiol. **11:**202–203.

Suzuki, H., T. Konno, Y. Numazaki, S. Kitaoka, T. Sato, A. Imai, F. Tazawa, T. Nakagomi, O. Nakagomi, and N. Ishida. 1984. Three different serotypes of human rotavirus determined using an interference test with coxsackievirus B1. J. Med. Virol. **13:**41–44.

Svensson, L., M. Grandien, and C.-A. Pettersson. 1983. Comparison of solid-phase immune electron microscopy by use of protein A with direct electron microscopy and enzyme-linked immunosorbent assay for detection of rotavirus in stool. J. Clin. Microbiol. **18**:1244–1249.

Taniguchi, K., T. Urasawa, S., Urasawa, and T. Yasuhara. 1984. Production of subgroup-specific monoclonal antibodies against human rotaviruses and their application to an enzyme-linked immunosorbent assay for subgroup determination. J. Med. Microbiol. **14**:115–125.

Taniguchi, K., S. Urasawa, and T. Urasawa. 1985. Preparation and characterization of neutralizing monoclonal antibodies with different reactivity patterns to human rotaviruses. J. Gen. Virol. **66**:1045–1053.

Theil, K. W., E. H. Bohl, and A. G. Agnes. 1977. Cell culture propagation of porcine rotavirus (reovirus-like agent). Am. J. Vet. Res. **38**:1765–1768.

Theil, K. W., C. M. McClosky, L. J. Saif, D. R. Redman, E. H. Bohl, D. D. Hancock, E. M. Kohler, and P. D. Moorhead. 1981. A rapid, simple method for preparing rotaviral double-stranded ribonucleic acid for analysis by polyacrylamide gel electrophoresis. J. Clin. Microbiol. **14**:273–280.

Thouless, M. E., A. S. Bryden, T. H. Flewett, G. N. Woode, J. C. Bridger, D. R. Snodgrass, and J. A. Herring. 1977. Serological relationships between rotaviruses from different species as studied by complement fixation and neutralization. Arch. Virol. **53**:287–294.

Thouless, M. E., G. M. Beards, and T. H. Flewett. 1982. Serotyping and subgrouping of rotavirus strains by the ELISA test. Arch. Virol. **73**:219–230.

Tsuchie, H., K. Shimase, I. Tamura, O. Kurimura, E. Kaneto, T. Katsumoto, M. Ito, and T. Kurimura. 1983. Comparison of enzyme-linked immunosorbent assay, electron microscopy, and reversed passive hemagglutination for detection of human rotavirus in stool specimens. Biken J. **26**:87–92.

Tufvesson, B., and T. Johnsson. 1976. Immunoelectroosmophoresis for detection of reolike virus: methodology and comparison with electron microscopy. Acta Pathol. Microbiol. Scand. B **84**:225–228.

Urasawa, T., S. Urasawa, and K. Taniguchi. 1981. Sequential passages of human rotavirus in MA-104 cells. Microbiol. Immunol. **25**:1025–1035.

Urasawa, S., T. Urasawa, and K. Taniguchi. 1982. Three human rotavirus serotypes demonstrated by plaque neutralization of isolated strains. Infect. Immun. **38**:781–784.

Vesikari, T., H. K. Sarkkinen, and M. Mäki. 1981. Quantitative aspects of rotavirus excretion in childhood diarrhoea. Acta Paediatr. Scand. **70**:717–721.

Vesikari, T., E. Isolauri, E. D'Hondt, A. Delem, F. E. Andre, and G. Zissis. 1984. Protection of infants against rotavirus diarrhoea by RIT 4237 attenuated bovine rotavirus strain vaccine. Lancet **1**:977–981.

von Bonsdorff, C. H., T. Hovi, P. Makela, L. Hovi, and M. Tevalvoto-Aarn. 1976. Rotavirus associated with acute gastroenteritis in adults. Lancet **2**:423.

Wang, S., R. Cai, J. Chen, R. Li, and R. Jiang. 1985. Etiologic studies of the 1983 and 1984 outbreaks of epidemic diarrhea in Guangxi. Intervirology **24**:140–146.

Ward, C. W., T. C. Elleman, A. A. Azad, and M. L. Dyall-Smith. 1984. Nucleotide sequence of gene segment 9 encoding a nonstructural protein of UK bovine rotavirus. Virology **134**:249–253.

Ward, C. W., A. A. Azad, and M. L. Dyall-Smith. 1985. Structural homologies between RNA gene segments 10 and 11 from UK bovine, simian SA11, and human Wa rotaviruses. Virology **144**:328–336.

Ward, R. L., D. R. Knowlton, and M. J. Pierce. 1984.

Efficiency of human rotavirus propagation in cell culture. J. Clin. Microbiol. **19**:748–753.

Wenman, W. M., D. Hinde, S. Feltham, and M. Gurwith. 1979. Rotavirus infections in adults: results of a prospective family study. N. Engl. J. Med. **301**:303–306.

White, L., I. Perez, M. Perez, G. Urbina, H. B. Greenberg, A. Z. Kapikian, and J. Flores. 1984. Relative frequency of rotavirus subgroups 1 and 2 in Venezuelan children with gastroenteritis as assayed by monoclonal antibodies. J. Clin. Microbiol. **19**:516–520.

WHO Programme for Diarrhoeal Diseases Control. 1984. Nomenclature of human rotaviruses: designation of subgroups and serotypes. Bull. WHO **62**:501–503.

WHO Scientific Working Group. 1980. Rotavirus and other viral diarrhoeas. Bull. WHO **58**:183–198.

Wilsnack, R. E., J. H. Blackwell, and J. C. Parker. 1969. Identification of an agent of epizootic diarrhea of infant mice by immunofluorescent and complement fixation tests. Am. J. Vet. Res. **30**:1195–1209.

Woode, G. N. 1982. Rotaviruses in animals, p. 295–313. *In* D. A. J. Tyrrell and A. Z. Kapikian (ed.), Virus infections of the gastrointestinal tract. Marcel Dekker, New York.

Woode, G. N., J. Jones, and J. Bridger. 1975. Levels of colostral antibodies against neonatal calf diarrhoea virus. Vet. Rec. **97**:148–149.

Woode, G. N., J. C. Bridger, J. M. Jones, T. H. Flewett, A. S. Bryden, H. A. Davies, and G. B. B. White. 1976. Morphological and antigenic relationships between viruses (rotaviruses) from acute gastroenteritis of children, calves, piglets, mice, and foals. Infect. Immun. **14**:804–810.

Woode, G. N., N. E. Kelso, T. F. Simpson, S. K. Gaul, L. E. Evans, and L. Babiuk. 1983. Antigenic relationships among some bovine rotaviruses: serum neutralization and cross-protection in gnotobiotic calves. J. Clin. Microbiol. **18**:358–364.

Wyatt, R. G., C. A. Mebus, R. H. Yolken, A. R. Kalica, H. D. James, Jr., A. Z. Kapikian, and R. M. Chanock. 1979. Rotaviral immunity in gnotobiotic calves: heterologous resistance to human virus induced by bovine virus. Science **203**:548–550.

Wyatt, R. G., W. D. James, E. H. Bohl, K. W. Theil, L. J. Saif, A. R. Kalica, H. B. Greenberg, A. Z. Kapikian, and R. M. Chanock. 1980. Human rotavirus type 2: cultivation in vitro. Science **207**:189–191.

Wyatt, R. G., H. B. Greenberg, W. D. James, A. L. Pittman, A. R. Kalica, J. Flores, R. M. Chanock, and A. Z. Kapikian. 1982. Definition of human rotavirus serotypes by plaque reduction assay. Infect. Immun. **37**:110–115.

Wyatt, R. G., H. D. James, Jr., A. L. Pittman, Y. Hoshino, H. B. Greenberg, A. R. Kalica, J. Flores, and A. Z. Kapikian. 1983a. Direct isolation in cell culture of human rotaviruses and their characterization into four serotypes. J. Clin. Microbiol. **18**:310–317.

Wyatt, R. G., A. Z. Kapikian, and C. A. Mebus. 1983b. Induction of cross-reactive serum neutralizing antibody to human rotavirus in calves after in utero administration of bovine rotavirus. J. Clin. Microbiol. **18**:505–508.

Yolken, R. H. 1982. Enzyme immunoassays for detecting human rotavirus, p. 51–74. *In* D. A. J. Tyrrell and A. Z. Kapikian (ed.), Virus infections of the gastrointestinal tract. Marcel Dekker, New York.

Yolken, R. H., and F. J. Leister. 1981. Evaluation of enzyme immunoasssays for the detection of human rotavirus. J. Infect. Dis. **144**:379.

Yolken, R. H., and P. J. Stopa. 1979. Analysis of nonspecific reactions in enzyme-linked immunosorbent assay testing for human rotavirus. J. Clin. Microbiol. **10:**703–707.

Yolken, R. H., H. W. Kim, T. Clem, R. G. Wyatt, A. R. Kalica, R. M. Chanock, and A. Z. Kapikian. 1977. Enzyme linked immunosorbent assay (ELISA) for detection of human reovirus-like agent of infantile gastroenteritis. Lancet **2:**263–266.

Yolken, R. H., L. Mata, B. Garcia, J. J. Urrutia, R. G. Wyatt, R. M. Chanock, and A. Z. Kapikian. 1978a. Secretory antibody directed against rotavirus in human milk – measurement by means of enzyme-linked immunosorbent assay (ELISA). J. Pediatr. **93:**916–921.

Yolken, R. H., R. G. Wyatt, H. W. Kim, A. Z. Kapikian, and R. M. Chanock. 1978b. Immunological response to infection with human reovirus-like agent: measurement of anti-human reovirus-like agent immunoglobulin G and M levels by the method of enzyme-linked immunosorbent assay. Infect. Immun. **19:**540–546.

Yolken, R. H., R. G. Wyatt, G. Zissis, C. D. Brandt, W. J. Rodriguez, H. W. Kim, R. H. Parrott, J. J. Urrutia, L. Mata, H. B. Greenberg, A. Z. Kapikian, and R. M. Chanock. 1978c. Epidemiology of human rotavirus types 1 and 2 as studied by enzyme-linked immunosorbent assay. N. Engl. J. Med. **299:**1156–1161.

Yow, M. D., J. L. Melnick, R. J. Blattner, W. B. Stephenson, N. M. Robinson, and M. A. Burkhardt. 1970. The association of viruses and bacteria with infantile diarrhea. Am. J. Epidemiol. **92:**33–39.

Zissis, G., and J. P. Lambert. 1978. Different serotypes of human rotavirus. Lancet **1:**38–39.

Zissis, G., J. P. Lambert, and D. de Kegel. 1978. Routine diagnosis of human rotaviruses in stools. J. Clin. Pathol. **31:**175–178.

CHAPTER 22

Togaviridae and *Flaviviridae:* The Alphaviruses and Flaviviruses

CHARLES H. CALISHER and THOMAS P. MONATH

Diseases: Yellow fever, dengue, St. Louis encephalitis, Japanese encephalitis, Wesselsbron, tick-borne encephalitis, louping ill, Kyasanur Forest disease, other tick-borne hemorrhagic fevers, Murray Valley encephalitis, Rocio encephalitis, equine encephalitides (eastern, western, Venezuelan), chikungunya, o'nyong-nyong, Ross River, Mayaro, Sindbis, Ockelbo.

Etiologic Agents: Yellow fever, dengue-1, dengue-2, dengue-3, dengue-4, Central European encephalitis, St. Louis encephalitis, Japanese encephalitis, West Nile, Murray Valley encephalitis, Wesselbron, Ilheus, Rocio, Russian spring-summer encephalitis, Omsk hemorrhagic fever, louping ill, Kyasanur Forest disease, Powassan, eastern equine encephalitis, western equine encephalitis, Venezuelan equine encephalitis, Ross River, Mayaro, Sindbis, Ockelbo.

Source: Mosquitoes, ticks; Omsk hemorrhagic fever may be water-borne.

Clinical Manifestations: Fever, fever with rash, fever with rash and polyarthritis, fever with rash, myalgia, and arthralgia, hemorrhagic fever with shock, abortion, encephalitis.

Pathology: Disturbance of the integrity of microcirculation, with leakage of plasma and plasma proteins into extravascular spaces (hemorrhagic fevers); typical viral encephalitis.

Laboratory Diagnosis: Virus isolation, fourfold or greater increase or decrease in antibody in infected individuals, antigen detection or IgM antibody capture ELISA, neutralization, hemagglutination-inhibition, complement-fixation, indirect fluorescent antibody tests.

Epidemiology: Focally or widespread worldwide, dependent on distribution of virus, vectors, and vertebrate hosts.

Treatment: Symptomatic, immune plasma.

Prevention and Control: Prevention of bite by infected arthropod, insecticide spraying, vaccination.

Abbreviations for virus names: yellow fever (YF), louping ill (LI), West Nile (WN), Japanese encephalitis (JE), Russian spring-summer encephalitis (RSSE), St. Louis encephalitis (SLE), western equine encephalitis (WEE), eastern equine encephalitis (EEE), Venezuelan equine encephalitis (VEE), Sindbis (SIN), Semliki Forest (SF), Ilheus (ILH), Uganda S (UGS), Ntaya (NTA), dengue-1 (DEN-1), dengue-2 (DEN-2), Murray Valley encephalitis (MVE), Rocio (ROC), Central European encephalitis (CEE), Omsk hemorrhagic fever (OHF), Kyasanur Forest disease (KFD), Powassan (POW), chikungunya (CHIK), Ross River (RR), Mayaro (MAY), Getah (GET), Sagiyama (SAG), Bebaru (BEB).

Introduction

At the beginning of this century it was reported that YF was transmitted to humans by the bite of *Aedes aegypti* mosquitoes infected with that virus (Reed et al., 1983). That seminal work led to a concentration of efforts to eradicate this disease, which at the time was a considerable public health peril. In turn, this not only led to a greater understanding of YF and YF virus, but effected a significant expansion of our knowledge of other arthropod-borne viruses. Louping Ill, WN, JE, RSSE, and SLE viruses were all isolated prior to 1940 (Karabatsos, 1985). A worldwide effort was begun to isolate viruses from arthropods and to determine the role of these viruses in human and animal diseases; there was a need to determine whether the viruses being isolated were unique or were pieces in a larger epidemiologic puzzle. First, SLE, JE, and LI viruses were reported to be distinct from one another (Webster et al., 1935), but then it was shown that sera from patients who had recovered from JE neutralized not only the homologous virus but SLE virus as well, though not to the same extent (Webster, 1938). This indicated an antigenic relationship between the two viruses.

Smorodintseff (1940) showed by neutralization tests that RSSE and LI viruses were related to each other, but that JE virus, although related to both, was sufficiently distinct from them to be considered a "less complex" virus. Then Smithburn (1942), investigating the reactivities of antibodies produced in humans and monkeys infected with WN, SLE, and JE viruses, discovered even more complex serologic interrelationships.

It was not until the mid-1940s that a simpler tool, the complement-fixation (CF) test, was applied by Casals (1944) to the problem of antigenic interrelationships. He found that RSSE and LI viruses were more closely related to each other than they were to the similarly closely related JE, WN, and SLE viruses and suggested that each of these sets of viruses should be considered separate antigenic "complexes". Further, he demonstrated by CF that WEE virus was not related to any of these. Later, Sabin (1950) showed that two types of dengue viruses elicited the production of CF antibodies to YF, WN, and JE viruses in monkeys and that experimental infections of human volunteers with one of the dengue viruses elicited the production of antibody not only to the homologous virus but to YF, WN, and JE viruses as well.

In the years that followed, more viruses related to YF, JE, SLE, WN, RSSE, and LI viruses were isolated from mosquitoes, ticks, humans, and other mammals, and serologic surveys provided further evidence for cross-reactivity between antibodies to some but not other viruses. Development and application of the hemagglutination-inhibition (HI) test (Sabin and Buescher, 1950; Casals and Brown, 1954; Clarke and Casals, 1958) led to the concept of "serogroups" (sets of viruses that are antigenically related to one another). First, Casals and Brown (1954) demonstrated that 16 of the 21 viruses they studied belonged to one of two serogroups, Group A or Group B. Those in Group A included WEE, EEE, VEE, SIN, and SF viruses, whereas YF, JE, SLE, WN, RSSE, LI, ILH, UGS, NTA, DEN-1, and DEN-2 viruses belonged to the Group B viruses.

The other five viruses were poliovirus, rabies virus, or arboviruses not belonging to either Group A or Group B. It was later reported that 7 viruses belonged in Group A and 17 in Group B and that others either belonged to other serogroups or were not at that time classifiable on the basis of serologic testing (Casals, 1957). By the mid-1960s, 19 Group A viruses and 39 Group B viruses had been recognized (World Health Organization, 1967).

The International Committee on the Taxonomy of Viruses (ICTV), an organ of the International Union of Microbiology Societies, has been attempting to classify viruses (Matthews, 1982). The ICTV originally placed Group A and Group B arboviruses in the family *Togaviridae*, which also included rubella and hog cholera and related viruses, on the basis of molecular and morphologic characteristics. The ICTV has now recognized a reorganization of this family, based on additional molecular data regarding strategy of replication, and has separated viruses formerly placed in the family *Togaviridae* into two families, *Togaviridae* (genera *Alphavirus*, *Rubivirus*, *Pestivirus*, and *Arterivirus*) and *Flaviviridae* (genus *Flavivirus*) (Westaway et al., 1985a,b). Invertebrate hosts are not known to carry viruses of the genera *Rubivirus* (rubella virus), *Pestivirus* (mucosal disease-bovine virus diarrhea virus, hog cholera virus, and border disease virus), or *Arterivirus* (equine arteritis virus); viruses belonging to the latter two genera are not human pathogens. Because rubella is covered in Chapter 23 of this book, none of the viruses of these three genera will be addressed in this chapter.

The term "arbovirus", a contraction of the term "*arthropod-borne virus*, denotes a virus maintained in nature in a biological transmission cycle between susceptible vertebrate hosts (in which they replicate) and hematophagous arthropods (in which they also replicate). The generic term "arbovirus" generally has an ecologic connotation. In this chapter, mention of these viruses will employ universal taxonomic descriptions. Differences in antigenic, morphologic, biochemical, and genetic characteristics are used to separate the arboviruses into families, genera, serogroups, complexes, viruses, subtypes, and varieties, in an increasing order of relatedness. Most recently, molecular analyses have substantiated previous antigenic classification schemes, and a clearer view of the taxonomy of these viruses has emerged. The foresightedness of Casals was remarkable!

Family *Flaviviridae,* Genus *Flavivirus*

Flavivirus virions are spherical, 40 to 50 nm in diameter. They contain one molecule of single-stranded, positive-sense RNA, three structural proteins, and have a molecular weight about 4×10^6. One protein is the core protein (C) with a molecular weight of 13 to 16×10^3, another is a membrane-associated protein (M) with a molecular weight of 7 to 9×10^3, and the third is an envelope glycoprotein (E) with a molecular weight of 51 to 59×10^3. At present more than 70 viruses, subtypes, and varieties have been assigned to this genus (the former Group B arboviruses). Antigenic classification schemes have been proposed for the flaviviruses (de Madrid and Porterfield, 1974; Varelas-Wesley and Calisher, 1982). Simian hemorrhagic fever virus has been assigned to the family *Flaviviridae*, but information regarding it is insufficient, and it has not been placed in a genus. One antigenic classification for the flaviviruses is presented in Tables 1 and 2 (for definitions of the antigenic categories serogroup, complex, virus, subtype, and variety see Chapter 32, *Bunyaviridae*). Some of these viruses are transmitted principally by mosquitoes, some by ticks, and some have not been associated with an arthropod vector. When segregated by vector associations, flaviviruses generally exhibit distinct antigenic differences, possibly as a consequence of such associations (Chamberlain, 1980). Among the mosquito-borne flaviviruses are YF, WN, SLE, MVE, JE, ROC, and the four dengue viruses. The viruses that cause encephalitides are more closely related to one another antigenically than they are to the viruses that cause YF or dengue. So, too, YF and dengue viruses cause diseases distinct from each other and are antigenically quite distant from each other.

The tick-borne flaviviruses, including RSSE, CEE, OHF, and KFD viruses, are of considerable medical importance and concern from central Europe to eastern Siberia. The antigenic relationships of these viruses to the mosquito-borne flaviviruses are distant, but that relationship is significant phylogenetically. It may be that evolutionary pressures (e.g., geographic isolation, environmental selection, survival of most genetically fit) have created a divergence of virus–vector relationships, perhaps from a common origin. For example, WN virus replicates nearly as well in ticks as in its usual arthropod vectors, *Culex* species mosquitoes. Powassan, a tick-borne flavivirus isolated first and principally in Can-

TABLE 1. Classification of viruses of the family *Flaviviridae*, genus *Flavivirus* (Group B arboviruses)

Complex	Virus (subtype) ⟨variety⟩
Mosquito-borne viruses	
St. Louis encephalitis	St. Louis enc. ⟨3⟩, Alfuy, Japanese enc., Kokobera, Koutango,[a] Kunjin, Murray Valley enc., Stratford, Usutu (Usutu) ⟨Yaounde⟩, West Nile
Uganda S	Uganda S, Banzi, Bouboui, Edge Hill
Dengue	Dengue-1, dengue-2, dengue-3, dengue-4
Ntaya	Ntaya, Tembusu (Tembusu) ⟨Yokose⟩, Israel turkey meningoenc. (Israel turkey meningoenc.) ⟨Bagaza⟩
Tick-borne viruses	
Russian spring-summer encephalitis	Russian spring-summer enc.[b] (Russian spring-summer enc.) ⟨Omsk hemorrhagic fever⟩ ⟨Central European enc.⟩ ⟨Kumlinge⟩ ⟨Kyasanur Forest disease⟩ ⟨louping ill⟩ ⟨Langat⟩ ⟨Langat⟩ ⟨Negishi⟩ ⟨Carey Island⟩, Powassan, Karshi, Royal Farm
Tyuleniy	Tyuleniy, Saumarez Reef, Meaban
Vector-unassociated viruses	
Modoc	Modoc, Cowbone Ridge, Jutiapa, Sal Vieja, San Perlita (San Perlita) ⟨San Perlita⟩ ⟨MA387-72⟩
Rio Bravo	Rio Bravo, Apoi, Bukalasa bat, Dakar bat, Entebbe bat, Saboya
Flaviviruses not assigned to an antigenic complex	

Aroa	Phnom-Penh bat
Bussuquara	Rocio
Cacipacore	Sepik
Gadgets Gulley	Sokuluk
Ilheus	Spondweni
Jugra	Tamana bat
Kadam	Wesslesbron
Kedougou	yellow fever
Montana Myotis leukoencephalitis	Zika
Naranjal	

[a] Koutango virus has not been isolated from naturally infected mosquitoes.
[b] enc. = Encephalitis.

TABLE 2. Viruses of the family *Togaviridae*, genus *Alphavirus* (Group A arboviruses)

Complex	Virus (subtype) ⟨variety⟩
Eastern equine encephalitis	Eastern equine encephalitis
Middelburg	Middelburg
Ndumu	Ndumu
Semliki Forest	Semliki Forest, chikungunya (chikungunya) ⟨several⟩ (o'nyong-nyong), Getah (Getah) (Sagiyama) (Bebaru) (Ross River), Mayaro (Mayaro) (Una)
Venezuelan equine enc.	Venezuelan equine enc. (Venezuelan equine enc.[a]) ⟨A-B⟩ ⟨C⟩ ⟨D⟩ ⟨E⟩ ⟨F⟩, (Everglades), (Mucambo) ⟨Tonate⟩ ⟨71D-1252⟩, (Pixuna), (Cabassou), (AG80-663)
Western equine enc.	Western equine enc. (several), Y 62-33, Highlands J, Fort Morgan, Sindbis (Sindbis) (Babanki) (Ockelbo) (Whataroa) (Kyzylagach), Aura
Barmah Forest	Barmah Forest
(No complex assigned)	Zingilamo

[a] enc. = Encephalitis.

ada but also isolated in the northern United States and the Soviet Union, is widely divergent antigenically from the mosquito-borne flaviviruses, but has been isolated from adult *Anopheles hyrcanus* mosquitoes and from larvae of *Aedes togoi* mosquitoes in the Soviet Union.

The flaviviruses that have not been associated with an arthropod vector are not closely related antigenically to other flaviviruses, but they are connected to each other by virtue of host and geography (*Rodentia* in the New World; *Chiroptera* in the Old World). This set of viruses represents divergent evolution, possibly indicating relatively restricted spread (vertebrate to vertebrate), whereas the mosquito- and tick-borne viruses represent more moderate divergent evolution and relatively unlimited geographic spread. Given the rather discrete geographic foci and econiches in which the vector-unassociated flaviviruses are found, phylogenetic divergence reflected by antigenic dissimilarities is not surprising.

Relationships between the flaviviruses may be summarized as follows: a) viruses that are mosquito-borne are far more antigenically similar to one another than they are to the tick-borne or vector-unassociated flaviviruses, whereas b) viruses that are vector-unassociated are nearly as antigenically dissimilar from one another as they are from mosquito- and tick-borne flaviviruses. As an example, Koutango virus has not been associated with an arthropod vector in nature but has been transmitted to vertebrates by mosquitoes and passed transovarially in laboratory studies. This virus is closely related antigenically to the mosquito-borne flaviviruses, suggesting common ancestry.

Family *Togaviridae,* Genus *Alphavirus*

Alphavirus virions are spherical, 60 to 70 nm in diameter. They contain one molecule of single-stranded, positive-sense RNA, three structural proteins, and have a molecular weight 4×10^6. Two of the proteins are cell membrane-derived envelope glycoproteins (El and E2) with molecular weights of 50 to 59 × 10³, and one is a nonglycosylated capsid protein with a molecular weight of 30 to 34 × 10³ (Semliki Forest virus possesses three envelope glycoproteins). At present, more than 37 viruses, subtypes, and varieties have been assigned to this genus (Table 2); a classification scheme for these viruses has been published (Calisher et al., 1980). Alphaviruses have been isolated on six continents. All but Fort Morgan virus have been isolated from mosquitoes. This virus, which has been isolated only from the bird-nest bug *Oeciacus vicarius* and from passerine birds, is an example of a virus whose distribution is restricted by its vector. Alphaviruses of the EEE and WEE antigenic complexes appear to employ birds as principal vertebrate hosts in their maintenance cycles in nature; this probably accounts for their widespread distribution. Alternatively, subtypes of VEE virus appear to be restricted to small mammals in discrete enzootic foci and have been found only in the Americas. Semliki Forest complex members SF, CHIK, RR, and MAY viruses also are widely distributed; MAY virus has been associated with avian hosts. Two of the four antigenic subtypes of GET virus, SAG and BEB, appear to be geographically isolated, but little is known of their natural histories. Triniti virus may be a member of the family *Togaviridae*, but insufficient information regarding it is available, and it has not been placed in a genus.

Clinical Features

Flaviviruses

Several clinical syndromes are caused by infections with flaviviruses, including uncomplicated fever; fever with rash; fever with rash, myalgia, and arthralgia; hemorrhagic fever with shock; stillbirth and abortion (in domestic livestock); and encephalitis.

Representative examples of flaviviruses that have been associated with each disease category are presented in Table 3. Evidence from serologic surveys indicates that the ratio of inapparent to apparent infections is quite high, so that these viruses, even in situations in which the incidence of infection is high, only rarely cause disease. The flaviviruses that cause encephalitis (Table 3), for example, usually cause abortive infection characterized by fever with headache or other relatively benign signs. However, in those few individuals who develop full-blown infection, disease may be severe or fatal. The case-fatality rate may be 20% or more in the encephalitides, YF, and dengue shock syndrome.

YF, the dengue viruses, and two tick-borne flaviviruses (OHF and KFD) are associated with hemorrhagic fever syndrome. The common denominator in the clinical features of these infections is hemorrhagic diathesis and circulatory failure (hypotension, shock). Hepatic necrosis occurs, but only in YF to an extent that is clinicopathologically significant; patients with YF develop signs of severe hepatic dysfunction. The central event in the pathophysiology of dengue hemorrhagic fever/shock syndrome is a disturbance of the integrity of the microcirculation, with leakage of plasma and plasma proteins into the extravascular space, shock, and hemorrhage. Kyasanur Forest disease virus, in addition to causing hemorrhagic fever, frequently produces meningoencephalitis and thus displays some pathobiological similarities to its antigenic relative, RSSE virus.

Alphaviruses

As with the flaviviruses, certain alphaviruses can cause a variety of clinical syndromes (Table 4). The most severe, and therefore the most extensively studied, are the encephalitides, caused by EEE, WEE, and VEE viruses. Eastern equine encephalitis and VEE viruses and their antigenic relatives occur only in the Americas, but antigenic relatives of WEE virus also occur in Europe, Africa, Asia, and the South Pacific (Table 2). Humans and horses are clinically affected; humans are dead-end hosts for these three viruses, whereas horses are the principal viremic host for VEE virus. Horses may serve only occasionally as a source of mosquito infection by EEE virus. Human infections are most often self-limiting or subclinical, and the severe manifestations of encephalitis occur only infrequently in infected individuals. Rates of inapparent to apparent infections are age- and perhaps strain-dependent. Children appear to be at greater risk of encephalitis.

Encephalitis, whether caused by flaviviruses or alphaviruses, is an infection of the brain parenchyma and results in either localized or diffuse signs of cerebral dysfunction. Signs of meningeal irritation (meningoencephalitis) are also nearly always present, but may not be obvious in the very young, the very old, or the comatose patient. Inflammation of the leptomeninges may occur without evidence of brain dysfunction; this is termed aseptic meningitis. Onset of neurologic disease may be insidious, preceded by a period in which the patient has an influenza-like illness. Encephalitis may follow quite soon after the onset of this rather mild stage, or it may occur days or weeks later. Humans infected with tick-borne encephalitis viruses (flaviviruses) may experience a bimodal fever curve, with encephalitis appearing only after occurrence of the second febrile episode, days to weeks after the first.

All individuals with encephalitis experience some alteration in degree of consciousness. Generalized convulsions may occur, and damage to corticospinal tracts causes paresis, paralysis, hyperactive reflexes, and plantar extensor responses. In encephalitis due to SLE, JE, and other viruses, there is involvement of extrapyramidal structures, causing tremors and muscular rigidity. Cerebellar dysfunction leads to incoordination, dysmetria, and ataxic speech. Damage

TABLE 3. Flaviviruses associated with various clinical syndromes

Syndrome	Flavivirus
Fever	Dengue 1–4, West Nile, St. Louis encephalitis, Banzi, Bussuquara, Ilheus, Kunjin, Japanese enc., Murray Valley enc., Rio Bravo, Rocio, Sepik, Spondweni, Wesselsbron, yellow fever, Zika, tick-borne enc.,[a] Kyasanur Forest Disease, Omsk hemorrhagic fever, Powassan, Usutu
Fever with rash	Dengue 1–4, West Nile
Fever with rash, myalgia, and arthralgia	Dengue 1–4
Hemorrhagic fever with shock	Dengue 1–4, yellow fever, Kyasanur Forest disease, Omsk hemorrhagic fever
Abortion	Japanese enc. (pigs), Wesselsbron (sheep)
Encephalitis	St. Louis enc., West Nile, Ilheus, Japanese enc., Murray Valley enc., Rocio, Apoi, Rio Bravo, tick-borne enc., louping ill, Powassan, Negishi

[a] enc. = Encephalitis.

TABLE 4. Alphaviruses associated with illness in humans or equines

Syndrome	Alpha virus
Fever	Eastern equine enc.,[a] western equine enc., Venezuelan equine enc., Mucambo, Everglades, chikungunya, o'nyong-nyong, Mayaro, Ross River, Sindbis, Ockelbo, Semliki Forest
Fever with rash	Sindbis, Ockelbo
Fever with rash and poly-arthritis	Ross River
Fever with myalgia, arthralgia, and rash	Chikungunya, o'nyong-nyong, Mayaro
Encephalitis	Eastern equine enc., western equine enc., Venezuelan equine enc., Semliki Forest

[a] enc. = Encephalitis.

to brainstem nuclei or supranuclear tracts results in cranial nerve palsies; cardiovascular irregularity, urinary retention, and sialorrhea reflect autonomic disturbances. Confusion, memory defects, changes in speech, personality, and behavior, as well as pathologic reflexes indicate damage to the cerebral cortex, hypothalamus, thalamus, or temporal lobes. Inappropriate antidiuretic hormone secretion and hyperthermia indicate disturbance of the pituitary–hypothalamic axis. Spinal cord involvement may be indicated by lower motor neuron and sensory deficits, hyperreflexia, and paralysis of the bladder. Interference with respiratory functions, laryngeal paralysis, cardiac arrhythmia, and cerebral edema are potentially life-threatening complications. Surviving patients may suffer permanent neuropsychiatric or other damage.

Clinical laboratory findings are rather nonspecific but usually include a leukopenia. The cerebrospinal fluid contains predominantly polymorphonuclear cells early in the course of infection and lymphocytes later, with the cell count usually less than 500/mm^3. An exception to this is EEE virus infection, in which a more fluid polymorphonuclear reaction occurs. There may be a moderate elevation in cerebrospinal fluid protein, but glucose and lactate concentrations usually are normal. Changes in serum-enzyme levels indicate pathologic processes in the myocardium, liver, and skeletal muscle.

Diagnosis

Viremia and Antibody Responses

It is important to determine the specific etiologic agent quickly and accurately in instances of any arboviral disease, not only so that the patient can be treated appropriately, but so that public health officials can institute vector-control operations to limit further spread of the virus. Laboratory confirmation of a clinical diagnosis depends upon direct detection of antigen, virus isolation, or serologic tests, such as HI, CF, neutralization, fluorescent antibody and, more recently, enzyme-linked immunosorbent assays (ELISA). It is not often that one isolates flaviviruses that cause encephalitis from blood or spinal fluid taken in the acute stage of illness of infected individuals. Often the viremic stage has passed before the individual becomes ill. This is not the case with a few human pathogens, for example, dengue viruses, YF, KFD, and WN viruses, which may be consistently isolated during the first few days to a week after onset of disease. St. Louis encephalitis and JE viruses and antigens may be more often detected in brain collected postmortem. Venezuelan equine encephalitis complex viruses can be isolated from the blood of febrile individuals, but EEE and WEE viruses usually are isolated only from brain.

Antibody generally is not detectable until the end of the viremic phase. Virus–antibody complexes have been demonstrated in YF (Sarthou and Lhuillier, 1984) and dengue (Theofilopolous et al. 1976). In the case of YF, Sarthou and Lhuillier reported the dissociation of IgM antibody–YF virus complexes by use of a mild reducing agent (dithiothreitol), with subsequent isolation of virus that had been rendered noninfectious by antibody.

The IgM fraction of serum contains both neutralizing and HI antibodies appearing soon after onset, and its presence can serve as an indicator of recent infection (Burke et al., 1982; Monath et al., 1984b). IgG antibody appears later, containing both neutralizing and HI as well as CF antibodies. IgG antibodies to the flaviviruses probably persist for life, as evidenced by the persistence of HI and neutralizing antibodies and the absence of detectable IgM antibodies in late convalescent-phase serum. However, IgM antibodies to flaviviruses have been detected for long periods (Monath, 1971), and it appears that a minority subset of patients have prolonged IgM antibody responses, limiting somewhat the value of these assays as a measure of very recent infection. IgM antibodies in serum appear to be relatively type-specific (Westaway et al., 1974; Lhuillier and Sarthou, 1983;

Monath et al., 1984b), but complex- and serogroup-reactivity also are observed (Monath et al., 1984b). In the only study of sera from humans infected with various alphaviruses, IgM antibodies were antigenic complex-specific (Calisher et al., 1986b).

Measurement of IgM antibody in cerebrospinal fluid is an extremely useful method of serodiagnosis. Because IgM antibodies do not cross the blood-brain barrier, finding them in cerebrospinal fluid implies local antibody synthesis in response to central nervous system infection. Antibody-producing cells have been found in the cerebrospinal fluids of JE patients (Burke et al., 1982). Moreover, the titer of IgM antibody in cerebrospinal fluid is a prognostic indicator of JE (Burke et al., 1982) and possibly other flaviviral encephalitides.

HI antibody is broadly reactive among viruses of a serogroup, making this a useful test for preliminary screening. CF antibody is both more complex-specific and short-lived than HI antibody, and therefore is useful (as is IgM) for determining recent infections. Neither HI, CF, nor IgM antibody capture (MAC) ELISA are virus-specific. However, MAC ELISA is, at present and for the foreseeable future, the test of choice for making provisional serodiagnoses with single serum samples or with cerebrospinal fluids and is of greater or equal value to HI and CF tests when paired acute- and convalescent-phase serum samples are available.

Specimen Collection

After collection, whole blood, serum, cerebrospinal fluid, and tissue samples should be processed immediately or placed on dry ice ($-70°C$) or other suitable deep-freezing agent if virus isolation is to be attempted. Although this may not be such a critical issue for antigen detection, it appears sensible to ship and store specimens at low temperatures to prevent degradation of proteins. When serum is to be tested only for antibody, it can be shipped at ambient temperatures for brief periods, provided it has been collected aseptically and not subsequently contaminated with microorganisms. If transit time to the laboratory is longer than several days, refrigeration or addition of a small quantity of antibiotics is necessary for prevention of deterioration of the specimen; other chemical sterilants can be used, provided they will not interfere with the tests.

Virus Isolation and Identification

No single virus-isolation system is adequate for all flaviviruses and alphaviruses. As newer, more sensitive isolation systems (inoculation of mosquitoes in vivo, inoculation of arthropod cells in vitro) are in-creasingly employed, it becomes apparent that many virus strains and viruses have not been revealed because of the bias incurred by use of "traditional" systems, such as suckling mice and vertebrate cell cultures. For example, in initial isolation attempts, strains of yellow fever virus may cause cytopathic effects in C6/36 (*Aedes albopictus* mosquito) cell cultures, but may not kill suckling mice inoculated intracranially (Mendez et al., 1984).

Traditional methods of virus isolation are still firmly entrenched in many laboratories. Suckling mice have been used as laboratory hosts for amplifying virus in diagnostic specimens and from field-collected mosquitoes, ticks, and animal tissues. Mice are inoculated intracranially with clarified suspensions of clinical specimens or with macerated and clarified arthropod pools or animal tissues. Because suckling mice are available to almost all laboratories, this system holds certain advantages over others.

Nevertheless, mosquito cell cultures, particularly C6/36 (*Aedes albopictus*), AP-61 (*Aedes pseudoscutellaris*), TR-248 (*Toxorhynchites amboinensis*), increasingly are being used for virus isolation. These and other cell culture systems have the additional advantage of ease of containment and reduction of aerosols. Because they are highly stable and have growth optima at lower temperatures than do mammalian cells, cultures and mosquitoes may be taken to the field, inoculated with specimens, and returned to the laboratory weeks later, after amplification of the virus being sought.

For several viruses, for example, YF, dengue, and vesicular stomatitis, mosquito cell cultures are more sensitive than mice or mammalian cell culture systems for virus isolation. However, they have the disadvantage in some cases of not producing cytopathic effects, and thus require specific, secondary steps for recognition of the presence of virus in the culture. Intrathoracic inoculation of *Toxorhynchites* mosquitoes, which do not take blood meals but in which dengue and other viruses replicate, has also been used with sensitivity and safety for virus isolation (Rosen, 1981).

The classical procedure for isolating and identifying an arbovirus begins with inoculation of suckling mice or a cell culture system in which cytopathic effects or plaques develop. The isolate is characterized by testing its ability to pass through a filter that excludes bacteria and its sensitivity to lipid solvents such as ether, chloroform, or sodium deoxycholate (Theiler 1957). It is often useful to determine the pathogenicity of the agent and to determine end point titers in various laboratory animals and cell cultures. A crude alkaline extract or partially purified (sucrose–acetone-extracted) antigen is prepared for use in serologic tests (Clarke and Casals, 1958). The antigen is tested for its ability to agglutinate the erythrocytes of male domestic geese (*Anser cinereus*) and to react in CF tests (Clarke and Casals, 1958; Casey, 1965) with homologous antibody preparations. One then tests the antigen by HI or CF with a battery of antibody preparations, including a) those against viruses of various serogroups, b) those against viruses suspected as the etiologic agent of the disease, and c) those known to be present in the area in which the specimen was collected or in which the patient contracted the illness.

The best method for identifying a flavivirus or an alphavirus is one that is rapid, specific, and inexpensive. In some laboratories, electron microscopy can be used as an early step to identify the virus family to which the isolate belongs, and this can greatly facilitate subsequent characterization. The application of direct or indirect fluorescent microscopy (Gardner and McQuillan, 1980) with polyclonal or monoclonal (Henchel et al., 1983) antibodies can provide a rapid and simple means of virus identification. Because a complete battery of reagents is not yet available, this method is now used only for identifying certain viruses. Direct and indirect fluorescent antibody tests have been used to detect viral antigens in clinical specimens. Once the isolate is characterized to the level of serogroup or antigenic complex, neutralization tests are performed with antisera against individual viruses to confirm the identification. If necessary, antiserum is also prepared against the isolate and cross-tested against antigens of viruses in the serogroup to which it belongs. Most of the data regarding antigenic characterization of arboviruses have been generated with these tests, so that they will remain the standards by which newly isolated viruses are identified and against which new methods of characterization are to be judged. Newly developed reagents (such as monoclonal antibodies) and procedures (such as ELISA) will add significantly to our diagnostic ability and enable us to more fully characterize the epitopes and other antigenic moieties of viruses.

For early detection of flavivirus antigens in mosquito cell cultures inoculated with clinical materials, fluorescent antibody can provide identification of YF (Saluzzo et al., 1985), dengue (Halstead et al., 1983), and other viruses in fewer than 72 to 96 h, long before the appearance of cytopathic effects in these mosquito cells or mammalian cells or before the appearance of signs of illness in suckling mice. The indirect immunofluorescence test is also used to identify viral antigens in intrathoracically inoculated *Toxorhynchites* (Rosen and Gubler, 1974).

Monoclonal antibodies are available with group-specificity against all flaviviruses and all alphaviruses. This allows rapid assignment of an arbovirus isolate to a serogroup. In addition, monoclonal antibodies have been characterized which show com-plex-reactive as well as type-specific and even strain-specific reactivities against YF, JE, Kunjin, WN, MVE, DEN-1, DEN-2, DEN-3, DEN-4, CEE, both North and South American varieties of EEE, various subtypes (variants) of VEE including a vaccine strain (TC-83), SF, WEE, Highlands J, and SIN viruses. These monoclonal antibodies have been prepared against the viruses considered important either for epidemiologic purposes or for research. In the future, monoclonal antibodies undoubtedly will be prepared against other flaviviruses and alphaviruses, and a more extensive battery of reagents will be available for use in diagnostic laboratories. Given an appropriate battery of serogroup-specific, antigenic complex-specific, and type- and subtype-specific antibodies, a virus isolate can be quickly and specifically identified.

Direct Antigen Detection

Direct detection of viruses and antigens in clinical specimens can provide a rapid diagnosis but is less sensitive than methods that require replication of the virus and thus amplify the antigenic components. Rapid, early diagnosis has obvious benefits both in formulating a therapeutic plan for the individual patient and in formulating appropriate public health measures in the instances of a potentially epidemic disease. The more rapid, sensitive, simple, and specific the test, the better. For many years respiratory viruses have been identified directly in nasopharyngeal aspirates by detection of their antigens via immunofluorescence (Gardner and McQuillan, 1980) or by ELISA (Sarkkinen et al., 1981). More recently, enteric viruses have been used to develop rapid identification techniques, such as nucleic acid hybridization (Hyypia et al., 1984). With fluorescent antibody or immunoperoxidase techniques, both flaviviruses and alphaviruses may be detected directly in tissues from infected humans or animals. Frozen sections or impressions smears (as for rabies) are suitable for such tests. Sections (6 μm) are cut on a cryostat–microtome, placed on clean slides, and fixed for 15 min in cold ($-20°C$) acetone. Indirect fluorescent antibody (IFA) or immunoperoxidase tests are then performed with the most sensitive and specific immune reagents available. In the case of IFA tests, primary (antiviral) antibody is added to the section or smear and incubated at 37°C for 30 min in a humidified chamber; the slides are washed in phosphate-buffered saline (PBS), pH 7.2 to 7.4, air-dried, and stained with fluorescein-conjugated antibody to the species in which the primary antibody was made. Trypan blue may be added to the conjugate at a concentration of 1:2,000, as a counterstain. Optimal concentrations of both the first and second (conjugated) antibodies are determined by prior titrations with control antigens (virus-infected cell cultures or tissues). Controls include infected and uninfected tissues and cell cultures as well as infected tissues or cell cultures incubated with normal mouse ascitic fluid or other appropriate reagent. Stained slides are examined with a fluorescence microscope fitted with appropriate filters and objective. Test results are scored semiquantitatively according to the intensity of fluorescence; the percentage of fluorescing cells per field may be determined as well. Nonspecific reactivity may be reduced by prior adsorption of the primary and, if necessary, secondary antisera with normal tissue.

The IFA technique has been used for primary detection of JE (Kusano, 1966) and SLE (Gardner and Reyes, 1980) antigens in human brain tissue and of EEE antigen in horse brain tissue (Monath et al., 1981b). Immunocytochemical staining may also be applied to formalin-fixed tissues. For IFA staining of fixed tissues, tissues must first be treated with trypsin or protease to uncover antigenic determinants. The immunoperoxidase technique appears to be more satisfactory for direct antigen detection in fixed materials and has been used for the diagnosis of YF in liver tissue (de la Monte et al., 1983). The immunofluorescence technique requires skilled technicians, as well as expensive microscopes. These needs, the subjectivity of the results, and the lack of sensitivity of the method are additional drawbacks. For these reasons, ELISA and nucleic acid hybridization are increasingly used for direct detection of viral antigens in clinical specimens, in tissues from

wild vertebrates, and in mosquitoes, ticks, and other arthropods.

Detection of antigen in clinical specimens requires construction of a sensitive and specific immunoassay utilizing the most appropriate combination of capture and detecting antibodies, enzyme, and substrate, as well as optimal conditions of incubation (Fig. 1a,b,c,d). Development of such assays requires a research study, first with laboratory-cultivated virus and then with clinical samples. Capture antibodies generally are characterized by high avidity for antigen and may be polyclonal (Halonen et al., 1980; Hildreth et al., 1984), physically separated IgM antibodies of human (or other species) origin (Sarthou and Lhuillier, 1984), or monoclonal (Monath and Nystrom, 1984a). Capture antibody is added at optimal dilution (determined by titration, usually 1:500 to 1:2,000) in carbonate–bicarbonate buffer, pH 9.6, to the solid-phase support (for example, a polystyrene microtiter plate) and allowed to bind for 3 h at 22°C or 37°C, or overnight at 4°C. The plate is washed repeatedly with PBS containing 0.05% Tween 20, and the clinical sample is added; virus-positive and -negative controls are included in each assay. After optimal incubation (usually 16 h at 22 to 45°C, depending on the virus system) followed by washing, detecting antibody is added (double-sandwich method); detecting antibody may be polyclonal or monoclonal. When polyclonal or monoclonal antibody of the same species is used for both capture and detection, the detecting antibody must be conjugated to an enzyme (or radionuclide); if antibody prepared in another species is used, it should be followed

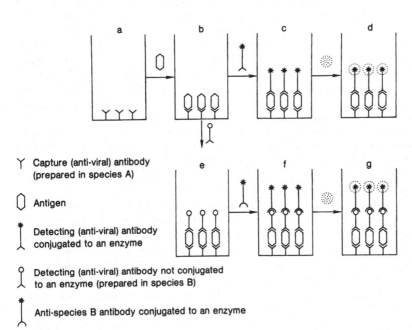

Y Capture (anti-viral) antibody
   (prepared in species A)

⊔ Antigen

Detecting (anti-viral) antibody
conjugated to an enzyme

Detecting (anti-viral) antibody not conjugated
to an enzyme (prepared in species B)

Anti-species B antibody conjugated to an enzyme

⋯ Substrate

FIG. 1. Double-sandwich antigen detection enzyme-linked immunosorbent assay.

by conjugated anti-species-detecting antibody (Fig. 1a,b,e,f,g). The color change is then recorded as optical density with a spectrophotometer. If incubations and washings between the additions of reagents were adequate, the results can be read as the ratios of test to control; usually, ratios greater than or equal to 2.0 are considered positive for IgM antibody. Should the optical density of the control (background) be exceedingly low, such ratios will be artificially high; common sense should be applied in these situations.

An advantage of this system is that there is no need for enzyme-conjugated antibodies for all the secondary antibodies used, and that for secondary antibodies one can use hyperimmune mouse ascitic fluids (Tikasingh et al., 1966), available in almost all arbovirus diagnostic laboratories and from the Centers for Disease Control (Atlanta, Georgia, or Ft. Collins, Colorado). The principal disadvantage of the double-sandwich ELISA is the need for an additional step and the possible increase in attendant nonspecific reactivity. We have not found the latter to be a problem, because the high titers of hyperimmune mouse ascitic fluids used as secondary antibodies allow the use of high dilutions, thereby diluting any nonspecifically reacting substances as well.

The double-sandwich method is generally not as sensitive, however, as the simultaneous-sandwich technique (Fig. 2). In this single incubation assay, the virus-containing sample and detecting antibody conjugate are added and incubated simultaneously (Monath et al., 1986a,b). If the simultaneous-sandwich method is used for enzyme immunoassays, the detecting antibody must be linked to biotin, because the horseradish peroxidase molecule is so large that it sterically interferes with virus–capture-antibody interaction. Following interaction of virus and biotinylated detecting antibody, the plate is washed,

and avidin-peroxidase is added at optimal dilution. The final step in either the double-sandwich or simultaneous-incubation assay is addition of enzyme substrate. Many substrates are available, and some show considerable promise, but we prefer 2.2'-azinodi[3-ethyl] benzthiazoline sulfonate. The color reaction may be stopped with 5% sodium dodecylsulfate after an appropriate incubation at room temperature in the dark. Usually this requires 10 to 20 min, the time necessary for maximal color generation in positive controls without appearance of color in negative controls.

Antibody Assays

Classical techniques for serologic diagnosis include the HI (Clarke and Casals, 1958), CF (Casey, 1965), neutralization (Lindsey et al., 1976), and IFA (Monath et al., 1981a) tests. The acceptable standard has been a fourfold or greater increase or decrease in antibody titers determined by these methods. For confirmation, these tests require multiple specimens, collected days to weeks apart. When rapidity of diagnosis is not critical or when infections with closely related viruses complicate interpretation of results, these tests are often invaluable. However, their lack of specificity creates a variety of problems in interpretation. For example, in areas hyperendemic for flaviviruses, broad cross-reactions may occur in HI, CF, and even neutralization tests, limiting their usefulness in defining the infecting agent (Monath et al., 1980). Moreover, prior infection results in a specific rise in antibodies to the heterologous (related) virus responsible for a remote infection, rather than in antibodies to the current infecting virus (Halstead, et al., 1983). Antibody cross-reactivity to closely related viruses also tends to broaden with time after

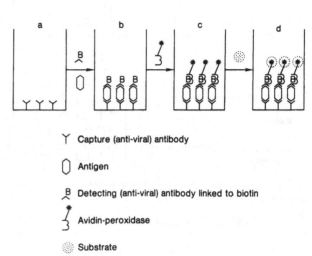

FIG. 2. Simultaneous-sandwich antigen detection enzyme-linked immunosorbent assay.

Y   Capture (anti-viral) antibody

◊   Antigen

B   Detecting (anti-viral) antibody linked to biotin

⌇   Avidin-peroxidase

⋮   Substrate

infection. In the case of primary infection, however, any one of the classical serologic methods applied to paired acute- and early convalescent-phase serum samples can provide a specific diagnosis.

The relatively recent adaptation and application of IgM antibody capture–enzyme-linked immunosorbent assay (MAC ELISA) to arbovirus serology has provided a tool for more rapid serodiagnostic testing and has been extensively studied in the case of flavivirus and alphavirus infections (Heinz et al., 1981; Burke et al.,1982; Lhuillier and Sarthou, 1983; Monath et al., 1984b; Calisher et al., 1985, 1986a,b,c). Unlike IgG antibody, which is detected later in infections and persists for years after a primary infection, IgM antibody is detected in serum soon after infection with arboviruses and, in most cases, does not persist in high titer. Presence of IgM antibodies, therefore, indicates recent infection in most instances. IgM antibodies are also quite specific, although cross-reactions remain an interpretive problem in some cases, particularly in distinguishing between infections with dengue serotypes. IgM antibodies may also have prognostic value, as shown by Burke et al. (1985) in the case of patients infected with JE virus. Patients who died had low levels of IgM antibody in the cerebrospinal fluid. The presence of IgM antibody in cerebrospinal fluid indicates neuroinvasion by the virus in question and provides important evidence in the etiology of encephalitis.

Briefly, MAC ELISA is performed as follows (Fig. 3a,b,c,d,e): wells of a microtiter plate are coated with (commercially available) antibody to human IgM. Patient serum or cerebrospinal fluid is then introduced, usually at a dilution of 1 : 100 for serum and 1 : 10 for cerebrospinal fluid. Density-gradient-purified virus, supernatant fluid from virus-infected cell cultures, or other viral antigen (such as sucrose–acetone-extracted antigen prepared from the brains of virus-infected suckling mice) and antiviral (secondary) antibody are then added. If the detecting antibody has been conjugated to an enzyme, the (commercially available) substrate is then introduced, and the intensity of the color change is recorded as optical density. Optical densities of positive serum samples are proportional to antibody titer (Monath et al., 1984b). Therefore, a single dilution of patient serum can be used to estimate changes in titers between paired acute- and convalescent-phase serum samples. Alternatively, positive serum can be diluted serially twofold and each dilution tested for positivity. In this way, titers comparable to those developed in HI, CF, and neutralization tests can be obtained. However, the presence of IgM antibody in a single serum or cerebrospinal fluid sample can be taken as presumptive evidence of recent infection by the virus used in the test.

A double-sandwich ELISA, with anti-species IgG third (detecting) antibody conjugated to an enzyme that reacts with the secondary (antiviral) antibody, also has been used with considerable success (Fig. 3,a,b,c,f,g,h). The inconvenience caused by having to take the extra step is largely overcome by being able to use a single anti-species conjugate, rather than one for each antiviral reagent. For alphaviruses and flaviviruses, this is accomplished by using serogroup-reactive, enzyme-conjugated, monoclonal an-

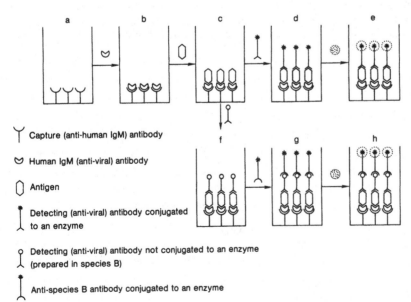

Y Capture (anti-human IgM) antibody

♡ Human IgM (anti-viral) antibody

◊ Antigen

⊥ Detecting (anti-viral) antibody conjugated to an enzyme

φ Detecting (anti-viral) antibody not conjugated to an enzyme (prepared in species B)

⊥ Anti-species B antibody conjugated to an enzyme

🌐 Substrate

Fig. 3. IgM antibody capture enzyme-linked immunosorbent assay.

tibodies. The individual needs of given situations and laboratories warrant tailoring the technique to specific purposes.

Rheumatoid factor, an IgM antibody directed against IgG, can cause false-positive results in immunoassays for IgM. In MAC ELISAs, all IgM antibodies, including rheumatoid factor, are captured. Rheumatoid factor captured in this way may result in binding of a) human IgG antibodies against the viral antigen used in the test or b) mouse (or other species) antiviral IgG antibodies used to detect antigen. Interference by rheumatoid factor cannot be controlled by use of control antigens, but specific tests for rheumatoid factor are commercially available. Additionally, rheumatoid factor can be removed from serum by various methods, which have been reviewed by Meurman (1983).

IgG antibody produced after infections with arboviruses persists for months, years, or even for the life of the individual. Therefore, as with HI and neutralizing antibodies, the presence of IgG antibody does not necessarily denote an active or recent arboviral infection, unlike IgM antibody. The fetus or neonate produces IgM, but not IgG, in response to infections in utero or shortly after birth. Thus, the presence of IgM in the fetus or neonate indicates active infection of the fetus or neonate; the relevance of this observation to diagnosing alphavirus and flavivirus infections is limited.

IgG antibody in serum or in cerebrospinal fluids may be determined as follows (Fig. 4): a) coat microtiter plate wells with about 350 to 500 ng of gradient-purified virus (Fig. 4a) or coat the wells with antibody (prepared in a nonhuman species) and then add virus or sucrose-acetone-extracted antigen or supernatant fluid from virus-infected cells (Fig. 4aa); b)

introduce the serum or cerebrospinal fluid suspected to contain IgG antibody, testing only a single dilution of serum or the first of a series of dilutions beginning with about 1:40; c) add anti-human IgG (which does not contain antibody to species (A) conjugated with an enzyme; d) add the substrate and record the optical density of the test and control sera. Again, ratios of test to control greater than or equal to 2.0 are considered positive. This IgG assay is performed relatively quickly, has the advantages of allowing precoating and storage of plates, and requires only a few steps for completion. It is somewhat less sensitive and specific than neutralization tests, but because it is performed relatively easily, it is used to rapidly detect antibody to a variety of viruses. A distinct disadvantage of this assay is the need for either purified virus or separate coating antibodies for each virus included in a panel. Also, because there is a need for paired acute- and convalescent-phase serum samples, it holds little advantage over the HI test.

In summary, we suggest using MAC ELISA for rapidly detecting antibody in situations of immediate importance, and neutralization, HI, IFA, or IgG assays for determining antibody in paired serum samples and for serosurveys. In individual instances in which IgM antibody has been shown to be present, HI and CF can be used as accessory tests, and neutralization tests can be used for definitive and confirmatory determinations of the infecting agent.

Interpretation of Serologic Data

All viruses of the genus *Flavivirus* are antigenically related to one another, as are all viruses of the genus *Alphavirus*. Antigen sharing provides a practical ad-

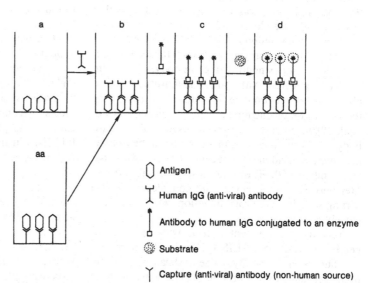

FIG. 4. IgG antibody detection enzyme-linked immunosorbent assay.

○ Antigen

Υ Human IgG (anti-viral) antibody

Ι Antibody to human IgG conjugated to an enzyme

▦ Substrate

Υ Capture (anti-viral) antibody (non-human source)

vantage, in that evidence for the presence of shared antigens can be used to place an isolate within a serogroup, thereby eliminating the possibility that the virus belongs to one of the many other serogroups in which arboviruses are found. Biological characterization cannot be used to place an isolate within a serogroup, and extensive and complex biochemical and electron microscopic methods are expensive and time-consuming. Therefore, detecting group-reactive antigen can be of value in serologic tests for antibodies to flaviviruses and alphaviruses.

When a human or other animal is infected with a virus, that animal produces type-specific, complex-specific, and group-specific antibodies, which are detected as a mixture in serum taken from the animal. Therefore, identification of group-specific antibody in a patient indicates infection with a member of that serogroup, identification of complex-specific antibody indicates infection with a member of a particular antigenic complex within that serogroup, and identification of type-specific antibody indicates infection with a particular virus (type).

The flavivirus hemagglutinin is contained in the envelope (E) glycoprotein (Trent, 1977), whereas the hemagglutinins of alphaviruses are contained in one of the two glycoproteins (E1 and E2); for example, for SIN virus in E1 and for VEE virus in E2 (reviewed by Roehrig, 1986). Irrespective of site, antibodies to all flaviviruses and alphaviruses inhibit hemagglutination by other members of those serogroups, so that the HI test can be used to place a virus in a serogroup.

The nucleocapsid antigen is responsible for the broad group- and antigenic complex-reactivity of both the flaviviruses and the alphaviruses (Dalrymple, 1972; Trent, 1977); this is one of the antigens that could participate in the CF reaction. Finding CF antibody to a particular virus in an individual indicates that that individual was recently infected with that virus or with one closely related to it. After dengue virus infections, individuals produce CF antibodies that persist for as many as 50 years (Halstead, 1974), or for life, as does neutralizing antibody. Certain individuals infected with other flaviviruses and alphaviruses never produce CF antibody or produce it too late to be of diagnostic value (Calisher and Poland, 1980). Nevertheless, the presence of CF antibody to a particular virus in a patient can be used as presumptive evidence of recent infection with that virus and, as with HI or other tests, a fourfold rise in titer between paired acute- and convalescent-phase serum samples confirms infection with that or a closely related virus.

Reactivity of a patient's serum with a given antigen by any test, be it ELISA, HI, CF, direct or indirect fluorescent antibody, or neutralization, is useful in the serodiagnosis of arbovirus infections. Because of the significant cross-reactivity of the flaviviruses, patient serum can be tested with the suspected etiologic agent of the infection and a few other viruses (flaviviruses and others). If the test detects antibody, it can be used to make a presumptive diagnosis of infection with that virus or a member of that serogroup, with some notable exceptions. First, antibody in a single serum merely indicates infection at some time in the past and, therefore, finding antibody in such a serum cannot be used to indicate any more than a presumptive diagnosis. Second, antibody in a single serum could be due to vaccination with YF, JE, or other flavivirus. IgM antibody detected by ELISA (as described above) usually confirms infection with the virus for which the specimen was tested or with a closely related virus. In epidemic situations ELISA-positive serum samples can be considered presumptive evidence of recent infections, but in all instances adequate epidemiologic information must be obtained to be secure in one's interpretation of results of serologic tests. Examples of the variety of serodiagnostic test results that can be obtained with flavivirus and alphavirus infections of humans are presented in Tables 5 and 6, respectively.

Case Reports

Patient F-1 (Table 5) was a 65-year-old female from Mississippi. Serum, collected on the 4th day after onset of encephalitic signs, was positive in MAC ELISA, HI, and neutralization tests for antibody to SLE virus. Only the MAC ELISA was specific, however, and the CF test was negative. Based on these results, a presumptive (and epidemiologically conclusive) serodiagnosis of SLE virus infection could be made. No advantage was conferred by performing similar tests with convalescent-phase serum, collected 18 days after onset. The MAC ELISA was no less specific, the HI test was no more specific, the CF test was uninterpretable (except to say that the patient had recently been infected with a flavivirus), and the neutralization test, although it revealed higher titers to SLE virus than to the other flaviviruses with which the serum was tested, nevertheless was not type-specific. Both HI and neutralization tests provided confirmatory evidence (fourfold rises in titer between paired serum samples) of recent infection with a flavivirus, most likely SLE virus.

Some areas outside the United States are hyperendemic for multiple flaviviruses. Residents of these areas, besides being at risk of exposure to more flaviviruses and to the possible complications brought about by sequential infections with such viruses, may respond to the latest in a series of flavivirus infections by producing broadly cross-re-

TABLE 5. Examples of serodiagnostic results obtained with serum samples from patients infected with flaviviruses

Patient	Days after onset	Test[b]	Virus[a]					
			SLE	JE	WN	MVE	YF	LGT
F-1	4	MAC ELISA	3.15	1.22	1.60	1.79	1.10	1.59
		HI	160	80	160	80	20	10
		CF	—	—	—	—	—	—
		N	160	20	NT	20	—	—
	18	MAC ELISA	2.98	1.15	1.21	1.37	1.38	1.01
		HI	1,280	320	320	1,280	40	40
		CF	16	8	8	8	8	—
		N	1,280	160	NT	160	40	10
			LGT	POW	SLE	YF	LAC	
F-2	14	HI	20	—	20	20	—	
		CF	32	—	—	—	—	
	18	HI	20	—	20	20	—	
		CF	32	—	—	—	—	
	35	HI	80	10	40	20	—	
		CF	256	16	—	—	—	
			YF	ZIKA	WN	UGS	USU	
F-3	14	IgM IFA	64	—	—	—	—	
		IgG IFA	128	—	8	—	—	
		HI	1,280	10	10	40	40	
		CF	8	—	—	—	—	
		N	640	—	—	10	—	
	42	IgM IFA	64	—	—	NT	NT	
		IgG IFA	128	—	—	—	—	
		HI	640	10	—	20	20	
		CF	32	—	—	—	—	
		N	640	—	—	—	—	
F-4	28	IgM IFA	128	—	—	8	8	
		IgG IFA	256	256	256	128	256	
		HI	1,280	1,280	640	1,280	1,280	
		CF	32	—	64	—	—	
		N	320	320	NT	160	320	
	77	IgM IFA	—	—	—	—	—	
		IgG IFA	512	128	128	16	32	
		HI	640	1,280	640	1,280	1,280	
		CF	512	—	8	—	—	
		N	80	320	—	10	80	

[a] SLE = St. Louis encephalitis; JE = Japanese encephalitis; WN = West Nile; MVE = Murray Valley encephalitis; POW = Powassan; YF = yellow fever; LGT = Langat; LAC = LaCrosse; ZIKA = Zika; UGS = Uganda S; USU = Usutu.
[b] IgM antibody capture–enzyme-linked immunosorbent assay (MAC ELISA) results given as ratio of optical densities with test and control antigens and serum at the lowest test dilution, in this instance 1 : 100 (greater than or equal to 2.0 considered positive). Results of other tests given as titers; − indicates <10 HI (hemagglutination-inhibition) or N (neutralization), <8 CF (complement-fixation), <8 IFA (indirect fluorescent antibody); NT indicates not tested.

active HI, CF, and neutralizing antibodies (Monath et al., 1980; Patient F-4, Table 5). It has not yet been proven, but it is unlikely that antibody detected under such circumstances by MAC ELISA is less specific than in the example provided here; thus, MAC ELISA is the test of choice in any circumstance in which a flavivirus is the suspected etiologic agent of human infection:

Patient F-2 (Table 5), a 4-year-old female from Ohio, became febrile while in central Europe. An engorged tick was found on her scalp; it was removed and her symptoms disappeared. Two weeks later, she returned to the United States and became ill again, this time with obvious signs of central nervous system disease. A tick-borne encephalitis virus was suspected as the etiologic agent of her illness, and a serum sample was serologically tested. Rather than using several tick-borne viruses known to be hazardous to laboratory workers, we used Langat virus as antigen. This virus occurs in southeast Asia, but not in Europe. It is related to CEE, POW of North America, and other viruses of the tick-borne complex of flaviviruses (Table 1). Results of HI and CF tests revealed that the patient had HI antibody to the four flaviviruses, but not to LaCrosse, a California serogroup bunyavirus. Serologic conversion, demonstrated to Langat virus, and stable titers to the other flaviviruses suggested recent infection by a flavivirus belonging to the tick-borne encephalitis complex. CF test results confirmed this; an eightfold rise in antibody to Langat and a lesser rise to POW virus was demonstrated between paired serum samples. Neutralization tests were not performed because of the hazard to laboratory workers. Extensive studies of IgM antibodies by Heinz et al. (1981) now have shown that ELISA for antibodies to CEE virus is a rapid and sensitive method for detecting such antibodies in serum samples and in cerebrospinal fluids from humans infected with this and related viruses.

Patient F-3 (Table 5), from The Gambia, West Africa, had a history of febrile illness with jaundice clinically compatible with YF. Diagnosis was made on the basis of fourfold rise in CF antibody between paired serum samples; antibody titers determined by other tests were stable, albeit high, and heterologous flavivirus reactivity was minimal or not detected. YF could be readily diagnosed from the fluorescent antibody test results, because high IgM and IgG antibody titers were present in the absence of heterologous cross-reactions. This is an example of primary YF virus infection, rather than infection by one flavivirus at some time after another.

Patient F-4 (Table 5), also from The Gambia, had a clinical history similar to that of patient F-3. Diagnosis was made on the basis of a 16-fold decrease in IgM antibody, 16-fold increase in CF antibody, fourfold decrease in neutralizing antibody, and elevated HI and IgG titers to YF virus. Extensive cross-reactions with other flaviviruses indicated that this was a clear instance of flaviviral superinfection, that is, infection with YF virus at some time after infection with another flavivirus.

Patient A-1 (Table 6) was a 2-year-old female from Florida. As has been shown previously (Calisher et al., 1986b), MAC ELISA was type-specific for infection with EEE virus, as were CF test results. HI and neutralization test results with both single and paired serum samples could be used as presumptive and confirmatory evidence, respectively, for recent infection with EEE virus.

Patient A-2 (Table 6), a 35-year-old female from Minnesota, was ill with fever and headache during an outbreak of WEE in that state. Again, the MAC ELISA was not type-specific for WEE virus; it reacted with SIN virus, a member of the same antigenic complex. Neutralization test results could be used to make an accurate serodiagnosis.

Patient A-3 (Table 6) was a 30-year-old male from Texas who became ill with fever and headache during the temporary incursion of VEE into that state in 1971. Serum samples from this patient and patient A-2 were stored at −20°C and tested by MAC ELISA more than 10 years after collection. Acute-phase serum, collected on the day of onset, did not react in any of the four tests. However, the convalescent-phase serum, collected 31 days after onset, was positive in MAC ELISA and in HI, CF, and neutralization tests for both VEE virus and the closely related Everglades virus. Titers were considerably higher to the former than to the latter. The MAC ELISA was the only test that did not provide evidence for reactivity with heterologous alphaviruses.

Patient A-4 (Table 6), was a male in Indonesia with fever during an outbreak of chikungunya there. MAC ELISA was positive to both CHIK virus and RR virus, a closely related alphavirus; however, as with MAC ELISA of the convalescent-phase serum of Patient A-3 with VEE and Everglades viruses, the serum of Patient A-4 reacted to much higher titer with the homologous (infecting) virus. That this was, in fact, an infection caused by CHIK virus could be determined from the neutralization test results and by the fact that the HI and CF tests were positive to CHIK and not to RR virus. Results of similar tests with sera from individuals infected with RR virus have given reciprocally similar results to these.

Epidemiology and Ecology

Because of the large number of alphaviruses and flaviviruses that have been implicated in diseases of humans and domesticated animals, selection of test methods for virus isolation and of antigens for use in diagnostic tests should be based on a thorough understanding of the clinical features and epidemiology of these viruses. The disease syndrome associated with infection can narrow the etiologic possibilities (Tables 3, 4, 7), but it should be emphasized that viruses that cause encephalitis or hemorrhagic fever also produce mild, nonspecific, systemic febrile illnesses. The geographic location of the patient's exposure should be considered in terms of virus distribution (Table 7), and diagnostic tests should be chosen and designed accordingly.

TABLE 6. Examples of serodiagnostic results obtained with serum samples from patients infected with alphaviruses

Patient	Days after onset	Test[b]	Virus[a]						
			EEE	WEE	SIN	VEE	EVE	CHIK	RR
A-1	1	MAC ELISA	4.21	1.51	1.31	0.89	1.14	NT	NT
		HI	40	—	—	—	—	—	—
		CF	—	—	NT	—	—	NT	NT
		N	20	—	NT	—	NT	NT	NT
	22	MAC ELISA	3.10	1.26	1.00	1.44	1.71	1.37	1.47
		HI	320	10	—	—	—	—	—
		CF	32	—	—	—	—	—	—
		N	10,240	20	NT	20	NT	NT	NT
A-2	3	MAC ELISA	1.39	1.79	1.82	1.03	1.61	1.45	1.63
		HI	—	—	—	—	—	—	—
		CF	—	—	—	—	NT	NT	NT
		N	—	—	—	—	NT	NT	NT
	10	MAC ELISA	1.57	3.89	2.15	1.28	1.46	1.23	1.08
		HI	—	40	20	—	—	—	—
		CF	—	8	—	—	NT	NT	NT
		N	—	160	20	—	NT	NT	NT
A-3	0	MAC ELISA	1.28	0.90	1.41	1.28	1.36	1.04	1.61
		HI	—	—	—	—	—	—	—
		CF	—	—	—	—	—	—	—
		N	—	—	NT	—	—	NT	—
	31	MAC ELISA	1.12	1.35	1.49	3.33	3.27	1.60	1.02
		HI	10	40	—	1,280	320	40	80
		CF	—	—	8	128	32	—	—
		N	—	—	—	2,560	160	—	10
A-4	5	MAC ELISA	1.82	1.61	1.20	1.47	1.18	5.52	3.39
		HI	—	—	—	—	—	40	—
		CF	—	—	NT	—	NT	8	—
		N	—	—	NT	—	NT	40	—
	19	MAC ELISA	1.72	1.34	1.05	1.22	1.66	4.62	2.82
		HI	—	—	NT	—	—	80	—
		CF	—	—	NT	—	—	16	—
		N	—	—	NT	—	NT	640	40

[a] EEE = eastern equine encephalitis; WEE = western equine encephalitis; SIN = Sindbis; VEE = Venezuelan equine encephalitis (variety IAB); EVE = Everglades (VEE type II); CHIK = chikungunya; RR = Ross River.
[b] See footnote [b], Table 5.

Season of the year also is important; alphaviruses and flaviviruses transmitted by mosquitoes in temperate areas cause human infections in the summer and early fall, during the period of vector activity. Tick-borne flavivirus infections usually occur during the spring and early summer. In tropical areas, infections generally coincide with the rainy season, the monthly distribution thus varying with latitude. Information about the occurrence of multiple cases in the same place and time, as well as association with disease in domesticated animals, will greatly assist in design of diagnostic studies. For example, an outbreak of fever, arthralgia, and rash in an African city is more likely to be due to dengue or chikungunya virus than to other flaviviruses or alphaviruses that have been associated only with individual or sporadic cases (for example, Zika, Spondweni, Banzi, or Wesselsbron viruses). The concurrence of cases of encephalitis in horses and humans suggests, depending on geographic area, a diagnosis of EEE, WEE, VEE, or JE.

Age, sex, and occupation are also significant risk factors for alphavirus and flavivirus infection. For example, an outbreak of encephalitis in the western United States could be due to WEE or SLE viruses, or both. Because children are more susceptible to severe encephalitis caused by WEE virus, cases in infants are more likely to be due to that virus; in contrast, the elderly are more prone to develop SLE.

Most alphaviruses and flaviviruses have unique transmission cycles involving specific vector arthropods and vertebrate hosts. The principal elements of these cycles are listed in Table 7.

TABLE 7. Clinical, epidemiologic, and ecological features of alphaviruses and flaviviruses

Genus	Virus	Clinical features	Geographic distribution
Alphavirus	Chikungunya	Fever, arthralgia, myalgia, rash	Africa, southern Asia, Philippines
	O'nyong nyong	Fever, arthralgia, myalgia, rash	East Africa
	Mayaro	Fever, arthralgia, myalgia, rash	Tropical South America
	Ross River	Fever, arthralgia, myalgia, rash	Australia, South Pacific
	Sindbis and related agents (e.g., Ockelbo)	Fever, arthralgia, myalgia, rash	Africa, Asia, Europe
	Eastern equine enc.	Encephalitis	North and South Americas, Caribbean
	Western equine enc.	Encephalitis	North and South Americas
	Venezuelan equine enc.[a]	Encephalitis	South America, Central Americas (enzootic subtype in Florida)
Flavivirus	Dengue-1, -2, -3, -4	Fever, arthralgia, myalgia, rash	Tropics worldwide
	West Nile	Fever, arthralgia, myalgia, rash	Africa, tropical Asia, Mediterranean region
	Banzi	Fever, arthralgia, myalgia, rash	Africa
	Bussuquara	Fever, arthralgia, myalgia, rash	Tropical South America
	Ilheus	Fever, arthralgia, myalgia, rash	Tropical South America
	Sepik	Fever, arthralgia, myalgia, rash	New Guinea
	Spondweni	Fever, arthralgia, myalgia, rash	Africa
	Wesselsbron	Fever, arthralgia, myalgia, rash	Africa
	Zika	Fever, arthralgia, myalgia, rash	Africa
	Usutu	Fever, arthralgia, myalgia, rash	Africa
	St. Louis enc.	Encephalitis	North, Central, South America
	Japanese enc.	Encephalitis	Asia
	Murray Valley enc. and Kunjin	Encephalitis	Australia and New Guinea
	Rocio	Encephalitis	Southeastern Brazil
	Yellow fever	Fever, jaundice, hemorrhage, and renal failure	Tropical Africa and Americas
	Kyasanur Forest disease	Fever, hemorrhage, encephalitis	Southwest India
	Omsk hemorrhagic fever	Fever, hemorrhage	Central U.S.S.R.
	Tick-borne enc.[b]	Encephalitis	U.S.S.R., Eastern Europe, Scandinavia, Finland
	Louping ill	Encephalitis	British Isles
	Powassan	Encephalitis	Canada, United States, U.S.S.R.

[a] Multiple antigenic subtypes and varieties, of which two (IAB, IC) are responsible for epizootic/epidemic disease; other (enzootic) viruses (e.g., subtypes ID, IE, II, IIIA, IIIB, IV) cause sporadic human infections.
[b] Multiple antigenic types and subtypes, including Russian spring-summer (Far Eastern) enc., Central European enc., and Kumlinge viruses.

TABLE 7. *Continued*

Epidemiologic features	Principal vector(s)	Principal hosts
Urban epidemics similar to dengue; enzootic cycle similar to yellow fever	*Aedes aegypti, Ae. africanus, Ae. furcifer-taylori*	Humans, monkeys
Single large rural outbreak in 1950s	*Anopheles* sp.	Humans
Sporadic cases, limited outbreaks associated with deforestation	*Haemagogus* sp.	Monkeys, ? rodents, birds
Annual epidemics (Australia), occasional introductions and epidemic spread (South Pacific)	*Ae. vigilax, Culex annulirostris, Ae. polynesiensis*	Large marsupials, humans
Sporadic causes, epidemics in nonendemic areas	*Culex* sp., *Culiseta* sp.	Wild birds
Periodic small equine epizootics and epidemics	*Cu. melanura, Cq. perturbans*	Wild birds
Periodic equine epizootics and epidemics	*Cx. tarsalis*	Wild birds
Periodic equine epizootics and epidemics	*Aedes, Mansonia, Culex* spp. (epizootic subtypes) *Cx.* (*Melanoconion*) (enzootic subtypes)	Equines (epizootic) Rodents (enzootic)
Urban epidemics	*Ae. aegypti*, other *Aedes* spp.	Humans
Endemic (tropics) with sporadic cases; summertime epidemics (Mediterranean, temperate South Africa)	*Culex* spp.	Wild birds
Sporadic cases	*Culex* spp.	Rodents
Sporadic cases	*Culex* (*Melanoconion*) spp.	Rodents
Sporadic cases	*Psorophora* sp., other species	Wild birds
Sporadic cases	*Armigeres, Ficalbia* sp.	?
Sporadic cases	*Aedes* sp., other species	?
Sporadic cases	*Aedes* sp.	? Sheep
Sporadic cases	*Ae. africanus*	Monkeys
Sporadic cases	*Culex* sp.	Wild birds
Periodic epidemics, North America; sporadic disease, tropical South America	*Culex tarsalis* (western U.S.); *Culex pipiens* (eastern U.S.); Multiple *Culex* sp (tropics)	Wild birds
Endemic (S.E. Asia); summertime epidemics (northern and central Asia)	*Culex tritaeniorhynchus*	Swine, birds
Periodic summertime epidemics	*Cx. annulirostris*	Mammals, birds
Singular epidemic	? *Psorophora, Aedes* sp.	Wild birds
Periodic epidemics (Africa); sporadic cases related to jungle exposure	*Ae. aegypti*, other *Aedes* sp., *Haemagogus* sp.	Monkeys, humans
Sporadic cases; occasional outbreaks related to deforestation	*Haemaphysalis* ticks	Rodents
Sporadic cases; winter outbreaks associated with trapping muskrats	*Dermacentor* ticks	Rodents
Sporadic cases with periodic high incidence; outbreaks associated with ingestion of infected raw milk	*Ixodes* ticks	Rodents, birds
Sporadic cases, especially acquired by direct contact with infected sheep or by tick bite	*Ixodes* ticks	Rodents, birds
Sporadic cases	*Ixodes* ticks	Small mammals

Prevention and Control

Identification of even one patient with an arthropod-borne disease known to undergo epidemic spread is extremely important in developing an appropriate prevention and control strategy, because such a patient serves as an indicator of a larger public health problem. Surveillance systems designed to detect virus transmission between vectors, wild and sentinel vertebrate hosts, and humans early in the season require collection and testing of samples. Positive cases should be reported immediately to the responsible public health agency.

Prevention and control of epidemics of mosquito-borne diseases generally is accomplished by reducing vector populations. Preventive strategies involve long-term programs aimed at limiting breeding of mosquitoes through source reduction and use of larvicides. The control of active virus transmission in preepidemic and epidemic situations requires use of space sprays to kill infected adult mosquitoes. Public education programs aimed at protecting against mosquito bites emphasize use of protective clothing, repellents, window screens, and bed nets.

Vaccines are available for a few viruses. Human immunization against YF, JE, and tick-borne encephalitis is relatively common; veterinary vaccines against EEE, WEE, VEE, and JE are available to protect horses. Use of VEE vaccine in horses, which are the principal natural viremic hosts implicated in infecting mosquito vectors, therefore provides protection against human disease. Similarly, immunization of swine is used to prevent human outbreaks of JE in Asia.

Vaccines against the equine encephalitides also are used to protect certain high-risk groups of humans, including laboratory workers. Indeed, in diagnostic laboratories in which work with EEE, WEE, VEE, YF, or JE viruses is anticipated, laboratory workers should be immunized, and then immunity should be ascertained by testing for neutralizing antibodies. For assistance in obtaining information regarding vaccines not otherwise available, contact the Division of Vector-Borne Viral Diseases, Centers for Disease Control, P.O. Box 2087, Fort Collins, Colorado 80522. Information regarding experimental vaccines for DEN-2, CHIK, KFD, and OHF also is available from this source.

Literature Cited

Burke, D. S., W. Lorsomrudee, C. J. Leake, C. H. Hoke, A. Nisalak, V. Chongswasdi, and T. Laorakpongse. 1985. Fatal outcome in Japanese encephalitis. Am. J. Trop. Med. Hyg. 34:1203–1209.

Burke, D. S., A. Nisalak, M. A. Ussery. 1982. Antibody capture immunoassay detection of Japanese encephalitis virus immunoglobulin M and G antibodies in cerebrospinal fluid. J. Clin. Microbiol. 16:1034–1042.

Calisher, C. H., V. P. Berardi, D. J. Muth, and E. E. Buff. 1986a. Specificity of immunoglobulin M and G antibody responses in humans infected with eastern and western equine encephalitis viruses: application to rapid serodiagnosis. J. Clin. Microbiol. 23:369–372.

Calisher, C. H., W. Brandt, J. Casals, R. E. Shope, R. B. Tesh, and M. E. Wiebe. 1980. Recommended antigenic classification of registered arboviruses. I. Togaviridae, alphaviruses. Intervirology 14:229–232.

Calisher, C. H., A. O. El-Kafrawi, M. I. Al.-D. Mahmud, A. P. A. Travassos da Rosa, C. R. Bartz, M. Brummer-Korvenkontio, S. Haksohusodo, and W. Suharyono. 1986b. Complex-specific immunoglobulin M antibody patterns in humans infected with alphaviruses. J. Clin. Microbiol. 23:155–159.

Calisher, C. H., M. I. Al.-D. Mahmud, A. O. El-Kafrawi, J. K. Emerson, and D. J. Muth. 1986c. Rapid and specific serodiagnosis of western equine encephalitis virus infection in horses. Am. J. Vet. Res. 47:1296–1299.

Calisher, C. H., O. Meurman, M. Brummer-Korvenkontio, P. E. Halonen, and D. J. Muth. 1985. Sensitive enzyme immunoassay for detecting immunoglobulin M antibodies to Sindbis virus and further evidence that Pogosta disease is caused by a western equine encephalitis complex virus. J. Clin. Microbiol. 22:566–571.

Calisher, C. H., and J. D. Poland. 1980. Laboratory diagnosis, p. 571–601. In T. P. Monath (ed.), St. Louis encephalitis. American Public Health Association, Washington, D.C.

Casals, J. 1944. Immunological relationships among central nervous system viruses. J. Exp. Med. 79:341–359.

Casals, J. 1957. The arthropod-borne group of animal viruses. Trans. N.Y. Acad. Sci. II. 19:219–235.

Casals, J., and L. V. Brown. 1954. Hemagglutination with arthropod-borne viruses. J. Exp. Med. 99:429–449.

Casey, H. L. 1965. Part II. Adaptation of LBCF method to micro technique. In Standardized diagnostic complement fixation method and adaptation to micro test. Public Health Monogr. 74:1–34.

Chamberlain, R. W. 1980. Epidemiology of arthropod-borne togaviruses: the role of arthropods as hosts and vectors and of vertebrate hosts in natural transmission cycles, p. 175–227. In R. W. Schlesinger (ed.), The togaviruses. Academic Press, Inc., New York.

Clarke, D. H., and J. Casals. 1958. Techniques for hemagglutination and hemagglutination-inhibition with arthropod-borne viruses. Am. J. Trop. Med. Hyg. 7:561–573.

Dalrymple, J. M. 1972. Biochemical and biophysical characteristics of Venezuelan equine encephalitis virus, p. 56–64. In Venezuelan encephalitis. Sci. Publ. No. 243. Pan American Health Organization, Washington, D.C.

de la Monte, S. M., A. L. Linhares, A. P. A. Travassos da Rosa, and F. P. Pinheiro. 1983. Immunoperoxidase detection of yellow fever virus after natural and experimental infections. Trop. Geogr. Med. 35:235–242.

de Madrid, A. T., and J. S. Porterfield. 1974. The flaviviruses (group B arboviruses): a cross-neutralization study. J. Gen. Virol. 23:91–96.

Gardner, J. J., and M. G. Reyes. 1980. Pathology, p. 551–569. In T. P. Monath (ed.) St. Louis encephalitis. American Public Health Association, Washington, D.C.

Gardner, P. S., and J. McQuillan. 1980. Applications of immunofluorescence, 2nd ed. Butterworth, London.

Halonen, P. E., H. Sarkkinen, P. Arstila, E. Hjertsson, and E. Torfason. 1980. Four-layer radioimmunoassay for detection of adenovirus in stool. J. Clin. Microbiol. 11:614–617.

Halstead, S. B. 1974. Etiologies of the experimental dengues of Siler and Simmons. Am. J. Trop. Med. Hyg. **23**:974–982.

Halstead, S. B., S. Rojanasuphot, and N. Sangkawibha. 1983. Original antigenic sin in dengue. Am. J. Trop. Med. Hyg. **32**:154–156.

Heinz, F. X., M. Roggendorf, H. Hofmann, C. Kunz, and F. Deinhardt. 1981. Comparison of two different enzyme immunoassays for detection of immunoglobulin M antibodies against tick-borne encephalitis virus in serum and cerebrospinal fluid. J. Clin. Microbiol. **14**:141–146.

Henchel, E. A., J. M. McCown, M. C. Seguin, M. K. Gentry, and W. E. Brandt. 1983. Rapid identification of dengue virus isolates by using monoclonal antibodies in an indirect immunofluorescence assay. Am. J. Trop. Med. Hyg. **32**:164–169.

Hildreth, S. W., B. J. Beatty, H. K. Maxfield, R. F. Gilfillan, and B. J. Rosenau. 1984. Detection of eastern equine encephalomyelitis virus and Highlands J virus antigens within mosquito pools by enzyme immunoassay (EIA). II. Retrospective field test of the EIA. Am. J. Trop. Med. Hyg. **33**:973–980.

Hyypia, T., P. Stalhandske, R. Vainionpaa, and U. Pettersson. 1984. Detection of enteroviruses by spot hybridization. J. Clin. Microbiol. **19**:436–438.

Karabatsos, N. (ed.). 1985. International catalogue of arboviruses including certain other viruses of vertebrates, 3rd ed. Am. Soc. Trop. Med. Hyg., San Antonio.

Kusano, N. 1966. Fluorescent antibody technique and Japanese encephalitis. Symposium on viral encephalitides in the Japanese, Tokyo, April 7, 1966. Psychiat. Neurol. Jap. **68**:300.

Lhuillier, M., and J.-L. Sarthou. 1983. Interet des IgM antiamariles dans de diagnostic et la surveillance epidemiologique de la fievre jaune. Ann. Virol. (Inst. Pasteur). **134E**:349–359.

Lindsey, H. S., C. H. Calisher, and J. H. Mathews. 1976. Serum dilution neutralization test for California group virus identification and serology. J. Clin. Microbiol. **4**:503–510.

Matthews, R. E. F. 1982. Classification and nomenclature of viruses. Fourth report of the International Committee on Taxonomy of Viruses. Intervirology **17**:1–199.

Mendez, M. R., C. H. Calisher, H. Kruger, F. Sipan, S. Sanchez, and J. S. Lazuick. 1984. A continuing focus of yellow fever in the Apurimac River valley, Ayacucho, Peru, and the first isolation of yellow fever virus in that country. Bull. Pan. Am. Health Org. **18**:172–179.

Meurman, O. 1983. Detection of antiviral IgM antibodies and its problems—a review, Curr. Top. Microbiol. Immunol. **104**:101–131.

Monath, T. P. 1971. Neutralizing antibody response in the major immunoglobulin classes to yellow fever 17D vaccination of humans. Am. J. Epidemiol. **93**:122–129.

Monath, T. P., R. B. Craven, D. J. Muth, C. J. Trautt, C. H. Calisher, and S. A. Fitzgerald. 1980. Limitations of the complement-fixation test for distinguishing naturally acquired from vaccine-induced yellow fever infection in flavivirus-hyperendemic areas. Am. J. Trop. Med. Hyg. **29**:624–634.

Monath, T. P., C. B. Cropp, D. J. Muth, and C. H. Calisher. 1981a. Indirect fluorescent antibody test for the diagnosis of yellow fever. Trans. Royal Soc. Trop. Med. Hyg. **75**:282–286.

Monath, T. P., L. J. Hill, N. V. Brown, C. B. Cropp, J. J. Schlesinger, J. F. Saluzzo, and J. R. Wands. 1986a. Sensitive and specific monoclonal immunoassay for detecting yellow fever virus in laboratory and clinical specimens. J. Clin. Microbiol. **23**:129–134.

Monath, T. P., R. G. McLean, C. B. Cropp, G. L. Parham, J. S. Lazuick, and C. H. Calisher. 1981b. Diagnosis of eastern equine encephalomyelitis by immunofluorescent staining of brain tissue. Am. J. Vet. Res. **42**:1418–1421.

Monath, T. P., and R. R. Nystrom. 1984a. Detection of yellow fever virus in serum by enzyme immunoassay. Am. J. Trop. Med. Hyg. **33**:151–157.

Monath, T. P., R. R. Nystrom, R. E. Bailey, C. H. Calisher, and D. J. Muth. 1984b. Immunoglobulin M antibody capture enzyme-linked immunosorbent assay for diagnosis of St. Louis encephalitis. J. Clin. Microbiol. **20**:784–790.

Monath, T. P., J. R. Wands, L. J. Hill, M. K. Gentry, and D. J. Gubler. 1986b. Multisite monoclonal immunoassay for dengue viruses: detection of viraemic human sera and interference by heterologous antibody. J. Gen. Virol. **67**:639–650.

Reed, W., J. Carroll, A. Agramonte, and J. W. Lazear. 1983. Classics in infectious diseases. The etiology of yellow fever: a preliminary note. Rev. Infect. Dis. **5**:1103–1111.

Roehrig, J. T. 1986. The use of monoclonal antibodies in studies of the structural proteins of togaviruses and flaviviruses, p. 251–278. *In* S. Schlesinger and M. J. Schlesinger (ed.), The Togaviridae and Flaviviridae. Plenum Publishing Corp., New York.

Rosen, L. 1981. The use of *Toxorhynchites* mosquitoes to detect and propagate dengue and other arboviruses. Am. J. Trop. Med. Hyg. **30**:177–183.

Rosen, L., and D. Gubler. 1974. The use of mosquitoes to detect and propagate dengue viruses. Am. J. Trop. Med. Hyg. **23**:1153–1160.

Sabin, A. B. 1950. The dengue group of viruses and its family relationships. Bacteriol. Rev. **14**:225–232.

Sabin, A. B., and E. L. Buescher. 1950. Unique physicochemical properties of Japanese B encephalitis virus hemagglutinin. Proc. Soc. Exp. Biol. Med. **74**:222–230.

Saluzzo, J. F., T. P. Monath, M. Cornet, V. Deubel, and J. P. Digoutte. 1985. Comparaison de differentes techniques pour la detection du virus de la fievre jaune dans les prelevements humains et les lots de moustiques: interet d'une methode rapide de diagnostic par ELISA. Ann. Inst. Pasteur/Virol. **136E**:115–129.

Sarkkinen, H. K., P. E. Halonen, P. P. Arstila, and A. A. Salmi. 1981. Detection of respiratory syncytial, parainfluenza Type 2, and adenovirus antigens by radioimmunoassay and enzyme immunoassay on nasopharyngeal specimens from children with acute respiratory disease. J. Clin. Microbiol. **13**:258–265.

Sarthou, J. F., and M. Lhuillier. 1984. Diagnostic immunoenzymatique rapide de la fievre jaune: detection directe dans le serum des malades de l'antigene amarile libre (Ag-YF) ou engage dans les immunscomplexes circulants (IgM-Ag-YF). Comm. Soc. Franc. Microbiol. Paris, February 2.

Smithburn, K. C. 1942. Differentiation of the West Nile virus from the viruses of St. Louis and Japanese B encephalitis. J. Immunol. **44**:25–31.

Smorodintseff, A. A. 1940. The spring-summer tick-borne encephalitis. (Synonyms: Forest Spring encephalitis). Arch. Gesamte Virusforsch. **1**:468–480.

Theiler, M. 1957. Action of sodium desoxycholate on arthropod-borne viruses. Proc. Soc. Exp. Biol. Med. **96**:380–382.

Theofilopoulos, A. N., W. E. Brandt, P. K. Russell, and F. T. Dixon. 1976. Replication of dengue-2 virus in cultured human lymphoblastoid cells and subpopulations of human peripheral leukocytes. J. Immunol. **117**:953–961.

Tikasingh, E. S., L. Spence, and W. G. Downs. 1966. The use of adjuvant and sarcoma 180 cells in the production of mouse hyperimmune ascitic fluids to arboviruses. Am. J. Trop. Med. Hyg. **15**:219–226.

Trent, D. W. 1977. Antigenic characterization of flavivirus structural proteins separated by isoelectric focusing. J. Virol. **22**:608–618.

Varelas-Wesley, I., and C. H. Calisher. 1982. Antigenic relationships of flaviviruses with undetermined arthropod-borne status. Am. J. Trop. Med. Hyg. **31**:1273–1284.

Webster, L. T. 1938. Japanese B encephalitis virus: its differentiation from St. Louis encephalitis virus and its relationship to louping ill virus. J. Exp. Med. **67**:609–618.

Webster, L. T., G. L. Fite, and A. D. Clow. 1935. Experimental studies on encephalitis. IV. Specific inactivation of virus by sera from persons exposed to encephalitis, St. Louis. J. Exp. Med. **62**:827–847.

Westaway, E. G., M. A. Brinton, S. Y. Gaidomovich, M. C. Horzinek, A. Igarashi, L. Kaariainen, D. K. Lvov, J. S. Porterfield, P. K. Russell, and D. W. Trent. 1985a. *Togaviridae*. Intervirology **24**:125–139.

Westaway, E. G., M. A. Brinton, S. Y. Gaidamovich, M. C. Horzinek, A. Igarashi, L. Kaariainen, D. K. Lvov, J. S. Porterfield, P. K. Russell, and D. W. Trent. 1985b. *Flaviviridae*. Intervirology **24**:183–192.

Westaway, E. G., A. J. Della-Porta, and B. M. Reedman. 1974. Specificity of IgM and IgG antibodies after challenge with antigenically related togaviruses. J. Immunol. **112**:656–663.

World Health Organization. 1967. Arboviruses and human disease. W.H.O. Tech. Rep. Ser. No. 369.

CHAPTER 23

Togaviridae: Rubella Virus

JENNIFER M. BEST and SIOBHAN O'SHEA

Disease: Rubella, congenital rubella.

Etiologic Agent: Rubella virus, genus *Rubivirus* of *Togaviridae.*

Clinical Manifestations: Asyptomatic infections are common; fever with rash; lymphadenitis; arthropathy and arthritis in adults. Congenital infection may result in fetal death and spontaneous abortion, live birth of severely malformed infant, an infant with minimal damage, or a healthy infant.

Pathology: Primary site of virus replication is the nasopharyngeal mucosa, followed by spread of virus to local lymph nodes and a viremia 7 days after infection. The appearance of a rash may be caused by viral antigen-antibody complexes. During pregnancy, the virus infects the placenta and is then disseminated to all fetal tissues.

Laboratory Diagnosis: IgG serology in paired serum specimens; IgM serology in a single serum. IgG serology is in large-scale use for immunity testing.

Epidemiology: Worldwide distribution; endemic in temperate climates, with seasonal peaks during spring and early summer. In countries with successful childhood vaccination programs, major epidemics have been prevented.

Treatment: No antiviral chemotherapeutic agent available.

Prevention and Control: Attenuated rubella vaccines administered subcutaneously are widely used, often in the form of combined measles-mumps-rubella vaccine; have greatly reduced total number of rubella cases and similar reduction has been observed in congenital rubella, the main target of vaccination programs.

Rubella virus was first isolated in cell cultures in 1962 by Weller and Neva (1962) and Parkman et al. (1962), but large-scale epidemiologic and diagnostic studies were possible only after introduction of the hemagglutination inhibition (HAI) test in 1967 by Stewart et al. (1967) and Halonen et al. (1967). Attenuated vaccines were developed by Parkman et al. (1966), Peetermans and Huygelen (1967), and Plotkin et al. (1967). Vaccines were first licensed in the United States in 1969 and the United Kingdom in 1970. In the United States, where all children are offered combined measles-mumps-rubella vaccine before school entry, the total number of rubella cases has been greatly reduced (Centers for Disease Control, 1987a). A similar reduction has been observed in cases of congenital rubella, the main target of vaccination programs. The total eradication of rubella is still in the distant future in many developed countries

(Banatvala, 1987; Taranger, 1982) and has not even begun in most developing countries.

Rubella immunity tests continue to be an important large-scale function of diagnostic laboratories. The assessment of pregnant women who have developed or been in contact with rubella-like illnesses is still one of the most challenging tasks in diagnostic virology. It requires high laboratory standards, expertise in IgM serology, and reliable reagents or kits.

The Virus

Classification

Rubella virus is a non-arthropodborne togavirus and the only member of the genus *Rubivirus* (Westaway et al., 1985). There is no antigenic relationship with other togaviruses or other virus groups (Horzinek, 1981; Mettler and Petrelli, 1968) and no significant homology in the major envelope proteins has been noted (Frey et al., 1986; Nakhasi et al., 1986).

Morphology and Antigens

The diameter of the pleomorphic rubella virus particle is 50 to 70 nm (mean 58 nm). The 30-nm core is covered by an envelope with surface projections of 5 to 6 nm (Holmes and Warburton, 1969; Smith and Hobbins, 1969). The exact symmetry of the nucleocapsid has not been established, but icosahedral symmetry has been suggested. The RNA is a single-stranded 11-kbp molecule of 38S to 40S, with a molecular weight of 3.3 to 3.5×10^3 kD (Hovi and Vaheri, 1970; Sedwick and Sokol, 1970).

Only a single type of rubella virus has been described. Major antigenic differences have not been observed between strains by kinetic and cross-hemagglutination-inhibition tests, but some differences have been observed in plaque morphology, cross-neutralization, and neutralization kinetic experiments, growth in rabbits, and using monoclonal antibodies (reviewed by Banatvala and Best, 1984).

Rubella virus has hemagglutinating (HA), complement-fixing (CF), precipitating, and platelet-aggregating antigens and a hemolysin that is associated with the HA antigen. The HA activity is associated with the spikes on the viral envelope. Hemagglutinating antigen is usually produced in infected BHK-21 and Vero cell cultures maintained in serum-free medium.

Erythrocytes of many species are agglutinated; the most commonly used are newborn chick, goose, and human O cells (Schmidt et al., 1971). Hemagglutination is pH-dependent, with the optimal pH 6.2 (Halonen et al., 1967). It is also Ca^{++} ion-dependent (Furukawa et al., 1967; Liebhaber, 1970). The CF activity of rubella virus is associated with at least three different structural proteins including the whole virion particle, a subunit of the virus envelope, and the ribonucleoprotein core (Schmidt and Lennette, 1966a; Vesikari, 1972). Complement-fixating antigens are produced by alkaline extraction of infected cells or by concentration of supernatant fluid (Halonen et al., 1976, Schmidt and Lennette, 1966a, b).

Immunoprecipitating activity is associated with two small-size antigens, "theta" and "iota," as first suggested by LeBouvier (1969). The theta antigen may consist of the same proteins on the virus envelope that have CF activity. The iota antigen is probably associated with the ribonucleoprotein core.

Platelet-aggregating antigens, identified by Pent-

tinen and Myllylä (1968), are associated with the envelope and the ribonucleoprotein core of the virion.

Rubella virus has three structural proteins, two envelope proteins (E1, E2a, and E2b), and the capsid protein C (Clarke et al., 1987; Oker-Blom et al., 1983; Oker-Blom, 1984; Toivonen et al., 1983; Vaheri and Hovi, 1972; Waxham and Wolinsky, 1985a). Antigenic and functional properties of these structural polypeptides have been studied with monoclonal antibodies. At least six antigenic sites have been identified in E1 (Gerna et al., 1987a; Waxham and Wolinsky, 1985b). Four sites are associated with hemagglutinating activity; two of these sites have hemolysin activity, and one is associated with neutralization. Another site is associated with neutralization, but not with hemagglutination. The E2 glycoprotein has at least one epitope associated with neutralization (Green and Dorsett, 1986).

Replication

Replication of rubella virus takes place in the cytoplasm and is remarkably similar to that of other members of the *Togaviridae*. First, a complementary negative strand 40S RNA is produced and serves as a template for synthesis of progeny 40S RNA and subgenomic 24S mRNA (Oker-Blom et al., 1984). The subgenomic mRNA encodes the 110 kD precursor of the three structural proteins (Oker-Blom et al., 1984; Vidgren et al., 1987; Waxham and Wolinsky, 1985b). E1 and E2 are envelope-associated glycoproteins, while the C protein is nonglycolysated and associated with genomic RNA with which it forms the nucleocapsid. E2 is in two forms, E2a and E2b, which differ in their degree of glycolysation (Bowden and Westaway, 1985; Waxham and Wolinsky, 1985a). Rubella virus matures by budding from host cell membranes, where the glycoprotein-containing envelope is formed.

Growth in Cell Cultures

A wide range of cell cultures can be used to grow rubella virus (Banatvala and Best, 1984; Herrmann, 1979). A cytopathic effect (CPE) is not usually observed in primary cell cultures. A CPE is produced in such continuous lines as RK13, SIRC, and Vero cells, but only if cell culture conditions are carefully controlled. Rubella virus can be detected in vervet monkey kidney cells by interference since it blocks the growth of viruses, such as enteroviruses (Parkman et al., 1964). The RK-13, SIRC, Vero, and primary vervet monkey kidney (VMK) cells are used for virus isolation.

BHK-21 and Vero cell cultures are used for the

production of the high titers of rubella virus required as antigens in antibody assays. The optimal growth requirements include the use of a virus strain, which has been adapted to the subline of cell cultures used.

Plaque assays have been developed in RK13, BHK-21, SIRC, and Vero cells. Carboxymethyl cellulose is often used in the overlay.

Pathogenicity for Animals

Primates, ferrets, rabbits, hamsters, and suckling mice can be experimentally infected (Herrmann, 1979), but teratogenic effects have not been consistently produced (reviewed by Banatvala and Best, 1984; Herrmann, 1979). Monkeys develop a subclinical infection, with viremia, virus excretion, and an antibody response (Oxford and Potter, 1971; Parkman et al., 1965) and have, therefore, been valuable models for vaccine testing.

Sensitivity to Chemical and Physical Agents

Rubella virus is readily inactivated by heat, detergents, and organic solvents (reviewed by Herrmann, 1979; Horzinek, 1981).

Pathogenesis and the Disease

Postnatally Acquired Infection

CLINICAL FEATURES

Rubella is transmitted mainly via respiratory secretions. It is not highly contagious, and many susceptible individuals escape infection during epidemics. The primary site of virus replication is the nasopharyngeal mucosa, followed by spread of the infection to local lymph nodes. Within 7 days of infection, a viremia occurs, which results in dissemination of the virus. Virus can be isolated from throat swabs, nasopharyngeal aspirates, and occasionally from urine, stools, synovial fluid, and cervical swabs (Green et al., 1965; Heggie, 1978; Heggie and Robbins, 1969).

The incubation period is 14 to 21 days, after which the rash and lymphadenopathy may appear. Characteristic lymphadenitis affecting cervical and occipital nodes can appear a few days before rash. The typical rubella rash is macular or maculopapular, initially appearing on face and rapidly spreading to the trunk, arms, and legs and fading within a few days. The rash may be present for only 1 day and may be unnoticed, especially in dark skinned individuals. It is similar to the rash caused by other viruses such as enteroviruses, human parvovirus B19, and some ar-

boviruses, which may also induce joint symptoms. Because of these difficulties in clinical diagnosis, a history of rubella is unreliable.

Patients with rubella are potentially infectious for 1 week before the onset of rash and 7 to 10 days thereafter. Viremia disappears at the time rubella antibodies appear.

Complications of rubella are infrequent, particularly in children. In adults, joint symptoms may appear soon after the appearance of rash; they may last for a few days or occasionally for more than a month. Postinfectious encephalitis occurs in about 1 in 10,000 rubella infections.

IMMUNE RESPONSES

Rubella antibodies develop rapidly after the onset of rash (Enders et al., 1985; Meurman, 1978; O'Shea et al., 1985), and IgG antibodies persist for life. Rubella-specific IgM antibody usually appears 1 to 5 days after the onset of rash, sometimes at the same time as IgG antibodies. IgM antibodies usually persist for 6 to 12 weeks (Enders et al., 1985; Enders and Knotek, 1986; Meurman et al., 1977), but occasionally for longer. The detection of rubella-specific IgA depends on the test used; it may persist for a number of years (Halonen et al., 1979; O'Shea et al., 1985). IgD and IgE antibodies develop rapidly after the onset of rash and persist for several months (Salonen et al., 1985). Studies on IgG subclasses have demonstrated that in acute rubella, there is consistently an IgG1 response (Skvaril and Schilt, 1984), and less frequently IgG3 and IgG4 responses (Linde, 1985).

REINFECTION

Reinfection is rare in those with naturally acquired immunity. It is almost always subclinical and can be detected by demonstrating an increase in IgG antibody in paired serum specimens. A weak and transient IgM response may also be detected when highly sensitive immunoassays are used (Cradock-Watson et al., 1985; Morgan-Capner et al., 1985). Experimental challenge studies have suggested that reinfection is more likely to occur in persons with vaccine-induced immunity, when antibody levels are lower (Horstmann et al., 1970; Davis et al., 1971; Harcourt et al., 1980; Fogel et al., 1978; O'Shea et al., 1983).

Current evidence suggests that subclinical reinfection, even with an IgM response, is not usually associated with fetal infection or damage (Cradock-Watson et al., 1985; Morgan-Capner et al., 1985). However, viremia must have occurred in reinfection with clinical symptoms (Forrest et al., 1972; Horstmann et al., 1969; Morgan-Capner et al., 1983a, 1984; Wilkins et al., 1972) and has been demonstrated after experimental challenge (Balfour et al., 1981; O'Shea et al., 1983; Schiff et al., 1985). Morgan-Capner (1986) has discussed those few cases of maternal reinfections that may have resulted in congenital rubella.

Congenitally Acquired Infection

CLINICAL FEATURES

The clinical features of the congenital rubella syndrome are listed in Table 1, and have been reviewed by Best and Banatvala (1987). In addition to the many and often severe defects seen in the newborn, some abnormalities, including hearing loss, may be detected months or years later. Likewise, some manifestations such as diabetes mellitus, glaucoma, and progressive panencephalitis, which resembles subacute sclerosing panencephalitis caused by measles virus, may develop some years later (reviewed by Burke et al., 1985; Hanshaw et al., 1985).

RISK TO THE FETUS

The risk of rubella in pregnancy varies with the gestational age at which maternal infection occurs. The outcome of rubella in utero may be fetal death and spontaneous abortion, live birth of a severely malformed infant, an infant with minimal damage, or a healthy infant with serologic evidence of congenital infection.

If the maternal infection occurs during the first 8 weeks of pregnancy, the rate of spontaneous abortion is about 20% (Siegel et al., 1971). During the first 12 weeks, the rate of fetal infection and rubella-in-duced defects in live births is 90 to 100%, by 16 weeks is has declined to 10 to 20% (Cooper et al., 1969; Munro et al., 1987; Peckman, 1972; Ueda et al., 1979). Maternal infection between the 17th and 20th week of gestation may occasionally result in hearing loss and retinopathy (Grillner et al., 1983; Peckham, 1972).

VIRUS EXCRETION AND IMMUNE RESPONSES

Infants with congenital rubella may excrete high titers of virus in nasopharyngeal and other secretions for months after birth (Cooper and Krugman, 1967) and contact with them is, therefore, a risk for susceptible pregnant women. The immunity of nursing staff of child-bearing age should be checked by reliable antibody assays before they take care of such infants. Virus can be isolated from all organs obtained at autopsy from infants who die with severe congenital rubella. Rubella-specific IgG and IgM antibodies develop in utero. However, serologic diagnosis of infants with congenital rubella requires sensitive and specific assays, expertise of clinical virologists, and often a series of serum specimens obtained from the mother and child. At birth, maternal IgG antibodies as well as fetal IgG and IgM are present in the newborn. The half-life of maternal IgG is about 3 weeks, and in infants without congenital rubella, maternal antibody usually disappears within 6 months depending, however, on the sensitivity of the antibody assay used and on the level of IgG antibody at birth. IgM antibody to rubella virus can usually be detected at birth, and detectable levels persist for several months (Cradock-Watson et al., 1979). In principle, the diagnosis of congenital rubella should be reliable

TABLE 1. Clinical features associated with congenitally acquired rubella

Category (according to Cooper, 1975)	Common	Uncommon
Transient	Low birth weight, thrombocytopenic purpura, hepatosplenomegaly, bone lesions	Cloudy cornea, hepatitis, generalized lymphadenopathy, hemolytic anemia, pneumonitis
Developmental	Sensorineural deafness, peripheral pulmonary arterial stenosis, mental retardation, central language defects, diabetes mellitus	Severe myopia, thyroiditis, hypothyroidism, growth hormone deficiency, late onset disease
Permanent	Sensorineural deafness, peripheral pulmonary stenosis, pulmonary valvular stenosis, patent ductus arteriosus, ventricular septal defect, retinopathy, cataract, microphthalmia, psychomotor retardation, cryptochidism, inguinal hernia, diabetes mellitus	Severe myopia, thyroid disorders, dermatoglyphic abnormalities, glaucoma, myocardial abnormalities

Reproduced with permission from Banatvala and Best (1984).

using sensitive IgM assays in the first 3 months of life, discussed under the section on Diagnosis of Congenitally Acquired Infection.

Immunoprecipitation and immunoblotting experiments have shown that antibody responses to the three structural proteins have different patterns of reactivity in congenital rubella than in postnatally acquired rubella (de Mazancourt et al., 1986; Katow and Sugiura, 1985). Cell-mediated immune responses in children with congenital rubella are depressed (Buimovici-Klein et al., 1979), natural killer cell activity is impaired (Fuccillo et al., 1974) and T-cell activities show persistent abnormalities (Rabinowe et al., 1986).

Laboratory Diagnosis

Virus Isolation

Virus isolation is seldom used for the diagnosis of postnatally acquired rubella because serologic techniques are more rapid and reliable. Virus isolation is of value, however, for confirming a diagnosis of congenitally acquired rubella and determining the duration of virus excretion in such infants. Virus isolation has also been employed to examine the products of conception after rubella in pregnancy and to assess the risk of reinfection and inadvertent vaccination of rubella-susceptible pregnant women.

SPECIMENS

Rubella virus can be isolated from respiratory secretions, blood, urine, and stools, as well as from cataract material, lens fluid, tears, cerebrospinal fluid, and autopsy material from cases of congenital rubella, and amniotic fluid, placenta, and fetal tissues obtained from spontaneous and therapeutic abortions. If specimens cannot be inoculated immediately into cell cultures, they should be stored at 4°C. Rubella virus is rarely isolated from specimens that have been transported at ambient temperature for more than 3 days.

PROCESSING OF SPECIMENS

Specimens are inoculated into at least three tube cultures of the cells employed and adsorbed onto the monolayer for 1 to 2 h at room temperature. Mechanically shaking swabs in transport medium for 20 min at 4°C prior to inoculation into cell cultures increases the amount of virus recovered (Harcourt et al., 1979). The processing of products of conception is described in detail by Thompson and Tobin (1970) and Herrmann (1979).

CELL CULTURES

Cell cultures used for virus isolation are RK13, SIRC, Vero, and primary VMK. Some batches of calf serum contain inhibitors that interfere with the growth of rubella virus in these cell cultures. It is, therefore, essential that serum used in both growth and maintenance media is prescreened. Sensitivity of continuous cell cultures should be periodically checked by titrating virus of known titer, since the sensitivity of such cell cultures may be reduced on multiple passage.

Rubella virus produces characteristic CPE in RK13 cells within 3 to 7 days of inoculation (Best and Banatvala, 1967; Pattison, 1982). A CPE will only be apparent if the cells are maintained under carefully controlled conditions using pretested serum. At least one passage is required to confirm the presence of virus because cellular material present in the specimen may produce nonspecific effects, which may be confused with CPE. A CPE may not always develop after primary inoculation, and further passage of cells and culture fluid is always recommended. At least two serial passages should be made before a specimen is considered negative.

The use of SIRC cells was first described by Leerhoy (1965). Rubella virus produces a characteristic CPE in these cells, but only if conditions are carefully controlled (Leerhoy, 1965, 1966).

A CPE is only produced in some sublines of Vero cells. Specimens should initially be inoculated into these cell cultures for 7 to 10 days. Cell cultures are then freeze-thawed and passed once before a final passage in Vero, RK13 or BHK21 cell cultures for identification by indirect immunofluorescence (IIF) (O'Shea et al., 1983). Alternatively, material may be passed into RK13 or SIRC cell cultures in which rubella virus can be identified by its characteristic CPE. The use of primary VMK cells has been described in detail by Herrmann (1979) and Parkman et al. (1964).

IDENTIFICATION OF ISOLATES

The identity of all isolates should be confirmed by neutralization or IIF using specific antisera or monoclonal antibodies (Mabs). A number of immunofluorescence techniques have been described (Pattison, 1982; Schmidt et al., 1978). The method described by O'Shea et al. (1983) has proved to be sensitive, simple to perform, and applicable to large numbers of specimens. Those laboratories that do not have a fluorescent microscope may prefer to use an immunoperoxidase technique employing horseradish peroxidase-conjugated anti-species immunoglobulin (Schmidt et al., 1978).

Antiserum to rubella virus can be produced in a wide variety of animals including rabbits, guinea-pigs, ferrets, pigs, horses, sheep, and monkeys. Preparation of immune serum has been described by Herrmann (1979). A satisfactory rubella antiserum should have a neutralizing titer greater than 32 and an HAI titer greater than 256. Hyperimmune sera are available from commercial sources. Mabs to rubella have been produced.

Detection of Viral Antigens and Nucleic Acid

Virus isolation has the disadvantage that it takes at least 2 weeks to obtain a result. A more rapid method would potentially be useful for the diagnosis of rubella in pregnancy and for the prenatal diagnosis of congenital infection. Preliminary experiments have shown that rubella antigens can be detected on leucocytes by flow cytometry (O'Shea et al., 1988) and rubella RNA by nucleic acid hybridization using a cDNA probe (Ho-Terry et al., 1988, Terry et al., 1986). Further evaluation of these techniques is required before they can be introduced into diagnostic laboratories.

Antibody Assays

PRODUCTION AND PURIFICATION OF RUBELLA VIRUS FOR SOLID-PHASE ANTIBODY ASSAYS

Bulk production of high titered (10^7 to 10^8 TCD50/ml) rubella virus requires highly standardized and carefully optimized conditions. An initial concentration of supernatant fluid by ultrafiltration followed by further concentration and at least partial purification of the virus by ultracentrifugation is usually required to provide antigen for IgG and IgM antibody assays. A method with some modifications used successfully for many years in the Department of Virology at the University of Turku (Meurman, 1978) is shown in Figure 1.

SCREENING ASSAYS

There has been an increasing demand for rubella antibody screening of adult women in order to identify susceptible individuals who should be offered vaccination. Several different techniques are available (Table 2). Most assays are commercially available in kit form, and commercially available reagents such as latex agglutination (LA), passive hemagglutination (PHA), and enzyme immunoassay (EIA) are usually employed in laboratories in the United States. Such techniques are preferred to HAI, which is more time-consuming, labor-intensive, and expensive and may sometimes give false positive results due to incomplete removal of beta-lipoprotein inhibitors of HA. Single radial hemolysis is the test of choice in many European laboratories (Champsaur et al., 1980; Grillner and Strannegard, 1976; Grillner et al., 1985; Kurtz et al., 1980; Morgan-Capner, 1983).

Single Radial Hemolysis

Single radial hemolysis (SRH) is sensitive, specific, reliable, and inexpensive (Vaananen and Vaheri, 1979). Although SRH plates are available commercially, the short shelf-life of the gels makes long distance distribution difficult. Therefore, SRH plates are usually prepared in the laboratory using commercially available reagents (Kurtz et al., 1980). Rubella antigen, red blood cells (RBCs), complement, and agarose must be carefully screened for suitability in the test. In the United Kingdom, a 15,000 iu/liter of standard serum is usually tested on each plate. Sera giving zones of hemolysis greater than 15,000 iu/liter are reported to contain rubella antibodies (Pattison, 1982). There is some variability in the quality of zones of hemolysis. Sera giving no test zone, a hazy zone, or a zone on the control plate should be retested by SRH and/or an alternative technique such as LA, PHA, EIA, or IIF. When a serum gives a zone of hemolysis on the control plate, a third test is required to confirm a positive result, or sera may be

TABLE 2. Characteristics of single radial hemolysis (SRH), enzyme immunoassay (EIA), latex agglutination (LA), immunofluorescence (IIF), and passive hemagglutination (PHA) tests for rubella antibody screening

Test	Labor intensive	Speed	Sensitivity	Capital cost	Comments
SRH	+	19 h	++	Minimal	Method of choice in many European laboratories
EIA	++	4 h	+++	Spectrophotometer	Print-out results; IgM antibodies can be tested simultaneously with the same method
LA	+	10 min	+++	Minimal	Reading of results subjective
PHA	+	30 min to 3 h	+++	Minimal	Reading of results subjective
IIF (FIAX)	++	2–3 h	+	Fluorimeter	Same system used for other antibodies

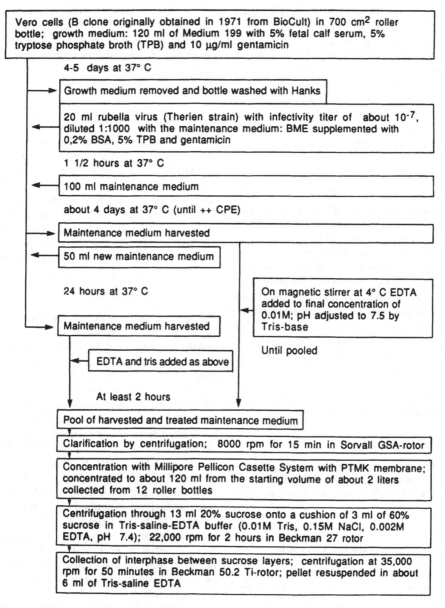

FIG. 1. The scheme of rubella antigen production, concentration, and purification for rubella-specific IgG and IgM antibody detection by enzyme immunoassay.

retested by SRH after adsorption with RBCs from the same species as used in the test. It may be necessary to send such sera to a reference laboratory.

Provided that a control plate without rubella antigen is employed to check for nonspecific hemolysis, false positive results (>15,000 iu/liter) should not be obtained. Vigilance is required, however, to prevent operator and transcription errors, especially when large numbers of sera are being tested. Very occasional sera may give equivocal or even negative results on repeat testing, even though they are positive when tested by an alternative technique. Such interference with lysis in SRH may be caused by a

"blocking factor" (Champsaur, 1983; Kangro et al., 1983). Hazy zones of hemolysis may also be indicative of recent rubella infection (Hedman et al., 1984; Morgan-Capner et al., 1982). The sensitivity of the SRH can be increased by enlarging the size of the serum well, double-filling the well, and reducing the concentration of RBCs employed (Morgan-Capner, 1983).

Enzymeimmunoassay and Radioimmunoassay

A number of commercial EIAs are available (Enders et al., 1985; Ferraro et al., 1987; Kleeman et al., 1983; Skendzel et al., 1983). Both types of assay can also be developed in the laboratory (Ferghani and Schmidt, 1979; Kalimo et al., 1976; Pattison, 1982; Sugishita et al., 1978; Vaheri and Salonen, 1980). Results of a typical EIA test are shown in Table 3. These techniques are sensitive, relatively simple to perform, and suitable for testing large numbers of sera. The EIA is more suitable than RIA for use in a diagnostic laboratory because of the longer shelf-life of enzyme-conjugated anti-human immunoglobulins and lack of radiation hazard.

Latex Agglutination

Latex agglutination tests are available commercially. Such tests have the advantage of speed and simplicity, since sera require no heat inactivation or dilution before testing and results can be obtained within 3 to 8 min of setting up a test (Kleeman et al., 1983; Meegan et al., 1982; Sever et al., 1983; Skendzel and Edson, 1986; Skendzel et al., 1983; Vaananen et al., 1985). It should be remembered that these assays are very sensitive, as they can detect as little as 5,000 iu/liter of rubella antibody. Such tests have the disadvantage that the reading of results by eye is subjective. Negative results should be confirmed by retesting the sera at a 1 : 10 and 1 : 100 dilution and/or testing by an alternative technique, since a prozone effect has been detected with occasional high titered sera (Freeman et al., 1983).

Passive Hemagglutination

These tests are also commercially available (Fayram et al., 1987; Kleeman et al., 1983; Meurman, 1978; Safford et al., 1985; Safford and Whittington, 1976;

Skendzel et al., 1983). Detailed protocols are provided with test kits and should be followed exactly. As with LA, PHA is sensitive and simple to perform and no heat inactivation or dilution of sera is required. Results are obtained 2 h or more after setting up the test; the reading of results is subjective.

Indirect Immunofluorescence

This test can be used to detect rubella antibodies (Cradock-Watson et al., 1972). Although it is too labor-intensive for routine use, it may be useful for "trouble-shooting." A commercial assay is available, employing rubella antigen on a dip-stick, the results being read automatically in a fluorimeter (Fayram et al., 1987; Kleeman et al., 1983). Rubella antibodies can also be detected by time-resolved fluoroimmunoassay (Meurman et al., 1982).

Interpretation of Results

The main problem with interpretation of tests for rubella immunity is what level of rubella antibody should be considered as indicative of full immunity. As discussed above, viremia may occur after reinfection, and this may happen more frequently among those with vaccine-induced immunity and low levels of antibody. In addition, since the sensitivity of screening assays varies, occasional sera may have antibodies detectable by some techniques but not by others (Balfour et al., 1981; Best et al., 1980; Buimovici-Klein et al., 1980; Kleeman et al., 1983; Morgan-Capner, 1983). Many commercial assays have not been standardized, and it is difficult to determine the level of antibody in international units. It is essential that "in-house" assays for rubella antibody are carefully standardized and adequate controls included in each test to ensure reliability of results. Nonspecialist laboratories should take particular care when interpreting results obtained with commercial kits.

Women with antibody levels less than 15,000 iu/liter detectable by such sensitive assays as EIA, LA, and PHA are usually reported to have "low level of rubella antibody, regard as susceptible." However, some women fail to produce antibody levels greater than 15,000 iu/liter even after repeated vaccinations. Many laboratories, therefore, consider women vaccinated more than once to be immune if antibodies are detectable even at low levels.

In the United Kingdom, a 15,000 iu/ml standard serum is available from the Public Health Laboratory Service for use in SRH tests. Other national standard sera may be available or can be made by standardization against an international standard (Clarke et al., 1975; Forsgren, 1985) in order to establish a suitable "cut-off" level.

TABLE 3. Results of a representative EIA test for rubella-specific IgG

Controls and test sera	$E100^a$	Conclusion[b]
15 iu/ml control serum	1.0	
Negative control serum	0.03	
Positive control serum	3.0	
Test sera		
1	0.01	Negative
2	4.0	Positive
3	2.1	Positive
4	0.2	Negative
5	0.7	Low titer
6	0.8	Low titer

^{a} $E100 = \dfrac{\text{Mean specific absorbance test serum}}{\text{Mean specific absorbance 15 iu/ml control serum}}$,

where the specific absorbance

$\qquad$ = absorbance on rubella antigen

$\qquad\qquad$ − absorbance on control antigen.

[b] Positive > 1.0; negative < 0.6; low titer 0.6 to 1.0.

Diagnostic Assays

Hemagglutination Inhibition. The critical step in HAI is removal of nonspecific inhibitors, identified as heat-stable serum lipoproteins (Haukenes, 1973), from the test sera. Several methods are available for removing these inhibitors, including kaolin treatment at pH 9.0, heparin-MnCl$_2$, and dextran-sulfate, but all methods occasionally fail to remove all nonspecific inhibitors (Haukenes and Blom, 1975; Kangro et al., 1983). Newborn chick, pigeon, and trypsinized human O cells are the RBCs most widely used. Red cell agglutinins must be removed from test sera by adsorbing with RBCs. All test variables (serum pretreatment, RBCs, antigen, diluent, and temperature) must be carefully controlled to provide a sensitive and reliable HAI test. Standardized reference methods were developed at the Centers for Disease Control, and have been described in detail elsewhere (Herrmann, 1979; Palmer et al., 1977).

SRH. This assay detects rubella antibodies of the IgG class, which develop more slowly than IgG antibodies measured by HAI, EIA, and RIA (Enders et al., 1985; Grillner and Strannegård, 1976; Strannegård et al., 1975). Recent primary rubella infection is suggested by detecting "soft hemolysis" in SRH (Hedman et al., 1984, 1986; Morgan-Capner et al., 1982). This specific pattern in SRH is not dependent on the presence of IgM antibody, but probably on low avidity IgG antibodies (Hedman and Seppälä, 1988).

EIA and RIA. Many laboratories now diagnose rubella by solid phase immunoassays (EIA and RIA). In experienced laboratories, using optimal antigen preparations the RIA and EIA may be one of the most sensitive methods for serologic diagnosis of rubella (Enders et al., 1985; Forghani and Schmidt, 1979; Kalimo et al., 1976; Vaheri and Salonen, 1980; Vejtorp et al., 1979; Voller and Bidwell, 1975). These methods are so sensitive that testing the serum specimens at several dilutions and expressing the test results in titers is inconvenient. Therefore, the results can be expressed in unit values that are calculated from the standard curve included in each test, which requires only a single dilution of the test serum (Kalimo et al., 1976; Meurman et al., 1982).

IgM Antibody Assays. The first rubella IgM assays were based on physical separation of IgM from IgG by fractionating serum specimens on sucrose gradients or by gel filtration (Best et al., 1969; Bürgin-Wolff et al., 1971; Vesikari and Vaheri, 1968), followed by HAI assay of the fractions. Solid phase immunoassays have now replaced these methods. In these assays, immunologic separation of IgM is achieved without any pretreatment of the test serum

specimens. Various assay principles have been used. In indirect assays, rubella antigen is bound to the solid phase and test serum at a single dilution or in a dilution series, it is incubated with solid phase antigen followed by labeled anti-human IgM (Kalimo et al., 1976; Kangro et al., 1978; Vejtorp et al., 1979; Ziegelmaier et al., 1981). These indirect EIAs are commercially available (Best et al., 1984; Chernesky et al., 1984; Enders and Knotek, 1986). The tests are subject to interference by IgM-class antiglobulins such as rheumatoid factor (Meurman and Ziola, 1978), which may produce false positive results. Sera should, therefore, be pretreated to remove IgM antiglobulins (Ziegelmaier et al., 1981) or with anti-human IgG to precipitate IgG (Enders and Knotek, 1986).

The most sensitive and specific techniques for routine diagnosis of postantal and congenital infections are the anti-μ-capture RIA (MACRIA) (Mortimer et al., 1981; Tedder et al., 1982) and the anti-μ-capture EIA (MACEIA) (Bellamy et al., 1985; Gerna et al., 1987b; Hodgson and Morgan-Capner, 1984; Kurtz and Malic, 1981). MACEIA are commercially available. Both techniques are based on labeled Mabs for the detection of antibody-bound rubella antigen. Rheumatoid factor does not usually interfere in these assays, but false positive results may occur in MACEIA if aggregated IgG in the enzyme-labeled antiserum complexes with bound anti-globulins (Kurtz and Malic, 1981). Sera with Epstein-Barr virus (EBV), cytomegalovirus, and parvovirus B19-specific IgM antibodies may give false positive results in anti-μ-capture assays (Cohen and Supran, 1987; Gerna et al., 1987b; Kurtz and Anderson, 1985; Morgan-Capner et al., 1983b) and heat-inactivated sera may give false positive results in MACRIA if RF is present (Ronald et al., 1985).

Erythrocytes can be used as indicators in anti-μ-capture assays. In the solid phase hemadsorption-inhibition test (HIT), erythrocytes are adsorbed on antibody-bound rubella antigen (Krech and Wilhelm, 1979; van der Logt et al., 1981). Enzyme labeled antigen as an indicator has also been employed (Bonfanti et al., 1985).

Characteristics of tests for rubella-specific IgM are shown in Table 4, and representative results of a MACRIA test in Table 5.

Since false positive results are possible with all tests for rubella-specific IgM antibody and low levels of rubella-specific IgM antibody may not always indicate a recent primary infection, a second specimen should be collected a few days later, and the paired serum specimens tested for IgG and IgM antibody. If a significant increase in antibody concentration is not demonstrated, retesting the specimens with low levels of IgM antibody by an alternative technique is often advisable. Interpretation of results requires vi-

TABLE 4. Characteristics of tests for rubella-specific IgM

Tests	Sensitivity	Speed (h)	Comments
Anti-μ-capture RIA (MACRIA)	+++++	<24	Sera containing heterophil antibodies and IgM to other
Anti-μ-capture EIA (MACEIA)	++++	<24	viruses and heated sera may give false-positive results. False-positives may occur in MACEIA due to aggregated IgG in enzyme-conjugated antiserum.
Indirect EIA	+++	7 to 24	Avoid interference by IgM antiglobulins (e.g., rheumatoid factor) by pretreatment of sera.
Sucrose density gradient fractionation and HI (long incubation)	++	48	Labor intensive
Hemadsorption immunosorbent technique (HIT)	++	24	Pretreatment of sera necessary Reading of results subjective

rologic and immunologic expertise. For instance, rubella-specific IgM antibody may develop as late as 10 days after the onset of rash and a second specimen may, therefore, need to be tested.

Serologic Assessment of Pregnant Women

Diagnosis of rubella during pregnancy is one of the most demanding tasks in clinical virology. It has been discussed by Best and Banatvala (1987) and Best and O'Shea (1988). False positive and false negative results must be avoided, and the laboratory procedures used must be tuned to the highest possible specificity and sensitivity. In addition to the laboratory data, it is essential to obtain detailed clinical information on the date of the onset of illness, the date and duration of exposure, vaccination status, other epidemiologic information and if possible specimens from the contact persons with suspected rubella. A firm virologic diagnosis of rubella or other virus infection should be established with a minimum of delay to help the physician and the patient if termination of pregnancy may need to be considered.

Paired serum specimens should be tested, with the first collected as soon as possible after the appearance of rash or after exposure, and the second 3 to 10 days later. A third specimen may be necessary if symptoms appear only after the second specimen has been collected. If the first specimen has been collected within 10 days of exposure, the results of the immunity tests dictate whether additional specimens are required. Seronegative women and women with low levels of antibody (< 15,000 iu/liter) should be tested at intervals for 4 weeks after the last exposure to a rubella-like illness.

If rubella-like symptoms have appeared, tests for IgG and IgM antibodies must be conducted on the first specimen, and the presence of IgM antibody reported without delay. If IgG is positive and IgM negative or borderline positive, or if the IgG is negative, paired serum specimens must be tested for IgG and IgM antibodies. If IgM antibody remains negative and there is no increase in IgG antibody, the data strongly suggest that there has been no recent rubella infection. However, the dates of specimen collections in relation to the symptoms or exposure should be evaluated in detail before issuing a report.

Unfortunately the serum specimen is often collected some weeks after rash or exposure. In these cases, the serologic diagnosis is difficult if not impossible, since negative IgM results on specimens collected 4 weeks or more after onset of symptoms may not rule out recent rubella infection (Enders and Knotek, 1986; Meurman, 1978). In the future, detection of low avidity IgG in serum (Hedman and Seppälä, 1988) or detection of rubella antigen or RNA sequences in biopsy specimens from chorionic villus (Ho-Terry et al., 1988) may provide additional diagnostic methods in these cases.

In many countries, rubella antibody screening is

TABLE 5. Representative results of anti-μ-capture RIA test (MACRIA) for rubella IgM antibody

Controls and test sera	Counts per minute	Result (au/ml)[b]	Conclusion
Control sera[a]			
Negative	525, 518, 500, 530		
1.0 au/ml	1,500, 1,550		
3.3 au/ml	3,100, 3,300		
10 au/ml	6,200, 6,300		
40 au/ml	15,000, 15,500		
Test sera			
1	659, 660	<1.0	Negative
2	1,020, 1,000	<1.0	Negative
3	15,0000, 15,100	35	Positive
4	4,000, 4,200	4.4	Positive
5	2,200, 2,107	1.8[c]	Equivocal

[a] Control sera were used to construct a calibration curve (counts per minute value versus au/ml). Rubella-specific IgM values for test sera were obtained from this curve.
[b] Arbitrary units of rubella-specific IgM.
[c] 1 to 3.3 au/ml considered equivocal.

included in the tests organized by antenatal clinics. These specimens, stored for a few months after delivery, provide valuable preexposure specimens if a patient is later exposed to rubella. If congenitally acquired infection is suspected after delivery, the stored maternal serum will be available for retesting with later specimens.

Diagnosis of Congenitally Acquired Infection

Postnatal. Congenitally-acquired rubella is diagnosed by:

1. detection of rubella IgM antibody in cord blood or serum samples obtained in infancy,
2. detection of a persistent rubella antibody response in the infant after 6 months of age, and
3. isolation of rubella virus from the infant.

The criteria of the Centers for Disease Control for classification of congenital rubella syndrome are shown in Table 6.

Prenatal. No firmly established methods are available for prenatal diagnosis of congenital rubella infection (Best and O'Shea, 1988). Such methods would be valuable in cases when maternal infection has occurred during the second trimester, maternal infection is suspected but not confirmed, and in cases of confirmed reinfection during the first trimester. The diagnostic methods used to date are isolation of virus from amniotic fluid, detection of RNA or viral antigens in biopsy specimens of chorionic villus or amniotic fluid (Ho-Terry et al., 1988; Terry et al., 1986), and IgM antibody assays on fetal blood (Daffos et al., 1984; Enders and Jonatha, 1987; Morgan-Capner et al., 1984b).

TABLE 6. Centers for Disease Control criteria for classification of congenital rubella syndrome (CRS) cases (Centers for Disease Control, 1986)

CRS Confirmed = defects present and one or more of the following:
   1) Rubella virus isolated.
   2) Rubella-specific IgM present.
   3) Rubella HAI titer in the infant persisting above and beyond that expected from passive transfer of maternal antibody (i.e., rubella HAI titer in the infant which does not fall off at the expected rate of one 2-fold dilution/month).

CRS Compatible = laboratory data insufficient for confirmation and any two complications listed in 1) or one from 1) and one from 2):
   1) Cataracts, congenital glaucoma (either or both count as one), congenital heart disease, loss of hearing, pigmentary retinopathy.
   2) Purpura, splenomegaly, jaundice, microcephaly, mental retardation, meningoencephalitis, radiolucent bone disease.

CRS Possible = some compatible clinical findings present, but not sufficient to fulfill the criteria for a compatible case.

Congenital rubella infection only = laboratory evidence of infection, but no defects present.

Stillbirths = stillbirths that are thought to be secondary to maternal rubella infection.

Not CRS = One or more of any of the following inconsistent laboratory findings in a child without evidence of an immunodeficiency disease:
   1) Rubella HAI titer absent in a child < 24 months.
   2) Rubella HAI titer absent in mother.
   3) Rubella HAI titer decline in an infant consistent with the normal decline of passively transferred maternal antibody after birth [see CRS confirmed 3)].

Epidemiology

Before the introduction of vaccination, rubella was endemic in temperate climates, with seasonal peaks during spring and early summer. Outbreaks involving school children and young adults living in institutions (including army training centers) were common. Seroepidemiologic studies demonstrated that in temperate climates, the rate of seropositivity increased with age, and among women of childbearing age, about 80 to 85% were immune (Hanshaw et al., 1985). Lower seropositive rates, however, exist in certain tropical island populations such as Hawaii (Sever, 1965). Puerto Rico (Greenberg et al., 1973), Trinidad, Jamaica, and Taiwan (Grayston et al., 1972, Harcourt et al., 1979).

Extensive childhood vaccination, often with combined measles-mumps-rubella vaccine, has completely changed the epidemiology of rubella in the United States. Since the licensing of rubella vaccines in 1969, no major rubella epidemics have occurred in

the United States, although minor local outbreaks and sporadic cases have occurred. In 1986, there were only 0.23 cases per 100,000 population (Centers for Disease Control, 1987a).

Prevention

The first rubella vaccine (HPV77) was developed at the National Institutes of Health in Bethesda, Maryland. Later, HPV77.DE5, Cendehill, RA27/3 and several vaccines in Japan were developed (Perkins, 1985). They are all attenuated vaccines administered subcutaneously, although RA27/3 can be administered intranasally. RA27/3 is now the most widely used vaccine. Virus is excreted up to 4 weeks after vaccination, but is not transmitted to susceptible contacts (Halstead and Diwan, 1971; Regamey et al., 1969).

Rubella vaccines induce an immune response in about 95% of susceptible vaccinees. Immune responses have been discussed in more detail by Best and O'Shea (1988). Vaccines produce minimal reactions in children, but arthropathy may develop in adults between 10 days and 4 weeks after vaccination (Banatvala, 1977; Perkins, 1985). Pregnancy is an absolute contraindication for rubella vaccination and should be avoided for 1 month after vaccination. At least 403 susceptible women who elected to go to term after inadvertent vaccination during pregnancy have now been followed-up (Best and O'Shea, 1988). None of their infants had defects compatible with congenital rubella, although 1.5% had evidence of congenital infection. It has been calculated that the theoretical maximum risk of rubella-induced major malformations among infants born to susceptible mothers is 1.4% (Centers for Disease Control, 1987b; Preblud et al., 1981).

Three types of vaccination policy have been employed in different countries: childhood vaccination with the aim of interrupting the transmission of rubella, first introduced in the United States in 1969 (Bart et al., 1985); selective vaccination for girls before childbearing age (e.g., in the United Kingdom) (Dudgeon, 1985); combined measles-mumps-rubella vaccination of all children at 1½ years of age and again at 6 to 12 years of age with the aim of eradicating these three diseases (e.g., in Finland, Norway, and Sweden) (Peltola et al., 1986; Rabo and Taranger, 1984). Many European countries that initially adopted a policy of selective vaccination have only recently introduced universal vaccination of children at about 15 months of age (e.g., Belgium, Finland, Federal Republic of Germany, Greece, Luxembourg, Norway, Spain, Sweden, and Switzerland). Selective vaccination programs have not affected the overall incidence of rubella (Tobin et al., 1985), and there is a continuing risk for those few remaining susceptible pregnant women. Therefore, in 1987, the United Kingdom decided to offer measles-mumps-rubella to both boys and girls at 15 months of age in order to reduce and ultimately eradicate rubella (Banatvala, 1987).

Literature Cited

Balfour, H. H., K. E. Groth, C. K. Edelman, J. M. Best, and J. E. Banatvala. 1981. Rubella viraemia and antibody responses after rubella vaccination and reimmunisation. Lancet 1:1078–1080.

Banatvala, J. E. 1977. Health of mother, fetus and neonate following maternal viral infections during pregnancy, p. 437–488. In C. R. Coid (ed.). Infections and pregnancy. Academic Press, London.

Banatvala, J. E. 1987. Measles must go and with it rubella. Br. Med. J. 295:2–3.

Banatvala, J. E., and J. M. Best. 1984. Rubella, p. 271–302. In F. Brown, and G. Wilson (ed.). Topley and Wilson's principles of bacteriology, virology and immunity, vol. 4, 7th edition. Edward Arnold, London.

Bart, K., W. A. Orenstein, S. R. Preblud, and A. R. Hinman. 1985. Universal immunization to interrupt rubella. Rev. Infect. Dis. 7(Suppl 1):S177–184.

Bellamy, K., J. Hodgson, P. S. Gardner, and P. Morgan-Capner. 1985. Public Health Laboratory Service IgM antibody capture enzyme-linked immunosorbent assay for detecting rubella specific IgM. J. Clin. Pathol. 38:1150–1154.

Best, J. M., and J. E. Banatvala. 1967. A comparison of RK13 vervet monkey kidney and patas monkey kidney cell cultures for the isolation of rubella virus. J. Hyg. Camb. 65:263–271.

Best, J. M., and J. E. Banatvala. 1987. Rubella, p. 315–353. In A. J. Zuckerman, J. E. Banatvala, and J. R. Pattison (ed.). Principles and practice of clinical virology. John Wiley & Sons, Chichester.

Best, J. M., J. E. Banatvala, and D. Watson. 1969. Serum IgM and IgG responses in postnatally acquired rubella. Lancet 2:65–69.

Best, J. M., G. C. Harcourt, A. Druce, S. J. Palmer, S. O'Shea, and J. E. Banatvala. 1980. Rubella immunity by four different techniques: results of challenge studies. J. Med. Virol. 5:239–247.

Best, J. M., and S. O'Shea. 1988. Rubella. In N. J. Schmidt and R. W. Emmons (ed.). Diagnostic procedures for viral, rickettsial and chlamydial infections, 6th ed. American Public Health Association, Washington D.C.

Best, J. M., S. J. Palmer, P. Morgan-Capner, and J. Hodgson. 1984. A comparison of rubazyme-M and MACRIA for the detection of rubella-specific IgM. J. Virol. Methods 8:99–109.

Bonfanti, C., O. Meurman, and P. Halonen. 1985. Detection of specific immunoglobulin M antibody to rubella virus by use of an enzyme-labelled antigen. J. Clin. Microbiol. 21:963–968.

Bowden, D. S., and E. G. Westaway. 1985. Changes in glycosylation of rubella virus envelope proteins during maturation. J. Gen. Virol. 66:201–206.

Buimovici-Klein, E., P. B. Lang, P. R. Ziring, R. Philip, and L. Z. Cooper. 1979. Impaired cell-mediated immune response in patients with congenital rubella: correlation with gestational age at time of infection. Pediatrics 64:620–626.

Buimovici-Klein, E., A. J. O'Beirne, S. J. Millian, and L. Z. Cooper. 1980. Low level rubella immunity detected by ELISA and specific lymphocyte transformation. Arch. Virol. 66:321–327.

Bürgin-Wolff, A., R. Hernandez, and M. Just. 1971. Separation of rubella IgM, IgA and IgG antibodies by gel filtration on agarose. Lancet 2:1278–1280.

Burke, J. P., A. R. Hinman, and S. Krugman. 1985. International symposium on prevention of congenital rubella infection. Rev. Infect. Dis. 7: Suppl 1.

Centers for Disease Control. 1986. Rubella and congenital rubella syndrome—United States, 1984–1985. Morbid. Mortal. Weekly Rep. 35:129–135.

Centers for Disease Control. 1987a. Rubella and congenital rubella—United States, 1984–1986. Morbid. Mortal. Weekly Rep. 36:664–675.

Centers for Disease Control. 1987b. Rubella vaccination during pregnancy—United States, 1971–1986. Morbid. Mortal. Weekly Rep. 36:457–461.

Champsaur, H. 1983. Loophole in rubella screening: a factor blocking radial haemolysis. Lancet 1:182–183.

Champsaur, H., E. Dussaix, and P. Tournier. 1980. Hemagglutination inhibition, single radial hemolysis and ELISA tests for the detection of IgG and IgM to rubella virus. J. Med. Virol. 5:273–286.

Chernesky, M. A., L. Wyman, J. B. Mahony, S. Castriciano, J. T. Unger, J. W. Safford, and P. S. Metzel.

1984. Clinical evaluation of the sensitivity and specificity of a commercially available enzyme immunoassay for detection of rubella virus-specific immunoglobulin M. J. Clin. Microbiol. **20:**400–404.

Clarke, M., J. A. Dudgeon, R. Ferris, G. Colinet, H. Tate, and F. T. Perkins. 1975. The British standard for antirubella serum. J. Biol. Stand. **3:**151–161.

Clarke, D. M., T. W. Loo, I. Hui, P. Chong, and S. Gilam. 1987. Nucleotide sequence and in vitro expression of rubella virus 24S subgenomic messenger RNA encoding the structural proteins E1, E2 and C. Nucleic Acids Res. **15:**3041–3057.

Cohen, B. J., and E. M. Supran. 1987. IgM serology for rubella and human parvovirus B19. Lancet **1:**393.

Cooper, L. Z. 1975. Congenital rubella in the United States. *In* S. Krugman and A. A. Gershon (ed.). Infections of the fetus and the newborn infant. Prog. Clin. Biol. Res. **3:**1–21. Alan R. Liss, New York.

Cooper, L. Z., and S. Krugman. 1967. Clinical manifestations of postnatal and congenital rubella. Arch. Ophthal. **77:**434–439.

Cooper, L. Z., B. Matters, J. K. Rosenblum, and S. Krugman. 1969. Experience with a modified rubella hemagglutination-inhibition antibody test. J. Am. Med. Assoc. **207:**89–93.

Cradock-Watson, J. E., M. D. Bourne, and E. M. Vandervelde. 1972. IgG, IgA, IgM responses in acute rubella determined by the immunofluorescent technique. J. Hyg. (Camb.) **70:**473–485.

Cradock-Watson, J. E., M. K. S. Ridehalgh, M. J. Anderson, and J. R. Pattison. 1985. Rubella reinfection and the fetus. Lancet **2:**1039.

Cradock-Watson, J. E., M. K. S. Ridehalgh, J. R. Pattison, M. J. Anderson, and H. O. Kangro. 1979. Comparison of immunofluorescence and radioimmunoassay for detecting IgM antibody in infants with the congenital rubella syndrome. J. Hyg. (Camb.) **83:**413–423.

Daffos, F., F. Forestier, L. Grangeot-Keros, M. C. Pavlovsky, P. Lebon, M. Chartier, and J. Pillot. 1984. Prenatal diagnosis of congenital rubella. Lancet **2:**1–3.

Davis, W. J., E. H. Larson, P. Simsarian, P. D. Parkman, and H. M. Meyer. 1971. A study of rubella immunity and resistance to infection. J. Am. Med. Assoc. **215:**600–608.

de Mazancourt, A., M. N. Waxham, J. C. Nicolas, and J. S. Wolinsky. 1986. Antibody response to the rubella virus structural proteins in infants with the congenital rubella syndrome. J. Med. Virol. **19:**111–122.

Dudgeon, J. A. 1985. Selective immunization: protection of the individual. Rev. Infect. Dis. **7(Suppl):**185–190.

Enders, G., and W. Jonatha. 1987. Prenatal diagnosis of intrauterine rubella. Infection **15:**1620–1640.

Enders, G., and F. Knotek. 1986. Detection of IgM antibodies against rubella virus: comparison of two indirect ELISAs and an anti-IgM capture immunoassay. J. Med. Virol. **19:**377–386.

Enders, G., F. Knotek, and U. Pacher. 1985. Comparison of various serological methods and diagnostic kits for the detection of acute, recent and previous rubella infection, vaccination and congenital infections. J. Med. Virol. **16:**219–232.

Fayram, S. L., S. Akin, S. L. Aarnaes, E. M. Peterson, and L. M. de la Maza. 1987. Determination of immune status in patients with low antibody titres for rubella virus. J. Clin. Microbiol. **25:**178–180.

Ferraro, M. J., W. M. Kallas, K. P. Welch, and A. Y. Lau. 1987. Comparison of a new, rapid enzyme immunoassay with a latex agglutination test for qualitative detection of rubella antibodies. J. Clin. Microbiol. **25:**1722–1724.

Fogel, A., Ch. B. Gerichter, B. Barnea, R. Handsher, and

E. Heeger. 1978. Response to experimental challenge in persons immunized with different rubella vaccines. J. Pediatr. **92:**26–29.

Forghani, B., and N. J. Schmidt. 1979. Antigen requirements, sensitivity and specificity of enzyme-linked immunoassays for measles and rubella viral antibodies. J. Clin. Microbiol. **9:**657–664.

Forrest, J., M. A. Menser, M. C. Honeyman, M. Stout, and A. M. Murphy. 1972. Clinical rubella eleven months after vaccination. Lancet **2:**399–400.

Forsgren, M. 1985. Standardization of techniques and reagents for the study of rubella antibody. Rev. Infect. Dis. **7(Suppl 1):**S129–132.

Freeman, S., L. Clark, and N. Dumas. 1983. Evaluation of a latex agglutination test for detection of antibodies to rubella virus in selected sera. J. Clin. Microbiol. **18:**197–198.

Frey, T. K., L. D. Marr, M. L. Hemphill, and G. Dominquez. 1986. Molecular cloning and sequencing of the region of the rubella virus genome coding for glycoprotein E1. Virology **154:**228–232.

Fuccillo, D. A., R. W. Steele, S. A. Hensen, M. E. Vincent, J. B. Hardy, and J. A. Bellanti. 1974. Impaired cellular immunity to rubella virus in congenital rubella. Infect. Immun. **9:**81–84.

Furukawa, T., S. A. Plotkin, W. D. Sedwick, and M. L. Profeta. 1967. Hemagglutinin of rubella virus. Nature **215:**172–173.

Gerna, G., M. G. Revello, M. Dovis, E. Petrucelli, G. Achilli, E. Percivalle, and M. Torsellini. 1987a. Synergistic neutralization of rubella virus by monoclonal antibodies to viral haemagglutinin. J. Gen. Virol. **68:**2007–2012.

Gerna, G., M. Zannino, M. G. Revello, E. Perruzzelli, and M. Dovis. 1987b. Development and evaluation of a capture enzyme-linked immunosorbent assay for determination of rubella immunoglobulin M using monoclonal antibodies. J. Clin. Microbiol. **25:**1033–1038.

Grayston, J. T., J. L. Gale, and R. H. Watten. 1972. The epidemiology of rubella in Taiwan. 1. Introduction and description of the 1957–58 epidemic. Int. J. Epidemiol. **1:**245–265.

Green, R. H., M. R. Balsamo, J. P. Giles, S. Krugman, and G. Mirick. 1965. Studies of the natural history and prevention of rubella. Am. J. Dis. Child. **110:**348–365.

Green, K. Y., and P. H. Dorsett. 1986. Rubella virus antigens: localization of epitopes involved in hemagglutination and neutralization by using monoclonal antibodies. J. Virol. **57:**893–898.

Greenberg, E. R., P. A. Blake, B. L. Cline, and K. L. Herrman. 1973. Rubella susceptibility among Puerto Rican mothers. Bol. Assoc. Med. P.R. **65:**259–261.

Grillner, L., M. Forsgren, B. Barr, M. Bottiger, L. Danielsson, and C. De Verdier. 1983. Outcome of rubella during pregnancy with special reference to the 17th–24th weeks of gestation. Scand. J. Infect. Dis. **15:**321–325.

Grillner, L., M. Forsgren, and E. Nordenfelt. 1985. Comparison between a commercial ELISA, rubazyme and hemolysis-in-gel test for determination of rubella antibodies. J. Virol. Methods **10:**111–115.

Grillner, L., and O. Strannegård. 1976. Evaluation of the hemolysis-in-gel test for the screening of rubella immunity and the demonstration of recent infection. J. Clin. Microbiol. **3:**86–90.

Halonen, P. E., H. L. Casey, J. A. Stewart, and A. D. Hall. 1976. Rubella complement fixing antigen prepared by alkaline extraction of virus grown in suspension culture of BHK-21 cells. Proc. Soc. Exp. Biol. Med. **125:**167–172.

Halonen, P., O. Meurman, M-T. Matikainen, E. Torfason, and H. Bennich. 1979. IgA antibody response in acute rubella determined by solid-phase radioimmunoassay. J. Hyg. (Camb.) **83**:69–75.

Halonen, P. E., J. M. Ryan, and J. A. Stewart. 1967. Rubella hemagglutinin prepared with alkaline extraction of virus grown in suspension culture of BHK-21 cells. Proc. Soc. Exp. Biol. Med. **125**:162–167.

Halstead, S. B., and A. R. Diwan. 1971. Failure to transmit rubella virus vaccine: a close contact study in adults. J. Am. Med. Assoc. **215**:634–636.

Hanshaw, J. B., J. A. Dudgeon, and W. C. Marshall. 1985. Viral diseases of the foetus and newborn, 2nd ed. WB Saunders and Co., Philadelphia.

Harcourt, G. C., J. M. Best, and J. E. Banatvala. 1979. HLA antigens and responses to rubella vaccination. J. Hyg. (Camb.) **83**:405–412.

Harcourt, G. C., J. M. Best, and J. E. Banatvala. 1980. Rubella-specific serum and nasopharyngeal antibodies in volunteers with naturally acquired and vaccine induced immunity following intranasal challenge. J. Infect. Dis. **142**:145–155.

Haukenes, G. 1973. A rubella hemagglutination inhibitor simulating antibody. Acta Pathol. Microbiol. Scand. **81**:719–723.

Haukenes, G., and H. Blom. 1975. False positive rubella hemagglutination inhibition reactions: occurrence and disclosure. Med. Microbiol. Immunol. **161**:99–106.

Hedman, K., K. Raiha, O. Meurman, and A. Vaheri. 1984. Altered hemolysis in single radial hemolysis from a single serum sample as an indicator of recent rubella virus infection. J. Med. Virol. **13**:323–330.

Hedman, K., E-M. Salonen, J. Keski-Oja, and K. Raiha. 1986. Single serum radial hemolysis to detect recent rubella virus infection. J. Infect. Dis. **154**:1018–1023.

Hedman, K., and I. Seppälä. 1988. Recent rubella virus infection indicated by low avidity of specific IgG. J. Clin. Immunol. **8**:1–8.

Heggie, A. D. 1978. Pathogenesis of the rubella exathem: distribution of rubella virus in the skin during rubella with and without rash. J. Infect. Dis. **137**:74–77.

Heggie, A. D., and F. C. Robbins. 1969. Natural rubella acquired after birth: clinical features and complications. Am. J. Dis. Child. **118**:12–17.

Herrmann, K. L. 1979. Rubella virus. In E. H. Lennette and N. J. Schmidt (ed.). Diagnostic procedures for viral, rickettsial and chlamydial infections, 5th ed. American Public Health Association, Washington D.C.

Hodgson, J., and P. Morgan-Capner. 1984. Evaluation of a commercial antibody capture enzyme immunoassay for the detection of rubella specific IgM. J. Clin. Pathol. **37**:573–577.

Holmes, I. H., M. C. Wark, and F. M. Warburton. 1969. Is rubella an arbovirus? ii Ultrastructural morphology and development. Virology **37**:15–25.

Horstmann, D. M., H. Liebhaber, G. L. LeBouvier, D. A. Rosenberg, and S. B. Halstead. 1970. Rubella: reinfection of vaccinated and naturally immune persons exposed in an epidemic. N. Engl. J. Med. **283**:771–778.

Horstmann, D. M., T. G. Pajot, and H. Liebhaber. 1969. Epidemiology of rubella subclinical infection and occurrence of reinfection. Am. J. Dis. Child. **118**:133–136.

Horzinek, M. C. 1981. Non-arthropod borne Togaviruses. Academic Press, London.

Ho-Terry, L., G. M. Terry, P. Londesborough, K. R. Rees, F. Wielaard, and A. Denissen. 1988. Diagnosis of fetal rubella infection by nucleic acid hybridization. J. Med. Virol. **24**:175–182.

Hovi, T., and A. Vaheri. 1970. Rubella virus-specific ribonucleic acids in infected BHK-21 cells. J. Gen. Virol. **6**:77–83.

Kalimo, K. O. K., O. H. Meurman, P. E. Halonen, B. Ziola, M. Viljanen, K. Granfors, and P. Toivanen. 1976. Solid-phase radioimmunoassay of rubella virus immunoglobulin G and immunoglobulin M antibodies. J. Clin. Microbiol. **4**:117–123.

Kangro, H. O., A. Campbell-Benzie, and R. B. Heath. 1983. Rubella screening. Lancet **1**:1278.

Kangro, H., J. R. Pattison, and R. B. Heath. 1978. The detection of rubella-specific IgM by radioimmunoassay. Br. J. Exp. Pathol. **59**:577–583.

Katow, S., and A. Sugiura. 1985. Antibody response to individual rubella virus proteins in congenital and other rubella virus infections. J. Clin. Microbiol. **21**:449–451.

Kleeman, K. T., D. J. Kiefer, and S. P. Halbert. 1983. Rubella antibodies detected by several commercial immunoassays in hemagglutination inhibition-negative sera. J. Clin. Microbiol. **18**:1131–1137.

Krech, U., and J. A. Wilhelm. 1979. A solid-phase immunosorbent technique for the rapid detection of rubella IgM by hemagglutination inhibition. J. Gen. Virol. **44**:281–286.

Kurtz, J. B., and M. J. Anderson. 1985. Cross-reactions in rubella- and parvovirus-specific IgM tests. Lancet **2**:1356.

Kurtz, J. B., and A. Malic. 1981. Rubella-specific IgM detected by an antibody capture assay/ELISA technique. J. Clin. Pathol. **34**:1392–1395.

Kurtz, J. B., P. P. Mortimer, P. R. Mortimer, P. Morgan-Capner, M. S. Shafi, and G. R. B. White. 1980. Rubella antibody measured by radial hemolysis: characteristics and performance of a simple screening method for use in diagnostic laboratories. J. Hyg. (Camb.) **84**:213–222.

LeBouvier, G. 1969. Physicochemical characteristics of rubella antigens theta and iota. Nature **221**:78–79.

Leerhoy, J. 1965. Cytopathic effect of rubella virus in a rabbit-cornea cell line. Science **149**:633–634.

Leerhoy, J. 1966. The influence of different media on cell morphology and rubella virus titre in a rabbit cornea cell line (SIRC). Arch. Ges. Virusforsch. **XIX**:210–220.

Liebhaber, H. 1970. Measurement of rubella antibody by hemagglutination inhibition. II. Characteristics of an improved HAI test employing a new method for the removal of non-immunoglobulin HA inhibitors from serum. J. Immunol. **104**:826–834.

Linde, A. G. 1985. Subclass distribution of rubella virus-specific immunoglobulin G. J. Clin. Microbiol. **21**:117–121.

Meegan, J. M., B. K. Evans, and D. M. Horstmann. 1982. Comparison of the latex agglutination test with the hemagglutination inhibition test, enzyme-linked immunosorbent assay and neutralization test for detection of antibodies to rubella virus. J. Clin. Microbiol. **16**:644–649.

Mettler, N. E., and R. L. Petrelli. 1968. Absence of antigenic cross-reactions between rubella virus and arboviruses. Virology **36**:503–504.

Meurman, O. H. 1978. Antibody responses in patients with rubella infection determined by passive hemagglutination, hemagglutination inhibition, complement fixation, and solid-phase radioimmunoassay tests. Infect. Immun. **19**:369–372.

Meurman, O. H., I. A. Hemmila, T. N. E. Lovgren, and P. E. Halonen. 1982. Time-resolved fluoroimmunoassay: a new test for rubella antibodies. J. Clin. Microbiol. **16**:920–925.

Meurman, O. H., M. K. Viljanen, and K. Granfors. 1977. Solid phase radioimmunoassay of rubella virus immunoglobulin M antibodies: comparison with sucrose density gradient centrifugation test. J. Clin. Microbiol. **5**:257–262.

Meurman, O. H., and B. R. Ziola. 1978. IgM-class rheumatoid factor interference in the solid-phase radioimmunoassay of rubella-specific IgM antibodies. J. Clin. Pathol. **31:**483–487.

Morgan-Capner, P. 1983. The detection of rubella-specific antibody. Microbiol. Digest **1:**6–11.

Morgan-Capner, P. 1986. Does rubella reinfection matter? p. 50–62. *In* P. P. Mortimer (ed.). Public health virology, 12 reports. Public Health Laboratory Service, London.

Morgan-Capner, P., C. Burgess, and S. Fisher-Hoch. 1982. Radial hemolysis for the detection of rubella antibody in acute postnatal rubella. J. Hyg. (Camb.) **89:**311–320.

Morgan-Capner, P., C. Burgess, R. M. Ireland, and J. C. Sharp. 1983a. Clinically apparent rubella reinfection with a detectable rubella-specific IgM response. Br. Med. J. **286:**1616.

Morgan-Capner, P., J. Hodgson, M. H. Hambling, C. Dulake, T. J. Coleman, P. A. Boswell, R. P. Watkins, J. Booth, H. Stern, J. M. Best, and J. E. Banatvala. 1985. Detection of rubella-specific IgM in subclinical rubella reinfection in pregnancy. Lancet **1:**244–246.

Morgan-Capner, P., J. Hodgson, J. Sellwood, and J. Tippett. 1984a. Clinically apparent rubella reinfection. J. Infect. **9:**97–100.

Morgan-Capner, P., C. H. Rodeck, K. Nicolaides, and J. E. Cradock-Watson. 1984b. Prenatal diagnosis of rubella. Lancet **2:**343.

Morgan-Capner, P., R. S. Tedder, and J. E. Mace. 1983b. Rubella-specific IgM reactivity in sera from cases of infectious mononucleosis. J. Hyg. (Camb.) **90:**407–413.

Mortimer, P. P., R. S. Tedder, M. H. Hambling, M. S. Shafi, F. Burkhardt, and U. Schilt. 1981. Antibody capture radioimmunoassay for anti-rubella IgM. J. Hyg. (Camb.) **86:**139–153.

Munro, N. D., S. Sheppard, R. W. Smithells, H. Holzel, and G. Jones. 1987. Temporal relations between maternal rubella and congenital defects. Lancet **2:**201–204.

Nakhasi, H. L., B. C. Meyer, and T-Y. Liu. 1986. Rubella virus cDNA: sequence and expression of E1 envelope protein. J. Biol. Chem. **261:**1616–1621.

Oker-Blom, C., N. Kalkkinen, L. Kääriäinen and R. F. Pettersson. 1983. Rubella virus contains one capsid protein and three envelope glycoproteins, E1, E2a and E2b. J. Virol. **46:**964–973.

Oker-Blom, C., I. Ulmanen, L. Kaariainen, and R. F. Pettersson. 1984. Rubella virus 40S genome RNA-specifies: a 24S subgenomic mRNA that codes for a precursor to structural proteins. J. Virol. **49:**403–408.

Oker-Blom, C. 1984. The gene order for rubella virus structural proteins is NH$_2$—C—E$_2$—E$_1$—COOH. J. Virol. **51:**354–358.

O'Shea, S., J. M. Best, and J. E. Banatvala. 1983. Viremia, virus excretion and antibody responses after challenge in volunteers with low levels of antibody to rubella virus. J. Infect. Dis. **148:**639–647.

O'Shea, S., D. Mutton, and J. M. Best. 1988. In vivo expression of rubella antigens on human leucocytes: detection by flow cytometry. J. Med. Virol. (in press).

O'Shea, S., J. M. Best, J. E. Banatvala, and W. M. Shepherd. 1985. Development and persistence of class-specific antibodies in the serum and nasopharyngeal washings of rubella vaccines. J. Infect. Dis. **151:**89–98.

Oxford, J. S., and C. W. Potter. 1971. Persistent rubella virus infection in laboratory animals. Arch. Ges. Virusforsch. **34:**75–81.

Palmer, D. F., J. J. Cavallaro, and K. L. Herrmann. 1977. A procedural guide to the performance of rubella hemagglutination-inhibition tests. Centers for Disease Control, Immunology Series, no. 2 revised. Atlanta, Georgia.

Parkman, P. D., E. L. Buescher, and M. S. Artenstein.

1962. Recovery of rubella virus from army recruits. Proc. Soc. Exp. Biol. Med. **111:**225–230.

Parkman, P. D., E. L. Buescher, M. S. Artenstein, J. M. McCowan, F. K. Mundon, and A. D. Druzd. 1964. Studies of rubella. 1. Properties of the virus. J. Immunol. **93:**595–607.

Parkman, P. D., H. M. Meyer, B. L. Kirschstein, and H. E. Hopps. 1966. Attenuated rubella virus. I. Development and laboratory characterization. N. Engl. J. Med. **275:**569–574.

Parkman, P. D., P. E. Phillips, R. L. Kirschstein, and H. M. Meyer. 1965. Experimental rubella virus infection in the rhesus monkey. J. Immunol. **95:**743–752.

Pattison, J. R. (ed.). 1982. Laboratory investigation of rubella. Public Health Laboratory Service, Monograph Series no. 16. HMSO, London.

Peckham, C. S. 1972. Clinical and laboratory study of children exposed in utero to maternal rubella. Arch. Dis. Child. **47:**571–577.

Peetermans, J., and C. Huygelen. 1967. Attenuation of rubella virus by serial passage in primary rabbit kidney cell cultures. I. Growth characteristics in vitro and production of experimental vaccines at different passage levels. Arch. Ges. Virusforsch. **21:**134–143.

Peltola, H., V. Karanko, T. Kurki, V. Hukkanen, M. Virtanen, K. Penttinen, M. Nissinen, and D. P. Heinonen. 1986. Rapid effect on endemic measles, mumps and rubella of nationwide vaccination programme in Finland. Lancet **1:**137–139.

Penttinen, K., and G. Myllylä. 1968. Interaction of human blood platelets, viruses and antibodies. 1. Platelet aggregation test with microequipment. Ann. Med. Exp. Biol. Fenn. **46:**188–192.

Perkins, F. T. 1985. Licensed vaccines. Rev. Infect. Dis. **7(Suppl 1):**S73–76.

Plotkin, S. A., J. Farquhar, M. Katz, and T. H. Ingalls. 1967. A new attenuated rubella virus grown in human fibroblasts: evidence for reduced nasopharyngeal excretion. Am. J. Epidemiol. **86:**468–477.

Preblud, S. R., H. C. Stetler, J. A. Frank, W. L. Greaves, A. R. Hinman, and K. L. Herrmann. 1981. Fetal risk associated with rubella vaccine. J. Am. Med. Assoc. **246:**1413–1417.

Rabinowe, S. L., K. L. George, R. Loughlin, S. J. Stuart, and G. S. Eisenbarth. 1986. Congenital rubella: monoclonal antibody-defined T cell abnormalities in young adults. Am. J. Med. **81:**779–782.

Rabo, E., and J. Taranger. 1984. Scandinavian model for eliminating measles, mumps and rubella. Br. Med. J. **289:**1402–1404.

Regamey, R. H., A. deBarbieri, W. Hennessen, D. Ikic, and F. T. Perkins (ed.). 1969. International symposium on rubella vaccines, London, 1968. Symposium Series Immunobiol. Standard. Karger, Basel/New York.

Ronalds, C. J., A. E. Hardiman, P. C. A. Grint, and H. O. Kangro. 1985. Rubella-specific IgM determination on heat-treated sera. Lancet **2:**1071–1072.

Safford, J. W., G. G. Abbott, and C. M. Deimler. 1985. Evaluation of a rapid passive hemagglutination assay for anti-rubella antibody: comparison to hemagglutination inhibition and a vaccine challenge study. J. Med. Virol. **17:**229–236.

Safford, J. W. Jr., and R. Whittington. 1976. A passive hemagglutination assay for detecting rubella antibody. Fed. Proc. **35:**813.

Salonen, E-M., T. Hovi, O. Meurman, T. Vesikari, and A. Vaheri. 1985. Kinetics of specific IgA, IgD, IgE, IgG and IgM antibody responses in rubella. J. Med. Virol. **16:**1–9.

Schiff, G. M., B. C. Young, G. M. Stefanovic, E. F. Stamler, D. R. Knowlton, B. J. Grundy, and P. H. Dorsett. 1985. Challenge with rubella virus after loss of detectable vaccine-induced antibody. Rev. Infect. Dis. **7(Suppl 1):**S157–163.

Schmidt, N. J., J. Dennis, and E. H. Lennette. 1971. Rubella virus hemagglutination with a wide variety of erythrocyte species. Appl. Microbiol. **22:**469–470.

Schmidt, N. J., J. Dennis, and E. H. Lennette. 1978. Comparison of immunofluorescence and immunoperoxidase staining for identification of rubella virus isolates. J. Clin. Microbiol. **7:**576–583.

Schmidt, N. J., and E. H. Lennette. 1966a. Rubella complement-fixing antigens derived from the fluid and cellular phases of infected BHK-21 cells: extraction of cell-associated antigen with alkaline buffers. J. Immunol. **97:**815–821.

Schmidt, N. J., and E. H. Lennette. 1966b. The complement-fixing antigen of rubella virus. Proc. Soc. Exp. Biol. Med. **121:**243–250.

Sedwick, W. D., and F. Sokol. 1970. Nucleic acid of rubella virus and its replication in hamster kidney cells. J. Virol. **5:**478–489.

Sever, J. L. 1965. Rubella antibody among pregnant women in Hawaii. Am. J. Obstet. Gynecol. **92:**1006–1008.

Sever, J. L., N. R. Tzan, I. C. Shekarchi, and D. L. Madden. 1983. Rapid latex agglutination test for rubella antibody. J. Clin. Microbiol. **17:**52–54.

Siegel, M., H. T. Fuerst, and V. F. Guinee. 1971. Rubella epidemicity and embryopathy. Am. J. Dis. Child. **121:**469–473.

Skendzel, L. P., and D. C. Edson. 1986. Latex agglutination test for rubella antibodies: report based on data from the College of American Pathologists surveys, 1983 to 1985. J. Clin. Microbiol. **24:**333–335.

Skendzel, L. P., K. R. Wilcox, and D. C. Edson. 1983. Evaluation of assays for the detection of antibodies to rubella: a report based on data from the College of American Pathologists surveys of 1982. Am. J. Clin. Pathol. **80(Suppl):**594–598.

Skvaril, F., and V. Schilt. 1984. Characterisation of the subclass and light chain types of IgG antibodies to rubella. Clin. Exp. Immunol. **55:**671–676.

Smith, K. O., and T. E. Hobbins. 1969. Physical characteristics of rubella virus. J. Immunol. **102:**1016–1023.

Strannegård, O., L. Grillner, and I-M. Lindberg. 1975. Hemolysis-in-gel test for the demonstration of antibodies to rubella virus. J. Clin. Microbiol. **1:**491–494.

Stewart, G. L., P. D. Parkman, H. E. Hopps, R. D. Douglas, J. P. Hamilton, and H. M. Meyer. 1967. Rubella virus hemagglutination-inhibition test. N. Engl. J. Med. **276:**554–557.

Sugishita, C., S. O'Shea, J. M. Best, and J. E. Banatvala. 1978. Rubella serology by solid-phase radioimmunoassay: its potential for screening programmes. Clin. Exp. Immunol. **31:**50–54.

Taranger, J. 1982. Vaccination programme for eradication of measles, mumps and rubella. Lancet **1:**915–916.

Tedder, R. S., J. L. Yao, and M. J. Anderson. 1982. The production of monoclonal antibodies to rubella haemagglutinin and their use in antibody-capture assays for rubella-specific IgM. J. Hyg. (Camb.) **88:**335–350.

Terry, G. M., L. Ho-Terry, R. C. Warren, C. H. Rodeck, A. Cohen, and K. R. Rees. 1986. First trimester prenatal diagnosis of congenital rubella: a laboratory investigation. Br. Med. J. **282:**930–933.

Thompson, K. M., and J. O'H. Tobin. 1970. Isolation of rubella virus from abortion material. Br. Med. J. **2:**264–266.

Tobin, J. O'H., S. Sheppard, R. W. Smithells, A. Milton, N. Noah, and D. Reid. 1985. Rubella in the United Kingdom 1970–1983. Rev. Infect. Dis. **7(Suppl 1):**S47–52.

Toivonen, V., R. Vainionpaa, A. Salmi, and T. Hyypia. 1983. Glycopolypeptides of rubella virus. Arch. Virol. **77:**91–95.

Ueda, K., Y. Nishida, K. Oshima, and T. H. Shepherd. 1979. Congenital rubella syndrome: correlation of gestational age at time of maternal rubella with type of defect. J. Pediatr. **94:**763–765.

Vaananen, P., V-M. Haiva, P. Koskela, and O. Meurman. 1985. Comparison of a simple latex agglutination test with hemolysis-in-gel, hemagglutination inhibition and radioimmunoassay for detection of rubella virus antibodies. J. Clin. Microbiol. **21:**793–795.

Vaananen, P., and A. Vaheri. 1979. Hemolysis-in-gel test in immunity surveys and diagnosis of rubella. J. Med. Virol. **3:**245–252.

Vaheri, A., and T. Hovi. 1972. Structural proteins and subunits of rubella virus. J. Virol. **9:**10–16.

Vaheri, A., and E. M. Salonen. 1980. Evaluation of solid-phase enzyme-immunoassay procedure in immunity surveys and diagnosis of rubella. J. Med. Virol. **5:**171–181.

van der Logt, J. T. M., A. M. van Loon, and J. van der Veen. 1981. Hemadsorption immunosorbent technique for determination of rubella immunoglobulin M antibody. J. Clin. Microbiol. **13:**410–415.

Vejtorp, M., E. Fanoe, and J. Leerhoy. 1979. Diagnosis of postnatal rubella by the enzyme-linked immunosorbent assay for rubella IgM and IgG antibodies. Acta Pathol. Microbiol. Scand. **87:**155–160.

Vesikari, T. 1972. Immune response in rubella infection. Scand. J. Infect. Dis. **(Suppl 4):**1–42.

Vesikari, T., and A. Vaheri. 1968. Rubella: a method for rapid diagnosis of a recent infection by demonstration of the IgM antibodies. Br. Med. J. **1:**221–223.

Vidgren, G., K. Takkinen, N. Kalkkinen, L. Kaariainen, and R. F. Pettersson. 1987. Nucleotide sequence of the genes coding for the membrane glycoproteins E1 and E2 of rubella virus. J. Gen. Virol. **68:**2347–2357.

Voller, A., and D. E. Bidwell. 1975. A simple method for detecting antibodies to rubella. Br. J. Exp. Pathol. **56:**338–339.

Waxham, N. M., and J. S. Wolinsky. 1985a. A model of the structural organization of rubella virions. Rev. Infect. Dis. **7:(Suppl 1):**S133–139.

Waxham, N. M., and J. S. Wolinsky. 1985b. Detailed immunologic analysis of the structural polypeptides of rubella virus using monoclonal antibodies. Virology **143:**153–165.

Weller, T. H., and F. A. Neva. 1962. Propagation in tissue culture of cytopathic agents from patients with rubella-like illness. Proc. Soc. Exp. Biol. Med. **111:**215–225.

Westaway, E. G., M. A. Brinton, S. Y. Gaidamovich, M. C. Horzinek, A. Igarashi, L. Kääriäinen, D. K. Lvov, J. S. Porterfield, R. K. Russell, and Trent, D. W. 1985. Togaviridae. Intervirol. **24:**125–139.

Wilkins, J., J. M. Leedom, M. A. Salvatore, and B. Portnoy. 1972. Clinical rubella with arthritis resulting from reinfection. Ann. Intern. Med. **77:**930–932.

Ziegelmaier, R., F. Behrens, and G. Enders. 1981. Class-specific determination of antibodies against cytomegalovirus (CMV) and rubella virus by ELISA. J. Biol. Standards **9:**23–33.

CHAPTER 24

Coronaviridae: The Coronaviruses

JOHN C. HIERHOLZER and GREGORY A. TANNOCK

Disease: Respiratory illness; possibly infant gastroenteritis.

Etiologic Agents: HCV 229E, OC43; HECV.

Source: Respiratory secretions/fomites; presumably fecal for HECV.

Clinical Manifestations: Upper respiratory illness, coryza, fever; rare lower respiratory tract disease; possible gastroenteritis.

Pathology: Typical viral upper respiratory infection with inclusions.

Laboratory Diagnosis: Indirect fluorescent antibody and enzyme immunoassay on direct respiratory secretions; electron microscopy and immune electron microscopy on stool specimens; hemagglutination inhibition, standard serum neutralization, and enzyme immunoassay serologic tests are best.

Epidemiology: Worldwide.

Treatment: None.

Prevention and Control: None.

Description of Disease

Human coronaviruses (HCV) were discovered in the 1960s during the period of active searching for agents of the common cold. The first evidence that a new group of viruses was involved in human respiratory illness came from Salisbury, England in 1962, when volunteers who were given respiratory secretions developed colds; however, no virus or bacteria could be isolated in the conventional systems then in use (Kendall et al., 1962). Further evidence was obtained, although not recognized at the time, when antibodies to mouse hepatitis viruses (MHV) were found in human sera (Hartley et al., 1964). Shortly thereafter, viruses from the Salisbury study that were distinct from other known respiratory pathogens were grown in human fetal tracheal organ culture, in which viral replication was evidenced by the reduction of ciliary activity in the epithelial cells (Tyrrell and Bynoe, 1965, 1966). Further proof of the involvement of these new viruses in respiratory disease was obtained epidemiologically and by volunteer transmission studies (Bradburne et al., 1967; Tyrrell and Bynoe, 1966).

During this time, another new respiratory virus was recovered in HEK cells from a group of medical students in Chicago (Hamre and Procknow, 1966), and additional viral strains, unrelated to those discussed previously, were isolated in tracheal organ cultures from adults in Washington, D.C. (McIntosh et al., 1967b). The latter were quickly adapted to suckling mouse brain and were shown to be morphologically identical to MHV (McIntosh et al., 1967a). In 1967, all of these new viruses were recognized as morphologically identical to each other and to avian infectious bronchitis viruses, yet distinct from all other known viruses (Almeida and Tyrrell, 1967; Becker et al., 1967; Hamre et al., 1967; McIntosh et al., 1967a; Tyrrell and Almeida, 1967). Because all were characterized by a "corona" of spikes projecting from the membrane, they were termed *coronaviruses*. Thus was established a new and distinct group of viruses, with representative strains already known to infect mice, humans, and chickens (Tyrrell et al., 1968).

Classification of these viruses within the family *Coronaviridae* by the International Committee on Taxonomy of Viruses (ICTV) was made according to similarities in their fine structure as revealed by electron microscopy and, to a lesser extent, their biologic and antigenic properties (Almeida and Waterson, 1970; Davies and Macnaughton, 1979; Oshiro, 1973; Tyrrell et al., 1975, 1978). The most recent definition of *Coronaviridae* (Third Report of the Coronavirus Study Group, Vertebrate Virus Subcommittee, ICTV) still places primary emphasis on morphology, followed by unique structural, replicative, antigenic, and biologic properties (Siddell et al., 1983a). These unique features are the focus of this chapter, with particular emphasis on the laboratory diagnosis of HCV infections.

The Viruses

Recognized Strains

Of the many coronaviruses putatively associated with human disease, only two are well established in the literature. Progress in identifying new strains has been slow because of their fastidious growth requirements in culture, where culture is possible, and their general physical lability. Furthermore, the occurrence of similar, enveloped-like particles in specimens from nonill control patients has often complicated the establishment of a definite etiology.

The first four HCVs identified—strains B814, 229E, OC43, and 692—were associated with upper respiratory disease (Table 1). Strain B814 (Tyrrell and Bynoe, 1965) was subsequently lost in the laboratory and so cannot be compared with later strains. Strain 229E, isolated from a specimen collected from a patient with upper respiratory infection in January 1962 and selected as the prototype of five strains isolated at that time (Hamre and Procknow, 1966), has been found worldwide in numerous serologic and epidemiologic studies (Bradburne and Somerset, 1972; Cavallaro and Monto, 1970; Hamre and Beem, 1972; Isaacs et al., 1983; Kapikian et al., 1969; Kaye and Dowdle, 1975; Larson et al., 1980; Wenzel et al., 1974). Strain 229E, although usually associated with mild to moderately severe colds in children and adults, can cause tonsillitis, otitis, and pneumonia in young children (Isaacs et al., 1983; McIntosh et al., 1970b, 1974). Currently, strain 229E is considered to represent a subgroup of antigenically related isolates (Macnaughton et al., 1981b; Monto and Lim, 1974; Reed, 1984).

Strain OC43, from a nasopharyngeal wash collected in January 1966, was selected as the prototype of six "IBV-like" strains isolated in fetal tracheal organ culture (McIntosh et al., 1967b). It also has been associated with upper respiratory illness worldwide (Bradburne and Somerset, 1972; Hovi et al., 1979; Isaacs et al., 1983; Kaye et al., 1971; Monto and Lim, 1974; Riski et al., 1974; Wenzel et al., 1974). Strain OC43 has been associated with sporadic cases of bronchitis, pneumonia, and possibly central nervous system disease, pericarditis, pancreatitis, and gastrointestinal disease. However, unlike 229E, these infections can occur in adults as well as children (Isaacs et al., 1983; McIntosh et al., 1970b, 1974; Riski and Hovi, 1980; Wenzel et al., 1974). Like 229E, strain OC43 also has been shown to represent a group of related but heterogeneous strains (Macnaughton et al., 1981b; McIntosh et al., 1967a, 1970b).

Strain 692 was shown to be a coronavirus and unrelated to either 229E or OC43 by immune electron microscopy (Kapikian et al., 1973). It was detected in organ culture harvests of a nasopharyngeal wash collected from a 29-year-old man in January 1966. No further laboratory or epidemiologic work on this virus has been reported.

Tettnang virus was, for a short time, another coronavirus associated with human respiratory disease. Tettnang was isolated in 1978 from the cerebrospinal fluid of a 1-year-old girl with rhinitis, pharyngitis, and mild encephalitis (Malkova et al., 1980). However, the virus was recovered only in suckling mouse brain in single mice on second passage, and exhibited a high degree of cross-reactivity with MHV and HCV OC43 antisera by indirect fluorescent-antibody and complement fixation tests. Also, anti-Tettnang serum, prepared in mice, reacted with MHV and OC43 antigens. Therefore, this isolate appeared to be reactivated MHV or at least heavily contaminated

TABLE 1. Coronaviruses in human respiratory disease

Strain	Specimen	Year	Patient	Illness	Original reference	Status
B814	NS/NW	1960	Youth	Common cold	Tyrrell and Bynoe (1965)	1[a]
229E	NPS	1962	Young adult	Minor URI[b]	Hamre and Procknow (1966)	2
OC43	NPW	1966	Young adult	URI	McIntosh et al. (1967b)	2
692	NPW	1966	29, male	URI	Kapikian et al. (1973)	3

[a] 1 = Strain lost in laboratory; 2 = 229E-like subgroup of strains and OC43-like subgroup of strains are well established virologically and epidemiologically (see text); 3 = definitive studies on viral characterization or etiologic association not yet reported.
[b] URI = Upper respiratory infection.

with MHV. A subsequent study by Bardos et al. (1980) in fact confirmed Tettnang as MHV.

Putative Strains

Many tentative associations have been reported between coronavirus infection and other diseases, such as hepatitis, nephropathy, and multiple sclerosis, and even more attempts have been made to link coronaviruses to necrotizing enterocolitis, gastroenteri-

tis, and other human diarrheal diseases (Table 2). All of these diseases will be discussed in detail. However, it is important to keep in perspective that as of 1988, an etiology has been established only for the common cold coronaviruses HCV 229E and OC43, which have been isolated and thoroughly characterized in the laboratory.

Coronavirus-like particles (CVLPs) were found in sera from a number of patients with chronic active hepatitis who were negative for Australia-SH antigen (Ackermann et al., 1974; Holmes et al., 1970;

TABLE 2. Observations of coronaviruslike particles in humans

Country	Season	Age	Disease	Direct EM[a] P/T	Direct EM[a] Specimen	Culture[a] P/T	Culture[a] Serologic confirmation	Reference
England		Adults	Hepatitis	1/2	Serum		Yes	Zuckerman et al. (1970)
United States		Adults	Hepatitis	2/2	Serum			Holmes et al. (1970)
England		Adults	Var. liver	+	Serum			Wright (1972)
Canada		Adults	Hepatitis	11	Serum			Ackerman et al. (1974)
Yugoslavia		All	Nephropathy	7	Kidney		Yes	Apostolov et al. (1975)
Romania		All	Nephropathy	+	Kidney			Georgescu et al. (1978)
United States		33 yr	MS	1	Brain			Tanaka et al. (1976)
England		Young adult	Gastroenteritis	6/9	Stool	0/6		Caul et al. (1975)
	Fall	Young adult	Gastroenteritis	>1	Stool	1/1	Yes	Caul and Clarke (1975)
	Summer	Young adult	Gastroenteritis	Many	Stool			Caul and Clarke (1975)
	Winter	Young adult	Homosexual	8/23	Stool			Riordan et al. (1986)
	Winter	Young adult	AIDS	1/1	Stool			Riordan et al. (1986)
India		Adults	Sprue	14/16	Stool	0/14		Mathan et al. (1975)
		Adults	None	27/29	Stool			Mathan et al. (1975)
		Children	None	10/12	Stool	0/10		Mathan et al. (1975)
Australia	All	Infants	Gastroenteritis	18/94	Stool			Moore et al. (1977)
	All		None	10/65	Stool			Moore et al. (1977)
		Children	Gastroenteritis	37/55	Stool			Schnagl et al. (1978)
	Fall	Children	Gastroenteritis	157/537	Stool		Yes	Schnagl et al. (1978, 1986)
	Fall	Children	None	145/226	Stool		Yes	Schnagl et al. (1978, 1986)
France	All	Children	Gastroenteritis	25/190	Stool	4/4		Peigue et al. (1978)
	Summer	Infants	NNEC	30	Stool	1	Yes	Sureau et al. (1980)
	Winter	Infants	NNEC	23/32	Stool	2	Yes	Chany et al. (1982)
	Winter	Infants	Diarrhea	2/12	Stool			Chany et al. (1982)
	Winter	Infants	None	3/47	Stool		No	Chany et al. (1982)
	Winter	Children	Diarrhea	16/19	Stool		Yes	Chany et al. (1982)
	Winter	Adults	None	17/75	Stool		Yes	Chany et al. (1982)
	Summer	Infants	NNEC	8	Stool			Caldera and Badoual (1982)
Germany	Fall, winter, spring	Infants	Diarrhea	15/24	Stool	1	Yes	Maass and Baumeister (1983)
	Fall, winter, spring	Infants	Malaise	5/8	Stool			Maass and Baumeister (1983)
	Fall, winter, spring	Adults	Gastroenteritis	7/116	Stool			Maass and Baumeister (1983)
	Fall, winter, spring	Adults	None	16/265	Stool			Maass and Baumeister (1983)
Italy		Infants	Gastroenteritis	34/208	Stool		Yes	Gerna et al. (1984)
		Infants	None	3/182	Stool			Gerna et al. (1984)
United States	Fall	Infants	Gastroenteritis	32/88	Stool		Yes	Vaucher et al. (1982)
	Winter	Children	Diarrhea	17/38	Stool		Yes	Vaucher et al. (1982)
	Fall, winter	Children	Diarrhea	49/126	Stool			Mortensen et al. (1985)
	Winter	Infants	NNEC	7/15	Stool	2/15	Yes	Resta et al. (1985)
Gabon	Fall, spring	Children	Diarrhea	60/156	Stool			Sitbon (1985)
	Fall, spring	Children	None	75/115	Stool			Sitbon (1985)

EM = electron microscopy; P/T = number of positive specimens over the total number tested, where given; MS = multiple sclerosis; NNEC = neonatal necrotizing enterocolitis.

Wright, 1972; Zuckerman et al., 1970). These particles are occasionally seen in sera from nonill humans and monkeys, suggesting that, at best, coronaviruses are only rarely hepatotrophic in humans (unpublished observations).

In addition to liver disease, coronaviruses may occasionally be involved in human kidney disease. Endemic (Balkan) nephropathy, a slow degenerative kidney disease affecting villagers in close contact with swine herds in Yugoslavia, Bulgaria, and Romania, was first described in the late 1950s, but only recently was associated with porcine coronaviruses. Coronavirus-like particles were seen throughout the nephron in seven patients with endemic nephropathy, and only those families involved in pig husbandry developed the disease (Apostolov et al., 1975). However, attempts to detect swine coronavirus antibodies in the sera of patients with nephropathy using a local HEV antigen were unsuccessful, suggesting that the etiologic agent is unrelated to HEV or its antigenic cousin, transmissible gastroenteritis virus (Georgescu et al., 1978).

Coronavirus-like particles were seen in active lesions in brain tissue from one patient with multiple sclerosis (Tanaka et al., 1976). Later, two coronaviruses (SD and SK) were isolated in suckling mouse brain from the brain tissue of two other patients with multiple sclerosis (Burks et al., 1980). These SD and SK viruses cross-reacted antigenically with OC43, and suitable caution had been exercised regarding the possibility of reactivating latent MHV in the host mice (Gerdes et al., 1981b). Nonetheless, SD and SK viruses were shown to possess 90% of the MHV-A59 RNA genome (Weiss, 1983). Furthermore, neither SK antigen nor OC43 RNA could be detected directly in brain tissue from these patients (Burks et al., 1984; Sorensen et al., 1986). Thus, these isolates are now considered to be adventitious murine contaminants. In an independent study, levels of coronavirus antibodies in patients with multiple sclerosis were identical to those in control groups (Leinikki et al., 1981). No solid evidence currently exists to associate coronaviruses with multiple sclerosis in humans.

Numerous groups have reported an association between CVLPs in stool specimens and human diarrheal disease (reviewed by Macnaughton and Davies, 1981; Resta et al., 1985), but few report an association between the development of serum antibody to CVLPs and recovery from infection (Schnagl et al., 1986). For simplicity, and in anticipation that an etiology will be established, the CVLPs in stool specimens are presently called *human enteric coronaviruses* (HECV). In this respect, there is a parallel with the human enteric adenoviruses that were denoted as a subgroup long before they were actually isolated in the laboratory. Human enteric cor-

onaviruses have been sought and tentatively identified by electron microscopy in various parts of the world. These studies will be summarized here in some detail to draw together common findings and emphasize the need for further research.

The earliest reported outbreaks were in England in 1965 (Weston), 1971 (Bristol), and 1975 (Somerset), in all of which CVLPs were seen in fecal specimens from young adults with gastroenteritis (Caul and Clarke, 1975; Caul et al., 1975). In the Weston outbreak in 23 patients, most had vomiting, 50% had diarrhea, and a few had fever or pharyngitis; CVLPs were seen in six specimens (Caul et al., 1975). In the Bristol outbreak among hospital nurses, a coronavirus was propagated in human fetal intestinal organ culture from a specimen in which large numbers of CVLPs were detected by electron microscopy. This virus, from patient X, could also grow to low titer in primary human embryonic kidney (HEK) cell cultures. Detection in both HEK and the organ cultures was by indirect immunofluorescence with convalescent serum from the patient (Caul and Clarke, 1975). Further ultrastructural studies of Patient X virus in intestinal organ cultures have been reported (Caul and Egglestone, 1977), but the virus has not yet been adapted to more available culture systems. In a later study (Riordan et al., 1986), CVLPs were seen without any clinical illness in 35% of male homosexuals in Manchester.

In Vellore, India, CVLPs were seen in fecal samples from many patients with epidemic tropical sprue (Mathan and Mathan, 1978; Mathan et al., 1975). In one patient with chronic tropical sprue, CVLPs were excreted for 8 months and biopsies of the jejunum revealed coronavirus vesicles. Coronavirus-like particles were also found in fecal specimens from a high percentage of apparently healthy persons in rural areas, but were not found in neonates.

Similar findings were also made in southern and northern Australia. Coronavirus-like particles were found in four relatively closed communities around Adelaide, consisting of Caucasian infants with gastroenteritis, native (Aboriginal) infants with gastroenteritis, healthy Vietnamese refugees, and institutionalized healthy children (Moore et al., 1977). They were also found in many communities of Western Australia, not only among Caucasian and Aborigines children, but also in dogs kept by the Aborigines (Schnagl et al., 1978). In this study, the particles were equally prevalent in children with or without diarrhea. Coronavirus-like particles increased in frequency with increasing age, reaching a high percentage of the adult population. They were rarely associated with symptomatic disease in adults. As in the Indian study, the incidence of CVLPs in stools of nonill persons was higher in rural areas and lower socioeconomic groups. An association between disease and the occurrence of antibody to CVLPs has been found in Australia, suggesting that CVLPs are infectious agents that may be related to chronic diarrhea in certain Australian populations (Schnagl et al., 1986, 1987).

Human enteric coronaviruses have been detected in Europe since 1978. Peigue et al. (1978) reported finding CVLPs in stool specimens from 25 of 190 children hospitalized with gastroenteritis in the Clermont-Ferrand region of central France over a 1-year period. Laporte's laboratory cultivated coronaviruses from four of these specimens in HRT-18 human rectal tumor adenocarcinoma cells. Growth was monitored by negative-stain electron microscopy (Laporte and Bobulesco, 1981; Sureau et al., 1980), but after 10 passages, the viruses could no longer be detected. They may have become contaminated with NCDCV bovine coronavirus (Patel et al., 1982).

Chany et al. (1982) reported two outbreaks of neonatal necrotizing enterocolitis (NNEC) that occurred in a large hospital in Paris. The first, from March to May 1979, was not studied by electron microscopy and no unusual microbial agents were isolated by routine procedures. In the second, from September 1979 to March 1980, 32 of 58 infants developed NNEC and 23 of these were positive for CVLPs by electron microscopy. Coronavirus was apparently isolated in cell culture from two of these specimens, but further details have not been forthcoming. By including babies born in a second maternity hospital in the same section of Paris, three nonill control infants and two infants with diarrhea were CVLP positive. In addition, 17 adult contacts of the infants with NNEC and 16 children with acute diarrhea in a local nursery were CVLP positive; most were antibody positive as well, suggesting a community-wide outbreak of CVLPs during this period. Serologic response to the CVLPs was documented by immune electron microscopy in 7 of 10 infants with NNEC, whereas none of 15 controls had CVLP antibody (Chany et al., 1982). In a 1981 outbreak of NNEC in one of these hospitals, CVLPs were observed in the stools of eight infants (Caldera and Badoual, 1982).

In a longitudinal study in Munster, West Germany during 1980 and 1981, Maass and Baumeister (1983) found CVLPs in the stools of infants with gastroenteritis and in healthy adults throughout the year, although there was a greater incidence in the winter and spring months. Gerna et al. (1984) reported CVLPs in 34 infants with gastroenteritis and in 3 agematched controls in Pavia, Italy, and demonstrated a significant bilateral cross-reaction between HCV OC43 and two strains of HECV. The antigenic relatedness was shown in one direction by hemagglutination inhibition and serum neutralization tests with OC43 virus versus sera from sick children, control patients, and mice and guinea pigs immunized with purified HECV, and in the other direction by immune electron microscopy with purified HECV and reference antisera to OC43. Additional relationships between HECV and OC43 were shown in subsequent studies (Battaglia et al., 1987; Gerna et al., 1985).

Coronavirus-like particles have frequently been observed in the western half of the United States. Vaucher et al. (1982) found CVLPs in 49 children, including 32 neonates, with diarrhea or gastroenteritis in Tucson, Arizona. Most of the illnesses occurred during the winter of 1979. Seroconversion, measured by immune electron microscopy, was noted in several children. The CVLPs recovered in this study did not exhibit antigenic cross-reactions with HCV OC43 or 229E (Mortensen et al., 1985; Vaucher et al., 1982). Resta et al. (1985) found CVLPs in seven stool specimens during an outbreak of NNEC in Dallas. Coronaviruses, designated A14 and C14, were isolated in human fetal intestinal organ cultures of two of the specimens and were successfully passaged at least 14 times. Preliminary data have shown that these isolates possess typical coronavirus polypeptides and antigens, and induced a specific serologic response. Further work may show the relationship of these strains to those previously described from other outbreaks of gastrointestinal disease. In a separate study (Rettig and Altshuler, 1985), CVLPs were seen in intestinal contents and within epithelial cells of the ileum in a fatal case of severe enteritis in Oklahoma.

In one study reported from Africa, peaks of CVLPs were detected during the rainy seasons (February to May and October to December) in equatorial Gabon (Sitbon, 1985). In this study, healthy children had a higher incidence of CVLPs (75 of 115 specimens) than did children with diarrhea (60 of 156 specimens). As in earlier studies in India and Australia, the incidence of CVLPs increased with age and in older individuals was clearly unrelated to illness. Coronavirus-like particles were found in diverse population groups in Gabon, as well as in domestic goats and dogs and in captive primates.

Definitive proof of the association between CVLPs and human gastrointestinal disease has been difficult to establish. Coronavirus-like particles have proved extraordinarily difficult to cultivate in cell or organ culture and most of those isolated thus far have been lost during subsequent passage. Also, CVLPs have frequently been observed in the stools of healthy individuals (see Table 2), and similar particles have been detected in the stools of both healthy and diarrheic nonhuman primates and other animals (see Table 3). Thus, the CVLPs in general may just be passenger particles tolerated by the host, with no role in disease. On the other hand, the ubiquity of coronaviruses and their etiology in a variety of well-described nephritic, neurologic, and enteric diseases

TABLE 3. Diversity of disease and organ tropism in animal coronaviruses

Abbreviation	Name	Host	Central nervous system	Eye	Lymphatic	Respiratory	Heart	Abdominal viscera	Gastrointestinal	Genitourinary	Other
IBV	Avian infectious bronchitis virus	Chick				+				+	
TGEV	Transmissible gastroenteritis virus	Pig			+	+			+		
HEV	Hemagglutinating encephalomyelitis virus	Pig	+			+			+		
PECV	CV777 porcine epidemic diarrhea virus	Pig							+		
MHV[a]	Mouse hepatitis virus	Mouse	+		+	+		+	+	+	
FIPV	Feline infectious peritonitis virus	Cat	+	+	+	+		+	+	+	
FECV	Feline enteric coronavirus	Cat			±	+			+		
SDAV	Sialodacryoadenitis virus	Rat		+	+	+					
RCV	Rat coronavirus	Rat			±	+					
TCV	Bluecomb disease virus	Turkey	±			+		±	+	±	
BCV (or NCDCV)	Neonatal calf diarrhea virus	Cattle							±		
BECV	Bovine enteric coronavirus	Cattle							+		
CCV	Acute enteritis coronavirus 1-71	Dog							+		
CECV	Canine enteric coronavirus	Dog							+		
RbCV	Cardiomyopathy virus	Rabbit		+	+	+	+	±		+	+
RbECV	Rabbit enteric coronavirus	Rabbit				+			+		
—	Puffinosis virus	Sea birds		+		+					
EECV	Equine enteric coronavirus	Horse							+		
OECV	Ovine enteric coronavirus	Sheep							+		
SECV	Simian enteric coronavirus	Monkey							+		+

[a] Includes serotypes 1, 2, 3, 4 (JHM), A59, and S (including LIVIM), and possibly others less well characterized (Wege et al., 1982).

of animals (see following section) suggests that the list of human diseases in Tables 1 and 2 is far from complete.

Animal Strains

In recent years, many coronaviruses of animals have been implicated in severe or fatal disease, especially in their young (Table 3). These viruses include infectious bronchitis viruses (IBV) (Beaudette and Hudson, 1937; Dawson and Gough, 1971); transmissible gastroenteritis virus (TGEV) of pigs (Doyle and Hutchings, 1946; Kodama et al., 1981); mouse hepatitis viruses (MHV) (Cheever et al., 1949; Rowe et al., 1963); lethal intestinal virus of infant mice (LIVIM) (Kraft, 1962), actually a substrain of MHV-S (Hierholzer et al., 1979); hemagglutinating encephalomyelitis virus (HEV) of pigs (Andries and Pensaert, 1980; Greig et al., 1962); feline infectious peritonitis virus (FIPV) (Pedersen et al., 1981a; Wolfe and Griesemer, 1966); sialodacryoadenitis virus (SDAV) of rats (Bhatt et al., 1977; Hirano et al., 1986; Jonas et al., 1969); rat coronavirus (RCV) (Parker et al., 1970); turkey enteritis (bluecomb disease, TBDV) virus (TCV) (Adams et al., 1970; Panigrahy et al., 1973); neonatal calf diarrhea virus (NCDCV or BCV) (Stair et al., 1972; Tektoff et al., 1983); acute enteritis coronavirus 1-71 (CCV) of dogs (Keenan et al., 1976); rabbit infectious cardiomyopathy virus or Stockholm agent or pleural effusion disease agent (RbCV) (Osterhaus et al., 1982; Small et al., 1979); and puffinosis virus of seabirds (Nuttall and Harrap, 1982). In addition, enteric CVLPs have been reported from many domestic and laboratory animals and may be associated with diarrhea. These animals include cattle (McNulty et al., 1975), horses (Bass and Sharpee, 1975), sheep (Tzipori et al., 1978), dogs (Schnagl and Holmes, 1978), pigs (porcine CV777 epidemic diarrhea virus, PEDV or PECV) (Ducatelle et al., 1981; Pensaert and deBouck, 1978); monkeys (Caul and Egglestone, 1979; Smith et al., 1982), rabbits (contagious diarrheal disease agent, RbECV) (Lapierre et al., 1980; Osterhaus et al., 1982), and cats (Hoshino and Scott, 1980; Pedersen et al., 1981b, 1984; Stoddart et al., 1984).

All of these viruses cause respiratory or enteric disease, or both, and some are also responsible for central nervous system, nephritic, conjunctival, or generalized disease. The animal viruses are, therefore, important as models for HCV disease, not only because their pathogenicity is so diverse, but because their antigenic composition and replication characteristics provide insight into the nature of the HCV. These properties have been the subject of frequent reviews (Bradburne, 1970; Bradburne and Tyr-

rell, 1971; Estola, 1970; McIntosh, 1974; Monto, 1974; Robb and Bond, 1979; Siddell et al., 1983b; Sturman and Holmes, 1983; Wege et al., 1982).

Physicochemical Properties and Morphology

Despite similarities based on morphology, HCV strains 229E and OC43 differ considerably in their host cell susceptibility, physical stability, and antigenic properties. These viruses are large (diameter, 80 to 200 nm) and pleomorphic, with a buoyant density of about 1.18 g/ml. Their unique morphology is characterized by the presence of clublike projections up to 20-nm long protruding from a lipid-containing envelope. They also possess an internal helical ribonucleoprotein, which has been observed as a long strand 1 to 2 nm in diameter (Caul et al., 1979; Davies et al., 1981; Kennedy and Johnson-Lussenburg, 1976) or as a helix condensed into coillike structures 10 to 20 nm in diameter (Macnaughton et al., 1978).

The HCV genetic material consists of singlestranded RNA with a molecular weight of approximately 6×10^6 daltons, which is the largest for any RNA virus (Macnaughton and Madge, 1978; Tannock and Hierholzer, 1977). An earlier study with OC43 purified from infected suckling mouse brain suggested that its RNA genome was fragile and could readily be disrupted with heat or organic solvents (Tannock and Hierholzer, 1977). However, more recent work with strain 229E grown in cell culture reveals no such lability (Macnaughton and Madge, 1978), and it seems likely that the instability observed for OC43 was associated with extensive nicking of the RNA genome during growth in the suckling mouse brain. Virion RNAs prepared from various animal coronaviruses after growth in cell culture have similarly been shown to be stable (Lai and Stohlman, 1978; Schochetman et al., 1977; Tannock, 1973). Human coronaviral RNA, like that of other coronaviruses, contains polyadenylate residues located at the 3' terminus, but does not appear to contain an RNA transcriptase in the outer coat of the virion (Macnaughton and Madge, 1978; Tannock and Hierholzer, 1978). Coronaviridae are, therefore, positive-stranded viruses, with their virion RNA having the same polarity as that of their intracellular viral messenger RNAs.

The structural protein compositions of HCV OC43 and 229E differ little from those reported for other coronaviruses (Cavanaugh et al., 1986b,c; Garwes and Reynolds, 1981; Hierholzer, 1976; Hierholzer et al., 1972, 1981; Hogue and Brian, 1986; King and Brian, 1982; Macnaughton, 1980, 1981; Robbins et al., 1986; Stern and Sefton, 1982; Stern et al., 1982; Sturman, 1977; Sturman and Holmes, 1977, 1983; Sturman et al., 1980; Sugiyama et al., 1986;

Wege et al., 1979, 1982; Wesley and Woods, 1986). The HCV proteins comprise four major groups of polypeptides, which are summarized in Table 4. The high-molecular-weight peplomeric glycoprotein, P, ranges from 160 to 200 kilodaltons (kDa) in various studies. It constitutes the principal antigen detected by neutralization tests and contains host cell receptors and fusion and cell-mediated immunity (CMI) activities. This glycoprotein is considered to be a dimer of two dissimilar proteins (P1 and P2) weighing 106 and 91 kDa, respectively, which in tetrameric form constitute the peplomer or spike. Monoclonal and polyclonal antibodies to the spike proteins of MHV block virus-mediated cell fusion and neutralize infectivity (Collins et al., 1982; Holmes et al., 1984; Nakanaga et al., 1986). Similar results were found for IBV and TGEV (Cavanagh and Davis, 1986; Cavanagh et al., 1986a; Delmas et al., 1986; Laude et al., 1986; Jimenez et al., 1986). These proteins in avian IBV are not dissociable with 2-mercaptoethanol (Cavanagh, 1983b, 1984) and, therefore, do not appear to be held together by disulfide bonds.

The hemagglutinin protein, H, is a 60- to 66-kDa glycoprotein (the 130-kDa monomer is seen in nonreducing gels) found in the peplomers of HCV OC43 and 229E as well as in hemagglutinating animal coronaviruses. The phosphorylated 47- to 55-kDa nucleoprotein, N, is the core antigen. The family of proteins having molecular weights of approximately 40, 27, 24, and 20 kDa have identical polypeptide structures but varying degrees of glycosylation; they constitute the matrix, M, protein that bridges the double-shelled envelope.

Minor virion polypeptides with molecular weights of 107, 92, and 39 kDa, which had previously been reported (Hierholzer, 1976; Kemp et al., 1984c; Stern et al., 1982; Sturman et al., 1980), are now known from tryptic digest analyses and radioimmunoprecipitation to represent P1, P2, and M1, respectively. The other matrix glycoproteins of HCV, M2, M3, and M4, are thought by analogy with murine and avian coronaviruses to provide a link between the nucleocapsid and envelope and span the lipid bilayer in virion assembly (Rottier et al., 1984; Sturman et al., 1980). Glycosylation by o-linked oligosaccharides probably takes place in the Golgi apparatus, and the matrix glycoprotein then influences both the formation of the envelope and budding by the virus into the endoplasmic reticulum (Rottier et al., 1984). The nature of the sugar moieties and the function of coronavirus glycoproteins have been thoroughly reviewed elsewhere (Cavanagh, 1983a,b; Frana et al., 1985; Holmes et al., 1981, 1984; Sturman and Holmes, 1983, 1984, 1985; Sturman et al., 1985).

Strain OC43 readily agglutinates chicken, rat, mouse, vervet monkey, and human group O erythrocytes (Kaye and Dowdle, 1969), a property useful for the detection of antibody by hemagglutination inhibition. Furthermore, attached virus can be readily eluted from erythrocytes by gentle warming, and thus adsorption and elution are an efficient means of partially purifying the virus. The chemical basis for elution is unknown, but does not appear to involve neuraminidase activity (Hierholzer et al., 1972; Pokorny et al., 1975; Sheboldov et al., 1973; Zak-

TABLE 4. Polypeptide composition of the human respiratory coronaviruses

Protein designation[a]		Glycosylation	Molecular weight[b] (kilodaltons)	Location	Function	Assigned mRNA
Functional	Other					
P	E2	++	186 106 (P1) 91 (P2)	Peplomer	SN, CF, fusion activity; binds to cell receptors; induces cell-mediated cytotoxicity	3
H	gp65	++	63	Peplomer	Hemagglutinin	?
N	N	--	50 40 (M1)	Internal core	Ribonucleoprotein, phosphorylated	7
M	E1	++	25 27 (M2) 24 (M3) 20 (M4)	Envelope	Matrix or transmembrane protein	6

[a] Functional designation (P = peplomer; H = hemagglutinin; N = nucleoprotein; M = matrix) taken from Hogue and Brian (1986); other designations (e.g., envelope 2 and 1) taken from Sturman and Holmes (1977) and Lai (1987).

[b] P exists predominantly in the 186-kDa form in either reducing or nonreducing gels, and therefore P1 and P2 are "minor"; H is found predominantly in hemagglutinating coronaviruses; and exists as a 130-kDa dimer in nonreducing gels or as the 63-kDa monomer in reducing gels; N may also be seen as a 160- to 165-kDa trimer or aggregate in nonreducing gels; M constitutes a family of identical polypeptides with varying degrees of glycosylation, with M1 being a minor component and M2–4 present in approximately equal amounts. Molecular weights are the means of published reports on HCV (Hierholzer, 1976; Hierholzer et al., 1972; Hogue and Brian, 1986; Hogue et al., 1984; Kemp et al., 1984c; Macnaughton, 1980; Schmidt and Kenny, 1982).

stelskaya et al., 1972a). Human coronavirus OC43 also effects a nonspecific or "false" hemadsorption with rat or mouse erythrocytes when the virus is grown in BSC-1, MK, HEK, or WI38 cell cultures, a property of potential usefulness in neutralization tests (Bucknall et al., 1972a; Kapikian et al., 1972). The nonspecific HAd is due to the presence of a high-density lipoprotein inhibitor in fetal calf serum (Bucknall et al., 1972a).

Human coronavirus 229E does not hemagglutinate, even when grown or concentrated to high infectivity titers (Hierholzer, 1976) or after treatment with trypsin, as is required for IBV hemagglutination (unpublished observations). Like many other nonhemagglutinating viruses, however, it does adhere to tannic acid-treated sheep red blood cells to make an indirect hemagglutination antigen that is the basis of a convenient IHA serologic test (Kaye et al., 1972). Data on the viral stability for HCV OC43 and 229E were variable; OC43 is more stable than 229E both at 33°C and 37°C and at low pH, but both viruses have the same ultraviolet inactivation rates (Bucknall et al., 1972b).

Replication

Several reviews on the molecular features of coronavirus replication have been derived mainly from studies on animal coronaviruses, especially the mouse hepatitis viruses (Lai, 1987; Siddell, 1983; Sturman and Holmes, 1983). Replication begins with viropexis similar to that in a number of other viral groups, and has been demonstrated for HCV 229E infecting diploid fibroblast cells (Patterson and Macnaughton, 1981).

The principal unique features of coronavirus replication are summarized in Fig. 1. Human coronaviral RNA is large and polyadenylated, and there is no evidence of an RNA polymerase (transcriptase) within the virion outer coat (Macnaughton and Madge, 1978; Tannock and Hierholzer, 1978). The isolated RNAs of other coronaviruses have, in addition, been shown to be infectious (Schochetman et al., 1977; Wege et al., 1978) and to possess a 5' methylated cap structure (Budzilowicz et al., 1985; Lai et al., 1982). Therefore, the polarity of viral RNA replication is the same as that of intracellular viral mRNAs. Replication occurs within the cytoplasm, where up to seven intracellular viral RNA species can be detected. All have common 3' terminal ends of varying size, all copied from the same negative-stranded template to produce a so-called 3' coterminal "nested set." The nested-set scheme allows the independent synthesis of each intracellular RNA, except for the largest RNA, without a processing step from a large precursor molecule. This type of RNA transcription is unique to coronaviruses and involves RNA-dependent RNA polymerase(s), which are coded for by the virion RNA (Brown et al., 1986; Dennis and Brian, 1981; Kapke and Brian, 1986; Lai, 1987; Lai et al., 1981; Leibowitz et al., 1982a; Mahy et al., 1983; Sawicki and Sawicki, 1986; Stern and Kennedy, 1980). Brayton et al. (1982) noted two separate polymerase functions, one occurring at 1 h and the other at 6 h postinfection. The early polymerase may be responsible for the synthe-

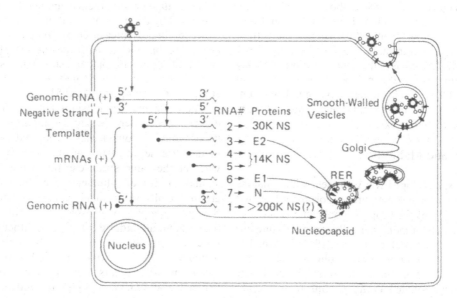

FIG. 1. Nested-set hypothesis for coronavirus RNA replication (Sturman and Holmes, 1983). (Reprinted by permission of the author and Academic Press, Inc.)

sis of the negative-stranded template and later for the synthesis of the positive-stranded RNAs. Oligonucleotide mapping studies have revealed the presence of common sequences at the 5' termini of most MHV A59 RNAs, that are translocated from genomic RNA and thus constitute a leader sequence (Lai and Stohlman, 1978; Lai, 1987).

The role of the cell nucleus in replication is uncertain. Hierholzer (1976) showed that replication of HCV 229E in human embryonic lung fibroblast cells was susceptible to actinomycin D (AMD), a study later confirmed with 229E grown in L132 cells (Kennedy and Johnson-Lussenburg, 1978). An earlier report indicated that replication of IBV required nuclear function and host transcriptional factors (Evans and Simpson, 1980). However, more recent studies indicate that yields of MHV strains A59 or JHM are unaffected by treatment of cells with AMD or the DNA inhibitor a-amanatin, and that growth can take place in enucleated cells (Brayton et al., 1981; Mahy et al., 1983; Wilhelmsen et al., 1981). This area therefore remains controversial.

Studies of viral-specific proteins present in infected cells have proved difficult because there is little inhibition of cell macromolecular synthesis during infection. However, the translation products of several subgenomic RNAs of MHV have been determined after cell-free synthesis in *Xenopus* oocytes or rabbit reticulocytes. The largest RNA (RNA 1) is believed to code for RNA-dependent RNA polymerase(s), RNA 2 for a 35K nonstructural protein, RNA 3 for the peplomeric surface glycoprotein, RNA 4 for a 14K nonstructural protein, RNA 5 for a 10K nonstructural protein, RNA 6 for the M or transmembrane protein, and RNA 7 for the nucleocapsid protein (Jacobs et al., 1986; Leibowitz et al., 1982b; Rottier et al., 1981; Siddell, 1983; Skinner and Siddell, 1985; Skinner et al., 1985; Stern and Sefton, 1984). RNA 1 is translated initially as a large 250K polypeptide, which is subsequently proteolytically cleaved into 220 and 28K polypeptides (Denison and Perlman, 1986). Such studies have not been conducted with HCV, but a similar mechanism of replication is assumed to exist.

Assembly and Maturation

Unlike most enveloped RNA viruses, which mature by budding at the cell membrane, coronaviruses bud into the lumen of the rough endoplasmic reticulum, where assembly of monomeric glycoproteins and nucleoproteins occurs (Becker et al., 1967; Bucknall et al., 1972a; Hamre et al., 1967; Holmes et al., 1984; Oshiro et al., 1971). During this time, the peplomer protein is glycosylated as the glycoproteins migrate to the Golgi apparatus (see Fig. 1). The assembled particles then appear in single-walled vesicles that migrate to the cell membrane, where they are probably released by exocytosis (Sturman and Holmes, 1983).

During replication of HCV 229E in human embryonic lung fibroblasts, roughly circular electronlucent structures with dense limiting membranes, sometimes empty but frequently containing netlike beaded strands, can be seen in the cytoplasm (Kemp et al., 1984b). These structures vary in size from 300 to 900 nm in diameter and appear approximately 6 h postinfection (i.e., at the initiation of viral protein synthesis, but before virion assembly). Formation of these electronlucent structures was not inhibited by treatment of infected cells with an inhibitor of glycosylation (tunicamycin) or of glycoprotein transport (monensin). These structures, which were noted in suckling mouse brain infected with HCV OC43 and a number of other coronaviruses, appear to be an early feature of replication for the *Coronaviridae* (Kemp et al., 1984a,b).

Antigenic Composition

Soon after the discovery of coronaviruses, significant antigenic relationships among strains from different animal species became apparent, along with significant antigenic divergence among strains within the same animal species. Antigenic variation has been noted within strains of IBV, as measured by neutralization, IFA, and enzyme immunoassay tests (Dawson and Gough, 1971; Snyder and Marquardt, 1984), in strains of TGEV differentiated by monoclonal antibodies (Laude et al., 1986), and in strains of HCV 229E and OC43 distinguished by enzyme immunoassays (Macnaughton et al., 1981b; Reed, 1984). Antigenic divergence among MHV strains, first noted by the marked variations in pathogenicity, has recently been measured by complement fixation, neutralization, and radial hemolysis tests (Hierholzer et al., 1979) and by plaque reduction tests and radioimmunoprecipitation (Childs et al., 1983; Dalziel et al., 1986; Fleming et al., 1983, 1986).

The antigenic relationships among viruses from different species emphasize the importance of studies of animal coronaviruses for an understanding of human disease. In particular, it is important to know whether the same strain can infect more than one animal species or whether the coronaviruses as a group actually share major antigenic determinants. Several examples of cross-species infections have been noted, all under experimental rather than natural conditions. In one study, newborn pigs were infected with TGEV, FIPV, and CCV (Woods et al., 1981). In others, cats were infected with TGEV (Reynolds and Garwes, 1979) and rodents were infected with BCV (Akashi et al., 1981). In all of these studies, the infected animals developed typical clinical disease, gut pathology, and serum antibody responses, suggesting that infection in multiple species can occur.

There is little question that the coronaviruses may share at least one major antigenic determinant, which is probably located on the peplomers. The original observation of Hartley et al. (1964) that antibodies in human sera (now known to be anti-OC43) reacted with MHV has been confirmed by a number of independent studies (Leinikki et al., 1981; McIntosh et al., 1969, 1970b). Also, animal hyperimmune antisera against HCV OC43 and various MHV strains have been shown to contain high levels of cross-reactive antibodies in both directions (Bradburne, 1970; Gerdes et al., 1981a,b; Hierholzer et al., 1979; Hogue et al., 1984; Macnaughton, 1981). The cross-reaction between OC43 and MHV-3 is variable with different isolates of OC43 virus (Macnaughton et al., 1981b).

Human convalescent sera and animal hyperimmune antisera have also been shown to react with BCV of cattle (Cereda et al., 1986; Gerna et al., 1981, 1982; Hogue et al., 1984; Kaye et al., 1975; Storz and Rott, 1981), HEV of pigs (Georgescu et al., 1978; Kaye et al., 1977), and the IBV and puffinosis strains of birds (Macnaughton, 1981; McIntosh et al., 1969; Miller and Yates, 1968; Nuttall and Harrap, 1982). All of these relationships have been determined with HCV OC43, are high titered, and, for the most part, are reciprocal.

Studies with HCV 229E have demonstrated bilateral cross-reactions with TGEV and FIPV (Pedersen et al., 1978), with rabbit cardiomyopathy virus (RbCV) (Small et al., 1979), and with MHV-3 (Hasony and Macnaughton, 1982). However, the cross-reaction between 229E and MHV-3 is a false one, apparently caused by the adherence of bovine serum components from culture media; the cross-reaction is removed by absorption of antisera with bovine serum or by growing antigen in serum-free cultures (Kraaijeveld et al., 1980a).

No antigenic relationship has been demonstrated between HCV OC43 and 229E (McIntosh, 1974; Robb and Bond, 1979; Schmidt, 1984; Schmidt and Kenny, 1981). Furthermore, each serotype represents a group of antigenically similar strains (Macnaughton et al., 1981b; Reed, 1984). The more recently described CVLPs seen in infant and adult stool specimens, or HECV, may well represent more than one serotype, at least one of which cross-reacts with HCV OC43 (Gerna et al., 1984; Macnaughton and Davies, 1981; Schnagl et al., 1986).

In addition to important bilateral relationships between the HCV and mammalian and avian coronaviruses, many antigenic relationships exist among the animal viruses themselves. The TGEV of pigs cross-reacts with the CCV of dogs (Garwes and

Reynolds, 1981; Horzinek et al., 1982; Norman et al., 1970), the FIPV of cats (Horzinek and Osterhaus, 1979; Pedersen et al., 1981b, 1984; Reynolds et al., 1977), and others (Horzinek et al., 1982; Pedersen et al., 1978; Pike and Garwes, 1979). The HEV of pigs cross-reacts with the BCV of cattle (Sato et al., 1980). The swine PECV (CV777) is antigenically related to several other coronaviruses (Pensaert et al., 1981). RbECV cross-reacts both with IBV and TGEV (Descoteaux et al., 1985). Many cross-reactions have been shown among and between the canine and feline coronaviruses (Evermann et al., 1981; Horzinek et al., 1982; Pedersen et al., 1978, 1981a,b, 1984; Tupper et al., 1987). The puffinosis coronavirus is related to MHV, RCV, SDAV, and HEV in addition to OC43 as noted above (Nuttall and Harrap, 1982); FIPV is antigenically related to FECV (Pedersen et al., 1981b, 1984); and RCV cross-reacts with SDAV (Bhatt et al., 1977). Many of these relationships have been reviewed in detail elsewhere (Bradburne, 1970; Bradburne and Tyrrell, 1971; McIntosh, 1974; Robb and Bond, 1979; Siddell et al., 1983b; Sturman and Holmes, 1983; Wege et al., 1982).

A phylogenetic tree summarizing these relationships has been constructed for the better-studied coronaviruses (Fig. 2) (Brian et al., 1984). Starting from an ancestor that will probably never be identified, the coronaviruses formed two major branches in avians and two major branches in mammals, each with several subbranches. These antigenic lineages were determined by reciprocal serum neutralization, IFA, enzyme immunoassay, and radioimmunoprecipitation tests carried out by different laboratories and summarized by Wege et al. (1982) and Brian et al. (1984). Undoubtedly, more relationships will be documented as additional strains are compared in reciprocal serologic tests. Altogether, the intricate picture of antigenic-relatedness among the human and animal coronaviruses will remain a major focus of research because of its impact on the biology and pathogenesis of these viruses.

The serologic cross-reactions observed among coronaviruses are related to antigenic similarities on the spike or peplomer component. Three to six virion antigens have been detected in studies with HCV grown in cell culture, and three have been found in OC43 grown in suckling mouse brain. No soluble antigens have been found in cell culture supernatant fluids. The 186-kDa glycoprotein of the human strains is the predominant protein in the peplomer (Table 4) and contains the hemagglutinating, complement-fixing, and neutralizing activities (Hierholzer, 1976; Hierholzer et al., 1972; Kaye et al., 1970; Macnaughton et al., 1981a; Pokorny et al., 1975; Schmidt, 1984; Schmidt and Kenny, 1981, 1982; Yaseen and Johnson-Lussenburg, 1981). Macnaughton

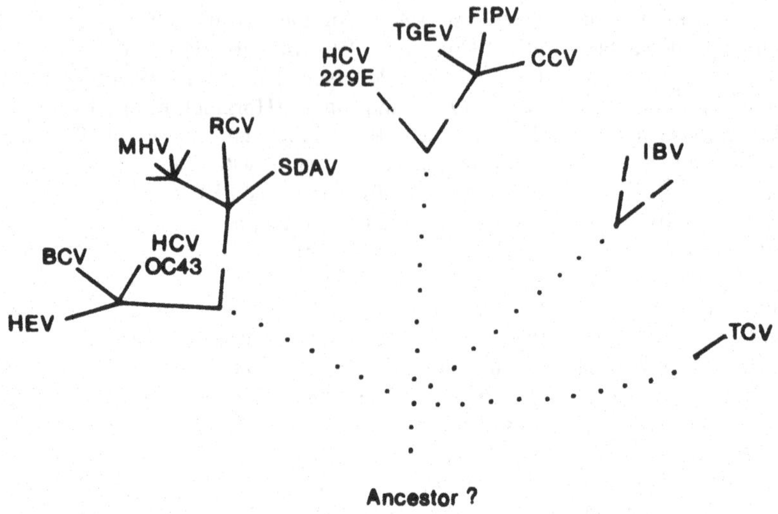

Fig. 2. Phylogenetic tree showing antigenic interrelationships among the human and animal coronaviruses (Brian et al., 1984). (Reprinted by permission of the author and publisher from Proceedings of the 4th International Symposium on Neonatal Diarrhea, October 3–5, 1983, University of Saskatchewan, p. 100–115, Veterinary Infectious Disease Organisation, Saskatchewan, Canada.)

et al. (1981a) showed that human volunteers experimentally infected with HCV 229E produced most antibody to the P antigen and little antibody to N or M antigens, as determined by antigen-specific enzyme immunoassays. This finding was consistent with P antibody having neutralizing and protective functions. Most naturally acquired human antibody to HCV infection also is to the P antigen (Schmidt, 1984). The same observations have been made with many animal coronaviruses, in which hemagglutinating and neutralizing activities were shown to reside in the largest glycoprotein, which was predominant in the intact peplomer (Cavanagh et al., 1984; Jimenez et al., 1986; King et al., 1985). Direct evidence in support of these findings came from bromelain-treated HCV, in which totally despiked particles were devoid of biologic activity (Hierholzer, 1976; Hierholzer et al., 1972; Schmidt and Kenny, 1982), and from monoclonal antibodies to the large glycoprotein of TGEV, which were shown to fully neutralize the virus (Jimenez et al., 1986).

The other major antigens are the N or ribonucleoprotein core and M or transmembrane glycoprotein, both of which are detectable by immunoelectrophoresis and immunodiffusion tests (Schmidt, 1984; Yaseen and Johnson-Lussenburg, 1981). Antibodies to the N and M antigens are detectable in human convalescent sera and probably react to some extent in complement-fixation tests, but they are not involved in hemagglutination inhibition and serum neutralization tests (Schmidt, 1984). Furthermore, the human antibody response to the P, N, and M antigens of 229E and OC43 is type-specific; that is, no cross-reactions were observed between the viruses by antigen-specific enzyme immunoassay, immunoelectrophoresis, or complement fixation tests.

Genetics

Because of technical difficulties in growing human coronaviruses to workable titers in cell culture, genetic studies have generally been precluded. Persistent infections with HCV 229E have been demonstrated in a variety of cell cultures, in particular L132 cells derived from human embryonic lung (but containing HeLa markers), and their relevance to human infection has been widely discussed (Chaloner-Larsson and Johnson-Lussenburg, 1981; Sturman and Holmes, 1983). Persistent and latent infections of MHV in mice are well known to cause sweeping epidemics of encephalomyelitis, hepatitis, and gastroenteritis in laboratory mouse colonies (Lamontagne and Dupuy, 1984; Robb and Bond, 1979) and equally well known as the source of contaminating viruses recovered after inoculation of mice with unrelated materials (Bardos et al., 1980; Weiss, 1983).

In addition to latency, MHV in mice has been subjected to critical analyses of the genetic variation found within the MHV subgroup of strains. The MHV strain 4 (JHM) is highly neurovirulent, strains 2, 3, and A59 are slightly neurovirulent, and strain 1 is not neurovirulent at all. Using T_1-RNAse-resistant oligonucleotide mapping, Wege et al. (1981) found that the nucleotide sequence of JHM differed significantly from that of the other four strains. Lai et al. (1983) found that the oligonucleotide fingerprints of MHV strains causing bone marrow necrosis and diarrhea were similar to those of the A59 strain, but were distinct from the fingerprints of other MHV strains, and suggested that this sequence divergence occurred naturally during persistent infection. Furthermore, mutations in the E2 peplomer antigen selected by monoclonal antibodies reveal that certain epitopes on E2 are the site of neurovirulence in MHV-4 (Dalziel et al., 1986). "Virulence groups" have also been shown with TGEV (Kodama et al., 1981), but as yet no similar data are available for HCV.

Some temperature-sensitive mutants of MHV-A59 and MHV-JHM have been made and used to study different aspects of the replication of the murine coronaviruses (Robb and Bond, 1979). Leibowitz et al. (1982a) found seven overlapping complementation groups in their genetic analysis of 37 temperature-sensitive (*ts*) mutants of JHM. Plaque-size variants and *ts* mutants of A59 were recovered from persistently infected Balb/c-3T3 cells (Holmes and Behnke, 1981). Also using *ts* mutants of A59, Makino et al. (1986) and Keck et al. (1988) showed that the genomes of different MHV strains could recombine at a high frequency, with five distinct recombination sites identified between the parent strains.

Reactions to Physical and Chemical Agents

A study by Kaye et al. (1970) showed that HCV OC43 could be inactivated by heating at 56°C for 30 min, a procedure that did not reduce its hemagglutinin activity. Treatment with trypsin, ether, Tween-80, sodium desoxycholate, sodium lauryl sulfate, *b*-propriolactone, and Triton X-100 also abolished infectivity and caused a reduction in hemagglutinin activity according to the time of incubation (Kaye et al., 1970; Schmidt and Kenny, 1981; Zakstelskaya et al., 1972a). Bucknall et al. (1972b) studied the kinetics of inactivation of strain 229E and a strain of OC43 that had been adapted to monkey kidney cells by continuous passage. They showed that in the presence of 0.2% bovine plasma albumin, strain 229E was more labile. When the rate of ultraviolet inactivation for either virus was determined in the presence of 2% fetal calf serum, multihit infectivity inactivation kinetics were observed due to the capacity of the virions to clump. However, in the presence of 0.2% bovine plasma albumin, one-hit kinetics similar to those obtained with influenza A were obtained.

Human coronavirus 229E is similar to OC43 in its sensitivity to lipid solvents and detergents (Schmidt and Kenny, 1981; Yaseen and Johnson-Lussenburg, 1981). Hierholzer (1976) showed that treatment of strain 229E with chloroform (5% for 10 min), heat (50°C for 1 h at pH 7.0 in the presence or absence of 1 M $MgCl_2$), or acid (pH 3, 4 h, 23°C) completely abolished infectivity. In the same study, depending on the multiplicity of infection, peak viral titers were obtained by 22 to 24 h postinfection during growth in human embryonic lung diploid fibroblast (HELF) cultures at 35°C. Thereafter, a precipitous decline in infectivity (4 to 5 log_{10} $TCID_{50}$) occurred, which coin-

cided with the development of an extensive cytopathic effect (CPE). Similar growth curves have been reported in L132 and HeLa cells (Bradburne, 1972; Bradburne and Tyrrell, 1969; Chaloner-Larsson and Johnson-Lussenburg, 1981), in WI38 cells (Hamre et al., 1967), and in FT and RD cells (Schmidt et al., 1979), except that the decline in infectious viral titers that peaked by 1 day of growth was less when lower incubation temperatures (namely, 33°C) were used. Strain 229E is therefore a relatively difficult virus to passage, which may explain why few diagnostic laboratories report its isolation in cell culture.

Pathogenesis, Pathophysiology, and Pathology

In early morphogenetic studies with OC43, the human respiratory coronaviruses were thought to have a clear tropism for ciliated epithelial cells because of their effect on these cells in tracheal organ cultures and their failure to grow in standard cell monolayers. Their cell tropism and their growth at 33°C indicated that they replicate in the surface layer of the upper respiratory tract. In in vivo studies, 229E and OC43 established superficial infection in nasal tissue within 3 or 4 days after inoculation of human volunteers with nasal drops; the infection was documented by cytoplasmic immunofluorescence in the cells obtained by nasal wash and serologic evaluation (McIntosh et al., 1978). Other volunteer studies have shown production of high titers of B814 virus (Tyrrell and Bynoe, 1965), 229E virus (Callow, 1985; Kraaijeveld et al., 1980b; Larson et al., 1980), and OC43 virus (Larson et al., 1980) in nasal secretions within 2 to 5 days of inoculation by intranasal drops. These volunteers also produced measurable antibody responses to whole virus (Bradburne and Somerset, 1972; Kraaijeveld et al., 1980b; Macnaughton et al., 1981b; Reed, 1984) and to the P, N, and M antigenic components (Macnaughton et al., 1981a). In addition, a specific IgA humoral antibody response was shown in nasal secretions from 229E-infected volunteers (Callow, 1985).

In HECV infection, CVLPs were abundant in ileum epithelial cells obtained at autopsy from an infant with enteritis (Rettig and Altshuler, 1985). Another report described biopsies of jejunal tissue obtained during a case of chronic tropical sprue, in which the cytopathologic findings consisted of viral-filled vesicles similar to coronavirus vesicles found in experimentally infected animals (Mathan and Mathan, 1978). These vesicles are one of the prominent features of intestinal coronavirus infection in birds and mammals (Bridger et al., 1978; Ducatelle and Hoorens 1984; Haelterman, 1972; Hierholzer et al.,

1979; Mebus et al., 1973; Mengeling et al., 1972; Pomeroy et al., 1978).

The pathologic descriptions of coronaviral infections in animals are extensive because coronaviruses infect such a diversity of organ systems in these species (see Table 3). Coronaviral pathogenesis has been particularly well studied in laboratory, domestic, and food-source animals because of the great economic impact of these infections. (Boyle et al., 1987; Dalziel et al., 1986; Ducatelle and Hoorens, 1984; Fishman et al., 1985; Fleming et al., 1986; Keljo et al., 1987; Nguyen et al., 1986; Pearson and Mims, 1985; Reynolds et al., 1985; Robb and Bond, 1979; Siddell et al., 1983; Sorensen et al., 1984; Van Berlo et al., 1986; Wege et al., 1982; Wilson et al., 1986).

Clinical Features

MODE OF TRANSMISSION

HCV is transmitted by the respiratory route via aerosols, droplets, and probably fomites. As discussed in the previous section, experimentally induced infection in volunteers was achieved by inoculating B814, 229E, and OC43 strains into the nose in the form of filtered drops. HECV is presumably transmitted by the fecal-oral route. Animal coronaviruses are infectious by the fecal-oral route and probably also by the respiratory route, except for the enteric coronaviruses (CVLPs). Because of the difficulty in recovering HCV or HECV in cell culture, no definitive transmission data are available. However, the epidemiologic behavior of HCV in the few outbreaks studied suggests that HCV is transmitted by aerosol and by large droplet, thus allowing rapid spread in the community (Cavallaro and Monto, 1970; Kaye and Dowdle, 1975; Macnaughton, 1982; Monto and Lim, 1974).

Animal reservoirs or vectors do not appear to have a role in HCV or HECV transmission because most coronaviruses are species-specific in their natural environments. The laboratory-induced infections of certain animals with coronaviruses from unrelated animal species are noted earlier in this chapter.

INCUBATION AND INFECTIOUS PERIODS

The incubation period of coronavirus colds has been studied in volunteers and ranges from 2 to 4 days. Virus is shed from the time symptoms begin and for 1 to 4 days afterward. In one study, some volunteers shed 229E virus longer than 5 days, which appeared to be correlated with decreased nasal IgA production (Callow, 1985). In another study, some children with recurrent upper and lower respiratory illness appeared to shed 229E or OC43 for up to several months (Macnaughton et al., 1983). HECV is appar-

ently excreted in the stool for several months at a time, with little correlation with illness (Chany et al., 1982; Gerna et al., 1984; Maass and Baumeister, 1983; Mathan et al., 1975; Moore et al., 1977; Mortensen et al., 1985; Schnagl et al., 1986; Sitbon, 1985; Vaucher et al., 1982).

Symptoms, Signs, and Clinical Course

Human coronavirus is associated with upper respiratory illness and common colds of mild to moderately severe intensity like those typical of rhinoviruses. Experimentally induced B814 colds were of similar intensity as those caused by 229E in human volunteers, and the same appears to be true in natural infections (Bradburne and Somerset, 1972; McIntosh et al., 1978; Tyrrell and Bynoe, 1965). Studies of human volunteers, peaks of cases during community outbreaks, and serodiagnosed individual infections, all showed coryza, rhinitis, and profuse watery nasal discharge as the prominent findings (Bradburne and Somerset, 1972; Hendley et al., 1972; Kraaijeveld et al., 1980b; McIntosh et al., 1978; Reed, 1984). In serologic studies, approximately 30% of patients are asymptomatic; the symptomatic patients report coryza (as much as 100%), sore throat or pharyngitis, cough or wheezing, headache, fever, cervical adenitis, and gastrointestinal symptoms (abdominal pain or diarrhea), in decreasing order (Cavallaro and Monto, 1970; Kapikian et al., 1969; Kaye and Dowdle, 1975; Larson et al., 1980). Adults reported more headache than children. Otherwise, the prevalence of symptoms was the same, regardless of age.

For 229E, the incidence of lower respiratory illness (including croup, bronchitis, and pneumonia) ranges from very rare in infants to 50% in preschool children to scattered in adults (Isaacs et al., 1983; McIntosh et al., 1970a,b, 1974; Wenzel et al., 1974). Tonsillitis and otitis media also have been reported occasionally. Persons with OC43-related illnesses have considerably more cough and sore throat and a higher incidence of lower respiratory symptoms than do those with 229E infections (Kaye et al., 1971; McIntosh et al., 1974; Riski and Hovi, 1980; Wenzel et al., 1974). As with 229E, headache, fever, diarrhea, and other symptoms are reported occasionally.

In volunteers, HCV colds last from 6 to 8 days and occasionally as long as 20 days, with generally no symptoms other than coryza and sore throat. Therefore, the symptoms reported in the serologic studies cited above must be viewed with caution; these studies are limited by the serologic methods used and by the lack of evidence that dual viral infections might be causing the additional symptoms.

Human enteric coronavirus appears to be related to human gastrointestinal disease, but not to respiratory or central nervous system disease. It has been associated with gastroenteritis, diarrhea, and necrotizing enterocolitis (see Table 2), and in some studies with vomiting and fever. However, because of the extraordinary difficulties in identifying CVLPs as HECV, developing specific serologic assays, or cultivating the virus, any association of coronavirus with human enteritis must remain an open possibility at best.

Complications

Complications from the human respiratory coronaviruses have not been proved. This finding is significant because they have been actively sought based on animal disease studies. The serologic studies that implicate HCV, especially OC43, in scattered cases of pneumonia, encephalitis, gastroenteritis, and the like have not been confirmed by isolation or antigen detection and are subject to other interpretations.

Diagnosis

Despite recent advances in enzyme immunoassay and other technologies, the laboratory diagnosis of

HCV or HECV infections is still developing. This predicament arises from the general inability of laboratories to work with coronaviruses and their lessened interest due to the mild nature of proven human coronaviral disease. Isolation of HCV or HECV from either respiratory or enteric specimens is almost never accomplished and, therefore, is rarely attempted. The time-honored serologic tests are not very sensitive, although they have been used to generate a base of epidemiologic data. The procedures reported for direct detection of virus and viral antigens and for viral isolation and serology are described here in detail (Table 5).

Specimen Collection

As with any virus, the type of specimen and the manner of collection and storage depend on the laboratory methods anticipated. Nasal swabs are the easiest specimens to collect for respiratory viruses and are also the best specimens for the respiratory coronaviruses. Comparative data in one study on direct detection by enzyme immunoassay (EIA) showed that 34% of nasal swabs, 18% of throat swabs, and 18% of nasopharyngeal aspirates were positive for HCV, mostly 229E (Macnaughton et al., 1983). For nasal swabs, urogenital calginate swabs are inserted

TABLE 5. Laboratory diagnosis of human coronavirus infections

Concept	Test method	Nasal specimens (HCV)[a]				Stool HECV
		B814	229E	OC43	692	
Direct viral detection						
Electron (EM) and immunoelectron (IEM) microscopy					−	+
Direct antigen detection						
Indirect fluorescent antibody (IFA)			+	+		
Enzyme immunoassay (EIA)			+	+		
Viral isolation in organ cultures						
Identified by cilia, interference, EM/IEM		+	+	+	+	+
Identified by IFA, SRH, immunoblot						+
Identified by CF, SN			+			
Viral isolation in cell cultures						
Identified by EM/IEM, CPE, CF, SN		+	+			
Serology (IgG)						
Immunoelectron microscopy (IEM)						+
Serum neutralization (SN)			+	+		
Complement fixation (CF)			+	+		
Hemagglutination inhibition (HI)				+		
Indirect hemagglutination (IHA)			+			
Single radial hemolysis (SRH or HIG)			+	+		
Indirect fluorescent antibody (IFA)			+	+		+
Immune-adherence hemagglutination (IAHA)			+			
Solid-phase radioimmunoassay (RIA)				+		
Immunoperoxidase assay (IPA)			+	+		
Enzyme immunoassay (EIA or ELISA)			+	+		

[a] HCV = Human coronavirus; HECV = human enteric coronavirus.

into the nasal passages, gently rotated to absorb mucus and cells, and then vigorously twirled into 2 ml of transport medium (such as tryptose phosphate broth with 0.5% gelatin, veal heart infusion broth, or Trypticase soy broth [BBL Microbiology Systems, Cockeysville, Md.]), with or without antibiotics. Throat swabs can be obtained with cotton-tipped wooden applicators in the usual fashion. Nasopharyngeal aspirates are collected with a neonatal mucus extractor and mucus trap to which transport medium is added (Isaacs et al., 1983). Because of the lability of coronaviruses, specimens for indirect fluorescent antibody (IFA) tests or culture should be placed on wet ice and transported to the laboratory for immediate testing; specimens for other procedures may be frozen on dry ice and stored in ultracold (less than −60°C) freezers until testing.

Nasal washings may be preferred for organ culture (Larson et al., 1980) and are clearly preferred for IFA to obtain a suitable number of intact epithelial cells. Nasal washings can be obtained by washing the nasal cavity with 10 to 15 ml of normal saline and then mixing the wash with an equal volume of nutrient broth, or by instilling up to 10 ml of PBS or normal saline in the nostrils and collecting the expelled fluids (Kapikian et al., 1973; McIntosh et al., 1978). Specimens should be collected within 2 days of onset of symptoms. Fluids are kept cold and quickly processed for IFA by low-speed centrifugation (1,000 × g, 4°C, 10 min), dropping the cells resuspended in phosphate-buffered saline (PBS) onto slides, and fixing in cold acetone in usual fashion.

Nasal swabs are the preferred specimen for isolation of HCV in organ or cell cultures, although nasopharyngeal aspirates or washings have been used. Nasal swabs and washes stored at −60°C were the source of B814 and OC43 (McIntosh et al., 1967b; Tyrrell and Bynoe, 1965). Nasal washes yielded several strains of 229E (Larson et al., 1980; Reed, 1984). As described previously, all specimens for viral isolation must be collected early in the illness, kept cold between collection and storage, and stored at low temperatures. For processing for viral isolation, the specimens are thawed, treated with antibiotics, spun lightly to remove cell debris and bacteria, and inoculated onto the organ cultures or cell monolayers. Inoculum volumes of 0.2 to 0.5 ml per tube are used and adsorbed onto the cells for 1 h at ambient temperature before a fortified maintenance medium is added. Cultures are best incubated at 33°C on roller or rocker platforms. Under optimal conditions, freshly collected specimens should be inoculated as soon as possible to reduce loss during storage; however, "bedside inoculation" has not been reported for HCV, probably because the patients are rarely hospitalized.

Stool specimens for the direct detection of HECV or for viral isolation must be collected within 2 days

of onset of abdominal symptoms. Specimens are handled as described previously between collection and processing. For processing, a 10 to 20% suspension is made in distilled water, PBS, Eagle minimal essential medium (MEM), or similar medium, shaken vigorously, and clarified by intermediate-speed centrifugation (2,000 to 4,000 × g, 4°C, 30 min). The extract may be used as is for electron or immunoelectron microscopy or can be treated with antibiotics for attempts at culture (Maass and Baumeister, 1983; Sitbon, 1985; Vaucher et al., 1982).

Serum specimens for serologic assays are collected during the acute and convalescent phases of illness in usual fashion. A 3-week interval between serum collections is preferred to assure peak levels of convalescent antibody.

No special safety precautions are required for HCV procedures. Coronaviruses are Class II agents, causing minimal disease, and thus only commonsense safeguards are indicated. Clinical specimens of unknown etiology, however, should routinely be processed under a laminar flow hood by persons wearing gloves and mask and being careful to avoid autoinoculation via aerosolized droplets or finger-to-eye or mouth transmission. Serum specimens should be presumed to contain hepatitis B virus, so that appropriate safety precautions for this very stable pathogen will be taken automatically.

Direct Viral Detection

Coronaviruses have been consistently sought by electron microscopic (EM) observation of stool specimens or tissue sections and by immune electron microscopy (IEM) of specimens reacted with antibody. Both are direct tests, involving no culture or excessive manipulations. Successful EM and IEM have resulted both from the large size of the virion, which enables a threshold of 4 or 5 logs of virus to be visible, and from the large numbers of particles often present in HECV-positive stools or in tissue sections containing virus-packed vesicles.

Direct EM of clarified 10 to 20% stool extracts or of gently homogenized organ culture tissue fragments is conducted by negative staining (see Table 2). The technique is variable, but in general consists of staining with 1.6 to 3% potassium phosphotungstic acid (PTA) at pH 6, 6.5, or 7 and placing approximately 10 μl on 300- or 400-mesh carbon–Formvar-coated copper grids (Chany et al., 1982; Hierholzer et al., 1979; Maass and Baumeister, 1983; Mortensen et al., 1985; Tyrrell and Almeida, 1967; Vaucher et al., 1982). Grids can be loaded either by the droplet method or by the agar diffusion pseudoreplica technique. Additional variations include ammonium molybdate as stain (Schnagl et al., 1978) and carbon–collodion-coated copper grids with 4% PTA, pH 6.5 (Sitbon, 1985).

Alternatively, the HECV in clarified stool extracts can be pelleted by ultracentrifugation (50,000 × g, 2 h), transferred to grids, and negatively stained with 1.5% PTA, pH 6.5 (Caul and Egglestone, 1977). Virus in tissue culture harvests and in organ culture supernatants can be examined after concentration in the same way. In one study, 229E-infected organ culture supernatants were clarified and the virus pelleted at 100,000 × g for 1 h, resuspended in PBS, and applied to parlodion–carbon grids for negative staining (Larson et al., 1980).

In addition, HCV and HECV can be directly sought by thin-section EM on organ culture explants and infected tissues. In one study, sections were prepared for EM by fixing with 2.5% glutaraldehyde in 0.1 M cacodylate buffer and then postfixing for 1 h with 1% osmium tetroxide in the same buffer (Caul and Egglestone, 1977). The tissues were then dehydrated in a graded ethanol series and embedded in araldite for thin sectioning. The ultrathin sections were finally stained with uranyl acetate and lead citrate.

Immune electron microscopy enhances EM by aggregating viral particles with specific antibody. For HCV organ culture harvests, IEM was conducted by mixing 0.1 ml of harvest with 0.1 ml of a 1 : 20 convalescent serum dilution, incubating at ambient temperature, and pelleting the virus–antibody complex at 15,000 × g for 2 h. The pellet was resuspended in 0.1 ml of distilled water and stained with 3% PTA, pH 7.2, on 400-mesh carbon–Formvar-coated grids (Kapikian et al., 1973). For HECV stool extracts, IEM has been carried out by mixing 1 drop of extract with 1 drop of heat-inactivated (56°C, 30 min) convalescent serum at low dilutions (1 : 1 to 1 : 10) in PBS, usually at 23°C for 1 h. Grids are prepared and stained as for direct EM (Mortensen et al., 1985; Resta et al., 1985; Schnagl et al., 1986; Vaucher et al., 1982). Alternatively, HECV was sought by IEM after the virus–antibody complex was concentrated by ultracentrifugation (Chany et al., 1982; Maass and Baumeister, 1983). A generally applicable IEM technique is to incubate the virus with a 1 : 15 dilution of heat-inactivated antibody at 37°C for 2 h, followed by overnight incubation at 4°C and examination by pseudoreplica technique. The viral sample is dropped onto a 2% agarose block and allowed to air-dry, Formvar is added to the surface, and the film is floated off into the stain (2.5% PTA, pH 6.5) and onto a 400-mesh copper grid. Serum controls for IEM should be human or animal antisera to other agents.

Direct Antigen Detection

Detection of coronavirus directly in respiratory specimens has been conducted by indirect fluorescent-antibody (IFA) and enzyme immunoassay (EIA or ELISA). In the IFA study (McIntosh et al., 1978), nasal washings or NPS specimens were subjected to low-speed centrifugation, and the epithelial cells resuspended in a small volume of PBS for dropping onto glass slides. The slides were air-dried, fixed with cold acetone, and reacted with OC43 and 229E rabbit antisera that had been absorbed with normal host cells. After suitable incubation and washing, fluorescein-conjugated anti-rabbit globulin was added as the indicator. The slides were again washed and then read in usual fashion. Proper controls, such as cultures of other common respiratory viruses and normal host cells, showed that the coronavirus sera were specific. For both viruses, nasal epithelial cells exhibited bright green, particulate fluorescence in the cytoplasm. Some cross-reaction was observed between OC43 virus and 229E antiserum, but was not judged to be a significant problem.

In the EIA study (Macnaughton et al., 1983), purified HCV 229E and HECV CV-Paris (used as a cross-reactive antigen for OC43) were prepared to format the test system and evaluate type-specific rabbit antisera. Nasal specimens were collected and immediately frozen on dry ice for transport to the laboratory, where they were stored at −70°C until tested. For the EIA, specimens were diluted in carbonate buffer and adsorbed to polystyrene plates in 0.2 ml volumes by overnight incubation at ambient temperature. The plates were washed four times in a PBS/0.05% Tween 20/0.02% azide buffer, and 0.2 ml of type-specific rabbit antisera were added to respective wells. After a 4 h incubation at ambient temperature, the plates were again washed four times, and alkaline phosphatase-conjugated, anti-rabbit IgG was added at 0.2 ml per well. The plates were incubated overnight at ambient temperature, washed as before, reacted with 0.2 ml of substrate (0.1% p-nitrophenylphosphate in 10% diethanolamine buffer/0.02% azide/0.01% $MgCl_2$, pH 9.8), and finally read for absorbance at 405 nm. A 1 : 200 dilution of nasal swab and a 1 : 20 dilution of rabbit antiserum provided the best combination for both viral types. This direct detection EIA was subsequently applied to a prospective epidemiologic study with good results (Isaacs et al., 1983), and will probably by very useful in future studies because of the increased popularity and sensitivity of EIA.

Direct Viral Nucleic Acid Detection·

Methods for the direct detection of viral nucleic acids have been described recently for some viruses and mycoplasmas and may become a major tool in rapid diagnosis. Biotinylated RNA probes, in particular, may be able to detect small amounts of virus in clinical specimens. To date, however, these procedures have not been applied to HCV or HECV.

TABLE 6. Growth of human coronavirus and human enteric coronavirus in tissue and cell cultures

Cell type	HCV				HECV
	B814	229E	OC43	692	
For direct isolation					
Human embryonic nasal, tracheal organ culture	+	+	+	+	
Human embryonic intestinal organ culture					+
Primary, secondary human embryonic kidney (HEK)	−	+	−	−	−
Human embryonic lung diploid fibroblast (WI38)	−	+	−	−	−
Continuous human embryonic lung diploid fibroblast (MRC-c)		+			
Continuous human embryonic lung epithelium (LI32)	+	+			−
Human embryonic intestine diploid fibroblast (MA177)		+		−	−
For adaptation					
Suckling mouse brain (SMB)	−	−	+	−	−
Human embryonic lung diploid fibroblast (WI38, HEL, WD, RU-1, MRC-5, HELF, MA-321)	−	+	±	−	−
Human embryonic intestine diploid fibroblast (MA177)	−	+	−	−	−
Continuous human embryonic lung diploid fibroblast (MRC-c)		+	+		
Continuous human embryonic lung epithelium (LI32)	+	+	−	−	−
Human fetal tonsil diploid fibroblast (FT)	+	+	+	−	−
Human embryonal heteroploid rhabdomyosarcoma (RD)	+	+	+	−	−
Primary rhesus, African green monkey kidney (MK)	−	−	+	−	−
Continuous green monkey kidney epithelioid (BSC-1)	−	±	+	−	−
Rhesus monkey kidney epithelial line (LLC-MK$_2$)			±		
Human peritoneal macrophages		+			

[a] HCV = Human coronavirus; HECV = human enteric coronavirus.

Viral Isolation, Adaptation to Cell Cultures, and Identification

Only one of the human coronaviruses—either the respiratory HCV or the gastrointestinal HECV—can be propagated in cell cultures in the usual fashion (Table 6). As discussed previously, the HCV 229E group of strains can be isolated and serially passaged in human cell cultures and has therefore been used in genetic and biophysical studies. The other HCVs have been isolated only in organ cultures; some have been adapted to cell monolayers with varying degrees of success. Many attempts to isolate the HECV have resulted in putative growth, at least for a few subpassages, in human embryonic intestinal organ culture. It is clear, then, that standard viral isolation procedures are not as relevant to HCV and HECV studies as are more specialized procedures.

Originally HCV B814 was isolated in organ cultures prepared from the tracheas of 14- to 22-week human embryos (Tyrrell and Bynoe, 1965). Tissue fragments were planted in plastic dishes, ciliated side up, and immersed in medium 199. The cultures were incubated at 33°C with daily changes of medium. The explants were inoculated with 0.3 ml of nasal specimens, dripped onto the ciliary surface, and maintained for up to 10 days. Viral replication was seen by three methods: cessation of ciliary activity; viral

interference tests with Sendai, echo-11, or parainfluenza-3 challenge viruses; and production of colds in human volunteers. Growth was consistently observed in human tracheal organ cultures, but not in ferret tracheal organ cultures (Tyrrell and Bynoe, 1965). Subsequently, organ cultures were gently homogenized and examined by negative-stain EM for evidence of viral growth (Tyrrell and Almeida, 1967). B814 was later isolated from nasal washes directly in L132 (continuous human embryonic lung epithelial) cells, but with only minimal cytopathology (Bradburne and Tyrrell, 1969). The B814 from L132 cells was adapted to human fetal tonsil diploid fibroblast (FT) and human embryonal heteroploid rhabdomyosarcoma (RD) cells and to plaque assay in these cells (Schmidt et al., 1979).

Organ cultures were also used to isolate six strains of HCV (OC16, 37, 38, 43, 44, and 48) in a study of adult upper respiratory infection in 1966 (McIntosh et al., 1967b). Tracheas were obtained from 5- to 9-month fetuses and stored in cold Hanks balanced salt solution with 10% fetal calf serum for 2 to 48 h. Then, the tracheas were cut into 2- to 3-mm squares and placed in petri dishes with Leibovitz medium–0.2% BSA. After inoculating with 0.2 ml of a nasopharyngeal wash specimen, cultures were incubated at 33°C on a rocker platform with daily changes of medium. Viral replication was evidenced by cessation of ciliary movement and by negative-stain EM. These isolates did not grow in conventional cell cultures, but two of them, OC38 and OC43, were readily adapted to suckling mouse brain (SMB) (McIntosh et al., 1967a). Adaptation occurred on the first passage at 11 to 15 days after intracranial inoculation; encephalitic symptoms developed after progressively shorter times until all mice died within 60 h postinfection.

Subsequently, OC38 and OC43 were adapted to low-level growth in both rhesus and vervet primary monkey kidney (MK) cell monolayers (Bruckova et al., 1970; McIntosh et al., 1970b). Adaptation to rhesus MK was slightly faster from SMB-grown virus (two passages) than from organ culture-grown virus (three passages). The CPE observed was focal and somewhat syncytial, with gradual spread throughout the cell monolayer. From African green MK cells, OC43 was further adapted to MA-321 cells (Gerna et al., 1980). From rhesus or vervet MK cells, both OC38 and OC43 were further adapted to BSC-1 (continuous African green monkey kidney epithelioid) cells. In BSC-1, the CPE appeared on first passage and was less syncytial, but still involved the entire monolayer by 10 to 14 days after inoculation. All cultures were maintained under a fortified medium at 33°C on roller drums. From BSC-1 cultures, OC38 and OC43 were further adapted to FT and RD cells with production of definitive CPE and high infectivity titers (Schmidt and Kenny, 1982; Schmidt et al., 1979). In addition, OC43 was adapted from BSC-1 to LLC-MK$_2$, a rhesus monkey kidney epithelial line (Monto and Rhodes, 1977).

Other strains of OC43-like virus (GI, HO, RO) were isolated in tracheal organ cultures and in human volunteers and subsequently passed in both organ culture and MRC-c continuous human embryonic lung diploid fibroblast line (Larson et al., 1980). These are the most recent isolations of OC43 virus reported.

Plaque assays for OC38 and OC43 virus have been carried out in FT and RD cells (Schmidt and Kenny, 1981, 1982; Schmidt et al., 1979). Virus has also been identified in the cultures by EM, CPE, IFA, neutralization tests, and complement fixation.

Strain 229E was first isolated in secondary human embryonic kidney (HEK) cells and adapted to human embryonic lung diploid fibroblast cells (WD, HEL, WI38). The cytopathic effect was described as slow and stringy after 6 to 10 days of incubation at 33°C on roller drums (Hamre and Procknow, 1966; Hamre et al., 1967). Other strains of 229E have been isolated in MA177 human embryonic intestine diploid fibroblast cell cultures (Kapikian et al., 1969); in WI38, L132, and primary HEK cells and human embryonic tracheal organ cultures (Bradburne, 1969, 1972; Bradburne and Tyrrell, 1969); in WI38 cells

(McIntosh et al., 1974); in human embryonic nasal organ culture (Larson et al., 1980); and in MRC-c cell culture (Reed, 1984). All isolations were made under fortified medium, such as Eagle MEM, Leibovitz, or medium 199, with 2% fetal calf serum, in roller cultures at 33°C. The 229E-like strains isolated by Larson et al. (1980) in organ culture or in volunteers (strains AD, PA, PR, TO, KI) were passed both in organ culture and in MRC-c cells for further study.

Adaptation to other cells has been easily accomplished. Human coronavirus 229E was adapted from WI38 cultures to FT and RD cells (Schmidt and Kenny, 1982; Schmidt et al., 1979); from nasal organ cultures to MRC-c and MRC-5 cells (Larson et al., 1980; Reed, 1984); and from MRC-c cells to human macrophage cells (Patterson and Macnaughton, 1982). Again, incubation at 33°C in roller cultures appeared to be critical.

Plaque assays for 229E have been successful in a number of systems, which have been useful for viral replication studies. Plaque production was reported in various human embryonic lung diploid fibroblast cells (Hamre et al., 1967; Macnaughton et al., 1980), in L132 cells (Bradburne, 1972; Bradburne and Tyrrell, 1969; Chaloner-Larsson and Johnson-Lussenburg, 1981), and in FT and RD cells (Schmidt and Kenny, 1981, 1982; Schmidt et al., 1979). Details of cell concentrations, overlay mediums, incubation times, and stains used for visualization are given in the reports cited.

Virus can be identified in cell culture by EM, CF, and IFA (Hamre et al., 1967; Kapikian et al., 1969; Macnaughton et al., 1980), by plaque reduction neutralization assays (Bradburne, 1972; Bradburne and Tyrrell, 1969; Macnaughton et al., 1980; Schmidt et al., 1979), and by fluorescent focus assay (Macnaughton et al., 1980). In peritoneal macrophage culture, 229E was identified by an infectious center assay read by IFA (Patterson and Macnaughton, 1982). Cytopathic effect was also a valuable indicator of viral growth in most cell systems. As described in the section Antigenic Composition, coronaviruses do not produce excess proteins or soluble antigens during culture, which might amplify an assay.

Strain 692 replicated in human embryonic tracheal organ culture, but without inhibiting ciliary motion, was seen by IEM of the culture sediment when mixed with the patient's convalescent serum (Kapikian et al., 1973). This virus could not be grown in cell cultures used successfully for 229E.

The HECV has been propagated to a limited extent in human embryonic intestinal organ culture (Caul and Clarke, 1975; Caul and Egglestone, 1977; Resta et al., 1985). As with HCV in organ cultures, growth in virus was evidenced by destruction of the villous epithelium and by EM or IEM of organ culture fluids and cell sediments. Intestinal organ cultures are prepared from small (2 × 2 mm) pieces of small intestine from 5- to 8-month fetuses and are maintained under enriched medium, such as Leibovitz L-15 with 0.4% BSA, at pH 6.5 to 6.8. Rocker cultures at 35 to 37°C were optimal. Trypsin at a final concentration of 5 μg/ml appeared to enhance infectivity (Resta et al., 1985). Identification of virus was accomplished by EM, IEM, and IFA (Caul and Clarke, 1975; Resta et al., 1985) and by single radial hemolysis, immunoblot, and Western blot (Resta et al., 1985).

Antibody Assays

Serologic tests have provided all of the clinical and epidemiologic information about HCV. These range from immunodiffusion (ID), immunoelectrophoresis (IE), IEM, and IFA for simple detection of antibody (Schmidt, 1984), to broadly applicable neutralization and CF tests, to highly sensitive EIA tests in various formats. All of these tests presumably measure IgG antibody; no early or IgM serologies have been reported for HCV or HECV. Because the peplomers of HCV interact with animal erythrocytes, HI tests for OC43 and IHA tests for 229E are readily available. Most serologic tests for HCV require the intact virion as antigen because the coronaviruses do not produce soluble antigens during their replication cycle. The various serologic tests used for HCV diagnosis are outlined in Table 5 and described in detail below. For all serologic tests, serum specimens, generally at a 1:4 starting dilution, are routinely heat-inactivated at 56°C for 30 min.

IMMUNE ELECTRON MICROSCOPY

Antibodies to HCV and HECV have been detected in patient sera by IEM, as have viruses in stool specimens. Serologic IEM aims to detect the presence, not the titer, of an antibody that can bind to virus and produce small clumps of viral particles. The procedure involves mixing a 1:30 to 1:100 dilution of "antigen," incubating for a minimum of 2 h, and then examining under the electron microscope. The antigen may consist of stool suspension, tissue homogenate, or cell or organ culture supernatant. Stool extracts are prepared by making an approximating 10% suspension in veal heart infusion broth, PBS, or similar medium, shaking vigorously with glass beads, and then clarifying by centrifugation at 5,000 × g for 30 min (Chany et al., 1982; Gerna et al., 1984; Maass and Baumeister, 1983; Mortensen et al., 1985; Schnagl et al., 1986; Vaucher et al., 1982). This force will not bring down coronaviruses (Kaye et al., 1970). Tissue homogenates are prepared by gently grinding with mortar and pestle or glass homogenizers and clarifying as above. Cell and organ culture

supernatant fluids (the maintenance medium bathing the cells during viral culture) are used as antigen without dilution, after clarifying by centrifugation (Kapikian et al., 1973; Resta et al., 1985). The serum–antigen mixtures are incubated for at least 1 h at ambient temperature; in our laboratory, we prefer 2 h at 37°C and overnight at 4°C. Pseudoreplica grids are then prepared, negatively stained with PTA, and examined by EM as described previously.

SERUM NEUTRALIZATION

The standard serum neutralization (SN) test has been common to all studies of human and animal coronaviruses, both to identify isolates and to measure specific antibody response in the host species and heterotypic antibody responses in other species. In coronaviral research, as with so many other viral groups, the SN test is still regarded as the "gold standard" to which other tests are compared.

For 229E, conventional SN and plaque-reduction SN tests have been used. Antibody titers of 1:8 to 1:64 were considered positive titers in tube cultures of WI38, HEL, MA177, and RU-1 fibroblast cells (Cavallaro and Monto, 1970; Hamre and Procknow, 1966; Kapikian et al., 1969; Kaye et al., 1972; Monto and Lim, 1974; Monto and Rhodes, 1977). These tests used an infectious antigen that was freeze–thawed once and clarified by light centrifugation. After titration for viral endpoint, 30 to 300 $TCID_{50}$s of the virus per 0.1 ml were mixed with an equal volume of serial twofold dilutions of serum and incubated for 1 h at ambient temperature. The mixtures were then inoculated at 0.2 ml per tube, adsorbed at 35°C for 1 h, and overlaid with 1 ml of a fortified maintenance medium. The tests were read for inhibition of CPE after 4 to 10 days of incubation at 33°C on a roller drum. A positive titer was any value of 1:8 or higher; a seroconversion was a fourfold or greater increase in titer between the acute- and convalescent-phase sera.

A similar macro-SN test in HEK cells was described by Miyazaki et al. (1971). In that study, 32 $TCID_{50}$s of virus were mixed with serial serum dilutions and adsorbed onto the cell monolayers for 2 h. After maintenance medium was added, the cultures were incubated at 33°C on roller drums and read at 7 days.

Micro-SN tests for 229E antibody have been described with WI38, MRC-c, FT, and C-16 cells in standard 96-well, flat-bottomed microtiter plates (Callow, 1985; Gerna et al., 1978; Kraaijeveld et al., 1980b; Reed, 1984; Schmidt, 1984; Schmidt and Kenny, 1981). In a typical test in FT cells (Schmidt and Kenny, 1981), 300 $TCID_{50}$s of virus in 0.025 ml were mixed with 0.025 ml of twofold serum dilutions and incubated at ambient temperature for 30 min.

Then, 16,000 FT cells in 0.05 ml of MEM–5% fetal
calf serum growth medium were added per well and
the plates incubated at 33°C in a 2.5% CO_2 atmosphere. The plates were read for CPE and cell staining with crystal violet when the viral controls and
back-titration showed 100 $TCID_{50}$s. In this and similar micro-SN tests, the tests are readable by 5 days
and titers of 1 : 8 to 1 : 128 are commonly obtained.

Serum neutralization tests for OC43 antibodies
were conducted in suckling mouse brain at first, using as antigen a 10% SMB suspension in PBS or
Veronal-buffered diluent, clarified at 600 × g for 20
min (Kaye and Dowdle, 1969). After adaptation to
cell culture, macro-SN tests were carried out in
BSC-1 cell monolayers and read by hemadsorption
after 4 days of incubation on rollers at 33°C (Monto
and Rhodes, 1977). Micro-SN tests in FT cells were
as described previously (Schmidt, 1984; Schmidt and
Kenny, 1981). In all three systems, titers of 1 : 4 to
1 : 128 were commonly obtained.

Plaque-reduction SN tests have been used for serologic studies with 229E and OC43. For 229E, tests
were conducted with L132 cells and endpoints read
as 90% reduction in plaques (Bradburne and Somerset, 1972; Bradburne and Tyrrell, 1969). For OC43,
tests were carried out with MA-321 cells and serum
endpoints read by an immunoperoxidase assay
(Gerna et al., 1980). Serum titers of 1 : 20 to 1 : 2560
were obtained in the OC43 test.

COMPLEMENT FIXATION

As for other viral antibodies, the microtiter complement fixation (CF) test for coronaviruses is highly
reproducible, but it is not sensitive (Kaye et al.,
1969). It was used extensively for earlier epidemiologic studies because both 229E and OC43 virus were
includable in one test, often along with many other
antigens in the so-called respiratory battery (Kapikian et al., 1969; Tyrrell and Bynoe, 1965). The antibody measured by CF appears to be relatively short-
lived, which further reduces the sensitivity of the test
(Cavallaro and Monto, 1970).

In most studies, the antigen for 229E was a crude,
clarified fibroblast cell harvest, prepared by three
freeze–thaw cycles of WI38, HELF, or RU-1 cultures after 3 days of incubation followed by low-
speed centrifugation (Cavallaro and Monto, 1970;
Gerna et al., 1978; Hendley et al., 1972; Kaye et al.,
1972; McIntosh et al., 1970a,b, 1978; Monto and
Lim, 1974; Monto and Rhodes, 1977). In other studies, antigens were prepared in L132 cells (McIntosh
et al., 1974), sometimes followed by purification and
concentration by Sepharose 4B chromatography
(Bradburne and Somerset, 1972), or were prepared
as purified virus from RD cell cultures (Schmidt,
1984; Schmidt and Kenny, 1981). Serum titers of 1 : 8

to 1 : 64 were considered positive titers (Gerna et al.,
1978; Hamre and Procknow, 1966; Kapikian et al.,
1969; Schmidt and Kenny, 1981; Wenzel et al.,
1974).

The antigen for OC43 was a 10 to 20% SMB suspension in veronal-buffered diluent, clarified at
1000 × g for 20 min (Hierholzer and Tannock, 1977;
Hierholzer et al., 1979; Hovi et al., 1979; Kaye and
Dowdle, 1969; Kaye et al., 1971; McIntosh et al.,
1970a,b, 1978; Monto and Lim, 1974; Monto and
Rhodes, 1977; Riski and Hovi, 1980). Alternatively,
Sepharose-purified virus from SMB or virus purified
from RD cells has been used in other studies (Bradburne and Somerset, 1972; Schmidt, 1984; Schmidt
and Kenny, 1981). Serum titers of 1 : 8 to 1 : 128 were
considered positive for past exposure to OC43 virus.

HEMAGGLUTINATION

After OC43 was adapted to SMB—the key to success with this virus—Kaye and Dowdle (1969)
showed that virus grown to high titer in SMB could
agglutinate human, vervet monkey, chicken, rat, and
mouse erythrocytes. This HA activity was different
from that of myxoviruses because elution did not
occur, a finding later confirmed by the failure to detect neuraminidase (Hierholzer et al., 1972; Zakstelskaya et al., 1972a). The property of hemagglutination was immediately utilized in hemagglutination
inhibition (HI) tests for serum antibody (Kaye and
Dowdle, 1969). The HI test was practical, high-titered antigen was easily prepared as a clarified 10%
SMB suspension in PBS, and HI was more sensitive
than CF or SN in detecting seroconversions (Kaye
and Dowdle, 1969; Kaye et al., 1971; Monto and
Lim, 1974).

The standardized microtiter HA/HI procedure
calls for treatment of serum by heat inactivation
only, utilization of 4 HA units of antigen/0.025 ml
and 0.5% chicken red blood cell suspension, and incubation of the test at ambient temperature
(Hierholzer et al., 1969). Virtually all OC43 HI tests
reported have used the SMB antigen with chicken
erythrocytes (Bradburne and Somerset, 1972;
Hendley et al., 1972; Hierholzer and Tannock, 1977;
Hovi et al., 1979; Kaye and Dowdle, 1969; McIntosh
et al., 1974; Monto and Lim, 1974; Monto and
Rhodes, 1977; Reed, 1984; Riski and Estola, 1974;
Riski and Hovi, 1980; Wenzel et al., 1974; Zakstelskaya et al., 1972b). Serum titers commonly run
from 1 : 10 to 1 : 640, but probably only titers above
1 : 20 are positive (Gerna et al., 1980; Hendley et al.,
1972; Hierholzer and Tannock, 1977; Kaye and Dowdle, 1969; Riski and Estola, 1974). In one study,
OC43 HA antigen was prepared as purified, concentrated virus from RD cells and gave serum HI titers
of 1 : 8 to 1 : 64 (Schmidt and Kenny, 1981). Some-

times, human serum specimens have been found to contain an inhibitor (possibly the same high-density lipoprotein factor responsible for false hemadsorption, as described in the section Physicochemical Properties and Morphology) which can be removed by treating the serum with phospholipase C (Gerna et al., 1980). In some systems, this treatment has been necessary to avoid false-positive titers in HI tests and in plaque-reduction SN tests (Gerna et al., 1980; Hovi et al., 1979).

INDIRECT HEMAGGLUTINATION

Human coronavirus 229E grown in cell cultures does not hemagglutinate under any condition and has not been adaptable to SMB despite repeated attempts by many laboratories. However, 229E can sensitize glutaraldehyde-fixed, tannic-acid-treated sheep erythrocytes to form the basis of an indirect hemagglutination (IHA) test (Kaye et al., 1972). The IHA antibodies range from 1:10 to 1:5120 and are significantly higher than the CF or SN antibodies found in the same sera (Gerna et al., 1978; Hierholzer and Tannock, 1977; Kaye and Dowdle, 1975; Kaye et al., 1972).

SINGLE RADIAL HEMOLYSIS

The single radial hemolysis (SRH) test was developed for coronaviruses because it had broad applicability in many types of studies, much like the CF test (Hierholzer and Tannock, 1977). For the SRH test, sheep erythrocytes were washed and stabilized with 0.0073% glutaraldehyde. Then, to use the binding properties of the chromic cation, a 25% erythrocyte suspension was mixed with a high concentration of purified virus in the presence of 0.0016% aged chromic chloride. The reaction was stopped with phosphate–saline, and finally the treated, rewashed cells were mixed with complement and agarose at 45°C to prepare a gel on a microscope slide. The final mix consisted of 1% agarose, 0.1% azide, 5% reconstituted complement, and 0.82% treated cells. Wells 2 mm in diameter were loaded with 5 μl of serum dilution, incubated overnight at 4°C for diffusion of antibody and fixation of complement, and then incubated for 1 day at 37°C for development of zones of hemolysis. The diameter of the hemolytic zone was linearly related to antibody concentration as determined by CF, SN, HI, and IHA serologic tests (Hierholzer and Tannock, 1977). In a separate study, antibody to HECV in infant sera that was quantitated by SRH also appeared to correlate with antibody detected by IEM and Western blot analyses (Resta et al., 1985).

A similar hemolysis-in-gel test was described by Riski et al. (1977). This test also correlated well with HI titers in serosurveys (Hovi et al., 1979; Riski and Hovi, 1980).

INDIRECT FLUORESCENT ANTIBODY TEST

Monto and Rhodes (1977) described serologic indirect fluorescent-antibody (IFA) test for 229E and OC43 antibodies. For 229E, WI38 cells were grown on Leighton tube cover slips and infected with 229E by adsorption of virus to monolayers for 2 h at 37°C. Maintenance medium (MEM–2% fetal calf serum) was added and the cultures incubated at 34°C for 2 to 3 days. Then, the cover slips were rinsed, fixed in acetone, and stored at 4°C until tested. For OC43, LLC-MK$_2$ cells were grown on Leighton cover slips and infected with BSC-1 adapted virus at a high multiplicity. Virus was adsorbed for 2 h at 37°C, maintenance medium (medium 199–1% horse serum) was added, and the cultures were incubated at 34°C for 14 to 16 days, when CPE appeared. The cover slips were then washed and fixed as described above.

For the IFA test, dilutions of human sera were incubated with the cover slips for 1 h at 37°C. The cover slips were washed well, incubated with fluorescein-conjugated anti-human globulin, rewashed, and then counterstained with Eriochrome black. The IFA test was specific for both viruses and was most likely to be positive if the patient's seroconversion was detectable by more than one other test (Monto and Rhodes, 1977).

IMMUNE-ADHERENCE HEMAGGLUTINATION

An immune-adherence hemagglutination (IAHA) test was described by Gerna et al. (1978) as a sensitive test for 229E antibodies. Human serum dilutions were added in 0.025-ml amounts to microtiter V plates. Human coronavirus 229E antigen, prepared in RU-1 or WI38 cells, was added in 0.025 ml amounts and incubated at 37°C for 1 h. Then, 0.025 ml of an optimal dilution of complement, determined as for the CF test, was added and the plates were incubated at 37°C for 40 min, followed by 0.025 ml of 0.3% dithiothreitol in 0.04 M EDTA–Veronal buffer. Finally, 0.025 ml of 0.4% human O erythrocytes were added and the plates were incubated at ambient temperature for 1 h. Hemagglutination patterns were read as nonagglutinated cells (negative) to agglutinated cells (positive). The diluent throughout the test was gelatin–Veronal buffer as used in CF. The IAHA test performed in this manner appeared to be more sensitive than IHA, SN, or CF. Positive titers ranged from 1:16 to 1:512 (Gerna et al., 1978).

RADIOIMMUNOASSAY

A solid-phase radioimmunoassay (RIA) for OC43 antibodies has been described by Hovi et al. (1979). The OC43 virus, diluted in PBS, was adsorbed onto polystyrene beads, 6.4 mm in diameter and at a concentration of 6 μg per bead, during overnight incubation at ambient temperature. After air-drying, the beads were reacted with serum dilutions at 37°C for 1 h. The beads were then washed twice, ^{125}I-labeled anti-human IgG was added for another hour at 37°C, and the beads were finally washed and assayed in a gamma counter. The RIA test was decidedly more sensitive than SRH, HI, or CF (Hovi et al., 1979).

IMMUNOPEROXIDASE ASSAY

Gerna et al. (1979, 1980) described an immunoperoxidase assay (IPA) for OC43 antibodies that was useful in reading plaque assays. The IPA test was conducted on microcultures of primary African green monkey kidney cells or of human embryonic lung diploid fibroblast cells (MA-321). The cell cultures were grown under a fortified medium, but without serum. The cells were fixed with absolute ethanol when the monolayers showed 50% infected cells. Dilutions of human acute- and convalescent-phase sera were added to the fixed cells and incubated at 37°C for 60 min. The cultures were washed three times in PBS and then reacted with peroxidase-conjugated anti-human IgG. The test was developed histochemically. Optimal dilutions of the reagents, including those for the histochemical detection of the peroxidase, were predetermined by titrations of reference mouse immune ascitic fluid to OC43, followed by peroxidase-conjugated sheep anti-mouse IgG serum. In the serologic IPA, positive titers ranged from 1:40 to 1:640, similar to those obtained by HI (Gerna et al., 1980).

ENZYME IMMUNOASSAY

The enzyme immunoassay (EIA) or enzyme-linked immunosorbent assay (ELISA) is the most sensitive and versatile test described to date for HCV. In the procedure described by Kraaijeveld et al. (1980b) for HCV 229E, the viral antigen was grown in MRC-c cells and the test carried out in flat-bottomed, polystyrene microtiter plates. The plates were prewashed four times with Dulbecco PBS, and the wells were then coated with 0.2 ml of optimal antigen dilution in 0.1 M carbonate buffer, pH 9.6, during overnight incubation at ambient temperature. The plates were washed four times with PBST (PBS–0.05% Tween 20–0.02% azide), shaken dry, and used without further treatment. For the EIA test, 0.2 ml of serum dilutions in PBST were added to wells in duplicate and incubated 4 h at ambient temperature. The plates were again washed four times with PBST and shaken dry. Then, 0.2 ml of conjugate (anti-human IgG–alkaline phosphatase) was added per well and incubated overnight at ambient temperature. After another washing series, 0.2 ml of substrate (0.1% sodium *p*-nitrophenyl phosphate–0.01% magnesium chloride–0.02% sodium azide, in 10% diethanolamine buffer, pH 9.8) was added per well. The reaction was developed for 30 min and was then stopped with

0.05 ml of 3 M NaOH per well. Absorbance values were read at 405 nm.

The EIA test for 229E was more sensitive than a micro-SN test in MRC-c cells for detecting infections in volunteers and in patients with common colds (Kraaijeveld et al., 1980b; Macnaughton et al., 1981a, b). The same procedure was applied to OC43 using a 10% SMB antigen (Macnaughton et al., 1981b) or a cross-reacting CV-Paris antigen grown in HRT-18 cells (Macnaughton, 1982). In these studies, the EIA was used to document type-specific infection with 229E and OC43, to confirm the distinction between these two viruses, and to measure the frequency of HCV infections in populations (Macnaughton, 1982; Macnaughton et al., 1981a, b). Other EIA tests have been described using purified, concentrated 229E and OC43 viruses grown in RD cells (Schmidt, 1984; Schmidt et al., 1986) or clarified 229E virus grown in C-16 cells (Callow, 1985). The choice of antigen may be decided by each laboratory, but EIA is the test of choice for HCV serology because it can measure antibody levels to both coronaviruses at the same time and with high sensitivity.

Interpretation of Serologic Data and Limitations of Tests

Serologic tests such as IEM, ID, and IE are useful only to detect the presence or absence of HCV or HECV antibody because the sera must be used at low dilutions (1:1 to 1:15). Other tests (e.g., IFA, CF, SN, and HI) do quantitate the antibody and may be easily performed, but are not sensitive for low levels of antibody. Still others (e.g., plaque-reduction SN, SRH, and EIA) are more complex, but are much more sensitive for serodiagnosis and serosurveys.

Proper controls must be built into all serologic tests. These include known positive and negative sera to check the specificity of the antigen, known positive and negative antigen or cell culture controls to check the sensitivity of the test, and reagent blanks to allow calculation of background values. When the test is performed properly and all controls give the expected values, a serodiagnosis is possible by demonstrating a fourfold or greater rise in antibody titer between the acute- and convalescent-phase sera. It is critical that both sera in a pair be tested at the same time to obviate the day-to-day variation found in any test system.

Due to the ubiquity and reinfection rate of HCV and HECV, paired sera are required for serodiagnosis. The acute-phase serum should be drawn within 7 days of onset of symptoms. The convalescent-phase serum should be drawn approximately 3 to 4 weeks after onset so that peak titers by any test procedure are realized. This is true for all of the serologic tests described above. The high reinfection rate observed with coronaviruses implies that homotypic or heterotypic anamnestic antibody responses might confuse the serodiagnosis of a patient's illness. This possibility does exist, but as yet data are insufficient to evaluate the extent of the problem.

Serosurveys can be carried out on single serum specimens to measure the prevalence of antibody in various populations. These single titers will indicate past (or current) infection, but cannot be interpreted beyond this finding. The relatively low serum titers obtained with the tests used for HCV might suggest that specific HCV antibodies are not long-lived or that HCV does not elicit a dramatic serum antibody response at all. These questions warrant further research, particularly with regard to reinfection throughout life.

The antibody titer of a serum is recorded as the highest serum dilution that inhibits the property or effect of the antigen tested, when the antigen backtitration confirms that a standard dose was employed in the test. The property is different for each test, such as infectious viral dose, fixing complement, agglutinating red blood cells, cytoplasmic fluorescence, or optical density readings. Accordingly, the definition of serum endpoint dilution (titer) is an integral part of and is unique to each test. Definitions of titers and of acceptable background values are found in the test descriptions cited.

The choice of which serologic test or tests to perform is often dependent on what is most efficient for the laboratory. It is obviously preferable to use available tests, if they will suffice, rather than setting up new ones. The other factor in choosing a test is sensitivity. For HCV 229E, the order of sensitivity (from most to least sensitive) is EIA, IAHA, SRH, micro-SN, IHA, tube SN, CF, IFA, IE, ID, and IEM. For HCV OC43, the order of sensitivity is EIA, RIA, plaque-reduction SN, SRH, IPA, micro-SN, HI, tube SN, CF, IFA, IE, ID, and IEM. Results from any test procedure should be interpreted in consideration of the patient's clinical history (i.e., to make a serodiagnosis, the results should be used in conjunction with information available from clinical evaluation, other diagnostic tests, and the epidemiology of the virus).

Epidemiology and Natural History

The respiratory HCV 229E and OC43 are considered from various studies to account for 5 to 35% of infections of the upper respiratory tract and are therefore a major cause of common colds worldwide. Significant levels of specific antibody have been found in all age groups in the United States (Hamre and Beem, 1972; Hendley et al., 1972; Kaye and Dowdle, 1975;

Kaye et al., 1971; McIntosh et al., 1970a,b; Monto and Lim, 1974; Schmidt et al., 1986); Japan (Miyazaki et al., 1971); England (Bradburne and Somerset, 1972; Isaacs et al., 1983; Macnaughton, 1982); Russia (Zakstelskaya et al., 1972b); Finland (Hovi et al., 1979; Riski and Hovi, 1974) Brazil, and Italy (Gerna et al., 1978). Antibody prevalence for both HCV types ranges from 6 to 37% in the less than 1-year-old age group, to 54 to 80% in the 1- to 5-year-old age group, to 100% of persons infected after age 5. Between ages 21 and 50, only 85% of persons tested had antibody, suggesting that fewer reinfections occurred during adulthood.

Seasonal and annual fluctuations in HCV outbreaks were noted in many studies. All documented epidemics of HCV-related upper respiratory illnesses occurred during the winter and spring seasons. Outbreaks with both 229E and OC43 follow a 2- to 3-year cycle, with winter-season peaks of infections occurring annually (Kaye and Dowdle, 1975; Kaye et al., 1971; McIntosh, 1974; Monto, 1974). In one study in a childrens' home, outbreaks of 229E-related illness alternated with those caused by OC43; this observation, however, could have been by chance (Kaye and Dowdle, 1975).

The rate of asymptomatic HCV infection appears to be high, as one might assume from the generally mild nature of HCV-related colds. In one study in which this factor could be measured, 55% of adults infected with 229E, as determined by appropriate serologic tests, reported no symptoms at all (Cavallaro and Monto, 1970). In the volunteer studies described previously, approximately 30% of persons given intranasal drops with 229E or OC43 failed to develop any symptoms, although the virus could generally be demonstrated in their nasal washings and increases in antibody found in their sera. Approximately 50% of children infected with 229E or OC43, as documented by seroconversions, did not have symptoms of respiratory illness (Kaye and Dowdle, 1975; Kaye et al., 1971).

The nonprotective effects of serum antibodies and the apparently high reinfection rates of HCV 229E and OC43 are related to the asymptomatic infection rate (Monto, 1974). Preexisting HI antibody to OC43 was found in one-third of the children who seroconverted to OC43, and this antibody did not appear to play a role in modifying the severity of subsequent OC43-related illness (Kaye et al., 1971). In an identical manner, preexisting IHA antibody was found in one-third of children with 229E seroconversions, and this antibody had no obvious ameliorating effect on the respiratory illness caused by the virus (Kaye and Dowdle, 1975).

Hamre and Beem (1972) showed that the frequency of increases in SN titer to 229E was inversely proportional to preexisting levels of SN antibody, suggesting that this antibody possessed some protective effect. However, in other studies, up to 81% of adults infected with OC43 possessed prior SN antibody (Monto and Lim, 1974), and preexisting SN, CF, HI, or EIA antibodies to 229E or OC43 failed to protect children or adults against homotypic reinfection (Callow, 1985; Cavallaro and Monto, 1970; Hendley et al., 1972; Isaacs et al., 1983; Macnaughton, 1982; Reed, 1984; Schmidt et al., 1986). In fact, it was not uncommon to find two distinct HCV infections in adults within a single year, even in the presence of detectable levels of antibody. Thus, the relationship of circulating SN or other antibodies to modification of HCV infection is not clear at present. In addition, because of the frequency of reinfections by both viruses throughout life, it has not been possible to determine the natural persistence of the different antibodies elicited by HCV infection.

Because HCV infections involve the surface of the upper respiratory tract, it is likely that secretory IgA antibody is important in protection. Callow (1985) has shown that locally produced IgA does indeed protect from 229E infection and also shortens the period of virus shedding. However, this antibody is short-lived and probably has little effect on the natural spread of the respiratory coronavirus in communities or families.

The epidemiology of the HECV is a fertile field. Because the CVLPs have been found in many animal species in addition to humans, and are not clearly related to any disease, little is known beyond their observation by EM and IEM. There are no diagnostic or serologic tests that are specific for HECV and no HECV strains available with which to construct serologic assays.

Prevention and Control

At this time, control of HCV infections is neither possible nor necessary. The respiratory HCVs cause only mild to moderately severe colds in volunteers; the more severe illness seen in a low percentage of patients by serologic tests has not been confirmed by more direct antigen or isolation tests. More serotypes possibly exist, and the importance of HCV 692 has yet to be defined. Furthermore, the frequency of re-infection observed with these agents is high and indicates that IgG antibody is not protective. Environmental control of infection to minimize the spread of virus by droplets and fomites may be useful, as it would for any respiratory virus. However, such control has rarely been practical or possible in school or home settings.

In a study of prophylactic control of HCV (Turner et al., 1986), 55 volunteers were given recombinant interferon intranasally for 15 days and exposed di-

rectly to HCV by intranasal inoculation on the 8th day. In the placebo group, 73% of the volunteers developed colds, compared with 41% in the interferon-treated group. The interferon also reduced the severity of the cold symptoms and shortened the duration of the colds.

The HECV probably will have the greatest need for control once its role in infant gastroenteritis is ascertained. At present, the HECV requires proof of existence, relevance to disease, and comprehensive virology before its role as a pathogen in diarrhea can even be assessed.

Literature Cited

Ackerman, H. W., G. Cherchel, J. P. Valet, J. Matte, S. Moorjani, and R. Higgins. 1974. Expériences sur la nature de particules trouvées dans des cas d'hépatite virale: type coronavirus, antigène Australia et particules de Dane. Can. J. Microbiol. 20:193–203.

Adams, N. R., R. A. Ball, and M. S. Hofstad. 1970. Intestinal lesions in transmissible enteritis of turkeys. Avian Dis. 14:392–399.

Akashi, H., Y. Inaba, Y. Miura, K. Sato, S. Tokuhisa, M. Asagi, and Y. Hayashi. 1981. Propagation of the Kakegawa strain of bovine coronavirus in suckling mice, rats and hamsters. Arch. Virol. 67:367–370.

Almeida, J. D., and D. A. J. Tyrrell. 1967. The morphology of three previously uncharacterized human respiratory viruses that grow in organ culture. J. Gen. Virol. 1:175–178.

Almeida, J. D., and A. P. Waterson. 1970. Some implications of a morphologically oriented classification of viruses. Arch. Gesamte Virusforsch. 32:66–72.

Andries, K., and M. B. Pensaert. 1980. Virus isolation and immunofluorescence in different organs of pigs infected with hemagglutinating encephalomyelitis virus. Am. J. Vet. Res. 41:215–218.

Apostolov, K., P. Spasić, and N. Bojanić. 1975. Evidence of a viral aetiology in endemic (Balkan) nephropathy. Lancet 2:1271–1273.

Bárdoš, V., V. Schwanzer, and J. Peško. 1980. Identification of Tettnang virus ('possible arbovirus') as mouse hepatitis virus. Intervirology 13:275–283.

Bass, E. P., and R. L. Sharpee. 1975. Coronavirus and gastroenteritis in foals. Lancet 2:822.

Battaglia, M., N. Passarani, A. di Matteo, and G. Gerna. 1987. Human enteric coronaviruses: further characterization and immunoblotting of viral proteins. J. Infect Dis 155:140–143.

Beaudette, F. R., and C. B. Hudson. 1937. Cultivation of the virus of infectious bronchitis. J. Am. Vet. Med. Assoc. 90:51–60.

Becker, W. B., K. McIntosh, J. H. Dees, and R. M. Chanock. 1967. Morphogenesis of avian infectious bronchitis virus and a related human virus (strain 229E). J. Virol. 1:1019–1027.

Bhatt, P. N., R. O. Jacoby, and A. M. Jonas. 1977. Respiratory infection in mice with sialodacryoadenitis virus, a coronavirus of rats. Infect. Immun. 18:823–827.

Boyle, J. F., D. G. Weismiller, and K. V. Holmes. 1987. Genetic resistance to mouse hepatitis virus correlates with absence of virus-binding activity on target tissues. J. Virol. 61:185–189.

Bradburne, A. F. 1969. Sensitivity of L132 cells to some "new" respiratory viruses. Nature (London) **221**:85–86.

Bradburne, A. F. 1970. Antigenic relationships amongst coronaviruses. Arch. Gesamte Virusforsch. **31**:352–364.

Bradburne, A. F. 1972. An investigation of the replication of coronaviruses in suspension cultures of L132 cells. Arch. Gesamte Virusforsch. **37**:297–307.

Bradburne, A. F., M. L. Bynoe, and D. A. J. Tyrrell. 1967. Effects of a "new" human respiratory virus in volunteers. Br. Med. J. **3**:767–769.

Bradburne, A. F., and B. A. Somerset. 1972. Coronavirus antibody titers in sera of healthy adults and experimentally infected volunteers. J. Hyg. **70**:235–244.

Bradburne, A. F., and D. A. J. Tyrrell. 1969. The propagation of "coronaviruses" in tissue culture. Arch. Virol. **28**:133–150.

Bradburne, A. F., and D. A. J. Tyrrell. 1971. Coronaviruses of man. Prog. Med. Virol. **13**:373–403.

Brayton, P. R., R. G. Ganges, S. A. Stohlman. 1981. Host cell nuclear function and murine hepatitis virus replication. J. Gen. Virol. **56**:457–460.

Brayton, P. R., M. M. Lai, C. D. Patton, and S. A. Stohlman. 1982. Characterization of two RNA polymerase activities induced by mouse hepatitis virus. J. Virol. **42**:847–853.

Brian, D. A., B. Hogue, W. Lapps, B. Potts, and P. Kapke. 1984. Comparative structure of coronaviruses. *In:* Proc. 4th Intntl Symp on Neonatal Diarrhea, p. 100–115; Vet Inf Dis Org, Saskatoon, Sask., Canada.

Bridger, J. C., E. O. Caul, and S. I. Egglestone. 1978. Replication of an enteric bovine coronavirus in intestinal organ cultures. Arch. Virol. **57**:43–51.

Brown, T. D., M. E. Boursnell, M. M. Binns, and F. M. Tomley. 1986. Cloning and sequencing of 5' terminal sequences from avian infectious bronchitis virus genomic RNA. J. Gen. Virol. **67**:221–228.

Brucková, M., K. McIntosh, A. Z. Kapikian, and R. M. Chanock. 1970. The adaptation of two human coronavirus strains (OC38 and OC43) to growth in cell monolayers. Proc. Soc. Exp. Biol. Med. **135**:431–435.

Bucknall, R. A., A. R. Kalica, and R. M. Chanock. 1972a. Intracellular development and mechanism of hemadsorption of a human coronavirus, OC43. Proc. Soc. Exp. Biol. Med. **139**:811–817.

Bucknall, R. A., L. M. King, A. Z. Kapikian, and R. M. Chanock. 1972b. Studies with human coronaviruses. II. Some properties of strains 229E and OC43. Proc. Soc. Exp. Biol. Med. **139**:722–727.

Budzilowicz, C. J., S. P. Wilczynski, and S. R. Weiss. 1985. Three intergenic regions of coronavirus MHV strain A59 genome RNA contain a common nucleotide sequence that is homologous to the 3' end of the viral mRNA leader sequence. J. Virol. **53**:834–840.

Burks, J. S., B. L. DeVald, J. C. Gerdes, I. T. McNally, and M. C. Kemp. 1984. Failure to detect coronavirus SK antigen in multiple sclerosis brain tissue by autoradiography. Adv. Exp. Med. Biol. **173**:393–394.

Burks, J. S., B. L. DeVald, L. D. Jankovsky, and J. C. Gerdes. 1980. Two coronaviruses isolated from central nervous system tissue of two multiple sclerosis patients. Science **209**:933–934.

Caldera, R., and J. Badoual. 1982. Role of coronaviruses and anaerobic bacteria in necrotizing enterocolitis of the newborn: systematic study in a neonatology unit. Nouv. Presse Med. **11**:1949–1950.

Callow, K. A. 1985. Effect of specific humoral immunity and some non-specific factors on resistance of volunteers to respiratory coronavirus infection. J. Hyg. **95**:173–189.

Caul, E. O., C. R. Ashley, M. Ferguson, and S. I. Egglestone. 1979. Preliminary studies on the isolation of coronavirus 229E nucleocapsids. FEMS Microbiol. Lett. **5**:101–105.

Caul, E. O., and S. K. R. Clarke. 1975. Coronovirus propagated from patient with non-bacterial gastroenteritis. Lancet **2**:953–954.

Caul, E. O., and S. I. Egglestone. 1977. Further studies on human enteric coronaviruses. Arch. Virol. **54**:107–117.

Caul, E. O., and S. I. Egglestone. 1979. Coronavirus-like particles present in simian faeces. Vet. Rec. **104**:168–169.

Caul, E. O., W. K. Paver, and S. K. Clarke. 1975. Coronavirus particles in faeces from patients with gastroenteritis. Lancet **1**:1192.

Cavallaro, J. J., and A. S. Monto. 1970. Community-wide outbreak of infection with a 229E-like coronavirus in Tecumseh, Mich. J. Infect. Dis. **122**:272–279.

Cavanagh, D. 1983a. Coronavirus IBV glycopolypeptides: size of their polypeptide moieties and nature of their oligosaccharides. J. Gen. Virol. **64**:1187–1191.

Cavanagh, D. 1983b. Coronavirus IBV: structural characterization of the spike protein. J. Gen. Virol. **64**:2577–2583.

Cavanagh, D. 1984. Structural characterization of IBV glycoproteins. Adv. Exp. Med. Biol. **173**:95–108.

Cavanagh, D., J. H. Darbyshire, P. Davis, and R. W. Peters. 1984. Induction of humoral neutralizing and haemagglutination-inhibiting antibody by the spike protein of avian infectious bronchitis virus. Avian Pathol. **13**:573–583.

Cavanagh, D., and P. J. Davis. 1986. Coronavirus IBV: removal of spike glycopolypeptide S1 by urea abolishes infectivity and haemagglutination but not attachment to cells. J. Gen. Virol. **67**:1443–1448.

Cavanagh, D., P. J. Davis, J. H. Darbyshire, and R. W. Peters. 1986a. Coronavirus IBV: virus retaining spike glycopolypeptide S2 but not S1 is unable to induce virus-neutralizing or haemagglutination-inhibiting antibody, or induce chicken tracheal protection. J. Gen. Virol. **67**:1435–1442.

Cavanagh, D., P. J. Davis, and D. J. Pappin. 1986b. Coronavirus IBV glycopolypeptides: locational studies using proteases and saponin, a membrane permeabilizer. Virus Res. **4**:145–156.

Cavanagh, D., P. J. Davis, D. J. Pappin, M. M. Binns, M. E. Boursnell, and T. D. Brown. 1986c. Coronavirus IBV: partial amino terminal sequencing of spike polypeptide S2 identifies the sequence Arg-Arg-Phe-Arg-Arg at the cleavage site of the spike precursor propolypeptide of IBV strains Beaudette and M41. Virus Res. **4**:133–143.

Cereda, P. M., L. Pagani, and E. Romero. 1986. Prevalence of antibody to human coronaviruses 229E, OC43 and neonatal calf diarrhea coronavirus (NCDCV) in patients of northern Italy. Eur. J. Epidemiol. **2**:112–117.

Chaloner-Larsson, G., and C. M. Johnson-Lussenburg. 1981. Establishment and maintenance of a persistent infection of L132 cells by human coronavirus strain 229E. Arch. Virol. **69**:117–130.

Chany, C., O. Moscovici, P. Lebon, and S. Rousset. 1982. Association of coronavirus infection with neonatal necrotizing enterocolitis. Pediatrics **69**:209–214.

Cheever, F. S., J. B. Daniels, A. M. Pappenheimer, and O. T. Bailey. 1949. A murine virus (JHM) causing disseminated encephalomyelitis with extensive destruction of myelin. I. Isolation and biological properties of the virus. J. Exp. Med. **90**:181–194.

Childs, J. C., S. A. Stohlman, L. Kingsford, and R. Russell.

1983. Antigenic relationships of murine coronaviruses. Arch. Virol. **78:**81–87.

Collins, A. R., R. L. Knobler, H. Powell, and M. J. Buchmeier. 1982. Monoclonal antibodies to murine hepatitis virus 4 (strain JHM) define the viral glycoprotein responsible for attachment and cell-cell fusion. Virology **119:**358–371.

Dalziel, R. G., P. W. Lampert, P. J. Talbot, and M. J. Buchmeier. 1986. Site-specific alteration of murine hepatitis virus type 4 peplomer glycoprotein E2 results in reduced neurovirulence. J. Virol. **59:**463–471.

Davies, H. A., R. R. Dourmashkin, and M. R. Macnaughton. 1981. Ribonucleoprotein of avian infectious bronchitis virus. J. Gen. Virol. **53:**67–74.

Davies, H. A., and M. R. Macnaughton. 1979. Comparison of the morphology of three coronaviruses. Arch. Virol. **59:**25–33.

Dawson, P. S., and R. E. Gough. 1971. Antigenic variation in strains of avian infectious bronchitis virus. Arch. Virol. **34:**32–39.

Delmas, B., J. Gelfi, and H. Laude. 1986. Antigenic structure of transmissible gastroenteritis virus. II. Domains in the peplomer glycoprotein. J. Gen. Virol. **67:**1405–1418.

Denison, M. R., and S. Perlman. 1986. Translation and processing of mouse hepatitis virus virion RNA in a cell-free system. J. Virol. **60:**12–18.

Dennis, D. E., and D. A. Brian. 1981. Coronavirus cell-associated RNA-dependent RNA polymerase. Adv. Exp. Med. Biol. **142:**155–170.

Descoteaux, J. P., G. Lussier, L. Berthiaume, R. Alain, C. Seguin, and M. Trudel. 1985. An enteric coronavirus of the rabbit: detection by immunoelectron microscopy and identification of structural polypeptides. Arch. Virol. **84:**241–250.

Doyle, L. P., and L. M. Hutchings. 1946. A transmissible gastroenteritis in pigs. J. Am. Vet. Assoc. **108:**257–259.

Ducatelle, R., W. Coussement, M. B. Pensaert, P. Debouck, and J. Hoorens. 1981. In vitro morphogenesis of a new porcine enteric coronavirus, CV777. Arch. Virol. **68:**35–44.

Ducatelle, R., and J. Hoorens. 1984. Significance of lysosomes in the morphogenesis of coronaviruses. Arch. Virol. **79:**1–12.

Estola, T. 1970. Coronaviruses, a new group of animal RNA viruses. Avian Dis. **14:**330–336.

Evans, M. R., and R. W. Simpson. 1980. The coronavirus avian infectious bronchitis virus requires the cell nucleus and host transcriptional factors. Virology **105:**582–591.

Evermann, J. F., L. Baumgartener, R. L. Ott, E. V. Davis, and A. J. McKeirnan. 1981. Characterization of a feline infectious peritonitis virus isolate. Vet. Pathol. **18:**256–265.

Fishman, P. S., J. S. Gass, P. T. Swoveland, E. Lavi, M. K. Highkin, and S. R. Weiss. 1985. Infection of the basal ganglia by a murine coronavirus. Science **229:**877–879.

Fleming, J. O., S. A. Stohlman, R. C. Harmon, M. M. C. Lai, J. A. Frelinger, and L. P. Weiner. 1983. Antigenic relationships of murine coronaviruses: analysis using monoclonal antibodies to JHM (MHV-4) virus. Virology **131:**296–307.

Fleming, J. O., M. D. Trousdale, F. A. K. El-Zaatari, S. A. Stohlman, and L. P. Weiner. 1986. Pathogenicity of antigenic variants of murine coronavirus JHM selected with monoclonal antibodies. J. Virol. **58:**869–875.

Frana, M. F., J. N. Behnke, L. S. Sturman, and K. V. Holmes. 1985. Proteolytic cleavage of the E2 glycoprotein of murine coronavirus: host-dependent differences in proteolytic cleavage and cell fusion. J. Virol. **56:**912–920.

Garwes, D. J., and D. J. Reynolds. 1981. The polypeptide structure of canine coronavirus and its relationship to porcine transmissible gastroenteritis virus. J. Gen. Virol. **52:**153–158.

Georgescu, L., P. Diosi, I. Butiu, L. Plavosin, and G. Herzog. 1978. Porcine coronavirus antibodies in endemic (Balkan) nephropathy. Lancet **1:**163.

Gerdes, J. C., L. D. Jankovsky, B. L. de Vald, I. Klein, and J. S. Burks. 1981a. Antigenic relationships of coronaviruses detectable by plaque neutralization, competitive enzyme linked immunosorbent assay, and immunoprecipitation. Adv. Exp. Med. Biol. **142:**29–41.

Gerdes, J. C., I. Klein, B. L. Devald, and J. S. Burks. 1981b. Coronavirus isolates SK and SD from multiple sclerosis patients are serologically related to murine coronaviruses A59 and JHM and human coronavirus OC43 but not to human coronavirus 229E. J. Virol. **38:**231–238.

Gerna, G., G. Achilli, E. Cattaneo, and P. Cereda. 1978. Determination of coronavirus 229E antibody by an immune-adherence hemagglutination method. J. Med. Virol. **2:**215–224.

Gerna, G., M. Battaglia, P. M. Cereda, and N. Passarani. 1982. Reactivity of human coronavirus OC43 and neonatal calf diarrhoea coronavirus membrane-associated antigens. J. Gen. Virol. **60:**385–390.

Gerna, G., E. Cattaneo, P. M. Cereda, M. G. Revelo, and G. Achilli. 1980. Human coronavirus OC43 serum inhibitor and neutralizing antibody by a new plaque-reduction assay. Proc. Soc. Exp. Biol. Med. **163:**360–366.

Gerna, G., P. M. Cereda, M. G. Revello, E. Cattaneo, M. Battaglia, and M. T. Gerna. 1981. Antigenic and biological relationships between human coronavirus OC43 and neonatal calf diarrhoea coronavirus. J. Gen. Virol. **54:**91–102.

Gerna, G., P. M. Cereda, M. G. Revello, M. Torsellini-Gerna, and J. Costa. 1979. A rapid microneutralization test for antibody determination and serodiagnosis of human coronavirus OC43 infections. Microbiologica **2:**331–344.

Gerna, G., N. Passarini, M. Battaglia, and E. G. Rondanelli. 1985. Human enteric coronaviruses: antigenic relatedness to human coronavirus OC43 and possible etiologic role in viral gastroenteritis. J. Infect. Dis. **151:**796–803.

Gerna, G., N. Passarani, P. M. Cereda, and M. Battaglia. 1984. Antigenic relatedness of human enteric coronavirus strains to human coronavirus OC43: a preliminary report. J. Infect. Dis. **150:**618–619.

Greig, A. S., D. Mitchell, A. H. Corner, G. L. Bannister, E. B. Meads, and R. J. Julian. 1962. A hemagglutinating virus producing encephalomyelitis in baby pigs. Can. J. Comp. Med. **26:**49–56.

Haelterman, E. O. 1972. On the pathogenesis of transmissible gastroenteritis of swine. J. Am. Vet. Med. Assoc. **160:**534–540.

Hamre, D., and M. Beem. 1972. Virologic studies of acute respiratory disease in young adults. V. Coronavirus 229E infections during six years of surveillance. Am. J. Epidemiol. **96:**94–106.

Hamre, D., D. A. Kindig, and J. Mann. 1967. Growth and intracellular development of a new respiratory virus. J. Virol. **1:**810–816.

Hamre, D., and J. J. Procknow. 1966. A new virus isolated from the human respiratory tract. Proc. Soc. Exp. Biol. Med. **121:**190–193.

Hartley, J. W., W. P. Rowe, H. H. Bloom, and H. C. Turner. 1964. Antibodies to mouse hepatitis viruses in human sera. Proc. Soc. Exp. Biol. Med. **115**:414–418.

Hasony, H. J., and M. R. Macnaughton. 1982. Serological relationships of the subcomponents of human coronavirus strain 229E and mouse hepatitis virus strain 3. J. Gen. Virol. **58**:449–452.

Hendley, J. O., H. B. Fishburne, and J. M. Gwaltney. 1972. Coronavirus infections in working adults: eight-year study with 229E and OC43. Am. Rev. Resp. Dis. **105**:805–811.

Hierholzer, J. C. 1976. Purification and biophysical properties of human coronavirus 229E. Virology **75**:155–165.

Hierholzer, J. C., J. R. Broderson, and F. A. Murphy. 1979. New strain of mouse hepatitis virus as the cause of lethal enteritis in infant mice. Infect. Immun. **24**:508–522.

Hierholzer, J. C., M. C. Kemp, and G. A. Tannock. 1981. The RNA and proteins of human coronaviruses. Adv. Exp. Med. Biol. **142**:43–67.

Hierholzer, J. C., E. L. Palmer, S. G. Whitfield, H. S. Kaye, and W. R. Dowdle. 1972. Protein composition of coronavirus OC43. Virology **48**:516–527.

Hierholzer, J. C., M. T. Suggs, and E. C. Hall. 1969. Standardized viral hemagglutination and hemagglutination-inhibition tests. II. Description and statistical evaluation. Appl. Microbiol. **18**:824–833.

Hierholzer, J. C., and G. A. Tannock. 1977. Quantitation of antibody to non-hemagglutinating viruses by single radial hemolysis: serological test for human coronaviruses. J. Clin. Microbiol. **5**:613–620.

Hirano, N., H. Takamaru, K. Ono, T. Murakami, and K. Fujiwara. 1986. Replication of sialodacryoadenitis virus of rat in LBC cell culture: Brief report. Arch Virol **88**:121–125.

Hogue, B. G., and D. A. Brian. 1986. Structural proteins of human respiratory coronavirus OC43. Virus Res. **5**:131–144.

Hogue, B. G., B. King, and D. A. Brian. 1984. Antigenic relationships among proteins of bovine coronavirus, human respiratory coronavirus OC43, and mouse hepatitis coronavirus A59. J. Virol. **51**:384–388.

Holmes, A. W., F. Deinhardt, W. Harris, F. Ball, and G. Cline. 1970. Coronaviruses and viral hepatitis. J. Clin. Invest. **49**:45a.

Holmes, K. V., and J. N. Behnke. 1981. Evolution of a coronavirus during persistent infection *in vitro*. Adv. Exp. Med. Biol. **142**:287–299.

Holmes, K. V., E. W. Doller, and L. S. Sturman. 1981. Tunicamycin resistant glycosylation of a coronavirus glycoprotein: demonstration of a novel type of viral glycoprotein. Virology **115**:334–344.

Holmes, K. V., M. F. Frana, S. G. Robbins, and L. S. Sturman. 1984. Coronavirus maturation. Adv. Exp. Med. Biol. **173**:37–52.

Horzinek, M. C., H. Lutz, and N. C. Pedersen. 1982. Antigenic relationships among homologous structural polypeptides of porcine, feline, and canine coronaviruses. Infect. Immun. **37**:1148–1155.

Horzinek, M. C., and A. D. Osterhaus. 1979. The virology and pathogenesis of feline infectious peritonitis. Arch. Virol. **59**:1–15.

Hoshino, Y., and F. W. Scott. 1980. Coronavirus-like particles in the feces of normal cats. Arch. Virol. **63**:147–152.

Hovi, T., H. Kainulainen, B. Ziola, and A. Salmi. 1979. OC43 strain-related coronavirus antibodies in different age groups. J. Med. Virol. **3**:313–320.

Isaacs, D., D. Flowers, J. R. Clarke, H. B. Valman, and M. R. Macnaughton. 1983. Epidemiology of coronavirus respiratory infections. Arch. Dis. Child. **58**:500–503.

Jacobs, L., B. A. M. vander Zeijst, and M. C. Horzinek. 1986. Characterization and translation of transmissible gastroenteritis virus mRNAs. J. Virol. **57**:1010–1015.

Jiménez, G., I. Correa, M. P. Melgosa, M. J. Bullido, and L. Enjuanes. 1986. Critical epitopes in transmissible gastroenteritis virus neutralization. J. Virol. **60**:131–139.

Jonas, A. M., J. Craft, L. Black, P. N. Bhatt, and D. Hilding. 1969. Sialodacryoadenitis in the rat: a light and electron microscopic study. Arch. Pathol. **88**:613–622.

Kapikian, A. Z., H. D. James, S. J. Kelly, J. H. Dees, H. C. Turner, K. McIntosh, H. W. Kim, R. H. Parrott, M. M. Vincent, and R. M. Chanock. 1969. Isolation from a man of "avian infectious bronchitis virus-like" viruses (corona-viruses*) similar to 229E virus with some epidemiological observations. J. Infect. Dis. **119**:282–290.

Kapikian, A. Z., H. D. James, S. J. Kelly, L. M. King, A. L. Vaughn, and R. M. Chanock. 1972. Hemadsorption by coronavirus strain OC43. Proc. Soc. Exp. Biol. Med. **139**:179–186.

Kapikian, A. Z., H. D. James, S. J. Kelly, and A. L. Vaughn. 1973. Detection of coronavirus strain 692 by immune electron microscopy. Infect. Immun. **7**:111–116.

Kapke, P. A., and D. A. Brian. 1986. Sequence analysis of the porcine transmissible gastroenteritis coronavirus nucleocapsid protein gene. Virology **151**:41–49.

Kaye, H. S., and W. R. Dowdle. 1969. Some characteristics of hemagglutination of certain strains of "IBV-like" virus. J. Infect. Dis. **120**:576–581.

Kaye, H. S., and W. R. Dowdle. 1975. Seroepidemiologic survey of coronavirus (strain 229E) infections in a population of children. Am. J. Epidemiol. **101**:238–244.

Kaye, H. S., J. C. Hierholzer, and W. R. Dowdle. 1970. Purification and further characterization of an "IBV-like virus" (coronavirus). Proc. Soc. Exp. Biol. Med. **135**:457–463.

Kaye, H. S., H. B. Marsh, and W. R. Dowdle. 1971. Seroepidemiologic survey of coronavirus (strain OC43) related infections in a children's population. Am. J. Epidemiol. **94**:43–49.

Kaye, H. S., S. B. Ong, and W. R. Dowdle. 1972. Detection of coronavirus 229E antibody by indirect hemagglutination. Appl. Microbiol. **24**:703–707.

Kaye, H. S., W. B. Yarbrough, and C. J. Reed. 1975. Calf diarrhea coronavirus. Lancet **2**:509.

Kaye, H. S., W. B. Yarbrough, C. J. Reed, and A. K. Harrison. 1977. Antigenic relationship between human coronavirus strain OC-43 and hemagglutinating encephalomyelitis virus strain 67N of swine: antibody responses in human and animal sera. J. Infect. Dis. **135**:201–209.

Keck, J. G., L. H. Soe, S. Makino, S. A. Stohlman, and M. M. Lai. 1988. RNA recombination between fusion-positive MHV-A59 and fusion-negative MHV-2. J. Virol. **62**:1989–1998.

Keenan, K. P., H. R. Jervis, R. H. Marchwicki, and L. N. Binn. 1976. Intestinal infection of neonatal dogs with canine coronavirus 1-71: studies by virologic, histologic, histochemical, and immunofluorescent techniques. Am. J. Vet. Res. **37**:247–256.

Keljo, D. J., K. J. Bloch, M. Bloch, M. Arighi, and J. R. Hamilton. 1987. In vivo intestinal uptake of immunoreactive bovine albumin in piglet enteritis. J. Pediatr. Gastroenterol. Nutr. **6**:135–140.

Kemp, M. C., A. Harrison, J. C. Hierholzer, and J. S. Burks. 1984a. Assembly of 229E virions in human embryonic lung fibroblasts and effects of inhibition of glycosylation and glycoprotein transport on this process. Adv. Exp. Med. Biol. **173:**149–150.

Kemp, M. C., A. Harrison, J. C. Hierholzer, and J. S. Burks. 1984b. Electron lucent structures induced by coronaviruses. Adv. Exp. Med. Biol. **173:**153–154.

Kemp, M. C., J. C. Hierholzer, A. Harrison, and J. S. Burks. 1984c. Characterization of viral proteins synthesized in 229E infected cells and effect(s) of inhibition of glycosylation and glycoprotein transport. Adv. Exp. Med. Biol. **173:**65–77.

Kendall, E. J. C., M. L. Bynoe, and D. A. Tyrrell. 1962. Virus isolations from common colds occurring in a residential school. Br. Med. J. **2:**82–86.

Kennedy, D. A., and C. M. Johnson-Lussenburg. 1976. Isolation and morphology of the internal component of human coronavirus strain 229E. Intervirology **6:**197–206.

Kennedy, D. A., and C. M. Johnson-Lussenburg. 1978. Inhibition of coronavirus 229E replication by actinomycin D. J. Virol. **29:**401–404.

King, B., and D. A. Brian. 1982. Bovine coronavirus structural proteins. J. Virol. **42:**700–707.

King, B., B. J. Potts, and D. A. Brian. 1985. Bovine coronavirus hemagglutinin protein. Virus Res. **2:**53–59.

Kodama, Y., M. Ogata, and Y. Shimizu. 1981. Serum immunoglobulin A antibody response in swine infected with transmissible gastroenteritis virus, as determined by indirect immunoperoxidase antibody test. Am. J. Vet. Res. **42:**437–442.

Kraaijeveld, C. A., M. H. Madge, and M. R. Macnaughton. 1980a. Enzyme-linked immunosorbent assay for coronaviruses HCV229E and MHV3. J. Gen. Virol. **49:**83–89.

Kraaijeveld, C. A., S. E. Reed, and M. R. Macnaughton. 1980b. Enzyme-linked immunosorbent assay for detection of antibody in volunteers experimentally infected with human coronavirus strain 229E. J. Clin. Microbiol. **12:**493–497.

Kraft, L. M. 1962. An apparently new lethal virus disease of infant mice. Science **137:**282–283.

Lai, M. M. 1987. Replication of coronavirus RNA. *In* E. Domingo, J. Holland, and P. Ahlquist (ed.), RNA genetics. Vol. I: RNA-directed Virus Replication. pp. 115–136. Boca Raton, FL, CRC Reviews, Inc.

Lai, M. M., P. R. Brayton, R. C. Armen, C. D. Patton, C. Pugh, and S. A. Stohlman. 1981. Mouse hepatitis virus A59: mRNA structure and genetic localization of the sequence divergence from hepatotropic strain MHV-3. J. Virol. **39:**823–834.

Lai, M. M., J. O. Fleming, S. A. Stohlman, and K. Fujiwara. 1983. Genetic heterogeneity of murine coronaviruses. Arch. Virol. **78:**167–175.

Lai, M. M., C. D. Patton, and S. A. Stohlman. 1982. Replication of mouse hepatitis virus: negative-stranded RNA and replicative form RNA are of genome length. J. Virol. **44:**487–492.

Lai, M. M., and S. A. Stohlman. 1978. RNA of mouse hepatitis virus. J. Virol. **26:**236–242.

Lamontagne, L., and J. M. Dupuy. 1984. Persistent *in vitro* infection with mouse hepatitis virus 3. Adv. Exp. Med. Biol. **173:**315–326.

Lapierre, J., G. Marsolais, P. Pilon, and J. P. Descoteaux. 1980. Preliminary report on the observation of a coronavirus in the intestine of the laboratory rabbit. Can. J. Microbiol. **26:**1204–1208.

Laporte, J., and P. Bobulesco. 1981. Growth of human and canine enteric coronavirus in a highly susceptible cell line—HRT-18. Perspect. Virol. **11:**189–194.

Larson, H. E., S. E. Reed, and D. A. J. Tyrrell. 1980. Isolation of rhinoviruses and coronaviruses from 38 colds in adults. J. Med. Virol. **5:**221–230.

Laude, H., J. M. Chapsal, J. Gelfi, S. Labiau, and J. Grosolaude. 1986. Antigenic structure of transmissible gastroenteritis virus. I. Properties of monoclonal antibodies directed against virion proteins. J. Gen. Virol. **67:**119–130.

Leibowitz, J. L., J. R. DeVries, and M. V. Haspel. 1982a. Genetic analysis of murine hepatitis strain JHM. J. Virol. **42:**1080–1087.

Leibowitz, J. L., S. R. Weiss, E. Paavola, and C. W. Bond. 1982b. Cell-free translation of murine coronavirus RNA. J. Virol. **43:**905–913.

Leinikki, P. O., K. V. Holmes, I. Shekarchi, M. Iivanainen, D. Madden, and J. L. Sever. 1981. Coronavirus antibodies in patients with multiple sclerosis. Adv. Exp. Med. Biol. **142:**323–326.

Maass, G., and H. G. Baumeister. 1983. Coronavirus-like particles as etiological agents of acute non-bacterial gastroenteritis in humans. Dev. Biol. Standards **53:**319–324.

Macnaughton, M. R. 1980. The polypeptides of human and mouse coronaviruses. Arch. Virol. **63:**75–80.

Macnaughton, M. R. 1981. Structural and antigenic relationships between human, murine, and avian coronaviruses. Adv. Exp. Med. Biol. **142:**19–28.

Macnaughton, M. R. 1982. Occurrence and frequency of coronavirus infections in humans as determined by enzyme-linked immunosorbent assay. Infect. Immun. **38:**419–423.

Macnaughton, M. R., and H. A. Davies. 1981. Human enteric coronaviruses: brief review. Arch. Virol. **70:**301–314.

Macnaughton, M. R., H. A. Davies, and M. V. Nermut. 1978. Ribonucleoprotein-like structures from coronavirus particles. J. Gen. Virol. **39:**545–549.

Macnaughton, M. R., D. Flowers, and D. Isaacs. 1983. Diagnosis of human coronavirus infections in children using enzyme-linked immunosorbent assay. J. Med. Virol. **11:**319–326.

Macnaughton, M. R., H. J. Hasony, M. H. Madge, and S. E. Reed. 1981a. Antibody to virus components in volunteers experimentally infected with human coronavirus 229E group. Infect. Immun. **31:**845–849.

Macnaughton, M. R., and M. H. Madge. 1978. The genome of human coronavirus strain 229E. J. Gen. Virol. **39:**497–504.

Macnaughton, M. R., M. H. Madge, and S. E. Reed. 1981b. Two antigenic groups of human coronaviruses detected by using enzyme-linked immunosorbent assay. Infect. Immun. **33:**734–737.

Macnaughton, M. R., B. J. Thomas, H. A. Davies, and S. Patterson. 1980. Infectivity of human coronavirus strain 229E. J. Clin. Microbiol. **12:**462–468.

Mahy, B. W., S. Siddell, H. Wege, and V. Ter Meulen. 1983. RNA-dependent RNA polymerase activity in murine coronavirus-infected cells. J. Gen. Virol. **64:**103–111.

Makino, S., J. G. Keck, S. A. Stohlman, and M. M. Lai. 1986. High-frequency RNA recombination of murine coronaviruses. J. Virol. **57:**729–737.

Málková, D., J. Holubová, J. M. Kolman, F. Lobkovic, L. Pohlreichová, and L. Zikmundová. 1980. Isolation of Tettnang coronavirus from man. Acta Virol. **24:**363–366.

Mathan, M., and V. I. Mathan. 1978. Coronaviruses and

tropical sprue in southern India (Abstract). *In* 4th International Congress of Virology, The Hague, p. 469.

Mathan, M., V. I. Mathan, S. P. Swaminathan, S. Yesudoss, and S. J. Baker. 1975. Pleomorphic virus-like particles in human faeces. Lancet 1:1068–1069.

McIntosh, K. 1974. Coronaviruses: a comparative review. Curr. Top. Microbiol. Immunol. 63:85–129.

McIntosh, K., W. B. Becker, and R. M. Chanock. 1967a. Growth in suckling-mouse brain of "IBV-like" viruses from patients with upper respiratory tract disease. Proc. Natl. Acad. Sci. USA 58:2268–2273.

McIntosh, K., M. Brůčková, A. Z. Kapikian, R. M. Chanock, and H. Turner. 1970a. Studies on new virus isolates recovered in tracheal organ culture. Ann. NY Acad. Sci. 174:983–989.

McIntosh, K., R. K. Chao, H. E. Krause, R. Wasil, H. E. Mocega, and M. A. Mufson. 1974. Coronavirus infection in acute lower respiratory tract disease of infants. J. Infect. Dis. 130:502–507.

McIntosh, K., J. H. Dees, W. B. Becker, A. Z. Kapikian, and R. M. Chanock. 1967b. Recovery in tracheal organ cultures of novel viruses from patients with respiratory disease. Proc. Natl. Acad. Sci. 57:933–940.

McIntosh, K., A. Z. Kapikian, K. A. Hardison, J. W. Hartley, and R. M. Chanock. 1969. Antigenic relationships among the coronaviruses of man and between human and animal coronaviruses. J. Immunol. 102:1109–1118.

McIntosh, K., A. Z. Kapikian, H. C. Turner, J. W. Hartley, R. H. Parrott, and R. M. Chanock. 1970b. Seroepidemiologic studies of coronavirus infection in adults and children. Am. J. Epidemiol. 91:585–592.

McIntosh, K., J. McQuillin, S. E. Reed, and P. S. Gardner. 1978. Diagnosis of human coronavirus infection by immunofluorescence: method and application to respiratory disease in hospitalized children. J. Med. Virol. 2:341–346.

McNulty, M. S., W. L. Curran, and J. B. McFerran. 1975. Virus-like particles in calves' faeces. Lancet 2:78–79.

Mebus, C. A., E. L. Stair, M. B. Rhodes, and M. J. Twiehaus. 1973. Pathology of neonatal calf diarrhea induced by coronavirus-like agent. Vet. Pathol. 10:45–64.

Mengeling, W. L., A. D. Boothe, and A. E. Ritchie. 1972. Characteristics of a coronavirus (strain 67N) of pigs. Am. J. Vet. Res. 33:297–308.

Miller, L. T., and V. J. Yates. 1968. Neutralization of infectious bronchitis virus by human sera. Am. J. Epidemiol. 88:406–409.

Miyazaki, K., A. Tsunoda, M. Kumasaka, and N. Ishida. 1971. Presence of neutralizing antibody against the 229E strain of coronaviruses in the sera of residents. Jpn. J. Microbiol. 15:276–277.

Monto, A. S. 1974. Coronaviruses. Yale J. Biol. Med. 47:234–251.

Monto, A. S., and S. K. Lim. 1974. The tecumseh study of respiratory illness. VI. Frequency of and relationship between outbreaks of coronavirus infection. J. Infect. Dis. 129:271–276.

Monto, A. S., and L. M. Rhodes. 1977. Detection of coronavirus infection of man by immunofluorescence. Proc. Soc. Exp. Biol. Med. 155:143–148.

Moore, B. W., M. Hewish, and P. Lee. 1977. Are coronaviruses associated with gastroenteritis? Austral. Soc. Microbiol. Abstracts 1977:52.

Mortensen, M. L., C. G. Ray, C. M. Payne, A. D. Friedman, L. L. Minnich, and C. Rousseau. 1985. Coronaviruslike particles in human gastrointestinal disease. Am. J. Dis. Child. 139:928–934.

Nakanaga, K., K. Yamanouchi, and K. Fujiwara. 1986.

Protective effect of monoclonal antibodies on lethal mouse hepatitis virus infection in mice. J. Virol. 59:168–171.

Nguyen, T. D., E. Bottreau, S. Bernard, I. Lantier, and J. M. Aynaud. 1986. Neutralizing secretory IgA and IgG do not inhibit attachment of transmissible gastroenteritis virus. J. Gen. Virol. 67:939–943.

Norman, J. O., A. W. McClurkin, and S. L. Stark. 1970. Transmissible gastroenteritis (TGE) of swine: canine serum antibodies against an associated virus. Can. J. Comp. Med. 34:115–117.

Nuttall, P. A., and K. A. Harrap. 1982. Isolation of a coronavirus during studies on puffinosis, a disease of the Manx Shearwater (*Puffinus puffinus*). Arch. Virol. 73:1–14.

Oshiro, L. S. 1973. Coronaviruses, p. 331–343. *In* A. J. Dalton and Haguenau (ed.), Ultrastructure of animal viruses and bacteriophages: an atlas. Academic Press, Inc., New York.

Oshiro, L. S., J. H. Schieble, and E. H. Lennette. 1971. Electron microscopic studies of coronavirus. J. Gen. Virol. 12:161–168.

Osterhaus, A. D., J. S. Teppema, and G. Van Steenis. 1982. Coronavirus-like particles in laboratory rabbits with different syndromes in the Netherlands. Lab. Anim. Sci. 32:663–665.

Panigrahy, B., S. A. Naqi, and C. F. Hall. 1973. Isolation and characterization of viruses associated with transmissible enteritis (bluecomb) of turkeys. Avian Dis. 17:430–438.

Parker, J. C., S. S. Cross, and W. P. Rowe. 1970. Rat coronavirus (RCV): A prevalent, naturally-occurring pneumotropic virus of rats. Arch. Gesamte Virusforsch. 31:293–302.

Patel, J. R., H. A. Davies, N. Edington, J. Laporte, and M. R. Macnaughton. 1982. Infection of a calf with the enteric coronavirus strain Paris. Arch. Virol. 73:319–328.

Patterson, S., and M. R. Macnaughton. 1981. The distribution of human coronavirus strain 229E on the surface of human diploid cells. J. Gen. Virol. 53:267–274.

Patterson, S., and M. R. Macnaughton. 1982. Replication of human respiratory coronavirus strain 229E in human macrophages. J. Gen. Virol. 60:307–314.

Pearson, J., and C. A. Mims. 1985. Differential susceptibility of cultured neural cells to the human coronavirus OC43. J. Virol. 53:1016–1019.

Pedersen, N. C., J. W. Black, J. F. Boyle, J. F. Evermann, A. J. McKeirnan, and R. L. Ott. 1984. Pathogenic differences between various feline coronavirus isolates. Adv. Exp. Med. Biol. 173:365–380.

Pedersen, N. C., J. F. Boyle, and K. Floyd. 1981a. Infection studies in kittens, using feline infectious peritonitis virus propagated in cell culture. Am. J. Vet. Res. 42:363–367.

Pedersen, N. C., J. F. Boyle, and K. Floyd. 1981b. An enteric coronavirus infection of cats and its relationship to feline infectious peritonitis. Am. J. Vet. Res. 42:368–377.

Pedersen, N. C., J. Ward, and W. L. Mengeling. 1978. Antigenic relationship of the feline infectious peritonitis virus to coronaviruses of other species. Arch. Virol. 58:45–53.

Peigue, H., M. Beytout-Monghal, H. Laveran, and M. Bourges. 1978. Coronavirus et "astrovirus" observés dans les selles d'enfants atteints de gastro-entérites. Ann. Microbiol. 129B:101–106.

Pensaert, M. B., and P. Debouck. 1978. A new coronavirus-like particle associated with diarrhea in swine. Arch. Virol. 58:243–247.

Pensaert, M. B., P. Debouck, and D. J. Reynolds. 1981. An immunoelectron microscopic and immunofluorescent study on the antigenic relationship between the coronavirus-like agent, CV 777, and several coronaviruses. Arch. Virol. **68:**45–52.

Pike, B. V., and D. J. Garwes. 1979. The neutralization of transmissible gastroenteritis virus by normal heterotypic serum. J. Gen. Virol. **42:**279–288.

Pokorný, J., M. Brǔčková, and M. Rýc. 1975. Biophysical properties of coronavirus strain OC-43. Acta. Virol. **19:**137–142.

Pomeroy, K. A., B. L. Patel, C. T. Larsen, and B. S. Pomeroy. 1978. Combined immunofluorescence and transmission electron microscopic studies of sequential intestinal samples from turkey embryos and poults infected with turkey enteritis coronavirus. Am. J. Vet. Res. **39:**1348–1354.

Reed, S. E. 1984. The behavior of recent isolates of human respiratory coronavirus in vitro and in volunteers: evidence of heterogeneity among 229E-related strains. J. Med. Virol. **13:**179–192.

Resta, S., J. P. Luby, C. R. Rosenfeld, and J. D. Siegel. 1985. Isolation and propagation of a human enteric coronavirus. Science **229:**978–981.

Rettig, P. J., and G. P. Altshuler. 1985. Fatal gastroenteritis associated with coronaviruslike particles. Am. J. Dis. Child. **139:**245–248.

Reynolds, D. J., T. G. Debney, G. A. Hall, L. H. Thomas, and K. R. Parsons. 1985. Studies on the relationship between coronaviruses from the intestinal and respiratory tracts of calves. Arch. Virol. **85:**71–83.

Reynolds, D. J., and D. J. Garwes. 1979. Virus isolation and serum antibody responses after infection of cats with transmissible gastroenteritis virus. Arch. Virol. **60:**161–166.

Reynolds, D. J., D. J. Garwes, and C. J. Gaskell. 1977. Detection of transmissible gastroenteritis virus neutralizing antibody in cats. Arch. Virol. **55:**77–86.

Riordan, T., A. Curry, and M. N. Bhattacharyya. 1986. Enteric coronavirus in symptomless homosexuals [letter]. J. Clin. Pathol. **39:**1159–1160.

Riski, H., and T. Estola. 1974. Occurrence of antibodies to human coronavirus OC43 in Finland. Scand. J. Infect. Dis. **6:**325–327.

Riski, H., and T. Hovi. 1980. Coronavirus infections of man associated with diseases other than the common cold. J. Med. Virol. **6:**259–265.

Riski, H., T. Hovi, P. Väänänen, and K. Penttinen. 1977. Antibodies to human coronavirus OC43 measured by radial hemolysis in gel. Scand. J. Infect. Dis. **9:**75–77.

Robb, J. A., and C. W. Bond. 1979. Coronaviridae. Comp. Virol. **14:**193–247.

Robbins, S. G., M. F. Frana, J. J. McGowan, J. F. Boyle, and K. V. Holmes. 1986. RNA-binding proteins of coronavirus MHV: detection of monomeric and multimeric N protein with an RNA overlay-protein blot assay. Virology **30:**402–410.

Rottier, P., D. Brandenburg, J. Armstrong, B. Van der Zeijst, and G. Warren. 1984. *In vitro* assembly of the murine coronavirus membrane protein. Adv. Exp. Med. Biol. **173:**53–64.

Rottier, P. J., W. J. Spaan, M. C. Horzinek, and B. A. Van der Zeijst. 1981. Translation of three mouse hepatitis virus strain A59 subgenomic RNAs in *Xenopus laevis* oocytes. J. Virol. **38:**20–26.

Rowe, W. P., J. W. Hartley, and W. I. Capps. 1963. Mouse hepatitis virus infection as a highly contagious, prevalent, enteric infection of mice. Proc. Soc. Exp. Biol. Med. **112:**161–165.

Sato, K., Y. Inaba, and M. Matumoto. 1980. Serological relation between calf diarrhea coronavirus and hemagglutinating encephalomyelitis virus. Arch. Virol. **66:**157–160.

Sawicki, S. G., and D. L. Sawicki. 1986. Coronavirus minus-strand RNA synthesis and effect of cycloheximide on coronavirus RNA synthesis. J. Virol. **57:**328–334.

Schmidt, O. W. 1984. Antigenic characterization of human coronavirus 229E and OC43 by enzyme-linked immunosorbent assay. J. Clin. Microbiol. **20:**175–180.

Schmidt, O. W., I. D. Allan, M. K. Cooney, H. M. Foy, and J. P. Fox. 1986. Rises in titers of antibody to human coronaviruses OC43 and 229E in Seattle families during 1975–1979. Am. J. Epidemiol. **123:**862–868.

Schmidt, O. W., M. K. Cooney, and G. E. Kenny. 1979. Plaque assay and improved yield of human coronaviruses in a human rhabdomyosarcoma cell line. J. Clin. Microbiol. **9:**722–728.

Schmidt, O. W., and G. E. Kenny. 1981. Immunogenicity and antigenicity of human coronaviruses 229E and OC-43. Infect. Immun. **32:**1000–1006.

Schmidt, O. W., and G. E. Kenny. 1982. Polypeptides and functions of antigens from human coronaviruses 229E and OC43. Infect. Immun. **35:**515–522.

Schnagl, R. D., S. Brookes, S. Medvedee, and F. Morey. 1987. Characteristics of Australian human enteric coronavirus-like particles: comparison with human respiratory coronavirus 229E and duodenal brush border vesicles. Arch. Virol. **97:**309–323.

Schnagl, R. D., T. Greco, and F. Morey. 1986. Antibody prevalence to human enteric coronavirus-like particles and indications of antigenic differences between particles from different areas. Arch. Virol. **87:**331–337.

Schnagl, R. D., and I. H. Holmes. 1978. Coronavirus-like particles in stools from dogs from some country areas of Australia. Vet. Rec. **102:**528–529.

Schnagl, R. D., I. H. Holmes, and E. M. Mackay-Scollay. 1978. Coronavirus-like particles in aboriginals and non-aboriginals in Western Australia. Med. J. Austral. **1:**307–311.

Schochetman, G., R. H. Stevens, and R. W. Simpson. 1977. Presence of infectious polyadenylated RNA in the coronavirus avian bronchitis virus. Virology **77:**772–782.

Sheboldov, A. V., L. Y. Zakstelskaya, and V. M. Zhdanov. 1973. Sedimentation and density characteristics of coronavirus. Vopr. Virusol. **1:**59–64.

Siddell, S. 1983. Coronavirus JHM: coding assigments of subgenomic mRNAs. J. Gen. Virol. **64:**113–125.

Siddell, S., R. Anderson, D. Cavanagh, K. Fujiwara, H. D. Klenk, M. R. Macnaughton, M. Pensaert, S. A. Stohlman, L. Sturman, and B. A. Van der Zeijst. 1983a. Coronaviridae. Intervirology **20:**181–189.

Siddell, S., H. Wege, and V. ter Meulen. 1983b. The biology of coronaviruses. J. Gen. Virol. **64:**761–776.

Sitbon, M. 1985. Human-enteric-coronavirus-like particles (CVLP) with different epidemiological characteristics. J. Med. Virol. **16:**67–76.

Skinner, M. A., D. Ebner, and S. G. Siddell. 1985. Coronavirus MHV-JHM mRNA5 has a sequence arrangement which potentially allows translation of a second, downstream open reading frame. J. Gen. Virol. **66:**581–592.

Skinner, M. A., and S. G. Siddell. 1985. Coding sequence of coronavirus MHV-JHM mRNA4. J. Gen. Virol. **66:**593–596.

Small, J. D., L. Aurelian, R. A. Squire, J. D. Strandberg, E. C. Melby, T. B. Turner, and B. Newman. 1979. Rabbit cardiomyopathy: associated with a virus antigenically related to human coronavirus strain 229E. Am. J. Pathol. **95:**709–729.

Smith, G. C., T. L. Lester, R. L. Heberling, and S. S. Kalter. 1982. Coronavirus-like particles in nonhuman primate feces. Arch. Virol. **72:**105–112.

Snyder, D. B., and W. W. Marquardt. 1984. Use of monoclonal antibodies to assess antigenic relationships of avian infectious bronchitis virus serotypes in the United States. Adv. Exp. Med. Biol. **173:**109–113.

Sorensen, O., S. Beushausen, S. Puchalski, S. Cheley, R. Anderson, M. Coulter-Mackie, and S. Dales. 1984. *In vivo* and *in vitro* models of demyelinating diseases. VIII. Genetic, immunologic, and cellular influences on JHM virus infection of rats. Adv. Exp. Med. Biol. **173:**279–298.

Sorensen, O., A. Collins, W. Flintoff, G. Ebers, and S. Dales. 1986. Probing for the human coronavirus OC43 in multiple sclerosis. Neurology **36:**1604–1606.

Stair, E. L., M. B. Rhodes, R. G. White, and C. A. Mebus. 1972. Neonatal calf diarrhea: purification and electron microscopy of a coronavirus-like agent. Am. J. Vet. Res. **33:**1147–1156.

Stern, D. F., L. Burgess, and B. M. Sefton. 1982. Structural analysis of virion proteins of the avian coronavirus infectious bronchitis virus. J. Virol. **42:**208–219.

Stern, D. F., and S. I. T. Kennedy. 1980. Coronavirus multiplication strategy II. Mapping the avian infectious bronchitis virus intracellular RNA species to the genome. J. Virol. **36:**440–449.

Stern, D. F., and B. M. Sefton. 1982. Coronavirus proteins: structure and function of the oligosaccharides of the avian infectious bronchitis virus glycoproteins. J. Virol. **44:**804–812.

Stern, D. F., and B. M. Sefton. 1984. Coronavirus multiplication: locations of genes for virion proteins on the avian infectious bronchitis virus genome. J. Virol. **50:**22–29.

Stoddart, C. A., J. E. Barlough, and F. .W. Scott. 1984. Experimental studies of a coronavirus and coronavirus-like agent in a barrier-maintained feline breeding colony. Arch. Virol. **79:**85–94.

Storz, J., and R. Rott. 1981. Reactivity of antibodies in human serum with antigens of an enteropathogenic bovine coronavirus. Med. Microbiol. Immunol. **169:**169–178.

Sturman, L. S. 1977. Characterization of a coronavirus. I. Structural proteins: effects of preparative conditions on the migration of protein in polyacrylamide gels. Virology **77:**637–649.

Sturman, L. S., and K. V. Holmes. 1977. Characterization of a coronavirus. II. Glycoproteins of the viral envelope: tryptic peptide analysis. Virology **77:**650–660.

Sturman, L., and K. Holmes. 1985. The novel glycoproteins of coronaviruses. Trends Biochem. Sci. **10:**17–20.

Sturman, L. S., and K. V. Holmes. 1983. The molecular biology of coronaviruses. Adv. Virus. Res. **28:**35–112.

Sturman, L. S., and K. V. Holmes. 1984. Proteolytic cleavage of peplomeric glycoprotein E2 of MHV yields two 90K submits and activates cell fusion. Adv. Exp. Med. Biol. **173:**25–35.

Sturman, L. S., K. V. Holmes, and J. Behnke. 1980. Isolation of coronavirus envelope glycoproteins and interaction with the viral nucleocapsid. J. Virol. **33:**449–462.

Sturman, L. S., C. S. Ricard, and K. V. Holmes. 1985. Proteolytic cleavage of the E2 glycoprotein of murine coronavirus: activation of cell-fusing activity of virions by trypsin and separation of two different 90K cleavage fragments. J. Virol. **56:**904–911.

Sugiyama, K., K. Ishikawa, and N. Fukunara. 1986. Structural polypeptides of the murine coronavirus DVIM. Arch. Virol. **89:**245–254.

Sureau, C., C. Amiel-Tison, O. Moscovici, P. Lebon, J. Laporte, and C. Chany. 1980. Une épidémie d'enterocolitis ulcéronécrosantes en maternité, arguments en faveur de son origine virale. Bull. Acad. Natl. Méd. **164:**286–293.

Tanaka, R., Y. Iwasaki, and H. Koprowski. 1976. Intracisternal virus-like particles in brain of a multiple sclerosis patient. J. Neurol. Sci. **28:**121–126.

Tannock, G. A. 1973. The nucleic acid of infectious bronchitis virus. Arch. Gesamte Virusforsch. **43:**259–271.

Tannock, G. A., and J. C. Hierholzer. 1977. The RNA of human coronavirus OC-43. Virology **78:**500–510.

Tannock, G. A., and J. C. Hierholzer. 1978. Presence of genomic polyadeylate and absence of detectable virion transcriptase in human coronavirus OC-43. J. Gen. Virol. **39:**29–39.

Tektoff, J., M. Dauvergne, M. Duraflour, and J. P. Soulebot. 1983. Propagation of bovine coronavirus on Vero cell line: electron microscopic studies. Dev. Biol. Stand. **53:**299–310.

Tupper, G. T., J. F. Evermann, R. G. Russell, and M. E. Thouless. 1987. Antigenic and biological diversity of feline coronavirus: feline infectious peritonitis and feline enteritis virus. Arch. Virol. **96:**29–38.

Turner, R. B., A. Felton, K. Kosak, D. K. Kelsey, and C. K. Meschievitz. 1986. Prevention of experimental coronavirus colds with intranasal alpha-2b interferon. J. Infect. Dis. **154:**443–447.

Tyrrell, D. A., D. J. Alexander, J. D. Almeida, C. H. Cunningham, B. C. Easterday, D. J. Garwes, J. C. Hierholzer, A. Kapikian, M. R. Macnaughton, and K. McIntosh. 1978. Coronaviridae: second report. Intervirology **10:**321–328.

Tyrrell, D. A. J., and J. D. Almeida. 1967. Direct electron-microscopy of organ cultures for the detection and characterization of viruses. Arch. Gesamte Virusforsch. **22:**417–425.

Tyrrell, D. A. J., J. D. Almeida, D. M. Berry, C. H. Cunningham, D. Hamre, M. S. Hofstad, L. Mallucci, and K. McIntosh. 1968. Coronaviruses. Nature (London) **220:**650.

Tyrrell, D. A. J., J. D. Almeida, C. H. Cunningham, W. R. Dowdle, M. S. Hofstad, K. McIntosh, M. Tajima, L. Y. Zakstelskaya, B. C. Easterday, A. Kapikian, and R. W. Bingham. 1975. Coronaviridae. Intervirology **5:**76–82.

Tyrrell, D. A. J., and M. L. Bynoe. 1965. Cultivation of a novel type of common-cold virus in organ cultures. Br. Med. J. **1:**1467–1470.

Tyrrell, D. A. J., and M. L. Bynoe. 1966. Cultivation of viruses from a high proportion of patients with colds. Lancet **1:**76–77.

Tzipori, S., M. Smith, T. Makin, and C. McCaughan. 1978. Enteric coronavirus-like particles in sheep. Austral. Vet. J. **54:**320–321.

Van Berlo, M. F., G. Wolswijk, J. Calafat, M. J. Koolen, M. C. Horzinek, and B. A. Van der Zeijst. 1986. Restricted replication of mouse hepatitis virus A59 in primary mouse brain astrocytes correlates with reduced pathogenicity. J. Virol. **58:**426–433.

Vaucher, Y. E., C. G. Ray, L. L. Minnich, C. M. Payne, D. Beck, and P. Lowe. 1982. Pleomorphic, enveloped, virus-like particles associated with gastrointestinal illness in neonates. J. Infect. Dis. **145:**27–36.

Wege, H., A. Müller, and V. ter Meulen. 1978. Genomic RNA of the murine coronavirus JHM. J. Gen. Virol. **41:**217–228.

Wege, H., S. Siddell, and V. ter Meulen. 1982. The biology and pathogenesis of coronaviruses. Curr. Top. Microbiol. Immunol. **99:**165–200.

Wege, H., J. R. Stephenson, M. Koga, H. Wege, and V. ter Meulen. 1981. Genetic variation of neurotropic and non-neurotropic murine coronaviruses. J. Gen. Virol. **54:**67–74.

Wege, H., H. Wege, K. Nagashima, and V. ter Meulen. 1979. Structural polypeptides of the murine coronavirus JHM. J. Gen. Virol. **42:**37–48.

Weiss, S. R. 1983. Coronaviruses SD and SK share extensive nucleotide homology with murine coronavirus MHV A-59, more than that shared between human and murine coronaviruses. Virology **126:**669–677.

Wenzel, R. P., J. O. Hendley, J. A. Davies, and J. M. Gwaltney. 1974. Coronavirus infections in military recruits. Three-year study with coronavirus strains OC43 and 229E. Am. Rev. Respir. Dis. **109:**621–624.

Wesley, R. D., and R. D. Woods. 1986. Identification of a 17000 molecular weight antigenic polypeptide in transmissible gastroenteritis virus-infected cells. J. Gen. Virol. **67:**1419–1425.

Wilhelmsen, K. C., J. L. Leibowitz, C. W. Bond, and J. A. Robb. 1981. The replication of murine coronaviruses in enucleated cells. Virology **110:**225–230.

Wilson, G. A., S. Beushausen, and S. Dales. 1986. In vivo and in vitro models of demyelinating diseases. XV. Differentiation influences the regulation of coronavirus in-fection in primary explants of mouse CNS. Virology **151:**253–264.

Wolfe, L. G., and R. A. Griesemer. 1966. Feline infectious peritonitis. Pathol. Vet. **3:**225–270.

Woods, R. D., N. F. Cheville, and J. E. Gallagher. 1981. Lesions in the small intestine of newborn pigs inoculated with porcine, feline, and canine coronaviruses. Am. J. Vet. Res. **42:**1163–1169.

Wright, R. 1972. Chronic hepatitis. Br. Med. Bull. **28:**120–124.

Yaseen, S. A., and C. M. Johnson-Lussenburg. 1981. Antigenic studies on coronavirus. I. Identification of the structural antigens of human coronavirus, strain 229E. Can. J. Microbiol. **27:**334–342.

Zakstelskaja, L. Y., A. V. Sheboldov, and E. V. Molibog. 1972a. Some aspects of interaction between coronaviruses OC38 and OC43 and erythrocytes. Vestn. Akad. Med. Nauk SSSR **27:**40–43.

Zakstelskaja, L. Y., A. V. Sheboldov, V. J. Vasilieva, and L. J. Alekseenkova. 1972b. Occurrence of antibody to coronaviruses in sera of people living in the USSR. Vopr. Virusol. **17:**161–165.

Zuckerman, A. J., P. E. Taylor, and J. D. Almeida. 1970. Presence of particles other than the Australian-SH antigen in a case of chronic active hepatitis with cirrhosis. Br. Med. J. **1:**262–264.

CHAPTER 25

Paramyxoviridae: The Parainfluenza Viruses

MONICA GRANDIEN

Disease: Respiratory tract infections.

Etiologic Agents: Parainfluenza virus types 1, 2, 3, 4A, and 4B.

Source: Infected humans disseminate virus by contact or by droplets generated by coughs or sneezes.

Clinical Manifestations: A variety of syndromes ranging from a mild upper respiratory illness without fever to severe pneumonia; the infection is the most frequent cause of infantile croup (laryngotracheitis) and second to RS virus, the most common cause of lower respiratory tract infections (bronchiolitis and pneumonia) in young children; in older children and adults, reinfections occur frequently, usually with mild upper respiratory symptoms.

Pathology: Infection of cells in the respiratory tract leads to epithelial cell death and inflammatory reaction in the upper and lower respiratory tracts, causing rhinitis, pharyngitis, laryngotracheitis (croup), tracheobronchitis, bronchiolitis, and pneumonia.

Laboratory Diagnosis: Besides virus isolation (LLC-MK2 cells), detection of virus or viral antigens directly in clinical specimens (nasopharyngeal secretions) by immunofluorescence or solid phase immunoassays as ELISA gives a type-specific diagnosis; the serologic diagnosis is complicated by occurrence of heterotypic antibody responses to the different parainfluenza virus types and to mumps; detection of type-specific IgM has been described.

Epidemiology: The viruses occur worldwide and are isolated in 6 to 13% of children with acute respiratory infections.

Treatment: No specific treatment available; in experimental studies, ribavarin, a nucleoside analogue, has been reported to have some effect in human infections.

Prevention and Control: No vaccine available.

Introduction

The human parainfluenza viruses were discovered during the years 1956 to 1960. They were found by tissue culture isolation and detected because of their hemadsorption properties or by developed cytopathic effects (Beale et al., 1958; Chanock, 1956; Chanock et al., 1958; Johnson et al., 1960). Three years earlier, a parainfluenza virus, the Sendai virus, had been isolated from mice (Kuroya and Ishida, 1953). Originally, this virus was described as a human pathogen but was shown later to be a murine virus.

The viruses cause acute respiratory tract illness, mainly in infancy and childhood. They play a significant role in both upper and lower respiratory tract infections (Chanock et al., 1963; Gardner et al., 1971) and are, second to respiratory syncytial (RS) virus, the most important cause of serious respiratory infection in infants and young children. Reinfections occur at all ages, usually with symptoms much less pronounced.

Together with mumps and the avian Newcastle disease virus (NDV), the parainfluenza viruses form the genus paramyxovirus belonging to the family *Paramyxoviridae,* which also includes two other genera—*morbillivirus* and *pneumovirus* (Kingsbury et al., 1978). Morphologically, the paramyxoviridae resemble the family *Orthomyxoviridae* (the influenza viruses) in many respects; they differ from the

Orthomyxoviridae mainly by their helical nucleocapsid structure, which has a diameter of 18 nm—twice that of influenza virus. The different cellular sites for antigen accumulation and the differences in genetics also support the separation into different families (Waterson, 1962; Kingsbury, 1985).

Much of the knowledge regarding the parainfluenza viruses is based on the work of R. M. Chanock and his co-workers, who reported the first discovery of a human parainfluenza virus and described the basic properties of the viruses as well as their symptomatology and epidemiology (Chanock and McIntosh, 1985). Other early reports used serological means for epidemiological studies (Chanock et al., 1960) and reported the antigenic relationship between parainfluenza and mumps viruses (Gardner 1957).

The Viruses

There are four types of human parainfluenza virus: types 1, 2, 3, and 4 (Kingsbury et al., 1978). Type 4 virus has been divided into two subtypes, 4A and 4B (Canchola et al., 1964). Although the virus types share antigens, they are all serologically distinct, as are the two type 4 subtypes. Animal parainfluenza viruses have been characterized as related to the human types but with individual antigens allowing their recognition. Sendai virus isolated from mice has been included as a subtype into parainfluenza type 1 virus, Simian virus 5 (SV5), infecting monkeys and dogs, into type 2, and a bovine virus, SF4, into type 3. Whether the animal strains also infect humans is less clear. In addition to the report on human isolates of Sendai virus (Kuroya and Ishida, 1953), which has been questioned, a few reports have been published on SV5 isolates from humans.

Morphology and Physicochemical Properties

The parainfluenza viruses are enveloped pleomorphic viruses with a diameter of 120 to 300 nm. Glycoprotein spikes penetrate a lipid-bilayer membrane and protrude with one end from the envelope while the other end reaches beneath the membrane (Hiebert et al., 1985a; Lyles, 1979). The nucleic acid consisting of one piece of single-stranded RNA is incorporated in the nucleocapsid, which is coiled inside the envelope. When released, the nucleocapsid streches out into a rod of about 1 μm length (Compans and Choppin, 1967). The nucleocapsid unit is relatively large, and, observed by electron microscopy, it appears as a rod with a serrated edge (Kingsbury, 1985). The helical construction of the 18-nm-wide nucleocapsid produces an empty central core which allows the negative stain used for electron microscopy to penetrate into the helix.

The viruses have hemagglutination, neuraminidase, cell fusion, and hemolysis activities, all located to the surface glycoproteins, which can be seen by electron microscopy as spikes. While the hemagglutination and neuraminidase activities reside at different sites on the same glycoprotein molecule (HN), the fusion activity is located on a separate fusion (F) glycoprotein (Örvell and Grandien, 1982; Örvell and Norrby, 1977; Portner, 1981; Scheid et al., 1978; Yewdell & Gerhard, 1982; compare Fig. 1). The F protein produced in infected cells lyses membranes of other cells which leads to cell fusion and the formation of cell syncytia. This is a phenomenon characteristic for all paramyxovirus infections.

The molecular weight of the viruses are 5 to 7 $\times$ 10^8 daltons—greater for pleomorphic multiploid virions and less for defective virions. The sedimentation coefficient has been determined to 1,000S and the buoyancy density to 1.18 to 1.25 g/cm^3, depending on virus and buoyant medium. The content of RNA is 0.5 to 1%, protein 70%, lipid 20 to 25%, and

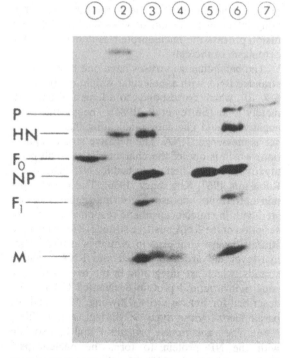

FIG. 1. Radioimmune precipitation assay with mouse ascites materials reacting with different Sendai virus proteins. The following materials were included (the positions in the gel are given in parentheses): [^{35}S]methionine-labeled Sendai virions grown in the presence of 0.3 μg trypsin per milliliter medium (3, 6); monoclonal antibodies directed against F (1), HN (2), M (4), NP (5), and P (7). Labeled cell-associated viral materials were used in tests with anti-F and anti-M antibodies and labeled extracellular virions in tests with anti-HN, anti-NP, and anti-P antibodies. The upper band seen in lane 2 probably represents dimers of the HN molecule. (From Örvell and Grandien, 1982; with permission.)

carbohydrate 6%, as determined for the type species, NDV (Kingsbury et al., 1978).

Viral Replication

All myxoviruses have a special affinity for mucopolysaccharides and glycoproteins—especially for sialic acid–containing receptors on cell surfaces (Markwell et al., 1981). The HN glycoprotein, one of the two types of envelope spikes, adsorbs via its hemagglutinin to the host cell neuraminic acid receptors and thus mediates virus attachment. The subsequent fusion of the virion envelope with the outer cell membrane is mediated by the other spike, the F protein.

The F protein's cell-fusing capacity develops only after the inactive precursor F_0, containing two glycopeptides linked by disulfide bonds, has been cleaved by the host cells' proteolytic enzymes into two subunits, F_1 and F_2 (Choppin and Scheid, 1980; Scheid and Choppin, 1977). The F_1 polypeptide is anchored in the envelope and plays the important role of initiating infection by fusion of virus and cell. By this fusion process the nucleoprotein is deposited in the cytoplasm of the cell.

The parainfluenza viruses have one large, single-stranded RNA with a molecular weight of about 5 to 6×10^6, which corresponds to a length of approximately 15 kb (Storey et al., 1984). The genome has a negative strand character, indicating that it cannot act as messenger RNA. The negative strand replication, which is one of the characteristics of all paramyxoviruses, was first described for NDV (Bratt and Robinson, 1967; Kingsbury, 1966). The virions contain an RNA polymerase, transcriptase that is associated with the nucleocapsid and responsible for transcription of the RNA positive strands. These positive strands govern viral protein synthesis at the ribosomal level and are also copied into RNA negative strands, which are integrated in the new virions. A small nonstructural protein designated C has been described for Sendai virus (Dowling, 1983) and for parainfluenza virus type 3 (Spriggs and Collins, 1986). The transcriptase proteins P and L associate with the NP protein to form the nucleocapsid (Portner and Murti, 1985). The P protein undergoes phosphorylization before entering the nucleocapsid (Lamb and Choppin, 1977). The virion thus carries its own transcriptase, which is a prerequisite for starting replication in new cells.

For assembly of the virion envelope, the synthesized glycoproteins accumulate at the cell plasma membrane. The transportation has been described for *Rhabdoviridae,* another virus family characterized by negative strand replication: the glycoproteins are transported via small vesicles from the endothelial reticulum via the Golgi apparatus to the plasma membrane (Rothman and Fine, 1980). The precise details of this transportation in the case of the parainfluenza viruses are not known. The glycoproteins consist of a hydrophobic transmembrane portion and are glycosylated at the hydrophilic part extruding from the surface of the cell.

The assembly of the virions is completed by the budding of the nucleocapsid through the cell plasma membrane studded with inserted glycoproteins. The role of the matrix (M) protein lining the envelope on the inner surface in this process is not yet fully clear. By electron microscopy the M protein of Sendai virus was found to be distributed in *clusters* along the entire length of the nucleocapsid (Portner and Murti, 1985). Investigations with peroxidase-labeled monoclonal antibodies revealed that M protein could not be seen at the infected cell surface except in budding viral particles (Kristensson and Örvell, 1983). The M protein in rhabdovirus maturation has been found to mediate the in vitro binding of membrane structures to the ribonucleoprotein (Ogden et al., 1986). It plays an important role in maintaining the structure of the virus. The neuraminidase activity, a property present in all parainfluenza viruses, prevents aggregation of the newly formed viruses by cleaving the binding of viral hemagglutinin to host cell sialic acid.

Occasionally the F protein can mediate fusion with adjacent cell membranes, which results in production of polycaryocytes. This fusion has been called "fusion from within" (Gallaher and Bratt, 1974) and results in the formation of syncytia. Using temperature sensitive mutants (ts) which express only the membrane-associated hemagglutinin at nonpermissive temperatures (39 to 40°C), cell fusion resulting in extensive polycaryocytes can be obtained even in the absence of any other virus constituents. Thus, the extension of the cytopathogenic effect can extend much further than the site of infection—such lesion has also been reported with HIV in lymphocytes and in the brain. When such VSV-ts mutants are injected i.c. in mice, they develop a spongiform encephalitis, simply due to the diffusion of envelope glycoproteins along the neurous (Chany et al., 1987). Because of its capacity to induce interferon Sendai virus has been used for large scale production of interferon in human buffy coat leukocytes.

Antigenic Composition and Genetics

Six major virus-coded structural proteins have been characterized in the parainfluenza virions. Their molecular weights have been found to vary to some extent and are given here for type 3 virus in kilodaltons, kDa (Elango et al., 1986; Spriggs and Collins, 1986). The glycoproteins HN (mol wt 69–72 kDa) and F (60–63 kDa) constitute the two surface antigens inserted into the lipid membrane; the M protein (35–40 kDa) lines the interior of the membrane, and the NP (66–68 kDa), P (83–90 kDa), and L (200 kDa) proteins together form the capsid of the virus. All of the virion structural proteins are also detected in extract from infected cells (Wechsler et al., 1985). Molecular weights of the structural proteins of other parainfluenza virus types differ to a certain extent from given figures (Cowley and Barry, 1983). A nonstructural protein, the C protein (22 kDa) has also been described for Sendai virus (Dethlefsen and Kolakofsky, 1983; Lamb et al., 1976) as well as for parainfluenza type 3 virus (Sanchez and Banerjee, 1985a; Spriggs and Collins, 1986). A gene coding for a small, highly hydrophobic polypetide SH has recently been identified in SV5 (Hiebert et al., 1985b). The cloning and gene assignment of mRNAs gives further information of type 3 virus (Elango et al., 1986; Sanchez and Banerjee, 1985b).

Antibodies to HN and F both are important for protection against infection (Merz et al., 1981), inhibiting attachment and fusion of virus to the host cell. This has been confirmed in an animal model in which passive immunization of mice with monoclonal antibodies against the F protein, and against the HN protein, protected mice against experimental Sendai virus infection (Örvell and Grandien, 1982). A thorough knowledge of the antigen determinants specific for one virus or shared between several of the parainfluenza viruses has been gained by use of monoclonal antibodies directed to different proteins of the virus types.

The *HN glycoproteins* possess multiple distinct antigenic determinants. On the basis of serological reactivity, at least four epitopes can be determined on the HN molecule of Sendai virus, and three or four topographically distinct domains have been defined as being involved in the hemagglutination activity (Örvell and Grandien, 1982; Portner, 1984; Totzawa et al., 1986). In NDV HN protein, there are two or three nonoverlapping regions involved, one of which is conserved among strains, while the others undergo antigenic variation (Iorio et al., 1986; Nishikawa et al., 1983). In the human parainfluenza viruses a number of unique epitopes have also been found on the HN molecule as reported for virus type 1 (Yewdell and Gerhard, 1982) and type 3 (Coelingh et al., 1986; Rydbeck et al., 1986; Van Wyke Coelingh et al., 1985). These form at least three distinct antigenic sites and two overbridging sites, which are all involved in hemagglutination and virus infectivity. Three more sites could be defined but have no known biological activity (Coelingh et al., 1986; Rydbeck et al., 1986). Epitopes at three sites are conserved among a wide range of human isolates. In other epitopes, antigenic variation occurs, not resulting from the accumulation of mutations over time

but representing genetic heterogeneity within the virus population (Van Wyke Coelingh et al., 1985).

The antigenic composition of the *F protein* has been further elucidated by the discovery of three distinct antigenic sites on Sendai virus (Örvell and Grandien, 1982) and at least two on parainfluenza type 3 virus (Rydbeck et al., 1986). The *M protein* characterized for parainfluenza type 3 virus seems to possess six epitopes representing at least four nonoverlapping regions (Rydbeck et al., 1986). The *NP protein* of parainfluenza virus type 3 has been defined to have at least six epitopes representing five nonoverlapping sites (Rydbeck et al., 1986).

Although the polypeptide profiles of animal and human parainfluenza viruses belonging to the same type show pronounced similarities, antigenic differences exist. The cross-reactions between type 1 viruses, the murine (Sendai) and the human parainfluenza virus type 1, have been characterized in animal experiments (Van der Veen and Sonderkamp, 1965). Reciprocal cross-reactions were seen by complement fixation but not by hemagglutination inhibition tests.

Antigenic relationships between NP as well as HN proteins of simian parainfluenza virus, SV5, and human parainfluenza type 2 virus have been documented with monoclonal antibodies, which were also used to define type-specific epitopes (Goswami and Russell, 1983).

Antigens shared between bovine and human parainfluenza type 3 viruses have been found by hemagglutination inhibition and neutralization tests. Although the HN glycoprotein of the two viruses differs to a major extent, they share at least one common epitope (Ray and Compans, 1986).

Immunological cross-reactions between the different parainfluenza virus types and between parainfluenza viruses and mumps virus are known to occur in human as well as in animal sera (Chanock, 1979; Gardner, 1957; Julkunen, 1984; Meurman et al., 1982; Ukkonen et al., 1980; Van der Veen and Sonderkamp, 1965). Individual epitopes of the different structural components involved in this antigenic relationship have been defined by monoclonal antibodies (Örvell et al., 1986; Ray and Compans, 1986). Thus, a reciprocal immunological relationship residing in the HN and NP proteins exists between type 1 (Sendai) and type 3 viruses (Goswami and Russell, 1983; Örvell et al., 1986), possibly involving also the M protein of the two viruses (Örvell et al., 1986). Parainfluenza type 2 virus has occasionally been found to raise a heterologous antibody response in humans (Ukkonen et al., 1980) as well as in animals (Van der Veen and Sonderkamp, 1962); the antigenic relationship to other parainfluenza viruses or to mumps has so far not been confirmed by use of monoclonal antibodies.

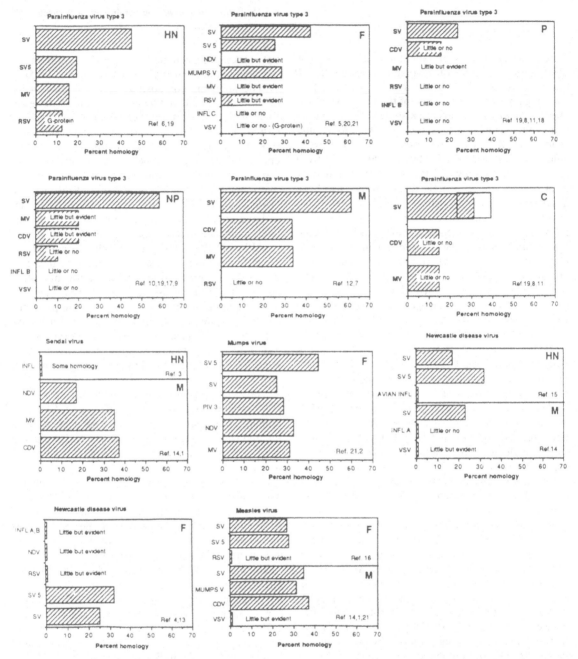

FIG. 2. Alignment of the amino acid sequence of some parainfluenza virus proteins with related myxoviruses and with vesicular stomatitis virus. Percent homology is graphically shown for each parainfluenza virus type 3 (PIV3) protein. HN = hemagglutinin-neuraminidase; F = fusion; P = phospho-; NP = nucleocapsid; M = matrix; C = small nonstructural protein compared to similar proteins of other viruses: Sendai virus (SV), simian virus 5 (SV5), Newcastle disease virus (NDV), mumps virus (mumps V), measles virus (MV), canine distemper virus (CDS), respiratory syncytial virus (RSV), influenza virus (Inflv) and vesicular stomatitis virus (VSV). Unfilled boxes denote uncertainty. The information included in this figure was collected by Dr. Robert Rydbeck, Karolinska Institute, Stockholm, Sweden from the following literature: *Ref:* 1. Bellini, W., et al., 1986. J. Virol. **58**:408; 2. Blumberg, B. et al., 1985. J. Gen. Virol. **66**:317; 3. Blumberg, B., et al., 1985. Cell **41**:269;

4. Chambers, P., et al., 1986. J. Gen. Virol. **67**:2685; 5. Côté, M.-J., et al., 1987. J. Gen. Virol. **68**:1003; 6. Elango, N., et al., 1986. J. Virol. **57**:481; 7. Galinski, M. S., et al., 1987. Virol. **157**:24; 8. Galinski, M. S., et al., 1986. Virol. **155**:46; 9. Galinski, M. S., et al., 1986. Virol. **149**:139; 10. Jambou, R. C., et al., 1986. J. Gen. Virol. **67**:2543; 11. Luk, D. et al., 1986. Virol. **153**:318; 12. Luk, D., et al., 1987. Virol. **156**:189; 13. McGinnes, L. W., et al., 1986. Virus Res. **5**:343; 14. McGinnes, L. W. and T. G. Morrison. 1987. Virol. **156**:221; 15. Millar, N. S., et al., 1986. J. Gen. Virol. **67**:1917; 16. Richardson, C., et al., 1986. Virol. **155**:508; 17. Sanchez, A., et al., 1986. Virol. **152**:171; 18. Satake, M., et al., 1984. J. Virol. **52**:991; 19. Spriggs, M. K. and P. L. Collins. 1986. J. Gen. Virol. **67**:2705; 20. Spriggs, M. K., et al., 1986. Virol. **152**:241; 21. Waxham, N., et al., 1987. Virol. **159**:381.

The two subtypes of parainfluenza virus type 4, 4A and 4B, are closely related and share NP antigens. Differences in neutralization and hemadsorption inhibition reactions indicate that they can be recognized by distinct HN proteins.

Although a reciprocal immunological relationship exists between mumps and Sendai virus NP proteins, a one-way cross-reaction is found between monoclonal antibodies to mumps P and F proteins and Sendai virus as well as between Sendai virus HN antibodies and mumps virus HN protein (Örvell et al., 1986). Parainfluenza type 3 virus seems to be antigenically less related to mumps virus (Örvell et al., 1986). The high degree of homology between similar proteins of related viruses is shown in Figure 2. Although the three-dimensional structures of the proteins are not known, the homology of the amino acid sequences indicates possible cross-reactions. Recent work on the amino acid homology among the paramyxoviruses shows that the HN proteins of SV_5, NDV and mumps virus are closely related to each other but distinct from Sendai virus and parainfluenza virus type 3 HN proteins. (Kövamees and Elango, 1988, personal communication). It appears that L (Fig. 2; Blumberg, B., personal communication), F, and M proteins are the most conserved proteins of the paramyxoviridae.

The six parainfluenza virus genes, coding for seven proteins, are located after each other within one piece of viral RNA. Their precise order has been given for parainfluenza type 3 virus and is from the 3' end of the genome: NP–(P+C)–M–F–HN–L (Spriggs and Collins, 1986). The order of genes is the same for NDV (Chambers et al., 1986; Wilde et al., 1986). The nucleotide sequences signaling start and termination of transcription of the different mRNAs have been defined as well as the nuclease sensitive intergenic regions (spacer sequences) (Amesse and Kingsbury, 1982; Kingsbury, 1985).

Mutation occurs in the parainfluenza viruses as well as in other single-strand RNA genomes lacking the error-correcting capability that double-strand genomes possess. Temperature-sensitive mutants have been described. When mutations occur in HN and F proteins, the virions may, however, become less viable, as the functional or structural character may change.

Interfering incomplete forms of parainfluenza virus may arise spontaneously during replication. One type of Sendai virus incomplete particles has genomes that are extensively deleted internally, their RNA species retaining only parts of the NP and L genes together with the terminal sequences. (Amesse et al., 1982). In other particles, an altered nucleotide sequence of the copy-back type has been found (Re and Kingsbury, 1986).

Reactions to Physical and Chemical Agents

The parainfluenza viruses are relatively unstable, when frozen and thawed, for example, they need additional proteins for protection. Serum proteins in media protect the virus to some extent from heat inactivation at temperatures of +37°C or higher. Addition of 0.5% bovine serum albumin or 5% chicken serum to the suspending medium prevents loss of infectivity during storage at −60°C for years (Chanock, 1979). The viruses are inactivated by exposure to lipid solvents such as 20% ether, to formaldehyde, to detergents, and to low pH (pH 3).

Pathogenesis, Pathophysiology, and Pathology

The infectivity of the parainfluenza viruses depends on the activation through cleavage of the fusion protein, F_0, and on the availability of proteases for the cleavage. The capacity of host cells to affect such cleavage may partly explain the localization of the virus infection in the human respiratory tract (Rott, 1979).

The spread of infection in tissue culture cells can occur either through released progeny virus adsorbing to new host cells scattered in the tissue culture or through fusion of the infected cell with neighboring cells. Some virus strains induce syncytium formation, leading to cell death; this is a characteristic feature of parainfluenza type 2 and 3 infections in human cell culture lines. No syncytium formation has been described in fatal human cases unless the affected child was severely immune-deficient (Aherne et al., 1970; Delage et al., 1979; Downham et al., 1975; Jarvis et al., 1979). Other strains may cause a persistent productive infection, which does not kill the cell (Choppin, 1964; Holmes and Choppin, 1966).

In humans, the viruses infect cells not only in the mucus membranes of the nose and throat but also in the lower respiratory tract, leading to epithelial cell death and inflammatory reactions. The different types vary in virulence. Whereas type 3 is the most virulent of the parainfluenza viruses, type 4 (A and B) seldom causes serious illness. Larynx and trachea may be involved, resulting in laryngotracheobronchitis (croup) in children. This syndrome is mainly associated with parainfluenza type 1 and 2 infections (Kim et al., 1961). Type 1 extends more often than type 2 infection to the bronchi, causing atelectasis and pneumonia (Parrot et al., 1959). Croup is more seldom seen in parainfluenza type 3 infections, which affect respiratory tract cells lining the small air passages and cause bronchitis or bronchiolitis and pneumonia. The similarity between the clinical manifestations of RS virus and parainfluenza virus type 3 in early infancy has been noted (Gardner et al., 1971).

Infections with parainfluenza viruses induce immunity, which, however, is of short duration. Reinfections occur frequently, usually with milder symptoms or asymptomatically (Hall et al., 1971). Parainfluenza virus type 3 has been shown to give reinfections within 9 months (Chanock et al., 1963).

The secretory immune response appears 7 to 10 days after onset of symptoms. The neutralizing activity in nasopharyngeal secretions reaches its peak 2 to 4 weeks after onset and is associated with secretory IgA. Whether this IgA alone or in association with a local cell-mediated immune response is responsible for recovery and immunity is not quite clear (Yanagihara and McIntosh, 1980).

As in RS virus infections, maternal antibodies to parainfluenza virus or sensitizing antibodies after a previous parainfluenza virus infection have been suggested to participate in the pathogenesis of severe bronchiolitis or pneumonia in early infancy. In an animal model this could not be demonstrated: moderate levels of passive antibodies or antibodies after a previous infection did not enhance development of pneumonia or pneumonitis in hamsters after intranasal infection by parainfluenza type 3 virus. However, an interesting phenomenon was noted: at reinfection, hamsters had significantly greater infiltrates if they had experienced primary infections in the presence of passively given antibodies than if they had primary infections without passive antibodies. Whether this depended on modified humoral antibody response caused by the presence of antibodies in the primary infection or on other parameters, such as a cellular immune response, could not be concluded (Glezen and Fernald, 1976). It is evident, however, that primary infections during the first years of life result in more severe illness than do reinfections.

An association has been shown between the formation of virus-specific IgE and the release of histamine in nasopharyngeal secretions leading to the development of croup during infection. This suggests a role of immunological mechanisms in the pathogenesis of severe forms of respiratory illness in parainfluenza virus infections (Welliver et al., 1982).

Viremia seems not to be common in parainfluenza virus infections of man, although it has been described for type 3 virus (Gross et al., 1973). Interferon, which may be important for inhibition of infection in the early stages of disease, is produced in nasal secretions in 30% of children infected with parainfluenza virus, compared to only 4% of children with RS virus infections (Hall et al., 1978).

Clinical Features

Mode of Transmission

Parainfluenza viruses spread readily to susceptible persons by contact with infected individuals or by droplets generated by coughs or sneezes. Although symptoms are less pronounced in reinfections, reinfected individuals can also transmit the disease (Mufson et al., 1973). In wards and institutions, infections with type 1 or 2 have been reported to spread to 40 to 70% of susceptible individuals, and infections with type 3 to all susceptible (Chanock et al., 1963).

Period of Infectivity

In primary infections, the period of type 3 virus shedding varies between 3 and 10 days; in reinfections, the period is usually shorter. Prolonged shedding is seen in immunosuppressed children and children with leukemia as well as in adults with chronic respiratory disease (Gross et al., 1973). Normal children shedding virus for 3 to 4 weeks have been observed (Chanock et al., 1963; Frank et al., 1981).

Incubation Period

The incubation period has been studied in adults with experimental infections and was found to be for type 1, 1 to 7 days, and for type 3, 2 to 6 days (Kapikian et al., 1961; Tyrrell et al., 1959). In institutional outbreaks, the incubation period in children for type 3 virus was 2 to 4 days (Chanock et al., 1963).

Symptoms

The four types of parainfluenza virus all cause acute respiratory infections. Parainfluenza type 1 and 3 viruses have been isolated from patients with acute parotitis (Bloom et al., 1961; Zollar and Mufson, 1970), but this seems to be exceptionally rare.

In children, primary infections usually produce cough, rhinorrhea, sore throat, and fever for 2 or 3 days (Parrot et al., 1959; Chanock et al., 1963). Severe illness affecting the lower respiratory tract occurs occasionally, especially during the first year of life and in immunosuppressed children. After 6 years of age, infections of the lower respiratory tract are rare.

Acute laryngotracheobronchitis (croup) is a common manifestation of infection with parainfluenza virus in children. The propensity of the different types to produce croup symptoms in children with lower respiratory tract infections has been reported from the United States to be about 60% for type 1 and 2 viruses and about 30% for type 3 virus (Denny et al., 1983).

The incidence of croup in various age groups has a peak in the 1-year-old group, with an attack rate of 15 children per 1,000 (Foy et al., 1973). About one-third of the cases with croup in children have been associated with parainfluenza virus infections. In an English report, 65% of isolated strains belonged to type 1, 30% to type 2, and 15% to type 3 virus (Gardner et al., 1971). In the developing countries, croup appears to be less common (Institute of Medicine, 1986).

Type 1 virus has been shown to cause bronchitis or pneumonia in about a quarter of infected children. In type 2 infections, croup often is the dominating symptom.

Parainfluenza virus type 3 infection is, second only to RS virus infection, the most common cause of lower respiratory tract infection in children below 6 months of age. The clinical picture is similar to that caused by RS virus with severe bronchiolitis and pneumonia. Although croup occurs, it is not a frequent feature in type 3 infections.

Infection with type 4 virus seems to be more often associated with mild symptoms than infections with the other types. However, serious illness has been seen in immunosuppressed children.

Reinfections in adults and older children are mostly of mild character (Bloom et al., 1961). Adenopathy is rare in parainfluenza infections (Parrot et al., 1959).

Laboratory Diagnosis

Parainfluenza virus infections can be diagnosed by specific identification of virus or viral antigens either directly in the clinical specimen or in tissue culture cells after virus isolation. Serological diagnosis is based on demonstration of rising antibody titers or on appearance of specific IgM. Although demonstration of secretory IgA may sometimes be of diagnostic value, a serum IgA antibody response does not seem to be significant. While conventional virus isolation and the demonstration of an antibody titer rise in convalescent-phase serum are slow methods, modern techniques with detection of virus or viral antigen directly in the clinical specimen permit an early and rapid diagnosis useful for clinical decisions (Gardner and McQuillin, 1980). Such techniques may also be used for epidemiological screenings (Ørstavik et al., 1984). If possible, confirmation of results obtained by antigen detection techniques should be made by virus isolation until proficiency in rapid diagnosis can be ensured.

Specimen Collection

Specimens for virus identification are taken from the respiratory tract, preferably within the first 3 or 4 days of illness. Only exceptionally are diagnostic efforts rewarding if specimens are taken more than 6 or 7 days after onset of illness.

During collection and transportation of specimens for virus isolation, it is important to preserve the viability of the virus. When collecting material for identification of virus directly in the specimen, avoidance of dilution (for immunoassays) and rupture of cells in the secretions (for immunofluorescence) may be of crucial importance; transportation per se is less crucial for the latter types of specimens.

Ideal material for all these investigations is the nasopharyngeal secretion (NPS), which is obtained by aspiration or suction, although for virus isolation nasal or cough swabs may also be used. The NPS gives a higher yield of parainfluenza virus than other material from the respiratory tract (Downham et al., 1974). Collection of NPS has been described in detail by Gardner and McQuillin (1980). The secretions are collected by suction through a sterile polythene feeding tube (size 8) into a plastic mucus extractor (see also Chapter 1). After collection, not more than 0.5 ml fluid should be used to rinse the tube if the specimen is to be used for detection of virus by immunoassays.

Sputum, tracheal secretions, tracheal or bronchial washings, or autopsy material (e.g., lung tissue as such or as biopsies) can also be used for virus diagnosis.

Detection of Viral Antigens Directly in the Clinical Specimen

Although no methods have been available for diagnosis of parainfluenza virus infections by detection of virus particles as such, reliable methods have been developed for detection of viral antigens. By use of specific antibodies, viral antigens are detected in the specimen by examination of infected cells by immunofluorescence staining or by investigation of the secretion by immunoassay. The main advantage of these methods over conventional virus diagnosis by virus isolation or serological investigation is that a diagnosis is obtained within a few hours and at a time when the patient is still ill. Furthermore, transportation and storage of specimens are facilitated, as no precautions have to be taken to keep the virus infectious. Viral antigens in the specimens are not affected by transportation at ambient temperature.

IMMUNOFLUORESCENCE STAINING FOR DETECTION OF PARAINFLUENZA VIRAL ANTIGENS

Immunofluorescence can be used for detection of viral antigens in cells present in the clinical specimen. These cells must be separated from the secretion and freed from mucus before they are applied on microscope slides for staining. The technique of preparation is of importance for the reliability of the fluorescence diagnosis. It has been described in detail for nasopharyngeal secretions (NPS) (Gardner and McQuillin, 1980). The cells are pelleted by centrifugation, and the supernatant is removed (for virus isolation).

The cell deposit is suspended in 0.5 ml PBS and gently pipetted with a Pasteur pipette until the deposit is broken up. A further 0.5 ml of PBS is added, and the suspension is pipetted to wash the cells further. This is repeated several times with additional 0.5-ml volumes of PBS until the tube contains 8 to 10 ml of fluid. After another centrifugation, the pelleted cells are suspended in a few drops of PBS and spread on small, previously etched areas on microscope slides, then air-dried and fixed in acetone for 10 min at 4°C (Gardner and McQuillin, 1980) or for 5 min at room temperature.

A simplified method using undiluted aspirated secretions for the preparation of cell deposits on microscope slides has been described for detection of RS virus as well as of parainfluenza virus type 3 (Ånestad, 1985). The resulting nonspecific staining of mucus, which may obscure the recognition of weak positive specimens, is only partly overcome by use of highly specific reagents such as monoclonal antibodies.

Areas with cell deposits fixed in acetone are investigated by the direct or indirect immunofluorescence technique for the presence of the different parainfluenza virus types.

Test Procedure and Results

For immunofluorescence diagnosis in clinical specimens, the indirect method has been recommended whenever the presence of several different viruses is investigated. The reason for this is to avoid standardization of several conjugates by using only one antispecies conjugate (Gardner and McQuillin, 1980).

The appearance of fluorescence in the infected cell varies somewhat for the different parainfluenza virus types. This can be seen in infected tissue culture cells as well as in respiratory epithelial specimen cells with animal hyperimmune sera (Fig. 3). Type 1 virus, in addition to granular fluorescence, produces rough strands or fluorescence with a floccular appearance in the cytoplasm of the infected cell, whereas type 3 gives mainly coarse granular fluorescence with inclusions of irregular size. Type 2-infected cells exhibit granular fluorescence and occasionally large, single, rounded inclusions. In diagnostic situations, knowledge of the antigen distribution pattern in virus-infected cells provides an additional confirmation of the diagnosis. This is especially valuable in parainfluenza virus immunofluorescence diagnosis, where good reagents may be difficult to obtain.

The use of monoclonal antibodies for identifica-

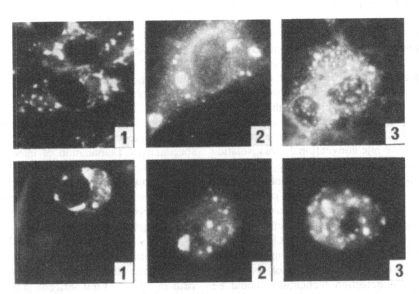

FIG. 3. Immunofluorescence of cells infected with parainfluenza virus types 1, 2, and 3. Viral antigens are detected in the cytoplasm of the infected cells by use of monospecific animal hyperimmune sera. Top row shows infected tissue culture cells; lower row shows infected cells in clinical specimens (nasopharyngeal secretions).

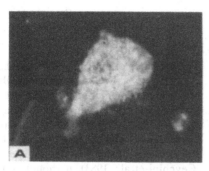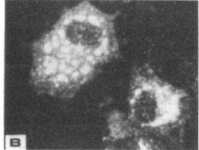

FIG. 4. Immunofluorescence pattern of parainfluenza type 1–infected cells as observed with monoclonal antibodies to the HN (A) and the NP (B) proteins.

tion of the viruses may change the appearance of fluorescence depending on the localization of the different viral antigens in the infected cells. Antibodies directed against epitopes on HN or F of the parainfluenza virus give a fine granular fluorescence covering the surface of cells—even if they are fixed in acetone. With antibodies against NP, P, or M, irregular cytoplasmic inclusions can be seen, and the borders of the cells are not clearly delineated (Fig. 4).

For parainfluenza virus type 1 (Sendai), the appearance of the different components has been analyzed by tracing them with monoclonal antibodies in immunofluorescence experiments at different times after infection. At $7\frac{1}{2}$ h postinfection, HN and F antigens could both be seen with a thin, fine granular pattern evenly distributed in the cytoplasm, and cell borders were clearly delineated. At 16 h postinfection, when the amount of antigen in the cytoplasm had increased, the empty nucleus was covered by the cytoplasm fluorescence and could usually not be seen. Antibodies against NP and P, which at all times of the replication cycle gave identical immunofluorescence patterns, first ($7\frac{1}{2}$ h postinfection) stained antigens with a fine floccular distribution, and later (9 h postinfection) with a coarse floccular distribution. During the following hours the coarse floccular pattern increased in the central part of the cell. A number of irregular inclusions appeared, which later (40 h postinfection) were enlarged and could be seen mainly around and near the nucleus. Likewise, antibodies directed against M showed that M first appeared as fine dots and later as large masses of antigen near the nucleus (Örvell and Grandien, 1982).

Parainfluenza virus–infected cells in the human specimen represent all stages of infection. The use of monoclonal antibodies directed against only one viral epitope may suffice to identify cells in a late state of infection. Cells in early infection give only faint fluorescence, and here a combination of monoclonal antibodies directed against several viral antigens offers advantages in diagnostic work.

Practical Applications

Immunofluorescence diagnosis of the parainfluenza virus infection is based on some early work on infected tissue culture cells. This work described the localization of intracellular antigen of Sendai virus (Traver et al., 1960) and parainfluenza virus type 3 (Sominina et al., 1967). Fluorescence of antigen was reported to occur at an early stage in the nucleoli and to be rapidly transferred to the cytoplasm of the infected cell (Sominina et al., 1967). Antigen could be detected in the cells sooner by immunofluorescence than by hemadsorption and much earlier than virus could be found by hemagglutination by the culture medium (Fedova and Zelenkova, 1969).

Rapid diagnosis was first performed on tissue culture cells incubated with clinical specimens. The cells were scraped off from the culture tubes and prepared as cell deposits on microscope glass slides and stained by immunofluorescence with specific antisera. Alternatively, virus was passaged from the hemadsorption-positive tissue culture into tubes containing cover slip cell cultures that were incubated for 1 to 3 days. When hemadsorption again appeared, the slides were fixed and stained for presence of viral antigens of the four parainfluenza virus types as well as of other respiratory viruses (Fedova et al., 1969).

Gardner and co-workers were the first to report immunofluorescence diagnosis directly in clinical specimens taken from children with parainfluenza viral infections. The results of a $2\frac{1}{2}$-year study period showed agreement between isolation results and immunofluorescence diagnosis for 53 of 55 specimens including infections with type 1, 2, and 3 viruses. In only two of the 25 parainfluenza type 1 isolation-positive specimens could the diagnosis not be established by immunofluorescence (Gardner et al., 1971). Other studies were not as successful, and for parainfluenza type 3, for example, the rate of success was reported to be 75% out of isolation positive specimens.

Gardner and co-workers extended their studies on diagnosis of parainfluenza viruses in children.

Results of investigations of NPS specimens for parainfluenza virus types 1, 2, and 3 by immunofluorescence and by virus isolation showed copositivity for
388 of 412 specimens—94.2%. Failure to identify virus by immunofluorescence was due either to a poor
specimen or to inadequate preparation of the specimen. Specimens giving a positive parainfluenza virus
diagnosis by immunofluorescence without yielding
virus on isolation were all collected from patients
after day 5 of illness. The authors conclude that at
this time, that is, in the convalescent stage of illness,
virus isolation may fail while immunofluorescence is
still positive, and that in later stages of infection,
immunofluorescence is the most sensitive of the two
methods (Gardner and McQuillin, 1980).

A 3-year study performed at six laboratories in
different European countries was organized 1978 to
1981 by the European Group for Rapid Viral Diagnosis. The aim was to promote the use of immunofluorescence for diagnosis of certain respiratory viral
infections and to study the epidemiology of these
infections in children by the technique. In addition to
RS and influenza A viruses, the presence of parainfluenza virus types 1 and 3 was also investigated in
specimens from infants and children below 6 years of
age admitted to the hospital with an acute respiratory
infection (Ørstavik et al., 1984). In 1,927 (25%) of the
7,716 NPS specimens, one of the four mentioned viruses was diagnosed. While RS and influenza A virus
was found in 19 and 1.6%, respectively, parainfluenza types 1 and 3 were found each in a proportion
varying from 0.3 to 4% at the six laboratories. Interesting epidemiological observations were made (see
Epidemiology).

Parainfluenza type 4 can also be diagnosed by immunofluorescence directly in clinical specimens
(Gardner and McQuillin, 1980). Antisera have been
prepared to each of the two subtypes, 4A and 4B.
Although the two sera showed certain heterologous
reactions with each other, they could be used for
diagnostic work after absorptions.

With great experience and after many years'
work, Gardner and McQuillin (1980) concluded that
the immunofluorescence technique could be used not
only for identification of parainfluenza virus types 1,
2, 3, 4A, and 4B in preference to neutralization tests
after isolation in tissue culture, but also for rapid
diagnosis of parainfluenza virus infections directly in
clinical specimens.

Immunofluorescence for detection of the parainfluenza viruses has also been used by other investigators. Wong and co-workers (1982) could detect
viral antigens in exfoliated cells by immunofluorescence for intervals of 9 to 15 days in infected patients. The successful use of monoclonal antibodies
has been reported for diagnosis of type 3 virus by the
same technique (Waner et al., 1985).

IMMUNOPEROXIDASE STAINING FOR DETECTION OF PARAINFLUENZA VIRAL ANTIGENS

The presence of endogenous peroxidase in nasopharyngeal cells as well as in red and white blood cells
has made the immunoperoxidase staining technique
less suitable for diagnosis of respiratory infections
(Gardner et al., 1978). If promises from a report using different technical procedures for the detection of
RS virus directly in the clinical specimen hold true
(Cevenini et al., 1983), it should be possible to use
also immunoperoxidase staining for diagnosis of the
parainfluenza viruses.

IMMUNOASSAYS FOR DETECTION OF PARAINFLUENZA VIRAL ANTIGENS

Sensitive immunoassays have been described for the
type-specific diagnosis of parainfluenza virus in human specimens (Grandien et al., 1985; Sarkkinen et
al., 1981a,b). Nasopharyngeal secretions (NPS) collected by suction from the nasopharynx (see earlier)
should be used for the investigation, as they contain
more virus and viral antigens than throat washings
and nasal or cough swabs. Unnecessary dilution
should be avoided during collection and processing
of the specimens.

Processing of Specimens for Immunoassays

The original NPS is diluted 1:2 in PBS containing
20% inactivated fetal calf serum, 2% Tween 20, and
10^{-4} M merthiolate and sonicated for 1 to 3 min to
homogenize the specimen. The specimen is then further diluted in the same buffer for test by the immunoassay (Sarkkinen et al., 1981b).

Test Procedure and Results

Use of immunoassays for type-specific diagnosis of
the parainfluenza virus types 1, 2, and 3 in NPS was
first reported from Finland (Sarkkinen et al.,
1981a,b; see also elsewhere in this volume). Polystyrene beads or separate wells in microtiter plates were
coated with guinea pig immunoglobulins to types 1,
2, and 3 virus. After addition of specimens and incubation overnight at 37°C, rabbit antiviral immunoglobulins were added. After 1 h incubation, horseradish peroxidase or ^{125}I-labeled antibodies to rabbit
Ig were added. For the enzyme assay, orthophenyl-
ene diamine was used as substrate. Specimens were
diagnosed as positive provided the confirmatory test,
performed as a blocking assay, indicated specific
binding.

The four-layer enzyme immunoassay (EIA) and
the radioimmunoassay (RIA) were found to possess

equal sensitivity when used with the same antiviral reagents (Sarkkinen et al., 1981b). The sensitivity of the assays varied between 1 and 10 ng viral antigen detected per milliliter for type 1 and 3 assays and between 10 and 30 ng/ml for the type 2 assay. In another type 3 ELISA (Grandien et al., 1985), other viral reagents were used, but the test was mainly constructed along the same principle as the described Finnish assay. The cutoff level was calculated as three times background activity, which corresponded to absorbance values of ≤0.1 and the test detected down to 10 ng viral proteins per test well (0.1 ml). A test protocol from this ELISA is given in Table 1.

Neither of the two studies reported any difficulties with cross-reactions in the antigen detection assays, and both obtained type-specific results.

Practical Applications

In a screening of specimens from 174 children with acute respiratory infections admitted to the hospital in Turku, Finland, 28 specimens were diagnosed by

TABLE 1. ELISA for parainfluenza virus type 3 diagnosis

| Antigen titration (ng/ml) | OD 450 nm | |
	NP	F
2,048	843	888
1,024	782	822
512	553	563
256	335	339
128	207	211
64	115	136
32	105	95
16	64	78
8	47	49
4	40	39
Buffer	32	30

| Nasopharyngeal secretions specimen number | OD 450 nm | |
	NP	F
1,118/85	772	173
1,439/85	1,104	648
1,261/85	1,231	375
125/85	414	54
1,808/85	19	8
1,932/85	969	437
2,068/85	39	53
2,315/85	454	225
1,807/85	37	20
Buffer	15	10

Antigen detection by monoclonal antibodies against the nucleocapsid (NP) and fusion (F) antigens in the fluid phase; guinea pig antiviral Ig on the solid phase.

immunofluorescence as positives: 4, 4, and 20 with parainfluenza types 1, 2, and 3, respectively. The immunoassay diagnosed all but one type 1 and one type 3 virus. Furthermore, four additional specimens were found positive by the assay, one type 1 and three type 3 infections (Sarkkinen et al., 1981b). A total agreement was seen between results obtained by ELISA and indirect immunofluorescence for the six parainfluenza type 3 positives found among the 376 specimens tested (Grandien et al., 1985).

During 3½ years' use of immunoassays for routine diagnosis of respiratory infections, the Turku laboratory reports investigation of 8,298 specimens. Parainfluenza virus types 1, 2, and 3 were diagnosed in 90 (1%), 52 (0.6%), and 172 (2%) cases, respectively (Sarkkinen, 1985).

The introduction of monoclonal antibodies, which can be used in the described immunoassays, will considerably increase the sensitivity and reduce the time of performance of the assay.

The time-resolved fluoroimmunoassay (TR-FIA) described as a highly sensitive test based on use of fluorescent probes (Soini and Kojola, 1983) can be used for type-specific detection of the parainfluenza viruses (Halonen et al., 1983). Employment of monoclonal antibodies and the reduction of background activity by the test principle itself allow a one-step procedure (specimen and monoclonal antibodies are mixed and incubated together) with a 1- to 2-h performance time.

Reagents for Antigen Detection

Production of reagents for the detection of type-specific parainfluenza virus antigens by immunofluorescence is time-consuming and difficult. By use of pools of carefully selected monoclonal antibodies, cross-reactions to other types can be eliminated, and the nonspecific reactions, notorious for immunofluorescence reagents, can be overcome (Waner et al., 1985).

For immunoassays, high-titer specific antisera were produced by Sarkkinen and co-workers (1981a, b) in guinea pigs and rabbits by using intradermal immunization with purified virus proteins in incomplete Freund's adjuvant. Intradermal boosters were given 3 and 6 weeks after the first injection. Immunoglobulin fractions were prepared by precipitation of serum with an 18% (wt/vol) final concentration of sodium sulfate followed by chromatography on a Sephadex G-25 column. In another study (Grandien et al., 1985), high-titer antisera were obtained to parainfluenza type 3 virus by intranasal infection of guinea pigs followed by intradermal injections. Monoclonal antibodies are used also for the solid-phase immunoassays, mainly in the fluid phase (Table 1).

Virus Isolation and Identification

ISOLATION IN TISSUE CULTURE

In addition to primary human embryonic and monkey kidney cell cultures, the continuous cell line LLC-MK2 provides a sensitive system for isolation of the parainfluenza viruses. Since the supply of suitable human cells is limited, most work with the parainfluenza viruses has been performed with primary monkey kidney cells. Rhesus and cynomolgus monkey kidney cultures have been found equally sensitive for isolation of type 1, 2, and 3 viruses (Chanock, 1979). For isolation of type 4 virus, however, rhesus monkey cells provide a more sensitive system than cynomolgus kidney cultures (Canchola et al. 1964)

The supply of monkey kidney is limited, and monkey breeding is expensive. The continuous cell lines, being easily available and less expensive, have therefore been investigated for their sensitivity to parainfluenza virus growth. The rhesus monkey kidney continuous cell line LLC-MK2 with trypsin-supplemented medium has been found to be a satisfactory alternative (Frank et al., 1979). In titration experiments, LLC-MK2 was found to have equal sensitivity to rhesus and cynomolgus kidney primary cultures. For reisolation from clinical specimens of type 1 and 3 viruses, the LLC-MK2 cells were more efficient than the primary monkey cells. Reisolation of type 2 virus seemed to be more successful in primary cynomolgus cultures than in LLC-MK2 cells; isolation rates obtained were 62 and 35%, respectively. There were, however, few type 2 isolates in the study. The authors summarize that for reisolation of the parainfluenza viruses, LLC-MK2 was superior to the primary rhesus and similar to the primary cynomolgus monkey kidney cells and was thus an acceptable tissue alternative to primary monkey kidney.

The Madin-Darby canine kidney (MDCK) cell line is less suitable for isolation of parainfluenza virus, although it is efficient for isolation of influenza virus strains. HeLa and Vero cells are not sensitive enough for virus isolation but can be used, as can MA104 cells, for serial passages of virus.

Specimens are inoculated into tissue culture without delay and incubated at temperatures from 33 to 36°C. Rotation during incubation increases the sensitivity of the tissue culture to type 4 virus (Canchola et al., 1964). The primary renal monkey cultures are best grown with 10% fetal calf serum. As the recovery of influenza virus may be inhibited by the presence of serum, the monkey kidney culture may be maintained in Eagle's minimal essential medium without serum. *Medium* for the continuous rhesus kidney culture LLC-MK2 has to be supplemented with 2 μg trypsin per milliliter of medium.

The parainfluenza viruses do not always produce enough cytopathic effect (CPE) to permit recognition in monkey kidney cultures during isolation (Chanock et al., 1963). Even after several passages, type 1 virus may cause only minimal destruction of the culture cells. Type 2 virus, which produces irregular syncytia, may be more easily recognized, particularly when the cells fall away from the cell culture, giving a "Swiss cheese" appearance. Type 3 virus, which like type 1 causes poorly recognizable CPE, may after passage give destruction of the cell sheet. In human cell lines, type 2 and type 3 virus have been described to cause syncytia (Chanock et al., 1963).

Primary monkey kidney cells are frequently contaminated by SV5, a simian parainfluenza virus type 2, which has biologic activities similar to the human parainfluenza viruses. The virus is antigenically related to the human type 2 virus. By addition of 0.2% SV5 rabbit immune serum to the medium, growth of this contaminant is eliminated (Chanock, 1979). By the use of LLC-MK2 cells, the risk of naturally occurring SV5 contamination is eliminated.

Hemadsorption Test

Since the parainfluenza viruses seldom show clear CPE in the tissue culture, the property of infected cells to adsorb red blood cells (RBC) is used for recognition of infection (Vogel and Shelokov, 1957). The cultures are examined for hemadsorption when CPE is seen or suspected; otherwise every 5 days. After the test, medium and RBC are replaced with fresh medium to prevent nonspecific adsorption. For the hemadsorption test, guinea pig (or human) RBC are prepared as a 10% suspension. The RBC can be stored for a few weeks and are washed once a week in saline. A suitable volume of the suspension is diluted in isotonic saline to a concentration of 0.4% before each test. To tissue-culture tubes with 1 to 1.5 ml fluid medium 0.2 ml of the RBC solution is added. The tubes are subsequently incubated at 4°C in a horizontal position for 30 min. If no hemadsorption occurs, the tubes are incubated at room temperature or 25 to 37°C, which is required for an optimal hemadsorption to cells infected with type 4 virus. Hemadsorptions caused by type 1 and 3 virus are reversed at this temperature.

Nonspecifically adsorbed RBC are dislodged by gently rocking or rotating the tubes before the examination. Each tube is read in a microscope at a magnification of at least 60×. Strong hemadsorption is easily seen; weak reactions may be questionable, especially as nonspecific hemadsorption often occurs. If after another 3 or 4 days' incubation repeated investigations still give questionable results, a passage to fresh tissue culture is advisable. Use of noninfected cultures from the same lot of cells as controls in the test facilitates the interpretation of test results.

If the growth of parainfluenza virus results in

large amounts of free virus in the medium, the added RBC are agglutinated. This may interfere with the hemadsorption, and when agglutination is seen, the cultures should be washed several times before addition of RBC (Chanock, 1979).

Identification of Virus

Isolates can be identified by immunofluorescence of infected cells, by hemadsorption inhibition performed in the tissue culture tube, by hemagglutination inhibition or neutralization tests performed with the tissue culture medium, or by CF tests of the infected cells.

Immunofluorescence offers a rapid, relatively cheap, and little work-demanding method for identification of isolates (Gardner and McQuillin, 1980; McIntosh and Clark, 1985). The neutralization test can be used for identification of all four types of parainfluenza virus; the hemadsorption inhibition test may be less suitable for identification of type 4 virus, since the virus is sensitive for inhibitors present in rabbit and chicken sera (Chanock, 1979).

Immunofluorescence

Tissue cultures in tubes showing hemadsorption are investigated with immunofluorescence before the CPE is extensive. The RBC are removed by repeated washings. A few drops of fresh PBS is added to the tube, and the tissue culture is scraped down from the walls of the tube. The cells are gently dispersed by pipetting with a Pasteur pipette. Drops of cell suspension are applied to etched areas on microscope slides and dried at room temperature before fixation for 5 min in acetone. The immunofluorescence staining is performed with specific antisera, preferably prepared in only one animal species to avoid use of several antispecies Ig FITC conjugates. The introduction of Mab in diagnostic work facilitates the otherwise difficult production of reagents for diagnostic use.

Hemadsorption Inhibition Test

Isolates are identified by sera that prevent an observed hemadsorption (Chanock, 1979). The sera are treated with receptor-destroying enzyme (RDE), inactivated, and diluted in physiological saline. The cultures are washed twice with balanced salt solution, BSS. Then 0.2 ml of RDE-treated serum diluted 1:10 and 0.6 ml BSS are added to the tube and incubated with the culture for 20 to 30 min at room temperature. Subsequently, RBC are added for the hemadsorption test.

GROWTH IN EMBRYONATED EGGS

Embryonated eggs can be used for growth of some parainfluenza virus strains, but they may not provide a sensitive enough system for isolation of naturally occurring virus strains, although some human strains of type 2 virus have been recovered in eggs. Though several strains of type 1 virus have been adapted to growth in the allantoic cavity, most strains of type 2 virus multiply only in the amniotic fluid and produce insufficient amounts of virus to cause agglutination of RBC (Chanock, 1979). However, SV5, closely related to the human parainfluenza type 2 virus, has been successfully cultivated to high titers in the allantoic cavity (Espmark, 1965).

For inoculation into the amniotic cavity 9- to 11-day-old embryonated eggs are used. Maximum virus titers are obtained after incubation for 3 to 7 days at 35°C.

INFECTION OF ANIMALS

Hamsters and guinea pigs can be infected intranasally by type 1, 2, and 3 viruses without signs of disease.

Antibody Assays

The solid-phase immunoassays and the complement fixation, hemagglutination inhibition, and neutralization tests can all be used to measure specific antibodies to the parainfluenza virus types. The serological diagnosis of an acute parainfluenza infection is based on the detection either of a significant rise of antibody titers in acute and convalescent phase sera or of virus-specific IgM. However, the antibody response in humans is not always type-specific. Heterotypic antibody responses to other parainfluenza virus types and mumps often occur. Sometimes the induced antibody response may not be strong enough to be recognized by laboratory tests.

SOLID-PHASE IMMUNOSORBENT ASSAY

Many laboratories are in the process of introducing the enzyme-linked immunosorbent assay (ELISA) for estimation of the specific antibody response to the individual virus types (Julkunen, 1984; Ukkonen et al., 1980). Purified or partly purified virions are prepared from tissue culture media or from allantoic fluids and disrupted for use as antigens in the test. The test can be performed in microtiter plates, as described elsewhere in this volume. Infected fixed monolayer cell cultures have also been used as solid-phase coupled antigen for ELISA (Bishai and Galli, 1978).

ELISA has been shown to be more sensitive than other serological methods for the diagnosis of parainfluenza virus infections (Julkunen, 1984). This may be partly explained by the sensitivity of the assay itself, which allows diagnosis of parainfluenza virus infections even when the antibody response is weak owing to reinfection or a limited viral replication in the respiratory tract. The measuring of antibodies of only one immunoglobulin class at a time also facilitates the detection of significant IgG titer rises, since IgG and IgM activity, appearing around the same time in serum, may cause high titers already in the acute-phase serum when measured together.

Cross-reactions, which are known to interfere with the serological parainfluenza diagnosis, are seen also in the ELISA and may even be more troublesome than in other tests, for example, CF. In virus type 1 and 3 infections, antibody titer rises are often seen to both type 1 and type 3 antigens. Occasionally, titer rises occur to all three types or to mumps virus.

In an ELISA report on the use of purified nucleocapsid and purified glycoprotein (surface components) parainfluenza antigens, antibody titer rises were frequently seen in type 1 and/or type 3 CF verified infections to both types 1 and 3 glycoprotein and nucleocapsid antigens but not to parainfluenza type 2 or mumps virus. Type 2 CF-diagnosed infections gave few heterologous titer rises by ELISA, but they did occur. Mumps induced heterologous antibody rises in several patients, mainly to the glycoproteins of the parainfluenza viruses and occasionally also to the nucleocapsid. Whereas an ELISA employing parainfluenza glycoprotein antigens detected homologous titer rises in nearly all serum pairs with antibody titer rises in the CF test, the HI test did not (Julkunen, 1984).

COMPLEMENT FIXATION (CF) TEST

Paramyxovirus antibodies must not be present in the guinea pig serum used as the source of complement, and they should be excluded by control tests. Antigens can be prepared to each of the parainfluenza virus types from infected monkey kidney cell cultures, preferably continuous cell lines. Type 2 and 3 viruses replicate well and yield large amounts of antigen in, for example, Ma 104 roller flask cell cultures. After 5 or 6 days' incubation, the antigen is harvested; and the cell fraction is pelleted, sonicated, and clarified by centrifugation. The tissue culture medium is concentrated 8 to 10 times. The clarified cell fraction antigen and the concentrated tissue culture medium are separately tested for their content of viral antigen. If titers are good, they are pooled for the final CF antigen, which can reach titers of 64 to 128. Type 4 virus cultures are harvested after 6 to 10 days for further concentration (Chanock, 1979).

Type 1 virus and certain strains of type 2 can be propagated also in embryonated eggs for production of CF antigen. Ten-day-old embryonated eggs are inoculated by the allantoic route, and after 3 days' incubation at 35°C the allantoic fluid is harvested. Virus antigen is also present in the chorioallantoic membrane, which for type 1 virus can be used as CF antigen after homogenization in a blender followed by clarification.

The CF test is used in many laboratories for routine diagnosis of parainfluenza virus infections. The test detects titer rises in serious illness in children. In mild or inapparent infections, however, this may not always be the case, and the CF antibody response to type 1 virus is especially poor (Chanock et al., 1960). Likewise, reinfection in adults may not result in a significant CF titer rise. On the other hand, heterotypic antibody titer rises often occur. This in addition to the frequently seen titer rises to the parainfluenza viruses after mumps, especially seen in adults, hampers the type-specific diagnosis.

HEMAGGLUTINATION INHIBITION (HI) TEST

Sera for testing should be treated with receptor-destroying enzyme (RDE) to remove nonspecific inhibitors of the hemagglutinin. Human sera may also contain agglutinins to the guinea pig RBC used in the test. Each serum should be checked for agglutinins, which can be removed by absorption with 5% guinea pig RBC before the RDE treatment.

Sensitive tests are obtained if an optimal agglutination temperature is used. The HI test does not regularly diagnose significant titer rises in parainfluenza virus infections and may not have the same sensitivity as the ELISA. Like other serological tests, it detects heterotypic antibody responses, making the type-specific diagnosis less reliable.

NEUTRALIZATION TEST

Neutralization tests can be performed either in tissue culture tubes or in microtiter plates, as described elsewhere in this volume. Infectious virus is prepared in monkey kidney tissue cultures. Bovine albumin (0.5%) can be used as stabilizer of the virus. Sera are heated (56°C for 30 min), except when tests are performed for detection of type 2 antibodies, where lower levels of antibodies may be obtained after heating of sera from children (Chanock, 1979). For the test, virus diluted in BSS and 0.5% gelatin is mixed with serum and incubated for 1 h at room temperature. Residual virus infectivity of the serum-virus mixture is shown in inoculated monkey kidney cells by testing the culture for presence of hemadsorption activity.

The technique is demanding and not suitable for diagnosis on a larger scale. It is also not especially sensitive for type-specific diagnosis in children. In adults, the presence of neutralizing antibodies in the acute-phase serum may jeopardize the detection of a titer rise.

DETECTION OF TYPE-SPECIFIC IgM

A hemadsorption immunosorbent technique (HIT) has been developed for detection of parainfluenza virus type-specific IgM. The test has been reported to facilitate diagnosis of infections with the different parainfluenza virus types by detecting specific IgM in acute-phase sera in 71% of patients with parainfluenza virus infections (Van der Logt et al., 1982, 1985). Seventy-two patients had IgM to one type, 32 to two types (especially types 1 and 3), and six to three types of the parainfluenza viruses.

In the test, goat antibodies to human IgM (μ-chain) coated on U-shaped microtiter plates capture IgM from the test serum. If virus-specific IgM is present, it binds the added virus, which in turn is detected by its reaction with guinea pig RBC. A positive result is seen as an even layer of red blood cells covering the bottom of the well.

The solid-phase immunocapture test (SPIT), based on the same test principle as HIT, has been used to confirm the sensitivity of the technique. Concomitant reactions to at least two parainfluenza type-specific antigens were reported in 28% of the sera investigated by the SPIT. However, half of them had higher titers to one single type than to the two others (Roussel et al., 1984).

So far, other solid-phase techniques for IgM detection in parainfluenza virus infections seem to be less successful. No specific IgM could be found by an indirect ELISA employing egg-grown Sendai virus as antigen. The same method detected mumps-specific IgM in sera from all patients with a verified mumps virus infection (Ukkonen et al., 1980).

Interpretation of Laboratory Data

Detection of a parainfluenza virus or its viral antigens in the human respiratory tract generally defines the etiological agent of a respiratory infection. Although a prolonged carrier state has been described in immunosuppression, the parainfluenza viruses are not usually excreted for long periods of time, and their detection is correlated to an acute infection.

The quality of specimens is of the utmost importance for diagnosis, and knowledge of the sampling technique and transportation conditions is a prerequisite for correct interpretation of obtained results. Results of virus isolation and antigen detection tests must also be correlated to time after infection, remembering that reinfections usually have a shorter virus excretion period than primary infections.

In cases where a parainfluenza virus has been recovered from the patient, serological investigation adds little of diagnostic value. In other cases, serology has a certain value. However, heterotypic antibody reactions often occur either as a result of cross-reactions based on common antigens or as an anamnestic antibody response to other parainfluenza virus types and to mumps virus. Sometimes not even a rise in antibody titers to only *one* virus allows a type-specific diagnosis to be made with certainty, as homologous antibody titer rises are occasionally absent.

Primary infections in infants and small children usually result in detectable type-specific antibody responses, provided methods sensitive enough are used. In *type 1 infections,* heterotypic antibody responses are seen mainly to type 3 but also to type 2 virus. In a study of 41 patients with type 1 virus infections diagnosed by virus isolation, 16 (39%) had only a homotypic CF antibody response. Besides a homotypic antibody response, 13 (31%) also had heterotypic responses to type 3 virus, and two (4%) to type 2 virus. Three patients had only heterotypic responses (Parrot et al., 1962). Antibody titer rises are not always seen in type 1 virus reinfections. Experimental reinfections have been shown often to result in heterotypic antibody titer rises to type 2 and 3 viruses (Cook et al., 1959).

In infants and small children, *type 2 virus infection* induces homotypic antibodies and occasionally also heterotypic antibodies to type 3 (Kim et al., 1961). Reinfections are usually followed by a homotypic antibody titer rise, although there are exceptions.

There is usually a type-specific antibody response in primary *type 3 virus infections* in infants and children. Of 45 patients with parainfluenza type 3 infections, 33 (73%) had a homotypic antibody titer rise, sometimes with a heterotypic rise to type 1 virus. No heterotypic titer rise was seen to type 2 virus (Jackson and Muldoon, 1975). (Antibody titer rises to type 3 virus are not unusual after infection with types 1 and 2.) After type 3 reinfections, heterotypic antibody rises to other parainfluenza virus types occur.

Mumps is closely related to the parainfluenza viruses, and cross-reactions frequently occur with these viruses. In a study of 47 patients with mumps, 10 had significant HI and/or CF antibody titer rises also to parainfluenza virus type 3, four to type 2, one to the human type 1 (HA2), and 20 to the murine type 1 (Sendai) virus (Lennette et al., 1963). Heterologous titer rises of type 4 antibodies in mumps have been reported to occur, mainly in adults (Johnson et al., 1960).

Epidemiology and Natural History

The acute respiratory infection is the most frequent type of illness in infancy and childhood. The parainfluenza viruses play important roles as etiologic agents, and they cause a wide range of respiratory illnesses. Second to the RS virus, they are the most important cause of lower respiratory tract infection in young children, and they are the leading cause of viral croup, which usually occurs in children 1–2 to 5 years of age. In older children and adults, the viruses cause also reinfections, commonly affecting the upper respiratory tract.

The overall rate of parainfluenza virus infections varies in different reports. In a longitudinal study of families with young children in Seattle, the rate was 44 infections per 100 person-years as estimated by serology (Fox and Hall, 1980). This study also confirmed that the infections were most frequent among preschool children 2 to 5 years of age (59 per 100 person-years), less frequent during the first year of life (46 per 100 person-years), and least frequent in older children and adults (40 per 100 person-years).

In a 5-year investigation of children admitted to the hospital with acute respiratory infections, Chanock and co-workers (1963) found parainfluenza viruses associated with 6% of all acute respiratory infections. In other studies, the proportion of parainfluenza virus infections has been reported to be around 9% in children with acute respiratory infections in England (Downham et al., 1974; Gardner et al., 1971).

The parainfluenza viruses are wide-spread, and they infect most persons during childhood. Children have usually experienced infection with type 3 virus during the first 2 years of life, and by 4 years of age 80% of the children have antibodies to parainfluenza type 3. Infections with types 1 and 2 viruses mainly occur later in childhood than infections with type 3 virus, and antibodies develop somewhat later. At the age of 5 years, 74 and 59% of the children in Washington, D.C., possessed antibodies to types 1 and 2, respectively (Parrot et al., 1962).

Evidence of infection with parainfluenza virus has been obtained in most parts of the world, including tropical areas (Clyde and Denny, 1983). An exception is the absence of antibodies to types 1 and 2 reported in children in remote Indian tribes in South America (Glezen et al., 1982).

Earlier reported serological data suggested that the distribution of viral respiratory disease agents in the tropics would be similar to that seen in temperate zones (Monto and Johnson, 1968). The percentage of parainfluenza virus occurring among viruses isolated from children with acute respiratory infection has been reported from Rio de Janeiro to be 6.6% and from Kuala Lumpur, West Bengal, Uganda, and Singapore 10 to 13% (Sutmoller et al., 1983).

The acute respiratory infection is also a leading cause of death in developing countries, especially among malnourished children. More than 4 million children below the age of 5 years, die annually of pneumonia, representing about 30% of the 14 million deaths of children occurring each year in the developing world (Berman and McIntosh, 1985). Together with RS virus, *H. influenzae*, and *S. pneumoniae*, the parainfluenza viruses constitute one of the four main causes of child deaths in acute respiratory infections. Thus, the parainfluenza viruses have been assumed to cause 5.5% of deaths in children due to noninfluenza acute respiratory infections (Institute of Medicine, 1985). This would mean a total mortality of 116,000 cases occurring each year in children below 5 years of age and an additional 8,000 cases in ages 5 to 14 years (Institute of Medicine, 1986).

The number of deaths has been calculated by presuming a case fatality rate of 10% for severe cases of the disease. The distribution of less severe episodes has been reported from the United States, where 50 mild and 10 moderate episodes were recorded for each severe case of parainfluenza infection.

However, there is no precise information on rates of mortality due to the various viral pathogens in the developing world. Isolation rates in respiratory disease may not give a true picture of the cause of mortality. Application of modern techniques for antigen detection directly in clinical specimens, circumventing the use of the difficult tissue culture technique for isolation of viruses, may contribute to more precise information.

The epidemic pattern of the parainfluenza virus types is not quite clear. Type 1, which has been reported to be endemic for long periods of time, has been noted to occasionally assume an epidemic pattern with outbreaks every two seasons (Brandt et al., 1974; Tyeryar et al., 1978). Epidemics have occurred mainly in the fall.

Type 2 has been found to have a distinct pattern with high attack rates producing minor respiratory illness in young children, as observed in surveillance studies (Glezen et al., 1982). The tendency of type 2 to produce mainly mild upper respiratory tract infections explains why it is seldom reported in studies performed on children admitted to the hospital, except in viral croup, to which it has a clear association. The outbreaks have mainly occurred during the fall and winter months.

Type 3 has been observed to occur endemically, although small outbreaks without predictable periodicity have been noted. Such outbreaks have also been seen in periods when other viruses are epidemic (Glezen et al., 1982). A development from an endemic to an epidemic pattern has been observed with activity peaking in the spring (Glezen et al., 1984). In an epidemiological study performed in six European countries during 1978 to 1981, two different epidemic patterns were seen. Whereas the virus predominated as a cause of summer illness with considerable outbreaks in Newcastle, England, it was most often detected during the colder months of the year in the other countries (Ørstavik et al., 1984). The reason for this may be differences among substrains of the virus. The difference in epidemic behavior has been observed earlier, when type 3 infections in a 14-year surveillance of illness in general practice in England were found to be more severe during the summer months and during the later years of the surveillance. A geographical discontinuity between summer and winter isolates gave additional support to the suggestion of the existence of two subtypes of type 3 virus (Hope-Simpson, 1981).

No definitive epidemic pattern has been described for type 4 viruses. They are only sporadically reported, which may be explained by the mild respiratory symptoms they cause (Tyrrell and Bynoe, 1969). Antibodies have been reported to develop in early childhood (Gardner, 1969). The viruses seem also to occur in children admitted to the hospital with febrile convulsions. This was seen in 7 of 16 children in whom type 4 (A and B) viruses were identified in 151 children with parainfluenza virus infections, as seen in a 2-year study (Downham et al., 1974). The type 4 viruses accounted for 8.3% of the total parainfluenza isolates in a hospital in northern England over a period of 8 years, and in children with debilitating illness, they caused infections serious enough for hospitalization (Gardner and McQuillin, 1980).

In open-ward situations, respiratory viruses represent a potential source of infection that may spread to other patients. This has been described not only for RS virus but also for parainfluenza virus type 3. One-fifth of 197 uninfected children in Chicago acquired type 3 infection during hospitalization (Mufson et al., 1973).

Reinfections, usually with less severe symptoms, occur in older children and adults. In a series of three outbreaks of type 3 infection in a nursery, 17% of children infected during an outbreak were reinfected during a subsequent outbreak, although the interval between the first and last outbreak was only 9 months (Chanock et al., 1963).

Control and Prevention

Because of their important roles as pathogens in children during the first year of life, the parainfluenza viruses have been considered possible vaccine candidates (Institute of Medicine, 1986). After a natural parainfluenza virus infection, general as well as mucosal immunity appears, conferring short-lived immunity. With time, the protection declines and reinfections occur.

The degree of host immunity has been shown to a certain extent to be related to the level of neutralizing antibodies. Although the secretory immunoglobulin A is responsible for the main virus neutralization capacity after a natural infection (Smith et al.,

1966), immunoglobulin G antibodies, which at high concentrations transude from serum to mucosal membranes, also contribute to protection.

The antigen specificity of the parainfluenza antibodies is of utmost importance for protection. Antibodies to the hemagglutinin-neuraminidase protein (HN) as well as to the fusion protein (F) have the ability to neutralize infectivity. The HN antibody prevents adsorption of virus to the cell, and the F antibody inhibits the spread of virus by cell fusion. The requirements of antibodies to the F protein for complete prevention of disease have been documented (Merz et al., 1980), and a discordance between virus neutralization and protection caused by insufficient amount of F antibodies has been demonstrated (Yanagihara and McIntosh, 1980). Anti-F activity can only be found in sera with high levels of neutralizing antibodies. This has been confirmed by a radioimmunoprecipitation assay (RIPA) showing that F antibodies after a natural infection rarely reached levels equal to those of HN antibodies (Kasel et al., 1984).

Although no immune pathological amplification has been seen during infection with parainfluenza virus after earlier immunizations with Formalin-inactivated virus, the vaccine's capacity to elicit F antibodies plays a role in protection. Earlier vaccines, which did not induce F antibodies, failed to confer immunity (Fulginiti et al., 1969); later vaccines, which have been shown to elicit antibodies in animals to both HN and F proteins, provide complete protection from challenge infection (Ray et al., 1985). Likewise, a component (subunit) vaccine prepared from the two different glycoproteins of parainfluenza type 3 virus induced a high antibody response and provided immunity in lambs against the infection (Morein et al., 1983). So far, no reports have been given on the use of genetic engineering technology for production of component (subunit) vaccine or on the synthesis of parainfluenza virus proteins for vaccine use.

The use of a parainfluenza vaccine would be to prevent serious disease and mortality caused by the infection. It can be assumed that the viruses contribute to a considerable extent to deaths in acute respiratory disease in children in the developing countries (see above).

Parainfluenza virus infections are serious during the first year of life. While the parainfluenza virus types 1 and 2 in temperate climates mainly affect children after the age of 6 months, type 3, like RS virus, causes severe illness in infants after the first month. Thus a future vaccine will have to be given as early as possible during infancy (at 6 weeks) in order to influence type 3 infections. Boosters should be given to maintain a level of antibodies high enough to prevent severe illness later in childhood.

Many nucleoside analogues have been synthesized during the past decades in the search for antiviral agents. Ribavarin, which is metabolized intracellularly to phosphate derivates inhibiting viral replication, has been documented to exhibit activity in vitro against parainfluenza as well as against many other RNA viruses (Gilbert and Knight, 1986). It has been reported to increase survival in parainfluenza virus–infected rodents (Chang and Heel, 1981), and recent reports indicate clinical efficacy also in human infections (Gelfand et al., 1983).

Literature Cited

Aherne, W., T. Bird, S. D. M. Court, P. S. Gardner, and J. McQuillin. 1970. Pathological changes in virus infections of the lower respiratory tract in children. J. Clin. Pathol. 23:7–18.

Amesse, L. S., and D. W. Kingsbury. 1982. Sendai virus gene sequences identified by oligonucleotide mapping. Virology 118:8–16.

Amesse, L. S., C. L. Pridgen, and D. W. Kingsbury. 1982. Sendai virus DI RNA species with conserved virus genome termini and extensive internal deletions. Virology 118:17–27.

Ånestad, G. 1985. Surveillance of respiratory viral infections by rapid immunofluorescence diagnosis, with emphasis on the epidemiological development of respiratory syncytial virus infections. J. Hyg. (Camb.) 94:349–356.

Beale, A. J., D. L. McLeod, W. Stackiw, and A. J. Rhodes. 1958. Isolation of cytopathogenic agents from the respiratory tract in acute laryngotracheobronchitis. Br. Med. J. 1:302–303.

Berman, S., and K. McIntosh. 1985. Selective primary health care: strategies for control of disease in the developing world. XXI. Acute respiratory infections. Rev. Infect. Dis. 7:674–691.

Bishai, F. R., and R. Galli. 1978. Enzyme-linked immunosorbent assay for detection of antibodies to influenza A and B and parainfluenza type 1 in sera of patients. J. Clin. Microbiol. 8:648–656.

Bloom, H. H., K. M. Johnson, R. Jacobsen, and R. M. Chanock. 1961. Recovery of parainfluenza viruses from adults with upper respiratory illness. Am. J. Hyg. 74:50–59.

Brandt, C. D., H. W. Kim, R. M. Chanock, and R. S. Parrott. 1974. Parainfluenza virus epidemiology. Pediatr. Res. 8:422.

Bratt, M. A., and W. S. Robinson. 1967. Ribonucleic acid synthesis in cells infected with Newcastle disease virus. J. Mol. Biol. 23:1–21.

Canchola, J., A. J. Vargosko, H. W. Kim, R. H. Parrott, E. Christmas, B. Jeffries, and R. M. Chanock. 1964. Antigenic variation among newly isolated strains of parainfluenza type 4 virus. Am. J. Hyg. 79:357–364.

Cevenini, R., M. Donati, A. Moroni, L. Franchi, and F. Rumpianesi. 1983. Rapid immunoperoxidase assay for detection of respiratory syncytial virus in nasopharyngeal secretions. J. Clin. Microbiol. 18:947–949.

Chambers, P., N. S. Millar, R. W. Bingham, and P. T. Emmerson. 1986. Molecular cloning of complementary DNA to Newcastle disease virus, and nucleotide sequence analysis of the junction between the genes encoding the haemagglutinin-neuraminidase and the large protein. J. Gen. Virol. 67:475–486.

Chang, T.-W., and R. C. Heel. 1981. Ribavarin and inosiplex: a review of their present status in viral diseases. Drugs 22:111–128.

Chanock, R. M. 1956. Association of a new type of cytopathogenic myxovirus with infantile croup. J. Exp. Med. 104:555–576.

Chanock, R. M. 1979. Parainfluenza viruses, p. 611–632. In E. H. Lennette and N. J. Schmidt (eds.), Diagnostic procedures for viral, rickettsial and chlamydial infections. American Public Health Association, Washington.

Chanock, R. M., and K. McIntosh. 1985. Parainfluenza viruses, p. 1241–1253. In N. Fields et al. (eds). Virology. Raven Press, New York.

Chanock, R. M., R. H. Parrott, K. Cook, B. E. Andrews, J. A. Bell, T. Reichelderfer, A. Z. Kapikian, F. M. Mastrota, and R. J. Huebner. 1958. Newly recognized myxoviruses from children with respiratory disease. N. Engl. J. Med. 258: 207–213.

Chanock, R. M., D. C. Wong, R. J. Huebner, and J. A. Bell. 1960. Serologic response of individuals infected with parainfluenza viruses. Am. J. Publ. Health 50:1858–1865.

Chanock, R. M., R. H. Parrott, K. M. Johnson, A. Z. Kapikian, and J. A. Bell. 1963. Myxoviruses: parainfluenza. Am. Rev. Respir. Dis. 88:152–166.

Chany, C., F. Chany-Fournier, and O. Robain. 1987. Cell fusion in viral diseases. Nature 326:250.

Choppin, P. W. 1964. Multiplication of a myxovirus (SV5) with minimal cytopathic effects and without interference. Virology 23:224–233.

Choppin, P. W., and A. Scheid. 1980. The role of viral glycoproteins in adsorption, penetration, and pathogenicity of viruses. Rev. Infect. Dis. 2:40–61.

Clyde, W. A., and F. W. Denny. (eds.). 1983. Workshop on acute respiratory diseases among children of the world. Pediatr. Res. 17:1023–1076.

Coelingh, K. J., C. C. Winter, B. R. Murphy, J. M. Rice, P. C. Kimball, R. A. Olmsted, and P. L. Collins. 1986. Conserved epitopes on the hemagglutinin-neuraminidase proteins of human and bovine parainfluenza type 3 viruses: nucleotide sequence analysis of variants selected with monoclonal antibodies. J. Virol. 60:90–96.

Compans, R. W., and P. W. Choppin. 1967. Isolation and properties of the helical nucleocapsid of the parainfluenza virus SV5. Proc. Natl. Acad. Sci. USA 57:949–956.

Cook, M. K., B. E. Andrews, H. H. Fox, H. C. Turner, W. D. James, and R. M. Chanock. 1959. Antigenic relations among the newer paramyxoviruses. Am. J. Hyg. 69:250–264.

Cowley, J. A., and R. D. Barry. 1983. Characterization of human parainfluenza viruses. I. The structural proteins of parainfluenza virus 2 and their synthesis in infected cells. J. Gen. Virol. 64:2117–2125.

Delage, G., P. Brochu, M. Pelletier, G. Gasmin, and N. Lapointe. 1979. Giant-cell pneumonia caused by parainfluenza virus. J. Pediatr. 94:426–429.

Denny, F. W., T. F. Murphy, W. A. Clyde, Jr., A. M. Collier, and F. W. Henderson. 1983. Croup: an 11-year study in a pediatric practice. Pediatrics 71:871–876.

Dethlefsen, L., and D. Kolakofsky. 1983. In vitro synthesis of the nonstructural C protein of Sendai virus. J. Virol. 46:321–324.

Dowling, P. C., C. Giorgi, L. Roux, L. A. Dethlefsen, M. E. Galantowicz, B. M. Blumberg, and D. Kolakofsky. 1983. Molecular cloning of the 3' proximal third of the Sendai virus genome. Proc. Natl. Acad. Sci. USA 80:5213–5216.

Downham, M. A. P. S., J. McQuillin, and P. S. Gardner.

1974. Diagnosis and clinical significance of parainfluenza virus infections in children. Arch. Dis. Child. **49:**8–15.

Downham, M. A. P. S., P. S. Gardner, J. McQuillin, and J. A. J. Ferris. 1975. Role of respiratory viruses in childhood mortality. Br. Med. J. **1:**235–239.

Elango, N., J. E. Coligan, R. C. Jambou, and S. Venkatesan. 1986. Human parainfluenza type 3 virus hemagglutinin-neuraminidase glycoprotein: nucleotide sequence of mRNA and limited amino acid sequence of the purified protein. J. Virol. **57:**481–489.

Espmark, Å. 1965. Hemagglutinating and plaque forming properties of two substrains of Simian parainfluenza virus SV5. Arch. Ges. Virusforsch. **17:**374–378.

Fedova, D., I. Pecenkova-Plachtova, and B. Tumova. 1969. Application of the fluorescent antibody method in the diagnosis of *M. parainfluenzae.* II. Identification of newly isolated *M. parainfluenzae* strains in monkey kidney tissue cultures. J. Hyg. Epidemiol. Microbiol. Immunol. **13:**181–190.

Fedova, D., and L. Zelenkova. 1969. Use of the fluorescent antibody method in the diagnosis of *M. parainfluenzae.* I. Proliferation of *M. parainfluenzae* types 1, 2 and 3 in monkey kidney tissue cultures. J. Hyg. Epidemiol. Microbiol. Immunol. **13:**13–23.

Fox, J. P., and C. E. Hall. 1980. Infections with other respiratory pathogens: influenza, parainfluenza, mumps and respiratory syncytial virus; mycoplasma pneumoniae, p. 335–381. *In* J. P. Fox and C. E. Hall (eds.), Viruses in families. PSG Publishing Company, Littleton, MA.

Foy, H. M., M. K. Cooney, and A. J. Maletzky. 1973. Incidence and etiology of pneumonia, croup and bronchiolitis in preschool children belonging to a prepaid medical care group over a four year period. Am. J. Epidemiol. **97:**80–102.

Frank, A. L., R. B. Couch, C. A. Griffis, and B. D. Baxter. 1979. Comparison of different tissue cultures for isolation and quantitation of influenza and parainfluenza viruses. J. Clin. Microbiol. **10:**32–36.

Frank, A. L., L. H. Taber, C. R. Wells, J. M. Wells, W. P. Glezen, and A. Paredes. 1981. Patterns of shedding of myxoviruses and paramyxoviruses in children. J. Infect. Dis. **144:**433–441.

Fulginiti, V. A., J. J. Eller, O. F. Sieber, J. W. Joyner, M. Minamitani, and G. Meiklejohn. 1969. Respiratory virus immunization. I. A field trial of two inactivated respiratory virus vaccines; an aqueous trivalent parainfluenza virus vaccine and an alum-precipitated respiratory syncytial virus vaccine. Am. J. Epidemiol. **89:**435–448.

Gallaher, W. R., and M. A. Bratt. 1974. Conditional dependence of fusion from within and other cell membrane alterations by Newcastle disease virus. J. Virol. **14:**813–820.

Gardner, P. S. 1957. Serological evidense of infection with Sendai virus in England. Br. Med. J. **1:**1143–1145.

Gardner, P. S., and J. McQuillin (eds.). 1980. Rapid virus diagnosis, 2d ed. Butterworths, London.

Gardner, P. S., J. McQuillin, R. McGuckin, and R. K. Ditchburn. 1971. Observations on clinical and immunofluorescent diagnosis of parainfluenza virus infections. Br. Med. J. **2:**7–12.

Gardner, P. S., M. Grandien, and J. McQuillin. 1978. Comparison of immunofluorescence and immunoperoxidase: Stockholm and Newcastle upon Tyne. Bull. WHO **56:**105–110.

Gardner, S. D. 1969. The isolation of parainfluenza 4 subtypes A and B in England and serological studies in their prevalence. J. Hyg. (Camb.) **67:**545–550.

Gelfand, E. W., D. McCurdy, C. P. Rao, and P. J. Middle-

ton. 1983. Ribavarin treatment of viral pneumonitis in severe combined immunodeficiency disease. Lancet **2:**732–733.

Gilbert, B. E., and V. Knight. 1986. Biochemistry and clinical applications of ribavarin. Antimicrob. Agents Chemother. **30:**201–205.

Glezen, W. P., and G. W. Fernald. 1976. Effect of passive antibody on parainfluenza virus type 3 pneumonia in hamsters. Infect. Immun. **14:**212–216.

Glezen, W. P., F. A. Loda, and F. W. Denny. 1982. Parainfluenza viruses, p. 441–454. *In* A. S. Evans (ed.), Viral infections of humans, epidemiology and control, 2d ed. Plenum, New York.

Glezen, W. P., A. L. Frank, L. H. Taber, and J. A. Kasel. 1984. Parainfluenza virus type 3: seasonality and risk of infection and reinfection in young children. J. Infect. Dis. **150:**851–857.

Goswami, K. K. A., and W. C. Russell. 1983. Monoclonal antibodies against human paramyxovirus type 3 and against SV5 virus: preparation and preliminary characterization. J. Gen. Virol. **64:**1663–1672.

Grandien, M., C.-A. Pettersson, P. S. Gardner, A. Linde, and A. Stanton. 1985. Rapid viral diagnosis of acute respiratory infections: comparison of enzyme-linked immunosorbent assay and the immunofluorescence technique for detection of viral antigens in nasopharyngeal secretions. J. Clin. Microbiol. **22:**757–760.

Gross, P. A., R. H. Green, and M. G. McCrea Curnen. 1973. Persistent infection with parainfluenza type 3 virus in man. Am. Rev. Respir. Dis. **108:**894–898.

Hall, C. E., C. D. Brandt, T. E. Frothingham, I. Spigland, M. K. Cooney, and J. P. Fox. 1971. The virus watch program: a continuing surveillance of viral infections in metropolitan New York families. IX. A comparison of infections with several respiratory pathogens in New York and New Orleans families. Am. J. Epidemiol. **94:**367–385.

Hall, C. B., R. G. Douglas, Jr., R. L. Simons, and J. M. Geiman. 1978. Interferon production in children with respiratory syncytial, influenza, and parainfluenza virus infections. J. Pediatr. **93:**28–32.

Halonen, P., O. Meurman, T. Lövgren, I. Hemmilä, and E. Soini. 1983. Detection of viral antigens by time-resolved fluoroimmunoassay, pp. 133–146. *In* P. A. Bachmann (ed.), New developments in diagnostic virology. Springer, New York.

Hiebert, S. W., R. G. Paterson, and R. A. Lamb. 1985a. Hemagglutinin-neuraminidase protein of the paramyxovirus simian virus 5: nucleotide sequence of the mRNA predicts an N-terminal membrane anchor. J. Virol. **54:**1–6.

Hiebert, S. W., R. G. Paterson, and R. A. Lamb. 1985b. Identification and predicted sequence of a previously unrecognized small hydrophobic protein, SH, of the paramyxovirus simian virus 5. J. Virol. **55:**744–751.

Holmes, K. V., and P. W. Choppin. 1966. On the role of the response of the cell membrane in determining virus virulence. Contrasting effects of the parainfluenza virus SV5 in two cell types. J. Exp. Med. **124:**501–520.

Hope-Simpson, R. E. 1981. Parainfluenza virus infections in the Cirencester survey: seasonal and other characteristics. J. Hyg. (Camb.) **87:**393–406.

Institute of Medicine. 1986. New vaccine development establishing priorities. *In* Diseases of importance in developing countries, Vol. II. National Academy Press, Washington.

Iorio, R. M., J. B. Borgman, R. L. Glickman, and M. A. Bratt. 1986. Genetic variation within a neutralizing domain on the haemagglutinin-neuraminidase glycoprotein of Newcastle disease virus. J. Gen. Virol. **67:**1393–1403.

Jackson, G. G., and R. L. Muldoon. 1985. Parainfluenza viruses. *In* Viruses causing common respiratory infections in man. University Chicago Press, Chicago.

Jarvis, W. R., P. J. Middleton, and E. W. Gelfand. 1979. Parainfluenza pneumonia in severe combined immunodeficiency disease. J. Pediatr. **94:**423–425.

Johnson, K. M., R. M. Chanock, M. K. Cook, and R. J. Huebner. 1960. Studies of a new human hemadsorption virus. I. Isolation, properties and characterization. Am. J. Hyg. **71:**81–92.

Julkunen, I. 1984. Serological diagnosis of parainfluenza virus infections by enzyme immunoassay with special emphasis on purity of viral antigens. J. Med. Virol. **14:**177–187.

Kapikian, A. Z., R. M. Chanock, T. E. Reichelderfer, T. G. Ward, R. J. Huebner, and J. A. Bell. 1961. Inoculation of human volunteers with parainfluenza virus type 3. JAMA **178:**537–541.

Kasel, J. A., A. L. Frank, W. A. Keitel, L. H. Taber, and W. P. Glezen. 1984. Acquisition of serum antibodies to specific viral glycoproteins parainfluenza virus 3 in children. J. Virol. **52:**828–832.

Kim, H. W., A. J. Vargosko, R. M. Chanock, and R. H. Parrott. 1961. Parainfluenza 2 (CA) virus: etiologic association with croup. Pediatrics **28:**614–621.

Kingsbury, D. W. 1966. Newcastle disease virus RNA. II. Preferential synthesis of RNA complementary to parenteral viral RNA by chick embryo cells. J. Mol. Biol. **18:**204–214.

Kingsbury, D. W. 1985. Orthomyxo- and paramyxoviruses and their replication, p. 1157–1178. *In* B. N. Fields et al. (eds.), Virology. Raven Press, New York.

Kingsbury, D. W., M. A. Bratt, P. W. Choppin, R. P. Hanson, Y. Hosaka, V. ter Meulen, E. Norrby, W. Plowright, R. Rott, and W. H. Wunner. 1978. Paramyxoviridae. Intervirology **10:**137–152.

Kristensson, K., and C. Örvell. 1983. Cellular localization of five structural proteins of Sendai virus studied with peroxidase-labelled Fab fragments of monoclonal antibodies. J. Gen. Virol. **64:**1673–1678.

Kuroya, N., and M. Ishida. 1953. Newborn virus pneumonitis (type Sendai). II. Report: the isolation of a new virus possessing hemagglutinin activity. Yokohama Med. Bull. **4:**217–233.

Lamb, R. A., and P. W. Choppin. 1977. The synthesis of Sendai virus polypeptides in infected cells. III. Phosphorylation of polypeptides. Virology **81:**382–397.

Lamb, R. A., B. W. J. Mahy, and P. W. Choppin. 1976. The synthesis of Sendai virus polypeptides in infected cells. Virology **69:**116–131.

Lennette, E. H., F. W. Jensen, R. W. Guenther, and R. L. Magoffin. 1963. Serologic responses to parainfluenza viruses in patients with mumps virus infection. J. Lab. Clin. Med. **61:**780–788.

Lyles, D. S. 1979. Glycoproteins of Sendai virus are transmembrane proteins. Proc. Natl. Acad. Sci. USA **76:**5621–5625.

Markwell, M. A. K., L. Svennerholm, and J. C. Paulson. 1981. Specific gangliosides function as host cell receptors for Sendai virus. Proc. Natl. Acad. Sci. USA **78:**5406–5410.

McIntosh, K., and J. C. Clark. 1985. Parainfluenza and respiratory syncytial viruses, p. 763–768. *In* E. H. Lennette (ed.), Manual of clinical microbiology, 4th ed. American Society for Microbiology, Washington.

Merz, D. C., A. Scheid, and P. W. Choppin. 1980. Importance of antibodies to the fusion glycoprotein of paramyxoviruses in the prevention of spread of infection. J. Exp. Med. **151:**275–288.

Merz, D. C., A. Scheid, and P. W. Choppin. 1981. Immunological studies of the functions of paramyxovirus glycoproteins. Virology **109:**94–105.

Meurman, O., P. Hänninen, R. V. Krishna, and T. Ziegler. 1982. Determination of IgG and IgM class antibodies to mumps virus by solid phase enzyme immunoassay. J. Virol. Methods **4:**249–257.

Monto, A. S., and K. M. Johnson. 1968. Respiratory infections in the American tropics. Am. J. Trop. Med. Hyg. **17:**867–874.

Morein, B., M. Sharp, B. Sundquist, and K. Simons. 1983. Protein subunit vaccines of parainfluenza type 3 virus: immunogenic effect in lambs and mice. J. Gen. Virol. **64:**1557–1569.

Mufson, M. A., H. E. Mocega, and H. E. Krause. 1973. Acquisition of parainfluenza 3 virus infection by hospitalized children. I. Frequencies, rates, and temporal data. J. Infect. Dis. **128:**141–147.

Nishikawa, K., S. Isomura, S. Suzuki, E. Watanabe, M. Hamaguchi, T. Yoshida, and Y. Nagai. 1983. Monoclonal antibodies to the HN glycoprotein of Newcastle disease virus. Biological characterization and use for strain comparisons. Virology **130:**318–330.

Ogden, J. R., R. Pal, and R. R. Wagner. 1986. Mapping regions of the matrix protein of vesicular stomatitis virus which bind to ribonucleocapsids, liposomes, and monoclonal antibodies. J. Virol. **58:**860–868.

Ørstavik, I., M. Grandien, P. Halonen, P. Arstila, C. H. Mordhorst, A. Hornsleth, T. Popow-Kraupp, J. McQuillin, P. S. Gardner, J. Almeida, F. Bricout, and A. Marques. 1984. Viral diagnoses using the rapid immunofluorescence technique and epidemiological implications of acute respiratory infections among children in different European countries. Bull. WHO **62:**307–313.

Örvell, C., and M. Grandien. 1982. The effects of monoclonal antibodies on biologic activities of structural proteins of Sendai virus. J. Immunol. **129:**2779–2787.

Örvell, C., and E. Norrby 1977. Immunologic properties of purified Sendai virus glycoproteins. J. Immunol. **119:**1882–1887.

Örvell, C., R. Rydbeck, and A. Löve. 1986. Immunological relationships between mumps virus and parainfluenza viruses studied with monoclonal antibodies. J. Gen. Virol. **67:**1–11.

Parrot, R. H., A. Vargosko, A. Luckey, H. W. Kim, C. Cumming, and R. Chanock. 1959. Clinical features of infection with hemadsorption viruses. N. Engl. J. Med. **260:**729–738.

Parrot, R. H., A. J. Vargosko, H. W. Kim, J. A. Bell, and R. M. Chanock, III. 1962. Myxoviruses: parainfluenza. Am. J. Publ. Health **52:**907–917.

Portner, A. 1981. The HN glycoprotein of Sendai virus: analysis of site(s) involved in hemagglutinating and neuraminidase activities. Virology **115:**375–384.

Portner, A. 1984. Monoclonal antibodies as probes of the antigenic structure and functions of Sendai virus, p. 345–350. *In* R. V. Compans and D. H. L. Bishop (eds.), Segmented negative strand viruses. Academic Press, Orlando, FL.

Portner, A., and K. G. Murti. 1985. Sendai virus nucleocapsid: localization of NP, P and M Proteins by immune electron microscopy. Virus Res. (Suppl. 1):46.

Ray, R., and R. W. Compans. 1986. Monoclonal antibodies reveal extensive antigenic differences between the hemagglutinin-neuraminidase glycoproteins of human and bovine parainfluenza 3 viruses. Virology **148:**232–236.

Ray, R., V. E. Brown, and R. W. Compans. 1985. Glycoproteins of human parainfluenza virus type 3: characterization and evaluation as a subunit vaccine. J. Infect. Dis. **152**:1219–1230.

Re, G. G., and D. W. Kingsbury. 1986. Nucleotide sequences that affect replicative and transcriptional efficiencies of Sendai virus deletion mutants. J. Virol. **58**:578–582.

Rothman, J. E., and R. E. Fine. 1980. Coated vesicles transport newly synthesized membrane glycoproteins from endoplasmic reticulum to plasma membrane in two successive stages. Proc. Natl. Acad. Sci. USA **77**:780–784.

Rott, R. 1979. Molecular basis of infectivity and pathogenicity of myxovirus. Arch. Virol. **59**:285–298.

Roussel, C., G. Duverlie, P. Daniel, and G. Desmet. 1984. Application of the solid phase immunocaptation test to paramyxovirus parainfluenzae and mumps infection diagnosis. Ann. Virol. **135**:269–276.

Rydbeck, R., C. Örvell, A. Löve, and E. Norrby. 1986. Characterization of four parainfluenza virus type 3 proteins by use of monoclonal antibodies. J. Gen. Virol. **67**:1531–1542.

Sanchez, A., and A. K. Banerjee. 1985a. Studies on human parainfluenza virus 3: characterization of the structural proteins and in vitro synthesized proteins coded by mRNAs isolated from infected cells. Virology **143**:45–54.

Sanchez, A., and A. K. Banerjee. 1985b. Clone and gene assignment of mRNAs of human parainfluenza virus 3. Virology **147**:177–186.

Sarkkinen, H. K. 1985. Respiratory viral antigen detection by solid-phase immunoassays: comparison with other diagnostic tests and application for routine diagnosis, p. 327–337. In K.-O. Habermehl (ed.), Rapid methods and automation in microbiology and immunology. Springer-Verlag, Berlin.

Sarkkinen, H. K., P. E. Halonen, P. P. Arstila, and A. A. Salmi. 1981a. Detection of respiratory syncytial, parainfluenza type 2, and adenovirus antigens by radioimmunoassay and enzyme immunoassay on nasopharyngeal specimens from children with acute respiratory disease. J. Clin. Microbiol. **13**:258–265.

Sarkkinen, H. K., P. E. Halonen, and A. A. Salmi. 1981b. Type-specific detection of parainfluenza viruses by enzyme immunoassay and radioimmunoassay in nasopharyngeal specimens of patients with acute respiratory disease. J. Gen. Virol. **56**:49–57.

Scheid, A., and P. W. Choppin. 1977. Two disulfide-linked polypeptide chains constitute the active F protein of paramyxoviruses. Virology **80**:54–66.

Scheid, A., M. C. Graves, S. M. Silver, and P. W. Choppin. 1978. Studies on the structure and function of paramyxovirus glycoproteins, p. 181–193. In W. J. Mahy and R. D. Barry (eds.), Negative strand viruses and the host cell. Academic Press, London.

Smith, C. B., R. H. Purcell, J. Bellanti, and R. M. Chanock. 1966. Protective effect of antibodies to parainfluenza type 1 virus. N. Engl. J. Med. **275**:1145–1152.

Soini, E., and H. Kojola. 1983. Time-resolved fluorometer for lanthanine chelates—a new generation of nonisotopic immunoassays. Clin. Chem. **29**:65–68.

Sominina, A. A., Y. N. Zubzhitsky, and A. A. Smorodintsev. 1967. Dynamics of immunofluorescence and infectivity of parainfluenza 3 virus in susceptible cells. Acta Virol. **11**:424–431.

Spriggs, M. K., and P. L. Collins. 1986. Human parainfluenza virus type 3: messenger RNAs, polypeptide coding assignments, intergenic sequences, and genetic map. J. Virol. **59**:646–654.

Storey, D. G., K. Dimock, and C. Y. Kang. 1984. Structural characterization of viral proteins and genomic RNA of human parainfluenza virus 3. J. Virol. **52**:761–766.

Sutmoller, F., J. P. Nascimento, J. R. S. Chaves, W. Ferreira, and M. S. Pereira. 1983. Viral etiology of acute respiratory diseases in Rio de Janeiro; first two years of a longitudinal study. Bull. WHO **61**:845–852.

Totzawa, H., H. Komatsu, K. Ohkata, T. Nakajima, M. Watanabe, Y. Tanaka, and H. Arifuku. 1986. Neutralizing activity of the antibodies against two kinds of envelope glycoproteins of Sendai virus. Arch. Virol. **91**:145–161.

Traver, M. I., R. L. Northrop, and D. L. Walker. 1960. Site of intracellular antigen production by myxoviruses. Proc. Soc. Exp. Biol. **104**:268–273.

Tyeryar, F. J., L. S. Richardson, and R. B. Belshe. 1978. From the National Institutes of Health: report of a workshop on respiratory syncytial virus and parainfluenza viruses. J. Infect. Dis. **137**:835–846.

Tyrrell, D. A. J., and M. L. Bynoe. 1969. Studies on parainfluenza type 2 and 4 viruses obtained from patients with common colds. Br. Med. J. **1**:471–474.

Tyrrell, D. A., M. L. Bynoe, K. Birkum Petersen, R. N. P. Sutton, and M. Pereira. 1959. Inoculation of human volunteers with parainfluenza viruses types 1 and 3 (Ha2 and HA1). Br. Med. J. **2**:909–911.

Ukkonen, P., O. Väisänen, and K. Penttinen. 1980. Enzyme-linked immunosorbent assay for mumps and parainfluenza type 1 immunoglobulin G and immunoglobulin M antibodies. J. Clin. Microbiol. **11**:319–323.

Van der Logt, J. T. M., A. M. van Loon, and J. van der Veen. 1982. Detection of parainfluenza IgM antibody by hemadsorption immunosorbent technique. J. Med. Virol. **10**:213–221.

Van der Logt, J. T. M., A. M. van Loon, F. W. A. Heessen, and J. van der Veen. 1985. Diagnosis of parainfluenza virus infection in children and older patients by detection of specific IgM antibody. J. Med. Virol. **16**:191–199.

Van der Veen, J., and H. J. A. Sonderkamp. 1965. Secondary antibody response of guinea pigs to parainfluenza and mumps viruses. Arch. Ges. Virusforsch. **15**:721–734.

Van Wyke Coelingh, K. L., C. Winter, and B. R. Murphy. 1985. Antigenic variation in the hemagglutinin-neuraminidase protein of human parainfluenza type 3 virus. Virology **143**:569–582.

Vogel, J., and A. Shelokov. 1957. Adsorption-hemagglutination test for influenza virus in monkey kidney tissue culture. Science **126**:358–359.

Waner, J. L., N. J. Whitehurst, T. Downs, and D. G. Graves. 1985. Production of monoclonal antibodies against parainfluenza 3 virus and their use in diagnosis by immunofluorescence. J. Clin. Microbiol. **22**:535–538.

Waterson, A. P. 1962. Two kinds of myxoviruses. Nature **193**:1163–1164.

Wechsler, S. L., D. M. Lambert, M. S. Galinski, and M. W. Pons. 1985. Intracellular synthesis of human parainfluenza type 3 virus-specific polypeptides. J. Virol. **54**:661–664.

Welliver, R. C., D. T. Wong, M. Middleton, Jr., M. Sun, N. McCarthy, and P. L. Ogra. 1982. Role of parainfluenza virus-specific IgE in pathogenesis of croup and wheezing subsequent to infection. J. Pediatr. **101**:889–896.

Wilde, A., C. McQuain, and T. Morrison. 1986. Identification of the sequence content of four polycistronic transcripts synthesized in Newcastle disease virus infected cells. Virus Res. **5:**77–95.

Wong, D. T., R. C. Welliver, K. R. Riddlesberger, M. S. Sun, and P. L. Ogra. 1982. Rapid diagnosis of parainfluenza virus infection in children. J. Clin. Microbiol. **16:**164–167.

Yanagihara, R., and K. McIntosh. 1980. Secretory immunological response in infants and children to parainfluenza virus types 1 and 2. Infect. Immun. **30:**23–28.

Yewdell, J., and W. Gerhard. 1982. Delineation of four antigenic sites on a paramyxovirus glycoprotein via which monoclonal antibodies mediate distinct antiviral activities. J. Immunol. **128:**2670–2675.

Zollar, L. W., and M. A. Mufson. 1970. Acute parotitis associated with parainfluenza 3 virus infection. Am. J. Dis. Child. **119:**147–148.

CHAPTER 26

Paramyxoviridae: Mumps Virus

CLAES ÖRVELL

Disease: Mumps.

Etiologic Agent: Mumps virus.

Source: Infected humans spread the virus by droplet infection of infected saliva to nonimmune contacts.

Clinical Manifestations: Generalized systemic infection with fever, swelling of parotid glands, and frequent involvement of the central nervous system.

Pathology: Primary infection of ductal epitial cells in salivary glands, resulting in cell death and an inflammatory response; pleocytosis in the cerebrospinal fluid in the majority of patients as a sign of virus multiplication in the central nervous systems.

Laboratory Diagnosis: Detection of virus or viral antigens directly in clinical specimens (saliva and cerebrospinal fluid by isolation of the virus in tissue culture and immunofluorescence); serological diagnosis is made by demonstration of IgM antibodies in a single serum sample or by the demonstration of significant titer increase in IgG antibodies in two consecutive serum samples from the patient.

Epidemiology: Worldwide disease; spread of infection from an acutely infected individual to a nonimmune person; a certain population density required for continued spread of infection.

Treatment: No specific treatment available.

Prevention and Control: The virus can be effectively controlled by vaccination with live attenuated mumps virus, which will result in a long-lasting immunity. Synthetic subcomponent vaccines are not yet available.

Introduction

A contagious disease probably caused by mumps virus was described by Hippocrates in the 5th century B.C. The characteristics of the disease were moderate fever of short duration and unilateral or bilateral enlargement of parotid glands. In some patients, painful swelling of one or both testes occurred. About 200 years ago central nervous system (CNS) involvement was first described as a clinical feature of the disease. Evidence that the illness was caused by a virus came from experiments published in 1934 (Johnson & Goodpasture, 1934). They could induce a mumps-like illness in rhesus monkeys with filtered parotid secretions from patients with mumps. After it was demonstrated that mumps virus could be iso-lated and propagated in eggs (Habel, 1945), a number of physical properties of the virus could be determined (Cantell, 1961). Biological activities of the virus were also demonstrated, that is, hemagglutination, hemolysis and neuraminidase activity (Burnet et al., 1946; Hirst, 1950; Levens and Enders, 1945; Morgan et al., 1948).

Mumps virus can effectively be controlled and prevented by live, attenuated mumps virus vaccines. The use of such vaccines for mass vaccination of susceptible children has drastically reduced the incidence of reported mumps virus cases in the United States. In many other countries, vaccination against mumps virus has been limited to selected groups of risk patients, and therefore mumps will probably remain a common childhood disease. A number of review articles on the virus have been written in recent years (Norrby, 1985; Tolpin and Schauf, 1984; Wolinsky and Server, 1985).

The Virus

Physicochemical Properties and Morphology

Mumps virus is a member of the family *Paramyxoviridae*, genus *Paramyxovirus* (Kingsbury et al., 1978). The virus contains one piece of single, negative-stranded RNA which is complementary in base sequence to the positive-sense, viral messenger RNA (Kingsbury et al., 1978). The RNA is associated with the nucleocapsid protein (NP) to form a helical structure, the nucleocapsid. A protein with enzymatic activity, the RNA-dependent RNA polymerase, is located in the nucleocapsid structure. The nucleocapsid has a unit length of 1 μm and a diameter of 17 nm. The coiled nucleocapsid is surrounded by a lipid-containing envelope, which gives the virion a diameter of 120 to 200 nm. The envelope is composed of the matrix (M) protein and two surface glycoproteins, the hemagglutinin-neuraminidase (HN) and the fusion (F) protein. The latter two proteins make up two different kinds of projections that protrude 12 to 15 nm from the virion surface.

The virus infectivity is sensitive to heat (60°C for 30 min will destroy infectivity), formalin treatment and Tween 80 and ether treatment. Formalin and Tween 80 and ether selectively destroy the hemolytic activity without destroying the hemagglutinating (HA) activity (Norrby and Penttinen, 1978). Trypsin, chymotrypsin, and pronase treatment of the virus removes the HA activity by a selective removal of the HN protein without affecting the F protein (Härfast et al., 1980, Örvell, 1978b).

Replication

The replication of mumps virus has not been studied in great detail. Like other members of the paramyxovirus family, mumps virus contains a negative-strand RNA genome that is transcribed intracellularly to form positive-strand RNA, which directs the synthesis of viral proteins. The polymerase enzyme in the first transcription is the RNA-dependent RNA polymerase of the virus. The positive-strand RNA molecules that are formed code for viral proteins and also serve as templates in the replication of the negative-strand virus genome. The gene order of the genes coding for the different structural proteins of the virus has been established in mumps virus as in other paramyxoviruses (Elango et al., unpublished data). The gene coding for the nucleocapsid protein has a position at the 3' end of the virus genome,

followed by genes for the phospho (P) and M proteins and the envelope glycoproteins. A gene coding for a large (L) protein is located at the 5' end of the genome in paramyxoviruses. In addition, a small gene has been identified, located between the genes coding for the envelope glycoproteins. This gene codes for a small hydrophobic (SH) protein (Elango et al., unpublished data).

A 50S negative-strand viral RNA molecule has been identified in the virus and in infected cells (East and Kingsbury, 1971). Shorter mRNA molecules in infected cells have not been studied in great detail. In one study nine different mRNA molecules were identified in mumps virus-infected cells (Rima and Martin, 1980). The time-related synthesis of mumps virus polypeptides in infected cells has been studied by different groups (Herrler and Compans, 1982; Merz et al., 1983; Naruse et al., 1981; Rima et al., 1980). The same proteins that can be identified in purified virions are also found in infected cells. The first proteins to become detectable in infected cells are the internal proteins of the mature virus, NP and P (Herrler and Compans, 1982; Naruse et al., 1981). The final step of maturation is budding of the virus from the cell membrane at about 20 h postinfection (Naruse et al., 1981). A difference in the maturation of mumps virus and other viruses of the genus paramyxovirus genus *Paramyxoviridae* concerns the F protein. In all members of the genus, including mumps virus, the F protein is built up by two disulfide-linked glycoprotein molecules (F_1 and F_2), which are generated from a larger precursor molecule (F_0). In mumps virus the larger precursor molecule is always processed to the active $F_{1,2}$ complex (Merz et al., 1983), whereas in other paramyxoviruses the F_0 precursor molecule is cleaved only under certain conditions (Nagai and Klenk, 1977; Scheid & Choppin, 1974).

Antigenic Structure

Five major structural proteins, the HN, F, NP, P, and M, have been identified in the virus (Table 1; Fig. 1). Mumps virus has somewhat larger F, NP, and M proteins compared with other paramyxoviruses (McCarthy and Johnson, 1980; Örvell, 1978b; Rima et al., 1980). In one study (McCarthy and Johnson, 1980) the molecular weights of the major structural proteins of five different mumps virus strains were compared. The size of the HN protein in the different strains was either 75K or 80K. This variability in size is probably due to differences in the glycosylation pattern of the HN protein. The molecular weights of other structural proteins were similar in the five mumps virus strains examined. As in other paramyxoviruses, *nonstructural protein(s)* has been identified in mumps virus. Actin, not coded by the

TABLE 1. Structural proteins of mumps virus: comparison of molecular weights with corresponding proteins of two other representative paramyxoviruses[a]

	Relative molecular weight (kDa)				
Designation	Mumps virus	Sendai virus	NDV[b]	Location	Function
L (large)	200			Nucleocapsid	Not known; possibly in transcriptive complex
HN (hemagglutinin-neuraminidase)	75–80	71	74	Surface spike	Adsorption to cells
NP (nucleocapsid)	68–73	57	53	Nucleocapsid	Protecting RNA
F (fusion)	65–74			Surface spike	Fusion of cells, hemolysis, virus entry
F_1	58–61	49	53	Larger disulfide-linked monomer of F	
F_2	10–16			Smaller disulfide-linked monomer of F	
P (phospho)	45–47	80		Nucleocapsid	In transcriptive complex, presumably enzymatic function
Actin	43	43	43	Not known	Not known
M (membrane or matrix)	39–42	36	36	Inside virion envelope	Virion assembly

[a] Molecular weight comparison with Sendai virus and NDV from Örvell (1978b).
[b] NDV = Newcastle disease virus.

virus, derived from the host cell, is a structural protein of mumps virus (Naruse et al., 1981; Örvell, 1978b). Ultrastructurally it has been shown that actin filaments are found in close association with nucleocapsids of paramyxoviruses (Bohn et al., 1986). Actin is probably involved in virus budding (Bohn et al., 1986; Tyrrell and Ehrnst, 1979).

The two surface glycoproteins, HN and F, form the target for antibodies that can maintain protective immunity to the virus. In analogy with other members of the paramyxovirus genus, both hemagglutination and neuraminidase activity are associated with the HN protein (Jensik and Silver, 1976; Örvell, 1976b, 1978a, 1978b). Recent studies have demonstrated that monoclonal antibodies directed against distinct structural parts of the HN glycoprotein can

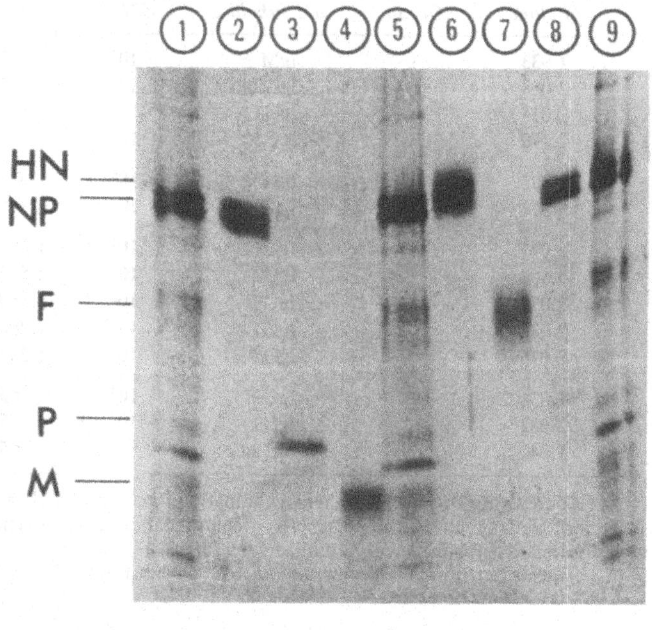

FIG. 1. Radioimmune precipitation assay (RIPA) with mouse ascites material reacting with different mumps virus proteins from the work of Örvell (1984). Materials included (the positions in the gel are given in parentheses): [^{35}S]methionine-labeled purified mumps virions (1, 5); monoclonal antibodies directed against NP (2, 8); P (3); M (4); HN (6); and F (7). Note that monoclonal antibodies against NP precipitate a double band (best seen in position 8), which both represent the nucleocapsid protein (Naruse et al., 1981). Band seen between P and M in purified virions is actin (Naruse et al., 1981).

inhibit different biological activities of the virus (Örvell, 1984; Tsurudome et al., 1986). In one study (Örvell, 1984), monoclonal antibodies directed against the HN glycoprotein of mumps virus could be divided into four different groups on the basis of their serologic reactivity (Tables 2 and 3). The first group of monoclonal antibodies could not inhibit any biological activity. The second group consisted of two monoclonal antibodies that could inhibit hemolytic (HL) activity, but not hemagglutination (HA) or neuraminidase (NA) activity. The third group of antibodies blocked HA, NA, and HL activity. The fourth group included monoclonal antibodies that could block NA, but not HA, activity of the virus. The spatial relationships between the different antibodies binding to the HN protein are presented in Table 3. Antibodies belonging to the second, third, and fourth groups had neutralization test (NT) activity, the antibodies with hemagglutination-inhibiting activity (group III) exhibited the highest titers of NT activity. It is not yet known whether antibodies binding to the HN protein with ability to block HL and cell-fusing activity (group II) block these activities directly by inhibiting corresponding activities of the HN protein, or indirectly by blocking the F protein (Örvell, 1984; Tsurudome et al., 1986). The grouping and serological characterization of monoclonal antibodies exhibit pronounced similarities with monoclonal antibodies against the HN protein of Sendai virus, another representative member of paramyxoviruses (Örvell and Grandien, 1982). The results show that HA and NA

activity are located on topographically distinct parts of the HN protein.

As yet few cloning data on the amino acid sequence of the proteins of different strains of mumps virus have been presented. Instead, monoclonal antibodies directed against different structural proteins of the virus have been used to compare strains (Örvell, 1984; Rydbeck et al., 1986; Wolinsky et al., 1985). Among the surface glycoproteins, the antigenic structure of the HN glycoprotein varies in different strains, whereas the F glycoprotein appears relatively well conserved (Örvell, 1984; Rydbeck et al., 1986; Wolinsky et al., 1985). Antibodies against the F protein can block hemolysis of the virus but have little or no demonstrable neutralizing activity (Örvell, 1976b, 1984; Wolinsky et al., 1985). In passive protection experiments, a monoclonal antibody against the fusion protein of mumps virus was found to confer marked protection in mumps virus-induced encephalitis in hamsters (Löve et al., 1986b). The monoclonal antibody used could not neutralize the infectivity of the virus, but was found to almost totally prevent extensive brain necrosis in the animals. In cross-neutralization experiments of rabbit hyperimmune sera immunized with a Swedish isolate (SBL 1) and the RW strain of mumps virus (Örvell 1984), approximately 10 to 20% of the homologous NT titers was obtained in tests with the heterologous virus; this reflects a pronounced functional antigenic difference between the HN proteins of these two strains (Örvell, unpublished data). The M protein ap-

TABLE 2. Determination of different antibody activities in mouse ascites material from 14 clones directed against the HN protein of mumps virus (Örvell, 1984)

Designation of clone	Group	Antibody class and subclass	Antibody titers in different serologic tests[a]						
			ELISA	HI	NI[b]	NI[c]	HLI	NT	NE
1.933	I	IgM	10^4	<5	<4	<4	<10	<4	<4
2.068		IgG2a	10^4	<5	<4	<4	<10	<4	<4
2.015		IgG1	10^5	<5	<4	<4	<10	<4	<4
2.170		IgG1	10^5	<5	<4	<4	<10	<4	<4
1.992	II	IgG2a	10^5	<5	<4	<4	1.280	16	512
5.374		IgG1	10^5	<5	<4	<4	1.280	16	1.024
2.041	III	IgG1	10^6	1.280	256	<4	1.280	512	32.000
5.500		IgG2a	10^6	2.560	256	<4	2.560	1.024	32.000
2.072		IgG1	10^6	2.560	256	256	2.560	512	32.000
2.073		IgG1	10^5	1.280	128	64	640	256	16.000
2.075		IgG1	10^5	1.280	128	128	1.280	256	16.000
2.034	IV	IgG1	10^5	<5	64	32	320	64	1.024
2.082		IgG1	10^5	<5	16	16	320	32	256
5,342		IgG2a	10^5	<5	64	64	640	128	1.024

[a] Abbreviations used: HI = hemagglutination inhibition; HLI = hemolysis inhibition; NI = neuraminidase inhibition; NT = neutralization test; NE = neutralization enhancement in the presence of anti-gamma globulin.
[b] NI tests performed with fetuin as substrate.
[c] NI tests performed with neuraminlactose as substrate.

TABLE 3. Competition experiment by ELISA with mouse ascites material from 14 clones producing antibodies against the HN protein of mumps virus in tests with 13 peroxidase-conjugated clones of similar specificity (Örvell, 1984)

Designation of clone	Group	Designation of peroxidase-conjugated clone												
		1.933	2.015	2.170	1.992	5.374	2.041	5.500	2.072	2.073	2.075	2.034	2.082	5.342
1.933	I	+[a](10^5)[b]	−[c]	−	−	−	−	−	−	−	−	−	−	−
2.068		−	−	−	−	−	−	−	−	−	−	−	−	−
2.015		−	+(10^3)	+(10^3)	−	−	−	−	−	−	−	−	−	−
2.170		−	+(10^1)	+(10^1)	−	−	−	−	−	−	−	−	−	−
1.992	II	+(10^2)	−	−	+(10^4)	+(10^4)	−	−	−	−	−	−	−	−
5.374		−	−	−	+(10^4)	+(10^4)	−	−	−	−	−	−	−	−
2.041	III	−	−	−	−	−	+(10^4)	+(10^4)	−	−	−	−	−	−
5.500		−	−	−	−	−	+(10^4)	+(10^4)	−	−	−	−	−	−
2.072		−	−	−	−	−	−	−	+(10^4)	+(10^4)	−	−	−	−
2.073		−	−	−	−	−	−	−	+(10^4)	+(10^4)	−	−	−	−
2.075		−	−	−	−	−	−	−	+(10^4)	+(10^4)	+(10^3)	−	−	−
2.034	IV	+(10^2)	−	−	−	−	−	−	−	−	−	+(10^3)	+(10^3)	+(10^3)
2.082		+(10^2)	−	−	−	−	−	−	−	−	−	+(10^2)	+(10^3)	+(10^3)
5.342		+(10^1)	−	−	−	−	−	−	−	−	−	+(10^3)	+(10^3)	+(10^3)

[a] + = Inhibition of binding of peroxidase-conjugated clone.
[b] Titer of inhibition with inhibiting clone.
[c] − = No inhibition of binding of peroxidase-conjugated clone.

pears to be well conserved in different strains of mumps virus (Örvell, 1984; Rydbeck et al., 1986). Of the internal proteins, the NP protein appears to vary to a greater extent than the P protein (Rydbeck et al., 1986).

The stability of different antigenic epitopes of mumps virus during passaging of the virus in vivo has been studied by Löve et al. (1986a). The neurotropic Kilham strain of mumps virus was serially passaged in newborn hamster brains. The virus was passaged with short (4 to 5 days) or long (10 to 12 days) intervals between inoculation and harvest. After 10 and 8 passages, respectively, two viral variants were isolated that differed in antigenic characteristics. With a collection of monoclonal antibodies directed against the structural proteins of mumps virus, marked differences between the two variants were found in the NP protein, and also some changes in the HN and P proteins. These results show that the virus can be subjected to antigenic modulation in vivo and possibly also in humans. An example of antigenic modulation and its effects on biologic and pathogenetic characteristics of the virus has been described by Löve et al. (1985b). One monoclonal antibody directed against the HN protein of the virus was used to generate four mutants of the neurotropic Kilham strain. One of the four mutants showed significantly lower neurovirulence in suckling hamsters than the parental strain.

The antigenic relationships between mumps virus and other paramyxoviruses have been studied in immunoprecipitation, enzyme-linked immunosorbent assay (ELISA) and immunofluorescence tests with both polyclonal and monoclonal antibodies (Julkunen, 1984; Örvell et al., 1986; Ray and Compans, 1986). A reciprocal antigenic relationship was found between mumps virus and Sendai virus. No antigenic relationship between mumps virus and parainfluenza virus type 2, 3 or Newcastle disease virus could be demonstrated (Örvell et al., 1986).

Pathogenesis, Pathophysiology, and Pathology

Careful clinical observations of mumps-infected patients and studies on experimental infections of humans and animals have provided information on the pathogenesis of mumps infection. The virus is spread between individuals via the respiratory route. Mumps virus infection may be subclinical (Mortimer, 1978). A systemic, generalized infection occurs and, although not always present, unilateral or bilateral parotitis is the first and most common clinical symptom. Viral replication takes place in the ductal cells of the gland. As a result of active virus replication and death of ductal epithelial cells, an inflamma-

tory response will be evoked, consisting of accumulation of lymphocytes and macrophages. An interstitial edema located around the ducti gives rise to swelling of the gland. Because of tissue damage, the enzyme amylase leaks out from cells into the blood stream. This pathologic process can give a significant elevation of serum amylase levels. Other organs that may become secondarily involved are the testes, ovaries, and, more seldom, pancreas. Peripheral nerves, eye, inner ear, and other organs may also become affected. Pleocytosis in the cerebrospinal fluid is found in more than 50% of patients with mumps virus infection as a sign of virus multiplication in the central nervous system (CNS), but clinical signs of CNS involvement are much less often recognized. CNS involvement is not correlated with occurrence of parotitis (Kilham, 1949). It is not known how the virus is spread to the CNS. Studies on newborn hamsters that were infected parenterally have shown that the spread of virus occurs by passage of infected mononuclear cells across the endothelium of plexus choroideus to the epithelial cells of plexus choroideus in that experimental system (Wolinsky et al., 1976). Another possibility for spread may be by direct passage of virus across the endothelium of plexus choroideus. In experimental infections of hamsters, it has been shown that the neuropathogenic capacity of different mumps virus strains is directly related to the fusion ability of the strain in question and indirectly related to the neuraminidase activity (Merz and Wolinsky, 1981). Neurologic symptoms of mumps infection of the CNS in humans are usually mild and limited to symptoms attributable to meningeal irritation. Symptoms of encephalitis, such as cranial nerve dysfunction, are more uncommon. The prognosis of mumps meningitis and meningoencephalitis is good, and as a rule the patients recover without long-lasting disabilities. In a few patients more serious neuropathologic effects, such as deafness and obstructive hydrocephalus, may follow (Everberg, 1957; Spataro et al., 1976; Thompson, 1979). The persistency of mumps virus in the CNS of mice and hamsters has been studied after experimental infection. In surviving hamsters inoculated as newborns either intracerebrally with a nonneuroadapted strain or intraperitoneally with a neuroadapted strain, viral antigen was demonstrable 50 days after the start of infection (Wolinsky and Stroop, 1978). Although viral nucleocapsids were found in ependymal cells, virus budding did not occur (Wolinsky, 1977). Newborn mice and hamsters respond differently to intracerebral injection of a neuroadapted strain (Kristensson et al., 1984). In hamsters a productive infection with expression in the neurons of all structural proteins will occur. In newborn mice, intracerebral inoculation of the virus resulted in a nonproductive infection of neurons. An expression of nucleocapsid and phospho protein antigens could be traced with monoclonal antibodies, but no expression of envelope antigens could be detected with monoclonal antibodies directed against HN, F, and M proteins (Kristensson et al., 1984). The difference between hamster and mice may be due to the fact that mouse neurons exert a host-cell restriction on virus maturation (Löve et al., 1987). Chronic mumps virus infection in the CNS of one human case has been described by Finnish workers (Vaheri et al., 1982). The clinical condition of the patient deteriorated over several years, and there were signs of chronic neurogenic lesions without active denervation. A pronounced antibody response developed in cerebrospinal fluid (CSF). Antibodies directed against all virus structural proteins could be identified.

Another chronic disease probably caused by mumps virus is a generalized, chronic, idiopathic, inflammatory myopathy, inclusion body myositis (IBM). The disease occurs in elderly persons (in approximately the sixth decade of life), but also in young adults. It is characterized by a slowly progressive, painless, muscular weakness of distal muscles to a greater extent than of proximal muscles. There are no signs of immunologic abnormality in these patients. Both intranuclear and intracytoplasmic microtubular inclusions were described from muscle biopsy specimens in 1967 (Chou, 1967). The morphology of the microtubular filaments was similar to nucleocapsids of paramyxoviruses. Immunoperoxidase screening for viral antigens has demonstrated that the inclusions stain with antibodies against mumps virus, but not with antibodies against other paramyxoviruses or influenza (Chou, 1986).

After puberty, involvement of the testicles is common in mumps virus infection. Viral multiplication in the affected testicle will result in lymphocytic infiltration and in local edema that give rise to swelling and pain of the affected testicle. Involvement of the pancreas may also occur as a result of mumps virus infection. The condition is often difficult to diagnose (Craighead, 1975). Mumps virus infection of the pancreas has been reported to be associated with onset of juvenile diabetes (Kahana and Berant, 1967; Messaritakis et al., 1966; Witte and Schanzer, 1968). Interestingly, it has been shown that human beta cells can support the growth of mumps virus and also that cytopathic effects of cells will develop as a result of virus replication (Prince et al., 1978). However, a causal relationship between mumps virus infection and diabetes mellitus has not been firmly established. Other conditions of little clinical importance that have been reported in association with mumps virus infection include myocarditis and a transient polyarticular arthritis (Bengtsson and Orndahl, 1954; Gordon and Lauter, 1982). If mumps-infected persons

are subjected to electrocardiography (EKG), a high proportion have EKG abnormalities, but the myocarditis as revealed by EKG is seldom symptomatic.

Clinical Features

Transmission of mumps virus is by means of droplet infection of infected saliva. It has not been excluded that virus secreted in the urine may spread the disease. The incubation period for mumps is usually 18 to 21 days but may extend from 12 to 35 days. Virus can be recovered in the saliva up to 6 days before the first clinical symptoms (Henle et al., 1948b). The excretion of virus in saliva continues up to 5 days after the onset of clinical symptoms, and after that time local antimumps IgA antibodies can be demonstrated (Chiba et al., 1973). The presence of virus in urine from infected individuals has been demonstrated up to 2 weeks after the start of illness (Utz et al., 1957, 1958).

Symptoms in the early phase of the disease are fever, anorexia, malaise and myalgia. Swelling of the parotid gland becomes apparent 1 to 7 days after onset of prodromal symptoms. The swollen gland can cause pain described as earache by the patient. The swelling is usually bilateral, although swelling may occur earlier in one gland than the other. Within 48 h the parotid gland reaches its maximal swelling, then the organ gradually returns to its normal size within 10 days. Parotitis may be the only clinical symptom of mumps virus infection but may also be absent. About 30% of all mumps infections are subclinical. More seldom, in 10% of patients, swelling of the sublingual and submaxillary salivary glands occurs. In uncomplicated mumps infection fever generally lasts 2 to 4 days. The peripheral white cell count is between 4,000 and 16,000 per mm^3 with a relative increase of lymphocytes, but lymphocytopenia may occur. Serum amylase concentration is often elevated when parotitis is present. An increase of cells in CSF is found in 50% of cases. Complications of mumps infection may involve a variety of organs. Orchitis is found in 25% of men and adolescent boys, but it is very rare before puberty. In most cases only one testis is involved. Orchitis usually develops within a few days after onset of parotitis. In severe cases there may be high fever, delirium, vomiting, and backache. The testis may become hard and extremely tender, with enlargement up to 5 times its normal size. Pain is a pronounced symptom. The ovary is involved in 5% of women with mumps. In severe oophoritis the patient is feverish, with large, tender ovaries. Pancreatitis occurs in 7% of cases, and when it is present the patient usually has swelling of the parotid gland. A rise in serum amylase is not specific for pancreatitis, since it occurs in connection with parotitis. The most common clinical signs are vomiting and epigastric pain. Meningoencephalitis is found in 10% of patients with mumps. It may be the only symptom of mumps, or it may develop at any time of mumps parotitis. When the patient develops meningoencephalitis, there is a rise in temperature, headache, vomiting, and stiff neck and back. The acute phase of mumps meningoencephalitis usually subsides within a few days. Convulsions are uncommon. There is pleocytosis in the CSF, usually with dominance of lymphocytes. The protein content is moderately elevated, but the sugar content is normal. The meningoencephalitis of mumps cannot be differentiated from other forms of viral meningitis clinically. Mumps virus infection seldom involves the eyes, but when it does, iritis and optic neuritis will develop. One-sided or two-sided deafness may occur during mumps parotitis or meningoencephalitis as a result of damage to the organ of Corti in the cochlea. Deafness is sometimes associated with tinnitus or Meniere's syndrome.

Diagnosis

Specimen Collection

The diagnosis of mumps can be made by demonstration of either virus or viral antigen in materials obtained from the area around Stensen's duct of the parotid gland, saliva, and urine. Saliva may be collected by a suction device or by swabbing, preferably from the area around the orifice of Stensen's duct. The swabs are placed in a tube containing Hanks balanced salt solution with a protein preservative and antibiotics. In cases where the CNS is involved in mumps infection, the virus can also be isolated from the cerebrospinal fluid. The virus should be isolated from the patient's saliva as early as possible, up to 5 days after the onset of clinical symptoms (Chiba et al., 1973). From the urine, the period of possible positive isolation is from 6 days before to 15 days after the onset of parotitis (Utz et al., 1958, 1964). The virus is relatively thermolabile and, in cases where isolation of virus is attempted, the specimen should be kept at 4°C for as short a time as possible before inoculation into tissue-culture cells. A way of increasing the frequency of positive isolation from urine and CSF is by means of concentration by centrifugation. The specimen should be centrifuged at $10,000 \times g$ for 1 h, and the pellet should be suspended in one-tenth of the original volume in Hanks balanced salt solution (BSS) containing 2% serum (Utz et al., 1957).

Paired serum samples should be used when serologic diagnosis is attempted, and the first serum should be collected as early as possible after the start of symptoms. The second serum sample should be collected 2 to 3 weeks after the start of symptoms. The diagnosis of mumps infection can be made on a single serum specimen collected by the 5th day of illness and up to 3 months or longer by demonstration of mumps-specific IgM antibodies (Glikmann et al., 1986; Sakata et al., 1985).

Direct Antigen Detection and Viral Nucleic Acid Detection

Direct or indirect fluorescence has been used for demonstration of paramyxovirus antigens in cells from the nasopharynx of patients with upper respiratory tract disease. Few studies have been performed to try to trace mumps virus antigen in cells from the respiratory tract or infected tissues. Immunoperoxidase staining with polyclonal sera has been reported for tracing of mumps virus antigen in human muscle tissue (Chou, 1986). Monoclonal antibodies directed against mumps virus (Örvell, 1984; Tsurudome et al., 1986; Wolinsky et al., 1985) can be expected to stain different structural components of the virus in tissues. For this purpose these monoclonal antibodies have been used successfully in animal experimental systems (Kristensson et al., 1984; Löve et al., 1985a).

Solid-phase immunoassays, that is, enzyme-linked immunosorbent assay (ELISA) and radioimmunoassay (RIA) have been used for demonstration of different viral antigens in the nasopharynx (for review, see Sarkkinen, 1985). These tests have not yet been applied to mumps virus antigen detection. Viral nucleic acid detection in cells from the nasopharynx or infected tissues by nucleic acid hybridization has not been reported.

Virus Isolation and Identification

For isolation of the virus, primary cell cultures and continuous cell lines from monkey kidneys; chicken embryo fibroblast cell cultures; human fibroblast cultures; and continuous cell lines, such as Hep-2 and HeLa cells; and embryonated hen eggs have been used. Human embryonic kidney cells have also been used. Primary monkey kidney cells and human embryonic kidney cells are considered to give the best results. Primary cell cultures or continuous cell lines are seeded in standard tissue culture tubes in 1 ml of growth medium containing 10% inactivated fetal calf serum and in a concentration to allow a monolayer of cells to be formed within 3 to 5 days at 37°C. A commonly used standard medium is Eagle's minimal essential medium supplemented with pyruvate and 0.5% freshly added glutamine and with addition of 3 ml of 7.5% sodium bicarbonate per 100 ml, 100 IU of penicillin per ml, and 100 μg of streptomycin per ml. When a monolayer of cells has formed, maintenance medium of the same composition but containing 3% fetal calf serum is added. Cultures can now be inoculated with 0.2 ml of the specimen used for isolation. The chance of successful isolation depends on how early the specimen is taken after the onset of symptoms. In mumps meningitis patients, virus can be obtained from the CSF within 8 to 9 days from the onset of symptoms of meningitis. From the CSF the virus has been isolated in 50% or fewer of cases (Kilham, 1949; Wolontis and Bjorvatn, 1973).

The tissue culture tubes, which are kept at stationary position or slowly rotating at 37°C, should be examined daily for the appearance of cytopathic effects (CPE) and compared with uninoculated control cells. It is essential that the cells be maintained in a good condition, and if necessary the medium must be replaced. Cytopathic effects, such as giant cell formation and rounding of cells, do not occur regularly with all isolates, and therefore the cultures should always be examined with a hemadsorption test after 7 days of incubation. It should be noted that cells, especially those from monkey kidney, may contain latent hemadsorbing viruses. Both inoculated and several control cultures should, therefore, be tested for comparison. The hemadsorption test is carried out in the following way: The cell culture medium is removed, and 0.2 ml of a cooled 1% chicken or guinea pig erythrocyte or human group 0 erythrocyte suspension in Veronal-buffered saline (VBS) is added to the cell sheet and kept in contact with the cell monolayer during the incubation period at 4°C for 45 min. After the incubation period the erythrocytes are poured off, and the cell cultures are washed 5 times with 2-ml volumes of cold, buffered saline before examination in the light microscope. A positive hemadsorption indicates presence of virus. Hemadsorption is a more sensitive indicator for the presence of virus than CPE. Provided that the erythrocyte suspension used in the test is sterile, the cell cultures can be used to passage the virus to other cell cultures or be subjected to further incubation.

Embryonated hen eggs can also be used for isolation of virus. The amniotic cavity of 7- to 8-day-old embryonated eggs is inoculated with 0.2 ml of the specimen. The eggs are incubated for 7 days at 37°C and then, after storage of the eggs at 4°C overnight, the amniotic fluid is collected. The presence of virus in the amniotic fluid can be examined by the hemagglutination (HA) test. Volumes of 0.05 ml of amniotic fluid, serially diluted in twofold dilutions, are incubated with 0.05 ml of a 1% suspension of chick, guinea pig, or human 0 erythrocytes in VBS and in-cubated at 4°C for 60 min before examination for hemagglutination. Amniotic fluid from uninoculated eggs is included in the test as a negative control, and an erythrocyte control with erythrocytes and buffer alone is also included in the test.

The virus isolates obtained from the tissue culture tubes or embryonated eggs can be identified in different ways, that is, by hemadsorption-inhibition, neutralization test, immunofluorescence staining, and hemagglutination-inhibition. In the hemadsorption-inhibition test, the medium from infected cell cultures is poured off, and the cell sheets are washed with VBS. Appropriately diluted antimumps serum, which inhibits hemadsorption of mumps virus, is added. Two other infected cultures receive only VBS. After incubation at room temperature or at 37°C for 1 h, the hemadsorption test is performed as described above.

In the neutralization test (NT), material from infected cultures is collected and clarified by centrifugation at 1,500 rpm (200 × *g*) for 10 min. The infectivity titer of the material must then be determined prior to use in NT. Thereafter, 10 to 50 TCID$_{50}$ of the virus is mixed with appropriately diluted antimumps- or antibody-negative reference serum. After incubation at room temperature for 60 min, equal portions of the mixtures are added to several cell cultures and incubated at 37°C. If the antimumps serum can inhibit the appearance of CPE in the test cultures compared with the control cultures, the isolate has been identified as mumps virus.

A rapid and practical identification of mumps virus isolates can be accomplished by immunofluorescence staining of the primary isolate grown on cells on a glass slide inserted into the tissue culture tube. This immunofluorescence antibody technique allows a direct identification of the isolate by staining with specific reagents, preferably monoclonal antibodies against mumps virus. Virus isolated from the allantoic cavity or tissue culture tubes can be identified by a hemagglutination-inhibition (HI) test or by demonstration of mumps virus complement-fixing (CF) antigen in the allantoic fluid. Monoclonal antibodies blocking hemagglutination of mumps virus can be used in HI tests, but in CF tests the use of monoclonal antibodies is of limited practical importance.

Serologic Diagnosis

Determination of a rise in specific IgG antibodies against mumps virus or the demonstration of IgM antibodies in sera of patients is a common way to establish the diagnosis of mumps. The requirements for serologic methods are that they should be sensitive, specific, and simple to perform. Comparison of

IgG titers on paired serum samples—with the first sample taken at the start of clinical illness, and the second sample taken 2 to 3 weeks later—usually demonstrates a significant titer rise but carries the drawback that the diagnosis is established late, when the patient has recovered from the infection. In such cases the (serologic) diagnosis is established too late to have influence on the doctor's management of the patient. Older tests used in serology, the NT, HI, and CF tests, have their drawbacks. The neutralization test is specific, and NT titers are found in all sera from patients 3 weeks after the start of clinical symptoms of mumps. However, the NT is cumbersome, time-consuming and therefore not suitable for routine serology (Buynak et al., 1967). The HI and CF tests are simple to perform but are relatively insensitive. Therefore, these tests are not suitable when low levels of antibodies are present, that is, for screening of immunity to mumps or for demonstration of seroconversion after vaccination. Cross-reacting antibodies against other paramyxoviruses may be a source of erroneous results (Meurman et al., 1982; Nicolai-Scholten et al., 1980). It should be pointed out that broader cross-reactivities may be found in patient sera owing to an anamnestic response (Meurman et al., 1982) than are found in monospecific sera as a result of a single immunization in experimental animals (Julkunen, 1984; Örvell et al., 1986). That an anamnestic response is involved is strengthened by the fact that in patient sera the frequency of heterologous antibody rises correlate closely with the age of the patient (Lennette et al., 1963). If nonspecific inhibitors are present, the HI test may be inaccurate for determination of low titers of antibodies. A more recently introduced test in mumps virus serology is the hemolysis-in-gel (HIG) test (Grillner and Blomberg 1976; Väänänen et al., 1976). An advantage of this test over the HI test is that it is less influenced by nonspecific inhibitors in sera. The test has a moderate sensitivity, and, in comparison with the NT, both false-negative and false-positive results have been reported (Christenson et al., 1983; Grillner and Blomberg, 1976). In recent years the ELISA has been introduced for determination of mumps-specific IgG and IgM antibodies. The ELISA has a higher sensitivity than the CF, HI, and HIG tests and to date appears to be the method of choice for demonstration of antibodies to mumps virus (Glikmann and Mordhorst, 1986; Nicolai-Scholten et al., 1980; Ukkonen et al., 1980). A newer development of the ELISA technique for demonstration of IgM antibodies against mumps virus is a direct immunoglobulin M antibody-capture enzyme immunoassay with enzyme-conjugated antigen (Gut et al., 1985; Sakata et al., 1985) or with antigen and enzyme-conjugated antibodies (Glikmann et al., 1986). Subclass determination of antimumps IgG antibodies by ELISA has recently been shown to be of value in the serologic diagnosis of mumps (Linde et al., 1987).

ELISA

The first development of the ELISA technique for serologic diagnosis of mumps virus infection was an indirect IgG and IgM assay (Nicolai-Scholten et al., 1980; Ukkonen et al., 1980). Mumps virus grown in embryonated eggs and purified from chorioallantoic fluid by ultracentrifugation should be used to coat plastic plates. A solution with a protein content of 10 to 20 μg/ml of purified virions diluted in 0.05 M carbonate–bicarbonate buffer, pH 9.6, is used to coat plastic plates in a volume of 0.1 ml per well. The plates should be incubated at room temperature overnight or at 37°C for 3 h before use. A catch-up ELISA for demonstration of IgG antibodies has been described (Glikmann and Mordhorst, 1986). In this test, antimumps antibodies are coated to the plates and used to bind mumps antigen for use in the test. After being coated, the plates should be washed 3 times with phosphate-buffered saline (PBS) with 0.05% Tween 20. Sera are diluted in serial two-, four-, five-, or tenfold dilutions in the same buffer with 1% bovine serum albumin, and 0.1 ml of each serum dilution is added to a well and incubated at 37°C for 1 to 2 h. After that the plates are washed again and appropriately diluted, enzyme-conjugated antihuman IgG or IgM antibodies are added in 0.1 ml. After a further incubation at 37°C for 2 h (some laboratories prefer overnight incubation), the plates are washed and the corresponding substrate is added. The colorimetric reaction is read spectrophotometrically after a certain time. The titer is expressed as the highest dilution of serum that gives a certain absorbance value, which has been defined in the test situation. The determination of IgM antibodies by addition of conjugated antihuman IgM antibodies to the test allows the diagnosis to be made by analysis of a single serum specimen. Antibodies of IgM class do not cross-react with other paramyxoviruses (Nicolai-Scholten et al., 1980; Meurman et al., 1982).

A newer development for determination of IgM antibodies is the direct immunoglobulin M antibody capture technique. In this test antihuman IgM antibodies at a certain dilution are used to coat the wells of the plastic plates. After washing of the plates, serial dilutions of the sera of patients are added to the plates. If IgM antibodies are present in the serum, they will bind to the anti-IgM antibodies coated on the plates. After further incubation and washing, enzyme-labeled, purified mumps virions are added. Finally, after the last incubation and washing, the appropriate substrate is added to the reaction. The IgM capture technique is more specific than the indirect

ELISA for IgM determination. The former test does not give false-positive reactions for sera containing rheumatoid factor or false-negative reactions for sera containing high titers of IgG antibodies that can be found in the indirect test (Gut et al., 1985; Sakata et al., 1985). The sensitivity of the test may be increased by addition of unconjugated antigen, followed by addition of enzyme-conjugated antimumps serum (Glikmann et al., 1986).

An indirect ELISA for determination of IgG subclass antibodies has recently been developed by Linde et al. (1987). Purified viral antigens were coated on plastic plates. In the first step the patient sera were added to the wells at a certain dilution. The second step involved addition of four enzyme-labeled monoclonal antibodies directed against each of the four human IgG subclasses, IgG1, IgG2, IgG3, and IgG4. Finally, in the last step substrate was added. In that work it was demonstrated that substantial levels of IgG3 to mumps NP were present only in current infection. In the future, subclass determination of mumps IgG antibodies may be of value in serologic diagnosis.

HEMOLYSIS IN GEL (HIG) TEST

The principle of the test is as follows. Sheep or chicken erythrocytes are washed and mixed with mumps virus antigen (preferably purified virions) at a final concentration of 320 to 640 hemagglutinating units per 25 μl. A solution of $CrCl_3$ is added to the mixture to allow coupling of the antigen to the erythrocytes. After coupling, the virus-erythrocytes are washed and suspended in agarose containing fresh guinea pig serum. The agarose solution is then poured into plastic petri dishes placed on a level surface. After solidification of the gel mixture on the plastic surface, holes with a diameter of 3 mm are punched in the gel, and 5 μl of patient serum is applied to each well. The diffusing antibodies form a zone of hemolysis owing to the action of complement on the erythrocytes. The diameter of the zone correlates with the antibody titer of serum.

HEMAGGLUTINATION INHIBITION (HI) TEST

Heat-inactivated sera are diluted in serial twofold dilutions in 25 μl VBS in microplates. If nonspecific inhibitors are contained in sera, they must be removed. They may be removed by treatment of sera with acid-washed kaolin or with the receptor-destroying enzyme of Vibrio cholera, or periodate treatment can be used. In the periodate method, 0.15 ml of freshly prepared 0.1 M NaIO$_4$ is added to 0.5 ml serum. After an incubation period at 37°C for 30 min, 0.15 ml of 40% glucose solution is added, and the serum is adjusted to a 1 : 2 solution by addition of buffered saline solution. Nonspecific agglutinins may be removed by absorption of serum with packed 10%

erythrocytes to be used in the test for 30 min at room temperature. Mumps virus antigen diluted to contain 4 hemagglutinating (HA) units in a volume of 25 μl is added to each serum dilution. In order to increase the HA titer of the antigen and the sensitivity of the test, the mumps virus antigen should be treated with Tween 80 and ether prior to use (Buynak et al., 1967). A 0.1-volume of a 1.25% solution of Tween 80 is added to the mumps antigen, mixed carefully, and incubated at room temperature for 5 min. Then an equal volume of anesthetic ether is added, and the mixture is shaken in an ice bath for 15 min. After centrifugation at 200 g for 10 min, the aqueous phase is recovered, and residual ether is removed by bubbling nitrogen gas through the preparation. Virus and serum mixture in the HI test is incubated at room temperature for 1 h. Serum, virus, and erythrocyte controls are included in the test. A 0.5 or 1% suspension of chick, guinea pig, monkey, or human 0 erythrocytes in VBS is added in a volume of 0.05 ml. The plates are shaken well, and the erythrocytes are allowed to settle for 60 to 90 min at 4°C or room temperature before the test is read. Agglutination will make the erythrocytes form a shield of cells along the sides and bottom of the cup, whereas nonagglutinated erythrocytes will form a small central button with a smooth margin. When there is partial hemagglutination, the central button is surrounded by a margin of serrated appearance. The inverted value of the highest serum dilution that inhibits hemagglutination is considered the titer of the serum. It may be necessary to tilt the plate to decide the titer with low serum dilutions. Nonagglutinated erythrocytes will stream towards a vertical position at the low side of the cups. A means of increasing the sensitivity of the test is the addition of appropriately diluted anti-human immunoglobulin to the virus-serum mixtures for 1 or 2 h prior to the addition of erythrocytes. In this way antibodies directed against the F or HN protein blocking hemolysis but not hemagglutination in the absence of antigamma globulin may become detectable (Örvell, 1976b, 1984; Tsurudome et al., 1986).

HEMOLYSIS INHIBITION (HLI) AND MIXED HEMADSORPTION (MH) TESTS

These tests measure antibodies against the envelope glycoproteins of the virus. The hemolysis inhibition test measures antibodies against both the F protein and the HN protein of the virus (Örvell, 1976b, 1978a, 1984). The test may be carried out as follows (Norrby and Gollmar, 1972; Örvell, 1976a).

Serial twofold dilutions of serum in 0.2 ml of phosphate-buffered saline (PBS) are mixed with an equal volume of purified virions. The test tubes are shaken well, and after incubation at room temperature for 1 h, a 0.1 ml of 10% erythrocyte suspension of washed monkey or human 0 cells is added to each tube. The mixtures are incubated at 37°C for 3 h, and shaken intermittently during the incubation period. After that the erythrocytes are removed by low-speed centrifugation. The optical density values of the supernatants are read spectrophotometrically at 540 nm. The virus material included in the test is diluted to give an optical density of 0.3 to 0.5 in control tubes containing PBS instead of serum. The highest dilution of serum that reduces the specific hemolysis by 50% or more is considered the end point titer.

The MH technique has been described by Espmark and Fagraeus (1965; Fagraeus et al., 1965) and has been applied for testing of antibodies against mumps (Norrby et al., 1977; Örvell, 1978a). Monolayer cultures of permissive cells are infected with mumps virus, preferably egg-grown virus at a multiplicity of infection of 1 to 10. Two or three days later the cultures are used in the test. The monolayer is then covered by an agarose overlay, and 20 μl of serum is added to filter paper disks of 5 mm diameter, which are placed on the agar overlay. The antibodies diffuse radially and bind to the surface of the infected cells. After 3 days of incubation at room temperature, the agar layer is removed, and an indicator cell system consisting of sheep erythrocytes covered with human antisheep erythrocytes is added to the cultures. Antihuman antibodies from another species are added in order to bind the antibody-coated erythrocytes to antibodies bound on the infected cell surface. A zone of erythrocytes will form, and the diameter of the zone is proportional to the antibody content of the serum. Experiments with monospecific sera directed against either HN or F have shown that the MH test measures antibodies against both these components (Norrby et al., 1977; Örvell, 1978a).

COMPLEMENT FIXATION (CF) TEST

Whole virus antigen prepared by three cycles of freezing and thawing of infected cells may be used in the CF test. Such antigen preparations contain all the antigens of the virus. Mumps virus antigen can be divided into S (soluble) or V (viral) antigens (Henle et al., 1948a). The S antigen probably represents nucleocapsid antigens and the V antigen viral glycoproteins. After infection there is a displacement in time between the appearance of antibodies against S and V antigen in the sera from patients (see below). The CF test is performed as described in earlier textbooks (Palmer et al., 1977).

FLUORESCENT ANTIBODY (FA) TEST

The fluorescent antibody technique has little practical use in routine serological work. However, due to

an anticipated more frequent use of monoclonal antibodies in the future, the technique can be expected to be of value for comparing different viral strains antigenically (Mufson et al., 1985; Örvell and Norrby, 1985; Rydbeck et al., 1986). Therefore, the technique has a potential application in epidemiological surveys, that is, to study the circulation of different viral strains in the population and to study the change of epitopes of different structural components of the virus in strains isolated in different years. As mentioned earlier, the test may also be used for identification of virus isolates. In the FA technique the cells to be infected with virus are preferably grown on microscope slides inserted in Leighton tubes or in petri dishes in a CO_2 incubator. The virus-infected cells are air dried and fixed in cold acetone (-20°C) for 10 min at room temperature. The indirect FA test is performed with murine ascites containing monoclonal antibodies when available, or with paired human sera or animal hyperimmune sera. An advantage with the use of monoclonal sera is their specificity, potency, and reproducibility. However, care must be taken to ascertain that the epitope with which the monoclonal antibody reacts is present in all strains. In certain situations it may be necessary to use a mixture of different monoclonal antibodies.

NEUTRALIZATION TEST (NT) AND OTHER LESS COMMONLY USED TECHNIQUES FOR DETERMINATION OF MUMPS VIRUS ANTIBODIES

The neutralization test is a somewhat cumbersome technique for determination of antibodies against mumps virus. Heat-inactivated sera (56°C, 30 min) are titrated in serial twofold dilutions in tissue culture medium. To each dilution is added an equal volume of mumps virus containing 60 to 200 $TCID_{50}$ in 0.1 ml. The virus-antibody mixtures are incubated at room temperature for 1 h and then at 4°C overnight. Cell cultures are then inoculated with 0.1 ml of the mixtures. Cytopathic changes are read after 7 days of incubation at 37°C. The highest dilution of serum that inhibits the multiplication of virus is the end point titer. Other techniques not commonly used include the immunodiffusion technique, which allows the separation of different antigen–antibody systems (Örvell, 1978a); imprint immunofixation technique, which can be used to identify mumps virus-specific antibody in fractionated immunoglobulin (Vandvik et al., 1982); and radioimmune precipitation assay, which allows identification of the structural components that react with an antibody (Fig. 1; Örvell, 1984; Rydbeck et al., 1986).

Interpretation of Laboratory Data·

The only situation in which it has been reported possible to isolate mumps virus from a patient is during current infection, and, therefore, isolation of mumps virus is diagnostic. With the development of ELISAs for demonstration of viral antigens (Sarkkinen, 1985) and application of this test to an acute infection, isolation of the virus may become superfluous in the future. Also, it will be possible to study the different chronic infections in more detail, especially since monoclonal antibodies will allow selective identification of individual structural components.

Mumps virus infection is more often diagnosed serologically than by isolation of virus or by demonstration of antigen. A fourfold increase in the titers between the acute and convalescent serum is diagnostic. There is good correlation between the CF test and ELISA IgG (Popow-Kraupp, 1981; Ukkonen et al., 1980). The ELISA is more sensitive and has been reported to give a higher percentage of positive diagnosis than the CF test, the catch-up IgG test being more specific and sensitive than the indirect IgG test (Glikmann and Mordhorst, 1986).

The ELISA for demonstration of IgM antibodies has a high specificity and is superior to the other tests, since this test requires only one serum sample to be analyzed. The demonstration of IgM antibodies against mumps in a single serum sample is diagnostic. IgM antibodies are present from 1 to 5 days after the start of illness and persist for 3 to 5 months or longer (Glikmann et al., 1986; Gut et al., 1985; Sakata et al., 1985; Ukkonen et al., 1980). A false-positive reaction in the indirect IgM test may be due to the presence of rheumatoid factor in the serum, and false-negative reactions may be due to high titers of IgG antibodies. The IgM capture ELISA is the method of choice since this test is not influenced by the presence in serum of rheumatoid factor or high titers of IgG antibodies (Gut et al., 1985; Sakata et al., 1985). Another marker for current infection is the demonstration in serum of IgG3 subclass antibodies against mumps. These antibodies can be determined by a subclass ELISA (Linde et al., 1987). These antibodies appear a few days after the appearance of IgM antibodies.

In mumps meningitis patients, both IgG and IgM antibodies can be demonstrated in the CSF by ELISA, but some patients with moderate to high IgM antibody titers in serum lack demonstrable IgM antibody titers in CSF (Forsberg et al., 1986; Ukkonen et al., 1981b). The titer of antibodies in the CSF reaches its maximum 1 to 2 weeks after debut of symptoms of meningitis. Evidence for an intrathecal synthesis of antibodies in the CNS in mumps meningitis patients is more frequently demonstrated by comparing the serum/liquor ELISA IgG titers than the IgM titers (Forsberg et al., 1986; Ukkonen et al., 1981b). By the sensitive IgG ELISA, 80 to 90% of patients with mumps meningitis produce intrathecally synthesized antibodies (Forsberg et al., 1986; Ukkonen et al., 1981b).

It has been known for a long time that antibodies directed against the V and S antigens of mumps virus are formed at different times after infection (Henle et al., 1948a). Antibodies against S antigen can be detected earlier than antibodies against V antigen. The CF, HIG, and ELISA tests measure antibodies against nucleocapsid antigens (S antigen), and the HI, HLI, MH and neutralization tests measure antibodies against virion envelope antigens (V antigen). Maximal titers in the former tests are reached within 3 weeks after the start of illness (Ukkonen et al., 1981a), whereas antibodies in the latter tests reach their maximum 1 to 2 weeks later. Therefore, if the first serum is collected late after infection, the chances of obtaining a significant titer increase is greater by applying the HI and neutralization tests. By application of these tests, a certain risk for determination of cross-reacting antibodies directed against other paramyxoviruses exists. This risk is eliminated by the determination of IgM antibodies, since IgM antibodies directed against mumps do not appear to cross-react with other paramyxoviruses (Meurman et al., 1982; Nicolai-Scholten et al., 1980; Ukkonen et al., 1980). IgM antibodies appearing early after infection have been shown to be directed predominantly against the nucleocapsid component of the virus, but a small reaction with the membrane antigen could also be demonstrated (Glikmann and Mordhorst, 1986).

For determination of immunity to mumps, the ELISA IgG is suitable. The HIG test has also been used for screening of mumps immunity, but this test appears to be less suitable inasmuch as both false-negative and false-positive results in comparison with NT titers have been found (Grillner and Blomberg, 1976; Christenson et al., 1983). Although the NT is specific, it is too cumbersome for routine determination of mumps immunity. In selected cases, that is, if the screening of immunity by the IgG ELISA gives equivocal results, the sera should be subjected to the NT.

Epidemiology and Natural History

Mumps virus infection is a worldwide disease that occurs only in humans, and the spread of the virus is from an acutely infected individual to another nonimmune person. Mumps virus requires a population of 300,000 people in order for continued spread of the virus to occur. Populations exceeding this size have existed for only 5,000 years. There is some evidence that Sendai virus in mice, which has existed for a longer time than mumps virus, may be the ancestor virus from which mumps virus has evolved. This possibility is based on the fact that there exists a close antigenic relationship between Sendai virus and mumps virus (Julkunen, 1984; Örvell et al., 1986). The incidence of mumps virus infection in a population should preferably be studied serologically, since a large proportion of mumps infections are subclinical (Levitt et al., 1970; Mortimer, 1978). Immunity to mumps virus in a dense urban population in the Netherlands has been reported (Wagenvoort et al., 1980). The peak incidence of mumps virus infection was found to be at 4 to 6 years of age, and at 6 years of age more than 50% of the children had developed antibodies to mumps virus, 90% of children had acquired antibodies before the age of 14 years. In adults, 95% had antibodies to mumps virus. Analogous results were reported in a study from the United Kingdom (Mortimer, 1978) and also by the Centers for Disease Control, Atlanta, Georgia, in 1972 (CDC Report, 1972). Immunity to mumps virus after natural infection is long-lasting. There are no reports of reinfections with mumps virus, but cross-reactivity with antibodies to other paramyxoviruses in human sera makes it difficult to study the possibility of reinfection with mumps virus.

New York City, with an unvaccinated urban population studied from 1935 to 1972, showed that mumps virus occurred in epidemics every year, primarily in the first 6 months of the year (Yorke and London, 1973). The seasonal variation was due primarily to increased contacts from the gathering of children in schools (London and Yorke, 1973). Although in some years there were fewer cases of mumps than in other years, there were no pronounced biannual fluctuations, as observed with measles virus in the same study.

After 20 years of active immunization in the United States, natural mumps virus infection has now become an exceedingly rare disease in this country (CDC Report, 1982).

Control and Prevention

Attempts to prevent the spread of mumps virus by isolation of infected individuals are not effective, since there is a high incidence of asymptomatic infections and because the virus is shed for several days before the start of clinical symptoms. The use of hyperimmune gamma globulin can reduce the susceptibility if given before exposure to the virus (Copelovici et al., 1979), but has no protective effects on the symptoms or complications if given after exposure to the virus (Pollock, 1969). Vaccination with both killed and live mumps vaccines has been used in order to control the virus. Immunization with virus killed by formalin inactivation was found to provide protection against challenge with live virus already in the 1940s (Enders, 1946). Formalin-inactivated vaccines were then prepared and tested in field trials (Habel, 1951; Henle et al., 1951; Stokes et al., 1946). The vaccine provided protection against symptomatic mumps virus infection in the majority of susceptible individuals, but a long-lasting immunity could not be maintained without repeated booster injections. Formalin-killed mumps virus vaccines were, therefore, considered impractical for clinical use. Instead, attenuated live vaccines have come into use. The Jeryl Lynn B strain of mumps virus was originally obtained as an egg isolate from a girl with mumps infection. The isolate was attenuated by repeated passaging in embryonated eggs and then propagated in chick-embryo-cell cultures before use. Another live mumps strain used for vaccination is the Urabe-Ann 9 strain. Immunization with the Jeryl Lynn vaccine strain rarely causes any symptoms of mumps infection or adverse effects in susceptible children (Ennis et al., 1969). Administration of the vaccine is contraindicated in immunocompromised persons and also to pregnant women because of a potential teratogenic effect of the virus.

The most sensitive tests for demonstration of seroconversion are the ELISA and neutralization tests. With these tests seroconversion rates between 90 and 100% have been demonstrated after vaccination (Christenson et al., 1983; Popow-Kraupp et al., 1982), and the vaccine has been reported to give complete (Brunell et al., 1969; Weibel et al., 1972) or almost complete protection (Hilleman et al., 1967). The mumps virus vaccine is often combined with attenuated measles and rubella virus vaccines. Combination of mumps virus with measles or rubella virus or with both measles and rubella virus does not result in a lower antibody response to mumps virus (Weibel et al., 1980). Antibody persistence against mumps virus could be demonstrated in all 43 children immunized with three different combinations of mumps virus together with measles and/or rubella virus 7 to 10 years after vaccination (Weibel et al., 1980). The use of attenuated mumps virus vaccines has been widespread in the United States, where more than 40 million doses have been administered to susceptible individuals during the last 20 years. This vaccination program has resulted in an approximately 50-times reduction of the reported incidence of mumps in the United States. Common practice in the United States is to administer attenuated mumps virus in a combined vaccine together with attenuated measles and rubella virus to children of about 15 months of age, at a time when passively transferred maternal antibodies have disappeared. In other countries, mumps virus vaccine was used for a long time more selectively and given to persons at high risk for contracting the disease or with difficult complications. It is uncertain whether vaccine given at the time of exposure to the virus can protect against the disease, but it is not contraindicated. In case the individual does not become ill with mumps as a result of that exposure, protection against subsequent exposures to the virus will be attained.

The question why formalin-inactivated mumps virus does not result in a stable, long-lasting immunity has been addressed by Norrby and Penttinen (1978). They compared the antibody response in individuals vaccinated with either formalin-killed or live, attenuated virus. The overall antibody responses, measured by HIG, HI, and HLI tests were of a similar magnitude in the two groups (Norrby and Penttinen, 1978; Penttinen et al., 1979). In these sera, antibodies blocking hemagglutination were removed by adsorption of sera with Tween 80 and ether. After this procedure, antibodies that could react in the HLI test were found only in sera from persons who had received the live vaccine. These findings were interpreted to mean that both Tween 80–ether treatment and formalin treatment destroy an antigenic structure of the F protein responsible for induction of HLI antibodies.

Killed vaccine does not protect against mumps virus when given within 3 days after exposure (Meyer et al., 1966). However, killed mumps virus vaccine may have a potential value for boosting immunity (Ilonen et al., 1984; Penttinen et al., 1979), and in immunocompromised, susceptible patients it is the only alternative.

In the future, with the help of monoclonal antibodies and knowledge of the protein sequence and three-dimensional configuration of the envelope proteins, common immunogenic domains eliciting protective antibodies may be identified. This knowledge may eventually result in subcomponent, synthetic vaccines for use in humans.

Literature Cited

Bengtsson, E., and G. Orndahl. 1954. Complications of mumps with special reference to the incidence of myocarditis. Acta Med. Scand. **149**:381–388.

Bohn, W., G. Rutter, H. Hohenberg, K. Mannweiler, and P. Nobis. 1986. Involvement of actin filaments in budding of measles virus; studies on cytoskeleton of infected cells. Virology **149**:91–106.

Brunell, P. A., A. Brickman, and S. Steinberg. 1969. Evaluation of a live attenuated mumps vaccine (Jeryl Lynn) with observation on the optimal tissue for testing serologic responses. Am. J. Dis. Child. **118**:435–440.

Burnet, F. M., J. F. McCrea, and J. D. Stone. 1946. Modification of human red cells by virus action. I. The receptor gradient for virus action in human red cells. Br. J. Exp. Pathol. **27**:228–236.

Buynak, E. B., J. E. Whitman, R. R. Roehm, D. H. Morton, G. P. Lampson, and M. R. Hilleman. 1967. Comparison of neutralization and hemagglutination-inhibition techniques for measuring mumps antibody. Proc. Soc. Exp. Biol. Med. **125**:1068–1071.

Cantell, K. 1961. Mumps virus. Adv. Virus Res. **8**:123–164.

CDC (Centers for Disease Control). 1972. Mumps surveillance. Report no. 2.

CDC (Centers for Disease Control). 1982. Annual summary 1981. Reported morbidity and mortality in the United States. Morbid. Mortal. Weekly Rep. **30**.

Chiba, Y., K. Horino, M. Umetsu, Y. Wataya, S. Chiba, and T. Nakao. 1973. Virus excretion and antibody responses in saliva in natural mumps. Tohoku J. Exp. Med. **111**:229–238.

Chou, S. M. 1967. Myxovirus-like structures in a case of human chronic polymyositis. Science **158**:1453–1455.

Chou, S. M. 1986. Inclusion body myositis: A chronic persistent mump myositis? Hum. Pathol. **17**:765–777.

Christenson, B., L. Heller, and M. Böttiger. 1983. The immunizing effect and reactogenicity of two live attenuated mumps virus vaccines in Swedish schoolchildren. J. Biol. Stand. **11**:323–331.

Copelovici, Y., D. Streelovici, A. Cristea, V. Tudor, and V. Armasu. 1979. Data on the efficiency of specific antimumps immunoglobulin in the prevention of mumps and its complications. Rev. Roum. Med. Virol. **30**:171–177.

Craighead, J. E. 1975. The role of viruses in the pathogenesis of pancreatic disease and diabetes mellitus. Prog. Med. Virol. **19**:161–214.

East, J. C., and D. W. Kingsbury. 1971. Mumps virus replication in chick embryo lung cells: properties of ribonucleic acids in virions and infected cells. J. Virol. **8**:161–173.

Enders, J. F. 1946. Mumps: techniques of laboratory diagnosis tests for susceptibility, and experiments on specific prophylaxis. J. Pediatr. **29**:129–142.

Ennis, F. A., R. D. Douglas, H. E. Hopps, H. M. Meyer, E. R. Brown, T. E. Hobbins, and F. C. Biehusen. 1969. Clinical studies with virulent and attenuated mumps viruses. Am. J. Epidemiol. **89**:176–183.

Espmark, J. Å., and A. Fagraeus. 1965. Identification of the species of origin of cells by mixed hemadsorption: a mixed antiglobulin reaction applied to monolayer cell cultures. J. Immunol. **94**:530–537.

Everberg, G. 1957. Deafness following mumps. Acta Otolaryngol. **48**:397–403.

Fagraeus, A., J. Å. Espmark, and J. Jonsson. 1965. Mixed haemadsorption: a mixed antiglobulin reaction applied to antigens on a glass surface. Immunology **9**:161–175.

Forsberg, P., A. Frydén, H. Link, and C. Örvell. 1986. Viral IgM and IgG antibody synthesis within the central nervous system in mumps meningitis. Acta Neurol. Scand. **73**:372–380.

Glikmann, G., and C.-H. Mordhorst. 1986. Serological diagnosis of mumps and parainfluenza type 1 virus infections by enzyme immunoassay, with a comparison of two different approaches for detection of mumps IgG antibodies. Acta Pathol. Microbiol. Scand. Sect. C **94**:157–166.

Glikmann, G., M. Pedersen, and C.-H. Mordhorst. 1986. Detection of specific immunoglobulin M to mumps virus in serum and cerebrospinal fluid samples from patients with acute mumps infection, using an antibody-capture enzyme immunoassay. Acta Pathol. Microbiol. Scand. Sect. C **94**:145–156.

Gordon, S. C., and C. B. Lauter. 1982. Mumps arthritis; unusual presentation as adult Still's disease. Ann. Intern. Med. **97**:45–47.

Grillner, L., and J. Blomberg. 1976. Hemolysis-in-gel and neutralization tests for determination of antibodies to mumps virus. J. Clin. Microbiol. 4:11–15.

Gut, J.-P., C. Spiess, S. Schmitt, and A. Kirn. 1985. Rapid diagnosis of acute mumps infection by a direct immunoglobulin M antibody capture enzyme immunoassay with labeled antigen. J. Clin. Microbiol. 21:346–352.

Habel, K. 1945. Cultivation of mumps virus in the developing chick embryo and its application to studies of immunity to mumps in man. Public Health Rep. 60:201–212.

Habel, K. 1951. Vaccination of human beings against mumps: vaccine administration at the start of an epidemic. I. Incidence and severity of mumps in vaccinated and control groups. Am. J. Hyg. 54:295–311.

Härfast, B., C. Örvell, A. Alsheikhly, T. Andersson, P. Perlmann, and E. Norrby. 1980. The role of viral glycoproteins in mumps virus dependent lymphocyte mediated cytotoxicity in vitro. Scand. J. Immunol. 11:391–400.

Henle, G., W. J. Bashe, J. S. Burgoon, C. F. Burgoon, G. R. Hunt, and W. Henle. 1951. Studies on the prevention of mumps. III. The effect of subcutaneous injection of inactivated mumps virus vaccines. J. Immunol. 66:561–577.

Henle, G., S. Harris, and W. Henle. 1948a. The reactivity of various human sera with mumps complement-fixation antigens. J. Exp. Med. 88:133–147.

Henle, G., W. Henle, K. K. Wendell, and P. Rosenberg. 1948b. Isolation of mumps virus from human beings with induced apparent or inapparent infections. J. Exp. Med. 88:223–232.

Herrler, G., and R. W. Compans. 1982. Synthesis of mumps virus polypeptides in Vero infected cells. Virology 119:430–438.

Hilleman, M. R., R. E. Weibel, E. B. Buynak, J. Stokes, and J. E. Whitman. 1967. Live, attenuated mumps virus vaccine. 4. Protective efficacy as measured in a field evaluation. N. Engl. J. Med. 276:252–257.

Hirst, G. K. 1950. Receptor destruction by viruses of the mumps–NDV–influenza group. J. Exp. Med. 91:161–175.

Ilonen, J., A. Salmi, H. Tuokko, E. Herva, and K. Penttinen. 1984. Immune responses to live attenuated and inactivated mumps virus vaccines in seronegative and seropositive young adult males. J. Med. Virol. 13:331–338.

Jensik, S. C., and S. Silver. 1976. Polypeptides of mumps virus. J. Virol. 17:363–373.

Johnson, C. D., and W. E. Goodpasture. 1934. An investigation of the etiology of mumps. J. Exp. Med. 59:1–19.

Julkunen, J. 1984. Serological diagnosis of parainfluenza virus infections by enzyme immunoassay with special emphasis on purity of viral antigens. J. Med. Virol. 14:177–187.

Kahana, D., and M. Berant. 1967. Diabetes in an infant following inapparent mumps. Clin. Pediatr. 6:124–125.

Kilham, L. 1949. Mumps meningoencephalitis with and without parotitis. Am. J. Dis. Child. 78:324–333.

Kingsbury, D. W., M. A. Bratt, P. W. Choppin, E. P. Hanson, Y. Hosaka, V. ter Meulen, E. Norrby, W. Plowright, R. Rott, and W. H. Wunner. 1978. Paramyxoviridae. Intervirology 10:137–152.

Kristensson, K., C. Örvell, G. Malm, and E. Norrby. 1984. Mumps virus infection of the developing mouse brain—appearance of structural virus proteins demonstrated with monoclonal antibodies. J. Neuropathol. Exp. Neurol. 43:131–140.

Lennette, E. H., F. W. Jensen, R. W. Guenther, and R. L. Magoffin. 1963. Serologic responses to parainfluenza viruses in patients with mumps virus infection. J. Lab. Clin. Med. 61:780–788.

Levens, J. H., and J. F. Enders. 1945. The hemagglutinative properties of amniotic fluid from embryonated eggs infected with mumps virus. Science 102:117–120.

Levitt, L. P., D. H. Mahoney, H. L. Casey, and J. O. Bond. 1970. Mumps virus in general population. A seroepidemiologic study. Am. J. Dis. Child. 120:134–138.

Linde, A., M. Granström, and C. Örvell. 1987. Serodiagnosis of mumps infection and mumps immunity: a comparison of Ig class and IgG subclass ELISA:s and microneutralization assay. J. Clin. Microbiol. 25:1653–1658.

London, W. P., and J. A. Yorke. 1973. Recurrent outbreaks of measles, chickenpox and mumps. I. Seasonal variation in contract rates. Am. J. Epidemiol. 98:453–468.

Löve, A., G. Malm, R. Rydbeck, E. Norrby, and K. Kristensson. 1985a. Developmental disturbances in the hamster retina caused by a mutant of mumps virus. Dev. Neurosci. 7:65–72.

Löve, A., R. Rydbeck, K. Kristensson, C. Örvell, and E. Norrby. 1985b. Hemagglutinin-neuraminidase glycoprotein as a determinant of pathogenicity in mumps virus hamster encephalitis: analysis of mutants selected with monoclonal antibodies. J. Virol. 53:67–74.

Löve, A., R. Rydbeck, Å. Ljungdahl, K. Kristensson, and E. Norrby. 1986a. Selection of mutants of mumps virus with altered structure and pathogenicity by passage in vivo. Microb. Pathogen. 1:149–158.

Löve, A., R. Rydbeck, G. Utter, C. Örvell, K. Kristensson, and E. Norrby. 1986b. Monoclonal antibodies against the fusion protein are protective in necrotizing mumps meningoencephalitis. J. Virol. 58:220–222.

Löve, A., T. Andersson, E. Norrby, and K. Kristensson. 1987. Mumps virus infection of dissociated rodent spinal ganglia in vitro. Expression and disappearance of viral structural proteins from neurons. J. Gen. Virol. 68:1755–1759.

McCarthy, M., and R. T. Johnson. 1980. A comparison of the structural polypeptides of five strains of mumps virus. J. Gen. Virol. 46:15–27.

Merz, D. C., A. C. Server, M. N. Waxham, and J. S. Wolinsky. 1983. Biosynthesis of mumps virus F glycoprotein: nonfusing strains efficiently cleave the F glycoprotein precursor. J. Gen. Virol. 64:1457–1465.

Merz, D. C., and J. S. Wolinsky. 1981. Biochemical features of mumps virus neuraminidases and their relationship with pathogenicity. Virology 114:218–227.

Messaritakis, J., C. Karabula, C. Kattamis, and N. Matsaniotis. 1966. Diabetes following mumps in sibs. Ann. Paediatr. Basel 207:236–246.

Meurman, O., P. Hänninen, R. V. Krishna, and T. Ziegler. 1982. Determination of IgG- and IgM-class antibodies to mumps virus by solid-phase enzyme immunoassay. J. Virol. Methods 4:249–257.

Meyer, M. B., W. C. Stifler, and J. M. Joseph. 1966. Evaluation of mumps vaccine given after exposure to mumps, with special reference to exposed adults. Pediatrics 37:304–315.

Morgan, H. R., J. F. Enders, and P. F. Wagley. 1948. A hemolysin associated with the mumps virus. J. Exp. Med. 88:503–514.

Mortimer, P. P. 1978. Mumps prophylaxis in the light of a new test for antibody. Br. Med. J. 2:1523–1524.

Mufson, M. A., C. Örvell, B. Rafnar, and E. Norrby. 1985. Two distinct subtypes of human respiratory syncytical virus. J. Gen. Virol. 66:2111–2124.

Nagai, Y., and H. D. Klenk. 1977. Activation of precursors of both glycoproteins of Newcastle disease virus by proteolytic cleavage. Virology 77:125–134.

Naruse, H., Y. Nagai, T. Yoshida, M. Hamaguchi, T. Matsumoto, S. Isomura, and S. Suzuki. 1981. The polypeptides of mumps virus and their synthesis in infected chick embryo cells. Virology **112:**119–130.

Nicolai-Scholten, M. E., R. Ziegelmaier, F. Behrens, and W. Höpken. 1980. The enzyme-linked immunosorbent assay (ELISA) for determination of IgG and IgM antibodies after infection with mumps virus. Med. Microbiol. Immunol. **168:**81–90.

Norrby, E. 1985. Mumps virus, p. 774–778. *In* H. Lennette, A. Balows, W. J. Hausler, and H. J. Shadomy, (ed.), Manual of clinical virology. American Society for Microbiology, Washington, D.C.

Norrby, E., and Y. Gollmar. 1972. Appearance and persistence of antibodies against different virus components after regular measles infections. Infect. Immun. **6:**240–247.

Norrby, E., M. Grandien, and C. Örvell. 1977. New tests for characterization of mumps virus antibodies: hemolysis inhibition, single radial immunodiffusion with immobilized virions and mixed hemadsorption. J. Clin. Microbiol. **5:**346–352.

Norrby, E. and K. Penttinen. 1978. Differences in antibodies to the surface components of mumps virus after immunization with formalin-inactivated and live virus vaccines. J. Infect. Dis. **138:**672–676.

Örvell, C. 1976a. Identification of paramyxovirus specific haemolysis-inhibiting antibodies separate from· haemagglutinating-inhibiting and neuraminidase-inhibiting antibodies. 1. Sendai virus haemolysis-inhibiting antibodies. Acta Pathol. Microbiol. Scand. Sect. B **84:**441–450.

Örvell, C. 1976b. Identification of paramyxovirus-specific haemolysis-inhibiting antibodies separate from haemagglutinating-inhibiting and neuraminidase-inhibiting antibodies. 2. NDV and mumps virus haemolysis-inhibiting antibodies. Acta Pathol. Microbiol. Scand. Sect. B **84:**451–457.

Örvell, C. 1978a. Immunological properties of purified mumps virus glycoproteins. J. Gen. Virol. **41:**517–526.

Örvell, C. 1978b. Structural polypeptides of mumps virus. J. Gen. Virol. **41:**527–539.

Örvell, C. 1984. The reactions of monoclonal antibodies with structural proteins of mumps virus. J. Immunol. **129:**2622–2629.

Örvell, C., and M. Grandien. 1982. The effects of monoclonal antibodies on biological activities of structural proteins of Sendai virus. J. Immunol. **129:**2779–2787.

Örvell, C., and E. Norrby. 1985. Antigenic structure of paramyxoviruses, p. 241–264. *In* A. R. Neurath and M. H. V. van Regenmortel (ed.), Immunochemistry of viruses—the basis for serodiagnosis and vaccines. Elsevier, Amsterdam.

Örvell, C., R. Rydbeck, and A. Löve. 1986. Immunological relationships between mumps virus and parainfluenza viruses studied with monoclonal antibodies. J. Gen. Virol. **67:**1929–1939.

Palmer, D. F., L. Kaufman, W. Kaplan, and J. J. Cavallaro. 1977. Serodiagnosis of mycotic diseases. Charles C. Thomas, Publisher, Springfield, Illinois.

Penttinen, K., E. Helle, and E. Norrby. 1979. Differences in antibody response induced by formaldehyde inactivated and live mumps vaccines. Dev. Biol. Stand. **43:**265–268.

Pollock, T. M. 1969. Human immunoglobulin in prophylaxis. Br. Med. Bull. **25:**202–207.

Popow-Kraupp, T. 1981. Enzyme-linked immunosorbent assay (ELISA) for mumps virus antibodies. J. Virol. Methods **8:**79–88.

Popow-Kraupp, T., C. Kunz, and H. Hofmann. 1982. Antibody response after application of a new live attenuated mumps vaccine (Pariorix) measured by enzyme-linked immunosorbent assay (ELISA). J. Med. Virol. **10:**119–129.

Prince, G. A., A. B. Jenson, L. C. Billups, and A. L. Notkins. 1978. Infection of human pancreatic beta cell cultures with mumps virus. Nature (London) **271:**158–161.

Ray, R., and R. W. Compans. 1986. Monoclonal antibodies reveal extensive antigenic differences between the hemagglutinin-neuraminidase glycoprotein of human and bovine parainfluenza 3 virus. Virology **148:**232–236.

Rima, B. K., and S. J. Martin. 1980. Messenger RNA synthesis in mumps virus infected cells. Biochem. Soc. Trans. **8:**441–442.

Rima, B. K., M. W. Roberts, W. D. McAdam, and S. J. Martin. 1980. Polypeptide synthesis in mumps virus-infected cells. J. Gen. Virol. **46:**501–505.

Rydbeck, R., A. Löve, C. Örvell, and E. Norrby. 1986. Antigenic variation of envelope and internal proteins of mumps virus strains detected with monoclonal antibodies. J. Gen. Virol. **67:**281–287.

Sakata, H., M. Tsurudome, M. Hishiyama, Y. Ito, and A. Sugiura. 1985. Enzyme-linked immunosorbent assay for mumps IgM antibody: comparison of IgM capture and indirect IgM assay. J. Virol. Methods **12:**303–311.

Sarkkinen, H. 1985. Respiratory viral antigen detection by solid-phase immunoassays: comparison with other diagnostic tests and application for routine diagnosis, p. 329–337. *In* K.-O. Habermehl (ed.), Rapid methods and automation in microbiology and immunology. Springer-Verlag, New York.

Scheid, A., and P. W. Choppin. 1974. Identification of biological activities of paramyxovirus glycoproteins. Activation of cell fusion, hemolysis and infectivity by proteolytic cleavage of an inactive precursor protein of Sendai virus. Virology **57:**475–490.

Spataro, R. F., S. R. Lin, F. A. Horner, C. B. Hall, and J. V. McDonald. 1976. Aqueductal stenosis and hydrocephalus. Rare sequela of mumps virus infection. Neuroradiology **12:**11–13.

Stokes, J., J. F. Enders, E. P. Maris, and L. W. Kane. 1946. Immunity in mumps. VI. Experiments on the vaccination of human beings with formalinized mumps virus. J. Exp. Med. **84:**407–428.

Thompson, J. A. 1979. Mumps: a case of acquired aqueductal stenosis. J. Pediatr. **94:**923–924.

Tolpin, M. D., and V. Schauf. 1984. Mumps virus, p. 311–331. *In* R. B. Belshe (ed.), Textbook of human virology. PSG Publishing Company, John Right Publisher, Bristol, England.

Tsurudome, M., A. Yamada, M. Hishiyama, and Y. Ito. 1986. Monoclonal antibodies against the glycoproteins of mumps virus; fusion inhibition by anti-HN monoclonal antibody. J. Gen. Virol. **67:**2259–2265.

Tyrrell, D. L. J., and A. Ehrnst. 1979. Transmembrane communication in cells chronically infected with measles virus. J. Cell Biol. **81:**396–402.

Ukkonen, P., M. L. Granström, and K. Penttinen. 1981a. Mumps-specific immunoglobulin M and G antibodies in natural mumps infection as measured by enzyme-linked immunosorbent assay. J. Med. Virol. **8:**257–265.

Ukkonen, P., M. L. Granström, J. Räsänen, E.-M. Salonen, and K. Penttinen. 1981b. Local production of mumps IgG and IgM antibodies in the cerebrospinal fluid of meningitis patients. J. Med. Virol. **8:**257–265.

Ukkonen, P., O. Väisänen, and K. Penttinen. 1980. Enzyme-linked immunosorbent assay for mumps and parainfluenza type 1 immunoglobulin G and immunoglobulin M antibodies. J. Clin. Microbiol. 11:319–323.

Utz, J. P., V. N. Houk, and D. W. Alling. 1964. Clinical and laboratory studies on mumps. IV. Viruria and abnormal renal function. N. Engl. J. Med. 270:1283–1286.

Utz, J. P., J. A. Kasel, H. G. Cramblett, C. F. Szwed, and R. H. Parrott. 1957. Clinical and laboratory studies of mumps. I. Laboratory diagnosis by tissue-culture techniques. N. Engl. J. Med. 257:497–502.

Utz, J. P., C. F. Szwed, and J. A. Kasel. 1958. Clinical and laboratory studies of mumps. II. Detection and duration of excretion of virus in urine. Proc. Soc. Exp. Biol. Med. 99:259–261.

Väänänen, P., T. Hovi, E. P. Helle, and K. Penttinen. 1976. Determination of mumps and influenza antibodies by haemolysis-in-gel. Arch. Virol. 52:91–99.

Vaheri, A., J. Julkunen, and M. L. Koskiniemi. 1982. Chronic encephalomyelitis with specific increase in intrathecal mumps antibodies. Lancet 25:685–688.

Vandvik, B., R. E. Nilsen, F. Vartdal, and E. Norrby. 1982. Mumps meningitis: specific and non-specific antibody responses in the central nervous system. Acta Neurol. Scand. 65:468–487.

Wagenvoort, J. H. T., M. Harmsen, B. J. K. Boutar-Trouw, C. A. Kraaijeveld, and K. C. Winkler. 1980. Epidemiology of mumps in the Netherlands. J. Hyg. 85:313–326.

Weibel, R. E., E. B. Buynak, A. A. McLean, R. R. Roehm, and M. R. Hilleman. 1980. Persistence of antibody in human subjects for 7 to 10 years following administration of combined live attenuated measles, mumps and rubella vaccines (40967). Proc. Soc. Exp. Biol. Med. 165:260–263.

Weibel, R. E., E. B. Buynak, and J. Stokes. 1972. Measurement of immunity following live mumps (5 years), measles (3 years), and rubella (2½ years) virus vaccines. Pediatrics 49:334–341.

Witte, C. L., and B. Schanzer. 1968. Pancreatitis due to mumps. J. Am. Med. Assoc. 203:1068–1069.

Wolinsky, J. S. 1977. Mumps virus-induced hydrocephalus in hamsters. Ultrastructure of the chronic infection. Lab. Invest. 37:229–236.

Wolinsky, J. S., T. Klassen, and J. R. Bahringer. 1976. Persistence of neuroadapted mumps virus in brains of newborn hamsters after intraperitoneal inoculation. J. Infect. Dis. 133:260–267.

Wolinsky, J. S., and A. C. Server. 1985. Mumps virus, p. 1255–1284. In B. N. Fields (ed.), Virology. Raven Press, New York.

Wolinsky, J. S., and W. G. Stroop. 1978. Virulence and persistence of three prototype strains of mumps virus in newborn hamsters. Arch. Virol. 57:355–359.

Wolinsky, J. S., M. N. Waxham, and A. C. Server. 1985. Protective effects of glycoprotein-specific monoclonal antibodies on the course of experimental mumps virus meningoencephalitis. J. Virol. 53:727–734.

Wolontis, S., and B. Bjorvatn. 1973. Mumps meningoencephalitis in Stockholm, November 1964–July 1971. II. Isolation attempts from the cerebrospinal fluid in a hospitalized group. Scand. J. Infect. Dis. 5:261–271.

Yorke, J. A., and W. P. London. 1973. Recurrent outbreaks of measles, chickenpox and mumps. II. Systematic differences in contact rates and stochastic effects. Am. J. Epidemiol. 98:469–480.

CHAPTER 27

Paramyxoviridae: Measles Virus

ERLING NORRBY

Disease: Measles.

Etiologic Agent: Measles virus; a single serotype.

Source: Acutely infected individuals; contact or inhalation.

Clinical Manifestations: Abrupt onset of disease with sneezing, running nose, cough, redness of eyes, and rapidly rising fever; 2 to 3 days later, a maculopapular exanthema develops; complications in the form of otitis, pneumonia, and encephalitis may occur; a progressive encephalitis is a rare late complication.

Pathology: Massive involvement of the lymphatic system and different membranes; the rash is an immunopathologic reaction causing damage to fine blood vessels.

Laboratory Diagnosis: Primarily serologic; antibody rises in paired sera or direct identification of virus-specific IgM.

Epidemiology: Worldwide; requires a population of about 300,000 people to allow continuous circulation.

Treatment: No specific treatment; symptomatic interventions.

Prevention and Control: Live attenuated vaccine.

Introduction

Measles is an acute febrile illness with characteristic clinical features (Modlin, 1984; Norrby, 1985). One prominent symptom is a generalized rash, which normally concludes the sequence of evolution of different symptoms. Because of the highly contagious nature of this community-related disease, it was identified as an important exanthematous disease already in the 10th century. Today the disease is readily diagnosed without the aid of laboratory assistance. However, the extensive use of effective live vaccine in some industrialized countries has forced the disease to retreat into trickling occurrence. Under these conditions, the opportunities for physicians to accumulate clinical acquaintance with the disease are limited. This circumstance, as well as the awareness of the potential occurrence of disease suppressed to a varying extent in individuals a long time after vaccination, has caused an increased need for accurate diagnosis of acute disease as well as for evaluation of immunity after infection or vaccination.

Even though the importance of measles in many industrialized countries has been reduced as a result of vaccination, the disease still ranks as the number one killer of young children in the majority of developing countries. The severe form of the disease in these countries probably reflects the simultaneous occurrence of other infections, which may harm the cell-mediated immunity required for recuperation from disease. In rare cases, severe forms of measles are seen also in industrialized countries in children with impaired, cell-mediated immunity as a consequence of, e.g., genetic defects or the occurrence of tumors in the lymphatic system. In terms of medical importance of the disease, it should be mentioned that not only acute but also progressive late disease in the brain, subacute sclerosing panencephalitis (SSPE), is seen (ter Meulen and Carter, 1984). Although for the individual this fatal disease is a tragic event, in terms of numbers it does not play an important role.

The viral etiology of measles was defined by early transmission experiments with monkeys. The main breakthrough in experimentation with the virus was the demonstration, more than four decades later, that the virus could be propagated in primary monolayer cultures of primate kidney cells. The distinct cytopathic changes of the virus in these cultures, later demonstrated also in certain heteroploid cell lines, allowed studies of the virus and virus–cell interactions. Results of these studies have led to the classification of measles virus as a member of the family *Paramyxoviridae*, genus *Morbillivirus* (Kingsbury et al., 1978).

The Virus

Morphology and Structural Components

Like other paramyxoviruses, measles virions are spherical, enveloped particles with a centrally located, helical nucleocapsid. The diameter of the pleomorphic particles varies between 100 and 250 nm, with a mean value of about 150 nm. There are six virus-specific proteins participating in the formation of virions, three of these proteins occur in the envelope, and the other three are associated with the internal structures. The envelope contains a bimolecular lipid layer of cellular origin with one of the virus envelope proteins, the matrix (M) protein, located on its inside. The other two components are transmembranous proteins forming 9- to 15-nm-long radial projections, peplomers (Varsanyi et al., 1984). One kind of peplomer is an oligomer of a protein with a capacity to anchor the virus particles to specific receptors on cells. Since similar receptors occur also on erythrocytes of Old World monkeys, these cells are agglutinated by measles virions or envelope fragments, hence the term hemagglutinin (H) for this protein. The other kind of peplomer is composed of a protein that, after proteolytic cleavage, is endowed with a capacity to fuse membranes. Thus, fusion of the virus envelope and the cytoplasmic membrane following the attachment of particles to the surface of cells allows penetration of nucleocapsid structures into the cytoplasm, initiating the events of replication. A consequence of the fusion of nucleated cells can be the formation of multinucleated giant cells. Such syncytia are one hallmark of measles virus cytopathology. Fusion events between virus particles and the membrane of erythrocytes results in hemolysis. The name of the component with these biological activities is fusion (F) protein.

Replication

The replicative events of measles virus are characteristic of those of paramyxoviruses. The negative-stranded linear RNA is transcribed by the nucleocap-sid-associated polymerase. This enzyme initiates the transcription at the 3' end of the RNA molecule and directs the synthesis of consecutively appearing mRNA molecules. Because the enzyme has only one attachment point and transcribes the whole genome RNA by skidding through intergenic regions, the consecutive occurrence of genes can be deduced from inactivation studies. The tentative gene order determined by such studies has then been consolidated by nucleotide sequencing of essentially the whole viral genome. The order of genes in terms of their products is nucleoprotein (NP), phosphoprotein (P), M, F and H genes and finally a gene directing the synthesis of a large (L) protein. The P gene also gives rise to a nonstructural protein named C. The virion RNA serves as a template not only for production of mRNA, but, in addition, for replication of intact RNA via a plus-strand intermediate. As a consequence of accumulation of new genomic RNA and the different structural proteins, virion morphogenesis can take place. This occurs at the cytoplasmic membrane via a budding off process. Prior to the appearance of the virus transmembranous proteins at the cell surface, they have been modified by attachment of N-linked carbohydrate chains of cellular origin.

The release of virus particles from a single cell occurs over an extended period of time. This time varies in length from a few hours, if the cell succumbs rapidly to the virus-induced cytopathology, to an unlimited time if the infection takes on chronic, steady-state characteristics. There are many possibilities for development of chronic infections such as emergence of temperature-sensitive mutants and interference by defective particles. In most cases, chronic infections are nonproductive, that is, no or only small amounts of infectious virus are released, but this is not always the case. In acute as well as many chronically infected cell lines, the cytoplasmic membrane contains virus-specific antigens. In virus-infected cell cultures this is apparent by the potential for attachment of monkey erythrocytes to the cell surface, *hemadsorption*. Occasionally, persistent infections in vivo may be caused by highly defective virus variants that do not express any viral antigens at the cell surface.

Antigenic Composition

All the six different structural components of the virus can serve as antigens as well as immunogens. The possibilities to detect antigen–antibody reactions involving individual components are dependent on the design of the serological test employed. Tests in which whole virus material is employed as antigen do not distinguish reactions with individual components. However, a test like ELISA can be made component specific by certain arrangements. For this purpose, either monoclonal antibody (MAb) is used for anchoring of an individual virus component from a lysate onto a plate, or a purified virus component is prepared and labeled (Salonen et al., 1986). In addition, it is possible to distinguish different antigenic sites on a single protein by use of a competition ELISA (Mäkelä et al., 1987). The most efficient way of identifying antigen–antibody interactions involving individual components is to use the Western blot (WB) or radioimmune precipitation assay (RIPA). The WB test has a distinct disadvantage in that the virus components are denatured and separated by SDS–polyacrylamide gel electrophoresis (PAGE) prior to their interaction with antibodies. In contrast, RIPA is based on the dissociation of isotope-labeled virus components by use of a mild lysis buffer followed by reaction with antibodies. Immune complexes are then recovered by use of *staphylococcus* A protein, separated by SDS–PAGE and identified by autoradiography. The limitations of the RIPA are that even the mild lysis buffer used may denature some highly sensitive epitopes and that different proteins may be labeled to varying extents with the isotope-tagged amino acid used, usually [^{35}S]methionine. The latter problem may be overcome by use of an alternate amino acid.

In the case of measles virus, the capacity of different virus components to react with antibodies has been identified only to a limited extent by use of WB tests (Hankins and Black, 1986). Immune complexes involving H, P, N, and M proteins but not the F protein were discernible. The RIPA has been used extensively, and all structural virus components are readily identified by this assay (Norrby et al., 1981) with the exception of the L protein. This protein is seen only occasionally (Graves, 1981), and, furthermore, its identity is difficult to determine, since aggregates of other virus components may occur in the same region of the electropherogram. Since the P protein shows the highest degree of susceptibility to proteolysis, special precautions, such as short-time labeling or purification of virions for use as antigen, have to be made to allow its identification.

Murine MAbs have been produced against all structural components except L (Norrby et al., 1982). The availability of sets of MAbs against different components has allowed a dissection of the occurrence of different antigenic sites. For this purpose, competition ELISAs are used (Carter et al., 1982, 1983; Sheshberadaran et al., 1983). In the case of neutralizing epitopes, a further distinction of antigenic site specificity can be achieved by characterization of non-neutralizable virus variants. Also, comparison of MAb reactivity with different virus strains may allow subdivision of MAb reactivity, possibly to the level of identification of individual epitopes.

Measles virus is monotypic. Comparison of virus strains by use of convalescent sera or hyperimmune sera generally shows homology. Minor antigenic differences have been observed in neutralization tests comparing laboratory strains with fresh isolates from cases of acute measles (Albrecht et al., 1981) and SSPE (Payne and Baublis, 1973). There is no evidence that measles infections can occur twice owing to the existence of antigenic variants. Further, live measles virus vaccines appear to provide a universal protection against this disease. The two surface glycoproteins H and F represent the prime target in immune protection reactions (Örvell and Norrby, 1985). The neutralizing antibodies determined by in vitro assays react almost exclusively with the H protein. However, low-titers, of neutralizing antibodies specific for the F protein were identified in human gamma globulin preparations after removal of all hemagglutination-inhibiting (HI) antibodies by absorption with Tween 80 and ether-treated measles virus antigens (Norrby and Gollmar, 1975), and in rabbit hyperimmune sera against purified F protein (Varsanyi et al., 1984). Although neutralization in vitro primarily depends on antibodies to the H protein, protection against infection in vivo also requires F specific immunity. This was borne out by the experiences with the use of inactivated vaccines (Norrby et al., 1975). The formalin and Tween 80–ether treatments used to inactivate virus in these products were found to destroy critical immunogenic parts of the F protein. The vaccines induced the formation of readily measurable titers of HI and consequently also neutralizing antibodies, but these did not suffice to provide protection against disease except possibly when they occurred in high titers. In addition, some immunized children developed an atypical form of disease upon exposure to wild virus. This form of measles was interpreted to reflect immune pathological events resulting from the concomitant occurrence of actively replicating virus and a preexisting anti-H immunity. It would appear that, if a future inactivated vaccine should ever be developed, it would have to include both H and F immunogens.

The surface glycoproteins of measles virus have been found to show a high antigenic stability, reflecting the monotypic nature of the virus. This has been demonstrated by characterizing the reactions between sets of component-specific monoclonal antibodies and collections of measles virus strains. No variations in the occurrence of epitopes on the F protein of different strains of virus have been found. In fact, there appears to be an evolutionary conservation of the F protein in members of the morbillivirus genus (Sheshberadaran et al., 1986). It was found that immunization of dogs with inactivated measles virus gave an infection-permissive immunity against exposure to virulent distemper virus; this was interpreted as reflecting an immune response to the F protein (Appel et al., 1984). The potential of an isolated anti-F immunity to give protection against canine distemper disease was later demonstrated by immunization with distemper virus F protein isolated by chromatography on a column containing measles virus-specific F MAb (Norrby et al., 1986).

Minor variations in the occurrence of H epitopes in different spontaneously occurring strains of measles virus have been shown. Further, strains of measles virus lacking specific epitopes were selected by propagation of the virus in the presence of neutralizing MAb. In one study (Sheshberadaran and Norrby, 1986), consecutive exclusion of 8 different neutralizing epitopes on the H protein was achieved. Interestingly, one of the 8 epitopes remained on the H protein in the selected variant, but had changed from a neutralizing to a nonneutralizing site. Neither the spontaneously occurring small variations nor the more extensive alterations seen in variants in the latter study appeared to influence the overall antigenic properties of the H protein, since no variation was found in the capacity of convalescent sera or hyperimmune sera to neutralize the different virus strains or variants. Some variations were also seen in the occurrence of different epitopes on internal components of virus strains. With the reagents employed, only occasional variations were seen in the nucleoprotein, whereas variations in epitopes on the M protein were a more common event (Sheshberadaran et al., 1983).

Reactions to Physical and Chemical Agents

Virion infectivity is a labile property. The half-life at 37°C in a protein-containing medium is about 2 h; however, it is possible to stabilize the infectivity of the virus by inclusion of certain additives in the suspension. This is done to increase the stability of the live measles vaccine after it has been dissolved from its frozen-dried state. Removal of infectivity without interference with hemolytic (cell-fusing) and hemagglutinating activities can be obtained by heating freeze-dried virus material at 60°C for 30 min. Selective destruction of hemolytic activity but retention of hemagglutinating activity is seen after formalin or Tween 80–ether treatment, as mentioned above. Conversely, treatment with trypsin removes H peplomers from the virus envelope but leaves F peplomers intact. However, the resulting particles lacking the H attachment components do not exhibit hemolytic or cell-fusion activities, since the expression of this biological activity is contingent upon the anchoring of virus particles to membranes. The relative sensitivity of different biological activities to in-

activating agents in general is infectivity > hemolytic activity > hemagglutinating activity > antigen activity.

Pathogenesis and Pathology

The events of virus replication and spread during the incubation time preceding the emergence of disease are conjectural. It is assumed that after local virus replication in the upper respiratory or alimentary tract, or possibly the conjunctival epithelium, there is a lymphatic and later viremic spread of virus (Kempe and Fulginiti, 1965). The viremic spread may be accentuated by virus replication in selected cells in the reticuloendothelial system and in certain populations of stimulated lymphocytes. In unstimulated lymphocytes, virus replication generally is restricted (McChesney and Oldstone, 1987). Such cells may serve as vehicles for dissemination of the virus infection and as a source of virus persistence. If peripheral blood cells are collected from an individual during the prodromal phase of measles and stimulated with a mitogen, for example, phytohemagglutinin, an activation of virus replication and formation of syncytia are seen (Osunkoya et al. 1974). During the disseminated stage of the disease there is a massive involvement of the lymphatic system, as evidenced by the occurrence of leukopenia with relative lymphocytosis and impairment of cell-mediated immune phenomena, as illustrated by extinction of the delayed hypersensitivity reaction to tuberculin for weeks after recovery from the infection. When prodromal symptoms emerge, the infection is already widely disseminated. It includes different epithelial membranes in conjunctivae and in respiratory, alimentary, and urinary tracts, smaller blood vessels, the lymphatic system, and frequently also the central nervous system (CNS). Enteropathological changes appear to be a particular problem in developing countries, and the infection in this compartment must derive from hematogenous spread of virus.

The uncomplicated acute disease is curtailed after 3 to 5 days by the development of immune reactions. This event is signaled by the appearance of a rash, which is a consequence of the interaction between virus-infected endothelial cells and immune T cells. In cases of dysfunction of T cells, no rash is seen. This is a serious situation, since it permits a relentless progression of the disease, frequently resulting in a fatal outcome. This complicated form of measles is seen in children with inherited or acquired (for example, leukemic children treated with cytotoxic drugs), defective, cell-mediated immunity. One characteristic feature of this disseminated form of disease is the appearance of giant cell pneumonia, reflecting unrestricted virus replication in the lung tissue. Encephalitis resulting from active virus replication in brain tissue is also observed. This form of CNS involvement should be distinguished from two other forms of complication involving the brain that appear to occur under conditions of normal immune functions.

The first of these two complications is postinfectious encephalitis, which occurs with a frequency of about 1 in 1000 cases of measles and develops on the average at 5 to 7 days after appearance of the rash. Involvement of the brain in connection with measles appears to occur much more often than the frequency of postinfectious encephalitis would indicate. Pleocytosis in the cerebrospinal fluid is seen in about 10% of all cases, and electroencephalographic changes are seen in about 50% of children with measles. The postinfectious measles encephalitis has been speculated to have an autoimmune etiology. This speculation is based on the findings that it develops some days after termination of the acute disease, that infectious virus only rarely has been identified in the brain of these cases, and, finally, that the pathological changes encompasses perivascular lymphocytic infiltration and demyelination.

The additional form of CNS involvement in measles is the late complication, subacute sclerosing panencephalitis (SSPE) (ter Meulen and Carter, 1984). This is a rare event, occurring in about 1 in 300,000 cases of measles. The progressive encephalitis in SSPE derives from activation of a defective cell-associated infection. This infection spreads by direct cell-to-cell transmission, and there is no evidence of maturation of virus particles at the surface of cells. Probably this surface is not antigenically modified, thus allowing infected cells to escape the surveillance of the immune defense system.

The facts that virus may remain to produce the late infection SSPE and that a generalized case of measles induces a life-long immunity against this disease have raised questions about the efficacy of clearance of virus. By use of the in situ hybridization technique, measles virus RNA has been demonstrated in peripheral blood lymphocytes and in brain cells in cases of SSPE (Fournier et al., 1985) and in bone cells in Paget's disease (Baslé et al., 1986). Evidence was also given for the persistence of virus genetic material in peripheral mononuclear cells of healthy individuals (Fournier et al., 1985). At this stage it is an open question whether intermittent limited replication of activated endogenous virus provides a source of antigen for maintenance of the life-long immunity. The profile of antibody persistence with time does not support this contention.

No unique pathological features can be distinguished in the catarrhal inflammation in measles. Throughout the body, hyperplastic lymphoid tissue may be found, and multinucleated giant cells occur in lymph nodes, tonsils, adenoids, spleen, and appendix and also in epithelial tissue. The most extensive fusion of cells is seen in patients with giant-cell pneumonia. Acidophilic inclusions may occur in the cytoplasm and nuclei of both individual and fused infected cells. Intranuclear inclusions are a sign of a long-standing virus infection. Thus they occur in neurons and glial cells in the brain in SSPE, an observation that led to the identification of the etiology of this disease. Perivascular inflammation and demyelination is seen in the brain of patients with postinfectious encephalitis, whereas in SSPE the encephalitis is more generalized and involves both the white and the gray matter. Demyelination is seen only at an advanced stage of SSPE in a few cases.

Clinical Features

Initiation of a measles infection occurs by spread of virus to the respiratory tract, the mouth and pharynx, and perhaps the conjunctivae. The source of infection is always an immediate contact with a diseased individual. Besides airborne transmission, transfer of virus by direct contact probably also plays an important role. The incubation time is 9 to 11 days, but after the parenteral administration of the live vaccine it is 2 to 3 days shorter. The prodromal symptoms are mostly catarrhal. There is an abrupt onset of disease with sneezing, running nose, cough, redness of eyes, and a rapidly raising fever. After 2 to 4 days 1- to 3-mm pale, bluish-white spots in erythematous areolas frequently can be observed in the buccal mucosa. These are the Koplik's spots pathognomonic for measles. Two to 3 days after onset of the disease the fever may subside somewhat, only to rise again when a rash develops 1 to 2 days later. The rash is a characteristic macular or maculopapular exanthema, which may become confluent. The rash initially appears on the forehead and behind the ears, and within 2 days it spreads first over the trunk and later the limbs. The fading rash leaves a brownish discoloration, which resolves by fine desquamation during about 10 days. If there is a complicating thrombocytopenia, a more hemorrhagic form of rash may be seen.

The involvement of the lymphatic system usually results in enlarged cervical lymph nodes and possibly splenomegaly. Occasionally a lymphoid inflammatory change in the appendix may signal appendicitis. Bacterial complications in the form of otitis media and pneumonia are not uncommon.

The period of infectiousness extends from the time of appearance of the very first prodromal symptoms until a few days after the development of rash. Virus has been isolated even at somewhat later times from urine and from stool. The latter isolation was made from cases of measles in developing countries, in which for unknown reasons the patients are particularly prone to develop enteritis. The involvement of the enteric tract in these children may cause a protein-losing enteropathy, leading to an extension of the time until complete recovery.

In complicated forms of measles, symptoms of pneumonia and encephalitis may develop. In cases of acute, postinfectious encephalitis fever, convulsions and coma may be seen. The prognosis is relatively good, but sequelae are seen. The progressive form of encephalitis in immunodeficient individuals is life-threatening. Also, SSPE generally has a fatal outcome, but in this case the onset of disease is more insidious. Symptoms develop over months, eventually to reflect extensive loss of cerebral cortex functions. In the majority of cases there is a focal retinitis leading to blindness.

Diagnosis

Collection and Storage of Specimens

Samples for virus isolation or for the detection of viral antigens include blood (primarily leukocytes), nasopharyngeal and conjunctival secretions, urine and, as appropriate, stool, and, in special circumstances, skin and brain biopsies. Isolation of virus from blood and mucosal secretion may be successful from the beginning of the prodromal phase until 1 or 2 days after the appearance of rash. The earlier the sample is collected, the better. Infectious virus may be present in urine (preferably sediment) and stool at later times, but the efficacy of such late virus isolation has not been defined. Material to be used for virus isolation should be kept at 4°C. It should not be frozen. Samples should be mixed with a balanced salt solution or medium containing 2 to 10% of serum except in the case of blood material. Transport times should be minimized. This is particularly urgent in cases in which explant cultures are to be established, as in the case of brain biopsies.

Special considerations apply to specimens for rapid diagnosis by immunofluorescence. Preferably, smears of cells should be fixed as early as possible in cold (−20°C) acetone for 10 min at room temperature. It is a considerable advantage if the steps for preparation of cell smears and fixation can be made in the clinical setting. Transport of fixed samples to the laboratory should be made at 4°C, and storage in the latter milieu should be at −20°C.

Serological analysis preferably is performed with paired serum samples, and the initial specimen should be obtained as early as possible. The rise of antibody titers starts around the time of appearance of rash. Single specimens collected within 10 to 20 days after the onset of the rash may be useful for identification of virus-specific immunoglobulin M (IgM). Demonstration of a local production of antibodies within the CNS can be made by examination of simultaneously collected serum and cerebrospinal fluid samples. The use of dried whole blood absorbed by filter paper (Brody et al., 1963) may offer advantages when samples are to be collected and transported under primitive conditions. Samples to be examined for antibody activity should be stored at −20°C.

Direct Virus Detection

Electron microscopy potentially could be used for identification of paramyxovirus particles in samples from measles patients. This technique has not been applied for the purpose of establishing a diagnosis of measles, as can be judged from the absence of published data. Possibly new filtration techniques could be used to concentrate particles in, for instance, urine or cerebrospinal flow, allowing virion identification by transmission (negative contrasting) or scanning electron microscopy (J. Andersson, personal communication). For a type-specific identification of particles, immune electron microscopy would need to be performed.

Direct Identification of Viral Antigen

During the phase of measles catarrhal symptoms, cells in nasopharyngeal secretions and urine can be examined. Detection of giant cells and inclusion bodies, possibly both cytoplasmic and intranuclear, may help in the diagnosis. For detection of specific antigen in cells, immunofluorescence (IF) or immunoperoxidase methods need to be used. Measles virus antigen has been demonstrated in cells collected from the nasopharyngeal secretion of acutely infected patients (Fulton and Middleton, 1975; McQuillin et al., 1976). Antigen-containing cells were found in samples collected between 4 days before and 6 days or even 10 days after onset of rash. Identification of measles cases by the IF technique showed a high degree of correlation with clinical and serological data. Because of the infrequent use of virus isolation techniques, no estimate is available on the comparative capacity of different methods for identification of virus or viral products. Successful IF examination of cells in urine specimens has also been described, and the same techniques can also be applied to biopsy material. Detection of measles virus antigen in skin biopsies from cases of acute measles (Olding-Stenkvist and Bjorvatn, 1976) and in brain biopsies from cases of SSPE (Norrby et al., 1985; ter Meulen and Carter, 1984) has been described.

Either the direct or indirect IF technique may be used. In the former case, high-titer human sera can be employed, but since the latter technique offers certain practical advantages, hyperimmunization reagents from experimental animals generally are used. These reagents may be polyclonal, hyperimmune sera and ascites or tissue culture supernatants containing MAbs. A very distinct and specific fluorescence is obtained with MAbs specific for the NP and P proteins. These antigens occur in abundance both in acutely and persistently infected cells. Furthermore the antigenic epitopes detected by available MAbs do not appear to vary to any large extent between virus strains. Still it would seem advisable to use a mixture of two MAbs against different epitopes on either of the two proteins, or one MAb against each protein. For experimental purposes it may be of interest also to identify different envelope antigens in persistent infections by IF by use of selected MAbs. This may provide information on the state of defectiveness of virus–cell interactions (see Norrby et al., 1985).

Measles viral antigens in tissues can also be identified by other immunobiological methods. Immunoultrastructural studies were performed with SSPE brain material (Jenis et al., 1973), and chronically infected hamster brains were examined by use of the immunoperoxidase technique (Johnson and Swoveland, 1977). In these two investigations it was found that IgG accumulates in the infected tissue, an observation that previously was made also in the abovementioned IF studies.

Cell-associated IgG may cause a potential limitation in examination of cells in secretions or tissue materials from patients, since it may represent blocking antibodies deriving from body fluids. In order to remove such complexed Ig treatment with chaotropic ions (ter Meulen et al., 1969) or acid buffers (Fulton and Middleton, 1975) have been used. However, such a treatment does not appear to be required, since at least a fraction of intracellular antigen should escape antibody blocking. The use of heterospecies antibodies in IF when employing the indirect technique would appear advisable to avoid nonspecific staining of tissue-associated patient Ig.

Potentially, antigen detection ELISA may be applied to cellular materials or secretions from patients with measles. An immunoassay for measles virus nucleocapsid antigen has been presented (Salmi and Lund, 1984). Threshold sensitivities of the assays were 1 to 10 ng antigen per ml. When antigen-antibody complexes were examined, difficulties were encountered in dissociating this complex without extensive antigen destruction. It should be remembered that in some phases of pathogenetic events, immune complexes occur naturally (Ziola et al., 1983).

Direct Viral Nucleic Acid Detection

The majority of measles virus genes have been cloned and replicated by use of recombinant DNA technique. Such DNA has been used for experimental purposes (Baslé et al., 1986; Fournier et al., 1985; Haase et al., 1981). This technique has demonstrated measles virus-specific RNA in lymphocytes, brain cells, and osteocytes in diseased individuals. Peripheral blood cells in seropositive individuals also contained viral RNA: 0.1 to 5% cells in adults, and 10 to 15% cells in children (Fournier et al., 1985). The significance of these remarkably high figures remains to be determined. Once specificity criteria have been defined, nucleic acid hybridization techniques may also have application as a diagnostic method.

Virus Isolation and Identification

Isolation of virus is not conventionally used as a technique to identify measles infections. The reasons for this are that the procedure is time-consuming and technically complex and that the results are unpredictable. This can be exemplified by a recent study (Sakaguchi et al., 1986) in which throat swabs collected shortly after onset of rash in 20 patients yielded two virus strains. Both these strains produced minimal amounts of cell-free virus during early passages.

When attempts to isolate virus are considered desirable, primary human or simian cells should be employed. Primary human kidney cells have been used most extensively, but primary human amnion cells were also found to be useful. Primary simian cells of different species origin were also used, but they have the inherent disadvantage of potential contamination with indigenous viruses. Human diploid cells and monkey cell lines such as Vero or BSC cells may offer a satisfactory alternative, but data are not available on comparative susceptibilities. Virus can be isolated in cell cultures from nasopharyngeal secretions up to 1 to 2 days after onset of rash (Ruckle and Rogers, 1957) and from the buffy coat of blood and urine 1 or 2 days longer (Gresser and Katz, 1960).

Brain materials from postinfectious measles encephalitis generally do not allow virus isolation. However, explant cultures of brain tissue from patients with SSPE may carry a persistent, nonproductive infection, which can be demonstrated by IF or hemadsorption techniques (Katz and Koprowski, 1973; Wechsler and Meissner, 1982). A production of extracellular infectious virus can be achieved in a small fraction of cases by cocultivation with cells susceptible to the virus.

Antibody Assays

As already discussed, different kinds of tests show varying capacity to discriminate the immune response to different virus components. Tests employing whole-virus antigen and a general indicator system to identify the antigen–antibody complexes formed demonstrate the sum of reactions of all specific antibodies. In this category belong complement fixation (CF) tests, radioimmune assays (RIA) and ELISA. With time the latter tests have become of primary use. They offer the advantage of higher sensitivity. In addition they provide an opportunity, by choice of the appropriate indicator system, for identification of different classes of antibodies: IgG, IgM, IgA, IgD, and IgE.

Serological tests based on inhibition of a defined biological activity of the virus demonstrate antibodies to one or a few structural components. Hemagglutination-inhibition (HI) tests selectively indicate antibodies to the hemagglutinin, hemolysis-inhibition tests (HLI) demonstrate antibodies to the hemolysin, but also to the hemagglutinin, since the hemolytic (fusion) activity of the virus requires attachment of particles prior to membrane fusion, and the neutralization test, finally, measures antibodies to the hemagglutinin and, to a minor extent, to the fusion factor.

Tests such as Western blot and RIPA, in which different structural components are separated prior to reaction with different antibodies, were also discussed above. It might be appropriate in this context to summarize antibody responses to different virus polypeptides prior to a discussion of different tests.

The dominant part of the antibody response involves the nucleocapsid antigen (Graves et al., 1984; Norrby et al., 1981). The intensity and duration of the antibody response to the P antigen probably is also pronounced. This antibody response is easily underestimated in the RIPA but may be more correctly illustrated by results of Western blot tests (Hankins and Black, 1986). The antibody responses to the peplomer glycoproteins are of particular relevance, since they are involved in immune protection. Antibodies to these components emerge somewhat later than antibodies to the internal components of the virus after the acute infection (Graves et al., 1984). In fact, there may also be asynchrony in the appearance of antibodies to different epitopes on the hemagglutinin (Kramer and Cremer, 1984a). The antibody response to the M protein has been a particular focus of interest, since it was proposed that a deficient synthesis of M antigen and hence a poor antibody response to this protein would be a special feature of SSPE (Hall et al., 1979). As a consequence of the availability of additional data, this hypothesis has become modified. First, there is a synthesis of M antigen in brain cells in at least some patients with SSPE (Norrby et al., 1985) and secondly a deficient M antibody response is not unique to patients with SSPE. In fact the antibody response to M antigen after the acute infection is weak and in many cases transient (Graves et al., 1984; Norrby et al., 1981). Besides SSPE, other disease conditions are associated with accentuated titers of measles antibodies. Weakly, but significantly increased antibody titers are found in patients with multiple sclerosis (Norrby, 1978), and markedly increased titers have reproducibly been encountered in two autoimmune disorders, systemic lupus erythematosus and active chronic hepatitis (Laitinen and Vaheri, 1974; Triger et al., 1976). High antibody titers were also encountered in individuals immunized with inactivated vaccines, who developed atypical measles (Graves et al., 1984; Norrby et al., 1981). In fact, the most pronounced antibody responses to the M protein have been seen in sera collected from cases of atypical measles.

PREPARATION OF ANTIGEN MATERIAL FOR SEROLOGIC TESTS

The requirements for antigens to be used in different tests vary extensively. In neutralization tests performed by plaque inhibition, the virus–cell system should allow a rapid development of distinct and sizable plaques, whereas for many other tests the main aim is to allow accumulation of large amounts of antigen, sometimes expected to express a particular biologic activity.

The yield of antigen varies considerably in different virus–cell systems. Various primate cell lines such as Vero and HeLa cells provide useful systems for propagation of measles virus. If the goal is to obtain large plaques in the test, a nonautointerfering virus strain, preferably taken through 1 to 3 consecutive plaque isolations, which procedure gives rapid growth with syncytia formation as its main hallmark, should be used. If, instead, the purpose is antigen accumulation, a more slowly growing virus strain giving preferentially strand-cell-forming cytopathic effects should be used. In the latter case, cultures are harvested when advanced cytopathic effects have developed, if extracellularly accumulated antigen is going to be used. Alternatively, the cultures may be harvested about 1 day earlier by discarding the medium and scraping off the cells in an appropriate buffer to a 10% vol/vol suspension. Disruption of cells can be achieved by freezing and thawing, treatment with ultrasonics, or both. In some cases detergent-containing buffers are used to dissociate virus components. Materials should be clarified at low speed in a table centrifuge before use to remove particulate material.

SPECIAL REQUIREMENTS FOR DIFFERENT TESTS

CF tests, RIA, and ELISA specific for measles antibodies are performed by standard methods and therefore require no particular consideration. The importance of using a rapidly growing virus in plaque-neutralization tests was already mentioned. In order to maintain cultures in a satisfactory condition to obtain large plaques, relatively more durable cell monolayers such as those provided by Vero cells and a CO_2-independent HEPES buffer may be used. In HLI and HI tests there is a need for accessibility of erythrocytes from Old World monkeys, preferentially green monkey (*Cercopitecus aethiops*). The HLI test is straightforward (Norrby and Gollmar, 1972, 1975), but requires availability of antigen preparations which, under standardized conditions, give extinction values at 540 nm exceeding 0.5 to 0.6. Sometimes concentration of tissue culture fluid material is necessary, but the use of cell pack material can also be considered. In order to allow comparison between tests it is necessary to store the antigen preparations at $-70°C$.

For increase of the sensitivity of HI tests, the antigen preparations should be disrupted by Tween 80 and ether treatment (Norrby, 1962). This treatment gives a four- to tenfold increase of hemagglutinin titer, hence allowing the saving of antigen, and increases the sensitivity of the test with a factor of 2 to 4 because of the small size of the antigen. Testing of sera at low dilution may require removal of nonspecific agglutinins and inhibitors. The agglutinins are readily removed by adsorption with packed erythrocytes, and the inhibitors can be eliminated by treatment with kaolin.

THE EVOLUTION OF ELISA FOR MEASLES ANTIBODY DETECTION AND PERMUTATIONS OF THIS ASSAY CURRENTLY CONSIDERED FOR USE

The first development from the traditional CF test for overall antibody detection was the introduction of indirect RIAs with whole-virus antigen (Arstila et al., 1977; Vuorimaa et al., 1978) or purified virus components (Moore et al., 1978). The RIA tests were within a short time exchanged for indirect ELISA, which originally was used with antigen attached to the solid phase (Forghani and Schmidt, 1979; Kahane et al., 1979; Kleiman et al., 1981; Voller and Bidwell, 1976). The next development was the introduction of capturing antibodies to be used for coating of the solid phase. This allowed more effective and specific identification of virus-specific immunoglobulins of different classes (Forghani et al., 1983; Kramer and Cremer, 1984b; Pedersen et al., 1982; Tuokko and Salmi, 1983; see also below). Recently one additional modification was introduced. Instead of the indirect technique with four layers—antihuman Ig for capture, test serum specimen, viral antigen, and enzyme-labeled antiserum—the direct three-layer technique employing enzyme-labeled antigen was used (Salonen et al., 1986). Specificity of reactions in the latter test was found to be dependent on the purity of the viral antigen preparations. Some nonspecific results were observed when infected-cell-lysate antigen was used.

COMPARATIVE SENSITIVITY OF DIFFERENT TESTS FOR MEASLES VIRUS-SPECIFIC ANTIBODIES

It is difficult to make precise statements about the relative sensitivity of different kinds of serological tests, since the procedures used in individual laboratories may introduce variations. In general terms, the HI test, when split antigen is used, is more sensitive than the CF test and about equally as sensitive as the neutralization test. Other tests, such as the HLI test, have a sensitivity equal to or in some instances higher than the HI test (Neumann et al., 1985), whereas the IF or the immunoperoxidase tests (Shani et al., 1981) used for antibody determination may have a sensitivity slightly inferior to or equal to the HI test. The ELISA antibody titers always exceed those of other tests (Cremer et al., 1985; Forghani and Schmidt, 1979; Kahane et al., 1979; Neumann et al., 1985; Weigle et al., 1984), except RIAs, which are not in practical use. The order of magnitude of the difference varies with the form of expression of ELISA antibody titers, but may be of the order of 10 and 100. In spite of this, the predictive value of the ELISA test for evaluation of immunity to measles is not higher than, for example, plaque neutralization and HI tests.

SELECTIVE TESTS FOR DIFFERENT CLASSES OF ANTIBODIES: THE USE OF IgM ANTIBODY DETECTION FOR SINGLE-SAMPLE SEROLOGIC DIAGNOSIS

There have been extensive studies of measles virus-specific IgM antibodies as a means of providing a diagnosis of an ongoing infection. Only limited studies have concerned IgA antibodies, and no characterization of virus-specific IgE has been published, although it has been noted that the total IgE concentration is higher than normal in patients with measles (Shalit et al., 1984). The finding of measles virus-specific IgA in respiratory secretions has been interpreted to reflect an ongoing infection (Friedman et al., 1983), and absence of serum IgA was suggested to indicate susceptibility to subclinical reinfection (Pedersen et al., 1986). There is a single report (Luster et al., 1976) showing measles virus-specific IgD antibodies in patients with SSPE. It was speculated that this class of antibodies would have a special role in pathogenesis. It is apparent that the conditions of synthesis of measles virus-specific Ig of class-characteristics other than IgG and IgM deserve further attention.

The evolution of IgM-specific measles antibody assays is as follows: The first studies employed sucrose density gradient centrifugation (Schluederberg, 1965) or DEAE-cellulose column chromatography (Tikhinova et al., 1973) to separate IgG and IgM and measure the distribution of HI antibodies. When class-specific anti-Ig became available, this was used in selective IF studies (Connolly et al., 1971) and RIAs (Arstila et al., 1977; Kiessling et al., 1977; Vuorimaa et al., 1978). Eventually the RIAs were exchanged for the more convenient ELISA (Forghani et al., 1983; Kleiman et al., 1983; Lievens and Brunell, 1986; Pedersen et al., 1982; Tuokko, 1984; Tuokko and Salmi, 1983). The tests were evaluated for their capacity to identify acute infections and also chronic infections, namely, SSPE. Three major problems emerged during the development of tests. One problem was that the concentration of virus-specific IgM might be underestimated or remain undetected if specific IgG simultaneously were present in the sample. The second problem was that the presence of rheumatoid factor, that is, IgM anti-IgG, could give false-positive reactions. Finally, the third problem related to the fact that many viral infections elicit a production of IgM autoantibodies against the host cell cytoskeleton or nuclear structures. Ways of circumventing these problems will be discussed in relation to IgM antibody determination in acute and persistent measles virus infections.

Studies of measles virus-specific IgM in acute infections by use of ELISA was initially performed by use of the indirect method (Tuokko and Salmi, 1983), but this technique was exchanged in an early phase for the reverse -μ-"capture"-technique (Forghani et al., Kleiman et al., 1983; Pedersen et al., 1982). This modification allowed a selective focusing on only IgM throughout all steps of the procedure, which had the important consequence that interference by IgM rheumatoid factor could be avoided (Tuokko, 1984). The level of false-positive reactions was 1.5% in 238 sera collected after various other viral infections,

whereas sera from healthy individuals did not give nonspecific reactions. The IgM "capture" technology takes care not only of the potentially false-positive reactions with rheumatoid factor, but also eliminates the problem of competition for antigen of IgG and IgM of related specificity. However, it does not solve the problem caused by IgM-specific autoantibodies reacting with different cellular components. Only by using purified viral antigens in the test would it be possible to guarantee that the reactions observed reflect the presence of only virus-specific IgM. One report using directly labeled crude or purified antigen was recently published (Salonen et al., 1986). The sensitivity of this direct ELISA was found to be slightly higher than that of the indirect ELISA.

The time for detection of specific IgM after acute measles obviously is dependent on the sensitivity of the test employed. For practical purposes it is important to identify the time of development of IgM in relation to the pathogenic events. The rash generally is used as the reference time point. As expected, specific IgM antibodies detected by ELISA in most cases are already present at the time of the rash (Forghani et al., 1983; Lievens and Brunell, 1986; Tuokko and Salmi, 1983), but occasional negative samples are seen. The duration of the IgM antibody response measured by the ELISA technique is 30 to 60 days. With the more sensitive RIA, antibodies of the IgM class may be detected for another month (Vuorimaa et al., 1978).

The attempts to characterize measles virus IgM in the chronic infection SSPE by immunofluorescence studies have given variable results. Measles virus IgM antibodies were reported to occur in all (Chinmei and Szu-chin, 1977; Connolly et al., 1971; Kiessling et al., 1977), in some (Mehta et al., 1977; Thomson et al., 1975), and in none (Massaro et al., 1978; Najera et al., 1972) of the SSPE patients analyzed in individual studies. To some extent these variable results may have a technical background, but a particular complication again probably has been caused by IgM-class rheumatoid factor. In order to exclude interference by this factor, some studies included pretreatment of samples with aggregated human IgG (Thomson et al., 1975; Ziola et al., 1979) or separation of IgG and IgM in samples (Mehta et al., 1977). The conclusion from the most recent application of RIAs (Ziola et al., 1986) shows that measles virus IgM antibodies do not represent a useful marker for SSPE even though there was a tendency for their synthesis at the early phase of the disease. In contrast, a recent application of the μ-capture ELISA technique (Chiodi et al., 1986) showed virus-specific IgM to be present in all 6 SSPE patients included in the study. A local production of specific IgM in the CNS was also identified in the latter study. Further studies are needed to resolve this conflicting issue. By use of the same ELISA technique, no measles virus-specific IgM antibodies were seen in sera and CSF from patients with multiple sclerosis (Chiodi et al., 1987). By application of another technique, measles virus-specific IgM was described to be present in sera from patients with active chronic hepatitis (Christie et al., 1984), another etiologically enigmatic disease in which the involvement of measles virus has been discussed. This observation needs to be confirmed by use of a different approach.

Interpretation of Laboratory Data

As already mentioned, the acute form of measles shows such distinct clinical features that there is no demand for laboratory diagnosis. The clinical virologic laboratory becomes involved when the primary infection is of an atypical nature and when the virus gives chronic infections. Aspects on interpretation of the different assays were already given in connection with the presentation of the individual tests. The following summarizing remarks can be given.

Virus isolation is not a convenient and reproducible laboratory test for diagnosis of measles. Identification of virus-infected cells or the presence of viral antigens or nucleic acid may be more useful. However, the tests for identification of viral antigens or nucleic acid need to be further elaborated. Direct identification of the presence of virus is of particular importance in the characterization of the chronic infection of SSPE. However, this requires the performance of a brain biopsy, since reproducible identification of antigen in cells outside the CNS has not been made.

Serologic diagnosis can be readily performed with paired serum samples by use of modern methodology. Determination of virus-specific IgM in sera collected immediately after the rash allows demonstration of an ongoing infection in almost all cases. The use of IgM antibody determination for identification of SSPE needs to be further evaluated. However, the characterization of specific IgG antibodies in this disease reveals not only markedly increased serum titers of antibodies to internal structural components and more variably to envelope components, but in particular a local production of antibodies within the CNS. A local production of antibodies in the CNS may also be encountered in cases of multiple sclerosis, but this production is more limited than in SSPE.

Epidemiology and Natural History

In nonimmune individuals, subclinical infections with measles are rare. However, low-grade passive or active immunity may mitigate symptoms or allow a restricted virus replication without clinical symptoms.

The generalized measles infection causes an extensive mobilization of the immune defense system. A continued synthesis of antibodies can be demonstrated throughout life in essentially all individuals even in the absence of reexposure. Seroepidemiologic surveys therefore allow a determination of the age-specific distribution of infections in the society. In developing countries, most measles infections occur before the age of 5 years, and a considerable fraction of these cases occur already during the first year of life, although overt measles before the age of 6 months is rare owing to the protection provided by maternal antibodies. In these countries measles is the most important cause of death between 1 and 5 years. Death from measles is a rare event in industrialized countries. The reasons for this can be that the infections occur at higher ages, normally 5 to 10 years, in children under better general health conditions, including fewer concomitant infections, which may upset the balance in the immune system.

Measles causes epidemics every 2 to 5 years in nonimmunized populations, with each epidemic lasting 3 to 4 months (Black, 1976). In industrialized countries, the epidemic peaks usually occur during late winter and early spring. This epidemic pattern has been extensively changed as a consequence of the general use of live vaccines. It has been seen that for establishment of an effective herd immunity, more than 90% of a population needs to be immune. The spread of measles in a susceptible community is rapid and occurs exclusively as a chain of infections between acutely diseased individuals. The persistent infection SSPE is not a source of infectious virus. Further, there is no animal reservoir of virus. As a consequence of these facts, measles requires a population exceeding 300,000 people in order to sustain the continued presence of the disease. Populations of this critical size did not exist prior to the development of the River Valley civilization in the Middle East. This indicates that measles appeared as a disease about 4500 years ago (Black, 1966). A possible source of virus may have been domesticated animals. There is some evidence that rinderpest virus may be the archvirus in the *Morbillivirus* genus of *Paramyxoviruses,* from which, first, canine distemper and, later, measles virus have evolved (Sheshberadaran et al., 1986).

The epidemiology of SSPE is unique (Jabbour et al., 1972). More than half of all cases have had their preceding regular measles infection before the age of 2 years. The time interval between the acute disease and development of SSPE is 5 to 6 years. The disease is 3 times more common in boys than in girls.

Control and Prevention

There is no direct treatment of measles, but immunoglobulin can be administered up to 5 days after exposure as a late prophylaxis. Since the duration of immunity after a mitigated infection is uncertain, a supplementary vaccination needs to be given. This should be provided 3 to 6 months later. If the exposure to measles is identified early, passive immune prophylaxis can be substituted for live vaccine; if it is given within 2 to 3 days after exposure, the replication of vaccine virus takes preference over replication of wild virus (Ruuskanen et al., 1978).

The live measles vaccine has been documented to be highly effective (Hinman et al., 1982; Krugman, 1977). The preferable age of vaccination is 18 months in industrialized countries. Because of the young age at which measles occurs in developing countries, the vaccine has to be given earlier; the age of 9 months for vaccination, recommended by the World Health Organization, is a compromise between the attempt to prevent as large a fraction of early measles as possible and the desire to have a high conversion rate after vaccination. Since the conversion rate under those conditions is about 60 to 70%, a follow-up vaccination about 9 months later is recommended. The need for a second dose of vaccine is small when the vaccine is given at 18 months, since the conversion rate exceeds 90% and the duration of immunity after vaccination in practical terms is lifelong. Still some countries advocate the use of a second dose of vaccine, usually administered at the age of 12 to 13 years. The main intent of such a program is to maintain the herd immunity on a level high enough to preclude circulation of the virus.

Vaccine virus does not spread from immunized individuals. Reactions to vaccine usually are not activity-limiting. In about 10 to 30% of the vaccinees, mild symptoms are seen, primarily fever and sometimes also a rash. Severe complications are rare. Acute encephalitis essentially does not occur, and the frequency of SSPE is reduced by a factor of at least 10 (Modlin et al., 1977). Vaccine should not be given to T-cell-deficient individuals, since a progressive, potentially lethal infection may ensue, nor to pregnant women, although there is no evidence for teratogenic effects.

Antibody titers after vaccination are lower than after regular measles. Still, the remaining antibody titers can be detected in the majority of vaccinees 15 years after immunization. Exposure to wild virus may boost antibody titers via subclinical infections. However, the immunity appears to be durable even in the absence of circulation of wild virus in the community.

It follows from the data discussed here that measles is an eradicable disease. However, it appears that global eradication should be a goal only in the long-term perspective, whereas regional elimination of disease can be established as a more realistic, attainable goal.

Literature Cited

Albrecht, P., K. Herrmann, and G. R. Burns. 1981. Role of virus strain in conventional and enhanced measles plaque neutralization test. J. Virol. Methods 3:251–260.

Appel, M. J. G., W. R. Shek, H. Sheshberadaran, and E. Norrby. 1984. Inactivated and heterotypic live virus vaccines induce incomplete immunity to canine distemper. Arch. Virol. 82:73–82.

Arstila, P., T. Vuorimaa, K. Kalimo, P. Halonen, M. Viljanen, K. Granfors, and P. Toivanen. 1977. A solid phase radioimmunoassay for IgG and IgM antibodies against measles virus. J. Gen. Virol. 34:167–176.

Baslé, M. F., J. G. Fournier, S. Rozenblatt, A. Rebel, and M. Bouteille. 1986. Measles virus RNA detected in Paget's disease bone tissue by in situ hybridization. J. Gen. Virol. 67:907–913.

Black, F. L. 1966. Measles endemicity in insular populations: critical community size and its evolutionary implications. J. Theoret. Biol. 11:207–211.

Black, F. L. 1976. Measles, p. 297–316. In A. S. Evans (ed.), Viral infections of humans. Epidemiology and control. John Wiley & Sons, London.

Brody, J. A., R. McAlister, R. Haseley, and P. Lee. 1963. Use of dried whole blood collected on filter paper disks in adenovirus complement fixation and measles hemagglutination inhibition tests. Am. J. Epidemiol. 92:854–857.

Carter, M. J., M. M. Willcocks, S. Löffler, and V. ter Meulen. 1982. Relationships between monoclonal antibody-binding sites on the measles virus haemagglutinin. J. Gen. Virol. 63:113–120.

Carter, M. J., M. M. Willcocks, S. Löffler, and V. ter Meulen. 1983. Comparison of lytic and persistent measles virus matrix protein by competition radioimmunoassay. J. Gen. Virol. 64:1801–1805.

Chin-Mei, Y., and W. Szu-chin. 1977. Subacute sclerosing panencephalitis: clinical and immunological investigation of two cases. Chin. Med. J. 3:419–422.

Chiodi, F., V.-A. Sundqvist, H. Link, and E. Norrby. 1987. Viral IgM antibodies in serum and cerebrospinal fluid in patients with multiple sclerosis and controls. Acta Neurol. Scand. 475:201–208.

Chiodi, F., V.-A. Sundqvist, E. Norrby, M. Mavra, and H. Link. 1986. Measles IgM antibodies in cerebrospinal fluid and serum in subacute sclerosing panencephalitis. J. Med. Virol. 18:149–158.

Christie, K. E., C. Endresen, and G. Haukenes. 1984. IgM antibodies in sera from patients with chronic active hepatitis reacting with the measles virus matrix protein and nucleoprotein. J. Med. Virol. 14:149–157.

Connolly, J. H., M. Haire, and D. S. M. Hadden. 1971. Measles immunoglobulins in subacute sclerosing panencephalitis. Br. Med. J. 1:23–25.

Cremer, N. E., C. K. Cossen, G. Shell, J. Diggs, D. Gallo, and N. J. Schmidt. 1985. Enzyme immunoassay versus plaque neutralization and other methods for determination of immune status to measles and varicella-zoster viruses and versus complement fixation for serodiagnosis of infection with these viruses. J. Clin. Microbiol. 21:869–874.

Forghani, B., and N. J. Schmidt. 1979. Antigen requirements, sensitivity and specificity of enzyme immunoassays for measles and rubella viral antibodies. J. Clin. Microbiol. **9:**657–664.

Forghani, M. B., C. M. Myoraku, and N. J. Schmidt. 1983. Use of monoclonal antibodies to human immunoglobulin M in "capture" assays for measles and rubella immunoglobulin M. J. Clin. Microbiol. **18:**652–657.

Fournier, J. G., M. Tardieu, P. Lebon, O. Robain, G. Ponsot, S. Rozenblatt, and M. Bouteille. 1985. Detection of measles virus RNA in lymphocytes from peripheral blood and brain perivascular infiltrates of patients with subacute sclerosing panencephalitis. N. Engl. J. Med. **313:**910–915.

Friedman, M., I. Hadari, V. Goldstein, and I. Sarov. 1983. Virus specific secretory IgA antibodies as a means of rapid diagnosis of measles and mumps infection. Isr. J. Med. Sci. **19:**881–884.

Fulton, R. E., and P. J. Middleton. 1975. Immunofluorescence in diagnosis of measles infections in children. J. Pediatr. **86:**17–22.

Graves, M. C. 1981. Measles virus polypeptides in infected cells studied by immune precipitation and one dimensional peptide mapping. J. Virol. **38:**224–230.

Graves, M., D. E. Griffin, R. T. Johnson, R. L. Hirsch, I. Lindo De Soriano, S. Roedenbeck, and A. Vaisberg. 1984. Development of antibody to measles virus polypeptides during complicated and uncomplicated measles virus infections. J. Virol. **49:**409–412.

Gresser, I., and S. L. Katz. 1962. Isolation of measles virus from urine. N. Engl. J. Med. **263:**452–454.

Haase, A. T., P. Ventura, C. J. Gibbs, and W. W. Tourtellotte. 1981. Measles virus nucleotide sequences: detection by hybridization in situ. Science **212:**672–675.

Hall, W. W., R. A. Lamb, and P. W. Choppin. 1979. Measles and subacute sclerosing panencephalitis virus proteins. Lack of antibodies to the M protein in patients with subacute sclerosing panencephalitis. Proc. Natl. Acad. Sci. USA **76:**2047–2051.

Hankins, R. W., and F. L. Black. 1986. Western blot analysis of measles virus antibody in normal persons and in patients with multiple sclerosis, subacute sclerosing panencephalitis or atypical measles. J. Clin. Microbiol. **24:**324–329.

Hinman, R. A., L. D. Eddine, D. C. Kirby, A. W. Orenstein, H. R. Bernier, M. P. Turner, Jr., and B. A. Block. 1982. Progress in measles elimination. J. Am. Med. Assoc. **247:**1592–1595.

Jabbour, J., D. Duenas, J. L. Sever, H. M. Krebs, and L. Horta-Barbosa. 1972. Epidemiology of subacute sclerosing panencephalitis. J. Am. Med. Assoc. **220:**959–962.

Jenis, E. H., M. R. Knieser, P. A. Rothouse, G. E. Jensen, and R. M. Scott. 1973. Subacute sclerosing panencephalitis. Immunoultrastructural localization of measles-virus antigen. Arch. Pathol. **95:**81–89.

Johnson, K. P., and P. Swoveland. 1977. Measles antigen distribution in brains of chronically infected hamsters. An immunoperoxidase study of experimental subacute sclerosing panencephalitis. Lab. Invest. **37:**459–465.

Kahane, S., V. Goldstein, and I. Sarov. 1979. Detection of IgG antibodies specific for measles virus by enzyme linked immunosorbent assay (ELISA). Intervirology **12:**39–46.

Katz, M., and H. Koprowski. 1973. The significance of failure to isolate infected viruses in cases of SSPE. Arch. Gesamte Virusforsch. **41:**390–393.

Kempe, C. H., and V. A. Fulginiti. 1965. The pathogenesis of measles virus infection. Arch. Gesamte Virusforsch. **16:**103–128.

Kiessling, W. R., W. W. Hall, L. L. Yung, and V. ter

Meulen. 1977. Measles-virus-specific immunoglobulin-M response in subacute sclerosing panencephalitis. Lancet **i:**324–327.

Kingsbury, D. W., M. A. Bratt, P. W. Choppin, R. P. Hansen, Y. Hosaka, V. ter Meulen, E. Norrby, W. Plowright, R. Rott, and W. H. Wunner. 1978. *Paramyxoviridae.* Intervirology **10:**137–152.

Kleiman, B. M., C. K. L. Blackburn, S. E. Zimmerman, and M. L. V. French. 1981. Comparison of enzyme linked immunosorbent assay for acute measles with hemagglutination inhibition, complement fixation and fluorescent antibody methods. J. Clin. Microbiol. **14:**147–152.

Kleiman, M. B., C. K. L. Blackburn, S. E. Zimmerman, M. L. V. French, and L. J. Wheat. 1983. Rapid diagnosis of measles using enzyme-linked immunosorbent assay for measles immunoglobulin M. Diagn. Microbiol. Infect. Dis. **1:**205–213.

Kramer, S. M., and N. E. Cremer. 1984a. Biological activity of a monoclonal antibody to a measles virus haemagglutinin epitope detected late in infection. J. Gen. Virol. **65:**577–583.

Kramer, S. M., and N. E. Cremer. 1984b. Detection of IgG antibodies to measles virus using monoclonal antibodies for antigen capture in enzyme immunoassay. J. Virol. Methods **8:**255–263.

Krugman, S. 1977. Present status of measles and rubella immunization in the United States: a medical progress report. J. Pediatr. **90:**1–13.

Laitinen, O., and A. Vaheri. 1974. Very high measles and rubella virus antibody titers associated with hepatitis, systemic lupus erythematous and infectious mononucleosis. Lancet **i:**194–198.

Lievens, A. W., and P. A. Brunell. 1986. Specific immunoglobulin M enzyme-linked immunosorbent assay for confirming the diagnosis of measles. J. Clin. Microbiol. **24:**391–394.

Luster, M. I., R. C. Armen, J. V. Hallum, and G. A. Leslie. 1976. Measles virus specific IgD antibodies in patients with subacute sclerosing panencephalitis. Proc. Natl. Acad. Sci. USA **73:**1297–1299.

Mäkelä, M., E. Norrby, and A. Salmi. 1987. Measurement of polypeptide- and antigenic site-specific antibodies to measles virus using a competitive enzyme immunoassay (EIA). J. Virol. Methods **16:**65–74.

Massaro, A. R., A. M. Agliano, and R. Grillo. 1978. Immunoglobulin M specific for measles in serum and cerebrospinal fluid of patients with multiple sclerosis and other neurological diseases. J. Neurol. **217:**191–194.

McChesney, M. B., and M. B. A. Oldstone. 1987. Viruses perturb lymphocyte functions: selected principles characterizing virus-induced immunosuppression. Annu. Rev. Immunol. **5:**279–304.

McQuillin, J., T. M. Bell, P. S. Gardner, and M. A. P. S. Downham. 1976. Application of immunofluorescence to a study of measles. Arch. Dis. Child. **51:**411–420.

Mehta, P. D., A. Kane, and H. Thormar. 1977. Quantitation of measles virus specific immunoglobulins in serum, CSF and brain extract from patients with subacute sclerosing panencephalitis. J. Immunol. **118:**2254–2261.

Modlin, J. F. 1984. Measles virus, p. 333–360. In R. E. Belshe (ed.), Textbook of human virology, PSG Publishing Co. Inc., Littleton, Massachusetts.

Modlin, J. F., J. T. Jabbour, J. J. Witte, and N. A. Halsey. 1977. Epidemiologic studies of measles vaccine and subacute sclerosing panencephalitis. Pediatrics **59:**505–513.

Moore, P. M. E., E. C. Hayes, S. E. Miller, L. L. Wright, C. E. Machamer, and H. J. Zweerink. 1978. Measles virus nucleocapsids: Large scale purification and use in radioimmunoassays. Infect. Immun. **20:**842–846.

538　E. Norrby

Najera, R., A. G. Saiz, I. Herrera, and L. Valenciano. 1972. Serological and tissue culture observations from cases of subacute sclerosing panencephalitis. Ann. Inst. Pasteur 123:565–570.

Neumann, P. W., J. M. Weber, A. G. Jessamine, and M. W. O'Shaughnessy. 1985. Comparison of measles antihemolysis test, enzyme-linked immunosorbent assay, and hemagglutination inhibition test with neutralization test for determination of immune status. J. Clin. Microbiol. 22:296–298.

Norrby, E. 1962. Hemagglutination by measles virus. 4. A simple procedure for production of high potency antigen for hemagglutination-inhibition (HI) tests. Proc. Soc. Exp. Biol. Med. 111:814–818.

Norrby, E. 1978. Viral antibodies in multiple sclerosis. Prog. Med. Virol. 24:1–39.

Norrby, E. 1985. Measles, p. 1305–1321. In: B. N. Fields et al. (ed.), Virology. Raven Press, New York.

Norrby, E., S. N. Chen, T. Togashi, H. Sheshberadaran, and K. P. Johnson. 1982. Five measles virus antigens demonstrated by use of mouse hybridoma antibodies in productively infected tissue culture cells. Arch. Virol. 71:1–11.

Norrby, E., G. Enders-Ruckle, and V. ter Meulen. 1975. Differences in the appearance of antibodies to structural components of measles virus after immunization with inactivated and live virus. J. Infect. Dis. 132:262–269.

Norrby, E., and Y. Gollmar. 1972. Appearance and persistence of antibodies against different virus components after regular measles infections. Infect. Immun. 6:240–247.

Norrby, E., and Y. Gollmar. 1975. Identification of measles virus-specific hemolysis-inhibiting antibodies separate from hemagglutinating-inhibiting antibodies. Infect. Immun. 11:231–239.

Norrby, E., K. Kristensson, W. J. Brzosko, and J. G. Kapsenberg. 1985. Measles virus matrix protein detected by immune fluorescence with monoclonal antibodies in the brain of patients with subacute sclerosing panencephalitis. J. Virol. 56:337–340.

Norrby, E., C. Örvell, B. Vandvik, and D. J. Cherry. 1981. Antibodies against measles virus polypeptides in different disease conditions. Infect. Immun. 34:718–724.

Norrby, E., G. Utter, C. Örvell, and M. J. G. Appel. 1986. Protection against canine distemper virus in dogs after immunization with isolated fusion protein. J. Virol. 58:536–541.

Olding-Stenkvist, E., and B. Bjorvatn. 1976. Rapid detection of measles virus in skin rashes by immunofluorescence. J. Infect. Dis. 134:463–469.

Örvell, C., and E. Norrby. 1985. Antigenic structure of paramyxoviruses, pp. 241–264. In: M. H. V. van Regenmortel and A. R. Neurath (ed.), Immunochemistry of viruses. The basis for serodiagnosis and vaccine. Elsevier Biomedical Press, Amsterdam.

Osunkoya, B. O., A. R. Cooke, O. Ayeni, and T. A. Adejumo. 1974. Studies on leucocyte cultures in measles. I. Lymphocyte transformation and giant cell formation in leucocyte cultures from clinical cases of measles. Arch. Gesamte Virusforsch. 44:313–322.

Payne, F. E., and J. V. Baublis. 1973. Decreased reactivity of SSPE strains of measles virus with antibody. J. Infect. Dis. 127:505–511.

Pedersen, I. R., A. Antonsdottir, T. Evald, and C. H. Mordhorst. 1982. Detection of measles IgM antibodies by enzyme linked immunosorbent assay (ELISA). Acta Pathol. Microbiol. Immunol. Scand. 90:153–160.

Pedersen, I. B. R., C. H. Mordhorst, T. Evald, and H. von Magnus. 1986. Long-term antibody response after mea-

sles vaccination in an isolated arctic society in Greenland. Vaccine 4:173–178.

Ruckle, G., and K. D. Rogers. 1957. Studies with measles virus II. J. Immunol. 78:341–352.

Ruuskanen, O., T. T. Salmi, and P. Halonen. 1978. Measles vaccination after exposure to natural measles. J. Pediatr. 93:43–46.

Sakaguchi, M., Y. Yoshikawa, K. Yamanouchi, K. Takeda, and T. Sato. 1986. Characteristics of fresh isolates of wild measles virus. Jpn. J. Exp. Med. 56:61–67.

Salmi, A., and G. Lund. 1984. Immunoassays for measles virus nucleocapsid antigen: effect of antigen–antibody complexes. J. Gen. Virol. 65:1655–1663.

Salonen, J., R. Vainionpää, and P. Halonen. 1986. Assay of measles virus IgM and IgG class antibodies by use of peroxidase labelled viral antigens. Arch. Virol. 91:93–106.

Schlueberg, A. 1965. Immune globulin in human viral infections. Nature 205:1232–1233.

Shalit, M., Z. Ackerman, S. Wollner, A. Morag, and Y. Levo. 1984. Immunoglobulin E response during measles. Int. Arch. Allergy Appl. Immunol. 75:84–86.

Shani, L., H. Haikin, and I. Sarov. 1981. A rapid immunoperoxidase assay for determination of IgG antibodies to measles virus. J. Immunol. Methods 40:359–365.

Sheshberadaran, H., S-N. Chen, and E. Norrby. 1983. Monoclonal antibodies against five structural components of measles virus. 1. Characterization of antigenic determinants on nine strains of measles virus. Virology 128:341–358.

Sheshberadaran, H., and E. Norrby. 1986. Characterization of epitopes on the measles virus hemagglutinin. Virology 152:58–65.

Sheshberadaran, H., E. Norrby, K. C. McCullough, W. C. Carpenter, and C. Örvell. 1986. The antigenic relationship between measles, canine distemper and rinderpest viruses studied with monoclonal antibodies. J. Gen. Virol. 67:1381–1392.

ter Meulen, V. T., and M. J. Carter. 1984. Measles virus persistency and disease. Prog. Med. Virol. 30:44–61.

ter Meulen, V., G. Enders-Ruckle, D. Müller, and G. Joppich. 1969. Immunhistological, microscopical and neurochemical studies on encephalitides. III. Subacute progressive panencephalitis, virological and immunhistological studies. Acta Neuropath. 12:244–259.

Thomson, D., J. H. Connolly, B. O. Underwood, and F. Brown. 1975. A study of immunoglobulin M antibody to measles, canine distemper and rinderpest viruses in sera of patients with subacute sclerosing panencephalitis. J. Clin. Pathol. 28:543–546.

Tikhinova, N. T., T. M. Khrometskaya, N. V. Khochec, M. P. Streltsova, and T. P. Nesterova. 1973. Study of the physico-chemical nature of anti-measles antibody in children who had typical measles. Vopr. Virusol. 1:27–32.

Triger, D. R., T. R. Gamlen, E. Paraskevas, R. S. Lloyd, and R. Wright. 1976. Measles antibodies and autoantibodies in autoimmune disorder. Clin. Exp. Immunol. 24:407–414.

Tuokko, H. 1984. Comparison of non-specific reactivity in indirect and reverse immunoassays for measles and mumps immunoglobulin M antibodies. J. Clin. Microbiol. 20:972–979.

Tuokko, H., and A. Salmi. 1983. Detection of IgM antibodies to measles virus by enzyme immunoassay. Med. Microbiol. Immunol. 171:187–198.

Varsanyi, T. M., G. Utter, and E. Norrby. 1984. Purification, morphology and antigenic characterization of measles virus envelope components. J. Gen. Virol. **65**:355–366.

Voller, A., and D. E. Bidwell. 1976. Enzyme-immunoassays for antibodies in measles, cytomegalovirus infections after rubella vaccination. Br. J. Exp. Pathol. **57**:243–247.

Vuorimaa, T. O., P. P. Arstila, B. R. Ziola, A. A. Salmi, P. T. Hänninen, and P. E. Halonen. 1978. Solid-phase radioimmunoassay determination of virus-specific IgM antibody levels in a follow-up of patients with naturally acquired measles infection. J. Med. Virol. **2**:271–278.

Wechsler, S. L., and H. C. Meissner. 1982. Measles and SSPE viruses: similarities and differences. Prog. Med. Virol. **28**:65–95.

Weigle, K. A., M. D. Murphy, and P. A. Bunnell. 1984. Enzyme-linked immunosorbent assay for evaluation of

immunity to measles virus. J. Clin. Microbiol. **19**:376–379.

Ziola, B., P. Halonen, and G. Enders. 1986. Synthesis of measles virus-specific IgM antibodies and IgM-class rheumatoid factor in relation to clinical onset of subacute sclerosing panencephalitis. J. Med. Virol. **18**:51–59.

Ziola, B., A. Salmi, M. Panelius, P. Halonen, and B. Friis. 1979. Measles virus specific IgM antibodies and IgM class rheumatoid factor in serum and cerebrospinal fluid of subacute sclerosing panencephalitis patients. Clin. Immunol. Immunopathol. **13**:462–474.

Ziola, B., G. Lund, O. Meurman, and A. Salmi. 1983. Circulating immune complexes in patients with acute measles and rubella virus infections. Infect. Immun. **41**:578–583.

CHAPTER 28

Paramyxoviridae: Respiratory Syncytial Virus

LARRY J. ANDERSON

Disease: Acute upper and lower respiratory tract illness.

Etiologic Agent: Respiratory syncytial virus (RSV). There are at least two groups or subtypes of RSV.

Source: Respiratory secretions of humans.

Clinical Manifestations: Acute onset of fever, cough, rhinorrhea, and nasal congestion. In infants and young children, often progresses to pneumonia, bronchitis, or bronchiolitis.

Pathology: Disease is limited to the upper and lower respiratory tract except in persons with compromised immune systems.

Laboratory Diagnosis: RSV is isolated from or its antigens detected in respiratory secretions. It can also be diagnosed by a ≥ four-fold rise in neutralizing complement-fixing, IgG, IgA, or IgM antibodies.

Epidemiology: Present worldwide and occurs as yearly outbreaks lasting 2 to 5 months. These outbreaks occur during the winter or early spring in temperate climates and cause serious disease primarily in infants and young children.

Treatment: Aerosolized ribavirin.

Prevention and Control: No vaccine has been developed for RSV, but there are measures that can and should be instituted to control nosocomial transmission. These measures include contact isolation for RSV-infected patients, use of gowns when contact with secretions is likely, strict attention to good handwashing practices, and cohorting patients and staff during outbreaks.

Introduction

In 1955, a new virus was isolated from a chimpanzee during an outbreak of upper respiratory tract illness in a colony of 20 chimpanzees (Morris et al., 1956). Fourteen became ill, and 18 developed an antibody response to the virus. Three chimpanzees were challenged with the virus and two developed respiratory illness, had an antibody response to the virus, and were culture-positive for the virus. This virus was called chimpanzee coryza agent. One year later, similar viruses were isolated from children with acute lower respiratory tract illness in Baltimore (Chanock and Finberg, 1957; Chanock et al., 1957). Two of these viruses, the Long and Snyder strains, were indistinguishable from the chimpanzee coryza agent by cross-neutralization studies. Based on their association with respiratory illness and their ability to in-

duce syncytia in cells in tissue culture, respiratory syncytial virus (RSV) was proposed as the name of this group of viruses (Chanock and Finberg, 1957). Numerous studies have since shown that RSV is the leading cause of lower respiratory tract illness in infants and young children worldwide (Beem et al, 1960; Berman et al., 1983; Chanock et al, 1961, 1967; Doggett, 1965; Foy et al., 1973a; Glezen et al., 1971; Kim et al., 1973; Pan American Health Organization, 1982; Suto et al., 1965; Tyrrell, 1963) and is responsible for repeated attacks of acute respiratory illness throughout life (Hall et al., 1976b; Morales et al., 1983).

RSV is a pleomorphic, enveloped virus 100 to 350 nm in diameter, which is morphologically similar to the paramyxoviruses in the family *Paramyxoviridae* (Bachi and Howe, 1973; Blath and Norrby, 1967; Joncas et al., 1969a; Kalica et al., 1973; Norrby et al., 1970). Like the paramyxoviruses, it causes formation of syncytial cells in tissue culture. However, it differs from these viruses in that (1) the diameter of its nucleocapsid, 12 to 15 nm, is shorter than that of the paramyxoviruses, 17 to 18 nm (Berthiaume et al., 1974; Norrby et al., 1970); (2) the molecular weights of its major nucleocapsid (N) and phosphoprotein protein (P) are smaller (Levine, 1977; Wunner and Pringle, 1976); and (3) it does not cause hemagglutination (Richman et al., 1971). Differences between RSV and the paramyxoviruses also exist with respect to the number and characteristics of its proteins, number and order of its genes, and its intergenic sequences (Collins et al., 1984a; Satake et al., 1985; Spriggs and Collins, 1986; Wertz et al., 1985). These differences suggest that RSV should be placed in a separate genus, the pneumovirus genus, within the family *Paramyxoviridae* (Kingsbury et al., 1978). Other viruses within this genus include the pneumonia virus of mice, which is antigenically unrelated to RSV (Berthiaume et al., 1974; Cash et al., 1977; Compans et al., 1967); bovine respiratory syncytial virus, which is antigenically related to RSV (Cash et al., 1977; Paccaud and Jacquier, 1970), and caprine respiratory syncytial virus (Lehmkuhl et al., 1980). Bovine RSV can be differentiated from human RSV on the basis of its differential susceptibility to cell culture, but the relationship of the caprine virus to the bovine or human RSV has not been determined. Several studies have demonstrated naturally occurring antibodies against RSV in dogs, sheep, and cats (Lamontagne et al., 1985; Lungren et al., 1969; Pringle and Cross, 1978; Richardson-Wyatt et al., 1981), suggesting that viruses related to RSV, but as yet unrecognized, may be endemic to these species or that infection of these species by human RSV, bovine RSV, or caprine RSV can occur.

The Virus

Morphology and Biochemical Characteristics

RSV is a single-stranded negative-sense RNA virus. The linear RNA has a sedimentation coefficient of 50S and a molecular weight of about 5×10^6 daltons; it encodes for 10 proteins (Collins et al., 1984a; Huang and Wertz, 1982; Lambert et al., 1980). The virus is pleomorphic with particles ranging from 100 to 350 nm in diameter; occasional elongated particles are seen which may be up to 2,500 nm in length (Fig. 1). The virus particles are often associated with the cellular membrane and have clublike projections or spikes of 12 to 15 nm on their surface. The nucleocapsid is helical with a diameter of 12 to 15 nm and has a periodicity of 65 to 70 Å which appears as a

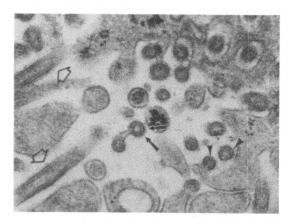

FIG. 1. RSV in infected HEp-2 cells. Arrow points to surface projections or spikes. Arrowheads point to nucleocapsids in cross-section, and open arrows point to elongated virus particles (magnification ×79,800). (Photograph courtesy of Cynthia Sporborg, Centers for Disease Control, Atlanta.)

herringbonelike structure by electron microscopy (Armstrong et al., 1962; Bloth and Norrby, 1967; Joncas et al., 1969a; Norrby et al., 1970; Zakstelskaya et al., 1967). RSV has neither hemagglutinin nor neuraminidase activity and is ether-, acid-, and heat-labile (Bennett and Hamre, 1962; Chanock et al., 1957; Hambling, 1964; Morris et al., 1956; Richman et al., 1971). Hambling (1964) demonstrated that 90% of its infectivity is lost after 2 days at 25°C, 3 days at 4°C, 4 weeks at −30°C, and 12 weeks at −65°C.

Replication

The steps involved in the replication of RSV are shown schematically in Figure 2. During adsorption of the virus to a susceptible cell, 50% of the virus has been shown to be attached within 30 min to 2 h and 80% within 2 to 4 h (Bennett and Hamre, 1962; Coates et al., 1966; Levine and Hamilton, 1960). The virus then enters the cell, and after uncoating, the viral RNA serves as a template for the production of 10 positive-sense messenger RNA species and a full-length, positive-sense complementary RNA (Collins et al., 1984a). Protein synthesis is not required for the synthesis of messenger RNA but is required for the synthesis of complementary RNA (Huang and Wertz, 1982). The messenger RNAs serve as the template for translation of viral proteins; the full-length, complementary RNA serves as a template for transcription of virion RNA.

All steps in replication of RSV occur within the cytoplasm, and viral antigens first become evident by immunofluoresent staining about 9 h postinfection (Bennett and Hamre, 1962; Follett et al., 1975; Lam-

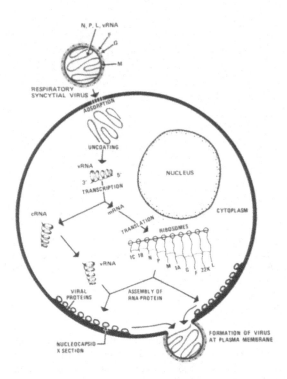

FIG. 2. Schematic representation of the replication of RSV. 1C and 1B are nonstructural proteins, N is the major nucleocapsid protein, P is the phosphoprotein, M is the membrane protein, 1A is presumed to be a structural protein, G is the large glycoprotein, F is the fusion protein, 22K is a structural protein whose location in the virion has not been determined, and L is a large protein presumed to be the polymerase. vRNA = viral RNA; mRNA = messenger RNA; cRNA = complementary RNA.

bert et al., 1980; Norrby et al., 1970). UV inactivation studies have shown that the messenger RNAs are produced sequentially from a single promoter site beginning at the 3' end of the viral RNA (Dickens et al., 1984). Unlike the paramyxoviruses, intergenic regions within RSV RNA are highly variable in length (1 to 52 nucleotides) and have no conserved sequences except for an adenosine at the end of each sequence (Collins et al., 1986). The first messenger RNA encodes for the nonstructural protein, 1C, followed by messenger RNAs that encode for the 1B, N, P, M, 1A, G, F, 22K, and L proteins, respectively (Table 1) (Collins et al., 1984a; 1986; Collins and Wertz, 1983; Huang and Wertz, 1983; Venkatesan et al., 1983). Three additional reading frames that overlap the 10 noted above have been identified—one in the sequence for the P protein of the Edinburgh strain of RSV (Lambden, 1985), one for the M protein (Satake and Venkatesan, 1984), and one for the 22K protein (Elango et al., 1985b) but no viral proteins have been found that correspond to these sequences. Two of the proteins, the F and G proteins,

are glycosylated after translation, and one, the P protein, is phosphorylated (Cash et al., 1979; Levine, 1977; Wunner and Pringle 1976). Nucleocapsids, consisting of viral RNA, N, P, and probably L (Huang et al., 1985; Peeples and Levine, 1979); accumulate within inclusion bodies in the cytoplasm that are visible by immunofluorescent staining within 10 to 24 h after infection (Kisch et al., 1962). At about the same time, projections of viral proteins appear on the surface of the cell, and infectious virions bud through the cell wall incorporating part of the cell membrane into their envelope (Berthiaume et al., 1974; Norrby et al., 1970). Infectious virus can first be detected in cell culture 11 to 13 h after infection (Bennett and Hamre, 1962; Levine and Hamilton, 1969) and has a nucleocapsid surrounded by an envelope which includes host cell membrane, M protein on the interior aspect of the envelope, and external projections of F and G proteins. The location of the 22K and 1A structural proteins has not been clearly established.

RSV replicates in a variety of tissue culture systems, including human heteroploid and diploid cell lines, primary monkey kidney cells, and continuous monkey cell lines (Jordan, 1962; Kisch et al., 1962). It can replicate in several animal species including subhuman primates, cattle, chinchillas, cotton rats, ferrets, guinea pigs, sheep, mice, and mink (Belshe et al., 1977; Cavallaro et al., 1967; Coates and Chanock, 1962; Lehmkuhl and Cutlip, 1979; Prince and Porter, 1976; Pricet et al., 1979a,b; Thomas et al., 1984b; Wright et al., 1970). Chimpanzees and owl monkeys develop clinical illness. Mice and cotton rats have been used most commonly in studies of RSV infection.

Proteins

The initial studies of the proteins of RSV were hampered by difficulties in purifying the virus and labeling the G protein. These studies identified between six and eight virus-specific proteins (Bernstein and Hruska, 1981; Dubovi, 1982; Fernie and Gerin, 1982; Lambert and Pons, 1983; Levine, 1977; Pringle et al., 1981; Wunner and Pringle, 1976). Recent studies have identified 10 proteins specific for RSV (Collins et al., 1984a, 1986; Huang et al., 1985) and the corresponding gene sequences for all but the L protein (Table 1). The first two proteins encoded by the genome are 1C and 1B, which are probably nonstructural proteins with molecular weights of 11,000 to 18,000 daltons by gel electrophoresis. Gene sequencing studies indicate that they are composed of 139 and 124 amino acids, respectively (Collins and Wertz, 1985b; Elango et al., 1985b; Huang et al., 1985). The next protein encoded by the RSV genome

TABLE 1. Proteins of respiratory syncytial virus

| Protein | Molecular weight by | | Location in virion | Function |
	PAGE	Sequence[a]		
1C	11–18K	15,567	Nonstructural	Unknown
1B	11–18K	14,674	Nonstructural	Unknown
N	40–45K	42,600	Nucleocapsid	Structural protein of nucleocapsid
P	31–37K	27,150	Nucleocapsid	?Part of the polymerase complex
M	27–29K	28,717	Envelope	Unknown
1A	9.5K	7,536	Unknown	Unknown
G	79–92K	32,587	Envelope	?Attachment to cell
F	66–70K	63,453	Envelope	Fusion, neutralization
22K	22–25K	22,156	Unknown	Unknown
L	160–200K	Unknown	Nucleocapsid	?Polymerase

[a] Calculated from the amino acid sequence deduced from the gene sequence.

is the N protein with an apparent molecular weight of 40,000 to 45,000 daltons and 391 amino acids (Collins et al., 1985; Elango and Venkatesan, 1983). It is rich in basic amino acids and is the major protein associated with the nucleocapsid (Cash et al., 1979; Huang et al., 1985; Levine, 1977; Wunner and Pringle, 1976). Monoclonal antibodies against the N protein stain the intracytoplasmic inclusion bodies in an immunofluorescent assay (IFA) (Cote et al., 1981; Walsh and Hruska, 1983b). The P protein is also associated with the nucleocapsid and has an apparent molecular weight of 31,000 to 37,000 daltons and consists of 241 amino acids. It is phosphorylated and relatively acidic (Cash et al., 1979; Dubovi, 1982; Huang et al., 1985; Lambden, 1985; Levine, 1977; Satake et al., 1984; Wunner and Pringle, 1976). Monoclonal antibodies against the P protein stain intracytoplasmic inclusion bodies (Gimenez et al., 1984; Walsh and Hruska, 1983b). When the gene that encodes the P protein of the A2 and Edinburgh strains of RSV were compared, only 18 changes were noted in the 726-base coding sequence. These changes predicted amino acid changes at positions 73 and 171 and a short second open reading frame in the Edinburgh but not the A2 strain (Lambden, 1985; Satake et al., 1984). The second open reading frame in the Edinburgh strain could encode a protein of 65 amino acids. The M protein, which is associated with the surface of the virion, has a molecular weight of 27,000 to 28,000 daltons and is composed of 256 amino acids (Cash et al., 1979; Huang et al., 1985; Levine, 1977; Satake and Venkatesan, 1984; Wunner and Pringle, 1976). From the deduced amino acid composition, M should be hydrophobic and basic. A second open reading frame begins at the 3′ end of the messenger RNA and could encode a protein of 75 amino acids.

The 1A protein has a molecular weight of 9,500 daltons and is composed of 64 amino acids (Collins

and Wertz, 1985a; Collins et al., 1984a; Huang et al., 1985). Low levels of the 1A protein have been found in purified virions, which suggests that it is a structural protein, but its location and function in the virion is unknown.

The G protein is one of two glycosylated proteins, has a molecular weight of 79,000 to 92,000 daltons, is composed of 298 amino acids, and is located on the surface of the virion (Bernstein and Hruska, 1981; Cash et al., 1979; Fernie and Gerin, 1982; Huang et al., 1985; Levine, 1977; Satake et al., 1985; Wertz et al., 1985; Wunner and Pringle, 1976). The difference between the molecular weight of its 298 amino acids and that obtained by gel electrophoresis (> 50,000 daltons) probably represents a high rate of glycosylation of its amino acids (Fernie et al., 1985; Gruber and Levine, 1985a; Satake et al., 1985; Wertz et al., 1985). Threonine and serine, which can serve as sites for O-linkage of oligosaccharides, constitute 30% of the amino acids of the G protein and studies with inhibitors of glycosylation suggest a high content of O-linked oligosaccharides and a lower content of N-linked oligosaccharides. Monoclonal antibodies against the G protein give a surface pattern by IFA (Fernie et al., 1982; Walsh and Hruska, 1983b). Since trypsin removes the G protein, it probably projects from the surface of the virion as a component of the spikes seen by electron microscopy (Gruber and Levine, 1983; Peeples and Levine, 1979; Ueba, 1980). Purified G protein has been shown to attach to cell surfaces, indicating that it may have a role in attachment of the virion to the cell surface analogous to the hemagglutinin protein of the parainfluenza viruses (Fernie and Gerin, 1982; Walsh et al., 1984b). Only a few monoclonal antibodies against G have been shown to neutralize the virus in vitro and complement was required to achieve neutralization (Anderson et al., 1986; Fernie et al., 1982; Taylor et al., 1984a; Walsh and Hruska, 1983b). However, polyclonal serum against purified G protein neutralized RSV without the addition of complement (Walsh et al., 1984b), and monoclonal antibodies against G when administered to cotton rats or BALB/c mice before challenge significantly reduced the titer of RSV in the lung (Taylor et al., 1984a; Walsh et al., 1984b).

The F protein is the other glycosylated protein. It has a molecular weight of 66,000 to 70,000 daltons, consists of 574 amino acids, and is associated with the surface of the virion (Bernstein and Hruska, 1981; Cash et al., 1979; Collins et al., 1984b; Elango et al., 1985; Fernie and Gerin, 1982; Gruber and Levine, 1983; 1985a,b; Huang et al., 1985a; Lambert and Pons, 1983). The initial translation product of the F protein is glycosylated with N-linked oligosaccharides, sulfated, and then cleaved into the disulfide-linked proteins F1 (molecular weight of 43,000 to 56,000 daltons) and F2 (molecular weight of 17,000 to 26,000 daltons). It is not yet clear whether the fully processed F protein requires further cleavage by cellular enzymes for its fusion activity, as has been reported for the F protein of the parainfluenza viruses (Dubovi et al., 1983; Spring and Tolpin, 1983). Monoclonal antibodies against the F protein give a surface pattern of fluorescence (Fernie et al., 1982; Walsh and Hruska, 1983b). The F protein is probably a component of the spikes that project from the surface of the virion (Peeples and Levine, 1979; Ueba, 1980). Polyclonal serum against purified F protein and monoclonal antibodies against F inhibit cell fusion, neutralize RSV in vitro, and reduce the titer of virus in the lungs of experimentally infected BALB/c mice and cotton rats (Anderson et al, 1986; Samson et al., 1986; Taylor et al., 1984; Walsh and Hruska, 1983b; Walsh et al., 1985).

The 22K protein has a molecular weight of 22,000 to 25,000 and is composed of 194 amino acids (Dubovi et al., 1982; Elango et al., 1985b; Huang et al., 1985; Peeples and Levine, 1979; Pringle et al., 1981; Wunner and Pringle, 1976). Its location on the virion has not been clearly determined; one report suggests it is associated with the nucleocapsid (Peeples and Levine, 1979), and another suggests it is associated with the envelope (Huang et al., 1985). A second open reading frame was identified at the 3' end of the messenger RNA encoding the 22K protein which is capable of encoding for a peptide of 90 amino acids.

The last messenger RNA encoded by the genome has a molecular weight of 2,500,000 and is the only one large enough to encode the L protein. The L protein is associated with the nucleocapsid and has a molecular weight as determined by gel electrophoresis of 160,000 to 200,000 daltons (Cash et al., 1977; Collins et al., 1984a; Fernie and Gerin, 1982; Huang et al., 1985; Peeples and Levine, 1979; Pringle et al., 1981). Its size and association with the nucleocapsid

suggest it is the viral, RNA-dependent RNA polymerase.

Strains

Shortly after its discovery in 1956, differences in isolates of RSV were identified by neutralization studies (Beem, 1967; Coates and Chanock, 1962; Coates et al., 1963, 1966; Doggett and Taylor-Robinson, 1965; Hierholzer and Hirsch, 1979; Prince et al., 1985c; Suto et al., 1965; Wulff et al., 1964). Differences between strains have also been identified from the electrophoretic profiles of some viral proteins (Cash et al., 1977; Gimenez et al., 1986), the peptide maps of the N protein (Ward et al., 1984), and the reaction of monoclonal antibodies against the N and P proteins (Gimenez et al., 1984; 1986; Ward et al., 1984). Recently, two major groups of RSV strains with antigenic differences on the N, G, and F proteins were identified with monoclonal antibodies (Anderson et al., 1985b). Group 1 is represented by the Long strain (Chanock et al., 1957), and group 2 by the 18537 strain (Chanock et al., 1963). Isolates from these two groups, recovered over more than 20 years from several continents, have relatively stable reaction patterns when tested against a panel of monoclonal antibodies (Anderson et al., 1985b). Subsequent studies using other panels of monoclonal antibodies have also identified two groups of RSV strains (designated by some as subtype A and B) (Akerlind and Norrby, 1986; Gimenez et al., 1986; Hendry et al., 1986c; Mufson et al., 1985). Results for the same strains tested in several studies suggest that the different panels of monoclonal antibodies identify the same two groups and that group 1 equals subtype A and group 2 equals subtype B; but further study is needed to confirm the identity of these two groupings. Antigenic differences between the two groups are greatest on the G protein, where differences were noted on four of five epitopes in one study and five of six in another (Anderson et al., 1986; Mufson et al., 1985). On the F, N, P, M and 22K proteins, 2 of 5, 1 of 6, 1 of 2, 1 of 6, and 1 of 2 epitopes respectively, were different (Akerlind and Norrby, 1986; Anderson et al., 1986; Gimenez et al., 1986; Mufson et al., 1985). The clinical and epidemiologic importance of the antigenic differences between the two groups has yet to be determined. Studies in Boston have shown that both groups are commonly isolated and that they can be isolated in the same community simultaneously but with different epidemiologic patterns (Hendry et al., 1987). In immunized animals, three- to 32-fold differences in neutralizing antibody titers induced by the two groups of isolates have been reported (Coates and Chanock, 1962; Coates et al., 1963; Doggett and Taylor-Robinson, 1965; Prince et al., 1985c).

The human antibody response appears to be broad enough to neutralize representatives from both groups (Beem, 1967; Suto et al., 1965), but it has not been determined if differences in the titer of neutralizing antibody induced by the two groups provides different levels of protection from lower respiratory tract disease. In one study, group 1 infections were less effective at inducing anti–G protein antibodies against group 2 strains than against group 1 strains, and vice versa (Hendry et al., 1986a).

In addition to the two major groups of RSV strains, a number of antigenically distinct variants within the groups can be identified by monoclonal antibodies (Anderson et al., 1985b; Finger et al., 1987; Hendry et al., 1986c). Identification of variants can be used to study patterns of spread of RSV and the effectiveness of infection control practices.

Pathogenesis and Pathology

From studies conducted during outbreaks and with adult volunteer studies, virus is shed from the upper respiratory tract starting 3 to 5 days after infection and persisting for an average of 7 to 8 days after onset of illness in infants and for a shorter time in adults (Frank et al., 1981; Hall et al., 1975a, 1976a, 1983a,b; Johnson et al., 1961; Sterner et al., 1966; Taber et al., 1983). Involvement of the lower respiratory tract usually follows onset of upper respiratory tract symptoms by several days. From studies of a limited number of patients, the two major lower respiratory tract syndromes, pneumonia and bronchiolitis, appear to have different pathologic features (Aherne et al., 1970). In patients who died from bronchiolitis, most of the pathologic changes were found in the bronchioles, where there was necrosis of the bronchiolar epithelium, round-cell infiltration, and edema of the submucosal and adventitial tissues. Mucus and cellular debris had collected in the lumen of bronchioles and presumably obstructed the lumen, which led to air trapping. The alveoli were usually spared, and only a few cells were positive for RSV by IFA. In patients who died from RSV pneumonia, the pathologic changes were more diffuse and included mononuclear infiltration of the subepithelial tissues of the bronchi, bronchioles, and interalveolar walls. In extensive disease there was epithelial cell necrosis, with sloughed cells present in the bronchioles and alveoli. In contrast to bronchiolitis, numerous cells were positive for RSV by IFA. Organ systems outside the respiratory tract have become infected in persons with compromised immune systems (Fishaut et al., 1980; Johnson et al., 1982) and RSV antigen has been found in circulating mononuclear leukocytes in normal persons with RSV infection (Domurat et al., 1985).

Several features of RSV disease suggest that its pathogenesis is different from that of other respiratory viruses (McIntosh and Fishaut, 1980). For example, bronchiolitis and pneumonia represent two ends of a spectrum of disease. In pneumonia the pathology can be explained by the cytopathic effect of the virus, but in bronchiolitis it can not and other factors are probably involved. Other distinctive features of RSV disease include (1) the peak incidence of disease in children 2 to 8 months of age despite the presence of neutralizing maternal antibody (Parrott et al., 1973), (2) the low incidence of disease in infants < 1 month old (Parrott et al., 1973), (3) the observation that natural infection neither prevents reinfection nor necessarily alters the severity of subsequent infection (Fernald et al., 1983; Glezen et al., 1986; Henderson et al., 1979); and (4) that RSV bronchiolitis has clinical features similar to asthma, an allergic disease. Furthermore, children <2 years of age vaccinated with a Formalin-inactivated RSV vaccine experienced more severe disease following a natural RSV infection than did the unvaccinated controls (Chin et al., 1969; Fulginiti et al., 1969; Kapikian et al., 1969; Kim et al., 1969b). These observations have led to speculation that the host's immune response not only provides protection from disease but in some instances enhances the disease process. Thus to understand the pathogenesis of RSV disease and develop a vaccine, we must understand the host's immune response to infection (Chanock et al., 1970).

Humoral Immunity

There is increasing evidence that humoral antibody contributes to protection from lower but not upper respiratory tract infection. High titers of neutralizing or IgG antibody in infants correlate with protection against lower respiratory tract disease, although in some instances lower respiratory tract disease can occur despite high titers of antibody (Glezen et al., 1981; 1986; Lamprecht et al., 1976; Ogilvie et al., 1981). In one study, the rate and severity of recurrent infections in children decreased as the titer of neutralizing antibody increased (Fernald et al., 1983).

Animal studies support the importance of humoral antibody in protecting against RSV lower respiratory tract disease. In cotton rats, neutralizing antibody, acquired either actively by immunization or passively by maternal transfer or injection, leads to a decrease in the titer of virus in the lung postchallenge, and the higher the titer of neutralizing antibody the greater the decrease in virus (Prince et al., 1983; 1985a,b; Wong and Ogra, 1986). In these studies, antibody did not protect from infection in nasal tissues. Similarly in adult humans the titer of serum-neutralizing antibody did not correlate with protection from upper respiratory tract disease (Johnson et al., 1961; Mills et al., 1971).

In recent years, groups have attempted to determine what proteins of the virus are most important in inducing protective antibodies. Immunoprecipitation tests show that infants develop antibodies against the G, F, N, and M proteins after infection (Vainionpaa et al., 1985; Ward et al., 1983). The response to G was weakest. In other studies, infants < 8 months old developed poorer anti-F, anti-G, and neutralizing antibody responses than did children >8 months of age, and the decreased response to F appeared to be related to age, whereas that to G appeared to be related to the presence of maternal anti-RSV antibodies (Murphy et al., 1986a,b). It is possible that some of the differences in the response to the G protein could be accounted for by differences in the infecting strain. Wagner et al. (1986) determined that after primary infection the anti-F immunoglobulin response was equally divided between the IgG1 and IgG3 antibodies and that the anti-G immunoglobulin response was composed mostly of the IgG3 antibodies plus some IgG1. In adults, there is a high IgG2 antibody response to the G protein consistent with a polysacharide antigen as the immunogen (Wagner et al., 1987). Animal studies with recombinant vaccinia viruses expressing the F and G proteins demonstrated that the highest titer of neutralizing antibody was elicited by an F recombinant which also conferred the greatest degree of protection from challenge with RSV (Ball et al., 1986; Elango et al., 1986; Olisted et al., 1986; Stott et al., 1986; Wertz et al., 1987). These studies suggest that F is more important than G in inducing a protective immune response but do not distinguish between the contributions of humoral antibody, local antibody, and cell mediated immunity.

Local Immunity

The role that local antibody plays in RSV disease is less clear than that of humoral antibody. Neutralizing activity and IgA, IgG, and IgM has been found in the acute-phase respiratory tract secretions of infants hospitalized with pneumonia or bronchiolitis (Kim et al., 1969a; McIntosh et al., 1978; 1979; Scott and Gardner, 1970; 1974). The most abundant immunoglobin is IgA, but IgG and IgM antibodies can also be found in these secretions, and all three can be found on RSV-infected cells before they are found free in secretions (McIntosh et al., 1979). However, McIntosh et al. (1978) found that neutralizing activity was not correlated with levels of IgA in the secretions and concluded that at least some of the neutralizing activity in secretions is not attributable to antibodies. The appearance of IgA, but not the presence of neutralizing activity, was temporally correlated with clearance of the virus from the secretions. Neutralizing activity in respiratory secretions before infection did correlate with decreased replication of virus in an adult volunteer study (Mills et al., 1971).

Several groups have postulated that local antibody might participate in the disease process through the formation of immune complexes that stimulate the local inflammatory process, leading to additional tissue damage. Several observations support this hypothesis: (1) complement components have been identified on RSV-infected cells in respiratory secretions; (2) RSV-infected cells can activate complement in vitro; (3) RSV antibody-antigen complexes can activate neutrophils and induce the release of mediators of inflammation in vitro; and (4) neutrophils attach better to RSV-infected cells in the presence of immune serum and release mediators of inflammation after attachment (Edwards et al., 1986; Faden et al., 1983, 1984; Kaul et al., 1982a,c; Smith et al., 1981). It has also been postulated that local anti-RSV IgE antibody develops after a priming infection, and the anti-RSV IgE causes a local type I hypersensitivity reaction after a second infection, leading to bronchiolitis (Gardner et al., 1970a). Epidemiologic data, however, suggest that prior infection is not a significant predisposing factor for bronchiolitis (Brandt et al., 1973), but some studies suggest that IgE may be important. IgE has been found on RSV-infected nasopharyngeal cells, and it persists longest in patients with RSV-associated bronchiolitis or asthma. Cell-free IgE and histamine have been found more often in RSV-infected patients with wheezing than in those without wheezing and RSV IgE has been found in secretions but not in serum (Welliver et al., 1980a, 1981, 1985). Although the studies on local antibody have provided some interesting results, their role in the pathogenesis of RSV disease is unknown.

Interferon

Interferon has also been studied in relation to RSV infections. Hall et al. (1978a) reported that fewer children with RSV than with influenza or parainfluenza infection had interferon activity in their respiratory tract secretions, and McIntosh et al. (1978) found interferon in low concentrations in respiratory tract secretions with RSV infection. It is possible that these studies found low levels of interferon with RSV infection because they measured the wrong type of interferon (Bell et al., 1983a).

Cell-Mediated Immunity

Cell-mediated immunity in RSV infections is the least understood component of the immune response. Several measures of cell-mediated immunity have been studied in humans following RSV infection, including antibody-dependent cellular cytotoxicity, lymphocyte transformation, and cytotoxic T-cell activity (Cranage and Gardner, 1980; Paul et al., 1982b; Kim et al., 1976; Meguro et al., 1979; Schauf et al., 1979; Scott et al., 1977, 1978, 1981, 1984; Welliver et al., 1979). These studies suggest a role for cell-mediated immunity in not only resolving infection but also in contributing to the disease. Recent studies with recombinant vaccinia viruses have demonstrated that target cells infected with vaccinia N or F but not G were recognized by human and mouse cytotoxic T cells (Bangham and McMichael, 1986; Bangham et al., 1986). In animal studies, evidence of antibody-dependent cellular cytotoxicity, lymphocyte transformation, and cytoxic T-cell activity has been demonstrated in association with RSV infection and immunization (Bangham and Askonas, 1986; Bangham et al., 1985; Koff et al., 1983; Sun et al., 1983a,b). In these studies, cytotoxic T-cell activity was temporally associated with clearance of virus, and cytotoxic T cells reacted with both homologous and heterologous strains of RSV. In mice, monoclonal antibodies against the F and G protein given before challenge inhibited the cytotoxic T-cell response to RSV infection, an effect that could be transferred to offspring (Bangham, 1986). Although these studies have not defined the role of cell-mediated immunity in the pathogenesis of RSV infection, observations of RSV infection in patients with altered immune systems suggest that cell-mediated immunity is important in clearing infection. Patients with immune deficiencies associated with factors ranging from steroid therapy to severe combined immune deficiency syndrome shed virus longer and experience more severe illness and a higher mortality rate than control patients (Craft et al., 1979; Delage et al., 1984; Fishaut et al., 1980; Hall et al., 1986; Milner et al., 1985; Spelman and Stanley, 1983). Studies of immunosuppressed cotton rats and cattle have similarly demonstrated prolonged excretion of virus and increased severity of disease (Johnson et al., 1982; Thomas et al., 1984a; Wong et al., 1985).

Recent serologic studies in cotton rats and humans administered the formalin-inactivated vaccine have shed some light on why vaccine recipients experienced more severe disease with natural infection than unvaccinated persons. The lessons from these trials are not necessarily applicable to natural infection, though. Animals challenged with RSV after receiving a Formalin-inactivated vaccine similar to the one used in the clinical trials had an increase in the number of neutrophils and lymphocytes in the lung, had lower titers of neutralizing antibody, and had similar titers of anti-F and anti-G IgG antibodies than controls (Prince et al., 1986). Vaccinated children much higher ratio of anti-F to neutralizing antibodies bodies than those who were naturally infected with RSV (Murphy et al., 1986c). Formalin inactivation of the virus appeared to selectively alter the epitopes responsible for the development of neutralizing RSV antibodies which led to an immune response that contributed to the disease. Similar findings have been noted with Formalin-inactivated measles vaccine (Fulginiti et al., 1967; Norrby et al., 1975). An increased cell-mediated immune response, as measured by lymphocyte transformation, was reported in recipients of the inactivated vaccine but the contribution of this response to the disease is not clear (Kim et al., 1976).

Clinical Characteristics of Infection

Volunteer studies and outbreak investigations have demonstrated that illness follows infection by 2 to 5 days (Johnson et al., 1961; Sterner et al., 1966; Wright et al., 1971). Most infections lead to illness that can range from a mild upper respiratory illness to a life-threatening lower respiratory tract illness (Chanock et al., 1961; Cooney et al., 1975; Hall et al., 1976b; Johnson et al., 1961; Kim et al., 1973b; McClelland et al., 1961). Severe disease is most common in infants and young children, among whom it is the single most important lower respiratory tract pathogen. In this age group, illness is characterized by fever (often over 102°F), cough, a rhinitis in 70 to 90% of cases, pharyngitis in about 50%, and gastrointestinal symptoms, conjunctivitis, or otitis media in 10 to 30% (Hall et al., 1976b; Moss et al., 1963; Reilly et al., 1961). From population-based studies 10 to 40% of RSV-infected children develop lower respiratory tract disease (Beem, 1969; Glezen et al., 1986; Hall et al., 1976b; Henderson et al., 1979; Lee et al., 1973; Loda et al., 1972; Kapikian et al., 1961; 1969; Sterner et al., 1966), and between 1 in 50 and 1 in 200 children <1 year old are hospitalized with RSV lower respiratory tract disease each year (Belshe et al., 1983; Glezen et al., 1986; Kim et al., 1973a; Martin et al., 1978; Sims et al., 1976). Most children have an acute respiratory illness for several days before developing the severe tachypnea and wheezing that lead to hospitalization. At the time of hospitalization, they are often hypoxemic and may have become afebrile. Most children are well enough to be discharged by 7 days after admission (Hall et al., 1975a, 1979a; Parrott et al., 1973; Reilly et al., 1961).

The clinical illness associated with RSV is age-dependent. Infants < 1 month old are less likely than those > 1 month old to have lower respiratory tract disease and more likely to have nonspecific symptoms or apnea (Hall et al., 1979b; Neligan et al., 1970). Older children and adults usually have symptoms of a common cold, which can be moderately severe, but they uncommonly develop pneumonia (Cooney et al., 1975; Hall et al., 1975b, 1978; Kravetz et al., 1961; Monto and Lim, 1971). When exposed to RSV by infected family members at home or infected children in the hospital, 30 to 40% of adults may become infected. In adults, especially the elderly, RSV can reappear as a significant lower respiratory tract pathogen and has been associated with outbreaks of serious respiratory illness in nursing homes (Fransen et al., 1967; Hart, 1984; Mather et al., 1980; Morales et al., 1983) and among older children and adults in an institution for the mentally retarded (Finger et al., 1987).

RSV is an important pathogen of otitis media in young children and can infect the middle ear directly or lead to predisposition to bacterial superinfection (Chonmaitree et al., 1986; Henderson et al., 1982; Klein et al., 1982; Sarkkinen et al., 1985). It may also lead to predisposition to chronic lung disease. Several studies demonstrated that children who have had RSV bronchiolitis are more likely to wheeze with subsequent respiratory infections and have residual pulmonary function abnormalities 8 to 10 years after their episode of bronchiolitis than those who have not (Hall et al., 1984; Mok and Simpson, 1982; Pullan and Hey, 1982; Sims et al., 1978). It is unclear if the children who had RSV bronchiolitis were predisposed to developing bronchiolitis and later pulmonary abnormalities or if the RSV infection itself leads to the later pulmonary abnormalities. One study found an association between history of childhood respiratory trouble and chronic lung disease in adults (Burrows et al., 1977). RSV has also been associated with apnea, meningitis and myelitis, myocarditis, rash illnesses, and sudden infant death syndrome (Anas et al., 1982; Bairan et al., 1974; Berkovich and Kibrick, 1964; Bruhn et al., 1977. Cappel et al., 1975; Giles and Gohd 1976; Wallace and Zealley, 1970).

Diagnosis

Virus Detection

SPECIMEN COLLECTION

The first step in performing an adequate diagnostic test is the collection of an appropriate specimen. RSV is detected in respiratory tract secretions, and two types of specimens— nasopharyngeal aspirates and nasal washes—give good results. Several studies have demonstrated that nasal wash or nasopharyngeal aspirate specimens are 2 to 3 times more sensitive than swab specimens for detecting RSV (Hall and Douglas, 1975; McIntosh et al., 1982; Sturdy et al., 1969; Truehaft et al., 1985).

The nasal wash is collected by placing a 1-oz ta-

pered rubber bulb with 5 to 7 ml of sterile saline solution into and occluding the nostril and then squeezing and releasing the bulb (Hall and Douglas, 1975). The nasal wash contents in the bulb are emptied into a container with 2 ml of transport medium. A nasopharyngeal aspirate is collected with a No. 5 to No. 8 French pediatric feeding tube attached to a collection trap and suction. The tube is placed through the nostril into the nasopharynx, and suction is applied with < 26 lb/in^2 vacuum (Gardner and McQuillin, 1980). This should be repeated with the other nostril. If other sources of suction are not available a hand-held manual suction pump (Nalge Company, Rochester, NY) works well. Swab specimens should be collected by thoroughly swabbing the nostril or nasopharynx and breaking the swab into a container containing transport medium.

The transport medium can be any of several physiologic salt solutions—for example, phosphate-buffered saline (PBS) with a protein stabilizer such as 0.5% gelatin or 0.5% bovine serum albumin (Hanks balanced salts solution has been reported to decrease the absorbance in one RSV EIA [Hendry and McIntosh, 1982]). Once collected the specimen should immediately be placed on wet ice for transport to the laboratory. A recent study demonstrated that bedside inoculation of a specimen onto tissue culture does not improve isolation rates over immediately transporting the specimen on ice to the laboratory (Bromberg et al., 1984). Specimens for isolation and immunofluorescent studies should be transported at 4°C, but specimens for enzyme immunoassays (EIA) can be transported at room temperature. Secretions remaining in the tube of the nasopharyngeal aspirate may need to be suctioned into the trap with 1 to 2 ml of sterile PBS.

Since the titer of RSV and other respiratory viruses decreases with freezing, thawing, and increased time of storage at 4°C, it is best to inoculate cells for isolation as soon as possible. If the specimen is to be stored for more than 3 or 4 days before inoculation, it is advisable to quick-freeze it and store it at −70°C. Long-term storage of specimens at −20°C can decrease their reactivity by EIA as well as their titer of infectious virus. Further preparation of the specimen is test-dependent and will be discussed in the following sections. All clinical specimens should be considered infectious for a variety of agents, not just those of diagnostic importance. Therefore, it is appropriate to handle all specimens as if they contained hepatitis B virus or human immune deficiency virus (HIV).

Direct Detection

Little has been written about direct detection of RSV by electron microscopy (Doane et al., 1969; Joncas et al., 1969b). Therefore, it is not possible to compare the sensitivity and specificity of this method with isolation or rapid antigen detection methods. However, the similarity in morphology between RSV and the parainfluenza viruses, the amount of cellular debris present in respiratory secretions (which can be confused with viruslike particles), and the low titer of virus in some specimens make this technique impractical compared to other rapid diagnostic tests.

Immunofluorescence Assay (IFA)

Detection of RSV antigens in tissue culture by IFA was first reported in 1962 (Bennett and Hamre, 1962; Kisch et al., 1962). IFA was then used for rapid identification of isolates of RSV by Schieble et al. (1965) and for direct detection of RSV in respiratory secretions by two groups in England in 1968 (Gray et al., 1968; McQuillin and Gardner, 1968). In these early studies and in most of the subsequent studies, 80 to 95% of specimens that were positive by isolation were also positive by IFA, and 5 to 10% of specimens that were negative by isolation were positive by IFA (Anestad et al., 1983; Fulton and Middleton, 1974; Gardner and McQuillin, 1980; Kaul et al., 1978; Lauer, 1982; Minnich and Ray, 1980, 1982; Treuhaft et al., 1985).

Gardner et al. (1970b) examined sequential specimens from patients whose initial specimen was positive by isolation and IFA and found that specimens collected 4 or more days after the initial positive specimen often gave fluorescence that was less bright, had human immunoglobins coating RSV-infected cells, and were more likely to be positive by IFA than by isolation. These results suggest that virus may be present but neutralized by antibody and that isolation should not necessarily be considered the most accurate test. Thus some IFA-positive, isolation-negative specimens are probably true positives. Isolation is also more sensitive to specimen handling; thus some IFA-positive, isolation-negative specimens can result from improper handling or storage. Several groups have recently compared EIA with IFA and found that they give similar results, though IFA tended to be slightly more sensitive than EIA (Chao et al., 1979; Flanders et al., 1986; Grandien et al., 1985; Korppi et al., 1986; McIntosh et al., 1982; Popow-Kraupp et al., 1986; Sarkkinen et al., 1981; Swenson and Kaplan, 1986).

Although the sensitivity of IFA has not been improved since the early studies in the 1960s, the availability and quality of reagents have greatly improved. Initially, polyclonal sera were used for IFA but now monoclonal antibodies have been developed, and some have been incorporated into diagnostic kits. Monoclonal antibodies can be chosen that will react with all the known RSV strains, and several groups have reported very good results for RSV IFA in clinical specimens (Bell et al., 1983b; Cheeseman et al., 1986; Freke et al., 1986; Freymuth et al., 1986; Kao et al., 1984; Kim et al., 1983; Pothier et al., 1985; Routledge et al., 1985). Monoclonal antibodies will probably become the standard reagent for RSV IFA.

Specimens for immunofluorescence should be prepared by adding 0.5 ml of PBS, vortexing, and centrifuging at low speed for 10 min. The supernatant can then be used to inoculate tissue culture cells for isolation studies. The pellet can be resuspended in 0.5 cc of PBS, centrifuged at low speed for 10 min, resuspended in a few drops of PBS, and spotted onto microscope slides. A spot for each antiserum plus one for a control antiserum should be made for each specimen. If the antisera are of high quality and come from one species of animal, the antiserum for one virus can be used as the control for another virus. Each test should also include RSV-infected and -uninfected cells as positive and negative controls. The cell suspension should be air-dried on the slide and fixed with ice-cold acetone for 10 min.

In the direct test, the spots are overlaid with the fluorescein-conjugated antiserum, incubated in a humidified 37°C incubator for 30 min, soaked three times in PBS for 10 min, and rinsed with distilled water. After air-drying, the slides are covered with a cover slip and mounting medium and viewed under a fluorescent microscope. In the indirect test, the spots are overlaid with the antivirus serum, incubated in a humidified 37°C incubator for 30 min, soaked three times in PBS for 10 min, and air-dried. Next, the spots are overlaid with fluorescein-conjugated antispecies sera, again incubated in a humidified 37°C incubator for 30 min, soaked three times in PBS for 10 min, rinsed with distilled water, and air-dried. After drying, the slides are covered with mounting media and a cover slip and viewed under a fluorescent microscope. Commercial kits should be used according to the manufacturer's instructions.

The first step in reading the slide is to evaluate the quality of the specimen by counting the number of cells. A satisfactory specimen should have about 200 cells on the slide or eight cells per high-power field. RSV-specific fluorescence is cell-associated and is characterized by particulate fluorescence on the cell surface or inclusion bodylike fluorescence in the cell cytoplasm. Monoclonal antibodies against the surface glycoproteins, F and G, give the surface pattern, and monoclonal antibodies against the nucleocapsid proteins, N and P, give the cytoplasmic pattern of fluorescence. Polyclonal sera made against whole virus or a mixture of the appropriate monoclonal antibodies gives a combination of the two patterns. Interpretation of IFA in clinical specimens requires a skilled microscopist (see also Chapter 4).

Enzyme Immunoassay (EIA)

Since the development of EIA in the early 1970s, several groups have applied EIA to the detection of RSV antigens in clinical specimens. The EIA has given results comparable to those with IFA, though some investigators have found it slightly less sensitive than IFA (Chao et al., 1979; Flanders et al., 1986; Grandien et al., 1985; McIntosh et al., 1982; Popow-Kraupp et al., 1986; Sarkkinen et al., 1981; Zrein et al., 1986). As with IFA, polyclonal serum was used initially, and monoclonal antibodies have recently been developed and incorporated into commercial EIA kits (Bromberg et al., 1985; Lauer et al., 1985; Swenson and Kaplan, 1986). The principal advantages of EIA over IFA are (1) it is easier to automate, (2) the results can be quantitated and require less skill to interpret, (3) it requires less technical skill to perform, and (4) specimens can be transported under less stringent conditions. One group demonstrated no loss in EIA activity when specimens were shipped by mail at room temperature or when they underwent a freeze-thaw (Hendry and McIntosh, 1982; McIntosh et al., 1982). One disadvantage of the EIA when compared with IFA is that the quality of the specimen cannot be evaluated as it can with IFA by counting the number of cells on the slide.

Several methods have been used to prepare specimens for EIA. The first step is to break up the mucus by aspirating the specimen up and down in a Pasteur pipette, sonicating the specimen, vortexing the specimen, or adding N-acetyl cysteine to the specimen. Since N-acetyl cysteine may decrease the ability of the G protein to react in EIA, one of the other procedures is probably better (M. Hendry, personnal communication). Sonication with a detergent has been reported to increase the EIA signal, presumably by releasing antigens from the cells or by dispersing the antigens and increasing the likelihood that they will react with the capture antibody (Hendry et al., 1986; Meurman et al., 1984b; Sarkkinen et al., 1981). Although sonication may increase the EIA signal, it does not necessarily increase the number of specimens that are positive. Whole secretions give higher EIA absorbance than do either washed cells or secretions from which the cells have been removed by centrifugation (Hendry et al., 1986b; Popow-Kraupp et al., 1986; Sarkkinen et al., 1981). One of several practical methods for preparing specimens is to dilute the original specimen 1:1 in PBS with 0.5% gelatin and 0.15% Tween 20 (PBS-GT) and vortex the specimen before testing by EIA.

Most RSV EIAs have been constructed as double-antibody sandwich assays. The capture antibody is immunoglobulin that is purified from serum by ammonium sulfate precipitation (Hebert, 1974; Hebert et al., 1973). The ammonium sulfate precipitation increases the amount of the desired antibody in the protein that is adsorbed to the solid phase. An EIA format that has worked well in our laboratory is constructed as outlined below. The capture antibody is diluted in 0.01 M carbonate buffer at pH 9.6 (others have used PBS at pH 7.2) and adsorbed to the solid phase by incubating overnight at 4°C. Several brands of 96-well microtiter plates have been shown to give good results as the solid phase (Shekarchi et al., 1984). Twelve-well microtiter strips can be also be used, and they make it easier to adapt the test to varying numbers of specimens. Next, the capture antibody is aspirated from the wells, and PBS with 0.5% gelatin is added to the wells and incubated for 30 min; after washing with PBS plus 0.05% Tween 20 (PBS-T), 0.075 ml of the specimen is added and incubated for 1.5 h at 37°C (some groups have reported an increase in sensitivity by incubating overnight at room temperature). After washing, the detector antibody, diluted in PBS-T with 0.5% gelatin (PBS-GT), is added and incubated for 1.5 h at 37°C; after washing, 0.075 ml of peroxidase-conjugated, antispecies antibody (the species of the detector antibody) diluted in PBS-GT is added and incubated for 1 h at 37°C; and after washing, 0.125 ml of substrate, 0.4 mg/ml O-phenylenediamine dihydrochloride, and 0.15% hydrogen peroxidase in citrate phosphate buffer at pH 5.5 is added and incubated for 30 to 45 min. The reaction is stopped with 0.025 ml of 3.5 M HCl, and absorbance is read at 490 nm. O-phenylenediamine is a potential carcinogen and therefore should be handled with caution. A variety of other enzymes and substrates can be used (Kurstak, 1985).

To minimize the potential for false-negative or false-positive results, a positive and negative test control (RSV and cell control antigen, respectively) and a negative specimen control (the specimen reacted against control serum as either capture or detector antibody) should be included. If the specimen is tested for other viruses by comparably constructed EIAs with high-quality antiserum from the same species, the EIA for one virus can be used as the specimen control for the other. A variety of methods have been used to determine negative and positive cutoff values. Some groups have used 3 standard deviations above the mean of 6 to 10 negative control specimens as a cutoff value. Others have used a P/N value (where P is the specimen against the RSV antiserum and N is the specimen against control serum) that is 3 standard deviations above the mean P/N value for negative-control specimens. The cutoff value needs to be independently determined for each EIA. An example of EIA results for clinical specimens is given in Table 2.

The quality of the EIA depends on the quality of the reagents. Good enzyme-conjugated antispecies

TABLE 2. Example of data for RSV antigen detection EIA[a]

Specimen[b]	Abs[c]	P–N	P/N
1	0.331	0.309	15.05
2	1.465	1.421	33.30
3	0.093	0.078	6.20
4	0.560	0.545	37.33
5	0.037	0.011	1.42
6	0.051	0.017	1.50
7	0.023	0.007	1.44
8	0.016	−0.009	0.64

[a] The EIA is an indirect antigen capture EIA as described by Hendry et al. (1982).
[b] Specimens 1 to 4 are positive, and specimens 5 to 8 are negative.
[c] Abs = Absorbance at 490 nm. P is the absorbance for the specimen reacted against bovine anti-RSV serum, and N is the absorbance for the specimen reacted against control bovine serum. A specimen with a P–N > 0.05 and P/N > 2.5 was considered positive These values were ≥ 3 SD above the respective means for negative specimens.

antibodies are available from several commercial sources. However, satisfactory antisera for the capture and detector antibody are not as readily available. Good results can be achieved with antiserum prepared from purified nucleocapsid (Sarkkinen et al., 1981) or from proteins purified by crossed immunoelectrophoresis (Hornsleth et al., 1981b) or from antiserum available commercially (Hendry and McIntosh, 1982). To develop a good antiserum for EIA, purified virus or viral proteins should be used, and purification techniques have been described in the literature (Fernie and Gerin, 1980; Hornsleth et al., 1982; Levine, 1977; Sarkkinen et al., 1981; Senterfit and Baldridge, 1974; Ueba, 1978; Wunner and Pringle, 1976).

There are several variations on the RSV EIA described above. Hornsleth et al. (1981a) developed a competition EIA and later reported the use of a conjugated detector antibody in a direct EIA (Hornsleth et al., 1982). The same group also biotinylated the detector antibody and used avidin peroxidase as the conjugate and found this EIA to be more sensitive than the direct EIA (Hornsleth et al., 1986). Since biotinylation of purified immunoglobulin can be accomplished with ease, it is an attractive method for labeling antibodies, including monoclonal antibodies (Anderson et al., 1986).

Radioimmunoassay (RIA) and Other Immunoassays

RIA has also been used to detect RSV (Ehrlicher et al., 1984; Sarkkinen et al., 1981). The tests are con-

structed in a fashion similar to the EIAs except that polystyrene beads are used for the solid phase, ^{125}I-labeled antispecies antibody is used in place of the conjugate, and radioactivity is measured instead of absorbance. The sensitivity and specificity of RIA are similar to those of EIA. Time-resolved fluoroimmunoassay (TR-FIA) has also been used to detect RSV antigens and is reported to detect a lower concentration of RSV antigens but not a higher percentage of positive clinical specimens than EIA in one study (Halonen et al., 1985). In the TR-FIA, both the capture and the detector antibodies are monoclonal antibodies. The detector antibody is labeled with europium, and, after it has reacted with the antigen, it is released into solution to form a fluorescent europium chelate, and fluorescence is measured with a time-lapse fluorometer. Since monoclonal antibodies are used as both the capture and detector antibodies, this assay eliminates the possibility of nonspecific reactions with the enzyme-conjugated antispecies antibody. An all-monoclonal antibody assay similar to the TR-FIA can be constructed for EIA by biotinylating the detector monoclonal antibody. Cranage et al. (1981) described a reverse passive hemagglutination test for RSV antigens, which appeared to give results similar to those from IFA and EIA.

Hybridization Assays

Since antigen detection assays have had such good sensitivity and specificity, there has been little incentive to develop nucleic acid probes for diagnosing RSV infections. Van Dyke (1986) reported preliminary results with a radiolabeled, cDNA probe that is complementary to the N protein gene of RSV and reacts with both groups of RSV strains. The probe technique was less sensitive than IFA and EIA for detecting RSV in clinical specimens, but the specimens had been frozen, and the probe technique might have been more sensitive with fresh specimens. Since the development of nucleic acid probes for RSV is in its infancy, it is not clear what role they will play in diagnosis of RSV infections (see also Chapter 8).

Virus Isolation

A variety of cell lines have been evaluated for the isolation of RSV including HEp-2, primary monkey kidney, HeLa, human diploid fibroblast, BSC-1, CV-1, Vero, and MDCK (Anderson and Beem, 1966; Arens et al., 1986; Beem et al., 1960; Bennett and Hamre, 1962; Holzel et al., 1963; Jordan, 1962; Levine and Hamilton, 1969; Meguro et al., 1979; Trepanier et al., 1980). HEp-2 and HeLa cells have usually been the most sensitive cell lines, but there can be year-to-year variation in the sensitivity of a cell line to isolates. Primary monkey and human diploid fibroblast cell lines have also given good results in some laboratories. RSV grown in human diploid fibroblast cells give a less typical cytopathic effect (CPE), which takes longer to develop. A combination of several lines, for example HEp-2 and primary monkey kidney cells or HEp-2 and human diploid fibroblast cells, will increase isolation rates. There also can be differences in sensitivity between different lines of a particular cell type, a decrease in the sensitivity of a cell type following multiple passages, and differences in the sensitivity of cells grown in different media (Jordan, 1962; Tyrrell, 1963). Glutamine appears to be necessary for optimal growth of virus and development of CPE (Marquez and Hsiung, 1967). To maintain good isolation results, stocks of an RSV-sensitive cell line, at a sensitive passage level, should be frozen with periodic thawing for regrowing. A nonconfluent layer of actively dividing cells is more sensitive and gives better CPE than does a confluent layer (Pons et al., 1983; Treuhaft et al., 1985). The characteristic CPE of RSV is syncytia which increase in size followed by degeneration of the entire cell layer within several days. In human diploid fibroblast cells and in confluent layers of other cell lines, cell degeneration may be more evident than syncytia formation.

Clinical specimens that have not been diluted with transport medium that contains antibiotics should be incubated with a 1:1 dilution of maintenance medium with antibiotics for about 30 min to minimize the chance for bacterial overgrowth and then inoculated onto tissue culture. A 0.5 ml aliquot of the specimen should be added to tubes with the appropriate cell lines and allowed to adsorb to the cells for 1 h, and then 1 ml of maintenance medium, for example Eagle minimum essential medium with 2% heat-inactivated fetal calf serum and antibiotics, is added. Best results are achieved with cell cultures that are 50 to 75% confluent. CPE first becomes evident 3 to 7 days after inoculation for most clinical specimens, but about 10 to 15% of specimens will not develop CPE until a second or third passage. Material can be passaged by scraping the cells into the medium, inoculating 0.5 ml into a fresh tube, adsorbing the material for 1 h, and then adding maintenance medium. Some strains, such as RSV 18537, may develop CPE more slowly and give lower titers of virus than others (Coates et al., 1966).

Virus Identification

Identification of an isolate of RSV can be made presumptively on the basis of the characteristic CPE in a heteroploid cell line and the lack of hemadsorption in primary monkey kidney cells. The identification can be confirmed by IFA, EIA, complement fixation, or neutralization. IFA and EIA are the simplest of these techniques. For IFA, the cells are harvested at 2+ to 3+ CPE (CPE present in 25 to 75% of the cells) by decanting the medium and washing the cells once with PBS, scraping them from the tube, and suspending them in several drops of PBS. The cell suspension is placed onto a microscope slide, air-dried, fixed with ice-cold acetone for 10 min, and tested as outlined in the section above on immunofluorescence assay. For EIA, the cells and medium are harvested at 3+ to 4+ CPE, diluted 1:1 in PBS-GT, and tested as outlined in the section above on enzyme immunoassay.

To type the virus by neutralization, the cells and medium are harvested at 3+ to 4+ CPE, the virus is titrated in an infectivity assay (see below, "Virus Quantitation"), and approximately 100 $TCID_{50}$ of virus is incubated both with an equal volume of 10 to 20 units of heat-inactivated anti-RSV serum and with an equal volume of a comparable dilution of heat-inactivated normal serum. Next, 0.2 ml of each of the serum virus mixtures, containing approximately 100 $TCID_{50}$ of virus and 10 to 20 units of antiserum, is added to two tubes, adsorbed for 1 h, and overlaid with 1 ml of maintenance medium. The tubes are incubated at 36°C until the virus control gives 3+ to 4+ CPE. If the antiserum prevents replication of the virus—that is, blocks the CPE—then the virus is RSV. Since antiserum against group 1 strains may neutralize group 2 strains at a lower titer and vice versa (Coates et al., 1966), some isolates may not be typed as RSV by neutralization tests with only one antiserum.

For complement fixation, the cells and media are harvested at 3+ to 4+ CPE and used undiluted as an unknown antigen against negative and positive anti-RSV serum (see Chapter 5).

Virus Quantitation

Virus grown in tissue culture can be assayed for either the content of antigen or titer of infectious virus. The content of antigen can be determined by testing serial, half-log dilutions of virus and an appropriate control in an antigen-detection EIA as described in the section on EIA. The end point is the last dilution that gives an absorbance above the cutoff value for the EIA.

The titer of infectious virus can be determined by a plaque assay or tissue culture infectivity assay. In the plaque assay, the virus is adsorbed to cells, such as HEp-2, and grown in sterile petri dishes or 6- to 24-well tissue culture plates for 1 h at room temperature. The dishes or plates should be rotated periodically to prevent drying of the cell sheet. Next, the cells are washed three times with medium and then overlaid with a methyl cellulose or agarose solution and placed into a CO_2 incubator at 36°C. The methyl cellulose solution should contain 1% methyl cellulose (4,000 centipoises), maintenance medium, and sufficient $NaHCO_3$ to achieve the desired pH (Coates et al., 1966). A 2× concentration of the methylcellulose solution is made by adding the appropriate amount to near boiling, tissue culture-grade water with continuous stirring and then cooling with continuous stirring. When cooled, the overlay solution is made by adding 2× concentration of maintenance medium with stirring. The agarose solution contains 0.6 to 0.7% agarose, maintenance medium, and sufficient $NaHCO_3$ to achieve the desired pH. The agarose solution is made up at a 2× concentration which is solubilized in boiling, tissue culture-grade water and stored at 4°C until used. To make the overlay solution, the 2× concentration of agarose is heated until it becomes liquid and then is mixed 1:1 with a 2× concentration of maintenance medium at 43 to 45°C. After 3 to 5 days, the overlay is removed, the cells are fixed with 10% Formalin and stained with Giemsa stain or crystal violet, and the plaques are read macroscopically. The titer of infectious virus, in plaque-forming units per volume of inoculum, is equal to the number of plaques divided by the dilution of the inoculum.

In a tissue culture infectivity assay, serial dilutions of the virus are adsorbed to the cells in tubes (2 tubes per dilution) for 1 h, maintenance medium is added, and the tubes or plates are placed in a 36°C incubator. Alternatively, 0.1 ml of serial dilutions of virus (in maintenance media) is placed in wells of a 96-well tissue culture microtiter plate, and approximately 15,000 HEp-2 cells in 0.1 ml of maintenance medium are added. The plate is covered and placed into a 36°C, humidified CO_2 incubator. After 4 to 7 days, the tubes or wells are inspected for CPE, and the titer of virus is calculated by the method of Reed and Muench or Karber. Alternatively, the end point can be determined by EIA performed in the microtiter plate (Anderson et al., 1985a). When 3+ to 4+ CPE becomes evident in the lowest dilution, usually 4 to 7 days after inoculation, the medium is aspirated, the plates are washed, and the cells are fixed with an 80% acetone and 20% PBS mixture. An anti-RSV serum is added and incubated for 1 h; after washing, an enzyme-conjugated, antispecies antibody is added, and the substrate is added and absorbance is read. Absorbance that is 3 standard deviations above that of uninfected cells is indicative of virus replication, and the end point is determined by the method of Reed and Muench or Karber.

Comment

The standard methods to detect RSV infection—isolation, IFA, and EIA—have similar sensitivity and specificity, but each has distinct advantages and disadvantages. Virus isolation may identify viruses not included in antigen detection tests, but it is expensive and time-consuming to perform. IFA is rapid and particularly well suited for laboratories that test a limited number of specimens and have an experienced microscopist to read the test. IFA is easily adapted to varying numbers of specimens. EIA, RIA, and the newly developed TR-FIA are also rapid and can be automated to facilitate large-scale testing. These tests are particularly well suited to large diagnostic laboratories. The quantitative nature of the results simplifies their interpretation and eliminates the requirement for a skilled microscopist. Use of 12-well microtiter strips, instead of 96-well microtiter plates, makes it easier to adapt the test to different numbers of specimens. Monoclonal antibodies should be the reagents of choice, since they can provide an unlimited quantity of standardized, high-quality reagents. Polyclonal sera made from synthetic peptides or proteins expressed from cloned genes could have many of the advantages of monoclonal antibodies.

Antibody Assays

Several serologic tests have been developed to detect a range of RSV antibodies, including class- and subclass-specific antibodies and their protein specificity. When these assays are used to diagnose RSV infections, their sensitivity and specificity depend not only on the test but also on the age of the patient and the timing of specimen collection. For example, the sensitivity of detecting RSV infection serologically is lower in infants < 6 months old than in infants and children ≥ 6 months old (Table 3). EIAs for IgG antibodies appear to be the most sensitive of the antibody tests for diagnosing RSV infection in the ≤ 6-month age group. The time of the antibody response depends on the test (Hamparian et al., 1961; Hornsleth et al., 1984; Meurman et al., 1984a; Welliver et al., 1980b). In primary infection, the IgM antibody response occasionally occurs by 4 and usually by 5 to 10 days after onset of illness, peaks at 10 to 20 days, and persists for as short as 2 weeks or as long as 10 or more weeks. In most persons, the IgM antibody is no longer detected at 10 weeks. In second infections, the IgM antibody response is more rapid, reaches a higher titer, and persists for a longer time. In primary infections, IgG antibody increases by 5 to 10 days after onset of illness, reaches a peak titer by 20 to 30 days, and is often still present but at a lower

TABLE 3. Sensitivity of serologic tests for detecting RSV infection by age

	Percent of RSV-infected patients diagnosed			
Age (months)	IgG antibodies	IgM antibodies	CF[a] antibodies	N[b] antibodies
<6	63–75%	35–62%	15–48%	18–38%
≥6	87–100%	71%	78–83%	85–100%

[a] CF = complement fixing.
[b] N = neutralizing.
From Chanock et al., 1961; Cranage and Gardner, 1980; Kapikian et al., 1961; Parrott et al., 1973; Richardson et al., 1978; Welliver et al., 1980.

titer after 1 year. The neutralizing antibody response parallels that for IgG antibodies. As with IgM, the IgG and neutralizing antibodies increase more rapidly and reach a higher titer with reinfections.

The complement-fixing antibody response occurs later, at 10 to 20 days after primary infection, and peaks about 20 to 30 days after onset of illness. Complement-fixing antibodies are often negative by 1 year after a primary infection but on subsequent infections are present at low titer for longer periods of time (McClelland et al., 1961).

IgG or neutralization antibody tests detect between 60 and 100% of infections identified by virus isolation or antigen detection, and vice versa (Bruhn and Yeager, 1977; Gerna et al., 1980; Kapikian et al., 1961; Korppi et al., 1986; Meurman et al., 1984b; Welliver et al., 1980b). In adults, the antibody response to RSV infection is similar to that in children experiencing repeated infections (Johnson et al., 1961, 1962; Mathur et al., 1980; Morales et al., 1983; Wright et al., 1971). For detecting the prevalence of RSV antibodies and thus evidence of past infection, the RSV IgG and neutralizing antibody tests are the most sensitive. The titers are approximately 10 to 100 times higher for these tests than for complement-fixing antibody tests (Belshe et al., 1982b; Gerna et al., 1980; Hornsleth et al., 1984; Kapikian et al., 1961; Richardson et al., 1978b; Vauer et al., 1986).

Complement Fixation (CF) Test

The CF test has probably been the most widely used test for the serologic diagnosis of RSV infections. The test has been well standardized and gives good results in persons > 6 months old but is less sensitive than tests for IgG antibodies (see Chapter 5 for description of the CF test). The antigen used in the test is RSV grown in a cell line that produces a high titer of antigen, such as HEp-2 cells. At 4+ CPE, the RSV-infected cells and medium are harvested, the cell-associated virus is released by several cycles of freezing and thawing or by sonication, and the cellular debris is removed by low-speed centrifugation. Uninfected control cells should be grown and har-

vested at the same time in the identical fashion. Several studies have shown that using 8 instead of 2 to 4 units of antigen markedly increases the sensitivity of the RSV CF test, especially in infants and young children (Chanock et al., 1961; Jacobs and Peacock, 1970; Kapikian et al., 1961).

Neutralization Tests

Several variations on the neutralization test have been developed that use CPE, plaques, or immunoassays to detect replication of RSV. Tests based on CPE have worked well both in tubes and in 96-well tissue culture microtiter plates. The virus is first titrated in the system to be used. For a tube neutralization test, 100 $TCID_{50}$/0.1 ml of the virus is incubated for 2 h at room temperature with an equal volume of serial dilutions of the heat-inactivated test serum. Then 0.2 ml of the mixture is added to each tube, two tubes for each dilution of serum. The virus-antibody mixture is adsorbed to the cells for 1 h at 36°C, and 1 ml of maintenance medium is added. The tubes are capped and incubated in a 36°C incubator.

For a microneutralization test, 200 $TCID_{50}$/0.1 ml is incubated with an equal volume of serial dilutions of the heat-inactivated test serum for 2 h. Then 0.1 ml of the mixture is added to 2 to 4 wells per dilution. Alternatively, the test serum can be titrated directly in the microtiter plate, and 100 $TCID_{50}$ of virus can be added to the serum dilutions (final volume should be 0.1 ml). Next, 15,000 to 20,000 cells in 0.1 ml of maintenance medium is added, and the plate is covered and incubated in a 36°C humidified CO_2 incubator.

For both tests a virus control (virus-infected cells with no serum), cell control (uninfected cells), and dilutions of the virus inoculum as a back titration should be included with each test. When the virus control tubes or wells have developed 3+ to 4+ CPE, the test is read by grading the CPE in each tube or well. The CPE can be visualized more clearly if the cells are reacted with a 1:80,000 dilution of neutral red for 1 h (Chanock et al., 1961). The neutral red stains the healthy cells a deep red. The end point can be calculated by counting any well with definite CPE as positive and using the Reed and Muench or Karber method. The inoculum of virus should give a titer between 32 and 320 $TCID_{50}$ for the test to be acceptable. Some studies have demonstrated that the titer of neutralizing antibody increases when complement is added to the virus antibody mixture (Buynak et al., 1979; Gerna et al., 1980; Kaul et al., 1981a; Kim et al., 1971).

Alternatively, viral antigen can be determined by EIA or immunoperoxidase staining (Anderson et al., 1985a; Gerna et al., 1980). In the EIA, when the virus control wells have developed 3+ to 4+ CPE, the plates are washed three times with PBS-T and fixed with 80% acetone and 20% PBS. After air-drying, the plates can be tested immediately or stored for several months at −20°C and then tested. The EIA is performed as described in the section on virus identification, and wells with absorbance over 3 standard deviations above the mean absorbance of the control wells are considered positive for virus replication. The tissue culture EIA has the advantages that it eliminates the potential ambiguities of reading CPE and is easily automated.

For the immunoperoxidase test, 24 to 48 h after inoculating the cells, the plates are fixed with cold absolute ethanol for 20 min at room temperature and then stained immediately or stored frozen and stained later (Gerna et al., 1980). When stored at −80°C, the plates are stable for at least 6 months. The cells are stained by incubating with an anti-RSV antibody for 1 h at 37°C, washing three times with PBS, incubating with a peroxidase-conjugated antispecies antibody for 1 h at 37°C, washing again three times with PBS, and adding a precipitable substrate—for example, diaminobenzidine and H_2O_2. Wells with antigen-positive cells, stained cells, are considered to be positive for virus replication. The immunoperoxidase test eliminates the potential ambiguities of reading CPE and can be completed in a few days.

Immunoassays for Class- and Subclass-Specific Antibodies

Two types of assay systems have been developed to detect class- and subclass-specific antibodies against RSV. One system reacts the specimen with RSV antigens and assays specific immunoglobulins that reacted with the antigens. This system is probably more sensitive for detecting IgG antibodies. The second system captures the specific immunoglobulin, reacts the immunoglobulin with RSV antigens, and then assays the RSV antigens captured by the captured immunoglobulin. This system is better for detecting IgM, IgA, and IgE antibodies. It minimizes false-positive results for IgM caused by rheumatoid factor and false-negative results caused by the more abundant IgG-blocking antigenic sites (Duermeyer and Van der Veen, 1978; Naot and Remington, 1980).

The first method, with fluorescence as the detection system, has been used by several laboratories (Cranage and Gardner, 1980; Kaul et al., 1981b; McIntosh et al., 1979; Oglivie et al., 1981; Welliver et al., 1980b; 1981). The antigen for IFA can be prepared in several ways. Infected and uninfected cells can be grown in flasks, harvested, and dispersed with 0.25% trypsin (diluted 1:3 in versene), spotted on microscope slides, and fixed with ice-cold acetone for 10 min; or after the cells are dispersed with trypsin, they can be reacted in an IFA as live cells to detect surface antigens. Alternatively, infected and uninfected cells can be grown directly on the slides and fixed with ice-cold acetone. The cells prepared by one of these methods are reacted with dilutions (in PBS) of the specimen, washed with PBS, reacted with a fluorescein-conjugated, antiimmunoglobulin antibody, washed again with PBS, and fluorescence read as described in the section on immunofluorescence assay. The titer is the highest dilution of the specimen that gives specific fluorescence. Simply by choosing the appropriate conjugated antiserum, IgA, IgE, IgG and subclasses, and IgM antibodies can be measured.

Similar test systems have been developed with EIA (Belshe et al., 1982b; Hornsleth et al., 1981b, 1984, 1985a,b; Meurman et al., 1984a,b; Richardson et al., 1978; Vaur et al., 1986; Welliver et al., 1981). The antigens for EIA antibody assays have been prepared in several ways: (1) infected and uninfected cells can be grown in microtiter plates and fixed with an acetone/PBS mixture or absolute ethanol; (2) cells can be grown in flasks to 3+ CPE, collected, washed with PBS, resuspended in PBS, and adsorbed to the plates by desiccation for 24 h at 37°C (Vaur et al., 1986); (3) virus and control antigens, prepared by freezing and thawing tissue culture harvest material, can be diluted in carbonate buffer and adsorbed to the plates by an overnight incubation at 4°C (Richardson et al., 1978b); (4) virus and control antigens can be purified and concentrated by sucrose-gradient centrifugation and diluted in carbonate buffer and adsorbed to the plates by incubating overnight at 4°C (Welliver et al., 1981); and (5) virus and control antigens can be reacted with an anti-RSV capture antibody that has previously been adsorbed to the plates by an overnight incubation at 4°C (Hornsleth et al., 1981b). The serum specimen, diluted in PBS or PBS-GT, is then reacted with antigen, prepared by one of the methods described above, for 1 h at 36°C. After washing with PBS-T, the enzyme-conjugated antihuman immunoglobulin is added and incubated for 1 h at 36°C, and after washing the substrate is added and absorbance read at 490 nm or visually. The titer is the highest dilution that gives absorbance, or color, above background levels. As with the IFA test, the different classes and subclasses of antibodies can be measured by choosing the appropriate conjugated antiserum. An example of results for an RSV IgG and IgM EIA is given in Table 4. Note the discrepancy between the moderately high P-N value for specimen 1A and the low P/N value. This is accounted for by a moderately high absorbance value for the specimen control (N) and underlines the importance of the specimen control in minimizing false-positive results.

Gerna et al. (1980) developed an immunoperoxidase assay similar to the EIA. The antigen consists of infected and uninfected cells grown and fixed in microtiter plates, and the cells are stained with a precipitable substrate. This assay has been described in more detail in the section on neutralization tests.

For the antibody-capture EIA, the antigen can be prepared by growing RSV in HEp-2 cells and, at 3+ to 4+ CPE, harvesting the cells and medium (and an uninfected control), freezing and thawing several times or sonicating, and storing at −70°C until testing. The first step in the assay is to adsorb antihuman immunoglobulin (diluted in PBS or carbonate buffer) to the plates by incubating overnight at 4°C; after washing, the serum specimen diluted in PBS-GT is added and incubated for 1 h at 37°C. RSV or control antigen is then added and incubated for 1 h at 37°C,

and an anti-RSV antibody diluted in PBS-GT is added and incubated for 1 h at 37°C. After washing, an enzyme-conjugated antispecies antiserum diluted in PBS-GT is added and incubated for 1 h, and after washing the substrate is added. The titer of antibody is the highest dilution that gives absorbance, or color, above background level. The different classes and subclasses of antibodies can be measured by choosing the appropriate capture antibody.

Protein-Specific Antibody Assays

Several laboratories have developed tests to detect antibodies against the specific proteins of RSV. Immunoprecipitation and transblot electrophoresis can detect antibodies against all of the viral proteins in one test (Vainionpaa et al., 1985; Ward et al., 1983), and the EIA against purified proteins can detect antibodies against one protein in one test (Murphy et al., 1986b,c; Wagner et al., 1986, 1987). In the immunoprecipitation test (Kessler, 1975; Vainionpaa et al., 1985; Ward et al., 1983), the human serum is reacted with Formalin-fixed preparation of the Cowans strain of *Staphylococcus aureus* or protein A–Sepharose beads for 1 h, radiolabeled RSV proteins are added and reacted for 2 h, and the beads or *S. aureus* are washed and boiled with the solubilization buffer and electrophoresed on a polyacrylamide gel. The gel is dried, and the radioactive bands corresponding to the immunoprecipitated proteins are detected by overlaying the gel with X-ray film and developing the film. If the virus is labeled only with [^{35}S]methionine, the G protein will not be detected.

The proteins can also be labeled with biotin, immunoprecipitated as described above, transferred to nitrocellulose paper, and reacted with avidin peroxidase and a precipitable substrate (Anderson et al., 1984). The transblot electrophoresis test is run with unlabeled proteins separated by polyacrylamide gel electrophoresis and transferred electrophoretically to nitrocellulose paper (Hierholzer et al., 1984). The nitrocellulose is cut into 2- to 5-mm strips and reacted with PBS-T for 30 min to block unreacted sites on the paper, reacted for 2 h with the serum specimen diluted in PBS-T, and then, after washing, reacted with the peroxidase-conjugated antispecies antibody for 2 h. Color is developed by adding an insoluble substrate such as diaminobenzidine plus H_2O_2.

An EIA that detects antibody against the F and G proteins uses proteins purified by affinity chromatography (Walsh et al., 1984b, 1985). In this test (Murphy et al., 1981, 1986b), the affinity-purified F or G protein (100 ng/ml in carbonate buffer at pH 9.8) is incubated overnight at 4°C; after washing, the serum specimen diluted in PBS-GT (or PBS-T and 1% fetal

TABLE 4. Example of EIA[a] data for RSV IgG and IgM antibodies

Serum specimen[b]	IgG antibodies		IgM antibodies	
	P–N[c]	P/N	P–N	P/N
1A	0.234	1.5	0.039	1.6
2A	0.258	3.1	0.017	1.2
2C	0.896	8.8	0.073	1.7
3A	0.236	3.4	0.016	1.2
3C	1.197	18.8	0.176	3.4
4A	0.115	2.9	0.260	4.6
4C	0.152	3.6	0.115	2.5

[a] The EIA is a capture IgG and IgM assay as described in Finger et al. (1987).
[b] A is the acute-phase and C is the convalescent-phase serum specimen.
[c] P is the absorbance for the serum specimen reacted against RSV antigens, and N is the absorbance for the serum specimen reacted against control antigens. A serum specimen with a P–N > 0.08 and P/N > 2.0 was considered positive. These values were ≥ 3 SD above the respective means for negative serum specimens.

calf serum) is added and incubated for 1.5 h at 36°C; after washing, the enzyme-conjugated antiserum is added and incubated for 1 h; and after washing, the substrate is added, and absorbance is read. The titer of antibody is the last dilution that gives a positive absorbance reading. The enzyme-conjugated antiserum used determines which class or subclass of antibodies is measured.

Comment

The best serologic method for identifying an RSV infection is to detect a ≥4-fold rise in IgG antibodies between acute- and convalescent-phase sera. EIA is probably the easiest way to measure RSV IgG, because it provides a quantitative result and can be partially automated. Other serologic assays, such as complement fixation and IgM assays, are less sensitive, especially in children < 6 months old. When an acute-phase serum specimen is not available, an RSV IgM assay can be used to test for evidence of a recent infection. Since RSV IgM can remain positive for several months, a positive result does not confirm a current infection. The IgM assay is probably most useful for case/control epidemiologic studies, and the capture assay is probably the best of the possible test configurations. Newer methods to prepare antigens, such as expressing proteins from cloned genes or synthesizing peptides, could be as important for serologic assays as monoclonal antibodies are for antigen detection assays. These methods could theoretically provide an unlimited source of standardized, high-quality antigens.

Interpretation of Results

To interpret results of a test correctly, one must take into account its sensitivity, specificity, and the predictive value of a positive and negative result. These characteristics are influenced by the characteristics of the test and other factors, including the type of specimen collected, the timing of specimen collection, the age of the patient, the handling of the specimen, and the rate of true positives in the population at the time the specimen was collected. These other factors can be more important than those of the test. The cutoff value should be chosen carefully, since it is a major determinant of the sensitivity and specificity of the test.

Epidemiology

RSV occurs as yearly epidemics in the winter or spring in temperate regions, but it can have a different pattern in other regions (Brandt et al., 1973;

Denny and Clyde, 1986; De Silva and Hanlon, 1986; Foy et al., 1973b; Hillis et al., 1971; Kim et al., 1973a; Martin et al., 1978; Mufson et al., 1973; Spence and Barratt, 1968). The severity and timing of these epidemics vary within a community from year to year and between communities during the same year (Centers for Disease Control, 1984, 1986). A community outbreak of RSV lasts from 2 to 5 months and often has a sharp peak of illness for 1 or 2 months, when 40% or more of the illness occurs. Studies in families have shown that the virus spreads efficiently to members of all ages and that it is most often introduced into the family by children (Cooney et al., 1975; Hall et al., 1976b). The virus can be spread by close contact with infected persons or their respiratory tract secretions and by fomites, but not by airborne transmission over 6 feet (Hall and Douglas, 1981a). RSV can survive on environmental surfaces for up to 6 h and is more likely to cause infection when it contacts the mucosa of the nose or the eye than the mouth (Hall et al., 1980, 1981). Most RSV infections probably occur when someone inadvertently inoculates the mucosa of the eye or nose with hands contaminated with virus. Schools and day-care centers are likely to be important in the spread of the virus in the community.

Most hospitalizations for RSV disease occur in the 2- to 6-month-old child, and more occur in boys than girls (Glezen and Denny, 1973; Glezen et al., 1981; Kim et al., 1973a). It is estimated from studies of hospitalized infants and young children that RSV is responsible for 38 to 40% of cases of bronchiolitis, 25 to 39% of cases of pneumonia, 9% of cases of croup, and 12 to 23% of all acute respiratory illnesses (Caul et al., 1976; Jacobs et al., 1971; Kim et al., 1973a; McClelland et al., 1961; Mufson et al., 1973; Paisley et al., 1984; Sobeslavsky et al., 1970). During outbreaks of RSV, the percentage of patients hospitalized with lower respiratory tract illness attributable to RSV infection may climb to 60% (Kim et al., 1973a). It accounts for a smaller proportion of pediatric outpatient acute upper and lower respiratory illness (Horn et al., 1975). RSV is responsible for a similar percentage of inpatient and outpatient cases of lower respiratory tract illness, but outpatient RSV disease involves a higher proportion of older children (Denny and Clyde, 1986; Foy et al., 1973). About 50% of children are infected in the first year of life, and nearly all by the second year of life (Beem et al., 1964; Chanock and Finberg, 1957; Moss et al., 1963; Suto et al., 1965). Other risk factors for hospitalization include lower socioeconomic status, not being breast-fed, and smoking in the household (Downham et al., 1976; Glezen and Denny, 1973; Glezen et al., 1981; Hall et al., 1984; Pullan et al., 1980; Sims et al., 1978).

Mortality among infants and young children hospitalized with RSV lower respiratory tract disease has been reported to be between 0.5 and 3% (Clarke et al., 1978; Eriksson et al., 1983; Hall et al., 1986; Jacobs et al., 1971; Sims et al., 1976). The mortality rate can be much higher in patients with compromised cardiac, immune, or pulmonary systems. For example, RSV infection has been associated with mortality rates of 37% in hospitalized infants with congenital heart disease and 10% in hospitalized children with compromised immune systems (Hall et al., 1986; MacDonald et al., 1982). In one study, 36% of deaths from lower respiratory tract illness in infants and young children were associated with RSV infection (Gardner et al., 1967).

Prevention and Treatment

Considerable effort has been directed toward developing a vaccine for RSV, with little success to date (Tyeryar, 1983). The first Formalin-inactivated vaccine, developed in the 1960s, in four separate clinical trials, proved not only to be ineffective but to be associated with severe adverse reactions. Following exposure to natural infection, vaccinated children < 2 years old not only became infected as frequent as the control children but had more severe disease (Chin et al., 1969; Fulginiti et al., 1969; Kapikian et al., 1969; Kim et al., 1969). Several live virus vaccines have also been developed, but none has proved to be an acceptable vaccine candidate (Belshe et al., 1982a; Buynak et al., 1978, 1979; Gharpure et al., 1969; Hodes et al., 1974; McIntosh et al., 1974; Wright et al., 1971; 1976, 1982). New approaches to immunizing against RSV, which involve the use of vaccinia-recombinant viruses and purified F and G proteins, are being evaluated. It is not yet clear if they will be acceptable candidates for an RSV vaccine, but they are proving useful for studying the host response to infection.

One area in which prevention can and should be applied is in the hospital. RSV is an important nosocomial pathogen in pediatric hospitals, and nosocomial infection has been associated with prolongation of hospital stay and increased mortality (Ditchburn et al., 1971; Gardner et al., 1973; Hall et al., 1975b; Meissner et al., 1984; Mintz et al., 1979; Sims, 1981; Valenti et al., 1982; Wenzel et al., 1977). Hospital staff appear to be important in the transmission of RSV, both by inadvertently carrying infectious material from one patient to another and by becoming infected themselves and directly infecting patients. Infection control practices, including cohorting patients and staff, using gowns when contact with infectious secretions is likely, and paying strict attention to good handwashing practices, may be effective in preventing spread to patients (Garner and Simmons, 1983; Hall and Douglas, 1981a; Hall et al., 1978b; Murphy et al., 1981). Wearing goggles may prevent spread to staff (Agah et al., 1985). Nosocomial spread of RSV has also occurred in other settings and among older children, adults, and the elderly (Centers for Disease Control, 1977, 1985; Garvie and Gray, 1980; Hart, 1984; Marthur et al., 1980; Morales et al., 1983; Sorvillo et al., 1984).

Treatment has recently become available for RSV infections. In adult volunteer studies and clinical trials in infants and young children, aerosolized ribavirin decreased the titer of virus excreted and sped improvement in clinical symptoms and arterial oxygenation (Cockburn et al., 1986; Hall and McBride, 1986; Hall et al., 1983a,b, 1985; Taber et al., 1983). Intravenous gamma globulin has been used prophylactically, and nonspecific transfer factor has been used therapeutically (Gouyon et al., 1985), but further study is needed to assess their role in the management of patients with RSV infection.

Acknowledgment. The author thanks Novella Goss for her help in preparing this chapter.

Literature Cited

Agah, R., J. Cherry, A. Garakian, and M. Chapin. 1985. A study of the respiratory syncytial virus (RSV) infection rate in personnel caring for children with RSV infection in which the routine isolation procedure was compared with the routine procedure supplemented with the use of masks and goggles. Program and Abstracts of the 25th Interscience Conference on Antimicrobial Agents and Chemotherapy, p. 174.

Aherne, W., T. Bird, S. D. M. Court, P. S. Gardner, and J. McQuillin. 1970. Pathological changes in virus infections of the lower respiratory tract in children. J. Clin. Pathol. 23:7–18.

Akerlind, B., and E. Norrby. 1986. Occurrence of respiratory syncytial virus subtypes A and B strains in Sweden. J. Med. Virol. 19:241–247.

Anas, N., C. Boettrich, C. B. Hall, and J. G. Brooks. 1982. The association of apnea and respiratory syncytial virus infection in infants. J. Pediatr. 101:65–68.

Anderson, J. M., and M. O. Beem. 1966. Use of human diploid cell cultures for primary isolation of respiratory syncytial virus. Proc. Soc. Exp. Biol. Med. 121:205–209.

Anderson, L. J., R. A. Coombs, C. Tsou, and J. C. Hierholzer. 1984. Use of the biotin-avidin system to study the specificity of antibodies against respiratory syncytial virus. J. Clin. Microbiol. 19:934–936.

Anderson, L. J., J. C. Hierholzer, P. G. Bingham, and Y. O. Stone. 1985a. Microneutralization test for respiratory syncytial virus based on an enzyme immunoassay. J. Clin. Microbiol. 22:1050–1052.

Anderson, L. J., J. C. Hierholzer, C. Tsou, R. M. Hendry, B. F. Fernie, Y. Stone, and K. McIntosh. 1985b. Antigenic characterization of respiratory syncytial virus strains with monoclonal antibodies. J. Infect. Dis. 151:626–633.

Anderson, L. J., J. C. Hierholzer, Y. O. Stone, C. Tsou, and B. F. Fernie. 1986. Identification of epitopes on respiratory syncytial virus proteins by competitive binding immunoassay. J. Clin. Microbiol. **23:**475–480.

Anestad, G., N. Breivik, and T. Thoresen. 1983. Rapid diagnosis of respiratory syncytial virus and influenza A virus infections by immunofluorescence: experience with a simplified procedure for the preparation of cell smears from nasopharyngeal secretions. Acta Pathol. Microbiol. Immunol. Scand. Sect. B. **91:**267–271.

Arens, M. Q., E. M. Swierkosz, R. R. Schmidt, T. Armstrong, and K. A. Rivetna. 1986. Enhanced isolation of respiratory syncytial virus in cell culture. J. Clin. Microbiol. **23:**800–802.

Armstrong, J. A., H. G. Pereira, and R. C. Valentine. 1962. Morphology and development of respiratory syncytial virus in cell cultures. Nature **196:**1179–1181.

Assaad, F., and W. C. Cockburn. 1974. A seven-year study of WHO virus laboratory reports on respiratory viruses. Bull. WHO **51:**437–445.

Bachi, T., and C. Howe. 1973. Morphogenesis and ultrastructure of respiratory syncytial virus. J. Virol. **12:**1173–1180.

Bairan, A. C., J. D. Cherry, L. F. Fagan, and J. E. Codd, Jr. 1974. Complete heart block and respiratory syncytial virus infection. Am. J. Dis. Child. **127:**264–265.

Baker, J. C., R. E. Werdin, T. R. Ames, R. J. F. Markham, and V. L. Larson. 1986. Study on the etiologic role of bovine respiratory syncytial virus in pneumonia of dairy calves. J. Am. Vet. Med. Assoc. **189:**66–70.

Ball, L. A., K. K. Y. Young, K. Anderson, P. L. Collins, and G. W. Wertz. 1986. Expression of the major glycoprotein G of human respiratory syncytial virus from recombinant vaccinia virus vectors. Proc. Natl. Acad. Sci. USA **83:**246–250.

Bangham, C. R. M. 1986. Passively acquired antibodies to respiratory syncytial virus impair the secondary cytotoxic T-cell response in the neonatal mouse. Immunology **59:**37–41.

Bangham, C. R. M., and B. A. Askonas. 1986. Murine cytotoxic T cells specific to respiratory syncytial virus recognize different antigenic subtypes of the virus. J. Gen. Virol. **67:**623–629.

Bangham, C. R. M., and A. J. McMichael. 1986. Specific human cytotoxic T cells recognize B cell lines persistently infected with respiratory syncytial virus. Proc. Natl. Acad. Sci. USA **83:**9183–9187.

Bangham, C. R. M., M. J. Cannon, D. T. Karzon, and B. A. Askonas. 1985. Cytotoxic T-cell response to respiratory syncytial virus in mice. J. Virol. **56:**55–59.

Bangham, C. R. M., P. J. M. Openshaw, L. A. Ball, A. M. Q. King, G. W. Wertz, and B. A. Askonas. 1986. Human and murine cytotoxic T cells specific to respiratory syncytial virus recognize the viral nucleoprotein (N), but not the major glycoprotein (G), expressed by vaccinia virus recombinants. J. Immunol. **137:**3973–3977.

Barry, F., F. Cockburn, R. Cornall, J. F. Price, G. Sutherland, and A. Vardag. 1986. Ribavirin aerosol for acute bronchiolitis. Arch. Dis. Child. **61:**593–597.

Beem, M. 1967. Repeated infections with respiratory syncytial virus. J. Immunol. **98:**1115–1122.

Beem, M., F. H. Wright, D. Hamre, R. Egerer, and M. Oehme. 1960. Association of the chimpanzee coryza agent with acute respiratory disease in children. J. Am. Med. Assoc. **263:**523–530.

Beem, M., R. Egerer, and J. Anderson. 1964. Respiratory syncytial virus neutralizing antibodies in persons residing in Chicago, Illinois. Pediatrics **34:**761–770.

Bell, D. M., E. E. Walsh, J. F. Hruska, K. C. Schnabel, and C. B. Hall. 1983a. Rapid detection of respiratory syncytial virus with a monoclonal antibody. J. Clin. Microbiol. **17:**1099–1101.

Bell, D. M., N. J. Roberts, Jr., and C. B. Hall. 1983b. Different antiviral spectra of human macrophage interferon activities. Nature **305:**319–321.

Belshe, R. B., L. S. Richardson, W. T. London, D. L. Sly, J. H. Lorfeld, E. Camargo, D. A. Prevar, and R. M. Chanock. 1977. Experimental respiratory syncytial virus infection of four species of primates. J. Med. Virol. **1:**157–162.

Belshe, R. B., L. P. Van Voris, and M. A. Mufson. 1982a. Parenteral administration of live respiratory syncytial virus vaccine: results of a field trial. J. Infect. Dis. **145:**311–319.

Belshe, R. B., L. P. Van Voris, M. A. Mufson, E. B. Buynak, A. A. McLean, and M. A. Hilleman. 1982b. Comparison of enzyme-linked immunosorbent assay and neutralization techniques for measurement of antibody to respiratory syncytial virus: implications for parenteral immunization with live virus vaccine. Infect. Immun. **37:**160–165.

Belshe, R. B., L. P. Van Voris, and M. A. Mufson. 1983. Impact of viral respiratory diseases on infants and young children in a rural and urban area of southern West Virginia. Am. J. Epidemiol. **117:**467–474.

Bennett, C. R. Jr., and D. Hamre. 1962. Growth and serological characteristics of respiratory syncytial virus. J. Infect. Dis. **110:**8–16.

Berkovich, S., and S. Kibrick. 1964. Exanthem associated with respiratory syncytial virus infection. J. Pediatr. **65:**368–370.

Berkovich, S., and L. Taranko. 1964. Acute respiratory illness in the premature nursery associated with respiratory syncytial virus infections. Pediatrics **34:**753–760.

Berman, S., A. Duenas, A. Bedoya, V. Constain, S. Leon, I. Borrero, and J. Murphy. 1983. Acute lower respiratory tract illnesses in Cali, Colombia: a two-year ambulatory study. Pediatrics **71:**210–218.

Bernstein, J. M., and J. F. Hruska. 1981. Respiratory syncytial virus proteins: identification by immunoprecipitation. J. Virol. **38:**278–285.

Berthiaume, L., J. Joncas, G. Boulay, and V. Pavilanis. 1973. Serological evidence of respiratory syncytial virus infection in sheep. Vet. Rec. **93:**337–338.

Berthiaume, L., J. Joncas, and V. Pavilanis. 1974. Comparative structure, morphogenesis and biological characteristics of the respiratory syncytial (RS) virus and the pneumonia virus of mice (PVM). Arch. Ges. Virusforsch. **45:**39–51.

Bloth, B., and E. Norrby. 1967. Electron microscopic analysis of the internal component of a respiratory syncytial (RS) virus. Arch. Ges. Virusforsch. **21:**71–77.

Brandt, C. D., H. W. Kim, J. O. Arrobio, B. C. Jeffries, S. C. Wood, R. M. Chanock, and R. H. Parrott. 1973. Epidemiology of respiratory syncytial virus infection in Washington, D.C. III. Composite analysis of eleven consecutive yearly epidemics. Am. J. Epidemiol. **98:**355–364.

Bromberg, K., B. Daidone, L. Clarke, and M. F. Sierra. 1984. Comparison of immediate and delayed inoculation of HEp-2 cells for isolation of respiratory syncytial virus. J. Clin. Microbiol. **20:**123–124.

Bromberg, K., G. Tannis, B. Daidone, L. Clarke, and M. F. Sierra. 1985. Comparison of ortho respiratory syncytial virus enzyme-linked immunosorbent assay and HEp-2 cell culture. J. Clin. Microbiol. **22:**1071–1072.

Bruckova, M., L. Kunzova, Z. Jezkova, and J. Vocel. 1979. Incidence of RS virus infections in premature children's ward. J. Hyg. Epidemiol. Microbiol. Immunol. **23:**389–396.

Bruhn, F. W., and A. S. Yeager. 1977. Respiratory syncytial virus in early infancy. Am. J. Dis. Child. **131**:145–148.

Bruhn, F. W., S. T. Mokrohisky, and K. McIntosh. 1977. Apnea associated with respiratory syncytial virus infection in young infants. J. Pediatr. **90**:382–386.

Burrows, B., R. J. Knudson, and M. D. Lebowitz. 1977. The relationship of childhood respiratory illness to adult obstructive airway disease. Am. Rev. Respir. Dis. **115**:751–760.

Buynak, E. B., R. E. Weibel, A. A. McLean, and M. R. Hilleman. 1978. Live respiratory syncytial virus vaccine administered parenterally (40112). Proc. Soc. Exp. Biol. Med. **157**:636–642.

Buynak, E. B., R. E. Weibel, A. J. Carlson, A. A. McLean, and M. R. Hilleman. 1979. Further investigations of live respiratory syncytial virus vaccine administered parenterally (40433). Proc. Socl. Exp. Biol. Med. **160**:272–277.

Cappel, R., L. Thiry, and G. Clinet. 1975. Viral antibodies in the CSF after acute CNS infections. Arch. Neurol. **32**:629–631.

Carlsen, K. H., L. Orstavik, and K. Halvorsen. 1983. Viral infections of the respiratory tract in hospitalized children. Acta Paediatr. Scand. **72**:53–58.

Cash, P., W. H. Wunner, and C. R. Pringle. 1977. A comparison of the polypeptides of human and bovine respiratory syncytial viruses and murine pneumonia virus. Virology **82**:369–379.

Cash, P., C. R. Pringle, and C. M. Preston. 1979. The polypeptides of human respiratory syncytial virus: products of cell-free protein synthesis and post-translational modifications. Virology **92**:375–384.

Caul, E. O., D. K. Waller, S. K. R. Clarke, and B. D. Corner. 1976. A comparison of influenza and respiratory syncytial virus infections among infants admitted to hospital with acute respiratory infections. J. Hyg. **77**:383–392.

Cavallaro, J. J., H. F. Maassab, and G. D. Abrams. 1967. An immunofluorescent and histopathological study of respiratory syncytial (RS) virus encephalitis in suckling mice. Proc. Soc. Exp. Biol. Med. **124**:1059–1064.

Centers for Disease Control. 1977. Respiratory syncytial virus—Missouri. Morbid. Mortal. Weekly Rep. **26**:351.

Centers for Disease Control. 1984. Update: respiratory virus surveillance—United States, 1984. Morbid. Mortal. Weekly Rep. **33**:223–224.

Centers for Disease Control. 1986. Respiratory syncytial virus—Oklahoma. Morbid. Mortal. Weekly Rep. **35**:162–164.

Cevenini, R., M. Donati, A. Moroni, L. Franchi, and F. Rumpianesi. 1983. Rapid immunoperoxidase assay for detection of respiratory syncytial virus in nasopharyngeal secretions. J. Clin. Microbiol. **18**:947–949.

Chanock, R. M. 1970. Control of acute mycoplasmal and viral respiratory tract disease: the prospects of eventual successful immunoprophylaxis through vaccination are encouraging. Science **169**:248–256.

Chanock, R., and L. Finberg. 1957. Recovery from infants with respiratory illness of a virus related to chimpanzee coryza agent (CCA). II. Epidemiologic aspects of infection in infants and young children. Am. J. Hyg. **66**:291–300.

Chanock, R., B. Roizman, and R. Myers. 1957. Recovery from infants with respiratory illness of a virus related to chimpanzee coryza agent (CCA). I. Isolation, properties and characterization. Am. J. Hyg. **66**:281–290.

Chanock, R. M., H. W. Kim, A. J. Vargosko, A. Deleva, K. M. Johnson, C. Cumming, and R. H. Parrott. 1961. Respiratory syncytial virus. I. Virus recovery and other observations during 1960 outbreak of bronchiolitis, pneumonia, and minor respiratory diseases in children. J. Am. Med. Assoc. **176**:647–653.

Chanock, R., L. Chambon, W. Chang, F. G. Ferreira, P. Gharpure, L. Grant, J. Hatem, I. Imam, S. Kalra, K. Lim, J. Madalengoitia, L. Spence, P. Teng, and W. Ferreira. 1967. WHO respiratory disease survey in children: a serological study. Bull. WHO **37**:363–369.

Chao, R. K., M. Fishaut, J. D. Schwartzman, and K. McIntosh. 1979. Detection of respiratory syncytial virus in nasal secretions from infants by enzyme-linked immunosorbent assay. J. Infect. Dis. **139**:483–486.

Cheeseman, S. H., L. T. Pierik, D. Leombruno, K. E. Spinos, and K. McIntosh. 1986. Evaluation of a commercially available direct immunofluorescent staining reagent for the detection of respiratory syncytial virus in respiratory secretions. J. Clin. Microbiol. **24**:155–156.

Chin, J., R. L. Magoffin, L. A. Shearer, J. H. Schieble, and E. H. Lennette. 1969. Field evaluation of a respiratory syncytial virus vaccine and a trivalent parainfluenza virus vaccine in a pediatric population. Am. J. Epidemiol. **89**:449–463.

Chonmaitree, T., V. M. Howie, and A. L. Truant. 1986. Presence of respiratory viruses in middle ear fluids and nasal wash specimens from children with acute otitis media. Pediatrics **77**:698–702.

Chuan-Liang, K., K. McIntosh, B. Fernie, A. Talis, L. Pierik, and L. Anderson. 1984. Monoclonal antibodies for the rapid diagnosis of respiratory syncytial virus infection by immunofluorescence. Diagn. Microbiol. Infect. Dis. **2**:199–206.

Clarke, S. K. R., P. S. Gardner, P. M. Poole, H. Simpson, and J. O. Tobin. 1978. Respiratory syncytial virus infection: admissions to hospital in industrial, urban, and rural areas. Br. Med. J. **2**:796–798.

Coates, H. V., and R. M. Chanock. 1962. Experimental infection with respiratory syncytial virus in several species of animals. Am. J. Hyg. **76**:302–312.

Coates, H. V., L. Kendrick, and R. M. Chanock. 1963. Antigenic differences between two strains of respiratory syncytial virus. Proc. Soc. Exp. Biol. Med. **112**:958–964.

Coates, H. V., D. W. Alling, and R. M. Chanock. 1966. An antigenic analysis of respiratory syncytial virus isolates by a plaque reduction neutralization tests. Am. J. Epidemiol. **83**:299–313.

Collins, P. L., and G. W. Wertz. 1983. cDNA cloning and transcriptional mapping of nine polyadenylylated RNAs encoded by the genome of human respiratory syncytial virus. Proc. Natl. Acad. Sci. USA **80**:3208–3212.

Collins, P. L., and G. W. Wertz. 1985a. The 1A protein gene of human respiratory syncytial virus: nucleotide sequence of the mRNA and a related polycistronic transcript. Virology **141**:283–291.

Collins, P. L., and G. W. Wertz. 1985b. Nucleotide sequences of the 1B and 1C nonstructural protein mRNAs of human respiratory syncytial virus. Virology **143**:442–451.

Collins, P. L., Y. T. Huang, and G. W. Wertz. 1984a. Identification of a tenth mRNA of respiratory syncytial virus and assignment of polypeptides to the 10 viral genes. J. Virol. **49**:572–578.

Collins, P. L., Y. T. Huang, and G. W. Wertz. 1984b. Nucleotide sequence of the gene encoding the fusion (F) glycoprotein of human respiratory syncytial virus. Proc. Natl. Acad. Sci. USA **81**:7683–7687.

Collins, P. L., K. Anderson, S. J. Langer, and G. W. Wertz. 1985. Correct sequence for the major nucleocapsid protein mRNA of respiratory syncytial virus. Virology **146**:69–77.

Collins, P. L., L. E. Dickens, A. Buckler-White, R. A. Olmsted, M. K. Spriggs, E. Camargo, and K. V. W. Coelingh. 1986. Nucleotide sequences for the gene junctions of human respiratory syncytial virus reveal distinctive features of intergenic structure and gene order. Proc. Natl. Acad. Sci. USA **83:**4594–4598.

Compans, R. W., D. H. Harter, and P. W. Choppin. 1967. Studies on pneumonia virus of mice (PVM) in cell cultures. II. Structure and morphogenesis of the virus particle. J. Exp. Med. **126:**267–276.

Cooney, M. K., J. P. Fox, and C. E. Hall. 1975. The Seattle virus watch. VI. Observations of infections with and illness due to parainfluenza, mumps and respiratory syncytial viruses and *Mycoplasma pneumoniae*. Am. J. Epidemiol. **101:**532–551.

Cote, P. J. Jr., B. F. Fernie, E. C. Ford, J. W. Shih, and J. L. Gerin. 1981. Monoclonal antibodies to respiratory syncytial virus: detection of virus neutralization and other antigen-antibody systems using infected human and murine cells. J. Virol. Methods **3:**137–147.

Cradock-Watson, J. E., J. McQuillin, and P. S. Gardner. 1971. Rapid diagnosis of respiratory syncytial virus infection in children by the immunofluorescent technique. J. Clin. Pathol. **24:**308–312.

Craft, A. W., M. M. Reid, P. S. Gardner, E. Jackson, J. Kernahan, J. McQuillin, T. C. Noble, and W. Walker. 1979. Virus infections in children with acute lymphoblastic leukaemia. Arch. Dis. Child. **54:**755–759.

Cranage, M. P., and P. S. Gardner. 1980. Systemic cell-mediated and antibody responses in infants with respiratory syncytial virus infections. J. Med. Virol. **5:**161–170.

Cranage, M. P., E. J. Stott, J. Nagington, and R. R. A. Coombs. 1981. A reverse passive haemagglutination test for the detection of respiratory syncytial virus in nasal secretions from infants. J. Med. Virol. **8:**153–160.

D'Alessio, D., S. Williams, and E. C. Dick. 1970. Rapid detection and identification of respiratory viruses by direct immunofluorescence. Appl. Microbiol. **20:**233–239.

Delage, G., P. Brochu, L. Robillard, G. Jasmin, J. H. Joncas, and N. Lapointe. 1984. Giant cell pneumonia due to respiratory syncytial virus: occurrence in severe combined immunodeficiency syndrome. Arch. Pathol. Lab. Med. **108:**623–625.

Denny, F. W., and W. A. Clyde, Jr. 1986. Acute lower respiratory tract infections in nonhospitalized children. J. Pediatr. **108:**635–646.

De Silva, L. M., and M. G. Hanlon. 1986. Respiratory syncytial virus: a report of a 5-year study at a children's hospital. J. Med. Virol. **19:**299–305.

Dickens, L. E., P. L. Collins, and G. W. Wertz. 1984. Transcriptional mapping of human respiratory syncytial virus. J. Virol. **52:**364–369.

Ditchburn, R. K., J. McQuillin, P. S. Gardner, and S. D. M. Court. 1971. Respiratory syncytial virus in hospital cross-infection. Br. Med. J. **3:**671–673.

Doane, F. W., N. Anderson, A. Zbitnew, and A. J. Rhodes. 1969. Application of electron microscopy to the diagnosis of virus infections. Can. Med. Assoc. J. **100:**1043–1049.

Doggett, J. E. 1965. Antibodies to respiratory syncytial virus in human sera from different regions of the world. Bull. WHO **32:**849–853.

Doggett, J. E., and D. Taylor-Robinson. 1965. Serological studies with respiratory syncytial virus. Arch. Ges. Virusforsch. **15:**601–608.

Downham, M. A. P. S., R. Scott, D. G. Sims, J. K. G. Webb, and P. S. Gardner. 1976. Breast-feeding protects against respiratory syncytial virus infections. Br. Med. J. **2:**274–276.

Dubovi, E. J. 1982. Analysis of proteins synthesized in respiratory syncytial virus-infected cells. J. Virol. **42:**372–378.

Dubovi, E. J., J. D. Geratz, and R. R. Tidwell. 1983. Enhancement of respiratory syncytial virus-induced cytopathology by trypsin, thrombin, and plasmin. Infect. Immun. **40:**351–358.

Duermeyer, W., and J. Van der Veen. 1978. Specific detection of IgM-antibodies by ELISA, applied in hepatitis-A. Lancet **2:**684–685.

Domurat, F., N. J. Roberts, Jr., E. E. Walsh, and R. Dagan. 1985. Respiratory syncytial virus infection of human mononuclear leukocytes in vitro and in vivo. J. Infect. Dis. **152:**895–902.

Edwards, K. M., P. N. Snyder, and P. F. Wright. 1986. Complement activation by respiratory syncytial virus-infected cells. Arch. Virol. **88:**49–56.

Ehrlicher, L., H. G. Hoffmann, and K. O. Habermehl. 1984. Detection of respiratory virus antigens in nasopharyngeal secretions from patients with acute respiratory disease by radio-immunoassay and tissue culture isolation. Med. Microbiol. Immunol. **173:**37–44.

Elango, N., and S. Venkatesan. 1983. Amino acid sequence of human respiratory syncytial virus nucleocapsid protein. Nucleic Acids Res. **11:**5941–5951.

Elango, N., M. Satake, J. E. Coligan, E. Norrby, E. Camargo, and S. Venkatesan. 1985a. Respiratory syncytial virus fusion glycoprotein: nucleotide sequence of mRNA, identification of cleavage activation site and amino acid sequence of N-terminus of F_1 subunit. Nucleic Acids Res. **13:**1559–1574.

Elango, N., M. Satake, and S. Venkatesan. 1985b. mRNA sequence of three respiratory syncytial virus genes encoding two nonstructural proteins and a 22K structural protein. J. Virol. **55:**101–110.

Elango, N., G. A. Prince, B. R. Murphy, S. Venkatesan, R. M. Chanock, and B. Moss. 1986. Resistance of human respiratory syncytial virus (RSV) infection induced by immunization of cotton rats with a recombinant vaccinia virus expressing the RSV G glycoprotein. Proc. Natl. Acad. Sci. USA **83:**1906–1910.

Eriksson, M., M. Forsgren, S. Sjoberg, M. von Sydow, and S. Wolontis. 1983. Respiratory syncytial virus infection in young hospitalized children: identification of risk patients and prevention of nosocomial spread by rapid diagnosis. Acta Paediatr. Scand. **72:**47–51.

Faden, H., T. N. Kaul, and P. L. Ogra. 1983. Activation of oxidative and arachidonic acid metabolism in neutrophils by respiratory syncytial virus antibody complexes: possible role in disease. J. Infect. Dis. **148:**110–119.

Faden, H., J. J. Hong, and P. L. Ogra. 1984. Interaction of polymorphonuclear leukocytes and viruses in humans: adherence of polymorphonuclear leukocytes to respiratory syncytial virus-infected cells. J. Virol. **52:**16–23.

Faulkner, G. P., P. V. Shirodaria, E. A. C. Follett, and C. R. Pringle. 1976. Respiratory syncytial virus mutants and nuclear immunofluorescence. J. Virol. **20:**487–500.

Fernald, G. W., J. R. Almond, and F. W. Henderson. 1983. Cellular and humoral immunity in recurrent respiratory syncytial virus infections. Pediatr. Res. **17:**753–758.

Fernie, B. F., and J. L. Gerin. 1980. The stabilization and purification of respiratory syncytial virus using $MgSO_4$. Virology **106:**141–144.

Fernie, B. F., and J. L. Gerin. 1982. Immunochemical identification of viral and nonviral proteins of the respiratory syncytial virus virion. Infect. Immun. **37:**243–249.

Fernie, B. F., P. J. Cote, Jr., and J. L. Gerin. 1982. Classification of hybridomas to respiratory syncytial virus glycoproteins (41509). Proc. Soc. Exp. Biol. Med. 171:266–271.

Fernie, B. F., G. Dapolito, P. J. Cote, Jr., and J. L. Gerin. 1985. Kinetics of synthesis of respiratory syncytial virus glycoproteins. J. Gen. Virol. 66:1983–1990.

Finger, F., L. J. Anderson, R. C. Dicker, B. Harrison, R. Doan, A. Downing, and L. Corey. 1987. Epidemic respiratory syncytial virus infection in institutionalized young adults. J. Infect. Dis. 155:1335–1339.

Fishaut, M., D. Tubergen, and K. McIntosh. 1980. Cellular response to respiratory viruses with particular reference to children with disorders of cell-mediated immunity. J. Pediatr. 96:179–186.

Flanders, R. T., P. D. Lindsay, R. Chairez, T. A. Brawner, M. L. Kumar, P. D. Swenson, and K. Bromberg. 1986. Evaluation of clinical specimens for the presence of respiratory syncytial virus antigens using an enzyme immunoassay. J. Med. Virol. 19:1–9.

Follett, E. A. C., C. R. Pringle, and T. H. Pennington. 1975. Virus development in enucleate cells: echovirus, poliovirus, pseudorabies virus, reovirus, respiratory syncytial virus and semliki forest virus. J. Gen. Virol. 26:183–196.

Foy, H. M., M. K. Cooney, A. J. Maletzky, and J. T. Grayston. 1973a. Incidence and etiology of pneumonia, croup and bronchiolitis in preschool children belonging to a prepaid medical care group over a four-year period. Am. J. Epidemiol. 97:80–92.

Foy, H. M., M. K. Cooney, R. McMahan, and J. T. Grayston. 1973b. Viral and mycoplasmal pneumonia in a prepaid medical care group during an eight-year period. Am. J. Epidemiol. 97:93–102.

Frank, A. L., L. H. Taber, C. R. Wells, J. M. Wells, W. P. Glezen, and A. Paredes. 1981. Patterns of shedding of myxoviruses and paramyxoviruses in children. J. Infect. Dis. 144:433–441.

Fransen, H., G. Sterner, M. Forsgren, Z. Heigl, S. Wolontis, A. Svedmyr, and G. Tunevall. 1967. Acute lower respiratory illness in elderly patients with respiratory syncytial virus infection. Acta Med. Scand. 182:323–330.

Freke, A., E. J. Stott, A. P. C. H. Roome, and E. O. Caul. 1986. The detection of respiratory syncytial virus in nasopharyngeal aspirates: assessment, formulation, and evaluation of monoclonal antibodies as a diagnostic reagent. J. Med. Virol. 18:181–191.

Freymuth, F., M. Quibriac, J. Petitjean, M. L. Amiel, P. Pothier, A. Denis, and J. F. Duhamel. 1986. Comparison of two new tests for rapid diagnosis of respiratory syncytial virus infections by enzyme-linked immunosorbent assay and immunofluorescence techniques. J. Clin. Micrbiol. 24:1013–1016.

Friedman, A. D., S. H. Naqvi, M. Q. Arens, and M. A. Eyler. 1986. Value of rapid diagnosis of respiratory syncytial virus infection on management of small infants. Clin. Pediatr. 25:404–406.

Fulginiti, V. A., J. J. Eller, A. W. Downie, and C. H. Kempe. 1967. Altered reactivity to measles virus: atypical measles in children previously immunized with inactivated measles virus vaccines. J. Am. Med. Assoc. 202:1075–1079.

Fulginiti, V. A., J. J. Eller, O. F. Sieber, J. W. Joyner, M. Minamitani, and G. Meiklejohn. 1969. Respiratory virus immunization. I. A field of two inactivated respiratory virus vaccines: an aqueous trivalent parainfluenza virus vaccine and an alum-precipitated respiratory syncytial virus vaccine. Am. J. Epidemiol. 89:435–448.

Fulton, R. E., and P. J. Middleton. 1974. Comparison of

immunofluorescence and isolation techniques in the diagnosis of respiratory viral infections of children. Infect. Immun. 10:92–101.

Gala, C. L., C. B. Hall, K. C. Schnabel, P. H. Pincus, P. Blossom, S. W. Hildreth, R. F. Betts, and R. G. Douglas, Jr. 1986. The use of eye-nose goggles to control nosocomial respiratory syncytial virus infection. J. Am. Med. Assoc. 256:2706–2708.

Gardner, P. S., and J. McQuillin. 1968. Application of immunofluorescent antibody technique in rapid diagnosis of respiratory syncytial virus infection. Br. Med. J. 3:340–343.

Gardner, P. S., and J. McQuillin. 1978. The coating of respiratory syncytial (RS) virus-infected cells in the respiratory tract by immunoglobulins. J. Med. Virol. 2:165–173.

Gardner, P. S., and J. McQuillin. 1980. Rapid virus diagnosis: application of immunofluorescence, 2d ed., p. 110–123, Butterworth, Boston.

Gardner, P. S., D. C. Turk, W. A. Aherne, T. Bird, M. D. Holdaway, and S. D. M. Court. 1967. Deaths associated with respiratory tract infection in childhood. Br. Med. J. 4:316–320.

Gardner, P. S., J. McQuillin, and S. D. M. Court. 1970a. Speculation on pathogenesis in death from respiratory syncytial virus infection. Br. Med. J. 1:327–330.

Gardner, P. S., J. McQuillin, and R. McGuckin. 1970b. The late detection of respiratory syncytial virus in cells in respiratory tract by immunofluorescence. J. Hyg. (Lond.) 68:575–580.

Gardner, P. S., S. D. M. Court, J. T. Brocklebank, M. A. P. S. Downham, and D. Weightman. 1973. Virus cross-infection in paediatric wards. Br. Med. J. 2:571–575.

Gardner, P. S., M. Grandien, and J. McQuillin. 1978. Comparison of immunofluorescence and immunoperoxidase methods for viral diagnosis at a distance: a WHO collaborative study. Bull. WHO 56:105–110.

Garner, J. S., and B. P. Simmons. 1983. Guideline for isolation precautions in hospitals. Infect. Control (Suppl.) 4:245–325.

Garvie, D. G., and J. Gray. 1980. Outbreak of respiratory syncytial virus infection in the elderly. Br. Med. J. 281:1253–1254.

Gerna, G., E. Cattaneo, P. M. Cereda, M. G. Revello, and G. Achilli. 1980. Serodiagnosis of respiratory syncytial virus infections in infants and young children by the immunoperoxidase technique. J. Clin. Microbiol. 11:79–87.

Gharpure, M. A., P. F. Wright, and R. M. Chanock. 1969. Temperature-sensitive mutants of respiratory syncytial virus. J. Virol. 3:414–421.

Giles, T. D., and R. S. Gohd. 1976. Respiratory syncytial virus and heart disease: a report of two cases. J. Am. Med. Assoc. 236:1128–1130.

Gimenez, H. B., P. Cash, and W. I. Melvin. 1984. Monoclonal antibodies to human respiratory syncytial virus and their use in comparison of different virus isolates. J. Gen. Virol. 65:963–971.

Gimenez, H. B., N. Hardman, H. M. Keir, and P. Cash. 1986. Antigenic variation between human respiratory syncytial virus isolates. J. Gen. Virol. 67:863–870.

Glezen, W. P. 1983. Viral pneumonia as a cause and result of hospitalization. J. Infect. Dis. 147:765–770.

Glezen, W. P., and F. W. Denny. 1973. Epidemiology of acute lower respiratory disease in children. N. Engl. J. Med. 288:498–505.

Glezen, W. P., F. A. Loda, W. A. Clyde, Jr., R. J. Senior, C. I. Sheaffer, W. G. Conley, and F. W. Denny. 1971. Epidemiologic patterns of acute lower respiratory disease of children in a pediatric group practice. J. Pediatr. **78:**397–406.

Glezen, W. P., A. Paredes, J. E. Allison, L. H. Taber, and A. L. Frank. 1981. Risk of respiratory syncytial virus infection for infants from low-income families in relationship to age, sex, ethnic group, and maternal antibody level. J. Pediatr. **98:**708–715.

Glezen, W. P., L. H. Taber, A. L. Frank, and J. A. Kasel. 1986. Risk of primary infection and reinfection with respiratory syncytial virus. Am. J. Dis. Child. **140:**543–546.

Gouyon, J. B., P. Pothier, F. Guignier, H. Portier, H. P. Pujol, A. Kazmierczak, P. Chatelain, and M. Alison. 1985. Outbreak of respiratory syncytial virus in France. Eur. J. Clin. Microbiol. **4:**415–416.

Grandien, M., C. Pettersson, P. S. Gardner, A. Linde, and A. Stanton. 1985. Rapid viral diagnosis of acute respiratory infections: comparison of enzyme-linked immunosorbent assay and the immunofluorescence technique for detection of viral antigens in nasopharyngeal secretions. J. Clin. Microbiol. **22:**757–760.

Gray, K. G., D. E. MacFarlane, and R. G. Sommerville. 1968. Direct immunofluorescent identification of respiratory syncytial virus in throat swabs from children with respiratory illness. Lancet **1:**446–448.

Gruber, C., and S. Levine. 1983. Respiratory syncytial virus polypeptides. III. The envelope-associated proteins. J. Gen. Virol. **64:**825–832.

Gruber, C., and S. Levine. 1985a. Respiratory syncytial virus polypeptides. IV. The oligosaccharides of the glycoproteins. J. Gen. Virol. **66:**417–432.

Gruber, C., and S. Levine. 1985b. Respiratory syncytial virus polypeptides. V. The kinetics of glycoprotein synthesis. J. Gen. Virol. **66:**1241–1247.

Habermehl, K. O. 1986. Rapid diagnosis of respiratory virus infections in patients with acute respiratory disease. Diagn. Microbiol. Infect. Dis. **4:**17S–22S.

Hall, C. B. 1981. Nosocomial viral respiratory infections: perennial weeds on pediatric wards. Am. J. Med. **70:**670–676.

Hall, C. B. 1983. The nosocomial spread of respiratory syncytial viral infections. Annu. Rev. Med. **34:**311–319.

Hall, C. B., and R. G. Douglas, Jr. 1975. Clinically useful method for the isolation of respiratory syncytial virus. J. Infect. Dis. **131:**1–5.

Hall, C. B., and R. G. Douglas, Jr. 1981a. Nosocomial respiratory syncytial viral infections: should gowns and masks be used? Am. J. Dis. Child. **135:**512–515.

Hall, C. B., and R. G. Douglas, Jr. 1981b. Modes of transmission of respiratory syncytial virus. J. Pediatr. **99:**100–103.

Hall, C. B., and J. T. McBride. 1986. Ribavirin and respiratory syncytial virus. Am. J. Dis. Child. **140:**331–332.

Hall, C. B., R. G. Douglas, Jr., and J. M. Geiman. 1975a. Quantitative shedding patterns of respiratory syncytial virus in infants. J. Infect. Dis. **132:**151–156.

Hall, C. B., R. G. Douglas, Jr., J. M. Geiman, and M. K. Messner. 1975b. Nosocomial respiratory syncytial virus infections. N. Engl. J. Med. **293:**1343–1346.

Hall, C. B., R. G. Douglas, Jr., and J. M. Geiman. 1976a. Respiratory syncytial virus infections in infants: quantitation and duration of shedding. J. Pediatr. **89:**11–15.

Hall, C. B., J. M. Geiman, R. Biggar, D. I. Kotok, P. M. Hogan, and R. G. Douglas, Jr. 1976b. Respiratory syncytial virus infections within families. N. Engl. J. Med. **294:**414–419.

Hall, C. B., R. G. Douglas, Jr., R. L. Simons, and J. M. Geiman. 1978a. Interferon production in children with

respiratory syncytial, influenza, and parainfluenza virus infections. J. Pediatr. **93:**28–32.

Hall, C. B., J. M. Geiman, R. G. Douglas, Jr., and M. P. Meagher. 1978b. Control of nosocomial respiratory syncytial viral infections. Pediatrics **62:**728–732.

Hall, C. B., W. J. Hall, and D. M. Speers. 1979a. Clinical and physiological manifestations of bronchiolitis and pneumonia. Am. J. Dis. Child. **133:**798–802.

Hall, C. B., A. E. Kopelman, R. G. Douglas, Jr., J. M. Geiman, and M. P. Meagher. 1979b. Neonatal respiratory syncytial virus infection. N. Engl. J. Med. **300:**393–396.

Hall, C. B., R. G. Douglas, Jr., and J. M. Geiman. 1980. Possible transmission by fomites of respiratory syncytial virus. J. Infect. Dis. **141:**98–102.

Hall, C. B., R. G. Douglas, Jr., K. C. Schnabel, and J. M. Geiman. 1981. Infectivity of respiratory syncytial virus by various routes of inoculation. Infect. Immun. **33:**779–783.

Hall, C. B., J. T. McBride, E. E. Walsh, D. M. Bell, C. L. Gala, S. Hildreth, L. G. Ten Eyck, and W. J. Hall. 1983a. Aerosolized ribavirin treatment of infants with respiratory syncytial viral infection: a randomized double-blind study. N. Engl. J. Med. **308:**1443–1447.

Hall, C. B., E. E. Walsh, J. F. Hruska, R. F. Betts, and W. J. Hall. 1983b. Ribavirin treatment of experimental respiratory syncytial viral infection: a controlled double-blind study in young adults. J. Am. Med. Assoc. **249:**2666–2670.

Hall, C. B., W. J. Hall, C. L. Gala, F. B. MaGill, and J. P. Leddy. 1984. Long-term prospective study in children after respiratory syncytial virus infection. J. Pediatr. **105:**358–364.

Hall, C. B., J. T. McBride, C. L. Gala, S. W. Hildreth, and K. C. Schnabel. 1985. Ribavirin treatment of respiratory syncytial viral infection in infants with underlying cardiopulmonary disease. J. Am. Med. Assoc. **254:**3047–3051.

Hall, C. B., K. R. Powell, N. E. MacDonald, C. L. Gala, M. E. Menegus, S. C. Suffin, and H. J. Cohen. 1986. Respiratory syncytial viral infection in children with compromised immune function. N. Engl. J. Med. **315:**77–81.

Hall, W. J., C. B. Hall, and D. M. Speers. 1978. Respiratory syncytial virus infection in adults. Clinical, virologic, and serial pulmonary function studies. Ann. Intern. Med. **88:**203–205.

Halonen, P., G. Obert, and J. C. Hierholzer. 1985. Direct detection of viral antigens in respiratory infection by immunoassays: a four year experience and new developments. *In* L. M. de la Maza and E. M. Pearson (ed.), Medical virology IV. Lawrence Erlbaum Associates, Hillsdale, NJ.

Hambling, M. H. 1964. Survival of the respiratory syncytial virus during storage under various conditions. Br. J. Exp. Pathol. **45:**647–655.

Hamparian, V. V., A. Ketler, M. R. Hilleman, C. M. Reilly, L. McClelland, D. Cornfeld, and J. Stokes, Jr. 1961. Studies of acute respiratory illnesses caused by respiratory syncytial virus. 1. Laboratory findings in 109 cases. Proc. Soc. Exp. Biol. Med. **106:**717–722.

Hart, R. J. C. 1984. An outbreak of respiratory syncytial virus infection in an old people's home. J. Infect. **8:**259–261.

Harter, D. H., and P. W. Choppin. 1967. Studies on pneumonia virus of mice (PVM) in cell culture: replication in baby hamster kidney cells and properties of the virus. J. Exp. Med. **126:**251–266.

Hayle, A. J. 1986. Culture of respiratory syncytial virus

infected diploid bovine nasal mucosa cells on cytodex 3 microcarriers. Arch. Virol. **89**:81–88.

Hebert, G. A. 1974. Ammonium sulfate fractionation of sera: mouse, hamster, guinea pig, monkey, chimpanzee, swine, chicken, and cattle. Appl. Microbiol. **27**:389–393.

Hebert, G. A., P. L. Pelham, and B. Pittman. 1973. Determination of the optimal ammonium sulfate concentration for the fractionation of rabbit, sheep, horse, and goat antisera. Appl. Microbiol. **25**:26–36.

Henderson, F. W., A. M. Collier, W. A. Clyde, and F. W. Denny. 1979. Respiratory-syncytial-virus infections, reinfections and immunity. N. Engl. J. Med. **300**:530–534.

Henderson, F. W., A. M. Collier, M. A. Sanyal, J. M. Watkins, D. L. Fairclough, W. A. Clyde, and F. W. Denny. 1982. A longitudinal study of respiratory viruses and bacteria in the etiology of acute otitis media with effusion. N. Engl. J. Med. **306**:1377–1383.

Hendry, R. M., and K. McIntosh. 1982. Enzyme-linked immunosorbent assay for detection of respiratory syncytial virus infection: development and description. J. Clin. Microbiol. **16**:324–328.

Hendry, R. M., B. F. Fernie, L. J. Anderson, E. Godfrey, and K. McIntosh. 1985. Monoclonal capture antibody ELISA for respiratory syncytial virus: detection of individual viral antigens and determination of monoclonal antibody specificities. J. Immunol. Methods. **77**:247–258.

Hendry, R. M., J. C. Burns, G. A. Prince, W. Rodriquez, H. W. Kim, V. G. Hemming, P. J. Wright, B. S. Graham, E. E. Walsh, K. McIntosh, and B. R. Murphy. 1986a. Infant sera recognize antigenically distinct strains of respiratory syncytial virus (RSV). Pediatr. Res. **311A**:917.

Hendry, R. M., B. F. Fernie, L. J. Anderson, and K. McIntosh. 1987. Antigenic and epidemiologic analysis of distinct strains of respiratory syncytial virus from two successive community outbreaks 1983–1985, p. 397–403. *In* B. W. J. Mahy and D. Kolakofsky (ed.), Biology of negative strand viruses, Elsevier, Amsterdam.

Hendry, R. M., L. T. Pierik, and K. McIntosh. 1986b. Comparison of washed nasopharyngeal cells and whole nasal secretions for detection of respiratory syncytial virus antigens by enzyme-linked immunosorbent assay. J. Clin. Microbiol. **23**:383–384.

Hendry, R. M., A. L. Talis, E. Godfrey, L. J. Anderson, B. F. Fernie, and K. McIntosh. 1986c. Concurrent circulation of antigenically distinct strains of respiratory syncytial virus during community outbreaks. J. Infect. Dis. **153**:291–297.

Hierholzer, J. C., and M. S. Hirsch. 1979. Croup and pneumonia in human infants associated with a new strain of respiratory syncytial virus. J. Invest. Dis. **140**:826–828.

Hierholzer, J. C., R. A. Coombs, and L. J. Anderson. 1984. Spectrophotometric quantitation of peroxidase-stained protein bands following gel electrophoresis and the Western blot transfer technique with respiratory syncytial virus. J. Virol. Methods **8**:265–268.

Hillis, W. D., M. R. Cooper, F. B. Bang, A. K. Dey, and K. V. Shah. 1971. Respiratory syncytial virus infection in children in West Bengal. Indian J. Med. Res. **59**:1354–1364.

Hodes, D. S., H. W. Kim, R. H. Parrott, E. Camargo, and R. M. Chanock. 1974. Genetic alteration in a temperature-sensitive mutant of respiratory syncytial virus after replication in vivi (37972). Proc. Soc. Exp. Biol. Med. **145**:1158–1164.

Holzel, A., L. Parker, W. H. Patterson, L. L. R. White, K. M. Thompson, and J. O. Tobin. 1963. The isolation of respiratory syncytial virus from children with acute respiratory disease. Lancet **1**:295–300.

Horn, M. E. C., E. Brain, and I. Gregg. 1975. Respiratory viral infection in childhood. A survey in general practice, Roehampton 1967–1972. J. Hyg. **74**:157–168.

Hornsleth, A. 1966. Effect of inhibitors and respiratory factors on the growth of respiratory syncytial (RS) virus in HEp-2 cell cultures. Acta Pathol. Microbiol. Scand. **68**:293–304.

Hornsleth, A. 1969. Growth of respiratory syncytial (RS) virus in HEp-2 cell cultures. Acta Pathol. Microbiol. Scand. **76**:637–645.

Hornsleth, A., E. Brenoe, B. Friis, F. U. Knudsen, and P. Uldall. 1981a. Detection of respiratory syncytial virus in nasopharyngeal secretions by inhibition of enzyme-linked immunosorbent assay. J. Clin. Microbiol. **14**:510–515.

Hornsleth, A., P. C. Grauballe, B. Friss, J. Genner, and I. R. Pedersen. 1981b. Production of antiserum to respiratory syncytial virus polypeptides: application in enzyme-linked immunosorbent assay. J. Clin. Microbiol. **14**:501–509.

Hornsleth, A., B. Friis, P. Andersen, and E. Brenoe. 1982. Detection of respiratory syncytial virus in nasopharyngeal secretions by ELISA: comparison with fluorescent antibody technique. J. Med. Virol. **10**:273–281.

Hornsleth, A., B. Friis, P. C. Grauballe, and P. A. Krasilnikof. 1984. Detection by ELISA of IgA and IgM antibodies in secretion and IgM antibodies in serum in primary lower respiratory syncytial virus infection. J. Med. Virol. **13**:149–161.

Hornsleth, A., N. Bech-Thomsen, and B. Friis. 1985a. Detection by ELISA of IgG-subclass-specific antibodies in primary respiratory syncytial (RS) virus infections. J. Med. Virol. **16**:321–328.

Hornsleth, A., N. Bech-Thomsen, and B. Friis. 1985b. Detection of RS-virus IgG-subclass-specific antibodies: variation according to age in infants and small children and diagnostic value in RS-virus-infected small infants. J. Med. Virol. **16**:329–335.

Hornsleth, A., B. Friis, and P. A. Krasilnikof. 1986. Detection of respiratory syncytial virus in nasopharyngeal secretions by a biotin-avidin elisa more sensitive than the fluorescent antibody technique. J. Med. Virol. **18**:113–117.

Huang, Y. T., and G. W. Wertz. 1982. The genone of respiratory syncytial virus is a negative-stranded RNA that codes for at least seven mRNA species. J. Virol. **43**:150–157.

Huang, Y. T., and G. W. Wertz. 1983. Respiratory syncytial virus mRNA coding assignments. J. Virol. **46**:667–672.

Huang, Y. T., P. L. Collins, and G. W. Wertz. 1985. Characterization of the 10 proteins of human respiratory syncytial virus: identification of a fourth envelope-associated protein. Virus Res. **2**:157–173.

Jacobs, J. W., and D. B. Peacock. 1970. Differentiation of actively and passively acquired complement-fixing antibodies in infants with respiratory syncytial virus infection. J. Med. Microbiol. **3**:313–324.

Jacobs, J. W., D. B. Peacock, B. D. Corner, E. O. Caul, and S. K. R. Clarke. 1971. Respiratory syncytial and other viruses associated with respiratory disease in infants. Lancet **1**:871–876.

Johnson, K. M., R. M. Chanock, D. Rifkind, H. M. Kravetz, and V. Knight. 1961. Respiratory syncytial virus IV. Correlation of virus shedding, serologic response, and illness in adult volunteers. J. Am. Med. Assoc. **176**:663–667.

Johnson, K. M., H. H. Bloom, M. A. Mufson, and R. M.

Chanock. 1962. Natural reinfection of adults by respiratory syncytial virus: possible relation to mild upper respiratory disease. N. Engl. J. Med. **267:**68–72.

Johnson, R. A., G. A. Prince, S. C. Suffin, R. L. Horswood, and R. M. Chanock. 1982. Respiratory syncytial virus infection in cyclophosphamide-treated cotton rats. Infect. Immun. **37:**369–373.

Joncas, J., I. Berthiaume, and V. Pavilanis. 1969a. The structure of the respiratory syncytial virus. Virology **38:**493–496.

Joncas, J. H., L. Berthiaume, R. Williams, P. Beaudry, and V. Pavilanis. 1969b. Diagnosis of viral respiratory tract infections by electron microscopy. Lancet **1:**956–959.

Jones, P. D., and G. L. Ada. 1986. Influenza virus-specific antibody-secreting cells in the murine lung during primary influenza virus infection. J. Virol. **60:**614–619.

Jordan, W. S. Jr. 1962. Growth characteristics of respiratory syncytial virus. J. Immunol. **88:**581–590.

Kalica, A. R., P. F. Wright, F. M. Hetrick, and R. M. Chanock. 1973. Electron microscopic studies of respiratory syncytial temperature-sensitive mutants. Arch. Ges. Virusforsch. **41:**248–258.

Kao, C., K. McIntosh, B. Fernie, A. Talis, L. Pierik, and L. Anderson. 1984. Monoclonal antibodies for the rapid diagnosis of respiratory syncytial virus infection by immunofluorescence. Diagn. Microbiol. Infect. Dis. **2:**199–206.

Kapikian, A. Z., J. A. Bell, F. M. Mastrota, K. M. Johnson, R. J. Huebner, and R. M. Chanock. 1961. An outbreak of febrile illness and pneumonia associated with respiratory syncytial virus infection. Am. J. Hyg. **74:**234–248.

Kapikian, A. Z., R. H. Mitchell, R. M. Chanock, R. A. Shvedoff, and C. E. Stewart. 1969. An epidemiologic study of altered clinical reactivity to respiratory syncytial (RS) virus infection in children previously vaccinated with an inactivated RS virus vaccine. Am. J. Epidemiol. **89:**405–421.

Kasupski, G. J., and W.-D. Leers. 1983. Case report: presumed respiratory syncytial virus pneumonia in three immunocompromised adults. Am. J. Med. Sci. **285:**28–33.

Kaul, A., R. Scott, M. Gallagher, M. Scott, J. Clement, and P. L. Ogra. 1978. Respiratory syncytial virus infection: rapid diagnosis in children by use of indirect immunofluorescence. Am. J. Dis. Child. **132:**1088–1090.

Kaul, T. N., R. C. Welliver, and P. L. Ogra. 1981a. Comparison of fluorescent-antibody, neutralizing-antibody, and complement-enhanced neutralizing-antibody assays for detection of serum antibody to respiratory syncytial virus. J. Clin. Microbiol. **13:**957–962.

Kaul, T. N., R. C. Welliver, D. T. Wong, R. A. Udwadia, K. Riddlesberger, and P. L. Ogra. 1981b. Secretory antibody response to respiratory syncytial virus infection. Am. J. Dis. Child. **135:**1013–1016.

Kaul, T. N., H. S. Faden, E. Middleton, T. Fugitani, and P. L. Ogra. 1982a. Release of pharmacological mediators from human neutrophils by viral immune complexes: Effects on mucosal smooth muscle function and possible role in clinical disease. Pediatr. Res. **16:**224A.

Kaul, T. N., R. C. Welliver, and P. L. Ogra. 1982b. Development of antibody-dependent cell-mediated cytotoxicity in the respiratory tract after natural infection with respiratory syncytial virus. Infect. Immun. **37:**492–498.

Kaul, T. N., R. C. Welliver, and P. L. Ogra. 1982c. Appearance of complement components and immunoglobulins on nasopharyngeal epithelial cells following naturally acquired infection with respiratory syncytial virus. J. Med. Virol. **9:**149–158.

Kessler, S. W. 1975. Rapid isolation of antigens from cells with a staphylococcal protein A–antibody absorbent: parameters of the interaction of antibody-antigen complexes with protein A. J. Immunol. **115:**1617–1624.

Kim, H. W., J. A. Bellanti, J. O. Arrobio, J. Mills, C. D. Brandt, R. M. Chanock, and R. H. Parrott. 1969a. Respiratory syncytial virus neutralizing activity in nasal secretions following natural infection. Proc. Soc. Exp. Biol. Med. **131:**658–661.

Kim, H. W., J. G. Canchola, C. D. Brandt, G. Pyles, R. M. Chanock, K. Jensen, and R. H. Parrott. 1969b. Respiratory syncytial virus disease in infants despite prior administration of antigenic inactivated vaccine. Am. J. Epidemiol. **89:**422–434.

Kim, H. W., J. O. Arrobio, G. Pyles, C. D. Brandt, E. Camargo, R. M. Chanock, and R. H. Parrott. 1971. Clinical and immunological response of infants and children to administration of low-temperature adapted respiratory syncytial virus. Pediatrics **48:**745–755.

Kim, H. W., J. O. Arrobio, C. D. Brandt, B. C. Jeffries, G. Pyles, J. L. Reid, R. M. Chanock, and R. H. Parrott. 1973a. Epidemiology of respiratory syncytial virus infection in Washington, D.C. I. Importance of the virus in different respiratory tract disease syndromes and temporal distribution of infection. Am. J. Epidemiol. **98:**216–225.

Kim, H. W., J. O. Arrobio, C. D. Brant, P. Wright, D. Hodes, R. M. Chanock, and R. H. Parrott. 1973b. Safety and antigenicity of temperature sensitive (TS) mutant respiratory syncytial virus (RSV) in infants and children. Pediatrics **52:**56–63.

Kim, H. W., S. L. Leikin, J. Arrobio, C. D. Brandt, R. M. Chanock, and R. H. Parrott. 1976. Cell-mediated immunity to respiratory syncytial virus induced by inactivated vaccine or by infection. Pediatr. Res. **10:**75–78.

Kim, H. W., R. G. Wyatt, B. F. Fernie, C. D. Brandt, J. O. Arrobio, B. C. Jeffries, and R. H. Parrott. 1983. Respiratory syncytial virus detection by immunofluorescence in nasal secretions with monoclonal antibodies against selected surface and internal proteins. J. Clin. Microbiol. **18:**1399–1404.

Kimball, A. M., H. M. Foy, M. K. Cooney, I. D. Allan, M. Matlock, and J. J. Plorde. 1983. Isolation of respiratory syncytial and influenza viruses from the sputum of patients hospitalized with pneumonia. J. Infect. Dis. **147:**181–184.

Kingsbury, D. W., M. A. Bratt, P. W. Choppin, R. P. Hanson, Y. Hosaka, V. ter Meulen, E. Norrby, W. Plowright, R. Rott, and W. H. Wunner. 1978. Paramyxoviridae. Intervirology **10:**137–152.

Kisch, A. L., K. M. Johnson, and R. M. Chanock. 1962. Immunofluorescence with respiratory syncytial virus. Virology **16:**177–189.

Klein, B. S., F. R. Dollete, and R. H. Yolken. 1982. The role of respiratory syncytial virus and other viral pathogens in acute otitis media. J. Pediatr. **101:**16–20.

Koff, W. C., F. R. Caplan, S. Case, and S. B. Halstead. 1983. Cell-mediated immune response to respiratory syncytial virus infection in owl monkeys. Clin. Exp. Immunol. **53:**272–280.

Korppi, M., P. Halonen, M. Kleemola, and K. Launiala. 1986. Viral findings in children under the age of two years with expiratory difficulties. Acta Paediatr. Scand. **75:**457–464.

Kravetz, H. M., V. Knight, R. M. Chanock, J. A. Morris, K. M. Johnson, D. Rifkind, and J. P. Utz. III. 1961. Production of illness and clinical observation in adult volunteers. J. Am. Med. Assoc. **176:**657–663.

Kurstak, E. 1985. Progress in enzyme immunoassays: production of reagents, experimental design, and interpretation. Bull. WHO 63:793–811.

Lambden, P. R. 1985. Nucleotide sequence of the respiratory syncytial virus phosphoprotein gene. J. Gen. Virol. 66:1607–1612.

Lambert, D. M., and M. W. Pons. 1983. Respiratory syncytial virus glycoproteins. Virology 130:204–214.

Lambert, D. M., M. W. Pons, G. N. Mbuy, and K. Dorsch-Hasler. 1980. Nucleic acids of respiratory syncytial virus. J. Virol. 36:837–846.

Lamontagne, L., J. P. Descoteaux, and R. Roy. 1985. Epizootiological survey of parainfluenza-3, reovirus-3, respiratory syncytial and infectious bovine rhinotracheitis viral antibodies in sheep and goat flocks in Quebec. Can. J. Comp. Med. 49:424–428.

Lamprecht, C. L., H. E. Krause, and M. A. Mufson. 1981. Role of maternal antibody in pneumonia and bronchiolitis due to respiratory syncytial virus. J. Infect. Dis. 134:211–217.

Lauer, B. A. 1982. Comparison of virus culturing and immunofluorescence for rapid detection of respiratory syncytial virus in nasopharyngeal secretions: sensitivity and specificity. J. Clin. Microbiol. 16:411–412.

Lauer, B. A., H. A. Masters, C. G. Wren, and M. J. Levin. 1985. Rapid detection of respiratory syncytial virus in nasopharyngeal secretions by enzyme-linked immunosorbent assay. J. Clin. Microbiol. 22:782–785.

Lehmkuhl, H. D., M. H. Smith, and R. C. Cutlip. 1980. Morphogenesis and structure of caprine respiratory syncytial virus. Arch. Virol. 65:269–276.

Levine, S. 1977. Polypeptides of respiratory syncytial virus. J. Virol. 21:427–431.

Levine, S., and R. Hamilton. 1969. Kinetics of the respiratory syncytial virus growth cycle in HeLa cells. Arch. Ges. Virusforsch. 28:122–132.

Lundgren, D. L., M. G. Magnuson, and W. E. Clapper. 1969. A serological survey in dogs for antibody to human respiratory viruses. Lab. Anim. Care 19:352–359.

MacDonald, N. E., C. B. Hall, S. C. Suffin, C. Alexson, P. J. Harris, and J. A. Manning. 1982. Respiratory syncytial viral infection in infants with congenital heart disease. N. Engl. J. Med. 307:397–400.

Marquez, A., and G. D. Hsiung. 1967. Influence of glutamine on multiplication and cytopathic effect of respiratory syncytial virus. Proc. Soc. Exp. Biol. Med. 124:95–99.

Martin, A. J., P. S. Gardner, and J. McQuillin. 1978. Epidemiology of respiratory viral infection among paediatric inpatients over a six-year period in north-east England. Lancet 2:1035–1038.

Mathur, U., D. W. Bentley, and C. B. Hall. 1980. Concurrent respiratory syncytial virus and influenza A infections in the institutionalized elderly and chronically ill. Ann. Intern. Med. 93:49–52.

McClelland, L., M. R. Hilleman, V. V. Hamparian, A. Ketler, C. M. Reilly, D. Cornfeld, and J. Stokes, Jr. 1961. Studies of acute respiratory illnesses caused by respiratory syncytial virus. 2. Epidemiology and assessment of importance. N. Engl. J. Med. 264:1170–1175.

McIntosh, K., and J. M. Fishaut. 1980. Immunopathologic mechanisms in lower respiratory tract disease of infants due to respiratory syncytial virus. Prog. Med. Virol. 26:94–118.

McIntosh, K., A. M. Arbeter, M. K. Stahl, I. A. Orr, D. S. Hodes, and E. F. Ellis. 1974. Attenuated respiratory syncytial virus vaccines in asthmatic children. Pediatr. Res. 8:689–696.

McIntosh, K., H. B. Masters, I. Orr, R. K. Chao, and R. M. Barkin. 1978. The immunologic response to infection with respiratory syncytial virus in infants. J. Infect. Dis. 138:24–32.

McIntosh, K. J. McQuillin, and P. S. Gardner. 1979. Cell-free and cell-bound antibody in nasal secretions from infants with respiratory syncytial virus infection. Infect. Immun. 23:276–281.

McIntosh, K., R. M. Hendry, M. L. Fahnestock, and L. T. Pierik. 1982. Enzyme-linked immunosorbent assay for detection of respiratory syncytial virus infection: application of clinical samples. J. Clin. Microbiol. 16:329–333.

Meguro, H., M. Kervina, and P. F. Wright. 1979. Antibody-dependent cell-mediated cytotoxicity against cells infected with respiratory syncytial virus: characterization of in vitro and in vivo properties. J. Immunol. 122:2521–2526.

Meissner, H. C., S. A. Murray, M. A. Kiernan, D. R. Snydman, and K. McIntosh. 1984. A simultaneous outbreak of respiratory syncytial virus and parainfluenza virus type 3 in a newborn nursery. J. Pediatr. 104:680–684.

Meurman, O., O. Ruuskanen, H. Sarkkinen, P. Hanninen, and P. Halonen. 1984a. Immunoglobulin class-specific antibody response in respiratory syncytial virus infection measured by enzyme immunoassay. J. Med. Virol. 14:67–72.

Meurman, O., H. Sarkkinen, O. Ruuskanen, P. Hanninen, and P. Halonen. 1984b. Diagnosis of respiratory syncytial virus infection in children: comparison of viral antigen detection and serology. J. Med. Virol. 14:61–65.

Miller, H. R., H. Phipps, and E. Rossier. 1986. Reduction of nonspecific fluorescence in respiratory specimens by pretreatment with N-acetylcysteine. J. Clin. Microbiol. 24:470–471.

Mills, J. V., J. E. Van Kirk, P. F. Wright, and R. M. Chanock. 1971. Experimental respiratory syncytial virus infection of adults. J. Immunol. 107:123–130.

Mills, B. G., F. R. Singer, L. P. Weiner, and P. A. Holst. 1981. Immunohistological demonstration of respiratory syncytial virus antigens in Paget disease of bone. Proc. Natl. Acad. Sci. USA 78:1209–1213.

Milner, M. E., S. M. de la Monte, and G. M. Hutchins. 1985. Fatal respiratory syncytial virus infection in severe combined immunodeficiency syndrome. Am. J. Dis. Child. 139:1111–1114.

Minnich, L., and C. G. Ray. 1980. Comparison of direct immunofluorescent staining of clinical specimens for respiratory virus antigens with conventional isolation techniques. J. Clin. Microbiol. 12:391–394.

Minnich, L. L., and C. G. Ray. 1982. Comparison of direct and indirect immunofluorescence staining of clinical specimens for detection of respiratory syncytial virus antigen. J. Clin. Microbiol. 15:969–970.

Mintz, L., R. A. Ballard, S. H. Sniderman, R. S. Roth, and W. L. Drew. 1979. Nosocomial respiratory syncytial virus infections in an intensive care nursery: rapid diagnosis by direct immunofluorescence. Pediatrics 64:149–153.

Mok, J. Y. Q., and H. Simpson. 1982. Outcome of acute lower respiratory tract infection in infants: preliminary report of seven-year follow-up study. Br. Med. J. 285:333–337.

Monto, A. S., and S. K. Lim. 1971. The Tecumseh study of respiratory illness: Am. J. Epidemiol. 94:290–301.

Morales, F., M. A. Calder, J. M. Inglis, P. S. Murdoch, and J. Williamson. 1983. A study of respiratory infections in the elderly to assess the role of respiratory syncytial virus. J. Infect. 7:236–247.

Morris, J. A., R. E. Blount, Jr., and R. E. Savage. 1956. Recovery of cytopathogenic agent from chimpanzees with coryza (22538). Proc. Soc. Exp. Med. Biol. **92:**544–549.

Moss, P. D., M. O. Adams, and T. O. Tobin. 1963. Serological studies with respiratory syncytial virus. Lancet **1:**298–300.

Mufson, M. A., H. D. Levine, R. E. Wasil, H. E. Mocega-Gonzales, and H. E. Krause. 1973. Epidemiology of respiratory syncytial virus infection among infants and children in Chicago. Am. J. Epidemiol. **98:**88–95.

Mufson, M. A., C. Orvell, B. Rafnar, and E. Norrby. 1985. Two distinct subtypes of human respiratory syncytial virus. J. Gen. Virol. **66:**2111–2124.

Murphy, B. R., D. W. Alling, M. H. Snyder, E. E. Walsh, G. A. Prince, R. M. Chanock, V. G. Hemming, W. J. Rodriguez, H. W. Kim, B. S. Graham, and P. F. Wright. 1986a. Effect of age and preexisting antibody on serum antibody response of infants and children to the F and G glycoproteins during respiratory syncytial virus infection. J. Clin. Microbiol. **24:**894–898.

Murphy, B. R., B. S. Graham, G. A. Prince, E. E. Walsh, R. M. Chanock, D. T. Karzon, and P. F. Wright. 1986b. Serum and nasal-wash immunoglobulin G and A antibody response of infants and children to respiratory syncytial virus F and G glycoproteins following primary infection. J. Clin. Microbiol. **23:**1009–1014.

Murphy, B. R., G. A. Prince, E. E. Walsh, H. W. Kim, R. H. Parrott, V. G. Hemming, W. J. Rodriguez, and R. M. Chanock. 1986c. Dissociation between serum neutralizing and glycoprotein antibody responses of infants and children who received inactivated respiratory syncytial virus vaccine. J. Clin. Microbiol. **24:**197–202.

Murphy, D., J. K. Todd, R. K. Chao, I. Orr, and K. McIntosh. 1981. The use of gowns and masks to control respiratory illness in pediatric hospital personnel. J. Pediatr. **99:**746–750.

Naot, Y., and J. S. Remington. 1980. An enzyme-linked immunosorbent assay for detection of IgM antibodies to *Toxoplasma gondii:* use for diagnosis of acute acquired toxoplasmosis. J. Infect. Dis. **142:**757–766.

Neligan, G. A., H. Steiner, P. S. Gardner, and J. McQuillin. 1970. Respiratory syncytial virus infection of the newborn. Br. Med. J. **3:**146–147.

Norrby, E., H. Marusyk, and C. Orvell. 1970. Morphogenesis of respiratory syncytial virus in a green monkey kidney cell line (Vero). J. Virol. **6:**237–242.

Norrby, E., G. Enders-Ruckle, and V. ter Meulen. 1975. Differences in the appearance of antibodies to structural components of measles virus after immunization with inactivation and live virus. J. Infect. Dis. **132:**262–269.

Ogilvie, M. M., A. S. Vathenen, M. Radford, J. Codd, and S. Key. 1981. Maternal antibody and respiratory syncytial virus infection in infancy. J. Med. Virol. **7:**263–271.

Olmsted, R. A., N. Elango, G. A. Prince, B. R. Murphy, P. R. Johnson, B. Moss, R. M. Chanock, and P. L. Collins. 1986. Expression of the F glycoprotein of respiratory syncytial virus by a recombinant vaccinia virus: comparison of the individual contributions of the F and G glycoproteins to host immunity. Proc. Natl. Acad. Sci. USA **83:**7462–7466.

Orstavik, I., M. Grandien, P. Halonen, P. Arstila, C. H. Mordhorst, A. Hornsleth, T. Popow-Kraupp, J. McQuillin, P. S. Gardner, J. Almeida, F. Bricout, and A. Marques. 1984. Viral diagnoses using the rapid immunofluorescence technique and epidemiological implications of acute respiratory infections among children in different European countries. Bull. WHO **62:**307–313.

PHLS. 1983. Communicable disease surveillance center, respiratory syncytial virus infection in the elderly 1976–82. Br. Med. J. **287:**1618–1619.

Paccaud, M. F., and C. L. Jacquier. 1970. A respiratory syncytial virus of bovine origin. Arch. Ges. Virusforsch. **30:**327–342.

Paisley, J. W., B. A. Lauer, K. McIntosh, M. P. Glode, J. Schachter, and C. Rumack. 1984. Pathogens associated with acute lower respiratory tract infection in young children. Pediatr. Infect. Dis. **3:**14–19.

Pan American Health Organization. 1982. Acute respiratory infections in children. Pan American Health Organization, Washington, D.C., 1982.

Parrott, R. H., A. Vargosko, A. Luckey, H. W. Kim, C. Cumming, R. Chanock. 1959. Clinical features of infection with hemadsorption viruses. N. Engl. J. Med. **260:**731–738.

Parrott, R. H., A. J. Vargosko, H. W. Kim, C. Cumming, H. Turner, R. J. Huebner, and R. M. Chanock. 1961. Respiratory syncytial virus. II. Serologic studies over a 34-month period of children with bronchiolitis, pneumonia, and minor respiratory diseases. J. Am. Med. Assoc. **176:**653–657.

Parrott, R. H., H. W. Kim, J. O. Arrobio, D. S. Hodes, B. R. Murphy, C. D. Brandt, E. Camargo, and R. M. Chanock. 1973. Epidemiology of respiratory syncytial virus infection in Washington, D.C. II. Infection and disease with respect to age, immunologic status, race and sex. Am. J. Epidemiol. **98:**289–300.

Peeples, M., and S. Levine. 1979. Respiratory syncytial virus polypeptides: their location in the virion. Virology **95:**137–145.

Peeples, M., and S. Levine. 1980. Metabolic requirements for the maturation of respiratory syncytial virus. J. Gen. Virol. **50:**81–88.

Pons, M. W., A. L. Lambert, D. M. Lambert, and O. M. Rochovansky. 1983. Improvement of respiratory syncytial virus replication in actively growing HEp-2 Cells. J. Virol. Methods **7:**217–221.

Popow-Kraupp, T., G. Kern, G. Binder, W. Tuma, M. Kundi, and C. Knuz. 1986. Detection of respiratory syncytial virus in nasopharyngeal secretions by enzyme-linked immunosorbent assay, indirect immunofluorescence, and virus isolation: a comparative study. J. Med. Virol. **19:**123–134.

Pothier, P., J. C., Nicolas, P. G. De Saint Maur, S. Ghim, A. Kazmierczak, and F. Bricout. 1985. Monoclonal antibodies against respiratory syncytial virus and their use for rapid detection of virus in nasopharyngeal secretions. J. Clin. Microbiol. **21:**286–287.

Prince, G. A., and D. D. Porter. 1976. The pathogenesis of respiratory syncytial virus infection in infant ferrets. Am. J. Pathol. **82:**339–352.

Prince, G. A., A. B. Jenson, R. L. Horswood, E. Camargo, and R. M. Chanock. 1978. The pathogenesis of respiratory syncytial virus infection in cotton rats. Am. J. Pathol. **93:**771–792.

Prince, G. A., R. L. Horswood, J. Berndt, S. C. Suffin, and R. M. Chanock. 1979a. Respiratory syncytial virus infection in inbred mice. Infect. Immun. **26:**764–766.

Prince, G. A., S. C. Suffin, D. A. Prevar, E. Camargo, D. L. Sly, W. T. London, and R. M. Chanock. 1979b. Respiratory syncytial virus infection in owl monkeys: viral shedding, immunological response, and associated illness caused by wild-type virus and two temperature-sensitive mutants. Infect. Immun. **26:**1009–1013.

Prince, G. A., R. L. Horswood, E. Camargo, S. C. Suffin, and R. M. Chanock. 1982. Parental immunization with live respiratory syncytial virus is blocked in seropositive cotton rats. Infect. Immun. **37:**1074–1078.

Prince, G. A., R. L. Horswood, E. Camargo, D. Koenig, and R. M. Chanock. 1983. Mechanisms of immunity to respiratory syncytial virus in cotton rats. Infect. Immun. **42:**81–87.

Prince, G. A., V. G. Hemming, R. L. Horswood, and R. M. Chanock. 1985a. Immunoprophylaxis and immunotherapy of respiratory syncytial virus infection in the cotton rat. Virus Res. **3:**193–206.

Prince, G. A., R. L. Horswood, and R. M. Chanock. 1985b. Quantitative aspects of passive immunity to respiratory syncytial virus infection in infant cotton rats. J. Virol. **55:**517–520.

Prince, G. A., R. L. Horswood, D. W. Koenig, and R. M. Chanock. 1985c. Antigenic analysis of a putative new strain of respiratory syncytial virus. J. Infect. Dis. **151:**634–637.

Prince, G. A., A. B. Jenson, V. G. Hemming, B. R. Murphy, E. E. Walsh, R. L. Horswood, and R. M. Chanock. 1986. Enhancement of respiratory syncytial virus pulmonary pathology in cotton rats by prior intramuscular inoculation of Formalin-inactivated virus. J. Virol. **57:**721–728.

Pringle, C. R., and A. Cross. 1978. Neutralization of respiratory syncytial virus by cat serum. Nature **276:**501–502.

Pringle, C. R., P. V. Shirodaria, P. Cash, D. J. Chiswell, and P. Malloy. 1978. Initiation and maintenance of persistent infection by respiratory syncytial virus. J. Virol. **28:**199–211.

Pringle, C. R., P. V. Shirodaria, H. B. Gimenez, and S. Levine. 1981. Antigen and polypeptide synthesis by temperature-sensitive mutants of respiratory syncytial virus. J. Gen. Virol. **54:**173–183.

Pullan, C. R., and E. N. Hey. 1982. Wheezing, asthma, and pulmonary dysfunction 10 years after infection with respiratory syncytial virus in infancy. Br. Med. J. **284:**1665–1669.

Pullan, C. R., G. L. Toms, A. J. Martin, P. S. Gardner, J. K. G. Webb, and D. R. Appleton. 1980. Breast-feeding and respiratory syncytial virus infection. Br. Med. J. **281:**1034–1036.

Reilly, C. M., J. Stokes, Jr., L. McClelland, D. Cornfeld, V. V. Hamparian, A. Ketler, and M. R. Hilleman. 1961. Studies of acute respiratory illnesses caused by respiratory syncytial virus. 3. Clinical and laboratory findings. N. Engl. J. Med. **264:**1176–1182.

Richardson, L. S., R. B. Belshe, W. T. London, D. L. Sly, D. A. Prevar, E. Camargo, and R. M. Chanock. 1978a. Evaluation of five temperature-sensitive mutants of respiratory syncytial virus in primates. I. Viral shedding, immunologic response, and associated illness. J. Med. Virol. **3:**91–100.

Richardson, L. S., R. H. Yolken, R. B. Belshe, E. Camargo, H. W. Kim, and R. M. Chanock. 1978b. Enzyme-linked immunosorbent assay for measurement of serological response to respiratory syncytial virus infection. Infect. Immun. **20:**660–664.

Richardson-Wyatt, L. S., R. B. Belshe, W. T. London, D. L. Sly, E. Camargo, and R. M. Chanock. 1981. Respiratory syncytial virus antibodies in nonhuman primates and domestic animals. Lab. Anim. Sci. **31:**413–415.

Richman, A. V., and N. M. Tauraso. 1971. Growth of respiratory syncytial virus in suspension cell culture. Appl. Microbiol. **22:**1123–1125.

Richman, A. V., F. A. Pedreia, and N. M. Tauraso. 1971. Attempts to demonstrate hemagglutination and hemadsorption by respiratory syncytial virus. Appl. Microbiol. **21:**1099–1100.

Ross, C. A. C., E. J. Stott, S. McMichael, and I. A. Crowther. 1964. Problems of laboratory diagnosis of respiratory syncytial virus infection in childhood. Arch. Ges. Virusforsch. **14:**553–562.

Routledge, E. G., J. McQuillin, A. C. R. Samson, and G. L. Toms. 1985. The development of monoclonal antibodies to respiratory syncytial virus and their use in diagnosis by indirect immunofluorescence. J. Med. Virol. **15:**305–320.

Samson, A. C. R., M. M. Willcocks, F. G. Routledge, L. A. Morgan, and G. L. Toms. 1986. A neutralizing monoclonal antibody to respiratory syncytial virus which binds to both F_1 and F_2 components of the fusion protein. J. Gen. Virol. **67:**1479–1483.

Sarkkinen, H. K., P. E. Halonen, P. P. Arstila, and A. A. Salmi. 1981. Detection of respiratory syncytial, parainfluenza type 2, and adenovirus antigens by radioimmunoassay and enzyme immunoassay on nasopharyngeal specimens from children with acute respiratory disease. J. Clin. Microbiol. **13:**258–265.

Sarkkinen, H., O. Ruuskanen, O. Meurman, H. Puhakka, E. Virolainen, and J. Eskola. 1985. Identification of respiratory virus antigens in middle ear fluids of children with acute otitis media. J. Infect. Dis. **151:**444–448.

Satake, M., and S. Venkatesan. 1984. Nucleotide sequence of the gene encoding respiratory syncytial virus matrix protein. J. Virol. **50:**92–99.

Satake, M., N. Elango, and S. Venkatesan. 1984. Sequence analysis of the respiratory syncytial virus phosphoprotein gene. J. Virol. **52:**991–994.

Satake, M., J. E. Coligan, N. Elango, E. Norrby, and S. Venkatesan. 1985. Respiratory syncytial virus envelope glycoprotein (G) has a novel structure. Nucleic Acids Res. **13:**7795–7812.

Schauf, V., C. Purcell, M. Mizen, and S. Mizen. 1979. Lymphocyte transformation in response to antigens of respiratory syncytial virus. Proc. Soc. Exp. Biol. Med. **161:**564–569.

Schieble, J. H., A. Kase, and E. H. Lennette. 1967. Fluorescent cell counting as an essay method for respiratory syncytial virus. J. Virol. **1:**494–499.

Scott, R., and P. S. Gardner. 1970. Respiratory syncytial virus neutralizing activity in nasopharyngeal secretions. J. Hyg. **68:**581–588.

Scott, R., and P. S. Gardner. 1974. The local antibody response to R.S. virus infection in the respiratory tract. J. Hyg. **72:**111–120.

Scott, R., M. O. De Landazuri, P. S. Gardner, and J. J. T. Owen. 1977. Human antibody-dependent cell-mediated cytotoxicity against target cells infected with respiratory syncytial virus. Clin. Exp. Immunol. **28:**19–26.

Scott, R., A. Kaul, M. Scott, Y. Chiba, and P. L. Ogra. 1978. Development of in vitro correlates of cell-mediated immunity to respiratory syncytial virus infection in humans. J. Infect. Dis. **137:**810–817.

Scott, R., M. Scott, and G. L. Toms. 1981. Cellular and antibody response to respiratory syncytial (RS) virus in human colostrum, maternal blood, and cord blood. J. Med. Virol. **8:**55–66.

Scott, R., C. R. Pullan, M. Scott, and J. McQuillin. 1984. Cell-mediated immunity in respiratory syncytial virus disease. J. Med. Virol. **13:**105–114.

Senterfit, L. B., and P. B. Baldridge. 1974. Separation and concentration of respiratory syncytial virus (RSV) antigens. J. Immunol. Methods **4:**349–367.

Shekarchi, I. C., J. L. Sever, Y. J. Lee, G. Castellano, and D. L. Madden. 1984. Evaluation of various plastic microtiter plates with measles, toxoplasma, and gamma globulin antigens in enzyme-linked immunosorbent assays. J. Clin. Microbiol. **19:**89–96.

Sieber, O. F., M. L. Wilska, and R. Riggin. 1976. Elevated nitroblue tetrazolium dye reduction test response in acute viral respiratory disease. Pediatrics **58:**122–124.

Sims, D. G. 1981. A two year prospective study of hospital-acquired respiratory virus infection on paediatric wards. J. Hyg. **86:**335–342.

Sims, D. G., M. A. P. S. Downham, J. McQuillan, and P. S. Gardner. 1976. Respiratory syncytial virus infection in north-east England. Br. Med. J. **2:**1095–1098.

Sims, D. G., M. A. P. S. Downham, P. S. Gardner, J. K. G. Webb, and D. Weightman. 1978. Study of 8-year-old children with a history of respiratory syncytial virus bronchiolitis in infancy. Br. Med. J. **1:**11–14.

Smith, T. F., K. McIntosh, M. Fishaut, and P. M. Henson. 1981. Activation of complement by cells infected with respiratory syncytial virus. Infect. Immun. **33:**43–48.

Sobeslavsky, O., M. Bruckova, L. Kunzova, K. Voitechovsky, and L. Syrucek. 1970. Adenovirus, RS virus and M. pneumoniae infections in young population of Prague in 1962–1967. J. Hyg. Epidemiol. Microbiol. **14:**350–359.

Sorvillo, F. J., S. F. Huie, M. A. Strassburg, A. Butsumyo, W. X. Shandera, and S. L. Fannin. 1984. An outbreak of respiratory syncytial virus pneumonia in a nursing home for the elderly. J. Infect. **9:**252–256.

Spelman, D. W., and P. A. Stanley. 1983. Respiratory syncytial virus pneumonitis in adults. Med. J. Aust. **1:**430–431.

Spence, L., and N. Barratt. 1968. Respiratory syncytial virus associated with acute respiratory infections in Trinidadian patients. Am. J. Epidemiol. **88:**257–266.

Spriggs, M. K., and P. L. Collins. 1986. Human parainfluenza virus type 3: messenger RNAs, polypeptide coding assignments, intergenic sequences, and genetic map. J. Virol. **59:**646–654.

Spring, S. B., and M. D. Tolpin. 1983. Enzymatic cleavage of a glycoprotein of respiratory syncytial virus. Arch. Virol. **76:**359–363.

Sterner, G., S. Wolontis, B. Bloth, and G. de Hevesky. 1966. Respiratory syncytial virus: an outbreak of acute respiratory illness in a home for infants. Acta Paediatr. Scand. **55:**273–279.

Stott, E. J., L. A. Ball, K. K. Young, J. Furze, and G. W. Wertz. 1986. Human respiratory syncytial virus glycoprotein G expressed from a recombinant vaccinia virus vector protects mice against live-virus challenge. J. Virol. **60:**607–613.

Sturdy, P. M., J. McQuillin, and P. S. Gardner. 1969. A comparative study of methods for the diagnosis of respiratory virus infections in childhood. J. Hyg. **67:**659–670.

Sun, C., P. R. Wyde, and V. Knight. 1983a. Correlation of cytotoxic activity in lungs to recovery of normal and gamma-irradiated cotton rats from respiratory syncytial virus infection. Am. Rev. Respir. Dis. **128:**668–672.

Sun, C., P. R. Wyde, S. Z. Wilson, and V. Knight. 1983b. Cell-mediated cytotoxic responses in lungs of cotton rats infected with respiratory syncytial virus. Am. Rev. Respir. Dis. **127:**460–464.

Suto, T., N. Yano, M. Ikeda, M. Miyamoto, S. Takai, S. Shigeta, Y. Hinuma, and N. Ishida. 1965. Respiratory syncytial virus infection and its serologic epidemiology. Am. J. Epidemiol. **82:**211–224.

Swenson, P. D., and M. H. Kaplan. 1986. Rapid detection of respiratory syncytial virus in nasopharyngeal aspirates by a commercial enzyme immunoassay. J. Clin. Microbiol. **23:**485–488.

Taber, L. H., V. Knight, B. E. Gilbert, H. W. McClung, S. Z. Wilson, H. J. Norton, J. M. Thurson, W. H. Gordon, R. L. Atmar, and W. R. Schlaudt. 1983. Ribarvirin aerosol treatment of bronchiolitis associated with respiratory syncytial virus infection in infants. Pediatrics **72:**613–618.

Taylor, G., E. J. Stott, M. Bew, B. Fernie, P. J. Cote, A. P. Collins, M. Hughes, and J. Jebbett. 1984a. Monoclonal

antibodies protect against respiratory syncytial virus infection in mice. Immunology **52:**137–142.

Taylor, G., E. J. Stott, M. Hughes, and A. P. Collins. 1984b. Respiratory syncytial virus infection in mice. Infect. Immun. **43:**649–655.

Thomas, L. H., E. J. Stott, A. P. Collins, S. Crouch, and J. Jebbett. 1984a. Infection of gnotobiotic calves with a bovine and human isolate of respiratory syncytial virus. Modification of the response by dexamethasone. Arch. Virol. **79:**67–77.

Thomas, L. H., E. J. Stott, A. P. Collins, and J. Jebbett. 1984b. Experimental pneumonia in gnotobiotic calves produced by respiratory syncytial virus. Br. J. Exp. Pathol. **65:**19–28.

Tidwell, R. R., J. D. Geratz, and E. J. Dubovi. 1983. Aromatic amidines: comparison of their ability to block respiratory syncytial virus induces cell fusion and to inhibit plasmin, urokinase, thrombin, and trypsin. J. Med. Chem. **28:**294–298.

Trepanier, P., P. Payment, and M. Trudel. 1980. A simple and rapid microassay for the titration of human respiratory syncytial virus. J. Virol. Methods **1:**343–347.

Trepanier, P., P. Payment, and M. Trudel. 1981. Concentration of human respiratory syncytial virus using ammonium sulfate, polyethylene glycol or hollow fiber ultrafiltration. J. Virol. **3:**201–211.

Trepanier, P., P. Payment, and M. Trudel. 1983. Modified immunoprecipitation procedure for the identification of human respiratory syncytial virus polypeptides. J. Virol. Methods **7:**149–154.

Treuhaft, M. W., J. M. Soukup, and B. J. Sullivan. 1985. Practical recommendations for the detection of pediatric respiratory syncytial virus infections. J. Clin. Microbiol. **22:**270–273.

Tyeryar, F. J. 1983. Report of a workshop on respiratory syncytial virus and parainfluenza viruses. J. Infect. Dis. **148:**588–598.

Tyrrell, D. A. J. 1963. Discovering and defining the etiology of acute respiratory viral disease. Am. Rev. Respir. Dis. **88:**77–84.

Ueba, O. 1978. Respiratory syncytial virus. I. Concentration and purification of the infectious virus. Acta Med. Okayama **32:**266–272.

Ueba, O. 1980. Respiratory syncytial virus. II. Isolation and morphology of the glycoproteins. Acta Med. Okayama **34:**245–254.

Urquhart, G. E. D., and G. H. Walker. 1972. Immunofluorescence for routine diagnosis of respiratory syncytial virus infection. J. Clin. Pathol. **25:**843–845.

Vainionpaa, R., O. Meurman, and H. Sarkkinen. 1985. Antibody response to respiratory syncytial virus structural proteins in children with acute respiratory syncytial virus infection. J. Virol. **53:**976–979.

Valenti, W. M., T. A. Clarke, C. B. Hall, M. A. Menegus, and D. L. Shapiro. 1982. Concurrent outbreaks of rhinovirus and respiratory syncytial virus in an intensive care nursery: epidemiology and associated risk factors. J. Pediatr. **100:**722–726.

Van Dyke, R. B. 1986. Detection of respiratory syncytial virus (RSV) in clinical samples by nucleic acid hybridization. Program and abstracts of the Twenty-sixth Interscience Conference on Antimicrobial Agents and Chemotherapy, p. 83.

Vaur, L., H. Agut, A. Garbarg-Chenon, P. G. de Saint-Maur, J. C. Nicolas, and F. Bricout. 1986. Simplified enzyme-linked immunosorbent assay for specific antibodies to respiratory syncytial virus. J. Clin. Microbiol. **24:**596–599.

Venkatesan, S., N. Elango, R. M. Chanock. 1983. Construction and characterization of cDNA clones for four respiratory syncytial viral genes. Proc. Natl. Acad. Sci. USA **80**:1280–1284.

Wagner, D. K., B. S. Graham, P. F. Wright, E. E. Walsh, H. W. Kim, C. B. Reimer, D. L. Nelson, R. M. Chanock, and B. R. Murphy. 1986. The serum IgG antibody subclass responses to respiratory syncytial virus F and G glycoproteins following primary infection. J. Clin. Microbiol. **24**:304–306.

Wagner, D. K., D. L. Nelson, E. E. Walsh, C. B. Reimer, F. W. Henderson, and B. R. Murphy. 1987. Differential IgG antibody subclass titers to respiratory syncytial virus F and G glycoproteins in adults. J. Clin. Microbiol. **25**:748–750.

Wallace, S. J., H. Zealley. 1970. Neurological, electroencephalographic, and virological findings in febrile children. Arch. Dis. Child. **45**:611–623.

Walsh, E. E., and J. F. Hruska. 1983a. Identification of the virus-specific proteins of respiratory syncytial virus temperature-sensitive mutants by immunoprecipitation (41546). Proc. Soc. Exp. Biol. Med. **172**:202–206.

Walsh, E. E., and J. Hruska. 1983b. Monoclonal antibodies to respiratory syncytial virus proteins: identification of the fusion protein. J. Virol. **47**:171–177.

Walsh, E. E., J. J. Schlesinger, and M. W. Brandriss. 1984a. Protection from respiratory syncytial virus infection in cotton rats by passive transfer of monoclonal antibodies. Infect. Immun. **43**:756–758.

Walsh, E. E., J. J. Schlesinger, and M. W. Brandriss. 1984b. Purification and characterization of GP90, one of the envelope glycoproteins of respiratory syncytial virus. J. Gen. Virol. **65**:761–767.

Walsh, E. E., M. W. Brandriss, and J. J. Schlesinger. 1985. Purification and characterization of the respiratory syncytial virus fusion protein. J. Gen. Virol. **66**:409–415.

Walsh, E. E., P. J. Cote, B. F. Fernie, J. J. Schlesinger, and M. W. Brandriss. 1986. Analysis of the respiratory syncytial virus fusion protein using monoclonal and polyclonal antibodies. J. Gen. Virol. **67**:505–513.

Waner, J. L., N. J. Whitehurst, S. Jonas, L. Wall, and H. Shalaby. 1986. Isolation of viruses from specimens submitted for direct immunofluorescence test for respiratory syncytial virus. J. Pediatr. **108**:249–250.

Ward, K. A., P. R. Lambden, M. M. Ogilvie, and P. J. Watt. 1983. Antibodies to respiratory syncytial virus polypeptides and their significance in human infection. J. Gen. Virol. **64**:1867–1876.

Ward, K. A., J. S. Everson, P. R. Lambden, and P. J. Watt. 1984. Antigenic and structural variation in the major nucleocapsid protein of respiratory syncytial virus. J. Gen. Virol. **65**:1749–1757.

Welliver, R. C., A. Kaul, and P. L. Ogra. 1979. Cell-mediated immune response to respiratory syncytial virus infection: relationship to the development of reactive airway disease. J. Pediatr. **94**:370–375.

Welliver, R. C., T. N. Kaul, and P. L. Ogra. 1980a. The appearance of cell-bound IgE in respiratory-tract epithelium after respiratory-syncytial-virus infection. N. Engl. J. Med. **303**:1198–1202.

Welliver, R. C., T. N. Kaul, T. I. Putnam, M. Sun, K. Riddlesberger, and P. L. Ogra. 1980b. The antibody response to primary and secondary infection with respiratory syncytial virus: kinetics of class-specific responses. J. Pediatr. **96**:808–813.

Welliver, R. C., M. Sun, D. Rinaldo, and P. L. Ogra. 1985. Respiratory syncytial virus-specific IgE responses following infection: evidence for a predominantly mucosal response. Pediatr. Res. **19**:420–424.

Welliver, R. C., D. T. Wong, M. Sun, E. Middleton, Jr., R. S. Vaughan, and P. L. Ogra. 1981. The development of respiratory syncytial virus-specific IgE and the release of histamine in nasopharyngeal secretions after infection. N. Engl. J. Med. **305**:841–846.

Wenzel, R. P., E. C. Deal, and J. O. Hendley. 1977. Hospital-acquired viral respiratory illness on a pediatric ward. Pediatrics **60**:367–371.

Wertz, G. W., P. L. Collins, Y. Huang, C. Gruber, S. Levine, and L. A. Ball. 1985. Nucleotide sequence of the G protein gene of human respiratory syncytial virus reveals an unusual type of viral membrane protein. Proc. Natl. Acad. Sci. USA **82**:4075–4079.

Wertz, G. W., E. J. Stott, K. K. Y. Young, K. Anderson, and L. A. Ball. 1987. Expression of the fusion protein of human respiratory syncytial virus from recombinant vaccinia virus vectors and protection of vaccinated mice. J. Virol. **61**:293–301.

Wong, D. T., and P. L. Ogra. 1986. Neonatal respiratory syncytial virus infection: role of transplacentally and breast milk-acquired antibodies. J. Virol. **57**:1203–1206.

Wong, D. T., M. Rosenband, K. Hovey, and P. L. Ogra. 1985. Respiratory syncytial virus infection in immunosuppressed animals: implications in human infection. J. Med. Virol. **17**:359–370.

Wright, P. F., W. G. Woodend, and R. M. Chanock. 1970. Temperature-sensitive mutants of respiratory syncytial virus: in-vivo studies in hamsters. J. Infect. Dis. **122**:501–512.

Wright, P. F., J. V. Mills, and R. M. Chanock. 1971. Evaluation of a temperature-sensitive mutant of respiratory syncytial virus in adults. J. Infect. Dis. **124**:505–511.

Wright, P. F., M. A. Gharpure, D. S. Hodes, and R. M. Chanock. 1973. Genetic studies of respiratory syncytial virus temperature-sensitive mutants. Arch. Virusforsch. **41**:238–247.

Wright, P. F., T. Shinozaki, W. Fleet, S. H. Sell, J. Thompson, and D. T. Karzon. 1976. Evaluation of a live, attenuated respiratory syncytial virus vaccine in infants. J. Pediatr. **88**:931–936.

Wright, P. F., R. B. Belshe, H. W. Kim, L. P. Van Voris, and R. M. Chanock. 1982. Administration of a highly attenuated, live respiratory syncytial virus vaccine to adults and children. Infect. Immun. **37**:397–400.

Wulff, H., P. Kidd, and H. A. Wenner. 1964. Respiratory syncytial virus: observations on antigenic heterogeneity. Proc. Soc. Exp. Biol. Med. **115**:240–243.

Wunner, W. H., and C. R. Pringle. 1976. Respiratory syncytial virus proteins. Virology **73**:228–243.

Zakstelskaya, L. Y., J. D. Almeida, and C. M. P. Bradstreet. 1967. The morphological microscope technique. Acta Virol. **11**:420–423.

Zrein, M., G. Obert, and M. H. V. van Regenmortel. 1986. Use of egg-yolk antibody for detection of respiratory syncytial virus in nasal secretions by ELISA. Arch. Virol. **90**:197–206.

CHAPTER 29

Rhabdoviridae: Rabies and Vesicular Stomatitis Viruses

PIERRE SUREAU, MONIQUE LAFON, and GEORGE M. BAER

Diseases: Rabies, hydrophobia (rage, tollwut, la rabia, mal de cadeiras); vesicular stomatitis.

Etiologic Agents: Lyssaviruses include lyssavirus serotype 1 (rabies virus), lyssavirus serotype 2 (Lagos bat virus), lyssavirus serotype 3 (Mokola virus), lyssavirus serotype 4 (Duvenhage virus); vesicular stomatitis viruses.

Source: Rabies—dogs, foxes, skunks, raccoons, insectivorous bats, vampire bats; vesicular stomatitis virus—cattle, horses, pigs.

Clinical Manifestations: Rabies—encephalitis, often with intermittent hyperexcitability, hydrophobia and aerophobia, following a prodromal period with tingling at the bite site; paralysis, usually ascending (Landry), coma, and death. Vesicular stomatitis—tongue vesicles and hoof (coronary band) and snout lesions in cattle, horses, and pigs.

Laboratory Diagnosis: Rabies—fluorescence of brain impressions by the fluorescent antibody technique. Isolation of virus in brain or saliva by mouse inoculation or by neuroblastoma cell culture. Vesicular stomatitis—complement fixation testing of antigen in vesicular fluid, with confirmation by fluorescent antibody and neutralization.

Epidemiology: Rabies—worldwide, except in a few countries, mostly islands such as Australia, England, and Japan. A disease of dogs in Asia, Africa, and Latin America, but of wild animals (foxes, skunks, raccoons, and insectivorous bats) in Europe, Canada, and the United States. Vampire bat rabies extends from northern Mexico to northern Argentina. Vesicular stomatitis—limited to the western hemisphere, endemic in the tropical and subtropical countries, and sporadically extends, with outbreaks, to temperate areas.

Treatment: None.

Prevention and Control: Rabies—prexposure vaccination of dogs and cats; local wound treatment and postexposure vaccination of humans; preexposure vaccination of humans at special risk. Vesicular stomatitis—awaits clarification of the epidemiology of the disease.

The family *Rhabdoviridae* (the rhabdoviruses) encompasses more than 100 viruses of vertebrates, invertebrates, and plants; the virions of all have a distinctive bullet-shaped morphology. Important human pathogens occur in two subgroups of the family, the genus *Lyssavirus,* of which rabies and the rabies-like viruses are members, and the genus *Vesiculovirus,* containing the vesicular stomatitis viruses. Only the lyssaviruses and the vesiculoviruses will be discussed in this chapter.

Lyssaviruses

Introduction

Canine rabies and its transmission to humans is mentioned in the Eshnunna code, 23rd Century BC (Theodorides, 1986; Tierkel, 1975a). The Greeks called rabies Lyssa and the human disease Hydrophobia. Democritus and Aristoteles described canine rabies in the 4th and 5th Centuries BC (Steele, 1975).

Canine rabies remains a serious health problem in developing countries, with rabid dogs causing many human exposures and deaths (WHO Survey, 1984–1985). Many human cases go unreported, as shown in a study done between 1962 and 1966 in which serially accessed human cadavers from Cali, Colombia, were examined, with 2% being diagnosed in retrospect as rabies deaths. However, no diagnosis of rabies had been made there, and the authors concluded that "this alarming percentage of rabies cases . . . is a clear reflection of the situation as it exists in our community, and which we are sure is no different from that in the rest of the country. It may be that our figures are higher because of our care in finding cases . . . We believe that a careful examination of pathological material in other teaching hospitals would show that the situation in Cali really is no different from that in the rest of the country" (Sanmartín et al., 1967). Canine rabies has only been controlled in the developed countries of Europe, the United States, and Canada. In these countries wildlife cases continue to be routinely reported (Centers for Disease Control, 1986; Schneider, 1985).

The relatively few human rabies cases in the developed world stand in sharp contrast to the continuing number of cases in developing countries; in India alone an estimated 40,000 to 50,000 human cases occur annually (Sehgal and Bhatia, 1985), and comparable figures have been cited for Thailand (Warrell, 1987) and the Philippines (Beran et al., 1972).

Until 1885, when Pasteur introduced rabies vaccination, many different treatments were recommended for people bitten by rabid dogs, including cautery with a red-hot iron (an effective technique used in the Middle Ages). Before the end of the 19th Century there was no laboratory diagnosis for rabies. Even the dog that bit the first patient vaccinated by Pasteur (Joseph Meister, on 6 July, 1885) was only considered rabid because its stomach was "full of hay, straw and fragments of wood" (Pasteur, 1885). Meister was vaccinated with 14 daily doses of vaccine prepared from rabbit spinal cords. The treatment has now been refined to include the injection of human (or equine) rabies immune globulin followed by five injections of a potent human diploid vaccine.

Galtier (1879) was the first person to transmit rabies by the subcutaneous inoculation of rabbits, but Pasteur found that intracerebral inoculation of rabbits with the brain tissue from rabid dogs was the most reliable technique for virus isolation and serial passage (Pasteur, 1881). During the beginning of the 20th Century, the intracerebral inoculation of rabbits or guinea pigs was commonly used to diagnose rabies. Rabbits were considered safer than guinea pigs because they always developed paralytic rabies (Remlinger, 1917), even though guinea pigs appeared

to be more susceptible to the virus (Marie, 1930). Rats also were found to be susceptible, but they generally developed furious rabies (Galli-Valerio, 1906). Later, Chinese and Syrian hamsters were both found to be susceptible to street rabies virus (Yen, 1936), but the best isolation procedure discovered was the intracerebral inoculation of mice (Webster and Dawson, 1935). Over many years, mouse inoculation proved to be accurate, rapid, and the least expensive virus isolation and identification method (Leach, 1938; Sulkin and Wilett, 1939).

Canine rabies control efforts only began 35 years after humans were first vaccinated (Umeno and Doi, 1921), but the steep drop in the incidence of dog rabies only began after World War II when effective canine vaccines came into use as part of comprehensive control programs (Tierkel, 1975b). Where dog rabies has been controlled wildlife cases exceed canine cases by a ratio of 50:1 or greater; foxes, skunks, raccoons, and insectivorous bats are primarily affected. In Latin America, a unique epidemiologic cycle exists in which the bites of rabid vampire bats cause hundreds of thousands of cattle deaths annually (Acha, 1967).

The genus *Lyssavirus* includes Duvenhage virus (named after a man who died after being bitten by a sick bat in Transvaal, South Africa), Mokola virus (first isolated from shrews near Ibadan, Nigeria), and Lagos bat virus (from frugivorous bats on Lagos Island, Nigeria) (Tignor and Shope, 1972). Common rabies vaccines appear to protect against all these viruses except Mokola. A list of miscellaneous rhabdoviruses is given in Table 1.

Physicochemical Properties and Morphology

The lyssaviruses and the vesiculoviruses are approximately 75 nm wide, 180 nm long, and consist of a lipid-containing bilayer envelope with glycoprotein projections surrounding a helically wound nucleocapsid. This helically wound nucleocapsid gives the viruses their distinctive bullet-shaped morphology. The nucleocapsid contains a single linear molecule of negative-sense single-stranded RNA, size 13 to 16 kb. The viruses contain five proteins (terms and molecular mass values for vesicular stomatitis Indiana virus), including a transcriptase (L, 190K), a nucleoprotein (N, 50K), a matrix protein (M, 20–30K), a protein associated with the transcriptase (NS, 40–50K), and a glycoprotein (G, 69K) that forms the surface projections and contains the epitopes against which neutralizing antibodies are directed. The chemical composition and properties of rabies and numerous other rhabdoviruses have been described (Banerjee, 1987; McSharry, 1979).

The main feature of rabies infection is neurotropism, presumably due to a specific acetylcholine re-

TABLE 1. Unclassified members of the *Rhabdoviridae* and their geographic distribution

Virus	Species from which virus has been isolated	Geographic distribution
Bovine ephemeral fever	Cattle, culicoides	Africa, Australia
Kimberley	Cattle, mosquito	Australia
Adelaide River	Cattle	Australia
Berrimah	Cattle	Australia
		Japan
		Middle East
Marco	Lizards	Brazil
Kern Canyon	Bats	California
Klamath	Rodents	California
Barur	Bats, mosquitoes, rodents, ticks	Kenya, India
Flanders	Birds, mosquitoes	North America
Hart Park	Birds, mosquitoes	North America
Mosqueiro	Mosquitoes	Brazil
Mount Elgon bat	Bats	Kenya
Joinjakaka	Mosquitoes	New Guinea
Navarro	Birds	Colombia
Le Dantec	Humans	Senegal
Keuraliba	Gerbils	Senegal
Almpiwar	Skink	Australia
Aruac	Mosquitoes	Trinidad
Boteke	Mosquitoes	Central African Republic
Gray Lodge	Mosquitoes	California, USA
Inhangapi	Spiny rat	Brazil
Kolongo	Birds	Central African Republic
Nkolbisson	Mosquitoes	Cameroon
Parry Creek	Cattle	Australia
Kwatta	Mosquitoes	Surinam
BeAn 157575	Birds	Brazil
Mossuril	Birds, mosquitoes	Central and Southern Africa
Sawgrass	Ticks	Florida
Connecticut	Ticks	Connecticut
New Minto	Ticks	Alaska
Sena Madureira	Lizard	Brazil
Bangoran	Bird	Central African Republic
Charleville	Lizard, mosquitoes	Australia
Cuiaba	Toad	Brazil
Kamese	Mosquitoes	Uganda; Central Africa
Chaco	Lizards	Brazil
S-1643	Mosquitoes	Sarawak
BFN-3187	Mosquitoes	California
Oita 296	Bats	Japan
Kununurra	Mosquitoes	Australia
Tibrogargan	Mosquitoes	Australia
Egtved (viral hemorrhagic septicemia)	Trout	Europe
Sigma	Fruit fly	France
Infectious hematopoietic necrosis	Salmon, trout	United States
Spring viremia of carp	Carp	Europe
Pike fry disease	Pike	Europe
Rhabdovirus of eels	Eels	Japan
Rhabdovirus of grass carp	Grass carp	
Rhabdovirus of *entamoeba*	*Entamoeba sp.*	
Rhabdovirus of blue crab	crab	Canada

Reprinted by permission from Brown, F. and J. Crick. 1979. Natural history of the rhabdoviruses of vertebrates and invertebrates. *In* Rhabdoviruses, Vol. I, D. H. L. Bishop, (ed), CRC Press, Boca Raton, Florida; and Calisher, C. H. and N. Karabatsos. Arbovirus serogroups: definition and geographic distributions. *In* Epidemiology of arthropod-borne viral diseases. T. P. Monath (ed), CRC Press, Boca Raton, Florida.

ceptor on neurons (Lentz et al., 1982). Further evidence for this specificity of this receptor is the close homology between the amino acid sequence of the rabies virus glycoprotein and a snake venom neurotoxin directed at the same acetylcholine receptor (Lentz et al., 1984). Because rabies has a wide host range in vitro (Clark and Wiktor, 1972; Wiktor and Clark, 1975) an alternate receptor binding action has been suggested (Reagan and Wunner, 1984; Wunner et al., 1984).

The rabies virus receptor detected on fibroblasts appears to share many characteristics with phosphatidylserine, a major vesicular stomatitis virus (VSV) receptor (Schlegel et al., 1983).

The viral portion that first attaches to cells appears to be the external glycoprotein (Perrin et al., 1982; Reagan and Wunner, 1984); attachment apparently occurs at the flat end of the virion (Iwasaki et al., 1973).

After its attachment to the host cell membrane, the virus can penetrate into the cytoplasm by (1) fusion of the virus and cell membranes and direct nucleocapsid injection into the cytoplasm, or (2) pinocytosis, in which the virus is endocytosed and passes into endosomes. Electron micrographs show that both mechanisms occur (Iwasaki, 1973).

Replication

After rabies virus attaches to cells, the virus penetrates the cytoplasm and is uncoated. Its replication strategy is similar to that of other negative-stranded RNA viruses: The virion RNA is transcribed into five monocistronic mRNAs by virus transcriptase; these mRNAs are then translated, forming the viral proteins. Although virus replication is entirely cytoplasmic, the nucleus may be involved because virus maturation is reduced in enucleated cells (Wiktor and Koprowski, 1974).

The duration of the rabies virus growth cycle depends on the virus strain, the host cell, and the multiplicity of infection. There is little cytopathic effect or depression of cellular DNA, RNA, or protein synthesis (Madore and England, 1977).

By analogy with VSV, it is assumed in rabies infection that a full-length positive strand of RNA is produced and assembled into nucleocapsids, which serve as a template for the production of negative-stranded genomic progeny RNA.

With VSV the viral polymerase enzyme is thought to switch from transcription to replication when enough N protein has been synthesized to protect the newly formed genomic RNA (Blumberg et al., 1981). Virus maturation is initiated as soon as nucleocapsid strands containing genomic RNA are available. Virus particles form on plasma membranes or endoplasmic reticulum membranes modified by the insertion of glycoprotein. Truncated defective interfering particles, having the same structure but incomplete RNA, are formed in the same way. These particles have a low infectivity and interfere with infectious virus application (Kawai et al. 1975).

Antigenic Composition and Genetics

ANTIGENIC STRUCTURE OF RABIES GLYCOPROTEIN

The rabies virus glycoprotein is the major immunity-inducing antigen. Virus neutralizing antibodies that develop after rabies vaccination are directed against the virus glycoprotein (Cox et al., 1977; Crick and Brown, 1969, 1970; Wiktor et al., 1973). The glycoprotein is also a target for immune T-lymphocytes, as shown in proliferative T-lymphocyte assays (Macfarlan et al., 1984), and in cytotoxic assays (Wiktor et al., 1984).

Monoclonal neutralizing antibodies have been used to delineate epitopes on the glycoprotein. There are three independent epitopes on ERA (Ellen, Rocketwicki: Abelseth) and CVS (Challenge Virus Standard) rabies strains and two additional sites on the ERA strain (Lafon et al., 1983, 1984).

ANTIGENIC STRUCTURE OF STREET RABIES AND RELATED VIRUSES

The nucleocapsid proteins of all lyssaviruses share common epitopes as shown by cross-reactivity of polyclonal antinucleocapsid antibodies, but quantitative differences exist (Schneider et al., 1973; Tignor et al., 1977). Monoclonal nucleocapsid antibodies (Wiktor and Koprowski, 1978) have resolved three epitopes on the N protein and two on the NS; all of these epitopes are common to all laboratory rabies strains except high egg passage (HEP) (Lafon and Wiktor, 1985). Rabies-related viruses (Mokola, Lagos bat, and Duvenhage), however, exhibit major differences in their N protein reactivity.

Selected N and NS monoclonal antibodies have helped differentiate numerous rabies and rabies-related viruses (Wiktor et al., 1980) such as street rabies isolates from Canada (Charlton et al., 1982), Europe, Africa, Asia (Schneider et al., 1985; Sureau et al., 1983), and the United States (Koprowski et al., 1985; Rupprecht and Wiktor, 1984). The differences in the nucleocapsid epitopes are powerful markers for identification of viral strains and geographic variants (Rupprecht and Wiktor, 1984; Rollin and Sureau, 1984; Smith et al., 1984; Sureau et al., 1983).

The glycoproteins of rabies and rabies-related viruses differ considerably by virus neutralization and cross-protection tests in mice (Koprowski et al., 1985; Tignor and Shope, 1972; Schneider et al., 1973; Schneider et al., 1982). On the basis of these serologic tests, the lyssaviruses have been grouped into four serotypes by the World Health Organization (WHO, 1973): serotype 1, rabies laboratory strains; serotype 2, Lagos bat; serotype 3, Mokola; serotype 4, Duvenhage. The use of glycoprotein monoclones has helped explain the failure of rabies vaccine to protect against Mokola virus challenge (Foggin, 1983; Koprowski et al., 1985). Mokola virus is not neutralized by the 42 monoclonal antibodies that neutralize rabies virus strains (Wiktor et al., 1983), nor is there any cross-reactivity between Mokola and rabies viruses by cytotoxic T-lymphocyte assay (Koprowski et al., 1985).

Pathogenesis, Pathophysiology, and Pathology of Rabies Virus

After it penetrates the skin, rabies virus first replicates in muscle cells (Murphy and Bauer, 1974), where it lies quiescent for most of the prolonged and variable incubation period (Baer and Cleary, 1972). The virus then reaches the peripheral nerves via the neuromuscular junctions and moves, apparently passively, along the peripheral axons and the spinal ganglia to the spinal cord, and then rapidly to the brain. During this period no antibodies are detectable nor are there any clinical signs of disease. After critical changes in the brain occur, several clinical signs appear, including (in humans) parasthesia at the bite site, nuchal rigidity, incoordination, and behavioral changes (including alternating periods of extreme excitability and calm).

In experimentally infected animals, virus is found in the spinal cord and brain at least a week before central nervous system signs are noted (Baer et al., 1968). The clinical signs of rabies appear to follow pathophysiologic changes in neuronal functions caused by infection (Hattwick and Gregg, 1975). Most humans with rabies die before antibodies appear in the cerebrospinal fluid. The deleterious effect of antibody has been clearly shown in a mouse study in which antibodies were suppressed by the administration of an immunosuppressive agent (Cytoxan, Mead Johnson, Evansville, Indiana). Rabid mice remained clinically normal for weeks despite having very high levels of virus in the brain during that period (Smith et al., 1982); immune serum when injected caused death almost immediately.

Lesions in rabies are limited, with congestion of the meningeal vessels being the only common gross finding. Almost all the known histologic changes in rabies were first described between 1870 and 1900; these include perivascular cuffing, Babes' nodules, neuronophagia, and in some cases modest neuronal necrosis. Extensive descriptions of the pathologic changes in rabies have been published (Perl, 1975).

Clinical Features

Mode of Transmission

Human rabies almost always follows the bite or scratch of a rabid animal or, very rarely, licks on the mucous membranes (Leach and Johnson, 1940). Exceptional cases of infection by aerosol have been reported, either in bat caves (Constantine, 1967) or in the laboratory (Tillotson et al., 1977; Winkler et al., 1973). There have been four unexpected cases of human-to-human transmission of rabies via corneal transplants (Galian et al., 1980; Houff et al., 1979; Thongcharoen et al., 1981). There are no other documented cases of human-to-human rabies transmission (Anderson et al., 1984; Marder, 1986; Remington et al., 1985).

Incubation Period

The incubation period usually varies between 20 to 60 days. Virus transmitted by bites on the head may result in appreciably shortened incubation periods (Baltazard and Ghoddsi, 1954). Incubation periods of less than 15 days or more than a year are unusual (Hattwick and Gregg, 1975). There are no signs or symptoms during the incubation period.

Period of Infectivity

Although rabies virus is often found in the salivary glands and saliva of humans with rabies (Leach and Johnson, 1940), there is no documented case of human-to-human transmission by bite. The peripheral distribution of virus may reach the cornea, the tissue involved in the four known cases of human-to-human transmission. It is not known how long the virus can be excreted in human saliva before symptoms appear, although this period ranges from 0 to 14 days in dogs (Fekadu et al., 1982).

Clinical Course

The first signs and symptoms noted are usually at the bite site and include pruritis, neuritis, and paresthesia. These signs start 1 to 4 days before further neurologic manifestations appear. Prodromal neurologic signs include anxiety, insomnia, malaise, headache, and fever; they are seen 2 to 10 days before the acute neurologic phase (Hattwick and Gregg, 1975).

The two clinical forms observed are the typical "furious rabies," in which hyperactivity is dominant, and "paralytic rabies," in which paralysis and paraplegia cloud the diagnosis.

In furious rabies, hyperactivity may occur sponta-neously or be caused by light, noise, or clinical ex-amination; hyperactive episodes consist of agitation, seizures, aggressiveness, or even biting. A typical reaction is hydrophobia, characterized by painful pharyngeal spasms when the patient attempts to drink, or even at the sight of a glass of water. Such spasms also may be provoked by a simple draft of air (aerophobia). The patient is relatively lucid but anx-ious between hyperactive episodes. Difficulty in swallowing saliva produces foaming at the mouth and continuous spitting. After a few days, paralytic symptoms appear and the mental status gradually de-teriorates from confusion to stupor, coma, and death.

Paralytic signs predominate in paralytic rabies. The paralysis is of the ascending (Landry) type, with paraplegia extending to quadriplegia, followed by re-spiratory paralysis, encephalitis, coma, and death (Hattwick and Gregg, 1975). Such paralysis was ob-served in two patients not diagnosed as rabid who contributed corneas for transplantation (Galian et al., 1980; Houff et al., 1979). Rabies is rarely diag-nosed in people who die after paralytic symptoms in tropical countries because adequate laboratory con-firmation is rare (Sanmartín et al., 1967; Thong-charoen et al., 1981).

Human rabies is almost always rapidly lethal; in a recent study of 177 cases in India, 93% died within 5 days of onset (Lakhanpal and Sharma, 1985). When intensive supportive treatment is applied, however, prolonged survival has been reported: 64 days in one case (Rubin et al., 1970) and 133 in another (Emmons et al., 1973). To date there have been only three doc-umented human recoveries from rabies; two in pa-tients who had received postexposure treatment (Hattwick et al., 1972; Porras et al., 1976) and one in a preimmunized laboratory worker infected by aero-sol (Tillotson et al., 1977).

Complications

The complications of human rabies have been re-viewed by Hattwick and Gregg (1975). They are nu-merous and may involve many major organ systems. Neurologic complications include cerebral edema, hydrocephalus, and seizures; pulmonary complica-tions include hypoxia, respiratory arrest, and pneu-monia; cardiac complications include arrhythmias, congestive heart failure, hypotension, and cardiac arrest; miscellaneous complications include second-ary bacterial infections, hypothermia, hyperpyrexia, and gastrointestinal bleeding. Despite the intensive treatments used to resolve these complications in the past, no unvaccinated humans have survived rabies (Lennette and Emmons, 1971; Rubin et al., 1970). In

many fatal cases with prolonged survival, hypoxia may be a partial cause of the neurologic damage.

Epidemiology and Natural History

The epidemiology of rabies depends very much on the geographic area and animal species involved (Baer, 1984). The global incidence of rabies can be divided into three general categories. A few coun-tries are free from rabies, including Australia, New Zealand, Japan, England, Spain, Portugal, most of the Scandinavian countries, and almost all of the small Caribbean islands. Another limited group of countries have mostly wildlife rabies, either in foxes (most European countries and Canada) or skunks and raccoons (the United States). The remaining countries—almost all of Asia, Africa, and Latin America—are affected with endemic dog rabies just as they have been for centuries.

The epidemiology of rabies in various species has been described (Baer, 1975). Rabies in terrestrial ani-mals is almost always transmitted by bite; the bitten animals usually sicken after an incubation period of 2 to 8 weeks and then transmit the disease to other animals by bite. The popular belief that almost all people bitten by rabid dogs die is not true: in studies of bitten persons in the years before the Pasteur treatment was adopted, "only" 15 to 20% died (Babes, 1912), indicating the relative resistance of humans to rabies. However, tens of thousands of persons die of rabies every year, and the number does not appear to be decreasing. Almost all human rabies deaths come from dog bites (WHO Rabies Survey, 1984–1985, PAHO 1984), and the dying per-sons have rarely received any postexposure treat-ment.

Prevention, Prophylaxis, and Control

Rabies is best controlled by eliminating the virus from dog populations by vaccinating owned dogs and removing stray animals. This has been accomplished in numerous rabies-free islands such as England, Ja-pan, and Taiwan. The disease in dogs also has been reduced to almost nil in some large land areas such as western Europe, Canada, and the United States, but various wildlife species continue to spread the dis-ease in these places.

In countries where it has not been possible to con-trol the disease in dogs, many human exposures still occur and many people receive the postexposure prophylaxis recommended by the World Health Or-ganization. Most persons exposed in developing countries, however, receive treatment with vaccine only.

Laboratory Diagnosis

The histologic diagnosis of rabies dates from the end of the last century, when alterations in the nervous system were described (Babes, 1892; Van Gehuchten, 1900). Negri's (1903) description of the pathognomonic cytoplasmic inclusions of street rabies virus infection was a decisive discovery. More recently, electron microscopy has helped in the study of Negri bodies for diagnosing the disease (Matsumoto, 1963; Miyamoto and Matsumoto, 1965).

Specimen Collection

Street rabies virus in saliva, in salivary glands, and in the central nervous system is highly infectious and should be considered hazardous (CDC-NIH Biosafety in Microbiological and Biomedical Laboratories, 1984). All precautions should be taken to avoid human exposure when collecting specimens and performing diagnostic procedures.

Preexposure Immunization

Preexposure rabies immunization is an absolute prerequisite for all laboratory staff. Immunization consists of three injections of tissue culture vaccine on days 0, 7, and 28 by the subcutaneous route (1.0 ml) or by the intradermal route (0.1 ml). The titers of persons vaccinated should be checked regularly, and a minimum level of 0.5 IU/ml should be maintained. Booster injections are given every 2 years (or as necessary to maintain a minimum titer) as long as the person remains at risk (WHO, 1984a). In case of accidental exposure (such as a finger cut), the affected area should be disinfected, and two booster doses of vaccine should be given (WHO, 1984a).

Immunization is an efficient measure against the risk of laboratory exposure, but should only be considered one step, along with other safety precautions described below.

It must be noted that there are two instances in which preexposure immunization might not be protective: exposure to one of the African rabies-related viruses, such as Mokola, against which rabies vaccines are not effective and accidental exposure to infectious aerosols. In fact, inhalation has been incriminated in two laboratory accidents not even due to street rabies virus; the first case occurred in a veterinarian who homogenized goat brains infected with fixed rabies virus in a "kitchen-type" blender (Winkler et al., 1973); the second case occurred in a laboratory worker who sprayed a suspension of modified live virus South Alabama Dufferin (SAD) in a pharmaceutical manufacturing machine (Tillotson et al., 1977).

Packaging and Shipment of the Specimens

The procedures, described in detail by Tierkel (1973), may be summarized as follows: the head of the animal suspected of being rabid (or the whole body for small species) is placed in a heavy gauge plastic bag, securely closed by a knot; this bag is then placed in a large double-walled outer plastic bag that has been partly filled with crushed ice. This bag is closed with a knot and placed inside an insulated polystyrene box protected by a carton box or carefully wrapped in thick paper and sealed with tape. The contents of the package must be clearly indicated "specimen for rabies diagnosis" and should be shipped by express freight via car, rail, or air to the authorized diagnostic laboratory. The specimen may be frozen for long-distance shipment. When refrigeration is not available, small specimens may be preserved in 50% glycerol-saline in screw-cap glass or plastic bottles, also carefully wrapped.

Necropsy and Removal of Animal Brains

The opening of the skull and removal of the brain is the most risky part of the necropsy procedure. Gowns, plastic aprons, heavy rubber gloves, and face shields should always be worn. The most simple tools are the best, such as a butcher's knife, hammer, and strong cutting blades; electric circular saws are less suitable as they may generate aerosols.

In laboratories, necropsies should be performed inside a biologic safety cabinet (Trimarchi et al., 1979). The grinding of brains or salivary glands for inoculation should also be done inside a biological safety cabinet (Trimarchi et al., 1979). The use of electric blenders or homogenizers must be prohibited outside the safety cabinet.

Direct Virus Detection

Electron microscopy of central nervous system specimens may permit direct detection of rabies virus particles. This is not a routine diagnostic procedure in animals but has been used for the diagnosis and pathologic study of human rabies. (Baer et al., 1982; Conomy et al., 1977; DeBrito et al., 1973; Garcia-Tamayo et al., 1972; Gonzalez-Angulo et al., 1970; Lamas et al., 1980; Larraza et al., 1978; Morecki and Zimmerman, 1969; Sung et al., 1976.)

In this technique, fresh specimens of the central nervous system, cut into approximately 1-mm^3 blocks, are placed in glutaraldehyde for 2 h at 4°C, fixed for 30 min in 1% phosphate-buffered osmium tetroxide, dehydrated in ethanol, and embedded in an araldite-epon mixture. Negri bodies can be detected in thick sections stained with toluidine blue, as can viral matrices of various sizes, shapes, and numbers in neurons. Thin sections stained with uranyl and lead acetate show typical rabies virions, bullet-shaped in longitudinal sections (approximately 75 nm in diameter and 180 nm in length) or circular or elliptical in transverse sections, budding from cytoplasmic membranes. Occasional tubular structures of variable diameter and length (Matsumoto, 1963) may also be seen.

Direct Antigen Detection

Direct antigen detection, the best technique for rapid rabies diagnosis, can be done by the fluorescent, immunoperoxidase, or rapid rabies enzyme immunodiagnosis techniques. All require potent hyperimmune serum for conjugate preparation.

IMMUNE GLOBULIN

The virus used for immunization should be a Pasteur fixed virus, either CVS or PV/PM. The antigen may be brain suspension (Dean and Abelseth, 1973) or, preferably, virus grown in tissue culture, especially in BHK-21 cells (Kissling, 1975). The antigen should be inactivated with β-propiolactone at a concentration of 1/4000, pH 7.2 to 7.4, for 2 h at 37°C. The ideal antigen, however, is a purified nucleocapsid preparation that induces specific antinucleocapsid antibodies, because fluorescent rabies antigen is mostly nucleocapsid. The technique is described by Sokol (1973): BHK-21 cells infected with PV virus are disrupted in distilled water and homogenized, the nucleocapsid-containing cell extract is centrifuged in cesium chloride, and the nucleocapsids are banded twice (1.32 g/ml) and inactivated with β-propiolactone.

Several inoculation schedules may be used for immunization. One is to inject hamsters with four weekly intraperitoneal doses of inactivated antigen, followed by two booster injections; the immunized hamsters are bled 2 weeks after the last booster (Dean and Abelseth, 1973). Another method is to inject rabbits with four intramuscular weekly doses of 200 μg of nucleocapsid proteins in 500 μl of Freund's adjuvant, one every week, followed by two booster injections. The immunized rabbits are bled 1 week after the second booster, and the antibody level is determined. Immunized rabbits can be kept for 1 year with monthly bleedings (30 ml) if booster injections are given (Atanasiu et al., 1974).

The immune globulin fraction of the serum should be used for conjugation. The globulin is precipitated by repeated ammonium sulfate treatment (Dean and Abelseth, 1973) or collected by chromatography on an ion exchange cellulose (DEAE Sephadex A50,

Pharmacia, Inc., Piscataway, NJ) column (Atanasiu et al., 1974; Joustra and Lundgren, 1969).

Monoclonal antibodies against nucleocapsid antigens are useful in diagnosing and identifying rabies and rabies-related viruses (Wiktor and Koprowski, 1978; Wiktor, et al., 1980). Such antibodies may be used in the indirect immunofluorescent technique (using an antimouse fluorescent conjugate) or by direct conjugation to fluorescein isothiocyanate (FITC) for the direct test. For actual laboratory diagnosis, several monoclonal antibodies are pooled. At least one pool is now commercially available (Durham et al., 1986).

FLUORESCENT ANTIBODY (FA) TEST

The greatest improvement in the rapid diagnosis of rabies was the introduction of fluorescent antibody (FA) staining of rabies antigen in brains and salivary glands about 30 years ago (Goldwasser and Kissling, 1958; Goldwasser et al., 1959). Years of experience in laboratories around the world (Atanasiu et al., 1970; Lennette et al., 1971; McQueen et al., 1960; Wachendorfer, 1967) have confirmed it as "the most accurate microscopic test . . . available for the diagnosis of rabies" (Dean and Abelseth, 1973). High-quality conjugates currently available and the recent addition of epifluorescence microscopic equipment have further improved the technique.

To prepare conjugate, the purified immune globulin is conjugated with FITC to a final protein concentration of 20 mg/ml and a FITC/protein ratio of 1:100. The unattached FITC is eliminated by filtration of the conjugate on a Sephadex G50 (Millipore Corp., Bedford, MA) column in phosphate buffer, pH 7.2. The conjugate is then filtered through a 0.22 μm Millipore filter (Atanasiu et al., 1974). The concentrated conjugate is distributed in convenient aliquots and stored frozen or lyophilized to avoid loss of potency. Each batch of conjugate must be titrated to determine its optimal dilution for diagnosis. In the titration, twofold dilutions of conjugate are dropped on impressions of positive brain material. The highest dilution giving excellent fluorescence (++++) and little or no background fluorescence is the dilution to use for calculating the working dilution, which should be four times this concentration. Each laboratory must determine its own working dilution, as results vary with differing fluorescent microscopes in different laboratories.

A concentrated, nonabsorbed conjugate should be used for that titration because the working dilution may fall and prediluted preparations might not be concentrated enough (Larghi et al., 1986). There rarely is nonspecific staining when conjugate is used on infected tissue cultures, but some may be noted when infected brain tissue impressions are stained; it can be reduced or eliminated by absorbing the conjugate with a suspension of normal brain. The specificity of the FA staining can be checked by absorbing the conjugate with a suspension of rabies-infected brains, which diminishes (or eliminates) specific staining in duplicate diagnostic specimens. These absorptions are done by mixing two types of brain suspension with conjugate: four volumes of either normal or infected (CVS) mouse brain (20% weight/volume) suspension in phosphate-buffered saline (PBS), pH 7.2–7.4, are mixed with one volume each of undiluted conjugate. After a 1-h absorption at room temperature the mixtures are centrifuged for 10 min at 1,000 *g* and distributed in aliquots for a weeks work, then stored frozen until used (Dean and Abelseth, 1973).

In most instances brain specimens are submitted for the FA. Several areas of the brain should be examined, including brainstem, cerebellum, and hippocampus (Ammon's horn); the thalamus and the cerebral cortex may be included if deemed appropriate. Only the brainstem and Ammon's horn are routinely examined in laboratories where many specimens are examined daily. The most adequate samples for FA testing are fresh refrigerated tissues; frozen material, shipped in dry ice, also may be used. Glycerol-preserved specimens must be thoroughly rinsed in PBS and blotted on filter paper before impressions are made.

The mucus in salivary gland material makes it difficult to prepare proper impressions; the glands should be blotted with filter paper so that they will stick to glass slides. An alternate procedure is to mince the gland and then crush small pieces onto glass slides. Brain samples fixed in formalin may also be submitted for FA, although the fluorescence is reduced and special techniques must be used. In a recent case of human rabies, a diagnosis was made 2 months postmortem by FA of formalin-fixed brain tissue (Geeslin et al., 1985).

For better FA results, formalin-fixed specimens can be treated with trypsin (Barnard and Voges, 1982; Umoh and Blenden, 1981). Formalin-fixed brain pieces are homogenized in five volumes of PBS, the suspension is centrifuged for 5 min at 1,000 *g*, and the supernatant is discarded. The sediment is resuspended at the initial volume in 0.25% trypsin solution containing 0.3% sodium citrate and 0.6% sodium chloride, and the pH is adjusted to 7.8. After trypsin digestion for 30 min at 37°C, the suspension is centrifuged, the sediment is washed once in PBS, and the smears are prepared on slides that have been air-dried and fixed in acetone. This technique has proven useful in field conditions (Barrat, 1985; Umoh et al., 1985).

The FAT also may be used in rabies-infected tissues embedded in paraffin (Fischman, 1969; Sainte-

Marie, 1962); one can enhance FA staining in formalin-fixed, paraffin-embedded tissues by trypsin digestion (Johnson et al., 1980; Swoveland and Johnson, 1979) or by pepsin and trypsin digestion (Reid et al., 1983).

Fluorescent antibody testing also has been used on specimens taken during the clinical phase of the disease. Immunofluorescence may be performed on cells in the sediment of centrifuged cerebrospinal fluid as well as the sediment of centrifuged saliva. Skin biopsy specimens from the nape of the neck (hair follicles should be included) have been examined on several occasions (Bryceson et al., 1975; Duhme et al., 1983; Zeidner et al., 1984). Corneal impressions (Schneider, 1969) may be examined for the clinical diagnosis in humans (Cifuentes et al., 1971; Cowherd et al., 1979; Larghi et al., 1973; Koch et al., 1975). On rare occasions, brain biopsy specimens have been examined to help diagnose human cases (Duhme et al., 1983; Swanson et al., 1984). Fluorescent antibodies staining of skin biopsy specimens has been successful in dogs, skunks (Fuh and Blenden, 1971; Smith et al., 1972), cats, cattle, and mongoose (Blenden, 1974).

The details of FA staining have been described (Dean and Abelseth, 1973; Kissling, 1975). Smears are air-dried and fixed in acetone at −20°C for 1 h. Longer fixation (4 h or overnight) is not harmful but not convenient for rapid diagnosis; fixation for shorter periods may give reduced fluorescence (Larghi and Jimenez, 1971; Segre and Zothner, 1985). It must be remembered that acetone-fixed preparations may still contain viable virus (Fischman and Ward, 1969) and should be handled with care. Acetone fixation at 37°C for 4 h or at 50°C for 30 min will inactivate the virus without reducing fluorescence (White and Chappell, 1982). After fixation, the slides are air-dried and the impressions are covered with conjugate and incubated in a humidified chamber for 30 min at 37°C. Evans blue may be added to the conjugate as a counterstain at a final concentration of 1 : 2000. After staining, the slides are rinsed twice in PBS, pH 7.2 to 7.4, for 10 min each, preferably with stirring, and mounted under a coverslip with 90% glycerol, pH 8.6. The stability of monoclonal conjugates depends on the pH of the mounting medium; it is important to use a mounting medium of pH 8.5 or higher (Durham et al., 1986).

Several methods may be used to control the sensitivity and the specificity of FA. One procedure is to make two smears of the same specimen on each slide; one is stained with conjugate absorbed with normal mouse brain, the other with conjugate absorbed with rabid mouse brain. A positive specimen will only react with the first (normal mouse brain absorbed) conjugate. Another method is to use only conjugate absorbed with normal mouse brain, previously tested in positive and negative control slides.

Any time field specimens are tested, control slides, including positive brain impressions and normal brain impressions, must be stained along with the diagnostic specimens. The field specimens are examined only when both positive and negative controls give satisfactory results. The rabies antigen stains as a brilliant fluorescent apple-green showing various shapes and sizes; some oval or round inclusions are seen that stain brilliantly at their periphery. The background is gray or bluish. If Evans blue is used as a counterstain, the background is dark red in contrast to the apple-green fluorescence.

The FA appears to be the most suitable technique for rabies diagnosis when well-preserved specimens are examined from several areas of the brain. At least 99% of the rabies-infected brains submitted for diagnosis are quickly diagnosed (Kissling, 1975). The FA may even be positive in specimens in which the virus has dropped in titer during transport or when specimens have been formalin fixed. Positive FA may also be seen in decomposed or putrid brain specimens not suitable for mouse inoculation. In any case, no technique is completely reliable when inadequate specimens are submitted, and the laboratory should simply state "Sample Unsuitable for Examination."

IMMUNOPEROXIDASE TECHNIQUE

The immunoperoxidase technique has been used for the direct detection of rabies antigen in brain smears of rabid animals (Atanasiu et al., 1971), but has not been adopted for routine diagnosis. The advantage of the technique is that it does not require the use of fluorescent microscope equipment. The peroxidase-stained preparations are examined under an ordinary light microscope, and rabies antigen inclusions of various sizes and shapes appear yellowish-brown (Atanasiu et al., 1974). The immunoperoxidase technique was compared with the fluorescent antibody test in 500 field specimens and found to be as specific but slightly less sensitive, but needed longer preparation time and was more difficult to perform than FA (Genovese and Andral, 1978).

RAPID RABIES ENZYME IMMUNODIAGNOSIS

This is a new test developed for rabies antigen detection in brain and salivary gland suspensions. It combines the immunocapture of rabies antigen by an antinucleocapsid immune globulin with the detection of the captured antigen by the same immune globulin conjugated to peroxidase (Perrin et al., 1986). To perform the technique, enzyme-linked immunosorbent assay (ELISA) microplates are sensitized by adding 1 μg/well of purified antinucleocapsid IgG diluted in carbonate buffer (0.05 M, pH 9.6). Suspensions (20%) of the suspect brain specimens are prepared in PBS and clarified by low-speed centrifugation. After the plates are washed twice in PBS-Tween solution, the supernatant from each specimen is placed in two wells (200 μl/well); positive and negative controls are included. The plates are incubated for 1 h at 37°C, and then washed six times in PBS-Tween. The peroxidase conjugate is added (200 μl/well). After addition of the substrate and peroxide, the reaction is allowed to develop in the dark for 30 min at room temperature. The reaction is stopped by adding 50 μl/well of 4 N sulfuric acid. Rabies nucleocapsid antigen appears yellow; the negative specimens are colorless. The color also may be measured by reading the optical density in a photometer at 492 nm.

This technique appears to be as specific and almost as sensitive as the FA. In some instances, it has given positive results with putrid brain specimens unsuitable for FA and with specimens inactivated by heat during transport that were unsuitable for virus isolation. It can also be used with glycerol-preserved specimens, but not with formalin-fixed brain tissue.

Direct Viral Nucleic Acid Detection

Rabies virus nucleic acid may be extracted from virus-infected brain specimens and detected by hybridization with ^{32}P-labeled specific cDNA probes. One gram of brain is homogenized in a solution of 1 ml phenol and 1 ml extraction solution (1% NP40, 1% SDS, 50 μg dextran sulfate in water). This step preferentially extracts the intracellular rabies virus nucleic acid. After centrifugation, the aqueous phase is collected, and the rabies virus nucleic acid is concentrated with ethanol, washed twice in 70% ethanol, and redissolved in 200 μl of distilled water. Each sample is diluted 1 : 10 and 1 : 100. One microliter of each dilution is deposited on a nylon membrane (Highbond-N, Amersham, Inc., Arlington Heights, IL), dried, and irradiated with ultraviolet light for 3 min to fix the rabies virus nucleic acid on the membrane. As a first step in hybridization, cDNA probes, complementary to the first 6,000 nucleotides of the rabies genome, are ^{32}P-labeled with a multiprime DNA labeling system. The nylon membranes are incubated overnight at 52°C with the labeled probes in hybridization solution (50% formamide, 5 SSC, 2 Denhart, 0.1% SDS, 200 μg/ml salmon sperm DNA). The membranes are then washed twice, at 45°C for 20 min in the first solution (2 SSC, 0.1% SDS), then at 22°C for 20 min in the second solution (0.1 SSC, 0.1% SDS). Hybridized, labeled probes are detected by autoradiography.

The cDNA probes specifically hybridize the viral rabies nucleic acid extracted from rabies-infected brains; no hybridization occurs with the rabies virus nucleic acid extracted from normal brains. The rabies virus nucleic acid can be extracted within 24 h after death with no nucleic acid degradation. This technique may be useful for research purposes to study rabies pathogenesis but has not been evaluated for the diagnosis of human or animal rabies in diagnostic laboratories.

Virus Isolation and Identification

Virus isolation confirms the direct detection of virus, viral antigen, or viral nucleic acid or is used for further studies of the virus isolates. Such studies may include characterization of antigenic variants (by animal species and geographic origin), identification of other members of the *Lyssavirus,* neutralization tests for virus identification, and cross-protection tests in experimentally vaccinated mice.

Virus isolation is usually attempted from postmortem specimens. It also may be attempted, most often in humans, from specimens (saliva, sputum, tears, cerebrospinal fluid [CSF], or brain biopsy specimen) taken during the clinical disease (Duhme et al., 1983; Swanson et al., 1984; Zeidner et al., 1984).

Fresh specimens must be refrigerated. Virus isolation may be by animal inoculation or inoculation of cell culture.

Animal Inoculation

The most common procedure for virus isolation is the intracerebral inoculation of the white Swiss mouse, recommended by the World Health Organization (WHO) and described in detail by Koprowski (1973). The mice should be from a breeding colony free of latent murine viruses or, preferably, from a specific pathogen-free colony. Sick animals should not be used. The white Swiss mouse is the animal of choice, but all mouse strains appear equally susceptible to the intracerebral inoculation of rabies virus (Johnson and Leach, 1940; Lodmell and Ewalt, 1985). Weanling mice, 21 to 28 days old (11 to 16 g), or suckling mice under 3 days old are preferred (Bagnarolli et al., 1970; Nilsson and Sugay, 1966; Pilon Moron et al., 1967). A minimum of six weanling mice should be inoculated with each specimen; with suckling mice a litter of 10 is convenient. The intracerebral inoculation (0.03 ml in weanling mice, 0.02 ml in newborn) requires the use of 1/4-ml glass syringes or 1-ml plastic tuberculin syringes, fitted with 26 to 27-gauge needles. Weanling mice may be anesthetized with ether; this is not necessary for suckling mice. Inoculated mice are checked daily for clinical signs. Records are kept using symbols such as: R = ruffled hair, H = humped, P = paralysis, Pr = prostrate, D = dead. Deaths the first 3 days after inoculation are usually due to inoculation trauma or bacterial infection, and those mice are discarded. Mice are observed for 28 days, as paralysis usually occurs between 8 and 15 days in weanling mice and 5 to 10 days in suckling mice. The clinical signs observed in the inoculated mice should be confirmed by FA of the brains or, in laboratories where FA is not available, by Negri body detection, histologic examination, or virus identification with an antirabies reference serum (Johnson, 1973). Diagnosis may be made more quickly if one mouse is killed every day or every other day, beginning on the 5th postinoculation day for weanling and on the 3rd day with suckling mice, and the brains examined by FA. By so doing a positive result may be obtained in 7 days in weanling mice and in 4 days with suckling mice (Pilon-Moron et al., 1967).

Mouse inoculation should be performed inside a biologic safety cabinet, and the cages should then be housed inside a negative pressure isolator equipped with HEPA filters. If such equipment is not available each cage should be protected with a filter paper cover and placed in a room with negative pressure and restricted access.

Cell Culture Inoculation

Although mouse inoculation is still widely used to confirm direct diagnosis, it has the drawback of delaying results for several days. Isolation in tissue culture, however, usually permits isolation in less than 48 h. Human rabies treatment often depends on laboratory results.

The first isolation of street rabies virus was done in primary hamster kidney cells by Kissling (1958). Later, Larghi recommended the use of the BHK-21 cell line for virus isolation from saliva (Larghi et al., 1975), a technique routinely used for the isolation of street virus since 1980 (Rudd et al., 1980). However, mouse neuroblastoma cells have been shown to be even more susceptible and more convenient for diagnosis (Portnoi et al., 1982; Rudd and Trimarchi, 1983; Smith et al., 1978; Umoh and Blenden, 1983); their use provides results in 2 days.

Routine tissue culture isolation for rabies diagnosis was begun at the Diagnostic Laboratory of the Rabies Unit at the Institut Pasteur, Paris, in 1981 (Portnoi et al., 1982), and about 4,000 specimens per year have been tested since then. The initially published technique has been modified and is now performed as follows: a 20% suspension (weight/volume) of the brain (or salivary gland) material is prepared in Dulbecco (D-MEM) tissue culture medium supplemented with 30% calf serum and antibiotics, the suspension is centrifuged at 3,000 rpm for 30 min at 4°C, and the supernatant is used for inoculation. The neuroblastoma cells (Neuro 2a; American Type Culture Collection CCL131) are maintained in plastic tissue-culture flasks in D-MEM supplemented with 10% calf serum, 0.3% tryptose phosphate broth, 30 mg/l L-glutamine, and antibiotics, and subcultured every 3 days at a ratio of 1:5. For virus inoculation, the cells are trypsinized and resuspended at a concentration of 200,000 cell/ml in D-MEM; 400 μl of suspension (about 80,000 cells) is placed in each chamber of Lab-Tek (Miles Labs., Naperville, IL) tissue culture 8-chamber slides, and the slides are kept at 37°C for 30 min in the CO_2 incubator to let the cells attach to the slide. At that time, 50 μl of supernatant from each specimen is inoculated per chamber; positive and negative controls are always included. The inoculated chambers are incubated overnight, and the next morning (after a 20 h incubation), the medium in the chambers is removed and the cells are air-dried inside a vertical laminar flow cabinet. After the plastic chambers are removed, the slides are fixed in cold acetone and stained by the FAT. Positive specimens exhibit specific cytoplasmic fluorescence.

Cell culture for virus isolation should be carried out in a biologic safety cabinet.

VIRUS IDENTIFICATION

Isolates obtained by mouse inoculation may be further identified by virus neutralization in mice (Johnson, 1973). A 20% brain suspension is prepared from the second mouse passage; serial 10-fold dilutions of the suspension are made and mixed with equal volumes of the immune and normal serum and incubated at 37°C for 1 h. Each mixture is then injected intracerebrally in groups of five mice each, which are then observed for 21 days. The LD_{50} of each series is calculated, and the identity of the isolate is confirmed if the neutralization index (titer in normal serum minus titer in immune serum) is greater than 2 logarithms.

The FAT (or the immunoperoxidase technique) can also be used to detect antigen in the brains of inoculated mice or in infected cell cultures.

Polyclonal conjugates give positive results with all lyssaviruses. Monoclonal antibodies may be needed in particular epizootiologic circumstances, such as occurred, for instance, in an outbreak of "atypical rabies" in several cats and a rabies-vaccinated dog in Zimbabwe (Foggin, 1982). The fluorescence noted in the brain smears of those animals was unusually weak, and analysis by monoclonal antibodies showed that the deaths were due to Mokola virus (Foggin, 1983; Wiktor et al., 1984a).

In fact, the production of monoclonal antibodies against rabies virus less than 10 years ago (Wiktor and Koprowski, 1978) completely changed the way rabies viruses were identified and opened up fascinating new approaches to this subject. The use of monoclonal antibodies that specifically recognize antigenic determinants of the nucleocapsid (NC) and glycoprotein (G) of rabies viruses now makes possible the detection of virus variants and the differentiation of rabies virus from the rabies-related viruses (Wiktor et al., 1980). In addition, field isolates can be characterized by species and by geographic origin with specific monoclonal antibodies (Smith et al., 1984; Sureau et al., 1983). The anti-NC monoclonal antibodies are used in an indirect fluorescent antibody test on brain smears of the original animal, on mouse passage or in cell culture. The anti-G monoclonal antibodies are used in serum neutralization tests, usually in cell cultures with adapted strains. The characterization of the antigenic profile of rabies virus isolates has become an important part of the laboratory diagnosis of rabies. A recent interesting example is the identification of Duvenhage virus in bats from northern Europe (Mollgard, 1985; Schneider, 1982; Schneider, 1985).

Antibody Assays

Detecting rabies-neutralizing antibodies in humans or animals may be helpful in diagnosing rabies infection and determining the immunity of vaccinated persons. The diagnosis of several recent cases of human rabies was made by detecting a rise in antibody titer in the serum and the cerebrospinal fluid (Devriendt and Staroukine, 1982; Duhme et al., 1983; Kerton et al., 1979; Moler et al., 1983; Swanson et al., 1984; Tillotson et al., 1977); in other cases, however, no antibodies were detected (Cowherd et al., 1979; Zeidner et al., 1984).

Neutralizing antibodies were first measured by mouse neutralization tests in mice (Webster and Dawson, 1935; Webster, 1936), but during the past few decades several other techniques have been used. These include the inhibition of the cytopathic effect in cell cultures (Kissling and Reese, 1963), the immune cytolysis test (Wiktor, et al., 1968), and the plaque assay (Wiktor and Clark, 1973). Other serologic techniques used to titrate rabies antibodies include the mouse serum neutralization test (Atanasiu, 1973), the indirect fluorescent antibody technique (Thomas et al., 1963), the hemagglutination inhibition test (Halonen et al., 1968), the passive hemagglutination test (Gough and Dierks, 1971), the complement fixation test (Kuwert, 1973), the counter immunoelectrophoresis test (Diaz and Varela-Diaz, 1977), ELISA (Atanasiu and Perrin, 1979), the radioimmunoassay (Wiktor et al., 1972), and an immunoelectron microscopy test (Chaudhary et al., 1979). However, the serologic technique most widely used is the rapid fluorescent focus inhibition test (RFFIT) (Smith et al., 1973).

As originally described, the RFFIT is performed in Lab-Tek tissue culture 8-chamber slides; fivefold dilutions of each serum are made in the chambers and 0.1 ml of the proper dilution of challenge virus standard (CVS) is added to 0.1 ml final volume of all serum dilutions. A reference serum of known titer is included in each test. After incubation at 37°C for 90 min, the BHK-21 cell suspension (pretreated with DEAE-dextran 10 μg/ml) containing 100,000 cells/0.2 ml of growth medium is added to each chamber. After a 24-h incubation, the slides are rinsed, fixed in acetone, and stained with FA. The slides are examined with a fluorescent microscope at low magnification (160×); 20 fields are observed. If fewer than 50% of the fields have fluorescent foci, neutralizing antibody is indicated in the serum dilution examined. The titer is calculated by comparison with the reference serum and is expressed in international units per milliliter (IU/ml). An index system for reporting RFFIT titers has recently been proposed (Brown et al., 1984).

The RFFIT also may be performed in 96-well flat-bottomed microplates (Zalan et al., 1979). In this technique the use of HEPES buffer eliminates the need for a CO_2 incubator, the substitution of acetone with formalin provides a better fixation of the cells, and the examination of only eight microscopic fields is sufficient for satisfactory results. To avoid the cumbersome and time-consuming microscopic examination and to eliminate personal bias in the reading, the FA staining can be replaced by an immunoenzymatic titration of the nonneutralized rabies antigen in the cells with an automatic photometer reading (Perrin et al., 1985).

The enzyme-linked immunosorbent assay (ELISA) does not require cell cultures and provides results in a few hours. It is performed in 96-well microplates coated with whole purified virus or with purified glycoprotein (Atanasiu and Perrin, 1979; Perrin et al., 1986). Antibodies bound to the rabies antigen are detected with peroxidase-conjugated protein A, and the reaction is revealed with H_2O_2-ortho-phenylene-diamine. The optical density is measured by photometer at 492 nm, and the antibody titer calculated by comparison with a reference serum. The use of peroxidase-conjugated antihuman IgM makes possible the early detection of IgM antibodies in clinical human rabies (Atanasiu et al., 1978; Savy and Atanasiu, 1978).

Vesiculoviruses

Vesicular stomatitis (VS) is a disease of cattle, horses, swine, and humans. According to Hanson (1952), it was first reported in horses in the United States in 1821. An early description of what appears to be a similar disease was made during the US Civil War in 1862 (McClelland, 1862): "The artillery and cavalry required large numbers [of horses] to cover losses sustained in battle, and on the march and by diseases. Both of these arms were deficient when they left Washington in September 1862. A most violent and destructive disease made its appearance at this time which put nearly 4,000 animals out of service. Horses reported perfectly well one day would be dead lame the next, and it was difficult to foresee when it would end or what number would cover the loss. They were attacked in hoof and tongues. No one seemed able to account for the appearance of the disease. Animals kept at rest would recover in time but could not be worked."

A virus causing vesicular disease in horses and presumed to be vesicular stomatitis virus was temporarily introduced to France in 1916 after its apparent spread from the United States during World War I (Vigel, 1916), but except for this episode vesicular stomatitis virus–New Jersey (VSV-NJ) and vesicular stomatitis virus–Indiana (VSV-IND) are limited to the western hemisphere.

Serious outbreaks of VSV may occur in cattle (Ellis and Kendall, 1964), but the disease is usually self-limiting. Symptoms in cattle, horses, and swine include vesicles on the tongue, teats, or, less often, the coronary band and interdigital spaces of the hoofs. These signs of illness, however, are similar to those seen in cattle and pigs infected with foot and mouth disease (FMD) virus, a far more serious disease of livestock. It is, therefore, critical to distinguish the two diseases so that control measures can be taken when FMD is present. In humans, VS is an influenza-like disease, with fever, muscular aches, stiffness, and occasional blisters on the lips (Fields and Hawkins, 1967); inapparent infections are common (Brody et al., 1967; Tesh et al., 1969).

Although both New Jersey and Indiana serotypes have caused numerous epizootics in tropical America (Johnson et al., 1969) and in the United States (Hanson, 1981), the New Jersey strain is the more common. In contrast to the limited areas where VS outbreaks have occurred in the United States, VS-IND and VS-NJ viruses are enzootic in tropical America.

The Viruses

Viruses causing VS are members of the genus *Vesiculovirus*. This genus contains one serogroup composed of five antigenic complexes: the VS-Indiana complex, including VS-Indiana, VS-Alagoas, Carajas, Cocal, Maraba, and Perinet viruses; the VS-New Jersey complex, including VS-New Jersey virus; the Piry complex, including Piry virus; the Chandipura complex, including Chandipura virus; and the Isfahan complex, including Isfahan, Jurona, La Joya, Yug Bogdanovac, and Calchaqui viruses.

Only a few of these viruses cause VS. The prototype virus of the serogroup, VS-Indiana virus, was isolated from the tongue epithelium of a calf shipped to Indiana from Kansas City in 1925 (Cotton, 1927). VS-IND occurs in most countries of North, Central, and South America (Karabatsos, 1985). Cocal virus, also a member of the VS-Indiana antigenic complex, has been isolated from rice rats (*Oryzomys laticeps velutinus*) in Trinidad and from other rodents, a horse, mosquitoes, and mites (*Gigantolaelaps* sp.) in Trinidad, Brazil, and Argentina. Vesicular stomatitis Alagoas was isolated from a mule with vesicular lesions in Brazil (Andrade, 1974). Vesicular stomatitis virus–New Jersey was first isolated in 1926 (Cotton, 1927). The prototype strain has been lost. The 1952 Georgia isolate is the Hazelhurst strain, isolated from vesicles on the snout of a domestic pig (Karstad and Hanson, 1958). The virus has been isolated in the United States and Central and South America from cattle, horses, swine, and laboratory workers, as well as from mosquitoes, *Culicoides* sp., and various species of house flies, stable flies, and other flies.

Piry virus has been isolated in Brazil from a Philander opossum (cited in Karabatsos, 1985). Chandipura virus has been isolated in India from humans with febrile illnesses (and from sandflies), and in Nigeria from hedgehogs (*Atelerix* sp.). Viruses of the Isfahan complex have been isolated from phlebotomine flies, ticks, mosquitoes, and sandflies. Thus, not only must VS viruses be distinguished from foot and mouth disease (FMD) virus, they must also be distinguished from each other.

Pathogenesis, Pathophysiology, and Pathology of Vesicular Stomatitis Virus

Francy et al. (1987) concluded that insect vectors transmitted virus from animal to animal—that the insects became infected when feeding near the eyes and mouth lesions of infected cattle, because fluids from vesicular lesions are known to contain high concentrations of virus. Because viremia in VS has rarely been demonstrated in wild animals (Geleta and Holbrook, 1961; Jonkers et al., 1964) or humans (Fellowes et al., 1955; Gaidamovich et al., 1966), the only source of infectivity appears to be infected vesicles containing high levels of virus (Francy et al., 1987).

Clinical Features

Mode of Transmission

Vesicular stomatitis is enzootic in certain tropical and subtropical areas of the Americas and is probably spread by a variety of insect vectors. Although clinical disease has been seen only in horses, cattle, pigs, and humans, serologic studies have shown that antibodies to the virus are widespread in wild animals (Tesh et al., 1969). In the summer months, epizootics spread north from the enzootic areas into the United States and Canada, or southward into temperate South America. Massive epizootics are sometimes reported (Jenney, 1967). The disease usually disappears shortly after the first frost, suggesting that it is transmitted by an arthropod vector. However, during the large 1982–1983 epizootic of VSV-NJ in the western United States, virus transmission continued long after the first killing frost (Walton et al., 1987). Other common epidemiologic characteristics are the rapidity of spread and the spread from natural water sources.

Incubation Period

The incubation period in cattle, horses, and swine varies from 2 to 7 days (Chow and McNutt, 1953; Mohler, 1930; Shahan et al., 1946). In humans infected in the laboratory or the field, fever and chills occur about 30 h after exposure (Hanson et al., 1950; Johnson et al., 1966).

Period of Infectivity

As no viremia has been detected in pathogenesis studies of the disease, it is thought that infected vesicles on the tongue or lips of infected animals are the only source of virus for transmission. Although these vesicles are known to contain high levels of virus when they form (Francy et al., 1987), it is not known how long the animals may be infective before vesicles appear.

Clinical Course

In cattle, VS usually occurs as sporadic cases in endemic areas. Outbreaks occur during summer months, especially during the rainy seasons in tropical areas. The lesions in pastured cattle are usually in the oral cavity. In milk cows teat lesions are common, perhaps because of the minor irritation involved in milking, but oral lesions occur also. Vesicles on the coronary band or interdigital spaces of the hoof are occasionally noted (Brody et al., 1967;

Gibbons, 1961). Because the fever is transient and minor ulcerations rather than vesicles may be present, the classic febrile disease with vesicles may not even be noted by veterinarians. The vesicles that do appear are found on the tongue, gums, or lips, and result in excess salivation. The mouth vesicles often break and the mucosa sloughs, exposing raw, red underlying tissue. Anorexia is then common. Sloughing of the necrotic epithelium may even occur during examination of the mouth (Gibbons, 1961). Animals with early vesicles are only found when all animals in a herd are carefully examined, making it possible to identify all the animals that have clinical signs of the disease. Recovery is usually rapid.

Subclinical cases are much more common than clinical ones. In a recent outbreak 60 of 1,341 cattle (4.5%) in one ranch had clinical signs of disease, but 69 of 102 studied (67.6%) had neutralizing antibodies (Webb et al., 1987). Lesions in horses are similar to those in cattle except that hoof (i.e., coronary band) lesions are apparently as common as mouth lesions (Heinz, 1945).

Swine may have lesions on the snout (Chow and McNutt, 1953; Shahan et al., 1946) as well as vesicles on the lips, the coronary band, and the interdigital space. The erosive stage lasts approximately 1 week.

Complications

Anorexia and excessive salivation are usually self-limiting complications resulting from the severe tongue lesions of VS, with occasional sloughing (Gibbons, 1961). When teat lesions predominate in milk cows (Acree et al., 1964; Brandly et al., 1951; Laverman et al., 1962), high attack rates are often observed. In one outbreak in Alabama 105 of 109 cows showed major teat lesions, with raw, eroded areas, resulting in 40 cases of severe mastitis (Ellis and Kendall, 1964). Such cows are nearly impossible to milk (Ellis and Kendall, 1964; Heinz, 1945). Cattle may slough large areas of the muzzle area (Brandly et al., 1951) as well as become almost immobile because of the severe lesions at the coronary (hoof) band. Horses may develop severe lameness because of the coronary band lesions (Hanson, 1952). Humans usually develop an influenza-like disease with malaise, myalgia, headache, and eye and chest pains (Fields and Hawkins, 1967; Hanson et al., 1950).

Epidemiology and Natural History

Studies of a recent outbreak of VSV-NJ in the United States revealed interesting findings on the possible transmission route. The first case occurred in May 1982 in Arizona; the disease eventually spread to 13 western and midwestern states (Walton et al., 1987) via the Rio Grande and Colorado River valleys. In entomologic studies carried out in Colorado in August 1982, the virus was isolated from houseflies and other nonhematophagous diptera. Two pools of unengorged blackflies (*Simulidae*) were also found infected; one of these had a high virus titer, suggesting biologic transmission had occurred (Francy et al., 1987). It still has not been shown, however, whether houseflies, often the most abundant type of insect during those outbreaks, can become infected feeding on these lesions. The authors state that:

Epidemiologic observations on one ranch suggest that insects are important in animal-to-animal transmission. Horses in separate stalls with no common feed or water source presented with clinical disease. Direct contact could almost certainly be eliminated as a means of spread . . . (Francy et al., 1987).

Yet the final conclusion drawn by the same authors was that: ". . . the [epizootiology of VS] remains an enigma."

Prevention, Prophylaxis, and Control

As Yuill (1981) has pointed out, "effective and economical control of VS is difficult without a clear understanding of the epidemiology of these infections." Both live and inactivated vaccines have been used; they appear to be effective in dairy cattle and showed no side effects (Correa, 1964; Lauerman and Hanson, 1963).

Various sanitary practices help to limit the spread of the disease in dairy cattle, including washing hands and machines thoroughly and separating infected animals from healthy ones. Normal veterinary care should be given to cattle with teat lesions.

Diagnosis

Specimen Collection

According to recent reports (Carbrey, 1984), some minor modifications and improvements have been made in diagnostic techniques since the diagnosis of VS moved from the animal room to the laboratory bench. The procedures, however, are basically the same as those originally developed by vesicular disease researchers in the 1940s and 1950s. Virus isolation can be made from throat swabs, mouth swabs, vesicle fluids, or epithelial tissue from ruptured vesicles (Meyer et al., 1960). Human infections are common in virus research workers and in veterinarians examining infected cattle (Reif et al., 1987). All persons working with infected animals should wear gloves and goggles at all times and be cognizant of the danger of direct contact with infected tissue.

Direct Virus Detection

Techniques for direct antigen detection and confirmation by virus isolation and identification are the techniques preferred for rapid virus diagnosis in infected animals, as discussed by Carbrey (1984), and direct virus detection is not usually performed.

Direct Antigen Detection

COMPLEMENT FIXATION

In 1984, Carbrey described the rapid diagnostic procedure used at the National Veterinary Services Laboratory, USDA, Ames, Iowa:

The first stage of the diagnostic procedure is rapid and presumptive with the use of complement fixation (CF) to detect antigen in vesicular fluid and tissue by the tissue CF test and antibody in serum by the serum CF test. The fluorescent antibody virus isolation technique and the neutralization test serve as definitive backup techniques requiring 18 to 24 hours and 2 to 3 days, respectively. The procedures are rigidly standardized, and we do not encourage creativity in their application.

The CF test is carried out largely according to procedures first described in 1958 (Jenney and Mott, 1958); numerous minor changes have been made. In the early days of CF testing, it was soon possible to differentiate VSV-NJ from VSV-IND (Jenney and Mott, 1963). Vesicular stomatitis virus from vesicular fluids can be rapidly typed by direct complement-fixation (CF) tests. The ease with which such tests are performed make them the most commonly used tests for typing.

Vesicular epithelium is triturated in Eagle's minimum essential medium (MEM), pH 7.2, to prepare a 20% suspension. The suspension is clarified by centrifugation for 15 min at $1500 \times g$ and 4°C. A preliminary complement titration is made to determine the degree of anticomplementary effect of the antigen. The minimal amount of complement that produces 100% hemolysis is multiplied by a factor of 1.7 to determine the amount to be used in the test with that antigen. The antigen is tested against VSV-NJ and VSV-IND subtypes 1 to 3 (Indiana, Cocal, and Brazil). Control antigens include VSV-NJ, VSV-IND, and normal tissue. The typing antiserum is placed in vertical rows of plastic U-bottom microtitration plates, and the tissue antigens are placed in duplicate horizontal rows. Appropriate controls are prepared. Complement is added to each well, and the plates are agitated and incubated for 1 h at 37°C. Sensitized erythrocytes are then added, and incubation at 37°C

is repeated for 30 min. After centrifugation, the test is read.

FLUORESCENT ANTIBODY

The first fluorescent antibody (FA) method for detecting VSV was developed in 1960 (Hopkins and Jenney, 1962). Hyperimmune guinea pig serum was conjugated with fluorescein isothiocyanate and used to detect viral antigen in swine kidney cell cultures infected with VSV, usually from tongue lesions. Multivalent antiserum against several viruses, including VSV, has also been prepared (Mengeling and Van der Maaten, 1971) to screen isolates, permitting a rapid diagnosis that excludes similar viruses.

Cell cultures are propagated on coverslips in Leighton tubes and inoculated with specimen tissue suspension. After incubation, the coverslip cultures are stained with conjugates prepared from VSV-NJ and VSV-IND antisera produced in guinea pigs. To obtain optimum staining, it is necessary to use serotype-specific conjugates. Primary embryonic bovine kidney (EBK) and primary embryonic swine kidney (ESK) cell cultures are prepared in Leighton tubes with coverslips. The clarified suspension as prepared for the tissue CF is injected into the Leighton tubes after the medium is discarded. Probang specimens are diluted in an equal volume of medium and centrifuged at $1500 \times g$ for 15 min at 4°C, and the sediment is triturated in a TenBroeck grinder. A 20% reconstitute is injected into the cell cultures. The suspension is incubated in contact with the cell sheet for 1 h at 34°C, and then the cell sheet is washed with medium. Growth medium supplemented with 2.5% equine serum that is free of bovine enterovirus antibodies is placed in the culture tubes, and the cultures are incubated overnight. The EBK cells are held at 37°C and the ESK cells at 34°C. After 18 to 24 h, some of the coverslips are removed, fixed with acetone, and stained with the anti-VSV-NJ and anti-VSV-IND conjugates. Virus is detected in plaques of green-fluorescing cells containing VS viral antigen. Specific fluorescence in virus-infected cells is detected after only a few hours' incubation.

The remaining cultures are observed daily for cytopathic effects and stained with VSV-NJ and VSV-IND conjugates as appropriate. If no cytopathic effect is observed after 7 days, the remaining coverslips are stained and examined.

Direct Viral Nucleic Acid Detection

Direct viral nucleic acid detection has not been used to diagnose VS; other methods are preferable unless it were shown to have value in diagnosing the disease because of its rapidity or sensitivity.

Virus Isolation and Identification

Virus can usually be isolated directly from vesicles of infected field animals or, during field investigations, from diptera and other insects that may carry the virus. Preliminary identification of the virus is by CF, but neutralization tests are used for confirmation.

LABORATORY ANIMALS

Guinea pigs are the laboratory animals most susceptible to VS by footpad scarification (Cotton, 1926; Olitsky et al., 1927; Wagner, 1931). This is an important isolation technique (along with tissue culture), and most useful for differentiating VS from foot-and-mouth virus.

Death occurs in young mice inoculated with VSV by any route, but only after intracerebral or intranasal inoculation of adult mice (Clark, 1979). Mice are also susceptible by the intranasal, subcutaneous, and oral routes, but less so than to intracerebral inoculation. Only newborn mice are susceptible to intraperitoneal inoculation.

Hamsters are highly susceptible to VSV by either intracerebral or intranasal inoculation (Cotton, 1926), as are ferrets (Kowalczyk and Brandley, 1954) and rhesus and cynomolgus monkeys (Olitsky et al., 1934).

EMBRYONATING EGGS

Vesicular somatitis virus readily grows on the chorioallantoic membrane of embryonating eggs (Burnet and Galloway, 1934); 7-day embryos are more susceptible than are 10-day embryos. These embryos have been used to produce viral antigen for the complement fixation test and as a host for neutralization testing.

TISSUE CULTURE

In 1955, two laboratories (Bachrach et al., 1955; Sellers, 1955) reported that VSV could be propagated in tissue culture with an accompanying cytopathic effect. Almost all cell cultures have since been shown to be susceptible to VSV, with high virus yields and a cytopathic effect in many (McClain and Hackett, 1958; see also the reviews by Clark, 1979; Fellowes et al., 1956; Karstad and Hanson, 1958). In a study of a recent outbreak (Francy et al., 1987), many more isolates were obtained with the highly susceptible C6/36 (*Aedes albopictus mosquito*) cells than with other cell culture systems (Table 2). The replication of VSV is very rapid in permissive cells; newly released virus is detectable after 2 h, and maximum yields are obtained at 10 to 12 h. Vesicular stomatitis virus replication occurs over a wide range of temperatures.

TABLE 2. Isolations of vesicular stomatitis New Jersey virus from Diptera in seven cell culture systems during studies along the Rocky Mountain front range, Colorado, 1982

Test system	Number of isolations
C6/36 only[a]	20
Vero only	0
DE only	1
C6/36 and DE	3
C6/36 and Vero	0
DE and Vero	0
C6/36, Vero, and DE	10
Total	34

[a] C6/36 cultures were blind passed after 3 days incubation into Vero cells for virus assay.
Reprinted with permission from Francy, D. B., et al. 1987. Am..J. Trop. Med. Hyg.

Antibody Assays

Numerous antibody assays have been developed to diagnose VS. The most commonly used are CF and neutralization (N).

COMPLEMENT FIXATION

The CF test was first used to detect serum antibodies from infected horses (Rice and McKercher, 1954). Antigen was prepared in chick embryos and CF antibody was detected in horses 6 to 8 days after infection. Maximum titers were found at 10 to 14 days, remained constant for the next 7 to 10 days, and then declined. In some early studies in swine (Geleta and Holbrook, 1961), circulating CF antibodies were detected 5 days after exposure to virus, with the highest concentration at 12 to 18 days. Within 2 to 4 months, few pigs had residual CF titers.

During the early days of CF testing (late 1950s) VSV antigen was prepared from virus from vesicular lesions, but now supernatant fluid from infected cell cultures or infected mouse brain is used to prepare high-titered, standardized virus stocks.

NEUTRALIZATION TEST

The neutralization test was developed by Sorensen et al. in 1958 after VSV had been adapted to tissue culture. When the fluorescent antibody technique is used after virus isolation, the virus can be identified in 24 to 72 h.

Neutralizing (N) antibody appears in the serum of infected cattle and swine as early as CF antibody is found (Geleta and Holbrook, 1961, Holbrook, 1962). Earlier in vivo neutralization tests have been described in a review of VSV (Yuill, 1981). An efficient

588   P. Sureau et al.

Vero cell plaque-reduction test can be used for diagnosis (Earley et al., 1967), as can microneutralization tests on monolayers (Rosenthal and Sheckmeister, 1971) and metabolic inhibition tests using bovine kidney cells (Castañeda and Hanson, 1966). Antibodies are usually detected about 7 days after inoculation (or infection), and highest titers are found 2 to 4 weeks later.

ENZYME-LINKED IMMUNOSORBENT ASSAY

An ELISA was recently developed (Vernon and Webb, 1985) for detecting IgM antibodies in serum of cattle and horses. It was first used in animals infected experimentally and naturally with VSV-NJ. Neutralizing antibody and ELISA titers were comparable when IgM was present, but neutralization titers persisted long after IgM titers disappeared. Complement fixation and indirect fluorescent antibody titers correlate closely with the ELISA titer for IgM. This is a rapid and practical method that can readily be used in the field.

Interpretation of Laboratory Results

Serodiagnosis of VSV infection is difficult because antibody varies in persistence, depending on the test used. Neutralizing antibody titers in cattle may persist for years (Geleta and Holbrook, 1961). Because the IgM antibody capture ELISA detects only IgM, it may be used to determine infection in early stages in animals and thus to determine the approximate date of illness and to clarify the epidemiology of the disease.

Neutralization titers of 8 and 16 are considered equivocal. Neutralization at 32 or greater is considered positive for previous infection with VSV. When paired acute- and convalescent-phase serum samples are tested, a fourfold or greater increase in titer is considered indicative of recent infection. Pigs and cattle with titers of 8, 16, and rarely 32 have been found without any supporting history of clinical VS or having inhabited VS-endemic areas. When these "nonspecific" neutralization reactions are found during tests for export, the animals are not accepted by the importing country; however, these reactions may be reduced, especially in swine, by withholding food from the animals for 24 h before collecting the blood.

In one study (Castañeda and Hanson, 1966), the resistance of cattle to VSV-NJ virus challenge was directly related to the levels of antibody measured by modified complement fixation (Boulanger, 1960), serum protection testing, colorimetric neutralization (Kuns, 1959), and chicken embryo neutralization (Vadlamudi and Hanson, 1963).

Literature Cited

Acha, P. N. 1967. Epidemiology of paralytic bovine rabies and bat rabies. Bull. Off. Int. Epizoot. **67:**343–382.

Acree, J. A., D. R. Hodgson, and R. W. Page. 1964. Epizootic Indiana vesicular stomatitis in Southwestern U.S. Proc., U.S. Livestock, Sanit. Assoc. **68:**375.

Anderson, L. J., L. P. Williams, and J. B. Layde. 1984. Nosocomial rabies: investigation of contacts of human rabies cases associated with a corneal transplant. Am. J. Pub. Health. **74:**370–372.

Andrade, C. M. 1974. Thesis, Instituto de Microbiologia, Universidad Federal de Rio de Janeiro, Brazil.

Atanasiu, P., A. Gamet, J. C. Guillon, P. Lepine, J. C. Levaditi, H. Tsiang, and A. Vallée. 1970. Possibilités et limites de l'immunofluorescence dans le diagnostic de la rage au laboratoire. Pathol. Biol. **19:**207–212.

Atanasiu, P., and P. Perrin. 1979. Microméthode immunoenzymatique de titrage des anticorps antirabiques: utilisation de la glycoprotéine rabique et de la protéine A conjugée à la peroxydase. Ann. Microbiol. (Inst. Pasteur). **130A:**257–268.

Atanasiu, P., P. Perrin, S. Favre, C. Chevallier, and H. Tsiang. 1974. Immunofluorescence and immunoperoxidase in the diagnosis of rabies, p. 141–155. In E. Kurstak, R. Morisset (ed), Viral imunodiagnosis. Academic Press Inc., New York, San Francisco, London.

Atanasiu, P., V. Savy, and C. Gibert. 1978. Rapid immunoenzymatic technique for titration of rabies antibodies IgG and IgM. Results Med. Microbiol. Immunol. **166:**201–208.

Babes, V. 1892. Sur certains caractères des lésions histologiques de la rage. Ann. Inst. Pasteur. **4:**209–223.

Babes, V. 1912. Traité de la rage, p. 81–119. Bailliére et Fils, Paris.

Bachrach, H., J. J. Callis, and W. R. Hess, 1955. The growth and cytopathogenicity of vesicular stomatitis virus in tissue culture. J. Immunol. **75:**186–191.

Baer, G. M. 1975. The natural history of rabies, Vol. II. Academic Press Inc., New York.

Baer, G. M. 1984. Control of rabies infection in animals and humans, p. 79–92. In E. Kurstak (ed), Control of virus diseases. Marcel Dekker, Inc., New York and Basel.

Baer, G. M., and W. F. Cleary. 1972. A model in mice for the pathogenesis and treatment of rabies. J. Infect. Dis. **125:**520–527.

Baer, G. M., J. H. Shaddock, S. A. Houff, A. K. Harrison, and J. J. Gardner. 1982. Human rabies transmitted by corneal transplant. Arch. Neurol. **39:**103–107.

Baer, G. M., T. R. Shantha, and G. H. Bourne. 1968. The pathogenesis of street rabies virus in rats. Bull. W.H.O. **38:**119–125.

Bagnarolli, R. A., O. P. Larghi, and N. Marchevsky. Susceptibilidad de ratones lactantes y adultos al virus rabico demostrada por inmunofluorescencia. Bol. Ofic. Sanit. Panamericana. **68:**388–392.

Baltazard, M., and M. Ghodssi. 1954. Prevention of human rabies—treatment of persons bitten by rabid wolves in Iran. Bull. W.H.O. **10:**797–803.

Banerjee, A. K. 1987. Transcription and replication of rhabdoviruses. Microbiol. Rev. **51:**66–87.

Barnard, B. J. H., and S. F. Voges. 1982. A simple technique for the rapid diagnosis of rabies in formalin-preserved brain. Onderstepoort J. Vet. Res. **49:**193–194.

Barrat, J. 1985. Rabies diagnosis in formalin-preserved brain specimens treated with trypsin. Rabies Info. Exch. **12:**5–15.

Beran, G. W., A. P. Nocete, O. Elvina, S. B. Gregorio, R. R. Moreno, J. C. Nakao, G. A. Burchette, H. L.

Canizares, and F. F. Macasaet. 1972. Epidemiological and control studies on rabies in the Philippines. Southeast Asian J. Trop. Med. Pub. Hlth. **3:**433.

Blenden, D. C. 1974. Diagnosis of rabies in various species by immunofluorescent staining of skin. J. Am. Vet. Med. Assoc. **165:**735.

Blumberg, B. M., M. Leppert, and D. Kolakofsky. 1981. Interaction of VSV leader RNA and nucleocapsid protein may control VSV genome replication. Cell **23:**837–845.

Boulanger, P. 1960. Technique of a modified direct complement fixation test for viral antibodies in heat inactivated cattle serum. Can. J. Comp. Med. **24:**262.

Brandly, C. A., R. P. Hanson, and T. L. Chow. 1951. Vesicular stomatitis, with particular reference to the 1949 Wisconsin epizootic, Proc. 88th Annual Mtr., Am. Vet. Med. Assoc. **61:**61–67 Milwaukee, WI.

Brody, J. A., G. F. Fischer, and P. H. Peralta. 1967. Vesicular stomatitis virus in Panama. Human serologic patterns in cattle raising area. Am. J. Epidemiol. **86:**158.

Brown, F., and J. Crick. 1979. Natural history of the rhabdoviruses of vertebrates and invertebrates, p. 1–22. *In* D. H. L. Bishop (ed), Rhabdoviruses, Vol. I. CRC Press, Boca Raton, FL.

Brown, J., D. W. Dreesen, J. S. Smith, and J. W. Sumner. 1984. A proposed index system for reporting rabies fluorescent focus inhibition test titres. J. Biol. Standard **12:**233–240.

Bryceson, A. D. M., B. M. Greenwood, D. A. Warrell, N. McD. Davidson, H. M. Pope, J. H. Lawrie, H. J. Barnes, W. E. Bailie, and G. E. Wilcox. 1975. Demonstration during life of rabies antigen in humans. J. Infect. Dis. **131:**71–74.

Burnet, F. M., and I. A. Galloway. 1934. The propagation of the virus of vesicular stomatitis in the chorio-allantoic membranes of the developing hen's egg. Br. J. Exp. Pathol. **15:**105–113.

Carbrey, E. A. 1984. Laboratory diagnosis of vesicular stomatitis, p. 446–456. Proceedings of an International Conference on Vesicular Stomatitis, Mexico City, 24–27 September.

Castañeda, J., and R. P. Hanson. 1966. Complement-fixing antibodies as a measure of immunity of cattle to the virus of vesicular stomatitis New Jersey. Am. J. Vet. Res. **27:**963.

Centers for Disease Control. 1986. Rabies surveillance annual summary 1985. Centers for Disease Control, Atlanta, GA.

Centers for Disease Control, Atlanta, Ga, and National Institutes of Health, Bethesda, Md. 1984. Public Health Service, U.S. Department of Health and Human Services, CDC-NIH Biosafety in Microbiological and Biomedical Laboratories. U.S. Government Printing Office, March.

Charlton, K. M., G. A. Casey, D. W. Boucher, and T. J. Wiktor. 1982. Antigenic variants of rabies virus. Comp. Immunol. Infect. Dis. **5:**113–115.

Chaudhary, R. K., H. C. Cho, and M. T. Monette. 1979. Measurement of antibodies to rabies virus by the immunoelectron microscopy. Can. J. Microbiol. **25:**1209–1211.

Chow, T. L., and S. H. McNutt. 1953. Pathological changes of experimental vesicular stomatitis of swine. Am. J. Vet. Res. **14:**420.

Cifuentes, E., E. Calderon, and G. Bijlenga. 1971. Rabies in a child diagnosed by a new intravitam method. The cornea test. J. Trop. Med. Hyg. **74:**23–25.

Clark, H. F. 1979. Systems for assay and growth of rhabdoviruses. *In* Bishop, D. H. L. (ed.), Rhabdoviruses, Vol. I. CRC Press, Boca Raton, FL.

Clark, H. F., and T. J. Wiktor. 1972. Rabies virus, p. 177–182. *In* S. A. Plotkin (ed.), Strains of human viruses. S. Karger, Basel.

Conomy, J. P., A. Leibovitz, W. McCombs, and J. Stinson. 1977. Airborne rabies encephalitis: demonstration of rabies virus in the human central nervous system. Neurology **27:**67–69.

Constantine, G. D. 1967. Rabies transmission by air in bat caves. (U.S. DHEW/PH N°1617), U.S. Government Printing Office, Washington, D.C.

Correa, W. M. 1964. Prophylaxis of vesicular stomatitis: A field trial in Guatemalan dairy cattle. Am. J. Vet. Res. **25:**1300–1302.

Cotton, W. E. 1926. Vesicular stomatitis in its relation to the diagnosis of foot-and-mouth disease. J. Am. Vet. Med. Assoc. **69:**313.

Cotton, W. E. 1927. Studies on vesicular stomatitis virus. Vet. Med. **22:**169.

Cowherd, H. J., S. Reeves, and Riegler. 1979. Human rabies—Kentucky. Morbid. Mortal. Weekly Rep. **28:**590.

Cox, J. H., B. Dietzschold, and L. G. Schneider. 1977. Rabies virus glycoprotein. II. Biological and serological characterization. Infect. Immun. **16:**743–759.

Crick, J., and F. Brown. 1969. Viral subunits for rabies vaccination. Nature **222**(No. 5188):92.

Dean, D. J., and M. K. Abelseth. 1973. The fluorescent antibody test, p. 73–84. *In* M. M. Kaplan, H. Koprowski (ed.), Laboratory techniques in rabies. World Health Organization, Geneva.

DeBrito, T., M. F. Araujo, and A. Tiriba. 1973. Ultrastructure of the Negri body in human rabies. J. Neurol. Sci. **20:**363–372.

Devriendt, J., and M. Staroukine. 1982. Human rabies—Rwanda. Morbid. Mortal. Weekly Rep. **31:**135.

Diaz, A. M. O., and V. M. Varela-Diaz. 1977. The counter immunoelectrophoresis test for detection of antibodies to rabies virus. Ann. Microbiol. (Inst. Pasteur). **128A:**331–337.

Duhme, D., D. Butman, and S. Aoki. 1983. Imported human rabies. Morbid. Mortal. Weekly Rep. **32:**78–85.

Durham, T. M., J. S. Smith, and F. L. Reid. 1986. Stability of immunofluorescence reactions produced by polyclonal and monoclonal antibody conjugates for rabies virus. J. Clin. Microbiol. **24:**301–303.

Earley, E., P. H. Peralta, and K. M. Johnson. 1967. A plaque neutralization method for arboviruses. Proc. Soc. Exp. Biol. Med. **125:**741.

Eichorn, A. 1917. Vesicular stomatitis in cattle. Am. J. Vet. Med. **12:**162–170.

Ellis, E. M., and H. E. Kendall. 1964. The public health and economic effect of vesicular stomatitis in a herd of dairy cattle. J. Am. Vet. Med. Assoc. **144:**377.

Emmons, R. W., L. L. Leonard, F. De Gennaro Jr., E. S. Protax, P. L. Bazaley, S. T. Giammona, and K. Sturckow. 1973. A case of human rabies with prolonged survival. Intervirology **1:**60–72.

Fekadu, M., J. H. Shaddock, and G. M. Baer. 1982. Excretion of rabies virus in the saliva of dogs. J. Infect. Dis. **145:**715–719.

Fellowes, O. N., G. T. Dimpoullos, and J. J. Callis. 1955. Isolation of vesicular stomatitis virus from an infected laboratory worker. Am. J. Vet. Res. **16:**623.

Fellowes, O. N., G. T. Dimpoullos, J. Tessler, W. R. Hess, T. H. Vardaman, and J. J. Callis. 1956. Comparative titrations of vesicular stomatitis virus in various animal species and in tissue culture. Am. J. Vet. Res. **17:**799.

Fields, B. N., and K. Hawkins. 1967. Human infection with the virus of vesicular stomatitis during an epizootic. N. Engl. J. Med. **277:**988.

Fischman, H. R. 1969. Infectivity of fixed impression

smears prepared from rabies virus-infected brain. Am. Vet. Med. **30**:2205–2208.

Foggin, C. M. 1982. Atypical rabies virus in cats and a dog in Zimbabwe. Vet. Record **110**:338.

Foggin, C. M. 1983. Mokola virus infection in cats and a dog in Zimbabwe. Vet. Record **113**:115.

Francy, D. B., G. G. Moore, G. C. Smith, W. L. Jakob, S. A. Taylor, and C. H. Calisher. 1987. Epizootic vesicular stomatitis in Colorado, 1982. Isolation of virus from insects collected along the Northern Colorado Rocky Mountain front range, 1986. Am. J. Trop. Med. Hyg. (in press).

Fuh, T. H., and D. C. Blenden. 1971. Biopsy technique for the early clinical diagnosis of rabies, p. 176. *In* American Society for Microbiology, Bacteriological Proceedings.

Gaidamovich, S. Ya., V. N. Uvarov, and A. A. Alekseeva. 1966. Isolation of vesicular stomatitis virus from a patient. Vop. Virusol. **11**:77.

Galian, A., J. M. Guerin, M. Lamotte, Y. Le Charpentier, J. Mikol, J. B. Dureux, M. Gerard, E. De Lavergne, P. Atanasiu, P. Ravisse, and P. Sureau. 1980. Human-to-human transmission of rabies via corneal transplant. France. Morbid. Mortal. Weekly Rep. **29**:25–26.

Galli-Valerio, B. 1906. Recherches experimentales sur la rage des rats avec observations sur la rage du surmulot, de la souris et du mulot. Zbl. Bakt. **40**:318–331.

Galtier, P. V. 1879. Etudes sur la rage. Rage du lapin. C. R. Acad. Sci. (Paris) **89**:444–446.

Garcia-Tamayo, J., A. Avila-Mayor, and E. Anzola-Perez. 1972. Rabies virus neuronitis in humans. Arch. Pathol. **94**:11–15.

Geeslin, B. B., B. B. Trotter, and D. Armstrong. 1985. Human rabies diagnosed 2 months postmortem—Texas. Morbid. Mortal. Weekly Rep. **34**:700–706.

Geleta, J. N., and A. A. Holbrook. 1961. Vesticular stomatitis—patterns of complement fixating and serum-neutralizing antibodies in serum of convalescent cattle and horses. Am. J. Vet. Res. **22**:713.

Genovese, M. A., and L. Andral. 1978. Comparaison de deux techniques utilisees pour le diagnostic de la rage: l'immunofluorescence et l'immunoperoxydase. Rec. Med. Vet. **154**:667–671.

Gibbons, W. J. 1961. An outbreak of vesicular stomatitis. Mod. Vet. Practice **42**:30.

Goldwasser, R. A., and R. E. Kissling. 1958. Fluorescent antibody staining of street and fixed rabies virus antigens. Proc. Soc. Exper. Biol. Med. **98**:219–223.

Goldwasser, R. A., R. E. Kissling, T. R. Carski, and T. S. Hosty. 1959. Fluorescent antibody staining of rabies virus antigens in the salivary glands of rabid animals. Bull. W.H.O. **20**:579–588.

Gonzalez-Angulo, A., H. Marquez-Monter, and A. Feria-Velasco. 1970. The ultrastructure of Negri bodies in Purkinje neurons in human rabies. Neurology **20**:323–328.

Gough, P. M., and R. S. Dierks. 1971. Passive haemagglutination test for antibodies against rabies. Bull. W.H.O. **45**:741–745.

Halonen, P. E., F. A. Murphy, B. N. Fields, and D. R. Reese. 1968. Haemagglutinin of rabies and some other bullet-shaped viruses. Proc. Soc. Exper. Biol. Med. **127**:1037–1042.

Hanson, R. P., A. F. Rasmussen, C. A. Brandley, and J. W. Brown. 1950. Human infection with the virus of vesicular stomatitis. J. Lab. Clin. Med. **36**:754.

Hanson, R. P. 1952. The natural history of vesicular stomatitis. Bacteriol. Rev. **16**:179.

Hanson, R. P. 1981. Vesicular stomatitis, p. 2517. *In* E. P. J. Gibbs (ed.), Virus diseases of food animals, a world geography of epidemiology and control. Academic Press, Inc., New York.

Hattwick, M. A. W., and M. B. Gregg. 1975. The disease in man, p. 281–304. *In* G. M. Baer (ed.), The natural history of rabies, Vol. II. Academic Press, Inc., New York.

Hattwick, M. A. W., L. Corey, and W. B. Creech. 1976. Clinical use of human globulin immune to rabies virus. J. Infect. Dis. **133**:A266–A272.

Hattwick, M. A. W., T. T. Weis, J. Stechschulte, G. M. Baer, and M. B. Gregg. 1972. Recovery from rabies. A case report. Ann. Intern. Med. **76**:931–942.

Heinz, E. 1945. Vesicular stomatitis in cattle and horses in Colorado. North Am. Vet. **26**:726.

Hopkins, S. F., and G. C. Jenney. 1962. The fluorescent antibody technique applied to vesicular stomatitis virus. Am. J. Vet. Res. **23**:603.

Houff, S. A., R. C. Burton, and R. W. Wilson. 1979. Human-to-human transmission of rabies by corneal transplant. N. Engl. J. Med. **300**:603–604.

Iwasaki, Y., T. J. Wiktor, and H. Koprowski. 1973. Early events of rabies virus replication in tissue cultures. An electron microscopic study. Lab. Invest. **28**:142–148.

Jenney, E. W. 1967. Vesicular stomatitis in the United States, 1963–1967. Proc. U.S. Livestock Sanit. Assoc. **71**:371.

Jenney, E. W., and L. O. Mott. 1958. Serologic studies with VSV. I. Typing of vesicular stomatitis by complement fixation. Am. J. Vet. Res. **19**:993.

Jenney, E. W., and L. O. Mott. 1963. Serologic studies with VSV. II. Typing of vesicular stomatitis convalescent serum by direct complement fixation. Am. J. Vet. Res. **24**:874–879.

Johnson, H. N. 1973. The virus neutralization index test in mice, p. 94–97. *In* M. M. Kaplan, H. Koprowski (eds.), Laboratory techniques in rabies. World Health Organization, Geneva.

Johnson, H. N., and C. N. Leach. 1940. Comparative susceptibility of different strains of mice to rabies virus. Am. J. Hyg. **32**:38–45.

Johnson, K. M., R. E. Tesh, and P. H. Peralta. 1969. Epidemiology of vesicular stomatitis virus: some new data and a hypothesis for transmission. J. Am. Vet. Med. Assoc. **155**:2033.

Johnson, K. M., J. B. Vogel, and P. H. Peralta. 1966. Clinical serological response to laboratory-acquired human infection by Indiana type vesicular stomatitis virus. Am. J. Trop. Med. Hyg. **15**:244.

Johnson, K. P., P. T. Swoveland, and R. W. Emmons. 1980. Diagnosis of rabies by immunofluorescence in trypsin-treated histologic sections. J. Am. Med. Assoc. **244**:41–43.

Jonkers, A. H., L. Spence, C. A. Cookwell, and J. J. Thornton. 1964. Laboratory studies with wild rodents and viruses native to Trinidad. Am. J. Trop. Med. Hyg. **13**:613.

Joustra, M., and H. Lundgren. 1969. Preparation of freeze-dried monomeric and immunochemically pure IgG by a rapid and reproducible chromatographic technique, p. 511–515. *In* H. Peeters (ed.), Protides of the biological fluids. Proceedings of the 17th Coloquium, Bruges.

Karabatsos, N. (ed.). 1985. International catalogue of arboviruses, 3rd ed., Amer. Soc. Trop. Med. Hyg.

Karstad, L., and R. D. Hanson. 1958. Primary isolation and comparative titrations of the field strains of vesicular stomatitis virus in chicken embryos, hogs, and mice. Am. J. Vet. Res. **19**:233.

Kawai, A., S. Matsumoto, and K. Tanabe. 1975. Characterization of rabies viruses recovered from persistently infected BHK cells. Virology **67**:520–533.

Kerton, L., S. Schwartz, E. M. Clearver, et al. 1979. Human rabies. Oklahoma. Morbid. Mortal. Weekly Rep. **28:**476–481.

Kissling, R. E. 1958. Growth of rabies virus in non-nervous tissue culture. Proc. Soc. Exper. Biol. Med. **98:**223–225.

Kissling, R. E., and D. R. Reese. 1963. Antirabies vaccine of tissue culture origin. J. Immunol. **91:**362–368.

Kissling, R. E. 1975. The fluorescent antibody test in rabies, p. 401–416. *In* G. M. Baer (ed.), The natural history of rabies. Academic Press, Inc., 1975. New York, San Francisco, London.

Koch, F. J., J. W. Sagartz, D. E. Davidson, and K. Lawhaswasdi. 1975. Diagnosis of human rabies by the cornea test. Am. J. Clin. Pathol. **63:**509–515.

Koprowski, H. 1973. The mouse inoculation test, p. 85–93. *In* M. M. Kaplan and H. Koprowski (ed.), Laboratory techniques in rabies. World Health Organization, Geneva.

Koprowski, H., T. J. Wiktor, and M. K. Abelseth. 1985. Cross-reactivity and cross-protection: rabies variants and rabies-related viruses, p. 30–39. *In* E. Kuwert, C. Merieux, H. Koprowski, and K. Bogel (ed.), Rabies in the tropics. Springer-Verlag, Berlin.

Kowalczyk, T., and C. A. Brandley. 1954. Experimental infection of dogs, ferrets, chinchillas, and hamsters with vesicular stomatitis virus. Am. J. Vet. Res. **15:**98.

Kuns, M. L. 1959. The epizootiology of vesicular stomatitis in middle America. PhD thesis, Univ. of Wisconsin, Madison.

Kuwert, E. 1973. The complement fixation test, p. 124–134. *In* M. M. Kaplan, H. Koprowski (ed.), Laboratory techniques in rabies. World Health Organization, Geneva.

Lafon, M., T. J. Wiktor, and R. I. Macfarlan. 1983. Antigenic sites of the CVS rabies virus glycoprotein. Analysis with monoclonal antibodies. J. Gen. Virol. **64:**843–851.

Lafon, M., J. Ideler, and W. H. Wunner. 1984. Investigations of antigenic structure of the rabies virus glycoprotein by monoclonal antibodies, p. 219–225. Development in Biological Standardization Joint WHO/AIBS Symposium on the Standardization on Rabies Virus, Vol. 57. S. Karger, Basel.

Lafon, M., and T. J. Wiktor. 1985. Antigenic sites on the ERA rabies virus nucleoprotein and non-structural protein. J. Gen. Virol. **66:**2125–2133.

Lakhanpal, U. and R. C. Sharma. 1985. An epidemiological study of 177 cases of human rabies. Int. J. Epidemiol. **14:**614–617.

Lamas, C. C., A. J. Martinez, and R. Baraff. 1980. Rabies encephaloradiculomyelitis: case report. Acta Neuropathol. **51:**245–247.

Larghi, O. P., L. Gonzalez, and J. R. Held. 1973. Evaluation of the corneal test as a laboratory method for rabies diagnosis. Appl. Microbiol. **25:**181–189.

Larghi, O. P., and Ch. E. Jimenez. 1971. Methods for accelerating the fluorescent antibody test for rabies diagnosis. Appl. Microbiol. **21:**611–613.

Larghi, O. P., A. E. Nerel, L. Lazaro, and V. L. Savy. 1975. Sensitivity of BHK-21 cells supplemented with diethylaminoethyl-dextran for detection of street rabies virus in saliva samples. J. Clin. Microbiol. **1:**243–245.

Larghi, O. P., O. Oliva–Pinheiro, and E. Gonzalez-Luarca. 1986. Evaluacion de un conjugado antirabico por titulacion en diferentes microscopios fluorescentes. Rev. Inst. Med. Trop. Saò Paulo **28:**2–5.

Larraza, H. O., J. E. Olvera-Rabiela, and S. Poucell-Lopez. 1978. Rabia humana: estudio clinico y anatomo-patologico de 52 casos. Patologia **16:**59–76.

Lauerman, L. H., and R. P. Hanson. 1963. Field trial vaccination against vesicular stomatitis in Panama. Proc. U.S. Livestock Sanit. Assoc. **67:**483–490.

Lauerman, L. H., M. L. Kuns, and R. P. Hanson. 1962. Field trial of live virus vaccination procedure for prevention of vesicula stomatitis in dairy cattle. I. Preliminary immune response. Proc. U.S. Livestock Sanit. Assoc. **66:**365.

Leach, C. N. 1938. Comparative methods of diagnosis of rabies in animals. Am. J. Pub. Health. **28:**162–166.

Leach, C. N., and H. N. Johnson. 1940. Human rabies with special reference to virus distribution and titer. Am. J. Trop. Med. **20:**335–340.

Lennette, E. H., and R. W. Emmons. 1971. The laboratory diagnosis of rabies: review and perspective, p. 77–90. *In* Y. Nagano, F. M. Davenport, (ed.), Rabies. University of Tokyo Press. Tokyo.

Lentz, T. L., T. G. Burrage, and A. L. Smith. 1982. Is the acetylcholine receptor a rabies virus receptor? Science **215:**182–184.

Lentz, T. L., P. T. Wilson, E. Hawrot, and D. W. Speicher. 1984. Amino acid sequence similarity between rabies virus glycoprotein and make venom curaremimetic neurotoxins. Science **226:**847–848.

Lodmell, D. L., and L. C. Ewalt. 1985. Pathogenesis of street rabies virus infections in resistant and susceptible strains of mice. J. Virol. **55:**788–795.

Madore, H. P., and J. M. England. 1977. Rabies virus protein synthesis in infected BHK-21 cells. J. Virol. **22:**102–112.

Marder, G. 1986. Human rabies transmission: the substitution of sense to panic. J. Am. Med. Assoc. **255:**321.

Marie, A. C. 1930. Sensibilité du cobaye au virus rabique. CR. Acad. Sci. (Paris) **103:**868–869.

Matsumoto, S. 1963. Electron microscope studies of rabies virus in mouse brains. J. Cell. Biol. **19:**565–591.

McClain, M. E., and A. J. Hackett. 1958. A comparative study of the growth of vesicular stomatitis virus in five tissue culture systems. J. Immunol. **80:**356–361.

McClellan, G. B.1862. Report on the organization of the army of the Potomac and of its campaigns in Virginia and Maryland, July 26, 1861 to November 1862. The Times Steam Printing House, Chicago.

McFarlan, R. I., B. Dietzschold, T. J. Wiktor, M. Kiel, R. Houghten, R. A. Lerner, J. G. Sutcliffe, and H. Koprowski. 1984. T. cell responses to cloned rabies virus glycoprotein and to synthetic peptides. **133(5):**2748–2752.

McSharry, J. J. 1979. The lipid envelope and chemical composition of rhabdoviruses, p. 107–117. *In* D. H. L. Bishop (ed.), Rhabdoviruses, Vol. 1. CRC Press, Boca Raton, Florida.

McQueen, J. L., A. L. Lewis, and N. J. Schneider. 1960. Rabies diagnosis by fluorescent antibody. Its evaluation in a public health laboratory. Am. J. Public Health **50:**1743–1752.

Mengeling, W. L., and M. J. Van der Maaten. 1971. Identification of selected animal viruses with fluorescent antibodies prepared from multivalent antiserums. Am. J. Vet. Res. **32:**1825.

Meyer, N. L., W. M. Moulton, E. W. Jenney, and R. J. Rodgers. 1960. Outbreaks of vesicular stomatitis in Oklahoma and Texas. Proc. U.S. Livestock Sanit. Assoc. **64:**324.

Miyamoto, H., and S. Matsumoto. 1965. The nature of the Negri body. J. Cell. Biol. **27:**677–682.

Mohler, J. R. 1930. Vesicular stomatitis in horses and cattle. Bull. No. 662. U.S. Dept. Agriculture, Washington, D.C.

Moler, F. W., B. C. Johnson, and E. Daniel. 1983. Human rabies. Michigan. Morbid. Mortal. Weekly Rep. **32:**159.

Mollgard, S. 1985. Bat-rabies in Denmark. Rabies Bull. Eur. **4:**11.

Morecki, R., and H. M. Zimmerman. 1969. Human rabies encephalitis: fine structure study of cytoplasmic inclusions. Arch. Neurol. **20:**599–604.

Murphy, F. A., and S. P. Bauer. 1974. Early street rabies virus infection in striated muscle and later progression to the central nervous system. Intervirology **3:**256–268.

Negri, A. 1903. Contribution allo studio dell'etiologica della rabbia. Bol. Soc. Med. Chir. Pavia. **88:**114.

Nilsson, M. R., and W. Sugay. 1966. O uso de camundongos lactantes no diagnostico da raiva. Arq. Inst. Biol. Sao Paulo **33:**47–48.

Olitsky, P. K., H. R. Cox, and J. T. Syverton. 1934. Comparative studies on the viruses of vesicular stomatitis and equine encephalomyelitis. J. Exp. Med. **59:**159.

Olitsky, P. K., J. Traum, and H. W. Schoening. 1927. Comparative studies on vesicular stomatitis and foot-and-mouth disease. J. Am. Vet. Med. Assoc. **70:**147.

Pasteur, L. 1881. Sur la rage. CR. Acad. Sci. (Paris) **92:**1259–1260.

Pasteur, L. 1885. Méthode pour prévenir la rage aprè morsure. CR. Acad. Sci. (Paris) **101:**765–773.

Perl, D. P. 1975. The pathology of rabies in the central nervous system, p. 236–242. *In* G. M. Baer (ed.), The natural history of rabies, Vol. 1. Academic Press, New York.

Perrin, P., M. Lafon, P. Versmisse, and P. Sureau. 1985. Application d'une methode immunoenzymatique au titrage des anticorps neutralisants en culture cellulaire. J. Biol. Standard. **13:**35–42.

Perrin, P., D. Portnoi, and P. Sureau. 1982. Etude de l'adsorption et de la pénétration du virus rabique: interactions avec les cellules BHK-21 et des membranes artificielles. Ann. Virol. (Inst. Pasteur) **133E:**403–422.

Perrin, P., P. E. Rollin, and P. Sureau. 1986. A rapid rabies enzyme immunodiagnosis (RREID): a useful and simple technique for the routine diagnosis of rabies. J. Biol. Standard **14:**217–222.

Perrin, P., P. Versmisse, and J. F. Delagneau. 1986. The influence of the type of immunosorbent on rabies antibody EIA; advantages of purified glycoprotein over whole virus. J. Biol. Stand. **14:**95–102.

Pilon Moron, E., J. Vincent, P. Sureau, and R. Néel. 1967. Diagnostic rapide de la rage par l'inoculation du cerveau et de la glande sous-maxillaire aux souriceaux et par l'immunofluorescence. Arch. Inst. Algérie. **45:**5–10.

Porras, C., J. J. Barboza, and E. Fuenzalida. 1976. Recovery from rabies in man. Intern. Med. **85:**44–48.

Portnoi, D., S. Favre, and P. Sureau. 1982. Use of neuroblastoma cells (MNB) for the isolation of street rabies virus from field specimens. Rabies Info. Exch. **6:**336.

Reagan, K. J., and W. H. Wunner. 1984. Early interactions of rabies virus with cell surface receptors, p. 387–392. *In* D. H. L. Bishop and R. W. Compans (ed.), Nonsegmented negative strain viruses. Academic Press, New York.

Reid, F. L., N. H. Hall, and J. S. Smith. 1983. Increased immunofluorescence staining of rabies-infected, formalin-fixed brain tissue after pepsin and trypsin digestion. J. Clin. Microbiol. **18:**968–971.

Reif, J. S., P. A. Webb, T. P. Monath, J. K. Emerson, J. D. Poland, G. E. Kemp, and G. Cholas. 1987. Epizootic vesicular stomatitis in Colorado, 1982. III. Infection in occupational risk groups. Am. J. Trop. Med. Hyg. **36:**187.

Remington, P. L., TY. Shope, and J. Andrews. 1985. A recommended approach to the evaluation of human rabies exposure in an acute-care hospital. J. Am. Med. Assoc. **254:**67–69.

Remlinger, P. 1917. Comparaison de l'inoculation du virus rabique au lapin et au cobaye. CR. Soc. Biol. **80:**670–672.

Rice, C. E., and P. D. McKercher. 1954. Studies of the complement-fixation reaction in virus systems. VI. In vesicular stomatitis in horses, cattle, and swine. J. Immunol. **73:**309–317.

Rollin, P. E., and P. Sureau. 1984. Monoclonal antibodies as a tool for rabies epidemiological studies. International Symposium on Monoclonal antibodies: Standardization of their characterization and use, Paris, France, 1983. Dev. Biol. Stand. **57:**193–197.

Rosenthal, L. J., and I. L. Sheckmeister. 1971. Comparison of microtiter procedures with the plaque technique for assay of VSV. Appl. Microbiol. **21:**400.

Rubin, R. H., L. Sullivan, and R. Summers. 1970. A case of human rabies in Kansas. Epidemiologic, clinical and laboratory considerations. J. Infect. Dis. **122:**318–322.

Rudd, R. J., C. V. Trimarchi, and M. K. Abelseth. 1980. Tissue culture technique for routine isolation of street strain rabies virus. J. Clin. Microbiol. **12:**590–593.

Rudd, R. J., and C. V. Trimarchi. 1983. BHK-21 vs neuroblastoma cells lines for the isolation of street rabies virus. Rabies Info. Exch. **8:**38–40.

Rupprecht, C. E., and T. J. Wiktor. 1987. Antigenic variants of rabies virus in Pennsylvania Wildlife. *In* Fischman, H. (ed.), North American Symposium on Rabies in Wildlife (in press).

Sainte-Marie, G. 1962. A paraffin embedding technique for studies employing immunofluorescence. J. Histochem. **10:**250–256.

Sanmartín, C., P. Correa, A. Dueñas, and N. Muñoz. 1967. Algunas consideraciones sobre 42 casos humanos de rabia, p. 155–161. Memorias del Primer Seminario Nacional Sobre Rabia, Medellin, Colombia, July 26–29.

Savy, V., and P. Atanasiu. 1978. Rapid immunoenzymatic technique for titration of rabies antibodies IgG and IgM. Dev. Biol. Stand. **40:**247–253.

Schlegel, R., T. S. Tralka, M. C. Willinggham, and I. Pastan. 1983. Inhibition of VSV binding and infectivity by phosphatidyl serine. Is phosphatidyl serine a VSV binding site. Cell **32:**639–646.

Schneider, L. G. 1969. The cornea test: a new method for the intravitam diagnosis of rabies. Zbl. Vet. Med. B **16:**24–31.

Schneider, L. G. 1982. Antigenic variants of rabies virus. Comp. Immunol. Microbiol. Infect. Dis. **5:**101–107.

Schneider, L. G., B. J. H. Barnard. and H. P. Schneider. 1985. Applications of monoclonal antibodies for epidemiological investigations and vaccination studies. 1 African viruses, p. 47–59. *In* E. Kuwert, C. Mérieux, H. Koprowski, and K. Bögel (ed.), Rabies in the tropics. Springer-Verlag, Berlin.

Schneider, L. G., B. Dietzschold, and R. E. Dierks. 1973. Rabies group-specific ribonucleoprotein antigen and a test system for grouping and typing of rhabdoviruses. J. Virol. **11:**748–755.

Segre, L., and A. E. Zothner. 1985. Instantaneous fluorescent antibody staining of rabies virus antigen. Medicina **44:**529.

Sehgal, S., and R. Bhatia. 1985. Rabies: current status and proposed control programmes in India, p. 11. Proceedings of Workshop on Rabies Surveillance and Control held at National Institute of Communicable Diseases, Delhi, India, 15–17 October.

Sellers, R. F. 1955. Growth and titration of foot and mouth disease and vesicular stomatitis in kidney monolayer tissue culture. Nature **176:**547–549.

Shahan, M. S., A. H. Frank, and L. O. Mott. 1946. Studies of vesicular stomatitis with special reference to a virus of swine origin. J. Am. Vet. Med. Assoc. **108**:5.

Smith, J. S., C. L. McClelland, F. L. Reid, and G. M. Baer. 1982. Dual role of the immune response in street rabies-virus infection of mice. Infect. Immunol. **35**:213–221.

Smith, J. S., J. W. Sumner, and L. F. Roumillat. 1984. Antigenic characteristics of isolates associated with a new epizootic of raccoon rabies in the United States. J. Infect. Dis. **149**:769–774.

Smith, A. L., C. H. Tignor, R. W. Emmons, and J. D. Woodie. 1978. Isolation of field rabies virus strains in CER and murine neuroblastoma cell cultures. Intervirology **9**:359–361.

Smith, J. S., P. A. Yager, and G. M. Baer. 1973. A rapid reproducible test for determining rabies neutralizing antibody. Bull. W.H.O. **48**:335–341.

Smith, J. S., P. A. Yager, and G. M. Baer. 1973. A rapid tissue culture test for determining rabies neutralizing antibody, p. 354–357. *In* M. M. Kaplan, H. Koprowski (ed.), Laboratory techniques in rabies. World Health Organization, Geneva.

Smith, W. B., D. C. Blenden, T. H. Fuh, and L. Hiler. 1972. Diagnosis of rabies by immunofluorescent staining of frozen sections of skin. J. Am. Vet. Med. Assoc. **1961**:1495–1501.

Sokol, F. 1973. Purification of rabies virus and isolation of its components, p. 165–178. *In* M. M. Kaplan, H. Koprowski, (ed.), Laboratory techniques in rabies. World Health Organization, Geneva.

Sorensen, D. K., T. L. Chow, T. Kowalczyk, R. P. Hanson, and C. A. Brandly. 1958. Persistence in cattle of serum-neutralizing antibodies of vesicular stomatitis virus. Am. J. Vet. Res. **19**:74–77.

Steele, J. H. 1975. History of rabies, p. 1–29. *In* G. M. Baer (ed.), The natural history of rabies, Vol. 1. Academic Press, Inc., New York.

Sulkin, S. E., and J. C. Wilett. 1939. A comparative study of the mouse and guinea pig inoculation methods in the diagnosis of rabies. Am. J. Pub. Hlth. **29**:921–926.

Sung, J. H., M. Mayano, and A. R. Mastri. 1976. A case of human rabies and ultrastructure of the Negri body. J. Neuropathol. Exp. Neurol. **35**:541–559.

Sureau, P., P. Rollin, and T. J. Wiktor. 1983. Epidemiologic analysis of antigenic variations of street rabies virus: detection by monoclonal antibodies. Am. J. Epidemiol. **117**:605–609.

Swanson, D., R. Feigin, and L. Tanney. 1984. Human rabies—Texas. **33**:469–470.

Swoveland, P. T., and K. P. Johnson. 1979. Enhancement of fluorescent antibody staining of viral antigens in formalin-fixed tissue by trypsin digestion. J. Infect. Dis. **140**:758–764.

Tesh, R. B., P. H. Peralta, and K. M. Johnson. 1969. Ecologic studies of vesicular stomatitis virus. I. Prevalence of infection among animals and humans living in an area of endemic VSV activity. Am. J. Epidemiol. **90**:255.

Théodoridès, J. 1986. Histoire de la rage. Masson, Paris, New York, Barcelona, Milan, Mexico, Sao Paulo.

Thomas, J. B. 1975. The serum neutralization, indirect fluorescent antibody, and rapid fluorescent focus inhibition test, p. 417–433. *In* G. M. Baer (ed.), The natural history of rabies, Vol. I. Academic Press, New York, San Francisco, London.

Thomas, J. B., R. K. Sikes, and A. S. Richer. 1963. Evaluation of indirect fluorescent antibody technique for detection of rabies antibody in human sera. J. Immunol. **91**:721–723.

Thongcharoen, P., C. Wasi, and S. Sirikavin. 1981. Human-

to-human transmission of rabies via cornea transplant—Thailand. Morbid. Mortal. Weekly Rep. **30**:473–474.

Tierkel, E. S. 1973. Shipment of specimens and techniques for preparation of animal tissues, p. 29–40. *In* M. M. Kaplan, H. Koprowski (ed.), Laboratory techniques in rabies. World Health Organization, Geneva.

Tierkel, E. S., 1975a. Canine rabies, p. 123–137. *In* G. M. Baer (ed.), The natural history of rabies, Vol. II. Academic Press, Inc., New York, San Francisco, London.

Tierkel, E. S. 1975b. Control of urban rabies, p. 189–201. *In* G. M. Baer (ed.), The natural history of rabies, Vol. 11. Academic Press, Inc., New York.

Tignor, G. H., F. A. Murphy, and H. F. Clark. 1977. Duvenhage virus: morphological biochemical, histopathological and antigenic relationships to the rabies serogroup. J. Gen. Virol. **37**:595–611.

Tignor, G. H., and R. E. Shope. 1972. Vaccination and challenge of mice with viruses of the rabies serogroup. J. Infect. Dis. **125**:322–324.

Tillotson, J. R., D. Axelrod, and D. O. Lyman. 1977. Rabies in a laboratory worker—New York. Morbid. Mortal. Weekly Rep. **26**:183–184.

Trimarchi, C. V., R. J. Rudd, and M. K. Abelseth. 1979. Animal isolators and necropsy hoods in the rabies laboratory. Rabies Info. Exch. **1**:30–33.

Umeno, S., and Doi, Y. 1921. A study in the antirabic inoculation of dogs. Kitasato Arch. Exp. Med. **4**:89–108.

Umoh, J. U., and D. C. Blenden. 1981. Immunofluorescent staining of rabies virus antigen in formalin-fixed tissue after treatment with trypsin. Bull. W.H.O. **59**:737–744.

Umoh, J. U., and D. C. Blenden. 1983. Comparison of primary skunk brain and kidney and raccoon kidney cells with established cell line for isolation and propagation of street rabies virus. Infect. Immun. **41**:1370–1372.

Umoh, J. U., C. D. Ezeokoli, and A. E. J. Okoh. 1985. Immunofluorescent staining of trypsinized formalin-fixed brain smears for rabies antigen: results compared with those obtained by standard methods for 221 suspect animal cases in Nigeria. J. Hyg. Camb. **94**:129–134.

Vadlamudi S., and R. P. Hanson. 1963. The neutralization test for vesicular stomatitis virus in chicken embryos and tissue culture. Cornell Vet. **53**:16.

Van Gehuchten, A., and C. Nelis. 1900. Les lésions histologiques de la rage chez les animaux et chez l'homme. Bull. Acad. Roy. Med. Belg. **14**:31–36.

Vernon, S. D., and P. A. Webb. 1985. Recent VSV infection detected by immunoglobulin M antibody capture enzyme-linked immunosorbent assay. J. Clin. Microbiol. **22**:582.

Vigel, F. 1916. Sur la stomatite erosive. Bull. Soc. Cent. Med. Vet. **69**:59–61.

Wachendorfer, G. 1967. Tollwutdiagnostik beim Tier mit Hilfe der Immunofluoreszens. Wien Tierarztl. Mschr. **54**:451.

Wagner, K. 1931. Infection and immunity of vesicular stomatitis in guinea pigs. Vet. Med. Kansas City, Missouri **26**:3881.

Walton, T. E., P. E. Webb, W. L. Kramer, G. C. Smith, T. Davis, F. R. Holbrook, C. G. Moore, T. Schiefer, R. H. Jones, W. L. Jakob, and G. Janney. 1987. Epizootic vesicular stomatitis in Colorado, 1982: epidemiologic and endomologic studies on the western slope, 1986. Am. J. Trop. Med. Hyg. **36**:176.

Warrell, D. A. 1988. Human rabies and its prevention: overview. Proceedings of "Research Towards Rabies Prevention." An international symposium presented by The Fogarty International Center, National Institutes of Health, Washington, D.C., November 3–5, 1986 (to be published in *Reviews of Infectious Diseases,* 1988).

Webb, P. A., T. P. Monath, J. S. Reif, G. C. Smith, G. E. Kemp, J. S. Lazuick, and T. E. Walton. 1987. Epizootic vesicular stomatitis in Colorado, 1982: epidemiologic studies among the Northern Colorado front range. Am. J. Trop. Med. Hyg. **36:**193.

Webster, L. T. 1936. Diagnostic and immunological tests of rabies in mice. Am. J. Public Health **26:**1207–1210.

Webster, L. T., and J. R. Dawson Jr. 1935. Early diagnosis of rabies by mouse inoculation. Measurement of humoral immunity to rabies by mouse protection test. Proc. Soc. Exp. Biol. Med. **32:**570–573.

White, L. A., and W. A. Chapell. 1982. Inactivation of rabies virus in reagents used for the fluorescent rabies antibody test. J. Clin. Microbiol. **16:**253–256.

Wiktor, T. J., and H. F. Clark. 1973. Application of the plaque assay technique to the study of rabies virus-neutralizing antibody interactions. Ann. Microbiol. (Inst. Pasteur) **124A:**271–282.

Wiktor, T. J., and H. F. Clark. 1975. Growth of rabies virus in cell culture, p. 155–179. *In* G. M. Baer (ed.). The natural history of rabies, Vol. 1, Academic Press, New York.

Wiktor, T. J., A. Flamand, A. H. Koprowski. 1980. Use of monoclonal antibodies in diagnosis of rabies virus infection and differentiation of rabies and rabies related viruses. J. Virol. Methods **1:**33–46.

Wiktor, T. J., and H. Koprowski. 1974. Rhabdovirus replication in enucleated host cells. J. Virol. **14:**300–306.

Wiktor, T. J., and H. Koprowski. 1978. Monoclonal antibodies against rabies virus produced by somatic cell hybridization: detection of antigenic variants. Proc. Natl. Acad. Sci. USA **75:**3938–3942.

Wiktor, T. J., and H. Koprowski. 1980. Antigenic variants of rabies virus. J. Exp. Med. **152:**99–112.

Wiktor, T. J., H. Koprowski, and F. Dixon. 1972. Radioimmunoassay procedure for rabies binding antibodies. J. Immunol. **109:**464–470.

Wiktor, T. J., E. Kuwert, and H. Koprowski. 1968. Immune lysis of rabies virus infected cells. J. Immunol. **101:**1271–1282.

Wiktor, T. J., R. I. MacFarlan, C. M. Foggin, and H. Koprowski. 1984. Antigenic analysis of rabies and Mokola virus from Zimbabwe using monoclonal antibodies. Dev. Biol. Standard. **57:**199–211.

Wiktor, T. J., R. I. Macfarlan, and K. J. Reagan. 1984. Protection from rabies by a vaccinia virus recombinant containing the rabies virus glycoprotein gene. Proc. Natl. Acad. Sci. (Wash.) **81:**7194–7198.

Winkler, W. G., T. R. Fashinell, and L. Leffingwell. 1973. Airborne transmission in a laboratory worker. J. Am. Med. Assoc. **226:**1219–1221.

World Health Organization Expert Committee on Rabies. 1973. Sixth Report, Wld. Hlth. Org. Techn. Rep. Ser. No. 523, Geneva.

World Health Organization Expert Committee on Rabies. 1984. Seventh Report, Wld. Hlth. Org. Techn. Rep. Ser. 709, Geneva.

World Health Organization. 1987. World survey of rabies XXII (for years 1984/85), WHO/RABIES/87.198, World Health Organization, Geneva.

Wunner, W. H., and K. J. Reagan et al. 1984. Characterization of saturable binding sites for rabies virus. J. Virol. **50:**691–697.

Wunner, W. H., C. L. Smith, and M. Lafon et al. 1984. Comparative nucleotide sequence analysis of the G gene of antigenically altered rabies virus, p. 279–284. *In* D. H. L. Bishop and R. W. Compans (eds.). Nonsegmented negative strand viruses. Academic Press, San Diego.

Wunner, W. H., B. Dietzschold, C. L. Smith, M. Lafon, and E. Golub. 1985. Antigenic variants of CVS rabies virus with altered glycosylation-sites. Virology **140:**1–12.

Yelverton, E., S. Norton, J. F. Obijeski, and D. V. Goeddel. 1983. Rabies virus glycoprotein analogs: biosynthesis in Escherichia coli. Science **219:**614–620.

Yen, A. C. H. 1936. Experimental virus infections in chinese hamster. II. Susceptibility to street rabies virus. Proc. Soc. Exper. Biol. Med. **34:**648–651.

Yuill, T. M. 1981. Vesicular stomatitis. *In* J. H. Steele (ed.). CRC Handbook series in zoonoses: viral zoonoses. CRC Press, Boca Raton, FL.

Zalan, E., C. Wilson, and D. Pukitis. 1979. A microtest for the quantitation of rabies neutralizing antibodies. J. Biol. Stand. **7:**213–220.

Zeidner, D., A. Bowman, and J. Dennehy. 1984. Human rabies. Pennsylvania **33:**633–634.

CHAPTER 30

Filoviridae: Marburg and Ebola Viruses

MICHAEL P. KILEY

Disease: Marburg hemorrhagic fever, Ebola hemorrhagic fever.

Etiologic Agents: Marburg virus, Ebola-Sudan virus, Ebola-Zaire virus.

Source: Unknown.

Clinical Manifestations: Acute onset with headache, myalgia, and lethargy. Leads to gastrointestinal symptoms and rash. Serious cases involve severe hemorrhagic manifestations. Death follows blood loss and shock. Mortality can be near 90% in an outbreak.

Pathology: Virus is present in most organs, and lesions are found in many organs. Vascular collapse leading to shock may be due to endothelial damage.

Laboratory Diagnosis: Diagnosis is made by virus isolation from blood and other tissues or by a fourfold rise in indirect fluorescence antibody titer. Caution must be used when handling specimens as viruses are extremely virulent.

Epidemiology: Viruses are indigenous to Africa.

Treatment: None proven.

Prevention and Control: No vaccine is available. Control is by good barrier nursing techniques and proper use of syringes and needles.

Introduction

Marburg and Ebola viruses are the charter members of a new family of negative-stranded RNA viruses, the *Filoviridae* (Kiley et al., 1982). They both cause severe hemorrhagic fever and are indigenous to Africa. Marburg virus disease was first diagnosed in laboratory workers in Marburg, Germany, and Belgrade, Yugoslavia, in 1967. These workers had been exposed to tissues and blood from African green monkeys (*Cercopithecus aethiops*) imported from Uganda. There were 25 primary cases and six secondary cases in the outbreak. The seven patients who died all had primary cases (Siegert et al., 1967; Smith et al., 1967; Stojkovic et al., 1971). Three cases of Marburg virus disease occurred in South Africa in 1975 (Gear et al., 1975). The index patient died, whereas the two patients with secondary cases survived. The only other confirmed cases (two) of Marburg virus disease occurred in Kenya in 1980. The first patient acquired the disease in the field and

died in a hospital. The second patient, who survived, was the physician attending the first patient (Centers for Disease Control, 1980; Teepe et al., 1983). The source of the virus in the South African and Kenyan episodes is unknown as is the original source of the Marburg outbreak.

Hemorrhagic fever caused by Ebola virus was first recognized in Zaire and Sudan in 1976 (W.H.O., 1976a,b; Table 1). The two outbreaks occurred at about the same time and were initially thought to be caused by the same virus strain. Subsequent studies showed that the outbreaks were caused by two distinct Ebola virus strains (Buchmeier et al., 1983; Cox et al., 1983; Richman et al., 1983). In the Zaire outbreak, 280 of the 318 confirmed cases died. In Sudan there were 212 confirmed cases, 117 of which died. Many of the cases were the result of nosocomial transmission. At one mission hospital in Zaire, where all 17 staff members had contact with patients or instruments used for treating patients, 13 employees contracted the disease and 11 died. The most

TABLE 1. Summary of Marburg and Ebola virus outbreaks

Year	Virus	Location	Cases No.	Case-fatality ratio (%)
1967	Marburg[a]	W. Germany, Yugoslavia	31	22
1975	Marburg	South Africa (Zimbabwe)	3	33
1980	Marburg	Kenya	2	50
1976	Ebola	Zaire	318	88
1976	Ebola	Sudan	228	51
1979	Ebola	Sudan	34	65

[a] The source of the 1967 Marburg outbreak in Europe was infected monkeys from Uganda.

recent outbreak of Ebola virus disease occurred in Sudan in 1980, where there were 34 confirmed cases and 22 deaths (Baron et al., 1983).

The Viruses

Morphologic studies have shown that both Marburg and Ebola viruses are unique in structure. Both Marburg and Ebola virions, when obtained from clinical material or cell culture, have a uniform width of about 80 nm but vary in length from approximately 1 to 14 μm. Virions, especially Ebola virions, are extensively branched (Fig. 1). Marburg virions are also long and often are seen in the shape of the number "6" as are Ebola virions (Murphy et al., 1978; Peters et al., 1971).

When Marburg and Ebola virions are purified by rate zonal-sucrose gradient centrifugation, a more conventional virion structure is seen (Regnery et al., 1981). Under these conditions virions are bacilliform in shape (Fig. 2); they have a sedimentation coefficient of approximately 1400S. Each virion is of uniform length, 790 μm for Marburg and 970 μm for Ebola. Genome inactivation studies have demonstrated that each of these bacilliform particles contains one functional copy of virion RNA. Except for the differences in length, both virions seem to be very similar in morphology. The mechanism by which virions of defined length assemble into the long forms seen in nature is unknown.

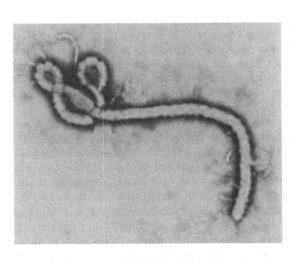

FIG. 1. Ebola virus. Unfixed diagnostic specimen from first Vero cell passage demonstrating branching and internal helix. Sodium phosphotungstate; magnification × 84,500 (courtesy of F. A. Murphy).

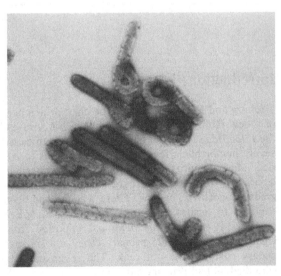

FIG. 2. Ebola virus. Glutaraldehyde-fixed virus particles harvested from a band obtained after centrifugation in a rate zonal sucrose gradient. Demonstrates bacilliform shape of the particles and uniform length of 970 nm for this virus. Sodium phosphotungstate; magnification × 68,700 (courtesy of R. L. Regnery and E. L. Palmer).

Virions contain a helical nucleocapsid surrounded by a lipoprotein unit membrane envelope. There are spikes on the virion surface approximately 7 nm long, spaced at approximately 10-nm intervals. The nucleocapsid structure consists of a dark, central space 20 nm wide surrounded by a helical capsid approximately 50 nm wide. The helical nucleocapsid has cross-striations with a periodicity of about 5 nm. Infected cells contain rod-shaped structures with a diameter of approximately 50 nm, which are thought to represent free viral nucleocapsids. Within the nucleocapsid is an axial channel of approximately 10 to 15 nm (Murphy et al., 1978; Peters et al., 1971).

The density of virions is approximately 1.14 g/ml, consistent with a structure containing RNA, protein, and lipid.

Virion RNA is single stranded with a molecular weight of 4.5×10^6 (Regnery et al., 1980). Virions contain seven proteins: one glycoprotein, three proteins associated with the viral ribonucleoprotein (RNP), and three proteins whose location within the virions have not yet been determined. The glycoprotein (125 Kds) forms the surface projections, NP (100 Kds) and VP30 (30 Kds) are nucleoproteins, and "VP35" (35 Kds) is loosely associated with the RNP, and is probably a replicative enzyme component. By analogy with other RNA viruses, the "L" protein (180 Kds) is probably the viral RNA polymerase, and "VP40" (40 Kds) and "VP24" (24 Kds) are probably membrane components.

The virion RNA of both Marburg and Ebola viruses is negative stranded and the viruses appear to replicate by means similar to other negative-stranded viruses (Sanchez and Kiley, 1987). Six monocistronic messenger RNA (mRNA) species have been identified and isolated from Ebola virus-infected cells. They program the synthesis of authentic viral proteins in an in vitro protein-synthesizing system. No mRNA for the "L" protein has yet been detected. RNA extracted from purified virions hybridizes with each of the mRNA species, thus confirming its negative-stranded nature.

Virus-infected cells contain prominent cytoplasmic inclusion bodies consisting of viral nucleocapsid masses. Budding of completed virions takes place at cell membranes where surface projections have been inserted.

Even though Marburg and Ebola virions are morphologically similar, possess similar-sized genomes, and have similar PAGE protein profiles, there are important differences between the two (Kiley et al., 1988). There are also differences between the Sudan and Zaire Ebola virus strains, which must be considered in diagnosis. Antibodies against antigens of one strain may react with antigens of the other strain.

The filoviruses are inactivated by a variety of detergents, including commercial phenolic detergents. The viruses are also inactivated by UV- and gamma-irradiation, 1% formalin, and beta-propiolactone (Bowen et al., 1969; Elliott et al., 1982).

Pathogenesis, Pathophysiology, and Pathology

Marburg and Ebola viruses cause severe hemorrhagic fever in humans and laboratory primates. In humans, the Zaire strain of Ebola virus carries the highest mortality when compared with the Sudan strain or Marburg virus. The virulence of Marburg virus is difficult to compare with the Ebola strains because Marburg virus-infected patients have been treated in modern medical facilities and Ebola patients, except for one, have been treated under difficult field conditions. In experimental infection of monkeys, both the Zaire strain of Ebola virus and Marburg virus kill all infected animals, whereas some animals survive Sudan virus infection.

In humans, the major pathologic changes seen in Marburg and Ebola virus infections are similar (Dietrich et al., 1978; Murphy, 1978). Virus is present in most organs of the host with the most striking lesions found in liver, spleen, and kidney. These lesions are characterized by moderate to severe focal hepatic necrosis, with remarkably little inflammatory response and virtually complete sparing of biliary architecture and function. Severe follicular necrosis of lymph nodes and spleen is found, sometimes with hyperplasia of the reticuloendothelial elements of these organs. In late stages of the disease, hemorrhage occurs in the gastrointestinal tract, pleural, pericardial, and peritoneal spaces and into the renal tubules with deposition of fibrin. These late manifestations have been attributed to disseminated intravascular coagulation (DIC). Interstial pneumonia, pancreatitis, orchitis, and iridocyclitis also have been described in human infections, and Marburg virus has been isolated from the anterior chamber of the eye late in the disease and from semen several months after recovery from the acute disease in one patient.

Recent experiments with Ebola-infected rhesus monkeys indicate that although DIC may be present during the later stages of infection, it in itself is not sufficient to account for the pathologic effects seen. No single organ shows sufficient histologic damage to account for the generalized vascular collapse seen in both diseases. It has been shown that dysfunction of endothelial cells and platelets may be associated with microcirculation failure and the ensuing coagulopathy. This leads to the sudden development of hypovolemic shock, multiple effusions, and multiple bleeding (Fisher-Hoch et al., 1985).

Clinical Features

No information is available about the origin of either Marburg or Ebola viruses nor about the exposure leading to the primary infection in any natural setting. All secondary cases have been either nosocomial or caused by intimate contact with a patient. One Marburg case was acquired by sexual contact more than 60 days after the original infection. The incubation period is between 4 and 16 days, during which the virus probably replicates in lymph nodes, spleen, and fixed-tissue macrophages of various organs. The period of acute febrile illness is marked by high viremia and the presence of virus in many visceral organs and tissues.

In the early stages of the disease, some of the symptoms are nonspecific; clinical diagnosis is difficult until the later stages. Onset is rapid with severe frontal headache, fever, and general weakness. Myalgia, especially centered in the lower back, is often present, and lethargy and an "expressionless face" are almost always present. On about the third day of the disease, patients develop a severe watery diarrhea, abdominal pain and cramping, and nausea and vomiting. Many patients develop a maculopapular rash that begins on the face and trunk and spreads centrifugally to the arms and legs between day 5 and 7. This rash is difficult to see on dark skin. Three or 4 days later desquamation may occur. There also may be pharyngitis, cough, conjunctivitis, and CNS abnormalities including lethargy, confusion, irritability, stupor, aggression, and signs of meningeal irritation. The severity of these disturbances often reflects the severity of the disease.

A high proportion of patients develop severe hemorrhagic manifestations, usually between day 5 and 7. Bleeding is often from multiple sites, with the gastrointestinal tract, lungs, and gingiva the most commonly involved. Bleeding and oropharyngeal lesions usually heralds a fatal outcome. Death occurs between day 7 and 16, with most on day 8 and 9; usually from shock with or without severe blood loss.

Virus may be present in the throat, blood, and urine for usually only the first week of the disease, but precautions should be observed with all patient samples. Virus may be present in seminal fluid for 2 months after infection (Fig. 3). High viremia is usually associated with a fatal outcome; virus may be recovered from all organs, where it can also be easily seen by electron microscopy. In survivors, there are usually few sequelae, although some CNS and personality disturbances have been reported (W.H.O., 1978a,b).

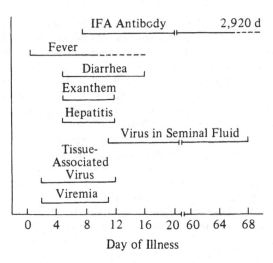

FIG. 3. Selected parameters in an idealized infection with Marburg or Ebola virus. The best times for collecting diagnostic specimens are included.

Epidemiology and Natural History

The source of infection in the first Marburg outbreak in 1967 were African green monkeys, *Cercopithicus aethiops,* imported from Uganda to Germany and Yugoslavia, but these animals were not the reservoir of the virus. Indeed, many species of monkeys become severely ill and die when experimentally infected and it is unlikely they could sustain transmission in nature. The index cases in the other outbreaks of Marburg hemorrhagic fever are known, but the source of infection is not. Neither the source of infection nor the index case are known for the three major Ebola outbreaks. In each outbreak, the initial patient spread the disease to close family members through intimate contact with the patient. After hospitalization of an infected person, the disease spread rapidly via contaminated needles and contact with blood. The disease subsided rapidly when barrier nursing procedures were instituted.

Data from several serosurveys, using indirect immunofluorescence (IFA), have suggested that Ebola antibody is present in a number of human and animal populations (Johnson et al., 1986; Saluzzo et al., 1981; van der Waals et al., 1986). Until the antibody detected in these surveys is proven to be virus specific, for example, by Western blot or radioimmune precipitation (RIP) techniques (Elliott et al., 1985), results obtained in these surveys must be considered tentative at best. To date the only antibody confirmed by Western blot or RIP have been found in serum of known patients from one of the disease outbreaks.

Ongoing studies of the genome structure of these viruses suggest that the viruses have evolved from a progenitor that also gave rise to families *Paramyxoviridae* and *Rhabdoviridae.* Members of all three families share many properties, including common sequences in the virion RNA that play a role in viral replication (Sanchez et al., submitted, J. Virol.).

Control and Prevention

There is currently no way to prevent infection with either Marburg or Ebola virus. Unlike several of the other viral hemorrhagic fevers, there is no proven method of postexposure prophylaxis either by immune plasma or antiviral drug. No antiviral drug has yet been shown to be effective, even in vitro, against Marburg or Ebola viruses. Experimental attempts to demonstrate passive protection by immune plasma in animals have failed. The viruses are not neutralizable in vitro. In experiments in which laboratory animals have been given virus antigens or inactivated whole virus vaccines, there has been no protection against subsequent challenge with live virus. Such protection experiments have been successful with Lassa virus, which also produces hemorrhagic fever, only if a replicating attenuated virus (Mopeia virus) or a vaccinia vector expressing Lassa antigen were used.

Many of the cases in each outbreak have been due to nosocomial infection, usually through contact with contaminated needles. Aerosol spread does not seem to play a role. In the known outbreaks, virus spread was halted rapidly when simple barrier nursing techniques were begun. Precautions consisted of gowns, gloves, masks, and isolation of patients.

Diagnosis

Because Ebola and Marburg viruses are highly virulent and because there are no vaccines or effective treatments, special precautions need to be taken when collecting clinical specimens. The most suitable specimens for virus isolation are venous blood and at autopsy liver, spleen, and heart tissues. When collecting samples, personnel should be acutely aware of the danger of accidental inoculation and of sprays, spills, and aerosols. Personnel should not attempt to replace the plastic needle guard on a used needle, but should discard the needle and syringe into a rigid plastic container containing 0.5% sodium hypochlorite. The container should then be autoclaved and discarded or incinerated. To avoid unnecessary exposure of laboratory personnel, blood specimens should not be centrifuged or separated. Gloves, mask, and gown should be worn while collecting clinical specimens.

Attempts to isolate virus must be conducted only in maximum containment laboratories (Biosafety

level 4 laboratories; Richardson and Barkley, 1984). All such samples should be considered infectious. Specimens should be collected in strong, tightly sealable screw-capped containers, preferably made of polypropylene. This container should be dipped in or swabbed with a disinfectant (0.5% sodium hypochlorite). Specimens should then be double bagged and surface disinfected before they are taken from the patient's room. In general, specimens for shipment should be handled as follows (Centers for Disease Control, 1983):

1. Place the specimen in a watertight, primary container (screw-capped plastic test tube or vial) and seal the cap with tape and label with patients name and date.
2. Wrap the primary container in sufficient absorbent material (paper towels or tissue) to absorb the entire contents in case of leakage.
3. Place the primary container in a durable, watertight secondary container such as a screw-capped metal mailing tube or sealed metal can. Multiple primary containers may be placed in the secondary containers if the total volume does not exceed 50 ml. Any excess space in the secondary container should be filled with absorbent material.
4. On the outside of the secondary container, place the specimen data forms, letters, and other information identifying the specimen.
5. Place the secondary mailing container and specimen information in an outer mailing tube or box. This should be mailed with appropriate labels and permits, as necessary.

Laboratories with Biosafety level 4 facilities are listed at the end of this chapter. Arrangements must be made with a representative of the maximum containment laboratory before specimens are shipped.

Viremia in Marburg and Ebola virus infections is usually high enough that virus can be visualized by electron microscopy of acute-phase serum samples (Simpson, 1977). Postmortem liver specimens, obtained with a biopsy needle, should be divided in half; one half should be placed in buffered formalin for histopathologic and electron microscopic studies and the other frozen at $-70°C$ for virus isolation and immunofluorescent studies. Virus nucleocapsids and virions are often observed by electron microscopy of liver sections (Bowen et al., 1977; Johnson et al., 1977; Pattyn et al., 1977).

Viral antigens in human tissues can be detected by the indirect fluorescent antibody (IFA) test using either polyclonal human convalescent-phase serum, animal serum, or monoclonal antibodies. Care must be taken in choosing the test antibody and antigen because some Ebola-Sudan antibody does not cross-react with Ebola-Zaire antigen and vice versa.

Positive diagnosis of infection may be accomplished in as little as 3 days by inoculating Vero cell cultures with clinical material. Cell cultures are screened daily by IFA. Some virus strains, especially Ebola isolates from Sudan, are difficult to isolate in cell culture (Baron et al., 1983); in these cases intraperitoneal inoculation of guinea pigs may be necessary. Guinea pigs are monitored for fever and illness; blind passage may be necessary. Blind passage in Vero cells also may be effective. Once isolated, these viruses usually grow well in Vero cells; cell culture materials may be used for characterizing the virus further and for preparing viral antigen.

Serum antibody levels are routinely determined by either IFA (Johnson et al., 1981) or ELISA. Radioimmunoassays for Marburg and Ebola antibodies also have been developed (Richman et al., 1983). The antigen used for most assays is virus-infected cells inactivated by gamma-irradiation (Elliott et al., 1982). Antigen is prepared by infecting Vero E6 cells with virus and harvesting cells when approximately 90% of the cells are IFA-positive. These infected cells are mixed with uninfected cells to give a final ratio of about 1 : 5. The cells are spotted onto glass slides, air dried, acetone fixed, and gamma-irradiated (Johnson et al., 1981). Serum from patients with suspected cases to be used for antibody studies should be inactivated by either heat (60°C, 60 min) or gamma-irradiation.

A fourfold rise in IFA antibody titer between acute- and convalescent-phase serum samples indicates a recent infection with these viruses. A reciprocal IFA titer of > 256 and the presence of virus specific IgM antibodies are considered positive indication of infection in one serum specimen. A high cutoff titer is required especially for Ebola virus, because the viral antigen produces nonspecific reactions. For example, one U.S. laboratory worker, with no exposure to Ebola virus, has had an IFA titer of 64 for more than 15 years.

Specific reactivity of serum that is positive by IFA or ELISA can be confirmed by a Western blot reaction with purified virus or radioimmune precipitation with purified virus or infected cell lysates (Elliott et al., 1985). If a suspect case is not part of a large outbreak, care must be taken to ensure that the antibody reported is specific for Marburg or Ebola virus. This is accomplished by Western blot analysis or radioimmune precipitation. These tests are performed only at specified laboratories.

Isolation of virus from acute-phase serum samples is the most reliable method of diagnosis. New isolates are characterized by reaction with positive control polyclonal serum and by analyzing virion RNA and protein. Monoclonal antibodies also can be used to characterize new isolates.

Literature Cited

Baron, R. C., J. B. McCormick, and O. A. Zubeir. 1983. Ebola virus disease in southern Sudan: hospital dissemination and intrafamilial spread. Bull. W.H.O. **61**:997–1003.

Bowen, E. T. W., D. I. H. Simpson, W. F. Bright, I. Zlotnik, and D. M. R. Howard. 1969. Vervet monkey disease: studies on some physical and chemical properties of the causative agent. Br. J. Exp. Pathol. **50**:400–407.

Bowen, E. T. W., G. Lloyd, W. J. Harris, G. S. Platt, A. Baskerville, and E. E. Vella. 1977. Viral hemorrhagic fever in southern Sudan and northern Zaire, preliminary studies on the aetiological agent. Lancet **1**:571–573.

Buchmeier, M. J., R. U. DeFries, J. B. McCormick, and M. P. Kiley. 1983. Comparative analysis of the structural polypeptides of Ebola viruses from Sudan and Zaire. J. Infect. Dis. **147**:276–281.

Centers for Disease Control. 1980. Marburg virus disease–Kenya. Morbid. Mortal. Weekly Rep. **29**:145–146.

Centers for Disease Control. 1983. Viral hemorrhagic fever: initial management of suspected and confirmed cases. Morbid. Mortal. Weekly Rep. **32(suppl)**:275–395.

Cox, N. J., J. B. McCormick, K. M. Johnson, and M. P. Kiley. 1983. Evidence for two subtypes of Ebola virus based on oligonucleotide mapping of RNA. J. Infect. Dis. **147**:272–275.

Dietrich, M., H. H. Schumacher, D. Peters, and J. Knobloch. 1978. Human pathology of Ebola (Maridi) virus infection in the Sudan. p. 37–42. In Pattyn S. R. (ed.), Ebola virus haemorrhagic fever. Elsevier/North Holland, Amsterdam, New York.

Elliott, L. H., J. B.McCormick, and K. M. Johnson. 1982. Inactivation of Lassa, Marburg, and Ebola viruses by gamma irradiation. J. Clin. Microbiol. **16**:704–708.

Elliott, L. H., M. P. Kiley, and J. B. McCormick. 1985. Descriptive analysis of Ebola virus proteins. Virology **147**:169–176.

Fisher-Hoch, S. P., G. S. Platt, G. H. Neild, T. Southee, A. Baskerville, R. T. Raymond, G. Lloyd, and D. I. H. Simpson. 1985. Pathophysiology of shock and hemorrhage in a fulminating viral infection (Ebola). J. Infect. Dis. **152**:887–894.

Gear, J. S. S., G. A. Cassell, A. J. Gear, B. Trappler, L. Clausen, A. M. Meyers, M. C. Kew, T. H. Bothwell, R. M. Sher, G. B. Miller, J. Schneider, H. J. Koornhof, E. D. Gomperts, M. Isaacson, and J. H. S. Gear. 1975. Outbreak of Marburg virus disease in Johannesburg. Br. Med. J. **4**:489–493.

Johnson, B. K., D. Ocheng, S. Oogo, L. G. Gitau, C. Wambui, A. Gichogo, D. Libondo, P. M. Tukei, and E. J. Johnson. 1986. Seasonal variation in antibodies against Ebola virus in Kenyan fever patients. Lancet **1**:1160.

Johnson, K. M., P. A. Webb, J. V. Lange, and F. A. Murphy. 1977. Isolation and partial characterization of a new virus causing acute haemorrhagic fever in Zaire. Lancet **1**:569–571.

Johnson, K. M., L. H. Elliott, and D. L. Heymann. 1981. Preparation of polyvalent viral immunofluorescent intracellular antigens and use in human serosurveys. J. Clin. Microbiol. **14**:527–529.

Kiley, M. P., E. T. W. Bowen, G. A. Eddy, M. Isaacson, K. M. Johnson, J. B. McCormick, F. A. Murphy, S. R. Pattyn, D. Peters, O. W. Prozesky, R. L. Regnery, D. I. H. Simpson, W. Slenczka, P. Sureau, G. van der Groen, P. A. Webb, and H. Wulff. 1982. Filoviridae: a taxonomic home for Marburg and Ebola viruses? Intervirology **1824**:32.

Kiley, M. P., N. J. Cox, L. H. Elliott, A. Sanchez, R. Defries, M. J. Buchmeier, D. D. Richman, and J. B. McCormick. 1988. Physicochemical properties of Marburg virus: evidence for three distinct virus strains and their relationship to Ebola virus. J. Gen. Virol., in press.

Murphy, F. A. 1978. Pathology of Ebola virus infection. p. 43–60. *In:* Pattyn S. R. (ed.), Ebola virus haemorrhagic fever, Elsevier/North Holland Biomedical Press, Amsterdam.

Murphy, F. A., G. van der Groen, S. G. Whitfield, and J. V. Lange. 1978. Ebola and Marburg virus morphology and taxonomy, p. 61–84. *In* Pattyn S. R. (ed.), Ebola virus haemorrhagic fever, Elsevier/North Holland Biomedical Press, Amsterdam.

Pattyn, S., G. van der Groen, W. Jacob, P. Piot, and G. Courteille. 1977. Isolation of Marburg-like virus from a case of haemorrhagic fever in Zaire. Lancet 1:573–574.

Peters, D., G. Muller, and W. Slenczka. 1971. Morphology, development and classification of the Marburg virus, p. 68–83. *In* G. A. Martini and R. Siegert (eds.), Marburg virus disease. Springer-Verlag, New York.

Regnery, R. L., K. M. Johnson, and M. P. Kiley. 1980. Virion nucleic acid of Ebola virus. J. Virol. 36:465–469.

Regnery, R. L., K. M. Johnson, and M. P. Kiley. 1981. Marburg and Ebola viruses: possible members of a new group of negative strand viruses, p. 971–977. *In* D. H. L. Bishop and R. W. Compans (eds.), The replication of negative strand viruses. Elsevier/North Holland, Inc., New York.

Richardson, J. H., and W. E. Barkley (eds.). 1984. Biosafety in microbiological and biomedical laboratories: US Department of Health and Human Services (HHS publication no. (CDC)86–8395).

Richman, D. D., P. H. Cleveland, J. B. McCormick, and K. M. Johnson. 1983. Antigenic analysis of strains of Ebola virus: identification of two Ebola virus serotypes. J. Infect. Dis. 147:268–271.

Saluzzo, J. F., J. P. Gonzalez, A. J. Georges, and K. M. Johnson. 1981. Antibodies against the Marburg virus among human populations in the southeastern Central African Republic. C. R. Acad. Sci. (Paris) 292:29–31.

Sanchez, A., and M. P. Kiley. 1987. Identification and analysis of Ebola virus messenger RNA. Virology 157:414–420.

Siegert, R., H. L. Shu, W. Slenczka, D. Peters, and G. Muller. 1967. Zur atiologie einer unbekannten, von affen ausgegangenen menschlichen infektionskrankheit. Dt. Med. Wschr. 92:2341–2343.

Simpson, D. I. H. 1977. Marburg and Ebola virus infections: a guide for their diagnosis, management, and control. WHO offset publication no. 36, World Health Organization, Geneva.

Smith, C. E. G., D. I. H. Simpson, E. T. W. Bowen, and I. Zlotnik. 1967. Fatal human disease from vervet monkeys. Lancet 2:1119–1121.

Stojkovic, Lj., M. Bordjoski, A. Gligic, and Z. Stefanovic. 1971. Two cases of Cercopithicus monkeys-associated haemorrhagic fever, p. 24–33. *In* G. A. Martini and R. Siegert (eds.), Marburg virus disease. Springer-Verlag, New York.

Teepe, R. G., B. K. Johnson, D. Ocheng, A. Gichogo, A. Langatt, A. Ngindu, M. Kiley, K. M. Johnson, and J. B. McCormick. 1983. A probable case of Ebola virus haemorrhagic fever in Kenya. East Afr. Med. J. 60:718–722.

van der Waals, F. J., K. L. Pomeroy, J. Goudsmit, D. M. Asher, and D. C. Gajdusek. 1986. Hemorrhagic fever virus infections in an isolated rainforest area of central Liberia. Limitations of the indirect immunofluorescence slide test for antibody screening in Africa. Trop. Geogr. Med. 38:209–214.

Webb, P. A., K. M. Johnson, H. Wulff, and J. V. Lange. 1978. Some observations on the properties of Ebola virus, p. 91–94. *In* S. R. Pattyn (ed.), Ebola virus haemorrhagic fever. Elsevier/North Holland Biomedical Press, Amsterdam.

W.H.O. International Study Team Report. 1976. Ebola haemorrhagic fever in Sudan, 1976. Bull. W.H.O. 56:247–270.

W.H.O. International Study Team Report. 1976. Ebola haemorrhagic fever in Zaire, 1976. Bull. W.H.O. 56:271–293.

Wulff, H., W. Slenczka, and J. H. S. Gear. 1978. Early detection of antigen and estimation of virus yield in specimens from patients with Marburg virus disease. Bull. W.H.O. 56:633–639.

Appendix

Locations of Maximum Containment Laboratories

Special Pathogens Unit
Public Health Laboratory Service
61 Colindale Avenue
London NW9 5DF
England

Special Pathogens Branch
Division of Viral Diseases
Center for Infectious Disease
Centers for Disease Control
Atlanta, Georgia 30333

CHAPTER 31

Orthomyxoviridae: The Influenza Viruses

ALAN KENDAL and MAURICE W. HARMON

Disease: Influenza, "flu," "grippe."

Etiologic Agents: Influenza type A (subtypes HlNl, H2N2, H3N2), type B, and type C.

Source: Human to human transmission. Occasionally transmitted from swine. Animal reservoirs may harbor viruses which, through genetic reassortant, become infectious for humans.

Clinical Manifestations: Rapid onset of malaise, fever, and myalgia, usually with nonproductive cough or sore throat. May often be subclinical.

Pathology: Cell necrosis and sloughing of ciliated columnar epithelium of the upper and lower respiratory tract. Complications include primary viral pneumonia and secondary bacterial pneumonia. Occasionally other complications such as myositis, myocarditis, and encephalitis occur.

Laboratory Diagnosis: Virus isolation from upper or lower respiratory tract. Antibody titer rise when acute-phase and convalescent-phase serum are tested by hemagglutination-inhibition, neutralization, enzyme immunoassay, or complement fixation.

Epidemiology: Worldwide. Seasonal in temperate climates; November to April in Northern Hemisphere and May to October in Southern Hemisphere. May be endemic in tropical areas.

Treatment: Oral amantadine or rimantadine if administered within 48 h of onset for influenza A only. Experimentally, aerosolized ribavirin for types A and B. Aspirin (salicylates) should be avoided in children younger than 19 years to reduce risk of Reye's syndrome.

Prevention and Control: Inactivated vaccine (many countries) or live attenuated vaccine (USSR), amantadine and rimantadine. In institutions, vaccine efficacy is influenced by vaccination rate.

Introduction and Taxonomic Considerations

Epidemics of respiratory disease believed caused by influenza viruses have been recognized for centuries, being characterized by the rapid involvement of persons of all ages throughout entire communities (Beveridge, 1977). Scientific analysis of such epidemics may have begun with William Farr's reports of increases in mortality during epidemics in England in the 19th Century. Observations at that time highlighted reasons why influenza is of such concern, namely the sharp rise in deaths due to the direct effects of influenza illness and pneumonia, as well as an associated increase in deaths officially attributed

to other underlying conditions such as heart disease. Excess mortality remains the hallmark of most contemporary epidemics (Lui and Kendal, 1987).

Although it is now known that an influenza virus was isolated first from an avian source in about 1902, even during the great human influenza pandemic of 1918 the causative agent of human illness was not available for study. The isolation of an influenza virus from humans did not occur until 1933, and for the next few years it proved possible, by passage of the virus in laboratory animals and neutralization tests in animals, to move into the era when variation among viruses could be recognized. Before long it was found that influenza virus could be propagated in hens' eggs, and could be examined by hemagglutination tests, greatly simplifying laboratory work. In this same period, the identification of influenza virus as a natural pathogen in swine, and the involvement of an antigenically related virus as a cause of the 1918 pandemic, established a clear link between enzootic influenza and human influenza (see Hoyle, 1968 for descriptions of this early period of study).

As human epidemic viruses continued to be isolated and compared with each other, the first subdivisions of the viruses into different serogroups began. Thus, in 1940 the existence of two quite unrelated groups of viruses led to the names influenza A and B viruses. In 1947 it was believed that a new form of influenza A virus was occurring because vaccines prepared with an earlier isolate, and used in the armed forces, proved to be ineffective against the 1947 epidemic strain. This was designated "A1," to distinguish it from the earlier "A0" viruses (Francis et al., 1947). Also witnessed during 1947 was the isolation of an unusual form of virus that shared many, but not all, biologic properties of influenza viruses; this isolate was the prototype for the "influenza C" viruses (Hirst, 1950). Asian influenza in 1957 (United States Public Health Service, 1960) and Hong Kong influenza in 1968 (World Health Organization, 1969) were each found to represent new antigenic forms of influenza A virus from humans when tested in hemagglutination assays.

Although most attention was focused on the hemagglutinating component of the virus, Gottschalk (1957) pioneered work that identified the receptor-destroying property of influenza viruses to be an enzyme neuraminidase (sialidase). This finding altered the perception of viruses as being totally inert outside of their host. Antigenic analysis of this enzymatic component (Rafelson et al., 1963), followed by the first understanding of the genetic independence of the influenza A components responsible for hemagglutination and neuraminidase activities (Laver and Kilbourne, 1966), necessitated a further revision of virus names to define both these antigens. Continuing studies with animal influenza viruses showed that surface antigens similar to those found in human influenza isolates were present in viruses from diverse animal species (Pereira et al., 1965; Hinshaw and Webster, 1982). Other studies using genetic comparisons questioned the validity of the distinction between pre- and post-1947 isolates and the distinction between certain human and animal virus hemagglutinin genes (reviewed by Scholtissek, 1983). A further revision of influenza nomenclature followed, which is currently in use (World Health Organization, 1980). Thus, the present taxonomic description of influenza viruses has arisen in several discrete phases, each clearly related to the introduction of a new technology, and its application by scientists who are fascinated by the interrelationships of influenza viruses from different epidemic or enzootic episodes.

Influenza viruses are now recognized to comprise a major genus, which includes influenza A and B viruses, and a probable second genus for influenza C viruses (International Committee on Taxonomy of Viruses, 1982). Influenza A viruses are subdivided both according to the subtype of their hemagglutinin, and also of their neuraminidase antigens. Thirteen hemagglutinin subtypes and 9 neuraminidase subtypes have been recognized for influenza A isolates. These are designated sequentially (Hl-H13; N1-N9), regardless of the host from which the virus was isolated. Further variation of a more gradual kind exists within each subtype. From time to time particular isolates are designated by the World Health Organization as antigenic reference strains for these more subtle variants which are of epidemiologic significance. Nomenclature takes into account, therefore, the virus type, location of isolation, serial number from that location, year of isolation, and antigenic properties of hemagglutinin and neuraminidase antigens, as for example with the recent epidemic strain A/Philippines/2/82(H3N2). For influenza A isolates from animals, the species from which the virus was isolated is also given as in A/Seal/Massachusetts/1/70 (H7N7). In the case of influenza B or C viruses antigenic subtypes are not relevant, and designations such as B/USSR/100/83 or C/Ann Arbor/1/50 are used.

The above comments are intended to establish for the reader both an understanding of the rationale behind the terminology used to describe influenza viruses, as well as some feeling for the significance attached to understanding the variation and interrelationships of influenza isolates. Clearly the role of the diagnostic laboratory is central to the development of data about the occurrence of different strains. A further role of the laboratory, which is rapidly developing increasing importance, is to provide diagnostic results that are of immediate practical value in preventing or treating disease. Methods relevant to both these long- and short-term objectives are described below, following brief descriptions of the virus and the disease it causes. For more extensive background reading, useful texts are Kendal and Patriarca (1986), Kilbourne (1987), Palese and Kingsbury (1983), and Stuart-Harris et al., (1985).

Influenza A and B Viruses

Structural Components and Nonstructural Proteins

Virions of influenza are either filamentous or approximately spherical, the latter being about 100 nm in diameter (Fig. 1). The envelope of the virions contains a lipid bilayer derived from the host cell, modi-

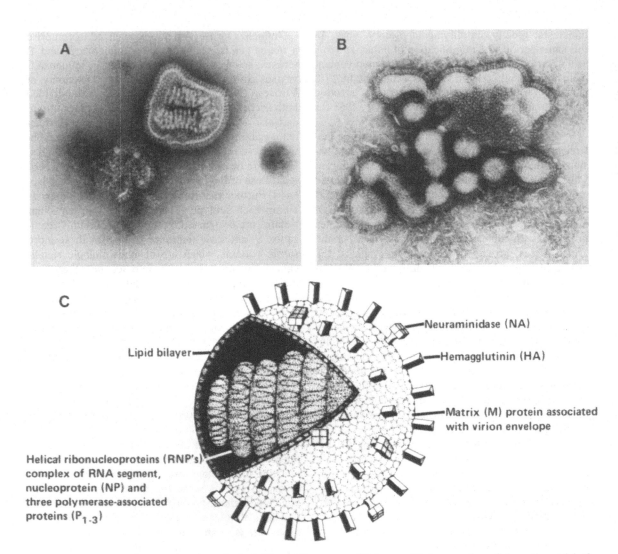

FIG. 1. Electron micrographs showing spherical (A) and filamentous (B) forms of influenza virus. Schematic model of influenza type A or B virion (C).

fied by the presence of two glycoproteins on the outside, hemagglutinin (HA) and neuraminidase (NA), and a matrix (M) protein on the inside. The envelope encloses the nucleocapsid, which for an infectious virion exists as eight segments of single-stranded RNA associated with nucleoprotein (NP) and three polymerase proteins designated as "acidic" or "basic" according to their electrophoretic mobility (PA, PB1, and PB2). Influenza A and B viruses synthesize in infected cells two other proteins designated as nonstructural proteins 1 and 2 (NS1 and NS2). In addition, both influenza A and B viruses recently have been found to synthesize an extra protein. These proteins are coded by different genes in influenza A and B viruses, representing one of the few biochemical distinctions between the two serotypes. However, both proteins have similar molecular structures, and appear to be membrane-spanning proteins. They are designated M2 and NB proteins, respectively, for the A and B viruses. The presence of these proteins in virions is not established.

The hemagglutinin and neuraminidase have been extensively studied to attempt to understand the evolution of new variants and the molecular basis for their biologic activities (see Webster, et al., 1983 for review and specific references). Both are glycoproteins, and the presence of carbohydrate side chains appears essential for correct processing of the molecules. Hemagglutinin molecules are trimers of three identical polypeptides which contain intra- but not inter-chain disulfide bonds. Each HA polypeptide chain is synthesized as a precursor molecule, with a hydrophobic N-terminal "leader" sequence attaching the polypeptide to intracellular membranes as it is processed to accept carbohydrate side chains. They are then modified by trimming and extension reactions. The leader sequence is ultimately cleaved, and the final HA molecules are left extending from the outside of the infected cells' plasma membranes, or virion envelopes after virus maturation. They are attached by a membrane spanning hydrophobic region at their C-terminal end. The HA molecule's shape is known from x-ray crystallography to consist of a stalk extending from the membrane, with a globular head at the distal region. The receptor binding site is located in the globular head, with one site in each of the three polypeptides of the trimer (Fig. 2).

Neuraminidase molecules are composed of four identical polypeptide subunits, with many intrachain disulfide bonds and interchain bonds also present. Unlike the HA, the NA is attached to the plasma membrane of infected cells, or virion envelopes, by a hydrophobic membrane-spanning region at the molecule's N-terminal end. The molecules have a fibrous stalk region extending from the membrane, and four enzyme-active sites are present in a tetrameric symmetric head at the molecule's distal end.

Antigenic Properties

Both the nucleoprotein and the matrix protein are type specific. Although certain highly specific monoclonal antibodies can differentiate these antigens between some strains, diagnostic antisera produced in animals will recognize all influenza A or B viruses according to reaction of their NP and M proteins. Monoclonal antibodies reactive with all type A or all type B viruses have been prepared as new type-specific diagnostic reagents. Although the same may be the case for polymerase proteins and NSl and NS2 proteins (as well as M2 and NB proteins), experimental results are lacking at present.

The division of influenza A HA and NA proteins into antigenic subtypes is based primarily on the use of gel immunoprecipitation tests with hyperimmune sera specific for purified isolated proteins. All members of a given subtype are expected to be precipitated in an Ouchterlony test by antiserum to one member of the subtype, whereas cross-reaction between subtypes is not expected to occur. Supporting data are obtained from studies of gene homology, and from studies about the ability of one member of a subtype to prime a host so that antibodies to that strain will be boosted when the host is immunized with another member of the subtype ("the Doctrine of Original Antigenic Sin") (Davenport et al., 1953).

Within each subtype antigenic variation is readily recognized when viruses from different epidemic periods are compared (Table 1). This phenomenon is called antigenic drift, and comparison of the amino

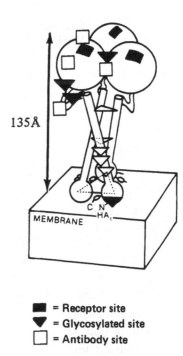

= Receptor site
= Glycosylated site
= Antibody site

FIG. 2. Schematic model of influenza virus hemagglutinin.

acid sequence changes between successive antigenic variants, when plotted on the three-dimensional models of the HA or NA molecules, have provided insight about the location of important antigenic domains on the proteins. (Also see section below on evolution.) For the HA of influenza A viruses, there are probably four or five major antigenic areas recognized, two close to the tip of the HA molecule, two somewhat lower down the side of the molecule, and one that involves areas in the interface between the three identical polypeptide chains that comprise each HA protein molecule (Fig. 2; Both et al., 1983). Variation in the sequences of amino acids near the molecule's tip (particularly the area near the receptor binding pocket) are most likely to cause significant antigenic variation between strains within a subtype (Nakajima et al., 1983). Variation in this region of the HA, or in the receptor binding site itself, also seem to occur when the same virus isolate undergoes passage in the laboratory that may alter the nature of its interaction with nonspecific inhibitors of hemagglutination, or with cell membrane receptors (Rogers et al., 1983; Robertson et al., 1985). The epidemiologic significance of this latter type of variation is not fully understood, whereas the evolution of epidemiologically successful new field strains clearly involves the occurrence and maintenance of several mutations, usually within more than one antigenic area of the molecule (Both et al., 1983; Raymond et al., 1986; Wiley et al., 1981). Less clear information is available about the variation of NA, but antigenic do-

TABLE 1. Results of hemagglutination-inhibition tests that demonstrate antigenic drift of influenza type A (H3N2) viruses from 1968 to 1986

Reference antigen	Postinfection ferret sera								
	A/Hong Kong/8/68	A/England/42/72	A/Port Chalmers/1/73	A/Victoria/3/75	A/Texas/1/77	A/Bangkok/1/79	A/Philippines/2/82	A/Mississippi/1/85	A/Leningrad/360/86
A/Hong Kong/8/68	320	320	<	<	<	<	<	<	<
A/England/42/72	80	320	80	40	<	<	<	40	<
A/Port Chalmers/1/73	80	160	320	80	40	<	<	<	<
A/Victoria/3/75	80	160	320	640	160	40	160	160	<
A/Texas/1/77	<	40	160	160	1280	320	320	320	160
A/Bangkok/1/79	320	80	320	320	1280	2560	1280	640	160
A/Philippines/2/82	<	<	<	<	80	40	320	160	80
A/Mississippi/1/85	<		80	40	160	160	320	640	320
A/Leningrad/360/86	<	<	<	<	80	20	320	320	640

< = less than 20.

mains can be mapped by the same techniques as applied to the HA, as the three-dimensional structure of NA is also known (Coleman et al., 1983).

Enzymic and Other Biologic Activities

The influenza HA has at least two biologic activities. One is to bind to sialic acid-containing receptors on the surface of target cells, initiating the events of the replicative cycle. At least two chemical specificities are known to exist for this binding, because some strains can preferentially bind to receptors containing sialic acid bound through a 2-6 glycosidic linkage (Rogers et al., 1983), whereas variants exist that preferentially recognize sialic acid bound through 2-3 glycosidic linkages. It is possible that this chemical specificity can play a role in determining the host range of the virus in some cases, as well as affecting the virus' sensitivity to inhibition with naturally occurring sialoglycoproteins ("nonspecific inhibitors"), which interfere with in vitro hemagglutination assays.

The second biologic activity of the HA is the induction of membrane fusion. For fusion to occur, the virion HA must be proteolytically cleaved, at a point near the base of the HA which is inserted into the virion membrane. Intramolecular disulfide bonds maintain attachment of the HA region distal to the membrane anchor, but a conformational change in the molecule occurs upon cleavage which is essential for the fusion activity (Skehel et al., 1982). Virions grown in host cells that do not contain a protease of appropriate specificity to cleave the HA of the infecting virus are not infectious unless an effective protease is resent in the extracellular environment (Klenk et al., 1975; Lazarowitz and Choppin, 1975). Avian influenza viruses of the H7 subtype are known to exist that have altered amino acid sequences around the HA cleavage site that renders them more susceptible to cleavage by common proteases, for example by addition of an extra lysine or arginine that creates a site sensitive to host proteases. These strains of virus can infect multiple tissues in their natural host and are usually highly pathogenic in chickens. Conversely, variants of virulent strains can arise which by indirect interactions of the cleavage site can reduce its susceptibility to cleavage, and hence reduce virulence (Rott and Klenk, 1986).

A practical consequence of this is the recommended use of trypsin in cell maintenance medium when isolating or propagating influenza A viruses, to maximize the yield of infectious virus (Tobita et al., 1975).

During the replication of influenza virus, it is believed that membrane fusion occurs in the presence of cleaved HA only at slightly acid pH (e.g., about 5.5), as found in lysosomes (Skehel et al., 1982). Thus, it is probable that after virions attach to cell membranes they are taken up into the cells in lysosomal-type vesicles ("coated pits"), which through ion-pump mechanisms reduce the environmental pH until a point is reached at which cleaved HA undergoes the necessary nonformational change to induce fusion of virion and vesicle membranes. This releases the virion core from the vesicle, that is into the cell's cytoplasm.

Influenza A and B, but not influenza C viruses (Kendal, 1975), possess an enzyme neuraminidase, which cleaves certain sialic acid molecules from the terminal alpha-o-glycosidic linkage on glycoproteins and gangliosides (Drzeniek, 1972; Gottschalk, 1957). In so doing, the receptor for influenza HA is destroyed. The biologic outcome(s) of this enzymatic modification may include inactivation of nonspecific inhibitors, prevention of superinfection of a cell already infected by one virion with a successive one, or facilitating release of newly formed virions from infected cells. The relative importance of these possible roles is not clear.

The PA and PB proteins are responsible for several enzymic functions necessary for RNA transcription and replication. Protein PB2 binds to 5'-methylguanosine-(capped) RNA fragments about 8 to 15 nucleotides long, generated from the 5' ends of heterogeneous nuclear RNA in the host cell. A second viral P protein (PB1) is believed to then initiate virion RNA transcription using the 3' ends of the host cell RNA fragments as a primer, and to elongate the RNA chain into almost full length copies of the viral RNA segments. Replication requires a slightly different strategy for RNA polymerase activity (see below). It should be emphasized that although biochemical and genetic studies have implicated proteins as essential for specific biologic activities, each activity is likely to require the integrated functioning of a complex of all three P proteins and the NP protein, as well as perhaps other components which may regulate the processes described.

Nucleic Acid and Replication Strategy

Influenza A and B viruses contain nucleic acid which is complementary to that of the virus messenger RNAs. Both the 5' and 3' ends of the virion RNA contain conserved sequences: each segment has a sequence 5'-AGUAGAAACAAGC for type A strains, and an identical or closely related sequence 5'-AGUAGUAACAA is found for the first 11 nucleotides in type B strains. The 3' ends are U-rich, for example 3'-UCGUUUUCGUCC in many type A virus genes. Sizes range from about 900 to 2400 nucleotides for the individual gene segments.

To direct synthesis of virus-specific proteins, most genes are transcribed to produce a complementary (messenger) sense RNA strand, with a short 5'-capped region derived from host cell RNA as described, and a sequence containing nearly all of the virus-specific sequences except for a short region at the 3' end which is not transcribed. The messenger RNAs are polyadenylated. A few mRNAs are generated by a procedure that involves splicing two segments of complementary RNA in such a way that the 5' and 3' sections correspond to sequences that would be in different reading frames had the virion RNA been transcribed continuously (Lamb, 1983). For RNA segment 7 and 8 of influenza A and B viruses, this permits each segment to code for two distinct proteins, sharing only a few amino acids at their N terminus. RNA 8 of influenza B virus is similar. For influenza B viruses, RNA segment 6 which codes for the NA gene also encodes a separate small protein NB read in a separate reading frame from that encoding the NA protein (Shaw et al., 1983).

Although synthesis of mRNA begins in the nucleus of the infected cells where primers are generated, as described, protein synthesis occurs in the cytoplasm. By as yet unknown mechanisms, control is exerted over RNA synthesis so that as infection proceeds, the type of complementary RNA synthesized changes from slightly short, polyadenylated mRNA to exact full length copies of the RNA found in virions. These full length complementary strands of RNA then act as templates for RNA replication, which results in synthesis of multiple copies of the RNA identical to those in infectious virions. Migration of NP antigen from the cytoplasm to the nucleus occurs early during replication, for unknown reasons.

Inhibitors of Influenza Replication

Some nonspecific glycoprotein inhibitors react with influenza HA which binds their sialic acid, mimicking the normal function of binding to cell receptors (Krizanova and Rathova, 1969). Neutralization probably occurs when the binding of the nonspecific inhibitor occurs with a very high affinity, and when the sialic acid on the inhibitor is relatively poorly released by the viral neuraminidase.

Amantadine hydrochloride, and its analogue rimantadine, are effective in vitro inhibitors of many influenza A virus strains, and are also effective clinically in prophylaxis or therapy of natural infection with recent influenza A epidemic strains (Dolin et al., 1982; Van Voris et al., 1981; Younkin et al., 1983). These drugs interfere with virus replication at a step after attachment of virus to host cells, but before RNA replication occurs. Genetic and biochemical analysis indicates that there may be at least two primary modes of action of the drugs. At high concentrations the drugs may nonspecifically affect the function of the HA molecule, whereas at lower concentrations they may affect function of the small M2 protein (Hay et al., 1985). In the former case the drugs' action may be to raise the pH of the vesicles containing virions that have been engulfed at the cells plasma membrane, to a level at which the pH-dependent conformational change in viral HA necessary for membrane fusion does not occur. Specific inhibition of influenza B viruses by the drugs is not known to occur, which may be related to the lack of an M2 protein in type B influenza viruses.

Drugs that interfere with proteolytic cleavage of influenza HA also have been considered as agents to prevent the HA developing fusion activity, and thereby reducing spread of infectious virus (Richardson et al., 1980; Zhirnov et al., 1985).

Actinomycin D, which inhibits the cellular RNA polymerase responsible for production of nuclear heterogeneous RNA, can inhibit replication of influenza viruses in vitro if present within the first few hours of infection (Barry et al., 1962), because the capped primers needed to initiate synthesis of viral mRNA cannot then be produced. Alpha-amanitin, a specific inhibitor of host cell RNA polymerase II, has a similar effect (Rott and Scholtissek, 1970). Though of use for research, the effects of these inhibitors are generally directed at DNA-dependent RNA synthesis, and have no clinical application.

Ribavirin, a nucleic acid analog, can inhibit replication of influenza A or B viruses in vitro, possibly through relatively nonspecific interference with the overall nucleic acid synthetic processes in cells (Stuart-Harris et al., 1985). Clinically, the drug has a modest therapeutic efficacy against natural infection with influenza A or B viruses, but must be administered through an aerosol for long periods to achieve results, without toxicity (Gilbert et al., 1986).

Interferon was first detected as a result of studies indicating that cells infected with one strain of influenza produced a substance that could render cells resistant to infection. Influenza viruses are less sensitive to interferon in vitro than many other RNA viruses, and studies in humans have been unable at present to identify a clinically acceptable regimen for practical use against influenza, although some efficacy has been seen on occasion.

Inhibitors of protein glycosylation can be successfully used in vitro to block the intracellular processing of the HA and NA glycoproteins. Tunicamycin, for example, acts to prevent the transfer of small carbohydrate chains onto protein. In the absence of carbohydrate, biologically active HA and NA are not formed and transported to the cells plasma membrane. The effect is not specific for influenza however, and clinical use has not been considered.

Evolution

Like other RNA viruses, it is expected on theoretical grounds that influenza viruses will undergo a higher rate of mutation than DNA viruses, due to the lack of RNA proofreading enzymes. However, the rapid appearance of new influenza variants should not be ascribed to an even more unusually high *mutation* rate as opposed to a high *evolutionary* rate. This presumably results from the conservation of mutations by factors which select for both viability and antigenic variation, and may be a function of the segmented single-stranded RNA genome, as well as other replication characteristics (Kendal et al., 1985). Consistent with this is the preliminary observation, based on limited comparisons of nucleic acid sequences, that the rate of evolution of genes coding for HA antigens averages about 0.7 to 0.8% base changes per year, whereas genes coding for proteins not believed to be exposed to antigenic pressure evolve at about 0.2 to 0.4% base changes per year.

Such extrapolations assume that evolution/mutation occurs at a fairly constant rate over time. However, comparisons of oligonucleotide maps of total viral RNA of type A(H1N1) viruses of humans suggest that, in fact, there may be a succession of years when "mutational drift" occurs at a constant level, followed by a year when a much larger number of mutations occur in association with appearance of a new epidemic variant (Kendal and Cox, 1985).

Additionally, genetic variation can occur in influenza viruses through reassortment of genes between two different viruses simultaneously infecting the same host. The genome of "Hong Kong" influenza that arose in 1968 is well documented to contain 7 RNA segments virtually indistinguishable from those of H2N2 "Asian" influenza viruses from 1967, and an H3 HA presumably derived by gene reassortant with another, unknown virus (Scholtissek, 1983). Because viruses with related H3 HA circulate in horses and ducks, it is generally held that an animal virus was the source of the gene coding for the Hong Kong virus HA.

In 1978 reassortment must have occurred between the newly reappeared type A(H1N1) virus and the still circulating type A(H3N2) strains. The outcome was a set of viruses, widely distributed in the world, which contained H1N1 antigens, and usually M and NS proteins of H1N1 origin, but other genes from type A(H3N2) strains. Eventually, "these reassortant H1N1" viruses disappeared, leaving "pure" H1N1 and H3N2 strains continuing to circulate until the present time (Nakajima et al., 1981).

Application of rapid nucleic acid sequencing techniques also has shown that the evolution of those influenza A and B virus HA genes which have been adequately studied can be traced in a continuous path over many years. Thus, new epidemic strains around the world are continuously being derived by evolution from common intermediate precursors, even though there is microheterogeneity in sequence between isolates circulating at the same time (Both et al., 1983; Raymond et al., 1986). Paradoxically, therefore, influenza viruses are both very variable, in terms of longitudinal evolution, but highly stable, in the conservation of the fundamental genetic composition of successful epidemic strains.

From sequencing limited portions of the HA of all known HA subtypes, a dendrogram of closest relationships was derived (Air, 1981). In this scheme, for example, H1 and H2 are more closely related to each other than either are to H3 HA. Nevertheless, the exact mechanism by which different HA subtypes evolved is unknown.

Influenza C Viruses

In many respects influenza C viruses resemble influenza A and B viruses, but significant differences do exist. Morphologic studies, for example, indicate that it is often (but not always) possible to detect a reticular pattern on parts of the envelope of negatively stained influenza C viruses, and that the diameter of the virion nucleocapsids is almost 9 nm, about one-third greater than for influenza A or B (Apostolov et al., 1970; Martin et al., 1977). Thus far, 7 segments of RNA have been detected in influenza C viruses, and only one of these is believed to code for a glycoprotein, with other segments codings for putative polymerase proteins, a matrix protein, and at least one nonstructural protein (Meier-Ewert et al., 1978; Meier-Ewert et al., 1981).

Because influenza C viruses agglutinate red blood cells, are capable of causing cell membrane fusion, and have receptor-destroying activity, it is assumed that all these biologic activities are caused by the virus' one glycoprotein species, in contrast to influenza A or B viruses, or paramyxoviruses. The chemical specificity of the receptor for influenza C virus recently has been determined as the 9-0-acetyl derivative of sialic acid, and the receptor-destroying enzyme is an esterase that liberates the 9-0-actyl group, without cleaving the glycosidic bond between sialic acid and the penultimate sugar (Herrler et al., 1985). This represents another significant biochemical difference between influenza C virus and influenza A or B viruses. There are, however, similarities in the terminal sequences of influenza C nucleic acid segments and those of influenza A or B viruses, and the strategy of replication may be generally similar. Antigenic drift occurs to a limited extent in the genome

of influenza C virus, but the overall evolution of the virus does not appear to be one of a single path, rather than multiple parallel or interconnected paths (Buonagurio et al., 1985). This may be related to the epidemiology of the virus, which is essentially one of a pediatric, mild respiratory pathogen only occasionally affecting adults (Homma, 1986). The virus is believed to infect pigs in China (Guo and Desselberger, 1984), may infect dogs (Homma, 1986), but other natural animal hosts have not been recognized.

Influenza in Animals

Studies of influenza in animals and humans have been mutually beneficial in extending knowledge of the virus (Hinshaw and Webster, 1982). The natural host range of influenza is known to include horses, swine and mink on land, seals and whales in water, and ducks, turkeys, chickens, quail, gulls, terns, and occasionally other species in the air. Transmission chains have been postulated to include virus shed into water, particularly by ducks and sea-going birds.

Only two antigenic types of viruses are known to infect horses (H7N7, originally called "equine 1," and H3N8, originally called "equine 2"). Disease is often severe, and equine influenza is a significant veterinary problem for the horse-racing industry.

Traditional "swine influenza virus" is a member of the H1N1 subtype, is endemic in pigs in the United States, and after many years of apparent absence has been recognized recently in several countries in Asia and Europe, usually linked to the importation of swine from the United States. Two other variants of H1N1 viruses have been found in swine, and shown by antigenic and molecular analyses to be similar to an avian H1N1 virus or contemporary strains of human H1N1 viruses, respectively. The avian H1N1-like virus has caused outbreaks in swine in parts of Germany, Belgium, and France in recent years.

After the appearance of the human pandemic of "Hong Kong" (H3N2) influenza in 1968, it was rapidly found that similar viruses were infecting swine in Taiwan and the United Kingdom. Numerous investigators have found continuous evidence of infections of swine with these and later human variants of H3N2 influenza. Apparently pockets of "old H3N2" viruses have been maintained in swine after such strains were displaced by newer variants in humans (Shortridge et al., 1977). Disease in swine is generally mild, may be complicated by coinfecting bacteria, but does retard development of swine to an extent that can be commercially significant.

The recognition of infections of mink with fulminant disease was recently observed (Okazaki et al.,

1983), as has also been the case with seals (Lang et al., 1981). The nature of influenza disease in whales is unclear. In these species, molecular and antigenic analysis suggest the causative viruses were derived from subtypes naturally found in birds (Hinshaw et al., 1986).

The spectrum of diseases of avian influenza virus in birds ranges from subclinical (the customary situation with ducks) to a disseminated, rapidly fatal disease (as in "fowl plague virus" outbreaks in chickens). Only H5 and H7 HAs, which have an unusual cleavage site, have been associated with highly pathogenic avian influenza viruses, but avirulent H5- and H7-containing strains also circulate and are, in fact, probably more common. It is clear that congregation of birds (as in feral ducks), the ability of many avian influenza viruses to infect and survive in the duck intestinal tract, and the usual lack of pathogenicity provide excellent situations for newer genetic combinations of avian influenza to be generated by reassortment.

Occasionally, apparently, the gene composition produced, possibly with further evolution, has unusual host range (e.g., as in the described isolates of influenza from minks, seals, and whales) and/or pathogenicity (e.g., as in "fowl plague virus," or the situation observed with seals and minks).

Influenza in Humans

Clinical Symptoms

Infection of humans with influenza virus is often subclinical, or produces very mild symptoms (hence the statement "I've never had flu" from persons with serologic evidence of infection with one or more virus types or strains)! Although nausea, vomiting, and diarrhea are often observed in outbreaks in school children during influenza epidemics, the "classical" symptoms of influenza in children, adolescents, or adults are the rapid onset (i.e., over about 12 h or less) of malaise, feverishness, and myalgia, usually with nonproductive cough or sore throat. These symptoms generally last for several days, and are often severe, requiring absence from work or school and bedrest. It is not uncommon for malaise and cough to persist into the second week after, onset.

Complications of influenza include exacerbation of underlying chronic illness or pneumonia, which may be due to spread of virus to the lower respiratory tract, secondary bacterial infections, or both. In an average epidemic in the United States, up to about 50,000 excess deaths may occur in association with influenza. Of these deaths, 80 to 90% are in persons greater than 65 years of age (Lui and Kendal, 1987). Several hundred thousand excess hospitalizations for pneumonia and influenza also occur primarily in older persons but in young adults or children too (Barker, 1986). In very young infants, severe influenza resembles other severe respiratory virus infections, with bronchiolitis, febrile convulsions, and occasionally encephalitis (Glezen et al., 1980). Case reports suggest other types of complications can arise, including otitis media, myositis, myocarditis, toxic shock syndrome, and Reye syndrome. A major risk factor for the latter is the use of salicylate-containing medications, which are therefore contraindicated for use in children and adolescents during epidemics of influenza (or chicken pox) (Halpin et al., 1982).

Epidemiology

Outbreaks of influenza are occurring at one or more locations in the world at any point in time. In temperate countries and in the Northern Hemisphere the "influenza season" is from approximately November to April, and in the Southern Hemisphere from May to October. The pattern in tropical regions is different, as influenza may be endemic, with epidemics of increased activity occurring more than once per year (Fig. 3).

The basic cause underlying influenza epidemics is the continual emergence of strains that vary antigenically from past strains, so that they can bypass individual immunity and cause illness in persons of all ages (Glezen et al., 1986; Kendal, 1987). Such epidemics are normally preceded in each community by sporadic cases, most of which go unrecognized, and occasionally focal outbreaks, over a 1 to 3-month period, before more widespread illness becomes apparent (Marine et al., 1976). Illness rates peak over a few weeks, and then decline. In some years epidemics are caused predominantly by one virus type or subtype, but in other years more than one virus circulates widely, either simultaneously or sequentially (Kendal et al., 1979). Experiences in different regions within one country, as well as between different countries, often vary in one season, but over a period of several years are fairly equivalent. This applies not only to the predominant virus strains, but also to the level of impact they cause.

Pandemic influenza occurs at irregular and unpredictable intervals when an antigenic subtype of influenza A appears that is either new or has not circulated for many years. The best documented pandemics are in 1918, 1957, 1968, and 1977. In 1918, about 20 million persons worldwide are estimated to have died (500,000 in the United States alone) from infections with a virus believed to be of H1N1 sub-

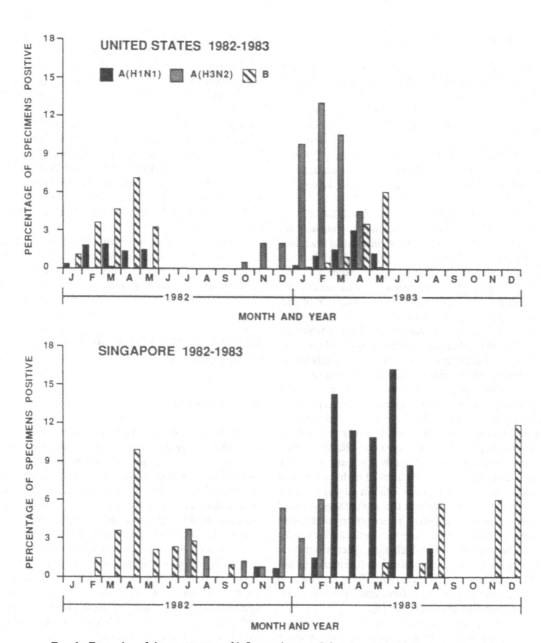

FIG. 3. Examples of the occurrence of influenza by month in temperate and tropical regions.

type, closely related to swine influenza strains. H1N1 strains continued to circulate until 1956, evolving major variants within the subtype. Asian influenza, caused by H2N2 virus, occurred in 1957; Hong Kong influenza, caused by H3N2 virus, occurred in 1968; and H1N1 virus reappeared in 1977. It is known that in 1957 and in 1968 the pandemic strains totally displaced the previous subtype, which rapidly disappeared in humans. The reappearance in 1977 was unusual, however, in two ways. Firstly, the pandemic strain was virtually identical to strains prevalent in 1950 (Kendal et al., 1978), and very few illnesses were observed in persons born before about the mid-1950s, presumably a result of immunity from prior infections with related H1N1 strains (Kendal et al., 1979). Secondly, H1N1 viruses have cocirculated with H3N2 strains, at least up to now, continuing to affect almost exclusively children and younger adults, whereas H3N2 influenza (and recent type B strains) have caused severe illnesses and excess mortality in older adults (Kendal et al., 1979; Lui and Kendal, 1987).

Prevention and Treatment

Experience with H1N1 influenza in 1977 and thereafter illustrates the potential for natural infection to induce almost solid long-lasting immunity against illness from related strains. The potential for success in the prevention of influenza through the use of vaccines is supported by such observations, but tempered by the knowledge that it is customary for variant strains of virus to appear that can partially or completely overcome immunity induced by vaccine strains.

In the United States only inactivated (killed) vaccines are used, whereas live attenuated vaccines are the subject of research. Live intranasal vaccines have been used for many years, however, in mass vaccination programs in the USSR (Zhdanov, 1986). When antigenically appropriate vaccine strains are used, either type of vaccine appears capable of preventing illness in 50 to 70% of those vaccinated, higher values usually being observed in the best controlled studies where the effects of non-influenza illness can be accounted for during data analysis. In institutions where either exposure or transmission risks may be higher than normal, or where the population may have naturally impaired immunity or host defenses (e.g., homes for the elderly or chronically ill) protection against infection is lower (e.g., about 35%), but high protection against complications leading to hospitalization and death often are seen (Patriarca et al., 1985). Achievement of high vaccination rates (e.g., more than about 75%) in closed or semi-closed populations permits the advantages of herd immunity to be exploited in preventing outbreaks (Patriarca et al., 1986). Live vaccines can induce local secretory antibody which may offer short-term advantages in preventing infection (Clements et al., 1986). Practical experience with live vaccines in terms of infectiousness, safety, and protective efficacy in high-risk groups is lacking.

Antiviral drugs that may be taken orally and have a 70 to 90% prophylactic efficacy against influenza A infections are amantadine and, its analog, rimantadine (Dolin et al., 1982). Experience has indicated that rimantadine may be slightly safer, in terms of CNS side effects, than amantadine (Dolin et al., 1982) and probably is well tolerated by the elderly (Patriarca et al., 1984). Considering, however, that the optimal dosages may vary and are imprecisely determined thus far for different age groups and persons with different underlying diseases, it is important to continue to search for information on these points for both drugs to prevent premature conclusions. Likewise, although in normally healthy college students both drugs have shown therapeutic efficacy when given in the first 1 to 2 days after illness onset (Younkin et al., 1983, Van Voris et al., 1981),

similar data are only now beginning to be sought in controlled studies in high-risk groups.

Ribavirin, a totally different drug, has the advantage over amantadine and rimantadine of being effective against influenza A and B viruses. However, it is a more toxic drug, having a known hematologic interaction. It is not generally considered effective against influenza when given orally, but rather must be administered by almost continuous aerosol treatment. Therapeutic effectiveness is not known except for normally healthy college students (Gilbert et al., 1986). In the United States, only amantadine presently is approved by the FDA for use against influenza.

No other measures have been proven to be specifically effective against influenza viruses, although treatment of symptoms with "over the counter" medications is commonplace.

Collection and Processing of Specimens

Virus Isolation

Cultures taken within the first 3 days of illness are most likely to yield a virus isolate. A swab culture of the throat and nasal passage is the most practical method of collecting the specimen for influenza virus isolation from acutely ill outpatients. Higher isolation rates are likely to be obtained from nasal washes, nasal aspirates, or broncho-alveolar lavages which should, if possible, be collected from hospitalized patients where the success and speed of diagnosis are most critical for patient management. Throat and nasal swabs are taken as follows: the pharynx is vigorously swabbed with a cotton-tipped applicator, the secretions are eluted from the cotton, and the applicator stick is broken off in a vial of transport medium. A wire nasopharyngeal swab is passed into the nostril parallel with the palate and gently rotated; the secretions are then eluted from the swab in the same vial of transport medium containing the throat swab. Nasal washes are collected by instilling 3 ml of transport medium alternately in each nostril while the patient sits with head tipped back and closes the airway by beginning to say "car." Fluid is then collected in a cup by tipping the head forward and blowing through the nose. A suitable transport medium is veal infusion broth or tryptose broth (pH 7 to 7.2) containing 0.5% gelatin or bovine serum albumin (pentex crystalline, preferred).

Specimens should be transported to the laboratory on wet ice and maintained at 4°C before inoculation. The isolation rate is essentially unchanged when specimens are held at 4°C for up to 4 days (Baxter et al., 1977). If inoculation is not possible within 4 days, specimens should be frozen to −70°C immediately after collection, and maintained in a mechanical freezer or on dry ice. If dry ice is used, vials should be tightly sealed to prevent entry of CO_2, which will inactivate viral infectivity by lowering the pH. Before inoculation, a 10 × solution of antibiotics should be added to specimens to yield a final concentration of 800 U of penicillin and 400 μg streptomycin, or 50 μg of gentamycin sulfate per ml. Specimens should be agitated and then centrifuged at 1500 × g for 15 min at 4°C to remove particulate debris and bacteria.

Postmortem specimens from the respiratory tract should be obtained as soon after death as possible to avoid loss of virus viability and excessive bacterial growth. Swab cultures, secretions such as tracheal aspirates, or blocks of hemorrhagic lung tissue should be collected aseptically. Virus isolation may be attempted from heart muscle in cases of myocarditis, and skeletal muscle occasionally has been reported also as a site of influenza virus; however, better isolation rates will be obtained with specimens from the respiratory tract collected early in the illness.

Immunofluorescent Examination of Cells

Secretions containing respiratory epithelial cells may be aspirated from the nasopharynx through the nose with a fine plastic catheter connected to a mucus trap and a suction machine. Suction also can be applied using a syringe if a suction machine is not available. Cells are sedimented by centrifugation for 10 min at 4°C and resuspended in phosphate-buffered saline (PBS) by gentle pipetting. Visible mucus is removed, and the cells are again sedimented by centrifugation. The cells are resuspended and spotted onto glass slides, allowed to dry, and fixed in acetone at 4°C for 10 min. Slides may be held at 4°C for 48 h before staining, but for longer term storage slides should be placed in slide boxes which are placed in plastic bags and held at −70°C.

A sufficient number of cells may sometimes be obtained from pharyngeal and nasal swabs. The pharynx is swabbed vigorously until the patient coughs or gags, and with a second swab the nasal mucosa is rubbed thoroughly; both swabs are placed in a single vial of transport medium and the cells are eluted by pipetting the transport medium over the swabs and by vigorous mixing. Smears are prepared as above. Tracheal and bronchial secretions obtained by lavage procedures or at autopsy may be processed in a similar manner. Frozen sections of lung tissue also may be examined by immunofluorescent staining. Immunofluorescence techniques have been described in detail (Gardner and McQuillan, 1974).

Specimens for Immunoassay Procedures

Specimens are collected as above for immunofluorescence examination, but cells, which are often rich with antigen, are not separated. The entire specimen is diluted 2.5-fold with PBS containing 20% inactivated fetal calf serum, 2% Tween 20, and 0.004% merthiolate, and sonicated for 1 to 3 min with a cell disruptor (Branson Sonic Power Co., Shelton, CT) to solubilize the mucus. The specimens are then further diluted (5-fold final) in the same buffer and tested by enzyme immunoassay (EIA) or time-resolved fluorescent immunoassay (TR-FIA) (Sarkkinen et al., 1981; Halonen et al., 1983). The specimens may be sent to the laboratory without prior cooling and can be stored for 2 days at room temperature without significantly decreasing the antigen titer (Halonen et al., 1983).

Electron Microscopy

The demonstration of influenza virus in respiratory secretions by electron microscopy has been reported (Joncas et al., 1969); however, other respiratory viruses are more commonly detected in respiratory secretions by this method (Doane et al., 1969). Electron microscopy can be helpful in the identification of hemagglutinating isolates that are difficult to identify as influenza in serologic tests as when antigenic shift occurs.

Serologic Diagnosis

An acute-phase serum specimen should be collected within 7 days of onset of illness, and a convalescent-phase serum specimen should be collected between the 14th and 21st days after onset of illness. From 1 to 3 ml of serum should be removed aseptically from clotted blood and placed in a tightly closed screw-cap vial. Both acute and convalescent specimens should be submitted and tested together. Serum specimens may be stored frozen at −20°C or unfrozen at 4°C. However, long-term storage at 4°C may allow some bacterial growth to occur. Shipping serum frozen on dry ice reduces leakage and the risk of bacterial growth. When long-distance shipping without refrigeration is required, sodium azide may be added to a final concentration of 0.1% to prevent bacterial growth.

Laboratory Diagnosis

Only isolation of influenza virus (or detection of influenza antigen in an appropriate clinical specimen) and/or detection of a rise in antibody specific for one or more influenza virus antigens, provides acceptable laboratory diagnosis. Virus isolation and antigen detection is preferable in terms of speed, and although traditional methods took up to several weeks, new approaches and techniques permit quite sensitive and highly specific diagnosis within a day or two. Full procedures for technical aspects of undertaking laboratory diagnosis of influenza are available from the Centers for Disease Control (Kendal et al., 1982).

Isolation and Identification

Influenza virus types A, B, and C may be isolated in embryonated hen's eggs. Embryos 10 to 11 days old are optimal for isolation of influenza types A and B, whereas embryos 7 to 8 days old are optimal for influenza type C. After inoculation, eggs are incubated at 33 to 34°C for 2 to 3 days for influenza types A and B or 5 days for influenza type C. Although influenza types A and B can often be recovered from both the allantoic and amniotic cavities, this is not always the case. Therefore, the most efficient method of primary isolation in eggs is to inoculate and harvest both cavities. Influenza type C, however, can be isolated only from the amniotic cavity. At times the virus may be recovered in harvests from eggs inoculated at 10 to 11 days (rather than 7 to 8 days), and incubated for 2 to 3 days, as for types A and B.

Influenza types A and B also may be isolated in susceptible tissue culture, traditionally primary Rhesus monkey kidney cells. The use of proteolytic enzymes, particularly trypsin, has extended the host range to include MDCK cells (Tobita et al., 1975) with sensitivity equal to Rhesus monkey kidney cells (Frank et al., 1979). Primary monkey kidney cells will support the growth of a wider range of viruses than MDCK cells. This may be an advantage for laboratories interested in the whole spectrum of respiratory viruses. Associated with the use of these cells, however, is the possible presence of adventitious hemagglutinating or hemadsorbing viruses of simian origin. Adequate controls of uninfected cells always must be carried out to detect this possibility. Alternative cells used for isolating influenza A or B viruses include primary cynomolgus monkey kidney cells, and possibly the continuous line of LLCMK2 cells (Frank et al., 1979). Influenza type A can be isolated in both eggs and tissue culture but, in general, tissue culture is more sensitive. With influenza type B, however, the isolation rate is much higher in tissue

culture (Monto et al., 1981). Because of variations in growth characteristics from year to year in the strains of viruses circulating, maximal sensitivity of isolation will be obtained if specimens are inoculated into both eggs and tissue culture. This may not be practical for all laboratories because eggs are not widely used for studies of viruses other than influenza, and because they require specialized incubators for maintenance before inoculation. Conversely, eggs may be essential in regions where adequate facilities for tissue culture do not exist, particularly in developing nations.

Tissue culture cells always should be washed with serum-free medium before inoculation of influenza as serum used for growth and maintenance of cells may contain nonspecific inhibitors that will neutralize influenza virus infectivity. Incubation should be at 33 to 34°C., as many wild strains of influenza are temperature sensitive. A roller drum may increase the rate or speed of isolation, but is not essential.

When harvesting egg and tissue culture fluids it is usually unknown whether they contain virus, the exception being when tissue culture cells show significant CPE. Thus, fluids should be handled as if they contain high levels of infectious virus. Furthermore, inoculation of new clinical specimens and harvesting of previously inoculated eggs and tissue culture should be done in separate areas and at separate times. Cross-contamination of clinical specimens with laboratory-adapted strains of influenza being grown for reagent production, "quality control," or research purposes is a well known occurrence. All laboratory staff should be made aware of this problem and procedures instituted to minimize its occurrence (Budnick et al., 1984; Cox et al., 1986).

After incubation, eggs should be examined with a candler to assure that they are viable and not grossly contaminated with bacteria. Dead or contaminated eggs should be discarded without harvesting because the presence of bacteria can cause false-positive hemagglutination. During the course of harvesting fluids, additional cases of bacterial or fungal contamination may be observed that were not detected by candling or microscopic examination (tissue culture cells). Any such fluids should be discarded.

Traditionally, influenza viruses isolated in tissue culture are detected by hemadsorption of guinea pig red blood cells added to tissue culture tubes when cytopathic effects are observed, or 7 days postinoculation in the absence of CPE. For egg harvests, hemagglutination is performed with guinea pig or human type O cells (0.4%), or chicken red blood cells (0.5%). The latter have a more rapid settling time, but may be less sensitive. Only chicken or human type O cells will agglutinate influenza type C virus. Figure 4 is a general scheme for influenza isolation, based on these traditional approaches.

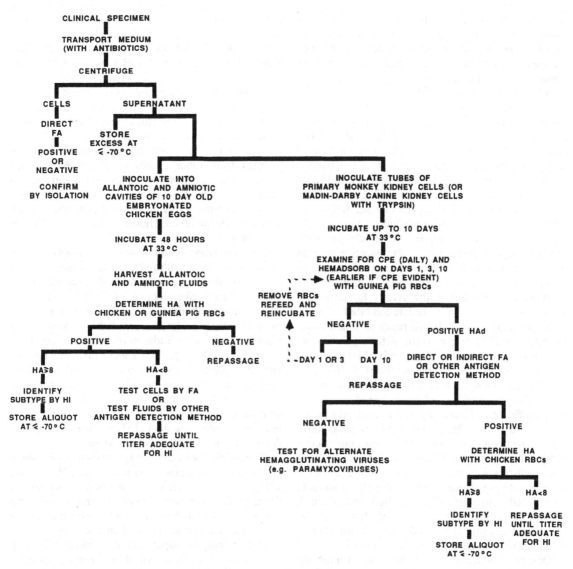

FIG. 4. Schematic for influenza virus isolation. FA = Fluorescent antibody; HA = hemagglutination; HI = hemagglutination inhibition; RBC = red blood cell; CPE = cytopathic effect; HAd = hemadsorption.

A newer approach, which particularly lends itself to rapid detection of influenza in outbreaks or in community surveillance programs, is to detect virus antigen in cell cultures 1 to 3 days after inoculation by early hemadsorption (Minnich and Ray, 1987), fluorescence microscopy, or immunoassay (see below).

Influenza isolates can be identified by the hemagglutination inhibition (HI) test using antisera to current virus strains. Because influenza viruses undergo continual antigenic drift, appropriate antisera are usually not commercially available. The Centers for Disease Control makes available each year chicken antisera to current strains, as well as appropriate HA

antigens through State Health Laboratories and World Health Organization (WHO) National Influenza Centers.

To perform the HI test in microtiter plates, the unknown isolate and control HA antigens must be adjusted to contain 4 HA units per 25 μl. The HA titer of each antigen is divided by 8 to obtain the correct dilution factor. Once these dilutions are made, each antigen must be backtitered to assure they contain 4 HA units.

Many influenza isolates are sensitive to factors in serum which nonspecifically inhibit agglutination. These inhibitors can usually be removed by treatment with receptor-destroying enzyme (RDE) of *Vi-*

brio cholerae. Sensitivity to inhibitors varies with virus strains, however, and at times treatment with trypsin, periodate, or kaolin is required. In addition, the serum may need to be adsorbed with red cells to remove nonspecific agglutinins.

After treatment of serum, twofold dilutions of antisera (generally 1 : 10 through 1 : 2560), are prepared in microtiter plates in 25 μl volumes. The control and unknown antigens adjusted to contain 4 HA units are added and incubated, followed by red blood cells. The HI titer is the reciprocal of the highest dilution of serum that completely inhibits agglutination of the red blood cells by virus. Where the HI test is correctly performed, each virus isolate will be inhibited by serum to only one influenza type or subtype and each control antigen will be inhibited only by homologous antiserum. False-negative results (i.e., isolate fails to be inhibited by any antisera) may be due to significant antigenic drift of the isolate or a virus with an extremely low avidity for antibody. False-positive results (i.e., isolate inhibited by more than one sera) may be due to residual nonantibody inhibitors of HA, or a mixture of hemagglutinating viruses occurring as a result of a mixed infection (or as a result of laboratory contamination); the possibility that the "viral" HA is actually due to bacterial contamination should also be considered.

Identification of influenza viruses also may be accomplished by the fluorescent antibody procedure. This is most commonly done with tissue culture cells but cells recovered from allantoic fluid are also suitable. Cells are scraped from tissue culture tubes (or centrifuged from allantoic fluid) and multiple aliquots are spotted onto glass slides, air dried, and fixed in acetone. Cells are then stained with fluorescein-labeled antibody (direct method) or by fluorescein-labeled antispecies globulin (indirect method).

One positive feature of the fluorescent antibody method is that rapid identification of a number of common respiratory viruses can be made using a single tube of cells. The choice of staining procedures (direct or indirect) at present is largely dependent on reagent supply. As direct conjugates to a complete battery of common respiratory viruses becomes available, it is likely that the direct FA test will become more common.

Antisera from almost any species of animal can be used in this test if the serum is type specific, that is, reacts with NP or M antigens. Immunization of animals with purified HA, however, can render the fluorescent antibody test subtype specific. Monoclonal antibodies have been developed that show subtype-specific staining (Schmidt et al., 1982). In the interest of providing long-term, stable reagents for influenza diagnosis, the CDC has developed type-specific monoclonal antibodies directed to the NP and M proteins of influenza A, and in the case of influenza B,

antibody directed toward constant determinants in the HA as well as NP (Walls et al., 1986a). These reagents have been made available through State Health Departments and WHO National Influenza Centers. Their distribution by reagent manufacturers is in progress. The reagents will recognize all type A or type B virus-infected cells regardless of antigenic drift or shift and are useful for rapid culture confirmation of isolates. Culture confirmation with these reagents has been done by FA staining on cells scraped from tubes (Walls et al., 1986a) on cell monolayers grown on coverslips and maintained in shell vials (Espy et al, 1986), or by enzyme immunoassay or time-resolved fluoroimmunoassay of supernatants from infected tissue culture cells (Walls et al., 1986a, b).

For the enzyme immunoassay procedure, harvests from tissue cultures inoculated with clinical specimens are added to Immulon II microtiter plates and incubated overnight at 4°C. After washing, monoclonal antibody is added for 30 min, followed by goat antimouse IgG horseradish peroxidase conjugate for 1 h. After addition of substrate, color development in the appropriate wells identifies the isolate. Positive controls also must be used as well as uninoculated egg or tissue culture harvests from the same lot as the unknown isolate to rule out nonspecific reactions. This procedure has proved to be much more sensitive than HAI, particularly for type A (H1N1) viruses (Walls et al., 1986a).

Even higher sensitivity may be obtained in a time-resolved fluoroimmunoassay procedure (Walls et al., 1986b), and it is likely that as optimal conditions are developed, greater than 80% of specimens containing infectious virus can be identified by antigen detection of cell cultures within 1 day of inoculation.

Pilot testing of rapid culture confirmation with specimens mailed to a central laboratory resulted in about 40% being positive in 3 days at the peak of an epidemic (Reichelderfer et al., 1987). As monoclonal antibodies to the other common respiratory viruses are assembled and tested, and reagents made generally available, these types of procedures will become most practical for rapid culture confirmation in diagnostic laboratories.

Direct Detection of Antigen

Although rapid culture confirmation may provide results with high sensitivity in 1 to 3 days, direct detection of viral antigen in clinical material can be performed within hours. This may be particularly useful for hospitalized patients, because amantadine and rimantadine are effective for antiviral chemotherapy of influenza type A, but not type B. Rapid diagnosis also may help guide management of immunocompromised patients with influenza as well as to help control nosocomial spread. In addition, prompt diagnostic information is of value in educating physicians and informing patients. Inappropriate therapy can be eliminated as can the expense of unnecessary diagnostic tests. On a larger scale, rapid diagnosis of outbreaks can alert public health officials to the possible onset of an epidemic and encourage the early initiation of specific epidemiologic control measures.

The possibility of establishing rapid diagnosis by fluorescent antibody (FA) staining of influenza infected nasal epithelial cells was first demonstrated by Liu (1956). Other investigators have confirmed these initial findings for influenza and have shown the technique to be applicable to a number of other respiratory viruses (Gardner and McQuillan, 1974). Some laboratories have found the method to be as sensitive for influenza detection as isolation, whereas others have reported a significant increase in the number of virus identifications when isolation and direct detection by FA were applied (Daisy et al., 1979; Fulton and Middleton, 1974; Ray and Minnich, 1987; Shalit et al., 1985). An important variable in test results was the quality of clinical specimen, specifically the number of cells recovered. There is some evidence that FA may have a diagnostic advantage over isolation when specimens are taken in the late stages of infection, presumably due to the presence of antiviral antibody which neutralizes virus infectivity.

More widespread use of the FA procedure, as well as other rapid methods (see below) has been hampered by the nonavailability of antisera of acceptable quality. Antiserum must have a high titer for the homologous virus, and must be free of unwanted antibodies to host proteins or other agents. Some of these requirements for viral antisera have been met by immunization with virus produced in a homologous host system; by the use of highly purified virus for immunization; or by extensive absorption of antisera with host tissue to remove unwanted antibodies. All of these approaches have limitations, and inconsistencies in supply and quality of viral antisera have been major deterrents to the development of viral diagnostic capabilities in clinical microbiology laboratories.

The monoclonal antibodies developed by CDC also were evaluated for use in an indirect FA procedure using cells from nasopharyngeal secretions. The reagents gave good type-specific diagnostic results in patients with type A influenza (both H1N1 and H3N2 subtypes) and type B influenza (McQuillan et al., 1985; Shalit et al., 1986). It should be pointed out here that because there is no objective endpoint, diagnosis by immunofluorescence microscopy requires considerable experience and thoughtful interpretation to be a reliable procedure, even with monoclonal

antibodies. In the initial evaluation (McQuillan et al., 1985) a pool of 3 monoclonal antibodies were used in the influenza A pool. Careful observation by an experienced microscopist resulted in the detection of an unexpected reaction with normal cell nuclei, which led to the exclusion of one of the monoclonal antibodies from further use for this purpose.

Although the rapid technique most commonly used in respiratory virus diagnostic laboratories today, the FA technique is inefficient in its requirement for expertise in test performance and interpretation of results. Moreover, the method lacks an objective endpoint and is unlikely to be automated or even applied with ease to large numbers of samples. Immunoassays overcome these disadvantages of FA and have been applied for the detection of influenza antigens. The prototype immunoassays, radioimmunoassay, have been largely abandoned because of the hazards and expense of radioactive reagents.

Enzyme immunoassays for antigen detection have many formats, the most common of which uses a capture antibody bound to the solid phase, usually a microtiter well or plastic bead. Incubation with a clinical specimen results in the binding of viral antigen present in that specimen to the antibody and thus to the solid phase. The bound viral antigen is then detected with another antibody. In the direct method, the detection antibody is directly conjugated to an enzyme. In the indirect method, the detector antibody is not labeled, but is detected using an antispecies, IgG antibody that is conjugated to an enzyme. The bound enzyme is then detected and quantitated by the development of color or fluorescence when an appropriate substrate is added. These assays can be performed with automated machinery, the results read in seconds by specially designed spectrophotometers, and the data promptly collated and analyzed by computer.

An assay of this configuration that used polyclonal antisera and nasopharyngeal aspirates as the clinical specimen was as sensitive as FA staining of nasal epithelial cells (Sarkkinen et al., 1981). Although isolation data were not available in this study, similar assays using throat swabs and nasal washes as a source of clinical material were 50 to 75% as sensitive as isolation depending upon whether a colorigenic (Harmon and Pawlik, 1982), or fluorogenic (Harmon, et al., 1983) substrate was used. Other investigators using both an enzyme-linked fluorescence assay (Yolken and Stopa, 1979) and an ultrasensitive enzymatic radioimmunoassay (Harris et al., 1979) demonstrated with clinical samples that assays with sensitivity comparable to isolation could be developed (Berg et al., 1980). That level of sensitivity, however, required the use of two assays, one of which used a radioactive substrate requiring separation of unreacted from reacted substrate.

As mentioned, a significant deterrent to immunoassay development was the lack of high-quality reagents. The successful application of the monoclonal antibodies developed at CDC in other assay procedures led to their examination in a direct antigen detection system that uses europium chelate as a label in a time-resolved fluorescent immunoassay (Fig. 5). A fluorometer that generates quantitative data with high statistical precision with only a 1-second counting time had been developed which takes advantage of the unique properties of europium fluorescence (Hemmila et al., 1984; Soini, 1984). Comparison of the developed time-resolved fluorescent immunoassay (TR-FIA) with highly sensitive serodiagnostic assays showed that 85% of patients who had serologic evidence of influenza infection had detectable levels of antigen in their nasopharyngeal secretions, with 95% specificity (Walls et al., 1986b). When compared with results of a previously described EIA for antigen detection, the TR-FIA detected 20% more positive nasopharyngeal aspirate specimens among patients with serologic evidence of infection. Because serologic diagnosis of influenza infection is generally more sensitive than virus isolation for confirmation, the results show the TR-FIA to be highly efficient for antigen detection in nasopharyngeal aspirate specimens, without being nonspecific.

Detection of Viral Enzyme and Nucleic Acid

The approaches to viral diagnosis, for the most part, use the structural proteins of viruses which are usually detected and identified on the basis of their antigenicity, that is, by antibody. However, other viral components, specifically viral enzymes and nucleic acids, also can be used for virus detection and identification.

All influenza A and B viruses possess a surface glycoprotein that has neuraminidase activity. Although neuraminidases from different strains of influenza vary structurally and antigenically, they all catalyze the same reaction, the breakage of the alpha-glycosidic bond of N-acetyl-neuraminic acid. A sensitive substrate, 4-methylumbelliferyl-alpha-ketoside of N-acetyl-D-neuraminic acid, has been developed. The substrate has little fluorescence itself but when cleaved by neuraminidase to yield free 4-methylumbelliferone produces intense fluorescence detectable with a simple filter fluorometer. This technique has been applied to the direct detection of influenza (Yolken et al., 1980). However, distinguishing influenza neuraminidase activity from that due to bacterial and mammalian neuraminidases may pose some specificity problems.

Detection of the nucleic acid component of influenza is a method that has considerable potential. It almost certainly will be possible to identify nucleotide sequences that are unique to the influenza genome and that therefore constitute probes with a high degree of specificity. The high avidity of complementary nucleic acid strands for each other should permit hybridization assays that are highly sensitive. Molecular cloning will enable the production of virtually unlimited quantities of probes identified as being suitable for the use intended. This could include broadly reactive type A or type B probes (NP or M gene sequences), or more narrow probes specific for Hl or H3 HA gene sequences, for example.

Detection of nucleic acid on nitrocellulose filters is fairly simple, at least for DNA. A very stable molecule, DNA resists exposure to 0.3 molar NaOH at 60°C, conditions that can be used to free DNA in clinical material from proteins, RNA, and membranes. The resulting denatured single-stranded DNA can then be applied to nitrocellulose filters by blotting or filtration, and baked onto the filter. The result is a solid-phase system containing immobilized DNA derived from a clinical specimen. Labeled virus-specific probes can then be hybridized to the DNA derived from clinical material to detect and identify any virus-specific DNA that is present.

Before this method can be used to any extent for detection and identification of influenza in the diagnostic laboratory, two major problems must be overcome. The first concerns the conditions for extracting single-stranded RNA (ssRNA) from clinical material. The NaOH extraction methods used for DNA hydrolyze ssRNA. In addition, respiratory secretions, from which influenza RNA will be extracted, contain high levels of ribonuclease activity which degrades RNA as soon as it is extracted from virions or host cells.

Richman et al., (1984) reported using ribonuclease inhibitors to modify existing RNA extraction techniques, which permitted the extraction of influenza RNA from respiratory secretions and subsequent detection with labeled probes. However, details of the procedure were not provided in the review article.

The second problem facing nucleic acid detection systems is the nature of the label. Probes are usually labeled by incorporation of ^{32}P-labeled deoxynucleoside triphosphates. The advantage of probes labeled with ^{32}P nucleotides is extreme sensitivity due to the high level of specific activity of the probe. These probes, however, are usable for only 2 weeks and pose a radiation hazard. The approach being taken to overcome this problem is to use probes with biotinylated deoxyuridine triphosphate which can be incorporated as efficiently as ^{32}P-labeled deoxyuridine triphosphate. The resulting biotin-labeled DNA probe can then be used to detect nucleic acid bound to nitrocellulose by adding enzyme-labeled avidin,

One Wash CAPTURE TR-FIA

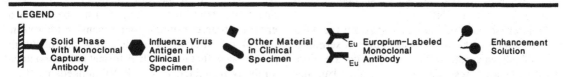

FIG. 5. Schematic illustration of the TR-FIA procedure. Reprinted with permission from Walls H. H., K. H. Johannson, M. W. Harmon, P. E. Halonen, and A. P. Kendal. 1986. Time-resolved fluoroimmunoassay with monoclonal antibodies for rapid diagnosis of influenza infections. J. Clin. Microbiol. **24:**907–912.

which binds firmly to biotin. The enzyme label then permits colorimetric measurement of bound nucleic acid and also may reduce assay time by eliminating the need for autoradiographic detection of the signal. Such assays, however, have not yet reached the point of practical application to influenza.

Serologic Diagnosis

Serodiagnosis of influenza by conventional methods such as hemagglutination inhibition (HI), complement fixation (CF), and neutralization is based upon demonstrating a significant rise (generally fourfold) in antibody titer to a given viral antigen over the course of the patients' illness. Because of the delay involved in obtaining a convalescent-phase specimen, serologic diagnosis is retrospective, which limits its usefulness in many clinical situations. However, in addition to being more economical than virus isolation, serodiagnosis has other advantages such as establishing the diagnosis of influenza when isolation attempts fail due to the short period of virus excretion.

Currently recommended procedures for routine serodiagnostic use are the HI and CF tests, although solid-phase immunoassays may displace these tests in the near future. The HI has the advantage of relative simplicity and economy and fulfills most of the requirements for an ideal test. Results of HI testing often approximate the neutralization test, most likely because HI measures antibody which prevents an essential biologic function of the virus, that is, attachment of HA to cells which is the first step in the process of viral infection. Therefore, the concentration of serum HI antibody is related to immunity to influenza infection. Although there is no HI titer at which protection is assured, titers of 32 or 40 are generally believed to be the lowest levels associated with significant protection, and protection is more likely with higher antibody levels. Because HI measures, for the most part, antibody to the HA it is a subtype specific test for influenza serodiagnosis. The disadvantages of the HI test include the need to inactivate nonspecific serum inhibitors and variation in test sensitivity according to the avidity for antibody of the particular influenza strains prevalent. The sensitivity of the HI test can be increased by ether treatment of the antigen (Isaacs et al., 1952). Ether-treated antigen is usually needed for testing influenza B strains (Monto and Maassab, 1981), but test specificity is reduced (Kendal and Cate, 1983). Use of tissue culture grown, rather than egg grown, antigen may also increase the sensitivity of the HI test when contemporary strains have low avidity for antibody in human sera (Turner et al., 1982).

The second widely used test for serologic diagnosis of influenza is CF, which uses the type-specific nucleoprotein antigen of influenza A or B virus. By using the NP antigen, the CF test does not distinguish between infections with different HA subtypes of influenza A. The CF test is often used because of its wide applicability in the virus laboratory for diagnosis of other infectious agents. Unlike the HI test, the type-specific CF test is not influenced by either antigenic variability of prevalent strains or by non-specific inhibitors. The results of CF and HI testing do not always correlate. The relative effectiveness of each test depends upon a number of variables, such as the previous antigenic experience of the individual, the appropriateness of the strain used in the HI test, and the interval between collection of acute and convalescent sera. For these reasons use of both CF and HI tests increases sensitivity of diagnosis. The CF test may be particularly useful when new virus subtypes appear that may induce low (primary) antibody responses to HA, but larger secondary antibody responses to the NP antigen.

Solid-phase enzyme immunoassays are the most recent addition to the methods available for serodiagnosis of influenza infection. These assays are more sensitive than CF tests (Hammond et al., 1980; Madore et al., 1983) and in many cases more sensitive than HI (Murphy et al., 1980, 1981). More important, however, they offer greater flexibility in the measurement of antibody to the various proteins of influenza. The test can measure the HA-specific antibody response, which correlates with resistance to infection, by coating the solid phase with purified strain-specific HA protein (Murphy et al., 1981). Coating the solid phase with whole virus will result in the measurement of antibody to both type and strain-specific antigens (Hammond et al., 1980; Madore et al., 1983; Murphy et al., 1980). In addition to flexibility, immunoassay is the only test that combines high sensitivity, specificity, and large throughput capabilities, and in which the specific antibody response in various antibody classes or isotypes can be conveniently measured. Isotype-specific tests are particularly important for influenza (and respiratory viruses in general) because of the high frequency and magnitude of the IgA response in serum and nasal washes (Burlington et al., 1983; Clements and Murphy, 1986; Johnson et al., 1986; Murphy et al., 1982).

In an effort to promote reproducible and standardizable tests for influenza antibody detection, the performance of well-characterized monoclonal antibodies specific for various human immunoglobulin isotypes has been evaluated in an HA-specific EIA (Harmon et al., 1986). The antibody titers and number of rises measured with these monoclonal antibodies correlated highly with results using polyclonal rabbit antisera to immunoglobulin classes that have been extensively used for this purpose (Murphy et al., 1981). Although polyclonal antisera are available, these reagents are difficult to produce and lot-to-lot variations in titer and specificity occur. The availability of well-characterized monoclonal antibodies to human immunoglobulin isotypes (Reimer

et al., 1984) will provide consistency and specificity not possible with polyclonal reagents.

In a further effort to modernize serologic testing for influenza antibody, we recently demonstrated the feasibility of using the NP antigen of influenza A, produced by recombinant DNA techniques, and monoclonal antibody to human IgG, in an immunoassay designed to replace the CF test (Harmon et al., to be published). The assay detected more four-fold titer rises, maintained specificity, and was easier to standardize than the CF test.

The neutralization test is perhaps the most sensitive and specific of the available serodiagnostic tests for influenza. It unequivocally correlates with protection because it measures neutralization of viral infectivity. However, in the past it has been considered too time consuming and expensive for routine use. Modification of the test to a rapid microneutralization assay that uses an EIA format and monoclonal antibody to influenza A or B NP antigens has been accomplished (Harmon et al.) The following is a brief description of the assay. Sera are diluted in microtiter wells, and approximately 100 TCID 50 of the relevant virus strain is added. After 1 h, MDCK cells are added at a concentration sufficient to form a monolayer. After overnight incubation, the cells are fixed with 80% acetone–20% PBS. Virus growth is detected by adding the appropriate NP specific monoclonal antibody followed by goat antimouse IgG conjugated to peroxidase and the substrate. The plates are then read in an EIA reader. High optical density (OD) readings should be observed in the virus control wells, and low OD readings in serum-containing wells is correlated with neutralizing antibody, which inhibited viral infectivity. The test provides objective and quantitative data and is completed in approximately 24 h. In addition, antisera from virtually any species of animal can be tested with the same reagents.

Other tests which have been used for serodiagnosis include single radial hemolysis, radial immunodiffusion, passive hemagglutination, and enhanced hemagglutination inhibition. These have been reviewed in (Kendal, 1982). In theory, serodiagnosis by measurement of antibodies to neuraminidase (Aymard-Henry et al., 1973) or matrix protein (Kahn et al., 1982) is possible, but such tests are not practical for routine use and tend to have low sensitivity.

Serodiagnosis of Outbreaks Using Batches of Single Sera

In outoreak situations when influenza is suspected, a rapid presumptive diagnosis can often be made by examining single serum specimens from a group of selected individuals. Sera are collected from 10 or

TABLE 2. Antibody prevalence to influenza A/Taiwan/1/86 and A/Chile/1/83 among residents of Palau after outbreaks in July, 1986

	A/Chile/1/83	A/Taiwan/1/86
Number of serum pairs	101	101
GMT S1	20	8
GMT S2	43	21
Number with fourfold titer increase	29	32
Concurrent fourfold increase to both viruses (%)	93	84

more patients who are in the acute stage of the disease and from the same number of age-matched cohorts who experienced the same symptoms 10 or more days earlier. All sera are tested simultaneously for antibody titers by HI, CF, or neutralization tests. If the outbreak was caused by influenza, the geometric mean antibody titer for type A or B influenza should be significantly higher (by the *t* test) in the latter group than in the former (Grist et al., 1961; Kendal and Dowdle, 1986). If the rise in antibody titer is fourfold or greater, the difference may be considered significant without resorting to statistical analysis. A diagnosis made on this basis should be confirmed if possible by conventional methods of virus isolation and serological diagnosis with paired sera. It should be pointed out that the comparison must be made between antibody titers of sera from two groups of individuals with recent influenza-like illness, rather than comparing antibody titers of sera from one group of convalescent individuals and from another group without any recent influenza-like illness. The reason is that the group without recent illness may contain a large proportion of persons with sufficiently high antibody titers to have prevented recent infection or may have had an asymptomatic infection.

Interpretation of Diagnostic Results

The laboratory diagnosis of influenza is based upon the isolation and identification of the virus, the demonstration of a fourfold or greater increase in antibody titer, or both. The identity of the isolate should be reported only as to type, unless tests have been performed with hemagglutinin- or neuraminidase-specific antisera to characterize the antigens fully. If the virus was identified with influenza A HA-specific antisera, then the corresponding subtype of the neuraminidase is usually inferred, because thus far epidemic isolates of human influenza A have almost never been found to have exchanged HA and NA antigens even when two subtypes (H1N1 and H3N2) cocirculate.

In the absence of virus isolation, caution should be exercised in the interpretation of serologic results.

A fourfold or greater increase in antibody titer measured in the CF test with NP antigen is interpreted only to mean infection, or vaccination, with type A or type B influenza virus. Because the NP antigen is identical for all viruses of a given type, this technique does not provide information identifying the subtype causing the disease.

Results of HI tests on patients' sera also should be interpreted with caution. The specific antigen used for diagnostic serology does not necessarily identify the infecting virus (Table 2). Anamnestic responses frequently occur, depending on the previous immunologic experience of the patient. Antibody response to an earlier antigen may be greater than to the current infecting virus (Table 2). For maximal diagnostic efficiency, antigens that closely resemble currently prevalent strains and antigens of past prevalent strains may both be included in the HI test.

In the interpretation of antibody data from immunoassays, the nature of the antigen added to the solid phase must be known. If the antigen included one or both of the internal proteins (NP or M), then the comments relative to the CF test are applicable. Likewise, if pure HA was the solid-phase antigen then interpretation would be analogous to an HI test. Also, because immunoassays are very sensitive, the possibility exists of measuring antibody to contaminating proteins, such as NA in a "pure" HA preparation. Therefore, the purity of antigen preparations added to the solid phase should be documented to facilitate interpretation of the resulting data.

Literature Cited

Air, G. M. 1981. Sequence relationships among the haemagglutinin genes of 12 subtypes of influenza A virus. Proc. Natl. Acad. Sci. USA **78:**7639–7643.

Apostolov, K., T. H. Flewett, and A. P. Kendal. 1970. Morphology of influenza A, B, C and infectious bronchitis virus (IBV) virions and their replication, p. 3–26. *In* R. D. Barry and B. W. J. Mahy (eds.), The biology of large RNA viruses, Academic Press Inc., New York.

Arden, N. H., P. A. Patriarca, and A. P. Kendal. 1986. Experiences in the use and efficacy of inactivated influenza vaccine in nursing homes, p. 155–168. *In* A. P. Kendal, P. A. Patriarca (eds.), Options for the control of influenza, Alan R. Liss, New York.

Aymard-Henry, M., M. T. Coleman, W. R. Dowdle, W. G. Laver, G. C. Schild, and R. G. Webster. 1973. Influenza-virus neuraminidase and neuraminidase-inhibition test procedures. Bull. W.H.O. 48:199–202.

Barker, W. H. 1986. Excess pneumonia and influenza associated hospitalization during influenza A epidemics in the US, 1970–1978, p. 75–87. In A. P. Kendal, P. A. Patriarca (eds.), Options for the control of influenza, Alan R. Liss, New York.

Barry, R. D., D. R. Ives, and J. G. Cruickshank. 1962. Participation of deoxyribonucleic acid in the multiplication of influenza virus. Nature 194:1139–1140.

Baxter, B. D., R. B. Couch, S. B. Greenberg, and J. A. Kasel. 1977. Maintenance of viability and comparison of identification methods for influenza and other respiratory viruses of humans. J. Clin. Microbiol. 6:19–22.

Berg, R. A., R. H. Yolken, S. I. Rennard, R. Dolin, B. R. Murphy, and S. E. Straus. 1980. New enzyme immunoassays for measurement of influenza A/Victoria/3/75 virus in nasal washes. Lancet ii:851–853.

Beveridge, W. B. 1977. Influenza: the last great plague. Heineman, London.

Both, G. W., M. J. Sleigh, N. J. Cox, and A. P. Kendal. 1983. Antigenic drift in influenza virus H3 haemagglutinin from 1968 to 1980: multiple evolutionary pathways and sequential amino acid changes at key antigenic sites. J. Virol. 48:52–60.

Budnick, L. D., M. E. Moll, H. F. Hall, J. M. Mann, and A. P. Kendal. 1984. A pseudo-outbreak of influenza A associated with use of laboratory stock strain. Am. J. Public Health 74:607–609.

Buonagurio, D. A., S. Nakada, U. Desselberger, M. Krystal, and P. Palese. 1985. Noncumulative sequence changes in the haemagglutinin genes of influenza C virus isolates. Virology 146:221–232.

Burlington, B. D., M. L. Clements, G. Meiklejohn, M. Phelan, and B. R. Murphy. 1983. Haemagglutinin-specific antibody responses in immunoglobulin G, A, and M isotypes as measured by enzyme-linked immunosorbent assay after primary or secondary infection of humans with influenza A virus. Infect. Immun. 41:540–545.

Clements, M. L., R. F. Betts, E. L. Tierney, and B. R. Murphy. 1986. Comparison of inactivated and live influenza A virus vaccines, p. 255–269. In A. P. Kendal, P. A. Patriarca (eds.), Options for the control of influenza, Alan R. Liss, New York.

Clements, M. L., and B. R. Murphy. 1986. Development and persistence of local and systemic antibody responses in adults given live attenuated or inactivated influenza A virus vaccine. J. Clin. Microbiol. 23:66–72.

Coleman, P. M., J. N. Varghese, and W. G. Laver. 1983. Structure of the catalytic and antigenic sites in influenza virus neuraminidase. Nature 303:41–44.

Couch, R. B., J. M. Quarles, T. R. Cate, and J. M. Zahradnik. 1986. Clinical trials with live cold-reassortant influenza virus vaccines, p. 223–241. In A. P. Kendal, P. A. Patriarca (eds.), Options for the control of influenza, Alan R. Liss, New York.

Cox, N. J., S. Nakajima, R. Black, and A. P. Kendal. 1986. Oligonucleotide mapping of viral ribonucleic acid as an aid in identifying laboratory contaminants of influenza virus. Diagn. Microbiol. Infect. Dis. 4:231–239.

Daisy, J. A., F. S. Lief, and H. M. Friedman. 1979. Rapid diagnosis of influenza A infection by direct immunofluorescence of nasopharyngeal aspirates in adults. J. Clin. Microbiol. 9:688–692.

Davenport, F. M., A. V. Hennessy, and T. Francis, Jr. 1953. Epidemiologic and immunologic significance of age distribution of antibody to antigenic variants of influenza virus. J. Exp. Med. 98:641–656.

Doane, F. W., N. Anderson, A. Zbitnew, and A. J. Rhodes. 1969. Application of electron microscopy to the diagnosis of virus infections. Can. Med. Assoc. J. 100:1043–1049.

Drzeniek, R. 1972. Viral and bacterial neuraminidases, p. 35–74. In Current topics in microbiology and immunology, Vol. 59, Springer-Verlag, New York.

Dolin, R., R. C. Reichman, H. P. Madore, R. Maynard, P. N. Linton, and J. Webber-Jones. 1982. A controlled trial of amantadine and rimantadine in the prophylaxis of influenza A infection. N. Engl. J. Med. 307:580–584.

Espy, M. J., T. F. Smith, M. W. Harmon, and A. P. Kendal. 1986. Rapid detection of influenza virus by shell vial assay with monoclonal antibodies. J. Clin. Microbiol. 24:677–679.

Francis, T., J. E. Salk, and J. J. Quilligan. 1947. Experience with vaccination against influenza in the spring of 1947. Am. J. Public Health 37:1013–1016.

Frank, A. L., R. B. Couch, C. H. Griffis, and B. D. Baxter. 1979. Comparison of different tissue cultures for isolation and quantitation of influenza and parainfluenza viruses. J. Clin. Microbiol. 10:32–36.

Fulton, R. E., and P. J. Middleton. 1974. Comparison of immunofluorescence and isolation techniques in the diagnosis of respiratory viral infections of children. Infect. Immun. 10:92–101.

Gardner, P. S., and J. McQuillan. 1974. Rapid viral diagnosis. Application of immunofluorescence. Butterworth & Co., Ltd., London.

Gilbert, B. E., S. Z. Wilson, and V. Knight. 1986. Ribavirin aerosol treatment of influenza virus infections, p. 343–356. In A. P. Kendal, and P. A. Patriarca (eds.), Options for the control of influenza, Alan R. Liss, New York.

Glezen, W. P., A. Paredes, and L. H. Taber. 1980. Influenza in children. Relationship to other respiratory agents. J. Am. Med. Assoc. 243:1345–1349.

Glezen, W. P., H. R. Six, A. L. Frank, L. H. Taber, D. M. Perrotta, and M. Decker. 1986. Impact of epidemics upon communities and families, p. 63–73. In A. P. Kendal and P. A. Patriarca (eds.), Options for the control of influenza, Alan R. Liss, New York.

Gottschalk, A. 1957. Neuraminidase: the specific enzyme of influenza virus and Vibrio cholerae. Biochem. Biophys. Acta 23:645–646.

Grist, N. R., J. Kerr, and A. Isaacs. 1961. Rapid serological diagnosis of an outbreak of influenza. Br. Med. J. 2:431.

Guo, Y., and V. Desselberger. 1984. Genome analysis of influenza C viruses isolated in 1981/82 from pigs in China. J. Gen. Virol. 65:1857–1872.

Halonen, P., O. Meurman, T. Lovgren, I. Hemmila, and E. Soini. 1983. Detection of viral antigens by time-resolved fluoroimmunoassay. Curr. Top. Microbiol. Immunol. 104:133–146.

Halpin, T. J., F. J. Holtzhaver, R. J. Campbell, L. J. Hall, A. Correa-Villasenor, R. Lanese, J. Rice, and E. S. Hurwitz. 1982. Reye's syndrome and medication use. J. Am. Med. Assoc. 248:687–691.

Hammond, G. W., S. J. Smith, and G. R. Noble. 1980. Sensitivity and specificity of enzyme immunoassay for serodiagnosis of influenza A virus infections. J. Infect. Dis. 141:644–651.

Harmon, M. W., and K. M. Pawlik. 1982. Enzyme immunoassay for direct detection of influenza type A and adenovirus antigens in clinical specimens. J. Clin. Microbiol. 15:5–11.

Harmon, M. W., D. J. Phillips, C. B. Reimer, and A. P. Kendal. 1986. Isotype-specific enzyme immunoassay for influenza antibody with monoclonal antibodies to human immunoglobulins. J. Clin. Microbiol. 24:913–916.

Harmon, M. W., L. L. Russo, and S. Z. Wilson. 1983.

Sensitive enzyme immunoassay with B-D-Galactosidase-Fab conjugate for detection of type A influenza virus antigen in clinical specimens. J. Clin. Microbiol. **17:**305–311.

Harris, C. C., R. H. Yolken, H. Krokan, and I. C. Hsu. 1979. Ultrasensitive enzymatic radioimmunoassay: application to detection of cholera toxin and rotavirus. Proc. Natl. Acad. Sci. USA **76:**5336–5339.

Hay, A. J., A. J. Wolstenholme, J. J. Skehel, and M. H. Smith. 1985. The molecular basis of the specific anti-influenza action of amantadine. EMBO J. **4:**3021–3024.

Hemmila, I. S., V. M. Dakubu, H. Siitari Mukkala, and T. Lovgren. 1984. Europium as a label in time-resolved immunofluorometric assays. Anal. Biochem. **137:**335–343.

Herrler, G., R. Rott, H. D. Klenk, H. P. Muller, A. K. Shukla, and R. Schaver. 1985. The receptor-destroying enzyme of influenza C virus is neuraminate-O-acetylestesase. EMBO J. **4:**1503–1506.

Hinshaw, V. S., W. J. Bean, J. Geraci, P. Fiorelli, G. Early, and R. G. Webster. 1986. Characterization of two influenza A viruses from a pilot whale. J. Virol. **58:**655–656.

Hinshaw, V. S., and R. G. Webster. 1982. The natural history of influenza A viruses, p. 79–104. *In* A. S. Beare (ed.), Basic and applied influenza research, CRC Press, Boca Raton, FL.

Hirst, G. K. 1950. The relationship of the receptors of a new strain of virus to those of the mumps-NDV-influenza group. J. Exp. Med. **91:**177–184.

Homma, M. 1986. Epidemiologic characteristics of type C influenza, p. 125–138. *In* A. P. Kendal and P. A. Patriarca (eds.), Options for the control of influenza, Alan R. Liss, New York.

Hoyle, L. 1968. The influenza viruses, Springer-Verlag, New York.

Huang, R. T. C., R. Rott, and H. D. Klenk. 1981. Influenza viruses cause hemolysis and fusion of cells. Virology **110:**243–247.

Hurwitz, E. S., D. B. Nelson, C. Davis, D. Morens, and L. B. Schonberger. 1982. National surveillance for Reye syndrome: a five-year review. Pediatrics **70:**895–900.

International Committee on Taxonomy of Viruses: 1982. Fourth report on classification and nomenclature of viruses. R. E. F. Matthews (eds.), Intervirology **17:**1–199.

Isaacs, A., A. W. Gledhill, and C. H. Andrews. 1952. Influenza A viruses. Laboratory studies with special reference to European outbreak of 1950–1951. Bull. WHO **6:**287–315.

Johnson, P. R., S. Feldman, J. M. Thompson, J. D. Mahoney, and P. F. Wright. 1986. Immunity to influenza A virus infection in young children: a comparison of natural infection, live cold-adapted vaccine, and inactivated vaccine. J. Infect. Dis. **154:**121–127.

Joncas, J. H., L. Berthiaume, R. Williams, P. Beaudry, and V. Pavilanis. 1969. Diagnosis of viral respiratory infections by electron microscopy. Lancet **i:**956–959.

Kendal, A. P., N. H. Arden, and P. A. Patriarca. 1985. Influenza viruses, new strategy against an old enemy, p. 377–408. *In* L. M. de la Maza and E. M. Peterson (eds.), Medical virology IV, Lawrence Erlbaum Associates, Hillsdale, NJ.

Kendal, A. P., and T. R. Cate. 1983. Increased sensitivity and reduced specificity of haemagglutination inhibition tests with ether-treated influenza B/Singapore/222/79. J. Clin. Microbiol. **18:**930–934.

Kendal, A. P., and N. J. Cox. 1985. Forecasting the epidemic potential of influenza virus variants based on their molecular properties. Vaccine **3**(suppl):263–266.

Kendal, A. P., and W. R. Dowdle. 1986. Influenza virus,

p. 515–520. *In* N. R. Rose, H. Friedman, and J. L. Fahey (eds.), Manual of clinical laboratory immunology (3rd ed.). American Society for Microbiology, Washington, D.C.

Kendal, A. P., J. M. Joseph, and G. Kobayashi. 1979. Laboratory-based surveillance of influenza virus in the United States during the winter of 1977–1978. 1. Periods of prevalence of H1N1 and H3N2 influenza A strains, their relative rates of isolation in different age groups, and detection of antigenic variants. Am. J. Epidemiol. **110:**449–461.

Kendal, A. P., G. R. Noble, J. J. Skehel, and W. R. Dowdle. 1978. Antigenic similarity of influenza A (H1N1) viruses from epidemics in 1977–1978 to "Scandinavian" strains isolated in epidemics of 1950–1951. Virology **89:**632–636.

Kendal, A. P., and P. A. Patriarca (eds.). 1986. Options for the control of influenza, Alan R. Liss, New York.

Kendal, A. P., J. J. Skehel, and M. S. Pereira. 1982. Concepts and procedures for laboratory-based influenza surveillance. U.S. Department of Health and Human Services, Public Health Service, Centers for Disease Control, Atlanta, GA.

Kendal, A. P. 1987. Epidemiologic implications of changes in the influenza virus genome. Am. J. Med. **82**(suppl. 6A):4–14.

Kendal, A. P. 1982. Newer techniques in antigenic analysis with influenza viruses, p. 51–78. *In* A. S. Beare (ed.), Basic and applied influenza research. CRC Press, Inc., Boca Raton, FL.

Kendal, A. P. 1975. A comparison of "influenza C" with prototype myxoviruses: receptor-destroying activity (neuraminidase) and structural polypeptides. Virology **65:**87–99.

Khan, M. W., D. J. Bucher, A. K. Kous, G. Kalish, H. Smith, and E. D. Kilbourne. 1982. Detection of antibodies to influenza virus M protein by an enzyme-linked immunosorbent assay. J. Clin. Microbiol. **16:**813–820.

Kilbourne, E. D. 1987. Influenza, Plenum, New York.

Klenk, H. D., R. Rott, M. Orlich, and J. Blodorn. 1975. Activation of influenza A viruses by trypsin treatment. Virology **68:**426–439.

Krizanova, D., and V. Rathova. 1969. Serum inhibitors of myxoviruses, p. 125. *In* Current topics in microbiology and immunology, Vol. 47, Springer-Verlag, New York.

Lamb, R. A. 1983. The influenza virus RNA segments and their encoded proteins, p. 21–69. *In* P. Palese and D. W. Kingsbury (eds.), Genetics of influenza viruses, Springer-Verlag, New York.

Lang, G., A. Gagnon, and J. R. Geraci. 1981. Isolation of an influenza A virus from seals. Arch. Virol. **68:**189–195.

Laver, W. G., and E. D. Kilbourne. 1966. Identification in a recombinant influenza virus of structural proteins derived from both parents. Virology **30:**493–501.

Lazarowitz, S. G., and P. W. Choppin. 1975. Enhancement of infectivity of influenza A and B viruses by proteolytic cleavage of the haemagglutinin polypeptide. Virology **68:**440–454.

Liu, C. 1956. Rapid diagnosis of human influenza infection from nasal smears by means of fluorescein-labeled antibody. Proc. Soc. Exp. Biol. Med. **92:**883–887.

Lui, K. J., and A. P. Kendal. 1987. Impact of influenza epidemics on mortality in the USA from October 1972 to May 1985. Am. J. Public Health **77:**712–716.

Madore, H. P., R. C. Reichman, and R. Dolin. 1983. Serum antibody responses in naturally occurring influenza A virus infection determined by enzyme-linked immunosorbent assay, haemagglutination-inhibition, and complement fixation. J. Clin. Microbiol. **18:**1345–1350.

Marine, W. M., J. E. McGowan, and J. E. Thomas. 1976. Influenza detection: a prospective comparison of surveillance methods and analysis of isolates. Am. J. Epidemiol. **104:**248–255.

Martin, M. L., E. L. Palmer, and A. P. Kendal. 1977. Lack of characteristic hexagonal surface structure on a newly isolated influenza C virus. J. Clin. Microbiol. **6:**84–86.

McQuillan, J., R. C. Madeley, and A. P. Kendal. 1985. Monoclonal antibodies for the rapid diagnosis of influenza A and B virus infections by immunofluorescence. Lancet **ii:**911–914.

Meier-Ewert, H., R. W. Compans, D. H. L. Bishop, and G. Herrler. 1978. Molecular analysis of influenza C virus, p. 127–133. In B. W. J. Mahy and R. D. Barry (eds.), Negative strand viruses and the host cell. Academic Press, Inc., New York.

Meier-Ewert, H., A. Nagele, G. Herrler, S. Basak, and R. W. Compans. 1981. Analysis of influenza C virus structural proteins and identification of a virion RNA polymerase, p. 173–180. In D. H. L. Bishop and R. W. Compans (eds.), Replication of negative strand viruses, Elsevier, New York.

Minnich, L. L., and G. C. Ray. 1987. Early testing of cell cultures for detection of hemadsorbing viruses. J. Clin. Microbiol. **25:**421–422.

Monto, A. S., J. S. Koopman, and I. M. Longini, Jr. 1985. Tecumseh study of illness XIII. Influenza infection and disease, 1976–1981. Am. J. Epidemiol. **121:**811–822.

Monto, A. S., H. F. Maassab, and E. R. Bryan. 1981. Relative efficacy of embryonated eggs and cell culture for isolation of contemporary influenza viruses. J. Clin. Microbiol. **13:**233–235.

Monto, A. S., and H. F. Maassab. 1981. Ether treatment of type B influenza virus antigen for the haemagglutination inhibition test. J. Clin. Microbiol. **13:**54–57.

Murphy, B. R., D. L. Nelson, P. F. Wright, E. L. Tierney, M. A. Phelan, and R. M. Chanock. 1982. Secretory and systemic immunological response in children infected with live attenuated influenza A virus vaccines. Infect. Immun. **36:**1102–1108.

Murphy, B. R., E. L. Tierney, B. A. Barbour, R. H. Yolken, D. W. Alling, H. P. Holley, R. E. Mayner, and R. M. Chanock. 1980. Use of the enzyme-linked immunosorbent assay to detect serum antibody responses of volunteers who received attenuated influenza A virus vaccines. Infect. Immun. **29:**342–347.

Murphy, B. R., M. A. Phelan, D. L. Nelson, R. Yarchoan, D. Tierney, W. Alling, and R. M. Chanock. 1981. Haemagglutinin-specific enzyme-linked immunosorbent assay for antibodies to influenza A and B viruses. J. Clin. Microbiol. **13:**554–560.

Nakajima, S., K. Nakajima, and A. P. Kendal. 1983. Identification of the binding sites to monoclonal antibodies on A/USSR/90/77 (H1N1) haemagglutinin and their involvement in antigenic drift in H1N1 influenza viruses. Virology **131:**116–127.

Nakajima, S., N. J. Cox, and A. P. Kendal. 1981. Antigenic and genomic analyses of influenza A (H1N1) viruses from different regions of the world, February 1978 to March 1980. Infect. Immun. **32:**287–294.

Nakajima, K., S. Nakajima, K. F. Shortridge, and A. P. Kendal. 1982. Further genetic evidence for maintenance of early Hong Kong-like influenza A (H3N2) strains in swine until 1976. Virology **116:**562–572.

Okazaki, K., R. Yanagawa, and H. Kida. 1983. Contact infection of mink with 5 subtypes of avian influenza virus. Arch. Virol. **77:**265–269.

Palese, P., and D. W. Kingsburg. (eds.). 1983. Genetics of influenza viruses. Springer–Verlag, New York.

Patriarca, P. A., J. A. Weber, and R. A. Parker. 1985. Efficacy of influenza vaccine in nursing homes. Reduction in illness and complications during an influenza A (H3N2) epidemic. J. Am. Med. Assoc. **253:**1136–1139.

Patriarca, P. A., J. A. Weber, and R. A. Parker. 1986. Risk factors for outbreaks of influenza in nursing homes. A case-control study. Am. J. Epidemiol. **124:**114–119.

Patriarca, P. A., N. A. Kater, A. P. Kendal, D. J. Bregman, J. D. Smith, and R. K. Sikes. 1984. Safety of prolonged administration of rimantadine hydrochloride in the prophylaxis of influenza A virus infections in nursing homes. Antimicrob. Agents Chemother. **26:**101–103.

Pereira, H. G., B. Tumova, and V. G. Law. 1965. Avian influenza A viruses. Bull. W.H.O. **32:**855–860.

Rafelson, M. E., M. Schneir, and V. W. Wilson, Jr. 1963. Studies on the neuraminidase of influenza virus. II. Additional properties of the enzymes from the Asian and PR8 strains. Arch. Biochem. Biophys. **103:**424–430.

Ray, G. C., and L. L. Minnich. 1987. Efficiency of immunofluorescence for rapid detection of common respiratory viruses. J. Clin. Microbiol. **25:**355–357.

Raymond, F. L., A. J. Caton, N. J. Cox, A. P. Kendal, and G. G. Brownlee. 1986. The antigenicity and evolution of influenza H1 haemagglutinin from 1950–1957 and 1977–1983: two pathways from one gene. Virology **148:**275–287.

Reichelderfer, P. S., K. A. Kappus, and A. P. Kendal. 1987. Economical laboratory support system for influenza virus surveillance. J. Clin. Microbiol. **25:**947–948.

Reimer, C. B., D. J. Phillips, C. H. Aloisio, D. D. Moore, G. G. Galland, T. W. Wells, C. M. Black, and J. S. McDougal. 1984. Evaluation of thirty-one mouse monoclonal antibodies to human IgG epitopes. Hybridoma **3:**263–275.

Richardson, C. D., A. Scheid, and P. W. Choppin. 1980. Specific inhibition of paramyxovirus and myxovirus replication by oligopeptides with amino acid sequences similar to those at the N-terminal of the F, or HA2 viral polypeptides. Virology **105:**205–222.

Richman, D. D., D. H. Cleveland, D. C. Redfield, M. N. Oxman, and G. M. Wahl. 1984. Rapid viral diagnosis. J. Infect. Dis. **149:**298–310.

Robertson, J. S., C. W. Naeve, R. G. Webster, J. S. Bootman, R. Newman, and G. Schild. 1985. Alterations in the haemagglutinin associated with adaptation of influenza B virus to growth in eggs. Virology **143:**166–174.

Rogers, G. N., J. C. Paulson, R. S. Daniels, J. J. Skehel, I. A. Wilson, and D. C. Wiley. 1983. Single amino acid substitutions in the influenza haemagglutinin change the specificity of receptor binding. Nature **304:**76–78.

Rott, R., and H. D. Klenk. 1986. Pathogenicity of influenza virus in model systems, p. 53–62. In A. P. Kendal and P. A. Patriarca, (eds.), Options for the control of influenza, Alan R. Liss, New York.

Rott, R., and C. Scholtissek. 1970. Specific inhibition of influenza replication by a-amanitin. Nature **228:**56.

Sarkkinen, H. K., P. E. Halonen, and A. A. Salmi. 1981. Detection of influenza A virus by radioimmunoassay and enzyme-immunoassay from nasopharyngeal specimens. J. Med. Virol. **7:**213–220.

Schmidt, N. J., M. Ota, D. Gallo, and V. L. Fox. 1982. Monoclonal antibodies for rapid, strain-specific identification of influenza virus isolates. J. Clin. Microbiol. **16:**763–765.

Scholtissek, C., W. Rohde, V. vopn Hoyningen, and R. Rott. 1978. On the origin of the human influenza virus subtypes H2N2. Virology **87:**13–20.

Scholtissek, C. 1983. Enetic relatedness of influenza viruses, p. 99–126. In P. Palese and D. W. Kingsbury (eds.), Genetics of influenza viruses, Springer-Verlag, New York.

Shalit, I., P. A. McKee, H. Beauchamp, and J. L. Waner. 1985. Comparison of polyclonal antiserum versus monoclonal antibodies for the rapid diagnosis of influenza A virus infections by immunofluorescence in clinical specimens. J. Clin. Microbiol. **22:**877–879.

Shaw, M. W., P. W. Choppin, and R. A. Lamb. 1983. A previously unrecognized influenza B virus glycoprotein from a bicistronic mRNA that also encodes the viral neuraminidase. Proc. Natl. Acad. Sci. USA **80:**4879–4883.

Shortridge, K. F., R. G. Webster, C. K. Butterfield, and C. H. Campbell. 1977. Persistence of Hong Kong influenza virus variants in pigs. Science **196:**1454–1455.

Skehel, J. J., P. M. Bayley, E. M. Brown, S. R. Martin, M. D. Waterfield, J. M. White, I. A. Wilson, and D. C. Wiley. 1982. Changes in the conformation of influenza virus haemagglutinin at the pH optimum of virus-mediated membrane fusion. Proc. Nat. Acad. Sci. USA **79:**968–972.

Soini, E. 1984. Pulsed light, time-resolved fluorometric immunoassay, p. 197–208. *In* C. A. Bizollon (ed.), Monoclonal antibodies and new trends in immunoassays, Elsevier Science Publishers, New York.

Stuart-Harris, C. H., G. C. Schild, and J. S. Oxford. 1985. Influenza. The viruses and the diseases. Edward Arnold, Baltimore, MD.

Stuart-Harris, C. N., G. C. Schild, and J. S. Oxford. 1985. The chemoprophylaxis and chemotherapy of influenza, p. 222–243. *In* Influenza. The viruses and the disease. Edward Arnold, Baltimore, MD.

Tobita, K., A. Sugiura, C. Enomoto, and M. Furuyama. 1975. Plaque assay and primary isolation of influenza A viruses in an established line of canine kidney cells (MDCK) in the presence of trypsin. Med. Microbiol. Immunol. **162:**9–14.

Turner, R., J. L. Lathey, L. P. Van Voris, and R. B. Belshe. 1982. Serological diagnosis of influenza B virus infection: comparison of an enzyme-linked immunosorbent assay and the haemagglutination inhibition test. J. Clin. Microbiol. **15:**824–829.

United States Public Health Service. 1961. International conference on Asian influenza. Am. Rev. Respir. Dis. **83:**part 2.

Van Voris, L. P., R. F. Betts, F. G. Hayden, W. H. Christmas, and R. G. Douglas, Jr. 1981. Successful treatment of naturally occurring influenza A/USSR/90/77 H1N1. J. Am. Med. Assoc. **245:**1128–1131.

Walls, H. H., M. W. Harmon, J. J. Slagle, C. Stocksdale, and A. P. Kendal. 1986a. Characterization and evaluation of monoclonal antibodies developed for typing influenza A and influenza B viruses. J. Clin. Microbiol. **23:**240–245.

Walls, H. H., K. H. Johannson, M. W. Harmon, P. E. Halonen, and A. P. Kendal. 1986b. Time-resolved fluoroimmunoassay with monoclonal antibodies for rapid diagnosis of influenza infections. J. Clin. Microbiol. **24:**907–912.

Webster, R. G., W. G. Laver, and G. M. Air. 1983. Antigenic variation among type A influenza viruses, p. 127–168. *In* P. Palese and D. W. Kingsbury (eds.), Genetics of influenza viruses, Springer-Verlag, New York.

Wiley, D. C., I. A. Wilson, and J. J. Skehel. 1981. Structural identification of the antibody-binding sites of Hong Kong influenza haemagglutinin and their involvement in antigenic variation. Nature **289:**373–378.

World Health Organization. 1980. A revision of the system of nomenclature for influenza viruses: a WHO memorandum. Bull. WHO **58:**585–591.

World Health Organization. 1969. International Conference on Hong Kong influenza. Bull. WHO **41:**335–748.

Yolken, R. H., and P. J. Stopa. 1979. Enzyme-linked fluorescence assay: ultrasensitive solid-phase assay for detection of human rotavirus. J. Clin. Microbiol. **10:**317–321.

Yolken, R. H., V. M. Torsch, R. Berg, B. R. Murphy, and Y. C. Lee. 1980. Fluorometric assay for measurement of viral neuraminidase. Application to the rapid detection of influenza virus in nasal wash specimens. J. Infect. Dis. **142:**516–523.

Younkin, S. W., R. F. Betts, F. R. Roth, and R. G. Douglas, Jr. 1983. Reduction in fever and symptoms in young adults with influenza A/Brazil/78 H1N1 infection after treatment with aspirin or amantadine. Antimicrob. Agents Chemother. **23:**577–582.

Zhdanov, V. M. 1986. Live influenza vaccines in USSR: development of studies and practical application, p. 193–205. *In* A. P. Kendal and P. A. Patriarca (eds.), Options for the control of influenza, Alan R. Liss, New York.

Zhirnov, O. P., A. V. Ovcharenko, and A. G. Bukrinskaya. 1985. Suppression of influenza virus replication in infected mice by protease inhibitors. J. Gen. Virol. **65:**191–196.

CHAPTER 32

Bunyaviridae: The Bunyaviruses

CHARLES H. CALISHER and ROBERT E. SHOPE

Diseases: Hemorrhagic fever with renal syndrome, Rift Valley fever, Crimean-Congo hemorrhagic fever, Nairobi sheep disease, Oropouche, LaCrosse encephalitis, Akabane, sandfly fever, and undifferentiated fevers.

Etiologic Agents: Hantaan, Rift Valley fever, Crimean-Congo hemorrhagic fever, Nairobi sheep disease, Oropouche, LaCrosse, Akabane, sandfly Naples, sandfly Sicilian, and other viruses (virus distribution related to distribution of arthropod vector and mammalian reservoirs.

Source: Infected mosquitoes, ticks, or mammals, depending on virus.

Clinical Manifestations: Undifferentiated fevers, fever and rash, hemorrhagic fever, hemorrhagic fever with renal syndrome, encephalitis, optic retinitis, teratogenesis.

Pathology: Dependent on etiologic agent, but rather typical of specific infection (i.e., encephalitis, fever, rash, etc.).

Laboratory Diagnosis: Virus isolation and serologic tests for antibody (IgM antibody capture ELISA, neutralization).

Epidemiology: Diseases occur when humans or animals are in contact with infected vertebrates or when fed on by infected arthropods in specific geographic areas.

Treatment: Symptomatic.

Prevention and Control: Prevention of contact (removal from area, insecticides, screening, rodent-proofing) between susceptible human or animal and source of virus (infected arthropod or vertebrate), vaccination.

Introduction

During studies of yellow fever in Uganda in 1943, Smithburn et al., (1946) isolated Bunyamwera virus from pooled *Aedes* sp. mosquitoes. Casals and Whitman (1960) subsequently demonstrated that this virus is related antigenically to Wyeomyia (Colombia), Cache Valley (Utah), and Kairi (Trinidad) viruses. Relationships among these viruses were based largely on studies of their complement-fixing (CF) antigens. Further studies of these as well as Germiston (South Africa), Batai (Malaysia), Guaroa (Colombia), and other related viruses from Brazil showed that although closely related by CF, these viruses are distinguishable by neutralization (N) tests. As additional isolations of arboviruses were made, workers at The Rockefeller Foundation detected relationships between and among many of the viruses. For each two or more viruses shown to be antigenically related (usually by hemagglutination-inhibition [HI] or CF tests), a virus serogroup was established. Within about a 10-year period, not only had the Bunyamwera serogroup been recognized, but the groups C, Guama, California, Capim, Anopheles A, Simbu, Bwamba, Patois, Koongol, Tete, and other serogroups also were distinguished (Casals and Whitman, 1961; Whitman and Casals, 1961; Whitman and Shope, 1962; Theiler and Downs, 1973).

As the number of serogroups increased, a need to provide some order became apparent and definitions were proposed for classification terms. We use these terms as defined herein. A *serogroup* is two or more viruses, distinct from each other by quantitative serologic criteria (fourfold or greater differences between homologous and heterologous titers of both sera) in one or more tests, but related to another virus of the serogroup by some serologic method. Arboviruses that are very closely related, but distinct from each other, constitute an antigenic *complex*. Individual agents, antigenically related but easily separable (fourfold or greater differences between homologous and heterologous titers of both sera) by the N test, are considered *viruses* or *types*. *Subtypes* are virus isolates separable from the type virus by at least a fourfold difference between the homologous and heterologous titers of one, but not both, of the two sera tested. *Varieties* are those isolates differentiable only by the application of special tests or reagents (kinetic HI, monoclonal antibody assays, and the like). In present practice, the first discovered virus of a newly recognized serogroup lends its name to the antigenic cluster.

Through the efforts of Casals and others of The Rockefeller Foundation, low-level, often one-way relationships between individual viruses in two or more serogroups were demonstrated. For example, as determined by CF tests, Guaroa virus is a member of the Bunyamwera serogroup. However, it is related to certain members of the California serogroup by both HI and N tests. Because of the intergroup cross-reactivity among members of Group C, Guama, Capim, California, and Bunyamwera serogroups, Casals suggested the establishment of what he called the Bunyamwera supergroup (Casals, 1963). Subsequent electron microscopic studies supported and extended Casals' concept, in that the viruses of the supergroup could not be distinguished by size, morphology, or morphogenesis in infected cells (Murphy et al., 1973). Other viruses, belonging to other serogroups, such as the phlebotomus fever and Uukuniemi serogroup viruses, morphologically resembled the supergroup members, but antigen sharing could not be demonstrated among members of these serogroups.

The term "arbovirus" denotes a virus maintained in nature by a biologic (propagative) transmission cycle between susceptible vertebrate hosts and hematophagous arthropods. The International Catalogue of Arboviruses lists more than 500 viruses, most with antigenic relationships to others, but some that are antigenically, taxonomically, and/or ecologically distinct (Karabatsos, 1985). Many of the viruses registered in this catalogue do not meet the definition of an arbovirus, some because information is lacking regarding transmission by arthropods and other catalogued viruses because they certainly are not arthropod-borne.

The term "arbovirus" is used here to denote an ecologic description, but in keeping with newer classification, these viruses will be referred to by universal taxonomic descriptions. Differences in antigenic, morphologic, biochemical, and genetic characteristics are used to separate the arboviruses into families, genera, serogroups, complexes, viruses, subtypes, and varieties in an increasing order of relatedness. Most recently, molecular analyses have substantiated the previous antigenic classification schemes, and a clear view of the taxonomy of the arboviruses has emerged.

The Viruses

Members of the family *Bunyaviridae* have certain characteristics in common: single-stranded RNA, three RNA segments whose ends are hydrogen-bonded so that the molecules are circular; and spherical or oval enveloped 90 to 100-nm diameter virions. All members of the family so far tested are acid (pH 3), lipid solvents (ether, chloroform), and detergent (sodium deoxycholate) labile; heat (56° for 15 to 30 min) labile; and formalin, 70% ethanol, 5% iodine, and ultraviolet (UV) irradiation labile (Karabatsos, 1985). Molecular weight of viruses in the family is 300 to 400 × 10^6. Sedimentation coefficients of representative members are between 400 and 500 S and bouyant densities are about 1.18 to 1.20 g/ml (Obijeski and Murphy, 1977). Uukuniemi virus has been determined to contain about 2% RNA, 58% protein, 33% lipid, and 7% carbohydrate (Obijeski and Murphy, 1977), but the composition of most other members of the family is not known.

The virions consist of a unit membrane envelope with what have been termed "fuzzy" (bunyaviruses) or "ordered" (uukuviruses) projections surrounding a rather unstructured interior from which a helical, 2.5-nm wide nucleocapsid can be extracted (von Bonsdorff et al., 1969; Murphy et al., 1973). Virus particles contain a transcriptase enzyme. Constituent synthesis takes place in the cytoplasm, and morphogenesis occurs without prior core formation, by budding directly into the Golgi complex and vesicles of infected cells (Murphy et al., 1973). No enzymatic function has been associated with the envelope glycoproteins of members of the family. No evidence has been obtained for reassortment of genes between members of the different genera within the family. However, reassortment has been detected between viruses belonging to the same genus and closely related antigenically, that is restricted to RNA segment reassortment between closely related viruses. Naturally occurring reassortants have been obtained from genotype analyses of field isolates (Klimas et al., 1981; Ushijima et al., 1981).

At least four genera of viruses have been distinguished within the family Bunyaviridae: *Bunyavirus, Phlebovirus, Nairovirus,* and *Uukuvirus* (Bishop et al., 1980). Recently, a fifth genus, *Hantavirus,* has been proposed for this family of biologically diverse viruses (Schmaljohn and Dalrymple, 1983). The prominent role of The Rockefeller Foundation workers and the remarkably perceptive predictions of Casals should not be understated. All subsequent genetic studies have borne out Casals' hypothesis that viruses shown to be interrelated by studies of their antigens are related genetically. Thus, the *Bunyavirus* genus is composed of members of the former Bunyamwera supergroup as well as more recently isolated viruses shown to be antigenically related to one or more viruses within the supergroup (Table 1); the *Phlebovirus* genus includes all members of the Phlebotomus fever serogroup (Table 2); the *Nairovirus* genus is composed of viruses belonging to at least six serogroups (Table 3); the *Uukuvirus*

TABLE 1. Viruses of the family *Bunyaviridae*, genus *Bunyavirus*

Complex	Virus (subtype) ⟨variety⟩
	(Bunyamwera Serogroup)
Bunyamwera	Bunyamwera, Batai, Calovo, Birao, Bozo, Cache Valley (Cache Valley) ⟨Cache Valley⟩ ⟨Tlacotalpan⟩ (Maguari) ⟨Maguari⟩ ⟨CbaAr426⟩ ⟨AG83-1746⟩ (Playas) (Xingu), Germiston, Ilesha, Lokern, Mboke, Ngari, Northway, Santa Rosa, Shokwe, Tensaw
Kairi	Kairi
Main Drain	Main Drain
Wyeomyia	Wyeomyia, Anhembi (Anhembi) (Iaco) (Macaua) (Sororoca) (Taiassui) (BeAr328208)
	(Anopheles A Serogroup)
Anopheles A	Anopheles A, Las Maloyas, Lukuni, Trombetas, CoAr3624, ColAn57389
Tacaiuma	Tacaiuma (Tacaiuma) ⟨SPAr2317⟩ ⟨SPAr2317⟩ ⟨Virgin River⟩ (H-32580), CoAr1071 (CoAr1071) ⟨CoAr1071⟩ ⟨CoAr3627⟩
	(Anopheles B Serogroup)
Anopheles B	Anopheles B, Boraceia
	(Bwamba Serogroup)
Bwamba	Bwamba (Bwamba) (Pongola)
	(Group C Serogroup)
Caraparu	Caraparu (Caraparu) ⟨2⟩ (Ossa), Apeu, Bruconha, Vinces
Madrid	Madrid
Marituba	Marituba (Marituba) (Murutucu) (Restan), Nepuyo (Nepuyo) ⟨Nepuyo⟩ ⟨63U11⟩ (Gumbo Limbo)
Oriboca	Oriboca (Oriboca) (Itaqui)
	(California Serogroup)
California enc.	California enc. (California enc.) (Inkoo) (LaCrosse) ⟨snowshoe hare⟩ (San Angelo) (Tahyna) ⟨Tahyna⟩ ⟨Lumbo⟩
Melao	Melao (Melao) ⟨Melao⟩ ⟨AG83-497⟩ (Jamestown Canyon) ⟨Jamestown Canyon⟩ ⟨Jerry Slough⟩ ⟨South River⟩ (Keystone) (Serra do Navio)
trivitattus	trivitattus
Guaroa	Guaroa
	(Capim Serogroup)
Capim	Capim
Guajara	Guajara (Guajara) ⟨Guajara⟩ ⟨GU71U350⟩
BushBush	BushBush (BushBush) (Benfica) ⟨GU71U344⟩, Juan Diaz
Acara	Acara, Moriche
Benevides	Benevides
	(Gamboa Serogroup)
Gamboa	Gamboa, Pueblo Viejo (75V-2621)
Alajuela	Alajuela, San Juan (San Juan) (78V-2441) (75V-2374)
	(Guama Serogroup)
Guama	Guama, Ananindeua, Mahogany Hammock, Moju
Bertioga	Bertioga, Cananeia, Guaratuba, Itimirim, Mirim
Bimiti	Bimiti
Catu	Catu
Timboteua	Timboteua
	(Koongol Serogroup)
Koongol	Koongol, Wongal
	(Minatitlan Serogroup)
Minatitlan	Minatitlan, Palestina

TABLE 1. *Continued*

Complex	Virus (subtype) ⟨variety⟩
	(Olifantsvlei Serogroup)
Olifantsvlei	Olifantsvlei (Olifantsvlei) ⟨Olifantsvlei⟩ ⟨Bobia⟩
Botambi	Botambi
	(Patois Serogroup)
Patois	Patois, Babahoyo, Shark River, Abras
Zegla	Zegla, Pahayokee
	(Simbu Serogroup)
Simbu	Simbu
Akabane	Akabane, Yaba-7
Manzanilla	Manzanilla (Manzanilla) (Ingwavuma) (Inini) (Mermet), Buttonwillow, Nola, Oropouche (Oropouche) (Facey's Paddock) (Utinga) (Utive), Sabo (Sabo) (Tinaroo)
Sathuperi	Sathuperi (Sathuperi) (Douglas)
Shamonda	Shamonda, Sango (Sango) (Peaton)
Shuni	Shuni, Aino (Aino) ⟨Aino⟩ ⟨Kaikalur⟩
Thimiri	Thimiri
	(Tete Serogroup)
Tete	Tete (Tete) (Bahig) (Matruh) (Tsuruse), Batama
	(Turlock Serogroup)
Turlock	Turlock, Umbre, Lednice
M'Poko	M'Poko (M'Poko) (Yaba-1)
	(No Serogroup Assigned)
	Kaeng Khoi

TABLE 2. Viruses of the family *Bunyaviridae*, genus *Phlebovirus*

Complex	Virus (subtype)
	(Phlebotomus Fever Serogroup)
Sandfly fever Naples	Sandfly fever Naples, Karimabad, Tehran, Toscana
Bujaru	Bujaru, Munguba
Rift Valley fever	Rift Valley fever (Rift Valley fever) (Belterra), Icoaraci
Candiru	Candiru, Alenquer, Itaituba (Itaituba) (Oriximina), Nique, Turuna
Punta Toro	Punta Toro (Punta Toro) (Buenaventura)
Frijoles	Frijoles, Joa
Chilibre	Chilibre, Cacao
Salehabad	Salehabad, Arbia
(no complex assigned)	Sandfly fever Sicilian, Aguacate, Anhanga, Arumowot, Caimito, Chagres, Corfou, Gabek Forest, Gordil, Itaporanga, Pacui, Rio Grande, Saint-Floris, Urucuri

TABLE 3. Viruses of the family *Bunyaviridae*, genus *Nairovirus*

Complex	Virus (subtype)
	(Crimean-Congo Hemorrhagic Fever Serogroup)
Crimean-Congo hemorrhagic fever	Crimean-Congo hemorrhagic fever, Hazara, Khasan
	(Dera Ghazi Khan Serogroup)
Dera Ghazi Khan	Dera Ghazi Khan, Abu Hammad, Abu Mina, Kao Shuan, Pathum Thani, Pretoria
	(Hughes Serogroup)
Hughes	Hughes, Farallon, Fraser Point, Punta Salinas, Raza, Sapphire II, Soldado, Zirqa
	(Nairobi Sheep Disease Serogroup)
Nairobi sheep disease	Nairobi sheep disease, Dugbe
	(Qalyub Serogroup)
Qalyub	Qalyub, Bandia, Omo
	(Sakhalin Serogroup)
Sakhalin	Sakhalin (Tillamook), Kachemak Bay, Clo Mor, Avalon, Taggert

TABLE 4. Viruses of the family *Bunyaviridae*, genus *Uukuvirus*

Complex	Virus (subtype)
	(Uukuniemi Serogroup)
Uukuniemi	Uukuniemi (Uukuniemi) (Oceanside), Grand Arbaud, Manawa, Murre, Ponteves, Precarious Point, Zaliv Terpeniya, EgAn-1825-61, Fin V-707, UK FT/254

TABLE 5. Viruses of the family *Bunyaviridae*, genus *Hantavirus*

Complex	Virus (subtype) (variety)
	(Hantaan Serogroup)
Hantaan	Hantaan, Seoul (Seoul) (Tchoupitoulas) (Girard Point) (Sapporo Rat), Prospect Hill
Puumala	Puumala

genus includes all members of the Uukuniemi sero-group (Table 4), and the *Hantavirus* genus includes all members of the Hantaan serogroup (Table 5). Thirty-five other viruses possess morphologic, mor-phogenetic, or other properties in common with vi-ruses of the family *Bunyaviridae*. However, these 35 have not been characterized sufficiently to warrant placement in one of the recognized genera; they are listed in Table 6.

Among the nearly 300 viruses that have been as-signed to the family *Bunyaviridae* are those causing Rift Valley fever (Daubney et al., 1931), phleboto-mus fever (Taussig, 1905), Crimean-Congo hemor-rhagic fever (Hoogstraal, 1979), hemorrhagic fever with renal syndrome (Gajdusek, 1953), and Nairobi sheep disease (Montgomery, 1917). Other members of the family have been associated primarily with encephalitis (*Bunyavirus*, California serogroup) or with febrile diseases (*Bunyavirus*, serogroups C, Guama, Bunyamwera, Bwamba, Tataguine, Simbu). Akabane virus is teratogenic in sheep and cattle, causing arthrogryposis and anencephaly (Kurogi et al., 1976). Of course, most of the viruses in the fam-ily have not been associated with any illness of hu-mans or livestock. Those that do cause disease, spe-cifically Rift Valley fever, the phlebotomus fevers,

TABLE 6. Viruses of the family *Bunyaviridae*, genus unassigned

Complex	Virus (subtype)
	(Bakau Serogroup)
Bakau	Bakau, Ketapang
	(Bhanja Serogroup)
Bhanja	Bhanja, Kismayo
	(Kaisodi Serogroup)
Kaisodi	Kaisodi, Silverwater, Lanjan
	(Mapputta Serogroup)
Mapputta	Mapputta, Maprik (Maprik) (GanGan), Trubanaman
	(Matariya Serogroup)
Matariya	Matariya, Burg el Arab, Garba
	(Nyando Serogroup)
Nyando	Nyando, Eretmapodites-147
	(Resistencia Serogroup)
Resistencia	Resistencia, Barranqueras, Antequera
	(Upolu Serogroup)
Upolu	Upolu, Aransas Bay
	(No Serogroup Assigned)
	Bangui, Belmont, Bobaya, Caddo Canyons, Enseada, Kowanyama, Lone Star, Pacora, Razdan, Sunday Canyon, Tamdy, Tataguine, Wanowrie, Witwatersrand

Crimean-Congo hemorrhagic fever, hemorrhagic fever with renal syndrome and similar diseases caused by related viruses, arthrogryposis, and Nairobi sheep disease, are pathogens of great epidemiologic and economic significance.

A case in point is Rift Valley fever (genus *Phlebovirus*). Until 1977 this disease was limited geographically to sub-Saharan Africa where it circulated in an enzootic-epizootic cycle. Before 1977 only four human fatalities had been ascribed to this disease, although human infections were observed during some epizootics, and the disease was described as a self-limited dengue-like illness when it occurred in laboratory workers, field investigators, or animal handlers. In 1977 an explosive epidemic of the disease was demonstrated in the Nile Valley and the Nile Delta areas of Egypt, involving more than 200,000 humans, with 600 deaths (Meegan, 1979). Widespread and severe morbidity and excessive mortality attributed to Rift Valley fever virus was seen in a variety of domestic animals and livestock (Ali and Kamel, 1978).

Another example is a tick-borne hemorrhagic fever, which has been recognized in Central Asia since the 12th Century and later in the Balkans (Gajdusek, 1953). This disease was seen in the Crimea shortly after World War II when farmers and soldiers clearing and cultivating land in this war-devastated region became sick. The disease was called Crimean hemorrhagic fever, and more than 200 cases were documented in that outbreak. It is known to occur from southern Africa, throughout sub-Saharan Africa (Congo hemorrhagic fever), eastern Europe, the Middle East, and Asia. Its geographic distribution parallels roughly the distribution of the principal arthropod vector of Crimean-Congo hemorrhagic fever virus (genus *Nairovirus*), ticks of the genus *Hyalomma*. Although the total number of recognized cases is not large and the widespread geographic distribution of cases is continuous but focal, the significance of this disease in regard to agricultural workers, nosocomial infections, and military personnel makes it noteworthy.

Hemorrhagic fever with renal syndrome (HFRS), caused by members of the Hantaan serogroup (genus *Hantavirus*), actually is a complex of diseases. Occurring in Europe and Asia, it is known by the names nephropathia epidemica, epidemic hemorrhagic fever, Korean hemorrhagic fever, or simply "mild" or "severe" HFRS. The disease was known in ancient China and thousands of cases occurred in military personnel in Korea during the early 1950s. However, it is only recently that the etiologic agents have been isolated and specific diagnoses could be made. The essentially worldwide distribution of the hantaviruses is now causing a flurry of scientific activity.

The classification of diseases caused by viruses belonging to the family *Bunyaviridae* roughly parallels, with a few exceptions, the classification of the viruses causing these diseases (Table 7). Bunya-

TABLE 7. Some clinical syndromes and their associated viruses in the family *Bunyaviridae*

Undifferentiated fever (human)	
Africa	Germiston, Shuni, Nairobi sheep disease, Bhanja, Dugbe, Kasokero, sandfly fevers
Europe	Bhanja, sandfly fevers
Asia	Shuni, Nairobi Sheep disease, Bhanja
Americas	Wyeomyia, Guaroa, Apeu, Caraparu, Itaqui, Madrid, Marituba, Murutucu, Oriboca, Ossa, Guama, Catu, Oropouche
Fever and rash (human)	
Africa	Bunyamwera, Ilesha, Bwamba, Tataguine, Bangui
Hemorrhagic fever (human)	
Africa	Rift Valley fever (also causes hepatitis and abortion in sheep and cattle), Crimean-Congo hemorrhagic fever
Europe	Crimean-Congo hemorrhagic fever
Asia	Crimean-Congo hemorrhagic fever
Hemorrhagic fever with renal syndrome (human)	
Europe	Puumala
Asia	Hantaan, Seoul
Encephalitis (human)	
Africa	Rift Valley fever
North America	LaCrosse, California encephalitis, snowshoe hare, Jamestown Canyon
Optic retinitis (human)	
Africa	Rift Valley fever
Teratogenesis (sheep and cattle)	
Australia, Asia, Israel	Akabane

viruses generally cause either encephalitis with fevers or fevers with rash in people; phleboviruses cause fevers, hemorrhagic fevers, or encephalitis; nairoviruses cause a variety of clinical illnesses ranging from undifferentiated fevers to hemorrhagic fevers; hantaviruses cause mild to severe hemorrhagic fever with renal involvement; and uukuviruses are not known to cause human illness.

Family *Bunyaviridae,* Genus *Bunyavirus*

Members of the genus *Bunyavirus* (Table 1) possess negative-sense RNA replication strategy. Total molecular weight of the RNA segments is 4.78 to 5.9 $\times$ 10^6. The 3' terminal sequence is UCAUCACAUG. Two of the four structural proteins are glycosylated (G1 and G2) with molecular weights of 108 to 120 and 29 to 41 $\times$ 10^3, respectively. Another is a nucleocapsid protein (N), molecular weight 19 to 25 $\times$ 10^3. A minor large protein (L) with a molecular weight of 145 to 200 $\times$ 10^3 also has been recognized. Within this genus, 16 serogroups, containing more than 150 viruses, are known. The bunyaviruses are found worldwide; are transmitted by mosquitoes and culicoids; have as their principal vertebrate hosts rodents and other small mammals, primates, birds, or ungulates; and usually exist in silent sylvatic transmission cycles. At least 30 of these viruses or their subtypes and varieties have been reported to cause disease in humans or animals of veterinary importance and three (Oropouche, LaCrosse, and Bwamba viruses) have caused epidemics in humans.

Bunyamwera serogroup viruses have not been reported from Australia, and only Batai virus has been isolated in Asia and Europe, probably because birds and migrating mammals are not involved in the mosquito–rodent cycles of these viruses in nature. Bunyamwera serogroup viruses are commonly isolated in the Americas and Africa (Karabatsos, 1985). The relative insularity of the North American Bunyamwera serogroup viruses may in some way be related to the distributions of their principal vertebrate hosts and the competence of their vectors. It is interesting to speculate that Batai and Northway viruses may represent phylogenetic links between Bunyamwera serogroup viruses in Africa and in North America. Rabbits, for example, are viremic or produce antibody after inoculation with Batai, Northway, Tensaw, or Cache Valley viruses, whereas horses and perhaps hares are not susceptible to these viruses (Karabatsos, 1985). Two other Bunyamwera serogroup viruses from North America, Lokern, and Main Drain viruses, replicate well in hares and have been isolated most frequently from *Culicoides* sp., not mosquitoes.

Restriction of a virus to one specific vertebrate-vector pairing with defined geographic distribution may lead to natural isolation and genetic stability in divergent evolution. African Bunyamwera serogroup viruses have been found in distinct or overlapping geographic areas and ecosystems, but the South American members appear to coexist in horizontally or vertically contiguous, but not identical, ecosystems. One factor influencing the separate maintenance of sympatric, closely related serotypes may be differences in vector susceptibility. Woodall (1979) has suggested that the group C bunyavirus Itaqui is transmitted mainly by *Culex vomerifer,* a species apparently resistant to infection by Oriboca virus. Apeu and Marituba, also group C viruses, have been isolated from marsupials but not rodents, whereas Caraparu and Murutucu viruses, present in the same area of Brazil, have been isolated frequently from rodents. Cross-protection tests in monkeys indicate that immunity to one group C virus confers protection against another. Thus, coexistence of group C viruses is probably explained by their separate mosquito–small mammal cycles. The application of more sophisticated analyses will be necessary if we are to confirm the reasons for the coexistence of such closely related viruses in certain areas.

Like the Bunyamwera serogroup members, bunyaviruses of the California serogroup are transmitted between small mammals by mosquitoes, principally of the genus *Aedes.* The vector and host relationships of each virus appear to be quite restricted, possibly as a consequence of transovarial transmission in the mosquito (Watts et al., 1974). Thus, California serogroup viruses are geographically distributed in relation to the range of their vectors and hosts (Sudia et al., 1971). For example, Keystone virus has been isolated from rabbits and from *Sigmodon hispidus* (cotton rats) from Georgia and Florida. Because *S. hispidus* is only infrequently found outside the southeastern United States, association of these natural hosts with the implicated principal mosquito vectors *Aedes atlanticus* and *Aedes infirmatus* limits the distribution of Keystone virus. With other mosquitoes and other mammalian hosts, similar associations could be shown for other members of the California serogroup. Such postulations have been made for all of the North American members of this serogroup (Sudia et al., 1971).

The California and group C viruses are, in many respects, quite similar biologically to those of the Bunyamwera serogroup; mammal-feeding mosquitoes transmit virus to small mammal hosts within geographic foci determined largely by the distribution and limited movements of the vertebrate hosts. Potentially competitive serotypes are excluded by natural, selective disadvantages imposed by vector-host restriction and ability to induce cross-protection. RNA segment reassortment, when it occurs, takes place only between closely related serotypes (Bishop and Shope, 1979). Shope and Causey (1962) have shown that six of the group C bunyaviruses form three indistinguishable antigen pairs in CF tests. Karabatsos and Shope (1979) extended these studies and suggested that, because the CF antigen common to the members of the pairs is not an antigen shared by all the members of the serogroup, "pairing" might have resulted from natural genetic recombination.

The absence from Australia of known Bunyamwera and California serogroup viruses, for which placental mammals are the principal vertebrate hosts, may indicate that these viruses arose somewhat later than the drift of the Australian continent, perhaps fewer than 10 million years ago.

Simbu, Tete, and Turlock serogroup viruses have been isolated from resident and migrating birds, which may account for their relatively worldwide, hemisphere-wide, or continent-wide distributions. Other bunyaviruses are confined to single continents because they replicate in mosquitoes and rodents, marsupials, bats, and other mammals, but not in birds. Gamboa serogroup viruses are an example of a third type. They replicate in and are transovarially transmitted by *Aedeomyia squamipennis* mosquitoes, which feed principally on birds. However, the geographic distribution of these viruses appears to be limited to the distribution of the arthropod vector, not the vertebrate host.

Family *Bunyaviridae,* Genus *Phlebovirus*

Members of the genus *Phlebovirus* possess an ambisense RNA replication strategy. Total molecular weight of the RNA segments is 5.1 to 5.8 × 10^6. The 3' terminal sequence is UGUGUUUCG. Two of the four structural proteins are glycosylated (G1 and G2) with molecular weights of 55 to 70 and 50 to 60 × 10^3, respectively. Another is a nucleocapsid protein (N), molecular weight 20 to 30 × 10^3. A minor large protein (L) with molecular weight of 145 to 200 × 10^3 also has been recognized.

A single serogroup, the Phlebotomus fever serogroup, constitutes this genus. Of the 37 members, none occur in Australia, one has been found in the United States, three occur only in Europe, three only in Asia, five only in Africa, and two in Africa, Asia, and Europe (Table 2). The last, sandfly fever Sicilian and sandfly fever Naples viruses, have been responsible for epidemics in these areas. Rift Valley fever virus, thus far limited to the African continent, is widespread there and, as mentioned, has caused extensive and serious epidemics and epizootics. Many

phleboviruses occur focally in South America or Central America, suggesting a relationship between arboreal or ground-dwelling mammalian hosts and virus distribution, or simple restriction of distribution in parallel to the restricted distribution of the arthropod vector.

Family *Bunyaviridae,* Genus *Nairovirus*

RNA replication strategy of viruses belonging to the genus *Nairovirus* is unknown. However, total molecular weight of the three RNA segments is 6.2 to 7.5 × 10^6. The 3' terminal sequence is AGAGUUUCU. Two of the four structural proteins are glycosylated (G1 and G2) with molecular weights of 72 to 84 and 30 to 40 × 10^3, respectively. Another is a nucleocapsid protein (N), molecular weight 48 to 54 × 10^3. A minor large protein (L) with a molecular weight of 145 to 200 × 10^3 also has been recognized.

Six serogroups, containing 28 viruses and subtypes, comprise this genus (Table 3). All the nairoviruses have been isolated from ticks; Dugbe virus and the Ganjam strain of Nairobi sheep disease viruses of the Nairobi sheep disease serogroup also have been isolated from culicine mosquitoes; strains of Dugbe, Nairobi sheep disease, and Crimean-Congo hemorrhagic fever viruses have been obtained from *Culicoides* sp. as well. With certain notable exceptions (Nairobi sheep disease and Crimean-Congo hemorrhagic fever viruses), little is known of the vertebrate hosts of the nairoviruses; however, at least one member each of the Sakhalin and Hughes serogroups have been isolated from seabirds, probably accounting for the relatively widespread distributions of these viruses.

Family *Bunyaviridae,* Genus *Uukuvirus*

RNA replication strategy of viruses belonging to the genus *Uukuvirus* is unknown. However, total molecular weight of the three RNA segments is 3.4 to 4.4 × 10^6. The 3' terminal sequence is UGUGUUUCUGGAG. Two of the four structural proteins are glycosylated (G1 and G2) with molecular weights of 70 to 75 and 65 to 70 × 10^3, respectively. Another is a nucleocapsid protein (N), molecular weight 20 to 25 × 10^3. A minor large protein (L) with a molecular weight of 180 to 200 × 10^3 also has been recognized.

The uukuviruses have been recovered primarily from ticks and use birds as their principal vertebrate hosts. None of these viruses is known to cause disease in humans or livestock. The genus contains one serogroup with 11 members (Table 4). With the exception of Uukuniemi virus, which is distributed throughout Europe, viruses belonging to this genus (serogroup) appear to be focally distributed in areas of Europe, Asia, Africa, North America, and Australia (Macquarie Island), where shorebirds and sea birds nest and otherwise congregate.

Family *Bunyaviridae*, Genus *Hantavirus*

Members of the proposed genus *Hantavirus* possess negative-sense RNA replication strategy. Total molecular weight of the RNA segments is about 4.5×10^6. The 3' terminal sequence is AUCAUCAUCUG. Two of the four structural proteins are glycosylated (G1 and G2) with molecular weights of 68 to 72 and 54 to 60 $\times 10^3$, respectively. Another is a nucleocapsid protein (N), molecular weight 50 to 53 $\times 10^3$. A minor large protein with a molecular weight of about 200×10^3 also has been recognized.

At present, only one serogroup comprises this proposed genus. The seven Hantaan serogroup viruses, relatively recently described, have been found in Asia, North America, South America, or Europe (Table 5). It is possible that one or more members of this serogroup will be found in Australia and that many more members of the serogroup will be isolated in the near future. Information currently available suggests that the host associations of the hantaviruses are species specific. Hantaan virus has been recovered from *Apodemus* sp. field mice; Seoul, Tchoupitoulas, Girard Point, and Sapporo Rat viruses from rats; Prospect Hill virus from *Microtus* sp. meadow voles; and Puumala virus from *Clethrionomys* sp. bank voles. Several hantavirus isolates have not been definitively typed, so these apparent host and geographic associations may or may not be borne out.

Bunyavirus-Like Viruses

Thirty-five viruses have morphologic or molecular characteristics in common with members of the family *Bunyaviridae;* usually the common denominator is morphology. In the absence of adequate molecular and genetic studies or antigenic relatedness with a recognized member of any of the serogroups within the family, these bunyavirus-like viruses have been placed provisionally within the family but are denoted only as "possible members of the family." Included among these are 21 viruses belonging to 8 serogroups and 14 ungrouped viruses (Table 6). Viruses of the Bhanja, Kaisodi, and Upolu serogroups are principally tick-borne, whereas those of the Ba-

kau, Mapputta, Nyando, and Resistencia serogroups appear to be principally mosquito-borne. Matariya serogroup viruses have been isolated only from birds. Knowledge of the geographic distributions of all these viruses is limited, as is that of the ungrouped mosquito-borne and ungrouped tick-borne bunyavirus-like viruses. Certain of these viruses are known to be pathogenic to humans and have widespread geographic distributions, thus studies of the taxonomy and classifications of these viruses have potential medical importance.

Pathogenesis and Pathology

The pathogenetic mechanisms and accompanying pathology of viruses belonging to the family *Bunyaviridae* can best be discussed by clinical classification (febrile illness with or without rash, encephalitides, and hemorrhagic fevers), rather than by virus classification (genera). Most viruses in the family cause febrile illnesses or no disease at all in humans. It is only the rare or exotic member of the family that causes hemorrhagic fever or other life-threatening syndromes; however, it is precisely this potential severity that creates medical and political concern and prompts efforts toward prevention and treatment.

LaCrosse virus (genus *Bunyavirus*, California serogroup) may be the most well-studied member of the family, insofar as virus ecology, disease etiology, and pathogenesis are concerned (Calisher and Thompson, 1983). In the two fatal human cases studied, neuronal and glial damage, perivascular cuffing of capillaries and venules, and cerebral edema were noted on pathologic examination. Although not quantitatively different from pathologic findings in other viral encephalitides, their distribution may be distinctive, being most focal in cortical gray matter of frontal, temporal, and parietal lobes, basal nuclei, midbrain, and pons, with other regions spared (Kalfayan, 1983). Brain biopsy in a single case showed congestion, margination of polymorphonuclear leukocytes, focal neuronal necrosis, and endothelial cell swelling (Balfour et al., 1973).

The pathogenesis of encephalitis, that is viremia followed by antibody production and encephalitis, is significant to the success of early diagnosis where antibody (IgM) is detected in a single specimen at admission to hospital. Rift Valley fever is usually an uncomplicated, temporarily prostrating febrile illness in humans, producing hemorrhagic manifestations, encephalitis, or eye lesions in a small proportion of persons. The most prominent pathologic finding in fatal cases of Rift Valley fever with hemorrhagic manifestations is hepatic necrosis, the probable cause of hepatic failure often seen in patients who are dying in shock, with jaundice, gastrointestinal and mucous membrane bleeding, and disseminated vascular damage.

Crimean-Congo hemorrhagic fever virus has not been shown to cause hemorrhagic fever in laboratory animals. In humans the primary pathophysiologic event appears to be leakage of plasma and erythrocytes through the vascular endothelium (Karmysheva et al., 1973). At autopsy, histopathologic findings include edema, focal necrosis and hemorrhage, and vascular congestion of the heart, brain, and liver.

In hemorrhagic fever with renal syndrome, vascular instability is the basic lesion. Capillary leakage occurs without inflammation but with significant loss of protein colloid. Autopsy findings include gross hemorrhagic necrosis of the kidney, pituitary, and right auricle. Microscopic changes found in the kidneys of patients who die with this disease depend on the stage of illness at death (Oliver and MacDowell, 1957). Those who die in the late febrile or early hypotensive stages show congestion in the subcortical medullary vessels without obvious damage to the renal tubules and without obstructive changes in the tubules. In the hypotensive stage, findings include intertubular congestion in the corticomedulla and swelling of the proximal convolutions with hydration and vacuolation of the epithelial cells. Later in the course of disease, progressive damage to the tubules is observed. The lumens are compressed and filled with desquamated cells and hyaline material. During the oliguric stage, hemorrhage develops in the congested zones and tubular necrosis occurs with coagulated proteins in the lumens of tubules. In patients who die during the diuretic stage, fibrosis and epithelial proliferation have occurred (Shope, 1985).

Clinical Features

The mechanism of transmission of viruses of the family *Bunyaviridae* is intimately related to the arthropod or vertebrate from which the virus is acquired. That is, the individual becomes infected by the bite of an infected arthropod or by direct or aerosol contact with virus excreted by a reservoir vertebrate. Depending on how successful the arthropod vector has been in feeding (subcutaneous introduction of virus while probing for a capillary or capillary introduction when feeding successfully), or the viremic titer in the vertebrate serving as the source of virus, virus replication is initiated as soon as the virus comes in contact with susceptible cells in the host. After an incubation period of 3 to 6 days in bunyavirus, phlebovirus, and probably nairovirus infections, to as long as 5 weeks in hantavirus infections.

Many bunyaviruses and phleboviruses cause undifferentiated febrile diseases with or without rash

(Table 7). The illness typically has an abrupt onset with accompanying chills and fever. Headache, with or without photophobia; retroorbital pain; myalgia; arthralgia; asthenia; nausea; and other signs and symptoms, including vomiting, diarrhea, or constipation, abdominal pain, upper respiratory distress, or pulmonary infiltrates with or without sore throat and cough, may occur. Over the duration of the typical acute illness, lymphadenopathy, conjunctival injection, and abdominal tenderness also may occur. Biphasic febrile episodes may be observed and leukopenia, leukocytosis, or normal leukocyte counts may be seen. Although the normal course of illness is 2 to 4 days to a week, convalescence may require several days but residua do not occur. Rash, occurring in infections with certain Bunyamwera, Bwamba, Simbu, and Nairobi sheep disease serogroup viruses and in infections due to Tataguine virus is typically maculopapular, appearing after the onset of illness, lasting 1 to 3 days and occurring most commonly on the trunk.

Persons infected with most of the bunyaviruses, phleboviruses, and nairoviruses can be considered infective only as a source of virus for a subsequently feeding, uninfected arthropod; as mentioned, Rift Valley fever virus, Crimean-Congo hemorrhagic fever virus, and probably other members of the family *Bunyaviridae* as well, can be transmitted by aerosol. The viremic period is about 2 or 3 days in individuals infected with bunyaviruses, phleboviruses, and probably most nairoviruses, but viremia persists for as long as 7 to 10 days in subjects with Crimean-Congo hemorrhagic fever. Infection with hemorrhagic fever with renal syndrome virus, acquired by aerosol or by direct contact with fomites or excreta from infected animals, may lead to viremia, but the long incubation period precludes detection.

The serious nature of these diseases and the evidence that body fluids from patients suffering from them may be highly infectious, suggests that immediate supportive care must be given them and that patient isolation measures are indicated, particularly with Crimean-Congo hemorrhagic fever patients. This is not the case with mild 3-day fevers, but until and unless a confirmed laboratory diagnosis is available, such persons should be watched carefully to see that they do not develop more serious complications.

Diagnosis

Dynamics of Viremia and Antibody Responses

In geographic areas where febrile illnesses are common and laboratory support for clinical observations is poor, it may be that only in special cases or in epidemics are laboratories called on to isolate viruses or to provide serodiagnostic support. Bunyaviruses and phleboviruses are isolated most often from blood samples collected from febrile humans. LaCrosse virus (California serogroup) has been isolated only twice from humans (brain), although serologic evidence suggests that this is the most common arboviral cause of pediatric encephalitis in the United States. Thus, viruses of the family *Bunyaviridae* are isolated from encephalitic humans rarely or only with great effort. One reason for this is that the viremic stage in bunyavirus infections is brief, generally no more than 1 to 3 days after onset. Often the viremic stage has passed by the time CNS symptoms have appeared. Virus is readily isolated from patients with fever and with fever and rash. During the first 3 days of illness, bunyavirus viremia titers may be very high in these patients. Figure 1 presents a hypothetical model of viremia and antibody responses in a person infected with a typical bunyavirus. The initial antibody response appears at the end of the viremic period; this may be coincidental or may serve to quench the viremia. In either case, virus (antigen)-antibody complexing may make both virus isolation and antibody determinations more difficult. The IgM fraction of the serum contains both neutralizing and HI antibodies early after infection, and its presence can serve as an indicator of recent infection (Niklasson et al., 1984; Calisher et al., 1986a, b, c). Later, IgG antibody appears, con-

taining both neutralizing and HI as well as CF antibodies; IgG antibody to the bunyaviruses appears to persist for years, perhaps for the life of the individual (Calisher et al., 1986a). In summary, the initial response to infection is by production of IgM antibody, then IgG antibody. As the titer and presence of antiviral IgM antibody wanes, antiviral IgG antibody predominates, until antibody can be detected only in the IgG fraction.

Antiviral antibody contained in the IgM fraction of serum appears to be antigenic complex-specific but not virus-specific in alphavirus (Calisher et al., 1986b) and flavivirus (Monath et al., 1984) infections but may be only serogroup-specific in infections caused by bunyaviruses (Calisher et al., 1986a). Nevertheless, assays for IgM class antibodies are valuable because their presence provides evidence that the patient had been infected recently.

Laboratory Safety

The dynamics of both the viremia and antibody responses provide opportunities for virus isolation and for serodiagnosis. When attempting virus isolations, prime consideration should be given to safety. When handling specimens assumed to contain viruses such as Rift Valley fever, Crimean-Congo hemorrhagic fever, or other class 4 and 5 pathogens, the possibility of laboratory-acquired infection should be of concern. Transmission to humans by aerosol was demonstrated in six Egyptian laboratory workers attending the slaughter of a sheep with Rift Valley fever virus infection (Hoogstraal et al., 1979) and in agricultural workers slaughtering Rift Valley fever virus-infected sheep (Shope, 1985). Hospital-based outbreaks of Crimean-Congo hemorrhagic fever have been reported from the Soviet Union (Kulagin et al., 1962), and infection with hemorrhagic fever with renal syndrome (Hantaan) virus was recorded in more than 100 laboratory workers in Moscow, workers who apparently were infected by aerosols generated by field-collected rodents (Shope at al., 1985). Workers handling laboratory rats as well have been infected with Hantaan or a related hantavirus (Umenai et al., 1979).

Most other members of the family *Bunyaviridae* may be handled with less stringent precautions, yet with relative safety, but it should be kept in mind that under certain circumstances, many are potential pathogens.

Wearing rubber or plastic gloves and a laboratory coat and using a laminar flow hood and aseptic technique should be sufficient for handling class 1 or 2 viruses. For viruses that pose special hazards to laboratory workers (class 3), special conditions are required for containment. For viruses that pose ex-

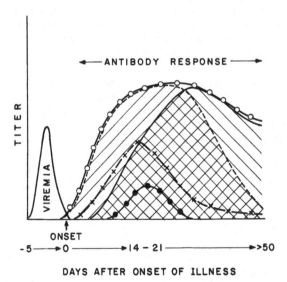

FIG. 1. A hypothetical model of viremia and antibody responses in a person infected with a typical bunyavirus. O—O—O = neutralizing antibody; ×—×—× = hemagglutination-inhibiting antibody; ●—●—● = complement-fixing antibody; ⫶⫶⫶⫶⫶ = IgM class antibody; ⌒⌒⌒⌒ = IgG class antibody.

treme hazards to laboratory workers or that may cause epidemic disease (class 4), conditions of maximum containment are needed. It is always best to direct questions regarding problem or potentially hazardous specimens to a specialty laboratory. In all instances, exposure to aerosols should be minimized or eliminated by using closed systems for grinding tissues, using blunt-ended needles for collecting and transferring liquids, and providing appropriate ventilation systems for filtering or otherwise scrubbing or cleaning exiting air. Personnel showers, disinfecting for individual decontamination, and the use of germ-free-type isolators (for class 4 viruses) are recommended. Specimens in both the field and the laboratory should be handled under the assumption that virus is present and potentially dangerous.

Specimen Collection

After collection, whole blood, serum, or tissue samples should be processed immediately or placed on dry ice (−70°C) or another suitable deep-freezing agent if virus isolation is to be attempted with the specimen. Although for antigen detection, this may not be such a critical issue, it appears sensible to ship and store specimens for this purpose at low temperatures to prevent further degradation of proteins. When sera are to be tested for antibody only, they can be shipped and stored at ambient temperatures, unless they are contaminated with microorganisms or will be in transit for long periods.

Virus Isolation and Identification

Because there is such a wide variety of individual characteristics of viruses within the family *Bunyaviridae,* no single virus isolation system suffices and suitable isolation systems have not yet been developed for all viruses of the family. In general, viruses of the genus *Bunyavirus* may be isolated in suckling mice inoculated intracranially with clarified suspensions of clinical specimens or with macerated and clarified arthropod pools. However, because essentially all the bunyaviruses were first found by using a suckling mouse isolation system, it follows that attempted isolation in other systems, including cell cultures derived from arthropods or mammals, may yield viruses not yet discovered. Because suckling mice are available to essentially all laboratories, particularly those that perform rabies virus isolations, this system holds certain advantages over others. Nevertheless, little loss in sensitivity and perhaps some increase in sensitivity may result from using mosquito cells grown in vitro, such as C6/36 (*Aedes albopictus*) or AP-61 (*Aedes pseudoscutellaris*) cell lines, for virus isolation. These hold the additional

advantage of ease of containment and reduction of aerosols. Inoculation of *Toxorhynchites amboinensis* mosquitoes, which do not take blood meals but in which dengue and other viruses replicate, also have been used with sensitivity and safety (Rosen, 1981).

For more than 30 years the classical procedure for identifying an arbovirus was to: (1) isolate the virus by inoculation of suckling mice; (2) perform a modicum of characterization (filterability, sensitivity to lipid solvents (Theiler, 1957), determination of replication and titers in various laboratory hosts such as animals and cell cultures, and so on); (3) prepare a sucrose-acetone extracted antigen (Clarke and Casals, 1958); (4) test that antigen for its ability to agglutinate the erythrocyte of male, domestic geese (*Anser cinereus*) and its ability to react in CF tests (Clarke and Casals, 1958; Casey, 1965) with homologous antibody preparations; and then (5) test by either HI or CF with a battery of antibody preparations. Although we now have a greater understanding of the structure and function of arbovirus virions and newer methods of testing have been introduced, the basic techniques used to identify and to antigenically characterize arboviruses have not and probably will not change in the foreseeable future. Because the G1 and G2 glycoproteins of most members of the family *Bunyaviridae* function in virus attachment to cells and play a role in HI and neutralization of these viruses and because the nucleocapsid protein participates in CF tests, much can be learned about a particular virus by studying the participation of these antigens in relatively simple laboratory tests. Also, because most of the antigenic data already accumulated have been generated using these tests they will, therefore, remain the standards by which newly isolated viruses are judged. This does not mean that newly devised tests are to be minimized. On the contrary, any additional tests that expand our ability to characterize fully the epitopes and other antigenic moieties of viruses are of great use. In this chapter, we limit the discussion of virus identification procedures to those recently developed and shown useful and the tried and tested techniques that have been so reliable and functional.

As with other arboviruses, hemagglutination by arboviruses is pH dependent. That is, virus adsorbs to indicator erythrocytes only over a narrow, usually low (pH 5.9 to 6.3, but sometimes as high as 6.8 to 7.4) pH range. Hemagglutinins for some viruses of the family *Bunyaviridae* have been prepared with difficulty or not at all, so this procedure is not as useful for their identification as for arboviruses of other families. Nevertheless for certain viruses, such as for the California and Phlebotomus fever serogroup viruses, they are quite useful. Using a battery of polyclonal hyperimmune mouse ascitic fluids (Tikasingh et al., 1966), the hemagglutinin is then tested for reactivity; a resulting positive reaction (HI) is then taken to indicate that the virus is antigenically related to one or more of the viruses used to prepare the polyvalent antibody. This allows one to place an isolate in a serogroup and to perform additional tests that provide more discrimination, that is, identification to antigenic complex or to type.

Complement-fixing antigens of viruses belonging to the family *Bunyaviridae* (except for some phleboviruses) are more serogroup-, or at least, complex-specific than are those belonging to other families of arboviruses because sharing of nucleocapsid protein is more extensive among the *Bunyaviridae*. This holds both advantages and disadvantages for virus identification because the CF test can be used to place an isolate in a serogroup, but specific identification of the isolate still remains to be done (Calisher et al., 1981). Nevertheless, because the known and registered number of viruses in the Catalogue of Arboviruses has surpassed 500 (Karabatsos, 1985), it is of considerable benefit to have techniques such as HI and CF available to reduce the possibilities.

Neutralization tests, whether performed in cell cultures by serum dilution-plaque reduction (Lindsey et al., 1976; Hunt and Calisher, 1979) or in mice by virus or serum dilution are the definitive tests for identification of virus types and subtypes. Virus dilution-constant serum neutralization test results provide a quantitative indication of the power of a serum sample to neutralize a virus, whereas constant virus-serum dilution tests provide a quantitative indication of the potency of that serum. The virus dilution-constant serum technique may be closer to the natural situation (that is, how much virus will this serum sample neutralize?); however, the serum dilution neutralization technique is more sensitive diagnostically (i.e., how much can the serum be diluted and still neutralize a given quantity of virus?) and the latter is most often used to determine significant differences in neutralizing antibody titers between paired sera. When a virus isolate is tested for its ability to be neutralized by antibody prepared against a reference virus, the titer of that antibody preparation is compared with its homologous titer; differences between the two titers are taken to indicate differences between the viruses. Monoclonal antibodies are increasingly being produced and used for virus typing and subtyping (Gonzalez-Scarano et al., 1983). As hybridomas are selected for epitope specificity, they are being used to prepare reagents useful not only for virus characterization but for practical and immediate applications, such as virus identifications during epidemic situations.

In summation, a virus isolate is tested for its ability to agglutinate the erythrocytes from male domestic geese. If found positive, the hemagglutinin is tested at its pH optimum for inhibition by one or more of a battery of immune reagents. Whether or not the antigen hemagglutinates, it is tested by CF against a battery of immune reagents; fewer reagents are necessary if preliminary (HI) tests give indication of the serogroup to which the isolate belongs. Serum dilution-plaque reduction neutralization tests are then performed to identify the virus to type, and epitope-specific monoclonal antibodies are then used for further, subtypic identification. A combination of monoclonal antibodies and oligonucleotide fingerprint mapping studies (Klimas et al., 1981) can then be applied to the isolate for determining genetic and epidemiologic origins.

Direct Antigen Detection

The relatively recent wide acceptance of enzyme-linked immunosorbent assays (ELISA) for detecting many viruses, including arboviruses, has proven a boon to clinical virologists, not to speak of the patient. Treatment of patients infected with a variety of viruses can be instituted as soon as laboratory tests indicate that such treatment would be useful. Judicious nontreatment, with supportive care only, can also be indicated by such tests. Therefore, it is in the interests of the patient, the attending physician, and the laboratorian to be able to determine with some certainty the etiologic agent of infection caused by members of the family *Bunyaviridae*. The more rapid, sensitive, and specific the test, the better for all concerned. For many years, respiratory viruses have been identified directly in nasopharyngeal aspirates. More recently, enteric viruses, including rotaviruses, have served as models for developing rapid identification techniques. Immunofluorescence has proved to be a relatively rapid, sensitive, and specific tool for direct identification of virus antigens in clinical specimens, without the need for an intervening amplification in a laboratory host (Gardner and McQuillin, 1980). Although offering such an advantage, immunofluorescence tests require highly skilled, precise, and accurate technicians, and expensive microscopes. These needs and the subjectivity of the results are drawbacks in addition to the absence of appropriate clinical specimens containing large quantities of virus in infections caused by viruses of the family *Bunyaviridae*.

Therefore, the ELISA is increasingly used for direct detection of viral antigens in clinical specimens as well as in tissues from wild vertebrates and in mosquitoes. ELISA systems have been developed for detecting *Bunyaviridae* members LaCrosse virus in mosquitoes (Hildreth et al., 1982) and in mice (Beaty et al., 1982) and Rift Valley fever virus in experimental animals, vaccines, and infected mosquitoes (Niklasson et al., 1983). In brief, material suspected to contain whole virus or viral antigen is diluted in pH 9.6 carbonate-bicarbonate coating buffer and introduced to wells of à microtiter plate and incubated at room temperature overnight or for 3 hours at 37°C. Incubation of all subsequent reagents is usually done at either 37°C for 1 h or at 4°C for 18 h. In direct ELISA's specific immune serum, or another antibody-containing reagent, conjugated to an enzyme such as peroxidase or alkaline phosphatase, and then substrate for this enzyme, are added sequentially. Diluent for all antibodies and antigens except the coating buffer is phosphate-buffered saline (pH 7.4) containing 5% heat-inactivated fetal bovine serum, 0.5% Tween 20, and 40 μg of merthiolate per ml. Sufficient washing of the wells after adding each reagent ensures that substances not attaching to specific binding sites on the previously added reagents will be flushed out. An adequate washing solution is phosphate-buffered saline containing 0.1% Tween 20. If antigen is present to react with enzyme-conjugated antibody, the substrate will be reduced (change from a colorless to a translucent material in solution), and the optical density of the resulting colored solution can be read either mechanically (spectrophotometer) or estimated by eye.

Indirect (sandwich) ELISAs use an antigen-capturing antibody for coating the wells. This antibody should be sufficiently avid and potent to capture the very small quantities of antigen that are in tissues, viremic sera, or arthropod tissues; either polyclonal (Halonen et al., 1980) or monoclonal antibodies may be used. To the antiviral antibody-coated wells are added, sequentially, the specimen to be tested for viral antigen, a secondary (detecting) antibody conjugated with enzyme, and a substrate for that enzyme. As an alternative to the sandwich ELISA, a double sandwich ELISA may be used. This is simply a variant of the sandwich ELISA in that the secondary antibody is not enzyme-conjugated, necessitating a third (detecting) antibody that is enzyme conjugated. This detecting antibody is directed against the antiviral secondary antibody (antispecies). An advantage of this method is the obviation of a need for enzyme-conjugated antibodies for all the viruses tested; principal disadvantages are the extra step (time) required and the possible increase in attendant nonspecific reactivity.

Viral antigens may be detected directly in cells by use of antibodies conjugated to fluorescing dyes, such as fluorescein isothiocyanate. When viral antigens are exposed to these antibodies, the fluorescent dye is immobilized at the site of antigen-antibody complexing and may be seen through a fluorescence microscope (Gardner and McQuillin, 1980). Because clinical specimens from patients infected with most members of the family *Bunyaviridae*, are, with notable exceptions, either devoid of virus or contain in-

sufficient quantities of virus to be detected by this method, the direct fluorescent antibody technique, and its counterpart, the indirect fluorescent antibody technique (virus-infected cells, to which antiviral antibody, fluorescein isothiocyanate-conjugated antispecies antibody are added, sequentially) are not routinely used for identifying arboviral antigens in clinical specimens. The exceptions alluded to above are Rift Valley fever, hemorrhagic fever with renal syndrome, and Crimean-Congo hemorrhagic fever. In these and possibly other infections caused by members of the family *Bunyaviridae*, virus often is detected directly in the tissues of patients, domestic animals, or wild rodents infected with the virus. Because the hantaviruses cause no cytopathology in cells infected with them, fluorescent antibody techniques have become invaluable for making an early diagnosis relevant to the course of illness of patients infected with these viruses.

Direct Detection of Viral Nucleic Acid

Recent advances in recombinant DNA technology have made it possible to use nucleic acid probes for diagnosing viral infections and for detecting viral genetic material in tissue from other sources. In situ hybridization permits detection of viral nucleic acid in infected cells. In these assays, virus-specific nucleic acids are bound to a solid phase and then detected with radiolabeled DNA probes (Hyypia et al., 1984) or RNA probes. This technique could provide a method for rapid and exquisitely specific identification of viral RNA in tissues from patients, reservoir hosts, and arthropod vectors of viruses in the family *Bunyaviridae;* however, it is not yet refined or widely accepted for use with arboviruses due to low sensitivity and the need for radiolabeled probes. It has the further potential advantage of identifying specific sequences in viral genetic material and will undoubtedly not only function in the future as a rapid diagnostic method but will also serve as an epidemiologic tool.

Antibody Assays

In the past, fourfold or greater increases or decreases in HI, CF, neutralizing, or other antibody titers have been the basis for serodiagnostic confirmation of arboviral infections. However, these tests require multiple specimens, collected days to weeks apart. When rapidity of diagnosis is not a critical issue, or when infection with closely related viruses confounds interpretation of results, these tests are often invaluable. However, their lack of specificity also may create a variety of diagnostic problems. For example, in areas hyperendemic for flaviviruses, vaccination with yellow fever virus, coincidental with the circulation of the related West Nile, Zika, Uganda S, Spondweni, dengue-1, Ntaya, and wild-type yellow fever viruses limits the usefulness of both the ordinarily broadly cross-reactive HI tests as well as the usually more specific CF test (Monath et al., 1980). Viruses of the same serogroups in the family *Bunyaviridae* (California, C, Patois, Guama, Capim, phlebotomus fever, Uukuniemi, Hantaan, to name a few) may occur sympatrically. Therefore, antibody to one virus in a serogroup does not provide definitive evidence for infection with that virus. When rapidity of serodiagnosis is needed, other techniques must be used. The relatively recent adaptation of antibody capture ELISAs for IgM antibody to arboviruses has provided such a tool (Heinz et al., 1981; Jamnback et al., 1982; Burke et al., 1982; Niklasson et al., 1984; Monath et al., 1984; Calisher et al., 1985a; Calisher et al., 1985b; Calisher et al., 1986a; Calisher et al., 1986b; Calisher et al., 1986c). Unlike IgG antibody, which is produced later in infections, IgM antibody is produced early in arboviral infections and does not persist in high titer; its presence, therefore, is indicative of recent infection.

Briefly, the IgM antibody capture ELISA is performed as follows: Wells of a microtiter plate are coated with antibody to human IgM. Patient serum or cerebrospinal fluid is then introduced, usually at a dilution of 1:100 for serum or 1:10 for cerebrospinal fluid. Density gradient-purified virus, supernatant fluid from virus-infected cell cultures, or other viral antigen such as sucrose-acetone extracted antigen prepared from the brains of virus-infected suckling mice, and antiviral (secondary) antibody are then added. If the secondary antibody has been conjugated to an enzyme, substrate is then introduced and the intensity of the color change recorded as optical density, using a spectrophotometer. If incubations and washings between the addition of each reagent were sufficient, the results can be read as the ratio of test to control; usually ratios greater than or equal to 2.0 are considered positive for IgM antibody. Sera can be tested at single dilutions or diluted serially twofold and each dilution tested.

A double sandwich ELISA, using an antispecies IgG third (detecting) antibody conjugated to an enzyme that reacts with the secondary (antiviral) antibody also can be used. As was mentioned in the section on antigen detection, such a system has advantages and disadvantages. However, when testing only a few or a large number of sera for IgM antibody to many viruses the double sandwich ELISA is preferred, because enzyme-conjugated antibody preparations are not needed for each of the antiviral reagents.

The presence of IgM antibody in a single serum specimen or spinal fluid specimen can be taken as essential evidence for recent infection by the virus with which the test was done. Not enough is known yet about the reactivity of IgM antibody produced during infections with viruses of the family *Bunyaviridae*. It may be that, as in infections with alphaviruses and flaviviruses, IgM class antibody produced after infection with viruses of the family *Bunyaviridae* are complex-specific but are not type (virus)-specific. Furthermore, recent evidence suggests that IgM antibody persists in cerebrospinal fluid and serum for months, years, and perhaps for life after central nervous system involvement in Japanese encephalitis virus infections of humans (Burke et al., 1985); should this prove to be the case for members of the family *Bunyaviridae* as well (IgM to LaCrosse virus has been shown to persist [Jamnback et al., 1982]), the IgM antibody capture ELISA, as applied to these viruses must be considered somewhat less confirmatory than it now appears.

IgG antibody produced after infections with arboviruses persists for months or years. Therefore, its presence does not necessarily denote an active or recent viral infection, contrary to the significance of the presence of IgM antibody. The fetus or newborn produces IgM but not IgG in response to infections in utero or shortly after birth. Thus, the presence of IgG in the fetus or newborn indicates passive reception of IgG across the placenta; IgM cannot cross the placenta. IgM and IgG are produced after vaccination with live, including attenuated, viruses.

For estimating antibody prevalence in populations, HI, indirect fluorescent antibody, and ELISA IgG, or neutralization assays are the most useful; CF and IgM assays are often the least useful, although CF is quite useful in specific instances, such as for determining antibody to the dengue viruses.

IgG antibody in sera or spinal fluids may be determined as follows: (1) coat wells of a microtiter plate with about 350 to 500 ng of gradient-purified virus; an alternative is to coat the wells with antibody (prepared in species A) and then add sucrose-acetone extracted antigen or supernatant fluid from virus infected cells; (2) introduce the clinical specimen suspected to contain IgG antibody, testing only a single dilution of serum or the first of a series of dilutions beginning at 1:40; (3) add antihuman IgG (which does not contain antibody to species A conjugated with an enzyme; and (4) add the substrate and record the optical density of the test and control sera. Again, ratios of test/control greater than or equal to 2.0 are considered positive. This IgG antibody assay is performed relatively quickly and has the distinct advantages of allowing precoating and storage of plates as well as requiring few steps for performance. It is somewhat less sensitive and specific than neutralization tests but, because it is relatively easy to perform and is a binding assay, it can be used to rapidly detect antibody to a variety of viruses. A distinct disadvantage of this test is the need for either purified virus for every virus used for determining antibody and the need for separate coating antibodies to each of these viruses.

In summary, the IgM antibody capture ELISA is suggested for use in rapidly detecting antibody, and IgG assays for use in determining antibody in paired sera and for serosurveys. HI and CF tests can be used to supplement these tests and neutralization tests used for definitive and confirmatory determinations of the infecting agent.

Interpretation of Serologic Data

Viruses of the family *Bunyaviridae* differ in their antigenic relationships within and between the various serogroups, that is, certain viruses are very closely related and others are very distantly related. This is reflected not only in cross-tests between viruses but in antibody responses in humans and other animals infected with these viruses. As mentioned, infection with more than one member of a serogroup may or may not confer protection to heterologous members of that serogroup but almost assuredly will create serodiagnostic problems. Therefore, the combined use of IgM and IgG antibody assays as well as HI, CF, and neutralization tests may be required for final, presumptive determination of the etiologic agent of infection. Table 8 presents test results with sera from three persons naturally infected with viruses of the California, C, and Phlebotomus fever serogroup bunyaviruses. Note that a combination of tests was needed to make a presumptive determination of the infecting virus. Although inhibition of hemagglutinin of a virus by patient serum is indicative of past infection with a member of the serogroup to which the virus belongs, close antigenic relationships between viruses prevents specific determination of the infecting agent by HI tests (Table 8, case A, La Crosse and snowshoe hare viruses). Where viruses share CF antigen but not hemagglutinins (Oriboca and Murutucu viruses, case B), HI and neutralization tests are valuable. Where HI and CF titers are low, neutralization tests are valuable. Increasingly, it will likely be found that IgM assays are a valuable addition to the present diagnostic armamentarium and may very well replace the more classical tests in many instances.

In summary, the best method for serodiagnosis is the one that works best for the individual laboratory performing the tests. Labor intensiveness and overall costs are always considerations, but if testing of a serum sample is worth doing at all, it is worth doing with the best available tests. For serodiagnosis of arbovirus infections, using an inadequate test is worse than not testing at all, because false-negative results will bias the diagnostic perception and perhaps allow an invasive diagnostic procedure that is unwarranted (Calisher and Bailey, 1981). Routine use of an insensitive CF test, an IgG assay on a single serum, an HI test in a single serum from a vaccinated person, tests with poor antigens, or otherwise uncontrolled assays and inappropriate tests are misleading and are to be avoided. The World Health Organization Centres for Arbovirus Research and Reference at the Centers for Disease Control, Fort Collins, Colorado and at Yale University, New Haven, Connecticut, are available to provide advice and reagents to laboratories establishing diagnostic procedures for arboviruses.

Epidemiology and Natural History

The terms "arthropod-borne" and "rodent-borne" refer to the major biologic phenomena involved in the cyclical transmission of certain viruses. In the natural setting, these viruses are maintained in invertebrate-to-vertebrate-to-invertebrate or vertebrate-to-vertebrate cycles, usually not involving humans. Although Rift Valley fever and Crimean-Congo hemorrhagic fever viruses have been known to be transmitted by aerosols and infected body fluids, in most instances transmission is a complex interaction of virus, vector, and host. The bunyaviruses, nairoviruses, phleboviruses, and uukuviruses are transmitted from infected arthropods (mosquitoes, ticks, phlebotomine flies, and ticks, respectively) to susceptible vertebrates. After an incubation period, the vertebrate becomes viremic and is capable of infecting an uninfected arthropod that happens to feed on it. If the vector is genetically competent both to support replication and to transmit virus, and if the now-infected arthropod feeds on a susceptible vertebrate host, the natural cycle is continued. This general concept holds true where biologic, not simply mechanical, transmission occurs. The following factors come into play as regards vector competence: virus and vector genetics, virus concentration in the infected vertebrate host that is the source of virus, ambient temperature, autogeny, salivary gland and gut barriers, and other less clearly understood factors. Certain arthropods may serve as maintenance or amplification vectors. Vectors that could maintain the virus would have low infection thresholds, minimum evidence of barriers to disseminated infection of the arthropod, and high transmission efficiency. Vectors that could amplify the virus would have high

TABLE 8. Results of antibody determination with sera from humans presumably infected with bunyaviruses LaCrosse (California serogroup), Oriboca (Group C), or Toscana (Phlebotomus fever serogroup) viruses

Case	Serum no.	Days after onset		Antigen[a] used in test						
				LAC	SSH	CE	SAN	KEY	JC	TVT
A	1	2	(HI)	−[b]	−	−	−	−	−	−
			(CF)	−	−	−	−	−	−	−
			(IgM)	2.1	2.2	2.4	2.0	−	2.0	−
			(IgG)	−	−	−	−	−	−	−
			(N)	10	−	−	−	−	−	−
	2	15	(HI)	160	80	40	40	20	−	−
			(CF)	8	8	−	−	−	−	−
			(IgM)	4.2	4.5	3.9	3.2	2.7	3.3	−
			(IgG)	+[c]	+	+	+	+	+	+
			(N)	640	160	40	80	40	10	−
	3	130	(HI)	20	20	20	10	−	10	−
			(CF)	8	16	8	−	−	−	−
			(IgM)	−	−	−	−	−	−	−
			(IgG)	+	+	+	+	+	+	+
			(N)	160	320	40	40	20	−	−
				ORI	MUR	ITQ	APEU	CAR	MTB	NEP
B	1	0	(HI)	−	−	−	−	−	−	−
			(CF)	−	−	−	−	−	−	−
			(IgM)	2.8	−	−	−	−	−	−
			(IgG)	−	−	−	−	−	−	−
			(N)	40	−	−	−	−	−	−
	2	23	(HI)	160	−	−	−	−	−	−
			(CF)	64	128	8	8	−	−	−
			(IgM)	4.7	−	−	−	−	−	−
			(IgG)	+	+	+	+	−	−	−
			(N)	320	20	−	−	10	−	−
				TOS	SFN	SFS				
C	1	2	(HI)	−	−	−				
			(CF)	8	−	−				
			(IgM)	12.1	11.2	4.2				
			(IgG)	−	−	−				
			(N)	−	−	−				
	2	87	(HI)	20	−	−				
			(CF)	8	8	−				
			(IgM)	7.6	3.6	2.8				
			(IgG)	+	+	+				
			(N)	640	−	−				

[a] Abbreviations of virus names are: La Crosse (LAC), snowshoe hare (SSH), California encephalitis (CE), San Angelo (SAN), Keystone (KEY), Jamestown Canyon (JC), Trivitattus (TVT); Oriboca (ORI), Murutucu (MUR), Itaqui (ITQ), Apeu (Apeu), Caraparu (CAR), Marituba (MTB), Nepuyo (NEP), Toscana (TOS), sandfly fever Naples (SFN), and sandfly fever Sicilian (SFS).

[b] Signifies <10 HI or neutralization; <8 CF; ratio of optical densities of test:control sera <2.0 at the lowest dilution tested by ELISA for IgM (1:100) or IgG (1:40) antibodies.

[c] Signifies ratio of optical densities of test:control sera equal to or greater than 2.0 at the lowest dilution (1:40) tested by ELISA for IgG antibodies.

infection thresholds and poor rates of bite transmission (often compensated for by extremely high seasonal population densities and feeding on a variety of vertebrate host species). Many members of the family *Bunyaviridae* that are arthropod transmitted have been shown to be transovarially transmitted in the arthropod. Mosquito-borne, tick-borne, and phlebotomine fly-borne members are among those that have this capacity. Such a mechanism allows the virus to be maintained without intervening feedings of the arthropod on viremic vertebrates.

Vertebrate hosts of these viruses also play a critical role in their maintenance and amplification. Replication of the virus in the vertebrate must attain high enough titer to serve as a source of virus for the host-feeding arthropod. In addition, the size of the population of susceptible vertebrates must be sufficient to provide relative assurance of contact with the arthropod, and population turnover must be adequate to provide a continual number of susceptible hosts. The vertebrate host becomes another source of amplification in epidemic situations, because one infected vertebrate can serve to infect many arthropods. Except for the Bunyamwera, Simbu, certain of the California, and Nairobi sheep disease serogroup viruses, most of the viruses of the family *Bunyaviridae* are transmitted to small mammals. Among the vertebrate hosts of the viruses of this family are rodents, camels, and birds, attesting to the remarkable diversity and adaptability of the viruses.

The hantaviruses are exceptions to the rule of the family, in that they appear to be transmitted only directly from mammal to mammal. Thus, maintenance and transmission of these viruses occur in the same host. Chronic infection of the vertebrate host, usually rodents, is characterized by urinary and salivary excretion and persistence of virus in noncytolytic, immunocompetent infection. The epidemiology of the hantaviruses is, therefore, limited by the geographic and ecologic distributions of the rodent host. Rodent genetics undoubtedly play a role in the induction of chronic infection, but this is, as yet, not understood.

When humans impinge upon the natural ecosystem of the virus, infection can result. Viruses causing mild febrile illnesses, or no illnesses at all in humans, may go unnoticed. But when epidemics or epizootics of life-threatening proportions occur in agricultural workers or livestock, in forest workers or forest animals of esthetic importance, or in military personnel, it is only with disastrous consequences that such occurrences are ignored. Human social and behavioral patterns impact upon the epidemiology of all the viruses of the family *Bunyaviridae*, so understanding such patterns can assist in controlling the spread of the viruses, or at least in lessening the probability of acquiring them.

Prevention and Control

The basic tenet of controlling any disease is to prevent its occurrence. To control the prevalence of viruses of the family *Bunyaviridae*, it is necessary first to understand the ecology of the disease: persistence in interepidemic periods, mechanisms of amplification of the virus, and mechanisms of transmission to the affected host. For mosquito-borne, tick-borne, and other arthropod-borne viruses, a thorough understanding of the maintenance and amplification cycles of the viruses is the sine qua non of both prevention and control.

First and foremost, contact between the susceptible human or animal and the vector must be reduced. This can be done by either removing the target vertebrate from contact with the maintenance and amplification vectors or by removing the arthropods from contact with the target vertebrate. An alternative is to remove the intermediate vertebrate host from contact with the arthropod vector. As this is more easily said than done, a more realistic target is to reduce contacts between arthropod, intermediate vertebrate host, and susceptible human or other animal hosts. This can be effected by mosquito- or tick-control operations, using larvacide or adulticide techniques, either on a large (spraying of insecticides from airplanes) or a small (spraying of insecticides from backpack units) scale. Public education projects with adequate, on-going publicity and legal means are also often used alternatives to direct intervention. Ongoing clinical, virologic, serologic, and histologic arthropod and rodent surveillance techniques should be used to ascertain the presence and prevalence of these viruses.

For preventing contact of mosquito, culicoid, and phlebotomine fly vectors with humans, the use of netting and window and door screening is recommended. Removing and subsequently destroying mosquito breeding sites (source reduction), providing piped water, maintaining liquid waste systems, and managing irrigation systems have been successful ways of reducing populations of both maintenance and amplification vectors.

Protection from tick infestations can be effected by personal protection (wearing clothes impregnated with suitable repellents), dipping cattle, sheep, dogs, and other vertebrate hosts, and treating camp sites with acaricides. These measures have all been used with success.

Preventive measures taken against commensal rodents include rodent-proofing habitations and food-storage areas, proper disposal of domestic and public wastes, and using permanent anticoagulant baiting stations to depress rodent populations.

If measures taken to control arthropod vectors and rodent hosts of viruses are not effective or are not possible, given economic, political, or geographic considerations, vaccines hold great potential for disease prevention. A Rift Valley fever vaccine for use in humans at risk and vaccines for sheep and cattle have been developed. Given that viruses of the family *Bunyaviridae* possess segmented genomes, it may be possible to construct live attenuated vaccines for any of these viruses. The G1 and G2 glycoproteins, appear to be responsible for major virulence characteristics, as well as for inducing neutralizing antibodies. If the middle-sized RNA segment of all *Bunyaviridae* codes for the G1 and G2 glycoproteins, then such live attenuated vaccines can be produced for Rift Valley fever, Crimean-Congo hemorrhagic fever, hemorrhagic fever with renal syndrome, or any other virus affecting humans.

Literature Cited

Ali, A. M. M., and S. Kamel. 1978. Epidemiology of Rift Valley fever in domestic animals in Egypt. J. Egypt. Publ. Hlth. Assoc. **53**:255–263.

Balfour, Jr., H. H., R. A. Siem, H. Bauer, and P. G. Quie. 1973. California arbovirus (La Crosse) infections: clinical and laboratory findings in 66 children with meningoencephalitis. Pediatrics **52**:680–691.

Beaty, B. J., S. W. Hildreth, D. C. Blenden, and J. Casals. 1982. Detection of La Crosse (California encephalitis) virus antigen in mouse skin samples. Am. J. Vet. Res. **43**:684–687.

Bishop, D. H. L., and R. E. Shope. 1979. Bunyaviridae, p. 1–156. *In* H. Fraenkel-Conrat, R. A. Wagner, (ed.), Comprehensive virology, Plenum Press, New York.

Bishop, D. H. L., C. H. Calisher, J. Casals, M. P. Chumakov, S. Ya. Gaidamovich, C. Hannoun, D. K. Lvov, I. D. Marshall, N. Oker-Blom, R. F. Pettersson, J. S. Porterfield, P. K. Russell, R. E. Shope, and E. G. Westaway. 1980. Bunyaviridae, Intervirology **14**:125–143.

Burke, D. S., A. Nisalak, and M. A. Ussery. 1982. Antibody capture immunoassay detection of Japanese encephalitis virus immunoglobulin M and G antibodies in cerebrospinal fluid. J. Clin. Microbiol. **16**:1034–1042.

Burke, D. S., W. Lorsomrudee, C. J. Leake, C. H. Hoke, A. Nisalak, V. Chongswasdi, and T. Laorakpongse. 1985. Fatal outcome in Japanese encephalitis. Am. J. Trop. Med. Hyg. **34**:1203–1209.

Calisher, C. H., and R. E. Bailey. 1981. Serodiagnosis of La Crosse virus infections in humans. J. Clin. Microbiol. **13**:344–350.

Calisher, C. H., T. P. Monath, N. Karabatsos, and D. W. Trent. 1981. Arbovirus subtyping: applications to epidemiologic studies, availability of reagents, and testing services. Am. J. Epidemiol. **114**:619–631.

Calisher, C. H., and W. H. Thompson (ed.). 1983. California serogroup viruses. Prog. Clin. Biol. Res. **123**:1–399.

Calisher, C. H., J. D. Poland, S. B. Calisher, and L. Warmoth. 1985a. Diagnosis of Colorado tick fever virus infection by enzyme immunoassays for immunoglobulin M and G antibodies. J. Clin. Microbiol. **22**:84–88.

Calisher, C. H., O. Meurman, M. Brummer-Korvenkontio, P. E. Halonen, and D. J. Muth. 1985b. Sensitive enzyme immunoassay for detecting immunoglobulin M antibodies to Sindbis virus and further evidence that Pogosta

disease is caused by a western equine encephalitis complex virus. J. Clin. Microbiol. **22**:566–571.

Calisher, C. H., C. I. Pretzman, D. J. Muth, M. A. Parsons, and E. D. Peterson. 1986a. Serodiagnosis of La Crosse virus infections in humans by detection of immunoglobulin M class antibodies. J. Clin. Microbiol. **23**:667–671.

Calisher, C. H., A. O. El-Kafrawi, M. I. Mahmud, Al-D. Travassos, A. P. A. da Rosa, C. R. Bartz, M. Brummer-Korvenkontio, S. Haksohusodo, and W. Suharyono. 1986b. Complex-specific immunoglobulin M antibody patterns in humans infected with alphaviruses. J. Clin. Microbiol. **23**:155–159.

Calisher, C. H., V. P. Berardi, D. J. Muth, and E. E. Buff. 1986c. Specificity of immunoglobulin M and G antibody responses in humans infected with eastern and western equine encephalitis viruses: application to rapid serodiagnosis. J. Clin. Microbiol. **23**:369–372.

Casals, J., and L. Whitman. 1960. A new antigenic group of arthropod-borne viruses. The Bunyamwera group. Am. J. Trop. Med. Hyg. **9**:73–77.

Casals, J., and L. Whitman. 1961. Group C. A new serological group of hitherto undescribed arthropod-borne viruses. Immunological studies. Am. J. Trop. Med. Hyg. **10**:250–258.

Casals, J. 1963. New developments in the classification of arthropod-borne animal viruses. Anais Microbiol. **11**:13–34.

Casey, H. L. 1965. Part II. Adaptation of LBCF method to micro technique. Standardized diagnostic complement fixation method and adaptation to micro test. Publ. Hlth. Monogr. **74**:1–34.

Clarke, D. H., and J. Casals. 1958. Techniques for hemagglutination and hemagglutination-inhibition wth arthropod-borne viruses. Am. J. Trop. Med. Hyg. **7**:561–573.

Daubney, R., J. R. Hudson, and P. C. Garnham. 1931. Enzootic hepatitis or Rift Valley fever: an undescribed disease of sheep, cattle, and man from East Africa. J. Pathol. Bacteriol. **34**:545–579.

Gajdusek, D. C. Acute infectious hemorrhagic fevers and mycotoxicoses in the Union of Soviet Socialist Republics. Med. and Sci. Publ. No. 2, Army Med. Serv. Grad. Sch.; Walter Reed Army Medical Center, Washington, D. C.

Gardner, P. S., and J. McQuillin. 1980. Applications of immunofluorescence (2nd ed.), Butterworth, London.

Gonzalez-Scarano, F., R. E. Shope, C. H. Calisher, and N. Nathanson. 1983. Monoclonal antibodies against the G1 and nucleocapsid proteins of LaCrosse and Tahyna viruses, p. 145–156. *In* C. H. Calisher and W. H. Thompson (ed.), California serogroup viruses. Prog. Clin. Biol. Res., Vol. 123, Alan R. Liss, Inc., New York.

Halonen, P. E., H. Sarkkinen, P. Arstila, E. Hjertsson, and E. Torfason. 1980. Four-layer radioimmunoassay for detection of adenovirus in stool. J. Clin. Microbiol. **11**:614–617.

Heinz, F. X., M. Roggendorf, H. Hormann, C. Kunz, and F. Deinhardt. 1981. Comparison of two different enzyme immunoassays for detection of immunoglobulin M antibodies against tick-borne encephalitis virus in serum and cerebrospinal fluid. J. Clin. Microbiol. **14**:141–146.

Hildreth, S. W., B. J. Beaty, J. M. Meegan, C. L. Frazier, and R. E. Shope. 1982. Detection of La Crosse arbovirus antigen in mosquito pools: application of chromogenic and fluorogenic enzyme immunoassay systems. J. Clin. Microbiol. **15**:879–884.

Hoogstraal, H. 1979. The epidemiology of tick-borne Crimean-Congo hemorrhagic fever in Asia, Europe, and Africa. J. Med. Entomol. **15**:307–417.

Hoogstraal, H., J. M. Meegan, G. M. Khalil, and F. K.

Adham, 1979. The Rift Valley fever epizootic in Egypt 1977–1978. 2. Ecological and entomological studies. Trans. Royal Soc. Trop. Med. Hyg. **73:**624–629.

Hunt, A. R., and C. H. Calisher. 1979. Relationships of Bunyamwera group viruses by neutralization. Am. J. Trop. Med. Hyg. **28:**740–749.

Hyypia, T., P. Stalhandske, R. Väinionpää, and U. Pettersson. 1984. Detection of enteroviruses by spot hybridization. J. Clin. Microbiol. **19:**436–438.

Jamnback, T. L., B. J. Beaty, S. W. Hildreth, K. L. Brown, and C. B. Gunderson. 1982. Capture immunoglobulin M system for rapid diagnosis of La Crosse (California encephalitis) virus infections. J. Clin. Microbiol. **16:**577–580.

Kalfayan, B. 1983. Pathology of La Crosse virus infections in humans, p. 179–186. *In* C. H. Calisher, W. H. Thompson, (ed.), California serogroup viruses. Prog. Clin. Biol. Res., Vol. 123, Alan R. Liss, Inc., New York.

Karabatsos, N., and R. E. Shope. 1979. Cross-reactive and type-specific complement-fixing structures of Oriboca virions. J. Med. Virol. **3:**167–176.

Karabatsos, N. (ed.). 1985. International catalogue of arboviruses including certain other viruses of vertebrates (3rd ed.), Am. Soc. Trop. Med. Hyg., San Antonio.

Karmysheva, V. Ya, E. V. Leshchinskaya, A. M. Butenko, A. P. Savinov, and A. F. Gusarev. 1973. Results of some laboratory and clinical–morphological investigations of Crimean hemorrhagic fever. Arkh. Patol. **35:**17–22.

Klimas, R. A., W. H. Thompson, C. H. Calisher, G. G. Clark, P. R. Grimstad, and D. H. L. Bishop. 1981. Genotypic varieties of La Crosse virus isolated from different geographic regions of the continental United States and evidence for a naturally occurring intertypic recombinant La Crosse virus. Am. J. Epidemiol. **114:**112–131.

Kulagin, S. M., N. I. Fedorova, and E. S. Ketiladze. 1962. Laboratornaia vspyshka gemorragicheskoi likhoradki s pochechnym sindromom (kilinko-epidemiologicheskaia charakteristika) [A laboratory outbreak of hemorrhagic fever with renal syndrome (clinical-epidemiological characteristics)]. Zh. Mikrobiol. Epidemiol. Immunobiol. **33:**121–126.

Kurogi, H., Y. Inaba, E. Takahashi, K. Sato, T. Omori, Y. Miura, Y. Goto, Y. Fujiwara, Y. Hatano, K. Kodama, S. Fukuyama, N. Sasaki, and M. Matumoto. 1976. Epizootic congenital arthrogryposis-hydranencephaly syndrome in cattle: isolation of Akabane virus from affected fetuses. Arch. Virol. **51:**67–74.

Lindsey, H. S., C. H. Calisher, and J. H. Mathews. 1976. Serum dilution neutralization test for California group virus identification and serology. J. Clin. Microbiol. **4:**503–510.

Meegan, J. M. 1979. The Rift Valley fever epizootic in Egypt 1977–1978. 1. Description of the epizootic and virological studies. Trans. Royal Soc. Trop. Med. Hyg. **73:**618–623.

Monath, T. P., R. B. Craven, D. J. Muth, C. J. Trautt, C. H. Calisher, and S. A. Fitzgerald. 1980. Limitations of the complement-fixation test for distinguishing naturally acquired from vaccine-induced yellow fever infection in flavivirus-hyperendemic areas. Am. J. Trop. Med. Hyg. **29:**624–634.

Monath, T. P., R. R. Nystrom, R. E. Bailey, C. H. Calisher, and D. J. Muth. 1984. Immunoglobulin M antibody capture enzyme-linked immunosorbent assay for diagnosis of St. Louis encephalitis. J. Clin. Microbiol. **20:**784–790.

Montgomery, E. 1917. On a tick-borne gastroenteritis of sheep and goats occurring in British East Africa. J. Comp. Pathol. Therap. **30:**28–57.

Murphy, F. A., A. K. Harrison, and S. G. Whitfield. 1973. Bunyaviridae: morphologic and morphogenetic similarities of Bunyamwera supergroup viruses and several other arthropod-borne viruses. Intervirology **1:**297–316.

Niklasson, B., M. Grandien, C. J. Peters, and T. P. Gargan II. 1983. Detection of Rift Valley fever virus antigen by enzyme-linked immunosorbent assay. J. Clin. Microbiol. **17:**1026–1031.

Niklasson, B., C. J. Peters, M. Grandien, and O. Wood. 1984. Detection of human immunoglobulins G and M antibodies to Rift Valley fever virus by enzyme-linked immunosorbent assay. J. Clin. Microbiol. **19:**225–229.

Obijeski, J. F., and F. A. Murphy. 1977. Bunyaviridae: recent biochemical developments. J. Gen. Virol. **37:**1–14.

Oliver, J., and M. MacDowell. 1957. The renal lesion in epidemic hemorrhagic fever. J. Clin. Invest. **36:**99–222.

Rosen, L. 1981. The use of *Toxorhynchites* mosquitoes to detect and propagate dengue and other arboviruses. Am. J. Trop. Med. Hyg. **30:**177–183.

Schmaljohn, C. S., and J. M. Dalrymple. 1983. Analysis of Hantaan virus RNA: evidence for a new genus of Bunyaviridae. Virology **131:**482–491.

Shope, R. E., and O. R. Causey. 1962. Further studies on the serological relationships of group C arthropod-borne viruses and the application of these relationships to rapid identification of types. Am. J. Trop. Med. Hyg. **11:**283–290.

Shope, R. E. 1985. Bunyaviruses, p. 1055–1082. *In* B. N. Fields (ed.), Virology. Raven Press, New York.

Smithburn, K. C., A. J. Haddow, and A. F. Mahaffy. 1946. Neurotropic virus isolated from *Aedes* mosquitoes caught in Semliki Forest. Am. J. Trop. Med. Hyg. **26:**189–208.

Sudia, W. D., V. F. Newhouse, C. H. Calisher, and R. W. Chamberlain. 1971. California group arboviruses: isolation from mosquitoes in North America. Mosq. News. **31:**576–600.

Taussig, S. 1905. Die Hundskrankheit, endemischer Magankatarrh in der Herzegowina. Wien Klin. Wochenschr. **18:**129–136, 163–169.

Theiler, M. 1957. Action of sodium desoxycholate on arthropod-borne viruses. Proc. Soc. Exper. Biol. Med. **96:**380–382.

Theiler, M., and W. G. Downs. (compilers and editors). 1973. The arthropod-borne viruses of vertebrates. Yale University Press, New Haven, CT.

Tikasingh, E. S., L. Spence, and W. G. Downs. 1966. The use of adjuvant and sarcoma 180 cells in the production of mouse hyperimmune ascitic fluids to arboviruses. Am. J. Trop. Med. Hyg. **15:**219–226.

Umenai, T., H. W. Lee, P. W. Lee, T. Saito, T. Toyoda, M. Hongo, K. Hoshinaga, T. Nobunaga, T. Horiuchi, and N. Ishida. 1979. Koren hemorrhagic fever in staff in an animal laboratory. Lancet **1:**1314–1316.

Ushijima, H., C. M. Clerx-van Haaster, and D. H. L. Bishop. 1981. Analyses of the Patois group bunyaviruses: evidence for naturally occurring recombinant bunyaviruses and existence of viral coded nonstructural proteins induced in bunyavirus infected cells. Virology **110:**318–332.

von Bonsdorff, C. H., P. Saikku, and N. Oker-Blom. 1969. The inner structure of Uukuniemi and two Bunyamwere supergroup arboviruses. Virology **39:**342–344.

Watts, D. M., W. H. Thompson, T. M. Yuill, G. R. DeFoliart, and R. P. Hanson. 1974. Overwintering of La

Crosse virus in *Aedes triseriatus*. Am. J. Trop. Med. Hyg. **23**:694–700.

Whitman, L., and J. Casals. 1961. The Guama group: a new serological group of hitherto undescribed viruses. Immunological studies. Am. J. Trop. Med. Hyg. **10**:259–263.

Whitman, L., and R. E. Shope. 1962. The California complex of arthropod-borne viruses and its relationship to the Bunyamwera group through Guaroa virus. Am. J. Trop. Med. Hyg. **11**:691–696.

Woodall, J. P. 1979. Transmission of Group C arboviruses (Bunyaviridae), p. 123–128. *In* E. Kurstak, (ed.), Academic Press, New York. Proc. Second Intl. Symp. on Arctic Arboviruses.

CHAPTER 33

Arenaviridae: The Arenaviruses

JOSEPH B. McCORMICK

Diseases: Lymphocytic choriomeningitis (LCM), Lassa fever, Argentine hemorrhagic fever (AHF), Bolivian hemorrhagic fever (BHF).

Etiologic Agents: Lymphocytic choriomeningitis virus, Lassa virus, Junin virus, Machupo virus.

Source: Infected urine of various rodent species, and person to person transmission for Lassa virus.

Clinical Manifestations: Febrile illness with systemic symptoms usually including headache, myalgias, and weakness. Meningitis occurs with LCM; sore throat and gastrointestinal symptoms are common with Lassa fever. Hemorrhage and late neurologic disease are common in AHF and BHF. Infections range from asymptomatic for all to fatal in some cases of AHF, BHF, and Lassa fever.

Pathogenesis: Generalized virus infections. Each virus probably somewhat different. LCM results from immune response and specifically T-cell destruction of infected cells in the central nervous system. Other diseases involve dysfunction of the hemostatic system and endothelium, resulting in bleeding and shock.

Laboratory Diagnosis: Isolation of virus from blood, cerebrospinal fluid or tissues, or demonstration of four-fold increase in virus-specific IgG antibodies, or high titered IgG antibody ($\geq$256) with virus-specific IgM.

Epidemiology: LCM occurs in all of the Americas, Europe, and Asia. Lassa fever in West Africa, AFH in a circumscribed area south of Buenos Aires, BHF in small area of Bolivia. All result from contact with infected rodent urine, usually through contamination of cuts and scratches on skin.

Treatment: Convalescent plasma for AHF, ribavirin for Lassa fever. Both therapies might work with BHF, but no data available. Symptomatic therapy for LCM.

Prevention and Control: Avoid contact with infected rodent urine. Barrier nursing practices in hospitals prevent nosocomial transmission of Lassa virus. Vaccines are under development, but not yet available.

The Viruses

History and Classification

Lymphocytic choriomeningitis (LCM) virus, the first identified arenavirus, was isolated by Lillie and Armstrong in 1933 from the cerebrospinal fluid of a patient suspected of having St. Louis encephalitis (Armstrong and Lillie, 1934). The virus was again isolated in 1935 from patients with aseptic meningitis (Rivers and Scott, 1935) and finally by Traub in 1935 from laboratory mice. More than 20 years passed before other members of this taxon were identified: Junin virus was isolated from patients with Argentine hemorrhagic fever (Parodi et al., 1958), and Machupo virus from patients with Bolivian hemorrhagic fever (Johnson et al., 1965). Several other arenaviruses were also obtained from rodent species in South America, but none has yet been associated with human illness (Webb et al., 1970; Pinheiro et al., 1966). Lassa virus was the first arenavirus isolated from Africa in 1969 (Buckley et al., 1970), the same year in which Murphy proposed a new taxon, based primarily on morphologic identity between these geographically diverse viruses; this group was given the name *Arenaviridae* (Murphy et al., 1969). (Arena was taken from the Latin word for sand because host–cell ribosomes appear like grains of sand in virus particles, as seen by electron microscopy.)

There are now 13 members of the *Arenaviridae*, which may be divided roughly into two geographic groups, Old and New World *Arenaviridae* (Table 1). Only four of the viruses—LCM, Lassa, Junin, and Machupo—are pathogenic for humans. The arenaviruses from each geographic area are antigenically related to other viruses from the same area (Wulff et al., 1978), and there are also some antigenic relationships between the Old and New World arenaviruses (Buchmeier et al., 1980). But the difficulty in establishing consistent plaque-reduction neutralization assays for many of these viruses has made definitive relationships difficult to determine. No effort has yet been made to create serogroups within the *Arenaviridae*, but a more detailed analysis of arenaviruses with the tools of molecular biology and monoclonal antibodies should lead to a more precise definition of the relationships between various members.

Physical, Chemical, and Morphologic Features

The arenaviruses are virtually indistinguishable from one another morphologically (Murphy et al., 1969). They are enveloped, pleomorphic, membrane viruses ranging in diameter from 50 to 300 nm, with a mean diameter of 110 to 130 nm. The virion density in sucrose is 1.17 g/cm³. They contain two segments of single-stranded RNA tightly associated with a nucleocapsid protein of 65,000 to 72,000 molecular weight. This is enclosed in a membrane consisting of two glycosylated proteins of molecular weights of about 34,000 to 44,000 and 54,000 to 72,000; the latter has been reported to be absent in some viruses (Matthews, 1982). A protein of about 200,000 molecular weight, found in the virion, is presumed to be the RNA polymerase. The characteristic granules noted in electron micrographs of arenaviruses have been shown to be host-cell ribosomes, which have no known role in replication of the virion (Rawls et al., 1975).

The genome of arenaviruses consists of two segments of single-stranded RNA, a large strand of negative-sense RNA of molecular weight 2.0 to 3.2×10^6 that codes for the RNA-dependent viral RNA polymerase. No RNA has yet been completely cloned and sequenced. The second segment or small RNA (sRNA) is an ambisense single strand of about 1.1 to 1.6×10^6 molecular weight, which encodes for the glycoprotein precursor (GPC) and for the nucleoprotein (NP) (Auperin et al., 1984, 1986). The NP and GPC genes are encoded in nonoverlapping reading frames with origins at the 3' and 5' ends of the molecule, respectively. Thus, the N gene is encoded by the 5' half of the viral complementary RNA sequence corresponding to the 3' half of the viral RNA molecule. The GPC gene is encoded by the 5' half of the viral RNA molecule. The GPC genes of Lassa and LCM viruses share many common sequences, espe-

TABLE 1. Members of the *Arenaviridae*

Virus	*Location*	*Primary host*
Old World		
Lymphocytic choriomeningitis virus	Americas, Europe, Asia	*Mus musculus*
Lassa virus	West Africa	*Mastomy natalensis*
Mopeia virus	Southern Africa	*Mastomy erythrolucus*
Mobala	Central Africa	*Praomys jacksoni*
New World		
Junin virus	Argentina	*Calomys sp.*
		Akodon asarae
Machupo	Bolivia	*Calomys callosus*
Amapari	Brazil	*Oryzomys goeldi*
		Neacomys guianae
Flexal	Northern Brazil	
Parana	Paraguay	*Oryzomys buccinatus*
Tamiami	USA	*Sigmodon hispideis*
Pichinde	Colombia	*Oryzomys albigularis*
Latino	Bolivia	*Calomys calosus*
Tacaribe	Jamaica	

TABLE 2. Estimated sizes of viral structural proteins of selected arenaviruses

Virus	Nucleocapsid proteins (NP)	Glycoprotein precursor (GPC)	Glycoproteins	
			GP1	GP2
LCM	63,000	62,250[a]	54,000	35,000
Lassa	61,000	55,820[a]	45,000	38,000
Pichinde	72,000		72,000	34,000
Junin	64,000		52,000	38,000
Tacaribe	68,000		42,000	

[a] Nonglycosylated form.

cially in the GPC portions (Auperin et al., 1984). The genes are separated by a hairpin structure that is a noncoding intergenic region which may act as an RNA transcription terminator. It is postulated that this coding mechanism allows independent control over the production of N and GPC, which may play a central role in the ability of the viruses to produce persistent infection (Auperin et al., 1984). The details of how this may occur, however, are not yet clear.

The nucleoprotein associated with the viral nucleocapsid is presumed to protect the RNA from intracellular damage and is known to be the antigen associated with the complement fixation (CF) reaction. The GPC is cleaved into two proteins, both of which are glycosylated, but the specific roles of the two glycosylated proteins (GP1 and GP2) are unproven, though they are believed to be associated with attachment to and penetration of the cell membrane.

Several studies of epitopic mapping of mouse monoclonal antibodies have shown conserved epitopes between arenaviruses, with the highest conservation between viruses from neighboring geographic areas but with some conservation of epitopes even between Old and New World arenaviruses. For example, an epitope on the GP2 of African arenaviruses is also found on many of the South American arenaviruses. Studies of LCM have similarly demonstrated epitopes on the GP2 and NP of LCM that are shared by Mopeia virus from southern Africa, and a GP2 epitope from Lassa virus is broadly reactive with many arenaviruses from South America and Africa (Table 2).

Clinical Characteristics

Old World Arenavirus Diseases

LYMPHOCYTIC CHORIOMENINGITIS

Lymphocytic choriomeningitis virus causes an acute febrile disease in humans, often associated with an aseptic meningitis. Although rarely fatal, the disease can be quite severe, resulting in hospitalization and a prolonged convalescence. Mild and asymptomatic infections also occur; the severity of illness may depend on both dose and route of infection, although no conclusive studies are available. The virus is transmitted to humans from several species of rodent (see Epidemiology and Ecology), and illness follows an incubation period of 1 to 3 weeks. There is no evidence of person-to-person transmission.

The typical disease begins with fever, malaise, weakness, headache, and myalgia. The headache is often retro-orbital, and the myalgia is most severe in the lumbar region. Anorexia, nausea, and dizziness are also common. As many as 50% of patients may have any or all of several manifestations including sore throat, vomiting, and arthralgias, with chest pain and pneumonitis occurring less frequently. Alopecia and orchitis have also occurred with LCM infection. Physical examination shows pharyngeal inflammation, usually without exudate, and in more severely ill patients meningeal signs and even signs of encephalopathy occur. The white blood cell count is often 3,000 or less, with a mild thrombocytopenia. Cerebrospinal fluid from patients with meningeal signs contains several hundred white cells, predominantly lymphocytes (>80%), with mildly increased protein and occasionally low sugar levels. Convalescence is prolonged, with persistent fatigue, somnolence, and dizziness.

LASSA FEVER

Lassa virus also causes an acute febrile disease, but this may progress to shock, respiratory distress, encephalopathy, and death. The disease is acquired by humans from the rodent *Mastomys natalensis,* but can also be transmitted from person to person (see Epidemiology and Ecology).

Lassa fever begins insidiously after 7 to 18 days of incubation (McCormick et al., 1987b). The early symptoms are fever, weakness, and malaise. More than 50% of patients complain of joint pain and lumbar pain by the 3rd or 4th day of illness. More than 60% of patients develop a nonproductive cough, also by the 3rd or 4th day of illness. Severe headache, usually frontal, and a very painful sore throat occur in a majority of patients by the 4th or 5th day of illness. Many patients develop a severe retrosternal chest pain on the 3rd to 5th day of illness. Vomiting or diarrhea associated with cramping abdominal pain occurs in 50% of Lassa fever patients. Prostration develops between days 6 and 8 of illness in 25% to 30% of hospitalized Lassa fever patients. On physical examination the respiratory rate, temperature, and pulse rate are elevated, and the blood pressure may be low. Conjunctivitis is seen in about 30% of patients. Pharyngitis is seen in 70% of patients, with more than 50% also having tonsillar exudates. Bleeding is seen in 15% to 20% of patients, limited primarily to the mucosal surfaces. Edema of the face and neck is uncommon, but is associated with more severe disease. Pneumonitis and pleural and pericardial effusions occur in some patients. The abdomen is tender in 50% of patients. Neurologic signs consist of unilateral or bilateral deafness in as many as 20% of all infections, and, in severe disease, diffuse encephalopathy with or without general seizures in a minority of patients. The mean white count in Lassa fever patients is normal (about 6,000/mm), with a relative lymphopenia. Thrombocytopenia is mild when it occurs, but platelet function is abnormal. Proteinuria occurs in two-thirds of patients. Lassa fever is fatal in 15 to 20% of untreated hospitalized patients.

Ribavirin, a guanosine analog, is effective in treating acute Lassa fever (MCCormick et al., 1986a). A five- to tenfold decrease in the case–fatality ratio was demonstrated in patients treated with ribavirin compared with untreated patients when therapy was given within the first 6 days of illness. A smaller, but still significant, decrease in fatality was also seen in patients treated later in illness. The demonstration of effective antiviral therapy is a further stimulus to develop and employ quantitative rapid laboratory diagnosis of arenaviral diseases.

New World Arenavirus Diseases

ARGENTINE AND BOLIVIAN HEMORRHAGIC FEVERS

The clinical aspects of Argentine hemorrhagic fever (AHF) and Bolivian hemorrhagic fever (BHF) are sufficiently alike to warrant a joint description (Maiztegui, 1975; Mercado, 1977). These diseases also have an insidious onset of malaise, fever, general myalgia, and anorexia. Nausea and vomiting commonly occur after 2 or 3 days of illness. Temperature is high, reaching 40°C or above with little variation. Unlike LCM and Lassa fever, AHF and BHF do not usually lead to respiratory symptoms and sore throat. On physical examination, patients often have conjunctivitis, but unlike LCM and Lassa fever, erythema of the face, neck, and thorax are common. There may be a pharyngeal enanthem. Relative bradycardia is often observed.

The second stage of AHF and BHF begins between the 4th and 6th days of illness. Unlike Lassa fever and LCM, the South American diseases are frequently associated with hemorrhagic manifestations that herald the second stage of illness, most commonly epistaxis and/or hematemesis. The bleeding may be from mucosal surfaces or into the skin, since petechiae and hemorrhagic rash are observed in AHF and BHF (but rarely in LCM and Lassa fever). The loss of blood is rarely sufficient to be fatal. Fifty percent of AHF and BHF patients also manifest neurologic symptoms, such as tremors of the hands and tongue, during the second stage of illness (Biquard et al., 1977), signs not observed in Lassa fever or LCM. These manifestations may progress in some patients to delirium, oculogyrus, and strabismus. Between 50 and 70% of patients with BHF and AFH experience hypotension between the 6th and 10th days of illness. As in Lassa fever, it is the hypotension, which becomes irreversible in some patients, that accounts for the majority of deaths from AFH and BHF. In severe illness the bleeding and shock may persist from 2 to 4 days. Leukopenia and thrombocytopenia (which are much more pronounced in AHF and BHF than in Lassa fever), hemoconcentration, and high urinary protein excretion are common features of the acute diseases.

Similar to Lassa fever and LCM, the third or convalescent stage of AHF and BHF is long (3 to 6 weeks) and characterized by weakness, weight loss, autonomic instability, and occasionally allopecia.

Convalescent-phase plasma has been proven an effective therapy for Argentine hemorrhagic fever (Maiztegui et al., 1979). Patients treated with plasma containing neutralizing antibodies in the first 8 days of illness with AHF had a 1% case fatality compared with 16% in untreated patients or those treated later in illness. The efficacy of the plasma depends on the titer of neutralizing antibody (Enria et al., 1986). Similar therapy for Lassa fever with convalescent-phase plasma has not been effective (McCormick et al., 1986a).

Pathogenesis and Pathology

The most common site of initiation of infection for the arenaviruses is not yet known, although it seems likely to be cuts and abrasions in the skin; however, under appropriate circumstances infection may be initiated in mucous membranes (for example, contact with infected rodent colonies) (Gregg, 1975). After the initiation of infection, all of the arenaviruses appear to produce generalized multiorgan infections, especially of the reticuluendothelial system and the lymph nodes (Walker et al., 1982; Elsner et al., 1973). From this point, however, the pathogenesis and pathology of the various diseases diverge.

Old World Arenavirus

LYMPHOCYTIC CHORIOMENINGITIS

There are no published descriptions of the pathology of LCM infection in humans. In animal studies, the leptomeninges have been densely infected with lymphocytic cells, with little involvement of the brain parynchema. An interstitial pneumonitis has also been described in nonhuman primate studies post mortem. However, one must be very cautious about extrapolating these observations to human disease. In humans, presumably LCM virus invades the meninges, followed by lymphocytic infiltration causing swelling and inflammation.

LASSA FEVER

Necropsy studies of Lassa fever patients have shown few gross pathologic changes. The most frequently and consistently observed microscopic lesions in fatal Lassa fever are focal necrosis of the liver, adrenal glands, and spleen (McCormick et al., 1986b; Walker et al., 1982). The liver damage in 19 cases showed a wide range in the degree of hepatocytic necrosis. The liver appears to go through cellular injury, necrosis, and regeneration, with any or all present at death. Although a vigorous macrophage response occurs, little if any cellular inflammatory response with lymphocytic infiltrate has been observed. However, none of the 19 fatal cases studied exhibited sufficient

hepatic damage to implicate hepatic failure as a cause of death. Similarly, moderate splenic necrosis has been consistently observed, primarily involving the marginal zone of the periarteriolar lymphocytic sheath (Walker et al., 1982). Diffuse focal adrenocortical cellular necrosis was observed in three cases; it was most prominent in the zona fasciculata but involved no more than 10% of the cells. Although high virus titers occur in other organs, such as brain, ovary, pancreas, uterus, and placenta, no significant lesions have been found. Thus, few clues as to the pathophysiology of Lassa fever are found in standard pathologic studies.

Whatever the final pathway to disease may be, it has clearly been demonstrated for Lassa fever that the outcome and severity of disease are related to the level of viremia (McCormick et al., 1986a; Johnson et al., 1987). In fatal cases, virus titers are significantly higher throughout illness, compared with those in nonfatal illness (Fig. 1).

Despite being categorized as a hemorrhagic fever, Lassa fever does not produce dramatic alterations in measures of coagulation (Fisher-Hoch et al., 1987; Lange et al., 1985). Platelet and fibrinogen turnover are normal, and there is no increase in fibrinogen breakdown products. Thus, disseminated intravascular coagulation is not a significant component in the course of fatal Lassa fever (Fisher-Hoch et al., 1987). Profound abnormalities in platelet function (despite adequate numbers of circulating platelets) and in endothelial function have been observed in primates and humans infected by Lassa virus (Fisher-Hoch et al., 1987). These abnormalities appear to be mediated by, or at least related to, disturbances in arachidonic acid metabolism in platelets and endothelium. Thus, a marked decrease in prosta-

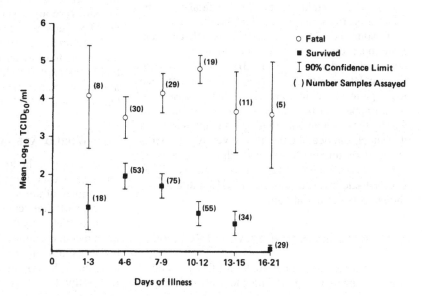

FIG. 1. Virema as related to outcome in Lassa fever.

cyclin production by endothelium has been measured in primates infected by Lassa virus. These disturbances are associated with the massive capillary leakage seen in severe disease and with the bleeding that results from poorly functioning platelets. One can only speculate about possible mediators of these disturbances, such as effectors released from neutrophils and macrophages, since definitive studies are lacking.

Argentine Hemorrhagic Fever and Bolivian Hemorrhagic Fever

Argentine hemorrhagic fever and BHF are classic hemorrhagic diseases and have very similar pathologic features, some of which differ from Lassa fever (Elsner, 1977). Unlike patients with Lassa fever, patients with AHF manifest a skin rash and petechiae. Gross examination of organs at necropsy also show petechiae on the organ surfaces. Bleeding ulcerations of the digestive tract have been described, although exsanguinating bleeding with hemoperitoneum has not been described.

Microscopic examination shows a general alteration in endothelial cells with a mild edema of the vascular walls, with capillary swelling, and perivascular hemorrhage. Large areas of intra-alveolar or bronchial hemorrhage are often seen with no evidence of inflammatory process. In half of the cases pneumonia, necrotizing bronchitis or pulmonary emboli were observed. Hemorrhage and a lymphocytic infiltrate were observed in the pericardium, occasionally with interstitial myocarditis. The lymph nodes are enlarged and congested with reticular-cell hyperplasia. Splenic hemorrhage is common. Medullary congestion with pericapsular and pelvic hemorrhages are commonly observed. Acute tubular necrosis is observed in about half of the fatal cases, but adrenal necrosis has not been reported.

Despite the different degrees hemorrhagic manifestations, there are sufficient similarities in the course of the disease in AHF, BHF, and Lassa fever to speculate that there exists a similar pathophysiologic pathway underlying all of the diseases. All would appear to be primarily diseases severely affecting the vascular endothelium, leading to irreversible shock and death in the most severely ill patients. It seems reasonable to speculate that the events described in the studies of Lassa fever, and even Ebola hemorrhagic fever, also occur in AHF and BHF, although proof is not at hand.

Immune Response to Arenavirus Infections

The very large body of literature on LCM virus infection in animals is a good indication of the complex relationships among factors such as host genetics, virus genetics, route of inoculation, age of the host, and titer of the virus inoculum. All of these factors play a role in arenavirus infection in small animals. The details of these and many other aspects of arenavirus immunology are summarized in a recent review (Oldstone, 1987).

The classical arenavirus model, that of LCM in the mouse, illustrates the central role of the cellular immune response in both virus clearance and disease. In this model, for a given route of infection and virus strain, it is the timing of infection that is critical. Infection in utero or within about 5 days after birth results in the establishment of persistent infection (Hotchin and Weigard, 1961; Riviere et al., 1986; Zinkernagel et al., 1985). A humoral immune response accompanies persistent infection, but there is no evidence of an effective cellular immune response. Systemic illness does not result. However, infection after the first week of life usually results in a humoral and cellular immune response and ensuing classical LCM disease in mice, whereas intraperitoneal infection is cleared within 2 to 3 weeks without evidence of disease. The meningitis appears to be the direct effect of virus-specific, cytotoxic T cells on infected cells, since ablation of the cytotoxic T-cell response results in persistent infection and no disease (Cole et al., 1972).

In newborn rodents, infection with other arenaviruses is chronic and may result in disease. However, infection of adult rodents (either a laboratory mouse or the natural virus host) does not result in a disease similar to that seen from LCM virus infection. The immune response elicited by these other arenaviruses in feral or laboratory rodents is incompletely studied. The viruses produce a humoral immune response; however, the presence of neutralizing antibodies remains variable, and the cellular immune response has not been adequately characterized for arenaviruses other than murine LCM virus.

The immune response to arenavirus infections in the accidental human host varies significantly from that in the natural rodent host. However, the human immune response also varies considerably between Old and New World arenaviruses.

OLD WORLD ARENAVIRUSES

Most studies of the immune response to LCM in humans have used the CF test and have reported the appearance of antibody as early as the first week of illness, with titers culminating in 40 to 60 days (Smadel and Wall, 1940). No data are available on the human cellular immune response to LCM. The antibody response to Lassa fever occurs early in the disease, with IgG antibodies in 46% of patients and IgM antibody in 59% by the 6th day of illness; both

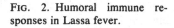

FIG. 2. Humoral immune responses in Lassa fever.

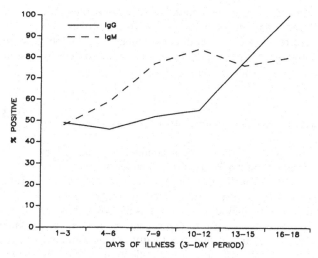

are found in 100% of patients by the 16th day of illness (Cossio et al., 1979) (Fig. 2). The IgG response includes antibodies to all structural proteins, as demonstrated in humans as well as primate studies (Mitchell et al., 1987). However, neutralizing antibodies are not found in the serum of most patients recovering from Lassa fever. In a minority of patients, an elevated log neutralization index (≥ 2.0) can be demonstrated with late convalescent-phase serum (Jahrling and Peters, 1984). However, neutralizing antibodies cannot be demonstrated by a constant virus–varying-serum-dilution test and are not sufficient to eliminate virus during acute infection. In more than 25% of patients, virus may be isolated in high titer in the face of high antibody levels.

Attempts to measure a cell-mediated immune (CMI) response to Lassa virus in humans and primates have been limited. In one study, primates with fatal Lassa infection did not show a lymphocyte transformation response (LTR) to inactivated Lassa virus, and no cytotoxic T-cell killing of infected syngeneic target cells was demonstrated (Mitchell et al., 1987). However, during the acute infection there was a profound lymphopenia, suggesting the possibility of immunosuppression during the acute illness. It has been possible to demonstrate an LTR in a convalescent Lassa fever patient. Thus, rather strong circumstantial evidence suggests that recovery from acute Lassa fever is CMI dependent. While passive administration of concentrated convalescent-phase plasma to monkeys has prevented infection (Jahrling et al., 1984), the vaccination of monkeys with killed vaccine followed by a strong anemestic antibody response was insufficient to protect the monkeys from fatal Lassa infection (Mitchell et al., 1987). These and human data indicate that a humoral immune response to natural infection is not adequate for viral clearance and probably not sufficient to prevent primary infection.

NEW WORLD ARENAVIRUSES

Studies of the antibody response to Junin and Machupo virus infections in humans indicate that the CF antibody was usually detectable 3 to 4 weeks after the onset of illness and that it was measurable for only a few months (McKenzie et al., 1965). Neutralizing antibody appears at approximately the same time as CF antibody but is measurable for at least several years (Webb et al., 1969). In contrast to Lassa fever, the appearance of antibody in Junin infection coincides with disappearance of virus in the blood. More recently, the use of indirect fluorescent antibody tests has shown that antibodies may be detected at the end of the second week in illness, still later than the antibodies elicited by Lassa virus infection (Cossio et al., 1979). No data are available on the protein specificity of the immune response, but studies of monoclonal antibodies suggest that antibodies to the glycoprotein-1 are responsible for neutralization. There are no data on an antigen-specific, cell-mediated immune response to either of these infections.

Laboratory Diagnosis of Arenavirus Infections

The laboratory diagnosis of arenavirus infections rests on the demonstration of a fourfold rise in antibody titer, or the isolation of virus, or demonstration of high-titered IgG antibody, and virus-specific IgM antibody in association with compatible clinical disease. Since three of the four human diseases are caused by viruses requiring laboratory biosafety level 4, any manipulation of live virus in tissue culture requires this level of containment. Fortunately, antigens may be prepared and inactivated so that the serologic diagnosis may be performed without such safety precautions (Elliott et al., 1982).

Specimen Collection, Storage, and Transfer

Routine isolation may be accomplished from serum specimens taken from acutely ill patients (Johnson et al., 1987). Specimens should be drawn, preferably in a vacuum tube system to minimize risk of contamination, and the blood allowed to clot at room temperature. The specimen is centrifuged at low speeds (1,000 × g) for 10 min in a tightly closed container (for example, vacuum blood collection tube). The serum is then removed by pipette and placed into a sealable vial (plastic screw cap is preferable) for storage or transfer. Urine should be mixed with an equal amount of bovine serum albumin, pH 7.4, before freezing. Other fluids should be frozen undiluted. All of these specimens will keep best if they are frozen at −70°C or lower as soon as possible.

Virus isolation may be attempted from serum specimens, throat swabs, urine, cerebrospinal fluid, breast milk, or other (for example, joint) fluid. Throat swabs should be obtained from the posterior pharynx and tonsillar fossae, and the swab should be placed in Eagle medium with 2.75% BSA buffered to pH 7.4 for storage or transport. Specimens for virus isolation are placed at −70°C for virus isolation. If the specimen cannot be stored at −70°C, the virus will survive several days at −20°C; however, with loss of virus titer. Isolation may also be accom-

plished from whole blood in the event it is impossible to safely separate serum. In this case, the whole blood should be stored in the same manner as the serum. Antibody may be stored for several weeks at 4°C, or for years at −20°C. Tissue specimens for virus isolation should be minced to 1-mm cubes in a petri dish in Eagle minimum essential medium (EMEM) (pH 7.4) with 2.75% bovine serum albumin (BSA), placed in a vial with EMEM, and frozen at −70°C.

Inactivation and Disinfection of Arenaviruses

Noninactivated materials derived from Junin, Machupo, and Lassa viruses must be kept in a biosafety level (BSL) 4 laboratory at all times, and those from most LCM strains kept in BSL 3. For work with diagnostic or other reagents derived from these viruses at a BSL 1 or lower, the reagents must be inactivated. Antigenic properties are best conserved by inactivation with gamma radiation (Elliott et al., 1982). Heat inactivation, β-propiolactone, formalin, and UV radiation are also effective; however, antigenic characteristics are less well conserved by these methods (Mitchell and McCormick, 1984). Extraction of virion RNA by phenol is sufficient to completely inactivate the virus, as is treatment with detergents during the isolation and purification of viral proteins. Curves defining the quantity of virus inactivated by a given dose of radiation are illustrated in Figure 3.

Disinfection of arenaviruses can be effectively accomplished by washing with 0.5% phenol in detergent (for example, Lysol), by 0.5% hypochlorite solution, or by formaldehyde or paracetic acid.

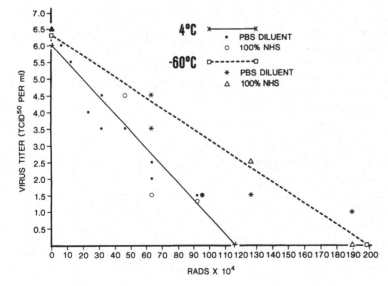

FIG. 3. Inactivation of Lassa virus by gamma irradiation.

Virus Isolation Methods

Isolation of arenaviruses may be best accomplished by use of cell culture techniques. Alternatively, suckling mice may be used. Primary arenavirus isolation may be performed with Vero cells or baby hamster kidney cells (BHK), but the most sensitive cell line for primary isolation of all the arenaviruses may be the E6 clone of Vero cells (Johnson et al., 1987). Cells are grown in growth medium until confluent (usually 24 to 48 h of growth). Roller tubes or a microtiter culture system may be used. Virus material (0.1 ml of either serum or other material such as 10% tissue suspension) is added to 0.9 ml of maintenance medium in roller tubes or 24-well microtiter plates, and the cells are incubated at 37°C with the caps tightly closed or in 5% CO_2 with loosened caps (panels must be in 5% CO_2). Material that may be toxic to the cell culture should be removed by washing after 1 h of incubation, and fresh EMEM should be placed over the monolayer. For routine isolation, the cells should be observed for cytopathic effect (CPE) (it may not always be present) and may be harvested after 7 days. Virus presence can be assayed by several techniques including immunofluorescent antibody (IFA), enzyme-linked immunosorbent assay (ELISA), or reverse passive hemagglutination (RPHA) (Johnson et al., 1987). While some arenaviruses produce CPE, not all do, especially on the primary isolation. Consequently, absence of CPE is not a definitive criterion for absence of virus. For a more rapid diagnosis, cells may be sampled after 48 to 72 h and assayed by IFA for presence of antigen. Alternatively, a plaque method may be used, in which case the first overlay of agar is made at the time of infection, and the second agar overlay made at 5 to 7 days; the plaques can be read in 24 to 48 h. For titration of virus samples, four- or tenfold dilutions of the original material are made and assayed as described. However, viruses that give little CPE may plaque poorly.

The simplest method of detecting an arenavirus in cell culture is a direct fluorescent antibody (FA) test with a broadly reacting monoclonal or polyvalent antibody on infected cells fixed on a microscopic slide. Alternatively, ELISA capture, ELISA antigen-detection system, or radioimmunoassay (RIA) may be used.

Suckling mice and guinea pigs have also been used for virus isolation. However, these are much less preferable because of the expense, safety considerations, and complexity involved. All such work must be done in a BSL 4 laboratory. For mice, 0.03 ml of undiluted material is placed in the brain of a 1- to 3-day-old mouse. The mice are killed after 7 to 9 days, and the brain material is harvested. A 10% suspension of the material may be titrated in cell culture, or the material can be examined directly for presence of virus by ELISA, or the IFA test can be performed on tissue sections. When guinea pigs are used, a young animal (4 to 6 weeks old) should be inoculated intraperitoneally with 1 ml of diagnostic material. The animal should be followed daily for fever, and the virus harvested from blood and organs on the 3rd or 4th day of fever.

Serologic Tests

IMMUNOFLUORESCENT ANTIBODY TEST (IFAT)

The IFAT is commonly performed on infected cell monolayers that have been inactivated (if an arenavirus pathogenic for humans), preferably by gamma irradiation (Elliott et al., 1982), but also by several other methods, including heat, formalin, and β-propiolactone (Mitchell and McCormick, 1984) (see section on Inactivation and Disinfection above). Cell suspensions are prepared by inoculating cell monolayers with a multiplicity of infection (MOI) of 0.01. Cells may be harvested from 3 days (for LCM virus) to 5 days (for Lassa virus) or longer, depending on the particular arenavirus. After attaining 50 to 100% cellular infection, the cells are harvested by trypsin or glass beads, washed three times in phosphate-buffered saline (PBS) at pH 7.4, and assayed for the percentage of infected cells by IFAT. These cells are then mixed with uninfected cells to obtain a suspension of 50% infected cells, which are then inactivated. Aliquots of cells may be placed into cell-freeze medium and stored at −20°C or −70°C.

When antigen slides are desired, an aliquot is thawed, and a 10-μl drop of cell suspension is placed on each well of a microscope slide covered with a perforated layer of Teflon (Johnson et al., 1981). The slides are thoroughly dried and then fixed in cold acetone for 10 min. For pathogenic viruses, the slides may be subjected at this point to gamma-irradiation or β-propiolactone for inactivation if the cell suspension was not previously inactivated (Elliott et al., 1982; Mitchell and McCormick, 1984). The slides may then be stored at −20°C for 6 months or at −70°C for several years.

The IFAT is performed by placing 20 μl of the test sample in each well of the dry slide. The slides are incubated in a moist chamber for 30 min at room temperature. They are then rinsed in distilled water and washed for 10 min in PBS and thoroughly dried. The conjugated antiserum is placed in each well, and after incubation for 1 h, the slides are rinsed in distilled water and washed in PBS for 10 min. Excess water is drained, and cover slips are mounted with phosphate-buffered glycerine, pH 8.4. Best results are obtained with an epi-illuminated fluorescent microscope with at least a 50-watt mercury or xenon lamp. Test results are based on the presence or absence of antigen-specific fluorescence compared with a known positive serum. The proportion of fluorescent cells must also be compared with the known percentage of infected and uninfected cells placed on the slide (for example, if 75% of cells fluoresce when only 50% are known to be infected, then background or nonspecific fluorescence is occurring). In most laboratories, fluorescence must be present at a dilution of at least 1 : 16 before the serum is declared to contain virus-specific antibody, but such conventions differ from laboratory to laboratory. Many laboratories also use a subjective system of grading the brightness of the fluorescence from 1 to 4. This is semiquantitative at best, since a test is considered positive at any of those levels, but it can be useful for estimating titrations since the intensity of fluorescence diminishes with dilution of the serum. In any event, each laboratory should have its own positive control serum, which should form the basis for comparison with reactions by unknown sera. The illustrations in Figure 4 are representative of specified anti-IgG conjugated serum. Anti-IgM is more difficult to read because it gives a fluorescent rim around positive cells instead of staining the entire cytoplasm.

ENZYME-LINKED IMMUNOSORBENT ASSAY

The ELISA test is presently the best candidate for replacing the IFAT. Its advantages are the elimination of the need for a fluorescent microscope and the greater degree of quantification. On the other hand, the morphologic information from IFAT can be helpful, since antibodies to predominantly a single viral protein often give characteristic morphology (Fig. 4). Comparisons of IFAT and ELISA have not thus far shown one to be more sensitive or specific than the other for the arenaviruses. Furthermore, the preparation of the IFAT reagents is generally less complicated than that for the ELISA reagents. Finally, background, nonspecific reactions may be easier to detect by IFAT than by ELISA since the control cells are mixed with the positive cells in a single well. A radioimmunoassay can be used in some laboratories, but the need for radioisotopes will limit its general usefulness.

The ELISA test may be performed with purified virus, but more commonly with infected cells as target antigens. Infected cells as described above are diluted from 1 to 5×10^5/ml. Of the cell suspension in carbonate buffer, pH 9.3, 100 μl are placed in each well of a 96-well plate. For most serum specimens to be tested, especially those that have been

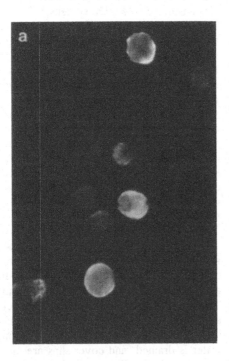 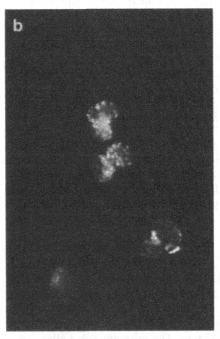

FIG. 4a and b. Immunofluorescence patterns of Lassa-infected Vero E-6 cells on human IgG.

stored frozen for long periods or that are from populations with high levels of immune complexes, such as found in many developing countries, use of uninfected control cells is recommended. The plates are tightly covered and incubated overnight at 4°C. The plates are rinsed three times with Tween 20* buffer. Then 0.1 μliter of test serum, which has been diluted in PBS with 0.5% gelatin and 15% Tween 20 buffer, is added to each well, and the plates are incubated for 1 h at 37°C. They are washed three times with Tween 20 buffer, and conjugated antiserum is added to each well, followed by 1 h of incubation at 37°C. The plates are washed three times as above, and appropriate substrate is added, depending on the enzyme system used. Plates are incubated for 30 to 60 min and then analyzed, preferably by spectrophotometer, but they may be read by eye if a spectrophotometer is not available. Such readings by the naked eye must be interpreted very conservatively, however, and only those sera that clearly react with antigen-containing wells and without evidence of background on uninfected cells may be recorded as positive. There are several methods of defining a positive test, including a specific optical density (OD) reading, or a positive serum may be defined as that which yields both

$$\text{OD of test} - \text{OD of control} > 0.05$$

and

$$\frac{\text{OD of test}}{\text{OD of control}} \geq 1.5$$

Sera that may contain high levels of immune complexes should each be reacted with infected and uninfected cells, since background reaction varies from serum to serum. Moreover, using only one "negative control" for all sera may produce many false positive tests. When high levels of reactivity are observed for infected and uninfected cells for a particular serum, then further dilution may be required to detect an antigen-specific reaction, with the risk of losing some sensitivity. Improved reagents from genetically engineered antigen-production systems may produce highly purified antigens, which may help in avoiding some of the difficulties of false positive reactions. However, many of these reactions are truly nonspecific and may occur regardless of the protein in the well. The ELISA may emerge as the preferable system, particularly if single arenavirus protein-specific reagents for the ELISA are developed.

NEUTRALIZATION TEST

The neutralization test is well established for some arenaviruses (Junin, Machupo), but its value is less certain for others. The test is performed for Junin

and Machupo viruses by the constant-virus–varying-serum-dilution method; 24-well panels or Labtek* slides can be used. Target cells in growth medium (EMEM with 5% fetal calf serum [FCS]) are grown in the wells of the plate or slide until confluent. One hundred microliters of serum at twice the final desired dilution in EMEM is mixed with an equal quantity of EMEM containing 100 PFU of virus. The mixture is incubated for 2 h at 37°C. The growth medium is removed from the cell layer, and the mixture of antibody and virus is placed over the cell layer. A layer of 1% agar is then placed over the cell layer. The cells are incubated at 37°C in 5% CO_2 for from 72 h to 5 days, depending on the virus. A second layer of agar containing neutral red dye is placed over the cell layer, and the cells are incubated for a further 24 h. An 80% reduction of plaques compared with control infection is considered the best measure of neutralization. Some laboratories use a 50% reduction, depending on the virus being evaluated. Neutralization tests may also be conducted in vivo with various animal models. These tests are much more cumbersome, particularly for the BSL 4 viruses. Nevertheless, guinea pigs and in some instances mice have been used for such evaluations.

IgM ANTIBODY ASSAY

Assays for IgM antibody may be reliably used for the arenaviruses (Johnson et al., 1987). Two assays have been used for detecting arenavirus-specific IgM antibodies, the IFAT and the capture ELISA system. The IFAT is conducted in exactly the same manner as the IFAT described above, except that anti-IgM antibodies conjugated to fluorescent dye are used in the last step of the assay. The assay is very specific, but its sensitivity depends on the amount of IgG antibody present; IgG antibody may compete for virus-binding sites and dilute the IgM antibodies. The IgM ELISA capture assay circumvents this problem by using anti-μ chain antibodies to capture only IgM antibodies from the serum being assayed. The disadvantage to the IgM capture system is that the arenaviruses, because they require BSL 4 procedures, must be grown, purified, and inactivated in order to supply antigen substrate for the capture system, whereas the IFAT uses only the same infected cells as used for the IgG assay. No satisfactory method of separating IgM from IgG antibodies in small serum samples has yet been demonstrated. Like the IgM capture assay, an IgG capture assay may also be performed.

---

* Use of trade names is for identification only and does not constitute endorsement by the Public Health Service or by the U.S. Department of Health and Human Services.

OTHER TESTS

The complement-fixation test was used in early investigations of human and animal arenavirus infections. However, the test is insensitive compared with the IFAT and is now rarely used; the reader is therefore referred elsewhere for details (Casals, 1975). The reverse passive hemagglutination and hemagglutination-inhibition tests have been used for Lassa virus and could be adapted to the other arenaviruses (Goldwasser et al., 1980). The RPHA test has been used primarily for antigen detection in tissue culture material and will be described in the Antigen Detection section. It appears to be less sensitive than the IFAT and ELISA tests for antibody detection and is therefore little used for that.

CELL-MEDIATED IMMUNITY ASSAYS

Assays for cell-mediated immune response may be the key to determining immunity to some arenaviruses. The simplest assay is the lymphocyte transformation test (LTT), for which peripheral white cells are collected in heparin-coated blood tubes. White cells are separated by centrifugation on a lymphocyte separation medium. At this stage, cells may be frozen in liquid nitrogen for later testing. For optimal preservation, the cells are placed in 1 ml of fetal calf serum (FCS), and then 1 ml of freeze medium (80% RPMI and 20% dimethyl sulfoxide) is added slowly with thorough mixing. The cells should be placed in an insulated container in vapor-phase nitrogen for 24 h before being submerged in liquid nitrogen. The LTT is performed by measuring ^{3}H thymidine after the cells are incubated in the presence of specific antigen compared with the amount of labeled thymidine incorporated by unstimulated cells. For Lassa fever, live antigen from cell culture is given 2,000 rads to limit replication of the cells before performing the LTT. An increase in incorporation greater than 20% above the level in unstimulated cells constitutes a positive response. Ideally, a cytotoxic T-cell assay needs to be performed. However, this assay presents considerable practical difficulties when applied to primates and humans, since the cells will not function satisfactorily after freezing, and syngeneic target cells are needed. These conditions are especially difficult to meet in areas endemic for arenavirus infections.

Antigen Detection

ANTIGEN-CAPTURE ELISA

The demonstration of effective therapy for AHF with immune plasma and Lassa fever with ribavirin has increased the practical value of a rapid diagnostic test for these diseases. An antigen-capture assay developed for Lassa virus appears to be moderately sensitive at least early in illness, although assay correlation with infectious virus levels declines later in the acute illness when more noninfectious antigen is present (Jahrling et al., 1985).

The assay is performed by placing Lassa virus immune monkey serum at a dilution of 1 : 200 carbonate buffer in wells of a 96-well microtiter plate and incubating for 1 h at 37°C. The plates are washed with Tween 20 buffer, and 0.1 ml of test serum undiluted or diluted in Tween 20 buffer with 3% BSA is placed in each well; the plates are incubated for 1 h at 37°C and then thoroughly washed as above.

To each well is added 0.1 mliter of virus-specific polyvalent guinea pig serum, and the plate is reincubated for 1 h. After washing, a third layer of rabbit anti-guinea pig immunoglobulin is added, and the plate is again incubated for 1 h and washed. Finally, swine anti-rabbit serum conjugated with alkaline phosphatase is added, followed by another 1-h incubation. Substrate is then added, and the reaction is read on a spectrophotometer after 20 min. An OD reading two standard deviations above the background of 30 normal serum sample readings is considered positive.

IN SITU ANTIGEN DETECTION

A second method of rapid antigen detection is the detection of viral protein by monoclonal antibodies in tissue imprints on a microscope slide. This method has been used for human post-mortem diagnosis as well as for detecting LCM virus and Lassa virus antigens in rodent tissues. The tissue to be used for the imprint is freshly removed and blotted on paper to clean off excess blood. The tissue may be a small piece from a necropsy, or it may be from a needle biopsy. Use of liver tissue has been most successful in detecting hemorrhagic fever viruses in both rodents and human post-mortem biopsies. The tissue is firmly placed against a thoroughly cleaned microscope slide and then lifted. Smearing or wiping motions are to be avoided, since they distort the cellular architecture. Four to six imprints may be made on a single slide, depending on the size of the tissue. Only four to six imprints should be made from a single tissue surface. For more imprints, the tissue surface should be recut. The slides are thoroughly dried and fixed for 10 min in acetone. They may be stored at −70°C for several months.

It appears that the nucleocapsid antigen is by far the most abundantly expressed antigen in infected cells during acute infection (McCormick et al., 1986b; Oldstone and Buchmeier, 1982). Consequently, the best success in antigen detection has resulted from the use of antinucleocapsid monoclonal antibodies as antigen probes. The monoclonal antibody is used at a dilution of 1 : 40 and is incubated on the cell imprint for 45 min at room temperature in a moist chamber. The slide is rinsed in distilled water and washed in PBS for 10 min. After it is dried, conjugated, anti-monoclonal antibody is placed on the imprints, and the slide is incubated for another 30 min. It is rinsed in distilled water and washed for 10 min in PBS. Excess water is allowed to run to the edge of the slide and blotted. Two drops of buffered glycerine are placed on the slide, and a cover slip is added.

Antigen may be found in single cells or in cell clusters, depending on the extent of infection. Efforts to detect antigen in conjunctival scrapings, cells from pharyngeal aspirates, and urinary sediment have not been successful.

A third method of virus detection is through the use of cDNA probes. Such probes are available for several arenaviruses (Auperin et al., 1984, 1986; Romanowski and Bishop, 1985); however, they have not been evaluated for diagnostic purposes.

Source of Reagents

Reagents for diagnosing and studying most arenaviruses are not available commercially. Many reagents are available from some noncommercial laboratories in limited quantities for use in research and diagnosis. These include antigens in the form of inactivated infected cells, positive antiserum, monoclonal antibodies, and cDNA probes. (Requests for such materials will be considered on an individual basis.) Inquiries regarding such reagents may be made to Special Pathgoens Branch, Division of Viral Diseases, Centers for Disease Control, Atlanta, Georgia 30333.

Epidemiology and Ecology

The epidemiology of all the arenaviruses is associated with the activity of the individual rodent species that are the host reservoirs for each of the viruses. The common feature is that all the rodent hosts are capable of persistent infection throughout life by their respective arenaviruses and chronically excrete the virus in their urine. However, the location of the contact between infected rodent and human target varies with each disease. Human intrusion into the rodents' ecologic niches is usually the determining factor for the frequency of human infection. That niche varies with the rodent species, and the degree of contact varies with the activities of the human population.

Old World Arenaviruses

LYMPHOCYTIC CHORIOMENINGITIS VIRUS

The epidemiology of LCM is associated with transmission from rodents in several different settings. LCM virus was originally associated with sporadic or natural infection from transmission of virus from feral rodents to humans. This source of human infection rarely results in an epidemic or even clusters of disease, since the frequency of contact between human and rodent excreta is rare. These cases usually occur in the winter when rodents seek shelter from cold temperatures in houses, thus increasing the probability of human contamination from rodent excreta. The mode of transmission is unknown, but would appear to be through direct contact with rodent urine deposited in the home, and infection through cuts or abrasions in the skin. The lack of clusters of these types of "natural cases" in families would suggest that aerosol transmission is unlikely. There are few studies of the prevalence of antibody to LCM virus in the general population. However, the data available suggest that the prevalence is probably below 0.1%.

The usual source of infection from LCM is through the contamination of pet rodents (Gregg, 1975), particularly hamsters and white mice, and through infection of laboratory personnel who unknowingly handle infected laboratory rodents. The use of cesarean deliveries of colonized rodents has been effective in reducing viral infections in many commercial rodent colonies; however, this technique is less effective for LCM virus, since the virus is also transmitted in utero and has little effect on the outcome of pregnancy or the health of the offspring. Thus, cohorting and screening colonies for antibody are the most effective method for detecting infection and reducing its spread in a rodent colony. Limiting the number of rodents that are imported for pets and for laboratory use has helped to reduce the frequency of this source of infection.

LASSA VIRUS

Lassa fever was first described in West Africa in the 1950s, although the virus was not isolated until 1969. Since that time a complex of antigenically related arenaviruses has been associated with several rodent species throughout Africa. However, the disease Lassa fever has only been shown to occur in West Africa, but with a wide geographic area from Northern Nigeria to Guinea, encompassing a population of perhaps 100 million. It has been estimated that more than 100,000 infections from Lassa virus may occur each year in West Africa, with several thousand deaths (McCormick et al., 1987a). The only known reservoir of Lassa virus in West Africa is *Mastomys natalensis,* one of the most commonly occurring rodents in Africa. At least two species of *Mastomys* occupy West Africa, and both have been found to harbor the virus (McCormick et al., 1987a). These rodents, especially the species with 32 chromosomes, are highly commensal with man. Studies of *Mastomys* have shown that their movement within a village is very limited and that their average life span is about 6 months. From 5% to as many as 70% of *Mastomys* have been found to be infected with virus in some village houses. The conclusion from systematic rodent and human studies is that most transmission takes place in and around the homes. This is reflected by the fact that infection and disease occur in all age groups and that antibody prevalence increases with age (McCormick et al., 1987a).

The death rate in systematically studied hospitalized patients with Lassa fever is 16%, very similar to that described for Junin and Machupo infections (McCormick et al., 1987b). The use of ribavirin therapy could cut this rate to 2 or 3% or less, on the basis of studies in West Africa (McCormick et al., 1986a). However, the death rate for all Lassa virus infections is more nearly 2 or 3%, since only 10 to 25% of infections result in a febrile illness. However, in endemic areas, from 5 to 15% of all febrile illness in a village may be caused by Lassa virus infection.

New World Arenaviruses

JUNIN VIRUS

Argentine hemorrhagic fever, caused by Junin virus, was recognized in the 1950s in the northwestern part of the Buenos Aires Province in Argentina, an area of very fertile farmland. The disease spread over the ensuing years from an area of 16,000 km^2 and a quarter of a million persons to an area greater than 100,000 km^2 and an affected population of over 1 million persons (Maiztegui et al., 1986). Since 1958 more than 20,000 cases of AHF have been reported. Before the effective therapy was used, the mortality rate was 15 to 20%; however, the routine use of convalescent-phase plasma has reduced the mortality rate to less than 3% (Enria et al., 1986).

The incubation period is 7 to 10 days. AHF is a seasonal disease with peak yearly incidence in May (Maiztegui et al., 1986). The disease is predominant in, but not limited to adult men, especially rural farm workers. The seasonality is associated with the corn harvest. Rodent populations in the field increase significantly during the harvesting season, and the number of field workers is also at a maximum during the harvest season. The reservoirs of Junin virus appear to be at least two rodents, *Calomys musculinus* and *Calomys laucha.* These rodents become chronically infected; the virus can be isolated throughout life from blood, and probably from urine.

MACHUPO VIRUS

The cause of Bolivian hemorrhagic fever, Machupo virus was first isolated in 1964. The virus is limited in its geographic distribution to a portion of the department of Beni in Bolivia. The only known reservoir for the virus is *Calomys callosus,* a cricetid rodent that is found in the highest density at the borders of tropical grassland and forest. The distribution of this rodent appears to include the eastern Bolivian plains, as well as northern Paraguay and adjacent areas of western Brazil. The disease was recognized in 1959, and by 1962 more than 1,000 cases had been identified in a confined area of two provinces, with a 22% case–fatality ratio. The largest known epidemic of BHF involving several hundred cases occurred in the town of San Joaquin in 1963 and 1964 (McKenzie et al., 1965). This outbreak occurred because of a marked increase in the *Calomys* population and the subsequent invasion of the homes in the town by these rodents. Since the *Calomys* appears to be capable of living both in the areas surrounding the towns as well as in the towns themselves, human contact may occur in both places. The most efficient transmission of virus would appear to occur in the domestic setting. There has not appeared to be any increase in the geographic areas affected by BHF in the last decade.

Prevention and Control

The key to prevention and control is either to interrupt the contact between infectious source and susceptible persons or to avoid disease in the event of infection. The ideal method of prevention for these rodent-borne diseases is to prevent contact between rodents and humans. The effectiveness of this has been admirably shown in the outbreaks of BHF in the 1960s when rodent control programs in the villages were highly successful in eliminating the epidemic situation (Kuns et al., 1965).

However, the prospects for rodent control in preventing AHF are not so bright. The human–rodent encounter resulting in AHF occurs during the crop harvests, and it is difficult to imagine how control of noncommensal feral rodents could be accomplished. The solution may be better protection of the agricultural worker from contact with rodent secretions and blood.

Similarly the control of rodents as a broad approach to preventing Lassa fever is not a realistic approach. The improvement of housing and food storage might reduce the domestic rodent population, but such changes are not easily made. Rodent trapping in an individual village where transmission is high has resulted in as much as an 80% reduction in the rate of virus transmission. However, such a program is applicable only in villages with exceptional transmission rates, and would certainly not be applicable to large areas.

Vaccine development is proceeding for both AHF and Lassa fever. A live attenuated vaccine is being developed against AHF. This vaccine is currently being safety tested in humans. For Lassa fever, a vaccina-vectored vaccine is currently going through initial safety and efficacy trials in primates.

In the event of identifiable exposure to Lassa virus in a hospital or laboratory setting, the prophylactic use of orally administered ribavirin is recommended.

Literature Cited

Armstrong, C., and R. D. Lillie. 1934. Experimental lymphocytic choriomeningitis of monkeys and mice produced by a virus encountered in studies of the 1933 St. Louis encephalitis epidemic. Public Health Rep. (Washington, D.C.) **49:**1019–1027.

Auperin, D. D., V. Romanowski, M. Galinski, and D. H. L. Bishop. 1984. Sequencing studies of Pichinde arenavirus S RNA indicate a novel coding strategy, and ambisense viral S RNA. J. Virol. **52:**897–904.

Auperin, D. D., D. R. Sasso, and J. G. McCormick. 1986. Nucleotide sequence of the glycoprotein gene and intergenic region of the Lassa virus S genome RNA. Virology **154:**155–167.

Biquard, H. A., D. A. Figini, M. J. Monteverde, and F. Alvarez. 1977. Manifestaciones neurologies de la fielue hemorragica Argentine. Medicina **37:**193–199.

Buchmeier, M. J., H. A. Lewicki, O. Tomori, and K. M. Johnson. 1980. Monoclonal antibodies to lymphocytic choriomeningitis virus react with pathogenic arenaviruses. Nature **288:**486–487.

Buckley, S. M., J. Casals, and W. G. Downs. 1970. Isolation and antigenic characterization of Lassavirus. Nature (London) **227:**174.

Casals, J. 1975. Arenaviruses. Yale J. Biol. Med. **48:**115–140.

Cole, G. A., N. Nathamm, and R. A. Pendergast. 1972. Requirement for theta-bearing cells in lymphocytic choriomeningitis virus-induced central nervous system disease. Nature (London) **238:**335–337.

Cossio, P. M., A. Rabinovich, J. I. Maiztegui, G. Carballal, M. D. Casanova, V. Ritacco, and R. M. Arana. 1979. Immunofluorescent anti-Junin virus antibodies in Argentine haemorrhagic fever. Intervirology **12:**26–31.

Elliott, L. E., J. B. McCormick, and K. M. Johnson. 1982. Inactivation of Lassa, Marburg and Ebola viruses by gamma irradiation. J. Clin. Microbiol. **16:**704–708.

Elsner, B. 1977. Anatomia pathologica de la fiebre hemorragica Argentina. Medicina **37:**200–204.

Elsner, B., E. Schwarz, O. G. Mando, J. Maiztegui, and A. Vilches. 1973. Pathology of 12 fatal cases of Argentine haemorrhagic fever. Am. J. Trop. Hyg. **22:**229–236.

Enria, D. A., S. G. Franco, A. Ambrosio, D. Vallejos, S. Lewis, and J. Maiztegui. 1986. Current status of the treatment of Argentine hemorrhagic fever. Med. Microbiol. Immunol. **175:**173–176.

Fisher-Hoch, S. P., S. W. Mitchell, D. R. Sasso, J. V. Lange, R. Ramsey, and J. B. McCormick. 1987. Physiologic and immunologic disturbances associated with shock in Lassa fever in a primate model. J. Infect. Dis. **155:**465–474.

Fisher-Hoch, S. P., G. S. Platt, G. H. Neild, T. Southern, A. Baskewille, R. T. Raymond, G. Lloyd, and D. I. H. Simpson. 1985. Pathophysiology of shock and hemorrhage in a fulminating viral infection (Ebola). J. Infect. Dis. **152:**887–894.

Goldwasser, R. A., L. H. Elliott, and K. M. Johnson. 1980. Preparations and use of erythrocyte–globulin conjugates to Lassa virus in reversed passive hemagglutination and inhibition. J. Clin. Microbiol. **11:**593–599.

Gregg, M. B. 1975. Recent outbreaks of lymphocytic choriomeningitis in the United States of America. Bull. W.H.O. **52:**549–553.

Hotchin, J., and H. Weigard. 1961. Studies of lymphocytic choriomeningitis virus in mice. I. The relationship between age at inoculation and outcome of infection. J. Immunol. **86:**392–400.

Jahrling, P. B., B. Nickolson, and J. B. McCormick. 1985. Early diagnosis of human Lassa fever by ELISA detection of antigen and antibody. Lancet **1:**250–252.

Jahrling, P. B., and C. J. Peters. 1984. Passive antibody therapy of Lassa fever in cynomolgus monkeys: importance of neutralizing antibody and Lassa virus strain. Infect. Immun. **44:**528–533.

Johnson, K. M., L. H. Elliott, and D. L. Hayman. 1981. Preparation of polyvalent viral immunofluorescent intracellular antigens and use in human serosurveys. J. Clin. Microbiol. **14:**527–529.

Johnson, K. M., J. B. McCormick, P. A. Webb, E. Smith, L. H. Elliott, and I. J. King. 1987. Lassa fever in Sierra Leone: clinical virology in hospitalized patients. J. Infect. Dis., in press.

Johnson, K. M., N. H. Wiebenga, R. B. MacKenzie, M. L. Juns, N. M. Tauraso, A. Shelokov, P. A. Webb, G. Justines, and H. K. Beye. 1965. Virus isolations from human cases of hemorrhagic fever in Bolivia. Proc. Soc. Exp. Biol. Med. **118:**113–118.

Kuns, M. L. 1965. Epidemiology of Machupo virus infection. II. Ecological and control studies of hemorrhagic fever. Am. J. Trop. Med. Hyg. **14:**813–816.

Lange, J. V., S. W. Mitchell, J. B. McCormick, D. H. Walker, B. L. Evatt, and R. R. Ramsey. 1985. Kinetic study of platelets and fibrinogen in Lassa virus-infected monkeys and early pathologic events in Mopeis virus-infected monkeys. Am. J. Trop. Med. Hyg. **34:**999–1007.

Maiztegui, J. I. 1975. Clinicaland epidemiological patterns of Argentine hemorrhagic fever. Bull. W.H.O. **55:**567–575.

Maiztegui, J. I., N. J. Fernandez, and A. J. de Damilano. 1979. Efficacy of immune plasma in treatment of Argentine haemorrhagic fever and association between treatment and a late neurological syndrome. Lancet **2:**1216–1217.

Maiztegui, J. I., M. Feuillade, and A. Briggiler. 1986. Progressive extension of the endemic area and changing incidence of Argentine hemorrhagic fever. Med. Microbiol. Immunol. **175:**142–152.

Marker, O., A. R. Thomsen, M. Volkert, B. L. Hansen, and I. H. Clemmenski. 1985. High dose survival in the lymphocytic choriomeningitis virus infection is accompanied by suppressed DTH but unaffected by T-cell cytotoxicity. Scand. J. Immunol. **21:**81–91.

Matthews, R. E. F. 1982. Classification and nomenclature of arenaviruses. Intervirology **4:**119–122.

McCormick, J. B., I. J. King, P. A. Webb, C. L. Schribner, R. B. Craven, K. M. Johnson, L. H. Elliott, and B. Williams. 1986a. Lassa fever: effective therapy with Ribavirin. N. Engl. J. Med. **314:**20–26.

McCormick, J. B., I. J. King, P. A. Webb, K. M. Johnson, R. O'Sullivan, E. S. Smith, and S. Trippel. 1987b. Lassa fever: a case-control study of the clinical diagnosis and course. J. Infect. Dis., in press.

McCormick, J. B., D. H. Walker, I. J. King, P. A. Webb, L. H. Elliott, S. G. Whitfield, and K. M. Johnson. 1986b. Lassa virus hepatitis: a study of fatal Lassa fever in humans. Am. J. Trop. Med. Hyg. 35:401–407.

McCormick, J. B., P. A. Webb, J. W. Krebbs, K. M. Johnson, and E. S. Smith. 1987a. A prospective study of the epidemiology and ecology of Lassa fever. J. Infect. Dis., in press.

McKenzie, R. B. 1965. Epidemiology of Machupo virus infection. I. Pattern of human infection, San Joaquin, Bolivia, 1962–1964. Am. J. Trop. Med. Hyg. 14:808–813.

McKenzie, R. B., P. A. Webb, and K. M. Johnson. 1965. Detection of complement-fixing antibody after Bolivian hemorrhagic fever, employing Machupo, Junin and Tacaribe virus antigens. Am. J. Trop. Med. Hyg. 14:1079–1084.

Mercado, R. 1977. Evolucion clinica y tratamiento de la fiebre hemorragica Boliviana. Medicina 37:216–218.

Mitchell, S. W., and J. B. McCormick. 1984. Physicochemical inactivation of Lassa, Ebola, and Marburg viruses and effect on clinical laboratory analyses. J. Clin. Microbiol. 20:486–489.

McCormick, J. B., S. W. Mitchell, D. R. Sasso, M. P. Kiley, and S. Ruo. 1987. Immune response to gamma-inactivated Lassa virus fails to prevent lethal infection in primates. Submitted for publication.

Murphy, F. A., P. A. Webb, K. M. Johnson, and S. G. Whitfield. 1969. Morphological comparison of Machupo with lymphocytic choriomeningitis virus: basis for a new taxonomic group. J. Virol. 4:535–541.

Oldstone, M. B. A. (ed.). 1987. Current topics in microbiology and immunology, Vol. 134 and 135. Springer-Verlag, Heidelberg.

Oldstone, M. B. A., and M. J. Buchmeier. 1982. Restricted expression of viral glycoprotein in cells of persistently infected mice. Nature (London) 300:360–362.

Parodi, A. A., J. D. Greenway, H. R. Rugiero, E. Rivero, M. Frigerio, J. M. de la Barrera, N. Mettler, F. Garzon, M. Boxaca, L. de Guerrero, and N. Nota. 1958. Sobre la etiologia del brote epidemico de Junin, Dia Med. 30:2300–2302.

Pinheiro, F. P., R. Shope, H. P. de Andrade, C. Bensabath, G. V. Cacios, and J. Casals. 1966. Amapari, a new virus of the Tacaribe group from rodents and mites of Amapa territory, Brazil. Proc. Soc. Exp. Biol. Med. 122:531–535.

Rawls, W. E., and M. Buchmeier. 1975. Arenavirus: purification and physicochemical nature. Bull. W.H.O. 52:393–399.

Rivers, T. M., and T. F. M. Scott. 1935. Meningitis in man caused by a filterable virus. Science 81:439–440.

Riviere, Y., P. J. Southern, R. Ahmsad, and M. C. Oldstone. 1986. Biology of cloned cytotoxic T lymphocytic specific for LCM virus. V. Recognition is restricted to gene products encoded by the viral sRNA segment. J. Immunol. 136:698–704.

Romanowski, V., and D. H. L. Bishop. 1985. Conserved sequences and coding of two strains of lymphocytic choriomeningitis virus (WE and ARM) and Pichinde arenavirus. Virus Res. 2:35–51.

Smadel, J. E., and M. J. Wall. 1940. A soluble antigen of lymphocytic choriomeningitis. III. Independence of antisoluble substance antibodies and neutralizing antibodies and the role of soluble antigen and inactive virus in immunity to infection. J. Exp. Med. 72:389–405.

Traub, E. 1935. A filterable virus recovered from white mice. Science 81:298–299.

Walker, D. H., J. B. McCormick, K. M. Johnson, P. A. Webb, G. Komba-Kono, L. H. Elliott, and J. J. Gardner. 1982. Pathologic and virologic study of fatal Lassa fever in man. Am. J. Pathol. 107:349–356.

Webb, P. A., K. M. Johnson, and R. B. MacKenzie. 1969. The measurement of specific antibodies in Bolivian haemorrhagic fever by neutralization of virus plaques. Proc. Soc. Exp. Biol. Med. 130:1013–1019.

Webb, P. A., K. M. Johnson, J. B. Hibbs, and M. L. Kuns. 1970. Parana, a new tacaribe complex virus from Paraguay. Arch. Gesamte Virusforsch. 32:379–388.

Weissenbacher, M. C., E. Edelmuth, M. J. Frigerio, C. E. Coto, and L. V. de Guerrero. 1980. Serological survey to detect subclinical Junin virus infection in laboratory personnel. J. Med. Virol. 6:223–226.

Wulff, H., J. V. Lange, and P. A. Webb. 1978. Interrelationships among arenaviruses measured by indirect immunofluorescence. Intervirology 9:344–350.

Zinkernagel, R. M., T. First, H. Hengartner, and A. Althage. 1985. Susceptibility to lymphocytic choriomeningitis virus isolates correlates directly with early and high cytotoxic T-cell activity, as well as footpad swelling reaction, and all these are regulated by H-2D. J. Exp. Med. 162(a):2125–2141.

CHAPTER 34

Retroviridae: Human T-Lymphotropic Virus-I (HTLV-I)/Adult T-Cell Leukemia Virus (ATLV)

NAOKI YAMAMOTO and YORIO HINUMA

Disease: Adult T-cell leukemia.

Etiologic Agent: Human T-cell leukemia virus type I/adult T-cell leukemia virus.

Clinical Manifestations: Lymphadenopathy, hepatomegaly, splenomegaly, cutaneous lesions without severe itching or excoriation, hypercalcemia, no mediastinal tumor, and several immune deficiencies.

Pathology: Appearance of abnormal T lymphocytes with lobulated, deformed nuclei in the blood and bone marrow.

Laboratory Diagnosis: Detection of serum antibody. Isolation of etiologic agent from peripheral blood. Identification of monoclonal integration of provirus DNA in the lymphocytes.

Epidemiology: Japan, especially in the southwest, Central Africa, and the Caribbean area.

Treatment: Resistant to current antileukemic agents, 2'-deoxy-coformycin may be useful.

Prevention and Control: Prevent infection with the causative agent through mother's milk, semen, and blood transfusion of infected individuals.

Introduction

Retroviruses are involved in many naturally occurring neoplasms in various animal species (Aaronson and Stephenson, 1976; Hardy et al., 1980). Their involvement in human neoplasias has been the subject of extensive research, but only recently was definitive evidence obtained for causal association of human retroviruses with a certain human disease. A type C retrovirus, named human T-cell leukemia virus (HTLV), was isolated in 1980 from cases of sporadic, cutaneous T-cell lymphoma-leukemia in the USA (Poiesz et al., 1980, 1981). Independently, a retrovirus designated as adult T-cell leukemia virus (ATLV) was obtained in 1981 from cases of endemic adult T-cell leukemia (ATL) in Japan (Hinuma et al., 1981; Yoshida et al., 1982). Later the two viruses were shown to be indistinguishable by molecular analysis (Watanabe et al., 1984). There is now evidence that this retrovirus is etiologically related only to ATL and not to other malignancies, including T-cell lymphoid malignancies other than ATL.

Adult T-cell leukemia, which was first described in 1977 as a new malignant entity (Takatsuki et al.,

1977), is endemic in Japan, especially in the southwest. Clinically, ATL is characterized by rapid progression, poor prognosis, skin lesions, and hypercalcemia (Uchiyama et al., 1977). Recently, several cases of a leukemia indistinguishable from ATL were also found in other parts of the world such as the Caribbean basin and Taiwan (Blattner et al., 1983; Catovsky et al., 1982; Hinuma et al., 1983; Kuo et al., 1985).

In 1981, an antigen detectable by indirect immunofluorescence (IF) was found in a T-cell line, MT-1, derived from peripheral blood lymphocytes (PBL) of a patient with ATL (Hinuma et al., 1981). Moreover, C-type retrovirus particles were detected in the same cell line by electron microscopy. The antigen, ATLA, reacted with sera from almost all ATL patients and also with sera from about 25% of the healthy adults tested in areas where ATL is endemic, but with very few sera of subjects from nonendemic areas (Hinuma et al., 1982b). Treatment of MT-1 cells with iododeoxyuridine (IUdR) increased their production of HTLV-I particles. Even larger amounts of HTLV-I and ATLA were detected in another T-cell line, MT-2 cells, established by cocultivation of normal cord-blood lymphocytes with leukemic cells from an ATL patient (Miyoshi et al., 1981). Furthermore, the antigens and virus particles were also shown to be induced when PBL from seropositive subjects were cultured in vitro, regardless of whether the subjects were ATL patients. Therefore, healthy seropositive subjects were considered to be HTLV-I carriers (Gotoh et al., 1982; Hinuma et al., 1982a; Miyoshi et al., 1982a). Immortalization of normal lymphocytes has been achieved quite easily by their cocultivation with X-irradiated MT-2 cells or primary PBL from seropositive individuals (Yamamoto et al., 1982b). In biochemical studies, monoclonal integration of HTLV-I proviral DNA was detected in fresh PBL from ATL patients, but not in PBL from healthy adults (Yoshida et al., 1982). These data strongly indicate that HTLV is involved in the leukemogenesis of ATL.

Animal retroviruses have contributed greatly to research on malignancies, for instance, the discovery of *onc* genes, the identification of promoter sequences for gene transcription, and immune prophylaxis. After the discovery of the first human retrovirus, it has been possible for us to use the same strategies as those used in research on animal viruses to study the mechanism, prevention, and therapy of human neoplasias.

The Virus

Physicochemical Properties and Morphology

Retroviruses (RNA tumor viruses) have been isolated from many and diverse vertebrate species, including man (Teich, 1982). All retroviruses share a similar genetic structure and a similar pathway for replication. Retrovirus genomes contain two subunits of 30 to 35S RNA, forming a 60 to 70S RNA complex and reverse transcriptase. Maturation of retroviruses is completed after budding from the membrane of host cells. Retroviruses can be classified on the basis of their morphology as type B, type C, and type D. In addition, some RNA tumor viruses are known to be transmitted through germ cells (endogenous retrovirus) (Coffin, 1982). HTLV-I appears to correspond to type C, but not to type B or type D retrovirus, on the basis of its appearance in electron micrographs of thin sections of ATL cell lines (Nakai et al., 1982; Poiesz et al., 1980). HTLV-I, however, seems to differ from known animal retroviruses in the following points. First, budding and completely electron-lucent forms of HTLV-I are rarely seen, in spite of the fact that numerous virus particles are found extracellularly in the MT-2 cell line. Second, the virus particles vary greatly in size, from about 50 to 150 nm in diameter. These characters suggest that there are some differences between HTLV-I and other known retroviruses in their steps of maturation or release. Upon short-term culture of PBL obtained from ATL patients and healthy virus carriers, type C virus particles indistinguishable from those in ATL cell lines were observed (Gotoh et al., 1982; Hinuma et al., 1982a). HTLV-I was shown to have a density of 1.152 to 1.16 g/cm^3 by sucrose-density gradient centrifugation (Poiesz et al., 1980; Yoshida et al., 1982).

Antigenic Composition and Genetics

The ATLA complex detected by immunofluorescence (IF) has been studied biochemically by immunoprecipitation with ATLA-reactive sera and lysates of HTLV-I-producing cells, precipitated HTLV-I, and culture supernatants free of cells and virus (Kalyanaraman et al. 1981; Schneider et al., 1984a,b; Yamamoto and Hinuma, 1982). The ATLA antigen complex was found to consist of three glycopolypeptides which are *env* gene products, gp46, gp67, and gp68; the viral core polypeptides p28, p24, p19, and p15, and about eight intracellular polypeptides of 40 to 70K, some of which are precursors of viral structural polypeptides. But the ATLA antigen complexes in different ATLA-positive cell lines were found to vary to some extent (Koyanagi et al., 1984; Sugamura et al., 1984). Although primary leukemic cells contain HTLV-I genomes, these antigens are detectable only after in vitro culture of the cells (Gotoh et al., 1982; Hinuma et al., 1982a).

An HTLV-I provirus integrated into fresh leukemic cells of an ATL patient was cloned, and its complete nucleotide sequence was determined (Seiki et al., 1983). The proviral genome of HTLV-I is composed of 9032 nucleotides, and the gene includes *gag, pol, env,* and *pX*. The *pX* sequence, where 4 proteins with molecular weights of 10,000 (10K) to 27K could be encoded, was implicated as being a transforming gene (*onc* gene) of HLTV-I. However, unlike all known *onc* genes in retroviruses, this region was not homologous to the proto-*onc* genes in normal cellular DNA. Thus HTLV-I was established as being a replication-competent retrovirus without a typical *onc* gene, although the function of the *pX* gene is not yet clear.

Analysis of DNA from fresh leukemic cells of ATL patients for HTLV-I provirus by Southern blotting showed that in all ATL cases examined, at least one HTLV-I genome was integrated into the leukemic cell DNA, but that its site of integration varied (Yamaguchi et al., 1984; Yoshida et al., 1984). These findings indicate that a model of activation of a specific cellular *onc* gene, such as that of c-*myc* by avian leukemia virus, was not applicable to HTLV-I leuke-mogenesis. Defective proviruses were found in fresh leukemic cells and shown to retain the *env, pX*, and 3'LTR regions, suggesting that these regions, and especially the *pX* region of the HTLV-I provirus, are important for initiation or maintenance of mono-clonal proliferation of ATL cells. In the following part of this section, the genes and gene products of HTLV-I are reviewed.

gag

As in other retroviruses, the *gag* gene of HTLV-I is located near the 5'-end of the viral RNA and encodes a 53,000 dalton polypeptide (p53). Pulse-chase exper-iments with anti-p19 and anti-p24 antibodies raised in rabbits or with anti-p19 mouse monoclonal antibody showed that p53 is a precursor polypeptide of the HTLV-I *gag* gene (Koyanagi et al., 1984; Schneider et al. 1984b; Tanaka et al., 1983). P53 is cleaved proteolytically to 3 lower-weight *gag* proteins (p19, p24, p15). P53 and p19 are phosphoproteins (Koba-yashi et al., 1984a) like many *gag* gene products of other retroviruses (Pal and Ray-Burman, 1975).

pol

The *pol* gene product of retroviruses is known to be reverse transcriptase. It is generally synthesized from only genome size RNA as a *gag-pol* product and is detected as an 180K to 200K dalton polyprotein in infected cells. The nucleotide sequence of HTLV-I suggests that the *pol* gene of HTLV-I could code for a 99K dalton polyprotein, but the *pol* gene product of HLTV-I has not yet been identified in HTLV-I-infected cells.

env

The glycosylated envelope proteins of retroviruses are synthesized from a subgenomic mRNA (Hay-ward, 1977). In HTLV-I, the *env* gene product is translated from a subgenomic 26S RNA (Kobayashi et al., 1984b; Yoshida et al., 1982). The *env* gene of HTLV-I has the capacity to code for a 54K dalton polypeptide that contains 5 possible sites for glyco-sylation (Seiki et al., 1983). Most ATLA-positive hu-man cell lines examined contain gp61 (Koyanagi et al., 1984; Sugamura et al., 1984). All these cell lines were also found to be positive for gp46, which could be concentrated by affinity chromatography from vi-rus- and cell-free culture medium (Yamamoto et al., 1982b). The gp61 precursor product of the *env* gene was processed to gp46 and gp20. In MT-2 cells, which produce much HTLV-I, gp68 was shown to be a major glycopolypeptide by immunoprecipitation with sera from ATL patients (Schneider et al., 1984b; Yamamoto and Hinuma, 1982; Yamamoto et al.,

1983a). Later, this polypeptide was shown to be a read-through product containing parts of both the *env* and *pX* regions from the subgenomic 20S RNA in MT-2 cells (Takeuchi et al., 1985).

pX

The sequence of HLTV-I in this unique *pX* region (*pX*I-IV) suggested 4 possible open reading frames with capacities to encode polypeptides with molecu-lar weights of 11K, 10K, 12K, and 27K daltons (Seiki et al., 1983). In studies with rabbit antibodies to syn-thetic oligopeptides specific to the predicted amino acid sequences of these proteins, a 40K polypeptide was recognized in HTLV-I-infected cells (Kiyokawa et al., 1984; Miwa et al., 1984). This 40K protein was also immunoprecipitated by serum of ATL patients; this indicates that the protein is expressed in vivo. The p28 protein is one of the major ATLA polypep-tides in MT-2 cells and has serine-specific phospho-kinase activity (Kobayashi et al., 1984a). It is en-coded by a 24S, defective, proviral genome that is composed of portions of the *gag* and *pX* genes. The gene product of *pX*IV was suggested to facilitate the transcription of viral and host genomes through *trans*-acting transcriptional activation (Sodroski et al., 1984). Thus the gene product of the *pX*IV region probably plays a crucial role in cellular transforma-tion.

HTLV/ATLV Subtypes

HTLV-II was isolated in 1982 from a cell line estab-lished from the spleen of a patient with a T-cell type of hairy cell leukemia (Kalyanaraman et al., 1982). This virus was related to, but distinct from, the pro-totype of HTLV-I. HTLV-II can infect both normal human T- and B-cells by coculture, but causes trans-formation only of T-cells (Chen et al., 1983). The complete nucleotide sequence of HTLV-II has been determined (Shimotohno et al., 1985). The HTLV-II provirus sequence is 8,952 base pairs (bp) long and consists of LTR and *gag, pol*, and *env* regions, like that of HTLV-I. In addition, HTLV-II contains 3 open reading frames between the *env* and 3'LTR. This is also very similar to the case in HTLV-I. Al-though HTLV-II is similar to HTLV-I biologically as well as biochemically, its isolation from cases of hairy T cell leukemia has been achieved only rarely. Thus the etiological role of the virus in a specific human disease is not obvious at present.

HIV, also known as lymphadenopathy-associated virus (LAV)/HTLV-III/ARV (Barre-Sinoussi et al., 1983; Levy et al., 1984; Popovic et al., 1984), is most probably a causative agent of the acquired immuno-deficiency syndrome (AIDS) (Centers for Disease Control, 1981; Gottlieb et al., 1981). HIV preferentially infects a subset of T-cells with the OKT-4[+] surface marker, and its reverse transcriptase shows preference for Mg^{2+} over Mn^{2+} for maximal activity. In these characters it appears to be somewhat like HTLV-I. However, HIV has cytopathic effects such as causing syncytium formation, but it does not cause cell transformation. HIV belongs to the lentivirus group of retroviruses, whereas HTLV-I belongs to the oncovirus group.

Viruses different from, but very similar to HTLV-I have been found in Old World monkeys (Hayami et al., 1983; Hunsmann et al., 1983; Miyoshi et al., 1982b; Yamamoto et al., 1983a). Infection of man with virus from monkeys or vice versa is very unlikely, as judged from the results of seroepidemiological and biochemical studies. But the viruses in monkeys provide useful models for use in studies on the origin and immunoprevention of HTLV-I.

Biology and Pathogenesis of HTLV-I

Cocultivation of normal lymphocytes from adults or newborn babies with HTLV-I-positive cell lines (e.g., MT-2) or primary leukemic cells of patients with ATL resulted in transformation and continuous growth of the recipient cells (Hoshino et al., 1983; Miyoshi et al., 1981; Popovic et al., 1983; Yamamoto et al., 1982b). All the cell lines established in this way gave positive reactions for ATLA and released HTLV-I into the medium, as revealed by electron microscopy. Most of these cell lines contained cells with the T-cell surface-markers Leu 1, Leu 3a, and Leu 4, which were very similar to those exhibited by leukemic cells of ATL patients. Although most of the cell lines established by coculture of normal lymphocytes with either primary leukemic cells or cell lines carrying HTLV-I are of T-cell lineage, the host range of this virus is very wide with respect to infectivity (Yamamoto and Hinuma, 1985); the virus can infect a variety of human and mammalian cells, such as those of rabbit, rat, monkey, and feline origin, by the coculture method. Moreover, it can also infect non-lymphoid cells, including fibroblasts, endothelial cells, lung cells, sarcoma cells, and cancer cells. Interestingly, after viral infection, only T-cells seem to be transformed. These data suggest that non-T-cells may play a role as a reservoir of the virus in vivo, although they do not become malignant. Indeed, continuous Epstein-Barr virus-genome-positive B cell lines expressing ATLA were established from PBL or lymph node biopsy specimens from ATL patients without the aid of interleukin 2 (IL-2) or cocultivation (Yamamoto et al., 1982a).

Several outstanding characteristics of ATL leukemogenesis by HTLV-I are known (Hunsmann and Hinuma, 1987; Yamamoto and Hinuma 1985): 1) random integration of provirus DNA of HTLV-I into DNA of the host cells; 2) monoclonal proliferation of leukemic cells; 3) the absence of viral proteins in tumor cells in vivo; 4) a long latency; and 5) a low frequency of occurrence (1/1,000 to 1/3,000), but high frequency of establishment of infected cells in vitro. These characteristics could be explained as follows. HTLV-I initially infects various types of cells, including T4, T8, and B cells, polyclonally. Expression of viral genomes, especially pX genomes, in T4 cells leads to proliferation of the cells. However, these cells may also express viral structural antigens on their surface and so be eliminated by the host immune surveillance system. This period of infection may last a long time, corresponding to the long latent period in healthy carriers. Then, selection of a cell carrying HTLV-I provirus but no viral antigens takes place for some as yet unknown reason, leading to monoclonal proliferation of leukemic cells. Although HTLV-I is essential for the occurrence of ATL, an additional factor(s) must be needed for selection of a single cell and initiation of its monoclonal growth from the population of cells that are polyclonally infected with HTLV-I. Thus, it appears that ATL leukemogenesis results from the interaction of HTLV-I with various additional factors in the environment.

Clinical Features

Mode of Transmission and Seroepidemiology

There are many healthy, seropositive individuals in endemic areas of ATL, their incidence apparently correlated with the frequency of occurrence of ATL (Hinuma et al., 1981; 1982b; Shimoyama et al., 1982). A nationwide survey of anti-ATLA in volunteer blood donors in Japan (Maeda et al., 1984) indicated that the incidence of seropositive donors was high in Kyushu (8%), with lower incidences of 0.3 to 1.2% in other areas. The incidence of seropositive donors was found to increase with age. Moreover, the virus is prevalent in certain families in ATL endemic areas (Miyoshi et al., 1982a; Tajima et al., 1982); when a mother of a family is seropositive, the chance of her children also being seropositive is greater than when the mother is seronegative. ATLA-positive lymphocytes can readily be detected by short-term culture of peripheral lymphocytes of seropositive individuals; this indicates that most of these persons are healthy carriers of HTLV-I. Vertical genetic transmission of viruses, known as endogenous retroviruses, is rather common in animals. However, this route of transmission has been ruled out in the case of HTLV-I. Several possibilities for mother-to-child infection either vertically or horizontally have been considered, including transplacental and intracervical infections and perinatal infections by HTLV-I-contaminated cervical secretions or breast milk. Of these, milk from virus-carrier mothers is considered to be the most probable, since virus-carrying lymphocytes can be detected in it (Nakano et al., 1986). Examination of married couples showed that the virus may also be transmitted horizontally from husband to wife. In this regard it is important to note that the semen from seropositive adults contains lymphocytes that are positive for HTLV-I provirus (Nakano et al., 1984).

Blood transfusion is also a very important route of horizontal transmission of HTLV-I, since, of several possible routes, this iatrogenic infection was found to be the major cause of increase in the frequency of affected persons (Okochi et al., 1984). Patients who received at least 1 unit of whole blood or packed erythrocytes or platelet concentrates from donors carrying antibodies to HTLV-I were shown to produce antibodies to ATLA. In contrast, no anti-ATLA was detected in any recipients of fresh frozen plasma prepared from seronegative donors or cell-free blood components even from seropositive donors.

Patients who had received whole blood or blood components containing cells derived from seropositive donors usually became positive for antibodies to HTLV-I within 50 days. This seroconversion was most probably due to primary infection with HTLV-I, because the recipients initially showed an IgM antibody response. The sequence of HTLV-I proviral DNA was also detected in the peripheral blood of these recipients (Sato and Okochi, 1986).

Individuals who showed seroconversion after blood transfusion did not show any symptoms indicative of acute infection with HTLV-I. In this regard, ATL seems to differ from AIDS, in which infectious mononucleosis-like symptoms have been observed as a result of acute viral infection.

Symptoms, Clinical Course, Complications

The incidence of ATL is slightly higher in men than women. The average age of the patients is 55 years. The most characteristic finding is rapid proliferation of abnormal T-lymphocytes with lobulated, deformed nuclei in the blood and bone marrow. Most of the patients have lymphadenopathy and hepatomegaly or splenomegaly, or both. No mediastinal tumors are observed. About one-third of the patients have various cutaneous lesions without severe itching or excoriation. Blood chemical analyses frequently show hypercalcemia, dysfunction of the liver, and hypoproteinemia. Remission is rarely achieved by common antileukemic chemotherapy and is frequently interrupted by lethal complications, presumably owing to deficiency of cell-mediated immunity. Several immune deficiencies are notable in ATL patients and virus-infected individuals (Takatsuki et al., 1986). These include general lymphadenopathy; chronic lung diseases (e.g., interstitial pneumonitis, bronchopneumonia, and lung fibrosis); opportunistic infections of the lung; cancer in various sites (e.g., the liver, stomach, skin, and vagina); M proteinemia; chronic renal failure; skin candidiasis; and some parasitic diseases such as strongyloididiasis. The median survival time is about 3 months from the onset of treatment. In Japan, more than 200 patients are estimated to die each year from ATL, particularly in endemic areas.

Adult T-cell leukemia is a clinical entity that includes various spectra of disease. It is subdivided into 5 groups (Takatsuki et al., 1986). 1) Acute ATL—this is the most common disease pattern; however, there may be no fundamental difference between this type and the crisis type described below, which develops from smoldering ATL. 2) Chronic ATL—some cases of this type of ATL were previously diagnosed as having T-cell chronic lymphatic leukemia. When such patients have anti-HTLV-I antibody as well as proviral DNA integrated into leukemia cells, they should be regarded as having chronic ATL. 3) Smoldering ATL—in this type of ATL, there is no increase in the leukocyte number, but abnormal lymphocytes are definitely present in the peripheral blood for a rather long time. 4) Crisis—acute conversion from chronic or smoldering ATL. 5) Lymphoma type ATL—in this type of ATL, no leukemic cells are detectable in the peripheral blood, and the prognosis is usually bad. Conversion from the lymphoma type to leukemia is not unusual. The proportions of these types of ATL are about 54% acute type, 17% chronic type, 11% smoldering type, 7% crisis type, and 11% lymphoma type.

In addition, it should be noted that 5 patients with clinically typical ATL without HTLV-I have recently been reported (Shimoyama et al., 1986). Thus, HTLV-I may not be involved in all patients with ATL: some factors other than HTLV-I may be able to cause a clinical picture indistinguishable from that of typical ATL without involvement of HTLV-I.

Diagnosis

Specimen Collection

Since HTLV-I virus can be isolated only by culture of the cells, whole blood or tissue samples are processed for separation of mononuclear cells (lymphocytes) by the Ficoll-Conray method. In the case of whole blood, heparinized blood is used for this purpose. For collection of mononuclear cells, samples should be separated immediately or kept at room temperature during transport. It appears sensible to ship and store specimens for this purpose at low temperature (below 10°C) to prevent decrease in the number of lymphocytes. Plasma can also be obtained for antibody tests before application of the blood to a Ficoll-Conray gradient to separate mononuclear cells. When sera are to be tested for antibody only, they can be shipped and stored at ambient temperatures, unless they are contaminated with microorganisms or will be in transit for a long period. Since the expression of viral genomes is almost totally switched off in vivo, the materials should be treated with the same caution as materials containing hepatitis B virus.

Detection of HTLV-I

There is evidence that HTLV-I genomes are expressed only slightly in vivo. This is probably owing to the in vivo selection of leukemic cells that do not express most viral information, especially membrane antigens, on which immune attack occurs. Thus, the strategies for detection of virus and virus antigens depend mainly on the expression and amplification of viral genomes after culture. However, proviral DNA can be detected directly in cells without their cultivation.

NUCLEIC ACID HYBRIDIZATION METHOD

Provirus DNA of HTLV-I integrated into cells is detectable with complementary DNA (cDNA) to the viral genome labeled with radioactive materials (Reitz et al., 1981, 1983; Yoshida et al., 1982). The Southern-blot and dot-blot methods can both be used for this purpose (Kafatos et al., 1979; Southern, 1975). For the Southern-blot procedure, lymphocyte preparations containing tumor cells are isolated from fresh peripheral blood or lymph nodes of patients by centrifugation through a Ficoll-Conray gradient. High-molecular-weight DNA is extracted from the cells by treating them with SDS—proteinase K and then phenol extraction. A provirus clone such as λ ATK-1, containing HTLV provirus, is used as a source of the probe for detecting the proviral sequence, since the total nucleotide sequence of this clone has been determined (Seiki et al., 1983). Cellular DNA is digested with restriction endonuclease (Eco-RI), separated by agarose gel electrophoresis, and transferred to a nitrocellulose membrane by a modification of the Southern procedure. The DNA is hybridized with the representative viral ^{32}P-DNA un-

der relatively stringent conditions in buffer containing 4× SSC (SSC: 0.15 M NaCl, 0.015 M sodium citrate, pH 7.5), 5× Denhardt solution, 100 μg/ml of sonicated and heat-denatured *Escherichia coli* DNA per ml, and 40 μg of poly(A)/ml at 65°C for 24 to 40 h. The filter is washed several times with 0.5× SSC at 65°C and then exposed to X-ray film. This procedure can be applied when 10^7 or more mononuclear cells from the peripheral blood of ATL patients are available, the number required depending on the number of provirus-positive cells in the blood. Since smaller numbers of virus-carrying cells are usually present in the healthy carriers of HTLV-I than in ATL patients, larger amounts of blood of carriers are usually necessary for the tests. At present this hybridization method is the only way to demonstrate HTLV-I in the cells directly. The biotin—avidin system can also be used in place of radiolabeled materials for detection of nucleic acid hybridization. However, this method can be carried out only in some laboratories, and the indirect ways described in the next section are generally used, because they are easier.

INDUCTION OF HTLV-I IN VITRO

Mononuclear cells (PBL) are separated by Ficoll-Conray gradient centrifugation from the peripheral blood of ATL patients or healthy virus-carriers. They are then cultured with or without normal PBL from cord blood or peripheral blood. Cells are also cultured either with or without interleukin 2 (IL-2). Cocultivation of provirus-containing cells with normal lymphocytes results in transformation and establishment of cell lines at high frequencies. For this, samples of freshly prepared PBL suspension (1×10^6 cells/ml) from ATL patients or healthy carriers are mixed with an equal volume of a suspension of normal PBL (1×10^6 cells/ml), and samples of the mixtures are cultured in plastic wells (Falcon model 3008). Control cultures consist of normal PBL alone. The cultures are examined at least once a week for cell transformation, which is defined as the appearance of scattered aggregations and subsequent increase in their size and number. Transformation is confirmed by observation of continuous growth of the cells during serial subcultivation. Without cocultivation, however, even leukemic cells from patients can usually grow only transiently. When transformed or growing, these cells are used for studies on the virus or viral information detected by immunofluorescence (IF), enzyme-linked immunosorbent assay (ELISA), Western blot (WB), radioimmunoprecipitation (RIP), electron microscopy (EM), or nucleic acid hybridization. Details of the immunological methods are described in the section on measurement of anti-HTLV-I antibodies.

Measurement of Anti-HTLV-I Antibodies

Immunofluorescence (IF) has been used as a reliable and standard method for measuring anti-HTLV-I antibodies. ELISA and passive particle agglutination (PA) (Ikeda et al., 1984) were developed for screening for anti-HTLV-I antibodies in donor blood. Kits for the ELISA and PA procedures are available commercially [EITEST ATL (Eisai Inc., Tokyo, Japan) and Serodia-ATLA (Fuji-rebio Inc., Tokyo, Japan), respectively]. IF, RIP, and WB have also been used as confirmatory tests.

IMMUNOFLUORESCENCE (IF)

As described in a previous section, this method has contributed greatly to the discovery of ATLV (HTLV-I), its association with the disease, and seroepidemiological studies of the virus (Hinuma et al., 1981; Maeda et al., 1984; Tajima and Hinuma, 1985). IF is very reliable and specific and can be used for both confirmation and screening of anti-ATLA antibodies.

As a target, MT-1 cells (Miyoshi et al., 1979) expressing ATLA in 2 to 5% of the total cell population are used widely, although other HTLV-I-carrying cells could also be used. The cells are smeared on a glass slide, dried thoroughly at room temperature for 1 hour, and then fixed with acetone at room temperature. Then, the slides are treated with test sera in phosphate-buffered saline (PBS) at 37°C for 30 min, washed with PBS, and treated with fluorescein isothiocyanate (FITC)-labeled rabbit anti-human IgG as a second antibody at 37°C for 30 min. The smear is then washed with PBS, covered with a cover glass, and examined in a fluorescence microscope. When the sera are positive for anti-ATLA antibodies, mono- or multinuclear giant cells show specific fluorescence. A great advantage in use of MT-1 cells for the IF procedure is this apparent morphological feature of positive cells. Moreover, it is easy to determine whether the test sera are positive for antibodies, since many of the MT-1 cells do not contain antigen and so serve as negative controls. The antibody titer is expressed as the reciprocal of the end dilution of the serum that gives positive fluorescence. FITC-labeled anti-human IgM can also be used as a second antibody to detect IgM antibody to HTLV-I.

ENZYME-LINKED IMMUNOSORBENT ASSAY (ELISA)

ELISA for ATLA antibody (Taguchi et al., 1983; Saxinger and Gallo, 1983) is based on the following principle. ATLA on a solid phase is treated with ATLA antibody; then ATLA antibody specifically combined with the ATLA on the solid phase is treated with enzyme-labeled anti-human IgG antibody; the resulting antigen–antibody complex is detected by the activity of the enzyme, and from the activity the amount of antibody can be determined. Antigens are prepared from extracts of MT-2 cells or other cells that produce much HTLV-I by treatment of the cells with detergent. The assay kit consists of antigen-coated cups, the reaction solution, the enzyme substrate, and solution for stopping the enzyme reaction. The assay procedure with microcups is as follows.

Standard Method

1. Take out the number of ATLA-coated cups necessary for the test and cool them by inserting them into the precooled cup holder.
2. Introduce 100 μl of reaction solution into each cup under cool conditions.
3. Place 20 μl of specimen in a cup. Introduce 20 μl volumes of reaction solution into 2 cups as controls.
4. Mix the liquid in the cups with a plate mixer or by manual agitation.
5. After incubation for 60 min at 37°C, cool immediately (the first reaction).
6. After removal of the liquid contents of the cups under cool conditions, carefully wash the inside of the cups with PBS and remove the wash solution completely.
7. Introduce 100 μl of enzyme-labeled antibody fluid into each cup under cool conditions.
8. Incubate for 60 min at 37°C for the reaction and then promptly cool.
9. Remove the liquid contents of the cups under cool conditions, wash the inside of the cups carefully with PBS, and remove the wash solution completely.
10. Introduce 100 μl of enzyme substrate solution into each cup under cool conditions.
11. Incubate for 30 min at 37°C and then promptly cool.
12. Add 100 μl of solution to each cup under cool conditions to stop the enzyme reaction.
13. Remove the liquid contents from the cups and measure their absorbance at 405 nm in a spectrophotometer with distilled water as a blank.

Rapid Method

Carry out the first, second, and third reactions at 40°C for 30 min. Follow the standard method for the rest of the procedure.

The results are evaluated as follows. In the standard method, the value obtained by adding 0.06 (or 0.39 when a Corona MTR-12 spectrophotometer is used for an EIA microplate) to the average absorbance value of the two controls provides a cut-off value. When the specimen shows an absorbance of more than the cut-off value, it is judged to be ATLA-antibody-positive; when it shows a value of less than the cut-off value, it is judged not to contain ATLA antibody. (An absorbance 2.5 times that of the ATLA antibody-negative blank can also be taken as the cut-off value). In the rapid method, the cut-off value is obtained by adding 0.05 (or 0.33 when a Corona MTP-12 spectrophotometer is used for an EIA microplate) to the average absorbance of the two controls. An absorbance 2.5 times that of the ATLA-antibody-negative blank can be taken as the cut-off value. Other details are the same as for the standard method.

PARTICLE AGGLUTINATION TEST (PA)

The reagent for the PA test is prepared from gelatin particle carrier sensitized with HTLV-I antigens (ATLA) on the principle that these sensitized particles are agglutinated by anti-ATLA antibody in serum or plasma (Ikeda et al., 1984; Kobayashi et al., 1988). Antigen is prepared by concentrating the culture fluid of a virus-producing cell line, subjecting it to sucrose-gradient centrifugation, collecting the virus fraction corresponding to a density of about 1.16 g/cm^3, and finally disrupting the purified HTLV-I with detergent. The PA assay kit consists of several reagents, including reconstituting solution, serum diluent, sensitized particles, unsensitized particles, and positive control serum. Plastic microplates are used for the PA test. The test is performed for either qualitative (screening) or quantitative purposes as follows.

Qualitative (Screening) Test Procedure

1. Place 1 drop (25 μl) of serum diluent in wells 1 through 3 with a calibrated pipette dropper.
2. Introduce 25 μl of test serum into the diluting loop of the diluter. Transfer this to well no. 1 and mix by rotation. Then move the diluter successively to wells no. 2 and 3, and repeat this procedure to obtain up to 2^n dilution.
3. Place 1 drop (25 μl) of unsensitized particles in well no. 2 and 1 drop (25 μl) of sensitized particles in well no. 3 with the droppers supplied in the kit.
4. Mix the contents of the wells thoroughly with a tray mixer (automatic rotary shaker). Then cover the plate and stand it on a level surface at room temperature (15 to 25°C) for 2 h. Read the patterns.

Quantitative Test Procedure

Quantitative analysis is advisable for confirmation of positive results on sera by the qualitative test.

1. Place 1 drop (25 μl) of serum diluent in wells no. 1 through 12 with a calibrated pipette dropper.
2. Add 25 μl of serum sample to well no. 1 with a micropipette. (Introduce the samples gently onto the surface of the serum diluent in the wells.)
3. Place the diluter in well no. 1 and rotate to stir. Then move the diluter to the other wells in succession and repeat this procedure to obtain up to 2^n dilution.
4. Place 1 drop (25 μl) of unsensitized particles in well no. 2 and 1 drop (25 μl) of sensitized particles in wells no. 3 through 12 with the droppers supplied in the kit.
5. Mix the contents of the wells thoroughly with a tray mixer (automatic rotary shaker). Then cover the plate and stand it on a level surface at room temperature for 2 h. Read the patterns.

In the quantitative test, a sample that shows a negative reaction with unsensitized particles (final serum dilution, 1:8) but shows agglutination with sensitized particles (final serum dilution, 1:16 or more) is regarded as giving a positive reaction. The antibody titer is determined as the highest serum dilution giving a positive pattern.

Sensitized and unsensitized particles should be reconstituted 30 min before the test. Furthermore, erythrocytes or other visible components present in the serum or plasma samples should be removed by centrifugation before the test to avoid nonspecific reactions. If a test sample induces agglutination with both unsensitized and sensitized particles, it should be retested after absorption with unsensitized particles; however, inactivation of serum samples is not necessary. This assay has the following advantages. 1) As a microtiter technique, the test procedure is very simple and is particularly suitable for mass-screening of test samples. 2) The test is rapid: results can be assessed by the naked eye after about 2 h. 3) The PA test involves the use of a newly developed artificial carrier, Fuji particles, that do not show the nonspecific agglutination usually observed with use of other carriers. 4) In the single PA test, IgM antibody as well as IgG antibody is detectable.

WESTERN BLOTTING (WB)

Viruses from the culture medium of HTLV-I-producing cells purified by sucrose-density gradient centrifugation are used after their solubilization with detergents. They are electrophoresed by SDS polyacrylamide gel and blotted onto nitrocellulose, where the antigen-antibody reaction is done. The sensitivity can be increased by several modifications, such as by use of radioisotope- or enzyme (i.e., peroxidase)-labeled second antibody or the biotin–avidin method. By this procedure IgG and IgM antibodies can be detected separately. Results are usually assessed in the presence or absence of antibodies against a viral core protein P19, P21, P24, or P28. In this procedure, proteins with high molecular weights are not transferred effectively. Moreover, the amounts of *env* gene products (e.g., gp46, gp62, and gp68) are usually small. Therefore, reactions against such products are generally identified by RIP as described in the next section.

RADIOIMMUNOPRECIPITATION (RIP)

Virus-producing cells are labeled with a radioisotope such as [^{35}S]methionine, [^{35}S]cysteine, or [^{3}H]leucine. The labeled cells or virus is solubilized, and the lysate is treated with serum with or without antibody to HTLV-I. The immunoprecipitate is bound to protein A–Sepharose, washed, and extracted. The extract is then separated on 8 to 16% SDS–PAGE, the gel is processed for fluorography before drying, and radioactive bands are located with sensitized X-ray films. Glycoproteins can be concentrated from cell- and virus-free culture supernatants by affinity chromatography on concanavalin A–Sepharose (Yamamoto et al., 1982c). Most, if not all ATL patients and healthy virus carriers have antibodies against three HTLV-I-specific glycoproteins, gp68, gp62, and pg46 (Yamamoto et al., 1983b). Some sera, however, show no reactivity with nonglycosylated core polypeptides (P28, P24, P19, and P15), and these polypeptides can be precipitated only by sera with rather high antibody titers against ATLA by IF. Therefore, antibodies to ATLA are probably directed mainly to these glycopolypeptides among the various HTLV-I-specific polypeptides in ATLA antibody-positive serum. This is reasonable, since antibodies to the viral envelope or its precursor may well be the first detectable antibodies after initial HTLV-I infection.

When purified viral polypeptides are available, RIP assay can also be performed without the use of PAGE. For this, a sample of radiolabeled purified protein (150,000 cpm), for example gp68, is mixed with diluted serum in a microtest tube and incubated for 15 h at 4°C. Then protein A–Sepharose is added to each tube, the mixtures stand for 30 min at 4°C, the Sepharose is washed three times with high salt extraction buffer, and the radioactivity in the tubes is determined (Yamamoto et al., 1983b).

Screening of Sera of Blood Donors by ELISA, PA, and IF, and Interpretation of Laboratory Data

As described previously, ATL is a chronic malignancy that has a long latent period after viral infection. The onset of leukemia probably results from the interaction of the virus with other factors. Thus, for control of ATL, it is very important to prevent viral infection itself. For preventing HTLV-I infection from transfused blood, virus-infected blood must be identified and excluded from the blood bank.

PROBLEMS ASSOCIATED WITH SCREENING FOR ANTI-HTLV-I ANTIBODY

Three procedures—PA, ELISA and IF—can be used to screen for HTLV-I antibody in the serum of donor blood, but each procedure has both advantages and disadvantages. For the PA test, a rather low dilution of serum, such as 1 : 16, is generally used. However, this could result in nonspecific agglutination due to undetermined factors and lead to misreading of the results. Moreover, the results for serum samples with low PA titers, such as 16 and 32, are not highly reproducible. However, all IF-positive samples are also positive by PA, and, therefore, PA may be suitable for screening sera of donor bloods. Since sera that are PA-positive but IF-negative usually have PA titers of 16 or 32, if these PA titers are regarded as negative, the coincidence rate of PA and IF is increased. But if this is done, some IF-positive samples with PA titers of 16 and 32 may be overlooked. Thus, for the purpose of screening donor bloods, it seems better to regard sera with PA titers of more than 16 as positive. This idea is also supported by the fact that some sera that are PA-positive but IF-negative have been shown to be positive for antibody to an *env* gene product, gp68 of HTLV-I, by RIP. However, it is also true that about 20% of the PA-positive, IF-negative samples tested were found to be negative even by RIP. Thus, it is important to study whether this discrepancy is caused by a nonspecific reaction, especially one due to impurity of the antigen preparations.

ELISA also has problems. In this method, the assessment is usually based on the degree of absorbance of negative control sera. Thus, results on samples that have an absorbance of about the cut-off value are not always reproducible, because the absorbance of the negative control sera themselves also differs somewhat from test to test.

The IF test is reliable and specific. But this procedure is not so suitable as PA or ELISA for screening large numbers of serum samples, such as those in blood centers. It sometimes provides false-negative results because it is less sensitive than RIP or PA. Moreover, it requires special expertise for reading of the results, and results for samples with low IF titers in particular may be misjudged. Furthermore, this procedure is not applicable to serum samples from patients with autoimmune diseases such as systemic lupus erythematodes, because these give various fluorescences that are not specific for HTLV-I. It is noteworthy that these sera also give a nonspecific reaction by the ELISA procedure.

The PA test can be used for detection of anti-HTLV-I antibody in sera from patients with autoimmune diseases. But heat-inactivation of the serum results in increase in its absorbance, and so in erroneous data. In experiments with sera from 2,316 blood donors, the average cut-off value in ELISA increased from 0.46 ± 0.15 in untreated controls to 0.86 ± 0.26 after their heat treatment, that is, a 1.9-fold increase of absorbance (Kobayashi et al., 1988). In this test, only 2 sera had an absorbance of more than 1.00 above the cut-off value initially, while the number increased to 29 after their heat treatment. This situation is not unique to HTLV-I ELISA, since it is rather commonly recognized with other viruses also (Anon, 1984). Thus, heat-inactivated sera are not appropriate for use in ELISA. In contrast, in the IF procedure, heat treatment does not increase the number of sera recorded as positive, and in fact is beneficial, since it usually decreases nonspecific fluorescence and so facilitates reading of the results.

Some serum samples appear positive only by PA or ELISA. However, no sera giving positive reactions in both PA and ELISA but a negative reaction in the IF test have been found. These facts suggest that the three assay systems may each show nonspecific reactions for different reasons. Thus, positive results must be confirmed by different procedures such as IF, WB, or RIP. When a large number of sera from blood donors are examined, the samples may be screened initially by a qualitative PA test, and then the positive sera may be examined by quantitative PA and ELISA. When results are positive by both PA and ELISA, the test serum should be regarded as ATLA-positive. When the serum shows an antibody titer by PA of more than 64, but is negative by ELISA, it should be treated with 2-mercaptoethanol (ME) for detection of IgM antibody (see next section), and the effect of this treatment on the PA reaction should be examined. This combination of PA and ELISA allows the screening of a large number of sera or antibody to HTLV-I with higher sensitivity and specificity than that achieved by the IF test.

IgM Antibody to HTLV-I

Among samples with high PA titers but undetectable levels of IF antibody, some were shown to have significantly decreased PA titers after 2-ME treatment. Since the IgM fractions of these sera after separation by high-pressure liquid chromatography still gave a positive reaction in the PA test, the PA method must detect both IgM antibody and IgG antibody. At the beginning of seroepidemiological studies on ATL by the IF procedure, only anti-ATLA IgG antibody was usually investigated, and scarcely any studies were carried out on IgM antibody. However, in the studies of Okochi et al. (1984), which clearly demonstrated the transmission of HTLV-I by blood transfusion, IgM antibody was found to be produced before IgG antibody. After development of the PA procedure as a screening method and during the step of evaluation of this method, PA was shown to detect more positive sera than the IF test and not to give false-negative results. One of the main reasons for this difference was that the PA method detects not only IgG antibody, but also IgM antibody. Detection of IgM antibody by the PA method with IgM fractions obtained by gel filtration of sera of patients who had a history of blood transfusion has also been reported (Wakasugi et al., 1986). Moreover, IgM antibody was detected in the sera of hemodialysis patients, hemophiliacs, and healthy carriers that have discrepant results in PA and IF tests (Hiroshige et al., 1986; Kobayashi et al., 1987); the sera tested had PA titers of more than 64, but undetectable levels of IF antibody.

Control and Prevention

The basic strategy to control any infectious disease is to prevent infection with the causative agent. Seroepidemiological studies showed that there are many anti-ATLA-positive adults among healthy residents in ATL-endemic areas. Furthermore, results strongly suggested that the familial clustering of anti-ATLA-positive individuals results from perinatal infection from mother to child or transmission from husband to wife by sexual contact. The main cause of HTLV-I infection from mother to child is suspected to be breast feeding. Attempts to block this route of infection have already started; namely, babies of seropositive mothers are fed artificial nutrients, not mother's milk. This type of approach for prevention of HTLV-I infection seems important. It will also be possible to block virus transmission from carrier husbands to their wives by conventional ways to prevent introduction of infected semen.

The spread of HTLV-I infection by blood transfusion can be prevented by screening donor blood for antiviral antibodies and discarding infected blood. In Japan there are thought to be about 1 million healthy carriers (Hinuma, 1985). Blood transfusion may be a main cause of increase in the number of infected persons, directly as well as indirectly through natural infection from seroconverted individuals to uninfected ones. For prevention of this, more than 8 million blood donors must be screened for anti-ATLA every year in Japan. Therefore, it is most important to develop a suitable method for mass screening of the blood. For this purpose the ELISA and PA methods have been developed, and blood screening should lead to a significant decrease of HTLV-I infection in recipients of transfused blood.

Several viral vaccines are very effective. The viral structural proteins, their precursors, and the genome structures encoding these polypeptides have been studied extensively. Viral glycoproteins encoded by the *env* gene of HTLV-I are present not only on the viral surface, but also on membranes of infected cells. Thus, they will be good targets for the immune surveillance system of the host. The efficacy of viral glycopolypeptides for prophylactic purposes has been clearly shown in the case of animal retroviruses such as feline and mouse leukemia viruses. Gene technology will be very helpful for challenge of this subject. *Env* gene products that elicit an immune response in the recipient can be produced in large quantities in *E. coli* or yeast. The purified proteins are then tested for reactivity with sera from ATL patients to show that they retain the natural antigenicities of the envelope glycoproteins. If this is shown, the fused envelope proteins are tested in laboratory animals, such as cynomolgus monkeys, for ability to induce resistance to challenge with HTLV-I. These steps have nearly been completed, and HTLV-I vaccination will become a reality in the near future. Possibly totally different problems will prove more difficult. For example, the question may arise before vaccination of who should be immunized? Another strategy of HTLV-I vaccination with vaccinia virus, into which DNA fragments of the *env* gene have been inserted, is also being examined (Shida, 1985).

Since multiplication of HTLV-I in vivo is poor, continuous expression of the viral genome in ATL cells is probably no longer required for maintenance of the transformed state. This conclusion suggests that antiviral drugs, such as inhibitors of reverse transcriptase, might not be therapeutically effective against ATL. Therefore, drugs that interfere with growth of ATL cells should be more effective. In this regard, it is interesting that 2′-deoxy-coformycin (DCF) has been reported to be effective in some T-acute lymphatic leukemia patients who were resistant to current antileukemic agents. DCF is known to be a strong inhibitor of the synthesis of adenosine deaminase (ADA), thus providing a theoretical basis for treatment of ATL. Conceivably, inhibition of ADA by DCF results in accumulation of the substrate deoxy-ATP, and the consequent inhibition of ribonucleotide reductase suppresses DNA synthesis in the cells. Recently this drug was found to be effective in one ATL case in The Netherlands (Daenen et al., 1984) and caused complete remission in 2 of 6 cases of ATL in Japan (Takatsuki et al., 1986). These

data appear to encourage and warrant further extensive studies.

Acknowledgments. This work was supported by grants-in-aid for Cancer Research from the Ministry of Education, Science, and Culture and the Ministry of Health and Welfare, Japan. We thank Mr. T. Yoshida and Dr. S. Kobayashi for helpful discussion and Miss C. Isida for preparation of the manuscript.

Literature Cited

Aaronson, S. A., and J. R. Stephenson. 1976. Endogenous type-C RNA viruses of mammalian cells. Biochim. Biophys. Acta **458**:323–354.

Anon, E. 1984. AIDS screening: false test results raise doubts. Nature (London) **312**:583.

Barre-Sinoussi, F., J. C. Chermann, F. Rey, M. T. Negeyre, S. Chamaret, J. Gruest, C. Dauguet, C. Axler-Blin, F. Vezinet-Brun, C. Rouzioux, W. Rozenbaum, and L. Montagnier. 1983. Isolation of a T-lymphotropic retrovirus from a patient at risk for acquired immune deficiency syndrome (AIDS). Science **220**:868–871.

Blattner, W. A., D. W. Blayney, M. Robert-Guroff, M. G. Sarangadharan, V. S. Kalyanaraman, P. S. Sarin, E. S. Jaffe, and R. C. Gallo. 1983. Epidemiology of human T-cell leukemia/lymphoma virus. J. Infect. Dis. **147**:406–416.

Catovsky, D., M. F. Greaves, M. Rose, D. A. G. Galton, A. W. G. Goolden, D. R. McClusky, J. M. White, I. Lampert, G. Bourikas, R. Ireland, A. I. Brownell, J. M. Bridges, W. A. Blattner, and R. C. Gallo. 1982. Adult T-cell lymphoma-leukemia in blacks from the West Indies. Lancet **1**:639–643.

Centers for Disease Control. 1981. Kaposi's sarcoma and pneumocystis pneumonia among homosexual men in New York City and California. Morbid. Mortal. Weekly Rep. **30**:305–308.

Chen, I. S. Y., S. G. Quen, and D. W. Golde. 1983. Human T-cell leukemia virus type II transforms normal human lymphocytes. Proc. Natl. Acad. Sci. USA **80**:7006–7009.

Coffin, J., 1982. Endogenous viruses, p. 1109–1204. *In* R. Weiss, N. Teich, H. Varmus, and J. Coffin (ed.), RNA tumor viruses. Cold Spring Harbor Laboratory, Cold Spring Harbor, New York.

Daenen, S., R. A. Rojer, J. W. Smit, M. R. Hais, and H. O. Nieweg. 1984. Successful chemotherapy with deoxycoformycin in adult T-cell lymphoma-leukaemia. Br. J. Haematol. **58**:723–727.

Gotoh, Y. I., K. Sugamura, and Y. Hinuma. 1982. Healthy carriers of a human retrovirus, adult T-cell leukemia virus (ATLV): demonstration by clonal nature of ATLV-carrying T-cells from peripheral blood. Proc. Natl. Acad. Sci. USA **79**:4780–4782.

Gottlieb, M. S., R. Schroff, H. M. Schanker, J. D. Weisman, T. P. T. Fan, R. A. Wolf, and A. Saxon. 1981. Pneumocystis carinii pneumonia and mucosal candidiasis in previously healthy homosexual men. Evidence of a new acquired cellular immunodeficiency. N. Engl. J. Med. **305**:1425–1431.

Hardy, W. D., A. J. McClelland, E. E. Zuckerman, H. W. Snyder, E. G. McEwen, D. P. Francis, and M. Essex.

1980. The immunology and epidemiology of FeLV non-producer feline lymphosarcomas, p. 677–697. M. Essex, G. Todaro, and H. zur Honsen (ed.). Viruses in naturally occurring cancers. Cold Spring Harbor Laboratory, Cold Spring Harbor, New York.

Hayami, M., K. Ishikawa, A. Komuro, Y. Kawamoto, K. Nozawa, K. Yamamoto, T. Ishida and Y. Hinuma. 1983. ATLV antibodies in cynomolgus monkeys in the wild. Lancet 2:620.

Hayward, W. S. 1980. Size and genetic content of viral RNAs in avian ooncovirus infected cells. J. Virol. 24:47–63.

Hinuma, Y. 1985. Natural history of the retrovirus associated with a human leukemia. BioEssays 3:205–209.

Hinuma, Y., T. Chosa, H. Komoda, I. Mori, M. Suzuki, K. Tajima, I. H. Pam, and M. Lee. 1983. Sporadic retrovirus (ATLV)-seropositive individuals outside Japan. Lancet 1:824–825.

Hinuma, Y., Y. Gotoh, K. Sugamura, K. Nagata, T. Goto, M. Nakai, N. Kamada, T. Matsumoto, and K. Kinoshita. 1982a. A retrovirus associated with human adult T-cell leukemia: in vitro activation. Gann 73:341–344.

Hinuma, Y., H. Komoda, T. Chosa, T. Kondo, M. Kohakura, T. Takenaka, M. Kikuchi, M. Ichimaru, K. Yunoki, I. Sato, R. Matusuo, Y. Takiuchi, H. Uchino, and M. Hanaoka. 1982b. Antibodies to adult T-cell leukemia-virus-associated antigen (ATLA) in sera from patients with ATL and controls in Japan: a nationwide sero-epidemiologic study. Int. J. Cancer 29:631–635.

Hinuma, Y., K. Nagata, M. Hanaoka, M. Mitsuoka, M. Nakai, T. Matsumoto, K. Kinoshita, S. Shirakawa, and I. Miyoshi. 1981. Adult T-cell leukemia: antigen in an ATL cell line and detection of antibodies to the antigen in human sera. Proc. Natl. Acad. Sci. USA 78:6476–6480.

Hiroshige, Y., T. Yoshida, S. Kobayashi, T. Matsui, and N. Yamamoto. 1986. Measurement of anti-ATLA antibodies: comparative studies with IF, ELISA and PA methods. Rinsho to Uirus 14:189–193 (in Japanese).

Hoshino, H., H. Esumi, M. Miwa, N. Shimoyama, K. Minato, K. Tobinai, N. Hirose, S. Watanabe, N. Imada, K. Kinoshita, S. Kamihira, M. Ichimaru, and T. Sugimura. 1983. Establishment and characterization of ten cell lines derived from patients with adult T-cell leukemia. Proc. Natl. Acad. Sci. USA 80:6061–6065.

Hunsmann, G., and Y. Hinuma. 1987. Human adult T-cell leukemia virus and its association with disease. Adv. Viral Oncol. 5:147–172.

Hunsmann, G., J. Schneider, J. Schmitt, and N. Yamamoto. 1983. Detection of serum antibodies to adult T-cell leukemia virus in non-human primates and in people from Africa. Int. J. Cancer 32:329–332.

Ikeda, M., R. Fujino, T. Matsui, T. Yoshida, H. Komoda, and I. Imai. 1984. A new agglutination test for serum antibodies to adult T-cell leukemia virus. Gann 75:845–848.

Kafatos, F. C., C. W. Jones, and A. Efstratiadis. 1979. Determination of nucleic acid sequence homologies and relative concentrations by a dot hybridization procedure. Nucleic Acids Res. 7:1541–1551.

Kalyanaraman, V. S., M. G. Sarangadharan, B. Poiesz, F. W. Ruscetti, and R. C. Gallo. 1981. Immunological properties of a type-C retrovirus isolated from cultured human T-lymphoma cells and comparison to other mammalian retroviruses. J. Virol. 38:906–915.

Kalyanaraman, V. S., M. G. Sarangadharan, M. Robert-Guroff, I. Miyoshi, D. Blayney, D. Golde, and R. C. Gallo. 1982. A new subtype of human T-cell leukemia virus (HTLV-III) associated with a T-cell variant of hairy cell leukemia. Science 218:572–573.

Kiyokawa, T. M. Seiki, K. Imagawa, F. Shimizu, and M. Yoshida. 1984. Identification of a protein (p40x) encoded by a unique sequence pX of human T-cell leukemia virus type I. Gann 75:747–751.

Kobayashi, N., Y. Koyanagi, N. Yamamoto, Y. Hinuma, H. Sato, K. Okochi, and M. Hatanaka. 1984a. 28,000-dalton polypeptide (p28) of adult T-cell leukemia virus has an associated protein kinase activity. J. Biol. Chem. 259:11162–11164.

Kobayashi, N., N. Yamamoto, Y. Koyanagi, J. Schneider, G. Hunsmann, and M. Hatanaka. 1984b. Translation of HTLV (human T-cell leukemia virus) RNA in a nuclease-treated rabbit reticulocyte system. EMBO J. 3:321–325.

Kobayashi, S., T. Yoshida, Y. Hiroshige, T. Matsui, and N. Yamamoto. 1988. Comparative studies of commercially available particle agglutination assay and enzyme-linked immunosorbent assay for screening of human T-cell leukemia virus type I antibodies in blood donors. J. Clin. Microbiol. 26:308–312.

Koyanagi, Y., Y. Hinuma, J. Schneider, T. Chosa, G. Hunsmann, N. Kobayashi, M. Hatanaka, and N. Yamamoto. 1984. Expression of HTLV-specific polypeptides in various human T-cell lines. Med. Microbiol. Immunol. 173:127–140.

Kuo, T., H. L. Chan, I. J. Su, T. Eimoto, Y. Maeda, M. Kukuchi, M. J. Chen, Y. Z. Kuan, W. J. Chen, C. F. Sun, L. Y. Shih, J. S. Chen, and M. Takeshita. 1985. Serological survey of antibodies to the adult T-cell leukemia virus-associated antigen (HTLV-A) in Taiwan. Int. J. Cancer 36:345–348.

Levy, J. A., A. D. Hoffman, S. M. Kramer, J. A. Landis, J. M. Shimabukuro, and L. S. Oshiro. 1984. Isolation of lymphocytopathic retroviruses from San Francisco patients with AIDS. Science 225:840–842.

Maeda, Y., M. Furukawa, Y. Takehara, K. Yoshimura, K. Miyamoto, T. Matsuura, Y. Morishima, K. Tajima, K. Okochi, and Y. Hinuma. 1984. Prevalence of possible adult T-cell leukemia virus-carriers among volunteer blood donors in Japan: a nationwide study. Int. J. Cancer 33:717–721.

Miwa, M., K. Shimotohno, H. Hoshino, M. Fujino, and T. Sugimura. 1984. Detection of pX proteins in human T-cell leukemia virus (HTLV)-infected cells by using antibody against peptide deduced from sequences of X-IV DNA of HTLV-I and Xc DNA of HTLV-II proviruses. Gann 75:752–755.

Miyoshi, I., I. Kubonishi, M. Sumida, S. Yoshimoto, S. Hiraki, T. Tsubota, H. Kobashi, M. Lai, T. Tanaka, I. Kimura, K. Miyamoto, and J. Sato. 1979. Characteristics of a leukemic T-cell line derived from adult T-cell leukemia. Jpn. J. Clin. Oncol. 9:485–494.

Miyoshi, I., I. Kubonishi, S. Yoshimoto, T. Akagi, Y. Ohtsuki, Y. Shiraishi, K. Nagata, and Y. Hinuma. 1981. Type C virus particles in a cord T-cell line derived by co-cultivating normal cord leukocytes and human leukaemic T-cells. Nature (London) 294:770–771.

Miyoshi, I., H. Taguchi, T. Fujishita, K. Niiya, T. Kitagawa, Y. Ohtsuki, and T. Akagi. 1982a. Asymptomatic type-C virus carriers in the family of an adult T-cell leukemia patient. Gann 73:332–333.

Miyoshi, I., S. Yoshimoto, M. Fujishita, H. Taguchi, I. Kubonishi, K. Niiya, and M. Minezawa. 1982b. Natural adult T-cell leukemia virus infection in Japanese monkeys. Lancet 2:658.

Nakai, M., T. Goto, H. Miyoshi, K. Sano, T. Chosa, and Y. Hinuma. 1982. Reactivity of sera from patients with adult T-cell leukemia (ATL) with type C virus particles associated with ATL cell lines: immunoelectron microscopic study. Gann **73:**511–513.

Nakano, S., Y. Ando, M. Ichijo, I. Moriyama, S. Saito, K. Sugamura, and Y. Hinuma. 1984. Search for possible routes for vertical and horizontal transmission of adult T-cell leukemia virus. Gann **75:**1044–1045.

Nakano, S., Y. Ando, K. Saito, I. Moriyama, M. Ichijo, T. Toyama, K. Sugamura, J. Imai, and Y. Hinuma. 1986. Primary infection of Japanese infants with adult T-cell leukaemia-associated retrovirus (ATLV): evidence for viral transmission from mothers to children. J. Infect. **12:**205–212.

Okochi, K., H. Sato, and Y. Hinuma. 1984. A retrospective study on transmission of adult T-cell leukemia virus by blood transfusion: seroconversion in recipients. Vox Sang. **46:**245–253.

Pal, B. K., and P. Ray-Burman. 1975. Phosphoproteins. Structural components of oncorna viruses. J. Virol. **15:**540–549.

Poiesz, B. J., F. W. Ruscetti, A. F. Gazdar, P. A. Bunn, J. D. Minna, and R. C. Gallo. 1980. Detection and isolation of type-C retrovirus particles from fresh and cultured lymphocytes of a patient with cutaneous T-cell lymphoma. Proc. Natl. Acad. Sci. USA **77:**7415–7419.

Poiesz, B. J., F. W. Ruscetti, M. S. Reitz, V. S. Kalyanaraman, and R. C. Gallo. 1981. Isolation of a new type C retrovirus (HTLV) in primary uncultured cells of a patient with Sezary T-cell leukemia. Nature (London) **294:**268–271.

Popovic, M., G. Lange-Wantzin, P. S. Sarin, D. Mann, and R. C. Gallo. 1983. Transformation of human umbilical cord blood T-cells by human T-cell leukemia/lymphoma virus. Proc. Natl. Acad. Sci. USA **80:**5402–5405.

Popovic, M., M. G. Sarangadharan, E. Read, and R. C. Gallo. 1984. Detection, isolation, and continuous production of cytopathic retroviruses (HTLV-III) from patients with AIDS and pre-AIDS. Science **224:**497–500.

Reitz, M. S., B. J. Poiesz, F. W. Ruscetti, and R. C. Gallo. 1981. Characterization and distribution of nucleic acid sequences of a novel type-C retrovirus isolated from neoplastic human T-lymphocytes. Proc. Natl. Acad. Sci. USA **78:**1887–1891.

Reitz, M. S., Jr., M. Popovic, B. F. Haynes, S. C. Clark, and R. C. Gallo. 1983. Relatedness by nucleic acid hybridization of new isolates of human T-cell leukemia-lymphoma virus (HTLV) and demonstration of provirus in uncultured leukemic blood cells. Virology **126:**668–672.

Sato, H., and K. Okochi. 1986. Transmission of human T-cell leukemia virus (HTLV-I) by blood transfusion: demonstration of proviral DNA in recipients' blood lymphocytes. Int. J. Cancer **37:**395–400.

Saxinger, C., and R. C. Gallo. 1983. Application of the indirect enzyme-linked immunosorbent assay microtest to the detection and surveillance of human T cell leukemia-lymphoma virus. Lab. Invest. **49:**371–377.

Schneider, J., N. Yamamoto, Y. Hinuma, and G. Hunsmann. 1984a. Sera from adult T-cell leukemia patients react with envelope and core plypeptides of adult T-cell leukemia virus. Virology **132:**1–11.

Schneider, J., N. Yamamoto, Y. Hinuma, and G. Hunsmann. 1984b. Precursor polypeptides of adult T-cell leukemia virus: detection with antisera against isolated polypeptides gp68, p24 and p19. J. Gen. Virol. **65:**2249–2258.

Seiki, M., S. Hattori, Y. Hirayama, and M. Yoshida. 1983. Human adult T-cell leukemia virus: complete nucleotide sequence of the provirus genome integrated in leukemia cell DNA. Proc. Natl. Acad. Sci. USA **80:**3618–3622.

Shida, H., 1985. Study on vaccinia virus hemagglutinin and attempt to develop vaccines against adult T-cell leukemia. Abstract for the 33rd Meeting of the Society of Japanese Virologists, p. 366–367.

Shimotohno, K., Y. Takahashi, N. Shimizu, T. Gojobori, D. W. Golde, I. S. Y. Chen, M. Miwa, and T. Sugimura. 1985. Complete nucleotide sequence of an infectious clone of human T-cell leukemia virus type II: an open reading frame for the protease gene. Proc. Natl. Acad. Sci. USA **82:**3101–3105.

Shimoyama, M., K. Minato, K. Tobinai, N. Horikoshi, T. Ibuka, K. Deura, T. Nagatani, Y. Ozaki, N. Inada, H. Komoda, and Y. Hinuma. 1982. Anti-ATLA (antibody to the adult T-cell leukemia cell-associated antigen)-positive hematologic malignancies in the Kanto district. Jpn. J. Clin. Oncol. **12:**109–116.

Shimoyama, M., Y. Kagami, K. Shimotohno, M. Miwa, K. Minato, K. Tobinai, K. Suemasu, and T. Sugimura. 1986. Adult T-cell leukemia/lymphoma not associated with human T-cell leukemia virus type I. Proc. Natl. Acad. Sci. USA **83:**4524–4528.

Sodroski, J. G., C. A. Rosen, and W. A. Haseltine. 1984. Trans-acting transcriptional activation of the long terminal repeat of human T lymphotropic viruses in infected cells. Science **225:**381–385.

Southern, E. M. 1975. Detection of specific sequence among DNA fragments separated by gel electrophoresis. J. Mol. Biol. **98:**503–517.

Sugamura, K., M. Fujii, M. Kannagi, M. Sakitani, M. Takeuchi, and Y. Hinuma. 1984. Cell surface phenotype and expression of viral antigens of various human cell lines carrying human T-cell leukemia virus. Int. J. Cancer **34:**221–228.

Taguchi, H., T. Sawada, M. Fujishita, T. Morimoto, K. Niiya, and I. Miyoshi. 1983. Enzyme-linked immunosorbent assay of antibodies to adult T-cell leukemia-associated antigens. Gann **74:**185–187.

Tajima, K., and Y. Hinuma. 1985. Epidemiological features of adult T-cell leukemia virus, p. 75–87. *In* G. Mathe and P. Reizenstein (ed.), Pathological aspects of cancer epidemiology. Pergamon Press, Inc., Elmsford, NY.

Tajima, K., S. Tominaga, T. Suchi, T. Kawagoe, H. Komoda, Y. Hinuma, T. Oda, and K. Fujita. 1982. Epidemiological analysis of the distribution of antibody to adult T-cell-leukemia-virus-associated antigen: possible horizontal transmission of adult T-cell leukemia virus. Gann **73:**893–901.

Takatsuki, K., T. Uchiyama, K. Sagawa, and J. Yodoi. 1977. Adult T-cell leukemia in Japan, p. 73–77. *In* S. Seno, F. Takaku, and S. Irino (ed.), Topics in hematology. Excerpta Medica, Amsterdam.

Takatsuki, K., K. Yamaguchi, T. Hattori, N. Aso, T. Chosa, T. Oda, T. Kiyokawa, and M. Matsuoka. 1986. Biochemistry of the disease—adult T-cell leukemia. Metabolism **23:**65–74 (in Japanese.)

Takeuchi, K., N. Kobayashi, S. Nam, N. Yamamoto, and M. Hatanaka. 1985. Molecular cloning of cDNA encoding gp68 of adult T-cell leukaemia associated antigen: evidence of expression of the pXIV region of human T-cell leukaemia virus. J. Gen. Virol. **66:**1825–1829.

Tanaka, Y., Y. Koyanagi, T. Chosa, N. Yamamoto, and Y. Hinuma. 1983. Monoclonal antibody reactive with both p28 and p19 of adult T-cell leukemia virus-specific polypeptides. Gann **74:**327–330.

Teich, N. 1982. Taxonomy of retroviruses. p. 25–208. *In* R. Weiss, N. Teich, H. Varmus, and J. Coffin (ed.), RNA tumor viruses. Cold Spring Harbor Laboratory, Cold Spring Harbor, New York.

Uchiyama, T., J. Yodoi, K. Sagawa, K. Takatsuki, and H. Uchino. 1977. Adult T-cell leukemia: clinical and hematologic features of 16 cases. Blood **50**:481–492.

Wakasugi, K., M. Sakamoto, and H. Nakamura. 1986. Change of ATLA antibodies in acute leukemia by blood transfusion. Igaku no Ayumi **136**:701–702 (in Japanese.)

Watanabe, T., M. Seiki, and M. Yoshida. 1984. HTLV type I (U.S. isolate) and ATLV (Japanese isolate) are the same species of human retrovirus. Virology **133**:238–241.

Yamaguchi, K., M. Seiki, M. Yoshida, H. Nishimura, F. Kawano, and K. Takatsuki. 1984. The detection of human T-cell leukemia virus proviral DNA and its application for classification and diagnosis of T cell malignancy. Blood **63**:1235–1240.

Yamamoto, N., and Y. Hinuma. 1982. Antigens in an adult T-cell leukemia virus-producer cell line: reactivity with human serum antibodies. Int. J. Cancer **30**:289–293.

Yamamoto, N., and Y. Hinuma. 1985. Viral aetiology of adult T-cell leukemia. J. Gen. Virol. **66**:1641–1660.

Yamamoto, N., T. Matsumoto, Y. Koyanagi, Y. Tanaka, and Y. Hinuma. 1982a. Unique cell lines harbouring both Epstein-Barr virus and other T-cell leukemia virus established from leukemia patients. Nature (London) **299**:267–269.

Yamamoto, N., M. Okada, M. Koyanagi, and Y. Hinuma. 1982b. Transformation of human leukocytes by cocultivation with an adult T-cell leukemia virus producer cell line. Science **217**:737–739.

Yamamoto, N., J. Schneider, Y. Hinuma, and G. Hunsmann. 1982c. Adult T-cell leukemia-associated antigen (ATLA): detection of a glycoprotein in cell- and virus-free supernatant. Z. Naturforsch. Sect. C **37**:731–732.

Yamamoto, N., Y. Hinuma, H. zur Hausen, J. Schneider, and G. Hunsmann. 1983a. African green monkeys are infected with adult T-cell leukaemia virus or a closely related agent. Lancet **1**:240–241.

Yamamoto, N., J. Schneider, Y. Koyanagi, Y. Hinuma, and G. Hunsmann. 1983b. Adult T-cell leukemia (ATL) virus-specific antibodies in ATL patients and healthy virus carriers. Int. J. Cancer **32**:281–287.

Yoshida, M., I. Miyoshi, and Y. Hinuma. 1982. Isolation and characterization of retrovirus from cell lines of human adult T-cell leukemia and its implication in the disease. Proc. Natl. Acad. Sci. USA **79**:2031–2035.

Yoshida, M., M. Seiki, K. Yamaguchi, and K. Takatsuki. 1984. Monoclonal integration of human T-cell leukemia provirus in all primary tumors of T-cell leukemia suggests causative role of human T-cell leukemia virus in the disease. Proc. Natl. Acad. Sci. USA **81**:2534–2537.

CHAPTER 35

Retroviridae: Human Immunodeficiency Viruses

JAY A. LEVY

Disease: Acute mononucleosis-like syndrome; AIDS-related complex (ARC); acquired immunodeficiency syndrome (AIDS).

Etiologic Agents: Human immunodeficiency virus 1 and 2.

Source: Humans; transmission by blood and blood products, sexual contact, and from infected mother to fetus and newborn.

Clinical Manifestations: Acute syndrome: fever, malaise, myalgia, arthralgia, headache, macular rash, and lymphadenopathy, which may occur in the first few months after infection, last for 1 to 3 weeks, and recur. ARC syndrome: persistent generalized lymphadenopathy, oral candidiasis, fever, and weight loss, occurring at varying times after infection. AIDS: Kaposi's sarcoma, malignancies (especially B-cell lymphomas), central nervous system disease, hairy leukoplakia, and nonspecific manifestations of immunosuppression, such as *Pneumocystis carinii* pneumonia, *Mycobacterium avium–intracellulare* pneumonia, toxoplasmosis, herpes zoster, chronic diarrhea, and cryptococcal meningitis, occurring at varying times after infection.

Pathology: Lymphopenia, especially depletion of CD 4+ cells; reactive hyperplasia in lymph nodes; encephalitis; vacuolar myelopathy; pathologic changes associated with opportunistic infections and malignancies.

Laboratory Diagnosis: Viral isolation from blood and tissues by culture using assays for viral reverse transcriptase activity or viral antigens. Antiviral antibody detection by enzyme-linked immunosorbent assay, indirect immunofluorescence, immunoblot (Western blot), and radioimmunoprecipitation methods.

Epidemiology: Worldwide occurrence, but concentrated in Central Africa, Haiti, the United States, and European countries. Major risk groups are homosexual and bisexual men, prostitutes, intravenous drug abusers, blood recipients, hemophiliacs, the sexual contacts of these groups, and newborn children born of infected mothers. The virus is not spread by casual contact.

Treatment: No curative therapy, but lengthened survival and clinical improvement in some patients with azidothymidine and some other experimental drugs. Primary treatment is directed at symptoms, opportunistic infections, and malignancies.

Prevention and Control: Serologic screening of blood and heat treatment of blood products are effective in removing risk from these sources. Education concerning safe sexual behavior and practices, including the use of condoms, is a key to comprehensive community-based prevention programs. Education of intravenous drug abusers, concerning the need to avoid sharing needles and syringes, is also very important for prevention programs.

Introduction

In 1981, the Centers for Disease Control (CDC) in Atlanta, Georgia, received several reports on the unusual occurrence of *Pneumocystis carinii* pneumonia in young men in the United States. This observation and subsequent reports of an increased incidence of Kaposi's sarcoma in previously healthy individuals led to the recognition of a new disease entity, acquired immunodeficiency syndrome (AIDS) (Jaffe et al., 1983). Further studies by several investigators indicated that this disease was also found in Haiti and Africa. The infection probably had been present in localized areas of Africa for at least several decades, but only in the late 1970s did substantial numbers of cases appear. In the United States, infection had been present since the late 1970s; the earliest retrospective evidence of disease dates from 1977. The initial epidemiologic data strongly suggested that an infectious agent was responsible for the disease and that the agent was transmitted by blood and by sexual contact.

In early 1983, a group from the Pasteur Institute (Barre-Sinoussi et al., 1983) reported the isolation of a retrovirus from a young homosexual man with lymphadenopathy. This virus, subsequently called lymphadenopathy-associated virus (LAV), proved to be different from the other human retroviruses known at the time, the virus associated with adult T-cell leukemia, termed human T-cell leukemia virus type I (HTLV-I) (Poiesz et al., 1980), and a virus isolated rarely from human leukocytes, HTLV-II. Later in 1983, retroviruses similar to LAV were isolated from individuals from San Francisco and other parts of the United States. These were called human T-cell lymphotropic virus type III (HTLV-III) (Gallo et al., 1984) and AIDS-associated retrovirus (ARV) (Levy et al., 1984). Biologic and molecular studies of the prototype viruses indicated that each of the research groups had isolated a virus with very similar characteristics; moreover, it became clear that the isolates represented a new class of human retrovirus (Rabson and Martin, 1985). The AIDS viruses differed from HTLV-I and HTLV-II in several ways: they replicated quickly and to high titer in cultures of T lymphocytes; they caused a distinct cytopathic effect in infected cell cultures; they were unable to transform cells; and their antigenic and molecular properties were unique.

In 1986, a subcommittee of the International Committee on Taxonomy of Viruses (ICTV) reviewed available data and concluded that the isolates represented a distinct new virus; it was recommended that the virus be named "human immunodeficiency virus" (HIV) (Coffin et al., 1986). The subcommittee suggested that strains of the virus be designated by use of a code with geographically in-

formative letters and sequential numbers, placed in brackets or as a subscript (e.g., for prototypes, HIV_{LAV-1}, $HIV_{HTLV-IIIB}$, HIV_{ARV-2}; for isolates, HIV_{SF-85} for the 85th isolate from San Francisco). The subcommittee also agreed that for some time the names of the initial prototypes (LAV, HTLV-III, ARV) could be used vernacularly until the new terminology became widespread.

As of mid-1988, more than 50,000 cases of AIDS have been reported in the United States, and it is expected that another 200,000 or more cases will be recorded by 1991. In some communities, such as San Francisco and New York City, AIDS has become the most important cause of death in young adult age groups. As of this writing, HIV has been associated with disease in 123 countries throughout the world.

In 1986, a major variant of HIV, now termed HIV-2, was isolated in West Africa from patients with AIDS and from seropositive, asymptomatic individuals (Clavel et al., 1986). The same virus has been detected in patients in Europe and Brazil. HIV-2 has some biologic characteristics that are similar to those of HIV-1, but differences in its molecular structure and antigenic properties clearly distinguish it from the original prototypic virus (Clavel et al., 1986; Guyader et al., 1987). There is some antigenic cross-reactivity between the polymerase and core proteins of HIV-1 and HIV-2, but the envelope proteins are distinct. Therefore, in most cases seroreactivity against HIV-2 can be detected using HIV-1 antigens, and vice versa. However, some individuals with low levels of antibody to HIV-2 or with antibodies only to the HIV-2 envelope proteins are negative when tested in current HIV-1-based serologic tests. Thus, it is clear that HIV-2 antigens must be incorporated into screening serologic tests, and when reactivity is found, sera must be subjected to specific confirmatory tests for HIV-2 antibodies.

The Human Immunodeficiency Viruses

Human Immunodeficiency Virus 1

The physicochemical, molecular, and biologic characteristics of HIV place it in the family *Retroviridae*, subfamily *Lentivirinae* (Table 1). There are three subfamilies of the family *Retroviridae: Oncovirinae* (human members, HTLV-1, HTLV-II), *Spumavirinae* (human foamy virus), and *Lentivirinae*, which in addition to HIV-1 and HIV-2 contains such important veterinary pathogens as visna/maedi virus of sheep, equine infectious anemia virus, caprine arthritis–encephalitis virus, a bovine lentivirus, and a feline lentivirus (Table 2).

TABLE 1. Lentivirus characteristics of human immunodeficiency viruses

Clinical
  1. Associated with a disease with a long latency period.
  2. Involvement of hematopoietic system.
  3. Involvement of nervous system.
  4. Morphology of virus particle: cylindric nucleoid.
Biological
  1. Cytopathic effects in infected cells (fusion; multinucleated cells).
  2. Accumulation of unintegrated circular and linear forms of proviral DNA in infected cells.
  3. Latent infection in some infected cells.
Molecular
  1. Large provirus size (9.7 kb).
  2. Primer binding site is tRNA^{lys}.
  3. Truncated *gag* gene: three processed *gag* proteins.
  4. Polymorphism: envelope region.

Retrovirus virions are spherical, measure 80 to 130 nm in diameter, and have a unique three-layered structure. Innermost is the genome–nucleocapsid complex, which is thought to have helical symmetry and is associated with reverse transcriptase molecules. This complex is enclosed within an icosahedral capsid, which in turn is surrounded by a host-cell membrane-derived envelope from which project glycoprotein peplomers. Virion RNA is a linear, non-segmented, single-stranded, positive-sense molecule of about 9 kb. The RNA molecule has a 3′ polyadenylated tail and a 5′ cap. Two identical copies of this RNA molecule are found in each virion; that is, the genome is diploid (for review, see Levy [1986]). The genome of all retroviruses contains three genes, each coding for two or more polypeptides. The *gag* gene (standing for *g*roup-specific *a*nti*g*en) encodes the virion core proteins, the *pol* gene encodes the reverse transcriptase (*pol*ymerase), and the *env* gene encodes the virion envelope peplomer (surface projection) proteins and the transmembrane protein. The RNA of some oncoviruses, but not lentiviruses

TABLE 2. Subfamilies of human retroviruses[a]

Oncovirinae
HTLV-1 (ATLV)
HTLV-II
Spumavirinae
Human foamy virus
Lentivirinae
HIV-1
HIV-2

[a] HTLV = Human T-cell leukemia virus; HIV = human immunodeficiency virus.

such as HIV, contains a fourth major gene, the viral oncogene (v-*onc*). The RNA of HIV-1 contains several other important genes that regulate the synthesis of structural proteins, including tat, art/trs, orf-A(sor), and orf-B(3′orf).

The major proteins of HIV are 1) the envelope (*env*) proteins, a precursor glycoprotein of M_r 160,000 (gp160) and its processed cleavage products the external surface, gp120 and the transmembrane protein gp41; 2) the polymerase (*pol*) proteins, p65, p53, the protease (p10), and p31 (the endonuclease); and 3) the core (*gag*) proteins, the precursor p55 and its processed products, p25, p16–18, p14, and p7 (p24). Recently, a precursor intracellular *gag* protein of M_r 40,000 has been identified (Evans et al., 1988). Antibodies to this protein may be confused with those reacting with gp41. Two proteins have been detected that are coded for by orf-A and orf-B (p23, p27) (Rabson and Martin, 1985).

The tat gene encodes a 14 kilodalton trans-acting protein that is essential for virus replication and enhances the production of the structural proteins, especially gp120. The art/trs gene product (p20) is also essential for structural protein production and viral replication (Sodroski et al., 1986). This protein appears to deregulate inhibitory regulatory functions. The orf-A gene product may be involved in virus assembly. The orf-B gene product is a down-regulator; mutants defective in this protein have greatly increased levels of virus replication.

The most unique gene product of the retrovirus genome is the reverse transcriptase. This enzyme permits the genomic RNA to be transcribed into a DNA copy and thus be in a form appropriate for integration into the chromosome of the infected cell.

Retroviruses first attach to the cell surface. The major receptor for HIV appears to be the CD4 antigen complex, detected by the OK-T4 (or Leu 3) monoclonal antibody (Dalgleish et al., 1984; Klatzmann et al., 1984a,b). The CD4 molecule has binding avidity for gp120 of HIV. After entry into the cell, the virion releases its RNA into the cytoplasm. The viral reverse transcriptase, acting as an RNA-dependent DNA-polymerase, then makes a DNA copy of the genomic RNA. The single-stranded DNA is made double stranded by the same enzyme, now acting as a DNA-dependent DNA-polymerase. This double-stranded DNA enters the nucleus, becomes circularized, and is then integrated into the host cell chromosome (for review, Levy, 1986). The integrated DNA (provirus) can then serve as template either for the production of mRNAs, which are translated into proteins, or for the production of genomic RNA for insertion into progeny virions. Some retroviruses, particularly the lentiviruses such as HIV, may be able to replicate without integration.

HIV can establish a latent infection with or without an initial productive phase (Folks et al., 1986); in this state, the integrated proviral DNA remains silent without transcription or expression of most viral proteins. Certain factors, however, can activate the virus, and convert the latent state into productive HIV infection (Folks et al., 1986; Levy et al., 1987).

As with other retroviruses, replicating HIV buds from the surface of infected cells, with or without cytopathology (Fig. 1). During this process, the virus can incorporate into its envelope some cellular proteins, particularly some of the major histocompatibility complex proteins. These nonviral proteins probably contribute to the false-positive serologic reactivities that are a particular problem in ELISA screening tests (see Interpretation of Tests).

Because of their lipid-membrane envelope, retroviruses are highly sensitive to polar solvents and detergents. Retroviruses are also very sensitive to iodophores, such as organic iodine compounds. The viruses are very labile when exposed to pH extremes (pH 2, pH 12). The viruses can be inactivated by ultraviolet irradiation, X-irradiation, or by heating in liquid solution at 56°C. For inactivation of virus in serum, prior to serologic testing, specimens are routinely heated at 56°C for 30 min.

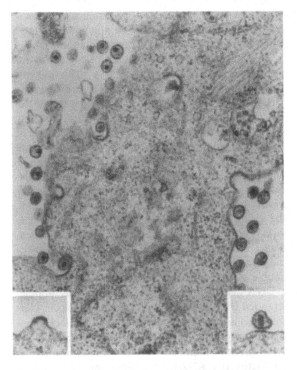

FIG. 1. The HIV$_{SF2}$ isolate (formerly called AIDS-associated retrovirus, ARV-2) replicating in a normal T cell. Magnification × 32,000. Note the cylindrically shaped nucleoid (insert, right, magnification × 54,000). (Electron photomicrograph by Lyndon S. Oshiro, California State Public Health Laboratory, Berkeley, Calif.).

Human Immunodeficiency Virus 2

The recently described relative of HIV-1, first isolated in West Africa and now called HIV-2, causes a similar spectrum of disease as the prototype virus. However, this virus can be distinguished by differences in its molecular structure, the size of its envelope proteins (gp140 and gp36), and its reactivity with particular antisera (Clavel et al., 1986; Kanki et al., 1986; Guyader et al., 1987). As noted above, core and polymerase proteins of HIV-1 and HIV-2 often cross-react, but the envelope proteins are quite distinct. Thus, some of the serologic tests that are routinely used to detect HIV-1 may not always detect HIV-2. The problem is particularly noted in serologic tests, such as the usual screening ELISA, in which there may not be much p25 or other cross-reactive proteins. Immunofluorescence assays, using HIV-1 infected cells as antigen substrate; immunoblot assays, using HIV-1 proteins as antigen substrate; and radioimmunoassays, using HIV-1 p25 as antigen substrate, do detect cross-reactive antibodies to this newly recognized virus in most sera. For confirmation, HIV-2-reactive antibodies must be distinguished from HIV-1-reactive antibodies either with immunoblots using the two viruses as substrates or with radioimmunoassays using HIV-1 gp41 and HIV-2 gp36. Recently, a synthetic peptide immunoassay has been described that can distinguish the two types of HIV (Gnann et al., 1987).

Pathogenesis and Pathology

The outcome of HIV infection reflects the effects of the virus on the immune system. The immunologic disorders that occur can lead to opportunistic infections, autoimmune syndromes, and malignancies. In general, the reduction of T-helper lymphocytes (T$_4$ lymphocytes) is the underlying cause of disease. In an immunonaive person, such as a child, the progression to AIDS is rapid (often within 6 months); in older persons, where the immune system has more residual competence, there may be a very long "incubation period," and the course of clinical disease may be intermittent or protracted. It has been estimated that the mean incubation period for development of the clinical syndrome, AIDS, in infected homosexual men is 60 to 84 months. What determines whether infection leads to clinical disease, whether the incubation period will be long or short, and whether the clinical course will be acute or protracted is multifactorial and largely unknown. Current estimates are that 30% of infected adults will develop AIDS within 7 years; another 20 to 35% will have symptoms of HIV infections.

Although HIV preferentially infects T-helper lymphocytes (Klatzmann et al., 1984a,b), macrophages/ monocytes, promonocytes, promyelocytes, and other leukocytes are susceptible as well (Levy et al., 1985d). The most common receptor for HIV appears to be one associated with the CD4 antigen complex found on many different cells (Dalgleish et al., 1984; Klatzmann et al., 1984a,b). The virus can also infect many different cells of the brain, especially macrophages, endothelial cells, astrocytes, and oligodendrocytes (Cheng-Mayer et al., 1987; Levy et al., 1987). Thus, the tropism of the virus appears to be more diverse than that of the other human retroviruses.

Infection of T-helper lymphocytes can lead to fusion of these cells with other cells expressing the CD4 antigen. These can include other T lymphocytes, B lymphocytes, and macrophages/monocytes. Fusion occurs most likely via the viral envelope protein and the multinucleated cells that formed eventually die. This kind of cytopathology, occurring in some cell culture systems within 3 to 7 days after infection, is one of the distinguishing characteristics of HIV infection (Fig. 2). Nevertheless, some infected T-helper lymphocytes can live for months without showing cytopathic effects (Hoxie et al., 1985). Infected macrophages rarely show cytopathology in vitro. Just how much cytopathology and cell death occur in vivo as a result of HIV infection is not known and is only estimable by studying serial leukocyte counts (and leukocyte/lymphocyte subset counts) and by evaluating lymphoid and other tissue pathology at biopsy and autopsy.

The infection of the central nervous system by HIV is a major aspect of the overall syndrome. Infection of the brain itself may be the major cause of the encephalopathy observed in many patients, but neurologic disease is also caused by infection of the supporting tissues of the brain, spinal cord, and peripheral nerves (Ho et al., 1985; Hollander and Levy, 1987; Levy et al., 1985c). Infection of astrocytes, endothelial cells, and macrophages may lead to a breakdown in the blood–brain barrier and penetration into the brain of opportunistic microorganisms and toxic substances. Direct infection of oligodendrocytes may be responsible for the myelin disarray and vacuolar myelopathy seen in some patients and the dementia seen in others (Levy et al., 1987). Certain aspects of dementia and myelopathy may result from infection-induced effects on nerve conduction rates. Finally, the malabsorption and chronic diarrhea observed in some individuals with HIV infection can reflect direct infection of the bowel by the virus (Nelson et al., 1988).

A major aspect of HIV pathogenesis is the polyclonal activation of B lymphocytes. This feature is particularly apparent in infected children. An enhanced production of immunoglobulins plays a key

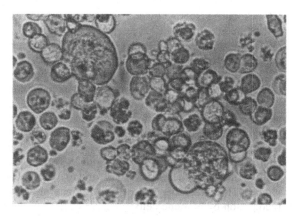

FIG. 2. Peripheral mononuclear cells infected with HIV. Note the balloon degeneration of the cells and formation of multinucleated cells. Magnification × 50.

role in the development of autoimmune sequelae that occur in some infected persons. Proliferation of B lymphocytes may also be pathogenetically involved in the eventual development of B-lymphocytic lymphomas in some patients. These B-lymphocytic lymphomas may also be the consequence of EB virus infection.

Kaposi's sarcoma, another malignancy importantly associated with HIV infection, is of unknown etiology. It may be the product of an aberrant, depressed, or enhanced immunologic responsiveness caused by HIV infection, but the nature of any such interactions with other etiologic factors is unknown. For example, an increased production of angiogenesis-promoting factors could be responsible for the endothelial cell proliferation characteristic of Kaposi's sarcoma (Levy and Ziegler, 1983). These possibilities require further study.

Although 95% of blood donors who have transmitted virus are culture positive, many have exhibited no clinical signs and very little laboratory evidence of the presence of the virus. However, such individuals have been capable of transmitting the virus throughout their lives. For example, in a few cases, AIDS has developed in recipients of blood transfusions when the donor was not only asymptomatic but also not viremic when assayed by the most sensitive viral isolation techniques. Thus, in individuals such as these donors, it would seem that HIV infection exists in some kind of suppressed state. Some persons seem able to maintain such a suppressed state of infection for a long time, perhaps for a lifetime. It is not known whether this state derives from an exceptional viral phenotype or an exceptional host response. In any case, the opposite situation, the person with easily detectable viremia, is of greater significance as a transmitter of virus and is at greater risk of developing disease.

In some infected individuals from whose blood

HIV cannot be isolated by conventional means, virus can be detected when CD8+ suppressor T lymphocytes are removed (Walker et al., 1986). Thus, this control over viral replication, could affect the progression of infection to clinical disease and the efficiency of viral transmission from one person to the next. When the virus exists in a latent state (Folks et al., 1986) or as a noncytopathic infection, as with macrophage infection, removal of infected cells by the immune system is essential. Such cells can be the source of virus for spread throughout the body, including the central nervous system, and to other individuals. However, this "silent" nature may prevent recognition by immune cells.

Clinical Features

HIV infection leads variably to a spectrum of diseases, including asymptomatic infection, an acute mononucleosislike syndrome, "AIDS-related complex" (ARC), and frank AIDS (Jaffe et al., 1983). In a considerable proportion of cases, where early observation is possible, distinctive features of acute infection are recognized. They include fever and a mononucleosislike illness, with lymphadenopathy, headache, fatigue, and sometimes myalgia and a macular rash (Cooper et al., 1985). This syndrome may last from 3 to 14 days; generally the patient recovers and remains asymptomatic for months to years. Occasionally, the syndrome can occur after 1 to 3 months. Seroconversion after acute infection usually takes place within 1 to 3 weeks, with a maximum delay of 6 or perhaps 12 months.

A majority of HIV-infected persons remain asymptomatic for 2 to 5 years. In time, many show the effects of immune disorders caused by the virus. Within 5 years, more than 25% of infected individuals develop signs of ARC. This complex of diseases includes persistent lymphadenopathy, wherein more than one lymph node is enlarged more than 2 cm and is present for at least 3 months. Histologically, these lymph nodes exhibit reactive hyperplasia (Abrams et al., 1984). Some patients with ARC have oral candidiasis; some may have chronic fever, intermittent diarrhea, weight loss (more than 10% of body weight), or recurrent herpes zoster. It is estimated that the chance of ARC patients progressing to AIDS is 2 to 4% per year.

About 20 to 30% of seropositive individuals develop AIDS in a 5-year period. They suffer from Kaposi's sarcoma, a malignancy of the endothelial lining cells of the lymphatic, *Pneumocystis carinii* pneumonia, chronic diarrhea often caused by cryptosporidia, cryptococcal meningitis, toxoplasmosis, unexplained encephalitis, encephalopathies and dementia, and B-lymphocytic lymphomas (particularly intracranial lymphomas). Recently, anal/rectal carci-nomas have been recognized at an increased rate in HIV-infected patients. In Africa, the most common presenting signs of HIV infection are chronic diarrhea of unknown origin, a wasting syndrome, resistant tuberculosis, aggressive Kaposi's sarcoma, and persistent herpes zoster (Quinn et al., 1986).

Some HIV-infected individuals also develop autoimmune disease, including immune thrombocytopenic purpura, neutropenia, and peripheral neuropathy. These diseases result from autoantibody production to normal cellular proteins. Antilymphocyte antibodies that might play a role in the depletion of lymphocytes have also been found (Stricker et al., 1987).

A new clinical entity, oral hairy leukoplakia, has been described in HIV infection (Greenspan et al., 1984). Its presence correlated with progression to AIDS within the ensuing 31 months in 83% of cases (Greenspan et al., 1987). Hairy leukoplakia is characterized by white plaques on the side or surface of the tongue that do not resolve with antifungal therapy. The cause appears to be direct infection of epithelial cells by EB virus, and perhaps by papillomaviruses.

Epidemiology and Natural History

HIV enters the body through blood transfusion, contaminated blood products, intravenous drug use involving contaminated needles and syringes, intrauterine and perinatal infection, and intimate sexual contact. It is not spread by casual contact or by arthropod vectors. Transmission occurs via virus-infected cells; little free virus can be found in the plasma phase of blood or in other body fluids (Levy et al., 1985a). HIV can be detected free in plasma and serum in about 30% of cases, but generally in low titers (10 to 50 infectious units/ml). The virus can be recovered from cerebrospinal fluid at somewhat higher frequency and titer, but it is not likely to be transmitted via this body fluid. The virus can be isolated in low titer from genital secretions, such as seminal and vaginal fluids (Levy et al., 1985a; Wofsy et al., 1986). In these secretions, however, HIV-infected cells appear to be the most important source of virus for transmission. On rare occasions, the virus has been found in saliva, both as free virus and virus associated with cells, but only in trace amounts. Thus, saliva is not an important source of contagion. Other fluids, including tears, ear fluids, urine, and bronchial secretions, have been found to contain virus, but not at significant titers.

The major route of sexual transmission of HIV is anal-receptive contact. By this route, the receptive partner is much more susceptible to infection than the insertive partner (Winkelstein et al., 1987). Direct viral infection of the mucosal epithelial cells of the bowel may occur, or virus can gain access to susceptible cells deep in supportive tissues via breaks in the mucosal lining of the bowel. Transmission of the virus by vaginal intercourse also occurs. However, the risk of infection through conventional sexual activity is not as great as that associated with anal–genital contact. Sexual transmission of HIV is enhanced in the presence of other sexually transmitted diseases, such as herpes simplex type 2 infection, syphillis, gonorrhea, and especially genital ulcer diseases, such as chancroid observed in Africa (Quinn et al., 1986).

The two major risk groups for HIV infection in the United States and European countries are homosexual and bisexual men (73% of all recorded cases in the United States) and intravenous drug abusers (17%). The latter is the major group of heterosexuals infected. However, with the increased spread of HIV in all risk groups, the number of heterosexuals infected by sexual contact with risk-group members has increased. Children born of seropositive mothers, most of whom in the United States have been infected via intravenous drug abuse, are also at particular risk.

The other risk groups for AIDS have been hemophiliacs and blood transfusion recipients. Fortunately, with the current serologic screening of blood before transfusion and the heat treatment of clotting factor concentrates, transmission of HIV to these groups has been virtually eliminated in the United States and European countries, and programs are being established in other parts of the world.

Prevention, Prophylaxis, and Control

At present there is no vaccine against HIV infection, and because of the nature of the virus and the infection, an effective vaccine seems a distant goal. The only means of preventing infection is avoiding contaminated blood, blood products, and sexual contact with HIV-infected persons. Education, aimed at behavior modification, appears to be the best means currently available for reducing the spread of HIV. Condoms are effective in blocking the passage of HIV (Conant et al., 1986) and are recommended for persons engaging in high-risk sexual activity. As noted previously, infection by blood and blood products has been virtually controlled in many countries through the use of serologic screening programs conducted in blood banks and by the heat treatment of factor VIII and factor IX concentrates used by hemophiliacs (Levy et al., 1985b). There have been rare reports of seroconversion in individuals receiving heat-treated products, but in these cases apparently the inactivation was not done properly.

Immunoglobulin preparations used in therapy and prophylaxis of other viral infections have been found to be free of infectious HIV. The methods used for their preparation inactivate HIV (Mitra et al., 1986).

Diagnosis

Specimen Collection

HIV can be detected in many body fluids, including plasma, serum, cerebrospinal fluid, saliva, tears, inner ear fluid, milk, urine, seminal fluid, vaginal secretions, and cervical secretions (Homsy et al., 1987; Levy et al., 1985a; Wofsy et al., 1986). These fluids can be collected as specimens for viral isolation or viral antigen detection by standard sterile techniques. To isolate virus from peripheral mononuclear cells, heparinized blood samples must be collected. The amount of virus present in most body fluids is small; nevertheless, caution should be taken when obtaining any clinical specimen. Gloves should be worn, and needles should be handled by safe procedures after collection of blood.

Direct and Indirect Viral Detection

Reverse Transcriptase Assay

The reverse transcriptase (RT) assay has been well described (Hoffman et al., 1985). It has been adapted to automated procedures suitable for running large numbers of assays and is available in commercial kits. For HIV, the assay employs magnesium instead of manganese as the divalent cation in the reaction mixture. Generally, RT assays are done on pelleted virus from 1 ml of culture fluid, although some laboratories use up to 3 ml for detecting evidence of virus in initial isolate cultures. One practical method of pelleting virus is to spin the sample for 2 h at full speed in a microfuge (10,000 rpm). Precipitation of virus with polyethylene glycol (PEG) can also be employed, but PEG can interfere with the RT assay (Hoffman et al., 1985).

Viral Antigen Enzyme-Linked Immunosorbent Assays

Procedures for detecting HIV antigen use either sandwich or competition ELISA techniques. The core antigen, p25, which occurs in relatively high abundance in virions and in infected cells, is generally the target of the assay (Goudsmit et al., 1986). This antigen is detected after treating the sample with detergent, which ruptures cells and intact virions. The sandwich technique is based on the capture of p25 by specific antibodies fixed onto a solid phase. The p25 is detected by addition of a second enzyme-labeled anti-25 antibody, and this reaction is measured by a subsequent enzymatic color reaction. The competition assay uses polyclonal antibody to HIV that is fixed to the solid phase (usually a microtiter plate) and an enzyme-labeled p25 recombinant protein as a competitor of the viral p25. This antigen assay can be adjusted to detect as little as 10 pg of viral protein per ml or about 100 infectious virions per ml. The procedure is also applicable for measuring viral antigens in cell lysates (Homsy et al., 1987).

INDIRECT IMMUNOFLUORESCENCE ASSAY

The presence of virus in peripheral mononuclear cell cultures or in other cells or tissues can be confirmed by the detection of viral antigens using specific antiserum in an indirect immunofluorescence assay (Kaminsky et al., 1985). The sera used to detect viral antigens must be prescreened to be certain they lack antibodies to normal cell antigens. Monoclonal antibodies to particular HIV proteins can be used alone or in mixtures to obtain broad or narrow strain specificity. Either fluorescein-labeled or preoxidase-labeled antibodies specific for the species of the globulins of the antiviral antibodies are used as the indicator. This procedure, employing HIV-infected cells as substrate, is also used to detect anti-HIV antibodies (see Detection of Antibodies).

IN SITU HYBRIDIZATION

Viral replication can be detected in cells by in situ hybridization (Harper et al., 1986) using a radioactive probe to the HIV genome. This procedure can detect small numbers of cells producing virus, but it cannot be used to estimate the titer of virus in the culture because of the varying influence of nonproductive infected cells. The method is not as sensitive for viral detection as viral isolation, where viral replication in cell culture serves to amplify greatly the initial viral content. In the latter (in principle), one infectious particle can be found, if cultures are held for a long time and subculturing is used to maximize amplification (Levy and Shimabukuro, 1985).

PLAQUE AND SYNCYTIUM FORMATION

Harada et al. (1985) described a procedure for detecting HIV by induction of plaques in MT-2 and MT-4 human T-cell lines. The procedure involves growing the cells as a monolayer on polylysine-treated plastic dishes. The virus is added, and the culture is overlaid with 0.6% agarose. In 5 to 7 days, plaques appear and can be quantitated. The assay only works with some HIV strains and is not particularly useful when working with primary isolates or low-passage strains. However, the method can be employed for neutralization assays (see Detection of Antibodies). Similarly, virus detection and assay systems have

been developed using syncytium formation as the indicator of infection. As with plaque formation, syncytium formation assays require particular cell lines and selected viral strains.

RADIOIMMUNOASSAY

Radioimmunoassay can be used in capture systems to detect HIV antigens, but has mostly been used to detect antibodies (see Detection of Antibodies).

Viral Isolation and Cultivation

VIRAL RECOVERY FROM CLINICAL SPECIMENS

The initial isolation of HIV was made by cultivating peripheral mononuclear cells from a person with lymphadenopathy syndrome (Barre-Sinoussi et al., 1983). In principle, HIV should be recoverable from the blood within a few days after initial infection. However, many variables influence the efficacy of viral isolation (e.g., CD8+ lymphocytes in primary peripheral mononuclear cell cultures are inhibitory) (Levy and Shimabukuro, 1985; Walker et al., 1986).

Primary viral isolation from patient material depends on the amplification of virus present in as few as 0.001% of leukocytes; therefore, an adequate amount of heparinized blood is needed (i.e., as little as 3 to 5 ml from children or as much as 30 ml from adults). Ideally, the blood specimen should be cultured within a few hours after venipuncture. Whole blood can be held for up to 36 h at room temperature without significant reduction in virus recovery, but if cooled to 4°C it is often difficult to later separate the buffy coat. Any blood sample to be kept for more than 36 h before analysis should be subjected to buffy-coat separation first. After the buffy coat is separated, leukocytes can be maintained in a viable state in serum-containing medium for 7 to 10 days before they are placed in culture.

For viral isolation, the buffy coat is separated from heparinized blood in Ficoll–Hypaque (F/H) gradients. In some laboratories, heparinized blood is first spun at 2,000 rpm for 10 min to separate the buffy coat. The buffy coat is then diluted in Hanks balanced salt solution (1 : 3, vol/vol) and overlaid on a F/H gradient (3 : 1, vol/vol). In this manner, the amount of F/H used is reduced and the separation is very efficient. At least three washings must be done before the separated cells are resuspended in RPMI 1640 medium with 10% fetal calf serum (prescreened to be certain it is not toxic to human lymphocytes), 2 mM glutamine, polybrene (1 to 2 μg/ml), and usually 10% human interleukin-2 (IL-2, T-cell growth factor). The IL-2 should be pretested for toxicity and for optimal concentration. This latter parameter can be assessed by titrating the IL-2 effect on cell division or labelled thymidine uptake. Recombinant IL-2 is generally not as effective for growing mononuclear cells as crude IL-2. The polyanion, polybrene, enhances viral infection and spread in culture.

In the initial isolation of HIV-1$_{LAV}$ (Barre-Sinoussi et al., 1983), the mitogen phytohemagglutinin (PHA), at concentrations necessary to induce optimal blastogenesis in T lymphocytes (usually 3 to 10 μg/ml), was added directly to cultures and left for 3 days. Subsequently, the cells were removed from the medium by centrifugation, washed, and resuspended in fresh RPMI 1640 medium.

In most laboratories, after 5 to 7 days, mononuclear cells from seronegative donors were added to primary isolation cultures (Levy and Shimabukuro, 1985). More efficient viral isolation has recently been achieved in the author's laboratory by leaving PHA out of initial cultures. Instead, leukocytes from seronegative persons that have been previously stimulated with PHA are added directly to the leukocytes from the person under study (1:1 to 3:1 ratio of normal:subject lymphocytes, vol/vol). The success of this approach may be due to avoidance of a PHA induction of suppressor T-lymphocyte activity that might reduce viral replication (Walker et al., 1986). In many laboratories, antibody to human α-interferon has been added to leukocyte culture medium (Barre-Sinoussi et al., 1983); this step, however, now appears unnecessary for conventional viral isolation.

Cultures are monitored for the presence of virus after 6 to 8 days, with medium changes and assays being done every 3 days thereafter. For detection of the presence of viral growth, culture fluid must be either filtered or subjected to centrifugation (10,000 rpm for 30 min) to remove the cell debris. Subsequently, the viral antigen reverse transcriptase assays can be conducted on the fluid, as described previously. To detect very small amounts of virus, or viruses that do not grow well in culture, it may be necessary to add fresh PHA-stimulated leukocytes from seronegative persons every 5 to 7 days for up to 40 days.

It can generally be assumed that all seropositive persons have virus present in their tissues, and most have virus present in their circulating leukocytes. As noted previously, in some cases the presence of CD8+ suppressor T cells may limit the recovery of virus when the total mononuclear leukocyte population is cultured (Walker et al., 1986). Nevertheless, by adding to the cultures PHA-stimulated leukocytes from seronegative persons, this CD8+ effect can often be abrogated.

Virus is easiest to isolate from patients showing symptoms of ARC or AIDS; in general, 80 to 90% will yield infectious virus in their blood. Asymptomatic seropositive individuals yield virus in 60 to 80% of cases, and manipulations of leukocyte cultures, including assays of separated CD4+ cells, can enhance this recovery dramatically. Body fluids, except for blood and cerebrospinal fluid, rarely contain free virus. When virus is present in these fluids, it may be at such low titer that detection may require more than a month of culture. The cells present in these body fluids are the major source of virus, and the yield of virus from these cells can be amplified by mixing them with PHA-stimulated leukocytes from seronegative persons.

The procedures used to detect infectious virus in saliva, genital secretions, and other body fluids are time consuming and expensive; therefore, they should be employed judiciously. It is not practical to employ viral isolation procedures merely to confirm contagion or transmissibility. Failure to isolate virus or detect viral antigen does not prove that the specimen is negative—the techniques are too insensitive. Rather, it is practical to assume that virus and virus-infected cells are present in all body fluids and tissues of all seropositive individuals. Infected individuals should be counseled accordingly. Titration of viral infectivity is very difficult.

Because there is no proof that the amount of virus present in the blood or other body fluids of infected individuals can be used to foretell transmissibility, clinical course, or prognosis, there is no reason to attempt virus titration procedures in the diagnostic laboratory setting. Even though there may be exceptional circumstances where viral isolation procedures might be warranted to answer specific clinical diagnostic questions, it is not practical to attempt to quantitate the virus. Research toward practical titration methods is under way in several laboratories.

CULTIVATION OF HUMAN IMMUNODEFICIENCY VIRUS

For viral cultivation in vitro, PHA-stimulated leukocytes from seronegative persons are plated in RPMI 1640 [supplemented with glutamine (2 mM), IL-2 (10%), and polybrene (2 μg/ml), and, if desired, antibiotics (100 units/ml penicillin, 100 μg/ml streptomycin)]. The seed virus stock is then inoculated. Absorption of the virus onto the cells for 30 to 60 min may enhance the efficiency of infection. The cells are centrifuged and placed in fresh medium every 3 days. The supernatant is assayed after 4 to 6 days for particle-associated reverse transcriptase or antigen production.

Once an isolate is identified, it can be passed to several different established human cell lines and maintained as a chronically infected culture (Levy et al., 1984). Some viral strains have been grown in different established T-lymphocyte lines, B-lymphocyte lines (those that contain the EB virus genome), macrophage lines (e.g., U937 cells), and established human glioma cell lines (Cheng-Mayer et al., 1987; Evans et al., 1987a; Levy et al., 1987). However, some viral isolates do not replicate at all in these established cell lines (Evans et al., 1987a). It has become common practice to make primary isolations by cultivating leukocytes from the infected person, as described previously, and then to adapt isolates to selected established cell lines. Passage of an isolate to an established cell line may require cell to cell contact; thus, addition of infected primary leukocytes to the cell lines may initially be necessary. (Evans et al., 1987a). As with primary leukocyte cultures, established human cell lines may undergo cytopathic changes and some cell death, but after a few days a subpopulation of cells usually emerges that is chronically infected and grows continuously.

Detection of Antibodies to Human Immunodeficiency Virus

Enzyme-Linked Immunosorbent Assay

One of the most sensitive methods for detecting antibodies to HIV is the indirect ELISA (Weiss et al., 1985b). This method, which has become standard for detecting antibodies to other infectious agents, is the basis for the most common HIV assay. The assay employs either ruptured virus (purified on sucrose density gradients) or infected cell lysate as antigen. A new "generation" of ELISAs has been developed that are based on recombinant-DNA-derived HIV proteins, such as the p25 core protein, the p31 polymerase protein, and the gp41 envelope protein. The HIV proteins are coated onto microtiter plates and reacted with the test serum. The serum is diluted to more than 1:50 to avoid nonspecific reactions. Subsequently, an enzyme-conjugated antihuman globulin is added and the bound fraction detected by the addition of an enzyme substrate. The resulting color reaction can be quantified by spectrophotometry. There are many variations of this assay, most involving proprietary conveniences that add to speed and simplicity of testing.

Indirect Immunofluorescence Assay

For indirect immunofluorescence assays, cells that have been chronically infected with the virus (such as established T-cell lines) are employed as substrate. These cells are best mixed 1:1 with uninfected cells to provide a known background of non-reacting cells as a built-in control (Kaminsky et al., 1985). Nonspecific cell staining can then be detected easily: if more than 50% of cells fluoresce. This nonspecificity can then be dealt with by other means. The cells are dried on slides, then fixed in acetone for 10 min, redried, and stored frozen for use. They can

be stored for up to 3 months at −70°C. The staining pattern for HIV is characteristic and can thus also be used to determine a true antibody reaction. The distribution of viral antigen may be both diffuse and reticular in the cytoplasm, localized at the periphery of the cell (crescent formation), or found in areas of high intensity as a result of the concentration of viral proteins in some parts of the cytoplasm of multinucleated cells (Kaminsky et al., 1985). Recent studies suggest that IFA may be the best assay for detecting the earliest response to HIV infection (Cooper et al., 1987). Both IgG and IgM reactions can be measured separately, but IgM assays will require further development for routine use. These assays have other merits as well: they are rapid and, where necessary, can be done in as little as 30 min. They lend themselves to situations where a limited number of tests are performed at one time.

Immunoblot (Western Blot)

For the immunoblot (Western blot) procedure, either purified viral or infected cell lysates are separated by polyacrylamide gel electrophoresis, then transferred to nitrocellulose paper by standard blotting procedures (Pan et al.,1987). The blot is then cut into narrow strips. Such strips, as well as complete kits containing a detector system, are available commercially. Each strip is incubated with a serum specimen (usually diluted 1:50), and following washing steps, any of several detector (anti-Ig reagents) systems are used. This process detects specific antibody binding to individual viral proteins (Fig. 3). If lysates of cells are employed, one can detect antibody binding to the precursor envelope protein, gp160, as well as to the precursor *gag* protein, p55. The precursor envelope protein, in particular, is not detected when purified viral preparations are used. Because some individuals may only have antibodies to this protein (Pan et al., 1987), there is advantage in using infected cell lysates as the immunoblot substrate.

Neutralization Assays

To detect antibodies that neutralize HIV, standard procedure can be employed using as the target cells PHA-stimulated leukocytes from seronegative persons. The virus used must be prescreened for its ability to induce a sufficient amount of reverse transcriptase activity or antigen (as detected by a p25 antigen assay or IFA). For neutralization studies with some isolates, established cell lines can be employed (Weiss et al., 1985a). Alternatively, for certain viral strains, a plaque inhibition or syncytium inhibition system can be used (see Viral Detection).

For a standard neutralization assay, the viral preparation (0.1 ml) is mixed 1:1 with the serum specimen (e.g., 0.1 ml), which has been prediluted

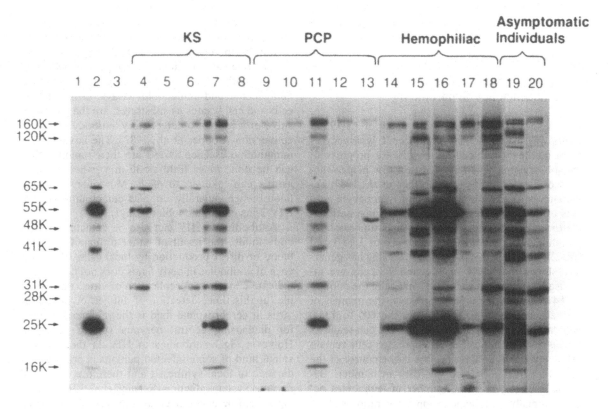

FIG. 3. Immunoblot analysis of sera from patients with Kaposi's sarcoma (KS) and *Pneumocystis carinii* pneumonia (PCP), and from asymptomatic, hemophiliac, and homosexual men. The viral proteins detected are to the envelope (gp160, gp120, gp41), the polymerase (p65, p31), and the core proteins (p55, p48, p25, p16). (Reprinted from Pan et al., 1987, with permission from the University of Chicago Press.)

1 : 10 or more to avoid nonspecific reactions. After 30 to 60 min at room temperature, the mixture is diluted 1 : 10 and added to the leukocyte culture. This second dilution step is important to avoid any cytotoxic effect of the antibody. The virus–cell mixtures are then cultured, as described previously, and viral production is detected by reverse transcriptase assay, IFA, or other antigen assay. It should be noted that viral isolates differ in their sensitivity to neutralization. Thus, neutralization assays should include either several different isolates or an isolate known to be highly neutralizable.

RADIOIMMUNOASSAY

To detect HIV antibodies, several dilutions of a test serum in buffer containing 0.01 M sodium phosphate, pH 7.5, 0.15 M NaCl, 0.1% Nonidet P-40 (NP-40), 0.01 M phenyl-methyl-sulfonyl-fluoride, and 2 mg/ml bovine serum albumin are incubated with ^{125}I-labeled p25 core polypeptides at 4°C for 16 h. Immune complexes are recovered by adding formalin-fixed *Staphylococcus aureus,* which are then washed in dilution buffer and assayed for bound radioactivity.

This procedure can also be used in a competition assay to determine the presence of virus by the presence of p25 in infected cells (Kanner et al., 1986).

RADIOIMMUNOPRECIPITATION ASSAY

Antibodies to HIV can also be detected by a radioimmunoprecipitation assay (RIPA) (Allan et al., 1985). The infected cells are labeled with ^{35}S-methionine or ^{35}S-cysteine by conventional procedures and lysed in RIPA buffer (10 mM Tris–Cl, pH 7.6; 150 mM NaCl, 1% DOC; 1% Triton X-100; 0.1% SDS; 1 mM PMSF). The labeled lysate is then precleared and reacted with the test serum overnight at 4°C; the antibody–antigen complexes are then precipitated by the addition of *Staphylococcus* protein A coupled to Sepharose Cl-4B. The precipitates are washed extensively with RIPA buffer, and then eluted in a sample buffer containing 2% SDS, 0.05 M 2-mercaptoethanol, 0.06 M Tris–Cl (pH 6.8), 10% glycerol, and 0.2% bromophenol blue by boiling at 100°C for 2 min. The samples are then analyzed by SDS–PAGE followed by autoradiography. The viral proteins are identified by their appropriate sizes. This same approach can be used to detect viral antigen with a known positive antiserum.

Lymphocyte Subsets in Acquired Immunodeficiency Syndrome

One of the earliest observations made of AIDS patients was that their helper/inducer T lymphocytes were markedly reduced as the disease progressed; this observation was reflected in a low helper/suppressor T-cell ratio (<1.0) (Gottlieb et al., 1987; Seligmann et al., 1984). Screening of lymphocyte subsets of infected persons, using monoclonal antibodies to antigen markers on helper and suppressor T lymphocytes (e.g., OKT4 and OKT8 or Leu-3 and Leu-2), has clinical value. Such screening has generally shown that early in infection T8 cells are increased, but there is very little change in the number of T4 cells. As infection progresses, the number of T4 cells decreases to well below the 600 to 1,000 cells/mm³ found in normal persons and can reach as low as 50 cells/mm³. The T-suppressor cells remain relatively stable until rather late in the course of the infection; then they also decrease in number. Because fluctuations in the number of leukocytes occurs normally, one helper/suppressor ratio or absolute T-cell count is not meaningful. However, repeated measurement of T-cell subsets does have prognostic value. These tests, now routinely performed in many laboratories, using flow cytometry (FACS) (Levy et al., 1985e), can be influenced by other infections and diseases. Therefore careful clinical interpretation of results is necessary.

Interpretation of Tests

When evaluating an individual for HIV infection, the presence of antibody should first be measured. In general, an ELISA can be used as an initial screening for antibody. If the patient has been recently infected, the antibody should be first detected within 2 to 4 weeks, usually by immunofluorescence or immunoblot. Testing for IgM antibodies may be helpful (Cooper et al., 1987), but the specificity of this response requires further development.

Confirmation of the presence of HIV antibody is necessary, particularly after ELISA screening, because of nonspecific reactions and the serious meaning of a positive test. As noted previously, false-positive reactions in the ELISA usually occur because of cellular proteins contaminating the viral preparations used as antigens and the natural presence in the serum of some individuals of antibodies to these proteins. In general, sera giving a positive ELISA are retested by a second ELISA. Then, if still positive, sera are confirmed by immunofluorescence or immunoblot, using appropriate control sera and cells.

Rarely, there will be an antibody-negative but virus-positive patient. Most of these patients will seroconvert within 1 to 6 months. In general, immunofluorescence and immunoblot assays, using infected cells and cell lysates as substrates, are the most sensitive procedures to detect early antibody and seroconversion (Cooper et al., 1987). The advantage of immunofluorescence assays are their rapidity; they can be done in as little as 30 min, whereas other assays can take up to 48 h. Moreover, immunofluorescence generally can also detect antibodies to HIV-2, as can immunoblot assays.

Antibodies to HIV can also be detected in most body fluids, the result of either diffusion from the blood or direct production by local cells. The presence of antibodies in body fluids does not necessarily indicate concomitant replication of virus. For example, at this time, detection of antigen or infectious virus in cerebrospinal fluid is the preferred method for diagnosing central nervous system infection. However, IgG antibodies to HIV in the cerebrospinal fluid of some infected persons is probably indicative of local synthesis, and therefore of central nervous system infection (Resnick et al., 1985). Testing should be encouraged as a means of developing confidence in this approach for confirming central nervous system infection.

The nature of the antibody response to HIV proteins can be potentially helpful in evaluating the prognosis of patients. In most acutely infected persons, the first antibodies produced are to the *gag* proteins, although some individuals initially develop antibodies to envelope proteins, particularly to gp160. As the disease progresses, antibodies to all of the viral proteins are made, particularly the polymerase proteins p65 and p31. When antibodies to the core and, later, the polymerase proteins decrease, a more advanced disease state is observed (Kalyanaraman et al., 1984; Lange et al., 1986; Pan et al., 1987; Steimer et al., 1986). This finding may reflect enhanced viral release in the blood and formation of immune complexes, or a decrease in specific antibody production. Longitudinal studies and further analyses of infected individuals are needed to clarify this point. It is noteworthy that neutralizing antibodies and antibodies to the envelope protein, gp160, may remain constant until the terminal stages of AIDS (Pan et al., 1987; Weiss et al., 1985a).

Laboratory Safety

Laboratories involved in HIV antibody testing should use Biosafety Level 2 standards and practices. These are the same standards recommended for handling all clinical specimens, such as specimens potentially contaminated with hepatitis B virus. The basic principle in these standards is to consider blood and other body fluids from *all* patients as infective. This approach eliminates the need for warning labels and avoids the confusion of having different specimens handled differently.

1. Specimens of blood and body fluids should be put in well-constructed containers with secure lids to prevent leakage during transport. Care should be taken when collecting specimens to avoid contaminating the outside of the container and paperwork accompanying specimens.
2. Persons processing blood and body fluid specimens in the laboratory should wear gloves. Masks and protective eyewear should be worn if mucous-membrane contact with specimens is anticipated. Gloves should be changed and hands washed after completion of specimen processing.
3. A biologic safety cabinet is not necessary for routine procedures such as microbiologic culturing, histologic processing, and clinical chemistry–hematology processing. However, a biologic safety cabinet should be used whenever procedures are done that generate droplets or aerosols, such as blending, sonicating, and vigorous mixing.
4. Mechanical pipetting devices should be used for manipulating all fluids in the laboratory.
5. Use of needles and syringes should be limited to situations in which there is no alternative.
6. Laboratory work surfaces should be decontaminated with an appropriate chemical germicide after a spill and when work is finished.
7. Contaminated materials used in laboratory tests should be decontaminated and disposed of in accordance with institutional policies for disposal of infective waste.
8. Diagnostic equipment that has been contaminated with blood or body fluids should be decontaminated and cleaned.
9. Persons should wash their hands after completing laboratory activities and should remove protective clothing before leaving the laboratory.

A biosafety cabinet should be used for viral isolation work, as well as gloves and a gown. This level of caution is still part of the standards of Biosafety Level 2. There are extremely low levels of virus in body fluids, including blood and genital secretions; therefore, inordinate concern about transmission via aerosol from primary specimens is unfounded. Activities such as growing research laboratory-scale amounts of HIV or related viruses, working with concentrated viral preparations, or conducting procedures that may produce droplets or aerosols should be performed in an appropriate Biosafety Level 3 facility.

The inactivation of HIV under clinical laboratory conditions has been well studied (Resnick et al., 1986; Spire et al., 1985). Although relatively stable in a dry or lyophilized state (Levy et al.,1985b), the virus is very sensitive to detergents, including soap, and can be eliminated rapidly by bleach (0.5% sodium hypochroite) and lipid solvents. Alcohol (more than 70%) or acetone–alcohol also inactivates the virus efficiently at room temperature or higher. As noted previously, irradiation and heating (in liquid solution, 56°C for 30 min) are also effective in inactivating the virus.

Literature Cited

Abrams, D. I., B. J. Lewis, J. H. Beckstead, C. A. Casavant, and W. L. Drew. 1984. Persistent diffuse lymphadenopathy in homosexual men: endpoint or prodrome? Ann. Intern. Med. 100:801–808.

Allan, J. A., J. E. Coligan, F. Barin, M. F. McLane, J. G. Sodroski, C. A. Rosen, W. A. Haseltine, T. H. Lee, and M. Essex. 1985. Major glycoprotein antigens that induce antibodies in AIDS patients are encoded by HTLV-III. Science 228:1091–1094.

Barré-Sinoussi, F., M. Nugeyre, C. Dauguet, E. Vilmer, C. Griscelli, F. Brun-Vezinet, C. Rouzioux, J. Gluckman, J. Chermann, and L. Montagnier. 1983. Isolation of a T-lymphotropic retrovirus from a patient at risk for acquired immune deficiency syndrome. Science 220:868–871.

Cheng-Mayer, C., J. T. Rutka, M. L. Rosenblum, T. McHugh, D. P. Stites, and J. A. Levy. 1987. The human immunodeficiency virus (HIV) can productively infect cultured human glial cells. Proc. Natl. Acad. Sci. USA 84:3526–3530.

Clavel, F., D. Guetard, F. Brun-Vezinet, S. Chamaret, M. A. Rey, M. O. Santos-Ferreira, A. G. Laurent, C. Dauguet, C. Katlama, and C. Rouzioux. 1986. Isolation of a new human retrovirus from West African patients with AIDS. Science 233:343–346.

Cóffin, J., A. Haase, J. A. Levy, L. Montagnier, S. Oroszlan, N. Teich, H. Temin, K. Toyoshima, H. Varmus, P. Vogt, and R. Weiss. 1986. Human immunodeficiency viruses (letter). Science 232:697.

Conant, M., D. Hardy, J. Sernatinger, D. Spicer, and J. A. Levy. 1986. Condoms prevent transmission of the AIDS-associated retrovirus. J. Am. Med. Assoc. 255:1706.

Cooper, D., A. A. Imrie, and R. Penny. 1987. Antibody response to human immunodeficiency virus after primary infection. J. Infect. Dis. 155:1113–1118.

Cooper, D. A., J. Gold, P. MacLean, B. Donovan, R. Finlayson, T. G. Barnes, H. M. Michelmore, P. Brooke, and R. Penny. 1985. Acute AIDS retrovirus infection. Lancet 1:537–540.

Dalgleish, A., P. Beverley, P. Clapham, D. Crawford, M. Greaves, and R. Weiss. 1984. The CD4 (T4) antigen is an essential component of the receptor for the AIDS retrovirus. Nature (London) 312:763–766.

Evans, L., T. McHugh, D. Stites, and J. A. Levy. 1987. Differential ability of human immunodeficiency virus isolates to productively infect human cells. J. Immunol. 138:3415–3418.

Evans, L. A., J. M. Homsy, J. W. Morrow, C. D. Sooy, and J. A. Levy. 1988. Human monoclonal antibodies directed against *gag* gene products of the human immunodeficiency virus. J. Immunol. **140**:941–943.

Folks, T., D. M. Powell, M. M. Lightfoote, S. Benn, M. A. Martin, and A. S. Fauci. 1986. Induction of HTLV-III/LAV from a non-virus producing T-cell line: implications for latency. Science **231**:600–602.

Gallo, R. C., S. Z. Salahuddin, M. Popovic, G. Shearer, M. Kaplan, B. Haynes, T. Palker, R. Redfield, J. Oleske, B. Safai, G. White, P. Foster, and P. Markham. 1984. Frequent detection and isolation of cytopathic retroviruses (HTLV-III) from patients with AIDS and at risk for AIDS. Science **224**:500–502.

Gnann, J. W., Jr., J. B. McCormick, S. Mitchell, J. A. Nelson, and B. A. Oldstone. 1987. Synthetic peptide immunoassay distinguishes HIV type 1 and HIV type 2 infections. Science **237**:1346–1349.

Gottlieb, M. S., R. Schroff, H. M. Schanker, J. D. Weisman, T. F. Peng, R. A. Wolf, and A. Saxon. 1981. *Pneumocystis carinii* pneumonia and mucosal candidiasis in previously healthy homosexual men. N. Engl. J. Med. **305**:1425–1430.

Goudsmit, J., D. A. Paul, and J. M. A. Lange. 1986. Expression of human immunodeficiency virus antigen (HIV-Ag) in serum and cerebrospinal fluid during acute and chronic infection. Lancet **2**:177–180.

Greenspan, D., J. S. Greenspan, M. Conant, V. Petersen, S. Silverman, Jr., and Y. de Souza. 1984. Oral "hairy" leucoplakia in male homosexuals: evidence of association with both papillomavirus and a herpes-group virus. Lancet **2**:831–834.

Greenspan, D., J. S. Greenspan, N. G. Hearst, L.-Z. Pan, M. A. Conant, D. I. Abrams, H. Hollander, and J. Levy. 1987. Relation of oral hairy leukoplakia to infection with HIV and the risk of developing AIDS. J. Infect. Dis. **155**:475–481.

Guyader, M., M. Emerman, P. Sonigo, F. Clavel, L. Montagnier, and M. Alizon. 1987. Genome organization and transactivation of the human immunodeficiency virus type 2. Nature (London) **326**:662–669.

Harada, S., Y. U. Koyanagi, and N. Yamamoto. 1985. Infection of HTLV-III/LAV in HTLV-I carrying cells MT-2 and MT-4 and application in a plaque assay. Science **229**:563–566.

Harper, M. E., L. M. Marselle, R. C. Gallo, and F. Wong-Staal. 1986. Detection of lymphocytes expressing human T-lymphotropic virus type III in lymph nodes and peripheral blood from infected individuals by *in situ* hybridization. Proc. Natl. Acad. Sci. USA **83**:772–776.

Ho, D., T. Rota, R. Schooley, J. Kaplan, J. Allan, J. Groopman, L. Resnick, D. Felsenstein, C. Andrews, and M. Hirsch. 1985. Isolation of HTLV-III from cerebrospinal fluid and neural tissues of patients with neurologic syndromes related to the acquired immunodeficiency syndrome. N. Engl. J. Med. **313**:1493–1503.

Hoffman, A. D., B. Banapour, and J. A. Levy. 1985. Characterization of the AIDS-associated retrovirus, reverse transcriptase and optimal conditions for its detection in virions. Virology **147**:326–335.

Hollander, H., and J. A. Levy. 1987. Neurologic abnormalities and human immunodeficiency virus recovery from cerebrospinal fluid. Ann. Intern. Med. **106**:692–695.

Homsy, J., G. Thomson-Honnebier, C. Cheng-Mayer, and J. A. Levy. 1988. Detection of human immunodeficiency virus (HIV) in body fluids and culture supernatants by sequential competition ELISA. J. Virol. Method. **19**:43–56.

Hoxie, J. A., B. S. Haggarty, J. L. Rackowski, N. Pilsbury, and J. A. Levy. 1985. Persistent noncytopathic infection of human lymphocytes with AIDS-associated retrovirus (ARV). Science **229**:1400–1402.

Jaffe, H., D. Bregman, and R. Selik. 1983. Acquired immune deficiency syndrome in the United States: the first 1,000 cases. J. Infect. Dis. **148**:339–345.

Kalyanaraman, V. S., C. D. Cabradella, J. P. Getchell, R. Narayanan, E. H. Braff, J. C. Chermann, T. Barre-Sinoussi, L. Montagnier, T. J. Spira, J. Paklan, D. Fishbein, H. W. Jaffe, J. W. Curran, and D. P. Francis. 1984. Antibodies to the core protein of lymphoadenopathy-associated virus (LAV) in patients with AIDS. Science **225**:321–322.

Kaminsky, L., T. McHugh, D. P. Stites, P. Volberding, G. Henle, W. Henle, and J. A. Levy. 1985. High prevalence of antibodies to AIDS-associated retroviruses (ARV) in acquired immune deficiency syndrome and related conditions and not in other disease states. Proc. Natl. Acad. Sci. USA **82**:5535–5539.

Kanki, P. J., S. M'Boup, D. Ricard, F. Borin, F. Denis, C. Boye, L. Sangare, K. Travers, M. Albaum, R. Marlink, J.-L. Romet-Lemonne, and M. Essex. 1987. Human T-lymphotropic virus type 4 and the human immunodeficiency virus in West Africa. Science **236**:827–831.

Kanner, S. B., C. Cheng-Mayer, R. B. Geffin, W. P. Parks, G. A. Belta, L. O. Arthur, K. P. Samuel, and R. S. Papas. 1986. Human retrovirus env and gag polypeptides: serologic assays to measure infection. J. Immunol. **137**:674–678.

Klatzmann, D., F. Barre-Sinoussi, M. Nugyre, C. Dauguet, E. Vilmer, C. Griscelli, F. Brun-Vezinet, C. Rouzioux, J. Gluckman, J. C. Chermann, and L. Montagnier. 1984a. Selective tropism of lymphadenopathy associated virus (LAV) for helper-inducer T lymphocytes. Science **225**:59–62.

Klatzmann, D., E. Champagne, S. Chamaret, J. Cruest, D. Guetard, T. Hercend, J.-C. Gluckman, and L. Montagnier. 1984b. T lymphocyte T4 molecule behaves as receptor for human retrovirus LAV. Nature (London) **312**:767–768.

Lange, J. M. A., R. A. Coutinho, W. J. A. Krone, L. F. Verdonck, S. A. Danner, J. Van der Noorda, and J. Goudsmit. 1986. Distinct IgG recognition pattern during progression of subclinical and clinical infection with lymphadenopathy associated virus/human T lymphotropic virus. Br. Med. J. **292**:228–230.

Levy, J. A. 1986. The multifaceted retrovirus. Cancer Res. **46**:5457–5468.

Levy, J. A., L. Evans, C. Cheng-Mayer, L.-Z. Pan, A. Lane, C. Staben, D. Dina, C. Wiley, and J. Nelson. 1987. The biologic and molecular properties of the AIDS-associated retrovirus that affect antiviral therapy. Ann. Inst. Pasteur **138**:101–111.

Levy, J. A., A. D. Hoffman, S. Kramer, J. Landis, J. Shimabukuro, L. Oshiro. 1984. Isolation of lymphocytopathic retroviruses from San Francisco patients with AIDS. Science **225**:840–842.

Levy, J. A., L. S. Kaminsky, W. J. W. Morrow, K. Steimer, P. Luciw, D. Dina, J. Hoxie, and L. Oshiro. 1985a. Infection by the retrovirus associated with the acquired immunodeficiency syndrome. Ann. Intern. Med. **103**:684–699.

Levy, J. A., G. A. Mitra, M. F. Wong, and M. Mozen. 1985b. Survival of AIDS-associated retrovirus (ARV) during factor VIII purification from plasma: inactivation by wet and dry heat procedures. Lancet **1**:1456–1457.

Levy, J. A., and J. Shimabukuro. 1985. Recovery of AIDS-associated retroviruses from patients with AIDS, related conditions, and clinically healthy individuals. J. Infect. Dis. **152**:734–738.

Levy, J. A., J. Shimabukuro, H. Hollander, J. Mills, and L. Kaminsky. 1985c. Isolation of AIDS-associated retroviruses (ARV) from cerebrospinal fluid and brain from patients with neurological findings. Lancet **2:**586–588.

Levy, J., J. Shimbukuro, T. McHugh, C. Casavant, D. Stites, and L. Oshiro. 1985d. AIDS-associated retroviruses (ARV) can productively infect other cells besides human T helper cells. Virology **147:**441–448.

Levy, J. A., L. H. Tobler, T. M. McHugh, C. H. Casavant, and D. P. Stites. 1985e. Long-term cultivation of T-cell subsets from patients with acquired immune deficiency syndrome. Clin. Immunol. Immunopathol. **35:**328–336.

Levy, J. A., and J. L. Ziegler. 1983. Acquired immune deficiency syndrome (AIDS) is an opportunistic infection and Kaposi's sarcoma results from secondary immune stimulation. Lancet **2:**78–81.

Mitra, G., M. F. Wong, M. M. Mozen, J. S. McDougal, and J. A. Levy. 1986. Elimination of infectious retroviruses during preparation of immunoglobulins. Transfusion **26:**394–397.

Nelson, J. A., C. A. Wiley, C. Reynolds-Kohler, C. E. Reese, W. Margaretten, and J. A. Levy. 1988. Human immunodeficiency virus detected in bowel epithelium from patients with gastrointestinal symptoms. Lancet **1:**259–262.

Pan, L.-Z., C. Cheng-Mayer, and J. A. Levy. 1987. Patterns of antibody response in individuals infected with the human immunodeficiency virus. J. Infect. Dis. **155:**626–632.

Poiesz, B. J., A. F. Ruscetti, P. A. Gazdar, P. A. Bunn, J. D. Minna, and R. C. Gallo. 1980. Detection and isolation of type C retrovirus particles from fresh and cultured lymphocytes of a patient with cutaneous T-cell lymphoma. Proc. Natl. Acad. Sci. USA **77:**7415–7419.

Quinn, T. C., J. M. Mann, J. W. Curran, and P. Piot. 1986. AIDS in Africa: an epidemiologic paradigm. Science **234:**955–963.

Rabson, A. B., and M. A. Martin. 1985. Molecular organization of the AIDS retrovirus. Cell **41:**477–480.

Resnick, L., F. diMarzo-Veronese, J. Schupbach, W. W. Tourtellotte, D. D. Ho, F. Muller, P. Shapshak, M. Vogt, J. E. Groopman, P. D. Markham, and R. C. Gallo. 1985. Intra-blood–brain-barrier synthesis of HTLV-III specific IgG in patients with neurologic symptoms associated with AIDS or AIDS-related complex. N. Engl. J. Med. **313:**1498–1503.

Resnick, L., K. Veren, S. Z. Salahuddin, S. Tondreau, and P. D. Markham. 1986. Stability and inactivation of HTLV-III/LAV under clinical and laboratory environments. J. Am. Med. Assoc. **255:**1887–1891.

Seligmann, M., L. Chess, J. L. Fahey, A. S. Fauci, P. J. Lachman, J. L'Age-Stehr, J. Ngu, A. J. Pinching, F. S. Rosen, T. J. Spira, and J. Wybran. 1984. AIDS—an immunologic reevaluation. N. Engl. J. Med. **311:**1286–1292.

Sodroski, J., W. C. Goh, C. Rosen, A. Dayton, E. Terwilliger, and W. Haseltine. 1986. A second post-transcriptional *trans*-activator gene required for HTLV-III replication. Nature (London) **321:**412–417.

Spire, B., F. Barré-Sinoussi, D. Dormont. 1985. Inactivation of lymphadenopathy-associated virus by heat, gamma rays, and ultraviolet light. Lancet **1:**188–189.

Steimer, K., J. Puma, M. Power, M. Powers, C. George-Nascimento, J. Stephans, J. Levy, R. Sanchez-Pescador, P. A. Luciw, P. Barr, and R. Hallewell. 1986. Differential antibody responses of individuals infected with AIDS-associated retroviruses surveyed using the viral core antigen p25^{gag} expressed in bacteria. Virology **150:**283–290.

Stricker, R. B., M. T. McHugh, D. J. Moody, W. J. W. Morrow, D. P. Stites, M. A. Shuman, and J. A. Levy. 1987. An AIDS-related cytotoxic autoantibody reacts with a specific antigen on stimulated CD4+ T cells. Nature (London) **327:**710–713.

Walker, C. M., D. J. Moody, D. P. Stites, and J. A. Levy. 1986. CD8+ lymphocytes can control HIV infection *in vitro* by suppressing virus replication. Science **234:**1563–1566.

Weiss, R. A., P. R. Clapham, P. R. Cheingson, A. G. Dagleish, C. A. Carne, I. V. D. Weller, and R. S. Tedder. 1985a. Neutralization of human T-lymphotropic virus type III by sera of AIDS and AIDS-risk patients. Nature (London) **316:**69–72.

Weiss, S., J. Goedert, M. Sarngadharan, A. Bodner, R. Biggar, J. Clark, R. Dodd, E. Geimann, J. Giron, M. Greene, M. Melbye, M. Popovic, M. Robert-Guroff, C. Saxinger, M. Simberkoff, D. Winn, R. Gallo, and W. Blattner. 1985b. Screening test for HTLV-III (AIDS agent) antibodies. J. Am. Med. Assoc. **253:**221–225.

Winkelstein, W., Jr., D. M. Lyman, N. Padian, R. Grant, M. Samual, J. A. Wiley, R. E. Anderson, W. Lang, J. Riggs, and J. A. Levy. 1987. Sexual practices and risk of infection by the human immunodeficiency virus: the San Francisco Men's Health Study. J. Am. Med. Assoc. **257:**321–325.

Wofsy, C. B., J. B. Cohen, L. B. Hauer, N. S. Padian, B. A. Mechaelis, L. A. Evans, and J. A. Levy. 1986. Isolation of the AIDS-associated retrovirus from vaginal and cervical secretions from women with antibodies to the virus. Lancet **1:**527–529.

CHAPTER 36

Picornaviridae: The Enteroviruses (Polioviruses, Coxsackieviruses, Echoviruses)

JACOBA G. KAPSENBERG

Diseases: Minor illness (all types); aseptic meningitis, exanthema (many types); poliomyelitis, polioencephalitis (some specific types); vesicular disease (some types); pleurodynia; (neonatal) carditis (some types); acute hemorrhagic conjunctivitis (two types).

Etiologic Agents: Poliovirus serotypes 1–3; Coxsackievirus A serotypes 1–17, 19–22, 24; Coxsackievirus B serotypes 1–6; echovirus serotypes 1–7, 9, 11–27, 29–33; human enterovirus serotypes 68–71.

Sources: Excreta of humans (i.e., fecal contamination, airborne droplets from the throat, eye discharge).

Clinical Manifestations: Most infections subclinical; acute undifferentiated febrile illness; upper respiratory infection; gastrointestinal symptoms; major illness involving different target organs with different serotypes (see diseases).

Pathology: Unspecified in most infections; pleocytosis in cerebrospinal fluid in meningitis cases and poliomyelitis; degeneration of motor neurons in the spinal cord in paralytic infections; inflammation and perhaps immunopathologic reactions in other target organs (skin, eye, heart, brain).

Laboratory Diagnosis: Virus isolation from feces and throat; from cerebrospinal fluid, ocular discharge, vesicle fluid, pericardial effusion in relation to symptoms; identification of the infecting serotype by neutralization.

Epidemiology: Most serotypes distributed worldwide, with a typical seasonal pattern of epidemic infections in summer and early autumn in temperate zones; polioviruses widely prevalent in countries with insufficient vaccination programs; epidemics of acute hemorrhagic conjunctivitis caused by enterovirus 70 and coxsackie virus A24 in Africa and southeast and south Asia.

Treatment: No specific treatment.

Prevention and Control: Good personal and nosocomial hygiene; vaccination against poliomyelitis by injection of concentrated vaccine containing high doses of inactivated virulent viruses of the three serotypes or by orally given attenuated virus strains; global eradication of polioviruses is possible.

Introduction

In the series "Portraits of Viruses: the Picornaviruses," one of the founding fathers of modern medical virology, J. L. Melnick, evoked a picture of what his students called "ancient-as-hell" virology. He described the fundamental transmission of the agent of poliomyelitis to monkeys in 1908 and the discovery that this virus was not primarily neurotropic, but was produced in and excreted from the intestinal tract (Melnick, 1983). In view of the dramatic increase of paralytic poliomyelitis, it was of paramount importance to find other methods for virus isolation and research. Inoculation of newborn mice with stool samples led in 1948 to the discovery of human enteric viruses, which could be distinguished as two groups (coxsackie A and B) according to the disease and pathology produced in these animals. The name coxsackie is reminiscent of the township where the first strains were isolated. As soon as the replication of poliovirus in tissue culture was recognized, first in explanted human embryonic material, later in trypsin-dispersed cells growing in a monolayer—both in conjunction with a cytopathic effect (Robbins, Enders, and Weller, 1950)—again numerous unknown viruses were found in stools from patients and also from healthy children. The development of cell-culture techniques was greatly advanced by the addition of the previously unknown antibiotics (penicillin and streptomycin) to sample extract and culture medium. The new viruses were called "enteric cytopathogenic human orphan" (ECHO) in 1955. Although most of the viruses are no longer orphan, that is, in search of disease, the name echovirus has been approved for this group.

With the growing knowledge of virus properties, a reshuffling of some viruses and serotypes was necessary. At this time poliovirus (types 1 to 3), coxsackie virus A (types 1 to 22, 24), coxsackie virus B (types 1 to 6), echovirus (types 1 to 9, 11 to 27, 29 to 33) and the later recognized enteroviruses (types 68 to 72) together form the 68 known human members of the genus *Enterovirus* of the family of *Picornaviridae* (Matthews, 1982). Enterovirus 72 is the cause of acute infectious hepatitis; it is only recently numbered among the enteroviruses and will be discussed elsewhere in this volume.

Infections with human enteroviruses are extremely common, but severe clinical manifestations are not the rule. A good example is the occurrence of paralytic poliomyelitis before the introduction of poliovaccine. Only 0.1 to 1% of infected persons would develop paralytic disease, whereas 90 to 95% would remain asymptomatic in spite of vigorous virus production in the intestinal tract, with virus titers of more than 10^6 TCID$_{50}$ per gram feces. The other 1 to 2% would suffer from meningitis, and 4 to 8% from a minor illness.

Almost all nonpolio enterovirus serotypes are associated with aseptic meningitis, frequently attracting medical attention during regional outbreaks. Occasionally paralytic disease simulating poliomyelitis is seen during an epidemic with a special virus type—for instance, with coxsackie virus A7 in 1952 in the USSR, or with enterovirus 71 in 1975 in Bulgaria and in 1978 in Hungary.

Other signs and symptoms associated with (groups of) enteroviruses are vesicular disease, exanthems, myalgia, myo- and pericarditis, respiratory disease, and enteritis. A special position is taken by enterovirus 70 and coxsackievirus A24. Both caused huge epidemics of acute hemorrhagic conjunctivitis throughout the world.

The family of *Picornaviridae* encompasses viruses of a small size (22 to 30 nm) without envelope, with a capsid structure consisting of 60 subunits, each comprising one molecule of each of the four capsid polypeptides. The virion contains one molecule of positive-sense (messenger type) ssRNA. The main difference of the enteroviruses from the other genus with human members in this family, the rhinoviruses, is from a laboratory viewpoint that enteroviruses are stable at acid pH (below 5 to 6), whereas the common cold viruses are not.

The Viruses

The list of human enteroviruses, given in Table 1, is composed to supply some information on serotype particulars relevant to laboratory diagnosis. The first and second columns tell the story of the difficulty to define species in virus taxonomy. Owing to more knowledge, the following numbers were omitted from the list: echo 10 and 28 (do not belong to the genus); coxsackie A23, echo 8, and echo 34 (later recognized by cross-neutralization respectively as echo 9, echo 1, and coxsackie A24 variants). Also coxsackie A18 should be canceled as a serotype, this virus is a variant of coxsackie A13.

The prototype strains should not be considered as representative strains. Most are just the earliest strains found, by neutralization tests, to be different from the other enteroviruses known at that time. Only strain Bastianni (echo 30) was selected as a true prototype from a number of variants because of its broader antigenicity (Melnick et al., 1963). Most prototypes are deposited in the American Type Culture Collection, and their original source and references are described in its catalogue (5th edition, 1986). Poliovirus strains used in oral vaccine production (Sabin strains) are:

LS-c, 2ab, derived from Mahoney (type 1)
P712, Ch, 2ab (type 2)
Leon, 12 a$_1$b, derived from Leon (type 3).

The virulent strains used in the inactivated, Salk-type poliovaccine are:

Mahoney (original isolation, 1941) (type 1)
MEF-1 (original isolation, 1942) (type 2)
Saukett (original isolation, 1950) (type 3).

References for the origin of the older enterovirus strains can be found in "Strains of Human Viruses" (Majer and Plotkin, 1972).

In the next columns of Table 1, special properties of the strains are listed: growth characteristics, typing problems, and miscellaneous information. Coxsackie A and B viruses were grouped for their ability to infect newborn mice. Most strains could be adapted to tissue culture except for coxsackie A1, 19, and 22. This is not similar to virus isolation. Only some coxsackie A viruses (types 9, 16, 21, and 24), but all coxsackie B viruses are better isolated in tis-

TABLE 1. Human enteroviruses

Species	Sero-type	Prototype strain	Special properties of strains				Occurrence in epidemics[d]
			Isolation, growth[a]	Typing problems[b,c]	Miscellaneous	Special symptoms	
Poliovirus	1	Brunhilde		+	Vaccine strains	Poliomyelitis	+
	2	Lansing		+	See text	Poliomyelitis	+
	3	Leon		+		Poliomyelitis	+
Coxsackievirus A	1	Tompkins	No TC adaptation				−
	2	Fleetwood				Herpangina	−
	3	Olson				Herpangina	+
	4	High Point				Herpangina	+
	5	Swartz				Herpangina	+
	6	Gdula				Herpangina	+
	7	Parker		+,C		Polioencephalitis	−
	8	Donovan				Herpangina	+
	9	Bozek	Good TC isolation			Meningitis	+
	10	Kowalik				Herpangina	−
	11	Belgium-1					−
	12	Texas-12					−
	13	Flores		++	CA18 included		−
	14	G-14					−
	15	G-9					+
	16	G-10	Good TC isolation	C		Hand-foot-mouth disease	−
	17	G-2					−
	19	NIH-8663	No TC adaptation	++			+
	20	IH-35					−
	21	Kuykendall	Good TC isolation			Respiratory disease	−
	22	Chulman	No TC adaptation				+
	24	Joseph	Good TC isolation	++	E34 included	Hemorrh. conjunctivitis	+
Coxsackievirus B	1	Conn.5	Good TC isolation	+		Bornholm; carditis	+
	2	Ohio-1	Good TC isolation	+		Bornholm; carditis	+
	3	Nancy	Good TC isolation	+		Bornholm; carditis	+
	4	JVB	Good TC isolation	+		Bornholm; carditis	+
	5	Faulkner	Good TC isolation		Related to Swine VDV	Bornholm; carditis	+
	6	Schmitt	Good TC isolation	++		Bornholm; carditis	−

		Growth in TC; mice[a]	Intratypic antigenic variability[b,c]		Disease	Epidemics described[d]
Echovirus						
1	Farouk		++	E8 included		−
2	Cornelis		+			−
3	Morrisey		++,C			+
4	Pesascek		+			+
5	Noyce		+			+
6	D'Amori					+
7	Wallace	Mice				+
9	Hill		++		Exanthem	+
11	Gregory		++		Respiratory disease	−
12	Travis					−
13	Del Carmen					−
14	Tow					+
15	CH96-51					+
16	Harrington		++		Exanthem	+
17	CHHE-29					+
18	Metcalf		+,C		Exanthem	−
19	Burke					+
20	JV-1		+			−
21	Farina					−
22	Harris	Difficult, slow		Different RNA	Infant diarrhea	+
23	Williamson	Difficult, slow		Different RNA		−
24	De Camp	Difficult, slow				+
25	JV-4					−
26	Coronel		++			−
27	Bacon		+			−
29	JV10		++			+
30	Bastianni		+			−
31	Caldwell		+			−
32	PR-10		+			+
33	Toluca-3					−
Human enterovirus						
68	Fermon					−
69	Toluca-1					+
70	J 670/71	Difficult	+		Hemorrh. conjunctivitis	+
71	BrCr	Difficult; mice	++,C		Hand-foot-mouth disease, polioencephalitis	+
72	CR 326	Difficult		Suggested prototype	Hepatitis A	+

[a] TC = Tissue culture; mice = some strains pathogenic for newborn mice.

[b] + = Intratypic antigenic variability (++ = pronounced).

[c] C = Some strains require chloroform treatment to release membrane-bound virions for neutralization.

[d] Epidemics described.

sue culture. On the other hand, strains of echovirus 9 and enterovirus 71 can be pathogenic for suckling mice with coxsackie A pathology. Other viruses are difficult to isolate and require blind passages or expert attention for the recognition of their cytopathic effect (echoviruses 22, 23, 24; and enterovirus 70, 71, 72).

Important typing problems are due to intratypic antigenic variability or insufficient neutralization, probably caused by aggregation with cellular membrane structures. These will be discussed later.

As mentioned in the introduction, all enteroviruses are associated with several clinical syndromes. Some serotypes, however, are more frequently isolated from patients with special diseases or symptoms. The picture given in Table 1 is intended for rapid orientation and of course, far from complete. The last column summarizes information about described occurrences in epidemics, be they extended or small. Physicians should keep in mind that many enterovirus serotypes circulate together in an unpredictable way. Only virus isolation can decide for certain whether a wave of similar diseases is caused by one or more viruses.

Physicochemical Properties and Morphology

Poliovirus is the best characterized of all enteroviruses. An immense amount of knowledge has been accumulated, extended even to the three-dimensional structure at 2.9 Å resolution (Hogle, Chow, and Filman, 1985). This virus, therefore, serves as representative for the genus. The other members are less well studied. Differences in some properties can be anticipated, but these do not preclude inclusion in the genus. Only with echoviruses type 22 and 23 does some doubt exist, because a different structure of the RNA genome of E22 and perhaps a different mode of replication of both types were discovered (Seal and Jamison, 1983)

Common features of *Picornaviridae* (selected from Koch and Koch, 1985; Matthews, 1982; Rueckert, 1985) are:

a. a small size (22 to 30 nm) with a slightly skewed spherical conformation as seen in electron micrographs, which fail to detect the fine structure of the surface
b. a capsid consisting of 60 subunits, called (mature) protomers, each comprising one molecule of each of the four major polypeptides (VP 1 to 3 of 22 to 40 kDa, VP4 of 5 to 10 kDa), which are formed by cleavage of one molecule of the structural protein precursor
c. within the capsid, one molecule of infectious, single-stranded, plus-sense RNA with a molecular weight of ± 2.5 × 10⁶ daltons, containing 7,000 to 8,000 bases. The 3′ terminus is polyadenylated; at the 5′ terminus, a small virus-encoded protein (VPg) is carried
d. a dry molecular weight of the virion of 8 to 9 × 10⁶ daltons
e. a structural icosahedral symmetry (5 : 3 : 2-fold cubic)

Crystals of poliovirus type 1, strain Mahoney, were analyzed by X-ray crystallographic methods (Hogle, Chow, and Filman 1985). This provided more insight into the three-dimensional structure of the capsid, its assembly, and the features of the outer surface. The investigators describe the last as a landscape: peaks at the fivefold axes are surrounded by broad valleys, and a broad plateau is formed at the threefold axes, separated by saddle surfaces at the twofold axes. The positions of VP1, VP2, and VP3 within the protomer are shown in Figure 1. The small VP4 has an internal position. By several methods, the distribution of antigenic sites on the surface of the virion were investigated. Most resided in prominent radial projections.

Features distinguishing enterovirus particles as a genus are:

a. stability at pH3 to 9.
b. buoyant density at about 1.34 g/cm³ in CsCl.
c. sedimentation coefficient, 155S.

Enteroviruses are not disrupted by detergents such as sodium deoxycholate and are insensitive to lipid solvents (ether, chloroform). Exposure to a temperature of 50°C inactivates them rapidly, but the

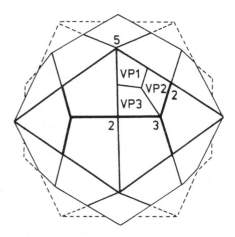

FIG. 1. Geometric representation of the exterior surface features of poliovirus, with the positions of VP1, VP2, and VP3 from the same protomer indicated. (Redrawn with permission from Hogle, Chow, and Filman, 1985, © AAAS.)

addition of molar magnesium chloride postpones inactivation for at least 1 hour and for weeks, months, or years at lower temperatures. This property of cationic stabilization was used in former years as one of the optional requirements for new human enterovirus candidates (Melnick et al., 1962). Of practical importance is the routine stabilization by molar $MgCl_2$ for cold storage of enterovirus strains and of the live attenuated poliovaccine.

Replication

The early concepts on enterovirus replication are rapidly being replenished or replaced by insights gained by molecular biology. Specialists on this subject should be consulted for the state of the art at this moment (for example, Koch and Koch, 1985, Rueckert, 1985). In the context of this book, however, a general view is sufficient, with emphasis on practical consequences. Many factors influence virus growth, connected with the serotype (even the special strain used in a certain laboratory), the host cells (again, those in use there and then), and the culture conditions.

Per definition, enteroviruses follow the rules of Picornavirus replication, which are extraordinarily well studied with poliovirus (Fig. 2). The first step is attachment of the virus to the cell. A specific site on the surface of the virion binds to a place called a receptor, drifting in the plasma membrane. Cells not expressing receptors for the virus in question are not susceptible to infection. Only when, under exceptional circumstances, infectious RNA is introduced into the cell, can replication go on. Some receptors are available in great numbers on the cells, others are not. Some receptors are shared by viruses, others seem to be type-specific. The detection of receptors explains to some extent tropism of viruses to certain cells, both in vivo and in vitro. The normal cellular function of virus receptors is not yet understood.

The concentration of the virus, the available number and mitotic phase of the cells, and environmental factors (such as pH, ionic composition of the medium, and temperature) determine the speed of attachment. Generally, most of the virus brought into contact with susceptible cells is attached within 60 min. This is the usual incubation time in experiments before further procedures are performed. During this period more receptors attach to the virion, which becomes more and more surrounded by the cell membrane and finally is engulfed in the cytoplasm, either free or in micropinocytotic vesicles. During this process, enterovirus capsid proteins undergo conformational changes, simultaneously with changes in antigenicity. The small protein VP4,

which acts as a sort of uncoating plug, is lost. Viral RNA is released, and from this moment the cell is infected, and its organization is being used for the replication of virus particles. The first translation of the virus RNA brings forth in about 15 min a polyprotein, which is cleaved to generate, among others, a protease and an RNA polymerase. Then replication proceeds with virus −RNA synthesis, followed by synthesis of more +RNA, part of which will be used as mRNA and part of which will be packed into virions. Capsid proteins are synthesized and assembled into immature subunits, containing VP1, VP3, and VP0, which at the final stage of virion assembly is cleaved into VP2 and VP4. Meanwhile, host protein and RNA synthesis are inhibited, and the structure of the living cell changes.

In poliovirus replication, capsid assembly and RNA synthesis are intimately associated with the smooth endoplasmic reticulum of a growing number of membrane-enclosed vesicles. Viral RNA with attached VPg is thought to be "sucked" into the nearly finished capsid. The last maturation step is the cleavage of VP0 into VP2 and VP4. The VP4 molecules might also be "sucked in" to take their internal position and to close the doors.

The first progeny virus can be found 3 to 4 h post infection. At the completion of the cycle at 6 to 8 h, a cell contains 25 to 100×10^3 virus particles, which are released from the lytic dying cell, some still enclosed in their membranous sacs.

The effects of virus replication on the host cell could be described by light and electron microscopic observations correlated with biochemical events, as Koch and Koch did (1985, page 209), but in the diagnostic laboratory the term cytopathic effect (CPE) is reserved for alterations in cellular morphology as seen in unstained, living cell cultures. Normal cells are stretched out on the glass or plastic surface and have intimate contact with one another. Enterovirus-infected cells begin to retract their peripheral parts and round up, become refractile, shrink, and get loose from the surface, sometimes in fragments. The experienced observer can predict from the speed of CPE development and the appearance of the cells, considering the cell type and the diagnostic procedure, whether an enterovirus is involved and, to some degree, to which group it belongs. In stained preparations the nucleus is crumpled and displaced to the periphery of the cell by an eosinophilic mass. The appearance of the nucleus in echovirus 22 and 23 infections is different, the chromatin is marginated, and the nucleus looks empty and not distorted (Fig. 3).

Synthesis of a mass of viral proteins is possible only with cells in a good nutritional condition. For an optimal virus harvest from a culture infected with a

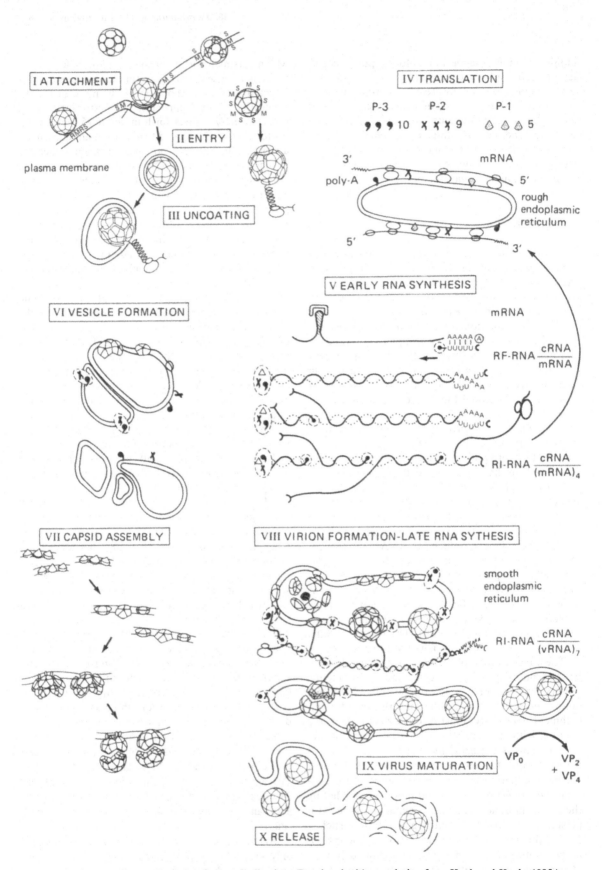

FIG. 2. Schematic overview of poliovirus replication. (Reprinted with permission from Koch and Koch, 1985.)

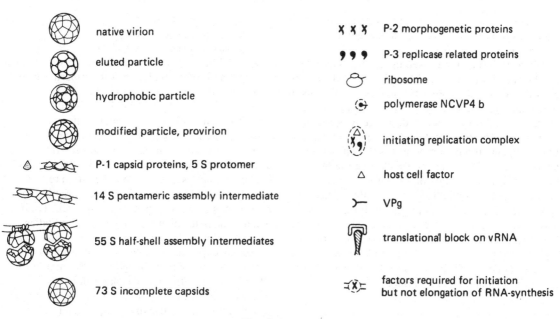

FIG. 2 *Continued.*

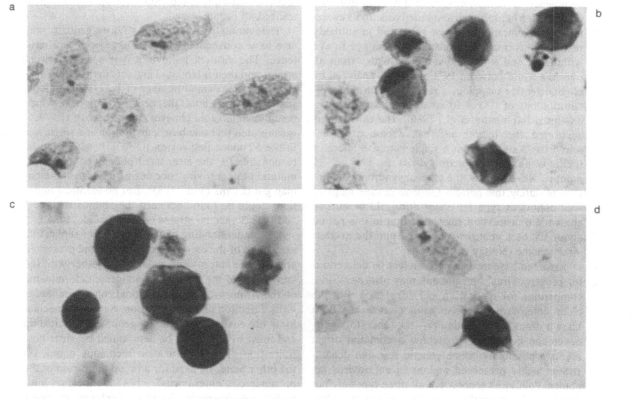

FIG. 3. Enterovirus cytopathic effects in tertiary monkey kidney cells stained after 2 days. (a) Cell control; (b) echovirus 6; (c) echovirus 22; and (d) enterovirus 71. The figure appears in full color at the beginning of this volume in Color Plate 1.

minimum of virus (which requires several replication cycles), changing of the medium at regular intervals is essential.

None of the cell types used in the virus laboratory for the detection of enteroviruses is identical with the cells that produce the enormous amounts of virus in the human intestinal tract. Virus growing in tissue culture may become adapted: progeny virus with properties of better attachment or speed of replication will gain an advantage. For isolation purposes, adaptation is often achieved by passing the first culture to fresh cells, sometimes before a cytopathic effect becomes visible. Some of the enterovirus serotypes are difficult to adapt and, therefore, to isolate. This is very much the case with enteroviruses 70 and 71. Strains of these viruses tend to stick to the cell type of the first isolation. This explains why these viruses, although occurring in large epidemics, are not readily isolated and identified.

Genetics and Antigenic Determinants

RNA virus genomes are extremely mutable, probably because they lack a proofreading system (Holland, 1984). This could readily be studied with the emergence of enterovirus 70 in 1969, a virus that causes acute hemorrhagic conjunctivitis. In its epidemiological evolution it is not subjected to antibody restriction as it can be transmitted from eye to eye within one and a half days. With isolates from all over the world between 1971 and 1981, analyzed by oligonucleotide mapping, a rapid and constant substitution rate of 1.83×10^{-3} base per year was demonstrated (Miyamura et al., 1986). The calculations supported the hypothesis that evolution of enterovirus 70 started from a single common ancestor (perhaps an animal picornavirus) in 1967 ± 15 months. Nevertheless, the clinical syndrome did not change during this period. Oligonucleotide maps of wild poliovirus type 1 isolates during 13 months of epidemic transmission suggested that in this period about 100 base changes were fixed into the average viral genome (Nottay et al., 1981).

Errors in base incorporation might be deleterious for progeny virus, but mutants may also profit in a competition for replication and adaptation. For the older enteroviruses—polio as a disease has been known from early antiquity—the ancestor virus never can be traced. Somehow a common origin seems a possibility, since genetic research demonstrates highly preserved regions in enterovirus genomes. Poliovirus genomes of all three types are homologous for more than 70% (Toyoda et al., 1984; Tracy and Smith, 1981). Probes containing the gene coding for the replicase at the 3' end and a piece of the 5' end of poliovirus 1 and coxsackie virus B3

have been shown to share enough sequences to hybridize to the RNA counterparts of many different enterovirus serotypes (Hyypiä et al., 1984; Rotbart et al., 1985). Only echovirus 22 escaped hybridization, giving further evidence of its disparity from the other enteroviruses (Rotbart et al., 1985).

The similarity also of the 5' end of different enterovirus genomes, which contains the code for the capsid proteins, suggests that their capsid "backbone" in its basic construction and relative protein arrangement is almost identical. The differences between the many serotypes, per definition revealed by virus neutralization, reside in the variable surface protrusions of the virion (Koch and Koch, 1985). By the use of neutralizing monoclonal antibodies, the antigenic determinants of poliovirus 1 and 3 were found on exposed, readily accessible loops on the surface of the virion (Hogle, Chow and Filman, 1985). Studies with clinical isolates of coxsackievirus B4, tested in a microneutralization assay with monoclonal antibodies, disclosed highly conserved epitopes present on all strains, also upon passage of the virus, beside moderately and poorly conserved epitopes disappearing or appearing upon passage in tissue culture (Prabhakar et al., 1985). In this sensitive assay a hyperimmune polyclonal serum did not detect the variation in individual epitopes, but it neutralized all isolates.

The variability of the enterovirus genome may also have consequences for pathogenicity and virulence. The virulent poliovirus type 1 and 3 parent strains and their attenuated mutants in the Sabin vaccine differ in a small number of base substitutions, scattered throughout the genome, some of which result in amino acid changes. In the case of type 3, it is suggested that one base substitution at a single site in the 5' noncoding region (C → U → C) may be responsible for the attenuated phenotype and backmutation to neurovirulence during passage in the human gut (Evans et al., 1985). But no identical mutations could be found between the attenuated type 1 and type 3 vaccine strains (Almond et al., 1984).

The understanding of the high rate of replicative errors and of the existence of variable and preserved areas in the enterovirus genomes, together with the knowledge of the location of antigenic determinants on the virion surface, has practical consequences for virus detection and identification. Serotype specificity is not easily detectable by genome hybridization and must by definition be determined by virus neutralization. The hybridization technique could, on the other hand, be used for a broad indication of the presence of "enterovirus" RNA. Monoclonal antibodies are too specific for virus identification unless they are selected by neutralization and recognize highly conserved epitopes.

The perspectives of the new information on genetics and molecular biology for vaccine production are not yet fully explored (Almond et al., 1984).

Reactions to Physical and Chemical Agents, with Emphasis on Inactivation and Disinfection

The enteroviruses are relatively resistant to many physical and chemical agents, owing to their compact structure and small size. Because they lack a lipid envelope, they are not destroyed by detergents and lipid solvents (such as ether or chloroform). They are also resistant to phenolic reagents. Thus, many of the commonly used disinfectants will have no effect on these viruses.

They are rapidly inactivated by UV irradiation, which causes cross-linking in the closely packaged RNA. Gamma-irradiation may also act via alteration of the RNA. Agents changing the conformational state of the proteins in the virus capsid destroy the ability to bind to the cellular receptor or may cause the release of VP4, whereupon the RNA can slip out of its confinement. Most disinfectants cause a denaturation of the proteins. In all cases, concentration of the agent and the duration of exposure are important, besides the presence of organic matter and other environmental factors (for instance, pH, the ionic nature of the medium, and the temperature). Enteroviruses do not behave as a group, but differ widely, for instance, in their susceptibility to chlorine (Engelbrecht et al., 1980).

In laboratory and hospital conditions, complete inactivation can be achieved by heating above 50°C (specimens, clothes) or submersion in 2% sodium hydroxide (disposables). Smooth surfaces, such as working tables and safety cabinets, can be treated by UV irradiation. Disinfectants containing free available chlorine, 70% ethanol, or 2% glutaraldehyde can be used on inanimate noncorrodible objects.

For delicate instruments, as yet no common rule can be given. Hand disinfection should be done routinely by thorough washing with 70% ethanol (Mahnel, 1983; Narang and Codd, 1983). Washing with water and soap is not effective (Schürmann and Eggers, 1985).

For the production of inactivated poliovaccine, formalin is used in a thoroughly controlled procedure that conserves the antigenic determinants of the virion.

Pathogenesis

With respect to pathogenesis, again poliovirus is the best studied of all enteroviruses, in both man and monkeys. The general pattern of the other serotypes is assumed to be the same. Differences are reflected in symptoms from the various target organs (skin, mucous membranes, meninges, or motor neurons in the central nervous system, heart, and perhaps muscle).

Portals of entry are epithelial cells of the mouth, nose, and throat, and occasionally of the eye. The primary, local infection includes replication of the virus in lymphoid tissue. A few days later, virus is found in the blood, free in the plasma and probably also in leucocytes (Prather et al., 1984). At that moment virus replication takes place in the small intestine, certainly in Peyer's patches, but there is no reason to believe that epithelial or other mucosal cells are not involved in the process, in view of the amounts of virus excreted in the feces. From these secondary sites, virus enters the bloodstream again and reaches the other target organs. Meninges can be affected, and infection of the skin can become visible as a rash or as vesicles. In special circumstances, virus can reach the central nervous system along the axons of peripheral nerves, whose endings are supposed to be injured by some intervention (such as injection or operation). This is known for poliomyelitis, even in vaccine-related cases, but probably injections also act as a provocation for the rare, severe neurological complications of acute hemorrhagic conjunctivitis (Kono et al., 1977).

In patients with an intact immune system, host responses start immediately after infection. The early appearance of neutralizing antibodies in the blood (from the 4th day on) may block the pathway of the virus to the central nervous system. Interferon is found in serum and cerebrospinal fluid as an indication of a nonimmunologic defense (Flowers and Scott, 1985). Cellular immune control is a complex series of events; for enterovirus infections it has been insufficiently explored. Inflammation accompanies the virus infection of nerve cells in poliomyelitis and the meningoencephalitis caused by enteroviruses, and it is recognized by the appearance of white blood cells in the cerebrospinal fluid. Persistent functional disorder and paralysis develop when neurons are destroyed by the infection, whereas temporarily paralysis, acute focal neurological signs, or depression of conscious level may be caused by the local immune response.

The primary infection of the nasopharynx and small intestine can cause a minor illness with fever, upper respiratory symptoms, nausea, vomiting, abdominal pain, and diarrhea. Apparent recovery is the rule, but in a small percentage a second attack follows after a few days; this is the major illness with involvement of the target organs. This diphasic illness is frequently seen with polio- and coxsackie B virus infections.

Except for persistent lameness in polio (like) disease, complete cure is usual. A chronic enterovirus infection, however, may develop in persons with agammaglobulinemia. If the virus persists in the meninges, this may lead to a finally fatal CNS syndrome, combined with dermatomyositis, with dissemination of virus throughout the body. It has been found that viruses, repeatedly isolated from these patients, acquire alterations in their genome, resulting in reversion to neurovirulence in the case of poliovaccine strains, or in changes in the antigenic make up (Hara et al., 1981; Kapsenberg et al., 1981). These patients have an intact cellular immunity. Therefore, termination of an enterovirus infection and ultimate recovery seem to be impossible without specific antibodies.

Clinical Features

Transmission

Enteroviruses are transmitted by the fecal-oral route and by infection of the nasopharynx via airborne droplets. Transmission is rapid in close communities with several young children or susceptible adolescents, such as households, crèches, schools, overcrowded settlements, or a recruit population. In the hospital setting, maternities are notorious for the spreading of viruses to the newborn, where especially infections with coxsackie B viruses and echovirus 11 are to be feared. The highly infectious discharge of acute hemorrhagic conjunctivitis, caused by enterovirus 70 or coxsackievirus A24, can be transferred from eye to eye by hands, sometimes with smooth surfaces as intermediary, by ophthalmological instruments, or tissues and the like. By way of sewage disposal, surface and recreational waters become contaminated. Since the majority of enterovirus infections pass unnoticed, the water route of transmission seems not to exist in the rich industrial world. In poorer countries, where water is used without proper treatment, this route certainly is a possibility. The same might be true for spreading of enteroviruses by ingestion (or inhalation) of aerosolized or dried sludge from sewage treatment plants or the consumption of contaminated shellfish.

Incubation

The incubation period of enterovirus infections can be divided into two parts: the illness at the site of the primary infection and the disease of the secondarily affected organs. For the infected eye (enterovirus 70, coxsackievirus A24), this period is approximately 24 h. Coxsackievirus A21 causes a common cold after from 36 h to 4 days. Signs of systemic infection

with fever, headache, muscle pains, or diarrhea may precede a major illness. No general incubation time for the development of the full disease can be given. It depends on the invasiveness of the epidemic strain in question, the target organ, and host factors. Exanthem and meningitis with or without encephalitis usually follow infection after ±10 days to 3 weeks. The paralytic phase of poliomyelitis can start abruptly a few days after the first signs of CNS involvement (pleocytosis, for instance). The onset of the neurological complications after enterovirus 70 conjunctivitis ranged from 0 to 120 days, usually 1 to 4 weeks (Kono et al., 1974; WHO, 1986).

In newborn infants, coxsackie B viruses seem to find their way to the myocardial tissue in 2 to 8 days after infection. In older patients a viral myopericarditis may follow the initial infection after a week or so. There is strong evidence of immunologic factors involved in the pathogenesis of cardiomyopathy after virus replication has been terminated (Woodruff, 1980). The onset of acute heart failure can sometimes be precipitated by other host factors, such as forced exercise or steroid therapy.

Period of Infectivity

Most enteroviruses are excreted from the throat during about 1 to 2 weeks and from the intestinal tract for many weeks, even months. This applies also to the completely inapparent infections and to the diseases in which only the secondary infection is noticed. Exceptions are enterovirus 70, which is infectious from the eye discharge during ± 1 week and which only occasionally has been found in late stool samples, and coxsackievirus A21, which in experiments with healthy volunteers was excreted from the nose and throat during a mean period of 16.5 days and which in half the cases was not isolated from the feces. At the end of the excretion period the amount of virus is, of course, diminished, and thus infectivity is less. Nevertheless, with virulent viruses the minimum infective dose for susceptible persons may be low, as low as a few tissue culture infectious doses (Ward and Akin, 1984). As the vesicles associated with some enterovirus infections contain virus, they might play a small role in disease transmission.

Symptoms, Signs, and Clinical Course

As already outlined above, enteroviruses cause several clinical syndromes, mostly not so severe as to need a definite laboratory diagnosis. Epidemics, however, and serious disease requiring hospitalization receive medical and public (!) attention. In this section, a few examples of typical disease will be given. When specimens for virological investigation are submitted together with clinical information of this kind, the medical virologist will immediately be alert for the shortest way to arrive at a laboratory diagnosis. For more clinical information, books on viral infectious diseases should be consulted.

PARALYTIC POLIOMYELITIS

In well-vaccinated countries this disease has become so rare that the diagnosis of the first case can be missed, because the doctors do not suspect its occurrence. A diagnosis of atypical Guillain-Barré polyradiculitis had been made in two cases during the outbreak of poliomyelitis in Finland in 1984 (where polio had been absent for 20 years), 2 months before the first isolation of wild poliovirus type 3 alerted the public health authorities (Kinnunen et al., 1986). In the type 1 epidemic of 1978 in The Netherlands, the first paralytic child, belonging to an unvaccinated religious community, was in a hospital ward without proper diagnosis for 4 weeks (Bijkerk, 1979). An informative review of the clinical criteria for the definite diagnosis of paralytic poliomyelitis and essential findings not compatible with that diagnosis is given by Sabin (1981). In a classical case with or without a preceding minor illness, an abrupt disease develops, with headache, vomiting, fever, a flushed face and fits of profuse sweating, followed by painful muscular spasms (often in the back and then compared to the pain of lumbago), remarkably symmetric and not correlated with the paralysis that appears a few days later. Often there are abdominal pains and constipation. The onset of flaccid paralysis is also abrupt and fully developed within 3 days, usually asymmetric and in a very variable distribution. After some time a limited regression occurs, and the ultimate damage to motor neurons and thus to muscle groups can be evaluated. The most life-threatening form is bulbar poliomyelitis, often with symptoms of diffuse encephalitis and pharyngeal paralysis. Conditions with paralysis of the spinal respiratory muscles also require immediate measures for assisted respiration.

ASEPTIC ENTEROVIRUS MENINGITIS

Meningitis is the "major" illness of almost all enterovirus infections. Its course is usually benign, although headache, fever, and general malaise may be severe. In typical cases a short prodomal phase and a latent period of 1 to 6 days is followed by fever and a severe frontal headache, nausea, and vomiting. Symptoms of meningeal irritation impel to investigation of the spinal fluid. A pleocytosis is found, initially polynuclear; later, lymphocytes predominate. Usually the cell count is not very high, but during an epidemic in 1956 caused by echovirus 9, counts of more than 5,000/mm^3 were noted. In these cases a maximum was found on about day 6, with a slow decline lasting 2 to 3 weeks. With several serotypes, a few days after the onset a rash is seen, more in children than in adults. The illness lasts about a week, but convalescence may be slow, fatigue and the need for sleep remaining for a further fortnight.

HAND-FOOT-AND-MOUTH DISEASE

The name hand-foot-and-mouth disease was given to a clinical entity, recognized as such when it occurred in an outbreak caused by coxsackievirus A16 (Alsop et al., 1960). Other coxsackie A viruses cause similar symptoms in sporadic cases. Enterovirus 71 was responsible for a large-scale epidemic in Japan in 1978, with more than 36,000 patients (Tagaya et al., 1981). The disease can cause some panic in a rural community, when people are not aware that foot-and-mouth-disease virus of cattle is not infectious for man.

Mostly patients are young children who may have had mild prodromata (malaise, headache, abdominal pains). Suddenly a painful stomatitis appears on the tongue, the buccal mucosa, and the soft palate, with slight fever. The lesions in the mouth are small bullae and superficial ulcers surrounded by a narrow zone of erythema. At the same time a maculopapular rash is seen on the hands, feet, and often the buttocks. Those on the hands and feet develop into bullae that contain a serous, later turbid fluid. They are not very painful. Histologically, in the epidermis ballooning degeneration of cells is seen, resulting in the formation of intraepidermal vesicles that unite to form subepidermal bullae. The epidermis becomes necrotic, and an inflammatory infiltrate appears. No intranuclear inclusions are seen, which are typical for herpes simplex and varicella/zoster (Berretty et al., 1983). The lesions regress within a week or two.

The brief febrile disease called herpangina is a minor illness of children. The onset is abrupt, with fever and sore throat. Greyish-white papules or small vesicles with a red areola can be seen in the throat and posterior part of the mouth, not elsewhere. This syndrome is associated with several coxsackie A viruses, especially the lower serotype numbers.

BORNHOLM DISEASE

During epidemics by coxsackie B virus types, many cases of a typical illness occur, called Bornholm disease (after a Danish island). The onset is abrupt, with severe pain localized in the diaphragm (= pleurodynia), radiating to the adjacent chest or abdomen. The patient has a fever and headache, and is sweating. These alarming symptoms can disappear and reappear several times. Mostly recovery is complete within 1 to 2 weeks. In adult cases, especially in males, pericarditis can occur. Coxsackie B virus can be isolated from pericardial fluid. Other complications of adult Bornholm disease are pleuritis and orchitis.

NEONATAL MYOCARDITIS

A sudden onset of lethargy and a greyish pallor in infants aged 5 to 10 days are ominous features of neonatal myocarditis caused by a coxsackie B virus, which is mostly acquired just before or at birth from the mother. When the patient develops tachycardia, dyspnea, and cyanosis, a grave prognosis is to be feared. The disease is often fatal, overwhelming many other organs (brain, liver, and pancreas). It is possible that coxsackie B serotypes 3 and 4 more often than other types are responsible for severe disease in newborn infants, although in these infections electrocardiograms also show abnormalities. A neonate surviving coxsackie B4 myocarditis showed the presence and subsequent resolution of myocardial calcifications and a left bundle-branch block, probably caused by such a lesion in the bundle of His (Barson et al., 1981).

MYOCARDIAL DISEASE OF ADULTS

In the course of a primary infection with coxsackievirus B, sometimes manifestations of myocarditis with typical ECG changes are found in adult people. Usually the disease is benign; death seldom occurs due to arrhythmia or congestive heart failure. There is also evidence that exceptionally the myocardial inflammation can become chronic or recurrent. In cases reviewed by Hirschman and Hammer (1974) and Woodruff (1980), the etiologic association of coxsackie B viruses with this type of disease was of moderate order. A role for an autoimmune response through cross-reactions between antibodies to epitopes of coxsackie B virus and myocard fiber (molecular mimicry) is still a matter of speculation, although the evidence for immunopathologic factors as primary cause of cardiac injury is supported by experiments in mice (Huber and Lodge, 1986; Notkins et al., 1984; Woodruff, 1980). Furthermore, in a few cases of probable coxsackie B virus myocarditis, antimyolemmal autoantibodies (detected by immunofluorescence and complement-dependent cytolysis on vital adult rat cardiocytes) could be adsorbed from patient serum by a commercial coxsackie B viral antigen (Maisch et al., 1982).

Findings of presumed specific coxsackie B immunofluorescence in myocardial tissue obtained at routine autopsies are controversial for several reasons and not yet confirmed (Woodruff, 1980). The same applies to the detection of presumed coxsackie B virus-specific RNA sequences in myocardial biopsy samples of 1 to 5 mg wet weight with a coxsackie B virus probe complementary to the 3' region of the genome of coxsackievirus B2 with the pi-

cornavirus replicase gene (Bowles et al., 1986). Reports of a high rate of occurrence of "coxsackie B virus specific" IgM in older patients with acute myocardial infarction must also be interpreted with caution (Hannington et al., 1986; Lau, 1986).

ACUTE HEMORRHAGIC CONJUNCTIVITIS (AHC)

This disease was first recognized in Ghana in 1969. It spread all over the world in the following years and is now endemic in Southeast and South Asia and parts of Africa. The onset is sudden, with swelling, redness, watering, and pain often in both eyes. The most characteristic symptom is subconjunctival hemorrhage, which varies from minute petechiae to large blotches of frank hemorrhage, accompanying conjunctival inflammation (Kono et al., 1972). This is a description of the disease caused by enterovirus 70, which, because of its replication in nonprimate cell cultures and the high percentage of neutralizing antibodies in animal sera, probably originated from an animal or insect picornavirus. Large epidemics in Singapore of a similar acute conjunctivitis but with less hemorrhage were caused by a variant of coxsackievirus A24 (Mirkovic et al., 1974).

Complete recovery ensues within 2 weeks; rarely, neurological features complicate infection with enterovirus 70. It is estimated that during an epidemic in India in 1981, a few thousand patients must have been paralyzed out of the millions of Indians who suffered AHC. The essential neurologic disease is a lower motor neuron paralysis, affecting mainly the lumbar cord, less often the lower cervical cord. Sometimes facial and trigeminal nerves are also affected. It is an acute, asymmetric, mainly proximal, hypotonic and areflexic paralysis. Persistent wasting of the muscles is the result (Anonymous, 1982; WHO, 1986).

Diagnosis

Specimen Collection

When an enterovirus infection is suspected, the first specimen to be collected is *feces*, because their virus content is high and the period of excretion is long. No time should be wasted in poliomyelitis-like cases; if no feces are available, a carefully taken rectal swab should be sent in transport medium immediately to the laboratory, with information on the probable clinical diagnosis. A long period of constipation diminishes the chance of virus isolation from swabs. Therefore, as soon as possible, a real stool sample should be sent from the patient and, in the case of poliomyelitis, also from the patient's household contacts. Whether fecal samples need to be cooled depends upon the transport time and temperature. A time of 3 to 4 days is acceptable for enteroviruses without cooling. In the laboratory the samples are placed in the refrigerator and, after processing, at $-20°C$.

The next specimen should be a *throat swab,* placed in a virus transport medium. As a consequence of the brief excretion time, virus isolation from this location enhances the probability that this virus is the cause of this disease. When an exanthem, vesicles, or ulcers are present, the lesions should be thoroughly swabbed. Transport medium provided by virus laboratories usually will be composed of a balanced salt solution at pH 7 containing a small amount of protective proteins (such as broth, gelatin, yeast extract, or lactalbumin hydrolysate) and antibiotics. The use of a transport medium has two advantages: the sample upon arrival needs only centrifugation, and more labile viruses such as mumps virus or herpes simplex virus (both sometimes possible agents in the differential diagnosis) remain protected. Never should a bacteriological transport medium or a metallic applicator be used. A sterile cotton swab on a plain wooden stick, which can easily be broken, is sufficient. It happens that commercial applicator sticks contain substances toxic for cell culture and thus cannot be used for virus isolation purposes. A problem is that sometimes the broken sticks are too long when placed in the vial, whereupon the screw cap is not safely closed and the material is spilled on shipment. This should of course be avoided.

Cerebrospinal fluid should be sent when the clinical condition necessitates a lumbar puncture. Special transport vials should be chosen, because the use of toxic vials or stoppers often prevents a normal isolation procedure in cell culture. In certain cases the CSF is also used for antibody determination. Therefore, the minimum amount is 2 to 3 ml, and the fluid should not be added to a transport medium.

Specimens of *ocular discharge* have to be taken according to the rules described for throat swabs, as early in the disease as possible. Information on the suspicion of an enterovirus conjunctivitis must accompany the sample, because the impact of this disease on an ophthalmologic clinic, for example, is tremendous, and the isolation of enterovirus 70 requires some time and special measures, such as incubation at a temperature of 33°C (Miyamura et al., 1974).

Other specimens, which are not routinely taken but depend upon the nature of the illness, such as *vesicle fluid, pericardial* or *pleural effusion, biopsy tissue,* require shipment in transport medium. In the case of *post mortem tissue* (brain, spinal cord, heart, liver, pancreas, and the like), a medium with broad-spectrum antibiotics is necessary to prevent bacterial growth.

Blood or serum is often sent for the determination of a specific antibody response. Diseases caused by enteroviruses often share symptoms with other viral ailments. Therefore, a general rule should be observed: send one sample as early as possible, followed by a second one some 10 days later.

It is possible to try virus isolation from blood. In a special study, viremia was detected on the day of hospitalization in 11 of 28 young children with proven enterovirus infection. Serum and washed mononuclear leukocytes were the best specimens for a positive result (Prather et al., 1984). Heparinized capillary blood samples obtained by finger prick may also be useful.

Sample Processing

In the laboratory, feces are mechanically processed into a 10 to 20% suspension in a flask containing glass beads and a balanced salt solution with antibiotics. If no other investigations have to be done with the sample, it is advantageous to add chloroform (0.5 ml/10 ml) to the mixture. This will destroy bacteria and perhaps remove cellular membranes still attached to enteroviruses. It should not be used when electron microscopy is needed for the detection of other virus types, and not when rotavirus has to be cultured. Though rotavirus does not contain lipids, its structure seems to be sensitive to chloroform because infectivity is reduced.

The fecal suspensions are clarified by centrifugation at speeds from 3,000 rpm for 1 hour to $\pm10,000$ rpm for a shorter time. The supernatants are much clearer with the use of chloroform than without. The chloroform sediments with the beads and need not be removed by separate means.

Enteroviruses can cause serious diseases, and samples from hospitalized patients may contain other pathogenic agents. Therefore, all handling should be done under good, but not too rigid, safety precautions. A laboratory worker must always keep in mind that shaking and centrifugation produce infectious aerosols.

Specimens shipped in transport medium will be centrifuged only to remove gross particles and threads from the swabs. Tissues are ground, preferably frozen, in a very cold, sterile mortar; seldom is addition of sterile sand necessary to make a good suspension, which is made up in a small amount of medium.

Direct Virus Detection

Direct methods are not of much practical use in the laboratory diagnosis of enterovirus disease. The viruses are small and must be concentrated to at least

10^6 to 10^7 particles per ml to be visualized in the electron microscope. Since the usual source of enterovirus will be a fecal specimen (and not the watery fluid that permits easy EM in many cases of gastroenteritis), the sample has to be homogenized, clarified, and concentrated by differential centrifugation. Addition of specific antibodies to obtain immunoclumping theoretically might better the chance to find virus. For this general purpose, namely, to fit to the many serotypes, pooled human γ-globulin could serve. But the titer of antibody against a certain virus can be so high that its morphology will be obscured. In the future, perhaps some of the small, round, structureless viruses, up to now detected only by EM in the feces of gastroenteritis patients, might become grouped among the enteroviruses and "stained" for typing by immuno-gold–electron microscopy with antibodies raised against that virus and conjugated to colloidal gold. In principle, the detection of just one virus particle decorated by gold grains would prove the presence of that particular virus. Only in this way would direct virus detection equal the sensitivity and surpass the speed of virus isolation, but at this moment this is pure speculation.

There are a few reports (some reviewed by Grist et al., 1978) about pseudocrystalline deposits resembling picornaviruses and presumed to be coxsackievirus aggregates, seen by EM in muscle biopsies taken from patients with chronic muscle disease (myopathia, idiopathic scoliosis, or polymyositis). By using a special technique, Green et al. (1979) gave evidence that these particles were not virus aggregates but crystalline glycogen molecules.

Direct Antigen Detection

Solid-phase immunoassays, if developed for the detection of enterovirus antigen in clinical specimens, have been used only for research purposes. The problem is the preparation of immunoreagents specific against so many serotypes, and the relatively small amount of virus antigen in the most probable specimen: stool. This material has its own pitfalls for the standardization of a reliable EIA technique, for instance, nonspecific reactions due to lipids or to immunoglobulin-binding products of bacteria.

Also, not much use has been made of techniques to detect antigen by immunofluorescence. Claims of finding coxsackie B viral antigen in cardiac tissue at routine autopsy and in some special heart diseases are not supported by other studies, including experimental heart disease induced by coxsackievirus B in mice (French et al., 1972; Grist et al., 1978; Woodruff, 1980).

A simple, commercially available, direct-antigen-detection test is most needed for the rapid laboratory diagnosis of acute hemorrhagic conjunctivitis caused by enterovirus 70, because this virus is so difficult to isolate. A move towards this goal is the indirect immunofluorescence test on smears from eye swabs, fixed by cold acetone, as described by Pal et al. (1983), and the development of a set of monoclonal antibodies against this virus that could be used by immunofluorescence and EIA techniques (Anderson et al., 1984).

Direct Viral Nucleic Acid Detection

Research on poliovirus genetics and replication has provided not only knowledge about its nucleotide sequence and genome products, but also tools for investigations by nucleid acid hybridization in human specimens. The first use of labeled cDNA complementary to poliovirus RNA for this purpose was reported by Kohne et al. (1981), who took up the old question of a possible etiological role for poliovirus in amyotrophic lateral sclerosis. The assay by itself was sensitive enough, the probes were capable of detecting the presence of about 3×10^4 poliovirus particles or about 1 virus genome/200 cells. The brain tissues did not show the presence of poliovirus RNA. The paper describes in detail the difficulties in detecting hybridizable RNA in tissues obtained after death. Even though the brain specimens were removed as soon as possible and stored at −70°C, a large fraction of the RNA was already highly degraded.

The cloning of cDNA fragments in plasmids of *Escherichia coli* enables the production of large amounts of probes containing sequences complementary to defined regions of the parent RNA. This approach is still in its research phase. Some fundamental facts have emerged from the first experiments with enteroviruses.

a. In the genomes of the three poliovirus serotypes 71% of the nucleotide sequences are common (Toyoda et al., 1984; Tracy and Smith 1981).
b. The noncoding terminal 5' end of coxsackie B3 genome, strain Nancy, is more or less shared with the other coxsackie B (CB) serotypes. CB 3 also shares at the 3' end region a well-conserved homology with the other CB serotypes (Tracy, 1984). In another investigation, a 4,500-base-pair cDNA clone derived from the 3' terminal part of CB 3 virus strain Nancy was sequenced and compared with the corresponding region of poliovirus type 1. For the replicase gene a homology of 68% was found (Stålhandske et al., 1984). This DNA was used as a probe for hybridization experiments with different enteroviruses in a spot hybridization test (Hyypiä et al., 1984). With tissue-

culture-grown virus strains of CB2, B3, B4, A9, echovirus 17 and 30, and poliovirus 3, remarkably good hybridization signals were obtained, owing to the highly conserved replicase gene of the picornavirus family. But the same test was not sensitive enough to detect the presence of RNA in more than 1 of 8 enterovirus-positive stool samples.

c. Three probes were made from cDNA clones of the poliovirus type 1 genome and used in hybridization experiments with the other poliovirus types, CA9, CB1, and echovirus 11. This revealed another highly conserved nucleotide sequence among the enteroviruses, mapping between bases 220 and 1,809 (Rotbart et al., 1984). Poliovirus type 1- and coxsackievirus B3-derived clones were tested for the detection of enterovirus added to cerebrospinal fluid. The patterns of hybridization produced led to the conclusion that any combination of two probes, one containing a 5' end and the other a part of the 3' end, with the replicase gene was capable of detecting RNA from the three polioviruses, CA virus 9 and 16, CB virus 1 and 6, and echovirus 2, 4, 6, and 11 (Rotbart et al., 1985).

The probes revealing common sequences among many serotypes of enteroviruses are represented in Figure 4 in relation to the location of the coding regions of the poliovirus genome. In principle, the whole range of serotypes now has to be screened by crosshybridization, because, as already shown, echovirus 22 did not react with the probes (Rotbart et al., 1985).

Certain parts of the enterovirus genomes will contain genes for serotype specificity. Because serotypes are defined by neutralization with polyclonal antisera and the antigenic sites, as discussed above, are found on small protruding loops on the virion surface and are related to the three-dimensional conformation of the capsid proteins, typing of enteroviruses by hybridization probably will be impossible.

Introduction of hybridization for clinical use will need standardization of the technique. Specimens still containing cultivable viruses will need special care before and during the preparation for hybridization. In the report on cerebrospinal fluid, several additions were recommended, of which formaldehyde certainly will destroy infectivity. The assays mentioned above differed in many particulars, for instance, type of membrane filters, temperature, and time of hybridization performance. But all probes were radioactively labeled, and hybridization was visualized by autoradiography. This is not a practical procedure for routine diagnosis, because the labels have a short half-life, the personnel must be specifically trained, and disposal is expensive. In the near future more nonradioactive labels will be used, for instance, biotin, coupled with a (strept)avidin detection system. This technique also has advantages for rapid in situ hybridization, which combines the visualization of a pathologic picture with location of the residing viral pathogen. When the sensitivity of viral RNA detection approaches that of virus isolation, hybridization could become a screening test to select patient samples to be cultivated further for isolation and identification.

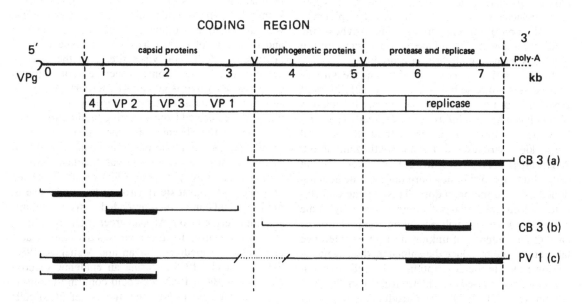

FIG. 4. Probes revealing enterovirus common sequences (bold lines). Data derived from (a) Hyypiä et al., 1984; (b) Rotbart et al., 1985; (c) Rotbart et al., 1984.

Virus Isolation and Identification

In the laboratory diagnosis of enterovirus diseases, the master role is played by virus isolation and identification. This provides solid information on the presence of a possibly pathogenic agent at the moment of the disease. The necessity of rapid testing is often exaggerated; most physicians will recognize an enterovirus disease, notably during the epidemic season, by its clinical picture and physical signs, and will avoid the needless use of antibacterial drugs. Speed is certainly urgent in the case of poliomyelitis. Here virus isolation and identification can be achieved within 24 h, provided that 1) a good fecal specimen taken early in the disease, 2) reaches the laboratory immediately, 3) is cultivated immediately in the most susceptible cells, 4) while its suspension is used as "virus" in a neutralization assay with specific antisera.

These four points also give the reasons why laboratory diagnosis seems to take so much time: 1) the feces have to be produced by the patient; 2) transport to the nearest virus laboratory usually takes place during normal working hours; 3) the laboratory has a practical working scheme for the administration and processing of many fecal samples and the inoculation of tissue cultures; a deviation will be accepted only when the clinical information indicates the urgency; and 4) although an enterovirus-specific CPE often is seen within a few days, the final typing of an isolate requires the presence of typing sera and a flawless technique. Early reporting (we have seen an enteroviruslike CPE) could be helpful to the clinician, especially if the specimen is cerebrospinal fluid, but would also distract attention from the final typing result, which gives clues on pathogenicity, epidemicity and so on of current virus strains. Anyhow, the medical virologist should have an open mind for the clinicians' exigency of early information.

In former times, recovery of coxsackie viruses necessitated a mouse breeding colony with litters of great size, for both isolation and typing. Inoculation of suckling mice, by different routes, really is a cumbersome and time-consuming method. Because all coxsackie B viruses and the most pathogenic group A serotypes (CA7, 9, 16, 21, 24) grow very well in cell culture, the use of newborn mice has been abandoned as a routine procedure. This has the unfortunate consequence that infections caused by some coxsackie A viruses will pass unrecognized. In the case of an epidemic of unknown origin, a reference laboratory will introduce the animal method again, as demonstrated in the description of the poliomyelitis-like epidemic caused by enterovirus 71 in Bulgaria (Chumakov et al., 1979). The introduction in the routine laboratory of a cell line (RD, see later) suitable for the propagation of 20 serotypes of coxsackie A

virus did not much enhance isolation of these virus types from clinical specimens, but had many other advantages (Bell and Cosgrove, 1980).

Enterovirus Isolation in Cell Culture

Because no single cell culture system is satisfactory for recovery of all naturally occurring enteroviruses, a diagnostic laboratory has to employ a combination of cell types. Limiting factors are workload and costs. Also, although most enterovirus serotypes have been detected in primary monkey kidney cells from Macaca species, nowadays objections are evident against the use of primary monkey cells, obtained from a diminishing supply of free-living monkeys, according to the Convention on International Trade in Endangered Species. Established monkey kidney cell lines often have lost sensitivity for many virus types. In the experience of the author, the problem can be solved by the use of tertiary monkey kidney cells, derived from a homogeneous unit process with microcarriers as developed for the large-scale cultivation of animal cells (van Wezel, 1985). The cell yield obtained from a pair of kidneys has improved considerably by the introduction of perfusion trypsinization and subcultivation on microcarriers. Cells from the secondary culture may be trypsinized from the carriers and frozen in liquid N_2 in medium with 5% dimethyl sulfoxide at a concentration of about 20×10^6 cells/ml. Ampoules containing enough cells can be used to start the weekly necessary monolayer cultures in tubes, flasks, and even microtiter wells. In our hands (unpublished data) monolayer cultures of tertiary cynomolgus monkey kidney cells (representing about 12 cell generations) were as sensitive as primary cells for the propagation and detection of enteroviruses with respect to appearance of CPE and virus titers. In the case of echovirus 22 and some other enteric viruses (reovirus, adenovirus 40 and 41), tertiary cells were superior. Further advantages of tertiary cells are that they can be derived from a captive-bred monkey colony, where the absence of some simian viruses is monitored, especially herpesvirus simiae, SV40, and foamy viruses (Osterhaus and van Steenis, 1981).

When laboratory diagnosis first of all is focused on polio and coxsackie B viruses, the single use of the BGM cell can be recommended. This continuous cell line, derived from African green monkey kidney, is highly sensitive for these viruses and almost acts as a selective medium (Dagan and Menegus, 1986; Schmidt et al., 1978). During an epidemic of coxsackievirus B4 in 1982, we could confirm the observation of the first authors that the use of BGM cells improved the speed of recovery of this serotype.

Human embryonic lung fibroblasts (HEL-cells)

are indispensable for the isolation of echoviruses. These cells are diploid and keep their original viral susceptibility during the 30 to 50 generations of a lifetime without transformation. A different origin might cause some diversity in viral susceptibility, but generally all polioviruses, most echoviruses (including echovirus 21), many coxsackie A virus types, and virtually no coxsackie B viruses can be recovered in HEL-cells. When fecal suspensions are clarified with chloroform and not more than 0.1 ml is inoculated per tube, the toxicity of the specimen will not be a major problem.

The role of RD cells in the enterovirus laboratory is peculiar. These cells are derived from a human rhabdomyosarcoma, but are epithelial-like. It is absolutely necessary to store a good supply in liquid N_2, because the viral susceptibility of the cells tends to diminish with prolonged cultivation, and a new start has to be made. Good confluent monolayers, for instance, tubes seeded with ± 100,000 cells/ml, are extremely susceptible to a wide variety of enteroviruses: complete CPE will develop in a few days. When no virus is present or not enough cells were seeded, the tubes will show much degeneration at the end of the normal observation time of 7 to 10 days, necessitating a "blind" passage, which mostly is inefficient. There are many, sometimes conflicting, reports on the virus range of RD cells. They are clearly susceptible for coxsackie A serotypes grown in suckling mice, enabling neutralization assays in cell culture, except for coxsackie Al, 19, and 22 (Schmidt et al., 1975). Direct isolation of A type coxsackieviruses has been disappointing, however, and blind passages must often be employed for that purpose. Other enteroviruses are recovered remarkably well in RD cells: echovirus 30 (Wecker, 1977); echovirus 33 (Druyts-Voets et al., 1985); echovirus 11, 12, 17, 19, and 20 (Dagan and Menegus, 1986); and echovirus 24 (Hamparian et al., 1985). RD cells usually fail to detect group B coxsackie-viruses and are slow in the recovery of echovirus 9.

The laboratory technician should be aware that several enterovirus types can vary significantly from year to year with respect to their host cell range and that cell lines propagated continuously may happen to change their viral susceptibility. Many older references on this subject mention cell lines of human origin, such as the HeLa cells or cells with other names but often with HeLa cell chromosomes. This HeLa contamination is indicated in the catalog of the American Type Culture Collection for cell lines. These cells have a narrow enterovirus range, mainly for polioviruses and especially coxsackievirus B type 3 and 5. Furthermore, antibodies should be absent in the serum used for the cell-culture medium. Bovine sera may contain inhibitors for the polioviruses, which probably are antibodies of the IgM class, and

in tropical areas also antibodies against enterovirus 70. Fetal serum and serum obtained from colostrum-deprived calves have to be checked before use in an enterovirus laboratory.

The best conditions for attachment of enterovirus from processed clinical specimens to cells are to add a rather small inoculum (feces, 1 to 3 drops, other specimens, about 0.5 ml) to a drained tube with a young monolayer and to incubate for 1 h at 37°C in a rolling apparatus, before maintenance medium is added. Only very toxic specimens have to be inoculated into the medium, which then is changed after 1 h or 1 day. Rolling is profitable for further cultivation, but not necessary. The cultures are if possible observed daily for CPEs. As soon as more than half of the cells are degenerated, the culture is placed at −20°C and, after thawing, passaged to a fresh culture of the same cell type and to RD cells. The observation time before a culture is estimated as negative depends on the cells and the specimen. HEL-cells can be observed for at least 14 days; stool specimens may be toxic, causing the cells to be virtually degenerated at the 10th day. For most enteroviruses a prolonged observation time is not necessary, even when very little virus is present. In our experience, the CPE of echovirus 22, 23, 24, and enterovirus 71 develops slowly; at a longer cultivation time virus loss occurs at medium changes and by thermal inactivation. Therefore, subpassaging has to be done at least before the 14th day. Of course the presence of a variety of other viruses, not belonging to the enterovirus genus, in clinical specimens has to be considered. Confusion is possible only with the CPE of rhinoviruses, isolated from respiratory specimens. They are distinguished by their inactivation at low pH.

Enterovirus Isolation in Newborn Mice

Although this method no longer has priority for routine diagnosis, an enterovirus laboratory should be acquainted with its principles. With the modern methods of mouse breeding and characterization, in theory the most susceptible mouse strain should be known, but for the coxsackie viruses of group A this is not true. It may be possible that all mice are susceptible to group A when inoculated on the first day of life. Certainly a minimal infective dose exists, which in the case of CA viruses growing well in cell culture is high, at least 10^3 $TCID_{50}$. On the contrary, many mouse strains have been compared for their susceptibility to group B coxsackieviruses, mostly with the intention to get a mouse model for virus induced myocardiopathy or diabetes. In this respect coxsackievirus susceptibility is age-dependent, species-dependent, and also differentially organ-dependent. For information about research on a murine model with coxsackieviruses for heart disease, the reader is referred to papers by Cao et al. (1984) and by Huber and Lodge (1986); the diabetic concept is discussed by Notkins and Yoon (1984). From an investigation to determine the effects of all six serotypes of coxsackie B in mouse strains with different genetic background, the BALB/c mouse came out as the most suitable for virus isolation purposes (Minnich and Ray, 1980). This study also underlined the great variability in tissue tropism by different virus strains within a given group B serotype.

A litter of newborn mice not older than 1 day is inoculated by 3 routes: intracerebrally, 0.02 ml; subcutaneously between the scapulae, 0.03 ml; or intraperitoneally, 0.05 ml. Mice are observed for 14 days. The disease caused by coxsackie A viruses is polymyositis. The animals show progressive flaccid paralysis due to muscle degeneration and die within 48 h. The disease caused by group B coxsackieviruses is variable. Symptoms distinguishing group B directly from group A result from encephalitis: tremors and spastic paralysis. Sometimes the mice will recover; in different strains (both of virus and mice) death occurs after a day or two. The definitive diagnosis (A or B) depends on histologic examination. Dead or ill and sacrificed mice are skinned and eviscerated. Soft organs, that is brain, heart, pancreas, and liver, are removed and fixed and embedded together. The mouse carcass is treated separately. By this procedure sections can be made through the limbs with attached muscles without damage to or loss of the soft organs.

Lesions seen in coxsackievirus group A infections are confined to the skeletal muscles. A generalized degeneration is found with swelling, loss of striation, and necrosis. Definitive pathology caused by the viruses of group B is meningoencephalomyelitis with loss of neurones, and necrosis with calcification in fat tissue, especially in the interscapular "brown" fat of the hibernation organ. Myocarditis can be found as small foci of myocyte necrosis with cellular infiltrates. Liver lesions consist of necrotic areas of hepatocytes with mononuclear infiltrates. In the pancreas, acinar cells degenerate; occasionally islets of Langerhans are involved. With careful examination small focal lesions might be found in muscles. The variable distribution of the histopathologic findings by mouse strain and group B serotype (and even virus strain) is unexplained (Minnich and Ray, 1980). But coxsackievirus B can grow in the heart cells of newborn mice without causing myocarditis (Cao et al., 1984). Passages of mouse isolates to mice or cell culture are made by suspensions from the carcass and the soft organs (brain or heart).

Enterovirus Identification

The empirical method to distinguish polio- from other enteroviruses by neutralization has lost nothing of its practical importance during decades of enterovirus research. No method is as sensitive for distinguishing between types as the binding of a specific antibody to a sequence or configuration of a few amino acid molecules at a virion surface, resulting in the loss of infectivity. One can only marvel at the expedience of a good standardization of simple laboratory conditions, such as pH, temperature, reaction time, and choice of cells. On the other hand, studies of the mechanism of neutralization have revealed that it is not a simple reaction and that several pathways may act together on a given population of virus particles. In the case of poliovirus, which of course has been analyzed in detail, the most probable pathway is the antibody-induced rearrangement of capsid proteins, which does not prevent virus attachment, but its uncoating. Also, aggregation of virus particles by antibody might lower the infectivity of the suspension (Mandel, 1985; Wetz et al., 1986). A rapid, specific and clear-cut identification of enterovirus isolates requires not only a standardized method, but also the availability of a diversity of typing antibodies. The mere fact that these antibodies are not for sale reflects the complexity of enterovirus identification rather than its commercial value. Some of the typing problems are inherent in the preparation of the antigen, others to the animal and the scheme used for immunization, and a different group to antigenic variation within the same serotype.

VIRAL ANTIGEN

To evoke neutralizing antibodies, the virus must present its type-specific antigenic determinants, which are preserved only on infectious virus. Thus, antiserum raised against a preparation that contains much thermally inactivated or incomplete virus (empty capsids) will include many nonneutralizing antibody molecules, which may interfere with its neutralizing potency. Since successful immunization needs high doses of antigen, also for the booster, a method for virus cultivation has to be chosen that results in high virus titers (if possible, more than 10^8 TCID$_{50}$/ml) and that reduces the necessity of vigorous and perhaps inactivating concentration procedures. High titers were reported for most echoviruses (except type 1 and 9) grown in RD cells (von Zeipel, 1980).

Infected tissue culture fluid always contains host cell material. This has to be removed as much as possible, both from the immunizing antigen and from the virus in the neutralization test. Antibodies against host proteins act by two ways: by giving a false-positive neutralization (often misunderstood as heterotypic cross-reaction) and by toxic effects on freshly trypsinized cells added in a microneutralization assay. Antibodies against serum components of the medium surprisingly have no effect.

Host cell fragments, presumably lipid membranes, attached to the virus may cause a bizarre phenomenon in some serotypes. Neutralizing antibodies seem not to be capable of acting on virus infectivity, although this is more obvious in some cells than in others. As soon as the material is removed by chloroform treatment (as published by Kapsenberg et al., 1979) or by filtration (adsorption to filter material), as recommended by Melnick (1979), the serum exhibits its potency. An example is presented in Table 2 for prototype strain Pesascek of echovirus 4, which is notorious for its poor neutralization. A serum was raised in a cynomolgus monkey, and the results of homologous titrations in primary monkey kidney, HEL and RD cells with or without chloroform treatment are given.

A better accessibility of the neutralization antigenic sites to antibody molecules is a reasonable explanation for the action of chloroform. The phenomenon has been found with strains of echovirus 4 and 18, coxsackievirus A7 and 16, enterovirus 71 (Kapsenberg et al., 1979), and with the hepatitis A virus enterovirus 72 (Lemon and Binn, 1985). Because chloroform treatment is so easy to perform, it should be used with routine enterovirus typing: shake 9 parts of a crude virus suspension with 1 part chloroform for 15 min at room temperature, centrifuge for 15 min at 1500 g, and collect the supernatant carefully. It is more difficult to explain the role of the host cells. RD cells apparently seem to "recognize" a virus–antibody complex as neutralized, whereas HEL and MK cells do so only partially. Investigations on the neutralizability of echovirus 4 (Kjellén and von Zeipel, 1984) and enterovirus 71 (Kjellén

1985) resulted in several findings and a hypothesis, which perhaps may be linked to the evidence of the chloroform treatment. The explanation could run as follows.

1. Within a virus suspension of certain strains a non-neutralizable fraction, perhaps an RNA variant, exists, which is bound to cellular membrane fragments.
2. This membrane-bound virus can be neutralized only when the virus is already attached to the host cell membrane.
3. A part of the antibody molecules present in a polyclonal hyperimmune serum reacts with hidden antigenic sites of the virus, which become accessible by the conformational changes of the capsid occurring in the process of attachment and penetration. This process is very much host-cell dependent. Kjellén demonstrated a different time course in RD versus GMK cells for the susceptibility of enterovirus 71 to antibodies during this process.

For the practice of enterovirus identification, the conclusion is: RD cells are at this moment the cells of choice both for the cultivation of the viral antigen and the performance of the neutralization test. Only when the virus does not grow well in these cells must another cell system be used.

IMMUNIZATION OF ANIMALS

Long before it was recognized that antibodies belong to different immunoglobulin classes which appear in a regular sequence after immunogenic stimulation, it was known from the production of therapeutic sera in horses that the highest titers were obtained when a period of months elapsed between the primary and the secondary course of injections. This principle was worked out for the immunization scheme of inactivated poliovaccine, where a booster dose after at least 6 months proved the most successful. When the need for large amounts of enterovirus reference antisera rose, studies in horses (who came out better than other large animals) proved the rewarding effect of a booster dose of antigen given 21 to 28 weeks after the primary course (Hampil et al., 1965). This is illustrated by the titers of horse serum produced against echovirus 4 (Pesascek), of necessity tested by the good, neutralizable strain of echovirus 4 Du Toit. The immunizing antigen had a low titer of 1.8×10^6 PFU/ml. At week 40 the neutralizing antibody titer had declined to below 50, but was elevated by a booster dose to a ceiling titer of about 6,400. We now know that the primary course gives rise only to IgG class antibody with a sufficiently high dose of antigen, and that it is the IgG memory which reacts at

TABLE 2. Homologous neutralization titer of hyperimmune monkey serum against echovirus 4 Pesascek with or without chloroform treatment in different cell types

Virus[a] treatment	pCMK	HEL	RD
None	20	160	40960
Chloroform	640	81920	81920

[a] Virus adapted to the cell strain of the test, treated before dilution to 100 TCD$_{50}$/0.1 ml.

restimulation with synthesis of antibodies to a high-titer plateau. But of course a booster reaction is also obtained for antibodies against host cell impurities in the antigen.

No comparative study has been made on the usefulness of small animals for the production of monovalent antisera. Perhaps monkeys make a better product than rabbits when immunized with monkey-kidney-cell-grown virus, because they do not have to make antibodies against a mass of foreign host proteins.

A general schedule for the preparation of typing antiserum is:

1. Preimmunization serum is taken and tested for the absence of antibodies for at least the virus in question.
2. Animals should receive intramuscularly a first dose of concentrated, purified, infectious virus, mixed with adjuvant, followed 14 days later by a dose without adjuvant at several sites, if possible also intraperitoneally.
3. After an interval of at least 8 weeks, a test bleeding is taken, and subsequently the animal receives a booster dose with or without adjuvant.
4. About 10 days after the last injection, the animal can be exsanguinated or bled on alternate weeks.

ANTIGENIC VARIATION

As already outlined above, the enterovirus RNA is subject to a high degree of replication errors, some of which may become expressed in the epitopes for the neutralization reaction. During natural transmission in a partly immune population, such antigenic variants may start a new wave of infections. The difference must be great to become recognized by a polyclonal hyperimmune antiserum in a routine constant virus–serum dilution neutralization assay. The relationship between two strains is called reciprocal when the homologous titer of the antiserum against each strain is higher than the homotypic titer. As long as the titer of the typing serum is sufficiently high and the serum is not too much diluted, the reciprocal variants will become neutralized. Another variation is the existence of "prime" strains, which present a broad spectrum of epitopes to the antibody production system, whereas others (and, as often happens, the prototype strain) are much smaller in the expression of neutralization sites. It can be seen from Table 3 that a typing serum produced with the prototype of echovirus 11 (isolated in 1953) was not of much use for the identification of isolates from 1959/1960 in The Netherlands.

As a matter of fact, prime strains should be chosen for the production of typing antiserum and become available by exchange and cooperation of reference laboratories. But a periodical review and

TABLE 3. Neutralization titers of echovirus 11 prototype and prime typing serum

Virus strain	Monkey antiserum against	
	E11 prototype	Strain 60-3590
E11 prototype	1280	10240
NL 59-2602[a]	<10	10240
NL 60-3590	<10	10240

[a] NL = The Netherlands.

replacement of reference strains has never been put into practice.

Other problems in neutralization tests caused by antigenic variation are more difficult to evade. Viral isolates within a certain serotype sometimes can be grouped according to the neutralizing titers of antisera prepared against representative group members. Homologous titers may even be much lower than homotypic titers. Often the method of the test and the cells used are important to produce evidence of the antigenic relationship. Echovirus type 16 and 27 belong to this category.

In Table 1, known typing problems caused by antigenic variability are indicated, the more severe by a double plus (++). Probably variation exists within more enterovirus serotypes. But for the majority of enterovirus isolates, a rapid and clear-cut identification is the rule.

A much more subtle intratypic variation, the differences between wild, vaccine, and vaccine-derived poliovirus strains, is discussed separately.

ENTEROVIRUS TYPING PROCEDURES

The importance of a standardized neutralization test was illustrated by a collaborative study on poliovirus antibody titration with 20 participating laboratories from 12 countries (Albrecht et al., 1984). Considerable differences in test results were found. The highest sensitivity and reproducibility were obtained by tests in microtiter trays with small virus doses (25 to 50 $TCID_{50}$ per well) and a long serum–virus incubation time (20 h at 36°C). Polioviruses are rapidly growing, stable viruses, easily neutralized without a special treatment. Such a long incubation time at elevated temperature is not recommended for unknown viruses, nor is the use of a very small, exact virus dose, because this has to be determined by a previous virus titration. For a routine typing procedure the virus dose can be estimated by experience from the rapidity of CPE development, and it is in this respect that a passage in RD cells might be favorable (von Zeipel, 1980). Only when difficulties occur, for instance, early breakthrough of CPE in a presumed neutralized test, must an exact dose of ± 100 TCID$_{50}$ be used.

As explained above, variability within serotypes is great. Therefore, typing serum should be polyclonal, preferably prepared with strains carrying a broad spectrum of neutralization epitopes. It should not be diluted too much. Talking about the number of necessary antibody "units" to be used in a test is not realistic when the serum is standardized only against the prototype strain. On the other hand, typing sera are precious reagents that have to be used sparingly. Therefore, a compromise has to be found. The advice not to type every single enterovirus isolate (Melnick and Wimberly, 1985) is, in the opinion of the author, a *testimonium paupertatis* and not in the interest of clinical virology.

It has been a great idea to pool monotypic horse reference sera in combinations which make identification of isolates possible by the pattern of pools neutralizing the virus. The first set of these LBM (Lim, Benyesh-Melnick) combination pools is recently newly prepared from frozen stocks and is, with the second set, available from the Microbiology and Immunology Support Services of the World Health Organization at Geneva (Melnick and Wimberly, 1985). However, the supply of the pools is limited. The sera have other limitations. Antibody against normal monkey kidney antigen is present, which interferes with a good reading of the test with some serotypes, especially in microtiter assays, when the cells are added as a fresh suspension to the virus–serum mixture. This last procedure enhances the sensitivity of the cells for the virus and is also convenient and economical. The use of only 50 antibody "units" will preclude the identification of variants (for example of coxsackievirus B3). Some serotypes are included in only one pool, and precisely among these are viruses with known typing problems by antigenic variation (echovirus 11 and 27). A mixture of virus types with almost equal titer, for example, poliovaccineviruses type 1 and 3, will not be typed.

In the author's laboratory the principle of pooling was reversed from serum to antigen. Horses were hyperimmunized with pools of antigen according to a pattern that allows identification of a virus by neutralization in two sera. By adsorption with human placental tissue, antibodies reacting with the cell culture were removed. Locally isolated prime strains of echovirus 9, 11, and 29 were used for immunization. Our first equine sera from 1968, permitting the typing of 20 echovirus serotypes and coxsackievirus A9, have recently been distributed into ampoules containing 0.5 ml. For the detection of polioviruses or coxsackie B virus, separate polio- and coxsackie B pools are included. A standard procedure for microneutralization has to be followed, with chloroform-treated virus, serum diluted 1:20, and a medium containing HEPES buffer. The plates can be sealed by tape and placed in an ordinary dry incubator. The plates are examined for viral CPE with an inverted microscope. The final reading should be done when the virus control shows complete cell degeneration, which in most cases takes 2 to 4 days. Slow viruses will be typed within 10 days, without change of medium. In our hands, the horse sera were of enormous value for the rapid typing of thousands of enteroviruses (information on availability at the author's address).

The original neutralization tests in culture tubes with monovalent typing serum (preferably adsorbed for anti-host cell antibody) are to be used when difficulties are encountered, for instance when the virus is growing slowly (echovirus 22, enterovirus 71) or a virus mixture is suspected because of late breakthrough.

Isolates made in suckling mice, presenting with coxsackie A pathology, should be passed to RD cells, which allows typing in a cell culture system. Typing by neutralization in litters of newborn mice is time-consuming and often not satisfactory and clearcut with field isolates (Melnick et al., 1977).

Intratypic Differentiation of Poliovirus Isolates

The oral vaccination with infectious but not virulent strains of the three types of poliovirus entailed the urgent need for the characterization of virus isolates in wild, vaccine, or vaccine-derived strains. Usually Sabin-vaccine strains excreted shortly after vaccination are completely similar to the vaccine, but, with prolonged passage in the human gut, differences occur. Presence or absence of neurovirulence as such can be associated with both wild and vaccine-derived strains. Much research has been done to arrive at a clear and reproducible assignment of strain origin. Methods advised by the World Health Organization are distinction by polyclonal, specifically adsorbed antisera and oligonucleotide mapping (Minor and Schild, 1981). Serologic tests demonstrate the presence of neutralization epitopes on a virus which are characteristic for the vaccine strains and are absent in wild strains of the same type. RNA fingerprinting shows characteristic vaccine-strain-oligonucleotide spots and, furthermore, contingent epidemiologic relationships between wild poliovirus strains.

SEROLOGIC DIFFERENTIATION

The original test as described by van Wezel and Hazendonk (1979) uses sera of rabbits immunized by two intravenous injections with an extremely high dose (10^{10} TCID$_{50}$) of purified infectious virus. The sera are adsorbed with a concentrated virus suspension of the heterologous strain of the same type, for instance anti LS-c,2ab serum by Mahoney virus, until only a strain-specific reaction remains in a microneutralization and double-diffusion test. The adsorbed sera can also be applied in an enzyme-linked immunosorbent assay (Osterhaus et al., 1983). The use of monoclonal antibodies for serodifferentiation by neutralization or EIA seems to be more tricky. An optimal panel of monoclonal antibodies has to be prepared because monoclonal antibodies are epitope-specific, and Sabin strains undergo modifications in the neutralization pattern during passage in humans (Crainic et al., 1983). Such panels are at this moment not available (Ferguson et al., 1986). The information obtained so far is more interesting for the study of natural poliovirus variation than for the urgent identification of poliovirus origin in a case of paralytic disease.

The EIA test is rapid and economical, and might be the method of choice when, from clinical or epidemiological evidence, the strains probably are derived from recently vaccinated individuals. In these cases a perfect fit can be obtained. Wells of a microtiter plate are coated with the IgG fraction of bovine monotypic antipolioserum. Then the virus strain to be typed is added. This can be ordinary culture-tube-grown virus, provided that a titer of at least 10^{-7} is obtained. The next step is the addition of the specific, adsorbed rabbit antiserum. Binding is demonstrated with enzyme-conjugated sheep anti-rabbit IgG, followed by the substrate and the reading of the optical density. As a control, the virus is also tested with a nonadsorbed rabbit serum; and the standard wild and vaccine strains are included in every series of tests. Higher extinction values with one of the specific adsorbed sera indicate antigenic relationship, which means that the virus is Sabin-like (SL) or non-Sabin-like (NSL). Intermediate strains may react with both specific sera to high extinction value. A high value with the nonadsorbed, type-specific serum and virtually no reaction with the adsorbed sera means that the test virus cannot be characterized. This happens quite often with type 2 strains, which by other means usually prove to be SL strains.

The neutralization test is performed as a standardized titration of the adsorbed sera with a virus dose of 100 $TCID_{50}$ in microtiter plates. The best results are obtained when the incubation time of the antiserum–poliovirus mixture is long, 4 h at 37°C. A suspension of secondary or tertiary monkey kidney cells is added. Microscopic reading of CPE is recorded at days 3 and 5. With the standard wild and vaccine strains, only homologous neutralization will be found; with other wild and vaccine-derived strains the difference in neutralization titer is usually clear-

cut for the origin. One of the advantages of the microneutralization test is that a mixture of serotypes, as is often excreted by persons who have been vaccinated, can be recognized by virus breakthrough. Few intermediate strains (neutralized by both sera) are found. The anti-Sabin type 2 serum had to be adsorbed with another wild-type 2 virus to become truly Sabin-specific (van Wezel and Hazendonk, 1979). Examples of the intratypic differentiation of polioviruses with cross-adsorbed antisera are presented in Table 4. Even in experienced hands, neutralization titers and extinction values of the standard controls may fluctuate, also because of batch differences in serum preparations.

It must be mentioned here that the tests indicate only the presence of antigenic sites similar to the vaccine-specific antigens. Wild viruses may contain by chance an identical epitope, but this surely is an exception.

OLIGONUCLEOTIDE MAPPING OF VIRAL RNA

Digestion of virion RNA with ribonuclease T1, which cleaves specifically at guanosine residues, results in many small oligonucleotides and a group of larger fragments with 12 or more base residues, representing 10 to 15% of the genome. When radioactively labeled fragments are resolved by two-dimensional electrophoresis, the large, structurally unique oligonucleotides yield a characteristic pattern of 50 to 60 spots: a fingerprint. Research with many poliovirus strains has demonstrated that even after "mutational drift," the epidemiologic relationship can be unambiguously determined (Kew and Nottay, 1984; Nottay et al., 1981). Vaccine and vaccine-derived strains and their wild-type parents display fingerprints which are similar, with small map differences. The molecular epidemiology of polioviruses has its limits in the rapid genetic change of the poliovirus genome. It is estimated that more than half of the spots on a map could be altered before genetic relatedness becomes unrecognizable (Minor and Schild, 1981). The method is technically demanding and has to be performed by specialists of this field.

Serodiagnosis

When enterovirus multiplies in the body, the host responds with antibody production against many antigenic determinants of the virus released from infected cells. Immunogenic determinants, eliciting a neutralization reaction, are serotype specific. Other epitopes may be type-specific, but not reacting with neutralizing antibodies, or cross-reactive with enteroviruses of the same group (polio, coxsackie B), or broader with many other types. After several infections with different serotypes (an occurrence that

TABLE 4. Examples of intratypic serodifferentiation of poliovirus isolates

Type	Strain[a]	Geographic origin[b]		SL[c]	NSL[d]	Interpretation
		Microneutralization by cross-adsorbed antisera				
1	83-14500	Zaire[p]		<2	8	NSL
	Mahoney			<2	128	
	LS-c,2ab			64	<2	
1	83-4198	River Rhine		<2	16	NSL
	Mahoney			<2	128	
	LS-c,2ab			512	<2	
1	83-19283	India		1024	<2	SL
	Mahoney			<2	256	
	LS-c,2ab			1024	<2	
1	K7	Kenya		128	<2	SL
	Mahoney			<2	32	
	LS-c,2ab			1024	<2	
2	83-19706	India		<2	8	NSL
	MEF-1			<2	16	
	P712, Ch,2ab			16	<2	
2	83-16197	Netherlands		<2	4	NSL
	MEF-1			<2	32	
	P712, Ch,2ab			16	<2	
2	83-20070	Colombia		16	<2	SL
	MEF-1			<2	16	
	P712, Ch,2ab			8	<2	
2	80-12990	River Rhine		2	4	Intermediate
	MEF-1			<2	128	
	P712, Ch,2ab			32	<2	
3	80-12013	Turkey[p]		<2	64	NSL
	Saukett			<2	256	
	Leon, 12a$_1$b			64	<2	
3	81-3804	River Rhine		<2	32	NSL
3	81-4304	River Rhine		8	<2	SL
	Saukett			<2	64	
	Leon, 12a$_1$b			64	<2	
3	83-13879	Korea		8	16	Intermediate
	Saukett			<2	128	
	Leon, 12a$_1$b			16	<2	
		EIA with cross-adsorbed antisera (extinction values)				
1	84-14341	Morocco[p]	1503	95	724	NSL
	Mahoney		1880	241	1589	
	LS-c,2ab		1497	1364	75	
1	85-4771	Colombia	1799	1547	644	Intermediate
	Mahoney		2000	140	1787	
	LS-c,2ab		2000	1849	247	
1	K 842	Kenya	1267	1378	267	SL
1	K 940	Kenya	1323	223	1089	NSL
	Mahoney		1400	222	1266	
	LS-c,2ab		1311	1267	100	
2	84-12137	Colombia	2000	1859	217	SL
2	84-10629	Korea	1150	225	118	Not possible
	MEF-1		1952	131	1364	
	P712,Ch,2ab		1586	1774	148	
2	84-15073	Scotland	1240	1444	324	SL
	MEF-1		2000	492	1979	
	P712,Ch,2ab		2000	1981	262	
3	K 228	Kenya	548	97	323	NSL?
	Saukett		1989	164	1046	
	Leon, 12a$_1$b		1485	892	176	
3	85-1470	Korea	1662	1363	713	Intermediate
3	85-1471	Korea	1754	1420	225	SL
	Saukett		1772	160	1315	
	Leon, 12a$_1$b		1507	1122	221	

[a] Strains isolated in The Netherlands.
[b] p = Paralytic case, recently arrived from other country. Other strains from adoptive children; special surveys; chance findings with minor illness.
[c] SL = Sabin-like.
[d] NSL = Non-Sabin-like.

is the rule with increasing age), a large amount of antibody to such crossreactive determinants is built up.

When the infecting serotype is known by virus isolation or presumed from epidemiologic evidence, a search for the serotypic neutralizing antibody response can be made with the use of good neutralizable virus. Special strains are necessary for echovirus 4 and enterovirus 71, and sometimes the response can be measured only by the epidemic strain. Since the serious disease, which requires a diagnosis, develops several days after the infection, it is often too late to detect a significant titer rise, but with young children the chances of thus establishing a temporal relationship between this viral infection and this illness by a neutralization test still are good. Moderate to high levels of neutralizing antibodies and/or declining titers should never be regarded as indicative of a recent infection.

The initial response is the production of IgM antibodies, some of which are neutralizing. Therefore, tests have been explored to detect serotype-specific, neutralizing IgM antibodies. Dissociation of IgM by a sulfhydryl reducing agent should reduce the titer of an early serum, but this test is unpleasant to perform because of the smell of the reagent. Alternatively, fractionation of the serum on a sucrose gradient or treatment by ion-exchange chromatography may be used to separate the IgM antibodies. The first technique requires ultracentrifugation and concentration of the IgM fraction to obtain measurable titers, and contamination by IgG is not excluded. The latter method was applied with success during a severe epidemic of acute CNS disease in 1978 caused by enterovirus 71 in Hungary (Nagy et al., 1982). The assay used with the IgM fraction was a radial-plaque neutralization, demonstrating protection of infected cells by zones around wells in an agar overlay, filled with 5 μl of treated or untreated serum. To ensure accurate measurement of the zones, monolayers were stained with Coomassie brilliant blue G-250 after fixation in 4% formaldehyde. A linear relationship between the average diameter of the zones and the logarithm of the serum dilution was established (Nagy and Takátsy, 1980). The method made an etiological diagnosis possible in many hundred cases, whereas isolations were positive with difficulty in only 44 patients.

The presence of neutralizing, type-specific enterovirus IgM antibodies may indicate infection by this particular virus, but heterotypic antibodies may be found without any relation to a recent infection. This is illustrated by a study of IgM fractions by microneutralization during an epidemic of echovirus 33 in France in 1982. Major IgM cross-reactivities were observed with other echoviruses, progressively increasing with the age of the subjects to ±40% in sera of adults over 40 years old (Pozzetto et al., 1986).

It may be presumed that inapparent primary infections of adults with relatively uncommon serotypes (such as E33 and E71) result in a true, specific, IgM antibody response to the infecting virus, whereas a heterotypic neutralization reaction is found because of immunologic memory. Then, of course, it must be assumed that IgM antibodies, though specifically neutralizing these other enteroviruses, do not indicate a recent infection. Alternatively, it can be assumed that with increasing age the percentage of auto-reactive, IgM-antibodies-excreting B-cells is growing, of which some probably will recognize epitopes of enteroviruses.

The picture of broadly cross-reactive antibodies of the IgM class, which complicate serodiagnosis, is still more evident in enzyme immunoassays with the more common serotypes as antigen. This type of test has many advantages, of which speed and relative ease of performance are most relevant to diagnostic purposes. If cross-reactions to common enterovirus structural antigens should develop in every case and not be present in a control group without disease, such a test would be attractive for a rapid preliminary diagnosis.

Perhaps in young children the presence of enterovirus-specific IgM would be indicative of a recent or current infection by any of the many serotypes, because monotypic titers tend to disappear after some time. But, in view of the large number of enterovirus infections in childhood, this indication would also become unreliable with increasing age. The most conflicting results were found with IgM antibody measured against coxsackie B viral antigens in adults with or without myocardial infarction. In a carefully controlled series, 13 to 15% of adults over 31 years of age gave positive results in the indirect EIA test with 5 different coxsackie B serotypes as antigen, whether presenting with cardiac disease or not (Hannington et al., 1986). Some sera with high levels of cross-reactive activity in the EIA–IgM test had no detectable IgM neutralizing antibody, or no neutralizing antibody in complete serum for several of the coxsackie B serotypes. In Epstein-Barr virus infections, which elicit a polyclonal B-cell stimulation, strong cross-reactive IgM reactions were often found.

In serologic tests with enteroviruses, the presentation of the antigens is important. Serotype-specific reactions, for example, neutralization, require the availability of the three-dimensional configuration as present on the virion surface. It has been shown by the immunoblot technique that cross-reactive IgG antibodies react only with epitopes of capsid protein VP1 probably not present on the surface of intact virus particles, and that IgM antibodies react with

VP1 and often more strongly with VP2 or VP3 (Reigel et al., 1985). In this technique the polypeptides are denatured to a linear configuration. Differences in cross-reactivities reported in investigations with human sera may be explained by the particular presentation of the antigens in the system used.

At this time it is uncertain whether adult IgM antibodies against enterovirus epitopes result at all from a continuous antigenic stimulus caused by repeated infections or by persisting virus somewhere in the body, or are in fact auto-antibodies against antigens perhaps released from damaged organs, for instance myopericardium as the investigations by Maisch et al. (1982) suggest.

EIA tests for IgM antibodies are afflicted with many technical and biological problems, as extensively reviewed by Meurman (1983). In the case of enterovirus IgM indirect tests with viral antigen bound to the solid phase or μ antibody capture assays require reagents which are not yet available. Workers in the field are calling for the development of a single, broadly reactive antigen and monoclonal antibodies, to improve specificity and lower the cost (Bell et al., 1986; McCartney et al., 1986).

In light of the difficulties as discussed above, positive IgM findings should be regarded with caution and should not be interpreted as, for example, a recent or persistent coxsackie B-virus-infection. These viruses are excreted during an acute infection in the feces for some time and are easily isolated. It is improbable that, in so many cases of myo/pericarditis or meningitis, virus isolation would have failed and IgM detection would have made a reliable diagnosis, as some recent reports suggest.

Serologic Diagnosis by Demonstration of Antibodies in Cerebrospinal Fluid

Involvement of the CNS by a viral infection may be accompanied by intrathecal antibody synthesis. Demonstration of specific viral antibodies in the cerebrospinal fluid thus can be diagnostic, provided that the blood-brain-barrier is intact. A reliable diagnostic procedure should include not only the titration of antibodies in CSF, but also in serum taken simultaneously, for the suspected and for another virus. A normal serum–CSF ratio is equal to or greater than 160; intrathecal antibody synthesis is indicated by a much lower ratio (serum titer : CSF titer, titers expressed as the reciprocal of the dilution). Supplementary evidence can be obtained by an elevated

$$\text{IgG index} = \frac{\text{CSF-IgG}}{\text{serum-IgG}} : \frac{\text{CSF-albumin}}{\text{serum-albumin}}$$

Estimation of neutralizing antibodies in CSF and serum was the only laboratory indication of the

neurovirulence of enterovirus 70, causing polio-like paralysis associated with acute hemorrhagic conjunctivitis in 1981 in Bombay, India, where all attempts at virus isolation failed (Wadia et al., 1983). In cases suspected of poliomyelitis, examination of the serum : CSF ratio could also be helpful. The documented, limited experience during an outbreak caused by type 3 in Finland, however, showed difficulties because the ratios versus the other poliotypes were also abnormal in patients with poliomyelitis and some others with polyradiculitis (Hovi et al., 1986).

Antibodies in Tears

An almost incredible immediate antibody response to enterovirus 70 infection of the eye provided a rapid diagnosis by the estimation of neutralizing antibodies in tear specimens collected during early onset of conjunctivitis (Yin-Murphy et al., 1985). Infectious virus particles were sedimented by ultracentrifugation and the supernatants heated for 10 min at 60°C. In microtitration plates the samples were titrated simultaneously against enterovirus 70 and coxsackievirus A24. Antibody titers ranged from 8 to 256. Because virus isolation of enterovirus 70 is difficult and patients often will not contribute both an acute and a convalescent serum sample, this unusual approach may be a fortunate alternative.

Epidemiology

Enterovirus infections may be endemic, epidemic, or sporadic, with differences of behavior occurring between geographic areas and with the lapse of time. In warm climates, prevalence of (inapparent) infections may be so great that within a short period almost all serotypes can be isolated from a small group of young children. In temperate zones a typical seasonal pattern is found; most viruses are isolated during summer and early autumn. Viruses isolated in spring sometimes may predict a high prevalence or epidemic spread during the following months. Where a reporting system exists and most isolates are typed, unpredictable changes in the predominance of different serotypes are found. This is beautifully illustrated in a report from a nationwide surveillance of infectious agents in Japan (Yamazaki et al., 1984) (Fig. 5). Serotypes known to occur in epidemic waves are indicated in Table 1. Many variables contribute to the epidemiology of enteroviruses, of which of course vaccination against poliomyelitis changed the epidemic incidence of polioviruses greatly. For more details the reader is referred to the chapter on enteroviruses by Melnick (1982).

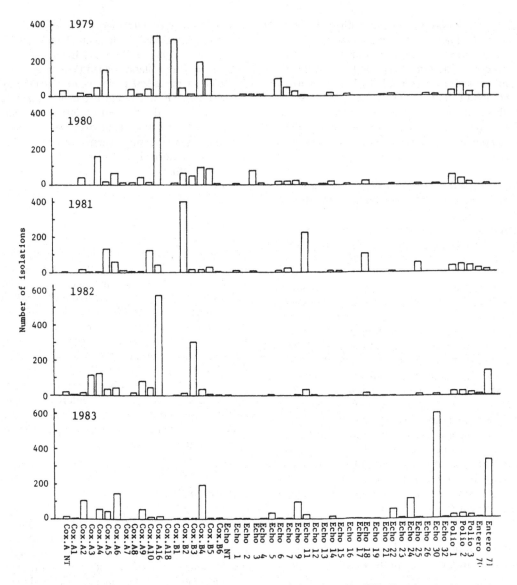

FIG. 5. Reported isolations of enteroviruses from humans in Japan, 1979–1983. (Reprinted with permission, Yamazaki et al., 1984.)

Changes in epidemic behavior can be anticipated when new variants arise by mutations in viral RNA, which alter virulence or antigenic make-up. A new coxsackie A24 variant was responsible for epidemics of conjunctivitis in Singapore, whereas enterovirus 70 probably gained human virulence by mutation of an animal or insect picornavirus. The exceptional epidemiology of these viruses is reviewed by Yin-Murphy (1984).

Prevention and Control

Good personal and nosocomial hygiene could prevent transmission of enteroviruses by the fecal-oral and (in special diseases) eye-to-eye route, but not infections transmitted through respiratory secretions. The large number of serotypes is prohibitive for the production of an all-in-one vaccine. It is, therefore, fortunate that the majority of infections pass without or with only a minor illness, and that not more than three serotypes cause the serious crippling disease poliomyelitis. Some outbreaks have occurred with diseases mimicking paralytic polio, caused by coxsackievirus A7, enterovirus 70 and 71. This proves that enteroviruses can acquire neurovirulence. Physicians (pediatricians, neurologists, orthopedic surgeons) and virologists should be on the alert whenever, in an area presumed to be free of poliomyelitis, an unexplained paralytic illness occurs. It could be a rare case of true polio, caused by wild or vaccine-associated poliovirus, or by a new variant of a normally more benign enterovirus.

At present poliomyelitis can be prevented by vaccines containing the three serotypes, either attenuated but infectious (OPV) or inactivated (IPV). The aim of vaccination is 1) to block the invasion of the CNS of an infected individual by the presence of neutralizing antibodies in the blood; 2) to diminish the spread of virus to susceptible persons in the community; 3) worldwide eradication of the virus as an ultimate goal, because in nature poliovirus has been found only in human hosts. Both vaccines have their merits in the prevention of paralytic poliomyelitis in individuals and have achieved almost complete reduction of the incidence in economically privileged countries, where a good vaccination strategy is maintained. The end came within sight by new developments in the production of IPV by biotechnologic methods. Poliovaccine with a high content of antigen can be combined with diphtheria–tetanus–pertussis vaccine and used in a simplified immunization schedule in economically underdeveloped countries. The oral vaccine in these countries has to cope with the high prevalence of nonpolio enteric infections, necessitating annual mass administration to young children without reference to the number of vaccinations they may have had previously. The Expanded Program on Immunization of the World Health Organization will make immunization services available for all children of the world by 1990. The early use of potent IPV combined with diphtheria–tetanus–pertussis vaccine will make this goal also economically feasible. The poliovirus antigens in IPV should at this moment be changed only in potency; replacement by viral subunits or polypeptides produced by modern technology has no priority. The careful inactivation of virulent polioviruses by formaldehyde has proved to be reliable, safe, and efficient with respect to the only important purpose: immunization causing the appearance of the appropriate neutralizing antibodies. A neutralizing response elicited in animals by viral subunits and polypeptides is not equivalent to protection of the human host and, therefore, this type of vaccine would require extensive field trials. With elimination in sight, depending only on a worldwide political will, no time should be lost. With the use of the present attenuated living vaccines, polioviruses will not be eradicated, but remain circulating in susceptible populations, with the danger of change to virulence by passage through the human gut. Persons with an abnormal immune function can excrete the virus for a lifetime. Eradication means that vaccination can finally be terminated because the virus is not longer present in its natural host in the whole world. The many aspects of poliomyelitis vaccination were discussed during an International Symposium on Poliomyelitis Control by experts from all parts of the globe. In his stimulating summary, Frederick C. Robbins said: "It would appear that a practical and feasible goal is worldwide control of paralytic poliomyelitis within this century, but global eradication as the ultimate goal should not be abandoned." (Robbins, 1984.)

Literature Cited

Albrecht, P., G. van Steenis, A. L. van Wezel, and J. Salk. 1984. Standardization of poliovirus neutralizing antibody tests. Rev. Infect. Dis. 6(suppl. 2):540–544.

Almond, J. W., G. Stanway, A. J. Cann, G. D. Westrop, D. M. A. Evans, M. Ferguson, P. D. Minor, M. Spitz, and G. C. Schild. 1984. New poliovirus vaccines: a molecular approach. Vaccine 2:177–184.

Alsop, J., T. H. Flewett, and J. R. Foster. 1960. "Hand, foot and mouth disease" in Birmingham in 1959. Br. Med. J. 10:1708–1711.

American Type Culture Collection. 1986. Catalogue of animal and plant viruses. American Type Culture Collection, Washington D.C.

Anderson, L. J., M. H. Hatch, M. R. Flemister, and G. E. Marchetti. 1984. Detection of enterovirus 70 with monoclonal antibodies. J. Clin. Microbiol. 20:405–408.

Anonymous. 1982. Neurovirulence of enterovirus 70. Lancet i:373–374.

Barson, W. J., J. Craenen, D. M. Hosier, R. L. Brawley, and M. D. Hilty. 1981. Survival following myocarditis and myocardial calcification associated with infection by Coxsackievirus B-4. Pediatrics 68:79–81.

Bell, E. J., and B. P. Cosgrove. 1980. Routine enterovirus diagnosis in a human rhabdomyosarcoma cell line. Bull. W.H.O. 58:423–428.

Bell, E. J., R. A. McCartney, D. Basquill, and A. K. R. Chandhuri. 1986. μ-Antibody capture ELISA for the rapid diagnosis of enterovirus infections in patients with aseptic meningitis. J. Med. Virol. 19:213–217.

Berretty, P. J. M., C. H. M. Hoek, J. C. Rademaker, and J. W. M. Tan-Go. 1983. De hand-, voet- en mondziekte. Ned. Tijdschr. Geneeskd. 127:1910–1913.

Bijkerk, H. 1979. Poliomyelitis epidemic in the Netherlands, 1978. Dev. Biol. Stand. 43:195–206.

Bowles, N. E., P. J. Richardson, E. G. J. Olsen, and L. C. Archard. 1986. Detection of coxsackie-B-virus-specific RNA sequences in myocardial biopsy samples from patients with myocarditis and dilated cardiomyopathy. Lancet i:1120–1123.

Cao, Y., D. P. Schnurr, and N. J. Schmidt. 1984. Differing cardiotropic and myocarditic properties of group B type 4 coxsackievirus strains. Arch. Virol. 80:119–130.

Chumakov, M., M. Voroshilova, L. Shindarov, I. Lavrova, L. Gracheva, G. Koroleva, S. Vasilenko, I. Brodvarova, M. Nikolova, S. Gyurova, M. Gacheva, G. Mitov, N. Ninov, E. Tsylka, I. Robinson, M. Frolova, V. Bashkirtsev, L. Martiyanova, and V. Rodin. 1979. Enterovirus 71 isolated from cases of epidemic poliomyelitis-like disease in Bulgaria. Arch. Virol. 60:329–340.

Crainic, R., P. Couillin, B. Blondel, N. Cabau, A. Boué, and F. Horodniceanu. 1983. Natural variation of poliovirus neutralization epitopes. Infect. Immun. 41:1217–1225.

Dagan, R., and M. A. Menegus. 1986. A combination of four cell types for rapid detection of enteroviruses in clinical specimens. J. Med. Virol. 19:219–228.

Druyts-Voets, E., F. Yane, E. Bosmans, J. Colaert, and J. Desmyter. 1985. Method for selecting optimal cells for enterovirus isolation as determined in an outbreak of

Echovirus type 33 meningitis. Eur. J. Clin. Microbiol. 4:331-334.

Engelbrecht, R. S., M. J. Weber, B. L. Salter, and C. A. Schmidt. 1980. Comparative inactivation of viruses by chlorine. Appl. Environ. Microbiol. 40:249-256.

Evans, D. M. A., G. Dunn, P. D. Minor, G. C. Schild, A. J. Cann, G. Stanway, J. W. Almond, K. Currey, and J. V. Maizel. 1985. Increased neurovirulence associated with a single nucleotide change in a noncoding region of the Sabin type 3 poliovaccine genome. Nature (London) 314:548-550.

Ferguson, M., D. I. Magrath, P. D. Minor, and G. C. Schild. 1986. WHO collaborative study on the use of monoclonal antibodies for the intratypic differentiation of poliovirus strains. Bull. W.H.O. 64:239-246.

Flowers, D., and G. M. Scott. 1985. How useful are serum and CSF interferon levels as a rapid diagnostic aid in virus infections? J. Med. Virol. 15:35-47.

French, M. L. V., N. J. Schmidt, R. W. Emmons, and E. H. Lennette. 1972. Immunofluorescence staining of group B coxsackieviruses. Appl. Microbiol. 23:54-61.

Green, R. J. L., J. N. Webb, and M. H. Maxwell. 1979. The nature of virus-like particles in the paraxial muscles of idiopathic scoliosis. J. Pathol. 129:9-10.

Grist, N. R., E. J. Bell, and F. Assaad. 1978. Enteroviruses in human disease. Prog. Med. Virol. 24:114-157.

Hamparian, V. V., A. C. Ottolenglis, and J. H. Hughes. 1985. Enteroviruses in sludge: multiyear experience with four wastewater treatment plants. Appl. Environ. Microbiol. 50:280-286.

Hampil, B., J. L. Melnick, C. Wallis, R. W. Brown, E. T. Braye, and R. R. Adams. 1965. Preparation of antiserum to enteroviruses in large animals. J. Immunol. 95:895-908.

Hannington, G., J. C. Booth, R. J. Bowes, and H. Stern. 1986. Coxsackie B virus-specific IgM antibody and myocardial infarction. J. Med. Microbiol. 21:287-291.

Hara, M., Y. Saito, T. Komatsu, H. Kodama, W. Abo, S. Chiba, and T. Nakao. 1981. Antigenic analysis of poliovirus isolated from a child with agammaglobulinemia and paralytic poliomyelitis after Sabin vaccine administration. Microbiol. Immunol. 25:905-913.

Hirschman, S. Z., and G. S. Hammer. 1974. Coxsackie virus myopericarditis, a microbiological and clinical review. Am. J. Cardiol. 34:224-232.

Hogle, J. M., M. Chow, and D. J. Filman. 1985. Three-dimensional structure of poliovirus at 2.9 Å resolution. Science 229:1358-1365.

Holland, J. J. 1984. Continuum of changes in RNA virus genomes, p. 137-143. In A. L. Notkins, and M. B. A. Oldstone (ed.), Concepts in viral pathogenesis. Springer-Verlag, New York.

Hovi, T., M. Stenvik, and E. Kinnunen. 1986. Diagnosis of poliomyelitis by demonstration of intrathecal synthesis of neutralizing antibodies. J. Infect. Dis. 153:998-999.

Huber, S. A., and P. A. Lodge. 1986. Coxsackievirus B-3 myocarditis. Identification of different pathogenic mechanisms in DBA/2 and Balb/c mice. Am. J. Pathol. 122:284-291.

Hyypiä, T., P. Stålhandske, R. Vainionpää, and U. Pettersson. 1984. Detection of enteroviruses by spot hybridization. J. Clin. Microbiol. 19:436-438.

Kapsenberg, J. G., R. A. Coutinho, A. G. Hazendonk, A. B. R. Ran, and A. L. van Wezel. 1981. Epidemiological implications of the isolations and intratypic serodifferentiation of poliovirus strains in The Netherlands. Dev. Biol. Stand. 47:293-301.

Kapsenberg, J. B., A. Ras, and J. Korte. 1979. Improvement of enterovirus neutralization by treatment with so-dium deoxycholate or chloroform. Intervirology 12:329-334.

Kew, O. M. and B. K. Nottay. 1984. Molecular epidemiology of poliovirus. Rev. Infect. Dis. 6(suppl. 2):499-504.

Kinnunen, E., T. Hovi, and M. Stenvik. 1986. Outbreak of poliomyelitis in Finland in 1984. Description of nine cases with persisting paralysis. Scand. J. Infect. Dis. 18:15-18.

Kjellén, L. 1985. A hypothesis accounting for the effect of the host cell on neutralization-resistant virus. J. Gen. Virol. 66:2279-2283.

Kjellén, L., and G. von Zeipel. 1984. Influence of host cells on the infectivity and neutralizability of echovirus 4 (Pesascek). Intervirology 22:32-40.

Koch, F., and G. Koch. 1985. The molecular biology of poliovirus. Springer-Verlag, New York.

Kohne, D. E., C. J. Gibbs, L. White, S. M. Tracy, W. Meinke, and R. A. Smith. 1981. Virus detection by nucleic acid hybridization: examination of normal and ALS tissues for the presence of poliovirus. J. Gen. Virol. 56:223-233.

Kono, R., K. Miyamura, E. Tajiri, A. Sasagawa, P. Phuapradit, N. Roongwithu, A. Vejjajiva, C. Jayavasu, P. Thongcharoen, C. Wasi, and P. Rodprassert. 1977. Virological and serological studies of neurological complications of acute hemorrhagic conjunctivitis in Thailand. J. Infect. Dis. 135:706-713.

Kono, R., K. Miyamura, E. Tajiri, S. Shiga, A. Sasagawa, P. F. Irani, S. M. Katrak, and N. H. Wadia. 1974. Neurologic complications associated with acute hemorrhagic conjunctivitis virus infection and its serologic confirmation. J. Infect. Dis. 129:590-593.

Kono, R., A. Sasagawa, K. Ishii, S. Sugiura, M. Ochi, H. Matsumiya, Y. Uchida, K. Kameyama, M. Kaneko, and N. Sakurai. 1972. Pandemic of new type of conjunctivitis. Lancet i:1191-1194.

Lau, R. C. H. 1986. Coxsackie B virus-specific IgM responses in coronary care unit patients. J. Med. Virol. 18:193-198.

Lemon, S. M., and L. N. Binn. 1985. Incomplete neutralization of hepatitis A virus in vitro due to lipid-associated virions. J. Gen. Virol. 66:2501-2505.

Mahnel, H. 1983. Desinfektion von Viren (Review). Zentralbl. Veterinarmed. Reihe B 30:81-96.

Maisch, B., R. Trostel-Soeder, E. Stechemesser, P. A. Berg, and K. Kochsiek. 1982. Diagnostic relevance of humoral and cell-mediated immune reactions in patients with acute viral myocarditis. Clin. Exp. Immunol. 48:533-545.

Majer, M., and S. A. Plotkin (ed.). 1972. Strains of human viruses. S. Karger, Basel.

Mandel, B. 1985. Virus neutralization, p. 53-70. In M. H. V. van Regenmortel and A. R. Neurath (ed.) Immunochemistry of viruses. The basis for serodiagnosis and vaccines. Elsevier Science Publishers B.V., Amsterdam.

Matthews, R. E. F. 1982. Classification and nomenclature of viruses (fourth report of the ICTV). Intervirology 17:1-199.

McCartney, R. A., J. E. Banatvala, and E. J. Bell. 1986. Routine use of μ-antibody-capture ELISA for the serological diagnosis of coxsackie B virus infections. J. Med. Virol. 19:205-212.

Melnick, J. L. 1982. Enteroviruses, p. 187-251. In A. S. Evans (ed.), Viral infections of humans. Epidemiology and control, 2nd ed. Plenum Medical Book Company, New York and London.

Melnick, J. L. 1983. Portraits of viruses: the picornaviruses. Intervirology 20:61-100.

Melnick, J. L., R. M. Chanock, H. Gelfand, W. McD. Hammon, R. J. Huebner, L. Rosen, A. J. Sabin, and H. A. Wenner. 1963. Picornaviruses: classification of nine new types. Science **141**:153–154.

Melnick, J. L., G. Dalldorf, J. F. Enders, H. M. Gelfand, W. McD. Hammon, R. J. Huebner, L. Rosen, A. B. Sabin, J. T. Syverton, and H. A. Wenner. 1962. Classification of human enteroviruses. Virology **16**:501–504.

Melnick, J. L., N. J. Schmidt, B. Hampil, and H. H. Ho. 1977. Lyophilized combination pools of enterovirus equine antisera: preparation and test procedures for the identification of field strains of 19 group A coxsackievirus serotypes. Intervirology **8**:172–181.

Melnick, J. L., H. A. Wenner, and C. A. Phillips. 1979. Enteroviruses, p. 471–534. *In* E. H. Lennette and N. J. Schmidt (ed.), Viral, rickettsial and chlamydial infections, 5th ed. American Public Health Association, Washington, D.C.

Melnick, J. L., and I. L. Wimberly. 1985. Lyophilized combination pools of enterovirus equine antisera: new LBM pools prepared from reserves stored frozen for two decades. Bull. W.H.O. **63**:543–550.

Meurman, O. 1983. Detection of antiviral IgM antibodies and its problems. A review. Curr. Top. Microbiol. Immunol. **104**:101–131.

Minnich, L. L., and C. G. Ray. 1980. Variable susceptibility of mice to group B coxsackievirus infections. J. Clin. Microbiol. **11**:73–75.

Minor, P. D., and G. C. Schild. 1981. Identification of the origin of poliovirus isolates. Lancet **ii**:968–970.

Mirkovic, R. R., N. J. Schmidt, M. Yin-Murphy, and J. L. Melnick. 1974. Enterovirus etiology of the 1970 Singapore epidemic of acute conjunctivitis. Intervirology **4**:119–127.

Miyamura, K., M. Tanimura, N. Takeda, R. Kono, and S. Yamazaki. 1986. Evolution of enterovirus 70 in nature: all isolates were recently derived from a common ancestor. Arch. Virol. **89**:1–14.

Miyamura, K., S. Yamazaki, E. Tajiri, and R. Kono. 1974. Growth characteristics of acute hemorrhagic conjunctivitis (AHC) virus in monkey kidney cells. 1. Effect of temperature on viral growth. Intervirology **4**:279–286.

Nagy, G., and S. Takátsy. 1980. Modified radial-plaque-neutralization test for demonstration of enterovirus neutralizing antibodies. Arch. Virol. **65**:89–98.

Nagy, G., S. Takátsy, E. Kukán, I. Mihály, and I. Dömök. 1982. Virological diagnosis of enterovirus 71 infections: experiences gained during an epidemic of acute CNS diseases in Hungary in 1978. Arch. Virol. **71**:217–227.

Narang, H. K., and A. A. Codd. 1983. Action of commonly used disinfectants against enteroviruses. J. Hosp. Infect. **4**:209–212.

Notkins, A. L., T. Onodera, and B. S. Prabhakar. 1984. Virus-induced autoimmunity, p. 210–215. *In* A. L. Notkins and M. B. A. Oldstone (ed.), Concepts in viral pathogenesis. Springer-Verlag, New York.

Notkins, A. L., and J. W. Yoon. 1984. Virus-induced diabetes mellitus, p. 241–247. *In* A. L. Notkins and M. B. A. Oldstone (ed.), Concepts in viral pathogenesis. Springer-Verlag, New York.

Nottay, B. K., O. M. Kew, M. H. Hatch, J. T. Heyward, and J. F. Obijeski. 1981. Molecular variation of type 1 vaccine-related and wild polioviruses during replication in humans. Virology **108**:405–423.

Osterhaus, A. D. M. E., and G. van Steenis. 1981. Virologic control of monkeys used for the production of poliomyelitis virus vaccine. Dev. Biol. Stand. **47**:157–161.

Osterhaus, A. D. M. E., A. L. van Wezel, A. G. Hazendonk, F. G. C. M. UytdeHaag, J. A. A. M. van Asten,

and G. van Steenis. 1983. Monoclonal antibodies to polioviruses. Comparison of intratypic strain differentiation of poliovirus type 1 using monoclonal antibodies versus cross-adsorbed antisera. Intervirology **20**:129–136.

Pal, S. R., J. Szücs, and J. L. Melnick. 1983. Rapid immunofluorescence diagnosis of acute hemorrhagic conjunctivitis caused by enterovirus 70. Intervirology **20**:19–22.

Pozzetto, B., J. C. Le Bihan, and O. G. Gaudin. 1986. Rapid diagnosis of echovirus 33 infection by neutralizing specific IgM antibody. J. Med. Virol. **18**:361–367.

Prabhakar, B. S., M. A. Menegus, and A. L. Notkins. 1985. Detection of conserved and nonconserved epitopes on coxsackievirus B4: frequency of antigenic change. Virology **146**:302–306.

Prather, S. L., R. Dagan, J. A. Jenista, and M. A. Menegus. 1984. The isolation of enteroviruses from blood: a comparison of four processing methods. J. Med. Virol. **14**:221–227.

Reigel, F., F. Burkhardt, and U. Schilt. 1985. Cross-reactions of immunoglobulin M and G antibodies with enterovirus-specific viral structural proteins. J. Hyg. **95**:469–481.

Robbins, F. C. 1984. International symposium on poliomyelitis control, summary and recommendations. Rev. Infect. Dis. (suppl. 2)**6**:S596–600.

Robbins, F. C., J. F. Enders, and T. H. Weller. 1950. Cytopathogenic effect of poliomyelitis viruses in vitro on human embryonic tissues. Proc. Soc. Exp. Biol. Med. **75**:370–374.

Rotbart, H. A., M. J. Levin, and L. P. Villareal. 1984. Use of subgenomic poliovirus DNA hybridization probes to detect the major subgroups of enteroviruses. J. Clin. Microbiol. **20**:1105–1108.

Rotbart, H. A., M. J. Levin, L. P. Villareal, S. M. Tracy, B. L. Semler, and E. Wimmer. 1985. Factors effecting the detection of enteroviruses in cerebrospinal fluid with coxsackievirus B3 and poliovirus 1 cDNA probes. J. Clin. Microbiol. **22**:220–224.

Rueckert, R. R. 1985. Picornaviruses and their replication, p. 705–738. *In* B. N. Fields et al. (ed.), Virology. Raven Press, New York.

Sabin, A. B. 1981. Paralytic poliomyelitis: old dogmas and new perspectives. Rev. Infect. Dis. **3**:543–564.

Schmidt, N. J., H. H. Ho, and E. H. Lennette. 1975. Propagation and isolation of group A coxsackieviruses in RD cells. J. Clin. Microbiol. **2**:183–185.

Schmidt, N. J., H. H. Ho, J. L. Riggs, and E. H. Lennette. 1978. Comparative sensitivity of various cell culture systems for isolation of viruses from wastewater and fecal samples. Appl. Environ. Microbiol. **36**:480–486.

Schürmann, W., and H. J. Eggers. 1985. An experimental study on the epidemiology of enteroviruses: water and soap washing of poliovirus 1-contaminated hands, its effectiveness and kinetics. Med. Microbiol. Immunol. **174**:221–236.

Seal, L. A., and R. M. Jamison. 1983. Evidence for secondary structure within the virion RNA of echovirus 22. J. Virol. **50**:641–644.

Stålhandske, P. O. K., M. Lindberg, and U. Pettersson. 1984. Replicase gene of coxsackievirus B3. J. Virol. **51**:742–746.

Tagaya, I., R. Takayama, and A. Hagiwara. 1981. A large-scale epidemic of hand, foot and mouth disease associated with enterovirus 71 infection in Japan in 1978. Jap. J. Med. Sci. Biol. **34**:191–196.

Toyoda, H., K. Michinori, Y. Kataoka, T. Suganama, T.

722   J. G. Kapsenberg

Omata, N. Imura, and A. Nomoto. 1984. Complete nucleotide sequence of all three poliovirus serotype genomes: implications for genetic relationship, gene function and antigenic determinants. J. Mol. Biol. **174**:561–585.

Tracy, S. 1984. A comparison of genomic homologies among the Coxsackie virus B group: use of fragments of the cloned Coxsackie virus B3 genome as probes. J. Gen. Virol. **65**:2167–2172.

Tracy, S., and R. A. Smith. 1981. A comparison of the genomes of polioviruses by cDNA: RNA hybridization. J. Gen. Virol. **55**:193–199.

van Wezel, A. L. 1985. Monolayer growth systems: homogeneous unit processes, p. 265–282. *In* R. E. Spier and J. B. Griffiths (ed). Animal cell biotechnology, vol. 1. Academic Press, London.

van Wezel, A. L., and A. G. Hazendonk. 1979. Intratypic serodifferentiation of poliomyelitis virus strains by strain-specific antisera. Intervirology **11**:2–8.

von Zeipel, G. 1980. Most echoviruses reach higher titers in RD than in GMK-AH1 cells. Arch. Virol. **63**:143–146.

Wadia, N. H., S. M. Katrak, V. P. Misra, P. N. Wadia, K. Miyamura, K. Hashimoto, T. Ogino, T. Hikiji, and R. Kono. 1983. Polio-like motor paralysis associated with acute hemorrhagic conjunctivitis in an outbreak in 1981

in Bombay, India: clinical and serologic studies. J. Infect. Dis. **147**:660–668.

Ward, R. L., and E. W. Akin. 1984. Minimum infective dose of animal viruses. Crit. Rev. Environ. Control **14**:297–310.

Wecker, I. 1977. The use of RD cells in the isolation of Echo type 30 virus from patients with aseptic meningitis. Med. Microbiol. Immunol. **163**:45–51.

Wetz, K., P. Willingmann, H. Zeichhardt, and K.-O. Habermehl. 1986. Neutralization of poliovirus by polyclonal antibodies requires binding of a single IgG molecule per virion. Arch. Virol. **91**:207–220.

W.H.O. 1986. Acute haemorrhagic conjunctivitis: neurological complications. Weekly Epidemiological Records **61**:8–10.

Woodruff, J. F. 1980. Viral myocarditis, a review. Am. J. Pathol. **101**:427–479.

Yamazaki, S. et al. 1984. Editorial committee of findings of infectious agents in Japan. Jap. J. Med. Sci. Biol. suppl., **37**:33.

Yin-Murphy, M. 1984. Acute hemorrhagic conjunctivitis. Prog. Med. Virol. **29**:33–44.

Yin-Murphy, M., N. Abdul Rahim, M. C. Phoon, Baharuddin-Ishak, and J. Howe. 1985. Early and rapid diagnosis of acute haemorrhagic conjunctivitis with tear specimens. Bull. W.H.O. **63**:705–709.

CHAPTER 37

Picornaviridae: Rhinoviruses—Common Cold Viruses

WIDAD AL-NAKIB and DAVID A. J. TYRRELL

Disease: Common cold.

Etiologic Agents: Rhinoviruses (over 115 serotypes).

Source: Infected individuals excreting virus in nasal secretions; airborne transmission over short distances; enter via nasal epithelium or conjunctiva; transmission via contaminated fingers and fomites also possible.

Clinical Manifestations: Rhinorrhea, nasal obstruction, some element of pharyngitis and cough; usually acute; subclinical infections occur.

Pathology: Rhinitis probably due mainly to direct cytocidal effect of virus replication in the nasal epithelium; edema and cellular infiltration occur; there is narrowing of the nasal cavities with excessive mucus secretion.

Laboratory Diagnosis: Virus isolation in cell or organ culture; seroconversion or four-fold or greater increases in neutralizing antibody; direct virus detection with cDNA probes is being clinically evaluated.

Epidemiology: Worldwide distribution with two main seasonal peaks, spring and early autumn; infection is most common in early life and generally decreases with increase in age.

Treatment: No specific treatment is yet available.

Prevention: No vaccine available, but intranasal interferon prophylaxis may be used in some circumstances.

Introduction

The term "cold" has generally been taken to mean a short mild illness in which the main symptoms involve the upper respiratory tract and in which nasal symptoms predominate. Colds have been recognized as a specific medical entity as early as 400 BC, and although it was suspected to be contagious, it was not until 1914 that Kruse provided experimental evidence for the infectious nature of colds by inoculating cell-free nasal material from persons with symptoms of colds into volunteers who later developed colds. However, evidence that colds are caused by a virus and not bacteria came from experiments by Dochez et al. (1930), who inoculated volunteers with bacteria-free filtrates from individuals with symptoms of colds (for details of historic events, see Gwaltney, 1982; Tyrrell, 1965).

Rhinoviruses were first cultured in vitro in 1953 in explants of human embryonic lungs (Andrewes et al., 1953). However, it was not until some years later that this work could be repeated when virus was shown to produce cytopathic effect in human embryo kidney cells (Tyrrell et al., 1960, 1962). In the meantime, two groups in the United States reported the isolation in monkey kidney cultures of a virus involved in upper respiratory illness, and this was subsequently designated rhinovirus 1A (Pelon et al., 1957; Price, 1956). Since then, other rhinovirus-sensitive cells have been found, and many rhinoviruses have been isolated, and in general all have been found to cause "common colds."

It is now clear that rhinoviruses are responsible for the majority of acute upper respiratory illnesses collectively known as the common cold. Coronavirus and to a much lesser extent adenoviruses, parainfluenza, enteroviruses, and respiratory syncytial viruses may also cause common colds and contribute to the total load of disease (Fig. 1). It has been estimated that acute respiratory illness accounts for over half of all acute disabling conditions in the United States and that one-third to one-half of all acute respiratory illnesses are caused by rhinoviruses (Couch, 1984). Furthermore, it has been estimated that common colds occur at rates of 2 to 5 per person per year (Couch, 1984; Fox et al., 1975, 1985). Common colds are estimated to cost billions of dollars each year in terms of lost working days, cold remedies, and analgesics (Couch, 1984). Despite their enormous impact on both morbidity and economic loss, there has been little progress in finding methods of control. However, over the past 2 years or so there have been some very important advances in rhinovirus research. Thus the genome of two human rhinoviruses has been cloned and sequenced (Callahan et al., 1985; Skern et al., 1985; Stanway et al., 1984), the major antigenic sites of one rhinovirus have been mapped using monoclonal antibodies (Sherry et al., 1986), and details of the fine structure of one, the same rhinovirus, human rhinovirus 14 (HRV-14) have been elucidated by X-ray crystallography techniques (Rossmann et al., 1985). Furthermore, a number of new procedures have recently been developed for the detection of rhinovirus nucleic acid (Al-Nakib et al., 1986), antigens (Dearden and Al-Nakib, 1987), and antibodies (Barclay and Al-Nakib, 1987). Although these procedures are still at the experimental stage, their future impact on the rapid diagnosis of rhinoviruses will be very important and hence very relevant to the subject of this book. In this chapter we shall omit further reference to other viruses that cause common colds which are dealt with elsewhere and concentrate on reviewing our present knowledge of rhinoviruses.

The Viruses

List of Viruses

On the basis of serum neutralization, over 115 different serotypes of human rhinoviruses have been identified (Cooper et al., 1978; Kapikian et al., 1971; Melnick, 1980) (Table 1). Although most of the human rhinoviruses are antigenically distinct, cross-relationships between them have been detected (Cooney et al., 1975). Indeed, Cooney et al., (1982) produced potent rabbit antisera to 90 of the classified rhinovirus serotypes and by systematic neutralization tests detected significant numbers of cross-reactions. Using heterotypic antibody reaction in both rabbit and guinea pig antirhinovirus sera, they defined 16 groups of rhinoviruses (Table 2).

Physiochemical Properties and Morphology

Rhinoviruses share with other members of the family picornaviridae a number of physical and biochemical characteristics. They are small RNA viruses, 28 to 34 nm in diameter with a relative molecular mass (M_r) of approximately 8.5×10^6, of which RNA accounts for some 30% by weight. The capsid has an icosahedral symmetry and is made up of 60 copies of each of four virus-coded structural proteins (VP1, VP2, VP3, and VP4) and encloses a single-stranded RNA genome of molecular weight of approximately 2.6×10^6 (Brown et al., 1970; Cooper et al., 1978;

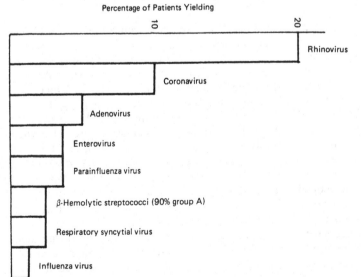

Percentage of Patients Yielding

10   20

Rhinovirus

Coronavirus

Adenovirus

Enterovirus

Parainfluenza virus

β-Hemolytic streptococci (90% group A)

Respiratory syncytial virus

Influenza virus

FIG. 1. Proportion of viruses and streptococci isolated from patients with common colds (from Tyrrell et al., 1979, with permission).

TABLE 1. List of human rhinoviruses

Rhinovirus number	Prototype strain	Rhinovirus number	Prototype strain
1A	Echo-28	58	21-CV 20
1B	B632	59	611-CV 35
2	HGP	60	2268-CV 37
3	FEB	61	6669-CV 39
4	16/60	62	1963M-CV 40
5	Norman	63	6360-CV 41
6	Thompson	64	6258-CV 44
7	68-CV 11	65	425-CV 47
8	MRH-CV 12	66	1983 CV 48
9	211-CV 13	67	1857-CV 51
10	204-CV 14	68	FO2-2317-Wood
11	1-CV 15	69	FO2-2513-Mitchinson
12	181-CV 16	70	FO2-2547-Treganza
13	353	71	SF365
14	1059	72	K2207
15	1734	73	107E
16	11757	74	328A
17	33342	75	328F
18	5986-CV 17	76	HOOO62
19	6072-CV 18	77	130-63
20	15-CV 19	78	2030-65
21	47-CV 21	79	101-1
22	127-CV 22	80	277G
23	5124-CV 24	81	483F2
24	5146-CV 25	82	03647
25	5426-CV 26	83	Baylor 7
26	5660-CV 27	84	432D
27	5870-CV 28	85	50-525-CV-54
28	6101-CV 29	86	121564-Johnson
29	5582-CV 30	87	FO2-3607-Corn
30	106F	88	CVD-01-0165-Dambrauskas
31	140F	89	41467-Gallo
32	363		
33	1200		
34	137-3		
35	164A		
36	342H		
37	151-1		
38	CH 79		
39	209		
40	1794		
41	56110		
42	56822		
43	58750		
44	71560		
45	Baylor 1		
46	Baylor 2		
47	Baylor 3		
48	1505		
49	8213		
50	A2 No. 58		
51	FO1-4081		
52	FO1-3772		
53	FO1-3928		
54	FO1-3774		
55	WIS 315E		
56	CH82		
57	CH47		

From Hamparian, 1979, with permission.

Medappa et al., 1971; Stott and Killington, 1972). The RNA is of positive polarity and has been shown to be polyadenylated at the 3′ end and is covalently linked to a small virus-encoded protein, VPg, at the 5′ terminus (Gauntt, 1980; Nair and Owens, 1974). In addition, rhinoviruses like other members of the picornaviridae family are ether-resistant. However, they differ from other picornaviruses in that they are acid-labile, having a relatively high bouyant density, and grow optimally at temperatures ranging from 33 to 35°C (Gwaltney, 1982).

Based on recent X-ray crystallography studies, it has been suggested that the 25-Å deep "canyons" found on each icosahedral face of HRV-14 are the sites for cell receptor binding (Rossmann et al.,1985). Using competition binding assays between pairs of serotypes, Abraham and Colonno (1984) have been able to show that rhinoviruses can be grouped in two receptor groups depending on their binding to the

TABLE 2. Antigenic grouping of rhinoviruses into 16 groups based on their cross-relationship as revealed by cross-neutralization using both rabbit and guinea pig anti-rhinovirus sera[a]

Group	Prototype of group	Rhinovirus types included[b]
I	1A	1B
II	2	**49**
III	3	(79), 6, **14**
IV	9	**32, 67**
V	11	15, 40, 74, 76
VI	12	**78**
VII	13	41
VIII	22	61
IX	29	5, 17, (19, 21, 30), **42**, (43), **44**, 70
X	36	50, (18), **58, 89**
XI	38	60
XII	39	54
XIII	(48)	(55)
XIV	(56)	(57)
XV	(59)	(63), (85)
XVI	66	77

[a] From Cooney et al. (1972), with permission.
[b] **Boldface** indicates a reciprocal cross-reaction with the group prototype. Parentheses indicate heterotypic antibody seen only in reference guinea pig anti-rhinovirus sera.

Replication

The initial step in the replication of rhinoviruses is the attachment of virus particles to the host cell via receptors. Picornaviruses are very specific with regard to the cells they infect, and this specificity is to a large extent determined by the type of receptor present on the host cell surface. It has been esti-

cellular receptors for virus attachment. Thus, the large majority of rhinoviruses studied, 20 (84%) of 24 (HRV-3, 5, 9, 11, 12, 14, 15, 17, 22, 32, 36, 39, 41, 51, 58, 59, 60, 66, 67, 89) appear to share the same receptor (in HeLa cells), while the remaining four (16%) (HRV-1A, 2, 44, 49) share another receptor, and these are distinct from those used by other picornaviruses. Although this study was confined to only 24 serotypes, the authors have chosen these viruses at random and therefore concluded that a wider selection of unstudied rhinoviruses was unlikely to alter the pattern of this relationship (Abraham and Colonno, 1984). Indeed, in a later study the same workers reported that a mouse monoclonal antibody was isolated that had the precise specificity predicted by the competition binding study and this antibody protected HeLa cells from infection by 78 of 88 human rhinovirus serotypes (Colonno et al., 1986).

mated that there are about 10^4 receptor sites per cell on HeLa and WI-38 cells for rhinovirus serotypes 2 and 14 (Butterworth et al., 1976; Lonberg-Holm and Korant, 1972; Macnaughton, 1982) and greater than 10^6 for human rhinovirus 1A and 18 on another line of human fibroblasts (Macnaughton, 1982; Medrano and Green, 1973). As mentioned earlier, human rhinoviruses can be divided into two receptor groups; the majority share one receptor, but a minority attach to a different receptor on a human cell line (Abraham and Colonno, 1984; Colonno et al.,1986). The binding of rhinoviruses to the receptor is enhanced by the presence of a high concentration of $MgCl_2$ and is temperature-dependent; the rate of attachment increases with increase in temperature from 7 to 37°C (Butterworth et al., 1976; Macnaughton, 1982). It has been suggested that more than one receptor is required for proper virus attachment and that the increase in temperature assists in the movement of the receptor subunits in the membrane lipids to achieve proper orientation for virus attachment (Butterworth et al., 1976). Following binding to the host cell receptors, rhinoviruses undergo conformational changes to what are termed A particles. In this process they lose one of the viral capsid proteins, namely VP4, and thus although they still contain RNA, they have lost their infectivity to a great extent. This "eclipsed" stage of the virus multiplicative cycle is particularly temperature-dependent, more so than the attachment of the virus to the host cell, and may represent the first stage of virus uncoating. Then follows the formation of B particles (empty capsids), which have lost their viral RNA. It has been estimated for HRV-2 that the process of viral penetration takes some 40 to 60 min. Following uncoating of rhinoviruses, the infecting virion RNA (plus strand) is first translated to produce an RNA-dependent RNA polymerase(s), which is required to replicate the genomic RNA (plus strand) to produce complementary virus RNA (minus strands) and later for the minus strands to replicate to produce progeny plus strands, which will ultimately become incorporated into progeny virus particles. Thus as in the replication of other picornavirus RNA, both the parent plus strand and later the minus RNA strands act as templates for the synthesis of their respective complementary strands (Butterworth et al., 1976; Macnaughton, 1982).

Rhinovirus proteins, like those of other picornaviruses, arise by proteolytic cleavage of a polyprotein. Thus the entire genome is first translated, translation being initiated at a point at or near the 5' end of the genome. The polyprotein is usually cleaved during translation. Initially, the polyprotein is cleaved into three species (VP0, VP1, and VP3), and later VP0 undergoes further cleavage to form VP2 and VP4; this last cleavage step is thought to occur only after the RNA is secured inside the immature capsid (Macnaughton, 1982). Details of the probable cleavage sites in relation to the nucleotide sequence of the genome have recently been published (Stanway et al., 1984). It is thought that cleavage allows some new contacts to occur between the various polypeptides in order that the particle become locked together (Baltimore, 1985). Maturation of virus has been shown by electron microscopy to occur in the cytoplasm of the infected cells some 16 to 18 h after infection following which particles may be seen in the form of hexagonal or rectangular crystals (Butterworth et al., 1976; Macnaughton, 1982). High concentrations of magnesium ions are thought to enhance the crystallization of rhinovirus particles (Blough et al., 1969).

Antigenic Composition and Genetics

As mentioned above, over 115 different serotypes of human rhinoviruses have so far been identified by neutralization (Melnick, 1980). However, recently, the antigenic sites on the human rhinovirus 14 capsid have been mapped using mutants of HRV-14 selected for resistance to individual neutralizing antibodies (Sherry and Rueckert, 1985; Sherry et al., 1986). These studies have now been integrated with X-ray crystallography studies (Rossmann et al., 1985). This indicates that there are four major immunogenic neutralization sites (Rossmann et al., 1985; Sherry et al., 1986). The three-dimensional structure data show that these are localized into four distinct areas on the surface of the viral capsid (Nim-IA, Nim-IB, Nim-II, and Nim-III) and that the residues all face outward toward the viral exterior (Rossmann et al., 1985). Nim-IA and IB are located on VP1, and Nim-II and III are located on VP2 and VP3, respectively (Rossmann et al., 1985). It is interesting to note that despite the 60-fold equivalence of each potential binding site on the virus, it is suggested that as few as four neutralizing antibody molecules would be required to confer "neutralization" of the virus particle. This binding of antibody with virus often results in conformational changes in the virus structure (Rossmann et al., 1985).

Recently published sequence data on the HRV-14 and 2 genomes (Callahan et al., 1985; Skern et al., 1985; Stanway et al., 1984) indicate that the human rhinovirus genome is a single strand of RNA of positive polarity comprising some 7,200 nucleotides. The first 624 nucleotides from the 5' end of the genome do not appear to code for any amino acid. This noncoding region is followed by a long open reading frame of some 6,537 nucleotides (6,450 for HRV-2; Skern et al., 1985), which comprises some 90% of the genome and terminates in another noncoding region, this time a 3' end of 47 nucleotides. The viral RNA is covalently linked to a small virus-encoded protein, VPG, at the 5' end and is polyadenylated at the 3' end. Details of the genomic structure of rhinoviruses and its relation to the published poliovirus genomes have been well reviewed elsewhere (Callahan et al.,1985; Skern, 1985; Stanway et al., 1984).

Effect of Physical and Chemical Agents

Heat

Rhinoviruses are inactivated on heating with apparent first-order kinetics. They are relatively stable at temperatures ranging from 24 to 37°C. Calculation of the activation energy and tests for the presence of infectious RNA indicate that the lower temperatures inactivate these viruses by damaging the viral RNA rather than the viral capsid proteins (Dimmock, 1967; Gauntt and Griffith, 1974; Jackson and Muldoon, 1973; Killington et al., 1977). Exposure to temperatures such as 50 or 56°C inactivates viruses rapidly by damage to peptides. When rhinoviruses like polioviruses are inactivated, C-type (empty shell) virus particles are produced, and these have a wider serological reactivities than the native or D-type particles. The presence of 1 M $MgCl_2$ appears to stabilize some but not all rhinoviruses when exposed to temperatures of 50°C for 1 h. In contrast, enteroviruses are completely stabilized when exposed to heat in the presence of 1 M $MgCl_2$ (Dimmock and Tyrrell, 1962, 1964; Ketler et al., 1962).

Ultraviolet Irradiation

The half-life of rhinoviruses when exposed to standard UV irradiation ranged from 3 to 5 sec, and no infectious particles could be detected after 90 sec of exposure (Hamparian, 1979).

pH

Unlike other picornaviruses, rhinoviruses are rapidly inactivated at pH values of below 6, particularly at below pH 3, although there are significant differences in the sensitivity of different rhinoviruses to exposure to pH values between 5.0 and 6.0 (Butterworth et al., 1976; Hughes et al., 1973). This property of inactivation at low pH remains the simplest criterion to distinguish the rhinoviruses from the enteroviruses.

Lipid Solvents and Detergents

Rhinoviruses, like other picornaviruses, are resistant to treatment with lipid solvents such as ether, ethanol, fluorocarbon, or chloroform (Jackson and Muldoon, 1973; Stott and Killington, 1972). Furthermore, detergents such nonidet P40, sodium dodecyl sulfate (SDS), or sodium deoxycholate do not appear to have any effect on the virus, although there are some data that suggest that SDS can inactivate a human bovine rhinovirus (Ide and Darbyshire, 1972).

PROTEOLYTIC ENZYMES

Trypsin and other proteolytic enzymes can affect the infectivity of some rhinoviruses while others do not appear to be affected (Ide and Darbyshire, 1972; Kisch et al., 1964).

VIRUCIDAL SUBSTANCES

Recently a number of substances have been shown to have virucidal activity and hence inactivate rhinoviruses. These include 2% glutaric acid (Hayden et al., 1984) and a combination of citric acid, malic acid, and sodium lauryl sulfate (Dick et al., 1986; Hayden et al., 1985b). The application of these substances in the interruption of rhinovirus transmission will be reviewed in more detail later, when control of rhinovirus infection is discussed.

OTHER AGENTS

Rhinoviruses have been shown to be inactivated by a number of other chemical agents such as hydrogen peroxide (Mentel and Schmidt, 1974; Mentel et al., 1977), iodine (Carter et al., 1980), and 2 M urea, which produces empty or C-type particles (Lonberg-Holm and Yin, 1973).

Pathogenesis, Pathophysiology, and Pathology

The primary site of rhinovirus infection is thought to be the nasal epithelial cells. Viruses attach to specific receptors in the nasal epithelial cells and begin to initiate infection. Within 24 h, virus in small quantities can be recovered from nasal washings of volunteers challenged with a rhinovirus. The virus concentration then rises to reach a peak between the second and third day and then gradually declines to undetectable levels by the fourth to fifth day after infection. Figure 2 shows the typical course of virus shedding in a volunteer infected with HRV-9.

The incubation period of a rhinovirus infection is approximately 2 to 3 days, and peak virus excretion often corresponds with the acute phase of rhinitis. During the acute phase of illness, most of the nasal epithelial cells become infected, as shown by biopsy studies (Douglas, 1970). Virus is thought to spread directly from cell to cell until most of the nasal epithelium becomes infected. The mechanism of rhinitis is still unknown but is thought to be due to direct

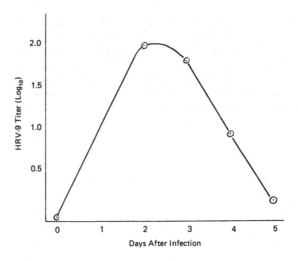

FIG. 2. Typical course of virus excretion by a volunteer infected with human rhinovirus 9 (HRV-9).

cytocidal effect of virus multiplication on the nasal epithelial cells. During the acute phase of rhinitis, there is marked edema with infiltration of neutrophils, lymphocytes, plasma cells, and eosinophils. The nasal cavities become narrowed with excessive secretory activity.

Our understanding of the pathophysiology following acute rhinovirus infection is still very limited, but recent work, including some of our own volunteer studies, indicates that a significantly prolonged nasal mucociliary clearances, reduced ciliary beat frequency, and loss of large areas of ciliated epithelium occur in volunteers infected with rhinovirus and who are symptomatic (Carson et al., 1985; Pedersen et al., 1983; Sakakura et al., 1973; Wilson et al., 1987). Scanning electron microscopy of bovine tracheal organ cultures infected with a bovine rhinovirus also showed extrusion of large numbers of ciliated cells from the epithelium (Reed and Boyde, 1972).

Although rhinovirus infections are usually confined to the upper respiratory tract, particularly the nasal mucosa, there are reports of lower respiratory tract infection, particularly the association of rhinovirus infections with the exacerbation of chronic bronchitis and asthma in children (Cherry et al.,1967; Craighead et al., 1969; Gregg, 1983; Horn et al., 1979; Minor et al., 1974, 1976; Wulff et al., 1969). However, the definite role of rhinoviruses in lower respiratory tract infection is still unclear, and further direct evidence of virus multiplication in the lower respiratory tract is required, since recovery of virus from sputum does not in itself indicate unequivocally that virus actually multiplies in the lower respiratory tract, but on occasions more virus has been found in sputum than in nasal specimens.

Immunology

The study of the local and serum immune response to rhinovirus infection and protection against reinfection has so far been limited mainly to the measurement of neutralizing antibodies. As yet, the role of the local and circulating cell-mediated immunological responses in limiting a rhinovirus infection and/or providing resistance to virus challenge has not been clearly defined. However, a very recent study showed that circulating lymphocytes, particularly the T4$^+$ population, are important in the progression of rhinovirus infection and severity of symptoms of the common cold. Thus, following challenge, significant decreases in the lymphocyte population were observed among infected persons, and this was associated with increase in the severity of symptoms. In addition, persons who had the greatest decrease in total lymphocyte count also shed virus more frequently (Levandowski et al., 1986). It was suggested that selective removal of T lymphocytes from the peripheral blood to the nasopharyngeal site of rhinovirus infection together with the release of neutrophils from the marginated pool or hematopoietic tissues may have taken place in response to the inflammatory process. Alternatively, rhinovirus infection may have inhibited the proliferation or increased the destruction of lymphocytes (Levandowski et al., 1986). However, further studies are required to elucidate the role of the local and circulating cell-mediated immune responses in limiting and/or preventing a rhinovirus infection.

Despite two decades of study, the role of serum and local antibodies in influencing the course of a rhinovirus infection and protecting against reinfection is still not very clear (for review see Gwaltney, 1982). However, it appears that resistance to challenge with a rhinovirus is generally associated with the presence of high levels of secretory antibody, most probably of the IgA class and/or of serum-neutralizing antibodies (for review see Dick and Chesney, 1981). Nevertheless, volunteers with high prechallenge serum-neutralizing antibodies of up to 1:256 have been infected when a relatively high challenge dose of a rhinovirus has been given (Cate et al., 1966). In addition, we have found that a high proportion of our volunteers who are challenged with smaller doses of virus, even though they have serum-neutralizing antibody titers of >1:8, actually do get infected whereas others with no prechallenge serum-neutralizing antibodies to the infecting virus may resist infection. Thus, the presence of serum-neutralizing antibodies does not always correlate with protection against reinfection.

In order that we can obtain a better understanding of the local and serum immunoglobulin response following infection, we have recently developed an en-zyme-linked-immunosorbent assay (ELISA) and applied this in the study of the immune response. The data we have so far obtained, although preliminary, indicate that resistance to infection is associated with the presence of rhinovirus-specific IgG and IgA in the serum and specific IgA in nasopharyngeal secretions (Barclay and Al-Nakib, 1987; Barclay et al., unpublished data). However, further detailed studies are required to elucidate in depth the role of the immune response in the protection against reinfection by both the homologous and heterologous rhinoviruses, and these are now in progress in our laboratory.

Clinical Features

Mode of Transmission

The peak excretion of rhinoviruses corresponds with the acute phase of the illness which is within 2 or 3 days after infection. Thereafter, virus excretion rapidly declines, although in some volunteer experiments it has been found that it may persist for as long as 2 weeks after infection but at very low levels (Douglas et al., 1966). The virus has been found to be present in particularly high concentrations in the nasal mucus secretions rather than in pharyngeal secretions or saliva (Gwaltney, 1982; Hendley et al., 1973). In studies of natural infections, a high rate of transmission was more likely to occur after long exposure, whereas after short exposures (15 min to 3 days), the transmission rates were low or variable (D'Alessio et al., 1984; Fox et al., 1975; Gwaltney, 1982). Furthermore, close contact (e.g., among married couples) and crowding appear to increase transmission of rhinoviruses (D'Alessio et al., 1976; Holmes et al., 1976). Nevertheless, low transmission rates have been observed in settings such as nursery schools (Beem, 1969), offices (Gwaltney et al., 1968), and even families (Dick and Chesney, 1981).

Several workers have failed to obtain evidence for airborne transmission (Couch et al., 1970; Gwaltney et al., 1978), although a recent report suggested that rhinoviruses may indeed also be transmitted by aerosol (Dick et al., 1986). Rather, it was mainly due to contact of contaminated fingers with the conjunctiva and the nares (Couch et al., 1970; Gwaltney et al., 1978). In recent volunteer experiments, it was found that the saliva, lips, and even the external nares of infected volunteers contained very small concentrations of virus, and it was therefore concluded that only a modest rate of transmission could be achieved by kissing (D'Alessio et al., 1984). It thus appears that the infected mucus secretions contaminating hands, fingers, and fomites and then the direct inoculation of virus through the conjunctiva and/or nasal epithelium are the most important route of transmission. There is some evidence that only small concentration of virus would be required to initiate an infection in an individual provided that the virus successfully reaches the appropriate portal of entry (Gwaltney, 1982; Tyrrell et al., 1962).

Clinical Course

The common cold is characterized by rhinorrhea, nasal obstruction, some element of pharyngitis, and cough. Fever and systemic reaction are infrequent. The length of the illness is normally 7 days, with a peak of symptoms and signs between the second and third days. The acute phase of rhinitis usually corresponds with peak virus excretion. In most individuals, symptoms subside by the eighth day, although in some patients symptoms may persist for 2 weeks or even as long as 1 month (Douglas, 1970; Gwaltney, 1982). As mentioned earlier, there is some evidence that rhinoviruses may be involved in lower-respiratory-tract complications and may precipitate asthmatic attacks and exacerbate chronic bronchitis, especially among children (Gregg, 1983; Hilleman et al., 1963; Horn and Gregg, 1973; Horn et al., 1979; Minor et al., 1974, 1976; Stott et al., 1968).

Diagnosis

Specimen Collection

The most appropriate specimen for the isolation of rhinoviruses is that of nasopharyngeal washings. It is our experience and that of others that both nasal and pharyngeal swabs are less satisfactory than nasopharyngeal washings. If, however, nasopharyngeal washings cannot be obtained for any reason, it is recommended that nasal swabs be taken rather than pharyngeal swabs. However, it is important to emphasize that specimens must be collected as soon as possible after patients present with symptoms, since the peak of virus excretion is reached during the first 2 to 3 days after onset of symptoms and then declines considerably. Like other virological specimens, those for rhinovirus isolations must be transported to the laboratory as soon as possible, preferably on ice and using a virus transport medium (see below). At the laboratory, specimens should be inoculated directly into the appropriate cell culture system. However, should this not be possible, they can be stored at −70°C in transport medium.

Nasal washing procedure. The patient is asked to sit with his or her head tilted slightly backward while approximately 1 ml of sterile phosphate-buffered saline (PBS) is instilled into one of the nostrils. The patient is asked to lean forward, and the washing then drips from the nostril into a sterile petri dish, aided if necessary by a gentle nose blow. The same procedure is then repeated in the other nostril until each nostril has been washed with approximately 5 ml of PBS. The washings are collected into a sterile container bottle, and an equal volume of sterile nutrient broth (5 ml) is added. Penicillin (1,000 U/ml), streptomycin (1,000 μg/ml), and amphotericin (10 μg/ml) should also be normally included in the transport medium to prevent bacterial and fungal growth. We also normally include phenol red (in a final concentration of 0.001%) in the PBS nasal wash to monitor any acidity which may affect the rate of rhinovirus isolation.

Virus Isolation and Identification

Rhinoviruses in clinical specimens are best isolated using a sensitive cell culture system. A number of cell cultures are susceptible to human rhinovirus multiplication. These include human embryonic kidney cells (HEK); human embryonic lung fibroblasts (HEL); especially known virus-sensitive strains such as WI-26, WI-38, and MRC-5; and some continuous cell lines such as Ohio HeLa cells. Although many laboratories use fibroblast cells for the isolation of rhinoviruses, we routinely use Ohio HeLa cell lines for this purpose. Our initial evaluation of this cell line suggested that it is sensitive to a large number of human rhinovirus serotypes, provides a good replicative medium for rhinoviruses with good yield, produces consistent results, and is easy to handle for routine purposes. Following initial inoculation of Ohio HeLa cells with rhinoviruses (for details see below), cytopathic effects (CPE) begin to form as a single "focus" or sometimes multiple "foci" in the monolayer, some as early as 24 to 48 h but generally by the 4th day of incubation. In the majority of cases, this CPE begins to spread in the infected monolayer over the next 2 to 3 days, producing more characteristic foci until 50 to 75% of the cell layer is affected, and generally by the 8th day after inoculation all the cells are rounded and many are detached. The amount of cell damage also depends on the rhinovirus serotype and its concentration in the original clinical specimen; sometimes only one focus or two foci of infection are seen, and these do not seem to spread to other areas of the cell layer. We normally do not go beyond 8 days of observation, since cells (including uninfected cells) generally begin to age, round up, and become difficult to distinguish from a genuine CPE. Furthermore, if there is only one focus or few foci of CPE, these can easily be overlooked or in some instances overgrown by uninfected cells. The CPE produced by rhinoviruses is indistinguishable from that produced by other picornaviruses and is generally characterized by rounding of cells.

The CPE produced by rhinoviruses in fibroblasts is often much clearer to read than that seen in HeLa cells, since cell rounding in fibroblasts is more easily recognized. For the same reason, CPE is often detected earlier in fibroblasts than in HeLa cells (i.e., within 24 to 48 h) (for review of rhinovirus isolation in HEK or HEL cells, see Hamparian, 1979).

Recently Geist and Hayden (1985) compared the sensitivity of two strains of fibroblasts—a strain of MRC-5 human embryonic lung fibroblast cells, and the Cooney strain of human fetal tonsil cells—for the isolation of rhinoviruses from clinical specimens. Ninety-seven percent of specimens were positive when isolated in the Cooney fetal tonsil cells compared to only 74% when isolation was attempted in MRC-5 cells. Furthermore, rhinovirus replication in the fetal tonsil cells reduced the mean time to cytopathic effect development. It was therefore suggested that fetal tonsil cells are superior to MRC-5 for the isolation of rhinoviruses (Geist and Hayden, 1985).

ISOLATION OF RHINOVIRUSES IN HEK AND HEL CELLS

Cells are seeded at a concentration of 2.5×10^5 cells/ml using tubes (in a final volume of 1 ml/tube), three tubes per specimen in growth medium (10% calf serum plus 0.5% lactalbumin hydrolysate in Hanks balanced salt solution (BSS) containing 0.03% sodium bicarbonate, penicillin (100 U/ml), streptomycin (100 μg/ml), and mycostatin (20 U/ml); alternatively, minimal essential medium (MEM) in BSS is used with 10% fetal bovine serum, 0.035% sodium bicarbonate, and glutamine, with appropriate antibiotics of choice. Tubes are shaken and incubated at 37°C in a stationary position. When the cell layer in each tube has reached confluence, usually within 48 h, the growth medium is removed, and the clinical specimen (0.2 ml) is inoculated into each of three tubes. Maintenance medium, which is similar in composition to the growth medium except that the concentration of calf serum or fetal bovine serum is reduced to 2 to 5%, is added to a final volume of 1 ml per tube (i.e., 0.8 ml). Tubes are incubated at 33°C on a roller drum (with the pH maintained at about 7) for 24 h in the first instance when cells are examined under the microscope for cell cytotoxicity and reincubated and then examined for the appearance of rhinovirus-specific CPE on the 4th day and given a final reading on the 8th day after inoculation. Tubes can, of course, be examined at more frequent intervals if desired. With fibroblasts it is generally possible to maintain the cultures for periods of 2 weeks or more. However, once the cytopathic effect is obvi-

ous or has ceased to progress, it is advisable to harvest the tubes in order to prevent the loss of virus by inactivation, which can take place if incubation is continued too long. Specimens that show little or no CPE may be passaged again. This will enable the virus when present in low concentration or ill adapted to grow in cell culture to replicate further; this may reduce the number of "false" negative specimens. Furthermore, further passage(s) are needed to give an ill-adapted virus opportunity to improve its yield so that it can be further characterized and identified.

ISOLATION OF RHINOVIRUSES IN OHIO HELA CELLS

Cells are seeded at a concentration of 2×10^5 cells/ml using tubes (in a final volume of 1 ml/tube), three tubes per specimen in growth medium (10% newborn calf serum in Eagle's basal medium, with 0.088% sodium bicarbonate, penicillin (100 units/ml), streptomycin (100 μg/ml), and gentamycin (50 μg/ml), pH maintained around 7). Tubes are shaken and incubated at 37°C in a stationary position. When the cell layer in each tube has reached confluence, usually within 48 h, the growth medium is removed, and the clinical specimen (0.2 ml) is inoculated into each of three tubes. Maintenance medium (2% of heat-inactivated fetal calf serum in Eagle's basal medium, 0.088% sodium bicarbonate, gentamycin (50 μg/ml), tryptose phosphate broth (0.13%), magnesium chloride (30 mM), amphotericin B (2.5 μg/ml), and neomycin (1 μg/ml) is added to a final volume of 1 ml per tube (i.e., 0.8 ml). Tubes are incubated at 33°C on a roller drum (with the pH maintained at about 7) for 24 h in the first instance at which time the cells are examined under the microscope for cell cytotoxicity and reincubated and examined for the appearance of specific rhinovirus CPE by the 4th day and given a final reading on the 8th day after inoculation. We do not normally read tubes further than the 8th day, since, as was mentioned earlier, the general state of the HeLa cells tends to deteriorate with time, and general rounding of cells begins to be observed. As with fibroblast cultures, specimens showing little or no CPE are repassaged in fresh Ohio HeLa cells in order to exclude false-negative specimens and to improve the growth characteristics if necessary. This is also important in order that the isolated virus can be further characterized and identified.

ISOLATION OF RHINOVIRUSES IN ORGAN CULTURES

Human embryonic nasal epithelium (HENE) or tracheal (HET) cultures are often useful in the isolation and study of rhinoviruses, especially those that are still not well adapted for growth in tissue cultures. Furthermore, they are also useful in the evaluation of antirhinovirus compounds that have already been found to be active in tissue cultures but are to be evaluated further prior to human volunteer experiments, especially since both HENE and HET provide a system very close to that of the human nasal epithelium in vivo.

The procedure for preparation of organ culture and virus propagation in this system as used in our laboratory is as follows. Embryonic nasal epithelium and trachea are dissected under sterile conditions. Small pieces ($\sim$ 1 cm^2) of nasal epithelium with the underlying cartilage of septum or turbinate are cut very carefully, and each piece is placed in a tube in 1 ml maintenance medium (MEM), supplemented with 0.2% bovine plasma albumin (BPA), 100 U/ml penicillin, and 100 μg/ml streptomycin. Similarly, small rings of the trachea are carefully cut, and one or two rings of HET are placed in a tube with 1 ml MEM as described earlier for HENE. Tubes are incubated at 33°C in a roller drum and examined under the microscope daily for ciliary activity. After 24 to 48 h, the cilia in the tracheal cultures should be beating regularly, and the medium in both types of organ cultures should become acidic, indicating that the cultures are viable. The medium is then removed, and virus (0.1 ml) is added with fresh MEM (0.9 ml) to give a final volume of 1 ml per tube. Tubes are incubated at 33°C in a roller drum and examined daily for ciliary activity and change of pH. Organ cultures are normally kept for 5 to 7 days, and medium is harvested every 2 or 3 days. Harvests are pooled and either passaged further in organ cultures and/or in tissue cultures, or the virus is identified as described later. If rhinovirus is growing well, ciliary activity in the tracheal cultures ceases after 5 to 7 days.

IDENTIFICATION OF RHINOVIRUSES

It is important that virus isolates from patients with signs and symptoms of the common cold be identified as rhinovirus. There are a number of stages in the identification of a rhinovirus. These are as follows:

1. Recognition of a characteristic CPE. Rhinoviruses produce a CPE in HEL and HEK fibroblasts and in HeLa cells that is characterized by rounding of cells very similar to that seen following the multiplication of other picornaviruses such as polio-, coxsackie-, and echoviruses. However, this can usually be readily distinguished from the rounding produced by viruses such as herpes simplex and adenoviruses.
2. Demonstration of resistance to lipid solvents. Rhinoviruses, like other picornaviruses, are not sensitive to lipid solvents such as chloroform or ether. Mix 1 ml of isolate with 0.1 ml of chloroform, and shake vigorously for 10 min then centri-

fuge at 1,000 g for 5 min. Collect the aqueous phase, and titrate in an appropriate cell culture the chloroform-treated specimen in parallel with one that has not been treated, and look for loss of infectivity. Other viruses that are not sensitive to chloroform treatment, such as a picornavirus, and viruses that are sensitive to treatment, such as a herpes simplex, can be included as controls. Rhinovirus and picornavirus titers should not be affected by chloroform treatment; the titer should not be reduced by more than 1 log$_{10}$ dilution. Herpes simplex virus, on the other hand, is enveloped and hence is totally inactivated and normally fails to grow at the highest concentration tested.

3. Test for temperature sensitivity. Rhinoviruses, unlike other picornaviruses, grow poorly at 37°C or higher temperatures and grow best at 33°C. Therefore, they are temperature-sensitive relative to the other picornaviruses. Dilutions of virus can be inoculated into tube cultures and incubated at 33 or 37°C and examined daily. CPE will occur earlier and to higher titer at 33°C than at 37°C.
4. Test for pH stability. Rhinoviruses, unlike other picornaviruses such as polio-, coxsackie-, and echoviruses, are sensitive to low-pH treatment, such as pH 3. Expose 0.2 ml of isolate to 1.8 ml of sterile HEPES buffer at pH 3 and another one at pH 7 for 1 h at 4°C. Then adjust the pH of the samples to 7 with N/10 NaoH, and titrate both samples in Ohio HeLa cells. Include a known rhinovirus as a positive control and an enterovirus such as coxsackie A21 as a negative control. If the isolate is a rhinovirus, it should be inactivated by treatment at low pH; hence the titer should be reduced by at least 2 log$_{10}$ dilutions. Similarly, the control rhinovirus should also not grow under these conditions, whereas the control picornavirus should not be affected by acid treatment.

The above procedures, if conducted, should be sufficient to provide the investigator with evidence that the specimen contains a rhinovirus.

However, specialized laboratories with access to specific antisera to the various rhinovirus serotypes may embark on a more extensive investigation—that of identification by serological typing of the virus by neutralization. However, this may involve the repassage of the isolate a number of times to improve virus titer followed by an extensive cross-neutralization assay to identify the virus using intersecting or combinatorial pools of guinea pig, rabbit, or goat hyperimmune sera. However, a simple procedure for typing, as used in our laboratory, is as follows.

Using tissue culture microtiter plate, mix in duplicate 50 μl of undiluted, 1 : 10 dilution and 1 : 100 dilution of virus with an equal volume (50 μl) of an appropriate dilution of an antirhinovirus serum (depending on the neutralizing antibody titer of that serum; for an unknown isolate, some 10 to 20 intersecting or combinatorial sera pools may be required for the investigation). Shake the plates, and incubate at room temperature for 1 h. Virus and serum dilution are prepared in maintenance medium. Add 100 μl of Ohio HeLa cell suspension at 3.0×10^5 cells/ml to each well of a microtiter plate. Include a control titration of the unknown virus in maintenance medium, no antiserum, antiserum controls with no virus, and cell controls with no virus or antiserum only MEM. Incubate in a box containing 5% CO_2 at 33°C for 4 days. Examine the wells that contained virus and antisera under the microscope for inhibition of viral CPE, and compare with control wells. A specimen is typed if it had at least 10 $TCID_{50}$ of virus and was neutralized when preincubated with 20 units of an antiserum known to be free of heterologous cross-reactivity. The isolate should show CPE when preincubated with the other neutralizing antisera or when incubated with MEM only. The typing should be confirmed by a repeated test with 10 to 20 units of monospecific serum.

It is possible that, in the near future, rhinovirus isolates will be identified more rapidly by either ELISA or cDNA probes as described in the later sections, especially if these become more widely available.

Direct Antigen Detection

Because of the great number of rhinovirus serotypes, it has been difficult to develop serological assays for the direct detection of rhinovirus antigens. It would require a large number of monovalent sera to be able to detect different viruses. Furthermore, early attempts to detect rhinoviruses in nasopharyngeal smears by the indirect immunofluorescence assay provided only poor correlation with virus isolation (Dreizin et al., 1971, 1975). However, studies on cross-immunization of rabbits with different serotypes have demonstrated some broader antigenic relationships (Cooney et al., 1975). Furthermore, Cooney et al. (1982) prepared potent antisera in rabbits to the 90 classified rhinoviruses and were able to show significant numbers of cross-relationships among rhinoviruses and thus were able to group rhinoviruses on the basis of heterotypic antibody in both rabbit and guinea pig antirhinovirus sera into 16 groups. Indeed, we, too, have recently found using an ELISA system that there is considerable serological cross-reactivity between rhinoviruses (Dearden and Al-Nakib, 1987). This together with the fact that different rhinoviruses may share a C-type cross-reactivity following treatment with 2 M urea or low pH, it

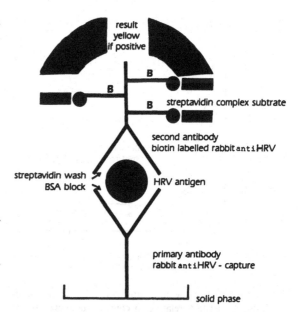

Fig. 3. Principle of the B-A ELISA system for rhinovirus antigen detection (B = biotin; BSA = bovine serum albumin; ONPG = orthonitrophenyl-β-galactoside; HRV = human rhinovirus).

is conceivable that it would be possible to produce sera with much broader seroreactivity than those now available, or alternatively, a selected panel of antisera known to react with a number of serotypes could be pooled or combined and then be utilized in a serological assay such as ELISA that will detect rhinovirus antigens directly. Such a procedure is likely, if shown to have a reasonable level of sensitivity, to be rapid, simple, and widely available. Furthermore, it will allow a specific diagnosis of rhinoviruses to be achieved within a few hours. This will be a prerequisite in the future should antirhinovirus chemotherapy become available and is indicated in persons who are compromised and in whom a rhinovirus infection may present with complications. Indeed, the majority of synthetic antirhinovirus compounds have very specific activity not only against rhinoviruses but, in the majority of cases, against specific serotypes. This therefore further emphasizes the importance of rapid diagnosis of rhinoviruses.

We have recently developed a capture biotin-avidin ELISA (B-A ELISA) system for rhinovirus antigen detection using a biotin-labeled serological probe to detect the captured virus and a streptavidin-β-galactosidase preformed complex to detect the biotin-labeled antibody probe reacted with the captured virus. The principle of this system is illustrated diagramatically in Fig. 3. However, an interesting feature of this B-A ELISA system is that the captur-

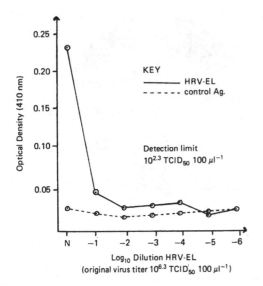

FIG. 4. B-A ELISA for rhinovirus detection. A titration of HRV-EL by ELISA relative to control uninfected tissue culture fluids (cutoff value taken as 1.5 × mean optical density readings obtained with control uninfected tissue culture fluids).

ing antibody and the biotin-labeled detecting antibody are the same. Therefore, this obviates any cross-reactivity between reagents in the system, a problem often seen in most ELISA systems. Furthermore, our second layer the streptavidin-β-galactosidase complex (Amersham) react via biotin-streptavidin bridges. The binding between biotin and streptavidin is extremely strong (binding constant 10^{15}). In addition, although endogenous β-galactosidases exist in human tissue, these have different characteristics and are not active under the conditions employed for assaying β-galactosidase from *E. coli* used in the ELISA system. Hence, the cumulative effect of these factors is that the B-A ELISA for rhinovirus has an extremely low background. This low background or "noise" therefore allows the detection of small quantities of viral antigen or "signal." Figure 4 shows a typical rhinovirus antigen titration by this B-A ELISA and that concentrations as low as $10^{2.3}$TCID$_{50}$/ml of rhinovirus can be detected. Furthermore, we found this system to provide extremely reproducible results.

Initially we used human rhinovirus EL, HRV-EL (an untyped virus), as a model. We then developed another B-A ELISA for HRV-2 on the same principle. Interestingly, we found that when we investigated a total of 57 human rhinoviruses in each system, the HRV-EL antiserum detected 34 of 57 (59.6%) rhinoviruses and the HRV-2 system detected 32 of 57 (56%). These two serological probes together detected 47 of 57 (82.5%) of the rhinoviruses

investigated, whereas control respiratory viruses such as influenza A and B, coronavirus 229E, and parainfluenza viruses did not react at all in either system. These preliminary data therefore suggest that an ELISA system to detect a large number of rhinoviruses is quite possible and may provide a rapid (24 to 48 h) and sensitive assay for rhinovirus antigen detection. Clinical evaluation is under way at the moment in our laboratory. Details of these ELISAs and data obtained with them will be published in due course (Dearden and Al-Nakib, 1987).

Direct Viral Nucleic Acid Detection

As an alternative or complementary procedure to the rapid detection of rhinovirus antigens, we have recently developed a cDNA:RNA hybridization system to detect viral RNA. Details of this system are in Al-Nakib et al. (1986).

Basically an M13 template comprising the first 800 nucleotides from the 5' end noncoding region of human rhinovirus 14 has been constructed (Al-Nakib et al., 1987a), and this was used to generate a cDNA probe (Fig. 5). This ^{32}P-labeled probe was then used to detect rhinovirus RNA. Viral RNA was extracted using a phenol/chloroform mixture following sodium dodecyl sulfate and proteinase K treatment. Some interesting data have emerged that may have important implications in the diagnosis of rhinoviruses. For example, the probe from the 5' end of the HRV-14 genome cross-hybridized with 54 of 56 (96.4%) of the human rhinoviruses investigated (Al-Nakib et al., 1986). However, there was a considerable variation in the strength of the hybridization signal obtained depending presumably on the amount of homology between these viruses in the region of the genome represented by the probe (Fig. 6). In reconstruction experiments we have been able to detect a $10^{2.8}$TCID$_{50}$/ml of virus in nasal washings (Al-Nakib et al., 1986). The sensitivity of the system might be further enhanced by including probes from more than one region of the genome. For the detection of many serotypes of rhinoviruses we could include a mixture of probes prepared from a careful selection of viruses and/or decrease the stringency of hybridization—for example, reduce the concentration of formamide in the hybridization mixture from 40 to 20%. The present test can be completed within 48 h if probes of high specific activity are used, and therefore it is more rapid than virus isolation. We are now investigating this system in depth, evaluating a number of rhinovirus probes both singly and in combination and also the condition most suitable for the rapid extraction of viral RNA. We hope to evaluate this procedure clinically using nasal washing material from volunteers who were excreting virus. However,

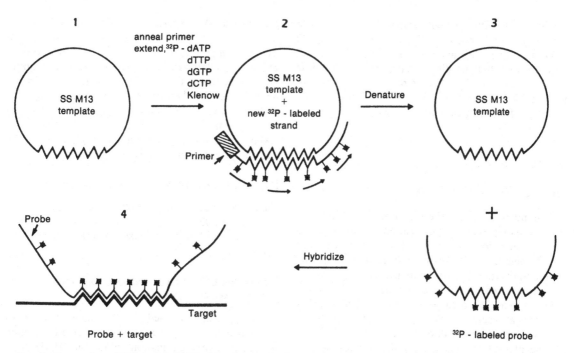

FIG. 5. Generation of ^{32}P-labeled cDNA rhinovirus probe from an M13 template using primer extension method (from Al-Nakib et al., 1987a, with permission).

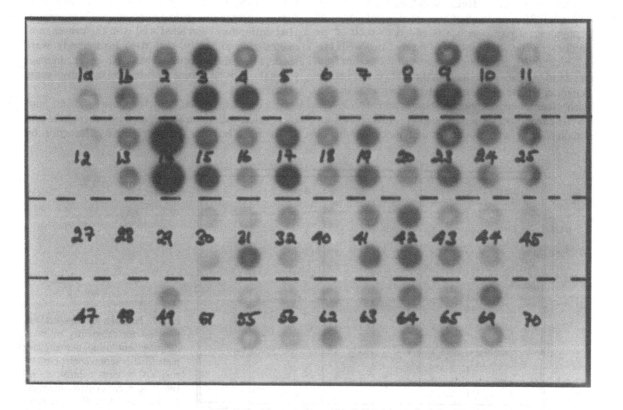

FIG. 6. Detection of human rhinoviruses by cDNA : RNA hybridization using ^{32}P-labeled probes from the 5' end of HRV-14 genome (from Al-Nakib et al., 1986, with permission).

we feel that if this system is likely to have a wide application, these probes will need to be labeled with nonradioactive reagents such as biotin. Nevertheless, we feel that the direct detection of rhinovirus RNA is now possible, that it represents a new approach to the rapid diagnosis of rhinoviruses, and that with further modification it will have a much wider application.

Antibody Assays

NEUTRALIZATION ASSAYS

The standard assay for the measurement of antibodies to rhinoviruses is neutralization. A variety of virus neutralization assays have been described, such as tube neutralization, plaque reduction test, and microneutralization. We routinely use microneutralization procedures, since these provide a simple, practicable system suited to routine measurements. The system is also sensitive and reproducible. The procedure we employ is as follows.

The plan of the microneutralization test is presented in Fig. 7. To each well of the microtiter plate 50 μl of HeLa maintenance medium with 5% fetal calf serum is added except for the cell control wells to which 100 μl of HeLa maintenance medium is added. Fifty microliters of heat-inactivated acute and convalescent phase serum is added to each of the first two wells, and another to the last two wells (serum controls). Doubling dilutions of the serum from 1 : 2 to 1 : 128 are then made in the plate using a multichannel pipette. Fifty microliters of virus at the appropriate required concentration is added to all the wells except the serum and cell controls. Plates are shaken using a microshaker and incubated in the presence of 5% CO_2 (using a plastic box) for 1 h at room temperature. One hundred microliters of freshly stripped Ohio HeLa cells at a concentration of 3×10^5 cells/ml in 5% fetal calf serum and maintenance medium is added to each of the wells in the plate. Plates are then incubated at 33°C for approximately 5 days in 5% CO_2 and examined microscopically for inhibition of CPE. The titer of the serum is taken as the highest dilution that completely inhibited the production of CPE by the virus.

RAPID MICRONEUTRALIZATION ASSAY

This is basically similar to that described above except that a 200-fold higher concentration of virus is added to each well. The result of this assay is normally obtained by 50 h.

HEMAGGLUTINATION INHIBITION AND COMPLEMENT FIXATION ASSAYS

Other procedures used in the measurement of rhinovirus antibody such as complement fixation (CF) and hemagglutination inhibition (HAI) tests have been described. In the limited number of studies conducted on the relationships between HAI antibody and infection, a good correlation was obtained between the presence of these antibodies in the serum and protection against reinfection. Furthermore, HAI antibodies correlated well with the presence of neutralizing antibody, and rises in antibody were specific for the infecting virus. However, hemagglutination is a property shared by only some rhinoviruses.

In contrast, CF antibody responses showed poor correlation with neutralizing antibody responses and had a much broader C-type reactivity. Infection by

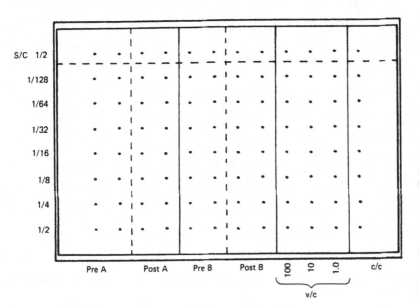

FIG. 7. Plan of a microneutralization test for rhinovirus antibody measurement using a microtiter plate (S/C = serum control, no virus; V/C = virus control, no serum; C/C = cell control, no serum or virus; Pre A or B = preinfection serum from patient A or B; Post A or B = postinfection serum from patient A or B).

one virus elicited a CF antibody response that cross-reacted with other rhinoviruses, and hence it was clear that the CF test will not be practical for serological diagnosis of rhinovirus (for review see Hamparian, 1979).

ELISA

We have recently developed a capture ELISA system to measure class-specific immunoglobulins both in the serum and in nasopharyngeal secretions. Figure 8 shows the principle of the ELISA system for rhinovirus antibody measurement. Preliminary results suggest that the amount of virus-specific IgA in the serum correlates more accurately with neutralizing antibody than does IgG. Thus, sera from infected volunteers who showed rises in neutralizing antibody also showed rises in serum rhinovirus-specific IgA. However, further experience with this new assay is required, and in due time a full and detailed clinical evaluation of this new ELISA will be conducted. However, these preliminary results are encouraging, especially since such an ELISA system is simple and rapid. Preliminary results also indicate that it is extremely sensitive—for example, detecting immunoglobulin responses by assay of unconcentrated nasal washing. Details of these preliminary findings have recently been published (Barclay and Al-Nakib, 1987).

Interpretation of Laboratory Data

As is the case with most other virological diagnosis, the isolation of a rhinovirus from a person with symptoms and signs of the common cold clearly indicates a positive diagnosis of a rhinovirus infection. However, it is essential that the virus be confirmed as a rhinovirus as described in the previous section including, if possible, typing of isolates. Seroconversion or fourfold or greater rises in neutralizing antibody in paired sera also indicates a recent infection with a rhinovirus. However, it would not be possible to detect a rise in neutralizing antibody in a patient unless a virus had been isolated from the patient or a close associate such as a member of the family. Blind screening with even a proportion of all the known serotypes is just wasteful and impractical. Because of this, an ELISA system with a broader rhinovirus serological reactivity may be able to overcome this problem and detect rises in antibody (i.e., probably of the IgA class) to most serotypes. Like virus isolation, the direct detection of viral antigens or viral RNA (when this becomes possible in the future) in nasopharyngeal secretions from patients with symptoms and signs of the common cold will also indicate a recent infection with a rhinovirus.

The presence of high levels of an antibody in the

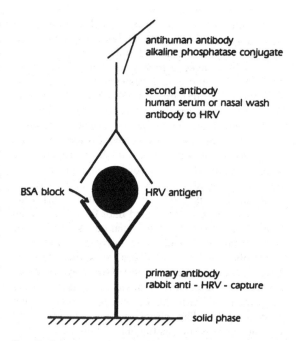

FIG. 8. Principle of the ELISA system for rhinovirus antibody detection (BSA = bovine serum albumin; HRV = human rhinovirus).

serum specific to a particular rhinovirus serotype (e.g., by neutralization) generally indicates protection against illness and possibly to reinfection by that serotype. However, data obtained from a number of studies indicate that although the majority of rhinovirus infections are symptomatic, the ratio of apparent to inapparent infections is about 3:1 (Dick and Chesney, 1981; Gwaltney, 1982), and in volunteers at this unit the ratio is almost 1:1.

Epidemiology and Natural History

Since over 115 different serotypes of rhinovirus have so far been identified and infection with one serotype does not confer immunity to another, rhinovirus infections are very common. It is estimated from studies in the United States that on average the incidence of rhinovirus infection is about 0.5 per person per year (Gwaltney, 1982). True rates of infection are probably much higher, since current procedures for the detection of rhinovirus infection are not optimal for sensitivity and are not widely available, so a true estimate of the rate of infection cannot be obtained. However, it has been shown that 2 to 5 episodes of common colds in a person per year is not unusual (Couch, 1984). Infection is most common in early life, and incidence declines with increase in age probably because of the higher frequency of antibody among adults who have experienced infection by a larger number of rhinoviruses. Indeed, studies on the distribution of neutralizing antibodies among different age groups in the United States indicate that sera from children aged 2 to 4 years neutralized about 10% of the rhinovirus investigated (a total of 56 rhinoviruses, rhinovirus 1A-55) whereas sera from adults 30 to 40 years of age neutralized just over half of the rhinoviruses investigated (Dick and Chesney, 1981; Gwaltney, 1982). However, it is expected that measures of antibody prevalence by more sensitive assays such as ELISA will reveal a much higher proportion of individuals with antibody to rhinoviruses.

Rhinovirus infection is distributed worldwide (Taylor-Robinson, 1965), and data from the United States suggest two main seasonal peaks—one major peak during late summer and early autumn (August to October), and a second peak in spring (April or May) (Dick and Chesney, 1981; Gwaltney, 1982). In the tropics, the peak of rhinovirus infection appears to correspond with the rainy season (Monto and Johnson, 1968). The school, nurseries, families, and confined groups such as the military provide the major sites for the spread of rhinoviruses in any community. It has long been accepted that a number of serotypes can circulate simultaneously within a community. However, whether new serotypes emerge frequently owing to mutation, recombination, or simple antigenic drift is still not clear; the answer to that question would require specific studies using some of the newly developed technologies such as T1-oligonucleotide mapping or partial genomic sequencing.

Control and Prevention

It has been clear from the early days of rhinovirus research that control of rhinovirus infection by vaccination is unlikely to prove successful in view of the great number of rhinoviruses and the fact that naturally acquired infection with one or several rhinoviruses does not appear to confer immunity and resistance to reinfection by other serotypes. Nevertheless, early attempts to induce local and systemic immunity using inactivated monovalent or decavalent vaccines (Andrewes et al., 1965; Douglas and Couch, 1972; Hamory et al, 1975), or live-attenuated vaccines (Draper et al., 1969) have not been satisfactory. Therefore, attention has since focused on prevention or control by chemotherapy. Although many antirhinovirus compounds have been synthesized and although those recently studied show very high specific activity in vitro (minimal inhibitory concentrations in the region of 0.003 μg/ml) (Al-Nakib and Tyrrell, 1987), none of the earlier compounds have been successful when evaluated clinically. Thus, Enviroxime [2-amino-1-(isopropyl sulfonyl)-6-benzimidazolephenylketoneoxime], when given intra-

nasally and orally, reduced both virus shedding and clinical illness by 33% relative to placebo (Phillpotts et al., 1981); however, when given intranasally, it had no effect (Hayden and Gwaltney, 1982; Levandowski et al., 1982). Other compounds such as 4',6-dichloroflavan, when given orally (Phillpotts et al., 1983a) or intranasally (Al-Nakib et al., 1987b), gave a reduction in illness of only 28% and 15%, respectively. Furthermore, two other antirhinovirus compounds—a chalcone, Ro-09-0415 given orally (Phillpotts et al., 1984a), and a new synthetic compound, 2-[(1,5,10,10-a-tetrahydro-^{3}H-thiazolo[3,4b]isoquinolin-3 ylindene)-amino]-4-thiazoleacetic acid (S) designated 44-081 R.P. (Zerial et al., 1985) given intranasally—gave a reduction in illness of 17% and 0%, respectively. Therefore, to date none of the antirhinovirus compounds evaluated in vivo have had any significant effect. The reason for this is still unclear despite the fact that most regimes were designed to deliver sufficient antirhinovirus activity.

In contrast, various prophylactic trials with different interferons have shown considerable reduction in illness relative to placebo. For example, a partially purified human leukocyte interferon, [IFN-α (le)], when given prophylactically (total dose of 14MU) reduced illness by 80% (Merigan et al., 1973), whereas purified IFN-α (le) and interferon alpha-2 (rIFN-alfa-2b) (total dose of 90 MU) gave a 100% reduction in illness relative to placebo (Scott et al., 1982a,b). Other interferons have also been successful in preventing experimental common colds. Thus, recent studies by Samo et al., (1983) showed that a recombinant leukocyte A interferon, rIFN-αA (rIFN-alfa-2a), when given at a total dose of 40 MU, reduced illness by 89% relative to placebo. Interestingly Hayden and Gwaltney (1983) showed that these differences in the percentage rate of reduction of rate of illness depended not so much on the total interferon dose given as on the schedule of medication. Thus, rIFN-alfa-2b, when given at a dosage of 4 × 11.4 MU per day for 4 days (a total dose of 118.2 MU), gave 100% reduction in illness, whereas the same interferon, when given as a single dose of 42.8 MU for 5 days (a total of 212 MU), resulted in only 75% reduction in illness. The significance of dosage schedule and timing have also been well documented by Phillpotts et al., (1983b).

Lymphoblastoid interferon (IFN-alfa-N1) (Phillpotts et al., 1984b) and a recombinant interferon beta and (rIFN-β) (Higgins et al., 1986) also produced a significant reduction in the rate of rhinovirus illness 100% and 66%, respectively. Prophylaxis with most interferons also reduced the rate of virus shedding, and with some, the frequency of seroconversion or antibody rises. Field studies in the United States and Australia indicate that alpha-2 interferon (IFN-alfa-2b) could significantly prevent rhinovirus infection in the family setting when 5 MU was applied daily for 7 days by contacts who started medication within 48 h of the onset of a cold in a family member (Douglas et al., 1986; Hayden et al., 1986). These promising data on the effects of interferons in vivo are therefore in contrast with data obtained with synthetic antirhinovirus compounds. Nevertheless, the use of interferon as a prophylactic agent is still very limited, since prolonged administration results in local inflammation, manifested by blood-stained secretions and nasal mucosal abnormalities (Hayden et al., 1985a). As yet, none of the antirhinovirus agents so far available have had any significant therapeutic effect.

An interesting new concept is the interruption of rhinovirus transmission from an infected person excreting virus to a susceptible contact with virucidal agents. It has been postulated that since rhinovirus is commonly recovered from the hands of persons with symptomatic colds and virus transmission may occur by nasal and/or conjunctival inoculation, transmission may be interrupted by using virucidal agents in the form of lotions or in handkerchiefs. Recent studies by Hayden et al., (1984) indicated that hand lotion containing 2% glutaric acid was significantly more effective than placebo in inactivating rhinovirus serotype 2, although glutaric acid was not effective against all rhinovirus serotypes. They later showed that paper tissues treated with a combination of citric acid, malic acid, and sodium lauryl sulfate significantly reduced contamination of the hands of the user and reduced experimental rhinovirus infection rates when compared to controls who used no tissues. Surprisingly, however, these workers found that placebo nasal tissues were also effective in blocking viral transmission and raised the question of whether virucidal treatment of tissues is actually required for interrupting virus transmission (Hayden et al., 1985b,c). More recently, Dick et al. (1986) described the successful use of paper handkerchiefs impregnated with 9.1% citric acid, 4.5% malic acid, and 1.8% sodium lauryl sulfate in interrupting rhinovirus transmission among participants of a poker game some of whom had rhinovirus colds and were shedding virus. In these experiments, none of the recipients of the virucidal tissues caught colds, whereas 42 to 75% of the recipients of cotton handkerchiefs without virucide did. Field studies are in progress at the moment, and it remains to be seen whether this approach to the interruption of colds would also prove effective in a natural setting.

Although the above review does suggest some encouraging trends regarding prevention of rhinovirus infection with interferons and also possibly the interruption of rhinovirus spread from infected individuals to close susceptible contacts, we still do not have the ideal prophylactic agent and certainly no treatment for the common cold. Control by vaccination may yet be possible, although a useful vaccine must still be a long way off. Recent advances in our understanding of the viral genome and details of the virus structure including the mapping of the major antigenic sites will be combined in the near future with a better understanding of the immune response to rhinovirus infection. It is conceivable that this will lead to progress in developing a vaccine perhaps based on synthetic peptides mimicking selected antigenic sites.

Addendum

While this chapter was going to press, some significant advances were reported.

There was international agreement to define and number additional serotypes, bringing the total to 100 (Hamparian et al., 1987). The major receptor for rhinoviruses was isolated in a high state of purity (Tomassini and Colonno, 1986). There was further progress in defining the relative importance of the airborne and contact routes of infections in an experimental setting in which finger and fomite transmissions were prevented mechanically, transmission continued at the same rate (Dick et al., 1987), while the use of virucidal tissues in a family setting had no significant effect on rates of spread of colds (Farr et al., 1988).

Synthetic oligonucleotide probes have now been made, which hybridize well with virtually all rhinoviruses tested and are being evaluated for clinical diagnosis (Bruce et al., 1988). Evaluation in a group of volunteers shows that the ELISA test is highly sensitive and specific in detecting rhinovirus infection in such volunteers (Al-Nakib et al., unpublished data).

Literature Cited

Abraham, G., and R. J. Colonno. 1984. Many rhinovirus serotypes share the same cellular receptor. J. Virol. **51:**340–345.

Al-Nakib, W., G. Stanway, M. Forsyth, P. J. Hughes, J. W. Almond, and D. A. J. Tyrrell. 1987a. Rhinoviruses detection by cDNA : RNA hybridization, p. 487–495. *In* M. A. Brinton and R. R. Rueckert (ed.), Positive strand RNA viruses. UCLA Symposia on molecular and cellular biology. New series, Volume 54, Alan R. Liss, New York.

Al-Nakib, W., G. Stanway, M. Forsyth, P. J. Hughes, J. W. Almond, and D. A. J. Tyrrell. 1986. Detection of human rhinoviruses and their molecular relationship using cDNA probes. J. Med. Virol. **20:**289–296.

Al-Nakib, W., J. Willman, P. G. Higgins, D. A. J. Tyrrell, W. M. Shepherd, and D. S. Freestone. 1987b. Failure of intranasally administered 4'-6-dichloroflavan to protect against rhinovirus infection in man. Arch. Virol. **92:**255–260.

Al-Nakib, W., and D. A. J. Tyrrell. 1987. A "new" generation of more potent synthetic antirhinovirus compounds: comparison of their MICs and their synergistic interactions. Antiviral Res. 8:179–188.

Andrewes, C. H., D. M. Chaproneiri, A. E. H. Gompels, H. G. Pereira, and A. T. Roden. 1953. Propagation of common cold virus in tissue cultures. Lancet 1:546–547.

Andrewes, C., D. A. J. Tyrrell, P. B. Stones, A. J. Beale, R. D. Andrews, D. G. Edwards, A. P. Goffe, J. E. Doggett, R. F. Homer, R. S. Crespi, and E. M. B. Clements. 1965. Prevention of colds by vaccination against a rhinovirus: a report by the Scientific Committee on Common Cold vaccines. Br. Med. J. 1:1344–1349.

Baltimore, D. 1985. Picornaviruses are no longer black boxes. Science 229:1366–1367.

Barclay, W. S., and Al-Nakib, W. 1987. An ELISA for the detection of rhinovirus specific antibody in serum and nasal secretion. J. Virol Meth. 15:53–64.

Beem, M. O. 1969. Acute respiratory illness in nursery school children: a longitudinal study of the occurrence of illness and respiratory viruses. Am. J. Epidemiol. 90:30–44.

Blough, H. A., J. M. Tiffany, G. Gordon, and M. Fiala. 1969. The effect of magnesium on the intracellular crystallization of rhinovirus. Virology 38:694–698.

Brown, F., J. F. E. Newman, and E. J. Stott. 1970. Molecular weight of rhinovirus ribonucleic acid. J. Gen. Virol. 8:145–148.

Bruce, C. B., W. Al-Nakib, D. A. J. Tyrrell, and J. W. Almond. 1988. Synthetic oligonucleotides as diagnostic probes for rhinoviruses. Lancet 2:53.

Butterworth, R. E., R. R. Grunert, B. D. Korant, K. Longberg-Holm, and F. H. Yin. 1976. Replication of rhinoviruses. Arch. Virol. 51:169–189.

Callahan, P. L., S. Mizutani, and R. J. Colonno. 1985. Molecular cloning and complete sequence determination of RNA genome of human rhinovirus type 14. Proc. Natl. Acad. Sci. USA 82:732–736.

Carter, C. H., J. O. Hendley, L. A. Mika, and J. M. Gwaltney. 1980. Rhinovirus inactivation by aqueous iodine in vitro and on skin. Proc. Soc. Exp. Biol. Med. 165:380–383.

Carson, J. L., A. M. Collier, and S. S. Hu. 1985. Acquired ciliary defects in nasal epithelium of children with acute viral upper respiratory infections. N. Engl. J. Med. 312:463–468.

Cate, T. R., R. D. Rossen, R. G. Douglas, W. T. Butler, and R. B. Couch. 1966. The role of nasal secretion and serum antibody in the rhinovirus common cold. Am. J. Epidemiol. 84:352–363.

Cherry, J. D., J. A. Diddams, and E. C. Dick. 1967. Rhinovirus infections in hospitalized children: provocative bacterial interrelationships. Arch. Environ. Health 14:390–396.

Colonno, R. J., P. L. Callahan, and W. J. Long. 1986. Isolation of a monoclonal antibody that blocks attachment of the major group of human rhinoviruses. J. Virol. 57:7–12.

Cooney, M. K., J. A. Wise, G. E. Kenny, and J. P. Fox. 1975. Broad antigenic relationships among rhinovirus serotypes revealed by cross-immunization of rabbits with different serotypes. J. Immunol. 114:635–639.

Cooney, M. K., J. P. Fox, and G. E. Kenny. 1982. Antigenic groupings of 90 rhinovirus serotypes. Infect. Immun. 37:642–647.

Cooper, P. D., V. I. Agol, H. L. Bachrach, F. Brown, Y. Ghendon, A. J. Gibbs, J. H. Gillespie, K. Longberg-Holm, B. Mandel, J. L. Melnick, S. B. Mohanty, R. C. Povey, R. R. Rueckert, F. L. Schaffer, and D. A. J. Tyrrell. 1978. Picornaviridae: second report. Intervirology 10:165–180.

Couch, R. B. 1984. The common cold: control? J. Infect. Dis. 150:167–173.

Couch, R. B., R. G. Douglas, Jr., K. M. Lindgren, P. J. Gerone, and V. Knight. 1970. Airborne transmission of respiratory infection with coxsackievirus A type 21. Am. J. Epidemiol. 91:78–86.

Craighead, J. E., J. Meier, and M. H. Cooley. 1969. Pulmonary infection due to rhinovirus type 13. N. Engl. J. Med. 281:1403–1404.

D'Alessio, D. J., J. A. Peterson, C. R. Dick, and E. C. Dick. 1976. Transmission of experimental rhinovirus colds in volunteer married couples. J. Infect. Dis. 133:28–36.

D'Alessio, D. J., C. K. Meschievitz, J. A. Peterson, C. R. Dick, and E. C. Dick. 1984. Short-duration exposure and the transmission of rhinoviral colds. J. Infect. Dis. 150:189–194.

Dearden, C. J., and W. Al-Nakib. 1987. Direct detection of rhinoviruses by an enzyme-linked immunosorbent assay. J. Med. Virol. 23:179–189.

Dick, E. C., and P. J. Chesney. 1981. Rhinoviruses. p. 1167–1186. In R. D. Feigin and J. D. Cherry (ed.). Textbook of paediatric infectious diseases, Vol. II. W. B. Saunders, Philadelphia.

Dick, E. C., L. C. Jennings, K. A. Mink, et al. 1987. Aerosol transmission of rhinovirus colds. J. Infect. Dis. 156:442–448.

Dick, E. C., K. A. Mink, T. A. Demke, and S. L. Inhorn. 1986. Aerosol transmission of rhinovirus infection. In Proceedings of the 86th Annual Meeting of the American Society for Microbiology, March, 1986.

Dimmock, N. J., and D. A. J. Tyrrell. 1962. Physiochemical properties of some viruses isolated from common colds (rhinoviruses). Lancet 2:536–537.

Dimmock, N. J., and D. A. J. Tyrrell. 1964. Some physiochemical properties of rhinoviruses. Br. J. Exp. Pathol. 45:271–280.

Dimmock, N. J. 1967. Differences between the thermal inactivation of picornaviruses at "high" and "low" temperatures. Virology 31:338–353.

Dochez, A. R., G. S. Shibley, and K. C. Mills. 1930. Studies in the common cold. IV. Experimental transmission of the common cold to anthropoid apes and human beings by means of a filtrable agent. J. Exp. Med. 52:701–716.

Douglas, R. G., T. R. Cate, P. J. Gerone, and R. B. Couch. 1966. Quantative rhinovirus shedding patterns in volunteers. Am. Rev. Respir. Dis. 94:159–167.

Douglas, R. G., Jr. 1970. Pathogenesis of rhinovirus common colds in human volunteers. Ann. Otol. Rhinol. Laryngol. 79:563–571.

Douglas, R. G., Jr., and R. B. Couch. 1972. Parenteral inactivated rhinovirus vaccine: minimal protective effect. Proc. Soc. Exp. Biol. Med. 139:899–902.

Douglas, R. M., B. W. Moore, H. B. Miles, L. M. Davis, N. M. H. Graham, P. Ryan, D. A. Warswick, and J. K. Albrecht. 1986. Prophylactic efficacy of intranasal alpha-2 interferon against rhinovirus infections in the family setting. N. Engl. J. Med. 314:65–70.

Draper, C. C., E. J. Stott, and M. Whitaker. 1969. Oral administration of rhinoviruses grown at different temperatures in human diploid cells. Arch. Virusforsch. 28:93–96.

Dreizin, R. S., E. M. Vikhnovich, N. M. Borovkova, and T. I. Ponomareva. 1971. The use of indirect fluorescent antibody technique in studies on reproduction of rhinoviruses and for the detection of rhinoviral antigen in materials from patients with acute respiratory diseases and conjectivitides. Acta Virol. 15:520.

Dreizin, R. S., N. M. Borovkova, T. I. Ponomareva, E. M. Vikhnovich, A. A. Kheinitis, and T. V. Leichinskaya.

1975. Diagnosis of rhinovirus-infections by virological and immunofluorescent methods. Acta Virol. **19:**413–418.

Farr, B. M., J. O. Hendley, D. L. Kaiser, et al. 1988. Two randomized controlled trials of virucidal nasal tissues in the prevention of natural upper respiratory tract infection. Am. J. Epid. (*In press*).

Fox, J. P., M. K. Cooney, and C. E. Hall. 1975. The Seattle virus watch. V. Epidemiologic observations of rhinovirus infections. 1965–1969, in families with young children. Am. J. Epidemiol. **101:**122–143.

Fox, J. P., M. K. Cooney, C. E. Hall, and H. M. Foy. 1985. Rhinoviruses in Seattle families. Am. J. Epidemiol. **122:**830–846.

Gauntt, C. J. 1980. Fragility of the rhinovirus type 14 genome to incubation at 60°. Intervirology **13:**7–15.

Gauntt, C. J., and M. M. Griffith. 1974. Fragmentation of RNA in virus particles of rhinovirus type 14. J. Virol. **13:**762–764.

Geist, F. C., and F. G. Hayden. 1985. Comparative susceptibilities of strain MRC-5 human embryonic lung fibroblast cells and the Cooney strain of human foetal tonsil cells for isolation of rhinoviruses from clinical specimens. J. Clin. Microb. **22:**455–456.

Gregg, I. 1983. Provocation of airflow limitation by viral infections: implication for treatment. Eur. J. Respir. Dis. Suppl. **128:**369–379.

Gwaltney, J. M. 1982. Rhinoviruses. p. 491–517. *In* A. S. Evans (ed.), Viral infections of humans: epidemiology and control, 2nd ed., Plenum, New York.

Gwaltney, J. M., Jr., J. O. Hendley, G. Simon, and W. S. Jordan, Jr. 1968. Rhinovirus infections in an industrial population. III. Number and prevalence of serotypes. Am. J. Epidemiol. **87:**158–166.

Gwaltney, J. M., Jr., P. B. Moskalski, and J. O. Hendley. 1978. Hand-to-hand transmission of rhinovirus colds. Ann. Intern. Med. **88:**463–467.

Hamory, B. H., V. V. Hamparian, R. M. Conant, and J. M. Gwaltney, Jr. 1975. Human responses to two decavalent rhinovirus vaccines. J. Infect. Dis. **132:**623–629.

Hamparian, V. V. 1979. Rhinoviruses. p. 535–575. *In* E. H. Lennette and N. J. Schmidt (ed.), Diagnostic procedures for viral, rickettsial and chlamydial infections, 5th ed. American Public Health Association. New York.

Hamparian, V. V., R. J. Colonno, M. K. Cooney, E. C. Dick, et al. 1987. A collaborative report: rhinoviruses—extension of the numbering system from 89 to 100. Virology **159:**191–192.

Hayden, G. F., and J. M. Gwaltney, Jr. 1982. Prophylactic activity of intranasal enviroxime against experimentally induced rhinovirus type 39 infection. Antimicrob. Agents Chemother. **21:**892–897.

Hayden, G. F., and J. M. Gwaltney, Jr. 1983. Intranasal interferon α2 for prevention of rhinovirus infection and illness. J. Infect. Dis. **148:**543–550.

Hayden, G. F., D. Deforest, J. O. Hendley, and J. M. Gwaltney, Jr. 1984. Inactivation of rhinovirus on human fingers by virucidal activity of glutaric acid. Antimicrob. Agents Chemother. **26:**928–929.

Hayden, G. F., J. M. Gwaltney, Jr., and M. E. Johns. 1985a. Prophylactic efficacy and tolerance of low-dose intranasal interferon-alpha2 in natural respiratory viral infections. Antiviral Res. **5:**111–116.

Hayden, G. F., J. M. Gwaltney, Jr., D. F. Thacker, and J. O. Hendley. 1985b. Rhinovirus inactivation by nasal tissues treated with virucide. Antiviral. Res. **5:**103–109.

Hayden, G. F., J. O. Hendley, and J. M. Gwaltney, Jr. 1985c. The effect of placebo and virucidal paper handkerchiefs on viral contamination of the hand and transmission of experimental rhinoviral infection. J. Infect. Dis. **152:**403–407.

Hayden, G. F., J. K. Albrecht, D. L. Kaiser, and J. M. Gwaltney, Jr. 1986. Prevention of natural colds by contact prophylaxis with intranasal alpha-2 interferon. N. Engl. J. Med. **314:**71–75.

Hendley, J. O., R. P. Wenzel, and J. M. Gwaltney, Jr. 1973. Transmission of rhinovirus colds by self-inoculation. N. Engl. J. Med. **288:**1361–1364.

Higgins, P. G., W. Al-Nakib, J. Willman, and D. A. J. Tyrrell. 1986. Interferon β_{ser} as prophylaxis against experimental rhinovirus infection in volunteers. J. Interf. Res. **6:**153–159.

Holmes, M. J., S. E. Reed, E. J. Stott, and D. A. J. Tyrrell. 1976. Studies of experimental rhinovirus type 2 infections in polar isolations and in England. J. Hyg. (Lond.) **76:**379–393.

Horn, M. E. C., and I. Gregg. 1973. Role of viral infection and host factors in acute episodes of asthma and chronic bronchitis. Chest **64:**44–48.

Horn, M. E. C., E. Brain, I. Gregg, J. M. Inglis, S. J. Yealland, and P. Taylor. 1979. Respiratory viral infection and wheezy bronchitis in childhood. Thorax **34:**23–28.

Hughes, J. H., D. C. Thomas, and V. V. Hamparian. 1973. Acid lability of rhinovirus type 4: effect of pH, time and temperature. Proc. Soc. Exp. Biol. Med. **144:**555–560.

Ide, P. R., and J. H. Darbyshire. 1972. Studies on a rhinovirus of bovine origin. I. Growth in vitro of strain RS 3X. Arch. Ges. Virusforsch. **36:**166–176.

Jackson, G. G., and R. L. Muldoon. 1973. Viruses causing common respiratory infections in man. J. Infect. Dis. **127:**328–355.

Kapikian, A. Z., R. M. Conant, V. V. Hamparian, R. M. Chanock, E. C. Dick, J. M. Gwaltney, D. Hamre, W. S. Jordan, G. E. Kenny, E. H. Lennette, J. L. Melnick, W. J. Mogabgab, C. A. Phillips, J. H. Schieble, E. J. Stott, and D. A. J. Tyrrell. 1971. Collaborative report: rhinoviruses—extension of numbering system. Virology **43:**524–526.

Ketler, A., V. V. Hamparian, and M. R. Hilleman. 1962. Characterization and classification of ECHO 28-rhinovirus-coryzavirus agents. Proc. Soc. Exp. Biol. Med. **110:**821–831.

Killington, R. A., E. J. Stott, and D. Lee. 1977. The effect of temperature on the synthesis of rhinovirus type 2 RNA. J. Gen. Virol. **36:**403–411.

Kisch, A. L., P. A. Webb, and K. M. Johnson. 1964. Further properties of five newly recognised picornaviruses (rhinoviruses). Am. J. Hyg. **79:**125–133.

Levandowski, R. A., C. T. Pachucki, M. Rubenis, and G. G. Jackson. 1982. Topical enviroxime against rhinovirus infection. Antimicrob. Agents Chemother. **22:**1004–1007.

Levandowski, R. A., D. W. Ou, and G. G. Jackson. 1986. Acute-phase decrease of T-lymphocyte subsets in rhinovirus infection. J. Infect. Dis. **153:**743–748.

Longberg-Holm, K., and B. D. Korant. 1972. Early interaction of rhinoviruses with host cells. J. Virol. **9:**29–40.

Longberg-Holm, K., and F. H. Yin. 1973. Antigenic determinants of infective and inactivated human rhinovirus type 2. J. Virol. **12:**114–123.

Macnaughton, M. R. 1982. The structure and replication of rhinoviruses. Curr. Top. Microbiol. Immunol. **97:**1–26.

Medappa, K. C., C. McLean, and R. R. Rueckert. 1971. On the structure of rhinovirus 1A. Virology **44:**259–270.

Medrano, L., and H. Green. 1973. Picornavirus receptors and picornavirus multiplication in human-mouse hybrid cell lines. Virology **54:**515–524.

Melnick, J. L. 1980. Taxonomy of viruses. Prog. Med. Virol. **26:**214–232.

Mentel, R., R. Schirrmacher, A. Kewitsch, R. S. Drezin, and J. Schmidt. 1977. Inactivation of viruses with hydrogen peroxide. Vopr. Virusol. 6:731–733.

Mentel, R., and J. Schmidt. 1974. Experiments in chemical inactivation of rhinoviruses and coronaviruses. Z. Gesamte. Hyg. 20:530–533.

Merigan, T. C., R. E. Reed, T. S. Hall, and D. A. J. Tyrrell. 1973. Inhibition of respiratory virus infection by locally applied interferon. Lancet 2:563–567.

Minor, T. E., E. C. Dick, A. N. DeMeo, J. J. Ouellette, M. Cohen, and C. E. Reed. 1974. Viruses as precipitants of asthmatic attacks in children. J. Am. Med. Assoc. 227:292–298.

Minor, T. E., E. C. Dick, J. W. Baker, J. J. Ouellette, M. Cohen, and C. E. Reed. 1976. Rhinovirus and influenza type A infections as precipitants of asthma. Am. Rev. Respir. Dis. 113:149–153.

Monto, A. S., and K. M. Johnson. 1968. Respiratory infections in the American tropics. Am. J. Trop. Med. Hyg. 17:867.

Nair, C. N., and M. J. Owens. 1974. Preliminary observations pertaining to polyadenylation of rhinovirus RNA. J. Virol. 13:535–537.

Pedersen, M., Y. Sakakura, B. Wintler, S. Brofeld, and N. Mygind. 1983. Nasal mucociliary transport, number of ciliated cells, and beating pattern in naturally acquired cold. Eur. J. Respir. Dis. 64(Suppl. 128):355–364.

Pelon, W., W. J. Mogabgab, I. A. Phillips, and W. E. Pierce. 1957. A cytopathogenic agent isolated from naval recruits with mild respiratory illness. Proc. Soc. Exp. Biol. Med. 94:262–267.

Phillpotts, R. J., R. W. Jones, D. C. DeLong, S. E. Reed, J. W. Wallace, and D. A. J. Tyrrell. 1981. The activity of enviroxime against rhinovirus infection in man. Lancet 2:1342–1344.

Phillpotts, R. J., J. Wallace, D. A. J. Tyrrell, D. S. Freestone, and W. M. Shepherd. 1983a. Failure of oral 4′,6-dichloroflavan to protect against rhinovirus infection in man. Arch. Virol. 75:115–121.

Phillpotts, R. J., G. M. Scott, P. G. Higgins, J. Wallace, D. A. J. Tyrrell, and C. L. Gauci. 1983b. An effective dosage regimen for prophylaxis against rhinovirus infection by intranasal administration of HuIFN-α2. Antiviral Res. 3:121–136.

Phillpotts, R. J., P. G. Higgins, J. S. Willman, D. A. J. Tyrrell, and I. Lenox-Smith. 1984a. Evaluation of the antirhinovirus Ro 09-0415 given orally to volunteers. J. Antimicrob. Chemother. 14:403–409.

Phillpotts, R. J., P. G. Higgins, J. S. Willman, D. A. J. Tyrrell, D. S. Freestone, and W. M. Shepherd. 1984b. Intranasal lymphoblastoid interferon ("Wellferon") prophylaxis against rhinovirus and influenza virus in volunteers. J. Interf. Res. 4:535–541.

Price, W. H. 1956. The isolation of a new virus associated with respiratory clinical disease in humans. Proc. Natl. Acad. Sci. USA 42:892–896.

Reed, S. E., and A. Boyde. 1972. Organ cultures of respiratory epithelium infected with rhinovirus and parainfluenza virus studied in a scanning electron microscope. Infect. Immun. 6:68–76.

Rossmann, M. G., E. Arnold, J. W. Erickson, E. A. Frankenberger, J. P. Griffith, H. J. Hecht, J. E. Johnson, G. Kamer, M. Luo, A. G. Mosser, R. R. Rueckert, B. Sherry, and G. Vriend. 1985. Structure of a human common cold virus and functional relationship to other picornaviruses. Nature 312:145–153.

Sakakura, Y., Y. Sasaki, R. B. Hornick, Y. Togo, A. R. Schwartz, H. N. Wagner, and D. F. Proctor. 1973. Mucociliary function during experimentally induced rhinovirus infection in man. Ann. Otol. Laryngol. 82:203–211.

Samo, T. C., S. B. Greenberg, R. B. Couch, J. Quarles, P. E. Johnson, S. Hook, and M. E. Harmon. 1983. Efficacy and tolerance of intranasally applied recombinant leukocyte A interferon in normal volunteers. J. Infect. Dis. 148:535–542.

Scott, G. M., R. J. Phillpotts, J. Wallace, D. S. Secher, K. Cantell, and D. A. J. Tyrrell. 1982a. Purified interferon as protection against rhinovirus infection. Br. Med. J. 284:1822–1825.

Scott, G. M., R. J. Phillpotts, J. Wallace, C. L. Gauci, J. Greiner, and D. A. J. Tyrrell. 1982b. Prevention of rhinovirus colds by human interferon alpha-2 from Escherichia coli. Lancet 2:186–188.

Sherry, B., and R. R. Rueckert. 1985. Evidence of at least two dominant neutralization antigens on human rhinovirus 14. J. Virol. 53:137–143.

Sherry, B., A. G. Mosser, R. J. Colonno, and R. R. Rueckert. 1986. Use of monoclonal antibodies to identify four neutralization immunogens on a common cold picornavirus, human rhinovirus 14. J. Virol. 57:246–257.

Skern, T., W. Sommergruber, D. Blaas, P. Gruendler, F. Fraundorfer, C. Pieler, I. Fogy, and E. Kuechler. 1985. Human rhinovirus 2: complete nucleotide sequence and proteolytic processing signals in the capsid protein region. Nucleic Acids Res. 13:2111–2126.

Stanway, G., P. J. Hughes, R. C. Mountford, P. D. Minor, and J. W. Almond. 1984. The complete nucleotide sequence of a common cold virus: human rhinovirus 14. Nucleic Acids Res. 12:7859–7877.

Stott, E. J., and R. A. Killington. 1972. Rhinoviruses. Annu. Rev. Microbiol. 26:503–524.

Stott, E. J., N. R. Grist, and M. B. Eadie. 1968. Rhinovirus infections in chronic bronchitis: isolation of eight possibly new rhinovirus serotypes. J. Med. Microbiol. 1:109–118.

Taylor-Robinson, D. 1965. Respiratory virus antibodies in human sera from different regions of the world. Bull. WHO 32:833–847.

Tomassini, J. E., and R. J. Colonno. 1986. Isolation of a receptor protein involved in attachment of human rhinoviruses. J. Virol. 58:290–295.

Tyrrell, D. A. J. 1965. Common cold and related diseases. Edward Arnold, London.

Tyrrell, D. A. J., and R. Parsons. 1960. Some virus isolations from common colds. III. Cytopathic effects in tissue cultures. Lancet 1:239–242.

Tyrrell, D. A. J., M. L. Bynoe, F. E. Buckland, and L. Hayflick. 1962. The cultivation in human embryo cells of a virus (D.C.) causing colds in man. Lancet 2:320–322.

Tyrrell, D. A. J., I. Phillips, C. S. Goodwin, and R. Blower. 1979. Microbial disease: the use of the laboratory in diagnosis, therapy and control. Edward Arnold, London.

Wilson, R., E. Alton, A. Rutman, P. Higgins, W. Al-Nakib, D. M. Geddes, D. A. J. Tyrrell, and P. J. Cole. 1987. Upper respiratory tract viral infection and mucociliary clearance. Eur. J. Respir. Dis. 70:272–279.

Wulff, H., G. R. Noble, J. E. Maynard, E. Feltz, J. F. Poland, and T. D. Y. Chin. 1969. An outbreak of respiratory infection in children associated with rhinovirus types 16 and 29. Am. J. Epidemiol. 90:304–311.

Zerial, A., G. H. Werner, R. J. Phillpotts, J. S. Willman, P. G. Higgins, and D. A. J. Tyrrell. 1985. Studies on 44 081 R.P., a new antirhinovirus compound in cell cultures and in volunteers. Antimicrob. Agents Chemother. 27:846–850.

CHAPTER 38

Picornaviridae: Hepatitis A Virus

ARIE J. ZUCKERMAN

Disease: Hepatitis A virus.

Etiologic Agent: Human fecal material.

Clinical Manifestations: Asymptomatic and subclinical infections are common in children; jaundice increases with age and may progress rarely to fulminant hepatitis.

Pathology: Parenchymal hepatic cell necrosis and histiocytic periportal inflammation.

Laboratory Diagnosis: Hepatitis A antibody of the IgM class usually detected at the onset of illness and persists for about 10 weeks.

Epidemiology: Hepatitis A occurs endemically in all parts of the world.

Treatment: There is no specific treatment.

Prevention and Control: Control of infection is by simple hygienic measures and the sanitary disposal of excreta; passive immunization with normal human immunoglobulin can be used for prophylaxis; vaccines are under development.

Introduction

Many viruses may infect the liver of animals and humans, producing severe hepatitis. In humans, the general term *viral hepatitis* refers to infections caused by at least five different viruses: hepatitis A (infectious or epidemic hepatitis); hepatitis B (serum hepatitis); hepatitis D (the δ agent); non-A, non-B hepatitis, which appears to be caused by more than one virus; and an epidemic form of non-A hepatitis. Viral hepatitis is recognized as a major public health problem occurring endemically in all parts of the world.

Acute viral hepatitis is a generalized infection, with emphasis on inflammation of the liver. The clinical picture of the infection varies in its presentation from asymptomatic or subclinical infection, mild gastrointestinal symptoms and the anicteric form of the disease, acute illness with jaundice, and severe prolonged jaundice to acute fulminant hepatitis. Hepatitis B and coinfection with the δ agent and non-A, non-B hepatitis may be associated with a persistent carrier state; these forms of infection may progress to chronic liver disease, which may be severe. There

is now compelling evidence of an etiologic association between hepatitis B virus and hepatocellular carcinoma, one of the 10 most common malignant tumors worldwide.

Hepatitis A, hepatitis B, and hepatitis D viruses have been characterized, and these infections can now be differentiated by sensitive and specific laboratory tests for the antigens and antibodies associated with these infectious agents. Non-A, non-B hepatitis is at present the most common type of posttransfusion hepatitis in some areas and an important cause of sporadic hepatitis in adults, although virologic criteria and specific laboratory tests are not yet available.

Hepatitis A occurs endemically in all parts of the world, but the exact incidence is difficult to estimate because of the high proportion of asymptomatic and anicteric infections, differences in surveillance, and differing patterns of disease. Serologic surveys have shown that although the prevalence of hepatitis A in industrialized countries, particularly in northern Europe, North America, and Australia, is decreasing, the infection is almost universal in most countries.

The Nature of Hepatitis A Virus

Morphology and Physicochemical Properties

In 1973, small cubic virus particles (Figs. 1 to 3) measuring 27 nm in diameter were identified by immune electron microscopy in fecal extracts obtained during the early acute phase of illness from adult volunteers infected orally or by the parenteral route with the MS-1 strain of hepatitis A virus. The mature virion bands at a buoyant density of 1.32 to 1.34 g/ml in cesium chloride and with a sedimentation coefficient in sucrose of 156 to 160 s. A much smaller proportion

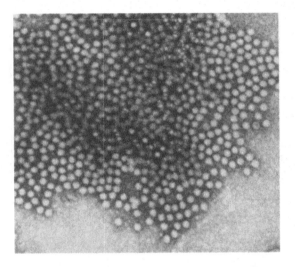

FIG. 1. Electron micrograph showing hepatitis A virus in fecal extracts during the late incubation period of the infection. Magnification ×70,000. (From a series by A. Thornton and A. J. Zuckerman.)

of infectious virions band at 1.42 to 1.46 and at about 1.27 g/ml CsCl.

The availability of viral antigen permitted the identification of specific antibody and the development of serologic tests for hepatitis A. Human hepatitis A has been transmitted experimentally to certain species of marmosets and to chimpanzees free of hepatitis A antibody. The infection in nonhuman primates is mild, often subclinical, and always anicteric.

Large numbers of virus particles are found during the incubation period in experimental infection in chimpanzees, beginning as early as 9 days after exposure, and viral shedding usually continues until peak elevation of serum aminotransferases is attained (Dienstag et al., 1975; Thornton et al., 1977). Similar observations have been made in the course of experimental and natural infections in humans. The virus is also detected during the acute phase of illness, but the number of infectious viral virus particles decreases rapidly after the onset of clinical jaundice. Prolonged excretion of virus and a persistent carrier state have not been demonstrated. It should be noted that in chimpanzees and humans, antibody to hepatitis A virus is found in the serum during the late incubation period of the infection, coinciding approximately with the onset of rising serum aminotransferase levels.

Hepatitis A virus is unenveloped, containing a linear genome of single-stranded RNA (32 to 35 s) approximately 7,500 nucleotides in length that codes for three major polypeptides with molecular weights of 21,000 to 23,000, 24,000 to 26,000, and 30,000 to 33,000. A fourth polypeptide with a molecular weight of about 14,000 has been reported, but not confirmed; sequence analysis of the viral genome indicates that this protein, if present at all, is exception-

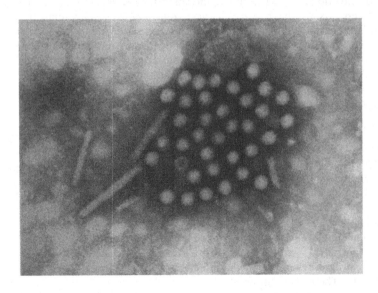

FIG. 2. Morphology of hepatitis A virus, showing both "full" particles and an "empty" particle measuring 25 to 27 nm in diameter. Magnification ×400,000. (From a series by A. Thornton and A. J. Zuckerman.)

ally small, with a molecular weight of about 3,000. No information is currently available on the arrangement of the individual polypeptides on the viral capsid. The organization of the genome of hepatitis A virus is similar to that of the picornavirus, and it has been classified as enterovirus type 72 (Melnick, 1982), although there are substantial differences between hepatitis A and poliovirus, the prototype enterovirus.

The genome of hepatitis A virus has a positive polarity. Cloning and sequencing data indicate a polyadenylic tract at the 3' terminus and a small, covalently bound protein at the 5' terminus. The available sequences do not provide evidence for the existence of a poly(C) tract in the vicinity of the 5' terminus, which is characteristic of picornaviruses.

A single open reading frame extends from nucleotides 710 to 750 at the 5' terminus to about 60 nucleotides in advance of the 3' terminal poly(A) tract. This sequence can encode a protein with a molecular weight of about 250,000. The predicted amino acid sequence compared with analogous regions of other picornaviruses suggests that the 5' region of the genome codes for the three major structural viral proteins (VP), including the presumed fourth small VP4. A polymerase is probably coded in the 3' region. Dipeptide cleavage sites present in poliovirus are not found in hepatitis A virus; however, a detailed description of the posttranslational processing of hepatitis A viral proteins is not yet available (Lemon, 1985).

The virus is ether resistant, stable at pH 3.0, and relatively resistant to inactivation by heat. Hepatitis A virus is partially inactivated by heat at 60°C for 1 h, mostly inactivated at 60°C for 10 h, and inactivated at 100°C for 5 min. The virus is inactivated by ultraviolet irradiation and by treatment with a 1 : 4,000 concentration of formaldehyde solution at 37°C for 72 h. There is also evidence that hepatitis A virus is inactivated by chlorine at a concentration of 1 mg/liter for 30 min (Abb and Deinhardt, 1986) (Table 1).

Replication

The successful propagation of hepatitis A virus in 1979 in primary monolayer and explant cell cultures and in continuous cell strains of primate origin has been a major advance and opened the way to the preparation of hepatitis A vaccines. The viral antigen is detectable by immunofluorescence, radioimmunoassay, an indirect quantitiative autoradiographic plaque assay, and complementary DNA–RNA hybridization.

The virus, which does not induce cytopathic changes, replicates in several types of cell cultures of primate origin, tends to induce persistent infections in cell cultures, and remains largely cell associated in

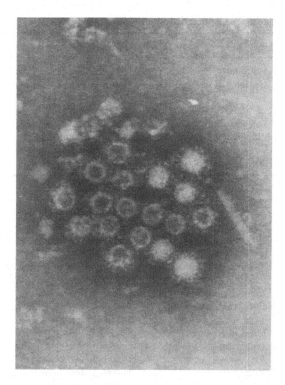

FIG. 3. Immune electron micrograph of an aggregate of hepatitis A virus particles heavily coated with antibody, giving the appearance of a "halo." Both full and empty particles are shown. Magnification ×400,000. (From a series by A. Thornton and A. J. Zuckerman.)

most infected cell cultures. However, primary isolation of wild virus is difficult, and several weeks elapse before intracellular antigen is detectable in the cytoplasm of infected cells. Isolation of the virus is not a practical diagnostic technique in routine laboratories. Adaptation occurs after passage with more rapid production of intracellular antigen and with higher final yields. Virus adapted to growth in cell culture may become attenuated (Provost et al., 1986).

The virus replicates in vivo in the liver, and there is only one report that replication may also occur in experimental infection in the mucosa of the small intestine (Karayiannis et al., 1986).

Antigenic Composition

Only one serotype of hepatitis A has been identified in experimentally infected human volunteers, in patients from different outbreaks of hepatitis A, and in naturally and experimentally infected chimpanzees. This serotype has also been confirmed by viral cross-neutralization tests (Lemon and Binn, 1983; Weitz and Siegel, 1984) and by the protective efficacy of

TABLE 1. Characteristics of hepatitis A virus[a]

General Characteristics	
Natural host	Humans
Experimental host	Marmosets (*Saguinus* spp., *Callithrix jacchus*), chimpanzees and other apes
Multiplication in vitro	Various cell cultures of human and nonhuman primate origin

Biophysical and Biochemical Characteristics	
Size	27 nm (range, 25–29 nm)
Capsid symmetry	Cubic
Virion	Unenveloped
Capsid assembly	Cytoplasm
Density in CsCl and sedimentation constants in sucrose gradients	1.34 g/cm³; 160 s (complete particles)
Nucleic acid	
Type	Single-stranded RNA
Molecular weight	1.9×10^6
Polypeptides[b]	
VP1	33,000–36,000 daltons
VP2	26,000–31,000 daltons
VP3	22,000–26,000 daltons
VP4	9,000–14,000 daltons
Stability	
Temperature	
4°C	Stable for weeks–several months of storage
−20°C to −70°C	Stable for years of storage
50°C (60 min)	Stable
60°C (60 min)	Mostly stable
60°C (10 h)	Mostly inactivated
100°C (5 min)	Inactivated
Ether (20%, 4°C, 24 h)	Stable
Acid (pH 3.0)	Stable
Formalin (1 : 4,000, 72 h, 37°C)	Inactivated (semipurified viral preparation)
Ultraviolet irradiation	Inactivated (semipurified viral preparation)
Chlorine (1 mg/liter, 30 min)	Inactivated

[a] Adapted from McCollum and Zuckerman (1981).
[b] VP = Viral protein.

pooled human immunoglobulin obtained from different geographic regions. However, strain-specific differences exist at least at the level of the nucleotide sequences of viral RNA from different strains of hepatitis A virus (Baroudy et al., 1985; Linemeyer et al., 1985; Najarian et al., 1985; Ticehurst et al., 1983).

The topographic nature of the viral surface antigen has not been fully defined, however, although recent evidence using monoclonal antibodies suggests that it is associated with the principal surface polypeptide, VP1 (Hughes et al., 1984). Lemon

(1985) considers that secondary or higher orders of protein structure may play essential roles at this antigenic site, as it has not been possible to detect this predominant antigen in viral preparations disrupted with detergent (Hughes et al., 1984).

The mechanisms underlying liver injury in hepatitis A are not understood. The initial noncytopathic phase, during which virus replicates and is released, is followed by decreased multiplication of the virus, and inflammatory cell infiltration suggests that immune mechanisms are involved in its pathogenesis.

Pathologic Changes in the Liver

Two features are constant in acute viral hepatitis: parenchymal cell necrosis and histiocytic periportal inflammation. Usually the reticulin framework of the liver is well preserved, except in some cases of massive and submassive necrosis. The pattern of histologic changes in the liver is essentially similar in hepatitis types A and B and consists of marked focal activation of sinusoidal lining cells; accumulations of lymphocytes in histiocytes within the parenchyma, often replacing hepatocytes lost by necrosis; mild diffuse hepatocytic changes with occasional coagulative necrosis in the form of acidophilic bodies; and focal regeneration and portal inflammatory reaction with alteration of bile ductules. The lesions in hepatitis A develop earlier and the duration of morphologic changes is shorter, whereas the lesions in hepatitis B linger, fluctuate, and regress slowly. There is some difference in the distribution of the lesions: in hepatitis A, the localization of parenchymal changes is predominantly periportal, whereas in hepatitis B the lesions are diffuse and tend to be accentuated around the hepatic vein tributaries, and streaks of focal necrosis may extend from portal tracts to hepatic vein tributaries. Viral antigen is localized in the cytoplasm of hepatocytes and Kupffer cells and can be readily detected by immunofluorescence. There is no evidence of persistence of the infection, and progression to chronic liver damage does not occur.

Immunology

The pathogenesis of liver damage suggests that immune mechanisms are involved, although little is known about the role of cell-mediated immune responses in hepatitis A (see also Karayiannis et al. [1986]). The humoral immune response is characteristic. Specific hepatitis A immunoglobulin (Ig)M antibody appears during the incubation period and rises rapidly in titer; it is almost always detectable by the time of onset of symptoms. IgG and IgA antibodies are also present at this stage. IgM antibody persists for up to 120 days (although the sensitivity of laboratory tests is usually adjusted to detect this class of antibody for up to 60 days). IgM antibodies possess some, but not complete, neutralizing activity. Hepatitis A IgG, a neutralizing antibody, also appears early in the course of infection and persists for many years, often for life. IgA antibody is detectable in fecal specimens collected during the convalescent phase of hepatitis A infection, but the role of this antibody in immunity is not known.

Clinical Features

Mode of Transmission

Hepatitis A virus is spread by the fecal–oral route, most commonly by person-to-person contact, and infection is particularly common in conditions of poor sanitation and overcrowding. Common source outbreaks result most frequently from fecal contamination of drinking water and food, but waterborne transmission is not a major factor in industrialized communities. On the other hand, many and an increasing number of food-borne outbreaks have been reported in developed countries. This finding can be attributed to the shedding of large amounts of virus in the feces during the incubation period of the illness in infected food handlers, and the source of the outbreak can often be traced to uncooked food or food that has been handled after cooking. The consumption of raw or inadequately cooked shellfish cultivated in polluted water is associated with a high risk of hepatitis A infection. However, although hepatitis A is common in developed countries, the infection occurs mainly in small clusters, for example, in nursery schools and in institutions for the mentally handicapped, often with only few identified cases. Hepatitis A is highly endemic in many tropical and subtropical areas, with the occasional occurrence of large epidemics. This infection is frequently acquired by travelers from areas of low endemicity to areas of high endemicity. Sexual transmission is common among homosexual men, particularly those who practice oral–anal contact.

Outbreaks of hepatitis A have also been described among handlers of newly captured nonhuman primates, particularly chimpanzees and other great apes. Hepatitis A has been reported to have been transmitted by blood and blood products on exceptionally rare occasions and very rarely by the parenteral route, although such infections have been achieved experimentally in volunteers and in susceptible nonhuman primates.

Incubation Period

The incubation period of hepatitis A is between 3 and 5 weeks, with a mean of 28 days. Subclinical and anicteric infections are common and although the disease has, in general, a low mortality, adult patients may be incapacitated for many weeks.

Age Incidence

All age groups are susceptible to hepatitis A. The highest incidence in the civilian population is observed in children of school age, but in many countries in northern Europe and North America most cases occur in adults. This shift in age incidence is similar to that found with poliomyelitis during and after the World War II, reflecting improvement in socioeconomic and hygienic conditions.

Seasonal Pattern

In temperate zones the characteristic seasonal trend is toward an increased incidence in the autumn and early winter months, falling progressively to a minimum during midsummer. More recently, however, the seasonal peak has been lost in some countries. In many tropical countries the incidence of infection tends to peak during the rainy season, with a lower incidence during dry periods.

Symptoms and Signs: Clinical Course

The clinical picture of hepatitis ranges from inapparent or subclinical infection, slight malaise, mild gastrointestinal symptoms, and the anicteric form of the disease to acute icteric illness, severe prolonged jaundice with or without cholestasis, and acute fulminant hepatitis. Hepatitis A does not lead to chronic liver disease.

Infection in children, particularly in those less than 5 years of age, is usually asymptomatic or anicteric, whereas in adults about 75% of infections are associated with jaundice. Nevertheless, the proportion of anicteric to icteric infections varies widely from outbreak to outbreak, and the ratio of patients without jaundice to those with jaundice varies from 2 or 3 to 1 to as many as 10 or more to 1.

It should be noted that in the individual case, the symptoms and signs of hepatitis A are indistinguishable from those caused by other types of viral hepatitis (Zuckerman, 1975).

Diagnosis of Hepatitis A Infection

Hepatitis A virus is excreted in the feces for 1 to 2 weeks before the onset of the disease; shedding of the virus is relatively brief and may have ceased by the time the patient seeks medical attention. The virus can be detected by immune electron microscopy

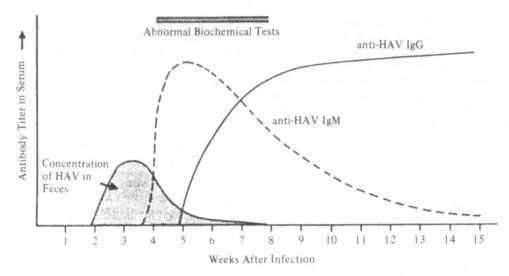

FIG. 4. Typical course of hepatitis A infection. (Reproduced with permission from McCollum and Zuckerman [1981].)

in approximately 50% of patients tested within 1 week of the onset of dark urine. Detection of virus falls to 10 to 25% in the second week and less than 10% in the third. Viral RNA and antigen may be detected by molecular hybridization. These techniques may be useful for epidemiologic studies and for detection of the source of infection and environmental pollution, including water supplies, contaminated shellfish, and food, but are not used for routine diagnosis. Isolation of the virus is not a practical diagnostic procedure; at present, however, immune electron microscopy has a role in distinguishing hepatitis A from epidemic non-A, non-B hepatitis.

Serum antibody against hepatitis A virus appears early in the course of the illness and titers rise rapidly. Confirming the diagnosis by demonstrating a fourfold or greater rise in antibody titer is often difficult, however, unless the patient is seen very early.

Serum antibody of the IgM class against hepatitis A virus is usually detected at the onset of illness and persists for about 10 weeks. This assay is the test of choice for the diagnosis of acute hepatitis A. Serum antibody of the IgG class appears more slowly, persists for many years, and indicates immunity to the disease (Fig. 4). Convenient tests include radioimmunoassay and enzyme-linked immunosorbent assay (Abb and Deinhardt, 1986; Zuckerman and Howard, 1979).

Epidemiology and Natural History

Hepatitis A occurs endemically in all parts of the world, but the exact incidence is difficult to establish because of the high proportion of asymptomatic and anicteric infections, differences in surveillance, and differing patterns of disease. The degree of underreporting, even in countries where notification is a statutory requirement, is known to be very high. Spread of the disease has been described previously under Mode of Transmission.

Serologic surveys, particularly among blood donors in industrialized countries, indicate that infection is related to age and socioeconomic status. Thus, in northern and western Europe, North America, and Australia, 5 to 20% of the population have antibody to hepatitis A by the age of 20 years and 40 to 50% by the age of 50 years. The prevalence is higher in central and southern European countries, whereas in most other countries up to 90% of the population is infected by the age of 10 years. With improvement in sanitation and socioeconomic status, the incidence of hepatitis A declines rapidly.

Control and Prevention

Control of the infection is difficult. As fecal shedding of the virus is at its highest during the late incubation period and prodromal phase of the illness, strict isolation of cases is not a useful control measure. Spread of hepatitis A is reduced by simple hygienic measures and the sanitary disposal of excreta.

Normal human immunoglobulin, containing at least 100 IU of antibody against hepatitis A virus per milliliter, given intramuscularly before exposure to the virus or early in the incubation period will prevent or attenuate clinical illness. The dosage should be at least 2 IU/kg of body weight, but in special cases such as pregnancy or in patients with liver dis-

TABLE 2. Passive immunization with normal immunoglobulin against hepatitis A virus for travelers to highly endemic areas

Weight (kg)	Dose of immunoglobulin required	
	< 3 months	> 3 months
< 25	50 IU (0.5 ml)	100 IU (1.0 ml)
25–30	100 IU (1.0 ml)	250 IU (2.5 ml)
> 50	200 IU (2.0 ml)	500 IU (5.0 ml)

ease the dosage may be doubled (Table 2). Immunoglobulin does not always prevent infection and excretion of hepatitis A virus, and inapparent or subclinical hepatitis may develop. The efficacy of passive immunization is based on the presence of hepatitis A antibody in the immunoglobulin; however, the minimum titer of antibody required for protection has not yet been established. Immunoglobulin is used most commonly for close personal contacts of patients with hepatitis A and for those exposed to contaminated food. Immunoglobulin has also been used effectively for controlling outbreaks in institutions, such as homes for the mentally handicapped, and in nursery schools. Prophylaxis with immunoglobulin is recommended for persons without hepatitis A antibody visiting highly endemic areas. After 6 months the administration of immunoglobulin to travelers should be repeated, unless it has been demonstrated that the recipient has developed his own hepatitis A antibodies. Live attenuated and killed hepatitis A vaccines are under clinical evaluation. Vaccines produced by recombinant DNA techniques are being developed.

Literature Cited

Abb, J., and F. Deinhardt. 1986. Hepatitis A, p. 303–319. *In* A. J. Zuckerman (ed.), Viral hepatitis. Clinics in tropical medicine and communicable diseases. The W. B. Saunders Co., London.

Baroudy, B. M., J. R. Ticehurst, T. A. Miele, J. V. Maizel, R. H. Purcell, and S. M. Feinstone. 1985. Sequence analysis of hepatitis A virus cDNA coding for capsid proteins and RNA polymerase. Proc. Natl. Acad. Sci. USA **82:**2143–2147.

Dienstag, J. L., S. M. Feinstone, R. H. Purcell, J. H. Hoofnagle, L. F. Barker, W. T. London, H. Popper, J. M. Peterson, and A. Z. Kapikian. 1975. Experimental infection of chimpanzees with hepatitis A virus. J. Infect. Dis. **132:**532–541.

Hughes, J. V., L. W. Stanton, J. E. Tomassini, W. J. Long, and E. M. Scolnick. 1984. Neutralizing monoclonal antibody to hepatitis A virus; partial localization of a neutralizing antigenic site. J. Virol. **52:**465–473.

Karayiannis, P., T. Jowett, M. Enticott, D. Moore, M. Pignatelli, F. Brenes, P. J. Scheuer, and H. C. Thomas. 1986. Hepatitis A virus replication in tamarins and host immune response in relation to pathogenesis of liver damage. J. Med. Virol. **18:**261–276.

Lemon, S. M. 1985. Type A viral hepatitis. N. Engl. J. Med. **313:**1059–1067.

Lemon, S. M., and L. N. Binn. 1983. Antigenic relatedness of two strains of hepatitis A virus determined by cross-neutralization. Infect. Immun. **42:**418–420.

Linemeyer, D. L., J. G. Menke, A. Martin-Gallardo, J. V. Hughes, A. Young, and S. W. Mitra. 1985. Molecular cloning and partial sequencing of hepatitis A viral cDNA. J. Virol. **54:**247–255.

McCollum, R. W., and A. J. Zuckerman. 1981. Viral hepatitis: Report on a WHO informal consultation. J. Med. Virol. **8:**1–29.

Melnick, J. L. 1982. Classification of hepatitis A virus as enterovirus type 72 and of hepatitis B as hepadenovirus type 1. Intervirology **18:**105–106.

Najarian, R., D. Caput, W. Gee, S. J. Potter, A. Renard, J. Merryweather, G. Van Nest, and D. Dina. 1985. Primary structure and gene organization of human hepatitis A virus. Proc. Natl. Acad. Sci. USA **82:**2627–2631.

Provost, P. J., R. P. Bishop, R. J. Gerety, M. R. Hilleman, W. J. McAleer, E. M. Scolnick, and C. E. Stevens. 1986. New findings in live, attenuated hepatitis A vaccine development. J. Med. Virol. **20:**165–175.

Thornton, A., K. N. Tsiquaye, and A. J. Zuckerman. 1977. Studies on human hepatitis A virus in chimpanzees. Br. J. Exp. Pathol. **58:**352–360.

Ticehurst, J. R., V. R. Racaniello, B. M. Baroudy, D. Baltimore, R. H. Purcell, and S. M. Feinstone. 1983. Molecular cloning and characterization of hepatitis A virus cDNA. Proc. Natl. Acad. Sci. USA **80:**5885–5889.

Weitz, M., and G. Siegel. 1984. Comparison of HAV strains isolated in far apart geographical regions (Abstract), p. 704. *In* G. N. Vyas, J. L. Dienstag, and J. H. Hoofnagle (ed.), Viral hepatitis and liver disease. Grune and Stratton, Orlando, Fla.

Zuckerman, A. J. 1975. Human viral hepatitis, 2nd ed., p. 92–108. North Holland/American Elsevier, Amsterdam.

Zuckerman, A. J., and C. R. Howard. 1979. Laboratory tests for hepatitis A and the nature of the virus, p. 157–166. *In* Hepatitis viruses of man. Academic Press, Inc. (London), Ltd., London.

CHAPTER 39

Hepadnaviridae: Hepatitis B Virus and the Delta Virus

S. A. LOCARNINI and I. D. GUST

Hepadnaviridae: Hepatitis B Virus

Disease: Hepatitis (jaundice, serum hepatitis).

Etiologic Agent: Hepatitis B virus (four major subtypes).

Source: In blood mainly, from about 1 to 2 months before onset of illness until about 1 to 2 months after onset. Can also be found in saliva, semen, sweat, and urine, but not in feces.

Clinical Manifestations: Incubation period, 50 to 180 days. Frequently asymptomatic. Prodrome of transient rashes and arthralgia with nausea, anorexia, and vomiting. Dark urine and yellow sclerae herald the onset of jaundice with pale stools and palpable liver. In about 5% of cases, chronic hepatitis B carriage follows the acute episode.

Pathology: Hepatitis B virus not cytocidal. Primary hepatocellular damage due to host's immune response. Hepatocellular damage particularly in centrilobular areas, with lobules and portal tracts infiltrated with mononuclear cells.

Laboratory Diagnosis: *Acute infection:* primary anti-HBc or anti-HBs response; high-titered anti-HBc IgM; positive tests for HBsAg or HBV DNA. *Chronic carriers* defined as viremia or antigenemic for more than 6 months.

Epidemiology: Spread of infection is via two patterns—1) *Low-prevalence areas:* 80% by parenteral inoculation of blood or blood products and about 20% nonpercutaneous (close contacts including venereal spread). 2) *High-prevalence areas:* perinatal transmission from infected mother to neonate at or shortly after delivery plays major role.

Treatment: 1) *Nonspecific:* symptomatic relief, bed rest, with low fat, high carbohydrate, and balanced protein diet. Vitamin K. 2) *Specific:* antiviral and immunomodulating agents controversial. Alpha-interferon, ara A, and acyclovir alone and in various combinations explored.

Prevention and Control: 1) *General measures:* exclude contact with HBV-infected blood and secretions; minimize needle-stick by scrupulous technique. 2) *Passive prophylaxis:* intramuscular injection of hepatitis B immune globulin within 7 days of exposure. 3) *Active prophylaxis:* sub-unit vaccine available; course of three intramuscular injections with booster every 5 years. 4) *Combined active-passive prophylaxis:* very effective with postexposure situations, especially interruption of perinatal mechanism of transmission.

Hepatitis Delta Virus

Disease: Delta hepatitis.

Etiologic Agent: Hepatitis delta virus, a defective virus completely dependent on HBV for its replicative functions.

Source: As for HBV.

Clinical Manifestations: Two patterns seen—1) *Coinfection of HDV with HBV:* results in an illness almost indistinguishable from a typical attack of acute hepatitis B. 2) *Superinfection with HDV in a HBV carrier:* presents as a clinical exacerbation or an episode of acute hepatitis in a chronic hepatitis B carrier. Severe sequelae common, including chronic active hepatitis, cirrhosis, and massive hepatic necrosis.

continued

continued

Pathology: HDV is directly cytopathic in HBV-infected cells. HDV replication significantly inhibits HBV replication.

Laboratory Diagnosis: Primary anti-HD response; presence of anti-HD IgM; positive test for HDAg or HDV RNA.

Epidemiology: As for HBV.

Treatment: Alpha-interferon demonstrates some benefit. Ribavirin suitable on theoretical grounds. Phosphonoformate beneficial in fulminant hepatitis D.

Prevention and Control: As for HBV.

Introduction

At least five distinct forms of viral hepatitis are recognized (hepatitis A, hepatitis B, delta hepatitis, and the parenterally and enterically transmitted forms of hepatitis non-A non-B). Although epidemic hepatitis was well documented in the 17th and 18th centuries, especially during warfare, this disease seems almost certainly to have been hepatitis A. The first indication of the existence of another form of hepatitis with a different mode of spread was probably the Bremen outbreak of 1833 (Lurmann, 1885), in which 191 shipyard workers developed hepatitis within 2 to 8 weeks of being vaccinated with a batch of smallpox vaccine containing human serum. No cases of hepatitis were seen among coworkers vaccinated with other batches of this vaccine. The disease, which became known as "serum hepatitis," was assumed to be due to a blood-borne infectious agent and regarded as something of a medical curiosity. Occasional outbreaks were seen in the 1920s among patients with sexually transmitted diseases (Stokes et al., 1926) and diabetics receiving injections from the same syringe (Flaum et al., 1926), and in the 1930s among children who had received injections of pooled convalescent serum as prophylaxis against measles (Propert, 1938) and adults who had received a batch of injections from vaccine that had been stabilized with human plasma (Findlay and MacCallum, 1937).

Although by the outbreak of World War II there was good evidence that two distinct forms of viral hepatitis (A and B) existed, these concepts were outside the mainstream of medical thought. It required massive epidemics of hepatitis amongst civilians and troops to drive this message home.

The importance of serum hepatitis (hepatitis B) was underlined by a massive outbreak among American soldiers who had been inoculated against yellow fever. The immunization program, which commenced in 1941, used a vaccine produced by the Rockefeller Foundation by growing infected cells in a medium supplemented with 1% human serum (collected from students, interns, and nurses of the Johns Hopkins School of Hygiene and Public Health). Mass immunization began in autumn 1941, and by February 1942 it became apparent that an epidemic of hepatitis was taking place among immunized troops. Altogether about 50,000 cases occurred, after incubation periods of 60 to 154 days (Havens, 1968). As a result of this outbreak and the high incidence of other forms of hepatitis among troops, both the British and American Governments established research groups to study hepatitis and develop control methods.

It soon became obvious that neither of the major forms of the disease (hepatitis A and B) could be transmitted to traditional laboratory animals or cultivated in embryonated eggs or the cell culture systems then available. For overcoming this barrier, a series of transmission studies were undertaken in human volunteers; these served to define the major epidemiological features of the diseases and the physical characteristics of the etiological agents (reviewed by MacCallum, 1972). These studies confirmed that hepatitis B was a blood-borne infection, which could be transmitted by inoculation of relatively small quantities of infected blood or plasma, and that it was serologically distinct from hepatitis A.

After the end of the war, these findings were amplified in a series of experiments carried out among mentally retarded children admitted to the Willowbrook State School in New York (Krugman et al., 1967). These studies not only clarified the duration of infectivity, but provided important insights into the natural history of hepatitis B infection and the first evidence that it could be controlled by immunization. During the course of these studies, two infectious pools of sera were established, which became known as MS-1 and MS-2. These pools were found to regularly transmit hepatitis with either a short (MS-1) or long (MS-2) incubation period, irrespective of whether they were administered orally or parenterally. It was concluded that the MS-1 pool contained the hepatitis A agent and the MS-2 pool the hepatitis B agent. These pools and serial sera collected from children who received them proved to be of critical importance in defining the role of putative etiologic agents.

In 1963, Barry Blumberg, an American geneticist, described a novel precipitating antigen in the serum of two Australian aborigines, which he called the "Australia antigen" (Blumberg et al., 1965). Subsequent studies showed that, whereas this antigen was relatively common among Pacific Islanders, it was rare in the United States except among patients with leukemia and institutionalized, mentally retarded children (Blumberg et al., 1967).

A possible association with hepatitis was noticed by accident when a technician in Blumberg's laboratory developed the disease and became briefly antigen-positive. A review of hospital patients with clinical diagnosis of viral hepatitis revealed that a high proportion were antigen-positive during the acute phase of their illness and that the antigen disappeared from the blood during convalescence. Although the association between Australia antigen and viral hepatitis was clear, what was not certain was whether this was a marker for all or for only one type of disease.

This dilemma was solved by the prospective studies of Okochi (Okochi and Murakami, 1968) in Tokyo and Prince in New York (Prince, 1968), both of whom had collected plasma from blood donors and recipients of large volumes of blood. When patients who had developed posttransfusion hepatitis were selected, they were found to become Australia antigen-positive prior to the onset of illness, with antigen levels reaching peak titer about the time of onset of biochemical hepatitis and declining thereafter—the pattern one would expect of an etiological agent. Furthermore, on each occasion, one of the donors was also antigen-positive; this suggests that posttransfusion hepatitis was acquired following transfusion of a unit of blood from a chronic carrier of the disease. These findings were rapidly confirmed by examination of stored sera from the Willowbrook studies. Australia antigen was found to be present in

the MS-2 serum pool and in children who received this pool, but not in the MS-1 pool or in children who received it (Giles et al., 1969). As a result of these studies, the Australia antigen was recognized to be a sensitive marker of the presence of the virus responsible for the long-incubation-period type of hepatitis (hepatitis type B).

Shortly thereafter, electron microscopy revealed that the antigen was present on the surface of viruslike particles that could be readily observed in the serum of patients with hepatitis or in chronic carriers (Bayer et al., 1968; Dane et al., 1970). It received a variety of names (hepatitis antigen [HAg], hepatitis-associated antigen [HAAg], serum hepatitis antigen [SHAg]) before a WHO Committee recommended terms now universally accepted (see Table 1). In this system, the antigen present on the surface of hepatitis B virus (formally known as the Australia antigen) is referred to as the hepatitis B surface antigen (HBsAg).

The discovery of the hepatitis B virus proved to be a landmark in our understanding of viral hepatitis. Within a short period, sensitive assays were developed for detecting current or past infection, and it became possible to define the epidemiology, mode of transmission, and natural history of the disease. To the surprise of many, hepatitis B was found not to be just a medical curiosity, but one of the most common and important infectious diseases of man, producing an alarming morbidity and mortality. In recent years advances in molecular biology have led to a detailed knowldge of the structure and replicative strategy of the virus and to the development of a range of vaccines, which now make it possible to control and perhaps even eradicate the disease.

It was initially thought that the hepatitis B virus was unique, but studies in several laboratories have revealed that it is a member of a group of morphologically similar animal viruses, which have a predispo-

TABLE 1. Nomenclature for viral hepatitis type B

Virus or antigen	Abbreviation	Definition
Hepatitis B virus	HBV	The 42-nm, double-shelled particle that consists of an 8-nm outer shell and a 27-nm inner core. The core contains a small, circular, partially double-stranded DNA molecule and DNA polymerase activity.
Hepatitis B surface antigen	HBsAg	The complex antigenic determinant that is found on the surface of HBV and on the 22-nm particles and tubular forms.
Hepatitis B core antigen	HBcAg	The antigenic specificity associated with the 27-nm core of HBV.
Hepatitis B e antigen	HBeAg	An antigenic determinant that is closely associated with the nucleocapsid of HBV.
Antibody to HBsAg, HBcAg, and HBeAg	Anti-HBs, anti-HBc, and anti-HBe	Specific antibodies produced in response to their respective antigenic determinants.

sition for the liver and are often associated with a carrier state and an increased risk of developing liver cancer.

In 1977 Mario Rizzetto and his colleagues in Turin (Rizzetto et al., 1977) described a new antigen/antibody system that could be detected in the hepatocytes of some chronic carriers of HBV. They referred to the new antigen as "delta." Extensive studies in Rizzetto's laboratory and at the National Institutes of Health have revealed that this antigen is associated with a virus-like particle—the delta virus (HDV)—which is capable of replicating only in the presence of HBV. Little is known about this intriguing agent, but the development of diagnostic tests now makes it possible to define its epidemiology more precisely and to study its replicative strategy.

The Viruses

Hepadnaviridae

TAXONOMIC CLASSIFICATION AND LIST OF VIRUSES

Since the discovery of hepatitis B surface antigen by Blumberg et al. (1965), the hepatitis B virus (Dane et al., 1970) has become one of the most studied human pathogens. HBV has a number of unique features that distinguish it from all other human viruses. These include a distinctive morphology and genome structure, secretion of excess particulate-envelope antigen into the bloodstream of the infected host, and a replicative strategy involving reverse transcription to produce virion DNA. Hepatitis B virus has three close relatives found in lower animal species, namely, Eastern woodchucks (Summers et al., 1978), ground squirrels (Marion et al., 1980) and Pekin ducks (Mason et al., 1980). These viruses share many of the properties that make human HBV distinct from all other viruses. Recently, the International Committee on Taxonomy of Viruses Study Group on the Nomenclature of Hepatitis Viruses (Gust et al., 1986) recommended that they be classified into a new family known as the *Hepadnaviridae*.

Woodchuck hepatitis B virus (WHBV) was discovered by Summers et al. in 1978 in the serum of members of a *Marmota monax* colony in the Philadelphia zoo. Animals in the colony were known to have an increased incidence of hepatitis and hepatocellular carcinoma. Ground squirrel hepatitis B virus (GSHBV) was discovered by Marion et al. in 1980 in the serum of wild *Spermophilis beecheyi* in Northern California. The duck hepatitis B virus (DHBV) was initially detected in the sera of domestic ducks from the People's Republic of China. These animals are known to have a high incidence of hepatocellular carcinoma (Zhou, 1980).

In summary, the hepadnaviruses are a family of hepatotropic viruses that tend to cause persistent infection, and all but GSHBV have been associated with the development of hepatocellular carcinoma. Except for DHBV (Tuttleman et al., 1986), they have not yet been propagated in tissue culture. All four viruses have similar morphologic features, and some share antigenic properties; all share a taxonomic position within the same genus of the *Hepadnaviridae* family (Gust et al., 1986; Marion and Robinson, 1983).

PHYSICOCHEMICAL PROPERTIES AND MORPHOLOGY

The structure of HBV and the location of the major antigens is shown in Figure 1; the standard nomenclature used to describe the virus and its components parts is listed in Table 1. When examined under the electron microscope, sera from patients with hepatitis B infection reveal three morphologically distinct entities in varying proportions (Fig. 2). The hepatitis B virion (or Dane particle) is spherical and has a diameter of approximately 42 nm. It is enclosed in an outer envelope approximately 8 nm in width, which is composed of protein, lipid, and carbohydrate, and has on its surface a mosaic of antigens known as the hepatitis B surface antigens (HBsAg's). The outer envelope surrounds a 27-nm, electron-dense, hexagonal inner core, which bears another antigen known as the hepatitis B core antigen (HBcAg). Inside the core particle is a genome of circular, partially double-stranded DNA and an endogenous DNA polymerase. A third viral antigen, the hepatitis B e antigen (HBeAg), is located within the nucleocapsid.

In addition to the mature virions, sera from patients infected with HBV contain large numbers of spherical 22-nm particles and tubular forms that are

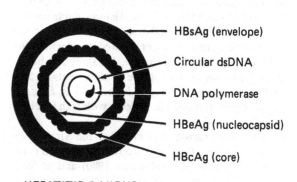

HEPATITIS B VIRUS

FIG. 1. The structure and components of the hepatitis B virus.

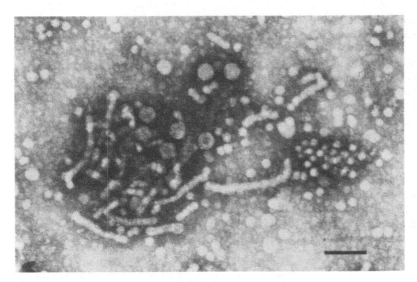

FIG. 2. The three types of particles visualized in the serum of a patient with HBeAg-positive chronic hepatitis B: 42-nm hepatitis B virions (Dane particles), 22-nm spherical forms, and the long filamentous forms. Negative stain with phosphotungstic acid. Bar = 100 nm. (Electron micrograph courtesy of Dr. John Marshall.)

composed entirely of HBsAg. These 22-nm-spherical and tubular particles represent excess viral coat protein and are immunogenic, but not infectious. They are often in excess of intact HBV particles by as many as a million to one (Robinson and Lutwick, 1976). Peak levels of HBsAg achieved during acute or chronic hepatitis can be as high as 200 to 500 μg/ml serum, which represents a range of 10^{13} to 10^{14} HBsAg particles/ml. In contrast, the peak number of infectious HBV particles is rarely greater than 10^8 per ml (Hoofnagle and Schafer, 1986).

Mature HBV particles have a buoyant density of 1.28 g/cm³ in CsCl, while 22-nm particles have an average buoyant density of 1.20 g/cm³ in CsCl and 1.17 g/cm³ in sucrose, with a sedimentation coefficient that ranges from 39 to 54S. There is no reliable information regarding the sedimentation coefficient of HBV.

The envelope of the virion can be removed by treatment with nonionic detergents, releasing the 27-nm core particles (Almeida et al., 1971) that contain both HBcAg and HBeAg. HBeAg is a soluble protein of approximately 15 kilodaltons (15 kDa), which appears to be a cryptic antigen located on or within the core particle (Takahashi et al., 1983). It can be released from the core structure by detergents or proteolytic enzymes. HBeAg is found only in HBsAg-positive serum, usually in patients with high levels of circulating HBV.

The nucleocapsid also contains the viral DNA (Robinson et al., 1974) with a covalently bound polypeptide (Gerlich and Robinson, 1980) and at least two enzymes, an endogenous DNA polymerase (Kaplan et al., 1973; Robinson and Greenman, 1974) and a protein kinase (Albin and Robinson, 1980). HBV DNA is a partially double-stranded, circularized molecule which contains a variable, single-stranded region (Robinson, 1977) and is about 3,200 nucleotide bases in length (Robinson et al., 1974; Sattler and Robinson, 1979; Summers et al., 1975). During replication, endogenous DNA polymerase enzyme closes the single-stranded region, while the protein kinase phosphorylates the major core polypeptide (Feitelson et al., 1982b).

MOLECULAR VIROLOGY OF HEPATITIS B VIRUS

HBV DNA

The HBV genome has an unusual and characteristic structure (see Fig. 3) (Burrell et al., 1979). At least six complete nucleotide sequences have been reported, ranging in length from 3,182 to 3,221 base pairs (bp) (Fujiyama et al., 1983; Galibert et al., 1979; Ono et al., 1983; Pasek et al., 1979; Valenzuela et al., 1980). The restriction patterns within the same subtype are generally identical (Sninksy et al., 1979).

The genome of the virus is made up of two strands of DNA referred to as the long (L or minus) and short (S) strand (see Fig. 3). The L-strand is not a closed circle, but is interrupted by a nick at a unique site near the 5' end of the S-strand (Sattler and Robinson, 1979). The sequence near this nick can form a stable, hair-pin structure, which may function as an origin of DNA replication (see below) (Galibert et al., 1982). A polypeptide is covalently attached to the 5' end of the L-strand and may function as a primer for viral DNA synthesis (see below) (Gerlich and Robinson, 1980). The L-strand has a fixed length of approximately 3,200 nucleotides, while the S-strand has a large, incomplete region and is of variable length (Summers et al., 1975). Thus, HBV DNA has a single-stranded region that can vary in length in different virus particles from 10 to 50% of the genome length. The 5′ ends (or the cohesive region) of both strands are positioned near short (10 to 12 nucleotides) direct repeats, known as DR1 and DR2 (Tiollais et al., 1984).

The minus strand carries all of the protein-coding capacity of the virus (Tiollais et al., 1985). Nucleotide sequencing of this strand has revealed four open reading frames (ORF) delineated by a start and a stop codon, which essentially represent the coding regions for the viral proteins, accounting for most of the known functions of HBV (Galibert et al., 1979; Valenzuela et al., 1980). These four potential genes have been labeled "S", "C,", "P" and "X" (Tiollais et al., 1985; Moriarty et al., 1985). The four regions overlap with one another so that the minus strand is read one-and-a-half times (Fig. 3). This strategy enables HBV, the smallest of the known mammalian double-stranded DNA viruses, to increase its coding capacity (Gerber and Thung, 1985).

Pre-S and S Gene

The envelope of the virion contains three proteins known as the major, middle, and large proteins (Tiollais et al., 1985). The S gene encodes the major protein of HBsAg, which is composed of 226 amino acids and exists in both glycosylated (P1: 27 kDa) and non-glycosylated (P11: 24 kDa) forms (Gerin and Shih, 1978). The primary structure of these polypeptides suggests the presence of two hydrophobic regions separated by a hydrophilic stretch as seen in transmembrane proteins (Misharo et al., 1980; Tiollais et al., 1985). From this, a model of the viral envelope has been proposed in which P1 and P11 form a dimeric subunit (49 kDa) that is embedded in a lipid bilayer (Tiollais et al., 1981).

There are two other short codons in front of the S gene, either of which can initiate synthesis of larger proteins. These regions are known as pre-S1 and pre-S2, and collectively as pre-S. If synthesis of HBsAg begins from the first start codon, a large protein of HBsAg (400 amino acids) is produced that contains information coded by pre-S1, pre-S2, and S and has a molecular weight of 39 kDa (if glycosylated, 42 kDa). If synthesis is encoded from the second start codon, another (middle) protein (281 amino acids) is produced, which contains pre-S2 and S information and has a molecular weight of 33 kDa (if glycosylated, 36 kDa) (Table 2).

The envelope of complete HB virions consists of 300 to 400 major protein (24 kDa) molecules and 40 to 80 middle (33 kDa) and large (39 kDa) molecules (Heermann et al., 1984). The protein composition of the tubular forms is identical with that of the complete HBV, whereas the composition of the 22-nm particles is significantly different (Table 3). In the serum of chronic carriers in which active viral repli-

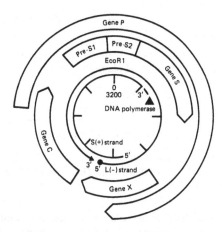

FIG. 3. Physical and genetic map of HBV DNA. Numbering of the nucleotides is shown on the inner circles, starting from the single Eco R1 restriction enzyme site as frame of reference (0/3,200). The broad arrows surrounding the genome represent the four open reading frames of the long minus strand [L(−)] showing gene S with its two pre-S regions and genes X, C, and P. [Modified from Gerber and Thung (1985); Tiollais et al. (1985).]

cation is occurring, these 22 nm particles contain both the major and middle proteins in the same proportion as HBV, but only 1% of the large protein. In the absence of viral replication, the 22-nm particles consist mostly of the major protein, with less than 1% of the middle protein and no large protein at all (Tiollais et al., 1985). The rates of synthesis of the three envelope proteins of HBsAg are probably regulated at the transcriptional level and are a direct reflection of the state of integration of the HBV DNA in hepatocytes, that is, episomal versus chromosomal (see Table 3).

The significance of the pre-S region has only recently become apparent. Intact HBV particles are enriched in the pre-S sequences (Hansson and Purcell, 1979; Imai et al., 1979; Neurath et al., 1985) containing middle (33 kDa) and large (39 kDa) pro-

TABLE 2. The gene products of HBV[a]

Gene	Protein	Mol. wt. (glycosylated) (kDa)
S	HBsAg : major	24 (27)
Pre-S2 + S	HBsAg : middle	33 (36)
Pre-S1 + Pre-S2 + S	HBsAg : large	39 (42)
C	HBcAg	15
X	HBxAg	16
P	DNA Polymerase	95

[a] Modified from Hoofnagle and Schafer (1986).

TABLE 3. Distribution of pre-S and S gene products in the three forms of HBsAg[a]

HBsAg found on	Major composition	Minor composition
HBV and filaments	Major protein (90%)	Middle protein (5%) Large protein (5%)
22-nm forms: when HBV DNA, episomal	Major protein (94%)	Middle protein (5%) Large protein (1%)
HBV DNA, chromosomal	Major protein (99%)	Middle protein (1%) Large protein (0%)

[a] Modified from Tiollais et al. (1985) and Heermann et al. (1984).

teins, and Milich et al. (1985) have demonstrated that the immune response to HBsAg containing these forms is significantly greater than for major forms alone. The pre-S region of HBsAg is quite sensitive to some of the chemical steps that are used to inactivate HBsAg during the production of commercial vaccines (Szmuness et al., 1980); this may explain the relatively poor immunogenicity of some HB vaccines (Neurath et al., 1985). In this regard, it may also be important to include the pre-S region in any cloned HBV DNA molecule used to produce recombinant HBsAg for vaccine production (Purcell and Gerin, 1985; Tiollais et al., 1985).

The pre-S region has another characteristic that may be of importance. Pre-S2 binds to aggregated human proteins, in particular to polyaggregated human serum albumin (pHSA) (Hansson and Purcell, 1979; Imai et al., 1979; Neurath et al., 1985). This region of HBsAg has been called the pHSA receptor, (Neurath and Strick, 1979) and it has been hypothesized that HBV attaches to and enters hepatocytes via this site (Pontisso et al., 1983). In support of this hypothesis is the finding that the pre-S2 pHSA-binding activity is found largely on intact HBV particles and that this activity is specific for human (or higher ape) proteins (Imai et al., 1979). The pHSA receptor activity on HBV particles can readily account for the species and organ specificity of HBV infection. Interestingly, antibodies against pre-S2 proteins have been detected in the sera of patients with acute HBV infection at early stages of the disease and may be involved in eliminating the virus from the infected individual (Neurath et al., 1985).

The C Gene

The C gene specifies the major polypeptide (p19) of the nucleocapsid (HBcAg) which has a molecular weight of 19 kDa and is 183 amino acids long. A possible 22 kDa precursor (214 amino acids) HBcAg polypeptide has recently been identified in the liver (Feitelson et al., 1982a). The 40 amino acid residues at the C-terminus of p19 are extremely rich in arginine, serine, and proline Such sequences resemble

protamine-like structures and may be involved in binding to the HBV DNA molecule (Tiollais et al., 1981). This polypeptide appears to be phosphorylated by the protein kinase associated with the core of Dane particles (Albin and Robinson, 1980).

Purified HBcAg may be converted by proteolytic cleavage into HBeAg (MacKay et al., 1981), a soluble protein with a molecular weight of 15.5 kDa (Imai et al., 1982). HBeAg is a hydrophilic polypeptide that probably binds the core structure to the viral envelope (Tiollais et al., 1985). The carboxy terminus of HBcAg with its DNA-binding properties is absent from HBeAg (Dreesman, 1984).

The P Gene

The third open reading frame of HBV DNA is P. This gene is a long sequence, covering almost 80% of the genome, which overlaps the other three genes and the ORF containing the S gene. The protein encoded by the P gene is not known, but from the sequence data it should be a histidine-rich protein, 832 to 845 amino acids long, with a molecular weight of 95 kDa. It has been postulated that this gene represents a virion DNA polymerase (Galibert et al., 1979; Pasek et al., 1979) that repairs the single-stranded region in the DNA to make it fully double-stranded.

On the basis of sequence homology with the polymerase genes of several retroviruses, the P gene appears to encode a reverse transcriptase (Toh et al., 1983), and reverse transcriptase activity has been associated with the endogenous DNA polymerase of hepadna viruses (Summers and Mason, 1982).

The X Gene

The final HBV gene, X, specifies a polypeptide (HBxAg) of 145 to 154 amino acids in length, with a molecular weight of 16 to 17 kDa (Galibert et al., 1979; Valenzuela et al., 1980). The X gene partially overlaps with the C gene and, like core, is interrupted by the nick in the L-strand (Gerber and Thung, 1985). The function of the protein is unknown, but it has been suggested that it represents the primer protein for synthesis of viral DNA and is mostly likely the protein covalently attached to the 5' end of the minus strand of HBV DNA (Robinson et al., 1984a). It may also prevent phosphorylation of this strand (Molnar-Kimber et al., 1983).

Recently, Moriarty et al. (1985) have detected a polypeptide (P28) in HBV-infected liver that reacts with antibodies against synthetic peptides corresponding to the X region. Antibodies against the translation product of this X region have also been detected in the sera of patients with primary hepatocellular carcinoma (Kay et al., 1985; Moriarty et al., 1985). However, its size (28 kDa) is larger than that predicted from the X-gene sequence (18 kDa).

Transcription of HBV DNA

Two major HBV-specific RNA species, containing poly A, have been detected in infected chimpanzee liver (Cattaneo et al., 1983): a larger-than-genome length (3.5 kilobases [kb]) RNA and a sub-genomic (2.1 kb) RNA. Both are transcripts from the large minus strand. The 3.5-kb RNA is produced from covalently closed, double-stranded DNA and processed only at the second run of transcription (Tiollais et al., 1985). No splicing event seems to exist for these two major HBV transcripts (Cattaneo et al., 1984).

The larger 3.5-kb RNA can probably encode all of the viral antigens and can also be used as a replicative intermediate, the pregenomic RNA. This species contains a terminal redundancy that encompasses DR1 (Seeger et al., 1986). The 3.5-kb RNA probably encodes the DNA polymerase and is the RNA messenger for the core protein (Pourcel et al., 1982). The small 2.1-kb RNA is the messenger for the major protein of the envelope (Cattaneo et al., 1983; Tiollais et al., 1985). From the studies of Pourcel et al., (1982), the mechanisms controlling the synthesis of the 2.1-kb RNA and 3.5-kb RNA are clearly different. They found that during active HBV replication both the 2.1- and 3.5-kb RNAs are present in approximately equal amounts, whereas the 2.1-kb RNA form dominated when the HBV entered its chromosomal (integrated) phase of replication.

In vitro cell-free transcription experiments have characterized other products. The L (−) strand directed the synthesis of two other RNA transcripts when RNA polymerase II was used (Rall et al., 1983). Both transcripts could be used for translation of the pre-S1 and pre-C regions (Tiollais et al., 1985). The S (+) strand directed the synthesis of a 700-bp RNA species dependent on RNA polymerase III (Standring et al., 1983). The initiation sites of these RNAs were separated by only 50 bp; this suggests some form of coordinate regulation of transcription of these two genes (Tiollais et al., 1985).

MECHANISM OF VIRAL REPLICATION

Despite the lack of a permissive tissue culture system (Zuckerman, 1975) and narrow host-range specificity, much is known about the replicative strategy of HBV (Summers and Mason, 1982).

Attachment, Penetration, Uncoating

The current hypothesis is that HBV uses polymerized human serum albumin as a bridge to attach to the hepatocyte surface membrane (Imai et al., 1979). The organ specificity of the HBsAg-associated receptors for human polyalbumin (Thung and Gerber, 1981, 1983a, 1984; Gilja et al., 1985) would explain the organ tropism and narrow host range of HBV. Following entry of HBV into the cell, fusion of the viral envelope with the membrane of the endocytic vacuole or uncoating of the virus by lysosomal enzymes results in release and "activation" of the viral nucleoprotein (Gerber and Thung, 1985).

Replication of HBV DNA

In addition to virion DNA and the two major classes of virus-specific RNA, two other forms of hepadnaviral nucleic acid have been detected in infected cells: superhelical, closed circular duplex DNA (Ruiz-Opazo et al., 1982; Weiser et al., 1983) and incomplete forms of viral DNA, consisting mainly of protein-linked minus strands alone, or paired with nascent plus strands, both representing probable intermediates in DNA synthesis (Tiollais et al., 1985). Summers and Mason (1982) have developed an elegant working model for the replicative cycle of HBV DNA based on observations of the replication of hepadnaviruses in the livers of ducks, ground-squirrels (Mason et al., 1982) and man (Fowler et al., 1984; Robinson et al., 1984a). It proceeds through two clear stages. In the first, the infecting DNA genome enters the nucleus and is converted to open circular, double-stranded DNA, then to supercoiled DNA, which serves as a template for the synthesis of viral RNA by host cell RNA polymerase II (Seeger et al., 1986). It is the L (−) strand which functions as the template for transcription, resulting in the production of the 3.5-kb RNA (the pregenome of HBV) as well as viral messenger RNA molecules (2.1-kb RNA). Multiple copies of the pregenome are synthesized from each L (−) strand. Viral mRNA is translated on host-cell ribosomes to produce HBsAg, HBcAg, HBxAg, and DNA polymerase. While this translational process is continuing, the pregenomic RNA and newly synthesized viral DNA polymerase (with endogenous reverse transcriptase activity), primer protein (HBsAg), and HBcAg are packaged into immature core structures in the cytoplasm (Buescher et al., 1985). This completes the first stage of HBV DNA replication (Fig. 4).

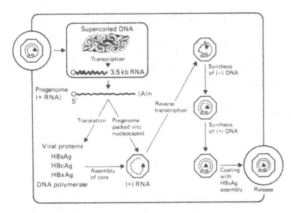

FIG. 4. Replication cycle of HBV. Following attachment, penetration, and uncoating, the virion DNA is repaired to complete double-stranded DNA and is then converted to supercoiled DNA. This viral DNA is transcribed, producing the RNA pregenome. Reverse transcription of this RNA pregenome produces the (−)DNA strand, and then initiation of synthesis of the (+)DNA strand completes DNA replication. The phases of reverse transcription and DNA strand replication occur in the viral core structures found in the cytoplasm. These structures are then coated with HBsAg and exported from the cell as complete HBV. [Modified from Tiollais et al. (1985).]

The second stage involves the reverse transcription of the plus-strand, viral pregenomic RNA into minus-strand DNA, with the simultaneous degradation of the RNA template by the RNase H activity of the HBV-specific DNA polymerase. Initiation of reverse transcription takes place at DR1. The DNA-linked protein covalently bound to the 5' end of the L (−) strand probably acts as the primer for minus-strand synthesis (Molnar-Kimber et al., 1984). A small RNA fragment of about 20 bp, probably derived from the 5' end of the pregenome, is thought to prime the synthesis of the plus strand at DR2 (Lieu et al., 1986). Complete synthesis of this strand is not necessary for coating of the nucleocapsids and export of virus particles, thus accounting for the single-stranded region of infectious hepatitis B virions (Fig. 4).

Assembly and Release of Mature HBV

The core particles are now coated by HBsAg and lipid-containing envelopes derived from the cell membrane and exported from the cell (Gerber et al., 1974; Gudat and Bianchi, 1977; Huang and Neurath, 1979; Kamimura et al., 1981; Yamada and Nakane, 1977). In situ hybridization of infected human hepatocytes suggests that viral replication occurs predominantly in the cytoplasm (Blum et al., 1984; Burrell et al., 1982, 1984). Little is known about the production of HBcAg and the formation of core particles. Likewise, the intracellular biosynthesis, processing, and secretion of HBsAg have not been clearly defined.

HBV DNA is not synthesized by the usual semiconservative mechanism of DNA replication, but rather is synthesized asymmetrically from an RNA intermediate, much as occurs in the replication of retroviruses (Varmus and Swanstrom, 1982). This and the significant sequence homology of the reverse transcriptase gene of retroviruses and the P gene of HBV suggest a common origin of the hepadnaviruses and the retroviruses (Toh et al., 1983), in particular the murine leukemic/sarcoma type C retroviruses (Miller and Robinson, 1986).

Compelling evidence for the successful in vitro propagation of DHBV has been reported recently (Tuttleman et al., 1986), describing a reproducible system for infecting primary duck hepatocytes and demonstrating de novo viral DNA synthesis, virus-specific protein production, and release of infectious virus into the medium. The system described will permit detailed analysis of the replicative strategy of DHBV in vitro, as well as providing a framework for testing various models of hepadnavirus replication.

ANTIGENIC COMPOSITION

Hepatitis B surface antigen is antigenically complex, and at least five major specificities may be found on HBsAg particles (Robinson, 1985). A group-specific determinant, designated "a", is common to all HBsAg particles, together with one of each of two pairs of subtype determinants "d" or "y" and "w" or "r", that are for the most part mutually exclusive and behave like alleles (Bancroft et al., 1972; Le Bouvier, 1971).

Eight HBsAg subtypes—namely, *ayw*, *ayw₂*, *ayw₃*, *ayw₄*, *ayr*, *adw₂*, *adw₄*, and *adr*—have been identified (Courouce et al., 1975). The predominant subtypes found in North America, Europe, and Africa are *adw* and *ayw*. Subtype *adr* prevails in parts of Southeast Asia and the Far East, along with *adw* and *ayw*. Subtype *ayr* occurs much less frequently in most of the world, but has been identified in a few isolated populations in Oceania (Courouce et al., 1975; Nishioka et al., 1975). HBV subtypes do not change after infection; however, occasionally isolated cases from the Far East with unusual combinations of HBsAg subtype determinants (e.g., *awr*, *adwr*, *adyw*, *adyr*, and *adywr*) have been reported (Courouce et al., 1975). The subtype determinants in these unusual cases are found on the same particles, suggesting that phenotypic mixing or unusual genetic recombinants have formed during mixed infections (Robinson, 1985).

Subtype determinants are specified by the viral genome and not by the host, and, in this way, HBsAg subtypes have provided very useful markers for epidemiological studies of the spread of virus in populations and in individual cases of transmission.

REACTIONS TO PHYSICAL AND CHEMICAL AGENTS

The stability of HBV does not always coincide with that of HBsAg, that is, immunogenicity and antigenicity are often retained after exposure to ether, acid, and heating, whereas infectivity is usually lost, especially if the concentration of HBV is not excessively high (Hollinger and Dienstag, 1985).

HBV has been shown to retain infectivity for humans when stored in serum for at least 6 months at 30°C (Hollinger and Dienstag, 1985), and for 15 years when frozen at −20°C. The virus also remains viable after being dried and stored at 25°C for at least 1 week (Bond et al., 1981).

When HBV is heated to 60°C for up to 4 h, infectivity for human volunteers may still remain (Murray and DieFenback, 1953). Heating of serum albumin for 10 h at 60°C renders it free of infectious virus under ordinary circumstances (Gellis et at., 1948). However, if the concentration of HBV is excessively high, inactivation may be incomplete (Shikata et al., 1978; Soulier et al., 1972). Infectivity in serum is partially or completely destroyed at 98°C after 1 min (Krugman et al., 1970) or 20 min (Robinson, 1985), respectively. Infection has also been destroyed after 10 h by dry heat at 160°C (Hollinger and Dienstag, 1985).

HBsAg is stable at pH 2.4 for at least 6 h and stable to exposure to ether and heat (98°C for 1 min; 60°C for 10 h), but HBV infectivity is lost under each of these conditions. Combined β-propiolactone/ultraviolet irradiation has been reported to reduce the titer of infectious HBV in human plasma by about a million-fold (Prince et al., 1983).

Each of the following disinfectants has been shown to destroy HBV infectivity in chimpanzees (Bond et al., 1983): sodium hypochlorite with 500 mg of free available chlorine per liter; cidex CX-250, a 2% aqueous glutaraldehyde solution, pH 8.4; sparicidin, pH 7.9, containing 0.12% glutaraldehyde and 0.44% phenol; 70% isopropyl alcohol; and Wescodyne diluted 1:213 and containing 84 mg of available iodine per liter.

HEPADNAVIRUS BIOLOGY: AN OVERVIEW AND COMPARISON WITH HBV

The four known hepadnaviruses have the following common properties: 3.0 to 3.2-kb circular, partially duplex DNA genome with covalently bound proteins at the 5′ end of the minus strand; common virion structure and morphology; the presence of virion-associated polymerase activity; production of excess surface antigen in the form of particles lacking nucleic acid; hepatotropism.

Although these viruses appear to be evolutionarily related, there are significant differences in their ultrastructure and biology, in particular the following

The Intracellular Localization of the Viral Component

Immunohistochemical staining of HBV-infected hepatocytes has revealed HBsAg in the cytoplasm or on the plasma membrane, or on both, and HBcAg mainly in the nuclei and less in the cytoplasm or on the plasma membrane. In contrast the core and surface antigens of the WHBV and DHBV can be detected by immunofluorescence only in the cytoplasm of infected hepatocytes (Halpern et al., 1983; Mason et al., 1984; Robinson et al., 1984a).

The Severity of the Hepatitis and the Development of Primary Hepatocellular Carcinoma (PHC)

In woodchucks, experimental or natural infection results in a spectrum of disease from mild, transient hepatitis to severe, chronic active hepatitis (Frommel et al., 1984), and PHC (Popper et al., 1981). The incidence of PHC is extremely high, with up to one-third of infected animals per year succumbing (this is 300 times greater than the rate observed in carriers of HBsAg). Although infected woodchucks do not develop cirrhosis, these animals offer the best available system for studying the oncogenic potential of hepadnaviruses.

By contrast to the situation in woodchucks, infection of ground squirrels with GSHBV results in mild or anicteric illness, and to date only a single case of PHC has been observed (Marion et al., 1983b). The natural routes of transmission of WHBV and GSHBV are unclear, but horizontal infection can be readily demonstrated in the laboratory by inoculation of viremic serum.

More than 50% of domestic ducks in China and up to 10% of Pekin ducks in the United States are persistently infected with DHBV, which may cause hepatitis, cirrhosis, and PHC (Marion et al., 1984; O'Connell et al., 1983; Omata et al., 1983; Robinson et al., 1984b). In infected flocks, transmission appears to be largely transovarial, and infection can be readily achieved in the laboratory by inoculation of embryonated eggs. Interestingly, the susceptibility of ducklings to infection declines rapidly in the first few weeks of life. Also, DHBV is the first hepadnavirus to be successfully propagated in vitro (Tuttleman et al., 1986).

Organization of the DHBV Genome

The DHBV genome is smaller (3.0 kb), and there is very little nucleotide sequence homology between it and the other hepadnaviruses (Siddigvi et al., 1981). The absence of an intergenic region between the X and the C genes results in one large ORF, the expression of which yields a larger polypeptide (37 kDa) of the DHBV core antigen (Mandart et al., 1984). This contrasts with the major polypeptides derived from the core and surface antigens of the other two animal hepadnaviruses, which are similar to those of HBV (Feitelson et al., 1981; Marion et al., 1983a).

Relationship of Structural Antigens

There is no serological cross-reactivity between DHBV and any other hepadnavirus, while the antigens of HBV, GSHBV, and WHBV are related, with the cross-reactivity between core antigens being greater (Werner et al., 1979) than between the surface antigens (Feitelson et al., 1981; Gerlich et al., 1980; Werner et al., 1979). Not surprisingly, the amino acid sequences of the C gene products (deduced from the nucleotide sequences that have been determined for all four viruses) are more highly conserved than those of the S gene products. Robinson et al. (1984a) has calculated 70% to 80% amino acid homology between GSHBV and WHBV 40% between WHBV or GSHBV and HBV, and significantly less between the mammalian hepatitis viruses and DHBV.

The Hepatitis Delta Virus

Introduction

The hepatitis delta virus (HDV) is unique among virus-like agents. It is a defective virus that requires active helper functions from the hepatitis B virus (HBV) for its replication (Rizzetto, 1983). Infection with HDV, therefore, may occur in patients with acute or chronic HBV infections and in the latter is often associated with severe and progressive liver disease, including chronic active hepatitis and cirrhosis. HDV has also been implicated in cases of fulminant hepatitis (Arico et al., 1985; Govindarajan et al., 1984; Jacobson et al., 1985), occasionally in epidemic form. Hepatitis D virus has been experimentally transmitted to HBV-infected chimpanzees (Rizzetto et al., 1980a, 1980b) and also to Eastern woodchucks chronically infected with WHBV (Ponzetto et al., 1984a).

The hepatitis delta virus is endemic among the HBV carrier population in some parts of Italy and has also been identified in many other parts of the world where it occurs as a result either of coinfection with HBV or superinfection of existing carriers of HBV (Rizzetto, 1983). High-risk groups include intravenous drug users and hemophiliacs who have received clotting factors derived from pooled blood donations (Rizzetto, 1983). HDV has not yet been classified, and its taxonomic position is far from clear.

Physicochemical Properties and Morphology

The hepatitis delta agent (see Fig. 5) consists of a 35- to 37-nm particle resembling a large HBsAg spherical particle and sharing the HBsAg coat and antigenic specificities of the strain of HBV that is also present. Its genetic material is composed of a small piece of single-stranded RNA (5.5×10^5 Da) surrounded by a protein, the delta antigen, which lacks any morphologic form. HDV has a buoyant density of 1.25 gm/cm^3 and a sedimentation coefficient intermediate between those of the 22-nm spherical forms of HBsAg and intact HBV particles.

Delta antigen can be detected in the nuclei of HBV-infected hepatocytes during the late incubation period and early acute phase of infection (Rizzetto, 1983). HW replicates efficiently, and the synthesis of HBV components is usually suppressed during HDV replication. Serum HBsAg and HBeAg levels and DNA polymerase activity fall to low and even undetectable levels, and intrahepatic markers of HBV synthesis, especially HBcAg, may not be detectable (Rizzetto, 1983). Also, HDV can be more efficiently packaged and secreted than HBV and is usually accompanied by a 1,000-fold excess of HBsAg (Bonino et al., 1986). During an acute HDV infection, the acute phase serum can have an estimated titer of 10^9 to 10^{10} HDV-associated particles, with only a 10-fold excess of HBsAg *not* associated with HDAg (Bonino et al., 1986). Although HDAg is present in the serum during the incubation period and acute phase of the illness, it is coated by HBsAg and can be detected only after the particles are pelleted by ultracentrifugation and disrupted with detergent.

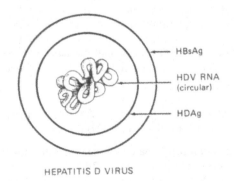

HEPATITIS D VIRUS

FIG. 5. Viral components of the hepatitis delta virus. (Modified from Deinhardt and Gust, 1982.)

MOLECULAR BIOLOGY OF HDV

HDAg

Preliminary characterization of HDAg has established that it is sensitive to proteases and high pH, and in guanidine behaves like a protein of approximately 68 kDa (Bonino et al., 1984; Rizzetto et al., 1980a). Further analysis of HDAg by denaturing electrophoresis and Western blot techniques and polyclonal anti-HD has revealed that the HDV particle contains antigens of 24 kDa and 27 kDa (Bergmann and Gerin, 1986). Bonino et al. (1986) have also reported similar HDV-specific polypeptides with molecular weights of 27 kDa and 29 kDa respectively.

The HBV-Derived Envelope of HDV

In an elegant study, Bonino et al., (1986) analyzed the envelope of HDV to determine if it resembled that of complete HBV or was more similar to the 22-nm spherical forms of HBsAg (see Table 2). Analysis of the HBsAg envelope of HDV revealed 95% major protein, 5% middle protein, and 1% large protein; these values are closer to the composition of the 22-nm particle than to that of the complete virion (Stibbe and Gerlich, 1982) (Tables 2 and 3).

HDV RNA

Hepatitis delta virus RNA of HDV is single-stranded, contains 1,700 nucleotides, and has considerable secondary structure. It is not polyadenylated and exists within HDV as a covalently closed, circular RNA molecule (Kos et al., 1986; Wang et al., 1986).

HDV virions contain only genomic HDV RNA species, whereas analysis of total RNA extracted from HDV-infected livers contains both genomic and antigenomic HDV-specific RNA (Chen et al., 1986; Wang et al., 1986). It is most likely that HDAg polypeptides are translation products expressed by the antigenomic RNA species found only in HDV-infected liver (R.H. Purcell, personal communication). Kos et al., (1986) recently demonstrated that HDV is the first animal virus known to possess a circular RNA genome. Since circular RNAs are unable to attach to eukaryotic ribosomes, HDV is probably a negative-stranded RNA virus. Interestingly, Kos et al. (1986) were also able to demonstrate considerable similarity of their HDV cDNA sequences to two plant viruses: tobacco mosaic virus (TMV) and the cassava latent virus (CLV). Wang et al. (1986) extended these observations by noting certain similarities of HDV RNA to a novel class of agents—the viroid-like agents from plants. These investigators have identified two short sequences, one of five nucleotides and the other of six, that exist in virtually all viroids as well as in HDV; in addition, there were two longer regions that had 65% homology with the central conserved region of all viroid molecules. Recently, Dimter-Gottlieb (1986) noted structural similarities between viroids and group I introns (a class of molecule found in ribosomal RNA genes and mitochondrial messenger RNA genes), which have the ability to self-splice, that is, to excise themselves and ligate the parental strands in the absence of enzymes, releasing a small circular RNA molecule. Whether or not HDV and plant viroids are escaped introns as well as the significance of the homologies between HDV and plant viruses, will be settled only by more detailed sequence comparisons and further characterization of the agent (Lewin, 1986).

Pathology of Hepatitis B

Immunopathology of Acute and Chronic Hepatitis B

INTRODUCTION

The clinical and histological presentation of viral hepatitis is determined by both viral and host factor (Mims, 1974). The primary hepatocellular damage can be due to direct viral cytotoxicity, immune destruction, or both. The outcome of the viral infection depends largely on efficient elimination of circulating and cell-associated viral antigens. In the normally immune, competent host, this will result in a self-limited inflammation which eliminates the viral infection.

The cytopathic potential of virus plays a crucial role in the initial presentation and course of the disease. It is now generally accepted that HBV itself is not directly cytopathic (Gerber and Thung, 1985). In the absence of an immune response, noncytotoxic viruses will not induce liver cell damage to a clinically detectable level. Therefore, in hepatitis B infection the grade and quality of the hepatic lesion will be due largely to immunologically mediated cell damage. Accordingly, a noncytotoxic infection like HBV may be established without clinical illness and persist beyond the incubation period as long as no primary immune response is mounted. In contrast, the liver cell damage and associated inflammation will be high and effective in hosts with normal immunologic responsiveness, but will become gradually lowered as a specific immune deficiency develops. Thus, viral replication will be increasingly facilitated as the process of immune elimination is decreased (Bianchi and Gudat, 1983).

Host-determined factors are extremely important in the pathology of hepatitis B, and the several defense systems of the body will significantly influence the course of hepatitis B infection. In particular, the well-established mechanisms for the elimination of virus-infected cells by B and T lymphocytes, including complement-mediated and antibody-mediated cytotoxicity, respectively, as well as the role of the cytotoxic T lymphocyte (Levy and Chisari, 1981) and the natural killer cell in the overall antiviral defense system, are important (White and Fenner, 1986). The interferon system in its classical cell-protective action, its involvement in the inhibition of the growth of virus-transformed cells, plus enhancement of cell-surface-antigen expression, and profound immune-regulating effects, with macrophage and natural killer-cell activation, also plays an important role (Bianchi and Gudat, 1983). Finally, the histocompatability-linked antigen (HLA) system is intricately involved in the control of the immune system, particularly the T lymphocyte in its role for effective elimination of viral infections (Eddleston and Williams, 1979).

GENERAL HISTOPATHOLOGY OF ACUTE AND CHRONIC HEPATITIS B

Acute Hepatitis

Four presentations are commonly seen (Bianchi and Gudat, 1983).

1. *Acute hepatitis with spotty necrosis* (classical lobular hepatitis), which consists of a diffuse and evenly distributed intralobular lymphohistiocytic inflammation associated with spotty necrosis, creating a picture of lobular disarray.
2. *Acute hepatitis with piecemeal necrosis,* consisting of periportal piecemeal necrosis, added to the histological features of acute hepatitis with spotty necrosis.
3. *Acute hepatitis with bridging (confluent) necrosis,* where, in addition to the features of acute spotty necrotic hepatitis, confluent necrosis bridges neighboring hepatic venules (central–central bridging) or central vein areas and portal tracts (central–portal bridging).
4. *Acute hepatitis with massive necrosis,* consisting of extensive confluent necrosis involving whole conventional liver lobules (panlobular, multilobular necrosis) is seen in a substantial part of the liver. Its clinical correlate is fulminant hepatitis.

In man, HBV infection results in acute hepatitis, where all of the above degrees of severity may be encountered, but acute hepatitis with piecemeal necrosis is the most common finding (Bianchi et al., 1971). In the chimpanzee, HBV infection results in acute hepatitis with different lobular changes and tends to run a more prolonged course than hepatitis A infection (Popper et al., 1980).

In most cases of hepatitis B studied, no viral antigens are detectable in the liver at the height of the disease. If viral antigen expression is demonstrable late in the course of the illness, this can be taken to be an indicator of impending chronicity (Bianchi and Gudat, 1983).

Chronic Hepatitis

In chronic hepatitis type B, two histological patterns are commonly seen (Bianchi and Gudat, 1983):

1. *Chronic active (aggressive) hepatitis.* In this presentation, portal and periportal inflammation is seen with erosion of the parenchymal limiting plate, piecemeal necrosis, and fibrosis. This is a precirrhotic lesion and is usually accompanied by a focal expression of HBcAg, together with plasma membrane expression of HBsAg. A minority of cases have a more generalized HBcAg expression. An inverse relationship between the amount of HBcAg expression in tissue and the level of the inflammatory process has been described (Bianchi and Gudat 1983). This group of patients are often HBV DNA positive.
2. *Chronic persistent hepatitis.* This lesion involves a predominantly lymphocytic inflammation restricted to enlarged portal tracts, leaving the parenchymal limiting plate preserved. It can be associated with the generalized focal or an HBcAg-free type of viral expression. A shift from one type to another with concomitant change of the inflammatory activity occurs with considerable frequency. These patients have a large number of hepatocytes expressing intracytoplasmic, but not plasma-membrane HBsAg and are usually asymptomatic HBsAg carriers who are HBV-DNA negative and anti-HBc-positive.

To summarize, few viral components are detected in acute and fulminant hepatitis B, although HBcAg and HBsAg may be demonstrable in the early or prenecrotic periods. Variable expression of HBsAg and HBcAg is usually seen in chronic active hepatitis B, whereas large numbers of HBsAg-positive hepatocytes are observed in inactive cirrhosis and in healthy carriers with chronic persistent hepatitis (Gudat et al., 1985).

IMMUNOPATHOLOGY OF HEPATITIS B

Recent studies have demonstrated the presence of HBcAg on the hepatocyte surface (Trevisan et al., 1982) and indicate that T lymphocytes are cytotoxic for hepatocytes with plasma membrane HBcAg or HBeAg, but not for the hepatocytes with membrane

HBsAg or IgG deposits (Naumov et al., 1984). Apparently HBcAg, together with class I HLA antigens on the cell surface, represent the principal target antigen for cytotoxic T lymphocytes in hepatitis B, and hepatocytes with active HBV replication will be lysed by these effector cells (Mondelli and Eddleston, 1984; Thomas and Lok, 1984). Antiviral antibodies (including anti-HBc) could bind to the cell surface and mask full viral antigen expression on infected hepatocytes, which then may escape immune clearance and support persistence of HBV. Therefore, variations in the outcome of hepatitis B will be related to a number of interrelating factors, including immunoregulatory cells, plus the factors that influence the cytotoxic T lymphocyte, the expression of antigens on the hepatocyte plasma membrane, and also the production of anti-HBc, which may block the immune attack by the effector cells (Thomas et al., 1984).

Thus, the immune response of the patient determines the expression pattern of HBsAg and HBcAg in liver tissue and the type, severity, and course of hepatitis. The second factor of importance in the pathogenesis of HBV-induced hepatocellular injury is viral replication.

THE STATE OF HBV DNA

Recent applications of molecular biology to the study of the natural history of hepatitis B have demonstrated that HBV DNA exists in either a freely replicating (episomal) form or an integrated (chromosomal) form. This analysis has led investigators (Shafritz and Rogler, 1984) to the hypothesis that HBV infection may proceed through an early replicative (permissive) stage and a later nonreplicative (nonpermissive) stage (Hoofnagle and Alter, 1984).

During acute infection and the initial period of chronic infection, HBV is present in episomal (free or extrachromosomal) form and replicates in the hepatocyte with complete transcription of the genome, resulting in production of intact virions, HBV DNA, DNA polymerase, HBsAg, HBcAg, and HBeAg. This replicative phase is characterized by high infectivity and progression of liver disease (Hoofnagle and Alter, 1984).

Depending on the host-immune response and several other antiviral factors, this phase may be followed after a period of more than 6 months by a nonreplicative phase of infection characterized by integrated (chromosomal) HBV DNA. In some patients integration may be observed earlier, during acute or fulminant hepatitis (Brechot et al., 1982). The integrated viral genome may be defective, resulting in incomplete viral gene transcription, mainly of the S gene, and little or no production of HBcAg,

HBeAg, DNA polymerase, and infectious virus (Brechot et al., 1981). Whereas the S gene can be transcribed from integrated sequences, freely replicating viral genomes seem to be necessary for C and P gene expression. This nonreplicative phase is usually associated with the markers of anti-HBc in the serum, decreased infectivity, and remission of disease activity in the liver, which may show chronic, persistent hepatitis or inactive cirrhosis.

RELATION OF HBV TO HEPATOCARCINOGENESIS

Several lines of evidence—namely epidemiologic, molecular, and histologic—strongly suggest that HBV is causally related to primary hepatocellular carcinoma (PHC) (Gerber et al., 1983; Popper et al., 1982). These data (Szmuness, 1978) include a strong correlation between the distribution of HBV carriers and the incidence of PHC, high frequency of serum markers of current or past HBV infection in patients with PHC, high risk of chronic carriers of HBV developing PHC, time sequence of HBV infection and subsequent development of PHC, family clusters of HBV carriers and PHC in which the mother is HBsAg-positive and transmission is presumed to have occurred perinatally, and the presence of HBV antigens in tumor tissue and cultured cells derived from PHCC (Gerber and Thung, 1985).

A number of cell lines have been derived from chronic carriers of HBV with PHC, of which the Alexander cell line is the prototype. This line secretes HBsAg, HBeAg, and occasionally HBcAg, and when injected into nude mice produces solid hepatocellular carcinoma (HCC) nodules (Alexander et al., 1978). Wen et al., (1985) have demonstrated a hepatitis B virus-associated nuclear antigen (HNA) in such cell lines, which resemble the neoantigens associated with viral transformation by DNA tumor viruses (T antigen).

Further evidence supporting the relationship between chronic HBV infection and PHC comes from studies in animal models and the detection of integrated HBV DNA in the genome of hepatocellular carcinoma cells. With rare exception, HBV DNA sequences can be detected in all HBsAg-positive PHCC patients (Gerber and Thung, 1985).

MECHANISM OF INTEGRATION OF HBV DNA

Integration of HBV DNA appears to be a nonspecific process (Tiollais et al., 1984). An analysis of viral DNA and flanking cellular sequences has revealed that viral DNA sequences are randomly integrated into the cell genome of both human (Koshy et al., 1983; Shafritz and Rogler, 1984) and woodchuck (Ogston et al., 1982) tumors, and extensive rearrangement, including deletions, partial duplications, and inverted duplications, have been observed (Mizusawa et al., 1985; Tiollais et al., 1984).

Based on the localization and the nucleotide sequences of host viral functions, two models for integration of HBV DNA have been proposed. The first model, described by Koshy et al., (1983), includes integration through the single-stranded region of the HBV DNA, close to the 5' end of the short strand or through the nick on the L(−) strand. This cohesive end contains a TATA sequence (Valenzuela et al., 1980) and shows gene promotor activity (Chakraborty et al., 1981a, 1981b. The second model, proposed by Tiollais et al., (1984), includes integration through the 11 bp direct repeats, with both DR1 and DR2 being targets of integration (Dejean et al., 1985). To date rearrangement of viral and cellular sequences has been observed in PHC, but not in chronic infection, and may therefore represent a unique characteristic of PHC (Shafritz and Rogler, 1984).

Integration appears to be a critical event in viral oncogenesis, but the exact mechanism of HBV-directed carcinogenesis is still unclear. There is no indication that the cellular sequences flanking viral inserts contain cellular oncogene sequences or that their expression is altered via the "promotor insertion" mechanism (Hayward et al., 1981). Neither is there direct evidence that HBV carries its own oncogene (Robinson et al., 1984b) or that HBV infection results in mutations of cellular genes that directly result in transformation, the "hit and run" mechanism (Galloway and McDougall, 1983).

Hepadnaviruses in Cells Other than Hepatocytes

Hepatotropism is considered one of the characteristic features of HBV (Gust et al., 1986). However, recent studies have demonstrated the presence of HBV antigens and DNA, including replicating forms, in cells other than hepatocytes. For example, HBsAg, HBcAg, or HBeAg has been found deposited, usually in the form of immune complexes, in extrahepatic tissue such as glomeruli, blood-vessel walls, synovium, skeletal and cardiac muscle, and lymphoid tissue (Gerber and Thung, 1985). In this form, it is usually associated with immune complex manifestation of diseases such as serum sickness, polyarteritis nodosa, glomerulonephritis, arthritis and essential mixed cryoglobulinemia (Dienstag, 1981).

More recently, replication-specific forms of viral DNA have been demonstrated in pancreas and kidney in naturally and experimentally infected Pekin ducks with DHBV-specific DNA probes (Halpern et al., 1984). Surface and core antigens of DHBV were detected in bile-duct epithelium of liver, proximal tubular epithelium of kidney, and acinar and islet cells (both α and β islets) of the pancreas (Halpern et al., 1983; Mason et al., 1984).

In man, integrated HBV DNA sequences have been found in kidney, pancreas, and skin (Brechot et al., 1984; Dejean et al., 1984, and evidence of viral multiplication occurring in pancreatic cells has been reported (Brechot et al., 1984), supporting the previous demonstration of HBcAg and HBsAg in pancreatic acinar cells of some HBV carriers (Hohenberger, 1984; Robinson et al., 1982). HBV DNA has also been demonstrated by in situ hybridization in bile duct epithelium, endothelial and smooth muscle cells, and spleen (Blum et al., 1983), as well as in the DNA extracted from skin and Kaposi's sarcoma tissue (Siddiqui, 1983) of HBV-infected patients.

In blood mononuclear cells of HBsAg carriers, free viral DNA has been detected in HBeAg-positive cases (Pontisso et al., 1984) and integrated viral DNA detected in anti-HBe-positive cases (Brechot et al., 1984), extending previous reports of the detection of extrachromosomal HBV DNA, HBsAg, HBcAg, and HBV particles in a cloned, marrow-derived lymphoblastoid cell line (Elfassi et al., 1984; Romet-Lemoune et al., 1983).

Cloned hepadna-virus genomes have been directly transfected into chimpanzee, duck, or ground squirrel liver, resulting in infectious progeny (Seeger et al., 1985; Sprengel et al., 1984; Will et al., 1982). These observations indicate that the host-cell range for hepadna viruses is more extensive than previously thought and raise challenging questions regarding the attachment and penetration of hepadna viruses to cells and the role of the cell-membrane-receptor interaction with virus-specified, receptor-binding proteins in this process.

Modulation by the Hepatitis Delta Virus

The most extensively studied model for interference with an ongoing HBV infection is superinfection or coinfection with HDV. The HDV can be demonstrated in liver biopsies mainly within the hepatocyte nucleus (Rizzetto, 1983), and less frequently in the cytoplasm (Rizzetto et al., 1977). Thung and Gerber (1983b) have shown that the delta antigen is released into the cytoplasm and then enveloped with the HBsAg to produce mature 37-nm particles, which are released into the blood. Interestingly, HDV infection is accompanied by a reduction in HBV replication and a corresponding decline in the level of various markers of HB infection (Hadziyannis et al., 1985).

The mechanism of hepatocyte injury by HDV is not known but is suspected to be due to a direct viral cytopathic effect. Thus, infection with HDV usually leads to acute hepatitis, which can occasionally progress to massive hepatic necrosis (Govindarajan et al., 1984) and rapid progression of chronic hepatitis (Rizzetto et al., 1983). In outbreaks of viral hepatitis in Venezuela (Popper et al., 1983) and Colombia, HDAg was demonstrated in hepatocyte nuclei in scattered cells in the focal necrotic stage, then in isolated cells in the massive necrotic stage, and also in many cells during the transition to cirrhosis; this suggested the development of persistent HDV infection (Hadler et al., 1984).

Clinical Features

Mode of Transmission

In developed countries, hepatitis B has traditionally been a hazard of blood transfusion, the use of shared syringes, and exposure to infected blood. For these reasons high rates of infection occur in certain occupational groups such as dentists, staff and patients in dialysis units, biochemists, and hematologists. Transmission can occur in the family setting and tends to be related to the degree of crowding of the household and the intimacy of each person to the case or carrier. It may occur as a result of accidental percutaneous inoculation following the use of shared razors, toothbrushes, bath brushes, or towels, or by close contact.

Hepatitis B virus has been detected in a variety of body secretions and excretions, including saliva, semen, and vaginal fluid, so that infection may be transmitted by kissing or by sexual intercourse. Increased rates of infection are found among people with multiple sexual partners, with the highest rates being observed among prostitutes and male homosexuals. Hepatitis B does not appear to be transmitted by the fecal–oral route, and urine is probably not infectious unless contaminated with blood. There is no convincing evidence that airborne infections occur.

The high prevalence of hepatitis B infection in many tropical countries led to the suggestion that mosquitoes and other bloodsucking insects might be important in transmission of the disease; however, there is no convincing evidence to support this suggestion.

Transmission from chronic carrier mothers to their babies appears to be the single most important factor in determining the prevalence of HBV infection in some areas. The risk of infection depends upon the proportion of HBeAg-positive carrier mothers, which may be as high as 40% in some countries. Although HBV can infect the fetus in utero, this rarely happens, and most infections appear to occur at birth owing to a leak of maternal blood into the baby's circulation, ingestion, or accidental inoculation.

Two different patterns of infection are recognized. In populations in which HBV is hyperendemic, infection is usually acquired early in life. The highest infection rates and carrier rates are seen among children and young adults. The carrier rate declines with increasing age, as does the prevalence of specific antibody.

In populations in which HBV is relatively uncommon, the majority of infections and the peak prevalence of HBsAg and specific antibody are in the 15 to 29-year-old group. The highest rates of infection are found among groups who have a high risk of contact with blood or blood products. Prominent among these are health workers (particularly laboratory staff, dentists, and surgeons), patients and staff in institutions for the mentally retarded, intravenous drug users, and male homosexuals.

Incubation Period

Studies in human volunteers which were carried out in the 1940s, 1950s and 1960s helped to define some of the major features of hepatitis B, such as its incubation period, the major mode of transmission, and the existence of a chronic carrier state. It was not until specific diagnostic tests became available, however, that a complete picture of the clinical features of the disease emerged.

The disease has a variable incubation period, ranging from 35 to 150 days, which seems to be determined mainly by the dose of virus encountered and the mode of inoculation. In general, the larger the dose, the shorter the incubation period (e.g. posttransfusion HB generally has a shorter incubation period than disease acquired as a result of a needlestick, while studies in volunteers have shown that the incubation period of infections acquired orally is longer than when the same dose is given by injection). Host factors also appear to play a role, since children or adults inoculated by the same dose of HBV may have incubation periods differing by weeks to months.

Duration of Infectivity

People who are infected with HBV are potentially infectious while the virus is circulating in their blood. In uncomplicated infections, the period of infectivity may precede the onset of jaundice by several weeks and persist for 6 to 8 weeks, even after liver function tests have returned to normal. While HBsAg may be detected in the blood for prolonged periods, infectivity is greatest during the late incubation period and early acute phase of the disease when viral replication is at its maximum (see Fig. 7).

Symptoms and Signs

The most commonly recognized sign of hepatitis is the onset of jaundice. This is commonly preceded by a prodrome which may last for several days or even a few weeks and which is characterized by fever, anorexia, malaise, nausea, and sometimes vomiting. The disease may be accompanied by upper abdominal discomfort, and weight loss is common. The first objective sign of illness is usually the onset of dark urine, which is followed soon thereafter by increasing jaundice, recognized clinically by a yellowing of the skin and sclera. With the onset of jaundice, symptoms begin to subside, and the patient regains his/her appetite and begins to feel better.

The clinical diagnosis is confirmed by a biochemical assessment of liver function supplemented by tests for HBsAg and perhaps anti-HBc IgM. The most commonly utilized biochemical assays are serum bilirubin (total and direct) and serum alanine aminotransferase (ALT) or aspartate aminotransferase (AST) levels. Other markers that may be useful include serum proteins and measurement of clotting factors.

Peak abnormalities of liver function usually coincide with the onset of clinical illness and decline rapidly thereafter, usually returning to normal within 4 to 6 weeks. Occasionally patients may develop fulminant hepatitis. This complication seems to be largely dose-dependent, and in countries in which routine screening of blood donors is practiced, this form of the disease has almost disappeared.

There is no speciific management for patients with viral hepatitis, therapy being limited to making the patient comfortable and treating any troublesome symptoms (e.g., pruritis). While the benefit of bed rest is largely unproven, it is generally recommended. Hospitalization is generally unnecessary unless the patient is severely ill or unable to care for himself/herself. This is usually determined by the patient's wishes, and there is no evidence but some widely held beliefs, for example, of the deleterious effect of fat or the need to abstain from alcohol for a prolonged period.

Complications

A small proportion of subjects infected with HBV develop fulminant hepatitis. Under these circumstances the patient develops a high fever and rapidly increasing jaundice, which is followed by the onset of encephalopathy, coma, and death. Mortality is closely correlated with age, being highest in patients over the age of 45.

While a variety of heroic measures, such as exchange transfusion, plasmaphoresis, continuous liver perfusion, etc. have been advocated for the management of fulminant hepatitis, there is no evidence that any of these is of value. In some studies; the combined use of lactulose and neomycin have been found to be of some benefit, especially in the encephalopathic patient.

A variable proportion of patients infected with HB become chronic carriers of HbsAg, and some of these develop chronic liver disease (chronic persistent hepatitis, chronic active hepatitis, cirrhosis) or primary hepatocellular carcinoma. The clinical presentation, investigation and management of these conditions is beyond the scope of this chapter.

Diagnosis

Hepatitis B

SPECIMEN COLLECTION AND BIOSAFETY

Hepatitis B virus is transmitted through direct inoculation of contaminated blood or blood products through the skin or mucous membranes or through intimate contact. HBV is highly infectious, and titers as high as 10^8 infectious units per ml have been documented in inoculation studies in human volunteers (Barker and Murray, 1971). HBV is also transmissible by nonpercutaneous means, but the virus is less infectious by these routes. Therefore, precautions consist mainly of preventing exposure of skin and mucous membranes to potentially infectious material. Additional precautions should be undertaken when procedures are employed that might generate aerosols of highly infectious material; for example, the purification and concentration of antigens or the homogenization of infected liver tissue.

Laboratory personnel should regard all specimens as potentially infectious. Gowns and disposable gloves should be worn and hand-washing procedures strictly enforced. Mouth pipetting and smoking, eating, or drinking in the laboratory should be strictly prohibited. The use of hypodermic needles and other sharp objects should be minimized, and, where their use is necessary, they should be disposed of in such a way as to prevent inadvertent needlestick injuries. For example, all sharps should be placed in metal cans and then autoclaved before being-discarded; disposable plastic ware can be collected into heat-resistant bags and then autoclaved. Recyclable materials can be placed in discard pans and autoclaved at 121°C for 40 min. Work areas can be decontaminated with either 10% formaldehyde/glutaraldehyde solutions or 0.5% active sodium hypochlorite (prepared fresh each week). Serum samples should be collected from all personnel at regular intervals (a minimum of 6 months) and tested for biochemical and serological evidence of HBV infection. Additional biosafety recommendations can be found in the literature (Bond et al., 1977).

The various HBV serological markers are generally stable, and sera containing antibodies to HBV antigens can be stored for at least 10 days at 37°C and almost indefinitely at −20°C. However, repetitive freezing and thawing can lead to substantial losses in titer. If testing is to take place within 5 to 7 days, then serum samples can be stored at 4 to 8°C. After this time, samples should be frozen, preferably at −20°C.

No special storage procedures are required if bacterial contamination is minimal. Some bacterial enzymes can destroy HBsAg, and bacterial contamination can make serum anti-complementary and introduce new antigens, which can interfere with serological diagnosis. If a bacteriostatic agent must be added to the samples, then 0.1% (wt/vol) sodium azide or 50 μg/ml gentamicin sulfate to final concentration is suitable. The use of severely hemolyzed blood or blood containing anticoagulant can occasionally result in false-positive immunoassay responses.

If serum samples ne d to be transported, they should be shipped frozen in dry ice and in doubly-sealed containers, as stipulated by the Interstate Quarantine Regulations (Code of Federal Regulations, Title 42, Part 72–25 Etiologic Agents) or equivalent code of safety practice.

DIRECT INDICATORS OF DISTURBED LIVER-CELL INTEGRITY: LIVER-FUNCTION TESTS

Liver-function tests refer to a group of biochemical investigations (usually bilirubin and plasma protein levels and measurement of certain liver enzymes) useful in confirming that the liver is diseased. In viral hepatitis, serum bilirubin levels reflect the degree of jaundice, and repeated estimations can be useful in following the progress of disease. Another useful test is serum transferase activity: alanine aminotransferase (ALT) and aspartate aminotransferase (AST) are both present in the cytosol of hepatocytes, with AST also present in mitochondria. Even before jaundice develops, serum transferase levels exceeding 400 IU/ml are not uncommonly found. These enzymes are of principal value in the diagnosis of acute hepatitis and in differentiating hepatocellular from obstructive jaundice. They are of no prognostic value in either acute or chronic liver disease.

DIRECT DETECTION OF HBV AND ITS ANTIGENS

Immunofluorescence, immunoperoxidase, biotin-avidin staining, and electron microscopy have been used to examine specimens (mainly liver biopsy or fresh autopsy material) and serum samples for the presence of HB antigens and particles (Huang et al., 1976; Hsu et al., 1981a, 1981b). Liver biopsy is a simple and usually safe procedure in the hands of an experienced clinician. The procedure is useful in the diagnosis of chronic hepatitis particularly in separating persistent from active forms and in establishing a diagnosis of cirrhosis. It is not required in acute hepatitis because the diagnosis can readily be made on other grounds. These techniques and procedures are not applicable to rapid, large-scale screening of HBV infections by clinical laboratories, but they have been invaluable research tools in elucidating the origin of the various antigens within infected cells (Gerber and Thung, 1985).

Immunofluorescent staining of cryostat sections is superior to the use of paraffin sections, but cellular outlines are better resolved in uniformly cut paraffin sections than in frozen sections (Hollinger and Dienstag, 1985). Reduction of nonspecific, background staining of formalin-fixed sections can be achieved by trypsin digestion of tissue before staining (Huang et al., 1976), and frozen sections can be treated for 5 min with 0.5 M glycine-HCl buffer, pH 1.2, before specific antibody is added. The biotin-avidin system offers a number of advantages over other immunodiagnostic clinical techniques, including the standard immunoperoxidase method (Hsu et al., 1981a, 1981b)—in particular, increased sensitivity with reduced background staining, and the ability to study both formalin-fixed and frozen sections, which can also be examined by light microscopy.

Negative staining of sera containing HBsAg reveals the presence of three distinct and characteristic morphological entities (Figs. 1 and 2). It was Dane et al., (1970) who first suggested that the double-shelled 42-nm particles represented HBV, and the small particles and tubular forms of the antigen were noninfectious, surplus virus-coat material. Electron microscopy is a useful and rapid diagnostic technique provided the sample contains large numbers of particles, but is expensive and time-consuming. The sensitivity of the technique can be increased 10-fold by the addition of specific anti-HBs.

IN VITRO ISOLATION AND ANIMAL MODELS

Despite evidence for limited replication of HBV in some cells and organ cultures (Zuckerman, 1975), serial propagation has not been accomplished. The recent success of Tuttleman et al. (1986) in propagating DHBV in primary duck embryo hepatocytes will almost certainly renew attempts to propagate HBV under similar conditions.

Chimpanzees and gibbons are highly susceptible to experimental infection with human HBV, but are not routinely available (Barker et al., 1975). In both species, the pattern of infection is similar to that observed in humans except that the disease is milder. Rhesus monkeys and woolly monkeys are also susceptible to HBV infections, but virus adaptation appears to be necessary for successful studies in these animals. Nevertheless, these higher primate investigations have and will continue to play an important role in inactivation studies and disinfection kinetics (Bond et al., 1983), as well as in monitoring the safety of vaccines and studying the pathogenesis and immunopathology of the disease.

DIRECT VIRAL NUCLEIC ACID DETECTION

HBV-Associated DNA Polymerase

Hepatitis B virus contains a virion-associated DNA polymerase (Kaplan et al., 1973), which is an excellent marker of active viral replication (Alberti et al., 1979; Robinson, 1975). This test is not included in most routine diagnostic evaluations, because reliable assays are not yet commercially available, nor do the results add substantially to the management of most patients. However, serial determinations of DNA polymerase and HBV DNA (if available, see below) can be extremely useful in monitoring the therapeutic effect of antiviral compounds used for treatment of chronic hepatitis B.

If DNA polymerase assays are performed, it is important to confirm that the reactivity is associated with HBV, either by immunoprecipitation studies (with anti-HBs before Nonidet P-40 detergent treatment and anti-HBc after the addition of Nonidet P-40) (Robinson, 1975), or by demonstrating the unique properties of the HBV DNA polymerase (Lin et al., 1983) compared with the other contaminating human DNA polymerases, alpha, beta, or gamma (Weissbach, 1975).

HBV DNA

Analysis of serum or biopsy specimens for HBV DNA sequences by molecular hybridization techniques (Hadziyannis et al., 1983) is the most sensitive method of detecting HBV replication (Brechot et al., 1981) and is capable of detecting as little as 0.2 to 0.5 picograms of DNA (Lieberman et al., 1983).

In patients with acute hepatitis B, HBV DNA is detected only transiently (Krogsgaard et al., 1985), whereas it may be persistent in patients with chronic hepatitis B. The presence of HBeAg and HBV DNA in the serum is evidence of ongoing replication (Bonino et al., 1981). Although HBeAg is more easily measured, serum HBV DNA levels are currently the most sensitive serological markers of infectivity (Berninger et al., 1982; Brechot et al., 1982). The most widely used assay is dot-blot hybridization (Lieberman et al., 1983; Scotto et al., 1983), with complete genome-length copies of HBV as the DNA probe (Pasek et al., 1979). The test serum sample is spotted onto a nitrocellulose filter sheet (Shafritz et al., 1982), viral particles are disrupted, and DNA is denatured by alkaline treatment and digested with proteinase K. The DNA is fixed onto the filter by baking at 80°C for 2 h, and bound DNA is then detected by hybridization using a cloned HBV DNA probe labeled with [^{32}P]. After washing, positives are identified by autoradiography.

Hepatitis B virus DNA can be further characterized by restriction enzyme digestion and Southern blot hybridization (Bonino et al., 1981; Shafritz and Kew, 1981). The viral DNA is extracted from partially purified HBV and redissolved in Tris–EDTA buffer prior to digestion with various restriction enzymes (Eco R1, Hind III) (Bonino et al., 1981). The digested sample is then electrophoresed through a 0.8 to 1.2% horizontal agarose slab gel, and the DNA fragments are transferred to nitrocellulose paper and processed for filter hybridization by the method of Southern (1975). Hybridization and autoradiography are performed under the same conditions as dot-blot analysis.

Recent improvements in technology include the introduction of filtration-hybridization methods (Morace et al., 1985), which allow rapid definition of HBV infectivity in sera. This method is easy to perform and, most importantly, can be applied to large-scale clinical studies and epidemiologic investigations. Likewise, analysis of HBV DNA has been qualitative, based on optical densities of autoradiographic spots, with scanning densitometry. Fagan et al. (1985) has modified the technique and used [^{32}P]scintillation Cerenkov counting to quantitate HBV DNA in serum, by using serial standards of cloned HBV DNA for direct comparisons. The use of biotin-labeled DNA probes for identifying viral DNA in human tissues (Brigati et al., 1983; Langer et al., 1981; Negro et al., 1985) is a further extension of this technology. This application has several advantages over the use of radioactive probes for either filter or in situ hybridization. This assay is rapid and avoids the risks of radiation, it can be used to detect various antigens (HBsAg, HBcAg, and so on) in the specimen, and it can be used on formalin-fixed and paraffin-embedded tissue, thus enabling cells containing virus DNA to be localized easily within specimens. This is also suitable for radioactive DNA probes (Burrell et al., 1984; Gowans et al., 1981), but resolution is often poor.

Bacterial contamination has been postulated as a possible explanation for the existence of HBV DNA in HBsAg-negative serum (Sherlock and Thomas, 1983), and a recent publication by Diegutis et al. (1986) is the first report of a false-positive result with the HBV DNA dot hybridization assay. These workers were able to show that the reactivity was due to the presence of sequences in the serum which reacted with residual bacterial plasmid vector sequences in the DNA probe. This problem would be overcome by using vector free probes as well as sub-genomic viral probes. Despite this small false-positive rate, these highly sensitive assays for HBV DNA will be of particular value in the lower range of DNA polymerase positivity and should provide additional information regarding HBV infectivity, as well as providing a rationale for the duration and outcome of antiviral therapies.

SEROLOGIC IDENTIFICATION OF HEPATITIS B ANTIGENS AND ANTIBODIES

Introduction

A wide range of serological methods has been developed for detection of HB antigens and their specific antibodies (Table 4). These include agarose gel diffusion, counter-immunoelectrophoresis, rheophoresis, complement fixation, latex agglutination, hemagglutination, immune electron microscopy, enzyme immunoassay, and radioimmunoassay. Each method offers certain advantages and disadvantages and differs in sensitivity, specificity, ease of performance, and cost.

In the United States, all commercially available reagents for detection of hepatitis B virus-specific antigens and antibodies are subject to Federal license and are accompanied by detailed directions for use.

A list of licensed manufacturers of the reagents for various test methods for hepatitis B and hepatitis delta can be obtained by corresponding with the Director, Bureau of Biologics, Food and Drug Administration, 8800 Rockville Pike, Bethesda, Maryland, 20014.

PRODUCTION OF HEPATITIS B-SPECIFIC ANTIGENS AND ANTIBODIES

The widespread availability of sensitive and specific commercial kits for the detection of both HBV-specific antigens and antibodies has greatly reduced the need to prepare highly purified antigens or antibody of high specificity, affinity, and avidity.

Situations do arise when independent preparation may be desirable, for example, subtyping new isolates of HBV. Detailed descriptions of the preparation of purified HBsAg and HBcAg are given elsewhere (Feinstone et al., 1979; Hollinger and Dienstag, 1980).

Anti-HBs of sufficient titer for use in the serological tests described below can be obtained from multiply transfused patients (hemophiliacs and thalassemics) or from patients previously infected with HBV who develop an anamnestic response following infusion of blood or blood products, or from people who have received several doses of HB vaccine. The major advantage of human (or chimpanzee) serum is that it is free from antibodies to normal human serum contaminants.

Techniques for immunizing goats, guinea pigs, and rabbits and a detailed account of the preparation of anti-HBs in seronegative guinea pigs can be found in Hollinger and Dienstag (1980). However, virtually all of these preparations have detectable levels of antibody to human serum proteins, especially human serum albumin. These contaminating antibodies can

TABLE 4. Methods for detection of HBsAg and anti-HBs[a]

Classification (relative sensitivity)	Assay methods	Time to complete (h)
First generation (least sensitivity, × 1)	Agar gel diffusion	24–72
Second generation (intermediate sensitivity, × 5–10)	Counter immunoelectrophoresis	1–2
	Rheophoresis	24–72
	Complement fixation	18
Third generation (most sensitivity, × 50–100)	Latex agglutination	1
	Hemagglutination	3–6
	Enzyme immunoassay	4–24
	Radioimmunoassay	3–24
Research laboratory	Immune electron microscopy	
	Immunofluorescence microscopy	
	DNA hybridization	
	DNA polymerase assay	

[a] Modified from Feinstone et al. (1979) and Hollinger and Dienstag (1985).

be effectively removed by absorption with glutaraldehyde-cross-linked polymers of normal human serum (Locarnini et al., 1978).

For the preparation of subtyping reagents, rabbits and guinea pigs are generally preferred to goats. This is because antibody with excellent antigen-precipitating capacity (anti-*a*) is readily prepared in goats. Thus, goat antiserum is usually better for screening than for subtyping. For subtyping, specific IgG containing anti-*d* or anti-*y* can be obtained by affinity chromatography techniques in which the HBV-specific antiserum is passed through a column containing the heterologous HBsAg subtype. Anti-*y* (and presumably anti-*w*) can be subsequently recovered from the column by elution with 0.1 *M* glycine, pH 2.8.

Several groups of investigators have produced and characterized high-affinity monoclonal IgM and IgG anti-HBs and used them to develop sensitive monoclonal radioimmunoassays (M-RIAs) for the detection of HBsAg-associated determinants in serum (Shafritz et al., 1982; Wands and Zurawski, 1981). Wands et al. (1984) have successfully used these monoclonal antibodies in various clinical studies, and more recently commercial companies have produced a number of monoclonal antibodies against both the group *a* determinant and subtypes *ad* and *ay*.

Agarose Gel Diffusion (AGD)

This technique is the least sensitive method available for detection of HBsAg and HBeAg (Magnius, 1975). Its main advantages are that it is simple, cheap, and allows different subtypes to be distinguished. While largely superseded in developed countries, this technique is still widely used in some developing countries.

Sera containing HBsAg and anti-HBs are allowed to diffuse radially from adjacent wells cut in low-concentration (0.6%) agarose gels. When optimal proportions of antigen and antibody are attained, precipitin lines, which are visible with the naked eye, are formed. Protein stains may be useful to increase sensitivity; weak precipitin lines can be reinforced by placing reference HBsAg-positive sera in wells adjacent to the specimen. This same reinforcement pattern can be used for subtyping antigens where monospecific anti-HBs containing only anti-*d* or anti-*y* antibody can be substituted for the reference antiserum. Slides are observed for a "spur" signifying different antigenic determinants (Hollinger and Dienstag, 1980).

Counterimmunoelectrophoresis (CIEP)

Unlike radial gel diffusion, the reactants in a CIEP test are driven together in the agarose matrix under the influence of an electric field. The HBsAg has an isoelectric point between 4.4 and 5.2; thus, in an alkaline (pH 8.2) environment, the antigen is negatively charged and migrates toward the anode. Conversely, IgG, being closer to its isoelectric point, moves toward the cathode by electroendosmosis, and precipitin lines form when optimal concentrations of the antigen and antibody occur (Hollinger and Dienstag, 1980).

Because of the greater physical effects of electrophoresis, an agarose medium of 1% is usually required. CIEP results in a tenfold increase in sensitivity and a much faster reaction time than AGD. Sensitivity and speed of reaction can be increased by using a discontinuous buffer system that enhances the movement of acidic proteins towards the anode and globulins towards the cathode (Wallis and Melnick, 1971).

Both specificity and sensitivity depend on the use of potent precipitating sera, rendered free from antibodies to normal human serum by prior adsorption. False-positive results are uncommon, but false-negative results due to the prozone phenomenon can occur in the region of HBsAg excess. This is minimized by diluting the reagent antibody in normal, homologous, whole serum or its globulin fraction rather than in buffer (Hollinger and Dienstag, 1980).

Rheophoresis

Rheophoresis uses the same principle as agar gel diffusion but achieves greater sensitivity. Rheophoresis relies on continuous evaporation of water molecules through a central hole placed directly over either an antigen or antibody well (Jambazian and Halper, 1972). Protein solutions, which are now placed in peripheral wells, are transported under the influence of hydrodynamic forces to the central area of dehydration through the flow of a low-ionic-strength (0.01 *M* Tris-HCl, pH 7.6) buffer, which is placed external to the agarose.

The sensitivity of rheophoresis is equivalent to that of CIEP, and this method has been particularly valuable in the study of HBsAg subtypes.

Complement Fixation (CF)

The CF test is commonly used in viral serology laboratories and is slightly more sensitive than CIEP or rheophoresis (Purcell et al., 1970). A potent and specific CF antiserum of human or animal origin is a prerequisite, as many precipitating antisera are not satisfactory for use in CF. The CF test has also been used for the detection of anti-HBc (Hoofnagle et al., 1973). This technique detects anti-HBc following most clinical acute HBV infections and in patients with chronic hepatitis B infection.

The major disadvantage of the CF test is the relatively large quantity of antigen required compared with other serological tests, such as RIA and EIA. Furthermore, the relative complexity of the method and the frequency of anti-complementary reactions in test sera, not uncommon in the acute phase of hepatitis, limit the usefulness of this procedure.

Latex Agglutination

The reverse passive latex agglutination test is the least sensitive of the "third generation" tests but has several major advantages, including speed, simplicity, and long shelf life (approximately 5 months). These factors make this test useful for emergency situations that do not allow adequate time for an evaluation of the sample by other procedures (Leach and Ruck, 1971).

False-negative reactions are quite low (about 5%), but false-positive reactions are not uncommon (15 to 20% in some series). The cause of these reactions includes the presence of autoimmune antibodies, rheumatoid factor and related reactivities, heterophile antibodies, lipemic serum, albumin-to-globulin-ratio disturbances, electrolyte imbalances, pH abnormalities, and various drug metabolites (Hollinger and Dienstag, 1985).

The false-positive rate can be cut by 50% if the serum is heat-inactivated and the rheumatoid factor absorbed out (Zalan et al., 1973). All other agglutination reactions should be confirmed by a blocking test. Confirmation of weak positive reactions by an alternative method of at least equivalent or preferably greater sensitivity and specificity is essential.

Hemagglutination (HA)

Erythrocytes are not naturally agglutinated by HBsAg. However, cells can be treated chemically to allow coating with anti-HBs (reversed passive hemagglutination) or with HBsAg (passive hemagglutination). A number of different types of erythrocytes have been used, including turkey, sheep, and human group O cells, and detailed methods of preparation and conjugation have been described by Hollinger et al. (1971).

Major advantages of HA tests (Juji and Yokochi, 1969) are conservation of sera (less than 10 μl required), the use of simple equipment that can be partly automated, rapid completion (1 to 3 h), simplicity, and ease of quantification (Hirata et al., 1973). Disadvantages include a relatively large number of nonspecific reactions and the need for personnel experienced in hemagglutination techniques. False-positive reactions frequently occur at dilutions below 1 in 8, thereby reducing the sensitivity accordingly. Appropriate controls must be incorporated in each test run and should include heat inactivation and adsorption of the sera with uncoated erythro-cytes (to remove antibodies against ruminant IgG and Forssman antigen), use of control erythrocytes coated with normal immunoglobulin from the same species, and removal of rheumatoid factor-like reactivity from the test serum samples. Confirmatory hemagglutination-inhibition tests should always be attempted (Hollinger and Dienstag, 1985).

Radioimmunoassay (RIA)

The technique of RIA is still the most sensitive and specific method available for detecting the various serological markers of hepatitis B (HBsAg, HBeAg, anti-HBs, anti-HBc, anti-HBe). This assay has excellent precision, is easy to perform, and has unexcelled specificity and sensitivity. Limiting factors include the need for a gamma counter, the relatively short shelf life of the radiolabeled reagents, and the need for facilities to store and dispense radiochemicals.

The radioimmunoassay principle was developed during the late 1960s and introduced into clinical medicine in the early 1970s. Essentially two methods are used, the solid-phase sandwich RIA technique (SPRIA) and the double-antibody RIA procedure.

The double-antibody RIA test (a radioimmunoprecipitation test or RIP) is a useful research tool that measures the primary interaction between antigen and antibody and can be used to study the kinetics of this interaction (Hollinger et al., 1971). The RIP is a highly sensitive, specific, and reproducible assay that uses a fixed amount of [^{125}I]labeled, highly purified HBsAg to compete with HBsAg in the test sample for a finite quantity of anti-HBs (first antibody). The antigen–antibody complexes that form are precipitated by an anti-IgG (the second antibody). The results are recorded as decreased binding of labeled HBsAg to anti-HBs (as revealed by a reduced number of counts in the coprecipitate), which is proportional to the concentration of unlabeled HBsAg present in the test serum sample (Hollinger and Dienstag, 1980).

HBsAg. The solid-phase sandwich radioimmunoassay (SPRIA) for HBsAg detection is comparable in sensitivity to the double-antibody RIA method (Wilde, 1970). The SPRIA is more convenient to use and has become the principal system used by most commercially available hepatitis B RIA tests and is capable of detecting HBsAg to a level of 0.1 μg/ml.

The solid phase used in these assays includes Sepharose, polystyrene or polyethylene beads or pearls, polystyrene tubes, and controlled pore glass beads. The principle of the SPRIA is that HBsAg in the test sample forms an immunologic complex with the anti-HBs preabsorbed onto the solid phase and is detected by the addition of [^{125}I]labeled anti-HBs. The labeled antibody attaches to any bound HBsAg particles, forming an antibody–antigen–antibody sandwich, and can be detected by use of a gamma-counter. The test protocol usually requires that at least three negative and two positive control sera be included with each test batch. The mean radioactivity (P) of the test sample (expressed as counts per minute [cpm]) is divided by the mean radioactivity of the negative control sample (N), and a P/N ratio is calculated. Samples with P/N ratios greater than 2.1 are generally considered to contain HBsAg. All positive tests are confirmed in a neutralization reaction with reference anti-HBs. Test samples must be confirmed in this way before the result can be reported as positive (see below).

Within the constraints of the assay, there is a direct correlation between the final cpm and the concentration of HBsAg particles in the test sample, particularly when low levels (0.1 to 1.0 μg/ml) of antigen are used. In sera containing more than 1 μg/ml of HBsAg, all the antibody sites are saturated, and a plateau response is seen. As some HBsAg-positive carriers have concentrations of HBsAg in excess of 100 μg/ml, serial dilutions are required to permit quantitation.

The specificity and sensitivity of the SPRIA is usually excellent. Repeatedly positive reactions that cannot be confirmed as positive for HBsAg are highly unusual (less than 0.1%). Confirmation of positive results may be achieved by the use of another licensed HBsAg assay or by performance of a blocking or inhibition assay to determine whether unlabeled anti-HBs will specifically inhibit the reaction. The detection of anti-HBc in the absence of anti-HBs will also corroborate a positive HBsAg result (Hollinger and Dienstag, 1985).

The negative control sample must be comparable to the test sample if an erroneous interpretation is to be avoided. For example, when testing for HBsAg in saliva or semen, "normal" saliva or semen should be used as negative controls. In general, protein-deficient specimens like cerebrospinal fluid and recalcified plasma result in higher background levels.

In urgent cases, the SPRIA can be modified to provide a result within 1 h of receipt of the sample. The sensitivity of the shortened assay is only slightly less than that of the standard protocol, and usually the number of true positive specimens likely to be missed is relatively small. In this situation, the P/N ratio will often fall into the 1.5 to 2.1 range, and so the cut-off can be reduced to more than 1.5 times the mean negative control. Because this increased sensitivity will also increase the number of false-positive reactions, verification of such a positive result by the normal procedure is essential.

Anti-HBs. The solid-phase RIA test for the detection of anti-HBs is identical in principle with that for

HBsAg, except that the solid phase (usually a polystyrene bead) is coated with HBsAg. After the specimen is added to the bead, radiolabeled HBsAg is used to detect anti-HBs bound to the fixed HBsAg. Specimens are counted as above for HBsAg detection, and P/N ratios greater than 2.1 are considered positive.

With the introduction of the hepatitis B immunization programs, the need for improved accuracy and precision in anti-HBs testing has become apparent. Hollinger et al. (1981) have modified the RIA test described above, so that the results can be expressed in milli-international units per milliliter (mIU/ml), with the World Health Organization Reference Preparation for hepatitis B immunoglobulin (HBIG: lot 26/1/77) as reference standard. This preparation has been designated to contain 125 mIU/ml. To express the results of a sample in mIU/ml, the following formula is used (Hollinger and Dienstag 1985):

$$\frac{\text{Sample cpm} - \text{mean negative control cpm}}{\text{Reference control cpm} - \text{mean negative control cpm}}$$

The lower limit of detection of anti-HBs with currently available commercial kits (SPRIA) in 0.7 mIU/ml.

Anti-HBc. Most tests for anti-HBc are based on a competitive binding assay rather than the direct measurement described above. In a competitive assay, the test sample competes with a constant amount of [^{125}I]labeled anti-HBc (of human origin) for a limited number of binding sites on a solid phase precoated with HBcAg. Because of the nature of the test, the proportion of [^{125}I]labeled anti-HBc bound to the solid phase is inversely proportional to the concentration of anti-HBc in the test specimen, that is, a low test cpm indicates a positive result. Thus, a specimen is positive for anti-HBc if the cpm is less than the cut-off value (Hollinger and Dienstag, 1985). The cut-off value is determined by the formula:

$$\frac{P + N}{2}$$

where P is the cpm of the mean positive control and N is the cpm of the mean negative control.

HBeAg and Anti-HBe. Commercial RIA kits are also available for HBeAg and anti-HBe. The HBeAg test uses the sandwich principle to measure HBeAg in serum or plasma. The solid-phase is coated with anti-HBe, and after addition of the test sample and incubation, [^{125}I]labeled anti-HBe is added. A positive result is indicated by a P/N ratio >2.1 times the mean of the negative control.

The test for anti-HBe is a modified, competitive binding assay with the same reagents that are required for detection of HBeAg. The anti-HBe in the test serum competes with the anti-HBe-coated solid phase for a standardized amount of HBeAg. If the test sample contains anti-HBe, less HBeAg will bind to the solid phase, and, therefore, less [^{125}I]labeled anti-HBe will be bound when the sandwich is completed. In general, the greater the amount of anti-HBe in the specimen, the lower the binding will be. The amount of bound radioactivity is compared with the cut-off value, which is determined by dividing the sum of the mean negative control and the mean positive control by two.

In general, competitive assays based on the principles described above are only sensitive and specific when the class and species of antibody are the same. For example, IgM will not compete as effectively as does IgG for the same antigenic determinants. This is the reason that competitive assays are unreliable for the detection of IgM-specific antibodies, prompting the introduction of such antibody capture assays (see below) to measure this class of immunoglobulin.

Enzyme Immunoassay (EIA)

Adaptation of the technique of EIA to hepatitis B serology has been accomplished by preparing conjugates of appropriate antibody with enzymes such as horseradish peroxidase or alkaline phosphatase. The principle of the test is the same as for the radioimmunoassays described above, except that the final indicator system is an enzyme-substrate-reaction product that can be read visually or in a spectrophotometer.

A wide range of commercial kits is now available for detection of HBsAg, HBeAg, and anti-HBe by EIA (Halbert and Anken, 1977; Wolters et al., 1977). As in the SPRIA tests, appropriate positive and negative controls are included to ensure the reliability and specificity of the test. In EIA the final readings must be made within 2.0 h of the addition of acid (or base) that stops the enzyme–substrate reaction from proceeding further. Care should be taken to avoid compounds like sodium azide in wash solutions, since they act as inhibitors/poisons of most enzyme systems.

Assays for IgM-specific anti-HBc by EIA are also available. The most valuable tests are based on the antibody capture principle described originally for IgM antibody to hepatitis A virus (Duermeyer et al., 1979). The test sample is added to a polystyrene bead on which antibody specific for human IgM (μ-chain specific) has been absorbed, thus capturing all the IgM present in the test sample. The assay is completed by addition of HBcAg, followed by anti-HBc conjugated with horseradish peroxidase. The absorbance is proportional to the quantity of IgM-specific anti-HBc present in the patient's serum. A cut-off value of 0.25 times the sum of the mean of the posi-

tive control plus the negative control mean is determined. Test specimens giving absorbance values equal to or greater than this cut-off value are considered positive for IgM-specific anti-HBc. The test usually has an overnight incubation step, but can be modified in an emergency situation to give a result on the same day as the specimen is received in the laboratory.

The EIA is about as sensitive as RIA (Locarnini et al., 1978) but is superior to reverse passive hemagglutination and reverse passive latex agglutination. Recent improvements in the test have reduced the number of repeatable, false-positive reactions. However, confirmatory testing with known negative and positive sera is necessary (Hollinger and Dienstag, 1985).

INTERPRETATION OF LABORATORY DATA

From the preceding discussion, it is apparent that there is a complex system of diagnostic markers available for investigation of patients infected with HBV. Many studies have reported clinical correlations with the appearance, persistence, or disappearance of these various HBV parameters, antigens and their corresponding antibodies. Differentiation between an acute infection and a virus carrier state, with or without chronic hepatitis and with different degrees of direct viral replicative activity, is important. Such an assessment is based on the determination of these markers in association with the patient's history, general clinical data, clinical chemistry, and, when possible, liver immunohistology obtained through liver biopsy specimens (Bianchi and Gudat, 1983). A simplified scheme of these determinations and their interpretation are presented in Table 5 and will provide a framework for the discussion to follow.

Following infection with HBV, both humoral and cellular immune mechanisms are stimulated. The most frequent response to HBV infection is a transient, subclinical infection followed by clearance of virus, production of antibody, and apparent permanent immunity (see Fig. 6). This presentation accounts for the finding of antibody to HBV markers in an individual who denied a previous history of jaundice or hepatitis (Hoofnagle and Schafer, 1986). Patients with subclinical infections usually produce high levels of anti-HBs, which last for many years.

The three principal forms of clinical expression of infection with HBV are: acute hepatitis B, chronic hepatitis B, and the asymptomatic ("healthy") HBsAg carrier state.

Acute Hepatitis Type B: The "Typical" Case

Approximately 30% of adults infected with HBV (Fig. 6) develop acute hepatitis with jaundice, liver

TABLE 5. Serodiagnostic profiles in viral hepatitis

HBsAg	Anti-HAV IgM	Anti-HBc		Anti-HD IgM	Interpretation
		Total	IgM (titer)		
−	+	−		−	Acute hepatitis A
+	−	−		−	Prodromal acute hepatitis B
+	−	+	+ (high)	−	Acute hepatitis B
−	−	+	+ (high)	−	Acute hepatitis B
+	−	+	+ (low) or −	−	Chronic hepatitis B
+	+	+	+ (low) or −	−	Recent acute hepatitis A in a HBsAg carrier
−	−	−		−	Possible non-A, non-B hepatitis. Need to exclude EBV, CMV, HSV, toxic and drug-induced liver disease
+	−	+/−	−	+ (Usually persistent)	Superinfection with HDV
+	−	+	+	+ (Usually brief)	Coinfection with HDV

enzyme elevations, and symptoms (Hoofnagle et al., 1978). The serological and clinical course of a typical case of acute type B hepatitis is shown in Figure. 7. The incubation period between exposure and the onset of symptoms (usually defined as the onset of dark urine) ranges from 1 to 6 months. Acute hepatitis B infection can be divided into an early, replicative phase and a late, nonreplicative phase.

In the early, replicative phase, which includes the incubation period, HBsAg first appears in the blood, followed by HBeAg, HBV DNA, and DNA polymerase. When serum aminotransferase (alanine aminotransferase, ALT) activity begins to increase and symptoms first appear, the levels of HBV and HBsAg are usually at their peak or starting to decrease. By the stage that most patients present to their physicians, more than half are negative for serum HBV DNA and DNA polymerase (Alberti et al.,

1979; Krogsgaard et al., 1985). These markers of active viral replication disappear within 1 to 8 weeks of the onset of symptoms, heralding the second, nonreplicative phase, during which HBsAg is still present but there is no evidence of continuing viral replication. Figure 7 shows that HBsAg can usually be detected 2 to 4 weeks before the ALT level becomes abnormal, and 3 to 5 weeks before the patient develops symptoms or becomes jaundiced. HBsAg usually remains detectable in serum through the clinical illness, disappearing in convalescence.

In some patients, HBsAg can still be detected months after the onset of illness. This is usually because HBsAg is cleared slowly from blood, with a half-life of approximately 8 days (Frosner and Franco, 1984), so that it may require months for the high levels of HBsAg achieved during acute hepatitis

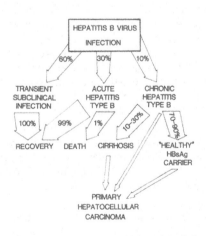

FIG. 6. Outcome of acute hepatitis B virus infections in adults. [Modified from Hoofnagle and Schafer (1986).]

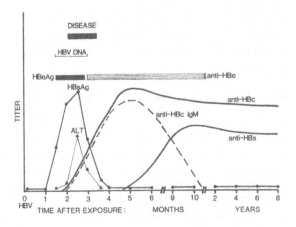

FIG. 7. Diagrammatic representation detailing the serologic responses following a "typical" acute hepatitis B virus infection.

B to reach undetectable levels. If there is concern or anxiety about the patient becoming a chronic hepatitis B carrier, then quantitative determination of HBsAg in serial sera is a useful guide. Nielsen et al. (1981) demonstrated that persistence of HBsAg in acute hepatitis B for more than 13 weeks is always associated with the development of chronic infection, whereas those patients who cleared HBsAg from their circulation by this time recovered completely. Frosner and Franco (1984) have claimed that monitoring the actual concentration of HBsAg in patients with acute hepatitis B is useful in predicting who will become chronic carriers. In patients who do not become carriers, a significant decrease in HBsAg concentration occurs within 12 days in 90% of patients, within 16 days in 98%, and in all patients by 20 days after the onset of symptoms. By contrast, in those patients who become chronic carriers, no significant decrease in HBsAg is observed.

HBeAg is usually found only in the presence of high titers of HBsAg and disappears at or soon after the peak in serum enzymes (Aikawa et al., 1978). The disappearance of HBeAg and the development of anti-HBe is considered a good prognostic marker (Hoofnagle et al., 1978).

The antibody responses in acute hepatitis B are complex (Fig. 7). Anti-HBc occurs shortly before or at the time of the onset of the illness (Hoofnagle et al., 1978) and is invariably present by the time jaundice is apparent (Krugman et al., 1974). The initial response is primarily IgM (Lemon et al., 1981); however, with recovery, IgG antibody increases while the anti-HBc IgM titers decline and usually disappear within 3 to 12 months. For this reason, anti-HBc IgM has become a useful method of diagnosing acute hepatitis B (Chau et al., 1983), especially in countries in which HB is hyperendemic and the presence of HBsAg alone does not establish the diagnosis. Very high titers are usually detected in the early to mid-phase of HBs antigenemia as well as in the "window" period of acute hepatitis B (that time between disappearance of HBsAg and first demonstration of anti-HBs). All patients with acute hepatitis type B produce anti-HBc, and this antibody probably persists for life (Hoofnagle and Schafer, 1986).

Anti-HBe is the next antibody to become detected (Aikawa et al., 1978). It usually appears when HBeAg becomes negative. Anti-HBe, unlike anti-HBc, rarely reaches high titers in acute hepatitis and often disappears within a few months or years. As a generalization, disappearance of HBeAg and development of anti-HBe are good prognostic markers (Hoofnagle et al., 1978), but do not guarantee complete clearance of HBV (Bonino et al., 1981; Scotto et al., 1983). Thus, it has been suggested (Gowans, 1986) that the HBeAg/anti-HBe testing system (Alter et al., 1976; Toug et al., 1977) be replaced with HBV

DNA detection and HBsAg quantitative determinations.

Anti-HBs generally appears during recovery, after HBsAg has been cleared. There is often a window period between the disappearance of detectable HBsAg and the appearance of anti-HBs (Hoofnagle et al., 1978). The appearance of anti-HBs is considered the marker of immunity; however, subclinical infections have been shown to develop in some patients who were known to be anti-HBs-positive. This uncommon occurrence could have been due to low (and not protective) concentrations of anti-HBs (Robinson, 1985) or to the presence of a subtype-specific anti-HBs response with subsequent infection with a different HBV subtype (Robinson, 1985).

Acute Hepatitis Type B: The Variants

A small proportion of patients with acute hepatitis B fail to produce anti-HBs despite the fact that they clear HBsAg and recover normally (Hoofnagle et al., 1978). A similar phenomenon has been observed following large-scale hepatitis B immunization (Szmuness et al., 1980). Whether these patients have a higher population of suppressor lymphocytes (Ts T11+, HNK-1, and T8+) that alters their normal immune response to HBV (Nowicki et al., 1985) or in some way are histocompatibility (MHC/HLA)-restricted (Walker et al., 1981) awaits further clarification.

In some series, up to 10% of patients with acute hepatitis B demonstrate atypical serological responses. These patients usually have a heightened and rapid clearance of HBV and are often HBsAg-negative when they present to their physicians (Hoofnagle et al., 1978). Typically, these patients rapidly develop substantial titers of anti-HBs without a window phase. Occasionally in patients with mild or fulminant hepatitis (Shimuzu et al., 1983) HBeAg may persist for a few weeks longer than HBsAg, yielding the unusual serological pattern of HBeAg without HBsAg (Hoofnagle and Schafer, 1986). This group of patients can easily be misdiagnosed as non-A non-B hepatitis, but testing for anti-HBc IgM will establish the correct diagnosis (Kryger et al., 1982a; Lemon et al., 1981).

Acute Hepatitis Type B: Summary

In a typical HBV infection, the following serological markers appear sequentially in serum: HBsAg, HBeAg, anti-HBc, anti-HBe, and then anti-HBs. Acute hepatitis B is usually diagnosed on the basis of a transient HBs-antigenemia in the acute phase of the illness; or a rising level of anti-HBc or a seroconversion to anti-HBc antibody between acute and convalescent serum samples; or the presence of a high titer of anti-HBc IgM in a acute-phase serum samples; or a seroconversion to anti-HBs antibody between acute and convalescent serum specimens; or a combination of the above.

Chronic Hepatitis Type B

Of adults with HBV infection, 5 to 10% do not eliminate HBsAg and go on to become chronic carriers of the virus (Hoofnagle et al., 1978; Schulman et al., 1980). The carrier state is usually defined as the presence of HBsAg in serum for a period of 6 months or longer (Fig. 8).

The initial pattern of HBV markers in patients who progress to chronic type B hepatitis is identical with that of patients with acute, self-limited disease (Hoofnagle and Schafer, 1986); however, in contrast to these cases, HBsAg titers do not decline as transaminase levels fall (Fig. 8). The persistence of HBV DNA, DNA polymerase, and high titers of HBeAg into the period of ALT elevations should suggest that chronic infection has become established (Aikawa et al., 1978; Alberti et al., 1979; Krogsgaard et al., 1985; Krugman et al., 1979).

Chronic hepatitis B infection can also be divided into two phases, the viral replicative phase and the nonreplicative phase:

Viral Replicative Phase. In the first phase, active viral replication occurs and is recognized by the detection of HBV DNA (Bonino et al., 1981; Brechot et al., 1981; Scotto et al., 1983), DNA polymerase (Alberti et al., 1979), HBeAg (Hoofnagle and Schafer, 1986), and anti-HBc IgM (Kryger et al., 1981; Sjogren and Hoofnagle, 1985). During this period, nuclear HBcAg can be identified in infected liver cells, and HBV DNA is episomal and not integrated (Bre-

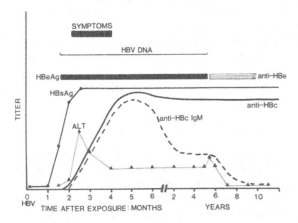

FIG. 8. Diagrammatic representation detailing the serologic responses commonly seen in chronic type B hepatitis. Also shown is the eventual loss of HBeAg and resultant seroconversion to anti-HBe 6 years after exposure to HBV.

chot et al., 1981; Hadziyannis et al., 1983; Shafritz et al., 1981).

The level of replication of HBV varies from carrier to carrier and seems to depend upon a number of factors, such as the age at which infection occurs. Although the amount of HBsAg produced by chronic carriers may be extremely high, the ratio of mature (Dane) particles to 22-nm spheres and tubules varies widely. The higher the proportion of mature virions, the more infectious the serum. While the simplest marker of infectivity is detection of HBcAg in the serum, the presence of HBV DNA is a more accurate and reliable marker (Brechot et al., 1981; Bonino et al., 1981; Pontisso et al., 1985).

Recently Hadziyannis et al. (1983) have produced evidence that the presence of HBV DNA is also useful for predicting which patients will develop chronic hepatitis. The HBV DNA assay can be used to monitor the clearance of HBV and has recently been shown to be extremely useful in diagnosing reactivation of HBV (see below). Where its value is doubtful is in fulminant hepatitis, where Scotto et al. (1983) failed to detect HBV DNA in the sera of HBeAg-positive patients with very high ALT values. In these situations, anti-HBc IgM testing will establish the diagnosis (see below).

The antibody response in chronic hepatitis B infection is marked by the development of prolonged high titers of anti-HBc and the absence of anti-HBs. The initial anti-HBc response is largely IgM (Fig. 8), which declines to low levels after several months (Fig. 8). The titer of anti-HBc IgM may be of use in distinguishing between acute and chronic hepatitis B infections (Chau et al., 1983). In chronic carriers of HBV, the level of anti-HBc IgM tends to correlate with the severity of the accompanying disease (Kryger et al., 1981; Sjogren and Hoofnagle, 1985). Further characterization of the anti-HBc IgM response in chronic hepatitis B has revealed that it is a 7 to 8S IgM molecule (Sjogren and Lemon, 1983; Tsuda et al., 1984), whereas in acute, self-limited hepatitis anti-HBc IgM exists essentially as the conventional 19S moiety.

Nonreplicative Phase. Chronic hepatitis B infection may alter with time (Fig. 8). For example, chronic carriers of HBsAg lose HBeAg and develop anti-HBe at a rate of 3 to 25% per annum, depending on the population studied, with an overall rate of at least 50% of patients in longitudinal studies (Liaw et al., 1983; Realdi et al., 1980; Sjogren and Hoofnagle, 1985). The loss of HBeAg is usually accompanied by the disappearance of serum HBV DNA and DNA polymerase activity, a lowering of serum ALT levels, and a marked improvement in histology (Hoofnagle and Alter, 1984). Interestingly, there is often a brief flare-up of disease just before seroconversion (see Fig. 8) (Liaw et al., 1983). This process heralds the second (nonreplicative) phase of the carrier state, in which HBV DNA is integrated into the host genome (Brechot et al., 1981; Kam et al., 1982).

In this phase HBsAg can usually be detected in the serum, although there is evidence that viral replication has stopped. The persistence of HBsAg is probably due to synthesis of HBsAg from molecules of HBV DNA that have become integrated into host chromosomal DNA (Brechot et al., 1981). No HBcAg, HBeAg, or HBxAg is synthesized or detected during this stage, probably because the mode of integration splits the genome in the region of HBcAg and HBxAg (Dejean et al., 1985). Integration of HBV DNA occurs commonly during the course of chronic HB infection and is probably an essential step in the development of primary hepatocellular carcinoma (Shafritz et al., 1981).

The persistence of HBsAg in the serum without evidence of viral replication or chronic liver disease is referred to as the "healthy" (asymptomatic) HBsAg carrier state. Chu et al., (1985) have postulated that chronic hepatitis B infection proceeds through three stages, not two. The initial replicative phase is characterized by high concentrations of HBV DNA and HBeAg in the serum and the absence of significant histological abnormalities. This is followed by an intermediate phase, in which either HBeAg or anti-HBe is present, HBV DNA continues to be detectable but in low concentrations, and liver biopsies show features of chronic liver disease. Ultimately a nonreplicative phase occurs, in which anti-HBe is detected, serum HBV DNA cannot be identified, and there is no histological evidence of continuing disease even though HBV DNA is integrated into the hepatocyte genome. This view accommodates the clinical diversity seen in the chronic phase of HBV infection and also presents the data in a way that reflects the natural history of the disease.

The final stage of the carrier state, disappearance of HBsAg, occurs at a rate of approximately 1 to 2% per year (Alward et al., 1985) and accounts for the steady decrease in the prevalence of carriers of HBsAg with increasing age (Szmuness et al., 1978). Subjects who lose HBsAg generally develop anti-HBs and normal liver tests (Lindsay et al., 1981), although some patients may be left with a degree of portal fibrosis or even post necrotic cirrhosis (Omata et al., 1978). The possibility that patients may develop sequelae of chronic hepatitis B infection despite the loss of HBsAg has been raised by several groups of investigators, who have detected low levels of HBsAg, HBcAg, or HBV DNA in the serum or liver of patients who are HBsAg-negative by conventional assays (Brechot et al., 1985; Wands et al., 1982). These findings have not been confirmed, and it

remains to be proved that their disease is due to HBV infection and not to infection with another agent (Feinstone and Hoofnagle, 1984).

Chronic Hepatitis Type B: Summary. During chronic HB infection, HBsAg and anti-HBc are present, and the patient can be either HBeAg- or anti-HBe-positive. Asymptomatic carriers of HBsAg are generally HBeAg negative and have low or undetectable levels of anti-HBc IgM. HBV DNA is rarely detected in serum samples. In patients with chronic "active" hepatitis, HBcAg is usually present and HBV DNA readily detected in serum. Anti-HBc titers are often high, and anti-HBc IgM usually persists.

The persistence of high titers for HBsAg, HBV DNA, and HBeAg correlates with an increased risk of the patient developing chronic liver disease. Conversely, the persistence of low titers of HBsAg, the absence of HBV DNA, and presence of anti-HBe tend to correlate with more benign infections, such as the asymptomatic carrier state or chronic persistent hepatitis.

Hepatitis D Virus (Delta Virus)

DETECTION OF HDAG

Immunofluorescence (IF) and immunoperoxidase (IP) staining are useful techniques for detecting HDAg in liver tissue from patients or experimentally infected animals. These techniques are restricted to research laboratories and are not suitable for use as screening tests in clinical laboratories. Antibodies (which can be conjugated with fluorescein or peroxidase) are usually obtained from the serum of a patient with high titers of anti-HD and a low level of anti-HBc. Although it is possible to reduce or eliminate the anti-HBc reactivity by simple dilution, suitable controls such as an anti-HD-negative anti-HBc-positive serum should be used to confirm the specificity of any reaction detected.

HDAg is located primarily in hepatocyte nuclei (Rizzetto et al., 1977) but can occasionally be detected in the cytoplasm as well. For detection of HDAg by IF or IP, frozen sections are preferable; however, it possible to detect the antigen in stored, paraffin-embedded tissue, usually after digestion of the section with trypsin or pronase (Crivelli et al., 1983).

DETECTION OF HDV RNA

The development of cloned cDNA probes (Smedile et al., 1984) has provided a mechanism for the detection of HDV RNA in serum by dot-blot hybridization (Northern) analysis. The principles of DNA–RNA hybridization are essentially the same as the Southern hybridization for HBV DNA described in earlier sections. Denniston et al. (1986) have used this technique to identify HDV RNA in the serum of patients with either acute or chronic HDV infection and demonstrated much higher (50 to 500-fold) levels in patients with acute infections. In an assessment of the clinical usefulness of this assay, Smedile et al. (1984) found HDV RNA in the serum of 61% of patients during the acute phase of their illness, but never during the recovery phase. The presence of HDV RNA correlated with other markers of active HDV replication (including raised titers of anti-HD IgM).

SEROLOGIC IDENTIFICATION OF DELTA ANTIGEN AND ANTIBODY

Antibodies to HDAg have usually been detected and quantitated by RIA or EIA (Crivelli et al., 1981; Rizzetto et al., 1980d). The major limitation to these techniques has been a shortage of HDAg. The recent development of commercial immunoassays has resulted in a significant increase in the number of diagnostic laboratories performing these tests.

Preparation and Purification of HDAg and Anti-HD Reagents

Tests for anti-HD require a reliable source of HDAg, which is usually extracted from aliquots of HDAg-positive human (postmortem) or chimpanzee livers. Occasionally HDAg-positive serum can be used (Dimitrakakis et al., 1984). More recently, HDAg-positive woodchuck liver has become available and provides a reliable source of antigen. Liver tissue (usually 2.0 g) is finely chopped and washed in cold PBS, then extracted with strong dissociating agents such as 6 M guanidine HCl or 8 M urea. The sample is clarified, dialyzed, and then diluted for use after assay for HDAg activity (Dimitrakakis et al., 1986b). When liver tissue is obtained from experimentally infected chimpanzees or woodchucks prior to the appearance of anti-HD, HDAg can be harvested by simple aqueous extraction.

Serum containing anti-HD can be collected from patients or experimentally infected animals with either acute or chronic delta infection. As mentioned above, anti-HBc reactivity is invariably present in anti-HD-positive serum samples. Fortunately, it is often low in those samples that are high in anti-HD activity. Residual anti-HBc activity can be easily diluted out, and HBsAg is removed by ultracentrifugation. The IgG component of the anti-HD-positive serum specimens can be readily prepared by processing the sample on protein-A Sepharose or by ion exchange chromatography (Dimitrakakis et al., 1984).

To detect IgM-specific anti-HD, capture assays based on the methods of Hansson et al., (1982) have

been most commonly used. A commercially prepared goat anti-human IgM or rabbit anti-human IgM (μ-chain-specific) is usually quite suitable for adaptation in this assay (see below).

Radioimmunoassay

HDAg. Delta antigen can be detected by a solid-phase sandwich RLA, as described by Rizzetto et al., (1980d). The test is based on the binding of HDAg in serum to anti-HD adsorbed to a solid phase (polyvinyl microtiter plate). The HDAg bound to the plate is then detected by incubation with [^{125}I]labeled, IgG-purified anti-HD. After washing, each sample is measured in a gamma counter and is considered positive for HDAg if the P/N value is >2.1. Because HDAg is enclosed within an HBsAg encapsidated virion, specimens must be treated with detergent (0.6% Nonidet P40, 0.5% Tween-80 or deoxycholate) prior to testing. This SPRIA for HDAg can be easily modified for detection of anti-HD in a blocking RLA (Rizzett et al., 1980d) (see below).

Anti-HD Blocking RIA. A standardized quantity of HDAg is added to either wells or beads coated with anti-HD. Test serum or plasma samples containing unlabeled anti-HD are added, and time is allowed for any antibody present to bind to the HDAg particles. A constant amount of the [^{125}I]labeled anti-HD (standardized probe) is then added, and, after washing, the residual bound radioactivity is measured in a gamma counter. When compared with the negative controls, sera that blocked 50% or more of the binding of radiolabeled antibody are considered positive for anti-HD (Rizzetto et al., 1980d; Dimitrakakis et al., 1984).

Competitive RIA. A commercial RIA for anti-HD has recently become available (Abbott Anti-Delta, Abbott Laboratories). The test is a competitive binding RLA in which anti-HD in the test serum competes with [^{125}I]labeled human anti-HD for woodchuck liver-derived HDAg coating a solid-phase polystyrene bead. After appropriate incubation and washing steps, the net cpm (minus background counts) for the particular test sample are compared with a calculated cut-off (0.4 of the mean negative control counts [N] plus 0.6 of the mean positive control counts [P]). Samples with counts equal to or below the cut-off range are considered positive. Validity of the test requires that, for each run, the ratio of the mean-negative-control counts to the mean-positive-control counts should be >4.0. The titer of anti-HD can be determined by testing serial dilutions of serum or plasma and selecting the dilution that yields counts closest to but not greater than the cut-off value, as the final anti-HD titer (Hollinger and Dienstag, 1985).

Enzyme Immunoassay

HDAg and Anti-HD. HDAg can now be measured and investigated by a commercially available enzyme immunoassay (Deltassays, Nochtech, Dublin, Ireland), which employs serum as the source of delta antigen (Shattock and Morgan, 1984). While the principle of the assay is essentially the same as for the SPRIA, several workers have found it to be more sensitive for the detection of HDAg (Buti et al., 1986; Shattock et al., 1983). Crivelli et al. (1981) have described an EIA for detection of anti-HD, which has a configuration similar to that described for the commercial RIA test.

Anti-HD IgM. Acute delta hepatitis infection (either coinfection with acute HBV infection or acute hepatitis delta superinfection of an HBV-positive carrier) is accompanied by an early anti-HD response predominantly of the IgM class. Detection of anti-HD IgM is usually performed as described by Hansson et al. (1982) and Smedile et al. (1982), using a capture solid-phase RIA for anti-HD IgM. Briefly, antibody to human IgM (μ-chain-specific) is bound to a solid phase, and the test serum is added. If the test serum contains anti-HD IgM, it will bind to the solid phase; when the HDAg is added subsequently, it binds to the IgM anti-HD. The sandwich is completed by addition of [^{125}I]labeled anti-HD (IgG fraction). The assay conditions should be optimized with respect to both the capture antibody (goat or rabbit anti-human IgM, μ-chain-specific) and the standardized HDAg used as a probe (Dimitrakakis et al., 1986b). Specimens are considered positive for anti-HD IgM when the number of counts per minute (cpm) is at least 2.1 times greater than the mean cpm of the negative controls. Standardized positive controls should be used, so that the end-point dilution of the standard positive control for anti-HD IgM is equivalent to that reported by Smedile et al., 1983).

Specific IgM tests based on the competitive assay principle are not so sensitive or reliable as the μ-capture immunoassay. This is because anti-HD IgM does not compete as efficiently as anti-HD IgG (the labeled probe) for the same antigenic determinants on HDV.

INTERPRETATION OF LABORATORY DATA

There are several features of HDV infection that need to be considered when interpreting laboratory data. First, the humoral antibody response is relatively poor when compared with that for HAV or HBV. Anti-HD can be detected only by repeated testing over several weeks. The presence of anti-HD IgM is the single most useful diagnostic marker and often provides crucial prognostic information. Waning levels of anti-HD IgM confirm resolution of HDV infection, while persistence indicates that the infection has become chronic. In self-limited disease, anti-HD IgM is transient, with IgG antibody developing in convalescence. In chronic HDV infection, a brisk IgM antibody response is seen in the acute phase with persistence of both IgM and IgG anti-HD in the convalescent period (Aragona et al., 1987). Also, the role of other markers of hepatitis B virus infection need to be evaluated. Although anti-HBc IgM does not distinguish absolutely between acute and chronic HBV infection, its presence is a reliable indicator of recent infection, while its absence is a reliable indicator of infection in the remote past. Therefore, in simultaneous acute HBV and HDV infection (coinfection mode), anti-HBc IgM will be detected, whereas in acute delta hepatitis infection (superinfection mode) in chronic carriers of HBsAg, anti-HBc will be primarily of the IgG class.

Coinfection Mode

In coinfection the clinical and biochemical features are usually indistinguishable from those of HBV infection alone (Rizzetto, 1983). The presence of delta hepatitis infection can be identified by demonstrating the presence of HDAg in the serum or in liver biopsies or, more practically, by detecting a rise in titer of anti-HD or anti-HD IgM. Diagnosis of acute coinfection with HBV and HDV can be difficult, because the anti-HD response may be only of the IgM class and transient, without subsequent appearance of anti-HD IgM (Smedile et al., 1982). In acute infection, HDAg has been detected by both RIA and EIA (Crivelli et al., 1981; Rizzetto et al., 1980d). Recent studies by Shattock et al., (1983) and Buti et al. (1986), using the same ELA method (Deltassays, Nochtech, Dublin, Ireland), have demonstrated that HDAg can be detected in the serum within 2 weeks of the onset of symptoms, but disappears rapidly (Fig. 9).

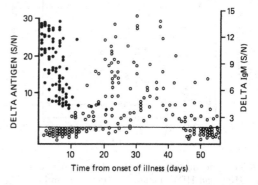

FIG. 9. Detection of delta antigen (●) and anti-delta IgM (○) in serial sera collected from 56 patients coinfected with HBV and HDV. S/N = Specimen activity/mean activity of negative control sera. [Reprinted from Dimitrakakis et al. (1986b), with permission.]

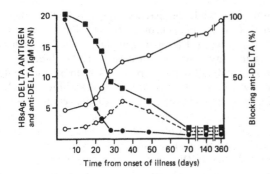

FIG. 10. Detection of anti-delta IgM in a patient coinfected with HBV and HDV. Pattern 1: Transient, anti-delta IgM response followed by development and persistence of blocking antibody. Delta antigen (●), anti-delta IgM (○— —○), blocking anti-delta (○———○), HBsAg (■). S/N = Specimen activity/mean activity of negative control sera. [Reprinted from Dimitrakakis et al. (1986b), with permission.]

Dimitrakakis et al., (1986b) demonstrated two different antibody patterns in patients coinfected with HBV and HDV (see Figs. 10 and 11). The first pattern is characterized by a transient, anti-HD IgM response, with subsequent development of long-lasting blocking anti-HD; the second pattern, by a transient, anti-HD IgM response and failure to develop blocking anti-HD (see below for discussion).

Superinfection

In contrast to patients with acute HBV infection, patients with chronic HBV infection can support

HDV replication indefinitely. Acute HDV infection becomes chronic when it is superimposed on an underlying chronic HBV infection. In these cases, the delta superinfection appears as a clinical exacerbation or an episode of acute hepatitis in someone already chronically infected with HBV.

The serologic response to HDV superinfection is usually characterized by the early appearance of HDAg, which is rapidly replaced by anti-HD IgM. As in coinfection, delta antigen is detected transiently or not at all in patients presenting early in their illness and usually is accompanied by a significant decrease in HBsAg levels.

Again, Dimitrakakis et al., (1986b) have recognized two distinct patterns of antibody response in patients superinfected with HDV (Figs. 12 and 13). The first pattern was seen in patients whose liver function tests returned to normal and was characterized by stationary low levels of anti-HD IgM and the development and persistence of high titers of blocking anti-HD. The second pattern was characterized by fluctuating levels of anti-HD IgM and biochemical evidence of continuing liver damage. The implication is that recrudescence of activity of HDV is associated with exacerbation of liver disease and a rise in specific IgM levels.

It is often possible to distinguish between coinfection and superinfection with HDV on the basis of the duration of the anti-HD IgM response. In coinfection, the response is brief (lasting only 2 to 3 weeks) and usually at a low level, while in superinfection it is usually prolonged and at a high level (Dimitrakakis et al., 1986b). These investigators have also demonstrated that, in patients coinfected with HDV and

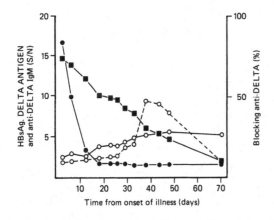

FIG. 11. Detection of anti-delta IgM in a patient coinfected with HBV and HDV. Pattern 2: Transient, anti-delta IgM response and failure to develop blocking antibody. Delta antigen (●), anti-delta IgM (○— —○), blocking anti-delta (○———○), HBsAg (■). S/N = Specimen activity/mean activity of negative control sera. [Reprinted from Dimitrakakis et al. (1986b), with permission.]

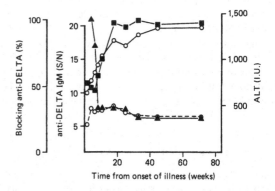

FIG. 12. Detection of anti-delta IgM in a chronic carrier of HBsAg superinfected with HDV. Pattern 1: Low levels of anti-delta IgM and persistent high levels of blocking antibody. Anti-delta IgM (○— —○), blocking anti-delta (○———○), HBsAg (■), ALT international units (▲). S/N = Specimen activity/mean activity of negative control sera. [Reprinted from Dimitrakakis et al. (1986b), with permission.]

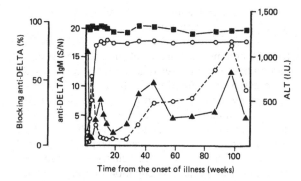

FIG. 13. Detection of anti-delta IgM in a chronic carrier of HBsAg superinfected with HDV. Pattern 2: Fluctuating levels of anti-delta IgM and ALT and persistent high levels of blocking antibody. Anti-delta IgM (O— —O), blocking anti-delta (O———O), HBsAg (■), ALT international units (▲). S/N = Specimen activity/mean activity of negative control sera. [Reprinted from Dimitrakakis et al. (1986b), with permission.]

HBV, only 50% develop blocking anti-HD. Therefore, epidemiological studies based on this test alone are likely to underestimate the prevalence of coinfection. The prevalence of coinfection is most accurately assessed by testing specimens of sera, collected from patients early in the course of their illness for anti-HD IgM.

The detection anti-HD IgM may be of value to the clinicians managing chronic carriers of HBsAg. In patients with progressive liver disease, anti-HD IgM tends to persist at high titer for many years, whereas in those patients whose disease improves or resolves it declines or disappears (Farci et al., 1986).

Serodiagnosis of Acute Viral Hepatitis: An Overview

The serodiagnosis of acute viral hepatitis is best made by testing for four serological markers: anti-HAV IgM, HBsAg, anti-HBc IgM, and anti-HD IgM. Sensitive and specific immunoassays for each of these markers are commercially available, the pattern of results of which will usually provide a serological diagnosis (Table 5).

Acute type A hepatitis can be diagnosed by the finding of IgM anti-HAV (Locarnini et al., 1979) in acute-phase sera. Since IgM anti-HAV is usually present at the onset of clinical illness, repeat testing during convalescence is not usually necessary. One rare problem with this assay is that it may remain positive for as long as a year after clinical illness, and in patients with two episodes of hepatitis in a 12-month period, IgM anti-HAV results may be unreliable (Hoofnagle and Schafer, 1986).

Acute type B hepatitis can usually be diagnosed by the finding of HBsAg in serum of a patient with acute hepatitis. However, confirmation that the disease is acute rather than chronic can sometimes require that the serum also be tested for IgM anti-HBc. A commercial immunoassay that detects only high titers of IgM anti-HBc is available (Chau et al., 1983) and has reasonable specificity for acute type B hepatitis. Unfortunately, IgM anti-HBc is not a perfect marker for acute hepatitis B. In some cases IgM anti-HBc can be detected only after 1 to 2 weeks of illness, while in other instances IgM anti-HBc is never detected (Hoofnagle and Schafer, 1986). If high-titer IgM anti-HBc is present, however, this is helpful in establishing the diagnosis of acute hepatitis B.

Since up to 10% of patients with acute hepatitis B can present to the physician when HBsAg is no longer detected, IgM anti-HBc testing prevents these cases from being mistakenly diagnosed as due to non-A, non-B hepatitis (Kryger et al., 1981,1982a, 1982b; Shimizu et al.,1983). Thus, IgM anti-HBc testing is not essential for serodiagnosis of acute hepatitis, but it can provide helpful information in establishing that HBsAg-positive cases are acute rather than chronic hepatitis and that HBsAg-negative cases are due to a resolving hepatitis B rather than to non-A, non-B hepatitis.

Patients with acute hepatitis who have HBsAg without IgM anti-HBc in serum present a difficult diagnostic problem (Table 5) (Hoofnagle and Schafer, 1986). In these patients, the acute liver disease may be type A hepatitis (IgM anti-HAV); delta hepatitis (anti-HDIgM); non-A, non-B hepatitis; or even a nonviral form of acute liver injury (such as drug-induced liver disease). Thus, a negative IgM anti-HBc result suggests that HBsAg is present as a result of chronic HBV infection and that a superimposed liver disease accounts for the acute symptoms and signs of hepatitis.

Another cause of acute hepatitis in a chronic HBsAg carrier is reactivation of chronic type B hepatitis (Davis and Hoofnagle, 1985; Davis et al., 1984). In reactivation, a patient with mild chronic hepatitis B or in the healthy carrier state redevelops markers of hepatitis B virus replication (HBV DNA, DNA polymerase, and, in some situations, HBeAg). In association with this return to a replicative state of HBV infection, there is usually a return of the hepatitis. Some cases of reactivation are severe and can mimic acute viral hepatitis.

Acute delta hepatitis used to be a difficult serologic diagnosis (Rizzetto, 1983; Smedile et al., 1982). However, application of an anti-HD IgM assay will resolve most situations. Delta coinfection will present as acute hepatitis B; the patient will be positive for both HBsAg and IgM anti-HBc, but in addition should also have transient titers of anti-HD IgM.

Occasionally, delta coinfection may result in two peaks of serum aminotransferase activity (Rizzetto, 1983).

Delta superinfection will usually present as an acute hepatitis in a known chronic HBsAg carrier: the patient will be positive for HBsAg and have persistently elevated levels of anti-HD IgM, but be anti-HBc IgM negative. The diagnosis of delta infection should be considered particularly if the patient is an intravenous drug user or a hemophiliac, or if the hepatitis is severe.

The absence of IgM anti-HAV, HBsAg, and IgM anti-HBc is the serologic pattern found in cases of non-A, non-B hepatitis (Dienstag, 1983a, 1983b). Non-A, non-B is a diagnosis of exclusion, and HAV and HBV are not the only viruses that need to be excluded (Hoofnagle and Schafer, 1986). Acute mononucleosis; CMV infection; primary or secondary syphilis; cholangitis; acute, drug-induced liver injury; and autoimmune, chronic, active hepatitis also need to be considered.

Epidemiology and Natural History

Hepatitis B Infection

Hepatitis B differs from most other viral infections in that some people fail to eliminate the virus and become chronic carriers; this results in a permanent reservoir of infection in the community. The risk of becoming a chronic carrier is greatest if infection is acquired in the first years of life. Babies who are infected at or shortly after birth, for example, have a greater than 85% chance of becoming carriers (Stevens et al., 1985). The probability that infection will be followed by chronic carriage of the virus declines during the first few years of life, and from the age of 5 is between 5 and 10%.

Several studies have demonstrated that prolonged carriage of HBV is associated with an increased risk of developing chronic active hepatitis, cirrhosis, and primary hepatocellular carcinoma (Dudley et al., 1972; Redeker, 1975). Since there are more than 200 million chronic carriers of HBsAg in the world, it is estimated that more than 40 million will die of the long-term sequelae of what is now a preventable and potentially eradicable disease.

GLOBAL DISTRIBUTION

Evidence of infection with hepatitis B has been obtained in every population studied, irrespective of geographical location, ethnic origin, or stage of development. It is widespread among Eskimos in Northern Canada, the Polynesians of the South Pacific, and the Bushmen of Africa, as well as being a common cause of acute disease in cities such as London, New York, and Rome.

Since subclinical infections with HBV are extremely common, estimates of the prevalence of infection based upon cases of acute hepatitis B grossly underestimate its importance. In the United States, as in many developed countries, clinical hepatitis B is most commonly seen among intravenous drug users (Levine and Payne, 1960), male homosexuals (Kryger et al., 1982b), contacts (especially sexual partners) of people who are acutely or chronically infected, and certain occupational groups whose work brings them into contact with blood or blood products (Nest, 1984). In Europe the incidence of hepatitis B ranges between 4 and 30/100,000 per year, with the lowest figures reported from the Netherlands, Scandinavia, and the United Kingdom; intermediate figures from Switzerland and Germany; and the highest incidence in Italy and Greece (Deinhardt and Zuckerman, 1985). Figures from the Eastern Mediterranean region and Southeast Asia are relatively rare, but reports from Egypt and Hong Kong suggest an incidence of 20 to 30/100,000 per year. A great deal of information is available on th prevalence of serological markers of hepatitis B infection in different populations. Although many of the data cannot be compared directly (because different markers were sought, different techniques used, and different populations studied), certain generalizations can be made.

The world can be broadly divided into three regions according to the prevalence of infection with hepatitis B virus. At one end of the scale are many countries located in Southeast Asia, sub-Saharan Africa, and the Western Pacific, in which infection is almost universal. In these populations infection is acquired early in life and is usually subclinical. Carrier rates are in excess of 5% and may reach 20% or even higher in some populations (Krugman, 1983), and the remainder of the population has serological evidence of past infection manifested by the detection of anti-HBs, anti-HBc, or both. In most of these countries early childhood infection and high chronic carrier rates are associated with an increased incidence of chronic liver disease and primary hepatocellular carinoma.

At the other extreme are Scandinavia, the United Kingdom, several of the countries of Western Europe, the United States, Canada, Australia and New Zealand, in which the prevalence of infection is low. In these countries, less than 10% of the population have serological evidence of past infection with hepatitis B virus, and the carrier rate is about 0.1%. Between these two extremes are a large number of countries such as those in South America, Northern Africa, the Middle East, and southern and eastern Europe, in which the prevalence of infection is quite high. In these countries 1 to 5% of the population may be chronic carriers of HBsAg, and the prevalence of antibodies ranges from 20% to 40%.

While these general patterns exist, they oversimplify what is often a very complicated epidemiological situation. Many countries are not racially homogeneous, so that, while the predominant pattern may be as described, the prevalence of infection may vary within different groups of the population. For example, on the basis of studies of healthy, first-time blood donors, Australia is usually quoted as a country in which the prevalence of chronic carriers is 0.1% (Beal, 1976). If these figures were extrapolated to the whole country, they would produce a pool of about 15,000 chronic carriers. This calculation assumes that blood donors are a typical cross-section of the community, which they are not, since a number of groups, such as migrants and aboriginals (in whom the prevalence of infection with hepatitis B virus is high, are under-represented. Approximately 11% of the Australian population are migrants from Europe or Asia, and a further 1% are aboriginals or Torres Strait Islanders. Since the prevalence of HBsAg in these groups is 2.8%, 15%, and 26% respectively, hepatitis B infection is rather more common than realized, and the actual number of chronic carriers is probably in the vicinity of 150,000 (Gust, 1984).

Patterns of infection with hepatitis B virus may change dramatically owing to changing social or sexual mores (intravenous drug use, promiscuity), environmental changes (screening of blood donors, large-scale use of vaccine) or migration. Introduction of families from areas with a high prevalence of infection into communities in which this infection is rare may increase the risk of transmission and the incidence of disease.

The risk of transmission depends on the intimacy of the contact. Two interesting and different epidemiological situations exist in the Western Pacific region, in Fiji and New Zealand. Fiji has a population of about 600,000 people, of whom half are Melanesian and half Indian (the descendants of laborers introduced at the turn of the century). Hepatitis B infection is common among Melanesians, and the carrier rate is high. By contrast, the carrier rate and infection rate among the Indian population is low. The two groups lead essentially separate lives, with separate housing, separate schools, separate political parties and little intermarriage; few opportunities exist for transmission of infection (Gust, 1984).

By contrast, extensive transmission of infection has been documented between Maori and non-Maori children in some parts of New Zealand. In this country about 85% of the population are of Anglo-Saxon origin; the remainder are Maori or Pacific Islanders, principally from Niue and the Cook Islands. Several studies in mixed communities have demonstrated that horizontal transmission of HBV from Maori to non-Maori children is common, with up to 60% of adolescent Caucasian children showing serological evidence of past infection (Milne et al., 1985).

MODE OF TRANSMISSION

Traditionally, hepatitis B has been thought of as a disease associated with transfusion of contaminated blood or blood products, sharing of contaminated needles and syringes, and, to a lesser extent, promiscuity. While it is true that these factors are (or have been) important in the etiology of clinical hepatitis in developed countries, they represent the tip of an iceberg of infection, most of which is subclinical and most of which is acquired by intimate contact.

On a global scale, early childhood infections are the most important, not only because they are the most common, but because they are likely to be followed by a chronic carrier state. Two different patterns of the transmission occur in Asia and Africa, the regions in which hepatitis B infection has the highest incidence. In China, Taiwan, and Hong Kong, a high proportion of infections are acquired in the perinatal period, usually by transmission from a carrier mother. In these countries, 5 to 15% of women of childbearing age are chronic carriers of HBsAg, and a high proportion (more than 50%) are HBe antigen (HBeAg)-positive, a condition associated with a high risk of transmission (Chin et al., 1981; Stevens et al., 1985). By contrast, in sub-Saharan Africa, although childhood infection is common, transmission appears to be largely horizontal (Basalamah et al., 1984). In these countries, although the prevalence of HBsAg among women of childbearing age is high, a relatively low proportion are HBeAg-positive. This fact and the high prevalence of HBeAg among infected children probably account for the low rate of vertical and the high rate of horizontal transmission.

In developing countries in which hepatitis B is hyperendemic, children who escape infection in the first 5 years of life usually acquire it during their first few years at school. The precise mode of transmission is unknown, but since HBV is found in the blood, saliva, and serous exudates (Karayiannis et al., 1985), it is likely that direct contact of the fluids with the damaged skin or mucosal surfaces is responsible. Although tattooing and ritual scarification provide an opportunity to transmit blood-borne infections, these activities usually occur at or after puberty, by which time hepatitis B transmission has usually occurred.

In developing countries that utilize unscreened blood for transfusion the risk of acquiring posttransfusion hepatitis B depends on the prevalence of HBsAg and anti-HBs in the population. In countries in which most adults have been infected, posttransfusion hepatitis is essentially unknown. In other countries such as Burma, where the prevalence of infection is moderate, up to 10% of patients receiving transfusions develop hepatitis B (Khin Maung Tan, 1985, personal communication).

Although WHO has recommended screening all blood donors for HBsAg, this practice is largely restricted to developed countries. In developing countries it is either not employed or limited to a few large urban centers.

With the licensing and wide-scale use of vaccines against hepatitis B, the epidemiology patterns are likely to change dramatically in the next few years. Hepatitis B vaccine is available throughout Asia and is being produced in China, Korea, and Japan. Several countries have embarked on programs to immunize all newborn babies, and China is aiming at 85% coverage by 1995.

In developed countries, certain groups within the community are at higher risk of acquiring HBV infection. The prevalence of infection among intravenous drug users who share needles and syringes is high, and it is significantly higher in homosexual men than in heterosexual men (Kryger et al., 1982b). Residents of institutions for the mentally retarded have a high risk of infection (Williamson et al., 1982), while health-care workers have about a four times greater chance of acquiring HBV infection than the rest of the population (Denes et al., 1978; Mosley et al., 1975). Measures such as screening of blood donors, segregation of infected dialysis patients, improved laboratory safety, and selective use of hepatitis B vaccine have resulted in major changes in the patterns of infection (Alter et al., 1984), although little impact has been made to date on transmission among intravenous drug users and male homosexuals.

Hepatitis Delta Infection

GLOBAL DISTRIBUTION

When first recognized, delta infection was thought to be restricted to the Italian population, with small pockets in other parts of Europe (Rizzetto et al., 1980c). The delta agent has now been recognized in many parts of the world, although it appears to have an irregular geographic distribution (Rizzetto, 1983; Dimitrakakis and Gùst, 1984). Infection is endemic in the Balkan peninsula and the eastern Mediterranean, in the Middle East, some countries of northern Africa, the Amazon basin of South America, and some Pacific Islands. Elsewhere, for example in central and northern Europe, the United States, and Australia, delta infection appears to be a relatively new phenomenon, which is restricted to intravenous drug

users and their sexual contacts (Raimondo et al., 1982).

Coinfection

Several studies of patients admitted to hospitals with acute hepatitis B have been reported. At Fairfield Hospital, Melbourne, stored sera are available from every patient with acute hepatitis B admitted since 1971. Of 1,929 patients admitted between January 1971 and July 1985, 1,136 (59%) were intravenous drug users. Although 166 patients showed serologic evidence of coinfection with the delta agent, 159 of these were intravenous drug abusers; thus, the overall prevalence of coinfection among drug users was 14% compared with less than 1% among non-drug users. When panels of stored sera were examined, delta infection was shown to have been introduced into the population in the early 1970s and to have become steadily more prevalent.

A study from Malmo, Sweden, in which 600 patients with acute hepatitis B and nearly 200 chronic carriers of HBsAg were tested for markers of delta infection, showed that intravenous drug users constituted the vast majority of cases (Hansson et al., 1982). In Malmo it appears that the delta agent was introduced into the drug user population in about 1973 and that by 1981, 73% of intravenous drug users with chronic HBsAg carriage had evidence of concomitant delta infection.

In the United Kingdom the prevalence of delta markers among intravenous drug users with hepatitis B infection has been relatively stable since 1976 and ranges from 13% to 26%, depending on the city studied (Ponzetto et al., 1983; Weller et al., 1983). Despite the association of hepatitis B infection with the homosexual community, the prevalence of delta markers among this group is still exceedingly low.

In the United States, delta infection has been identified most commonly in intravenous drug users and in hemophiliacs, and studies suggest that it has been present since 1972 (Ponzetto et al., 1984b). In 1983–1984 an outbreak of hepatitis that occurred among intravenous drug users in Massachusetts appears to have been due to coinfection of HBV and HDV (U.S. Department of Health and Human Services, 1984). In total, of 50 cases of acute hepatitis B were reported, with 6 deaths, giving a case fatality rate of 12%. Delta markers were not sought in every instance, but 3 of 4 patients tested prior to death had circulating anti-delta IgM, and in 4 of 22 patients with non-fulminant hepatitis, anti-delta IgM was detected.

There are few data on the prevalence of coinfection in many of the most populous areas of the world. Recently we tested sera from more than 200 patients with hepatitis B admitted to hospitals in Japan, the

Philippines, Singapore, and Malaysia. Evidence of coinfection was obtained on only one occasion (Dimitrakakis et al., 1986a).

Superinfection

Accurate data on the incidence of superinfection are difficult to obtain, since specific laboratory tests are unavailable in most countries. Such data as exist are difficult to compare because they have been obtained from a variety of different groups. Many of the published data come from retrospective studies of sera collected from healthy chronic carriers of HBsAg (or carriers whose health status is unknown (Rizzetto et al., 1980c). Since superinfection with the delta agent carries a very high risk that the patient will develop chronic liver disease, surveys that concentrate on healthy carriers will underestimate the importance of delta infection in the community. For example, we recently tested 300 chronic carriers of HBsAg with normal-liver-function tests and a similar number with abnormal-liver-function tests for delta markers. Whereas no markers were detected in the former group, approximately 20% of carriers with abnormal-liver-function tests had evidence of superinfection with HDV.

Despite limitations in the data, certain patterns are beginning to emerge. Delta infection appears to be endemic in southern Italy (where 50 to 60% of carriers of HBsAg have antibodies to delta antigen) and parts of the Balkan peninsula, North Africa, the Middle East, Nauru, Niue, Western Samoa, and American Samoa (Dimitrakakis and Gust, 1984; Rizzetto, 1983; Rizzetto et al., 1980c). Despite the high prevalence of hepatitis B infection, superinfection with the delta agent appears to be rare in Japan, Singapore, and the Philippines. A low level of infection has been reported from Taiwan, restricted largely to intravenous drug users (Chen et al., 1984).

Superinfection with HDV is common among intravenous drug users in many countries and among haemophiliacs, with isolated reports of its occurrence in male homosexuals and institutionalized, mentally retarded children.

Mode of Transmission

The routes of transmission of HDV appear to be similar to those of HBV. The direct inoculation of infected blood or blood products represents the most effective mode of transmission and is responsible for infection among intravenous drug users and hemophiliacs. Nonparenteral spread via mucosal exposure to infectious body fluids and transfer of infected serum from open skin wounds may account for the endemic situation in southern Europe and parts of Africa and the Middle East.

The risk of developing delta infection with blood that has been screened for HbsAg is very low. On the other hand, infection of HBsAg-positive hemophiliacs (treated with coagulation factors prepared from pooled plasma obtained from several thousand donors) has been documented. The prevalence of delta infection in these individuals approaches 50% in Italy, Germany, and the United States (Rizzetto, 1983; Rizzetto et al., 1980c, 1982) and appears to be an important cause of chronic liver disease in this group.

Since delta infection requires the ongoing replication of HBV, vertical transmission of delta can occur only if HBV is also transmitted. Most pregnant women with delta infection are not HBeAg-positive and are, therefore, relatively unlikely to transmit HBV infection to their babies (Alessandro et al., 1983). In the rare instance in which a pregnant woman is a carrier of HBsAg, is HBeAg-positive and infected with the delta agent, vertical transmission of the delta agent can be prevented by administration of hepatitis B immunoglobulin (HBIG) and hepatitis B vaccine to the baby shortly after birth.

Control and Prevention

A number of methods of interrupting transmission of HBV have been developed; these include education of people at high risk of infection about the mode of transmission of the virus and about measures that can be taken to reduce the risk, screening blood donors for HBsAg and excluding known carriers, passive prophylaxis with high-titered hepatitis B immunoglobulin, and active immunization.

Environmental Factors

Education of health-care workers has had a dramatic effect in reducing the incidence of HB infection in situations in which contamination of the environment with blood or other body fluids is common, for example, dialysis and obstetric units, autopsy rooms, and clinical pathology laboratories.

Detailed guidelines have been developed for the safe handling of patients and their specimens and for cleaning and sterilizing instruments or other equipment that has become contaminated. Interested readers are referred to several authoritative documents.

Passive Immunization

There is good evidence that preparations of human immunoglobulin with measurable levels of anti-HBs are effective in preventing hepatitis B when given prior to exposure.

Szmuness et al. (1974) demonstrated that preparations with anti-HBs titers as low as 1 : 16 (by PHA) significantly reduce the frequency of infections among institutionalized children. Since that time, several manufacturers have produced lots of hepatitis B immune globulin (HBIG) with anti-HBs titers of 1 : 100,000 (RIA) or greater. Where these preparations have been used in subjects at high risk of infection, significant protection has been observed (Seef et al., 1975).

The most common clinical indication for the use of HBIG is postexposure prophylaxis either of a health-care worker who has been inadvertently exposed to infection (by percutaneous inoculation, oral ingestion, or contact with mucus membranes) (see Seef et al., 1975), sexual partners of patients with acute hepatitis B, or babies born to infected mothers. Several studies have attempted to define the benefits of HBIG in each of the situations, and each has shown that HBIG significantly reduces the risk of infection (reviewed by Hollinger and Melnick, 1985). The timing of administration appears to be critical. The best results are obtained when HBIG is administered within 24 h of exposure, with efficacy declining rapidly over the next 48 h.

Prevention of perinatal transmission of HBV from carrier mothers to their babies is an extremely important issue, since a high proportion of infections acquired in this manner are likely to result in development of the chronic carrier state.

In controlled trials, HBIG (0.5 ml) has been shown to be highly effective in preventing infection, provided that it is administered within 24 h of birth and additional doses are given at 3 and 6 months (Stevens et al., 1982). Alternative dose schedules and routes of administration have also been shown to be effective.

HBIG is a relatively expensive preparation and in limited supply. To date, its use has been restricted largely to developed countries, usually for the treatment of health-care workers who have been exposed to infection or for babies born to carrier mothers. In developed countries, the need for this material has been greatly reduced because of the availability of safe, effective vaccines.

Active Immunization

The need for a vaccine that would prevent infection and thus the development of the carrier state is obvious. Most of the viral vaccines currently available (e.g., polio, measles, mumps, and rubella) are produced from virus that has been adapted to growth in cell culture. Hepatitis B has never been open to this approach, because the virus has never been isolated in cell culture.

The first attempt to produce a vaccine was made by Krugman and his colleagues in 1971 by the simple technique of injecting volunteers with serum known to contain HBV that had been heated to 98°C for 1 min. This preparation failed to produce disease in recipients and provided partial protection against challenge with live virus (Krugman and Giles, 1973). As it was clear that such crude preparations would not meet the approval of licensing authorities, several groups, notably Gerin and Purcell at the National Institutes of Health, Maupas and his colleagues at the Pasteur Institute, and Hilleman and his colleagues at the Merck Institute for Therapeutic Research, began to develop more highly purified and chemically defined vaccines. They collected blood from chronic carriers and sought to purify the excess HBsAg associated with the 22-nm particles from the mature virions and other particulate matter. It proved relatively easy to do this, by purification procedures based on the size and buoyant density of the particles (usually several cycles of ultracentrifugation).

As it was recognized that large pools of plasma would inevitably contain other blood-borne viruses, a series of measures (pepsin at low pH, 8 M urea and formalin per liter) were introduced which were known to be capable of inactivating members of all groups of vertebrate viruses, including the slow viruses. In view of the subsequent emergence of the acquired immune deficiency syndrome (AIDS) as a major public health problem, this was an extremely prudent and far-sighted approach. The purification and inactivation procedures to which the virus is submitted are so drastic that its antigenicity is considerably reduced, necessitating the use of an alum adjuvant and a 3-dose schedule.

The safety, antigenicity, and efficacy of the vaccine was established in chimpanzees before it was tried cautiously in man. When the optimal-dose schedule had been ascertained, it was evaluated in a large-scale, placebo-controlled, double-blind study in a group of sexually active homosexual men in New York. This study (Szmuness et al., 1980) demonstrated that the vaccine was well tolerated, antigenic, and effective. More than 95% of recipients developed antibody after 3 doses of vaccine and were fully protected against the virus. There was also evidence of protection among people who had received 1 or 2 doses; this suggested that the vaccine might be valuable for postexposure prophylaxis.

Hepatitis B vaccines were first licensed in 1981 and are now available in many developed and some developing countries. Several million doses have been administered without serious side effects being reported.

Vaccine Use

Because of the high cost of development, current hepatitis B vaccines are quite expensive (from $40 to $120 per person, depending on the manufacturer) and are thus unsuited for mass immunization programs.

A variety of strategies have been developed for use of the vaccine, depending upon the prevalence of infection, the age at which it is usually acquired, and the funds available (Stevens et al., 1980).

In developed countries in which the prevalence of HB infection is low, the major priorities are (a) reduction in the reservoir of chronic carriers (by immunizing babies born to carrier mothers and children at high risk of infection); and (b) protection of adults at increased risk of developing disease (e.g., regular sexual partners of infected subjects and health-care workers who have regular contact with blood, body fluids, or tissues).

By contrast, in hyperendemic areas, most infections are acquired early in life, usually from an infected mother or other household member. These children have a high risk of becoming chronic carriers and of developing chronic sequelae. Under these circumstances, the only policy likely to lead to a major reduction in morbidity and mortality would be immunization of all babies.

In a second classical study in Taiwan, Beasley and Stevens (1978) demonstrated that infection is frequently transmitted by carrier mothers to their babies at or shortly after birth and that transmission can be interrupted by combined active/passive immunization.

Armed with this information and conscious of the extent of the problem in developing countries, the World Health Organization (WHO) has launched a global, viral-hepatitis-control program aimed at stimulating local production of low-cost vaccine, developing adequate delivery systems, and integrating hepatitis B immunization into the Expanded Program of Immunization.

To date, the major emphasis of this program has fallen on the Western Pacific Region, a vast area extending from Mongolia to New Zealand and from French Polynesia to Kampuchea. A Regional Task Force has been established, which has sought to influence Health Ministries to give high priority to the program, and has provided expert assistance in vaccine production, training of support staff, and development of quality control programs. WHO has provided some seed money and assistance in mobilizing resources from donor agencies. The results have been very gratifying.

Whereas in 1982 the only vaccine available in the region was that purchased from Merck, Sharp and Dohme or the Pasteur Institute, the vaccine is now being produced in nine centers (China, 4; Japan, 3;

Korea, 2), while Singapore and Taiwan have entered into agreements to produce vaccines under license in China. WHO-trained staff are expected to produce 5 million doses of vaccine this year, and 12 million doses next year. If these targets are met, it should be possible to immunize every newborn baby in China before the end of the decade.

Despite these achievements, plasma-derived vaccines are likely to remain relatively expensive because of the lengthy purification procedure and the costly safety testing. It is clear that other approaches are needed (Lerner et al., 1981; Skolnick et al., 1985; Smith et al., 1983).

Literature Cited

Aikawa, T., H. Sairenjii, and S. Furuta. 1978. Seroconversion from hepatitis Be antigen to anti-HBe in acute hepatitis B virus infection. N. Engl. J. Med. **298:**439–441.

Alberti, A., S. Diana, A. L. W. F. Eddleston, and R. Williams. 1979. Changes in hepatitis B virus DNA polymerase in relation to the outcome of acute hepatitis type B. Gut **20:**190–195.

Albin, C., and W. S. Robinson. 1980. Protein kinase activity in hepatitis B virus. J. Virol. **34:**297–302.

Alessandro, Z. R., T. Elisabetta, F. Pierino, and E. Magliano. 1983. Vertical transmission of the HBV-associated delta agent, p. 127–132. *In* G. Verne, F. Bonino, and M. Rizzetto, (ed.), Viral hepatitis and delta infection. R. Allan, New York.

Alexander, J., G. Macnab, and R. Saunders. 1978. Studies on *in vitro* production of HBsAg by a human hepatoma cell line, p. 103–120. *In* M. Pollard (ed.), Perspectives in virology, vol. 10. Raven Press, New York.

Almeida, J. D., D. Rubenstein, and E. J. Scott. 1971. New antigen-antibody system in Australia antigen positive hepatitis. Lancet **2:**1225–1227.

Alter, H. J., J. B. Seeff, P. M. Kaplan, V. J. McAuliffe, E. C. Wright, J. L. Gerin, R. H. Purcell, P. V. Holland, and H. J. Zimmerman. 1976. Type B hepatitis: the infectivity of blood positive for e antigen and DNA polymerase after accidental needlestick exposure. N. Engl. J. Med. **295:**909–913.

Alter, M. J., M. S. Favero, and J. E. Maynard. 1984. Haemodialysis-associated hepatitis in the United States (abstract), p. 636. *In* G. N. Vyas, J. L. Dienstag, and J. H. Hoofnagle (ed.), Viral hepatitis and liver disease. Grune and Stratton, Orlando.

Alward, W. L. M., B. J. McMahon, D. B. Hall, W. L. Heyward, D. R. Francis, and T. R. Bender. 1985. The long term serological course of asymptomatic hepatitis B virus carriers and the development of primary hepatocellular carcinoma. J. Infect. Dis. **151:**604–609.

Aragona, M., S. Macagno, F. Caredda, O. Crivelli, C. Lavarini, E. Marcan, P. Farci, R. H. Purcell, and M. Rizzetto. 1987. Serological response to the hepatitis delta virus in hepatitis D. Lancet **1:**478–480.

Arico, S., M. Aragona, M. Rizzetto, F. Caredda, A. Zanetti, G. Marinucci, S. Diana, P. Farci, M. Arnone, N. Caporaso, A. Ascione, P. Dentico, G. Pustore, G. Raimondo, and A. Craxi. 1985. Clinical significance of antibody to the hepatitis delta virus in symptomless HBsAg carriers. Lancet **2:**356–358.

Bancroft, W. H., F. K. Mundon, and P. K. Russell. 1972.

Detection of additional antigenic determinants of hepatitis B antigen. J. Immunol. **109**:842–848.

Barker, L. F., J. E. Maynard, R. H. Purcell, J. H. Hoofnagle, K. R. Berquist, and W. T. London. 1975. Viral hepatitis, type B, in experimental animals. Am. J. Med. Sci. **270**:189–195.

Barker, L. F., and R. Murray. 1971. Relationship of virus dose to incubation time of clinical hepatitis and time of appearance of hepatitis-associated antigen. Am. J. Med. Sci. **263**:27–33.

Basalamah, A. H., F. Serebour, and E. Kazim. 1984. Materno-foetal transmission of hepatitis B in Saudi Arabia. J. Infect. **8**:200–204.

Bayer, M. E., B. S. Blumberg, and B. Werner. 1968. Particles associated with Australia antigen in the sera of patients with leukaemia, Down's syndrome and hepatitis. Nature (London) **218**:1057–1059.

Beal, R. W. 1976. Hepatitis B antigen screening in South Australia and Northern Territory blood donors. Pathology **8**:21–27.

Beasley, R. P., and C. E. Stevens. 1978. Vertical transmission of HBV and interruption with globulin, p. 333–345. *In* G. N. Vyas, S. N. Cohen, and R. Schmid (ed.), Viral hepatitis. The Franklin Institute Press, Philadelphia.

Bergmann, K. F., and J. L. Gerin. 1986. Antigens of hepatitis delta virus in the liver and serum of humans and animals. J. Infect. Dis. **154**:702–706.

Berninger, M., M. Hammer, B. Hoyer, and J. L. Gerin. 1982. An assay for the detection of the DNA genome of hepatitis B virus in serum. J. Med. Virol. **9**:57–68.

Bianchi, L., J. DeGroote, V. J. Desmet, P. Gedigk, G. Korb, H. Popper, H. Poulsen, P. J. Schever, M. Schmid, H. Thaler, and W. Wepler. 1971. Morphological criteria in viral hepatitis. Lancet **1**:333–337.

Bianchi, L., and F. G. Gudat. 1983. Histo- and immunopathology of viral hepatitis, p. 335–382. *In* F. Deinhardt and G. Deinhardt (ed.) Viral hepatitis: laboratory and clinical science. Marcel Dekker, Basel.

Blum, H. E., A. T. Haase, and G. N. Vyas. 1984. Molecular pathogenesis of HBV infection: simultaneous detection of viral DNA and antigens in paraffin-embedded liver sections. Lancet **2**:771–775.

Blum, H. E., L. Stowring, and A. Figus. 1983. Detection of hepatitis B virus DNA in hepatocytes, bile duct epithelium and vascular elements by *in situ* hybridization. Proc. Natl. Acad. Sci. USA. **80**:6685–6688.

Blumberg, B. S., H. J. Alter, and A. Visnich. 1965. A "new" antigen in leukaemia sera. J. Am. Med. Assoc. **191**:541–546.

Blumberg, B. S., B. J. S. Gerstley, D. A. Hungerford, W. T. London, and A. J. Sutnick. 1967. A serum antigen (Australia antigen) in Downs syndrome, leukaemia and hepatitis. Ann. Intern. Med. **66**:924–931.

Bond, W. W., M. S. Favero, J. J. Petersen, and J. W. Ebert. 1983. Inactivation of hepatitis B virus by intermediate-to-high level disinfectant chemicals. J. Clin. Microbiol. **18**:535–538.

Bond, W. W., M. S. Favero, J. J. Petersen, C. R. Gravelle, J. W. Ebert, and J. E. Maynard. 1981. Survival of hepatitis B virus after drying and storage for one week. Lancet **1**:550–551.

Bond, W. W., N. J. Petersen, and M. S. Favero. 1977. Viral hepatitis B: aspects of environmental control. Health Lab. Sci. **14**:235–252.

Bonino, F., K. H. Heerman, M. Rizzetto, and W. H. Gerlich. 1986. Hepatitis delta virus: protein composition of delta antigen and its hepatitis B virus-derived envelope. J. Virol. **58**:945–950.

Bonino, F., B. Hoyer, E. J. Ford, J. W. K. Shih, R. H. Purcell, and J. L. Gerin. 1981. The delta agent: HBsAg particles with delta antigen and RNA in the serum of an HBV carrier. Hepatology **1**:127–131.

Bonino, F., B. Hoyer, J. Nelson, R. Engle, G. Verme, and J. L. Gerin. 1981. Hepatitis B virus DNA in the sera of HBsAg carriers: a marker of active hepatitis B virus replication in the liver. Hepatology **1**:386–391.

Bonino, F., B. Hoyer, J. W. K. Shih, M. Rizzetto, R. H. Purcell, and J. L. Gerin. 1984. Delta hepatitis agent: structural and antigenic properties of the delta-associated particle. Infect. Immun. **43**:1000–1005.

Brechot, C., P. Degos, C. Lugassy, V. Thiers, S. Zafrani, D. Franco, H. Bismuth, C. Trepo, J-P. Benhamou, J. Wands, K. Isselbacher, P. Tiollais, and P. Berthelot. 1985. Hepatitis B virus DNA in patients with chronic liver disease and negative tests for hepatitis B surface antigen. N. Engl. J. Med. **312**:270–276.

Brechot, C., M. Hadchouel, J. Scotto, F. Degos, P. Charnay, C. Trepo, and P. Tiollais. 1981. Detection of hepatitis B virus DNA in liver and serum: a direct appraisal of the chronic carrier state. Lancet **2**:765–768.

Brechot, C., C. Lugassy, A. Dejean, P. Pontisso, V. Thiers, P. Bertholet, and P. Tiollais. 1984. Hepatitis B virus DNA in infected human tissues, p. 395–410. *In* G. N. Vyas, J. L. Dienstag, and J. H. Hoofnagle, (ed.), Viral hepatitis and liver disease. Grune and Stratton, New York.

Brechot, C., C. Pourcel, M. Hadchouel, A. Dejean, A. Louise, J. Scotto, and P. Tiollais. 1982. State of hepatitis B virus DNA in liver diseases. Hepatology **2** (Suppl.): 27–34.

Brigati, D. J., D. Myerson, J. J. Leary, B. Spalholz, S. Z. Travis, C. K. Y. Fong, G. D. Hsiung, and D. C. Ward. 1983. Detection of viral genomes in cultured cells and paraffin-embedded tissue sections using biotin-labelled hybridization probes. Virology **126**:32–50.

Buescher, M., W. Reiser, H. Will, and H. Schaller. 1985. Characterization of transcripts and the potential RNA pregenome of duck hepatitis B virus: implication for replication by reverse transcription. Cell **40**:717–724.

Burrell, C. J., E. J. Gowans, A. R. Jilbert, J. R. Lake, and B. P. Marmion. 1982. Hepatitis B virus DNA detection by *in situ* cytohybridization: implications for viral replication strategy and pathogenesis of chronic hepatitis. Hepatology **2** (Suppl.): 85–91.

Burrell, C. J., E. J. Gowans, R. Rowland, P. Hall, A. R. Jilbert, and B. P. Marmion. 1984. Correlation between liver histology and markers of hepatitis B virus replication in infected patients. Hepatology **4**:20–24.

Burrell, C. J., P. MacKay, P. J. Greenaway, P. H. Hofschneider, and K. Murray. 1979. Expression in *Escherichia coli* of hepatitis B virus DNA sequences cloned in plasmid pBR 322. Nature (London) **279**:43–47.

Buti, M., R. Esteban, R. Jardi, J-I. Esteban, and J. Guardia. 1986. Serological diagnosis of acute delta hepatitis. J. Med. Virol. **18**:81–85.

Cattaneo, R., H. Will, N. Hernandez, and H. Schaller. 1983. Signals regulating hepatitis B surface antigen transcription. Nature (London) **305**:336–338.

Cattaneo, R., H. Will, and H. Schaller. 1984. Hepatitis B virus transcription in infected liver. EMBO J. **3**:2191–2196.

Chakraborty, P. R., N. Ruiz-Opazo, and D. A. Shafritz. 1981a. Transcription of human HBV core antigen gene sequences in an *in vitro* HeLa cellular extract. Virology **111**:647–652.

Chakraborty, P. R., N. Ruiz-Opazo, D. Shouval, and D. A. Shafritz. 1981b. Identification of integrated hepatitis B virus DNA and expression of viral DNA in an HBsAg producing human hepatocellular carcinoma cell line. Nature (London) **286**:532–533.

Chau, K. H., M. P. Hargie, R. H. Decker, I. K. Mushawar, and L. R. Overby. 1983. Serodiagnosis of recent hepatitis B infection by IgM anti-HBc. Hepatology **3:**142–149.

Chen, D. S., M. Y. Lai, and J. L. Sung. 1984. Delta agent infection in patients with chronic liver diseases and hepatocellular carcinoma—an infrequent finding in Taiwan. Hepatology **4:**502–503.

Chen, P. J., G. Kaplana, J. Goldberg, W. Mason, B. Werner, J. L. Gerin, and J. Taylor. 1986. Structure and replication of the genome of the hepatitis delta virus. Proc. Natl. Acad. Sci. USA **83:**8774–8778.

Chin, K. C., A. K. R. Chaudhuri, and E. A. C. Follett. 1981. Materno-foetal transmission of HBsAg by asymptomatic carrier mothers in Glasgow. J. Infect. **3:**246–252.

Chu, C-M., P. Karayiannis, M. J. F. Fowler, J. Monjardino, Y-F. Liaw, and H. C. Thomas. 1985. Natural history of chronic hepatitis B virus infection in Taiwan: studies of hepatitis B virus DNA in serum. Hepatology **5:**431–434.

Courouce, A. M., P. V. Holland, J. Y. Muller, and J. P. Soulier. 1975. HBsAg antigen subtypes: Proceedings of the International Workshop on HBs antigen subtypes. Bibl. Haematologica **42:**1–158.

Crivelli, O., M. Rizzetto, C. Lavarini, A. Smedile, and J. L. Gerin. 1981. Enzyme-linked immunosorbent assay for detection of antibody to the hepatitis B surface antigen-associated delta antigen. J. Clin. Microbiol. **14:**173–177.

Crivelli, O., J. W. K. Shih, and M. Rizzetto. 1983. Methods for detection of the delta antigen and antibody in liver and serum, p. 121–126. *In* G. Verne, F. Bonino, and M. Rizzetto (ed.), Viral hepatitis and delta infection. Alan R. Liss, New York.

Dane, D. S., C. H. Cameron, and M. Briggs. 1970. Viruslike particles in serum of patients with Australia antigen associated hepatitis. Lancet **1:**695–698.

Davis, G. L., and J. H. Hoofnagle. 1985. Reactivation of chronic type B hepatitis presenting as acute viral hepatitis. Ann. Intern. Med. **102:**762–765.

Davis, G. L., J. H. Hoofnagle, and J. F. Waggoner. 1984. Spontaneous reactivation of chronic hepatitis B virus infection. Gastroenterology **86:**230–235.

Deinhardt, F., and I. D. Gust. 1982. Viral hepatitis. Bull. W.H.O. **60:**661–691.

Deinhardt, F., and A. J. Zuckerman. 1985. Immunization against hepatitis B: report of a WHO meeting on viral hepatitis in Europe. J. Med. Virol. **17:**209–217.

Dejean, A., C. Lugassy, and E. S. Zafrani. 1984. Detection of hepatitis B virus DNA in pancreas, kidney and skin of two human carriers of the virus. J. Gen. Virol. **65:**651–655.

Dejean, A., P. Sonigo, S. Wain-Hobson, and P. Tiollais. 1985. Specific hepatitis B virus integration in hepatocellular carcinoma DNA through a viral 11-base-pair direct repeat. Proc. Natl. Acad. Sci. USA **81:**5350–5354.

Denes, A. E., J. L. Smith, and J. E. Maynard. 1978. Hepatitis B virus infection in physicians. Results of a nationwide seroepidemiologic survey. J. Am. Med. Assoc. **239:**210–212.

Denniston, K. J., B. H. Hoyer, A. Smedile, F. V. Wells, J. Nelson, and J. L. Gerin. 1986. Cloned fragments of the hepatitis delta virus RNA genome: sequence and diagnostic applications. Science **232:**873–875.

Diegutis, P. S., E. Keirnan, L. Burnett, B. N. Nightingale, and Y. Cossart. 1986. False-positive results with hepatitis B virus DNA dot-hybridization in hepatitis B surface antigen-negative specimens. J. Clin. Microbiol. **23:**797–799.

Dienstag, J. L. 1981. Hepatitis B as an immune complex disease. Semin. Liver Dis. **1:**45–49.

Dienstag, J. L. 1983a. Non-A, non-B hepatitis. I. Recognition, epidemiology and clinical features. Gastroenterology **85:**439–462.

Dienstag, J. L. 1983b. Non-A, non-B hepatitis. II. Experimental transmission, putative virus agents and markers, and prevention. Gastroenterology, **85:**743–768.

Dimitrakakis, M., A. G. Coulepis, R. C. Pringle, and I. D. Gust. 1984. Detection of delta infection using reagents obtained from the serum of patients infected with HBV. J. Virol. Methods **8:**331–334.

Dimitrakakis, M., S. Crowe, and I. D. Gust. 1986a. Prevalence of delta infection in the Western Pacific region. J. Med. Virol. **20:**305–311.

Dimitrakakis, M., and I. D. Gust. 1984. Delta infection in Pacific islanders. Med. J. Aust. **2:**197.

Dimitrakakis, M., M. J. Waters, A. Wootton, and I. D. Gust. 1986b. Detection of IgM antibodies to delta antigen after coinfection and superinfection with the delta virus. J. Med. Virol. **20:**305–311.

Dimter-Gottlieb, G. 1986. Viroids and virusoids are related to group I introns. Proc. Natl. Acad. Sci. USA **83:**6250.

Dreesman, G. R. 1984. Molecular immunology of hepatitis antigens, p. 583–586. *In* G. N. Vyas, J. L. Dienstag, and J. H. Hoofnagle. Viral hepatitis and liver disease. Grune and Stratton, New York.

Dudley, F. J., P. J. Schever, and S. Sherlock. 1972. Natural history of hepatitis B-associated antigen-positive chronic liver disease. Lancet **ii:**1388–1391.

Duermeyer, W., F. Wielaard, and J. van der Veen. 1979. A new principle for the detection of specific IgM antibodies applied in an ELISA for hepatitis A. J. Med. Virol. **4:**25–32.

Eddleston, A. L. W. F., and R. Williams. 1979. HL-A system and liver disease, p. 285–299. *In* H. Popper and F. Schaffner (ed.), Progress in liver diseases, vol. 6. Grune and Stratton, New York.

Elfassi, E., J.-L. Romet-Lemoune, M. Essex, M. Frances-McLane, and W. A. Haseltine. 1984. Evidence of extrachromosomal forms of HBV DNA in a bone marrow culture obtained from a patient recently infected with HBV. Proc. Natl. Acad. Sci. USA **81:**3526–3528.

Fagan, E. A., P. Guarner, S. D. K. Perera, R. Trowbridge, N. Rolando, F. Davison, and R. Williams. 1985. Quantitation of hepatitis B virus DNA (HBV DNA) in serum using the sport hybridization technique and scintillation counting. J. Virol. Methods **12:**251–262.

Farci, P., J. L. Gerin, M. Aragona, I. Lindsey, O. Crivelli, A. Balestrieri, A. Smedile, H. C. Thomas, and M. Rizzetto. 1986. Diagnostic and prognostic significance of the IgM antibody to the hepatitis delta virus. J. Am. Med. Assoc. **255:**1443–1446.

Feinstone, S. M., L. F. Barker, and R. H. Purcell. 1979. Hepatitis A and B, p. 879–925. *In* E. H. Lennette and N. J. Schmidt (ed.), Diagnostic procedures for viral, rickettsial and chlamydial infections, 5th ed. American Public Health Association, Washington, D.C.

Feinstone, S. M., and J. H. Hoofnagle. 1984. Non-A, maybe-B hepatitis. N. Engl. J. Med. **311:**185–189.

Feitelson, M. A., P. L. Marion, and W. S. Robinson. 1981. Antigenic and structural relationships of the surface antigens of hepatitis B virus, ground squirrel hepatitis virus, and woodchuck hepatitis virus. J. Virol. **39:**447–454.

Feitelson, M. A., P. L. Marion, and W. S. Robinson. 1982a. Core particles of HBV and GSHV. I. Relationship between HBcAg and GSHcAg associated polypeptides by SDS-PAGE and tryptic peptide mapping. J. Virol. **43:**687–696.

Feitelson, M. A., P. L. Marion, and W. S. Robinson. 1982b. Core particles on HBV and GSHV. II. Characterization of the protein kinase reaction associated with ground-squirrel hepatitis and hepatitis B virus. J. Virol. **43**:741–748.

Findlay, G. M., and F. O. MacCallum. 1937. Note on acute hepatitis and yellow fever immunization. Trans. R. Soc. Trop. Med. Hyg. **31**:297.

Flaum, A., H. Malmros, and E. Persson. 1926. Eine nosocomiale Ikterus-Epidemie. Beta. Med. Scand. Suppl. **16**:544.

Fowler, M. J. F., J. Monjardino, K. N. Tsiquaye, A. J. Zuckerman, and H. C. Thomas. 1984. The mechanism of replication of hepatitis B virus: evidence of asymmetric replication of the two DNA strands. J. Med. Virol. **13**:83–91.

Frommel, D., D. Crevat, L. Vitvitsky, C. Pichoud, O. Hanitz, M. Chevalier, J.-A. Grimaud, J. Lindberg, and C. G. Tolpo. 1984. Immunopathologic aspects of woodchuck hepatitis. Am. J. Pathol. **115**:125–134.

Frosner, G. G., and E. Franco. 1984. New developments of open questions in the serology of hepatitis A and B, p. 487–496. *In* G. N. Vyas, J. L. Dienstag, and J. H. Hoofnagle (ed.), Viral hepatitis and liver disease. Grune and Stratton, New York.

Fujiyama, A., A. Miyanohara, C. Nozaki, T. Yoneyama, N. Ohtomo, and K. Matsubara. 1983. Cloning and structural analysis of hepatitis B virus DNAs, subtype adr. Nucleic Acids Res. **II**:4601–4610.

Galibert, F., T. N. Chen, and E. Mandart. 1982. Nucleotide sequence of a cloned woodchuck hepatitis virus genome: comparison with the hepatitis B virus sequence. J. Virol. **41**:51–65.

Galibert, F., E. Mandart, F. Fitoussi, P. Tiollais, and P. Charnay. 1979. Nucleotide sequence of hepatitis B virus genome (subtype ayw) cloned in *E. coli*. Nature (London) **281**:646–650.

Galloway, D. A., and J. K. McDougall. 1983. The oncogenic potential of herpes simplex viruses: evidence for a "hit and run" mechanism. Nature (London) **302**:21–24.

Gellis, S. J., J. R. Neefe, J. Stokes, Jr., L. E. Strong, C. E. Janeway, and G. Scatchard. 1948. Chemical, clinical and immunological studies on the products of human plasma fractionation. XXXVI. Inactivation of the virus of homologous serum hepatitis in solution of normal serum albumin by means of heat. J. Clin. Invest. **27**:239–244.

Gerber, M. A., S. Hadziyannis, C. Vissoulis, F. Schaffner, F. Paronetto, and H. Popper. 1974. Electron microscopy and immunoelectron microscopy of cytoplasmic hepatitis B antigen in hepatocytes. Am. J. Pathol. **75**:489–502.

Gerber, M. A., and S. N. Thung. 1985. Biology of disease: molecular and cellular pathology of hepatitis B. Lab. Invest. **52**:572–590.

Gerber, M. A., S. N. Thung, and H. Popper. 1983. Viral hepatitis and hepatocellular carcinoma, p. 317–334. *In* F. Deinhardt and J. Deinhardt (ed.), Viral hepatitis: laboratory and clinical sciences. Marcel Dekker, Inc., Basel.

Gerin, J. L., and J. W. K. Shih. 1978. Structure of HBsAg and HBcAg, p. 147–158. *In* G. N. Vyas, S. N. Cohen, and R. Schmid (ed.), Viral hepatitis. The Franklin Institute Press, Philadelphia.

Gerlich, W. H., M. A. Feitelson, P. L. Marion, and W. S. Robinson. 1980. Structural relationships between the surface antigens of ground squirrel hepatitis virus and human hepatitis B virus. J. Virol. **36**:787–795.

Gerlich, W. H., and W. J. Robinson. 1980. Hepatitis B virus contains protein attached to the 5′ terminus of its complete DNA strand. Cell **31**:801–809.

Giles, J. P., H. E. McCollum, L. W. Berndtson, and S.

Krugman. 1969. Viral hepatitis Australia: SH antigen and Willowbrook MS2 strain. N. Engl. J. Med. **281**:119–122.

Gilja, B., E. Kasambalides, R. S. Bressler, S. N. Thung, and M. A. Gerber. 1985. Secretion of polyalbumin receptors *in vitro*. J. Med. Virol. **15**:335–341.

Govindarajan, S., K. P. Chin, A. G. Redeker, and R. L. Peters. 1984. Fulminant B viral hepatitis: role of delta agent. Gastroenterology **86**:1417–1420.

Gowans, E. J. 1986. Relationship between HBeAg and HBV DNA in patients with acute and persistent hepatitis B infection. Med. J. Aust. **145**:439–441.

Gowans, E. J., C. J. Burrell, A. R. Jilbert, and B. P. Marmion. 1981. Detection of hepatitis B virus DNA sequences in infected hepatocytes by *in situ* hybridization. J. Med. Virol. **8**:67–78.

Gudat, F., and L. Bianchi. 1977. Evidence for phasic sequences in HbcAg formation and cell membrane-directed flow of core particles in chronic hepatitis B. Gastroenterology **73**:1194–1197.

Gudat, F., L. Bianchi, W. Sonnabend, G. Thiel, W. Aenishaenslin, and G. A. Stalder. 1975. Pattern of core and surface expression in liver tissue reflects state of specific immune response in hepatitis B. Lab. Invest. **32**:1–9.

Gust, I. D. 1984. The epidemiology of viral hepatitis, p. 415–421. *In* G. N. Vyas, J. L. Dienstag, and J. H. Hoofnagle, (ed.), Viral hepatitis and liver disease. Grune and Stratton, Orlando.

Gust, I. D., C. J. Burrell, A. G. Coulepis, W. S. Robinson, and A. J. Zuckerman. 1986. Toxonomic classification of human hepatitis B virus. Intervirology **25**:14–29.

Hadler, S. C., M. DeMonzon, A. Ponzetto, E. Anzola, D. Rivero, A. Mondolfi, A. Bracho, D. Francis, M. Gerber, S. Thung, J. L. Gerin, J. Maynard, H. Popper, and R. Purcell. 1984. Delta virus infection and severe hepatitis. An epidemic in the Yucpa Indians of Venezuela. Ann. Int. Med. **100**:339–344.

Hadziyannis, S. J., H. M. Lieberman, G. G. Karvountzis, and D. A. Shafritz. 1983. Analysis of liver disease, nuclear HBcAg, viral replication, and hepatitis B virus DNA in liver and serum of HBeAg versus anti-HBe positive carriers of hepatitis B virus. Hepatology **3**:656–662.

Hadziyannis, S. J., M. Sherman, H. M. Lieberman, and D. A. Shafritz. 1985. Liver disease activity and hepatitis B virus replication in chronic delta antigen-positive hepatitis B virus carriers. Hepatology **5**:544–547.

Halbert, S. P., and M. Anken. 1977. Detection of hepatitis B surface antigen (HBsAg) with use of alkaline phosphatase-labelled antibody to HBsAg. J. Infect. Dis. **136**:318–323.

Halpern, M. S., J. Egan, W. S. Mason, and J. M. England. 1984. Viral antigen in endocrine cells of the pancreatic islets and adrenal cortex of Pekin ducks infected with DHBV. Virus. Res. **1**:213–224.

Halpern, M. S., J. M. England, and D. T. Deery. 1983. Viral nucleic acid synthesis and antigen accumulation in pancreas and kidney of Pekin ducks infected with duck hepatitis B virus. Proc. Natl. Acad. Sci. USA **80**:4865–4869.

Hansson, B. G., T. Moestrup, A. Widell, and E. Nordenfelt. 1982. Infection with delta agent in Sweden: introduction of a new hepatitis agent. J. Infect. Dis. **146**:472–478.

Hansson, B. G., and R. H. Purcell. 1979. Sites that bind polymerized albumin on hepatitis B surface antigen particles: detection by radioimmunoassay. Infect. Immun. **26**:125–130.

Havens, W. P. Jr. 1968. Viral hepatitis, chap. 13. *In* R. S.

Anderson, (ed.) Internal medicine in World War II, vol. 3. Office of the Surgeon General, Department of the Army, Washington, D.C.

Hayward, W. S., B. G. Neel, and S. M. Astrin. 1981. Activation of a cellular *onc* gene by promoter insertion in ALV-induced lymphoid leukosis. Nature (London) **290:**475–480.

Heermann, K. H., U. Goldmann, W. Schwartz, T. Seyffarth, H. Baumgarten, and W. H. Gerlich. 1984. Large surface proteins of HBV containing the pre-S sequence. J. Virol. **52:**396–402.

Hirata, A. A., A. J. Emerick, and W. F. Boley. 1973. Hepatitis B virus antigen detection by reverse passive haemagglutination. Proc. Soc. Exp. Biol. Med. **143:**761–766.

Hohenberger, P. 1984. Detection of HBsAg in the pancreas in cases on pancreatic carcinoma. Hepatogastroenterology **31:**239–241.

Hollinger, F. B., E. Adam, D. Heiberg, and J. L. Melnick. 1981. Response to hepatitis B vaccine in a young adult population, p. 451–466. *In* W. Szmuness, H. J. Alter, and J. E. Maynard, (ed.) Viral hepatitis, 1981 International Symposium. The Franklin Institute Press, Philadelphia.

Hollinger, F. B., and Dienstag, J. L. 1980. Hepatitis viruses, p. 899–921. *In* E. H. Lennette, A. Balows, W. J. Hausler, and J. P. Truant. (ed.) Manual of clinical microbiology, 3rd ed. American Society for Microbiology, Washington, D.C.

Hollinger, F. B., and J. L. Dienstag. 1985. Hepatitis viruses, p. 813–835. *In* E. H. Lennette, A. Balows, W. J. Hausler, Jr., and H. J. Shadomy, (ed.) Manual of clinical microbiology. 4th ed. American Society for Microbiology, Washington, D.C.

Hollinger, F. B., and J. L. Melnick. 1985. Prevention and control of hepatitis B infection, p. 165–203. *In* F. B. Hollinger, J. L. Melnick, and W. S. Robinson (ed.), Viral hepatitis. Raven Press, New York.

Hollinger, F. B., V. Vorndam, and G. R. Dreesman. 1971. Assay of Australia antigen and antibody employing double-antibody and solid-phase radioimmunoassay techniques and comparison with the passive haemagglutination methods. J. Immunol. **107:**1099–1111.

Hoofnagle, J. H., and H. J. Alter. 1984. Chronic viral hepatitis, p. 97–113. *In* G. N. Vyas, J. L. Dienstag, and J. H. Hoofnagle, (ed.), Viral hepatitis and liver disease. Grune and Stratton, Orlando, Florida.

Hoofnagle, J. H., R. J. Gerety, and L. F. Barker. 1973. Antibody to hepatitis B virus core in man. Lancet **2:**869–873.

Hoofnagle, J. H., and D. F. Schafer. 1986. Serologic markers of hepatitis B virus infection. Semin. Liver Dis. **6:**1–10.

Hoofnagle, J. H., L. B. Seeff, Z. B. Bales, R. J. Gerety, and E. Tabor. 1978. Serological responses in hepatitis B, p. 219–242. *In* G. N. Vyas, S. N. Cohen, and R. Schmid, (ed.), Viral hepatitis. Franklin Institute Press, Philadelphia.

Hsu, S. M., L. Raine, and H. Fanger. 1981a. A comparative study of the peroxidase-antiperoxidase method and an avidin-biotin complex method for studying polypeptide hormones with radioimmunoassay antibodies. Am. J. Clin. Pathol. **75:**734–738.

Hsu, S. M., L. Raine, and H. Fanger. 1981b. Use of avidin-biotin-peroxidase complex (ABC) in immunoperoxidase techniques: a comparison between ABC and unlabelled antibody (PAP) procedures. J. Histochem. Cytochem. **29:**577–580.

Huang, S., H. Minassian, and J. D. More. 1976. Application of immunofluorescent staining on paraffin sections

improved by trypsin digestion. Lab. Invest. **35:**383–390.

Huang, S. N., and A. R. Neurath. 1979. Immunohistologic demonstration of hepatitis B viral antigens in liver with reference to its significance in liver injury. Lab. Invest. **40:**1–17.

Imai, M., M. Nomura, T. Gotanda, T. Sano, K. Tachibana, H. Miyamoto, K. Takahashi, G. Toyama, Y. Miyakawa, and M. Mayumi. 1982. Demonstration of two distinct antigenic determinants on HBeAg by monoclonal antibodies. J. Immunol. **128:**69–72.

Imai, M., Y. Yanase, T. Nojiri, Y. Miyakawa, and M. Mayumi. 1979. A receptor for polymerized human and chimpanzee albumins on hepatitis B virus particles occurring with HBeAg. Gastroenterology **76:**242–247.

Jacobson, I. M., J. L. Dienstag, B. G. Werner, D. B. Brettler, P. H. Levine, and I. K. Musahawar. 1985. Epidemiology and clinical impact of hepatitis D virus (delta) infection. Hepatology **5:**188–191.

Jambazian, A., and J. C. Halper. 1972. Rheophoresis: a sensitive immuno-diffusion method for detection of hepatitis associated antigen. Proc. Soc. Exp. Biol. Med. **140:**560–564.

Juji, T., and T. Yokochi. 1969. Haemagglutination technique with erythrocytes coated with specific antibody for detection of Australia antigen. Jpn. J. Exp. Med. **39:**615–619.

Kam, W., L. Rall, E. Smuckler, R. Schmid, and W. Rutter. 1982. Hepatitis B viral DNA in liver and serum of asymptomatic carriers. Proc. Natl. Acad. Sci. USA **79:**7522–7526.

Kamimura, T., A. Yoshikawa, F. Ichida, and H. Sasaki. 1981. Electron microscopic studies of Dane particles in hepatocytes with special reference to intracellular development of Dane particles and their relation to HBeAg. Hepatology **1:**392–397.

Kaplan, P. N., R. L. Greenman, J. L. Gerin, R. H. Purcell, and W. L. Robinson. 1973. DNA polymerase associated with human hepatitis B antigen. J. Virol. **12:**995–1005.

Karayiannis, P., D. Novick, A. S. Lok, M. J. F. Fowler, J. Monjardino, and H. C. Thomas. 1985. Hepatitis B virus DNA in saliva, urine, and seminal fluid of carriers of hepatitis Be antigen. Br. Med. J. **290:**1853–1854.

Kay, A., E. Mandert, C. Trepo, and G. Galibert. 1985. The HBV HBx gene expressed in *E. coli* is recognized by sera from hepatitis patients. EMBO J. **4:**1287–1292.

Kos, A., R. Dijkeura, A. C. Arnberg, P. H. Van der Meide, and H. Schellekeus. 1986. The hepatitis delta (δ) virus possesses a circular RNA. Nature (London) **323:**558–560.

Koshy, R., S. Koch, A. Freytag von Loringhoven, R. Kahmann, K. Murray, and P. H. Hofschneider. 1983. Integration of hepatitis B virus DNA: evidence for integration in the single-stranded gap. Cell **34:**215–223.

Krogsgaard, K., P. Kryger, J. Aldershuile, P. Andersson, C. Brechot, and The Copenhagen Hepatitis Acuta Programme. 1985. Hepatitis B virus DNA in serum from patients with acute hepatitis B. Hepatology **5:**10–13.

Krugman, S. 1983. The development of a vaccine against hepatitis B. Scand. J. Infect. Dis. **38** (suppl.): 9–12.

Krugman, S., and J. P. Giles. 1973. Viral hepatitis, type B (MS2 strain): further observations of natural history and prevention. N. Engl. J. Med. **288:**755–760.

Krugman, S., J. P. Giles, and J. Hammond. 1967. Infectious hepatitis: evidence for two distinctive clinical, epidemiological and immunological types of infection. J. Am. Med. Assoc. **200:**365–369.

Krugman, S., J. P. Giles, and J. Hammond. 1970. Hepatitis virus: effect of heat infectivity and antigenicity of MS-1 and MS-2 strains. J. Infect. Dis. **122:**432–436.

Krugman, S., J. P. Giles, and J. Hammond. 1971. Viral hepatitis, type B (MS2 strain): studies in active immunization. J. Am. Med. Assoc. **217**:41–45.

Krugman, S., J. H. Hoofnagle, R. J. Gerety, P. M. Kaplan, and J. L. Gerin. 1974. Viral hepatitis, type B: DNA polymerase activity and antibody to hepatitis B core antigen. N. Engl. J. Med. **290**:1331–1335.

Krugman, S., L. R. Overby, I. K. Mushahwar, C. Ling, G. G. Frosner, and F. Deinhardt. 1979. Viral hepatitis, type B: studies on natural history and prevention reexamined. N. Engl. Med. J. **300**:101–106.

Kryger, P., J. Aldershuile, L. R. Mathieson, J. O. Nielsen, and The Copenhagen Hepatitis Acuta Programme. 1982a. Acute type B hepatitis among HBsAg negative patients detected by anti-HBc IgM. Hepatology **2**:50–53.

Kryger, P., L. R. Mathieson, J. Aldershuile, J. O. Nielsen, and The Copenhagen Hepatitis Acuta Programme. 1981. Presence and meaning of anti-HBc IgM as determined by ELISA in patients with acute type B hepatitis and healthy HBsAg carriers. Hepatology **1**:233–237.

Kryger, P., N. S. Pederson, L. Mathieson, and J. O. Nielsen. 1982b. Increased risk of infection with hepatitis A and B viruses in men with a history of syphilis. J. Infect. Dis. **145**:23–26.

Langer, P. R., A. A. Waldrop, and D. C. Ward. 1981. Enzymatic synthesis of biotin-labelled polynucleotides: novel nucleic acid affinity probe. Proc. Natl. Acad. Sci. USA **78**:6633–6637.

Leach, J. M., and B. J. Ruck. 1971. Detection of hepatitis associated antigen by the latex agglutination test. Br. Med. J. **4**:597–598.

LeBouvier, G. L. 1971. The heterogenicity of Australia antigen. J. Infect. Dis. **123**:671–675.

Lemon, S. M., N. L. Gates, T. E. Simms, and W. H. Bancroft. 1981. IgM antibody to hepatitis B core antigen as a diagnostic parameter of acute infection with hepatitis B virus. J. Infect. Dis. **143**:803–809.

Lerner, R. A., N. Green, H. Alexander, F. T. Liu, J. G. Sutcliffe, and T. M. Shinnick. 1981. Chemically synthesized peptides predicted from the nucleotide sequence of the hepatitis B virus genome elicit antibodies reactive with the native envelope protein of Dane particles. Proc. Natl. Acad. Sci. USA **78**:3403–3407.

Levine, R. A., and M. A. Payne. 1960. Homologous serum hepatitis in youthful heroin users. Ann. Intern. Med. **53**:164–178.

Levy, G. A., and F. V. Chisari. 1981. The immuno pathogenesis of chronic HBV induced liver disease. Semin. Immunopathol. **3**:439–459.

Lewin, R. 1986. New class of animal virus found in virulent form of human hepatitis. Science **234**:423–424.

Liaw, Y.-F., C.-M. Chu, I.-J. Su, M.-J. Huang, D.-Y. Lin, and C.-S. Chang-Chien. 1983. Clinical and histological events preceding hepatitis Be antigen seroconversion in chronic type B hepatitis. Gastroenterology **84**:216–219.

Lieberman, H. M., D. R. Labrecque, M. C. Kew, S. J. Hadziyannis, and D. A. Shafritz. 1983. Detection of hepatitis B virus DNA directly in human serum by a simplified molecular hybridization test: comparison to HBeAg/anti-HBe status in HBsAg carriers. Hepatology **3**:285–291.

Lieu, J.-M., C. E. Aldrich, and W. S. Mason. 1986. Evidence that a capped oligoribonucleotide is the primer for duck hepatitis B virus plus-strand DNA synthesis. J. Virol. **57**:229–236.

Lin, H.-J., J. P.-W. Kwan, P.-C. Wu, and W. Chak. 1983. Phosphonoformic acid-inhibitable nucleotide incorporation as a measure of hepatitis B viral DNA polymerase activity. J. Med. Virol. **12**:61–70.

Lindsay, K. L., A. G. Redeker, and M. Ashcavai. 1981. Delayed HBsAg clearance in chronic hepatitis B viral infection. Hepatology **1**:586–589.

Locarnani, S. A., A. G. Coulepis, A. M. Stratton, J. Kaldor, and I. D. Gust. 1979. Solid-phase enzyme-linked immunosorbent assay for detection of hepatitis A-specific immunoglobulin M. J. Clin. Microbiol. **9**:459–465.

Locarnini, S. A., S. M. Garland, N. I. Lehmann, R. C. Pringle, and I. D. Gust. 1978. Solid-phase enzyme-linked immunosorbent assay for detection of hepatitis A virus. J. Clin. Microbiol. **8**:277–282.

Lurmann, A. 1885. II. Eine Icterusepidemic. Berl. Klin. Wchnschr. **22**:20.

MacCallum, F. O. 1972. Early studies on viral hepatitis. Br. Med. Bull. **28**:105–108.

Mackay, P., J. Lees, and K. Murray. 1981. The conversion of hepatitis B core antigen synthesized in *E. coli* into e antigen. J. Med. Virol. **8**:237–243.

Magnius, L. O. 1975. Characterization of new antigen-antibody system associated with hepatitis B. Clin. Exp. Immunol. **20**:209–216.

Mandart, E., A. Kay, and F. Galibert. 1984. Nucleotide sequence of a cloned duck hepatitis B virus genome: comparison with woodchuck and human hepatitis B virus sequences. J. Virol. **49**:782–792.

Marion, P. L., S. S. Knight, M. A. Feitelson, L. S. Oshiro, and W. S. Robinson. 1983a. Major polypeptide of duck hepatitis B surface antigen particles. J. Virol. **48**:534–541.

Marion, P. L., S. S. Knight, B.-K. Ho, Y.-Y. Guo, W. S. Robinson, and H. Popper. 1984. Liver disease associated with DHBV infection of domestic ducks. Proc. Natl. Acad. Sci. USA **81**:898–902.

Marion, P. L., S. S. Knight, F. H. Salazar, H. Popper, and W. S. Robinson. 1983b. Ground squirrel hepatitis virus infection. Hepatology **3**:519–527.

Marion, P. L., L. Oshiro, D. C. Regnery, G. H. Scullard, and W. S. Robinson. 1980. A virus in Beechey ground squirrels that is related to hepatitis B virus of humans. Proc. Natl. Acad. Sci. USA **77**:2941–2945.

Marion, P. L., and W. Robinson. 1983. Hepadnaviruses. Hepatitis B and related viruses. Curr. Top. Microbiol. Immunol. **105**:99–121.

Mason, W. S., C. Aldrich, and J. Summers. 1982. Asymmetric replication of duck hepatitis B virus DNA in liver cells: free minus-strand DNA. Proc. Natl. Acad. Sci. USA **79**:3997–4001.

Mason, W. S., J. Newbold, G. Seal, C. E. Aldrich, L. Coates, J. M. England, J. Summers, and M. S. Halpern. 1984. Expression of duck hepatitis B virus in congenitally and experimentally infected Pekin ducks, p. 443–450. *In* G. N. Vyas, J. L. Dienstag, and J. H. Hoofnagle, (ed.), Viral hepatitis and liver disease. Grune and Stratton, New York.

Mason, W. S., G. Seal, and J. Summers. 1980. Virus of Pekin ducks with structural and biological relatedness to human hepatitis B virus. J. Virol. **36**:829–836.

Milich, D. R., G. B. Thornton, A. R. Neurath, S. B. Kent, M.-L. Michel, P. Tiollais, and F. U. Chisari. 1985. Enhanced immunogenicity of the pre-S region of hepatitis B surface antigen. Science **228**:1195–1199.

Miller, R. H., and W. S. Robinson. 1986. Common evolutionary origin of hepatitis B virus and retroviruses. Proc. Natl. Acad. Sci. USA **83**:2531–2535.

Milne, A., G. K. Allwood, C. D. Moyes, N. E. Pearce, and C. R. Lucas. 1985. The prevalence of hepatitis B infections in a multiracial New Zealand community. N. Z. Med. J. **98**:529–531.

Mims, C. A. 1974. Factors in the mechanism of persistence of viral infections. Progr. Med. Virol. **18:**1–15.

Misharo, S., M. Imai, K. Takahashi, A. Machida, T. Gotanda, Y. Miyakawa, and M. Mayumi. 1980. A 49,000-dalton polypeptide bearing all the antigenic determinant and full immunogenicity of 22nm hepatitis B surface antigen particles. J. Immunol. **124:**1589–1593.

Mizusawa, H., M. Taira, K. Yaginuma, M. Kobayashi, E. Yoshida, and K. Koike. 1985. Inversely repeating integrated HBV DNA and cellular flanking sequences in the human hepatoma-derived cell line husp. Proc. Natl. Acad. Sci. USA **82:**208–212.

Molnar-Kimber, K. L., J. Summers, and W. S. Mason. 1984. Mapping of the cohesive overlap of DHBV DNA and of the site of the initiation of reverse transcription. J. Virol. **51:**181–191.

Molnar-Kimber, K., J. Summers, J. M. Taylor, and W. S. Mason. 1983. Protein covalently bound to minus-strand DNA intermediates of duck hepatitis B virus. J. Virol. **45:**165–172.

Mondelli, M., and A. L. W. F. Eddleston. 1984. Mechanism of liver cell injury in acute and chronic hepatitis B. Semin. Liver Dis. **4:**47–51.

Morace, G., K. Von Der Helm, and F. Deinhardt. 1985. Detection of hepatitis B virus DNA in serum by a rapid filtration-hybridization assay. J. Virol. Methods **12:**235–242.

Moriarty, A. M., H. Alexander, R. A. Lerner, and S. B. Thornton. 1985. Antibodies to peptides detect new hepatitis B antigen: serological correlation with hepatocellular carcinoma. Science **227:**429–432.

Mosley, J. W., V. M. Edwards, and G. Casey. 1975. Hepatitis B virus infection in dentist. N. Engl. J. Med. **293:**729–734.

Murray, R., and W. D. DieFenbach. 1953. Effect of heat on the agent of homologous serum hepatitis. Proc. Soc. Exp. Biol. Med. **84:**230–231.

Naumov, N. V., M. Mondelli, G. J. M. Alexander, R. S. Tedder, A. L. W. F. Eddleston, and R. Williams. 1984. Relationship between expression of hepatitis B virus antigens in isolated hepatocytes and autologous lymphocyte cytotoxicity in patients with chronic hepatitis B virus infection. Hepatology **4:**63–68.

Negro, F., M. Berninger, E. Chiaberge, P. Gugliotta, G. Bussolati, G. C. Actis, M. Rizzetto, and F. Bonino. 1985. Detection of HBV-DNA by *in situ* hybridization using a biotin-labelled probe. J. Med. Virol. **15:**373–382.

Neurath, A. R., S. B. H. Kent, N. Strick, P. Taylor, and C. E. Stevens. 1985. Hepatitis B virus contains pre-S gene encoded domains. Nature (London) **31:**154–156.

Neurath, A. R., and N. Strick. 1979. Radioimmunoassay for albumin-binding sites associated with HBsAg: correlation of results with the presence of e-antigen in serum. Intervirology **11:**128–132.

Nielsen, J. O., O. Dietrichson, and P. Elling. 1981. Incidence and meaning of persistence of Australia antigen in patients with acute viral hepatitis: development of chronic hepatitis. New Engl. J. Med. **285:**1157–1160.

Nishioka, K., A. G. Levin, and M. J. Simons. 1975. Hepatitis B antigen, antigen subytpes, and hepatitis B antibody in normal subjects and patients with liver disease. Bull. W.H.O. **52:**293–300.

Nowicki, M. J., M. J. Tong, and R. E. Bohman. 1985. Alterations in the immune response of non responders to the hepatitis B vaccine. J. Infect. Dis. **152:**1245–1248.

O'Connell, A. P., M. K. Urban, and W. T. London. 1983. Naturally occurring infection of Pekin duck embryos by duck hepatitis B virus. Proc. Natl. Acad. Sci. USA **80:**1703–1706.

Ogston, C. W., G. J. Jonak, C. E. Rogler, S. M. Astrin, and

J. Summers. 1982. Cloning and structural analysis of integrated woodchuck hepatitis virus sequences from hepatocellular carcinomas of woodchucks. Cell **29:**385–394.

Okochi, K., and S. Murakami. 1968. Observations on Australia antigen in Japanese. Vox Sang. **15:**374–379.

Omata, M., A. Afroukedis, C. T. Liew, M. Ashcavai, and R. L. Peters. 1978. Comparison of serum hepatitis B surface antigen (HBsAg) and serum anti-core with tissue HBsAg and hepatitis B core antigen (HBcAg). Gastroenterology **75:**1003–1009.

Omata, M., K. Uchiumi, Y. Ho, O. Yokosuka, J. Mori, K. Terao, Y. Wei-Fa, A. P. O'Connell, W. T. London, and K. Okuda. 1983. Duck hepatitis B virus and liver diseases. Gastroenterology **85:**260–267.

Ono, Y., H. Onada, R. Sasanda, K. Igarashi, Y. Sugino, and K. Nishioka. 1983. The complete nucleotide sequence of the cloned hepatitis B virus DNA: subtypes *adr* and *adw*. Nucleus Acids Res. **11:**1747–1757.

Pasek, M., T. Goto, W. Gilbert, B. Zink, H. Schaller, P. MacKay, G. Leadbetter, and K. Murray. 1979. Hepatitis B virus genes and their expression in *E. coli*. Nature (London) **282:**575–579.

Pontisso, P., A. Alberti, F. Bortolotti, and G. Realdi. 1983. Virus-associated receptors of polymerized human serum albumin in acute and chronic hepatitis B virus infection. Gastroenterology **84:**220–226.

Pontisso, P., L. Chemello, G. Fattovich, A. Alberti, G. Realdi, and C. Brechot. 1985. Relationship between HBcAg in serum and liver and HBV replication in patients with HBsAg-positive chronic liver disease. J. Med. Virol. **17:**145–152.

Pontisso, P., M. C. Poon, P. Tiollais, and C. Brechot. 1984. Detection of hepatitis B virus DNA in mononuclear blood cells. Br. Med. J. **228:**1563–1566.

Ponzetto, A., B. J. Cohen, E. M. Vandervelde, and P. P. Mortimer. 1983. Delta antigen in Britain. Lancet **i:**1141–1142.

Ponzetto, A., P. J. Cote, H. Popper, B. H. Hoyer, W. T. London, E. C. Ford, F. Bonino, R. H. Purcell, and J. L. Gerin. 1984a. Transmission of the hepatitis B virus-associated S agent to the eastern woodchuck. Proc. Natl. Acad. Sci. USA **81:**2208–2212.

Ponzetto, A., L. B. Seeff, Z. Buskell-Bales, and the Veterans Administration Hepatitis Co-operative Study Group. 1984b. Hepatitis B markers in United States drug addicts with special emphasis on the delta hepatitis virus. Hepatology **4:**1111–1115.

Popper, H., J. L. Dienstag, S. M. Feinstone, H. J. Alter, and R. H. Purcell. 1980. The pathology of viral hepatitis in chimpanzees. Virchows Arch. [Pathol. Anat.] **387:**91–106.

Popper, H., M. A. Gerber, and S. N. Thung. 1982. The relation of hepatocellular carcinoma to infection with hepatitis B and related viruses in man and animals. Hepatology Suppl **2:**1–9.

Popper, H., J. W. K. Shih, J. L. Gerin, D. C. Wong, B. H. Hoyer, W. T. London, D. L. Sly, and R. H. Purcell. 1981. Woodchuck hepatitis and hepatocellular carcinoma: correlation of histologic with virologic observations. Hepatology **1:**91–98.

Popper, H., S. N. Thung, M. A. Gerber, S. C. Hadler, M. Le Monzon, A. Ponzetto, E. Anzola, D. Rivera, A. Mondolfi, A. Bracho, D. P. Francis, J. L. Gerin, J. E. Maynard, and R. H. Purcell. 1983. Histologic studies of severe delta agent infection in Venezuelan Indians. Hepatology **3:**906–912.

Pourcel, C., A. Louise, M. Gervais, N. Chenciner, M-F. Dubois, and P. Tiollais. 1982. Transcription of the hepatitis B surface antigen gene in mouse cells transformed with cloned viral DNA. J. Virol. **42**:100–105.

Prince, A. M. 1968. An antigen detected in the blood during the incubation period of serum hepatitis. Proc. Natl. Acad. Sci. USA **60**:814–821.

Prince, A. M., W. Stephan, and B. Brotman. 1983. β-Propiolactone/ultraviolet irradiation: a review of its effectiveness for inactivation of virus in blood derivates. Rev. Infect. Dis. **5**:92–107.

Propert, S. A. 1938. Hepatitis after prophylactic serum. Br. Med. J. **2**:677–679.

Purcell, R. H., and J. L. Gerin. 1985. Prospects for second and third generation hepatitis B vaccines. Hepatology **5**:159–163.

Purcell, R. H., G. L. Gerin, P. V. Holland, W. L. Cline, and R. M. Chanock. 1970. Preparation and characterization of complement-fixing hepatitis-associated antigen and antiserum. J. Infect. Dis. **121**:222–226.

Raimondo, G., A. Smedile, L. Gallo, A. Balbo, A. Ponzetto, and M. Rizzetto. 1982. Multicentre study of prevalence of HBV associated delta infection and liver disease in drug addicts. Lancet **1**:249–251.

Rall, L. B., D. N. Standring, O. Laub, and W. J. Rutter. 1983. Transcription of HBV by RNA polymerase II. Mol. Cell. Biol. **3**:1766–1773.

Realdi, G., A. Alberti, M. Rugge, F. Bortolotti, A. M. Rigoli, F. Tremolada, and A. Ruol. 1980. Seroconversion from hepatitis B e antigen to anti-HBe in chronic hepatitis B virus infection. Gastroenterology **79**:195–199.

Redeker, A. G. 1975. Viral hepatitis: clinical aspects. Am. J. Med. Sci. **270**:9–16.

Rizzetto, M. 1983. The delta agent. Hepatology **3**:729–737.

Rizzetto, M., M. G. Canese, S. Arico, O. Crivelli, C. Trepo, F. Bonino, and G. Verme. 1977. Immunofluorescence detection of new antigen-antibody system (δ/anti-δ) associated to hepatitis B virus in liver and in serum of HBsAg carriers. Gut, **18**:997–1003.

Rizzetto, M., M. G. Canese, J. L. Gerin, W. T. London, O. L. Sly, and R. H. Purcell. 1980a. Transmission of the hepatitis B virus-associated delta antigen to chimpanzees. J. Infect. Dis. **141**:590–602.

Rizzetto, M., B. Hoyer, M. G. Canese, J. W. K. Shih, R. H. Purcell, and J. L. Gerin. 1980b. Delta agent: the association of delta antigen with hepatitis B surface antigen and ribonucleic acid in the serum of delta-infected chimpanzees. Proc. Natl. Acad. Sci. USA **77**:6124–6128.

Rizzetto, M., C. Morello, P. M. Mannucci, D. J. Gocke, J. A. Spero, J. H. Lewis, D. H. van Thiel, C. Scaroni, and F. Peyretti. 1982. Delta infection and liver disease in haemophiliac carriers of hepatitis B surface antigen. J. Infect. Dis. **145**:18–22.

Rizzetto, M., R. H. Purcell, and J. L. Gerin. 1980c. Epidemiology of HBV-associated delta agent: geographical distribution of anti-delta and prevalence in polytransfused HBsAg carriers. Lancet **1**:1215–1218.

Rizzetto, M., J. W. Shih, and J. L. Gerin. 1980d. The hepatitis B virus-associated δ antigen: isolation from liver, development of solid-phase radioimmunoassays for δ antigen and anti-δ and partial characterization of δ antigen. J. Immunol. **125**:318–324.

Rizzetto, M., G. Verme, S. Recchia, F. Bonino, P. Farci, S. Arico, R. Calzia, A. Picciotto, M. Colombo, and H. Popper. 1983. Chronic hepatitis in carriers of HBsAg with intra hepatic expression of the δ antigen. Ann. Int. Med. **98**:437–441.

Robinson, W. S. 1975. DNA and DNA polymerase in the core of the Dane particle of hepatitis B. Am. J. Med. Sci. **270**:151–159.

Robinson, W. S. 1977. The genome of hepatitis B virus. Ann. Rev. Microbiol. **31**:357–377.

Robinson, W. S. 1985. Hepatitis B virus, p. 1384–1406. In B. N. Fields, D. M. Knipe, R. M. Chanock, J. L. Melnick, B. Roizman, and R. E. Shope (ed.), Virology. Raven Press, New York.

Robinson, W. S., D. A. Clayton, and R. L. Greenman. 1974. DNA of a human hepatitis B virus candidate. J. Virol. **14**:384–391.

Robinson, W. S., and R. L. Greenman. 1974. DNA polymerase in the core of the human hepatitis B virus candidate. J. Virol. **13**:1231–1236.

Robinson, W. S., and L. I. Lutwick. 1976. Hepatitis B virus: a cause of persistent virus infection in man, p. 787–811. In D. Baltimore, A. Huang, and C. F. Fox (ed.), Animal virology. Academic Press, New York.

Robinson, W. S., P. L. Marion, M. Feitelson, and A. Siddiqui. 1982. The hepadna virus group: hepatitis B and related viruses, p. 57–68. In W. Szmuness, H. J. Alter, and J. W. Maynard, (ed.), Viral hepatitis—1981 International Symposium, Franklin Institute Press, Philadelphia.

Robinson, W. S., P. L. Marion, and R. H. Miller. 1984a. The hepadna viruses of animals. Semin. Liver Dis. **4**:347–351.

Robinson, W. S., R. H. Miller, L. Klote, P. L. Marion, and S. C. Lee. 1984b. Hepatitis B virus and hepatocellular carcinoma, p. 245–264. In G. N. Vyas, J. L. Dienstag, and J. H. Hoofnagle, (ed.), Viral hepatitis and liver disease. Grune and Stratton, New York.

Romet-Lemoune, J.-L., M. F. McLane, E. Elfassi, W. A. Haseltine, J. Azocar, and M. Essex. 1983. Hepatitis B virus infection in cultured human lymphoblastoid cells. Science **221**:667–669.

Ruiz-Opazo, N., P. R. Chakraborty, and D. A. Shafritz. 1982. Evidence for supercoiled hepatitis B virus DNA in chimpanzee liver and serum Dane particles: possible implications in persistent HBV infection. Cell **29**:129–138.

Sattler, F. R., and W. S. Robinson. 1979. Hepatitis B viral DNA molecules have cohesive ends. J. Virol. **32**:226–233.

Schulman, A. N., N. D. Fagan, M. Brezing, H. Silver, A. Nitzze, D. Morton, and G. L. Gitnick. 1980. HBe-antigen in the course and prognosis of hepatitis B infection: a prospective study. Gastroenterology **78**:253–258.

Scotto, J., M. Hadhovel, C. Hery, J. Yvart, P. Tiollais, and C. Brechot. 1983. Detection of hepatitis B virus DNA in serum by a simple spot hybridization technique: comparison with results for other viral markers. Hepatology **3**:279–284.

Seeff, L. B., H. J. Zimmerman, E. C. Wright, B. F. Felsher, J. D. Finkelstein, P. Garcia-Pout, H. B. Greenless, A. A. Dietz, J. Hamilton, R. S. Koff, C. M. Leevy, T. Kiernan, C. H. Tamburro, E. R. Schiff, Z. Vlahcevic, R. Zemel, D. S. Zimmon, and N. Nath. 1975. Efficacy of hepatitis B immune serum globulin after accidental exposure. Lancet **2**:939–941.

Seeger, C., D. Ganem, and H. E. Vermus. 1984. The cloned genome of ground squirrel hepatitis virus is infectious in the animal. Proc. Natl. Acad. Sci. USA **81**:5849–5852.

Seeger, C., D. Ganem, and H. E. Varmus. 1986. Biochemical and genetic evidence for the hepatitis B virus replication strategy. Science **232**:477–484.

Shafritz, D. A., and M. C. Kew. 1981. Identification of integrated hepatitis B virus DNA sequences in human hepatocellular carcinomas. Hepatology **1**:1–8.

Shafritz, D. A., H. M. Lieberman, K. J. Isselbacher, and J. R. Wands. 1982. Monoclonal radioimmunoassays for HBsAg: demonstration of HBV DNA or related sequences in serum and viral epitopes in immune complexes. Proc. Natl. Acad. Sci. USA **79**:5675–5679.

Shafritz, D. A., C. E. Rogler. 1984. Molecular characterization of viral forms observed in persistent hepatitis infections, chronic liver disease and hepatocellular carcinoma in woodchucks and humans, p. 225–244. *In* G. N. Vyas, J. L. Dienstag, and J. H. Hoofnagle, (ed.), Viral hepatitis and liver disease. Grune and Stratton, New York.

Shafritz, D. A., D. Shouval, H. Sherman, S. Hadziyannis, and M. Kew. 1981. Integration of hepatitis B virus DNA into the genome of liver cells in chronic liver disease and hepatocellular carcinoma. N. Engl. J. Med. **305**:1067–1073.

Shattock, A. G., and B. M. Morgan. 1984. Sensitive enzyme immunoassay for the detection of delta antigen and anti-delta, using serum as the delta antigen source. J. Med. Virol. **13**:73–82.

Shattock, A. G., B. M. Morgan, J. Peutherer, M. J. Inglis, J. F. Fielding, and M. G. Kelly. 1983. High incidence of delta antigen in serum. Lancet **2**:104–105.

Sherlock, S., and H. C. Thomas. 1983. Hepatitis B virus infection: the impact of molecular biology. Hepatology **3**:455–456.

Shikata, T., T. Karasawa, K. Abe, T. Takahashi, M. Mayumi, and T. Oda. 1978. Incomplete inactivation of hepatitis B virus after heat treatment at 60°C for 10 hours. J. Infect. Dis. **138**:242–244.

Shimuzu, M., M. Ohyama, Y. Takahashi, K. Udo, M. Kojima, M. Kametani, F. Tsuda, E. Takai, Y. Miyakawa, and M. Mayumi. 1983. Immunoglobulin M antibody against hepatitis B core antigen for the diagnosis of fulminant type B hepatitis. Gastroenterology **84**:604–610.

Siddiqui, A. 1983. Hepatitis B virus DNA in Kaposi sarcoma. Proc. Natl. Acad. Sci. USA **80**:4861–4864.

Siddiqui, A., P. L. Marion, and W. S. Robinson. 1981. Ground squirrel hepatitis virus DNA; molecular cloning and comparison with hepatitis B virus DNA. J. Virol. **38**:393–397.

Sjogren, M., and J. H. Hoofnagle. 1985. Immunoglobulin M antibody to hepatitis B core antigen in patients with chronic type B hepatitis. Gastroenterology **89**:252–258.

Sjogren, M. H., and S. M. Lemon. 1983. Low molecular weight IgM antibody to hepatitis B core antigen in chronic hepatitis B virus infection. J. Infect. Dis. **148**:445–451.

Skolnick, E. M., A. A. McLean, D. West, W. J. McAlleer, W. J. Miller, and E. B. Buynak. 1985. Clinical evaluation in healthy adults of a hepatitis B vaccine made by recombinant DNA. J. Am. Med. Assoc. **251**:2812–2815.

Smedile, A., C. Lavarini, O. Crivelli, G. Raimondo, M. Fassone, and M. Rizzetto. 1982. Radioimmunoassay detection of IgM antibodies to the HBV-associated delta (δ) antigen: clinical significance in δ infection. J. Med. Virol. **9**:131–138.

Smedile, A., C. Lavarini, P. Farci, S. Arico, G. Marinucci, P. Dentico, G. Giuliani, A. Gargnel, C. U. Blanco, and M. Rizzetto. 1983. Epidemiologic patterns of infection with hepatitis B virus associated delta agent in Italy. Am. J. Epidemiol. **117**:223–229.

Smedile, A., M. Rizzetto, F. Bonino, J. Gerin, and B. Hoyer. 1984. Serum delta-associated RNA (DAR) in chronic HBV carriers infected with the delta agent, p. 613–614. *In* G. N. Vyas, J. L. Dienstag, and J. F. Hoofnagle, (ed.), Viral hepatitis and liver disease. Grune and Stratton, New York.

Smith, G. L., M. Mackett, and B. Moss. 1983. Infectious vaccinia virus recombinants that express hepatitis B surface antigen. Nature (London) **302**:490–495.

Sninsky, J. J., A. Siddiqui, W. S. Robinson, and S. N. Cohen. 1979. Cloning and endonuclease mapping of the hepatitis B virus genome. Nature **279**:346–348.

Soulier, J. P., C. Blatix, A. M. Courouce, D. Benamon-Djiane, P. Amouch, and J. Drouet. 1972. Prevention of virus B hepatitis (SH virus). Am. J. Dis. Child. **123**:429–434.

Southern, E. M. 1975. Detection of specific sequences among DNA fragments separated by gel electrophoresis. J. Mol. Biol. **98**:503–517.

Sprengel, R., C. Kuhn, C. Manso, and H. Will. 1984. Cloned duck hepatitis B virus DNA is infectious in Pekin ducks. J. Virol. **52**:932–937.

Standring, P. N., L. B. Rall, O. Laub, and W. J. Rutter. 1983. Hepatitis B virus encodes an RNA polymerase III transcript. Mol. Cell. Biol. **3**:1774–1782.

Stevens, C. E., R. P. Beasley, C.-C. Lin, L.-Y. Huang, T.-S. Sun, F.-J. Hsieh, K.-Y. Wang, and W. Szmuness. 1982. Perinatal hepatitis B virus infection: use of hepatitis B immune globulin, p. 527–535. *In* W. Szmuness, H. J. Alter, and J. E. Maynard (ed.), Viral Hepatitis—1981 International Symposium. Franklin Institute Press, Philadelphia.

Stevens, C. E., W. Szmuness, A. I. Goodman, S. A. Weseley, and M. Fotino. 1980. Hepatitis B vaccine: immune responses in haemodialysis patients. Lancet **2**:1211–1213.

Stevens, C. E., P. T. Toy, M. J. Tong, P. E. Taylor, G. N. Vyas, P. V. Nair, M. Gudavalli, and S. Krugman. 1985. Perinatal hepatitis B virus transmission in the United States. J. Am. Med. Assoc. **253**:1740–1745.

Stibbe, W., and W. H. Gerlich. 1982. Structural relationships between minor and major proteins of hepatitis B surface antigen. J. Virol. **46**:626–628.

Stokes, J. H., R. Ruedemann, Jr., and W. S. Lemon. 1926. Epidemic infectious jaundice and its relation to the therapy of syphilis. Arch. Int. Med. **26**:521–525.

Summers, J., and W. S. Mason. 1982. Replication of the genome of a hepatitis B-like virus by reverse transcription of an RNA intermediate. Cell **29**:403–415.

Summers, J., J. M. Smolec, and R. Snyder. 1978. A virus similar to human hepatitis B virus associated with hepatitis and hepatoma in woodchucks. Proc. Natl. Acad. Sci. USA **75**:4533–4537.

Summers, J. A., A. O'Connell, and L. Millman. 1975. Genome of hepatitis B virus: restriction enzyme cleavage and structure of DNA extracted from Dane particles. Proc. Natl. Acad. Sci. USA **72**:4597–4601.

Szmuness, W. 1978. Hepatocellular carcinoma and the hepatitis B virus: evidence for a causal association. Prog. Med. Virol. **24**:40–69.

Szmuness, W., E. J. Harley, H. Ikram, and C. E. Stevens. 1978. Sociodemographic aspects of the epidemiology of hepatitis B, p. 297–320. *In* G. N. Vyas, S. N. Cohen, and R. Schmid. (ed.), Viral hepatitis. Franklin Institute Press, Philadelphia.

Szmuness, W., A. Prince, M. Goodman, C. Ehrich, R. Pick, and M. Ansari. 1974. Hepatitis B immune serum globulin in prevention of nonparenterally transmitted hepatitis B. N. Engl. J. Med. **290**:701–706.

Szmuness, W., C. E. Stevens, E. J. Harley, E. A. Zang, W. R. Oleszko, D. C. William, R. Sadovsky, J. M. Morrison, and A. Kellner. 1980. Hepatitis B vaccine: demonstration of efficacy in a controlled clinical trial in a high-risk population in the United States. N. Engl. J. Med. **303**:833–841.

Takahashi, K., A. Machida, G. Funatsu, M. Nomura, S. Usuda, S. Aoyagi, K. Tachibaua, H. Miyamoto, M. Imai, T. Nakamura, Y. Miyakawa, and M. Mayumi.

1983. Immunochemical structure of hepatitis B e antigen in the serum. J. Immunol. **130**:2903–2907.

Thomas, H. C., and A. S. F. Lok. 1984. The immunopathology of autoimmune and hepatitis B virus-induced chronic hepatitis. Semin. Liver Dis. **4**:36–41.

Thomas, H. C., M. Pignatelli, A. Goodall, J. Waters, P. Karayiannis, and D. Brown. 1984. Immunologic mechanisms of cell lysis in hepatitis B virus infection, p. 167–177. In G. N. Vyas, J. L. Dientstag, and J. H. Hoofnagle (ed.), Viral hepatitis and liver disease. Grune and Stratton, New York.

Thung, S. N., and M. A. Gerber. 1981. HBsAg-associated albumin receptors and antialbumin antibodies in sera of patients with liver disease. Gastroenterology **80**:260–264.

Thung, S. N., and M. A. Gerber. 1983a. Albumin binding sites on human hepatocytes. Liver **3**:290–294.

Thung, S. N., and M. A. Gerber. 1983b. Immunohistochemical study of delta antigen in an American metropolitan population. Liver **3**:392–397.

Thung, S. N., and M. A. Gerber. 1984. Polyalbumin receptors: their role in the attachment of hepatitis B virus to hepatocytes. Semin. Liver Dis. **4**:69–73.

Tiollais, P., P. Charnay, G. N. Vyas. 1981. Biology of hepatitis B virus. Science **213**:406–411.

Tiollais, P., A. Dejean, C. Brechot, M.-L. Michel, P. Sonigo, and S. Wain-Hobson. 1984. Structure of hepatitis B virus DNA, p. 49–65. In G. N. Vyas, J. L. Dienstag, and J. H. Hoofnagle, (ed.), Viral hepatitis and liver disease. Grune and Stratton, New York.

Tiollais, P., C. Pourcel, and A. Dejean. 1985. The hepatitis B virus. Nature (London) **317**:489–495.

Toh, H., H. Hayashida, and T. Miyata. 1983. Sequence homology between retroviral reverse transcriptase and putative polymerase of hepatitis B virus and cauliflower mosaic virus. Nature (London) **305**:827–828.

Tong, M. J., D. Stevenson, and I. Gordon. 1977. Correlation of e antigen, DNA polymerase activity and Dane particles in chronic benign and chronic active type B hepatitis B infection. J. Infect. Dis. **135**:980–984.

Trevisan, A., G. Realdi, A. Alberti, G. Ongaro, E. Pornaro, and R. Meliconi. 1982. Core antigen-specific immunogloublin G bound to the liver cell membrane in chronic hepatitis B. Gastroenterology **82**:218–222.

Tsuda, F., S. Naito, E. Takai, Y. Akahane, S. Furuta, Y. Miyakawa, and M. Mayumi. 1984. Low molecular weight (7S) immunoglobulin M antibody against hepatitis B core antigen in the serum for differentiating acute from persistent hepatitis B virus infection. Gastroenterology **87**:159–164.

Tuttleman, J. S., J. C. Pugh, and J. W. Summers. 1986. In vitro experimental infection of primary duck hepatocytic cultures with duck hepatitis B virus. J. Virol. **58**:17–25.

U.S. Department of Health and Human Services. 1984. Morbidity and Mortality Weekly Report. **33**(35):493–494.

Valenzuela, P., M. Quiroga, J. Zaldivar, P. Gray, and W. J. Rutter. 1980. The nucleotide sequence of the hepatitis B genome and the identification of the major viral genes, p. 57–70. In B. Fields, R. Jaenisch, and C. F. Fox (ed.), Animal virus genetics. Academic Press, New York.

Varmus, H., and R. Swanstrom. 1982. Replication of retroviruses, p. 369–512. In R. Weiss, N. Teich, H. Varmus, and J. Coffin, (ed.), RNA Tumour Viruses. Cold Spring Harbor Press, Cold Spring Harbor, New York.

Walker, M., W. Szmuness, C. Stevens, and P. Rubinstein. 1981. Genetics of anti-HBs responsiveness I. HLA-DR7 and non-responsiveness to hepatitis vaccination. Transfusion **21**:601.

Wallis, C., and J. L. Melnick. 1971. Enhanced detection of Australia antigen in serum hepatitis patients by discontinuous counter-immunoelectrophoresis. Appl. Microbiol. **21**:867–869.

Wands, J. R., E. Ben-Porath, and M.-A. Wong. 1984. Monoclonal antibodies and hepatitis B: a new perspective using highly sensitive and specific radioimmunoassays, p. 543–559. In G. N. Vyas, J. L. Dienstag and J. H. Hoofnagle (ed.), Viral hepatitis and liver disease. Grune and Stratton, New York.

Wands, J. R., and V. R. Zurawski. 1981. High affinity monoclonal antibodies to hepatitis B surface antigen (HBsAg) produced by somatic cell hybrids. Gastroenterology **80**:225–232.

Wang, K.-S., Q.-L. Choo, A. J. Weiner, J.-H. Ou, R. C. Najarian, R. M. Thayer, G. T. Mullenbach, K. J. Denniston, J. L. Gerin, and M. Houghton. 1986. Structure, sequence and expression of the hepatitis delta (δ) viral genome. Nature (London) **323**:508–514.

Weiser, B., D. Ganem, C. Seeger, and H. E. Varmus. 1983. Closed circular viral DNA and asymmetrical heterogeneous forms in livers from animals infected with GSHV. J. Virol. **48**:1–9.

Weissbach, A. 1975. Vertebrate DNA polymerases. Cell **5**:101–108.

Weller, I. V. D., P. Karayiannis, A. S. F. Lok, L. Montano, M. Bamber, H. C. Thomas, and S. Sherlock. 1983. Significance of delta agent infection in chronic hepatitis B infection: a study on British carriers. Gut **24**:1061–1063.

Wen, Y.-M., K. Mitamura, and B. Merchant. 1983. Nuclear antigen detected in hepatoma cell lines containing integrated hepatitis B virus DNA. Infect. Immun. **39**:1361–1367.

Werner, B. G., J. M. Smolec, R. Snyder, and J. Summers. 1979. Serological relationship of woodchuck hepatitis virus to human hepatitis B virus. J. Virol. **32**:314–322.

West, D. J. 1984. The risk of hepatitis B infection among health professionals in the United States: a review. Am. J. Med. Sci. **287**:26–33.

White, D. A., and F. F. Fenner. 1986. Medical virology, 3rd ed. Academic Press, New York.

Wilde, L. 1970. Solid phase antigen-antibody systems, p. 199–206. In K. E. Kirkham and W. M. Hunter, (ed.), Radioimmunoassay methods. E. & S. Livingston, Edinburgh.

Will, H., R. Cattaneo, H. G. Koch, G. Darai, H. Schaller, H. Schellekens, M. C. A. van Eerd Pan, and F. Deinhardt. 1982. Cloned HBV DNA causes hepatitis in chimpanzees. Nature (London) **299**:740–742.

Williamson, H. G., N. I. Lehmann, M. Dimitrakakis, D. L. B. Sharma, and I. D. Gust. 1982. A longitudinal study of hepatitis infection in an institution for the mentally retarded. Aust. N. Z. J. Med. **12**:30–34.

Wolters, G., L. P. C. Kuijpers, J. Kacaki, and A. H. W. M. Shuurs. 1977. Enzyme-linked immunosorbent assay for hepatitis B surface antigen. J. Infect. Dis. **136** (Suppl.):311–317.

Yamada, G., and P. L. Nakane. 1977. Hepatitis B core and surface antigen in liver tissue. Lab. Invest. **36**:649–659.

Zalan, E., C. Wilson, and N. A. Labzoffsky. 1973. Elimination of non-specific reaction in latex agglutination test for the detection of hepatitis-associated antigen. Arch. Gesamte Virusforsch. **40**:171–175.

Zhou, Y.-Z. 1980. A virus possibly associated with hepatitis and hepatoma in ducks. Shanghai Inst. Biochem. Acad. Sinica **3**:641–644.

Zuckerman, A. J. 1975. Tissue and organ culture studies of hepatitis B virus. Am. J. Med. Sci. **270**:197–204.

CHAPTER 40

Unclassified: Non-A, Non-B Hepatitis

ARIE J. ZUCKERMAN

Disease: Non-A hepatitis, non-B hepatitis.

Etiologic Agents: Diagnosis based on exclusion; the viral agents causing the parenterally transmitted forms of the infection have not been identified; the enterically transmitted epidemic form appears to be caused by a calici-like virus which has not yet been fully characterized.

Clinical Manifestations: Clinically similar to other types of acute viral hepatitis; the epidemic form is associated more frequently with cholestasis and is severe in pregnancy, causing high mortality.

Pathology: As for other forms of viral hepatitis.

Laboratory Diagnosis: By exclusion of hepatitis A, B, delta hepatitis, and hepatitis caused by cytomegalovirus and Epstein-Barr virus.

Epidemiology: Epidemiologically, the parenteral forms are identical to hepatitis B. The parenteral types are distributed worldwide; the epidemic form is common in the subcontinent of India, the Middle East and North Africa.

Prevention and Control: The epidemic form is spread by the fecal-oral route, usually by contamination of water supplies and food by sewage.

Treatment: There is no specific treatment.

Introduction

The identification in recent years of specific viral markers of hepatitis A, hepatitis B, and delta hepatitis (hepatitis D) led to the development of sensitive laboratory tests for the diagnosis of these infections. In addition, the specific diagnoses of hepatitis types A, B, and D revealed a previously unrecognized form of hepatitis that is clearly unrelated to any of these three types. Results obtained from several surveys of post-transfusion hepatitis in the United States (Hollinger, 1984) and elsewhere provided strong epidemiologic evidence of "guilt by association" of an infection of the liver referred to as non-A, non-B hepatitis. This is now the most common form of hepatitis occurring after blood transfusion in some areas of the world. More direct evidence of at least two transmissible agents in non-A, non-B hepatitis has come from experimental transmission to chimpanzees. Studies have also shown that this infection occurs in hemodialysis and other specialized units,

that it occurs in a sporadic form in the general population, that it can be transmitted by therapeutic plasma components, and that a prolonged carrier state in the blood may occur. There is also evidence that the parenterally transmitted infection, like hepatitis type B, may progress to chronic liver disease.

Another, more recently described form of epidemic non-A, non-B hepatitis (a better term would be epidemic non-A hepatitis), which resembles hepatitis A but is serologically distinct, has been reported from the subcontinent of India, central and southeast Asia, the Middle East, and North Africa and in travelers returning from these regions. The infection is acute, self-limiting, and occurs predominantly in young adults. The incubation period is 30 to 40 days. It is more severe in pregnant women, in whom it is associated with high mortality, especially during the last trimester of pregnancy. The infection is spread by the ingestion of contaminated water and probably by food, but secondary cases appear to be uncommon.

There is preliminary evidence that the virus morphologically (but not serologically) resembles hepatitis A virus, measuring approximately 27 nm in diameter.

The extensive literature on non-A, non-B hepatitis has been exhaustively reviewed elsewhere (Dienstag et al., 1977; Dienstag, 1983a,b; Fagan and Williams, 1984; Gerety and Iwarson, 1986; Tabor, 1985; Zuckerman and Howard, 1979).

Epidemiology

Parenterally Transmitted Non-A, Non-B Hepatitis

Non-A, non-B hepatitis has been found in every country in which it has been sought and shares a number of features with hepatitis B. This form of hepatitis has been most commonly recognized as a complication of blood transfusion, and in countries where all blood donations are screened for hepatitis B surface antigen by very sensitive techniques, non-A, non-B hepatitis may account for as many as 90% of all cases of post-transfusion hepatitis. Outbreaks of non-A, non-B hepatitis have also been reported after the administration of blood-clotting factors VIII and IX. Non-A, non-B hepatitis has occurred in hemodialysis and other specialized units, among drug addicts, and after accidental inoculation with contaminated needles and other sharp objects, and occasionally maternal to infant transmission has been reported (reviewed by Dienstag, 1977, 1983a,b; Fagan and Williams, 1984).

In several countries, a significant number of cases are not associated with transfusion, and such sporadic cases of non-A, non-B hepatitis account for 10 to 25% of all adult patients with recognized viral hepatitis. The route of infection or the source of infection cannot be identified in many of these patients. Males predominate, as is the case with hepatitis B. Differences from the epidemiology of hepatitis B include lack of evidence of transmission of infection in the family setting, and no evidence of transmission by sexual contact, either heterosexual or homosexual.

Although in general the illness is mild and often subclinical or anicteric, severe hepatitis with jaundice does occur, and the infection is a significant cause of fulminant hepatitis. There is considerable evidence that the infection may be followed in many patients, and in experimentally infected chimpanzees, by prolonged viremia and the development of a persistent carrier state. Studies of the histopathologic sequelae of acute non-A, non-B hepatitis infection revealed that chronic liver damage, which may

be severe, may occur in as many as 40 to 50% of the patients.

Clinical, epidemiologic, and experimental studies in several laboratories indicate that non-A, non-B hepatitis may be caused by two and possibly more than two infectious agents. Clinical evidence is based on the observation of multiple attacks of hepatitis in individual patients. Epidemiologically, short incubation (2 to 5 weeks) and long incubation (5 to 10 weeks or longer) forms of non-A, non-B hepatitis have been described. The incubation period, however, does not appear to be a reliable index for differentiating between the two non-A, non-B types of hepatitis, and it is likely that differences in the incubation period represent differences in the infective dose. Experimental evidence for the existence of at least two distinct non-A, non-B hepatitis viruses has been obtained from cross-challenge experimental transmission studies in chimpanzees, but final confirmation must await the development of specific laboratory tests and the identification and characterization of the virus(es) (Bradley et al., 1980; Hollinger et al., 1980; Tabor, 1985; Tsiquaye and Zuckerman, 1979; Yoshizawa et al., 1981).

Sporadic Form of Non-A, Non-B Hepatitis

In the course of epidemiologic investigation of hepatitis in Costa Rica, Villarejos et al. (1975) examined sera from 103 patients for hepatitis A antibody, hepatitis B surface antigen and surface antibody, for antibodies to cytomegalovirus, and for infectious mononucleosis. Infection with these viruses was excluded in 11 patients with hepatitis who had not been transfused and in whom there was no indication of exposure to hepatotoxic agents, including alcohol and narcotic drugs. The evidence indicated person-to-person transmission, the illnesses were as often subclinical as clinical, and all the patients were previously diagnosed on epidemiologic, clinical, and biochemical grounds as probably suffering from hepatitis A. However, the infection in these patients was evidently non-A, non-B hepatitis, as was the case in a subsequent small study. Mosley et al. (1977) examined, for serologic evidence of infection with hepatitis A, hepatitis B, cytomegalovirus, and Epstein-Barr (EB) virus, sera from 13 patients who were readmitted to a hospital in Los Angeles with repeated attacks of acute hepatitis. These patients had a total of 30 episodes of acute viral hepatitis; 12 of the 13 patients had evidence that hepatitis B was the cause of one of the multiple episodes of acute hepatitis, and only 2 patients had evidence that one of their infections was hepatitis A. One explanation for multiple attacks of acute hepatitis would be clinical recurrence of chronic infection, but this is un-likely because each episode was a distinct departure from good health and was accompanied by typical prodromal symptoms of acute viral hepatitis. In addition, all liver biopsies in subsequent episodes had the features of acute hepatitis without evidence of underlying chronic lesions. Furthermore, there was no statistically significant change in serologic indexes of hepatitis A or B during non-A, non-B episodes, and the serologically identifiable episodes did not necessarily precede the unidentified infections.

Finally, multiple attacks were most commonly observed in drug addicts using the parenteral route. Dienstag et al. (1977) investigated the cause of non-B acute hepatitis in 45 patients admitted to a hospital in Los Angeles; 20 patients were diagnosed by exclusion of non-A, non-B hepatitis. Clinically, the infection could not be distinguished from hepatitis A, but the predominance of women among the non-A, non-B group was striking. Stewart et al. (1978) and Farrow et al. (1981) reported the results of an investigation of the relative importance of non-A, non-B hepatitis in probably the most extensive and comprehensive survey of viral hepatitis carried out in a total population group between 1972 and 1975 in three West London boroughs with a population of 682,695. Of 368 patients with clinical hepatitis, 48 (13%) were attributed to non-A, non-B hepatitis. The illness was milder than hepatitis B, as judged by the duration of jaundice and peak serum bilirubin and alanine aminotransferase levels. The ratio of men to women was 1.4 to 1, but there was an excess of women in their 20s, most of whom were single. Only one had received blood, and none was a drug addict. Other surveys in other countries yielded essentially similar results, although the range of non-A, non-B hepatitis was 10 to 25%.

therapy. However, this proliferation differs from a normal physiologic response, or that of drug detoxification in that it is accompanied by an increase in complexity and formation of narrow, occasionally branching tubules. These arise from expanded elements of SER which, especially when the affected membrane surrounds mitochondria or other cytoplasmic organelles, appear to be formed by loss of ribosomal particles from contiguous areas of RER. The diameter of these tubules is between 15 and 25 nm, and they vary in electron density, some being as dense as adjacent glycogen granules, of which there appears to be a normal distribution.

Where cross-sections of the tubules are discernible, the electron density is seen to surround a less dense, central core or lumen, and where tubules are very close together, they merge into one another. In any one section, tubules may present as parallel, electron-dense lines approximately 20 nm thick and up to 2 μ in length, or as circlets of varying diameter. In some cells, four or five concentric circlets are seen together, with parallel lengths or the tubules flanked by smooth or rough endoplasmic reticulum (Figs. 1 to 3).

The whole complex of bizarre profiles, in which whole lengths of apparently normal endoplasmic reticulum are interwoven with varying lengths of electron-dense tubules, makes interpretation of any one isolated area on its own virtually impossible. In many sections, the tubules exhibit regular alternating bands of electron opacity and lucidity, with a periodicity of approximately 20 nm. The earliest indication of SER abnormality was seen 7 days after infection, and 13 weeks after infection similar abnormalities were still present, which in one chimpanzee was 12 weeks after serum enzyme levels had returned to

Ultrastructural Changes in the Liver in Experimental Non-A, Non-B Hepatitis

Cytoplasmic and nuclear ultrastructural abnormalities occur in experimental non-A, non-B hepatitis in chimpanzees.

Characteristic cytoplasmic changes have been recorded by Jackson et al. (1979), Shimizu et al. (1979), and Tsiquaye et al. (1980). They are associated with smooth endoplasmic reticulum (SER) rather than rough endoplasmic reticulum (RER), although the SER involved might have been formed from RER through loss of ribosomes. The first deviation from normality is a marked proliferation of SER similar to that known to accompany starvation or barbiturate

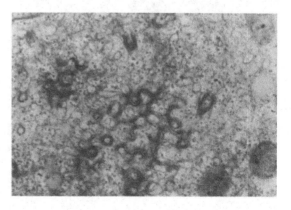

FIG. 1. Attaching curved membranes resulting from close opposition of two cisternae of smooth endoplasmic reticulum. Magnification ×30,000.

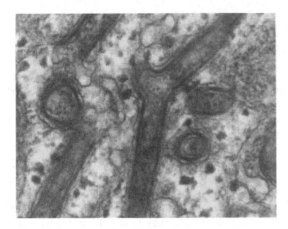

FIG. 2. Abnormal cytoplasmic structures in the hepatocytes during early experimental acute non-A, non-B hepatitis. Many of these structures have double membranes separated by electron-dense material. The tubular structures contain ribosomal material. Branching of the tubules may be a prominent feature. Magnification ×30,000.

their normal limits. Because of the wide variation in chromatin distribution and granular content of normal nuclear material, it is difficult to attribute specific early changes to a non-A, non-B hepatitis agent. However, as the serum alanine aminotransferase levels increased, it was noted that many hepatocyte nuclei within the biopsy specimens showed a generalized reduction in chromatin content, most marked around the periphery just within the nuclear membrane. There was, at the same time, an overall rarefaction of the nuclear granular matrix.

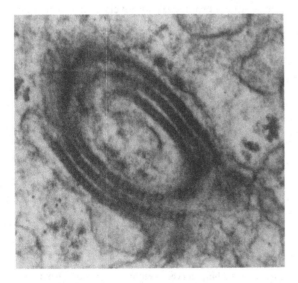

FIG. 3. Tubular structure with a regular surface periodicity. The dark-band interval measures approximately 17 nm. Magnification ×200,000.

This initial change was followed, or accompanied, by a condensation and clumping of the residual chromatin into one, two, or possibly three irregularly shaped masses with patchy electron density. Within some of these masses, as well as within limited areas between the chromatin masses and the nuclear envelope, there were groups of irregularly shaped, small, electron-dense particles ranging in diameter from 15 to 20 nm. Although virus-like in appearance, the variation in particle size and their irregularity tend to negate this interpretation, which appears to be nonspecific (McCaul et al., 1982b,c).

Another pathologic change noted by Bradley et al. (1980) is the presence of crystalline structures consisting of particles measuring 25 to 30 nm in diameter in the cytoplasm of endothelial cells of the hepatic sinusoids and Kupffer cells during the acute phase of infection. Individual particles within crystalline arrays were examined by the Markham rotation technique (Fig. 4) and found to possess an outer structure with 16 to 18 divisions (McCaul et al., 1982a). However, reinforcement of the periodical structures was not as well defined as in other objects with symmetric patterns, and it was concluded that these crystalline structures probably reflect host cell response to infection.

Such definitive and characteristic ultrastructural changes have not been found in human non-A, non-B hepatitis, although a number of ultrastructural alterations have been described (see, for example, Anderson et al., 1982; Watanabe et al., 1984).

The Viruses

1. The virus(es) causing the parenterally transmitted form of non-A, non-B hepatitis have not been identified (although numerous particles have been described), and their mode of replication and antigenic composition remain unknown despite intensive efforts in many laboratories throughout the world. There is no homology between hepatitis B virus and the non-A, non-B agents (Fowler et al., 1983; Prince et al., 1982).

2. Although the virus causing epidemic non-A hepatitis has not been fully characterized, preliminary studies indicate that the virus is excreted in feces (Khuroo, 1980); it morphologically resembles hepatitis A virus as a nonenveloped particle measuring 27 nm in diameter. The buoyant density of the particles in cesium chloride is 1.35 g/cm. These virus particles are antigenically distinct from hepatitis A virus, and immune aggregates are formed and detected by immune electron microscopy after the addition of convalescent serum, but not with

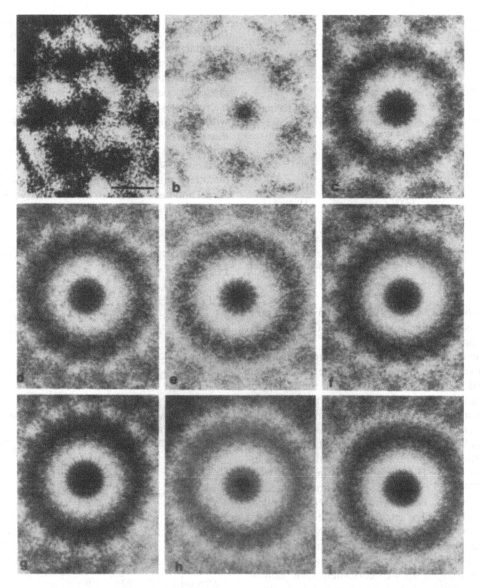

FIG. 4. Application of the Markham rotation technique to an individual unit of the crystalline structure shown in panel a.

convalescent serum from patients with serologically confirmed hepatitis A (Balayan et al., 1983; Sreenivasan et al., 1984).

The virus has not been cultivated in culture. The infection has been transmitted to *Macaca cynomolgous,* but not to chimpanzees. Results of transmission to species of marmosets are irregular, and further studies are required.

The physicochemical properties of the parenterally transmitted non-A, non-B hepatitis viruses are as follows. Bradley et al. (1980) prepared pellets by ultracentrifugation of plasma from a non-A, non-B chimpanzee carrier and from the liver of another chimpanzee during the acute phase of infection. Ex-

traction of the pellets with 20 or 83% chloroform did not inactivate the infectious agent, whereas a second, chloroform-sensitive non-A, non-B agent was identified. The chloroform-sensitive virus passes through an 80-nm filter, has a sedimentation coefficient of 200 to 280S and a buoyant density in cesium chloride of 1.24 g/cm. The H strain of non-A, non-B hepatitis was also reported to be sensitive to 10% vol/vol chloroform by Feinstone et al. (1983) and by Zuckerman, Tsiquaye, Schellekens, and Murray (1986, unpublished observations).

One of the agents found in non-A, non-B hepatitis has been inactivated by formalin 1 : 2,000 at 37°C for 72 to 96 h (see Tabor, 1985). Heating at 100°C for

TABLE 1. Is non-A, non-B hepatitis caused by a retrovirus?

	Non-A, non-B hepatitis	Retroviruses
Reverse transcriptase particles	+ 60–85 nm 85–90 nm	+ 80–120 nm
Density in sucrose	1.14 g/ml	1.16 g/ml
Antigenic cross-reactivity	Glycoprotein (gp)	gp HIV
EM changes	Tubular structures Liver Nonhuman primates	HIV: Tubular structures Lymphocytes Spleen

Similarities between non-A, non-B hepatitis and retroviruses.

5 min (Yoshizawa et al., 1981) and possibly at 60°C for 10 h may be sufficient to inactivate these virus(es) (Gerety and Iwarson, 1986).

3. Reverse transcriptase activity in human and chimpanzee phase non-A, non-B hepatitis has been described. Seto et al. (1984) reported the finding of reverse transcriptase in association with the parenterally transmitted forms of non-A, non-B hepatitis. Peak reverse transcriptase activity was found at the 1.14 g/ml fraction in a sucrose gradient, and inoculation of this fraction into susceptible chimpanzees induced infection. The enzyme is distinguished from cellular DNA polymerases, and it is not inhibited by actinomycin D (which inhibits DNA-dependent DNA synthesis). The enzyme is sensitive to RNase with endogenous template primer. These results suggest that these viruses bear similarities to retroviruses. Seto and Gerety (1985) also found a glycoprotein antigen to be associated with non-A, non-B hepatitis, with cross-reactivity with disrupted HIV (LAV/HTLV III) and with a protein derived from HIV. The similarities between the parenteral form of non-A, non-B hepatitis and HIV are summarized in Table 1. Hallam (1985), Khan and Hollinger (1986), and Itoh et al. (1986) were unable to confirm the finding of reverse transcriptase activity in sera from patients and chimpanzees with acute non-A, non-B hepatitis. Confirmation is also awaited of the description of a membrane-coated virus replicating in chimpanzee liver cell cultures after inoculation with an infectious non-A, non-B hepatitis serum (Prince et al., 1984) and of retrovirus-like particles in liver biopsies from patients with non-A, non-B hepatitis (Iwarson et al., 1985).

4. A monoclonal antibody, named 48-1, associated with non-A, non-B hepatitis was produced in a chimpanzee lymphoblastoid cell line established by

in vitro transformation with Epstein-Barr virus (Shimizu et al., 1985). This IgM antibody produced granular immunofluorescent staining in the cytoplasm of infected chimpanzees in almost every hepatocyte. Immunoperoxidase staining and electron microscopy confirmed that this antibody reacted with microtubules and microtubular aggregates, but not with the cytoplasmic tubular structures. Positive immunofluorescence was also obtained in a number of patients with acute non-A, non-B hepatitis. It is not yet known whether the antibody is directed against the infectious agent or a host antigen induced by the agent, and further data on the specificity and reactivity of monoclonal 48-1 are awaited.

Surrogate Tests for Non-A, Non-B Hepatitis in Donor Blood

In the absence of specific serologic tests for identifying non-A, non-B hepatitis, several "nonspecific" tests have been recommended for screening of units of blood.

Two large studies were conducted to assess the role of anti-hepatitis B surface antibody (HBs) de-

TABLE 2. Transfusion-Transmitted Viruses Study (1974–1979, United States)

Center	%	Hepatitis B	Non-A, non-B hepatitis
1. Recipients	429	1.(0.2%)	34 (7.9%)
Controls	414	0	9 (2.2%)
2. Recipients	506	2 (0.4%)	22 (4.3%)
Controls	522	1 (0.2%)	12 (2.3%)
3. Recipients	392	8 (2.1%)	68 (17.4%)
Controls	370	0	17 (4.6%)
4. Recipients	206	4 (1.9%)	32 (15.5%)
Controls	282	0	8 (2.8%)
Total			
Recipients	1,533	15 (1.0%)	156 (10.2%)
Controls	1,588	1 (0.06%)	46 (2.9%)

[a] The risk of non-A, non-B hepatitis in hospital patients was 2 to 6 times greater for transfused patients.
[b] 72% (range 47 to 82%) of the risk of hepatitis could be attributed to blood transfusion.
[c] The lowest attack rate was with volunteer blood obtained primarily from middle class white donors. Most of the blood used in hospitals with the highest attack rates was obtained from low socioeconomic groups including family and friends of patients, and until 1976 from paid blood donors.
[d] It is emphasized that in virtually every study in the United States where volunteer and commercial blood were compared directly in the same population of recipients, the risk of hepatitis in recipients of paid donors was 2.5 times higher. For example, among 2,079 patients in five studies in the United States (1970–1978), the risk of hepatitis in those receiving only volunteer blood was 6.9% compared with 27.5% risk in those receiving paid donor blood.

TABLE 3. Transfusion-Transmitted Viruses Study (United States)

Volume of transfusion	Risk of hepatitis
1 unit	6.9%
2–3 units	10.5%
4–5 units	11.0%
6–15 units	12.0%

The lack of linear relationship between volume of transfusion and risk of hepatitis, observed previously in other studies, is believed to be due in part to secondary neutralization of the virus by specific antibodies present in some of the units.

Adjustment of the effect of alanine aminotransferase (ALT) level in the donor in this study showed that the highest attack rates of 37 to 42% were in recipients of at least 1 unit of blood with ALT > 45 IU/liter attack, and rates of 5 to 8% with ALT < 45 IU/liter.

Thus, the effect of the volume of transfusion is associated only indirectly with the development of post-transfusion hepatitis.

tected in blood donor units in the subsequent development of non-A, non-B hepatitis (Conrad et al., 1977). Although the studies did show a higher incidence of hepatitis in recipients of anti-HBs-positive blood, it has been reported that it was not related to the presence of anti-HBs per se, but to the higher frequency of anti-HBs in commercial blood. Others, however, failed to confirm the association between anti-HBs in donor blood and the increased risk of non-A, non-B hepatitis in recipients.

The Transfusion Transmitted Viruses (TTV) Study Group proposed that units of blood that were positive for anti-HBc were associated with a two- to threefold greater risk of non-A, non-B hepatitis in recipients than were units without anti-HBc (Tables 2 and 3) (Stevens et al., 1984). This was confirmed more recently by Deloris et al. (1986), who calculated that by excluding anti-HBc-positive donors, 43% of non-A, non-B cases may be prevented, with a donor loss of only 4%.

The nonspecific indicator that has received most attention is serum aminotransferase levels in blood donors. Several studies have shown that the risk of non-A, non-B post-transfusion hepatitis is directly related to the serum alanine aminotransferase (ALT) level of the donor (Aach et al., 1981). However, it was concluded that exclusion of blood units with serum ALT levels of 53 IU/liter or higher would prevent 29% of post-transfusion hepatitis, with a loss of only 1.6% of donor units. This method is thus better than screening for anti-HBc, since the corrected effi-

cacy of anti-HBc as a screening test was slightly less than that of ALT and the number of blood units lost would be twice those if ALT were used. However, the sensitivity of the test for ALT is only 26%, and despite the high specificity, the predictive value is only 42%. Thus, almost two out of three units of blood with an elevated ALT level will not transmit non-A, non-B hepatitis. ALT levels vary with age, sex, alcohol use, and geographical region and would, therefore, not be useful as a surrogate marker of non-A, non-B hepatitis.

Prospective Studies on the Risk of Hepatitis in Recipients of Blood Screened for Hepatitis B Surface Antigen

Three prospective studies in the United States and one in Japan (1975 to 1979), using volunteer blood screened for hepatitis B surface antigen (HBsAg), revealed an incidence of hepatitis ranging from 7 to 12%. In another prospective study in the U.S. Army, the incidence of posttransfusion hepatitis was 17%. In each of these studies, over 90% of cases of hepatitis were attributed to non-A, non-B hepatitis.

Prospective studies of post-transfusion hepatitis since 1980, conducted primarily in patients undergoing open heart surgery in The Netherlands, Sweden, Australia, Spain, and Italy, showed an incidence ranging from 3% in Australia to 44% in Japan. The proportion of cases attributed to non-A, non-B hepatitis ranged from 78 to 100%.

Literature Cited

Aach, R. D., W. Szmuness, J. W. Mosley, F. B. Hollinger, R. A. Kahn, C. E. Stevens, V. M. Edwards, and J. Werch. 1981. Serum alanine aminotransferase of donors in relation to the risk of non-A, non-B hepatitis in recipients: the transfusion-transmitted viruses study. N. Engl. J. Med. 304:989–994.

Anderson, M., I. M. Murray-Lyon, J. C. Coleman, T. F. McCaul, R. G. Bird, G. Tovey, and A. J. Zuckerman. 1982. Non-A, non-B hepatitis: a case report. J. Med. Virol. 9:217–229.

Balayan, M. S., A. G. Andjaparidze, S. S. Savinskaya, E. S. Ketiladze, D. V. Braginsky, A. P. Savinov, and V. F. Poleschuk. 1983. Evidence for a virus in non-A, non-B hepatitis transmitted via the faecal-oral route. Intervirology 20:23–31.

Bradley, D. W., J. E. Maynard, E. H. Cook, J. W. Ebert, C. R. Gravelle, K. N. Tsiquaye, H. Kessler, A. J. Zuckerman, M. F. Miller, C.-M. Ling, and L. R. Overby. 1980. Non-A, non-B hepatitis in experimentally infected chimpanzees: cross-challenge and electron microscopic studies. J. Med. Virol. 6:185–201.

Bradley, D. W., J. E. Maynard, H. Popper, E. H. Cook, J. W. Ebert, K. A. McCaustland, C. A. Schable, and

H. A. Fields. 1983. Post-transfusion non-A, non-B hepatitis: physicochemical properties of two distinct agents. J. Infect. Dis. **148**:254–265.

Conrad, M. E., R. G. Knodell, E. L. Bradley, Jr., E. P. Flannery, and A. L. Ginsberg. 1977. Risk factors in transmission of non-A, non-B post-transfusion hepatitis: the role of hepatitis B antibody in donor blood. Transfusion **17**:579–585.

Deloris, E., M. T. Koziol, P. V. Holland, D. W. Alling, J. C. Melpolder, R. E. Solomon, R. H. Purcell, L. M. Hudson, F. J. Shoup, H. Krakauer, and H. J. Aeter. 1986. Antibody to HBcAg as a paradoxical marker for non-A, non-B hepatitis agents in donated blood. Ann. Intern. Med. **104**:488–495.

Dienstag, J. L. 1983a. Non-A, non-B hepatitis. I. Recognition, epidemiology, and clinical features. Gastroenterology **85**:439–462.

Dienstag, J. L. 1983b. Non-A, non-B heptitis. II. Experimental transmission, putative virus agents and markers, and prevention. Gastroenterology **85**:743–768.

Dienstag, J. L., A. Alaama, J. W. Mosley, A. G. Redeker, and R. H. Purcell. 1977. Etiology of sporadic hepatitis B surface antigen-negative hepatitis. Ann. Intern. Med. **87**:1–6.

Fagan, E. A., and R. Williams. 1984. Non-A, non-B hepatitis, p. 314–335. In A. J. Zuckerman (ed.), Infectious agents as causes of liver disease, Seminars in liver disease, vol. 4.

Farrow, L. J., J. S. Stewart, H. Stern, R. E. Clifford, H. G. Smith, and A. J. Zuckerman. 1981. Non-A, non-B hepatitis in West London. Lancet **1**:982–984.

Feinstone, S. M., K. B. Mihalik, T. Kamimura, H. J. Alter, W. T. London, and R. H. Purcell. 1983. Inactivation of hepatitis B virus and non-A, non-B hepatitis by chloroform. Infect. Immun. **41**:816–821.

Fowler, M. J. F., J. Monjardino, I. V. Weller, M. Bamber, P. Karayiannis, A. J. Zuckerman, and H. C. Thomas. 1983. Failure to detect nucleic acid homology between some non-A, non-B viruses and hepatitis B virus DNA. J. Med. Virol. **12**:205–213.

Gerety, R. J., and S. A. Iwarson. 1986. Non-A, non-B hepatitis, p. 441–458. In A. J. Zuckerman (ed.), Viral hepatitis. Clinics in tropical medicine and communicable diseases. London: W. B. Saunders.

Hallam, N. F. 1985. Non-A, non-B hepatitis: reverse transcriptase? Lancet **2**:665.

Hollinger, F. B. 1984. Prevention of post-transfusion hepatitis, p. 319–337. In G. N. Vyas, J. L. Dienstag, and J. H. Hoofnagle (ed.), Viral hepatitis and liver disease. New York: Grune and Stratton.

Hollinger, F. B., J. W. Mosley, W. Szmuness, R. D. Aach, R. L. Peters, and C. Stevens. 1980. Transfusion-transmitted viruses study: experimental evidence for two non-A, non-B hepatitis agents. J. Infect. Dis. **142**:400–407.

Itoh, Y., S. Iwakiri, K. Kitajima, T. Gotanda, M. Miyaki, Y. Miyakawa, and M. Mayumi. 1986. Lack of detectable reverse transcriptase activity in human and chimpanzee sera with a high infectivity for non-A, non-B hepatitis. J. Gen. Virol. **67**:777–779.

Iwarson, S., Z. Schaff, B. Seto, G. Norkrans, and R. J. Gerety. 1985. Retrovirus-like particles in hepatocytes of patients with transfusion-acquired non-A, non-B hepatitis. J. Med. Virol. **16**:37–45.

Jackson, D., E. Tabor, and R. J. Gerety. 1979. Acute non-A, non-B hepatitis: specific ultrastructural alterations in endoplasmic reticulum of infected hepatocytes. Lancet **1**:1249–1250.

Khan, N. C., and F. B. Hollinger. 1986. Non-A, non-B hepatitis agent. Lancet **1**:41.

Khuroo, M. S. 1980. Study of an epidemic of non-A, non-B hepatitis. Possibility of another human hepatitis virus distinct from post-transfusion non-A, non-B type. Am. J. Med. **68**:818–824.

McCaul, T. F., K. N. Tsiquaye, G. Tovey, W. Duermeyer, and A. J. Zuckerman. 1982a. Examination of crystalline arrays in non-A, non-B hepatitis. J. Med. Virol. **9**:185–188.

McCaul, T. F., K. N. Tsiquaye, G. Tovey, C. Hames, X-Z. Lu, D. N. Hogben, and A. J. Zuckerman. 1982b. A study of ultrastructural alterations in experimental non-A, non-B hepatitis by electron beam analysis. Br. J. Exp. Pathol. **63**:325–329.

McCaul, T. F., K. N. Tsiquaye, G. Tovey, C. Hames, and A. J. Zuckerman. 1982c. Application of electron microscopy to the study of structural changes in the liver in non-A, non-B hepatitis. J. Virol. Methods **4**:87–106.

Mosley, J. W., A. G. Redeker, S. M. Feinstone, and R. H. Purcell. 1977. Multiple hepatitis viruses in multiple attacks of acute viral hepatitis. N. Engl. J. Med. **296**:75–78.

Prince, A. M., C. Brechot, P. Charnay, B. Brotman, L. Richardson, and P. Tiollais. 1982. Absence of detectable HBV-like DNA sequences in chimpanzee liver infected with non-A, non-B hepatitis virus(es): a preliminary report, p. 657–658. In W. Szmuness, H. J. Alter, and J. E. Maynard (ed.), International symposium on viral hepatitis 1981. Franklin Institute Press.

Prince, A. M., T. Huima, B. A. A. Williams, L. Bardina, and B. Brotman. 1984. Isolation of a virus from chimpanzee liver cell cultures inoculated with sera containing the agent of non-A, non-B hepatitis. Lancet **2**:1071–1074.

Seto, B., W. G. Coleman, Jr., S. Iwarson, and R. J. Gerety. 1984. Detection of reverse transcriptase activity in association with the non-A, non-B hepatitis agent(s). Lancet **2**:941–943.

Seto, B., and R. J. Gerety. 1985. A glycoprotein associated with the non-A, non-B hepatitis agent(s): isolation and immunoreactivity. Proc. Natl. Acad. Sci. USA **82**:4934–4938.

Shimizu, Y. K., S. M. Feinstone, R. H. Purcell, H. J. Alter, and W. T. London. 1979. Non-A, non-B hepatitis: ultrastrutural evidence for two agents in experimentally infected chimpanzees. Science **205**:197–200.

Shimizu, Y., M. Oomura, K. Abe, M. Uno, E. Yamada, Y. Ono, and T. Shikata. 1985. Production of antibody associated with non-A, non-B hepatitis in a chimpanzee lymphoblastoid cell line established by in vitro transformation with Epstein-Barr virus. Proc. Natl. Acad. Sci. USA **82**:2138–2142.

Sreenivasan, M. A., V. A. Arankalle, A. Sehgal, and K. M. Pavri. 1984. Non-A, non-B epidemic hepatitis: visualization of virus-like particles in the stool by immune electron microscopy. J. Gen. Virol. **65**:1005–1007.

Stevens, C. E., R. D. Aach, F. B. Hollinger, J. W. Mosley, W. Szmuness, R. Kahn, J. Werch, and V. Edwards. 1984. Hepatitis B virus antibody in blood donors and the occurrence of non-A, non-B hepatitis in transfusion recipients: an analysis of the transfusion-transmitted viruses study. Ann. Intern. Med. **101**:733–738.

Stewart, J. S., L. J. Farrow, R. E. Clifford, S. G. S. Lamb, N. F. Coghill, R. L. Lindon, I. M. Sanderson, P. A. Dodd, H. G. Smith, J. W. Preece, and A. J. Zuckerman. 1978. A three year survey of viral hepatitis in West London. Quart. J. Med. **47**:365–384.

Tabor, E. 1985. The three viruses of non-A, non-B hepatitis. Lancet. **1**:743–745.

Tsiquaye, K. N., R. G. Bird, G. Tovey, J. Wyke, R. Williams, and A. J. Zuckerman. 1980. Further evidence of cellular changes associated with non-A, non-B hepatitis. J. Med. Virol. **5**:63–71.

Tsiquaye, K. N., and A. J. Zuckerman. 1979. New human hepatitis virus. Lancet **1**:1135–1136.

Villarejos, V. M., K. A. Visona, C. A. Eduarte, P. J. Provost, and M. R. Hilleman. 1975. Evidence for viral hepatitis other than type A or type B among persons in Costa Rica. N. Engl. J. Med. **293**:1350–1352.

Watanabe, S., R. Reddy, L. Jeffers, G. M. Dickinson, M. O'Connell, and E. R. Schiff. 1984. Electron microscopic evidence of non-A, non-B hepatitis markers and virus-like particles in immunocompromised humans. Hepatology **4**:628–632.

Yoshizawa, H., Y. Itoh, S. Iwakiri, K. Kitajima, A. Tanaka, T. Nojiri, Y. Miyakawa, and M. Mayumi. 1981. Demonstration of two different types of non-A, non-B hepatitis by reinjection and cross-challenge studies in chimpanzees. Gastroenterology **81**:107–113.

Zuckerman, A. J., and C. R. Howard. 1979. Hepatitis viruses of man. London: Academic Press.

CHAPTER 41

Unclassified Viruses and *Caliciviridae:* Other Viruses Associated with Gastroenteritis

CHARLES RICHARD MADELEY

Disease: Diarrhea and vomiting, gastroenteropathy, "gastroenteritis," "winter vomiting," etc.

Etiologic Agents: Astrovirus, calicivirus, Norwalk virus, small round virus(es) (SRVs), small round structured virus(es) (SRSVs), coronavirus(es).

Sources: Humans, directly (fecal, oral, droplet?) or indirectly (for example, by ingesting shellfish); at present animal sources seem unlikely, but cannot be excluded.

Clinical Manifestations: Diarrhea and vomiting, with or without mild fever, of short duration. May be endemic, epidemic, or originate from food-poisoning.

Pathology: Little known, but by analogy with other viruses, it is likely to include acute villous destruction and repair from undamaged crypts.

Laboratory Diagnosis: Demonstration of virus or viral antigen in feces by negative contrast electron microscopy or antibody-based tests (for Norwalk virus). No routine antibody tests are available, but when sought, antibody has been widely present.

Epidemiology: Endemic and/or epidemic throughout the world; more common in poorer and overcrowded areas. Some or perhaps all viruses transmitted in contaminated food (mostly raw shellfish) or water. Astroviruses occur most commonly in children younger than 5 years. Other viruses occur at any age.

Treatment: No specific treatment. Symptomatic, including oral rehydration, supplemented by parenteral fluids if necessary.

Prevention and Control: None other than care in the consumption of raw shellfish and maintenance of clean water supplies. No vaccines available.

For many years, viruses were presumed to be among the causes of diarrhea and vomiting, particularly in children, but whenever they were sought by culture, the yield of positives was very low. Since the early 1970s when electron microscopy (EM) was first applied to stool extracts, a lengthening list of hitherto unrecognized viruses has been compiled. The list now includes rotaviruses, adenoviruses, astroviruses, caliciviruses, coronaviruses, Norwalk virus, other small round viruses (SRVs), small round structured viruses (SRSVs), and possibly toroviruses (Berne- and Breda-like particles). The only properties possessed by all are an association with gastrointestinal disease and a failure to grow in routine cell cultures. This has severely limited progress in characterizing them, although strains of rotaviruses (Chapter 22) and adenoviruses (Chapter 15) have now been serially propagated. Consequently, more is known about them, as these chapters show.

This chapter covers the remainder and, because it covers a miscellany of viruses with only these properties in common, it will be more incomplete than others and, probably, more untidy.

Description of Disease

All the viruses covered in this chapter infect the gut and are potential inducers of diarrhea with or without vomiting, although the severity of the resultant disease and whether diarrhea or vomiting predominates varies from virus to virus and, possibly, from patient to patient. Excretion of virus by apparently normal individuals is not uncommon, and the reason or reasons for this variability are unknown. Astroviruses are associated with endemic diarrhea most often in infants, but the other viruses may affect all ages. This point is discussed further below.

These disease presentations have led to ill-defined terms such as "winter vomiting disease," but they are not confined to one season and probably occur throughout the year. Such viral infections have been found throughout the world wherever they have been sought, although severity may vary according to local conditions.

The disease is mild and self-limiting, with rapid recovery when adequate hydration is maintained. Not surprisingly, such illness is more serious in hotter, more overcrowded locales with less adequate medical facilities.

Communicability and Transmissibility

Virus is detected in fecal extracts by EM, which indicates excretion of considerable quantities since EM is a far from sensitive technique. Hence there will be plenty of opportunity for fecal-oral transmission to occur. To date, only Norwalk virus has been found in vomitus (Greenberg et al., 1979). However, it is possible that droplet spread to and from the oropharynx occurs even though no direct evidence has been obtained.

The incubation period is probably short. Volunteer studies have been attempted only with astrovirus, Norwalk and calicivirus (see below).

Etiologic Agents

The agents described in this chapter are all viruslike in appearance and their identity as viruses is not seriously doubted (with the exception of some coronaviruses). Nonetheless, evidence for formal verification as viruses of human origin has not yet been obtained for most of them. For convenience, they will be referred to as viruses, but some uncertainties remain.

The viruses discussed in this chapter are listed in Table 1 with some of their properties and are illustrated in Figure 1. Individual examples of SRVs and SRSVs are listed in Table 2, and are referred to by either the name of the location of the outbreak or the vehicle of transmission. They may be antigenically distinct, but the difficulties of putting them into groups is discussed under Laboratory Diagnosis. It will be immediately clear that they have little in common other than a predilection for the gut in whose cells there is circumstantial evidence for replication.

Sources

All may be recovered from fecal extracts. The difficulties in growing them mean that, in 1987, there are no recognized prototype strains. Those investigators who have published the details of specific viruses in Table 2 may be willing to provide material if approached, but this should not be assumed.

For the majority, viruses of similar morphology have been recovered from domestic animals, but these are usually antigenically distinct. Hence no parallel behavior in humans as the natural host or in laboratory tests should be assumed. Equally, the ability or failure to cross species barriers also cannot be predicted, although the latter will have been attempted (if only in newborn mice) knowingly or unknowingly for the majority. No evidence for breaking species barriers has been found.

Clinical Manifestations

All the viruses are presumed to grow in and damage the cells of the gut. This leads to a disruption of normal function, which will usually present to the clinician as diarrhea with or without vomiting. Neither the diarrhea nor the vomiting is likely to be very copious, as would be found with cholera for example, but the fluid loss may become insidiously severe if hydration is not monitored and may eventually lead to circulatory collapse and death. This is an avoidable complication.

The stools may be greenish and foul-smelling, but variations have been noted (for example, the whitish stool called *hakuri* in Japan [Konno et al., 1975]). Claims to be able to distinguish types of diarrhea associated with any specific virus have not been substantiated, and there are probably no genuine clues. Much will depend on individuals and the type of food consumed. This is particularly true for adults who will constitute many of the patients.

Vomiting is a less constant feature. It is probably more frequently a feature of Norwalk virus infections and outbreaks of food-poisoning, but its presence or absence is a poor indicator of the type of virus likely to be present.

As indicated in Table 3, some of the viruses are more likely to be involved in one of three epidemiologic presentations. Pyrexia is variable, but generally low grade (Dolin, 1985). When present, it is usually brief and rarely exceeds 38.5°C. Malaise, headache, nausea, and abdominal discomfort are all features that are variably present. The incubation period has not been defined for all the viruses, but appears to be in the range of 2 to 4 days, which is long for most forms of bacterial food-poisoning. However, there is some evidence to suggest that the incubation period

TABLE 1. Viruses (other than rotavirus and adenovirus) associated with diarrhea

Virus	Approx. size (nm)	Appearance[a]	Nucleic acid	Polypeptides	Buoyant density in CsCl (g/ml)[b]	Resistant to lipid solvents	Animal parallels[c]	No. of serotypes
Astrovirus	25 to 30	Smooth outline, spherical, 5- or 6-pointed surface star on 10%, possible surface projections	ssRNA (positive sense)	Possibly 4 (36,5 K, 34 K, 33 K, and 32 K)	1.35–1.37	Resistant	Several species[d]	5?
Calicivirus	29 to 40	Spiky or ill-defined outline, proportion show the typical 2-, 3-, and 5-fold symmetry of icosahedra, round or oval surface hollows (32 in number?)	ssRNA (positive sense?)	1 major[e] 60–70 K	1.38–1.39	Resistant	Several species	5
Coronavirus	Pleomorphic (>100)	Pleomorphic envelope studded with pinlike projections in one or two tiers, vary between strains but consistent for each in length and distribution, no nucleocapsid seen	ssRNA?[f]	—[g]	—	—	Morphologically different[h]	—

Norwalk	29 to 35	Spiky or ill-defined outline, no clearly defined surface substructure or other unequivocal recognition features (SRSV)	ssRNA[j]	1 major 55 K[e]	1.38–1.39	—	None known	2?
Parvovirus[k]	20 to 26	Plain spheres but may have hexagonal outline (SRV)	DNA?	—	1.38–1.40	—	Several species	—
SRVs[l]	25 to 40	Plain featureless spheres	—	—	1.38–1.40 and —	—	—	—
SRSVs[m]	30 to 35	Spherical particles with spiky ill-defined outline, no clearly defined surface substructure	—	—	1.36–1.41 and —	—	—	—
Toroviruses	Pleomorphic (>100)	Torus or kidney-shaped cores inside an envelope with an ill-defined surface fringe	ssRNA (positive)	—	—	Sensitive	Horses, calves	—

[a] As seen in the electron microscope.
[b] Figures will be slightly different in other salts.
[c] Morphologically but antigenically distinct in cases found in one or more animal species.
[d] Associated with diarrhea.
[e] Possibly two or three minor polypeptides.
[f] By analogy with other coronaviruses.
[g] — No information available at present.
[h] Animal viruses differ in important details of their surface projections.
[j] Not DNA as previously reported.
[k] By morphology; otherwise unconfirmed.
[l] Small round viruses.
[m] Small round structured viruses.

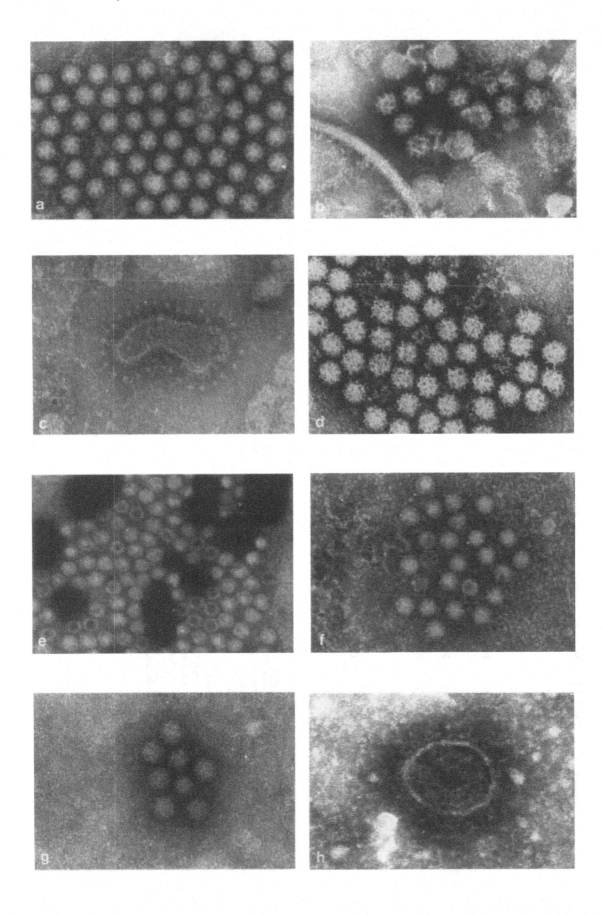

TABLE 2. Relationship among various small round viruses from outbreaks of gastroenteritis

	Agent	Relationship	Community affected	Country	Reference
Small round viruses (SRVs) (Fig. 1f)	Wollan	Related	School	UK	Paver et al., 1973
	Ditchling		School	UK	Appleton et al., 1977
	Cockle	Distinct	Widespread food-borne outbreak	UK	Appleton and Pereira, 1977
	Parramatta	Distinct	School	Australia	Christopher et al., 1978
Small round structured viruses (Fig. 1g)	Norwalk	Related	School	USA	Thornhill et al., 1977
	Montgomery County		Family	USA	
	Hawaii	Distinct	Family	USA	
	Otofuke	Related to each other, but	Institute for mentally retarded	Japan	Taniguchi et al., 1979
	Sapporo[a]	distinct from Norwalk and Hawaii	Children's home	Japan	Kogasaka et al., 1981
	Taunton	Not compared with other agents	Hospital	UK	Caul et al., 1979
	Amulree	Distinct from Norwalk and Hawaii	Hospital	UK	Appleton and Gostling, unpublished 1982
	Snow Mountain	Distinct from Norwalk and Hawaii	Resort camp	USA	Dolin et al., 1982

[a] Sapporo agent is also unrelated to caliciviruses observed in other outbreaks in the same children's home. (Adapted from Appleton, 1987.)

is dose-related, and viruses should not be ruled out because the period is too short.

The illness normally will last only 2 to 3 days, and it is reasonable to doubt whether a prolonged episode in an otherwise normal subject is due entirely to any virus found in the stool several days after onset. Virus excretion is also normally short, but children may continue to excrete enteroviruses and adeno-

viruses asymptomatically for several weeks (Bell et al., 1961; Kidd et al., 1982). This does not appear to happen commonly with the other gut viruses, and hospitalized patients found to have a fecal virus identified several days after admission are likely to have acquired it from nosocomial spread in the hospital ward.

Immunocompromised patients, whether naturally

◁ FIG. 1. Negative contrast electron micrographs of viruses seen in stool extracts. (a) Astrovirus. Note star-shaped surface morphology on some of the particles that have an entire edge. (b) Calicivirus. Note surface hollows that are round or oval in outline and filled with the negative stain. Note also the "spiky" outline (compare with astrovirus). (c) Coronavirus. The surface of the envelope is studded with pin-like projections of uniform length. Only the "head" of each projection is resolved clearly. The projections (peplomers) vary in appearance between strains, with some having two tiers. (d) Norwalk virus. This preparation was mixed with an acute serum, but the clump is probably a naturally occurring one and there is no sign of antibody being present. The virus bears some similarities to calicivirus (b), but any surface hollows are not so clearly visualized. (e) Parvovirus. Small, probably icosahedral viruses, often with a hexagonal outline and with a high proportion of stain-penetrated "empty" particles. (f) Small round viruses (SRVs). These are larger than the parvovirus (e), and this clump contains a few empties, although the proportion is usually lower. (g) Small round structured viruses (SRSVs). These viruses are generally similar to calicivirus (b) and Norwalk virus (d), and lack clearly distinguishing features, but their "hairiness" is not due to attached antibody. (h) Breda-like virus. This is a larger enveloped particle with a characteristic toroidal internal component (? nucleocapsid). The "doughnut" shape is not a complete circle, and C-shapes are common. All micrographs printed at 200,000× final magnification, and all preparations negatively contrasted with neutral potassium phosphotungstate. (d) By courtesy of Dr. A. Z. Kapikian and the editors of the *Journal of Virology*. (h) By courtesy of Mr. G. M. Beards.

TABLE 3. Types of epidemiologic behavior

Type	Viruses and age groups likely to be involved	Vehicle of spread
Endemic (outbreaks rare but not unknown)	Astrovirus, coronavirus (prolonged excretion possible), SRVs, SRSVs, torovirus Mostly in young children	None known
Epidemic (uncommon except in outbreaks likely to have a common source)	Calicivirus, Norwalk virus, SRSVs Children of all ages and adults	None known
Food poisoning (clearly common source, usually related to an easily identified event)	SRVs, SRSVs (astrovirus) Mostly in adults	Uncooked food, especially shellfish, less commonly salad bars

immunodeficient or immunosuppressed, frequently have viral respiratory tract infections. This is not paralleled in the gut. Immunodeficient patients may have frequent, multiple, and/or persistent infections that probably contribute to their eventual death, but immunosuppressed patients do not appear to be similarly susceptible. Even children being treated for cancer or leukemia do not have episodes of diarrhea attributable to viruses (they may have loose stools as a consequence of their drug therapy, but microorganisms are rarely implicated) (Madeley, Craft, Wilkinson, and Laidler, unpublished observations). When gut infections do occur, these children recover as quickly as their normal counterparts. Some severe episodes have been reported (Townsend et al., 1982), but these are very rare compared with the large numbers of patients undergoing treatment.

Pathology

With a generally nonfatal disease, details of the pathologic changes in children are scanty. In some cases (Norwalk, for instance), volunteer studies in adults have included gut biopsies (Agus et al., 1973), but most of the available data on the other viruses have been obtained from animal experiments. These should not be extrapolated to humans in toto, but there is no reason to doubt that there is considerable similarity.

The animal experimental work has been well reviewed by Hall (1987), and he himself has contributed substantially to the morphologic studies. In general, the viruses usually infect the cells covering the intestinal villi, leading to atrophy and stunting. This reduces their functional area and the number of cells available to contribute to digestion and absorption. Not surprisingly, this causes a malabsorption, which

can be documented by assaying the enzymes involved and fluid transport across the mucosal surface. However, most studies have concentrated on rotaviruses, and far less is known of the functional lesion (and the cells involved) with the other viruses. The evidence from thin section EM and labeled antibody staining suggests that different viruses may invade different parts of the villus and may induce different functional lesions as a result. Data from Hall (1987) to make this point are summarized in Figure 2.

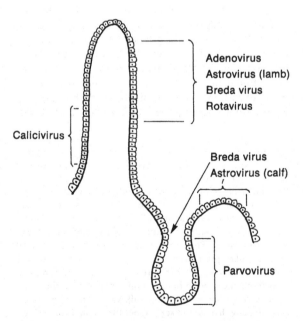

FIG. 2. Schematic view of intestinal villus and crypt system, indicating zones preferentially invaded by various gut viruses. These are based on veterinary experiments and may not be paralleled exactly in human patients. Adapted from Hall (1987).

If applicable to humans, this would suggest that dual or multiple infections, which are not uncommon, could be mutually potentiating. So far no clear evidence for this has been found, but this aspect deserves further investigation.

Villi stunted by the sloughing of the distal part after infection are normally restored by new enterocytes migrating up the villus from the crypts. This is the normal process to replace the considerable number of cells lost from the tips of the villi every day, but one which is accelerated after infection. Most viruses infect directly the cells on the villi, but an exception is provided by the parvovirus(es) which, in animals, may produce a very severe and often fatal infection. In this case, the targets are the crypt cells, possibly because these are the most actively dividing cells. This also leads to stunting with fusing of the damaged villi, but by destroying the source of new enterocytes to replace those lost, the effect is more profound and longer-lasting. Not surprisingly, this diarrhea is very difficult to treat, even if fluid balance is maintained.

There is some doubt whether the infection is terminated because the new and immature enterocytes are refractory to infection by the viruses still present in the lumen or whether they are protected by locally produced IgA antibody. Even with rotaviruses, this query is unresolved, and there is much less information about the other viruses. With the parvoviruses that are thought to infect dividing and therefore immature cells preferentially, the refractoriness argument is less convincing, and they do not seem to cause as severe an infection in humans as in kittens and puppies. Here also, the majority of infections occur in adults whose crypt cells may be less vulnerable.

To complicate the situation still further, the role of antibody, and immunity generally is unclear with these viruses. Previous infection with Norwalk virus in adult volunteers does not necessarily protect them from further symptomatic infection. Immunity in the gut is either very short-lived or irrelevant. Volunteers who developed illness, however, did seroconvert, although only half were susceptible in the first place. These, paradoxically, had lower titers of serum antibody before and after challenge and possession of antibody appeared to be a risk factor (Parrino et al., 1977).

With astroviruses, volunteers developed antibody after challenge (Kurtz et al., 1979) as occurs with most viral infections. The volunteers were not rechallenged, although there was evidence that preexisting antibody gave some protection. Calicivirus challenge of adult volunteers by the nasal/oral route induced an illness similar to natural infection, but possibly more mild, as might be expected with older subjects (Cubitt, 1987).

At present, it is impossible to draw any general conclusions over immunity. Nothing is known about the role of immunity in infections by and resistance to most of the stool viruses. Interferon may be important, but has yet to be investigated.

Laboratory Diagnosis

Much of the investigation of these viruses has been done by research workers exploring their role in disease. Because the episodes of disease associated with each are unpredictable, there has been no incentive to develop widely used tests as found with other viruses. Consequently, the availability of reagents has been determined on a personal contact basis and reagents are not commercially available. Electron microscopy remains the only technique for recognizing their presence although, as will become apparent, its value has clear limits.

Electron Microscopy

METHODS

The essence of preparation is to deliver as many virus particles as possible to the microscope grid as free as possible of contaminating proteins, salt, and other debris. Competent laboratories develop their own techniques, and a search of the literature will show a variety of alternatives. Newcomers should try several and settle on one that suits their methods of working. It would be difficult to prove that one method is better than another.

Only the negative contrast technique ("negative staining") is applicable to screening considerable numbers of fecal extracts, and provides the most characteristic image of most viruses. If large numbers of stool extracts are to be examined, the more elaborate preparation techniques should be avoided. Keep it simple—there is no evidence that more complex methods are any more sensitive than less complex methods. The method outlined next is used in the author's laboratory. It is simple and works satisfactorily.

Steps in preparation are:

1. Make an aproximate 10% suspension of the feces in 5 to 10 ml of phosphate buffered saline (PBS) by shaking by hand an appropriate quantity. If none of the extract is to be used for inoculating cell cultures, distilled water may be used instead of PBS.
2. Clarify in a bench centrifuge, preferably at 4°C if the room temperature is more than 18°C, at 3,000 rev · min⁻¹ (about 1,500 g) for 15 min.
3. Take the supernatant (which may appear crystal

clear or cloudy, but which should be free of visible aggregates) and, after removing a small amount (<1 ml) for culture, centrifuge the remainder in an ultracentrifuge at an average force of 100,000 g for 1 h. This is considerably more than is needed to sediment all known viruses, but it ensures that no significant amounts are left in the supernatant.

4. Pour off the supernatant into a suitable disinfectant (for example, hypochlorite) and, while it is held inverted, plug the mouth of the centrifuge tube with a tissue to ensure that all the supernatant is removed. Leave inverted to drain for 5 to 10 min.

5. Resuspend the pellet in one or two drops of diluent (0.1% bacitracin in distilled or deionized water) using a pasteur pipette. The bacitracin is present only as a wetting agent to ensure spreading of the extract over the whole grid. Its value is as a small pure polypeptide, not as an antibiotic.

6. Dilute the resuspended material to light opalescence in the diluent by successive addition with the pasteur pipette on dental wax, waxed paper, microscope slide, or a similar impervious and nonwetting surface. It is usually necessary to dilute most pellets 10- to 20-fold, thereby apparently negating the concentration from the centrifugation. In fact, the concentration is probably less important than the removal of soluble proteins and salt from the specimen.

7. Mix the diluted specimen with an equal volume of negative stain. The best routine stain is still probably 2 to 3% phosphotungstic acid adjusted to a neutral pH. Others such as uranyl salts (acetate and formate), silicotungstate, ammonium molybdate, and methylamine tungstate at various pH values are preferred by some microscopists.

8. A drop of the specimen/stain mixture is put on a carbon-formvar-coated grid. The surplus is drawn off with the torn edge of a piece of filter paper.

9. The grid is allowed to dry and is then examined in the electron microscope.

This procedure leaves the virus "live" on the grid. With the majority of viruses, inactivation is probably unnecessary, and the methods available are either unreliable or liable to alter the virus morphology unacceptably. Where dangerous pathogens are likely to be present (very rare in stool extracts), prudence or laboratory safety rules may require some form of inactivation to be used. The microscopist must consider the implications of the method used and be careful not to assume that the grid is totally safe if there is no way to assess this with a virus that cannot be grown in cell culture.

Space does not permit a full discussion of the problems of finding virus that is often not distributed evenly over the grid. It is obvious that a careful search is needed for viruses that may be present only in small numbers. The use of one routine and high (about 50,000×), screening magnification is less obvious, but the programming of the observer's memory is very important in improving the chances of finding virus. It is also important to remember to look for a second virus—dual infections are common enough for the extra vigilance to be rewarded.

RESULTS

With the exception of the enveloped viruses (coronaviruses and toroviruses), all the viruses sought here are in the size range 20 to 35 nm, which is appreciably smaller than rotaviruses and adenoviruses, both of which are easily recognizable as individual particles and are often present in stools in very large numbers. In contrast, only astroviruses among the viruses listed in Table 1 are likely to be present in similarly high concentrations. The other viruses will have to be sought with care and persistence, and can easily be missed by the inexperienced microscopist.

Only caliciviruses and astroviruses of the small spherical viruses possess positive identification features, reviewed by Madeley (1979). Both may exhibit surface stars, but the central hole in the star of David of the calicivirus is both diagnostic and unmistakable when seen (Fig. 1). Six-pointed stars on astroviruses never have this central stain-filled hole, but typical particles of each virus may be found only after careful searching and the operator should not expect easy certainty. Indeed, it may be easier to be certain only from a high quality micrograph, particularly with astroviruses.

The need to consult a micrograph means that representative negatives should be taken of all positive specimens and that they should be of high technical quality (in focus, and without drift or astigmatism) and taken on an electron microscope that has been calibrated at the magnification to be used.

Other small spherical viruses should be called small round viruses (SRVs) or small round structured viruses (SRSVs) depending on whether they have smooth entire edges and a featureless presenting surface (SRVs) or a hairy or feathery outline with ill-defined threadlike structures on the presenting surface (SRSVs). By EM alone, no further identification is possible, and further information can only be gained by the use of antiserum to react with the virus in immune electron microscopy (IEM), solid phase IEM (SPIEM, "flypaper," or Derrick technique), or other antibody-based tests (immune adherence hemagglutination [IAHA], etc.). In the absence of suitable validated antisera, the investigator is left with only a resemblance to some of the "named" SRVs and SRSVs but no certain identification. Care must be exercised not to overinterpret EM appearance. Similarity is not identity.

Coronaviruses and toroviruses present a different problem. Fecal extracts contain considerable amounts of membraneous debris, and distinguishing this from virus requires both judgment and experience. Recognition features include a very regular pin-shaped surface fringe that may be two-layered (coronaviruses) and the presence of a horseshoe-shaped or doughnut-shaped internal component (toroviruses) (see Fig. 1).

The problems of EM recognition are more fully explored by Doane and Anderson (1987), Madeley (1988) and Madeley and Field (1988), and the microscopist is referred to these books.

Other Techniques

Alternative techniques to EM, which is expensive, time-consuming, and extravagant in its use of trained personnel (and insensitive with undertrained microscopists), are needed for routine diagnosis. Some, as discussed next, have been explored, but development has been handicapped by shortage of starting material. Without good stocks of virus, antisera cannot be prepared nor tests evaluated. These difficulties, combined with the limited availability of diagnostic EM, have diminished the perceived importance of most of these viruses.

The reported techniques are listed by virus in Table 4, and are additional to direct EM on stool extracts. None so far has become widely used, and use has been confined almost exclusively to the laboratory in which they were developed. Details may be found in the original reports. There is not enough space to repeat them here, but some general points can be made.

As already indicated, these viruses have proved to be difficult to grow both for diagnosis and to provide further material. Reports of the growth of astroviruses (Lee and Kurtz, 1981) and caliciviruses (Cubitt, 1987) have been published, but have yet to be confirmed. If there is a general underlying reason for this failure, it remains to be found. Even the use of fetal gut organ culture (although as close to the natural habitat as can be achieved in vitro) has failed to provide the answer.

Antibody tests that have been developed have mostly been used to investigate the appearance of antibody after infection and for antibody surveys. They have shown that seroconversion will normally follow infection, although possession of antibody may not necessarily protect (see, for example, Norwalk virus: Parrino et al., 1977).

Serologic surveys have been reported for astroviruses (Kurtz and Lee, 1978), caliciviruses (Sakuma et al., 1981), and Norwalk (Kapikian et al., 1978), although the numbers studied in the first two reports were small. As with rotaviruses, acquisition of anti-

TABLE 4. Alternative methods to electron microscopy for detecting virus or antibody

Virus	Detection of virus	Detection of antibody
Astrovirus	Growth in HEK[a] cells and trypsin (Lee and Kurtz, 1981)	Immunofluorescence (Kurtz and Lee, 1978)
Calicivirus	Growth in dolphin kidney cell line (NBL-10) and HEK cells (Cubitt, 1987)	IEM[b] (Cubitt et al., 1979)
Coronavirus	Growth in foetal gut organ culture (Caul and Egglestone, 1977)	—
Norwalk virus	IEM[b] (Kapikian et al., 1972), RIA[c] (Greenberg and Kapikian, 1978)	IAHA[d] (Kapikian et al., 1978)
SRVs SRSVs Toroviruses	—	—

[a] Human embryo kidney.
[b] Immune electron microscopy.
[c] Radioimmunoassay.
[d] Immune adherence hemagglutination.

body to astrovirus and calicivirus appears to start early, and by the age of 5 years the majority have antibody, although only a few will have had a recorded episode of significant diarrhea and/or vomiting. Hence the significance of finding antibody to either of these viruses in serum may have to await tests to detect specific IgM.

In contrast, antibody to Norwalk virus is acquired more slowly, reaching a level of 50% of the U.S. population only in middle age. Consequently, a search for seroconversion is likely to be more rewarding with this virus.

Epidemiology

Viruses whose habitat is the gut are likely to spread best in the poorer overcrowded tropical parts of the world, as has already been demonstrated with enteroviruses. In fact, the evidence is patchy. Rotaviruses and adenoviruses appear to be ubiquitous, but the prevalence data for the other viruses are more uneven so far. This may be due to genuine differences in distribution or to variations in the sensitivity of the techniques used.

As already indicated in Table 3, two types of spread can be found. Endemic spread means that cases of infection are common, but outbreaks are unusual. Of the viruses discussed in this chapter, astroviruses and coronaviruses mostly show endemic spread, while Norwalk virus and calicivirus show epidemic spread. The type of spread of the other viruses is more uncertain, mainly because of the difficulties in making positive identifications. However, all may be found from time to time in individual patients, and an epidemic has to be "seeded" by a few starter endemic cases unless a vehicle such as food or water is involved.

Both SRVs and SRSVs have been involved in outbreaks of food-poisoning, mostly through ingestion of uncooked shellfish. These animals are filter-feeders, and the source of the virus is thought to be raw sewage released into the sea. The shellfish filter out the virus and concentrate it to an infecting dose. Waterborne outbreaks of rotavirus have been recorded, which means that there has been substantial contamination of drinking water with sewage because the virus will not multiply outside the body. The concentrations of some viruses can reach very high levels in feces, but this is rarely so with SRVs and SRSVs. Hence the question of whether they can spread through contaminated water remains unanswered at present, although this has been suggested for one outbreak (Koopman et al., 1982).

The peak incidence for astrovirus infections is probably less than 5 yr of age and may be about 6 mon. Adult volunteers showed comparatively low susceptibility (Kurtz et al., 1979), but may not have been as fully vulnerable as infants. The other viruses may occur at any age.

With antibody surveys suggesting considerable underreporting, much of the epidemiology of these viruses remains unrecorded. Volunteer studies in adults whose level of susceptibility is difficult to estimate accurately reveal little of their true pathogenic potential. Experiments on babies and children are unethical. Nonetheless, the adult volunteer studies with Norwalk virus reported by Parrino et al. (1977) showed some very unexpected results. The results are summarized in Figure 3 and show that half (6 of 12) of the volunteers became ill after the initial inoculation; these same 6 were again ill after the second challenge 27 to 42 mon later and 1 was ill a third time after a third challenge 8 wk after the second challenge. These results suggest that prior exposure does not apparently protect and this is supported by antibody studies in which 10 of 13 subjects with pre-existing antibody became ill, while 17 of 25 subjects without antibody did not (Blacklow et al., 1979). Such results would have important implications for a vaccine for Norwalk virus if one was proposed, but the apparent impossibility of growing the virus makes this unlikely unless genetic engineering provides an alternative.

Whether the failure to induce immunity to Norwalk virus is a general phenomenon is not known at present. To date, volunteer studies with other viruses have been confined to astroviruses (Kurtz et al., 1979), and the volunteers in that study were not rechallenged.

In addition to the study by Parrino and his colleagues, a number of other Norwalklike strains (Montgomery County, Hawaii, Snow Mountain,

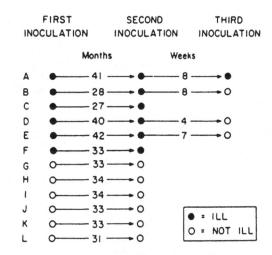

FIG. 3. Results of challenge experiments in adult volunteers with Norwalk virus. Twelve subjects (*A to L*) were challenged initially and six (*filled circles*) were ill. They were rechallenged 27 to 41 months later, and four were challenged a third time 4 to 8 weeks later. (Reprinted with permission from Parrino et al., 1977.)

etc.) have been used in volunteers and have confirmed the ability of stool extracts containing them to induce vomiting and diarrhea (Dolin et al., 1975, 1982; Wyatt et al., 1974).

Typical coronaviruses cause diarrhea in animals. A good example is the virus of transmissible gastroenteritis of swine (Garwes, 1982), but there is no evidence that these animal viruses can infect man. Similarly, human strains do not appear to infect animals, and the same probably holds true for toroviruses. The Breda and Berne strains of the latter infect calves and horses, respectively, but are distinct from the candidate human strains described by Beards et al. (1984).

Little is known about the epidemiology of human fecal coronaviruses. They have been associated with necrotizing enterocolitis in infants in France (Chany et al., 1982), but their role (if any) in causation is unclear and has not been confirmed elsewhere. Prolonged excretion over several months has been reported from England (Clarke et al., 1979) and Australia (Schnagl et al., 1978), thereby casting doubt on their role as inducers of clinical disease.

Treatment

The main problem caused by these gut infections is dehydration. This is not usually serious, but may rapidly become serious in the tropics and with some individuals elsewhere. If it is kept under control, the illness should resolve spontaneously. With most patients, oral rehydration is the treatment of choice. It may have to be supplemented by parenteral fluids, although this should not be necessary if the patient is adequately monitored and the oral rehydration fluid (ORF) has the correct composition (WHO/UNICEF, 1985).

There is no specific treatment for any of these viruses. No chemotherapeutic agent has been used, and with infections that are naturally and rapidly self-limiting, it would be difficult to recommend anything but the safest drugs for an illness whose occurrence is unpredictable and largely over by the time the virus is identified. Nevertheless, because diarrhea may have many causes, it is important to identify any putative causes. Accurate and rapid diagnosis should be an important part of managing the patient. At present, the parts of the world with most cases are also the parts least likely to have the facilities for diagnosis.

Prevention and Control

The logical way to prevent infection with a group of viruses spread feco-orally is by adequate sewage disposal and provision of clean water. To do so would probably limit spread of these viruses to some extent, but is unlikely to provide a complete answer since they also spread successfully in highly developed communities as well as in hospitals. The usual advice about scrupulous care in hygiene (hand-washing, etc) will cut down spread, but will not abolish it, possibly because spread by droplets may take place.

Spread through contaminated shellfish is directly related to their being eaten raw. The viruses are unlikely to survive the usual cooking routines, but few of those who enjoy shellfish will seriously contemplate modifying the flavor even if viruses contribute to it!

If adequate cooking is ruled out, the alternative is to allow time for the molluscs to purge themselves in clean water before they are released for consumption. The number of outbreaks reported throughout the world is testimony to the difficulties in pursuing this policy successfully.

The principles of control are clear, but implementing them in practice is difficult. With the variety of viruses involved and with their incidence being for the most part unpredictable (except perhaps for astroviruses), the prospects for control by vaccine(s) are small. As already discussed, above under Epidemiology, even prior infection with wild Norwalk virus does not appear to enhance natural resistance although half the volunteers challenged were refractory on two occasions. The reasons for this refractoriness are unknown and would be worth exploring in search of ways of potentiating it. There seems few scientific reasons for developing vaccines and no good reason to expect that they would work.

Literature Cited

Agus, S. G., R. Dolin, R. G. Wyatt, A. J. Tousimis, and R. S. Northrup. 1973. Acute infectious nonbacterial gastroenteritis: Intestinal histopathology. Ann. Intern. Med. **79**:18–25.

Appleton, H. 1987. Small round viruses: Classification and role in food-borne infections, p. 108–119. *In* G. Bock and J. Whelan, ed. Novel diarrhoea viruses. John Wiley and Sons, Chichester, England.

Appleton, H., M. Buckley, B. T. Thom, J. L. Cotton, and S. Henderson. 1977. Virus-like particles in winter vomiting disease. Lancet **i**:409–411.

Appleton, H., and M. S. Pereira. 1977. A possible virus aetiology in outbreaks food-poisoning from cockles. Lancet **i**:780–781.

Beards, G. M., C. Hall, J. Green, T. H. Flewett, F. Lamouliatte, and P. DuPasquier. 1984. An enveloped virus in stools of children and adutls with gastroenteritis that resembles the Breda virus of calves. Lancet **ii**:1050–1052.

Bell, J. A., R. J. Huebner, L. Rosen, W. P. Rowe, R. M. Cole, F. M. Mastrota, T. M. Floyd, R. M. Chanock, and R. A. Shvedoff. 1961. Illness and microbial experience of nursery children at Junior Village. Am. J. Hyg. **74**:267–292.

Blacklow, N. R., G. Cukor, M. K. Bedigian, P. Echeverria, H. B. Greenberg, D. S. Schreiber, and J. S. Trier. 1979.

Immune response and prevalence of antibody to Norwalk enteritis virus as determined by radioimmunoassay. J. Clin. Microbiol. **10**:903–909.

Caul, E. O., C. R. Ashley, and J. V. S. Pether. 1979. Norwalk-like particles in epidemic gastroenteritis in the UK. Lancet **ii**:1292.

Caul, E. O., and S. I. Egglestone. 1977. Further studies on human enteric coronaviruses. Arch. Virol. **54**:107–117.

Chany, C., O. Moscovici, P. Lebon, and S. Rousset. 1982. Association of coronavirus infection with neonatal necrotizing enterocolitis. Pediatrics **69**:209–214.

Christopher, P. J., G. S. Grohmann, R. H. Millsom, and A. M. Murphy. 1978. Parvovirus gastroenteritis. A new entity for Australia. Med. J. Aust. **i**:121–124.

Clarke, S. K. R., E. O. Caul, and S. I. Egglestone. 1979. The human enteric coronaviruses. Postgrad. Med. J. **55**:135–142.

Cubitt, W. D. 1987. The candidate caliciviruses, p. 126–138. *In* G. Bock and J. Whelan, ed. Novel diarrhoea viruses. John Wiley and Sons, Chichester, England.

Cubitt, W. D., D. A. McSwiggan, and W. Moore. 1979. Winter vomiting disease caused by calicivirus. J. Clin. Pathol. **32**:786–793.

Doane, F. W., and N. Anderson. 1987. Electron microscopy in diagnostic virology. A practical guide and atlas, p. 178. Cambridge University Press, Cambridge, England.

Dolin, R. 1985. Viral gastroenteritis, p. 1734–1735. *In* J. B. Wyngaarden, and L. H. Smith, ed. Textbook of Medicine, 17th Ed. W.B. Saunders Co., Philadelphia.

Dolin, R., A. G. Levy, R. G. Wyatt, T. S. Thornhill, and J. D. Gardner. 1975. Viral gastroenteritis induced by the Hawaii agent. Jejunal histopathology and serologic response. Am. J. Med. **59**:761–768.

Dolin, R., R. C. Reichman, K. D. Roessner, T. S. Tralka, R. T. Schooley, W. Gary, and D. Morens. 1982. Detection by immune electron microscopy of the Snow Mountain agent of acute viral gastroenteritis. J. Infect. Dis. **146**:184–189.

Garwes, D. J. 1982. Coronaviruses in animals, p. 315–359. *In* D. A. J. Tyrrell and A. Z. Kapikian, ed. Virus infections of the gastrointestinal tract. Marcel Dekker, New York.

Greenberg, H. B., R. G. Wyatt, and A. Z. Kapikian. 1979. Norwalk virus in vomitus. Lancet **i**:55.

Greenberg, H. B., and A. Z. Kapikian. 1978. Detection of Norwalk agent antibody and antigen by solid phase radioimmunoassay and immune adherence hemagglutination assay. J. Am. Vet. Med. Assoc. **173**:620–623.

Hall, G. A. 1987. Comparative pathology of infection by novel diarrhoea viruses, p. 192–207. *In* G. Bock and J. Whelan, ed. Novel diarrhoea viruses. John Wiley and Sons, Chichester, England.

Kapikian, A. Z., R. G. Wyatt, R. Dolin, T. S. Thornhill, A. R. Kalica, and R. M. Chanock. 1972. Visualization by immune electron microscopy of a 27 nm particle associated with acute infectious nonbacterial gastroenteritis. J. Virol. **10**:1075–1081.

Kapikian, A. Z., H. B. Greenberg, W. L. Cline, A. R. Kalica, R. G. Wyatt, H. D. James, N. L. Lloyd, R. M. Chanock, R. W. Ryder, and H. W. Kim. 1978. Prevalence of antibody to the Norwalk agent by a newly developed immune adherence hemagglutination assay. J. Med. Virol. **2**:281–294.

Kidd, A. H., B. P. Cosgrove, R. A. Brown, and C. R. Madeley. 1982. Faecal adenoviruses from Glasgow babies. Studies on culture and identity. J. Hyg. Camb. **88**:463–474.

Kogasaka, R., S. I. Nakamura, S. Chiba, Y. Sakuma, H. Terashima, T. Yokoyama, and T. Nakao. 1981. The 33- to 39-nm virus-like particles tentatively designated as Sapporo agent associated with an outbreak of gastroenteritis. J. Med. Virol. **8**:187–193.

Konno, T., H. Suzuki, and N. Ishida. 1975. Reovirus-like agent in Japanese infants with gastroenteritis. Lancet **i**:918–919.

Koopman, J. S., E. A. Eckert, H. B. Greenberg, B. C. Strohm, R. E. Isaacson, and A. S. Monto. 1982. Norwalk virus enteric illness acquired by swimming exposure. Am. J. Epidemiol. **115**:173–177.

Kurtz, J. R., and T. W. Lee. 1978. Astrovirus gastroenteritis. Age distribution of antibody. Med. Microbiol. Immunol. **166**:227–230.

Kurtz, J. B., T. W. Lee, J. W. Craig, and S. E. Reed. 1979. Astrovirus infection in volunteers. J. Med. Virol. **3**:221–230.

Lee, T. W., and J. B. Kurtz. 1981. Serial propagation of astrovirus in tissue culture with the aid of trypsin. J. Gen. Virol. **57**:421–424.

Madeley, C. R. 1979. A comparison of the features of astroviruses and caliciviruses seen in samples of faeces by electron microscopy. J. Infect. Dis. **139**:519–524.

Madeley, C. R. 1988. The search for viruses by negative contrast. *In:* D. W. Henderson, J. A. Papadimitriou, and D. Spagnolo, ed. Diagnostic ultrastructure of non-neoplastic diseases. Churchill Livingstone, Edinburgh. In press.

Madeley, C. R., and A. M. Field. 1988. Virus morphology. 2nd ed. Churchill Livingstone, Edinburgh.

Parrino, J. A., D. S. Schreiber, J. S. Trier, A. Z. Kapikian, and N. R. Blacklow. 1977. Clinical immunity in acute gastroenteritis caused by Norwalk agent. N. Engl. J. Med. **297**:86–89.

Paver, W. K., E. O. Caul, C. R. Ashley, and S. K. R. Clarke. 1973. A small virus in human faeces. Lancet **i**:237–240.

Sakuma, Y., S. Chiba, R. Kogasaka, H. Terashima, S. Nakamura, K. Horino, and T. Nakao. 1981. Prevalence of antibody to human calicivirus in general population of northern Japan. J. Med. Virol. **8**:221–225.

Schnagl, R. D., I. H. Holmes, and E. M. Mackay-Scollay. 1978. Coronavirus-like particles in aboriginals and non-aboriginals in western Australia. Med. J. Aust. **i**:307–309.

Taniguchi, K., S. Urasawa, and T. Urasawa. 1979. Virus-like particles 35 to 40 nm, associated with an institutional outbreak of acute gastroenteritis in adults. J. Clin. Microbiol. **10**:730–736.

Thornhill, T. S., R. G. Wyatt, A. R. Kalica, R. Dolin, R. M. Chanock, and A. Z. Kapikian. 1977. Detection by immune electron microscopy of 26 to 27 nm viruslike particles associated with two family outbreaks of gastroenteritis. J. Infect. Dis. **135**:20–27.

Townsend, T. R., E. A. Bolyard, R. H. Yolken, W. E. Beschorner, C. A. Bishop, W. H. Burns, G. W. Santos, and R. Saral. 1982. Outbreak of coxsackie A1 gastroenteritis: a complication of bone-marrow transplantation. Lancet **i**:820–823.

WHO/UNICEF. 1985. The management of diarrhoea and use of oral rehydration therapy. 2nd Ed. World Health Organization, Geneva.

Wyatt, R. G., R. Dolin, N. R. Blacklow, H. L. duPont, R. F. Buscho, T. S. Thornhill, A. Z. Kapikian, and R. M. Chanock. 1974. Comparison of three agents of acute infectious nonbacterial gastroenteritis by cross-challenge in volunteers. J. Infect. Dis. **124**:709–714.

CHAPTER 42

Unclassified: Spongiform Encephalopathies

PATRICIA A. MERZ and HENRYK M. WISNIEWSKI

Disease: Scrapie of sheep and goats, transmissible mink encephalopathy, chronic wasting disease of captive mule deer and elk, and the human diseases kuru, Creutzfeldt-Jakob, and Gerstmann-Straussler syndrome.

Etiologic Agent: Nature unknown.

Source: Unknown.

Clinical Manifestations: Insidious onset with neurologic abnormalities progressing to dementia, listlessness, emaciation, and death.

Pathology: Limited to one organ, the brain; all cases exhibit status spongiosus, neuronal loss, and astrogliosis. The occurrence of amyloid plaques is observed in kuru and Gerstmann-Straussler syndrome and a few cases of Creutzfeldt-Jakob disease.

Laboratory Diagnosis: Primarily based on clinical signs, postmortem neuropathology, transmission studies, and presence of SAF and PrP in brain extracts.

Epidemiology: Scrapie and Creutzfeldt-Jakob disease occur worldwide. The other diseases are focal and all can occur in a familial manner.

Treatment: None known.

Prevention and Control: None known.

Introduction

In this chapter, we turn our attention to one of the continuing enigmas in the transmissible diseases—the spongiform encephalopathies, or unconventional slow viral diseases. Part of the enigma surrounding these diseases stems from the nature of the infectious agents, which are still unknown and unclassified. We present the current information and discuss some of the speculation and uncertainty that continue to surround the etiological agents of these diseases. Over the past 6 years, workers in the area of unconventional slow viral diseases have for the first time had morphological and biochemical tools with which to pursue these elusive agents. Although these tools have opened up new areas for investigation, they have yet to settle the issue of the nature of the agents. It is anticipated that this field will continue to be very active over the coming years, hopefully leading to further understanding of the nature of the infectious agents and perhaps resulting in some form of prevention for these diseases.

What are the spongiform encephalopathies? The spongiform encephalopathies comprise the following natural diseases: scrapie of sheep and goats; transmissible mink encephalopathy of mink; chronic wasting disease of captive mule deer and elk; and kuru, Creutzfeldt-Jakob disease (CJD), and Gerstmann-Straussler syndrome (GSS) in man (Gajdusek, 1985). Scrapie has been known for over two centuries in Europe, where it has wreaked economic havoc with sheep breeders (Journal of the House of Commons, 1755; Parry, 1983). Transmissible mink encephalopathy and potentially chronic wasting disease of captive mule deer and elk are thought to have arisen from the ingestion of scrapie-contaminated sheep tissue. Since transmissible mink encephalopathy has not occurred on mink ranches in the United States since 1963 and chronic wasting disease of captive mule deer and elk has only been known since 1980 (Burger and Hartsough, 1965; Hartsough and Burger, 1965; Marsh and Hanson, 1979; Williams and Young, 1980, 1982), we will not discuss these diseases further. CJD was first reported early in this century by the two researchers for whom the disease is named (Creutzfeldt, 1920, 1921; Jakob, 1921). GSS was first reported in 1936 (Gerstmann et al., 1936), and kuru was first reported in 1957 (Gajdusek and Zigas, 1957, 1959). The transmissible nature of kuru, CJD, and GSS has been known for less than 30 years (Gajdusek et al., 1966; Gibbs et al., 1968; Masters et al., 1981b).

The spongiform encephalopathies fulfill the criteria for slow infections first described by Sigurdsson many years ago (Sigurdsson, 1954). Slow infections are characterized by long incubation periods (on the order of ⅓ of the host's life-span), a clinical and pathological picture that is limited to one organ—in this case, the central nervous system (CNS)—and an invariably progressive, degenerative, and fatal course.

The mystique of these diseases is a consequence of the unusual nature of the etiologic agents. Although the detailed nature of these agents remains unknown and unclassified, it is clear that they represent a class of new and unusual transmissible agents. These are the first human progressive, degenerative, lethal, CNS diseases which were shown caused by a transmissible agent. This discovery has raised hopes of understanding other chronic, progressive CNS diseases of man. It should also be pointed out that scrapie, kuru, and GSS were originally thought to be purely hereditary diseases with an autosomal-dominant mode of inheritance (Claridge, 1795; Gajdusek and Zigas, 1959; Gerstman et al., 1936; Parry, 1983; Seitelburger, 1962). It is expected that an understanding of the nature of these agents will reveal new mechanisms of CNS disease and dysfunction and will open up new areas of biology and molecular biology.

Another impetus to the study of these diseases is the possible relationship to other human CNS diseases, such as Alzheimer's disease, whose incidence is predicted to increase dramatically in the coming years (Wisniewski et al., 1983) but is not known to be transmissible. The spongiform encephalopathies and Alzheimer's disease share several similarities: onset in middle to late years; primary pathology limited to the CNS; amyloid deposition in the brain; and sporadic and familial patterns of occurrence (Wisniewski et al., 1975). An understanding of the pathogenesis of the spongiform encephalopathies may well assist in understanding basic mechanisms operational in Alzheimer's disease and other degenerative CNS diseases of unknown etiology.

The Agents

The most studied model of the spongiform encephalopathies is scrapie. This is largely based on the ease of transmission of scrapie to laboratory animals (i.e.,

hamsters and mice) and the safety of working with nonhuman pathogens. Taken together, the agents causing the spongiform encephalopathies appear to share similar biochemical, morphological, biophysical, and inactivation properties. On the other hand, there are strain differences based on host range and incubation period, neurological signs and symptoms, and the distribution, severity, and kinds of neuropathological lesions. Over 15 strains of scrapie agent have been identified in experimental systems owing to the outstanding work of Alan Dickinson and co-workers (Bruce and Dickinson, 1979; Dickinson, 1976; Dickinson and Outram, 1979; Fraser and Dickinson, 1973). It has been suggested that the spongiform encephalopathy agents share a common ancestor from which evolved different agents with their individual biological properties (Gajdusek, 1985).

The agents are best known for their unusual properties, particularly their resistance to inactivation. This is related, at least in part, to their association with cellular membranes, which has also led to great difficulties in isolation and purification (Clark and Millson, 1976a; Gibbons and Hunter, 1967; Manuelidis and Manuelidis, 1983; Millson et al., 1971, 1976). The agents are unusually resistant to inactivation by 10% Formalin, wet or dry heat up to 100°C, ultraviolet irradiation, and X rays (Alper and Haig, 1968; Alper et al., 1966, 1967; Bellinger-Kawahara et al., 1987; Brown, 1982; Brown et al., 1986a; Carp et al., 1985a, 1985b; Court and Cathala, 1983; Dickinson and Outram, 1983; Dickinson and Taylor, 1978; Gajdusek, 1977, 1985; Gajdusek et al., 1976; Gibbs et al., 1978; Gordon, 1946; Haig et al., 1969; Latarjet, 1979; Latarjet et al., 1970; Millson et al., 1976; Rohwer, 1983, 1984b,c). These agents are also highly resistant to chemical inactivation by such compounds as B-propiolactone, nucleases, psoralens, and lipases (Brown, 1986a; Carp et al., 1985b; Dickinson and Taylor, 1978; Diener et al., 1982; Gajdusek, 1977, 1985; Kingsbury et al., 1983b; McKinley et al., 1981, 1983b; Millson and Manning, 1979; Millson et al., 1976; Prusiner et al., 1987; Rohwer, 1984b; Ward et al., 1974).

They are, however, susceptible to certain proteases and strong denaturing compounds such as phenol, chlorox, and sodium hydroxide (Brown et al., 1986a; Carp et al., 1985b; Diener et al., 1982; Gajdusek, 1977, 1985; Kingsbury, 1983b; Millson and Manning, 1979; Millson et al., 1976; Prusiner et al., 1981, 1987; Rohwer, 1984b; Ward et al., 1974). The inactivation profile has suggested an essential protein component as part of the infectious agent (Carp et al., 1985b; Cho, 1980a,b; Gajdusek, 1985; Lax et al., 1983; McKinley et al., 1981, 1983a). Despite this potential for the generation of an immune response, the agents fail to induce an antibody reaction by the host or affect the immune system in any manner.

Although the agents replicate to some degree in

tissue culture systems (Asher et al., 1979; Clark, 1979; Clark and Haig, 1970; Clark and Millson, 1976b; Gibson et al., 1972; Markovits et al., 1983; Rubenstein et al., 1984), there is no cytopathic effect or indication of the presence of the agents other than the induction of disease in vivo. The only known marker for the presence of these agents is infectivity. Most of the conventional virological markers or detection systems have not been applicable to these diseases.

A key breakthrough in the study of these diseases was the discovery of an abnormal fibril, scrapie-associated fibrils (SAF), consistently present in cellular extracts from diseased brains of the spongiform encephalopathies (Table 1) (Merz et al., 1981, 1983c, 1984a). SAF do not resemble conventional viruses but are similar to the pathological fibrils known as amyloids, which are observed in many human diseases (Merz, 1983a,b). The next advance was the identification of a 27- to 30-kDa protease-resistant protein (PrP) in infectious fractions (Bolton et al., 1982). In experiments designed to isolate infectivity from scrapie-infected hamsters, copurification of SAF, PrP, and infectivity was achieved (Diringer et al., 1983a,b; Prusiner et al., 1983). It now appears that the protein component of SAF is PrP (Barry et al., 1985; Bode et al., 1985; Merz et al., 1987a) and that both SAF and PrP are uniquely associated with spongiform encephalopathies (Bendheim et al., 1985; Bockman et al., 1985; Bode et al., 1985; Brown et al., 1986b; Gibson et al., 1987; Merz et al., 1981, 1983a,c, 1984a). The PrP 27- to 30-kDa polypeptide is a glycosylated cleavage product of a larger host glycoprotein of 33 to 35 kDa (Bolton et al., 1985; Chesebro et al., 1985; Hope et al., 1986; Manuelidis et al., 1985; Merz et al., 1987; Multhaup et al., 1985; Oesch et al., 1985; Prusiner et al., 1984b; Robakis et al., 1986a; Rubenstein et al., 1986). Infectivity has been associated with SAF and the presence of both forms of the protein (27–30 and 33–35 kDa) (Diringer et al.,

1983a; Kascsak et al., 1985; Manuelidis et al., 1985; Prusiner et al., 1983a). However, it is still not known whether SAF represent a form of the agent or if infectivity is merely trapped among these structures during purification.

A third advance has been the production of both polyclonal and monoclonal antibodies to natural polypeptides and synthetic peptides of PrP (Barry and Prusiner, 1986; Bendheim et al., 1984; Cho, 1986; Diringer et al., 1984a; Kascsak et al., 1986, 1987a; Shinagawa et al., 1986; Takahashi et al., 1986). These reagents have provided a specific and sensitive means to further isolate and characterize SAF and PrP.

Three main hypotheses are currently entertained as models of the etiologic agents for these diseases. Although some evidence exists that can support each model, there is no definitive proof supporting any one model. The *prion hypothesis* suggests, similar to older theories, that protein alone is sufficient for infectivity (Alper et al., 1978; German and Marsh, 1983; Griffith, 1967; Prusiner, 1982). Posttranslational or conformational alterations in PrP, the major component of the infectious agent, could potentially generate disease and perpetuate similar changes in newly generated agent. The *virus hypothesis* suggests that these diseases are caused by a conventional viruslike agent whose unusual properties have beclouded its detection (Braig et al., 1985; Czub et al., 1986a,b; Manuelidis and Manuelidis, 1986; Manuelidis et al., 1987; Rohwer, 1983, 1984b,c). If this hypothesis proves to be true, obviously neither SAF nor PrP are likely to be associated with this agent. In the *virino hypothesis*, favored by our group, the informational molecule is a nucleic acid surrounded and protected by a host-coded protein (Bruce and Dickinson, 1987; Dickinson and Fraser, 1979; Dickinson and Outram, 1983; Kascsak et al., 1985; Kimberlin, 1982b). Prp is an obvious candidate for this protein molecule, which when assembled into SAF forms a morphological entity that protects the nucleic acid. This model does not rule out the possibility of other host proteins' being able to assume this same protective role. The characteristics of each agent would be controlled by the nucleic acid component. An argument against this hypothesis has been the inability to detect this nucleic acid molecule (Borras and Gibbs, 1985; Borras et al., 1982; Hunter et al., 1976; Manuelidis et al., 1980; Marsh et al., 1974; Merz et al., 1986a; Prusiner 1984a,c). Readers are urged to consult recent reviews for more detail on the conceptual differences between these models (Carp et al., 1985a,b; Gajdusek, 1985; Harris, 1985; Hope and Kimberlin, 1987; Johnson, 1982; Kimberlin, 1982a,b, 1984, 1986a; McKinley and Prusiner, 1986; Prusiner, 1982, 1984a,c; Prusiner et al., 1987; Robertson et al., 1985; Rohwer, 1984a; Traub, 1983).

TABLE 1. Observation of SAF and PrP in spongiform encephalopathies

	Natural	Transmitted
Creutzfeldt-Jakob	Human	Monkeys Guinea pigs Hamsters Mice
Kuru	Human	Monkeys
Gerstmann-Straussler	Human	
Scrapie	Sheep	Sheep Monkeys Hamsters Mice
Chronic wasting disease	Mule deer Elk	Mule deer

Scrapie

Partial characterization of the scrapie agents has led to a greater appreciation of their unique and unusual properties. In addition, it has provided a basis for understanding some of the perplexing features of the much less well defined human diseases such as Gerstmann-Straussler syndrome, an autosomal-dominant transmissible genetic disease.

Early in the study of experimental scrapie in mice, it appeared that the disease showed a bewildering spectrum of variation in incubation period (e.g., 130 to 500 days), clinical signs and course, and the distribution and kinds of neuropathological lesions (e.g., the presence of large numbers of amyloid plaques in some infections but not in others). In a series of painstaking studies that has spanned more than two decades, Alan Dickinson and his colleagues have shown that this variation is generated by the specific interaction of the biological properties of each strain of scrapie agent and the host species. When one systematically varied such factors as the source of the scrapie isolate, strain of host mouse, dose, and route of infection, it became clear that scrapie agent strains exhibit several heritable features (Bruce and Dickinson, 1985, 1987; Bruce and Fraser, 1982; Bruce et al., 1976; Carp et al., 1984, 1987; Dickinson, 1975; Dickinson and Fraser, 1977, 1979; Dickinson and Meikle, 1969b, 1971; Dickinson and Outram, 1979, 1983; Dickinson et al., 1969, 1984; Fraser, 1976, 1979; Fraser and Dickinson, 1973; Kimberlin and Walker, 1977, 1978a,b, 1982, 1983).

Incubation periods are (1) precisely predictable (e.g., a mean incubation period of 220 days with a standard error of the mean of as little as 1 or 2 days), and (2) determined by a host autosomal gene called *Sinc* (scrapie *inc*ubation period) whose two alleles s7 and p7 give shorter and prolonged incubation periods

for scrapie strain ME7 respectively and exhibit no dominance with this particular strain (Table 2) (Dickinson and Fraser, 1977, 1979; Dickinson and Meikle, 1969, 1971; Dickinson et al., 1969a). Similarly, the kinds of neuropathological lesions and their severity and distribution within the CNS are so reproducible as to enable one to identify individual agent strains on the basis of their "lesion profiles" (Fraser, 1976, 1979; Fraser and Dickinson, 1973a).

The biggest obstacle to the further definition of the transmissible encephalopathy agents in general and the scrapie agents in particular has been the difficulty in their purification, which has only been accomplished in the past few years. This intractability has in large measure been a consequence of the lack of suitable markers other than LD_{50} titrations, the tight association of infectivity with cell membranes, and the ease with which the infectious agents aggregate. In the natural and transmitted cases of scrapie, the highest titers of infectivity are associated with the brain, followed by the lymphoreticular system (Clark and Haig, 1971b; Clark and Millson, 1976a; Dickinson and Outram, 1979; Fraser and Dickinson, 1978; Hadlow et al., 1974, 1979; Kimberlin, 1981; Kimberlin and Walker, 1979; Scott and Dickinson, 1985; Sigurdsson, 1954). The specific cell type(s) responsible for agent replication, however, are not known.

The detection of scrapie strains in original isolates from sheep were made on the basis of the incubation period differences observed in the two genotypes (s7s7 or p7p7) of mice (Dickinson, 1976; Dickinson and Fraser, 1979). The strains of scrapie can be said to exhibit three classes of stability (Bruce and Dickinson, 1979, 1985a,b, 1987; Bruce and Fraser, 1982; Dickinson and Fraser, 1979; Dickinson and Outram, 1983; Dickinson et al., 1984; Scott and Dickinson, 1985). The three stability classes into which scrapie

TABLE 2. Scrapie strain characteristics

| Scrapie strain | Host species | Incubation period | | Neuropathology |
		Genotype	Length	
ME7	Mice	s7s7	153 ± (1)[a]	Gray matter vacuolation with slight white matter changes,
		p7p7	300 ± (3)	moderate amyloid plaque formation in p7p7 mice
139A	Mice	s7s7	125 ± (1)	Gray and white matter vacuolation, no plaque formation
		p7p7	163 ± (1)	
22L	Mice	s7s7	130 ± (1)	High-level cerebellar vacuolation, no plaque formation
		p7p7	200 ± (1)	
87V	Mice	s7s7	467 ± (21)	Gray matter vacuolation, high plaque formation in s7s7 and
		p7p7	311 ± (5)	p7p7 mice
22A	Mice	s7s7	356 ± (4)	Gray matter vacuolation, plaque formation in p7p7
		p7p7	180 ± (2)	
263K	Hamsters		60 ± (5)	Low vacuolation, low plaque formation

[a] Mean ± standard error (in days).

strains fall are (1) those strains (e.g., ME7) that maintain their biological properties irrespective of the genotype of mouse through which they have been passaged; (2) those strains (e.g., 22A) that exhibit defined biological properties in one *Sinc* genotype but if passaged in a different genotype will exhibit gradual and consistent changes in the length of the incubation period and lesion profile over several passages (these characteristic changes can be interpreted as host-permitted selection of strains that arose by accumulated point mutations): and (3) those agents that exhibit instability as a single discontinuity in a single mouse genotype, observed as a sudden large change in properties in contrast to the more gradual change seen in (2). All strains exhibiting this type of change originate from scrapie strains possessing a high incidence of amyloid plaque deposition in both s7s7 and p7p7 mice. The new strain that is selected in s7s7 mice has lost the ability to cause amyloid plaque formation. A mutational event has been postulated for this observation, since once established, the new scrapie strain is stable on further passage, and the original amyloidogenic strain is lost (Bruce and Dickinson, 1987).

SAF isolated from several different scrapie strain-host combinations exhibit strain-specific differences in their morphological, biochemical, and immunological properties plus differences in protease sensitivity which, in some cases, can be attributed to differences in the agent strain. (Kascsak et al., 1985, 1986; Rubenstein et al., 1986). The scrapie strain differences in SAF do not appear to be solely explainable on the basis of alterations in the host-coded glycoprotein which is their major component but may reflect agent-specific events or agent-specific influences generated as part of the infectious process. A scrapie agent-specific informational molecule is unknown at this time, but the biological evidence supports the existence of a genome which is an integral part of the agent similar to conventional viruses and all other living organisms. As described above, evidence exists to support mutational events within two classes of scrapie strains, and this plus all the other evidence on different scrapie strains is most easily understood in terms of an informational molecule that is a nucleic acid.

Kuru

Kuru is a progressive, degenerative, lethal CNS disease of unknown origin. Its occurrence is limited to the Fore tribe in the New Guinea highlands. It has never been seen elsewhere, and at its height it was of epidemic proportions, accounting for the majority of the deaths in this population (Gajdusek, 1977; Gajdusek and Zigas, 1957, 1959). The infectious nature of

this disease was demonstrated by Carlton Gajdusek and co-workers following inoculation of primates with brain tissue from kuru cases (Gajdusek et al., 1966). These studies were initiated after the similarity between the neuropathology of kuru and scrapie, a known transmissible disease, was pointed out in a letter to *Lancet* by William J. Hadlow (Hadlow, 1959). Kuru agents are included in the spongiform encephalopathies on the basis of their neuropathology, transmissibility, long incubation period, lack of immune response, and resistance to inactivation. Because the agents have been assayed only in higher primates, a more detailed description of their characteristics is lacking. SAF and PrP have been detected in extracts of human and transmitted cases (Brown et al., 1986b; Merz et al., 1984a). It is possible that kuru will ultimately turn out to be a variant of CJD to which the Fore people were particularly susceptible.

Creutzfeldt-Jakob Disease

Second only to scrapie, CJD is the most studied of the spongiform encephalopathies. With the realization that kuru and CJD exhibited a similar neuropathological picture, Clarence J. Gibbs Jr. and co-workers succeeded in transmitting CJD to primates (Gibbs et al., 1968). Similar to transmission studies on natural scrapie with rodents, over 90% of all CJD cases can be transmitted to primates, and the characterization of these agents has been performed on the transmitted isolates (Gajdusek, 1985). The inactivation characteristics of CJD agents are similar to those of scrapie (Brown et al., 1982, 1986a; Gajdusek, 1985; Gajdusek et al., 1976; Gibbs et al., 1978; Latarjet, 1979). Ample evidence exists for the presence of multiple strains among different isolates, but because of the inherent problems of long-term studies with primates, the characterization of these isolates has not progressed to the stage of scrapie strains (Gibbs et al., 1979; Tateishi et al., 1983). The recent demonstration of the susceptibility of laboratory rodents to primary CJD isolates will undoubtedly lead to further characterization of individual CJD strains.

SAF have been observed in all natural cases so far examined, and PrP have been detected in 75% of the human cases (Bendheim et al., 1985; Bockman et al., 1985; Bode et al., 1985; Brown et al., 1986b; Merz et al., 1983c, 1984a). Several studies have shown variation in silver stain and Western blot profiles of PrP associated with different natural CJD cases. Variation has also been seen in the neuropathological profile of different isolates (Manuelidis, 1985a; Masters and Richardson, 1978; Mizutani et al., 1981; Oppenheimer, 1975; Park et al., 1980; Tateishi et al., 1983). Analogies to findings in experimental scrapie suggest that there are also different strains of the CJD agent.

About 10 to 15% of the reported CJD cases are familial in origin, indicating the potential for *Sinc*like incubation-period genotypes being present in the human population (Asher et al., 1983; Cathala et al., 1980; Ferber et al., 1974; Haltia et al., 1979; Masters et al., 1979, 1981a).

Gerstmann-Straussler Syndrome

Gerstmann-Straussler syndrome is a rare, progressive, degenerative, lethal disease with an autosomal-dominant pattern of inheritance observed in families (Baker et al., 1985; Boellaard and Schlote, 1980; Gerstmann et al., 1936; Masters et al., 1981b; Schumm et al., 1981; Seitelberger, 1962). Inoculation of animals with brain suspensions from several GSS cases has transmitted a CJD-like disease to either primates, marmosets, or rodents (Baker et al., 1985; Masters et al., 1981b; Tateishi et al., 1983).

As stated earlier, natural scrapie in sheep was for many years considered to be a genetic, inheritable disease. The clear demonstration of its transmissible nature was incompatible with a genetic origin for that disease. Before the demonstration of its transmissible nature, the best explanation for the epidemiology and incidence of kuru was as a genetic disease. With GSS, we are again faced with a disease whose incidence within families follows a clear autosomal-dominant pattern of inheritance, but, like scrapie and kuru, GSS has also been shown to be a transmissible disease. One possible explanation for the incidence pattern of GSS within a family is contact transmission of a particular strain of CJD agent into a specific genetic background present in these families.

Again, SAF have been detected in extracts in GSS-affected human brains, and PrP have been identified on Western blots (Brown et al., 1986b; Merz et al., 1983c). No other information is available on the GSS agents at the present time, since the transmissible nature of this disease has been demonstrated only recently.

Pathogenesis and Pathology

The spongiform encephalopathies exhibit pathology only in the brain. There is a spectrum of pathological changes including neuronal loss, vacuolar changes in gray matter and white matter, reactive astrogliosis, and, in certain agent-host combinations, occurrence of amyloid plaques (Beck and Daniel, 1979; Beck et al., 1964; Fraser, 1976, 1978; Klatzo et al., 1959; Lampert et al., 1971, 1972; Marsh and Kimberlin, 1975; Masters and Richardson, 1978; Masters et al., 1984; Mizutani and Shiraki, 1985). The prominence of the vacuolar changes in the neuronal perikarya

and processes has given this class of diseases its name, the spongiform encephalopathies. The distribution of the vacuolar changes in specific areas of the scrapie-infected brain (i.e., the lesion profile) has aided in the differentiation of scrapie strains. The vacuolar changes are observed mainly in the gray matter, but several scrapie strains and some CJD cases exhibit vacuolation of the white matter in addition to or instead of gray matter changes (Fraser, 1979; Macchi et al., 1984; Mizutani et al., 1981; Park et al., 1980). The reactive astrogliosis is easily demonstrated by Cajal staining or by immunocytochemical methods (Fig. 1). No inflammatory reaction has been detected in the affected brains, again reflecting the apparent lack of an immune response to the infecting agent.

Scrapie

Neuropathological examination of natural scrapie cases has revealed that some animals at the clinical stages of disease exhibit no spongiform changes in the brain (Dickinson, 1976a). In contrast, other animals displaying similar clinical signs can exhibit a wide range in the degree and localization of the vacuolar changes (Bertrand et al., 1937; Dickinson, 1976a; Fraser, 1976). Reactive astrogliosis is a common pathological feature of all the cases. The pathogenesis of brain infection is unknown, but the agent is widespread in the lymphoreticular system of sheep for several years prior to the presence of infectious agent in the CNS (Hadlow et al., 1974, 1979, 1982). Extensive CNS agent replication is followed by the progressive development of clinical signs eventually leading to death. Infectious agent is not detected in the body fluids of the infected sheep. Sheep are thought to be naturally infected through contact with infected ewes or contact with the agent on surfaces when the sheep are crowded together in wintering pens (Brotherson et al., 1968; Dickinson, 1976; Dickinson et al., 1974; Hourrigan et al., 1979; Kimberlin, 1981; Palsson, 1979; Pattison, et al., 1974).

During the early experiments in transmitting scrapie to sheep and mice, a variable incubation period was observed in animals infected with one isolate, $SSBP/_1$, of scrapie. Host genes controlling the incubation period of scrapie were identified, and selective inbreeding for these genes over many years has yielded two genotypes of sheep and two genotypes of mice exhibiting precise, reproducible incubation times with a specific scrapie strain—ME7 (Dickinson, 1976; Dickinson and Meikle, 1969; Dickinson et al., 1969; Kimberlin, 1979; Scott and Dickinson, 1985). Two alleles of the mouse gene (which is called *Sinc*) control the incubation period of all scrapie strains in mice, either short (s7s7) or long (p7p7) depending on the particular scrapie strain. An allele analogous to *Sinc*, called *Prn-i*, has recently been identified in NZW/LACJ and I/Ln strains of mice (Carlson et al., 1986).

Experimental infection of the different mouse genotypes has led to the identification of a large number of scrapie strains exhibiting different biological properties. To date, most inbred strains of mice have been shown to have the short incubation genotype, s7s7 (Carp et al., 1985a). Only four strains of mice are known that carry the long incubation period genotype, p7p7—VM/DK, IM/DK, MB/DK, and I/Ln (Bruce and Dickinson 1985; Bruce et al., 1976; Carp et al., 1987).

Studies using a cDNA clone encoding PrP have revealed restriction fragment length polymorphisms (RFLP) associated with the incubation period gene (Carlson et al., 1986). An Xba 1 fragment of 3.8 kb segregates with s7s7, and a 5.5-kb Xba 1 fragment segregates with p7p7. Thus, the *Sinc* and PrP genes appear to be closely linked, probably within 7 centimorgans of each other. *Sinc* controls the overall pathogenesis of scrapie but details are not known of precisely where it acts. But the key question is whether or not the PrP and *Sinc* gene are the same. Possible roles for these genes in scrapie replication involve processing or posttranslational modifications of PrP and/or receptor recognition sites on either agent or susceptible cells. It is anticipated that more information on the mapping of PrP gene (localized on mouse chromosome 2 and human chromosome 20) and its relationship to *Sinc* will be needed before conclusions can be drawn about the presence of similar genetic factors in humans and their role in sporadic and familial cases of spongiform encephalopathies (Liao et al., 1986; Robakis et al., 1986b; Sparkes et al., 1986). If the PrP gene controls the incubation period of disease, then it might allow agent strain typing in humans and sheep and even help in the prevention and control of at least the sheep disease.

To date, pathogenesis studies with the mouse (139A) and hamster (263K) models indicate that following peripheral infection, the agent travels by potentially neural spread from the spleen and visceral lymph nodes to the spinal cord and then spreads anteriorly in the brain (Kimberlin and Walker, 1980, 1982, 1983, 1986). Agent is also spread along the optic nerve and tract into the brain following intraocular inoculation (Buyukmihci et al., 1983; Fraser, 1981; Fraser and Dickinson, 1985). Infection of the sciatic nerve has also shown that it is possible to infect peripheral nerves (Kimberlin et al. 1983). The mode of spread has implications for the natural human and sheep diseases.

It is known that once the agent is present in the host, a precise clock is set in motion which is depen-

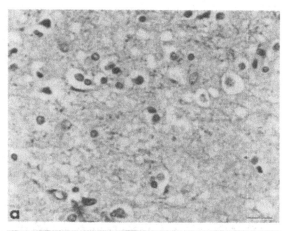
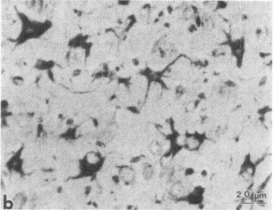

FIG. 1. The neuropathology of the spongiform encephalopathies includes (a) vacuolation and (b) astrogliosis of the CNS. Formalin-fixed paraffin sections from a CJD brain (a) stained with hematoxylin and eosin and (b) immunochemically stained with rabbit antiglial fibrillary acidic protein. Bar represents 20 μm.

dent on host genetics, strain of agent, and the dose and route of inoculation (Outram, 1976; Scott and Dickinson, 1985). The dynamics of agent replication and pathogenesis develop with this clocklike precision. In some cases, mice heterozygous for the *Sinc* gene have been shown to have incubation periods for certain strains of agent that are longer than the lifespan of the mouse (Collis and Kimberlin, 1985; Dickinson et al., 1975, 1976). The titer of infectious agent can attain levels of 10^{7-8} LD_{50}/g of brain and 10^{5-6} LD_{50}/g of spleen for many of the mouse scrapie strains; scrapie strain 263K in hamsters reaches 10^{9-10} LD_{50}/g of brain and 10^{5-6} LD_{50}/g of spleen (Kimberlin and Walker, 1986b, Carp, personal communication).

Like the natural forms of the disease, the neuropathology of scrapie in the mouse or hamster is characterized by reactive astrogliosis and vacuolar changes in the gray or white matter (Fraser, 1979;

Marsh and Kimberlin, 1975; Masters et al., 1984). In addition, amyloid plaques are consistently observed in certain specific scrapie agent-host strain combinations (Bendheim et al., 1984; Bruce and Fraser, 1975; Fraser and Bruce, 1973; Wisniewski et al., 1975, 1981). (Figs. 2 and 3). Plaques are also seen in the initial transmission of natural scrapie to mice following inoculation with a high infectious dose (Fraser, 1983).

The amyloid deposits are extracellular accumulations of fibers assembled from B-pleated cleavage products of the 33- to 35-kDa protein (Barry and Prusiner, 1986; Bendheim et al., 1984; De Armond et al., 1985; Kimberlin, 1986; Kitamoto et al., 1986; Merz et al., 1986a). Like other amyloids, they exhibit green-red birefringence when strained with Congo red. The fibers are long and 4 to 8 nm in diameter. Experimental scrapie constitutes the only model for the formation of amyloid plaques in the central nervous system. The number of plaques and their distribution

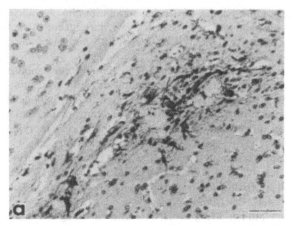
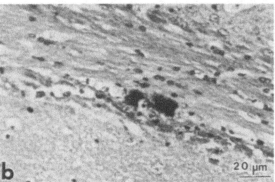

FIG. 3. Light micrographs of amyloid plaques in the corpus callosum of a mouse injected with scrapie strain 87V and reacted with (a) rabbit antiglial fibrillary acidic protein, illustrating the reactive astrocytosis in the vicinity of the amyloid plaques; and (b) rabbit anti-ME7-PrP antiserum, illustrating the relationship of amyloid plaques to the SAF-PrP protein. Bar represents 20 μm.

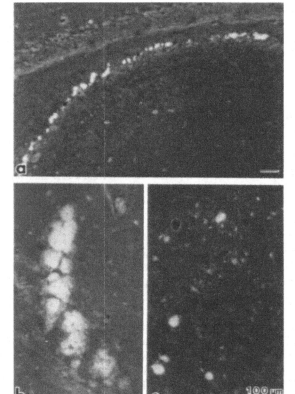

FIG. 2. Fluorescence micrographs of amyloid plaques observed in some of the spongiform encephalopathies. Formalin-fixed paraffin sections of brains from mice of the p7p7 genotype injected with scrapie strain 87V and stained with thioflavin S. This illustrates the distribution of amyloid plaques following IC inoculation: (a) corpus callosum at low magnification and (b) high magnification, and (c) cortex. Bar represents 100 μm. (Courtesy of Dr. R. C. Moretz.)

within the CNS are controlled by several variables: host genotype, strain of agent, dose, and route of infection (Bruce and Dickinson, 1985, 1987; Bruce et al., 1976; Carp et al., 1987; Merz et al., 1987; Wisniewski et al., 1981).

The other major degenerative CNS disease with amyloid deposition is Alzheimer's. It is now clear that Alzheimer and scrapie amyloids are composed of different proteins (Masters et al., 1985a, b; Merz et al., 1986b; Roberts et al., 1986; Wisniewski et al., 1986). However, understanding the mechanism of amyloid plaque deposition in the scrapie model systems should provide insight into the pathogenetic mechanism by which amyloid deposits are formed in human Alzheimer's disease.

Different biochemical mechanisms have to be postulated for SAF formation and amyloid deposition, since (1) not all scrapie- or CJD-infected natural or experimental hosts display extracellular amyloid deposition, but all infected hosts have SAF and PrP, and (2) amyloid fibrils are found in situ, whereas SAF have only been observed in detergent-extracted brain (and spleen) homogenates (Merz et al., 1981, 1983c, 1984a,b; Rubenstein et al., 1986, 1987; Shinagawa et al., 1986).

Kuru

In kuru, the majority of the infectivity has been detected in the brain (Gajdusek, 1977, 1985). However, very low amounts of infectivity have been detected in spleens, which is in contrast to the comparatively high titers present in spleens of scrapie-infected animals (Clark and Haig, 1971b; Dickinson and Fraser, 1972; Fraser and Dickinson, 1970; Hadlow et al., 1979, 1982; Kimberlin and Walker, 1978b, 1979, 1980, 1986). As with scrapie, infectivity has not been associated with body fluids.

The pathology in the natural and transmitted cases of kuru are similar in that both exhibit astrogliosis and spongiform changes. In only one respect is there a difference: amyloid plaques are detected in the cerebellum of 75% of the natural cases, whereas no amyloid deposits have been found in experimental kuru cases (Beck and Daniel, 1979; Beck et al., 1966, 1969, 1973; Klatzo et al., 1959; Lampert, 1972).

In fact, even in the experimental scrapie systems, no model is known in which amyloid plaques are deposited only in the cerebellum. An explanation for the deposition of amyloid plaques in the natural kuru cases could include a genetic predisposition in the Fore tribe, the long incubation period and/or continual contact infection with the agent during the peak of the epidemic. Experimental animals may fail to display amyloid deposition, because they lack the proper genetic background, or a nonamyloidogenic strain of kuru present in the initial inocula may be selectively enriched during passage in animals.

Creutzfeldt-Jakob Disease

The pathogenesis of CJD has been more extensively studied than that of kuru but not to be the same extent as experimental scrapie. The majority of CJD cases occur with a worldwide incidence of 1 per million population (Brown, 1980; Chatelain et al., 1981; Gajdusek, 1977, 1985; Masters et al., 1979; Matthews, 1975; Will et al., 1986). About 10 to 15% of the cases are familial (Asher et al., 1983; Cathala et al., 1980; Ferber et al., 1974; Haltia et al., 1979; Masters et al., 1979, 1981a; Prusiner and Hadlow, 1979).

Transmission studies have shown that high levels of infectivity are present in the brain, and only low levels of infectivity are found occasionally in lymph nodes, lung, liver, kidney, and white blood cells (Gajdusek, 1977, 1985; Manuelidis, 1975; Manuelidis et al., 1976, 1977a, 1978b, 1985; Tateishi et al., 1983). Body fluids are negative except that a viremia has been reported in infected guinea pigs and mice (Kuroda et al., 1983; Manuelidis et al., 1978a). Thus, the distribution of infectivity is somewhat different from that observed in scrapie.

The spleens of CJD cases have been consistently negative in transmission experiments with primates but not in transmission studies with rodents (Gajdusek, 1977, 1985; Tateishi et al., 1983). The inability to easily detect CJD infectivity in human spleens following infection of primates must reflect a species barrier effect, since infection of rodents with spleen tissue has transmitted disease. One strain of experimental scrapie may mimic the human CJD strains. Scrapie strain 263K in hamsters may mimic the CJD agents in that it exhibits a low-level viremia (Diringer, 1984b) and replicates to titers of 10^{9-10} LD_{50}/g of brain whereas the spleen contains only 10^5 LD_{50}/g.

The genotype of the host and/or strain of agent can influence the expression of the disease in experimental systems and may also be operational with human disease. SAF and PrP have been observed in all postmortem brain samples from CJD patients, but no information is available from CJD patients in an earlier clinical phase.

The pathology of CJD is similar to those of the other spongiform encephalopathies in that neuronal loss, vacuolar changes, and astrogliosis are prominent (Oppenheimer, 1975). Amyloid plaques are observed in the cerebellum of 10 to 15% of CJD human cases. Experimental transmission of CJD to rodents reveals differences in the localization of the vacuolar changes in the brain which appear to be reproducible on further passage (Manuelidis, 1975; Manuelidis et al., 1976, 1977a, 1978b; Tateishi et al., 1983). These variations in pathology could be interpreted as strain variations of the agents causing CJD, similar to those previously described for the scrapie mouse systems.

In contrast to primates, in which amyloid plaques have never been seen after inoculation with human CJD tissue, plaques have been observed in the brains of some rodents inoculated with human CJD tissue, even in cases in which plaques were not part of the human neuropathology (Hikita et al., 1985; Tateishi et al., 1979, 1981, 1983). A similar phenomenon occurs with the primary passage of sheep scrapie into rodents. In this instance it has been shown that plaque formation in both the primary and subsequent passages is contingent on inoculation with very high doses of infectivity. When diluted inocula are used, plaque formation is irrevocably lost. One explanation offered for this phenomenon is that the original isolate contains a mixture of agents one of which is an amyloidogenic strain that is diluted out at low doses (Dickinson et al., 1984).

Gerstmann-Straussler Syndrome

GSS, a rare disease with an autosomal-dominant pattern of inheritance, has recently been added to the transmissible spongiform encephalopathies (Baker et al., 1985; Masters et al., 1981b; Tateishi et al., 1983). It was surprising to find that this disease was transmissible although it is comparable to the patterns that occur in high incidence scrapie outbreaks. The autosomal-dominant pattern of inheritance had indicated that this was a genetic disease. Initial infectivity studies have been performed, and primates, marmosets, hamsters, mice, and rats have since all developed a CJD-like disease on primary passage from a few GSS specimens.

Again, the pathological changes in the human CNS tissue are characterized by neuronal loss, vacuolar changes, and reactive astrogliosis. However, a striking finding is the predominance of amyloid plaques throughout the brain in contrast to the cerebellar localization seen with kuru and CJD in humans (Boellaard and Schlote, 1980; Masters et al., 1981b; Schumm et al., 1981; Seitelberger, 1962). These amyloid plaques are of the kuru type and do not exhibit the neuritic dystrophy and degeneration seen with the amyloid plaques of Alzheimer's disease (Gerstmann et al., 1936; Masters et al., 1981b). Like kuru and CJD, inoculation of GSS tissue into primates and rodents results in a disease that is identical to transmitted kuru or CJD but no amyloid plaques are observed.

The presence of amyloid plaques in GSS and the lack of them on transmission may reflect certain requirements for amyloid deposition. A particular genotype present in the family may be required, and a specific strain of agent may only be able to generate amyloid plaques in certain species. There may also be selection of a nonamyloidogenic strain of agent by the experimental host. GSS brain samples have never been inoculated into a p7p7 strain of mice for analysis of production of disease and amyloid plaque deposition.

The mechanism by which this syndrome is naturally transmitted is unknown. It is tempting to consider that the GSS families have a genetic predisposition for susceptibility to agent and/or for formation of amyloid deposition following infection.

Clinical Features

Scrapie

The natural mode of transmission of scrapie in sheep is unknown, but the outlines are reasonably well established. Natural scrapie seems to be spread both horizontally and vertically (Kimberlin, 1976, 1981; Parry, 1983; Prusiner and Hadlow, 1979). Vertical transmission is thought to occur from contact with infected placentas or in utero, since over 80% of the scrapie cases occur in the offspring of infected ewes (Dickinson et al., 1974; Hadlow et al., 1979; Hourrigan, 1979; Patterson et al., 1974). Infected ewes give birth during the preclinical stage of scrapie disease, which results in the majority of their offspring being infected. The offspring can then subsequently spread scrapie both vertically/or maternally and horizontally through the flock.

The classic example of horizontal transmission of scrapie occurred in Iceland and Scotland. In Iceland, a pastureland used by a scrapie-affected flock was left vacant for 3 years (Palsson, 1979). New sheep from scrapie-free flocks were transferred to this pastureland. Over the next few years, some of these sheep became infected with scrapie even though the source flocks remained scrapie-free. Survival of infectivity under these conditions could be related to the remarkable resistance of these agents to complete physical and chemical inactivation. The only known reservoir for scrapie is in sheep and goats. So cases could have been due to the persistence of infection, from this reservoir, in the environment. There could have been an intermediate animal reservoir, but no evidence for this exists at the present time.

The second example of horizontal spread of scrapie in sheep flocks involved Scottish Blackface sheep (Dickinson et al., 1974). The Scottish Blackface source flocks of more than 18,000 animals had been free of scrapie for over 30 years. Seventy-five Scottish Blackface sheep were placed in "field contact" with Suffolk sheep in which scrapie was endemic. Twenty-eight percent of these Scottish Blackface sheep developed scrapie over the next several years, but the source flocks remained free of disease.

In a number of natural scrapie outbreaks, the spread of scrapie appears to coincide with the introduction of new breeding stock to scrapie-free flocks (Parry, 1983). The new stock may have been contaminated with the infectious agent. It is also possible that the genetic composition of the new breeding stock introduced a new genotype into subsequent offspring, making them more susceptible. If scrapie is endemic in the area but unable to produce clinical illness because of a particular genotype in the host animals, this genetic change may allow agent replication and the emergence of clinical disease. Once established within a flock, the disease is self-perpetuating through infection by the maternal, transplacental, and oral route and through skin wounds that come into contact with contaminated fence posts (Brotherson et al., 1968; Gibbs et al., 1980; Hadlow, 1982;

Hourrigan, 1979; Kimberlin, 1981; Patterson et al., 1974).

In contrast, experimental scrapie in rodents exhibits no vertical transmission and will only transmit horizontally through scarification or cannibalism of dead animals (Amyx et al., 1981; Carp et al., 1981; Clark and Haig, 1971a). Thus, the experimental system does not completely mimic the conditions of the natural infection.

Two forms of natural scrapie in sheep have been observed—a nervousness with puritus or incoordination of the hind legs as the presenting symptoms. Clinical disease is usually manifest in sheep between 2 and 5 years of age. The clinical manifestations of this disease have not changed very much in the 200 years of observation (Besnoit, 1899; Cauvet, 1854; Claridge, 1795; Dickinson, 1976; Journal of the House of Commons, 1755; Kimberlin, 1981; Palsson, 1979; Wilson, 1950). Good shepherds can invariably pick out sheep with scrapie while the animals are still in an early stage of the disease process as evidenced by subtle changes in behavior or temperament, such as a slight apprehensiveness and general restlessness. The course progresses fairly rapidly over the next 3 to 6 months to pruritis, debilitation, rubbing against fixed objects (origin of the name scrapie), rapid movements of the tongue and lips, grinding of teeth, biting of feet and limbs, locomotor incoordination, loss of wool, loss of weight with no change in appetite or food intake, and finally death (Palsson, 1979; Parry, 1983). Experimental scrapie in the mouse usually presents with incoordination, reduced activity, failure to thrive, and wasting, which results in death (Carp et al., 1985a; Chandler, 1961; Morris and Gajdusek, 1963).

During the early history of natural scrapie, arguments raged over the cause of this disease—genetic or infectious. Similar to today's differing viewpoints on the nature of the spongiform encephalopathy agents, the proponents could present evidence to support their views. The hereditary concept was supported by the occurrence of scrapie in the progeny of certain rams and not others and the use of these rams for excessive inbreeding of the sheep breeds (Claridge, 1795). The infectious concept was supported by outbreaks of scrapie following the introduction of new sheep breeds into flocks, primarily Spanish merinos (Girard, 1829, 1830).

The transmissible nature of scrapie disease was first convincingly documented by transmissions into goats and sheep in 1936 (Cuille and Chelle, 1936). This was confirmed soon after by these and other investigators (Chelle, 1942; Gordon, 1946; Wilson, 1950). Working with transmissions of scrapie into sheep and goats, these same investigators demonstrated characteristics of scrapie disease that are now taken for granted by workers in this field. They

were as follows: all routes of inoculation were effective (intracerebral, epidural, intraocular, or subcutaneous), and the incubation period was dependent upon the route of inoculation, with shorter incubation periods following IC inoculation and longer incubation periods following subcutaneous injection.

The unusual physical properties of scrapie agents were beautifully illustrated in an iatrogenic mass inoculation of scrapie agent (Gordon, 1946). A louping-ill virus preparation of a 15% sheep brain, spinal cord, and spleen suspension was treated with Formalin to prepare a louping-ill vaccine. Sheep were inoculated subcutaneously with this vaccine. Approximately 10% of the recipient animals developed scrapie around 15 months after inoculation. The vaccine, unknowingly prepared with tissue from scrapie-infected animals and very effective against louping-ill virus, clearly demonstrated the Formalin resistance of scrapie agents. With the demonstration by Chandler in 1961 that scrapie could be transmitted to laboratory rodents, the death knell was rung for scrapie as purely a genetic disease (Chandler, 1961, 1963; Chandler and Fisher, 1963; Eklund et al., 1963; Morris and Gajdusek, 1963). As we have seen, a similar conflict is still present in the field of spongiform encephalopathies. GSS, a disease that clearly has an autosomal-dominant mode of inheritance, has been transmitted to experimental animals where it causes a CJD-like disease.

Experimental scrapie in mice and hamsters has served as the model system to examine the relationship between the route of infection, dose of agent, and pathogenesis of scrapie disease (Bruce and Fraser, 1981; Carp et al., 1985a; Cole and Kimberlin, 1985; Czub et al., 1986; Dickinson and Fraser, 1977; Fraser, 1981; Fraser and Dickinson, 1978, 1985; Kimberlin and Walker, 1978b, 1979, 1983, 1986; Merz et al., 1983a, 1984b; Mould et al., 1967; Scott and Dickinson, 1985). Intracerebral inoculation of 10-fold dilutions of infected tissue homogenates into rodents yield incubation periods whose length is dose-dependent. Infectivity is detected in the brain from midway through the incubation period in s7s7 mice, with clinical signs appearing during the latter part of the incubation period. The duration of the clinical phase is very short, on the order of 15% of the incubation period. In mice with the s7s7 genotype, infectivity can be detected from the initial IC inoculation through clinical signs, but in mice with the p7p7 genotype the accumulation of scrapie infectivity in the brain can be delayed for as long as 300 days. Peripheral inoculation of scrapie infectivity yields a longer incubation period than the IC route for the same strain of scrapie and a delayed appearance of infectivity in the brain, with maximum titers being detected in close proximity to the appearance of clinical signs. Subcutaneous inoculation of scrapie strain 139A into Compton white mice requires approximately a 20,000-fold higher dose of agent than the IC route to cause disease.

In contrast to the other organs, the brain is composed of functionally integrated groups of specialized cell types with complex cell-cell interactions. The specialized function of each cell type confers on that cell specific properties and characteristics that can determine selective susceptibility to various noxious stimuli and infectious agents. Clinical symptoms often reflect the damage to a particular subpopulation of these specialized brain cells. "Pathoclysis" refers to the selective vulnerability of groups of cells to damaging agents (toxins or viruses) that may be accompanied by morphological lesions in a given area (Vogt, 1925). The idea of clinical target areas was introduced in scrapie to account for the discrepancy between the widespread extent of morphological lesions (e.g., vacuolation), the apparent targeting of neurological symptoms and the differences in the duration of the replication phase of the agents in the brain detected after an 1C or peripheral route of inoculation (Kimberlin and Walker, 1983, 1986b).

The incubation period after an IC injection is much shorter than after peripheral injection. However, the length of time between when the agent replicated or accumulated in the brain and when clinical signs occurred was much shorter after peripheral injection than after IC injection. This indicates that following a peripheral inoculation the agent more readily reaches the life-supporting areas of the brain.

The mode of spread of a scrapie agent following peripheral inoculation appears to be from the spleen and visceral lymph nodes through the splenic nerve to the spinal cord, the cerebellum, and finally the cortical centers. An intraspinal injection of scrapie has a shorter incubation period than an IC injection, supporting the view that the critical areas concerned with the disease process are in the brainstem (Cole and Kimberlin, 1985; Kimberlin and Walker, 1979, 1980, 1983; Kimberlin et al., 1987).

Stereotaxic injections into different areas of the CNS have confirmed the results with intraspinal inoculation and have led to significant differences in clinical symptoms and/or incubation periods for a particular scrapie strain (Gorde et al., 1982; Kim et al., 1987a,b). In addition, different scrapie strains seem to differ in biological properties which can be correlated with specific areas of the brain. In SJL mice infected with scrapie strain ME7, obesity is observed as a clinical symptom, whereas some other scrapie strains do not elicit this clinical symptom (Carp et al., 1984). Stereotaxic injection of ME7 into the hypothalamus of SJL mice causes two effects: an earlier appearance and increased severity of obesity, and shorter incubation periods (Kim et al., 1987b). Scrapie strain 22L distinguishes itself neuropatho-

logically with high levels of vacuolation of the white matter in the cerebellum following IC injection (Fraser, 1979). Stereotaxic injection of 22L into the cerebellum, but not the caudate nucleus, substantia nigra, cerebral cortex, or thalamus, causes a dramatic reduction in the incubation period for this scrapie strain (Kim et al., 1987a).

The findings with experimental scrapie that the route of injection influences the dynamics of the development of clinical symptoms may be relevant to the differences seen in patients with kuru, CJD, and GSS. An example of its applicability is seen with the iatrogenic transmissions of CJD by two different routes of inoculation, CNS and peripheral. CNS inoculation causes widespread status spongiosus in the brain and neurological symptoms with dementia within 1 to 2 years following inoculation. In contrast, the peripheral inoculation with CJD-contaminated human pituitary growth hormone has revealed CJD presenting as a cerebellar ataxia with no dementia. One patient from this group, who had died from other causes, was found to exhibit classic status spongiosus in only two sites, the basal ganglia and cerebellum (Brown et al., 1987). This is remarkably similar to that observed following peripheral inoculation of animals with scapie agent with the first lesions being observed in the cerebellum (Cole and Kimberlin, 1985).

Kuru

This disease was endemic in the Fore tribe of New Guinea (Gajdusek and Zigas, 1957). The mode of transmission has been established as contact with infected brains during the preparation of the deceased body for ritualistic cannibalism (Glass, 1963; Gajdusek, 1977, 1985; Gibbs et al., 1980). Women and children would prepare the body for the rite, during which time there was continual contact with infected tissue. Brains were given to the women. It can easily be seen how the agent could infect the women and children through natural cuts and by wiping agent-containing tissue debris in the eyes or nose. As a result of this ceremonial practice, the affected population was predominantly women and young children of both sexes (Alpers, 1968, 1979; Gajdusek, 1977, 1985). As ritual cannibalism ceased, the number of kuru cases has dropped, and those now occurring can be traced to cannibalistic acts that took place 20 to 30 years ago (Klitzman et al., 1985). This indicates that natural kuru can have an incubation period of more than 30 years. The minimum incubation period has been derived from the young children who have become ill with the disease at 5 years of age. Infectivity has not been detected in placentas of kuru-affected women, nor has kuru developed in children of kuru mothers who were not part of the "cannabalistic rites," so vertical transmission does not seem to be a natural route of infection in this disease (Gajdusek, 1985).

Experimental kuru in primates has a minimum incubation period of 1.5 years following an intracerebral inoculation with brain tissue (Gajdusek, 1977, 1985; Gajdusek et al., 1966). Vertical transmission of kuru has not been observed with experimentally infected animals (Amyx et al., 1981).

Kuru presents with early insidious signs as an unsurety or clumsiness in walking and balance, which progresses until death (Alpers, 1979; Gajdusek, 1977, 1985; Gajdusek and Zigas, 1957, 1959; Hornabrook, 1979). The clinical course is very uniform, in contrast to the wide spectrum of clinical signs observed in CJD. The initial clumsiness becomes progressively worse, leading to a bedridden state, with excessive contraction of synergistic muscles, slurred speech leading to muteness, and difficulty in swallowing, which contributes to the emaciation and feebleness at the terminal stages of the disease. In juveniles, a more rapid and progressive course is observed, with more varied clinical signs (Hornbrook, 1979). Dementia is not a leading symptom in young people affected with kuru.

Creutzfeldt-Jakob Disease

The natural mode of transmission of this sporadic disease is unknown. Its uniform occurrence around the world at a constant level indicates a worldwide distribution of the CJD agent(s) (Gajdusek, 1977, 1985; Masters et al., 1979). The rarity of this disease indicates that several factors or events may be required to transmit disease. A high infectious dose of agent may be required to achieve clinical disease, a mutational event in the agent may convert a nonneuroinvasive agent to a neuroinvasive one and/or the genetic makeup of the individual may determine whether a person is susceptible to a given strain or agent. The mode of infection in familial cases is likewise unknown.

In the experimental CJD system, maternal or vertical transmission of CJD has not been obsered in studies with offspring of infected primates (Amyx et al., 1981; Manuelidis and Manuelidis, 1979). Likewise, a woman iatrogenically infected with a CJD-contaminated electrode delivered a normal, healthy baby while affected with CJD. Only time will tell if the child will go on to develop CJD (Bernoulli et al., 1977).

The more worrisome means of transmission with CJD is iatrogenic. This has occurred with contaminated corneas, electrodes, inadequately sterilized surgical equipment, human pituitary growth hor-

mone preparations, and dura mater transplants (Bernoulli et al., 1977; Brown, 1987; Brown et al., 1985a,b; Duffy et al., 1974; Gajdusek, 1977, 1985; Gibbs et al., 1985; Hintz et al., 1985; Koch et al., 1985; Manuelidis et al., 1977b; Prickard et al., 1987; Tintner et al., 1986; Weller et al., 1986; Will and Matthews, 1982). The tremendous increase in medical devices and use of natural products has increased the risk of iatrogenic tranmission due to inadvertent contamination with infected tissue.

There is no method available to detect persons who may be incubating the disease. Once the agent is introduced and accumulates or replicates in the brain, the disease is fatal. Extreme caution is needed to ensure that adequate decontamination procedures are used and that all donors have been carefully screened. Of critical importance is the ability to identify those people who may harbor the agent prior to onset of clinical disease. A prospective study of families with familial CJD could lead to the identification of preclinical markers. Transmission studies, SAF isolation, and/or PrP identification from lymph nodes and placentas following birth of offspring and/or monitoring cerebrospinal fluid from family members for novel polypeptides by two-dimensional gel electrophoresis are procedures that have a diagnostic value and importance in understanding the mode of transmission of the disease.

The minimum incubation period for the CJD agents depends on the initial site and dose of infection and has been derived from the cases of iatrogenic transmission by inoculation into the CNS with CJD-contaminated material. In these cases, disease has appeared 16 to 28 months after introduction of the agent into the CNS by either corneal transplant or contaminated instruments. The incubation period of CJD in experimental primates is 15 to 20 months following an IC inoculation of a 10% brain suspension (Gajdusek, 1977, 1985; Gibbs et al., 1968). Even though the incubation periods for the iatrogenic human CJD cases and the experimental CJD disease in chimpanzees are very similar, the dose received by the inoculated humans is presumably very low. The infectious dose required to develop disease in humans may be very different from that required to develop the disease in chimpanzees. The incubation period in sporadic cases is unknown. A person may be replicating agent subclinically and thereby posing a potential danger.

Injection of people with CJD-contaminated human pituitary growth hormone preparations caused clinical disease within 4 to 23 years after inoculation (Brown et al., 1985b; Gibbs et al., 1985; Hintz et al., 1985; Koch et al., 1985; Tintner et al., 1986; Weller et al., 1986). Experimental infection of chimpanzees by a similar peripheral route of inoculation has an incubation period of 5 to 10 years (Brown, 1987). The iatrogenic transmissions with human pituitary growth hormone preparations by peripheral inoculation have revealed differences in the incubation period and clinical picture of CJD compared to the CNS inoculation route. In these cases, CJD presents with cerebellar signs and very little dementia, symptoms remarkably similar to kuru. At the moment it is not known whether this reflects the age of the affected patient (approximately 20 years of age), a particular strain of the CJD agent, or differences in the mode of spread of the CJD agent to the brain.

In contrast to the other spongiform encephalopathies, CJD presents a varied progressive, clinical presentation, even though all patients exhibit terminal sequelae of mutism, listlessness, emaciation, and finally death. The variety of presenting symptoms and duration of disease may reflect many of the factors discussed earlier: infection with difference strains of agent, infection through different portals of entry, different host genetic susceptibility, or a mixture of all of the above.

TYPICAL CJD

In 30% of the cases is there a prodromal phase of vague complaints. Then definite neurological signs and symptoms develop in which there may be sensory, motor, or mental deterioration initially, along with myoclonus and abnormal EEG tracings. From this stage the course of the disease is rapid, so the patient is beridden within a few months and expires usually in less than 6 months. The average age of incidence is in the 6th decade of life with an average duration of clinical signs from 2 months to 2 years (Bernoulli et al., 1979; Court and Cathala, 1983; Creutzfeldt, 1920, 1921; Gajdusek, 1977, 1985; Jakob, 1921; Masters and Gajdusek, 1982; Masters et al., 1979, 1981a; Mizutani and Shiraki, 1985; Prusiner and Hadlow, 1979; Tateishi et al. 1981, 1983).

Familial CJD presents a similar clinical picture as that seen in typical CJD, but the age of onset is earlier (46 ± 6 years) and the duration of illness is much longer (12 to 120 months). The distinction between familial CJD and familial Alzheimer's disease is often difficult to make and requires postmortem neuropathological examination (Asher et al., 1983; Ball, 1980; Cathala, 1980; Ferber et al., 1974; Flament-Durand and Couck, 1979; Gajdusek, 1985; Haltia et al., 1979; Masters et al., 1981a; Seitelberger, 1962).

ATYPICAL CJD

Here the age of onset may vary. Instead of clustering in the 6th decade, CJD may appear in the 3rd decade of life. Instead of a clinical course of under 6 months, the duration of illness may be much longer, and when dementia is a major clinical sign, the patient may also be diagnosed as having Alzheimer's disease. A patient with atypical CJD may be diagnosed and then expire within 1 month. This short clinical course suggests that the patient may have an acute encephalitis; however, the results of all laboratory tests will be in the normal range (Bernoulli et al., 1979; Brown et al., 1979, 1984; Masters et al., 1979; Mizutani and Shiraki, 1985; Packer et al., 1980; Park et al., 1980; Schoene, 1981). As stated earlier, iatrogenic transmissions of CJD may present very differently in the early stages. Because of these variations in the clinical picture, postmortem examinations are needed to confirm diagnoses. The incidence of inapparent CJD infections in the general population is unknown, since infected people may expire from other causes before clinical disease has appeared.

Gerstmann-Straussler Syndrome

Little is known about the incubation period of this disease in humans, but experimental transmissions in spider monkeys, squirrel monkeys, and marmosets yield a 1- to 2-year incubation period following intracerebral inoculation with a 10% brain suspension (Baker et al., 1985; Masters et al., 1981b; Tateishi et al., 1983). As with familial CJD, GSS exhibits an autosomal-dominant mode of inheritance with the appearance of disease around 48 years, which is earlier than sporadic CJD, and with a longer duration of disease (13 to 132 months). The GSS patient presents with primarily cerebellar and pyramidal signs and dementia. These progress until the patient becomes bedridden and terminally ill (Gerstmann et al., 1936; Kuguhara et al., 1985; Masters et al., 1981b; Schumm et al., 1981; Seitelberger, 1962).

To date, about 50% of GSS cases have successfully transmitted a CJD-like disease when GSS tissue has been inoculated into nonhuman hosts. Such a high incidence of a coincidental infection is unlikely. This raises the questions whether GSS is a genetic or an infectious disease. As stated earlier, both scrapie and kuru were thought to be genetic diseases until both were clearly demonstrated to be transmissible. Thus, the GSS families have become another test system in which to investigate the prevailing concepts of the nature of the agents causing spongiform encephalopathies. The hereditary concept can be considered to be the "prion" theory in which outbreaks of GSS in families originate as a mutation in the germ plasm, which is continued through the succeeding generations. The infectious concept is embodied in the virus or virino theories in which the disease is transmitted maternally and/or by contact transmission to the offspring in a similar manner to scrapie. The GSS families provide a rare opportunity to study the human genes that influence susceptibility to spongiform encephalopathies. Genes that control scrapie incubation periods have been identified in sheep and mice. An analogous mechanism may be operational in GSS families. Experiments are in progress in our laboratories to examine polymorphisms which segregate with the eventual development of clinical disease within GSS families. PrP gene polymorphisms are of particular interest considering the possible relationship between *Sinc* and PrP genes in the experimental mouse system (Carlson et al., 1986). A low-frequency Pvu 11 polymorphism has been detected in the human PrP gene (Wu et al., 1987). These studies will provide additional information on agent-host interactions within the spongiform encephalopathies.

Diagnosis

The spongiform encephalopathies present difficulties in diagnosis, because most of the routine viral diagnostic procedures cannot be used. Since the infectious agent has not been characterized, techniques used to identify either viral-specific protein or nucleic acid cannot be utilized. The host does not generate a specific immune response to infection at either the cellular or the humoral level. Therefore, the serological tests that are part of the repertory of most diagnosis laboratories do not apply. These agents replicate poorly, at best, in tissue culture systems and produce no demonstrable cytopathic effect (Asher et al., 1979; Clark, 1979; Clark and Haig, 1970; Clark and Millson, 1976b; Gibson et al., 1972; Kuroda et al., 1983; Markovits et al., 1983; Oleszak et al., 1986; Rubenstein et al., 1984). Cell fusion in tissue culture has been induced by brain extracts from patients with CJD (Kidson et al., 1978; Moreau-Dubois et al., 1979). Although this is an interesting phenomenon, the assay does not appear to have the specificity necessary for a diagnostic test.

Until recently the only assay for the presence of these agents is the transmission of disease to test animals. However, it can take 1 to 2 years or more before animals manifest clinical signs. Such transmission studies are obviously very time-consuming and costly, and they require special facilities to maintain and contain the infected animals. These restrictions result in transmission studies being performed only rarely and only in specialized laboratories. At present, a tentative diagnosis in humans can be made on the basis of the clinical signs and characteristic electroencephalogram changes of paroxymal bursts of high-voltage slow waves (Burger et al., 1972; Chiofala et al., 1980; Court and Cathala, 1983; Gajdusek, 1985; Prusiner and Hadlow, 1979). The

diagnosis can be strengthened with the demonstration of degenerative vacuolar changes and astrogliosis in biopsied or autopsied CNS tissue. At this point, even if all of the above criteria indicate a spongiform encephalopathy disease, definitive diagnosis can only be made following positive transmission of the disease to susceptible animals.

Transmission studies are also not without their problems. These agents can exhibit a profound species barrier in primary transmission to experimental animals. Transmission studies using chimpanzees and squirrel monkeys have been fairly successful (75 to 90% of the suspected CJD cases transmit disease), but they are very costly and time-consuming, with incubation periods of up to 10 years. CJD has been transmitted to a variety of laboratory animals including mice, rats, hamsters, and guinea pigs, with an incubation period of 1 to 2 years on primary transmission (Ferber et al., 1974; Gajdusek, 1977, 1985; Gibbs et al., 1968, 1980; Manuelidis, 1975, 1985; Manuelidis et al., 1977a, 1978b; Tateishi et al., 1979, 1981, 1983). The incidence of transmission to laboratory rodents appears to be equal to that seen in nonhuman primates (Tateishi et al., 1983). Because of the species barrier mentioned earlier, incubation times are generally reduced by one-half to one-third on second passage in homologous species (Manuelidis et al., 1976; Tateishi et al., 1983). The intracerebral route of inoculation is the most rapid means of transmitting the disease in all species, but other routes have also proved successful (Gajdusek, 1977, 1985; Gibbs et al., 1980).

Over the past 4 years, progress has been made in developing new markers of the diseases which may help in establishing a diagnostic laboratory test. SAF have been detected in extracts of organs containing high levels of infectivity (Merz et al., 1981, 1983c, 1984a,b). The SAF, as stated earlier, are composed of proteins derived from a normal host glycoprotein of 33 to 35 kDa which can be cleaved with proteinase K (PK) to yield protease-resistant forms of the protein. Since the protein in normal, noninfected individuals is completely sensitive to proteolysis, the presence of PK-resistant proteins is indicative of the disease (Bendheim et al., 1985; Bockman et al., 1985; Bode et al., 1985; Brown et al., 1986b; Gibbs et al., 1985; Kascsak et al., 1986; Rubenstein et al., 1986). Polyclonal and monoclonal antibodies have been raised to these PrP which have been used to detect PrP by immunological means (Barry and Prusiner, 1986; Barry et al., 1985; Bendheim et al., 1984; Cho, 1986; Diringer et al., 1984; Kascsak et al., 1986, 1987a,b; Shinagawa et al., 1986; Takahashi et al., 1986). We will now discuss the potential use of SAF, PrP, and antibodies generated to them in aiding in the diagnosis of these diseases.

Isolation Procedures

Several rapid purification methods are available that utilize both the aggregation properties of the SAF and the protease resistance of their component PrP (Brown et al., 1986b; Hilmert and Diringer, 1984; Hope et al., 1986; Kascsak et al., 1987a,b; Manuelidis et al., 1985). One must be cautious, since SAF and PrP derived from different strains of scrapie and CJD have been shown to possess different sensitivities to PK degradation (Kascsak et al., 1985; Manuelidis et al., 1985). SAF from one scrapie strain (139A) have been shown to be highly susceptible to proteolytic digestion. In unknown cases, it is best to examine material with and without PK treatment to guard against false-negative results when using PK treatment alone. A drawback of the procedures based on aggregation is that false negatives can also be seen if the number of fibrils are not sufficient to form aggregates large enough to be isolated in these procedures.

Nonproteinase K Subcellular Fractionation

As low as 0.250 g of brain tissue may be processed in this procedure (Merz et al., 1981, 1983c, 1984a; Rubenstein et al., 1986; Somerville et al., 1986). After gentle homogenization, the nuclei are removed with low-speed centrifugation. A crude mitochondrial synaptosomal preparation is recovered as a pellet. The membranes are lysed with the detergent octylglucoside, and the sample is layered on a step sucrose gradient and centrifuged overnight. A broad band forms at the interface of the 1.4 and 1.8 M sucrose, which is analyzed by electron microscopy (EM) and Western blotting (Fig. 4). This procedure does not depend on the ability of SAF to aggregate, nor does it depend on the use of proteases. However, it may be helpful to pretreat the samples with PK prior to Western blotting (Merz et al., 1987a; Rubenstein et al., 1986). Both the normal and disease forms of the 33- to 35-kDa protein are isolated in this procedure, such that conversion of the 33- to 35-kDa protein to the lower-molecular-weight forms is necessary for definitive diagnosis. This procedure has the advantage of detecting low numbers of SAF that may be present preclinically (Merz et al., 1983a, 1984b). But the procedure requires multiple steps and experience in subcellular fractionation, and, as stated earlier, the normal protein is also isolated.

SAF Isolation Dependent on Aggregation of Fibrils

Several modifications of a basic procedure are available (Gibbs et al., 1985; Hilmert and Diringer, 1984; Hope et al., 1986; Kascsak et al., 1987b; Manuelidis

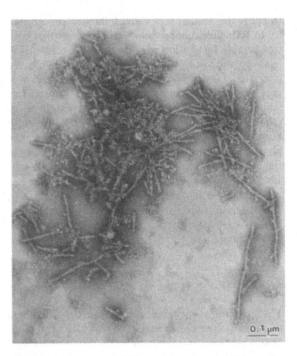

FIG. 4. Electron micrograph of SAF isolated by nonproteinase K procedure from the brains of mice injected with scrapie strain 139A and stained with sodium phosphotungstate. Proteinase K was not used in this procedure. Bar represents 0.1 μm.

et al., 1985). The initial quantity of the brain tissue sample can be as low as 0.1 g, which is homogenized with a Tissumizer or Dounce homogenizer as a 20% sarkosyl homogenate. Centrifugation in a Beckman airfuge has proved very successful when small amounts of tissue are being used (0.1 to 0.2 g). Larger amounts of tissue can be processed in a standard high-speed ultracentrifuge. The sarkosyl extract is pelleted, washed in high salt, and repelleted. The pellet contains SAF, which are then analyzed as coded specimens for both SAF by EM and PrP by Western blotting (Fig. 5). The entire analysis can be performed in 24 hr when the airfuge procedure is employed. The advantages of this procedure are the small amount of starting tissue, the ease with which a fraction can be prepared and analyzed without using proteolytic enzymes, and the absence of the normal host form of this protein. Any immunological reactivity with PrP antisera is, therefore, indicative of a positive diagnosis. The drawback of this procedure is the potential for false negatives if there are low quantities of SAF.

The nonproteinase K subcellular fractionation procedure can be used as an epidemiological tool for analysis of the distribution of SAF in other tissues of affected and non-affected hosts (Rubenstein et al., 1986). The SAF isolation procedure and variations of

FIG. 5. Electron micrographs of purified SAF isolated from the brains of hamsters injected with scrapie strain 263K. Proteinase K and extensive sonication were used in this procedure. SAF are stained with uranyl acetate. Bar represents 0.1 μm.

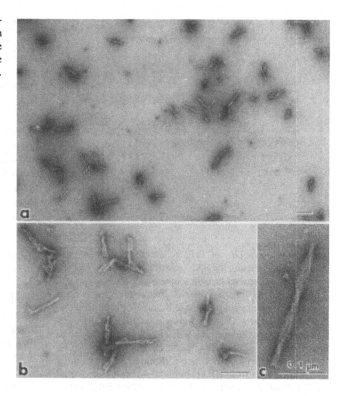

it have failed to yield SAF or PrP from spleens of experimental scrapie in the mouse or hamster (Czub et al., 1986a; Merz, unpublished observations). In addition to their use with CNS tissue, these isolation procedures could also be used on sheep placentas, an organ that may be important in the transmission of natural scrapie. This process may alert the shepherds to the presence of scrapie prior to overt clinical disease in the field. As stated earlier, similar analyses can be performed in a prospective study of human placentas or lymph nodes in known familial CJD and GSS families.

Analysis

ELECTRON MICROSCOPY

Second only to infectivity studies, detection of SAF by negative-stain electron microscopy remains the most sensitive diagnostic assay (Merz et al., 1984a; Rubenstein et al., 1987). An aliquot of the sample obtained from one of the above procedures can be analyzed for the presence of SAF by routine viral diagnostic EM. This can be done in most well-equipped viral diagnostic laboratories. The specimen is placed on Formvar carbon or carbon-coated grids, negatively stained with 3% sodium phosphotungstate, pH 7.2, and observed at EM magnifications of ×18,000 to ×36,000. After the microscopist has

gained a little experience in distinguishing SAF from normal CNS structures, the analysis is performed with little difficulty, since the fibrils will be easily recognizable (Figs. 5 and 6). Observing a maximum of 10 squares is enough for the SAF isolation procedure, whereas several duplicate grids may have to be observed for the human diseases using the nonproteinase K. All samples are coded, and positive and negative controls are included in every analysis.

Thus far, all suspected biopsy (3) and autopsy (12) CJD cases examined with the above two procedures have been positive for SAF. In contrast to human material, 50% to 100% of natural scrapie sheep cases have been positive for the presence of SAF, probably owing to autolysis of the brain specimen (Gibson et al., 1987; Rubenstein et al., 1987). CSF and whole blood from CJD cases have been negative for both SAF and PrP (Merz, unpublished observations). This is not surprising, since very low titers of CJD infectivity have been detected in human blood, and only 16% of the CSF has been infectious when inoculated into primates (Gajdusek, 1985).

PROTEINS

In conjunction with morphological analysis of the samples, PrP proteins can be separated by electrophoresis (Laemmli, 1979) and detected by Western blot analysis (Towbin, 1979). As indicated earlier, the crude samples can be analyzed with or without

WESTERN BLOT

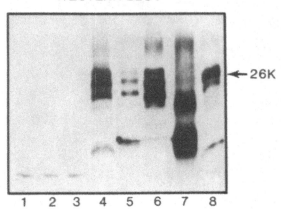

FIG. 6. Western blot of PrP isolated from human brain tissue prepared using the SAF isolation procedure. Homogenates were prepared in 10% sarcosyl and SAF isolated by differential centrifugation using a Beckman airfuge. Samples were treated with proteinase K and electrophoresed on 12% Laemmli gels. Proteins were electrophoretically transferred to nitrocellulose and Western blotted using monoclonal antibody 263K–3F4. Lane designations: 1, 2, 3—preparations from Alzheimer's disease patients; 4, 5, 6—preparations from CJD patients; 7—preparation from a GSS patient; 8—positive control preparation from scrapie strain 263K-infected hamsters. Note the variety in the Western blot profiles in the spongiform encephalopathy cases (lanes 4 to 8), potentially reflecting different strains of the spongiform encephalopathy agents infecting each case. 26K indicates the position of the 26- to 30-kDa PrP protein.

PK treatment for determination and analysis of PrP. To date, we and others have been able to detect PrP in brain samples from 75 to 100% of the CJD cases analyzed and 25% of the sheep cases with natural scrapie (Bendheim et al., 1985; Bockman et al., 1985; Brown et al., 1986b; Rubenstein et al., 1987). One of the most encouraging aspects of this work has been the absence of false positives (Fig. 6). Well-characterized antisera and monoclonals to PrP are used to detect antigen by means of ELISA assay, Western blots, or dot blots (Barry and Prusiner, 1986; Kascsak et al., 1986, 1987a). It is envisioned that newer and more sensitive tests will become available as further research is performed on these diseases.

It is encouraging that using biopsy or autopsy CNS tissue a diagnostic test has been developed that offers an indication that one is dealing with a disease caused by unconventional agents. The analyses can be performed within 12 to 24 h. Previously, postmortem neuropathology, often taking 2 weeks to 3 months, and transmission studies (1.5 to 10 years) were the only mechanisms available. With the increase in the geriatric population, this is a valuable asset for hospitals to use for the differential diagnosis of CNS disease, thereby detecting potential biohazards, and preventing iatrogenic transmissions.

Interpretation of Data

As stated earlier, SAF and PrP have never been seen in other diseases of viral, degenerative, or toxic origin or in normal controls and are, therefore, diagnostic markers for the spongiform encephalopathies (Brown et al., 1986b; Merz et al., 1984a). At present, the analysis for SAF by EM or PrP by Western blot should be referred to a specialty laboratory with expertise in these areas.

Other Tests

The protein pattern of CSF has recently been analyzed by silver staining following two-dimensional gel electrophoresis as a potential diagnostic test for degenerative CNS diseases (Harrington et al., 1986). Two proteins (26 kDa, isoelectric point (PI) 5.2, and 29 kDa, PI 5.1) were present in all CJD patients and in 5 of 10 patients with herpes simplex encephalitis but not in any of the control groups. These proteins are not related to PrP, since they fail to react with PrP-specific antisera. Their identity and origin are unknown, but their presence may aid in the diagnosis of CJD. CSF from kuru cases were negative, but this may have been due to storage conditions. CSF from GSS patients or their family members have not been analyzed. Since it is fairly easy to differentiate between CJD and herpes simplex encephalitis on a clinical basis, the lack of specificity to CJD may not cause a problem in diagnosis. Because of the expertise required for this analysis, all tests should be referred to a specialty laboratory.

Biosafety

The spongiform encephalopathy agents can be transmitted, but no evidence exists that the agents are contagious to humans by the common portals of entry. The fatal nature of these diseases and the harsh conditions required for total decontamination of the agents mandate that strong preventive measures be taken to protect hospital patients and laboratory personnel from the human spongiform encephalopathy agents (American Association of Tissue Banks, 1984; Asher et al., 1986; Barringer et al., 1980; Chatigny and Prusiner, 1980; Gajdusek, 1977, 1985; Rosenberg et al., 1986). Processing of tissue must be performed under P2 containment conditions, but even more stringent conditions are suggested. Gloves, masks, and gowns are worn at all times. All work is performed in specially designed laminar flow hoods (total exhaust, HEPA filters, prefilters), and disposable laboratory utensils are required. A potential problem is the generation of aerosols during sonication, which is a required step in certain agent-isolation procedures.

Iatrogenic transmissions due to inadequate inactivation procedures are a problem in hospitals. Because of the lethality of these diseases, with no known remissions or cures, the effective inactivation of infectivity is essential. The current recommended decontamination procedures require the following: All contaminated equipment is immersed in 1.0 NaOH or undiluted bleach (Clororx) for several hours, followed by sterilization for 1 h at 15 psi and a temperature setting of 132°C, followed by standard washing and resterilization. Since these agents survive Formalin fixation, separate microtome knives should be used in the histopathology laboratories for the cutting of suspected CJD tissue.

With the increase in medical use of natural products as evidenced by the iatrogenic transmission of CJD from isolated human pituitary growth hormone and dura mater grafting, precaution should be taken in the selection of donor tissue in order to minimize potential iatrogenic transmissions. Tissues, particularly those of CNS origin, should only be taken from individuals with well-documented clinical and family histories. Careful selection of tissue at this level forms the major barrier against inadvertent transmission of disease. Many of the concepts and procedures described in this chapter should help in this regard.

Epidemiology

As stated throughout this chapter, two epidemiological patterns are observed with the spongiform encephalopathy diseases—focal or worldwide.

Both scrapie and sporadic CJD occur with a worldwide distribution (Brown, 1980; Chatelain et al., 1981; Court and Cathala, 1983; Gajdusek, 1977, 1985; Hourrigan et al., 1979; Masters et al., 1979; Parry, 1983; Prusiner and Hadlow, 1979; Will et al., 1986). Scrapie has been eradicated in Australia and New Zealand by slaughter of affected sheep flocks and restriction on importation of new breeding stocks, but it remains endemic in portions of North America, Europe, Iceland, and India.

Sporadic CJD occurs with an incidence of 1 per million worldwide, irrespective of the presence of scrapie in sheep. As stated earlier, 10 to 15% of the CJD cases are familial in origin and have an autosomal-dominant pattern of inheritance. CJD does not seem to be contagious, since it does not spread easily within a family.

Transmissible mink encephalopathy arose in captive mink from the consumption of scrapie-contaminated sheep carcasses and has been totally controlled by prevention (Burger and Hartsough, 1965; Hartsough and Burger, 1965; Marsh and Hanson, 1979; Marsh and Kimberlin, 1975; Marsh et al., 1974). Chronic wasting disease of captive mule deer and elk is endemic in captive herds and is thought to have occurred by similar mechanisms as transmissible mink encephalopathy (Williams and Young, 1980, 1982). The incidence of this disease in wild mule deer and elk is unknown.

Kuru has a focal incidence in the Fore tribe of the New Guinea highlands and is now disappearing since the cessation of ritual cannibalism. Gerstmann-Straussler syndrome is entirely familial, and sporadic cases have not been observed.

Treatment, Prevention, and Control

There is no known treatment for any of the spongiform encephalopathies except supportive therapy. No remissions or cures are known; the diseases are invariably fatal. With the human disease, CJD, iatrogenic transmissions can be prevented when adequate precautions are taken; however, no mechanisms are known to prevent and control the natural incidence of CJD, since the nature of the etiological agent and the mode of infection are unknown.

Scrapie is controlled by eradication programs in the United States. New outbreaks are likely to occur over the next several years as a consequence of budgetary restrictions curtailing the eradication program. Another factor, as illustrated previously, is the difficulty in effectively eliminating scrapie from an area once it has become endemic.

What does the future hold for the spongiform encephalopathies of animals and man? At the moment, prevention remains the only means for combating these diseases. Early clinical diagnosis combined with newly formulated laboratory diagnostic tests will assist in curtailing the spread of disease. Further studies on pathogenesis and the molecular biology of the infectious agents should provide insight into additional strategies to more effectively prevent disease and perhaps lead to a means of intervention and treatment of these conditions. These remain a unique class of transmissible diseases. An understanding of these diseases will provide information on basic mechanisms of infectious agent-host interactions within the central nervous system.

Acknowledgments. The authors gratefully acknowledge the superb secretarial assistance of Lucille F. Olsen and the patience and expertise of our collaborators Richard I. Carp, Richard J. Kascsak, Richard

Rubenstein, Monika Wrzolek, and George S. Merz. This work has been partially supported by the Public Health Service grants NS21349 and AG04220 from the National Institutes of Health.

Addendum

Bovine Spongiform Encephalopathy

A new disease, bovine spongiform encephalopathy, has recently been added to the list of spongiform encephalopathy diseases (Wells et al., 1987). It first appeared in herds of English Friesan/Holstein cattle aged 3 to 6 years. The clinical symptoms have an insidious onset, presenting with behavior changes and neurologic signs that progress to incoordination of gait, debility, and death. The pathology is restricted to the brain, with neuropil and neuronal vacuolation in the gray matter of the brain stem. SAF have been isolated from brain tissue of the affected cattle, but not from unaffected cattle brain. On the basis of the sporadic, slowly progressive, clinical neurologic signs, vacuolation in brain tissue, and isolation of SAF, this disease of cattle has been termed bovine spongiform encephalopathy. At present, the cause of this disease is not known. It may reflect: 1) a genetic shift in the cattle population such that a scrapie-like disease could emerge; 2) a common contamination with a scrapie-like agent similar to a transmissible mink encephalopathy outbreak; 3) a mutation in the scrapie-like agents such that the agent could now cause disease in cattle; or 4) any or all of the above. Epidemiologic studies, analysis of genetic data, and experimental transmission studies have all been initiated.

Unconventional Agents

Readers are urged to consult the most recent compilation of known facts, theories and discussion pertaining to the unconventional agents, which is the result of a symposium held by the CIBA Foundation in London from June 30 to July 2, 1987, and recently published.

Literature Cited

Alper, T., D. A. Haig, and M. C. Clark. 1966. The exceptionally small size of the scrapie agent. Biochem. Biophys. Res. Commun. **22**:278–284.

Alper, T., W. A. Cramp, D. A. Haig, and M. C. Clark. 1967. Does the agent of scrapie replicate without nucleic acid? Nature **214**:764–766.

Alper, T., and D. A. Haig. 1968. Protection by anoxia of the scrapie agent and some DNA and RNA viruses irradiated as dry preparations. J. Gen. Virol. **3**:157–166.

Alper, T., D. A. Haig, and M. C. Clark. 1978. The scrapie agent: evidence against its dependence for replication on intrinsic nucleic acid. J. Gen. Virol. **41**:503–516.

Alpers, M. P. 1968, Kuru: implications of its transmissibility for the interpretation of its changing epidemiologic pattern. p. 234–251. *In* Central nervous system, Vol. 9. International Academy of Pathology, Baltimore.

Alpers, M. P. 1979. Epidemiology and ecology of kuru. p. 67–90. *In* S. B. Prusiner, and W. J. Hadlow (ed.), Slow transmissible diseases of the nervous system, Vol. 1. Academic Press, New York.

American Association of Tissue Banks. 1984. Standards for tissue banking. American Association of Tissue Banks, Arlington, VA.

Amyx, H. L., C. J. Gibbs, Jr., D. C. Gajdusek, and W. E. Greer. 1981. Absence of vertical transmission of subacute spongiform viral encephalopathies in experimental primates. Proc. Soc. Exp. Biol. Med. **166**:469–471.

Asher, D. M., R. T. Yanagihara, N. G. Rogers, C. J. Gibbs, Jr., and D. C. Gajdusek. 1979. Studies of the viruses of spongiform encephalopathies in cell culture. p. 235–242. *In* S. B. Prusiner and W. J. Hadlow (ed.), Slow transmissible diseases of the nervous system, Vol. 2. Academic Press, New York.

Asher, D. M., C. L. Masters, D. C. Gajdusek, and C. J. Gibbs, Jr. 1983. Familiar spongiform encephalopathies. p. 273–291. *In* S. S. Kety, L. P. Rowland, R. L. Sidman, and S. W. Mattkysse (ed.), Genetics of neurological and psychiatric disorders. Raven Press, New York.

Asher, D. M., C. J. Gibbs, Jr., and D. C. Gajdusek. 1986. Slow viral infections: safe handling of the agents of subacute spongiform encephalopathies. p. 59–71. *In* B. M. Miller (ed.), Laboratory safety: principles and practices. American Society for Microbiology, Washington, D.C.

Baker, H. F., R. M. Ridley, and T. J. Crow. 1985. Experimental transmission of an autosomal dominant spongiform encephalopathy: does the infectious asgent originate in the human genome? Br. Med. J. **291**:299–302.

Ball, M. J. 1980. Features of Creutzfeldt-Jakob disease in brains of patients with familial dementia of Alzheimer type. Can. J. Neurol. Sci. **7**:51–57.

Barringer, J. P., D. C. Gajudsek, C. J. Gibbs, Jr., C. L. Masters, W. E. Stein, and R. D. Terry. 1980. Transmissible dementias: current problems in tissue handling. Neurology **30**:302–303.

Barry, R. A., and S. B. Prusiner. 1986. Monoclonal antibodies to the cellular and scrapie prion protein. J. Infect. Dis. **154**:518–521.

Barry, R. A., M. P. McKinley, P. E. Bendheim, G. K. Lewis, S. J. De Armond, and S. B. Prusiner. 1985. Antibodies to the scrapie protein decorate prion rods. J. Immunol. **135**:603–613.

Beck, E., P. M. Daniel, and H. B. Parry. 1964. Degeneration of the cerebellar and hypothalamic neurohypophysial systems in sheep with scrapie; and its relationship to human system degenerations. Brain **87**:153–176.

Beck, E., P. M. Daniel, M. Alpers, D. C. Gajdusek, and C. J. Gibbs, Jr. 1966. Experimental "kuru" in chimpanzees: a pathological report. Lancet **2**:1056–1059.

Beck, E., P. M. Daniel, W. B. Matthews, D. L. Stevens, M. P. Alpers, D. M. Asher, D. C. Gajdusek, and C. J. Gibbs, Jr. 1969. Creutzfeldt-Jakob disease: the neuropathology of a transmission experiment. Brain **92**:699–716.

Beck, E., P. M. Daniel, D. C. Gajdusek, and C. J. Gibbs, Jr. 1973. Experimental kuru in the chimpanzee: a neuropathological study. Brain **96**: 441–442.

Beck, E., and P. M. Daniel, 1979. Kuru and Creutzfeldt-Jakob disease: neuropathological lesions and their significance. p. 253–270. In S. B. Prusiner and W. J. Hadlow (ed.), Slow transmissible diseases of the nervous system, Vol. 1. Academic Press, New York.

Bellinger-Kawahara, C., J. E. Cleaver, T. O. Diener, and S. B. Prusiner. 1987. Purified scrapie prions resist inactivation by UV irradiation. J. Virol. 61:159–166.

Bendheim, P. E., R. A. Barry, S. J. De Armond, D. P. Stites, and S. B. Prusiner. 1984. Antibodies to a scrapie prion protein. Nature 310:418–421.

Bendheim, P. E., J. M. Bockman, M. P. McKinley, D. T. Kingsbury, and S. B. Prusiner. 1985. Scrapie and Creutzfeldt-Jakob disease prion proteins share physical properties and antigenic determinants. Proc. Natl. Acad. Sci. USA 82:997–1001.

Bernoulli, C. C., J. Siegfried, G. Baumgartner, F. Regli, T. Rabinowicz, D. C. Gajdusek, and C. J. Gibbs, Jr. 1977. Danger of accidental person to person transmission of Creutzfeldt-Jakob disease by surgery. Lancet 1:478–479.

Bernoulli, C. C., C. L. Masters, D. C. Gajdusek, C. J. Gibbs, Jr., and J. O. Harris. 1979. Early clinical features of Creutzfeldt-Jakob disease (subacute spongiform encephalopathy). p. 229–252. In S. B. Prusiner and W. J. Hadlow (ed.), Slow transmissible diseases of the nervous system, Vol. 1. Academic Press, New York.

Bertrand, J., H. Carre, and F. Leucam. 1937. La tremblante du mouton (recherches histo-pathologiques). Ann. Anat. Pathol. 14:565–586.

Besnoit, C. 1899. La tremblante ou nevrite peripherique enzootique du mouton. Rev. Vet. Toulouse 24:265–277, 333–343.

Bock, G. and J. Marsh (ed.) Novel infectious agents and the central nervous system. CIBA Foundation Symposium 135. John Wiley and Sons, Chichester.

Bockman, J. M., D. T. Kingsbury, M. P. McKinley, P. E. Bendheim, and S. B. Prusiner. 1985. Creutzfeldt-Jakob disease protein in human brains. N. Engl. J. Med. 312:73–78.

Bode, L., M. Pocchiari, H. Gelderbloom, and H. Diringer. 1985. Characterization of antisera against scrapie-associated fibrils (SAF) from affected hamster and cross reactivity with SAF from scrapie-affected mice and patients with Creutzfeldt-Jakob disease. J. Gen. Virol. 66:2471–2478.

Boellaard, J. W., and W. Schlote. 1980. Subakute spongiforme Encephalopathie mit multiformer Plaquebildung. "Eigenartige familiar-hereditare Krankheit des Zentralnervensystems [spino-cerebellare Atrophie mit Demenz, Plaques und plaqueahnlichen Ablagerungen im Klein- und Grosshirn" (Gerstmann, Straussler, Scheinker)]. Acta Neuropathol. (Berl.) 49:205–212.

Bolton, D. C., M. P. McKinley, and S. B. Prusiner. 1982. Identification of a protein that purifies with the scrapie prion. Science 218:1309–1311.

Bolton, D. C., R. K. Meyer, and S. B. Prusiner. 1985. Scrapie PrP 27-30 is a sialoglycoprotein. J. Virol. 53:596–606.

Borras, M. T., D. T. Kingsbury, D. C. Gajdusek, and C. J. Gibbs, Jr. 1982. Inability to transmit scrapie by transfection of mouse embryo cells in vitro. J. Gen. Virol. 58:263–271.

Borras, T., and C. J. Gibbs, Jr. 1986. Molecular hybridization studies of scrapie brain nucleic acids. 1. Search for specific DNA sequences. Arch. Virol. 88:67–78.

Braig, H. R., and H. Diringer. 1985. Scrapie: concept of a virus-induced amyloidosis of the brain. EMBO J. 4:2309–2312.

Brotherson, J. G., C. C. Renwick, J. T. Stamp, I. Zlotnick, and I. H. Patterson. 1968. Spread of scrapie by contact to goats and sheep. J. Comp. Pathol. 8:9–17.

Brown, P. 1980. An epidemiologic critique of Creutzfeldt-Jakob disease. Epidemiol. Rev. 2:113–135.

Brown, P. 1985. Virus sterility for human growth hormone. Lancet 2:729–730.

Brown, P., 1988. The clinical neurology and epidemiology of Creutzfeldt-Jakob disease, with special reference to iatiogenic cases. In: Novel infectious agents and the central nervous system. CIBA Foundation Symposium 135. In press.

Brown, P., F. Cathala, D. Sadowsky, and D. C. Gajdusek. 1979. Creutzfeldt-Jakob disease in France. III. Clinical characteristics of 124 consecutive verified cases during the decade 1968–1977. Ann. Neurol. 6:430–437.

Brown, P., R. G. Rohwer, E. M. Green, and D. C. Gajdusek. 1982. Effect of chemicals, heat and histopathologic processing on high infectivity hamster-adapted scrapie virus. J. Infect. Dis. 145:683–687.

Brown, P., P. Rogers-Johnson, D. C. Gajdusek, and C. J. Gibbs, Jr. 1984. Creutzfeldt-Jakob disease of long duration: clinical pathological characteristics, transmissibility and differential diagnosis. Ann. Neurol. 16:295–304.

Brown, P., D. C. Gajdusek, C. J. Gibbs, Jr., and D. M. Asher. 1985. Potential epidemic of Creutzfeldt-Jakob disease from human growth hormone therapy. N. Engl. J. Med. 313:728–731.

Brown, P., R. G. Rohwer, and D. C. Gajdusek. 1986a. Newer data on the inactivation of scrapie virus or Creutzfeldt-Jakob disease virus in brain tissue. J. Infect. Dis. 153:1145–1148.

Brown, P., M. Coker-Vann, K. Pomeray, M. Franko, D. M. Asher, C. J. Gibbs, Jr., and D. C. Gajdusek. 1986b. Diagnosis of Creutzfeldt-Jakob disease by Western blot identification of marker protein in human brain tissue. N. Engl. J. Med. 314:547–551.

Bruce, M. E. 1985. Agent replication dynamics in a long incubation period model of mouse scrapie. J. Gen. Virol. 66:2517–2522.

Bruce, M. E., and A. G. Dickinson. 1979. Biological stability of different classes of scrapie agent. p. 71–86. In S. B. Prusiner, and W. J. Hadlow (ed.), Slow transmissible diseases of the nervous system, Vol. 1. Academic Press, New York.

Bruce, M. E., and A. G. Dickinson. 1985. Genetic control of amyloid plaque production and incubation period in scrapie-infected mice. J. Neuropathol. Exp. Neurol. 44:285–294.

Bruce, M. E., and A. G. Dickinson. 1987. Biological evidence that scrapie agent has an independent genome. J. Gen. Virol. 68:79–89.

Bruce, M. E., and H. Fraser. 1975. Amyloid plaques in the brain of mice infected with scrapie: morphological variation and staining properties. Neuropathol. Appl. Neurobiol. 1:189–202.

Bruce, M. E., and H. Fraser. 1981. Effect of route of infection on the frequency and distribution of cerebral amyloid plaques in scrapie mice. Neuropathol. Appl. Neurobiol. 7:289–298.

Bruce, M. E., and H. Fraser. 1982. Focal and asymmetrical vacuolar lesions in the brains of mice infected with certain strains of scrapie. Acta Neuropathol. (Berl.) 58:133–140.

Bruce, M. E., A. G. Dickinson, and H. Fraser. 1976. Cerebral amyloidosis in scrapie in the mouse; effect of agent strain and mouse genotype. Neuropathol. Appl. Neurobiol. 2:471–478.

Burger, D., and G. R. Hartsough. 1965. Encephalopathy of

mink. II. Experimental and natural transmission. J. Infect. Dis. **115**:393–399.

Burger, L. J., A. J. Rowan, and E. S. Goldensohn. 1972. Creutzfeldt-Jakob disease: an electroencephalographic study. Arch. Neurol. **26**:428–433.

Buyukmihci, N., F. Goering-Harmon, and R. F. Marsh. 1983. Neural pathogenesis of experimental scrapie after intraocular inoculation of hamsters. Exp. Neurol. **81**:396–406.

Carlson, G. A., D. T. Kingsbury, P. A. Goodman, S. Coleman, S. T. Marshall, S. De Armond, D. Westaway, and S. B. Prusiner. 1986. Prion protein and scrapie incubation time genes are linked. Cell **46**:503–511.

Carp, R. I. 1981. Transmission of scrapie by the oral route: effect of gingival scarification. Lancet **1**:120–171.

Carp, R. I., S. M. Callahan, E. A. Sersen, and R. C. Moretz, 1984. Preclinical changes in weight of scrapie-infected mice as a function of scrapie agent-mouse strain combination. Intervirology **21**:61–69.

Carp, R. I., P. A. Merz, R. C. Moretz, R. A. Somerville, S. M. Callahan, and H. M. Wisniewski. 1985a. Biological properties of scrapie, an unconventional slow virus. p. 425–463. *In* K. Maramorosch, and J. J. McKelvey (ed.), Subviral pathogens of plants and animals: viroids and prions. Academic Press, New York.

Carp, R. I., P. A. Merz, R. J. Kascsak, G. S. Merz, and H. M. Wisniewski. 1985b. Nature of the scrapie agent: current status of facts and hypotheses. J. Gen. Virol. **67**:1357–1368.

Carp, R. I., R. C. Moretz, M. Natelli, and A. G. Dickinson. 1987. Genetic control of scrapie: incubation period and plaque formation in I mice. J. Gen. Virol. **68**:401–407.

Cathala, F., J. Chatelain, P. Brown, M. Dumas, and D. C. Gajdusek. 1980. Familial Creutzfeldt-Jakob disease: autosomal dominance in 14 members over 3 generations. J. Neurol. Sci. **47**:343–351.

Cauvet, J. 1854. Sur la tremblante. J. Vet. Midi **7**:441–448.

Chandler, R. L. 1961. Encephalopathy in mice produced by inoculation with scrapie brain material. Lancet **1**:1378–1379.

Chandler, R. L. 1963. Experimental scrapie in the mouse. Res. Vet. Sci. **4**:276–285.

Chandler, R. L., and J. Fisher. 1963. Experimental transmission of scrapie to rats. Lancet **2**:1165.

Chatelain, J., F. Cathala, P. Brown, S. Ra-Harrison, L. Court, and D. C. Gajdusek. 1981. Epidemiologic comparisons between Creutzfeldt-Jakob disease and scrapie in France during the 12-year period 1968–1979. J. Neurol. Sci. **51**:329–337.

Chatigny, M. A., and S. B. Prusiner. 1980. Biohazards of investigations on the transmissible spongiform encephalopathies. Rev. Infect. Dis. **2**:713–724.

Chelle, P. L. 1942. Un cas de tremblante chez la chevre. Bull. Acad. Vet. Fr. **15**:294–295.

Chesebro, B., R. Race, K. Wehrly, J. Nishio, M. Bloom, D. Lechner, S. Bergstrom, K. Robbins, L. Mayer, J. M. Keith, C. Garon, and A. Haase. 1985. Identification of scrapie prion protein specific mRNA in scrapie-infected and uninfected brain. Nature **315**:331–333.

Chiofala, N., A. Fuentes, and S. Galvez. 1980. Serial EEG findings in 27 cases of Creutzfeldt-Jakob disease. Arch. Neurol. **37**:143–145.

Cho, H. J. 1980a. Requirement of a protein component for scrapie infectivity. Intervirology **14**:213–216.

Cho, H. J. 1980b. Inactivation of the scrapie agent by pronase. Can. J. Comp. Med. **47**:313–320.

Cho, H. J. 1986. Antibody to scrapie-associated fibril protein identifies a cellular antigen. J. Gen. Virol. **67**:243–252.

Claridge, J. Extract from a general view of agriculture in the county of Dorset; with observations on the means of its improvement. Bath Papers 7:66–73; 1795.

Clark, M. C. 1979. Infection of cell cultures with scrapie agent. p. 225–233. *In* S. B. Prusiner and W. J. Hadlow, (ed.), Slow transmissibile diseases of the nervous system, Vol. 2. Academic Press, New York.

Clark, M. C., and D. A. Haig. 1970. Multiplication of scrapie agent in cell culture. Res. Vet. Sci. **11**:500.

Clark, M. C., and D. A. Haig. 1971a. An attempt to determine whether maternal transmission of scrapie occurs in mice. Br. Vet. J. **127**:32–34.

Clark, M. C. and D. A. Haig. 1971b. Multiplication of scrapie agent in mouse spleen. Res. Vet. Sci. **12**:195–197.

Clark, M. C., and G. C. Millson. 1976a. The membrane localization of scrapie infectivity. J. Gen. Virol. **31**:441–445.

Clark, M. C., and G. C. Millson. 1976b. Infection of a cell line of mouse L fibroblasts with scrapie agent. Nature **262**:144–145.

Cole, S., and R. H. Kimberlin. 1985. Pathogenesis of mouse scrapie: dynamics of vacuolation in brain and spinal cord after intraperitoneal infection. Neuropathol. Appl. Neurobiol. **11**:213–227.

Collis, S. C., and R. H. Kimberlin. 1985. Long-term persistence of scrapie infection in mouse spleens in the absence of clinical disease. FEMS Microbiol. Lett. **29**:111–114.

Court, L., and F. Cathala. (ed.). 1983. Virus non conventionals et affections du systeme neveure central, Vol. 1, p. 528. Masson, Paris.

Creutzfeldt, H. G. 1920. Uber eine eigenartige herdformige Erkrankung des Zentralnervensystems. Z. Gesamte Neurol. **57**:1–18.

Creutzfeldt, H. G. 1921. Uber eine eigenartige herdformige Erkrankung des Zentralnervensystems. p. 1–48. *In* F. Nissl and A. Alzheimer (ed.), Histologische und Histopathologische Arbeiten uber die Grosshirnrinde, Vol. 6. Gustav Fischer, Jena, G. D. R.

Cuille, J., and P. L. Chelle. 1936. La maladie dite tremblante du mouton est-elle inoculable. C. R. Acad. Sci. Paris **205**:1552–1554.

Czub, M., N. R. Braig, H. Blade, and H. Diringer, 1986a. The major protein of SAF is absent from spleen and thus not an essential part of the scrapie agent. Arch. Virol. **91**:383–386.

Czub, M., H. R. Braig, and H. Diringer. 1986b. Pathogenesis of scrapie: study of the temporal development of clinical symptoms, of infectivity titers and scrapie associated fibrils in brains of hamsters infected intraperitoneally. J. Gen. Virol. **67**:2005–2009.

De Armond, S. J., M. P. McKinley, R. A. Barry, M. B. Braunfeld, J. R. McColloch, and S. B. Prusiner. 1985. Identification of prion amyloid filaments in scrapie-infected brain. Cell **41**:221–235.

Dickinson, A. G. 1975. Host-pathogen interactions in scrapie. Genetics **79**:387–395.

Dickinson, A. G. 1976. Scrapie in sheep and goats. p. 209–241. *In* R. H. Kimberlin (ed.), Slow virus diseases of animals and man. North-Holland, Amsterdam.

Dickinson, A. G., and H. Fraser. 1972. Scrapie: effect of Dh gene on the incubation period of extraneurally injected agent. Heredity **29**:91–93.

Dickinson, A. G., and H. Fraser. 1977. Scrapie pathogenesis in inbred mice: an assessment of host control and response involving many strains of agent. p. 3–14. *In* V. ter Meulen and M. Katz (ed.), Slow virus infections of the central nervous system. Springer-Verlag, New York.

Dickinson, A. G., and H. Fraser. 1979. An assessment of the genetics of scrapie in sheep and mice. p. 367–385. *In* S. B. Prusiner, and W. J. Hadlow, (ed.), Slow transmissible diseases of the nervous system, Vol. 1. Academic Press, New York.

Dickinson, A. G., and V. M. H. Meikle. 1969. A comparison of some biological characteristics of the mouse-passaged scrapie agents, 22A and ME7. Genet. Res. **13:**213–225.

Dickinson, A. G., and V. M. H. Meikle, 1971. Host-genotype and agent effects in scrapie incubation: change in allelic interaction with different strains of agent. Mol. Gen. Genet. **112:**73–79.

Dickinson, A. G., and G. Outram. 1979. The scrapie replication-site hypothesis and its implication for pathogenesis. p. 13–31. *In* S. B. Prusiner, and W. J. Hadlow (ed.), Slow transmissible diseases in the nervous system, Vol. 2. Academic Press, New York.

Dickinson, A. G., and G. W. Outram. 1983. Operational limitations in the characterization of the infective units of scrapie. p. 3–16. *In* L. A. Court, and F. Cathala, (ed.), Virus non conventionnels et affections du systeme nerveux central. Masson, Paris.

Dickinson, A. G., and D. M. Taylor. 1978. Resistance of scrapie agent to decontamination. N. Engl. J. Med. **299:**1413–1414.

Dickinson A. G., V. M. H. Meikle, and H. Fraser. 1969. Genetical control of the concentration of ME7 scrapie agent in the brain of mice. J. Comp. Pathol. **79:**15–22.

Dickinson, A. G., J. T. Stamp, and C. C. Renwick. 1974. Maternal and lateral transmission of scrapie in sheep. J. Comp. Pathol. **84:**19–25.

Dickinson, A. G., H. Fraser, L. McConnell, G. W. Outram, D. I. Sales, and D. M. Taylor. 1975. Extraneural competition between different scrapie agent leading to loss of infectivity. Nature **253:**556.

Dickinson, A. G., H. Fraser, and G. W. Outram. 1976. Scrapie incubation time can exceed natural lifespan. Nature **256:**732–733.

Dickinson, A. G., M. E. Bruce, H. Fraser, and R. H. Kimberlin. 1984. Scrapie strain differences: the implication of stability and mutation. p. 105–118. *In* J. Tateishi (ed.), Proceedings of Workshop on Slow Transmissible Diseases. Japanese Ministry of Health and Welfare, Tokyo.

Diener, T. O., M. P. McKinley, and S. B. Prusiner. 1982. Viroids and prions. Proc. Natl. Acad. Sci. USA **79:**5220–5224.

Diringer, H. 1984. Sustained viremia in experimental hamster scrapie. Arch. Virol. **82:**105–109.

Diringer, H., H. Gelderblom, H. Hilmert, M. Ozel, C. Edelbluth, and R. H. Kimberlin. 1983a. Scrapie infectivity, fibrils and low molecular weight protein. Nature **306:**476–478.

Diringer, H., H. Hilmert, D. Simon, E. Werner, and B. Ehlers. 1983b. Toward purification of the scrapie agent. Eur. J. Biochem. **134:**555–560.

Diringer, H, H. C. Rahn, and L. Bode. 1984. Antibodies to protein of scrapie associated fibrils. Lancet **2:**345.

Duffy, P., J. Wolf, G. Collins, A. Devoe, B. Steeten, and D. Cowen. 1974. Possible person to person transmission of Creutzfeldt-Jakob disease. N. Engl. J. Med. **290:**692–693.

Eklund, C. M., W. J. Hadlow, and R. C. Kennedy. 1963. Some properties of the scrapie agent and its behavior in mice. Proc. Soc. Exp. Biol. Med. **112:**974–979.

Ferber, R. A., S. L. Wiesenfeld, R. P. Roos, A. R. Borowich, C. J. Gibbs, Jr., and D. C. Gajdusek. 1974. Familial Creutzfeldt-Jakob disease: transmission of the familial disease to primates. p. 308–380. *In* A. Subirana,

J. M. Espadaler, and E. H. Burrows (ed.), International Congress Series 319. Excerpta Medica, Amsterdam.

Flament-Durand, J., and A. M. Couck. 1979. Spongiform alterations in brain biopsies of presenile dementia. Acta Neuropathol. (Berl.) **46:**159–162.

Fraser, H. 1976. The pathology of natural and experimental scrapie. p. 267–305. *In* R. H. Kimberlin (ed.), Slow virus diseases of animals and man. North-Holland, Amsterdam.

Fraser, H. 1979. Neuropathology of scrapie, the precision of the lesions and their diversity. p. 387–406. *In* S. B. Prusiner. and W. J. Hadlow (ed.), Slow transmissible diseases of the nervous system, Vol. 1. Academic Press, New York.

Fraser, H. 1981. Neuronal spread of scrapie agent and targeting of lesions within the retino-tectal pathway. Nature **294:**149–150.

Fraser, H. 1983. A survey of primary transmissions of scrapie and rida to mice. p. 34–46. *In* L. A. Court, and F. Cathala (ed.), Virus non conventionnels et affections du systeme nerveux central. Masson, Paris.

Fraser, H., and M. E. Bruce. 1973. Argyrophlic plaques in mice inoculated with scrapie from particular sources. Lancet **1:**617–618.

Fraser, H., and A. G. Dickinson. 1970. Pathogenesis of scrapie in the mouse: the role of the spleen. Nature **226:**462–463.

Fraser, H., and A. G. Dickinson. 1973. Scrapie in mice: agent-strain differences in the distribution and intensity of grey matter vacuolation. J. Comp. Pathol. **83:**29–40.

Fraser, H., and A. G. Dickinson. 1978. Studies of the lymphoreticular system in the pathogenesis of scrapie: role of spleen and thymus. J. Comp. Pathol. **88:**563–573.

Fraser, H., and A. G. Dickinson. 1985. Targeting of scrapie lesions and spread of agent via the retino-tectal projection. Brain Res. **346:**32–41.

Gajdusek, D. C. 1977. Unconventional viruses and the origin and disappearance of kuru. Science **197:**943–960.

Gajdusek, D. C. 1985. Unconventional viruses causing subacute spongiform encephalopathies. p. 1519–1557. *In* B. N. Fields (ed.), Virology. Raven Press, New York.

Gajdusek, D. C., and V. Zigas. 1957. Degenerative disease of the central nervous system in New Guinea. The endemic occurrence of "kuru" in the native population. N. Engl. J. Med. **257:**974–978.

Gajdusek, D. C., and V. Zigas. 1959. Kuru: clinical, pathological and epidemiological study of an acute progressive degenerative disease of the central nervous system among natives of the eastern highlands of New Guinea. Am. J. Med. **26:**442–469.

Gajdusek, D. C., C. J. Gibbs, Jr., and M. Alpers. 1966. Experimental transmission of a kuru-like syndrome to chimpanzees. Nature **209:**794–796.

Gajdusek, D. C., C. J. Gibbs, Jr., G. Collins, and R. Traub. 1976. Survival of Creutzfeldt-Jakob disease in Formol fixed brain tissue. N. Engl. J. Med. **294:**553.

German, T. L., and R. F. Marsh. 1983. The scrapie agent: a unique self-replicating pathogen. p. 111–121. *In* F. E. Hahn (ed.), Progress in molecular and subcellular biology, Vol. 8. Springer-Verlag, Berlin.

Gerstmann, J., E. Straussler, and J. Scheinker. 1936. Uber eine eigenartige hereditarfamiliare Eikrankungen des Zentralnervensystems. Z Neurol. **159:**736–762.

Gibbons, R. A., and G. D. Hunter. 1967. Nature of the scrapie agent. Nature **215:**1041–1043.

Gibbs, C. J. Jr., D. C. Gajdusek, D. M. Asher, M. P. Alpers, E. Beck, P. M. Daniel, and W. B. Matthews. 1968. Creutzfeldt-Jakob disease (spongiform encephalopathy): transmission to the chimpanzee. Science **161:**388–389.

Gibbs, C. J. Jr., D. C. Gajdusek, and R. Latarjet. 1978. Unusual resistance to UV and ionizing radiation of the viruses of kuru, Creutzfeldt-Jakob disease and scrapie (unconventional viruses). Proc. Natl. Acad. Sci. USA 75:6268–6270.

Gibbs, C. J. Jr., D. C. Gajdusek, and H. Amyx. 1979. Strain variation in the viruses of Creutzfeldt-Jakob disease and kuru. p. 87–110. In S. B. Prusiner, and W. J. Hadlow (ed.), Slow transmissible diseases of the nervous system, Vol. 2. Academic Press, New York.

Gibbs, C. J. Jr., H. L. Amyx, A. Bacote, C. L. Masters, and D. C. Gajdusek. 1980. Oral transmission of kuru, Creutzfeldt-Jakob disease and scrapie to non-human primates. J. Infect. Dis. 142:205–208.

Gibbs, C. J. Jr., A. Joy, R. Heffner, M. Franko, M. Miyaszki, D. M. Asher, J. E. Parisi, P. B. Brown, and D. C. Gajdusek. 1985. Clinical and pathological features and laboratory confirmation of Creutzfeldt-Jakob disease in a recipient of pituitary-derived human growth hormone. N. Engl. J. Med. 313:734–738.

Gibson, P. E., T. M. Bell, and E. J. Field. 1972. Failure of the scrapie agent to replicate in L51784 mouse leukaemic cells. Res. Vet. Sci. 13:95–96.

Gibson, P. H., R. A. Somerville, H. Fraser, J. D. Foster, and R. H. Kimberlin. 1987. Scrapie associated fibrils in the diagnosis of scrapie in sheep. Vet. Res. Feb. 7, 125–127.

Girard, J. 1829, 1830. Notice sur quelques maladies peu connues des betes a laine. Rec. Med. Vet. 6:674–683; 7:26–39, 65–76.

Glass, R. M. 1963. Cannabilism in the Kuru region. Department of Public Health, Papua, New Guinea.

Gorde, J. M., J. Tamalet, M. Toga, and J. Bert. 1982. Changes in the nigrostriatal system following microinjection of an unconventional agent. Brain Res. 240:87–93.

Gordon, W. S. 1946. Louping-ill, tick borne fever and scrapie. Vet. Res. 58:516–525.

Griffith, J. S. 1967. Self-replication and scrapie. Nature 215:1043–1044.

Hadlow, W. J. 1959. Scrapie and kuru. Lancet 2:289–290.

Hadlow, W. J., C. M. Eklund, R. C. Kennedy, T. A. Jackson, H. W. Whitford, and C. C. Boyle. 1974. Course of experimental scrapie virus in the goat. J. Infect. Dis. 129:559–567.

Hadlow, W. J., R. E. Race, C. Kennedy, C., and C. M. Eklund. 1979. Natural infection of sheep with scrapie virus. p. 3–12. In S. B. Prusiner and W. J. Hadlow (ed.), Slow transmissible diseases of the nervous system, Vol. 2. Academic Press, New York.

Hadlow, W. J., R. C. Kennedy, and R. E. Race. 1982. Natural infection of Suffolk sheep with scrapie virus. J. Infect. Dis. 146:657–664.

Haig, D. A., M. C. Clark, E. Blum, and T. Alper. 1969. Further studies on the inactivation of the scrapie agent by ultraviolet light. J. Gen. Virol. 5:455–457.

Haltia, M., J. Kovanen, H. van Crevel, G. Bots, G. Th. A. M. Bots, and I. Stefanko. 1979. Familial Creutzfeldt-Jakob disease. J. Neurol. Sci. 42:381–389.

Harrington, M. G., C. R. Merril, D. M. Asher, and D. C. Gajdusek. 1986. Abnormal proteins in the cerebrospinal fluid of patients with Creutzfeldt-Jakob disease. N. Engl. J. Med. 315:279–283.

Harris, T. 1985. Persistently puzzling prions. Nature 315:278.

Hartsough, G. R., and D. Burger. 1965. Encephalopathy in mink. I. Epizoological and clinical observations. J. Infect. Dis. 115:387–392.

Hikita, K., J. Tateishi, and H. Nagara. 1985. Morphogenesis of amyloid plaques in mice with Creutzfeldt-Jakob disease. Acta Neuropathol. (Berl.) 68:138–144.

Hilmert, H., and H. Diringer. 1984. A rapid and efficient method to enich SAF-protein from scrapie brains of hamsters. Biosci. Rep. 4:165–170.

Hintz, R. M., M. MacGillivay, A. Joy, and R. Tintner. 1985. Fatal degenerative neurological disease in patients who received pituitary derived human growth hormone. Morbid. Mortal. Weekly Rep. 34:359–366.

Hope, J., and R. H. Kimberlin. 1987. The molecular biology of scrapie: the last two years. TINS 10:149–151.

Hope, J., L. J. D. Morton, C. F. Faraquhar, G. Multhaup, K. Beyreuther, and R. H. Kimberlin. 1986. The major polypeptide of scrapie-associated fibrils (SAF) has the same size, charge distribution and N-terminal protein sequence as predicted for the normal brain protein (PrP). EMBO J. 5:2591–2597.

Hornabrook, R. W. 1979. Kuru and clinical neurology. p. 37–66. In S. B. Prusiner and W. J. Hadlow (ed.), Slow transmissible diseases of the nervous system, Vol. 1. Academic Press, New York.

Hourrigan, J., A. Klingsporn, W. W. Clark, and M. deCamp. 1979. Epidemiology of scrapie in the United States. p. 331–356. In S. B. Prusiner, and W. J. Hadlow (ed.), Slow transmissible diseases of the nervous system, Vol. 1. Academic Press, New York.

Hunter, G. D., S. C. Collis, G. C. Millson, and R. H. Kimberlin. 1976. Search for scrapie-specific RNA and attempts to detect an infectious DNA or RNA. J. Gen. Virol. 32:157–162.

Jakob, A. 1921. Uber eigenartige Erkankungen des Zentralnervensystems mit bemerkensweitem anatomischen Befunde (spastische Pseudosclerose-Encephalomyelopathie mit disseminierten Degenerationsheiden). Dtsch. Z. Nervenheilkd. 70:132–146.

Johnson, R. T. 1982. The novel nature of scrapie. TINS 5:413–415.

Journal of the House of Commons. 1755. 27:87.

Kascsak, R. J., R. Rubenstein, P. A. Merz, R. I. Carp, H. M. Wisniewski, and H. Diringer. 1985. Biochemical differences among scrapie associated fibrils support the biological diversity of scrapie agents. J. Gen. Virol. 66:1715–1722.

Kascsak, R. J., R. Rubenstein, P. A. Merz, R. I. Carp, N. K. Robakis, and H. M. Wisniewski. 1986. Immunological comparison of scrapie associated fibrils isolated from animals infected with four different scrapie strains. J. Virol. 59:676–683.

Kascsak, R. J., R. Rubenstein, P. A. Merz, M. Tomma-DeMasi, R. Fersko, R. I. Carp, H. M. Wisniewski, and H. Diringer. 1987a. Mouse polyclonal and monoclonal antibody to SAF (PrP) protein. J. Virol. (Dec. 1987).

Kascsak, R. J., R. Rubenstein, P. A. Merz, R. I. Carp, and P. Brown. 1987b. Monoclonal antibody for detection of human unconventional slow virus diseases. (Submitted.)

Kidson, C., M. C. Moreau, D. M. Asher, P .W. Brown, H. G. Coon, D. C. Gajdusek, and C. J. Gibbs, Jr. 1978. Cell fusion induced by scrapie and Creutzfeldt-Jakob virus-infected preparations. Proc. Natl. Acad. Sci. USA 75:2969–2971.

Kim, Y. S., R. I. Carp, S. M. Callahan, and H. M. Wisniewski. 1987a. Clinical course of three scrapie strains in mice injected stereotaxically in different brain regions. J. Gen. Virol. 68:695–702.

Kim, Y. S., R. I. Carp, S. M. Callahan, and H. M. Wisniewski. 1987b. Scrapie-induced obesity in mice. J. Infect. Dis. 156:402–405.

Kimberlin, R. H. (ed). 1976. Slow virus diseases of animals and man. Elsevier, New York.

Kimberlin, R. H. 1979. Aetiology and genetic control of natural scrapie. Nature 278:303–304.

Kimberlin, R. H. 1981. Scrapie. Br. Vet. J. 137:105–112.

Kimberlin, R. H. 1982a. Reflections on the nature of scrapie agent. TIBS 7:392–394.

Kimberlin, R. H. 1982b. Scrapie agent: prions or virinos? Nature 297:107–108.

Kimberlin, R. H. 1984. Scrapie: the diseases and the infectious agent. TIBS 7:312–316.

Kimberlin, R. H. 1986. Scrapie: how much do we really understand? Neuropathol. Appl. Neurobiol. 12:131–147.

Kimberlin, R. H., S. Cole, and C. A. Walker. 1987. Pathogenesis of scrapie is faster when infection is intraspinal instead of intracerebral. Microbiol. Pathogen. 2:405–415.

Kimberlin, R. H., S. M. Hall, and C. A. Walker. 1983. Pathogenesis of mouse scrapie: evidence for direct neural spread of infection to the CNS after injection of sciatic nerve. J. Neurol. Sci. 61:315–325.

Kimberlin, R. H., and C. A. Walker. 1977. Characteristics of a short incubation model of scrapie in the golden hamster. J. Gen. Virol. 34:295–304.

Kimberlin, R. H., and C. A. Walker. 1978a. Evidence that the transmission of one source of scrapie agent to hamsters involves separation of agent strains from a mixture. J. Gen. Virol. 39:487–496.

Kimberlin, R. H., and C. A. Walker. 1978b. Pathogenesis of mouse scrapie: effect of route of inoculation on infectivity titers and dose-response curves. J. Comp. Pathol. 88:39–47.

Kimberlin, R. H., and C. A. Walker. 1979. Pathogenesis of mouse scrapie: dynamics of agent replication in spleen, spinal cord and brain. J. Comp. Pathol. 89:551–562.

Kimberlin, R. H., and C. A. Walker. 1980. Pathogenesis of mouse scrapie: evidence for neural spread of infection to the CNS. J. Gen. Virol. 51:183–187.

Kimberlin, R. H., and C. A. Walker. 1982. Pathogenesis of mouse scrapie: patterns of agent replication in different parts of the CNS following intraperitoneal infection. J. R. Soc. Med. 75:618–624.

Kimberlin, R. H., and C. A. Walker. 1983. Invasion of the CNS by scrapie agent and its spread to different parts of the brain. p. 17–33. In L. A. Court and F. Cathala (ed.), Virus non conventionnels et affections du systeme nerveux central. Masson, Paris.

Kimberlin, R. H., and C. A. Walker. 1986. Pathogenesis of scrapie (strain 263K) in hamsters infected intracerebrally, intraperitoneally or intraocularly. J. Gen. Virol. 67: 255–263.

Kingsbury, D. T., H. L. Amyx, and C. J. Gibbs, Jr. 1983. Biophysical properties of the Creutzfeldt-Jakob disease agent. p. 125–137. In L. Court and F. Cathala (ed.), Virus non-conventionnels et affections du systeme nerveux central. Masson, Paris.

Kitamoto, T., J. Tateishi, T. Tashima, I. Takeshita, R. A. Barry, S. J. DeArmond, and S. B. Prusiner. 1986. Amyloid plaques in Creutzfeldt-Jakob disease stain with prion protein antibodies. Ann. Neuol. 20:204–208.

Klatzo, I., D. C. Gajdusek, and V. Zigas. 1959. Pathology of kuru. Lab. Invest. 8:799–847.

Klitzman, R. L., M. P. Alpers, and D. C. Gajdusek. 1985. The natural incubation period of kuru and the episodes of transmission in three clusters of patients. Neuroepidemiology 3:3–20.

Koch, T. K., B. O. Berg, S. J. DeArmond, and R. F. Gravina. 1985. Creutzfeldt-Jakob disease in a young adult with idiopathic hypopituitarism: possible relation to the administration of cadaveric human growth hormone. N. Engl. J. Med. 313:731–733.

Kuguhara, S., I. Kanazawa, H. Sasaki, T. Nakanishi, and K. Shimamura. 1983. Gerstmann-Straussler-Scheinker's disease. Ann. Neurol. 14:216–225.

Kuroda, Y., C. J. Gibbs, Jr., H. L. Amyx, and D. C. Gajdusek. 1983. Creutzfeldt-Jakob disease in mice: persistent viremia and preferential replication of virus in low-density lymphocytes. Infect. Immun. 41:159–161.

Laemmli, U. K. 1970. Cleavage of structural proteins during the assembly of the head of bacteriophage T4. Nature 277:680–685.

Lampert, P. W., D. C. Gajdusek, and C. J. Gibbs, Jr. 1971. Experimental spongiform encephalopathy (Creutzfeldt-Jakob disease) in chimpanzees. J. Neuropathol. Exp. Neurol. 30:20–32.

Lampert, P. W., D. C. Gajdusek, and C. J. Gibbs, Jr. 1972. Subacute spongiform virus encephalopathies: scrapie, kuru and Creutzfeldt-Jakob disease. Am. J. Pathol. 68:626–646.

Latarjet, R. 1979. Inactivation of the agents of scrapie, Creutzfeldt-Jakob disease and kuru by radiations. p. 387–408. In S. B. Prusiner and W. J. Hadlow (ed.), Slow transmissible diseases of the nervous system, Vol. 2. Academic Press, New York.

Latarjet, R., B. Muel, D. A. Haig, M. C. Clark, and T. Alper. 1970. Inactivation of the scrapie agent by near monochromatic ultraviolet light. Nature 227:1341–1343.

Lax, A.J., G. C. Millson, and E. J. Manning. 1983. Involvement of protein in scrapie agent infectivity. Res. Vet. Sci. 34:155–158.

Liao, Y. C., R. V. Lebo, G. A. Clawson, and E. A. Smuckler. 1986. Human prion protein cDNA: molecular cloning, chromosomal mapping and biological implications. Science 233:364–367.

Macchi, G., H. L. Abbamondi, G. Di Trapani, and A. Ibriccoli. 1984. On the white matter lesions of the Creutzfeldt-Jakob disease. J. Neurol. Sci. 63:197–206.

Manuelidis, E. E. 1975. Transmission of Creutzfeldt-Jakob disease from man to guinea pig. Science 190:571–572.

Manuelidis, E. E. 1985. Presidential address: Creutzfeldt-Jakob disease. J. Neuropathol. Exp. Neurol. 44:1–17.

Manuelidis, E. E., and L. Manuelidis. 1979. Experiments on maternal transmission of Creutzfeldt-Jakob disease in guinea pigs. Proc. Soc. Exp. Biol. Med. 160:233–236; 1979.

Manuelidis, E. E., J. Kim, J. N. Angelo, and L. Manuelidis. 1976. Serial propagation of Creutzfeldt-Jakob disease in guinea pigs. Proc. Natl. Acad. Sci. USA 73:223–227.

Manuelidis, E. E., J. N. Angelo, E. J. Gorgacz, and L. Manuelidis. 1977a. Transmission of Creutzfeldt-Jakob disease to Syrian hamster. Lancet 1:479.

Manuelidis, E. E., J. N. Angelo, E. J. Gorgacz, J. H. Kim, and L. Manuelidis. 1977b. Experimental Creutzfeldt-Jakob disease transmitted via the eye with infected cornea. N. Engl. J. Med. 296:1334–1336.

Manuelidis, E. E., E. J. Gorgacz, and L. Manuelidis. 1978a. Viremia in experimental Creutzfeldt-Jakob disease in guinea pigs. Proc. Soc. Exp. Biol. Med. 160:233–236.

Manuelidis, E. E., E. J. Gorgacz, and L. Manuelidis. 1978b. Transmission of Creutzfeldt-Jakob disease with scrapie-like syndromes to mice. Nature 271:778–779.

Manuelidis, E. E., J. H. Kim, J. R. Mericangas, and L. Manuelidis. 1985. Transmission to animals of Creutzfeldt-Jakob disease from human blood. Lancet 2:896.

Manuelidis, L., and E. E. Manuelidis. 1980. Search for specific DNA's in Creutzfeldt-Jakob infectious brain fractions using "nick translation." Virology 109:435–443.

Manuelidis, L., and E. E. Manuelidis. 1983. Fractionation and infectivity studies in Creutzfeldt-Jakob disease. p.

399–412. *In* R. Katzman (ed.), Biological aspects of Alzheimer's disease, Vol. 15. Cold Spring Harbor Laboratory, Cold Spring Harbor, NY.

Manuelidis, L., and E. E. Manuelidis. 1986. Recent developments in scrapie and Creutzfeldt-Jakob disease. Prog. Med. Virol. **33**:78–98.

Manuelidis, L., S. Valley, S., and E. E. Manuelidis. 1985. Specific proteins associated with Creutzfeldt-Jakob disease and scrapie share antigenic and carbohydrate determinants. Proc. Natl. Acad. Sci. USA **82**:4263–4267.

Manuelidis, L., T. Sklaviadis, and E. E. Manuelidis. 1987. Evidence suggesting that PrP is not the infectious agent in Creutzfeldt-Jakob disease. EMBO J. **6**:341–347.

Markovits, P., C. Dauthville, D. Dormont, L. Dianoux, and R. Latarjet. 1983. in vitro propagation of the scrapie agent. 1. Transformation of mouse glia and neuroblastoma cells after infection with the mouse-adapted strain C-506. Acta Neuropthol. (Berl.) **60**:75–80.

Marsh, R. F., and R. P. Hanson. 1979. On the origin of transmissible mink encephalopathy. p. 451–460. *In* S. B. Prusiner, and W. J. Hadlow (ed.), Slow transmissible diseases of the nervous system, Vol. 1. Academic Press, New York.

Marsh, R. F., and R. H. Kimberlin. 1975. Comparison of scrapie and transmissible mink encephalopathy in hamsters. II. Clinical signs, pathology, and pathogenesis. J. Infect. Dis. **131**:104–110.

Marsh, R. F., J. S. Semancik, K. C. Medappa, R. P. Hanson, and R. R. Rueckert. 1974. Scrapie and transmissible mink encephalopathy: search for infectious nucleic acid. J. Virol. **13**:993–996.

Masters, C. L., and D. C. Gajdusek. 1982. The spectrum of Creutzfeldt-Jakob disease and the virus-induced subacute spongiform encephalopathies. p. 139–163. *In* W. T. Smith and J. D. Cavanagh (ed.), Recent advances in neuropathology, Vol. 9. Churchill Livingstone, Edinburgh.

Masters, C. L., and E. P. Richardson, Jr. 1978. Subacute spongiform encephalopathy (Creutzfeldt-Jakob disease): the nature and progression of spongiform change. Brain **101**:333–344.

Masters, C. L., J. O. Harris, D. C. Gajdusek, C. J. Gibbs, Jr., C. Bernoulli, and D. M. Asher. 1979. Creutzfeldt-Jakob disease: patterns of worldwide occurrence and the significance of familial and sporadic clustering. Ann. Neurol. **5**:177–188.

Masters, C. L., D. C. Gajdusek, and C. J. Gibbs, Jr. 1981a. The familial occurrence of Creutzfeldt-Jakob disease and Alzheimer's disease. Brain **104**:535–558.

Masters, C. L., D. C. Gajdusek, and C. J. Gibbs, Jr. 1981b. Creutzfeldt-Jakob disease virus isolations from the Gerstmann-Straussler syndrome. Brain **104**:559–588.

Masters, C. L., R. G. Rohwer, M. C. Franko, P. Brown, and D. C. Gajdusek. 1984. The sequential development of spongiform change and gliosis of scrapie in the golden Syrian hamster. J. Neuropathol. Exp. Neurol. **43**:242–252.

Masters, C. L., G. Simms, N. A. Weinman, G. Multhaup, B. L. McDonald, and K. Beyreuther. 1985a. Amyloid plaque core protein in Alzheimer disease and Down Syndrome. Proc. Natl. Acad. Sci. USA **82**:4242–4249.

Masters, C. L., G. Multhaup, G. Simms, J. Pottgiesser, R. N. Martins, and K. Beyreuther. 1985b. Neuronal origin of a cerebral amyloid: neurofibrillary tangles of Alzheimer's disease contains the same protein as the amyloid plaque cores and blood vessels. EMBO J. **4**:2759–2763.

Matthews, W. B. 1975. Epidemiology of Creutzfeldt-Jakob disease in England and Wales. J. Neurol. Neurosurg. Psychiatry **38**:210–213.

McKinley, M. P., and S. B. Prusiner. 1986. Biology and structure of scrapie prions. Int. Rev. Neurobiol. **28**:1–57.

McKinley, M. P., F. R. Masiarz, and S. B. Prusiner. 1981. Reversible chemical modification of scrapie agent. Science **214**:1259–1260.

McKinley, M. P., D. C. Bolton, and S. B. Prusiner. 1983a. A protease resistant protein is a structural component of the scrapie prion. Cell **35**:57–62.

McKinley, M. P., F. R. Masiarz, S. T. Isaacs, J. E. Hearst, and S. B. Prusiner. 1983b. Resistance of the scrapie agent to inactivation by psoralens. Photochem. Photobiol. **37**:539–545.

Merz, P. A., R. A. Somerville, H. M. Wisniewski, and K. Iqbal. 1981. Abnormal fibrils from scrapie-infected brains. Acta Neurolpathol. (Berl.) **54**:63–74.

Merz, P. A., R. A. Somerville, and H. M. Wisniewski. 1983a. Abnormal fibrils in scrapie and senile dementia of the Alzheimer type. p. 259–281. *In* L. A. Court and F. Cathala (ed.), Virus non-conventionnels et affections du systeme nerveux central. Masson, Paris.

Merz, P. A., H. M. Wisniewski, R. A. Somerville, S. A. Bobin, K. Iqbal and C. L. Masters. 1983b. Ultrastructure of amyloid fibrils from neuritic and amyloid plaques. Acta Neuropathol. (Berl.) **60**:113–124.

Merz, P. A., R. A. Somerville, H. M. Wisniewski, L. Manuelidis, and E. E. Manuelidis. 1983c. Scrapie associated fibrils in Creutzfeldt-Jakob disease. Nature **306**:474–476.

Merz, P. A., R. G. Rohwer, R. J. Kascsak, H. M. Wisniewski, R. A. Somerville, C. J. Gibbs, Jr., and D. C. Gajdusek. 1984a. Infection-specific particle from the unconventional slow virus diseases. Science **225**:437–440.

Merz, P. A., R. J. Kascsak, R. Rubenstein, R. I. Carp, and H. M. Wisniewski. 1984b. Variations in SAF from different scrapie agents. p. 137–145. *In* J. Tateishi (ed.), Proceedings of Workshop on Slow Transmissible Diseases. Japanese Ministry of Health and Welfare, Tokyo.

Merz, P. A., R. J. Kascsak, N. K. Robakis, R. Rubenstein, and H. M. Wisniewski. 1986a. Recent studies on SAF. p. 91–94. *In* A. Bignami, L. Bolis, and D. C. Gajdusek (ed.), Discussions in neurosciences, Vol. 3. Fondation pour l'Etude du Systeme Nerveux Central et Peripherique, Amsterdam.

Merz, P. A., H. M. Wisniewski, R. Rubenstein, and R. J. Kascsak. 1986b. Immunological studies on paired helical filaments of Alzheimer's disease. p. 58–68. *In* A. Bignami, L. Bolis, and D. C. Gajdusek (ed.), Discussions in Neurosciences, Vol. 3. Fondation pour l'Etude du Systeme Nerveux Central et Peripherique, Amsterdam.

Merz, P. A., R. J. Kascsak, R. Rubenstein, R. I. Carp, and H. M. Wisniewski. 1987a. Antisera to scrapie-associated fibril protein and prion protein decorate scrapie-associated fibrils. J. Virol. **61**:42–49.

Merz, P. A., R. J. Kascsak, N. Goller, R. Rubenstein, R. I. Carp, A. Vorbrodt, and H. M. Wisniewski. 1987b. Immunocytochemistry of polyclonal and monoclonal antibodies to SAF polypeptides. (Submitted.)

Millson, G. C., and E. J. Manning. 1979. The effect of selected detergents on scrapie infectivity. p. 409–424. *In* S. B. Prusiner and W. J. Hadlow (ed.), Slow transmissible diseases of the nervous system, Vol. 2. Academic Press, New York.

Millson, G. C., G. D. Hunter, and R. H. Kimberlin. 1971. An experimental examination of the scrapie agent in cell membrane mixtures. II. The association of scrapie activity with membrane fractions. J. Comp. Pathol. **81**:255–265.

Millson, G. C., G. D. Hunter, and R. H. Kimberlin. 1976.

The physico-chemical nature of scrapie agent, p. 243–266. *In* R. H. Kimberlin (ed.), Slow virus diseases of animals and man. North-Holland, Amsterdam.

Mizutani, T., and H. Shiraki. (ed.). 1985. Clinicopathological aspects of Creutzfeldt-Jakob disease. Elsevier, Amsterdam.

Mizutani, T., A. Okumura, M. Oda, and H. Shiraki. 1981. Panencephalopathic type of Creutzfeldt-Jakob disease: primary involvement of the cerebral white matter. J. Neurol. Neurosurg. Psychiatry **44**:103–115.

Moreau-Dubois, M. C., P. Brown, and D. C. Gajdusek. 1979. Comparisons of cell-fusing activity of brain suspensions from patients with Creutzfeldt-Jakob disease and other degenerative neurological diseases. Proc. Natl. Acad. Sci. USA **76**:5365–5367.

Morris, J. A., and D. C. Gajdusek. 1963. Encephalopathy in mice following inoculation of scrapie sheep brain. Nature **197**:1084–1086.

Mould, D. L., A. M. Dawson, and W. Smith. 1967. Determination of the dosage-response curve of mice inoculated with scrapie. J. Comp. Pathol. **77**:387–391.

Multhaup, G., H. Diringer, H. Helmert, H. Prinz, J. Heukeshoven, and K. Beyreuther. 1985. The protein component of scrapie-associated fibrils is a glycosylated low molecular weight protein. EMBO J. **4**:1495–1501.

Oesch, B., D. Westaway, M. Walchli, M. P. McKinley, S. B. Kent, R. Aebersold, R. A. Barry, A. Tempst, D. B. Teplow, L. E. Hand, S. B. Prusiner, and C. Weissmann. 1985. A cellular gene encodes scrapie PrP 27–30 protein. Cell **40**:735–746.

Oleszak, E. L., L. Manuelidis, and E. E. Manuelidis. 1986. In vitro transformation elicited by Creutzfeldt-Jakob-infected brain material. J. Neuropathol. Exp. Neurol. **45**:489–502.

Oppenheimer, D. R. 1975. Pathology of transmissible and degenerative diseases of the nervous system. p. 161–174. *In* L. S. Illis (ed.), Viral diseases of the central nervous system. Williams and Wilkins, Baltimore.

Outram, G. 1976. The pathogenesis of scrapie in mice. p. 325–357. *In* R. H. Kimberlin (ed.), Slow virus diseases of animals and man. North-Holland, Amsterdam.

Packer, R. J., D. R. Cornblath, N. K. Gonatas, L. A. Bruno, and A. K. Ashbury. 1980. Creutzfeldt-Jakob disease in a 20 year old woman. Neurology **30**:492–496.

Palsson, P. A. 1979. Rida (scrapie) in Iceland and its epidemiology. p. 357–366. *In* S. B. Prusiner and W. J. Hadlow (ed.), Slow transmissible diseases of the nervous system, Vol. 1. Academic Press, New York.

Park, T. S., G. M. Kleinman, and E. P. Richardson. 1980. Creutzfeldt-Jakob disease with extensive degeneration of the white matter. Acta Neuropathol. (Berl.) **52**:239–242.

Parry, H. B. 1983. Scrapie disease in sheep. Academic Press, New York.

Patterson, I. H., M. N. Hoare, J. N. Jebben, and W. A. Watson. 1974. Further observations on the production of scrapie in sheep by oral dosing with foetal membranes from scrapie-affected sheep. Br. Vet. J. **130**:65–67.

Pritchard, J., V. Thadani, R. Kalb, and E. E. Manuelidis. 1987. Rapidly progressive dementia in a patient who received a cadaveric dura mater graft. Morbid. Mort. Weekly Rep. **36**:49–55.

Prusiner, S. B. 1982. Novel proteinaceous infectious particles cause scrapie. Science **216**:136–144.

Prusiner, S. B. 1984a. Prions-novel infectious pathogens. Adv. Virus Res. **29**:1–56.

Prusiner, S. B. 1984b. Prions. Sci. Am. **251**:50–59.

Prusiner, S. B., and W. J. Hadlow (ed.). 1979. Slow transmissible diseases of the nervous system, Vols. 1 and 2. Academic Press, New York.

Prusiner, S. B., D. F. Groth, M. P. McKinley, S. P. Cochran, K. A. Bowman, and K. C. Kasper. 1981. Thiocyanate and hydroxyl ions inactivate the scrapie agents. Proc. Natl. Acad. Sci. USA **78**:4606–4610.

Prusiner, S. B., M. P. McKinley, K. A. Bowman, D. C. Bolton, P. E. Bendheim, D. F. Groth, and G. G. Glenner. 1983. Scrapie prions aggregate to form amyloid-like birefringent rods. Cell **35**:349–358.

Prusiner, S. B., D. F. Groth, D. C. Bolton, S. B. Kent, and L. E. Hand. 1984. Purification and structural studies of a major scrapie prion protein. Cell **38**:127–134.

Prusiner, S. B., R. Gabezin, and M. P. McKinley. On the biology of prions. Acta Neuropathol. (Berl) **72**:299–314; 1987.

Robakis, N. K., P. R. Sawh, G. C. Wolfe, R. Rubenstein, R. I. Carp, and M. A. Innis. 1986a. Isolation of a cDNA clone encoding the leader peptide of prion protein and expression of the homologous gene in various tissues. Proc. Natl. Acad. Sci. USA **83**:6377–6381.

Robakis, N. K., E. A. Devine-Gage, E. C. Jenkins, R. J. Kascsak, W. T. Brown, M. S. Krawczun, and W. P. Silverman. 1986b. Localization of a human gene homologous to the PrP gene in the p arm of chromosome 20 and detection of PrP-related antigens in normal human brain. Biochem. Biophys. Res. Commun. **140**:758–765.

Roberts, G. W., R. Lofthouse, R. Brown, T. J. Crow, R. A. Barry, and S. B. Prusiner. 1986. Prion protein immunoreactivity in human transmissible dementias. N. Engl. J. Med. **315**:1231–1233.

Robertson, H. D., A. D. Branch, and J. E. Dahlberg. 1985. Focusing on the nature of the scrapie agent. Cell **40**:725–727.

Rohwer, R. G. 1983. Scrapie inactivation kinetics: an explanation for scrapie apparent resistance to inactivation, a reevaluation of estimates of its small size. p. 84–113. *In* L. A. Court and F. Cathala (ed.), Virus non conventionnels et affections du systeme nerveux central. Masson, Paris.

Rohwer, R. G. 1984a. Scrapie-associated fibrils. Lancet **2**:36.

Rohwer, R. G. 1984b. Scrapie infectious agent is virus-like in size and susceptibility to inactivation. Nature **308**:658–662.

Rohwer, R. G. 1984c. Virus-like sensitivity of the scrapie agent to heat inactivation. Science **223**:600–602.

Rosenberg, R. N., C. L. White, P. Brown, D. C. Gajdusek, J. J. Volpe, J. Posner, and P. J. Dyck. 1986. Precautions in handling tissues, fluids and other contaminated materials from patients with documented or suspected Creutzfeldt-Jakob disease. Ann. Neurol. **19**:75–77.

Rubenstein, R., R. I. Carp, and S. M. Callahan. 1984. In vitro replication of scrapie agent in a neuronal model: infection of PC 12 cells. J. Gen. Virol. **65**:2191–2198.

Rubenstein, R., R. J. Kascsak, P. A. Merz, M. C. Papini, R. I. Carp, N. K. Robakis, and H. M. Wisniewski. 1986. Detection of scrapie-associated fibril (SAF) proteins using anti-SAF antibody in non-purified tissue preparations. J. Gen. Virol. **67**:671–681.

Rubenstein, R., P. A. Merz, R. J. Kascsak, R. I. Carp, C. L. Scalici, C. L. Fama, and H. M. Wisniewski. 1987. Detection of scrapie associated fibrils (SAF) and SAF proteins from scrapie affected sheep. J. Infect. Dis. **156**:36–42.

Schoene, M. C., C. L. Masters, C. J. Gibbs, Jr., D. C. Gajdusek, H. R. Tyler, and G. J. Dammin. 1981. Transmissible spongiform encephalopathy (Creutzfeldt-Jakob disease). Atypical clinical and pathological findings. Arch. Neurol. **38**:473–477.

Schumm, F., J. W. Boellaard, W. Schlote, and M. Stohr. 1981. Morbus Gerstmann-Straussler-Scheinker Familie Sch.—ein Bericht uber drei Kranke. Arch. Psychiatry Neurol. Sci. **230**:179–196.

Scott, J. R., and A. G. Dickinson. 1985. A view of dementia from the standpoint of unconventional infections. Gerontol Biomed. Acta **1**:127–150.

Seitelberger, F. 1962. Eigenartige familiar-hereditare Krankheit des Zentralnervensystems in eine neiderosterreichischen Sippe. (Zugleich ein Beitrag zur vergleichenden Neuropathologie des Kuru.) Wiener Klin. Wochenschr. **74**:687–691.

Shinagawa, M., E. Munekata, S. Doe, K. Takahashi, H. Gato, and G. Sato. 1986. Immunoreactivity of a synthetic pentadecapeptide corresponding to the N-terminal region of the scrapie prion protein. J. Gen. Virol. **67**:1745–1750.

Sigurdsson, S. B. 1954. Rida, a chronic encephalitis of sheep with general remarks on infections which develop slowly and some of their special characteristics. Br. Vet. J. **110**:341–354.

Somerville, R. A., P. A. Merz, and R. I. Carp. 1986. Partial copurification of scrapie associated fibrils and scrapie infectivity. Intervirology **25**:48–55.

Sparkes, R. S., M. Simon, V. H. Cohn, R. E. K. Fournier, J. Lem, I. Klisak, C. Heinzmann, C. Blatt, M. Lucero, T. Mohandos, S. J. DeArmond, D. Westaway, S. B. Prusiner, and L. P. Weiner. 1986. Assignment of the human and mouse prion protein genes to homologous chromosomes. Proc. Natl. Acad. Sci. USA **83**:7358–7362.

Takahashi, K., M. Shinagawa, S. Doi, S. Sasaki, H. Goto, and G. Sato. 1986. Purification of scrapie agent from infected animal brains and raising of antibodies to the purified fraction. Microbiol. Immunol. **30**:123–131.

Tateishi, J., Y. Sato, M. Koga, M. Ohta, and Y. Kuroiwa. 1979. A transmissible variant of Creutzfeldt-Jakob disease with kuru plaques. p. 175–185. *In* S. B. Prusiner and W. J. Hadlow (ed.), Slow transmissible diseases of the nervous system, Vol. 2. Academic Press, New York.

Tateishi, J., H. Doi, Y. Sato, M. Suetsugu, K. Ishii, and Y. Kuroiwa. 1981. Experimental transmission of human subacute spongiform encephalopathy to small rodents. III. Further transmission from three patients and distribution patterns of lesions in mice. Acta Neuropathol. (Berl.) **53**:161–163.

Tateishi, J., Y. Sato, and M. Ohta. 1983. Creutzfeldt-Jakob disease in humans and laboratory animals. p. 195–221. *In* H. M. Zimmerman (ed.), Progress in neuropathology, Vol. 5. Raven Press, New York.

Tintner, R., P. Brown, E. T. Hedley-White, E. B. Rappaport, C. P. Piccardo, and D. C. Gajdusek. 1986. Neuropathologic verification of Creutzfeldt-Jakob disease in the exhumed American recipient of human pituitary

growth hormone: epidemiologic and pathogenetic implications. Neurology **36**:932–936.

Towbin, H., T. Staehelin, and J. Gordon. 1979. Electrophoretic transfer of proteins from polyacrylamide gels to nitrocellulose sheets: procedure and some applications. Proc. Natl. Acad. Sci. USA **76**:4350–4354.

Traub, R. D. 1983. Recent data and hypotheses on Creutzfeldt-Jakob disease. Adv. Neurol. **38**:149–164.

Vogt, O. 1925. Zur Pathoklisenlekie. Arch. Psychiatr. Nervenki. **73**:740.

Ward, R. L., D. D. Porter, and J. G. Stevens. 1974. Nature of the scrapie agent: evidence against a viroid. J. Virol. **14**:1099–1103.

Weller, R. O., P. V. Steart, and J. D. Powell-Jackson. 1986. Pathology of Creutzfeldt-Jakob disease associated with pituitary-derived human growth hormone administration. Neuropathol. Appl. Neurobiol. **12**:117–129.

Wells, G. A. H., A. C. Scott, C. T. Johnson, R. F. Gunning, R. D. Hancock, M. Jeffrey, M. Dawson, and R. Bradley. 1987. A novel progressive spongiform encephalopathy in cattle. Vet. Rec. **121**:419–420.

Will, R. G., and W. B. Matthews. 1982. Evidence for case-to-case transmission of Creutzfeldt-Jakob disease. J. Neurol. Neurosurg. Psychiatry **45**:235–238.

Will, R. G., W. B. Matthews, P. G. Smith, and C. Hudson. 1986. A retrospective study of Creutzfeldt-Jakob disease in England and Wales 1970–1979. II. Epidemiology. J. Neurol. Neurosurg. Psychiatry **49**:749–755.

Williams, E. S., and S. Young. 1980. Chronic wasting disease of captive mule deer: a spongiform encephalopathy. J. Wildl. Dis. **16**:89–98.

Williams, E. S., and S. Young. 1982. Spongiform encephalopathy of Rocky Mountain elk. J. Wildl. Dis. **18**:465–471.

Wilson, D. R., R. D. Anderson, and W. Smith. 1950. Studies in scrapie. J. Comp. Pathol. **60**:267–282.

Wisniewski, H. M., M. E. Bruce, and H. Fraser. 1975. Infectious etiology of neuritic (senile) plaques in mice. Science **190**:1108–1109.

Wisniewski, H. M., R. C. Moretz, and A. S. Lossinsky. 1981. Evidence for induction of localized amyloid deposits and neuritic plaques by an infectious agent. Ann. Neurol. **10**:517–522.

Wisniewski, H. M., G. S. Merz, P. A. Merz, G. Y. Wen, and K. Iqbal. 1983. Neurofibrillary tangles and paired helical filaments in Alzheimer's disease. p. 196–221. *In* C. A. Marrota (ed.), Neurofilaments. University of Minnesota Press, Minneapolis.

Wisniewski, H. M., K. Iqbal, I. Grundke-Iqbal, R. Rubenstein, G. Y. Wen, P. A. Merz, R. Kascsak, and K. Kristensson. 1986. Amyloid in Alzheimer's disease and unconventional viral infections. Int. Symp. Dementia Amyloid Neuropathol. Suppl. **3**:87–94.

Wu, Y., W. T. Brown, N. K. Robakis, C. Dobkin, E. Divine-Gage, P. A. Merz, and H. M. Wisniewski. 1987. A Pvu II. RFLP detected in the human prion protein (PrP) gene. Nucleic Acids Res. **15**:3191.

CHAPTER 43

Chlamydiaceae: The Chlamydiae

JULIUS SCHACHTER

Disease: Trachoma, inclusion conjunctivitis, paratrachoma, urethritis, cervicitis, epididymitis, salpingitis, lymphogranuloma venereum, psittacosis, and atypical pneumonia.

Etiologic Agents: *Chlamydia trachomatis*—trachoma biovar (serovars A-K), lymphogranuloma venereum biovar (serovars L1, L2, and L3); *C. psittaci*—multiple biovars and serovars and the TWAR strains.

Source: *C. trachomatis*—person to person transfer, no known animal reservoir; *C. psittaci*—usually acquired from exposure to infected avians, occasionally from mammals. TWAR strain appears to occur only in humans.

Clinical Manifestations: *C. trachomatis*—trachoma, keratoconjunctivitis, urethritis, cervicitis, epididymitis, acute salpingitis, inguinal adenopathy (may be acute, chronic, or subclinical), infant pneumonia; *C. psittaci*—pneumonia or febrile toxic condition.

Pathology: *C. trachomatis*—trachoma biovar involves squamocolumnar/columnar epithelial cells; LGV biovar involves reticuloendothelial cells; *C. psittaci*—systemic infection, particularly mononuclear cells in lungs, liver, and spleen.

Laboratory Diagnosis: *C. trachomatis*—isolation from involved site, antigen detection; *C. psittaci*—isolation from respiratory tract or blood or fourfold rise in antibody.

Epidemiology: *C. trachomatis*, trachoma biovar—sexually transmitted infections worldwide, trachoma is a particular problem in developing countries, especially those in North and sub-Sahara Africa and Southeast Asia; lymphogranuloma venereum biovar is worldwide in distribution, more endemic in certain areas of Africa and Asia; *C. psittaci*—occupational disease in poultry industry, exposure to pet birds; TWAR strain—sporadic cases or outbreaks among humans.

Treatment: Tetracycline is the drug of choice for all chlamydial infections, erythromycin for pregnant women and young children.

Prevention and Control: *C. trachomatis*, trachoma biovar—mass treatment with topical antibiotics and endemic area reduces likelihood of blindness. Perinatal infections can be prevented by treatment of pregnant women; *C. psittaci*—psittacosis from pet birds may be controlled by chemoprophylaxis of exotic birds. No control available for other chlamydial infections.

Introduction

Chlamydia trachomatis

Trachoma, one of the oldest recognized human diseases, was described in Egyptian papyrus and in ancient Chinese writings. The causative agent, *Chlamydia trachomatis*, was first seen in stained conjunctival scrapings. The diagnostic intracytoplasmic inclusions were found first in specimens from experimentally infected subhuman primates and then from humans (Halberstaedter and von Prowazek, 1907). Shortly thereafter, the neonatal form of *C. trachomatis* conjunctivitis (inclusion conjunctivitis of the newborn [ICN] or inclusion blennorrhea) and the related genital tract infections were recognized

(Lindner, 1911). The agent, however, was not isolated until 1957, when Chinese workers recovered the organism by inoculating conjunctival scrapings from trachoma patients into the yolk sac of embryonated hens' eggs (T'ang et al., 1957). Interest in *Chlamydia trachomatis* grew with increased awareness of its role in genital tract disease (Grayston and Wang, 1975; Schachter, 1978). Improved laboratory methodology—particularly tissue culture isolation procedures and the valuable epidemiologic tool, the microimmunofluorescence (Micro-IF) test for measuring antibodies to *C. trachomatis*—led to the elucidation of the wide clinical spectrum of this organism (Gordon and Quan, 1965; Wang and Grayston 1970). A more invasive biovar of *C. trachomatis* causes the systemic sexually transmitted disease, lymphogranuloma venereum (LGV). This condition was first described in the late 1700s, and the causative agent was isolated in 1929 (Siegel, 1962).

C. trachomatis biovars infecting humans have no known animal hosts. *C. psittaci* is extremely common in avian species and lower mammals (Meyer, 1967). The human disease caused by this organism, psittacosis, was first recognized in the latter part of the 19th Century and came to worldwide attention during the pandemic of 1929 to 1930, when the organism was first isolated. The causative agent, *C. psittaci,* is not part of the human flora, being acquired only through incidental contact with an infected bird or, rarely, a mammal.

Chlamydia psittaci

Psittacosis was first described in Switzerland in the 1870s (Schachter and Dawson, 1978). Outbreaks of the disease were reported from a number of countries in Europe and the association with exotic or psittacine birds was recognized. The pandemic of 1929 to 1930 brought much attention to the disease because of the approximate 20% fatality rates seen in the preantibiotic era. In the 1950s the importance of *C. psittaci* infections (ornithosis) in poultry was recognized, and human psittacosis was described as an important occupational hazard to workers in poultry processing plants (Meyer, 1965).

Human psittacosis is a zoonosis, usually contracted from exposure to an infected avian species. *C. psittaci* is ubiquitous among avian species, and infection in the birds is usually of the intestinal tract. The organism is shed in the feces, contaminates the environment, and is spread by aerosol.

C. psittaci also is common in domestic mammals. In some parts of the world these infections have important economic consequences, as *C. psittaci* is a cause of a number of systemic and debilitating diseases in domestic mammals and most importantly, can cause abortions (Storz, 1971). Human chlamydial infections resulting from exposure to infected domestic mammals are known (McKinlay et al., 1985), but seem to be relatively uncommon.

During trachoma studies performed in Taiwan and Iran, some *C. psittaci* strains were recovered from conjunctival swabs. Seroepidemiologic studies have suggested that infections with these strains (currently designated as TWAR) are common in many parts of the world (Forsey et al., 1986; Grayston et al., 1986). Age-specific prevalence rates suggest that transmission occurs in childhood and peaks early in adult life. These TWAR strains appear to be circulating among humans without an avian reservoir. They have been associated with respiratory disease.

Classification

Chlamydiae are presently placed in their own order, the *Chlamydiales*, family *Chlamydiaceae*, with one genus, *Chlamydia* (Moulder et al., 1984). There are two species, *C. trachomatis* and *C. psittaci*. *C. trachomatis* includes the organisms causing trachoma, inclusion conjunctivitis, lymphogranuloma venereum (LGV), other sexually transmitted infections, and some rodent pneumonia strains. The mouse pneumonia agent, lymphogranuloma venereum agent, and the trachoma strains are representative of the three biovars in this species. *C. trachomatis* strains are sensitive to the action of sulfonamides and produce a glycogen-like material within the inclusion vacuole, which stains with iodine. *C. psittaci* strains infect many avian species and mammals, producing the diseases psittacosis, ornithosis, feline pneumonitis, bovine abortion, and so on (Page, 1972; Storz, 1971). They are resistant to the action of sulfonamides and produce inclusions that do not stain with iodine. There is virtually no DNA homology between the species. On the basis of DNA homology, the TWAR strains probably will be placed in a new species.

The Organism

List of Strains

Within *C. trachomatis* the chlamydiae are usually referred to by serovar or biovar. It is clear that *C. psittaci* is a much more heterogeneous species, and many of the organisms in this species are designated by the host and disease with which they have been associated. Thus, there are strains called bovine abortion, sheep arthritis, ovine pneumonitis, feline pneumonitis, and the like. This unhappy state will persist until better taxonomic tools are developed.

Physicochemical Properties

Chlamydiae are bacteria, not viruses (Moulder, 1966). It is traditional in manuals dealing with diagnostic methods to include chlamydiae (and rickettsiae) in with the viruses because the diagnostic methods for chlamydial infection are more related to those for viruses than for bacteria. This similarity in methodology is because chlamydiae are obligate intracellular parasites. In common with the viruses, they cannot be cultured on artificial media and require living host cells to support their replication. They also have a complicated life cycle. The infectious particle, called the elementary body, is approximately 350 nm in diameter. It is responsible for spread of infection from cell to cell and host to host. Once inside the cell the particle changes to a noninfectious reticulate body that is approximately 850 nm in diameter. Toward the end of the growth cycle the metabolically active and noninfectious reticulate bodies change back to elementary body form. Both chlamydial particles contain RNA and DNA. DNA of *C. trachomatis* has a quanine plus cytosine content of about 45%; in *C. psittaci* this proportion is about 41%. The organism is about 40 to 50% lipid, 35% protein, and there are small amounts (1 to 2%) of carbohydrate. Chlamydiae have a cell wall structure that is in some ways similar to that of gram-negative bacteria, although the peptidoglycan layer appears to be lacking. The rigidity of the elementary body is maintained by disulfide cross-linking among several cysteine-rich outer membrane proteins. The major outer membrane protein (MOMP) is generally in the molecular weight range of 39,000 to 45,000 daltons and comprises approximately 60% of the weight of the outer membrane (Caldwell et al., 1981). Chlamydiae also contain lipopolysaccharide (LPS) (Nurminen et al., 1983).

Replication

Chlamydiae are endocytosed by susceptible host cells (Byrne and Moulder, 1978). The phagocytic process is directly influenced by the chlamydiae, and ingestion of organisms is specifically enhanced. After attachment, apparently at specific sites on the surface of the cell, the elementary body enters the cell in a phagosome where the entire growth cycle is completed. The chlamydiae prevent phagolysosomal fusion. Once the elementary body (EB diameter, 0.25 to 0.35 nm) has entered the cell, it reorganizes into a reticulate particle (initial body), which is larger (0.5 to 1 nm) and richer in RNA. After approximately 8 h, the initial body begins dividing by binary fission. Approximately 18 to 24 h after infection, these initial bodies start to become elementary bodies by a poorly understood reorganization or condensation process. The elementary bodies are then released to initiate another cycle of infection. The elementary bodies are specifically adapted for extracellular survival and are the infectious form, whereas the intracellular metabolically active and replicating form, the initial body, does not survive well outside the host cell and seems adapted for an intracellular milieu.

Antigenic Composition and Genetics

All chlamydiae share a common complement-fixing genus-specific antigen. This antigen is a lipopolysaccharide. It is soluble in ether, stable to boiling, and sensitive to periodate. It is antigenically similar to the LPS of *Acinetobacter calcoaceticus* and Re mutants of *Salmonella* spp. (Nurminen et al., 1983). A ketodeoxyoctanoic acid may be the reactive moiety, although other group antigens with a hexose as a reactive group, also appear to exist (Dhir et al., 1972; Stuart and MacDonald, 1982). Protein antigens, of genus, species, and subspecies specificity, also have been identified (Schachter and Caldwell, 1980).

C. trachomatis may be divided by the microimmunofluorescence test into 15 serovars (Grayston and Wang, 1975). These serovars fall into two major complexes (the B and C complexes). The endemic trachoma strains are usually found within the A, B, Ba, and C serovars, whereas the D through K serovars are more commonly associated with sexually transmitted infections. The lymphogranuloma venereum biovars form another group of three serovars, Ll, L2 and L3, which fall within the B complex of *C. trachomatis* and are broadly cross-reactive. *C. psittaci* strains have many serovars, and no practical method for identifying them has yet been devised. The mammalian strains have been divided into two major groups on the basis of plaque reduction assays with hyperimmune antisera, but the same test has detected many serovars among the avian *C. psittaci* isolates (Schachter et al., 1975; Banks et al., 1970). Differences among these strains can also be demonstrated by Micro-IF (Eb et al., 1986).

The specific antigens have never been isolated in sufficient quantity to be well studied. Some studies suggest that they are protein molecules of approximately 29,000 to 30,000 molecular weight (Sacks & MacDonald, 1979). However, much of the serovar reactivity (as well as the species-specific reactions) are due to epitopes on MOMP (Caldwell and Schachter, 1982).

Chlamydiae have a number of antigenic properties that may be assayed. These include mouse toxicity (intravenous inoculation of large quantities of viable *Chlamydia* results in rapid death in mice and this can be neutralized by specific antisera), hemagglutinin for murine or certain fowl erythrocytes, and soluble complement fixation (CF) antigens. Except for the CF test, none of these tests are routinely used in serodiagnosis.

Reactions to Physical and Chemical Agents

In general, *C. trachomatis* is a much more fragile organism than *C. psittaci*. *C. trachomatis* will not survive dessication, whereas *C. psittaci* can remain infectious in litter contaminated with infected feces for months. Both organisms are heat labile and will be killed at 56°C for 30 min. These organisms are sensitive to common disinfectants and can be inactivated with formalin (0.5% for 24 h at 4°C) for preparation of experimental vaccines.

Pathogenesis, Pathophysiology, and Pathology

There is little information on the molecular basis of chlamydial pathogenicity. Studies in cell culture systems have identified a number of virulence factors that appear to be important in establishing infection (Moulder, 1985; Schachter and Caldwell, 1980). These include the ability to recognize specific attachment sites, the ability to induce phagocytosis, and the ability to avoid phagolysosomal fusion.

Members of the trachoma biovar of *C. trachomatis* are essentially parasites of columnar epithelial cells and appear to cause disease in most anatomic sites where these cells are found. These organisms are primarily pathogens of the conjunctivae and genital tract where these cells predominate. The organisms also can be important pathogens in the lower gastrointestinal tract and respiratory tract. The LGV biovar has a broader cell range, involving lymphoid and endothelial cells. It is more invasive and apparently capable of causing more tissue destruction as a result of the infectious process; late stages of LGV are often characterized by scar formation.

Because the trachoma biovar does not appear to be capable of infecting enough cells to cause the tissue damage that appears in some of the diseases, immunologic mechanisms for pathogenesis have been postulated (Grayston and Wang, 1975). More severe disease is often seen in secondary infection and in reinfection with heterologous biovars. It has been shown that a soluble extract from chlamydial particles can induce marked conjunctival inflamma-

tion in animals sensitized by a prior infection (Watkins et al., 1986). Thus, it seems that much of the pathogenesis of the trachoma biovar disease results from local hypersensitivity reactions.

C. psittaci is capable of infecting a wide variety of cells and can cause damage in most anatomic sites. The clinical manifestations of LGV and human *C. psittaci* infection are protean.

Clinical Features

Psittacosis

Psittacosis is the name used for the human infection with *C. psittaci*. In most cases, exposure to birds can be documented. Infection is contracted as a result of environmental contamination by fecal shedding from infected birds. For humans, the route of infection is usually respiratory and the disease may have a respiratory component or can present as a generalized toxic condition (Schachter and Dawson, 1978). Atypical pneumonia is a common presentation. The incubation period is usually between 7 to 14 days, although a much wider range is recognized. Prodrome is relatively nonspecific. Infections are often subclinical and may be mild, resembling a common cold or mild influenzal attack, but severe pneumonitis may occur. Overt clinical disease is almost always accompanied by fever and severe headache. Cough, when present, is usually nonproductive, but x-rays show extensive pneumonic processes. Hepatosplenomegaly is common. Person-to-person transmission is uncommon. Since the introduction of tetracycline therapy for psittacosis, fatalities are rare. Before antibiotic therapy the case fatality rate was quite high (> 20%) for infected individuals, with most fatalities seen in those above the age of 50 years. Treatment with tetracycline (1 g daily for 21 days) is almost always successful, although clinical response may not be rapid and recovery may be prolonged. Many complications are recognized. Prominent among these are meningoencephalitis, myocarditis, and hepatitis.

Lymphogranuloma Venereum

This venereal disease is caused by three serovars within the species *C. trachomatis* (Grayston and Wang, 1975). These organisms are also biovars, as they are more invasive and the disease is a systemic one. The LGV organism may be recovered from genital ulcers and also may be present in the cervix or urethra in an asymptomatic form. It is transmitted by sexual activity.

Lymphogranuloma venereum is usually described

in three stages (Schachter and Osoba, 1983; Siegel, 1962). The primary stage involves a superficial lesion, usually on the external genitalia, which develops 1 to 2 weeks after exposure. These lesions are painless and may be vesicular or ulcerative. In tropical areas ulcerative disease is more severe. The secondary stage, occurring a week or more after the primary stage, is characterized by inguinal lymphadenopathy. This is the typical stage for presentation of men with LGV who come to clinic because of enlarged inguinal lymph nodes (bubos). Fever and chills commonly occur at this stage. Women do not usually present with this form of the disease and the sex ratio is often greater than 10:1 males to females. If untreated, the disease may progress to the tertiary stage (often called the anogenital syndrome), which is characterized by inflammation and scarring of the genitalia and anorectal canal. Fistulae, strictures, and wide tissue destruction may occur. Because women often have no obvious early manifestations of LGV, they often first seek medical attention for tertiary disease.

Although the primary sites involved usually are within the genitourinary tract, if the organism is implanted elsewhere, local disease may occur. For example, if the eye is the site of infection, a form of Parinaud's oculoglandular syndrome may develop. Implantation within the oral cavity may result in cervical lymphadenopathy, and the like. Lymphogranuloma venereum proctocolitis is not uncommon among male homosexuals (Bolan et al., 1982; Quinn et al., 1981). Systemic complications, including hepatitis, pneumonia, arthritis, and meningoencephalitis, are recognized. Treatment is usually with tetracycline or sulfonamides, and prolonged therapy is often required. Response may be variable.

Trachoma

Trachoma is a chronic keratoconjunctivitis caused by *C. trachomatis* (Jones, 1975; Schachter and Dawson, 1978). It may begin as a mucopurulent conjunctivitis and is often complicated by secondary bacterial infection. Marked follicular reaction and papillary hypertrophy develop. In the hyperendemic area most active disease is seen in young children. As the follicular reaction resolves, some focal necrosis may occur and scarring of the upper conjunctivae can develop. Over time, these scars contract and cause an inturning of the upper eyelids, so that the eyelashes abrade the cornea. These lesions (trichiasis and entropion) represent the blinding lesions of trachoma. It takes many years for sufficient contraction of scars to occur to cause the lid distortion and thus, blindness is generally seen more than 25 to 30 years after the peak of active inflammatory processes. Mild cases of trachoma rarely lead to visual loss.

The main reservoir for infection is young children with severe inflammatory disease. Trachoma is a family disease and is typically spread from child to child by direct personal contact. Flies can act as mechanical vectors, as they feed on ocular discharges. Occasional adults may have active disease or may be inapparent shedders and the likely source of infection for some children in the household.

Trachoma may be effectively treated with systemic antibiotics. Tetracycline is the drug of choice, although sulfonamides and erythromycin have been used. In the endemic area, however, systemic therapy is difficult to manage and most trachoma control programs are based on mass topical treatment with topical tetracycline. This is usually done on an intermittent basis (given for 6 consecutive days, twice a year). This approach is not aimed at curing the infection but at reducing the severity of the disease to prevent blinding complications. For those with scarred lids, surgical intervention is available to prevent corneal damage.

Inclusion Conjunctivitis

Inclusion conjunctivitis in adults or in newborn infants is usually a relatively mild, self-limited conjunctivitis resulting from inoculation of conjunctivae with genital tract discharges. It is also called paratrachoma. Adults probably acquire the infection during sexual activity or by hand-to-eye transmission, whereas newborns acquire the organism during the birth process (Schachter and Grossman, 1981). The incubation period is typically between 6 and 19 days. In adults the disease is follicular, but infants having less developed lymphoid response, usually have a mucopurulent conjunctivitis. Severe disease with a poor prognosis occurs uncommonly.

Adults may be treated with systemic tetracycline (1 g daily for 21 days) and infants are usually treated with erythromycin (40 mg/kg for 7 to 10 days).

Genital Tract Infections

C. trachomatis is currently considered to be the most common sexually transmitted pathogen (Oriel and Ridgway, 1982; Schachter, 1978; Schachter et al., 1975). In men it is a major cause of nongonococcal and postgonococcal urethritis and epididymitis. In women the organism may cause a mucopurulent endocervicitis, urethritis, and endometritis. Chlamydiae have been shown to be important causes of acute salpingitis and are responsible for many of the complications of this condition. Thus, the organism is now known to be an important cause of Fitz-Hugh-Curtis syndrome (perihepatitis), tubal factor infertility, and ectopic pregnancy. Rectal infections and proctitis can occur in either sex. Mild and subclinical infections with these organisms are common. The organism may persist in the genital tract for years in the absence of antimicrobial therapy. Treatment is usually with tetracyclines (2 g daily for at least 7 days).

Neonatal infections

Although acute, self-limited conjunctivitis caused by *C. trachomatis* was first described early in this century, it was only in the 1970s that the wider spectrum of disease in the infant was recognized (Schachter and Grossman, 1981). Infected newborn infants are at risk of developing pneumonia (Schachter et al., 1986). Chlamydial pneumonia is a characteristic syndrome, with a chronic afebrile course and a staccato cough (Beem and Saxon, 1977). Laboratory clues may include eosinophilia and elevated immunoglobulins. Serous otitis media is reported as a complication of the pneumonia and may occur independently. Infants with chlamydia pneumonia may subsequently develop chronic respiratory problems (Weiss et al., 1986). Severe rhinitis has also been associated with chlamydial infection in the newborn (Cohen et al., 1982). Gastrointestinal tract infections occur but no clinical manifestations in the intestinal tract have been described (Schachter et al., 1979).

Clinical Spectrum of *C. psittaci* Disease in Animals

MAMMALS

C. psittaci strains are common pathogens of lower mammals where they can cause a variety of diseases (Storz, 1971). Some of these conditions are economically important. Among the conditions attributed to *Chlamydia* are arthritis, pneumonitis, abortion, enteritis, conjunctivitis, and encephalitis. Many of these infections have a common pathway in that the organism is found within the intestinal tract and shed in the feces. Inapparent infections are common.

AVIAN SPECIES

C. psittaci infection in birds is usually an intestinal tract infection, although respiratory components may exist (Meyer, 1967; Page, 1972). The disease, which may be subclinical, when clinically manifest is typified as a chronic wasting process which may be fatal. Diarrhea is common.

Feral birds are apparently a common source of infection for poultry as they may be asymptomatic shedders and contaminate the environment in which the poultry are being reared (Grimes et al., 1979).

TWAR Infection

Very little is known about the clinical spectrum of TWAR infection. The organism has been implicated as an important cause of atypical pneumonia in young adults (Grayston et al., 1986) It has been found in association with severe and fatal pneumonias in older individuals with underlying disease (Marrie et al., 1987). Seroprevalence studies indicate high rates of exposure to this organism, suggesting that either asymptomatic or very mild infections are common.

Diagnosis

Biosafety Considerations

C. trachomatis is not considered to be a particularly dangerous pathogen to handle in the laboratory. There have been a number of laboratory infections, usually manifested as follicular conjunctivitis. However, the LGV biovar is a more invasive organism, and severe cases of pneumonia have occurred when research workers were exposed to aerosols created by laboratory procedures such as sonication (Bernstein et al., 1984). *C. psittaci* must be considered as a potentially dangerous organism to handle in the laboratory. For many years it was a major cause of laboratory infections. These usually resulted from exposure to aerosols, but the stability of the organism is a potential problem. The organism should not be handled in laboratories without appropriate containment facilities.

Infected Sites and Specimen Collection

For cytologic studies, impression smears of involved tissues or scrapings of involved epithelial cell sites should be appropriately fixed (cold acetone or methanol for immunofluorescence and methanol for Giemsa stain).

For most *C. trachomatis* infections of humans, it is imperative that samples be collected from the involved epithelial cell sites by vigorous swabbing or scraping. Purulent discharges are inadequate and should be cleaned from the site before sampling. Appropriate sites include the conjunctiva for trachoma inclusion conjunctivitis and the anterior urethra (several centimeters into the urethra), or the cervix (within the endocervical canal) for genital infection. Because these strains appear to infect only columnar cells, cervical specimens must be collected at the transitional zone or within the os. The organism also can infect the urethra of the female; it may improve recovery rates if another sample is collected from the urethra and sent to the laboratory for testing in the same tube with the cervical sample. For women with salpingitis, the samples may be collected by needle aspiration of the involved fallopian tube or endometrial specimens may yield the agent (Mardh et al., 1977; Schachter et al., 1986). Rectal mucosa, the nasopharynx, and the throat also may be sampled. For infants with pneumonia, swabs may be collected from the posterior nasopharynx or the throat, although nasopharyngeal or tracheobronchial aspirates collected by intubation appear to be a superior source of agent.

For *C. psittaci* infection in humans the specimens include sputum and blood specimens in classical psittacosis and throat swabs for the TWAR strains. *C. psittaci* has been recovered from a variety of involved anatomic sites sampled by biopsy or at necropsy. For all culture methods the sites sampled are likely to be contaminated and the specimen should be collected into a medium which contains appropriate antibiotics to remove unwanted bacteria. Chlamydiae are resistant to the action of streptomycin, gentamycin, vancomycin, and fungicides such as nystatin or fungizone, and these antimicrobials are often included into the transport media.

PROCESSING SPECIMENS FOR THE ISOLATION OF *CHLAMYDIAE*

For practical purposes only three experimental systems need be considered for the isolation of chlamydiae. First, all known chlamydiae grow in the yolk sac of the embryonated hen's egg. Second, with centrifugation of the inoculum, it appears that all chlamydiae (with some variability) will grow in tissue culture; psittacosis and LGV agents are capable of serial growth in tissue culture without centrifugation. Third, the psittacosis agents will grow in mice after intracerebral (ic), intraperitoneal (ip), and intranasal (in) inoculation, and LGV agents after intracerebral and intranasal inoculation, although the latter route is rarely used. Mice are of no use in recovering trachoma biovars.

The specimens to be tested include ocular and genital tract epithelial cell scrapings for trachoma biovars, bubo pus, and genital tract specimens for LGV agents, and blood, sputum, and biopsy tissue specimens for psittacosis.

General guidelines for handling specimens are listed below. The diluents and antibiotics used to control bacterial contamination will differ with the isolation system being used, resulting in some minor variations. For yolk sac procedures a suitable collection medium is nutrient broth containing streptomycin (2.5 mg/ml), neomycin (0.5 mg/ml), and nystatin (100 U/ml). For isolation in tissue culture, a useful collection medium is complete cell culture medium containing gentamicin (10 μg/ml), vancomycin (100 μg/ml), and fungizone (4 μg/ml). Either diluent may

be used for mice. Other tissue culture collection media have been used, such as modifications of the sucrose-phosphate buffer solutions originally developed for rickettsiae (Darougar et al., 1971). These, however, are slightly more toxic to the cells. Fresh samples are preferred, but frozen material ($-60°C$) is acceptable.

Ocular and Genital Tract Specimens

For maximal results it is imperative that adequate specimens be collected. Because the object is to obtain a representative sample of epithelial cells, the specimens must be collected with some vigor. With genital tract specimens the samples must be obtained (either by swabbing or scraping) from the transitional zone of the cervix or the endourethra (2 to 4 cm from the meatus). Culture of discharges or of urine is inadequate. The material is inoculated into yolk sac or tissue culture.

Bubo Pus

Grind the viscous material. Suspend in nutrient broth or tissue culture medium to at least 20% of weight. Even when pus is not viscous, dilution is advisable. If the bubo is not fluctuant, sterile saline may be injected and then aspirated for isolation attempts. Test for bacterial contaminants, treat with antibiotics, and inoculate the material into mice intracerebrally, into eggs by the yolk sac route, or into cell cultures.

Blood

If there is a clot, grind it and add beef heart broth or tissue culture medium to make a 10% suspension. Inoculate directly into eggs or mice. For cell culture it is advisable to inoculate with several further dilutions.

Sputum or Throat Washings

Sputum is cultured for bacteria on blood agar plates. To prepare the emulsion, suspend sputum, depending on its consistency in 2 to 10 times its volume of sterile antibiotic-containing broth (pH, 7.2 to 7.4) or tissue culture medium; emulsify thoroughly by shaking with glass beads in a sterile, tightly stoppered container. Inoculate into the isolation system after 1 to 2 h of treatment with antibiotics at room temperature. It may be advisable to centrifuge extracts for 20 to 30 min at 100 × g to remove coarse material.

Fecal Samples

Cloacal or rectal swabs, the droppings from caged birds, or fecal pellets are suspended in antibiotic broth or medium. The suspension (approximately 20%) is shaken thoroughly. After centrifugation at

$300 \times g$ for 10 min, the supernatant fluid is removed. It may be further diluted (1 : 2 and 1 : 20) with antibiotic solution and held for 1 h at room temperature before inoculation into tissue culture, yolk sac, or mice. More concentrated material may be used for ip inoculation of mice.

Another method has been recommended by Storz and associates (1965) for testing fecal samples collected from sheep, cattle, or other mammals. Immediately after collection, the samples are taken to the laboratory, ground, and brought to a 10% suspension in Earle's balanced salt solution containing 0.5 mg of streptomycin/ml. The samples are centrifuged for 30 min at $1800 \times g$. Supernatant fluid is gently withdrawn, mixed with an equal amount of fresh diluent, and centrifuged again. This procedure is repeated once more. About 3 ml of supernatant fluid is then withdrawn, and 0.5 ml each of 1 : 40 and 1 : 400 dilutions is inoculated into yolk sacs of 7-day-old chicken embryos. Appropriate bacteriologic controls are also prepared.

Tissues

Frozen tissue is thawed in a refrigerator at about 4°C for 18 to 24 h. The specimen is weighed, minced with sterile scissors, and ground to a paste with mortar and pestle or homogenizer.

After the tissue has been ground thoroughly, the volume of antibiotic-containing diluent required to make a 10 to 20% emulsion is added and the suspension is thoroughly mixed. Plain nutrient broth (pH, 7.2 to 7.4) may be used if the tissue is bacteriologically sterile. For tissue culture, antibiotic-containing collection medium is used, and 10^{-1} and 10^{-2} dilutions are inoculated.

Direct Detection by Microscopy

Because chlamydiae are large enough to be seen by light microscopy and the intracytoplasmic inclusions are pathognomonic, much of the diagnostic methodology depends on microscopic indentification of the organism.

MACCHIAVELLO'S STAIN

This stain is used on impression smears from animal tissue and yolk sac.

Modified Macchiavello's Stain

This stain is prepared as follows:

Stock solutions

Basic fuchsin   0.25 g in 100 ml double-distilled
water

Citric acid   0.5 g in 200 ml double-distilled
water
Methylene blue   1.0 g in 100 ml double-distilled
water

Prepare citric acid solution fresh daily.

After drying in air, the smear or impression preparation is fixed by heat. The basic fuchsin solution, first passed through filter paper in a small funnel, is dropped onto the film and left for 5 min before being quickly drained off. The slide is first washed in tap water and then dipped for a few seconds in the citric acid solution which is best held in a coplin jar. The slide is then washed thoroughly with tap water and stained with 1% methylene blue for 20 to 30 s; it is washed again in tap water and dried.

The citric acid solution must be fresh. Exposure to citric acid for more than a few seconds decolorizes the chlamydiae, and they all stain blue. In a properly prepared slide, most EBs stain red against blue background.

Gimenez Modification of the Macchiavello Technique

Stock solutions

1. 10% (wt/vol) basic fuchsin in
   95% ethanol   100 ml
   4% (wt/vol) aqueous phenol   250 ml
   distilled water   650 ml
2. 0.1 M sodium PBS, pH 7.45 (mix 3.5 ml of 0.2 M NaH_2PO_4, 15.5 ml of 0.2 M Na_2HPO_4, and aqueous malachite green oxalate)
3. 0.8% aqueous malachite green oxalate

To prepare a working solution of carbol fuchsin, mix 4 ml of stock solution with 10 ml of buffer (pH, 7.45); filter immediately and filter again before each staining. The working solution remains satisfactory for about 40 h.

A very thin air-dried smear (heat fixation is not necessary for cytologic reasons but should be used for safety) is covered with the filtered carbol basic fuchsin working solution for 1 to 2 min. After a thorough washing in tap water, it is covered with the malachite green solution for 6 to 9 s and washed again in tap water. The slides are finally dried with absorbent paper. EBs stain red, and the background stains greenish.

IODINE STAINING TECHNIQUE

Scrapings are air dried, fixed in absolute methanol, and stained with Lugol iodine or 5% iodine in 10% potassium iodide for 3 to 5 min. Slides are examined as wet mounts. The matrix of inclusions may appear as a reddish-brown mass recognizable under low magnification. The slides may be decolorized with methanol and restained with Giemsa stain. This technique is the least sensitive cytologic procedure. It can be used to detect *C. trachomatis* inclusions in conjunctival scrapings (Rice, 1936), but it is not recommended for use with clinical specimens. Its speed and simplicity have made it the popular test for examining *C. trachomatis*-infected cell cultures. *C. psittaci* inclusions do not stain by this method.

GIEMSA STAINING TECHNIQUE

The smear or scraping is air dried, fixed with absolute methanol for at least 5 min, and dried again. It is then covered with the diluted Giemsa stain (freshly prepared the same day) for 1 h. Longer staining periods (1 to 5 h) may be preferable with heavy tissue culture monolayers. The slide is then rinsed rapidly in 95% ethyl alcohol to remove excess dye and to enhance differentiation; it is then dried and examined microscopically. Elementary bodies stain reddish purple. Visualization of the characteristic perinuclear elementary body inclusions is diagnostic. The initial bodies are more basophilic, staining bluish, as do most bacteria. Sensitivity and contrast of the method for detecting *C. trachomatis* inclusions in cell culture may be improved by darkfield examination. Inclusions appear lemon yellow and stand out against the dark background.

FLUORESCENT ANTIBODY TECHNIQUE

Most of the published experience with immunofluorescent procedures to detect typical *C. trachomatis* inclusions within epithelial cells used polyclonal antibody in either direct or indirect fluorescent antibody procedures (Hanna, 1968). These procedures were more sensitive than other cytologic methods (Giemsa or iodine). There were no commercial sources for antisera, and laboratories had to prepare their own reagents. The antisera were usually standardized against infected tissue culture monolayers to determine appropriate working dilutions. Trained microscopists were required to identify typical chlamydial intracytoplasmic inclusions. The procedure is less sensitive than isolation of the agent in cell culture.

More recently, fluorescein-conjugated monoclonal antibodies have been made available (Stephens et al., 1982). They can be used to identify *C. trachomatis* inclusions in cell culture (Stamm et al., 1983). The test may be applied directly to clinical specimens (Tam et al., 1984). In the direct test the elementary bodies, rather than inclusions, are visualized. This test is more sensitive than other cytologic means of diagnosing chlamydial infection, but it is less sensitive than culture. Although much of the published literature indicates a sensitivity on the order of 90% as compared with culture, when optimal culture techniques are used the procedure is proba-

bly only 60 to 70% sensitive (Lipkin et al., 1986; Schachter, 1985). It, however, still offers a nonculture method of diagnosing chlamydial infection in areas where cell culture is not available or where transportation problems preclude appropriate handling of specimens to maintain viability.

This test requires a trained microscopist who can distinguish between fluorescing chlamydial particles and nonspecific fluorescence. Several configurations of this test are commercially available, and monoclonal antibodies are available against the major outer membrane protein or against the lipopolysaccharide. Monoclonals to the LPS will stain either *C. trachomatis* or *C. psittaci,* but the quality of the fluorescence is somewhat mitigated by uneven distribution of LPS on the chlamydial particle. The anti-MOMP monoclonals are prepared against *C. trachomatis* and are species specific and, thus, will not stain *C. psittaci.* The quality of fluorescence is better because MOMP is evenly distributed on the chlamydial particle and if *C. trachomatis* is being sought, the anti-MOMP monoclonals are to be preferred. This procedure offers the possibility of rapid diagnosis, as the technique takes approximately 30 min to perform.

Direct Antigen Detection

In addition to the direct fluorescent antibody methods described, enzyme immunoassays are available for the detection of chlamydial antigens. The tests currently available include either monoclonal antibodies or polyclonal antibodies to the LPS and, thus, theoretically, could detect either *C. psittaci* or *C. trachomatis.* They have not been evaluated in diagnosis of infections with *C. psittaci.* For *C. trachomatis* they appear to be slightly more sensitive and slightly less specific than direct fluorescent antibody methods (Chernesky et al., 1986; Schachter, 1987). The technique takes approximately 4 h to perform and is suitable for batch processing and will allow a laboratory to test many specimens.

Nucleic Acid Detection

Currently, detection of nucleic acid in clinical specimens is a research technique. Tests are available for detection of chromosomal DNA or detection of a cryptic plasmid which is common to all *C. trachomatis* human pathogens (Horn et al., 1986). An advantage of the latter test is that multiple copies of the plasmid are present in each elementary body (Palmer and Falkow, 1986). Only radiolabeled probes have been shown to have sufficient sensitivity for use with most specimens and this has restricted their application in routine diagnostic testing.

Isolation

All chlamydiae appear to grow in the yolk sac of embryonated hen's eggs. Many *C. psittaci* strains and the lymphogranuloma venereum biovar of *C. trachomatis* can be cultured in mice or other laboratory animals. However, these procedures are relatively slow and will not provide an etiologic diagnosis in an appropriate time frame to be clinically relevant. A number of different cell lines have been used to support the growth of chlamydiae. It does not appear that any single cell line has a marked superiority to others, as successful studies have been performed using monkey kidney cells, HeLa cells, L-cells, and McCoy cells among others.

ISOLATION IN CELL CULTURE

The recommended procedure for primary isolation of chlamydiae is in cell culture. The most common technique involves inoculation of clinical specimens into cycloheximide-treated McCoy cells (Ripa and Mardh, 1977). The basic principle involves centrifugation of the inoculum into the cell monolayer at approximately 2,800 × g for 1 h, incubation of monolayers for 48 to 72 h, and then staining. Iodine is used for *C. trachomatis* to detect the glycogen-positive inclusions. Fluorescent antibody staining may allow earlier detection of the inclusion (Thomas et al., 1977). Use of fluorescein-conjugated monoclonal antibodies represents the most sensitive method for detecting *C. trachomatis* inclusions in cell culture (Stamm et al., 1983). The procedure requires more attention to staining than the iodine technique and is more costly. For *C. psittaci* the inclusions can be demonstrated by use of genus-specific monoclonal antibodies or by the Giemsa stain.

McCoy cells are plated onto 13-mm coverslips contained in 15-mm diameter (1 dram) disposable glass vials. Cell concentration (approximately 1×10^5 to 2×10^5) is selected to give a light, confluent monolayer after 24 to 48 h of incubation at 37°C. The cells should, for optimal results, be used within 24 to 72 h after reaching confluency. If the laboratory is only passing cells on a sporadic basis, they may then be held at room temperature or in a low (2%) serum medium for up to 2 weeks before inoculation.

The clinical specimens should be shaken with glass beads before inoculation. This is safer and more convenient than sonication. Standard inoculation procedure involves removing medium from the cell monolayer and replacing it with the inoculum in a volume of 0.1 to 1 ml. The specimen is then centrifuged onto the cell monolayer at approximately 3,000 × g at 35°C for 1 h. The vials should be held at 35°C for 2 h before the cells are washed or the medium is changed to medium containing 1 to 2 μg of cycloheximide per ml (this must be titrated for each batch). The cells are then incubated at 35°C for 48 to 72 h, after which one coverslip is examined for inclusions by use of iodine, Giemsa, or immunofluorescence staining. The use of immunofluorescence can speed up the process, as inclusions can clearly be seen (although smaller) at 24 h postinfection, but this requires availability of immunologic reagents and uses more difficult microscopic procedures. Giemsa stain is more sensitive than iodine stain, but the microscopic evaluation is more difficult. Slide reading can be facilitated by examining the Giemsa-stained coverslip by darkfield rather than brightfield microscopy (Darougar et al., 1971). The iodine stain is the simplest procedure and the one most commonly used, although it is less sensitive than either of the other two.

If passage of positive material or blind passage of negative material is desired, the material should be passed at 72 to 96 h postinoculation. The cell monolayer is disrupted by shaking with glass beads on a Vortex mixer; the material is treated by low-speed centrifugation to remove cell debris, and the supernatant is inoculated as above. For symptomatic patients 90% of specimens positive for *C. trachomatis* are inclusion positive in the first passage. In screening asymptomatic patients, who often have less agent, more (30 to 40%) of the positive specimens became positive with passage.

With trachoma, inclusion conjunctivitis, and the genital tract infections, the technique is as described earlier. In LGV, the aspirated bubo pus is diluted (10^{-1} and 1^{-2}) before inoculation. Second passages are always made because detritus from the inoculum may make it difficult to read the slides. For many *C. psittaci* isolation attempts it may be convenient to lengthen the incubation period to 5 to 10 days before examining the coverslips for inclusions. These organisms do not require mechanical assistance for cell-to-cell infection (Schachter et al., 1978).

For laboratories processing large numbers of specimens, it may be convenient to use flat-bottomed 96-well microtiter plates rather than vials (Yoder et al., 1981). Cells are plated onto coverslips or directly onto the plates. Processing and incubation will be as mentioned, but microscopy will be modified to use either long-working objectives or inverted microscopes. This procedure is less sensitive than the vial technique but offers considerable savings in terms of reagents and time and may be suitable for settings where mostly symptomatic patients are being screened. These patients usually yield higher numbers of chlamydiae and this minimizes the impact of the decreased sensitivity of the test.

ISOLATION IN MICE

The mice to be used should be proven susceptible to chlamydiae because there are some genetic variations in this regard. Mice should be obtained from a colony shown to be free of latent chlamydial infection. There have been at least seven reports of subclinical chlamydial infections in mouse colonies. These agents have been identified as *C. psittaci* as well as a *C. trachomatis* biovar, and some have been viscerotropic, whereas others were pneumotropic. These infections were revealed by persistent blind passage of "normal" mouse tissue.

Intraperitoneal Injection

This is used only for *C. psittaci*. Most chlamydiae from psittacines or turkeys will be lethal after inoculation by the ip route; those from pigeons, chickens, some turkeys, or ducks may produce significantly enlarged spleens and ascitic fluid but do not regularly cause death.

Inoculate 0.5 ml of the prepared 10 or 20% sterile emulsion. Virulent material from parrots, parakeets, humans, and some turkeys, injected by the ip route in this amount causes death of the mouse in 3 to 30 days.

If mice die within 2 or 3 days, little that is abnormal can be seen with the naked eye; spleen and liver may look normal in size and architecture. Some animals may show signs of vascular damage. Quite characteristic, and often the only sign, is a bloated duodenum covered with a thin viscous exudate. In some animals the surface of the liver and intestines may be moist and covered with a thin, sticky exudate that contains abundant endothelial cells packed with chlamydial particles. Macchiavello's stain is used with impression smears.

When death occurs within 5 to 15 days, the spleen is enlarged, and early necrotic lesions of the liver can be seen. Microscopically, hemorrhages and necrosis are common in the liver; the phagocytic cells of liver and spleen may be packed with chlamydiae. The abdominal cavity may be filled with stringy, turbid, fibrinous exudate.

If animals survive until day 21, they should be sacrificed, and further blind passage of emulsions of their spleen and liver made. The general rule is that if chlamydiae are not found by the third passage, they cannot be isolated no matter how many more passages are made. Mice that recover and are sacrificed 3 weeks after infection have few gross lesions. In general, the intestines are slightly distended and pale. Exudate may be present in the abdominal cavity. The spleen is conspicuously enlarged, the liver friable and mottled, and the kidneys grayish. EBs are sparse in tissue smears, but animal passage has shown that they may exist as long as 300 days after initial infection. Most survivors have infection immunity.

This technique offers the advantages of simplicity, reliability, and large inocula. If it is desired, the animals may receive multiple (at daily intervals) inoculations from the original specimen. In addition, the mice may "filter" out bacteria that have not been controlled by antibotics or centrifugation and dilution.

Intracranial Injection

Tissue specimens from humans, sterile exudates from the pericardial or air sacs of birds, or peritoneal fluid and suspensions prepared from infected mice may be safely infected by this route, which may furnish excellent specimens for rapid histologic diagnosis. Lymphogranuloma venereum specimens are also inoculated by the ic route. Inoculate 0.03 ml of the treated emulsion. Somnolence and paralysis often develop within 24 to 48 h, and death follows within 3 to 5 days with highly virulent material. Blind passage is performed at 10 days.

This route has the advantage of not involving the respiratory tract, precluding the possibility of activating latent mouse pneumonitis. Smears are made from the dura teem with chlamydiae. A relatively fast and sensitive method for isolating psittacosis and ornithosis agents, this technique is somewhat less effective with LGV. This route of inoculation suffers a disadvantage in terms of small volume of inoculum and in the susceptibility of the mice to bacteria that may contaminate the specimen.

Intranasal Instillation

Instill 0.03 to 0.05 ml of a 10% tissue suspension, with the mouse under light anesthesia (ether is suitable). If the material inoculated is virulent or if isolates have been established, signs of infection-hunched posture, apathy, and increasingly labored respiration develop rapidly, and death follows within 2 to 20 days. Bacterial contamination must be ruled out. In typical successful isolation atempts, death may take place between days 8 and 16 if the agent is present in high concentration. However, with less virulent material all symptoms may gradually disappear; in such cases blind passage should be performed 21 days after inoculation. Blind passage is usually required. Segments or entire lobes of the lung may be extensively consolidated. Discrete foci of pneumonia are manifested as limiting infective dilutions are approached. These areas, which are gray, almost translucent, and 1 to 3 mm in diameter.

Fewer EBs are seen in smears from lungs infected for more than 10 days, and there may be difficulty finding them in old lesions. Repassage may furnish excellent material for microscopy.

Yolk Sac Isolation Technique

Clinical specimens are collected in an appropriate antibiotic broth. The specimen is held for 1 h at room temperature before inoculation of 0.25 ml into the yolk sac, using a 3.2-cm 22-gauge needle. Before inoculation, the fertile eggs are incubated at 38.5 to 39°C in a moist atmosphere. The eggs to be used must be obtained from a flock fed an antibiotic-free diet. They should be free from *Mycoplasma*. When 7 days old, embryonated hen eggs are candled for viability, the location of air sacs is marked and the shell is painted with tincture of iodine, and a hole is gently punched. The specimen is inoculated at a slight angle away from the embryo; three or four eggs are labeled with a pencil or marking pen. After inoculation, the shell is again swabbed with iodine and the hole is sealed (with glue or tape). The eggs are then incubated in a moist environment at 35°C and candled daily for 13 days. Eggs that die in the first 3 days after inoculation are discarded.

The yolk sacs of eggs dying thereafter are harvested. This procedure entails painting the shell with iodine, cracking and removing the shell over the air sac, dissecting the shell and chorioallantoic membranes away, and removing the yolk sac with forceps. Excess yolk material may be stripped away. It is important that all instruments be sterile and that fresh instruments be used for each specimen. Impression smears are made and stained (Gimenez or the modified Macchiavello method). Sterility tests are performed on yolk sac with thioglycolate broth. If the embryos are still viable 13 days postinoculation, the eggs are chilled for several hours and yolk sacs are harvested, ground in nutrient broth, and centrifuged lightly. The supernatant is passed to another group of four 7-day-old embryonated hen eggs (1 ml of 50% yolk sac per egg). After two blind passages, attempts are terminated as negative.

The generally acceptable criteria for positive isolation are the finding of elementary bodies in the impression smears, serally transmissable egg mortality, the presence of group antigen in the yolk sac, and the absence of contaminating bacteria.

Identification

Because most laboratories will be using tissue culture isolation systems, the basic procedure for identification of chlamydiae involves demonstration of typical intracytoplasmic inclusions by appropriate (Giemsa or iodine) staining procedures. However, in laboratories initiating work with chlamydiae, it would be prudent to use at least one other parameter for identification of chlamydiae. Fluorescent-antibody staining provides both a morphologic and an immunologic identification. In yolk sac or mouse isolation procedures one can use heavily infected material to prepare a CF antigen.

C. trachomatis strains may be serotyped by the Micro-IF technique (Wang and Grayston, 1970). For this procedure antisera are produced by intravenous inoculation of mice at day 0, a booster is given at day 7, and exsanguination occurs at day 11. The mouse antiserum is then tested in a titration against all serotypes, as well as the immunizing agent, and the serotype is identified presumptively by the pattern of reactivity and finally by appropriate box titration with the appropriate prototype serovar. Undoubtedly serovar-specific monoclonal antibodies will become available for typing.

Antibody Assays

Although many different serologic tests are available, only the following two are widely used in diagnosis.

COMPLEMENT FIXATION (CF)

The CF test may be performed in either the tube system or the microtiter system. Reagents should be standardized in the tube system, regardless of which test system will be used. The microtiter systems are most useful in screening large numbers of sera, but it is preferable to retest all positive sera in the tube system. Occasionally, sera giving titers in the 1:4 to 1:8 range in the microtiter system are positive at 1:16 (the significant level) in the tube system. The microtiter system uses standard plates and volumes one-tenth of those used in the tube test. The CF test is performed on serum specimens heated at 56°C for 30 min (preferably acute and convalescent paired sera are tested together). In each test a positive control serum of high titer is included together with a known negative serum. The reagents for the CF test are standardized by the Kolmer technique and include special buffered saline, group antigen, antigen (normal yolk sac) control, the positive serum, the negative serum, guinea pig complement, rabbit antisheep hemolysin, and sheep erythrocytes. (The guinea pig complement should be carefully tested for chlamydial antibodies because many herds are enzootically infected with chlamydiae, the guinea pig inclusion conjunctivitis agent.) The hemolytic system is titrated and the complement unitage is determined. The standard units used in the test are 4 U of antigen and 2 exact U of complement. The test may be performed either using a water bath at 37°C for 2 h or using overnight incubation at 4°C, the former being preferable. Doubling dilutions of the serum (from 1:2) are made in a 0.25 ml volume of saline. The antigen is added at 4 U (0.25 ml), and 2 exact U of complement (0.5 ml) is added. Standard reagent controls are always included. The normal yolk sac control is used at the same dilution as the group antigen. The tubes are shaken well and incubated in a water bath at 37°C for 2 h. Then 0.5 ml of sensitized sheep erythrocytes is added and the tubes are placed in a water bath for 1 h, after which they are read for hemolysis on a 1+ to 4+ scale (roughly equivalent to 25 to 100% inhibition of erythrocyte lysis). The endpoint of the serum is considered the highest dilution inhibiting at least 50% (2+) hemolysis after a complete inhibition of hemolysis has been observed. It is good practice to shake the tubes to resuspend the settled cells, refrigerate them overnight, and recheck the results the next morning.

All reagents are available commercially, except for high-titered group antigen. This may be prepared as follows. Yolk sacs of 7-day embryonated eggs are inoculated with any *Chlamydia* (the psittacosis isolate 6BC has long been used) at a dose estimated to result in death of about 50% of inoculated eggs in 5 to 7 days. Eggs are candled daily, and those dying early are discarded. When the 50% death endpoint is approached, the remaining eggs (recently dead or live) are refrigerated for 3 to 24 h. The yolk sacs are then harvested. If examination of random samples shows large numbers of particles, the yolk sacs are pooled. This preparation may be stored at −20°C until further processing. The yolk sacs are ground in a mortar with sterile sand. Beef heart broth (pH, 7.0) is added to make a 20% suspension, and the material is cultured to determine if it is free of bacterial contamination. The suspension is placed in a flask containing sterile glass beads and stored at 4°C for 3 to 6 weeks with daily shaking. It is then centrifuged at approximately 500 × *g* to remove coarse particles, transferred to a heavy sterile flask, and steamed at 100°C or immersed in boiling water for 30 min. After it has cooled, liquified phenol is added to 0.5%. The antigen should then be refrigerated for at least 1 week before being used. It is stable for at least 1 year if not contaminated and should have an antigen titer of 1:256 or greater. A similar preparation from uninfected yolk sacs must be included as one of the controls.

Microimmunofluorescence (Micro-IF)

The Micro-IF test is usually performed against chlamydial organisms grown in yolk sac. Tissue culture-grown agent may be used, but it may be necessary to concentrate the elementary bodies and to add some normal yolk sac to improve contrast for microscopy. The individual yolk sacs are selected for elementary body richness and pretitrated to give an even distribution of particles. It is generally found that a 1 to 3% yolk sac suspension (PBS, pH, 7.0) is satisfactory. Formalin treatment renders the suspension noninfectious. The antigens may be stored as frozen aliquots; after thawing, they are well mixed on a vortex mixer before use. Antigen dots are placed on a slide in a specific pattern, with separate markings with a pen used for each antigen. Each cluster of dots

includes all the antigenic types to be tested. The antigen dots are air dried and fixed on slides with acetone (15 min at room temperature). Slides may be stored frozen. When thawed for use, they may sweat, but they can be conveniently dried (as can the original antigen dots) with the cool air flow of a hair dryer. The slides have serial dilutions of serum (or tears or exudate) placed on the different antigen clusters. The clusters of dots are placed sufficiently separated to avoid the running of the serum from cluster to cluster. After the serum dilutions have been added, the slides are incubated for 0.5 to 1 h in a moist chamber at 37°C. They are then placed in a buffered saline wash for 5 min, followed by a second 5-min wash. The slides are then dried and stained with fluorescein-conjugated antihuman globulin. Conjugates are pretitrated in a known positive system to determine appropriate working dilutions. This reagent may be prepared against any class of globulin being considered (IgA or secretory piece for secretions, IgG or IgM). Counterstains such as bovine serum albumin conjugated with rhodamine may be included. The slides are then washed twice again, dried, and examined by standard fluorescence microscopy. Use of a monocular tube is recommended to allow greater precision in determining fluorescence of individual elementary body particles. The endpoints are read as the dilution giving bright fluorescence clearly associated with the well-distributed elementary bodies throughout the antigen dot. Identification of the type-specific response is based on dilution differences reflected in the endpoints for different prototype antigens.

For each run of either CF or Micro-IF, known positive and negative sera should always be included. These sera should always duplicate their titers as previously observed within the experimental (±1 dilution) error of the system.

Interpretation of Laboratory Data

The most widely used serologic test for diagnosing chlamydial infections is the complement fixation (CF) test. This is useful in diagnosing psittacosis, in which paired sera often show fourfold or greater increases in titer. It may also be useful in diagnosing LGV, in which single-point titers greater than 1:64 are highly supportive of this clinical diagnosis. With LGV it is difficult to demonstrate rising titers because the nature of the disease results in the patient being seen by the physician after the acute stage. Any titer above 1:16 is considered significant evidence of exposure to chlamydiae. The CF test is not particularly useful in diagnosing trachoma inclusion conjunctivitis or the related genital tract infections, and it plays no role in diagnosing neonatal chlamydial infections. It seems to be useful in TWAR infections.

The Micro-IF method is a much more sensitive procedure for measuring antichlamydial antibodies. If appropriate antigens are included it may be used in diagnosing psittacosis, in which paired sera will show rising immunoglobulin G (IgG) titers (and often IgM antibody). With LGV it is again difficult to demonstrate rising titers, but in active cases usually have relatively high levels of IgM antibody (>1:32) and IgG levels ≥1:2,000. Trachoma, inclusion conjunctivitis, and the genital tract infections may be diagnosed by the Micro-IF technique if appropriately timed paired acute and convalescent sera can be obtained. However, it is often difficult to demonstrate rising antibody titers, particularly in sexually active populations. Many of these individuals will be seen for chronic or repeat infections. The background rate of seroreactors in venereal disease clinics is ≥60%, making it particularly difficult to demonstrate seroconversion. In general, first attacks of chlamydial urethritis have been regularly associated with seroconversion (Bowie et al., 1977). Individuals with systemic infection (epididymitis, salpingitis) usually have much higher antibody levels than those with superficial infections, and women tend to have higher antibody levels than men.

Serology is particularly useful in diagnosing chlamydial pneumonia in neonates. In this case, high levels of IgM antibody are regularly found in association with disease (Schachter et al., 1982). IgG antibodies are less useful because the infants are being seen at a time when they have considerable levels of circulating maternal IgG as all these infections are acquired from the infected mother, who is almost always seropositive. It takes between 6 and 9 months for maternal antichlamydial antibodies to disappear. Infants older than that age may be tested for determination of prevalence of chlamydial infection without fear of confounding effects of maternal antibody. Infants with inclusion conjunctivitis or respiratory tract carriage of *Chlamydia* without pneumonia usually have very low levels of IgM antibodies. Thus, a single titer ≥1:32 may support the diagnosis of chlamydial pneumonia.

The Micro-IF technique uses many serotypes of chlamydiae and the procedure as simplified by Wang is recommended (Wang et al., 1975). Because the IgM antibody responses tend to be markedly specific, the use of single broadly reacting antigens will miss at least 15 to 25% of the infections that can be proven to be due to *Chlamydia* by other procedures, or that would be positive by a multiple antigen Micro-IF. The single-antigen tests may involve either yolk sac suspensions of agent or identification of fluorescent inclusions in tissue monolayers (Richmond and Caul, 1977; Thomas et al., 1976). Serotypes

within the DEL serogroup are commonly chosen for this purpose. Inclusion fluorescence assays will measure genus-specific reactions also because of the LPS in the inclusion.

Monotypic A seroreactions, at least in industrialized countries, are liable to be spurious. Long-term longitudinal studies on infants suggest that the appearance of antibodies against type A (and to a lesser extent the cross-reacting CJI serotypes) may appear in response to nonchlamydial antigenic stimulus (Schachter et al., 1982). These antibodies are usually transient and do not result in the persistent high levels of IgG antibodies that usually follow chlamydial infections.

Enzyme immunoassay (EIA) techniques have been described which measure antichlamydial antibodies (Finn et al., 1983). Most of these procedures have been successful in measuring IgG antibody, albeit often less sensitive than the Micro-IF test. They have been less successful in measuring IgM antibody. Some tests are commercially available. There is inadequate published experience with those tests to recommend them. It is likely that they will be less sensitive than the Micro-IF and will miss some C complex reactors and cannot be readily applied to IgM antibody. The procedure may be of some use in selected instances and for serosurveys in laboratories where Micro-IF techniques are not available.

Epidemiology and Natural History

Psittacosis

This infection is usually spread by the aerosol route and human infection almost always occurs as a result of exposure to infected avian species (Page, 1972; Schachter and Dawson, 1978). Most infections are attributed to exposure to pet birds, usually exotic psittacine varieties or parakeets. It is likely that infection is acquired by aerosol from infective droppings. The human disease is also recognized as an occupational hazard within the poultry industry, particularly to workers in turkey processing plants. Although human-to-human transmission has been described, particularly in some very severe outbreaks of pneumonitis due to *C. psittaci*, such transmission is relatively uncommon.

Lymphogranuloma Venereum

Lymphogranuloma venereum is a sexually transmitted disease (STD) of worldwide distribution (Schachter and Osoba, 1983). Although relatively uncommon in certain of the western industrialized countries and once considered to be a tropical disease, it is now recognized to occur throughout the world. It is more important in some developing countries, where it is responsible for up to 2 to 6% of STD clinic visits. One feature of the disease which is underrecognized, is the importance of this organism as a contributor to proctocolitis in homosexual men (Bolan et al., 1982; Quinn et al., 1981).

Trachoma

Trachoma is still considered one of the world's leading causes of preventable blindness (Schachter and Dawson, 1978; Jones, 1975). It is a major public health problem in many developing countries, particularly those in North Africa and subSaharan Africa and Southeastern Asia where blinding trachoma still exists. In these settings the infection is holoendemic with virtually all members of affected communities acquiring the infection before they are 2 years of age. It is a disease of families and of poverty, being found at highest rates and in the most severe form among the poorest portions of the society. Trachoma responds markedly to improving socioeconomic conditions and has disappeared from some countries as a result of improved standard of living, rather than specific antitrachoma measures.

Genital Tract Infections and Inclusion Conjunctivitis

These infections are sexually transmitted and the proportions of society most affected are those at highest risk for other sexually transmitted agents (Grayston and Wang, 1975; Oriel and Ridgway, 1982; Schachter, 1978). Numerous surveys have shown that members of lower socioeconomic strata, particularly teenagers, have the highest prevalence of *C. trachomatis* infection (Schachter et al., 1983). It is estimated that in excess of 3 million chlamydial infections occur in the United States each year (Banks et al., 1970). To place this in proper perspective, a similar estimation of gonococcal infections places that number at 2 million per annum.

Control and Prevention

C. trachomatis

TRACHOMA

Trachoma is a disease of poverty in developing countries. There are no control measures better than improvement in standard of living. To prevent blinding complications in the hyperendemic areas, the World Health Organization recommends intermittent topi-

cal therapy with 1% tetracycline ointment. In the typical setting this is given to all school children or young children in a village once or twice daily for 6 consecutive days on a twice yearly schedule. This will not cure trachoma, but it will reduce the severity of the disease and thus prevent blinding complications. Corrective lid surgery is available to prevent development of blindness in those who already have lid deformity.

SEXUALLY TRANSMITTED INFECTIONS

It is likely that screening programs for asymptomatic sexually active individuals will yield higher rates of infection and be more cost effective for *C. trachomatis* infection than they are for gonococcal infection. Routine treatment of all gonococcal infections in heterosexuals with regimens effective against *C. trachomatis* will have some impact on the reservoir. However, there are no specific chlamydial control programs in effect. A program of screening pregnant women and erythromycin treatment for those found to be infected will prevent perinatal chlamydial infection (Schachter et al., 1986). Use of topical erythromycin or tetracycline for ocular prophylaxis in the newborn will prevent inclusion conjunctivitis, but not pneumonia (Laga et al., 1986).

C. psittaci

PSITTACOSIS

Chemoprophylaxis of exotic birds in quarantine stations or before they are introduced into commerce will reduce the infectious load. Tetracycline containing feeds can be effective in eradicating chlamydial infection in these species (Arnstein et al., 1968). Thus, infected birds are not released into commerce. This can minimize the human psittacosis resulting from this source. *C. psittaci* in poultry is often an inapparent infection and thus is difficult to control in the poultry industry. When clinical manifestations of chlamydial infection occur in these birds they can be treated by incorporation of tetracyclines into food.

Literature Cited

Arnstein, P., B. Eddie, and K. F. Meyer. 1968. Control of psittacosis by group chemotherapy and infected parrots. Am. J. Vet. Res. 29:2213–2227.

Banks, J., M. Eddie, M. Sung, N. Sugg, J. Schachter, and K. F. Meyer. 1970. Plaque reduction technique for demonstrating neutralizing antibodies for *Chlamydia*. Infect. Immun. 2:443–447.

Beem, M. O., and E. M. Saxon. 1977. Respiratory tract colonization and a distinctive pneumonia syndrome in infants infected with *Chlamydia trachomatis*. N. Engl. J. Med. 293:306–310.

Bernstein, D. I., T. Hubbard, W. Wenman, B. L. Johnson, Jr., K. K. Holmes, H. Liebhaber, J. Schachter, R. Barnes, and M. A. Lovett. 1984. Mediastinal and supraclavicular lymphadenitis and pneumonitis due to *Chlamydia trachomatis* serovars L1 and L2. N. Engl. J. Med. **311**:1543–1546.

Bolan, R. K., M. Sands, J. Schachter, R. C. Miner, and W. L. Drew. 1982. Lymphogranuloma venereum and acute ulcerative proctitis. Am. J. Med. **72**:703–706.

Bowie, W. R., S-P. Wang, E. R. Alexander, J. Floyd, P. Forsyth, H. Pollock, J-S. Tin, T. Buchanan, and K. K. Holmes. 1977. Etiology of nongonococcal urethritis: evidence for *Chlamydia trachomatis* and *Ureaplasma urealyticum*. J. Clin. Invest. **59**:735–742.

Byrne, G. I., and J. W. Moulder. 1978. Parasite-specified phagocytosis of *Chlamydia psittaci* and *Chlamydia trachomatis* by L and HeLa cells. Infect. Immun. **19**:598–606.

Caldwell, H. D., J. Kromhout, and J. Schachter. 1981. Purification and partial characterization of the major outer membrane protein of *Chlamydia trachomatis*. Infect. Immun. **31**:1161–1176.

Caldwell, H. D., and J. Schachter. 1982. Antigenic analysis of the major outer membrane protein of *Chlamydia* spp. Infect. Immun. **35**:1024–1031.

Chernesky, M. A., J. B. Mahony, S. Castrciano, M. Mores, I. O. Stewart, S. J. Landis, W. Seidelman, C. Leman, and E. J. Sargeant. 1986. Detection of *Chlamydia trachomatis* antigens by enzyme immunoassay and immunofluorescence in genital specimens from symptomatic and asymptomatic men and women. J. Infect. Dis. **154**:141–148.

Cohen, S. D., P. H. Azimi, and J. Schachter. 1982. *Chlamydia trachomatis* associated with severe rhinitis and apneic episodes in a one-month-old infant. Clin. Pediatr. **21**:498–499.

Darougar, S. J., R. Kinnison, and B. R. Jones. 1971. Simplified irradiated McCoy cell culture for isolation of chlamydiae, p. 63–70. *In* R. L. Nichols (ed.), Trachoma and related disorders caused by chlamydial agents. Excerpta Medica, Amsterdam.

Dhir, S. P., S. Hakomori, G. E. Kenny, and J. T. Grayston. 1972. Immunochemical studies on chlamydial group antigen (presence of a 2-keto-3-deoxycarbohydrate as immunodominant group). J. Immunol. **109**:116–122.

Eb, F., J. Orfila, A. Milon, and M. F. Geral. 1986. Microimmunofluorescence typing of *Chlamydia psittaci*: epidemiologic interest. Ann. Instit. Pasteur/Microbiol. **137B**:77.

Finn, M. P., A. Ohlin, and J. Schachter. 1983. Enzyme-linked immunosorbent assay for immunoglobulin G and M antibodies to *Chlamydia trachomatis* in human sera. J. Clin. Microbiol. **17**:848–852.

Forsey, T., S. Darougar, and J. D. Treharne. 1986. Prevalence in human being of antibodies to Chlamydia IOL-207, an atypical strain of *Chlamydia*. J. Infect. **12**:145–152.

Gordon, F. B., and A. L. Quan. 1965. Isolation of the trachoma agent in cell culture. Proc. Soc. Exp. Biol. Med. **118**:354–359.

Grayston, J. T., C-C. Kuo, S-P. Wang, and J. Altman. 1986. A new *Chlamydia psittaci* strain, TWAR, isolated in acute respiratory tract infections. N. Engl. J. Med. **315**:161–168.

Grayston, J. T., and S-P. Wang. 1975. New knowledge of chlamydiae and the diseases they cause. J. Infect. Dis. **132**:87–105.

Grimes, J., K. J. Owens, and J. R. Singer. 1979. Experimental transmission of *Chlamydia psittaci* to turkeys from wild birds. Avian Dis. **23**:915.

Halberstaedter, L., and S. von Prowazek. 1907. Zur atiologie des trachoms. Dtsch. Med. Wochenschr. **33**:1285–1287.

Hanna, L. 1968. An evaluation of the fluorescent antibody technique in the diagnosis of trachoma and inclusion conjunctivitis. Rev. Int. Trach. **4**:345–359.

Horn, J. E., T. Quinn, M. Hammer, L. Palmer and S. Falkow. 1986. Use of nucleic acid probes for the detection of sexually transmitted infectious agents. Diagn. Microbiol. Infect. Dis. **4**:101S–109S.

Jones, B. R. 1975. Prevention of blindness from trachoma. Trans. Ophthalmol. Soc. UK **95**:16–33.

Laga, M., H. Nsanze, F. A. Plummer, J. O. Ndinya-Achola, R. C. Brunham, and P. Piot. 1986. Comparison of tetracycline and silver nitrate for the prophylaxis of chlamydial and gonococcal ophthalmia neonatorum, p. 301–304. *In* D. Oriel, G. Ridgway, J. Schachter, D. Taylor-Robinson, and M. Ward (ed.), Chlamydial infections. Cambridge University Press, Cambridge.

Lindner, K. 1911. Gonoblennorrhoe, einschlussblennorrhoe, und trachoma. Graefe's Arch. Ophthalmol. **78**:380.

Lipkin, E. S., J. V. Moncada, M-A. Shafer, T. E. Wilson, and J. Schachter. 1986. Comparison of monoclonal antibody staining and culture in diagnosing cervical chlamydial infection. J. Clin. Microbiol. **23**:114–117.

Mardh, P-A., T. Ripa, L. Svensson, and L. Westrom. 1977. *Chlamydia trachomatis* infection in patients with acute salpingitis. N. Engl. J. Med. **296**:1377–1379.

Marrie, T. J., J. T. Grayston, S-P Wang, C-C. Kuo. 1987. Pneumonia associated with the TWAR strain of *Chlamydia*. Ann. Intern. Med. **106**:507–511.

McKinlay, A. W., N. White, D. Buxton, J. M. Inglis, F. W. A. Johnson, J. B. Kurtz, and R. P. Brettle. 1985. Severe *Chlamydia psittaci* sepsis in pregnancy. Quart. J. Med. **57**:689–696.

Meyer, K. F. 1965. Ornithosis, p. 675. *In* H. E. Biester and L. H. Schwarte (ed.), Diseases of poultry. 5th ed., Iowa State University Press, Ames.

Meyer, K. F. 1967. The host spectrum of psittacosis-lymphogranuloma venereum (PL) agents. Am. J. Ophthalmol. **63**:1225–1246.

Moulder, J. W. 1985. Comparative biology of intracellular parasitism. Microbiol. Rev. **49**:298–337.

Moulder, J. W. 1966. The relation of the psittacosis group (chlamydiae) to bacteria and viruses. Ann. Rev. Microbiol. **20**:107–130.

Moulder, J. W., T. P. Hatch, C-C. Kuo, J. Schachter, and J. Storz. 1984. Order II. Chlamydiales Storz and Page 1971, 334, p. 729–739. *In* N. R. Krieg and J. G. Holt (ed.), Bergey's manual of systematic bacteriology. Williams & Wilkins, Baltimore, Maryland.

Nurminen, M., M. Leinonen, P. Saikku, and P. H. Makela. 1983. The genus-specific antigen of *Chlamydia*: resemblance to the lipopolysaccharide of enteric bacteria. Science **220**:1279–1281.

Oriel, J. D., and G. L. Ridgway. 1982. Genital infection by Chlamydia trachomatis, Edward Arnold Ltd., London.

Page, L. A. 1972. Chlamydiosis (ornithosis), p. 414–417. *In* M. S. Hofstad (ed.), Diseases of poultry. Iowa State University Press, Ames.

Palmer, L., and S. Falkow. 1986. A common plasmid of *Chlamydia trachomatis*. Plasmid **16**:52–62.

Quinn, T. C., S. E. Goodell, E. Mkrtichian, M. D. Schuffler, S-P. Wang, W. E. Stamm, and K. K. Holmes. 1981. *Chlamydia trachomatis* proctitis. N. Engl. J. Med. **305**:195–200.

Rice, C. E. 1936. Carbohydrate matrix of epithelial cell inclusion in trachoma. Am. J. Ophthalmol. **19**:1–8.

Richmond, S. J., and E. O. Caul. 1977. Single-antigen indirect immunofluorescence test for screening venereal disease clinical populations for chlamydial antibodies, p. 259–265. *In* K. K. Holmes and D. Hobson (ed.), Nongonococcal urethritis and related infections. American Society for Microbiology, Washington, D.C.

Ripa, K. T., and P-A. Mardh. 1977. Cultivation of *Chlamydia trachomatis* in cycloheximide-treated McCoy cells. J. Clin. Microbiol. **6:**328–331.

Sacks, D. L., and A. B. MacDonald. 1979. Isolation of a type-specific antigen from *Chlamydia trachomatis* by sodium dodecyl sulfate polyacrylamide gel electrophoresis. J. Immunol. **122:**136–139.

Schachter, J. 1978. Chlamydial infections. N. Engl. J. Med. **298:**428–435, 490–495, 540–549.

Schachter, J. 1985. Immunodiagnosis of sexually transmitted diseases. Yale J. Biol. Med. **58:**443–452.

Schachter, J. 1987. Laboratory studies in Trachoma. *In* Transactions of the World Congress on the Cornea III. Raven Press, Inc., New York.

Schachter, J., J. Banks, N. Sugg, M. Sung, J. Storz, and K. F. Meyer. 1975. Serotyping of *Chlamydia:* isolates of bovine origin. Infect. Immun. **11:**904–907.

Schachter, J., and H. D. Caldwell. 1980. Chlamydiae. Ann. Rev. Microbiol. **34:**285–309.

Schachter, J., and C. R. Dawson. 1978. Human chlamydial infections. PSG Publishing, Littleton.

Schachter, J., and M. Grossman. 1981. Chlamydial infections. Ann. Rev. Med. **32:**45–61.

Schachter, J., M. Grossman, and P. H. Azimi. 1982. Serology of *Chlamydia trachomatis* in infants. J. Infect. Dis. **146:**530–535.

Schachter, J., M. Grossman, J. Holt, R. Sweet, and S. Spector. 1979. Infection with *Chlamydia trachomatis:* involvement of multiple anatomic sites in neonates. J. Infect. Dis. **139:**232–234.

Schachter, J., M. Grossman, R. L. Sweet, J. Holt, C. Jordan, and E. Bishop. 1986. Prospective study of perinatal transmission of *Chlamydia trachomatis.* J. Am. Med. Assoc. **255:**3374–3377.

Schachter, J., L. Hanna, E. C. Hill, S. Massad, C. W. Sheppard, J. E. Conte, Jr., S. N. Cohen, and K. F. Meyer. 1975. Are chlamydial infections the most prevalent venereal disease? J. Am. Med. Assoc. **231:**1252–1255.

Schachter, J., and A. Osoba. 1983. O. Lymphogranuloma venereum. Br. Med. Bull. **39:**151–154.

Schachter, J., E. Stoner, and J. Moncada. 1983. Screening for chlamydial infections in women attending family planning clinics: evaluations of presumptive indicators for therapy. West J. Med. **138:**375–379.

Schachter, J., N. Sugg, and M. Sung. 1978. Psittacosis: the reservoir persists. J. Infect. Dis. **137:**44–49.

Schachter, J. 1980. Chlamydiae, p. 700. *In* N. R. Rose and H. Friedman (ed.), Manual of clinical immunology. American Society for Microbiology, Washington, D.C.

Schachter, J., R. L. Sweet, M. Grossman, D. Landers, M. Robbie, and E. Bishop. 1986. Experience with the routine use of erythromycin for chlamydial infections in pregnancy. N. Engl. J. Med. **314:**276–279.

Siegel, M. M. 1962. Lymphogranuloma Venereum, p. 21. *In* Epidemiological, clinical, surgical and therapeutic aspects based on a study in the Caribbean. University of Miami Press, Coral Gables, Florida.

Stamm, W. E., M. Tam, M. Koester, and L. Cles. 1983. Detection of *Chlamydia trachomatis* inclusions in McCoy cell cultures with fluorescein-conjugated monoclonal antibodies. J. Clin. Microbiol. **17:**666–668.

Stuart, E. S., and A. B. MacDonald. 1982. Isolation of a possible group antigenic determinant of *Chlamydia trachomatis,* p. 57–60. *In* P-A. Mardh, K. K. Holmes, J. D. Oriel, P. Piot, and J. Schachter (ed.), Chlamydial infections. Elsevier Biomedical Press, Amsterdam.

Stephens, R. S., M. R. Tam, C-C. Kuo, and R. C. Nowinski. 1982. Monoclonal antibodies to *Chlamydia trachomatis:* antibody specificities and antigen characterization. J. Immunol. **128:**1083–1089.

Storz, J. 1971. Chlamydia and chlamydia-induced diseases, Charles C. Thomas, Springfield.

Storz, J., J. L. Shupe, M. E. Marriott, and W. R. Thornley. 1965. Polyarthritis of lambs induced experimentally by a psittacosis agent. J. Infect. Dis. **115:**9–18.

Tam, M. R., W. E. Stamm, H. H. Handsfield, R. Stephens, C-C. Kuo, K. K. Holmes, K. Ditzenberger, M. Crieger, and R. C. Nowinski. 1984. Culture-independent diagnosis of *Chlamydia trachomatis* using monoclonal antibodies. N. Engl. J. Med. **310:**1146–1150.

T'ang, F. F., H-L. Chang, Y-T. Huang, and K-C. Wang. 1957. Trachoma virus in chick embryo. Natl. Med. J. China **43:**81–86.

Thomas, B. J., R. T. Evans, G. R. Hutchinson, and D. Taylor-Robinson. 1977. Early detection of chlamydial inclusions combining the use of cycloheximide-treated McCoy cells and immunofluorescence staining. J. Clin. Microbiol. **6:**285–292.

Thomas, B. J., P. Reeve, and J. D. Oriel. 1976. Simplified serological test for antibodies to *Chlamydia trachomatis.* J. Clin. Microbiol. **4:**6–10.

Wang, S-P., and J. T. Grayston. 1970. Immunologic relationship between genital TRIC, lymphogranuloma venereum, and related organisms in a new microtiter indirect immunofluorescence test. Am. J. Ophthalmol. **70:**367–374.

Wang, S-P., J. T. Grayston, E. R. Alexander, and K. K. Holmes. 1975. Simplified microimmunofluorescence test with trachoma-lymphogranuloma venereum (*Chlamydia trachomatis*) antigens for use as a screening test for antibody. J. Clin. Microbiol. **1:**250–255.

Watkins, N. G., W. J. Hadlow, A. B. Moos, and H. D. Caldwell. 1986. Ocular delayed hypersensitivity: a pathogenic mechanism of chlamydial conjunctivitis in guinea pigs. Proc. Natl. Acad. Sci. USA **83:**7480–7484.

Weiss, S. G., R. W. Newcomb, and M. O. Beem. 1986. Pulmonary assessment of children after chlamydial pneumonia of infancy. J. Pediatr. **108:**659–664.

Yoder, B. L., W. E. Stamm, M. C. Koester, and E. R. Alexander. 1981. Microtest procedure for isolation of *Chlamydia trachomatis.* J. Clin. Microbiol. **13:**1036–1039.

CHAPTER 44

Rickettsiaceae: The Rickettsiae

JOSEPH E. McDADE and DANIEL B. FISHBEIN

Diseases: 1) Rocky Mountain spotted fever, 2) Boutonneuse fever, 3) rickettsialpox, 4) epidemic typhus, 5) murine typhus, 6) scrub Typhus, 7) trench fever, and 8) Q-fever.

Etiologic Agents: 1) *Rickettsia rickettsii,* 2) *Rickettsia conorii,* 3) *Rickettsia akari,* 4) *Rickettsia prowazekii,* 5) *Rickettsia typhi,* 6) *Rickettsia tsutsugamushi,* 7) *Rochalimaea quintana,* and 8) *Coxiella burnetii.*

Source: 1) and 2) bites of infected ticks, 3) infected mites, 4) and 7) infected body lice, 5) infected fleas, 6) infected chiggers, and 8) contracted from domestic animals by inhalation of infectious aerosols.

Clinical Manifestations: Rickettsial diseases are characterized by the sudden onset of fever, headache, myalgias, and arthralgias. Rash frequently accompanies all rickettsioses except Q-fever; the type of rash and its distribution help to distinguish the various diseases. Pulmonary and gastrointestinal symptoms are common in most rickettsioses.

Pathology: Information about Q-fever, rickettsialpox, and trench fever is limited. The other rickettsiae have a tropism for the endothelial cells of the capillaries and the principal damage is to the vascular endothelium. Gross and microscopic lesions may occur in the brain, kidneys, lungs, and heart.

Laboratory Diagnosis: Indirect fluorescent antibody tests are available for all diseases, except trench fever. Centers for Disease Control considers titers ≥ 256 diagnostic for Q-fever and titers ≥ 64 to be diagnostic for the other rickettsioses, but there is variability in diagnostic titers among laboratories. Isolation is useful but hazardous.

Epidemiology: 1) Limited to the western hemisphere; 2) occurs in Africa, the Mediterranean countries, and elsewhere in the eastern hemisphere; 3) Usually occurs as small outbreaks in urban tenements; 4) localized primarily in the highland areas of Africa, Central America, and South America, disease occurs mostly in outbreaks; 5) exists worldwide in areas where *Rattus* and the oriental rat flea (*Xenopsylla cheopis*) coexist; 6) endemic in the western Pacific countries, and reinfections occur due to the presence of multiple serotypes; 7) occurred in troops in both World Wars, but has been studied little since then, and *R. quintana* has been isolated in Mexico; and 8) occurs worldwide both sporadically and in outbreak situations, and it is an occupational disease among abattoir workers.

Treatment: All rickettsiae are susceptible to tetracyclines or chloramphenicol. A combination of tetracycline with another drug is often useful for the treatment of Q-fever endocarditis.

Prevention and Control: Vaccines are generally unavailable. Avoidance of vector-infested areas or vector removal can reduce the risk of all rickettsioses except Q-fever. Proper disposal of infected animal products can reduce the risk of Q-fever.

Introduction

The rickettsiae are gram-negative obligate intracellular bacteria; at least 10 species are pathogenic for humans. They are grouped into three genera, *Rickettsia, Coxiella,* and *Rochalimaea* (Weiss and Moulder, 1984). Every species is associated with an arthropod vector (lice, ticks, fleas, or mites) at some stage in its life cycle (Table 1). Except for louse-borne typhus rickettsiae (*Rickettsia prowazekii*), humans are accidental hosts in the life cycles of these microorganisms. All rickettsiae are included in the family *Rickettsiaceae* in the order *Rickettsiales.*

Rickettsial diseases have played an important role in the history of western civilization. For example, epidemics of louse-borne typhus were documented as early as 420 B.C. during the Peloponnesian War (Snyder, 1973), and have periodically decimated cities and ravaged armies from then to the present. Millions of cases of louse-borne typhus occurred among troops and prisoners during both World Wars, with fatalities due to typhus infection equalling or surpassing battlefield casualties (Snyder, 1973). Thousands of cases of louse-borne typhus are still reported each year, mostly in the remote highland areas of Africa and South America. The global incidence of typhus is greatly underreported, however, because of the lack of adequate surveillance systems in endemic areas.

Scrub typhus was well known to oriental missionaries centuries ago (Audy, 1968). Tens of thousands of cases occur annually in China, Japan, Indonesia, Southeast Asia, Australia, and various islands in the western Pacific (Traub and Wisseman, 1974). Antibody prevalence rates for scrub typhus exceed 60% in some populations (Brown et al., 1976). Fatality rates approaching 10% are common in untreated patients.

Various species of tick-borne rickettsiae are endemic throughout the world and cause mild to severe disease in indigenous populations (Hoogstraal, 1981). Included in this category are Rocky Mountain spotted fever (RMSF), Boutonneuse fever (Marseilles fever, Indian tick typhus, African tick typhus), Siberian tick typhus, and Queensland tick typhus (Table 1). An average of 1,000 cases of Rocky Mountain spotted fever has been reported annually in recent years in the United States alone (Fishbein et al., 1984). Tick-borne rickettsiae have lived in close association with their acarine and mammalian hosts for millennia, often with disastrous results (Marchette and Stiller, 1982). For example, RMSF ravaged the early settlers in Montana and Idaho (Wilson and Chowning, 1904).

Q-fever was not reported as a distinct clinical entity until 1937 (Derrick, 1937), but it subsequently has been documented in at least 51 countries throughout the world. Sporadic cases and outbreaks of Q-fever have occurred on several continents, and chronic Q-fever infections are documented with increasing frequency. Q-fever is a serious occupational disease of abattoir workers (Marmion et al., 1984).

Flea-borne (murine) typhus occurs commonly in many underdeveloped countries but is seldom reported (Traub et al., 1978). Other rickettsial diseases include rickettsialpox (Brettman et al., 1981; Huebner et al., 1946; Rose, 1948) and trench fever (Vinson, 1973). Incidence data for the last two diseases are generally unavailable, although both undoubtedly occur regularly in endemic areas. Major epidemics of trench fever were reported during World War I and II (Vinson, 1973).

Rickettsiae

Antigenic Characterization

The pathogenic rickettsiae have been subdivided into five groups of antigenically related microorganisms. This antigenic grouping follows the generic separation, except that the genus *Rickettsia* is subdivided into spotted fever, typhus, and scrub typhus serogroups. DNA-to-DNA hybridization studies of the rickettsiae performed to date (Myers and Wisseman, 1980, 1982) have confirmed that the species that comprise the serogroups are related.

The spotted fever and typhus groups include multiple species, whereas the scrub typhus group consists of different strains of only one species, *Rickettsiae tsutsugamushi* (Table 1). Convalescent-phase serum samples from patients infected with any given species of *Rickettsia* react with other species in the same serogroup, but they usually do not react with species in the other serogroups. Samples of convalescent-phase sera from patients with Rocky Mountain spotted fever or murine typhus occasionally react with typhus- and spotted fever-group rickettsiae, respectively (Ormsbee et al., 1978b), but these cross-reactions are the exception rather than the rule. Immune sera prepared in guinea pigs and most other species of animal are also group reactive. Species within the spotted fever and typhus serogroups can be distinguished with immune mouse sera (Philip et al., 1978), but such differentiations are only performed by some reference laboratories.

The species within the various serogroups have other properties in common, including vector association, optimal growth temperatures, and pathogenicity for animals (Weiss and Moulder, 1984). Because of the many common features of member species, the five different serogroups of rickettsiae are sometimes referred to as biogroups (Table 1).

TABLE 1. Outline of pathogenic rickettsiae

Serogroup	Member species	Human disease and synonyms	Distribution	Natural cycle of infection		
				Arthropod vector	Mammalian host	Transmission to humans
Spotted fever	R. rickettsii	Rocky Mtn. spotted fever, fiebre manchada	Western hemisphere	Ixodid ticks	Ticks/small mammals	Tick bite
	R. sibirica	Siberian tick typhus	Siberia, Mongolia	Ixodid ticks	Ticks/wild and domestic animals	Tick bite
	R. conorii	Boutonneuse fever, South African tick-bite fever, Kenya tick typhus, Indian tick typhus	Mediterranean countries, India, Africa	Ixodid ticks	Ticks/rodents, dogs	Tick bite
	R. australis	Queensland tick typhus	Australia	Ixodid ticks	Ticks/wild and domestic animals	Tick bite
	R. akari	Rickettsialpox	United States, Russia, Korea	Gamasid mites	Mite/mice, voles?	Mite bite
Typhus	R. prowazekii	Epidemic typhus, louse-borne typhus	Primarily highland areas of South America, Africa	Body lice	Humans	Infected louse feces into broken skin or mucous membranes
		Recrudescent typhus, Brill-Zinsser disease	Worldwide	None: recrudescent illness		Reactivation of latent infection years after contracting louse-borne typhus
		Flying squirrel-associated typhus fever	United States	Flying squirrel lice	Flying squirrels	Contract with flying squirrels (Glaucomys volans)
	R. typhi (-R. mooseri)	Murine typhus, endemic typhus	Worldwide	Fleas	Rodents	Infected flea feces into broken skin or mucous membranes
Scrub typhus	R. tsutsugamushi (multiple serotypes)	Scrub typhus, tsutsugamushi disease	Asia, Australia, Pacific Islands	Chiggers (trombiculid mite)	Chiggers/rodents?	Chigger bite
Q-fever	Coxiella burnetii	Q-fever	Virtually ubiquitous	Ticks?	Ticks/mammals	Inhalation of infectious aerosols
Trench fever	Rochalimaea quintana	Trench fever, 5-day fever	Uncertain—previously reported in Mexico, Europe, North Africa, Middle East	Body lice	Humans	Infected louse feces into skin or mucous membranes

Morphology, Replication, Physicochemical Properties, and Stability

RICKETTSIA

Members of the genus *Rickettsia* have morphologic and biochemical properties typical of gram-negative bacteria. They are short, rod-shaped or coccobacillary organisms, ranging from 0.8 to 2.0 μm long and 0.3 to 0.5 μm wide. They multiply by transverse binary fission (Ris and Fox, 1949). Elongated forms frequently develop under suboptimal growth conditions in which cell division is impaired. *Rickettsia* do not have flagella.

The *Rickettsia* have a five-layered outer envelope complex (Silverman and Wisseman, 1978; Silverman et al., 1978). A microcapsular layer is attached to the outer leaflet, and it is covered by a polysaccharide slime layer. The morphology of *Rickettsia tsutsugamushi* is distinct among the *Rickettsia* in that its outer leaflet is much thicker and its inner leaflet is much thinner than the corresponding leaflets in other rickettsial species. Lipopolysaccharide (LPS) has been detected in *Rickettsia* species (Schramek et al., 1976), as has the LPS precursor 2-keto-3-deoxyoctulosonic acid (Smith and Winkler, 1979). Ultrastructural observations of various species of *Rickettsia* also have revealed the presence of chromosomal structures and ribosomes (Ris and Fox, 1949; Silverman et al., 1978). All *Rickettsia* are exceptionally unstable outside the host cell and are readily inactivated at temperatures $\geq 56°C$ and by standard disinfectants.

Rickettsia enter host cells by induced phagocytosis (Walker and Winkler, 1978). Entry depends on both the metabolic activity of the rickettsiae and the phagocytic ability of the host cell (Winkler, 1982). The rickettsiae escape the phagosome, however, and multiply in the cytoplasm of the host cell. Wisseman and Waddell (1975) showed that *R. prowazekii* grew exponentially in the cytoplasm of irradiated chick embryo cells with a generation time of approximately 9 h. Almost 100 organisms accumulated in infected cells before they ruptured several days later.

The infection cycle of *R. rickettsii*, however, is somewhat more complex. Wisseman et al. (1976) found that after initial uptake by the host cells, *R. rickettsii* remained stable in the cytoplasm for several hours; some rickettsiae then moved bilaterally across the cell membrane and infected adjacent cells. *R. rickettsii* occasionally migrated unidirectionally across the nuclear membrane and infected cell nuclei. Several days after infection, the host cells degenerated and lysed, even though relatively few rickettsiae were present in the cytoplasm. In the few instances where growth occurred in the nucleus, compact masses of rickettsiae were observed. Other species of spotted fever-group rickettsiae can also proliferate in the nucleus.

The infection cycle of *R. tsutsugamushi* differs substantially from that of spotted fever- and typhus-group rickettsiae. Growth of *R. tsutsugamushi* occurs primarily in the perinuclear region; mature microorganisms are extruded from the host cell and surrounded by host cell membranes (Rikihisa and Ito, 1980). Hase (1985) concluded from his ultrastructural studies that *R. tsutsugamushi* is assembled de novo in the host cell cytoplasm, but that hypothesis has not been evaluated by others.

Although members of the genus *Rickettsia* grow only in the presence of eukaryotic host cells, they possess considerable synthetic capabilities and generate their own ATP via the tricarboxylic acid cycle (Weiss and Moulder, 1984). Additionally, *Rickettsia* possess an ADP/ATP translocator system that allows them to exchange their own ADP with the ATP in the host cell (Winkler, 1976). Presumably the rickettsiae use host cell ATP as long as it is available and produce their own supply when the host's becomes depleted. The precise regulatory mechanism for these events is currently under investigation (Krause et al., 1985; Phibbs and Winkler, 1982).

COXIELLA

The genus *Coxiella* is composed of a single species, *C. burnetii*. It is a short, rod-shaped microorganism, 0.2 to 0.4 μm wide and 0.4 to 1.0 μm long. Like the *Rickettsia*, *C. burnetii* is an obligate intracellular bacterium that enters the host cell by phagocytosis. Unlike *Rickettsia*, however, *C. burnetii* does not escape from the phagosome. Instead, it multiplies efficiently there (by binary fission) until the host cells eventually lyse and release microorganisms into the environment. The low pH optimum (4.5) of *C. burnetii* enzymes evidently allows this microorganism to survive and multiply in the phagosome (Hackstadt and Williams, 1981).

C. burnetii is resistant to inactivation by physical and chemical treatment (see Bernard et al., 1982). It can survive for a year or more when attached to wool or other fomites, but it is incompletely inactivated when held at 63°C for 30 min. *C. burnetii* is only partially inactivated when exposed to 1% formalin or phenol for 24 h. It can be inactivated by 0.05% hypochlorite, 5% H_2O_2, or a 1 : 100 dilution of Lysol.* Some attribute the stability of *C. burnetii* to the presence of an endospore (McCaul and Williams, 1981), but this has not yet been confirmed.

---

* Use of trade names is for identification only and does not constitute endorsement by the Public Health Service or by the United States Department of Health and Human Services.

C. burnetii displays an antigenic phase variation that is unique among the rickettsiae. The organism exists in two phases (I and II), analogous to the smooth to rough transition that occurs in some species of enteric bacteria. *C. burnetii* is in antigenic phase I in nature, but it changes to phase II after multiple passages in tissue cultures or embryonated eggs. The phase I antigen is most likely a polysaccharide component of *Coxiella* lipopolysaccharide, and the phase I to phase II transition probably occurs when one or more carbohydrate components from the LPS moiety are no longer synthesized (Hackstadt et al., 1985).

C. burnetii is also unique among the rickettsiae in that it contains plasmids (Mallavia et al., 1984; Samuel et al., 1983, 1985). Although their function is unknown, the presence or absence of plasmids does not affect the antigenic phase.

ROCHALIMAEA

Rochalimaea quintana is morphologically similar to the *Rickettsia* species. It differs from all other members of the family *Rickettsiaceae* in that it can be cultivated on cell-free media (Vinson and Fuller, 1961). A second species of *Rochalimaea* (*R. vinsonii*) has been described, but is not thought to be a human pathogen. Because knowledge of the epidemiology, biologic properties, and natural history of *R. quintana* is quite incomplete, this organism will not be discussed further. Additional information on this genus can be found elsewhere (Weiss and Moulder, 1984).

Pathogenesis, Pathology, and Pathophysiology

General

Members of the genus *Rickettsia* display a unique tropism for the endothelial cells of the microcirculatory system, especially the capillaries, and the consequent damage to the vasculature is the basis for similarities in the pathogenetic, pathologic, and pathophysiologic features of rickettsial diseases (Walker and Mattern, 1980). Information about the pathologic aspects of acute *Coxiella* infection is limited because Q-fever is rarely fatal and few specimens are available for pathologic examination. Available data indicate that this infection is very different from those caused by members of the genus *Rickettsia*. Thus, the pathology of diseases caused by *Coxiella* and *Rickettsia* are discussed separately.

Rickettsia

Rickettsia are transmitted by the bite of infected ticks and mites or by contamination of the skin or mucous membranes with louse or flea feces. Regardless of their mode of entry, rickettsiae invade the capillary vascular endothelium and frequently cause focal swelling and a mixed inflammatory infiltration at the site of inoculation. Rickettsiae then proliferate locally at the site of infection. A result is the primary lesion or eschar that frequently appears during the incubation period of scrub typhus, rickettsialpox, and Boutonneuse fever. In scrub typhus, the eschar begins intraepidermally and extends in all directions as the rickettsiae destroy adjacent tissue. At the edge of the suppuration and necrosis, the veins show evidence of internal damage and cellular infiltration, and it may be here that the vasculitis begins (Allen and Spitz, 1945). Host factors as well as the depth of injection of both *R. tsutsugamushi* and *R. conorii* are important in determining whether an eschar will form, but not whether generalized infection results (Allen and Spitz, 1945). The absence of an eschar in a substantial proportion of patients with scrub typhus (and other rickettsial diseases as well) militates against the pathogenic importance of the eschar.

After local proliferation in the endothelial cells, the rickettsiae become disseminated in the vasculature of many organ systems. The route of this dissemination has not been elucidated, but is presumed to be hematogenous or lymphangitic. Focal involvement of small blood vessels of the skin and other organs is seen, and this correlates closely with many of the clinical manifestations of typhus- and spotted fever-group infections. The disseminated lesions of rickettsial infections directly reflect rickettsia-induced focal swelling and proliferation of endothelial cells as well as the acute inflammatory infiltration (Walker and Mattern, 1980; Walker and Gear, 1985). In vitro studies suggest that "balance between rickettsial replication and rickettsia-induced host cell damage" permits a substantial increase in the number of microorganisms before the host cells are destroyed (Wisseman, 1986). The primary cellular lesion results from dilation and destruction of intracellular membranes, particularly the rough endoplasmic reticulum (Silverman, 1984).

As the endothelial cells die, necrosis of the intima and media of the blood vessels leads to the formation of hyaline thrombi composed of fibrin and cellular debris. These hyaline thrombi cause microinfarcts and extravasation of blood, and they are manifested grossly by the petechial lesions that are the hallmark of rickettsial disease. Gross and microscopic lesions are found in the brain, kidneys, lungs, and heart. The clinical consequences of infarcts in these tissues are much greater than those of the vasculitis.

Unlike other rickettsiae, *R. rickettsii* is not confined to capillary endothelium; it also invades and destroys vascular smooth muscle cells and the vascular endothelium of larger vessels. *R. rickettsii* spreads centripetally in the vascular system to the arterioles and veins (Walker and Mattern, 1980). These and other aspects of the clinical and pathologic spectrum of RMSF are unique among rickettsial infections and are related to the ability of *R. rickettsii* to migrate freely from cell to cell.

Typhus- and spotted fever-group infections may cause interstitial pneumonitis; there is considerable variation in the frequency of this finding. In one comparative pathologic study, the frequency and severity of pulmonary involvement were greatest in scrub typhus, intermediate in epidemic typhus, and least in RMSF (Allen and Spitz, 1945). In fatal cases of RMSF the distribution of *R. rickettsii* in lung tissue parallels observed pulmonary vasculitis, suggesting that rickettsial invasion of the pulmonary microcirculation causes the interstitial pneumonitis of RMSF (Walker et al., 1980a). The consequences of vasculitis, vascular damage, and increased permeability include alveolar septal congestion and interstitial edema; alveolar edema, fibrin formation, macrophage accumulation, and hemorrhage; and interlobar, septal, and pleural effusion (Walker et al., 1980a). The pulmonary dysfunctions (e.g., coughing, pneumonia, and pulmonary edema) that sometimes accompany RMSF are the clinical consequence of this involvement (Donohue, 1980).

Lesions in the central nervous system (CNS) play a prominent part in the clinical complications of RMSF and typhus. All portions of the brain and spinal cord may be involved, but the rickettsiae display a special proclivity for the midbrain and inferior olivary nucleus. Focal proliferations of endothelial and neuroglial (predominantly oligodendroglial) cells have been called "typhus nodules" (Allen and Spitz, 1945). (Once thought to be characteristic of rickettsial involvement of the CNS, "typhus nodules" have since been found in Chagas' disease, malaria, toxoplasmosis, typhoid fever, and many arthropod-borne encephalitides.) Other pathologic manifestations of CNS involvement in epidemic and scrub typhus include mononuclear cell meningitis, perivascular cuffing of arteries, focal hemorrhage, and degeneration of ganglion cells (Allen and Spitz, 1945). Unlike epidemic and scrub typhus, the neurologic findings of RMSF are the result of microinfarcts in the CNS and frequently involve the white matter (Allen and Spitz, 1945; Harrell, 1949).

A histologically similar interstitial myocarditis is found universally in postmortem heart specimens from patients with epidemic typhus, scrub typhus, and RMSF. In RMSF, heart lesions are patchy and consist primarily of a mixed mononuclear cell infiltrate. As in the lungs, the location of coronary vasculitis in RMSF correlates well with the localization of *R. rickettsii* in cardiac vessels, as demonstrated by specific immunofluorescence (Walker et al., 1980b).

A discrepancy is found between the severity of myocarditis and the severity of cardiac insufficiency. The hypotension that often accompanies RMSF (Walker and Mattern, 1980; Walker et al., 1980b) and typhus is not associated with left ventricular dilation; this strongly suggests that myocarditis is not involved in this complication. Shock is primarily attributable to an increase in capillary permeability that leads to a decrease in intravascular volume and systemic vascular resistance. In epidemic typhus, interstitial myocarditis, epicarditis, and mural endocarditis spare the heart valves (National Research Council, 1953) and are secondary to capillary injury. Cardiac involvement in epidemic typhus is generally infiltrative and primarily involves small blood vessels. As in RMSF, there are few well-documented clinical correlates of cardiac involvement in epidemic typhus.

The urinary tract is commonly involved in RMSF and typhus, but this infrequently leads to clinical consequences; the remote effects of decreased intravascular volume are primarily responsible for the abnormalities of renal function. Although the nodular proliferative reactions seen in the CNS and skin are uncommon in the kidneys, multifocal perivascular interstitial nephritis is frequent. The azotemia found in RMSF and in typhus is usually prerenal, at least early in the course of infection, and reflects intravascular volume depletion. Acute renal failure in RMSF is probably due to the acute tubular necrosis that results from systemic hypotension. Acute diffuse glomerulonephritis was described in pathologic studies of the RMSF patients conducted in the 1940s (Allen and Spitz, 1945), but more recent studies failed to detect this abnormality (Walker and Mattern, 1979, 1980).

In epidemic typhus, scrub typhus, and RMSF, histopathologic studies have shown the testes, epididymis, scrotal skin, and adrenals also are involved. Lesions attributable to typhus also have been found in most parts of the alimentary tract (with the possible exception of the colon and salivary glands), but there is general agreement that the gastrointestinal tract is neither frequently nor markedly involved (National Research Council, 1953), except in fatal cases of RMSF (Randall and Walker, 1984). Glossal muscles are almost always involved. Swollen Kupffer cells of the liver are prominent in typhus, and phagocytosis of erythrocytes and inflammatory cells is frequently noted in the spleen.

Coxiella

Microscopically the pathology of Q-fever resembles that of bacterial pneumonia. There is a severe interalveolar, focally necrotizing hemorrhagic pneumonia that is patchy in distribution, involves primarily the alveolar lining cells, and is associated with a necrotizing bronchitis and bronchiolitis. In one case report, the exudate was a nonfibrillary eosinophilic coagulum containing few cells (Whittick, 1950); in another, it was fibrinocellular with substantial numbers of histiocytes, lymphocytes, and plasma cells (Urso, 1975). *C. burnetii* is found in the histiocytes of the alveolar exudate. Histiocytic hyperplasia is found in the mediastinal lymph nodes, spleen, and adrenals.

Hepatocellular damage in acute Q-fever is not widespread; the serum transaminase levels generally are moderately elevated. In a series of 18 liver biopsies in patients with Q-fever, almost all showed granulomatous changes of the lobules; portal areas were only occasionally involved (Srigley et al., 1985). The granulomas consisted of a nondistinctive focal histiocytic and mixed inflammatory cell infiltrate with varying numbers of multinucleated giant cells. In 13 of the 18 biopsy specimens, lipid granulomas also were present. In only seven cases was distinctive lesion present; this was a lipid granuloma with a ring of eosinophilic material. Nongranulomatous changes, present in 11 cases, most commonly included mild to moderate mononuclear infiltrates in the portal areas and focal fatty change. In the late stages of disease, there is often necrosis in the center of the granulomas. Morphologic abnormalities may persist much longer than the clinical illness.

Clinical Features

Mode of Transmission

Spotted fever-group rickettsiae are present in the salivary secretions of ticks and mites and are injected into the body when the ectoparasites feed on humans. Typhus-group rickettsiae, on the other hand, enter the body when broken skin is contaminated by the infected feces of lice or fleas. *C. burnetii* organisms are usually transmitted when the host inhales infectious aerosols. They enter the host's cells by passive phagocytosis, preferentially infecting histiocytes and Kupffer cells (Baca and Paretsky, 1983).

Incubation Period

The incubation period of rickettsial diseases is generally from 7 to 21 days but can range from 1 to 32 days (Table 2). Q-fever has the longest incubation period, usually 18 to 21 days. Variability in the incubation periods might be due to differences in inoculum size, because variations in dose affected the incubation periods when human volunteers and animals were experimentally infected with *R. rickettsii, R. typhi,* and *C. burnetii* (Dupont et al., 1973; Ley et al., 1952; Snyder, 1965; Tigertt and Benenson, 1956). In rickettsialpox, where exposures to the mite vector often occur over a number of days, exact incubation periods are difficult to determine (Brettman et al., 1981).

Symptoms, Signs, and Clinical Course

Initial Symptoms

Most patients with rickettsial illnesses initially have nonspecific systemic symptoms and signs (Table 2). Fever and headache are most commonly reported, but chills, myalgias, arthralgias, malaise, and anorexia also are noted. Onset of disease is sudden in about half of the cases. The severity of symptoms and organ system involvement vary greatly, depending on the etiologic agent, host factors (especially age), inoculum size, and possibly strain differences.

Fever increases during the first week of illness, often reaching 104°F or higher. Nonspecific systemic symptoms and signs become common during the first week of illness. Mucous membrane involvement (e.g., conjunctivitis and pharyngitis) may be noted. In scrub typhus, early in the course of infection, regional lymphadenopathy proximal to the eschar is found in about 20% of the patients. Later, generalized lymphadenopathy that may be mistaken for mononucleosis is seen in about 80%.

Cutaneous Manifestations

Although rash is considered a hallmark of rickettsial infection, it usually follows systemic symptoms. Its absence should not rule out a possible rickettsial etiology, especially during the first week of illness. Rash is found during the first 3 days of clinical illness in only 50% of patients with RMSF (Helmick et al., 1984) and 10% of patients with scrub typhus (Blake et al., 1945). As many as two-thirds of the patients with scrub typhus (Berman and Kundin, 1973) and 10% of patients with RMSF (Helmick et al., 1984) never develop a rash, and, another 5% of RMSF patients do not develop rash until after the 10th day of illness. A rash may escape detection in dark-skinned individuals.

With these limitations in mind, the character and pattern of progression of the rash are useful in distinguishing the various rickettsial diseases. In RMSF, for example, the rash begins on the wrists and ankles in about 50% of the patients and on the palms and

TABLE 2. Clinical and laboratory features of rickettsial diseases[a]

	RMSF	Rickettsialpox	Epidemic typhus	Scrub typhus	Q-fever
Incubation period					
Usual, days	7	9–14	8–12	10–12	18–21
Range, days	1–14	7–24	5–15	4–18	8–32
Symptoms					
General					
Sudden onset	2–4+	4–5+	2+	2+	4+
Fever	5+	5+	5+	5+	5+
Chills	±	4+	1–4+	1–4+	3–4+
Myalgias	2–5+	4+	4+	2+	2–3+
Photophobia	1+	2+	2+		2+
Arthralgias	2+		1+	1+	1–3+
Lymphadenopathy	2+	1+		1–5+	
Conjunctivitis	2+	1+	3+	2–4+	3+
Pharyngitis	2+	1+	2–4+	2+	1–3+
Rash	3–4+	4+	2–4+	2–3+	±
Eschar	0	3–4+	0	2–3+	0
Palm and soles	2–5+	±–1+	±	2+	
Face		±	±	2+	0
Central nervous system					
Headache	3–5+	5+	5+	3–5+	4–5+
Abnormal CSF	1–2+		2+		±
Meningismus	1+		±	±	±
Seizures	±–1+				
Focal signs	±		1+		±
Coma	±–1+		1+		±
Decreased hearing	±–1+		2+	±–2+	
Cardiovascular system					
Cough	2+	±	5+	1–4+	2–4+
Rales	1+	±	1+		1+
Abnormal chest x-ray	±–2+		1+	1–2+	1–3+
Shock	±		±	±	
Gastrointestinal system					
Nausea or vomiting	2–4+	±	2–3+	2–3+	2–3+
Constipation	4+		2+		+
Abdominal pain	1–3+		1+	2+	±–1+
Splenomegaly	1–3+	+	1+	1–3+	±–2+
Hepatomegaly	1–2+			1+	±–3+
Diarrhea	±–1+			2+	±–1+
Jaundice	±–1+		±–1+		±
Routine laboratory tests					
Abnormal liver function	2+				5+
Azotemia	1+		3+		±
Thrombocytopenia	4+		2–4+	2+	±–2+

[a] Information was compiled from published reports as follows:
Rocky Mountain spotted fever (RMSF): Kapolowitz et al., 1981; Helmick et al., 1984.
Rickettsialpox: Greenberg et al., 1947; Rose, 1948; Bretman et al., 1981.
Epidemic typhus: Kamal and Messih, 1943; Dyer, 1944.
Scrub typhus: Blake et al., 1945; Sayan et al., 1946; Berman and Kundi, 1973.
Q fever: Clark et al., 1951; Powell, 1960; Spelman, 1982; Dupuis et al., 1985b.
[b] Percentage of patients with indicated characteristic: 0 = not reported to occur; ± = <1–5%; 1+ = 6–20%; 2+ = 21–40%; 3+ = 41–60%; 4+ = 61–80%; 5+ = 81–100%.

soles in another 25%. It is often macular and blanching during the early stages; later, the rash spreads centripetally and becomes petechial, ultimately involving the palms and soles in up to 80% of patients. The rash in Boutonneuse fever is very similar in timing and characteristics to that of RMSF, although Boutonneuse fever is almost always distinguishable epidemiologically and by the presence of the eschar that is found in two-thirds of patients (Raoult, 1986b).

The rash of rickettsialpox characteristically begins as sparse, discrete, maculopapular lesions on the face, trunk, and extremities and then becomes vesicular, but it usually spares the palms and soles. About 90% of rickettsialpox patients develop an eschar. Eschars begin as red papules and develop into painless, punched-out, shallow ulcers that characteristically are covered by a central, black-brown, crusted area.

In epidemic typhus, faint pink to red blanching macules appear first, usually on the anterior trunk and axillary folds on the 5th to 7th day of illness. Depending on the severity of the illness, the rash may remain on the upper thorax and abdomen, or it may spread rapidly to the extremities. Except in very ill patients, it spares the face, palms, and soles. In mild cases the lesions fade over the course of a few days; in more severe cases they become maculopapular and petechial and fade during the next few weeks.

The rash of scrub typhus is similar to that of epidemic typhus, beginning on the trunk, axillary folds, and proximal extremities as nonconfluent pink to red macules that fade on pressure. The initial lesions are larger than those seen in epidemic typhus. The rash spreads centrifugally, involving most of the body except the face, palms, and soles. It becomes maculopapular and petechial, then fixed and confluent, and in severe cases, hemorrhagic. It is evanescent or absent in 15 to 65% of cases in some areas (Berman and Kundin, 1973; Blake et al., 1945)

Organ System Involvement

Headache is a very common accompaniment of most rickettsial diseases. Photophobia is also common, but evidence of more serious CNS impairment (confusion, stupor, delirium, seizures, and coma) is found only in about 25% of patients with RMSF (Kaplowitz et al., 1981) and typhus (National Research Council, 1953) who receive proper treatment and is virtually never seen in the other rickettsial diseases. Abnormalities of the cerebrospinal fluid, especially pleocytosis and increased protein, tend to parallel other signs of CNS involvement. Mild pulmonary involvement, manifested by cough and infiltrates on the chest roentgenogram, is common in epidemic typhus and is found in about half of Q-fever and scrub typhus patients and in about one-third of patients with RMSF. However, pulmonary involvement is almost never found in rickettsialpox. Gastrointestinal symptoms, especially nausea or vomiting, are reported in about one-half of patients with RMSF (Kaplowitz et al., 1981) and in most other rickettsial diseases (except rickettsialpox). Constipation is reported by two-thirds of patients with RMSF and epidemic typhus. Diarrhea is reported by about 10% of patients with rickettsial diseases. Hepatomegaly and/or splenomegaly is found only in about 20% of patients with rickettsioses, except for Q-fever where hepatomegaly and hepatitis may dominate the clinical picture in as many as half of the patients.

The symptoms of murine typhus (Miller and Beeson, 1946) and flying squirrel-associated typhus fever (Duma et al., 1981; McDade et al., 1980) are similar but milder than those of epidemic typhus.

Course and Outcome

The course and outcome of rickettsial diseases are quite variable. The primary determinants are the specific infectious agent and the rapidity with which effective antibiotic treatment is initiated. Other factors include the patient's age and the virulence of the infecting strain (Hattwick et al., 1978; Helmick et al., 1984). Epidemics of louse-borne typhus have had attack rates as high as 96% and often have involved millions of people with fatality rates of 10 to 66% (Foster, 1981; Megaw, 1942). Poor nutrition and other underlying health problems undoubtedly contribute to the high fatality rates of some epidemics. The case fatality rate of RMSF was 23 to 70% in the preantibiotic era (Harrell, 1949; Wilson and Chowning, 1904). With the advent of antibiotics the fatality rates quickly fell below 10%, and it has remained at approximately 3 to 7% in the United States in the 1980s (Fishbein et al., 1984). In World War II, before tetracycline was available, the fatality rate for American soldiers with scrub typhus was 8.5%. Servicemen in Vietnam in the 1960s and 1970s with scrub typhus were treated with tetracycline, and no fatalities were attributable to this disease (Berman and Kundin, 1973). Boutonneuse fever is generally a much milder disease, but in a recent study in Marseilles, Raoult et al. (1986a) reported a fatality rate of 2% in hospitalized patients.

Infectivity

Patients with rickettsial disease develop rickettsemias of varying durations (Kaplowitz et al., 1981; Shirai et al., 1982b), at which time blood and tissues, including pathologic specimens obtained antemortem and postmortem, are potentially infectious.

However, there are no well-documented cases of human-to-human transmission of any rickettsial disease. In some Q-fever patients, however, the illness did appear to be attributable to human contagion (Holland et al., 1960; Whittick, 1950).

Clinical Laboratory Data

Routine laboratory tests are unlikely to be diagnostic for any rickettsial diseases with the possible exception of Q-fever. Mild abnormalities of liver function are found in the majority of patients with Q-fever (and therefore make this disease a part of the differential diagnosis of non-A, non-B hepatitis) and in about one-third of patients with RMSF and Boutonneuse fever. However, jaundice is rare in all these illnesses. Mild azotemia is common in RMSF but rare in Boutonneuse fever; severe abnormalities of renal function are seen frequently in epidemic typhus and occasionally in RMSF. Thrombocytopenia is observed in more than half of the patients with epidemic typhus, in about 40% of those with RMSF and Boutonneuse fever, and less commonly with scrub typhus and Q-fever. The white blood cell count is usually depressed or normal during the first week of illness, but later it may be elevated in any of the rickettsioses.

Recrudescence

Chronic Q-fever and recrudescent typhus (Brill-Zinsser disease) are illnesses that appear years after initial infection with *C. burnetii* and *R. prowazekii*, respectively. Chronic Q-fever is an uncommon illness that usually affects patients with preexisting valvular heart disease; it develops as a culture-negative endocarditis accompanied by liver function abnormalities and granulomatous hepatitis. Chronic Q-fever infections are frequently fatal (Ellis et al., 1982; Turck et al., 1976). It is perhaps noteworthy that plasmids from patients with Q-fever endocarditis are similar to each other but different from plasmids obtained from patients with acute, self-limited infection (Samuel et al., 1985). Whether the plasmids code for a factor that induces a chronic infection remains to be determined.

Brill-Zinsser disease is probably much more common than chronic Q-fever, but its incidence is difficult to ascertain. It occurs sporadically in persons who have lived in typhus-endemic areas and contracted primary cases of louse-borne typhus months to years before the onset of the recrudescent infection. The clinical manifestations of Brill-Zinsser disease are similar to, but milder than, those of primary louse-borne typhus: the duration of illness is shorter, temperature elevation is less sustained, rash is often absent, complications are less severe, and the case fatality rate is much lower (Snyder, 1965). However, Murray et al. (1950) reported that the clinical manifestations of Brill-Zinsser disease were almost indistinguishable from those of primary typhus infection.

Relapses also can occur in scrub typhus patients. The frequency of scrub typhus relapses are related inversely to the duration of therapy during the primary infections, but relapses rarely occur in more than 5% of the patients (Berman and Kundin, 1973; Smadel, 1949a).

Diagnosis

General

Techniques for the diagnosis of rickettsial diseases have improved considerably in recent years, but unfortunately they have not improved the outcome of most rickettsial infections. Attempts to directly detect rickettsiae or rickettsial antigens in clinical specimens have only been marginally successful, and even with the most sensitive serologic techniques one usually cannot detect rickettsial antibodies until a week or more after the onset of symptoms. Similarly, the lengthy generation time of rickettsiae usually preclude their isolation and identification in less than a week. These time frames are unsatisfactory because many rickettsioses have high fatality rates unless specific antibiotic therapy is administered within 3 to 5 days after the onset of illness.

Despite these shortcomings, techniques for the diagnosis of rickettsioses have inherent value and can contribute indirectly to a reduction in rickettsial morbidity and mortality when used as part of a detailed surveillance system. With these techniques, areas that are endemic and hyperendemic for rickettsial diseases can be identified, the index of suspicion can be raised, and diagnosis and treatment can be expedited accordingly.

Unfortunately, rickettsioses occur with the greatest frequency in remote areas of underdeveloped countries, where adequate facilities are usually unavailable. Rickettsial isolation attempts are impractical in such areas because of the lack of adequate biohazard containment facilities; newer antibody assays are precluded by the lack of necessary equipment and supplies. Retrospective serologic diagnosis at a central laboratory must usually suffice under such circumstances. In areas where all necessary facilities, equipment, and reagents are available, the diagnostic approach depends on the local prevalence of rickettsial diseases. For example, in areas of North Carolina, where the prevalence of RMSF is high, some hospitals and specialty laboratories attempt to detect rickettsiae directly in skin biopsies by fluorescent antibody techniques (see below). Such tests are performed infrequently in areas where RMSF is rare. Retrospective serologic diagnosis is used instead.

In fatal cases, direct fluorescent antibody tests of formalin-fixed paraffin-embedded tissues are quicker, safer, and much more convenient than rickettsial isolation for confirming the diagnosis. Rickettsial isolation is usually limited to situations in which no alternative is available for diagnosis, that is, when early acute-phase blood is the only specimen available for testing. Rickettsial isolation is also indicated when the isolate would be useful for epidemiologic or research purposes.

Specimen Collection

Rickettsiae are hazardous (Biosafety Level [BSL] 3) microorganisms that have been responsible for numerous laboratory infections over the years, usually in research laboratories where rickettsiae are propagated (Oster et al., 1977; Pike, 1976, 1979). In infected blood, rickettsiae are usually present at relatively low concentrations (100 viable organisms per ml) (DeShazo et al., 1976; Kaplowitz et al., 1983; Shirai et al., 1982b). Rickettsial titers in blood specimens begin to fall off dramatically after several hours at room temperature (Q-fever rickettsiae excepted), but viable organisms might still be present in blood kept at room temperature for several days. Processing freshly drawn blood to obtain serum does not pose a serious threat to the careful laboratory worker, provided that aerosolization is minimized and the clot is discarded carefully and then autoclaved. Surgical gloves should be worn when handling blood specimens. Tissues obtained postmortem contain far greater numbers of organisms, however, and require additional care in handling, including wearing surgical masks and back-fastening gowns.

Special care should be exercised when handling specimens containing Q-fever rickettsiae. *C. burnetii* is highly infectious, resists dessication and chemical inactivation, and is frequently shed in the urine and feces of infected animals (see Bernard et al., 1982). Although serum from suspected Q-fever patients present no unusual hazards when handled as described, infected tissues should be processed under strict BSL 3 conditions, and then only by highly qualified personnel (Richardson and Barkley, 1984).

Direct Detection of Rickettsiae

RAPID DIAGNOSTIC PROCEDURES

Rickettsiae can be detected in skin biopsies by direct fluorescent antibody tests. Unfortunately, however, the current procedure is not sufficiently sensitive for

definitive diagnosis. The technique is based on the assumption that most rickettsial infections cause focal vasculitis, and that rickettsiae are present in the blood vessels of these foci. Punch biopsies approximately 3 to 5 mm in diameter are obtained either from petechiae or from the center of a macule, embedded in polyethylene glycol, and frozen in a cryostat. The tissue is then sectioned, collected on glass slides, fixed in acetone, and tested with fluorescein isothiocyanate-conjugated antirickettsial sera (Walker et al., 1978). Evaluations to date (Fleisher et al., 1979; Raoult et al., 1984; Walker et al., 1978; Woodward et al., 1976) indicate that this procedure detects *R. rickettsii* and *R. conorii* in approximately 50% of the patients with RMSF or Boutonneuse fever. The reasons for such poor sensitivity are not known, although most false-negative results occurred with specimens from patients who had already begun specific antibiotic therapy (Raoult et al., 1984; Walker et al., 1978). The biopsy technique has not gained widespread acceptance, probably because of this lack of sensitivity and reproducibility (Linneman, 1980). However, despite its shortcomings, the general approach has potential value for the development of better rapid diagnostic procedures.

Rickettsiae are also present in circulating monocytes, but their relatively low titer precludes easy detection in infected blood. Buhles et al. (1975) and DeShazo et al. (1976) detected *R. rickettsii* in the monocytes of infected guinea pigs and Rhesus monkeys by direct fluorescent antibody tests, but only after the rickettsiae had been cultured in the monocytes for 3 to 6 days. This technique might be minimally sufficient for the early diagnosis of RMSF in humans, provided that the blood specimen is obtained immediately after the onset of illness and rickettsiae are detected in monocytes as soon as 3 days in culture. Otherwise, the diagnosis would be too late to allow prompt antibiotic therapy.

POSTMORTEM EVALUATIONS

Direct fluorescent antibody testing of tissues collected postmortem is a useful tool for the retrospective diagnosis of rickettsial diseases. This technique was first applied to the rickettsiae by Walker and Cain (1978), who detected rickettsia-like organisms in the kidney tissue of 7 of 10 patients who had died of suspected RMSF. In this technique sections of formalin-fixed, paraffin-embedded tissue are fixed on glass slides with glue and incubated in an oven at 60°C for 1 h. The sections are then deparaffinized in xylene and gradually rehydrated in a series of solutions from ethanol to distilled water. The rehydrated sections are digested with trypsin for 4 h, washed with distilled water and then with phosphate buffered-saline (PBS) pH 7.4. Fluorescein isothiocyanate-labeled antirickettsial sera are then added to the sections, and the sections are incubated for 30 min. They are washed again in PBS, rinsed in distilled water, overlayed with a glycerol solution, and examined under an ultraviolet microscope.

Morphologically distinct rickettsiae are readily detectable by this procedure. In our laboratory we have had best success with lung and spleen sections, although rickettsiae can also be seen in heart, liver, kidney, and brain tissues.

Direct Antigen Detection

Several decades ago the direct detection of rickettsial antigens in the blood and urine of typhus patients was reported. Unfortunately no subsequent reports have confirmed or refuted these early findings. Fleck (1947) and Fleck et al. (1960) concentrated urine specimens from epidemic and murine typhus patients by dialyzing them in gelatin. The concentrated urine specimens were subsequently tested for their ability to inhibit the agglutination of red blood cells sensitized either with extracts of rickettsial antigens or extracts of *Proteus* OX19, an organism that is antigenically related to the rickettsiae. Virtually all of the urine specimens collected from typhus patients during the 1st week of illness inhibited agglutination, whereas proportionately fewer of the specimens collected during the 2nd and 3rd week were inhibitory. Urine from normal persons or from patients with other febrile illnesses failed to inhibit agglutination. O'Connor and MacDonald (1950) similarly evaluated urine from scrub typhus patients, and although specimens from their patients also inhibited passive hemagglutination, the pattern of their results was sufficiently erratic to question the validity of their findings.

Smorodintzeff and Fradkina (1944) detected rickettsial antigens in the blood of 60% of epidemic typhus patients by complement fixation tests, when the sera were obtained very early in the course of illness. They reported similar success with a slide agglutination test developed for that purpose.

There have been no subsequent verifications of these early reports. We assume that others have attempted to repeat their observations but were unsuccessful. We have been unsuccessful in our attempts to detect rickettsial antigens in the urine of RMSF patients by a sensitive immunofluorometric assay system modified for that purpose, and tests of the urine of mice infected with *R. typhi* gave equivocal results (McDade et al., unpublished observations). Although it is possible that the results reported by early investigators were laboratory artifacts, this should not discourage the application of newer technology for the same purpose, particularly because early diagnosis of rickettsial diseases by direct antigen detection would greatly reduce morbidity.

Direct Rickettsial Nucleic Acid Detection

To date there has been only one report of the direct detection of rickettsiae in infected cells by nucleic acid hybridization technology. Regnery et al. (1986) cloned a 3.7-kilobase-pair DNA segment of *R. prowazekii* in plasmid pUC19. When radiolabeled and used in Southern blot analysis with *Bam*HI digests of the DNA from various *R. prowazekii* strains, this DNA segment distinguished flying squirrel isolates from human strains of European and African origin. Significantly, the probe also detected rickettsial DNA in *Bam*HI digests of cytoplasmic extracts of infected Vero cells, demonstrating its potential as a diagnostic probe for clinical specimens. However, the specificity of that probe for rickettsiae has not been evaluated. Additional progress in this area awaits the identification of DNA segments specific for the rickettsiae. Presumably DNA segments that code for specific rickettsial antigens will be the first diagnostic probes developed. If nucleic acid hybridization technology is to be useful for the routine diagnosis of rickettsial diseases, however, it must be able to detect as few as 100 copies of a given DNA segment. Only 100 to 1,000 viable rickettsiae are in 1 ml of infected blood, and there is no reason to believe that there are repeat segments on the rickettsial genome.

Rickettsial Isolation and Identification

PRIMARY ISOLATION

Two major factors dictate the general approach that must be followed in isolating rickettsiae from a clinical specimen. Like viruses, rickettsiae require eukaryotic host cells for their propagation. Unlike viruses, however, rickettsiae are susceptible to antibiotics, and therefore antibiotics cannot be used to suppress the growth of adventitious agents that might be present in the tissue specimen. Because tissues obtained at autopsy are usually contaminated with bacteria, selective propagation of rickettsiae in susceptible laboratory animals is usually a necessary first step for rickettsial isolation. Rickettsiae isolated from animals are then subcultured in tissue cultures or embryonated eggs to confirm their isolation and to rule out the presence of contaminants.* Specimens that are presumed to be free of contaminants (e.g., aseptically collected blood) can be injected directly into embryonated eggs or cell cultures for rickettsial isolation attempts, as appropriate.

---

* Because of the highly infectious nature of Q-fever rickettsiae, we do not recommend the passage of *C. burnetii* unless the isolate per se is needed. Isolation of Q-fever can be confirmed indirectly and more safely merely by demonstrating seroconversion in infected guinea pigs.

Differences in the pathogenicity of various rickettsiae for experimental animals dictate modifications of the basic isolation protocol. Guinea pigs are quite susceptible to *R. prowazekii*, *R. typhi*, and *R. rickettsii;* one organism of these species can initiate infection (Ormsbee et al., 1978a). However, guinea pigs are refractory to *R. tsutsugamushi*. Guinea pigs are also the animal of choice for the isolation of *Coxiella burnetii*. Male guinea pigs are specifically recommended for the isolation of spotted fever- and typhus-group rickettsiae because of the characteristic scrotal reaction that is elicited after inoculation of some rickettsial species (see below). Although meadow voles (*Microtus pennsylvanicus*) are also quite susceptible to spotted fever-group rickettsiae, the use of guinea pigs provides a practical advantage in that their body temperature can easily be monitored as an indicator of infection. Mice are the preferred hosts for the isolation of *R. akari* and *R. tsutsugamushi*. Because some inbred strains of mice are resistant to infection with *R. tsutsugamushi* (Groves et al., 1980), it is prudent to use outbred mice for all such isolation attempts.

Because rickettsae cannot survive long outside of living cells (Q-fever rickettsiae excepted), specimens for rickettsial isolation attempts must either be used immediately after they are obtained or stored at ≤−60°C until processing is begun. Groups of 4 guinea pigs for 8 mice, as appropriate, are recommended for the processing of each specimen. Individual animals are marked, and a preinoculation serum specimen is obtained from each animal as a control for serologic testing of surviving animals.

All clinical specimens for rickettsial isolation must be processed in a biological safety cabinet under appropriate biosafety conditions. A small (1 cm^2) piece of human tissue is triturated in sufficient PBS to form a 10% (weight/volume) suspension; the suspension is then injected intraperitoneally into the appropriate laboratory animal. An inoculum of 1 to 2 ml is recommended for guinea pigs, and 0.5 ml used for mice. All animals are then observed for signs of illness for 28 days. An aliquot of the original suspension should be stored at −60°C as reference material.

When guinea pigs are used for primary isolation, their rectal temperatures are recorded daily for 14 days. Because all rickettsiae have relatively long incubation periods, any fever (≥40°C) that occurs during the first 2 days is usually the result of bacterial peritonitis or the trauma associated with the inoculation. Fevers that occur from 3 to 14 days after inoculation are more likely to indicate rickettsial infection. Strains of *R. prowazekii* and *C. burnetii* usually induce fever in guinea pigs, without other symptoms, starting 5 to 12 days after infection. *R. typhi* has a somewhat shorter incubation period, and male guinea pigs infected with some strains of *R. typhi*

also develop scrotal swelling and edema. Rickettsiae of the spotted fever group have 3 to 10-day incubation periods in guinea pigs. Most strains of *R. rickettsii* also cause both swelling and necrosis of the scrotum.

Tissue specimens are collected aseptically from infected guinea pigs on the 2nd or 3rd day of fever, triturated in PBS, and subcultured in tissue cultures or embryonated eggs (see below). Infected spleen is the preferred tissue for passage, although liver, brain, and tunica vaginalis are also acceptable. At this stage in the isolation effort, one can make multiple impression smears of infected guinea pig tissues for direct microscopic examination. Selected smears are heat fixed and stained by the Giménez method (1964). If microscopic examination of the smears reveals the presence of numerous intracellular microorganisms, additional smears are fixed with acetone and tested with an appropriate fluorescein-labeled antirickettsial serum to determine the identity of the presumed isolate. Regardless of the outcome of such examinations, however, we recommend that suspected isolates be subcultured in tissue culture or embryonated eggs to confirm their identity and to rule out the presence of bacterial contaminants. An aliquot of the infected animal tissues should also be stored at −60°C in the event that these passaging attempts result in contaminated cultures.

When mice or *Microtus* are used for isolation attempts, animals are killed and their tissues harvested for passage when they develop overt signs of illness (lethargy, ruffled fur, and so on). Alternatively, two animals are killed on days 7, 10, and 14 after inoculation to obtain tissues for blind passage. Rickettsiae are usually found in the greatest numbers in the spleens of infected mice, although organisms can also be found in the peritoneal exudate, particularly in mice inoculated with scrub typhus rickettsiae. Mouse tissues are treated identically to those from guinea pigs.

Although most species of rickettsiae are pathogenic for guinea pigs or mice, some strains can cause inapparent infections in these animals. If the guinea pigs fail to develop fever during the observation period, one should obtain serum samples 28 days after inoculation and test them for the presence of rickettsial antibodies. Positive reactions indicate inapparent rickettsial infection, provided, of course, that negative results are obtained with the corresponding preinoculation sera. Similar tests should be performed on the sera from surviving mice.

SUBCULTURE OF ISOLATES

Although all rickettsiae will grow in embryonated hen eggs, there are some peculiarities that should be noted (Stoenner et al., 1962). First, members of the spotted fever group grow better in embryos that are approximately 5 days old at the time of inoculation, whereas other rickettsiae grow best in 6 to 7-day-old embryos. Second, the incubation period of spotted fever-group rickettsiae is 5 to 7 days compared with the 7 to 10-day incubation periods of the typhus, scrub typhus, and Q-fever rickettsiae. Third, members of the spotted fever group will continue to grow up to 48 h after the embryos have died. Thus, yolk sacs infected with spotted fever-group rickettsiae produce optimum yields of living organisms when harvested 24 h after the death of embryos.

Embryonated eggs of the required age are candled and infertile eggs are discarded. The top of each egg is disinfected with alcohol or tincture of iodine. Next, a 5% suspension of infected animal tissue, prepared in sterile PBS, is drawn into a 5 or 10-ml syringe fitted with a 20-gauge, 1.5-inch needle, and 0.5-ml aliquots of the suspension are inoculated into each of 12 embryonated eggs. The eggs are then sealed with paraffin or model airplane glue and incubated at 35°C. An aliquot of the inoculum is stored at −60°C as reference material, and another is spread onto appropriate bacteriologic media to check for contaminants.

Embryos that die during the first 3 days after inoculation should be discarded, because their deaths are usually due to trauma or bacterial contamination. Yolk sacs are harvested aseptically from embryos that die from days 4 through 10. A small piece of the harvested yolk sac is removed aseptically and prepared for direct microscopic observation and fluorescent antibody tests as indicated. Another small piece of yolk sac should be smeared on blood agar or other suitable bacteriologic media and incubated aerobically and anaerobically to check for bacterial contaminants. The remaining large fragment of each yolk sac is quick-frozen and stored at −60°C as prospective seed material.

If none of the embryos die during the 10-day incubation, or if the yolk sacs of dead embryos do not contain detectable rickettsiae, then the yolk sacs are harvested from all remaining live embryos for blind passage into additional embryonated eggs. A 20% suspension of several yolk sacs prepared in PBS is used as inoculum. Eggs are then monitored and yolk sacs harvested as described.

Although passage is embryonated eggs is the traditional method for propagating rickettsiae once they have been isolated in animals, cell cultures also have been used extensively and offer a convenient alternative for rickettsial propagation. The number of cell lines that have been used successfully for growing rickettsiae is too extensive to be listed here. A good example of the relatively uniform ability of rickettsiae to grow in cell cultures can be found in early studies by McDade et al. (1969, 1970), who showed

that spotted fever, typhus, scrub typhus, and Q-fever rickettsiae would all produce plaques in primary chick embryo cells. Later studies by Johnson and Pedersen (1978) demonstrated that several species of spotted fever-group rickettsiae would produce plaques in primary, diploid, and heteroploid cell lines. Thus, the preferred cell line for rickettsial cultivation is more likely to be dictated by the availability of the cell than by its ability to support the growth of rickettsie. One possible factor limiting the use of a given cell line is whether it can be maintained for several weeks in culture, because the incubation period of rickettsiae can be quite long, and growth of new isolates may be slow for a week or longer.

Infected animal tissues are homogenized thoroughly in sterile PBS to form a 5% (weight/volume) suspension. The growth medium is then decanted from several tissue culture flasks, and 0.1 to 0.3 ml of the infected cell suspension is added. The flasks are rocked to ensure that the inoculum covers the entire cell sheet, and then placed in a 35°C incubator for 30 min to allow infection to occur. Growth medium (without antibiotics) is added to the cultures, and they are incubated at 35°C. No other special growth conditions are necessary, although Kopmans-Gargantiel and Wisseman (1981) did show that *R. rickettsii, R. prowazekii,* and *R. typhi* (but not *R. tsutsugamushi*) grew better in an environment with 5% CO_2 than they did when exposed to the 0.2 to 0.3% CO_2 found in atmospheric air. The cell cultures should be monitored for 14 days.

We recommend that replicate cell cultures be inoculated, not just because of the potential hazards of bacterial contamination, but also to allow frequent sampling of the tissues for direct microscopic examination. Beginning on day 3 after inoculation, small pieces are scraped off the flask of a given culture with an inoculating loop or the equivalent, and smears are prepared and stained for rickettsiae as described.

In the event that rickettsiae cannot be detected in the cell cultures after 14 days, or if growth is so scanty as to preclude positive identification by fluorescent antibody testing, the cells are harvested and subcultured into additional flasks of the same cell line. Sampling and testing for rickettsiae in the second set of cell cultures should then be repeated.

Final confirmation of rickettsial isolation is accomplished when the isolate is morphologically similar to rickettsiae, grows intracellularly, fails to grow on bacteriologic media, stains red by the Giménez technique, and reacts with appropriate immune serum. Additional details concerning the procedures for rickettsial isolation can be found in the review by Weiss (1981).

Antibody Assays

Several serologic tests are currently available for the diagnosis of rickettsial diseases (Table 3). Rickettsial serology has improved in parallel with technological advances, beginning with complement fixation (CF) tests in the 1940s, continuing with indirect fluorescent antibody (IFA) tests in the 1960s, and progressing to sensitive enzyme (ELISA) and radioimmunoassay (RIA) systems of the present. Although many techniques have been used successfully for rickettsial serodiagnosis (Table 3), relatively few are used regularly by most laboratories. Some older tests lack sensitivity (CF, microagglutination, and Weil-Felix) (Berman and Kundin, 1973; Ormsbee et al., 1977; Newhouse et al., 1979; Philip et al., 1977) or specificity (Weil-Felix) (Hechemy et al., 1979), and reagents are not commercially available for most procedures. The Centers for Disease Control distributes IFA reagents for the diagnosis of RMSF, typhus, and Q-fever to qualified public health laboratories, but ELISA, RIA, IHA, or latex agglutination test reagents can only be acquired by special arrangement with reference laboratories.

Choosing the best serologic test is not easy. ELISA techniques, particularly immunoglobulin M (IgM) capture assays, are probably the most sensitive tests available for rickettsial diagnosis, but they require large quantities of purified antigens that are commercially unavailable. Although ELISA tests can readily screen multiple serum specimens, this is often not useful for rickettsial diseases, because, with the exception of epidemic typhus and Q-fever, rickettsioses usually occur sporadically at a relatively low frequency. However, the sensitivity of the ELISA test could be exploited and economy of antigen use maintained if the appropriate rickettsial antigens were included in a battery of antigens designed to test for several possible causes of febrile illnesses of unknown origin.

The IFA technique remains the most popular procedure for serodiagnosis of rickettsial diseases, primarily because of the ease and economy with which it can be incorporated into existing antibody screening systems. In addition, it readily detects IgM antibodies; this feature is quite valuable for serodiagnosis during the early stages of rickettsial infections (Dupuis et al., 1985a; Hunt et al., 1983; Philip et al., 1976). The IFA procedure for rickettsiae is similar to conventional techniques: inactivated yolk sac or tissue culture suspensions of rickettsiae are normally used as antigens (Newhouse et al., 1979; Philip et al., 1976). When infected yolk sacs are used as antigens, sera are diluted in a 3% suspension of normal yolk sac to adsorb antibodies to the egg substrate that might be present in the patient's serum.

Indirect fluorescent antibody tests are reasonably sensitive, although the subjectivity of endpoint determinations is a disadvantage. In addition, nonspecific reactivity at low serum dilutions confounds the interpretation of results (see below).

Evaluation of the indirect hemagglutination (IHA) test has been limited, but the test is apparently as sensitive as IFA (Kaplan and Schonberger, 1986). An erythrocyte sensitizing substance (ESS), obtained either from typhus- or spotted fever-group rickettsiae by alkali extraction, is adsorbed onto sheep or human group 0 erythrocytes, and the coated red blood cells are then used as antigens for simple agglutination tests (Chang, 1953; Chang et al., 1954). The respective ESSs exhibit group-specific antigenic reactivity. The convenience of the IHA technique earmarks it for a clinical setting, particularly in remote areas, although the relatively short shelf life (6 months) of sensitized red blood cells (Shirai et al., 1975) is a disadvantage.

The latex agglutination test has been used with success in several public health laboratories (Hechemy et al., 1983b). Latex spheres are coated with ESS, and the sensitized particles are used as antigens in an agglutination test. The simplicity of the latex agglutination test makes it more convenient than most other techniques, but because it is primarily an IgM assay, it lacks sensitivity for late convalescent-phase serum samples (Hechemy and Rubin, 1983). The ESS contains a limited repertoire of rickettsial antigens because of the method of its preparation. Some epitopes are also masked or destroyed when the ESS is adsorbed onto latex particles (Hechemy et al., 1983a), and so positive samples may be undetected as a result. Nonetheless, the convenience of the latex test makes it particularly useful in a clinical setting.

The latex technique or other simplified procedures (IHA, microagglutination) are recommended for that setting over the Weil-Felix test, which, unfortunately, is still used to diagnose rickettsial diseases. The Weil-Felix test became popular in the 1920s after it was observed that certain strains of *Proteus* would agglutinate early convalescent-phase sera from patients with rickettsial diseases due to shared antigens on the respective organisms (Weil and Felix, 1916). The test fell into disfavor years ago because it lacked both sensitivity and specificity (Berman and Kundin, 1973; Hechemy et al., 1979), but it has survived because of its convenience. Despite its simplicity, it does not provide either early or specific diagnosis of rickettsioses, and we recommend that its use be discontinued.

Other possible techniques for rickettsial diagnosis are complement fixation, radioimmunoassay, and microagglutination (MA). Radioimmunoassay has

Serologic test	Minimum positive titer	Time after onset diagnostic titers usually detected	Ability to measure antibody persistence	Comments	Selected references
Indirect immuno-fluorescence	Varies: 16 to 64, depending on the investigator	IgM positive 6–10 days; 2–3 weeks for IgG	IgM—10 weeks; IgG–avg 1 year; 40 months for total reversion in scrub typhus	Currently most widely used test; relatively sensitive and requires little antigen; can distinguish immunoglobulin isotypes	Berman & Kundin, 1973; Dupuis et al., 1985a; Hunt et al., 1983; Kaplan and Schonberger, 1986; Newhouse et al., 1979; Ormsbee et al., 1977; Philip et al., 1977; Saunders et al., 1980
Complement fixation	8 to 16, depending on investigator	3 to 4 weeks	≥1 year	Lacks sensitivity compared with IFA and ELISA, but is very specific; now used infrequently	Dupuis et al., 1980; Kaplan and Schonberger, 1986; Newhouse et al., 1979; Ormsbee et al., 1977; Philip et al., 1977
ELISA	50 for paper ELISA; or OD 0.25 > controls	≤1 week in many instances	Not thoroughly evaluated—1 year?	IgM capture assay promising for early diagnosis; paper ELISA may be suitable for field use	Crum et al., 1980; Dasch et al., 1979; Doller et al., 1984; Field et al., 1983; Halle et al., 1977; Herrmann et al., 1977
Latex agglutination	64	6–10 days	Approximately 8 weeks	Antigen preparation may mask some epitopes; lacks sensitivity for late convalescent phase sera	Hechemy et al., 1980, 1983a–c; Kaplan and Schonberger, 1986
Indirect hemagglutination	40 ?	1–2 weeks	Several months	Fixed red cells stable up to 6 months; sensitivity comparable to IFA, > than CF	Anacker et al., 1979; Kaplan and Schonberger, 1986; Philip et al., 1977; Shirai et al., 1975
Immunoperoxidase	20	7 days	Preliminary data suggest 1 year	Has had limited evaluations; may be useful in field situations	Raoult et al., 1985; Yamamoto and Minamishima, 1982
Microagglutination	≥8	6–10 days	At least several months; presume that agglutination is best with IgM-containing sera	Test is convenient, but uses considerable antigen; less sensitive than IFA; not compared with ELISA	Newhouse et al., 1979; Ormsbee et al., 1977; Philip et al., 1977
Radioimmunoassay	10 ?	7–10 days	Not thoroughly evaluated; presumably quite sensitive	IgM capture assay may be very useful for early diagnosis	Doller et al., 1984; Herrmann et al., 1977; Lackman et al., 1964
Weil-Felix	40–320, depending on investigator	1st–3rd week	Several months	Lacks both sensitivity and specificity	Berman and Kundin, 1983; Brown et al., 1983; Hechemy et al., 1983; Kaplan and Schnoberger, 1986; Ormsbee et al., 1977; Philip et al., 1977

not commonly been used for rickettsial diagnosis outside specialty laboratories. It is highly sensitive, but the potential hazards from exposure to isotopes and the requirement for expensive equipment are obvious disadvantages. It is unlikely that RIA will gain increased acceptance for rickettsial serodiagnosis in view of the proliferation of various enzyme assays and other new techniques of comparable sensitivity. The CF technique was the technique of choice for decades because of its extreme specificity, but the advent of newer techniques with increased sensitivity and adequate specificity has made the CF technique less attractive. Similarly, the relative expense and lack of sensitivity of the MA technique indicate that it cannot keep pace with newer technology.

Interpretation of Laboratory Data

With the exception of Q-fever, rickettsial diseases usually develop as febrile exanthemous illnesses with protean clinical manifestations. They must be distinguished from several viral and bacterial exanthems, including measles, meningococcemia, enterovirus infections, leptospirosis, rubella, and typhoid fever. The final diagnosis of rickettsioses is best made by laboratory testing, although clinical findings are also useful in suggesting a diagnosis.

Laboratory diagnosis of rickettsial diseases is based primarily on the results of serologic testing. Unfortunately, the interpretation of tests is confounded by the lack of specificity of available rickettsial reagents. Whole rickettsiae are used as antigens for most antibody assays, and because rickettsiae are antigenically related to other microorganisms, sera from healthy persons sometimes react at low titers with rickettsial antigens (Wilfert et al., 1985). These factors have necessitated the designation of minimum positive titers to confirm recent rickettsial infections (Table 3). The minimum titers for each procedure should not be considered absolute, however, because in many instances they were assigned rather arbitrarily and without adequate testing of control sera. Borderline titers obtained from single specimens collected either very early or very late in convalescence are the most difficult to interpret. Specific IgM testing of specimens collected early can confirm that a recent illness was rickettsial in origin, but it is not possible to determine if a borderline titer on a late convalescent-phase serum is due to a recent rickettsial illness. The best way to avoid problems is to collect acute- and convalescent-phase serum specimens approximately 2 to 3 weeks apart so that rising titers can be demonstrated.

The Centers for Disease Control has established criteria for positive response in the various serologic tests (Fishbein et al., 1984). A fourfold rise in titer

detected by any technique (Weil-Felix technique excepted) is considered evidence of rickettsial infection. Complement fixation titers of ≥ 16 in single serum specimens from patients with a clinically compatible illness are also considered positive. In the IFA test single titers of ≥ 64 are considered diagnostic for typhus- and spotted fever-group infections, whereas single IFA titers of 256 are minimal for confirmation of Q-fever. Although these titers are somewhat higher than those recommended by others (Table 3), they present no problems in identifying recent infections if serum collected at appropriate times is available for testing.

Because convalescent-phase serum from a patient infected with one species cross-reacts with all other species in that biogroup, identifying the species responsible for a rickettsial infection can be difficult. Although such identification is not essential to the proper treatment of individual patients, it can be important in outbreaks. For example, routine IFA tests cannot distinguish between epidemic and murine typhus infections, but the strategies for dealing with outbreaks of the two diseases are different. Widespread vaccination and delousing programs are needed during epidemics of louse-borne typhus, whereas rodent and flea control are indicated for the outbreaks of murine typhus. Expensive antibody absorption tests (Goldwasser and Shepard, 1959; McDade et al., 1980) or toxin neutralization tests (Hamilton, 1945) are still the only methods for differentiating between epidemic and murine typhus.

Determining the specific etiology of spotted fever-group infections is less difficult, because endemic areas for most species do not overlap. We have successfully used antibody absorption tests to determine the specific etiology of spotted fever-group infections (McDade et al., unpublished results), but such distinctions usually are unnecessary, and are used infrequently for epidemiologic purposes. An exception might be during outbreaks of rickettsialpox when verification of the specific etiology is necessary to determine proper rodent and other vector control measures. Even so, however, the clinical and epidemiologic features of rickettsialpox are sufficiently distinct from other spotted fever-group rickettsiae to virtually ensure specific diagnosis.

Interpretation of serologic results for the diagnosis of Q-fever is complicated by the antigenic phase variation of the etiologic agent, *C. burnetii*. In acute, self-limited Q-fever infections, antibodies to the phase II antigen appear first and dominate the humoral immune response, probably because of the inherently greater immunogenicity of the phase II antigen. In chronic Q-fever infections, however, phase I titers equal or exceed phase II titers, presumably because of the repeated antigenic stimulation with the phase I antigen. Peacock et al. (1983) observed that the ratio of phase II to phase I antibodies is a useful indicator for diagnosing acute versus chronic Q-fever. The ratios are >1 for primary Q-fever, ≥1 for granulomatous hepatitis, and ≤1 for endocarditis patients.

Multiple serotypes of *R. tsutsugamushi* can exist in endemic areas. Unless an appropriate combination of strains of *R. tsutsugamushi* are included as antigens, some infections could be undetected (Bourgeois et al., 1977; Shirai et al., 1982a).

Even though laboratory testing is critical to diagnosing rickettsial diseases, clinical findings can contribute, particularly if serologic results are equivocal. There are no true pathognomonic features for the rickettsioses, but eschars are useful indicators of rickettsialpox, Boutonneuse fever, and scrub typhus; the vesicular rash of rickettsialpox is unique among the rickettsioses, and the triad of fever, headache, and rash are useful indicators of RMSF.

Epidemiology and Natural History

Except for epidemic typhus and scrub typhus, the rickettsioses were not really recognized as distinct clinical entities until the 20th Century. Only when specific rickettsial serologic testing was introduced in the 1940s could the precise etiologies of various rickettsial diseases be determined with certainty. Thus, there is little information about the epidemiology of most rickettsioses before World War II. Even now, there is a paucity for surveillance for Boutonneuse fever, scrub typhus, rickettsialpox, or Q-fever; on a global scale, data on their prevalence, distribution, and movement are compiled primarily from scattered reports. In addition, studies of the natural history of the various rickettsiae have been uneven. For example, the natural history of *R. rickettsii* has been thoroughly evaluated (see McDade and Newhouse, 1986), but Boutonneuse fever and other tick-borne rickettsioses have been little studied and their natural history remain obscure.

Epidemic Typhus

Epidemic typhus has historically been one of the most devastating infectious diseases. Outbreaks have been reported on all continents except Australia. The last outbreak of louse-borne typhus in the United States occurred in 1921. Globally, louse-borne typhus remains endemic in mountainous regions of Mexico, Central and South America (especially Peru and Bolivia), Africa (Ethiopia, Rwanda, Burundi, and other countries that border the Sahara), and numerous countries of Asia (Benenson, 1985). The incidence of typhus is greatly underreported because of the lack of adequate surveillance systems in endemic areas.

In classic epidemic typhus, the human body louse, which spends its entire existence in the clothing of humans, becomes infected by feeding on a rickettsemic human. The rickettsiae multiply in the epithelial cells of the louse's midgut and are shed in its feces. Lice infected with *R. prowazekii* die within 14 days, but apparently their feces can remain infectious for several months if temperature and humidity are low (Weyer, 1961). Humans are infected when louse feces or crushed lice contaminate broken skin or the mucous membranes of the eyes or respiratory tract.

Sporadic cases of *R. prowazekii* infection have been reported in the United States in the past 10 years in association with flying squirrels (*Glaucomys volans*) (Duma et al., 1981; McDade et al., 1980). The flying squirrels are a zoonotic reservoir of *R. prowazekii* (Sonenshine et al., 1978). The exact mechanism by which flying squirrels transmit *R. prowazekii* to humans is unknown. Ectoparasitic lice (*Neohaematopinus sciuropteri* Osborn) serve as vectors among animals, but this species of lice is quite host specific and unlikely to parasitize humans (Sonenshine et al., 1978). It is possible, however, that inhalation of infectious feces from squirrel lice could initiate human infections. To date all cases of flying squirrel-associated typhus have occurred in the United States, and because human body lice are generally absent in the United States, no outbreaks of louse-borne typhus have resulted from these infections. Whether this or a similar zoonotic reservoir of *R. prowazekii* contributes to outbreaks of human louse-borne typhus in other countries remains to be determined. *R. prowazekii* isolated from flying squirrels multiplies very well in human body lice (Bozeman et al., 1981).

Murine Typhus

Like epidemic typhus, murine typhus has occurred since ancient times, but it was not distinguished from epidemic typhus until the 1920s. The geographic distribution is much broader than that of epidemic typhus. Murine typhus is endemic in at least some part of every continent except Antarctica, and occurs most commonly where *Rattus*, rat fleas, and humans are found together. The disease is usually transmitted to humans when crushed fleas or flea feces contact broken skin or mucous membranes. However, direct transmission of *R. typhi* by the bite of infected fleas was recently experimentally demonstrated (Farhang-Azad and Traub, 1985).

Rats, mice, and, possible other small mammals serve as the reservoir for murine typhus rickettsiae.

The tropical rat flea (*Xenopsylla cheopsis*) is the primary vector, but eight other species of fleas in seven genera also have been incriminated as vectors (Traub et al., 1978). The cat flea (*Ctenocephalides felis*) has considerable potential for transmitting murine typhus because it is found on so many species of mammals (Farhang-Azad et al., 1984). *R. typhi* multiplies in the cells of the flea gut, but does not kill the flea. The extent of the rickettsial multiplication depends on temperature (Farhang-Azad and Traub, 1985); this may explain the higher incidence of disease in warmer climates. The microorganism is shed in flea feces, where it can remain infectious for up to 100 days. Although rodents are considered the reservoir of *R. typhi* and fleas merely vectors, recent data indicate that transovarial transmission of *R. typhi* occurs to a limited extent in *Xenopsylla cheopis* (Farhang-Azad et al., 1985).

Scrub Typhus

Scrub typhus is widely distributed in central, eastern, and southeastern Asia, and the southeastern Pacific. Outbreaks of scrub typhus usually occur when susceptible humans enter an endemic area for military or vegetation-clearing operations. Scrub typhus accounts for a substantial proportion of febrile illnesses in some endemic areas (Brown et al., 1976).

Disease is transmitted to humans when they intrude into "typhus islands," the innumerable and sometimes very small (1 sq ft) foci where the agent (*R. tsutsugamushi*), the host/vector (*Leptotrombidium delienese*), and wild rats (*Rattus sp.*) coexist. The foci are found in such diverse settings as alpine meadows, disturbed rain forests, and seashores. Although in temperate zones disease incidence increases during the summer, more complex environmental factors influence incidence in the tropics. Chiggers (the larval stage of mites) are both vectors and reservoirs; females transmit *R. tsutsugamushi* transovarially to their progeny. Many species of murids, insectivores, marsupials, and domestic animals have been infected with *R. tsutsugamushi*. Small mammals have no role in the dissemination of rickettsiae; they only serve as hosts for the development of chiggers (Traub and Wisseman, 1974).

Spotted Fever Group

Tick-borne rickettsioses of the spotted fever group can be separated into two categories: RMSF in the western hemisphere and the various forms of tick typhus (Boutonneuse fever, Siberian tick typhus, Queensland tick typhus) in the eastern hemisphere. Rocky Mountain spotted fever is caused by a single species of rickettsiae (*R. rickettsii*), which is found in western Canada, the United States, Mexico, Central America, Colombia, and Brazil. The other types of tick typhus can be attributed to three different rickettsial species, each of which occurs over a broad but distinct geographic area: *R. conorii* (responsible for Boutonneuse fever) is found in the Mediteranean basin, including southern France and Italy, parts of Africa, the Middle East, India, and Pakistan; *R. sibirica* (Siberian tick typhus) occurs in a broad band that crosses the Soviet Union and parts of the Indo-Pakistan subcontinent; and *R. australis* (Queensland tick typhus) is found only in Eastern Australia.

Various species of ixodid ticks are both reservoirs and vectors of these rickettsiae. The microorganisms are maintained in the ticks for several generations by ovarial passage, but the rickettsiae may ultimately be deleterious to the tick host (see McDade and Newhouse, 1986). Many rodents and other mammals are susceptible to spotted fever-group rickettsiae, and amplify the infection in a particular tick-host ecosystem. Humans are infected by the ticks when they intrude on these cycles of infection (Hoogstraal, 1981; Linneman et al., 1973). Adult ticks are much more likely than larval or nymphal ticks to attach to humans, and are therefore more likely to transmit infection than are other tick stages. *R. akari* (responsible for rickettsialpox) infects mites that are ectoparasites of mice, presumably in a manner similar to that of tick-borne rickettsiae.

Q-Fever

Q-fever has been reported on all continents except Antarctica and is endemic in almost all countries except for certain ones in Scandinavia. However, even considering possible underdiagnosis, this disease is very rare in the United States.

Serologic evidence of Q-fever has been found in a large number of wild and domestic animals, but infection is almost always subclinical. The primary animal reservoir of *C. burnetii* varies from area to area. *C. burnetii* is carried by many species of tick, and although ticks may serve as a vector between animals, they probably do not transmit the disease to humans. Humans are usually infected when they inhale aerosols generated by infected livestock. Infection occurs most frequently in abbatoirs, sheep research facilities, dairies, and in animal husbandry operations: enormous numbers of organisms (10^9/g tissue) can be found in the products of conception.

Prevention, Control, and Treatment

The prevention and control of the rickettsial diseases are primarily accomplished by interrupting transmission of the infective agent from its vector or reservoir to humans. This can often be achieved by simply avoiding vector-infested areas (or, in the case of Q-fever, areas where infected animals are kept), by vector control, or by wearing protective clothing. At the present time immunization plays a minimal role in the prevention of rickettsial diseases, because, with the exceptions noted below, safe and effective vaccines are generally unavailable. In addition, with the possible exception of scrub typhus, chemoprophylaxis is generally both impractical and ineffective (Brezina, 1985; Wisseman and Ordonez, 1986).

Epidemic Typhus

Delousing remains the cornerstone of the control of epidemic typhus. A residual insecticide is applied to the patient and his contacts, their clothing, bedding, and living quarters. In facilities for washing clothes and bathing cannot be provided, patients should be isolated and their contacts kept under surveillance for 2 weeks. In an epidemic, the entire community may be treated. Residual insecticides including organochlorides (e.g., 10% DDT) were widely used in World War II. Increased resistance to these agents, however, has necessitated the use of the more expensive, but less efficient, organophosphates (e.g., 1% malathion) and the newer carbamates (Davidson, 1984). To choose the optimal insecticide, susceptibility testing may be required.

Immunization was widely used during World War II, and is still recommended by some organizations for persons working in close contact with infected patients in typhus-endemic areas (Benenson, 1985). However, commercial production of a killed epidemic typhus vaccine was discontinued in 1980 because of questions regarding the vaccine's efficacy and the need for safer production facilities (Centers for Disease Control, 1980). An attenuated vaccine prepared from the Madrid E strain of *R. prowazekii* is available under investigational new drug (IND) status in the United States, but is not routinely administered. No typhus cases are known to have occurred in an American traveler since 1950, and with reasonable sanitary measures the risk of typhus infection is extremely low. Moreover, tetracycline or chloramphenicol treatment is highly effective. Immunization is not currently required for entry into any country. Protective measures for persons occupationally at risk include wearing clothes that are either treated with repellent or storing them overnight with a dichorvos strip (NO Pest*) in an airtight bag.

---

* Use of trade names is for identification only and does not constitute endorsement by the Public Health Service or by the United States Department of Health and Human Services.

Murine Typhus

Murine typhus usually occurs sporadically or in small outbreaks. Persons can prevent infection by avoiding areas where flea-infested rats are found or by wearing repellent-treated clothing. Rat populations can be controlled by preventing their access to food. In murine typhus-endemic areas, vector control with insecticide powders should precede rodent control measures to prevent increased exposure of humans to fleas. No vaccine is available to murine typhus.

Scrub Typhus

The chigger vectors of scrub typhus are especially amenable to control because they often are found in distinct areas ("typhus islands") (Traub and Wisseman, 1974). These foci can be eliminated by treating the ground and vegetation with residual insecticides (of the organophosphate, organochlorine, or carbamate groups), reducing rodent populations, and destroying limited amounts of local vegetation. Persons who cannot avoid infested terrain should wear protective clothing, impregnate their clothing and bedding with a mitocide (e.g., benzyl benzoate, M-1960), and apply a mite repellent (e.g., diethyltoluamide, dibutylphosphate, benzyl benzoate) to exposed skin. Chemoprophylaxis should also be considered. In a controlled trial, the weekly administration of 200 mg doxycycline decreased the incidence of clinical illness but not of inapparent infection (Olson et al., 1980). Prophylaxis with chloramphenicol resulted in an unacceptable number of relapses when the drug was discontinued (Smadel et al., 1949b).

Spotted Fever Group

Rocky Mountain spotted fever and the other tick-borne rickettsioses (Boutonneuse fever, Queensland tick typhus, Siberian tick typhus) are prevented by avoiding tick-infested areas, carefully searching the body for ticks and removing them, wearing clothing designed to exclude ticks, and in very high-risk situations, wearing clothing impregnated with a tick repellent (e.g., N-N-butylacetanelide, diethyltoluamide) (Schreck et al., 1982). Integrated tick-control measures are generally not economically feasible (Hoch et al., 1971). Ticks should be removed from the body by grasping them with tweezers as closely as possible to the point of attachment and pulling slowly and steadily until they are dislodged (Needham, 1985). The tick-bite wound should then be washed with soap and water. Other methods, such as the application of nail polish remover or a lighted cigarette, do not work nearly as well. Because of the potential danger of becoming infected when crushing ticks, using fingers for this procedure is discouraged. The hands should always be washed thoroughly after removing a tick. Education about the early symptoms and signs of RMSF and the need for prompt empiric therapy if symptoms develop is also very important to control. An experimental vaccine for RMSF developed in the 1970s (DuPont et al., 1973) did not prove sufficiently protective and was not developed commercially.

Rickettsialpox is best prevented by the elimination of rodents, followed by the control of mites with residual acaricides. However, as the number of rodents decrease, mites are more likely to bite humans, and transmission may transiently increase. Improved garbage management may be responsible for a change in the predominant mite species and a decrease in the number of cases reported (Benenson, 1985).

Q-Fever

In the 1970s and 1980s, most reported outbreaks of Q-fever in the United States and other countries were associated with work-related exposures to sheep and their birth products. Therefore, prevention measures should be directed primarily at (1) avoiding exposures in institutions where sheep are used for research (Bernard et al., 1982), and (2) educating personnel in agricultural settings about disinfection and disposal of the products of conception. Vaccines made from killed organisms have been tested ever since *C. burnetii* was discovered (Anonymous, 1984; Marmion, 1967). A formalin-inactivated phase I vaccine, evaluated in a large comparative (but not randomized) study in Australian abattoir workers, provided complete protection to vaccinated subjects (Marmion et al., 1984). However, vaccination of individuals with pre-existing immunity has frequently caused erythema, induration, and sterile abscesses (Benenson, 1985; Lackman, 1962). Most of these adverse affects can be avoided if immune individuals are identified by skin tests before vaccination. One of the vaccines used in Australia is under IND status in the United States. Vaccination should be considered when contact with infected sheep is unavoidable.

Therapy

Prompt institution of effective antibiotic therapy against rickettsiae is the single most effective measure for preventing morbidity and mortality due to rickettsial diseases. Antirickettsial therapy improves the outcome of all rickettsioses, with the occasional

exception of fulminant or complicated cases of RMSF, epidemic typhus, and scrub typhus, where the illness is no longer susceptible to intervention (Woodward, 1959). If the illness is severe, the cardiac, pulmonary, renal, and central nervous systems should be assessed and additional measures instituted to prevent complications. Even in the less serious rickettsial diseases, such as Q-fever and rickettsialpox, treatment during the first few days appears to shorten the duration and lessen the severity of the illness (Powell et al., 1962; Rose, 1952).

Although they are rickettsiostatic rather than rickettsiacidal (Spicer et al., 1981; Wisseman and Ordonez, 1986), tetracyclines and chloramphenicol remain the only proven therapy for the rickettsial diseases. Clinical experience and limited comparative studies (Powell et al., 1962; Rose, 1952) suggest that the responses of patients treated with either antibiotic are similar, but others have found a slightly more rapid response with tetracycline than with chloramphenicol (Sheehy et al., 1973). Tetracyclines are also less likely to cause hematologic complications, but they may cause discoloration of teeth, hypoplasia of the enamel, and depression of skeletal growth in children; the extent of discoloration is directly related to the number of courses of tetracycline therapy received (Grossman et al., 1971). Therefore, chloramphenicol should be used for children under 8 years of age and for pregnant women.

There are no controlled data to suggest that any one tetracycline is superior to the others in terms of clinical response, but the high blood and tissue levels and long half-life of the lipophilic tetracyclines (minocycline, doxycycline) give them a theoretic advantage. In a small controlled study of epidemic typhus patients, the response to a single dose of 100 mg doxycycline was comparable to that of a 10-day course of chloramphenicol or tetracycline (Krause et al., 1975). Similar results were obtained in larger trials with doxycycline in scrub typhus patients (Brown et al., 1978; Olson et al., 1980). Although occasional relapses have been reported in patients receiving single-dose doxycycline therapy, no data suggest that relapses are more frequent with this therapy than with others. Doxycycline is excreted in the gastrointestinal tract and it is the tetracycline of choice in patients with renal failure (Whelton, 1978). Doxycycline may cause dental changes less frequently because it binds less with calcium than do other tetracyclines (Forti and Benincori, 1969), and it is preferred if tetracyclines must be used in children.

The response to therapy depends on the specific disease, severity of illness, and presence of other complicating factors (e.g., underlying illness, poor nutrition) at the time treatment is initiated. In general, improvement is rapid and dramatic, beginning about 24 h after initiation of therapy and continuing to complete defervescence in 1 to 4 days. Therapy is usually continued for 3 days after defervescence, but in more serious infections a minimum of 2 weeks of treatment is recommended.

Q-fever therapy requires separate consideration because of the unique characteristics of the chronic form of the disease and the antibiotic susceptibility of *C. burnetii.* Tetracycline is the only antibiotic with efficacy demonstrated in a controlled trial (Powell et al., 1962). Compared with no treatment, 3.0 g tetracycline followed by 0.5 g every 6 h shortened the duration of fever by about 2 days when treatment was started during the first 3 days of illness, but tetracycline was less effective when administration was begun later. Studies in embryonated chick eggs demonstrated that rifampin, trimethoprim, and the tetracyclines are effective against experimental Q-fever infection (Spicer et al., 1981). Erythromycin was not effective in this model, although apparent efficacy in humans treated with erythromycin alone has been reported (D'Angelo and Heatherington, 1979).

The poor prognosis of Q-fever endocarditis makes optimal management, including antibiotic therapy, essential. Experience with bacterial endocarditis suggests that combination therapy may be desirable, and clinical (Kimbrough et al., 1979; Turck et al., 1976) and experimental data (Spicer et al., 1981) suggest that in addition to a tetracycline, rifampin or the combination of trimethoprim and sulfamethoxazole are promising second drugs. Unfortunately, the small number of patients studied makes evaluation of the various therapeutic regimens difficult.

Literature Cited

Allen, A. C., and S. Spitz. 1945. A comparative study of the pathology of scrub typhus (tsutsugamushi disease) and other rickettsial diseases. Am. J. Pathol. **21**:603–680.

Anacker, R. L., R. N. Philip, L. A. Thomas, and E. A. Casper. 1979. Indirect hemagglutination test for detection of antibody for *Rickettsia rickettsii* in sera from humans and common laboratory animals. J. Clin. Microbiol. **10**:677–684.

Anonymous. 1984. Q fever: antigens and vaccines. Lancet **ii**:1435–1436.

Audy, J. R. 1968. Red mites and typhus. The Athlone Press, London.

Baca, O. G., and D. Paretsky. 1983. Q fever and *Coxiella burnetii:* a model for host-parasite interactions. Microbiol. Rev. **47**:127–149.

Benenson, A. S. (ed.) 1985. Control of communicable diseases in man, 14th ed. American Public Health Association, Washington, D.C.

Berman, S. J., and W. D. Kundin. 1973. Scrub typhus in South Vietnam. A study of 87 cases. Ann. Intern. Med. **79**:26–30.

Bernard, K. W., G. L. Parham, W. G. Winkler, C. G. Helmick. 1982. Q fever control measures: recommenda-

tions for research facilities using sheep. Infect. Control **3**:461–465.

Blake, F. G., K. F. Maxcy, J. F. Sadusk, Jr., G. M. Kohls, and E. J. Bell. 1945. Studies on tsutsugamushi disease (scrub typhus, mite borne-typhus) in New Guinea and adjacent islands: epidemiology, clinical observations, and etiology in the Dobadura area. Am. J. Hyg. **41**:243–396.

Bourgeois, A. L., J. G. Olson, C. M. Ho, R. C. Y. Fang, and P. F. D. Van Peenan. 1977. Epidemiological and serological study of scrub typhus among Chinese military in the Pescadores Islands of Taiwan. Trans. Roy. Soc. Trop. Med. Hyg. **71**:338–342.

Bozeman, F. M., D. E. Sonenshine, M. S. Williams, D. P. Chadwick, D. M. Lauer, and B. L. Elisberg. 1981. Experimental infection of ectoparasitic arthropods with *Rickettsia prowazekii* (GvF-16 strain) and transmission to flying squirrels. Am. J. Trop. Med. Hyg. **30**:253–263.

Brettman, L. R., S. Lewin, R. S. Holzman, W. D. Goldman, J. S. Marr, P. Kechijian, and R. Schinella. 1981. Rickettsialpox: report of an outbreak and a contemporary review. Medicine **60**:363–372.

Brezina, R. 1985. Diagnosis and control of rickettsial disease. Acta Virol. **29**:338–349.

Brown, G. W., D. M. Robinson, D. L. Huxsoll, T. S. Ng, K. J. Lim, and G. Sannasey. 1976. Scrub typhus: a common cause of illness in indigenous populations. Trans. Roy. Soc. Trop. Med. Hyg. **70**:444–448.

Brown, G. W., J. P. Saunders, S. Singh, D. L. Huxsoll, and A. Shirai. 1978. Single dose doxycycline therapy for scrub typhus. Trans. Roy. Soc. Trop. Med. Hyg. **72**:412–416.

Brown, G. W., A. Shirai, C. Rogers, and M. G. Groves. 1983. Diagnostic criteria for scrub typhus: probability values for immunofluorescent antibody and *Proteus* OXK agglutinin titers. Am. J. Trop. Med. Hyg. **32**:1101–1107.

Buhles, W. C., D. L. Huxsoll, G. Ruch, R. H. Kenyon, and B. L. Elisberg. 1975. Evaluation of primary blood monocyte and bone marrow cell cultures for the isolation of *Rickettsia rickettsii.* Infect. Immun. **12**:1457–1463.

Centers for Disease Control. 1980. Production of typhus vaccine discontinued in the United States. Morbid. Mortal. Weekly Rep. **29**:465.

Chang, R. S.-M., E. S. Murray, and J. C. Snyder. 1954. Erythrocyte-sensitizing substances from rickettsiae of the Rocky Mountain spotted fever group. J. Immunol. **73**:8–15.

Chang, S. 1953. A serologically active erythrocyte sensitizing substance from typhus rickettsiae. J. Immunol. **70**:212–214.

Clark, W. H., E. H. Lennette, O. C. Railsback, and M. S. Romer. 1951. Q Fever in California. Arch. Int. Med. **88**:155–167.

Crum, J. W., S. Hanchalay, and C. Eamsila. 1980. New paper enzyme-linked immunosorbent technique compared with microimmunofluorescence for detection of human serum antibodies for *Rickettsia tsutsugamushi.* J. Clin. Microbiol. **11**:584–588.

D'Angelo, L. J., and R. Heatherington. 1979. Q fever treated with erythromycin. Br. Med. J. **2**:305–306.

Dasch, G. A., S. Halle, and A. L. Bourgeois. 1979. Sensitive microplate enzyme-linked immunosorbent assay for detection of antibodies against the scrub typhus rickettsiae. J. Clin. Microbiol. **9**:38–48.

Davidson, G. 1984. Control of arthropods of medical importance, p. 923–940. *In* G. T. Strickland (ed.), Tropical medicine. The W. B. Saunders Co., Philadelphia.

DeShazo, R. D., J. R. Boyce, J. V. Osterman, and E. H. Stephenson. 1976. Early diagnosis of Rocky Mountain spotted fever. Use of primary monocyte culture technique. J. Am. Med. Assoc. **235**:1353–1355.

Derrick, E. H. 1937. 'Q' fever, a new fever entity: clinical features, diagnosis, and laboratory investigation. Med. J. Aust. **2**:281–299.

Döller, G., P. C. Döller, and H-J. Gerth. 1984. Early diagnosis of Q fever: detection of immunoglobulin M by radioimmunoassay and enzyme immunoassay. Eur. J. Clin. Microbiol. **3**:550–553.

Donohue, J. F. 1980. Lower respiratory tract involvement in Rocky Mountain spotted fever. Arch. Intern. Med. **140**:223–227.

Duma, R. J., D. E. Sonenshine, F. M. Bozeman, J. M. Veazy, Jr., B. L. Elisberg, D. P. Chadwick, N. I. Stocks, T. M. McGill, G. B. Miller, and J. N. MacCormack. 1981. Epidemic typhus in the United States associated with flying squirrels. J. Am. Med. Assoc. **25**:2318–2323.

DuPont, H. L., R. B. Hornick, A. T. Dawkins, G. G. Heiner, I. B. Fabrikant, C. L. Wisseman, Jr., and T. E. Woodward. 1973. Rocky Mountain spotted fever: a comparative study of the active immunity induced by inactivated and viable pathogenic *Rickettsia rickettsii*. J. Infect. Dis. **128**:340–344.

Dupuis, G., O. Péter, M. Peacock, W. Burgdorfer, and E. Haller. 1985a. Immunoglobulin responses in acute Q fever. J. Clin. Microbiol. **22**:484–487.

Dupuis, G., O. Péter, D. Pedroni, and J. Petite. 1985b. Aspects cliniques observes lors d'une epidemie de 415 cas de fievre Q. Schweiz. Med. Wschr. **115**:814–818.

Dyer, R. E. 1944. The rickettsial diseases. J. Am. Med. Assoc. **124**:1165–1168.

Ellis, M. E., C. C. Smith, and M. A. J. Moffat. 1982. Chronic or fatal Q-fever infection: a review of 16 patients seen in north-east Scotland (1967–80). Quart. J. Med., New Series **205**:54–66.

Farhang-Azad, A., R. Traub, M. Sofi, and C. L. Wisseman, Jr. 1984. Experimental murine typhus infection in the cat flea *Ctenocephalides felis* (Siphonaptera: pulicidae). J. Med. Entomol. **21**:675–680.

Farhang-Azad, A., and R. Traub. 1985. Transmission of murine typhus rickettsiae by *Xenopsylla cheopis*, with notes on experimental infection and effects of temperature. Am. J. Trop. Med. Hyg. **34**:555–563.

Farhang-Azad, A., S. Traub, and S. Baqar. 1985. Transovarial transmission of murine typhus rickettsiae in *Xenopsylla cheopis* fleas. Science **227**:543–545.

Field, P. R., J. G. Hunt, and A. M. Murphy. 1983. Detection and persistence of specific IgM antibodies to *Coxiella burnetii* by enzyme-linked immunosorbent assay: a comparison with immunofluorescence and complement fixation tests. J. Infect. Dis. **148**:477–487.

Fishbein, D. B., J. E. Kaplan, K. W. Bernard, and W. G. Winkler. 1984. Surveillance of Rocky Mountain spotted fever in the United States, 1981–1983. J. Infect. Dis. **150**:609–611.

Fleck, L. 1947. Specific antigenic substances in the urine of typhus patients. Texas Rep. Biol. Med. **5**:168–172.

Fleck, L., S. Porat, Z. Evenchik, and M. A. Klinberg. 1960. The renal excretion of specific microbial substances during the course of infection with murine typhus rickettsiae. Am. J. Hyg. **72**:351–361.

Fleisher, G., E. T. Lennette, and P. Honig. 1979. Diagnosis of Rocky Mountain spotted fever by immunofluorescent identification of *Rickettsia rickettsii* in skin biopsy tissue. J. Pediatr. **95**:63–65.

Forti, G., and C. Benincori. 1969. Doxycycline and the teeth. Lancet **i**:782.

Foster, G. M. 1981. Typhus disaster in the wake of war: the

American-Polish relief expedition, 1919–1920. Bull. Hist. Med. **55**:221–232.

Giménez, D. F. 1964. Staining rickettsiae in yolk-sac cultures. Stain Tech. **39**:135–140.

Goldwasser, R. A., and C. C. Shepard. 1959. Fluorescent antibody methods in the differentiation of murine and epidemic typhus sera; specificity changes resulting from previous immunization. J. Immunol. **82**:373–380.

Greenberg, M., O. Pellitteri, I. F. Klein, and R. J. Huebner. 1947. Rickettsialpox—a newly recognized rickettsial disease: II. Clinical observations. J. Am. Med. Assoc. **133**:901–906.

Grossman, E. R., A. Walchek, and H. Freedman. 1971. Tetracyclines and permanent teeth: the relation between dose and tooth color. Pediatrics **47**:567–570.

Groves, M. G., D. L. Rosenstreich, B. A. Taylor, and J. V. Osterman. 1980. Host defenses in experimental scrub typhus: mapping the gene that controls natural resistance in mice. J. Immunol. **125**:1395–1399.

Hackstadt, T., M. G. Peacock, P. J. Hitchock, and R. L. Cole. 1985. Lipopolysaccharide variation in *Coxiella burnetii*: intrastrain heterogeneity in structure and antigenicity. Infect. Immun. **48**:359–365.

Hackstadt, T., and J. C. Williams. 1981. Biochemical stratogem for obligate parasitism of eukaryotic cells by *Coxiella burnetii*. Proc. Natl. Acad. Sci. **78**:3240–3244.

Halle, S., G. A. Dasch, and E. Weiss. 1977. Sensitive enzyme-linked immunosorbent assay for detection of antibodies against typhus rickettsiae, *Rickettsia prowazekii* and *Rickettsia typhi*. J. Clin. Microbiol. **6**:101–110.

Hamilton, H. L. 1945. Specificity of toxic factors associated with epidemic and murine strains of typhus rickettsiae. Am. J. Trop. Med. Hyg. **25**:391–395.

Harrell, G. T. 1949. Rocky Mountain spotted fever. Medicine **28**:333–370.

Hase, T. 1985. Developmental sequence and surface membrane assembly of rickettsiae. Ann. Rev. Microbiol. **39**:69–88.

Hattwick, M. A. W., H. Retailliau, R. J. O'Brien, M. Slutzker, R. E. Fontaine, and B. Hanson. 1978. Fatal Rocky Mountain spotted fever. J. Am. Med. Assoc. **240**:1499–1503.

Hechemy, K. E., R. L. Anacker, N. L. Carlo, J. A. Fox, and H. A. Gaafar. 1983a. Absorption of *Rickettsia rickettsii* antibodies by *Rickettsia rickettsii* antigens in four diagnostic tests. J. Clin. Microbiol. **17**:445–449.

Hechemy, K. E., R. L. Anacker, R. N. Philip, K. T. Kleeman, J. N. MacCormack, S. J. Sasowski, and E. E. Michaelson. 1980. Detection of Rocky Mountain spotted fever antibodies by a latex agglutination test. J. Clin. Microbiol. **12**:144–150.

Hechemy, K. E., E. E. Michaelson, R. L. Anacker, M. Zdeb, and S. J. Sasowski. 1983b. Evaluation of latex-*Rickettsia rickettsii* test for Rocky Mountain spotted fever in 11 laboratories. J. Clin. Microbiol. **18**:938–946.

Hechemy, K. E., and B. B. Rubin. 1983. Latex *Rickettsia rickettsii* test reactivity in seropositive patients. J. Clin. Microbiol. **17**:489–492.

Hechemy, K. E., R. W. Stevens, S. Sasowski, E. E. Michaelson, E. A. Casper, and R. N. Philip. 1979. Discrepancies in Weil-Felix and microimmunofluorescence test results for Rocky Mountain spotted fever. J. Clin. Microbiol. **9**:292–293.

Helmick, C. G., K. W. Bernard, and L. J. D'Angelo. 1984. Rocky Mountain spotted fever: clinical, laboratory, and epidemiological features of 262 cases. J. Infect. Dis. **150**:480–488.

Herrmann, J. E., M. R. Hollingdale, M. F. Collins, and J. W. Vinson. 1977. Enzyme immunoassay and radioimmunoprecipitation tests for the detection of antibodies to *Rochalimaea* (*Rickettsia*) *quintana* (39655). Proc. Soc. Exp. Biol. Med. **154:**285–288.

Hoch, A. L., R. W. Barker, and J. A. Hair. 1971. Further observations on the control of lone star ticks (Acarina: Ixodidae) through integrated control procedures. J. Med. Entomol. **8:**731–734.

Holland, W. W., K. E. K. Rowson, C. E. D. Taylor, and A. B. Allen, M. Ffrench-Constant, and C. M. C. Smelt. 1960. Q fever in the R.A.F. in Great Britain in 1958. Br. Med. J. **1:**387–390.

Hoogstraal, H. 1981. Changing patterns of tickborne diseases in modern society. Ann. Rev. Entomol. **26:**75–99.

Huebner, R. J., P. Stamps, and C. Armstrong. 1946. Rickettsialpox—a newly recognized rickettsial disease: I. Isolation of the etiologic agent. Public Health Rep. **61:**1605–1614.

Hunt, J. G., P. R. Field, and A. M. Murphy. 1983. Immunoglobulin responses to *Coxiella burnetii* (Q fever): single-serum diagnosis of acute infection, using an immunofluorescence technique. Infect. Immun. **39:**977–981.

Johnson, J. W., and C. E. Pedersen, Jr. 1978. Plaque formation by strains of spotted fever rickettsiae in monolayer cultures of various cell types. J. Clin. Microbiol. **7:**389–391.

Kamal, A. M., and G. A. Messih. 1943. Typhus fever. Review of 11,410 cases. Symptomatology, laboratory investigations and treatment. J. Egypt Public Health Assoc. **1:**125–213.

Kaplan, J. E., and L. B. Schonberger. 1986. The sensitivity of various serologic tests in the diagnosis of Rocky Mountain spotted fever. Am. J. Trop. Med. Hyg. **35:**840–844.

Kaplowitz, L. G., J. J. Fischer, and P. F. Sparling. 1981. Rocky Mountain spotted fever: a clinical dilemma, p. 89–108. *In* J. S. Remington, M. N. Schwartz (ed.), Current clinical topics in infectious diseases, vol. 2. McGraw-Hill Book Co., New York.

Kaplowitz, L. G., J. V. Lange, J. J. Fischer, and D. H. Walker. 1983. Correlation of rickettsial titers, circulating endotoxin, and clinical features in Rocky Mountain spotted fever. Arch. Int. Med. **143:**1149–1151.

Kimbrough, R. C., III, R. A. Ormsbee, M. Peacock, R. Rogers, R. W. Bennetts, J. Raaf, A. Krause, and C. Gardner. 1979. Q fever endocarditis in the United States. Ann. Intern. Med. **91:**400–402.

Kopmans-Gargantiel, A. I., and C. L. Wisseman, Jr. 1981. Differential requirements for enriched atmospheric carbon dioxide content for intracellular growth in cell culture among selected members of the genus *Rickettsia.* Infect. Immun. **31:**1277–1280.

Krause, D. C., H. H. Winkler, and D. O. Wood. 1985. Cloning and expression of the *Rickettsia prowazekii* ADP/ATP translocator in *Escherichia coli.* Proc. Natl. Acad. Sci. USA **82:**3015–3019.

Krause, D. W., P. L. Perine, J. E. McDade, and S. Awoke. 1975. Treatment of louse-borne typhus fever with chloramphenicol, tetracycline or doxycycline. East Afr. Med. J. **52:**421–427.

Lackman, D. B., E. J. Bell, J. F. Bell, and E. G. Pickens. 1962. Intradermal sensitivity testing in man with a purified vaccine for Q fever. Am. J. Public Health **52:**87–91.

Lackman, D. B., G. Gilda, and R. N. Philip. 1964. Application of the radioisotope precipitation test to the study of Q fever in man. Health Lab. Sci. **1:**21–28.

Ley, H. L., J. E. Smadel, F. H. Diercks, and P. Y. Paterson. 1952. Immunization against scrub typhus. V. The infective dose of *Rickettsia tsutsugamushi* for men and mice. Am. J. Hyg. **56:**313–319.

Linneman, C. C., P. Jansen, and G. M. Schiff. 1973. Rocky Mountain spotted fever in Clermont County, Ohio. Am. J. Epidemiol. **97:**125–130.

Linnemann, C. C., Jr. 1980. Skin biopsy in diagnosis of Rocky Mountain spotted fever (letter). J. Pediatr. **96:**781–782.

Mallavia, L. P., J. E. Samuel, M. L. Kahn, L. S. Thomashow, and M. E. Frazier. 1984. *Coxiella burnetii* plasmid DNA. Microbiol. **1984:**293–296.

Marchette, N. J., and D. Stiller. 1982. Ecological relationships and evolution of the rickettsiae, vol. I. CRC Press, Boca Raton, Florida.

Marmion, B. P. 1967. Development of Q-fever vaccines, 1937 to 1967. Med. J. Aust. **2:**1074–1078.

Marmion, B. P., R. A. Ormsbee, M. Kyrkou, J. Wright, D. Worswick, S. Cameron, A. Esterman, B. Feery, W. Collins, and L. Hartman. 1984. Vaccine prophylaxis of abattoir-associated Q-fever. Lancet **ii:**1411–1414.

McCaul, T. F., and J. C. Williams. 1981. Developmental cycle of *Coxiella burnetii:* structure and morphogenesis of vegetative and sporogenic differentiations. J. Bacteriol. **147:**1063–1076.

McDade, J. E., and P. J. Gerone. 1970. Plaque assay for Q fever and scrub typhus rickettsiae. Appl. Microbiol. **19:**963–965.

McDade, J. E., and V. F. Newhouse. 1986. Natural history of *Rickettsia rickettii.* Ann. Rev. Microbiol. **40:**287–309.

McDade, J. E., C. C. Shepard, M. A. Redus, V. F. Newhouse, and D. J. Smith. 1980. Evidence of *Rickettsia prowazekii* infections in the United States. Am. J. Trop. Med. Hyg. **29:**277–284.

McDade, J. E., J. R. Stakebake, and P. J. Gerone. 1969. Plaque assay system for several species of *Rickettsia.* J. Bacteriol. **99:**910–912.

Megaw, J. W. D. 1942. Louse-borne typhus fever. Br. Med. J. **ii:**401–403; 433–435.

Miller, E. S., and P. B. Beeson. 1946. Murine typhus fever. Medicine **25:**1–15.

Murray, E. S., G. Baehr, G. Schwartzman, R. A. Mandelbaum, N. Rosenthal, J. C. Doane, L. B. Weiss, S. Cohen, and J. C. Snyder. 1950. Brills disease: I. clinical and laboratory diagnosis. J. Am. Med. Assoc. **142:**1059–1066.

Myers, W. F., and C. L. Wisseman, Jr. 1980. Genetic relatedness among the typhus group of rickettsiae. Int. J. System. Bacteriol. **30:**143–150.

Myers, W. F., and C. L. Wisseman, Jr. 1982. Genetic relatedness within the spotted fever biotype of the genus *Rickettsia.* Third National Meeting of the American Society for Rickettsiology and Rickettsial Diseases, Atlanta, Georgia, (abstract).

National Research Council, Division of Medical Sciences, Committee on Pathology. 1953. Pathology of epidemic typhus. Report of fatal cases studied by the United States of America Typhus Commission in Cairo, Egypt, during 1943–1945. Arch. Pathol. **56:**397–435, 512–553.

Needham, G. R. 1985. Evaluation of five popular methods for tick removal. Pediatrics **75:**997–1002.

Newhouse, V. F., C. C. Shepard, M. D. Redus, T. Tzianabos, and J. E. McDade. 1979. A comparison of the complement fixation, indirect fluorescent antibody and microagglutination tests for the serological diagnosis of rickettsial diseases. Am. J. Trop. Med. Hyg. **28:**387–395.

O'Connor, J. L., and J. M. MacDonald. 1950. Excretion of specific antigen in the urine in tsutsugamushi disease (scrub typhus). Br. J. Exp. Pathol. **31:**51–64.

Olson, J. G., A. L. Bourgeois, R. C. Y. Fang, J. C. Coolbaugh, and D. T. Dennis. 1980. Prevention of scrub typhus. Prophylactic administration of doxycycline in a randomized double blind trial. Am. J. Trop. Med. Hyg. **29**:989–997.

Ormsbee, R., M. Peacock, R. Philip, E. Casper, J. Plorde, T. Gabre-Kidan, and L. Wright. 1977. Serologic diagnosis of epidemic typhus fever. Am. J. Epidemiol. **105**:261–271.

Ormsbee, R., M. Peacock, R. Gerloff, G. Tallent, and D. Wike. 1978a. Limits of rickettsial infectivity. Infect. Immun. **19**:239–245.

Ormsbee, R., M. Peacock, R. Philip, E. Casper, J. Plorde, T. Gabre-Kidan, and L. Wright. 1978b. Antigenic relationships between the typhus and spotted fever groups of rickettsiae. Am. J. Epidemiol. **108**:53–59.

Oster, C. N., D. S. Burke, R. H. Kenyon, M. S. Ascher, P. Harber, and C. E. Pedersen, Jr. 1977. Laboratory-acquired Rocky Mountain spotted fever. N. Engl. J. Med. **297**:859–862.

Peacock, M. G., R. N. Philip, J. C. Williams, and R. S. Faulkner. 1983. Serological evaluation of Q fever in humans: enhanced phase I titers of immunoglobulins G and A are diagnostic for Q-fever endocarditis. Infect. Immunol. **41**:1089–1098.

Phibbs, P. V., Jr., and H. H. Winkler. 1982. Regulatory properties of citrate synthase from *Rickettsia prowazekii*. J. Bacteriol. **149**:718–725.

Philip, R. N., E. A. Casper, W. Burgdorfer, R. K. Gerloff, L. E. Hughes, and E. J. Bell. 1978. Serologic typing of rickettsiae of the spotted fever group by microimmunofluorescence. J. Immunol. **121**:1961–1968.

Philip, R. N., E. A. Casper, J. N. MacCormack, D. L. Sexton, L. A. Thomas, R. L. Anacker, W. Burgdorfer, and S. Vick. 1977. A comparison of serologic methods for diagnosis of Rocky Mountain spotted fever. Am. J. Epidemiol. **105**:56–67.

Philip, R. N., E. A. Casper, R. A. Ormsbee, M. G. Peacock, and W. Burgdorfer. 1976. Microimmunofluorescence test for the serological study of Rocky Mountain spotted fever and typhus. J. Clin. Microbiol. **3**:51–61.

Pike, R. M. 1976. Laboratory-associated infections: summary and analysis of 3,921 cases. Health Lab. Sci. **13**:105–114.

Pike, R. M. 1979. Laboratory-associated infections: incidence, fatalities, causes, and prevention. Ann. Rev. Microbiol. **33**:41–66.

Powell, O. 1960. "Q" fever: clinical features of 72 cases. Aust. Ann. Med. **9**:214–223.

Powell, O. W., K. P. Kennedy, M. McIver, and H. Silverstone. 1962. Tetracycline in the treatment of "Q" fever. Aust. Ann. Med. **11**:184–188.

Randall, M. B., and D. H. Walker. 1984. Rocky Mountain spotted fever. Gastrointestinal and pancreatic lesions and rickettsial infection. Arch. Pathol. Lab. Med. **108**:963–967.

Raoult, D., C. DeMicco, H. Chaudet, and J. Tamalet. 1985. Serological diagnosis of Mediterranean spotted fever by the immunoperoxidase reaction. Eur. J. Clin. Microbiol. **4**:441–442.

Raoult, D., C. DeMicco, H. Gallais, and M. Toga. 1984. Laboratory diagnosis of Mediterranean spotted fever by immunofluorescent demonstration of *Rickettsia conorii* in cutaneous lesions. J. Infect. Dis. **150**:145–148.

Raoult, D., P. Zuchelli, P. J. Weiller, C. Charrel, J. L. San Marco, H. Gallais, and P. Casanova. 1986a. Incidence, clinical observations and risk factors in severe form of Mediterranean spotted fever among patients admitted to hospital in Marseilles, 1983–1984. J. Infect. **12**:111–116.

Raoult, D., P. J. Weiller, A. Chagnon, H. Chaudet, H. Gallais, and P. Casanova. 1986b. Mediterranean spotted

fever: clinical, laboratory, and epidemiological features of 199 cases. Am. J. Trop. Med. Hyg. **35**:845–850.

Regnery, R. L., Y. F. Zhang, and C. L. Spruill. 1986. Flying squirrel-associated *Rickettsia prowazekii* (Epidemic typhus rickettsiae) characterized by a specific DNA fragment produced by restriction endonuclease digestion. J. Clin. Microbiol. **23**:189–191.

Richardson, J. H., and W. E. Barkley. 1984. Biosafety in microbiological and biomedical laboratories. U.S. Government Printing Office, Washington D.C.

Rikihisa, Y., and S. Ito. 1980. Localization of electron-dense tracers during entry of *Rickettsia tsutsugamushi* into polymorphonuclear leukocytes. Infect. Immun. **30**:231–243.

Ris, H., and J. P. Fox. 1949. The cytology of rickettsiae. J. Exp. Med. **89**:681–686.

Rose, H. M. 1948. Rickettsialpox. NY State J. Med. **48**:2266–2271.

Rose, H. M. 1952. The treatment of Rickettsialpox with antibiotics. Ann. NY Acad. Sci. **55**:1019–1026.

Samuel, J. E., M. Frazier, M. L. Kahn, L. S. Thomashow, and L. P. Mallavia. 1983. Isolation and characterization of a plasmid from phase I *Coxiella burnetii*. Infect. Immun. **41**:488–493.

Samuel, J. E., M. E. Frazier, and L. P. Mallavia. 1985. Correlation of plasmid type and disease caused by *Coxiella burnetii*. Infect. Immun. **49**:775–779.

Saunders, J. P., G. W. Brown, A. Shirai, and D. L. Huxsoll. 1980. The longevity of antibody to *Rickettsia tsutsugamushi* in patients with confirmed scrub typhus. Trans. Roy. Soc. Trop. Med. Hyg. **74**:253–257.

Sayan, J. J., H. S. Pond, J. S. Forrester, and F. C. Wood. 1946. Scrub typhus in Assam and Burma. A clinical study of 616 cases. Medicine **25**:155–214.

Schramek, S. R., R. Brezina, and I. V. Tarasevich. 1976. Isolation of a lipopolysaccharide antigen from *Rickettsia* species. Acta Virol. **20**:270.

Schreck, C. E., G. A. Mount, and D. A. Carlson. 1982. Wash and wear persistence of permethrin used as a clothing treatment for personal protection against the lone star tick (Acari: Ixodidae). J. Med. Entomol. **19**:143–146.

Sheehy, T. W., D. Hazlett, and R. E. Turk. 1973. Scrub typhus. A comparison of chloramphenicol and tetracycline in its treatment. Arch. Int. Med. **132**:77–80.

Shirai, A., J. C. Coolbaugh, E. Gan, T. C. Chan, D. L. Huxsoll, and M. G. Groves. 1982a. Serologic analysis of scrub typhus isolates from the Pescadores and Philippine Islands. Jpn. J. Med. Sci. Biol. **35**:255–259.

Shirai, A., J. W. Dietel, and J. V. Osterman. 1975. Indirect hemagglutination test for human antibody to typhus and spotted fever group rickettsiae. J. Clin. Microbiol. **2**:430–437.

Shirai, A., J. P. Saunders, A. L. Dohany, D. L. Huxoll, and M. G. Groves. 1982b. Transmission of scrub typhus to human volunteers by laboratory-reared chiggers. Jpn. J. Med. Sci. Biol. **35**:9–16.

Silverman, D. J. 1984. *Rickettsia rickettsii*-induced cellular injury of human vascular endothelium in vitro. Infect. Immun. **44**:545–553.

Silverman, D. J., and C. L. Wisseman, Jr. 1978. Comparative ultrastructural study on the cell envelopes of *Rickettsia prowazekii*, *Rickettsia rickettsii*, and *Rickettsia tsutsugamushi*. Infect. Immun. **21**:1020–1023.

Silverman, D. J., C. L. Wisseman, Jr., A. D. Waddell, and M. Jones. 1978. External layers of *Rickettsia prowazekii* and *Rickettsia rickettsii*: occurrence of a slime layer. Intect. Immun. **22**:233–246.

Smadel, J. E., T. E. Woodward, H. L. Ley, Jr., and R.

Lewthwaite. 1949a. Chloramphenicol (chloromycetin) in the treatment of tsutsugamushi disease (scrub typhus). J. Clin. Invest. **28:**1196–1215.

Smadel, J. E., R. Traub, H. L. Ley, Jr., C. B. Philip, T. E. Woodward, and R. Lewthwaite. 1949b. Chloramphenicol (chloromycetin) in the chemoprophylaxis of scrub typhus (tsutsugamushi disease). II. Results with volunteers exposed in hyperendemic areas. Am. J. Hyg. **50:**75–91.

Smith, D. K., and H. H. Winkler. 1979. Separation of inner and outer membranes of *Rickettsia prowazekii* and characterization of their polypeptide compositions. J. Bacteriol. **137:**963–971.

Smorodintzeff, A. A., and R. V. Fradkina. 1944. Slide agglutination test for rapid diagnosis of pre-eruptive typhus fever. Proc. Soc. Exp. Biol. **56:**93–94.

Snyder, J. C. 1973. The philosophy of disease control and the population explosion, p. 7–13. Proceedings of the International Symposium on the Control of Lice and Louse-Borne Diseases. Pan American Health Organization Scientific Publication No. 263.

Snyder, J. C. 1965. Typhus fever rickettsiae, p. 1059–1094. *In* F. L. Horsfall and I. Tamm (ed.), Viral and rickettsial infections of man. J. B. Lippincott Co., Philadelphia.

Sonenshine, D. E., F. M. Bozeman, M. S. Williams, S. A. Masiello, D. P. Chadwick, N. I. Stocks, D. M. Lauer, and B. L. Elisberg. 1978. Epizootiology of epidemic typhus (*Rickettsia prowazekii*) in flying squirrels. Am. J. Trop. Med. Hyg. **27:**339–349.

Spelman, D. W. 1982. Q fever. A study of 111 consecutive cases. Med. J. Aust. **1:**547–553.

Spicer, A. J., M. G. Peacock, and J. C. Williams. 1981. Effectiveness of several antibiotics in suppressing chick embryo lethality during experimental infections by *Coxiella burnetii, Rickettsia typhi,* and *R. rickettsii,* p. 375–383. *In* W. Burgdorfer and R. L. Anacker (ed.), Rickettsiae and rickettsial diseases. Academic Press, Inc., New York.

Srigley, J. R., W. R. Geddie, H. Vellend, N. Palmer, M. J. Phillips, W. R. Geddie, A. W. P. Van Nostrand, and V. D. Edwards. 1985. Q-fever. The liver and bone marrow pathology. Am. J. Surg. Pathol. **9:**752–758.

Stoenner, H. G., D. B. Lackman, and E. J. Bell. 1962. Factors affecting the growth of rickettsias of the spotted fever group in fertile hens' eggs. J. Infect. Dis. **110:**121–128.

Tigertt, W. D., and A. S. Benenson. 1956. Studies on Q fever in man. Trans. Am. Assoc. Phys. **69:**98–104.

Traub, R., and C. L. Wisseman. Jr. 1974. The ecology of chigger-borne rickettsiosis (scrub typhus). J. Med. Entomol. **11:**237–303.

Traub, R., C. L. Wisseman, Jr., and A. Farhang-Azad. 1978. The ecology of murine typhus-a critical review. Trop. Dis. Bull. **75:**237–317.

Turck, W. P. G., G. Howitt, L. A. Turnberg, H. Fox, M. Longson, M. B. Matthews, and R. Das Gupta. 1976. Chronic Q fever. Quart. J. Med. New Series **45:**193–217.

Urso, F. P. 1975. The pathologic findings in rickettsial pneumonia. Am. J. Clin. Pathol. **64:**335–342.

Vinson, J. W. 1973. Louse-borne diseases worldwide: trench fever, p. 76–79. Proceedings of the International Symposium on the Control of Lice and Louse-Borne Diseases. Pan American Health Organization Scientific Publication No. 263.

Vinson, J. W., and H. S. Fuller. 1961. Studies of trench fever: I. Propogation of rickettsia-like microorganisms from a patient's blood. Pathol. Microbiol. **24**(suppl.):152–166.

Walker, D. H., and B. G. Cain. 1978. A method for specific diagnosis of Rocky Mountain spotted fever on fixed, paraffin-embedded tissue by immunofluorescence. J. Infect. Dis. **137:**206–209.

Walker, D. H., B. G. Cain, and P. M. Olmstead. 1978. Laboratory diagnosis of Rocky Mountain spotted fever by immunofluorescent demonstration of *Rickettsia rickettsii* in cutaneous lesions. Am. J. Clin. Pathol. **69:**619–623.

Walker, D. H., and J. H. S. Gear. 1985. Correlation of the distribution of *Rickettsia conorii,* microscopic lesions, and clinical features in South African tick bite fever. Am. J. Trop. Med. Hyg. **34:**361–371.

Walker, D. H., C. G. Crawford, and B. G. Cain. 1980a. Rickettsial infection of the pulmonary microcirculation: The basis of interstitial pneumonitis in Rocky Mountain spotted fever. Human Pathol. **11:**263–272.

Walker, D. H., and W. D. Mattern. 1979. Acute renal failure in Rocky Mountain spotted fever. Arch. Intern. Med. **139:**443–448.

Walker, D. H., and W. D. Mattern. 1980. Rickettsial vasculitis. Am. Heart J. **100:**896–906.

Walker, D. H., C. E. Palleta, and B. G. Cain. 1980b. Pathogenesis of myocarditis in Rocky Mountain spotted fever. Arch. Pathol. Lab. Med. **104:**171–174.

Walker, T. S., and H. H. Winkler. 1978. Penetration of cultured mouse fibroblasts (L cells) by *Rickettsia prowazeki.* Infect. Immun. **22:**200–208.

Weil, E., and A. Felix. 1916. Zur serologischen Diagnose des Fleckfiebers. Wien. Klin. Wochenschr. **29:** 33–35.

Weiss, E. 1981. The family Rickettsiaceae: human pathogens, p. 2137–2160. *In* M. P. Starr, et al., (eds.), The prokaryotes. Spinger-Verlag KG, Berlin.

Weiss, E., and J. W. Moulder. 1984. Rickettsiales, p. 687–704. *In* N. R. Krieg, (ed.), Bergeys manual of systematic bacteriology, Vol. 1. The Williams & Wilkins Co., Baltimore, London.

Weyer, F. 1961. Zur Frage der widerstands fahigkeit von Rickettsien im Lausekot gegen physikalische Einflusse, insbesondere gegen Warme. Z. Tropenmed. Parasit. **12:**79–92.

Whelton, A. 1978. Tetracyclines in renal insufficiency: resolution of a therapeutic dilemma. Bull. NY Acad. Med. **54:**223–236.

Whittick, J. W. 1950. Necropsy findings in a case of Q fever in Britain. Br. Med. J. **1:**979–980.

Wilfert, C. M., J. N. MacCormack, K. Kleeman, R. N. Philip, E. Austin, V. Dickinson, and L. Turner. 1985. The prevalence of antibodies to *Rickettsia rickettsii* in an area endemic for Rocky Mountain spotted fever. J. Infect. Dis. **151:**823–831.

Wilson, L. B., and W. M. Chowning. 1904. Studies in *Pyroplasmosis hominis* ("spotted fever" or "tick fever" of the Rocky Mountains). J. Infect. Dis. **1:**31–57.

Winkler, H. 1982. Rickettsiae: intracytoplasmic life. ASM News **48:**184–187.

Winkler, H. H. 1976. Rickettsial permeability. An ADP-ATP transport system. J. Biol. Chem. **251:**389–396.

Wisseman, C. L., Jr. 1986. Selected observations on rickettsiae and their host cells. Acta. Virol. **30:**81–95.

Wisseman, C. L., Jr., E. A. Edlinger, A. D. Waddell, and M. R. Jones. 1976. Infection cycle of *Rickettsia rickettsii* in chicken embryo and L-929 cells in culture. Infect. Immun. **14:**1052–1064.

Wisseman, C. L., Jr., and S. V. Ordonez. 1986. Actions of antibiotics on *Rickettsia rickettsii.* J. Infect. Dis. **153:**626–628.

Wisseman, C. L., Jr., and A. D. Waddell. 1975. In vitro

studies on Rickettsia-host cell interactions: intracellular growth cycle of virulent and attenuated *Rickettsia prowazeki* in chick embryo cells in slide chamber cultures. Infect. Immun. **11:**1391–1401.

Woodward, T. E. 1959. Therapy of the rickettsial diseases with a discussion of chemoprophylaxis. Arch. Inst. Pasteur (Tunis) **35:**507–517.

Woodward, T. E., C. E. Pedersen, Jr., C. N. Oster, L. R. Bagley, J. Romberger, and M. J. Snyder. 1976. Prompt confirmation of Rocky Mountain spotted fever: identification of rickettsiae in skin tissues. J. Infect. Dis. **134:**297–301.

Yamamoto, S., and Y. Minamishima. 1982. Serodiagnosis of tsutsugamushi fever (scrub typhus) by the indirect immunoperoxidase technique. J. Clin. Microbiol. **15:**1128–1132.

Index to Volume II

Note: Virus *families* (suffix *-viridae*) refer to English plural vernacular forms, e.g., *Poxviridae, see also* Poxviruses.

Most virus *genera* (suffix *-virus*) are listed in plural vernacular form, e.g., *Cytomegalovirus* is listed as Cytomegaloviruses.

Virus *species* are listed in singular vernacular form, e.g., Orf virus.

Most *infections* are also listed, e.g., Orf.

Page numbers of chapter titles appear in **boldface** type.

Index to Volume I

Note: Page numbers of chapter titles appear in **boldface** type.

Printed in the United States
By Bookmasters